D1626121

OXFORD TEXTBOOK OF CRITICAL CARE

OXFORD MEDICAL PUBLICATIONS

EDITORS

ANDREW R. WEBB

Director of Intensive Care
University College London Hospitals
London, UK

MARC J. SHAPIRO

Professor of Surgery and Anesthesiology
Director, Trauma Service
St Louis University Health Sciences Center
St Louis, USA

MERVYN SINGER

Director and Reader in Intensive Care
The Bloomsbury Institute of Intensive Care Medicine
London, UK

PETER M. SUTER

Professor and Chief, Surgical Intensive Care
Hôpital Cantonal Universitaire de Genève
Genève, Switzerland

OXFORD TEXTBOOK OF CRITICAL CARE

Edited by

Andrew R. Webb
Marc J. Shapiro
Mervyn Singer
Peter M. Suter

Oxford New York Tokyo
OXFORD UNIVERSITY PRESS
1999

Oxford University Press, Great Clarendon Street, Oxford OX2 6DP

Oxford New York
Athens Auckland Bangkok Bogota Buenos Aires Calcutta
Cape Town Chennai Dar es Salaam Delhi Florence Hong Kong Istanbul
Karachi Kuala Lumpur Madrid Melbourne Mexico City Mumbai
Nairobi Paris São Paolo Singapore Taipei Tokyo Toronto Warsaw

and associated companies in
Berlin Ibadan

Oxford is a trade mark of Oxford University Press

Published in the United States
by Oxford University Press, Inc., New York

First published 1999

A catalogue record for this book
is available from the British Library

Library of Congress Cataloging in Publication Data
Oxford textbook of critical care / edited by Andrew R. Webb . . . [et al.]
(Oxford medical publications)
1. Critical care medicine. I. Webb, Andrew R. (Andrew Roy)
II. Series.
[DNLM: 1. Critical Care–methods. 2. Emergencies. 3. Intensive Care Units–organization & administration. WX 218098 1999]
RC86.7.094 1999 616.028–dc21 98-34573
ISBN 0 19 262737 6

Typeset by Dobbie Typesetting Limited, Tavistock, Devon

Printed in China

Preface

Few specialties are as exacting as critical care; the holistic complexity of the patient with multisystem dysfunction, the out-of-hours commitment, the often stressful and highly charged situations requiring considerable agility of brain and hand, and the continuing evolution (and occasional revolution) in perceived 'best practice' place major demands on intensive care unit staff.

Decision-making at the bedside should ideally be systematic, problem oriented, and based upon sound pathophysiological principles. Rapid access to up-to-date information is necessary, and this may need to be drawn from other medical or non-clinical disciplines. The *Oxford Textbook of Critical Care* is a single-volume major reference book of adult intensive care medicine aimed at fulfilling most of these requirements, albeit acknowledging that every single topic of importance to critical care medicine cannot possibly be covered in detail. Thus we decided that specialist monographs rather than comprehensive, detailed, and weighty reviews were more suited to our purpose. We also decided to avoid the specialty of pediatric critical care, recognizing key differences between the needs of and some of the disease processes in the critically ill child and adult. Notwithstanding these limitations, this is a state of the art book and should be used as a reference in the intensive care unit and the emergency department and also by all health care providers, including physicians, nurses, and respiratory therapists, who take care of critically ill patients.

Recognizing the needs of those working within the critical care environment, we have incorporated several important features. The book was written with CD-ROM in mind such that all text was 'tagged' at the outset for electronic cross-linking to allow rapid access to relevant information on the desktop PC. Access to electronic media is expanding rapidly; we envisage the CD-ROM version of the book as a central information resource within the critical care unit or library. Another advantage of publishing for both media is the disciplined organization required for the CD-ROM version. Thus the traditional chapter layout of a textbook was abandoned in favor of system-oriented sections. Each section is subdivided into short topics grouped within the section according to clinical problems, as we believe that the reader will often come to this book wishing to update on a specific clinical problem that matches an issue experienced at the bedside. Furthermore, this layout facilitates manageable printing of relevant searches from the CD-ROM. This editorial strategy constrained authors to write within a rigid framework for each chapter. Although we did not attempt to constrain individual styles, we hope that acceptable consistency has been achieved.

We acknowledge that there are often local, national, and international differences in philosophy and management strategy. Some of these differences are seemingly contradictory and it is often difficult for physicians in one country to assimilate information that is produced for another. We intended from the outset that the *Oxford Textbook of Critical Care* would be an international text. We have attempted to give a balanced view where international differences exist, and in many cases have sat squarely on the fence. We make no apology for this since we believe that the book should inform rather than dictate.

Many major textbooks take several years to come to fruition, in which time changes in theory and practice may have occurred. The idea for this textbook was conceived during late 1995, with publication just three years later. It was a mammoth task, co-ordinating the efforts of over 400 authors from all corners of the world, but the result is a textbook that we hope will remain current for longer. The thanks of the editors go to all those who have contributed to this project with a timeliness that has ensured the short production time, and to members of the staff of Oxford University Press for persuading us that this project was worth the blood, sweat, tears, and burnt midnight oil, and whose skill and support have been essential to the editorial and production process. Finally, the editors were saddened to hear of the deaths of Professor T. W. Smith and Dr P. B. Hazard since the submission of their contributions to the book.

A. R. Webb
M. J. Shapiro
M. Singer
P. M. Suter

Foreword

It is a paradox that this erudite textbook of almost 1400 pages with contributions from 450 authors from 20 countries deals with a subject which in many of those countries is not recognized as a specialty in its own right. It does seem unfortunate that a discipline as complex, subtle, and important as intensive care has clearly become, should still be regarded by the medical profession as a subspecialty of other longer-established disciplines. The fact that this book has involved so many authors from such a wide variety of backgrounds and interests indicates that intensive care has matured and will certainly play an increasingly pivotal role in the care and management of rising numbers of hospital patients.

This role will range from transporting critically ill patients between hospitals to advising colleagues in other specialties on the management of severe illnesses likely to be associated with a high mortality. The major role of intensive care specialists in treating patients is likely to increase rapidly as the population grows older, patients' expectations rise, and therapeutic options multiply. This is bound to lead to more demands on intensive care resources and staff. The processes involved in such an undertaking are very complex and require the intensive care specialist to have a wide range of skills. Not only do these specialists need extensive clinical expertise and significant managerial experience, but they must also keep abreast of the rapid technological developments taking place in therapeutics and equipment. This in turn may well accentuate the ethical dilemmas with which intensive care staff are often faced, resulting in the rising levels of stress so familiar to most of them. Ideally, an intensive care specialist needs to be a jack of all trades and master of most. In reality it is unlikely that any single individual can ever possess all these attributes, and this emphasizes that intensive care must be a multidisciplinary subject with close co-operation with many disciplines on a regular and frequent basis.

Any textbook that attempts to deal in a substantive way with these wide-ranging and complex issues is a massive undertaking, and I believe that the editors and authors of this book have succeeded admirably in completing this task. The editorial decision to limit the individual topics to not more than two or three thousand words is wise in that authors feel less pressurized and are more likely to produce a clear and focused contribution with an appropriate number of key references. This decision has, I believe, been amply justified.

Readers have the opportunity of using the book in several ways. They may use it as a learning tool in preparing for the various diplomas and examinations in intensive care that many countries have set up or are planning. Others might use the book to help with a diagnostic problem or to seek advice on a question of clinical management. The key references are helpful in directing readers towards the large body of literature that is now readily accessible electronically. The book not only deals with an enormous number of clinical issues but also covers other important matters outside the mainstream. These include ethics, the problems associated with research in the intensive care unit, planning and equipping the unit, and general planning and financial matters. Because the book is so comprehensive in its coverage, it should appeal to a wide audience ranging from doctors to nurses to technicians and students and perhaps even managers.

The reputation and power of this book will, I believe, be considerably enhanced by its availability as a CD-ROM. Many working in the intensive care environment are computer literate and will certainly welcome accessing such a substantial body of information at the touch of a button. Indeed, it is not outside the bounds of possibility that such instant access might in due course lead to a measurable improvement in clinical performance. The editors are to be congratulated for the vision with which they have produced this book. I know how hard they have worked, and this has resulted in what I believe to be a very important addition to the literature of intensive care medicine. Finally, I hope that this substantial contribution to the medical literature helps to convince some of our more doubting colleagues that intensive care has at last come of age.

E. D. Bennett
St George's Hospital Medical School
London, UK

Contents

3 The cardiovascular system

4 The gastrointestinal and hepatic systems

5 Nutrition

6 The renal system

7 The neurological system

8 The metabolic and endocrine systems

9 Poisoning

10 The hematological system

11 Trauma, burns, and physical disorders

12 Obstetric problems

13 Infection and systemic inflammation

14 Perioperative problems

15 Management of the critical care department

16 Monitoring equipment and techniques

17 Pharmacotherapeutics

18 Therapeutic equipment and techniques

Contributors

K. AHMED
Senior Registrar in Anaesthetics, The General Infirmary at Leeds, UK

P. J. D. ANDREWS
Consultant Anaesthetist and Part-Time Senior Lecturer, University of Edinburgh, UK

E. M. ANTMAN
Cardiovascular Division, Brigham and Women's Hospital, Harvard Medical School, Boston, Massachusetts, USA

D. O. C. ANUMBA
Research Fellow, Department of Obstetrics and Gynaecology, University of Newcastle upon Tyne, UK

A. I. ARIEFF
Chief, Geriatrics Research and Education, San Francisco VA Medical Center, and Professor of Medicine, University of California School of Medicine, San Francisco, California, USA

S. ARLOT
Chief of Endocrinology, Centre Hôpitalier Universitaire, Amiens, France

R. F. ARMSTRONG
Consultant in Anaesthesia and Intensive Care, University College London Hospitals, London, UK

A. ARTIGAS
Associate Professor, Physiology Unit, Biocellular and Physiology Department, Autonomous University of Barcelona, and Director, Intensive Care Service, Sabadell Hospital, Sabadell, Spain

C. J. ASH
Southwest Missouri State University, Springfield, Missouri, USA

M. AUBIER
Professor of Respiratory Medicine, Faculté Xavier Bichat, Paris, France

D. BALOGH
Professor of Anaesthesiology, Department of Anaesthesia, Leopold-Franzens-University of Innsbruck, Austria

S. J. Y. BARD
John Farman Intensive Care Unit, Addenbrooke's Hospital, Cambridge, UK

P. S. BARIE
Associate Professor of Surgery, Cornell University Medical College, and Director, Anne & Max A. Cohen Surgical Intensive Care Unit, New York Hospital, New York, USA

R. H. BARTLETT
Professor of Surgery, University of Michigan Hospitals, Ann Arbor, Michigan, USA

P. J. F. BASKETT
Consultant Anaesthetist, Frenchay Hospital, Bristol, UK

A. D. BAXTER
Department of Anesthesia, Ottawa General Hospital, Ottawa, Canada

H.G. BEGER
Professor, Department of General Surgery, University of Ulm, Germany

G. J. BELLINGAN
Lecturer in Intensive Care Medicine, University College London Hospitals, London, UK

R. BELLOMO
Intensive Care Unit, Austin & Repatriation Medical Centre, Heidelberg, Victoria, Australia

S. BELOUCIF
Department of Anesthesiology and Critical Care Medicine, Lariboisiere University, Paris, France

S. W. BENHAM
Nuffield Department of Anaesthetics and Intensive Therapy Unit, John Radcliffe Hospital, Oxford, UK

S. BENITO
Professor of Medicine, Servei Medicina Intensiva, Hospital de la Santa Creu i Sant Pau, Universitat Autonòma de Barcelona, Spain

W. M. BENNETT
Oregon Health Sciences University, Portland, Oregon, USA

A. BENZER
Professor of Anaesthesiology, Department of Anaesthesia, Leopold-Franzens-University of Innsbruck, Austria

J. BIERT
Department of Surgery, University Hospital Nijmegen, The Netherlands

D. BIHARI
Senior Registrar in Intensive Care, Department of Intensive Care, The St George Hospital, Kogarah, New South Wales, Australia

H. T. BILLY
Santa Barbara Community Health Care Services, Santa Barbara Cottage Hospital, Santa Barbara, California, USA

J. F. BION
Reader in Intensive Care Medicine, Birmingham University, and Honorary Consultant in Intensive Care Medicine, Queen Elizabeth Hospital, Birmingham, UK

G. BISHOP
Liverpool Hospital, Liverpool, New South Wales, Australia

C. BISMUTH
Professor of Medicine and Chief of Service of Intensive Care Unit, Fernand Widal Hospital, Paris, France

L. BLANCH
Section Head, Intensive Care Service, Sabadell Hospital, Sabadell, Spain

T. P. BLECK
Professor of Neurology, Neurological Surgery, and Internal Medicine, and Director, Neuroscience Intensive Care Unit, University of Virginia, Charlottesville, Virginia, USA

J.-M. BOLES
Professor of Intensive Care Medicine, Faculté de Médecine de Brest, and Head, Service de Réanimation Médicale et Urgences Médicales, CHU de la Cavale Blanche, Brest, France

C. F. BOLTON
Professor, Department of Clinical Neurological Sciences, University of Western Ontario, London, Ontario, Canada

J. V. BOOTH
Visiting Professor, Department of Anesthesiology, Duke University Medical Center, Durham, North Carolina, USA

D. C. BORGSTROM
St Paul-Ramsey Medical Center, St Paul, Minnesota, USA

S. W. BORRON
Visiting Researcher, Fernand Widal Hospital, University of Paris, France, and Clinical Associate Professor of Emergency Medicine, George Washington University, Washington, DC, USA

O. BOYD
Intensive Care Senior Registrar, St George's Hospital, London, UK

R. P. BOYER
University of Missouri Health Sciences Center, Columbia, Missouri, USA

P. G. BOYSEN
Professor and Chairman, Department of Anesthesiology, School of Medicine, University of North Carolina at Chapel Hill, North Carolina, USA

M. A. BRANTHWAITE
Barrister at Law, formerly Consultant Physician and Anaesthetist, Royal Brompton Hospital, London, UK

L. BRAZZI
Anesthesiology Staff, Department of Anesthesia and Intensive Care Medicine, Ospedale Maggiore, Milan, Italy

S. J. BRETT
Senior Registrar in Anaesthesia and Intensive Care, Royal Brompton Hospital, London, UK

M. BREZIS
Professor of Medicine and Chief Physician, Hadassah University Hospital–Mount Scopus, Jerusalem, Israel

S. BRIMIOULLE
Department of Intensive Care, Erasme University Hospital, Brussels, Belgium

A. BRINKMANN
Universitätsklinik für Anästhesiologie, Klinikum der Universität Ulm, Germany

A. BRISTOW
Consultant Anaesthetist, St Bartholomew's Hospital, London, UK

P. BRISTOW
Staff Intensivist, Liverpool Hospital, Liverpool, New South Wales, Australia

L. BROCHARD
Service Réanimation Médicale, Hôpital Henri Mondor, Creteil, France

L. J. BROERSE
Department of Nursing Affairs, University Hospital Groningen, The Netherlands

M. G. BROOK
Consultant Physician, Department of Genitourinary Medicine, Central Middlesex Hospital, London, UK

H. BURCHARDI
Department of Anaesthesiology, Emergency and Intensive Care Medicine, University Hospital, Göttingen, Germany

P. D. CAMERON
Senior Staff Specialist, Department of Intensive Care, Sir Charles Gairdner Hospital, Nedlands, Australia

A. J. CAMM
Department of Cardiological Sciences, St George's Hospital Medical School, London, UK

J. CARLET
Director, Intensive Care Unit, Hôpital Saint Joseph, Paris, France

P. A. CARLI
Professor of Anesthesiology and Critical Care, Hôpital Necker, Paris, France

I. CAVILL
Senior Lecturer in Haematology, University of Wales College of Medicine, Cardiff, UK

B. M. CHAIN
Department of Immunology, University College London, UK

H. R. CHAMPION
Research Professor of Surgery, University of Maryland, Baltimore, and Professor of Surgery, Uniformed Services University of the Health Sciences, Bethesda, Maryland, USA

M. CHAPMAN
Consultant Intensivist, Intensive Care Unit, Royal Adelaide Hospital, Adelaide, Australia

J.-C. CHEVROLET
Division of Medical Intensive Care, Hôpital Cantonal Universitaire, Geneva, Switzerland

P. L. CHIODINI
Consultant Parasitologist, Hospital for Tropical Diseases, London, and Honorary Senior Lecturer, London School of Hygiene and Tropical Medicine, London, UK

D. CHIUMELLO
Anesthesiology Staff, Department of Anesthesia and Intensive Care Medicine, Ospedale Maggiore, Milan, Italy

D. B. CLIFFORD
Professor of Neurology, Washington University School of Medicine, St Louis, Missouri, USA

J. COAKLEY
Consultant Physician in Intensive Care, St Bartholomew's Hospital, London, UK

H. COHEN
Senior Lecturer in Haematology, Imperial College School of Medicine, London, UK

S. L. COHEN
Consultant Physician, Intensive Care Unit, University College London Hospitals, London, UK

M. C. COLBRIDGE
Senior Information Specialist, Medical Toxicology Unit, Guy's and St Thomas's Hospital Trust, London, UK

P. J. COLE
Senior House Officer, John Farman Intensive Care Unit, Addenbrooke's Hospital, Cambridge, UK

A. D. COLLINS
Assistant Clinical Professor of Pediatrics, Tufts University School of Medicine, Springfield, Massachusetts, USA

J. B. CONE
Professor of Surgery and Head, Section of General Surgery, University of Arkansas for Medical Sciences, Little Rock, Arkansas, USA

G. CONTI
Clinical Chief, Intensive Care Unit, University of Rome 'La Sapienza', Rome, Italy

D. J. COOK
Associate Professor of Medicine and Clinical Epidemiology, McMaster University, Hamilton, Ontario, Canada

J. T. COOKE
Assistant Professor of Medicine, Cornell University Medical College, and Associate Director, Medical Intensive Care, New York Hospital, New York, USA

J. J. CORDINGLEY
Department of Intensive Care, Charing Cross Hospital, London, UK

M. J. COUSINS
Professor and Head, Department of Anaesthesia and Pain Management, Royal North Shore Hospital and University of Sydney, Australia

C.-P. CRIÉE
Kreiskrankenhaus, Bovenden-Lenglern, Germany

D. J. CULLEN
Chairman, Department of Anesthesiology, St Elizabeth's Medical Center, Boston, Massachusetts, USA

D. CUMMINS
Consultant Haematologist, Harefield Hospital, Harefield, UK

I. J. A. DAVIDSON
Surgical Director, Division of Renal Transplantation, Medical City Hospital, Dallas, Texas, USA

M. J. DAVIES
British Heart Foundation Professor of Cardiovascular Pathology, St George's Hospital Medical School, London, UK

N. M. DEARDEN
Consultant Anaesthetist, Leeds General Infirmary, Leeds, UK

G. DEBY-DUPONT
Department of Anaesthesia and Intensive Care Medicine, University of Liège, Belgium

M. DE SWIET
Consultant Physician, Queen Charlotte's and Chelsea Hospital for Women, London, UK

P. M. J .M. DE VRIES
Department of Pulmonary Medicine, Westfries Gasthuis, Hoorn, The Netherlands

R. DHAND
Associate Professor of Medicine, Stritch School of Medicine, Loyola University of Chicago, Maywood, Illinois, USA

G. J. DOBB
Intensive Care Unit, Royal Perth Hospital, Perth, Australia

J. DORCA
Chest Department, Hospital de Bellvitge, Barcelona, Spain

T. DUGERNIER
Director of the Intensive Care Unit, St Pierre Hospital, Ottignies, and Consultant Physician, Intensive Care Department, St Luc University Hospital, Brussels, Belgium

R. M. DURHAM
Associate Professor of Surgery and Director, Surgical Critical Care, St Louis University School of Medicine, St Louis, Missouri, USA

H. J. DWORKEN
Emeritus Professor of Medicine, Case Western Reserve University, Cleveland, Ohio, USA

N. B. EASTWOOD
Visiting Professor, Department of Anesthesiology, Duke University Medical Center, Durham, North Carolina, USA

P. ECKERT
Attending Physician, Medical Intensive Care, University Hospital, Lausanne, Switzerland

G. EDGE
Royal National Orthopaedic Hospital, Stanmore, UK

J. D. EDWARDS
Consultant Physician, University Hospital of South Manchester, UK

M. EL-EBIARY
UVIR-Servei de Pneumologia, Hospital Clínic, Barcelona, Spain

F. R. ELLIS
Professor of Anaesthesia, University of Leeds, Leeds, UK

H. EL-REFAEY
Senior Registrar, Academic Department of Obstetrics and Gynaecology, University College London Medical School, London, UK

D. C. EVANS
Clinical Director of Trauma Services, Montreal General Hospital, Montreal, Canada

T. W. EVANS
Professor of Intensive Care Medicine, Imperial College School of Medicine, University of London, and Consultant in Intensive Care Medicine, Royal Brompton Hospital, London, UK

B. K. EVETTS
Santa Barbara Community Health Care Services, Santa Barbara Cottage Hospital, Santa Barbara, California, USA

A. G. FABRIS
Department of Nephrology, City Hospital, Bassano del Grappa, Italy

K. J. FALKE
Professor, Virchow-Klinikum, Humboldt University, Berlin, Germany

M. E. FAYMONVILLE
Department of Anaesthesia and Intensive Care Medicine, University of Liège, Belgium

J. L. FELLAHI
Service de Réanimation Médicale, Hôpital Ambroise Paré, Boulogne, France

M. FENNELLY
Royal National Orthopaedic Hospital, Stanmore, UK

C. FERGUSON
Consultant Anaesthetist, St Bartholomew's Hospital, London, UK

H. M. FERNANDES
Research Associate in Neurosurgery, Department of Surgery (Neurosurgery), University of Newcastle upon Tyne, UK

R. FERNANDEZ
Clinical Chief, Intensive Care Department, Sabadell Hospital, Sabadell, Spain

J. M. FIELD
The Royal Hospital Haslar, Gosport, UK

M. FISHER
Department of Anaesthesia, University of Sydney, and Head, Intensive Therapy Unit, Royal North Shore Hospital, St Leonards, Australia

W. FITCH
Professor of Anaesthesia, University of Glasgow, UK

B. A. FOËX
Research Fellow, North Western Injury Research Centre, University of Manchester, UK

D. M. FORREST
Research Fellow, Critical Care Medicine, Department of Medicine, St Paul's Hospital, Vancouver, Canada

A. FOURNIER
Chief of Nephrology, Centre Hôpitalier Universitaire, Amiens, France

R. S. R. FOX
Consultant Anaesthetist, Royal National Orthopaedic Hospital, Stanmore, UK

P. FROST
Intensive Care Unit, Middlemore Hospital, Auckland, New Zealand

S. M. GABE
Research Fellow, Department of Gastroenterology and Nutrition, Central Middlesex Hospital, London, UK

T. J. GAN
Assistant Professor, Department of Anesthesiology, and Medical Director, Postanesthesia Care Unit, Duke University Medical Center, Durham, North Carolina, USA

R. M. GARDNER
Professor of Medical Informatics, University of Utah, and Co-Director of Medical Informatics, LDS Hospital, Salt Lake City, Utah, USA

C. S. GARRARD
Director of Intensive Care, John Radcliffe Hospital, Oxford, UK

A. GASPARETTO
Chairman, Institute of Anesthesiology, University of Rome 'La Sapienza', Rome, Italy

C. M. GASPERETTI
Assistant Professor of Medicine, Veterans Affairs Medical Center, Salem, Virginia, USA

L. GATTINONI
Professor of Anesthesia and Intensive Care Medicine, University of Milan, Italy

P. G. GAUGER
Critical Care Fellow, University of Michigan Hospitals, Ann Arbor, Michigan, USA

M. GAVALAS
Consultant in Emergency Medicine, Accident and Emergency Directorate, University College London Hospitals, London, UK

J. E. GILLIGAN
Director, Retrieval and Resuscitation, Royal Adelaide Hospital, Adelaide, Australia

A. E. S. GIMSON
Consultant Hepatologist, Addenbrooke's Hospital, Cambridge, UK

E. H. GLUCK
Chief, Medical Division, North Chicago VA Medical Center, Chicago, Illinois, USA

L. R. GOLDFRANK
Director, Department of Emergency Medicine, Bellevue Hospital and New York Medical Center, New York, USA

A. H. GOLDSTONE
Director of Clinical Haematology and Consultant Haematologist, Department of Haematology, University College London Hospitals, London, UK

J. C. GOLDSTONE
Senior Lecturer in Anaesthesia, University College London Hospitals, London, UK

R. J. A. GORIS
Chairman, Department of Surgery, University Hospital St Radboud, Nijmegen, The Netherlands

M. GRAHAM
Professor of Pathology, School of Medicine, St Louis University, and Chief Medical Examiner, City of St Louis, Missouri, USA

R. S. GRAHAM
Department of Neurological Surgery, West Hospital, Richmond, Virginia, USA

J. W. M. GREVE
Associate Professor of Surgery, Department of Surgery, University Hospital Maastricht, The Netherlands

B. P. GRIFFITH
Professor of Surgery and Chief, Cardiothoracic Surgery, University of Pittsburgh Medical Center, Pittsburgh, Pennsylvania, USA

G. K. GRIMBLE
Reader in Clinical Nutrition, Roehampton Institute, London, UK

A. B. J. GROENEVELD
Medical Intensive Care Unit, Free University Hospital, Amsterdam, The Netherlands

R. M. GROUNDS
Consultant in Anaesthesia and Intensive Care Medicine, St George's Hospital, London, UK

A. GÜNTHER
Department of Internal Medicine, Justus-Liebig University, Giessen, Germany

G. GUTIERREZ
Professor of Medicine, University of Texas Medical School, Houston, Texas, USA

O. P. HABLER
Institute for Surgical Research and Institute of Anesthesiology, University of Munich (LMU), Germany

U. HAGLUND
Professor of Surgery, Department of Surgery, Uppsala University Hospital, Uppsala, Sweden

P. J. HALSALL
Associate Specialist and Honorary Research Fellow in Anaesthesia, St James's University Hospital, Leeds, UK

J. HAMMOND
Professor of Surgery and Chief, Trauma/Surgical Critical Care, Robert Wood Johnson Medical School, New Brunswick, New Jersey, USA

K. K. HAMPTON
Senior Lecturer in Haematology, Royal Hallamshire Hospital, Sheffield, UK

C. W. HANSON
Center for Anesthesia Research, University Hospital of Pennsylvania, Philadelphia, Pennsylvania, USA

S. J. HARBARTH
Fellow, Infection Control Program, Geneva University Hospitals, Geneva, Switzerland

D. T. HARRINGTON
Chief, Burn Study Branch, US Army Insititute of Surgical Research, Fort Sam Houston, Texas, USA

C. N. HARRISON
Research Fellow, Department of Haematology, University College London Hospitals, London, UK

M. J. G. HARRISON
Professor in Clinical Neurology, University College London Medical School, London, UK

T. K. HARTUNG
Accident and Emergency Department, St Mary's Hospital, London, UK

W. HASIBEDER
Clinical Director, Department of General and Surgical Intensive Care Medicine, University of Innsbruck, Austria

F. HAWKER
Director, Intensive Care Unit, Cabrini Hospital, Malvern, Victoria, Australia

P. B. HAZARD (deceased)
Formerly Associate Clinical Professor of Medicine, University of Tennessee College of Medicine, Memphis, Tennessee, USA

S. HECKERS
Clinical Fellow, Department of Psychiatry, Massachusetts General Hospital, Boston, Massachusetts, USA

G. HEDENSTIERNA
Professor and Chairman, Department of Clinical Physiology, University Hospital, Uppsala, Sweden

M. HEMMER
Department of Anesthesiology and Surgical Intensive Care, Centre Hospitalier de Luxembourg, Luxembourg

C. P. H. HENEGHAN
Consultant Anaesthetist, Nevill Hall Hospital, Abergavenny, UK

J. A. HENRY
Professor of Accident and Emergency Medicine, Imperial College School of Medicine, St Mary's Hospital, London, UK

P. HEWITT
National Blood Service, London, UK

K. HILLMAN
Co-Director, Division of Critical Care, Liverpool Hospital, Liverpool, New South Wales, Australia

C. J. HINDS
Director of Intensive Care, St Bartholomew's Hospital, London, UK

N. P. HIRSCH
Consultant Anaesthetist and Honorary Senior Lecturer, Batten–Harris Intensive Therapy Unit, National Hospital for Neurology and Neurosurgery, London, UK

A. HOLE
Department of Anesthesiology, University Hospital of Trondheim, Norway

H. HON
Consultant Paediatrician, Intensive Care Unit, Department of Anaesthesia, University of the Witwatersrand, Baragwanath Hospital, Bertsham, Republic of South Africa

M. HOPLEY
Consultant Physician, Intensive Care Unit, Department of Anaesthesia, University of the Witwatersrand, Baragwanath Hospital, Bertsham, Republic of South Africa

H. M. HORST
Director, Surgical Critical Care, Henry Ford Hospital, Detroit, Michigan, USA

R. S. HOWARD
Consultant Neurologist and Honorary Senior Lecturer, Batten–Harris Intensive Therapy Unit, National Hospital for Neurology and Neurosurgery, London, UK

D. HUNT
Health Information Research Unit, McMaster University Health Sciences Centre, Hamilton, Ontario, Canada

G. IAPICHINO
Servizio Anestesia e Rianimazone, Ospedale San Paolo, Milan, Italy

J. P. ISBISTER
Head, Department of Haematology and Transfusion Medicine, Royal North Shore Hospital of Sydney, and Adjunct Professor, School of Biological Sciences, University of Technology, Sydney, Australia

R. M. JACKSON
Division of Pulmonary and Critical Care Medicine, University of Alabama at Birmingham, Alabama, USA

M. R. JACOBS
Professor of Pathology, Case Western Reserve University School of Medicine, Cleveland, USA

D. JACOBSEN
Head Physician and Professor, Clinic for Acute Medicine, Ullevål University Hospital, Oslo, Norway

A. JAEGER
Professor of Intensive Care, Hôpitaux Universitaires, Strasbourg, France

J. JALLO
Department of Neurosurgery, Temple University Hospital, Philadelphia, Pennsylvania, USA

F. JARDIN
Service de Réanimation Médicale, Hôpital Ambroise Paré, Boulogne, France

L. J. JARVIS
Head of Radiology, The Royal Hospital Haslar, Gosport, UK

K. N. JEEJEEBHOY
Professor of Medicine, University of Toronto, Canada

A. A. JEFFREY
Consultant Physician in Respiratory and Intensive Care Medicine, Northampton General Hospital, Northampton, UK

M. JEGERS
Free University Brussels (VUB), Belgium

D. C. JERWOOD
Morriston Hospital, Swansea, UK

P. JOLLIET
Assistant Head Physician, Division of Medical Intensive Care, Hôpital Cantonal Universitaire, Geneva, Switzerland

R. J. KAHN
Head, Department of Intensive Care, Erasme University Hospital, Brussels, Belgium

P. KALLIS
Consultant Cardiothoracic Surgeon, University College London Hospitals, London, UK

R. C. KARMY-JONES
Assistant Professor of Surgery, Division of Cardiothoracic Trauma, University of Washington Medical Center and Harbor View Medical Center, Seattle, Washington, USA

A. M. S. KASHMEERY
Professor of Physiology and Head of Biomedical Ethics Programme, Oxford Academy for Advanced Studies, Oxford, UK

E. S. KASSIS
Research Assistant, Thoracic Surgery, University of Pittsburgh Medical Center, Pittsburgh, Pennsylvania, USA

E. S. KAUFMAN
Senior Clinical Instructor, Metro Health Medical Center, Case Western Reserve University, Cleveland, Ohio, USA

C. A. KELLER
Professor, Division of Pulmonary Medicine, St Louis University School of Medicine, St Louis, Missouri, USA

M. J. KELLETT
Consultant Uroradiologist, Department of Uroradiology, St Peter's and Middlesex Hospital, London, UK

K. P. KELLY
Senior Registrar in Anaesthetics, Department of Anaesthetics, Royal Infirmary of Edinburgh, and Intensive Therapy Unit, Western General Hospital, Edinburgh, UK

B. F. KEOGH
Consultant in Anaesthesia and Intensive Care, Royal Brompton Hospital, London, UK

P. J. KESTENS
Professor of Surgery, University of Louvain Medical School, Brussels, Belgium

P. L. KHIMENKO
Instructor in Physiology, Department of Physiology, University of South Alabama College of Medicine, Mobile, Alabama, USA

J. KINGDOM
Division of Maternal–Fetal Medicine, Samuel Lunenfeld Research Institute, Mount Sinai Hospital, Toronto, Canada

O. C. KIRTON
Assistant Professor of Surgery, University of Miami School of Medicine, Miami, Florida, USA

H. KLEIN
Senior Registrar, Anaesthetics, Leeds General Infirmary, Leeds, UK

P. KLEPSTAD
Department of Anesthesiology, University Hospital of Trondheim, Norway

W. A. KNAUS
Professor and Chairman, Department of Health Evaluation Sciences, University of Virginia Medical School, Charlottesville, Virginia, USA

P. M. KOCHANEK
University of Pittsburgh, Pennsylvania, USA

K. KOYAMA
Assistant Professor, Institute of Clinical Medicine, University of Tsukuba, Japan

K. A. KRALOVICH
Assistant Professor of Surgery, Case Western Reserve University, and Senior Staff, Trauma and Critical Care Surgery, Henry Ford Hospital, Detroit, Michigan, USA

A. N. LAGGNER
Professor of Emergency Medicine, Vienna General Hospital, Vienna, Austria

M. LAMY
Head, Department of Anesthesia and Intensive Care Medicine, University of Liège, Belgium

S. E. LAPINSKY
Assistant Professor of Medicine, University of Toronto, Canada

P. F. LATERRE
Intensive Care Department, St Luc University Hospital, Brussels, Belgium

G. LAVERY
Intensive Care Consultant, Robert C. Gray Regional Intensive Care Unit, Royal Hospitals Trust, Belfast, UK

J.-R. LE GALL
Intensive Care Unit, Hôpital Saint Louis, Paris, France

T. S. LEARY
John Farman Intensive Care Unit, Addenbrooke's Hospital, Cambridge, UK

M. LEJAY
Hôpital Necker, Paris, France

S. LEMESHOW
Professor of Biostatistics, University of Massachusetts, Amherst, Massachusetts, USA

M. LEVIN
Professor of Paediatrics, Imperial College School of Medicine, London, UK

J. H. LEVY
Professor of Anesthesiology, Emory University School of Medicine, Atlanta, Georgia, USA

W. R. LEWIS
Senior Clinical Instructor, Metro Health Medical Center, Case Western Reserve University, Cleveland, Ohio, USA

P. LHEUREUX
Associate Clinical Director, Emergency Department, Erasme University Hospital, Brussels, Belgium

R. LIESNER
Camelia Botnar Laboratory, Great Ormond Street Hospital, London, UK

J. LIPMAN
Intensive Care Facility, Royal Brisbane Hospital, Brisbane, Australia

A. LISSONI
Anesthesiology Staff, Department of Anesthesia and Intensive Care Medicine, Ospedale Maggiore, Milan, Italy

R. A. LITTLE
Director, North Western Injury Research Centre, University of Manchester, UK

J. LLOYD
Senior Anaesthetic Registrar, St Bartholomew's Hospital, London, UK

L. LOH
Consultant Anaesthetist, Nuffield Department of Anaesthetics, Radcliffe Infirmary NHS Trust, Oxford, UK

K. LOWRY
Consultant Intensivist, Northern Ireland Regional Thoracic Surgical Department, Royal Victoria Hospital, Belfast, UK

J. D. LUKETICH
Assistant Professor, Section Head, Thoracic Surgery, University of Pittsburgh Medical Center, Pittsburgh, Pennsylvania, USA

D. C. MACALLAN
Division of Infectious Diseases, St George's Hospital Medical School, London, UK

J. R. MCEWAN
Senior Lecturer and Consultant Cardiologist, University College London Medical School, London, UK

J. A. MCGUIGAN
Consultant Thoracic Surgeon, Northern Ireland Regional Thoracic Surgical Department, Royal Victoria Hospital, Belfast, UK

S. J. MACHIN
Professor of Haematology, University College London Hospitals, London, UK

S. A. M. MCLEAN
International Bar Association Professor of Law and Ethics in Medicine, University of Glasgow, UK

M. MAGGIORINI
Department of Internal Medicine, University Hospital Zurich, Switzerland

A. MAIN
Leeds General Infirmary, Leeds, UK

E. MAJOR
Director of Intensive Therapy, Morriston Hospital, Swansea, UK

A. MALLICK
Specialist Registrar, Academic Unit of Anaesthesia, The General Infirmary at Leeds, UK

J. MANCEBO
Servei Medicina Intensiva, Hospital de Sant Pau–VAB, Barcelona, Spain

C. MANDEL
National Hospital for Neurology and Neurosurgery, London, UK

M. MANJI
Senior Registrar in Intensive Care Medicine, Queen Elizabeth Hospital, Birmingham, UK

J. J. MARINI
Professor of Medicine, University of Minnesota, Minneapolis-St Paul, Minnesota, USA

B. E. MARSHALL
Horatio C. Wood Professor of Anesthesia and Director, Center for Anesthesia Research, University Hospital of Pennsylvania Health System, Philadelphia, Pennsylvania, USA

J. C. MARSHALL
Director of Research, Medical/Surgical Intensive Care Unit, The Toronto Hospital, and Associate Professor of Surgery, University of Toronto, Canada

L. F. MARSHALL
Professor and Chair, Division of Neurological Surgery, University of California, San Diego, USA

A. MARSTON
Emeritus Consultant Surgeon, The Middlesex Hospital, London, UK

G. D. MARTICH
Assistant Professor, Anesthesiology and Critical Care Unit, and Associate Director, Cardiothoracic Intensive Care Unit, University of Pittsburgh Medical Center, Pittsburgh, Pennsylvania, USA

C. MARTIN
Professor of Anesthesia and Intensive Care, Marseilles Medical School, University of Marseilles, France

G. MAUGHAN
Resuscitation Training Officer, University College London Hospitals, London, UK

G. MAURER
Professor of Medicine and Director, Department of Cardiology, University of Vienna, Austria

J. E. MAZUSKI
Associate Professor of Surgery, St Louis University School of Medicine, St Louis, Missouri, USA

A. MEBAZAA
Assistant Professor, Department of Anesthesiology and Critical Care Medicine, Lariboisière University, Paris, France

P. G. MEDD
Department of Immunology, University College London, UK

G. U. MEDURI
Professor of Medicine, University of Tennessee, Memphis, Tennessee, USA

A. D. MENDELOW
Professor of Neurosurgery, Department of Surgery (Neurosurgery), University of Newcastle upon Tyne, UK

D. K. MENON
Lecturer in Anaesthesia, University of Cambridge, and Consultant, Neurosciences Critical Care Unit, Addenbrooke's Hospital, Cambridge, UK

K. MESSMER
Director, Institute for Surgical Research, University of Munich (LMU), Germany

S. W. METCALF
Specialist Registrar in Accident and Emergency Medicine, University College London Hospitals, London, UK

T. H. MILHORAT
Professor and Chairman, Department of Neurosurgery, State University of New York Health Science Center, Brooklyn, New York, USA

J. MILIC-EMILI
Professor of Physiology, McGill University, Montreal, Canada

D. R. MILLER
Associate Professor and Research Director, Department of Anesthesia, University of Ottawa, Canada

R. F. MILLER
Senior Lecturer and Consultant Physician, University College London Medical School, London, UK

A. R. MOLLOY
Senior Staff Specialist and Senior Lecturer, Department of Anaesthesia and Pain Management, Royal North Shore Hospital and University of Sydney, Australia

H. E. MONTGOMERY
Senior Registrar in Intensive Care, University College London Hospitals, London, UK

R. MOORE
Specialist Registrar in General Surgery, University College London Hospitals, London, UK

T. M. MOORE
Postdoctoral Fellow, Department of Pharmacology, University of South Alabama College of Medicine, Mobile, Alabama, USA

W. J. MOORE
Consultant Anaesthetist, Hereford County Hospital, Hereford, UK

B. MORGAN
The Richard Ivey Critical Care Trauma Center, London Health Sciences Center, London, Ontario, Canada

V. MORGAN
Regional Transplant Coordinator, North Thames NHS Executive, London, UK

A. G. MORRISON
Specialist Registrar in Anaesthetics, Western Infirmary, Glasgow, UK

N. MUTZ
Head, Division of General and Surgical Intensive Care Medicine, University of Innsbruck, Austria

M. G. MYTHEN
Consultant Anaesthetist, University College London Hospitals, London, UK

S. NADEL
Consultant in Pediatric Intensive Care, St Mary's Hospital, London, UK

R. K. NARAYAN
Department of Neurosurgery, Temple University Hospital, Philadelphia, Pennsylvania, USA

A. C. NATHWANI
Lecturer and Honorary Consultant, Department of Haematology, John Radcliffe Hospital, Oxford, UK

C. NELSON-PIERCY
Senior Registrar, Obstetric Medicine, Queen Charlotte's and Chelsea Hospital for Women, London, UK

M. NEVIN
Consultant in Anaesthesia and Intensive Care, Bristol Royal Infirmary, Bristol, UK

P. NEWMAN
Consultant in Intensive Care, St George's Hospital, London, UK

N. NICHOL
Department of Accident and Emergency Medicine, Royal Infirmary, Edinburgh, UK

M. W. NICOLLE
Assistant Professor of Neurology and Director, Myasthenia Gravis Clinic, University of Western Ontario, London, Ontario, Canada

R. NIRULA
Santa Barbara Community Health Care Services, Santa Barbara Cottage Hospital, Santa Barbara, California, USA

K. M. NOLAN
Intensive Care Unit, John Radcliffe Hospital, Oxford, UK

R. B. NOONE
Research Fellow in Surgery, Duke University Medical Center, Durham, North Carolina, USA

C. O'DONNELL
Specialist Registrar in Emergency Medicine, Whipps Cross Hospital, London, UK

T. E. OH
Professor and Chairman, Department of Anaesthesia and Intensive Care, Chinese University of Hong Kong, Shatin, Hong Kong

A. OLYAEI
Oregon Health Sciences University, Portland, Oregon, USA

L. OPPENHEIMER
Professor of Surgery, University of Manitoba, Winnipeg, Canada

J. J. ORO
Associate Professor of Neurological Surgery, University of Missouri School of Medicine, Columbia, Missouri, USA

E. OSCHATZ
Department of Emergency Medicine, Vienna General Hospital, Vienna, Austria

C. OZANNE
Consultant Haematologist, Guy's Hospital, London, UK

R. P. PACZYNSKI
Department of Neurology, Washington University School of Medicine, St Louis, Missouri, USA

M. G. A. PALAZZO
Department of Intensive Care, Charing Cross Hospital, London, UK

M. E. PALMER
Attending Physician, Emergency Services, New York City Poison Control Center, New York, USA

E. A. PANACEK
Associate Professor of Medicine, University of California Davis, Sacramento, California, USA

D. PAPPERT
Senior Staff, Klinik für Anaesthesiologie und operative Intensivmedizin, Virchow-Klinikum, Humboldt University, Berlin, Germany

G. R. PARK
Director, John Farman Intensive Care Unit, Addenbrooke's Hospital, Cambridge, UK

U. PATEL
Department of Diagnostic Radiology, St George's Hospital, London, UK

D. M. PAYEN
Professor and Chairman, Department of Anesthesiology and Critical Care Medicine, Lariboisière University, Paris, France

D. A. PAYNE
Executive Director, New York Regional Transplant Program Inc., New York, USA

S. PEAKE
Director, Intensive Care Unit, Lyell McEwin Hospital, Woodville, Australia

D. G. PENNEY
Professor of Physiology, Wayne State University School of Medicine, Detroit, and Director of Surgical Research, Providence Hospital, Southfield, Michigan, USA

J. L. PETERS
Consultant Surgeon, Princess Alexandra Hospital, Harlow, UK

C. A. PIANTADOSI
Professor of Medicine, Duke University Medical Center, Durham, North Carolina, USA

M. R. PINSKY
Professor of Anesthesiology, Critical Care Medicine and Medicine, Department of Anesthesiology and Critical Care Medicine, University of Pittsburgh Medical Center and Oakland VA Medical Center, Pittsburgh, Pennsylvania, USA

D. PITTET
Associate Professor of Medicine, and Chief, Infection Control Program, Geneva University Hospitals, Geneva, Switzerland

M. POEZE
Department of Surgery, University Hospital, Maastricht, The Netherlands

B. J. POLLARD
Professor of Anaesthesia, University of Manchester, UK

J. A. POLOGE
Manager of Research, Ohmeda, Louisville, Colorado, USA

G. PORENTA
Assistant Professor of Medicine, Department of Cardiology, University of Vienna, Austria

W. E. PORTER
Wayne State University School of Medicine, Detroit, Michigan, USA

F. POTIE
Staff Anesthetist, Marseilles Medical School, University of Marseilles, France

B. M. POWER
Staff Specialist in Intensive Care, Sir Charles Gairdner Hospital, Nedlands, Australia

U. B. S. PRAKASH
Scripps Professor of Medicine, Mayo Medical Center and Mayo Graduate School of Medicine, and Consultant in Pulmonary and Critical Care Medicine, Mayo Medical Center, Rochester, Minnesota, USA

B. A. PRUITT JR
Professor of Surgery, The University of Texas, Health Science Center at San Antonio, Texas, USA

P. L. RADERMACHER
Professor of Anesthesiology, Ulm University Hospital, Ulm, Germany

D. RADRIZZANI
IRCCS Ospedale Maggiore, Milan, Italy

G. RAMSAY
Associate Professor of Surgery, Chairman of Intensive Care, University Hospital, Maastricht, The Netherlands

V. M. RANIERI
Istituto di Anestesiologia e Rianimazione, Ospedale Policlinico, Bari, Italy

B. RAU
Department of General Surgery, University of Ulm, Germany

F. REGAN
National Blood Service, London, UK

J. REICHMANN
Department of Medicine, Hadassah University Hospital–Mount Scopus, Jerusalem, Israel

D. REIS MIRANDA
Department of Surgery, University Hospital Groningen, The Netherlands

J. RELLO
Staff Physician, Intensive Care Unit, Sabadell Hospital, Sabadell, Spain

M. S. REYNAERT
Professor of Medicine, Head of Department of Intensive Care and Emergency Medicine, St Luc University Hospital, Brussels, Belgium

A. RHODES
Senior Registrar in Intensive Care, St George's Hospital, London, UK

D. W. RICHTER
Zentrum Physiologie und Pathophysiologie, Georg-August-Universität, Göttingen, Germany

G. L. RIDGWAY
Honorary Senior Lecturer, University College London School of Medicine, and Consultant Microbiologist, University College London Hospitals, UK

S. A. RIDLEY
Consultant in Anaesthesia and Intensive Care, Elizabeth Fry (Intensive Care) Unit, Norfolk and Norwich Hospital, UK

J. E. RISDALL
Consultant Anaesthetist, The Royal Hospital Haslar, Gosport, UK

R. RITZ
Professor of Internal Medicine and Intensive Care, University Hospital, Basel, Switzerland

C. ROBERTSON
Department of Accident and Emergency Medicine, Royal Infirmary, Edinburgh, UK

R. D. ROBERTSON
Department of Surgery, University of Arkansas for Medical Sciences, Little Rock, Arkansas, USA

S. C. ROBSON
Professor of Fetal Medicine, Department of Obstetrics and Gynaecology, University of Newcastle upon Tyne, UK

J. ROCA
Associate Professor of Medicine, University of Barcelona, and Section Head, Servei de Pneumologia i Alergia Respatòria, Hospital Clínic, Barcelona, Spain

R. RODRIGUEZ-ROISIN
Professor of Medicine, University of Barcelona Medical School for Pulmonary Medicine, and Head of Respiratory Medicine Service, Servei de Pneumologia i Alergia Respatòria, Hospital Clínic, Barcelona, Spain

J.-A. ROMAND
Consultant, Surgical Intensive Care Unit, University Hospital of Geneva, Switzerland

C. RONCO
Department of Nephrology, St Bortolo Hospital, Vicenza, Italy

R. ROSSAINT
Virchow-Klinikum, Humboldt University, Berlin, Germany

D. ROSSER
Consultant and Honorary Senior Lecturer in Intensive Care Medicine, Queen Elizabeth Hospital, Birmingham, UK

H. U. ROTHEN
Department of Anesthesiology and Intensive Care, University Hospital, Bern, Switzerland

K. ROWAN
Director, Intensive Care National Audit and Research Centre (ICNARC), London, UK

R. RUGGIERO
Instructor, Division of Cardiothoracic Surgery, St Louis University School of Medicine, St Louis, Missouri, USA

J. A. RUSSELL
Chairman, Critical Care Medicine, Department of Medicine, St Paul's Hospital, Vancouver, Canada

D. W. RYAN
Director of Intensive Care, General Intensive Therapy Unit, Freeman Hospital, Newcastle upon Tyne, UK

J. M. RYAN
Leonard Cheshire Professor of Conflict Recovery, University College London Hospitals and University College London, UK

W. J. SACCO
Charles McC. Mathias Jr Study Center for Trauma and EMS, Baltimore, Maryland, USA

P. SAFAR
Distinguished Professor of Resuscitation Medicine, Safar Center for Resuscitation Research, University of Pittsburgh, Pennsylvania, USA

J. R. SAFFLE
Professor of Surgery, University of Utah Health Center, Salt Lake City, Utah, USA

M. SAIR
Garfield Weston Research Fellow, Unit of Critical Care, Imperial College School of Medicine, London, UK

J. SALMON
Consultant Physician and Head of Intensive Care, The Royal Hospital Haslar, Gosport, UK

I. C. SANDERSON
Department of Anesthesiology, Duke University Medical Center, Durham, North Carolina, USA

M. C. SAYERS
Clinical Associate Professor of Medicine–University of Virginia School of Medicine, Director of Coronary Care Unit and Cardiac Catheterization Laboratory, Carilion Roanoke Community Hospital, Roanoke, Virginia, USA

M. SCHÄFER
Universitätsklinikum Benjamin Franklin, Klinik für Anaesthesiologie und operative Intensivmedizin, Freie Universität Berlin, Germany

M.-D. SCHALLER
Chief Physician and Assistant Professor, Medical Intensive Care, University Hospital, Lausanne, Switzerland

M. SCHETZ
Department of Intensive Care Medicine, University Hospital Gasthuisberg, Leuven, Belgium

W. SCHOBERSBERGER
Department of Anesthesiology, University of Innsbruck, Austria

W. SEEGER
Professor of Internal Medicine, Department of Internal Medicine, Justus-Liebig University, Giessen, Germany

G. R. SERJEANT
Director, MRC Laboratories (Jamaica), University of the West Indies, Kingston, Jamaica

B. A. SHAPIRO
Professor of Clinical Anesthesia, Northwestern University Medical School, Chicago, Illinois, USA

M. J. SHAPIRO
Associate Professor of Surgery and Anesthesiology, Director, Trauma Service, Department of Surgery, St Louis Health Sciences Center, St Louis, Missouri, USA

D.V. SHATZ
Assistant Professor of Surgery, University of Miami School of Medicine, Miami, Florida, USA

W. C. SHOEMAKER
LAC + USC Medical Center, University of Southern California, Los Angeles, California, USA

W. J. SIBBALD
The Richard Ivey Critical Care Trauma Center, London Health Sciences Center, Victoria Campus, London, Ontario, Canada

A. SIELENKÄMPER
Research Fellow, Program in Critical Care, University of Western Ontario, London, Ontario, Canada

T. SIMILOWSKI
Service de Pneumologie et Réanimation, Groupe Hospitalier Pitié-Salpêtrière, Paris, France

S. G. SIMONSON
Associate in Medicine, Duke University Medical Center, Durham, North Carolina, USA

M. SINGER
Director and Reader in Intensive Care, Bloomsbury Institute of Intensive Care Medicine, University College London Medical School, London, UK

A. K. B. SLADE
Department of Cardiology, Treliske Hospital, Truro, UK

R. N. SLADEN
Professor and Vice-Chairman, Department of Anesthesiology, and Director, Cardiothoracic-Surgical Intensive Care Unit, College of Physicians & Surgeons of Columbia University, New York, USA

P. SLEIGHT
Professor Emeritus of Cardiovascular Medicine, John Radcliffe Hospital, Oxford, UK

A. S. SLUTSKY
Professor of Medicine and Surgery, and Biomedical Engineering Division Director, Respiratory Division, University of Toronto, Canada

M. SMITH
Consultant Neuroanaesthetist and Honorary Senior Lecturer in Anaesthesia, Tavistock Surgical ITU, The National Hospital for Neurology and Neurosurgery, London, UK

N. J. SMITH
University Hospital, Nottingham, UK

R. A. SMITH
Associate Professor, College of Medicine, University of South Florida, Tampa, Florida, USA

T. W. SMITH (deceased)
Formerly Chief, Cardiovascular Division, Brigham and Women's Hospital, Harvard Medical School, Boston, Massachusetts, USA

E. M. SPENCER
Consultant in Anaesthesia and Intensive Care, Gloucestershire Royal NHS Trust, Gloucester, UK

C. STEIN
Director, Universitätsklinikum Benjamin Franklin, Klinik für Anaesthesiologie und Operative Intensivmedizin, Freie Universität Berlin, Germany

T. STERN
Chief, Psychiatry Consultation Service, Massachusetts General Hospital, and Associate Professor of Psychiatry, Harvard Medical School, Boston, Massachusetts, USA

M. STERNBACH
Associate Professor of Medicine, McMaster University, and Senior Physician, St Joseph's Hospital, Hamilton, Ontario, Canada

J. M. STEVENS
Consultant Neurologist, National Hospital for Neurology and Neurosurgery and St Mary's Hospital, London, UK

C. P. STOUTENBEEK
Director of Intensive Care, Academic Medical Center, Amsterdam, The Netherlands

G. STRATENWERTH
Basel, Switzerland

W. STROBEL
Department of Internal Medicine, University Hospital, Basel, Switzerland

S. SUN
Associate Professor, Institute of Critical Care Medicine, University of Southern California School of Medicine, Palm Springs, California, USA

P. M. SUTER
Chief, Surgical Intensive Care Unit, University Hospital of Geneva, Switzerland

A. SWAMI
Consultant, Neurosciences Critical Care Unit, Addenbrooke's Hospital, Cambridge, UK

M. SYDOW
Associate Professor of Anesthesiology, Department of Anesthesiology, Georg-August-Universität Göttingen, Germany

J. E. SZALADOS
Clinical Assistant Professor, Department of Anesthesiology, School of Medicine, University of North Carolina at Chapel Hill, North Carolina, USA

D. TAGAN
Attending Physician, Medical Intensive Care, University Hospital, Lausanne, Switzerland

W. TANG
Associate Professor and Vice President, Institute of Critical Care Medicine, University of Southern California School of Medicine, Palm Springs, California, USA

A. E. TAYLOR
Professor and Chairman, Department of Physiology, University of South Alabama College of Medicine, Mobile, Alabama, USA

H. TEN CATE
Department of Internal Medicine, Slotervaartziekenhuis, and Center for Hemostasis, Thrombosis, Atherosclerosis and Inflammatory Research, Academic Medical Center, Amsterdam, The Netherlands

L. G. THIJS
Professor of Medicine and Director, Medical Intensive Care Unit, Free University Hospital, Amsterdam, The Netherlands

L. THOMACHOT
Staff Anesthetist, Marseilles Medical School, University of Marseilles, France

F. D. THOMPSON
Consultant Nephrologist, University College London Hospitals, London, UK

R. C. TILLEY
Registrar in Intensive Care, Department of Anaesthesia and Intensive Care, Royal Brompton Hospital, London, UK

J.-F. TIMSIT
Senior Physician, Intensive Care Unit, Hôpital Saint Joseph, Paris, France

M. J. TOBIN
Professor of Medicine, Stritch School of Medicine, Loyola University of Chicago, Maywood, Illinois, USA

A. TORRES
Director, UVIR–Servei de Pneumologia, Hospital Clínic, Barcelona, Spain

B. J. TORTELLA
UMDNJ-New Jersey Medical School, Newark, New Jersey, USA

E. M. TRINIDAD
Research Fellow, Department of Neurosurgery, State University of New York Health Science Center, Booklyn, New York, USA

M. TRYBA
Department of Anaesthesiology, Intensive Care Medicine and Pain Therapy, City Hospital Kassel, University Teaching Hospital, Kassel, Germany

J. G. TYBURSKI
Assistant Professor of Surgery, Wayne State University, and Assistant Director–SICU, Detroit Receiving Hospital, Detroit, Michigan, USA

M. ULRICH
Renal/Pancreas Transplant Department, Buffalo General Hospital, Buffalo, New York, USA

G. VAN DEN BERGHE
Department of Intensive Care Medicine, University Hospital Gasthuisberg, Leuven, Belgium

H. VAN GOOR
Department of Surgery, University Hospital St Radboud, Nijmegen, The Netherlands

E. C. M. VAN GORP
Department of Internal Medicine and Department of Infectious Disease, Academic Medical Centre, Amsterdam, The Netherlands

P. V. VAN HEERDEN
Staff Specialist, Department of Intensive Care, Sir Charles Gairdner Hospital, Nedlands, Australia

H. K. F. VAN SAENE
Department of Medical Microbiology, University of Liverpool, UK

R. VIALET
Staff Anesthetist, Marseilles Medical School, University of Marseilles, France

J.-L. VINCENT
Department of Intensive Care, Erasme University Hospital/Free University Hospital, Brussels, Belgium

L. Y. VITOU
Fellow, Critical Care Medicine, McMaster University, Hamilton, Ontario, Canada

M. VRANCKX
Emergency Department, Erasme University Hospital, Brussels, Belgium

J. W. WAGNER
Division of Trauma–SICU, Henry Ford Hospital, Detroit, Michigan, USA

H. G. WAKELING
Visiting Professor, Division of Critical Care, Duke University Medical Center, Durham, North Carolina, USA

P. G. M. WALLACE
Consultant in Anaesthesia and Intensive Care, Western Infirmary, Glasgow, UK

J. D. WARD
Professor of Pediatric Neurosurgery, West Hospital, Richmond, Virginia, USA

C. A. J. WARDROP
Senior Lecturer and Honorary Consultant in Haematology, University of Wales College of Medicine, Cardiff, UK

B. WATSON
Regional Transplant Coordinator, North Thames NHS Executive, London, UK

D. WATSON
Senior Lecturer in Intensive Care and Anaesthesia, St Bartholomew's and the Royal London School of Medicine and Dentistry, London, UK

K. WAXMAN
Director of Surgical Education, Santa Barbara Community Health Care Services, Santa Barbara Cottage Hospital, Santa Barbara, California, USA

A. R. WEBB
Director of Intensive Care, University College London Hospitals, London, UK

J. A. WEIGELT
University of Minnesota, St Paul, Minnesota, USA

M. H. WEIL
Distinguished University Professor and President, Institute of Critical Care Medicine, University of Southern California School of Medicine, Palm Springs, California, USA

J. WENDON
Senior Lecturer and Honorary Consultant Physician, Institute of Liver Studies, King's College London, UK

C. WERNER
Professor of Anesthesiology, Department of Anesthesiology, Technical University Munich, Germany

R. A. WHITE
Chief, Vascular Surgery, Associate Chair, Department of Surgery, and Professor of Surgery, UCLA School of Medicine, Harbor-UCLA Medical Center, Torrance, California, USA

R. WHITTA
Locum Consultant, Intensive Care and Anaesthetics, Whipps Cross Hospital, London, UK

C. J. M. WHITTY
Senior Registrar, Hospital for Tropical Diseases, London, and Honorary Lecturer, London School of Hygiene and Tropical Medicine, London, UK

S. M. WILLATTS
Director, Intensive Therapy Unit, Bristol Royal Infirmary, Bristol, UK

R. WILLIAMS
Director, Institute of Hepatology, University College London Medical School, London, UK

A. P. R. WILSON
Consultant Microbiologist, University College London Hospitals, UK

E. P. WILSON
Coastal Surgical Associates, Chesapeake, Virginia, USA

R. F. WILSON
Professor of Surgery, Wayne State University, and Chief, Surgical Services, Detroit Receiving Hospital, Detroit, Michigan, USA

R. G. WOOLFSON
Consultant Nephrologist, Institute of Urology and Nephrology, University College London Hospitals, London, UK

J. A. WORRELL
Vanderbilt University Medical Center, Nashville, Tennessee, USA

G. J. WRATHALL
Consultant in Anaesthesia and Intensive Care Medicine, Frenchay Hospital, Bristol, UK

G. B. YOUNG
Professor of Neurology, Department of Clinical Neurological Sciences, University of Western Ontario, London, Ontario, Canada

J. E. ZIMMERMAN
Professor of Anesthesiology and Medicine, and Director, Intensive Care Research, George Washington University Medical Center, Washington, DC, USA

R. ZÜRCHER ZENKLUSEN
Medical Intensive Care, University Hospital (Insel), Bern, Switzerland

B. ZWISSLER
Clinic of Anesthesiology, University of Munich, Klinikum Grosshadern, Munich, Germany

1 Resuscitation

1.1 Respiratory management

1.1.1 Basic airway management

J. M. Field and P. J. F. Baskett

Key messages

- Phonation, detection of breath sounds, particularly using a stethoscope applied over the upper trachea and both lung fields, and the feel of breath on the hand are readily observable clinical signs.
- Techniques of airway opening and maintenance include posture, chin lift, and jaw thrust.
- In patients with suspected or proven cervical spine injury it is recommended that the cervical spine be protected by manual in-line axial stabilization.
- Devices for airway maintenance during adult life support include the oropharyngeal airway, the nasopharyngeal airway, the laryngeal mask airway, the pharyngotracheal lumen airway, and the Combitube esophageal–tracheal airway.

Introduction

Respiration is the basis of all life. Without oxygen, cellular mechanisms fail, resulting in biochemical and physiological dysfunction which progresses to cell death. Without timely intervention to restore oxygen to the tissues, injury becomes irreversible. Oxygen molecules journey from the atmosphere to the cells by being drawn via a ventilatory mechanism through patent airways to the alveoli, where diffusion then occurs across the alveolar–capillary–erythrocyte membranes. Oxygen is transported around the body primarily as oxyhemoglobin using the circulatory system. The recognition of a compromised airway and the institution of techniques to restore and maintain that airway is fundamental to resuscitation.

Assessment of the airway

Just as the airway may be intact, partially obstructed, or completely obstructed, respiratory effort may be present, obtunded, or absent. While the techniques of airway management are essentially the same whether or not there is respiratory effort, in the latter group it must be assumed that the airway is compromised and basic airway maintenance techniques must be instituted in conjunction with ventilatory support.

In addition to assessment of the airway, the presence of ventilatory effort should also be monitored. Phonation, detection of breath sounds, particularly using a stethoscope applied over the upper trachea and both lung fields, and the feel of breath on the hand are readily observable clinical signs. In the presence of spontaneous ventilatory effort an obstructed airway may present with paradoxical chest movements. As the diaphragm contracts the abdominal contents descend, causing the girth to increase, while at the same time the chest wall collapses and a tracheal tug is evident as the more mobile tissues of the neck are drawn down towards the thoracic inlet. This alternates with relaxation of the diaphragm and contraction of the abdominal muscles, reversing the abdominal and thoracic movements.

The most common form of airway compromise is snoring, which results from loss of pharyngeal tone as a consequence of a reduced level of consciousness. This is an example of inspiratory stridor, which is characteristic of supraglottic obstruction. In contrast, airway obstruction below the larynx is usually more evident during the expiratory phase (wheezing). Upper airway obstruction may also be caused, or exacerbated, by excessive salivation, gastric contents, blood, or a foreign body in the pharynx or laryngeal opening. These may obstruct the airway directly or indirectly, by stimulating laryngeal spasm, and should always be considered, as should a history of maxillofacial or neck trauma.

Techniques of airway opening and maintenance

Posture

The classical 'sniffing the morning air' position improves the upper airway by tensing the soft palate, and by lifting the tongue off the posterior pharyngeal wall and the epiglottis from the laryngeal opening. The position is the result of a combination of flexion of the neck and extension of the head on the neck.

The three-quarters prone and prone positions will likewise improve the upper airway by the effect of gravity on the soft tissue structures at the base of the tongue. However, access to the airway in the prone position is compromised.

Chin lift and jaw thrust

The genioglossus muscle attaches the tongue to the mandible. Therefore maneuvers which displace the mandible anteriorly will also pull the tongue off the posterior pharyngeal wall. Upper airway patency can be improved further by lifting the chin forward using a finger and thumb. An alternative is to thrust the jaw forwards with a finger placed on either side just proximal to the angle of the mandible (Fig. 1).

It should be noted that many patients may have coincidental or associated nasal airway obstruction. Opening the mouth at the same time as applying the above techniques may further improve upper airway patency.

Cervical spine injury

In patients with suspected or proven cervical spine injury the above maneuvers may exaggerate any neurological insult. Therefore it is recommended that the cervical spine be protected by manual in-line axial stabilization, maintaining the head and neck in the neutral position (Fig. 2). Jaw thrust will probably be the optimal technique for improving the airway with the least likelihood of distracting the cervical spine.

Obstruction of the upper airway by foreign material

The emphasis of management of obstruction of the upper airway by foreign material will depend on the level of consciousness and the severity of airway obstruction. If air movement appears satisfactory, spontaneous respiratory effort and coughing should be encouraged as these will tend to clear any obstructing material. Supplementary oxygen therapy will help to compensate for mildly compromised alveolar ventilation. Back blows and abdominal thrusts (Heimlich maneuver) may help to dislodge impacted matter by augmenting forced expiration. Finger sweeps, using a gloved finger wrapped in a handkerchief or swab, should be reserved for those who do not have a gag reflex or the ability to bite. The optimal technique is the use of direct vision (usually with a laryngoscope) and suction for liquid or Magill's offset forceps for solid matter.

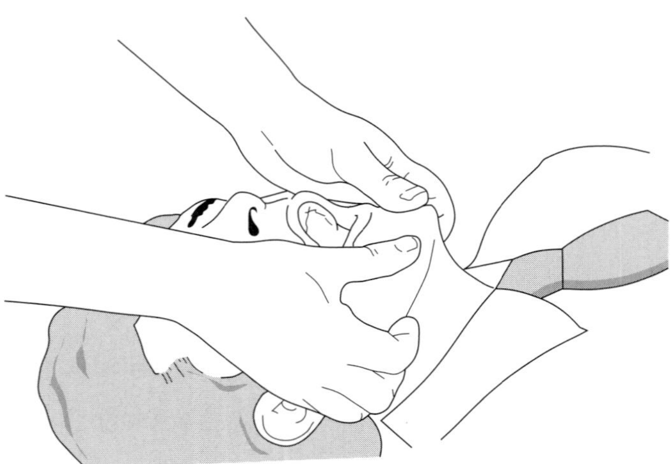

Fig. 1 Jaw thrust to open the airway.

Devices for airway maintenance during adult life support

The oropharyngeal airway

The oropharyngeal airway can serve three purposes.

1. Upper airway obstruction may be relieved by anterior displacement of the tongue.
2. Since it is an oral airway, it opens up an oral passage for air movement.
3. The incorporated bite block will protect the airway (and any other oral airway adjunct such as a laryngeal mask airway or oral tracheal tube) if masseter muscle tone recovers.

If the airway has been opened using the chin lift or jaw thrust techniques described above, insertion of an oropharyngeal airway may free up a pair of hands.

Clenched jaws, vulnerable dentition, and active bleeding in the oro/hypopharynx may preclude insertion of an oral airway. Active glossopharyngeal reflexes will cause the patient to reject the airway at the very least, and might even promote vomiting which would increase the risk of pulmonary aspiration of stomach contents. It is better to intubate these patients if ventilation is compromised or consider a nasopharyngeal airway.

The nasopharyngeal airway

The nasopharyngeal airway is inserted through the nose and advanced until the tip lies in the hypopharynx proximal to the epiglottis. Traumatic insertion is not unusual and bleeding into the pharynx may exacerbate airway compromise. However, it may be an invaluable airway adjunct in patients with clenched jaws, is an alternative to the oropharyngeal airway if the latter is contraindicated or of limited benefit, and is often better tolerated (once in place) than the oropharyngeal airway if glossopharyngeal reflexes are present. Midfacial fractures, base of skull fractures, and bilateral nasal

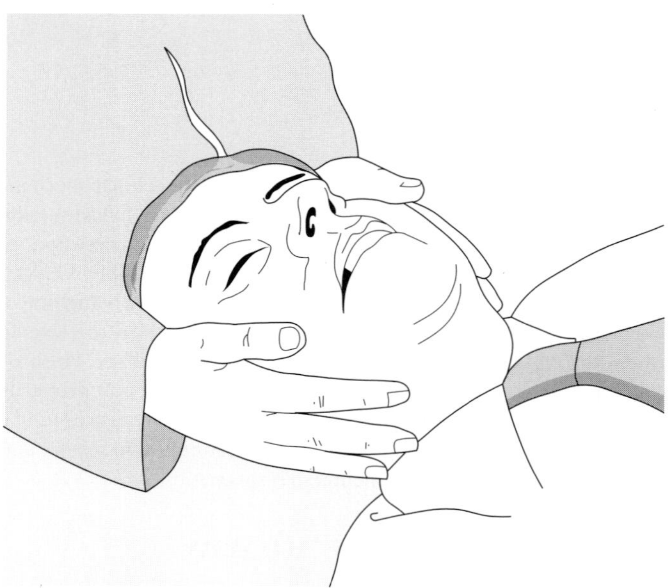

Fig. 2 Manual in-line stabilization of the head and neck.

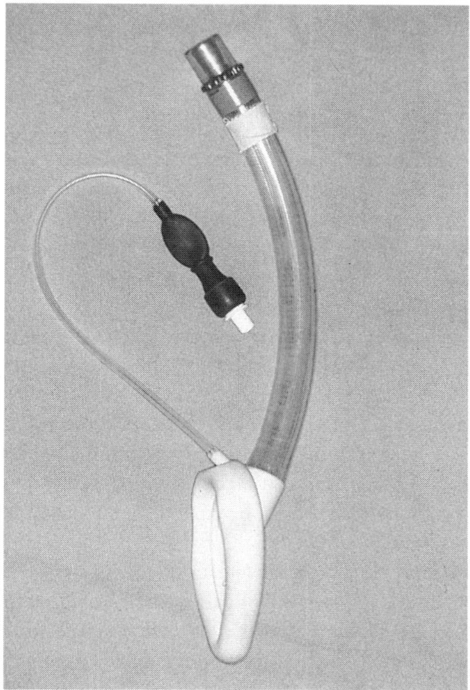

Fig. 3 The laryngeal mask airway.

passage obstruction or deformity would contraindicate use of the nasopharyngeal airway.

The laryngeal mask airway

The laryngeal mask airway (Fig. 3) consists of an inflating oval cuff that sits obliquely over the laryngeal inlet, lying from the base of the tongue anteriorly to the upper esophagus posteriorly. Arising proximally from this is a wide-bore tube which protrudes from the mouth and can be connected to standard airway circuitry. While it is not an alternative to tracheal intubation, the laryngeal mask airway provides a clear and secure airway through which gentle positive-pressure ventilation can be applied. Although airway protection is not guaranteed, pulmonary aspiration is very uncommon. It is relatively easy to insert with minimal training or equipment. Moreover, the laryngeal mask airway may be a satisfactory or even lifesaving alternative if tracheal intubation is difficult or potentially hazardous. The laryngeal mask airway is made in a variety of sizes to accommodate all patients from the neonate to the large adult (Table 1).

Rejection, straining, coughing, and laryngeal spasm may occur in patients with active reflexes. Incorrect placement and obstruction may result from folding of the tip of the cuff, folding down of the epiglottis at insertion, or rotation of the laryngeal mask airway. These can usually be corrected by careful reinsertion of the laryngeal mask airway using the manufacturer's recommended technique. Persistent leakage around the cuff may be due to incorrect insertion techniques or the use of an incorrectly sized laryngeal mask airway, inadequate cuff inflation, or excessive lung inflation pressure (which should not exceed 20 cmH_2O).

The laryngeal mask airway can also be used to facilitate difficult tracheal intubation.

Table 1 Sizing of the laryngeal mask airway

Patient	Laryngeal mask airway size
Neonate–1 year	1
Child aged 1–3 years	1½
Child aged 3–7 years	2
Child aged 7–10 years	2½
Child aged 10–small adult	3
Normal adult	4
Large adult	5

The pharyngotracheal lumen airway

In unconscious patients with absent glossopharyngeal and laryngeal reflexes the pharyngotracheal lumen airway can be used to provide an airway which may give protection from pulmonary aspiration similar to the laryngeal mask airway. It is introduced blindly into the mouth and consists of two tubes: the longer tube passes into the upper esophagus where its cuff is inflated to protect the laryngeal inlet against regurgitated gastric contents, while the shorter tube lies with its tip above the glottic opening where its large cuff obliterates the hypopharynx. Ventilation is through the short tube, allowing a separate gastric tube to be passed through the longer tube. If the long tube enters the trachea, it can then be used as an orotracheal tube. It is relatively difficult to introduce but is an alternative if tracheal intubation is difficult or precluded.

The Combitube esophageal–tracheal airway

The Combitube (Fig. 4), which is introduced blindly into the mouth, is a double-lumen tube with a small-volume distal cuff and a large-volume (100 ml) proximal cuff. One channel opens distally, while the other has a blind end with side-openings between the two cuffs. If the tube enters the trachea, the distal cuff is inflated and the distally opening channel is used as if for conventional orotracheal intubation. If the tube enters the esophagus, the distal cuff is inflated to protect against regurgitation and pulmonary aspiration, the proximal cuff is inflated to form an air-tight seal in the hypopharynx, and ventilation is via the blind-ended channel. Inflation pressures may be high owing to the small size of the side-openings.

Tracheal intubation

Tracheal intubation using the oral or, more rarely, the nasal route remains the gold standard for airway management.

Conclusions

Without an intact airway, further resuscitative measures, including ventilation, are usually futile. Recognition of a compromised, or potentially compromised, airway is crucial. The airway must be established early using the techniques and aids described. Experience is

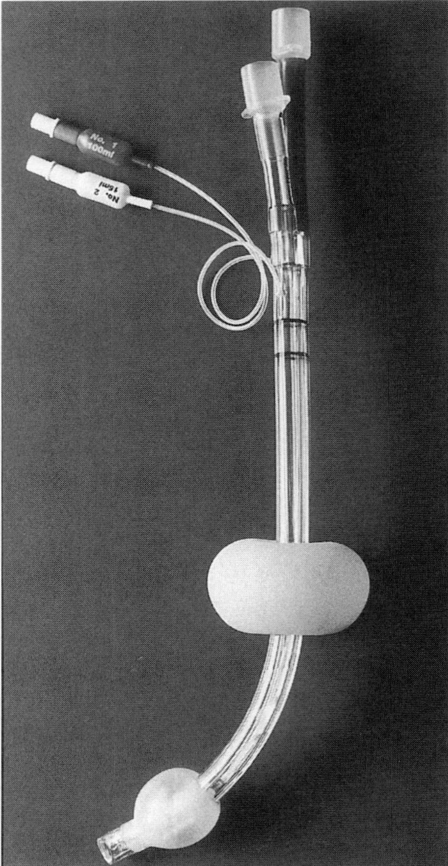

Fig. 4 The Combitube esophageal–tracheal airway.

required to master the use of adjuncts such as laryngeal mask airways and tracheal tubes. Once the airway has been established and oxygenation restored, its maintenance may require further intervention either with the devices described above or with more advanced techniques discussed elsewhere.

1.1.2 Indications for intubation

R. F. Armstrong

Key messages

- Intubation offers four advantages:
  - protection of the airway
  - relief of obstruction
  - tracheal toilet
  - enables mechanical ventilation.

Introduction

Indications for intubation are airway protection, access to secretions, bypassing obstruction, control of respiratory function, anesthesia, and transportation.

Airway protection

Protective laryngeal reflexes may be impaired if there is any reduction in level of consciousness. In the absence of an adequate cough reflex, aspiration of gastric contents or blood may contaminate the lung, or airway obstruction may occur leading to hypoxia and hypercarbia.

Head injury, brain tumor, cerebrovascular accident, drug overdose, epilepsy, or syncope are commonly associated with a failure of protective reflexes and thus a need for intubation. Occasionally bulbar palsy or neuromuscular weakness may result in impairment of airway control, necessitating intubation.

Loss of airway control is associated with a Glasgow Coma Scale score of 8 or less. Clinical signs of partial airway obstruction are noisy breathing or snoring. Total obstruction is soundless as no air traverses the larynx. Both partial and complete obstruction are associated with a characteristic paradoxical respiratory pattern in which the chest moves inward on inspiration with marked indrawing of the suprasternal notch. Signs of an obstructed airway may not be noticed by the casual observer.

Patients with weakness (e.g. Guillain–Barré syndrome or myasthena gravis), can be difficult to assess. A vital capacity below 15 ml/kg may be associated with ineffective cough and protective reflexes. If in doubt it is better to err on the cautious side and intubate.

Access to secretions

Retention of sputum occurs because the cough reflex is suppressed by pain or excessive sedation or the cough mechanism is inadequate. This may be because a glottic leak prevents the development of high pressures in the trachea prior to an explosive cough (e.g. bulbar palsy) or there is an inability to generate rapid gas flows (e.g. neuromuscular weakness). This situation may be encountered in intensive care unit (ICU) patients after prolonged intubation has rendered the larynx temporarily incompetent. As a result, pulmonary secretions accumulate in the respiratory tract. In these cases endotracheal intubation will allow protection of and access to the airway so that accumulated secretions can be removed. When respiratory function is otherwise satisfactory a minitracheostomy may be a better solution.

Bypassing obstruction

Trauma, foreign bodies, laryngotracheal inflammation, acute anaphylaxis, and inhalation of hot gases, chemicals, smoke, or steam may produce sufficient damage or swelling of the airway to create airway obstruction. Commonly, acute dyspnea develops with inspiratory and expiratory wheezing, dysphonia, and dysphagia. Once 50 per cent of the laryngeal inlet is obstructed by edema, stridor develops. If stridor is present, intubation or placement of a surgical airway becomes urgent. In the presence of facial, circumferential neck, or full-thickness head burns, intubation is recommended before airway obstruction develops.

Control of respiratory function

In the event of clinical respiratory distress or respiratory failure where immediate correction of the cause is not possible, intubation will be necessary as a prelude to ventilatory support. Common indications are a rising $P\text{a}CO_2$ sufficient to cause severe acidosis (pH < 7.2), a

Pao_2 less than 5.5 kPa (40 mmHg) on room air or less than 10 kPa (75 mmHg) on an Fio_2 of 0.4, or a vital capacity less than 15 ml/kg.

Anesthesia and transportation

When transfer to the operating theater is not practicable, anesthesia for surgical procedures may be performed in the ICU. Indications for intubation in these circumstances will include a full stomach, aspiration risk, morbid obesity, impaired respiratory function, or the need for positions incompatible with mask anesthesia.

Whenever transportation of the critically ill patient is contemplated (e.g. CT scanning), intubation and ventilation should be considered for the duration of the transfer. This is particularly important whenever respiratory function is marginal for patients unable to breathe lying flat and for patients normally on continuous positive airway pressure.

Intubation hazards

Intubation is not always easy, it has risks, and it contributes to anesthetic complications and death (Caplan *et al.* 1990). This applies not only to the operating room but also to the ICU. Table 1 helps to identify potentially difficult intubations.

Table 1 Factors associated with difficult or dangerous intubation

History of difficult airway
Vomiting/regurgitation risk
Obvious pathological/anatomical abnormalities
Poor head extension
Immobile temporomandibular joint
Visualization of posterior pharyngeal wall not possible
Thyromental distance <6.5 cm
Body mass index >30*
Contraindications to succinylcholine (suxamethonium) (e.g. burns, abdominal sepsis, neuropathy, hyperkalemia)

*Body mass index=weight (kg)/[height (m)]^2.
Modified with permission from King (1994).

References

Caplan, R.A., Posner, K.L., Ward, R.J., and Cheney, F.W. (1990). Adverse respiratory events in anesthesia: a closed claims analysis. *Anesthesiology*, 72, 828–33.

King, T.A. (1994). Emergency control of the airway. In *Recent advances in anesthesia and analgesia*, Vol. 18 (ed. R.S. Atkinson and A.P. Adams), p. 59. Churchill Livingstone, New York.

1.1.3 Restoring ventilation

R. F. Armstrong

Key messages

- Respiratory rate and patterns are important clinical signs of ventilatory failure.
- Stridor is a (too) late sign of airway obstruction.
- Bag-and-mask ventilation, insertion of a laryngeal mask airway, and cricothyrotomy are valuable manual skills in the intensive care unit.

Introduction

Failure of ventilation occurs as a result of a broad range of disorders: cardiorespiratory arrest; respiratory depression (drugs, head injury, obesity); respiratory obstruction (coma, infection, edema, foreign body, trauma); respiratory failure (infection, edema, bronchospasm, weakness).

Cardiorespiratory arrest

Once cardiac or respiratory arrest has occurred, early ventilation is a priority. Expired air ventilation is best performed with the aid of a specially designed pocket mask. In this type of system the rescuer holds the mask tightly to the patient's face and, using thumbs and forefingers, lifts the angles of the jaw to open the airway and blows into the inspiratory port. A unidirectional valve allows the patient's expired air to vent to the atmosphere.

Tidal volumes of 800 to 1000 ml, rates of 10 to 12 breaths/min, and inspiratory flow rates of 30 l/min are recommended for adults.

The risk of gastric inflation and regurgitation can be minimized by avoiding high inflation pressures, allowing full expiration to occur, and having an assistant apply cricoid pressure if necessary.

Although expired air resuscitation is an acceptable practice for the non-medical bystander, ventilation using a self-inflating bag and mask is more effective. Bag-and-mask systems comprise a self-inflating bag with an inlet valve at one end to entrain air–oxygen mixtures and a one-way patient valve at the other. This has a 15-mm connector which allows connection to a mask, endotracheal adaptor, endotracheal tube, laryngeal mask, or tracheostomy. At oxygen flow rates of 5 l/min through the inlet valve, concentrations of 40 to 50 per cent oxygen can be obtained. If higher inspired oxygen concentrations are needed, a reservoir bag can be attached to the inlet and oxygen flow rates of 8 to 10 l/min selected. When the reservoir bag is not used, caution is recommended at oxygen flow rates in excess of 5 l/min because overfilling of the self-inflating bag can occur causing the patient valve to stick in the inspiratory position. This allows expired air to re-enter the bag with consequent rebreathing of carbon dioxide. Positive end-expiratory pressure valves can be connected to the patient valve if necessary. During ventilation with this system, inspiratory–expiratory time (I:E) ratios of 1:3 are recommended with respiratory rates and chest excursion appropriate for patient size and age (Baskett 1989).

An alternative to bag-and-mask ventilation is the insertion of a laryngeal mask airway. Since its introduction (Brain 1983), the laryngeal mask airway has been used extensively and has proved easy and quick to insert. Ventilation of the lungs is reliably achieved, although controversy exists about the degree of protection against regurgitation. It can be invaluable in the intensive care unit (ICU) as an intermediate step between intubation and unsupported spontaneous breathing (Arosio and Conci 1995).

Respiratory depression

Slow or shallow breathing is the hallmark of respiratory depression. Respiratory rates below 10 breaths/min mandate medical assessment. Treatment with oxygen therapy should be instituted, and monitoring of oxygen saturations or arterial blood gases arranged. When respiratory rates are below 8 breaths/min, the cause of the depression should be reversed. In the case of opiates, the competitive opioid antagonist naloxone is used. As its effect following intravenous administration may only last for 1 h, additional doses may be necessary. Unfortunately this drug reverses analgesia. An alternative is doxapram (1–1.5 mg/kg) which increases respiratory rate and tidal volume without affecting analgesia.

Where pain control with opiates has been difficult to achieve, the clinician may opt not to treat low respiratory rates but, rather, reduce the opiate dose and await recovery. In these circumstances patient nursing and medical surveillance, maintenance of oxygen saturations above 90 per cent, and exclusion of significant carbon dioxide retention are critical.

Low tidal volumes are difficult to assess clinically. They can be measured using a respirometer (e.g. Wright's respirometer) and face mask, and should not be below 4 ml/kg. Persistent breathing at low tidal volumes will result in pulmonary atelectasis. In these circumstances continuous positive airway pressure (CPAP) will help maintain lung volume and $Pa\text{O}_2$ but will not reverse hypercarbia, and thus naloxone or doxapram may be necessary.

Occasionally respiratory depression may be the result of administration of oxygen concentrations which are too high. There is evidence that, in patients with long-standing hypercarbia, hypoxia becomes the predominant stimulus to ventilation via its effect on the carotid body. High inspired oxygen concentrations may then lead to respiratory depression. In these patients the use of controlled oxygen therapy with oxygen concentrations of 24 to 28 per cent is recommended, utilizing the steep gradient of the oxyhemoglobin dissociation curve to maximize oxygen content without reaching depressant levels of oxygen partial pressures.

Airway obstruction

Airway obstruction presents a major hazard in the comatose patient. The most common cause is the tongue, which slips backwards into the pharynx, blocking the airway. Characteristically, this produces noisy breathing or snoring. Laryngeal edema or tracheal obstruction is associated with inspiratory stridor. Although complete obstruction is silent, respiratory movements continue, often becoming more marked and labored, and characterized by a paradoxical 'seesaw' respiratory pattern in which the chest wall and suprasternal fossa are drawn inwards on inspiration.

Treatment of airway obstruction is by chin lift and elevation of the lower jaw, drawing the tongue out of the pharynx. In case of failure to relieve the obstruction or when injury to the cervical spine contraindicates neck extension, a jaw thrust technique is recommended. In this method the attendant pushes the lower jaw forward using pressure applied behind the angles of the mandible. If this is unsuccessful, insertion of an oral airway or a 6-mm nasopharyngeal tube may be needed.

When airway obstruction occurs at the level of the epiglottis, vocal cords, or trachea, intubation or a surgical airway is necessary. If intubation fails and oxygenation is threatened, catheter cricothyrotomy may be carried out using a 14G cannula introduced through the cricothyroid membrane. Once in place (air can be aspirated) and secured firmly, ventilation through the cannula can be carried out using a Sanders injector or high-frequency ventilator at rates of 40 to 50 breaths/min. There are other alternatives, such as connecting the male end of a 7.5-mm endotracheal tube adaptor to a plungerless 2-ml syringe which is then attached to the hub of the 14G cannula. Connection to an anesthetic breathing circuit then allows ventilation by either bag compression or use of the oxygen flush valve (Patel 1983) In these circumstances complete obstruction above the level of the cricothyrotomy may lead to barotrauma as expired gas cannot vent. This should be guarded against by insertion of another cannula or conversion to a wider-bore minitracheostomy or percutaneous tracheostomy as soon as control of oxygenation has been regained. While trying to overcome obstruction, it is essential that oxygen is presented to the airway at all times and that oxygen saturation and ECG monitoring are ongoing.

Respiratory failure

Failure to maintain oxygen saturations above 90 per cent or normal blood gases may dictate the need for respiratory support. Oxygen therapy of known concentration should be used initially. Once the $Pa\text{O}_2$ falls to less than 10 kPa on an FiO_2 of 0.4 or the respiratory rate exceeds 30 breaths/min, CPAP should be introduced, using a tightly fitting padded face mask with sufficiently high flows to maintain a positive pressure of initially 5 cmH_2O throughout the respiratory cycle. The stomach may have to be decompressed. Failure to reverse poor oxygenation, irreversible hypercarbia with a pH below 7.2, or further respiratory deterioration is an indication for intubation and mechanical ventilation.

Ventilator settings

After the ventilator function has been checked using a reservoir bag on the delivery tubing, the standard ventilator settings are as follows: tidal volume, 7 to 10 ml/kg; minute volume, 6 to 10 l/min; respiratory rate, 12 to 15 breaths/min; FiO_2, 1.0; peak inspiratory pressure, 40 cmH_2O; I:E ratio, 1:2; alarms on.

Once the patient is on the ventilator, arterial cannula insertion with blood gas measurement will allow further ventilatory manipulation (Schuster 1990). Persistent hypoxemia or high airway pressures should draw attention to the possibility of endobronchial intubation or pneumothorax.

References

Arosio, E.M. and Conci, F. (1995). Use of the laryngeal mask airway for respiratory distress in the intensive care unit. *Anaesthesia*, **50**, 635–6.

Baskett, P.J. (1989). *Resuscitation handbook*. J.B. Lippincott, Philadelphia, PA.

Brain, A.I.J. (1983). The laryngeal mask—a new concept in airway management. *British Journal of Anaesthesia*, **55**, 801–5.

Patel, R. (1983). Systems for transtracheal ventilation. *Anaesthesiology*, **59**, 165.

Schuster, D.P. (1990). A physiologic approach to initiating, maintaining, and withdrawing mechanical ventilatory support during acute respiratory failure. *American Journal of Medicine*, **88**, 268–78.

1.2 Circulatory management

1.2.1 Pathophysiology and causes of cardiac arrest

N. Nichol and C. Robertson

Key messages

- The most common predisposing factor in adult cardiac arrest is coronary heart disease.
- The initiating arrhythmias are most often ventricular fibrillation or pulseless ventricular tachycardia.
- In younger patients, asystole is relatively more common as an initiating rhythm and reflects differing etiologies.
- Correction of the precipitating factors is particularly important in electromechanical dissociation.

Epidemiological aspects

Cardiac arrest is the absence of cardiac pumping activity, characterized by the clinical features of pulselessness and loss of consciousness. It can be the common endpoint of many primary and/or secondary conditions affecting the heart.

Sudden unexpected cardiac arrest in adults has an incidence of 0.1 to 0.2 per cent in the general population, with 80 per cent of cases due to coronary heart disease. The incidence of cardiac arrest increases exponentially with age, reflecting the escalating frequency of coronary heart disease, but at all ages men are at greater risk than women. For coronary heart disease populations, hereditary factors producing hypertension, diabetes, lipid abnormalities, and male-pattern baldness predispose to cardiac arrest. Racial characteristics are important, with increased risks in colored populations, although the reasons for this are unclear. Cigarette smoking is the major acquired (and preventable) risk factor, although sedentary occupation and diet also play significant roles.

Although acute and chronic alcohol abuse is associated with an increased frequency of arrhythmias, moderate intake may have a protective effect on coronary heart disease and cardiac arrest by increasing 'protective' lipid concentrations. Other poorly understood factors may be involved in cardiac arrest. There is a circadian variation, with an excess of events occurring in the early morning (Muller and Mangel 1994). Mental and/or physical stress effected through autonomic activity, along with altered platelet activity, is suggested as a cause. Holter monitoring shows an increased incidence of ST depression (implying ischemia) between 12 midnight and 12 noon. There is some evidence to suggest that this can be prevented or modified with β-blockers and/or aspirin. A linkage between stress and cardiac arrest is popular in the media but the evidence is tenuous, although the 1994 Northridge earthquake reportedly caused an excess of sudden cardiac deaths, with a reduction in the following week, suggesting that emotional stress (independent of physical exertion) can be a triggering factor in predisposed individuals.

Causes of cardiac arrest that are unrelated to coronary heart disease include many disparate conditions, but drugs, trauma, infections, and cardiomyopathies are the principal causes. These conditions are relatively more common in younger patients and provoke cardiac arrest by producing increased susceptibility to malignant ventricular arrhythmias. Cocaine, a 'recreational' drug that increases myocardial sensitivity to catecholamines by blocking their presynaptic uptake, is one example.

Non-penetrating blunt myocardial injury, such as being struck on the chest by cricket balls, baseballs, or ice-hockey pucks, can lead to cardiac arrest without producing identifiable cardiac structural damage or in the absence of pre-existing pathology (commotio cordis). The etiology is believed to involve arrhythmias induced by relatively minor myocardial trauma occurring at an electrically vulnerable phase of ventricular excitability.

The cardiac causes of sudden death are listed in Table 1.

Electrophysiological characteristics of cardiac arrest

Cardiac arrest results either from the development of an arrhythmia which is incompatible with cardiac output, such as pulseless ventricular tachycardia, ventricular fibrillation, or asystole, or from a rhythm that is compatible with a cardiac output but in which no output occurs, termed electromechanical dissociation or pulseless electrical activity.

ECG monitoring suggests that the initiating arrhythmia in 75 per cent of cases of cardiac arrest is ventricular tachycardia or ventricular fibrillation. These malignant ventricular arrhythmias may be heralded by premature ventricular contractions or monomorphic ventricular tachycardia. With time ventricular tachycardia progresses to ventricular fibrillation and eventually to asystole, reflecting the consumption of myocardial high-energy phosphate stores, hypoxia, and progressive acidosis.

Asystole is a more common presenting rhythm in patients who are less than 17 years old, reflecting different underlying causes of cardiac arrest in this age group (Safranek *et al.* 1992). Asystole may also be provoked by sudden excessive vagal activity. Recognized precipitants include facial immersion in cold water, carotid sinus pressure, and stimulation of the peritoneum, cervix, or anus.

Table 1 Cardiac causes of sudden death

Coronary heart disease
Atherosclerotic
Ischemia/infarction
Injury
Congenital
Embolic
Vasculitic
Dynamic occlusion
Spasm/bridges
Hypertrophy
Primary
Hypertrophic cardiomyopathy
Secondary
Pressure/volume overload
Heart failure
Acute
Chronic
Congestive
Dilated cardiomyopathies
Myocarditis
Infective
Inflammatory
Infiltrative disorders
Neoplasms
Restrictive cardiomyopathiẽs
Valvular heart disease
Valve stenosis or obstruction (e.g. atrial myxoma)
Endocarditis
Non-infectious acute valvular regurgitation
Congenital heart disease
Presurgical or non-surgical
Following surgical correction
Electrophysiological disorders
Conduction system disease
Long QT syndromes
Ventricular fibrillation without overt structural heart disease
Accessory pathways
Pericardial tamponade

Reproduced with permission from Myerburg and Castellanos (1992).

The pathogenesis of cardiac arrhythmia initiation

The development of an arrhythmia is believed to result from the superimposition of a 'transient' factor on a pre-existing or static factor. However, an arrhythmic death may occur in the absence of clinically detectable cardiac disease. Only 20 per cent of cardiac arrest patients with coronary heart disease are subsequently shown to have developed an acute myocardial infarct with an acute plaque fissure causing platelet aggregation and subsequent thrombus formation. Transient myocardial ischemia–reperfusion causes the transient factor in the remainder, and in the presence of left ventricular hypertrophy or a previously healed myocardial infarction is particularly liable to be arrhythmogenic.

A model relating the transient and static factors is shown in Fig. 1. It incorporates the chance development of ventricular tachycardia or fibrillation due to randomly occurring premature ventricular contractions (electrogenic component) (Myerburg *et al.* 1989).

The cellular electrophysiological changes and the tendency for arrhythmias induced by acute ischemia develop within seconds. An initial peak relates to early cell injury with calcium and potassium shifts and developing electrochemical gradients as a direct result of ischemia. The second peak may be due to reperfusion injury or to different responses to ischemia in epicardial or endocardial cells, with the creation of an arrhythmogenic substrate. These changes promote disorders of myocardial electrical impulse conduction with local blocks and re-entry circuits.

Arrhythmias can result from disorders of electrical impulse formation (automaticity), disorders of electrical conduction, or a combination of both. After acute infarction, arrhythmias tend to be re-entrant in character in the first 30 min. Later, disorders of automaticity supervene, but after 3 days re-entrant arrhythmias again predominate.

The vulnerable period at which a premature ventricular contraction may trigger ventricular tachycardia or fibrillation is at the end

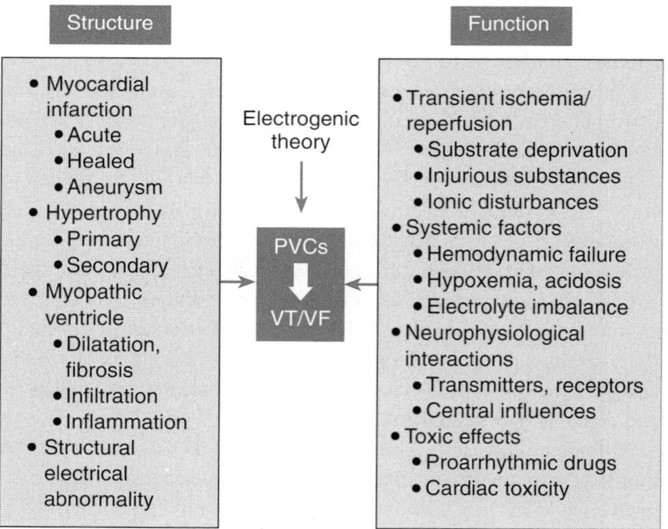

Fig. 1 Biological model of sudden cardiac death. Short- or long-term structural abnormalities and functional modulations interact to influence the propensity for premature ventricular contractions (PVCs) to initiate a ventricular tachycardia or fibrillation (VT/VF). The four major categories of structural abnormalities may be influenced by one or more functional events, as shown. (Reproduced with permission from Myerburg *et al.* (1989).)

(phase 3/4) of the myocardial action potential. In clinical practice this vulnerability is demonstrated by the association of the 'R on T' phenomenon with the development of ventricular arrhythmias.

Conduction abnormalities

Most ventricular tachyarrhythmias result from re-entrant phenomena. The normal progress of electrical impulses may be obstructed by an area of abnormal (commonly ischemic) myocardium exhibiting unidirectional block (Binah and Rosen 1992). The impulse travels round this zone and returns through an area of slow conduction, at which time repolarization has occurred and a sustained re-entrant rhythm can occur. If a single circuit with ordered re-entry is established, ventricular tachycardia results; if multiple circuits (random re-entry) are involved, ventricular fibrillation occurs (Fig. 2).

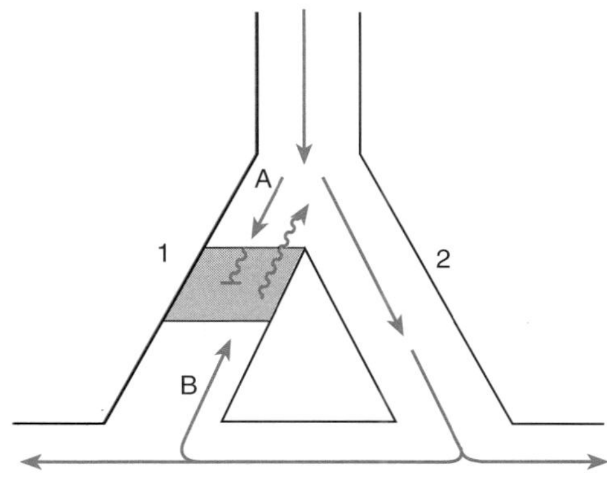

Fig. 2 The diagram shows a Purkinje fiber bundle dividing into two branches which are attached to ventricular muscle. The area at 1 (green shading) shows a zone of slow conduction and unidirectional block. The wave of depolarization travels normally down fiber 2 but cannot traverse fiber 1. The wave is conducted back up fiber 1 in a retrograde fashion, and may cause a circular or re-entrant rhythm if the proximal part of the loop is not refractory. (Reproduced from Binah and Rosen (1992).)

Abnormalities of automaticity

Every myocardial cell has intrinsic pacemaker activity (automaticity) with the potential to initiate electrical impulses. The varying rates at which the threshold potential is reached normally allows a co-ordinated progressive depolarization, beginning in the SA mode. Transient factors may alter the normal hierarchy of automaticity, either producing an accelerated rate (sinus tachycardia, escape rhythm, accelerated atrioventricular nodal rhythms) or triggering abnormal depolarizations (oscillations of phase 3/4 of the action potential may reach threshold level characterized by conditions producing the long QT syndrome (Table 2), causing early or late after-depolarizations).

Table 2 Prolonged repolarization syndromes associated with cardiac arrest

Congenital
Jervel and Lange–Nielson
Romano–Ward
Electrolyte disturbance
Hypokalemia
Hypomagnesemia
Hypocalcemia
Drugs and poisons
Quinidine (and other class Ia agents)
Amiodarone (and other class III agents)
Amitriptyline
Chlorpromazine, terfenadine
Erythromycin
Organophosphates
Miscellaneous
Bradycardia
Central nervous system disease
Mitral valve prolapse
Acute myocardial infarction
Prolonged fasting/liquid protein diets
Cardiac
Contusion
Concussion

Table 3 Causes of electromechanical dissociation

Respiratory insufficiency or arrest
Mechanical
Pump failure (myocardial infarction)
Circulatory obstruction (massive pulmonary embolism, myxoma)
Cardiac rupture
Hypovolemia
Pericardial tamponade
Tension pneumothorax
Metabolic
Hypocalcemia
Hypo/hyperkalemia
Hypothermia
Hypoglycemia
Severe acidosis
Drugs
Beta-blockers
Calcium-channel blockers
Digoxin

Early after-depolarizations are associated with class IA anti-arrhythmic drugs, acidosis, hypokalemia, hypoxia, and catecholamines. Delayed after-depolarizations are associated with digoxin therapy and postmyocardial infarction.

Electromechanical dissociation

In contrast with arrhythmic causes, electromechanical dissociation causing cardiac arrest has very different mechanisms. The electrical activity of the heart may be near normal, or at least compatible with cardiac output. However, the precipitating cause prevents this electrical activity from producing output.

The causes are summarized in Table 3. Since the primary defect is often unrelated to the myocardium, treatment efforts are directed to the specific correction, for example volume replacement for hypovolemia, decompression of tension pneumothorax, etc.

References

Binah, O. and Rosen, M.R. (1992). Mechanisms of ventricular arrhythmias. *Circulation*, **85** (Supplement 1), I-25–31.

Muller, J.E. and Mangel, B. (1994). Circadian variation and triggers of cardiovascular disease. *Cardiology*, **85** (Supplement 2), 3–10.

Myerburg, R.J. and Castellanos, A. (1992). In *Heart disease: a textbook of cardiovascular medicine* (4th edn) (ed. E. Braunwald), p. 756. W.B. Saunders, Philadelphia, PA.

Myerburg, R.J., Kessler, K.M., Bassett, A.L., and Castellanos, A. (1989). A biological approach to sudden cardiac death: structure, function and cause. *American Journal of Cardiology*, **63**, 1512–16.

Safranek, D.J., Eisenberg, M.S., and Larsen, M.P. (1992). The epidemiology of cardiac arrest in young adults. *Annals of Emergency Medicine*, **21**, 1102–6.

1.2.2 Cardiac massage

S. Sun and M. H. Weil

Key messages

- Techniques of conventional precordial compression generate a cardiac output that is typically less than 25 per cent of normal.
- The alternative interventions may be classified as modifications of chest compression, intravascular techniques, and open-chest cardiac massage.
- There is no persuasive evidence that modifications of chest compression, including interposed abdominal compression, vest cardiopulmonary resuscitation, active compression–decompression, and phased chest and abdominal compression–decompression, have improved the outcome of cardiac arrest.
- Although technically more demanding, intravascular methods, including ascending aortic balloon occlusion and extracorporeal circulation, and open-chest cardiac massage are currently hemodynamically more effective interventions.

Introduction

The primary management of global myocardial ischemia of cardiac arrest is re-establishment of myocardial and systemic blood flow. The discovery by Kouwenhoven *et al.* (1960) that myocardial and systemic blood flow could be generated by external compression of the chest ushered in the era of modern cardiopulmonary resuscitation. External chest compression became the standard of mechanical interventions intended to generate circulation during cardiac arrest.

Unfortunately, standard techniques of manual or mechanical external chest compression generate cardiac outputs of less than 30 per cent of normal; coronary blood flows are correspondingly reduced. Therefore alternative techniques have been investigated in an effort to increase cardiac output and both coronary and cerebral blood flows. Newly proposed interventions can be classified as either modifications of chest compression or the use of intravascular techniques. In addition, open-chest cardiac massage continues as an appropriate option. Mechanical interventions for maintaining systemic perfusion during cardiac arrest will be addressed in this chapter.

Chest compression

Conventional precordial compression

The standard technique of external precordial compression has changed little since it was described in the landmark paper of Kouwenhoven *et al.* (1960). External chest compressions are applied by the rescuer who places the heel of one hand over the lower half of the victim's sternum and the other hand on top of the first hand. Downward force displaces the sternum for a distance of 3.5 to 5 cm for adult patients. The compressions are repeated at a rate of 80 to 100/min with equal compression–relaxation intervals.

The efficacy of precordial compression progressively decreases with increasing duration of the resuscitation effort. This is coincident with a loss of myocardial compliance, deformation of the chest wall, fractured ribs, and fatigue on the part of the operator. The technique is maximally effective when cardiopulmonary resuscitation is initiated after only a brief interval of untreated cardiac arrest and during the initial 10 min of precordial compression.

The mechanism by which blood flow is generated during external precordial compression has been the subject of vigorous debate. Echocardiographic recordings now quite securely support the so-called cardiac pump theory. This theory holds that the heart is compressed and blood is actively ejected into both the pulmonary and systemic circulation during chest compression. Cardiac valve function is preserved so as to assure forward blood flow. When compression is released, corresponding declines in intrathoracic and intracardiac pressures allow for venous return and cardiac filling. Precordial compression is associated with closure of the mitral valve and opening of the aortic valve. Pressure gradients between the atria, ventricles, and aorta have also been documented under experimental conditions. These observations contrast with the thoracic pump theory which claims that chest compression increases intrathoracic pressure, and that this pressure is transmitted to the intrathoracic vasculature with extrusion of blood from the intrathoracic vessels and forward blood flow. Therefore this theory holds that the heart is a passive conduit, and that arterial–venous pressure gradients between intrathoracic and extrathoracic arteries and veins account for forward blood flow. The observation that patients who developed ventricular fibrillation during cardiac catheterization were effectively sustained by vigorous coughing, the so-called 'cough cardiopulmonary resuscitation', has been cited in support of the thoracic pump theory. The current consensus is that high-impulse compression does represent a cardiac pump mechanism, but that there is a thoracic component to the extent that the increase in intrathoracic pressure itself accounts for cardiac compression together with or independently of direct ventricular compression.

Injury to both thoracic and abdominal structures follows chest compression in as many as 60 per cent of victims. Most frequently the ribs and sternum are fractured, with occasional injuries occurring to the heart, great vessels, lungs, and liver. Other pathological findings include pulmonary edema, hemothorax, pneumothorax, and hemoperitoneum.

Interposed abdominal compression

Interposed abdominal compression is intended to increase venous return of blood from the viscera to the right atrium. At the same time, the concurrent compression of the abdominal aorta is intended to decrease infradiaphragmatic runoff of aortic blood flow. Consequently, venous return, forward blood flow into the aorta, and delayed runoff into the lower body preserve vital blood flow to the coronary and cerebral circuits. During the relaxation phase of precordial compression a second rescuer compresses the abdomen in the midline with a force equivalent to 200 mmHg at a site equidistant from the xyphoid process and the umbilicus (Fig. 1). Practical guidelines for securing the optimal compression force to the abdomen are still lacking. However, there is evidence that interposed abdominal compression may substantially increase the cardiac output and coronary perfusion pressure generated during precordial compression. In the setting of in-hospital cardiac arrest, this technique is reported to increase initial resuscitability and the number of neurologically intact patients discharged from the hospital (Sack *et al.* 1992). However, as yet there is no persuasive proof that interposed abdominal compression improves the outcome of cardiopulmonary resuscitation in settings of out-of-hospital cardiac arrest.

Complications of interposed abdominal compression are similar to those of external precordial compression. Contrary to expectation, traumatic injury to the abdominal viscera is not significantly increased.

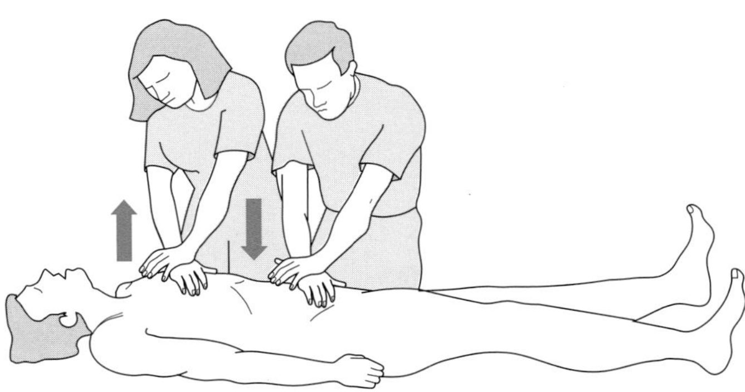

Fig. 1 Interposed abdominal compression.

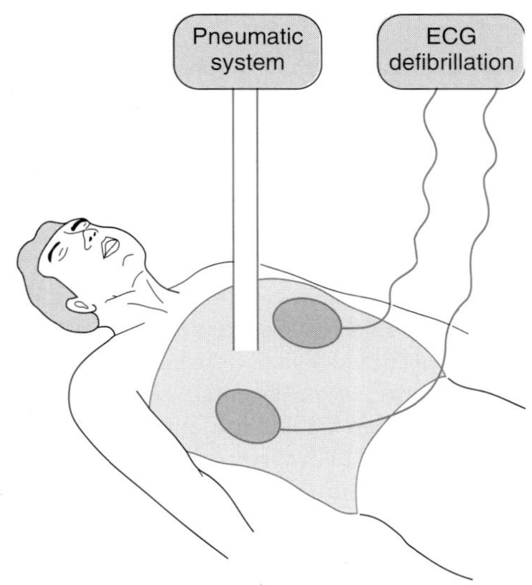

Fig. 2 Vest cardiopulmonary resuscitation.

Vest cardiopulmonary resuscitation

Vest cardiopulmonary resuscitation is based on concepts developed by a group of investigators at Johns Hopkins University in the mid-1980s. A vest, constructed similarly to a very large blood pressure cuff, is placed circumferentially around the chest (Fig. 2). It is inflated and deflated at a frequency of 60/min and to a pressure of 250 mmHg with a pneumatic pump. This technique significantly increases maximum aortic pressure and coronary perfusion pressure when compared with conventional precordial compression. However, neither increases in initial resuscitability nor 24-h survival have been achieved as yet. A comprehensive clinical trial is in process.

Vest cardiopulmonary resuscitation utilizes compression forces that are evenly applied circumferentially around the chest; the intrathoracic volume is decreased in proportion to the pressure applied. Trauma is potentially minimized because the compression forces are distributed circumferentially rather than localized to the sternum.

Only incomplete clinical data on vest cardiopulmonary resuscitation are currently available. As yet, no lower incidence of trauma or improved outcome compared with conventional precordial compression have been demonstrated. It may be uniquely useful in transport vehicles, elevators, and other constricted surroundings.

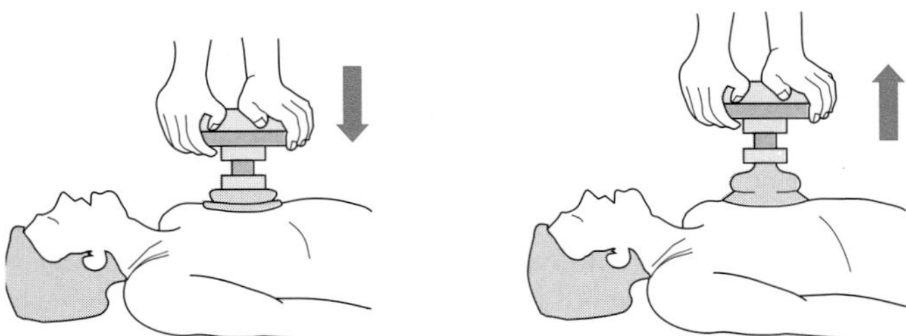

Fig. 3 Active compression–decompression.

Active chest compression–decompression

Perhaps the most acclaimed method of cardiopulmonary resuscitation is that of active compression and decompression using the equivalent of a sink plunger. The compression device consists of a manually operated plunger which is applied to the precordium with a suction cup. This allows for active compression and decompression of the chest (Fig. 3). Decompression actively re-expands the chest wall. The greater chest expansion during the relaxation phase of chest compression should increase the range of intrathoracic pressures and thereby augment pressure gradients, accounting for increased venous and aortic blood flow. In an initial report of randomized studies of victims of in-hospital cardiac arrest, active compression–decompression significantly increased the success of both initial resuscitation and 24-h survival. However, a more comprehensive clinical trial failed to confirm improvement in outcome (Schwab *et al.* 1995). Experimentally, the increases in coronary perfusion pressure produced by this method are quite moderate.

Except for a potentially greater incidence of injury to the skin, the complications of active compression–decompression are comparable with those of conventional precordial compression.

Phased chest and abdominal compression–decompression

Phased chest and abdominal compression–decompression is the newest alternative to conventional precordial compression. Experimental outcomes have only recently been reported (Tang *et al.* 1997). This technique provides for reciprocal compression and decompression of the chest and abdomen with a Lifestick™ resuscitator (Fig. 4) which resembles a manual seesaw. The precordium and abdomen are compressed at a rate of 60/min with compression forces of 120 lb (55 kg) and 50 lb (23 kg) respectively. The reciprocal seesaw operation involves a specified phase delay. This technique combines the advantage of active chest compression–decompression and interposed abdominal compression. Abdominal decompression serves to reduce both left and right ventricular afterloads with consequent increases in stroke volumes.

This technique has demonstrated impressive hemodynamic efficacy in animal studies. The coronary perfusion pressure generated was three times greater than that generated by conventional precordial compression. This was associated with a striking improvement in initial resuscitability and 72-h survival. Experimental studies have shown a significantly lower incidence of rib fractures and injuries to the thoracic or abdominal organs; human evaluation is ongoing. Hemodynamic efficacy has also been claimed in preliminary human trials.

Intravascular methods

Aortic balloon occlusion

Another option for cardiopulmonary resuscitation is the use of intermittent occlusion of the ascending aorta (Tang *et al.* 1993). When the ascending aorta is occluded by a balloon proximal to the innominate artery, the entire cardiac output generated by precordial compression is directed into the coronary circuit. Consequently, coronary blood flow is dramatically increased. The technique utilizes a balloon catheter which is advanced into the ascending aorta. The balloon is inflated for 30 s during each minute of precordial compression. This augments coronary perfusion pressure with typical increases from 12 to more than 90 mmHg. Even though cerebral blood flow ceases for 30 s of each minute, no neurological deficits have been seen.

Experimentally, the success of initial resuscitation attempts and 48-h survival are impressive. Other researchers have proposed continuous occlusion of the thoracic descending aorta. However, hemodynamic efficacy is lessened when the aorta is occluded at a more distal site. As yet, no confirmation of the applicability of balloon occlusion is available.

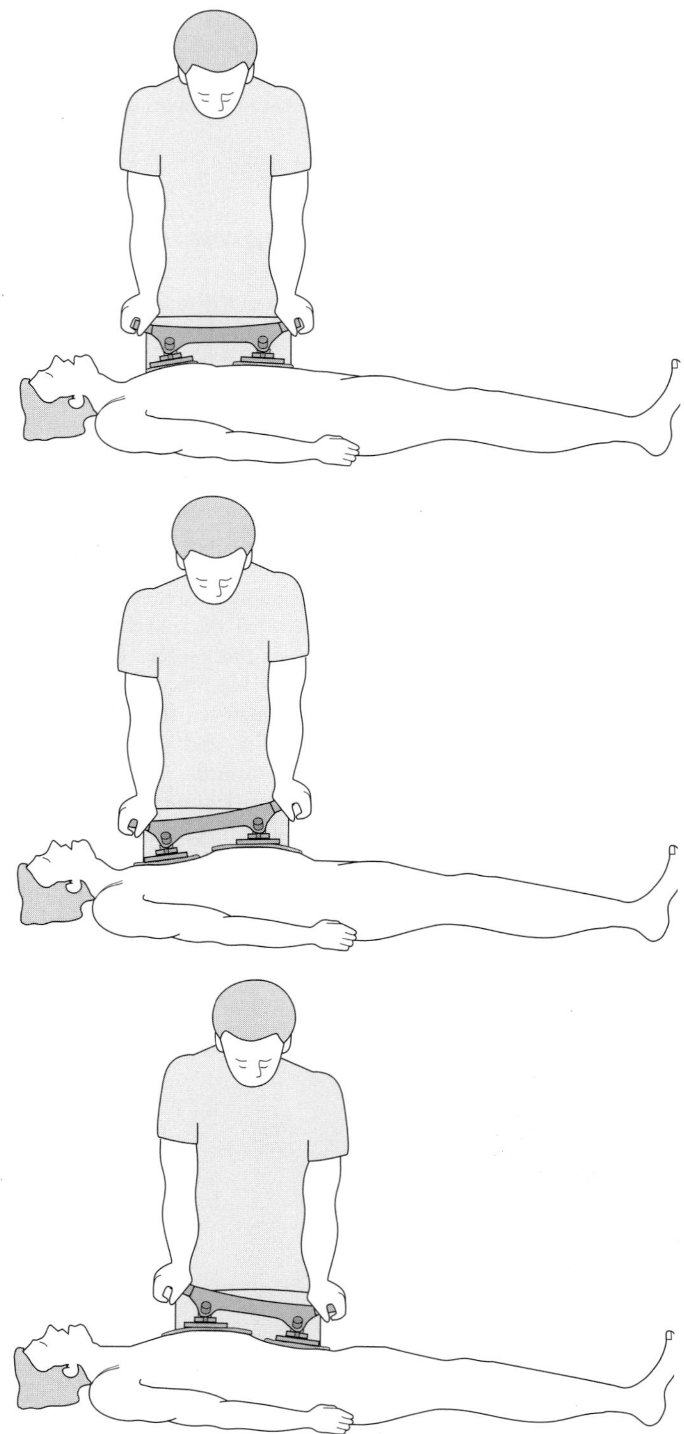

Fig. 4 Phased chest and abdominal compression–decompression.

Appropriate skill and significant time are required for aortic catheterization under the crisis conditions of cardiopulmonary resuscitation, and this produces a formidable constraint.

Extracorporeal circulation

Although technically more demanding, extracorporeal circulation is the most effective hemodynamic intervention for cardiac resuscitation available at present. However, it requires even more skill and greater complexity of instrumentation than balloon occlusion. Both arterial and venous cannulas are required, typically by percutaneous or surgical cannulation of the femoral artery and vein with an extracorporeal circuit from the patient to the membrane oxygenator and back to the patient using a centrifugal pump.

Extracorporeal circulation allows successful restoration of spontaneous circulation after 15 to 20 min of untreated cardiac arrest. Coronary perfusion pressure is almost immediately restored to normal levels. If applied sufficiently rapidly to victims who fail conventional methods, there may be an impressive improvement in outcome. The technique is confined to patients who sustain in-hospital cardiac arrest, particularly in operating room environments.

Complications include bleeding and vascular injury from cannulation.

Open-chest cardiac massage

Open-chest cardiac massage was first employed more than 120 years ago, and was the primary intervention after in-hospital cardiac arrest during the 1940s and 1950s. In the 1960s, however, non-invasive external precordial compression came into predominant use. Open-chest cardiac massage is now utilized only in settings of cardiac arrest caused by penetrating chest trauma, cardiac arrest after thoracotomy, or in victims who have major bony deformities of the chest that preclude successful chest compression.

An anterolateral thoracotomy is performed through the fourth or fifth intercostal space. After the chest is opened, the rescuer's open right hand, with the palm facing anteriorly, is advanced posteriorly behind the left ventricle. The palm of the hand compresses the heart against the posterior sternal surface at a frequency of 60/min. When compression is released the heart fills passively.

Open-chest cardiac massage is unequivocally more effective than external chest compression for increasing cardiac output, coronary perfusion, initial resuscitability, and survival. In fact, the aorta distal to the left subclavian can be cross-clamped to provide increased coronary and cerebral perfusion. However, complications from this procedure are significantly greater than those associated with external precordial compression. Traumatic injury to the heart, lungs, and surrounding tissues can occur. Thus, except in cases of penetrating torso trauma, no improvement in outcome has been demonstrated with open-chest cardiac massage.

Conclusion

Treatment of in- and out-of-hospital cardiac arrest with prompt re-establishment of blood flow to the coronary and systemic circuits is essential for improved outcome. At present there are no conclusive data demonstrating that any of the newer options have improved outcome in clinical settings.

By measuring such indices as end-tidal carbon dioxide, electrocardiographic ventricular fibrillation waveform voltage, and dominant frequencies, which correlate with cardiac output and coronary perfusion pressure during cardiopulmonary resuscitation, newer modes of resuscitation such as phased chest and abdominal compression–decompression may be shown to improve outcome.

References

Kouwenhoven, W.B., Jude, J.R., and Knickerbocker, G.G. (1960). Closed-chest cardiac massage. *Journal of the American Medical Association*, **173**, 1064–7.

Sack, J.B., Kasselbrenner, M.B., and Bregman, D. (1992). Survival from in-hospital cardiac arrest with interposed abdominal counterpulsation during cardiopulmonary resuscitation. *Journal of the American Medical Association*, **267**, 379–85.

Schwab, T.M., Callaham, M.L., Madsen, C.D., and Utecht, T.A. (1995). A randomized clinical trial of active compression–decompression CPR vs standard CPR in out-of-hospital cardiac arrest in two cities. *Journal of the American Medical Association*, **273**, 1261–8.

Tang, W., Weil, M.H., Noc, M., Sun, S.J., Gazmuri, R.J., and Bisera, J. (1993). Augmented efficacy of external CPR by intermittent occlusion of the ascending aorta. *Circulation*, **88**, 1916–21.

Tang, W., *et al.* (1997). Phased chest and abdominal compression–decompression: a new option for cardiopulmonary resuscitation. *Circulation*, **95**, 1335–40.

1.2.3 Therapeutic strategies during cardiac arrest

W. J. Moore and P. J. F. Baskett

Key messages

- Algorithms have been established for treating various life-threatening arrhythmias.
- Ventricular fibrillation and pulseless ventricular tachycardia are treated similarly by countershock.
- The differential causes of electromechanical dissociation must be considered.
- Asystole generally carries a poor prognosis.
- Adverse signs associated with bradycardia and with broad and narrow complex tachycardia need to be assessed as treatment varies.

Introduction

Cardiac arrest is defined as a sudden and complete loss of cardiac function. Death is virtually inevitable unless basic life support is initiated promptly. Advanced life support should be commenced as soon as possible. The therapeutic strategies used for managing the arrest are based largely on the electrocardiogram tracing. Algorithms have been developed by the European Resuscitation Council, the American Heart Association, the Australian Resuscitation Council, the Heart and Stroke Foundation of Canada, and the Resuscitation Council of Southern Africa (American Heart Association 1992; European Resuscitation Council 1992). In 1997 the International Liaison Committee on Resuscitation (ILCOR), composed of representatives from the above organizations, published Advisory

Statements which included the Universal Advanced Life Support Algorithm shown in Fig. 1 (International Liaison Committee on Resuscitation 1997).

Cardiac arrest may be due to ventricular fibrillation, asystole, or electromechanical dissociation (pulseless electrical activity).

Ventricular fibrillation

Ventricular fibrillation is the most common initial arrhythmia in cardiac arrest. It occasionally starts as pulseless ventricular tachycardia and then proceeds to ventricular fibrillation. Both rhythms can be treated identically. The most important therapeutic action for ventricular fibrillation is early defibrillation with d.c. countershock (Fig. 1). Survival after this rhythm decreases by 5 per cent per minute, even with basic life support. A precordial thump is still included in the algorithm as the first step in arrests witnessed by health-care professionals as it restores to sinus rhythm 40 per cent of patients with ventricular tachycardia and 2 per cent of patients in ventricular fibrillation. The risk of myocardial contusion and rib fracture is outweighed by the potential benefit of the rapid conversion of a pulseless arrhythmia to sinus rhythm. If the thump is unsuccessful, defibrillation should be performed immediately. One paddle should be applied below the outer half of the right clavicle, and the other just outside the position of the cardiac apex (ECG position lead V4–5). Good electrical contact must be assured by the use of electrode jelly pads and firm pressure. The energy sequence should be 200 J followed by 200 J and then 360 J. The first three shocks should be delivered within 30 to 45 s. A check of rhythm or pulse should be confirmed between each shock, but basic life support should be discontinued when automated external defibrillators are used between the first and third shock as this will only delay defibrillation. If there has been no success after the third shock, basic life support should be recommenced and, if not already done, the trachea should be intubated and intravenous access established. Basic life support should only be stopped

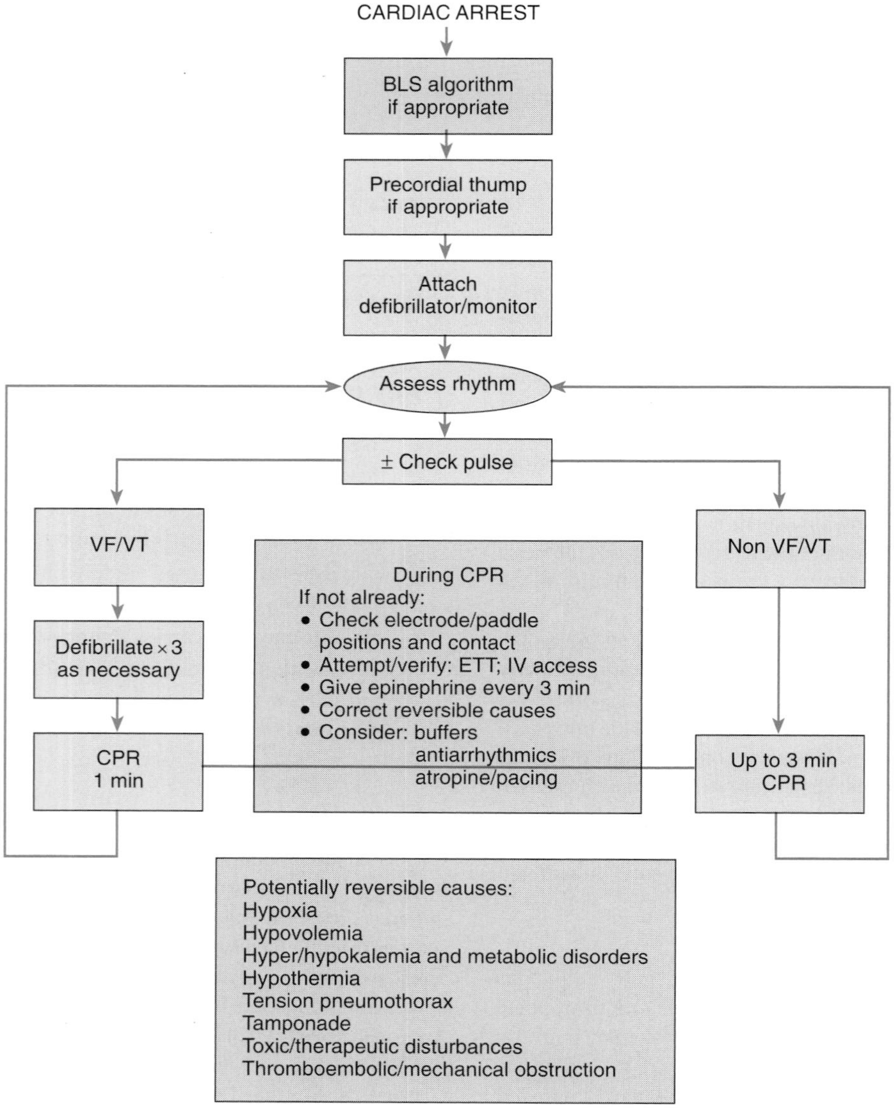

Fig. 1 The Universal Advanced Life Support Algorithm for the management of cardiac arrest in adults: VF, ventricular fibrillation; VT, ventricular tachyarrhythmia; CPR, cardiopulmonary resuscitation. Note that each successive step is based on the assumption that the one before has been unsuccessful. (Reproduced with permission of the Advanced Life Support Working Group of the International Liaison Committee on Resuscitation.)

for a maximum of 15 s to allow these procedures to be performed. Epinephrine (adrenaline) 1 mg is then given intravenously to increase the efficacy of basic life support and not as an adjuvant to defibrillation (Lindner and Koster 1992). A new set of countershocks at 360 J should follow as quickly as possible and should not be delayed if intubation or intravenous cannulation is difficult. Thereafter, the loop is repeated so that intravenous epinephrine 1 mg is given once every 3 min. After every three loops, other drugs may be appropriate. Sodium bicarbonate (50 ml of 8.4 per cent solution = 50 mmol) may be given if acidosis is suspected. It is better if acidosis is confirmed by arterial blood or central venous blood pH, bicarbonate, and base excess measurements. However, it must be remembered that intracellular acidosis bears little relationship to arterial or even mixed venous blood gas results. It is known that 8.4 per cent sodium bicarbonate is hyperosmolar and generates carbon dioxide that may give rise to a paradoxical intracellular acidosis as carbon dioxide can cross cell membranes very quickly. However, most clinicans will treat an arterial blood pH of less than 7.1 with 50 ml of 8.4 per cent sodium bicarbonate (note that if a mixed venous sample is taken, the pH will be approximately 0.3 units lower than that of the arterial sample).

The antiarrhythmic agents lidocaine (lignocaine), bretylium, and amiodarone may also be considered in this now desperate situation. Intravenous calcium, magnesium, and potassium may be of use in certain clinical situations when there is pre-existing or suspected electrolyte imbalance. Calcium has been implicated in ischemic tissue injury and should always be administered with caution. The number of loops used in the algorithm depends on clinical judgment; usually the resuscitation attempt should continue for 10 to 60 min and not be stopped while ventricular fibrillation exists. Persistent asystole is an indication that prospects of success are remote. Further resuscitation attempts should rarely last for more than 20 min unless hypothermia, near drowning, or drug intoxication are suspected of being associated with the arrest.

Asystole

A precordial thump may be useful in asystole as long as there has not already been a prolonged period of basic life support (Fig. 1). Recovery from asystole is unlikely unless there is either a trifasicular block (P waves should be seen on the electrocardiogram), the rhythm has evolved from an extreme bradycacrdia, or asystole is a transient sequel to defibrillation. The possibility of a mistaken diagnosis of asystole must also be remembered. Ventricular fibrillation may be mistaken for asystole if there is equipment failure, excess artefact, uncontrollable movement (e.g. in a moving ambulance or aircraft), or incorrect gain setting on the electrocardiogram. Thus, because ventricular fibrillation is much more likely to have a successful outcome, if it is suspected, it should be treated by an attempt at defibrillation first. If the waveform is asystole or very fine ventricular fibrillation, an automated defibrillator will not allow shocks to be given and time should not be wasted in persevering in these attempts. After excluding ventricular fibrillation, the patient's airway should be secured and intravenous access should be established. Intravenous epinephrine 1 mg should be given to enhance basic life support. Intravenous atropine 3 mg will completely block the vagus and is given once and once only to counter any excess vagal tone, although no proven benefit has been found in clinical studies. Undue delay in performing basic life support must be avoided. If there has been any recent electrical activity on the electrocardiogram, intravenous or transcutaneous pacing may be considered. In the absence of any electrical activity, further loops should be considered. High-dose (5 mg) intravenous epinephrine may be considered if there is no electrical activity after three loops. Recovery rarely occurs after 15 min of asystole.

Electromechanical dissociation

Electromechanical dissociation or pulseless electrical activity implies that there is electrical activity of the heart without mechanical activity. A definite diagnosis can rarely be made clinically because feeble contraction of the heart may produce no pulse, no detectable heartbeat, and no heart sounds, but the distinction is irrelevant since both absence of mechanical activity and undetectable mechanical activity carry an equally poor prognosis except when they are transient phenomena during a cardiac arrest or there is a specific remediable cause. Therefore search for and recognition of specific and correctable causes of the clinical picture of electromechanical dissociation is of prime importance (Fig. 1). If no evidence exists for any of the specific causes, cardiopulmonary resuscitation should be continued with the usual associated procedures of airway security, establishing venous access, and treatment with epinephrine. No recommendation based on sound scientific evidence can be made for the routine use of calcium salts or alkalizing agents, although they may be of value in some circumstances.

Periarrest arrhythmias

These arrhythmias can be categorized into three groups. There are a few differences of opinion between the American Heart Association and the European Resuscitation Council with regard to treatment of periarrest arrhythmias, but none are of major substance. However, for simplicity and clarity, the European Resuscitation Council algorithms will be discussed here.

Bradycardias and blocks

A heart rate that is either abnormally slow in absolute terms or is too slow to support an adequate circulation needs to be addressed. For example, a heart rate of 65 beats/min may be inappropriately slow for a patient who has developed a low output state. The recommendations for bradycardias and blocks depend on whether there is a recognizable appreciable risk of asystole (Fig. 2). Complete third-degree heart block with a narrow QRS is not in itself an indication for treatment because atrioventricular junctional ectopic pacemakers (with a narrow QRS) often provide an adequate and stable heart rate.

If asystole is believed to be a definite risk, the responder may wish to do no more than establish intravenous access and give atropine before seeking help from others with transvenous pacing skills. If a patient's condition is critical, placement of a transvenous ventricular pacing wire or external pacing may be appropriate. Isoproterenol (isoprenaline) 1 μg/min can be administered by infusing 2.5 mg of isoproterenol in 500 ml of a carrier solution with an infusion pump, initially at 0.2 ml/min. The dose may be increased rapidly. However, attention should be paid to the risk of precipitating or worsening ventricular arrhythmias and increasing myocardial oxygen consumption. Isoproterenol for infusion is

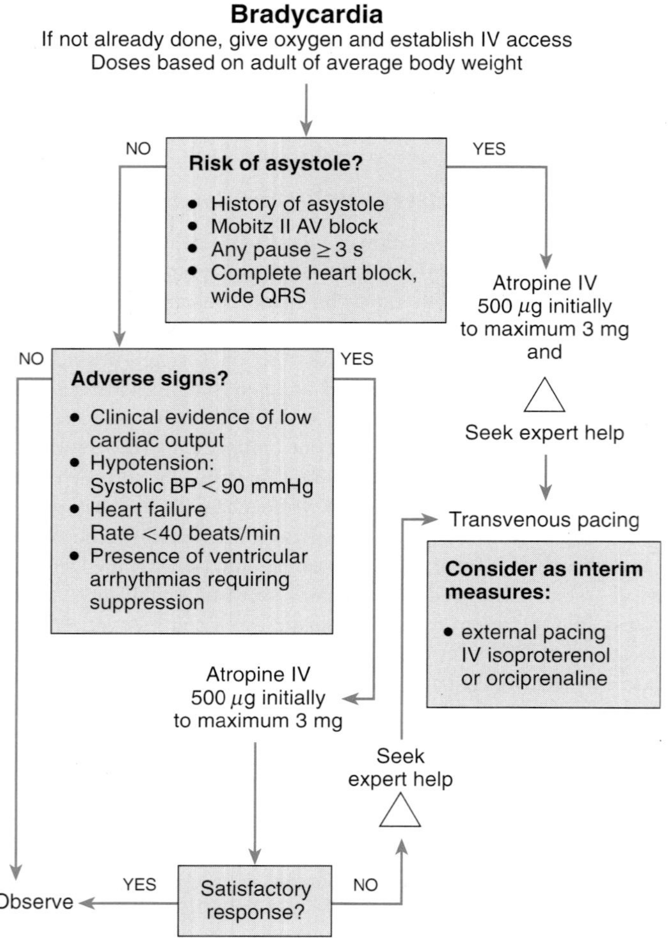

Fig. 2 Algorithm for bradycardia: AV, arteriovenous; BP, blood pressure.

not available in all European countries, and orciprenaline is an appropriate alternative.

If there is no perceived risk of asystole, adverse clinical signs including severe impairment of myocardial function, very slow heart rate, or the progression of emerging tachyarrhythmias that require suppression should be sought. Without such signs, observation may be all that is required; however if one or more signs are present, 500 μg of atropine should be administered initially by slow intravenous injection. Increments of 500 μg or 1 mg, up to a total dose of 3 mg, can be given at intervals. Higher doses are not beneficial and may produce unwanted effects. If there is a satisfactory response, only observation may be needed; however, failure to respond to atropine in the presence of adverse signs may require transvenous pacing.

Broad complex tachycardia

Although broad complex tachycardias may be supraventricular with aberrant intraventricular conduction, the distinction from sustained ventricular tachycardia may be difficult (Fig. 3). Little harm results if a supraventricular tachycardia is treated as ventricular in origin, but very serious consequences can follow from the converse assumption. The first determinant of management of a broad complex tachycardia is whether or not there is a palpable pulse. Pulseless ventricular tachycardia is a rhythm akin to ventricular fibrillation, and the patient will be unconscious due to inadequate cerebral perfusion. The treatment follows the guidelines for ventricular fibrillation (Fig. 1). If a palpable pulse is found, give oxygen, establish intravenous access, and look for adverse signs (Fig. 3).

With adverse signs, synchronized d.c. countershock is appropriate after sedation has been given. If this does not immediately resolve the dysrhythmia, lidocaine should be administered. If the plasma potassium concentration is known to be less than 3.6 mmol/l, particularly in the presence of recent infarction, infusions of potassium and magnesium should be administered while the patient is prepared for further cardioversion. Other pharmacological agents that might be considered for refractory cases include procainamide, flecainide, propafenone, bretylium tosylate, and amiodarone. Overdrive pacing may also be considered.

In the absence of adverse signs, lidocaine can be administered in conventional doses and, if the potassium level is known to be low, an infusion of potassium and magnesium is recommended to help prevent recurrent disturbances of rhythm. If lidocaine is ineffective, synchronized d.c. countershock should be considered, as for the symptomatic patient. For refractory cases without adverse signs, amiodarone should be given by slow intravenous injection followed by an infusion, with another attempt at synchronized cardioversion after a period of up to 1 h has elapsed to allow this antiarrhythmic drug to produce a powerful pharmacological effect.

Narrow complex tachycardia

Narrow complex tachycardia, which is a supraventricular tachycardia including atrial fibrillation, is less frequent and generally less hazardous than ventricular tachycardia, even though supraventricular tachycardias are recognized triggers of ventricular fibrillation (Hays *et al.* 1989).

As with all arrhythmias, oxygen should be administered and intravenous access secured. Performing vagal maneuvers such as a Valsalva or carotid sinus massage should be considered for treating supraventricular tachycardias. However, there are hazards that must be emphasized. Vigorous vagal maneuvers may cause sudden bradycardia or trigger ventricular fibrillation in the presence of acute ischemia or digitalis toxicity. Elderly patients are also vulnerable to plaque dislodgement with cerebrovascular complications following carotid massage.

The pharmacological treatment of choice for regular supraventricular tachycardia is adenosine (Garratt *et al.* 1992). Although this may cause unpleasant side-effects such as nausea, flushing, and chest discomfort, these are brief and well tolerated if a patient is informed of their nature and duration before the injection is given. The initial dose recommended in the algorithm is 3 mg. However, this will be effective in only a minority of cases, so that increasing increments every 1 or 2 min are required, with up to two injections of 12 mg if necessary. Adenosine is not available in some European countries but adenosine triphosphate is an alternative. If adenosine is not successful in establishing a satisfactory rhythm, or atrial fibrillation continues at a rate greater than 130 beats/min, therapy will depend upon whether or not adverse signs are present (Fig. 4). If they are present,

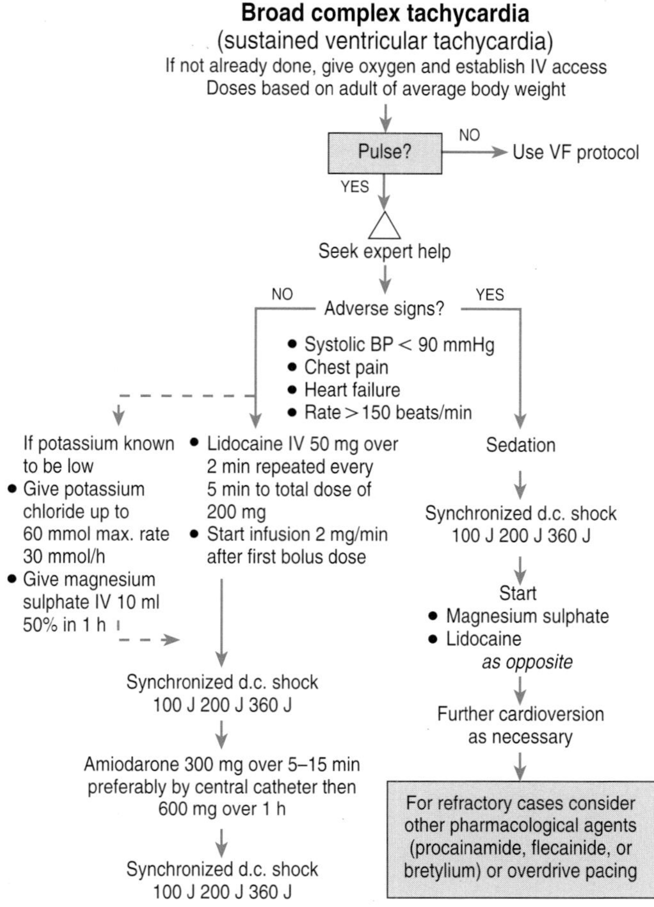

Fig. 3 Algorithm for broad complex tachycardia (sustained ventricular tachyardia): VF, ventricular fibrillation; BP, blood pressure.

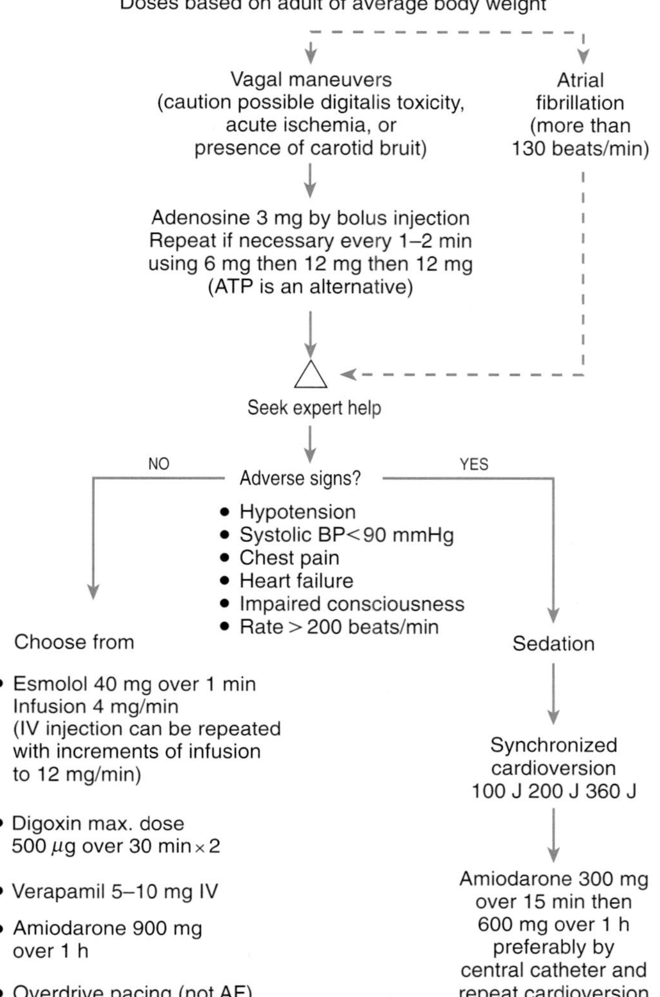

Fig. 4 Algorithm for narrow complex tachycardia (supraventicular tachycardia).

the treatment should consist of synchronized d.c. countershock after any necessary sedation. If this is unsuccessful, further shocks should be given after a slow intravenous injection of amiodarone followed by an infusion of amiodarone. If there is no perceived need for urgency, up to 1 h may elapse before further shocks are attempted.

In the absence of adverse signs, no single recommendation can be made because of different traditions of treatment between countries. Possible alternatives that might be considered include a short-acting β-blocker (e.g. esmolol), digoxin, verapamil, or amiodorone. Overdrive pacing may also be successful.

Verapamil, although widely used and a very successful agent, may be hazardous in certain circumstances. These include arrhythmias associated with Wolf–Parkinson–White syndrome, tachycardias that are ventricular and not supraventricular in origin, and some of the supraventricular arrhythmias of childhood. The potentially serious interaction between verapamil and β-blocking drugs should also be remembered; this is particularly dangerous if both drugs have been administered intravenously.

References

American Heart Association (1992). Guidelines for cardiopulmonary resuscitation and emergency cardiac care. *Journal of the American Medical Association*, **268**, 2199–241.

European Resuscitation Council (1992). *Guidelines for resuscitation*. European Resuscitation Council Secretariat, Antwerp.

Garratt, C.J., Malcolm, A.D., and Camm, A.J. (1992). Adenosine and cardiac arrhythmias. *British Medical Journal*, **305**, 3–4.

Hays, W., Lerman, B.B., and DiMarco, J.P. (1989). Nonventricular arrhythmias as precursors of ventricular fibrillation in patients with out-of-hospital cardiac arrest. *American Heart Journal*, **118**, 53–7.

International Liaison Committee on Resuscitation (1997). The Universal Advanced Life Support Algorithm. An Advisory Statement by the Advanced Life Support Working Group of the International Liaison Committee on Resuscitation. *Resuscitation*, **34**, 109–11.

Lindner, K.H. and Koster, R. (1992). Vasopressor drugs during cardiopulmonary resuscitation. *Resuscitation*, **24**, 147–53.

1.2.4 Resuscitating and protecting the brain

P. Safar and P. M. Kochanek

Key messages

- Cerebral resuscitation after cardiac arrest in patients should include reperfusion with a hypertensive bout, continued normotension, or mild hypertension; it might include mild hypothermia to a nasopharyngeal or tympanic membrane temperature of about 34 °C, as rapidly as possible, for 12 to 24 h.
- Cerebral resuscitation after severe brain trauma with coma in patients might include moderate hypothermia (30–32 °C) for 12 to 24 h.
- Hypothermia can be controlled by head–neck–trunk surface cooling or, more rapidly, by intraperitoneal or intravascular methods.
- Hyperthermia must be prevented and seizure activity suppressed.

Introduction and pathophysiology

Resuscitation of the brain after temporary complete global brain ischemic anoxia, as produced by prolonged normothermic cardiac arrest, is crucial. Pathophysiology and therapy are similar for the temporary incomplete global ischemic anoxia of asphyxia or severe shock, but are different for the permanent incomplete focal brain ischemia of stroke or for traumatic brain injury which can be multifocal and followed by lethal brain swelling. Measures effective for protection (pretreatment) and preservation (intra-insult treatment) may not be effective for resuscitation (to reverse the insult and support recovery).

Modern external cardiopulmonary resuscitation was established in the 1950s. However, by 1961 it was clear that a more comprehensive approach to resuscitation was needed, and the concept of cardiopulmonary–cerebral resuscitation was introduced, with a continuum from basic via advanced to prolonged life support (Safar and Bircher 1988). Secondary postarrest derangements in the brain and extracerebral organs — the post-resuscitation disease (Negovsky *et al.* 1983) — are extremely complex and multifactorial. Unfortunately, except for brain-oriented general extracerebral life support (Table 1), no treatment that prevents or reverses all the cascades involved in the development of postcardiac arrest brain damage has yet been statistically documented in humans. Since the 1970s, some potentially valuable therapeutic strategies have been reported in experimental animal models of cardiac arrest and other forms of cerebral ischemia by an increasing number of investigative groups (Koehler *et al.* 1996; Safar 1996). Until recently, documentation of benefit has been difficult because of a lack of reproducible large animal cardiac arrest outcome models. Translation into clinical practice has been problematic for many reasons, including the difficulty, cost, and unreliability of randomized clinical trials. There are numerous unknown or uncontrollable variables that influence cerebral outcome in cardiac arrest patients. A single specific cerebral resuscitation drug with minimal risk and a breakthrough effect may never be found. The multifactorial pathogenesis of the secondary derangements calls for multifaceted treatments (Safar 1996).

Normothermic cardiac arrest followed by restoration of spontaneous circulation within about 4 min (before exhaustion of energy charge in the brain) generally results in complete recovery of cerebral function within 3 to 7 days. During normothermic cardiac arrest of duration 10 to 20 min (with no blood flow), a combination of intraneuronal calcium loading, lactic acidosis, glutamate rise, increase in free fatty acids, hyperosmolality, metabolic silence, and membrane leakage sets the stage for reoxygenation injury. After cardiac arrest of duration 10 to 20 min, during and after reperfusion–reoxygenation, normal brain ATP, ion pump, tissue pH, and glutamate concentration are quickly restored. However, a combination of delayed, prolonged, and inhomogeneous cerebral hypoperfusion during rising cerebral

Table 1 Brain-oriented prolonged life support throughout coma

- Avoid even mild hypotension, hypoxemia, and hyperthermia
- After restoration of spontaneous circulation and a hypertensive bout, control normotension or mild hypertension, and normoxia and normocapnia, with controlled ventilation during coma
- Sedate with titrated benzodiazepine or barbiturate; control seizures with diazepam-phenytoin as needed
- Immobilize with partial neuromuscular blockade as needed
- Monitor core temperature (pulmonary, arterial, or rectal) and/or brain temperature (tympanic or nasopharyngeal)
- Explore feasibility of mild resuscitative cerebral hypothermia (34 °C), induced as early and as rapidly as feasible, and sustained for at least 12 h; the optimal duration of mild hypothermia is not known
- Use appropriate volume administration, pressors, and inotropic agents as indicated to treat shock
- Conduct hemodynamic monitoring to guide administration of drugs and fluids
- Keep hematocrit at 30%–35%; aim for normal values of plasma colloid osmotic pressure and serum electrolytes and osmolality
- Maintain fluid and acid–base balance and alimentation; control base deficit to below 5 mmol/l
- Administer fluids intravenously; do not use glucose in H_2O; initially use glucose 5% or 10% in NaCl 0.25% or 0.5% at about 50 ml/kg/24 h; then begin parenteral nutrition (TPN) and, later, enteral feeding as indicated
- Avoid iatrogenic hypoglycemia or severe hyperglycemia
- Use standard intensive care life support, with position supine horizontal; optional slight elevation of head; turn trunk side to side for atelectasis prophylaxis

O_2 demands (supply–demand mismatching), reoxygenation injury cascades triggered by free iron and free radicals, secondary calcium loading and excitotoxicity, and transient or prolonged extracerebral organ malfunction cause potentially preventable or treatable secondary derangements. The immediate multifocal cerebral no-reflow phenomenon seen with hypotensive reperfusion does not occur with hypertensive or normotensive reperfusion. Intracranial pressure increase due to brain swelling is not a problem after survivable cardiac arrest (but can be a major problem after focal brain ischemia or traumatic brain injury). After restoration of spontaneous circulation following prolonged cardiac arrest and a latent period of several hours, selectively vulnerable neurons (in the hippocampus, cerebellum, and neocortex) die over a period of 24 to 72 h (perhaps even longer) alongside surviving neurons. Why and how these cells die is still unclear, although triggering of programmed cell death (apoptosis) is a possibility.

Persistent unresponsiveness at 3 to 7 days after cardiac arrest and cardiopulmonary–cerebral resuscitation has been predictably followed by permanent severe brain damage, and therefore justifies consideration of discontinuation of life support. However, unresponsiveness for months or even years after traumatic brain injury has occasionally been followed by conscious survival. The mechanisms of the cerebral post-resuscitation syndrome are under intensive investigation. This may soon lead to additional novel therapies which would make those listed below more effective (Koehler *et al.* 1996; Safar 1996).

General brain-oriented life support

Minimizing arrest (no-flow) time and external cardiopulmonary resuscitation (low-flow) time with emergency control of airway, pulmonary oxygenation and ventilation, and high or normal perfusion pressures are basic requirements for cerebral recovery. This requires the earliest possible restoration of spontaneous circulation since the hypotension and hypoperfusion produced by standard external cardiopulmonary resuscitation is inadequate for cerebral resuscitation as the injured brain is very vulnerable. Therefore even mild postarrest hypotension, hypoxemia, and hyperthermia must be avoided. A spontaneous or induced hypertensive bout during or immediately after restoration of spontaneous circulation is associated with better cerebral outcome in dogs and humans. Titrated high-dose epinephrine (adrenaline) during cardiopulmonary resuscitation increases the chance of restoration of spontaneous circulation and of a spontaneous hypertensive bout. If it does not occur spontaneously, we recommend briefly producing a systolic arterial pressure of 150 to 200 mmHg immediately after restoration of spontaneous circulation, with a titrated intravenous infusion of epinephrine, norepinephrine, or dopamine, and continuing with controlled normotension (using fluids and titrated infusion of cardiovascular drugs). After prolonged cardiac arrest, buffer therapy for control of metabolic acidemia seems to be associated in dogs with better restoration of spontaneous circulation and better cerebral outcome. Brain-oriented prolonged life support should include controlled ventilation throughout coma. Pre- or intra-arrest hyperglycemia worsens cerebral lactic acidosis and augments postcardiac arrest brain damage. Postarrest moderate hyperglycemia plus insulin has been shown to be possibly beneficial for cerebral recovery in a rat model of cardiac arrest. Corticosteroid administration and osmotherapy are at best controversial.

Special cerebral resuscitation measures

Numerous drug treatments have been investigated in studies with short-term mechanism-oriented animal models, including outcome models of incomplete forebrain ischemia in rats, global brain ischemia in monkeys, and (clinically most relevant) cardiac arrest in dogs without a breakthrough effect. Intravenous barbiturate loading can lower increased intracranial pressure after traumatic brain injury and mitigate focal brain ischemia, but is risky after cardiac arrest. Calcium entry blockers (lidoflazine, nimodipine), given early after cardiac arrest or global brain ischemic anoxia and reperfusion, reduce brain damage in dogs and monkeys, and in subgroups of patients in randomized clinical trials. However, their use is controversial since they give no breakthrough effect and can produce postarrest hypotension or rearrest. Targeted agents for the prevention of calcium accumulation in neurons might become available in the future. After a normothermic cardiac arrest of duration 10 to 12 min in dogs, only measures promoting cerebral blood flow and mild hypothermia have reproducibly and significantly improved cerebral outcome.

In over 50 per cent of cases of sudden death outside hospital, standard external cardiopulmonary resuscitation with advanced life support has failed to achieve restoration of spontaneous circulation. In such cases, restoration of spontaneous circulation and promotion of cerebral blood flow, during and after reperfusion from prolonged cardiac arrest, might be achieved with open-chest cardiopulmonary resuscitation or (portable) emergency (closed-chest) cardiopulmonary bypass and a combination of early postcardiac arrest induced hypertension, mild hemodilution, and $P\text{aCO}_2$ control. Cerebral blood flow promotion can be achieved by creating the above-mentioned initial hypertensive bout followed by controlled mild hypertension for a few hours, mild hemodilution with colloid plasma substitute to a hematocrit of about 30 per cent, and $P\text{aCO}_2$ control. After admission to the intensive care unit (ICU), these measures to promote cerebral blood flow could be titrated against mixed cerebral venous $P\text{O}_2$ (superior jugular bulb) to achieve values of at least 30 mmHg. This variable would reflect normal average global cerebral oxygen delivery.

Since its introduction in the 1950s, therapeutic hypothermia has been used for cerebral protection/preservation in cardiac and neurological surgery, but not for resuscitation after normothermic cardiac arrest. The beneficial mechanism of hypothermia during and/or after ischemic or traumatic vital organ insults is multifactorial. The lower the temperature, the better is cerebral protection/preservation, but mild hypothermia (34–36 °C) seems to be more effective for post-cardiac arrest resuscitation than moderate (28–32 °C) or deep (15–25 °C) hypothermia. For hypothermia to be therapeutic, shivering must be prevented. Deep hypothermia causes cardiac arrest and requires cardiopulmonary bypass. Moderate hypothermia is compatible with spontaneous circulation but can cause management problems and arrhythmias, particularly when induced in unstable patients after cardiac arrest. Therefore resuscitative hypothermia was dormant for 30 years. In the late 1980s, the results obtained by us in dogs and by others in rats documented that even mild hypothermia had beneficial effects during or after ischemia. A series of cardiac arrest outcome studies in dogs demonstrated the protective and resuscitative effects of mild hypothermia, which is safe (Safar 1996).

Mild hypothermia should be induced as soon as possible during and after reperfusion, ideally starting during attempts to restore spontaneous circulation in the field, and be sustained for at least 12 h. Implementation of hypothermia requires monitoring of brain temperature as the tympanic membrane or nasopharyngeal temperature, and core temperature as the esophageal, central venous, pulmonary artery, or urinary bladder temperature. Mild hypothermia can be induced in many different ways. Methods available range from the slowest (head–neck–trunk surface cooling which could be started at the scene with cold packs) through intermediate (e.g. additional nasopharyngeal cold irrigation and gastric and intravenous cold loads) to rapid invasive brain cooling. The last of these requires a trained physician and includes intraperitoneal instillation of cold Ringer's solution, blood cooling with cardiopulmonary bypass or other measures, or intracarotid injection of cold Ringer's solution. Although achieving a tympanic membrane temperature of about 35 °C within 15 min of restoration of spontaneous circulation seems desirable, recent data from studies of rats and gerbils suggest that even mild cooling delayed several hours after reperfusion can be beneficial if sustained for 12 or 24 h.

Complete recovery after cardiac arrest (no-flow) of duration about 10 min is clinically important because the average 8-min response time of mobile ICU ambulance services cannot be reduced further. The most promising protocol so far for cerebral resuscitation has been demonstrated using a reproducible cardiac arrest outcome model in dogs (Safar *et al.* 1996); after 11 min of normothermic cardiac arrest, a combination of cerebral blood flow promotion and mild hypothermia (from 15 min to 12 h of reperfusion) achieved complete functional and near-complete histological cerebral recovery. Benefit from either treatment alone has been reported in several animal studies. A clinically feasible protocol based on this latest dog study is ready for clinical feasibility and side-effect trials inside and outside hospitals. Mild to moderate resuscitation hypothermia has also improved outcome after focal brain ischemia or severe hemorrhagic shock in rats and after brain trauma in rats, dogs, and humans.

Conclusion

The complex multifactorial nature of the cerebral post-resuscitation syndrome may obviate discovery of a breakthrough therapy. This does not justify therapeutic nihilism. Potentially neuron-saving measures in cardiac arrest patients, which are documented by outcome models in large animals, should undergo clinical trials to determine feasibility and side-effects. It will not be easy to apply novel cerebral resuscitation measures to clinically dead patients at the scene, in the ambulance, or in the hospital. Such clinical trials are justified because the brain is the target organ of cardiopulmonary–cerebral resuscitation.

References

Koehler, R.C., Eleff, S.M., and Traystman, R.J. (1996). Global neuronal ischemia and reperfusion. In *Cardiac arrest: the science and practice of resuscitation medicine* (ed. N. A. Paradis, H. R. Halperin, and R. M. Nowak), pp. 113–45. Williams & Wilkins, Philadelphia, PA.

Negovsky, V.A., Gurvitch, A.M., and Zolotokrylina, E.S. (ed.) (1983). *Postresuscitation disease*. Elsevier, Amsterdam.

Safar, P. (1996). Prevention and therapy of postresuscitation neurologic dysfunction and injury. In *Cardiac arrest: the science and practice of resuscitation medicine* (ed. N.A. Paradis, H.R. Halperin, and R.M. Nowak), pp. 859–87. Williams & Wilkins, Philadelphia, PA.

Safar, P. and Bircher, N.G. (ed.) (1988). *Cardiopulmonary–cerebral resuscitation: an introduction to resuscitation medicine* (3rd edn). W.B. Saunders, London.

Safar, P., *et al.* (1996). Improved cerebral resuscitation from cardiac arrest in dogs with mild hypothermia plus blood flow promotion. *Stroke*, **27**, 105–13.

1.2.5 Post-resuscitation management

W. Tang and M. H. Weil

Key messages

- There are two stages of cardiac resuscitation: the first is initial resuscitation with re-establishment of spontaneous circulation, and the second is management of post-resuscitation myocardial dysfunction.
- Post-resuscitation myocardial dysfunction includes mechanical dysfunction, characterized by decreased systolic and diastolic functions, and electrical dysfunction, characterized by potentially fatal ectopic ventricular rhythms.
- Doppler echocardiography has been a significant advance in monitoring the function of the heart, particularly in post-resuscitation patients.
- Mechanical circulatory support of the heart only serves as a bridge to more invasive interventions rather than definitive treatment.

Introduction

The concept of post-resuscitation disease as a unique and new nosological entity was introduced by Negovsky (1972). Until recently, however, the practice of cardiopulmonary resuscitation focused primarily on the success of its initial stage and therapeutic options by which a viable rhythm and spontaneous circulation would be restored. Even though the initial cardiopulmonary resuscitation success rate is approximately 39 per cent, with a range of 13 to 59 per cent across the United States, more than 60 per cent of these victims die within 24 h after initial successful resuscitation from cardiac arrest. Therefore cardiopulmonary resuscitation itself has a meaningful success rate for as few as 3 per cent of the 500 000 victims of unexpected cardiac arrest from diverse causes and at diverse sites in the United States each year. Both experimental and clinical studies have indicated that the very high post-resuscitation mortality, which averages more than 50 per cent and ranges from 44 to 71 per cent, may be related to myocardial dysfunction immediately following successful resuscitation.

Two discrete stages of cardiac resuscitation appear to exist. The first stage is that of initial resuscitation with re-establishment of a spontaneous rhythm and circulation. The second stage is that of post-resuscitation dysrhythmia and myocardial failure with increased risk of recurrent cardiac arrest (Tang *et al.* 1993).

Circulatory monitoring after successful cardiac resuscitation

The natural history of post-resuscitation myocardial dysfunction in human patients is not yet fully understood. However, the following hemodynamic abnormalities may indicate the presence of post-resuscitation myocardial dysfunction:

(1) increases in heart rate;

(2) decreases in arterial pressure and cardiac output;

(3) multiple ventricular arrhythmias;

(4) impaired myocardial work capability;

(5) acute ventricular dilation.

Routine hemodynamic and ECG monitoring, including the use of arterial cannulas, pulmonary artery flow-directed catheters, and echocardiographic studies, during the first 72 h following successful resuscitation are practical guides.

ECG monitoring is the standard practice in the measurement of heart rate and cardiac rhythm, with a 12-lead ECG preferred for monitoring post-resuscitation patients in the intensive care unit. Arrhythmias are the immediate cause of death during the first 72 h following initial successful resuscitation. Accordingly, continuous display of a standard electrocardiographic monitoring lead (II) is appropriate practice. The presence of arrhythmias and dysrhythmias such as premature ventricular complexes and supraventricular tachycardia requires clinical reassessment.

Continuous measurement of arterial pressure is recommended during the first 72 h after resuscitation. Arterial pressure reflects overall circulatory status but does not have diagnostic specificity.

The balloon-tipped flow-directed pulmonary artery catheter (Swan–Ganz catheter) allows for differentiation of cardiac failure from hypovolemia by evaluating filling pressures. Higher filling pressures are consistent with cardiac dysfunction. The catheter may be advanced through the subclavian, jugular, brachial, or femoral veins. Mean pulmonary artery pressure and pulmonary artery occlusion pressure may serve as indicators of left ventricular filling pressure, and a thermistor at the tip facilitates cardiac output measurements.

Chest radiography documents the position of endotracheal tubes and intravascular catheters, facilitates differential diagnosis of cardiogenic and non-cardiogenic pulmonary edema, and indicates changes in cardiac size. The effects of increased ventricular filling pressures may manifest with radiographic increases in pulmonary vascular congestion. Therefore serial chest radiographs are useful for the management of post-resuscitation patients.

Intermittent Doppler echocardiography has been a significant advance in monitoring the function of the heart, particularly in post-resuscitation patients. It provides quantitative assessment of both the structure and the function of the heart and great vessels by providing information on the systolic and diastolic functions of both left and right ventricles. Doppler echocardiography during the first 72 h after resuscitation allows for the early diagnosis of post-resuscitation myocardial dysfunction and the effects of therapy.

In addition, factors such as fever, anemia, acid–base balance, and electrolyte and endocrine abnormalities increase the heart rate and myocardial oxygen demand and lead to the further impairment of myocardial function following resuscitation. These factors should be identified and treated promptly.

Circulatory management after successful cardiac resuscitation

Pharmacological interventions

The goals of management of post-resuscitation myocardial dysfunction are (i) improved myocardial systolic and diastolic functions with increases in stroke volume and reduction of ventricular filling pressures, and (ii) control of arrhythmias.

The drugs recommended for the reversal of post-resuscitation myocardial systolic and diastolic dysfunction in the current guidelines for cardiopulmonary resuscitation (Anonymous 1992) are summarized in Table 1 together with dosage, rationale for use, and side-effects. They include inotropic agents (i.e. dobutamine and phosphodiesterase inhibitors), vasopressor agents (i.e. dopamine and levarterenol), and preload and afterload reducing agents (i.e. nitroglycerin (glyceryl trinitrate), nitroprusside, phosphodiesterase inhibitors, and angiotensin-converting enzyme inhibitors).

Dobutamine acts primarily on β_1- and β_2-adrenergic receptors. Its hemodynamic effects include increases in cardiac output and stroke volume, and decreases in systemic and pulmonary vascular resistance, and therefore improvement in systemic and coronary blood flow. Although dobutamine improves post-ischemic myocardial dysfunction, it has the potential of increasing myocardial oxygen consumption owing to its β-adrenergic actions. Therefore it may increase myocardial ischemic injury in settings in which coronary blood flow is critically reduced (Tang *et al.* 1995).

Phosphodiesterase inhibitors such as amrinone have a combination of inotropic and vasodilator effects. These agents increase the concentration of cyclic adenosine monophosphate (cAMP) and myocardial contractility. The therapeutic benefit in the setting of post-resuscitation myocardial dysfunction is not well established and thus amrinone serves as a third-line drug, to be used after other agents have failed.

Dopamine acts on α-adrenergic, β-adrenergic, and dopaminergic receptors. It also prompts norepinephrine (noradrenaline) release from cellular sites. Dopamine increases myocardial contractility, but α_1-adrenergic actions predominate at larger doses (>10 g/kg/min). This induced peripheral vasoconstriction accounts for increased afterload and results in increases in the workload on the heart and myocardial oxygen requirements. Therefore its use in the post-resuscitation setting is restricted to treatment of systemic hypotension. The same considerations apply to the more potent arterial vasoconstrictor, norepinephrine.

Dopexamine is a synthetic analog related to dopamine with intrinsic activity at both dopamine and β_2-adrenergic receptors. It has shown favorable hemodynamic effects in patients with severe congestive heart failure. The potential role of dopexamine in the setting of post-resuscitation myocardial dysfunction remains to be determined.

Table 1 Pharmacological agents available for the management of postresuscitation myocardial dysfunction

Agent	IV dose	Action	Effects: Vasoconstriction	Effects: Vasodilatation	Effects: Cardiac output	Side-effects
Dobutamine	2.5–15.0 μg/kg/min	Primarily β_1, some β_2 receptor agonist	+	++	++++	Ventricular and supraventricular arrhythmias
Amrinone	0.75 mg/kg loading, 5–15 μg/kg/min	Phosphodiesterase inhibitor	–	+++	+++	Arrhythmias, thrombocytopenia
Dopamine	2–20 μg/kg/min	α, β, and dopaminergic receptor agonist	++	++	++	Ventricular and supraventricular arrhythmias
Levarterenol	2–8 μg/min	Primarily α, some β_1 receptor agonist	++++	–	±	Dose-related hypertension, reflex bradycardia
Nitroglycerin	40–600 μg/min		–	++++	±	Tolerance develops within 72 h
Nitroprusside	1–10 μg/kg/min		–	++++	±	Thiocyanate toxicity
Lidocaine	1–4 mg/min	Sodium-channel blocker	–	±	±	Central nervous system toxicity
Adenosine	6–12 mg	Purinergic$_1$ receptor agonist	–	+	±	Heart block

Pharmacological interventions optimizing left ventricular afterload and preload remain the mainstay for the treatment of post-resuscitation myocardial dysfunction. Nitroglycerin and nitroprusside are the predominant vasodilator drugs in current use. At medium and high doses (5–10 g/kg/min) nitroglycerin relaxes both venous and arterial smooth muscle. The more selective coronary vasodilation produced by nitroglycerin also favors its use in the management of post-resuscitation myocardial dysfunction in which cardiac arrest was caused by ischemic heart disease. At low doses, nitroglycerin acts primarily as a venodilator which decreases preload. Intermittent treatment with nitroglycerin after 24 h mitigates tachyphylaxis which accounts for loss of hemodynamic efficacy.

Nitroprusside is a rapid acting venous and arteriolar smooth muscle dilator with a very short half-life. It may elevate plasma thiocyanate concentration, induce an anion gap acidosis, and lead to cyanide toxicity. However, its short-term use is not likely to be constrained by these side-effects, particularly if used in dosages below 10 g/kg/min. It is widely used because it is easily titrated in patients who have both arterial and pulmonary artery monitors.

In patients with coronary atherosclerosis, coronary vasodilation may reduce blood flow of partially occluded vessels (coronary steal). This may explain the increased frequency of angina in patients with ischemic heart disease despite the favorable hemodynamic response of nitroprusside.

The roles of other vasodilators such as angiotensin-converting enzyme inhibitors and calcium-channel blockers for the management of post-resuscitation myocardial dysfunction are less well defined. The relatively long half-lives of these drugs and the need for intravenous formulations limit their roles in post-resuscitation settings.

Lidocaine (lignocaine) remains the drug of choice for the management of post-resuscitation ventricular arrhythmia. Adenosine has emerged as a first-line drug for treatment of atrial tachyarrhythmia with a short duration of action and absence of prolonged systemic hypotension. When tachyarrhythmias lead to hemodynamic instability, prompt d.c. cardioversion (100–360 J) is preferred. The initial therapy for bradyarrhythmia is atropine and a temporary transvenous pacemaker if the bradycardia is persistent and symptomatic. Both hypokalemia and hypomagnesemia must be excluded as causes of post-ischemic arrhythmias and treated appropriately.

No data are currently available to guide the use of buffer agents, and particularly sodium bicarbonate, for the management of post-resuscitation myocardial dysfunction. Recent animal studies indicate that buffer agents administered during cardiopulmonary resuscitation may ameliorate post-resuscitation myocardial dysfunction. Carbon-dioxide-consuming buffers, such as trimethamine (TRIS) and Carbicarb, improve post-resuscitation survival after prolonged cardiac arrest (Sun *et al.* 1996).

Mechanical interventions

The rationale for mechanical support of the heart during myocardial dysfunction is to restore hemodynamic stability when pharmacological interventions fail. However, mechanical support of the heart serves as a bridge to more invasive interventions rather than definitive treatment.

Mechanical ventilation

The rationale for mechanical ventilator support in post-resuscitation myocardial dysfunction patients is to improve systemic oxygenation in settings of pulmonary edema by increasing functional residual capacity. This intervention typically reduces intrapulmonary shunting. Equally importantly, mechanical ventilation is also used to reduce excessive work of breathing. Minimal positive end-expiratory pressure should be applied, and it should be titrated so as to preclude hemodynamic decrements, particularly decreases in cardiac output.

Intra-aortic balloon counterpulsation

Intra-aortic balloon counterpulsation support is indicated for patients with post-resuscitation myocardial dysfunction refractory to pharmacological interventions, refractory angina with or without hypotension, or refractory malignant ventricular arrhythmias, and as a transitional option for patients who fail efforts to reduce cardiopulmonary bypass after surgical operations. The clinical benefit of prolonged intra-aortic balloon counterpulsation support is not securely documented. More often, the intra-aortic balloon counterpulsation allows for more definitive therapy including elective coronary angioplasty, coronary artery bypass surgery, or heart transplantation.

Other options

Left ventricular assist devices and extracorporeal membrane oxygenation are the most complex cardiac assist options. The left ventricular assist device is used in cardiac surgery after unsuccessful attempts are made to wean the patient from cardiopulmonary bypass. Extracorporeal membrane oxygenation has been used, with limited success, to treat refractive hypoxemia in adults and for respiratory distress syndrome.

References

Anonymous (1992). Guidelines for cardiopulmonary resuscitation and emergency cardiac care. III. Adult advanced life support. *Journal of the American Medical Association*, **268**, 2199–241.

Negovsky, V.A. (1972). The second step in resuscitation: the treatment of the post-resuscitation disease. *Resuscitation*, **1**, 1–7.

Sun, S.J., Weil, M.H., Tang, W., and Fukui, M. (1996). Effects of buffer agents on post-resuscitation myocardial dysfunction. *Critical Care Medicine*, **24**, 2035–41.

Tang, W., Weil, M.H., Sun, S.J., Gazmuri, R.J., and Bisera, J. (1993). Progressive myocardial dysfunction after cardiac resuscitation. *Critical Care Medicine*, **21**, 1046–50.

Tang, W., Weil, M.H., Sun, S.J., Noc, M., Yang, L., and Gazmuri, R.J. (1995). Epinephrine increases the severity of post-resuscitation myocardial dysfunction. *Circulation*, **92**, 3089–93.

1.3 Fluid management

1.3.1 Physiology of body fluids

P. L. Khimenko and A. E. Taylor

Key messages

- The fluid exchange between parenchymal cells and extracellular fluid is determined by the ionic imbalance across the cell membranes.
- The fluid exchange between plasma and interstitial space is related to the existing capillary pressure, the tissue fluid pressure, the oncotic (protein osmotic) pressure of plasma and tissue fluids, and the permeability of the capillary endothelial barrier.
- The forces operating across the capillary walls are almost balanced, and the small amount of fluid that enters the interstitium is removed by the lymphatics.
- Interstitial edema occurs when the capillary pressure is increased or the endothelial barrier is damaged and fluid containing protein enters the tissue in amounts that cannot be removed by the lymphatics, and excessive fluid accumulates in potential tissue spaces.
- Interestingly, when capillary pressure increases to 25 to 30 mmHg, little fluid enters the interstitium because tissue fluid pressure is increased, the protein osmotic pressure gradient acting across the capillary wall decreases, and lymph flow decreases. These changes, which oppose capillary filtration, are edema safety factors.

Introduction

In critically ill patients, adequate cardiac output and oxygen delivery must be provided to sustain life. This is not a simple task because plasma exchanges with the extracellular fluid and is influenced by events occurring at the capillary wall and within the endothelial cells. In addition, damage to epithelial cells causes loss of plasma volume into potential spaces. As different pathological conditions develop, such as in the inflammatory response associated with ischemia and reperfusion and in sepsis, various other systems are activated, and maintenance of proper fluid balance becomes difficult to achieve (Aukland and Reed 1993).

Fluid exchange between extracellular and intracellular spaces

Total body water in a normal 70-kg man is about 40 liters, of which 25 liters are in cells and 15 liters are in the extracellular volume. Plasma (3 liters) and the interstitial fluid (12 liters) comprise the extracellular volume. These fluids have identical crystalloid osmotic pressures and when isotonic saline is infused into the vascular system of normal patients, only about a fifth to a quarter of the volume remains in the plasma. If parenchymal cells are hypoxic and cannot maintain their normal ionic distributions, then extracellular fluid enters the cells. Potential fluid spaces, such as the subcutaneous and peritoneal spaces, can hold enormous volumes of extracellular fluid. Overexpansion of plasma volume disrupts epithelial barriers in the gastrointestinal tract and lungs, causing fluid loss into the intestine and the formation of alveolar edema in the lungs. These events often occur during attempts to increase blood flow and oxygen delivery in patients with multiorgan failure and can promote a vicious cycle. Thus the goal is to maintain an adequate cardiac output to provide optimal O_2 delivery in critically ill patients without promoting excessive tissue edema (Taylor *et al.* 1997).

Transvascular exchange

Structure of the capillary wall

Lungs, muscle, and other organs contain microvessels which form a continuous endothelial barrier. Fluid and small solutes cross this barrier through interendothelial junctions, but the larger proteins can only permeate by passing through either larger damaged junctions or large fused plasmalemmal vesicles. Fenestrated endothelial barriers are located in organs in which large amounts of transcapillary volume exchange occur, such as the gastrointestinal tract and renal capillaries. Continuous and fenestrated capillaries are very permeable to small solutes and water, but are relatively impermeable to plasma proteins since they contain numerous small pores of radius 4 to 5 nm and only a very few large pores of radius 20 nm. However, all plasma proteins leak into the interstitium in organs, except at the blood–brain barrier which is impermeable to all solutes, even small solutes which must cross by an active transport process. Capillaries with discontinuous barriers are found in liver sinusoids and have very large gaps in their walls (radius, 100–1000 nm) which allow even the largest plasma proteins to pass easily into the interstitial spaces.

Permeability and biophysical characteristics of the microvessels

The transcapillary volume flow across the endothelial barrier is described by the following equation which relates solvent flow J_{VC} to microvessel pressure P_c, interstitial fluid pressure P_T, and the oncotic (protein osmotic) pressures π_p and π_T of plasma and interstitial fluid respectively:

$$J_{VC} = K_{fc}[(P_c - P_T) - \sigma(\pi_P - \pi_T)]. \quad (1)$$

The filtration coefficient K_{fc} of the microvessel wall describes the permeability characteristics of the solvent. The reflection coefficient σ of the plasma proteins is a measure of the selectivity of the microvascular wall to plasma proteins. It is unity when the endothelial barrier is impermeable and zero when the endothelial barrier is freely permeable. The transcapillary filtration J_{VC} can be positive or negative depending on whether the exchange vessel is filtering or absorbing.

The movement of a solute between plasma and the interstitial spaces is described by the following solute flux equation developed by Taylor and Granger (1985) and Rippe and Haraldsson (1994):

$$J_{SC} = (1-\sigma)J_{VC}C_T + (x/e^x - 1)(PS\Delta C) \quad (2)$$

where C_p is the plasma concentration of the solute, ΔC is the difference between the plasma and tissue concentrations C_T of the solute, PS is the product of solute permeability and surface area and describes the permeability characteristics of the solute, and x describes the dynamics of solute exchange. The first term in eqn (2) is the solute movement associated with convective transport and the second term is the diffusional component of solute flux (Taylor and Granger 1985).

Transvascular fluid exchange

Tissue forces and lymph flow

Figure 1 is a schematic representation of a microvessel, its surrounding interstitium, and a small lymphatic vessel (Taylor 1996). The microcirculation normally filters a small amount of volume into the tissues, which then percolates through them and is removed by the lymphatics without any change in the interstitial volume. When microvessel pressure is increased, more fluid enters the tissues, increasing P_T, decreasing π_T, and increasing lymph flow, and buffers the increased pressure. However, when these forces can no longer change, observable edema develops. The ability of these forces and lymph flow to change and to oppose changes in microvascular pressure has been defined as edema safety factors (Aukland and Reed 1993; Taylor 1996). Table 1 shows normal forces measured in a variety of animal tissues. The sum of the forces $(P_c - P_T) - (\pi_p - \pi_T)$ in eqn (1) is small and usually positive, indicating that net transvascular filtration occurs in normal tissues. However, in organs that are always absorbing or filtering fluid, such as those found in the gastrointestinal tract and in the renal glomerular and peritubular capillaries, the imbalance in forces is large.

Edema safety factors

Edema forms in tissues when microvascular pressure increases and plasma protein concentration decreases, so that the endothelial barrier is damaged. Figure 2 illustrates the formation of edema when microvascular pressure increases. As microvascular pressure increases from about 7 to 25 mmHg, little fluid enters the tissues (curve labeled 'Pressure'). This occurs because, when P_T increases by 5 mmHg, the protein osmotic absorptive force $\sigma(\pi_p - \pi_T)$ increases by 10 mmHg and the lymphatic flow factor provides an additional safety factor of 3 mmHg. As microvascular pressure is increased to 30 mmHg, edema develops at a rapid rate which continues unabated at higher microvascular pressures. This rapid rate of fluid accumulation occurs because the safety factors can no longer change to oppose the elevated microvascular pressure. The curve labeled 'Decreased π' was obtained after π was lowered to half normal. Edema develops at lower microvessel pressures because $\sigma(\pi_p - \pi_T)$ cannot increase substantially when π_P is small. Observable edema does not develop in damaged microvessels until microvascular hydrostatic pressure is higher than that observed when only π_P is low. Obviously, observable edema does occur at lower pressures when the barrier is damaged, but does not occur until microvascular pressure is relatively high (curve labeled 'Permeability'). A recent analysis of this high lymph flow phenomenon concluded that compound(s) are released during the damaging process which cause the lymphatics to increase their pumping capability substantially (Taylor 1996).

Table 2 shows the percentage change in the edema safety factor in a variety of tissues after the microvascular pressure was increased by 20 mmHg, which does not produce observable edema.

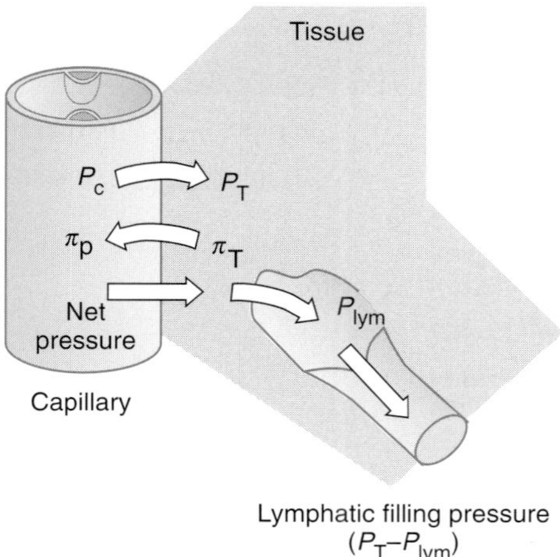

Fig. 1 Diagram of the capillary–tissue–lymphatic system. (Reproduced with permission from Taylor (1996).)

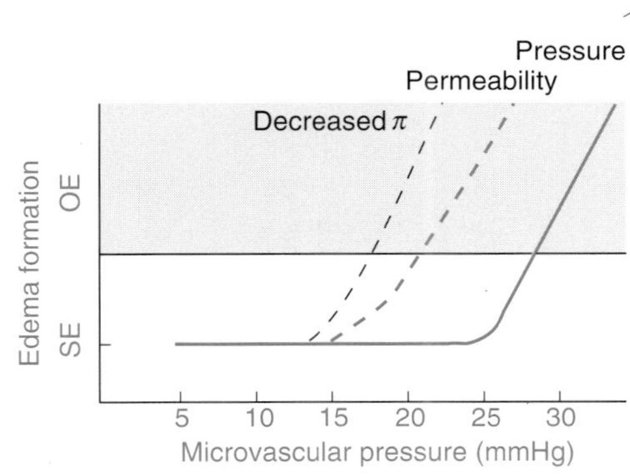

Fig. 2 Edema formation as a function of capillary pressure: OE, observable edema; SE, slight edema; green solid curve, normal lungs; black broken curve, decreased plasma proteins (π_p); green broken curve, damaged capillaries. (Reproduced with permission from Taylor (1996).)

Table 1 Starling and lymphatic flow forces in selected tissues

Tissue	P_c (mmHg)	P_T (mmHg)	π_p (mmHg)	π_T (mmHg)	LF (ml/min/100 g)	ΔP (mmHg)
Subcutaneous	13.0	−5	21	4	0.015	+1
Skeletal muscle	10.1	−3	20	8	0.005	+0.1
Intestine (normal)	16.0	2	23	10	0.08	+1
Intestine (absorbing)	16.0	3	23	5	0.10	−5
Liver	7.0	6	22	20	0.10	−1
Lung	7.0	−5	23	12	0.10	+1
Cardiac muscle	23.1	15	21	13	0.12	+0.1
Glomerular	50.0	15*	28	0*	2.0	+7
Renal peritubular	25.0	7	32	7	2.0	−7

LF, lymph flow.
ΔP is the sum of the forces $(P_c - P_T) - (\pi_p - \pi_T)$ and represents filtration when positive and absorption when negative. A zero sum indicates no filtration or absorption.
*Bowman capsule pressure rather than renal interstitial pressure.
Reproduced with permission from Taylor (1996).

Fluid therapy

No formula exists to predict how much fluid should be given to expand plasma volume in critically ill patients. Figure 3 shows the effect of increasing extracellular fluid volume on blood volume. The maximal blood volume that occurs for a given extracellular volume is a function of the existing microvascular pressures, the compliance of the interstitium, and the permeability of the microvascular barrier. When these factors are increased, a lower blood volume will always result for a given extracellular fluid volume (black broken curve). When all factors are decreased, a larger blood volume results (green broken curve). A higher compliance of the blood vessels will result in a larger blood volume relative to the extracellular fluid volume. The effective microvascular pressure can be estimated in most organs (Taylor 1996) and plasma oncotic pressure can be routinely measured in intensive care units, and so they can help to guide fluid replacement therapy. If, with a given mode of fluid replacement, the measured microvascular and plasma oncotic pressures are normal but the tissues are edematous, particularly in the lung, there is no doubt that the microvascular barrier has been damaged and fluid will leak out into the organ's interstitium. Effective patient therapy will help to stabilize cellular membrances, allowing for adequate resuscitation.

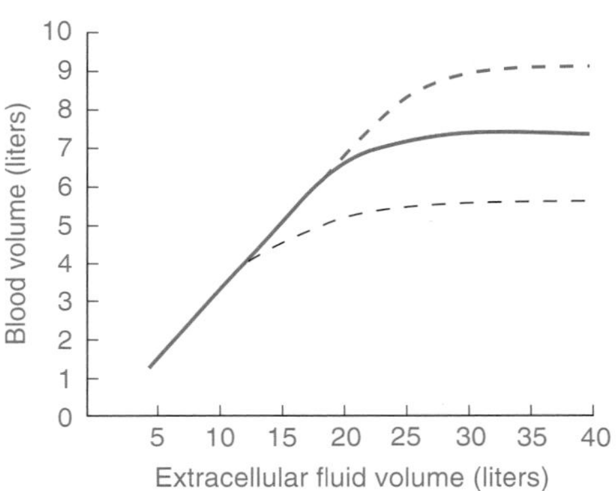

Fig. 3 Relationship between extracellular fluid volume and blood volume: green solid curve, normal conditions; green broken curve, decreased microvascular pressures, decreased interstitial compliance, and less permeable challenge; black broken curve, increased microvascular pressures, increased interstitial compliance, and more permeable challenge. (Reproduced with permission from Guyton *et al.* (1975).)

Table 2 Edema safety factors in selected tissues

Tissue	Increased $\sigma(\pi_p - \pi_T)$	Increased LF	Increased P_T
Lung	50	30	20
Hind paw	14	24	62
Small intestine	45	20	35
Colon	52	4	44
Liver	0	42	58
Heart	7	12	81

The values are shown as a percentage of the total safety factor measured when microvascular pressure was increased by 20 mmHg above control values.
LF, lymph flow.
Modified with permission from Taylor (1996).

References

Aukland, K. and Reed, R.K. (1993). Interstitial-lymphatic mechanisms in the control of extracellular fluid volume. *Physiological Reviews*, 73, 1–78.

Guyton, A.C., Taylor, A.E., and Granger, H.J. (1975). *Circulatory physiology: dynamics and control of body fluid*, pp. 18–52. W.B. Saunders, Philadelphia, PA.

Rippe, B. and Haraldsson, B. (1994). Transport of macromolecules across microvascular walls: the two pore theory. *Physiological Reviews*, **74**, 163–219.

Taylor, A.E. (1996). Microvascular fluid and solute exchange. In *Encyclopedia of human biology*, Vol. 5 (ed. R. Dulbecco), pp. 31–42. Academic Press, Orlando, FL.

Taylor, A.E. and Granger, D.N. (1985). Exchange of macromolecules across the microcirculation. In *Microcirculation handbook of physiology*, p. 467. American Physiological Society, Bethesda, MD.

Taylor, A.E., Khimenko, P.L., Moore, T.M., and Adkins, W.K. (1997). Fluid balance. In *The lung* (ed. R.G. Crystal *et al.*), pp. 113.1–18. Lippincott–Raven, Philadelphia, PA.

1.3.2 Therapeutic strategy for fluid replacement

M. J. Shapiro

Key messages

- Intravenous access should be established initially.
- Crystalloids are osmotically active in most instances and are available in many different types of solutions.
- Colloids are oncotically active and exist in many forms with specific indications for use.

Introduction

The issue on the type of fluids to be used for volume resuscitation has been debated for 50 years and still remains to be settled (Velanovich 1989). Research has focused on many fronts in trying to define the best type of fluid to be used for a particular circumstance, as there exist a wide range of clinical syndromes that can result in the shock state. The main objectives in shock therapy remain the same, independent of the etiology such as myocardial infarction, pulmonary embolism, traumatic bleeding, spinal cord injury, or sepsis: provide optimal oxygen delivery and nutrition to the cells in order to preserve their integrity. Fluid therapy remains a cornerstone for almost every form of shock, although the inciting insult must also be sought and dealt with.

Technical aspects of fluid therapy

After the airway and breathing have been addressed in the hypovolemic patient, circulation must be evaluated. Not only does this involve evaluation of the cardiac status, but also the cessation of life-threatening hemorrhage. Although in most major centers antecubital large-bore intravenous cannulas are placed, the ability to place 9 Fr introducers into central veins (subclavian, internal jugular, and femoral) has allowed aggressive resuscitation to proceed. Strategic location of access is crucial in that a patient with an obvious pneumothorax should have a central catheter placed on that side, although in massive hemothorax the catheter should be placed on the opposite side. In individuals with clinical conditions such as tense ascites, the rapid administration of fluids is best done centrally from above rather than via the femoral route which, owing to inferior vena caval compression, may limit resuscitation.

According to Poiseuille's law, the flow through a catheter depends on the radius to the fourth power, the length of the catheter, the viscosity of the fluid, and the pressure at which it is infused. Thus, large-bore short cannulas are best. The fluid viscosity affects flow rate in that cold packed red blood cells are more rapidly infused if they are diluted in an equal volume of normal saline and warmed. External devices are available to allow for rapid transfusions. In addition, these devices allow fluids to be warmed rapidly to body temperature, helping to ameliorate the hypothermia that frequently accompanies massive resuscitation.

Irrespective of the location, all catheters put in under urgent and emergency circumstances in the field or the emergency department should be changed within 24 h to decrease the incidence of infection and sepsis.

Crystalloids

Crystalloid solutions generally contain sodium as their major osmotically active particle. As shown in Table 1, there is marked variability in the amount of sodium and other electrolytes, pH, and osmolality with the different solutions available. Although the cost differential for the various types of crystalloids is small, there is a theoretical advantage to the utilization of lactated Ringer's solution, which has a more favorable pH and the advantage that the lactate is converted to bicarbonate by the liver. In situations where a metabolic acidosis may exist, this may be helpful in at least not exacerbating it.

The volume of distribution of crystalloids also varies, depending on what is utilized. For instance, 5 per cent glucose distributes throughout the total body water. If the rough estimation for a 70-kg individual is that two-thirds of the total body water is intracellular, then 667 ml/l goes into the cells and 333 ml/l is distributed extracellularly. Of this amount, three-quarters is interstitial and a quarter is intravascular. Thus, approximately 83 ml of each liter administered remains intravascular. In contrast, 0.9 per cent saline or lactated Ringer's solution remains in the extracellular space, so that administration of 1 liter will lead to 250 ml or 25 per cent going intravascularly, which is preferred. In a trauma situation lactated Ringer's solution or 0.9 per cent saline are the crystalloids of choice, whereas in other clinical conditions, such as renal failure or pulmonary edema, electrolyte values may guide the type, rate, and amount of fluid. In patients that are hypernatremic, a hypotonic solution may be utilized.

Interest in hypertonic saline is based on the fact that smaller amounts of fluid can be given with an equivalent or even augmentated improvement in intravascular volume. However, the problem arises that the sodium content may increase rapidly and markedly with hypertonic saline, leading to the serious complications associated with hypernatremia. The intravascular volume increase is for the most part transient.

Hypertonic fluids act as plasma volume expanders as fluid is translocated from the interstitial and intracellular compartments to the intravascular compartment. This can lead to an augmentation of cardiac output, mean arterial pressure, and other hemodynamic variables, and may actually lead to a decrease in intracranial pressure in patients where this is raised.

Table 1 Types of crystalloid

	Glucose (g/l)	Na^+ (mmol/l)	Cl^- (mmol/l)	K^+ (mmol/l)	Ca^{2+} (mmol/l)	Lactate (mmol/l)	Osmolality (mOsmol/kg)	pH	Cost ($/l)
Glucose 5%	50	0	0	0	0	0	252	4	0.81
Saline 0.2%	0	34	34	0	0	0	68	5	0.81
Saline 0.45%	0	77	77	0	0	0	154	5	0.83
Saline 0.9%	0	154	154	0	0	0	308	5	0.81
Lactated Ringer's solution	0	130	109	4	3	28	273	6.0–7.5	0.92
Saline 3%	0	513	513	0	0	0	1026	4.5–7.0	1.30
Saline 7.5%	0	1283	1283	0	0	0	2566	4.5–7.0	1.30

*St Louis University Pharmacy costs 1997.

The volume of crystalloid necessary to resuscitate patients adequately during hypovolemic hemorrhagic shock is three to seven times the blood loss. Advanced Trauma Life Support advocates the administration of 2 liters of crystalloid to adults initially as part of resuscitation. Burn patients follow one of a number of burn formulas for early therapy.

Peripheral edema is expected when crystalloids are used. The presence of edema in the interstitial compartment or soft tissues does not imply that intravascular repletion is complete. In fact, interstitial edema may lead to organ dysfunction if not corrected, and soft tissue edema may lead to extremity or abdominal compartment syndrome. The decrease in plasma oncotic pressure from crystalloid utilization may also contribute to pulmonary edema and acute respiratory distress syndrome. However, this is part of the ongoing controversy related to crystalloid use and remains to be proven.

Colloids

Colloids contain particles in suspension which are too large to pass through semipermeable membranes; thus they exert an oncotic pressure which draws fluid across the membrane into the intravascular space. By increasing intravascular oncotic pressure, fluid is drawn from the interstitial and intracellular spaces into the intravascular space. The oncotic pressure exerted by a specific fluid is dependent on the number of particles present in the plasma, their duration in the plasma, and the rate at which they are removed from the plasma.

Natural colloidal solutions are prepared from human plasma and include albumin, plasma protein fraction, and fresh frozen plasma. Plasma protein fraction is a mixture of proteins which comprises approximately 85 per cent albumin and 15 per cent other serum globulins. Since albumin is the primary component of plasma protein fraction, its volume expansion characteristics are similar to those of albumin.

As indicated in Table 2, two concentrations of albumin are generally utilized, a 5 per cent and a 25 per cent solution. The only major difference is the volume infused. As it is more concentrated, the 25 per cent solution, which usually comes in a 50-ml vial, is much more oncotically active and capable of drawing more fluids intravascularly than the 5 per cent solution. However, in practice the 25 per cent solution is diluted up to a liter total of 5 per cent. Albumin is isotonic and a major component (80 per cent) of the plasma oncotic pressure, with 40 per cent of albumin residing within the intravascular compartment and 60 per cent in the interstitial compartment. Equilibration occurs between the intravascular and extravascular compartments with a plasma half-life for albumin of about 16 h. However, in a pathological state, when membrane integrity is lost, albumin will leak out of the intravascular space into the interstitial and intracellular spaces, drawing water with it.

Solutions of albumin are heated to 60 °C for 10 h during processing to inactivate any viruses. Thus they are relatively risk free with respect to HIV and hepatitis. Although allergic reactions have been reported, with adverse reactions including urticaria, fever, and chills, they are rare.

The major artificial colloids in use are gelatins, starches, and glucose polymers. Owing to its effect on blood coagulation, including decreases in platelet adhesiveness, serum fibrinogen, and platelet factor 3, dextran should be avoided in the acute situation of hemorrhagic hypovolemic shock. Its utilization includes resuscitation of individuals who have microvascular thrombosis secondary to such conditions as polycythemia, since dextran reduces red cell aggregation.

Renal failure related to unrecognized hypovolemia may occur due to the osmotic diuresis induced by dextran along with a toxic effect on renal tubular cells. Anaphylactic reactions are rare, and dextran may result in a falsely elevated blood glucose level when certain methods of serological testing are used. The interference with cross-matching of blood due to *in vitro* aggregation of red cells is another reason for avoiding dextran when transfusions may be required. Gelatins and hydroxyethyl starch solutions form the mainstay of artificial colloid treatment worldwide.

The optimum hematocrit remains under debate (Lundsgaard-Hansen 1996). It has been suggested that patients over 40 undergoing elective procedures should have a hematocrit greater than 30 per cent unless silent myocardial ischemia has been excluded. However, the optimum level of transfusion also remains under debate, with concerns of viral transmission of hepatitis and HIV also needing to be considered (Greenwalt *et al.* 1988). Nonetheless, those patients who have large blood losses require blood replacement in a fastidious

Table 2 Types of colloid

	Na^+ (mmol/l)	Cl^- (mmol/l)	K^+ (mmol/l)	Ca^{2+} (mmol/l)	Osmolality (mOsm/kg)	pH	Cost* ($/l)
Albumin 5%	145	145	<2	0	290	6.9	154.08
Albumin 25%	145	145	<2	0	290	6.9	77.04/100 ml
Dextran 40	150	150	0	0	255	4.0	26.18
Dextran 70	150	150	0	0	309	5.0	17.90
Hydroxyethyl starch 6%	154	154	0	0	310	3.5–7.0	44.17

*St Louis University Pharmacy costs 1997.

fashion. This is one of the advantages of the rapid transfusion systems which also warm the blood as quickly as it is transfused, thus preventing hypothermia. Adverse effects of hypothermia include platelet dysfunction, reversible inhibition of thromboxane synthetase, and decreased clotting factor production. In addition, hypothermia increases hemoglobin oxygen affinity, causing a decrease in oxygen release to the tissues by shifting the oxyhemoglobin curve to the left. Initial urgent blood transfusions may be given using O-negative blood, with roughly 15 min required for typing and another 15 min for cross-matching. Coagulopathy frequently accompanies massive blood transfusions and is exacerbated by hypothermia. The trend for blood component therapy administration initially using packed red blood cells contributes to this coagulopathy since it is deficient in platelets, deficient in coagulation factors 5 and 8, and leads to a dilution of other platelet factors. This can be treated with the administration of fresh frozen plasma and platelet concentrates, although there is no consensus as to the best time to give or how much to give. In the face of surgical lesions (e.g. transected femoral artery), early operation remains the procedure of choice (Spence 1995).

Endpoint of resuscitation

The adequacy of resuscitation can be assessed in many different ways. Basic vital signs such as blood pressure, pulse, urine output, mentation, capillary refill, and skin profusion may guide clinical therapy. In the intensive care unit, use of a pulmonary artery catheter with the evaluation of hemodynamic variables can aid in resuscitation. Appropriate therapeutic goals depend on the clinical situation; a fluid challenge regimen allows optimization of the individual's filling pressures and/or stroke volume without prior assumption of fixed therapeutic endpoints.

The controversy continues

The choice of fluids remains controversial. Although most agree that crystalloids are effective in initial resuscitation of hemorrhagic shock from trauma, there may be other advantages to hypertonic solutions and investigations of combinations of hypertonic solutions, dextran, and hydroxyethyl starch are ongoing (Mattox *et al.* 1991). Once resuscitation is complete in a critically ill patient, hemodynamic variables and serology should be evaluated in order to determine whether crystalloids, colloids, or blood component therapy are necessary. Concerns may range from permeability defects, such as loss of integrity of the alveolar capillary membrane with subsequent transudation of large molecules with free water leading to the onset of acute respiratory distress syndrome, to the risk of viral transmission (Velanovich 1989). In these circumstances clinical evaluation coupled with invasive hemodynamic monitoring may help to guide fluid therapy.

References

Greenwalt, T.J. *et al.* (1988). Perioperative red blood cell transfusion. *Journal of the American Medical Association*, **260**, 2700–3.

Lundsgaard-Hansen, P. (1996). Safe hemoglobin or hematocrit levels in surgical patients. *World Journal of Surgery*, **20**, 1182–8.

Mattox, K.L. *et al.* (1991). Prehospital hypertonic saline/dextran infusion for post-traumatic hypotension. *Annals of Surgery*, **213**, 482–91.

Spence, R.K. (ed.) (1995). Blood management: surgical practice guidelines. *American Journal of Surgery*, **170**, Supplement 6A.

Velanovich, V. (1989). Crystalloid versus colloid fluid resuscitation: a meta-analysis of mortality. *Surgery*, **105**, 65–71.

1.3.3 Therapeutic goals

W. C. Shoemaker

Key messages

- Those patients who can obtain supranormal cardiac index, oxygen delivery, and oxygen consumption do best.
- Non-invasive technology allows continuous on-line display of cardiac function, pulmonary function, and tissue perfusion.

Introduction

Almost half of all deaths are due to acute illness and are associated with or caused by circulatory dysfunction, shock, and its sequela organ failure. Early recognition of shock and optimal initial management are the key to prevention of morbidity and mortality. It is extremely important to recognize shock as early as possible and to treat it while it is still easily reversible.

Conventional approach

The conventional clinical approach to shock is based on imprecise signs and symptoms such as cold clammy skin, altered mental status, weak thready pulse, and unstable vital signs. These subjective and observer-dependent findings are inadequate. They are seriously misleading for three reasons.

1. The signs and symptoms are the same for each etiological type of shock: they all have oliguria, tachycardia, and cardiovascular collapse.

2. The laboratory is not at all diagnostic.

3. The pathophysiology based on a one-dimensional description of each monitored variable fails to consider the interactions of the three major circulatory components: cardiac function, pulmonary function, and tissue perfusion. Tissue perfusion and oxygenation represents the important overall purpose of the circulation to supply body metabolism.

The most commonly monitored variables in hospital in-patients are the mean arterial pressure, heart rate, central venous pressure, and cardiac output. These are measured with the aim of using therapy to bring the values back into the normal range. In several thousand monitored values of patients who died, 76 per cent were restored to the normal range; nevertheless, these patients still went on to die (Shoemaker *et al.* 1988). Survivors had essentially the same values. Thus, something is wrong with the conventional approach to monitoring acutely ill or shocked patients. The wrong variables may be monitored, and normal values may be appropriate for normal healthy people but are not appropriate for critically ill patients, who require increased circulatory function because of their increased body metabolism.

Physiological measurements

In order to describe the time course of physiological events, hemodynamic and oxygen transport variables were monitored preoperatively, during surgery, and in the immediate postoperative period in 708 high-risk critically ill surgical patients. The temporal patterns of the survivors' values were compared with the patterns of those who died. The non-survivors were found to have relatively normal values for cardiac output, oxygen delivery DO_2, and oxygen consumption VO_2, while survivors had markedly increased values for these variables.

The physiology of the shocked state after high-risk surgery can be summarized in terms of oxygen delivery and metabolism. Oxygen delivery is the most direct measurement of circulatory function and the capacity of the circulation to compensate. Oxygen consumption is a direct measure of metabolic activity. It measures what the patient is actually consuming at the time, but not necessarily what is needed. On the basis of these key variables which predict outcome, it is hypothesized that inadequate oxygen delivery in the face of increased metabolic demands due to trauma, surgical operations, blood loss, anesthesia, and infection produce tissue hypoxia, organ dysfunction, organ failure, and death (Shoemaker *et al.* 1992*a*).

Indications for invasive monitoring

In the United States over 98 per cent of patients survive major surgery and 1.5 to 2 per cent die; as there are about 35 million major operations annually, over half a million patients die perioperatively each year. Seven per cent of patients subjected to surgery are defined as high risk (e.g. a history of prior cardiac infarction or stroke, extensive ablative cancer surgery, severe multiple trauma, massive hemorrhage, septic shock, respiratory failure, acute abdominal catastrophe, endstage vascular disease). Thirty-four per cent of these patients died.

Clinical decisions for each defect versus an overall therapeutic plan

In the traditional 'one at a time' approach, each monitored variable is identified, measured, and corrected. Often, simple solutions are applied to complex problems although the underlying pathophysiology is not understood. For example, if the pressure goes down, fluids are given; if that does not work, dopamine is given. Then the next deficiency is diagnosed, measured, and corrected, and so on. This results in an uncoordinated, disorganized, and sometimes contradictory therapeutic plan. Eventually all the appropriate therapy may be given, but not always at the right time, in the right doses, and in the right order.

An alternative is to use the physiological patterns of survivors as therapeutic goals. The survivors developed a pattern of supranormal cardiac index, DO_2, and VO_2. The hypothesis is that if high-risk patients are prophylactically driven to the survival pattern with early aggressive therapy that optimizes cardiac output, DO_2, and VO_2, there will be improved outcome.

In a prospective preoperatively randomized trial of this hypothesis, each high-risk patient was identified preoperatively and randomized to one of three groups:

(1) the central venous catheter;

(2) the pulmonary artery catheter with normal values as the goal of therapy;

(3) the pulmonary artery catheter protocol group with supranormal values as the goal.

If normal values were the goal, the central venous catheter was as effective as the pulmonary artery catheter. In contrast, the mortality of the pulmonary artery supranormal group was significantly reduced from 32 per cent to 4 per cent ($p < 0.02$), there was a significant reduction of 67 per cent in ventilator days, a reduction of 30 per cent in intensive care unit and hospital days, and a reduction of 25 per cent in the cost of treatment (Shoemaker *et al.* 1988).

Of interest was a group of high-risk patients who were not considered to be sick enough to need invasive monitoring. This 'non-randomized group' had the highest mortality and highest percentage of organ failure; ironically, 60 per cent had a pulmonary artery catheter placed postoperatively after they developed a life-threatening postoperative cardiorespiratory event. However, placement of the pulmonary artery catheter at this time did not improve the overall group mortality. This suggests that the pulmonary artery catheter can prevent but is not able to reverse lethal organ failure.

The cardiac index values of group 3 reached optimal values within 8 to 12 h postoperatively, while the values of group 1 remained at the preoperative levels. There were no significant differences in the other hemodynamic variables. Similarly, the oxygen transport values of group 1 were maintained at their preoperative normal values, while the $D\text{O}_2$ and $\dot{V}\text{O}_2$ values of group 3 reached their optimum goals in the first 12 h postoperatively. There were no significant differences in blood gases, hematocrit, or other oxygen transport variables.

The striking difference between the groups was the incidence of organ failure. Group 3 had only one patient with acute respiratory distress syndrome, while group 1 had significantly greater numbers of patients with organ failures. There were no significant differences between the groups in the incidence of complications not due to organ failure. The protocol did not protect against local mechanical or anatomic complications such as wound infection, dehiscence, and postoperative bleeding (Shoemaker *et al.* 1988). The importance of early therapy has been demonstrated in a meta-analysis. Seven randomized studies performed early showed improved outcome, but seven others performed late, after onset of acute respiratory distress syndrome or other organ failures, did not show improvement.

Oxygen debt as the underlying cause of organ failure and death

The basic underlying problem in all shock is tissue hypoxia which can be measured as the net cumulative oxygen debt. If this oxygen debt is prevented or corrected early on, there will be a reduction in organ failure and death. Oxygen debt can be determined directly in surgical patients by measuring oxygen consumption preoperatively and, after correcting for the effects of anesthesia and temperature, extrapolating this value to the intraoperative and immediate postoperative periods.

Oxygen debt due to reduction of $\dot{V}\text{O}_2$ to a level less than that needed by the body is the common denominator in all forms of shock. Oxygen debt from reduced $\dot{V}\text{O}_2$ is also the major determinant of outcome.

Invasive hemodynamic monitoring as a method of improving outcome

At least two-thirds of patients who die postoperatively do not die as a result of anatomical reasons but because of physiological problems that can be described, predicted, and prevented. Values observed in surviving critically ill high-risk patients should be considered as the goals of therapy. If these goals are achieved early, i.e. in the first 8 to 12 h postoperatively, there will be a reduction in morbidity and mortality.

Choice of fluid therapy

Invasive monitoring is extremely useful for evaluating objectively and scientifically the relative efficacy of alternative therapies in various clinical conditions. In a prospective random-order cross-over designed study, patients in early acute respiratory distress syndrome (defined as within 24 to 48 h after the diagnosis was established) were given 1.0 liters of Ringer's lactate and 100 ml of 25 per cent albumin. Each patient received each agent: half received Ringer's lactate first and then crossed over to albumin, and the other half received albumin first and then Ringer's lactate. Infusion of 1 liter of Ringer's lactate increased blood volume by only 200 ml, whereas infusion of 100 ml of 25 per cent albumin increased blood volume by an average of 450 ml. In essence, albumin did not leak but, rather, pulled 350 ml of interstitial water back into the plasma volume with a concomitant increase in cardiac index, colloid osmotic pressure, and O_2 consumption without worsening the $P(\text{A}-\text{a})\text{O}_2$ gradient or pulmonary shunt. Albumin also significantly improved both $D\text{O}_2$ and $\dot{V}\text{O}_2$, whereas Ringer's lactate only transiently increased $D\text{O}_2$ and actually decreased $\dot{V}\text{O}_2$.

Comprehensive non-invasive physiological monitoring

Non-invasive circulatory monitoring is attractive because it is easier, cheaper, and more feasible to use. Although previous bio-impedance instruments have been unreliable, new high-technology hardware and software innovations developed by the defense industry have greatly improved bio-impedance technology so that the data are more reliable.

Comparison of a new bio-impedance method (Renaissance Technologies, Newtown, PA) with the standard thermodilution technique in 842 paired measurements showed a close correlation ($r = 0.86$) in the ability to track and trend thermodilution changes (Shoemaker *et al.* 1994).

The new impedance cardiac output system can be combined with other non-invasive monitoring systems, including pulse oximetry to assess pulmonary function and transcutaneous oximetry to assess tissue oxygenation. These non-invasive monitors evaluate the interactions of the three circulatory components to identify, diagnose, and recommend treatment at an early stage when circulatory problems are easily reversible.

Irrespective of the precipitating event in shock, one should look at cardiac, pulmonary, and tissue perfusion in terms of the primary problem, pathophysiology, compensation mechanisms, decompensations, and therapy. This allows a more comprehensive approach to acute circulatory problems irrespective of the precipitating etiological event (cardiogenic, hemorrhagic, traumatic, septic, or postsurgical).

Conclusions

The pulmonary artery catheter has translated information previously only available in the cardiac catheterization laboratory to the bedside of patients in intensive care and has changed the approach to treating these patients. Similarly, non-invasive monitoring combined with advanced information systems can change the way that we treat severe acute illnesses as both these technologies are applied more broadly to improve analysis of interactions of multiple continuous functions as they unfold in time. In addition, intelligent programs will be used to individualize (i.e. customize) complex therapeutic algorithms according to clinical categories, comorbid states, demographic data, and time. They will be used to describe the postoperative trajectory of multiple variables and to predict the outcome of a given patient with known comorbid conditions in concert with the trajectories of a group of survivors and a group of non-survivors with the same comorbid conditions (Shoemaker *et al.* 1992*b*; Patil *et al.* 1996).

References

Patil, R.S., Adibi, J., and Shoemaker, W.C. (1996). Application of an artificial intelligence program to therapy of high risk surgical patients. *New Horizons*, **4**, 541–50.

Shoemaker, W.C., Appel, P.L., Kram, B., Waxman, K., and Lee, T.S. (1988). Prospective trial of supranormal values of survivors as therapeutic goals in high risk surgical patients. *Chest*, **94**, 1176–86.

Shoemaker, W.C., Appel, P.L., and Kram, H.B. (1992*a*). Role of oxygen debt in the development of organ failure, sepsis, and death in high-risk surgical patients. *Chest*, **102**, 208–15.

Shoemaker, W.C., Patil, R., Appel, P.L., and Kram, H.B. (1992*b*). Hemodynamic and oxygen transport patterns for outcome prediction, therapeutic goals, and clinical algorithms to improve outcome: feasibility of artificial intelligence to customize algorithms. *Chest*, **102**, 617S–25S.

Shoemaker, W.C., *et al.* (1994). Multicenter trial of a new thoracic electric bioimpedance device for cardiac output estimation. *Critical Care Medicine*, **22**, 1907–12.

1.3.4 The fluid challenge

A. R. Webb

Key messages

- An adequate circulating volume must be provided before considering other methods of circulatory support.
- The use of fluid to restore circulating volume overcomes many of the problems associated with inotropes in the hypovolemic patient.
- Smaller volumes of colloid than crystalloid are required for resuscitation with less contribution to edema.
- The volume required for correction of hypovolemia is determined by the hemodynamic effect of small aliquots of plasma substitute.

Introduction

The fluid challenge is a method of safely restoring circulating volume. Rather than using fixed hemodynamic endpoints, fluid is given in small aliquots with assessment of the dynamic hemodynamic response to each aliquot.

Fluids and circulatory support

Apart from the utilization of oxygen, correction of plasma volume deficit is the most critical intervention in shock. Early administration of fluid often corrects features of shock such as hypotension or oliguria. Hypovolemia must be treated urgently to avoid the serious complication of progressive organ failure. Indeed, milder degrees of hypovolemia are associated with the pathophysiological changes leading, ultimately, to organ failure or dysfunction.

An adequate circulating volume must be provided before considering other methods of circulatory support. This remains true in the patient with capillary leak, where some would choose to reduce the circulating volume in an attempt to avoid edema formation while supporting the circulation with inotropes ('dry and drive'). Figure 1 shows the effects of inotropes or colloid solutions on components of the Starling equation. The main effect of both inotropes and colloids is to increase capillary pressure; both therefore increase transcapillary fluid flux. The use of inotropic agents in hypovolemic patients often leads to tachycardia with an associated increase in myocardial oxygen demand. Inotropes also have various effects in different parts of the circulation depending on the balance of α, β, and dopaminergic stimulation. Since the endogenous stimulation of these receptors is not predictable for the critically ill individual, it follows that the balance of vasodilatation versus vasoconstriction is also not predictable for particular tissues, particularly where there is an endogenous response to hypovolemia. The result is an unpredictable maldistribution of flow.

The use of fluid to restore circulating volume overcomes many of the problems associated with inotropes in the hypovolemic patient. Increase in stroke volume is achieved without further increase in sympathetic drive, thus avoiding the problem of tachycardia. Indeed, in those who have a tachycardic response to hypovolemia, restoration of circulating volume would reduce the tachycardia and myocardial oxygen demand. Likewise, a reduction in peripheral vasoconstriction secondarily induced by hypovolemia will improve microcirculatory blood flow. Only after the circulating volume has been restored can the use of additional circulatory support be recommended to minimize problems associated with endogenous sympathetic drive.

Choice of fluid

The aim of a fluid challenge is to produce a small but significant (200 ml) and rapid increase in plasma volume. Colloid fluids are ideal, in that they give a reliable increase in plasma volume; crystalloids are rapidly lost from the circulation and larger volumes would be required to achieve an initial comparable increase of 200 ml in plasma volume. Furthermore, from Fig. 1 it can be seen that colloid fluids may increase the capillary oncotic pressure π_c and the reflection coefficient σ, thus increasing the intravascular retention of plasma.

Where blood loss is the cause of hypovolemia, it would seem logical to use blood as the replacement fluid. However, while whole blood would achieve an adequate increase in circulating volume, the high hematocrit of packed red cells will correct red cell volume with less effect on plasma volume. Therefore it is appropriate to replace the plasma volume deficit with a plasma substitute. Since the effect of colloid on the reflection coefficient is determined by the combination of the molecular size and the effective capillary pore size, a lower-

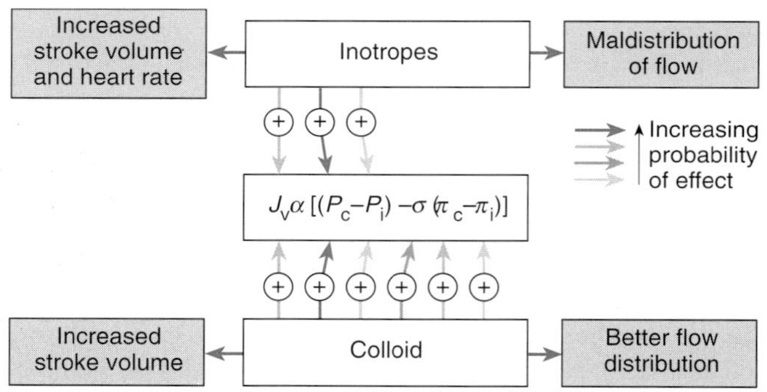

Fig. 1 Effects of colloid or inotropes in the hypovolemic patient. Transcapillary fluid flux J_v is increased by both inotropes and colloid since both mainly increase capillary hydrostatic pressure P_c. Colloids have the advantage of providing better microcirculatory flow, and they may increase intravascular oncotic pressure π_c and the reflection coefficient σ, thus increasing intravascular retention of fluid. Inotopes increase myocardial oxygen demand by increasing heart rate. Minor effects on interstitial pressures P_i and π_i are probably of little significance.

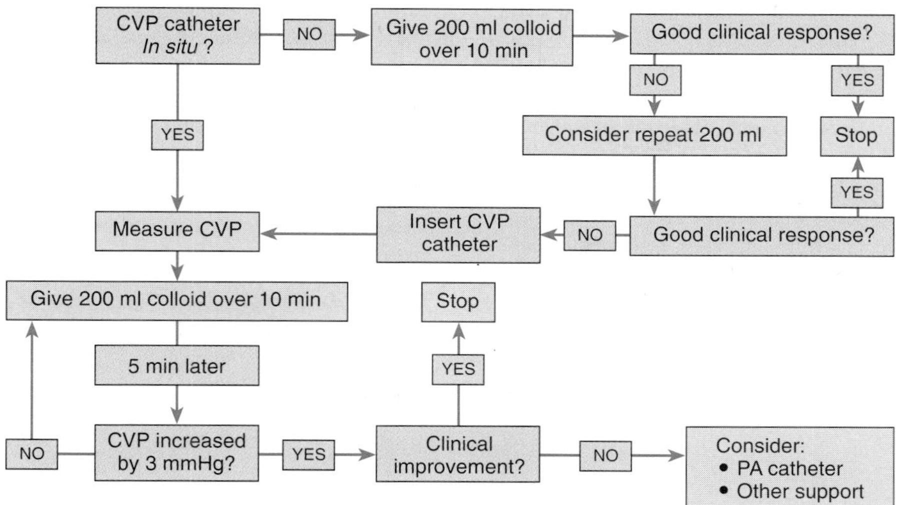

Fig. 2 Algorithm detailing the use of dynamic CVP responses to guide fluid therapy.

molecular-weight colloid (e.g. gelatin solution, low-molecular-weight hydroxyethyl starch, dextran) is recommended for short-term plasma volume expansion in simple hypovolemia and high-molecular-weight hydroxyethyl starch is recommended where there is a probability of capillary leak (Webb 1991).

Diagnosing hypovolemia

Clinical signs of hypovolemia (reduced skin turgor, low central venous pressure, oliguria, tachycardia, and hypotension) are late indicators. A high index of suspicion must be maintained; a normal heart rate, blood pressure, and central venous pressure (CVP) do not exclude hypovolemia (Weil *et al.* 1965), and the CVP is particularly unreliable in pulmonary vascular disease, right ventricular disease, isolated left ventricular failure, and valvular heart disease. The absolute CVP and pulmonary artery wedge pressure (PAWP) are also difficult to interpret since peripheral venoconstriction may maintain CVP despite hypovolemia; indeed, in patients with an intact sympathetic response to hypovolemia the CVP may fall in response to fluid (Baek *et al.* 1975). The response to a fluid challenge is the safest method of assessment.

Assessing the response to a fluid challenge

The response of CVP or PAWP should be monitored during a fluid challenge (Fig. 2). Fluid challenges should be repeated when the response suggests continuing hypovolemia. However, if such monitoring is not available it is reasonable to assess the clinical response to up to two fluid challenges (200 ml each).

The change in CVP after a 200-ml fluid challenge depends on the starting blood volume. A rise of 3 mmHg in CVP represents a significant increase and is probably indicative of an adequate circulating volume. However, a positive response may sometimes occur in the vasoconstricted patient with a lower blood volume. It is important to assess the clinical response in addition; if it is inadequate, it is appro-

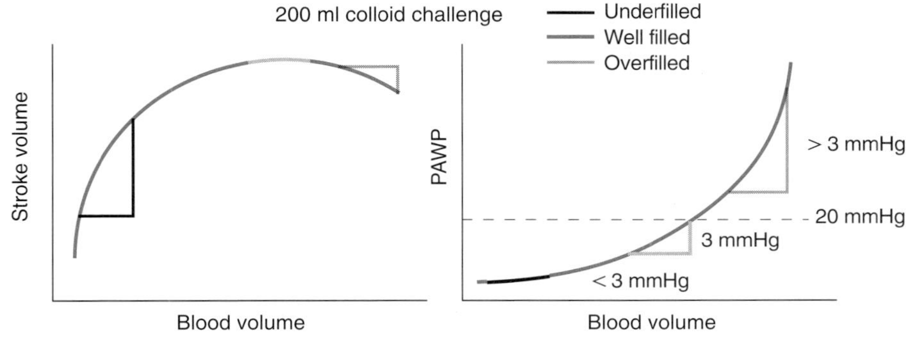

Fig. 3 The response of stroke volume and PAWP to small-volume fluid challenges. In the hypovolemic patient an increase in stroke volume with no significant rise in PAWP would be expected. In the optimally filled patient a rise in PAWP with no significant rise in stroke volume would be expected.

priate to monitor stroke volume and PAWP before further fluid challenges or considering further circulatory support.

In the inadequately filled left ventricle a fluid challenge will increase the stroke volume (Fig. 3). Failure to increase the stroke volume with a fluid challenge may represent an inadequate challenge, particularly if the PAWP fails to rise significantly (3 mmHg). This indicates that cardiac filling was inadequate and the fluid challenge should be repeated. Such a response may also be seen in right heart failure, pericardial tamponade, and mitral stenosis. It may be important to monitor stroke volume rather than cardiac output during a fluid challenge. If the heart rate falls appropriately in response to a fluid challenge, the cardiac output may not increase despite an increase in stroke volume.

References

Baek, S.M., Makabali, G.G., Bryan-Brown, C.W., Kusek, J.M., and Shoemaker, W.C. (1975). Plasma expansion in surgical patients with high central venous pressure (CVP); the relationship of blood volume to haematocrit, CVP, pulmonary wedge pressure, and cardiorespiratory changes. *Surgery*, **78**, 304–15.

Webb, A.R. (1991). The physical properties of plasma substitutes. *Clinical Intensive Care*, **1**, 58–61.

Weil, M.H., Shubin, H., and Rosoff, L. (1965). Fluid repletion in circulatory shock. *Journal of the American Medical Association*, **192**, 84–90.

2 The respiratory system

2.1 Normal physiology

2.1.1 The respiratory system

G. Hedenstierna and H. U. Rothen

Key messages

- In the healthy, quietly breathing subject, both ventilation V_A and perfusion Q increase down the lung.
- In addition, non-gravitational distribution of perfusion has been described.
- On average, $V_A/Q \approx 1$, with higher ratios in upper lung regions and lower ratios in lower lung regions.
- Important factors influencing V_A/Q are the gradient of the vertical pleural pressure, the shape of the thoracic cavity and the lung, forces exerted by abdominal organs, airway closure in dependent lung regions, gravitational forces, length of pulmonary arterial vessels, and cardiac output.

Introduction

In this chapter we focus on the distributions of pulmonary ventilation and blood flow under normal conditions, i.e. in the awake healthy subject. Aspects of the mechanical behavior of the lung will also be described, but only with the intention of explaining the distributions of ventilation and blood flow.

Distribution of ventilation

There are approximately 150 million alveoli in each lung, and even in the healthy subject inspired gas is not uniformly distributed between all alveoli. During quiet breathing ventilation increases down the lung, i.e. in the gravitational direction. In an upright subject the basal dependent regions are more ventilated than the apical regions, and in the supine position dorsal lung units are better ventilated than ventral regions (Milic-Emili 1991; Nunn 1993). In the lateral position the dependent lung receives more ventilation than the upper non-dependent lung.

Pressure–volume curve of the lung

During quiet breathing the distribution of inspired gas is governed mainly by the pressure–volume curve of the lung and the vertical pleural pressure gradient. The non-linear relationship between pressure and volume results in a larger volume increase for a given increment in pressure at low compared with high lung volumes. The pleural pressure increases down the pleural space by approximately 0.2 to 0.4 cmH_2O/cm distance. This results in a lower transpulmonary pressure (approximated by airway pressure minus pleural pressure) in dependent than in upper lung regions. Therefore dependent lung regions are positioned on the lower steeper part of the pressure–volume curve (Fig. 1). During inspiration the transpulmonary pressure is increased to an equal extent all over the lung; lower and upper lung regions move along the pressure–volume curve with similar pressure increments. Therefore the change in volume (or

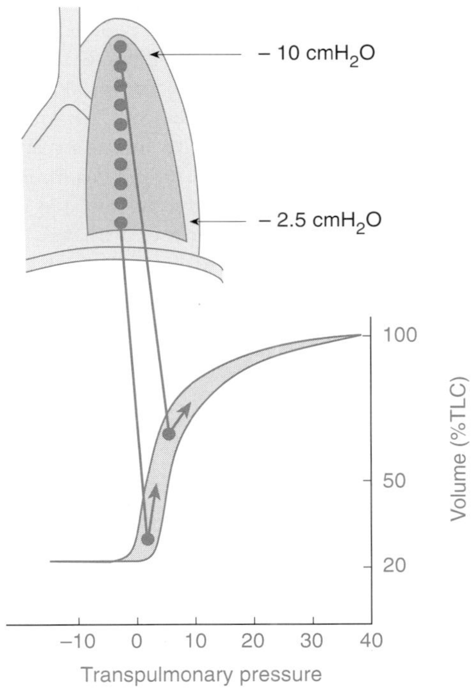

Fig. 1 Static pressure–volume curve of the total respiratory system in an upright subject: TLC, total lung capacity. The basal lung units are on a steeper part of the pressure–volume curve than the apical units. Arrows show the changes during quiet tidal breathing.

ventilation) will be larger in dependent regions of the lungs (Milic-Emili 1991; Nunn 1993).

The pleural pressure gradient is caused by the weight of the lung tissue. In addition, differences in the shape of the thoracic cavity and the lung, and the forces exerted by the abdominal organs on the diaphragm, may also affect the distribution of ventilation. The influence of the latter factor may explain why ventilation may be more uniform from top to bottom in the prone than in the supine position. It is of interest to note that ventilation appears to be evenly distributed in the lungs if gravitational forces are absent, as evidenced by measurements in short periods of weightlessness.

Airway closure

Both alveolar volume and airway dimensions are dependent on the transpulmonary pressure. Airway caliber decreases during expiration. Airways close during a deep expiration, when pleural and extramural airway pressures become positive and higher than luminal pressure (Milic-Emili 1991). Closure of airways during expiration will begin in the most dependent lung regions, owing to the pleural pressure gradient, and spread up the lung with deeper expiration. Thus airway closure will impede ventilation mostly in dependent lung regions.

Airway closure is smallest in subjects aged about 20 years, where it may be difficult to demonstrate even after expiration to residual volume. Above the age of about 65 years airway closure becomes more prominent and can be detected even above functional residual capacity (FRC) in the upright position. Closing capacity, which is the lung volume at which airways begin to close during expiration, appears to be independent of body position. Since the change in position from upright to supine reduces FRC by approximately 0.7 to 0.8 liters, airway closure will be seen more regularly above FRC in the supine position and can be expected at ages of about 50 years and older. Thus ventilation of dependent lung regions will decrease with increasing age. Since blood flows predominantly to dependent lung regions (see below), the phenomenon of airway closure seems to be an explanation of the well-known decrease in $P\text{aO}_2$ with age.

The increase in airway closure with increasing age is probably caused by a loss of elastic tissue in the lung. Thus the elastic recoil of the lung decreases, making the pleural pressure less negative. This also explains why FRC slowly increases with age (corrected for changes in body size) (Fig. 2). Further causes of an increase in closing capacity include a loss of elastic recoil due to lung disease (chronic bronchitis, emphysema) or an increase in airway tone

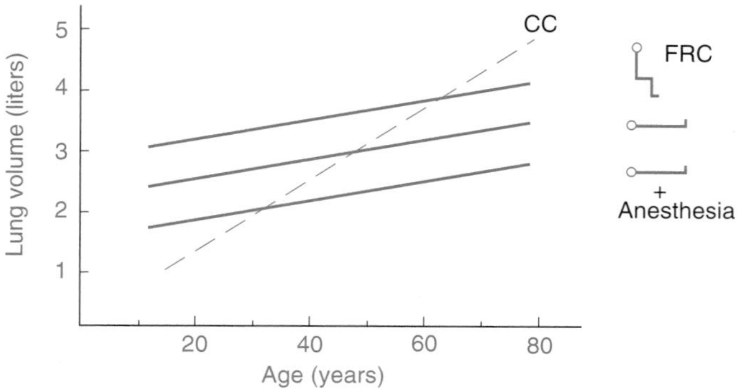

Fig. 2 Schematic diagram of the age dependence of FRC and closing capacity (CC) in the awake subject sitting, supine, and supine and anesthetized. See text for further explanations.

(asthma). Decrease in FRC can also be seen during anesthesia which may promote closure of airways and impede ventilation in dependent lung regions. Long-standing closure of airways will promote resorption atelectasis. While collapse of a lung unit will take several hours if it is filled with air, gas resorption and subsequent atelectasis may occur within 10 min on oxygen breathing. Breathing oxygen at very low lung volume, close to or at residual volume, further promotes atelectasis formation. Atelectasis is also a common finding in anesthetized subjects, and it can be hypothesized that any acute lung failure accompanying a small lung volume is sensitive to the composition of the respiratory gases. It is possible that the administration of oxygen and nitrous oxide contributes to the pathology of the lung as seen on CT scans.

Inspiratory flow resistance

Breathing at higher flow rates (e.g. during exercise) will affect the ventilation distribution because of an increasing influence of airway resistance. Owing to the vertical pleural pressure gradient, dependent airways are less dilated than those in upper lung units. Accordingly, resistance to gas flow will be higher in dependent lung regions and impede ventilation to these regions more than to upper regions. Thus the vertical distribution of ventilation within the lung becomes more uniform with increasing inspiratory flow rate (Milic-Emili 1991). Since the pulmonary perfusion is also more evenly distributed at a higher cardiac output (see below), ventilation–perfusion matching is approximately preserved compared with resting conditions. In addition, the more efficient use of the alveolar capillary bed during exercise increases the area available for diffusion of O_2 and CO_2.

The expiratory resistance may also influence the distribution of ventilation. Chronic obstructive lung disease (asthma, chronic bronchitis, emphysema) causes airway narrowing that may become apparent mainly or solely during expiration because of an expiratory muscle activity that compresses the lung and the airways. Lung regions that have not emptied completely because of an expiratory resistance exert an increased elastic recoil which will impede the inspiration of gas to those regions. It is not clear to what extent this has any bearing on acute lung disease requiring intensive care.

Distribution of pulmonary blood flow

Vertical distribution

Gravitational forces increase blood flow down the lung so that it is unevenly distributed. The lung has been described as being made up by a number of zones defined by the driving forces for blood flow (Hughes 1991; Nunn 1993). An upper lung region may exist where alveolar pressure exceeds arterial and venous pressures, and the alveoli compress the alveolar capillaries and obstruct their blood flow (zone I, no blood flow zone). Although the alveolar capillaries are closed and blood flow has ceased through them, a persistent tiny blood flow in zone I has been demonstrated. Histological studies of excised lung tissue have shown that this blood passes through corner vessels, i.e. vessels located in the junctions between alveolar septa. These vessels appear to be subject to forces other than those acting on the alveolar capillaries and they are kept patent even when in zone I. By using the multiple inert gas elimination technique (see below) it has also been demonstrated that this corner vessel blood flow participates in gas exchange, creating a distinct mode known as the high V_A/Q mode.

Further down the lung arterial pressure increases sufficiently to exceed alveolar pressure, although the latter is still higher than venous pressure. In this region (zone II) the driving pressure equals arterial pressure minus alveolar pressure. The effect on blood flow in this zone has variously been called the 'sluice', the 'waterfall phenomenon', or the 'Starling resistor'. In a simple model it can be seen that a constriction will develop at the downstream end of a collapsible tube surrounded by a pressure that is higher than venous pressure. The pressure inside the tube at the collapse point will then be equal to the external pressure. When the tube is completely collapsed, the higher arterial pressure will be transmitted to the collapse point and reopen the vessel. This may result in an unstable and fluttering vascular wall. Because of an increasing arterial pressure down this zone and a maintained alveolar pressure, blood flow increases down the zone.

Still further down the lung both arterial and venous pressures exceed alveolar pressure. Thus the driving force is arterial pressure minus venous pressure, similar to that in the systemic circulation. The pressure is constant down this zone (zone III), with hydrostatic pressure adding equally to both arterial and venous pressures. However, there is still some increase in blood flow down this zone, indicating a decrease in vascular resistance, presumably as a consequence of increasing dilatation of the alveolar capillaries. In the lowermost region of the lung, blood flow may be reduced (zone IV). This may be attributable to compression of extra-alveolar vessels brought about by increasing interstitial pressure.

Non-gravitational perfusion distribution

There also appears to be a non-homogeneous distribution of perfusion in non-gravitational planes, i.e. in the medial–lateral and apical–basal directions in the supine position. Thus more perfusion in central parts of the lung than in the periphery has been found during mechanical ventilation in the anesthetized dog, and a redistribution of blood flow towards the lateral and basal borders has been observed during ventilation with positive end-expiratory pressure. More recent studies of the perfusion distribution in an awake healthy man using single-photon-emission CT (see below) suggest a maximum blood flow in the middle of the lung, although different results and opinions have also been presented (Hughes 1991). It has been proposed that the decreasing blood flow towards the periphery of the lung is due to the increasing length of the pulmonary arterial vessels which increases the resistance to blood flow. Blood flow has also been fitted to a fractal model with the provocative conclusion that gravity is only a minor determinant of blood flow distribution in the lung. However, full agreement has not been reached on either the degree of non-gravitational inhomogeneity of perfusion or possible mechanisms behind such inhomogeneity.

Influence of cardiac output

When cardiac output is decreased by blood loss, flow is almost absent in upper regions of the lungs (Hughes 1991; Nunn 1993). This can be attributed to a decrease in pulmonary artery pressure while alveolar pressure remains unaffected. An increased dead-space, as measured by CO_2 elimination, has been seen in dogs under such circumstances, and this observation can be explained by abolished perfusion of upper lung regions (zone I).

In contrast, blood flow is proportionally more increased in apical than in basal lung regions after an increase in cardiac output. This redistribution of blood flow can be explained by the increase in pulmonary artery pressure that can be seen during exercise. It is also worth noting that the more even distribution of blood flow during exercise is accompanied by a more even ventilation distribution (see above) and an increase in the alveolar capillary membrane area available for gas transfer, thus facilitating gas exchange.

Influence of intrathoracic pressure

Increased intrathoracic pressure will impede the return to the right heart and therefore reduce cardiac output. It may also affect pulmonary vascular resistance, although this may be a less important modifier of cardiac output. However, it should be mentioned that pulmonary vascular resistance appears to be lowest at a lung volume close to FRC and increases regardless of whether lung volume is increased or reduced from that level. The diphasic dependence on lung volume can be explained by an increase in capillary resistance with increasing lung volume and an increase in extra-alveolar lung vessel resistance with decreasing lung volume.

Ventilation–perfusion ratio (V_A/Q)

If the distributions of ventilation and perfusion are known, V_A/Q can be estimated. The vertical distributions of ventilation and blood flow in a healthy subject, both awake and anesthetized, are summarized in the upper panels of Fig. 3. As discussed earlier, both ventilation and blood flow increase down the lung, with the fractional increase of blood flow being larger than that of ventilation. The average V_A/Q is approximately unity, with higher ratios in upper lung regions and lower ratios in lower lung regions. Thus V_A/Q will vary from approximately 0.5 to 5. V_A/Q can also be assessed without direct measurement of the spatial distributions of ventilation and blood flow.

Shunt and dead-space

An overall estimate of V_A/Q can be obtained by simple arterial blood gas analysis. A three-compartment analysis can be performed by recording the O_2 and CO_2 content in arterial and mixed venous blood and mixed expired gas. This enables the calculation of one compartment that is perfused but not ventilated (venous admixture), one compartment that is ventilated but not perfused (dead-space), and an ideal compartment that is both ventilated and perfused (Nunn 1993). However, venous admixture in this three-compartment model includes perfusion not only of non-ventilated lung tissue (true pulmonary shunt), but also of regions which are poorly ventilated or perfused in excess of their ventilation (low V_A/Q regions). Similarly, the dead-space compartment includes not only true dead-space (airways plus non-perfused alveoli) but also regions which are ventilated well in excess of their perfusion (high V_A/Q regions).

Multiple inert gas elimination technique

By using more than two gases with different solubilities in blood a better resolution of V_A/Q can be obtained, allowing a multicompartmental analysis. This can be achieved with the multiple inert gas elimination technique (Hedenstierna 1995). Ventilation and blood flow can be allocated to a number (e.g. 50) of hypothetical compartments ranging from shunt ($V_A/Q = 0$) through compartments with

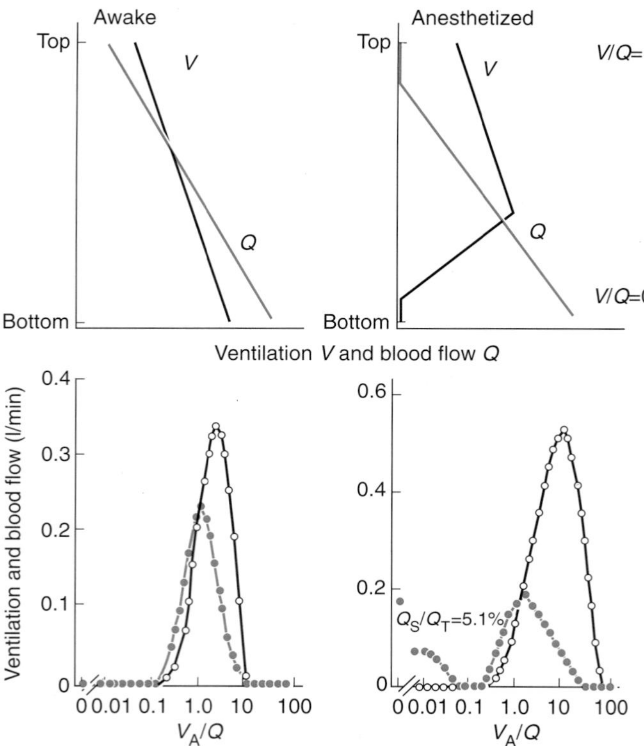

Fig. 3 Schematic ventilation V and perfusion Q distributions from top to bottom of the lung in a healthy subject when awake and anesthetized, as well as the ventilation–perfusion ratios V_A/Q that fit the vertical distributions of ventilation (open circles) and blood flow (solid circles). Note the increasing ventilation and blood flow down the lung in the awake subject, with a rather good match of V_A/Q at all vertical levels, which is also evidenced by the narrow distribution of V_A/Q ratios centered upon $V_A/Q = 1$. Anesthesia lowers FRC and causes collapse of dependent lung regions, and may promote airway closure, impeding ventilation in lower lung regions. Increased alveolar pressure, as may occur during mechanical ventilation, impedes perfusion of the uppermost lung regions. The ensuing V_A/Q distribution differs from the awake situation with a shunt ($V_A/Q = 0$) and perfusion of poorly ventilated lung regions ($V_A/Q < 0.1$). The poor perfusion in upper regions may also cause high V_A/Q ratios ($V_A/Q > 10$). It can be anticipated that much of the anesthesia-induced V_A/Q mismatch may also develop in acute respiratory failure.

low, normal, and high V_A/Q ratios to dead-space ($V_A/Q = \infty$). The procedure is based on the measurement of the retention and elimination of several (usually six) 'inert' gases (gases obeying Henry's law, i.e. showing a linear relationship between partial pressure and concentration in blood) with different solubilities in blood. The resulting V_A/Q distribution will depend on both gravitational and non-gravitational distributions of ventilation of blood flow but does not allow any spatial analysis. Rather, the V_A/Q distribution can be considered as a fingerprint of the lung's ability to transfer gas between alveoli and capillary blood. Examples of V_A/Q distributions in a healthy subject, both awake and during anesthesia, are shown in the lower panels of Fig. 3.

Isotope methods

Ventilation can be assessed by radioactive gas (e.g. xenon-133 or krypton-81m) or radiolabeled small particles (e.g. Technegas). Pulmonary blood flow is usually assessed with radiolabeled macro-aggregated albumin or microspheres (Diot *et al.* 1993). A typical application is the ventilation–perfusion scintigraphy used for the diagnosis of pulmonary emboli. A further development of the scintigraphic techniques has led to single-photon-emission CT where the gamma counter is moved around the body during the investigation and, if the condition under examination is stable, a two- or three-dimensional distribution of the isotope activity can be reconstructed.

Positron emission tomography can be used to estimate the lung blood flow, ventilation, ventilation–perfusion ratio, permeability to proteins, density of receptors, and many other biologically relevant aspects *in vivo*. However, this technique is an expensive research tool that is limited to a few research centers.

Conclusion

Both ventilation and perfusion are unevenly distributed, as they are influenced by gravitation with a preference for dependent lung regions. Pathological changes in acute respiratory failure are also distributed towards dependent lung regions. Any impairment of ventilation in dependent lung regions while blood flow is better preserved will have a considerable effect on the ability of the lung to oxygenate the blood. Therefore a knowledge of the basic normal physiology of the lung may guide in the understanding and treatment of such disease.

References

Diot, P., Baulieu, J.-L., and Lemarié, E. (1993). *Nuclear medicine and lung diseases*, p. 124. Springer-Verlag, Paris.

Hedenstierna, G. (1995). Ventilation–perfusion relationships during anaesthesia. *Thorax*, **50**, 85–91.

Hughes, J.M.B. (1991). Distribution of pulmonary blood flow. In *The lung: scientific foundations* (ed. R.G. Crystal *et al.*), pp. 1135–45. Raven Press, New York.

Milic-Emili, J. (1991). Topographical inequality of ventilation. In *The lung: scientific foundations* (ed. R.G. Crystal *et al.*), pp. 1043–51. Raven Press, New York.

Nunn, J.E. (1993). Distribution of pulmonary ventilation and perfusion. In *Nunn's applied respiratory physiology* (4th edn), pp. 156–84. Butterworth Heinemann, Oxford.

2.2 Upper airway obstruction

2.2.1 Airway control and upper airway obstruction

P. D. Cameron and P. V. van Heerden

Key messages

- Digital and suction clearance of the airway, supplemental oxygenation, optimal posturing, and endotracheal intubation may be lifesaving.
- Inspiratory stridor suggests upper airway obstruction.
- Patients with partial obstruction requiring additional diagnostic tests must be escorted by personnel skilled and equipped for management of an acute decompensation.
- Diminishing stridor, sweating, cyanosis, hypertension, arrhythmias, and coma are late and grave signs of airway obstruction.

Introduction

The upper airway includes the nose, mouth, pharynx, larynx, and external thoracic trachea. Obstruction presents to the intensivist in two main settings: acute patient presentation to the intensive care unit (e.g. facial trauma); associated with mechanical ventilation (e.g. secretions, airway edema).

Signs and symptoms of upper airway obstruction may present acutely (e.g. foreign body), insidiously (e.g. decline in conscious state), or progressively (e.g. laryngeal edema postextubation). Obstruction may prove rapidly life threatening and must be managed with vigilance and expedience.

Causes

Obstruction may be acute or chronic, and may be due to functional or mechanical causes occurring in the lumen, in the wall, or outside the lumen of the upper airway (Cameron 1990). Examples are summarized in Table 1.

Clinical presentation

Minor degrees of obstruction may manifest during increased respiratory rate and inspiratory flow, as may occur during exercise. Advanced signs of obstruction are as follows: extreme agitation and anxiety; inability to phonate; inspiratory stridor and gasping respirations; use of respiratory accessory muscles, nasal flaring, and suprasternal and intercostal recession; tracheal tug and paradoxical chest and abdominal wall movements; upright posture dependence.

Diminishing stridor, diaphoresis, cyanosis, hypertension, and evidence of severe hypoxemia and hypercarbia (loss of consciousness and cardiac arrhythmias) are late manifestations of obstruction and are grave prognostic signs.

Diagnosis

Clinical examination should occur concurrently with emergency management of the airway-obstructed patient, with resuscitation facilities immediately available. Evaluation includes the following.

1. Direct inspection of the oral cavity and oropharynx with clearance of any foreign material.

2. Auscultation over the larynx, trachea, and lung fields to exclude concurrent or alternate diagnoses (e.g. tension pneumothorax, lower airway obstruction). Inspiratory stridor suggests upper airway obstruction.

Table 1 Examples of the causes of upper airway obstruction

Acute
Infection (e.g. epiglottitis, diphtheria, abscess, tetanus)
Trauma (e.g. maxillofacial injury, airway burns, blood clot)
Foreign body (e.g. food bolus, surgical throat packs)
Loss of airway tone (e.g. residual muscle relaxation, depressed level of consciousness, polyradiculopathy, depressant drugs, myasthenia gravis)
Equipment related (e.g. occlusion of the endotracheal tube, postextubation glottic edema)
Allergic reactions
Laryngospasm (e.g. postanesthesia, hypocalcemia)
Following airway surgery
Chronic
Tumor (e.g. pharynx and larynx)
Scarring (e.g. postintubation stenosis)
Acute on chronic conditions
Infection (e.g. abscess formation in hypertrophic tonsils)
Hemorrhage (e.g. into a pre-existing tumor)
Trauma (e.g. laryngeal edema in an already scarred airway)
Inflammation (e.g. after radiation therapy to airway tumor)
Any of the above may present acutely, insidiously, or progressively

3. Palpation of the neck to identify possible tumor or hematoma.
4. Consideration of traumatized adjacent structures (e.g. major vessels, axial skeleton, facial bones).

Repeated clinical examinations are required to detect progressive or recurrent upper airway obstruction and its sequelae. Patients with partial obstruction requiring additional diagnostic tests must be escorted by personnel skilled and equipped for management of an acute decompensation.

A wide variety of indirect clinical signs may be detected during the initial examination or following relief of the upper airway obstruction (Table 2).

Special investigations

Special investigations are only possible when the obstruction is stable and non-life-threatening and will not delay or interfere with timely definitive management. Useful investigations defining the site and extent of obstruction include the following (Cameron 1990; Benumof 1991; Eagle 1992).

1. Visualization of the airway directly (laryngoscopy or tracheo-bronchsocopy via a rigid endoscope under local or general anesthesia) or via a fiber-optic endoscope.
2. Radiological evaluation of the airway: lateral cervical radiography or tomography; CT; magnetic resonance imaging with axial, coronal, and sagittal views; fluoroscopy may determine the dynamic component of obstruction.
3. Flow–volume loops may differentiate extra- from intrathoracic obstructions and fixed from dynamic/variable obstructions.
4. Assessment of adequacy of gas exchange by arterial blood gas analysis and pulse oximetry.
5. Polysomnography for suspected obstructive sleep apnea.
6. Primary pathology investigations for the various types of cervical tumors, tumor markers, angiography, etc.

Table 2 Indirect effects of upper airway obstruction

Decreased alveolar ventilation (hypercarbia)
Sympathetic nervous system
Increased endogenous catecholamines
Diaphoresis
Central nervous system
Anxiety
Increased cerebral blood flow
Progressive narcosis
Increased intracranial pressure
Cardiovascular system
Increased cardiac output
Hypertension
Tachycardia
Arrhythmias
Peripheral vasodilatation (wide pulse pressure)
Hypoxemia
Respiratory system
Increased respiratory drive
Tachypnea
Cardiovascular system
Systemic vasodilatation
Increased cardiac output
Tachycardia followed by bradycardia
Arrhythmias
ECG evidence of myocardial ischemia
Increased capillary permeability
Central nervous system
Decreased level of consciousness
Respiratory efforts against an obstructed upper airway
Surgical emphysema
Pneumothorax
Petechial hemorrhages (neck, trunk, conjunctivae, brain)
Post-obstructive pulmonary edema
Pulsus paradoxus
Increased work of breathing
Endocrine response
Increased level of stress hormones (cortisol)
Increased metabolic rate and blood sugar
Salt and water retention

Management

Management must include the following.

1. Vigilance and early recognition of the obstructed airway and its sequelae.
2. Supplemental oxygen.
3. Rapid clearance of the obstructed airway. Debris is removed manually, by suction, or by Magill forceps. Optimal positioning of the jaw, tongue, and neck is essential. The Heimlich maneuver (abdominal thrust) may prove lifesaving in the acute setting.
4. Establishment of intravenous access and a patent secure airway to permit adequate gas exchange and avoidance of aspiration of gastric and other debris.
5. Pursuit of an 'awake' technique (Benumof 1994) with avoidance of injudicious use of sedation.

These principles are applied at a rate and sequence consistent with the acuteness and severity of the obstruction. The patient with severe facial bone fractures and depressed level of consciousness requires rapid recognition of airway compromise, suctioning under direct vision, and intubation with a cuffed endotracheal tube within seconds to minutes. Less acute laryngeal edema patients with stridor require frequent assessment to detect deterioration and temporizing measures include parenteral corticosteroids and nebulized epinephrine (adrenaline) (2 ml 1:1000 epinephrine in 2 ml normal saline).

Morbidity and mortality occur under the following conditions.

1. There is failure to recognize obstruction, mostly in the acute setting (e.g. upper airway debris in a trauma patient) or when obstruction is insidious (e.g. slowly deteriorating level of consciousness and loss of compensatory mechanisms) or progressive (e.g. chemically or thermally injured upper airway or epiglottitis). The need for frequent evaluation of the patency and security of the airway cannot be overemphasized.

2. Management is inappropriate; for example, a laryngeal mask airway may be suitable for a fasted patient, pending recovery from muscle relaxants and narcosis, but not for an obtunded patient with copious blood in the pharynx secondary to a fractured larynx.

3. The chosen course of management fails; for example, endotracheal intubation fails owing to anatomical or technical difficulties. Failed intubation protocol is discussed elsewhere (Benumof 1991, 1994).

Alternatives to endotracheal intubation include the following.

1. Oro- or nasopharyngeal airway placement and protective posturing against pharyngeal secretion pooling.

2. Face mask and self-inflating resuscitation bag (for assisted ventilation).

3. Fiber-optic-guided intubation of the trachea.

4. Intubation over a guidewire or stylet introduced retrograde into the oropharynx via a cricothyroid puncture.

5. A laryngeal mask can be used as a temporizing measure.

6. Surgical access to the upper airway. Examples include the following.

   (a) Cricothyrotomy (open or percutaneous), which permits rapid placement of a small-bore tube or cannula into the trachea via the cricothyroid membrane to allow oxygen insufflation or jet ventilation (expiration must be unobstructed). Skill in this technique may be readily acquired by non-surgeons.

   (b) Percutaneous tracheostomy, which is an elective technique useful in avoiding expected postextubation stridor after a prolonged course of oral endotracheal intubation (van Heerden *et al.* 1996).

   (c) Surgical tracheostomy under local anesthesia when a difficult or impossible endotracheal intubation is predicted.

7. Extracorporeal membrane oxygenation may be planned when control of the upper airway is lost consequent upon a major elective surgical intervention.

Pharmacological adjunctive therapy includes the following.

1. Intravenous antihistamine, epinephrine, and corticosteroid therapy for anaphylactic or anaphylactoid swelling of upper airway mucosa.

2. Intravenous corticosteroids and nebulized epinephrine for postextubation laryngeal edema.

3. Antibiotics for epiglottitis or other local infections.

4. Fresh frozen plasma for swelling due to angio-edema or hematoma following coagulopathy.

5. Parasympathomimetic agents to counter excessive secretions.

Conclusions

Failure to recognize and correctly manage the obstructed upper airway may result in grave morbidity and mortality. Management is consistent with the acuity and severity of the obstruction, including only strategies familiar to the physician. Obstruction is recognized and managed by simple measures such as digital and suction clearance of the airway, supplemental oxygenation, optimal posturing, and endotracheal intubation. More advanced techniques should be undertaken by experienced personnel, if simple methods fail.

References

Benumof, J.L. (1991). Management of the difficult adult airway. *Anesthesiology*, **75**, 1087–1110.

Benumof, J.L. (1994). Management of the difficult airway: the ASA algorithm. In *1994 Annual refresher course lectures and clinical update program*, pp. 223/1–7. American Society of Anesthesiologists, Park Ridge, IL.

Cameron, P.D. (1990). Acute upper respiratory tract obstruction. In *Intensive care manual* (ed. T.E. Oh), pp. 144–9. Butterworths, Sydney.

Eagle, C.J. (1992). The compromised airway: recognition and management. *Canadian Journal of Anesthesia*, **39**, R40–6.

van Heerden, P.V., Webb, S.A.R., Power, B.M., and Thompson, W.R. (1996). Percutaneous dilational tracheostomy—a clinical study evaluating two systems. *Anaesthesia and Intensive Care*, **24**, 56–9.

2.2.2 Tracheobronchial obstruction

P. D. Cameron and B. M. Power

Key messages

- Recognize that obstruction is present.
- Define the site and nature of obstruction.
- Prepare to deal with an acute airway crisis.
- Perform procedures which allow 'elective' unhurried control of the airway if at all possible.
- Anticipate common complications.
- Avoid leaving inexperienced personnel with unstable airway patients in poorly equipped areas.

Anatomy

The adult trachea is approximately 11 cm long, with 3.5 cm located above the suprasternal notch. It extends from the lower border of the cricothyroid cartilage, at the level of the sixth cervical vertebra, to the carina which lies at the level of the upper border of the fifth thoracic vertebra posteriorly and the manubriosternal joint anteriorly, possibly extending caudally as far as the sixth thoracic vertebra during inspiration when erect (Dunne and Gillbe 1987). The trachea bifurcates into main bronchi, giving rise to lobar and segmental bronchi. The right main bronchus has a narrower angle of origin from the trachea and therefore is more prone to intraluminal obstruction by foreign bodies.

The trachea is U-shaped in cross-section and approximately 1.8 cm in diameter; its structure is maintained by supporting cartilaginous rings. Lesions may not produce symptoms until encroachment upon the lumen has reached 50 per cent. The trachea may be compressed

Table 1 Causes of tracheobronchial obstruction

Intrinsic (intraluminal, intramural)	Extramural
Blood clot, sputum, infection debris	Neck hematoma
Strictures	Thyroid tumors
Foreign bodies	Mediastinal lymph nodes and tumors
Tumor Primary Secondary invasive	Aneurysm
Granulomas and papillomas	
Tracheomalacia	
Vascular rings	
Acute trauma	
Disruption	
Mucosal tears, edema, burns	

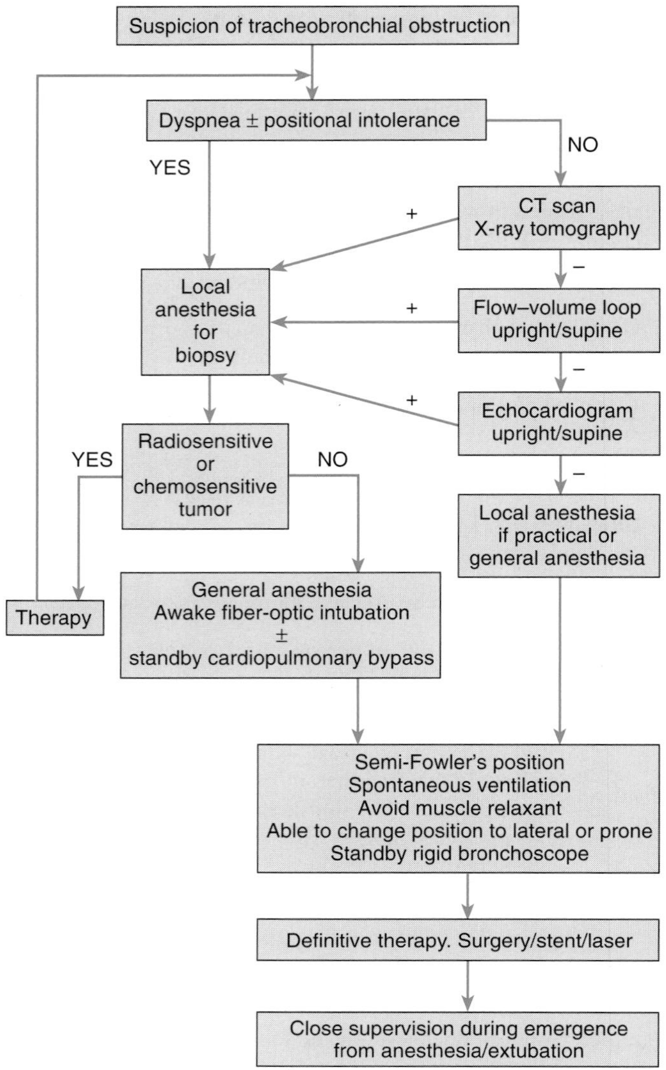

Fig. 1 Flowchart of a plan for preoperative evaluation of the patient with suspected tracheobronchial obstruction: + positive finding; – negative work-up. (Adapted with permission from Neuman *et al.* (1984).)

outside the thorax in the neck. The intrathoracic trachea and bronchi are exposed to changes in intrathoracic pressure and may collapse when intrathoracic pressure exceeds intraluminal pressure by approximately 50 cmH_2O.

Causes of tracheobronchial obstruction

The causes of tracheobronchial obstruction are listed in Table 1.

Strictures following previous prolonged tracheal intubation represent a common cause of obstruction. Primary tumors of the trachea (cylindromas) are uncommon and are outnumbered by invasive bronchial tumors (bronchogenic carcinoma, esophageal carcinoma).

Tracheomalacia should be considered in any patient with a large thyroid lesion. Obstructive symptoms may present postoperatively after extubation and when mucosal edema is progressive. Anterior mediastinal masses are commonly associated with airway compression and anesthetic planning is mandatory (Fig. 1).

Pathophysiology

Complete tracheal obstruction presents a life-threatening emergency. Bronchial obstruction may be better tolerated, but can prove life threatening in the presence of diminished cardiopulmonary reserve. Incomplete bronchial obstructions may be associated with hemodynamic collapse secondary to 'ball-valve' obstruction.

Airway collapse from partially obstructing tracheobronchial lesions can present during anesthesia induction, maintenance, or extubation. Patients may subsequently provide a history of respiratory symptoms, the significance of which was missed on initial evaluation.

Tracheal stenosis may be misdiagnosed as asthma; clues to its presence are recurrent attendances with failure to respond to bronchodilator therapy or with postural aggravation of symptoms. Some patients may present with recurrent respiratory arrest.

Patients with a known mediastinal mass or previous tracheostomy, or who have previously undergone prolonged ventilation, should be considered at risk of tracheal narrowing.

Specific investigations

Chest radiography

Examine for the presence of any masses, deviation of tracheobronchial structures, and complicating lung collapse.

Inspiratory and expiratory posterior–anterior films or fluoroscopy may demonstrate air trapping in the presence of endobronchial obstruction. In patients unable to perform these maneuvers, lateral radiographs with first the left side and then the right side up may demonstrate failure of the lung containing the foreign body to decompress when that lung is placed in the dependent (down) position.

Extrapulmonary gas shadows are indicative of airway disruption until proven otherwise.

Chest CT and magnetic resonance imaging

Chest CT and magnetic resonance imaging may be of use in further outlining masses and to determine tracheobronchial diameters. The patient with critical obstruction may be at risk during these studies.

Bronchoscopy and biopsy

Bronchoscopy and biopsy will help to define and diagnose intraluminal lesions and fixed areas of obstruction.

Flow–volume loop studies

Flow–volume loop studies performed in both the upright and the supine posture may help to distinguish between diffuse airway obstruction (e.g. chronic obstructive airways disease) and a lesion of the upper airway, larynx, or trachea.

Other radiological techniques

Barium swallows and angiography may occasionally be useful in outlining the anatomy of compressive lesions where the tissue of origin is known. Fluoroscopy may be used to demonstrate dynamic causes of airway obstruction, such as laryngomalacia, tracheomalacia, or bronchomalacia, which may be missed on static chest radiographs.

Emergency management

First aid measures

1. Administer supplemental oxygen.
2. Reassure the patient, and minimize anxiety and unnecessary straining.
3. Find the optimal posture. The supine position may decrease expansile forces on the trachea, precipitating complete obstruction. Such obstruction may be temporarily relieved by nursing in a lateral or even prone position, although the latter may make subsequent assessment and emergency management difficult.

Specific measures

Recognized obstruction is best managed by experienced staff with adequate equipment. Although the operating room may be the optimal location for management, a suitably stocked emergency airway trolley is necessary for acute management in other locations (Table 2).

The following procedures should be considered initially.

1. Release neck hematomas if postsurgical.
2. Perform the Heimlich maneuver if a foreign body is suspected.
3. Standby rigid bronchoscopy for removal of a foreign body.

If an airway cannot be maintained the following approaches are considered.

1. Laryngoscopy with elevation of laryngeal structures may provide tracheal support and reduce obstruction.
2. Perform awake intubation, bronchoscopically assisted, with a long thin endotracheal tube or catheter passed beyond obstruction. Small-diameter endotracheal tubes may not be long enough to bypass distal obstructions and adapted catheters may be required (Sutcliffe *et al.* 1995).
3. Muscle relaxants decrease chest wall muscle tone and remove the expansile tracheal forces of spontaneous ventilation. Although not absolutely contraindicated, paralysis prior to control of airway may be associated with total loss of airway control.
4. In upper tracheal lesions, jet ventilation via a transtracheal catheter may be possible provided that expiration is unimpeded.
5. Femoral vein–femoral artery partial cardiopulmonary bypass has been used in this group of patients.
6. In the presence of massive hemoptysis from a bronchial site or intrabronchial rupture of a lung abscess, urgent placement of a double lumen endobronchial tube may ‘isolate’ the lungs and prevent contamination of the contralateral lung prior to definitive treatment.

Continuing hypoxia or hemodynamic compromise after successful bypassing of obstruction should lead to consideration of the emergence of another complication such as postobstructive pulmonary edema, pneumothorax, or aspiration pneumonitis. Although a

Table 2 Difficult airway management trolley: useful items to supplement the intubation trolley

Important: Contents of a portable storage unit should be customized to meet the specific needs, preferences, and skills of the practitioner and health care facility

1. Rigid laryngoscope blades of alternative design and size from those routinely used
2. Endotracheal tubes of assorted size and length
3. Endotracheal tube guides; examples include (but are not limited to) semirigid stylets with or without a hollow core for jet ventilation or oxygen insufflation, light wands, and forceps designed to manipulate the distal portion of the endotracheal tube
4. Fiber-optic intubation equipment
5. Retrograde intubation equipment
6. At least one device suitable for emergency non-surgical airway ventilation; examples include a transtracheal jet ventilator, a hollow jet ventilation stylet, a laryngeal mask, and an esophageal–tracheal Combitube
7. Equipment suitable for emergency surgical airway access (e.g. cricothyrotomy)
8. An exhaled CO_2 detector

Additional equipment which may be useful for treatment of tracheal obstruction includes (Sutcliffe *et al.* 1995):
- Straightened pediatric RAE tubes (size 4.0)
- Other adult-length narrow-bore cannulas suitable for ventilation

Adapted with permission from Caplan *et al.* (1993).

small-size endotracheal catheter may allow acute crisis control, pending definitive treatment of obstruction the patient remains at risk from endotracheal tube kinking, luminal obstruction, suboptimal airway toilet, and difficulty in re-establishing spontaneous ventilation.

Elective management

The principles of acute management of tracheobronchial obstruction in intensive care overlap those of the anesthetic management of patients with known (potentially) obstructing lesions requiring diagnosis and surgery.

Algorithms for the anesthetic management of patients with mediastinal masses where respiratory obstruction is anticipated have been well described (Fig. 1)

Management at extubation

The patient with known tracheobronchial obstruction is also at risk at the time of extubation. Airway security in the presence of a high lesion may be provided by tracheostomy. In more distal lesions, the principles of management are as follows.

1. Minimize exacerbation of obstruction and swelling if possible.
2. Maintain spontaneous ventilation if possible.
3. Avoid anxiety and unnecessary straining.

Useful adjuncts at extubation may include the following.

1. Use of an endotracheal ventilation catheter (Cooper 1995).
2. Application of positive airway pressure by mask, i.e. continuous positive airway pressure and/or inspiratory pressure support.
3. Semisitting or sitting position.
4. When airway edema is a significant component, inhalation of epinephrine (adrenaline) 0.5 mg in 10 ml normal saline and/or hydrocortisone (intravenous) 1 to 2 mg/kg every 4 to 8 h can be considered.

References

Caplan, R.A., *et al.* (1993). Practice guidelines for management of the difficult airway. A report by the American Society of Anesthesiologists Task Force on Management of the Difficult Airway. *Anesthesiology*, **78**, 597–602.

Cooper, R.M. (1995). The use of an endotracheal ventilation catheter in the management of difficult extubations. *Canadian Journal of Anaesthesia*, **43**, 90–3.

Dunne, N.M. and Gillbe, C.E. (1987). Endobronchial apparatus and its intraoperative management. *Baillière's Clinical Anaesthesiology*, **1**, 79–98.

Neuman, G.G., Weingarten, A.E., Abramowitz, R.M., Kushins, L.G., Abramson, A.L., and Ladner, W. (1984). The anesthetic management of the patient with an anterior mediastinal mass. *Anesthesiology*, **60**, 144–7.

Sutcliffe, N., Remington, S.A.M., Ramsay, T.M., and Mason, C. (1995). Severe tracheal stenosis and operative delivery. *Anaesthesia*, **50**, 26–9.

2.3 The breathless patient

2.3.1 Control of breathing

H. Burchardi and D.W. Richter

Key messages

- The respiratory center controls ventilation in three phases: inspiration (I phase), postinspiration (PI phase, passive expiration), and stage 2 expiration (E2 phase, active expiration).
- Central chemoreceptors play a major role in the regulation of breathing through CO_2 pressure and pH. The ventilation response to $P\text{aco}_2$ is strongly affected by drugs (e.g. opioids, sedatives).

Introduction

The control system of the ventilatory pump is composed of an input unit for afferent information with central and peripheral sensors (chemoreceptors, mechanoreceptors, cortical influences), an integrator/controller (brainstem rhythm and pattern generator), an output unit (efferent pathways, motor nerves), and a motor (respiratory muscles), which moves the machinery (chest wall) (Fig. 1). The purpose is to adapt pulmonary ventilation to the needs of the body's metabolism. In this chapter we deal only with the central control and neuronal efferent and afferent aspects of this topic.

Respiratory rhythm

Ventilation of the lungs is controlled by a network of medullary respiratory neurons within the lower brainstem (Richter *et al.* 1992). In simple terms, this network is an active center that augments and diminishes inspiratory activity which is periodically interrupted during active expiration. The two phases of ventilatory mechanics are controlled by three neural phases: inspiration (I phase), postinspiration (PI phase, passive expiration), and stage 2 expiration (E2 phase, active expiration).

Rhythmic nervous output from the respiratory center is fed to spinal motoneurons and leads to periodic contraction of thoracic, abdominal, and diaphragmatic muscles. Breathing can be voluntarily interrupted (for speech, singing, sniffing, coughing, etc.); reflex interferences induce sneezing, swallowing, and reflex coughing. Changes in arterial blood gases may alter the pattern of respiration; for example voluntary hyperventilation is followed by periods of apnea.

Active expiratory phase activities can be totally blocked during rapid shallow breathing or panting.

Respiratory center

The bilateral respiratory center is localized within the pre-Bötzinger complex, which is located in the ventral medulla (Bianchi *et al.* 1995). This center produces rhythmic activity which to some extent is independent of peripheral feedback. The same region also contains the presumed central chemosensitivity. Medullary respiratory neurons are apparently not endogenously active in humans; their rhythmic activity seems to originate from synaptic interaction between neurons. Obviously, there is a separate gasping center within the caudal medulla.

Ventilatory pump

During inspiration, a ramp-like increase in the discharge of inspiratory spinal motoneurons contracts the diaphragm and inspiratory intercostal muscles and expands the thorax. Laryngeal abductor muscles are also activated during inspiration and dilate the larynx. Exhalation is generally driven by the stored recoil forces of lung tissue released during postinspiration when inspiratory movements have come to an end, but inspiratory muscles relax slowly. Thus exhalation occurs passively during this initial phase of exhalation (PI phase) without any active innervation of expiratory muscles. Lung volume is held by an actively controlled slow relaxation of the diaphragm during this phase. Expiratory airflow is additionally retarded by contraction of laryngeal adductor muscles (e.g. thyrohyoidic muscle) which increases the resistance to expiratory air flow.

Expiratory airflow can be reinforced by active contraction of expiratory intercostal and abdominal muscles (E2 phase). The dimensions of the upper airways, such as the pharynx, the larynx, and the bronchial tree, determine the flow resistance to air. They can be actively adjusted through cranial motoneurons which are synaptically coupled to the central respiratory network.

Common cardiorespiratory network

There are close interactions between the respiratory and cardiovascular control systems in the brainstem which are part of both respiratory and cardiovascular control. The structures are located in almost identical regions. The tight coupling between the two systems results in a matching of respiratory minute volume to cardiac output. Central respiratory activity can influence the performance of several basic cardiovascular reflexes (e.g. respiration-related variations in heart rate and blood pressure).

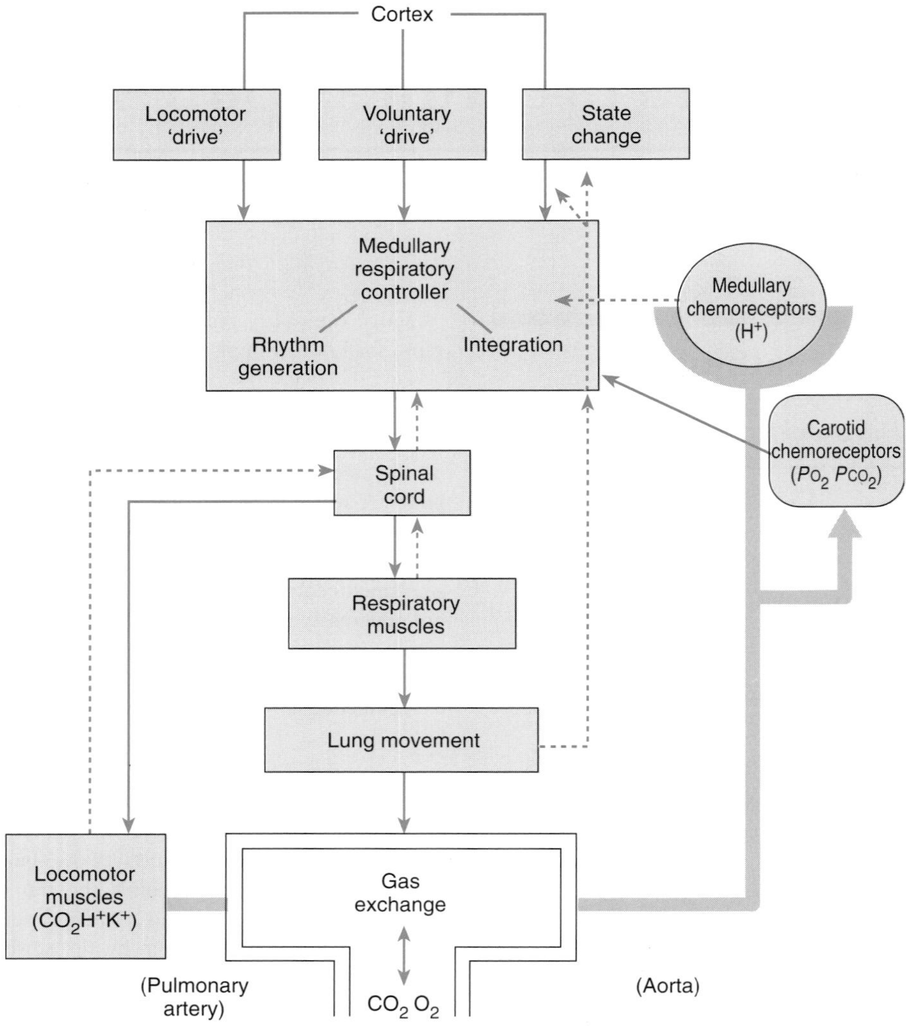

Fig. 1 Diagram of the respiratory control system showing the efferent nervous pathways (thin green line), the afferent nervous pathways (broken green line), and the humoral pathways (thick green line). (Reproduced with permission from Simon (1994).)

Reflex control of respiratory activity

Information about ventilatory mechanics, and also about chemical stimuli and temperature, is transmitted via afferent fibers running in the olfactory, trigeminal, glossopharyngeal, and vagal nerves. This provokes various reflexes, such as termination of inspiration (Hering–Breuer reflex), sneezing, aspiration, coughing, or vomiting, and behavioral reactions, such as sniffing and inhibition of motor activity during heavy working load.

Afferents from the glossopharyngeal and vagal nerves terminate in a specific relay nucleus known as the nucleus of the solitary tract (NTS) which contains the target neurons of afferents from 'pulmonary stretch' receptors (Hering–Breuer reflex). However, there are also direct projections of pulmonary afferents to pontine regions.

Reflexes from the upper respiratory tract (irritant receptors and mechanoreceptors) may cause bronchoconstriction, laryngospasm, and coughing. They can provoke serious complications during endotracheal intubation in sensitive individuals. Nociceptive receptors (J receptors) are activated by tissue damage, accumulation of interstitial fluid, and release of mediators resulting in rapid shallow breathing. This mechanism may cause dyspnea in pulmonary vascular congestion, lung edema, and pulmonary embolism.

Other sensory inputs (e.g. information of hyperinflation or deflation of the lungs from pulmonary stretch receptors) activate a variety of respiratory and non-respiratory NTS neurons for rapid adaptation (physiological mechanisms to turn off inspiration). Other NTS neurons also receive afferent inputs from arterial chemoreceptors and baroreceptors which can modulate the central respiratory network (e.g. respiratory depression during arterial hypertension).

The specific connectivity of all these relay neurons with the respiratory network is still unknown, but their axonal projections seem to be widespread.

Chemical control of respiration

The ultimate goal of respiration is to maintain adequate concentrations of oxygen, CO_2, and hydrogen ions in the body fluid. Excess CO_2 or hydrogen ions directly affect the respiratory center, whereas mild hypoxia only acts through the peripheral chemoreceptors. Changes in blood CO_2 play the greatest role in respiratory control (Nunn 1993) (see Fig. 1).

Central chemoreceptors

Central chemoreceptors (through the cerebrospinal fluid and cerebral blood vessels) play a major role in the regulation of breathing through the CO_2 pressure ($P\text{CO}_2$) and/or pH. An increase in $P\text{aCO}_2$ immediately stimulates ventilation. However, if an abnormal level of $P\text{CO}_2$ is maintained over several days, the pH of the cerebrospinal fluid returns to normal (compensatory bicarbonate shift). Thus changes in arterial CO_2 have a very potent acute effect on respiratory control, but only weak chronic effects after adaptation. Compensatory changes in cerebrospinal fluid bicarbonate and restoration of cerebrospinal fluid pH are important in chronic respiratory or metabolic acid–base imbalances (e.g. in patients with chronic obstructive pulmonary disease (COPD)).

Peripheral chemoreceptors

Stimuli through peripheral chemoreceptors (carotid and aortic bodies) are less dominant. They stimulate ventilation when $P\text{aO}_2$ falls below normal. They also react to hypoperfusion (enhanced respiration in response to hypotension). A fall in (alveolar) $P\text{AO}_2$ from 5.3 to 4.0 kPa (from 40 to 30 mmHg) increases ventilation by a factor of 1.5 to 1.7, whereas ventilation increases fourfold when blood pH decreases to 7.0 or tenfold when blood $P\text{aCO}_2$ increases by 50 per cent. This is because the powerful $P\text{aCO}_2$ and pH control mechanisms counteract the hypoxic stimulus ('braking' effect by compensatory mechanisms).

The stimulating effect of hypoxemia becomes much more significant when the compensatory effects of blood $P\text{aCO}_2$ and pH are ineffective, for example when the CO_2/pH regulation mechanisms become adapted to chronic hypoxemia (e.g. in COPD). However, in the case of combined respiratory pump and oxygenation failure the two mechanisms become additive. Ventilatory response to hypoxia is enhanced by exercise; however, it is virtually abolished during anesthesia or marked analgesia and sedation. During severe hypoxia the respiratory center is depressed.

The $P\text{aCO}_2$/ventilation response

There is a wide individual variation in the $P\text{aCO}_2$/ventilation response and it is strongly affected by drugs (e.g. opioids, sedatives). As $P\text{aCO}_2$ increases a point of maximum stimulation is reached at 13.3 to 26.7 kPa (100–200 mmHg), after which ventilatory stimulation decreases again and finally apnea may result. In metabolic acidosis (e.g. diabetic acidosis) the $P\text{aCO}_2$/ventilation response is shifted to the right, i.e. acidosis stimulates the respiratory center considerably for CO_2.

Clinical situations

Most patients with respiratory failure continue to make spontaneous efforts during mechanical ventilation unless they are under heavy sedation or neuromuscular blockade. Missing efforts and dissynchrony during assisted ventilation impair its effectiveness and increase the work of breathing. Mild hypocapnia during mechanical ventilation can reduce spontaneous respiratory drive and even totally inhibit inspiratory muscle activity. The end-tidal $P\text{CO}_2$ at which inspiratory muscle activity returns is known as the 'recruitment threshold' (Simon 1994). The recruitment threshold depends on the tidal volume and frequency (e.g. through the influence of sensory reflexes). Sudden cessation of continuous mechanical ventilation (even in awake normocapnic patients) can result in prolonged apnea. The mechanisms are not fully understood (it may be a 'memory' phenomenon).

Mild hypercapnia during weaning seems to be a physiological response to the additional resistive load (e.g. by the tube) even in patients with normal lungs (load-induced reduction in ventilation). Thus a 'normal' $P\text{aCO}_2$ is not strictly defended under additional load.

An increase in body temperature (e.g. fever) increases ventilation (rate as well as tidal volume) indirectly by the augmented metabolism and directly by raising the respiratory center activity.

References

Bianchi, A.L., Denavit-Saubié, M., and Champagnat, J. (1995). Central control of breathing in mammals: neuronal circuitry, membrane properties and neurotransmitters. *Physiological Reviews*, **75**, 1–45.

Nunn, J. (1993). *Nunn's applied respiratory physiology* (4th edn), pp. 97–109. Butterworth-Heinemann, Oxford.

Richter, D.W., Ballanyi, K., and Schwarzacher, S. (1992). Mechanisms of respiratory rhythm generation. *Current Opinion in Neurobiology*, **2**, 788–93.

Simon, P.M. (1994). Effect of mechanical ventilation on the control of breathing. In *Principles and practice of mechanical ventilation* (ed. M. J. Tobin), pp. 631–45. McGraw-Hill, New York.

2.3.2 Control of breathing: external influences and abnormalities

H. Burchardi and C.-P. Criée

Key messages

- Reflexes from the upper respiratory tract (irritant receptors, mechanoreceptors) may cause bronchoconstriction, laryngospasm, and coughing.
- Respiratory muscle fatigue is a loss of capacity of respiratory muscles to develop force and/or velocity under load. Failure in central motor output, neuronal transmission, and generating muscular force are possible mechanisms.
- Breathlessness is the recognition of an inappropriate relationship between respiratory work and total body work; it is a subjective symptom with many different factors involved. Adaptation allows a significant disability to be tolerated, which may result in underestimation of the actual risk.
- Hypercapnia only occurs when both the diaphragm and the intercostal muscles are involved. Under normal conditions, the paralysis of the diaphragm alone does not result in hypercapnia.

Effects of drugs and hormones on the respiratory response

Drugs can modify respiration by affecting the respiratory centers and their response to CO_2, but they may also influence peripheral receptors (by depression or stimulation) and neuromuscular junctions. Secondary effects may occur by modifying metabolism and CO_2 formation.

1. Central nervous depressants: narcotics, sedatives, and hypnotic drugs depress central respiratory activity to various degrees. Aggravating effects are seen when sedatives (e.g. barbiturates) are combined with opioids.

2. Central nervous stimulants: salicylates stimulate respiratory response to CO_2; aminophylline stimulates respiratory activity.

3. Anesthesia and sedation: apart from the direct effects of the drugs on the respiratory centers there are many secondary influences on ventilation. Central and peripheral chemosensitivity are depressed, as are the responses to hypoxia, hypercapnia, and exposure to cold (shivering is depressed). When metabolism is reduced, ventilation also decreases.

4. Hormones may alter respiration directly or indirectly. For example, insulin relieves respiratory stimulation in severe diabetes acidosis, thyroid hormone increases metabolism, CO_2 formation, and secondarily ventilation, and norepinephrine (noradrenaline) stimulates the CO_2 response.

Pathophysiological disturbance of the respiratory rhythm

Any reflex or direct perturbation of the primary oscillator or its interaction with the network has pathophysiological consequences.

1. Activation of inspiration phase activity by afferents from the larynx or pulmonary C fibers results in respiratory arrest (reflex postinspiratory apnea).

2. Prolonged inspiratory phases (apneusis) result whenever synaptic inhibition is reduced.

3. When synaptic inhibition is blocked (e.g. under hypoxia or ischemia), respiration stops during expiration, i.e. 'expiratory apnea'.

Cheyne–Stokes periodic breathing is a slow oscillation between hyper- and hypoventilation which occurs every 45 s to 3 min (Fig. 1). It may be caused by a delay in blood transport from the lungs to the brain as well as by a change in feedback gain of the respiratory center, resulting in a phase shift and an inadequate response to the respiratory stimulus. Its occurrence in severe brain damage or acute cerebral edema is an unfavorable sign.

Loads and sensations

Respiratory muscle fatigue

Respiratory muscle fatigue is a loss in the capacity of respiratory muscles to develop force and/or velocity under load, finally resulting in exhaustion or task failure which is reversible on rest (NHLBI Workshop 1990). Various mechanisms are possible: failure to generate force because of reduced central motor output (central fatigue), failure to transmit the neuronal message (transmission fatigue), or failure to generate force within the respiratory muscles (peripheral fatigue). About half the reduction in diaphragmatic force can be attributed to reduced central motor drive. However, it remains unclear whether this is a primary or a secondary mechanism (i.e. vital adaptation on a lower level).

Central fatigue is considered to be present when maximum voluntary contraction generates less force than maximum electrical

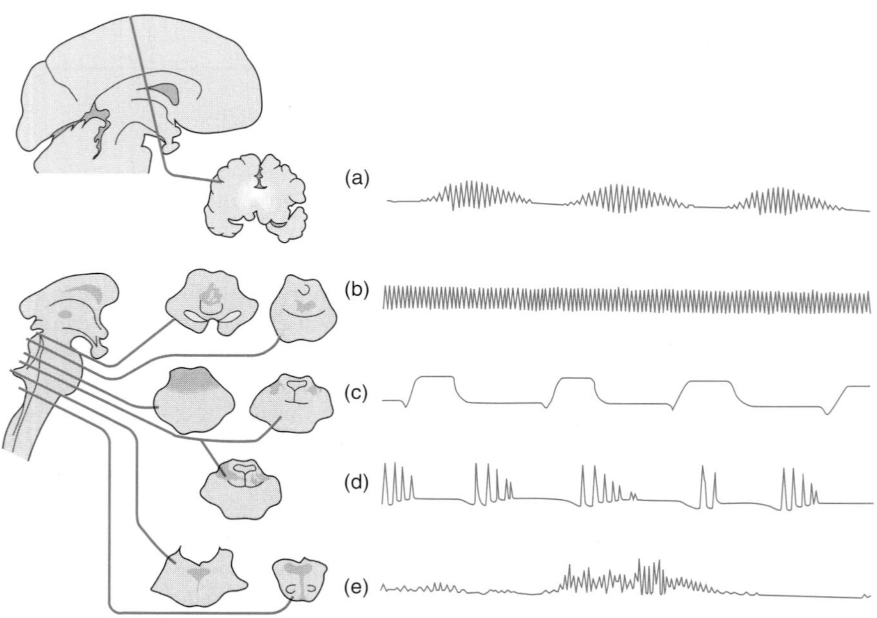

Fig. 1 Abnormal respiratory patterns associated with pathological lesions (shaded areas) at various levels of the brain (tracings by chest–abdomen pneumograph; inspiration reads up): (a) Cheyne–Stokes respiration; (b) central neurogenic hyperventilation; (c) apneusis; (d) cluster breathing; (e) ataxic breathing. (Reproduced with permission from Plum and Posner (1980).)

stimulation (diaphragmatic force tested by the twitch occlusion test). In contrast, the existence and relevance of transmission fatigue (i.e. neuromuscular function failure) is still questionable. Peripheral (i.e. muscular) fatigue is beyond the scope of this chapter.

Intermittent (nocturnal) non-invasive mechanical ventilation appears to be efficacious in some groups of patients with chronic respiratory failure. However, it is still not clear whether this benefit is due to 'resting' (unloading) fatigued muscles by the intermittent mechanical ventilation or to preventing the neural control system from adapting to high CO_2 levels.

Breathlessness and dyspnea

Breathlessness is the recognition of an inappropriate relationship between respiratory work and total body work (Burdon *et al.* 1994). It is a subjective symptom whose origin is not completely understood. In addition to the mechanics of breathing (i.e. work and lung volume), many other factors are involved. Perception depends on subjective experience: 'normal' (after exercise) or 'abnormal' (not related to adequate activity). The threshold of perception may vary. Some patients tolerate severe respiratory loads with little or no complaint of breathlessness; this may occur because of individual characteristics or because of temporal adaptation.

Temporal adaptation allows a significant disability to be tolerated with a reduction or absence of symptoms. This can be clinically dangerous, since it may result in underestimation of actual risk. About 25 per cent of the change in lung volume is necessary in acute asthma before any change in symptoms is noticed. Thus larger changes in background load are required in patients with abnormal pulmonary mechanics, i.e. patients with chronic background load (e.g. chronic inadequately treated asthma), before changes in sensation occur. However, if the change in background load is recent, smaller changes in sensation are noticed. There are considerable variations between individuals. Identification of the subgroup of patients who have impaired perception of breathlessness (e.g. external loading) is clinically important.

Asynchrony

Increased ventilatory drive in hyperinflated patients with chronic obstructive pulmonary disease activates the accessory respiratory muscles (from neck, rib cage, abdomen) and upper airway muscles and occasionally leads to chest–abdomen asynchrony (not necessarily a sign of fatigue). This can also be observed during unsuccessful weaning trials. Hyperinflation and excessive activation of neck inspiratory muscles are associated with tachypnea and the feeling of dyspnea. In this situation, reduction of tidal volume and a slight increase in $Pa\text{CO}_2$ can be regarded as a strategy of minimizing effort rather than a manifestation of fatigue. Respiratory drive ($P_{0.1}$) is increased in all situations of augmented load.

The reaction to external loading (external resistance) seems to be different to the reaction to internal loading (airway resistance). Under steady state conditions gas exchange is impaired only under large external loads, whereas during anesthesia mechanical loading rapidly leads to hypercapnia and hypoxemia.

Extrapulmonary respiratory failure

Respiratory center dysfunction

Decreased activity of the respiratory centers may be due to the following:

(1) structural damage (e.g. cerebral infarction, neoplasm, infection);

(2) central nervous system depressants (e.g. narcotics or barbiturates);

(3) metabolic abnormalities (e.g. metabolic alkalosis, hypothyroidism, starvation);

(4) loss of sensitivity to changes in $Pa\text{CO}_2$ (e.g. primary alveolar hypoventilation (Ondine's curse));

(5) periodic cessation of impulse formation during sleep (central sleep apnea).

Interruption of efferent neural pathways

Neural pathway interruption may be due to the following:

(1) spinal cord disease (e.g. trauma, neoplasm, hemorrhage, infection);

(2) anterior horn cell disease (e.g. amyotrophic lateral sclerosis, poliomyelitis);

(3) peripheral neuropathy (e.g. Landry–Guillain–Barré syndrome);

(4) neuromuscular junction disorders (e.g. myasthenia gravis, Eaton–Lambert syndrome, botulism, organophosphate poisoning, neuromuscular blocking, aminoglycoside antibiotics).

Hypercapnia only occurs when efferent nerves to both the diaphragm and the intercostal muscles are involved. Total paralysis of the diaphragm reduces vital capacity by about 50 per cent, but does not result in hypercapnia under normal conditions. Interruption of the efferent nervous pathways results in ventilatory restriction (decreased lung volume, decreased inspiratory and expiratory force). Clinically, impaired coughing increases the risk of aspiration, atelectasis, and pneumonia.

References

Burdon, J.G.W., Pain, M.C.F., Rubinfield, A.R., and Nana, A. (1994). Chronic lung disease and the perception of breathlessness: a clinical perspective. *European Respiratory Journal*, 7, 1342–9.

NBLBI Workshop (1990). Respiratory muscle fatigue: report of the Respiratory Muscle Fatigue Workshop Group. *American Review of Respiratory Diseases*, **142**, 474–90.

Plum, F. and Posner, J.B. (1980). *The diagnosis of stupor and coma* (3rd edn), p. 34. Davis, Philadelphia, PA

2.3.3 Causes of breathlessness

S. Benito

Key messages

- Breathlessness may be due to airflow limitation, lung parenchymal or extrapulmonary disease, or cardiovascular disease.
- Breathlessness can be classified according to pathophysiological mechanisms, for example increased ventilatory demands, decreased ventilatory capacity, or increased subjective sensitivity.

Introduction

Breathlessness is an extremely common and distressing symptom. It is characterized by four closely related sensations: tightness, excessive ventilation, excessive frequency of breathing, and difficulty in the act of breathing. These sensations cannot be isolated, and in some patients it is difficult to differentiate between them and other sensations referred to the chest.

Table 1 Principal mechanisms causing breathlessness

Increased ventilatory demands
Hypoxemia
Hypercapnia
Acidosis
Anemia
Exercise
Hyperthermia
Hypotension
Pulmonary hypertension
Decreased ventilatory capacity
Muscular weakness
Rib cage deformity
Low pulmonary volume
Stiff lung
Airflow obstruction
Increased subjective sensitivity
Anxiety
Depression
Hysteria

We do not yet know the basic physiological mechanisms causing breathlessness or which lung function test is the most appropriate for evaluating this distress. An excellent review of respiratory sensations has recently been published (Adams and Guz 1996).

The conscious sensation of discomfort associated with breathing is also called dyspnea. Strictly speaking, breathlessness is related more to a lack of satiaton and dyspnea to effort, but the terms are used interchangeably.

Assessment of symptoms

The perception of difficulty in breathing is normally measured clinically using the Borg modification scale (Burdon *et al.* 1982). This scale categorizes words describing increasing degrees of breathlessness, relating them to scores between zero and 10. When asked to score breathlessness, the patient chooses the number which most appropriately describes the sensation that he or she experiences.

A different approach to the quantification of breathlessness is related to measurement of the principal mechanisms involved, increased demands, and decreased reserves. The ratio of minute ventilation to maximum voluntary ventilation is a good indicator of the relationship between the patient's needs and his maximum capacity. Breathlessness can be estimated from the ratio of the inspiratory pressure needed for the actual tidal volume to the maximum inspiratory pressure that can be generated by the patient (Killian and Campbell 1995). We can also estimate respiratory demands by measuring the airway occlusion pressure at 0.1 s, and maximum capacities by determining the maximum inspiratory pressure. Finally, the forced expiratory volume at 1 s (FEV_1) is also considered a good indicator of breathlessness.

Causes of breathlessness

The causes of breathlessness are related to the mechanisms which induce ventilatory demands, the pathology which produces a decrease in ventilatory capacity, and the alterations which can produce an increase in the subjective sensitivity, the most common of which are listed in Table 1. The clinical pathologies most closely associated with breathlessness are caused by diseases of the airway, the lung, the pleura, and the cardiovascular system.

Airflow limitation

Breathlessness in asthma crises is one of the most distressing respiratory disorders. It has been clinically observed that some patients become very distressed with relatively little change in lung mechanics. The major mechanism occurring in asthma is increased elastic loading of the inspiratory muscle as a result of dynamic hyperinflation. The most objective changes contributing to breathlessness in acute bronchoconstriction are the decrease in inspiratory capacity, reflecting an increment in end-expiratory lung volume, and the change in esophageal pressure needed to produce the inspired tidal volume (Lougheed *et al.* 1993).

Breathlessness may be caused by upper airway obstruction due to a foreign body, an infection, or trauma or allergic reaction in the hypopharynx, tongue, or soft tissue in this area.

The mechanical abnormality in chronic airway obstruction is expiratory flow limitation related to dynamic airway compression and dynamic hyperinflation. However, dynamic airway compression plays an indirect role in the sensation of breathlessness. Exertional breathlessness mainly depends on ventilatory demands and reserves, but dynamic hyperinflation contributes significantly and explains the intersubject variability in the sensation of breathlessness (O'Donnell and Webb 1993). The disparity between mechanical impairment and the sensation of breathlessness confirms that other factors are involved in the generation of this sensation. The increase in minute ventilation is perceived differently by different patients, and it is difficult to determine whether this tolerance is an individual characteristic or a temporal adaptation to perception of respiratory load.

Pulmonary or pleural disease

A decrease in lung volume and hypoxemia are the main factors inducing acute tachypnea and dyspnea in patients with pneumonia, pneumothorax, or acute lung injury.

Cardiovascular disease

Patients with heart disease frequently present with the sensation of breathlessness in a variety of clinical circumstances, and the mechanism of this is poorly understood.

Acute heart failure may induce left ventricular failure leading to pulmonary edema, which is the probable cause of breathlessness. Ischemic heart disease can trigger transient left ventricular failure.

Different abnormalities may contribute to the sensation of breathlessness in chronic heart failure. Some patients present with bronchial hyper-responsiveness. Decreased lung compliance may contribute to breathlessness but does not seem to be a major determinant. Muscle fatigue, associated with low cardiac output, may also play a role.

Acute unexplained dyspnea may be a clinical sign of pulmonary embolism. Because of mismatching in the ventilation-to-perfusion ratio, even a low degree of obstruction produces hypoxemia which is responsible for the breathlessness. Greater obstruction may provoke pulmonary hypertension or cardiogenic shock, which also cause breathlessness.

References

Adams, L. and Guz, A. (1996). *Respiratory sensation*. Marcel Dekker, New York.

Burdon, J.G.W., Juniper, E.F., Killian, K.J., Hargreave, F.E., and Campbell, E.J.M. (1982). The perception of breathlessness in asthma. *American Review of Respiratory Disease*, **126**, 825–8.

Killian, K. and Campbell, E.J.M. (1995). Dyspnea. In *The thorax* (ed. C. Roussos), pp. 1709–47. Marcel Dekker, New York.

Lougheed, M.D., Lam, M., Forket, L., Webb, K.A., and O'Donnell, D.E. (1993). Breathlessness during acute bronchoconstriction in asthma: pathophysiologic mechanisms. *American Review of Respiratory Disease*, **148**, 1452–9.

O'Donnell, D.E. and Webb, K.A. (1993). Exertional breathlessness in patients with chronic airflow limitation. The role of lung hyperinflation. *American Review of Respiratory Disease*, **148**, 1351–7.

2.3.4 Therapeutic approach to breathlessness

S. Benito

Key messages

- Hypoxemia must be corrected to reduce ventilatory demand.
- Treatment aimed at improving respiratory muscle function is helpful.
- Some drugs may reduce breathlessness without detriment to pulmonary function.

Introduction

Treatment of breathlessness is efficacious when the cause is identified and can be modified. In the most frequent clinical situations of breathlessness, specific treatment normally provides little or no relief. However, it is mandatory to explore the possibility of alleviating the distress, specifically to find a therapeutic approach which decreases the sensation of breathlessness but does not cause changes in the mechanics and control of ventilation. Such therapy does not substitute specific treatment for the disease, but could play an important role in patient comfort, particularly if the main process is chronic and irreversible. In such cases, reducing the sensation of breathlessness can improve the quality of life.

In this discussion of treatment, an approach similar to that used in the discussion of the causes of breathlessness is also applicable. We analyze the procedures used to reduce the ventilatory demands and the interventions made to improve ventilatory capacity, and discuss management of the fear of breathlessness and its related subjective sensitivity. Pharmacological intervention is considered separately.

Reduction of ventilatory demands

Hypoxemia is a major stimulus to increasing ventilatory demands. This effect can be minimized or eliminated by supplementary oxygen. Normalizing blood oxygen, detected in the peripheral chemoreceptor, decreases the respiratory drive. In some patients, this reduction of minute ventilation may promote ventilatory failure, leading to an increase in arterial $P\text{CO}_2$.

Although the mechanisms of action are not completely understood, some interventions such as nasal gas flow through a cannula, chest wall vibration, and a flow of cold air onto the cheek may reduce breathlessness without causing a significant reduction in ventilation.

Improvement of ventilatory capacity

Therapeutic approaches to the improvement of respiratory muscle or ventilatory capacity in general cover a broad spectrum. As in other chronic diseases, the patient must understand the goals of therapy and therefore must be given adequate information and instruction. Correction of nutritional habits may be advisable. Reduction of obesity will improve mechanical efficiency and permit a better inspiratory displacement of the diaphragm. In other cases, counteracting malnutrion and its repercussions on muscle strength may be useful. Breathing training may also be valuable for some patients — relaxing the abdomen in the inspiratory phase with pursed-lip expiration in order to reduce the respiratory rate and increase the tidal volume.

During acute bronchoconstriction, dynamic hyperinflation contributes significantly to breathlessness. Despite its mechanism, which is related to the contraction of inspiratory muscles during expiration and prolonged expiratory time (Cormier *et al.* 1990), hyperinflation produces an inspiratory threshold load that must be overcome by the inspiratory effort in each breath. This threshold has been documented in assisted modes of mechanically ventilated patients, where applying a positive end-expiratory pressure (PEEP) lower than intrinsic PEEP improves the triggering of the ventilator.

It has been demonstrated that continuous positive airway pressure (CPAP) alleviates breathlessness in induced asthma, provided that it does not increase end-expiratory lung volume or minute ventilation. Optimal CPAP is associated with less effort, decreasing the inspiratory threshold load, the ratio of pressure needed for tidal volume generation to maximum capacity pressure, and the elastic work of breathing (Lougheed *et al.* 1995).

When chronic obstructive pulmonary disease patients with expiratory flow limitation and dynamic hyperinflation are mechanically ventilated because of acute ventilatory failure, the use of PEEP to counterbalance the intrinsic PEEP does not affect the respiratory mechanics, hemodynamic state, or gas exchange. When external PEEP exceeds internal PEEP, further hyperinflation and alteration of hemodynamics and gas exchange occurs (Ranieri *et al.* 1993) The reduction of the inspiratory threshold, due to intrinsic PEEP, by means of CPAP decreases the inspiratory work of breathing and breathlessness in patients with chronic obstructive pulmonary disease during weaning from mechanical ventilation (Petrof *et al.* 1990). In acute exacerbation of chronic lung disease, the use of non-invasive ventilation can obviate the need for intubation by improving ventilatory capacity.

Fear of breathlessness

Subjective sensitivity to the perception of breathlessness is influenced by the emotional state of the patient. As well as the limitation in activity caused by chronic illnesses, some patients present with depression or panic attacks related to their emotional state or previous experience in crises of acute breathlessness.

Pulmonary rehabilitation includes education about the respiratory system in both normal and disease states, functional limitations, muscle tension and relaxation, breathing techniques, panic control, and the role of stress and emotional factors in shortness of breath.

Pharmacological intervention

Some pharmacological agents can reduce breathlessness with no major adverse effects, and consequently may improve the quality of life. Such treatment is not used to substitute for specific therapy but should be considered because in many cases the clinical situation is irreversible.

Psychotropic drugs, specifically opioids, are the agents most frequently studied for the treatment of breathlessness. Their mechanisms of action are related to a reduction in ventilatory drive owing to a decrease in hypoxic and hypercapnic responses, and even more to an alteration of the central perception of breathlessness. Following morphine administration, there is a lower degree of breathlessness for a given level of ventilation.

Low-dose nebulized morphine is reported to improve breathlessness in patients with advanced lung disease. The small amounts of morphine administered in this way suggest that it has a local effect. This treatment has been suggested as palliative care for breathlessness in endstage situations.

Benzodiazepines have been considered to have a potential action in alleviating breathlessness. However, diazepam and alprazolam have proved ineffective in relieving this symptom. Buspirone is a non-benzodiazepine anxiolytic drug, and is better tolerated as it does not present sedative and anticholinergic actions. During buspirone treatment, chronic obstructive pulmonary disease patients show reduced anxiety and depression, and an improvement in exercise tolerance and sensations of dyspnea. A reduction in dyspnea has been observed using promethazine. It has been suggested that chlorpromazine could be useful for relieving breathlessness in dying patients.

Although obstruction is not reversible with bronchodilator treatment, it is possible that commonly used drugs may have an extrapulmonary action in reducing dyspnea. Theophylline improves diaphragmatic contraction, increases cardiac output, and reduces dyspnea, but its clinical significance has not been determined. Breathlessness is common in bronchial carcinoma related to airflow obstruction, and bronchodilator therapy may provide alleviation.

Final observations

When acutely ill patients are admitted to a hospital, treatment is generally directed towards the main disease rather than its symptoms. Patients may be considerably distressed and may not obtain the relief they hope for to lessen their discomfort. Treatment given may even worsen symptomatology and is only justifiable if there is the possibility that the patient can return to an acceptable quality of life if he or she survives the acute episode.

The discomfort of a breathless patient in hospital can often be considerably eased by such basic care as arranging the pillows in such a way that the body is supported and thus ventilation is enhanced. Pain control is not necessarily the most difficult problem in treating breathlessness in terminally ill patients. As fear often plays a major role in augmenting distress in these patients, they may often feel considerable relief if they are encouraged to talk about this and other things that may be upsetting them.

References

Cormier, Y., Lecours, R., and Legris, C. (1990). Mechanisms of hyperinflation in asthma. *European Respiratory Journal*, 3, 619–24.

Lougheed, M.D., Webb, K.A., and O'Donnell, D.E. (1995). Breathlessness during induced lung hyperinflation in asthma: the role of the inspiratory threshold load. *American Journal of Respiratory and Critical Care Medicine*, **152**, 911–20.

Petrof, B.J., Legaré, M., Goldberg, P., Milic-Emili, J., and Gottfried, S.B. (1990). Continuous positive airway pressure reduces work of breathing and dyspnea during weaning from mechanical ventilation in severe chronic obstructive pulmonary disease. *American Review of Respiratory Disease*, **141**, 281–9.

Ranieri, V.M., *et al.* (1993). Physiologic effects of positive end-expiratory pressure in patients with chronic obstructive pulmonary disease during acute ventilatory failure and controlled mechanical ventilation. *American Review of Respiratory Disease*, **147**, 5–13.

2.4 Acute respiratory failure

2.4.1 Hypoventilation

J. Milic-Emili and T. Similowski

Key messages

- In patients with severe abnormalities of the control system of breathing and of gas exchange, hypercapnia may result from relatively small increases in metabolic rate.
- The respiratory centers may be depressed directly or indirectly. In both instances, the increase in $Pa\text{CO}_2$ is due to decreased output of the respiratory centers.
- Causes of reduced respiratory muscle strength include neuromuscular disease, malnutrition, electrolyte abnormality, steroids, humoral mediators, thoracic deformity, deconditioning of respiratory muscles, peripheral fatigue, pulmonary hyperinflation, and rapid shallow breathing.

Introduction

The arterial $P\text{CO}_2$ ($Pa\text{CO}_2$) depends on the ratio of metabolic CO_2 production ($\dot{V}\text{CO}_2$) to alveolar ventilation ($\dot{V}_A$):

$$Pa\text{CO}_2 = k\dot{V}\text{CO}_2/\dot{V}_A \qquad (1)$$

where k is a constant. Equation (1) can also be expressed as

$$Pa\text{CO}_2 = \frac{k\dot{V}\text{CO}_2}{\dot{V}_E(1 - V_D/V_T)} \qquad (2)$$

where $\dot{V}_E$ is ventilation, V_D is the physiological dead-space, and V_T is tidal volume.

In normal individuals $Pa\text{CO}_2$ is maintained within a narrow range during most daily activities involving a variety of metabolic demands. Even more astonishing is the fact that most patients affected by respiratory diseases also exhibit a normal $Pa\text{CO}_2$. In some patients, however, the control system of breathing can be so disrupted that they lose the ability to maintain an adequate $\dot{V}_A$ for a given $\dot{V}\text{CO}_2$ with a resultant increase of $Pa\text{CO}_2$ above the normal limits (> 45 mmHg). This condition is defined by the synonyms hypoventilation, hypercapnia, or ventilatory failure.

Pathophysiology of ventilatory failure

Ventilatory failure may be acute or chronic. Both conditions are caused by the same general mechanisms which are shown schematically in Fig. 1, whose kernel is eqn (2).

Increased metabolic rate

Under normal conditions, any increase in metabolic CO_2 production is usually associated with a proportional increase in $\dot{V}_A$ such that $Pa\text{CO}_2$ is maintained within the normal limits. However, in patients with severe abnormalities of the control system of breathing and of gas exchange (e.g. severe airway obstruction), hypercapnia may result from relatively small increases in metabolic rate, such as in fever, sepsis, agitation, excessive weight, hyperthyroidism, increased work of breathing, and carbohydrate excess or increased amino acid content due to total parenteral nutrition.

Central depression

The respiratory centers may be depressed directly (e.g. central nervous system disease, drug overdose, anesthesia, metabolic alkalosis) or indirectly (e.g. obstructive sleep apnea syndrome, central fatigue of respiratory muscles). In both instances, the increase in $Pa\text{CO}_2$ is due to decreased output of the respiratory centers (i.e. decreased neural drive to the respiratory muscles) and hence decreased $\dot{V}_E$ and V_T (eqn (2)).

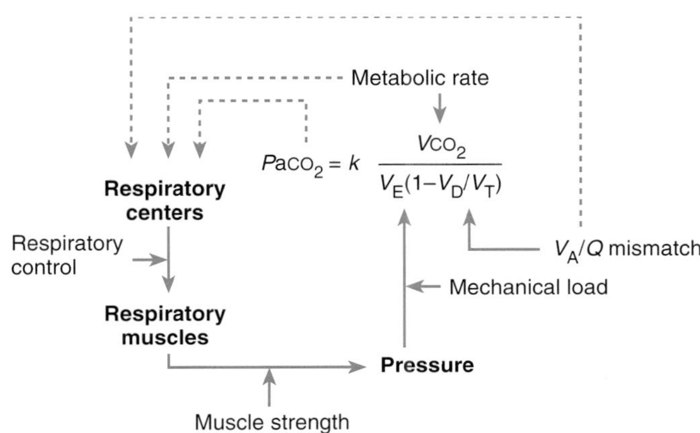

Fig. 1 Schematic diagram of the main mechanisms causing hypoventilation. (Reproduced with permission from Younes (1993).)

Reduced respiratory muscle strength

Causes of reduced respiratory muscle strength include the following.

1. Neuromuscular disease affecting the nerves (e.g. Guillain–Barré syndrome, diphtheria), anterior horn cells (e.g. amyotrophic lateral sclerosis, poliomyelitis), or respiratory muscles (e.g. myopathies).
2. Malnutrition.
3. Electrolyte abnormality (e.g. hypophosphatemia).
4. Steroids.
5. Humoral mediators (e.g. prostaglandins and oxygen free radicals).
6. Thoracic deformity and flail chest.
7. Deconditioning of respiratory muscles (e.g. prolonged mechanical ventilation).
8. Excessive inspiratory muscle activity which may lead to peripheral fatigue, i.e. the muscles become unable to continue to generate adequate pressure despite an appropriate central respiratory drive and intact chest wall. Such failure may activate feedback loops that modify the output of the respiratory centers in response to afferent signals from the fatiguing inspiratory muscles. An important part of the force decline during fatigue may be attributed to such an adaptation of the respiratory centers (central fatigue), reflecting a protective mechanism to prevent a catastrophic drop of intrinsic muscle strength.
9. Pulmonary hyperinflation, which may be due to increased elastic equilibrium volume of the total respiratory system (e.g. emphysema) or dynamic hyperinflation (see below).
10. Decreased respiratory muscle strength may result in hypoventilation due to decreased V_E and/or V_T (e.g. rapid shallow breathing) (eqn (2)).

Ventilation–perfusion inequality

Ventilation–perfusion inequality results in increased shunt and physiological dead-space. The latter promotes hypercapnia, particularly if associated with a reduction of V_T (eqn (2)). Indeed, the acute ventilatory failure of patients with chronic airway obstruction is characterized by increased V_D and rapid shallow breathing, resulting in a very high V_D/V_T ratio. Equipment dead-space also contributes to V_D. The ventilation–perfusion mismatch may lead to a hypoxic stimulation of V_E, which may in part offset the increase in $P\text{a}CO_2$ expected from the augmented V_D (e.g. pulmonary embolism).

Abnormal mechanical load

Elastic load

The elastic load is augmented in diseases characterized by decreased compliance of the lung and chest wall (i.e. restrictive diseases such as obesity, thoracic deformity, acute respiratory distress syndrome). The elastic load also increases when breathing takes place at high lung volumes, i.e. when the tidal volume encroaches into the flat part of the static volume–pressure curve of the total respiratory system. The latter may be due to mechanical ventilation with too high a positive end-expiratory pressure (PEEP) and/or too high a V_T.

Resistive load

The airway resistance may be increased by intrinsic lung disease (e.g. chronic obstructive lung disease) or by an added resistance (e.g. intubation with endotracheal tubes of small size).

Intrinsic PEEP

This load, which acts as an inspiratory threshold load, is probably the most important factor leading to acute ventilatory failure in patients with chronic obstructive lung disease because it is associated with decreased inspiratory muscle strength due to pulmonary hyperinflation, increased airway resistance and physiological V_D due to ventilation–perfusion mismatch, and increased metabolic rate due to increased work of breathing. In addition, the increased $P\text{a}CO_2$ and decreased pH may themselves reduce the inspiratory muscle strength. Furthermore, in chronic obstructive lung disease there is a decreased ventilatory response to CO_2.

Bibliography

Milic-Emili, J. (1982). Recent advances in clinical assessment of control of breathing. *Lung*, **160**, 1–17.

Whitelaw, W.A. and Derenne, J.-Ph. (1996). Control of respiration in acute respiratory failure in chronic obstructive pulmonary disease. In *Acute respiratory failure in chronic obstructive pulmonary disease* (ed. J.Ph. Derenne, W.A. Whitelaw, and T. Similowski), pp. 129–226. Marcel Dekker, New York.

Younes, M. (1993). Mechanisms of ventilatory failure. *Current Pulmonology*, **14**, 243–92.

2.4.2 Pathophysiology of acute respiratory distress syndrome and acute lung injury

G. Deby-Dupont and M. Lamy

Key messages

- The acute respiratory distress syndrome (ARDS) is characterized by tachypnea, severe hypoxemia, and the presence of diffuse bilateral pulmonary infiltrates (a reflection of modified pulmonary capillary permeability) in the absence of signs of left ventricular failure. According to a recent consensus definition, the level of hypoxemia distinguishes acute lung injury ($P\text{a}O_2/\text{F}IO_2 \leqslant 300$ mmHg (40 kPa)) from ARDS ($P\text{a}O_2/\text{F}IO_2 \leqslant 200$ mmHg (27 kPa)).

  The cause of ARDS may be intrathoracic or extrathoracic, but it is usually the earliest sign of a generalized increase in capillary permeability, often accompanying systemic multiple organ dysfunction.

- The pathophysiological mechanisms of ARDS involve complex interactions between activated cells and both humoral and cellular mediators, and the diffusion by these cells and mediators of an initially localized inflammatory reaction, leading to a systemic inflammatory reaction syndrome.
- The principal cells implicated in these reactions are phagocytic cells and cells of the vascular endothelium. Among the mediators involved are the cytokines, active oxygen species, and elements of the complement and coagulation cascades. There are no mediators or groups of mediators specific for ARDS; early identification in bronchoalveolar lavage fluid is more useful than identification in plasma.
- The absence of specific early markers of ARDS and the complex interactions between mediators and cells render therapy using inhibitors of mediators or activated cells difficult.
- Constant improvements in the clinical treatment of ARDS have decreased the mortality of this syndrome, in particular following trauma; new therapeutic approaches are being developed using molecular biological techniques and genetic engineering.

Overview: definition, causes, and mortality

The first description of a syndrome of acute respiratory distress following trauma or severe hypovolemic shock appeared in 1967; this report emphasized the presence of tachypnea, severe hypoxemia related to an intrapulmonary shunt, decreased lung compliance, and diffuse bilateral infiltrates. The definition of acute respiratory distress syndrome (ARDS) was initially based on clinical criteria. Use of Murray's scoring system followed, and a standardized definition was applied from 1994 onwards. Under this system, ARDS is the most severe form of acute lung injury. Table 1 summarizes the principal elements of this definition (Bernard *et al.* 1994).

Table 1 Major elements of the definition of acute lung injury and ARDS

Essential factors	Important secondary factors
Bilateral diffuse pulmonary infiltrates	Presence and type of assisted ventilation
No evidence of left ventricular failure or pulmonary capillary pressure $\leqslant 18$ mmHg	Other organ dysfunction(s)
Hypoxemia	Causes (direct or indirect)
$Pa_{O_2}/Fi_{O_2} \leqslant 300$ mmHg (40 kPa) for ALI	Associated pathologies
$Pa_{O_2}/Fi_{O_2} \leqslant 200$ mmHg (27 kPa) for ARDS	

ALI, acute lung injury.

The factors inciting ARDS can be categorized as either direct or indirect pulmonary insults (Table 2). Infections, trauma, and aspiration of gastric contents are responsible for 70 to 80 per cent of cases. The incidence increases in the presence of several etiological factors and/or concomitant pulmonary pathologies.

The mortality associated with ARDS has been approximately 50 per cent from the 1970s until recently, with individual studies reporting variations from 10 to 90 per cent (Krafft *et al.* 1996). These variations are apparently due to differences in underlying pathological processes, presence of infections, and existence of multiple organ failure, and are not related to the Pa_{O_2}/Fi_{O_2} ratio or the 'lung injury score'. However, recent studies appear to indicate decreases in mortality (Milberg *et al.* 1995), particularly for trauma-associated ARDS.

Mechanisms of ARDS and acute lung injury

The exact mechanisms responsible for the development of ARDS depend on whether the stimulus is direct or indirect, but always involve complex interactions between cellular elements and inflammatory mediators as well as cascade effects (Deby-Dupont *et al.* 1991). These mechanisms frequently lead to a 'snowballing' of the inflammatory reaction, with diffusion to other organs, often resulting in development of the multiple system organ failure syndrome. The precise nature of these mechanisms has not been fully elucidated for several reasons. First, animal models are at best imperfect reproductions of human ARDS. Second, clinical studies focus on the presence of mediators in plasma or bronchoalveolar lavage fluid. Lastly, investigation of the early phases of the syndrome is difficult in humans. Also, it is not understood why, for a given severity of clinical state, one patient develops complicated ARDS whereas another recovers uneventfully. Genetic factors, including a predisposition to increased synthesis of certain cytokines, are suspected in this clinical heterogeneity; early onset of decreased immune function may also explain these differences.

Mechanisms involved in direct pulmonary insults

Direct pulmonary lesions lead to damage to the lung epithelium (alveolar cells and surfactant) and activation of alveolar macrophages. Both these elements are capable of reciprocal interactions and amplification. Three phases can be distinguished: an exudative phase, with formation of edema, hemorrhage, and inflammation, a proliferative phase, with organization of the intraluminal exudate and the start of reparative processes, and a fibrotic phase, which is only present when regeneration is pathological. The final phase only concerns a limited number of patients.

The exudative phase (corresponding roughly to the first week) is characterized by alveolar and interstitial edema, swelling of capillaries, and intra-alveolar hemorrhage. This phase appears to be initiated by activation of macrophages (by micro-organisms and/or toxins) and direct injury of surfactant-producing and epithelial cells caused by these agents or by the initial trauma itself. The activated macrophages produce activated oxygen species, which in turn cause oxidative stress in neighboring cells, and liberate enzymes and peptide, protein, and lipid mediators. These mediators have multiple

Table 2 Main etiologies of ARDS

Thoracic ARDS	Extrathoracic ARDS	Thoracic and extrathoracic ARDS
Major etiologies		
Inhalation (gastric content, fumes, toxics)		Infections (bacteria, viruses, fungi, others) Trauma
Minor etiologies		
Near drowning Pulmonary emboli	Acute pancreatitis Severe shock, ischemia, reperfusion Massive transfusions Toxins Extracorporeal circulation	Burns

properties, including modification of the alveolocapillary membrane and chemoattraction for blood-borne leukocytes.

The inflammatory reaction rapidly reaches nearby endothelial cells, which react by producing new inflammatory mediators (cytokines, procoagulant factors, lipid mediators, vasoactive substances, etc.) and expressing adhesion receptors. These allow (or amplify) the binding of polymorphonuclear neutrophils which are marginated in the pulmonary vasculature, having been attracted by chemotactant agents released locally. This stimulation of adhesion, combined with the presence of chemoattractants and mediators capable of stimulating polymorphonuclear neutrophils, leads to diapedesis of neutrophils into the alveoli. This is followed by further expression of adhesion molecules and production of inflammatory mediators.

Concomitant with the passage of the polymorphonuclear neutrophils into the alveoli, there is an influx of protein- and mediator-rich fluid, as well as other cellular elements including lymphocytes, platelets, and erythrocytes. Bronchoalveolar lavage fluid collected early in the development of ARDS is the best evidence of processes occurring during this initial phase. This fluid is rich in polymorphonuclear neutrophils (up to 90 per cent of cells present), albumin, and inflammatory mediators such as cytokines, prostanoids (e.g. prostaglandin E_2), and leukotrienes. Products of complement activation, coagulation by-products, peptides, proteolytic and granulocyte enzymes (myeloperoxidase, elastase, collagenase, lactoferrin), and even active oxygen species (such as hydrogen peroxide) can also be detected in bronchoalveolar lavage fluid. These products are markers of the intra-alveolar activation of macrophages and neutrophils, which propagates the inflammatory reaction and leads to loss of surfactant and pneumocytes. This in turn leads to denudation of the basement membrane, so that the pulmonary interstitium comes into contact with the alveolar space. Cellular debris and hyaline membranes, which form during this initial phase, become adherent to the inside of the alveolus. Previously aerated lung becomes congested, edematous, and compacted, and obviously can no longer participate in gas exchange. Platelet and fibrin microthrombi together with leukocyte aggregates, both of which predispose to cellular hypoxia, form within the capillaries.

During the subsequent phases of proliferation and fibrosis, the inflammatory exudate begins to organize within the alveoli and interstial spaces, while the type II pneumocytes multiply, fill in denuded areas of the basal membrane, and differentiate. Under the influence of humoral factors liberated by various groups of cells (endothelial cells, platelets, macrophages, etc.), fibroblasts and myofibroblasts proliferate in the interstitium and the alveolar septae, and contribute to the reconstitution of functional alveoli. This reparative response is directed by several molecular species, including growth factors (tumor growth factor-α, epidermal growth factor, platelet-derived growth factor, etc.), as well as fibronectin, collagen fragments, elastin, and fibrin resulting from excessive enzymatic proteolysis.

In ARDS, these regulatory processes are often modified, with incomplete repair and/or excessive cellular proliferation. Thus fibroblasts can attach to the epithelial basement membrane and secrete matrix components which progressively fill the alveolar space and reduce pulmonary compliance. A fibrocellular proliferation can also be seen in the intima of the pulmonary vessels; this, in combination with the cellular aggregates and thrombi mentioned previously, increases the areas of hypoxia. Mesenchymal cells also proliferate abnormally, and intra-alveolar angiogenesis can also be seen. These processes contribute to alveolar fibrosis and loss of pulmonary function, and can lead to the late deaths described in ARDS. However, if the reparative processes are normally controlled, the patient recovers almost normal pulmonary function after several months.

Mechanisms involved in indirect pulmonary insults

The primary mechanism responsible for causing ARDS after extrathoracic injuries is the delivery to the lung of activated cells, inflammatory mediators, and microaggregates originating from a remote site. This form of ARDS constitutes a specific expression of a generalized inflammatory reaction, in which blood-borne cells (monocytes, polymorphonuclear neutrophils, platelets, erythrocytes) and vascular endothelial cells are the principal actors. After severe trauma, infection, or shock states with ischemia followed by reperfusion, cellular functions are activated *in situ*, initiating a localized inflammatory reaction. Phagocytic cells are stimulated by the trauma itself or by bacterial toxins, while endothelial cells are activated by both the initial lesion and the stimulated leukocytes. Under these circumstances, the usually antiaggregant and anticoagulant activity of the endothelium (via production of prostacyclin, thrombomodulin, etc.) becomes predominantly procoagulant (decreased production of thrombomodulin, and secretion of thromboplastin, plasminogen activator–inhibitor, and platelet activating factor). This activity predisposes to the development of disseminated intravascular coagulation (DIC), platelet consumption, and fibrin deposits. These cells release inflammatory mediators, such as cytokines and prostaglandins, as well as substances capable of modulating vascular tone (prostacyclin, endothelin, and nitric oxide (NO)). They activate the complement cascade by binding immune complexes, and express increasing numbers of adhesion receptors either by migration of preformed receptors to the cell surface or by *de novo* synthesis. These receptors facilitate neutrophil binding, allowing them to act locally on the surface of the endothelium, to migrate into the extravascular space, and to maintain and amplify the inflammatory reaction.

Localized activation of endothelial-bound neutrophils damages the vascular lining, alters capillary permeability, and leads to formation of cellular and tissue edema with passage of fluid and mediators into the alveoli. It also leads to modifications of the normal metabolic functions of the endothelium, such as decreased production of converting enzyme. The metabolism of active products is also decreased, leading to longer intravascular persistence of bradykinin (which causes vasodilation and increases vascular permeability), serotonin, prostaglandins E_1, E_2, and F_{2a} (which increase vascular permeability, modulate vascular tone, and cause platelet aggregation and bronchospasm), and norepinephrine (noradrenaline) which causes vasoconstriction.

Neutrophils are the second category of cells intimately involved in the development of ARDS after extrathoracic injury. Mediators derived from polymorphonuclear neutrophils and activated cells themselves are transported by the bloodstream from the damaged area or organ to the pulmonary capillaries. It should be remembered that these cells are found in large numbers in the lung under ordinary circumstances; three times as many are localized, but not activated, in the pulmonary circulation as are found in the general circulation. The transient peripheral neutropenia described early in the course of ARDS is believed to be due to sequestration of these cells in the lung, where they are slowed, trapped, and bound to endothelial adhesion receptors. This is followed by diapedesis into the interstitial and alveolar spaces, where they induce reactions identical to those seen with ARDS of thoracic origin. The demonstration of the existence of activated forms of oxygen in the bloodstream is difficult, but increased levels of protective enzymes such as superoxide dismutase and catalase are noted in infected patients who go on to develop ARDS.

Platelets also participate in the series of reactions leading to the development of ARDS. These cells are activated by endotoxins, complement fragments, and other mediators, such as platelet activating factor. They liberate mediators which include platelet activating factor itself, serotonin, platelet factor 4, etc. All of these probably contribute to amplification of the inflammatory reaction and the coagulation cascade, and participate in the formation of microaggregates; these aggregates can subsequently cause or worsen hypoxemia and DIC.

ARDS — the first marker of multiple organ failure

The alterations of capillary permeability, and the activation of phagocytic and endothelial cells, are not limited to the lungs, but are seen elsewhere in the organism following infections and shock states. They also occur in organs after successive cycles of ischemia and reperfusion, as a result of mitochondrial dysfunction. This leads to falls in ATP synthesis and increased levels of intracellular calcium. The result is cellular edema, activation of enzymes, and production of active oxygen species. The contribution of this mechanism (ischemia–reperfusion) depends on the given organ's sensitivity to hypoperfusion; however, some degree of altered capillary permeability and cellular activation is present in virtually all organs. The clinical expression of this lesion, and of the inflammatory reaction that accompanies it, can be discrete despite the organ's exquisite sensitivity (such as the pancreas), while the clinical syndrome of respiratory failure is readily apparent in the lung early in the disease process. Among the organs damaged first are the kidneys, the heart, and the brain (dysfunction of these systems is well described during and after shock states), and also the intestines, the liver, and the pancreas. Altered mucosal permeability of the digestive tract implies the possibility of translocation of bacteria, toxins, and active digestive enzymes. Failure of hepatic macrophages promotes penetration of micro-organisms via the portal circulation, decreased metabolism of inflammatory mediators and products of cellular destruction (such as hemoglobin), and reduced synthesis of normal proteins. Pancreatic lesions involve penetration of active digestive enzymes into the circulation (trypsin, elastase, lipase, amylase, etc.). Some of these enzymes (such as trypsin) can activate the complement and coagulation cascades, destroy proteins and peptides, and contribute to saturating endogenous antiproteases, with establishment of a protease–antiprotease imbalance.

Once localized inflammatory reactions become excessive and poorly controlled, they lead to dispersal of mediators towards other sites and to the systemic inflammatory response syndrome. This can progress to frank multiple organ failure, frequently complicated by sepsis. Under these circumstances, ARDS becomes the earliest and easiest to diagnose aspect of a simultaneous syndrome of widespread visceral failure, as opposed to an isolated clinical entity.

Mediators of ARDS and acute lung injury

Mediators of ARDS and acute lung injury, either of humoral origin or secreted by cells, trigger, spread, and sustain the inflammatory reaction and provoke effects via cascade mechanisms with multiple interactions (Demling 1995). Numerous mediators have been described, with the most recent being identified using molecular biological techniques and using bronchoalveolar lavage fluid; this biological substance allows processes occurring in the target organ itself to be studied early in the pathological pathway. It is useful, for ease of categorization, to classify these mediators according to their origin, i.e. humoral or cellular.

Humoral mediators

These substances originate primarily in the activation of the complement and coagulation–fibrinolytic cascades. The endotoxins, the acute phase reactant proteins (characteristic of the inflammatory reaction), the proteases released from damaged organs, and some peptides, proteins, and protein fragments (which appear in the blood during shock and inflammatory states) could also be added to this list. Complement degradation products, particularly C5a, are found in high concentrations in various pathological states associated with the development of ARDS, such as infections, extracorporeal circulation (used for cardiac surgery), acute pancreatitis, burns, and multiple trauma. Activation of the complement cascade, by either the classical or the alternative pathways, can occur early after trauma and infection. This system triggers the coagulation cascade and attracts and activates polymorphonuclear neutrophils (and thus spreads and amplifies the inflammatory reaction), but it is not the cause of ARDS; complement-split products are not specific markers of ARDS.

In ARDS, particularly in the presence of infection, the activation of the coagulation–fibrinolytic cascades is caused by complement fragments, endotoxins, and the original inciting trauma. Several elements tend to prove the activation of these systems, including demonstration of pulmonary sequestration of microthrombi, platelet aggregates, and fibrinogen, the presence of DIC, and unusually high plasma concentrations of active enzymes (thrombin, plasmin) and of abnormal kininogen and fibrin-split products.

Excessive proteolytic activity is responsible for the increased plasma concentrations of peptides with various activities (including immunosuppression) commonly seen in ARDS. These proteases also originate from the activation of various humoral cascades, but also from activated cells and damaged organs (particularly the pancreas) after lesions produced by hypoperfusion and the onset of the multiple organ failure syndrome. They rapidly saturate available endogenous antiproteases, such as antithrombin III and α_2-macroglobulin, and lead to an imbalance between active proteases and antiprotease defenses.

The presence of endotoxin in the bloodstream is related not only to infection, but also to translocation across an intestinal mucosal barrier altered by ischemia. However, this mechanism is controversial. Endotoxins activate not only monocytes and macrophages (with consequent production of cytokines), but also polymorphonuclear neutrophils and endothelial cells via complexes formed with binding proteins that are recognized by cellular receptors.

Cellular mediators

Most cellular mediators are produced by monocyte-macrophages, polymorphonuclear neutrophils, and endothelial cells, but they can also be derived from lymphocytes, platelets, alveolar cells, and fibroblasts.

Cytokines, secreted by macrophages and monocytes after stimulation by endotoxins, constitute a class of peptide molecules with effects on either the cells that produced them (autocrine effect) or more distant target cells (paracrine effect). These molecules are tumor necrosis factor-α and interleukins 1, 6, and 8 (IL-1, IL-6, and IL-8). The sequential release of the interleukins is consecutive to that of tumor necrosis factor-α. Their effects are mainly exerted on polymorphonuclear neutrophils, endothelial cells, and certain target organs such as the liver (where they stimulate production of acute phase reactant proteins). IL-8, which is produced by numerous cell types, is a chemoattractant, stimulates adhesion, and powerfully activates polymorphonuclear neutrophils. IL-12, which has recently been implicated in the pathogenesis of ARDS, acts by stimulating lymphocytes. High levels of tumor necrosis factor-α, IL-1, IL-6, and IL-8 are found in the plasma and bronchoalveolar lavage fluid during ARDS. Levels of tumor necrosis factor-α have been found to correlate with the incidence and intensity of the syndrome. A genetically determined variability in the capacity to produce cytokines, and therefore to develop ARDS, has also been invoked.

Among the lipid substances (prostanoids, leukotrienes, platelet activating factor) formed from phospholipids de-esterified by phospholipase A_2, thromboxane A_2 has been considered to be an early mediator with a role in the development of ARDS. In effect, its liberation precedes that of prostacyclin and the leukotrienes. None of the lipid mediators found in plasma or bronchoalveolar lavage fluid is considered to be specific to ARDS, but their varied actions (vasoconstriction, vasodilation, bronchoconstriction, chemotaxis, cellular activation) appear to confirm a key position in the generalization and maintenance of the inflammatory reaction seen during the syndrome.

Active oxygen species and free radicals (superoxide anion, hydrogen peroxide, the hydroxyl radical, hypochlorous acid, singlet oxygen) are produced by stimulated phagocytic cells. They are responsible for lipid peroxidation reactions and for molecular and tissue damage. They are also believed to be at the origin of oxidative stress and are considered to play an important role in the intracellular activation of nuclear transcription regulatory factors (such as NF B), the synthesis of messenger RNA, the induction of protein synthesis, and the triggering of apoptosis. Other sources of activated oxygen species are the mitochondria, the ischemic organs (by the action of xanthine oxidase), and free plasma hemoglobin. Their short half-lives make direct detection of these species difficult, but the time course of plasma concentrations of antioxidant molecules and

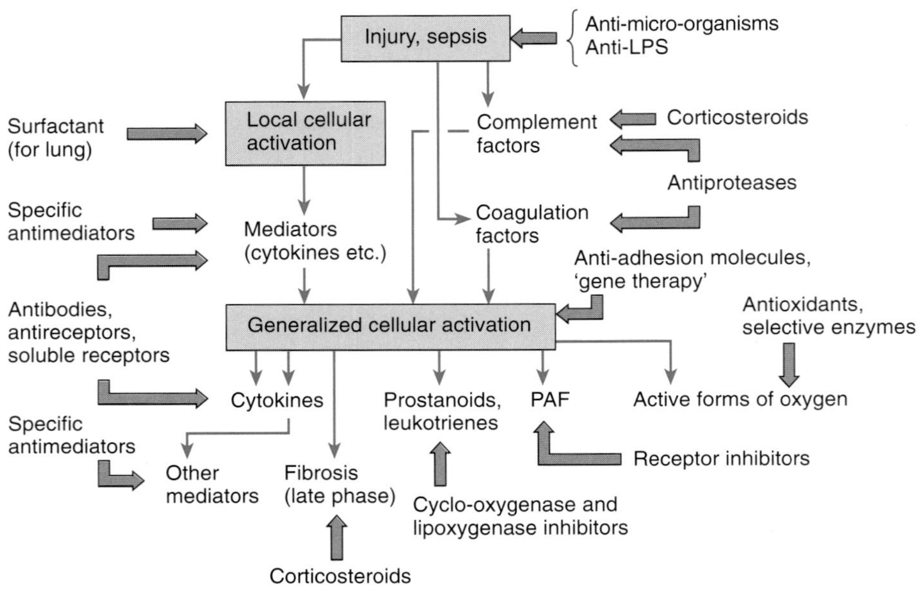

Fig. 1 Main pathways for pharmacological therapy in ARDS (→ indicates an inhibitory activity): LPS, lipopolysaccharides (endotoxins); PAF, platelet activating factor.

specific antiradical enzymes (such as superoxide dismutase, catalase, and glutathione peroxidase) indirectly suggests their presence.

The adhesion receptors found on cell surfaces play an important role by favoring cellular interactions. They belong to the family of the selectins and the integrins, and the immunoglobulin superfamily. Their expression is modulated during the inflammatory reaction by the influence of endotoxins and cytokines; it is increased by the action of stimulated polymorphonuclear neutrophils and by numerous signal molecules such as platelet activating factor and IL-8.

During shock states and inflammatory reactions, cells produce heat shock proteins. These peptides are chemotactic and immunosuppressive, and as yet their role is poorly characterized (protection from or propagation of the inflammatory reaction).

Treatment of ARDS

It would be useful to be able to define a specific early marker of ARDS, particularly if such a molecule belonged to one of the classes of mediators discussed above. This would allow early pharmacological treatment of the syndrome before or simultaneously with the appearance of the first symptoms. Such a role has been assigned to several mediators, often following the chronology of their discovery. Complement fragments, such as C5a, were initially believed to be the required marker. Subsequently, prostanoids and leukotrienes, coagulation proteins, etc. were thought to be the long-sought molecules. It is currently accepted that there is no marker specific for ARDS. Early studies of bronchoalveolar lavage fluid may provide a technique for the early rapid detection of the onset of ARDS, allowing early treatment.

Apart from specific causal therapy for ARDS and symptomatic supportive treatment of the respiratory failure and of the patient in general, the numerous attempts at pharmacological antimediator therapy have not been particularly successful (Fig. 1). This failure can be explained by the large number of mediators, their cascading interactions, the early generalization of the inflammatory reaction, and the development of multiple organ failure. The use of substances capable of inducing endogenous defenses, such as granulocyte colony-stimulating factor, granulocyte-monocyte colony-stimulating factor, and certain 'anti-inflammatory' cytokines, such as IL-4, IL-10, and IL-13, appears promising. Genetic therapy (by stimulation or repression of genes coding for molecules implicated in the pathogenesis of ARDS) is a possible future development.

Conclusions

ARDS is a complex clinical problem in which a disproportionate immunological response plays a major role. After 25 years, clinical treatment is showing slow steady improvement. However, it is still not possible to predict the impact of interventions on the clinical course of an individual patient because the fundamental mechanisms of the syndrome are only imperfectly understood.

References

Bernard, G.R., *et al.* (1994). Report of the American-European Consensus Conference on Acute Respiratory Distress Syndrome: definitions, mechanisms, relevant outcome and clinical trial coordination. *Journal of Critical Care*, **9**, 72–81.

Deby-Dupont, G., Lamy, M., Deby, C., Damas, P., and Faymonville, M.E. (1991). Adult respiratory distress syndrome: local or systemic disease. In *Perspectives in critical care* (ed. F.C. Cerra), pp. 57–83. Quality Medical, St Louis, MO.

Demling, R.H. (1995). The modern version of acute respiratory distress syndrome. *Annual Reviews of Medicine*, **46**, 193–202.

Krafft, P., *et al.* (1996). The acute respiratory distress syndrome: definitions, severity and clinical outcome. An analysis of 101 clinical investigations. *Intensive Care Medicine*, **22**, 519–29.

Milberg, J.A., Davis, D.R., Steinberg, K.P., and Hudson, L.D. (1995). Improved survival of patients with acute respiratory distress syndrome (ARDS): 1983–1993. *Journal of the American Medical Association*, **273**, 306–9.

2.4.3 Pulmonary mechanical dysfunction

J. Milic-Emili and T. Similowski

Key messages

- Dynamic hyperinflation exists whenever the duration of expiration is insufficient to allow the lungs to deflate to the relaxation volume prior to the next inspiration.
- When intrinsic positive end-expiratory pressure (PEEP) is present, onset of inspiratory muscle activity and inspiratory flow are not synchronous; inspiratory flow starts only when the pressure developed by the inspiratory muscles exceeds intrinsic PEEP.
- The aim of treatment of patients with continuous obstructive pulmonary disease who have respiratory failure should be reduction of respiratory frequency.
- Application of external PEEP during patient-triggered mechanical ventilation can counterbalance and reduce the inspiratory load imposed by intrinsic PEEP.

Introduction

Most normal subjects do not exhibit expiratory flow limitation even during maximal exercise. In contrast, patients with severe chronic obstructive pulmonary disease (COPD) often exhibit it even at rest, i.e. they exhale tidally along their maximal flow–volume curve. This promotes dynamic hyperinflation with a concomitant increase in work of breathing and impairment of inspiratory muscle function. This contributes to dyspnea and is probably the main cause of ventilatory failure in patients with COPD.

Dynamic hyperinflation

In normal subjects at rest, the end-expiratory lung volume (functional residual capacity (FRC)) corresponds to the relaxation volume (V_r) of the respiratory system, i.e. the lung volume at which the elastic recoil pressure of the respiratory system is zero. Pulmonary hyperinflation is defined as an increase of FRC above the predicted normal value. This may be due to increased V_r as a result of loss of elastic recoil (e.g. emphysema) or to dynamic pulmonary hyperinflation, which is said to be present when the FRC exceeds V_r. Dynamic hyperinflation exists whenever the duration of expiration is insufficient to allow the lungs to deflate to V_r prior to the next inspiration. This tends to occur under conditions in which expiratory flow is impeded (e.g. increased airway resistance) or when the expiratory time is reduced (e.g. increased breathing frequency). Expiratory flow may also be retarded by other mechanisms such as persistent contraction of the inspiratory muscles during expiration and expiratory narrowing of the glottic aperture. In COPD patients with acute ventilatory failure, dynamic hyperinflation is invariably present and is due to severe expiratory flow limitation.

Effects of dynamic hyperinflation on work of breathing and mechanical performance of inspiratory muscles

Figure 1 illustrates the pressure required to overcome the elastic recoil of the respiratory system for the same tidal volume (equal to 20 per cent of vital capacity (VC)) inhaled from V_r (equal to 34 per cent of VC) and from an end-expiratory lung volume increased to 67 per cent of VC. As shown by the hatched areas, the elastic work increases about fivefold when the breath is inhaled from 67 per cent VC (case B) relative to the breath taken from V_r (case A). Clearly, dynamic hyperinflation implies an increase of both elastic work and inspiratory muscle effort. Furthermore, as lung volume increases, there is a decreased effectiveness of the inspiratory muscles as pressure generators because the inspiratory muscle fibers become shorter (force–length relationship) and their geometrical arrangement changes. Thus, in COPD patients there is a vicious cycle: the inspiratory flow-resistive work is invariably increased due to airway obstruction but, more importantly, as a result of hyperinflation there is a concomitant increase in elastic work and impaired mechanical performance of the inspiratory muscles. With increased severity of COPD, a critical point is eventually reached at which the inspiratory muscles become fatigued.

Intrinsic positive end-expiratory pressure (PEEP)

Under normal conditions, the end-expiratory elastic recoil pressure of the respiratory system is zero (case A in Fig. 1). In this instance, as soon as the inspiratory muscles contract, the alveolar pressure becomes subatmospheric and gas flows into the lungs. When breathing takes place at lung volumes higher than V_r, the end-expiratory elastic recoil pressure is positive (15 cmH_2O in case B of Fig. 1). The elastic recoil pressure present at end-expiration has been termed auto-PEEP or intrinsic PEEP. When intrinsic PEEP is present, onset of inspiratory muscle activity and inspiratory flow are not synchronous; inspiratory flow starts only when the pressure developed by the inspiratory muscles exceeds intrinsic PEEP because only then does alveolar pressure becomes subatmospheric. In this respect, intrinsic PEEP acts as an inspiratory threshold load which increases the elastic work of breathing. As indicated above, this places a significant burden on the inspiratory muscles, which in hyperinflated COPD patients are operating under disadvantageous force–length conditions and abnormal thoracic geometry. Thus the presence of expiratory flow limitation during resting breathing is associated with hypercapnia and dyspnea.

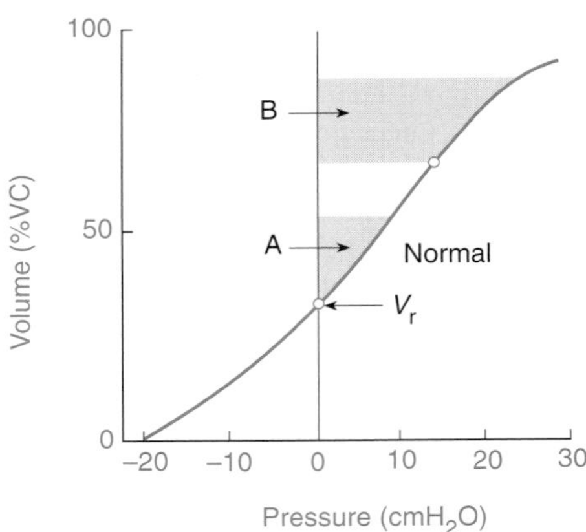

Fig. 1 Volume–pressure diagram of the relaxed respiratory system showing the increase in elastic work caused by dynamic hyperinflation: area A, elastic work for a breath that starts from the relaxation volume V_r; area B, elastic work for a similar breath to A but starting from a volume 29 per cent VC higher than V_r. In case B, the intrinsic PEEP is 15 cmH_2O, as indicated by the upper circle.

Dynamic hyperinflation and intrinsic PEEP in COPD patients with acute ventilatory failure

The highest values of intrinsic PEEP in stable COPD patients are 7 to 9 cmH_2O. Higher values have been reported in COPD patients with acute respiratory failure—up to 13 cmH_2O during spontaneous breathing and 22 cmH_2O during mechanical ventilation. Such high values of intrinsic PEEP have profound consequences on the energetics of breathing, as shown Fig. 2. Acute ventilatory failure in COPD patients is usually triggered by airway infection. As a result, there is an increase in airway resistance and expiratory flow limitation becomes more severe. The increased resistance causes increased resistive work of breathing and promotes dynamic hyperinflation. The latter is further exacerbated by the tachypnea which is invariably present in acutely ill COPD patients. Dynamic hyperinflation promotes an increase of elastic work of breathing due to intrinsic PEEP (Fig. 1). The increased resistive and elastic work of breathing, in association with the impaired inspiratory muscle performance, eventually leads to inspiratory muscle fatigue. As a result, the patient needs mechanical respiratory assistance.

Strategies to reduce the inspiratory load caused by intrinsic PEEP

As indicated in Fig. 2, the aim of treatment of COPD patients with respiratory failure should be to reduce respiratory frequency (and hence increase expiratory time) and decrease flow resistance. To the extent that tachypnea is due to fever and/or airway infection, an improvement of these variables by conventional treatment should be beneficial. Similarly, administration of bronchodilators may be useful in reducing both flow resistance and intrinsic PEEP. A less conventional but promising approach to dealing with intrinsic PEEP is the use of continuous positive airways pressure (CPAP). Indeed, CPAP has been found to reduce the magnitude of inspiratory muscle effort and work of breathing in stable patients with severe COPD and in patients with severe COPD during weaning from mechanical ventilation. This is related to a reduction in the inspiratory load imposed by

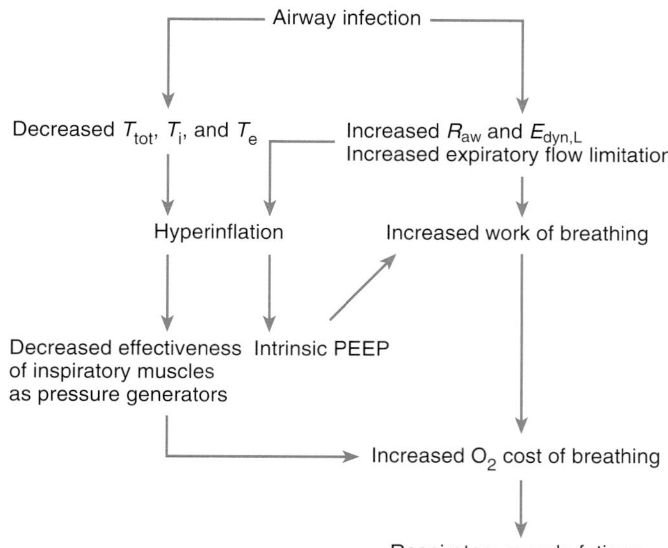

Fig. 2 Scheme of the pathophysiology causing acute ventilatory failure in COPD patients: T_{tot}, total breathing cycle duration; T_i, inspiratory time; T_e, expiratory time; R_{aw}, airway resistance; $E_{dyn,\,L}$, dynamic lung elastance.

intrinsic PEEP, and considerably improves patient–ventilator synchrony by decreasing the number of breaths where the patient fails to develop the sum of intrinsic PEEP and the triggering pressure which is required for the breath to be delivered. In addition, CPAP administered through a face or nasal mask may also be of therapeutic benefit during an acute exacerbation of severe COPD in the non-intubated patients. Conceivably, the early use of CPAP in this setting could preclude the need for intubation and mechanical ventilation in some COPD patients. Although application of CPAP may reduce considerably the inspiratory work of breathing and muscle effort, it does not seem to lead to a substantial increase in alveolar ventilation. To achieve an increase of alveolar ventilation, further respiratory assistance is required (e.g. pressure support). Finally, it should also be noted that application of external PEEP during patient-triggered mechanical ventilation can counterbalance and reduce the inspiratory load imposed by intrinsic PEEP.

Bibliography

Coussa, M.L., *et al.* (1993). Partitioning of work of breathing in mechanically ventilated COPD patients. *Journal of Applied Physiology*, **75**, 1711–19.

Gottfried, S.B. (1991). The role of PEEP in the mechanically ventilated COPD patient. In *Ventilatory failure* (ed. J.J. Marini and C. Roussos), pp. 392–418. Springer-Verlag, Berlin.

Guérin, C., *et al.* (1993). Lung and chest wall mechanics in mechanically ventilated COPD patients. *Journal of Applied Physiology*, **74**, 1570–80.

Pride, N.B. and Milic-Emili, J. (1995). Lung mechanics. In *Chronic obstructive pulmonary disease* (ed. P.M.A. Calverley and N.B. Pride), pp. 135–60. Chapman & Hall, London.

2.4.4 Cardiovascular interactions

M. R. Pinsky

Key messages

- The effects of ventilation on the circulation are due to changes in lung volume and intrathoracic pressure and the effort necessary to generate them.
- Spontaneous ventilatory efforts can occur in a subject on mechanical ventilation and will increase metabolic demands on the heart.
- Pulmonary vascular resistance is affected by lung volume changes.

  Small lung volumes, as occur with acute lung injury, increase pulmonary vascular resistance by the process of hypoxic pulmonary vasoconstriction.

  Overdistension will also increase pulmonary vascular resistance by alveolar vessel compression.
- Ventilatory strategies that restore lung volume toward normal will maintain low pulmonary vascular resistance, whereas those that overdistend the lung will decrease blood flow to the overdistended lung units, increasing dead-space ventilation and pulmonary vascular resistance.
- Changes in intrathoracic pressure alter the pressure gradients for both venous return and left ventricular ejection. These effects can be used to identify patients with normal or reduced cardiac functional reserve.

  Increasing intrathoracic pressure will decrease venous return and unload the left ventricle, decreasing intrathoracic blood volume. In normal conditions cardiac output will decrease, whereas in heart failure cardiac output will either remain constant or, if augmented by the afterload reduction effect, increase.

  Decreasing intrathoracic pressure will have the opposite effect and may precipitate cardiac failure and pulmonary edema in patients with borderline cardiac function, while augmenting cardiac output in normal subjects.

Introduction

Acute respiratory failure can directly alter cardiovascular function in a large number of seemingly unrelated ways. However, many of these effects are predictable from knowledge of cardiovascular function. The respiratory system and the cardiovascular systems are not separate but tightly integrated. The ultimate cardiovascular response to acute respiratory failure is dependent on the basal cardiovascular state of the subject, the type of respiratory dysfunction present, and the ventilatory pattern. Spontaneous inspiratory efforts during both acute bronchospasm and acute lung injury will induce markedly negative swings in intrathoracic pressure. Both underinflation and hyperinflation of the lung will alter pulmonary vascular resistance and heart–lung interactions, and increased work of breathing will stress the cardiovascular response to maintain and increase cardiac output to meet the increased oxygen demand ($\dot{V}O_2$). Furthermore, artificial ventilatory support will increase intrathoracic pressure during inspiration, in contradistinction to spontaneous ventilation which will decrease intrathoracic pressure for the same tidal breath. Heart–lung interactions involve four basic concepts

1. Inspiration increases lung volume above end-expiratory volume.
2. Spontaneous inspiration decreases intrathoracic pressure.
3. Positive-pressure ventilation increases intrathoracic pressure.
4. Ventilation is exercise: it consumes O_2 and produces CO_2.

Hemodynamic effects of changes in lung volume

Lung inflation alters autonomic tone and pulmonary vascular resistance; at high lung volumes, it compresses the heart in the cardiac fossa, limiting absolute cardiac volumes. The associated diaphragmatic dissent increases abdominal pressure and compresses the liver markedly, altering the flow characteristics of venous return. Each of these processes may predominate in determining the final cardiovascular state. Small tidal volumes (< 10 ml/kg) increase heart rate by vagal withdrawal (respiratory sinus arrhythmia), whereas larger tidal volumes (> 15 ml/kg) decrease heart rate, arterial tone, and cardiac contractility by sympathetic withdrawal (Butler 1983). These effects are probably only relevant in the diagnosis of dysautonomia and in the care of neonatal subjects where autonomic tone is high.

The major determinants of the hemodynamic response to increases in lung volume are mechanical in nature (Butler 1983). Lung inflation, independent of changes in intrathoracic pressure, primarily affects cardiac function and cardiac output by altering right ventricular preload and afterload and left ventricular preload. First, inspiration induces diaphragmatic dissent which may alter venous return. Venous return is a function of the ratio of the pressure difference between the right atrium and the systemic venous reservoirs to the resistance to venous return. Since a large proportion of the venous blood volume is in the abdomen, increases in intra-abdominal pressure will increase the venous pressure in the vascular spaces, augmenting venous blood flow (Fessler *et al.* 1992). However, diaphragmatic descent will also compress the liver, increasing hepatic outflow resistance and decreasing flow from the splanchnic venous reservoirs to the right heart. Complicating this further, inspiration will shift venous flow from high-resistance splanchnic circuits, which must drain through the liver, to low-resistance systemic venous circuits, making flow greater for the same driving pressure. Thus inspiration may increase, decrease, or have no effect on venous return depending on which of these factors is predominant. Inspiration will increase venous return in volume-overloaded states, whereas venous return will decrease in hypovolemic states and hepatic cirrhosis.

Right ventricular output is sensitive to changes in pulmonary outflow resistance. Alveolar collapse often occurs in acute lung injury states and is associated with increases in pulmonary vasomotor tone via hypoxic pulmonary vasoconstriction. Alveolar recruitment by restoring end-expiratory lung volume to functional residual capacity often reverses this process. Increasing lung volume above functional residual capacity also increases right ventricular outflow resistance (Butler 1983; Pinsky *et al.* 1985). However, this is due to progressive increases in transpulmonary pressure (airway pressure relative to intrathoracic pressure) associated with increasing lung volume. Since the heart and great vessels are located in the thorax and sense intrathoracic pressure as their surrounding pressure, increases in transpulmonary pressure such that it approaches pulmonary artery pressure will induce pulmonary vascular collapse (Butler 1983). Hyperinflation increases pulmonary artery pressure; therefore reversing hyperinflation, by prolonging expiration, reducing levels of positive-end expiratory pressure or tidal volumes, and bronchodilatation, may decrease pulmonary arterial pressure, thus improving right ventricular ejection.

Left ventricular end-diastolic volume (preload) can be altered by ventilation in three ways. First, since the right and left ventricular outputs are in series, changes in right ventricular preload must eventually alter left ventricular preload in the same direction. Second, because of ventricular interdependence, changes in right ventricular end-diastolic volume inversely change left ventricular diastolic compliance (Taylor *et al.* 1967). Ventricular interdependence is a major factor in altering left ventricular output during spontaneous ventilation when right ventricular end-diastolic volumes may vary widely from expiration (small) to inspiration (large). Third, increasing lung volume restricts absolute cardiac volume by direct compression of the heart (Butler 1983). As the lungs expand, the heart is compressed in the cardiac fossa and absolute biventricular volume is limited in a fashion analogous to cardiac tamponade.

Hemodynamic effects of changes in intrathoracic pressure

The heart within the thorax is a pressure chamber within a pressure chamber. Therefore changes in intrathoracic pressure will affect the pressure gradients for both systemic venous return to the right ventricle and systemic outflow from the left ventricle, independent of the heart itself (Buda *et al.* 1979; Pinsky *et al.* 1985). Increases in intrathoracic pressure, by both increasing right atrial pressure and decreasing transmural left ventricular systolic pressure, will reduce these pressure gradients, thereby decreasing intrathoracic blood volume. However, by the same argument, decreases in intrathoracic pressure will augment venous return and impede left ventricular ejection, thus increasing intrathoracic blood volume. Variations in right atrial pressure represent the major factor determining the fluctuation in pressure gradient for systemic venous return during ventilation (Fessler *et al.* 1992). Increases in intrathoracic pressure, as seen with positive-pressure ventilation or hyperinflation during spontaneous ventilation, decrease venous return, whereas decreases in intrathoracic pressure, as seen with spontaneous inspiration, increase venous return.

Spontaneous inspiratory efforts, by decreasing intrathoracic pressure, both increase lung volume and decrease right atrial pressure, accelerating blood flow into the right ventricle (Buda *et al.* 1979) and increasing pulmonary blood flow on the subsequent beat. Thus normal respiration-associated hemodynamic changes maximize ventilation–perfusion temporal matching because the spontaneous inspiration matches an increase in alveolar oxygen flux with an increase in pulmonary capillary flow. However, the augmentation of venous return is limited because, as intrathoracic pressure decreases below atmospheric pressure, the large systemic veins collapse as they enter the thorax, limiting maximal venous flow (Fessler *et al.* 1992). This 'flow limitation' is useful because intrathoracic pressure can decrease greatly with obstructive inspiratory efforts. If the increase in venous blood flow were unlimited, the right ventricle could become overdistended in volume and fail.

Left ventricular afterload or systolic wall tension is proportional to the product of transmural left ventricular pressure and left ventricular volume. Since increasing intrathoracic pressure will mechanically decrease transmural left ventricular pressure if arterial pressure is constant, increases in intrathoracic pressure unload the left ventricle, whereas decreases in intrathoracic pressure have the opposite effect (Buda *et al.* 1979). Thus, in ventricular failure states associated with fluid resuscitation, increases in intrathoracic pressure increase cardiac output (Buda *et al.* 1979; Pinsky *et al.* 1985).

Spontaneous ventilatory efforts performed against a resistive (bronchospasm) or elastic (acute lung injury) load, decrease left ventricular stroke volume via a complex mechanism collectively called pulsus paradoxus. Transient intraventricular septal shift into the left ventricular lumen combined with pericardial volume restraint decreases absolute left ventricular end-diastolic volume (Taylor *et al.* 1967; Butler 1983). Increases in left ventricular afterload (left ventricular pressure minus intrathoracic pressure) increase left ventricular end-systolic volume (Buda *et al.* 1983).

Sudden increases in intrathoracic pressure increase arterial pressure by an amount equal to the increase in intrathoracic pressure without changing aortic blood flow. However, if the increase in intrathoracic pressure is sustained, the decrease in systemic venous return induced will eventually decrease left ventricular output, thus decreasing arterial pressure (Buda *et al.* 1979). In the steady state, changes in intrathoracic pressure that result in altered cardiac output also alter peripheral vasomotor tone through baroreceptor mechanisms. Baroreceptor reflexes tend to keep systemic pressure (arterial pressure) and flow (cardiac output) constant. Thus, if intrathoracic pressure increased arterial pressure without changing transmural arterial pressure, the periphery would reflexly vasodilate to maintain a constant extrathoracic arterial pressure–flow relation (Pinsky *et al.* 1985). Since coronary perfusion pressure is not increased by intrathoracic-pressure-induced increases in arterial pressure, whereas mechanical constraint from the expanding lungs may obstruct coronary blood flow, coronary hypoperfusion from a combined coronary compression and a decrease in coronary perfusion pressure is a potential complication of increased intrathoracic pressure. Although increases in intrathoracic pressure should augment left ventricular ejection by decreasing left ventricular afterload, this effect should have limited therapeutic potential, just as all afterload-reducing therapies are limited by both the minimal end-systolic volume and the obligatory decrease in venous return. Thus the potential augmentation of left ventricular ejection by increasing intrathoracic pressure is limited under most conditions because increasing intrathoracic pressure, by reducing left ventricular ejection pressure, can only decrease end-systolic volume, which is usually already small and cannot decrease much more except in markedly dilated cardiomyopathies. However, the decrease in venous return associated with the increase in intrathoracic pressure can completely arrest venous blood flow.

Mechanically speaking, there is no difference between increasing intrathoracic pressure from a basal end-expiratory level and eliminating the negative end-inspiratory intrathoracic pressure swings seen in spontaneous ventilation. For many reasons, removing negative swings in intrathoracic pressure may be more clinically relevant than increasing intrathoracic pressure. First, many pulmonary diseases are associated with exaggerated decreases in intrathoracic pressure during inspiration. In restrictive lung disease states, such as interstitial fibrosis or acute hypoxemic respiratory failure, intrathoracic pressure must decrease substantially to generate a transpulmonary pressure that is large enough to ventilate the alveoli. Similarly, in obstructive diseases, such as upper airway obstruction or asthma, large decreases in intrathoracic pressure occur owing to increased resistance to inspiratory airflow. Second, exaggerated decreases in intrathoracic pressure require increased respiratory efforts that increase the work of breathing, taxing a potentially stressed circulation. Finally, the exaggerated decreases in

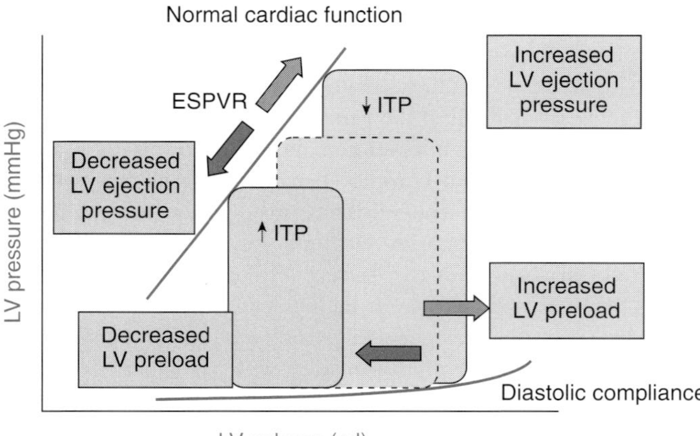

Fig. 1 The effect of increasing (green shading) and decreasing (gray shading) intrathoracic pressure (ITP) on the left ventricular (LV) relation with left ventricular contractility is normal. The slope of the left ventricular end-systolic pressure–volume relationship (ESPVR) is proportional to contractility. The slope of the diastolic left ventricular pressure–volume relationship defines diastolic compliance.

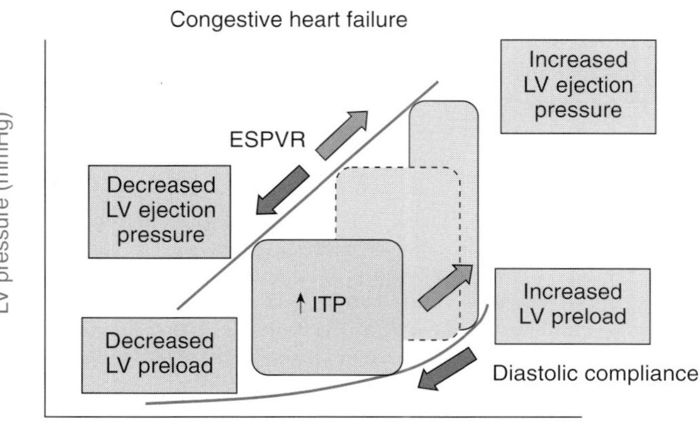

Fig. 2 The effect of increasing (green shading) and decreasing (gray shading) intrathoracic pressure (ITP) on the left ventricular (LV) relation in congestive heart failure when left ventricular contractility is reduced and intravascular volume is expanded. The slope of the left ventricular end-systolic pressure–volume relationship (ESPVR) is proportional to contractility. The slope of the diastolic left ventricular pressure–volume relationship defines diastolic compliance.

intrathoracic pressure can only increase venous blood flow so much before venous collapse limits blood flow. The level to which intrathoracic pressure must decrease to induce venous flow limitation is different in different circulatory conditions, but in most patients it occurs below an intrathoracic pressure of $-10\,cmH_2O$ (Butler 1983). Thus further decreases in intrathoracic pressure will only increase left ventricular afterload without increasing venous return. Accordingly, abolishing these markedly negative swings in intrathoracic pressure should disproportionally reduce left ventricular afterload more than venous return (left ventricular preload). These concepts of a differential effect of increasing and decreasing intrathoracic pressure on cardiac function are illustrated for both normal and failing hearts in Figs 1 and 2 using the left ventricular pressure–volume relationship during one cardiac cycle to interpose venous return (end-diastolic volume) and afterload (end-systolic volume). Using this logic, one would predict that, by endotracheally intubating and ventilating such patients, markedly negative swings in intrathoracic pressure can be abolished without any impairment in systemic venous return or any need to make the swings in intrathoracic pressure more positive. These interactions have important implications in the decision both to institute and to withdraw mechanical ventilatory support.

Ventilation as exercise

Spontaneous ventilatory efforts require muscular activity, consume O_2, and produce CO_2, and they represent a metabolic load on the cardiovascular system. Although ventilation normally requires less than 5 per cent of total O_2 delivery to meet its demand, the O_2 requirements may increase to 25 per cent or more of total O_2 delivery in lung disease states where the work of breathing is increased, such as pulmonary edema or bronchospasm. Furthermore, if cardiac output is limited, this level of activity (spontaneous ventilation) may not be possible even with additional cardiovascular support. The institution of mechanical ventilation for ventilatory and hypoxemic respiratory failure may reduce metabolic demand on the stressed cardiovascular system increasing $S\text{vO}_2$ for a constant cardiac output and $C\text{aO}_2$. Intubation and mechanical ventilation, when adjusted to the metabolic demands of the patient, may dramatically decrease the work of breathing, resulting in increased O_2 delivery to other vital organs and decreased serum lactic acid levels. Under conditions in which fixed right-to-left shunts exist, the obligatory increase in $S\text{vO}_2$ will result in an increase in $P\text{aO}_2$, despite no change in the ratio of shunt blood flow to cardiac output.

References

Buda, A.J., Pinsky, M.R., Ingels, N.B., Daughters, G.T., Stinson, E.B., and Alderman, E.L. (1979). Effect of intrathoracic pressure on left ventricular performance. *New England Journal of Medicine*, **301**, 453–9.

Butler, J. (1983). The heart is in good hands. *Circulation*, **67**, 1163–8.

Fessler, H.E., Brower, R.G., Wise, R.A., and Permutt, S. (1992). Effects of positive end-expiratory pressure on the canine venous return curve. *American Review of Respiratory Disease*, **146**, 4–10.

Pinsky, M.R., Matuschak, G.M., and Klain, M. (1985). Determinants of cardiac augmentation by increases in intrathoracic pressure. *Journal of Applied Physiology*, **58**, 1189–98.

Taylor, R.R., Covell, J.W., Sonnenblick, E.H., and Ross, J., Jr (1967). Dependence of ventricular distensibility on filling the opposite ventricle. *American Journal of Physiology*, **213**, 711–18.

2.4.5 Therapeutic strategy

M. Sair and T. W. Evans

Key messages

- Acute respiratory failure is the final common pathway of many pulmonary and extrapulmonary pathologies.
- Coexisting disease influences the assessment of acute respiratory failure. A pan-systemic approach is an important aspect of management.
- Augmentation of oxygenation and carbon dioxide clearance are therapeutic goals.
- Ventilation is indicated where acute respiratory failure is not responsive to continuous positive airways pressure (CPAP) or oxygen therapy alone, or where the use of CPAP is inappropriate.
- All modes of positive-pressure ventilation aim at maximizing alveolar recruitment by increasing mean airway pressure. Avoiding high peak-inspiratory pressures and large tidal volumes reduces the risk of barotrauma and volotrauma.
- Normal arterial blood gases may not be appropriate targets in severe acute respiratory failure.
- Patient posture may influence the degree of ventilation–perfusion mismatch
- Inhaled nitric oxide or nebulized prostacyclin may improve oxygenation and ameliorate pulmonary hypertension.
- Early CT scanning in acute respiratory failure indicates the nature and degree of pulmonary involvement and influences management strategy.

Introduction

Acute respiratory failure is not a single clinical entity but represents the final common pathway of many disparate precipitating causes. It is complicated by patient heterogeneity and variation in response to underlying diseases. Acute respiratory failure is a dynamic process, and frequent assessment of the patient and responses to therapy are required. Therapeutic intervention in acute respiratory failure should match acute and chronic changes in physiology with the nature and severity of the initiating insult. Detailed premorbid medical and social histories facilitate interpretation of acutely measured variables and the setting of appropriate therapeutic targets. As it is not possible to define acute respiratory failure in every clinical setting, a conceptual overview of available therapeutic options to management will be provided, with particular reference to acute lung injury and acute respiratory distress syndrome (ARDS). Secondly, a therapeutic strategy is proposed that might synthesize these interventions into a rational approach to management.

Background

The primary function of the alveolar capillary unit is to oxygenate pulmonary venous blood and remove carbon dioxide (CO_2).

Respiratory gases follow passive diffusion gradients, maintained by the cardiac and ventilatory pumps. In essence, the etiology of acute respiratory failure is often mixed, but is usually the consequence of mechanical pump failure and/or alveolar capillary dysfunction. Thus, hypoxemia results from hypoventilation, in which the underlying lungs are often normal, from shunt and membrane diffusion anomalies, and, most commonly, from ventilation–perfusion mismatch. ARDS is characterized by hypoxemia refractory to the provision of supplementing oxygen and a loss of lung compliance secondary to bilateral alveolar infiltration with non-cardiogenic edema and parenchymal consolidation. The consequences are ventilation–perfusion mismatching and a reduction in functional residual capacity. Strategies are aimed at supporting lung function during ARDS while resolution of lung injury occurs.

Therapeutic intervention

Oxygen therapy

Inspired oxygen concentration (FiO_2) is titrated to arterial oxygen tension (PaO_2) and arterial oxygen saturation (SaO_2). The risks of oxygen toxicity are reduced by administering the lowest concentration compatible with normoxemia or modest hypoxemia. In refractory acute respiratory failure, targets of $PaO_2 > 8$ kPa (60 mmHg) or $SaO_2 > 90$ per cent represent reasonable goals. Benefits depend on the degree of right-to-left shunt present in the pulmonary circulation. Large shunts (above 30 per cent) effectively nullify the effects of increasing FiO_2. In the spontaneously breathing patient, face-mask oxygen is administered from variable- or fixed-performance devices. The efficacy of variable-performance systems (e.g. Hudson or MC masks) depends on the oxygen flow and the respiratory pattern of the patient. As patient peak inspiratory flow rate can exceed 60 l/min, and thus the fresh gas delivery to such devices, entrainment of ambient air occurs. This results in variations in FiO_2 and rebreathing of dead-space gases. Fixed-performance systems (e.g. Ventimask) are preferable, as they rely on the Bernoulli principle to deliver fresh gas flow at a rate greater than patient peak inspiratory flow rate. The ratio of entrained air to oxygen determines FiO_2. By definition, acute lung injury and ARDS do not respond to oxygen therapy alone and further intervention is necessary.

Hemodynamic support

Cardiac function can be compromised in several ways, and the use of inotropic and vasoactive drugs is commonplace. Inotropic support of the circulation may be required because of other maneuvers, such as fluid restriction or the institution of positive-pressure ventilation. The effects of the condition itself, namely elevated pulmonary vascular resistance with consequent right ventricular strain and dysfunction, and the resulting leftward shift of the interventricular septum also impair left ventricular performance.

Fluid balance

The transudation of fluid across the alveolar capillary membrane is proportional to the mean pulmonary capillary pressure, as predicted by Starling's law. The reduction of hydrostatic forces in patients with ARDS reduces pulmonary edema formation and improves arterial oxygenation, but this must be balanced against the need to maintain cardiac output and adequate tissue perfusion. The circumspect use of diuretics and fluid restriction in patients with normal renal function does not appear to have long-term detrimental effects. In patients with coexisting renal failure, hemofiltration improves serum chemistry and produces the desired negative fluid balance.

Continuous positive airways pressure (CPAP)

The application of positive pressure in a closed breathing system through spontaneous breath cycles is known as CPAP; it increases mean airway pressure and may decrease the work of breathing by moving tidal ventilation to a more compliant part of the volume–pressure curve (Fig. 1). CPAP acts as a pneumatic splint, promoting alveolar re-expansion in areas of atelectasis, and may prevent the need for endotracheal intubation and conventional ventilation in certain patients. The hemodynamic effects of CPAP are dependent upon myocardial filling pressures and underlying cardiac function. Cardiac index improves in patients with left ventricular failure by reducing pre- and afterload, but falls in hypovolemic states. Mild lung injury may respond to CPAP, but tracheal intubation and mechanical ventilation are usually required.

Mechanical ventilation

The aims of mechanical ventilation are summarized in Table 1. Conventional mechanical ventilation is indicated where potentially reversible acute respiratory failure does not respond to the measures described above. A few patients with less severe lung injury respond to intermittent mandatory ventilation synchronized to their own spontaneous respiration. However, in practice, the majority of patients with established ARDS require continuous mandatory ventilation with varying amounts of positive end-expiratory pressure (PEEP). PEEP prevents repetitive opening and closing of recruitable alveoli and diminishes shear stress forces in the alveolar wall. It can be titrated against lung compliance in several ways to optimize alveolar recruitment (Tuxen 1994). Ventilatory modes are usually volume or pressure preset (Marini 1993).

Volume-preset ventilation

Traditional volume-preset modes of ventilation utilize constant inspiratory flow rates to produce supraphysiological tidal volumes. In theory, this diminishes dependent atelectasis and compensates

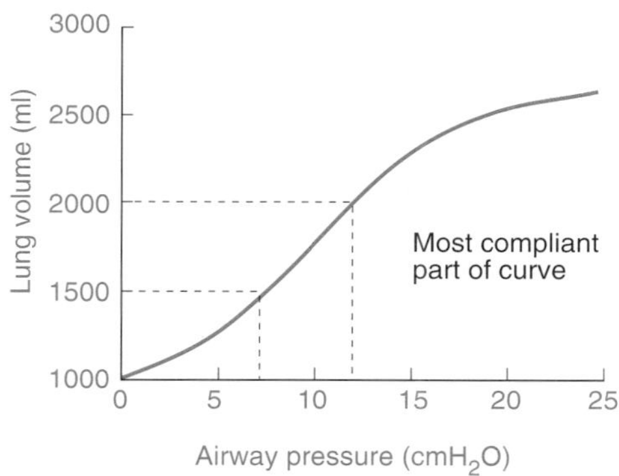

Fig. 1 Pulmonary pressure–volume relationship. Lung compliance (the change in lung volume per unit alteration in pressure) is improved with increasing lung volume. CPAP increases functional residual capacity to the compliant part of the curve, thus reducing the work of breathing.

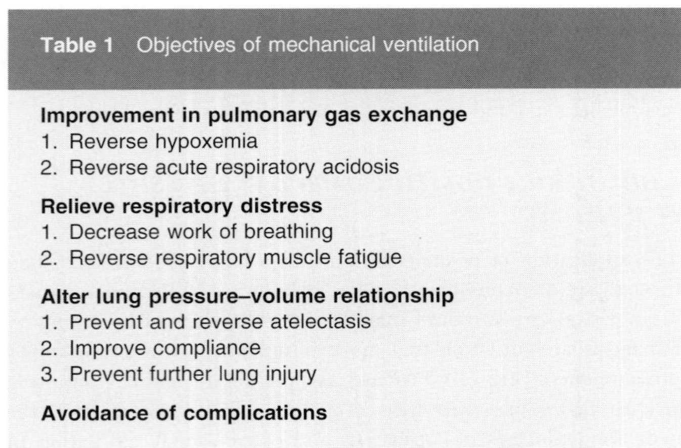

Table 1 Objectives of mechanical ventilation

Improvement in pulmonary gas exchange
1. Reverse hypoxemia
2. Reverse acute respiratory acidosis

Relieve respiratory distress
1. Decrease work of breathing
2. Reverse respiratory muscle fatigue

Alter lung pressure–volume relationship
1. Prevent and reverse atelectasis
2. Improve compliance
3. Prevent further lung injury

Avoidance of complications

Reproduced with permission from Tobin (1994).

for the increased dead-space asssociated with positive pressure ventilation. Mandatory respiratory rates are adjusted to maintain minute ventilation, and therefore $P\text{aCO}_2$, within the normal range. At higher respiratory rates, incomplete exhalation can result in unintentional PEEP as a consequence of air trapping; this phenomenon is known as intrinsic or auto-PEEP. The obligatory nature of control-mode ventilation precludes any patient effort and usually requires neuromuscular blockade. The later development of assist modes initiated by the patient's own efforts allows more patient–ventilator interaction, but their utility in the critically ill is dependent on trigger sensitivity and type. Applying an external source of PEEP to the ventilator circuit may improve pressure triggering in patients with significant degrees of intrinsic PEEP. The volume-preset strategy has been employed as a mode of choice in patients with normal lungs for many years, but in the presence of poorly compliant lungs it may result in unacceptably high peak inspiratory airway pressures.

Pressure-preset ventilation

ARDS has heterogeneous effects on lung parenchyma. Poorly compliant areas of diseased lung are contiguous with relatively normal alveoli. Aerated pulmonary tissue is usually compliant and ventilates normally, unlike consolidated lung. Therefore patients with ARDS lungs can be thought of as having small-volume lungs. Conventional intermittent positive-pressure ventilation with 'standard' tidal volumes of 10 to 15 ml/kg causes hyperinflation or 'volotrauma' in less affected areas and can result in iatrogenic lung injury. Advances in ventilator design and the recognition that iatrogenic pulmonary damage results from volotrauma and/or barotrauma secondary to high airway pressures have led to the concepts of lung protection and pressure-limited ventilation. As a result, volume-cycled ventilation has been largely superseded by pressure-preset modes of ventilation in the management of acute lung injury and ARDS. These methods employ rapid inflows of inspiratory gas in conjunction with a predetermined airway pressure limit. This results in a square pressure wave and an exponential deceleration of flow during the inspiration (Fig. 2). Pressure-preset ventilation increases mean alveolar pressure but reduces peak inspiratory pressure. Tidal volume cannot be adjusted directly, but is determined by preset pressure limits and pulmonary compliance.

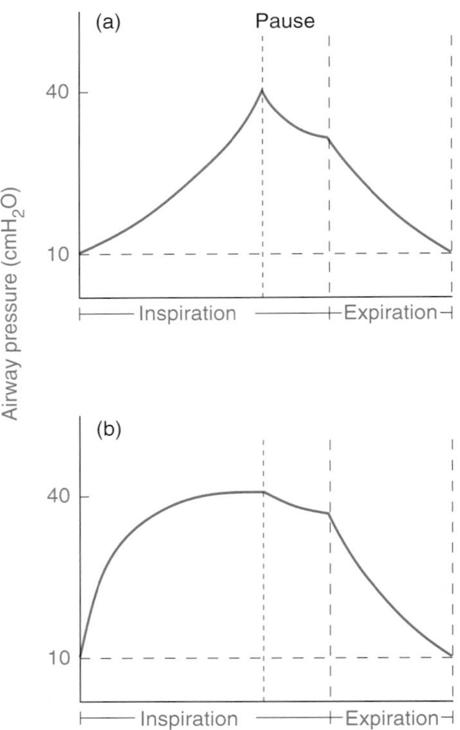

Fig. 2 Inverse ratio ventilation: airway pressure measured in (a) a volume-preset mode of ventilation (volume-controlled inverse ratio ventilation) and (b) with rapid inspiratory flow combined with a preset pressure limit (pressure-controlled inverse ratio ventilation). (Reproduced with permission from Keogh *et al.* (1990).)

Inverse ratio ventilation

The volume- and pressure-preset modes of ventilation can be manipulated further by adjusting cycling or 'duty' times. The inspiratory–expiratory time ratio (I:E ratio) of modern ventilators can be increased or reversed in order to sustain higher mean airway pressures throughout respiration and prevent alveolar collapse (Fig. 2). The prolonged inspiratory phase may also faciltate delivery of gases to alveoli with long time constants, but whether improvements in oxygenation are achieved without concomitant increases in auto-PEEP is a controversial issue. Volume-controlled inverse ratio ventilation delivers a preset tidal volume and provides predictable CO_2 clearance, but may result in significant air trapping and hemodynamic compromise, secondary to 'stacking' of mandatory tidal breaths and high peak inspiratory airway pressures. Therefore volume-controlled inverse ratio ventilation has been largely superseded by pressure-controlled inverse ratio ventilation, which delivers a constant preset inspiratory pressure throughout the designated inspiratory time. Although this maneuver avoids high peak pressures, the mean airway pressure, which determines oxygenation, is increased. Inverse ratio ventilation is uncomfortable for the conscious patient and requires adequate sedation and usually neuromuscular blockade.

High-frequency jet ventilation

The potential for a reduction in the incidence of pulmonary barotrauma encouraged the development of high-frequency low-tidal-volume techniques as alternative modes of ventilation.

Table 2 Terms used in extracorporeal membrane oxygenation

ECGE	Extracorporeal gas exchange, the blanket term for all techniques of extracorporeal respiratory support
ECMO	Extracorporeal membrane oxygenation: oxygenation is the prime objective of this technique, and CO_2 transfer occurs as a secondary effect
$ECCO_2R$	Extracorporeal CO_2 removal: the removal of CO_2 is the primary objective, thereby reducing the need for conventional mechanical ventilatory support
$P/TECCO_2R$	Extracorporeal CO_2 removal may be partial ($PECCO_2R$) or total ($TECCO_2R$), depending on the blood flow rate and the area of the gas-exchanging membrane

Humidified ventilator gases can be driven into the airway with a preset inspiratory time at frequencies varying from 1 to 10 Hz and still permit passive expiration. Tidal volumes are smaller than dead-space volume and gas exchange is believed to depend on convective forces. Ventilating the lungs near their natural resonant frequency (approximately 5 Hz) increases mean airway pressures and alveolar recruitment, reduces peak airway inspiratory pressure, and permits adequate CO_2 clearance (Swami and Keogh 1992). Increasing the I:E time and driving pressure augments tidal volume and results in improved functional residual capacity.

Permissive hypercapnia

The pursuit of normal arterial blood gases in patients with severe acute respiratory failure is not always appropriate. Improvements in oxygenation and CO_2 clearance can be offset by exacerbating lung injury. Acceptance of non-physiological targets may reduce this risk. Arterial CO_2 tension ($P\text{aCO}_2$) is gradually allowed to rise by reducing tidal volumes to 4 to 7 ml/kg and avoiding high peak airway inspiratory pressures. Permissive hypercapnia appears to be well tolerated, and the resulting respiratory acidosis is usually compensated for by renal retention of bicarbonate.

Posture

Ventilation–perfusion relationships in the injured lung can sometimes be improved by changes in patient posture. Evidence of a dependent distribution of edema, inflammation, or atelectasis on CT scanning may provide an indication for ventilating patients in the prone position, in an attempt to improve oxygenation by diverting blood flow to non-affected areas of lung. The gravitational nature of the edema means that improvements are often temporary, and frequent switching between the prone and supine positions may be necessary. The procedure is labor intensive and potentially hazardous in unstable patients.

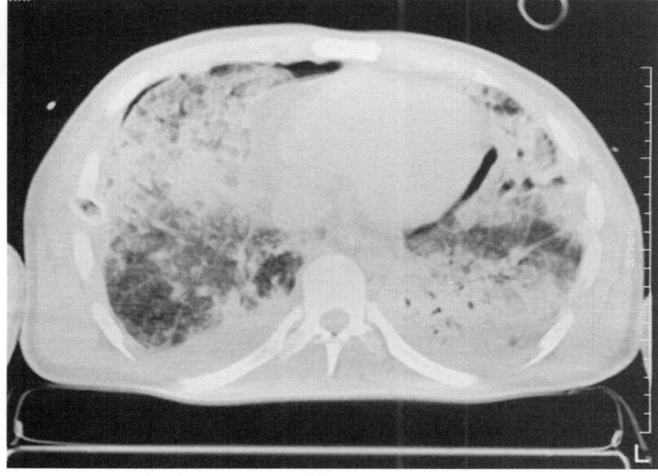

(a)

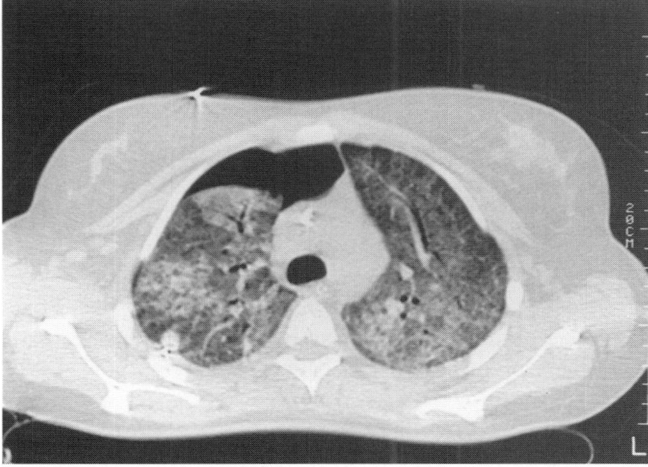

(b)

Fig. 3 (a) CT appearance of ARDS. The bilateral dependent consolidation and ground glass appearance are consistent with acute disease. These are accompanied bilaterally by small pneumothoraces. (b) CT demonstration of a large anterior pneumothorax not visible on a plain chest radiograph.

Artificial gas exchange membranes

The use of artificial membranes *ex vivo* to oxygenate and/or eliminate CO_2 from blood is collectively known as extracorporeal gas exchange (Table 2). Long-term extracorporeal membrane oxygenation (ECMO) for acute respiratory failure was first used in the 1970s. It is labor intensive, involving continuous venoarterial bypass, and is often complicated by bleeding diatheses. Enthusiasm for ECMO evaporated in the 1980s after a multicenter prospective clinical trial in the United States demonstrated no benefit in survival. Extracorporeal CO_2 removal ($ECCO_2R$) is a similar technique, but involves low flow and venovenous support combined with low-frequency positive-pressure ventilation. CO_2 clearance is then independent of ventilatory attempts to increase oxygenation and the lungs can be 'rested'. The method is popular in mainland Europe, but the results of the only randomized controlled trial conducted to date have not been encouraging and suggest that $ECCO_2R$ has no advantages over

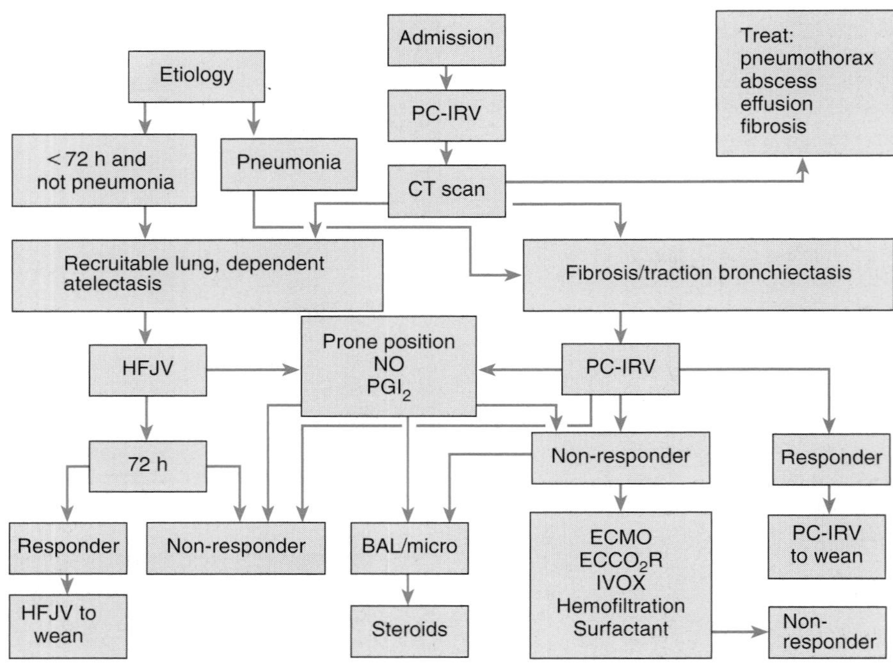

Fig. 4 Algorithm presenting a therapeutic strategy for managing acute respiratory failure secondary to acute respiratory distress syndrome: PC-IRV, pressure-controlled inverse ratio ventilation; HFJV, high-frequency jet ventilation; PGI2, prostacyclin; BAL, bronchoalveolar lavage; IVOX, intravascular oxygenator. The importance of early CT scanning is emphasized. Lung-protective modes of ventilation may reduce associated iatrogenic volutrauma. Patients with pneumonia or more than 72 h after the initiating insult do not usually respond to high-frequency ventilation.

conventional pressure-controlled inverse ratio ventilation in terms of reduction in mortality (Morris *et al.* 1994).

Pharmacological support

Attempts to modify the inflammatory process in ARDS with pharmacological agents have met with only limited success. Agents used have included non-steroidal anti-inflammatory drugs (e.g. ibuprofen), prostaglandins (alprostadil), thromboxane synthetase inhibitors (ketoconazole), antioxidants (*N*-acetylcysteine), and neutrophil activator inhibitors (e.g. pentoxifylline). Corticosteroids are beneficial in animal models, but clinical trials have been disappointing. High-dose steroids have a place in the treatment of patients with significant fibroproliferation or high bronchoalveolar eosinophil counts, where sepsis is excluded. Most progress has been made in the area of inhaled vasodilators used in an attempt to improve ventilation–perfusion relationships and decrease intrapulmonary shunting. This is achieved by increasing blood flow to ventilated lung units. Reducing pulmonary vascular resistance may also decrease pulmonary edema formation and therefore increase arterial oxygenation. Nitric oxide (NO) inhaled at concentrations from 4 to 40 ppm is a selective pulmonary vasodilator; its short half-life obviates effects on the systemic circulation. Although the results in individual patients are variable, delivery can be titrated against the response of the pulmonary circulation and $Pa\text{O}_2$. Rapid withdrawal of NO can result in severe rebound pulmonary hypertension and even cardiac arrest. NO is scavenged by red cell hemoglobin, which inactivates it, but methemoglobinemia is not a significant problem at the doses employed. Nebulized prostacyclin has similar vasodilating actions, but unlike NO may produce systemic effects. Surfactant replacement therapy does not have the same dramatic effects in adults compared with neonates in respiratory distress, and exogenous surfactant has not been demonstrated to benefit adult patients.

Therapeutic strategy

Treatment of patients with ARDS is primarily aimed at maintaining adequate oxygen delivery to vital organs and peripheral tissues, by means of respiratory and cardiovascular intervention, until lung injury resolves. Correction of anemia and the optimization of hemodynamic indices, with inotropes if necessary, may improve oxygen flux to systemic microvascular beds. Early assessment with CT indicates the extent and location of recruitable lung, and may also reveal loculated pneumothoraces or sources of infection (Fig. 3). The correction of exacerbating factors is a key aspect of therapeutic strategy. Eradication of sepsis with appropriate antibiotics and drainage, if indicated, is essential. Prudent attention to hydration, nutrition, electrolyte, and acid–base balance is important. Physiotherapy assists sputum clearance and lung re-expansion in established acute respiratory failure.

The application and integration of the above techniques into a therapeutic strategy is clearly a matter of individual experience as randomized prospective trials are lacking, and this has hampered the widespread acceptance of lung-protective approaches to ventilation. Therefore empirically based intervention algorithms are inevitably the product of anecdotal experience. Figure 4 illustrates one such approach to the management of refractory hypoxemia as a consequence of ARDS. The algorithm presented emphasizes the value of initial CT assessment and the importance of patient response to interventions. The recognition that deterioration of lung injury is often iatrogenic in origin has led to the adoption of pressure-controlled inverse ratio ventilation as our initial mode of ventilation in this group of patients. CT appearance and the etiology of any underlying process dictate further strategies to improve oxygenation, but individual patients combine distinct personal circumstances with precipitating illness, so that some flexibility in interpretation and management is required. A negative fluid balance is desirable and may produce improvements in gas exchange where edema is a prominent feature. Dependent atelectasis may be responsive to postural manipulations, but frequent switching between supine and prone positions may itself be hazardous. The use of pulmonary-selective vasodilators is commonplace, but their effects can be erratic and their impact on mortality is questionable. Pulsed steroids are administered to non-responders where there is evidence of fibroproliferation on CT scanning and no clinical or bronchoalveolar lavage evidence of sepsis. Several interventions suggested for use in non-responder groups remain unproven and await further scientific scrutiny. Therefore it is not possible to be dogmatic about these therapies, but their use in individual cases may be appropriate.

Conclusions

Mortality from ARDS may have fallen in the last decade. Advances in our understanding and the development of new approaches to treatment of lung injury are, in part, responsible. Acceptance of realistic physiological targets during refractory acute respiratory failure and lung-protective modes of ventilation are the cornerstones of these emerging techniques.

References

Keogh, B.F., *et al.* (1990). The management of adult respiratory distress syndrome. *British Journal of Hospital Medicine*, **43**, 26–34.

Marini, J.J. (1993). New options for the ventilatory management of acute lung injury. *New Horizons*, **1**, 489–503.

Morris, A.H., *et al.* (1994). Randomized clinical trial of pressure-controlled inverse ratio ventilation and extracorporeal CO_2 removal for adult respiratory distress syndrome. *American Journal of Respiratory and Critical Care Medicine*, **149**, 295–305.

Swami, A. and Keogh, B.F. (1992). The pulmonary physician and critical care. 2. The injured lung: conventional and novel respiratory therapy. *Thorax*, **47**, 555–62.

Tobin, M.J. (1994). Mechanical ventilation. *New England Journal of Medicine*, **330**, 1056–61.

Tuxen, D.V. (1994). Permissive hypercapnic ventilation. *American Journal of Respiratory and Critical Care Medicine*, **150**, 870–4.

2.5 Severe hypoxemia

2.5.1 Pathophysiology and causes

C.W. Hanson and B.E. Marshall

Key messages

- Hypoxemia is ubiquitous in the intensive care unit.
- Causes include hyperventilation, true shunt, diffusion impairment, ventilation–perfusion mismatch, decreased mixed venous oxygenation, and changes in the affinity of hemoglobin for oxygen binding.
- Diffusion impairment is not typically a significant cause of hypoxemia in the intensive care unit.
- Administration of pure oxygen does not correct hypoxemia due to pure shunt, whereas it does correct hypoxemia due to ventilation–perfusion mismatch.

Introduction

Hypoxemia is defined as deficient oxygenation of the blood, whereas hypoxia implies a problem with tissue oxygenation. Severe hypoxemia and hypoxia ($Pa\text{O}_2$ below about 4.5 kPa) results in anaerobic tissue metabolism and eventually cell death. Different organs are differentially sensitive to hypoxia; the cerebral cortex is most immediately dependent on oxygen and stops functioning after less than 1 min of deprivation, whereas the heart can survive for up to 5 min and skeletal muscle for approximately 2 h. Irreversible loss of function only occurs after four times the period of anoxia that results in cessation of function, and partial recovery can occur after intermediate anaerobic periods.

In the intensive care setting, it is useful to consider several discrete causes of severe hypoxemia: hypoventilation, true shunt, impaired diffusion, ventilation–perfusion inequality, and decreased mixed venous oxygen saturation. In reality, the pathophysiological processes commonly found in critically ill patients result in combinations of the above components. Acute respiratory distress syndrome, for example, affects both the pulmonary vasculature and the alveolar membrane, and therefore can cause both ventilation–perfusion ($\dot{V}/\dot{Q}$) mismatch and a barrier to diffusion. Each process is best understood using physiological principles (Nunn 1993; West 1998).

Hypoventilation

Hypoventilation occurs when the alveolar ventilation decreases relative to production of CO_2. The relationship between production and excretion is apparent from the following equation, which indicates that when alveolar ventilation ($\dot{V}_A$) falls and CO_2 production ($\dot{V}\text{CO}_2$) remains constant, the arterial $Pa\text{CO}_2$ must increase:

$$Pa\text{CO}_2 = (\dot{V}\text{CO}_2 / \dot{V}_A)K. \quad (1)$$

If it is assumed that the respiratory ratio R remains constant, the alveolar gas equation further demonstrates that, for each unit rise in arterial $P\text{CO}_2$, alveolar $P\text{O}_2$ decreases by an approximately constant amount:

$$PA\text{O}_2 = PI\text{O}_2 - (PA\text{O}_2/R) \quad (2)$$

where $PI\text{O}_2$ is the partial inspiratory oxygen tension. This equation also indicates that the hypoxemia of hypoventilation is readily overcome by administration of supplemental oxygen; inspiration of pure oxygen increases alveolar $P\text{O}_2$ to 102 kPa, while arterial $P\text{CO}_2$ rarely rises to levels above 14 kPa. The interrelationship of these variables is shown in Fig. 1.

True shunt

Shunt flow is blood that enters the arterial circulation without passing through ventilated lung. Most anatomical shunts are extrapulmonary, although arteriovenous malformations in the lung also cause shunt. Blood flowing through consolidated or atelectatic lung behaves as if it were shunt flow, although it can also be considered as one extreme of ventilation–perfusion mismatch. This is the most frequent cause of shunt encountered in clinical practice.

In venous blood, oxygen saturation is highest in the inferior vena cava, lower in the superior vena cava (which is most equivalent to

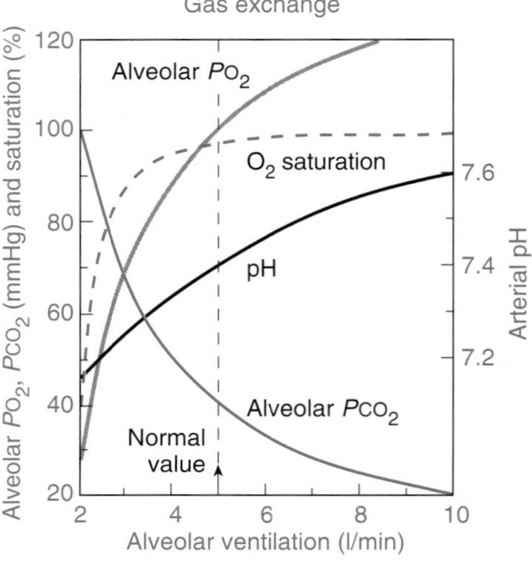

Fig. 1 Interrelationship between alveolar oxygen and carbon dioxide concentrations, pH, and oxygen saturation at varying minute ventilation. (Reproduced with permission from West (1998).)

mixed venous saturation), and lowest in the coronary sinus. Therefore the saturation of shunted blood depends on the level of the shunt.

Because of normal pressure gradients, blood ordinarily flows through congenital or acquired defects from the left side of the circulation to the right until pulmonary hypertension develops in response to chronic right-sided volume overload. At that point, pressures on the right may equal or exceed those on the left (in any given chamber) and venous blood enters the arterial circulation directly through the defect. A left-to-right shunt can also be acutely converted to a right-to-left shunt when pulmonary artery pressures rise abruptly. Pulmonary emboli due to air, fat, or clot, for example, cause acute increases in pulmonary artery pressure and can cause right-to-left shunting in susceptible patients such as those with a 'probe-patent' foramen ovale. Moreover, in patients with complex congenital heart disease, the pulmonary circulation is dynamically responsive to changes in the partial pressure of oxygen and CO_2, and a decrease in the former or an increase in the latter can result in life-threatening diversion of blood flow to the systemic circulation.

The shunt equation

$$\frac{Q_S}{Q_T} = \frac{C_C - C_a}{C_C - C\bar{V}} \qquad (3)$$

is used to calculate the degree of venous admixture, and implicitly assumes that all the cardiac output is divided into two parts, of which one passes through ventilated lung and the other (venous admixture) flows through unventilated lung or anatomical shunts (Wandrup 1995). Although convenient, this calculation erroneously suggests that all shunt blood is equivalent in saturation to that of mixed venous blood. Blood from the bronchial circulation, from the besian circulation, and from lung units with low ventilation–perfusion ratios all contribute to computed shunt. To be precise, the value derived for venous admixture represents the calculated quantity of blood, equivalent in saturation to mixed venous blood, which would be required to reduce the saturation of pulmonary end-capillary blood to the observed value.

Diffusion impairment

The true clinical significance of impaired oxygen diffusion is debatable, except during exercise and other high-output states. The diffusing capacity is defined as

$$DL_O = O_2\text{ uptake} / (PA_{O_2} - \text{mean pulmonary capillary } P_{O_2}). \qquad (4)$$

The capillary transit time, which equals the pulmonary capillary blood volume divided by the pulmonary blood flow (or cardiac output), determines the mean pulmonary capillary P_{O_2}. Therefore the factors that contribute to a decreased oxygen diffusion capacity are shorter capillary transit time, decreased area or increased thickness of the alveolar capillary membrane, or a reduction in the pulmonary capillary blood volume.

Increased cardiac output results in decreased contact time between the red cell and the alveolar interface or decreased capillary transit time. The normal mean transit time is approximately 0.75 s. However, it may decrease to 0.1 s during heavy exercise in normal subjects, and there is a wide distribution of values around the mean. Other hyperkinetic states, such as sepsis or thyrotoxicosis, also have short transit times. The rapid transit time and $\dot{V}/\dot{Q}$ mismatch characteristically found in septic patients contribute to the hypoxemia found in this population. Figure 2 shows the effect of shortened transit time on equilibration of oxygen tension between the alveolus and the red cell.

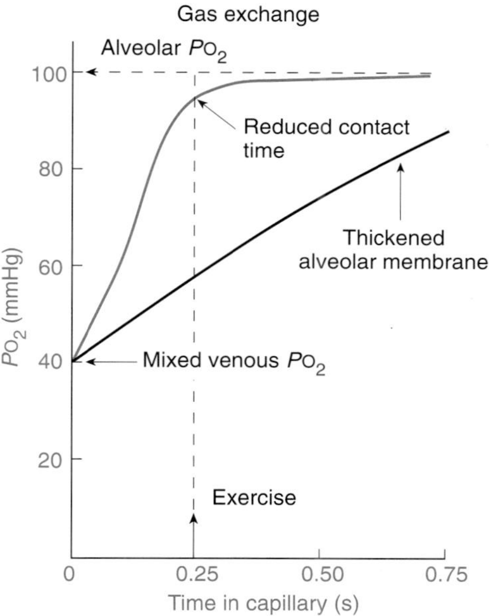

Fig. 2 Increased cardiac output (reduced contact time) and thickened alveolar membrane both affect diffusion of oxygen from the alveolus into the blood. (Reproduced with permission from West (1998).)

Ventilation–perfusion mismatch

$\dot{V}/\dot{Q}$ mismatch is responsible for most of the hypoxemia encountered in the intensive care environment, including hypoxemia in chronic obstructive pulmonary disease, pulmonary embolism, pulmonary edema, and interstitial lung disease (Rodriguez-Roisin and Wagner 1990). $\dot{V}/\dot{Q}$ mismatch is illustrated by comparing the two plots in Fig. 3, which shows a normal subject and a subject with $\dot{V}/\dot{Q}$ maldistribution. The $\dot{V}/\dot{Q}$ distribution for any lung unit must fall somewhere on the curve shown in Fig. 4 in a patient breathing room air. The curve for a specific patient will vary depending on the mixed venous partial pressure of oxygen and the oxygen concentration of the inspired gas.

As the degree of $\dot{V}/\dot{Q}$ maldistribution increases, the deleterious effect on oxygenation increases for two reasons. The first is that with $\dot{V}/\dot{Q}$ maldistribution, a greater percentage of the cardiac output passes through lung units with $\dot{V}/\dot{Q}$ ratios that are lower than normal (perfusion in excess of ventilation) so that less well saturated blood makes a greater contribution to total pulmonary blood flow. The second reason is related to the sigmoid shape of the hemoglobin dissociation curve; the oxygen content of blood from lung units with low $\dot{V}/\dot{Q}$ ratios exerts a greater effect on the mean saturation of blood flowing into the left side of the heart as demonstrated in Fig. 5.

Ventilation–perfusion relationships are primarily determined by regional perfusion characteristics, and hypoxic pulmonary vasoconstriction actively regulates $\dot{V}/\dot{Q}$ distribution (Marshall *et al.* 1994). In normal conditions, hypoxic pulmonary vasoconstriction acts to reduce blood flow to lung regions with low $\dot{V}/\dot{Q}$ ratios, and thereby improve the efficiency of gas exchange. Under certain pathological

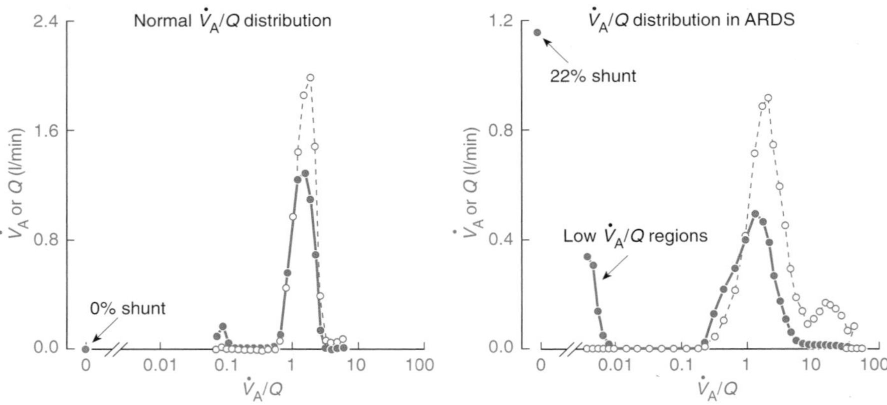

Fig. 3 Normal (left) and abnormal (right) distributions of ventilation (○) and perfusion (•) to lung units with differing $\dot{V}/Q$ ratios.

conditions, it is impaired or abolished by sepsis and trauma and by vasoactive mediators, such as prostaglandins, leukotrienes, cytokines, and platelet activating factor, or drugs, such as sodium nitroprusside and nitroglycerin, and blood flow to poorly ventilated lung increases with resulting hypoxemia.

Conversely, when the pulmonary vascular bed is partially obliterated (emphysema) or occluded (pulmonary embolism), pulmonary artery pressures can rise to the point that hypoxic pulmonary vasoconstriction is 'overcome' and is generally ineffective in redistributing blood flow. The failure of hypoxic pulmonary vasoconstriction to divert flow from atelectatic or consolidated lung results in $\dot{V}/Q$ mismatch and hypoxemia.

Finally, the combination of systemic hypoxemia and low mixed venous PO_2 is a potent stimulus for generalized hypoxic pulmonary vasoconstriction, causing increased pulmonary artery resistance and the potential for right heart failure. Pulmonary embolism and acute respiratory distress syndrome are two diseases in which this commonly occurs, and in this setting hypoxic pulmonary vasoconstriction is diffuse and maladaptive in contrast with its actions in pneumonia or atelectasis where it is regional and beneficial.

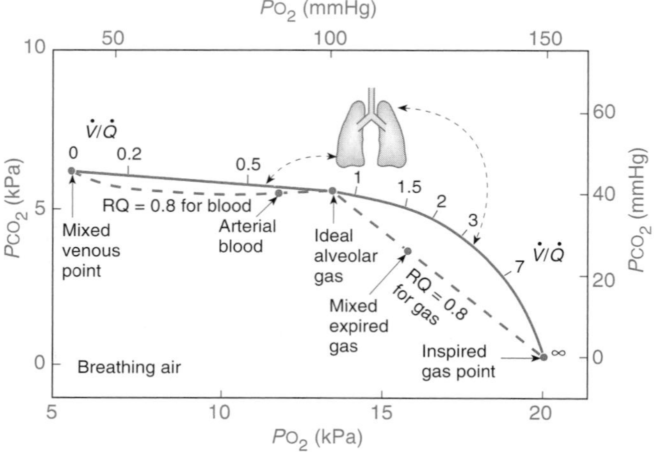

Fig. 4 The relationship between partial pressures of oxygen and carbon dioxide and ventilation–perfusion ratio at various points in the lung in a patient breathing room air: RQ, respiratory quotient. (Reproduced with permission from Nunn (1993).)

It is difficult to distinguish between the contributions of true shunt and those of $\dot{V}/Q$ maldistribution, and in many clinical situations the two effects are combined. However, analysis of the isoshunt diagram in Fig. 6 provides some general insights into the behavior of the two lesions. $\dot{V}/Q$ maldistribution results in hypoxemia because the spread of alveolar oxygen tensions is uneven. However, when breathing 100 per cent oxygen, the alveolar oxygen tension is almost homogeneous irrespective of the $\dot{V}/Q$ distribution and venous admixture is equal to true shunt. An estimate of the $\dot{V}/Q$ effect can be obtained by comparing the venous admixture when breathing 100 per cent oxygen with that at some lower fractional inspired oxygen concentration (FIO_2), and is approximately equal to the difference between the venous admixture on low FIO_2 and that on 100 per cent oxygen.

Decreased mixed venous oxygen saturation

While not an intuitively obvious or typically self-sufficient cause for severe hypoxemia, decreased mixed venous oxygen saturation acts synergistically with each of the previous lesions to worsen hypoxemia. Conversely, interventions which increase mixed venous oxygen saturation can be used to compensate for the other lesions.

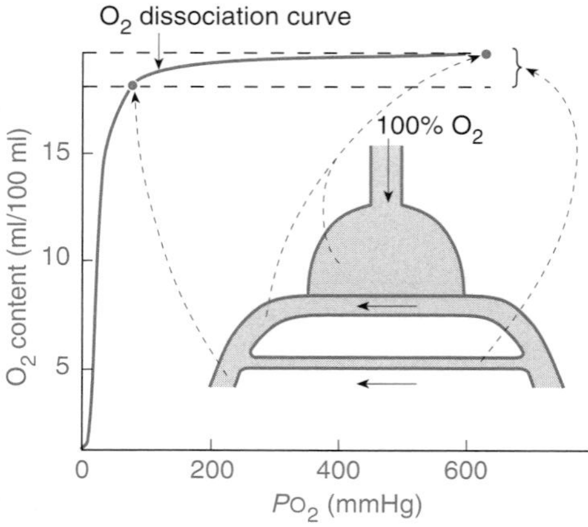

Fig. 5 Shunt while breathing pure oxygen. Note that the addition of a small amount of shunted blood significantly reduces the arterial partial pressure of oxygen. (Reproduced with permission from West (1998).)

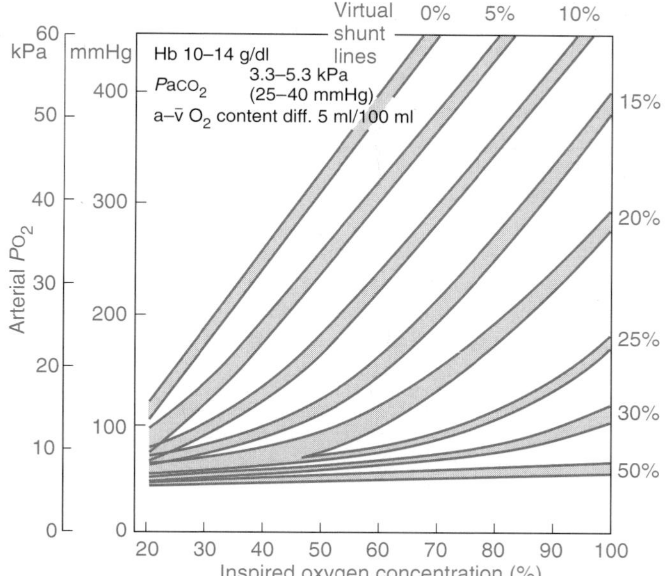

Fig. 6 Isoshunt diagram showing the percentage of shunt blood flow for different values of inspired and arterial oxygen concentration. (Reproduced with permission from Nunn (1993).)

The saturation of mixed venous blood influences oxygenation in the presence of pure shunt as well as $\dot{V}/Q$ mismatch. The saturation of the shunt blood flow is approximately equal to that of the mixed venous blood and, because of the sigmoid shape of the oxyhemoglobin dissociation curve, even a small amount of poorly saturated shunt flow significantly reduces the saturation of systemic blood. As the saturation of mixed venous blood decreases, the stimulus for hypoxic pulmonary vasoconstriction increases according to the formula

$$PSO_2 = PAO_2^{0.6} + P\bar{v}O_2^{0.4}. \quad (5)$$

It should be noted that as the stimulus for generalized hypoxic pulmonary vasoconstriction increases, pulmonary resistance rises and the flow through anatomical shunts may also increase, creating a self-reinforcing cycle. Decreases in mixed venous oxygen saturation interrelate with $\dot{V}/Q$ mismatch in a similar fashion.

Mixed venous oxygen saturation is determined by oxygen delivery and oxygen consumption. Oxygen delivery, in turn, is a function of hemoglobin quantity (Hb), arterial hemoglobin saturation, and cardiac output (CO):

$$O_2\text{ delivery} = 10 \times CO \times (1.38 \times Hb \times SaO_2). \quad (6)$$

Normal resting oxygen delivery for a 70-kg adult is approximately 1000 ml/min. Oxygen is consumed as blood circulates and normal resting oxygen consumption is about 250 ml/min. The combined blood from the superior vena cava, inferior vena cava, and coronary sinus normally has a saturation between 60 and 75 per cent. Normal resting oxygen delivery is approximately 16 ml/kg/min. 'Critical' oxygen delivery, which is the delivery below which consumption begins to decrease, is about half normal delivery or roughly 8 ml/kg/min. Therefore there is a substantial reserve. The fact that consumption remains constant over the range described implies that mixed venous oxygen decreases linearly over the same range.

Anemia, arterial hypoxemia, low cardiac output, and increased systemic oxygen consumption all contribute to reduced mixed venous oxygen saturation through their effects on oxygen delivery or consumption. However, it should be noted that the contributing factors have different relative 'weights' in terms of their impact on mixed venous oxygenation. A 2-l/min decrease in cardiac output may represent a 40 per cent reduction in oxygen delivery, while a decrease in arterial oxygen saturation from 100 to 90 per cent only decreases oxygen delivery by 10 per cent.

It is obvious that the combination of arterial desaturation and anemia or decreased cardiac output will have substantial effects on systemic oxygen delivery as well as mixed venous oxygen saturation. Under some circumstances, arterial hypoxemia and mixed venous hypoxemia can become self-reinforcing, as is evident from Fig. 7. In these conditions, oxygenation can be ameliorated by therapies aimed at improving mixed venous oxygenation, such as red cell transfusion, inotropic support, or sedation.

PO_2 at which hemoglobin is 50 per cent saturated (P_{50})

The hemoglobin dissociation curve is shifted to the left by hypothermia, alkalosis, or increased base deficit, and to the right by hyperthermia, acidosis, or base excess. These changes can have very significant effects; thus with a right shift of the curve, hemoglobin requires a greater alveolar oxygen tension to saturate as the blood passes through the lung, but conversely in the tissues the same right shift promotes greater unloading of oxygen at higher tensions. Because FIO_2 can generally be increased to compensate for the lung exchange, the tissue effects are more important. Acidosis may be beneficial and alkalosis detrimental to tissue oxygen delivery under critical circumstances. The above influences are caused by physicochemical changes in the hemoglobin molecule, and an additional influence is due to the presence of erythrocyte 2,3-diphosphoglycerate (2,3-DPG). Binding of this organic phosphate to hemoglobin reduces its affinity for oxygen, and therefore decreased 2,3-DPG (as in stored blood) causes a left shift of the dissociation curve. The latter effect is seldom important in clinical practice, except during massive transfusion.

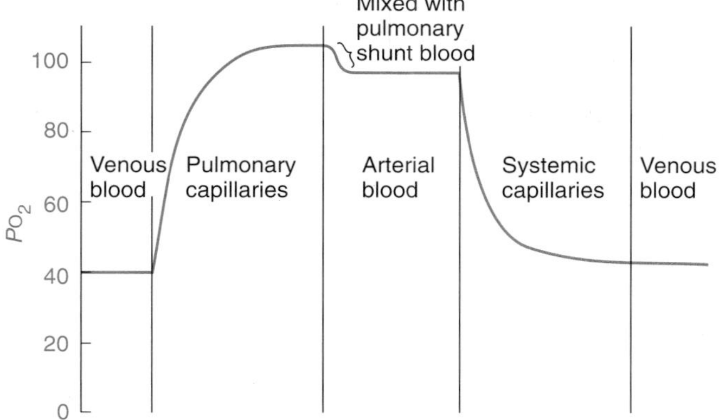

Fig. 7 The changes in oxygen concentration as blood passes through the lung. Note that if the mixed venous oxygen concentration is sufficiently low, it will affect all other oxygen concentrations. (Reproduced with permission from Guyton (1991).)

Diagnostic approach to hypoxemia

It is not always essential or practical to determine the etiology for hypoxemia in the intensive care environment, where a pragmatic approach is usually employed. In the majority of patients $\dot{V}/Q$ mismatch is the primary cause. Hypercarbic hypoxemia should be considered in the appropriate setting, such as the sedated or immediate postoperative patient with residual anesthesia and respiratory depression. Processes affecting the nerves or muscles of the respiratory tract also cause respiratory depression. These include diseases of the respiratory centers in the medulla (Ondine's curse), the spinal conducting pathways (cervical spine injury), the spinal cord (poliomyelitis), the peripheral nerves (Guillain–Barré syndrome, myasthenia gravis), the thoracic cage (crush injuries) and the airways (upper airway obstruction).

True diffusion block occurs rarely and is usually overwhelmed by other causes, so that it is ignored in diagnosing hypoxemia.

A value for the percentage of venous admixture can be calculated for both true shunt and $\dot{V}/Q$ mismatch, and it is often difficult to differentiate between the two at any given $F\text{IO}_2$. However, administration of 100 per cent oxygen allows the two processes to be distinguished. $\dot{V}/Q$ scatter, in the absence of true shunt, has a negligible effect on the alveolar–arterial oxygen gradient when pure oxygen is breathed. Therefore administration of higher concentrations of oxygen allows clearer differentiation between $\dot{V}/Q$ mismatch and true shunt using the isoshunt diagram (Fig. 6). However, concentrations of oxygen close to 100 per cent induce collapse of poorly ventilated alveoli and therefore increase shunt.

Finally, mixed venous oxygen desaturation can contribute to any of the processes already mentioned and should be considered in patients with .anemia, depressed cardiac output, or increased oxygen consumption. Pulmonary arterial catheterization is often employed in the treatment of severe hypoxemia in order to manage intravascular volume and the oxygen supply and demand ratio better.

References

Guyton, A.C. (1991). *Textbook of medical physiology* (8th edn). W.B. Saunders, Philadelphia, PA.

Marshall, B.E., Hanson, C.W., Frasch, F., and Marshall, C. (1994). Role of hypoxic pulmonary vasoconstriction in pulmonary gas exchange and blood flow distribution. 2 Pathophysiology. *Intensive Care Medicine*, **20**, 379–89.

Nunn, J.F. (1993). *Nunn's applied respiratory physiology* (4th edn). Butterworth Heinemann, Oxford.

Rodriguez-Roisin, R. and Wagner, P.D. (1990). Clinical relevance of ventilation–perfusion inequality determined by inert gas elimination. *European Respiratory Journal*, 3, 469–82.

Wandrup, J.H. (1995). Quantifying pulmonary oxygen transfer deficits in critically ill patients. *Acta Anaesthetica Scandinavica*, **107** (Supplement), 37–44.

West, J.B. (1998). *Pulmonary pathophysiology — the essentials* (4th edn). Williams & Wilkins, Baltimore, MD.

2.5.2 Therapeutic approach

L. Gattinoni and D. Chiumello

Key messages

- In the presence of severe hypoxemia search for the etiology and pathogenesis, and establish the dominant damage: consolidation in direct lung injury, compression atelectasis in indirect lung injury, and perfusion alterations in vascular disease.
- Immediately after intubation, open the lung by a recruitment maneuver. Consider that the transmural pressure for full opening is approximately 30 cmH_2O. Remember that in acute lung injury/acute respiratory distress syndrome (ALI/ARDS) caused by an indirect insult the chest wall elastance is higher than normal; consequently the airway pressure to reach a transmural pressure of 30 cmH_2O will be higher than in ALI/ARDS caused by a direct insult.
- Target the blood gases to the values that the patient is assumed to have had before the acute insult (consider age and previous respiratory diseases).
- Consider lung injury secondary to mechanical ventilation (fractional inspired oxygen ($F\text{IO}_2$), plateau pressure, tidal volume, intratidal collapse and decollapse), and remember that positive end-expiratory pressure (PEEP) seems to have a protective effect.
- Set the mechanical ventilation with PEEP adequate to keep open compression atelectasis and tidal volume adjusted according to the end-expiratory lung volume. Accept hypercapnia if plateau pressure (or transmural pressure) is excessively high. Always remember that a tidal volume that is too low may result in reabsorption atelectasis.

Introduction

Severe hypoxemia is still a major problem in the intensive care unit (ICU). Here, we will limit our discussion to the hypoxemia due to acute lung injury (ALI), of which the acute respiratory distress syndrome (ARDS) represents an extreme case, and we will describe the therapeutic approach in the early phase of ALI/ARDS. ALI and ARDS have recently been defined by the American–European Consensus Conference on ARDS (Bernard *et al.* 1994). Both are characterized by bilateral pulmonary infiltration, edema, and low compliance of the respiratory system. They are distinguished by blood gas values: ALI is defined by a $Pa\text{O}_2/F\text{IO}_2$ ratio below 300 mmHg (40 kPa), and ARDS by a $Pa\text{O}_2/F\text{IO}_2$ ratio below 200 mmHg (26.7 kPa), where $Pa\text{O}_2$ is the arterial oxygen tension and $F\text{IO}_2$ is the fractional inspired oxygen, irrespective of the level of positive end-expiratory pressure (PEEP) in use. However, this distinction is arbitrary, and the selected thresholds are not related to outcome.

We describe the approach to the individual patient, which is not substantially different in ALI and ARDS. The rational approach must include the following:

(1) understanding the etiology and pathogenesis;

(2) understanding the underlying pathology;

(3) tailoring the symptomatic respiratory treatment to the individual patient;

(4) general rules for setting mechanical ventilation.

Table 1 Etiologies associated with ALI/ARDS

Direct insult
Viral, bacterial, and fungal pneumonia
Lung contusion
Near drowning
Inhalation of toxic gases
Aspiration of gastric contents
Indirect insult
Septic, hemorrhagic, and cardiogenic shock
Drug ingestion
Uremia
Pancreatitis
Bowel infarction
Disseminated intravascular coagulation

Etiology and pathogenesis

The most common etiologies of ALI/ARDS, grouped according to the pathogenetic pathway (i.e. direct versus indirect insult), are summarized in Table 1. Traditionally, ALI/ARDS is considered to be a terminal event which is independent of the pathogenetic pathway. However, there is increasing experimental and clinical evidence that direct insult (via airways) and indirect insult (via bloodstream) may lead to different lung pathologies, with intra-alveolar damage being dominant if the insult is direct, as in diffuse pneumonia, and micro-vessel congestion and interstitial edema being dominant, with alveolar collapse, if the insult is indirect, as in peritonitis (Gattinoni *et al.* 1998).

The etiologies leading to direct or indirect insult account for more than 80 per cent of ALI/ARDS occurring in the ICU. However, ALI/ARDS may occur in another setting, i.e. pulmonary embolism, pulmonary thrombosis, or pulmonary vasculitis.

Thus, when confronted with severe hypoxemia, it is mandatory to identify the etiology from the history, clinical signs, and laboratory findings. Understanding the etiology allows the patient to be assigned to one of three main pathogenetic categories:

(1) direct insult (consolidation);

(2) indirect insult (edema/collapse);

(3) vascular disease (embolism, thrombosis, and vasculitis), frequently with a near-normal chest radiograph and severe hypoxemia.

Underlying pathology

Traditionally, ALI/ARDS was considered as a diffuse inflammation of the lung parenchyma. However, CT scanning has revealed that the alterations in the lung parenchyma are not homogeneously distributed, as the CT densities are more concentrated in the dependent lung regions. These observations led to the 'baby lung' model, which encompasses two concepts.

1. As little as 20 or 30 per cent of the normal lung volume has near-normal aeration.

2. The functional baby lung is located in the non-dependent regions of the lung.

Understanding of the functional anatomy of the ALI/ARDS lung has been improved further by quantitative regional CT scan analysis. In fact, the inflammatory edema is quite evenly distributed throughout the lung parenchyma, from sternum to vertebra. The increased lung weight ('wet sponge' model) leads to 'compression' of the most dependent regions, in which the increased densities are not due to greater edema but to reduced aeration as a result of 'compression atelectasis'.

The sponge model is valid for severe ALI/ARDS where the dominant pathology is represented by diffuse edema (increased lung weight) compression atelectasis. However, this model does not describe other ALI/ARDS conditions, in which the dominant underlying pathology is represented not by diffuse interstitial edema but by intra-alveolar lesions, as in diffuse pneumonia. For example, it has recently been shown that, in ALI/ARDS due to diffuse *Pneumocystis carinii* pneumonia, the CT densities are more homogeneously distributed than predicted by the sponge model. Although no specific studies have been performed, it can also be predicted that the sponge model does not describe early-phase pulmonary embolism, thrombosis, or vasculitis.

As the response to positive pressure may be different for intra-alveolar lesions (consolidation), compression ateletasis (collapse), or vascular disease, it is important to establish which is dominant. Unfortunately, the CT scan may not discriminate between collapse and consolidation. However, it may be possible to infer the dominant damage from the respiratory mechanics and imaging. The goal of the maneuvers described below is to establish, in a given patient, the potential for lung recruitment when applying positive pressure, i.e. to differentiate between consolidation and compression atelectasis.

Respiratory mechanics

One approach to characterizing the underlying pathology is the pressure–volume (P–V) curve, from which several variables can be derived (Gattinoni *et al.* 1987).

The initial slope (C_{start}) of the P–V curve, which is also called the initial compliance, gives an idea of the dimensions of the baby lung at atmospheric pressure. In an adult man, C_{start} values of 20 cmH$_2$O, 30 cmH$_2$O, 40 cmH$_2$O, etc. correspond to baby lung volumes of 20 per cent, 30 per cent, 40 per cent, etc. of the original healthy lung. The inflection point (or inflection zone) suggests the potential for recruitment (i.e. the presence of compression atelectasis) and indicates the pressure at which most of the recruitment may occur. The slope (C_{inf}) of the *P–V* curve after the inflection point may provide an estimate of the amount of recruitment (the steeper the *P–V* curve, the greater the recruitment). Finally, the upper inflection point indicates the pressure at which stretching of the pulmonary units becomes the dominant phenomenon.

The *P–V* curve may then allow an initial estimate of the underlying pathology. The presence of an inflection point suggests compression atelectasis, whereas its absence implies consolidation or vessel alterations. Unfortunately, the *P–V* curve may be misleading as the inflection point may be absent for a number of reasons other than lack of recruitment.

1. Recruitment proportional to the applied pressure results in a linear *P–V* curve.
2. A linear *P–V* curve may be obtained when one lung is overstretched and the other is being recruited.
3. As the *P–V* curve of the respiratory system is obtained from the *P–V* curves of the lung and the chest wall, an inflection point on the chest wall may lead to erroneous interpretation of the *P–V* curve of the total respiratory system.

Imaging

Another method of investigating the potential for recruitment in a given patient is to obtain CT scans of the lung at different pressures. Quantitative analysis of the images provides a precise estimate of the recruited lung at both end-inspiration and end-expiration. CT studies have shown clearly that recruitment occurs at the plateau pressure, and that PEEP just keeps open what has been previously recruited (Gattinoni *et al.* 1995). CT scanning has confirmed that the potential for recruitment differs between patients with ALI/ARDS of different pathologies.

Tailoring the symptomatic respiratory treatment to the individual patient

We shall not discuss the etiological treatment here. However, establishing the etiology is as urgent as selecting the respiratory treatment. The final outcome strongly depends on curing the disease leading to ALI/ARDS; respiratory treatment is only a method of 'buying time' to allow the cure to succeed.

Lung conditioning

Before setting the appropriate respiratory treatment, a recruitment maneuver is of major importance in every patient presenting with ALI/ARDS. In fact, regardless of the dominant damage underlying ALI/ARDS, it is likely that some degree of lung collapse is always present when the patient is referred to the ICU. The patient has probably had secretions, high respiratory frequency, and low tidal volume, i.e. several risk factors for reabsorption atelectasis, for hours or days before the admission. To achieve the goal of an 'open lung', airway patency must be assured (consider bronchoscopy) and the atelectasis recruited.

There are two types of atelectasis in ALI/ARDS.

1. Compression atelectasis, which develops at the end of each expiration, is due to the squeezing out of gases from the compressed pulmonary units. As some gas is left in the pulmonary units, the atelectasis is 'loose', and the transmural pressure (i.e. the difference between the airway pressure and the pleural pressure) required for opening is of the order of 10 to 20 cmH_2O.
2. In reabsorption atelectasis all gases have been reabsorbed so that the atelectasis is 'sticky', and the transmural pressure required for opening is between 30 and 35 cmH_2O.

The transmural pressure is a function of the airway pressure (P_{aw}) and of the elastances of the lung (E_L) and chest wall E_W:

$$\text{transmural pressure} = P_{aw}\frac{E_L}{E_L + E_W}.$$

Thus to achieve a transmural pressure of 30 cmH_2O when $E_L = E_W$, P_{aw} should be 60 cmH_2O. However, as E_L is high and E_W is normal in ALI/ARDS from direct insult, whereas E_W is greatly increased in ALI/ARDS from indirect insult, application of the same P_{aw} results in two different transmural pressures. Thus a higher P_{aw} is required in the recruitment maneuver for ALI/ARDS from indirect insult.

Moreover, the recruitment maneuver may be difficult when the compliance of the respiratory system is relatively good, as in moderate ALI, or when one lung has good compliance compared with the other. In these conditions it is difficult to achieve an adequate transmural pressure unless volumes of 2 liters or more are insufflated. Artificial alterations of the total respiratory system compliance (external compressions or positioning) may help to achieve the required transmural pressure without insufflating excessive volumes.

Target blood gases

The goal of respiratory support is an adequate gas exchange at the lowest iatrogenic cost. Traditionally, 'adequate' normal blood gases were considered to be a $Pa{O_2}$ of about 100 mmHg (13.3 kPa) and a $Pa{CO_2}$ of about 40 mmHg (5.3 kPa). However, age affects gas exchange even in normal individuals, so that $Pa{O_2}$ is usually given by

$$Pa{O_2} = 100\,\text{mmHg} - 0.3 \times \text{age (years)}$$

or

$$Pa{O_2} = 13.3\,\text{kPa} - 0.04 \times \text{age (years)}.$$

Moreover, it is obvious that it is nonsense to set a target of normal blood gases in a patient who, before the acute lung insult, had abnormal blood gas values because of pre-existing pulmonary disease. It follows that the target blood gases should be those that the patient is believed to have had before the acute event.

However, in the last few years the implementation of the philosophy of 'lung rest' has led to the acceptance of definitively abnormal blood gas values (particularly $Pa{CO_2}$, permissive hypercapnia) as a side-effect of a gentler treatment of diseased lungs. Although it may be reasonable to accept high $Pa{CO_2}$, we believe that $Pa{O_2}$ should be maintained at about 80 mmHg (10.7 kPa) (rather than 60 mmHg (8 kPa) as suggested by others) to avoid the risk of sudden deterioration of O_2 saturation.

Thus, in summary:

(1) target the blood gases at normal values (considering the age);

(2) adjust the target if the patient had previous pulmonary disease;

(3) accept elevated $Pa\text{CO}_2$ if the iatrogenic cost of ventilation is too high.

Iatrogenic cost of mechanical ventilation

Inspired oxygen fraction

The observations that a high FiO_2 (> 60 per cent) caused lung damage were mainly based on experimental studies in normal animals in which, in most cases, the delivered 100 per cent O_2 was not appropriately humidified. We do not believe that there is consistent evidence that a high FiO_2 is dangerous in ALI/ARDS, as there is worldwide experience of patients who have survived treatment for days or weeks at an FiO_2 of 100 per cent. It must also be remembered that, in ALI/ARDS, the lung surface exposed to high FiO_2 is greatly reduced compared with normal lung, as the collapsed and consolidated regions are relatively protected.

Plateau pressure

It has been suggested that a plateau pressure of 35 cmH_2O is the upper 'safe' threshold for mechanical ventilation (Slutsky 1994). However, the pressure *per se* is not dangerous (a diver may experience an alveolar pressure of several atmospheres!). The important factors are the transmural pressure and, probably, the pressure difference between end-expiration and end-inspiration (ΔP_{aw}). As discussed above, a plateau pressure of 35 cmH_2O may be associated with a wide range of transmural pressures depending, during paralysis and anesthesia, on the relationships between the lung and chest wall elastances, and in patients without paralysis (e.g. during pressure support ventilation), on the action of the respiratory muscles.

ΔP_{aw} may also play a role in the lung damage. Tidal 'stretching' with $\Delta P_{aw} = 35\ \text{cmH}_2\text{O}$ (starting from zero PEEP) is very different from that with $\Delta P_{aw} = 15\ \text{cmH}_2\text{O}$ (starting from a PEEP of 20 cmH_2O). Although no investigations have been performed in a clinical setting, there is consistent experimental evidence that a high ΔP_{aw} may have a deleterious effect on lung structure. Therefore the 'safe' pressure limit of 35 cmH_2O should be evaluated considering both the transmural pressure and ΔP_{aw}.

Tidal volume

Tidal volumes of 10 to 15 ml/kg were used for years in ALI/ARDS; however, after consistent experimental evidence that a 'high tidal volume' may damage the lung structures, it has now been reduced to between 4 and 8 ml/kg. Standardizing the tidal volume per kilogram of weight does not seem appropriate in ALI/ARDS, where the lung gas space (the baby lung) is not related to the body weight. As the risk of an inappropriately high tidal volume is related to 'stretching', i.e. the change in tension betweeen end-expiration and end-inspiration, the tidal volume should be related to the end-expiratory lung volume.

In a normal 70-kg man, 700 ml of tidal volume (i.e. 10 ml/kg) is delivered on an end-expiratory lung volume of approximately 2500 ml. This leads to a 28 per cent increase in end-expiratory lung volume. The same tidal volume in a 70-kg man with an end-expiratory lung volume of 700 ml (moderate to severe ARDS) leads to a 100 per cent increase in end-expiratory lung volume, which is equivalent to ventilating a normal man with a tidal volume of 2500 ml.

To avoid stretching we should maintain the ratio of tidal volume to end-expiratory lung volume within the physiological range (20–30 per cent). This is impossible in severe ARDS, even with permissive hypercapnia, because of the dimensions of the baby lung.

Intratidal collapse and decollapse

The final recognized risk factor for iatrogenic lung damage is the collapse and decollapse of the dependent lung regions within the tidal volume, which may damage the lung structures. This form of barotrauma is present when compression atelectasis is dominant. In fact, during inspiration, even a low plateau pressure (20–25 cmH_2O) is sufficient to open the dependent lung regions in which the 'loose' compression atelectasis is recruited. However, if the PEEP level is inadequate, the dependent lung regions collapse again at end-expiration. To avoid this, a PEEP sufficient to counteract the lung weight should be provided (Gattinoni *et al.* 1995) Unfortunately, increasing PEEP, for a given tidal volume, leads to an increase in plateau pressure. However, if high PEEP and decreased tidal volume are used, there is a risk of development of reabsorption atelectasis. If this occurs and a 'safe' plateau pressure (< 35 cmH_2O) is being used, the transmural pressure is not sufficient to reopen the atelectasis and a progressive decrease of oxygenation may occur. A possible role of 'sigh' should be re-evaluated.

Positive end-expiratory pressure

Rather than damaging the lungs, PEEP may have a protective role. If compression atelectasis is present, PEEP acts as a force counteracting the increased lung weight and keeping open the pulmonary units which will otherwise collapse at the end of expiration. Unfortunately, the level of PEEP necessary to keep the dependent lung regions open overinflates the non-dependent lung regions. Despite this, it is possible that the protective effect of PEEP is avoidance of intratidal collapse. Moreover, PEEP may protect the lung by decreasing ΔP_{aw} between inspiration and expiration, thus reducing intratidal stretching.

General rules for setting mechanical ventilation

We believe that the following rules should be adopted in the treatment of ALI/ARDS.

1. Collect an accurate history to establish the etiology and pathogenesis. This provides initial information about the underlying pathology.

2. Condition the lung immediately after intubation (recruitment), as described previously.

3. Assess the dominant underlying pathology in patients who are already anesthetized and paralyzed for intubation;
   (a) measure the respiratory mechanics (possibly including the 'pleural' pressure, by means of an esophageal balloon);

(b) obtain radiographs or CT scans at different airway pressures;

(c) perform imaging procedures for the detection of embolism/thrombosis if suspected.

4. Institute the type of mechanical ventilation with the most favorable cost–benefit ratio, aiming to 'open the lung and keep it open'.
   (a) After conditioning keep the lung open by adequate PEEP, remembering that the PEEP requirement in ALI/ARDS from indirect insult may be higher than that in ALI/ARDS from direct insult or vascular alterations. PEEP is mandatory to maintain recruitment when compression atelectasis is present. In consolidation and vascular alteration, PEEP may improve oxygenation by regional diversion of ventilation–perfusion. The inflection point on the *P–V* curve may help in setting PEEP. A PEEP trial is always recommended.

   (b) Try to maintain the plateau pressure below 35 cmH_2O, remembering that the transmural pressures for a given P_{aw} are lower in ALI/ARDS from indirect insult. In this case plateau pressures above 35 cmH_2O should not be dangerous.

   (c) After setting the pressures (plateau and PEEP), adjust Fio_2 to maintain Pao_2 around 80 mmHg (10.7 kPa). Accept hypercapnia and decrease tidal volume if the pressure set is too high. However, it should be remembered that an excessively low tidal volume may cause a slow deterioration of oxygenation through the development of reabsorption atelectasis.

5. Interventions other than mechanical ventilation should be considered, such as the prone position which is probably the most effective maneuver for regional recruitment and has been tested in a randomized trial.

References

Bernard, G.R., *et al.* (1994). The American–European Consensus Conference on ARDS: definitions, mechanisms, relevant outcomes, and clinical trial coordination. *American Journal of Respiratory and Critical Care Medicine*, **149**, 818–24.

Gattinoni, L., Pesenti, A., Avalli, L., Rossi, F., and Bombino, M. (1987). Pressure–volume curve of total respiratory system in acute respiratory failure: a computed tomographic scan study. *American Review of Respiratory Disease*, **136**, 730–6.

Gattinoni, L., Pelosi, P., Crotti, S., and Valenza, F. (1995). Effects of positive end expiratory pressure on regional distribution of tidal volume and recruitment in adult respiratory distress syndrome. *American Journal of Respiratory and Critical Care Medicine*, **155**, 1807–14.

Gattinoni, L., Pelosi, P., Suter, P.M., Pedoto, A., Vercesi, P., and Lissoni, A. (1998). *American Journal of Respiratory and Critical Care Medicine*, **159**, in press.

Slutsky, A.S. (1993). ACCP Consensus Conference: mechanical ventilation. *Chest*, **104**, 1833–59.

2.6 Aspiration and inhalation

2.6.1 Aspiration of gastric contents

G. J. Bellingan

Key messages

- Aspiration of gastric contents is generally preventable by good anesthetic practice and attention to detail on the intensive care unit.
- Mortality remains high in those requiring mechanical ventilation for more than 24 h.
- Aspiration of gastric contents with pH greater than 3 is not benign; food particles, even at neutral pH, are directly pathogenic to the airways and lung parenchyma.
- The presence of an endotracheal tube, even with cuff inflated, does not guarantee complete protection against aspiration

Incidence

The aspiration of gastric contents is a feared complication in the obtunded patient. The incidence is uncertain, as the diagnosis is not always clear and there is a significant incidence of subclinical aspiration. The incidence varies according to the underlying disorder; clinically significant aspiration occurs in approximately 1 in every 3900 cases with anesthesia for elective surgery and 1 in 895 with anesthesia for emergency surgery, but in up to 8 to 19 per cent of cases during non-anesthetic emergency intubations.

Risk factors

The major risk factor for gastric aspiration is a reduction in conscious level. Non-anesthetic causes of reduced consciousness include alcohol, drug overdose, head injury, metabolic (including diabetic) coma, and infective causes including meningitis. Factors predisposing to reduced gastric emptying or vomiting, such as gastroparesis in diabetics or vomiting associated with raised intracranial pressure, exacerbate the risk.

Some patients with normal conscious levels are at risk, including those with multiple sclerosis, motor neuron disease, Guillain–Barré and Miller–Fisher syndromes, and myasthenia gravis, and those with brainstem or motor disorders or laryngeal trauma affecting swallowing control. Indeed, videofluoroscopy demonstrates that a significant number of otherwise neurologically intact patients who have aspirated have evidence of oropharyngeal swallowing defects.

The other major group at risk of aspiration are those with reduced consciousness level due to general anesthesia. Here the risk of aspiration is greatly increased in those undergoing emergency rather than elective procedures and in those with certain predisposing conditions, particularly pregnancy where there is increased intra-abdominal pressure and delayed gastric emptying. Twenty-five per cent of maternal deaths associated with obstetric anesthesia can be attributed to the aspiration of gastric contents (in obstetric practice, this is also known as Mendelson's acid aspiration syndrome). Other disorders of gastric emptying, including achalasia, pyloric stenosis, and causes of ileus, are also risk factors. Aspiration during anesthesia occurs most frequently at the time of induction, but it can also occur with the endotracheal tube *in situ* or after extubation. In patients mechanically ventilated for a number of days, the same micro-organisms are commonly cultured from the stomach, pharynx, and tracheobronchial tree. This is due to chronic aspiration of gastric contents despite the presence of the cuffed endotracheal tube. The incidence of this aspiration is increased when patients are nursed supine rather than semirecumbent; the degree of aspiration is proportional to the time spent supine. This chronic aspiration is implicated in the pathogenesis of nosocomial pneumonia. Selective decontamination of the digestive tract can help to reduce the incidence of such nosocomial pneumonia but this has not translated into any benefit in mortality figures. Other factors that affect aspiration of contents in mechanically ventilated patients are the type of endotracheal cuff, with low-volume high-pressure cuffs allowing a greater degree of aspiration than high-volume low-pressure cuffs. Enteral feeding tubes are another factor related to the risk of aspiration pneumonia in mechanically ventilated patients. Although small-bore enteral feeding tubes (8 French) are believed to result in less reflux, there is dispute as to whether tube size is an important determinant of reflux. Nasojejunal rather than nasogastric placement of enteral feeding tubes certainly aids with emptying of gastric contents in the long-term enterally fed, but it is unclear whether this translates into any difference in the risk of aspiration. Certainly, up to 40 per cent of enteral feeding tubes become displaced, and in one study to determine the risk of use of oral contrast for CT in obtunded trauma patients where contrast (450 ml) was instilled via the nasogastric tube, the only case where contrast was found in the pulmonary tree was where the nasogastric tube was misplaced in the lung!

Gastric aspiration postextubation is due to laryngeal incompetence after prolonged intubation. In these patients there is often evidence of vocal cord dysfunction and temporary loss of sensorimotor function of the pharynx and larynx. Immediately post-extubation 33 per cent of intensive care patients swallowing contrast show some evidence of aspiration.

Table 1 Effective drugs for preoperative prophylaxis

Drug	Dose	Route	Time drug given preoperatively	Percentage of patients achieving adequate prophylaxis
Cimetidine	800 mg	Oral	8–12 h	85
	400 mg	IV	1 h	80
Ranitidine	300 mg	Oral	8–12 h	91
	50 mg	IV	30–45 min	85–98
Famotidine	40 mg	Oral	8–12 h	90–100
Omeprazole	40 mg	Oral	8–12 h	80–86
	40 mg	IV	30–45 min	90–95
Lansoprazole	30 mg	Oral	8–12 h	70–80
Sodium bicarbonate	20 ml (8.4%)	Oral	30 min	85
Sodium citrate	30 ml (0.3 M)	Oral	30 min	90

The percentage of patients achieving adequate prophylaxis when using drugs either orally or intravenously (IV) given at the times stated preoperatively is shown. A gastric aspirate pH of greater than 2.5 and volume of less than 25 ml is taken as adequate protection.

Pathophysiology

Aspiration is best viewed in three stages. Initially, the inhalation of food particles results directly in mechanical obstruction of the airways with distal collapse leading to increased shunt, loss of functional residual capacity, and an increase in the work of breathing. Later, in response to both acid fluid and particulate matter, there is a chemical injury response with bronchorrhea, bronchoconstriction, and mucosal edema. The associated reduction in airway defenses increases the risk of bacterial infection. Histology of airways of animals exposed to gastric contents demonstrates a complete loss of the ciliated epithelial lining during the first 2 days. This leads to vascular leak and increased airway sensitivity to histamine and neuropeptides, all reducing airway patency. Regeneration is rapid unless there is added aspiration of particulate matter, which significantly delays healing. Thus damage is due to both the acidic fluid and the particulate gastric contents. Gastric contents with pH below 2.5 and a volume of 25 to 50 ml have been shown to be far more damaging than an equal volume of a more alkaline fluid. Acid denatures surfactant and activates neutrophils, leading to atelectasis and augmentation of the pro-inflammatory responses. However, even at neutral pH the aspiration of particulate contents is damaging. A comparison of the pulmonary inflammatory potential of small gastric particles to that of acid fluid in rats showed a synergistic increase in alveolar capillary leak from the combination of particles and acid. Use of inert glass particles rather than gastric contents did not have any synergistic effect with acid saline. Lung biopsy of patients with aspiration pneumonia shows parenchymal lung damage with a granulomatous foreign body response and evidence later of fibroproliferation. The final stage of lung injury is due to the inflammatory response with release of tumor necrosis factor, interleukin 1, and other proinflammatory mediators including leukotrienes A_4 and B_4 and thromboxane A_2 which contribute to mucosal edema and bronchoconstriction. This process can develop into acute respiratory distress syndrome.

Clinical features

The clinical picture can vary very widely from subclinical aspiration to a fulminant picture with rapid death. In clinically apparent cases of aspiration, 12 per cent die within hours, 62 per cent have a rapid improvement over 4 to 5 days, and 26 per cent show an initial improvement but then deteriorate both clinically and radiographically; many, but not all, of these patients have evidence of bacterial superinfection. In patients under anesthesia, where the aspiration of gastric contents was diagnosed by the presence of bilious secretions or particulate matter in the tracheobronchial tree, 64 per cent of survivors did not develop any significant respiratory sequelae. The mortality in the 10 per cent requiring mechanical ventilation for more than 24 h was 50 per cent. Diagnosis is clear with a witnessed aspiration, but otherwise is made by having a high index of suspicion in the correct clinical setting with prompt onset of tachypnea and in many cases cough and wheeze. There may be fever and diffuse rales, and patients can be cyanotic; severe cases will be shocked or even apneic. Hypoxemia is usual, and there are often associated radiological changes on chest radiography. However, there are no characteristic radiological features, although bilateral lower-zone changes are probably the most common finding. Bronchoscopy has a useful place in both diagnosis and therapy.

Prevention

The main aim of treatment is prevention. This requires good anesthetic practice for those patients undergoing general anesthetic and attention to detail for patients in the intensive care unit (ICU). Patients at risk must be identified, including emergency cases, pregnant women, and diabetics. An experienced anesthetist is required for these cases, and regional anesthesia should be considered. For safe intubation patients need to be fasted; thus in emergencies rapid-sequence induction with correct use of cricoid pressure is required.

These patients must be preoxygenated, and given an appropriate sedative and muscle relaxant with the application of cricoid pressure from the moment of loss of consciousness until the endotracheal tube is confirmed to be correctly positioned with the cuff inflated. There is debate as to how long patients should be fasted before undergoing elective surgery. The accepted minimum time for solid food to be emptied from the stomach is 6 h. However, evidence initially obtained with children, and now extended to adults, suggests that only a 2-h fasting period is required for clear fluids, and there is no difference in the volume and acidity of gastric contents aspirated from fully fasted patients and those allowed fluids until 2 h preoperatively. This more relaxed fluid regime reduces thirst, improves compliance, reduces hypoglycemia, and allows a better recovery from the anesthetic. It remains essential to identify high-risk cases where standard fasting times will still be needed. The incidence of aspiration during general anesthesia in elective fasted patients where a laryngeal mask was used is negligible. Other approaches to prevention include use of awake intubation and the use of balloon-tipped nasogastric tubes to attempt to occlude the cardia.

Another major arm in prevention is the prophylactic use of antacids. A large number of different methods of alkalinizing the gastric contents have been tried and compared. The best is to use a non-particulate antacid, H_2 blocker, or proton pump inhibitor (the most common agents employed include cimetidine, ranitidine, or sodium citrate) in combination with an agent to promote emptying of gastric contents such as metoclopramide or cisapride. A number of agents given preoperatively have been shown both to increase effectively the pH and to reduce the volume of gastric contents (Table 1).

Unlike antacids, H_2 blockers and proton pump inhibitors significantly reduce the volume of gastric contents, and metoclopramide reduces this further. Oral ranitidine at 2 to 3.5 mg/kg is effective in children. Agents employed to promote gastric emptying are not always effective, particularly in diabetics where cisapride has no effect on the volume of gastric aspirates. It is essential to note the fallacy of the common belief that aspiration of less than 25 ml of gastric contents at a pH greater than 2.5 is benign. Many studies have demonstrated a direct pathogenic effect of food particles additional to that of any acid. Acute respiratory distress syndrome has been shown to develop with aspiration of gastric contents at a pH of 6.4 after use of Aludrox® (aluminum hydroxide). Thus alkalinization of gastric contents is helpful, but is not all that is needed.

In the ICU, similar preventive measures should apply for intubation of patients, many of whom will also have nasogastric tubes inserted. The correct positioning of these tubes is important and can be confirmed by aspiration of acid gastric juice ($pH < 4$ tested using litmus paper) combined with injection of air and auscultation over the epigastrium. In the absence of aspiration of acid fluid, a confirmatory chest radiograph is essential; auscultation alone is insufficient. Once nasogastric feeding has commenced, preventing gastric dilatation by regular aspiration of gastric contents to determine effectiveness of gastric emptying is essential. If gastric emptying is delayed, as demonstrated by regular gastric aspirates of more than 200 ml every 4 h, agents such as metoclopramide, cisapride, and erythromycin can be employed to promote it. Even a tracheostomy does not completely eliminate the risk of aspiration, as a significant number of patients aspirate some feed over a 48-h period. Finally, safe extubation is also an important factor in the ICU where patients can have an endotracheal tube in position for weeks. The patient must be able to cough and have an adequate gag reflex, the stomach contents must be aspirated, and the patient should be sitting up. The possibility of poor airway protection in the postextubation period must be borne in mind.

Treatment

If prevention fails, the immediate treatment is to place the patient in the head-down position on the right side and to suck clear the pharynx and larynx. Aggressive pulmonary toilet is lifesaving where there has been obstruction of major airways by food particles; fiberoptic bronchoscopy is of proven benefit, although rigid bronchoscopy is superior for large solids. High concentrations of inspired oxygen should be used and patients who cannot protect their airways should be intubated. Patients must be closely monitored but other management is symptomatic; colloids are used to maintain blood pressure and bronchodilators to treat bronchospasm. Steroids are not beneficial. They do not prevent the early alveolar capillary leak and, although they reduce fibroblast proliferation, they delay granulomatous healing. Their use does not improve mortality and is associated with an increased incidence of Gram-negative pneumonia. Antibiotics are important when the patient's clinical state has deteriorated or there is evidence of pulmonary infection. Their place immediately after aspiration is less clear. Most ICUs in the United Kingdom do not use them at this stage, but rely instead on surveillance microbiology and clinical assessment. The best choice of antibiotic if immediate therapy is instituted is probably clindamycin. Therapies directed against the host's own inflammatory cascade remain controversial, and remain completely unproven for treatment of aspiration pneumonitis.

Aspiration of gastric contents is preventable in most cases. With strict attention to detail throughout the patient's stay in the ICU, the incidence of this potentially fatal condition can be further reduced.

Bibliography

Britto, J. and Demling, R.H. (1993). Aspiration lung injury. *New Horizons*, **1**, 435–9.

Schwartz, D.J., Wynne, J.W., Gibbs, C.P., Hood, C.I., and Kuck, E.J. (1980). The pulmonary consequences of aspiration of gastric contents at pH values greater than 2.5. *American Review of Respiratory Disease*, **121**, 119–26.

Torres, A., *et al.* (1992). Pulmonary aspiration of gastric contents in patients receiving mechanical ventilation: the effect of body position. *Annals of Internal Medicine*,. **116**, 540–3.

Warner, M.A., Warner, M.E., and Weber, J.G. (1993). Clinical significance of pulmonary aspiration during the perioperative period. *Anesthesiology*, **78**, 56–62.

Wolfe, J.E., Bone, R.C., and Ruth, W.E. (1977). Effects of corticosteroids in the treatment of patients with gastric aspiration. *American Journal of Medicine*, **63**, 719–22.

2.6.2 Inhalation injury

D. T. Harrington and B. A. Pruitt Jr

Key messages

- Inhalation injury can substantially increase morbidity and mortality in burn patients.
- History and physical examination identify groups that are at risk of this injury.
- Diagnosis should be based on bronchoscopic findings or the results of xenon-133 scanning.
- Therapies include meticulous intensive care high-frequency interrupted-flow ventilation, and the use of permissive hypercapnia.

Introduction

There has been a sustained improvement in outcome following thermal injury as is evident in the progressive increase in the LD_{50} (the extent of burn in percentage of surface area associated with a 50 per cent mortality) that has occurred since 1950. Figure 1 shows this improvement for a 21-year-old patient, but during the period noted the LD_{50} for every age group has increased. Much of this improvement has resulted from the use of physiologically based fluid resuscitation, which essentially eliminates the occurrence of early renal failure and its high attendant mortality, and the development of effective topical antimicrobial creams, which control the microbial density of the burn wound and prevent invasive wound infection. Continued advances in our understanding of the pathophysiology of injury have led to further improvements in the care of the thermally injured. These include accurate diagnostic modalities and effective management techniques for inhalation injury, early excision and grafting of burn wounds, and the implementation of effective infection control policies.

Inhalation injury accompanies burn injury in up to 35 per cent of those admitted to burn centers and adds further physiological stress to an already burdened patient, resulting in higher morbidity and mortality. However, accurate diagnosis of the injury, understanding of its pathophysiology, and subsequent institution of targeted therapies have been instrumental in reducing the comorbid effect of inhalation injury following thermal injury (Shirani *et al.* 1987).

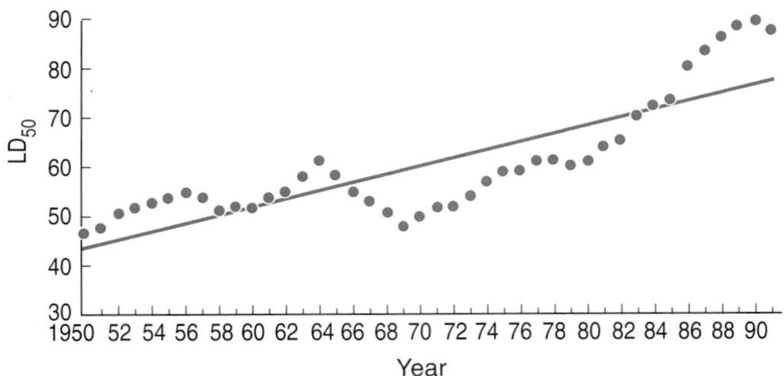

Fig. 1 The LD_{50} following thermal injury for a 21-year-old patient for the years 1950 to 1992.

Pathophysiology

Inhalation injury can be both thermal and chemical. Thermal injury is usually limited to the upper airway because of the low thermal capacity of heated air and the efficient heat exchange of the oropharynx. However, direct thermal injury to the trachea and bronchi can result following exposure to steam which can overwhelm the upper airway because of its higher thermal capacity. Chemical injury of the airway can be caused by irritants or cytotoxic chemicals that either adhere to the fine particles in smokes or become aerosolized as mists. The type and volume of these irritants generated by combustion can vary depending on the material burned, the temperature of the fire, and the amount of oxygen present in the fire environment. Injury can occur at all levels of the airway and is typically heterogeneous in nature, with some airways sustaining severe damage and others sustaining mild or no injury.

Even a short duration of exposure to highly reactive irritants can result in loss of cilia and superficial epithelial erosions of the tracheo-

Table 1 The positive predictive value (PPV) for presence of an inhalation injury based on the history of event and the physical examination is unacceptably low

	Inhalation		PPV (%)
	Yes	No	
Closed space			
Yes	272	144	65.4
No	101	541	
Facial burn			
Yes	348	356	49.4
No	25	330	

Data from Shirani *et al.* (1987).

bronchial tree. Injury activates the inflammatory cascade resulting in histologically evident inflammatory changes of the respiratory mucosa within 2 h of injury. More prolonged exposure results in epithelial necrosis and sloughing, exposing areas of bronchial and alveolar basement membrane. Such tissue injury evokes release of arachidonic acid metabolites and other vasoactive substances which may produce pulmonary hypertension and bronchospasm and cause mismatching of airflow and blood flow within the lung. Cast formation may occur with subsequent obstruction of bronchioles and atelectasis, which decrease compliance and impair oxygenation and ventilation.

A significant determinant of inhalation injury's effect on outcome is the extent of the accompanying burn. Smoke inhalation with minimal or no cutaneous burn is often limited to a chemical tracheobronchitis and does not significantly add to the morbidity and mortality of thermal injury. Patients with inhalation injury and a medium-sized burn have an increased fluid requirement during the resuscitative phase and often develop significant sequelae of their pulmonary injury. In these medium-sized burns inhalation injury can add up to 20 per cent to the expected mortality (Shirani *et al.* 1987). However, inhalation injury produces little appreciable increase in the already discouragingly high mortality of patients with extensive thermal injury.

Diagnosis

Several elements of the history and physical examination can alert the clinician to the possibility of an inhalation injury. Being burned in a confined space and having sensory impairment due to ethanol or drug ingestion, central nervous system trauma, or dementia place patients at high risk of this injury. Facial burns, singed nasal vibrissae, or burned eyelashes are commonly present in patients with inhalation injury. Even though the history of the event and physical examination of the patient can alert one to the possibility of inhalation injury, they are poor predictors of those patients who actually have that injury (Shirani *et al.* 1987) (Table 1). A chest radiograph, which is often normal in the acute setting, is also a poor indicator of inhalation injury.

The inaccuracy of history and physical examination in predicting inhalation injury has led to the use of early bronchoscopy and, if necessary in the face of a normal bronchoscopy, xenon-133 ventilation–perfusion lung scintiphotography. The accurate diagnosis of this condition allows these patients to receive therapies directed at reducing the comorbid effect of inhalation injury and allows for quantification of its effect on morbidity and mortality. Moreover, diagnostic bronchoscopy has the added benefit of identifying incipient upper airway obstruction due to edema elicited by the heat component of the smoke. Signs of upper airway edema upon initial fiber-optic examination warrant early intubation since edema and upper airway narrowing are typically progressive, usually peaking 12 to 24 h after injury. Carbonaceous debris and early changes of the respiratory epithelium such as hyperemia, edema, or superficial sloughing are consistent with inhalation injury. These findings often precede the clinical sequelae of inhalation injury such as hypoxia and declining pulmonary compliance. During the resuscitative phase early inflammatory changes may be absent owing to hypovolemia; therefore repeat bronchoscopy when the patient is 'fully resuscitated' will be more accurate.

A normal appearance of the upper airway on examination by fiber-optic bronchscopy does not always exclude inhalation injury. Distal airway and alveolar injury caused by finely particulate smokes which do not precipitate in the large-caliber airways can be missed by this modality. Therefore patients who are considered at risk for inhalation injury, yet have a negative bronchoscopic examination, should undergo xenon-133 scanning. Serial scintigraphs taken following intravenous injection of this radiolabeled inert gas document elimination via the lung. Retention of the isotope for more than 90 s after injection indicates injury of the small airways. Pre-existing chronic obstructive pulmonary disease or acute inflammatory processes such as bronchitis due to aspiration or infection can result in false-positive examinations. A combination of fiber-optic bronchoscopy with xenon-133 scanning detects inhalation injury with an accuracy to 93 per cent.

Treatment

Prompt removal of the patient from the fire environment minimizes airway damage. Thereafter, evaluation and control of the upper airway form the critical first step in the care of a thermally injured patient. Upper airway edema is often progressive; therefore early bronchoscopy and observation, in an intensive care setting, of patients thought to be at risk of inhalation injury are recommended. As noted above, signs of impending airway obstruction, such as stridor or significant edema noted at the time of bronchoscopy, warrant intubation. Treatment of these patients with nebulized racemic epinephrine (adrenaline) may afford a measure of temporary relief, but is not a definitive therapy and frequent monitoring for evidence of progressive airway compromise must be continued.

The presence of carbon monoxide (CO) poisoning, a non-inflammatory form of inhalation injury, should always be considered in those patients suspected of having inhalation injury. CO binds to hemoglobin 220 times more avidly than oxygen. It interferes with oxidative metabolism by decreasing the oxygen-carrying capacity of the blood, shifting the oxygen–hemoglobin dissociation curve to the left, and binding to cytochrome oxidase. For these reasons any patient suspected of inhalation injury should receive 100 per cent oxygen via a tight-fitting non-rebreathing mask until CO poisoning is excluded and the carboxyhemoglobin level is less than 10 per cent. The half-life of carboxyhemoglobin depends on the percentage concentration of oxygen delivered (FiO_2) (21 per cent FiO_2, $t_{1/2} = 250$ min; 100 per cent FiO_2 at 1 atm, $t_{1/2} = 40$–50 min; 100 per cent FiO_2 at 2 atm, $t_{1/2} = 27$ min). Although the 20-min advantage appears to justify the use of hyperbaric oxygen for CO poisoning, an alarming number of side-effects have been reported in patients receiving this therapy (Grube *et al.* 1988). Additionally, small hyperbaric chambers isolate patients from physician and nursing care during a critical phase of burn resuscitation and large chambers place the attendants at risk. Although hyperbaric oxygen therapy has some proponents, the accompanying risks and the lack of data confirming long-term benefit provide no support for its use outside prospective randomized trials.

Since there is no consensus on the frequency of cyanide poisoning as a clinically significant event, its treatment is also controversial. Cyanide, by virtue of its uncoupling of oxidative phosphorylation, is a known cellular toxin. In the patient with an unexplained severe acidosis, tachycardia, and tachypnea, yet a normal arterial oxygen content, cyanide poisoning may be present and treatment should be initiated. This consists of creating a cyanide sink in the form of a ferric ion on hemoglobin by the delivery of inhaled amyl nitrite or intravenous sodium nitrite. Further treatment includes supplying substrate, as thiosulfate, for the liver enzyme rhodanase which detoxifies cyanide to thiocyanate. Since treatment with nitrite compounds can cause profound hypotension and the true incidence of clinically significant cyanide poisoning is unknown, a reasonable clinical suspicion for the presence of cyanide poisoning is required before treatment with nitrite compounds is initiated. A third therapeutic agent, which has been shown to have few hemodynamic effects, is hydroxycobalamin. This vitamin combines with cyanide to form the inactive compound cyanocobalamin and is usually given intravenously as a 5-g dose, which should be doubled for patients with blood cyanide levels greater than 40 μmol/l.

Therapy for the inflammatory forms of inhalation injury is primarily supportive in nature. If further injury, such as barotrauma or intervening infection, is avoided, the epithelial injury and inflammatory response should resolve in 14 to 21 days. Delivery of humidified gas may prevent inspissation of airway secretions and should be standard therapy for these patients. Bronchospasm should be treated aggressively with inhaled bronchodilators. Except for those patients who are on corticosteroids for pre-existing medical conditions and those with intractable bronchospasm unresponsive to bronchodilators, steroid therapy should be avoided. Barotrauma can be minimized by ventilating at the lowest peak airway pressure and rate (rate–pressure product) that adequately ventilate the patient. Owing to the heterogeneous nature of inhalation injury, high airway pressures result in overdistention of normal lung units while the relatively non-compliant injured lung units remain underventilated. Barotrauma produced by high airway pressures causes increased histological damage to the lung parenchyma in animal models of inhalation injury. Maintaining a mild hypercapnia with the blood pH above 7.32 is well tolerated by most patients and may minimize barotrauma following acute lung damage caused by inhalation injury. For similar reasons, in the face of already high airway pressures, high levels of positive end-expiratory pressure should not be used to drive the oxygen saturation above 93 per cent, since this does not appreciably increase the oxygen content of the blood.

Following an airway and parenchymal injury such as inhalation injury, an ideal ventilator would apply low peak airway pressures, facilitate clearance of soot, sloughed mucosa, and secretions, and recruit collapsed airways. The volume diffusive respirator (VDR), which has been utilized by the authors for the mangement of burn patients with inhalation injury since the mid-1980s, has these abilities. It is a high-frequency interrupted-flow ventilator that superimposes oscillatory airflow on a pressure-controlled tidal volume. A retrospective analysis comparing the VDR ventilator with conventional volume ventilation in patients with similar distribution of age and burn sizes revealed a decreased rate of pneumonia and mortality in those patients ventilated with the VDR (Rue *et al.* 1993) (Fig. 2). The beneficial effects of this ventilator have been confirmed in other burn centers clinically and in a primate model of inhalation injury.

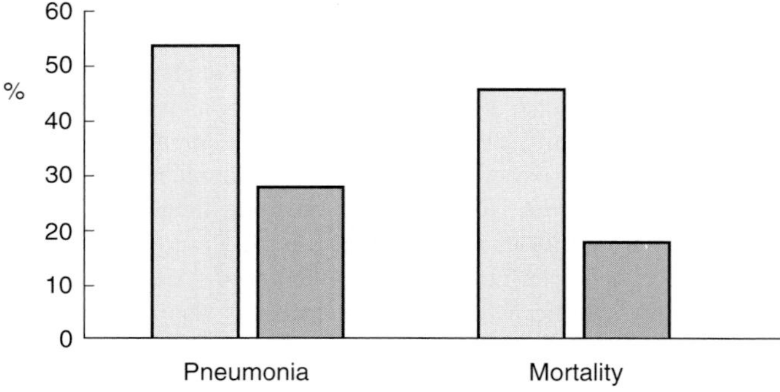

Fig. 2 Rates of pneumonia and mortality in 330 patients treated with conventional (green shading) or VDR (gray shading) ventilation. (Data from Rue *et al.* (1993).)

Airway and parenchymal injury are often not the only factors limiting ventilation following smoke inhalation injury; the burn eschar can also limit gas exchange. Burn injury to the trunk that is full thickness and circumferential may restrict chest wall expansion and result in the need for higher airway pressures to attain the same level of ventilation. Chest wall escharotomies, which release this constricting band, can be lifesaving. Escharotomies, which can be performed at the bedside because of the insensate nature of third-degree burns, are carried through the eschar and the superficial fascia only to that depth that will permit the cut edges of the eschar to separate. The incisions are made in the anterior axillary lines and, if the eschar extends onto the anterior abdominal wall and restricts motion of the costal margin, are connected transversely across the chest by a subcostal escharotomy incision.

Complications

Pulmonary infections following inhalation injury are common. The risk of developing pneumonia without evidence of inhalation injury is 8 to 10 per cent. With inhalation injury diagnosed by xenon scan only, the incidence of pneumonia increases to 20 per cent and, with bronchoscopic evidence of inhalation injury, it climbs to 50 per cent. Not surprisingly, increasing extent of burn also correlates with increased rates of pneumonia (Shirani *et al.* 1987). Once the diagnosis of pneumonia or tracheobronchitis is made, antibiotic therapy directed at the predominant organism in the sputum culture is begun. Antibiotics with a broad spectrum of activity are avoided unless required by the resistance pattern of the infecting organism.

Inhalation injury can also be the cause of mechanical complications. In those patients who require high airway pressures to maintain acceptable ventilation and oxygenation, barotrauma can result in pneumothorax, pneumomediastinum, subcutaneous emphysema, or pneumoperitoneum. The complications of prolonged intubation or tracheostomy, such as tracheal stenosis, tracheo-esophageal fistula, and laryngeal polyps, are very difficult to treat, and so prevention should be the first line of defense. Stabilization of the endotracheal or tracheostomy tube, maintenance of cuff pressures at a minimum to prevent air leak, and prompt weaning to extubation should be the standard of care for intubated patients. Tracheostomy has been reported to be required in 3 per cent of burn patients, and is indicated to improve endobronchial toilet in patients with tenacious secretions and for the comfort of patients who require long-term intubation. Although there is a wide range of opinion concerning the timing of tracheostomy, we consider tracheostomy, other than that required for early airway control, for any patient who will require continued endotracheal intubation after 2 weeks of mechanical ventilation.

Future therapies

The development of new clinical therapies and ongoing laboratory investigations promise further improvement in the treatment of inhalation injury. One new therapeutic intervention involves the treatment of severe inhalation injury with aerosolized heparin. Severe injury can denude the airway and lead to the formation of bloody casts consisting of inspissated mucus and hemorrhagic plugs that can be obstructive. In this setting, aerosolized heparin 10 000 IU, administered in nebulized form every 6 h, can reduce the catastrophic complication of occluded airways and respiratory decompensation (Cox *et al.* 1993). Pentoxifylline, another potential therapeutic agent, may improve the microcirculation through its hemorrheological and antithrombotic effects, and its inhibitory effect on cytokine release and leukocyte activation may offer relief from the hyperacute inflammatory response that follows inhalation injury. Platelet activating factor antagonists have also shown promise in animal models of inhalation injury by virtue of their ability to reduce the local and systemic responses evoked by the injury.

One of the mainstays of therapy following inhalation injury is to allow no further injury to occur. Repeated cycles of injury in the form of barotrauma or overdistension of lung units cause continued recruitment of inflammatory cells into the zone of injury and development of fibrosis through macrophage elaboration of basic fibroblast growth factor and other factors. Some investigators have evaluated intravenous CO_2 extraction as a means of decreasing the need for ventilation and its attendant high airway pressures in patients with severe physiological derangement. Another therapy that may offer relief from these repeated cycles of injury is perfluorocarbon-assisted gas exchange (PAGE). PAGE has been evaluated in many models of acute lung injury, and recently, in a porcine model of inhalation injury, transformed an inhalation injury with uniform lethality at 24 h into a survivable injury (Fitzpatrick *et al.* 1996).

References

Cox, C.S., Jr, Zwischenberger, J.B., Traber, D.L., Traber, L.D., Haque, A.K., and Herndon, D.N. (1993). Heparin improves oxygenation and minimizes barotrauma after severe smoke inhalation in an ovine model. *Surgery, Gynecology and Obstetrics*, **174**, 339–49.

Fitzpatrick, J.C., Jordan, B.S., Salman, N., Williams, J., Cioffi, W.G., Jr, and Pruitt, B.A., Jr (1997). The use of perfluorocarbon-assisted gas exchange to improve ventilation and decrease mortality after inhalation injury in a neonatal swine model. *Journal of Pediatric Surgery*, **32**, 192–7.

Grube, B.J., Marvin, J.A., and Heimbach, D.M. (1988). Therapeutic hyperbaric oxygen: help or hindrance in burn patients with carbon monoxide poisoning. *Journal of Burn Care and Rehabilitation*, **9**, 249–52.

Rue, L.W., Cioffi, W.G., Mason, A.D., Jr, McManus, W.F., and Pruitt, B.A., Jr (1993). Improved survival of burned patients with inhalation injury. *Archives of Surgery*, **128**, 772–80.

Shirani, K.Z., Pruitt, B.A., Jr, and Mason, A.D., Jr (1987). The influence of inhalation injury and pneumonia on burn mortality. *Annals of Surgery*, **205**, 82–7.

2.7 Acute bronchospasm

2.7.1 Pathophysiology and causes

M.-D. Schaller and D. Tagan

Key messages

- The onset of an asthma attack may be progressive over days or very sudden and unexpected, leading to respiratory arrest within minutes or a few hours (sudden asphyxic asthma). Bronchial obstruction results from mucosal inflammation, muscle spasm, and lumen occlusion by thick mucus.
- Sudden asphyxic asthma may have a distinct pathogenic mechanism and be attributable essentially to a bronchospasm.
- Expiratory air flow rate is reduced due to a decreased lung elastic recoil force, persistent contraction of inspiratory muscles, airways obstruction, and possible glottic aperture.
- Expiration time is so prolonged that inspiration begins before complete air exhalation, leading to progressive hyperinflation and positive end-expiratory pressure (auto-PEEP).
- Bronchial obstruction is associated with increased total lung capacity, functional residual capacity, and residual volume, and with decreased tidal volume and vital capacity.
- Deep negative intrapleural pressures developed during inspiration interfere with cardiovascular function; right ventricle preload and afterload are increased, left ventricle preload is depressed, and afterload is augmented. These alterations account for the development of pulsus paradoxus.

Introduction

Asthma is a disease characterized by episodes of bronchial obstruction, which are usually relieved by therapy. Occasionally, it may lead to acute respiratory failure or even death. In Western countries, asthma mortality is estimated as 1 to 8 per 100 000 annually. The tendency of a progressive increase in mortality observed in some countries and some low social classes in recent decades is a major concern. Causes of death have been attributed to underestimation of the severity of asthma by patients or relatives, poor compliance, low level of education, poverty, and psychiatric disorders. Bronchodilators have been implicated; overuse or abuse of β_2-agonists may provoke fatal arrhythmias or may delay seeking medical aid, although they may correlate with more severe asthma. Physician responsibility has sometimes been identified, including errors in diagnosis, underassessment of the severity of an attack, and inappropriate treatment. These preventable deaths occur in all age groups, and are particularly striking when considering the possible excellent long-term prognosis in such patients. Severe asthma remains a challenge in the intensive care unit (ICU), with a mortality of 1 to 3 per cent, and even of up to 40 per cent when mechanical ventilation is required. Therefore it is extremely important to recognize severe asthma attacks and status asthmaticus in order to admit such patients to the ICU.

Pathogenesis

When challenged with the appropriate stimulus, the asthmatic patient experiences bronchial obstruction. Trigger events may be an identified allergen (pollen, grass, dust, food, hymenoptera venom, or drugs) or non-specific factors such as cold, smoke irritants, air pollution, viral infection, or emotional stress. Airway narrowing results from smooth muscle spasm, mucosal inflammation, and/or occlusion of bronchial lumen by thick mucus. Involvement of these three components varies, with exclusive or preponderant bronchospasm being able to achieve lumen obstruction. Postmortem examination and biopsies have revealed infiltration of bronchial submucosa by eosinophils and mast cells, local denudation of mucosa, deposition of dense collagen beneath basal membrane, proliferation and hyperplasia of goblet cells, smooth muscle hypertrophy, and airways plugged by thick mucus (Figs 1 and 2). These tenacious secretions contain mucus, inflammatory exudate with plasma proteins, epithelial cells, and inflammatory cells (mainly eosinophils). Expectorations with a large eosinophil content are known as Charcot–Leyden crystals. However, in some asthma deaths the only pathological findings are empty airways and predominant neutrophil infiltration of bronchial walls. Histamine, leukotrienes, prostaglandins, thromboxane, and other vasoactive and bronchoconstrictive substances have been recovered from asthmatic bronchoalveolar lavage, suggesting a complex effect of numerous immunological and inflammatory mediators.

Following allergen exposure, B lymphocytes synthesize immunoglobin E (IgE) which is bound to mast cell surfaces preferentially but not exclusively; it also binds with lower affinity to other cells (eosinophils, macrophages, platelets) (Fig. 3). T lymphocytes play a major role in orchestrating immunoinflammatory response; mainly through the release of interleukins, they enhance IgE production by B lymphocytes, differentiate mast cells, and prime eosinophils. In the presence of allergens or other non-specific stimuli, mast cells degranulate and release histamine, prostaglandins, leukotrienes, and

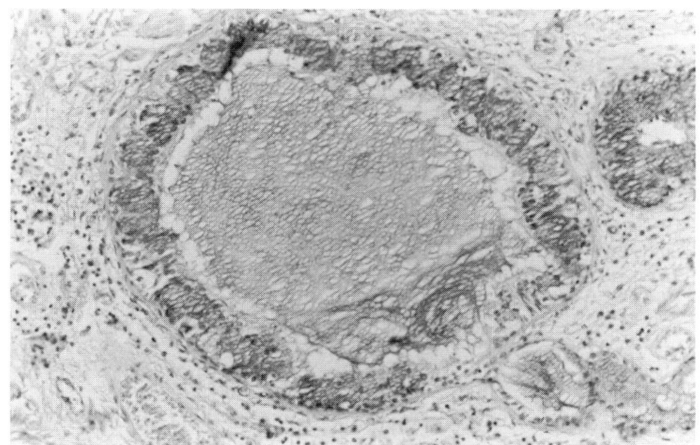

Fig. 1 Mucus plugging a dilated bronchiole surrounded by inflammatory cells. (Reproduced with permission from Schaller *et al.* (1996).)

tryptase, thus promoting bronchoconstriction, dilatation of blood vessels, increased permeability, and edema. They also modify the viscosity of mucus and enhance its production. This is known as immediate or early allergic reaction. Eosinophils release vasoactive and bronchoconstritive substances as well as cytotoxic factors. Major basic protein, peroxidase, and neurotoxin damage surrounding tissues, leading to epithelial shedding. They also depress ciliary function and hence clearance of mucus from the airways. Under these conditions,the unprotected nerve endings are easily stimulated. The ensuing parasympathetic activation causes bronchoconstriction through acetylcholine release. These neuronal reflexes may also be initiated by psychological stress without an identified external trigger factor. Stimulation of non-myelinated C-fiber endings in the bronchial epithelium is responsible for a neuronal reflex with local neuropeptides release. Thus substance P promotes vascular permeability, mucus secretion, and bronchoconstriction, whereas neurokinin A causes bronchoconstriction. Other cells and mediators have been implicated in the development of bronchial obstruction: alveolar macrophages with release of superoxide radicals, chemotactic factors, thromboxane, leukotrienes and platelet activating factor (PAF), platelets with cytokines and platelet activating factor,

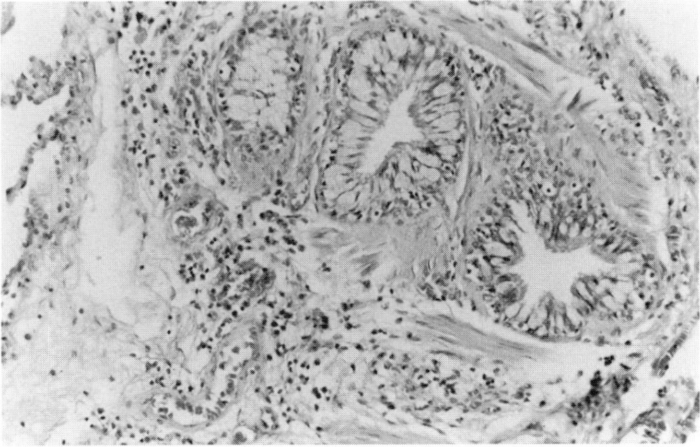

Fig. 2 Narrowing of the airway lumen by numerous hyperplastic goblet cells, hypertrophied smooth muscle, and inflammatory infiltrate. (Reproduced with permission from Schaller *et al.* (1996).)

damaged epithelial cells with decreased epithelial-derived bronchodilator factors, and lower mucosal protection.

These cellular, humoral, and neural mechanisms have all been implicated in the production of inflammation, edema, mucus accumulation, and muscle spasm. Autopsy findings following a prolonged and progressive asthma attack have confirmed extensive mucus plugging with marked airway inflammation. In addition, postmortem examinations of patients dying from a sudden asphyxic attack have shown empty airways and occasionally neutrophil infiltration, suggesting a predominant bronchospasm and distinct pathogenic mechanism in this setting.

Pathophysiology

Airway narrowing and closure are the most prominent pathophysiological abnormalities found in asthma. They are responsible for respiratory, cardiocirculatory, and gas exchange modifications (Fig. 4).

Respiratory consequences

Bronchial obstruction is associated with a progressive increase in lung volume. Expiration time is prolonged and inspiration begins before the exhalation of inspired gas is complete. A number of factors are responsible for the impairment of expiration flow rate. During expiration, there is a persistent contraction of inspiratory muscles promoting an outward motion of the thoracic cage, thus opposing expiration. There is a decreased lung recoil elastic force, the mechanism of which is not clearly understood. Moreover, resistance to flow is increased due to airway obstruction, sometimes associated with closure of the glottic aperture. This explains the use of expiratory muscles, as expiration is no longer passive. Bronchial obstruction is not uniform throughout the lungs; airways are completely closed, partially occluded, occasionally with check valve phenomena, or completely open. The ensuing prolonged expiration time in some areas leads to air gas trapping and progressive dilatation of the alveoli. This results in progressive hyperinflation.

In addition, there is an increased ventilatory drive associated with enhanced respiratory muscle activity and secondary to stimulation of pulmonary receptors of the vagus nerve. The resulting tachypnea interferes with a necessarily prolonged expiration and enhances further hyperinflation. However, an increase in lung volume is an advantage for the patient as it helps to keep open narrowed airways which would otherwise close at normal lung volumes. This is achieved by the radial traction exerted on the bronchial wall in this situation. With severe obstruction, there is an increase in total lung capacity (up to 100 per cent over control values), residual volume, and functional residual capacity (up to several-fold). Concomitantly, tidal volume and vital capacity decrease. All indices of expiratory flow rate are markedly reduced, as evidenced by the two more useful pulmonary function tests available at the bedside: the forced expiratory volume in 1 s (FEV_1) or the peak expiratory flow rate (PEFR).

Changes occur not only in lung volumes but also in airway pressure. Lung deflation is not completely achieved when the next inspiration begins. At the end of expiration, airway pressure does not return to zero or atmospheric pressure, but remains elevated, i.e. a positive end-expiratory pressure (PEEP), known as intrinsic PEEP or auto-PEEP, progressively develops.

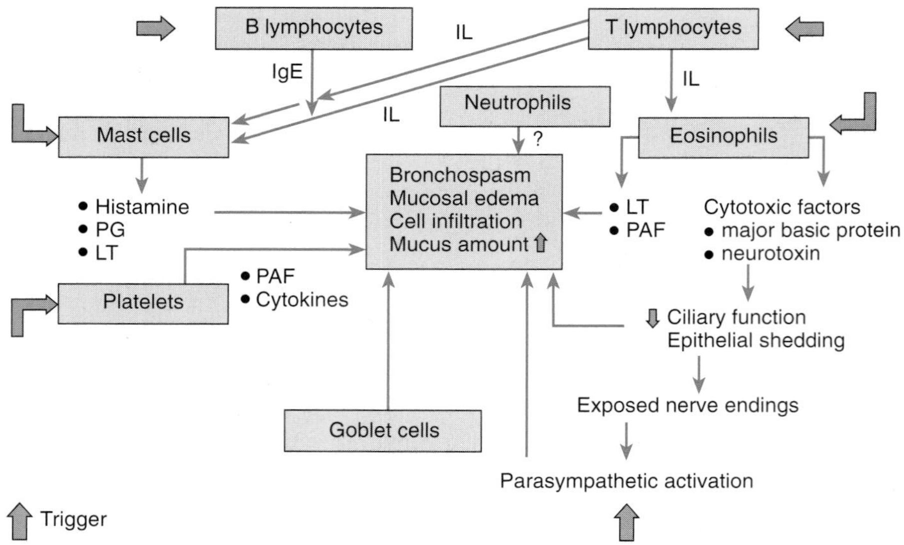

Fig. 3 Schematic representation of cell-derived mediator effects: IL, interleukins; PG, prostaglandins; LT, leukotrienes; PAF, platelet activating factor.

With the development of hyperinflation, auto-PEEP, and higher airways resistance, initiation of inspiratory flow requires greater force. Normal inspiration is achieved by inspiratory muscle work generating intrapleural pressures from –3 to –6 cmH_2O. During asthma, profound negative intrapleural pressures are necessary to obtain normal tidal volume. Expiration requires active muscle contraction. This results in increased work of breathing, with increased CO_2 production and O_2 consumption. Furthermore, increased minute ventilation, which is wasted in many areas with impaired ventilation–perfusion ratio, is stimulated. Dyspnea is a usual finding, and patients may be able to evaluate the degree of bronchial obstruction better than physicians. However, perception of breathlessness is reduced in some patient categories, particularly after severe attacks or in older asthmatics.

The diaphragm is flattened or even everted in the abdominal cavity. During inspiration, this leads to a paradoxical inward movement of the base of the thoracic cage associated with a tracheal tug concomitant with contractions of the sternocleidomastoid and intercostal muscles. The flattened and elongated diaphragm loses its efficacy. Muscle fatigue can develop, with the appearance of alternating abdominothoracic respiration, abdominal paradox, and a progressive decrease in respiratory rate.

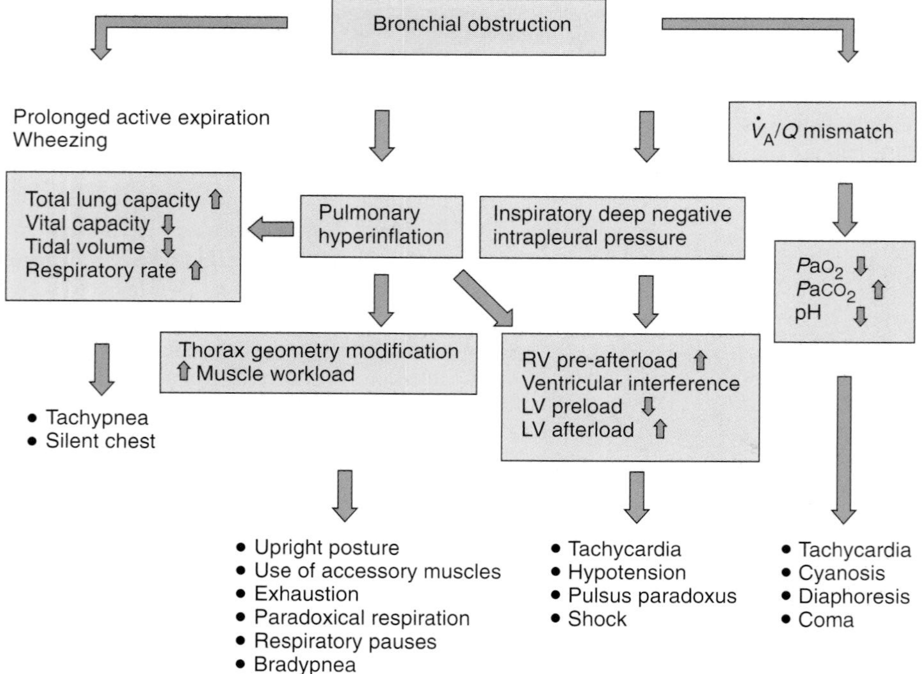

Fig. 4 Pathophysiological consequences of bronchial obstruction: RV, right ventricular; LV, left ventricular. (Reproduced with permission from Schaller *et al.* (1996).)

Cardiovascular consequences

The wide swings of intrathoracic pressure during respiratory cycles and hyperinflation modify the extramural pressure surrounding the heart and blood vessels and have serious circulatory consequences. In the presence of bronchial obstruction, intrapleural pressure may fall to –30 cmH$_2$O during inspiration and to –50 cmH$_2$O during forced inspiration. This profoundly negative pressure is directly transmitted to the right atrium and there is an enormous increase in venous return from the inferior vena cava. Consequently, right ventricular preload increases.

Hyperinflation may compress and stretch the capillary network, contributing to an increase in pulmonary arterial resistances. Intra-alveolar pressure is the extramural pressure for capillaries. It is positive in distended alveoli ventilated by partially occluded bronchi (auto-PEEP), profoundly negative in completely occluded areas, and approaches zero to –3 cmH$_2$O in normally ventilated alveoli. Globally, this results in a mean pressure value approaching zero, which is greater than intrapleural pressure. These phenomena result in an increase in afterload for the right ventricle which is surrounded by deep negative pressures, i.e. intrathoracic pressures. Hence the right ventricular ejection fraction falls, leading to increased end-diastolic volume. The right ventricle dilates and squeezes the left ventricle, since both are contained in a common inextensible pericardium. This effect, which is known as diastolic ventricular interference, occurs in parallel. Preload of the left ventricle is lower, owing to a reduced pulmonary blood flow following decreased right ventricular output, and produces ventricular interference in series. A decrease in the left ventricular compliance further impairs its filling. This is because of changes in the left ventricular configuration induced by compression of the right ventricle. The negative intrathoracic pressure augments the transmural aortic pressure, causing an increased left ventricular afterload. Thus left ventricular stroke volume decreases during inspiration, leading to the fall in systolic blood pressure known as pulsus paradoxus. This physiological inspiratory fall in systolic blood pressure can reach 10 mmHg. This phenomenon is markedly enhanced during an asthma attack, and pulsus paradoxus has been measured up to 50 mmHg. An inverse correlation between pulsus paradoxus and FEV$_1$ has been documented. However, this is not always reliable. Pulsus paradoxus may be weak in patients experiencing a very severe asthma attack when they are exhausted and unable to generate the required marked negative intrapleural pressures. It is probably a better reflection of pleural pressure swings than of spirometric indices of airflow obstruction.

The enormous increase in lung volumes with the concomitant development of PEEP may have a compressive effect on the heart. Moreover, venous return may be limited by the flattened diaphragm muscle, which compresses the inferior vena cava at the entrance to the thorax. Pulmonary circulation is also impaired by overdistended alveoli and associated high pressures. All these factors further reduce right and left ventricular preload and may impair cardiac output.

Gas exchange consequences

Distribution of ventilation is uneven due to different degrees of bronchial obstruction. Impairment of circulation may occur in overdistended areas. Some alveoli are poorly ventilated and normally perfused (low ventilation–perfusion ratio) and lead to hypoxemia. In completely obstructed areas, normal perfusion would constitute a true shunt. However, a small collateral ventilation of these alveoli accounts for the absence of true shunt. Other alveoli are overinflated, compress capillaries, and increase the dead-space. Evidence for the ensuing ventilation–perfusion imbalance is provided by blood gas analysis.

The increased minute ventilation following the stimulation of ventilatory drive reduces $P\text{aCO}_2$, creating respiratory alkalosis (Table 1, stages I and II). In more severe asthma, hypoxemia may develop associated with respiratory alkalosis. With progression of obstruction, there is a decrease in tidal volume and an increase in respiratory rate. $P\text{aCO}_2$ increases due to alveolar hypoventilation and concomitant increased CO$_2$ production, resulting in respiratory acidosis (stage IV). In severe forms, metabolic acidosis develops which can sometimes be attributed to a lactic origin. The explanation of this is still debated; it is probably due to tissue hypoxia, lactate production by respiratory muscles, and/or intracellular alkalosis.

Hypoxemia and hypercapnia account for cyanosis, diaphoresis, tachycardia, unconsciousness, and coma.

Clinical signs

Clinical signs are derived from the pathophysiological changes following bronchial obstruction (Fig. 4).

The patient experiencing a severe asthma attack remains seated, avoids speaking, and is dyspneic with profuse diaphoresis and cyanosis. The respiratory rate is increased (above 30 breaths/min). Prolonged and active expiration is covered by musical wheezes. The patient breathes with effort, using accessory muscles. Contraction of the sternocleidomastoid muscle, tracheal tug, and inward movement of the thoracic cage during inspiration may be noted. With fatigue,

Table 1 Arterial blood gas and pH values corresponding to various asthma severity stages

Stage		$P\text{aO}_2$	$P\text{aCO}_2$	pH
I	Mild	Normal or mild ⇓	Normal or ⇓	Normal or ⇑
II	Mild to moderate	⇓	⇓	⇑
III	Moderate to severe	⇓	Normal	Normal
IV	Severe	⇓ ⇓	⇑⇑	⇓ ⇓

there may appear a paradoxical respiration, bradypnea, and respiratory pauses. These signs or the appearance of a silent chest indicate imminent respiratory arrest. Complications such as pneumothorax or pneumomediastinum may occur any time.

The cardiopulmonary interaction promotes first a tachycardia in an attempt to maintain cardiac output despite reduced preload. A heart rate above 120 beats/min and pulsus paradoxus are severity criteria. Hypotension with more severe tachycardia and/or arrhythmias appear in more compromised stages. Cardiocirculatory collapse and shock are warning signs of impending cardiac or cardiorespiratory arrest, as are altered consciousness or coma reflecting impaired gas exchange.

If, at first sight, the patient is sitting, cyanotic, and sweating, most of the severity signs described above will be present.

Laboratory tests

Pulmonary function tests

Measurement of PEFR may occasionally provoke cardiorespiratory arrest in severe asthma. Indeed, in contrast with normal subjects, deep inspiration in asthmatics induces bronchoconstriction and may promote coughing. PEFR should not be measured in a patient experiencing severe asthma, i.e. unable to speak; it is below 100 l/min. However, repeated PEFR measurements are of value later, when major bronchial obstruction is relieved, to monitor response to therapy and stabilization. When PEFR values checked prior to hospital admission disclose unstable asthma with profound morning dips, sudden respiratory arrest is a real threat.

Chest radiography

Chest radiographs show what is expected from clinical examination, i.e lung hyperinflation, sometimes associated with small heart size. They should be obtained to exclude pneumothorax, pneumomediastinum, pneumonia, or other diseases mimicking bronchial obstruction.

Electrocardiography

ECG abnormalities are common and non-specific. They reveal tachycardia, arrhythmias, and right ventricular strain (right QRS axis deviation, repolarization abnormalities, P pulmonale, and partial right bundle branch block).

Arterial blood gas

Arterial blood gas monitoring is mandatory. Respiratory alkalosis with or without hypoxemia is expected first (Table 1, stages I–II). Normocapnia with hypoxemia reflects hypoventilation (stage III) and should alert the physician. The extreme situation favors the development of hypercapnic and possibly metabolic acidosis (stage IV).

Patterns of asthma

Asthma attack may progressively worsen over days, with more frequent, severe, and long-lasting episodes of bronchial obstruction, often associated with thick and viscid expectorations. A striking feature among clinical signs of severity is exhaustion.

In contrast, the attack may supervene suddenly in an apparently previously symptomless patient and develop in a fulminant way leading to severe respiratory failure in minutes or a few hours. This has been called sudden asphyxic asthma. Most of these patients are in coma, with a high incidence of respiratory arrest and cardiovascular instability. They develop marked mixed acidosis or extreme hypercapnia (with $Pa\text{CO}_2$ reaching values above 200 mmHg (26.7 kPa)). Response to therapy is usually very rapid, with fast normalization of $Pa\text{CO}_2$ and pulmonary function tests. In view of the scanty expectorations and rapid time course, which contrasts with other asthmatics experiencing slow-onset attack, bronchospasm may be the preponderant pathogenic mechanism in these patients.

Bibliography

Djukanovic, R., *et al.* (1990). Mucosal inflammation in asthma. *American Review of Respiratory Disease*, **142**, 434–57.

Ferrer, A., *et al.* (1993). Airway obstruction and ventilation–perfusion relationships in acute severe asthma. *American Review of Respiratory Disease*, **147**, 579–84.

Mountain, R.D., *et al.* (1990). Acid–base disturbances in acute asthma. *Chest*, **98**, 651–5.

Schaller, M.D., Tagan, D., and Perret, C. (1996). Asthme aigu grave. In *Encyclopédie médico-chirurgicale*, Vol. 36, *Anesthésie-réanimation*, p. 970-A-10. Elsevier, Paris.

Scharf, S.M. (1991). Cardiovascular effects of airways obstruction. *Lung*, **169**, 1–23.

2.7.2 Therapeutic approach to bronchospasm and asthma

K. M. Nolan and C. S. Garrard

Key messages

- The three basic mechanisms of the disease process of asthma are airway wall inflammation, smooth-muscle-mediated bronchoconstriction, and intraluminal mucus. Therefore therapy must include aggressive utilization of anti-inflammatory medications and bronchodilators.
- Maximal bronchodilator and supportive therapy should be provided under the highest level of observation and monitoring in an intensive care or high-dependency unit.
- Mechanical ventilation is indicated in the asthmatic who continues to deteriorate despite maximal medical therapy.
- The aim of mechanical ventilation is to maintain adequate arterial oxygenation and control acidosis while specific therapeutic measures reverse bronchospasm, reduce bronchial epithelium inflammation, mobilize secretions, and suppress airway hyper-reactivity.
- Deterioration despite optimal treatment, including oral steroids and accelerated use of inhaled β-agonists, identifies patients at higher risk for fatal asthma. A history of near-fatal asthma requiring mechanical ventilation is the greatest single predictor of subsequent asthma death.

Introduction

Only a decade ago, asthma was viewed as a disease in which bronchospasm was the principal component of airflow obstruction. Current paradigms emphasize the role of airway wall inflammation in the pathogenesis of this disease which, in conjunction with smooth-muscle-mediated bronchoconstriction and intraluminal mucus, results in airflow obstruction.

Drug treatment in asthma has changed during recent years, reflecting the increased awareness that asthma is not just a disease of reversible bronchoconstriction, and that the underlying inflammation occurring in the airways contributes in a major way to its pathological processes and progress.

There is an emerging consensus that airway inflammation is a consistent component of asthma, and its intensity relates to the severity of the disease. A therapeutic consequence of this concept is to encourage aggressive utilization of anti-inflammatory medications.

Clinical assessment

Medical history

Acute bronchial asthma is one of the most common medical emergencies encountered in clinical practice. The natural history of acute severe asthma varies from hours to weeks. It may result from two separate mechanisms; one develops rapidly over hours from a relatively normal state (an anaphylactoid type reaction), and the other, far more common, is a gradual deterioration against a background of chronic uncontrolled asthma. It is important to recognize the factors that identify patients at risk of death as well as those who may require mechanical ventilation. Risk factors of death from asthma include a history of intubation, hypercapnia, pneumomediastinum or pneumothorax, admission to hospital despite chronic oral steroid use, and medical non-compliance. Deterioration despite optimal treatment, including oral steroids and accelerated use of inhaled β-agonists, identifies patients at higher risk for fatal asthma. A history of near-fatal asthma requiring mechanical ventilation is the greatest single predictor of subsequent asthma death. Comorbid conditions such as coronary artery disease place patients at greater risk during severe asthma, and they may predispose to complications of drug treatment.

Physical examination

Clinical indicators of the severity of an attack should be assessed, namely ability to speak, inability to lie supine, diaphoresis, heart rate, respiratory rate, degree of arterial paradox (the fall in systolic blood pressure on inspiration), and peak expiratory flow rate (PEFR). A recent consensus statement by the British Thoracic Society (Woodhead 1993) has defined a severe asthma attack in terms of these variables (Table 1) and described features requiring admission to an intensive care unit (ICU) (Table 2).

Physical examination shows hyperinflation, accessory muscle use in both inspiration and expiration, and widespread rhonchi. The classic sign of wheezing correlates poorly with the degree of airflow limitation. Severely obstructed patients may have a silent chest if there is insufficient alveolar ventilation and airflow for wheezes to occur. A silent chest is another sign of imminent respiratory arrest.

The complications of acute asthma, including signs of pneumothorax, pneumomediastinum, and atelectasis (Table 3), should also be considered during the physical examination.

Table 1 Indicators of a severe asthma attack (statement by the British Thoracic Society)

Potentially life-threatening features
Unable to complete sentences in one breath
Respiratory rate ⩾25 breaths/min
Heart rate ⩾110 beats/min
PEFR ⩽50% of predicted normal or of best normal if known
PEFR <200 l/min if best normal not known
Arterial paradox >10 mmHg
Immediately life-threatening features
PEFR <33% of predicted normal or best
Silent chest, cyanosis, or feeble respiratory effort
Bradycardia or hypotension
Exhaustion, confusion, or coma

Monitoring and investigations

Patients with acute asthma should have arterial blood gases measured early in their hospital course; both the absolute values and changes in response to therapy are invaluable in assessing the potential need for mechanical ventilation. The most common pattern of blood gas abnormality is a combination of hypoxemia, hypocapnia, and respiratory alkalosis. Ventilation–perfusion abnormalities are common in severe asthma and maximal inspired oxygen concentrations should be used during resuscitation of the patient to achieve normal saturation and $Pa\text{O}_2$ values. A normal or increased $Pa\text{CO}_2$ has long been regarded as a danger sign, but the trend of repeated measurements is more important than one absolute value. Even patients presenting with gross hypercarbia can be managed without mechanical ventilation if they can be carefully observed in an ICU or high-dependency unit (HDU), and medical therapy produces rapid improvement. In contrast, a patient with a low $Pa\text{CO}_2$ which is increasing despite maximal medical therapy may require mechanical ventilation. Measurement of $Pa\text{CO}_2$ should not be considered in isolation but should be included in the overall assessment of the patient's clinical condition. Metabolic acidosis can occur in up to 28 per cent of cases and is usually associated with very severe bronchial narrowing. However, metabolic acidosis can be induced in acutely ill asthmatic patients by the excessive use of non-selective adrenergic agonists such as epinephrine (adrenaline).

Table 2 Indications for admission to intensive care (statement by the British Thoracic Society)

Deteriorating peak flow, worsening or persisting hypoxia, or hypercapnia
Exhaustion, feeble respiration, confusion, or drowsiness
Coma or respiratory arrest

Table 3 Complications of acute asthma
Pneumothorax
Pneumomediastinum
Pneumopericardium
Subcutaneous emphysema
Atelectasis
Mucus plugging
Electrolyte disturbances (hypokalemia, hypophosphatemia, hypomagnesemia)
Theophylline toxicity
Lactic acidosis

Investigations should include a chest radiograph to quantify the extent of hyperinflation and to exclude infection and manifestations of pulmonary barotrauma such as pneumothorax or pneumomediastinum. Arterial blood gases should be estimated regularly with the patient receiving oxygen. Electrolyte estimation may reveal hypokalemia secondary to β-agonist use or corticosteroids, and urea, creatinine, and full blood count are also useful. A 12-lead ECG may show right heart strain and arrhythmias, and help to exclude other pathologies that could masquerade as asthma such as left ventricular failure and pulmonary embolus.

ECG and pulse oximetry should be continuously monitored, and frequent estimations of respiratory rate, level of consciousness, blood pressure, temperature, and PEFR should be made. Regular examinations of the chest should evaluate air entry, accessory muscle use, and evidence of subcutaneous emphysema.

Therapeutic management

In acute severe asthma therapy must be directed at relief of bronchospasm and airway inflammation. Patients require continual assessment until the airway disease reverses. Intubation and mechanical ventilation may be required, and an understanding of the underlying abnormalities of lung mechanics and the interaction of the lungs and the circulation is necessary for successful management.

Oxygen

Virtually all patients with asthma have hypoxemia during acute exacerbations, and the more severe the obstruction, the lower the arterial oxygen tension becomes. It should be administered in high concentrations as there is little risk of promoting hypercapnia because respiratory drive is well maintained. The concentration of oxygen administered should be determined by pulse oximetry and regular arterial blood gases. Humidification of the inspired gas is desirable to lessen the chance of cold-induced bronchoconstriction and to minimize inspissated secretions.

Beta-sympathomimetic agents

Inhaled β_2-adrenergic agonists are by far the most effective bronchodilators in common use. These agents have a rapid onset of action and are indicated as first-line treatment for the short-term relief of bronchoconstriction and acute exacerbations of asthma symptoms. Commonly, initial treatment consists of 2.5 mg albuterol (salbutamol) (0.5 ml of a 0.5 per cent solution in 2.5 ml normal saline) by nebulization every 20 min for 1 h (three doses) followed by hourly treatments during the first few hours of therapy. Inhaled treatments can be given continuously to severely obstructed patients until an adequate clinical response is achieved or adverse side-effects (excessive tachycardia, arrhythmias, or tremor) limit further administration.

The means of administering aerosols does not appear to be important, and studies have shown that both compressor-driven hand-held nebulizers and multiple doses from a metered-dose inhaler equipped with a spacer will produce the same results in non-intubated patients.

Many questions remain regarding optimal delivery of inhaled β-agonists to intubated and mechanically ventilated patients. There is a consensus, however, that whether metered-dose inhalers or nebulizers are used, higher drug doses are required to achieve a physiological effect in intubated patients. Whichever device is used drug doses should be titrated to either a beneficial physiological response as judged by a fall in the peak airway pressure or to the development of toxic side-effects.

The available data do not support the routine use of intravenous infusion of β-agonists in the treatment of patients with severe asthma. Several studies have demonstrated that inhaled therapy is equal to or better than intravenous therapy in treating airflow obstruction, and is less likely to cause cardiac arrhythmias, hypokalemia, and lactic acidosis. One empirical approach is to start nebulized β-agonists as first-line treatment and then to use the intravenous form if the response to nebulized therapy is poor or the patient is moribund on admission. Whichever route is chosen, the dose should be determined by the response and side-effects, and current preference is for agents such as albuterol 5 mg or terbutaline 5 mg nebulized from solution. Each dose should be diluted with normal saline to a total volume of 4 to 5 ml.

Intravenous albuterol is given as an initial bolus of 200 μg over 10 min and then as a continuous infusion of 3 to 20 μg/min.

Recently it has been suggested that β-agonist use in chronic asthma may be detrimental. There is no evidence for this in acute severe asthma, and specific β-agonists remain the mainstay of immediate treatment in acute airway obstruction.

Corticosteroids

Chronic inflammation is central to the pathogenesis of asthma, and steroids are the most effective agents in the control of asthma. There are several studies demonstrating the effectiveness of steroids in acute severe asthma. However, the peak response for steroid action does not occur until 6 to 12 h after an intravenous dose; therefore steroids should be given early. Prednisolone 30 to 60 mg orally, hydrocortisone 200 mg intravenously, or methylprednisolone 60 to 125 mg intravenously should be given immediately and continued every 6 h during the initial 24 to 48 h of treatment depending on the patient's condition. Intravenous steroids should be continued until the life-threatening phase is considered to be over and then changed to decreasing oral doses.

Aminophylline

Intravenous aminophylline is usually regarded as a second-line drug because of its narrow therapeutic range and the high incidence of side-effects; the most serious of these are cardiac arrhythmias and seizures. The value of aminophylline in acute severe asthma is uncertain. Its mechanism of action is unclear; at therapeutic concentrations the inhibition of phosphodiesterase is minimal and it is a relatively weak bronchodilator, providing a third of the bronchodilatation of sympathomimetics. Randomized clinical trials comparing combinations of aminophylline and β-agonists have not demonstrated any additional benefits with aminophylline.

If other first-line treatment has been given and the patient is not responding, aminophylline may be used. Patients who have not received a theophylline preparation recently can be given a loading dose of 5 mg/kg over 30 min in a peripheral vein, followed by an infusion of 0.5 to 0.9 mg/kg/h. There are wide variations in the elimination of aminophylline, and serum levels should be checked within 6 h of intravenous loading to avoid toxic levels and to guide further dosing. There is an increased risk of toxicity in patients with congestive heart failure, liver failure, and concurrent use of a number of medications including ciprofloxacin, macrolide antibiotics, and cimetidine.

Anticholinergics

Anticholinergic agents block muscarinic receptors, inhibit vagal tone, and promote bronchodilation, but they are medium-potency bronchodilators and produce substantively less bronchodilatation than β-agonists. However, ipratropium bromide augments the bronchodilating effect of β-agonists in acute asthma. The consensus statement by the British Thoracic Society (Woodhead 1993) suggests that the addition of nebulized ipratropium is indicated only in patients whose asthma is very severe when first seen or who deteriorate or fail to improve rapidly when treated with the standard regime. Ipratropium should be withdrawn when patients are clearly responding.

Additional bronchodilators

Inhalation anesthetics

Halothane, enflurane, and isoflurane can acutely reduce peak airway pressure and $Paco_2$ in ventilated patients and can supplement continued β-agonists and systemic corticosteroids. Halothane (0.5–2.0 per cent) is effective in most patients, but has a low therapeutic ratio in the acidotic hypovolemic patient. In addition, use of halothane for more than 24 h may be associated with bromide toxicity. Isoflurane in similar concentrations is safer than and probably as effective as halothane. Adequate gas-scavenging facilities must be available if anesthetic gases are to be used in an ICU environment.

Intravenous anesthetics

Ketamine, an intravenous anesthetic with sedative, analgesic, anesthetic, and bronchodilating properties, has been used successfully for the emergency intubation of patients with severe asthma. The usual intubating dose is 1 to 2 mg/kg followed by an infusion of 10 to 40 μg/kg/min to maintain a bronchodilating and sedative effect. Ketamine increases catecholamine levels and directly relaxes bronchial smooth muscle.

Magnesium sulfate

Anecdotal uncontrolled experience has suggested that magnesium sulfate may offer useful bronchodilator properties but the mechanism is unknown. One possibility is that magnesium inhibits the calcium channels of airway smooth muscle, thus interfering in calcium-mediated smooth muscle contraction. It has been reported that intravenous administration of 1 g magnesium sulfate in the management of acute respiratory failure in a 72-year-old asthmatic produced sufficient improvement to avoid intubation. However, controlled studies have failed to confirm significant bronchodilator effect from the use of magnesium in acute asthma.

Heliox

Heliox is a mixture of helium and oxygen which is generally available with $He:O_2$ mixtures of 80:20, 70:30, and 60:40. Helium is a low-density gas which reduces airway resistance and decreases respiratory work. A significant decrease in pulsus paradoxus and a significant increase in peak flow has been observed in patients with acute asthma breathing heliox for 15 min compared with controls. A significant reduction in both peak airway pressures and $Paco_2$ has been noted in intubated patients within 20 min of ventilation with heliox mixtures. If heliox is used during mechanical ventilatory support, gas blenders and flowmeters must be recalibrated to this low-density gas to obtain accurate measures of oxygen concentration or tidal volume.

Nitric oxide

Nitric oxide (NO) is an important short-acting endogenous vasodilator and bronchodilator. As it is extremely lipophilic, inhaled NO may not only diffuse directly from the alveoli into vascular smooth muscle causing vasorelaxation but also diffuse through the bronchial epithelial barrier to reach airway smooth muscle and produce airway relaxation. One study concluded that NO inhaled at 80 ppm exerted a weak bronchodilatory effect in asthmatics but not in patients with chronic obstructive pulmonary disease.

Mechanical ventilation

Between 1 and 3 per cent of patients admitted to hospital with acute asthma require mechanical ventilation. Although it should not be undertaken lightly, delaying to the point of exhaustion and cardiorespiratory arrest may prove fatal. The guiding principle is that mechanical ventilation is indicated in the asthmatic who continues to deteriorate despite maximal medical therapy (Table 4).

Maximal therapy with inhaled or intravenous β-agonists, systemic corticosteroids, possibly combined with xanthine derivatives such as aminophylline, and anticholinergics must be continued up to and throughout the period of mechanical ventilation.

Table 4 Indications for mechanical ventilation in a patient on maximal medical therapy

Decreasing level of consciousness
Decreasing Pao_2
Increasing $Paco_2$
Increasing arterial paradox
Increasing heart rate
Increasing respiratory rate
Decreasing PEFR
Decreasing ability to converse

Intubation

Oral endotracheal intubation is the preferred technique. At the time of intubation, patients with severe asthma are often tachycardic, tachypneic, hypoxic, hypercarbic, acidotic, hypovolemic, and hypokalemic. Therefore intubation must be performed by experienced medical staff. Unless a difficult intubation is anticipated, intravenous induction with fentanyl, midazolam, propofol, and succinylcholine (suxamethonium), following preoxygenation with 100 per cent oxygen, is satisfactory. Alternatively, for potentially difficult intubations, inhalation induction with isoflurane in oxygen or an awake intubation with topical airway anesthesia (either blind nasal or with fiber optics) should safely secure the airway. The preferred and safest technique is usually the one with which the clinician is most familiar.

Sedation

Sedation is necessary for the ventilated asthmatic patient to improve comfort and to allow safe and effective mechanical ventilation. A combination of a benzodiazepine such as midazolam and a synthetic narcotic such as fentanyl provides sedation and depresses respiratory drive. Propofol meets many of the requirements for an ideal sedating drug in ICU.

Paralysis

In the early period following intubation and mechanical ventilation, muscle relaxants may be needed to supplement sedation to minimize peak inspiratory pressures. Vecuronium is recommended as it does not release histamine. Patients receiving prolonged non-depolarizing neuromuscular blockade over several days, particularly in combination with corticosteroids, may experience significant skeletal muscle wasting. This myopathy is usually fully reversible over weeks but may delay ventilator weaning.

Ventilators, modes of ventilation, and settings

The aim of mechanical ventilation is to maintain adequate arterial oxygenation and control acidosis while using specific therapeutic measures to reverse bronchospasm, reduce bronchial epithelium inflammation, mobilize secretions, and suppress airway hyper-reactivity.

The simplest type of control-mode ventilator (either volume- or pressure-cycled) will serve to support an asthmatic patient. However, the current generation of volume- and time-cycled pressure-limited ventilators, capable of several modes of ventilation, provide the flexibility and safety features which make them preferable to the less sophisticated machines. During the initial stages of stabilization of the patient following intubation, control-mode ventilation is ideal. As the patient recovers from sedation, synchronized intermittent mandatory ventilation may be possible. If pressure-limited ventilation is adopted, minimal tidal volume alarms must be set so as to avoid inadvertent (uncontrolled) hypoventilation.

The ventilator settings that must be selected include the tidal volume (V_T), respiratory rate, FiO_2, peak inspiratory flow, waveform, and ratio of inspiratory to expiratory time (I:E ratio). Recommended settings are shown in Table 5. Opposing philosophies have attached different weights to V_T selection, the relevance of peak inspiratory pressure and the necessity for its limitation, the rapid normalization of arterial blood gases, and the need for high inspiratory flow rates to maximize expiratory time. The strategy of controlled hypoventilation as a means of limiting complications was proposed by Darioli and Perret (1984). Their priorities were maintaining tissue oxygenation and limiting peak inspiratory pressures to less than 50 cmH_2O; this was achieved by reducing tidal volume, respiratory rate, and inspiratory flow rate. Present opinion is divided between high and low inspiratory flow rate ($\dot{V}_I$) in asthma. Several investigators (Darioli and Perret 1984; Bishop and Hillman 1993) recommend that a low $\dot{V}_I$ of 40 l/min should be used as one of the methods to lower peak inspiratory pressure and thus reduce the risk of barotrauma. However, Tuxen (1994) has shown that reducing $\dot{V}_I$ from 100 to 40 l/min reduces the peak inspiratory pressure but also decreases expiratory time, which in turn decreases lung emptying. This results in more overinflation of the lung which has been identified as one of the major factors in pulmonary barotrauma. The problem with delivering high flow rates and high peak inspiratory pressures to asthmatic lungs is that the gas would preferentially be applied to the more normal population of alveoli and airways, as the disease process is very heterogeneous.

Peak pressure probably determines the risk of barotrauma and is primarily determined by V_T and, in acute asthma, by the peak inspiratory flow. To minimize peak pressure (ideally below 40 cmH_2O), a V_T of 8 to 10 ml/kg and an inspiratory flow of 30 to 40 l/min might be chosen. Achieving an adequate alveolar ventilation then depends on setting the ventilator rate between such settings and invariably results in some compromise being accepted in terms of the I:E ratio which may approach 1:1.

An overlying principal that should be adopted is the use of controlled hypoventilation, whereby hypercapnia ($P\text{CO}_2 = 5$–15 kPa (37.5–112.5 mmHg)) is tolerated in the belief that it is less harmful than barotrauma. Indeed, hypocapnia should be avoided since it may produce an increase in airway resistance in the asthmatic patient. Recent studies of permissive hypercapnia in patients with acute respiratory distress system suggest that moderate levels of hypercapnia are associated with acceptable levels of respiratory acidosis and carry negligible risk. Unrestricted oxygen should be administered to the acute asthma patient during all phases of management. Ventilation–perfusion mismatch is typically very severe in the ventilated asthmatic patient and may be exaggerated further by the administration of oxygen. However, the overall benefit of an increase in $S\text{aO}_2$ outweighs this effect. Methods which attempt to correct ventilation–perfusion mismatch include bronchial lavage to remove mucus plugs and, controversially, the application of positive end-expiratory pressure (PEEP).

Table 5 Initial ventilator settings (adjust to maintain oxygenation and limit peak inspiratory pressure)

FiO_2=1.0 (reduce when $S\text{aO}_2$ known)
Low tidal volume (<10 ml/kg)
Low ventilator rate (<10 cycles/min)
Long expiratory time (I:E<1:2)
Peak inspiratory flow 30–40 l/min
Limit peak inspiratory pressure (<40 cmH_2O)
Minimal PEEP (<5 $cm H_2O$)

Table 6 Complications associated with mechanical ventilation in acute asthma (in order of frequency and relative urgency)

Hypotension
Pneumothorax
Cardiac arrhythmias
Displacement of endotracheal tube
Pneumonia
Sepsis syndrome
Ventricular failure
Gastrointestinal hemorrhage
Pulmonary embolism
Pneumomediastinum
Subcutaneous emphysema

Formal bronchoalveolar lavage using a double-lumen endotracheal tube is a procedure associated with significant periods of arterial desaturation which nevertheless have been followed by an improvement in oxygenation. Where single or multiple lobar collapse is suggested on chest radiography, fiber-optic bronchoscopy with the instillation of normal saline or *N*-acetylcysteine to aspirate mucus plugs can produce significant improvement in oxygenation and inflation pressures.

Complications of mechanical ventilation

The most common ventilation-related complications are listed in Table 6.

PEEP, CPAP, and auto-PEEP

Passive expiratory airflow dynamics and the necessity for long expiratory times have lead to the concept of 'auto-PEEP' and to consideration of the potential gains that might be achieved by the addition of PEEP or continuous positive airways pressure (CPAP). Although there is anecdotal evidence supporting the use of PEEP and CPAP, undesirable effects on the cardiovascular system, particularly in a volume-depleted patient, represent a significant hazard. If one of the major goals is to avoid overdistention of the alveoli and minimize the effects on the cardiovascular system, PEEP offers little or no benefit during controlled mechanical ventilation of asthma. If PEEP is required for purposes of oxygenation, its effect on lung volume must be assessed and compensated for by reductions in V_T or respiratory rate. In general, improved oxygenation, together with relief of mucus plugging, will follow reversal of bronchospasm.

Weaning

The reported duration of mechanical ventilation required in acute severe asthma varies considerably from an average of 12 h up to several days. It is essential that maximal bronchodilator therapy is applied during this period. Complications of mechanical ventilation, such as barotrauma and infection, tend to prolong the duration of ventilation and are associated with reduced survival in some series but not in others.

Reversal of bronchospasm is indicated by the absence of wheezing and a fall in inflation peak pressures and the calculated effective dynamic compliance $D(A-a)O_2$. Resumption of spontaneous ventilation following the withdrawal of relaxants and sedation is best accommodated by the use of synchronized intermittent mechanical ventilation, or by assist–control or pressure-support modes of ventilation. This is readily achieved with the current generation of mechanical ventilators. Extubation can often be managed with little difficulty, but occasionally the reduction of the level of sedation, the increasing level of awareness, and the presence of the endotracheal tube may precipitate further bronchospasm. Judicious use of a short-acting agent (e.g. propofol) or an inhalation agent (e.g. isoflurane) may allow extubation to be performed under light short-lived sedation. Following extubation, continued observation and aggressive therapy must be maintained to avoid relapse.

Mortality in asthma patients undergoing mechanical ventilation

Reported mortalities vary considerably from 38 per cent to zero. We have reviewed 511 cases reported over the past three decades and have identified a crude mortality of 11 to 21 per cent, including patients who have suffered brain injury from cardiorespiratory arrest prior to ventilatory support. Exclusion of these latter cases reduces mortality to 16 per cent. Death in asthmatics receiving mechanical ventilation appears to be the result of barotrauma, hypotension, cardiac arrhythmias, or sepsis.

Extracorporeal support

There are several reports of extracorporeal lung assist in severe asthma. Acute severe asthma is associated with such a low mortality with modern ICU management that the additional survival afforded by this technique is probably marginal at best or even harmful.

Outcome

An increase in mortality from asthma has been observed in several countries in recent years, despite the improvement in pathophysiological findings and the introduction of new effective therapeutic agents. The phenomenon is difficult to explain, but the causes of death and identification of high-risk patients have been widely studied. The patient who has previously required ICU admission and mechanical ventilation is at significant risk during subsequent exacerbations of asthma. Successful management of the asthmatic patient in intensive care requires high levels of observation and the timely and careful application of effective therapy.

References

Bishop, G.F. and Hillman, K.M. (1993). Acute severe asthma. *Intensive Care World*, **10**, 166–71.

Corbridge, T.C. and Hall, J.B. (1995). The assessment and management of adults with status asthmaticus. *American Journal of Respiratory and Critical Care Medicine*, **151**, 1296–1316.

Darioli, R. and Perret, C. (1984). Mechanical controlled hypoventilation in status asthmaticus. *American Review of Respiratory Disease*, **129**, 385–7

Tuxen, D.V. (1994). Permissive hypercapnic ventilation. *American Journal of Respiratory and Critical Care Medicine*, **150**, 870–4.

Woodhead, M. (1993). Guidelines on the management of asthma. Statement by the British Thoracic Society. *Thorax*, **48** (Supplement), S1–24.

2.8 Acute on chronic airflow limitation

2.8.1 Pathophysiology

R. Fernandez and A. Artigas

Key messages

- Infections, cardiac failure, and pulmonary embolism are the most common precipitating factors of acute respiratory failure in chronic obstructive pulmonary disease.
- Hypercapnia develops despite maintenance of minute ventilation because of worsening ventilation–perfusion ($\dot{V}/\dot{Q}$) mismatch.

Introduction

The evolution of chronic obstructive pulmonary disease (COPD) is characterized by episodes of acute respiratory failure. In the first stages of COPD, the transient impairment in lung function is only perceived as an increase in sputum, the appearance of wheezing, or shortness of breath. As the illness becomes more severe, the number and intensity of the symptoms associated with an episode of acute respiratory failure increase. In the most extreme cases, patients develop severe abnormalities in their blood gases and become unable to sustain spontaneous ventilation. In this scenario, the use of artificial ventilatory support is lifesaving and is the main indication for admission of such patients to the intensive care unit (ICU). Until recently, the most common way of supplying mechanical support was intubation and sedation, with respiratory muscle rest considered as an important aspect of treatment. With the development of non-invasive ventilation, maintenance of a spontaneous breathing pattern has become more important than total respiratory muscle rest. Therefore an understanding of the pathophysiology of acute respiratory failure is useful in the treatment of such patients.

The fact that acute respiratory failure in COPD patients is a severe illness with an associated mortality has been recognized for many years. Nevertheless, the reported mortality of 6 to 40 per cent shows large discrepancies because of the selection criteria used in different studies. Moreover, the main cause of death in these patients is rarely respiratory insufficiency but is coexisting disease, i.e. cardiac failure and infectious processes (Derenne *et al.* 1988). Some authors have suggested that acute respiratory failure in some varieties of COPD may have lower mortality; thus type B (blue bloaters) would have a better prognosis than type A (emphysematous), but again the different admission rates for each type of patient may have biased this conclusion.

Pathological studies of COPD patients dying during acute respiratory failure have been of little help in clarifying the pathophysiology. Most of the pathological findings have been related to the underlying disease or concurrent illnesses, such as pulmonary embolism, pulmonary edema, and pneumothorax (Derenne *et al.* 1988).

Precipitating factors

Factors commonly provoking acute respiratory failure in COPD patients are infections, cardiac failure, and pulmonary embolism; few episodes are due to an excess of sedatives or oxygen. Additionally, occurrences such as surgery or trauma, with long recovery times in the ICU, may result in acute respiratory failure in these patients (Curtis and Hudson 1994).

Infections

The exact role of infections as triggers of acute respiratory failure is difficult to establish because the respiratory tract in COPD patients is routinely colonized by micro-organisms. Moreover, some patients may have viable pathogens in their airways without symptoms of infection, while others may suffer severe infection by opportunistic micro-organisms.

One reason for this apparently erratic behavior is the inoculum effect. Pathogens with low virulence can result in severe infections if a large number arrive in the lower airways. Another important factor facilitating the growth of micro-organisms is a defective local immune system. Bacterial residence time is increased because of the thickening of the bronchial mucosa and reduced mucous clearance. These factors are enhanced by malnutrition, which is common in COPD patients.

In addition to the common respiratory pathogens (*Streptococcus pneumoniae*, *Hemophilus influenzae*, and *Mycoplasma pneumoniae*), *Moraxella catarrhalis* and viridans streptococci are increasingly being recognized as precipitants of acute respiratory failure (Curtis and Hudson 1994). Nevertheless, the lack of correlation between antibiotic treatment and resolution of acute respiratory failure suggests that viruses can also have an important responsibility in many episodes. Outbreaks of acute respiratory failure coincident with influenza epidemics are clear examples that viral infection is typically the key determinant. However, the difficulties in diagnosing active viral diseases preclude stronger statements.

Cardiovascular failure

Cardiac function is frequently compromised in COPD, mainly as cor pulmonale. This is characterized by right ventricle dysfunction caused by high impedance of the lung vasculature. In more advanced stages the right ventricle dilates, producing incompetence of the tricuspid valve which leads to an increase in right atrial pressure. Additionally, the change in the shape of the right ventricle shifts the interventricular septum into the left ventricle, reducing the left ventricular volume with associated impairment of performance.

Acute respiratory failure further damages cardiac function by additional mechanisms. First, hypoxemia increases pulmonary hypertension by enhancing the hypoxic vasoconstrictive effect. Second, when hyperinflation develops, some pulmonary vessels collapse, leading to a reduction in the vascular cross-sectional area that also impairs right ventricular impedance. Third, pleural pressure becomes more negative during inspiration and even positive during active expiration, translating these pressure swings to the transmural ventricular pressure. As a result, both right ventricular preload and afterload increase.

Gas exchange

Worsening of gas exchange is the initial factor of acute respiratory failure in COPD patients. Stable COPD patients show alterations in ventilation–perfusion ($\dot{V}/Q$) mismatch, mainly as low and high $\dot{V}/Q$ areas. When acute respiratory failure develops, this effect is increased because of increased resistance and hyperinflation. The former reduces ventilation in some zones of the lungs, which increases the number of low $\dot{V}/Q$ units. Conversely, hyperinflation reduces blood flow in distended alveoli, increasing the number of high $\dot{V}/Q$ units. These alterations in $\dot{V}/Q$ mismatch are the main reason for the increase in $P\text{aCO}_2$ despite maintenance of minute ventilation.

Worsening of $\dot{V}/Q$ mismatch is limited by hypoxic pulmonary vasoconstriction. This vascular reflex reduces blood flow to hypoventilated areas, redistributing flow to better ventilated areas. Administration of oxygen at high doses abolishes this reflex, leading to an impairment in $\dot{V}/Q$ mismatch and a further increase in $P\text{aCO}_2$.

Respiratory mechanics

Resistance

COPD patients are characterized by an increase in airway resistance which induces a reduction in the ability to generate airflow. The main feature of acute respiratory failure in COPD patients is a further reduction in airflow caused by an additional increase in airways resistance.

The interaction between inspiratory airflow and resistance is more complex. The resistance of the airways increases linearly with the flow, while parenchymal lung resistance increases at low flows. Thus the inspiratory flow pattern chosen by COPD patients during acute respiratory failure is a delicate compromise between airways and lung resistances. Moreover, lung volume also influences respiratory resistances. At high lung inflation volumes, lung resistance increases while airways resistance reduces, leading to a rise in total resistance with increasing inflation volume (Tantucci *et al.* 1991).

Elastance

The elastic component of the respiratory system is affected during acute respiratory failure only in cases of lung consolidation. Nevertheless, because the elastance in COPD patients is usually high, a slight reduction in elastance is unable to counterbalance the increase in resistance in order to maintain airflow. As a result, these patients are unable to expire to functional residual capacity and become hyperinflated. This volume excess helps to increase expiratory flow but represents an additional load for the inspiratory muscles, as will be discussed below. Because total lung capacity is fixed, the increase in end-expiratory volume leads to a reduction in inspiratory capacity. The only strategy left for increasing minute ventilation is not to increase tidal volume, but to increase respiratory rate. This pattern results in a further reduction in expiratory time and further hyperinflation.

Additionally, hyperinflation forces patients to breathe in a less compliant part of their pressure–volume curve. This impairment in the elastic properties of the lung further burdens the inspiratory muscles, but enhances expiratory flow by increasing pressure available for flow. In this way the problem of expiratory flow limitation must be compensated by inspiratory muscles.

Work of breathing

The work of breathing is closely related to the mechanical properties of the respiratory system and to the ventilatory pattern chosen by the patient. Classically, the work of breathing is divided into resistive and elastic works. Elevated airways resistance is the main component of the increased resistive work of breathing in stable COPD patients. When acute respiratory failure develops in these patients, the resistive work is further increased due to the impairment in airway resistance. Nevertheless, as hyperinflation appears and patients breathe in the less compliant part of the pressure–volume curve, the elastic work also increases. More importantly, dynamic hyperinflation maintains positive intrapulmonary pressure at the end of expiration (intrinsic PEEP or auto-PEEP). This forces the inspiratory muscles to perform excess work in early inspiration to overcome this threshold pressure. In some COPD patients this extra work induced by hyperinflation can exceed the resistive work.

Expiratory work is commonly not taken into account in normal subjects. However, in COPD patients expiration is active rather than passive. In this scenario, the contraction of the expiratory muscles helps to increase expiratory flow, but also increases the work of breathing. Moreover, hyperinflation also assists in developing expiratory work because the thorax becomes expanded and tries to return to the resting position.

Respiratory muscles

The respiratory muscles are responsible for maintaining ventilation despite factors affecting pulmonary mechanics. As a result, if the muscular reserve can afford increasing loads, patients will maintain ventilation and avoid respiratory insufficiency. Nevertheless, COPD patients are at a disadvantage in coping with extra loads for the following reasons.

First, chronic hyperinflation forces the inspiratory muscles to work in a less efficient position of their length–tension relationship. Similar to other skeletal muscles, respiratory muscles show a maximum ability to develop force when they are stretched to their maximum length, their force being reduced as far as their fibers shorten.

Second, hyperinflation changes the shape of the diaphragm, reducing the radius of its dome and making it flat and horizontal. This means that a given fiber shortening achieves lower volume displacement, reducing airflow and tidal volume. Moreover, flattening of the diaphragm reduces the apposition area near the costal insertions, which commonly plays an important role in ventilation (Duranti *et al.* 1995).

The third factor that reduces respiratory muscle performance is muscle weakness. This is defined as a reduction in the force developed in response to a given electrical excitation. Muscle weakness in COPD patients is multifactorial, but malnutrition is the main component. Malnutrition has been found in many stable COPD patients, and in an even higher proportion during acute respiratory failure. It is related to an imbalance between energy demand and intake. On one hand, COPD patients show higher basal metabolic rates related to an elevated work of breathing. On the other hand, the intake is commonly reduced by dyspnea and by the anorexia caused by concurrent illnesses. Moreover, chronicity is also important, as is evidenced by the correlation between the decline in forced expiratory volume in 1 s (FEV_1) and body weight. At present, the role of the deconditioning of respiratory muscles by a sedentary lifestyle as a key factor for weakness is under debate. Respiratory muscle training improves performance and dyspnea, but efficacy is lost when it is discontinued (Rochester 1991).

Respiratory muscle performance can be evaluated at the bedside by determining the maximal inspiratory pressure Pi_{max}, which is usually reduced in COPD patients.

Respiratory drive

The relation between the ability of the respiratory muscles to develop force and the effort made by these muscles is mainly controlled by the respiratory drive. It is elevated in stable COPD patients in proportion to the degree of pulmonary alteration. Moreover, in acute respiratory failure the respiratory drive increases to a similar level to that exhibited by normal subjects during heavy exercise, and is a good expression of the effort needed for ventilation (Derenne *et al.* 1988).

The common chemical ventilatory stimuli (i.e. oxygen, carbon dioxide, and pH) still work in COPD patients under acute respiratory failure, but at less sensitive thresholds. Because of the lack of accuracy of minute ventilation, tidal volume, and respiratory rate in describing respiratory drive in these patients, it is better evaluated by electromyography of the diaphragm and occlusion pressure $P_{0.1}$. These parameters have shown higher respiratory drive during acute respiratory failure in COPD patients, but minute ventilation remains constant due to a worsening of the mechanical properties of the respiratory system. This elevated respiratory drive accounts for the lack of efficacy of 'respiratory stimulants' in clinical studies.

In the most extreme cases, very high levels of $Paco_2$ cause obtundation and lead to a reduction in respiratory drive that heralds evolving respiratory arrest.

Conclusions

The pathophysiology of acute respiratory failure in COPD patients is triggered by a variety of precipitating factors, but mainly infections, cardiac failure, and pulmonary embolism. These elements worsen the already impaired $\dot{V}/Q$ mismatch, so that $Paco_2$ decreases despite maintenance of minute ventilation. The respiratory centers respond by increasing respiratory drive, but are unable to enhance ventilation because of worsened respiratory mechanics and reduced respiratory muscle performance.

References

Curtis, J.R. and Hudson, L.H. (1994). Emergent assessment and management of acute respiratory failure in COPD. *Clinics in Chest Medicine*, **15**, 481–500.

Derenne, J.P., Fleury, B., and Pariente, R. (1988). Acute respiratory failure of chronic obstructive pulmonary disease. *American Review of Respiratory Disease*, **138**, 1006–33.

Duranti, R., Misuri, G., Gorini, M., Goti, P., Gigliotti, F., and Scano, G. (1995). Mechanical loading and control of breathing in patients with severe chronic obstructive pulmonary disease. *Thorax*, **50**, 127–33.

Rochester, D.F. (1991). Respiratory muscle weakness, pattern of breathing, and CO_2 retention in chronic obstructive pulmonary disease. *American Review of Respiratory Disease*, **143**, 901–3.

Tantucci, C., Corbeil, C., Chassé, M., Braidy, J., Matar, N., and Milic-Emili, J. (1991). Flow resistance in patients with chronic obstructive pulmonary disease in acute respiratory failure. Effects of flow and volume. *American Review of Respiratory Disease*, **144**, 384–9.

2.8.2 Therapeutic strategy

P. Jolliet and J.-C. Chevrolet

Key messages

- Controlled oxygen therapy is used to increase Pao_2 to 8 kPa without carbonarcosis.
- Beta-2 agonists and ipratropium bromide are combined in aerosolized form.
- Corticosteroid is administered only to known 'responders' or patients already taking corticosteroids.
- Aminophylline, respiratory stimulants, and digitalis are of little use.
- Consider an early trial of non-invasive ventilation (pressure support preferred).
- In acute life-threatening situations (respiratory arrest, cardiovascular collapse) or deterioration despite treatment, resort to intubation and controlled mechanical ventilation.
- Pitfalls of mechanical ventilation are alkalosis, air trapping ('intrinsic' positive end-expiratory pressure), barotrauma, and difficult and protracted weaning.
- During difficult weaning, always consider factors contributing to respiratory muscle fatigue.

Conservative treatment

A detailed discussion is given by Derenne *et al.* (1988) and Ziment (1990).

Oxygen therapy

Patients with acute respiratory failure in the setting of chronic obstructive pulmonary disease frequently exhibit severe hypoxemia, with Pao_2 often decreasing below 6 kPa. Thus one of the therapeutic priorities should be to raise Pao_2 by O_2 therapy. However, aspects specific to this group of patients should be stressed to guide the optimal use of O_2. First, chronic obstructive pulmonary disease patients are chronically hypoxemic, and therefore develop counter-regulatory mechanisms such as increased hematocrit and enhanced O_2 extraction by the tissues. Hence the tolerance of these patients to acute hypoxemia is improved. Second, overzealous administration of O_2 can induce considerable worsening of hypercapnia through complex mechanisms involving altered ventilation–perfusion relationships, the Haldane effect, and a decrease in minute ventilation. This acute rise in $Paco_2$ can result in a further lowering of pH, as well as in a clinical entity known as 'carbonarcosis', presenting as agitation, drowsiness, or coma. Third, the lower range of Pao_2 discussed here corresponds to the steep portion of the hemoglobin dissociation curve. Hence even a modest gain in Pao_2 entails a substantial rise in arterial O_2 saturation and content. Consequently, the main goal of O_2 therapy should be to increase Pao_2 to approximately 8 kPa, which marks the end of the steep portion and the beginning of the flattened portion of the hemoglobin dissociation curve. To attain this objective, the inspired fraction of O_2 (Fio_2) should be carefully titrated upwards ('controlled oxygenation') by increments of 0.04 from an initial level of 0.24 or 0.28. Controlled oxygenation can be performed using nasal cannulas or Venturi masks. With the former, an increase in O_2 flow rate of 1 l/min raises Fio_2 by 0.04, provided that the patient's minute ventilation remains constant. With Venturi masks, Fio_2 is determined by the fixed relationship between O_2 and room air flow rates. The O_2 flow rate and resultant Fio_2 are usually indicated on interchangeable color-coded adapters fitted to the mask. Arterial blood gases should be assessed every 30 min. If Pao_2 remains under the target value, $Paco_2$ has not increased by more than 2 kPa, and the patient does not exhibit signs of carbonarcosis, Fio_2 should be increased to the next highest level. In severe cases of CO_2 retention after Fio_2 has been increased, Fio_2 must be lowered to the preceding level. Patients should not be returned to room air in the hope of accelerating the decrease in $Paco_2$, as severe hypoxemia due to increased CO_2 stores will result. If it is impossible to raise Pao_2 to the target levels without major CO_2 retention, non-invasive ventilation or intubation should be performed.

Bronchodilators

The major benefit expected from bronchodilators is a decrease in the reversible component of bronchial obstruction, leading to improved alveolar ventilation and gas exchange, and reduced work of breathing. The most commonly used bronchodilators are adrenergic β_2-receptor agonists, the anticholinergic drug ipratropium bromide, and methylxanthines.

Beta-2 agonists

Beta-2 agonists, such as salbutamol (albuterol) or terbutaline, can be given by oral, aerosolized, subcutaneous, or intravenous routes. There is no clear evidence that one agent is superior to the others in the acute setting. Long-acting preparations should be reserved for chronic treatment to avoid accumulation of repeated doses to toxic levels. Aerosols entail greater bronchodilatation than oral or subcutaneous routes, and should be preferred. Metered-dose inhalers or compressed air nebulizers can be used; the latter are a better choice if patient co-operation is limited. Aerosols should initially be administered every 4 h. The recommended doses for each nebulization are 1.25 to 2.5 mg for salbutamol and 2.5 to 5 mg for terbutaline. In patients incapable of performing aerosol therapy, the continuous intravenous route is preferable to subcutaneous injections because of the more rapid onset of action. A typical regimen for salbutamol is to begin the infusion at 0.07 mg/kg/min, and then to increase the rate up to a maximum of 0.3 mg/kg/min if severe obstruction persists. Side-effects of β_2-agonists are usually dose dependent, and include tremor, restlessness, tachycardia, arrhythmias, hypokalemia, worsening of ventilation–perfusion ratios, and peripheral vasodilatation (usually in high-dose intravenous administration).

Ipratropium bromide

The bronchodilating efficacy of ipratropium bromide is possibly superior to that of β_2-agonists in some chronic obstructive pulmonary disease patients. Furthermore, combining this drug with a β_2-agonist can result in an additive bronchodilator effect compared with ipratropium bromide alone. Ipratropium bromide is given by a metered-dose inhaler or a compressed air nebulizer. The usual dose for each nebulization is 0.125 mg, but up to 0.25 to 0.5 mg can be administered if obstruction is severe.

Aminophylline

The use of aminophylline in acute on chronic respiratory failure is controversial and should best be avoided. Indeed, its bronchodilator effect is inferior to that of β_2-agonists, and no benefit of combining the two agents has been demonstrated. Furthermore, even though aminophylline could increase diaphragmatic contractility in both normal humans and chronic obstructive pulmonary disease patients, there is no clear-cut evidence that it can prevent the respiratory muscle fatigue which often leads to mechanical ventilation. Finally, it has a narrow therapeutic margin with a substantial risk of arrhythmia, particularly if β_2-agonists are administered concomitantly.

Corticosteroids

Theoretical benefits derived from corticosteroids include a reduction in bronchial inflammation and hyper-responsiveness, as well as potentiation of the bronchodilating effect of β_2-agonists. However, only about 20 per cent of stable chronic obstructive pulmonary disease patients are 'responders', i.e. significantly improve their forced expiratory volume at 1 s (FEV_1) when receiving corticosteroids. Furthermore, published data on their efficacy in acute respiratory failure are scarce and controversial. Potential acute side-effects include gastric ulcer and bleeding, hypokalemia, sodium retention, and psychiatric disorders. Hence, we prefer to reserve these for patients who are known 'responders' and those who were taking corticosteroids prior to admission. When major bronchospasm is present, we use prednisolone at an initial dose of 1 mg/kg every 6 h during the first 2 days, tapering off over a few days to 1 mg/kg/day which is subsequently lowered according to the clinical course.

Other treatments

Antibiotics

Lower respiratory tract infections which cause the majority of episodes of acute respiratory failure are not always of bacterial origin. However, the bulk of evidence suggests that empirical antibiotic treatment directed against Gram-positive and Gram-negative organisms should be initiated when worsening dyspnea is accompanied by an increase in the purulence or production of sputum. Amoxicillin (amoxycillin) plus clavulanic acid or cefuroxime are effective against the pathogens most commonly identified in these patients (*Streptococcus pneumoniae*, *Hemophilus influenzae*, and *Moraxella catarrhalis*). However, the emergence of penicillin-resistant strains of *S. pneumoniae* warrants caution, and the choice of antibiotic must be guided by the geographical and epidemiological distribution of such strains and by local epidemiological data. Ceftriaxone is an alternative in this situation. When the chest radiograph suggests pneumonia, adding erythromycin in case of *Legionella pneumophilia* infection should be considered until sputum Gram stain and cultures are obtained.

Digitalis

Digitalis has no place in the treatment of cor pulmonale, and should only be used to treat acute supraventricular tachyarrhythmias such as atrial fibrillation.

Heparin

Either standard or low-molecular-weight heparin should be administered to prevent deep venous thrombosis. The dose should be 5000 units two to three times daily or 2500 units once daily respectively.

Nutrition

Many chronic obstructive pulmonary disease patients exhibit an increased resting metabolic rate, and nearly 50 per cent of those admitted to hospital with acute respiratory failure show some degree of protein calorie malnutrition. Nutritional support should aim to meet the raised metabolic needs of the acute episode, while a thorough assessment of the patient's nutritional status should guide objectives for the long term. Care should be taken not to administer large carbohydrate loads, as they lead to increased CO_2 production and O_2 consumption. Initially, patients should receive approximately 30 kcal/kg/day, not more than 40 per cent of which should consist of carbohydrates. The enteral route is preferred.

Respiratory stimulants

Hypercapnia in decompensated chronic obstructive pulmonary disease patients results from mechanisms such as increased dead-space ventilation, respiratory muscle fatigue, and increased CO_2 production in the presence of a considerable rise in the respiratory drive. Hence, agents leading to further increase in respiratory drive, such as doxapram or medroxyprogesterone, are not useful and can worsen dyspnea.

Mechanical ventilation

The main goal of mechanical ventilation is the support of gas exchange and fatigued respiratory muscles while the precipitating causes of acute respiratory failure are corrected. Before the advent of non-invasive ventilation, mechanical ventilation with endotracheal intubation was used for two categories of patients: those in life-threatening situations such as respiratory or cardiac arrest, extreme exhaustion, or major neurological alterations; those whose condition worsened despite conservative management. However, in the latter category, the early use of non-invasive ventilation can reduce the need for intubation from 75 to 25 per cent and possibly improve survival (Brochard *et al.* 1995). Therefore, since mechanical ventilation with intubation increases morbidity and mortality, a trial of non-invasive ventilation should be performed in patients with respiratory failure and intubation reserved for those unable to co-operate or whose condition continues to deteriorate.

Non-invasive ventilation

The term 'non-invasive ventilation' applies to various techniques not implying endotracheal intubation or tracheotomy, and includes negative- and positive-pressure ventilation. Positive-pressure ventilation is clearly preferred in the acute setting. Pressure support administered with a face mask or intermittent positive-pressure ventilation through a nasal mask are the two most widely used methods. The former is probably the technique of choice in chronic obstructive pulmonary disease patients, as patient–ventilator interaction is better tolerated than with the latter (Brochard *et al.* 1995). The initial level of pressure support should be around 20 cmH_2O, subsequently adjusted to patient tolerance and arterial blood gases. Trigger sensitivity should be maximum, to reduce inspiratory muscle work. The most frequent complications are patient discomfort and dyspnea, excessive air leaks at the junction between mask and face, gastric distension due to air ingestion, and pressure sores along the mask lining.

If pressure support fails, intermittent positive-pressure ventilation with a nasal mask should be attempted. The assist–control mode is preferable, with initial settings of 10 ml/kg tidal volume, 15 to 20 breaths/min respiratory rate, trigger level at maximum sensitivity, and an inspiratory–expiratory time ratio of 1:3. The complications are similar to those observed with pressure support, but air leaks around the nasal mask and pressure sores on the nose are added problems. Readily available masks can be unsuited to the patient's anatomy, but this difficulty can often be overcome by tailor-molding a mask from a special silicon paste.

Air trapping and intrinsic positive end-expiratory pressure (PEEP) due to incomplete emptying of the lung during expiration because of airflow obstruction is often present in these patients (Rossi *et al.* 1995). Hence, alveolar pressure is positive rather than atmospheric at the start of inspiration. Successful triggering of the delivery of either pressure support or a ventilator-controlled breath requires the generation of a spontaneous inspiratory flow, which implies that the alveolar pressure must become subatmospheric. More inspiratory muscle work is needed to achieve this if end-expiratory alveolar pressure is positive rather than atmospheric. Applying a PEEP of 5 to 10 cmH_2O can reduce the added work required to trigger the ventilator (Fig. 1).

At present, there are no clearly defined criteria for initiating non-invasive ventilatory support. Based on our experience, a patient whose condition is severe enough to warrant intensive care admission but who does not meet the criteria for immediate intubation discussed above should receive a trial of non-invasive ventilation.

Mechanical ventilation with intubation

The intensive care mortality in intubated severe chronic obstructive pulmonary disease ($FEV_1 < 1$ liter) patients varies between 10 and 30 per cent, with 1-year survival approximately 50 per cent. Prognostic criteria of a fatal outcome are scarce and their predictive ability is at best 70 to 80 per cent. Hence the decision to intubate or not must rest on patient history, in particular previous intubation with difficult weaning, and assessment of such difficult issues as quality of life before admission, nutritional state, age, and known FEV_1 in stable conditions. An endotracheal tube with an internal

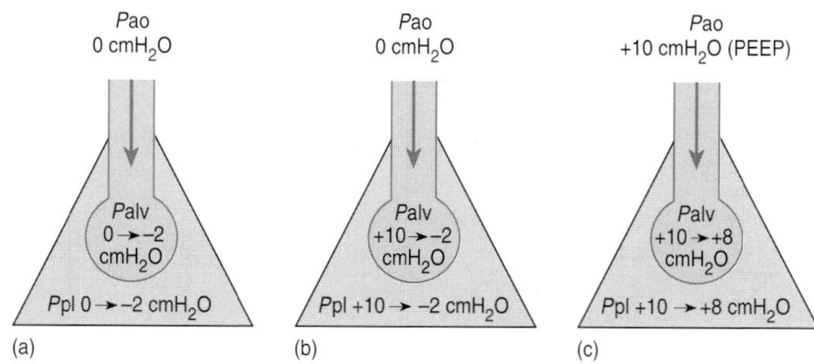

Fig. 1 Effects of intrinsic PEEP on respiratory muscle work required to trigger the ventilator. A 2-cmH_2O drop in alveolar pressure *P*alv and pleural pressure *P*pl below pressure at the mouth *P*ao is assumed to be necessary to induce an inspiratory flow sufficient for triggering. All pressures are expressed relative to atmospheric (0 cmH_2O). (a) In the absence of air trapping, *P*alv and *P*ao are atmospheric before inspiration. The 2-cmH_2O decrease in *P*pl and *P*alv generates the inspiratory flow. (b) Owing to air trapping, *P*alv is positive ('intrinsic PEEP'+ 10 cmH_2O) before inspiration. If *P*ao is atmospheric, *P*pl and *P*alv will have to decline by 12 cmH_2O to initiate an inspiratory flow. This will result in considerable added inspiratory muscle work. (c) If a PEEP equal to intrinsic PEEP is applied by the respirator, a drop of only 2 cmH_2O in *P*pl and *P*alv will be required for inspiration to start, thus reducing respiratory muscle work.

diameter of at least 8 mm reduces the work of breathing and allows flexible bronchoscopy to be performed if necessary. Most patients exhibit respiratory muscle fatigue at this stage. Controlled ventilatory modes are preferred for at least 48 h, or until the precipitating event is corrected; the goal is to rest the fatigued respiratory muscles before weaning can start. Sedation with a continuous infusion of a short-acting benzodiazepine, such as midazolam, combined with morphine is suitable. Muscle relaxants should be reserved for situations where ventilation is hampered by excessive rib-cage stiffness or respiratory muscle activity. Initially, a low respiratory rate (10 breaths/min) and tidal volume (8 ml/kg), a high inspiratory flow (60 l/min), and an inspiratory–expiratory time ratio of 1:3 should be used. There are two reasons for using this hypoventilation approach: first, the risk of air trapping is reduced; second, in the presence of increased bicarbonate levels often found in these patients, excessive alveolar ventilation resulting in normo- or hypocapnia can lead to severe alkalosis, with resultant arrhythmias or convulsions. Inspiratory pressures should be monitored to avoid barotrauma.

No consensus exists as to the optimal timing of tracheotomy in patients undergoing protracted ventilation. The low-pressure cuffs of modern endotracheal tubes cause much less damage than the older high-pressure type and can remain in place longer, thus avoiding the small added morbidity of tracheotomy. However, tracheotomy improves patient comfort by allowing eating, drinking, and phonation. Our policy is to perform tracheotomy after 21 days of mechanical ventilation.

The weaning process can be very difficult in chronic obstructive pulmonary disease patients, probably because of respiratory muscle fatigue. Many criteria have been published aimed at determining the optimal timing, clinical or laboratory parameters, and technique for weaning. However, the capacity for an individual patient to resume sustained breathing without any mechanical assistance ultimately depends on the outcome of trials of decreasing ventilatory support. As inspiratory muscle fatigue is probably central to the failure of these trials (Roussos and Zakynthinos 1996), factors that decrease strength and endurance or increase the load placed on these muscles should be identified and corrected (Table 1). Our approach, once the precipitating causes have been corrected, is to stop sedation while the patient is placed in the assist–control mode, followed by synchronized intermittent ventilation modes with pressure support levels of 30 cmH_2O or higher. Pressure support is then used for longer periods of time, with mechanical ventilation being reinstituted if signs suggesting respiratory muscle fatigue develop (respiratory rate of 30 breaths/min or more, intense accessory muscle activity, and acute rise in $Pa\text{CO}_2$ with decreasing pH). Subsequently, the level of pressure support is reduced. Once the patient can breathe at a pressure support level of 10 cmH_2O, a T-piece trial is performed. There is no consensus as to the optimum duration of such a trial before extubation is decided. Indeed, the shorter the trial, the greater is the risk of reintubation, and the longer the trial, the greater is the risk of respiratory muscle fatigue due to the increased work of non-assisted breathing through an endotracheal tube. The optimum time is probably between 2 and 8 h. Our approach is to extubate after at least 2 h of T-piece breathing. In the case of postextubation respiratory distress due to increased secretions, dyspnea, or patient anxiety, non-invasive ventilation can be temporarily useful to avoid reintubation. Patients should be monitored in the intensive care unit for 24 h following successful extubation.

Table 1 Main factors contributing to respiratory muscle fatigue during weaning

Increased respiratory drive	Decreased muscle capacity	Increased load
Pain	Atrophy	Bronchospasm
Anxiety, agitation	Malnutrition	Bronchial secretions
Fever	Hyperinflation	Pneumothorax
Metabolic acidosis	Acidosis	Atelectasis, pneumonia
Hypoxemia	Hypoxemia	Hyperinflation/intrinsic PEEP
Increased CO_2 production	Infection	Left ventricular failure
Excessive protein intake	PO_4^-, Mg^{2+}, K^+ depletion	Endotracheal tube resistance
	Severe anemia	Ventilator circuit resistance
	Insufficient level of pressure support	

References

Brochard, L., *et al.* (1995). Noninvasive ventilation for acute exacerbations of chronic obstructive pulmonary disease. *New England Journal of Medicine*, **333**, 817–22.

Derenne, J.P., Fleury, B., and Pariente, R. (1988). Acute respiratory failure of chronic obstructive pulmonary disease. *American Review of Respiratory Disease*, **138**, 1006–33.

Rossi, A., Polese, G., Brandi, G., and Conti, G. (1995). Intrinsic positive end-expiratory pressure. *Intensive Care Medicine*, **21**, 522–36.

Roussos, C. and Zakynthinos, S. (1996). Fatigue of the respiratory muscles. *Intensive Care Medicine*, **22**, 134–55.

Ziment, I. (1990). Pharmacologic therapy of obstructive airway disease. *Clinics in Chest Medicine*, **11**, 461–86.

2.9 Respiratory acidosis

2.9.1 Pathophysiology and causes

L. Gattinoni and A. Lissoni

Key messages

- Respiratory acidosis is a process in which the low pH is primarily due to a rise in $P\text{CO}_2$.
- In critically ill patients the CO_2 load may be greatly increased due to hypermetabolism, excess parenteral nutrition, or titration of HCO_3^- by fixed acids.
- The arterial $P\text{CO}_2$ for a given acid load is a function of alveolar ventilation and cardiac output.
- CO_2 is buffered by fixed buffers, i.e. hemoglobin, phosphates, and proteins.
- CO_2 buffering *in vivo* leads to an increase in HCO_3^- of 1 mmol/l for every increase of 10 mmHg (1.33 kPa) in $P\text{CO}_2$ above 40 mmHg (5.33 kPa).
- The decrease in pH stimulates the generation of more HCO_3^- in the tubular kidney cells by increased ammonium ion production. This mechanism leads to an increase in plasma HCO_3^- of approximately 3.5 mmol/l for every increase of 10 mmHg (1.33 kPa) in $P\text{CO}_2$ above 40 mmHg (5.33 kPa).
- Every step of acid–base regulation may be affected in critically ill patients: abnormal increase in acid load, decreased concentration of fixed buffers, altered cardiac output, inadequate alveolar ventilation, and impaired production of HCO_3^- in the presence of renal failure.
- Respiratory failure may orginate from changes in the central nervous system, the conductive system (airways), the lung parenchyma, the inspiratory muscles, or the chest wall.

Introduction

Although carbon dioxide (CO_2) is not an acid *per se*, it becomes a proton (H^+) donor when it is combined with water:

$$CO_2 + H_2O \Leftrightarrow H_2CO_3 \Leftrightarrow H^+ + HCO_3^-. \qquad (1)$$

When $P\text{CO}_2$ increases above its physiological levels (i.e. above 40 to 45 mmHg (5.33 to 6 kPa), corresponding to 1.2 to 1.35 mmol/l), acidemia usually develops. Thus respiratory acidosis can be defined as a process in which the tendency for the arterial pH to decrease is primarily due to an increase in $P\text{CO}_2$. However, it is important to note that, although this process is usually associated with acidemia, there are circumstances in which the pH may be maintained within near-physiological limits owing to the concurrent increase in bicarbonate HCO_3^-. In fact, the pH, which reflects the free H^+ activity in the blood, is a function of the ratio of HCO_3^- to CO_2 and not of their absolute values.

The CO_2 load

The amount of CO_2 produced metabolically in a normal male is approximately 200 to 250 ml/min, corresponding to an acid load of 13 000 to 16 000 mmol/day. However, because of increased metabolism, this load may be increased by 30 to 100 per cent depending on the basic disease (burns, infection, sepsis, etc.) (Long *et al.* 1979). An excessive caloric intake, particularly during parenteral nutrition, may also greatly increase the CO_2 production (Covelli *et al.* 1981).

Another possible source of CO_2 is provided by HCO_3^- when titrated by fixed acid, such as lactic acid (LAH):

$$\text{LAH} + HCO_3^- \rightarrow \text{LA}^- + CO_2 + H_2O. \qquad (2)$$

This mechanism is usually underestimated, but may be relevant in critically ill patients where metabolic acidosis frequently occurs. For example, in a closed system, a decrease in HCO_3^- of 1 mmol/l would cause an increase of 33.3 mmHg (4.44 kPa) in $P\text{CO}_2$ (the solubility coefficient of CO_2 is 0.003 ml/mmHg (0.023 ml/kPa)).

Excretion of CO_2

The acid load excreted by the lungs can be expressed as

$$\dot{V}\text{CO}_2 = \dot{V}_A \times K \times P\text{aCO}_2 \qquad (3)$$

where $\dot{V}\text{CO}_2$ is the CO_2 excreted, $\dot{V}_A$ is the alveolar ventilation, $P\text{aCO}_2$ is the partial pressure of CO_2 in arterial blood, and K is a factor converting pressure to volume fraction.

Equation (3) shows that, to maintain a normal $P\text{aCO}_2$ (and pH) when the volatile acid load $\dot{V}\text{CO}_2$ is increased, the alveolar ventilation must rise by the same proportion. If the alveolar ventilation cannot increase, the same acid load is eliminated at the cost of higher $P\text{aCO}_2$. In the steady state (i.e. unmodified $P\text{aCO}_2$), the $\dot{V}\text{CO}_2$ excreted by the lung equals the metabolic CO_2 produced.

However, CO_2 excretion is a function not only of alveolar ventilation, but also of cardiac output. According to Fick's law

$$\dot{V}\text{CO}_2 = Q(C\text{vCO}_2 - C\text{aCO}_2) \qquad (4)$$

where Q is the cardiac output, and $C\text{vCO}_2$ and $C\text{aCO}_2$ are respectively the CO_2 contents of mixed venous and arterial blood. Assuming, for simplicity, that the relationship between CO_2 content and partial pressure is linear in the range of physiological interest, and combining and rearranging eqns (3) and (4), it follows that

$$P\mathrm{a}CO_2 = \int \frac{Q}{\dot{V}_A + Q} P\bar{v}CO_2. \quad (5)$$

This equation shows that $P\mathrm{a}CO_2$ (which affects the acid–base status of the blood) for a given acid load and $P\bar{v}CO_2$ increase when cardiac output increases and decrease when alveolar ventilation increases. Equation (5) emphasizes the importance of the transport Q of the acid load from the tissues to the lungs. In fact, for a given acid load, which primarily affects $P\bar{v}CO_2$, if Q decreases and $\dot{V}_A$ remains constant, $P\mathrm{a}CO_2$ (and the arterial acid–base status) may be normal or even decrease, but $P\bar{v}CO_2$ will increase substantially, resulting in large $P\mathrm{a}CO_2 - P\bar{v}CO_2$ and pH gradients. In some circumstances, often encountered in critically ill patients (e.g. low flow states), arterial sampling may not give an accurate assessment of the acid–base status, and sampling of mixed venuous blood is mandatory (Weil *et al.* 1986).

In summary, the respiratory component of the acid–base status, i.e. $P\mathrm{a}CO_2$, is a function of the acid load, the alveolar ventilation, and the cardiac output.

Mechanism of the buffer system

Of the 15 000 to 30 000 mmol of protons delivered daily to the extracellular fluid, only 40 to 60 nmol/l are found free in the blood; the remainder are bound by the buffer system. Buffers, which are present in extracellular and intracellular fluids, are substances which are able to bind or release H^+, thus preventing large changes in free H^+ concentration. In general, buffers are weak acids and their ionized salts. At a given pH the relative concentrations of base (able to accept H^+) and acid (able to release H^+) are defined by the Henderson–Hasselbalch equation:

$$\mathrm{pH} = \mathrm{p}K + \log\frac{[A^-]}{[AH]} \quad (6)$$

where $[A^-]$ and $[AH]$ are the concentrations of dissociated base and undissociated acid respectively. The maximum efficiency of a given buffer occurs when pK = pH. In these conditions, the concentrations of A^- and AH are the same, and the buffer system is equally able to bind or release H^+. Also, the efficiency of a given buffer pair increases with increasing concentration.

The pK values and the concentrations of the various extracellular buffers are listed in Table 1. Three buffer systems (phosphates, proteins, and hemoglobin) have similar pK values (about 6.8). Thus they can be grouped together as a single buffer. The HCO_3^-–CO_2 buffer pair has a less favorable pK (6.1); however, it is the most important buffer pair in the body as it can be regulated independently by the lungs (CO_2 elimination) and by the kidneys (HCO_3^- reabsorption).

To understand the mechanism of buffering, it is convenient first to analyze the effects of an acid load in a closed system (e.g. venous blood before it reaches the lungs). When an acid load is delivered to a closed system, buffering is almost immediate. Quantitatively, it is interesting to know how a given acid load is shared by the different buffer pairs depending on their pK. The regulation of acid–base sharing between the different buffers is based on the following equilibrium:

$$\mathrm{pH} = 6.1 + \log\frac{[HCO_3^-]}{[CO_2]} = 6.8 + \log\frac{[Pr^-]}{[HPr]}$$

$$= 6.8 + \log\frac{[Hb^-]}{[HHb]} = 6.8 + \log\frac{[HPO_4^-]}{[H_2PO_4]} \quad (7)$$

where Pr^-, Pr, Hb^-, HHb, HPO_4^-, and H_2PO_4 are proteins, hemoglobin, and phosphates respectively in their dissociated and undissociated forms.

Since the pK values of proteins, hemoglobin, and phosphate are similar (although not exactly the same), these three systems can be grouped together (A^-–AH buffer pair). Thus the equilibrium simplifies to

$$\mathrm{pH} = 6.1 + \log\frac{[HCO_3^-]}{[CO_2]} = 6.8 + \log\frac{[A^-]}{[AH]}. \quad (8)$$

When normal values are substituted, the normal equilibrium can be written as

$$7.40 = 6.1 + \log\frac{24}{1.2}(\text{i.e. } PCO_2 = 40\,\text{mmHg}) = 6.8 + \log\frac{84.8}{21.2}. \quad (9)$$

Response to the CO_2 load

When there is an abnormal increase in the CO_2 load because of hypermetabolism, HCO_3^- titration, or CO_2 retention, the first response is buffering by the A^-–AH pairs, of which hemoglobin is quantitatively the main component under normal conditions. Titrating the A^-–AH buffer pairs with H_2CO_3 leads to an increase in HCO_3^-. The *in vivo* HCO_3^- increase due to this physiochemical mechanism is approximately 1 mmol/l HCO_3 for every increase of 10 mmHg (1.33 kPa) in PCO_2 above 40 mmHg (5.33 kPa) (Brackett *et al.* 1965; Weill *et al.*

Table 1 The blood buffers

Buffer pairs	Normal concentration (mmol/l)	pK	Base-to-acid ratio at pH 7.4
Bicarbonate–carbonic acid	25	6.1	$[HCO_3^-]/[CO_2] = 20/1$
Dibasic–monobasic phosphate	2	6.8	$[HPO_4^{2-}]/H_2PO_4^- = 4/1$
Proteins	14 (7 g/dl)	6.8	$Pr^-/HP = 4/1$
Hemoglobin	90 (15 g/dl)	6.8	$Hb^-/HHb = 4/1$

1986). As pH is determined by the ratio of [HCO_3^-] to [CO_2], the increase in HCO_3^- associated with the rise in $P\text{CO}_2$ limits the decrease in pH. However, both the increased $P\text{CO}_2$ and the decreased pH trigger the physiological control, which involves the central nervous system and the kidneys. The increased $P\text{CO}_2$ and the low pH stimulate the chemosensitive regions of the respiratory center in the brainstem and the peripheral chemoreceptors in the carotid bodies (Cunningham *et al.* 1986). The result is an increased respiratory drive which normally leads to increased alveolar ventilation.

This mechanism is effective when the CO_2 increase is due to hypermetabolism or HCO_3^- titration by fixed acid, with an intact pulmonary function. If the cause is primarly hypoventilation or the patient is receiving fixed mechanical ventilation, $P\text{CO}_2$ will rise until a new equilibrium is reached. The decreased pH in the tubular kidney cells stimulates the generation of NH_3 which combines with the H^+ secreted in lumen of the distal segments to form the ammonium ion NH_4^+ which is trapped in the lumen. As an OH^- ion is produced for each H^+ ion secreted (both derive from the dissociation of water catalyzed by carbonic anhydrase), new HCO_3^- is generated according to

$$OH^- + \text{cellular } CO_2 \rightarrow HCO_3^- \qquad (10)$$

and secreted in the peritubular capillaries. The generation of new HCO_3^- takes a few days to complete and leads to a further increase in HCO_3^- with a consequent rise in pH.

In vivo, in this chronic situation, an increase of 10 mmHg (1.33 kPa) in $P\text{CO}_2$ above 40 mmHg (5.33 kPa) leads to an increase in plasma HCO_3^- of approximately 3.5 mmol/l. The increase in HCO_3^- in respiratory failure is due to two mechanisms: buffering by A^-–AH buffer pairs, and generation of new HCO_3^- by tubular cells.

Acid–base regulation in critically ill patients

Every step of acid–base regulation may be affected in critically ill patients.

Table 2 The most common causes of ventilatory impairment

Central nervous system	Drugs, opiates, anesthetics Increased oxygenation in chronic hypercapnia Central sleep apnea
Chest wall mechanics	Kyphoscoliosis Obesity
Respiratory muscle impairment	Electrolyte disorders, paralysis, myasthenia, Guillain–Barré syndrome, etc.
Upper airway obstruction	Laryngospasm, foreign body inhalation
Alterations in lung parenchyma	Asthma, COPD, ARDS, pneumonia, pulmonary edema, pneumohemothorax, emphysema

COPD, chronic obstructive pulmonary disease; ARDS, acute respiratory distress syndrome.

1. The acid load may be altered by both the underlying disease (hypoperfusion, shock, increased metabolism, etc.) and therapeutic intervention (excessive parenteral feeding).

2. The buffer system is frequently abnormal in critically ill patients; decreased levels of hemoglobin, the most important component of the A^-–AH buffer pair, are not unusual. A decreased concentration of both A^-–AH and HCO_3^-–CO_2 buffer pairs implies a greater change in pH for a given acid load.

3. Transport of blood from the venous to the arterial side may also be affected (low-flow states). In these conditions the CO_2 clearance may be altered, resulting in a large difference between the acid–base status of the venous and arterial blood, with increases in the $P\text{CO}_2$ and pH gradients.

4. The physiological response to increased $P\text{CO}_2$ and decreased pH, i.e. increased ventilation, is usually impaired in critically ill patients for two main reasons: the underlying disease may affect the lung, thus preventing increased ventilation, and these patients are usually on mechanical ventilation. If volume-controlled ventilation is used, $P\text{CO}_2$ will rise as the patient (usually sedated and sometimes paralyzed) cannot change either the tidal volume or the respiratory rate. If pressure-support ventilation is used, the patient may respond to the increase in $P\text{CO}_2$ by triggering mechanical breathing more frequently. However, the depth of breath (i.e. the tidal volume) is often out of the patient's control. Thus it is obvious that mechanical ventilation has a deleterious effect on the physiological response to changes in acid–base status.

5. The physiological response to decreased pH in the renal tubular cells, in which HCO_3^- losses are replaced, may also be impaired in the critically ill, whose renal function is often affected. It is also important to realize that some forms of renal support, as hemofiltration, may lead to HCO_3^- losses, with consequent additional derangement of the acid–base balance.

In summary, it is important to remember that the physiological control of the acid–base equilibrium is often impaired in critically ill patients, and physicians must understand which mechanisms are altered so that an adequate substitute can be provided for the physiological control which has been lost.

Causes of acute and chronic respiratory acidosis

The causes of increased CO_2 may be excessive CO_2 load, hypermetabolism, HCO_3^- titration, or, more commonly, reduction of alveolar ventilation. An excessive CO_2 load rarely causes increased $P\text{CO}_2$ in spontaneously breathing subjects, as the CO_2 is a strong stimulus to ventilation. However, an extra CO_2 load may cause respiratory acidosis in mechanically ventilated patients.

The most common cause of increased $P\text{CO}_2$ is CO_2 retention which may be due to changes in any of the organs involved in normal ventilation: the central nervous system, the conductive system (airways), the lung parenchyma, the inspiratory muscles, and the chest wall.

The most common causes of CO_2 retention, which may be acute or chronic, are summarized in Table 2.

References

Brackett, N.C., Cohen, J.J., and Schwartz, W.B. (1965). Carbon dioxide titration curve of normal man: effect of increasing degrees of acute hypercapnia on acid–base equilibrium. *New England Journal of Medicine*, **272**, 6–10.

Covelli, H.D., Black, J.W., Olsen, M.S., and Beekman, J.F. (1981). Respiratory failure by high carbohydrate loads. *Annals of Internal Medicine*, **95**, 569–81.

Cunningham, D.J.C., Robbins, P.A., and Wolff, C.B. (1986). Integration of respiratory responses to changes in alveolar partial pressures of CO_2 and O_2 and in arterial pH. In *Handbook of physiology* (ed. A.P. Fishman, N.S. Cherniack, J.G. Widdicombe, and R.G. Stephen), pp. 475–528. American Physiological Society, Bethesda, MD.

Long, G.L., *et al*. (1979). Metabolic response to injury and illness: estimation of energy and protein needs from indirect calorimetry and nitrogen balance. *Journal of Parenteral and Enteral Nutrition*, **3**, 452–6.

Weil, M.H., Rackow, E.C., Trevino, R., Grundler, W., Falk, J.L., and Griffel, M.I. (1986). Difference in acid–base state between venous and arterial blood during cardiopulmonary resuscitation. *New England Journal of Medicine*, **315**, 153–6.

2.9.2 Therapeutic strategy

L. Gattinoni and A. Lissoni

Key messages

- If respiratory acidosis is present, identify the precipitating factors and assess whether the symptoms (hypoxemia, hypercapnia, and acidosis) are themselves life threatening.
- There are no threshold values for $Pa\text{O}_2$, $Pa\text{CO}_2$, and pH which dictate symptomatic correction. The indications for mechanical support are based on global clinical assessment (age, consciousness, cardiovascular stability, and trend of respiratory impairment).
- Assessment of the precipitating factors and the time required for their correction is of paramount importance in planning therapeutic strategy.
- The target blood gases during respiratory support must be established according to the previous respiratory conditions.
- Since hyperinflation of the lung, with its cardiovascular consequences, is the most harmful complication of mechanical ventilation in chronic respiratory acidosis, it is essential to set a small tidal volume associated with an adequate expiratory time.
- Non-invasive ventilation appears to be a promising treatment of chronic respiratory acidosis.
- Infusion of HCO_3^- to correct acidosis may be appropriate only if the decrease in pH is itself harmful, the infusion rate is titrated against the increase in $Pa\text{CO}_2$, the ventilation can be increased to clear the extra CO_2 generated, and the cardiac output is adequate. In these conditions HCO_3^- administration is equivalent to an acceleration of the natural process in the kidneys (if renal control is adequate).

Introduction

Hypercapnia with acidosis and with or without hypoxemia is common in the intensive care unit (ICU). There are many etiologies which may lead to respiratory acidosis. Some are immediately reversible, some can be corrected in hours or days, and some are irreversible. We will discuss the indications for respiratory support in these conditions, and how and to what degree hypercapnia, acidosis, and hypoxemia should be corrected. As mechanical ventilation may cause further complications, the disadvantages of introducing such a dangerous symptomatic treatment must be weighed against the benefits expected in relieving hypercapnia, acidosis, and hypoxemia. The therapeutic strategy is to design, for each patient, an appropriate plan of care, in which mechanical ventilation is probably the most important symptomatic treatment.

Approach to respiratory acidosis

In patients presenting with hypercapnia, acidosis, and hypoxemia, two lines of action are required:

(1) identification of the causes of respiratory acidosis, with particular focus on the correction of precipitating factors which may be reversible;

(2) treatment of the symptoms if they are themselves a possible cause of unfavorable outcome.

These two actions should be pursued together, as correction of the precipitating factors may lead to almost immediate resolution of the respiratory acidosis. However, the first goal in intensive care is maintenance of homeostasis, and correction of life-threatening symptoms is the priority. Thus the indications for symptomatic treatment will be discussed first.

Hypoxemia, hypercapnia, and acidosis

Hypoxemia

Hypoxemia is the most harmful symptom occurring during respiratory failure and usually needs correction. However, there is not a fixed threshold of 'harmful' $Pa\text{O}_2$, as it depends on previous history, age, and hemodynamic conditions. What is most important is the global clinical assessment, i.e. the presence of signs suggesting inadequate oxygenation such as increased autonomic outflow, impairment of the central nervous system, etc. During acute respiratory acidosis, most of symptoms caused by inadequate oxygenation may also be due to hypercapnia and acidosis. However, hypoxemia must be corrected first.

Hypoxemia may occur in the presence of either low or high partial alveolar oxygen pressure $PA\text{O}_2$. Low $PA\text{O}_2$ is unavoidable during CO_2 retention when the patient breathes room air since

$$FA\text{O}_2 = Fi\text{O}_2 - FA\text{CO}_2$$

where $FA\text{O}_2$ is the alveolar fraction of oxygen, $Fi\text{O}_2$ is the inspired fraction of oxygen, and $FA\text{CO}_2$ is the alveolar fraction of CO_2. This type of hypoxemia can easily be corrected by increasing $Fi\text{O}_2$. For example, when $PA\text{CO}_2 = 80\,\text{mmHg}$ (10.66 kPa) ($FA\text{CO}_2 = 11.2$), $FA\text{O}_2$ can be restored to normal by increasing $Fi\text{O}_2$ from 21 to 26.6 per cent.

Table 1 Relationship between Po_2 and Pco_2

	Normal Pco_2	Low Pco_2	High Pco_2
Normal Po_2	Normal	Pure hyperventilation if $Fio_2 = 21\%$ (stress + anxiety, metabolic acidosis); if $Fio_2 <$ 21%, the above conditions plus shunt	Ventilatory impairment while inhaling $Fio_2 > 21\%$
Low Po_2	Oxygenation impairment due to shunt plus relative ventilatory impairment	Oxygenation impairment due to shunt with physiological ventilatory response	Ventilatory impairment if $Fio_2 = 21\%$; oxygenation impairment due to shunt if $Fio_2 > 21\%$ associated with ventilatory impairment
High Po_2	$Fio_2 > 21\%$ without ventilatory impairment	Hyperventilation during inhalation of $Fio_2 > 21\%$	Ventilatory impairment while inhaling $Fio_2 > 21\%$

Hypoxemia with high FAo_2 is due to parenchymal disease, when part of the lung parenchyma is collapsed or consolidated, and represents true oxygenation impairment. Correction of this kind of hypoxemia (due to right to left shunt) is discussed elsewhere.

Various combinations of Po_2 and Pco_2 may occur, and their physiopathological meanings are listed in Table 1 (the conditions of high altitude or inhalation of hypoxic gas mixtures are excluded).

Hypercapnia and acidosis

As with Po_2, there is no threshold value of Pco_2 or pH which is 'harmful' *per se*. Many factors may condition the response to increased CO_2, such as the rate of increase in Pco_2 (acute or chronic), age, and cardiovascular conditions. If associated with normoxia, near-normal pH, consciousness, and hemodynamic stability, a high Pco_2 does not need any therapeutic intervention. The indications for mechanical assistance should be based on a global clinical assessment, considering the three main consequences of increased Pco_2, i.e. tissue acidosis, impairment of the central nervous system, and the cardiovascular response (Bidani *et al.* 1994).

Tissue acidosis

The immediate buffering of Pco_2 by fixed buffers (mainly hemoglobin) leads to an increase in HCO_3^- of about 1 mmol/l for every increase of 10 mmHg (1.33 kPa) in Pco_2.

As molecular CO_2 enters the cell membrane faster than HCO_3^-, it is generally believed that intracellular pH decreases more than extracellular pH. However, there is increasing evidence that the intracellular buffers limit tissue acidosis, and hypercapnia is well tolerated. There is no doubt that if cellular acidosis develops, cell function and viability are impaired.

Effects on the central nervous system

Increasing Pco_2 may have a severe effect on central nervous system activity. Experimentally, the brain excitability first decreases, then increases, with associated seizures, and finally decreases to anesthesia and coma. CO_2 is one of the major determinants of the cerebral vascular reactivity, both directly and indirectly (through pH changes), and acute hypercapnia may result in an increased cerebral blood flow and intracranial pressure.

Circulatory response to hypercapnia

The effect of hypercapnia on the cardiovascular system depends on the balance between the direct depressant effects of Pco_2 on heart and peripheral vascular smooth muscles and the increased plasma levels of epinephrine (adrenaline) and norepinephrine (noradrenaline) due to activation of the sympathetic nervous system. In normal conditions, the net result is an increase in cardiac output and a slight decrease in peripheral resistance. The arterial pressure tends to rise and the pulmonary artery pressure may increase substantially. It is important to remember that these reactions are observed in intact subjects. In patients given β-blockers, for example, hypotension and decreased cardiac output may be observed.

Summary

It is evident that the indications for the mechanical treatment of respiratory acidosis are based on three symptoms: hypoxemia, hypercapnia, and acidosis. However, no threshold values can be established, and only the history (chronic or acute derangement), the trends (the rate and type of deterioration), and the clinical signs (hemodynamic instability and level of consciousness) should be used to determine whether mechanical treatment is appropriate.

However, other factors should be taken into consideration (i.e. the precipitating factors and the time required for correction of acidosis) before mechanical support is planned.

Precipitating factors

The most common causes of respiratory acidosis and the time required for their correction are summarized in Table 2. They can be classified into three groups.

1. In this group the cause of hypercapnia can be removed easily. If hypoxemia can be corrected by supplemental oxygen administration, it is better, after removal of the precipitating factors, to wait for a spontaneous increase in alveolar ventilation. Hypercapnia

Table 2 Causes of respiratory acidosis

	Immediate reversibility	Reversibility within hours/days	Irreversible
Respiratory drive	Drugs (antidotes available) Nutritional insufficiency	Chronic loading Metabolic alkalosis Endocrine disturbances	Congenital
Airways	Secretions Foreign bodies Bronchospasm Airways apparatus	Asthma Bronchial stenosis	Terminal COPD
Muscles	Drugs (antidotes available)	Neuromyopathies Endocrine/electrolyte disorders Abdominal distension Hyperinflation	Quadriplegia Terminal neuromuscular disease
Chest wall		Flail chest	Kyphoscoliosis Thoracoplasty
Lung parenchyma	Pneumothorax Pleural effusion	Pulmonary edema Pulmonary embolism ARDS COPD exacerbation	Terminal obstructive and restrictive lung disease

ARDS, acute respiratory distress syndrome; COPD, chronic obstructive pulmonary disease.

does not require any treatment if associated with a stable pH, high HCO_3^-, and hemodynamic stability in a conscious patient.

2. In this group the correction of the precipitating factors will probably require hours or days. The need for mechanical ventilation should be determined on the basis of a global clinical assessment. In patients in whom hypercapnia is associated with clinical signs of severely increased work of the respiratory muscles, mechanical support should be introduced before the development of respiratory fatigue, which may lead to a sudden deterioration of $P\text{CO}_2$ and pH.

3. In the final group the cause cannot be corrected (e.g. late-stage neuromuscular disease). In most cases, the issue is more ethical than medical, and the therapeutic plan should be discussed with the patients and relatives (e.g. planning for home ventilation).

Assessment and correction of hypercapnia

Knowledge of the previous respiratory status is of paramount importance in determining the goal of respiratory support. A reasonable goal in a previous healthy subject could be normal blood gases ($P\text{aCO}_2 = 40$ mmHg (5.33 kPa), $P\text{aO}_2 > 80$ mmHg (10.66 kPa)). In a subject with previous chronic respiratory impairment, the goal should be the blood gas values present before the superimposed acute derangement (e.g. $P\text{aO}_2 = 60$ mmHg (8 kPa), $P\text{aCO}_2 = 50$ mmHg (6.66 kPa)). An accurate history is essential for differentiating chronic and acute respiratory impairment (i.e. to define the target blood gases). Measurement of the blood gases alone may be misleading. In fact, the normal increase in HCO_3^- of 1 mmol/l (acute conditions) or 3.5 mmol/l (chronic condition) for every increase of 10 mmHg (1.33 kPa) in $P\text{aCO}_2$ may be offset by the concurrent presence of metabolic acidosis which 'consumes' the HCO_3^-. Under these conditions the HCO_3^- level does not discriminate between acute and chronic respiratory acidosis.

In typical chronic respiratory acidosis, hypoxemia is usually corrected by increasing $F\text{IO}_2$. The risk of high $P\text{aO}_2$ in chronic patients breathing spontaneously (coma) has probably been overestimated (Aubier and Dombret 1993), and reasonable oxygenation is a mandatory target.

Since $P\text{aCO}_2 = \int \dot{V}\text{CO}_2/\dot{V}_A$, it is evident that it can be decreased by either decreasing $\dot{V}\text{CO}_2$ or increasing the alveolar ventilation $\dot{V}_A$. Methods of decreasing $\dot{V}\text{CO}_2$ include withdrawal of the excessive load of glucides delivered by parenteral nutrition, control of temperature in a hyperthermic patient, and artificial removal of part of $\dot{V}\text{CO}_2$ by extracorporeal methods such as dialysis or artificial lungs. However, the most efficient way of correcting hypercapnia is to increase alveolar ventilation, and this is usually achieved by mechanical ventilation. When the causes of hypercapnia are extrapulmonary, such as central nervous system or neuromuscular diseases, and the lung parenchyma is normal, mechanical ventilation causes no more problems than during general anesthesia in a normal patient.

Major problems may occur in patients in whom hypercapnia is due to dysfunction of the small airways (e.g. bronchospasm or asthma), parenchymal lesions (e.g emphysema), or diseases involving both airways and parenchyma (e.g. severe chronic obstructive pulmonary disease). If mechanical support of these patients is unavoidable, it is essential to avoid worsening hyperinflation which may lead to devastating hemodynamic consequences including cardiac tamponade. Small tidal volumes and prolonged expiratory time must be maintained when ventilating these patients, even at the cost of a relatively high $P\text{CO}_2$. The advantages of this setting have been shown in asthmatic patients (Darioli and Perret 1984) and suggested in other patient populations (Hikling *et al.* 1990). As a general rule, non-invasive ventilation is preferable to mechanical ventilation with intubation in the hypercapnic patient (Brochard *et al.* 1995). If the patient is hypercapnic, hyperinflated, and performing excessive respiratory work, the use of continuous positive airway pressure ventilation may decrease respiratory work and enable the patient to maintain spontaneous breathing.

Correction of respiratory acidosis with bicarbonate

When HCO_3^- is added to a closed system (e.g. venous blood) it is buffered by fixed buffer pairs A^-–AH (i.e. hemoglobin, proteins, and phosphates). This leads to generation of CO_2 , according to the following reaction:

$$HCO_3^- + AH \rightarrow H_2CO_3 + A^- \rightarrow CO_2 + H_2O + A^-.$$

However, not all the added HCO_3^- is transformed into CO_2. For example, if 10 mmol of HCO_3^- are added to a closed system, 9.56 mmol are unchanged and only 0.44 mmol are transformed to CO_2, so that $P\text{CO}_2$ increases from 40 to 54.5 mmHg (5.33 to 7.27 kPa). The concurrent increases in HCO_3^- and $P\text{CO}_2$ result in a change of pH from 7.40 to 7.41, which is clinically irrelevant.

This basic mechanism is important for the rational use of HCO_3^- in correcting acidosis. For example, if the patient is in circulatory arrest or a severe low-flow state, i.e. approaching a closed system, HCO_3^- is ineffective in correcting pH as both HCO_3^- and $P\text{CO}_2$ rise. The pH increases during HCO_3^- loading only if the newly generated CO_2 can be cleared out of the system by appropriate transport (cardiac output) and elimination (ventilation).

Under these conditions, administration of HCO_3^- is equivalent to acceleration of the process naturally occurring in the kidneys, in which new HCO_3^- is generated. Thus, when correcting the pH with HCO_3^- during CO_2 retention, the rate of HCO_3^- administration and the resetting of mechanical ventilation to clear the extra CO_2 derived from HCO_3^- titration must be carefully balanced. It is important to avoid sudden changes in the acid–base status, and pH correction using alkalinizing agents should be restricted to conditions in which the pH decrease is believed to be responsible for severe clinical deterioration. In most cases artificial $P\text{CO}_2$ control by mechanical ventilation and the natural kidney function (if present) is sufficient to restore an adequate acid–base balance.

References

Aubier, M. and Dombret, M.C. (1993). Acute exacerbation of chronic airflow obstruction. In *Pathophysiology: foundations of critical care* (ed. M.R. Pinsky and J.F.A. Dhainaut), pp. 427–46. Williams & Wilkins, Baltimore, MD.

Bidani, A., Tzounakis, A.E., Cardenas, V.J., and Zwischenberg, J.B. (1994). Permissive hypercapnia in acute respiratory failure. *Journal of the American Medical Association*, **272**, 957–62.

Brochard, L., *et al.* (1995). Noninvasive ventilation for acute exacerbations of chronic obstructive pulmonary disease. *New England Journal of Medicine*, **333**, 817–22.

Darioli, R. and Perret, C. (1984). Mechanical controlled hypoventilation in status asthmaticus. *American Review of Respiratory Disease*, **129**, 385–7.

Hikling, K.G., Henderson, S.J., and Jackson, R. (1990). Low mortality associated with low volume pressure limited ventilation with permissive hypercapnia in severe adult respiratory distress syndrome. *Intensive Care Medicine*, **16**, 372–7.

2.10 Respiratory alkalosis

2.10.1 Pathophysiology

J.-C. Chevrolet and R. Zürcher Zenklusen

Key messages

- Respiratory alkalosis is present when alveolar ventilation exceeds the amount necessary to eliminate the daily load of CO_2 produced by metabolic activity.
- Systemic and metabolic consequences are different in acute and chronic respiratory alkalosis.
- In acute respiratory alkalosis there is a decrease in $[HCO_3^-]$ of about 2 mmol/l for each decrease of 10 mmHg (1.3 kPa) in $Paco_2$.
- In chronic respiratory alkalosis there is a decrease in $[HCO_3^-]$ of about 5 mmol/l for each decrease of 10 mmHg (1.3 kPa) in $Paco_2$.
- Severe hypocapnia can induce brain hypoxia and elevated cerebral lactate output.

Definition

Respiratory alkalosis is a primary acid–base disorder which occurs when alveolar ventilation is increased to a level which exceeds the amount necessary to eliminate the daily load of carbon dioxide produced by the metabolic activity. This situation causes hypocapnia and is called hyperventilation. This should not be confused with tachypnea, i.e. an increased respiratory rate, which may or may not be associated with hyperventilation. When hyperventilation is present, carbon dioxide pressure is reduced in all body fluids. Alkalemia, i.e. a decrease in the proton concentration $[H^+]$ or an elevation of pH, will occur as a necessary consequence af alkalosis in all body fluid compartments unless some mechanism(s) of correction and/or compensation ensues.

Respiratory alkalosis can be explained using the alveolar gas equation

$$Paco_2 = \text{constant} \times \dot{V}co_2/\dot{V}_A \tag{1}$$

where $\dot{V}co_2$ is the carbon dioxide production, expressed in liters per minute, $\dot{V}_A$ is the alveolar ventilation at body temperature and ambient pressure, expressed in liters per minute, and the constant, which is given by

$$\frac{760 \times \text{body temperature (K)}}{273\ \text{K} \times \text{Hg}} = 863,$$

is introduced to normalize the units of $\dot{V}_A$ to $\dot{V}co_2$. The alveolar ventilation is given by

$$\dot{V}_A = f_b\,(V_T - V_D) \tag{2}$$

where f_b is the respiratory rate in breaths per minute, V_T is the tidal volume in liters, and V_D is the dead-space volume in liters. Consequently, when discussing the causes of respiratory alkalosis, one must systematically consider the different terms of both eqns (1) and (2).

Pathophysiology of hypocapnia

The body's response to hypocapnia results in a decrease in the extracellular proton concentration $[H^+]$ according to the Henderson–Hasselbalch equation

$$[H^+] = 24 \times Pco_2 / [HCO_3^-] \tag{3}$$

or

$$pH = 6.1 + \log [HCO_3^-]/0.0301Pco_2.$$

It is obvious from this equation that any decrease in $Paco_2$, as observed in primary respiratory alkalosis, leads to a reduction in $[H^+]$ and a consequent increase in pH, but that a subsequent lowering of $[HCO_3^-]$ will minimize the degree of alkalemia. The latter effect occurs *in vivo* by two different mechanisms which take place in two consecutive steps characterized by different time constants.

First, there is a rapid buffering whose maximum effect is attained after 10 min; $[H^+]$ leaves the intracellular compartment and combines with $[HCO_3^-]$ in the extracellular fluid. Intracellular buffers (proteins, hemoglobin, etc.) are the main sources of $[H^+]$ which leave the cells and enter the extracellular fluid. This can be described by the following equations for the intracellular fluid (ICF) and the extracellular fluid (ECF) respectively:

$$\text{H-buffers}_{ICF} \rightarrow H^+ + \text{buffers}_{ICF}^- \tag{4}$$

$$(H^+ + HCO_3^-)_{ECF} \rightarrow H_2CO_3 \rightarrow CO_2 + H_2O. \tag{5}$$

In addition, there may be intracellular lactic acid production due to alkalemia, which may also represent a source of H^+ ions aimed at buffering extracellular HCO_3^-. However, this last finding is not consistent and is controversial, and lactic acid production does not seem to play a significant role here even though acute respiratory alkalosis decreases lactate clearance by the liver in humans. Pragmatically, whatever the mechanism(s) causing some degree of hyperlactatemia in hyperventilation, plasma lactate is only moderately increased in this setting, i.e. its value is rarely higher than 3 to 4 mmol/l. Finally, the level of plasma chloride ions increases in acute respiratory alkalosis as a result of the chloride shift mechanism, in which ions are transferred across the red blood cell membrane (plasma chloride levels increase whereas plasma bicarbonate levels decrease) and all the ionic concentrations equilibrate at this new steady state in the extracellular fluid compartment. This causes a small decrease in [HCO_3^-]. The following rule of thumb can be used by intensive care clinicians: for each decrease of 10 mmHg (about 1.3 kPa) in $Pa{CO_2}$, [HCO_3^-] in the blood decreases by about 2 mmol/l. This buffering effect, although rapid, is not very efficient; it is only observed in acute respiratory alkalosis and is poorly suited to correcting alkalemia.

Secondly, if hyperventilation persists, plasma [HCO_3^-] decreases further due to renal compensation of the acid–base disorder. This mechanism begins to operate after 2 h of persistent hypocapnia and its maximum effect is attained after 2 to 3 days. Urinary [HCO_3^-] losses increase and simultaneously ammonium excretion in urine decreases, probably because of a signal mediated by the intracellular increase in [H^+] being sent to the tubular cells. Both the proximal and the distal nephrons are affected by changes in peritubular and intracellular $P{CO_2}$. For instance, bicarbonate recuperation in the proximal nephron dramatically decreases when peritubular $P{CO_2}$ is reduced in animal preparations. In addition, ammoniagenesis is inhibited by acute alkaline conditions in animals, although the exact sites of inhibition (proximal and/or distal tubules) and the ultimate regulation are incompletely understood at present. Taken together, these mechanisms promote a decrease in the daily [H^+] elimination and the extrapolation to humans seems to be valid.

Finally, the combined buffering processes lead to an average decrease of about 5 mmol/l of [HCO_3^-] for each reduction of 10 mmHg (1.3 kPa) in $P{CO_2}$ due to hyperventilation. This is another rule of thumb that may be useful for the clinician. Therefore, in chronic respiratory alkalosis, this combined buffering is more efficient for correcting the pH disturbance and reducing the magnitude of alkalemia. The occurrence of a mixed and more complicated acid–base disorder should be suspected when the observed reduction in plasma [HCO_3^-] is not consistent with the change in $P{CO_2}$.

Systemic and metabolic consequences of respiratory alkalosis

Electrolyte modifications occur in response to changes in blood $P{CO_2}$ (acidosis or alkalosis), and these modifications take place in opposite directions depending on the primary respiratory acid–base disorder. For instance, slight hyponatremia (a decrease of 2–4 mmol/l) has been observed in acute, but not chronic, respiratory alkalosis. This is probably due to increased water reabsorption in the proximal renal tubule associated with a distal enhancement of the action of ADH. Hyperchloremia of small magnitude, due to the chloride shift and to some degree of contraction of the extracellular space, has been observed. According to classical data, serum potassium decreases during the acute respiratory alkalosis. This is caused by a shift of this electrolyte into the cells with concomitant release of the H^+ ion (a potassium decrease of 0.1–0.4 mmol/l for each increase of 0.1 in pH). However, the increased catecholamine concentrations associated with voluntary hyperventilation may counteract these mechanisms, and the serum potassium may sometimes be slightly elevated (by about 0.3 mmol/l) in acute respiratory alkalosis. This effect is mediated by enhanced α-adrenergic activity and is partly counteracted by β-adrenergic stimulation. The serum potassium level remains normal in chronic respiratory alkalosis.

Hypophosphatemia (about 0.5–1.5 mg/dl or 0.3–0.8 mmol/l), as well as ionized hypocalcemia, is widely recognized with acute hypocapnia. This accounts in part for the well-known increased risk of tetany and seizures seen with hyperventilation. These changes are due to a shift of phosphate into the cells, probably mediated by the intracelluar alkalosis caused by hypocapnia. The increase in intracellular pH enhances glycolysis, leading to the formation of phosphorylated compounds, such as glucose 6-phosphate and fructose 1,6-diphosphate, and a decrease in free phosphate in both the cells and the extracellular fluid. This phenomenon does not seem to have clinical consequences *per se* and therefore should not need to be treated. However, the clinician should be aware of its existence in order to interpret the occurrence of hypophosphatemia. In animals, the parathyroid hormone (PTH) plasma level actually increases in

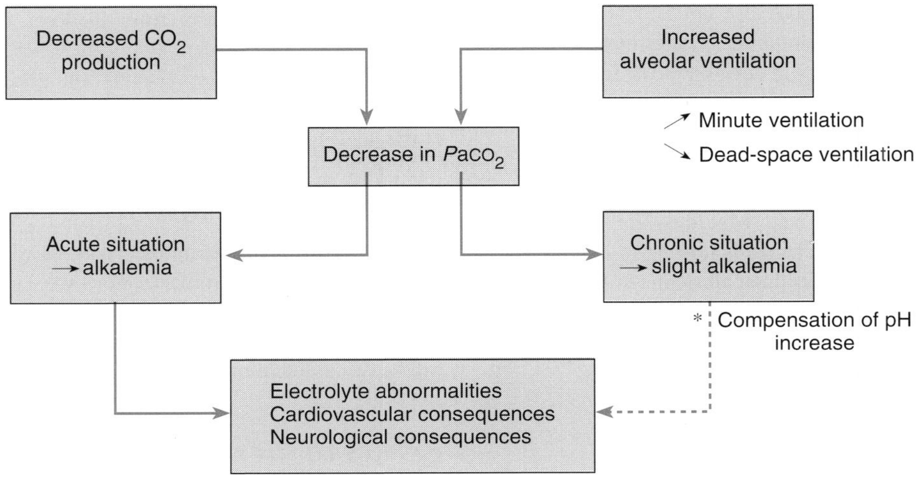

Fig. 1 Summary of the pathophysiology of respiratory alkalosis.

chronic respiratory alkalosis; this hormone has been shown to decrease proximal tubular bicarbonate resorption and to cause an increase in the transfer of phosphate from the extracellular to the intracellular compartment. However, in humans, chronic respiratory alkalosis induces some degree of renal PTH resistance, mediated by β-adrenergic receptors, which causes a slight hyperphosphatemia and hypocalcemia.

The cardiovascular system appears to be profoundly affected by respiratory alkalosis. In fact, both respiratory acidosis and respiratory alkalosis may cause a decrease in systemic vascular resistance mediated by a relaxation of the systemic arterial tone. However, these actions are partially counteracted by vasoconstriction mediated by the circulating catecholamines and by an increase of the activity of the sympathetic nervous system. The net clinical effect of respiratory alkalosis is a fall in peripheral vascular resistance and arterial blood pressure, with the magnitude of these effects varying according to the relative importances of these mechanisms. Generally, the decrease in blood pressure and systemic vascular resistance is slight. However, patients with a depressed central nervous system and/or under general anesthesia, particularly when passively hyperventilated by mechanical ventilation, may present major reductions in cardiac output and systemic blood pressure. In addition, these patients may be characterized by various levels of peripheral vascular resistance and different degrees of hyperlactatemia, particularly when in acute respiratory alkalemia. It is of paramount importance to be aware that some regional vascular beds react differently from the overall systemic circulation. One of these is the cerebral circulation. Cerebral blood flow decreases in acute hypocapnia, and, when $Pa\text{CO}_2$ is severely decreased, this may result in brain hypoxia and increased cerebral lactate output. In addition, acute respiratory alkalosis may be responsible for neurological disorders not only because of the low $P\text{CO}_2$, but also because of alkalemia, pH-induced shift of the oxyhemoglobin curve, and electrolyte abnormalities.

The function of the peripheral neuromuscular system is strongly modified by alkalemia. In essence, alkalemia, of either metabolic or respiratory origin, increases the binding of ionized calcium to its transporters, mainly serum albumin. Hence, for each pH increment of 0.1, ionized calcium will be reduced by about 0.2 mg/dl (or 0.05 mmol/l). The clinical consequences of this phenomenon depend not only on the level of serum ionized calcium, but also on the rate at which it decreases. Hence more severe symptoms (paresthesias, tetany, anxiety, cardiovascular changes, etc.) will be observed in acute respiratory alkalosis than in the setting of chronic hyperventilation.

Conclusions

Primary respiratory alkalosis is an acid–base disorder that can be caused by several very different mechanisms leading to hypocapnia. It results in an extracellular and intracellular alkalemia until compensatory mechanisms are fully operative. The consequences of this disorder are clearly dependent on the magnitude of the primary disorder and on the rapidity of the decrease in $P\text{CO}_2$ (Fig. 1).

Bibliography

Druml, W., Grimm, G., Laggner, A., Lenz, K., and Schneeweiss, B. (1991). Lactic acid kinetics in respiratory alkalosis. *Critical Care Medicine*, **19**, 1120–4.

Giesbisch, G., Berger, L., and Pitts, R. (1955). The extrarenal response to acute acid–base disturbances of respiratory origin. *Journal of Clinical Investigations*, **34**, 231–42.

Krapf, R., Jaeger, P., and Hulter, H. (1992). Chronic respiratory alkalosis induces renal PTH-resistance. *Kidney International*, **42**, 727–34.

Krapf, R., Caduff, P., Wagdi, P., Staubli, M., and Hulter, H. (1995). Plasma potassium response to acute respiratory alkalosis. *Kidney International*, **47**, 217–24.

Molony, D. and Jacobson, H. (1986). Respiratory acid–base disorders. In *Fluids and electrolytes* (ed. J. Kokko and R.Tannen). W.B. Saunders, Philadelphia, PA.

2.10.2 Causes and therapeutic strategy

J.-C. Chevrolet and R. Zürcher Zenklusen

Key messages

- Hyperventilation, manifested by hypocarbia, is very common in any kind of acute bronchial and/or lung disorder.
- The prognosis of the acid–base disorder depends strongly on its underlying cause.
- The main issues are the magnitude of the alkalemia and whether the actual disorder is acute or chronic.
- Respiratory alkalosis becomes life threatening when blood pH exceeds 7.55, and very life threatening when it exceeds 7.70, inducing seizures, coma, and cardiac arrest.
- When metabolic acidosis is corrected rapidly, superimposed respiratory and metabolic alkalosis occur.

Introduction

In considering the causes of respiratory alkalosis, reference should be made to the alveolar air equation which shows that respiratory alkalosis may be due to an increase in alvelolar ventilation and/or a reduction in the carbon dioxide output. When discussing the nonspecific treatment of respiratory alkalosis, the main issues are the magnitude of the alkalemia and whether the actual disorder is acute or chronic. The etiology of the alkalosis must be carefully considered when undertaking any treatment because the prognosis of the acid–base disorder depends strongly on its underlying cause.

Causes of respiratory alkalosis

The main causes of respiratory alkalosis which can be observed in clinical medicine are summarized in Table 1 according to the concepts on which the alveolar air equation is based. Some of these disorders are particularly relevant to the intensive care setting and therefore will be discussed in more detail.

Errors in the management of mechanical ventilation

There are basically two clinical circumstances which are sometimes observed in the intensive care unit when mechanically ventilated patients may suffer from iatrogenic hyperventilation: posthypercapnic

Table 1 Causes of respiratory alkalosis

Decrease in CO_2 output
Hypothermia
Hypothyroidism
Hypometabolism induced by drugs (i.e. barbiturates)
Increase in alveolar ventilation
Primary activation of the central nervous system
Central nervous system disorders (tumors, cardiovascular disorders, infections, trauma)
Drugs (salicylate, nicotine, almitrine, doxapram, methylxanthines, catecholamines, etc.)
Fever, heat stroke
SIRS, sepsis
Secondary stimulation of the central nervous system
Lung diseases (ARDS, pneumonia, bronchial asthma, pneumonia, pulmonary thromboembolism, lung edema, etc.)
Hypoxemia
Others (hypotension, anemia, high altitude, etc.)
Iatrogenic
Mechanical hyperventilation
Extrapulmonary loss of CO_2
Hemodialysis
Extracorporeal circulation

SIRS, systemic inflammatory response syndrome; ARDS, acute respiratory distress syndrome.

alkalosis, and the situation in which a patient with a reduced carbon dioxide output is 'normally' ventilated. The first issue is generally encountered in patients with chronic obstructive pulmonary disease or with chronic respiratory acidosis of other origin. The sudden introduction of efficient mechanical ventilation rapidly corrects hypercarbia, and severe alkalemia may occur if systemic bicarbonate plasma concentration is elevated due to renal compensation of chronic acidosis. Therefore slow correction of the hypercarbia, lasting at least 24 to 48 h and allowing a parallel decrease of plasma bicarbonate, is recommended in these patients. The second clinical issue is occasionally observed in hypothermic and/or hypometabolic patients (hypothyroidism, barbiturate intoxication, etc.) when they are mechanically ventilated using the usual ventilator setting, i.e. the respiratory rates and tidal volume generally recommended. Both these issues may lead to very severe hypocapnia if they are not rapidly recognized, and therefore they may cause neurological damage, cardiac arrhythmias, cardiovascular shock, and eventually death.

Rapid correction of metabolic acidosis

Another medical intervention aimed at altering the acid–base status may cause dangerous alkalemia. In metabolic acidosis, the normal compensatory response of the body, mediated by the central nervous system, is to increase minute ventilation in order to lower $P\text{aCO}_2$ and to increase the pH of body fluids. Because body fluid compartments equilibrate rapidly for CO_2, hypocapnia is also present in the cerebrospinal fluid whose pH is restored in a new steady state, characterized simultanously by low local $P\text{CO}_2$ and bicarbonate concentration. When exogenous bicarbonate is given rapidly in order to correct metabolic acidosis, CO_2 will be produced in the blood by the buffering of H^+ by HCO_3^-. Because CO_2 diffuses rapidly into the cerebrospinal fluid, whereas bicarbonate does not (there is a slow transport mechanism for this anion accross the cisternal epithelium), there may be a transient disequilibrium in pH, which may also decrease in regions of the brain close to the respiratory centers. Consequently, the respiratory drive increases even further, which leads to hyperventilation and alkalemia. In this clinical situation, superimposed respiratory and metabolic alkaloses occur, with the latter being caused by the administration of the buffer. This may be particularly dangerous in some patients, particularly in diabetic ketoacidosis. An increased risk of malignant cardiac arrhythmias and/or seizures has been described in this setting. Therefore, if possible, buffers should be given only when absolutely necessary and they should be administered slowly.

Salicylate poisoning

The acid–base disorder due to salicylate ingestion is very complex, particularly when large quantities of this drug are absorbed, for instance in suicide attempts. First, salicylate seems to stimulate the

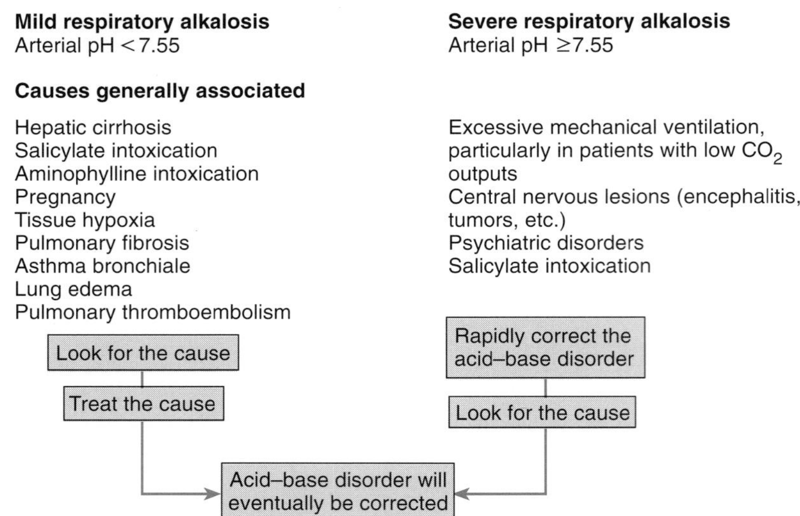

Fig. 1 Therapeutic strategy for respiratory alkalosis depending on the severity of arterial pH increase

cerebral respiratory centers directly and cause hyperventilation and respiratory alkalosis. However, in severe poisoning, salicylate can uncouple oxidative phosphorylation in an enormous number of cell systems, resulting in a metabolic lactic acidosis. Finally, in the most severe cases, hypoglycemia due to hepatic failure will ensue. Ketoacidosis will also occur because the lipid metabolism will be substituted as an energy source. The clinician must actively look for intoxication with this drug when a patient presents with an association of mixed acid–base disorder, such as alkalosis, elevated anion gap and osmolal gap acidosis, and lactic and ketotic acidosis. The only life-saving treatment is the rapid initiation of hemodialysis, and therefore no delay is permissible.

Hepatic failure

The pathophysiological mechanisms underlying hyperventilation in hepatic failure are rather obscure, but patients suffering from this disorder are frequently characterized by a low $Pa\text{CO}_2$. The magnitude of respiratory alkalosis is associated with a poor prognosis and, in severe hepatic dysfunction, $Pa\text{CO}_2$ may be as low as 30 mmHg (4 kPa). However, in the most severe cases, particularly when the patient presents with hepatic encephalopathy, the acid–base disorder becomes more complex and shows some degree of metabolic acidosis, which is probably due to the occurrence of a sepsis syndrome.

Respiratory alkalosis associated with bronchopulmonary disorders

Some degree of hyperventilation is associated with every acute bronchial and/or lung disorder (asthma, acute respiratory distress syndrome, pneumonia, lung edema, pulmonary thromboembolism, etc.). The causes of hyperventilation in these clinical situations are numerous. First, hypoxemia may play a role, because when $Pa\text{O}_2$ drops below approximately 65 mmHg (8 kPa), there is a significant stimulation of peripheral chemoreceptors sensing hypoxemia and hyperventilation ensues. However, even after the correction of hypoxemia, for instance by increasing the inspired oxygen fraction, many bronchopulmonary diseases remain characterized by hyperventilation. This is due to the stimulation of peripheral bronchial and/or lung chemoreceptors and mechanical receptors. From a pragmatic point of view, the intensive care physician should consider hyperventilation, manifested by hypocapnia, as a useful diagnostic signal providing two important items of information: first, that 'something is wrong within the respiratory system' and, second, that the respiratory muscles are still able to cope with the disorder, whatever it may be. The occurrence of normocapnia or, *a fortiori*, hypercapnia in this situation is of concern, because it strongly suggests that the inspiratory muscles are becoming fatigued.

Treatment of respiratory alkalosis

Generally speaking, respiratory alkalosis is less life threatening than respiratory acidosis, probably because most of the respiratory alkaloses observed in clinical situations are a consequence of chronic and relatively stable disorders. This is in striking contrast with the causes of many respiratory acidoses, which may be rapidly fatal when not corrected quickly. However, this may not be true in the intensive care setting, where some causes of very severe respiratory alkaloses can be found, as noted earlier.

The arterial pH often remains below 7.55 in respiratory alkalosis, so that symptoms are either mild or absent unless other factors, such as previous hypocalcemia, hypoxia, etc., are concomitantly present and may cause serious disorders, such as seizures, cardiac arrhythmias, etc. The clinical situation is completely different when the arterial pH exceeds 7.55: the acid–base disorder itself becomes very life threatening when blood pH exceeds 7.70. This degree of alkalemia is rarely tolerated for long periods and may lead to death by causing cardiovascular and/or neurological dysfunction, most frequently by inducing seizures and cardiac arrhythmias.

Therefore mild respiratory alkalosis (arterial pH 7.45–7.55) rarely necessitates a specific treatment. However, this acid–base disorder represents some sort of marker of an underlying disease that must itself be detected and treated. This will eventually correct the acid–base abnormality without the need for a specific treatment directed against the acid–base disorder. In contrast, severe respiratory alkalosis, characterized by an arterial pH exceeding 7.55, can lead to coma, cardiac arrhythmias, and death. In this situation, asterixis (flapping tremor) and myoclonus may sometimes precede coma and, when present, these symptoms prompt rapid and specific treatment in which alveolar ventilation is decreased by every available means in order to lower the arterial pH to below 7.50. When simple sedation is not sufficient to reach this goal quickly, the patient must be intubated and paralyzed in order to reduce the arterial pH. Therefore it may be necessary to use controlled mechanical ventilation until the primary disorder causing the severe alkalemia has been cured. This treatment must be undertaken rapidly before the occurrence of irreversible cardiac and/or cerebral damage, i.e. within a few minutes. Figure 1 shows the causes of respiratory alkalosis, listed according to the severity of the disturbance, and a simple therapeutic strategy for this acid–base disorder for both mild and severe pH increases.

Bibliography

Elisaf, M., Theodorou, J., Pappas, H., and Siamopoulos, K.C. (1993). Acid–base and electrolyte abnormalities in febrile patients with bactaeremia. *European Journal of Medicine*, **2**, 404–7.

Javaheri, S., Corbett, W., Wagner, K., and Adams, J.M. (1994). Quantitative cerebrospinal fluid acid–base balance in acute respiratory alkalosis. *American Journal of Respiratory and Critical Care Medicine*, **150**, 78–82.

Pierson, D.J. (1990). Complications associated with mechanical ventilation. *Critical Care Clinics*, **6**, 711–24.

Rose, B.D. (1989). *Clinical physiology of acid–base and electrolyte disorders* (3rd edn), pp. 580–8. McGraw-Hill, New York.

2.11 Failure to wean from mechanical ventilation

2.11.1 Pathophysiology of respiratory muscle dysfunction

M. Aubier

Key messages

- Inspiratory muscle energy demand is determined by the work of breathing and muscle strength.
- Muscle energy supply is determined by the blood flow and the energy stores of the muscle.
- Fatigue is due to an imbalance between energy supply and demand.

Introduction

The respiratory muscle can be considered as an engine in which fatigue occurs when the rate of energy consumption by the muscle is greater than the energy supplied by the blood. The factors predisposing to respiratory muscle dysfunction are those which increase the energy demands of the inspiratory muscles and which decrease the energy supplies to the respiratory muscles or the energy stores.

Factors determining inspiratory muscle energy demand

Work of breathing

The level of minute ventilation, the frequency of breathing, a decrease in lung compliance, or an increase in airway resistance increase the work of breathing and energy demands on the muscle. A hyperstimulated central respiratory drive also imposes an increased inspiratory muscle work of breathing.

Muscle strength

The greater the fraction of the maximum pressure that can be developed by the inspiratory muscles, the greater are their energy demands. This fraction can be increased by either increasing the pressure necessary to breath (e.g. increased airway resistance) or reducing the maximum force that the muscle can develop (hyperinflation, muscle atrophy). Hyperinflation, which is common in patients with chronic obstructive pulmonary disease (COPD), shortens the diaphragmatic fibers and thus reduces the maximum force of the diaphragm (maximum force is a function of initial fiber length).

Efficiency

For a given workload, the energy demands and oxygen cost of breathing increase as efficiency decreases. Efficiency decreases with airway resistance. In this condition, hyperinflation occurs, leading to a flatness of the diaphragm which, in turn, acts as a fixator and not as an agonist (quasi-isometric contraction). This requires energy expenditure without production of work.

Factors determining available inspiratory muscle energy

Energy supply

A decrease in energy supply to a muscle can be observed when muscle blood flow decreases (low cardiac output and increased intensity of the diaphragmatic contraction), the oxygen content of arterial blood decreases (anemia, hypoxemia), and the ability to extract sources of energy decreases.

Energy stores

Poor nutritional status, catabolic states (e.g. malignancy, sepsis), and depletion of glycogen by prolonged hyperventilation may all facilitate the occurrence of muscle fatigue.

Muscle blood flow and fatigue

The fact that the diaphragm is fatigued beyond a critical rate of tension strongly suggests an association between the rate of work, a limitation on the diaphragmatic capacity to extract or consume oxygen, and blood flow to the diaphragm. This association also implies that peripheral fatigue is a consequence of this mechanism.

The importance of blood flow as a cause of respiratory muscle fatigue was emphasized in a dog model in which breathing was unloaded and blood flow was limited by cardiogenic shock. Under these conditions, all spontaneously breathing animals died of respiratory arrest when the respiratory muscles failed to generate sufficient pressure to maintain alveolar ventilation despite an increase in phrenic nerve output. In this example, energy demand was unaffected; fatigue was due to a decrease in energy supply alone.

Respiratory failure may result from a failure of the lungs due to airway obstruction (status asthmaticus, chronic bronchitis, emphysema), parenchymal disease (cardiogenic or non-cardiogenic edema, severe infections, pneumonia, pulmonary fibrosis), or a failure of the pump. Failure of the respiratory muscles can result in acute respiratory failure.

Whether respiratory muscle fatigue exists during acute respiratory failure can be discussed with reference to the clinical conditions of acute respiratory failure.

Chronic obstructive pulmonary disease

In subjects with COPD in acute respiratory failure, some factors predispose to the occurrence of respiratory muscle fatigue. The respiratory centers are hyperstimulated, which increases the workload of the respiratory muscles. This workload is also increased by the high airway resistances. The efficiency of the respiratory muscles is impaired by hyperinflation and airway resistance. In the extreme case, the diaphragm contracts isometrically (i.e. with energy demand without production of work). Furthermore, acute respiratory acidosis decreases the contractility and endurance time of the diaphragm in humans and thus could lead to respiratory muscle fatigue.

The oxygen cost of breathing is increased in COPD, but it is unclear whether muscle blood flow is maintained at an adequate level with respect to muscle workload. It is known that muscle blood flow increases when contractile activity increases. However, beyond a critical level, mechanical compression of diaphragmatic vessels decreases blood flow. When the effects of cor pulmonale and right heart failure, which may impair cardiac output, are added to this factor, respiratory blood flow may decrease below the levels required to supply substrates and oxygen adequately.

In the literature, there is some argument for the existence of respiratory muscle fatigue in the setting of acute respiratory failure in patients with COPD. Paradoxical motion of the abdominal wall during the breathing cycle and thoracic alternans are frequently observed in these patients.

It has been demonstrated that the relaxation rate of the diaphragm is increased in the case of diaphragmatic fatigue and correlates closely with the modifications of the high-to-low ratio of the diaphragmatic electromyogram. It has also been shown, in 14 patients with COPD intubated and ventilated for acute respiratory failure, that the relaxation rate of the diaphragm just after intubation significantly decreases towards the range of normal values just before extubation. This attests to the fact that respiratory muscle fatigue probably exists in these patients at the onset of the acute respiratory failure and is reversed by mechanical ventilation. It appears from this study that respiratory muscle fatigue should be taken into account in the analysis of failure to wean patients from mechanical ventilation.

In another study of patients of the same type who were intubated and ventilated for acute respiratory failure, Murciano *et al.* (1988) showed that, at the onset of acute respiratory failure, the patients had high occlusion pressures (hyperstimulated respiratory centers) associated with a decreased high-to-low ratio of the diaphragmatic electromyogram, an index closely related to diaphragmatic fatigue. After a few days of mechanical ventilation, most of the patients were weaned successfully from the respirator, with a significant decrease in occlusion pressure and a significant increase in the high-to-low ratio. Nevertheless, some patients needed to be reventilated within 48 h of extubation. In these patients, the high-to-low ratio did not increase and occlusion pressure did not decrease. These individuals probably had persistent fatigue of the respiratory muscles, requiring a longer period of artificial ventilation.

Cardiogenic shock

Factors predisposing to respiratory muscle fatigue are present during cardiogenic shock with or without pulmonary edema. The work of breathing is increased in pulmonary edema (hyperventilation, stiff lungs resulting in an increased oxygen demand by respiratory muscles). Concomitantly, there is inadequate oxygen delivery to the respiratory muscles secondary to hypoxia and decreased cardiac output. Lactic acidosis, resulting from inadequate oxygen delivery to the tissues, further increases ventilation and respiration. Thus the work of breathing is increased, which could lead to respiratory muscle fatigue. This might explain the hypercapnia that is sometimes observed during acute pulmonary edema. Aubier *et al.* (1982) and Viires *et al.* (1983) provided evidence that the development of an imbalance between diaphragmatic metabolic demands and blood flow may result in diaphragmatic fatigue in this situation.

Septic shock

Sepsis is a frequent pathological event in which a sustained increase in protein catabolism has been reported in skeletal muscles. Skeletal muscle contractile proteins are used or consumed to provide amino acids. Furthermore, hyperventilation increases pulmonary resistance and decreases lung compliance, while increasing the work of breathing and the energy demand on the respiratory muscles. Decreases in blood pressure and cardiac output, which may occur in septic shock, decrease the energy supply to the respiratory muscles. In addition, the muscles are unable to extract oxygen. Thus fatigue of respiratory muscle is expected in septic shock. A model of endotoxic shock in spontaneously breathing dogs provided evidence of respiratory muscle fatigue. The effects of a 3-day pneumococcal infection on diaphragmatic strength and endurance capacity in an *in vivo* rat model have been studied. Inoculated rats were compared with control rats. Diaphragmatic strength was assessed by measuring the transdiaphragmatic pressure generated during electrical stimulation of the phrenic nerves at different frequencies. It was found that acute sepsis produced a decrease in diaphragmatic strength related to both impaired contractility and a reduction in diaphragmatic endurance capacity.

Therapeutic considerations

Restoring the balance between energy supply and demand

If fatigue results from the imbalance between energy supply and demand, the objectives of therapy must be to restore this balance. Energy supplies can be increased by treating hypoxemia and improving cardiac output, and the demands for energy can be decreased by treating bronchospasm or pulmonary edema. If the usual methods of restoring the balance between energy supply and demand fail, and clinical manifestations of inspiratory muscle fatigue persist, serious consideration must be given to resting the muscles by the use of artificial ventilation. Just how much rest is required is not known. Obviously, total rest is not required in normal subjects. If it were, recovery from experimental fatigue would not be possible. However, a body respirator that allows the diaphragm to rest may be useful for some patients. If these patients are in a chronic state of fatigue during quiet breathing, complete rest may restore their respiratory muscle function.

Recently, there has been a resurgence of interest in the delivery of positive-pressure ventilation with face or nasal masks to patients with COPD during acute respiratory failure. Compared with conventional treatment in patients who have an acute exacerbation of COPD, non-invasive ventilation may reduce the need for endotracheal intubation and decrease mortality (Brochard *et al.* 1990). Furthermore, once these patients with severe airflow obstruction are intubated, ventilator dependence may be a serious problem. The latter could be due, at least in part, to respiratory muscle atrophy, as suggested by recent experimental evidence (Le Bourdelles *et al.* 1994).

References

Aubier, M., *et al.*(1982). Respiratory muscle contribution to lactic acidosis in low cardiac output. *American Review of Respiratory Disease*, **121**, 648–52.

Brochard, L., Isabey, D., Piquet, J., Artigas, A., and Lemaire, F. (1990). Reversal of acute exacerbations of chronic obstructive lung disease by inspiratory assistance with a face mask. *New England Journal of Medicine*, **323**, 1523–30.

Le Bourdelles, G., Viires, N., Boczkowski, J., Seta, N., Pavlovic, D., and Aubier, M. (1994). Effects of mechanical ventilation on diaphragmatic contractile properties in rats. *American Journal of Respiratory Critical Care Medicine*, **149**, 1539–44.

Murciano, D., *et al.*(1988). Tracheal occlusion pressure: a simple index to monitor respiratory muscle fatigue during acute respiratory failure in COPD patients. *Annals of Internal Medicine*, **108**, 800–5.

Viires, N., *et al.* (1983). Regional blood flow distribution in dog during induced hypotension and low cardiac output, spontaneous breathing, VS arterial ventilation. *Journal of Clinical Investigation*, **72**, 935–47.

2.11.2 Pathophysiological cardiorespiratory interactions

M. R. Pinsky

Key messages

- Spontaneous ventilation is a form of cardiovascular stress.
- Spontaneous inspiratory efforts will, by decreasing intrathoracic pressure, augment systemic venous return, increasing right ventricular volumes.
- Spontaneous inspiratory efforts will, by decreasing intrathoracic pressure, increase the pressure gradient for left ventricular ejection, impeding left ventricular emptying. In heart failure states and if the swings in intrathoracic pressure are great, this effect can precipitate acute heart failure.
- In patients with obstructive lung disease, rapid spontaneous ventilation can induce marked hyperinflation which can cause hemodynamic compromise by (a) increasing pulmonary vascular resistance and (b) decreasing systemic venous return to the heart even if intrathoracic pressure decreases.

Introduction

The initiation of spontaneous ventilation in patients on mechanical ventilatory support (weaning) may induce markedly detrimental hemodynamic effects. Clinically relevant heart–lung interactions can be grouped as follows.

1. Ventilation is exercise: it consumes O_2 and produces CO_2, and thus may stress normal adaptive circulatory mechanisms.
2. Inspiration increases lung volume above end-expiratory volume.
3. Spontaneous inspiration decreases intrathoracic pressure.
4. Positive-pressure ventilation increases intrathoracic pressure.

When the cardiorespiratory unit is functioning normally, it balances the ever-varying and often conflicting metabolic demands by the body in a smooth and seemingly effortless fashion. However, during disease, marked cardiovascular dysfunction may occur as a result of two related but different reasons.

First, spontaneous ventilatory efforts require muscular effort and are a cardiovascular stress analogous to exercise. Spontaneous ventilatory efforts require increased respiratory muscle blood flow to meet their increased O_2 demands and produce a CO_2 load for the lungs. If the metabolic load required by spontaneous ventilation is greater than the ability of the cardiovascular system to handle it, the subject will be unable to breath spontaneously without developing acute cardiovascular collapse and/or acute respiratory failure. Thus failure to wean may be due to either inadequate cardiovascular reserve, which limits total blood flow, or ventilatory reserve, which limits alveolar ventilation. Cardiovascular compromise may occur if either the work of breathing is markedly increased or the cardiovascular reserve is compromised. Increases in the work of breathing occur due to increased resistive loads (acute bronchospasm in asthmatics, upper airway obstruction due to vocal cord paralysis and laryngeal edema) and elastic loads (restrictive lung diseases such as interstitial fibrosis as well as both tense abdominal acities and pulmonary edema), or decreased efficiency of the respiratory muscles (hyperinflation).

Second, ventilation by altering intrathoracic pressure and lung volume may both limit left ventricular filling (preload) and increase left ventricular ejection impedance (afterload) independent of any change in metabolic load. Although much attention has been focused on the relation between work of breathing, respiratory muscle fatigue, and ventilatory failure, there has been little interest in the cardiovascular effects of spontaneous ventilation in respiratory disease states and their subsequent effects on ventilation. This is surprising when one considers that most subjects with cardiovascular insufficiency who are not receiving mechanical ventilatory support die a respiratory death with apnea preceding cardiac standstill. Furthermore, failure to wean from mechanical ventilation is commonly associated with cardiovascular deterioration.

Ventilation as exercise

Spontaneous ventilatory efforts are induced by contraction of the respiratory muscles, including primarily the diaphragm and intercostal muscles. However, with increased ventilatory effort, abdominal wall muscles and muscles of the shoulder girdle also function as accessory respiratory muscles. Blood flow to these muscles is derived from several arterial sources whose absolute flow probably exceeds the maximum metabolic demand of exercising skeletal muscle under normal conditions (Roussos and Macklem 1982). Thus, under conditions of normal cardiovascular function, blood flow is not the limiting factor determining maximum ventilatory effort. Normally, spontaneous ventilation at rest requires less than 5 per cent of total O_2 delivery to meet its demand. However, in subjects with lung disease where the work of breathing is increased, such as pulmonary edema or bronchospasm, or with markedly increased minute ventilation requirements, such as breathing at high altitudes, the requirements for O_2 may increase to 25 or 30 per cent of total O_2 delivery, and maximum minute ventilation limits will define maximum exercise level. Furthermore, if cardiac output is limited, blood flow to other organs and to the respiratory muscles may be compromised, inducing both tissue hypoperfusion and lactic acidosis. Supporting the respiratory efforts of a subject with marginal cardiovascular reserve will relieve the metabolic demand on the cardiovascular system, reduce lactic acidosis, and increase mixed venous O_2 saturation $S\text{v}O_2$. Similarly, weaning from mechanical ventilation or spontaneous ventilatory efforts against an increased load or at an increased level will decrease $S\text{v}O_2$. Thus, in subjects with physiologically significant right to left shunts (>5 per cent), changes in mixed $S\text{v}O_2$ will induce directionally similar changes in arterial oxygenation. Therefore arterial oxygenation may decrease during weaning without any real change in gas exchange abnormalities.

Hemodynamic effects of changes in lung volume

Lung hyperinflation can occur during weaning from mechanical ventilation if expiratory time is inadequate. This scenario is common in patients with chronic obstructive pulmonary disease and acute bronchospasm. To the extent that respiratory rate increases during weaning trials, inadequate time of exhalation may occur. Acute hyperinflation may occur in patients with chronic obstructive pulmonary disease because of airway collapse during expiration. Hyperinflation can occur at any time in the management of the critically ill as long as inspiratory time is inadequate to allow complete exhalation. This may occur during weaning as a result of sensations of anxiety, respiratory distress, or increased work of breathing due to either respiratory apparatus resistance or intrinsic airway collapse. Hyperinflation has three important hemodynamic effects.

1. Hyperinflation places the respiratory muscles at a disadvantage, increasing the work cost of breathing (addressed above). However, if intrinsic positive end-expiratory pressure develops, the negative swings in intrathoracic pressure necessary to ventilate will become exaggerated and will alter cardiac function via changes in intrathoracic pressure, as described below.

2. Hyperinflation acts like tamponade to compress the heart within the cardiac fossa, decreasing biventricular end-diastolic volumes. Lung distension will compress both ventricles in the cardiac fossa (Marini *et al.* 1980). Because the chest wall and diaphragm can move away from the expanding lungs, whereas the heart is trapped within its cardiac fossa, juxtacardiac intrathoracic pressure may increase more than lateral chest-wall or diaphragmatic intrathoracic pressure. This decrease in 'apparent' left ventricular diastolic compliance when only intralumenal left ventricular pressure and left ventricular output are measured is lost when pericardial pressure is also measured, so that the left ventricular distending pressure can be calculated (intralumenal left ventricular pressure minus pericardial pressure). Lung compression of the heart is physiologically analogous to cardiac tamponade and is seen in subjects with severe airflow obstruction.

3. Hyperinflation increases pulmonary vascular resistance. Lung distension can increase pulmonary vascular resistance by overdistending aerated lung units and compressing alveolar vessels (Marini *et al.* 1980). Hyperinflation-induced increased pulmonary vascular resistance may be the most important processes impeding right ventricular ejection in subjects with chronic obstructive pulmonary disease.

Hemodynamic effects of changes in intrathoracic pressure

The heart within the thorax is a pressure chamber within a pressure chamber. Therefore changes in intrathoracic pressure will affect the pressure gradients for both systemic venous return to the right ventricle and systemic outflow from the left ventricle, independent of the heart itself. Decreases in intrathoracic pressure by increasing venous return and impeding left ventricular ejection will increase intrathoracic blood. Spontaneous inspiratory efforts, by decreasing intrathoracic pressure, both increase lung volume and decrease right atrial pressure (P_{ra}). The fall in P_{ra} accelerates blood flow into the right ventricle (Pinsky 1984). This increased venous return is transmitted to the pulmonary artery on the subsequent beat. Thus normal respiration-associated hemodynamic changes maximize ventilation–perfusion temporal matching because the spontaneous inspiration matches an increase in alveolar oxygen flux with an increase in pulmonary capillary flow. However, the maximum increase in venous blood flow is limited because, as P_{ra} becomes negative with respect to atmospheric pressure, the veins collapse as they enter the chest, limiting blood flow. Accordingly, maximum venous return is rapidly reached during spontaneous inspiration and further decreases in intrathoracic pressure do not produce any additional increase in venous return. Thus markedly negative swings in intrathoracic pressure, as may occur with spontaneous inspiratory efforts with upper airway obstruction, bronchospasm, or pulmonary edema, do not induce a massive increase in venous return which, potentially, could impair right ventricle function. Furthermore, since venous return is limited, the hemodynamic consequences of marked negative swings in intrathoracic pressure are not primarily due to increased venous return. However, decreases in intrathoracic pressure at a constant arterial pressure will increase left ventricular transmural pressure and thus increase left ventricular afterload, impeding left ventricular ejection (Buda *et al.* 1979). Spontaneous inspiration, by decreasing intrathoracic pressure, increases left ventricular afterload. This afterload effect will progressively increase as the swings in intrathoracic pressure become more negative until the ejection pressure is so high that no left ventricular ejection can occur.

To put both the venous return and left ventricular afterload processes together, loaded spontaneous ventilatory efforts decrease left ventricular stroke volume via a complex mechanism collectively called pulsus paradoxus. Transient intraventricular septal shift into the left ventricular lumen from right ventricular dilation plus pericardial volume restraint decrease absolute left ventricular end-diastolic volume. Increases in left ventricular afterload impede left ventricular ejection, increasing left ventricular end-systolic volume. Although the increase in left ventricular afterload can be explained by increases in left ventricular transmural pressure alone, increases in aortic input impedance and altered series contraction of the left ventricular myocardium have also been described during loaded spontaneous inspiration. However, these two additional effects appear to be mild.

Accordingly, weaning patients from positive-pressure ventilation, by allowing the return of decreases in intrathoracic pressure, may precipitate acute left ventricular failure and pulmonary edema in patients with borderline left ventricular function (Lemaire *et al.* 1988). In this regard, weaning can be seen as a form of cardiac stress testing because left ventricular loading invariably occurs in the transition from positive-pressure to spontaneous ventilation. Furthermore, this line of reasoning forms the basis for the assertion made above that cardiovascular impairment must play an important role in subjects who fail to wean from mechanical ventilatory support.

Clinical applications

The hemodynamic consequences of weaning from positive-pressure ventilation may be profound. In patients with markedly increased work of breathing, hypervolemia, or impaired left ventricular pump function, spontaneous ventilation may worsen cardiovascular function by adding an additional stress. Spontaneous ventilation will promote further cardiogenic pulmonary edema and hypoxemia. The re-institution of mechanical ventilatory support, by decreasing the work cost of breathing and abolishing the negative swings in intrathoracic pressure, can be lifesaving in this setting because of its ability to support the cardiovascular system while decreasing global O_2 consumption, independent of any beneficial effects that mechanical ventilation may have on gas exchange.

In patients with decreased pulmonary elastic recoil, increased pulmonary vascular resistance, hypovolemia, or airflow obstruction, the withdrawal of ventilatory support invariably increases intrathoracic blood volume and left ventricular afterload and can be used as a cardiovascular stress test to ascertain whether the patient's cardiovascular status is primarily determined by venous return of right ventricular ejection pressure. Thus the initiation and withdrawal of ventilatory support can be seen as a ventilatory probe into the determinants of cardiovascular homeostasis in the ventilator-dependent patient. Patients with stable but limited cardiac reserve may often not be weaned from mechanical ventilatory support unless supplemented by exogenous positive inotropes.

Perhaps the best documentation of cardiopulmonary interaction during weaning is the ventilator-dependent subject with chronic obstructive pulmonary disease. Weaning of such patients will tax the cardiovascular system. Many patients with severe chronic obstructive pulmonary disease and apparently adequate respiratory parameters for weaning may go into severe cardiogenic pulmonary edema during the weaning trial (Lemaire *et al.* 1988). Similarly, left ventricular ejection fraction may decrease in such patients during weaning trials but not during positive-pressure ventilation, suggesting that occult left ventricular failure is a common occurrence during the weaning of patients with chronic obstructive pulmonary disease from mechanical ventilation.

References

Buda, A.J., *et al.* (1979). Effect of intrathoracic pressure on left ventricular performance. *New England Journal of Medicine*, **301**, 453–9.

Lemaire, F., *et al.* (1988). Acute left ventricular dysfunction during unsuccessful weaning from mechanical ventilation. *Anesthesiology*, **69**, 171–9.

Marini, J.J., Culver, B.N., and Butler, J. (1980). Mechanical effect of lung distention with positive pressure on cardiac function. *American Review of Respiratory Disease*, **124**, 382–6.

Pinsky, M.R. (1984). Determinants of pulmonary arterial flow variation during respiration. *Journal of Applied Physiology*, **56**, 1237–45.

Roussos, C. and Macklem, P.T. (1982). The respiratory muscles. *New England Journal of Medicine*, **307**, 786–97.

2.11.3 Increased work of breathing

R. Dhand and M. J. Tobin

Key messages

- Work of breathing can be calculated from measurements of airway pressure and volume in a patient receiving controlled mechanical ventilation (passive ventilation).
- In a patient who is assisting the ventilator, calculation of work of breathing requires measurement of airway pressure, esophageal pressure, and volume.
- Work of breathing can be partitioned into various components, such as pulmonary flow-resistive work, elastic work, and expiratory work.
- Accurate measurements of work of breathing are difficult to obtain, and there are several limitations to their calculation and interpretation.
- Patients who fail to wean from the ventilator exhibit increased values of work of breathing, but the usefulness of measuring work of breathing in predicting failure from weaning has not been evaluated prospectively.

Introduction

Mechanical ventilation is commonly initiated because the work of breathing is increased and patients have a decreased capacity to perform this work. To wean a patient from the ventilator successfully it is important to correct the imbalance between the imposed respiratory load and the ability of the patient's respiratory muscles to perform the required work of breathing. Therefore measurements of work of breathing can provide important information about the pathophysiological nature of respiratory compromise in a ventilator-dependent patient. However, accurate measurements of respiratory work are difficult to obtain in ventilator-dependent patients (Tobin and Van de Graaff 1994).

Mechanical work is performed when a force moves its point of application through a distance. In mechanically ventilated patients, the mechanical work of breathing can be calculated by measuring the generation of intrathoracic pressure (P) due to contraction of the respiratory muscles (or a ventilator substituting for them) and the displacement of gas volume (V):

$$W = PV = \int P\mathrm{d}v.$$

The changes in pressure and volume can be analyzed graphically, and the area enclosed within a volume–pressure loop has the units of mechanical work. The various trans-structural pressures involved in overcoming different aspects of respiratory work include pressure difference across the lung (transpulmonary pressure) (P_{L}) and pressure difference across the respiratory system (P_{rs}):

$$P_{\mathrm{L}} = P_{\mathrm{aw}} - P_{\mathrm{pl}}$$

$$P_{\mathrm{rs}} = P_{\mathrm{aw}} - P_{\mathrm{bs}}$$

where P_{aw} is airway pressure, P_{pl} is pleural pressure, and P_{bs} is body surface pressure.

The respiratory muscles perform work against several different forces (Tobin and Van de Graaff 1994), including elastic forces developed in the tissue of the lungs and chest wall when a volume change occurs, flow-resistive forces resulting from flow of gas through the airways, viscoelastic forces resulting from stress adaptation within the lung and chest wall, and plastoelastic forces within thoracic tissues which are reflected in the difference in static elastic recoil of the lung and chest wall during inflation and deflation. Work required to overcome inertial forces, which depend on the mass of tissues and gases, is minimal and is usually ignored in calculating the work of breathing. The work required to overcome gravitational forces, to compress intrathoracic gas, and to distort the chest wall from its relaxed (passive) configuration makes a small contribution to the overall work of breathing. In addition, in ventilator-dependent patients work is performed to overcome the external applied impedances of many types, including resistive, elastic, and threshold loads.

Units of work

The most popular units of work are kilogram meters (kg m) and joules (J) (Tobin and Van de Graaff 1994). In general, 0.1 kg m approximates 1 J, and this can be visualized as the energy that is needed to move 1 liter through a pressure gradient of 10 cmH_2O.

Work performed by the ventilator

When a relaxed subject is passively inflated by a ventilator, the mechanical work necessary to inflate the total respiratory system can easily be determined by measuring inhaled volume and the airway pressure (P_{aw}). During passive ventilation, P_{aw} represents the pressure difference across the respiratory system (P_{rs}). Work performed on the elastic properties of the respiratory system can be shown graphically as a plot of P_{rs} against volume, as represented by the triangular area ABCA in Fig. 1. Airway pressure obtained at point C is determined under static or quasistatic conditions. If compliance is assumed to be constant, the elastic work of inspiration can be calculated as $(V_{\mathrm{T}} \times 0.5) \times$ pressure change; alternatively, it can be calculated as mean pressure × volume change.

Work performed in overcoming air flow resistance, and to some degree resistance due to movement of the chest wall and lung tissue, during passive inflation can be shown on a plot of volume versus P_{aw}; the inspiratory curve is bowed to the right of the compliance line, and the area ACDA represents the work of inspiration against air flow resistance (Fig. 2).

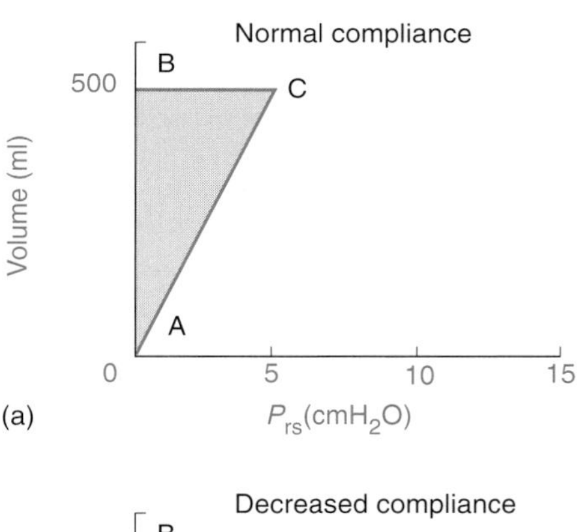

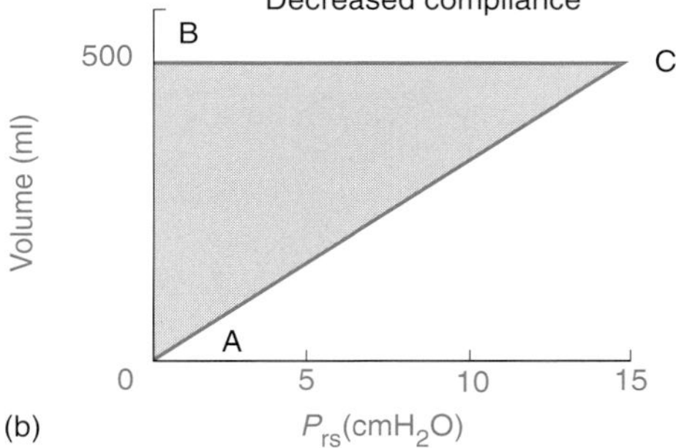

Fig. 1 Work performed in inflating the total respiratory system against elastic forces in (a) a normal subject and (b) a patient with decreased respiratory compliance. Work, represented by the triangular shaded area ABCA, is greater in the patient with decreased compliance. The slope of AC represents compliance of the respiratory system. (Reproduced with permission from Tobin and Van de Graaff (1994).)

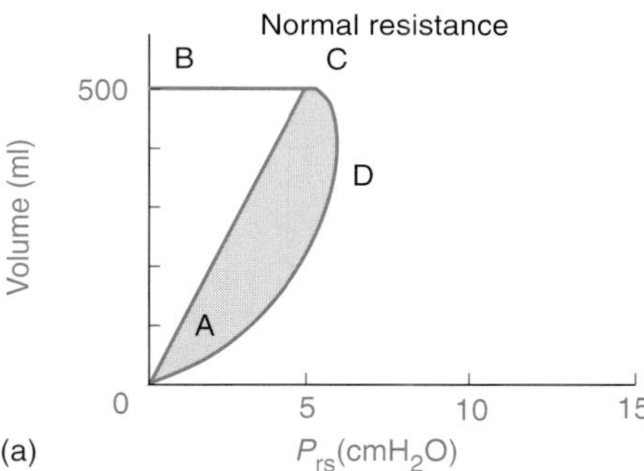

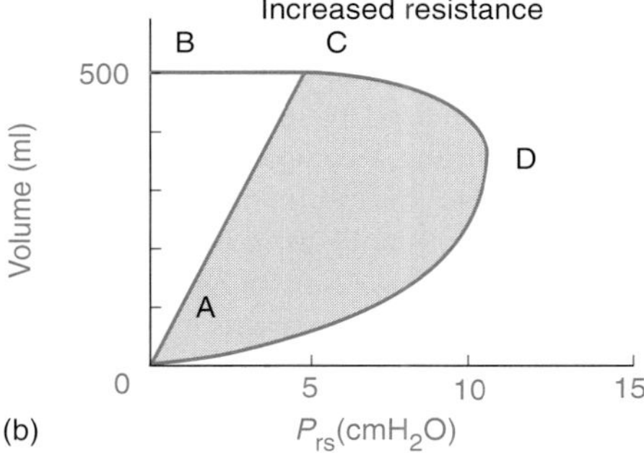

Fig. 2 Work performed in overcoming flow resistance during passive inflation is represented by the shaded area ACDA in (a) a normal subject and (b) a patient with increased resistance. Work performed against elastic forces is represented by the area ABCA. (Reproduced with permission from Tobin and Van de Graaff (1994).)

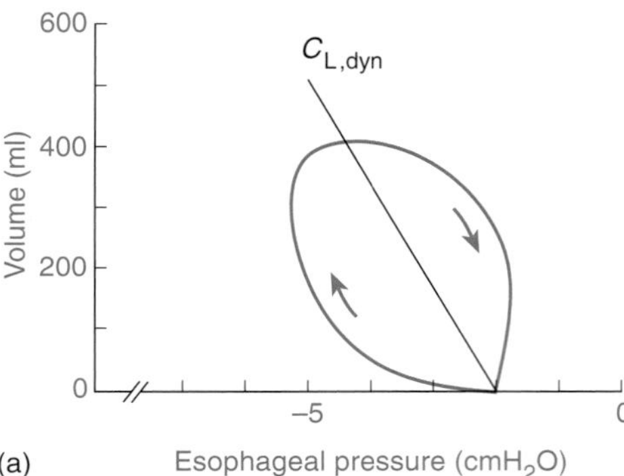

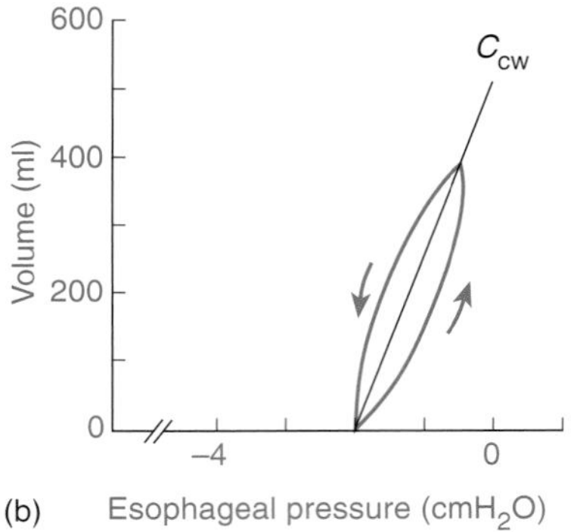

Fig. 3 A plot of esophageal pressure versus volume during (a) unassisted breathing and (b) passive ventilation. The $C_{L,dyn}$ line between the points of zero flow (the dynamic pulmonary compliance line) partitions resistive work into its inspiratory and expiratory components. Note the clockwise direction in which actively generated P_{es}–volume curves are formed. The C_{cw} line connecting the points of zero flow during passive ventilation is the static chest wall compliance line. Note the counterclockwise direction in which passively generated P_{es}–volume curves are formed. (Reproduced with permission from Tobin and Van de Graaff (1994).)

Work performed by the patient

Measurements based on airway pressure

In this method, it is assumed that similar amounts of work are required to inflate the respiratory system during controlled and assisted ventilation (Marini *et al.* 1985). The area enclosed within an inflation P_{rs}–volume curve is measured during controlled ventilation (Marini *et al.* 1985). Another curve is obtained with the patient assisting the ventilator under conditions of inspiratory flow, tidal volume, respiratory frequency, and end-expiratory pressure identical with those during controlled ventilation. The active work being performed by the patient's inspiratory muscles can be calculated by subtracting the area of the 'active' curve from that of the 'passive' curve (Marini *et al.* 1985).

Measurements based on esophageal pressure

Unless the ideal conditions defined above are met, recordings of P_{rs} are insufficient for measurement of respiratory work in a patient who is assisting the ventilator. Instead, an estimate of pleural pressure (P_{pl}) is required, which is usually achieved by recording esophageal pressure (P_{es}). The graphical approach to the analysis of P_{es}–volume loops introduced by Campbell allows work to be separated into several components (Tobin and Van De Graaff 1994).

Pulmonary flow-resistive work

During unassisted breathing, the total resistive work done on the lungs in each breath is obtained by integrating the area subtended by P_{es} and lung volume during a complete respiratory cycle (Fig. 3). Work is partitioned into its inspiratory and expiratory resistive components by drawing a line between points of zero flow, i.e. the dynamic pulmonary compliance line. In Fig. 3, the area enclosed by the ellipse to the left of this line represents inspiratory resistive work while the area to the right represents expiratory resistive work.

Elastic work

During unassisted breathing, work performed by the inspiratory muscles against elastic recoil of the lungs and chest wall can be calculated by constructing a Campbell diagram. This method requires measurement of the static compliance of the chest wall (C_{cw}), which can be closely estimated by recording P_{es} and volume during controlled ventilation while the respiratory muscles are completely relaxed (Fig. 3). The difference between P_{es} and P_{cw} indicates the pressure developed by the respiratory muscles in expanding the chest wall (P_{mus}).

The chest wall compliance line is fitted to the pressure–volume curve by passing it through the end-expiratory elastic recoil pressure of the chest wall, which coincides with P_{es} at zero flow (Fig. 4). The chest wall compliance line is generally assumed to be linear and is extrapolated to the tidal volume of each breath. Elastic inspiratory work is then measured as the area between the dynamic pulmonary compliance line ($C_{L,dyn}$) and the static chest wall compliance line (C_{cw}) within the tidal volume range (Fig. 4).

Expiratory work

During quiet breathing, expiration is passive and the work of breathing is usually performed entirely by the inspiratory muscles. However, when ventilatory demands increase or expiratory resistance is markedly increased, the expiratory muscles are recruited. Such expiratory muscle activity increases P_{es}, producing P_{es}–volume values to the right of the chest wall relaxation line and outside the elastic work area on a Campbell diagram, i.e. P_{es} is higher than the static elastic recoil pressure of the chest wall. This can be quantified

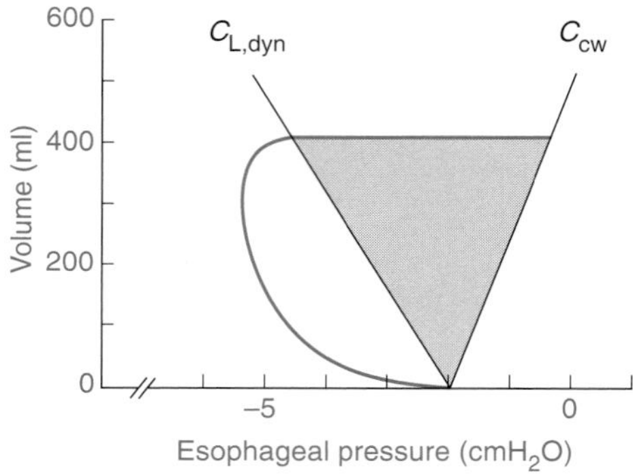

Fig. 4 Plot of esophageal pressure versus volume during unassisted ventilation. The static chest wall compliance line (C_{cw}) is fitted at the end-expiratory elastic recoil pressure of the chest wall. Inspiratory elastic work is calculated as the shaded area of the pressure–volume curve subtended by the C_{cw} and $C_{L,dyn}$ lines (see Fig. 3). (Reproduced with permission from Tobin and Van de Graaff (1994).)

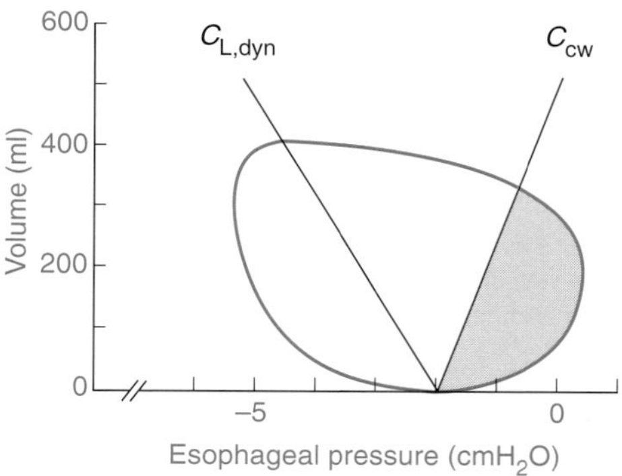

Fig. 5 Plot of esophageal pressure versus volume in a patient exhibiting expiratory muscle activity. The chest wall compliance line is measured and fitted to the pressure–volume curve as described in Figs 3 and 4. Expiratory work is quantified as the shaded area enclosed by the portion of the pressure–volume loop lying to the right of the C_{cw} line. (Reproduced with permission from Tobin and Van de Graaff (1994).)

as the area enclosed by the expiratory portion of the P_{es}–volume loop to the right of the chest wall relaxation line (Fig. 5).

Normal values of respiratory work

In the clinical setting, respiratory work should be described in terms of both work per minute and work per liter. Work per minute, also called power or work rate ($\dot{W}$), is calculated by multiplying the work of one breath by the corresponding respiratory frequency. Work per liter is calculated by dividing work per minute by minute ventilation. When breathing through the nose, the normal resting work of breathing per minute is approximately 3.9 J/min or 0.47 J/l. Work per liter closely reflects abnormalities in pulmonary mechanics (increased resistance, decreased compliance), whereas work per minute is more dependent on minute ventilation and is less well correlated with abnormalities in lung mechanics.

Increased respiratory work as a determinant of weaning outcome

A number of investigators have examined the utility of respiratory work measurements in predicting the outcome of a trial of weaning from mechanical ventilation. Fiastro *et al.* (1988) measured the work of spontaneous breathing in patients who were successfully weaned from mechanical ventilation and in patients who were ventilator dependent. Only the work being performed on the lungs was assessed. All the ventilator-dependent patients had work per minute values above 15.7 J/min and four (66 per cent) had work per liter values above 1.37 J/l (the remaining two patients each had work values of 1.08 J/l). Importantly, neither index alone (work per minute, work per liter) was as good at predicting weaning outcome as the combination of the two indices.

Henning *et al.* (1977) reported that patients remained ventilator dependent when the work rate was greater than 16.66 J/min, whereas spontaneous ventilation could usually be sustained if the work rate was 9.80 J/min or less. Again, these investigators measured work performed on the lungs only. Proctor and Woolson (1973) found that a work rate of 13.3 J/min provided the best separation of ventilator-dependent and ventilator-independent postoperative patients (657 measurements in 168 patients). However, this threshold value was associated with false-positive and false-negative rates of 14 per cent.

In conclusion, work of breathing is higher in ventilator-dependent than ventilator-independent patients, but the predictive value of work measurements as an index of weaning has not been prospectively determined. Furthermore, accurate measurements of work of breathing are difficult to perform, and there are several limitations to their calculation and interpretation (Tobin and Van de Graaff 1994). Whether work of breathing measurements provide information which cannot be obtained by simpler predictors of weaning outcome, such as the degree of rapid shallow breathing, is unclear. Therefore the ability of work of breathing measurements to enhance clinical decision-making needs to be examined in a prospective fashion.

References

Fiastro, J.F., Habib, M.P., Shon, B.Y., and Campbell, S.C. (1988). Comparison of standard weaning parameters and the mechanical work of breathing in mechanically ventilated patients. *Chest*, **94**, 232–8.

Henning, R.J., Shubin, H., and Weil, M.H. (1977). The measurement of the work of breathing for the clinical assessment of ventilator dependence. *Critical Care Medicine*, **5**, 264–8.

Marini, J.J., Capps, J.S., and Culver, B.H. (1985). The inspiratory work of breathing during assisted mechanical ventilation. *Chest*, **87**, 612–18.

Proctor, H.J. and Woolson, R. (1973). Prediction of respiratory muscle fatigue by measurements of the work of breathing. *Surgery, Gynecology and Obstetrics*, **136**, 367–70.

Tobin, M.J. and Van de Graaff, W.B. (1994). Monitoring of lung mechanics and work of breathing. In *Principles and practice of mechanical ventilation* (ed. M.J. Tobin), pp. 967–1003. McGraw-Hill, New York.

2.11.4 Causes of failure to wean

T. E. Oh

Key messages

- Failure to wean is due to an imbalance between respiratory demand and respiratory muscle capacity.
- Conditions which give rise to this situation include:
  - premature weaning with unresolved underlying illness;
  - poor respiratory muscle capacity from any etiology;
  - inadequate or inappropriate nutrition;
  - excessive breathing load imposed by circuits;
  - left ventricular dysfunction precipitated by weaning;
  - severe agitation.

Introduction

Weaning is the process whereby mechanical ventilation is gradually withdrawn and the patient resumes spontaneous breathing to allow eventual successful extubation. However, there is no universally accepted definition of successful weaning. Proposed definitions specify varying periods of sustained spontaneous breathing following discontinuation of mechanical ventilatory support, and some definitions require successful extubation as a condition. None of these definitions note the time at which the weaning process starts. With widespread use of patient-assisted ventilatory modes, it has become difficult to stipulate when mechanical ventilation changes from primary support to assistance in weaning. Hence the rate of weaning failure can range from 20 to 70 per cent, depending on the definition used for weaning success. Failure to wean is distinct from the need for continuing tracheal intubation, for example for airway protection.

Causes of failure to wean

Patients who repeatedly fail to wean pose considerable clinical and economic problems. In general, difficulties with weaning from mechanical ventilation arise when there is an imbalance between respiratory demand and respiratory muscle capacity. The major causes of failure to wean are listed in Table 1. Other causes are dyssynchrony between the patient and the ventilator due to inappropriate sedation or ventilator settings.

Table 1 Causes of difficult weaning

Unresolved underlying disease
Poor respiratory muscle capacity
Impaired central respiratory drive
Respiratory muscle atrophy and fatigue
Myopathies and neuropathies
Inadequate or inappropriate nutrition
Excessive respiratory load
Inspiratory load of breathing circuits
Intrinsic PEEP
Left ventricular dysfunction
Severe agitation

Unresolved underlying disease

Ongoing respiratory failure, fever, sepsis, multiple organ dysfunction, hemodynamic instability, and biochemical derangements contribute to inadequate oxygenation and gas exchange. Obviously, if weaning is initiated before the precipitating illness has resolved significantly, it is more likely to fail. Decreased level of consciousness secondary to the primary illness is common and will also jeopardize the success of weaning if it is started prematurely. Unfortunately, there are no clinical or laboratory indices which reliably define when a patient has recovered sufficiently from the underying condition to begin the weaning process.

Poor respiratory muscle capacity

Impaired central respiratory drive

Poor respiratory muscle capacity appears to be the principal cause of weaning failure. Residual effects of sedative agents will diminish spontaneous ventilation. However, if residual drug effects and premature weaning are excluded, a depressed central respiratory drive is rarely responsible for weaning failure (Slutsky 1994).

Respiratory muscle atrophy and fatigue

Difficulty with weaning is usually associated with patients who have received long-term mechanical ventilation, indicating an impairment of global respiratory muscle function with prolonged ventilation. The diaphragm is a major component of respiratory muscles, but its precise contribution in weaning failure is unknown; phrenic nerve function is usually satisfactory.

Concepts of disuse atrophy following prolonged ventilation and muscle fatigue during weaning are widely accepted. Increased work of breathing will lead to respiratory muscle fatigue when respiratory demands exceed capacity. During weaning, this is superimposed on some degree of muscle weakness and atrophy resulting from the period of mechanical ventilation. Unsuccessful weaning has been attributed to respiratory muscle fatigue, as electromyographic and transdiaphragmatic pressure measurements have demonstrated decreased diaphragmatic contractility and slowing of muscle relaxation rate (Goldstone *et al.* 1994). Nevertheless, the clinical importance of respiratory muscle fatigue remains unclear. Patients who fail a trial of weaning commonly have increased respiratory rates with low tidal volumes. This rapid shallow breathing pattern is seen as an adaptation to spare fatigued respiratory muscles. However, laboratory-induced fatigue (by resistive loading) does not result in rapid shallow breathing. Maximum inspiratory pressure (Pi_{max}) is a poor determinant of weaning outcome, and thus muscle strength *per se* may be assumed to be relatively unimportant in sustaining spontaneous breathing. Unfortunately, the pathophysiological mechanisms of respiratory muscle dysfunction, and the roles played by wasting, weakness, and fatigue, are poorly understood (Slutsky 1994). There are no satisfactory methods of measuring the different components of respiratory function in patients on ventilator support. Techniques to assess respiratory muscle function in the laboratory are not reliably applicable at the bedside. Respiratory muscle function is likely to be impaired in varying degrees, but the clinical significance of this dynamic continuum has not been investigated. Also, data on respiratory muscle function at the time when weaning is initiated are poor.

Myopathies and neuropathies

Of course, weaning difficulties are encountered in neuromuscular diseases (e.g. Guillain–Barré syndrome and motor neuron disease). Persistent paralysis in critically ill patients receiving mechanical ventilation has been reported following administration of corticosteroids or the non-depolarizing muscle relaxants vecuronium and pancuronium. The etiology is not clear, as reported features are inconsistent. Electrophysiological, serum creatine kinase, and muscle biopsy studies have revealed a neuropathic process in some afflicted patients and a myopathic process in others. An increased concentration of a vecuronium metabolite and renal dysfunction in patients who have received long-term vecuronium administration have been proposed as causes. The term 'critical illness polyneuropathy' has been used for patients reported to have unexplained persistent muscle weakness during weaning from mechanical ventilation. Again, the etiology and clinical picture are unclear, but the pathophysiological pattern is neurogenic and not myopathic (Raps *et al.* 1994). Axonal degeneration involving motor and sensory nerves is seen, and sepsis and multiple organ failure are commonly associated. The diagnosis is made clinically and is unrelated to blood concentrations of cytokines. These myoneuropathies may be manifestations of a common condition caused by toxic or metabolic disturbances following long-term mechanical ventilation. The contribution of neuromuscular diseases to weaning failure is probably underestimated.

Inadequate or inappropriate nutrition

Critically ill patients, including those on mechanical ventilation, are at risk of developing malnutrition. Unfortunately, the nutritional requirements of ventilated patients are difficult to study. Consequently, prospective randomized clinical trials to determine the role of nutritional support for ventilated critically ill patients are not available. Although the energy expenditure of critically ill patients receiving mechanical ventilation has been reported, data on energy requirements during weaning are lacking. Nonetheless, malnutrition affects muscle structure and function in humans. Since respiratory muscle endurance is important for meeting ventilatory demands during weaning, poor nutrition may contribute to weaning failure.

Although the optimal nutritional regimen for ventilated or weaning patients has not been determined, excessive administration of glucose may compromise successful weaning. Increased CO_2 production and hypercarbia result from increased glucose combustion and from lipogenesis, as excess glucose is converted to fat.

Excessive respiratory load

Inspiratory load of breathing circuits

Breathing difficulties caused by ventilators and circuits are often overlooked during weaning from mechanical ventilation. Additional work of breathing imposed by weaning circuits commonly contributes to weaning failure. Significant resistance to gas flow, resulting in increased inspiratory work, is imposed by endotracheal tubes, humidifiers, and ventilators. Circuit hoses have relatively small flow resistances as they have a large internal diameter (22 mm). From Poiseuille's law

$$\text{flow} = \frac{\Delta P r^4}{L},$$

where ΔP is the pressure gradient across the tube, r is the tube radius, and L is the tube length, small endotracheal tube sizes present very large resistances. *In vivo* endotracheal tube resistance may be greater, presumably due to kinking and intraluminal secretions.

Bubble or cascade water-bath humidifiers, in which gas is dispersed through heated water, have higher resistances than blow-by or pass-over types, in which gas passes over the water surface. Resistance is higher with low peak inspiratory flows, and, if used, bubble humidifiers contribute to weaning failure.

Ventilators use inspiratory demand valves to respond rapidly to changing spontaneous ventilatory efforts. Unfortunately, these demand valves add considerably to the work of breathing (Beydon *et al.* 1988), depending on the way that gas is delivered in response to an initiated spontaneous breath. Ventilators with pressurized gas reservoirs, for example Servo 900C (Siemens Elema AB, Solna, Sweden) and Engstrom Erica (Gambro Engstrom AB, Bromma, Sweden), produce a high gas flow response and can compensate to some extent for demand valve impedance. The high inspiratory loads imposed by the demand valves of some ventilators, particularly older models, are prominent but often unrecognized factors in weaning failure.

Continuous positive airways pressure (CPAP) circuits are available either incorporated in ventilators or as stand-alone units. Valveless circuits obviously impose less inspiratory work. The Bennett 7200a ventilator (Puritan-Bennett Corporation, California, USA) has a 'flow-by' facility for CPAP breathing which uses a separate continuous-flow circuit to bypass the demand valve. CPAP circuits which induce large swings in airway pressure during the respiratory cycle result in greater work of breathing. Weaning will also be severely compromised if breathing circuits do not provide sufficient fresh gas flow during spontaneous inspiration, particularly if the patient increases minute ventilation intermittently.

Intrinsic positive end-expiratory pressure (PEEP)

Patients with obstructive lung disease demonstrate some degree of expiratory flow limitation during weaning, a period when high respiratory rates, short expiratory times, and increased minute volumes are commonly seen. Dynamic hyperinflation results in an alveolar pressure that is greater than atmospheric pressure. This auto-PEEP or intrinsic PEEP imposes an inspiratory load, and the increased work of breathing will compromise successful weaning.

Left ventricular dysfunction

Failure to wean patients with heart or chronic lung disease may be due to the onset of left ventricular failure with cardiogenic pulmonary edema. With the resumption of spontaneous breathing, onset of respiratory distress is accompanied by an abrupt increase in pulmonary artery wedge pressure (Lemaire *et al.* 1988). This left ventricular dysfunction is due to acute myocardial ischemia and is associated with coronary artery disease. Myocardial scintographic studies have shown altered myocardial perfusion, left ventricular dilatation, and decreased left ventricular ejection fraction. Acute impairment of left ventricular function is probably a consequence of increased afterload due to an increase in systemic blood pressure and a change in intrathoracic pressure from positive to negative. The situation is worsened by the work imposed by poorly designed circuits and hypoxemia from any cause. Weaning will be difficult if the cardiopulmonary stress is not recognized.

Severe agitation and delirium

Weaning from mechanical ventilation is a time of physical and emotional stress for the patient. Some patients may be delirious or become extremely agitated, at a time when excessive use of sedative agents is avoided. The patient may experience significant pain. Weaning will not be successful if extreme delirium or agitation is poorly controlled, or if there is oversedation.

References

Beydon, L., Chasse, M., Harf, A., and Lemaire, F. (1988). Inspiratory work of breathing during spontaneous ventilation using demand valves and continuous flow systems. *American Review of Respiratory Disease*, **138**, 300–4.

Goldstone, J.C., Green, M., and Moxham, J. (1994). Maximum relaxation rate of the diaphragm during weaning from mechanical ventilation. *Thorax*, **49**, 54–60.

Lemaire, F., Teboul, J.-L., Cinotti, L., Giotto, G., Abrouk, F., and Steg, G. (1988). Acute left ventricular dysfunction during unsuccessful weaning from mechanical ventilation. *Anesthesiology*, **69**, 171–9.

Raps, E.C., Bird, S.J., and Hansen-Flaschen, J. (1994). Prolonged muscle weakness after neuromuscular blockade in the intensive care unit. *Critical Care Clinics*, **10**, 799–813.

Slutsky, A.S. (1994). Consensus conference on mechanical ventilation—January 28–30, 1993 at Northbrook, Illinois, USA. Part 2. *Intensive Care Medicine*, **20**, 150–62.

2.11.5 Therapeutic approach

T. E. Oh

Key messages

- The decision to start weaning is made on subjective clinical assessments rather than on outcome predictors.
- Conditions such as adequate recovery from the precipitating illness and hemodynamic stability must be met before weaning is started. Potential causes of weaning failure must be excluded.
- An organized approach should be planned for each patient.
- Common ventilatory modes used for weaning are T-tube breathing, intermittent mandatory ventilation, and pressure support ventilation. Use of low levels of continuous positive airways pressure is recommended.
- The clinician chooses the most appropriate mode. If used properly, there is no evidence to show that one mode is superior to the others. Ventilatory modes can be used in combination; if the patient is unsettled, another mode may be tried on an empirical basis.
- The pace of weaning and time to resume ventilatory support are decided on clinical grounds.
- Close observation, pulmonary care, and psychological reassurance are vital. Attention must be paid to pain relief, cardiac failure, anemia, electrolyte imbalance, and acid–base disorders.
- Appropriate nutrition avoids a high carbohydrate intake.
- The patient must be allowed sufficient rest and sleep. Overnight ventilation will help fatigued respiratory muscles to recover.
- Close observation extends beyond extubation or separation from the ventilator. Non-invasive ventilatory support can be useful in this period.

Introduction

Weaning from mechanical ventilation is instituted when the patient has recovered sufficiently to breathe spontaneously. Many patients can be weaned easily, but the process remains a challenge in those ventilated for prolonged periods. Patients who repeatedly fail to wean pose considerable clinical, economic, and ethical problems. Planning a rational approach is helpful (Knebel *et al.* 1994), taking into consideration the many interacting factors (Table 1).

Table 1 Associated factors that should be considered when planning weaning

Stages of weaning
Preweaning
Weaning process
Separation from ventilator
Decisions
When to initiate weaning
Weaning mode to use
Adjustments of ventilator
Use of CPAP, level of CPAP
Change weaning mode
Duration of weaning period, when to alternate ventilatory support
When to intervene to stop weaning if the patient is not coping
Assessment and monitoring
Clinical assessments
Adequate recovery from illness
Hemodynamic stability
No anemia, electrolyte imbalance, acid–base disorders
Vital signs
Ventilator settings
Respiratory variables
Oxygenation
Patient–ventilator dyssynchrony
State of respiratory muscles
Degree of fatigue
Oxygenation
Therapy
General supportive care
Reassurance, psychological support
Pain relief, appropriate sedation
Measures to encourage sleep
Physiotherapy
Nutrition
Respiratory care

Prediction of weaning outcome

The first fundamental problem in weaning is deciding when to initiate the process. With widespread use of patient-assisted ventilatory modes, it is difficult to stipulate when mechanical ventilation changes from primary support to assistance in weaning. Nevertheless, delayed discontinuation of mechanical ventilation or repeated failed attempts at extubation could be avoided if there were criteria to predict weaning success or failure. The accuracy of these predictors of weaning outcome (Table 2) are conflicting owing to differences in definitions, study methodologies, and cut-off values to separate success from failure. This is not surprising, given the many determinants of ventilator dependence. One proposed predictor, the ratio of frequency to tidal volume (Yang and Tobin 1991), is simple to measure and is an index of rapid shallow breathing. Nonetheless, there is no consistent evidence to support the usefulness of any set of criteria to hasten the process of weaning (Slutsky 1994). The decision to start weaning is still based on subjective bedside assessments, particularly during unsupported or minimally supported breathing. Rapid shallow breaths, use of accessory respiratory muscles, and prominent paradoxical movement of the abdomen and rib cage are signs of probable decompensation.

Table 2 Reported predictors of weaning outcome: cut-off variables suggesting weaning success

Standard weaning criteria
Respiratory frequency <35 breaths/min
Tidal volume ≥ 5 ml/kg
Vital capacity ≥ 10 ml/kg
Pi_{max} ≥ −25 cmH_2O
Pa_{O_2} ≥ 60 mmHg (on F_{IO_2}=0.6)
$P(A–a)O_2$ < 350 mmHg
Pa_{CO_2} < 60 mmHg
V_D/V_T < 0.6
Expired minute volume < 10 l/min
MVV > twice baseline minute volume
Work of breathing < 13 J/min
$\triangle V_{O_2}$ < 15% total body consumption
$P_{0.1}$ < 5 cmH_2O
Pi < 40% Pi_{max}
$Pi/Pi_{max} \times T_i/T_{tot}$ < 0.15
f/V_T < 105
CROP > 13
Pi/Pi_{max} < 0.3

CROP, integrated index of compliance, frequency, oxygenation, and airway pressure.
f/v_T, maximum voluntary ventilation.
MVV, maximum voluntary ventilation.
$P(A–a)O_2$, alveolar–arterial oxygen gradient.
$P_{0.1}$, mouth occlusion pressure.
Pi, inspiratory pressure.
Pi_{max}, maximum inspiratory pressure.
$Pi/Pi_{max} \times T_i/T_{tot}$, transdiaphragmatic inspiratory pressure: time index.
$\triangle V_{O_2}$, oxygen cost of breathing.
V_D/V_T, ratio of dead-space to tidal volume.

Ventilatory modes of weaning

Ventilatory modes of weaning are based on a progressive reduction in the contribution of the ventilator and a progressive increase in the patient's contribution to ventilation. While moderate muscle activity is desirable for successful weaning, the extent of muscle exertion required and the value of muscle training remain unknown. The ventilatory modes commonly used for weaning are T-tube breathing and the two patient-assisted modes, intermittent mandatory ventilation and pressure support ventilation. Mandatory minute volume ventilation, which incorporates intermittent mandatory ventilation or pressure support ventilation, can be classified as another weaning mode (see below).

Studies to determine the optimal weaning mode have been reported (see below), but the numbers of patients studied were small. Weaning protocols, definitions of weaning success/failure, and study populations were not standardized. Despite much enthusiasm for a particular mode, there is no consistent evidence to show that any mode is superior to the others when each mode is properly used. Indeed, modes are often used in combination, and the traditional mode, T-tube weaning, is still widely used.

T-tube circuit weaning

This weaning mode compels the patient to breathe spontaneously through a T-tube circuit after disconnection from the ventilator. Periods of unassisted breathing are gradually lengthened according to the patient's capability. The duration of each T-tube 'trial' and the integration and mode of alternating ventilator support are empirical. Assist–control ventilation is frequently used, and overventilation with respiratory alkalosis may occur. If that happens, the patient's P_{CO_2} and serum bicarbonate will be inappropriately low when T-tube weaning resumes. The circuit itself is simple and valveless, and imposes insignificant work of breathing. Unfortunately, the change from ventilator support to unassisted spontaneous breathing is an abrupt 'sink or swim' physiological adjustment, which some patients tolerate poorly. Esteban *et al.* (1995) recently reported that a once-daily T-piece trial of up to 2 h resulted in more rapid extubation than intermittent mandatory ventilation and pressure support ventilation.

Intermittent mandatory ventilation weaning

Intermittent mandatory ventilation avoids the drastic change to full unassisted breathing by allowing spontaneous breathing between preset ventilator-delivered breaths. The ventilator rate is progressively reduced as the patient's breathing ability improves, at a pace determined clinically. Advantages claimed for intermittent mandatory ventilation include better venous return via the thoracic pump mechanism and an ability to exercise respiratory muscles. Synchronized intermittent mandatory ventilation is a modification whereby the mechanical breaths are provided by patient-triggered ventilation. Advantages over intermittent mandatory ventilation are unproven. Intermittent mandatory ventilation may induce increased work of breathing due to the circuit demand valve and imprecise breath-to-breath co-ordination between ventilator and patient.

Pressure support ventilation weaning

Pressure support ventilation also allows a transition from mechanical to spontaneous ventilation. Each breath is initiated by the patient but is supported by a constant preset airway pressure. The pressure support ceases after a given fraction of inspiratory time, or when inspiratory flow falls below a predetermined fraction of the initial inspiratory flow. Expiration is passive. The pressure support is gradually decreased to a level that will compensate for circuit impedance (about 2–5 cmH_2O); however, effort is still required to open the demand valve. Brochard *et al.* (1994) found that pressure support ventilation was associated with a shorter duration of weaning and a higher success rate than T-tube breathing or intermittent mandatory ventilation.

Mandatory minute volume ventilation weaning

Mandatory minute volume ventilation is a general term used to describe a variety of ventilatory support modes that interact with the patient to deliver a predetermined minimum minute volume. The basic ventilator mode is intermittent mandatory ventilation or pressure support ventilation. Spontaneous breathing is monitored by the ventilator, which then self-adjusts to deliver the preset minute volume. Hence ventilator assistance automatically decreases as the patient's spontaneous efforts improve, i.e. the patient 'self-weans'. Mandatory minute volume ventilation was originally described with intermittent mandatory ventilation as the basic mode and minute volume as the targeted variable. Some newer ventilators use respiratory rate or tidal volume as the targeted variable, which is regulated by changing pressure support ventilation or flow pattern. Forms of mandatory minute volume ventilation are varied and can be confusing; they include volume-assured pressure support, volume support, and pressure-regulated volume control. The concept is appealing, but the technology for low-resistance rapid-response mandatory minute volume ventilation is complicated. Rapid shallow breathing may not be recognized, and the ventilator algorithm may induce non-physiological breathing patterns. Clinical experience is limited.

Airway pressure release ventilation and bi-positive airway pressure

These are newer ventilatory modes that have also been used for weaning. With airway pressure release ventilation, the ventilator applies a high continuous positive airways pressure (CPAP) and intermittently 'releases' the airway pressure to a lower level using a time-controlled release valve. The higher CPAP level and lung volume are re-established when the release valve is closed. The airway pressure release ventilation pressure versus time profile is the inverse of that for intermittent mandatory ventilation. Gas exchange, and thus ventilator support, is provided by lung deflation, which is superimposed on spontaneous CPAP breathing. Peak airway pressures never exceed the higher CPAP level.

Bi-positive airway pressure is similar to airway pressure release ventilation, except that the pressure release rate depends on the patient's spontaneous breathing rate (i.e. airway pressure is released every two to six spontaneous breaths). Better co-ordination between spontaneous and pressure release breaths may result. Pressure support ventilation can be added to augment spontaneous breathing. The role of airway pressure release ventilation and bi-positive airway pressure in weaning is unclear, and clinical experience is limited. Proportional assist ventilation is a reported new mode that augments spontaneous breathing by increasing pressure delivered to the airway in proportion to inspiratory effort. Both experience and availability of proportional assist ventilation remain limited.

Therapeutic approach

Time to initiate weaning

The weaning process is started when the patient's general condition is stable and the likelihood of success is judged to be good. Weaning is unlikely to be successful and must not be started if the conditions in Table 3 are not satisfied. Potential causes of weaning failure should be excluded. Patients with heart failure or chronic obstructive pulmonary disease may experience cardiopulmonary stress induced by weaning. This weaning-initiated left ventricular dysfunction should be recognized, and appropriate diuretic, vasodilator, or bronchodilator therapy given. The optimum level of arterial $P\text{CO}_2$ to maintain prior to weaning is unknown. Zealous correction of hypercarbia in patients with ventilatory dysfunction shortly before weaning may subsequently increase the patient's work of breathing. With patient-assisted ventilation, the patient will usually maintain a satisfactory $P\text{CO}_2$ level. Weaning is best started in the morning, when a full complement of staff is available.

Ventilatory strategy

The choice of weaning mode is determined by the attending clinician. Intermittent mandatory ventilation and pressure support ventilation are frequently used in combination. Both may be gradually decreased, or the level of one may be gradually decreased while the other is kept constant. One may be used as the predominant mode. For example, intermittent mandatory ventilation may be used and pressure support ventilation kept constant at a low level to overcome breathing circuit resistance. Alternatively, pressure support ventilation may be used with a constant low intermittent mandatory ventilation rate (e.g. 2–4 breaths/min). Mandatory minute volume ventilation, incorporating intermittent mandatory ventilation or pressure support ventilation, may be used if available. Despite proper use of a weaning mode, patients may sometimes feel more comfortable with another mode or combination mode. It seems reasonable to try another predominant mode on an empirical basis if the patient appears unsettled.

There are good reasons to introduce CPAP with whichever weaning mode chosen. The beneficial effects of CPAP stem from an increase in functional residual capacity and compliance, which potentially improve gas exchange and decrease work of breathing. Low levels of CPAP (e.g. below 5 cmH_2O) can reduce inspiratory work without increasing intrinsic positive end-expiratory pressure (auto-PEEP) and dynamic hyperinflation.

With all patient-assisted modes, triggering sensitivity and inspiratory flow rate must be adjusted to give the best co-ordination between patient and ventilator. Less effort is required for the patient to initiate inspiration with flow triggering (in ventilators with this facility) than with pressure triggering. Attention should be paid to the endotracheal tube size, humidifier and circuit resistance, equipment dead-space, demand valve, and fresh gas flow rate to minimize inspiratory work. Circuits which offer high resistances to air flow must be identified and rejected.

Whatever the mode or combination of modes used, unassisted breathing through a T tube or CPAP circuit can be the final stage of the weaning process. Bi-positive airway pressure can be tried as an intermediate stage between intermittent mandatory ventilation/pressure support ventilation and CPAP breathing for patients who are particularly difficult to wean. The CPAP level is gradually decreased to 2 cmH_2O, which approximates to intrinsic PEEP. Extubation at this low level of CPAP (without decreasing CPAP to zero) is proposed by some, as better effects on lung volume and oxygenation are claimed.

Table 3 Conditions necessary before weaning is initiated
Adequate resolution of the precipitating cause or causes of respiratory failure
Hemodynamic stability
Stable neurological status with an adequate central respiratory drive
Absence of residual paralyses or excessive sedation
Adequate resolution of fever and failure of other systems
Correction of anemia and electrolyte imbalance
Appropriate nutrition
Acceptable inspiratory loads imposed by breathing circuits

The optimal pace of weaning from mechanical ventilation is not established and has to be individually assessed. Respiratory muscles must not be overburdened, as fatigued muscles may require 10 h or more to recover. Relatively abrupt weaning may be applicable for patients with readily reversible forms of respiratory failure. However, graded weaning is necessary for sicker patients and those who recover more slowly. A more rapid weaning pace with earlier extubation must be weighed against a higher likelihood of reintubation. Adequate ventilatory support must be provided between weaning periods, particularly at night to permit sufficient sleep.

Supportive care

Close observation and good bedside care are essential during the weaning process. The patient should be positioned upright in bed or, if able to be moved out of bed, positioned sitting up in a chair. Mobilization is to be encouraged. Skilled physiotherapy, clearance of respiratory tract secretions, and monitoring of respiratory variables must be performed regularly. The use of pulse oximeters and CO_2 analyzers has reduced the need for frequent arterial blood gas sampling, but these tests should not be neglected. Stimulation of the respiratory center with analeptic agents is unlikely to be useful and cannot be recommended for routine use. Theophylline has been reported to improve diaphragmatic contractility, but this effect is small and evidence of clinical usefulness is lacking. The progress of the patient must be continually assessed. Criteria to terminate weaning do not exist. The attending nurse and doctor must appraise the level of fatigue that necessitates halting the weaning trial without waiting for absolute exhaustion to occur. Signs of cardiorespiratory distress must be recognized early.

Attention must be given to alleviating the patient's physical and emotional stress. Some patients may become extremely agitated during weaning. Pain may be present. Sedative agents and analgesics should be given cautiously. A continuous low-dose infusion of an opioid is often useful, supplemented if necessary by benzodiazepines. Regional local analgesic blocks can be applied where appropriate. Good communication, reassurance, and encouragement by staff (particularly to take periodic deep breaths) contribute to psychological support. A non-stressful environment and adequate rest and sleep between breathing trials should be part of the therapeutic aims.

Adequate nutrition must be given to provide energy for weaning, but excessive calories and high carbohydrate loads lead to excessive CO_2 production and lipogenesis, and must be avoided. For the majority of patients, 2000 to 3000 kcal/day will be appropriate, but the optimal nutritional regimen for weaning from mechanical ventilation is unknown.

Concurrent with providing nutrition, potential problems of sepsis, anemia, electrolyte imbalance, endocrine dysfunction, acid–base disorders, and unwanted effects of medications must be addressed. Potassium, calcium, magnesium, and phosphate are important minerals for muscle function, and deficiencies must be promptly corrected.

After extubation/separation from ventilator

Weaning from mechanical ventilation can occupy up to 41 per cent of the total time of ventilatory support. Nonetheless, extubation does not mark the end of the weaning process. The patient must continue to be closely monitored for 24 to 48 h following separation from the ventilator, when rapid deterioration can occur. Patients who were difficult to wean may benefit from continuing ventilatory support by non-invasive means after removal of their endotracheal tubes. Non-invasive ventilation or mask/nasal CPAP or bi-positive airway pressure may sometimes obviate the need for reintubation in those who initially deteriorate. There is no urgency to extubate patients with tracheostomies. Oxygen therapy, physiotherapy, airway toilet, and supportive care remain vitally important.

Other strategies

Some weaning techniques and strategies have been reported to facilitate or shorten the weaning process. They include respiratory muscle training, biofeedback, and use of an exclusive weaning team. The validity of these claims has not been universally accepted. Fundamental principles, such as close monitoring by staff and attention to the many factors that can adversely affect the process, are more important. An organized approach for each patient should be followed.

References

Brochard, L., *et al.* (1994). Comparison of three methods of gradual withdrawal from ventilatory support during weaning from mechanical ventilation. *American Journal of Respiratory and Critical Care Medicine*, **50**, 896–903.

Esteban, A., *et al.* (1995). A comparison of four methods of weaning patients from mechanical ventilation. *New England Journal of Medicine*, **332**, 345–50.

Knebel, A.R., Shekleton, M.E., Burns, S., Clochesy, J.M., Goodnough Hanneman, S.K., and Ingersoll, G.L. (1994). Weaning from mechanical ventilation: concept development. *American Journal of Critical Care*, **3**, 416–20.

Slutsky, A.S. (1994). Consensus conference on mechanical ventilation—January 28–30, 1993 at Northbrook, Illinois, USA. Part 2. *Intensive Care Medicine*, **20**, 150–62.

Yang, K.L. and Tobin, M.J. (1991). A prospective study of indexes predicting the outcome of trials of weaning from mechanical ventilation. *New England Journal of Medicine*, **324**, 1445–50.

2.12 Acute pneumonia

2.12.1 Pathophysiology

J. Rello and A. Torres

Key messages

- The majority of cases of nosocomial pneumonia are the result of aspiration of oropharyngeal and gastric contents to the lower airways.
- The development of pneumonia depends on interaction between bacterial virulence, innoculum size, and local lung defenses.
- Etiology in nosocomial pneumonia results from the selection of certain pathogens by local or systemic host factors, the virulence of the organism, the underlying disease of the patient, and exposure to antimicrobial agents.

Mechanisms of airway colonization and infection

The understanding of the pathogenesis of nosocomial pneumonia is absolutely necessary for the rational comprehension of risk factors involved with this intrahospital infection. The pathogenic scheme recently recommended by the Centers for Disease Control (Tablan *et al.* 1994) is shown in Fig. 1. 'Abnormal' colonization of the oropharynx and the gastric reservoir and the subsequent aspiration of their contents to lower airways in patients whose mechanical, cellular, and humoral defenses are impaired may lead to the development of nosocomial pneumonia (Table 1). The inoculation of micro-organisms to the oropharynx or lower airways is another related mechanism. An alternative mechanism may be the bacterial translocation of enteric bacteria and their products from ischemic gut to lung. Bacteremic lung metastasis is the pathogenic mechanism in some cases of nosocomial pneumonia (Ramphal 1994).

Abnormal oropharyngeal and gastric colonization

The majority of cases of nosocomial pneumonia are the result of aspiration of oropharyngeal and gastric contents to the lower airways. The 'abnormal' colonization of the oropharynx is a clear risk factor for the development of nosocomial pneumonia. Gram-negative bacillary colonization of the oropharynx increases from 35 per cent in moderately ill patients to 73 per cent in critically ill patients. Furthermore, the rate of nosocomial pneumonia is almost eight times higher in colonized compared with non-colonized patients. Intimal mechanisms for 'abnormal' oropharyngeal colonization are the release of oral proteases and the loss of fibronectin which is necessary to prevent the adherence of Gram-negative bacilli to the epithelial cells of the oropharyngeal mucosal surface. Antibiotic therapy seems to be one of the main mechanisms related to oropharyngeal colonization.

Transcolonization from contiguous structures to the oropharynx is an important feature in oropharyngeal colonization. Reservoirs such as dental plaque, sinuses, and nasal mucosa could be the initial reservoir for oropharyngeal colonization. The presence of sinusitis as a risk factor for pneumonia has been reported but is not widely appreciated. The presence of sinusitis has been linked to the nasal placement and duration of endotracheal and gastric intubation. Nosocomial pneumonia occurs significantly more frequently in the patients with maxillary sinusitis.

Gastric colonization as a reservoir for retrograde oropahryngeal colonization or direct tracheal colonization was detected some years ago. The stomach is usually sterile, but at alkaline pH (>4) may contain more than 100 million organisms/ml. Increases in gastric colonization may occur with advanced age, achlorhydria, alterations in gastric juice secretion, and particularly with treatment with antacids or H_2 blockers prescribed for stress bleeding prophylaxis in critically ill patients. Although still controversial, several randomized

Table 1 Pathogenesis

Mechanisms of airway colonization and infection
Oropharyngeal and gastric colonization
Aspiration to lower airways
Translocation
Inoculation of aerosols
Mechanisms of microbial clearance
Cellular defenses
Phagocytosis
Intracellular defense mechanisms
Activation of cellular defenses
The inflammatory cascade
Action of micro-organisms in immunological terms
Humoral defenses
Mechanical defenses

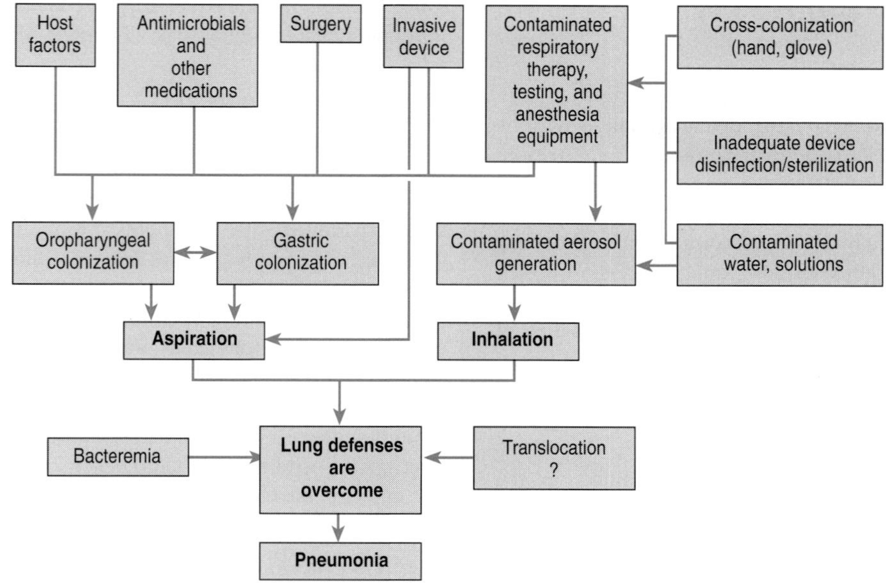

Fig. 1 Pathogenic scheme recommended by the Centers for Disease Control.

studies have shown that rates of nosocomial pneumonia, particularly in mechanically ventilated patients, are decreased when sucralfate is given rather than antacid or H_2 blockers. The absence of gastric alkalinization with sucralfate and its antibacterial effect may explain the protective effect of this drug against nosocomial pneumonia (Craven *et al.* 1994).

Another factor related to the existence of bacteria in the gastric cavity is the presence of bilirubin in gastric contents. Indeed, the presence and amount of bilirubin in the stomach is related to a decrease in duodenal motility.

Aspiration

Aspiration frequently occurs in healthy people during sleep and is one of the final steps in the development of nosocomial pneumonia. The amounts of aspiration and of bacterial inoculum are variables related to this pathogenic mechanism of nosocomial pneumonia. Overall, factors that promote alteration of conciousness favor aspiration. It is important to remember that even the presence of an endotracheal tube may not avoid the risk of aspiration between the tracheal tube cuff and the tracheal wall (Rello *et al.* 1997). Aspiration of gastric contents to lower airways is favored by the supine position and by the presence of a nasogastric tube.

Nasogastric intubation is a risk factor for nosocomial pneumonia since it indirectly favors several risk factors for pneumonia: oropharyngeal colonization, gastroesophageal reflux, bacterial migration, and sinusitis. Gastroesophageal reflux is a constant feature in mechanically ventilated patients with a nasogastric tube in place and potentially may increase oropharyngeal colonization. Reducing the size of nasogastric tubes would have a beneficial effect in reflux. Maintaining the patient in the upright position reduces the amount of aspiration of gastric contents to the lower airways.

An endotracheal tube 'bypasses' the natural mechanical host defenses and obviously facilitates the entry of bacteria into the lung. Also, leakage around the cuff allows the pooled secretions accumulated above the cuff to enter the trachea. Suction to remove subglottic secretions is an effective prophylactic mechanism for ventilator-associated pneumonia. The importance of careful airway management, including maintenance of adequate intracuff pressure, in reducing the risk of pneumonia should be emphasized (Rello *et al.* 1996).

Biofilm formation in endotracheal tubes and the dislodgement of these biofilms with bacteria to the lower airways is another risk factor suggested to be implicated in ventilator-associated pneumonia.

Bacterial translocation

Bacterial or toxin translocation from gut to lung is an alternative mechanism for the development of nosocomial pneumonia. Intramucosal gastric ischemia measured by tonometry was invariably present in a group of critically ill patients with nosocomial pneumonia and was also an independent risk factor in a multivariate analysis. More prospective investigations are needed to confirm that intramucosal gastric ischemia is a risk factor for the development of nosocomial pneumonia.

Aerosols or contamination of respiratory equipment

External colonization by aerosol inoculation or contamination of respiratory equipment is a well-known mechanism for respiratory nosocomial transmission. Among several types of equipment, nebulizers are the most dangerous since the size of particles produced easily bypasses the lower-airway defenses.

In mechanical ventilators, excessive manipulation of tubing and the formation of water condensate are risk factors for nosocomial pneumonia. Water humidifiers favor the formation of water condensate in ventilator tubing. Heat moisture exchangers that recycle exhaled heat and moisture eliminate water condensate.

Mechanisms of bacterial clearance

The development of pneumonia depends on interaction between bacterial innoculum size and local lung defenses. These may be altered by the underlying disease, comorbidities, malnutrition, and some therapies. Mechanical defenses prevent micro-organisms from entering the lower airway tract. If these defenses are bypassed in the normal host, viable organisms are eradicated from the lower respiratory tract by a very efficient and rapid process in which killing of micro-organisms by phagocytes is of great importance. However, uncontrolled infection may develop in the presence of highly virulent organisms, large microbial inoculums, or failure of cellular defenses to restrict microbial replication. In the normal host, local and systemic defenses can be further amplified by activation of an immune response mediated by cytokines that limits microbial invasion but may cause serious abnormalities in the lung or other organs.

Mechanical defenses

Mechanical defenses include filtration and humidification of inspired air in the upper airways, intact epiglottic and cough reflexes, tracheobronchial secretions, and mucociliary transport. Cigarette smoke alters both the ciliated epithelium and macrophage activity.

The presence of an endotracheal tube produces mucociliary dysfunction and a less effective cough. If mucociliary clearance is slow, respiratory tract mucus traps bacteria which, instead of being removed, proliferate and lead to both colonization and infection. In addition, if there is excessive mucus secretion, the bacterial receptors in secretions may promote colonization by providing organisms with more loci to bind within the airway.

Cellular defenses

Cell-mediated immunity is modulated by alveolar macrophages, T lymphocytes, and polymorphonuclear neutrophils. Alveolar macrophages account for 85 per cent of the total cellularity of the alveoli and are the most important defensive system against microbes in the lung. Macrophages interact with bacteria through the binding of molecules to specific membrane surface receptors.

The most important process during phagocytosis is oxidative burst. Non-oxidative mechanisms include the release of enzymes such as proteases, hydrolases, and lysozymes. Lymphocytes form 10 per cent of the alveolar cell population and produce soluble mediators that stimulate alveolar macrophages. This mechanism seems to be regulated by cytokines. In addition, alveolar macrophages are able to release several mediators (e.g. interleukin 8, C5a, leukotriene B4, and platelet activating factor) with potent chemotactic activity for polymorphonuclear bodies. These cells act mainly through phagocytosis and exert their antibacterial activity in a dose-dependent manner via the release of granular enzymes. They constitude the primary defense against pathogens such as *Staphylococcus aureus* or fungi. Finally, natural killer cells are lymphocytes responsible for immunological surveillance of viral infections.

Humoral immunity

B lymphocytes, immunoglobulins, and complement constitute the key points of humoral immunity. Structures on the microbial surface, such as endotoxin and certain carbohydrate domains, can activate complement components through the alternate pathway. In immune hosts, binding of specific immunoglobulin G or M molecules to the microbial surface initiates activation of the classical pathway. Both pathways of complement activation result in binding of C3b to the infectious agent, which can be recognized by the type 1 complement receptors on neutrophils and macrophages. In addition, local secretory immunoglobulin A promotes clumping and decreases the adherence of a variety of bacteria to the airway. It also protects the lung against certain viruses, and deficit of this substance predisposes to recurrent episodes.

Histopathological aspects

Postmortem studies of human ventilator-associated pneumonia (Fabregas and Torres 1996) have demostrated that pneumonia in critically ill patients is a diffuse polymicrobial process with a non-homogeneous distribution of micro-organisms, Furthermore, it is an ongoing dynamic structural process, showing different degrees of histological evolution, which are multifocal and predominate in lung-dependent zones. The association of pneumonia and diffuse alveolar damage is not an uncommon finding. The histological and microbiological characteristics of ventilator-associated pneumonia may be explicable by the unusual flow and volume patterns developed by mechanical ventilators which probably favor the distribution of bacteria from central to distal airways. Finally, it is important to note that there is a clear relationship between therapy and the decrease in lung burden.

References

Craven, D., Steger, K., and Duncan, R. (1994). Nosocomial pneumonia in critically ill and mechanically ventilated patients. In *Respiratory infections: a scientific basis for management* (ed. M.S. Niederman, G. Sarosi, and J. Glassroth), pp. 147–62. W.B. Saunders, Philadelphia, PA.

Fabregas, N. and Torres, A. (1996). New histopathological aspects of human ventilator-associated pneumonia. In *Yearbook of intensive care and emergency medicine* (ed. J.L. Vincent), pp. 520–30. Springer-Verlag, Berlin.

Ramphal, R. (1994). Pathogenesis of airway colonization. In *Respiratory infections: a scientific basis for management* (ed. M.S. Niederman, G. Sarosi, and J. Glassroth), pp. 45–56. W.B. Saunders, Philadelphia, PA.

Rello, J., Soñora, R., Jubert, P., Artigas, A., Rué, M., and Vallés, J. (1996). Pneumonia in intubated patients: role of respiratory airway care. *American Journal of Respiratory and Critical Care Medicine*, **154**, 111–15.

Rello, J., Cabello, H., and Torres, A. (1997). Epidemiology, risk and prognostic factors of nosocomial pneumonia. *European Respiratory Reviews*, in press.

Tablan, O.C., *et al.* (1994). Guidelines for prevention of nosocomial pneumonia. Part I. Issues on prevention of nosocomial pneumonia. *Infection Control and Hospital Epidemiology*, **15**, 588–625.

2.12.2 Causes and diagnosis of severe community-acquired pneumonia

J. Dorca and A. Torres

Key messages

- In community-acquired pneumonia, the need for admission to an intensive care unit (ICU) has been estimated to be between 3 and 5 per cent of total cases.
- The outcome of ICU-admitted patients with community-acquired pneumonia is poor, with most studies reporting a mortality near 50 per cent.
- Severe community-acquired pneumonia shows a characteristic etiological pattern different from that of milder cases. Although *Streptococcus pneumoniae* remains the most common pathogen, *Legionella pneumophila*, *Hemophilus influenzae*, enteric Gram-negative bacilli, and *Staphylococcus aureus* are not infrequent.
- The initial diagnostic approach to severe pneumonias should include sputum Gram stain and culture, two serial blood cultures, and acute phase serology for *Legionella* and atypical agents.
- If an invasive approach is considered, a transthoracic needle aspiration with an ultrathin needle is particularly indicated in non-ventilated pneumonias, whereas a bronchoscopic protected specimen brush should be preferred in patients receiving assisted ventilation.

Epidemiology

Pneumonia is a common infection in the community. The estimated incidence in developed countries has been reported to be between two and four cases per 1000 adult population. The rate of admission to hospital for community-acquired pneumonia appears to be more variable, with figures ranging from 22 to near 50 per cent. The need for admission to an intensive care unit (ICU) has been estimated to be between 3 and 5 per cent, although some studies report higher percentages. In general, elderly patients or those suffering chronic debilitating diseases are more prone to require hospital and, eventually, ICU admission. However, approximately one-third of patients admitted to hospital and the ICU are previously healthy individuals.

Community-acquired pneumonia is a relevant cause of mortality. Although the incidence of fatal outcome decreased dramatically after the introduction of antibiotics, the mortality in developed countries has remained essentially stable during the last 40 years, with a reported incidence of approximately 30 deaths per 100 000 population. Mortality is directly related to the severity of the episode. The mortality in cases treated in an ambulatory setting has been evaluated to be below 1 per cent, whereas in those cases needing hospital admission it is between 11.5 and 25 per cent, and several studies report a mortality of almost 50 per cent in ICU-admitted cases.

Various demographic and clinical factors have been related to the probability of a fatal outcome. Age has been shown to be independently associated with mortality, although its relative importance appears to be lower than other prognostic factors. Male gender may also be related to death probability, but to a low degree. Chronic debilitating diseases such as diabetes mellitus, chronic obstructive pulmonary disease, alcoholism, hepatic cirrhosis, chronic heart failure, renal insufficiency, neuralgic or mental disease, malignancy, and immunosuppression have been shown to be clearly related to mortality. As shown in Table 1, different clinical features at presentation, and also some laboratory and radiographic findings, have been associated with pneumonia outcome in univariate and multivariate studies. Finally, some etiologies, particularly *Staphylococcus aureus*, Gram-negative enteric rods, and *Legionella pneumophila*, present a higher probability of a fatal outcome when compared with other common etiological agents such as *Streptococcus pneumoniae*, *Mycoplasma pneumoniae*, or *Hemophilus influenzae*.

Etiology

Because of the limitations of diagnostic methods, the presence of a reliable etiological agent can be proved in only half of the community-acquired pneumonias affecting the general adult population. As shown in Table 2, the organisms identified most frequently are *Strep. pneumoniae*, *M. pneumoniae*, respiratory viruses, and *Chlamydia pneumoniae*. *Coxiella burnetii* has been reported to be prevalent in some rural areas. Other organisms, such as *L. pneumophila*, *H. influenzae*, Gram-negative aerobic bacilli, and *Staph. aureus*, are rare. In the case of severe community-acquired pneumonia, this etiological pattern shows some modifications (Table 2), as has been demon-

Table 1 Clinical features associated with unfavorable outcome of community-acquired pneumonia

Advanced age (>65 years) and/or coexisting chronic debilitating diseases
Physical signs on admission
Respiratory rate >30 breaths/min
Diastolic blood pressure ⩽60 mmHg or systolic blood pressure ⩽90 mmHg
Temperature >38.5 °C
Evidence of extrapulmonary sites of infection
Confusion and/or decreased level of consciousness
Laboratory findings
Severe leukocytosis ($>30\times10^9$/l) or leukopenia ($<4\times10^9$/l)
Respiratory failure: $Pa\text{O}_2<60$ mmHg (8.0 kPa) or $Pa\text{CO}_2>50$ mmHg (6.67 kPa) breathing room air
Abnormal renal function: serum creatinine >1.2 mg/dl (>106 μmol/l) or BUN >20 mg/dl
Anemia (hematocrit <30%)
Metabolic acidosis
Evidence of coagulation disorders
Radiographic findings
Multilobe involvement
Rapid radiographic progression
Evidence of cavitation
Presence of a significant pleural effusion

BUN, blood urea nitrogen.

strated in different studies using sophisticated or invasive methods. *Strep. pneumoniae* remains the most common pathogen, but the relative frequencies of *L. pneumophila*, *H. influenzae*, enteric Gram-negative bacilli, and *Staph. aureus* are substantially increased. In contrast, other etiologies such as *M. pneumoniae*, *C. pneumoniae*, and respiratory viruses become less relevant in this setting. *L. pneumophila* presents an important geographical and temporal variation. Occasionally, *Mycobacterium tuberculosis* may be responsible for severe community-acquired pneumonia.

The etiology of community-acquired pneumonia in patients from nursing homes has some peculiarities, particularly in those who are severely limited and handicapped. In such cases the etiological pattern may resemble that of nosocomial pneumonia, with enteric Gram-negative bacilli being the most common cause. Community-acquired pneumonia in patients who are immunosuppressed because of treatment for cancer or with long-term steroids, or in those with known HIV infection, may also be caused by a variety of opportunistic agents.

Table 2 Etiology of community-acquired pneumonia in the general population and in cases with severe presentation

Pathogens	General population	Severe presentation
Major	*Strep. pneumoniae* *M. pneumoniae* *C. pneumoniae* (?)	*Strep. pneumoniae* *L. pneumophila*
Less common	Viruses Other atypical bacteria *H. influenzae* *L. pneumophila*	*H. influenzae* Enteric Gram-negative bacilli
Rare	Enteric Gram-negative bacilli *Staph. aureus*	Atypical bacteria *Staph. aureus*

Diagnostic methods

The classic microbiological approach to community-acquired pneumonia is based on examination of easily obtainable samples (respiratory secretions, blood, or pleural fluid if available) for stains and culture. Serology has been the basic method for demonstrating the presence of atypical pathogens that are difficult to culture in non-specialized laboratories. The general effectiveness of this classic approach is far from optimal; only a minority of cases can be diagnosed during the episode, and obtaining a reliable diagnosis before antibiotic prescription is rare. Several new 'rapid' tests, generally based on immunological techniques, have been introduced to improve the yield of older methods. However, very few have satisfied a cost–benefit analysis. Molecular biology appears to be a promising alternative, but few tests are commercially available yet. Finally, more reliable respiratory samples can be obtained by invasive procedures such as bronchoscopy or direct transthoracic aspiration, but, owing to potential complications, their use is reserved for more severe cases.

Non-invasive procedures

Examination and culture of respiratory secretions

The sputum Gram stain has been the cornerstone of the microbiological investigation of community-acquired pneumonia for many years. Although its diagnostic efficacy is a matter of debate, it is a simple, rapid, and economical first step in patients not previously treated with antibiotics or with sputum or tracheal aspirate microbiological processing. The predominance of a particular bacterial morphology (Gram-positive diplococci, Gram-negative bacilli, Gram-negative coccobacilli, etc.) is a good indication of the most likely pathogen. In addition, the absence of bacteria in a sample with polymorphonuclear cells suggests *Legionella* or an atypical agent. Culture of respiratory secretions is of less value because contaminant flora can overgrow the real pathogens. Consequently, cultures are accepted only if they agree with a Gram stain. In practice, cultures are particularly useful for demonstrating obligate pathogens such as *L. pneumophila* or *M. tuberculosis*.

Pneumococcal antigen detection by counter-immunoelectrophoresis, latex agglutination, coagglutination, and enzyme immunoassay has been proposed in order to improve the sensitivity of Gram stain in pneumococcal pneumonia. However, these techniques have not been found to be cost effective, and consequently they are seldom carried out in clinical practice. Antigen detection for *H. influenzae* is only available for typable strains that seldom produce pneumonia in adults. Direct fluorescent antigen detection of *L. pneumophila* in sputum may be useful in selected cases although it is insensitive. The use of molecular biology techniques in sputum samples has recently been introduced for obligate pathogens such as *L. pneumophila*. Unfortunately, DNA probes were not very sensitive for detecting this organism.

Blood culture

Usually two blood cultures are routinely collected before antibiotic treatment is started. The sensitivity of this procedure is low (about 25 per cent in the case of pneumococcal pneumonia) although it is highly specific. The presence of bacteremia increases the risk of complications, and so the result of this procedure also has prognostic implications.

Pleural fluid

If there is significant pleural effusion, the pleural fluid should always be analyzed in order to exclude empyema. The diagnostic efficacy of pleural fluid Gram stain and culture is low. Detection of pneumococcal antigen in pleural fluid by latex agglutination could be a simple and effective diagnostic method when pleural effusion is present.

Urine

Detection of pneumococcal antigen in urine by counter-immunoelectrophoresis has an acceptable sensitivity. However, it is a cumbersome procedure. Detection of *L. pneumophila* serogroup 1 antigen in urine using latex agglutination or enzyme immunoassay is commercially available and seems to have an acceptable diagnostic efficacy.

Serology

Serological testing for atypical bacteria is of little help in the acute phase of pneumonia. Paired acute and convalescent serological testing usually provides a reliable retrospective diagnosis, useful only in epidemiological studies. The detection of IgM antibodies in acute *M. pneumoniae* pneumonia has proved to be a useful test. Serology for viruses has little practical use.

Invasive procedures

Transtracheal needle aspiration

Transtracheal needle aspiration, although still recommended in textbooks, is seldom performed owing to its poor tolerance. Although it has some lack of specificity, its sensitivity is very high. Minor complications have been reported occasionally. This procedure could be considered in non-intubated severe pneumonia without expectoration.

Bronchosocopic procedures

Among bronchoscopic methods, the protected specimen brush has been widely demonstrated to be highly sensitive and specific provided that the patient is not already on antibiotics. In this case, protected specimen brush can produce a false-negative result even after a brief antibiotic treatment (24 h). In practice, bronchoscopy and protected specimen brush are rarely carried out in non-intubated patients because of their complexity and the risk of precipitating the need for mechanical ventilation. Bronchoalveolar lavage has no advantage over protected specimen brush for diagnosing pyogenic bacterial infection; however, it is more useful for demonstrating opportunistic organisms such as *Pneumocystis carinii*. Often, this infection presents as a severe community-acquired pneumonia.

Transthoracic needle aspiration has good sensitivity and excellent specificity. Complications are minimal when the 'ultrathin' 25G needle is used, although mechanical ventilation is considered to be an absolute contraindication. Pneumococcal antigen detection by latex agglutination in the aspirated material substantially increases the sensitivity. Despite its simplicity, safety, and effectiveness, transthoracic needle aspiration with an ultrathin needle is seldom performed in practice.

Diagnostic approach

The initial diagnostic approach for severe pneumonias should include sputum Gram stain and cultures (including α-BCYE for *Legionella*), two serial blood cultures, acute phase serology for *Legionella* and atypical agents, and pleural fluid stains and cultures (aerobic and anaerobic) if pleural effusion is present (Table 3). If there are no contraindications, transthoracic needle aspiration could be considered in non-ventilated pneumonia (Gram stain, direct fluorescence antigen for *Legionella*, latex agglutination for pneumococcus, and cultures). If the patient is already receiving assisted ventilation, a sample of respiratory secretions for Gram stain should be obtained. Bronchoscopy could be carried out for protected specimen brush. Proceed to bronchoalveolar lavage if opportunistic organisms are suspected.

Table 3 Diagnostic approach to severe community-acquired pneumonia

Sputum Gram stain and culture (including α-BCYE medium)
Blood cultures
Serology for atypical organisms (acute phase)
Pleural fluid Gram stain and culture, LA for pneumococcal antigen*
Urine: CIE for pneumococcal antigen*, ELISA for *Legionella* antigen serogroup 1*
Invasive procedures
Non-ventilated patients
Transthoracic needle aspiration* in large consolidations
Transtracheal aspiration* in ill-defined consolidations
Ventilated patients
Tracheal aspirate Gram stain and culture (including α-BCYE medium)
Protected specimen brush*

LA, latex agglutination; CIE, counter-immunoelectrophoresis.
*Optional.

Finally, there is substantial debate about the best timing for the performance of invasive procedures. Some authors consider these diagnostic tools as the last step in non-responding cases. In contrast, others prefer to use them earlier, on the basis of determined prognostic factors, when the patient is not yet in a critical condition and could benefit from their result. The latter approach would have a greater impact on the outcome of pneumonia.

Bibliography

Campbell, G.D., *et al.* (1995). Hospital-acquired pneumonia in adults: diagnosis, assessement of severity, initial antimicrobial therapy and preventive strategies. *American Journal of Respiratory Care Medicine*, **153**, 1711–25.

Meduri, G.U. and Johanson, W.G. (ed.) (1992). International Consensus Conference on Clinical Investigation of Ventilator-associated Pneumonia. *Chest*, **102** (Supplement 1).

Niederman, M.S., *et al.* (1993). Guidelines for the initial management of adults with community-acquired pneumonia: diagnosis, assessment of severity and initial antimicrobial treatment. *American Review of Respiratory Disease*, **148**, 1418–26.

Torres, A. (1996). *Severe pulmonary infections*, Vols I and II. W.B. Saunders, Philadelphia, PA.

Torres, A. and Woodhead, M. (1996). *Pneumonia. European Respiratory Monograph 3*, Vol. 2. European Respiratory Society, Sheffield.

2.12.3 Causes and diagnosis of nosocomial pneumonia

J. Dorca and A. Torres

Key messages

- Nosocomial pneumonia accounts for approximately 10 to 15 per cent of all hospital-acquired infections.
- Ventilator-associated nosocomial pneumonia has the highest mortality of all hospital-acquired infections.
- Leading pathogens are the enteric aerobic Gram-negative bacilli and *Staphylococcus aureus*. Other organisms, such as *Streptococcus pneumoniae* or *Hemophilus influenzae*, are less common.
- Certain circumstances predict the etiology of hospital-acquired pneumonia: the severity of the episode; the presence of coexisting illness; previous antibiotic therapy; the duration of the period in hospital.
- In the case of nosocomial pneumonia affecting ventilated patients, simpler sampling methods such as a quantitative culture of endotracheal aspirates or distal samples obtained by 'blind' procedures are preferable.
- In non-ventilated nosocomial pneumonia, transthoracic or transtracheal aspiration is preferable to bronchoscopic procedures in order to obtain a reliable etiological diagnosis.

Epidemiology

Nosocomial pneumonia is one of the three major kinds of hospital-acquired infection (the others are urinary tract and surgical bound infections), and accounts for approximately 10 to 15 per cent of all infections in this setting. Its incidence has been estimated to be between five and ten cases per 1000 hospital admissions. However, the incidence among ventilated patients is much greater, and can present a risk as much as 20 times higher. In fact, pneumonia is the leading cause of nosocomial infection in this setting, with reported rates as high as 25 to 70 per cent. Although nosocomial pneumonia is less frequent in non-ventilated patients, it accounts for approximately one-third of total cases.

Nosocomial pneumonia is a severe complication and has the highest mortality of all hospital-acquired infections. This is particularly evident in ventilated patients, who present an attributable mortality of between a third and a half of all registered cases. Also, the economic cost of hospital-acquired pneumonia is high in terms of both treatment expense and increased duration of hospital stay.

Etiology

Hospital-acquired pneumonia has a characteristic etiological pattern reflecting the oropharyngeal colonization by specific flora suffered by patients after hospital admittance. Several factors, such as coexisting diseases, the length of hospital stay, some therapeutic procedures, and the occurrence of certain complications, modulate the timing and characteristics of airway colonization and the probability of developing pneumonia. As shown in Table 1, the bacterial pathogens most frequently associated with nosocomial pneumonia are enteric aerobic Gram-negative bacilli and *Staphylococcus aureus*. Other organisms such as *Streptococcus pneumoniae* or *Hemophilus influenzae* are less common, but a variety of other bacteria, such as streptococci and anaerobes, may be present. Opportunistic bacteria, viruses, and fungi are rarely seen, except in immunosuppressed patients. *Legionella pneumophila* is an endemic cause of pneumonia in some institutions. It is generally accepted that between a third and a half of cases present a polymicrobial etiology.

The accumulated experience of the circumstances that predict the etiology of hospital-acquired pneumonia has identified the key importance of four specific factors: the severity of the episode, the presence of coexisting illness, previous antibiotic therapy, and the length of hospital stay.

Severe presentation of nosocomial pneumonia may be due to the presence of specific circumstances or to particularly virulent causative organisms such as methicillin-resistant *Staph. aureus* or highly resistant Gram-negative bacteria (*Pseudomonas aeruginosa* or *Acinetobacter* species). Certain comorbidities or preceding antibiotic treatments may predispose to certain organisms (Table 2). Previous antibiotic treatments during hospital admittance clearly predispose to infection by more resistant organisms such as *Ps. aeruginosa*, *Acinetobacter* species, and methicillin-resistant *Staph. aureus*. Finally, the etiological pattern is also influenced by the length of hospital stay: in 'early-onset' pneumonias ($<$ 5 days) organisms such as *Staph. aureus*, *H. influenzae*, and *Strep. pneumoniae* are predominant, whereas in 'late-onset' pneumonias ($>$ 5 days) Gram-negative bacilli are much more common.

According to the official statement recently published by the American Thoracic Society, patients can be categorized in three groups.

1. Patients without risk factors who present with either mild to moderate nosocomial pneumonia occurring at any time during their stay in hospital or early-onset severe nosocomial pneumonia.

2. Patients with mild to moderate nosocomial pneumonia and risk factors, who develop pneumonia at any time.

Table 1 Etiology of nosocomial pneumonia

Mild to moderate pneumonia with no unusual risk factors, onset any time
Enteric Gram-negative bacilli
H. influenzae
Methicillin-sensitive *Staph. aureus*
Strep. pneumoniae
Severe pneumonia, early onset
H. influenzae
Methicillin-sensitive *Staph. aureus*
Strep. pneumoniae
Enteric Gram-negative bacilli
Severe pneumonia, late onset
Ps. aeruginosa
Acinetobacter species
Methicillin-resistant *Staph. aureus*

Table 2 Conditions that predispose to specific causes of nosocomial pneumonia

Condition	Cause
Recent abdominal surgery or witnessed aspiration	Anaerobes
Coma, head trauma, diabetes mellitus, renal failure	*Staph. aureus*
Cellular immunosuppression (corticoid treatment)	*L. pneumonophila*
Structural lung disease	*Ps. aeruginosa*
Prolonged ICU stay, prior antibiotics	*Ps. aeruginosa*, *Acinetobacter* species

3. Patients with severe early-onset nosocomial pneumonia but without risk factors, or patients with severe pneumonia occurring at any time and risk factors.

Diagnostic methods

The goals of microbiological studies in the diagnosis of nosocomial pneumonia are as follows.

1. To confirm the infectious origin of the clinical and radiographic features. The efficacy of diagnosis based on clinical and radiological features only is poor in terms of both sensitivity and specificity. Consequently, it is difficult to differentiate pneumonia from other non-infectious causes of febrile pneumonitis.
2. To identify the causative organisms. The etiological spectrum of nosocomial pneumonia is very broad and the diagnostic efficacy of routine diagnostic methods is poor. Although some predisposing factors may provide a guide to which pathogens are most likely to be found, reliable etiological diagnosis is usually obtained by means of specific diagnostic tests.
3. To establish the severity of the episode. Prognosis is partially dependent on the specific infecting organism.

Non-invasive procedures

Examination and culture of proximal airways secretions

Sputum or tracheal aspirate microbiological processing (Gram stain and cultures) is the simplest method of studying the etiology of nosocomial pneumonia. This procedure is considered to be very sensitive because the causative organisms involved in pneumonia are usually present in tracheobronchuial secretions. In contrast, its specificity is poor and it has a high rate of false-positive results (i.e. positive cultures in patients without pneumonia) owing to the difficulty of differentiating pneumonia from airways infection or colonization. Nevertheless, although this method is unsatisfactory for separating pneumonia from other causes of febrile lung infiltrates, it can be used to exclude some potential pathogens and, occasionally, can provide conclusive evidence about obligate pathogens (e.g. *L. pneumophila*) or some opportunistic agents.

Several recent studies have suggested that quantitative cultures of endotracheal aspirates could provide reliable information by substantially increasing the specificity of the procedure. According to these studies, true pathogenic organisms are usually cultured in concentrations higher than 10^6 colony-forming units.

Blood cultures

The sensitivity of blood cultures for diagnosing nosocomial pneumonia is low. Additionally, it is not uncommon for blood infection in critical patients to arise from a simultaneous extrapulmonary source. Owing to this lack of specificity, blood-cultured organisms should also be found in respiratory secretions before they can be considered as the causative organisms of pneumonia.

Pleural fluid examination and culture

In the event of significant parapneumonic pleural effusion, a pleural fluid sample should be stained and cultured to exclude empyema. If positive, these cultures provide extremely reliable information on the etiology of pneumonia. In practice, however, this procedure is of little help for demonstrating the causative organisms.

Serology

Serological studies are of little use in the diagnosis of nosocomial pneumonia. However, they can provide valuable epidemiological information for the identification of hospital-acquired *L. pneumophila* infections.

Urine

Urine cultures are used to exclude an extrapulmonary source of infection. Additionally, detection of *L. pneumophila* antigen could be useful in institutions where this etiology is common.

Invasive procedures

The poor performance of non-invasive methods, based on the examination and culture of samples which are easily obtainable by various microbiological techniques, has prompted the recovery of more representative samples by a variety of instrumental procedures. These invasive techniques (Table 3) enable samples to be obtained directly from, or close to, the site of infection, so that the yield of microbiological processing is substantially increased.

Bronchoscopic distal airway sampling

Because of its minimally invasive character, bronchoscopy has been considered the ideal method for studying lung infections invasively, providing rapid access to lower airways under direct visual control. However, the tracheobronchial tree is usually contaminated by upper respiratory secretions during the procedure. This problem has been solved by designing various protected devices. In addition, quantitative culture of recovered material has been introduced in order to differentiate between colonizing and pathogenic bacteria.

Two bronchoscopic procedures, protected specimen brush and bronchoalveolar lavage, have been widely used in the etiological diagnosis of ventilator-associated pneumonia. Recovered bacteria are considered to be pathogenic, thus confirming the presence of pneumonia, when cultured in concentrations equal to or higher than

Table 3 Invasive procedures in nosocomial pneumonia

Bronchoscopic distal airway sampling
Protected specimen brush
Bronchoalveolar lavage
Protected bronchoalveolar lavage
Non-bronchoscopic 'blind' distal airway sampling
Transthoracic needle aspiration
Transtracheal aspiration
Lung biopsy

10^3 colony-forming units per milliliter in the case of protected brush or 10^4 to 10^5 colony-forming units per milliliter in samples obtained by bronchoalveolar lavage. However, because antibiotics rapidly affect quantitative cultures, the validity of these thresholds in patients already on antibiotics has never been well established.

Bronchoscopic sampling procedures are easily carried out in ventilated patients and, although they are not innocuous, complications are infrequent. This is not the case in patients who are not carrying an orotracheal or a tracheostomy tube, where these techniques are poorly tolerated and could precipitate the need of ventilation. Further, bronchoscopy in this setting requires considerable experience.

Distal bronchial sampling by these bronchoscopic methods has proved to be very sensitive and specific compared with lung tissue cultures. It has also provided essential epidemiological data about the etiology of ventilator-associated pneumonia. Nevertheless, the role of quantitative bronchoscopic techniques in daily medical practice is a matter of controversy, mainly because no adequate cost–benefit or outcome studies have been performed.

Non-bronchoscopic distal airways sampling

Distal bronchial sampling can also be carried out by different non-bronchoscopic 'blind' protected devices, thus avoiding the need of bronchoscopy. The evident risk of sampling a non-consolidated area does not seem to be very relevant owing to the fact that ventilator-associated pneumonia is a multifocal infection, and causative organisms are usually present even in non-consolidated areas. Although the sensitivity of these 'blind' techniques is somewhat lower than that of bronchoscopic procedures, they are much simpler and, therefore, available at any time. So far, however, the precise indications of these 'blind' have not been established.

Transthoracic needle aspiration

Because of its simplicity and low cost, transthoracic needle aspiration could be a useful procedure for diagnosing pneumonia in non-ventilated nosocomial pneumonia. This procedure has proven to be sensitive and its specificity is excellent. Because the sample is seldom contaminated, its sensitivity could be enhanced by the use of rapid immunological methods and, eventually, molecular biology techniques. Although considered hazardous, transthoracic puncture using the 25G ultrathin needle appears to be a safe procedure, although mechanical ventilation is considered to be an absolute contraindication. Transthoracic needle aspiration has proved to have a significant therapeutical relevance in non-ventilated nosocomial pneumonia. The procedure could be indicated in the case of large or easily located consolidations.

Transtracheal aspiration

Transtracheal aspiration could also be indicated in non-ventilated nosocomial pneumonia. Although it is very sensitive, the specificity in hospital in-patients could be lower owing to the increased rate of tracheobronchial colonization by nosocomial flora.

Lung biopsy

Lung biopsy by means of thoracoscopy or thoracotomy is seldom indicated in the diagnosis of nosocomial pneumonia except for immunosuppressed patients.

Diagnostic approach

A variety of diagnostic strategies, such as the exclusive use of classical non-invasive procedures, the quantitative culture of endotracheal aspirates, and the use of quantitative invasive diagnostic 'blind' or bronchoscopic techniques, can be applied to nosocomial pneumonia

Table 4 Diagnostic approach to nosocomial pneumonia

Nosocomial pneumonia in ventilated patients
Non-invasive procedures (tracheal aspirate Gram stain, blood culture, pleural fluid examination and culture)
Quantitative culture of serial endotracheal aspirates
Bronchoscopic PSB or BAL when no response to empirical treatment
Nosocomial pneumonia in non-ventilated patients
Non-invasive procedures (sputum Gram stain, blood culture, pleural fluid examination and culture, serology for *L. pneumophila*, *Legionella* antigen in urine)
Contemplate invasive procedures
TNA in large consolidations and when there is no adequate response to antibiotic treatment

PSB, protected specimen brush; BAL, bronchoalveolar lavage; TNA, transthoracic needle aspiration.

affecting ventilated patients (Table 4). At present, the cost–benefit analysis of these different approaches and their potential influence on the outcome of pneumonia are a matter of controversy. Simpler methods such as a quantitative culture of endotracheal aspirates or distal samples obtained by 'blind' procedures, are generally preferable. Ideally, these procedures should be carried out before antibiotic treatment. More sophisticated bronchoscopic procedures, particularly protected specimen brush, could eventually be used for non-responding cases.

The diagnostic efficacy of non-invasive procedures in non-ventilated nosocomial pneumonia is very poor. Bronchoscopic procedures are usually not well tolerated by these patients and could precipitate the need of mechanical ventilation. Transthoracic and transtracheal aspiration could be a reasonable alternative for establishing a reliable etiological diagnosis in this setting.

Bibliography

Campbell, G.D., *et al.* (1995). Hospital-acquired pneumonia in adults: diagnosis, assessement of severity, initial antimicrobial therapy and preventive strategies. *American Journal of Respiratory and Critical Care Medicine*, **153**, 1711–25.

Meduri, G.U. and Johanson, W.G. (ed.) (1992). International Consensus Conference on Clinical Investigation of Ventilator-associated Pneumonia. *Chest*, **102** (Supplement 1).

Niederman, M.S., *et al.* (1993). Guidelines for the initial management of adults with community-acquired pneumonia: diagnosis, assessment of severity and initial antimicrobial treatment. *American Review of Respiratory Disease*, **148**, 1418–26.

Torres, A. (1996). *Severe pulmonary infections*, Vols I and II. W.B. Saunders, Philadelphia, PA.

Torres, A. and Woodhead, M. (1996). *Pneumonia. European Respiratory Monograph 3*, Vol. 2. European Respiratory Society, Sheffield.

2.12.4 Therapeutic approach

A. Torres and M. El-Ebiary

Key messages

- In severe community-acquired pneumonia, we recommend the association of a second- or third-generation cephalosporin in addition to erythromycin and rifampin (rifampicin).
- After effective therapy, an improvement in the clinical features of pneumonia should be seen in 48 to 72 h.
- When a patient is not responding or is deteriorating during empirical therapy, several complications and factors associated with the antibiotic, the organism, and the host should be considered.
- Duration of therapy should be individualized depending upon the severity of illness, the rapidity of clinical response, and the infecting agent.

Community-acquired pneumonia

Epidemiology and definition of severe community-acquired pneumonia

Community-acquired pneumonia remains an important cause of mortality in developed nations and accounts for 2.8 per cent of all hospital admissions. In the United States 50 000 deaths per year are attributed to it. It has been estimated that, annually, about 0.4 to 4 per 1000 individuals require hospitalization because of pneumonia. During the last 25 years, mortality has ranged from 4 to 24 per cent in adult patients admitted to hospital with community-acquired pneumonia. Severe community-acquired pneumonia is defined as a life-threatening pneumonia acquired in the community in non-immunocompromized patients requiring admission to an intensive care unit (ICU). There is little information about its frequency relative to the total number of pneumonias within and outside hospitals. However, epidemiological studies suggest that 3 to 5 per cent of cases of community-acquired pneumonia will be severe enough to require ICU admission. Most patients with severe community-acquired pneumonia have chronic underlying diseases such as chronic obstructive pulmonary disease or other forms of chronic respiratory failure. Other frequent underlying chronic conditions are alcoholism and diabetes mellitus. Over 30 per cent of cases involve previously healthy patients.

Etiology

Several studies report the etiology of severe community-acquired pneumonia. In most of them, *Streptococcus pneumoniae* and *Legionella pneumophila* are the two most frequent micro-organisms. However, *L. pneumophila* is much less frequent than *Strep. pneumoniae* in the overall population of community-acquired pneumonia, indicating that the former micro-organism produces more severe forms. *Hemophilus influenza*, other Gram-negative bacilli, and *Mycoplasma pneumoniae* are also frequently involved. *Staphylococcus* species, although infrequent in most series, were the second most frequent type of micro-organism involved in a recent large series of 299 cases of severe community-acquired pneumonia. Table 1 shows the different etiologies of severe community-acquired pneumonia in order of frequency.

Treatment

In view of the epidemiological data on severe community-acquired pneumonia we suggest the following empirical antibiotic regimens.

Table 1 Micro-organisms most commonly causing severe community-acquired pneumonia in order of frequency

Strep. pneumoniae
Legionella species
Aerobic Gram-negative bacilli
M. pneumoniae
H. influenzae
Staph. aureus

1. In healthy patients aged below 60 years, we recommend the association of a second-generation (cefuroxime) or third-generation (cefotaxime or ceftriaxone) cephalosporin in addition to erythromycin and rifampin (rifampicin) to cover *L. pneumophila*. In those countries with a high prevalence of penicillin-resistant *Strep. pneumoniae*, cefotaxime (2 g every 8 h) must be administered.

2. In patients older than 60 years or with an underlying chronic condition, we suggest adding an aminoglycoside (at least during the first 3 days of treatment) since the presence of Gram-negative bacilli, other than *H. influenzae*, is more likely and the combination of an aminoglycoside with a β-lactam can be of help in the case of bacteremia.

3. In patients with chronic bronchiectasis presenting with severe community-acquired pneumonia we have to consider the possibility of *Pseudomonas aeruginosa* so that the administration of antipseudomonal combinations (ceftazidime plus amikacin, imipenem, ciprofloxacin, etc.) is mandatory.

Obviously initial antibiotic regimens have to be modified in relation to microbiological findings in both rapid initial tests and cultures. The role of the new macrolides and new quinolones for severe community-acquired pneumonia has still to be assessed.

Duration of treatment

The duration of treatment has not yet been well established, but it seems reasonable to maintain antibiotic treatments for a period of 10 to 14 days. In confirmed cases of severe *L. pneumophila* it is advisable to administer antibiotics for 2 to 3 weeks. In the case of *Ps. aeruginosa* pneumonia the duration of treatment should be at least 2 weeks.

If empirical therapy is used, evaluation of clinical response is essential so that non-responding patients can be identified and appropriately evaluated. After effective therapy, an improvement in the clinical features of pneumonia should be seen in 48 to 72 h. Therefore therapy should not be changed in the first 72 h unless there is a marked clinical deterioration. Fever can persist for 2 to 4 days, and leukocytosis usually resolves by the fourth day. Radiographic evidence of deterioration is common during therapy and may have prognostic significance in the setting of severe pneumonia, when it is a predictor of a poor outcome.

Causes of non-response to initial antibiotic therapy

When a patient is not responding or is deteriorating during empirical therapy, several complications and factors associated with the antibiotic, the organism, and the host should be considered.

1. Empyema is a relatively frequent complication of pneumonia which explains the lack of response to initial antibiotic therapy in some cases. Particularly because of concern about empyema, any patient with an inadequate clinical response to therapy should have repeat chest radiography, and any pleural fluid should be sampled, cultured, and analyzed for cell count and chemistry. Other complications such as septic metastases can occur in as many as 10 per cent of patients with bacteremic pneumonia (arthritis, meningitis, pericarditis, etc.).

2. The etiological agent may be resistant to the drugs used in the initial empirical regimen. This is the case in penicillin-resistant *Strep. pneumoniae*, particularly in areas with reported high rates of resistance. Pneumococcal resistance to erythromycin and cephalosporins should also be considered.

3. The infecting pathogen may not be covered by the initial antibiotic selection. This is the case with *Staphylococcus aureus* or *Ps. aeruginosa* which are not optimally covered by the therapies outlined above.

4. Alternatively, the infection could be caused by an agent that is not responsive to antibiotics (e.g. a virus). Additional unusual pathogens could be involved in the etiology of community-acquired pneumonia. Such infections should be considered when clinical and radiographic findings persist, and the differential diagnosis includes tuberculosis, endemic fungal pneumonia, and *Pneumocystis carinii* pneumonia.

5. It is also possible that the patient does not have pneumonia, as some non-infectious processes can present with clinical features that mimic pneumonia. These include pulmonary embolism, congestive heart failure, obstructing bronchogenic carcinoma, and certain inflammatory lung diseases (e.g. bronchiolitis obliterans and organizing pneumonia, Wegener's granulomatosis, eosinophilic pneumonia).

Nosocomial pneumonia

Nosocomial pneumonia accounts for approximately 10 to 15 per cent of all hospital-acquired infections. It is currently the second leading cause of nosocomial infections, after urinary tract infections. The overall risk of acquiring nosocomial pneumonia is approximately 6.0 to 8.6 infections per 1000 admissions. The incidence of nosocomial pneumonia is greatly increased for all patients in ICUs, where respiratory infections have been reported to be the most frequent type of nosocomial infection. The reported incidence in ICUs ranges from 12 to 29 per cent. The risk to patients receiving mechanical ventilation is as much as 20 times greater, with rates as high as 25 to 70 per cent.

Etiology and initial empirical treatment

The spectrum of potential pathogens associated with nosocomial pneumonia differs from that of community-acquired pneumonia. The bacterial pathogens that are most frequently associated with nosocomial pneumonia are enteric Gram-negative bacilli and *Staph. aureus*, but accumulating data suggest that the etiology is polymicrobial in up to 50 per cent of mechanically ventilated patients.

The spectrum of potential pathogens can be identified by assessing several factors such as the severity of the episode, the presence of coexisting illness, the prior antibiotic therapy, and the length of time in hospital.

Knowledge of these factors allows patients to be separated into easily identifiable groups which form the basis for therapeutic decisions. In order to classify a patient appropriately, three questions must be asked.

1. Is the pneumonic process mild to moderate, or is it severe (Table 2)?

2. Are specific host or therapeutic factors predisposing to specific pathogens present?

Table 2 Definition of severe nosocomial pneumonia

Admission to the ICU

Respiratory failure, defined as the need for mechanical ventilation or the need for above 35% oxygen to maintain an arterial oxygen saturation above 90%

Rapid radiographic progression*, multilobar pneumonia, or cavitation of lung infiltrate

Evidence of severe sepsis with hypotension and/or endorgan dysfunction

- Shock (systolic blood pressure <90 mmHg or diastolic blood pressure <60 mmHg)
- Requirement of vasopressors for more than 4 h
- Urine output below 20 ml/h, or total urine output below 80 ml in 4 h
- Acute renal failure requiring dialysis

*Increase of more than 50% of the size opacities within 48 h.
Modified with permission from Campbell *et al.* (1996).

3. Is the pneumonia of early onset (occurring within less than 5 days of admission) or of late onset (occurring 5 days or more after admission)?

According to the American Thoracic Society official statement published recently, patients can be categorized into three groups.

1. Patients without risk factors who present with mild to moderate nosocomial pneumonia occurring at any time during their stay in hospital or with severe nosocomial pneumonia of early onset.
2. Patients with mild to moderate nosocomial pneumonia and risk factors who develop pneumonia at any time.
3. Patients with severe nosocomial pneumonia of early onset and no risk factors, or patients with severe pneumonia occurring at any time and risk factors.

The core organisms of group 1 are the enteric Gram-negative bacilli, *H. influenzae, Strep. pneumoniae,* and methicillin-sensitive *Staph. aureus.* These micro-organisms can be treated with second-generation or non-pseudomonal third-generation cephalosporins (Table 3). Combinations of β-lactams and β-lactamase inhibitors can be also used. In the case of penicillin allergy, fluoroquinolones, or clindamycin plus aztreonam can be given. Monotherapy is usually appropriate in this setting.

Patients from group 2 are susceptible to certain additional bacteria beyond the core organisms, such as anaerobes, *Staph. aureus, Legionella* species, and *Ps. aeruginosa*. These should be considered according to the presence or absence of specific risk factors for these organisms (Table 4). All of these patients should be treated for the core organisms as above, but usually require the addition of other antimicrobial agents to provide cover for other likely pathogens. For example, clindamycin or metronidazole are active against anaerobes and can be added to the core antibiotics in witnessed or suspected cases of gross aspiration, although a β-lactam plus β-lactamase inhibitor combination may be sufficient. Vancomycin should be added in cases with coma or head trauma until methicillin-resistant *Staph. aureus* (MRSA) is excluded. If *Legionella* infection is suspected (e.g. patients on high doses of steroids), erythromycin (with or without rifampin) should be administered.

Treatment of patients from group 3 should be directed against the core organisms as well as against more resistant and virulent Gram-negative bacilli such as *Acinetobacter* species and *Ps. aeruginosa*. These micro-organisms require the use of combination antimicrobial therapy. This can be achieved by an aminoglycoside or ciprofloxacin plus an anti-pseudomonal penicillin, a β-lactam plus β-lactamase inhibitor combination, ceftazidime or cefoperazone, imipenem, or aztreonam. In some cases it may be possible to complete therapy using a single agent. The decision to continue combination therapy can be made after treatment for 2 or 3 days based on the clinical response and microbial cultures. For instance, if *Ps. aeruginosa*, *Acinetobacter* species, *Enterobacter* species, or MRSA are not isolated, monotherapy can be used with confidence since there are data showing good efficacy with monotherapy in severe nosocomial pneumonia not caused by these micro-organisms. However, if one of the micro-organisms mentioned above is present, antibiotic combination therapy has to be continued to reduce mortality in bacteremic cases and to avoid the emergence of resistant mutants.

Table 3 Core organisms and core antibiotics in immunocompetent patients

Core organisms	Core antibiotics
Non-pseudomonal enteric Gram-negative bacilli	Cephalosporins
Enterobacter species	second-generation
Escherichia coli	non-pseudomonal third-generation
Klebsiella species	β-lactam+β-lactamase inhibitor
Proteus species	If allergic to penicillin:
Serratia marcescens	fluoroquinolone or clindamycin+aztreonam
H. influenzae	
Methicillin-sensitive *Staph. aureus*	
Strep. pneumoniae	

Modified with permission from Campbell *et al.* (1996).

Table 4 Specific risk factors for the emergence of certain micro-organisms

Risk factor	Agent
Witnessed gross aspiration	Anaerobic bacteria
Abdominal surgery	*Enterococcus* species, anaerobes
Coma, intravenous drug abuse	*Staph. aureus* (methicillin-sensitive)
Diabetes mellitus	*Staph. aureus* (methicillin-sensitive)
Chronic renal failure	*Staph. aureus* (methicillin-sensitive)
Corticosteroids	*Legionella* species, *Aspergillus* species
Prolonged hospital stay	*Ps. aeruginosa*, *Enterobacter* species, *Acinetobacter* species
Prolonged mechanical ventilation	*Ps. aeruginosa*, *Enterobacter* species, *Acinetobacter* species
Prior antibiotics	*Ps. aeruginosa*, *Enterobacter* species, *Acinetobacter* species
Structural lung diseases	*Ps. aeruginosa*
Tracheostomy	*Ps. aeruginosa*

There are several approaches to anti-pseudomonal therapy. The first is a combination of a β-lactam with an aminoglycoside, which in theory increases the synergism against *Ps. aeruginosa*. However, the existence of this synergism *in vivo* is not clear since aminoglycosides do not penetrate the pulmonary tissue well and are inactivated by the acidic environment. A second possibility is a combination of two β-lactam antibiotics, which may act by antagonism in some cases or even by induction of β-lactamases for one or both antibiotics. Finally, combination of ciprofloxacin with a β-lactam antibiotic may provide synergy, good parenchymal penetration, and lower toxicity compared with aminoglycosides.

Particular antibiotic issues when choosing empirical treatment of nosocomial pneumonia

In addition to the factors discussed above, the choice of antibiotic regimens in a specific hospital depends on the following.

1. The sensitivity pattern of the predominant flora in the unit: it is necessary to adjust antibiotic strategies according to the specific bacterial flora and sensitivity patterns of the ICU.
2. The type and degree of immunosuppression, and the type of antibiotics received either for prophylaxis or treatment, since these factors can modify the spectrum of pathogens.
3. Antibiotic-related aspects such as the ability to produce endotoxins and to select mutant strains. This issue is of particular importance and deserves careful study by the physicians.

Duration of treatment

Duration of therapy should be individualized depending upon the severity of illness, the rapidity of clinical response, and the infecting agent. No carefully controlled studies documenting duration of therapy have been reported, but antibiotics should be continued for a minimum of 14 to 21 days to reduce the chances of relapse when treating enteric Gram-negative bacilli or MRSA. In contrast, cure rates exceeding 95 per cent have been noted for nosocomial pneumonia caused by methicillin-sensitive *Staph. aureus*, *H. influenzae*, and *Strep. pneumoniae*. A 7- to 10-day course of treatment may be adequate for these pathogens.

Non-response to treatment

There are several possible explanations for a rapid deterioration or failure to improve. These include the possibility that the process is not pneumonia, or that certain host, bacterial, and therapeutic factors have not been considered. The non-infectious processes that may be mistakenly labeled as nosocomial pneumonia include atelectasis, congestive heart failure, pulmonary embolism with infarction, lung contusion in trauma patients, and chemical pneumonitis from aspiration. Patients with acute respiratory distress syndrome can have fibroproliferative diffuse alveolar damage, and mechanically ventilated patients can have pulmonary hemorrhage.

Host factors associated with non-response to treatment include underlying fatal conditions, superinfection, chronic obstructive pulmonary disease, and immunosuppression. Again, bacterial variables may also be associated with an adverse outcome of initial therapy. The infecting pathogen can be resistant at the outset to the chosen therapy or may acquire resistance during therapy, particularly if the agent is *Ps. aeruginosa*. Finally, pneumonia can be due to other pathogens (i.e. fungi).

Certain complications during therapy can also lead to failure to respond. Some patients with nosocomial pneumonia can have other sources of fever such as sinusitis, vascular-catheter-related infection, urinary tract infections, or pseudomembranous enterocolitis. Complications of the original pneumonia, including the development of lung abscess or empyema, can lead to failure.

Other measures for managing severe pneumonia

In some cases of severe pneumonia, the severe hypoxemia can only be adequately managed with oxygen administration and mechanical ventilation. Several therapeutic strategies, some of which are still under clinical investigation, have been suggested.

1. The administration of anti-inflammatory drugs such as aspirin or indomethacin may reverse the partial ablation of hypoxic pulmonary vasoconstriction that occurs in some cases of bacterial pneumonia owing to the local release of vasodilator substances such as prostacyclin. The administration of these drugs to improve hypoxemia in pneumonia is still under clinical investigation and remains controversial.

2. The aerosolization of prostacyclin (epoprostenol, prostaglandin I_2) may reduce pulmonary hypertension and shunt and improve arterial oxygenation in the absence of systemic effects in patients with severe pneumonia.

3. In cases of unilateral pneumonia, placing the patient in the lateral position (affected lung up) may increase $P\text{aO}_2$ by an average of 10 to 15 mmHg by improving the ventilation–perfusion mismatch. Specifically, there is an increase of blood flow to the well-ventilated areas.

4. Similarly, placing patients with refractory hypoxemia in the prone position has been shown to improve their oxygenation.

5. The inhalation of small amounts of nitric oxide (5–20 ppm) causes a selective arterial pulmonary vasodilation diverting blood flow to well-ventilated areas. The effect of nitric oxide in arterial oxygenation has been proved effective in acute respiratory distress syndrome patients with pulmonary hypertension. However, the beneficial effect in severe pneumonia is not confirmed and is still under clinical investigation.

Bibliography

Campbell, G.D., *et al.* (1996). Hospital acquired pneumonia in adults: diagnosis, assessment of severity, initial antimicrobial therapy, and preventive strategies. A consensus statement. *American Journal of Respiratory and Critical Care Medicine*, **153**, 1771–25.

Niederman, M.S., *et al.* (1993). Guidelines for the initial management of adults with community-acquired pneumonia: diagnosis, assessment of severity, and initial antimicrobial therapy. *American Review of Respiratory Disease*, **148**, 1418–26.

Torres, A. (ed.) (1996). Severe pulmonary infections I and II. *Seminars in Respiratory Infections*, **11**, 1–127.

Wunderink, R.G. (ed.) (1995). Pneumonia in the intensive care unit. *Clinics in Chest Medicine*, **16**, 1–223.

2.13 Viral pneumonitis and atypical pneumonia

2.13.1 Pathophysiology and causes

J.-F. Timsit and J. Carlet

Key messages

- *Legionella* is the most common pathogen causing life-threatening atypical pneumonia.
- Micro-organisms involved in atypical pneumonia are capable of:
  - resisting phagocytosis (*Mycoplasma pneumoniae*);
  - creating abnormal phagosomes which do not fuse with secondary lysosomes (*Legionella* species);
  - surviving within phagocytes by inhibiting phagolysosomal fusion (*Chlamydia pneumoniae*, *Mycobacterium tuberculosis*);
  - resisting lysosomal enzymes (*Coxiella burnetii, Mycobacterium tuberculosis*).
- The main defense against these infections is the cellular immune response.
- Essentially, any virus can produce any syndrome, but severe pneumonia in adults leading to admission to the intensive care unit is usually due to influenza and herpes viruses (varicella zoster virus).

Introduction

Atypical pneumonia and pulmonary tuberculosis account for 10 to 20 per cent of all cases of community-acquired pneumonia. The micro-organisms involved are capable of resisting cell phagocytosis or of surviving within phagocytes. Their pathogenesis involves mainly cellular immunity.

Mycoplasma pneumoniae pneumonia

Clinically apparent *M. pneumoniae* pneumonia accounts for 3 to 10 per cent of infected persons. However, severe diseases, often mimicking necrotizing bacterial pneumonia, have been described. Peribronchial and perivascular lymphoid infiltrates, and bronchitis and bronchiolitis with variable consolidation of alveoli, have been observed in cases that have gone to autopsy. Infection appears to be acquired via inhalation of infected material after exposure to an acutely ill coughing individual. The incubation period is 2 to 3 weeks.

M. pneumoniae is a short Gram-negative rod with a neuraminic acid receptor site at the end for attachment to the host cell membranes. *M. pneumoniae* escapes phagocytosis because of this specific attachment to the cell membrane. *M. pneumoniae* is believed to synthesize superoxide radicals and H_2O_2, which are toxic to mucosal cells, on the mucosa. Pathogenicity seems to require both properties.

Several classes of antibodies are produced during *M. pneumoniae* infection. Some are directed against the agent, but do not necessarily confer lifelong immunity as some reinfection can occur. The other are autoantibodies and include agglutinins to the brain, lung, smooth muscle, and cardiolipins. Cold agglutinins are found in half the cases of *M. pneumoniae* pneumonias at the beginning of the disease, and have been shown to be oligoclonal IgM autoantibodies directed against an altered I-antigen on the surface of erythrocytes of infected patients. A high titer of cold agglutinins can lead to erythrocyte destruction, presumably by complement activation, but rarely to significant hemolysis.

Many findings suggest that immune mechanisms, rather than actual direct infection, may play a role in the development of clinically apparent *M. pneumoniae* pneumonia as well as some of the extrapulmonary complications. First, *M. pneumoniae* is able to cause a wide variety of extrapulmonary symptoms but is never isolated from clinical material except sputum or nasopharyngeal secretions. Second, *M. pneumoniae* or its antigen have seldom been demonstrated in the lungs of patients with fatal pulmonary infection. Evidence has been reported to support the notion that repeated subclinical *M. pneumoniae* infections with consequent sensitization of T lymphocytes, and probably other components of the immune system (autoantibodies), may be necessary before manifestations such as pneumonitis occur. Third, antithymocyte globulin abrogates or diminishes the severity of experimental *M. pneumoniae* infection in animals, and corticosteroids have been used with some beneficial clinical effects in patients with severe *M. pneumoniae* pneumonia. Moreover, pneumonia failed to develop in a group of children with immunodeficiency syndromes who had acquired severe *M. pneumoniae* respiratory tract infections.

Chlamydial pneumonia

Patients infected with *Chlamydia pneumoniae* rarely need admission to the intensive care unit (ICU) and the mortality is low. The incubation period is not known precisely, but may be as long as 3 to 4 weeks.

C. pneumoniae are obligate intracellular parasites. The elementary bodies are highly condensed and therefore metabolically inactive. The lifecycle is initiated when an elementary body attaches to a susceptible epithelial cell. The elementary bodies differentiate intracellularly into reticulate bodies, which apparently inhibit phagolysosomal fusion, and multiply using high-energy phosphate compounds and some amino acids from the host cell. As they multiply, chlamydial antigens are released onto the surfaces of infected eukaryotic cells, stimulating the prduction of neutralizing antibodies by the host. They revert to elementary bodies before exocytosis or cell lysis.

Infection with *C. pneumoniae* leads to partial immunity. Reinfections in young adults do not manifest as pneumonia. However, in elderly patients, severe pneumonia appears to develop as a result of reinfection, probably because of a previous chlamydia-like sensitization. It is not known whether the consequent severity of such cases is due to reactivation of latent infection or to a hypersensitivy reaction to chlamydial organisms in the inflamed area.

Chlamydia psittaci

Infection with *C. psittaci* occurs after inhalation of dry bird excreta, although it can be contracted by handling contaminated plumage or tissues. The incubation period is between 6 and 15 days (rarely longer). The mechanisms of intracellular parasitism and host defenses are similar to those observed with other chlamydial infections.

Psittacosis may be responsible for severe pneumonia with myocarditis endocarditis and glomerulonephritis. After inhalation, the organism disseminates widely and may reach the reticuloendothelial cells of the liver and spleen. After replication in local mononuclear phagocytes, *C. psittaci* spreads hematogeneously to the lungs. The pulmonary lesion usually starts in the hilum and extends peripherally, mainly in the dependent lobes and segments. The initial events are characterized by a pulmonary capillary leak. Polynuclear cells predominate in the early stage, but they are soon replaced by mononuclear cells. The affected areas become consolidated, with a typical rubbery gelatinous consistency. Hemoptysis due to minor hemorrhage may subsequently occur. Non-specific splenic inflammation and reactive hepatitis have been described. Myocarditis is related to fatty degeneration and lymphocytic infiltration of the myocardium. In the kidney, hyaline glomerular capillary occlusions explain the occurrence of glomerulonephritis. Finally, fibrinous and gelatinous arachnoiditis around the central sylvian fissures has been described.

Coxiella burnetii (Q fever) pneumonia

Pneumonia is one manifestion of acute Q fever. Patients with Q fever pneumonia have fewer complications than those with mycoplasmal or legionella pneumonia. Aerosols from newborn animals, contaminated wool, or the placentas of infected mammals can infect humans. Infection occurs after an incubation period of 7 to 14 days. The first target cell is the alveolar macrophage.

The primary lysosome fuses with the phagosome. The acid pH activates the enzymes of *C. burnetii*, which multiplies and is able to undergo sporogenesis. Following cell lysis, spores are able to survive for a long time (more than a month on meat in cold storage, and more than 40 months in skim milk at room temperature). *C. burnetii* undergoes phase variation: the phase I form is virulent and the phase II form is avirulent. During the acute infection, host defenses consist of antibody mainly directed against phase II antigen and a cellular immune response resulting in immunity. However, the microorganism can still be isolated even after clinical cure.

Legionellosis

Legionnaires' disease is mainly caused by *Legionella pneumophila* serogroup 1. *L. pneumophila* serotypes 2 to 10, as well as other subspecies, have been associated with pneumonia. The Legionellaceae are pleiomorphic coccobacillary Gram-negative bacilli.

Legionellae are ubiquitous and survive in various, primarily aquatic, habitats. Plumbing fixtures, drinkable water, heat exchange units, cooling towers, air conditioners, and evaporative condensers may be contamined with legionellae, which are then disseminated via devices or circumstances that generate infective aerosols (such as cooling towers, air conditioners, evaporative condensers, industrial coolants, respiratory therapy, and devices such as showerheads and whirlpool spas). The incubation period is 2 to 10 days.

Once the pathogen has been inhaled, the severity of the disease seems to be related to the virulence of the organism, the innoculum, and the host's response. *L. pneumophila* serogroup 1 appears to be the most virulent as it is the most common cause of the disease, whereas all species are equally distributed in the environment.

Legionellae are able to resist phagocytosis in the absence of complement or specific antibody, and to resist antibody- and complement-mediated lysis. *L. pneumophila* is ingested via engulfment with a pseudopod and creates a sequestered intracellular niche in which it parasitizes cellular nutrients and hides from the toxic properties of lysosomes. Following phagocytosis, wild-type *L. pneumophila* multiplies within a specialized phagosome which does not fuse with secondary lysosomes. The organism then multiplies until the cell ruptures (Fig. 1). A variety of enzymes and toxins are produced which may contribute to the extrathoracic manifestation of the disease.

The quality of host response influences the course of the disease. Legionellae are not able to resist mucociliary clearance, and more than 60 per cent of those inhaled are cleared from the lung within 8 h in experimental models. This finding probably partly explains the increased risk of legionnaires' disease among cigarette smokers and alcoholics.

Once in the alveolar space, legionellae encounter the initial component of the phagocytic defense system, the alveolar macrophages, which are able to phagocytize *L. pneumophila* even in the absence of specific antibody but are unable to prevent the intracellular multiplication of the organism. Monocytes and polymorphonuclear cells are then recruited from the lung. Phagocytosis by human monocytes is complement mediated. The role of polymorphonuclear leukocytes in host defense against *Legionella* is unclear as neutropenic patients do not have an undue predilection for legionnaires' disease. *L. pneumophila* is able to resist killing by polymorphonuclear cells; however, intracellular replication is inhibited.

Legionellae that evade the mucociliary apparatus and survive attack by phagocytic cells are confronted by humoral and cell-mediated immunity. IgM, followed by IgG, is measurable in the first week of infection but the response may be delayed. Humoral immunity seems to play a modest role in host defense. Antibody does not promote killing of *L. pneumophila* by complement, promotes only limited

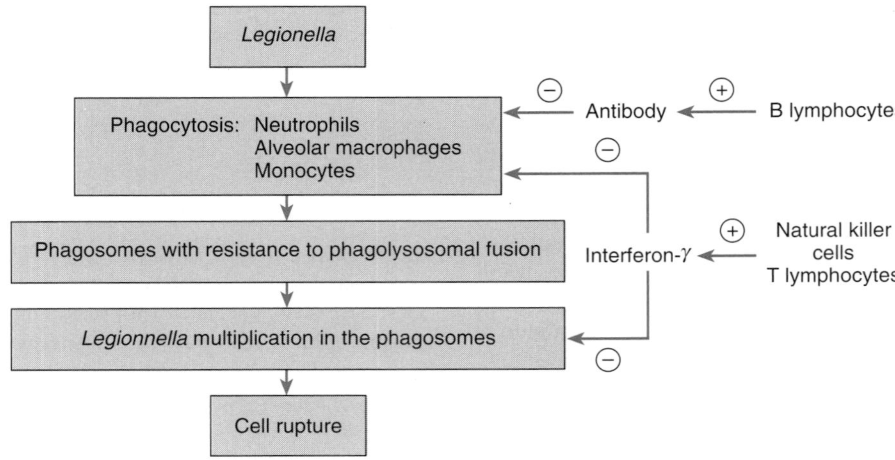

Fig. 1 Pathogenesis of *Legionella* infection.

destruction of *L. pneumophila* by phagocytes, and does not inhibit intracellular multiplication in monocytes or alveolar macrophages.

Viral pneumonitis

Viruses cause most pneumonia in infants and children, but viral pneumonia is uncommon in adults. Essentially, any virus can produce any syndrome, but severe pneumonia leading to ICU admission in adults is mainly due to influenza and herpes viruses, particularly varicella zoster virus. Hantavirus, which is known to cause hemorrhagic fever with renal syndrome, has been also responsible for multiple organ failure with non-cardiogenic pulmonary edema.

Pulmonary distress is due to viral-related necrosis of the trachea and bronchi (as in varicella pneumonitis), or to an interstitial alveolar inflammatory infiltrate associated with an interstitial intra-alveolar proteinaceous exudate, edema, intra-alveolar hemorrhage, hyaline membrane formation, and type 2 cell hyperplasia. These findings lead to diffuse alveolar damage similar to those observed in other types of acute respiratory distress syndrome.

Transmission can occur via the aerosolization of small particles, which have the capability of traveling distances greater than 1 m (influenza, coxsackie). Large droplets may account for a small number of close person-to-person transmissions. Direct hand-to-hand transmission may partially explain the transmission of rhinovirus and respiratory syncytial virus. Small particules and droplets infect the whole respiratory tract, including the alveolar spaces, whereas in the other mode of transmission the first target cells are in the upper respiratory tract.

The infectious cycle of most viruses lasts for 6 to 12 h and leads to cell death and the release of 100 virions per cell. The incubation period is generally short (1–3 days). After several generations and damage to a sufficient number of cells, lung function is impaired and signs and symptoms appear. This scheme suggests that the virus has a direct cytotoxic effect, but some interaction between host cells and the virus may explain some of the symptoms and the severity of the illness.

Some viruses such as influenza and adenovirus cause primary infection by antigenic variation. Other such as respiratory syncytial virus are able to reinfect the same individual. Finally, herpes viruses and adenovirus reactivate from a latent infected phase during a period of altered immunity.

The interaction between the virus and a target cell begins with attachment of the virus to the cell surface. This process is initiated by a random collision between a virus particle and the cell surface. Attachment may involve an interaction between specific proteins on the viral surface and specific receptors in the target cell membrane. Generally, receptors are molecules on the host cell surface which are designed to perform a variety of normal functions.

Pulmonary macrophages are the first cells involved in the defense against viral infections. The cell is capable of ingesting virus particles on the mucosal surface of the airways and transporting them from the lung on the mucociliary escalator. The interaction between virus and macrophage may depend on various factors: the permissiveness of the cell for viral replication, the presence of neutralizing antibody, and the local production of interferon-γ which increases the ability of macrophages to restrict viral replication.

This response is initiated locally by macrophages presenting viral antigen to lymphocytes located along the airway in the submucosa (bronchus-associated lymphoid tissue) which generate a local immune response at the mucosal level. A systemic immune response may also be generated when viremia occurs.

Viral-specific IgG, IgA, and IgM are produced, which are capable of neutralizing viral infectivity and initiating complement-mediated lysis of virus-infected cells by the alternative complement pathway. The cellular response to viral antigen includes natural killer cells and T lymphocytes. T lymphocytes produce various soluble mediators which recruit and activate monocytes and lymphocytes at the site of infection. They also produce substances such as interferon that directly inhibit viral replication.

Impairment of the chemotaxis of polymorphonuclear neutrophils has been found during influenza infection, and this may explain the increased bacterial adherence and the increased susceptibilty to bacterial infections. In addition, viral infection impairs the phagocytic functions of macrophages and depresses the chemotaxis of peripheral blood monocytes. CD14 is a surface glycoprotein present on macrophages, monocytes, and activated neutrophils which seems to increase the sensitivity of these cells to the endotoxin of Gram-negative bacteria. It has recently been suggested that cytomegalovirus lung infection decreases the defenses of human alveolar macrophages against Gram-negative bacteria by inhibiting the surface expression of CD14 and decreasing the release of soluble CD14.

Tuberculosis

The incidence of active tuberculosis in patients admitted to the ICU is increasing, particularly in young adult AIDS patients. The main form is miliary tuberculosis which leads to severe respiratory distress syndromes and multiple organ failures. Clinically active tuberculosis has been estimated to develop in only 5 to 15 per cent of those infected with the organism.

Very few infectious units contained in airborne droplets escape to the mucociliary apparatus and reach the terminal air space. The initial focus is almost always the subpleural space. The bacteria are ingested by alveolar macrophages. *Mycobacterium tuberculosis* inhabits phagosomes that seldom provoke lysosomal fusion. Even if lysosomal fusion occurs, the bacilli are able to resist lysosomal enzymes and can multiply and subsequently destroy the macrophage. Lymphocytes and monocytes are attracted. Monocytes differentiate into macrophages which ingest bacilli released from degenerate cells, and one or more areas of pneumonitis slowly develop. Infected macrophages are carried to regional lymph nodes via the lymphatics, but in the non-immune host they are not retained there and may spread throughout the body via the bloodstream. Tissues that favor retention and bacillary multiplication include most importantly the apical posterior areas of the lungs, but also the lymph nodes, kidneys, epiphyses of the long bones, vertebral bodies, and juxtapendymal meningeal areas adjacent to the subarachnod space. Before the development of hypersensitivity, microbial growth is virtually uninhibited. Generally, rapid destruction of the bacteria arrests the process. In a minority of cases antigen concentration in the primary complex (initial pulmonary focus and draining regional nodes) reaches a sufficient level, and the development of hypersensitivity results in necrosis and further fibrosis and calcification. In some cases, metastatic pulmonary apical foci of infection contain a sufficient number of bacteria, and necrosis ensues with the onset of hypersensitivity, producing a calcific deposit in which viable bacilli may persist in 50 per cent of cases. The primary complex may also progress in several ways. In children, large hilar and mediastinal lymph nodes may produce bronchial collapse with distal atelectasis or may erode into a bronchus. Primary infection may lead to a lymphohematogeneous dissemination hyperacute miliary tuberculosis in very young non-immune children or when the primary complex reaches the bloodstream via a large caseating metastatic focus in the wall of the pulmonary vein. Progressive hematogeneous dissemination may also lead to tuberculous meningitis within a few weeks. The subpleural parenchymal component of the primary infection may rupture in the pleural space, producing serofibrinous pleurisy.

Defense against infection requires 3 to 8 weeks and consists of tissue hypersensitivity and cellular immune response.

The pathological features of tuberculosis depend on the degree of hypersensitivity and the local concentration of antigen. Granuloma formation requires a high tissue hypersensitivity with a small antigen load. In contrast, when antigen load and degree of hypersensitivity are high, epithelioid cells and giant cells are sparse, lymphocytes, macrophages, and granulocytes are present in a less organized fashion, and there may be tissue necrosis. Necrosis tends to be incomplete, resulting in an acellular material called caseous necrosis. Caseous necrosis inhibits microbial multiplication but is unstable, particularly in the lung, where it tends to liquefy and discharge through the bronchial tree, producing a tuberculous cavity and providing conditions in which the bacterial population reaches a titer 10^5 to 10^6 times greater than in non-cavitary lesions. When the degree of hypersensitivity is very low, the tissue reaction may be non-specific with polymorphonuclear cells, mononuclear cells, and an enormous number of bacilli. This finding may explain the frequent presence of adenopathy, absence of cavitation, and the 50 per cent frequency of extrapulmonary tuberculosis seen in the late phase of HIV infection.

Bibliography

Cook, P.J. and Honeybourne, D. (1994). *Chlamydia pneumoniae*. *Journal of Antimicrobial Chemotherapy*, **34**, 859–73.

Haas, D.W. and Des Prez, R.M. (1995). *Mycobacterium* tuberculosis. In *Principles and practice of infectious diseases* (ed. G.L. Mandell, R.G. Douglas, and J.E. Bennett), pp. 2213–42. Churchill Livingstone, New York.

Macfarlane, J.T. and Miller, A.C. (1983). Psittacosis. *British Medical Bulletin*, **39**, 163–71.

Rose, R.M., Pinkston, P., O'Donnell, C., and Jensen, W.A. (1987). Viral infection of the lower respiratory tract. *Clinics in Chest Medicine*, 8, 405–17.

Skerrett, S.J. and Locksley, R.M. (1986). Legionellosis: ecology and pathogenesis. In *Respiratory infections* (ed. M.A. Sande, L.D.Hudson, and R.K. Root), pp. 161–90. Churchill Livingstone, New York.

2.13.2 Therapeutic approach

J.-F. Timsit and J. Carlet

Key messages

- Treatment of atypical pneumonia is mainly symptomatic, with adequate ventilation with end-expiratory pressure if necessary.
- Antibiotics active against intracellular pathogens must be retained in the cell and be able to express their activity in the intracellular medium where the bacteria are located.
- Legionellosis is by far the most common cause of life-threatening atypical pneumonia.
- Probabilistic treatment of community-acquired pneumonia must include treatment active against *Legionella* species, namely erythromycin or the fluoroquinolones.
- Most cases of viral pneumonia admitted to the intensive care unit require only symptomatic treatment.
- It must be remembered that bacterial superinfection with *Staphylococcus aureus*, *Staphylococcus pneumoniae*, or *Hemophilus influenzae* frequently accompanies viral pneumonia, particularly influenza pneumonia.

Introduction

In patients with severe community-acquired pneumonia, ineffectiveness of initial therapy is associated with a 12-fold increased risk of death. Atypical pneumonia accounts for 5 to 15 per cent of these patients. The treatment consists of symptomatic measures against respiratory failure and adapted antimicrobial therapy (Table 1).

Respiratory support

Nearly two-thirds of patients with severe community-acquired pneumonia require mechanical ventilation. In patients with moderate hypoxemia, continuous positive airway pressure (CPAP) via a face mask can be used to correct hypoxemia and to obviate mechanical ventilation. Positive airway pressure can reduce pulmonary shunt and improve perfusion inequality. When respiratory muscle fatigue occurs, mechanical ventilation is mandatory. For treatment of severe hypoxemia, other therapies should be used including adequate ventilation with permissive hypercapnia reducing the incidence of baro- or volotrauma, positioning, optimization of end-expiratory pressure, and inhaled nitric oxide.

Penetration, accumulation, and efficacy of intracellular antibacterial agents

In order to be active against an intracellular pathogen, an antibiotic must be able to penetrate the cell, to be retained by the cell, and to express its activity in the environment where the bacteria are located.

Uptake and release of antimicrobials

Marked intracellular accumulation of macrolide antibiotics has been observed in several types of culture cells. Thus the concentration of erythromycin in the macrophage may be twice to 10 times the extracellular concentration. Uptake of macrolides is non-saturable and usually rapid, but so is efflux. A marked recent exception is azithromycin, which, in contrast with other macrolides, has a slow penetration and a very high intracellular level (the ratio of intra- to extracellular level is greater than 400), contrasting with a low extracellular level and a slow release. Macrolides accumulate and distribute themselves in both lysosomes and the cytosol. Fluoroquinolones accumulate in phagocytic cells and are not associated with a specific organelle. The uptake is rapid, and the accumulated drugs are also quickly released. Accumulation of fluoroquinolones is similar to that of erythromycin (two- to eightfold) and is not influenced by the activation of macrophages. Very few data have been reported for other antibiotic families used in the treatment of intracellular infection. Rifampin (rifampicin) accumulates moderately in phagocytes (two- to threefold). Therefore it is possible that a high intracellular concentration does not reflect the effectiveness of an antimicrobial.

Antibiotic uptake may be influenced by other factors. For example, macrophages infected with *Legionella pneumophila* retain more pefloxacin than uninfected macrophages. It has been reported that interferon-γ and tumor necrosis factor enhance the uptake of azithromycin by macrophages.

Intracellular activity

Intracellular pH, oxygen content, and enzymatic activity within cell cytoplasm and phagocytic lysosome are different from those in serum. Any of these variables may render the antibacterial agent inactive, even if the agent and the bacteria are in the same intracellular location. However, the intracellular environment does not have too detrimental an effect on the antibacterial properties of macrolides and fluoroquinolones. Despite a relatively low intracellular concentration, rifampin is very active against intracellular bacteria. The most widely accepted hypothesis is that the activity of rifampin is somehow enhanced by the intracellular physicochemical (low pH) conditions at the site of infection.

Antibiotic–phagocyte interaction

Several antimicrobials can interfere directly with phagocyte function at different levels. However, available data gave conflicting results. An exception is the possible inhibition of neutrophil function by tetracycline which acts as a calcium chelator.

Table 1 Treatment of atypical pneumonia

	First choice	Alternative treatment	Duration (days)
C. pneumoniae	Doxycyclin 200 mg	Erythromycin 2 g	15–30
C. psitacci	Doxycyclin 200 mg	Erythromycin 2–3 g	10–21
C. burnetti	Doxycyclin 200 mg	Erythromycin+rifampin	15–21
M. pneumoniae	Erythromycin 2 g	Tetracycline 1 g	21
L. pneumophila	Erythromycin 4 g IV+rifampin 1200 mg	Fluoroquinolones	15–21
Probabilistic treatment*	Amoxicillin 3–6 g+erythromycin 3 g	Ceftriaxone 2 g (or cefotaxime 3–4 g)+fluoroquinolones	

IV, intravenous.

*Probabilistic treatment of severe community-acquired pneumonia in which atypical pneumonia is suspected.

Chlamydia pneumoniae

C. pneumoniae is susceptible to many antimicrobials *in vitro*, i.e. tetracycline, macrolides, and fluoroquinolones. The lowest minimum inhibitory concentrations are obtained with tetracycline, doxycycline, erythromycin, roxythromycin, and particularly clarithromycin and azithromycin which have excellent tissue and intracellular penetration. However, the results of *in vitro* tests are not good predictors of microbiological efficacy *in vivo*.

There are few data describing the response of *C. pneumoniae* to antibiotic treatment. In the first report, 1 g of erythromycin daily failed in most patients. Therefore the initial recommendation for treatment of *C. pneumoniae* pneumonia was 2 g of tetracycline daily for 7 to 10 days or 1 g daily for 21 days. The first-choice treatment remains doxycycline. However, persistence of *C. pneumoniae* positive culture after 2-week courses of erythromycin or 30-day courses of tetracycline or doxycycline have been reported. Pharmacokinetics and *in vitro* activity suggest that clarithromycin and azithromycin are the treatment of choice of *C. pneumoniae* infection; however the number of subjects in the studies reported are too small to reveal significant differences between antibiotic regimens.

Chlamydia psittaci

The recommended treatment for *C. psittaci* is tetracycline hydrochloride (2–3 g/day) or doxycycline (100 mg orally twice daily for 10–21 days). Erythromycin is the alternative treatment, although it appears to be less effective in severe cases.

Q fever

Doxycycline (100 mg twice daily for 15–21 days) or the association of erythromycin and rifampin are the recommended therapy for the treatment of *Coxiella burnetii* pneumonia.

Mycoplasma pneumoniae

Treatment with effective antimicrobials can shorten the illness. The effect of antimicrobials in extrapulmonary complications, which occur 2 to 3 weeks after respiratory infection, is debatable. The recommended treatment for *M. pneumoniae* is erythromycin (2 g/day) or tetracycline. The duration of therapy appears to be important. Although symptoms abate abruptly with therapy, relapses occur in up to 10 per cent of cases unless therapy is continued for 3 weeks.

Other antimicrobials such as azithromycin and clarithromycin have greater *in vitro* activity than the first-choice drugs and are well tolerated. However, there are few clinical data, particularly for the most severely affected patients. Fluoroquinolones such as ciprofloxacin and ofloxacin are effective, but both have minimum inhibitory concentrations greater than those for erythromycin.

Steroids have been used in conjunction with antimicrobials and appeared to be helpful in severe pneumonia and erythema multiforme, although there are no controlled studies supporting this therapy.

Legionellosis

The effectiveness of erythromycin has been validated by a large number of clinical studies in the last 15 years and seems to halve the fatality rate of legionnaires' disease compared with untreated cases. This first-choice drug has minimum inhibitory concentrations ranging from 0.125 to 0.5 g/ml. Its activity is rapidly reversed once the drug is removed. However, many clinical and bacteriological failures of erythromycin treatment have been reported, and may partly explain the fatality rates of 10 to 40 per cent under treatment, but no *in vitro* resistance has been reported.

The majority of the newer macrolides, such as josamycin, spiramycin, clarithromycin, and azithromycin, show better *in vitro* activity against *Legionnella* than erythromycin. The oral administration of 500 mg of clarithromycin twice daily has been associated with a 98 per cent success rate in an open study involving 46 patients with mild forms of the disease.

Rifampin is also very active against *Legionella*, with minimum inhibitory concentrations ranging from 0.001 to 0.125 g/ml. Rifampin penetrates intracellularly and reaches a high concentration in the lung. However, it should not be used as the sole therapy because of the theoretical possibility of emergence of resistance. There are no good clinical data demonstrating the superiority of combination therapy with rifampin and other agents over single-drug therapy without rifampin. However, anecdotal data show that the addition of rifampin to erythromycin therapy may improve outcome of immunosuppressed patients, and this treatment is preferred in severe patients with confirmed legionnaires' disease.

Tetracyclines are probably as effective as erythromycin. These agents were comparable with erythromycin in decreasing case fatality rate in the Philadelphia outbreak of 1976. Efficacy has been reported in patients in whom erythromycin therapy failed. However, failures of tetracycline treatment have also been described.

A wide variety of fluoroquinolones have been shown experimentally to be very active against *Legionnella*, achieving minimum inhibitory concentrations of less than 0.01 g/ml. These drugs also very active against intracellular *L. pneumophila,* always being more active than erythromycin. Many of the fluoroquinolones kill, rather than inhibit, the intracellular bacterium. The activity of pefloxacin has been shown to be superior to that of ofloxacin and ciprofloxacin, partly because of a postantibiotic effect. However, no large prospective data exist which support the inclusion of fluoroquinolones in the first-choice treatment regimen.

On the basis of experimental data and open clinical studies, a fluoroquinolone or one of the newer macrolide azalide antimicrobials are probably more effective than erythromycin. In the intensive care unit (ICU), treatment of confirmed severe legionellosis with a fluoroquinolone or a combination of erythromycin and rifampin is presumably preferable. The duration of therapy is 2 to 3 weeks.

Tuberculosis

Treatment of tuberculosis was extensively reviewed in a statement by the American Thoracic Society and the Center for Disease Control and Prevention in March 1993. In summary, a 6-month regimen consisting of isoniazid (5 mg/kg once daily) rifampin (10 mg/kg once daily), and pyrazinamide (15 mg/kg once daily) should be given for 2 months, followed by isoniazid and rifampin for 4 months for patients with fully susceptible organisms who adhere to treatment. In cases of non-conversion of cultures after the initial 2-month regimen, the two-drug regimen of the continuation phase should be extended for 6 to 7 months after cultures are negative. Ethambutol (15 mg/kg) should be included in the first regimen if there is a risk of drug resistance (i.e. previous antituberculous treatment, contact with a drug-resistant case, country with a high prevalence of drug resistance) until the results of drug susceptibility studies are available. Alternatively, a 9-month regimen with rifampin and isoniazid is acceptable for patients who cannot or should not take pyrazinamide. In cases of isoniazid resistance, a 12-month regimen with rifampin and ethambutol is recommended.

In adults, extrapulmonary tuberculosis is treated in a similar manner and with the same regimen as for pulmonary tuberculosis.

Most of the severe cases of tuberculosis requiring admission to the ICU have a disseminated form of the disease and frequently miliary tuberculosis. Steroids have been shown to be effective in preventing cardiac constriction from tuberculous pericarditis and in decreasing the neurological sequelae of all stages of tuberculous meningitis. Corticosteroids are frequently used for severe miliary tuberculosis with acute respiratory distress syndrome and are probably lifesaving.

Amikacin, quinolones, rifabutin, rifapentene, and clofazimine are potentially effective drugs against tuberculosis.

Viral pneumonias

Viral pneumonias which might lead to respiratory distress are mainly influenza, adenovirus, and varicella and cytomegalovirus infections. Most of the viral pneumonias admitted to the ICU only require symptomatic treatment. However, antiviral drugs are useful in some situations.

Amantadine hydrochloride has been shown to be useful for prophylaxis against influenza A infection. Anecdotal cases of improvement, when given early in the course of the disease, have been reported.

Intravenous acyclovir (aciclovir) has been used early in the onset of severe varicella zoster virus pneumonia and seems to reduce the respiratory rate and the duration of fever, and to improve oxygenation. It probably favors the healing of bronchial lesions.

The same treatment might be recommended for the rare cases of severe necrotizing herpes simplex virus pneumonia, although there are no published data, as acyclovir is highly active *in vitro* and is nontoxic. The dose for severe pneumonia due to varicella zoster virus is 500 mg/m^2 intravenously every 8 h for 5 to 10 days.

Ribavirin (tribavarin, Virazole) given in aerosolized form has been associated with a more rapid improvement of blood gases and decrease of viral shedding in respiratory syncytial virus bronchiolitis and pneumonia in children. Aerosol ribavirin has been used in influenza B infection in young adults in a placebo control trial and resulted in a more rapid defervescence, decrease in viral shedding, and improvement in clinical symptoms. Thus this agent might be useful in the treatment of severe influenza and respiratory syncytial virus pneumonia. The suggested dose is 6 g in 300 ml of water by continuous nebulization 18 to 24 h daily for 3 to 7 days. This difficult method of administration requires a closed system to prevent exposure of hospital personnel and this limits its usefulness.

Ganciclovir is used against cytomegalovirus and is effective in decreasing viral shedding from the lung of immunocompromised patients with pneumonia. However, mortality remains high under treatment, and some patients remain alive and well without treatment.

As was pointed out earlier, severe viral pneumonias require mainly symptomatic treatment and adequate ventilation with end-expiratory pressure if necessary. It must be remembered that bacterial superinfection with *Staphylococcus aureus*, *Staphylococcus pneumoniae*, or *Hemophilus influenzae* accompanying viral pneumonia is frequent, particularly with influenza pneumonia, and must be appropriately treated.

Empirical initial antibiotic treatment in severe community-acquired pneumonia

Except for some *Legionella* pneumonias, the diagnosis of atypical pneumonia is often made retrospectively on the basis of serology. Moreover, the pathogen responsible for pneumonia remains unknown in a third of cases.

The clinician confronted with severe pneumonia must target the empirical therapy to cover *Streptococcus pneumoniae* and some Gram-negative organisms. However, initial therapy must also include adequate coverage of atypical pneumonias such as *Legionella*.

If the Gram stain of proximal or distal bronchial samplings is negative, treatment is based on amoxicillin combined with erythromycin or fluroquinoles or on a third-generation cephalosporin combined with erythromycin.

Fluoroquinolones may be preferable to macrolides when methicillin-sensitive *Staph. aureus* or Gram-negative bacteria are also suspected.

Bibliography

Ad Hoc Committee of the Scientific Assembly on Microbiology, Tuberculosis, and Pulmonary Infections (1995). Treatment of tuberculosis and tuberculosis infection in adults and children. *Clinical Infectious Diseases*, **21**, 9–27.

Baum, S.G. (1995) *Mycoplasma pneumoniae* and atypical pneumonia. In *Principles and practice of infectious diseases* (ed. G.L. Mandell, R.G. Douglas, and J.E. Bennett), pp. 1704–13. Churchill Livingstone, New York.

Edelstein, P.H. (1995). Antimicrobial chemotherapy for legionnaires' disease: a review. *Clinical Infectious Diseases*, **21** (Supplement 3), S265–76.

Hammerschlag, M.R. (1994). Antimicrobial susceptibility and therapy of infections caused by *Chlamydia* pneumoniae. *Antimicrobial Agents and Chemotherapy*, **9**, 1873–8.

Kauffman, S. (1994). Viral pneumonia. In *Respiratory infections. Diagnosis and management* (ed. J.E. Pennington), pp. 515–32. Raven Press, New York.

2.14 Pneumothorax

2.14.1 Pathophysiology of pneumothorax

M. Sydow

Key messages

- Spontanous pneumothorax is generally due to subpleural blebs. It often occurs in chronic obstructive pulmonary disease, cystic fibrosis, and AIDS.
- Tension pneumothorax is associated with severe impairment of gas exchange and cardiac function.
- Barotrauma/pneumothorax occurs in 4 to 15 per cent of patients on mechanical ventilation, but in up to 60 per cent of patients with acute respiratory distress syndrome.
- Diagnostic methods include chest radiography and auscultation.

Introduction

Pneumothorax is defined as a pathological collection of extra-alveolar air in the pleural space between the visceral and parietal pleura. It is generally due to alveolar or airway injury, but can also be caused by penetrating injury of the chest wall. It can be classified as spontaneous pneumothorax or pneumothorax of traumatic and/or iatrogenic origin. Common causes and mechanisms of pneumothorax and other forms of extra-alveolar air are summarized in Table 1.

Owing to the retracting forces of the lung tissue, the pleural pressure is more negative than the alveolar or the ambient pressure. If the integrity of the alveolar wall and the visceral pleura or the thoracic wall and the parietal pleura is broken, the air follows the pressure gradient and causes pneumothorax. Subsequently, depending on the amount of gas in the pleural space, the ipsilateral lung collapses. Lung volume decreases and oxygenation deteriorates due to an increased intrapulmonary shunt. If the gas inflow is limited, the lung collapse is not progressive and no mediastinal shift will occur. Without further inflow the pleural air will be reabsorbed within a certain time. Uncomplicated pneumothorax is usually well tolerated in otherwise healthy individuals. However, in patients with decreased pulmonary reserve, such as those with severe chronic obstructive pulmonary disease or acute respiratory distress syndrome, deterioration of oxygenation can be critical with even a small pneumothorax.

Spontaneous pneumothorax

Spontaneous pneumothorax can be differentiated into primary and secondary types. Primary spontaneous pneumothorax is defined as occurring spontaneously in patients without underlying lung disease. However, visible subpleural blebs or apical subpleural emphysema have been demonstrated retrospectively in the majority of patients who underwent thoracotomy or diagnostic computer tomography. Smoking is the most likely risk factor for primary spontaneous pneumothorax. It is assumed to cause inflammation of the small airways with secondary increase of airway resistance and development of emphysematous cysts by air trapping.

Secondary spontaneous pneumothorax is a complication of many lung diseases, mainly chronic obstructive pulmonary disease, cystic fibrosis and, more recently, AIDS. In chronic obstructive pulmonary disease, airway inflammation increases airway resistance which results in hyperinflation with intrinsic positive end-expiratory pressure (PEEP). Subsequently, pulmonary emphysema and thinning of the lung parenchyma occur. If the transpleural pressure gradient exceeds a certain level because of a high intrinsic PEEP, the overdistended emphysematous bulla and visceral pleura rupture, causing pneumothorax. As severity of chronic obstructive pulmonary disease and level of intrinsic PEEP are closely related, the risk of secondary spontaneous pneumothorax is higher in patients with chronic obstructive pulmonary disease who are more critically ill. *Pneumocystis carinii* pneumonia, treatment with inhaled pentamidine, and cigarette smoking are risk factors for spontaneous pneumothorax in AIDS patients. The pathomechanism appears to be necrosis of subpleural lung cavities.

Tissue emphysema

Under conditions of positive alveolar or positive airway pressure, i.e. positive-pressure ventilation or dynamic hyperinflation, air can also dissect into surrounding tissues. This results in pulmonary interstitial emphysema (air cysts), mediastinal emphysema (pneumomediastinum), subcutaneous emphysema, pneumopericardium, pneumoperitoneum, or even systemic air embolism.

Harmful or life-threatening effects of extra-alveolar air depend mostly on their influence on global or regional circulation. Subcutaneous or mediastinal air usually does not compromise local blood flow because the loose connecting tissue is very compliant and allows further dissection of the air along fascial planes. Thus small amounts of systemic air, which might cause cerebral or myocardial air embolism, are more likely to be hazardous than a huge subcutaneous emphysema. Although sometimes impressive, subcutaneous emphysema and pneumomediastinum are not dangerous but, rather, are markers for an apparent air leak. In both cases the clinician's suspicion should be focused on the possibility of

Table 1 Common causes and pathways of extra-alveolar air

Cause	Pathway of the air	Extra-alveolar air
Emphysematous cysts	Lung cyst→pleural space	Pneumothorax (spontaneous)
Lung tissue necrosis (AIDS)	Alveoli→pleural space	Pneumothorax (spontaneous)
Blunt chest trauma	Alveoli/airway→lung tissue	Lung tissue emphysema
	Lung tissue→mediastinum	Mediastinal emphysema
	Mediastinum→pleural space/subcutaneous tissue	Pneumothorax (traumatic)/subcutaneous emphysema
Penetrating chest trauma	Alveoli→pleural space	Pneumothorax (traumatic)
	Ambient air→pleural space	Pneumothorax (traumatic)
Rib fractures	Alveoli→pleural space	Pneumothorax (traumatic)
	Pleural space→subcutaneous tissue	Subcutaneous emphysema
Attempted central line placement	Alveoli→pleural space	Pneumothorax (traumatic, iatrogenic)
Transthoracic needle aspiration	Alveoli→pleural space	Pneumothorax (traumatic, iatrogenic)
Transbronchial biopsy	Alveoli→pleural space	Pneumothorax (traumatic, iatrogenic)
Bronchoscopy	Alveoli/airway→lung tissue	Pulmonary interstitial emphysema
	Lung tissue→mediastinum	Mediastinal emphysema
	Mediastinum→pleural space/subcutaneous tissue	Subcutaneous emphysema/pneumothorax (traumatic, iatrogenic)
	Alveoli→pleural space	Pneumothorax (traumatic, iatrogenic)
Thoracic surgery	Alveoli/airway→lung tissue	Pulmonary interstitial emphysema
	Lung tissue→mediastinum	Mediastinal emphysema
	Mediastinum→pleural space/subcutaneous tissue	Subcutaneous emphysema/pneumothorax (traumatic, iatrogenic)
	Alveoli→pleural space	Pneumothorax (traumatic, iatrogenic)
Mechanical ventilation	Alveoli/airway→lung tissue	Pulmonary interstitial emphysema
	Lung tissue→mediastinum	Mediastinal emphysema
	Mediastinum→pleural space/subcutaneous tissue/peritoneum	Pneumothorax (traumatic, iatrogenic)/ subcutaneous emphysema/pneumoperitoneum
	Alveoli→pleural space	Pneumothorax (traumatic, iatrogenic)

an overlooked or threatening pneumothorax. Additionally, if air has dissected into the peritoneum, a visceral perforation must be excluded. The incidence of mediastinal emphysema in positive-pressure ventilation has been reported as up to 21 per cent (Gammon *et al.* 1992). Subcutaneous emphysema is a common complication in mechanically ventilated patients with penetrating or blunt chest trauma. It does not usually occur alone but subsequent to pneumothorax or mediastinal emphysema.

Tension pneumothorax

In the case of a one-way valve mechanism only a part of the inspired volume can be exhaled and the remaining volume is trapped in the pleural space. Thus the intrapleural gas accumulates and tension pneumothorax develops. Intrapleural pressure which exceeds ambient pressure throughout expiration is the specific characteristic of tension pneumothorax. Increased pleural pressure and volume results in a displacement of the mediastinal structures and the heart, and probably also in compression of the opposite lung. Venous return, and hence cardiac output, will be decreased by the increased intrathoracic pressure and/or by kinking of the vena cava due to the mediastinal shift. However, animal studies suggest that the major consequence of tension pneumothorax is not a decrease in cardiac output but severe hypoxemia due to increased intrapulmonary shunt (Gustman *et al.* 1983). These animal studies were conducted in spontaneous breathing sheep; thus inspiratory pleural pressure was lower than ambient pressure. Therefore venous return might be influenced less than during positive-pressure ventilation when pleural pressure exceeds ambient pressure even during inspiration. This is supported by observation of decreased cardiac output after the animals were artificially hyperventilated. Moreover, hemodynamic measurements during tension pneumothorax in a mechanically ventilated patient revealed a significant decrease of about 45 per cent in cardiac output and a reduction of $Pa\text{O}_2$ from 10.1 to 6.1 kPa (76 to 46 mmHg (Connolly 1993). Likewise, central venous pressure increased threefold, indicating venous congestion by the high intrathoracic pressure.

Tension pneumothorax is rare in spontaneous breathing individuals and is usually a complication during positive-pressure ventilation. However, significant tension pneumothorax can occur even during spontaneous breathing in patients with asthma or exacerbated chronic obstructive pulmonary disease with severe hyperinflation and intrinsic PEEP.

Pneumothorax and barotrauma during mechanical ventilation

Although direct pleural damage and the subsequent communication with ambient or pulmonary air can result from penetrating injury (e.g. rib fracture or accidental needlestick during line placement attempts), extra-alveolar air and pneumothorax from barotrauma during positive-pressure ventilation usually has a different pathway. Macklin and Macklin (1944) demonstrated that a pressure gradient between the bronchovascular sheath and the alveolus is a necessary precondition for alveolar wall rupture and development of pulmonary interstitial emphysema. Accordingly, during positive-pressure ventilation air dissects into the mediastinum primarily via the bronchovascular sheaths. From there it may penetrate the parietal pleura and produce pneumothorax or dissect via fascial planes into either the subcutaneous tissue or the peritoneum. However, in a recent study the predictive value of mediastinal air for subsequent pneumothorax was only 42 per cent in all intensive care patients studied and 50 per cent in patients with acute respiratory distress syndrome (Gammon *et al.* 1992).

The term barotrauma suggests pressure as the predominant factor for alveolar rupture. Indeed, several studies suggest a close relationship between high peak airway pressure and occurrence of pneumothorax or mediastinal emphysema (Gammon *et al.* 1992). However, in healthy individuals airway pressures above 200 cmH_2O can occur during coughing or maximum efforts without any deleterious consequence. Furthermore, experimentally induced high-pressure barotrauma during positive-pressure ventilation in animals was prevented by strapping the chest with belts (Dreyfuss *et al.* 1988). Chest belts have the same effect as actively contracting respiratory muscles during coughing. Both mechanisms decrease the chest wall compliance as well as the transpulmonary pressure and overdistension is avoided. Therefore current concepts support the hypothesis that, rather than high airway pressure, high-volume ventilation and alveolar overdistension as well as high transpulmonary pressure are the causative determinants for barotrauma during positive-pressure ventilation. Nevertheless, occurrence of pneumothorax seems to be most closely correlated with underlying disease, specifically with acute respiratory distress syndrome. The previously noted associations between airway pressures and barotrauma might be largely related to the occurrence of high airway pressures in acute respiratory distress syndrome. The incidence of pneumothorax during mechanical ventilation is between 4 and 15 per cent, while the incidence in acute respiratory distress syndrome can be up to 60 per cent (Gammon *et al.* 1992).

Bronchopleural fistula

A bronchopleural fistula is a connection between the airways and the pleural space. Bronchopleural fistulas cause pneumothorax in penetrating chest injury, after thoracic surgery, and in spontaneous pneumothorax as well as in most cases of iatrogenic pneumothorax (e.g. needlestick etc.). If a bronchopleural fistula persists during positive-pressure ventilation, tension pneumothorax develops in almost all cases. Drainage of a pneumothorax with a persistent bronchopleural fistula results in an air leak which can be detected by the continuous bubbling of air through the water seal of the suction system. A persisting bronchopleural fistula usually seals spontaneously within a few days. The incidence of persisting bronchopleural fistulas is higher during positive-pressure ventilation and in chronic obstructive pulmonary disease patients with intrinsic PEEP. Therapeutic measures to decrease the causative increased back-pressure increase the likelihood of sealing the air leak. However, this is a dilemma, particularly in mechanically ventilated acute respiratory distress syndrome patients who need a higher airway pressure to maintain oxygenation. Large bronchopleural fistulas occur particularly after thoracic trauma due to severe direct airway injury and during acute respiratory distress syndrome because of the inflamed fragile lung tissue and the application of high airway pressure. Major airway injury such as tracheal or main-stem rupture must be treated surgically, but lung surgery to seal large air leaks in acute respiratory distress syndrome is difficult and problematic because of the tissue fragility.

Diagnosis of pneumothorax and other forms of extra-alveolar air

Absent or decreased breath sounds on auscultation accompanied by a hypersonorous percussion sound on the affected side are usually present. Circulatory abnormalities depend on the size of the pneumothorax and include tachycardia, hyper- or hypotension, increased central venous pressure, and increased pulmonary artery pressure. Hypoxemia occurs only in large or tension pneumothorax and in patients with pre-existing lung disease. Patients with spontaneous pneumothorax usually have ipsilateral chest pain and sometimes have dyspnea. Mechanically ventilated patients present rather unspecific symptoms such as agitation and distress, labored breathing, and tachypnea. Tension pneumothorax must be considered in any sudden detoriation of circulation and/or oxygenation during mechanical ventilation. In volume-controlled ventilation the increasing pleural pressure by a tension pneumothorax results in an increase in the airway pressure (Connolly 1993). Thus the pressure gradient between the airways and the pleural space remains and maintains further gas flow into the pleura. Theoretically, this downgoing spiral is interrupted only by a pressure limit at the ventilator. However, an unexplained rise in airway pressure during volume-controlled ventilation or difficulty in ventilating the patient manually with a bag is often the first indicator of a tension pneumothorax. In contrast, during pressure-controlled ventilation a tension pneumothorax decreases the tidal volume while airway pressure is unchanged. Tissue crunching is caused by subcutaneous emphysema. The most probable reason for tissue crunching in the neck region is a pneumomediastinum. Hamman's sign (a crunching sound heard over the precordial region) can be auscultated in patients with pneumomediastinum and pneumopericardium.

In chest radiography, the typical radiological signs of pneumothorax include a lucent rim with the visceral pleura line, which is normally invisible, or the 'deep sulcus sign'. Radiographic signs of mediastinal emphysema are lucent lines along the border of the mediastinum and highlighting of the aortic knob. However, radiographic findings obtained with a portable chest X-ray apparatus in a supine patient are more subtle and may be more difficult to detect. A ventral pneumothorax is only visible on a lateral decubitus film or a chest CT. However, pneumothorax might be confused with large bullas in lung emphysema or by the occurrence of a skinfold which mimics a visceral pleural line. A skinfold can be identified by its unusual orientation, independent of anatomical structures, and positive lung tissue signs in the apparent pleural space. In contrast with the visceral pleura line, which is convex to the chest wall in pneumothorax, the bulla wall is concave. If pleural adhesions cause separation and atypical location of a pneumothorax, the differentiation to a giant bulla might be impossible; thus a chest CT is indicated. After central venous line placement, the possibility of a delayed pneumothorax must be considered. Routine chest radiography within 2 h of catheter placement might be normal, despite a slowly developing pneumothorax which can only be detected on a chest radiograph after 6 to 8 h.

Treatment of pneumothorax

A small pneumothorax of volume less than 20 per cent of the hemithorax in an asymptomatic patient can be treated conservatively. Only observation to avoid complications may be indicated. If room air (21 per cent oxygen) is in the pleural space, the daily expansion of the lung is about 1.25 per cent of the lung volume. Spontaneous resolution of the pneumothorax depends on the nitrogen gradient between the intrapleural gas and the surrounding tissue. Therefore an increase in the nitrogen gradient by breathing oxygen accelerates the reabsorption. A large pneumothorax (more than 20 per cent of the hemithoracic volume) should be treated either by needle aspiration or preferably by chest tube thoracostomy. Owing to the high recurrence rate, pleurodesis (with talc, tetracycline, fibrin glue, etc.) is recommended in spontaneous pneumothorax. Once a pneumothorax results in impaired oxygenation or circulatory symptoms the placement of a chest tube is mandatory. Reinflation of the lung can be improved by negative suction on the chest tube.

In mechanically ventilated patients, a pneumothorax must always be drained with a chest tube because of the high risk of developing tension pneumothorax. For the same reason it is advisable to place a chest drain prophylactically in patients with otherwise uncomplicated pneumothorax if general anesthesia and positive-pressure ventilation is planned.

Persistent large bronchopleural fistulas are frequently a problem during mechanical ventilation. Often a significant part of the inspired air leaves the lung via the fistula. CO_2 removal might be impaired if the bypassing air does not participate in gas exchange, but this can often be solved by increasing the inspired minute ventilation. However, a large air leak can result in a decrease of the mean intrapulmonary pressure and development of alveolar collapse, particularly in volume-controlled ventilation. Thus, pressure-controlled ventilation with a constant inspiratory airway pressure might be more advantageous. Although successful in infants, high-frequency ventilation in adults with large bronchopleural fistulas failed to improve gas exchange or to decrease gas flow through the fistula compared with conventional ventilation. Occasionally, in unilateral fistulas, independent lung ventilation facilitates individual adaptation of the ventilatory setting. While the intrapulmonary pressure in the lung with the fistula can be decreased to support sealing, gas exchange can be maintained by ventilation of the other lung.

References

Connolly, J.P. (1993). Hemodynamic measurements during a tension pneumothorax. *Critical Care Medicine*, **21**, 294–6.

Dreyfuss, D., Soler, P., Basset, G., and Saumon, G. (1988). High inflation pressure pulmonary edema. Respective effects of high airway pressure, high tidal volume, and positive end-expiratory pressure. *American Review of Respiratory Disease*, **137**, 1159–64.

Gammon, R.B., Shin, M.S., and Buchalter, S.E. (1992). Pulmonary barotrauma in mechanical ventilation. Patterns and risk factors. *Chest*, **102**, 568–72.

Gustman, P., Yerger, L., and Wanner, A. (1983). Immediate cardiovascular effects of tension pneumothorax . *American Review of Respiratory Disease*, **127**, 171–4.

Macklin, C.C. and Macklin, M.T. (1944). Malignant interstitial emphysema of the lungs and mediastinum as an important occult complication in many respiratory diseases and other conditions: an interpretation of the clinical literature in the light of laboratory experiment. *Medicine*, **23**, 281–352.

2.14.2 Assessment and management

R. C. Tilley and B. F. Keogh

Key messages

- Indications for drainage of a pneumothorax in a critical care patient are usually obvious.
- Any deterioration in hemodynamics or gas exchange in a ventilated patient should raise the possibility of pneumothorax.
- Clinical evidence of tension pneumothorax mandates immediate drainage.
- A pneumothorax suspected clinically, and confirmed radiologically, should similarly be drained.
- In less obvious circumstances, more definitive imaging, such as by CT scanning, may identify an occult collection.
- In certain circumstances, particularly in the presence of small iatrogenic pneumothoraces, a conservative approach may be used but must be accompanied by vigilant observation and recognition of the risk of rapid deterioration.

Introduction

Pneumothorax is a potentially life-threatening condition which may be difficult to diagnose accurately in critical care patients. Drainage indications depend on hemodynamics, lung architecture, cardiopulmonary reserve, and the nature of ventilatory support. The need for drainage in most situations is obvious. A more difficult clinical decision is to leave a pneumothorax untreated, which is an appropriate action in certain circumstances.

Pneumothoraces in spontaneously breathing patients

The traditional teaching that a pneumothorax that occupies less than 20 per cent of the hemithorax can be observed relates to patients, such as those with primary spontaneous pneumothorax, in whom pulmonary function is well preserved. Drainage indications in critical care patients depend much more on associated pulmonary pathology and cardiopulmonary reserve than on pneumothorax volume.

Small pneumothoraces are still usually well tolerated and can be observed. The onset of tachypnea, distress, discoordinate respiratory pattern, or hemodynamic compromise indicates the need for urgent definitive drainage. Pneumothoraces associated with recent chest trauma or significant obstructive airways disease should be drained as they are likely to increase in size.

Pneumothoraces in positive-pressure ventilation

Pneumothoraces secondary to underlying pulmonary pathology, trauma, or percutaneous procedures are always at risk of increasing in size and developing tension in positive-pressure ventilation. In general, all such collections should be drained. Exceptions may be small rims of air, thought to be insignificant, or small residual collections after a partially successful drainage procedure.

There is usually little debate on the need for drainage of a pneumothorax. Drainage should follow as soon as possible after identification of clinical signs of its significance. Delay in diagnosis or therapy may lead to catastrophic tension pneumothorax (Fig. 1). Suspicion of pneumothorax in a severely compromised patient mandates immediate speculative pleural space decompression. Although we do not encourage the insertion of small intercostal cannulas, preferring formal chest drain insertion, temporary cannulation may prove to be lifesaving. Treatment should be based on clinical signs. Mortality from tension pneumothorax is much lower (7 versus 31 per cent) if treatment is based on clinical signs rather than delayed until radiographic diagnosis is confirmed (Steier *et al.* 1974).

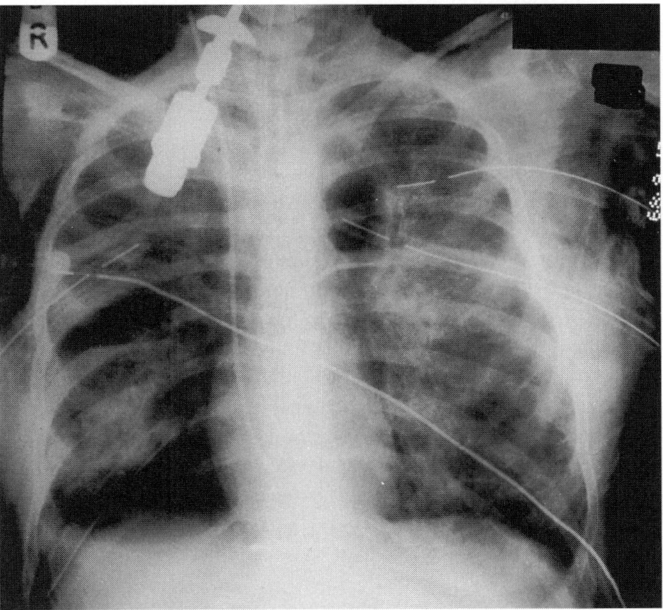

Fig. 1 Bilateral tension pneumothorax despite pleural drainage on both sides. Note the compressed mediastinal and cardiac structures. Bilateral subcutaneous emphysema is also present.

Investigation of the difficult pneumothorax

Many patients with acute respiratory failure suffer recurrent pulmonary air leaks. Loculation of pleural air may occur as a result of pleural inflammation, bleeding, or the presence of intercostal catheters. The lung may be tethered at many points, and localization of the pneumothorax, either clinically or with anteroposterior radiography, may not prove possible. In addition, pneumothorax distribution in supine patients may be atypical; anteromedial and subpulmonic collections were most common in one study (Tocino *et al.* 1985).

Patients with clinically significant pneumothoraces may already have several functioning intercostal catheters, and the site of new collections may be unclear. Lateral shoot-through decubitus films may be very helpful in guiding therapy. The improved definition of computerized phosphor-plate radiology has much increased the yield from lateral films in adults and even in obese patients. Radiographs from one such patient with acute respiratory distress syndrome (Fig. 2) show the anteroposterior radiograph, which suggested an anterior subpulmonary collection, to be misleading. The lateral film confirms a large posterior left-sided collection in an area which was previously densely consolidated.

High-definition CT scanning is the definitive imaging tool. Significant pneumothoraces may be completely missed on anteroposterior films. Fig. 3(a) shows a significant posterior collection which was not identified on the anteroposterior radiograph (Fig. 3(b)). The risk versus the benefit of transfer to the CT scanner is controversial, but in our experience the potential benefit in patients with refractory respiratory failure outweighs the logistic difficulties in safely transporting such patients.

Another unusual problem is the large pneumothorax not associated with ongoing air leaks. Figure 4(a) shows the CT scan of a patient with acute respiratory distress syndrome who, despite an excellent oxygenation response to nitric oxide therapy, exhibited deterioration of anteroposterior radiograph. A patent chest drain is seen in the left side. Direct pressure measurements of anterior air collections indicated no tension. An overenthusiastic reduction in mean airway pressure in response to selective vasodilator therapy had resulted in a loss of lung volume over several days. Aggressive lung volume maintenance was reinstituted, and Fig. 4(b) shows re-expansion, without further chest drains, 6 days later. In this case, lung volume expansion, rather than further pleural drainage, represented definitive treatment.

Pneumothoraces in poorly compliant lungs

Paradoxically, pneumothoraces which occur in the most severely compromised patients may be difficult to identify. Acute respiratory distress syndrome patients with static pulmonary compliance values below 15 ml/cmH_2O and intrapulmonary shunt above 50 per cent may show little change in already grossly disordered gas exchange when pneumothoraces occur. The lungs are extremely stiff and do not

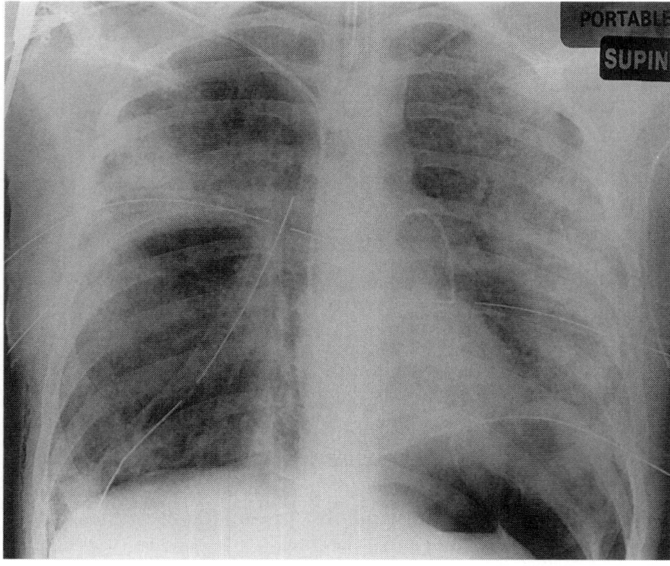

(a)

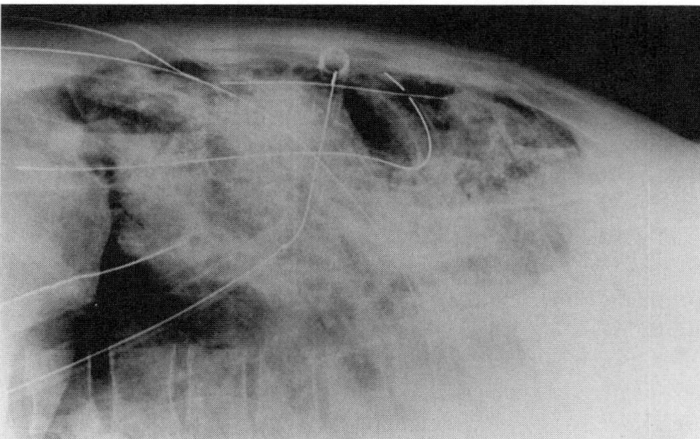

(b)

Fig. 2 The anteroposterior radiograph (a) suggests an anterior subpulmonary collection, whereas the lateral radiograph (b) shows a large left posterior collection.

Table 1 Early subtle signs of pneumothorax in patients with low pulmonary compliance
Decreased urine output
Increased central venous pressure
Tachycardia (may be masked by inotropes or disease process)
Pulsus paradoxus
Dampening of peripheral arterial saturation trace
Decreased cardiac index
Decreased mixed venous oxygen saturation with oximetric pulmonary artery catheter

collapse readily, and therefore the volume effects of relatively small collections may result in hemodynamic depression before gas exchange deteriorates. The early signs of pneumothorax in such patients may be subtle (Table 1), and most reflect hemodynamic changes. The non-uniformity of changes in commonly measured hemodynamic indices in the presence of tension pneumothorax has recently been highlighted (Beards and Lipman 1994)

Pressure-controlled ventilation is widely used in acute respiratory failure, particularly in acute respiratory distress syndrome. An early sign of pneumothorax when this form of pressure-limited ventilation is employed is the inevitable decrease in tidal and minute volume caused by the pneumothorax-induced decrease in pulmonary compliance.

Clinical dilemmas associated with pneumothoraces

Iatrogenic pneumothoraces

Small iatrogenic pneumothoraces in ventilated patients are inevitable in critical care. Typically, small air collections are an incidental or expected finding after difficult central venous cannulation. Subcutaneous emphysema may be palpable locally. Large ongoing leaks with obvious clinical signs clearly require intercostal drainage. Small and clinically insignificant pleural air collections due to needle laceration do not necessarily require drainage, as such lacerations may rapidly seal without further consequence. A repeat radiograph to determine pneumothorax evolution allows a reasoned decision in this situation. Vigilant observation in such patients is the key to avoiding extension of iatrogenic morbidity.

Pneumothoraces in ventilated asthmatics

The diagnosis of a pneumothorax in acute asthma may be extremely difficult. All ventilated asthmatics are at risk of pneumothorax, with a reported incidence of up to 20 per cent (Mansel *et al.* 1990). Patients may be moribund with extreme hyperinflation, and the clinical differentiation from pneumothorax may be extremely difficult. One discriminating point is that ventilated asthmatics usually exhibit normal arterial oxygen saturation at moderate F_{IO_2} (e.g. 0.5), whereas those with a significant pneumothorax may be hypoxic owing to shunt through associated collapsed lung. In addition, an extended period of apnea (>20 s) should see significant hemodynamic improvement in asthma, although not in tension pneumothorax. We believe that there is no indication for needle aspiration in asthma. If no pneumothorax is present before aspiration, it certainly will be following blind needling and will mandate subsequent pleural cavity drainage. If a pneumothorax is suspected, intercostal tube placement, often bilateral, is appropriate. The potential for direct damage to the hyperinflated lung, which will not collapse away when the pleural incision is made, must be recognized, and intercostal catheters must be inserted with great care.

The significance of subcutaneous emphysema

Air originating from damaged lung units may track back via the hilum to the mediastinum and be manifest only as subcutaneous emphysema in the neck. In this situation, no pneumothorax may be present. In critical care practice, these signs are usually associated

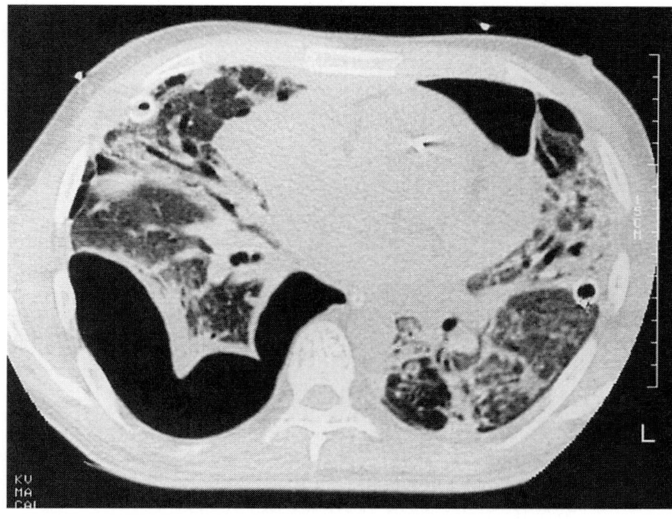

(a)

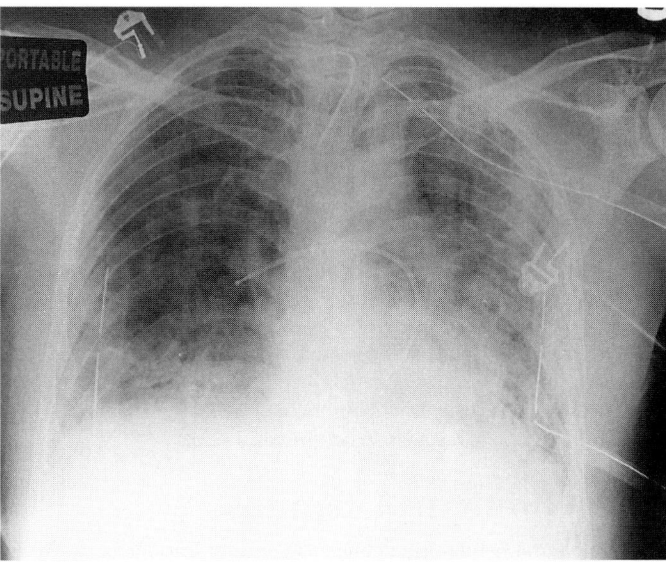

(b)

Fig. 3 Important posterior collection not identified on the anteroposterior film.

with pneumothorax or represent the precursor of more extensive pulmonary leaks in patients with predisposing pulmonary disease.

Subcutaneous emphysema, although cosmetically alarming, is usually a benign condition and does not, of itself, require treatment. However, massive surgical emphysema may render a patient difficult to intubate. It is reasonable to observe small areas of subcutaneous emphysema in the intubated patient if no pneumothorax can be identified. If subcutaneous air is extensive, and particularly if it is increasing, a CT scan may identify the origin. Alternatively, speculative bilateral pleural drainage may access an unrecognized air collection.

Subcutaneous emphysema of the chest wall, in the presence of a chest drain, may indicate that the drain is blocked and air is tracking around it or that the pneumothorax is incompletely drained. It may also occur if the proximal side-hole of the chest drain has migrated out of the pleural cavity.

Interhospital transfer of patients with severe respiratory failure

In patients undergoing positive-pressure ventilation, an attempt should be made to drain all pneumothoraces before interhospital transfer is undertaken. Strict limitation of peak inflation pressures or hyperinflation is impossible in such patients during transfer. Transport ventilators are often inadequate in poorly compliant patients, and manual ventilation, although necessary, may predispose to air leaks. Even the smallest pneumothorax should be drained, as the implications of its evolution during transport may be catastrophic.

Subcutaneous emphysema in such patients without obvious pneumothorax mandates bilateral prophylactic chest tube insertion. In pneumonia, acute lung injury, or other necrotizing lung disease, patients with poor compliance, who have been ventilated for several

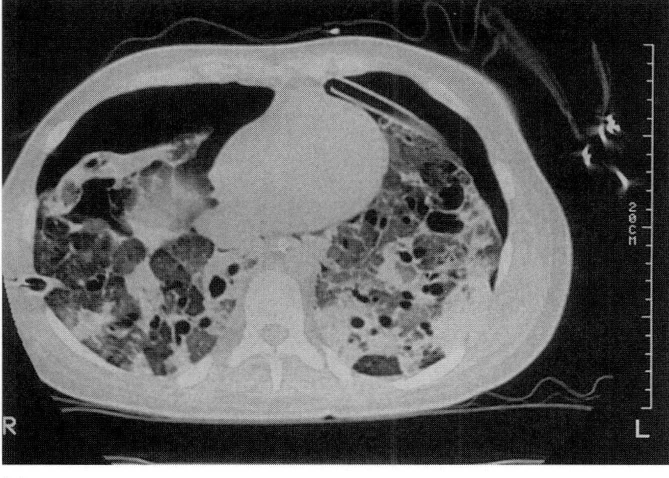

(a)

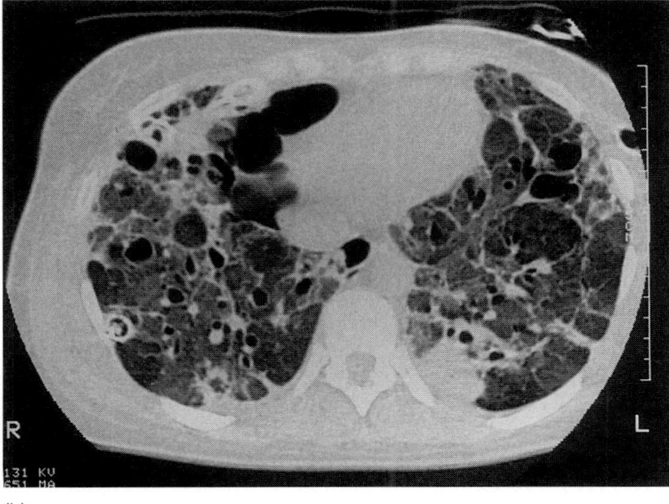

(b)

Fig. 4 CT scan of a patient with acute respiratory distress syndrome in whom rapid reduction in mean airway pressure resulted in loss of lung volume. Re-expansion was obtained by increasing inflation pressure, not by further pleural drainage.

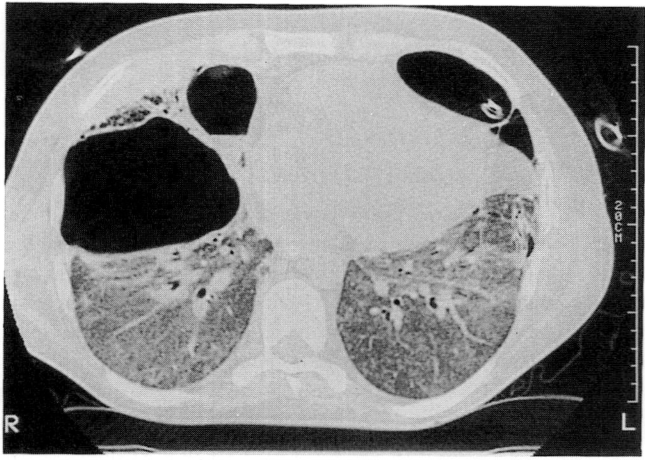

Fig. 5 Multiple residual collections following acute respiratory distress syndrome. Successful weaning was achieved without further drainage.

days and in whom pulmonary function is worsening, are at high risk of developing pneumothoraces in transit, and the insertion of bilateral prophylactic chest drains is reasonable and appropriate.

Pneumothoraces during weaning from ventilation

Patients with severe acute respiratory distress syndrome or pneumonia who exhibit recurrent air leaks during the acute illness are often left with intrathoracic air collections during the weaning phase. Extensive areas of pulmonary fibrosis, combined with lung tethering, result in eccentric collections of residual air. Such collections are often not under tension and have no active air leak. In addition, fibrotic lung cannot be expanded to fill the space. Figure 5 illustrates multiple residual collections following acute respiratory distress syndrome, in one of which, anteriorly, is a functioning chest drain. In our experience, there is a diminishing return in attempting to access all these collections, and weaning, although protracted, can be successfully achieved despite their presence.

Unrecognized tension pneumothorax

An unexpected cardiac arrest in a ventilated patient, in which electromechanical dissociation is observed, should immediately arouse suspicion of tension pneumothorax. Unsuspected large pneumothoraces are occasionally found at postmortem, although they may be secondary to resuscitative attempts rather than the cause of cardiac arrest.

References

Beards, S.C. and Lipman, J. (1994). Decreased cardiac index as an indicator of tension pneumothorax in the ventilated patient. *Anaesthesia*, **49**, 137–41.

Mansel, J.K., Stogner, S.W., Petrini, M.F., and Norman, J.R. (1990). Mechanical ventilation in patients with acute severe asthma. *American Journal of Medicine*, **89**, 42–8.

Steier, M., Ching, N., Roberts, E.B., and Nealon, T.F. (1974). Pneumothorax complicating continuous ventilatory support. *Journal of Thoracic and Cardiovascular Surgery*, **67**, 17–23.

Tocino, I.M., Miller, M.H., and Fairfax, W.R. (1985). Distribution of pneumothorax in the supine and semirecumbent position. *American Journal of Radiology*, **144**, 901–5.

2.14.3 Management of bronchial fistulas—pleural, esophageal, and aortic

E. S. Kassis and J. D. Luketich

Key messages

- Intrathoracic fistulas are rare and life-threatening events.
- Survival depends on rapid diagnosis and initiation of treatment.
- The definitive therapy for most intrathoracic fistulas is surgery.

Introduction

Fistulas arising within the thoracic cavity can result in significant morbidity and mortality. Fistulas can develop between different structures such as the tracheobronchial tree, the esophagus, the pleural space, major vascular structures, and skin.

Bronchopleural fistulas

Bronchopleural fistulas are defined as communications between the tracheobronchial tree and the pleural cavity. Bronchopleural fistulas are most frequently seen following pulmonary resection. Trauma, spontaneous pneumothorax, bacterial pneumonia, lung abscesses, and tuberculosis are also associated with bronchopleural fistulas.

The incidence of postresection bronchopleural fistulas has decreased in recent years with modern surgical technique, and the frequency is now 1.6 to 6.8 per cent (Vester *et al.* 1991; Asamura *et al.* 1992). Asamura and colleagues studied 52 cases of bronchopleural fistulas following 2359 pulmonary resections. Based on a multivariate analysis the authors concluded that pneumonectomies resulted in a higher incidence of bronchopleural fistulas than did lobectomies or wedge resections. They also found that residual carcinoma at the bronchial stump, preoperative radiation therapy, and diabetes were significant risk factors. Vester and colleagues found no correlation between postresectional bronchopleural fistulas and tumor stage, type of resection, and preoperative radiation or chemotherapy, although stapled closures resulted in fewer cases. Others have implicated a long bronchial stump, excessive peribronchial and paratracheal dissection, poor approximation of the mucosa, active infection of the resected lung, and tension at the suture lines as contributing factors in the development of bronchopleural fistulas.

Bronchopleural fistulas occurring within 10 days of resection are usually due to technical inadequacies. The presentation can be dramatic, with dyspnea, subcutaneous emphysema, and productive cough. Contaminated fluid may enter the airway from the postresectional space. The cough and airway contamination are accentuated with the patient in the lateral decubitus with the operated side up.

A somewhat delayed presentation may occur secondary to infection in the postresectional space. This can lead to rupture into the airway, causing severe contamination with resulting pneumonia. This infectious process may be related to technical problems of surgery such as devitalization of the bronchial stump or tension at the suture line.

The diagnosis of bronchopleural fistulas after pneumonectomy may be confirmed with a chest radiograph which demonstrates a decrease in the normal postoperative fluid level within the pneumonectomy space. Another technique used to confirm the diagnosis is the injection of methylene blue in the pleural cavity. If the dye appears in the sputum, the diagnosis is confirmed. CT scan may reveal the fistulous tract between the airway and the pleural space (Fig. 1). Bronchoscopy may also be used to detect defective suture lines in the airway.

Initial treatment is to provide adequate early drainage via tube thoracostomy and the administration of broad-spectrum antibiotics. Chest tube drainage should be performed immediately after diagnosis to drain the space and prevent contamination of the contralateral lung. Patients should be placed in the lateral decubitus position with the operated side down to minimize contamination of the tracheobronchial tree. If hemoptysis or massive expectoration is occurring, intubation of the contralateral bronchus can be lifesaving. If intubation is necessary, mechanical ventilatory indices such as airway pressure, tidal volume, positive pressure breaths, positive end-expiratory pressure, and inspiratory time must be limited.

If the interval between resection and diagnosis of bronchopleural fistulas is less than 7 to 10 days, immediate exploration and closure may be successful in over two-thirds of cases. The decision to perform immediate repair will be based on intraoperative findings, such as the extent of devitalized tissue and contamination. If contamination is too great to consider immediate closure, open drainage with a cutaneous extension in the form of a Clagett window can be performed. The bronchopleural opening can be closed at a later date and the cavity obliterated with omentum and/or a muscle flap.

Bronchopleural fistulas associated with pneumothorax

The most common cause of spontaneous pneumothorax is rupture of a subpleural bleb. These secondary pneumothoraces may result from bullous emphysematous disease, pneumonia, tuberculosis, lung abscesses, AIDS, pulmonary metastases, and positive-pressure ventilation. Preoperative radiation and chemotherapy are also risk factors. In the typical case of spontaneous pneumothorax, the patient is a young tall male. The only pathology present is generally microbullas of the apices of the upper lobe or superior segment of the lower lobe.

Spontaneous pneumothorax presents with sudden onset of chest pain, dyspnea, cough, and respiratory distress. Physical findings may include decreased breath sounds, hyper-resonance on percussion, and distended neck veins. If a pneumothorax is suspected, a chest radiograph should be obtained to confirm the diagnosis. Tension pneumothorax is suggested by a shift of the mediastinum and heart border into the contralateral chest. Occasionally a pneumothorax is not seen on plain film and may be diagnosed by a CT scan of the chest.

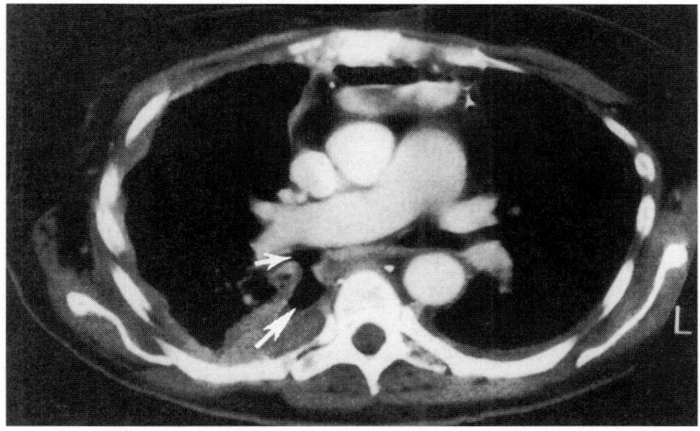

Fig. 1 CT scan demonstrating persistent bronchoesophagopleural fistula (small arrow) and loculated hydropneumothorax in the right posterior apex of the pleural space (large arrow).

The initial treatment of pneumothorax consists of chest tube placement. The position of the chest tube and expansion of the lung should be confirmed by chest radiograph after tube placement. The chest tube can be removed if complete lung expansion and cessation of the air leak has occurred. It is usually recommended that the chest tubes remain in place if the patient is on a ventilator with high peak inspiratory pressures. This acts as a safety valve in case of barotrauma.

The majority of pneumothoraces resolve following placement of a chest tube. Schoenberger *et al.* (1991) showed that if an air leak is present for more that 48 h, further chest tube suction for 8 to 10 days results in a minimal increase in pulmonary healing. Thus surgical intervention may be necessary after 4 to 5 days of persistent airleak following spontaneous pneumothorax. Other indications for surgical intervention include the second occurrence of spontaneous pneumothorax or a life-threatening event associated with the first spontaneous pneumothorax.

Surgical intervention is aimed at the excision of the offending bleb or bullas and mechanical abrasion to result in pleurodesis between the visceral and parietal pleura. In most instances, this can be performed videothoracoscopically.

Tracheo-esophageal fistula

Esophagorespiratory fistulas most commonly occur between the trachea and the esophagus and are referred to as tracheo-esophageal fistulas. They may occur secondary to malignancy, radiation therapy, chemotherapy, or photodynamic therapy.

Malignant tracheo-esophageal fistula is a devastating and often terminal complication occurring in a variety of malignancies. In the largest series of tracheo-esophageal fistulas, 207 malignant cases were reported (Burt *et al.* 1991). The most common tumors resulting in fistulas were esophageal (78 per cent), lung (16 per cent), and tracheal (2 per cent) carcinoma. The incidences of malignant tracheo-esophageal fistulas in patients with esophageal cancer and primary lung cancer are 4.5 per cent and 0.3 per cent respectively. The trachea is the most common site of fistulization, followed by the left and right mainstem bronchi.

In the series reported by Burt *et al.* (1991), 71 per cent, 11 per cent, and 9 per cent of patients had been treated with radiation therapy, resection, and chemotherapy respectively. Luketich *et al.* (1996*a*) reported a case of tracheo-esophageal fistula secondary to photodynamic therapy for malignant mesothelioma which was successfully treated by esophageal exclusion and gastric pull-up.

The majority of patients with malignant tracheo-esophageal fistulas are symptomatic. The most common symptoms are cough (56 per cent), aspiration (37 per cent), and fever (25 per cent). If tracheo-esophageal fistula is suspected, the diagnosis may be confirmed by chest radiography, CT scan of the chest (Fig. 2), barium swallow (Fig. 3), esophagoscopy, or bronchoscopy.

The survival of patients with malignant tracheo-esophageal fistulas primarily depends on the degree of pulmonary contamination and the performance status of the patient at the time of diagnosis. If significant pneumonia is present, most patients will die of respiratory failure within 30 days. Radiation therapy combined with esophageal bypass has resulted in the greatest survival advantage. Patients with minimal pulmonary involvement and good performance status should be considered for esophageal exclusion and gastric bypass.

Supportive care may be preferred in patients with pulmonary contamination or poor performance status. Immediate placement of a covered metallic stent may be effective palliation. As the time interval from diagnosis to treatment increases, the morbidity associated with tracheo-esophageal fistulas increases considerably.

Aortobronchial fistulas

Aortobronchial fistulas are a rare and life-threatening complication seen in pulmonary tuberculosis, staphylococcal pneumonia, bronchogenic lung cancer, aortic dissection without aneurysm, thoracic aortic aneurysm, and mycotic aneurysms, and after aortic valve replacement, repair of aortic coarctation, and unilateral lung transplantation.

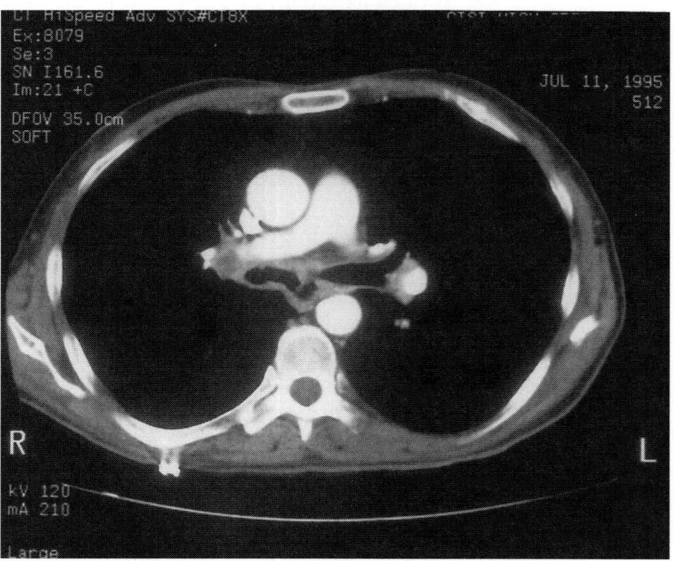

Fig. 2 CT scan demonstrating a fistulous tract between the esophagus and right mainstem bronchus. This fistula developed secondary to lymph node invasion from intrathoracic Hodgkin's disease.

Patients with aortobronchial fistula most commonly present with recurrent or massive hemoptysis. Intermittent hemoptysis is attributed to thrombus formation in the fistulous tract which stops the hemoptysis. With clot lysis or dislocation, a new episode of hemoptysis begins. Most aortobronchial fistulas are large enough to produce massive hemoptysis and immediate death. Patients may also present with cough, dyspnea, and chest pain, which are non-specific signs and are often misleading. Aortobronchial fistula should be considered in the differential diagnosis of hemoptysis, particularly in the presence of a thoracic aortic aneurysm.

Diagnosis should be confirmed with a dynamic CT scan or angiography. Bronchoscopy may result in the displacement of thrombus in the fistulous tract or perforation of the aneurysm, leading to exsanguination.

Surgical repair is the only definitive lifesaving therapy for aortobronchial fistula. Delay in the diagnosis and treatment results in a fatal outcome. Surgery is aimed at excision of the fistula and repair or replacement of the aorta. Sometimes resection of the adjacent lung is necessary.

Aorto-esophageal fistulas

Aorto-esophageal fistula is a rare and life-threatening complication. Primary aorto-esophageal fistula is usually caused by fistulization of an expanding atherosclerotic aneurysm into the esophagus or, less commonly, erosion of a reflux-associated esophageal ulcer into the aorta. Secondary aorto-esophageal fistula occurs after previous grafting of the descending aorta and represents a communication between the aorta and the esophagus, usually at one of the suture lines.

Patients with aorto-esophageal fistula present with massive hematemesis leading to exsanguination, often after a symptom-free interval. Patients may complain of a sentinel arterial hemorrhage, fever, and mid-thoracic pain. The definitive study to detect aorto-esophageal fistula is aortography. Other studies that may be useful include chest radiography and contrast studies of the esophagus.

Surgical repair is the only definitive lifesaving therapy for aorto-esophageal fistula. A delay in diagnosis is uniformly fatal. There have been seven reports of patients with secondary aorto-esophageal

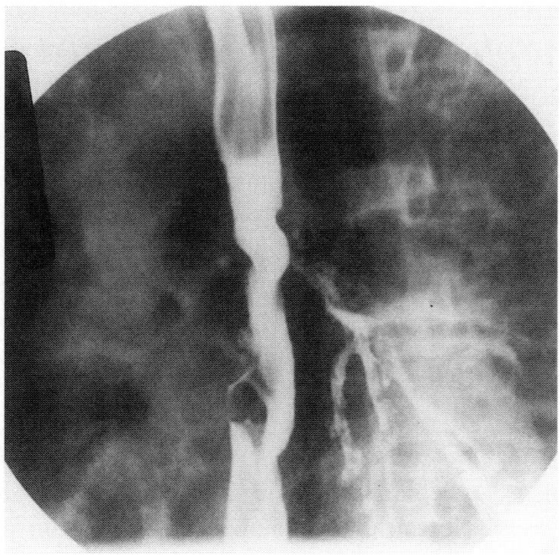

Fig. 3 Barium swallow of the patient in Fig. 2. Note the extravasation of contrast from the esophagus into the right mainstem bronchus.

fistula, with only one long-term survivor (Luketich *et al.* 1996*b*). This patient was successfully treated with an *in situ* polytetrafluoroethylene graft, esophagectomy, and substernal gastric bypass.

References

Asamura, H., Naruke, T., Tsuchiya, R., Goya, T., Kondo, H., and Suemasu, K. (1992). Bronchopleural fistulas associated with lung cancer operations. *Journal of Thoracic and Cardiovascular Surgery*, **104**, 1456.

Burt, M., *et al.* (1991). Malignant esophagorespiratory fistula: management options and survival. *Annals of Thoracic Surgery*, **52**, 1222–9.

Luketich, J.D., Westkaemper, J., Sommers, K.E., Ferson, P.F., Keenan, R.J., and Landreneau, R.J. (1996*a*). Bronchoesophagopleual fistula after photodynamic therapy for malignant mesothelioma. *Annals of Thoracic Surgery*, **62**, 283–4.

Luketich, J.D., *et al.* (1996*b*). Successful management of secondary aortoesophageal fistulas. *Annals of Thoracic Surgery*, **62**, 1852–4.

Schoenenberger, R.A., Haefeli, W.E., Weiss, P., and Ritz, R.F. (1991). Timing of invasive procedures in therapy for primary and secondary spontaneous pneumothorax. *Archives of Surgery*, **126**, 764–6.

Vester, S.R., Faber, L.P., Kittle, C.F., Warren, W.W., and Jensik, R.J. (1991). Bronchopleural fistula after stapled closure of the bronchus. *Annals of Thoracic Surgery*, **52**, 1253–8.

2.15 Pleural effusion

2.15.1 Pathophysiology of pleural effusion

H. T. Billy and K. Waxman

- The visceral and parietal pleura differ both anatomically and functionally. These differences determine the physiology of fluid transport across the pleural space.
- The gradient of partial pressure of gas between the parietal pleura and venous blood keeps the pleural space free of gas.
- Pleural effusions result when a breakdown in the balance between hydrostatic and cellular osmotic forces occurs.
- Upright chest radiography may detect pleural effusions larger than 300 ml. Decubitus views increase the sensitivity. Ultrasonography is more sensitive than chest radiography. Effusions as small as 2 ml may be detectable.
- Pleural fluid can be categorized as transudate or exudate by protein content, specific gravity, and pleural fluid to serum lactate dehydrogenase and protein ratios. Low pleural fluid glucose and low pH are characteristic of empyema.
- Evaluation of pleural white cell counts may help to differentiate bacterial infections from tuberculosis and neoplasm. Gram stains and cultures of pleural fluid are more reliable than sputum cultures for determining the etiology of an underlying pneumonia.

Anatomy and pathophysiology

Pleural effusion is a common development with pleural disorders and frequently indicates a significant underlying systemic disease. The parietal and visceral pleura comprise the serous membranes which line the pleural space. These membranes consist of a single layer of mesothelial cells 20 to 40 µm wide and 0.1 to 0.4 µm thick. Microvilli, which line the pleural surface, increase the functional surface area tremendously, thereby enhancing membrane transport and decreasing pleural friction. The pleura is composed of two membranes which differ both anatomically and functionally. The visceral pleura lines the lungs, mediastinum, diaphragm, and inner surface of the chest wall. It is thin, elastic, and tightly connected to the lung via fibrous extensions of connective tissue, and it reabsorbs fluid from the pleural space. In contrast, the parietal pleura is a thick membrane loosely attached to the underlying chest wall. Blood supplied by systemic capillaries to the parietal pleura creates a hydrostatic gradient which allows net movement of fluid from the parietal pleura into the pleural space. Oncotic pressures of visceral pleural fluid create the driving force which allows fluid reabsorption from the pleural space to the visceral pleura (as much as 5 to 10 l/day).

Pleural membranes are permeable to gas; however the pleural space remains gas free. This is due to a gradient between the partial pressure of gas in the pleural space and venous blood. Normal partial pressures of venous gas in the parietal pleura are as follows: oxygen, 40 mmHg (5.3 kPa); carbon dioxide, 46 mmHg (6.1 kPa); nitrogen, 673 mmHg (89.7 kPa); water, 47 mmHg (6.3 kPa). The total of these partial pressures (706 mmHg (94.1 kPa)) is approximately 50 mmHg (6.7 kPa) less than atmospheric pressure.

Pleural effusions result when there is a breakdown in the balance between hydrostatic and cellular osmotic forces. In the normal physiological state the balance between these forces creates a gradient where absorption of pleural fluid by the visceral pleura is favored. The rate of fluid formation and reabsorption depends on Starling forces, pleural lymphatics, and pleural surface area. Although this gradient allows large volumes of fluid flux into the pleural space, the visceral pleura reabsorbs almost all of the fluid. Only a small volume of pleural fluid (0.1–0.2 ml/kg body weight) is maintained. Pleural and systemic disease processes may result in a change in the normal balance and, as the equilibrium is shifted, accumulation of pleural fluid occurs.

Blood supply to the parietal pleura is systemic. Thus the mean capillary hydrostatic pressure in the parietal pleura is about 30 cmH_2O. The blood supply to the visceral pleura is pulmonary, with a mean capillary hydrostatic pressure of 6 to 10 cmH_2O. Lymphatic supply to the pleura is responsible for absorption of protein and particulate material. The pleural membranes have an effective lymphatic network which decreases the protein content of pleural fluid to about 100 mg/100 cm^3. This lymphatic drainage normally removes about 600 ml of fluid from the pleural space every day. Hydrostatic forces between the pleural space and parietal pleura are positive and exceed the oncotic forces. Thus there is a net flux of fluid from the chest wall into the pleural space. In contrast, the visceral pleura has a hydrostatic pressure which is less than the oncotic gradient and reabsorption of pleural fluid to the visceral pleura is favored.

Several mechanisms are responsible for breakdown of the normal physiological balance which keeps the pleural space free of fluid. An increase in hydrostatic pressure in the pulmonary capillaries decreases pleural fluid uptake. Inflammatory processes or neoplasms can increase capillary permeability resulting in a flux of fluid into the pleural space. Neoplastic cells, protein, and inflammatory products increase the colloid oncotic pressure of the plural fluid, decreasing the reabsorptive effect of the visceral pleura. Systemic hypoalbuminemia decreases pleural capillary osmotic pressure. The flux of fluid out of the parietal pleura is increased, while fluid uptake by the visceral pleura is decreased, leaving lymphatic reabsorption as the major route for fluid reabsorption. Obstruction of lymphatics decreases absorption of protein, cells, and particulate matter by the visceral pleura, which may result in pleural effusions.

Clinical manifestations

Pleural effusions occur as the result of significant pleural or systemic disorders (Table 1). Pleuritic pain, cough, and dyspnea are common findings in patients with pleural effusions. Pleural pain is caused by irritation of the intercostal sensory nerves supplying the parietal pleura. Inflammation of the parietal pleura results in pain localized to the overlying chest wall. The central diaphragm is innervated by the phrenic nerve, and inflammatory processes involving this region result in referred pain to the ipsilateral shoulder. The visceral pleura contains no somatic sensory nerves. Dyspnea is frequently present. Possible causes are decreased vital capacity, pleural splinting, and mechanical distortion of the lung and chest wall.

Careful physical examination may reveal decreased breath sounds and dullness to percussion. The presence of free pleural fluid may be confirmed by changes in percussion dullness with changes in patient position.

Table 1 Etiologies of pleural effusions

Congestive heart failure
Cirrhosis
Nephrotic syndrome
Malignancy
Bronchogenic carcinoma
Metastatic carcinoma
Mesothelioma
Infections
Bacterial pneumonia
Tuberculosis
Fungal
Parasitic
Viral
Pulmonary infarction
Pulmonary emboli
Systemic lupus erythematosus
Rheumatoid disease
Asbestos exposure
Dressler's syndrome
Meigs' syndrome
Subphrenic abscess
Pancreatitis
Radiation pneumonitis
Chylothorax
Yellow-nail syndrome
Trapped-lung syndrome
Ruptured amebic abscess
Superior vena cava syndrome
Status post upper abdominal surgery
Peritoneal dialysis
Drug-induced

Diagnosis of pleural effusions

Radiology

Upright posterior–anterior chest radiography is the most commonly used diagnostic modality for pleural effusions. Accumulation of pleural fluid occurs first under the lung at the base of the hemithorax. Pleural effusions greater than 300 ml will usually obliterate the costophrenic angle. Larger effusions will cause opacification of the hemithorax. Lateral decubitus views confirm the presence of free pleural fluid and are more sensitive than upright films for the detection of effusions. Decubitus views are sensitive in detecting small effusions; as little as 5 ml of fluid may be detected.

Ultrasonography is superior to radiography for detecting pleural effusions. As little as 2 ml of pleural fluid can be detected using ultrasonography (Gryminski *et al.* 1976). Ultrasound guidance is also useful and can be used to assist aspiration drainage of small or loculated effusions.

CT is also sensitive in detecting small pleural effusions. It is most useful for defining underlying parenchymal disease or pulmonary pathology. Distinguishing malignant from benign disease and evaluation of mediastinal adenopathy are better done using CT, while ultrasonography is superior for localizing free and loculated fluid.

Portable supine radiographs may be the only possible view in the critically ill patient unable to tolerate upright or decubitus views. Small effusions may be undetectable, or appear as a generalized haziness. Bilateral effusions may not be detected at all. Ultrasound and CT play an important role in these patients.

Thoracentesis

Although radiographic studies and a careful physical examination will usually confirm the presence of a pleural effusion, determining the etiology of the effusion often requires thoracentesis. Pleural fluid analysis categorizes effusions as either transudates or exudates. Transudates are effusions resulting from an imbalance in the Starling force. Systemic disease states, such as congestive heart failure, cirrhosis, and nephrotic syndrome, alter forces governing fluid exchange across pleural surfaces. Exudative effusions, in contrast, occur as the result of pleural membrane disease such as neoplastic and inflammatory processes. Abnormal protein clearance by pleural lymphatics and increased permeability of protein and fluid into the pleural space result in exudative effusions. Pleural fluid with a protein concentration of 3 g/dl and a specific gravity of 1.016 or more is usually considered to be an exudate. Additional tests of pleural fluid are of diagnostic value. Analysis of pleural fluid lactate

Table 2 Distinguishing values for exudative plural effusions

Protein	>3.0 g/dl
Specific gravity	>1.016
Lactate dehydrogenase	>200 IU
Fluid/serum lactate dehydrogenase	>0.6
Fluid/serum protein	>0.5

dehydrogenase and protein content will distinguish exudative from transudative effusions (Table 2). Lactate dehydrogenase levels above 200 IU, a pleural fluid to serum lactate dehydrogenase ratio above 0.6, and a pleural fluid to serum protein ratio above 0.5 distinguish exudative effusions (Light *et al.* 1972). Additional tests of pleural fluid are of diagnostic value for exudative effusions. Amylase, glucose, cell counts, and pH values do not help to distinguish between transudates and exudates.

Low pleural fluid glucose (below 60 mg/dl) or a pleural fluid to serum glucose ratio below 0.5 is typical of infectious effusions, particularly empyema. Tuberculous effusions may have a low glucose value. Rheumatoid effusions also have a characteristic low glucose level. Systemic lupus erythematosus, esophageal rupture, and 15 per cent of malignant effusions have low glucose levels. Low glucose levels are the result of increased glucose metabolism by bacteria and leukocytes which exceeds the rate of glucose transport into the pleural space.

Pleural fluid amylase is elevated above normal serum levels in effusions associated with pancreatitis and pseudocysts. Increased pleural amylase occurs in esophageal rupture and malignancy.

The presence of bacteria in the pleural fluid usually lowers the pleural pH. Thirty per cent of exudative effusions have a pH below 7.3, while transudates typically remain above 7.4. Accumulation of lactic acid by anaerobic glycolysis contributes to the lower pleural pH. Tube thoracostomy drainage should be considered for those effusions which present with gross purulence of effusions with a pH below 7.2. Effusions with pH above 7.2 will often respond to antibiotic therapy alone (Light *et al.* 1973). Pleural pH is useful for differentiating tuberculosis from malignant effusions: a pH above 7.4 suggests malignancy, while a pH below 7.3 suggests tuberculosis.

Pleural fluid white cell counts of more than 103/mm^3 are found in pleural effusions from 85 per cent of patients with pulmonary infection, 73 per cent with tuberculosis, 42 per cent with neoplasm, 27 per cent with cirrhosis, and 10 per cent with congestive heart failure (Paddock 1984). Predominance of neutrophils favors a diagnosis of bacterial infection, whereas predominance of lymphocytes favors tuberculosis or neoplasm. Red blood cells in the pleural fluid most frequently result from neoplasm. Other causes include congestive heart failure, infection, pulmonary infarction, trauma, and cirrhosis.

Gram stain and culture of pleural fluid are more reliable than sputum cultures for determining the etiology of the underlying pneumonia. Aerobic, anaerobic, and tuberculosis cultures should be prepared. However, tuberculosis cultures are positive in only 30 per cent of cases, even when pleural biopsy confirms the diagnosis of tuberculosis. Specific cultures for fungus and *Entamoeba histolytica* may also reveal the cause of effusions due to these organisms.

Hemothorax

Hemothorax with or without associated pneumothorax is a frequent result of chest trauma and may cause hemorrhagic shock due to massive blood loss into the pleural space. The shock state may be exacerbated by mediastinal shift and decreased.

The most important aspect of managing hemothorax is adequate chest tube drainage of the pleural space. With adequate chest tube drainage and re-expansion of the lung, bleeding will be controlled in 95 per cent of cases (Griffith *et al.* 1976). Only 5 per cent of chest trauma cases with hemothorax will require thoracotomy to control hemorrhage. If bleeding is more than 300 ml/h, surgery is indicated.

References

Griffith, G.L., *et al.* (1978). Acute traumatic hemothorax. *Annals of Thoracic Surgery*, **26**, 204–10.

Gryminski, J., Krabowka, P., and Lypaowicz, G. (1976). The diagnosis of pleural effusion by ultrasonic and radiologic techniques. *Chest*, **70**, 33–9.

Light, R.W., *et al.* (1972). Pleural effusions: the diagnostic separation of transudates and exudates. *Annals of Internal Medicine*, **77**, 507–12.

Light, R.W., *et al.* (1973). Diagnostic significance of pleural fluid pH and $P\text{CO}_2$. *Chest*, **64**, 591–8.

Paddock, F.K. (1984). The diagnostic significance of serous fluids in disease. *New England Journal of Medicine*, **233**, 1010–17.

2.15.2 Indications for drainage of pleural effusion

B. K. Evetts and K. Waxman

Key messages

- Indications for draining pleural effusions include an effusion suspected of being infected, a hemorrhagic effusion, and an effusion which is causing ventilatory compromise.
- Thoracentesis should not be attempted unless fluid can be clearly localized by chest radiography, ultrasound, or CT scan.
- It is important to determine whether an effusion is recurrent or of new onset, or if there is a history of trauma. Traumatic effusions should be assumed to be hemothoraces, and should be treated with tube thoracostomy.
- The treatment of effusion varies depending upon thoracentesis: thick and purulent empyema requires tube thoracostomy; thinner parapneumonic effusions with low protein content, high pH, and high glucose content can sometimes be treated with antibiotics alone; multiloculated pleural effusions require surgical drainage and decortication.
- Symptomatic recurrent pleural effusions may require pleurodesis via chest tube, video-assisted thoracoscopic surgery, or open thoracotomy.

Introduction

Pleural effusions are very significant since they are invariably secondary to an important underlying condition. Fluid collections within the pleural space are rarely apparent clinically or by radiography until they amount to about 500 ml. Indications for draining a pleural effusion emergently are usually limited to three settings: an effusion suspected of being infected, a suspected hemorrhagic effusion, and a high suspicion that the fluid in the pleural space is causing ventilatory compromise. Thoracentesis should not be attempted unless more than 1 cm of fluid layers out on lateral decubitus chest radiography or can be safely localized by other means. When considering drainage of pleural fluid, it is important to consider whether the diagnosis made from the fluid is going to change current treatment. This is important because thoracentesis performed on critically ill patients in the intensive care unit has significant risks. The critically ill patient may be too unstable to position properly or may be ventilator dependent. If the pleural fluid cannot be well localized, thoracentesis should not be attempted.

New-onset pleural effusion

When a pleural effusion is diagnosed, it is necessary to determine whether it is recurrent or of new onset, or if there is a history of trauma (Fig. 1). If there is a history of trauma a chest tube is indicated. The most important aspect of a new-onset pleural effusion is determination of the cause. This can usually be done by a thorough history, but diagnostic thoracentesis is often needed.

Based on findings from the thoracentesis, the treatment of the effusion may vary. If purulent fluid is aspirated at thoracentesis, a chest tube should be placed immediately to drain the empyema. However, a free-flowing parapneumonic effusion with a pleural fluid pH of 7.3 or above, glucose content above 60 mg/dl, and lactate dehydrogenase content below 1000 IU/l can usually be treated with antibiotics alone. If the parapneumonic effusion has pH below 7.1, glucose content below 40 mg/dl, and lactate dehydrogenase content above 1000 IU/l, early tube thoracostomy is advisable (Sahns 1993). In the patient with a multiloculated pleural effusion aggressive drainage and decortication is needed, usually via thoracotomy. If the fluid in the pleural space is secondary to congestive heart failure, medical treatment will usually suffice. However, if the patient is symptomatic, repeat thoracentesis may be necessary. If the effusion is determined to be sympathetic, treating the underlying cause usually resolves the effusion. Discomfort in patients with symptomatic new-onset malignant pleural effusions can often be relieved with thoracentesis. If thoracentesis is not successful, chest tube insertion and possible pleurodesis should be performed. If this is unsuccessful, the malignant effusion should be treated as a recurrent symptomatic effusion (Fig. 2).

Recurrent pleural effusion

The most frequent reason for draining recurrent pleural effusions is to relieve symptoms or improve ventilatory status. If the recurrent effusion of known cause is not symptomatic, observation is warranted. The majority of patients with large pleural effusions are

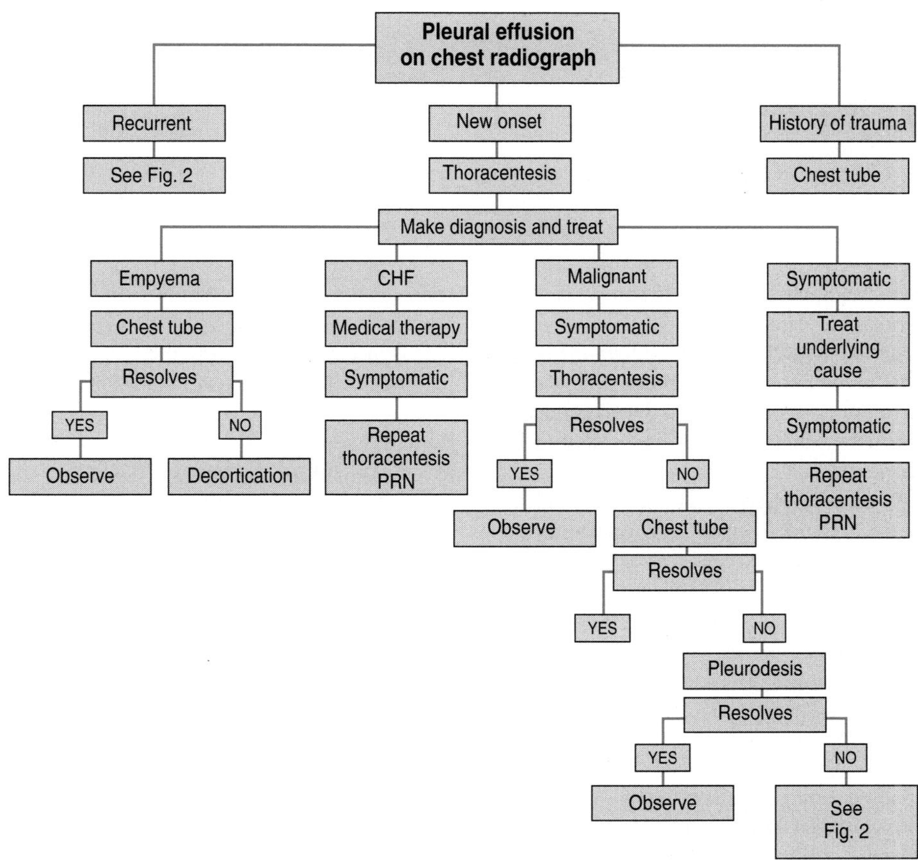

Fig. 1 Management of a new-onset pleural effusion: CHF, congestive heart falure; PRN, as required.

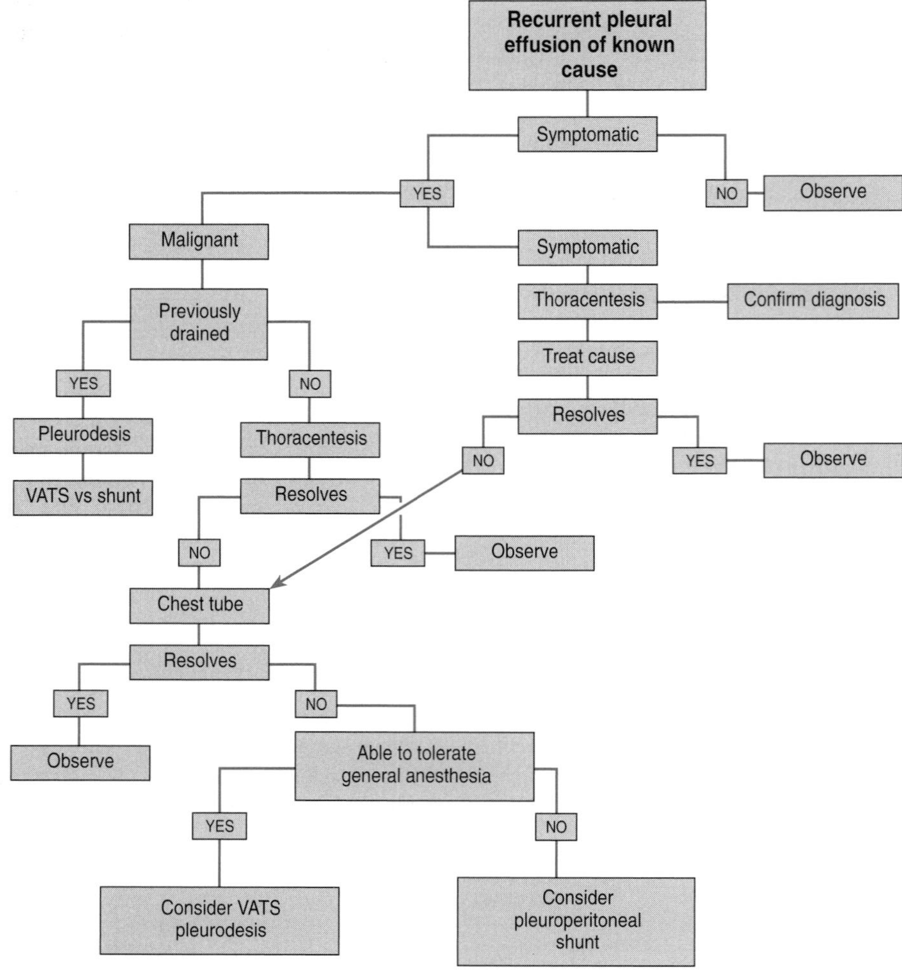

Fig. 2 Management of recurrent pleural effusion: VATS, video-assisted thorascopic surgery.

symptomatic and experience cough, chest pain, or dyspnea on exertion. It is important to know the cause of the effusion. If it is sympathetic, thoracentesis should be performed to confirm the diagnosis. Then treating the underlying medical problem will usually improve the effusion. If treatment does not improve the effusion and the patient remains symptomatic, chest tube insertion should be performed. If the symptomatic effusion remains problematic, video-assisted thoracic pleurodesis or pleuroperitoneal shunt should be considered, depending on the patient's overall medical condition (Fig. 2).

The most common cause of exudative pleural effusions in patients aged over 60 years is malignancy (Keller 1993). A malignant pleural effusion is usually an exudate (fluid protein, above 3 g/dl; pleural fluid to serum protein ratio, above 0.5; lactate dehydrogenase content, above 200 IU/l; pleural fluid to serum lactate dehydrogenase ratio, above 0.6) and hemorrhagic in appearance (Light 1983). In patients with malignant effusions that have previously been drained, pleurodesis should be attempted either via chest tube or by video-assisted thoracoscopic surgery. Repeat thoracentesis is a reasonable option in the patient with a previously undrained malignant effusion that recurs slowly. If this proves to be unsuccessful, tube thoracostomy with or without pleurodesis should be attempted. With the emerging use of video-assisted thoracoscopic surgery the management of malignant pleural effusions is changing. In those patients with recalcitrant effusions who can tolerate general anesthesia and single-lung ventilation, lysis of adhesion, decortication, and pleurodesis using video-assisted thoracoscopic surgery is the treatment of choice. In those patients in whom the lung remains trapped, a viable option is pleuroperitoneal shunt. In the patient who cannot tolerate general anesthesia, tube thoracostomy should be performed; if the lung can be fully expanded, pleurodesis will resolve the effusion in 66 to 92 per cent of patients, depending on the sclerosing agent used (Bayly 1978).

References

Bayly, T.C. (1978). Tetracycline and quinacrine in the control of malignant pleural effusion. *Cancer*, **41**, 1188–92.

Keller, S. (1993). Current and future treatment for malignant pleural effusion. *Chest*, **103**, 63S–7S.

Light, R.W. (1983). *Pleural diseases*. Lea & Febiger, Philadelphia, PA.

Sahns, S. (1993). Management of complicated parapneumonic effusion. *American Review of Respiratory Disease*, **148**, 813–17.

2.15.3 Drainage techniques

R. Nirula and K. Waxman

Key messages

- With relatively large effusions, thoracentesis can be performed using physical examination and chest radiography for landmarks. In smaller effusions, or those in which fluid is not easily drained, ultrasound is useful to guide insertion and placement of the pleural catheter.
- Post-thoracentesis chest radiography should be routine.
- The maximum volume of fluid which should be removed is 1.5 liters. Tube thoracostomy is most effective when performed early in the course of empyema. The use of fibrinolytic therapy may improve success rates of tube drainage of empyema.
- CT scans are useful for defining the anatomy and the extent of loculation in empyema, and may predict success rates of chest tube drainage.
- Pleurodesis can be performed via the chest tube, by video-assisted thoracoscopic surgery, or by open thoracotomy. Pleuroperitoneal shunts or indwelling catheters with implanted reservoirs may provide symptomatic relief in selected patients with malignant pleural effusions.

Introduction

Several techniques for draining pleural effusions are available to the clinician. Determining which is most appropriate for a given patient requires consideration of the therapeutic goals, the nature of the effusion, and the condition of the patient. The most widely used methods of drainage include thoracentesis, thoracostomy tube drainage, video-assisted thoracoscopic surgery, pleuroperitoneal shunting, and Tenckhoff catheter drainage.

Thoracentesis

Thoracentesis is usually performed under local anesthetic at the bedside. It is most commonly used for diagnostic purposes, but can be therapeutic. Traditionally, thoracentesis is performed using physical examination and chest radiography to identify landmarks for needle insertion into the effusion. However, thoracentesis can be difficult, even in experienced hands. In those instances where fluid is not easily drained or encountered using this technique, ultrasound can be used to document the presence and location of the effusion as well as to guide its drainage (Weingardt *et al.* 1994). Furthermore, in those cases where an attempt at thoracentesis fails, ultrasound can be useful in identifying the cause of the failed attempt (Weingardt *et al.* 1994).

Unguided thoracentesis may fail because there may, in fact, be an absence of pleural fluid, or the thoracentesis site is either below the diaphragm or above the effusion (Weingardt *et al.* 1994). Less frequently, the fluid may be loculated, in which case ultrasound guidance may be quite effective (Weingardt *et al.* 1994).

A successful thoracentesis depends upon optimal patient and clinician comfort. The most widely used and effective position is with the patient sitting with the head down on pillows over a table. The skin should be cleansed generously with antiseptic solution over the site of the effusion, and the needle should be inserted one intercostal space below the onset of percussion dullness and approximately 3 to 5 cm lateral to the spinal column. The skin, subcutaneous tissue, and parietal pleura should be anesthetized with local anesthetic (e.g. 1 per cent lidocaine (lignocaine) with epinephrine (adrenaline)) and a 'finder' needle then introduced directly over the rib to avoid injury to the neurovascular bundle coursing along the rib's inferior edge. Once pleural fluid is aspirated, the thoracentesis needle can be inserted following the same path taken by the 'finder' needle (Quigley 1995). A post-thoracentesis chest radiograph should be taken to exclude the presence of a pneumothorax secondary to the procedure. In those patients having a significant effusion requiring therapeutic thoracentesis, a volume of 1.5 liters is the maximum amount of fluid which should be removed to reduce the risk of re-expansion pulmonary edema (Quigley 1995).

While thoracentesis can provide relief of symptoms, in the case of malignant pleural effusions symptoms commonly recur as the effusion reaccumulates. In this instance, thoracentesis is primarily for diagnostic purposes. It is best not used repeatedly to drain these recurrent effusions because it is inconvenient, painful, and carries a risk of infection into the pleural space as well as a pneumothorax (Quigley 1995). Alternative methods of dealing with recurrent effusions are preferable.

Thoracostomy drainage

The use of chest tubes has been the standard modality once it has been decided to drain the pleural fluid. It can be effective for drainage of parapneumonic effusions, particularly if it is done early in the course of the disease process. Once loculations and viscous fluid have developed in the pleural space, the efficacy of chest tube drainage decreases and the likelihood that the lung becomes trapped or develops an empyema increases (Sahn 1993).

Barring inadequate chest tube positioning, the most common reason for failed chest tube drainage is tube obstruction by organized empyema with multiple loculations (Sahn 1993). The use of chest-tube-instilled fibrinolytics, such as urokinase, may improve drainage of such effusions if utilized early in the disease process before significant collagen is laid down within the pleural space (Sahn 1993).

In those instances where the effusion is not free flowing on lateral decubitus radiographs, a CT scan may be helpful in defining the nature of the effusion and whether it will be amenable to chest tube drainage. A single loculus with minimal pleural enhancement may be adequately drained via chest tube insertion as opposed to empyectomy with decortication.

Those patients who present with malignant pleural effusions can also be drained with a tube thoracostomy. If the lung re-expands to fill the chest cavity, then pleurodesis may be attempted with one of a number of agents. However, when the lung does not re-expand, pleurodesis is doomed to failure and the clinician must consider other alternatives, depending on the patient's condition (Keller 1993).

Video-assisted thoracoscopic surgery

For those patients who are able to tolerate a general anesthetic, video-assisted thoracoscopic surgery, with the intent to lyse adhesions, decorticate, and remove restraining or affected pleural tissue, may be performed. This may permit re-expansion of the lung refractory to chest tube drainage alone. Pleurodesis can then be performed and the chest tube subsequently removed (usually within 24 h of the procedure).

Pleuroperitoneal shunts and indwelling external catheters

Indwelling catheters provide a means of symptomatic relief in those patients with malignant pleural effusions who either cannot tolerate a general anesthetic and have a trapped lung, or have a trapped lung which is not amenable to video-assisted thoracoscopic surgery.

Pleuroperitoneal shunts consist of a pumping chamber that transports approximately 1 to 2 ml of pleural fluid with each compression. Pumping for 10-min intervals approximately four times daily is required for adequate drainage of most effusions. The entire system is intracorporeal, with the pump portion placed in the subcutaneous tissues. Fluid is pumped from the intrapleural space into the abdominal cavity where it is resorbed by the peritoneal cavity. The procedure can be performed under local anesthesia and there is minimal discomfort afterwards (Little *et al.* 1988). A pocket for the pump is made in the subcutaneous tissue from an incision in the lateral part of the inframammary crease. The pleural catheter is inserted at the superolateral aspect of the incision using the Seldinger technique. The abdominal catheter is then tunneled subcutaneously across the costal margin and directed downwards, into the peritoneal cavity, via a 2- to 3-cm skin incision carried down through the peritoneum (Little *et al.* 1988).

In approximately 10 per cent of patients the shunt will become occluded, necessitating a replacement (Little *et al.* 1988). The use of such a device is strictly palliative and should be reserved for those patients who have known metastatic disease, since there is concern with respect to pleuroperitoneal seeding (Little *et al.* 1988).

External indwelling catheter drainage is an alternative for palliative management of malignant pleural effusion. The catheter is placed in the intrapleural space and drains externally when effusions are symptomatic. Insertion is carried out with local anesthesia in the operating suite. The patient is placed in 30° of Trendelenburg, and local anesthetic is used to infiltrate the two incision sites made in the anterolateral chest. The incisions are approximately 1 cm in length and are made perpendicular to the direction of the ribs at approximately the sixth intercostal space. The intercostal membrane and pleural space are entered through the most lateral incision using a needle and then the catheter is inserted using the Seldinger technique. The remainder of the catheter is tunneled through the subcutaneous tissue towards the second incision, where it exits the body, and the catheter is adjusted such that the felt cuff sits underneath the skin at the medial incision. The lateral incision is closed completely and the medial incision is closed around the catheter which is sutured in place (Little *et al.* 1988).

The failure rate is reported to be less than that of the pleuroperitoneal shunt (Little *et al.* 1988). However, infection may occur at the exit site, requiring antibiotic therapy and possible removal.

The drainage technique used for a pleural effusion must be patient specific. Early drainage of parapneumonic effusions can best be achieved with tube thoracostomy, while more complex effusions with loculations often require decortication and lysis of adhesion to allow lung re-expansion and subsequent pleurodesis. The palliative management of malignant pleural effusions must be as painless and uncomplicated as possible in order to afford these patients improved quality of their remaining life.

References

Keller, S.M. (1993). Current and future therapy for malignant pleural effusion. *Chest*, **103**, 63S-7S.

Little, A.G., Kadowaki, M.H., Ferguson, M.K., Staszek, V.M., and Skinner, D.B. (1988) Pleural–peritoneal shunting. *Annals of Surgery*, **208**, 443–50.

Quigley, R.L. (1995). Thoracentesis and chest tube drainage. *Critical Care Clinics*, **11**, 111–26.

Sahn, S.A. (1993). Management of complicated parapneumonic effusions. *American Review of Respiratory Disease*, **148**, 813–17.

Weingardt, J.P., Guico, R.R., Nemcek, A.A., Li, Y.P., and Chiu, S.T. (1994). Ultrasound findings following failed, clinically directed thoracenteses. *Clinical Ultrasound*, **22**, 419–26.

2.15.4 Hemothorax

J. A. McGuigan

Key messages

- Spontaneous hemothorax is uncommon.
- Small-volume aspiration and analysis precede complete drainage.
- Hypovolemia is life threatening in acute large-volume drainage.
- Arrange for rapid transfusion prior to draining hemothorax.
- Acute massive hemothorax should be managed in an operating area before drainage, where time permits.
- Small-volume hemothorax in a stable patient may lead to a better outcome if managed conservatively rather than repeatedly aspirated or drained.
- Thoracoscopy should only be carried out by experienced personnel.
- Thoracotomy can add to morbidity if carried out by the inexperienced.
- A bag with a valve can be an alternative to an underwater-seal bottle.

Introduction

Hemothorax is defined as the presence of blood in the space between the visceral and parietal layers of the pleural cavity. This potential space can be filled by most of the patient's blood volume, and when it fills rapidly can lead to death from hypovolemia. When it develops slowly or is of smaller volume it may remain undetected, leading to adhesion formation and restriction of lung expansion.

Hemothoraces frequently follow blunt and penetrating trauma but also arise in other situations. The management varies from simple observation to emergency thoracotomy depending on factors apart from the volume of blood present. In well-resourced facilities selective thoracotomy or thoracoscopy can be the most appropriate therapy. Where resources are limited and experienced surgeons are not available, the optimal management is more likely to be simple drainage or aspiration. There is no uniform method for treating hemothoraces, and there is often variation in management techniques within individual units. In the following account basic principles are emphasized in the light of broad experience and well-established data.

The presence of blood in the pleural space without air or fluid in the absence of trauma is rare. It is surprising that even in hemophilia and other hemorrhagic diseases hemothorax is unusual. In the critical care environment the source of blood is the most important feature. The prognosis is closely related to the anatomical and physiological cause of the hemothorax rather than the volume of blood. Spontaneous hemothorax is uncommon, and at one time was associated with a mortality of around 15 per cent. When continuous excessive blood loss occurs there is usually retained clot and accumulation of blood within the thorax. In postoperative hemorrhage there is frequently a larger volume of clot within the pleural space than liquid blood in the drainage bottles. A small elderly female may have significant tension in a hemithorax containing 1 liter, while a large fit young man will have little tension in a hemithorax containing 4 liters. It is the physiological status of the patient which determines the urgency in management.

Mechanisms must be understood when dealing with hemothorax. Bleeding from a contained dissecting aortic aneurysm may leak into the pleural space. Hemorrhage from the abdominal cavity can present as hemothorax. The most common explanation for spontaneous presentation is in association with a pneumothorax from which air has been absorbed. Bleeding may occur synchronously with pneumothorax or may follow the absorption of air from the space. Bleeding following a long delay from surgery or insertion of a chest drain usually signifies erosion associated with infection and necrosis.

Pathophysiology

Blood in the pleural cavity usually remains liquid for some considerable time, particularly if it is a pure hemothorax. Hemopneumothoraces tend to undergo coagulation, and blood cannot be removed with a needle or drain. Large collections result in collapsed lung, and thrombus organization may lead to fibrothorax. Where feasible, significant collections should be removed as soon as possible. Blood does not usually form a complete coagulum in the pleural space. It is thought that physical agitation of the chest wall impairs the maturation of clots, and because of this liquefaction occurs. Relatively healthy lung collapses with bleeding, and the concept that a collection of blood can seal a point of hemorrhage is not supported by the evidence. Bleeding from the pulmonary circulation is at a lower pressure than the systemic circulation and tends to stop more rapidly. With systemic bleeding the local intrapleural pressure has to approximate to systemic blood pressure before bleeding will stop. Therefore clot formation is essential if the entire pleural space is not to be filled. Rapid hemorrhage and hematological abnormalities may prevent clot formation. Hemorrhage then only stops when the systemic blood pressure has fallen as a result of hypovolemia and hypotension, and the pleural cavity contains almost the entire circulating volume. While the right pleural space is 5 to 10 per cent larger than the left, large right hemothoraces have more severe effects on venous return.

Etiology

Hemothorax has a variety of causes. A distinction needs to be made between bloodstained pleural effusion and true hemothorax. When hemorrhage has occurred in the pleural space, there are changes in the blood's consistency when it is removed sometime later. Unless removal is immediate the hematocrit is not the same as the blood in the circulation. Various arbitrary hemoglobin levels have been quoted as the lower level for determining true hemothorax. Such levels are of doubtful value, but frequently hemothorax is regarded as a collection in the pleural space with a hemoglobin of more than 7 g/dl. The causes given in Tables 1 and 2 are those that may be considered when a pleural aspiration reveals blood. Where the patient is stable, an aspirate with a hemoglobin as high as the circulating blood may indicate that the aspirating needle has entered an intercostal vessel and the blood usually clots unless there is a coagulopathy.

Initial diagnosis of hemothorax

Rapid accumulation

Rapid accumulation of significant quantities of intrapleural blood usually presents as an increasing effusion on radiography or as hypovolemia. Only a small-volume aspiration should be undertaken initially from any large-volume effusion, unless the patient has severe respiratory distress. When a large hemothorax is found, blood should be grouped and cross-matched, intravenous access should be established, and the patient should be examined and monitored. If there is a high volume of drainage or rapid accumulation of blood following aspiration, the surgeons should be contacted again as exploration of the chest is often necessary.

Slow accumulation

The patient is less likely to be hypovolemic, but hemothorax may be detected following a diagnosis of respiratory distress. Unless there is severe or rapid deterioration of the respiratory status, the effusion should initially be managed as in a patient with rapid accumulation of a large-volume hemothorax. Bleeding which has been gradual can

Table 1 Common etiologies of hemothorax

Trauma
Thoracic surgery
Common etiologies in the critical care area
Following previous pleural drainage or aspiration
Cannulation of subclavian and jugular veins
Pulmonary catheter insertion and balloon inflation
Rib 'cough fracture'
Lung infarction following pulmonary embolism
Pneumothorax
Hemorrhagic disorders
Anticoagulants
Platelet deficiencies
Coagulation defects

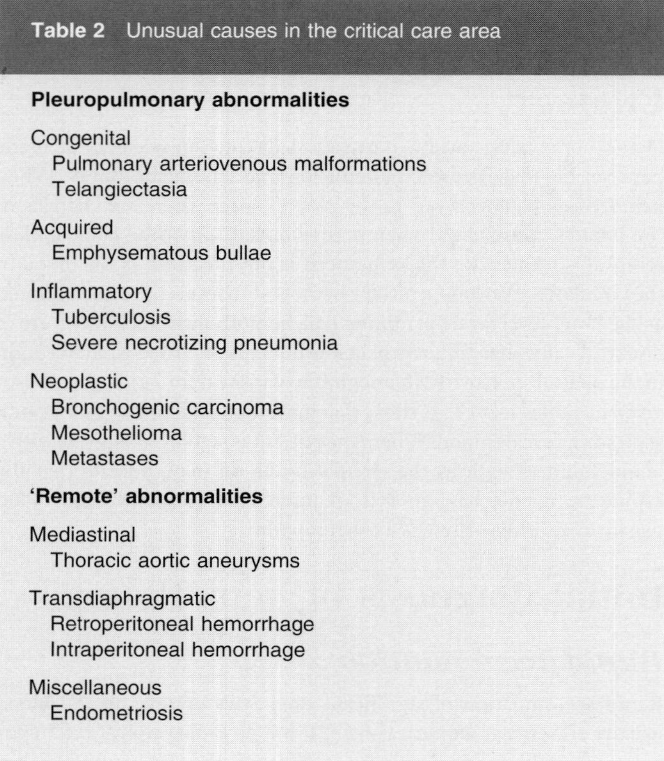

Table 2 Unusual causes in the critical care area

Pleuropulmonary abnormalities

Congenital
- Pulmonary arteriovenous malformations
- Telangiectasia

Acquired
- Emphysematous bullae

Inflammatory
- Tuberculosis
- Severe necrotizing pneumonia

Neoplastic
- Bronchogenic carcinoma
- Mesothelioma
- Metastases

'Remote' abnormalities

Mediastinal
- Thoracic aortic aneurysms

Transdiaphragmatic
- Retroperitoneal hemorrhage
- Intraperitoneal hemorrhage

Miscellaneous
- Endometriosis

become rapid when the pleura is drained, and again surgeons should be consulted and radiology arranged.

Small volume

Small effusions are managed without intervention or by aspiration. One of the most common causes of small-volume hemothorax is previous aspiration of an effusion. In a hemodynamically stable patient without respiratory compromise management should be by further aspiration or observation initially.

Investigations

Chest radiograph appearances of hemothorax must be distinguished from pulmonary parenchymal opacities. With large hemothoraces, physical examination will reveal dullness to percussion and diminished or absent breath sounds. Chest radiography will often appear normal where clinical signs are confined to the base. This is due to the necessity of taking chest radiographs supine in many critical care situations. Where possible, erect radiographs should be taken. The presence of fluid is more obvious and a rough estimate of volume can also be made (Fig. 1). Volumes of 500 ml or more have been drained from patients whose initial supine films show no obvious intrapleural fluid. Lateral decubitus views can also raise the diagnostic accuracy. If air is present, the estimate may be more accurate as the upper border of the fluid is thrown into sharper relief. Routine radiology cannot distinguish blood from other pleural effusions, apart from chylothorax which has a high fat content.

The diagnosis of hemothorax usually follows the identification of fluid on chest radiography. Changing the position of the patient will also determine whether the blood is free in the pleural space or is

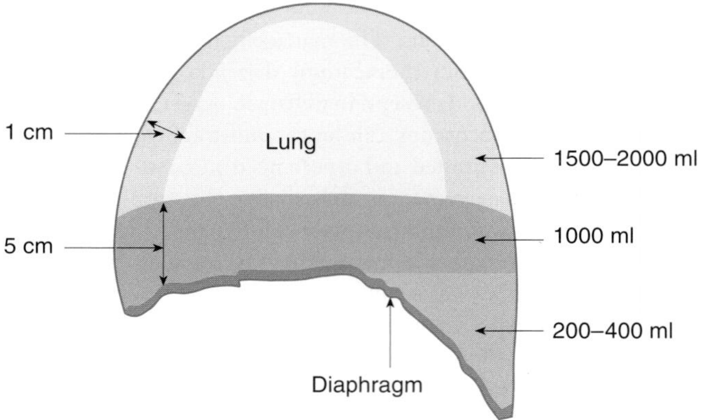

Fig. 1 Approximate volume of hemothorax on an erect chest radiograph in a patient of normal build: fills the costophrenic angle, 200–500 ml; air plus fluid level 5 cm over diaphragm, 1000 ml; separates the chest wall and diaphragm by 1 cm, 1500–2000 ml.

loculated. When available, CT scanning will help to identify the precise location of blood. Massive hemothorax is only diagnosed definitely when 2 liters drains from the chest.

The initial finding of a hemothorax in the non-traumatic patient is unexpected as other causes of intrapleural fluid are so much more common. Finding blood at pleural aspiration should not on its own be a cause of alarm. Blood removed from a probable hemothorax should be examined to assess whether clotting occurs in a bottle. The bloodstained fluid should be spun to establish its hematocrit, and bacteriology should be carried out to exclude infection. The blood pressure and pulse rate should be measured regularly.

If the patient is hemodynamically stable and not in respiratory failure, further radiological investigation should be undertaken if available. A CT scan with intravascular contrast enhancement may be carried out and the surgeons consulted. Magnetic resonance imaging may give some additional information, particularly in

Table 3 Suggested drainage procedures

Apparent hemothorax	
Small (< 500 ml)	None, or aspiration initially and tube drainage if rapid accumulation
Moderate (500 ml–1.5 liters)	Tube drainage and plan for thoracotomy if patient unstable or bleeding persists
Massive (> 1.5 liters)	Tube drainage inserted in the operating room and thoracotomy if patient is unstable or bleeding persists
Tension	Tube drainage in the operating room, followed by immediate thoracotomy; operating room set up for thoracotomy
Clotted	Elective minithoracotomy or thoracoscopy

distinguishing blood from other pleural effusions, and where available it may be appropriate to carry this out before aspiration.

Treatment

If no serious underlying vascular defect is detected, drainage of the pleural space can be undertaken using a chest drain or aspiration needle. Blood does not usually drain from the pleural space due to gravity alone but escapes with the re-expansion of the lung. Provided that loculation has not occurred, this process usually takes a short period of time unless there is continued intrapleural bleeding. Chest drainage should be through a tube large enough to allow patency and to permit some clots to be evacuated, but the pain that overlarge intercostal drains may cause should also be taken into account. Inadequate drainage is more common following needle aspiration than following intercostal tube insertion. Liquefied blood may be removed by needle aspiration. The use of an underwater seal is not essential. In many parts of the world sterile water can be difficult to obtain. A study of a drainage bag with a valve incorporated has shown good results (Graham *et al.* 1992). In the critical care area such bags simplify the environment around the patient. There is controversy over the necessity of controlling the rate of fluid removal. No good data are available. Rapid drainage of hemothoraces that have accumulated rapidly seems safe. However, it may be wiser to restrict drainage to 500 ml/h in those that have accumulated gradually with a relatively 'fixed' pleural space.

When patients remain well with hemothoraces, the blood is often broken down and gradually reabsorbed without complication. Fibrin can form and a fibrothorax can ensue, so that in critical care management it is best to remove blood by aspiration or chest tube insertion or, if this fails, by thoracoscopic division of loculi and large clots (McManus and McGuigan 1994). Where thoracoscopy is not available thoracotomy should be considered. From our trauma experience in Belfast, thoracotomy is best when carried out within 4 days of hemothorax formation or about 8 weeks after formation when an organized 'cortex' can be completely removed. Early thoracoscopy for hemothorax may save significant numbers of patients from more major procedures (Webb 1992). When surgery is carried out between these window periods, there are greater problems in dealing with edematous collapsed lung, further hemorrhage, and infection. When infection supervenes, tube drainage with the aid of a single short rib resection simplifies management. Only a small percentage will progress to require decortication. The management of hemothorax will vary according to the condition of the patient and the surgical and anesthetic skills available. Where experienced thoracic surgeons are available the added morbidity and mortality of a thoracotomy is low. Table 3 sets out a suggested management plan for this scenario. However, this is not a universal solution. Elsewhere, conservative management may be the safer method.

References

Graham, A.N.J., Cosgrove, A.P., Gibbons, J.R.P., and McGuigan, J.A. (1992). Randomized clinical trial of chest drainage systems. *Thorax*, **47**, 461–2.

McManus, K. and McGuigan, J.A. (1994). Minimally invasive therapy in thoracic injury. *Injury*, **25**, 609–14.

Webb, W.R. (1992). Diagnostic procedures for thoracic diseases. *Chest Surgery Clinics of North America*, **2**, 679–89.

2.16 Atelectasis and sputum retention

2.16.1 Pathophysiology of sputum retention

F. Potie and C. Martin

Key messages

- Postoperative diaphragmatic dysfunction is the major cause of alterations in respiratory mechanics and subsequent atelectasis.
- General anesthesia decreases total pulmonary compliance and favors the development of atelectasis in the dependent zones in the lung.
- Reduction in mucociliary clearance and cough contribute to the stagnation of bronchial secretions and favor atelectasis formation.
- Atelectasis is responsible for major changes in pulmonary gas exchange which often lead to severe hypoxemia.

Introduction

Atelectasis is defined as a collapse of the alveoli with a reduction in the quantity of intrapulmonary air. The mechanisms leading to atelectasis have been particularly well studied in patients on mechanical ventilation. They are multiple and involve respiratory function or mucociliary clearance or both. Indeed, the various elements responsible for atelectasis (surgery, posture, anesthetic agents, composition of inspired gases, etc.) target these functions. Atelectasis occurring during mechanical or spontaneous ventilation remains a frequent and severe pulmonary complication.

Pathogenic mechanisms

Alterations in ventilatory mechanics

The variations in respiratory mechanics that occur during the postoperative period have made it possible to identify the circumstances favoring atelectasis.

Atelectasis is essentially a restrictive syndrome with reduction of pulmonary volumes. Vital capacity is diminished by 40 to 60 per cent compared with the preoperative value after supraumbilical laparotomy, and by 20 to 40 per cent after subumbilical laparotomy. Tidal volume is diminished by 25 per cent and respiratory rate is increased by 20 per cent. This is accompanied by a 30 per cent reduction in functional residual capacity (FRC). These modifications are due to a marked diaphragmatic dyskinesia. This term includes all the diaphragmatic dysfunctions secondary to laparotomy or thoracic surgery.

Several possible causes of the mechanisms of diaphragmatic dyskinesia have been suggested. Decrease in diaphragmatic contractility after surgery seems to be unlikely, given the persistence of electromyographic contractility and the normal response secondary to bilateral phrenic contraction. A decrease in abdominal compliance after laparotomy is a possible cause, but is certainly not sufficient alone. The most probable mechanism is an inhibitory reflex of phrenic motor output, the origin of which is uncertain. Pain does not seem to be involved because a very satisfactory analgesia achieved using fentanyl administered by the peridural route does not improve diaphragmatic function (Simonneau *et al.* 1983).

Modes of ventilation and changes in the mechanical properties of the thoracic wall

Mechanical ventilation

It has been shown that the thoracopulmonary structure is less compliant under general anesthesia. The decrease in total compliance is due to the change in the elastic properties of the lung which would explain the occurrence of atelectasis. CT scanning has revealed that parenchymatous zones with increased density appear during the first few minutes of anesthesia. These dependent zones in the lung, whose extension is not influenced by the fraction of inspired oxygen, disappear during positive end-expiratory pressure (PEEP) of 10 cmH_2O (Brismar *et al.* 1985). These zones correspond more to atelectasis caused by compression than to that caused by adsorption. There is a significant correlation between these retention zones and the development of an intrapulmonary shunt (Brismar *et al.* 1985).

Spontaneous ventilation

Patients in the postoperative period ventilate at a greater respiratory rate and a smaller tidal volume. This type of ventilation causes alveolar collapse with reduction of FRC. When pulmonary volume falls below a certain value, the small bronchi collapse. Alveoli are then ventilated less well or not at all. The pulmonary volume at which the airways close is known as the closing volume. Its value is normally less that of the FRC in a standing subject aged less than 60 years. Atelectasis occurs when the FRC is less than the closing volume.

The FRC can be reduced by many other factors, including maintaining the patient in a supine or lateral position, obesity or abdominal distention, pain that limits thoracic expansion, postanesthesia alveolar hypoventilation, etc.

Stagnation of bronchial secretions

Physiological definition

Mucociliary clearance and cough are the principal defense mechanisms against the inhalation of particles and infectious agents. The reduction or disappearance of these mechanisms contributes to the stagnation of bronchial secretions, favoring the formation of atelectasis.

Mucociliary clearance

Physiologically, tracheal and bronchial secretions comprise 20 to 80 ml of seromucous liquid daily. Under normal conditions, these secretions are spontaneously drained from the small bronchi towards the trachea by the vibratile cilia in the bronchial epithelium. This is known as mucociliary clearance.

Depending on the acute or chronic pathologies in the respiratory tract, the volume and quality of the secretions will vary, ranging from very fluid and abundant to viscous and sticky. Certain pharmacological agents, which modify mucociliary clearance by acting on the viscoelastic properties of the mucus and/or the vibratile activity of the cilia (Table 1), have similar effects.

Secretions are then swallowed at the level of the pharynx. Coughing occurs at the level of the large bronchi in order to facilitate drainage and drive the secretions toward the larynx.

These defense mechanisms are altered during mechanical ventilation. The presence of an intubation or a tracheostomy tube short-circuits the upper airways and prevents evacuation of secretions. In addition, mechanical ventilation modifies the conditions for administration of inspired gases (temperature, humidity, oxygen concentration, pressure, etc.) which can affect mucociliary clearance, the only mechanism ensuring drainage of the peripheral bronchi, in the following ways.

1. The use of cold and dry gas reduces mucociliary clearance.
2. An increase in inspired fraction of oxygen Fio_2 reduces mucociliary clearance. It has been shown experimentally that it is totally inhibited in 30 min at $Fio_2 = 1$.
3. The cuff of the intubation tube blocks mucociliary clearance several centimeters above the carina.
4. Elevated insufflation pressures tend to crush and reverse the direction of movement of the vibratile cilia.
5. The dorsal decubitus position causes an accumulation of secretions.
6. Repeated tracheal aspirations produce abrasion of the tracheobronchial mucus, destroying the cilia over an extended surface.
7. Sedation and various anesthetic agents are probably responsible for a reduction in ciliary motility (see below).

Table 1 Pharmacological agents acting on mucociliary clearance

Agent	Action
Anticholinergic drugs	Depression
Local anesthetics	Depression
Opioids	Depression
Halothane	Depression
Thiopental (thiopentone)	Depression
Adrenergic agonists	Stimulation
Cholinergic drugs	Stimulation
Methylxanthines	Stimulation
Local antibiotics	Unknown
Corticosteroids	Unknown
Mucolytic agents	Unknown

Cough

Coughing is a complex phenomenon that is usually triggered by local 'irritation', making it possible to evacuate secretions from the central bronchi toward the larynx. It comprises an occlusion of the glottis, a sudden contraction of the abdominal muscles, a rapid elevation in airway pressure, and a sudden opening of the glottis. Under mechanical ventilation, this phenomenon is reduced because the glottis cannot close even if the resistance in the upper airways (intubation tube, ventilatory circuit) is sufficiently elevated to produce an increase in pressure. Coughing is suppressed during general anesthesia and can be significantly reduced by the administration of opioids. In a patient who is not intubated, diaphragmatic problems (medullar impairment, phrenic paralysis, myasthenia, Guillain–Barré syndrome, etc.), pain, or consciousness problems are responsible for 'ineffective' cough and tracheobronchial obstruction.

Airway obstructions

In addition to the accumulation of bronchial secretions, other obstacles in the tracheobronchial tree (inhalation of foreign bodies) can cause atelectasis.

Atelectasis by adsorption

In cases of incomplete bronchial obstruction, nitrogen in the inspired air tends to maintain alveolar stability. However, when high concentrations of oxygen are inspired, alveoli with a low ventilation-to-perfusion ratio can become unstable and collapse. In fact, since the total partial pressure in the alveolar gas is greater than that in the venous blood, the gas (rich in oxygen and poor in nitrogen) easily diffuses towards the venous blood and alveolar collapse rapidly occurs. Collapse is most likely to occur at the base of the lung where the parenchyma is less distended and the airways can be closed. However, this phenomenon of alveolar retention does not seem to be pathognomonic, as atelectasis by 'passive' compression also plays a role.

Passive atelectasis

This is the result of the compression of a segment or pulmonary lobe by an intra- or extrapulmonary mass. It can occur postoperatively (abdominal distention compressing the diaphragm) or in other pathological situations (total or partial pneumothorax and/or hemothorax, tumors).

Risk factors

General risk factors

1. Age is a risk factor for atelectasis for two reasons: increase in closure volume and reduction in cough reflex.
2. Atelectasis is a risk in obese individuals because the compression of the diaphragm reduces FRC and the closure volume then exceeds the FRC level. An ineffective cough and the supine position worsen the situation.
3. Smoking, through its related complications, plays a role in the occurrence of pulmonary conditions such as chronic obstructive pulmonary disease.
4. Pulmonary complications will occur three to four times more frequently in patients with a chronic respiratory condition than in those without such a pathology.
5. Dehydration with hyperosmolarity of viscous mucus has been reported as an etiological factor for atelectasis (Gaussorgues *et al.* 1988).

Risk factors linked to anesthetic agents

The importance of anesthetic agents in the occurrence of atelectasis has never been demonstrated in man. However, the action of anesthetic agents on mucociliary clearance has been demonstrated experimentally. Halothane, enflurane, and morphine reduce mucus transport in animals. Similarly, thiopental anesthesia increases the rigidity (elasticity and viscosity) of mucus in dog. However, there is no modification in the viscoelastic properties of mucus in patients submitted to general anesthesia. The action of anesthetic agents on mucociliary clearance is probably due more to a depression in ciliary activity than to a modification in mucus composition, and there is no difference between inhaled agents and intravenous anesthetics in the development of perianesthetic atelectasis (Rubin *et al.* 1990).

Risk factors linked to ventilation

Dry and cold inspired gases promote the rapid appearance of atelectasis by reducing mucociliary clearance. Hyperoxic mixtures cause the formation of atelectasis by adsorption a few minutes after the beginning of anesthesia to such a point that the wisdom of preoxygenation at $F\text{IO}_2 = 1$ during induction of anesthesia has been questioned (Rothen *et al.* 1995).

Risk factors linked to surgery

The type of surgery performed has an enormous influence on the rate of postoperative atelectasis. Upper abdominal and thoracic surgery are associated with a higher rate of respiratory complications than other types of surgery such as peripheral surgery. The type of incision also plays a role; transverse or subcostal incisions cause a postoperative restrictive syndrome that is less severe than that with median incisions. However, other studies have shown contradictory results for respiratory complications. Although the duration of anesthesia is implicitly linked to the type of surgery (with major surgery lasting longer), it is also a factor that favors postoperative respiratory complications including atelectasis.

Consequences of atelectasis

Effects on gas exchange

Intrapulmonary shunt

Regardless of whether atelectasis occurs under mechanical ventilation, it is responsible for changes in gas exchange which often lead to hypoxemia. This type of atelectasis (which is not visible on radiographs but appears on CT sections) causes abnormalities in the ventilation-to-perfusion ratio and creates an intrapulmonary shunt. This is defined as the passage of venous blood into the arterial system through the pulmonary capillaries with a low ventilation-to-perfusion ratio or limited diffusion. This phenomenon can become a true shunt in cases of total bronchial obstruction and hence total absence of ventilation. The degree of atelectasis is correlated with the intrapulmonary shunt and consequently the extent of hypoxemia.

Hypoxic vasoconstriction

Pulmonary vascular resistance is dependent on the oxygen pressure $P\text{O}_2$ in the alveolar gas. A reduction in $P\text{O}_2$ will cause a contraction of the smooth muscles in the walls of the arterioles in the hypoxic area. The existence of such areas may be due to insufficient ventilation caused by bronchial obstruction. The effect of this vasoconstriction is to divert blood flow from poorly ventilated to well ventilated areas.

Infectious consequences

The occurrence of atelectasis has classically been associated with the presence of infectious complications such as pneumonia. However, recent studies of postoperative atelectasis have found that this relationship is not significant when preventive treatment is prescribed.

Conclusions

Atelectasis is a complex and multifaceted phenomenon. It implies abnormalities in ventilatory mechanics, mucociliary clearance, and gas diffusion. The effects on hematosis are relatively frequent; they usually lead to a hypoxemia and can be responsible for other complications. A good understanding of the pathophysiology of atelectasis will make it possible to prescribe a curative and sometimes preventive treatment.

References

Brismar, B., Hedenstierna, G., Lundquist, H., Srandberg, A., Svenssson L., and Tokics, L. (1985). Pulmonary densities during anesthesia with muscular relaxation — a proposal of atelectasis. *Anesthesiology*, **62**, 422–8.

Gaussorgues *et al.* (1987). Etiologies des atélectasies lobaires aiguës sous ventilation mécanique. *Réanimation Soins Intensifs et Médecine d'Urgence*, **4**, 213–16.

Rothen, H.U., Sporre, B., Engberg, G., Wegenius, G., Reber. A., and Hedenstiernia, G. (1995). Prevention of atelectasis during general anesthesia. *Lancet*, **345**, 1387–91.

Rubin, B.K., Finegan, B., Ramirez, O., and King, M. (1990). General anesthesia does not alter the viscoelastic or transport properties of human respiratory mucus. *Chest*, **98**, 101–4.

Simonneau *et al.* (1983). Diaphragm dysfunction induced by upper abdominal surgery. *American Review of Respiratory Disease*, **128**, 899–903.

2.16.2 Conditioning of inhaled respiratory gases

L. Thomachot and C. Martin

Key messages

- Poor conditioning of the temperature and humidity of inspired gases is the source of an increase in morbidity, mortality, and cost of hospital stay.
- Modern heat and moisture exchangers (HMEs) provide adequate conditioning of inspired gases in terms of heat and water preservation.
- A heated humidifier may be preferable to an HME for hypothermic patients, when minute ventilation is 10 l/min or more, or when occlusion of the endotracheal tube occurs during ventilation with an HME.

Introduction

The importance of delivering warm humidified gas to patients ventilated through an endotracheal or tracheostomy tube is widely accepted. Mechanical ventilation with endotracheal intubation bypasses the upper airway and the normal heat- and moisture-exchanging process of inspired gases. A continuous loss of moisture and heat occurs, which predisposes patients to serious airway damage (Shelly *et al.* 1988)

The development of mechanical ventilation in the intensive care unit (ICU) has resulted in a number of complications which are the source of a major increase in morbidity, mortality, and the cost of hospital care. Some of these complications are due to the poor conditioning of the temperature and humidity of inspired gases (Table 1).

A fixed volume of gas at a given temperature can only contain a limited and precise amount of water vapor. The temperature of this gas determines its saturation point, i.e. the maximum concentration of water vapor that it can contain. The relative humidity (RH) is the humidity expressed as a percentage of the maximum water vapor that a gas can contain at a given temperature. Absolute humidity (AH) is the weight of water vapor that a unit volume of gas contains at a given temperature.

Techniques for regulating heat and moisture

Heat and moisture exchangers

The heat and moisture exchanger (HME) is located between the tracheal tube and the Y-piece of the ventilator (Fig. 1). It must be positioned above the tracheal tube so that the condensed water and the patient's secretions that are deposited in the filter do not increase the pressure drop (Table 2). Currently, HMEs are changed every 24 h in order to limit bacterial colonization, reduce the pressure drop, and restore the optimal conditioning capacities of the inspired gases.

Hydrophobic heat and moisture exchangers

The hydrophobic HME consists of a membrane punctured with pores of diameter approximately 0.2 μm. Gases and water vapor can pass through the membrane, but water in liquid form at the pressures normally used during mechanical ventilation cannot. The additional inspiratory resistances are very low. The hydrophobic HME functions like the human nose, capturing water vapor and energy from the expired gases and restoring them during the following

Table 1 Complications associated with poor conditioning of inspiratory gases

Complications associated with cold and dry gases	Complications due to overhumidification or hyperthermia of inspired gases
Epithelial cell disorders	Tracheal burning
Increased mucous viscosity	Alteration in surfactant
Restriction of the mucociliary function	Epithelial cell disorders
Decrease of surfactant production	Clinical consequences
Clinical consequences	Hyperthermia
Hypothermia	Increased resistance
Decreased compliance	Hyponatremia
Atelectasis	Edema
Ventilation–perfusion disorders	Decreased compliance
Hypoxemia	Atelectasis
	Ventilation–perfusion disorders
	Hypoxemia

Reproduced with permission from Thomachot *et al.* (1995).

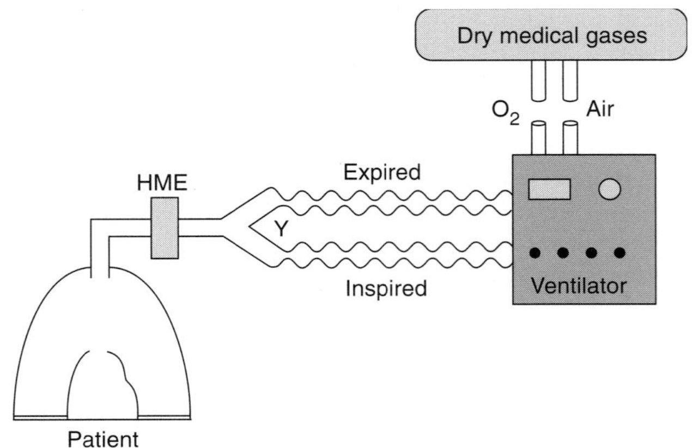

Fig. 1 Position of the HME in the ventilatory circuit. (Reproduced with permission from Thomachot *et al.* (1995).)

inspiration. Air is exhaled from the lungs at a temperature of 37 °C and 100 per cent RH (AH = 44 mg H_2O/l). If the exhaled air is cooled, both AH and the saturation point decrease, whereas RH remains at 100 per cent. Water vapor condenses on neighboring surfaces or forms very small droplets suspended in the gas. The expired air reaches the end of the tracheal tube at a temperature of approximately 33 °C. RH is still 100 per cent, but AH is only 36 mg H_2O/l. Thus 8 mg H_2O/l has condensed on the natural and artificial airways. When the expired air reaches the HME, water condenses on its surfaces, liberating the latent heat of the water vapor which heats the HME. The temperature of the gas leaving the HME is lower. If the expired gas leaves the HME at 20 °C, the AH is 17 mg H_2O/l. This leaves 19 mg of water vapor per liter of gas in the HME associated with the latent heat of vaporization of water. The higher the temperature difference between the two sides of the HME, the greater is the quantity of heat and moisture retained by the HME.

On the next inspiration, the HME transmits the heat and moisture stored during expiration, i.e. 19 mg H_2O/l. The tracheal tube or the tracheostomy provides 4 to 5 mg H_2O/l. In all, the HME and the tracheal tube provide approximately 25 mg H_2O/l. The patient should be able to provide the remaining moisture and heat from his bronchial tree.

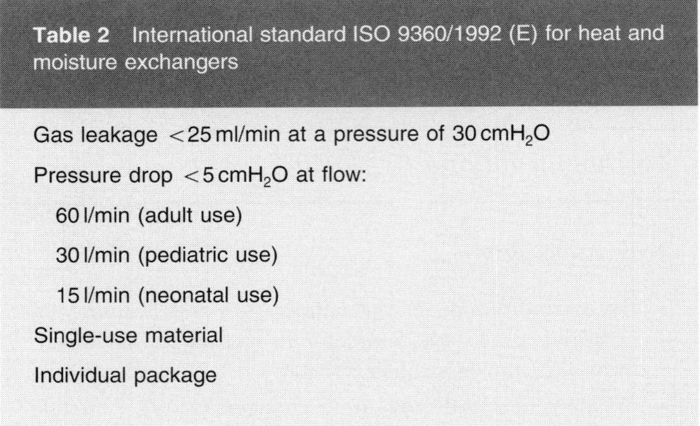

Table 2 International standard ISO 9360/1992 (E) for heat and moisture exchangers

Gas leakage < 25 ml/min at a pressure of 30 cmH_2O

Pressure drop < 5 cmH_2O at flow:

- 60 l/min (adult use)
- 30 l/min (pediatric use)
- 15 l/min (neonatal use)

Single-use material

Individual package

Data from International Organization for Standardization (1992).

Hygroscopic heat and moisture exchangers

Hygroscopic HMEs add the chemical conservation of water in a hygroscopic layer to the simple physical phenomenon of the hydrophobic HME (Fig. 2). The hygroscopic layer is composed of wool or foam coated with calcium, magnesium, or lithium chloride which retains moisture. On expiration, some of the water is condensed in the condenser element and some is retained in the hygroscopic layer and is released on inspiration without any energy requirement. The heat accumulated by the HME serves only to reheat the air and is no longer required to vaporize the water. The resistance of hygroscopic HMEs is low but can be increased in cases of excess water retention.

Advantages and disadvantages of heat and moisture exchangers

HMEs are also antibacterial filters. They have the advantage of being positioned at the end of the tracheal tube and therefore can filter inspired and expired gases. Pall® hydrophobic HMEs have an antibacterial efficacy of 99.9999 per cent. This means that only one out of 1 000 000 bacteria above the filter (*Serratia marcescens* at 0.5 µm in monodispersed form) will pass through it. Their cost is very competitive compared with other methods of conditioning inspired gases. Bronchopleural fistula and significant leaking around the tracheal tube are absolute contraindications for the use of this device. Hyperproduction of mucus and hypothermia are relative contraindications.

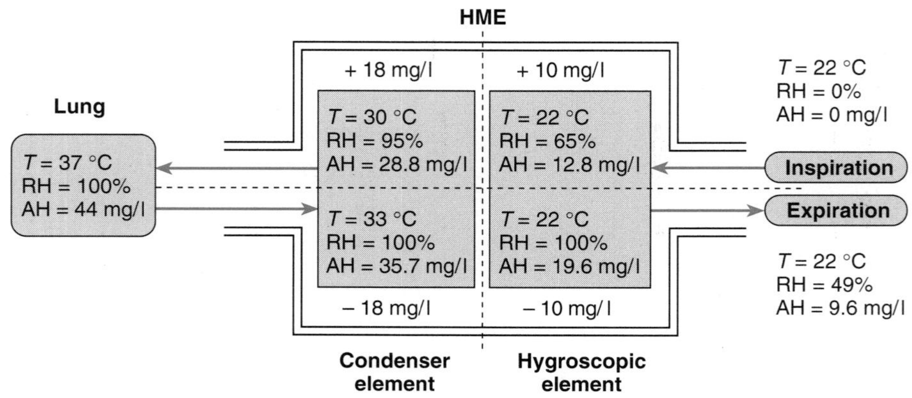

Fig. 2 Hygroscopic condenser humidifier. (Reproduced with permission from Thomachot *et al.* (1995).)

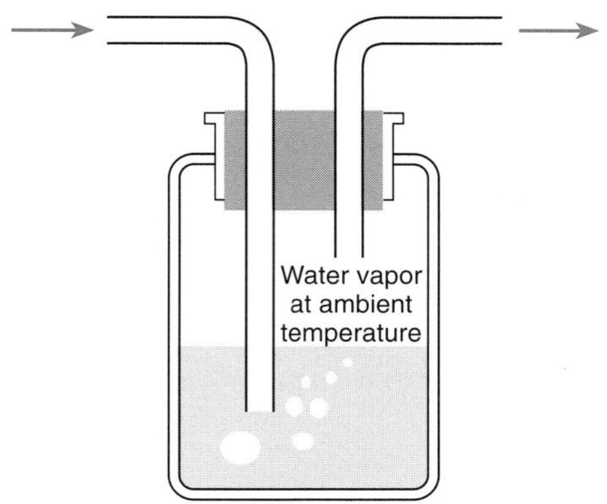

Fig. 3 Cold-water humidifier. (Reproduced with permission from Thomachot *et al.* (1995).)

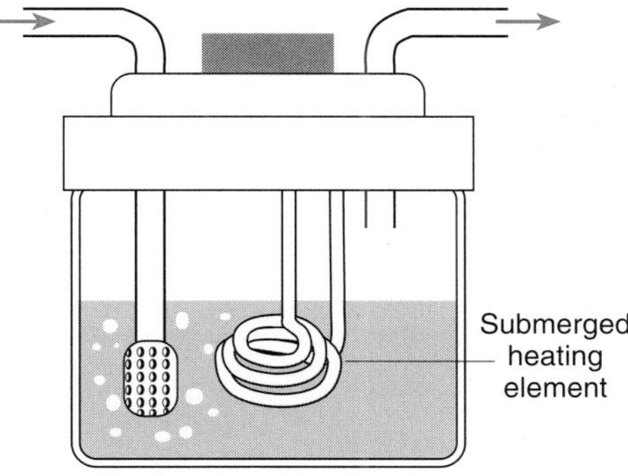

Fig. 4 Heated humidifiers. (Reproduced with permission from Thomachot *et al.* (1995).)

Heated and cold-water humidifiers

There are two types of cold-water humidifiers: air humidifiers and bubble humidifiers (Fig. 3). These devices have been gradually abandoned because of their poor performance in terms of heat preservation.

Heated humidifiers are now considered to be standard (Fig. 4). Water is supplied in vapor form. The device allows the delivery of gases that are totally saturated in water vapor. Many systems have a reservoir with an immersed heating element; others have an adjacent heating element or a heating chamber. The heated humidifier is located in the inspiration circuit between the ventilator and the patient. The most sophisticated models have a heating element in the inspiratory circuit up to the Y-piece. Therefore the inspired gases are maintained at a constant temperature in order to reduce the condensation of water vapor in the inspiratory circuit. A thermometer at the Y-piece controls the temperature of the inspired gases.

Staff using this type of device must be properly trained to check the electrical connections, the water level in the reservoir, the proper assembly of the circuits, and the aseptic filling of the water tank.

Nebulizers

Nebulizers produce a fine mist at room temperature. The humidification of inspired gases from these droplets depends on the RH. When the RH is above 80 per cent, the evaporation rate from the droplets decreases. Moreover, droplet diameter is of crucial importance. Droplets of diameter less than 1 µm can reach the alveoli. Only droplets with a diameter of 3 to 5 µm participate efficiently in the humidification of inspired gases.

There are three types of nebulizer:

(1) the centrifuge generator;

(2) the ultrasonic nebulizer (Fig. 5) in which a piezoelectric crystal vibrates at very high frequencies in order to produce a mist (droplet size, 2.8–4 µm);

(3) nebulizers in which gas is sprayed directly towards the end of the capillary tube and nebulizes the water aspired by the Bernoulli effect (droplet size, 0.5–3 µm).

Closed circuits

The conditions for humidifying and heating gases are comparable in a heated humidifier and a closed circuit with a fresh gas inflow of 0.5 l/min but adequate gas conditioning is achieved much later. The use of a higher fresh gas inflow reduces the level of humidification.

Humidification and temperature of inspired gases in intensive care unit patients

Under normal circumstances, it can be assumed that the temperature in the upper trachea ranges from 30 to 33 °C and RH is approximately 95 per cent, providing a water content of 30 mg H_2O/l. In mechanically ventilated patients, a gas temperature of 29 to 32 °C and 95 to 100 per cent RH should be adequate for inspired gases (AH, 25–30 mg H_2O/l). However, the optimum humidity of inspired gas for patients in the ICU is still a matter of controversy and the minimum acceptable level has not yet been established. It has been suggested that 23 to 33 mg H_2O/l is a desirable range with a tracheal temperature of 32 °C. Some authors have proposed higher tracheal temperatures (33–37 °C), leading to AH up to 44 mg H_2O/l. Literature values range from 17 to 44 mg H_2O/l. In man, during nose breathing of room air, temperature in the subglottic space is 32 °C, RH is 98 per cent, and

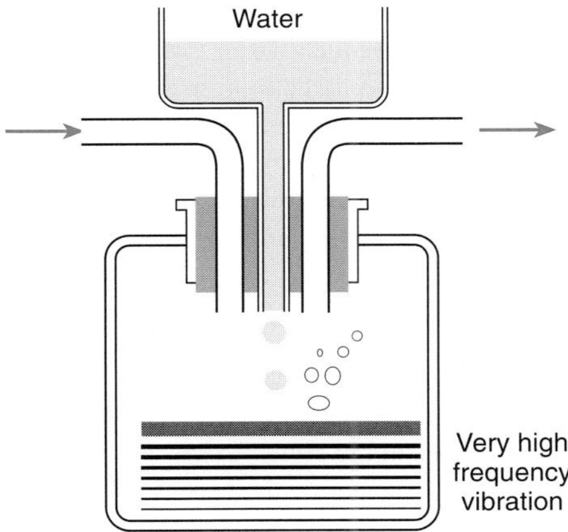

Fig. 5 Ultrasonic nebulizer. (Reproduced with permission from Thomachot *et al.* (1995).)

Table 3 Advantages and disadvantages of HMEs and heated humidifiers

	Advantages	Disadvantages
HME	No electrical connection	Added resistance
	Easy to handle	Clogging by sputum
	Decreased bacterial contamination of the circuit	Dead-space
	Inexpensive	Endotracheal tube occlusion if inspired gas is inadequately conditioned
	No workload	
HH	Control of humidification	Electrical connections
	Control of temperature	Electrical hazards
	Rewarming of hypothermic patients	Rain-out in ventilator tubings
		Temperature monitoring
		Bacterial contamination
		Overhumidification
		Hyperthermia, tracheal burning
		Connection errors
		Staff workload
		Cost

HH, heated humidifier.
Reproduced with permission from Thomachot *et al.* (1995).

the water content is 32.8 mg H_2O/l. A temperature of 33 °C and a water content of 25 mg H_2O/l were found in the larynx during nose breathing; at a point 9 cm below the vocal cords the temperature was 35 °C and the water content was 35 mg H_2O/l.

Currently, there is no routine method for continuous measurement of AH, RH, or tracheal temperature. If secretions become thick, the volume and frequency of instillations should be changed. A change in the fluidity of bronchial secretions, an increase in peak inspiratory pressure, and/or a flattening of pressure–volume curves are all signs of incorrect conditioning of inspired gases.

Incidents and accidents

Ventilation with cold and dry gases is responsible for several complications including hypothermia, water loss, epithelial and ciliary damage, atelectasis, and perturbations in gas exchange leading to hypoxia (Table 1).

In contrast, overhumidification or ventilation with gases that are too hot can cause hyperhydratation, tracheal burns, and atelectasis leading to hypoxia.

Incidents and accidents with heat and moisture exchangers

Some HMEs can cause occlusion of the tracheal tube (Martin *et al.* 1990, 1992). This catastrophic event is reported mostly with hydrophobic HMEs. The efficiency of these HMEs is significantly decreased when tidal volume increases, and the great majority of patients who have experienced tube occlusion were submitted to high minute ventilation (10 l/min or more). Water reduction in hydrophobic HMEs is a purely physical phenomenon, and increased tidal volume may reduce their performance compared with hygroscopic HMEs (Rathgeber *et al.* 1996) Use of an HME does not increase the levels of intrinsic positive end-expiratory pressure and dynamic hyperinflation.

Incidents and accidents with heated humidifiers

The absence of inspiratory temperature monitoring exposes the ventilated patient to a risk of tracheal burning or hyperhydratation. Moreover, bacterial contamination of the ventilator circuits occurs rapidly. The part of the circuit between the patient and the Y-piece is the most likely to become contaminated, after which contamination of the inspiratory and expiratory circuits occurs rapidly (33 per cent after 2 h, 64 per cent after 12 h, and 80 per cent after 24 h). Some teams place antibacterial filters between the ventilator and the inspiratory circuit and between the expiratory circuit and the ventilator. The tubes must be changed regularly (every 2–5 days), and ventilators must be sterilized between each patient.

Incidents and accidents with nebulizers

Nebulization of fine water droplets at room temperature causes cooling of the airways. Moreover, the nebulizer increases the volume of bronchial secretion and can promote hyperhydration in children (Table 1). Pulmonary resistance is increased in patients with chronic obstructive bronchopneumonia when nebulizers are used. Finally, aerosols can be colonized by bacteria which they then transport; this favors the emergence of nosocomial bronchopneumonia.

Conclusion

There are now a sufficient number of studies reporting that HMEs provide adequate conditioning of inspired gases. They are suitable for most ICU patients, and they provide adequate moisture and heat. The use of HMEs will also decrease staff workload. However, heated humidifiers may be preferable for hypothermic patients, when minute ventilation is 10 l/min or more, or when occlusion of the endotracheal tube occurs during ventilation with an HME. The advantages and disadvantages of HMEs and heated humidifiers are summarized in Table 3.

References

International Organization for Standards (1992). ISO 9360/1992(E). Anesthetic and respiratory equipment — heat and moisture exchangers for use in humidifying respired gases in humans, pp. 1–9.

Martin, C., Perrin, G., Gevaudan, M. J., Saux, P., and Gouin, F. (1990). Heat and moisture exchangers and vaporizing humidifiers in the intensive care unit. *Chest*, **97**, 144–9.

Martin, C., Papazian, L., Perrin, G., Bantz, P., and Gouin, F. (1992). Performance evaluation of three vaporizing humidifiers and two heat and moisture exchangers in patients with minute ventilation > 10 l/min. *Chest*, **102**, 1347–50.

Rathgeber, J., Züchner, K., and Burchardi, H. (1996). Conditioning of air in mechanically ventilated patients. In *Yearbook of intensive care and emergency medicine* (ed. J.-L. Vincent). Springer, Berlin.

Shelly, M.P., Lloyd, G.M., and Park, G.R. (1988). A review of the mechanisms and methods of humidification of inspired gases. *Intensive Care Medicine*, **14**, 1–9

Thomachot, L., Viviand, X., and Martin, C. (1995). Humidification et réchauffement des gaz inspirés. In *Principe de réanimation chirurgicale* (ed. J.-L. Pourriat and C. Martin), pp. 121–6. Arnette Blackwell, Paris.

2.16.3 Therapeutic approach

R. Vialet and C. Martin

Key messages

- Endotracheal aspiration is useful only in the presence of abundant tracheal secretion.
- Of the many techniques proposed to treat and prevent atelectasis, incentive spirometry should be encouraged.
- Adequate control of postoperative pain is of major importance if effective chest physiotherapy is to be performed.

Introduction

Studies have not been able to demonstrate the benefits of physiotherapy in terms of morbidity or to show the superiority of one technique over another (O'Donohue 1985). However, when physiotherapy techniques or mechanical ventilation are applied vigorously, the incidence of postoperative atelectasis and its associated morbidity are low. This justifies intensive care for patients at risk of atelectasis (Selsby and Jones 1990; Kacmarek 1995).

Analgesia

Effective analgesia must be provided to prevent parietal and pleural pain from affecting coughing, expectoration, and chest physiotherapy. All analgesic techniques may be used, but patient-controlled analgesia is preferred if possible. Analgesic drugs can be administered in combination.

Aerosols

Bronchodilator or 'mucokinetic' aerosols play an important role in the management of airway obstruction. Inhaled aerosolized medications are often preferable to oral or parental routes of administration because topical delivery gives direct benefits with limited toxic reactions. Aerosols are generated from a nebulizer solution, directly with a metered-dose inhaler, or with a dry powder inhaler. These devices deliver aerosols of pharmacologically active molecules to the lower airways.

The nebulizer and the metered-dose inhaler have identical efficacy when the patient is intubated. The metered-dose inhaler is the standard method for patients who are not intubated.

This technique is combined with physiotherapy in order to increase its effectiveness. It is used to relieve bronchospasm (β_2-agonists, anticholinergic agents) and also to optimize mucociliary clearance. *N*-Acetylcysteine may be useful for short-term treatment in lung diseases associated with mucous plugging.

Tracheobronchial aspiration

Endotracheal aspiration

Bronchial crackles indicate the presence of abundant tracheobronchial secretion. An increase in peak pressures during volume-controlled ventilation or a decrease in tidal volumes during controlled-pressure ventilation are also good indicators of bronchial sputum retention.

An aspiration catheter is inserted into the trachea. Once the catheter is positioned in the lower part of the bronchial tree, suction is applied. The tube is then slowly withdrawn over a period of 10 to 15 s with suction applied. The inspired oxygen fraction F_{IO_2} must be increased for several minutes before and after the procedure in order to reduce secondary hypoxia.

The diameter of the aspiration catheter must not exceed half the internal diameter of the endotracheal tube. Single-use flexible tubes are preferable. Instillations of 5 to 10 ml of isotonic saline solution are sometimes required when secretions are thick and tenacious.

Tracheobronchial ulcers can occur, and may bleed or become infected. The causes of these ulcers depend on the rigidity, the number of lateral perforations, the diameter of the aspiration catheter, the shape of the distal tip, and the frequency of tracheobronchial aspiration. Suction of over 200 mmHg exposes the patient to the risk of pulmonary atelectasis as well as invagination of tracheobronchial mucosa through the port of the aspiration catheter. Infection can be caused by introduction of exogenous bacteria into the trachea or contamination of the patient by his or her own flora. Severe hypoxia can occur when a patient is disconnected from the ventilator. There may also be cardiac arrhythmia or even cardiac arrest.

Nasotracheal or orotracheal aspiration

Nasotracheal aspiration is performed in non-intubated patients who cannot eliminate tracheobronchial secretions. A tracheal aspiration tube is passed through the nose or mouth. Passage of the tube triggers coughing which shows that the tube is positioned properly. This method is not without risk, and the patient must be fully co-operative.

Chest physiotherapy

Chest physiotherapy may be preventive or curative, and the indications must be strictly followed (Table 1).

Postural drainage therapy

Postural drainage therapy consists of positioning certain pulmonary segments higher than the carina. Secretions can then be eliminated easily. Posturing can also improve the ventilation-to-perfusion ratio and increase functional residual capacity. Some postures cause the patient major discomfort. Different postures (inclined, lateral decubitus) are used with aerosols, percussion, and cough stimulation.

In intensive care unit (ICU) patients on mechanical ventilation, changes in posture should be performed every 1 or 2 h and postural drainage every 4 to 6 h. The therapeutic indications must be reviewed every 24 h. Postural drainage can increase intracranial pressure.

Percussion and vibration

Manual or mechanical vibration of the thoracic wall can help to mobilize secretions. They can be used in ventilated, non-ventilated, conscious, or unconscious patients. They may induce hypoxemia or bronchospasm.

Directed cough

Coughing is a normal reflex that uses maximum forced expiration to clear irritants or bronchial excretions. A contraction of expiratory muscles suddenly increases intrathoracic pressure and concomitant opening of the glottis generates a very rapid expiratory airflow.

The anatomy of the bronchial tree has important repercussions on ventilatory flows which can be laminar or turbulent depending on the airways and bifurcations. As the bronchial walls are flexible, their

Table 1 Indications for different physiotherapy regimes

	Indications	Relative contraindications	Comments
Analgesia	All patients in pain		
Tracheobronchial aspiration	Intubated patient	Normal auscultation	After pulmonary auscultation
	Fully co-operative non-intubated patient		
Chest physiotherapy			
Posture change	Sedated patient		Every 1–2 h
	Unconscious patient		
	Neuromuscular disease		
Postural drainage			Every 4–6 h
Percussion and vibration	Obstruction of proximal airways	Bronchospasm	
Directed cough	Conscious and co-operative patient	Sedated patient Unconscious patient	Every 2–4 h
Incentive spirometry	Conscious patient	Vital capacity less than 10 ml/kg	The most widely used method with or without technical aid
Partial mechanical ventilation			
Positive airway pressure	Spontaneous ventilation		Every 1–6 h, pressure level not clearly defined (5–20 mmHg)
Intermittent positive pressure breathing	Treatment of atelectasis (if other methods have failed)		
	Alternative to intubation (in case of alveolar hypoventilation)		
Fiber-optic bronchoscopy	Treatment of atelectasis	Major hypoxia Hemodynamic instability	Insufflation can give better results than aspiration

diameters vary and the bronchi can collapse. If the viscosity of mucus increases, the flow at which it can be eliminated will be higher.

Directed cough is indicated only for conscious and co-operative patients and consists of coughing efforts with an open glottis (at low pulmonary volume to avoid dynamic compression of the airways) followed by ample ventilation. The physiotherapist can help the patient by applying phased pressure to the thorax and epigastric zone. The recommended rhythm is one session every 2 to 4 h.

This technique may be contraindicated in cases of active tuberculosis, intracranial hypertension, myocardial infarct or unstable angina, and unstable trauma of the spine, and in patients presenting a risk of pulmonary aspiration. This therapy also carries the risk of suture leakage after abdominal surgery.

Incentive spirometry

Incentive spirometry, which is also known as sustained maximal inspiration, is the most widely used method of bronchial hygiene therapy, but it requires the co-operation of the patient. It attempts to mimic the natural sighing or yawning that a healthy subject performs nine or ten times every hour. The patient is encouraged to take long slow deep breaths.

Maximum inspiration is facilitated by using an indicator (usually visual) that informs the patient that he or she has produced sufficient flow or insufflatory volume and maintained it for a sufficiently long period (at least 3 s). Of course, ample ventilation without technical aid is possible and can be as beneficial as incentive spirometry.

The contraindication is essentially a lack of co-operation. Incentive spirometry is not suitable for patients with a vital capacity less than 10 ml/kg.

Incentive spirometry should be superior to percussion and directed cough as a method of increasing mucus clearance and treating sputum retention, but, despite many studies, the superiority of this technique has not been clearly demonstrated (Bellet *et al.* 1995). However, it should be as effective as intermittent positive-pressure ventilation with less frequent undesirable effects. Incentive spirometry must not be confused with expiratory maneuvers (blow bottles) which can be harmful because they result in a decrease in pulmonary volumes.

Indications for chest physiotherapy techniques

Intermittant lateral turning is indicated for sedated patients and those presenting a neuromuscular disease, or hypoxemia in certain positions, when patients present risk factors for atelectasis or proven sputum retention, or when patients are intubated or tracheostomized.

Postural drainage is indicated to eliminate secretions in patients who have excessive production, ineffective cough, cystic fibrosis, or bronchiectasis, in patients who are intubated, and as treatment for atelectasis and sputum retention.

Percussion is effective when the proximal airways are obstructed. The indications are the same as for other physiotherapy techniques. Preventive or curative percussion can be of benefit to intubated or tracheostomized patients. Incentive spirometry is indicated for all clinical conditions with risk of atelectasis. It can also be used as therapy for minor atelectasis.

Postoperative chest physiotherapy does not provide any benefit when secretions are not abundant. Therefore it is only indicated postoperatively when the patient presents chronic bronchopulmonary pathology (inhaled bronchodilators can be used in combination with physiotherapy), when cough is ineffective, or when atelectasis or sputum retention is established.

In all cases, patients must be monitored (pulse oximeter) and their tolerance evaluated because some can become hypoxemic without warning. All the chest physiotherapy techniques must be reviewed regularly. Therapy should only be continued if there is improvement in expectoration, clinical state, blood gases, or on the chest radiograph.

Partial mechanical ventilation

These techniques are used in patients on spontaneous ventilation who are not usually intubated. However, in certain clinical conditions they can also be applied to intubated or tracheostomized patients. Patient co-operation is not always necessary, but tolerance is sometimes poor.

Positive airway pressure

Positive airway pressure (PAP) includes continuous positive airway pressure (CPAP) and expiratory positive airway pressure (EPAP).

Application of PAP increases intraluminal pressure and prevents the collapse of the bronchioles and alveoli. The increased pressure provides better conditions for the functioning of the diaphragm. The optimum level of PAP has not been clearly defined.

EPAP (exhalation against a threshold resistor) increases the work of the diaphragm, permitting the patient to recover normal diaphragmatic function more rapidly. When the transmural pressure gradient is increased, bronchiolar collapse is limited. Slow inspiratory flow provides better ventilation in the parts of the lung with a low time constant.

During CPAP therapy the patient breathes from a pressurized circuit against a threshold resistor that maintains constant airway pressure (from 1 to 15–20 cmH_2O) during both inspiration and expiration. The main aim is to increase the functional residual capacity.

The side-effects are hypercapnia, increase in intracranial pressure, hemodynamic failure, vomiting, pulmonary barotrauma, and patient discomfort. There are no absolute contraindications, but the indications should be carefully weighed in the following conditions: esophageal or cephalic surgery, chronic obstructive pulmonary disease, acute sinusitis, epistaxis or hemoptysis, nausea, tympanic rupture, and undrained pneumothorax. In the ICU, this therapy can be administered at intervals of 1 to 6 h and pressures of 5 to 20 mmHg.

These techniques are used to prevent or treat atelectasis and sputum retention in patients on spontaneous ventilation. The indication follows clinical examination revealing hypoxemia associated with bronchial obstruction, sputum retention that is not responsive to spontaneous or directed coughing, or radiological signs of atelectasis. The various clinical studies in the literature are often encouraging but have yet to clearly demonstrate the advantage of one technique over another. However, CPAP appears to be the best tolerated technique.

Intermittent positive-pressure breathing

The technique of intermittent positive-pressure breathing is used to provide short-term or intermittent mechanical ventilation for the purpose of increasing lung expansion. Intermittent positive-pressure breathing is not the technique of first choice when less invasive therapies can reliably meet clinical objectives. Positive-pressure ventilation can be used with volume- or pressure-limited ventilators.

There are numerous complications: increase in respiratory resistance, alveolar hyperinflation, pneumothorax, nosocomial infection, hypo- or hypercapnia, hypoxia or hyperoxia, aggravation of the ventilation-to-perfusion ratio, epistaxis or hemoptysis, mucus plugs if humidification is poor, gastric distention, and reduction of venous return. In cases of associated bronchospasm, clinical monitoring must be particularly careful. The only absolute contraindication is pneumothorax. However, the indications of intermittent positive-pressure breathing must be assessed more carefully in patients with emphysema, untreated tuberculosis, intracranical hypertension, hemodynamic instability, tracheo-esophageal fistula, recent surgery of the face, mouth, or esophagus, hemoptysis, nausea, regurgitation, and hiccups.

In the ICU, this therapy can be administered every 1 to 6 h. Its efficacy is evaluated as follows: increase of more than 25 per cent in the tidal volume and improvement of cough, chest radiograph, auscultation, and the well being of the patient.

The first indication for intermittent ventilation is for the curative treatment of atelectasis after failure of chest physiotherapy, incentive spirometry, deep breathing exercises, and posturing, or if the patient is not co-operative. The second indication is for prevention of obstruction during limited ventilation or for ineffective cough. It can be interesting to make use of functional respiratory explorations. The third indication is as an alternative to intubation during alveolar hypoventilation.

The therapeutic effects of intermittent positive pressure ventilation do not persist for more than 1 h after the treatment.

Fiber-optic bronchoscopy

Fiber-optic bronchoscopy is frequently proposed as an essential element in the treatment of atelectasis and sputum retention.

Fiber-optic bronchial aspiration

Fiber-optic bronchial aspiration is used to remove bronchial obstacles. It cannot always be performed, and may be dangerous when there is major hypoxia or hemodynamic instability. The effect of aspiration can be negative if it induces bronchial collapse or suction lesions that aggravate distal retention of secretions.

Fiber-optic bronchial insufflation

This technique consists of selective insufflation of the pulmonary atelectasis either through a balloon-tipped catheter or directly through a fiberscope. It can give better results than fiber-optic bronchial aspiration.

References

Bellet, P.S., Kalinyak, K.A., Shukla, R., Gelfand, M.J., and Rucknagel, D.L. (1995). Incentive spirometry to prevent acute pulmonary complications. *New England Journal of Medicine*, **333**, 699–703.

Kacmarek, R.M. (1995). Prophylactic bronchial hygiene following cardiac surgery: what is necessary? *Intensive Care Medicine*, **21**, 467–8.

O'Donohue, W.J. (1985). National survey of the usage of lung expansion modalities for the prevention and treatment of postoperative atelectasis following abdominal and thoracic surgery. *Chest*, **87**, 76–80.

Selsby, D. and Jones, J.G. (1990). Some physiological and clinical aspects of chest physiotherapy. *British Journal of Anaesthesia*, **64**, 621–31.

2.17 Hemoptysis

2.17.1 Pathophysiology and causes

R. Zürcher Zenklusen and P. Jolliet

Key messages

- Massive hemoptysis is present if blood loss is 600 ml/24 h or more and/or is life threatening.
- The most frequent causes of massive hemoptysis are tuberculosis, bronchiectasis, lung abscess, aspergillosis, and bronchial carcinoma.
- The source of bleeding is usually the bronchial circulation.
- Prognosis is closely related to the rate of bleeding.
- Published mortalities vary widely depending on inclusion criteria and cause.

Introduction

Massive hemoptysis is an uncommon and often dramatic emergency that mandates rapid stabilization and treatment. The definition of massive hemoptysis is not universally established, but the most widely accepted criterion is a blood loss of at least 600 ml/24 h. Massive hemoptysis is most often associated with inflammatory lung disease. The source of bleeding is usually the bronchial circulation and, rarely, the pulmonary circulation. Immediate mortality is primarily related to the rate of bleeding, regardless of the underlying disease.

Definition

Massive hemoptysis has been variably defined as blood loss exceeding 100 to 1000 ml/24 h, with the most widely accepted criterion being 600 ml/24 h (Cahill and Ingbar 1994). However, quantifying the amount of blood lost is often approximate. The functional definition of massive hemoptysis is bleeding into the tracheobronchial tree at a rate that poses a threat to life.

Etiology

Hemoptysis has been linked to over 100 disease states. Table 1 lists the major causes in relation to their incidence.

Tuberculosis continues to be the major cause of significant hemoptysis worldwide, but is relatively less common in industrialized countries than in the less developed parts of the world. It is the most common cavity-producing predisposing condition.

Bronchiectasis in the industrialized world has evolved from a common problem to an infrequent complication in patients with long-standing pulmonary disease and in those with inherited disorders, such as cystic fibrosis. The anatomical changes in the bronchial artery tree associated with bronchiectatic segments include bronchial artery hypertrophy, expansion of the dense peribronchial and submucosal bronchial arteriolar plexus, and augmentation of anastomoses with the pulmonary artery tree.

Hemoptysis occurs in 10 to 15 per cent of patients with primary lung abscess, of whom 20 to 50 per cent experience massive bleeding. The bleeding is probably directly related to the necrotizing effects of the primary infection on the lung parenchyma and vasculature. Cavities that lack adequate drainage have rapid progression of infection with an overwhelming inflammatory response that involves nearby pulmonary vessels. Rupture may occur as a consequence of direct infection or inflammation of the pulmonary vascular tree.

Fungal colonization of pre-existing cavities resulting from other diseases (tuberculosis, histoplasmosis, sarcoidosis, emphysema, or other chronic illness) produces a mycetoma. *Aspergillus* infection is

Table 1 Causes and incidence of massive hemoptysis

Tuberculosis (active or inactive)	30%–50%
Bronchiectasis	10%–20%
Lung abscess	10%
Mycetoma (e.g. aspergilloma)	10%
Bronchogenic carcinoma	5%–10%
Cystic fibrosis	5%
Cryptogenic	8%–15%
Miscellaneous*	
Iatrogenic (drug/procedure-related)	
Alveolar hemorrhage syndrome	
Cardiovascular disease (congenital/acquired)	
Bronchial adenoma	
Metastatic carcinoma	
Broncholithiasis	
Foreign-body aspiration	
Pulmonary embolism	
Autoimmune systemic disorder	
Dissecting aortic aneurysm	

*Incidence of each is less than 5% or single case reports.

the most common fungal disease leading to massive hemoptysis. The incidence of hemoptysis complicating mycetoma formation is 50 to 90 per cent, and that of massive hemoptysis is 5 to 25 per cent. Patients in the advanced stage of tuberculosis who are also alcoholics appear to be at particular risk of developing mycetomas. In contrast with the latter, invasive pulmonary fungal infections are infrequently associated with hemoptysis. The radiographic appearance of a mycetoma is that of an intracavitary mass separated from the surrounding cavity wall by a crescent of air. An intense inflammatory reaction accompanied by ectatic vascular growth (i.e. highly vascular granulation tissue) produces a thick-walled cavity. However, the precise etiology of hemorrhage is unclear. Theories include mechanical trauma of the vascular granulation tissue, produced by movement of the fungus ball in the cavity, vascular injury from *Aspergillus*-produced endotoxin, *Aspergillus*-related proteolytic acitivity, and vascular damage from a type III hypersensitivity reaction.

Bronchogenic carcinoma is one of the most common causes of non-massive hemoptysis owing to its high prevalence in the population; however, massive hemoptysis occurs in only about 3 per cent of lung cancer patients (Cahill and Ingbar 1994). Of the major malignant cell types, squamous cell carcinoma is far more likely to produce massive hemoptysis than are adenocarcinoma or small-cell carcinoma. Bronchial carcinoids frequently cause hemoptysis by virtue of their marked vascularity and endobronchial location. The most common primary sites of endobronchial metastatic tumors associated with hemoptysis include breast, colon, kidney, and melanoma. Tumors of the mediastinum, particularly esophageal carcinoma, may extend directly into the tracheobronchial tree, resulting in massive hemoptysis.

Hemoptysis occurs in approximately 60 per cent of patients with cystic fibrosis at some time in the course of their disease, and about 5 to 7 per cent experience massive bleeding; the latter is mainly due to extensive bronchiectasis, lung abscesses, or patchy bronchopneumonia.

Cryptogenic hemoptysis refers to hemoptysis in which the cause remains unclear despite a thorough diagnostic evaluation; this is the case in about 8 to15 per cent of patients with massive hemoptysis.

Iatrogenic hemoptysis is a potential complication of bronchoscopy, endobronchial or transbronchial biopsy, transthoracic needle biopsy, or pulmonary artery catheter use. Rupture of the pulmonary artery is the most catastrophic complication of pulmonary artery catheter use, with a mortality of at least 50 per cent.

Pathophysiology

The lungs have a dual vascular supply from both the pulmonary and bronchial circulations. The bronchial arteries originate from the aorta or intercostal arteries and provide nutrition to the pulmonary structures. They give rise to small penetrating arteries which form submucosal plexi supplying the bronchial mucosa. There are anastomoses between the bronchial and pulmonary circulations at the level of medium-sized arteries as well as of microvascular and precapillary vessels. Massive hemoptysis generally involves bleeding from the bronchial arteries or, less frequently, from a pulmonary artery when the pulmonary circulation is pathologically exposed to the high-pressure bronchial circulation through enlarged bronchopulmonary anastomoses (Sternbach and Varon 1995).

The response of the bronchial circulation to both acute and chronic disease involves vascular proliferation. Such 'neovascular' development can be seen as early as 4 to 5 days after an acute insult. This ability to proliferate distinguishes the bronchial from the pulmonary circulation. The bronchial circulation, which normally receives less than 1 per cent of the cardic output, can proliferate to the extent that as much as a third of the systemic blood flow passes through it (Deffebach *et al.* 1987).

Mechanisms by which massive hemorrhage is produced include the following (Sternbach and Varon 1995).

1. Various vascular alterations, such as aneurysm formation (e.g. 'Rasmussen's aneurysm' in tuberculosis patients), vasculitic reactions (Goodpasture's syndrome, systemic lupus erythematosus, idiopathic hemosiderosis), and pulmonary embolism.
2. Chronic parenchymal inflammation: erosion of bronchial and vascular walls, enlargement and proliferation of bronchial vessels, formation of anastomoses between bronchial and pulmonary circulation, and formation of bronchiectases or lung abscesses.
3. Broncholithiasis: erosion of a calcified lymph node into the tracheobronchial tree with disruption of peribronchial and submucosal bronchial vessels.
4. Vascular invasion by tumor.

Prognosis

There are no clinical or radiographic findings which can accurately predict the appearance of massive hemoptysis; mortality varies between the extremes of 0.9 and 85 per cent, with the lower rates in more recent series (Stoller 1992). Prognosis is closely related to the rate of bleeding rather than to the extent of pulmonary disease or the age of the patient. In a series of 59 patients, the death rate was six times higher in patients coughing more than 1000 ml of blood every 24 h (58 per cent) than in patients with smaller-volume hemoptysis (9 per cent), with an overall mortality of 34.5 per cent (Corey and Hla 1987). Clinical features that exclude patients as surgical candidates (e.g. inadequate pulmonary function, debilitated states, bilateral pulmonary bleeding sources, metastatic cancer) are also strongly associated with increased mortality, for example a mortality of 46 per cent in inoperable patients versus 16 per cent in operable patients in the series reported by Corey and Hla (1987). Appropriate interpretation of medical mortality among series requires comparison of mortality among patients deemed operable but managed medically. Thus the mortality in the surgical candidates managed medically in the series of Corey and Hla (1987) was 11 per cent compared with 29 per cent of all medically managed patients (including some inoperable patients). Obviously, the impact of the underlying disease is more crucial than the impact of surgery and medical treatment.

References

Cahill, B.C. and Ingbar, D.H. (1994). Massive hemoptysis. Assessment and management. *Clinics in Chest Medicine*, **20**, 147–68.

Corey, R. and Hla, R.M. (1987). Major and massive hemoptysis: reassessement of conservative management. *American Journal of Medical Sciences*, **294**, 301–9.

Deffebach, M.E., Charan, N.B., Lakshminarayan, S., and Butler, J. (1987). The bronchial circulation: small, but a vital attribute of the lung. *American Review of Respiratory Disease*, **135**, 463–81.

Sternbach, G. and Varon J. (1995) Massive hemoptysis. *Intensive Care World*, **12**, 74–9.

Stoller, K.J. (1992) Diagnosis and management of massive hemoptysis: a review. *Respiratory Care*, **37**, 564–81.

2.17.2 Therapeutic approach

R. Zürcher Zenklusen and P. Jolliet

Key messages

- The greatest immediate vital risk is asphyxia, prompting rapid airway control.
- Bronchoscopy is usually the single most valuable emergency diagnostic procedure.
- Bronchoscopic tamponade with a balloon cathether can prevent extensive alveolar flooding.
- Arteriography localizes the source of bleeding and allows treatment by embolization.
- Surgery should be limited to cases where arteriography is unavailable or fails.

Introduction

Optimum management involves a team approach including the intensive care specialist, the pulmonologist, the radiologist, the anesthetist, and the thoracic surgeon. The greatest immediate vital risk to the patient is asphyxia. Therefore control of the airway is the most important aspect of initial management. Bronchoscopy is usually the single most valuable diagnostic procedure. Medical treatment, including systemic vasopressors, topical treatment such as endobronchial tamponade, and/or intrabronchial instillation of a fibrin sealant, may help to stabilize the patient. In the majority of cases, definitive treatment will involve arteriography followed by embolization or, if embolization is not available or is unsuccessful, pulmonary resection if the patient has adequate pulmonary reserves and no major contraindications to surgery.

Diagnosis

Physical examination

The primary initial diagnostic aim in massive hemoptysis is to identify the source of bleeding. The clinician must confirm that the origin of the hemorrhage is, in fact, the respiratory tract below the level of the vocal cords rather than the gastrointestinal tract. Although the distinction is often obvious, it is not invariably so, inasmuch as vomited blood can spill into the tracheobronchial tree and induce coughing. This can also occur if the bleeding is of pharyngeal origin. In hematemesis the blood is generally darker, has an acidic pH, and may contain particles of partially digested food.

Once the pulmonary origin of bleeding is verified, attention should be focused on determining the side from which bleeding emanates. Patients themselves often cannot accurately localize the source of hemoptysis as being one lung or the other. Physical findings, such as localized rales, rhonchi, or wheezing are often not helpful, inasmuch as they may reflect pneumonitis or bronchial reaction to aspirated blood rather than denoting the bleeding site.

Chest radiography

A chest radiograph must be obtained in all cases of massive hemoptysis, as it sometimes provides clues to the etiology of hemorrhage such as a tuberculous cavern or an abscess. However, findings on chest radiographs are frequently diffuse and do not pinpoint the site of hemorrhage, being secondary to blood distribution throughout the tracheobronchial tree.

Bronchoscopy

Bronchoscopy is the initial procedure of choice for localizing the source of bleeding and enables the physician to choose between the various therapeutic options available. The timing of this procedure in massive hemoptysis has been somewhat controversial, but it is generally accepted that bronchoscopy during active bleeding significantly enhances the likelihood of diagnostic localization (Sternbach and Varon 1995). Because blood is distributed throughout the tracheobronchial tree, visualization of active bleeding, rather than clot, is essential for definite identification.

In the past, rigid bronchoscopy was advocated as the procedure of choice. Its advantages include improved suctioning capability, continuous airway control, and a larger lumen to clear the airway of large clots and debris as well as to introduce packing materials. However, the rigid instrument has a reduced visual range compared with the flexible bronchoscope; it also requires general anesthesia and so the procedure must be performed in the operation room. Generally, however, the patient needs immediate securing of the airways and cannot wait for the operation room to be prepared. Furthermore, widespread use of the rigid bronchoscope is declining, as fewer institutions currently provide training in this technique.

The flexible bronchoscope has an extensive visual range and possesses increased ability to visualize the upper-lobe bronchial orifices, a preferential bleeding site in hemoptysis. The flexible bronchoscope is tolerated more readily and for longer periods than the rigid instrument. Moreover, flexible bronchoscopy is a technique that can be performed under local anesthesia at the bedside. The drawbacks of flexible bronchoscopy include limited suctioning capacity and the ease with which the fiber-optic tip can become obscured by blood.

For all these reasons, patients are increasingly being investigated by flexible bronchoscopy, with or without intubation. On occasion, the two procedures can be used in combination, with passage of the flexible bronchoscope through the rigid instrument.

Arteriography

Selective arteriography can be used to assess the bronchial or pulmonary arterial circulation. Arteriography has a distinct advantage when the source of bleeding is peripheral to the bronchoscope's area of visualization and when bronchoscopy fails to localize the bleeding. Arteriographic findings suggestive of a bleeding vessel include parenchymal hypervascularity, vascular hypertrophy and/or tortuosity, capillary stasis, bronchopulmonary shunting, aneurysm formation, and thrombosis of vessels (Rabkin *et al.* 1987). Rarely, acute extravasation of contrast medium from the bronchial vasculature is seen when there is active bleeding during the procedure. Embolization of the bleeding vessel can often be performed and will be discussed below.

Computed tomography

Chest CT, particularly high-resolution CT, has largely replaced bronchography in the evaluation of stable patients with bronchiectasis. McGuiness *et al.* (1994) have shown its value in conjunction with bronchoscopy in assessing patients with non-massive hemoptysis and normal chest radiographs. However, as bronchogenic carcinoma is rarely the cause of massive hemoptysis, the additional information provided by chest CT is seldom relevant and does not justify its routine use in this situation.

Radionuclide scanning and bronchography

These two methods have been almost completely abandoned and are confined to situations where bronchoscopy, arteriography, and CT are unavailable or non-diagnostic.

Management

Because rapid and effective treatment of massive hemoptysis is critically important, the patient should be admitted to intensive care.

Airway control

Death from hemoptysis is not usually caused by blood loss, but rather from asphyxia produced by flooding of the airways and alveoli with aspirated blood. Therefore control of the airway is the most important aspect of initial management. Supplemental oxygen should be administered, and pulse oximetry or arterial blood gases monitored. The head of the bed should be lowered to promote drainage of blood from the airway. If the side of origin of the bleeding is known, the patient should be positioned with the affected side down to prevent spill-over into the non-involved lung.

If bleeding does not subside promptly, the patient should be intubated using an endotracheal tube with an internal diameter of at least 8 mm which is large enough to permit passage of suction catheters or a flexible bronchoscope. If the risk of asphyxiation is imminent and the side of bleeding is known, a standard endotracheal tube can be advanced into the main bronchus of the non-involved side and this side can be ventilated and protected from aspiration of blood from the contralateral side. This procedure can be performed blind or with the aid of a bronchoscope. The use of double-lumen endotracheal tubes (Carlens' or Robertshaw's tubes) has little advantage over selective main bronchial intubation. Moreover, their use is associated with possible complications, the need for frequent checking for positioning, and the potential for erosion of bronchial mucosa in the mechanically ventilated patient. Because of the small lumen of these tubes, suctioning of the airways is also difficult.

Mechanical ventilation is often needed because patients frequently have underlying lung disease with decreased lung reserves. Positive end-expiratory pressure has been advocated as a possible therapeutic modality, by attempting to accomplish tamponade of the bleeding vessel (particularly if it is a pulmonary artery) with positive intrathoracic pressures. However, care must be taken not to worsen hemodynamics or ventilation–perfusion mismatch in localized lung disease.

Hemodynamic stabilization

Two large-bore intravenous cannulas should be placed to allow fluid access and administration of medication. In a few patients, survival is threatened by the actual volume of blood lost through hemoptysis. Intravenous fluid should be administered commensurate with the vital signs. Routine laboratory analysis should include a hematocrit, platelet count, coagulation profile, arterial blood gases, and renal function studies. Blood should also be typed and cross-matched in case transfusion is required. Any coagulation defects should be corrected. If there is ongoing massive bleeding despite endobronchial management, vasopressors such as intravenous vasopressin (0.2–0.4 units/min as a continuous infusion), terlipressin (1–2 mg as a bolus every 6–8 h), or octreotide (50 μg intravenously as a bolus followed by 50 μg/h as a continuous infusion) can be used as a temporary measure.

Sedation and cough suppression

Cough suppression is controversial as it enhances aspiration. Antitussives should be reserved for patients with violent and persistent cough which appears to exacerbate hemoptysis.

Sedation with a benzodiazepine may help to alleviate the extreme anxiety that these patients often experience.

Topical and endobronchial management

Bronchial irrigation, vasoconstrictive agents, and topical coagulants

Cold saline lavage (4 °C), topical epinephrine (1:20 000), and topical thrombin and fibrinogen–thrombin solutions have been used with success in episodic case reports.

Endobronchial blockade and laser photocoagulation

Endobronchial balloon tamponade with embolectomy catheters (Fogarty) or pulmonary artery catheters have been used to control hemorrhage, to stabilize patients, and to treat patients with massive hemoptysis who are not considered suitable for embolization and/or surgery. The balloon remains inflated for 24 to 48 h at a maximum filling pressure between 30 and 50 mmHg (Jolliet *et al.* 1992). Laser photocoagulation has been used with success in patients with bleeding from airway neoplasms.

Arterial embolization

Bronchial embolization has become a mainstay in the treatment of life-threatening hemoptysis. Embolization is an attractive alternative to surgery in patients with bilateral disease, multiple bleeding sites, or borderline pulmonary reserve. In the largest available single series, acute cessation of bleeding was reported in 91 per cent of 360 patients undergoing bronchial embolization (Rabkin *et al.* 1987). A long-term success rate of 78 per cent was reported in the same series. Recurrent bleeding is most common if the underlying process is hypervascularization. The technical failure reported is 4 to 13 per cent. Although highly successful in controlling acute bleeding, embolization may be associated with complications such as spinal cord ischemia (in 5 per cent of patients the anterior spinal artery originates from the right bronchial artery) and vascular obstruction to other organs.

Fortunately, the frequency of 'ectopic' embolization in reported series is below 2 per cent.

Surgery

To the extent that surgery removes the bleeding lesion and the source vessels, resection is successful in arresting bleeding acutely as well as in preventing recurrence provided that the disease is localized to the resected segment and does not progress. All available series that compare surgical with medical therapy are observational; a randomized controlled trial of medical versus surgical therapy of massive hemoptysis has not yet been conducted. Older series are clearly in favor of surgical treatment, whereas more recent studies report lower mortality with medical management than with resection. Moreover, appropriate interpretation of medical mortality between series requires comparison of mortality rates among patients deemed operable but managed medically, since being considered inoperable itself suggests a poor prognosis (Corey and Hla 1987). Nowadays, surgery seems best reserved for operable patients when bronchial embolization is not available, technically impossible, or unsuccessful, when the volume of hemoptysis requires definitive initial treatment (i.e. it is more than 1000 ml/24 h or is associated with hemodynamic or major respiratory compromise), or when the cause of massive hemoptysis is unlikely to be arrested by bronchial embolization (pulmonary artery perforation by a pulmonary artery catheter or mycetoma with profuse systemic collateral supply).

Contraindications to surgery are bilateral pulmonary disease with inability to localize the bleeding site, unresectable cancer, hemoptysis secondary to valvular heart disease, bleeding diathesis, or diffuse vasculitic process.

References

Corey, R. and Hla, R.M. (1987). Major and massive hemoptysis: reassessement of conservative management. *American Journal of Medical Sciences*, **294**, 301–9.

Jolliet, P., Soccal, P., and Chevrolet, J.C. (1992). Control of massive hemoptysis by endobronchial tamponade with a pulmonary artery balloon catheter. *Critical Care Medicine*, **20**, 1730–2.

McGuiness, G., Beacher, J.R., Harkin, T.J., Garay, S.M., Rom, W.N., and Naidich, D.P. (1994). Hemoptysis: prospective high-resolution CT/bronchoscopic correlation. *Chest*, **105**, 1155–62.

Rabkin, J.E., Astafjev, V.I., Gothman, L.N., and Grigorjev, Y.G. (1987). Transcatheter embolization in the management of pulmonary hemorrhage. *Radiology*, **163**, 361–5.

Sternbach, G. and Varon, J. (1995). Massive hemoptysis. *Intensive Care World*, **12**, 74–9.

3 The cardiovascular system

3.1 Normal physiology

3.1.1 The cardiovascular system

B. A. Foëx and R. A. Little

Key messages

- The function of the systemic circulation is to transport oxygen and nutrients to tissues and remove the products of cellular metabolism via the cardiac output and the regional blood flow.
- Cardiac output is the product of heart rate, determined by the intrinsic electrical properties of the heart and extrinsic neural influences, and stroke volume, determined by the energy of contraction of the myocytes, the resistance against which blood is ejected. These are modulated by neural, humoral, and chemical factors.
- Blood flow is laminar in arteries, arterioles, venules, and veins. It is turbulent in the ventricles and single file in capillaries.
- Arterioles represent the major component of the peripheral resistance.

Introduction

The principal function of the cardiovascular system is to transport oxygen and nutrients to the tissues and remove the products of cellular metabolism. This is achieved in two ways: by diffusion; by convection or bulk flow.

Diffusion, which is a random process, is extremely slow and a more efficient means of transport is needed for distances greater than 0.1 μm.

Convection or bulk flow allows oxygen bound to hemoglobin to travel from the lungs to the most peripheral capillaries in about 30 s. Convection requires energy (a pump) and a complex system of conduits (the vascular tree). The heart is the pump or, rather, two pumps in series: the right atrium and ventricle provide the energy for convective transport to the lungs, and the left atrium and ventricle do the same for the systemic circulation. The function of the systemic circulation is to distribute cardiac output to all the tissues. Thus the cardiovascular system can be broken down into the pump function, i.e. the cardiac output, and regional blood flow.

In addition to its transport function, the cardiovascular system has a control or regulatory function. It provides the conduit for hormones to reach their target organs and it also regulates body temperature. We will not address these functions here, but will concentrate on the biomechanics of the pump function and the main principles of the circulation.

Cardiac output

Cardiac output is the volume of blood ejected by one ventricle during a period of 1 min. It depends on both heart rate and stroke volume. A 70-kg adult will have a stroke volume of 70 to 80 ml and a heart rate of 65 to 75 beats/min and so a cardiac output of approximately 5 l/min (75 ml × 70 beats/min). This may vary considerably.

Heart rate is determined both by the intrinsic electrical properties of the heart and cardiac muscle fibres and by extrinsic influences provided mostly by vagal and sympathetic innervation. Stroke volume is also determined by the intrinsic properties of cardiac muscle and by extrinsic influences (neural, hormonal, and chemical).

Two opposing factors determine stroke volume: the energy of contraction of the myocytes and the arterial pressure against which blood must be expelled. The energy of contraction depends on the following:

(1) the degree of stretch of the myocytes during diastole;

(2) the intrinsic strength of contraction (the contractility) of the myocytes for a given degree of stretch during diastole.

Contractility is largely influenced by extrinsic factors (neural, hormonal, and chemical).

The microscopic architecture of actin and myosin in skeletal and cardiac muscle is such that maximum contractile energy is only generated at a length of 2.2 to 2.3 μm. Under resting conditions, at the end of diastole, sarcomere lengths are suboptimal (usually below 2 μm). Stretching the sarcomeres will improve their contraction. Sarcomeres are stretched when ventricular filling is increased before systole. Thus force of contraction tends to increase with an increase in end-diastolic volume. This principle has become known as Starling's law of the heart. This can be represented graphically by a Starling curve, also known as a ventricular function curve (Fig. 1).

The ventricles are approximately spherical pumps. The internal pressure within a sphere is proportional to wall tension and inversely proportional to its radius (Laplace's law). As the radius increases the

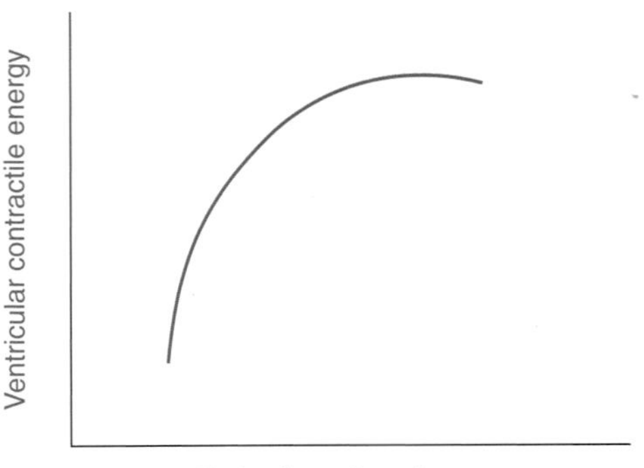

Fig. 1 Ventricular function curve (Starling curve). This is applied to any graph whose ordinate is a measure of contractile energy (e.g. stroke volume or stroke work) and whose abscissa is an index of resting fiber length (e.g. end-diastolic volume, filling pressure, or end-diastolic pressure).

curvature of the wall decreases. Less of the wall tension is directed towards the cavity and so less pressure is generated. This contradicts Starling's law of the heart. However, within the normal size range of the ventricles the improved performance of the myocytes induced by stretching them (and reducing the curvature of the sphere) outweighs the mechanical disadvantage of an enlarged ventricle. Beyond a certain limit pathological ventricular distension results in greatly reduced mechanical efficiency and the typical dilated 'failing' heart.

As well as ventricular filling (end-diastolic volume), arterial pressure is an important intrinsic determinant of stroke volume. This can be understood by examining the mechanics of cardiac contraction:

$$\text{work} = \text{force } (F) \times \text{distance } (L).$$

However, for a fluid force is the pressure applied over a given area. Therefore

$$\text{work} = \text{pressure} \times \text{area} \times \text{distance}.$$

Alternatively,

$$\text{work} = \text{pressure} \times \text{volume}.$$

In the case of ventricular contraction, pressure is actually the increase ΔP in ventricular pressure during systole. Therefore, for a given ventricular contraction, the equation can be rewritten in terms of stroke work (SW) and stroke volume (SV):

$$\text{SW} = \Delta P \times \text{SV}.$$

Any increase in ΔP must result in a reduction in stroke volume unless there is an increase in stroke work.

Preload and afterload are precise terms used in muscle physiology. Preload refers to the load used to stretch muscle before a contraction, and afterload is the tension at which the load is lifted by the contraction. Clinically, preload refers to the degree of ventricular filling (i.e. the degree of stretch of the myocytes during diastole). Afterload is the resistance against which blood is expelled (arterial pressure). Arterial pressure may also influence cardiac function via extrinsic mechanisms.

A number of reflexes are involved in the regulation of the cardiovascular system and their precise mechanisms are beyond the scope of this brief review. The baroreflex is important in regulating blood pressure and heart rate. It depends on baroreceptors, which are tonically active stretch receptors found in the aortic arch and carotid sinuses. Their afferents travel with the vagus nerve and the glossopharyngeal nerve respectively to the brainstem, terminating in the nucleus tractus solitarius. The efferent limb consists of both vagal and cardiac sympathetic nerves to the heart. An increase in arterial pressure will produce increased baroreceptor afferent activity and increased vagal efferent activity. This results in a bradycardia. At the same time there will be a reduction in sympathetic outflow to the heart. This causes a reduction in cardiac contractility and arterial pressure. Sympathetic outflow to the peripheral vasculature will also be reduced, causing a fall in peripheral vascular resistance and thus a fall in arterial pressure.

Contractility and inotropes

The intrinsic strength of contraction (contractility) of cardiac muscle fibers at a given length can be altered extrinsically by chemical agents known as inotropes (derived from *inos*, the Greek for strength).

Norepinephrine (noradrenaline), the most important natural inotrope, is released from cardiac sympathetic nerve endings in the ventricle wall. These nerves are part of the efferent limb of the baroreflex already discussed. Binding of norepinephrine to β_1-receptors on the myocytes results in an increase in intracellular calcium, and an increase in the proportion of cross-bridges between actin and myosin which are activated and therefore a greater contractile force.

This results in a more rapid increase in ventricular pressure during isovolumetric contraction and a higher arterial pressure. At the same time cardiac fiber shortening is enhanced and velocity of contraction is increased so that the ejection fraction increases. Although this will increase stroke volume initially, it will be limited by the increase in arterial pressure and the reduction in ventricular size (end-diastolic volume) inherent in any increase in ejection fraction.

Increases in stroke volume will only be maintained if there are simultaneous alterations in the peripheral circulation, such as a reduction in peripheral vascular resistance and therefore an attenuation of the increase in arterial pressure.

Positive inotropes, by increasing the strength of myocyte shortening, alter the ventricular function curve. This occurs in two ways: either the curve is shifted upwards or it becomes steeper. This alteration is dose dependent; the greater the inotropic stimulus, the greater is the effect on the ventricular function curve (Fig. 2).

Biomechanics of the circulation

The circulation is essentially a central pump distributing fluid to various organs and then recirculating it. The relationship between

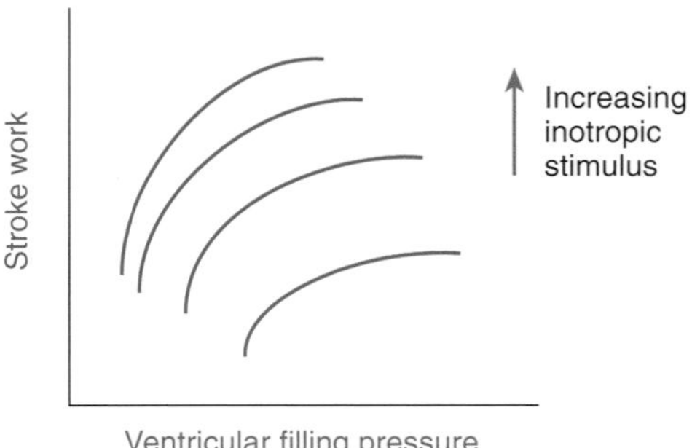

Fig. 2 Effect of increasing inotropic stimulation of the heart on the ventricular function curve.

flow, pressure, and resistance in a fluid circuit is similar to that in an electrical circuit:

$$\text{flow} = \text{pressure}/\text{resistance}.$$

In any part of the circulatory system flow will be determined by the effective perfusion pressure through that part of the system, i.e. the difference in pressure between the arterial and venous ends. Thus the law can be restated as

$$\text{flow} = (P_a - P_v)/R.$$

This is known as Darcy's law of flow. It can be applied to the whole circulation where flow equals cardiac output (CO) and the effective perfusion pressure is the difference between the mean arterial pressure (MAP) and the central venous pressure (CVP). Resistance is the total peripheral resistance, which is known as systemic vascular resistance (SVR). The equation then becomes

$$\text{CO} = (\text{MAP} - \text{CVP})/\text{SVR}.$$

Flow is expressed in terms of volume and time. In contrast, velocity is expressed in terms of distance and time. This distinction is important in considering the circulation, where the size of channels varies enormously from the aorta to the capillaries.

The mean velocity of a fluid is flow divided by the cross-sectional area of the channel. As blood passes from the aorta to the microvascular network, there is an enormous increase in the total cross-sectional area and therefore a progressive decrease in velocity, although overall flow remains the same.

Darcy's law of flow relates flow to pressure: blood flows from areas of high pressure to areas of low pressure. This is a simplification; consider the pressures involved when blood flows from the heart to the feet in an upright position.

Mean arterial pressure is typically 95 mmHg in the aorta, but in the foot it is about 180 mmHg (owing to the weight of the column of blood above the foot). However, blood does flow against this apparent gradient. A more general law of flow states that in the steady state there is flow from A to B if there is a difference in mechanical energy from A to B. This relationship was postulated by Bernoulli.

Mechanical energy is the sum of pressure energy, potential energy, and kinetic energy, whic are defined as follows:

(1) pressure energy is the product of pressure and volume (PV);

(2) potential energy is the capacity of a mass to do work in a gravitational field by virtue of its vertical height above a reference level, i.e. fluid mass (density × volume) × height × gravitational force;

(3) kinetic energy is the energy that a moving mass possesses due to its momentum (which increases in proportion to velocity squared).

Returning to the example, it is clear that blood in the aorta has more gravitational potential energy than blood in the foot. The total energy of aortic blood is greater than that in the foot; hence there is an energy gradient relative to the foot.

Kinetic energy, i.e. energy dependent on the momentum of flow, is particularly important in the venous side of the circulation. In the great veins blood velocity is similar to that in the aorta but pressure is much lower. Thus kinetic energy contributes more than pressure energy to the fluid energy in the great veins. When blood reaches the relaxed ventricle, the kinetic energy is almost zero. This gradient of kinetic energy from the veins to the ventricles contributes to ventricular filling.

In this brief description of the physical laws governing the flow of fluids it has been assumed that flow is constant. This is clearly not the case on the arterial side of the circulation where flow is pulsatile. The laws are still useful in considering mean flow, but slightly different rules apply for instantaneous flow. These are beyond the scope of this review.

Flow in blood vessels

Fluid does not flow as one uniform mass through a tube, and neither does blood flowing through the circulation. Blood flow varies according to the size of the vessel. In the arteries, arterioles, venules, and veins flow is laminar. Flow through the capillaries is single file. In the ventricles there is turbulent flow which allows mixing of saturated and unsaturated blood.

Laminar flow

Laminar flow occurs in cylindrical tubes. The fluid behaves as if it were a set of concentric shells or laminas moving at slightly different speeds. The lamina against the side of the tube is actually fixed to it by molecular cohesive forces and has zero velocity. The lamina next to it slides slowly past this non-slip lamina. Subsequent laminas all slip past each other and so gradually appear to flow faster than each other. At the entrance of a cylinder the laminas present a plug-like profile. Once flow is established the profile becomes parabolic (Fig. 3), with the central lamina experiencing the greatest velocity.

The situation is complicated in the case of blood as the fluid is actually a mixture of two phases: plasma and red cells. The velocity profile is blunted compared with that of water.

The movement of one lamina against another results in shearing. One effect of this is that red cells are displaced towards the centre of the column of fluid, resulting in 'axial flow', leaving a marginal layer of cell-deficient plasma along the vessel wall. This layer is only 2 to 4 μm thick, but in arterioles, which have a diameter of only 30 to 40 μm, it reduces the relative viscosity of blood. Lower viscosity means that less pressure is needed to perfuse the microvasculature and blood flow improves.

Single-file flow in capillaries

As the width of a red cell (8 μm) exceeds that of a capillary (5–6 μm), flow can only occur in single file with the cells slightly deformed. There is no possibility of laminar flow, and the cells and plasma move by 'bolus flow'. Red cell deformability is an essential prerequisite for bolus flow of this type. Thus the friction seen when lamina flows against lamina is eliminated, but there is still friction between

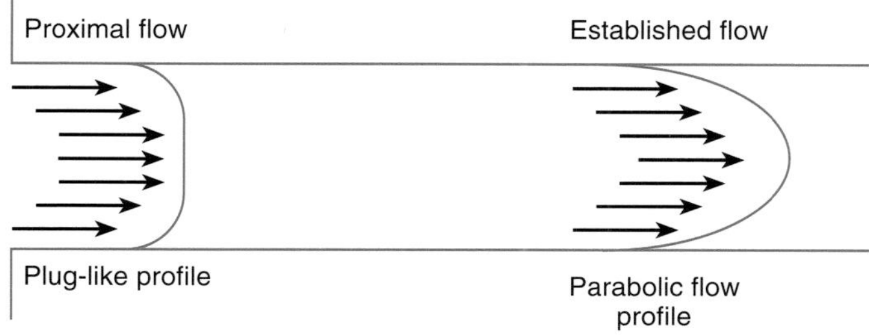

Fig. 3 Laminar flow in a cylinder illustrating the change from a plug-like flow profile to a parabolic flow profile.

the cells and the capillary wall. This is minimized by a thin film of plasma.

Polymorphonuclear leukocytes, which are larger, rounder, and stiffer than red cells, move more slowly through capillaries and so impair flow. In inflammation, adhesion of these cells to the endothelium can result in a very detrimental impairment of microvascular blood flow.

Turbulent flow

Darcy's law suggests that there is a linear relationship between flow and effective perfusion pressure. In practice this is not the case; above a certain critical pressure difference flow increases as the square root of pressure. This is because there has been a transition from laminar flow to turbulent flow in which swirling cross-currents dissipate some of the pressure energy of a fluid as heat. A number of variables promote turbulence: a high fluid velocity V; a large diameter D; a high fluid density ρ. However, an increase in viscosity η reduces turbulence. The relationship between these variables defines the Reynolds number (Re):

$$\mathrm{Re} = VD\rho/\eta.$$

The Reynolds number has practical implications; above a value of 2000 flow changes from laminar to turbulent. In normal situations turbulent flow is only seen in the ventricles and the left ventricular outflow tract. Anemia, which causes a reduction in viscosity, may result in turbulence, as may the presence of atheromatous plaques which disrupt the orderly arrangement of laminas.

Vascular resistance

Laminar flow is the characteristic flow pattern in all blood vessels except the ventricles and capillaries. The resistance to flow is the result of friction between the laminas rather than between the fluid and the vessel walls since the vessel is effectively lined by a non-slip lamina. Vessel geometry, by affecting fluid velocity, will affect vascular resistance. At higher fluid velocities there is more shearing between the laminas and so greater internal friction. As velocity is given by flow divided by cross-sectional area, the narrower the vessel the greater is the velocity and so the greater is the internal friction or vascular resistance.

Resistance to steady laminar flow along a straight tube is proportional to its length L and the fluid viscosity η, and inversely proportional to the tube radius to the power 4. Two constants are also involved so that the exact relationship is

$$R = 8\eta L/\pi r^4.$$

Using the definition of R from Darcy's law ($F = P_a - P_v/R$), we can rearrange the formula to give

$$F = (P_a - P_v)(\pi r^4/8\eta L)$$

which is known as Poiseuille's law of flow through a single tube.

The circulation is much more complex than a single tube. Vascular anatomy is such that vessels are arranged both in series and in parallel. When vessels are in series, the total resistance is the sum of the individual resistances. However, when they are in parallel (such as in a capillary bed) there is a summation of their conducting capacities. The net conductance NK is the product of the individual conductances K and the number N of tubes or capillaries. As the resistance R is the inverse of NK, then as N increases so resistance decreases. For a given effective perfusion pressure, the greater is the cross-sectional area of the capillary bed the greater is the flow.

When total peripheral resistance or systemic vascular resistance (SVR) is considered, it is apparent that the circulation appears as a series of tubes of different caliber all connected in series. Overall resistance will be the sum of these resistances. For each reduction in radius there is a greater reduction in r^4. As $R = 8\eta L/\pi r^4$, resistance increases dramatically.

The reduction in size of the vessels from the level of the aorta to that of the arterioles results in an enormous increase in resistance so that the arterioles represent the major component in SVR. Capillaries are less important, despite their small radius, because they occur in huge numbers, are very short (0.5 mm), and are arranged in parallel. Also, they are subject to single-file rather than laminar flow.

Arterioles are effectively the regulators of organ blood flow. They contain smooth muscle in their walls which is densely innervated by the sympathetic system. Vasoconstriction results in a reduction in arteriolar radius, an increase in resistance, and therefore a reduction in organ blood flow.

Bibliography

Ganong, W.F. (1995). *Review of medical physiology* (17th edn). Appleton and Lange, Norwalk, CT.

Levick, J.R. (1991). *An introduction to cardiovascular physiology*. Butterworths, London.

3.2 Acute chest pain

3.2.1 Causes and diagnosis of chest pain

J. R. McEwan

Key messages

- Cardiac ischemia (myocardial infarction or unstable angina) is a common cause of chest pain in men from middle life onwards.
- A coronary origin for chest pain in a premenopausal woman who has no family history of early ischemic heart disease is unusual, particularly if the cholesterol is normal and she does not smoke.
- Exclusion of acute myocardial infarction and unstable angina is a major goal, but this may require several hours of observation, serial ECGs, and measurement of cardiac enzymes.

The prevalence of coronary disease in Western males in late middle life and beyond is such that coronary artery disease is highly likely in any older man (over 55 years) who has any type of chest pain. The absence of ECG changes at the time of review may only indicate temporary resolution of the ischemia. At the other end of the spectrum, a non-smoking woman who is within 10 years of the menopause, who has no family history of ischemic heart disease, and has atypical pain is very unlikely to have significant coronary disease. Non-cardiac causes of chest pain have to be considered, but exclusion of unstable (pre-infarction) angina or myocardial infarction is the major goal. Short-lasting chest pain at rest in a patient of either gender, associated with any ECG abnormalities (e.g. transient ST depression, T inversion, or ST elevation), should be treated as unstable angina, and the patient should be admitted for treatment guided by local policy (i.e. conservative or investigative).

Acute chest pain is the presenting symptom of a number of life-threatening conditions, of which myocardial infarction is the most common. In up to 60 per cent of patients, myocardial infarction will not be recognized as such; in half of these it can be considered as silent, with no event recalled that could be associated with sudden coronary occlusion. Myocardial infarction is usually associated with acute injury potentials (ST segment elevation) seen on ECG. Sometimes, particularly in small inferior or evolving posterior myocardial infarctions, these ECG changes may be subtle and not immediately appreciated. Some young people, most often of Afro-Caribbean origin, have a high ST segment take-off which, in the presence of chest pain, may lead to confusion with myocardial infarction. The ST segment usually has a normal slope with a T wave that may appear somewhat peaked. Chest pain in these cases may be quite atypical, and the patient does not have the anxiety and apprehension which frequently accompany myocardial infarction. In these circumstances, there is time to consider carefully and repeat the ECG at 30-min intervals over the next hour or so. This will often reassure that there is no evolution of infarct-related changes. The benefits of thrombolysis are seen in those with ST segment elevation or left bundle branch block, and so there is time for procrastination when the picture is less clear.

Pulmonary embolism and aortic dissection can also present with profound symptoms. Myocardial infarction due to coronary dissection may also complicate aortic dissection. In the absence of an ECG typical of acute myocardial infarction in a patient presenting with acute chest pain, associated symptoms and signs and the results of other immediate investigations (e.g. chest radiography, blood gases, troponin T) will help in the differentiation (Table 1).

If there is a suspicion of aortic dissection, thrombolysis should be deferred. There is little to choose between spiral CT scanning, magnetic resonance imaging, experienced transesophageal echocardiography, and aortography with respect to both sensitivity and specificity of diagnosis of aortic dissection. The 24-h availability of one modality of investigation is the essential factor, as are facilities for safe escorted transfer of the patient to a referral center if such an investigation and surgical intervention are required. Pulmonary embolism can occur in patients at low risk of ischemic heart disease (e.g. premenopausal women). Any predisposing factor (obesity, immobility, oral contraceptive pill use, recent parturition or surgery, previous history of thromboembolic disease) should raise suspicion of this diagnosis.

When considering pericarditis, widespread convex ST segment elevation and posture-related changes in pain severity, as well as a sharp scratchy quality to the pain, are typical. The precordium should be auscultated with the patient in different positions (e.g. sitting forward and recumbent) in an attempt to elicit a rub. A two-dimensional echocardiogram may be useful, confirming normal left ventricular function and sometimes demonstrating a small amount of pericardial fluid.

The patient with pneumonia is often toxic and unwell, may have a fever, and may be hypoxemic. A raised neutrophil count may not always be seen, as overwhelming sepsis can suppress the acute marrow response.

Acute costochondritis is associated with exquisite chest tenderness, tightly localized to costochondal joints but often radiating along the intercostal nerve desmosomes.

Table 1 Differential diagnosis of chest pain

	Pain quality	Pain location	Associated symptoms
Acute myocardial infarction	Heavy, constricting, oppressive	Central to arms, neck, jaw	Sweating Pallor Apprehension
Aortic dissection	Sudden onset, often tearing	Beginning or radiating to back	Asymmetric pulses Aortic regurgitation History of hypertension or family history (Marfan's syndrome)
Massive pulmonary embolism	Oppressive	Central, but later sharp, stabbing over chest, worsening on inspiration	Dyspnea Hemoptysis Predisposition (recent operation, immobility, childbirth)
Pericarditis	Sharp, changes with posture and respiration	Central	Pericardial rub Fever
Pneumonia	Sharp, stabbing	Well-localized chest, sometimes shoulder, pain (diaphragm irritation)	Fever Toxic Dyspnea
Tietze's syndrome (viral costochondritis)	Sharp, worse with movement and respiration	Anterior chest wall	Exquisite tenderness over one or two costochondral junctions
Esophagitis and esophageal spasm	Burning or heavy	Retrosternal with radiation to throat	History of heartburn, dyspepsia; may be relieved or exacerbated by eating or drinking

Reflux esophagitis and esophageal spasm may present with or without a history of previous dyspepsia. In the absence of an ECG diagnostic of ischemia, it is worthwhile offering a liquid antacid, but the diagnosis is often retrospective and follows exclusion of ischemic heart disease. Many patients are admitted to a coronary care unit until lack of ECG, exercise testing, or enzyme evidence of ischemic heart disease allows discharge for further gastrointestinal investigation or a trial of a gastric acid secretion inhibitor such as omeprazole.

Bibliography

Braunwald, E. (1992). The history. In *Heart disease: a textbook of cardiovascular medicine*, Vol. 2 (4th edn) (ed. E. Braunwald), Chapter 38. W.B. Saunders, Philadelphia, PA.

Task Force of the Management of Acute Myocardial Infarction of the European Society of Cardiology (1996). Acute myocardial infarction: pre-hospital and in-hospital management. *European Heart Journal*, **17**, 43–63.

3.2.2 Diagnosis and management of angina pectoris

J. R. McEwan

Key messages

- Ischemic heart disease is so common in men from middle life onwards that all complaints of chest pain in this group should be considered to be cardiac in origin until proven otherwise.
- Unstable angina (chest pain at rest or with diminishing exercise tolerance) may progress to myocardial infarction. It is not yet clear whether conservative (medical) or invasive (endoluminal or surgical revascularization) management is best, but thrombolysis has no advantages.
- Non-cardiac surgery is stressful to all patients with coronary artery disease. Therefore careful preoperative assessment is merited in all patients, particularly those whose primary disease indicates high cardiac risk (e.g. peripheral vascular disease or surgery for lung cancer).

Diagnosis

Angina pectoris is pain or discomfort in the chest caused by ischemia of the myocardium. It is distinguished from the pain of acute myocardial infarction in that it is transient and not associated with myocardial necrosis. Pain which has the following characteristics is a highly sensitive marker of significant coronary disease, particularly in men:

(1) heavy or squeezing discomfort;

(2) indicated by clenched fist above the sternum or flat of hand in same position;

(3) radiates to throat and jaw and down the ulnar surface of one or both arms;

(4) associated with a sensation of breathlessness, tiredness, or faintness.

The following features are not usually associated with anginal pain:

(1) experienced away from the chest (epigastric discomfort is occasionally felt);

(2) lasting less than 10 s or more than 20 min (unless there is evolution to myocardial infarction in the latter case);

(3) located with one finger tip;

(4) described as a sharp or stabbing pain;

(5) precipitated by movement or deep inspiration;

(6) precipitated or relieved by eating.

During an episode of angina pectoris, the patient usually ceases activity and may wish to sit down and rest until the episode has passed. Occasionally, patients find that they are able to continue walking through the pain, which gradually dissipates and does not return even if exercise remains at a similar level ('walk-through' angina). Patients will sometimes report that their exercise tolerance is variable, but recent meals and cold weather can lower exercise tolerance before pain develops. Most typically, angina pectoris is predictable and the patient knows what sort of activity will precipitate an episode of chest discomfort.

Relief of angina is brought about by rest, but sublingual nitrates will usually relieve symptoms within 5 min. An absence of relief suggests either that the pain is not ischemic myocardial in origin, or that acute coronary occlusion and myocardial infarction is supervening.

A progressive or sudden change in exercise tolerance in a patient with previous stable angina pectoris is indicative of a precipitous change in the nature of the coronary obstruction from chronic atherosclerosis to plaque instability and transient near-occlusive thrombus. The term 'crescendo angina' is applied to this situation, although the term 'unstable angina' can cover either rapidly decreasing exercise tolerance or the development of angina pectoris at rest (see below).

On examination, the patient with angina may be completely normal or may reveal signs of the following risk factors associated with ischemic heart disease:

(1) hypercholesterolemia (xanthalasma, corneal arcus, xanthomata);

(2) smoking (nicotine staining of fingers, signs of obstructive airways disease);

(3) hypertension (signs of left ventricular hypertrophy, fundal changes of hypertension);

(4) peripheral vascular disease (absent pulses, bruit);

(5) diabetes mellitus (the major risk factor for women).

Angina pectoris may be the presenting feature of aortic stenosis. Indeed, this condition is a contraindication to exercise treadmill testing, and so auscultation of the precordium is mandatory before such an investigation.

Prinzmetal angina

Prinzmetal angina is a rare cause of chest pain, at rest or on provocation such as exercise or hyperventilation, in which spasm of the coronary artery at a site of severe or mild atheromatous narrowing causes intermittent occlusion and typical ST segment elevation on the ECG. The ischemia may provoke ventricular arrhythmias. Rapid relief with nitrates and calcium-channel blockers is common, but recurrent episodes may be troublesome.

Unstable angina

Unstable angina is defined as chest pain occurring on a background of diminishing exercise tolerance or at rest. The diagnosis requires previously identified coronary artery disease or the association of symptoms with transient ECG changes of ischemia. The pathology is recognized as that of an unstable plaque with superadded platelet thrombus, but thrombolytic therapy is of no proven benefit. Patients who present with unstable angina have an approximately 15 per cent chance of a further acute event (myocardial infarction or death) within the next year. Treatment is conservative with bed rest, oral aspirin, β-blockers, and calcium-channel blockers, and initially intravenous nitrates and heparin, until no further pain at rest is experienced. Alternatively, early investigation and endovascular or surgical revascularization may be required. The optimal treatment is as yet undefined, and so either approach is appropriate, depending on resources.

Investigations

A flowchart for the diagnosis and management of chest pain is shown in Fig. 1.

Non-invasive investigations

Non-invasive investigations include resting ECG, exercise tolerance testing, thallium perfusion scanning (exercise or pharmacological stress), stress echocardiogram, and isotope ventriculography. The investigation of angina can confirm the diagnosis and may also give some indication of disease severity and prognosis.

Electrocardiography

The resting ECG may be normal or show evidence of previous myocardial infarction in one or more territories. There may be evidence of hypertension, such as left ventricular hypertrophy. ECG changes may be subtle and non-specific, such as ST segment flattening in the lateral leads or T-wave inversion inferiorly.

Stress testing

In the presence of significant coronary stenosis, exercise or other chronotropic and inotropic stimuli (e.g. infusion of adrenergic

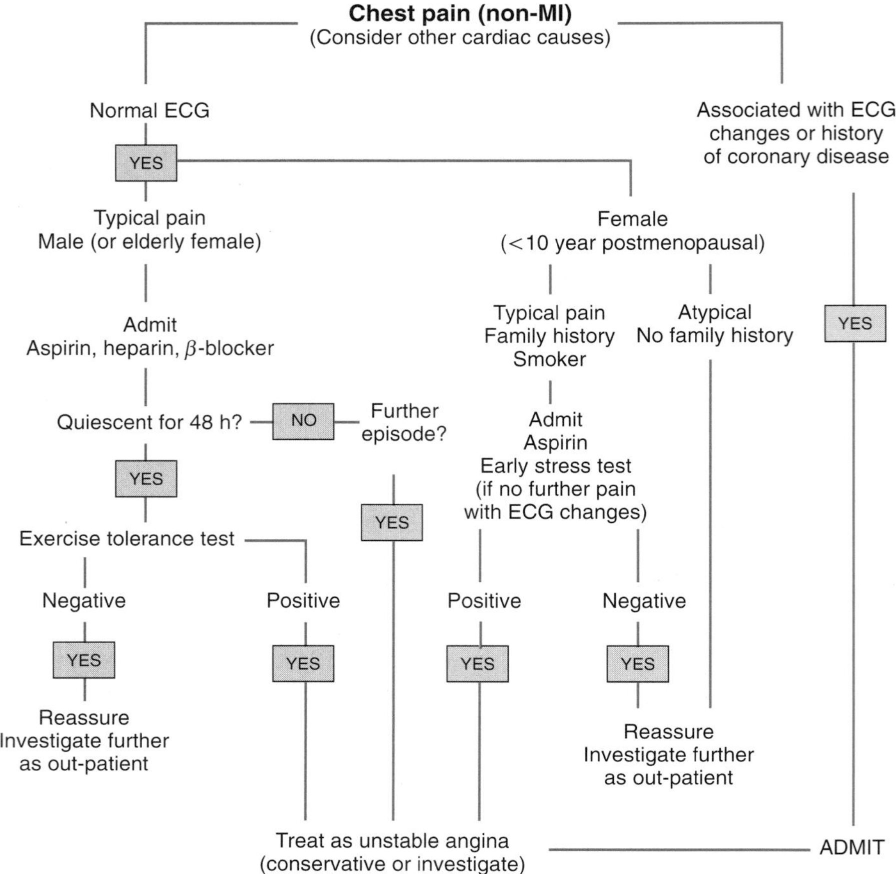

Fig. 1 Flowchart for diagnosis and management of chest pain: MI, myocardial infarction.

agonists such as dobutamine) will result in localized ischemia. This manifests itself as typical pain and also results in changes in the contractility of the ischemic myocardium, which can be detected by echocardiography (stress echo) or isotope ventriculography (multiple-gated acquisition (MUGA) scanning). ECG changes are repolarization abnormalities with ST segment depression. Thallium-201 is distributed in the myocardium in proportion to blood flow. Areas of infarction and scar tissue do not take up the isotope, and ischemic areas take up less than normal. The isotope is detected by a gamma camera.

Invasive investigations

Coronary angiography

Coronary angiography and left ventriculography can be carried out by a Seldinger technique from the femoral artery or by cut down to the brachial artery. Recommendations for coronary angiography include the following.

1. Angina pectoris present and exercise tolerance low, with ECG changes suggestive of severe proximal disease.
2. Symptoms intolerable despite maximal medical therapy.
3. Chest pain and ECG changes occurring at rest (unstable angina).
4. A requirement for valve surgery in a patient over the age of approximately 50 years (younger in men with a positive family history).
5. The diagnosis of chest pain is uncertain and there is frequent referral for admission or consultation.

Medical treatment of angina

Aspirin affects the outcome of stable angina pectoris, reducing the annual rate of acutely unstable events (unstable angina or myocardial infarction) by about 25 per cent. Aggressive cholesterol lowering with HMG Co-A reductase inhibitors (statins) or fibrates with diet has a similar benefit. All other treatments are symptomatic, although β-blockers improve early and late mortality when used postinfarction. We recommend a dose of 150 mg aspirin daily in stable angina pectoris.

Beta-blockers

Beta-blockers work by reducing heart rate and the inotropic effect of sympathetic stimulation, thereby decreasing myocardial oxygen demand. There is little to choose between the myriad of β-blockers available, although the more selective β_1-blockers reputedly have a better side-effect profile. Recent studies have demonstrated a beneficial effect of perioperative β-blockade in patients known to have ischemic heart disease undergoing non-cardiac surgery. A reduction in both acute ischemic episodes and death at 1 year was noted. Heart failure and asthma remain contraindications, but peripheral vascular disease will be of less consequence in these circumstances, at least for the immediate perioperative period.

Calcium-channel blockers

Calcium-channel blockers act by vasodilatation, reducing myocardial oxygen demand by pre- and afterload reduction; they also have some direct coronary dilating effects. They are usually used in conjunction with a β-blocker to reduce the side-effect of reflex sympathetic drive.

Nitrates

Organic nitrates such as isosorbide dinitrate and isosorbide mononitrate are available in many preparations. Headache may limit acceptability. Long-acting nitrates should be given in a regimen which allows a nitrate-free period at the time of day that fewest episodes of angina occur. The nitrate-free period reduces the development of tolerance. Nitroglycerin (glyceryl trinitrate) is still used sublingually as a spray or tablet to terminate episodes of angina or as prophylaxis before angina-provoking exertion.

Potassium-channel openers

Nicorandyl combines an effect on the ATP-sensitive K^+ channel with a nitrate-like effect; these act together to dilate coronary arteries.

Percutaneous transluminal coronary angioplasty and endoluminal stenting

Myocardial infarction

In the treatment of acute myocardial infarction, there is little to choose in outcome between early thrombolysis and percutaneous transluminal coronary angioplasty, although the resources required for the latter are greater. If there is no evidence of recanalization following thrombolysis of a patient having a large infarct, or if new pain and ECG changes suggest reocclusion late after thrombolysis, immediate angiography and, where appropriate, revascularization should be considered where resources are available. However, 'routine' angioplasty as an adjunct to thrombolysis is of no benefit.

Chronic angina

Arterial access for angioplasty is usually via the femoral artery, as a Judkins guide catheter for introduction of the guide wire and angioplasty balloon catheter is more stable than those used from a brachial approach. The procedure of angioplasty stretches the vessel and compresses the atheroma into the wall. In 30 per cent of patients an overexuberant headline response with neointimal hyperplasia renarrows the lumen and further angioplasty or bypass grafting may be required. Nevertheless, the procedure is popular with patients, with good relief of symptoms and a short hospital admission.

The delivery of an endovascular stent splints the vessel subjected to angioplasty. The lumen site will be greater, dissection flaps secured, and the likelihood of restenosis reduced. The technique is applicable only to larger vessels (3 mm) and patients require appropriate anticoagulation and antithrombotic treatment. The regimen is determined by the circumstances of the procedure. Elective stenting is preceded by ticlopidine (an antiplatelet drug) for 48 h and heparin administered around the time of the procedure. Ticlopidine is continued for 1 month with a full blood count at fortnightly intervals to identify any reversible neutropenia, which occurs as a complication in up to 2 per cent of patients. If the stent is placed in circumstances of an occlusive complication of angioplasty, or if thrombus is readily identified at the lesion site, abciximab, the platelet receptor glycoprotein (Gp IIb/IIIa) antagonist, is given first as a bolus and then by infusion in conjunction with carefully titrated heparin.

Referral for surgery

Symptomatic disease involving all three coronary arteries which is uncontrolled by medical therapy will be improved by coronary artery bypass grafting. Prognosis is unaltered unless there is co-existing left ventricular dysfunction (ejection fraction below 40 per cent). The increasing use of left internal mammary grafts may alter this balance in the future, as long-term results are better when this conduit can be used.

Left main stem stenosis above 30 per cent carries an adverse prognosis (1 year mortality of 20 per cent) which benefits from surgical revascularization. Proximal left anterior descending coronary artery stenosis above 70 per cent is a high-risk lesion, and there is little to choose between angioplasty and left internal mammary artery grafting. Stenting may prove more advantageous in the future.

Unstable multivessel coronary disease, which does not settle medically, merits early investigation and referral for surgery if other revascularization techniques are not applicable.

Patients with symptomatic disease of one or two vessels, which is unacceptable despite adequate medical therapy and unsuitable for percutaneous transluminal coronary angioplasty/stenting (or a failed attempt at these procedures), should be referred for surgical revascularization. The patient needs to understand the individual risks and benefits; mortality benefits are not seen, although there are good symptomatic benefits.

ECG evidence of acute ischemia in the ventilated/anesthetized patient

The incidence of perioperative (non-cardiac surgery) myocardial infarction may be up to 10 per cent in patients with known coronary disease. It may first manifest itself in the immediate period in the recovery room or intensive care unit as ECG changes of ischemia (ST elevation or depression). These changes may be due to acute coronary occlusion (probably related to platelet activation) or critical ischemia because of hypotensive anesthesia or a tachycardia in the presence of previously unrecognized severe coronary stenosis. With either infarction or severe ischemia, optimizing blood pressure and, if possible, infusion of nitrates to increase coronary dilatation is the first move.

Recent research has demonstrated the protective effect of β-blockers (intravenous β-blocker, e.g. atenolol 5 mg, with induction) preoperatively and thereafter.

Where β-blockade has not been carried out or is inadequate, as manifest by persisting tachycardia, further β-blockers can be given, initially intravenously. Contraindications are known asthma and heart failure. If ECG changes do not resolve within 1 h with this regimen, consideration should be given to cardiac catheterization and coronary angiography with a view to an immediate endovascular procedure such as angioplasty or stenting. The decision depends on the amount of territory deemed to be at risk and the hemodynamic upset associated with the ischemia.

A 'small' inferior infarct is best treated conservatively with only β-blockade and aspirin (300 mg and then 150 mg daily), whereas a large territory of ischemia across the anterior or anterolateral wall merits angiographic intervention. If necessary, aortic balloon counterpulsation can be used to support the circulation before, during, and after the investigation and intervention. Two-dimensional transthoracic echocardiography can help in the definition of the infarcting area because the ischemia is accompanied by hypo- or akinesis. Thrombolysis would normally be contraindicated post-surgery, but if the patient is ventilated and anesthetized for other reasons (e.g. respiratory disease), it should be given due consideration.

The demonstration of three-vessel disease in these circumstances may merit early bypass grafting as an alternative. An acute occlusion of one of three diseased vessels can be considered for angioplasty, although severe stenosis without occlusion is likely to be managed with referral for immediate or early bypass grafting.

Percutaneous transluminal coronary angioplasty/stenting versus coronary artery bypass grafting

The cost of percutaneous transluminal coronary angioplasty is slightly less than that of coronary artery bypass grafting, but this is offset by an increased requirement for further investigation and intervention. Survival data to 5 years is similar for both techniques in patients with two-vessel disease. Coronary artery bypass grafting is more advantageous In patients with diabetes or three-vessel disease,. The mortality for cardiogenic shock is 55 per cent, but recent evidence suggests that more aggressive support, investigation, and intervention (aortic balloon counterpulsation, angioplasty, or coronary artery bypass grafting) is associated with an improved outcome and lower mortality. If there is evidence to support the cause of cardiogenic shock being a single occlusion, such as a proximal left anterior descending occlusion, 'salvage' angioplasty should be considered where facilities permit.

Bibliography

Anderson, H.V., *et al.* for the TIMI IIIB Investigators. (1995). One year results of the Thrombolysis in Myocardial Infarction (TIMI) IIIB Clinical Trial. A randomized comparison of tissue-type plasminogen activator versus placebo and early invasive versus early conservative strategies in unstable angina and non-Q wave myocardial infarction. *Journal of the American College of Cardiology*, **26**, 1643–50.

Braunwald, E. and Sobel, B.E. (1992). Coronary blood flow and myocardial ischemia. In *Heart disease: a textbook of cardiovascular medicine*, Vol. 2 (4th edn) (ed. E. Braunwald), Chapter 38. W.B. Saunders, Philadelphia, PA.

Corr, L.A. (1996). The future of interventional cardiology. *Lancet*, **348**, 23–5.

Mamode, M., Cobbe, S., and Pollock, J.G. (1995). Infarcts after surgery (Editorial). *British Medical Journal*, **310**, 1215–16.

3.3 Acute myocardial infarction

3.3.1 Pathophysiology

M. J. Davies

Key messages

- Regional infarction is due to thrombosis in the subtending artery.
- Thrombi occur over plaques undergoing disruption.
- Plaques at risk of disruption have a high lipid and macrophage content.
- Cardiogenic shock is a complication of large infarcts.
- In cardiogenic shock further non-regional necrosis develops due to hypoperfusion.
- Cardiac rupture causes 10 per cent of infarct mortality.
- Transmural infarcts may expand and alter the long-term ventricular shape.
- Non-transmural infarcts do not expand.
- Non-transmural infarcts are associated with either prior collateral flow or re-establishment of antegrade flow within 6 h.

Introduction

The blood supply of the human myocardium is regional, i.e. each major branch of the coronary arteries supplies a discrete segment of the myocardium. In the normal human heart connections exist between adjacent coronary artery branches but are not functional as collaterals unless stenosis has caused a pre-existing pressure differential between the two systems. Myocardial infarction usually means a regional area of necrosis due to ischemia. Other forms of ischemic myocardial necrosis also occur.

Experimental model of regional infarction

If a major coronary artery branch is ligated in a dog a regional area of ischemia occurs (Jennings *et al.* 1995). The lateral margins of this area of myocardium at risk of infarction are determined instantly by the coronary anatomy. Necrosis begins in the subendocardial zone within 10 min and over hours spreads from the endocardium to the epicardial surface. If flow is restored within this 6-h time window, some myocardium can be saved and the infarction does not become transmural. Periods of occlusion lasting for more than 6 h produce a transmural regional infarction.

Human regional infarction

The arterial lesion

The dog model shows that a regional infarct is produced when flow is abolished in the artery subtending that region. This is also true of human infarction. The predominant cause of human arterial occlusion is thrombosis superimposed on pre-existing atherosclerosis. Rare causes are coronary spasm, either spontaneous or due to cocaine, coronary dissection, coronary emboli, and localized arteritis. About a quarter of major thrombi are due to superficial loss of the endothelium over a plaque with exposure of the underlying connective tissue matrix to platelets, leading to a thrombus which is attached to the luminal surface of the plaque. Plaque disruption is responsible for most thrombi (Davies 1990) and is a complication of plaques (American Heart Association types IV and Va) (Stary *et al.* 1995) which have a core of extracellular lipid separated from the arterial lumen by a cap of fibromuscular tissue. The core contains abundant macrophages expressing tissue factor. In disruption the plaque cap tears (Fig. 1), allowing blood from the lumen to enter the core where a platelet-rich thrombus forms. A second stage follows in which thrombus projects into but does not occlude the lumen; in the third stage thrombus occludes the lumen. Platelets are a major component of the first two stages; in the third stage the thrombus is rich in fibrin and enmeshed red cells. A range of factors may accelerate or inhibit this staged thrombotic response (Fig. 1), which is dynamic; intermittent occlusion is common (stuttering onset). Unstable angina is associated with non-occluding mural thrombus. Transient episodes of myocardial ischemia are caused by platelet emboli from the surface of the exposed mural thrombus impacting in the intramyocardial vascular bed. These microemboli cause microscopic foci of necrosis (Davies *et al.* 1986).

Plaques at high risk of disruption (i.e. vulnerable) can be angiographically invisible or cause any degree of stenosis. Total thrombotic occlusion of a coronary artery in which there is no prior stenosis, and therefore no collateral flow, is more likely to cause infarction than thrombosis at sites of previous high-grade stenosis. Previous angina also invokes preconditioning of the myocardium and limits infarction.

The myocardial lesion

In human infarction the occlusion can be sudden and persistent, but it is often slower in onset with intermittent occlusion occurring over some hours or even days. Mural thrombus is associated with platelet emboli and causes focal microscopic foci of necrosis of differing ages.

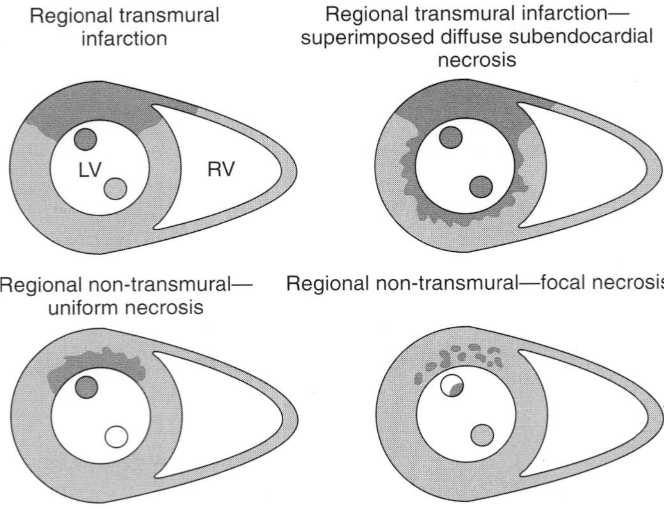

Fig. 1 Patterns of human myocardial infarction.

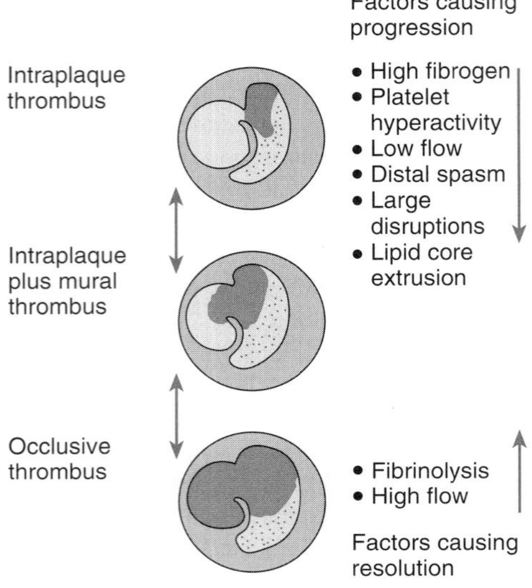

Fig. 2 Stages and factors influencing coronary thrombosis.

These different mechanisms lead to considerable variation in the structure of human infarcts (Fig. 2). The simplest form is a transmural infarct in which the tissue necrosis is uniform in age. Such infarcts are often associated with the appearance of Q waves in the ECG and have potentially serious consequences such as rupture. Regional infarcts are also formed by the coalescence of focal areas of necrosis of widely differing ages; such infarcts are more likely to be confined to the inner zone of the myocardium (non-transmural) and may not have Q waves.

Morphological changes in infarction

Within seconds of flow ceasing myocardial contraction stops. There is a switch to anaerobic glycolysis within the myocyte; lactic acid accumulates, and ATP and phosphocreatine fall. Irreversible damage is indicated by the loss of membrane integrity with loss of intracellular enzymes and by the influx of calcium ions (Jennings *et al.* 1995). Structural changes recognizable by microscopy occur after 8 h. For infarction to be recognized by a pathologist, the subject must survive for at least 8 h after the onset of arterial occlusion. In non-reperfused infarcts, the myocytes swell and lose all intracytoplasmic organization, and myofibrillary material forms amorphous masses. In reperfused infarcts the myocyte shows brightly eosinophilic cross-banding in the cytoplasm. This appearance (contraction band necrosis) is due to intense hypercontraction of the myofibrils owing to the influx of calcium ions and occurs only when blood supply is restored by fibrinolysis to myocardium which is undergoing infarction. Infarcted myocardial tissue invokes an inflammatory repair response. By 12 h small intramyocardial vessels contain polymorphs which also infiltrate the interstitial tissue. By 24 h myocytes have begun to disintegrate, and there is a florid macrophage response followed in turn by invasion of the infarct zone by fibroblasts and new capillaries. Collagen deposition begins by day 4 and continues for many weeks. The rate of repair depends on the original structure of the infarct. Non-transmural infarcts in which there are foci of surviving myocytes intermixed with foci of necrosis show very rapid fibrosis leading to a scar containing islands of surviving myocardium. Large areas of contiguous necrosis can only be organized by ingrowth of cells from the margins where there is viable tissue.

Complications of myocardial infarction

Early ventricular fibrillation is not related to infarct size. After 24 h ischemic myocytes have either undergone necrosis or recovered, and the risk of ventricular fibrillation diminishes. The majority of the other complications occur in transmural infarcts. Cardiogenic shock is directly related to the proportion of the left ventricular muscle mass which has been lost. Large infarcts are more likely to be anterior because the left anterior descending coronary aretery supplies more than 60 per cent of the myocardial mass.

Cardiogenic shock may lead to an extension of the area of necrosis (Fig. 2). Myocardial perfusion is dependent on the relation between aortic diastolic pressure and left ventricular cavity pressure. If aortic pressure falls and left ventricular diastolic pressure rises, subendocardial perfusion throughout the whole left ventricle falls; a circumferential zone of subendocardial necrosis occurs and the centers of the papillary muscles also undergo necrosis (Fig. 2). Patients enter a downward spiral in which falling myocardial perfusion causes new necrosis which in turn further depresses ventricular function.

Cardiac rupture, which is responsible for up to 10 per cent of infarct mortality, complicates transmural infarction alone. Rupture within the first 36 h is due to a tear developing at the margin of viable and non-viable myocardium. The tears are often slit-like and occur at a stage when the infarct is not readily apparent at autopsy. Later rupture is due to expansion of the infarct itself (Weisman and Healy 1987). In infarct expansion (Fig. 3), the necrotic tissue expands outward and the wall thins at the site. No increase in the volume of infarcted tissue occurs; it is purely a shape change and the ventricular cavity becomes asymmetric with one transverse axis greater than in a plane at right angles (Fig. 3). Expanding infarcts may rupture from day 3 onward and are associated with overlying pericarditis. Expansion is more common with large infarcts. The process of

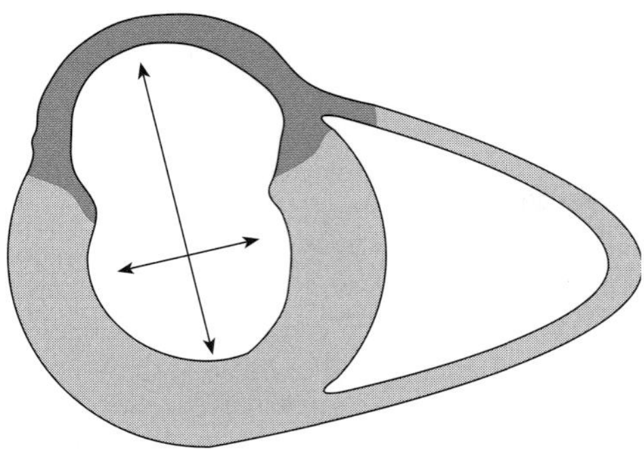

Fig. 3 Infarct expansion.

expansion begins within 24 h and continues for some days. As repair is initiated the expanded infarct is converted into a fibrous structure, permanently altering ventricular shape (remodeling) (Pfeffer and Braunwald 1990). The residual myocardium undergoes hypertrophy.

Ventricular aneurysms arise from severe degrees of acute infarct expansion and have a wide neck. A rarer form of aneurysm has a narrow entry into the left ventricle leading to a large external sac. These lesions follow a partial myocardial tear with a hematoma that develops beneath the visceral pericardium. These lesions are often referred to as 'pseudoaneurysms' because the outer wall is formed by pericardium rather than by the myocardium itself. Aneurysms which develop mural thrombus have a risk of systemic embolism. The factors which lead to thrombus with an aneurysm are not clear; apparently identical aneurysms may or may not be associated with thrombus in different individuals. Ventricular tachycardia is associated with the survival of subendocardial nests and strands of myocytes embedded in fibrous tissue at the margins of the aneurysm. These surviving myocytes form loops which may extend over several centimeters and act as a focus for re-entry.

Ventricular septal defects have the same characteristics as rupture of the left ventricular free wall. Papillary muscle rupture is not necessarily related to either transmural or large infarcts. Occlusion of the left marginal branch of the circumflex artery leads to very localized infarction of the anterolateral papillary muscle. Papillary muscle rupture also occurs in the setting of larger infarcts, and a combination of an expanding posteroseptal infarct, a ventricular septal rupture, and avulsion of the posteromedial papillary muscle occurs with proximal occlusions of dominant right coronary arteries. The magnitude of mitral regurgitation varies widely depending on whether avulsion occurs across the base of the papillary muscle or just in one subhead. Papillary muscle infarction without rupture occurs in up to 40 per cent of all acute infarcts, and subsequent fibrosis leading to either shortening or elongation is a factor in persistent mild mitral regurgitation.

References

Davies, M.J. (1990). A macro and micro view of coronary vascular insult in ischemic heart disease. *Circulation*, **82**, 1138–46.

Davies, M.J., Thomas, A.C., Knapman, P.A., and Hangartner, R. (1986). Intramyocardial platelet aggregation in patients with unstable angina suffering sudden ischemic cardiac death. *Circulation*, **73**, 418–27.

Jennings, R.B., Steenbergen, C., Jr, and Reimer, K.A. (1995). Myocardial ischemia and reperfusion. In *Cardiovascular pathology: clinicopathologic correlations and pathogenetic mechanisms* (ed. F.J. Schoen and M.A. Gimbrone Jr), pp. 47–80. Williams & Wilkins, Baltimore, MD.

Pfeffer, M.A. and Braunwald, E. (1990). Ventricular remodeling after myocardial infarction. Experimental observations and clinical implications. *Circulation*, **81**, 1161–72.

Stary, H.C., *et al.* (1995). A definition of advanced types of atherosclerotic lesions and a histological classification of atherosclerosis. *Circulation*, **92**, 1355–74.

Weisman, H.F. and Healy, B. (1987). Myocardial infarct expansion, infarct extension and reinfarction: pathophysiologic concepts. *Progress in Cardiovascular Disease*, **30**, 73–110.

3.3.2 Diagnosis

P. Sleight

Key messages

- Rapid diagnosis (<30 min) is important for deciding on the need for reperfusion.
- Beware of atypical presentation (painless; left heart failure; gastrointestinal symptoms), particularly in the elderly.
- Learn the ECG signs: ST elevation, terminal T-wave inversion, new Q waves, bundle branch block.
- Localized wall motion abnormalities on echocardiography are an early sign.
- Follow the evolution of cardiac enzyme release.

Introduction

The classical presentation of acute myocardial infarction is relatively simple to diagnose, but atypical presentation is common, particularly in the elderly and in those who are already ill or have undergone recent surgery. The earlier after onset that the patient is seen, the more difficult is the diagnosis.

Pain

Pain is the outstanding feature of acute myocardial infarction, but several studies have shown that up to a third may be painless, even on careful questioning. The pain is usually described as crushing or severe constriction, felt retrosternally, but radiating to the jaws or to both arms and wrists, particularly the left wrist. Sometimes ischemic pain is quite localized to the jaws and/or the left wrist, with no chest component. It is frequently felt below the left scapula, particularly in inferior or posterior myocardial infarction. Sometimes it is described as discomfort rather than pain, or as a heavy weight on the chest accompanied by dyspnea. Unfortunately, the pain of acute myocardial infarction may be mimicked by esophageal pain, since the afferent nerves input to the spinal cord at the same place. A glass of milk may be a useful therapeutic test when there are no other confirmatory features on the ECG.

The pain usually builds up over several minutes and is less acute and instantaneous in onset than the pain of aortic dissection, whose onset can often be timed to the second and which has a more tearing quality. The pain of myocardial infarction is usually felt by the patient to be life threatening or serious; this may help to differentiate it from the pain of pericarditis which is also distinguished by changes in posture or swallowing. Pleuritic pain is usually worse on breathing or coughing and is often sharper, stabbing, and unilateral. Finally, musculoskeletal or costochondral pain may have similar radiation; it is worthwhile attempting to 'spring' the ribs to see if pain is reproduced. It is obviously helpful if the patient has experienced angina before the myocardial infarction, but the character and site of myocardial infarction pain may not be the same as that with angina.

Dyspnea

Breathlessness is a very common feature of the early stages of myocardial infarction and may be the only clinical complaint. Sudden onset of breathlessness, particularly in the elderly, should always trigger suspicion of acute myocardial infarction. The reason for this is the increase in myocardial stiffness (diastolic dysfunction) which rapidly (within a few minutes) follows myocardial ischemia and equally rapidly resolves upon relief of ischemia. The left atrial and pulmonary artery wedge pressures rise rapidly, and so the lungs stiffen and may develop frank edema. The symptom has to be distinguished from other causes of sudden dyspnea: late-onset asthma, pneumothorax, pneumonia, or pulmonary embolism.

Unstable angina

Sudden worsening of previously stable angina should always be considered to be a warning of an impending myocardial infarction. It is usually diagnosed retrospectively when typical chest pain does not develop into myocardial infarction by ECG or enzyme criteria. However, this is somewhat semantic, since new-onset angina or sudden worsening of stable angina usually implies fresh thrombus in a coronary artery; more frequent blood sampling or ECGs may reveal evidence of cell death which would be missed on daily or twice-daily examination.

Nausea and vomiting

Nausea and vomiting are common early symptoms of acute myocardial infarction and, as a result, lead to misdiagnosis as a gastrointestinal upset or even an abdominal surgical emergency. It may be a pointer to a larger myocardial infarction since it is caused by the Bezold reflex which arises from stimulation of the receptors with c-fiber afferents which are present throughout ventricular muscle. These receptors can be excited by both mechanical and chemical stimuli. They are believed to monitor left ventricular contractility under normal conditions. They also give rise to considerable reflex autonomic disturbance as well as reflex vomiting.

Autonomic disturbances

The Bezold reflex also leads to reflex vagal activation and sympathetic withdrawal. Bradycardia and hypotension due to vasodilatation may lead to circulatory collapse and shock, particularly in large inferior infarction. This responds dramatically to intravenous atropine which blocks the vagal bradycardia and the vagally induced fall in left ventricular stroke volume. In some patients the picture is dominated by sympathetic overactivity as a result of the action of both carotid baroreceptor unloading due to the fall in blood pressure and stimulation of sympathetic afferents in the left ventricle. These autonomic effects can be very dramatic and usually occur in the first hour or two after the coronary occlusion.

The electrocardiogram

The ECG remains the mainstay of early diagnosis but can be quite difficult to interpret, particularly in the first hour of onset. The classical feature is ST elevation, which may be so marked as to mimic the widened QRS of bundle branch block. The height of the ST segment and the number of leads involved bears a clear relation to the amount of infarcted muscle and hence the prognosis for a particular patient. With extreme ST elevation it is useful to consider Prinzmetal or variant angina, and to observe the behavior after the use of nitrates, when alarming changes may revert to normal within a few minutes.

In practice, changes are usually less extreme, with minor ST elevation. It is then necessary to seek other supporting features, including terminal T-wave inversion giving a biphasic pattern. When seen in leads V2–V4, this usually indicates a critical and proximal stenosis of the left anterior descending artery. Inverted U waves are nearly always abnormal; together with minor ST elevation, this should always arouse suspicion of early myocardial infarction. Bundle branch block, particularly left bundle branch block, may mask the changes of acute myocardial infarction. In the setting of typical chest pain and history, left or right bundle branch block is usually indicative of a large infarction. Even when the age of the bundle branch block was unknown, the Fibrinolytic Therapy Trialists' overview showed clear a reduction of the high mortality in those given thrombolytic treatment compared with controls. The changes of acute myocardial infarction may also be hidden by ectopic rhythms, such as ventricular tachycardia or an idioventricular rhythm.

Q waves are considered to be pathological and indicative of infarction if they are wide (above 0.04 s) or deep (more than one-third the height of the following R wave). So-called 'septal' Q waves (from left to right activation of the septum) are, of course, normal in leads I, AVL, and V4–V6. Pathological Q waves usually develop over the first 24 h. They are absent in the early hours; in this case, lack of progression of the precordial R wave from V1 to V4 may be the only early indication of infarction. These changes are more significant if a prior ECG is available. ST depression and/or T-wave inversion do not exclude infarction. Widespread ST depression is sometimes due to subendocardial infarction, with a poor prognosis. However, ST depression and/or T-wave inversion can be a completely non-specific finding due to various other causes, such as digoxin, electrolyte abnormality, ventricular hypertrophy, anemia, thyroid disease, or cardiomyopathy. The Fibrinolytic Therapy Trialists' overview showed no benefit from fibrinolysis for ST depression, despite a 1-month mortality of approximately 20 per cent, when clearly many patients had sustained a myocardial infarction. However, the sample size of patients with ST depression was not large enough to be certain of this lack of benefit.

Biochemical markers

Estimation of enzymes released from necrotic myocardial cells has previously been used as confirmatory evidence of infarction, rather than to make the diagnosis on presentation, since the evolution of enzyme release is as important as the actual values recorded. Typically, in a patient seen within a few hours of infarction onset, the values rise and then fall over hours or days.

Most laboratories record serum glutamine oxaloacetic transaminase (SGOT) or serum aspartate aminotransferase (AST) and creatine kinase (CK). The CK-MB form is more useful as it is specific to heart muscle. These enzymes are usually present by 6 h from onset, peak at 24 h, and fall to normal by 3 to 4 days. The height of the peak (which may be missed by infrequent sampling) is a good guide to the amount of myocardial necrosis. Serum lactate dehydrogenase is released more slowly and persists for longer (2–4 days). The CK enzymes can be fractionated further, and so it is possible to date the time of infarction onset by comparing the ratio of MM to MB subforms.

Other more sophisticated enzyme tests measure serum myoglobin or troponin. These smaller molecules are released much earlier and can be used to diagnose infarction on admission when the ECG is equivocal. Bedside kits are now available. These may help to make a diagnosis of myocardial necrosis, particularly if the ECG is unhelpful because of previous infarctions or bundle branch block. However, these methods still involve some delay which may hinder proper administration of thrombolytic treatment. In practice, they have not yet proved clearly superior to clinical evaluation by the more traditional methods outlined above, except in the case of small infarctions when the value of reperfusion is less clear. Nevertheless, they may be useful as markers of early reperfusion or, more importantly, as markers of failure to reperfuse in the first 60 to 90 min, and so act as pointers for further intervention. Markers arising solely from the myocardium, such as the CK-MB and troponins, are quite specific for myocardial damage, whereas myoglobin and lactate dehydrogenase are non-specific and can be released, for example, by intramuscular injections and liver damage respectively. The sensitivity of these markers varies markedly, mainly depending on the appropriate timing of blood sampling. Troponins and myoglobin are released during the first few hours, AST and CK are released more slowly, and lactate dehydrogenase is somewhat delayed over 24 to 72 h. The sensitivity of these assays also depends on how frequently blood is sampled; once-daily sampling may miss the peak level.

Echocardiography

Whereas the newer enzyme methods are still the subject of intense research, immediate echocardiography can be of great and rapid help in equivocal cases. Immobility of a discrete area of the ventricular wall may be present before clear ECG signs develop. Recently developed quantitative measures of wall-motion abnormality may also indicate the need for early titration of angiotensin-converting enzyme inhibitors. Finally, in shocked patients, the echocardiogram may reveal some unexpected complications.

Radionuclide scanning

Most coronary care units do not have the facility for rapid scanning of left ventricular wall motion (from gated blood pool scans) or for thallium scanning to detect areas of non-uptake. These 'cold' areas only signify necrosis, and do not indicate whether this infarct is old or new. Pyrophosphate scans can detect recent infarction (as a 'hot' area), but they take too long to be useful clinically and so are used mainly for research purposes. In any case, none of these methods are sufficiently rapid or convenient for routine clinical use.

Bibliography

Fibrinolytic Therapy Trialists' (FTT) Collaborative Group (1994). Indications for fibrinolytic therapy in suspected acute myocardial infarction: collaborative overview of early mortality and major morbidity results from all randomised trials of more than 1000 patients. *Lancet*, **343**, 311–22.

Hamm, C.W. (1994). New serum markers for acute myocardial infarction. *New England Journal of Medicine*, **331**, 607–8.

Kennedy, R.L., Burton, A.M., Fraser, H.S., McStay, L.N., and Harrison, R.F. (1996). Early diagnosis of acute myocardial infarction using clinical and electrocardiographic data at presentation: derivation and evaluation of logistic regression models. *European Heart Journal*, **17**, 1181–91.

Sleight, P. (1981). Cardiac vomiting. *British Heart Journal*, **46**, 5–7.

Sleight, P. (1996). Essential hypertension. In *Oxford textbook of medicine*, Vol. II, Chapter 15.10.4, pp. 2331–49. Oxford University Press.

3.3.3 Management

P. Sleight

Key messages

- Rapid response and assessment is essential for effective treatment.
- Pain relief, oxygen, aspirin, and nitrates, with or without atropine as appropriate, should be given immediately (or pre-hospital).
- Thrombolysis is still the main reperfusion strategy, but encouraging results have been obtained with direct percutaneous transluminal coronary angioplasty.
- If transfer to the coronary care unit is not immediate, begin thrombolysis in the emergency room (or outside pre-hospital where appropriate).

Immediate treatment

Whoever sees the patient first, whether it be emergency room staff, the general practitioner, or paramedic, needs to relieve pain, generally with nitrates and an opiate (plus an antiemetic), give oxygen (since arterial desaturation is usual), and give half to one tablet of oral aspirin to be chewed for a rapid antiplatelet effect (since platelets are hyperaggregable in acute myocardial infarction). An ECG can be recorded while this is being instituted in order to make a decision about reperfusion therapy in the initial 15 to 30 min after arrival at the hospital. Telephone or radio transmission of the results of an ECG obtained outside hospital or in transit to the coronary care unit or emergency room can save valuable time.

Atropine

Profound autonomic disturbances often occur early in acute myocardial infarction. Reflex vagal overactivity slows the heart profoundly, causes vasodilatation, and reduces stroke volume. Intravenous atropine can rapidly reverse what otherwise may be diagnosed as cardiogenic shock due to massive ventricular damage.

Reperfusion strategies: percutaneous transluminal coronary angioplasty versus thrombolysis

The cornerstone of modern treatment is to restore blood flow to the ischemic myocardium in order to prevent cell death. Although this necrosis occurs within 20 to 30 min after coronary ligation in animals, it is now clear that in humans reperfusion can salvage the myocardium over a much longer time window, perhaps 3 h or more; this is probably due to a combination of incomplete or intermittent occlusion of an artery and/or collateral vessels which bypass the obstruction.

There is now solid evidence from the Fibrinolytic Therapy Trialists' collaboration that mortality is reduced by thrombolytic treatment initiated up to 12 or even 24 h after onset of acute myocardial infarction in patients with bundle branch block or ST elevation. Overall, about 30 lives can be saved per 1000 patients treated, with a downside of about one stroke due to cerebral bleeding or embolization. Immediate aspirin (160–300 mg) adds a further 20 to 25 lives saved with no downside. The earlier the treatment is started, the better the result; the 1-month mortality is halved by thrombolysis and aspirin started within 4 h of the onset of myocardial infarction.

Although the risks of thrombolysis increase with advancing age (above 55 years) and increasing systolic blood pressure (> 150 mmHg), older people have a much greater risk of dying and so a much greater absolute benefit from lysis. Thus age is certainly not a contraindication, unless other concomitant illness reduces the risk–benefit ratio.

More recently, interventional cardiologists have promoted direct percutaneous transluminal coronary angioplasty (PTCA) as an alternative to thrombolysis. Early trials comparing lysis with PTCA showed lower mortality, stroke, and reinfarction with PTCA. However, caution should be exercised before this is accepted as a routine treatment in centers where it is practical and rapidly available, as these small early trials involved expert-selected centers and, probably, selected patients also. Later data from the somewhat larger GUSTO-2 trial and unrandomized registry data from the United States and Germany are much more evenly balanced. Since many coronary thromboses occur on trivial non-stenotic plaques, it does not make sense to attempt to dilate these. Where the obstruction is mostly fresh thrombus, PTCA may achieve a good angiographic result at the expense of embolization of the distal microcirculation. Finally, direct PTCA, with 24-h rapid availability of a catheter laboratory, is simply not feasible for the vast majority of patients admitted to ordinary general hospitals. The inevitable delays in transfer to a tertiary center will probably outweigh any potential advantages from the greater coronary flow achieved with PTCA. Reocclusion and particularly restenosis (which occurs in about 50 per cent of patients by 6 months) are formidable unsolved problems. Stenting may be a temporary palliative measure, but the long-term outlook is not known in patients who may now survive for decades after an initial myocardial infarction. However, direct PTCA is certainly a viable alternative to lysis in expert centers, if delays can be avoided. It is particularly attractive when lysis is relatively contraindicated because of the risk of surgical or cerebral bleeding. Future research needs to focus on identifying patient characteristics which predict benefit for PTCA rather than thrombolysis and vice versa.

Which drug for thrombolysis: streptokinase or newer regimens?

As the battles between streptokinase and tissue plasminogen activator have shown, this is a highly commercial question both for manufacturers and for the health services who have to pay for the greatly increased costs of the newer drugs of which tissue plasminogen activator was the prototype. In Sweden during 1996 streptokinase was used in about two-thirds of acute myocardial infarction cases and tissue plasminogen activator in the remainder; however, the costs to the Swedish health care system of tissue plasminogen activator were about six to eight times that of streptokinase. Therefore one has to be quite sure of the strength of evidence in favor of newer drugs such as tissue plasminogen activator and its successors such as reteplase, staphylokinase, and bat tissue plasminogen activator. The basic problem is that small trials with surrogate endpoints on relatively few patients can show more rapid lysis of clot by 90 min, but cannot show the inevitable excess of rare but important complications such as cerebral hemorrhage. Even when randomized trials such as GUSTO-1 are carried out, and appear to show that these risks are outweighed by the benefits of earlier opening of the coronary artery by approximately 45 min, we cannot be sure of this benefit, particularly when other larger trials such as ISIS-3 or GISSI-2 showed no overall superiority.

The GUSTO-1 trial was an unblinded open trial which compared four regimens: streptokinase plus 25 000 units of subcutaneous heparin (as in ISIS-3), streptokinase plus full intravenous heparin, accelerated tissue plasminogen activator , and a 'combo' arm of streptokinase plus tissue plasminogen activator. As the results were presented, the two streptokinase arms combined were slightly but significantly less effective compared with the accelerated tissue plasminogen activator arm. Consequently, many physicians, particularly in the United States, use tissue plasminogen activator exclusively, at great extra financial cost and probably also at considerable medical cost in terms of increased incidence of stroke.

In Europe the general attitude differs; streptokinase is used for more than 80 per cent of cases of acute myocardial infarction, with tissue plasminogen activator reserved for patients who have had streptokinase or APSAC (anistreplase) previously, and so have antibodies to streptokinase, or for patients where the physician believes that more rapid lysis will outweigh the risks of tissue plasminogen activator (e.g. large infarcts in patients who present early).

Our local belief is that the mortality benefit presented in the GUSTO publication is less secure than generally believed, and that the benefit is exaggerated and more likely to amount to about three to four lives saved per thousand patients treated with tissue plasminogen activator compared with streptokinase rather than the published nine to eleven. If GUSTO is analyzed with each arm compared with the other three (as is usual in a four-arm trial), then the p value for the superiority of accelerated tissue plasminogen activator versus the ISIS-3 streptokinase regimen is only 0.04, i.e. the lower confidence interval is very close to zero. Bayesian analysis (which takes all the earlier evidence into account) suggests that the true result is indeed close to zero. However, the excess of cerebral hemorrhage seen with tissue plasminogen activator is highly consistent and statistically significant in all three trials (GUSTO, ISIS-3, and GISSI-2), so that a plausible final mortality reduction with tissue plasminogen activator might be nearer three to five per thousand treated, but with an excess stroke rate of three to four per thousand with only a marginal or zero net clinical benefit. However, this excess stroke rate is closely related to age (over 55 years) and initial systolic blood pressure (over 145 mmHg), presumably due to the presence of Charcot–Bouchard microaneurysms in the brain. In patients either aged over 75 years or with systolic blood pressure over 175 mmHg, the risk of cerebral hemorrhage with tissue plasminogen activator is five times that of streptokinase. Clearly, streptokinase is preferable for these patients. Tissue plasminogen activator may be superior for younger (below 55 years) normotensive patients with little risk of cerebral hemorrhage. However, since such patients are at low risk of death, the cost per life saved with tissue plasminogen activator compared with streptokinase will be very high, if not prohibitive.

Far more important than commercial arguments about which thrombolytic to use is the need to give thrombolytic therapy to more patients and to minimize delays in decision-making and treatment. A 'fast-track' approach where the emergency room or coronary care unit are notified of a likely myocardial infarction patient during transit has been shown to reduce the door-to-needle time dramatically. Similar very worthwhile savings can be achieved by out-of-hospital lysis by motivated local general practitioners, with or without fax or telephone transmission of the ECG to the coronary care unit.

Adjuvant therapy

Antithrombotic treatment

It is now realized that even during lytic therapy, or during spontaneous lysis from endogenous tissue plasminogen activator, fresh thrombus continues to be laid down and platelets are hyperaggregable. This increase in platelet stickiness is accentuated by thrombolytic drugs; this helps to explain the remarkable success of aspirin.

Aspirin

Aspirin 160 to 300 mg, with the first tablet crushed or chewed to speed absorption, has been shown to reduce short-term mortality by about 23 per cent in the ISIS-2 trial. Other data from the Anti-Platelet Trialists' Collaboration has shown a similar additional benefit if aspirin is continued for at least a few years. Unlike thrombolytics, there is no loss of benefit with time; therefore aspirin should be given to all patients with myocardial infarction. This does not mean that aspirin should be delayed, since all fibrinolytics, as well as the myocardial infarction itself, activate platelets.

Newer antiplatelet agents

Both antibodies developed against the platelet/fibrin IIb/IIIa adhesion receptors such as 7E3 (abcibcimax) and the integrelins reduce platelet adhesion by mechanisms which are different from the cyclo-oxygenase inhibition caused by aspirin. Preliminary studies in PTCA, unstable angina, and suspected myocardial infarction are very promising, although these studies are too small to assess any bleeding consequences. However, patients with von Willebrand's disease (who lack IIb/IIIa receptors) do not bleed, and so there is some hope from this approach to myocardial infarction.

Antithrombin agents: heparin or hirudins

Despite the widespread use of heparin in addition to aspirin and thrombolysis, there is no good evidence of mortality reduction with either streptokinase or tissue plasminogen activator. There is some evidence that the addition of heparin significantly increases major bleeding. Indirect comparison of ISIS-2 (where heparin was not used routinely) with GISSI-2 and ISIS-3 (where a dose of 25 000 units/day of subcutaneous heparin was randomly compared with no heparin) and GUSTO-1 (where intravenous heparin was randomly compared with 25 000 units of subcutaneous heparin in patients treated with streptokinase) showed no benefit and some possible harm as the heparin regimen became more aggressive. Initially, small studies with surrogate endpoints such as patency suggested that hirudin was superior to heparin and easier to control. However, three large trials of hirudin versus heparin were stopped prematurely because of unacceptable rates of cerebral hemorrhage. Further studies comparing low-dose hirudin with heparin showed no difference. The results of ongoing trials with an intermediate dose of hirudin are awaited. In keeping with the general thesis that increased patency with more aggressive therapy is accompanied by increased bleeding, these hirudin/heparin trials also used a more aggressive heparin regimen, and this was also associated with higher rates of cerebral hemorrhage.

In summary, the routine use of heparin in addition to aspirin and any thrombolytic is not proved, although there is some modest patency advantage for heparin added to tissue plasminogen activator. In the aspirin era any benefit of heparin is uncertain. In cases at special risk (e.g. large anterior myocardial infarction with mural thrombus, or low cardiac output and/or arrhythmia) the benefit of heparin added to aspirin may outweigh the risks, but this is unproven.

Beta-blockade

Two large trials in the prethrombolytic era demonstrated that intravenous β-blockade in myocardial infarction patients without contraindications reduced mortality, chiefly by preventing cardiac rupture. However, the patients entered in these trials were rather low risk, with a 1-week mortality of about 5 per cent. In addition, the reduction in mortality was only modest at about 15 per cent, and so the use of this treatment today is highly variable, ranging from 2 to 3 per cent of patients in the United Kingdom to 25 to 30 per cent in the United States and Australia. Another reason for the poor uptake of this proven treatment is that a few patients become hypotensive; although this does not seem dangerous (and indeed may be beneficial), it worries resident staff.

Oral β-blockade after myocardial infarction is used more frequently. Again, it cannot be given to all patients (perhaps 60 to 70 per cent), but it is well proven to reduce, in particular, sudden death and reinfarction.

Angiotensin-converting enzyme inhibition

This is the latest proven treatment in acute myocardial infarction. Earlier trials in heart failure patients had shown reduction in both mortality and progression of failure, with a probable reduction in reinfarction. This was generally thought to be related to the prevention of left ventricular dilatation and hypertrophy ('remodeling'). Later trials, such as SAVE, AIRE, and TRACE, tested captopril, ramipril, and trandolapril in patients with left ventricular dysfunction randomized 2 to 11 days after myocardial infarction. Clear benefit for angiotensin-converting enzyme inhibition was seen in each trial over the following months or years.

More recently, several large trials tested earlier angiotensin-converting enzyme inhibition in an unselected group of patients, where the treatment began on the first day of the myocardial infarction, once the patient was hemodynamically stable after treatment with thrombolysis and/or β-blockade. Angiotensin-converting enzyme inhibition, titrated carefully in patients with stable hemodynamics, was safe; mortality was reduced by about 5 per cent at 30 days by captopril or lisinopril. The reduction was proportionally similar in all subgroups of patients, and so the absolute benefit was related to the risk profile of the patient. In some patients at very low risk (e.g. after a small inferior myocardial infarction in younger subjects with no other risk factors for mortality), the side-effects, largely reversible hypotension, may outweigh the small benefit. However, it is clear from these trials that many people in Killip class I did receive benefit, and so it is reasonable to extend the indications for angiotensin-converting enzyme inhibition in myocardial infarction beyond the selected minority of patients with left ventricular dysfunction who were entered into the SAVE, AIRE, and TRACE studies. This is reinforced by the finding in GISSI-3 that most of the benefit was seen in the first seven days (even on days zero and one), and was due to a substantial reduction in death due to cardiac rupture or arrhythmia.

Nitrates

Although nitrates are widely used, the ISIS-4 and GISSI-3 studies showed that their routine use was safe but did not reduce mortality significantly. Their use should be restricted to patients with symptoms of continuing pain or left ventricular failure.

Magnesium

Despite the impressive, but marginally significant, reduction in mortality with intravenous magnesium which was seen in the LIMIT and earlier small trials, the 60 000-patient ISIS-4 trial showed a non-significant trend to harm. There is still controversy over these discrepant results, with some alleging that the magnesium in ISIS-4 should have been started earlier and before thrombolysis in those patients (approximately 90 per cent) who received non-trial thrombolytics. A new United States trial (MAGIC) will attempt to resolve this. Meanwhile, magnesium cannot at present be recommended as routine therapy for myocardial infarction. There were fewer arrhythmias in ISIS-4 in those patients allocated magnesium, but this expected benefit was offset by more shock and heart failure.

Treatment of complications

Heart failure

Heart failure occurs in about 20 to 30 per cent of all acute myocardial infarction and is associated with a much poorer long-term outlook. As indicated above, it is useful to use intravenous nitrates initially to reduce afterload. However, these rapidly become less effective because of tolerance. Angiotensin-converting enzyme inhibitors should be started cautiously, but only in patients with a systolic blood pressure above 100 mmHg. Diuretics, particularly intravenous furosemide (frusemide) or bumetanide are commonly given, but the potassium level and intravascular volume depletion should be watched carefully and avoided.

An urgent echocardiogram is essential in all cases to exclude surgical causes such as papillary muscle or chordal rupture, cardiac rupture, or cardiac tamponade. Even in the absence of these complications, an echocardiogram will reveal the extent of infarction.

A dobutamine stress echocardiogram may reveal the presence of viable but ischemic myocardium (hibernating) and indicate the need for urgent catheterization and reperfusion by PTCA or surgery. Other treatable causes of heart failure such as anemia or fluid overload should also be considered carefully before treatment which is only symptomatic is given.

Arrhythmia

Arrhythmia is common in acute myocardial infarction. Early randomized trials showed the benefit of routine intravenous lidocaine (lignocaine) in the prevention of ventricular fibrillation. However, a later meta-analysis showed an excess mortality in those allocated lidocaine, owing to an excess of asystole! Lidocaine is now reserved for symptomatic sustained ventricular tachycardia or resuscitated ventricular fibrillation. At the same time an alternative antiarrhythmic should be considered. The safest and best proven is a cardioselective β-blocker such as atenolol or metoprolol. Since this is not always effective or tolerated, many physicians go immediately to intravenous amiodarone, about 600 mg/day, followed by oral amiodarone, reducing rapidly to 400 mg orally or less. Once the first few days are over this can usually be stopped, unless ventricular tachycardia or ventricular fibrillation recurs. In this case the drug is continued orally at the lowest effective dose, often 100 to 200 mg/day, since side-effects are common (photosensitivity, bradycardia, thyroid dysfunction) or rare but often fatal (liver and lung toxicity).

Recent randomized trials (GEMICA, CAMIAT, EMIAT) have addressed the question of routine prophylaxis with amiodarone. Unfortunately, although there is clearly some benefit, this is almost outweighed by toxicity, and so it cannot be recommended in low-risk individuals. In addition, these trials were not mutually consistent; it is probable that the doses of amiodarone were too high and continued for too long. Nevertheless, clinical and anecdotal experience with the drug (at the lower doses given above) is reassuring. The drug is presently the most effective for established life-threatening arrhythmia, and it is remarkably well tolerated for short-term use in the coronary care unit, even in patients with overt left ventricular dysfunction.

Other antiarrhythmic agents, for example flecainide, encainide, sotalol (a β-blocker with class IV long-QT action) have generally been discarded because of adverse trial data in ischemic patients.

Cardiogenic shock

This condition, which is defined by hypotension, poor peripheral circulation, and oliguria, should be distinguished from the autonomic reflex vagal overactivity described above. True cardiogenic shock carries a high mortality (60–80 per cent) and should be investigated vigorously (echocardiogram) to exclude a surgical cause. If none is found, urgent revascularization should be considered. This is obviously only available in tertiary centers. In the usual coronary care unit thrombolysis should be considered, despite the often stated (but mistaken) belief that thrombolysis is ineffective. Trials show that thrombolysis reduces absolute mortality substantially, particularly with streptokinase (which because of its own hypotensive action in patients with normal blood pressure is often mistakenly withheld).

Surgical causes include mitral regurgitation due to chordal rupture, or cardiac rupture from the left to the right ventricle, or the left ventricle to the pericardial sac. Urgent intervention can be life saving; with tamponade, and hemopericardium seen on the echocardiogram, urgent pericardiocentesis is worth trying.

Continued pain or heart failure

When the symptoms of myocardial infarction fail to settle, or recur, it is important to consider urgent investigation. This usually involves cardiac catheterization followed by surgery or angioplasty, and therefore it is necessary to transfer the patient to a tertiary center. However, an increasing number of community hospitals are developing their own catheter laboratories without surgical back-up. This is a controversial issue, but in the author's opinion the benefits of local provision for PTCA and stenting will far outweigh the theoretical advantage of surgical cover, the need for which has declined rapidly since the availability of 'bail-out' stenting has increased.

Conclusions

The modern treatment of acute myocardial infarction is truly an area of critical care. Decisions have to be taken rapidly, usually by emergency room staff who are not cardiologists. The most important procedure is to have proper organization, training, and regular routine audit to ensure that policies are appropriate and are being properly implemented.

A fast-track triage, which begins in the ambulance, is best. It should be possible to assess the patient, carry out an ECG (and ideally also an echocardiogram), and treat the patient within 20 to 30 min. Time is muscle! Teamwork is essential, so that where complications are present or the diagnosis uncertain the cardiologist should be involved urgently. For most patients with myocardial infarction, contraindications to thrombolysis are relative. In the United Kingdom, the uptake of these proven treatments is patchy, although better than in many other countries, and could be improved with better organization, particularly in the emergency room and admissions area. Aspirin should be given to all as soon as possible, even at home, with the first dose chewed.

Finally, it is vital to pursue secondary prevention vigorously with proven treatments of aspirin, β-blockers, angiotensin-converting enzyme inhibitors where left ventricular dysfunction persists, and lipid lowering with diet and (probably) statins. Other lifestyle changes are important: quitting smoking, exercise, dietary sources of antioxidants, and moderate regular alcohol. Blood pressure lowering is important in hypertensive patients, but β-blockers and/or angiotensin-converting enzyme inhibitors will take care of the majority.

Bibliography

Antiplatelet Trialists' Collaboration (1994). Collaborative overview of randomised trials of antiplatelet therapy. II. Maintenance of vascular graft or arterial patency by antiplatelet therapy. *British Medical Journal*, **30**, 159–68.

Collins, R., Peto, R., Baigent, C., and Sleight, P. (1997). Aspirin, heparin and fibrinolytic therapy in suspected acute myocardial infarction. *New England Journal of Medicine*, **336**, 847–60.

Every, N.R., Parsons, L.S., Hlatky, M., Martin, J.S., and Weaver, M.D., for the Myocardial Infarction Triage and Intervention Investigators (1996). A comparison of thrombolytic therapy with primary coronary angioplasty for acute myocardial infarction. *New England Journal of Medicine*, **335**, 1253–60.

GUSTO Investigators (1993). An international randomized trial comparing four thrombolytic strategies for acute myocardial infarction. *New England Journal of Medicine*, **32**, 673–82.

3.4 Aortic dissection

3.4.1 Pathophysiology

J. Salmon

Key messages

- Dissection is most common in hypertensive males in their sixth and seventh decades.
- Nearly two-thirds of dissections arise in the ascending aorta.
- Without surgery, 40 per cent of patients with ascending aortic dissections will die within 24 h and only 30 per cent will survive for 1 week.

Introduction

Dissection of the aorta occurs when blood enters the aortic media and dissects the inner from the outer layers. There is usually a discrete proximal intimal tear, and there may be one or more re-entry tears distally. Dissections are regarded as acute during the first 14 days.

Anatomical considerations

The aortic media is composed of a series of concentric elastic lamellar plates, separated by smooth muscle and connective tissue. During development, in response to radial (compressive) and circumferential (tensile) stresses which are maximal at the intima and weakest at the adventitia, the inner lamellas thicken and smooth muscle cells develop more cell–cell contacts. Because of this structural modification, most of the stress-resistant properties are found in the inner two-thirds of the aortic media. There is a clear change at the junction between the middle and outer thirds of the media; this represents an area of weakness where tearing might be expected with exposure to high levels of stress and coincides with the observed plane of most aortic dissections.

The intimal tear is found most frequently approximately 4 cm above the aortic valve in the ascending aorta or just distal to the subclavian artery. The dissection usually propagates distally, but proximal extension from the ascending aorta may involve the aortic valve annulus, and tears distal to the subclavian artery may propagate in both directions to involve the aortic arch and ascending aorta as well as the descending aorta. The false channel usually extends 50 to 67 per cent around the aortic circumference, rarely involving the whole aortic wall. The outer wall of the false channel is thin and may rupture, particularly into the pericardium where the ascending aorta is unsupported, the pleural space (usually the left), or the abdomen. In the absence of frank rupture, blood may extravasate to form mediastinal or pericardial hematomas. In patients who survive the acute event, the false channel tends to become aneurysmal and may rupture subsequently. Occasionally, the channel may become thrombosed spontaneously. In a minority of patients there is a discrete re-entry tear so that the false channel produces a parallel aortic circulation, albeit not in communication with the aortic branch arteries.

Although an intimal tear is a universal finding in autopsy series, it may be difficult to identify clinically. The events leading to the intimal tear are poorly understood. It is uncertain whether the intimal tear is the primary event, or whether dissection is initiated by bleeding into the media from the vasa vasorum and the intimal tear follows. The major predisposing factors to dissection appear to be the level of shear stress to which the aorta is subjected and, in a proportion of cases, medial degeneration and structural abnormalities of the aortic wall.

With advancing age, elastin fragmentation, fibrosis, and medial necrosis lead to degeneration of the aortic lamellas. These changes are most pronounced in the ascending aorta and become less prominent as the aorta descends into the abdomen. These changes occur as part of the normal aging process but are more prominent in hypertensive patients. Pre-existing dilation of the aorta (as occurs in Marfan's syndrome) may also be relevant. Laplace's law relates wall stress to pressure and diameter: as the aorta dilates, so wall stress increases. Medial degeneration may result from structural abnormalities (e.g. Marfan's and Ehlers–Danlos syndromes) in which dissection may occur without pre-existing hypertension or, possibly, from ischemic necrosis due to occlusion of the vasa vasorum.

A large proportion of dissections take place in aortas without evidence of excessive degeneration. This is true particularly in patients where dissection follows instrumentation of the aorta. It appears that, whereas abnormalities of the aortic wall may make dissection more likely, once the intima and underlying media are damaged, dissections propagate rapidly due to entry of the blood into the media under pressure.

Predisposing causes

The majority of patients with dissection are hypertensive males between the ages of 40 and 70, usually in their sixth or seventh decades. It is more common in the black population than in Caucasians and is relatively rare in orientals. It is more common in Marfan's syndrome and occurs with increased frequency in Ehlers–Danlos, Turner's, and Noonan's syndromes and in patients with bicuspid aortic valves. Such patients tend to develop dissection at a younger age and frequently have no history of arterial hypertension. Rarer associations include pregnancy (usually in the third trimester),

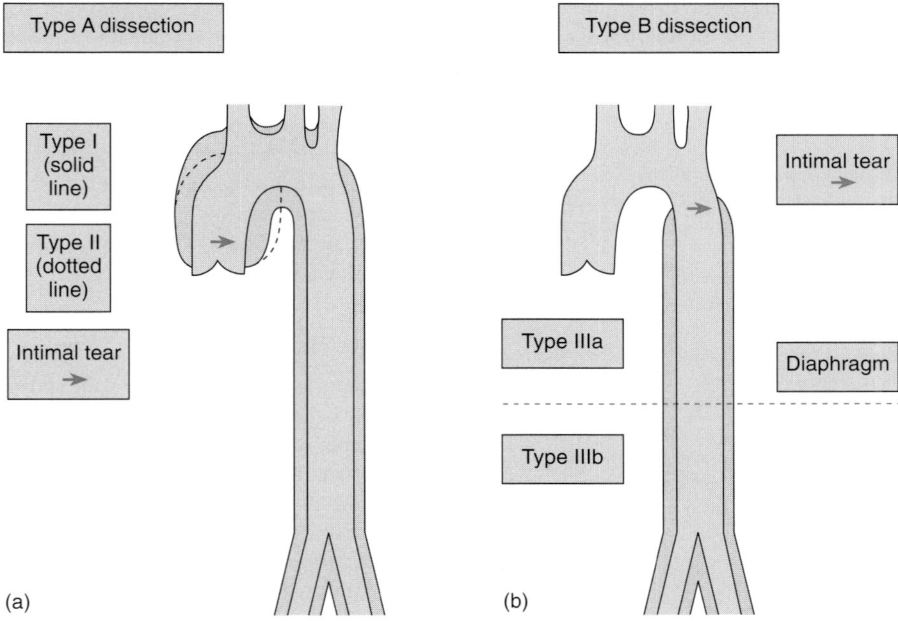

Fig. 1 Classification of aortic dissection.

vasculitis, and aortic coarctation. Dissection may occasionally follow trauma, aortic instrumentation, cocaine abuse, or aortotomy. Syphilis and atheroma lead to saccular aneurysms rather than dissection.

Site and classification of dissection

Sixty to sixty-five per cent of dissections originate in the ascending aorta, 5 to 10 per cent in the aortic arch, and 30 to 35 per cent in the first part of the descending aorta. Subsequent dissection usually occurs distally but may also extend proximally. The original De Bakey classification divides dissection into three types based on the extent of aortic involvement (not the site of the intimal tear, which may be difficult to identify) (Fig. 1).

I. This involves both the ascending and descending aorta and usually extends into the abdomen. Ten to fifteen per cent of such patients have an intimal tear distal to the left subclavian artery; in the remainder the origin is in the proximal aorta.

II. This is the least common type and involves the ascending aorta only, stopping just proximal to the innominate artery. This type of dissection is seen in Marfan's syndrome and may occasionally be found incidentally in operations for ascending aortic aneurysms.

III. This originates distal to the left subclavian artery and extends downwards. Type IIIa dissections are confined to the descending thoracic aorta only, whereas the more common type IIIb dissection extends into the abdominal aorta and may involve the iliac arteries. The majority of these aneurysms have an intimal tear just distal to the origin of the left subclavian artery. Type III dissections tend to occur in older patients.

The alternative Shumway (Stamford) classification divides dissection into two types only. This classification is less descriptive but is well suited to clinical decision-making (Fig. 1).

Type A. Involvement of the ascending aorta (De Bakey types I and II).

Type B. No involvement of the ascending aorta (De Bakey type III).

In autopsy series, type A dissections predominate in a ratio of approximately 2 to 1. In most clinical series, types A and B appear almost equally common, probably because type A dissections are more rapidly fatal and may pass unrecognized.

Consequences of dissection

The acute sequelae of dissection include aortic rupture (usually into the pericardium or pleura), disruption of the aortic valve apparatus leading to acute aortic regurgitation and left ventricular failure, and extension of the dissection into the branch walls of arteries originating from the aorta, leading to distal ischemia or infarction. This is more common in large arteries such as the innominate, subclavian, carotid, or renal arteries, but any artery may be affected. Involvement of intercostal arteries in descending aortic dissections may prejudice blood flow to the anterior spinal artery. Distal extension of type IIIb dissection involves the left iliac artery more commonly than the right. Involvement of branch arteries may lead to ischemia or infarction of the territory supplied.

Natural history following dissection

Without surgery, approximately 40 per cent of ascending aortic dissections are fatal within 24 h and only 30 per cent survive 1 week; 86 per cent of these deaths are due to aortic rupture. Ten per cent of patients will survive for 3 months. Survival is higher in dissections limited to the descending aorta: 10 to 20 per cent will die within the first 24 to 48 h, 75 per cent will survive for 1 month, and over 60 per cent will survive for 1 year. Descending aortic dissection tends to occur in an older age group, and much of the mortality is not directly related to the dissection.

Bibliography

Berry, C.L., Sosa-Melgarejo, J.A., and Greenwald, S.E. (1993). The relationship between wall tension, lamellar thickness, and intercellular junctions in the fetal and adult aorta: its relevance to the pathology of dissecting aneurysm. *Journal of Pathology*, **169**, 15–20.

Daily, P.O., Trueblood, H.W., Stinson, E.B., Wuerflein, R.D., and Shumway, N.E. (1970). Management of acute aortic dissection. *Annals of Thoracic Surgery*, **10**, 237–47.

De Bakey, M.E., Henly, W.S., Cooley, D.A., Morris, G.C., Crawford, E.S., and Beall, A.C. (1965). Surgical management of dissecting aneurysms of the aorta. *Journal of Thoracic and Cardiovascular Surgery*, **49**, 130–49.

Larson, E.W. and Edwards, W.D. (1984). Risk factors for aortic dissection: a necropsy study of 161 cases. *American Journal of Cardiology*, **53**, 849–55.

Leonard, J.C. and Hasleton, P.S. (1979). Dissecting aortic aneurysms: a clinicopathological study. *Quarterly Journal of Medicine*, **189**, 55–76.

3.4.2 Diagnosis

J. Salmon and L. J. Jarvis

Key messages

- Consider dissection in the differential diagnosis of all acute chest pain.
- Unequal pulses are only present in a minority of cases.
- CT scanning is the most convenient initial investigation.
- Transesophageal echocardiography is useful for refining the diagnosis of acute ascending aortic dissection. This can be performed at the bedside or in the operating room.
- Magnetic resonance imaging is the investigation of choice in descending aortic dissection.

Presentation and diagnosis

The clinical features are composed of pain due to the dissection itself, symptoms and signs of ischemia or infarction in arterial territories involved in the dissection, aortic valve involvement, and leak of extravasated blood into other structures, typically the pleura or pericardium, producing effusions and tamponade. The features are highly variable and may mimic a number of other conditions.

Although painless dissections do occur, dissection characteristically produces a severe tearing chest pain of abrupt onset which reaches an immediate crescendo and may be very poorly localized. The initial pain may mimic myocardial infarction but frequently extends beyond the usual boundaries of cardiac pain, often penetrating through to the back or radiating into the abdomen, particularly as the dissection progresses. Typically, dissections involving the ascending aorta produce anterior chest pain, while interscapular pain usually implies involvement of the descending aorta. Patients can frequently pinpoint the exact moment that it developed and sometimes relate it to a sudden movement or physical strain. Extension of the dissection may lead to remote ischemic pain or loss of function which may dominate the clinical picture. Severe aortic regurgitation may cause acute left ventricular failure. Patients may also present with collapse, loss of consciousness, or stroke if the dissection involves the innominate or left common carotid arteries. Similarly, separation of intercostal arteries from the aortic lumen may lead to paraplegia. Dissection of either coronary artery may cause sudden death.

Clinical findings

The patient is usually in severe pain and may appear shocked, but there may be no other obvious physical abnormality. Shock may be a consequence of hypovolemia, cardiac tamponade, or aortic regurgitation. Apparent 'shock', in which the patient is cold and clammy but has a normal or raised blood pressure, is described frequently. Absent or unequal pulses are said to be characteristic of the condition but are only found in a minority; approximately 25 per cent will have reduced pulses in one or both arms, and a similar percentage will have abnormal leg pulses. The right arm and the left leg are involved more commonly. Aortic regurgitation develops in over 50 per cent of patients with dissections of the ascending aorta but is initially severe only in a minority. Involvement of other major vessels may produce symptoms or signs suggestive of ischemia in the relevant arterial territory. Other findings may be of pleural or pericardial fluid due to leakage of blood, sometimes with cardiac tamponade. This may develop suddenly and catastrophically but, more commonly, is of gradual onset. Aortic rupture in type A dissections is almost invariably preceded by the development of a pericardial effusion. Patients presenting late frequently have a fever due to large quantities of extravasated blood; if there is associated aortic regurgitation, it is possible to make a mistaken diagnosis of endocarditis.

Dissection should be considered in the differential diagnosis of any patient with severe chest or upper abdominal pain. It is particularly important to consider the diagnosis in the initial assessment of patients with apparent myocardial infarction before giving thrombolytics. Important differential diagnoses include ischemic cardiac pain, pancreatitis, esophageal pain, and massive pulmonary embolus.

Ischemic cardiac pain

The pain of dissection is usually of very abrupt onset and frequently spreads beyond the usual borders of ischemic cardiac pain, but distinction between the two can be extremely difficult. In myocardial infarction there may be premonitory anginal symptoms; these are not a feature of dissection. Dissection should be considered in all patients with 'anginal' pain where the distribution of pain is atypical, if there is evidence of aortic valvar regurgitation, and particularly where there are no acute ECG changes and cardiac enzymes are normal. The presence of ECG and enzyme changes does not exclude dissection; myocardial ischemia or infarction may result from involvement of the coronary ostia in the dissection or extension of the dissection into the coronary arteries.

Pancreatitis

Dissection usually presents with chest pain in a known hypertensive with subsequent radiation into the upper abdomen and vomiting, whereas abdominal symptoms are usually the presenting feature in acute pancreatitis. Amylase may rise secondary to ischemia but usually to lower levels than seen in acute pancreatitis. If diagnostic doubt remains, CT scanning of the aorta and pancreas is probably the best method of distinguishing the two.

Esophageal pain

Esophageal spasm may be of similarly acute onset. However, the pain is usually more localized, less severe, and more remittent, and the patient appears less unwell.

Massive pulmonary embolus

This may present with poorly localized chest pain of sudden onset accompanied by shock. Associated clinical and ECG features will usually allow differentiation from acute dissection. If there is diagnostic doubt, contrast-enhanced CT scanning (particularly spiral CT) is appropriate.

Investigations

The ECG may be normal or may show ischemia or infarction in any territory owing to involvement of the coronary ostia. The chest radiograph shows mediastinal widening and enlargement of the aortic contour in up to 70 per cent of cases (Fig. 1); however, this may be difficult to appreciate on a portable film. Tracheal and esophageal displacement are common, and inward displacement of a calcified aortic rim more than 6 mm from the aortic border is a valuable feature which is found in 17 per cent of cases. Fluid in the pleural space, usually the left, may be apical or basal or may contribute to mediastinal widening. Although chest radiographs are now performed with decreasing frequency before thrombolysis, a film should be taken if there is any diagnostic doubt or if there are any features suggestive of dissection. However, dissection can occur when the chest radiograph is entirely normal.

CT, aortography, transesophageal echocardiography, and magnetic resonance imaging (MRI) have now evolved with complementary roles (Goldman *et al.* 1986; Erbel *et al.* 1990).

Echocardiography

Transthoracic echocardiography may show dissection of the aortic root, aortic regurgitation, and the presence of blood in the pericardium. Other findings supportive of a diagnosis of ascending aortic dissection include dilation or thickening of the aortic root and an absence of left ventricular dysfunction or regional wall motion abnormalities. Its major advantages are speed, non-invasion, and non-interference with treatment and monitoring. It has no value in the assessment of descending aortic dissections.

Transesophageal echocardiography is becoming the investigation of choice in unstable patients (Treasure and Raphael 1991) as it can be carried out at the bedside or in the operating theatre. It will usually allow accurate assessment of the origin of the dissection, its distal extent, and the state of the aortic valve.

CT scanning

Dynamic contrast-enhanced CT scanning is the preferred first investigation in most community hospitals. Diagnosis is much enhanced when spiral CT is available. The sensitivity ranges between 83 and 100 per cent and the specificity between 90 and 100 per cent for identification of dissection (Cigarroa *et al.* 1993). CT is useful for demonstrating the extent of dissection and the presence of hemopericardium, but can be prone to artefacts. It may be difficult to demonstrate the intimal flap or tear, limiting its value in planning surgery. In 70 per cent of cases it is possible to demonstrate contrast filling both true and false lumens separated by an intimal flap (Fig. 2). Other features suggesting dissection include differing opacification of the true and false lumens, compression of the true lumen by thrombosis in the false lumen, inward displacement of intimal calcification, and irregularity of the contrast-enhanced lumen. Blood may be seen in the aortic wall, peri-aortic tissues, or pericardium as high-attenuation material on unenhanced scans, but these features

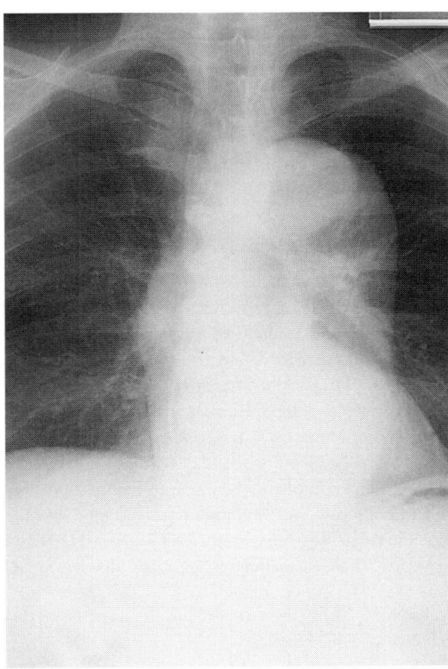

Fig. 1 Characteristic chest radiograph appearance of aortic dissection demonstrating mediastinal widening, aortic unfolding, and tracheal displacement.

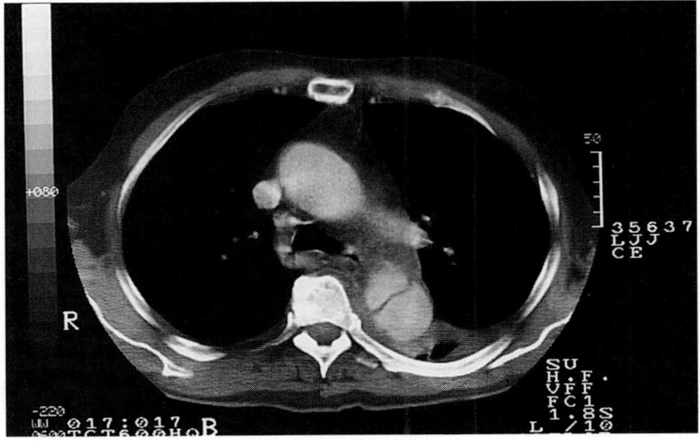

Fig. 2 Axial enhanced CT image in the same subject as Fig. 1, demonstrating the true and false lumens separated by the intimal tear.

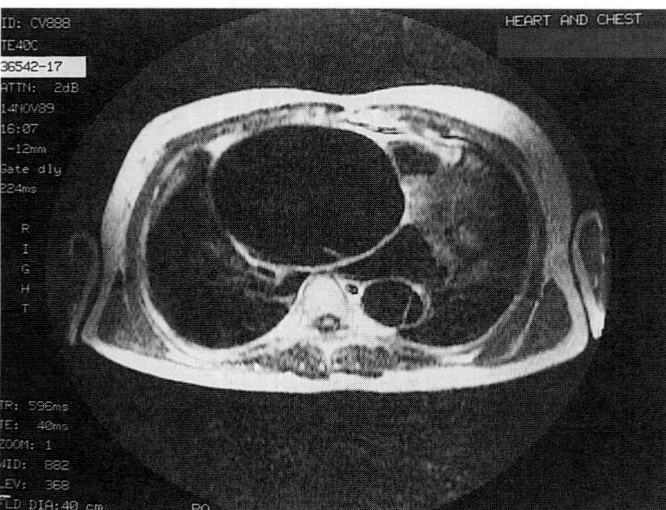

Fig. 3 Axial MRI of aortic arch dissection at a similar level but in a different patient to Fig. 2, demonstrating similar features. (Courtesy of The Royal Brompton Hospital, London, UK.)

may also be seen in aortic rupture. If there is strong clinical suspicion despite a negative CT scan, further investigation such as aortography is required.

Magnetic resonance imaging

MRI has the advantage that images can be obtained in any plane and intravenous contrast may not be required. It is difficult to image and monitor very sick patients in conventional closed-bore MRI scanners; however, the new designs of open scanner are considerably better suited. MRI shows the extent of dissection and allows visualization of the arch and other vessel involvement (Fig. 3). It will also show aortic regurgitation and rupture into the pericardium. MRI can be adapted to the specific clinical requirements, allowing detailed anatomical imaging with cardiac gated spin echo and gradient echo sequences, and directional blood flow assessment with phase contrast angiography and velocity mapping. As magnetic resonance technology advances, faster imaging techniques will improve its usefulness.

Aortography

Although largely superseded by transesophageal echocardiography and MRI, aortography remains useful in defining the proximal anatomy of a dissection found on CT scanning and should be considered if clinical suspicion of dissection is high but other investigations have been equivocal. It may require the use of upper limb arterial access to allow entry to the aorta. There is a risk of rupture of the weakened aorta during contrast injection. Although catheterization via the femoral artery will usually enter the true lumen, confirmation of catheter placement in the true lumen requires retrograde passage through the aortic valve. Aortography requires an experienced angiographer and the ability to obtain multiple angiographic projections. It is valuable in identifying involvement of branch vessels, demonstrating the intimal flap and unusual manifestations of dissection, for example rupture into the main pulmonary artery, the inferior vena cava, or a cardiac chamber. The most common cause of diagnostic error occurs when a tear has sealed over thrombosis of the false channel.

It is traditional practice to combine aortography with coronary arteriography. This is a procedure of questionable value. The procedure is technically difficult in a dissected aorta, is time consuming, and has a high false-negative rate. The vast majority of deaths following dissection of the ascending aorta are the result of aortic rupture, not coronary artery disease. The rationale for the procedure is that it allows assessment of the coronary arteries in a high-risk population who are about to undergo major cardiovascular surgery. Conversely, the diagnostic yield is low, and the procedure carries inherent risks and imposes extra delays before surgery during which time aortic rupture may occur.

Choice of investigation

In acute ascending aortic dissection, survival appears to be improved by rapid diagnosis followed as soon as possible by surgery (Cigarroa *et al.* 1991; Petasnick 1991). In most hospitals diagnosis can be made most rapidly using non-invasive methods. CT scanning or transesophageal echocardiography are usually the most convenient. If available, transesophageal echocardiography is the method of choice. Aortography should be used if it is more readily available, more reliable locally, or if the results of non-invasive investigations are equivocal. There is an increasing body of opinion that once the diagnosis of ascending aortic dissection has been established by any means, attempting to refine the diagnosis by further investigation should be avoided. The priority should be to transfer the patient to the operating room as soon as possible where, if necessary, further evaluation can be performed using transesophageal echocardiography.

MRI is probably the investigation of choice in stable patients with chronic dissections or in patients presenting acutely with no evidence of ascending aortic involvement who require further evaluation of suspected type III dissection.

References

Cigarroa, J.E., Issebacher, E.M., DeSanctis, R.W., and Eagle, K.A. (1993). Diagnostic imaging in the evaluation of suspected aortic dissection. *New England Journal of Medicine*, **328**, 35–43.

Erbel, R., *et al.* (1990). Diagnostic strategies in suspected aortic dissection: comparison of computed tomography, aortography and transesophageal echocardiography. *American Journal of Cardiac Imaging*, **4**, 157–72.

Goldman, A.P., Kotler, M.N., Scanlon, M.H., Ostrum, B., Parameswaran, R., and Parry, W.R. (1986). The complementary role of magnetic resonance imaging, Doppler echocardiography and computed tomography in the diagnosis of dissecting thoracic aneurysms. *American Heart Journal*, **111**, 970–81.

Petasnick, R. (1991). Radiologic evaluation of aortic dissection. *Radiology*, **180**, 297–305.

Treasure, T. and Raphael, M.J. (1991). Investigation of suspected dissection of the thoracic aorta. *Lancet*, **338**, 490–5.

3.4.3 Management

J. Salmon

Key messages

- Blood pressure should be reduced with both β-blockers and vasodilators, and not with vasodilators alone.
- Ascending aortic dissections should receive surgery as soon as possible.
- Descending aortic dissections should be managed medically provided that there are no complications.
- Lifelong follow-up and blood pressure control are essential in all dissections.

Introduction

Initial stabilization of all acute dissections is identical. Thereafter, type A dissections should be taken for surgery as rapidly as possible while type B dissections should be managed medically unless complications develop. Routine surgery in all patients with type B dissection has not yet been shown conclusively to improve overall mortality except in those with dissections secondary to Marfan's syndrome. Otherwise, surgery is indicated urgently in the presence of distal ischemia or potentially reversible complications. The best treatment for dissections originating in the aortic arch remains controversial. Surgery is difficult and is usually carried out under conditions of circulatory arrest and deep hypothermia; some authorities advocate initial medical treatment unless complications develop. Chronic dissections with evidence of developing complications should be treated in the same way as acute dissections; those with no complications may be treated less aggressively but require urgent control of blood pressure and close observation subsequently.

Suspected dissection

The immediate priorities are pain relief, control of blood pressure, and volume resuscitation, followed by appropriate imaging. Pain relief will require intravenous opiates, sometimes in high doses. The blood pressure should be reduced to normal or below; a target systolic blood pressure of 100 mmHg is usual, but higher pressures may have to be accepted if organ perfusion and urine output are inadequate. A combination of vasodilators and β-blockers should be used, with the aim of reducing not only blood pressure but also shear stress by decreasing dP/dt. This requires close monitoring of circulatory adequacy. Sodium nitroprusside by continuous infusion offers the most flexible method of blood pressure control. Beta-blockade is most conveniently achieved with infusions of labetalol or esmolol.

Definitive imaging should be performed as soon as possible. Close monitoring, blood pressure control, and analgesia must be maintained throughout. An experienced clinician, appropriately equipped, should accompany the patient at all times.

Proved dissection

If the patient presents to a community hospital without cardiac surgical facilities, arrangements should be made for emergency transfer to an appropriate unit. All patients with dissection should be moved to a cardiothoracic unit as soon as possible following initial stabilization, type A dissections to receive emergency surgery and type B dissections so that they are well placed to receive surgery should they deteriorate. The mode of transfer is not particularly important, but the patient should be monitored adequately, blood pressure control should be maintained throughout, appropriate staff and equipment should be carried, and sudden acceleration and deceleration should be avoided.

Medical management

Patients with type B dissections or chronic dissections without evidence of complications should be weaned gradually towards oral antihypertensive therapy; a combination of a β-blocker and a vasodilator such as nifedipine is appropriate. All patients (whether hypertensive or not) should receive β-adrenergic antagonists unless there are contraindications, in which case a calcium-channel antagonist such as diltiazem should be considered. The aim is to establish the patient on therapy with a low-normal resting blood pressure without postural symptoms or evidence of organ hypoperfusion. During the initial phases of medical treatment and during weaning the patient should be observed closely for evidence of developing complications.

Those in whom initial medical management is successful should be followed closely as many will require surgery in the ensuing months or years. Some will develop extension of the original dissection, while others may develop aneurysms of the thin-walled false channel which may rupture subsequently if the aorta is not repaired. Patients with chronic dissections have survived the period of greatest danger and surgery should be withheld unless complications develop.

Failure of medical management

Medical management should be considered to have failed if there is evidence of developing distal ischemia, organ failure, failure to maintain an adequate circulation, or aneurysm formation. Neurological dysfunction is a special case; paraplegia is usually permanent and so operation is not justified, whereas paraparesis may improve following operation. Stroke is not an absolute contraindication to operation; a small proportion of patients with hemiparesis or hemiplegia will improve significantly postoperatively.

Postoperative management

The postoperative management of dissection differs little from that of any postoperative cardiac surgical patient. In the majority of cases surgery is palliative; the intimal tear is repaired but the false channel remains. There is potential for further dissection and blood pressure control is particularly important (as in medical management). Maintenance of an adequate circulation can be difficult to achieve in patients with coexistent or consequent cardiac dysfunction. Traditionally, it has been taught that dissection is an absolute contraindication to the use of intra-aortic balloon pumps; however, there are case reports of their having been used *in extremis* to salvage patients who have had technically successful operations but who cannot maintain an adequate circulation postoperatively.

Follow-up

All patients should be followed up closely following dissection with regular CT scans and chest radiographs. Developing aneurysms should be treated in the normal way. Lifelong blood-pressure control is essential.

Dissection in pregnancy

Dissection is uncommon in pregnancy except in those with Marfan's syndrome, particularly those with aortic root dilation and progressive aortic regurgitation who may develop dissection at any stage of pregnancy. Those with no aortic regurgitation and an aortic root diameter below 4.5 cm have a better prognosis. Termination should be considered in the first trimester if there is evidence of progressive aortic root dilatation. Beta-blockade may reduce the risk of dissection and aortic root dilation but experience is limited.

Patients without Marfan's syndrome are most prone to dissection in the third trimester and the peripartum period. Diagnosis is best made by transesophageal echocardiography; treatment is by emergency blood pressure control with careful fetal monitoring and early surgery. An emergency Cesarean section will usually be necessary; however, the maternal and fetal mortalities remain very high.

Outcome

Most patients with dissection have another pathology which will contribute to a reduction in life expectancy. The short-term in-hospital mortality for type A dissections treated rapidly by surgery has fallen from 20 to 40 per cent to below 10 per cent in several series. The in-hospital mortality for type B dissections treated medically is approximately 20 per cent. The initial mortality for patients with type B dissections treated surgically is higher, largely because most such patients do not receive surgery unless severe complications develop. Although only 50 per cent will survive for a year, the excess mortality is almost all related to surgery; the slope of the survival curve for those leaving hospital is similar to that for those managed medically. Longer-term outcome is less clear; overall 10-year actuarial survival for all dissections is approximately 40 per cent, but this will probably improve as the results from recent series are analyzed. Operations for subsequent complications will be required in up to 23 per cent of patients who leave hospital. These include aneurysm formation and the development of severe aortic regurgitation. Of those who die subsequently, in nearly 30 per cent the cause will be development and rupture of an aneurysm, emphasizing the need for long-term follow-up of these patients.

Bibliography

Crawford, E.S., Svensson, L.G., Coselli, J.S., Safi, H.J., and Hess, K.R. (1988). Aortic dissection and dissecting aortic aneurysms. *Annals of Surgery*, **208**, 254–73.

DeSanctis, R.W., Doroghazi, R.M., Austen, W.G., and Buckley, M.J. (1987). Aortic dissection. *New England Journal of Medicine*, **317**, 1060–7.

Elefteriades, J.A., *et al.* (1992). Long-term experience with descending aortic dissection: the complication specific approach. *Annals of Thoracic Surgery*, **53**, 11–21.

Rizzo, R.J., *et al.* (1994). Rapid non-invasive diagnosis and surgical repair of acute ascending aortic dissection. *Journal of Thoracic and Cardiovascular Surgery*, **108**, 567–75.

Wheat, M.W., Harris, P.D., Malm, J.R., Kaiser, G., Bowman, F.O., and Palmer, R.F. (1969). Acute dissecting aneurysms of the aorta: treatment and results in 64 patients. *Journal of Thoracic and Cardiovascular Surgery*, **58**, 344–51.

3.5 The hypotensive patient

3.5.1 Pathophysiology of hypotension

A. Sielenkämper and W. J. Sibbald

Key messages

- Neural and humoral mechanisms keep the blood pressure fairly constant over a broad range of pressures. However, below a critical value (e.g. 50–60 mmHg for the brain), blood flow and oxygen delivery become dependent on perfusion pressures.
- The term hypotension describes a decrease in blood pressure which threatens organ perfusion and, eventually, oxygen delivery to the peripheral tissues.
- Decreased tissue oxygen availability may be followed by ischemia, cell death, and organ failure. It may also lead to 'priming' of sublethal injured cells, with release of second messengers and the possible consequence of a generalized inflammatory state following reperfusion of injured tissue with therapy.
- Consequences of hypotension for the 'vital' organ systems include cerebral and myocardial ischemia, particularly in the presence of coronary artery or cerebrovascular disease.
- While the final pathway of multiple organ failure complicating hypotension is rather uniform, the hemodynamic profile varies and is related to the underlying disease process.

Introduction

Depending on the circumstances, the term 'hypotension' may carry different meanings. Hypotension is classically used to describe a clinical scenario where the measured blood pressure is insufficient to support organ perfusion. It can also be viewed as the key clinical sign of 'shock'—a syndrome describing a situation in which tissue O_2 needs and availability are mismatched, i.e. either tissue O_2 availability is insufficient to match O_2 needs (anoxic, anemic, or circulatory anoxia), or the tissues are unable to utilize available O_2 (histiotoxic anoxia) (Table 1). The result of this mismatch between O_2 need and availability is cellular ischemia, a situation characterized by an intracellular O_2 deficit. This is followed by cellular damage that ranges, depending on the severity and duration of ischemia, from reversible alterations of cellular metabolic function to irreversible cell death. Arterial blood pressure is used as a surrogate marker of the adequacy of systemic perfusion and tissue O_2 delivery. For the purpose of this chapter, hypotension is discussed in the context of understanding the pathophysiology of both inadequate systemic perfusion and inadequate tissue O_2 delivery.

Table 1 Clinical criteria for the identification of shock

Evidence of decreased systemic perfusion
Hypotension: systolic blood pressure <90 mmHg or fall of 30–40 mmHg from previously recorded levels
Evidence of impaired perfusion of organ systems
Cerebral perfusion: restlessness/confusion
Renal blood flow: oliguria
Peripheral blood flow: cool mottled peripheries (except in septic shock where peripheries are warm)

Hypotension and tissue oxygen delivery

In health, O_2 is delivered in excess of tissue needs and systemic O_2 uptake is independent of delivery. When O_2 delivery falls, tissue oxygenation is initially maintained by increasing O_2 extraction. However, the ability to increase extraction is more efficient in some 'non-vital' organs (e.g. intestine, kidney) than in other 'vital' organs (e.g. heart, brain). Thus, when O_2 delivery falls because of decreased systemic flow and/or anemia, blood flow (and hence O_2 delivery) is redirected from non-vital to vital organs to maintain tissue oxygenation. However, if O_2 delivery continues to fall, the ability to compensate by redistributing blood flows and increasing O_2 extraction is eventually exceeded. Tissue O_2 uptake then becomes dependent on O_2 supply, a point referred to as the 'critical' O_2 delivery (Fig. 1(a)). Since an adequate O_2 delivery is critical to maintain cellular metabolic function, O_2 supply dependency, even if present only in selected organs, heralds the onset of cellular ischemia. If uncorrected, organ failure and death are inevitable consequences (Shoemaker *et al.* 1992).

Oxygen delivery is a function of arterial oxygen content (comprising hemoglobin concentration and O_2 saturation, and dissolved O_2) and blood flow:

$$O_2 \text{ delivery} = \text{flow} \times O_2 \text{ content}. \quad (1)$$

Blood flow, blood pressure, and vascular resistance are defined by the relationship

$$\text{blood pressure} = \text{flow} \times \text{resistance}. \quad (2)$$

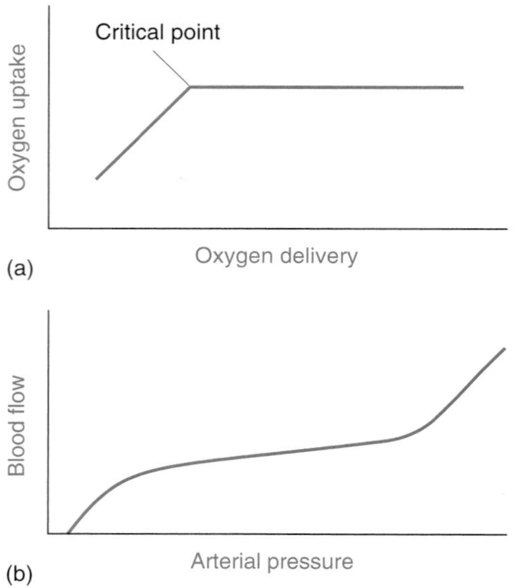

Fig. 1 (a) Relationship between O_2 delivery and O_2 uptake. Physiologically, O_2 uptake is not dependent on O_2 delivery. O_2 uptake becomes dependent on O_2 delivery, if O_2 delivery falls below the critical point. (b) Relationship between blood flow and arterial pressure (in skeletal muscle). Note that blood flow is well maintained over a wide range of arterial pressures.

This relationship implies that any fall in blood pressure eventually results in a decrease in O_2 delivery. However, the relationship of pressure to flow is non-linear; as blood pressure falls, local autoregulation (particularly in the vital organs) maintains organ blood flows (and hence O_2 delivery) across a broad range of perfusion pressures (Fig. 1(b)). Thus, autoregulation of cerebral blood flow usually maintains adequate oxygenation to the brain until the mean arterial blood pressure falls below 50 to 60 mmHg. Blood flow and organ O_2 delivery then become dependent on perfusing pressures—the lower the perfusion pressure, the lower the blood flow. Autoregulation of an organ's blood flow is locally mediated by release of factors which complicate ischemia (e.g. adenosine, CO_2, lactic acid) and activation of stretch receptors in the vascular smooth muscle of small vessels (Guyton and Hall 1996).

Hemodynamic determinants of hypotension

Equation (2) shows that blood pressure is dependent on blood flow and vascular resistance. Systemically, blood flow is referenced to the cardiac output (which, in turn, is distributed into the blood flow delivered to individual organs). Therefore hypotension may complicate a fall in either resistance or cardiac output. A depression in cardiac output can be related to its determinants:

$$\text{cardiac output} = \text{heart rate} \times \text{stroke volume} \qquad (3)$$

where stroke volume depends on ventricular preload, contractility, and afterload.

In this context, causes of hypotension can be classified. Briefly, some diseases cause hypotension primarily by affecting vascular resistance, while others predominantly depress cardiac output. Regardless of cause, a depression in blood pressure is accompanied by many autoregulatory responses, all of which attempt to maintain arterial perfusing pressures and local blood flows to prevent tissue ischemia (eqn (1) and Fig. 1(b)). Compensation involves all three major components of the circulation—the central circulation (the heart), the regional circulations (those supporting individual organs), and the microregional circulation (the capillary microcirculation)—and is dependent upon both neural and hormonal mechanisms.

The early response to hypotension is dominated by neural (sympathetic) activation which induces precapillary constriction, hence reducing capillary hydrostatic pressure. This mechanism acutely alters the balance of Starling forces and supports fluid redistribution from extravascular spaces into the capillary lumen, thereby augmenting intravascular fluid volume. This process, sometimes referred to as 'autotransfusion', helps to maintain (or even increase) cardiac output via the heart's Starling mechanism (the relationship between ventricular preload and cardiac output). Later responses to hypotension also attempt to maximize intravascular volume, including humoral mechanisms that increase sodium and water retention (i.e. stimulation of the renin–angiotension system, enhanced release of ADH, and activation of the ACTH–adrenocortical axis). In hypotension, sympathetic activation also augments myocardial contractility and heart rate, and promotes vasoconstriction of the arterial supply to non-vital organs. This latter response is the mechanism that facilitates redistribution of blood flow from non-vital organs to core vital organs. This autonomic response is mediated by decreased $Pa\text{O}_2$ or ischemia, although not by low blood pressure, and becomes extremely powerful at pressures below 60 mmHg (Guyton and Hall 1996).

Decompensation and cellular injury

If hypotension is not corrected promptly, the limited capability of the compensatory circulatory responses to support tissue oxygenation will be overwhelmed and an O_2 deficit may occur. Cellular ischemia induces alterations in cell function which may further exacerbate the circulatory response to the primary disease process. For example, a cellular O_2 deficit in prolonged hypotension is accompanied by anaerobic glycolysis and increased lactate production. Excess plasma lactate increases microvascular permeability, thereby promoting intravascular fluid loss. With ischemia, cellular ATP production is also depressed and the membrane's sodium pump fails, thereby further promoting intravascular fluid loss (in this case to the intracellular compartment) (Waxman 1996). Fluid shift away from the intravascular space reduces venous return and ventricular preload, thus worsening hypovolemia. As acidosis also depresses myocardial contractility, protracted hypotension leading to tissue ischemia is accompanied by further attenuation of stroke volume (myocardial depressant agents may also be released from ischemic tissues, e.g. the splanchnic circulation). This sequence is particularly deleterious in cardiogenic hypotension where stroke volume may be critically diminished already. Therefore the consequences of the metabolic processes induced by tissue ischemia are a further fall in blood pressure and O_2 delivery, and more tissue ischemia. The course of the primary disease process may then be modified by the consequences of the tissue ischemia—the concept of a vicious cycle which, if not prevented by rapid restoration of tissue oxygenation, hastens multiple organ failure and death.

Cell priming and reperfusion injury

Hypotension and shock, although capable of inducing tissue ischemia and organ failure directly, may also be antecedents of a generalized inflammatory state which complicates reperfusion during therapy. During the early phases of tissue ischemia (before irreversible ischemic cell injury), cell 'priming' occurs. This phenomenon is manifest only when the tissue is reperfused with restoration of blood pressure (Fig. 2). Cell priming by hypoxia leads to the release of second messengers (i.e. intracellular calcium, cyclic adenosine monophosphate, and reactive O_2 species) during the restoration of blood flow which, in turn, activate the inflammatory cascade. During reperfusion, leukocytes become entrapped in the previously ischemic microcirculation and release O_2 radicals and other inflammatory mediators locally and into the systemic circulation. Paradoxically, therefore, reperfusion following treatment of hypotension may also result in widespread tissue injury, the systemic inflammatory response syndrome, and multiple organ failure (Waxman 1996).

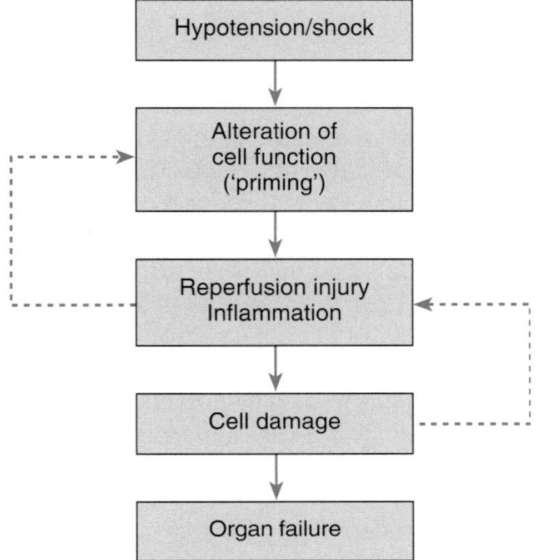

Fig. 2 Mechanisms of cellular and organ injury induced by hypotension and shock. Hypotension and/or hypoxia lead to cellular priming. Resuscitation may be followed by extensive mediator release into the vascular system and subsequent reperfusion injury and inflammation, resulting in further cellular damage and organ injury.

Pathways of hypotension for different disease processes

Although the final pathway of multiple organ failure complicating hypotension (or its treatment) is generally uniform regardless of the cause of the hypotension, the hemodynamic profile characterizing hypotension may vary depending on the disease processes (Fig. 3).

Myocardial ischemia

Myocardial ischemia leads to depressed ventricular contractility or systolic dysfunction. Cardiac output and blood pressure are depressed, while ventricular filling pressures frequently rise. As well as impairing systolic function, myocardial ischemia depresses ventricular compliance (i.e. diastolic dysfunction). This limits ventricular filling, thereby preventing full use of the heart's Starling mechanism to support cardiac output and function. Coronary perfusion may also be affected. Paradoxically, circulatory compensation, particularly elevated systemic resistance and tachycardia, may elevate heart work and myocardial O_2 needs at the very time that supply is threatened. Tachycardia may also impair diastolic coronary perfusion. Therefore autoregulatory compensation of cardiogenic hypotension may worsen myocardial O_2 balance, exacerbating myocardial ischemia.

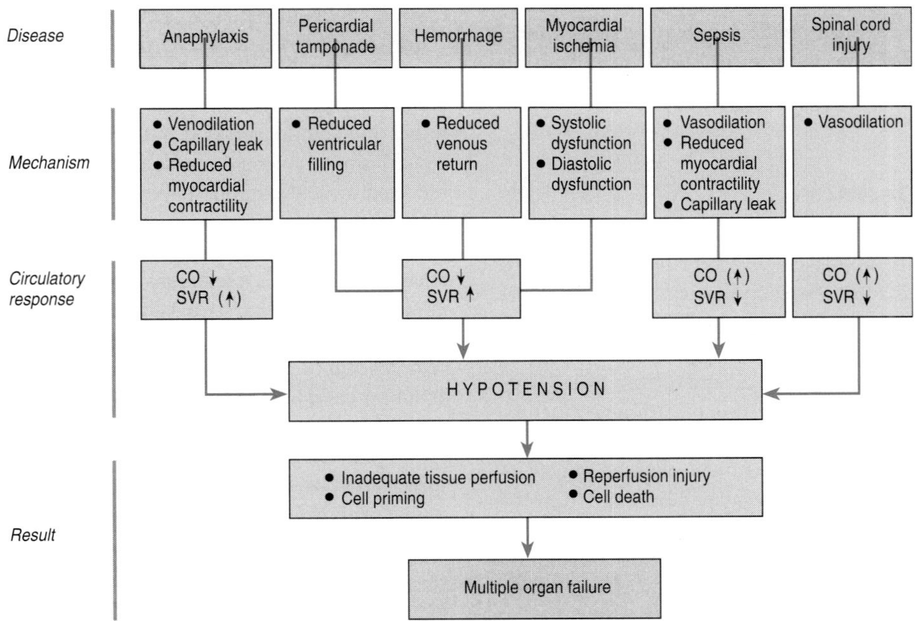

Fig. 3 Pathway and hemodynamic profile of different disease processes commonly leading to hypotension and multiple organ failure. Arrows indicate regular findings; arrows in parentheses represent facultative findings. CO, cardiac output; SVR, systemic vascular resistance.

Hemorrhage

Hemorrhage has two immediate sequelae which limit tissue oxygenation: depressed intravascular fluid volume and anemia. Hypotension in acute hemorrhage depresses venous return and, subsequently, cardiac output. In protracted shock, the metabolic consequences of tissue ischemia may lead to the release of myocardial depressant factor from the splanchnic circulation, thus depressing cardiac output further.

With activation of hemodynamic and metabolic autoregulatory responses, up to 20 to 30 per cent fluid loss may be compensated during hemorrhage without marked effects on blood pressure. If this volume is exceeded, or pre-existing diseases such as coronary heart disease or cardiac failure blunt the autoregulatory mechanisms, hypotension results and O_2 delivery falls.

Sepsis

The inflammatory process characterizing sepsis includes release of various mediators, most of which induce peripheral vasodilation and lead to increased microvascular permeability. Ventricular function is initially unaffected, and upregulation of cardiac output (which is necessary to maintain blood pressure in the presence of the vasodilatation typical of sepsis) is often sufficient to maintain perfusing pressures. Hypotension characterizing severe sepsis may occur when intravascular fluid volume is profoundly depressed (because of excessive venodilatation or microvascular fluid losses), myocardial depressant substances limit the ability to augment cardiac output, and/or the patient has underlying heart disease.

Sepsis is also characterized by microcirculatory injury, including depressed capillary perfusion and capillary dropout, both of which occur even without severe hypotension. With abnormalities in inter-organ distributions of systemic flow, these microcirculatory abnormalities could explain the inability to extract O_2 maximally in sepsis. A primary cellular defect may also exist whereby tissues are limited in their ability to utilize available O_2 — the concept of histiotoxic anoxia. However, this concept remains controversial and very difficult to investigate with existing tools. Any hypotension further exacerbates microcirculatory flow abnormalities and may cause the 'sludge' phenomenon, indicating severe microvascular plugging (Hinshaw 1996).

Other disease processes

Cardiac obstruction, due to pericardial tamponade or tension pneumothorax for example, can rapidly cause severe hypotension. The hemodynamic features of cardiac obstruction (for the left heart) are reduced diastolic compliance, reduced preload, and a decreased cardiac output.

Anaphylaxis refers to an antigen-provoked and IgE-mediated activation of mast cells and basophils with release of various mediators. Acute hypotension may complicate untreated anaphylaxis and is primarily the consequence of hypovolemia (due to elevated microvascular permeability and shifting of intravascular fluid away from the intravascular space), venodilatation with decreased venous return, and myocardial depression. Both cardiac filling pressures and cardiac output may be reduced.

An insufficient intravascular fluid volume is also the cause of neurogenic hypotension complicating spinal cord injury. Hypotension is related to an acute decrease in systemic vascular resistance caused by interruption of sympathetic nerves distal to the site of injury. Cardiac output is usually maintained or, in response to vasodilatation, increased. If the acute phase of neurogenic hypotension is bridged with adequate volume therapy and, when indicated, pressor agents, autoregulatory mechanisms will lead to a restoration of sufficient blood pressure within weeks or months in the majority of cases.

Effects of hypotension on organ systems

Vital organs

Cerebral ischemia occurs when blood flow, after loss of autoregulation, falls below 25 to 30 per cent of normal levels. An altered level of consciousness, ranging from mild confusion to coma, represents the most important clinical finding of cerebral ischemia. For the heart, the catecholamine release associated with hypotension not only supports myocardial contractility and induces tachycardia, but also increases oxygen demand and the risk of arrhythmia. Hypotension reduces coronary artery perfusion pressure and blood flow, and therefore may lead to myocardial ischemia, particularly in the presence of coronary heart disease or pre-existing cardiac dysfunction. Concomitant metabolic acidosis is followed by a decrease in the pH of myocardial cells and a decrease in contractility.

Non-vital organs

The respiratory system is commonly affected in severe hypoperfusion. Respiratory failure is often due to increased respiratory workload and impairment of diaphragmatic function. Hypoxemia and increased respiratory efforts associated with an increase in minute volume are typical; thus the O_2 needs of the respiratory muscles are increased at the very time that O_2 supply is limited. In these circumstances, type II respiratory failure may complicate hypotension. Delayed respiratory failure in hypotension may also be due to acute respiratory distress syndrome following injury to the lung's microcirculation. Hypotension also depresses renal blood flow and glomerular perfusion rate. The compensatory release of catecholamines and angiotensin II also decreases renal blood flow by inducing constriction of afferent arterioles. Ultimately, renal hypoperfusion is followed by acute tubular necrosis and renal failure. In severe hypotension, blood flow is directed away from the intestine and liver towards the heart and brain. Thus the liver and gut may be early sites of tissue damage. These organs are uniquely sensitive in reperfusion injury. In the liver, hypoxic damage leads to attenuated metabolic and synthetic function, with the latter including a decreased generation of proteins and coagulation factors. Gut injury is often followed by ileus or submucosal hemorrhage. Although it was previously assumed that inadequate gut perfusion may cause barrier failure, allowing translocation of bacteria and bacterial fragments into the circulation, it now appears that bacterial translocation does not cause bacteremia but is a source of activation of systemic inflammatory processes (Deitch *et al.* 1996).

References

Deitch, E.A., Rutan, R., and Waymack, J. P. (1996). Trauma, shock, and gut translocation. *New Horizons*, 4, 289–99.

Guyton, A.C. and Hall, J.E. (1996). *Textbook of medical physiology* (9th edn), pp. 199–220. W.B. Saunders, Philadelphia, PA.

Hinshaw, L.B. (1996). Sepsis/septic shock: participation of the microcirculation: an abbreviated review. *Critical Care Medicine*, **24**, 1072–8.

Shoemaker, W.C., Appel, P.L., and Kram, H.B. (1992). Role of oxygen debt in the development of organ failure sepsis, and death in high-risk surgical patients. *Chest*, **102**, 108–15.

Waxman, K. (1996). Shock: ischemia, reperfusion, and inflammation. *New Horizons*, **4**, 153–60.

3.5.2 Causes

A. Sielenkämper and W. J. Sibbald

Key messages

- The causes of hypotension can be categorized as cardiogenic, obstructive, hypovolemic, or distributive.
- The most frequent cause of cardiogenic hypotension is lack of functional myocardium due to coronary thrombosis. Other important causes are structural and mechanical abnormalities of the heart and electrical dysfunction.
- Obstructive hypotension reflects a hemodynamically relevant impediment of blood flow from either extravascular (extrinsic) or intravascular (intrinsic) compartments.
- Categorization of hypovolemic hypotension as being due to either erythrocyte or salt/water loss is important to direct replacement of the right type of fluid.
- Sepsis and the systemic inflammatory response syndrome are the most frequent causes of distributive hypotension and reflect the same pathophysiology, either with (sepsis) or without (systemic inflammatory response syndrome) infection.

Introduction

Hypotension complicates a variety of disease processes and can be classified according to its associated hemodynamic profile in one of four categories: cardiogenic, obstructive, hypovolemic, or distributive. The advantage of this classification, which was initially proposed by Weil, is that it creates a pathophysiological basis on which to develop rational treatment strategies.

When examining the hypotensive patient, it is important to remember that the cause of shock may be multifactorial. For example, if distributive shock occurs in an elderly patient with underlying coronary artery disease, the clinical manifestations may be a combination of both distributive and cardiogenic shock. Protracted cardiogenic shock may result in intravascular fluid loss and a hypovolemic component when microvascular dysfunction complicates the initial scenario. Septic shock may have a cardiogenic component when severe (secondary to myocardial depressant substances) or when complicating pre-existing ischemic heart disease; both may limit the anticipated increase in cardiac output otherwise typical of this condition.

Causes of hypotension

Cardiogenic

Hypotension due to cardiogenic causes occurs when cardiac output falls sufficiently low that increases in systemic resistance are inadequate to maintain perfusing arterial pressures. Cardiogenic hypotension (Table 1) may be a consequence of any one (or a combination) of the following: lack of functional myocardium, structural or mechanical abnormalities, or electrical dysfunction (conduction abnormalities, arrhythmia). However, by far the most frequent cause of cardiogenic hypotension is myocardial infarction associated with coronary thrombosis. In this context, more than 40 per cent of the left ventricle needs to be injured (acutely or chronically) for hypotension to be a complicating factor. This complication is usually associated with a high mortality. Patients with lesser degrees of ischemic myocardial necrosis may also develop hypotension with the coexistence of acute or chronic structural abnormalities (e.g. mitral insufficiency complicating papillary muscle dysfunction), intravascular volume contraction (e.g. the patient with pre-existing congestive heart failure and chronic diuretic use), and/or hemodynamically significant arrhythmias (e.g. atrioventricular block.).

Valvular dysfunction leading to shock secondary to mechanical difficulties may complicate almost any type of heart disease. It may be acute (e.g. mitral insufficiency complicating papillary muscle dysfunction, aortic insufficiency complicating endocarditis) or chronic (e.g. progressive aortic or mitral stenosis). Other structural causes of cardiogenic hypotension include ventricular septal defect and ventricular rupture complicating myocardial infarction. Blunt or penetrating injury resulting in traumatic myocardial injury is an infrequent cause of cardiogenic hypotension. In these cases, shock may be due to a loss of functional myocardium complicating ventricular contusion (the most frequent traumatic cause), arrhythmia, conduction abnormality, or valvular lesion (traumatic valvular rupture). Intravascular volume contraction (absolute or relative) is also an uncommon cause of hypotension in the patient with acute or chronic heart disease. However, this scenario is particularly frequent in the patient on long-term diuretic use who then receives vasodilators, narcotics, or positive-pressure ventilation in the acute management of myocardial infarction.

Table 1 Cardiogenic causes of hypotension

Myocardial
Myocardial necrosis following infarction
Myocarditis
Endstage congestive/ischemic cardiomyopathy
Traumatic myocardial contusion
Structural/mechanical
Ventricular septal defect
Valvular dysfunction
Papillary muscle rupture
Free-wall rupture
Ventricular aneurysm
Arrhythmia
Sinus bradycardia
Tachycardia
Atrioventricular blocks

Table 2 Obstructive causes of hypotension

Intrinsic
- Pulmonary embolism
- Atrial myxoma or thrombus
- Acute pulmonary hypertension

Extrinsic
- Pericardial tamponade
- Constrictive pericarditis
- Tension pneumothorax
- Obstructive intrathoracic tumors

Clinically, the patient with cardiogenic hypotension is diaphoretic and may present with tachycardia and severe pain when myocardial infarction is the cause. Pulmonary edema and cyanosis indicate severe cardiac failure, while the presence of a murmur is usually a sign of structural abnormality (e.g. acute ischemic lesions such as papillary muscle necrosis or rupture of the ventricular septum) but can also be a consequence of diseases that primarily affect structural components of the heart such as endocarditis.

Obstructive

Obstructive hypotension indicates obstruction to flow within the core circulation and may be categorized as either intrinsic or extrinsic (Table 2). Pulmonary embolism is the most important intrinsic cause and may be difficult to diagnose. The prevalence of potentially fatal acute pulmonary embolism among hospital in-patients is about 1 per cent. At autopsy, 14.6 per cent of patients had pulmonary embolism that may have contributed to their death. The most important cause of extrinsic obstruction is pericardial tamponade. This may follow blunt or penetrating chest trauma, cardiac surgery, and a variety of other chronic disease processes of which renal failure is probably the most common chronic cause seen in the intensive care unit.

Tamponade can be difficult to distinguish from right ventricular infarction as both may present with signs and symptoms of right ventricular failure (the inability to provide sufficient preload to the left ventricle to maintain cardiac output). It should be noted that hypotension occurring in endstage hypertrophic or restrictive cardiomyopathy should not be classified as primarily obstructive, as circulatory failure related to these diseases is usually related to cardiogenic mechanisms (i.e. arrhythmia).

The clinical appearance of the patient with obstructive hypotension is variable and related to the underlying disease. For example, patients with pulmonary embolism may show signs similar to those with myocardial infarction, including severe chest pain and diaphoreses, while a patient with saddle embolism may show signs of acute arterial occlusion of the lower extremities in combination with severe left ventricular failure.

Hypovolemia

When hypotension complicates hypovolemia, the intravascular fluid volume is depressed sufficiently for compensatory fluid mobilization from the extravascular to the intravascular space to be unable to maintain venous return, ventricular preload, and hence cardiac output.

Table 3 Hypovolemic causes of hypotension

External loss
- Acute hemorrhage
- Burns
- Dehydration
- Vomiting
- Diarrhea
- Polyuria
- Postoperative

Internal redistribution
- Thermal injury
- Trauma
- Capillary leak

Causes of hypovolemia are categorized as external, due to fluid loss, and internal, due to a redistribution of intravascular volume such that it is functionally unavailable to support cardiac preload (Table 3).

Hemorrhage, for example due to trauma or gastrointestinal bleeding, is the most common cause of hypovolemic hypotension due to external fluid loss. In this situation, acute losses up to 30 per cent of total blood volume (about 1600 ml blood in a 70-kg male) may be tolerated without marked decreases in blood pressure provided that circulatory compensation (sympathoadrenal mechanisms) is complete. Greater blood loss or blood loss in patients with inadequate circulatory compensation (i.e. in the presence of underlying ischemic heart disease which limits increases in cardiac output, or cord injury which limits augmented venous return and arterial vasoconstriction) is associated with hypotension and shock (Table 4). In the case of chronic dehydration, a loss of 6 to 10 per cent of the extracellular fluid volume is followed by a redistribution of blood volume to vital organs and an altered level of consciousness, while a fluid loss of 20 per cent or more of total body weight cannot be survived without immediate and aggressive intervention.

It is also important to categorize the cause of hypovolemia as due to either erythrocyte or salt/water loss. This categorization directs replacement of the right 'type' of fluid for emergency treatment and during further diagnostic procedures.

Table 4 Classification of hemorrhage

Blood loss (%)	Stage	Clinical findings
10–15	I	None
15–30	II	Tachycardia, normotension, anxiousness
30–40	III	Tachycardia, hypotension, oliguria, confusion
>40	IV	III+loss of consciousness, death

Table 5 Distributive causes of hypotension

Sepsis
- Bacterial
- Fungal
- Viral

Anaphylaxis
- IgE mediated
- Non-IgE mediated (anaphylactoid)

Neurogenic
- Spinal cord injury
- Spinal anesthesia

In patients with chronic hypovolemia, clinical findings other than hypotension typically include dry skin and tongue, loss of skin elasticity, and oliguria. Acute hypovolemia is accompanied by typical signs of shock such as tachycardia, vasoconstriction, cool peripheries, and confusion or loss of consciousness. Postoperative hypovolemia may often be masked by sympathoadrenal discharge due to post-operative pain. In these cases, the initiation of analgesia may be followed by a sudden decrease in blood pressure which reveals the need for immediate fluid therapy.

Distributive

Distributive hypotension refers to shock which complicates redistribution of intravascular fluid volume with decreases in arterial vascular tone, dilatation of the venous capacitance vessels, or both (Table 5). The most frequent cause of distributive hypotension is the systemic inflammatory response syndrome–sepsis continuum. Systemic inflammatory response syndrome follows acute insults such as massive reperfusion injury and pancreatitis, while sepsis reflects the same pathophysiology but complicates severe infection. Sepsis is associated with a mortality ranging from 35 to above 50 per cent. Distributive hypotension may also complicate anaphylaxis, neurogenic lesions (traumatic cord injury, spinal anesthesia), and thyroid storm (rare, but acute and life threatening).

In sepsis-related distributive hypotension, clinical findings generally include signs of decreased systemic resistance and a hyperdynamic circulatory response (e.g. warm skin, tachycardia, increased pulse pressure, and tachypnea). Both anaphylaxis and anaphylactoid reactions invariably cause flushing and tachycardia, but may also present with symptoms indicating an advanced form of an allergic reaction including urticaria, dyspnea, laryngeal edema, bronchospasm, or abdominal cramps and nausea.

Bibliography

Brun-Buisson, C., *et al.* (1995). Incidence, risk factors, and outcome of severe sepsis and septic shock in adults. *Journal of the American Medical Association*, **274**, 968–74.

Califf, R.M. and Bengtson, J.R. (1994). Cardiogenic shock. *New England Journal of Medicine*, **330**, 1724–30.

Gasche, Y., Pittet, D., and Suter, P.M. (1995). Outcome and prognostic factors in bacteremic sepsis. In *Clinical trials for the treatment of sepsis* (ed. W.J. Sibbald and J.-L. Vincent), pp. 35–51. Springer-Verlag, Berlin.

Peitzman, A.B., Billiar, T.R., Harbrecht, B.G., Kelly, E., Udekwu, A.O., and Simmons, R.L. (1995). Hemorrhagic shock. *Current Problems in Surgery*, **32**, 925–1002.

Stein, P.D. and Henry, J.W. (1995). Prevalence of acute pulmonary embolism among patients in a general hospital and at autopsy. *Chest*, **108**, 987–81.

3.5.3 Therapeutic strategy

A. Sielenkämper and W. J. Sibbald

Key messages

- Timely reversal of hypotension with maintenance of tissue oxygen delivery at levels consistent with tissue needs is critical to avoid life-threatening complications such as multiple organ failure.
- The key to adequate management is to treat at the same time that monitoring is being instituted and diagnostic procedures initiated.
- Restoration of systolic blood pressure to at least 90 mmHg with reversal of clinical signs of shock (indicating an adequate tissue oxygen availability) is the primary objective of emergency therapy.
- The two major therapeutic strategies in hypotension are normalization of intravascular volume by administration of fluids (both crystalloids and colloids) in the volume-depleted patient and pharmacological support of arterial tone and cardiac contractility.
- In cardiogenic hypotension due to infarction, further management objectives include limiting further myocardial injury, restoring ventricular function, and determining whether there is a mechanical defect that requires invasive management.
- Key issues in the septic patient with hypotension are source control (antibiotic therapy, eradication of the septic focus) and support of oxygen delivery to match the elevated metabolic needs typical of sepsis.

Introduction

Key goals in treating the hypotensive patient include normalization of arterial perfusing pressures and maintenance of tissue O_2 delivery at levels consistent with ongoing O_2 requirements. As the occurrence of primary multiple organ dysfunction syndrome increases in relation to the time taken to restore perfusing pressures and tissue oxygenation, timely reversal of hypotension is critical. Successful management also demands an organized and practiced approach; communicating clear objectives to the healthcare team and repeated evaluation of the management plan are crucial to ensuring successful outcomes. An added challenge is the need to treat at the same time that the precipitating cause of hypotension is being identified and controlled. Management will be discussed in the context of treatment periods which occur in sequence, namely ebb phase resuscitation, which includes restoring arterial pressures to either preinjury levels or levels consistent with adequate organ perfusion, followed by flow phase resuscitation, which necessitates supporting tissue O_2 delivery both to repay any O_2 debt created by the initial hypotensive episode and simultaneously to provide for ongoing tissue O_2 demands.

Diagnostic approach and monitoring

Monitoring is required to determine the adequacy of management designed to restore perfusing pressures and the underlying cause of the hypotension.

A primary assessment, including history and physical examination, must be undertaken in an ordered way. Inspection should be initially directed at pulse rate and quality, respiratory rate, skin circulation, and mental status. Hypotension is associated with a typical constellation of clinical signs with some differentiation according to the underlying cause (Table 1). It is important to remember that clinical findings may also be modified by pre-existing diseases. For example, an elderly or diabetic patient may not exhibit tachycardia because of a diminished cardiac response to sympathetic stimulation.

During ebb phase resuscitation the clinician also needs to decide on the monitoring and diagnostic approaches which facilitate early restoration of arterial perfusing pressures and to establish the cause of hypotension. Key to this early management phase is adequate monitoring of the perfusing pressures. Shock in the previously normotensive patient is defined as a systolic blood pressure above 90 mmHg, but, as the sphygmomanometer can be an unreliable tool in patients with shock and pre-existing vascular disease, confirmatory evidence by clinical examination is mandatory. Signs of clinically significant hypotension include a cool vasoconstricted periphery, oliguria (since skin and renal blood flows are redirected to the core circulation), and obtundation or restlessness (since cerebral blood flow is depressed). As a previously hypertensive patient is accustomed to elevated perfusing pressures, shock in this type of patient is defined when the blood pressure has fallen by at least 30 mmHg from previously recorded levels. We recommend inserting an arterial line (radial or femoral) if restoration of arterial perfusing pressures might be protracted and if this can be performed rapidly—time should not be spent trying to insert an arterial line to the exclusion of prompt management of the hypotension. An arterial line also provides access for rapid blood sampling or in-line biochemistry monitoring.

Other monitoring and diagnostic technologies that might be considered during ebb phase management are considered in Table 2. The principal objectives are to determine the adequacy of restoration of arterial perfusing pressures and to find clues to the underlying cause of the shock. Pulse oximetry and end-tidal CO_2 monitoring provide a non-invasive assessment of the integration of the respiratory and circulatory systems. However, in vasoconstricted shock pulse oximetry may be unreliable because capillary blood flows are impeded. End-tidal CO_2 monitoring has been used as a surrogate for circulatory function, particularly in cardiac arrest situations where restoration of core blood flows is accompanied by increases in end-tidal CO_2. Laboratory tests may provide diagnostic information about the etiology of hypotension and give an important baseline assessment of organ function. Leukocytosis or leukopenia can occur in sepsis, although in non-septic patients they may indicate a late stage of shock. Thrombocytopenia may be found in sepsis and after acute hemorrhage. Coagulation parameters need not be routinely measured unless there is clinical suspicion of a coagulopathy. Diagnostic blood sampling also provides an opportunity to perform early cross-matching for possible blood transfusion needs.

During flow phase resuscitation, the aim of monitoring is to determine whether O_2 delivery is adequately meeting the body's needs. Measuring arterial lactate may be useful in tracking the balance between whole-body O_2 supply and demand. Although many factors contribute to arterial lactate concentrations, the trend is useful over short periods of time as a surrogate marker of the adequacy of the whole-body O_2 supply–demand relationship. Thus, if lactate is increasing, the prevailing O_2 delivery is likely to be insufficient for ongoing metabolic needs. In this circumstance, treatments which increase O_2 delivery can be used as a diagnostic test: if the lactate falls after O_2 delivery is elevated, this is reasonable presumptive evidence of perturbation of the whole-body O_2 supply–demand balance. Ideally, this balance should be determined at the level of individual organs, since any test which represents the sum of all organs (as happens with arterial lactate) may obscure the presence of significant ischemia in individual organs. Gastric tonometry measures gastric intramucosal $P\text{CO}_2$ ($P\text{iCO}_2$); the difference between $P\text{aCO}_2$ and $P\text{iCO}_2$ is currently being evaluated as a surrogate of the adequacy of oxygenation in the gastrointestinal circulation. Urine $P\text{O}_2$ measurements have also been proposed to monitor the O_2 supply–demand balance in the core circulation.

The next issue in flow phase resuscitation is to determine how to improve O_2 delivery further when indicated. Figure 1 shows a treatment algorithm for increasing systemic O_2 delivery. Crucial to circulatory management is the assessment of the adequacy of ventricular preload. If it is low, fluid infusion to augment intravascular volume is appropriate. If it is high, inotropes and/or dilators are

Table 1 Clinical signs common to three frequent causes of hypotension

	Myocardial ischemia	Hemorrhage	Sepsis
Skin	Cool, vasoconstricted, cyanosis	Cool, vasoconstricted, cyanosis	Warm, vasodilated
Pulse	Weak, irregular	Weak, fast	Bounding
Neck veins	Distended	Flat	Flat
Auscultation			
Heart	Gallop rhythm, murmur	Tachycardia	Tachycardia
Lungs	Respiratory crepitations	None	None
Additional signs	Chest pain, fear	Apparent source of blood loss	Fever, septic focus

Table 2 Monitoring the hypotensive patient

Methods	Diagnostic marker
Oscilloscope	Heart rate
Intra-arterial catheter	Blood pressure, blood gases
Pulse oximeter	Arterial O_2 saturation
Capnometry	End-tidal CO_2
Central venous catheter	Central venous pressure
Radiology	Chest (and other)
Urethral/suprapubic catheter	Urine output
Biochemistry	Arterial lactate, WBC, hemoglobin, platelets, PT/PTT, electrolytes, Ca, Mg, creatinine, urea
Right heart catheter	Cardiopulmonary profile (CO, PAWP, So_2, derived variables)
Additional methods (dependent on availability)	
Gastric tonometer	Gastric pH_i, Pco_2 gap
Advanced pulmonary artery catheters	Right ventricular ejection fraction, continuous cardiac output

WBC, white blood count; PT, prothrombin time; PTT, partial thromboplastin time; PAWP, pulmonary artery wedge pressure.

Table 3 Specific diagnostic techniques

Assumed etiology	Diagnostic methods	Comment
Cardiogenic Myocardial Structural Arrhythmias	ECG (12-lead) Chest radiography Enzymes (CK, CKMB, LDH) Echocardiography	Rule out diseases that may imitate cardiogenic origin (e.g. pulmonary embolism, pancreatitis)
Hypovolemia External fluid loss Internal redistribution	Ultrasonography Endoscopy Colloid osmotic pressure Hematocrit Angiography CT and MRI scans	Choose methods appropriate to assumed cause of hypovolemia (bleeding? chronic dehydration?)
Obstructive Intrinsic Extrinsic	Echocardiography Angiography Perfusion scans CT and MRI scans	Acute cardiac tamponade may often require immediate surgery
Distributive sepsis	Cultures (e.g. blood, urine) Gram stains Ultrasonography CT scan Bronchoscopy	Apply ACCP–SCCM guidelines for diagnosis of sepsis—find and treat the septic focus

CK, creatinine phosphokinase; CKMB, creatine kinase–myocardial isoenzyme; LDH, lactate dehydrogenase; ACCP, American College of Chest Physicians; SCCM, Society of Critical Care Medicine.

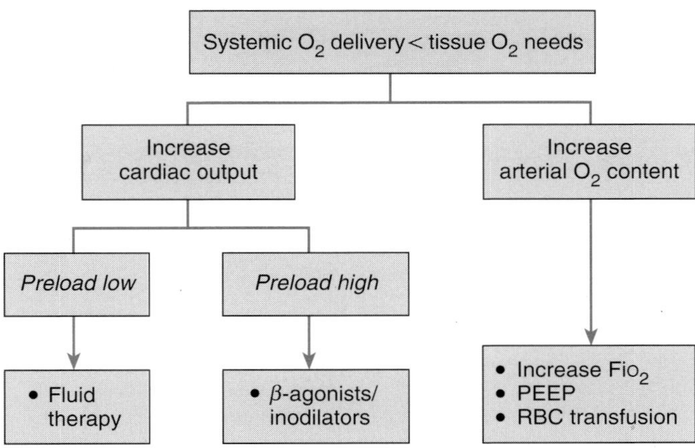

Fig. 1 General treatment algorithm for increasing systemic O_2 delivery during flow phase resuscitation (after restoration of blood pressure) in patients with inadequate O_2 delivery. PEEP, positive end-expiratory pressure; RBC, red blood cell.

more appropriate as fluid therapy may promote pulmonary edema, thereby further depressing the arterial O_2 content. At this point the clinician must decide about the usefulness of right heart catheterization. If uncertainty remains regarding the status of ventricular preload, and hence whether fluid or inotropic therapy is the most appropriate, we recommend pulmonary artery catheter insertion. As no clinical trial has satisfactorily demonstrated that information provided by the pulmonary artery catheter leads to improved patient outcome, this recommendation must be regarded as opinion only.

When early evaluation of the patient does not identify the cause of hypotension, specific diagnostic procedures must be considered (Table 3).

Treatment of hypotension in the ebb phase

Treatment requires both manual and intellectual dexterity. The history must be identified and vital organ function assessed clinically while therapy and essential monitoring are initiated simultaneously. The first goal in resuscitation is restoration of arterial perfusing pressures to levels consistent with reversing the numerical and clinical signs of shock, i.e. increasing systolic blood pressure to at least 90 mmHg in previously normotensive patients, and to levels not less than 10 to 15 per cent below pre-illness pressures in previously hypertensive patients. Achieving acceptable perfusing pressures is often accompanied by reversal of the signs of shock — improved peripheral skin temperatures and urine output, and conscious state.

A successful approach to ebb phase resuscitation is best ensured by following a pragmatic regimen, namely the ABC (airway, breathing,

Table 4 Emergency approach to ebb phase resuscitation

1. Secure airway, ensure adequate oxygenation
   Consider intubation
   Administer oxygen (maximal FiO_2)
   Assisted or controlled ventilation
2. Restore blood pressure and cardiac output

Ask question: Can I use fluids?

If the answer is 'Yes' (i.e. hypovolemia, sepsis, anaphylaxis)
   Establish large-bore venous access sites
   Intravenous fluids, combined colloids/crystalloids
   Add inotropes/vasopressors if indicated

If the answer is 'No' (e.g. cardiogenic hypotension accompanied by pulmonary edema)
   Support myocardial pump function with inotropes and vasopressors
   Treat pulmonary edema (e.g. diuretics, positive end-expiratory pressure)
   Consider treatment of arrhythmias (if present)

circulation) of resuscitation, as developed for adult cardiac life support (Table 4).

1. Secure the airway: most hypotensive patients will benefit from early intubation to protect against aspiration of gastric contents or blood and to facilitate appropriate ventilation and oxygenation.
2. Breathing: an elevated respiratory workload is common in hypotensive patients, increasing O_2 needs of the respiratory muscles at the very time that O_2 delivery is impaired. Controlled ventilation will discount the work of breathing, thereby allowing a redistribution of any limited systemic O_2 delivery to other core organs. To maximize diffusive O_2 transport in the lungs, we recommend starting with an inspiratory O_2 fraction (FiO_2) of 1.0 which can later be adjusted to maintain PaO_2 values above 70 mmHg (9.3 kPa) or an SpO_2 above 98 per cent. Without intubation in spontaneously breathing patients, most high-flow O_2 devices fail to provide an FiO_2 above 0.5. In contrast, an FiO_2 of 0.9 to 1.0 is easily obtained in the intubated patient.
3. Circulation: hypotension often signals a fall in O_2 delivery that may lead to irreversible dysfunction of vital organs. Long resuscitation times increase the risk of developing multiple organ dysfunction syndrome and reduce the likelihood of survival. Thus timely restoration of the arterial perfusing pressures is crucial. Once blood pressure has been restored to levels consistent with goals previously defined (i.e. systolic blood pressure above 90 mmHg and return of peripheral perfusion and urine output), time can be more carefully apportioned to further diagnostic evaluation to manage the cause of the hypotension, repay any O_2 debt that occurred during the ebb phase, and optimize O_2 delivery to match ongoing needs.

Two general approaches are used for restoring arterial perfusing pressures during ebb phase resuscitation: increasing cardiac preload; increasing arterial tone and cardiac contractility. Fluid restores intravascular volume in the volume-depleted patient, optimizing left

Table 5 Specific pharmacological agents commonly used in hypotensive patients

Drug	Predominant effect		Dose	Clinical use
	Cardiac	Peripheral		
Dobutamine	+Heart rate (β_1) +++Contractility (β_1, α)	+Arterial dilation (β_2)	1–10 μg/kg/min	Cardiogenic, septic, and obstructive hypotension
Dopamine	++Heart rate (β_1) ++Contractility (β_1, α)	++Arterial constriction (α) +++Renal, mesenteric, cerebral, coronary artery dilation (DA_1, DA_2)	1–15 μg/kg/min ('renal' dose 1–4 μg/kg/min)	All forms of hypotension (dose dependent)
Epinephrine	+++Heart rate (β_1) +++Contractility (β_1, α)	++Arterial constriction (α)	1–10 μg/min (increase transiently if no effect)	Anaphylaxis, refractory hypotension
Norepinephrine	++Heart rate (β_1) ++Contractility (β_1, α)	+++Arterial constriction (α)	2–20 μg/kg/min	Septic and refractory hypotension
Amrinone	+Heart rate (β_1) +++Contractility	+++Arterial dilation	0.75 mg/kg over 5 min once or twice, then 5–15 μg/kg/min	Cardiogenic hypotension
Nitroprusside	No direct effect on pump function	+++Arterial dilation ++Venodilation	0.5–8 μg/kg/min	Cardiogenic, septic, and obstructive hypotension
Nitroglycerine	No direct effect on pump function	+++Venodilation +Arterial dilation	5–66 μg/min (no maximum dose)	Cardiogenic, septic, and obstructive hypotension

ventricular preload and thereby elevating cardiac output according to the Starling hypothesis. As fluid therapy is potentially less injurious than exogenous sympathomimetics for improving cardiac output and blood pressure, an important first question is to determine whether fluids can be used. When history and clinical signs suggest intravascular volume contraction, a fluid challenge is appropriate while clinical signs (blood pressure, heart rate, etc.) are closely monitored. If the patient has cardiogenic pulmonary edema or hypotension does not respond quickly to fluid challenges, exogenous sympathomimetics are recommended (Table 5). As the initial therapeutic objective in ebb phase resuscitation is blood pressure elevation, α-agonists such as norepinephrine (noradrenaline) and epinephrine (adrenaline) are preferable. Sympathomimetics with predominately β or dopaminergic effects are not recommended, as the vasodilation that they induce may be greater than can be compensated for by an increase in cardiac output, in which case hypotension will be accentuated.

The use of bicarbonate in the hypotensive acidotic patient is controversial. CO_2 excretion may be impaired (e.g. due to respiratory failure) and tissue CO_2 increased due to tissue hypoperfusion. Under such circumstances bicarbonate therapy may cause an additional load of CO_2, resulting in further intracellular acidosis. Alternatively, a pH below 7.15 to 7.2 may limit the response to pharmacological treatment, particularly on β-receptors. Our practice is to use bicarbonate therapy when the arterial pH is below 7.2 and hypotension persists despite adequate fluid therapy.

Treatment of hypotension in the flow phase

After restoring blood pressure, therapy is directed at maintaining perfusing pressures and optimizing tissue O_2 availability. This requires a number of steps, including determining which component of the O_2 delivery equation should be improved to pay off any O_2 debt and match delivery with ongoing needs, and initiating a management (diagnostic and treatment) plan which adequately manages the primary cause of the hypotension.

Immediately post-shock, most patients have an elevated O_2 requirement because of the need to repay the O_2 debt that developed previously. Subsequently, traumatic and septic shock patients usually demonstrate ongoing elevations in tissue O_2 requirement since the metabolic rate is elevated by the healing processes in trauma and the generalized inflammation typical of sepsis. These patients may then need a systemic O_2 delivery above the normal range (> 600 ml/min/m^2). Increasing O_2 delivery is accomplished by keeping $P\text{ao}_2$ above 70 mmHg (9.3 kPa), transfusing packed red cells to correct anemia, and increasing cardiac output with fluid therapy, inotropes, and/or dilators. A right heart catheter may now be necessary to guide management. Left ventricular filling is optimized by additional fluids. If the pulmonary artery wedge pressure is above 15 to 18 mmHg, O_2 delivery is increased using vasodilators and inotropes. An increase in $S\text{vo}_2$ and a decrease in arterial lactate can be used to determine the efficacy of therapy.

Fluid therapy

During flow phase resuscitation, fluid therapy maintains intravascular volume in patients where initial management has already demonstrated success (i.e. ongoing losses as in hemorrhage). This maintains the preload levels necessary to support cardiac output and systemic O_2 delivery. Crystalloids and/or colloids are used for this objective. Fluid therapy also repletes the intravascular fluid lost to the extravascular compartment which characterizes the physiological adaptation to hypotension. Given the characteristics of extravascular fluid, crystalloids (i.e. Ringer's lactate) are preferred.

Each fluid prescription must be followed by reassessment to determine whether the therapeutic objective is being met. Where cardiogenic causes are suspected, the use of fluid to increase cardiac output is undertaken with particular caution with fluid challenges given in small steps (i.e. 50–100 ml colloid). In contrast, fluid challenges after hemorrhage are typically more aggressive (250–500 ml).

Debate persists as to the 'best' fluid for both emergent resuscitation and ongoing maintenance of intravascular volume in critically ill patients. With emergency resuscitation of intravascular volume, i.e. in hypovolemic and vasodilated septic shock, we recommend a balanced combination of colloids (e.g. albumin, Pentastarch) and isotonic crystalloids. Some practitioners recommend hypertonic saline as part of the early management of hypovolemic shock. Smaller doses of colloids and hypertonic solutions increase intravascular volume faster than larger volumes of isotonic crystalloids. They also facilitate movement of interstitial fluid into the circulation, making them more efficient resuscitation agents. However, attention should be paid to restoring extravascular volume during subsequent flow phase resuscitation. The most commonly used crystalloid solutions are Ringer's lactate and normal saline. Excessive use of normal saline may result in a hyperchloremic metabolic acidosis. Electrolyte solutions are the most commonly used fluids for maintenance therapy, although colloids are recommended if the colloid osmotic pressure remains excessively depressed.

Transfusion therapy

Current guidelines regarding red cell transfusions require consideration of the rate of blood loss, the hematocrit level, and hemodynamic parameters such as mean arterial pressure, cardiac index, and mixed venous oxygen saturation. The old 10/30 rule, i.e. an arbitrary decision to transfuse patients when the hemoglobin is below 10 g/dl (or hematocrit below 0.30), is now regarded as inappropriate because hemoglobin concentration is a poor clinical indicator of tissue oxygen delivery. Transfusion is rarely indicated when the hemoglobin concentration exceeds 10 g/dl and is almost always indicated when it is less than 6 g/dl, particularly when the anemia is acute. The decision as to whether intermediate hemoglobin concentrations (6–10 g/dl) justify red blood cell transfusions should be based on the patient's risk of developing complications arising from inadequate oxygenation. Those with transfusion needs at a relatively high hemoglobin level include patients with or at high risk of developing myocardial ischemia and those with cerebral vascular or peripheral circulatory disease.

Except in life-threatening situations with ongoing massive hemorrhage, intravascular volume should be normalized by crystalloid and colloid solutions before transfusion decisions are made. When transfusion is initiated, a 'unit-by-unit basis' approach is mandatory, with evaluation after each unit. Transfusion of one unit of blood in a non-bleeding adult increases hematocrit by approximately 3 per cent and hemoglobin concentration by 1 g/dl.

Pharmacological therapy

If hypotension is unresponsive to fluid therapy, or fluids cannot be used, pharmacological interventions with inotropes or vasodilators are indicated. In some patients, this approach may be further justified to increase vascular tone for a transient period parallel to the initiation of fluid therapy to prevent irreversible lesions of momentarily hypoperfused core organs. Table 5 summarizes the pharmacological profile and specific indications for drugs most commonly used in treating hypotension.

Sympathomimetic drugs are administered to increase vasoconstrictor tone by their effect on peripheral α-receptors and/or to increase heart rate and myocardial contractility by their effects on cardiac β_1-receptors and α-receptors. Phosphodiesterase inhibitors such as amrinone improve cardiac contractility by increasing cyclic AMP production but they also cause vasodilation. Once blood pressure has been restored to levels adequate to sustain perfusion of vital organs, vasodilator therapy may enhance cardiac output by reducing the ventricular afterload. Vasodilators may also be useful to decrease pulmonary artery wedge pressure in the presence of hydrostatic (cardiac) pulmonary edema.

Table 6 Therapeutic management of hypovolemia

Establish two large venous access sites
Intravenous fluids, combined colloids/crystalloids
Draw blood
- Biochemistry (hemoglobin (Hb), hematocrit, platelets, electrolytes, creatinine, lactate, coagulation)
- Blood gases
- Cross-match

Maintain $Pa\text{O}_2$ above 70 mmHg (9.3 kPa)
Consider blood component therapy (massive bleeding? Hb < 6 g/dl? Hb < 10 g/dl plus risk for inadequate oxygenation? coagulopathy?)
- Packed red blood cells
- Fresh frozen plasma
- Platelets

Pharmacological therapy
- Inotropes
- Correct severe acidosis

If the initial treatment is insufficient (i.e. persisting hypotension, no fall in lactate) consider further options
- Right heart catheter to guide the therapy
- Endoscopic control of bleeding
- Early surgery
- Rapid infusion system

Treatment issues in common causes of hypotension

Hypovolemic hypotension

Management objectives include limiting further fluid loss (source control) and restoring intravascular and extravascular fluid volumes (Table 6). To facilitate aggressive fluid therapy, at least two large venous cannulas are inserted. Placement of a sheath introducer into a major jugular or subclavian vessel also allows quick insertion of a right heart catheter if necessary.

Beyond the basic principles of resuscitation, pulmonary gas exchange, fluid and red blood cell transfusion therapy, and pharmacological interventions should also be optimized. Adequate monitoring and assessment of clinical and diagnostic markers of organ function are essential. Situations with extremely high and protracted blood loss can sometimes be controlled only by use of a high-rate rapid infusion device. In class III and IV hemorrhage, and after additional hemodilution due to fluid therapy, the infusion of asanguineous fluids alone is often ineffective in accomplishing resuscitation goals.

Infusion of blood components, rather than whole-blood preparations, is now standard practice. Red blood cell transfusions are used to maintain a sufficient O_2 transport capacity, while fresh frozen plasma and platelets are given to ensure adequate coagulation. The decision to use fresh frozen plasma or platelets is never drawn from biochemical markers of coagulation or the platelet concentration alone but is taken in conjunction with clinical evidence of insufficient coagulation.

Patients with hypovolemic hypotension sometimes fail to restore perfusing pressures and O_2 delivery despite aggressive fluid resuscitation. The clinician must consider whether there is ongoing, yet unrecognized, fluid loss or other causes of 'refractory' hypotension. There may also be coexistent cardiac disease (acute or chronic) which limits the appropriate responses to intravascular volume restoration. When blood pressure has not been restored with aggressive fluid resuscitation, sympathomimetics with predominantly vasoconstricting actions may be employed until the definitive cause of this 'unresponsive' shock state has been established and managed.

Cardiogenic hypotension

Management objectives include limiting further myocardial injury (in the presence of acute infarction), restoring myocardial pump function and thus improving cardiac output, and determining whether any mechanical defect amenable to an invasive intervention exists (e.g. valve replacement, angioplasty, or surgical revascularization).

Intubation and assisted ventilation are early considerations as such patients may die from respiratory fatigue because of the imbalance between O_2 supply to the respiratory muscles and the increased needs associated with a greater respiratory workload.

Circulatory management is facilitated by reviewing the patient according to the Forrester classification, which uses the presence of pulmonary congestion (both radiographic and auscultatory) and peripheral hypoperfusion (as diagnosed by decreased skin temperature, confusion, or oliguria in conjunction with either hypotension or tachycardia) (Table 7). This approach conveniently provides information about prognosis and directs early management (Fig. 2). Inherent in this approach is the use of the pulmonary artery catheter to measure left ventricular filling pressures and cardiac output. Although pulmonary artery wedge pressure is not a direct measure of cardiac preload, it does provide sufficient information to allow intelligent use of fluid in the management of cardiogenic hypotension.

Other key issues are pain relief, arrhythmia management, and treatment of pulmonary edema. Pain relief, for example by repetitive intravenous administration of small doses of morphine, will be accompanied by a reduction of myocardial O_2 consumption because endogenous catecholamine secretion will be reduced. Correcting

Table 7 Correlative classification of clinical and hemodynamic function after acute myocardial infarction

Subset	CI (l/min/m²)	PAWP (mmHg)
I No pulmonary congestion or peripheral hypoperfusion	⩾2.2	⩽18
II Isolated pulmonary congestion	⩾2.2	>18
III Isolated peripheral hypoperfusion	<2.2	⩽18
IV Both pulmonary congestion and hypoperfusion	<2.2	>18

CI, cardiac index; PAWP, pulmonary artery wedge pressure.
The classification is based on the evaluation of clinical and hemodynamic data of 200 patients with myocardial infarction.

CARDIOGENIC HYPOTENSION

↓

Cardiovascular stabilization
- Maintain $Pa\text{O}_2$ above 70 mmHg (9.3 kPa)
- Relieve pain
- Correct electrolyte abnormalities
- Treat arrhythmias
- Fluid therapy if no signs of pulmonary edema
- Vasoconstrictors in patients with Forrester III (isolated hypoperfusion) and Forrester IV (hypoperfusion + pulmonary edema)
- Dobutamine, diuretics if signs of pulmonary edema

↓

If no immediate response
- Hemodynamic monitoring
- Further diagnostic evaluation

↓

Differentiated pharmacological therapy
- Optimize preload:
  (1) PAWP <16–18 mmHg: colloidal fluid challenge
  (2) PAWP high + edema: nitroglycerin, diuretics
- To increase contractility: inotropes
- To decrease afterload: nitroglycerin, nitroprusside
- If pH<7.2, consider correction of acidosis

↓

No effect: refractory hypotension
- Additional fluid, increase inotrope dosage
- Add norepinephrine, epinephrine
- Intra-aortic balloon counterpulsation
- Reconsider surgical intervention (PTCA, bypass?)
- Consider ventricular assist device

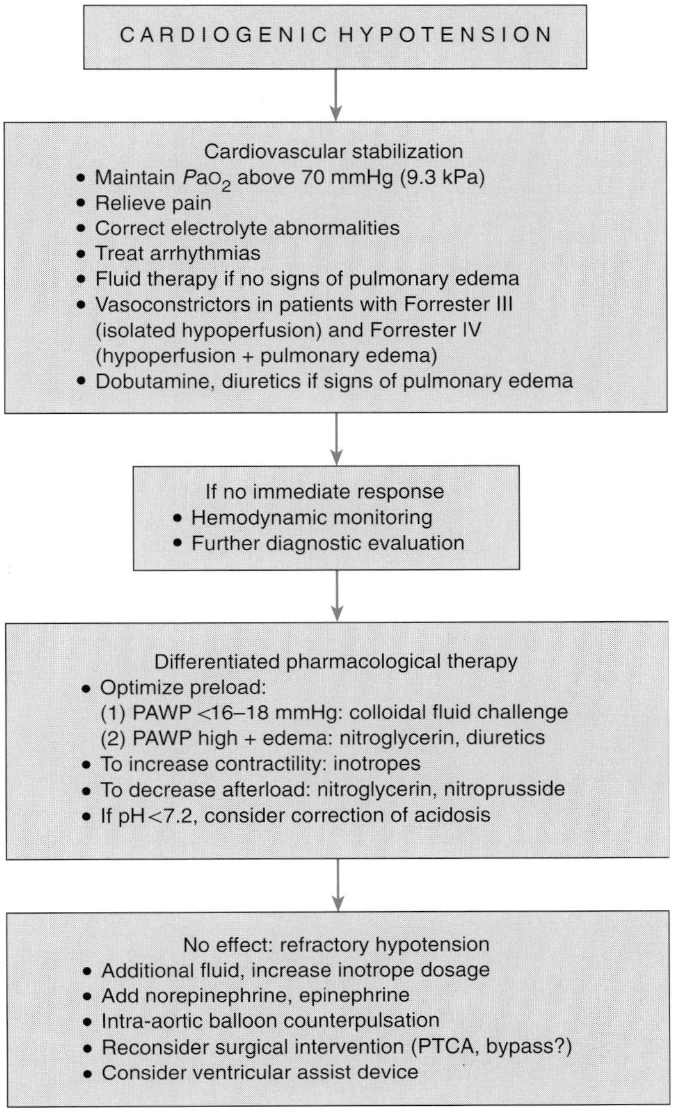

Fig. 2 Management of the patient with inadequate oxygen delivery. PTCA, percutaneous transluminal coronary angioplasty.

atrial arrhythmias may improve ventricular filling (with restoration of an atrial kick), thereby augmenting cardiac output. Correcting ventricular arrhythmias will also restore the cardiac output reserve. Primary issues to consider in the presence of arrhythmias include electrolyte abnormalities, pain, or arterial deoxygenation. Pacing may need to be considered for significant bradycardia. Continuous infusion of vasodilators plus diuretics may be needed for pulmonary edema.

Percutaneous insertion of an intra-aortic balloon counterpulsation may be considered for refractory cardiogenic hypotension to improve coronary blood flow and decrease left ventricular impedance (and thus afterload). Surgical intervention such as implantation of a ventricular assist device for bridging until a possible heart transplantation is only experimental at this stage. Mechanical abnormalities should also be excluded. Acute revascularization (thrombolysis, angioplasty, coronary artery bypass grafting) may be indicated. Other correctable lesions include acute valvular insufficiency (particularly mitral), ventricular rupture, and tamponade.

Septic hypotension

Management objectives (Table 8) include source control of the infectious process, by antibiotic therapy and drainage where appropriate, and ensuring that O_2 delivery is supported to match the typically elevated tissue metabolic needs.

Early localization and eradication of the inflammatory focus underlying the septic state is vital. Empirical antibiotic therapy is commenced after obtaining appropriate cultures. Although no clinical trial has revealed the optimal regimen in septic shock, a conservative approach includes a third-generation cephalosporin plus an aminoglycoside. If history and clinical examination suggest an anaerobic infection, then metronidazole or clindamycin should be added. Antibiotic therapy must be changed according to the susceptibility of specific organisms found by culture and sensitivity testing.

Initial therapy in sepsis also follows basic principles of resuscitation. Hypotension is often associated with hypovolemia and therefore may respond well to fluid therapy. Early red blood cell transfusion may be considered to maintain a hemoglobin of 100 to 130 g/l. Infusion of red cells stored for more than 15 days may produce splanchnic ischemia. Since vascular reactivity and myocardial contractility are impaired in sepsis, α- and β-agonists may be considered in the early treatment of circulatory compromise.

Table 8 Therapeutic management of the hypotensive patient with sepsis

Maintain $Pa\text{O}_2$ above 70 mmHg
Supportive therapies to increase O_2 delivery (>600 ml/min/m²)
  Fluids (colloids/crystalloids)
  Inotropes and vasopressors
  Red blood cell transfusion (Hb 100–130 g/l)

Antibiotic therapy
  Start with empirical therapy (third-generation cephalosporin+ aminoglygoside, +metronidazole or clindamycin if anaerobes are suspected)
  If cultures are positive, specific therapy

Consider correcting acidosis if pH ⩽ 7.2

After restoring normotension, therapy is directed at optimizing O_2 delivery to the tissues as this syndrome is a distributive disorder and a 'normal' blood pressure alone does not ensure adequacy of nutritive organ blood flow. Septic patients have elevated metabolic needs and an O_2 extraction deficit; therefore high normal levels of O_2 delivery ($> 600\,ml/min/m^2$) are considered beneficial. However, there is considerable controversy about the concept of augmenting O_2 delivery to 'supranormal' values (cardiac index above $4.5\,l/min/m^2$); there is no clear evidence that this improves patient outcome. Consequently, O_2 delivery may be supported towards a level thought to be appropriate for the patient's metabolic needs withot compromising organ function (e.g. increased myocardial O_2 consumption due to excessive doses of inotrope). This level of delivery may often be higher in a young patient than in an elderly patient with pre-existent cardiovascular morbidity. If the pulmonary artery wedge pressure is low, fluids may be beneficial; if it is high, vasodilators such as nitroglycerine may improve cardiac output.

A decrease in arterial lactate may be taken as a sign of response to therapy, since elevated arterial lactates have been associated with decreased survival from sepsis. The mixed venous oxygen saturation is maintained above 50 per cent to avoid the onset of regional O_2 debts in the face of a decreased ability of the tissues to extract O_2. Acidosis resulting in an arterial pH of less than 7.2 may indicate ongoing anaerobic metabolism.

At present it is only possible to offer supportive treatment for septic-specific lesions such as circulatory failure or impaired cellular O_2 utilization. New strategies currently under investigation, such as therapies directed at scavenging oxygen radicals, endotoxin, or other mediators of inflammation, may provide additional benefit.

Refractory hypotension

Beyond the consideration of more invasive therapeutic strategies, refractory hypotension should always lead to a re-evaluation of the diagnosis. In patients with trauma, pneumothorax or pericardial obstruction frequently remain unrecognized in the first few hours after the incident, but they may prevent (often concomitant to aggravation of the process) restoration of circulatory function. Addison's disease may lead to unresponsiveness of the circulation to standard therapies of hypotension and requires corticosteroid therapy.

Finally, the hypotensive condition may have caused organ failure (single or multiple), thus initiating further pathological processes such as gut ischemia, ischemia of the brain, or secondary myocardial depression, all of which are followed by additional circulatory failure.

Bibliography

American Heart Association (1992). Guidelines for cardiopulmonary resuscitation and emergency cardiac care, Parts I–III. *Journal of the American Medical Association*, **268**, 2171–241.

American Society of Anesthesiologists (1996). Practice guidelines for blood component therapy. *Anesthesiology*, **84**, 732–47.

Edwards, J.D. (1993). Management of septic shock. *British Medical Journal*, **306**, 1661–4.

Forrester, J.S., Diamond, G.A., and Swan, H.J.C. (1977). Correlative classification of clinical and hemodynamic function after acute myocardial infarction. *American Journal of Cardiology*, **39**, 137–45.

Goodnough, L.T. and Despotis, G.J. (1995). Establishing practice guidelines for surgical blood management. *American Journal of Surgery*, **170**, 16–20S.

3.6 Cardiac failure

3.6.1 Pathophysiology of myocardial failure and the systemic circulatory response

M. Singer

Key messages

- Ischemic heart disease and long-standing hypertension are the most common causes of heart failure, although numerous other causes (structural, electrical, pharmacological, metabolic, or inflammatory) should also be considered.
- Clinical manifestations do not necessarily correlate with the degree of ventricular dysfunction.
- Systolic and/or diastolic dysfunction may be present; systolic dysfunction usually predominates.
- Pulmonary edema is not the primary reason for dyspnea and is a poor guide to intravascular volume status.
- Compensatory mechanisms (within the heart and/or periphery) may prove paradoxically disadvantageous on ventricular stroke work and stroke volume.

Pathophysiology

Heart failure can be defined as the inability of the heart to meet the body's oxygen requirements. This definition is pertinent when considering treatment, as reducing demand is an important concept in patient management. Increases in ventricular filling pressures and/or volumes are usual, although not consistent, accompaniments.

Etiology

Although high-output failure states exist (e.g. sepsis, hyperthyroidism), the usual situation is pump failure leading to a low cardiac output. Ischemic heart disease and long-standing hypertension are the most common causes of myocardial failure, although other etiologies affecting muscle function (e.g. cardiomyopathy), valvular function (e.g. stenosis, endocarditis), electrical activity (e.g. bradycardia, tachyarrhythmia), or ventricular filling (e.g. pericardial tamponade) should also be considered (Table 1). Atheromatous coronary artery disease leads to either progressive myocardial ischemia and damage or, alternatively, an abrupt deterioration such as that seen following plaque and thrombus formation in acute myocardial infarction. A further scenario is acute decompensation where a chronically dysfunctional heart is subjected to an undue degree of stress, for example following infection or hemorrhage.

Table 1 Causes of heart failure

Loss of myocardial contractility
Partial, e.g. coronary artery disease (focal infarction/ischemia)
Generalized, e.g. widespread atheroma, dilated cardiomyopathy, myocarditis, malnutrition, drugs/poisons (e.g. β-blockers, alcohol), sepsis, metabolic (e.g. hypocalcemia, hypophosphatemia)
Overload
Pressure, e.g. hypertension, aortic stenosis, aortic coarctation
Volume, e.g. excess intravascular fluid, oligoanuric renal failure, valvular regurgitation (NB endocarditis), congenital left-to-right shunts, high output states (e.g. beri-beri, hyperthyroidism)
Abnormal filling
Mitral stenosis, atrial myxoma, restrictive cardiomyopathy, pericardial tamponade, constrictive pericarditis, pulmonary embolus, arrhythmias

The critically ill patient may be admitted to the intensive care unit with *de novo* heart failure or it may develop from a combination of factors including hypoperfusion, sepsis (where myocardial function is often depressed despite a normal or even high cardiac output), arrhythmias, drugs with undesired negative inotropic properties (e.g. most antiarrhythmics), metabolic abnormalities such as hypocalcemia and hypophosphatemia, pericardial effusion, and endocarditis.

Although an intracellular acidosis is negatively inotropic, myocardial dysfunction is often wrongly ascribed to extracellular (arterial) acidosis. Bicarbonate and equimolar saline have been compared in cardiac failure patients and no benefit was found from arterial pH correction. Patients with severe diabetic ketoacidosis rarely show any evidence of coexisting failure.

Another important distinction, albeit clinically difficult to assess, is to distinguish between a 'hibernating' and a 'stunned' myocardium. In the former, myocardial ischemia is ongoing, whereas in the stunned myocardium, which often follows an ischemia–reperfusion injury, blood flow is (almost) fully restored. Both retain a myocardial reserve and have reversible contractile dysfunction. However, with hibernation, an increase in contractile function is at the expense of metabolic recovery and treatment should aim to restore flow to the

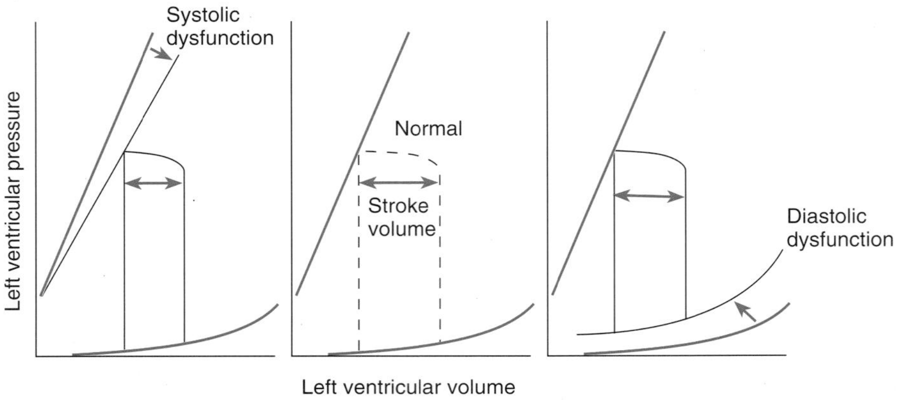

Fig. 1 Effects of systolic and diastolic dysfunction on the ventricular pressure–volume curve

hypoperfused tissue. Myocardial stunning often requires no treatment and usually recovers spontaneously although, if the patient is compromised, inotropic stimulation can be used without inducing further myocardial damage.

Cellular dysfunction

At the cellular level there appear to be abnormalities in cytosolic calcium handling, myofilament sensitivity to calcium, myocyte energetics, and phosphodiesterase and adenyl cyclase production/activity. Other changes include down-regulation of both β_1- and β_2-adrenergic receptors and possible uncoupling of these β-adrenoreceptors from adenylate cyclase. Alterations in the activity and/or number of myocyte enzymes and transport channels involved in the excitation–contraction coupling mechanism are also implicated in many of the above processes.

Clinical features

Clinical manifestations need not necessarily be apparent for a considerable degree of ventricular dysfunction to be present. Symptoms are usually manifest as breathlessness and fatigue resulting in limited exercise capacity (systolic dysfunction) and/or peripheral and pulmonary edema with orthopnea (diastolic dysfunction). However, severe pump failure will produce symptoms at rest, with evidence of impaired organ perfusion (oliguria, confusion, drowsiness) and/or lung congestion. Cardiogenic shock refers to the situation where oxygen supply by the heart is inadequate to prevent significant tissue hypoxia. This term is usually applied in error to describe a coexisting hypotensive state which need not be present.

Ventricular dysfunction

Systolic and diastolic dysfunction often coexist, although the systolic component usually predominates (Fig. 1). This relates to impaired ventricular contractility and a defective ability of the myofibrils to shorten against a load with a consequent decrease in ejection fraction. The ventricle lies on a flatter left ventricular function curve (Fig. 2); consequently, the ability to increase stroke volume by an increase in preload is consequently impaired. The ventricle is particularly sensitive to increases in afterload.

Diastolic dysfunction occurs despite a relatively normal ejection fraction and ventricular size. The fundamental process underlying this syndrome is a progressive decrease in ventricular compliance, resulting in marked increases in filling pressures at relatively normal end-diastolic volumes (Fig. 1). This abnormal pressure–volume relationship is related to changes in active relaxation during the initial phases of diastole and/or passive compliance during the slow-filling phase. Depending on the ventricle affected, rises in pressure will produce systemic and/or pulmonary venous congestion and interstitial edema. Compared with systolic dysfunction, there appears to be a lower sensitivity to increases in afterload and compensatory neuroendocrine activation is probably less marked. With decreases in

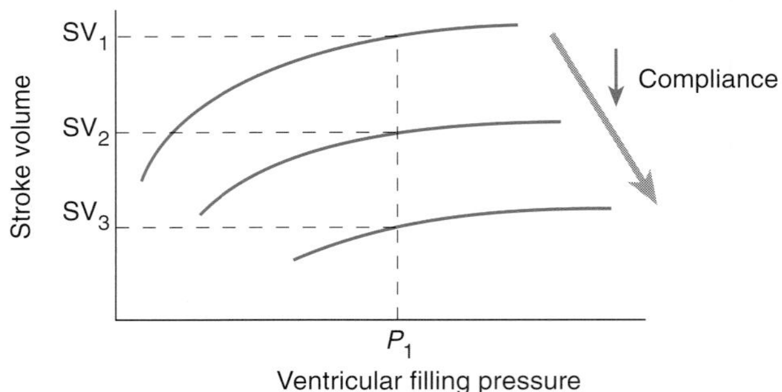

Fig. 2 Effect of progressive reductions in ventricular compliance on ventricular function curves. If blood pressure remains constant, a similar end-diastolic pressure will generate progressively lower stroke volumes as compliance decreases.

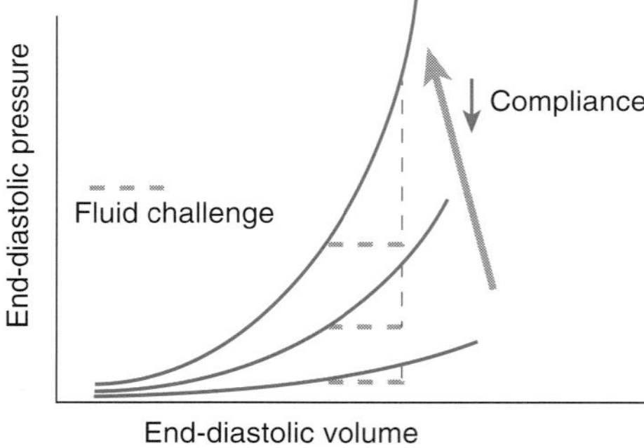

Fig. 3 Ventricular end-diastolic pressure–volume relationship. As compliance decreases, an increase in end-diastolic volume by a fluid challenge will generate a much higher increase in end-diastolic pressure. This can lead to venous congestion and edema.

compliance there is a progressively limited ability to respond positively to a fluid challenge (Fig. 3).

Diastolic dysfunction is more common in the elderly (up to 40 per cent); etiologies range from recurrent ischemia to hypertrophic cardiomyopathy, hypertension, and aortic stenosis. The response to standard failure therapy is generally worse in patients with predominant diastolic dysfunction.

One or both ventricles may be primarily affected by the underlying disease process. While symptoms and signs of systolic dysfunction are similar for both left and right ventricles, diastolic dysfunction affecting the left heart produces pulmonary edema and, for the right heart, hepatic congestion and dependent peripheral edema. A diseased ventricle will often impact upon the functioning ventricle (e.g. septal shift, increased pulmonary vascular resistance).

Pulmonary edema

Cardiogenic pulmonary edema formation occurs as a consequence of raised pulmonary venous pressures leading to increases in hydrostatic pressures. Increased pulmonary capillary permeability has been reported after myocardial infarction. Removal of interstitial edema is predominantly via the lymphatics. In chronic failure states, lymphatic clearance levels are raised considerably; however, in the acute situation, the lymphatics may take several days to reduce extravascular lung water and improve the radiographic appearance. The presence of pulmonary edema bears no relationship to intravascular volume and such patients should not be automatically assumed to be fluid overloaded.

Interstitial edema is erroneously considered to be the predominant reason for dyspnea in heart failure. Other factors include attempted respiratory compensation for the concurrent metabolic acidosis, respiratory muscle fatigue secondary to hypoperfusion, arterial hypoxemia arising from a low $S\text{v}O_2$, anxiety, and/or pain, and pulmonary venous engorgement leading to activation of afferent stretch receptors.

Compensatory mechanisms

Heart

In low-output failure the cardiac output is often maintained by increases in heart rate, chamber size, and muscle mass. Stroke volume usually falls unless severe bradycardia is present. Acute compensation occurs via sympathetic activation of heart rate. However, over days to weeks, progressive ventricular dilatation increases end-diastolic volume, thereby helping to maintain stroke volume in the face of a falling ejection fraction. By the Laplace law, dilatation will increase systolic wall stress; this is partially offset by a remodeling process in which functioning myofibers hypertrophy. The resultant increase in muscular mass permits an increase in cardiac work, although not at the expense of wall stress. Hypertrophy is a slow compensatory process which will nevertheless adversely affect diastolic function, ventricular compliance, and the end-diastolic pressure–volume relationship.

Ventriculoarterial coupling

The left ventricle can be considered an elastic chamber which periodically increases its volume elastance V_e to a value equal to the slope of the relationship between end-systolic pressure and end-systolic volume. The arterial load property is measured as arterial elastance E_a, i.e. the slope of the relationship between arterial end-systolic pressure P_{es} and stroke volume. The end-systolic elastance E_{es} varies in response to ventricular contractility, while P_{es} varies inversely with stroke volume SV for a given end-diastolic volume V_{ed}:

$$P_{es} = E_{es}(V_{ed} - \text{SV} - V_0)$$

$$= E_{es}(V_{es} - V_0)$$

where V_0 is a constant reference volume.

The ventricle produces maximal external work to the arterial load when the ventricular and arterial elastances are equalized. In normal subjects, the ventricular elastance is nearly twice as large as arterial elastance, thereby affording maximal efficiency. However, in moderate heart failure, the two elastance values are almost identical, thereby maximizing stroke work from a given end-diastolic volume but at the expense of work efficiency. Finally, in severe failure, ventricular elastance is less than half arterial elastance values, resulting in increased potential energy and an inability to maintain either stroke work or work efficiency properly.

Peripheral

In response to a falling cardiac output, neuroendocrine activation leads to vasoconstriction and, in the longer term, salt and water retention. Vasoconstriction is mediated via sympathetic and renin–angiotensin pathways, redistributing blood flow away from cutaneous, splanchnic, and muscular beds to more 'vital' organs such as brain and heart. Venous tone also increases, enhancing venous return. While vasoconstriction is an appropriate response to hemorrhage and hypovolemia, it is often deleterious in heart failure as myocardial work and function are adversely affected. The body initially attempts to compensate by raising circulating levels of atrial natriuretic peptide and adrenomedullin which enhance natriuresis and vasodilatation. However, over time, salt and water retention will predominate. This is primarily mediated by increased secretion of aldosterone and, to a lesser extent, antidiuretic hormone.

The negative consequences of these compensatory mechanisms include (i) a progressive decrease in renal blood flow and the potential development of renal dysfunction, (ii) pulmonary and systemic interstitial edema which may worsen oxygenation and symptoms, and (iii) a further fall in ventricular compliance and an increase in cardiac work, placing greater strain upon an already damaged heart.

Thus increased ventricular stiffness will result from pathophysiological processes occurring either as a direct consequence of heart failure or secondary to the body's compensatory mechanisms. Ischemia, ventricular hypertrophy, excessive afterload or preload, or an inadequate preload with compensatory vasoconstriction will all contribute further, as will misguided treatment regimens such as excessive diuresis. For the same end-diastolic pressure, a less compliant ventricle on a flatter ventricular function curve will generate a much lower stroke volume (Fig. 2). This underlines the potentially catastrophic consequences of afterload increases, the therapeutic importance of optimizing compliance, and the problems inherent in an over-reliance on pressure measurements (central venous pressure, pulmonary artery wedge pressure) to the exclusion of volume or flow. It further illustrates how a vicious downward spiral is actually facilitated by the body's attempts at correction.

Bibliography

Bogaert, M.G. and Fraeyman, N. (1991). Receptor function in heart failure. *American Journal of Medicine*, **90** (Supplement 5B), 10S–13S.

Gaasch, W.H. (1994). Diagnosis and treatment of heart failure based on left ventricular systolic or diastolic dysfunction. *Journal of the American Medical Association*, **271**, 1276–80.

Goldsmith, S.R. and Dick, C. (1993). Differentiating systolic from diastolic heart failure: pathophysiologic and therapeutic considerations. *American Journal of Medicine*, **95**, 645–55.

Sasayama, S. and Asanoi, H. (1991). Coupling between the heart and arterial system in heart failure. *American Journal of Medicine*, **90** (Supplement 5B), 14S–18S.

Schulz, R. and Heusch, G. (1994) Characterization of hibernating and stunned myocardium. *Herz*, **19**, 189–203.

3.6.2 Therapeutic strategy

M. Singer

Key messages

- Management is aimed at improvement of forward flow with restoration/maintenance of adequate organ perfusion and perfusion pressures, and relief of symptoms.
- The oxygen supply–demand balance should be optimized within both the myocardium and the body as a whole. Therapy should be aimed at both decreasing demand and increasing supply.
- Considerable improvement in outcome can be obtained with an aggressive regimen of myocardial revascularization and resting.

Introduction

The 5-year mortality from heart failure is approximately 50 per cent and directly correlates with both the severity of symptoms and exercise capacity. However, acute cardiogenic shock generally carries a hospital mortality of the order of 80 to 90 per cent, although recent data suggest considerable improvement with an aggressive regimen of myocardial revascularization and resting.

The fundamental therapeutic principles of heart failure management comprise improvement of forward flow with restoration/maintenance of adequate organ perfusion, and relief of symptoms. These symptoms arise as a consequence of a cardiac output/oxygen delivery that is insufficient to meet the body's needs, raised pulmonary venous pressures, and the body's attempt to compensate for low flow by peripheral vasoconstriction.

Appropriate management requires sound appreciation of the underlying pathophysiology, awareness of the actions and potential side-effects of each therapeutic intervention, and a level of monitoring and investigation sophisticated enough to assess disease severity and the effectiveness (or otherwise) of any treatment being given. Where possible, consideration of previous comorbid factors and chronic symptomatology should guide how aggressive intervention should be. However, these must be based on documented fact rather than hearsay or supposition. The patient should always be given the benefit of the doubt.

Principles of management

The aim is to maintain an adequate oxygen supply–demand balance within both the myocardium and the body as a whole. A decrease in cardiac output relative to the body's oxygen requirements is manifest as symptoms of poor forward flow, for example weakness, fatigue, oligoanuria, confusion, drowsiness, and tachypnea secondary to metabolic acidosis. Thus therapy should be aimed at both increasing supply and decreasing demand (Table 1).

Reducing demand

Reducing both myocardial and total body oxygen requirements is beneficial to the failing heart by lowering the cardiac work demanded of an already compromised and dysfunctional myocardium. This can be achieved pharmacologically by optimizing preload and afterload, and

Table 1 Achieving an adequate oxygen supply–demand balance

Decreasing the demand

Sedation, pain relief, calming (e.g. opiates)

Mechanical ventilation—permits heavier sedation and removes the work of breathing

Reduction in pyrexia (e.g. by antipyretics, tepid sponging)

Increasing the supply

Oxygen

FIO_2 0.6–1.0 to maintain (near) complete arterial oxygen saturation

Blood transfusion

The optimal figure is still contentious. A hemoglobin level of 10–12 g/dl is a reasonable compromise between oxygen carriage and microcirculatory flow

Optimization of preload and afterload

If fluid overloaded (rare): venodilatation+diuretics to optimal stroke volume

If hypovolemic: fluid challenge to optimal stroke volume

If vasoconstriction persists (SVR > 1400 dyn s/cm^5): arterial dilatation through a combination of dilator+fluid may be necessary

Inotrope administration

Inotropes increase cardiac work and place additional stress on an already diseased heart so should ideally be used after preload/afterload optimization when an inadequate organ perfusion and/or blood pressure remains

Mechanical ventilation+PEEP

Positive-pressure ventilation+additional PEEP will often augment cardiac output in heart failure states by preload and left ventricular afterload reduction

Intra-aortic balloon counterpulsation

Improves forward flow, augments coronary perfusion, and reduces afterload

Correction of any anatomical abnormalities (e.g. valve replacement, pericardiocentesis)

thus ventricular compliance. Such measures include appropriately monitored fluid challenges, vasodilators, and, in the relatively unusual situation of intravascular volume overload, diuretics. Mechanical support devices are also effective; these include early mechanical ventilation, which rests the heart by reducing preload and afterload, maintaining normoxemia, and removing the work of breathing; 30 to 40 per cent of cardiac output may be required purely to support the work of breathing in a dyspneic subject. Respiratory muscle fatigue leading to hypercapnia is a prelude to cardiorespiratory collapse.

Total body oxygen consumption can be reduced further by alleviation of pain or anxiety. This is achieved by psychological support in addition to adequate analgesia (usually opiates which also have intrinsic vasodilating and anxiolytic properties).

Increasing supply

Apart from reducing cardiac work, optimizing preload and afterload also enhances forward flow. Additional augmentation can be achieved by inotropic agents, although usually at the expense of increased cardiac work. Mechanical supports such as intra-aortic balloon counterpulsation and ventricular assist devices can also increase output.

Achieving adequate perfusion pressure

Despite improving flow, organ function may still remain compromised by an inadequate perfusion pressure. Inotropes, intra-aortic balloon counterpulsation, and, for high-output states, vasopressors may be used alone or in combination to elevate perfusion pressure. The ideal perfusion pressure varies between individuals. In general, patients who were not previously hypertensive can often cope with mean systemic blood pressures in the range 60 to 70 mmHg (or lower), while long-term hypertensives often require higher mean pressures (e.g. 70–80 mmHg). The drawbacks of an excessive blood pressure are an increase in cardiac work and a decreased output.

Relief of symptoms

1. Tachypnea results from pain, anxiety, metabolic acidosis, hypoxemia, and engorgement of the pulmonary vasculature. Treatment to provide analgesia and anxiolysis, improvement of organ perfusion, and reduction in pulmonary venous pressures, with supplemental high-flow oxygen administration, will often improve symptoms dramatically. Non-invasive ventilatory support, for example continuous positive airways pressure (CPAP), may also be useful.

2. Pain relief is provided by adequate analgesia and improved coronary perfusion.

3. Symptoms of poor forward flow (confusion, oliguria, fatigue, respiratory failure) are relieved by improving organ perfusion.

Investigation and monitoring

While 'first-aid' therapeutic measures are being implemented, continuous ECG and pulse oximetry monitoring and frequent sphygmomanometric blood pressure measurements should be commenced. Early and prompt invasive monitoring of cardiovascular function should be instituted, not only in the severely ill patient but also in those patients who are either deteriorating or responding poorly to initial therapy. This usually consists of arterial cannulation for continuous measurement of systemic blood pressure and intermittent blood gas analysis, bladder catheterization for hourly measurement of urine output, and pulmonary artery catheterization for measurement of stroke volume, intracardiac pressures, and mixed venous oxygen saturation, and calculation of systemic and pulmonary vascular resistances. Newer flow monitoring techniques such as aortic Doppler ultrasonography may also play a useful role.

Urgent investigations include the following: 12-lead (and sometimes right-sided) electrocardiography to diagnose any rhythm disturbances, evidence of infarction (new or old), or ongoing ischemia (ST-segment depression, T-wave inversion); cardiac enzymes to indicate recent infarction and perhaps prompt urgent thrombolysis with or without angioplasty or surgical revascularization; echocardiography to detect structural defects including heart chamber and wall dimension abnormalities, valvular disorders, septal defects, pericardial effusion, and segmental wall motion irregularities; chest radiography to seek abnormalities suggestive of specific disorders such as pneumothorax, aortic aneurysm, and pulmonary embolus.

Markers of severity

Bedside diagnosis of left ventricular failure is based on history, symptoms, and clinical signs. Right heart failure may present without chest signs but with evidence of poor organ perfusion and, over time, venous congestion (e.g. peripheral edema, hepatomegaly).

A greater degree of heart failure is suggested by progressive dyspnea, obtundation, hypotension, oliguria, and cyanosis. Biochemical markers of severity include a metabolic acidosis with associated hyperlactatemia. A low stroke volume and mixed venous oxygen saturation below 60 per cent are additional indicators of poor cardiac performance and inability to meet tissue oxygen demands. Pulmonary artery wedge pressures and central venous pressures are not necessarily elevated.

Treatment

Heart failure should be treated swiftly. The presence of any marker of severity as outlined above should prompt a more aggressive and interventionist approach to both monitoring and therapy. Etiology can usually be ascertained rapidly from history, examination, and investigations including electrocardiography, chest radiography, echocardiography, and biochemistry (cardiac enzymes). Specific diagnoses warrant specific treatments, for example thrombolysis for myocardial infarction (unless contraindicated), antiarrhythmics or cardioversion for any compromising arrhythmia, pericardiocentesis for tamponade, and valve replacement for significant valvular stenosis.

First aid

General treatment should follow standard ABC resuscitation guidelines. High-flow high-concentration oxygen should be administered. Addition of CPAP may be considered to improve oxygenation and aid hemodynamics. Early mechanical ventilation should be instituted promptly in severe cases for the reasons described previously.

Diuretic therapy is usually inappropriate; the vast majority of patients presenting in acute heart failure, and even many with acute-on-chronic failure, are hypovolemic on hospital admission as a consequence of sweating, vomiting, not drinking, and third-space fluid shifts. The beneficial immediate vasodilating effect of a loop diuretic is quickly negated by an often brisk diuresis. This further exacerbates any existing hypovolemia for which the body attempts to compensate by additional vasoconstriction, thereby placing an extra load on the heart which results in further reductions in cardiac output.

Vasodilatation is thus desirable and can be rapidly achieved, resulting in symptomatic relief, by four sublingual puffs of glyceryl trinitrate (nitroglycerin) spray followed by an intravenous nitrate infusion titrated to optimal effect. A patient presenting with low-output failure and hypotension may be unable to tolerate nitrates unless hypovolemia has been corrected and obstruction to flow excluded (e.g. pericardial tamponade). Myocardial function will remain compromised until an adequate perfusion pressure and flow are achieved. This may necessitate the 'blind' use of an inotrope such as epinephrine (adrenaline) until more sophisticated monitoring can be introduced.

Directed therapy

This requires instrumentation for monitoring as described previously. Suggested targets, many of which indicate adequate organ perfusion, are given in Table 2. No target figures are given for cardiac output, oxygen delivery, or oxygen consumption, as individual variations in the oxygen supply–demand balance are such that, while some patients may remain undertreated, many others will have their myocardium needlessly driven. Newer techniques such as gastric tonometry or tissue oxygen tension measurement require further validation before being recommended for general clinical usage.

Pharmacological support

The choice of therapeutic agent is dictated by therapeutic benefit, familiarity, and desired pharmacological action. If a drug proves unpredictably detrimental, it should either be substituted by another agent that achieves the required effect or any hitherto unrecognized problem should be dealt with appropriately. Dobutamine may be used as an illustrative agent for both circumstances: in some patients it may induce excessive vasodilatation and hypotension and require substitution with epinephrine, whereas generation of tachycardia or tachyarrhythmias implies possible concurrent hypovolemia requiring volume administration. Familiarity with the drugs being used and awareness of both their actions and potential side-effects are crucial.

1. Oxygen should be administered to ensure (near) complete arterial oxygenation.

2. Pain secondary to ongoing myocardial ischemia is best treated by nitrates, although opiates are useful for their additional anxiolytic and vasodilating properties.

3. Vasodilator agents such as glyceryl trinitrate, hydralazine, and sodium nitroprusside all have effects on both preload and afterload. Significant falls in blood pressure with small doses of the above imply either hypovolemia or flow obstruction. Nitrate tolerance develops after 24 h, with increasing doses required to maintain effect. Cyanide accumulation may develop after 24 h with sodium nitroprusside usage. It is this author's preference to commence angiotensin-converting enzyme inhibitors such as captopril at this time, provided that blood pressure and lack of contraindications permit, in rapidly escalating dosage while slowly weaning the original vasodilator.

4. Judicious fluid challenges are often appropriate as the heart failure patient often arrives in the intensive care unit in a hypovolemic state. The endpoint is no rise in stroke volume to a 100- to 200-ml fluid challenge. In the absence of flow monitoring, a rise in central venous or wedge pressure of 3 mmHg or more following a fluid challenge is suggestive of an adequately filled intravascular compartment.

5. Loop diuretics are often inappropriate and detrimental in the first instance for the reasons described earlier. However, they may be indicated in fluid-overload situations, when the patient is on chronic diuretic therapy, and when symptoms or oliguria persist after adequate loading with fluid and angiotensin-converting enzyme inhibitors.

6. Inotropes are indicated where low pressure and/or low output result in inadequate organ function. Epinephrine usually produces consistent rises in both. Inotropes with associated vasodilator activity, including dobutamine, dopexamine, enoximone, and milrinone, may all usefully improve output although they can occasionally produce excessive falls in blood pressure.

7. Pressors are rarely indicated, although norepinephrine (noradrenaline) is occasionally needed to generate a blood pressure sufficient to perfuse the organs, in particular the heart, to generate an adequate output. These agents should never be used without flow monitoring as they will more frequently compromise flow in the heart failure patient.

Table 2 Therapeutic endpoints

Adequate blood pressure to support adequate organ perfusion
Optimization of preload and afterload to maximize stroke volume
Mixed venous oxygen saturation $\geqslant$60%
Urine output $\geqslant$0.5 ml/kg/h
Removal of any metabolic acidosis secondary to lactic acid production
Removal of any evidence of persisting ischemia (seen on ECG)

Mechanical support

For the reasons described above, positive-pressure ventilation with or without positive end-expiratory pressure (PEEP), CPAP, intra-aortic balloon counterpulsation, and venticular assist devices may be useful adjuncts to treatment. In particular, mechanical ventilation is a much underestimated support mode.

Bibliography

Aubier, M., Trippenbach, T., and Roussos, C. (1981). Respiratory muscle fatigue during cardiogenic shock. *Journal of Applied Physiology*, **51**, 499–508.

Eltchaninoff, H., Simpfendorfer, C., Franco, I., Raymond, R.E., Casale, P.N., and Whitlow, P.L. (1995). Early and 1-year survival rates in acute myocardial infarction complicated by cardiogenic shock: a retrospective study comparing coronary angioplasty with medical treatment. *American Heart Journal*, **130**, 459–64.

Gaasch, W.H. (1994). Diagnosis and treatment of heart failure based on left ventricular systolic or diastolic dysfunction. *Journal of the American Medical Association*, **271**, 1276–80.

Goldsmith, S.R. and Dick, C. (1993). Differentiating systolic from diastolic heart failure: pathophysiologic and therapeutic considerations. *American Journal of Medicine*, **95**, 645–55.

Northridge, D. (1996). Frusemide or nitrates for acute heart failure? *Lancet*, **347**, 667–8.

3.7 Tachyarrhythmias

3.7.1 Causes and diagnosis

M. C. Sayers and C. M. Gasperetti

Key messages

- Look for P waves on the ECG to help diagnose the arrhythmia.
- With rapid rates, use calipers to measure RR intervals as irregular rhythms may appear to be regular.
- A wide complex tachycardia in an older patient, particularly with a history of heart disease, is usually ventricular tachycardia.

Introduction

Both supraventricular and ventricular tachycardia are frequently seen in the critically ill. If the patient is hemodynamically stable, a 12-lead ECG should be obtained initially and compared with a baseline recording. A supraventricular tachycardia is present if the QRS complex is narrow (less than 0.12 s). If it is 0.12 s or more, ventricular tachycardia must be differentiated from supraventricular tachycardia with aberrancy, provided that an underlying bundle branch block is not present. The ECG should then be scanned for P waves (best seen in leads II and V_1) and their relationship to the QRS complex. This relationship, together with the configuration of the P wave, may provide useful information, as may Lewis leads, i.e. placing the right and left arm leads along the sternal borders and monitoring lead I.

Supraventricular tachyarrhythmia

Sinus tachycardia

Sinus tachycardia is defined as a sinus rate above 100 beats/min. The P waves have a normal appearance, although they can be difficult to visualize with a rapid rate. Since this is an automatic focus, carotid sinus massage may slow the rate, allowing the P waves to become more prominent. There are many causes of sinus tachycardia, including pain, anxiety, fever, hypoxemia, volume depletion, thyrotoxicosis, drugs, heart failure, ischemia, hypotension, anemia, pulmonary embolus, and shock.

Sinus node re-entry tachycardia

This is an uncommon cause of supraventricular tachyarrhythmia, usually occurring in patients with underlying heart disease. The re-entry circuit is within the sinus node and so it will be identical to a sinus tachycardia. Vagal maneuvers may help to differentiate them (Table 1). Sinus node re-entry tachycardia is sudden in onset and termination.

Atrial premature contractions

Atrial premature contractions are also due to enhanced automaticity and are usually distinguished by a premature P wave that differs in configuration from the sinus P wave. Depending on where the focus occurs, the PR interval may be the same, longer, or shorter than the normal PR interval. If the atrial focus occurs in the low atrium, the PR interval will be shorter and the P wave will be inverted in the inferior leads as depolarization occurs in a retrograde fashion. The QRS complex should remain normal unless the atrial premature contraction occurs during the relative or effective refractory period of the previous beat. During the effective refractory period, the atrial premature contraction will be blocked and only the P wave can be identified unless it is obscured by the previous T wave. The pause that occurs is less than compensatory owing to resetting of the sinus node. When the atrial premature contraction occurs during the relative effective refractory period, the QRS will be wide and usually of right bundle block pattern as this area may not yet have recovered.

Atrioventricular nodal re-entry tachycardia

Nodal tachycardia, which is the most common cause of paroxysmal supraventricular tachyarrhythmia, usually occurs at rates of 100 to 180 beats/min. Although it is usually seen in young patients who have no cardiac disease, it may occur at any age and be associated with digitalis toxicity, pulmonary embolus, myocardial infarction, rheumatic heart disease, and chronic obstructive pulmonary disease. The sequence is usually initiated by a premature atrial beat which enters the re-entry loop in the atrioventricular node, activating both atrium and ventricle (Fig. 1). The P wave is not usually visible, and the QRS complex is identical to that occurring during normal sinus rhythm. The rhythm is also regular, which helps to differentiate it from atrial fibrillation or atrial flutter with varying atrioventricular block (Fig. 2). It is important to distinguish this rhythm from sinus tachycardia, which can be more difficult. Increasing vagal activity by carotid sinus massage may be helpful (Table 1).

Atrioventricular reciprocating tachycardia

Atrioventricular reciprocating tachycardia, which is another common form of paroxysmal supraventricular tachyarrhythmia, may also occur in young people who have no structural heart disease. An accessory

Table 1 Differentiation of types of supraventricular tachycardia

Type of supraventricular tachycardia	Ventricular rate (beats/min)	Effect of vagal stimulation	ECG	Mechanism
Sinus tachycardia	100–160	Gradual slowing Rate resumes when stimulation is removed	Normal P waves	Automatic
Sinus node	100–160	No effect or abrupt termination	Normal P waves	Re-entry
AV nodal tachycardia	150–200	No effect or abrupt termination	P wave usually not seen	Re-entry
AV reciprocating tachycardia	150–250	No effect or abrupt termination	P wave may not be identified or may follow QRS and be inverted in II, III, aVF	Re-entry
Automatic atrial tachycardia	100–160	No effect or may increase AV block	Abnormal P wave; may be AV block	Automatic
Intra-atrial entry	100–160	No effect or may increase AV block	Abnormal P wave; may be inverted in II, III, aVF	Re-entry without SA or AV involvement
Multifocal atrial tachycardia	100–160	No effect or may increase AV block	Variable P wave At least 3 separate foci Irregular ventricular response	Automatic
Atrial fibrillation	130–180	Increases AV block with slowing of ventricular response	Chaotic fibrillatory waves Irregular ventricular response	Presumed re-entry
Atrial flutter	140–320	Increases AV block with slowing of ventricular response	'Sawtooth' P waves; may be inverted in II, III, aVF	Re-entry
Non-paroxysmal junctional tachycardia	70–130	No effect or slows the rate	May be AV dissociation or inverted T waves in II, III, aVF	Automatic
Paroxysmal junctional tachycardia	150–200	No effect or abrupt termination	P wave usually not seen or inverted in II, III, aVF	Re-entry

AV, atrioventricular; SA, sinoatrial.

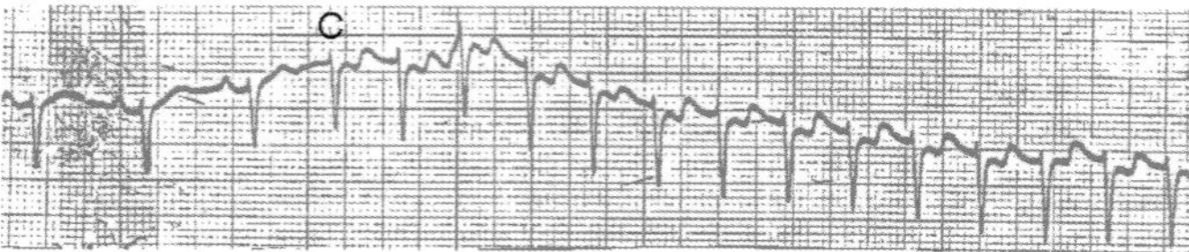

Fig. 1 Atrial premature contraction initiating supraventricular tachycardia.

pathway is required for completion of the circuit. The Wolff–Parkinson–White syndrome, which involves conduction of impulses from atrium to ventricle through the Kent bundle, is the most common pre-excitation syndrome. Classically, the baseline ECG shows a short PR interval with a delta wave and a broad QRS complex as the ventricle is depolarized prematurely. However, this

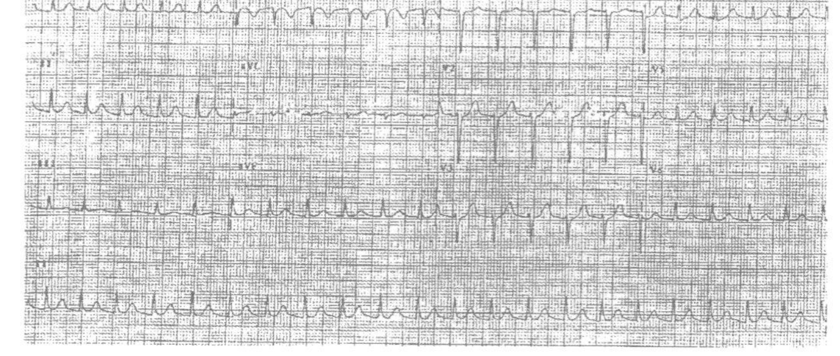

Fig. 2 Atrioventricular nodal tachycardia (rate, 130 beats/min). P waves are not visualized.

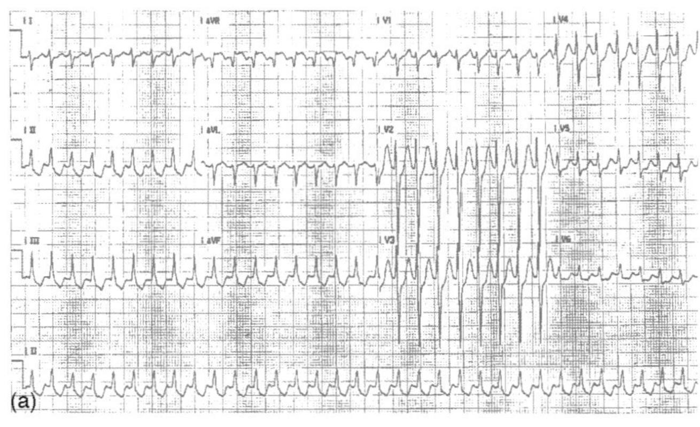

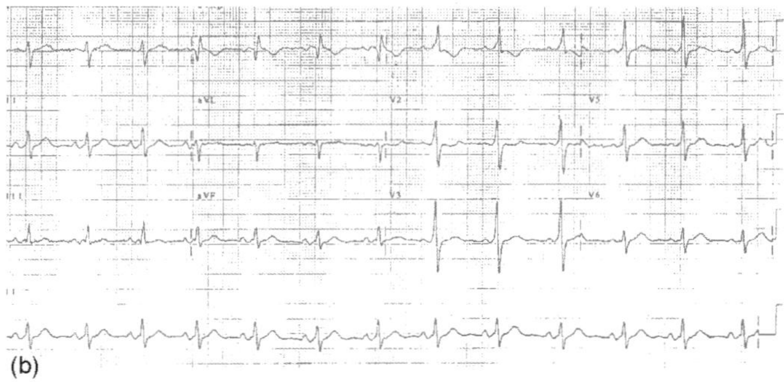

Fig. 3 (a) Wolff–Parkinson–White syndrome with paroxysmal supraventricular tachycardia. Notice that the rate is over 200 beats/min.

may be inconsistent and the QRS complex may be normal. Medication, stress, or exercise may increase conduction through the Kent fibers. Patients with accessory pathways are prone to paroxysmal supraventricular tachycardia as well as atrial fibrillation and atrial flutter (Fig. 3). Paroxysmal supraventricular tachyarrhythmia is usually antegrade over the atrioventricular node and retrograde through the accessory pathway, thereby giving a normal QRS complex. However, in approximately 10 per cent of cases, antegrade conduction is over the accessory pathway with retrograde conduction through the atrioventricular node, resulting in a wide complex tachycardia that may be difficult to distinguish from ventricular tachycardia (Table 2). Distinguishing atrioventricular nodal re-entry tachycardia from reciprocating tachycardia using a bypass tract can be difficult, but it is usually not critical because both may respond to the same treatment. Rates above 200 beats/min suggest that an accessory pathway may be involved.

Intra-atrial re-entry tachycardia

Intra-atrial re-entry tachycardia involves a circuit in the atrium and does not involve the atrioventricular node. Like paroxysmal atrial tachycardia, it is frequently initiated by an atrial premature contraction, establishing the re-entry circuit. QRS morphology is the same as seen in sinus rhythm; however, the P wave has a different morphology. As with other re-entry arrhythmias, onset and termination are sudden, but carotid sinus massage will not cause termination since neither sinoatrial nor atrioventricular nodes are involved.

Table 2 Differentiation of wide complex supraventricular tachycardia from ventricular tachycardia

	Favoring SVT with aberrancy	**Favoring ventricular tachycardia**
Rate	100–250 beats/min; may be greater than 170 beats/min	100–180 beats/min; usually less than 170 beats/min
Rhythm	Regular except A-fib, or atrial flutter with varying AV block or MAT	May demonstrate slight irregularity especially with captured beats
QRS pattern	Usually RBBB-V_1, triphasic (rsR) V_6–RS ratio usually >1 LBBB, V_1–rS or QS, V_6–R	Usually RBBB-V_1, monophasic V_6–RS ratio may be <1 LBBB, V_6–QS or qR
QRS duration	Less than 0.14 s	0.14 s or greater
QRS axis	Usually normal	Abnormal, frequently shifted to left
QRS configuration	Usually initial deflection same as normal QRS	Bizarre complex May have positive deflection in all precordial leads
AV conduction	If seen, P wave associated with QRS complex	AV dissociation May see retrograde P waves, fusion beats
Physical examination (except A-fib and MAT)	Cannon A waves, none or will be regular Regular pulse Constant first heart sound	Irregular Cannon A waves Varying peripheral pulse Varying first heart sound

AV, atrioventricular; SVT, supraventricular tachycardia; A-fib, atrial fibrillation; MAT, multifocal atrial tachycardia; LBBB, left bundle branch block; RBBB, right bundle branch block.

Non-paroxysmal atrial tachycardia

Non-paroxysmal atrial tachycardia, which is also called automatic tachycardia, is caused by enhanced atrial automaticity as an atrial focus depolarizes faster than the sinus node and becomes the pacemaker. It may be seen with toxic effects of digitalis, chronic obstructive pulmonary disease, heart failure, pneumonia with hypoxemia, certain drugs such as aminophylline, and metabolic derangements. The atrial rate is usually between 110 and 130 beats/min, and there may be a 2:1 or 3:1 atrioventricular block, particularly if digoxin is involved. As with other ectopic atrial rhythms, the P wave differs from that seen during normal sinus rhythm.

Multifocal tachycardia

Multifocal tachycardia is a very rapid irregular rhythm that can be difficult to differentiate from atrial fibrillation. Finding the P wave is very helpful as this rhythm is associated with distinct P waves that vary in morphology (Fig. 4). As with non-paroxysmal atrial tachycardia, the rate is usually 110 to 130 beats/min and is due to enhanced automaticity. It also has the same precipitating factors (e.g. pulmonary disease, hypoxemia, heart failure, acidosis, and theophylline toxicity).

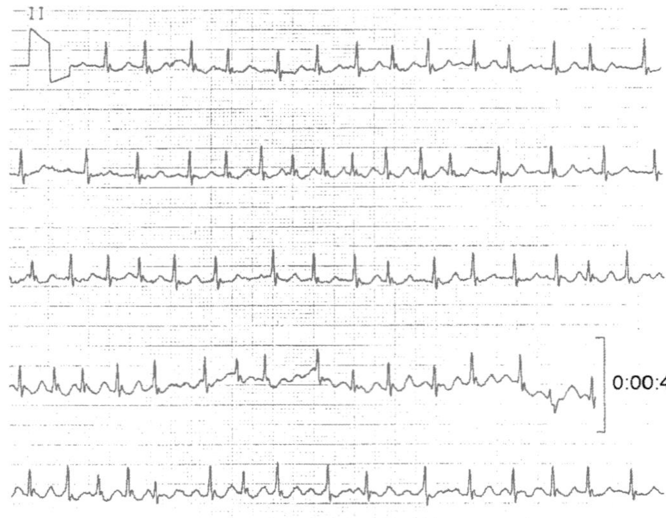

Fig. 4 Multifocal atrial tachycardia. Notice P waves of different morphology.

Atrial fibrillation

Atrial fibrillation, which is a common condition, occasionally occurs in normal individuals without evidence of structural heart disease, although is usually associated with abnormalities such as hypertensive or valvular heart disease, coronary artery disease, cardiomyopathies, pulmonary embolism, pericarditis, myocarditis, alcohol abuse, thyrotoxicosis, and chronic obstructive pulmonary disease. When seen in the setting of an acute anterior myocardial infarction, atrial fibrillation may be associated with higher morbidity and mortality as it may indicate larger areas of myocardial damage. There may be associated pericarditis, higher intra-atrial pressures, or atrial infarction.

Atrial fibrillation is completely irregular on the ECG, and P waves are not present. There is usually an undulating rhythm at a rate of 600 beats/min. Physical examination may be helpful by demonstrating fibrillatory jugular venous waves and a variable first heart sound. Even though this rhythm is highly irregular, it is believed to be re-entry in origin with the occurrence of a random sequence of depolarization. In patients without atrioventricular node disease, the ventricular response is usually 120 to 180 beats/min. The QRS complex is narrow unless aberrant ventricular conduction or a bundle branch block is present. Occasionally, Ashman's phenomenon will occur, particularly with a rapid ventricular response. This is seen when a short RR interval is preceded by a long RR interval and is usually of a right bundle branch block configuration. The long RR interval increases the refractory period of the following beat, thus accounting for the aberrancy (Fig. 5) The QRS may also be broadened in patients who have Wolff–Parkinson–White syndrome when depolarization is conducted down the accessory bypass tract. This rapid conduction can lead to ventricular rates in excess of 300 beats/min and can degenerate into ventricular fibrillation.

Atrial fibrillation with aberrancy can be difficult to distinguish from ventricular tachycardia. However, carefully plotting the RR intervals will be helpful because atrial fibrillation is grossly irregular while ventricular tachycardia is usually regular, although subtle irregularities may be seen.

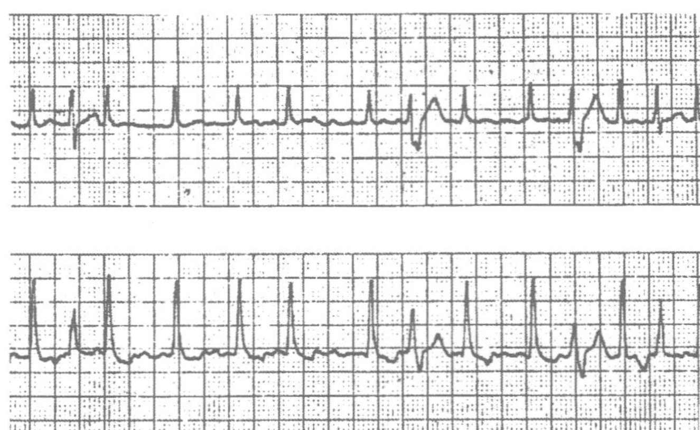

Fig. 5 Atrial fibrillation with Ashman's phenomenon.

Atrial flutter

Atrial flutter, which is another example of an entry tachycardia not involving the atrioventricular node, produces an extremely rapid atrial rate as the re-entry circuit involves the atrial walls. Underlying heart disease such as coronary artery or valvular disease, primary myocardial disease, or pericarditis almost always exists. This arrhythmia may also be associated with thyrotoxicosis, chronic obstructive pulmonary disease, pulmonary embolus, and alcohol consumption. Atrial rates are usually between 250 and 350 beats/min; in the absence of atrioventricular node disease there is usually a 2:1 block with a ventricular response of approximately 150 beats/min. As with atrial fibrillation, patients with Wolff–Parkinson–White syndrome may conduct down the accessory bypass tract and have a much faster rate which could degenerate into ventricular fibrillation. Conduction through the accessory bypass tract produces a wide complex tachycardia that may be difficult to distinguish from ventricular tachycardia. The ECG typically shows sawtooth flutter waves best seen in leads II or V_1 (Fig. 6). Carotid sinus massage may increase the atrioventricular block transiently to make the flutter waves more obvious.

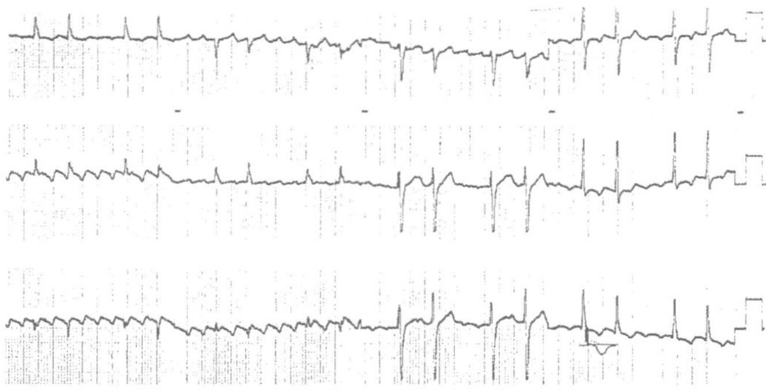

Fig. 6 Atrial flutter. Notice the sawtooth flutter waves best seen in the inferior leads.

Junctional tachyarrhythmias

Non-paroxysmal junctional tachycardia

The atrioventricular junction usually fires at 40 to 60 beats/min, but can become the pacemaker if the sinus node fails. When the rate exceeds 60 beats/min, by definition this becomes a junctional tachycardia. This may be seen in patients who have chronic obstructive pulmonary disease or structural heart disease, particularly after cardiac surgery. It may also be seen with digitalis toxicity following myocardial infarction, rheumatic heart disease, or myocarditis. Atrioventricular dissociation may occur as the atria continue to depolarize from the sinus node and the ventricle is depolarized from the junctional tachycardia. Alternatively, there may be retrograde depolarization of the atrium with inverted P waves in leads II, III, and aVF. The PR interval is usually less than 0.12 s.

Paroxysmal junctional tachycardia

This may occur and is identical to paroxysmal supraventricular tachyarrhythmia in rate and configuration.

Ventricular tachyarrhythmias

Ventricular premature complexes

Ventricular premature complexes arise below the bundle of His, prematurely depolarizing the ventricular myocardium and resulting in wide complexes with no preceding P wave. These beats are usually blocked in the atrioventricular node and there is no retrograde depolarization of the atrium. The sinus node continues to fire at its normal rate, thereby resulting in a fully compensatory pause (twice the normal RR interval). This pause does not occur with atrial premature contractions and, when present, helps to establish the diagnosis of ventricular premature complex. Occasionally, however, the ventricular premature complex does penetrate the atrioventricular node in a retrograde fashion and may reset the sinus node, resulting in a pause that is less than fully compensatory. Ventricular premature complexes may be monomorphic or polymorphic, and there is usually a constant interval between the sinus complex and the ventricular premature complex. This is referred to as fixed coupling and supports the concept that most are due to re-entry. If the re-entry circuit continues, more complexes may occur with pairs known as ventricular couplets and three or more sequential ventricular premature complexes resulting in ventricular tachycardia. When every other complex is a ventricular premature complex, the pattern is known as ventricular bigeminy; with every third it is known as trigeminy, and with every fourth it is known as quadrigeminy.

Ventricular premature complexes may occur in patients without cardiac or pulmonary disease and do not need to be treated. However, the detection of frequent ventricular premature complexes in an individual may warrant a thorough cardiovascular evaluation. If myocardial infarction has occurred, the tolerance for ventricular ectopy is decreased. Other causes include digitalis toxicity, cardiomyopathy, valvular heart disease, myocarditis, congenital heart disease, long QT syndrome, hypoxemia, hypokalemia, hypomagnesemia, tricyclic antidepressants, antiarrhythmic medication, theophylline toxicity, and other drugs.

Parasystole

Parasystole represents an automatic focus that fires at a regular fixed interval. The surrounding tissue will not be depolarized if the impulse occurs during the ventricular refractory period or if it is blocked. The intervals that occur will be multiples of the regular rate of the ectopic focus. There is no fixed coupling in this situation since the focus is independent. Occasionally, fusion beats may be seen.

Accelerated idioventricular rhythm

Accelerated idioventricular rhythm is a non-paroxysmal ventricular tachycardia with rates usually between 70 and 100 beats/min. It involves an automatic focus and is usually seen in the setting of an acute inferior myocardial infarction with sinus bradycardia, but may be also associated with digitalis toxicity.

Ventricular tachycardia

Ventricular tachycardia is defined as the occurrence of three or more consecutive ventricular premature complexes at a rate of 100 to 200 beats/min. It is considered to be sustained if the rhythm lasts for more than 30 s and non-sustained if the rhythm lasts for less than 30 s. It is occasionally difficult to differentiate ventricular tachycardia from supraventricular tachycardia with aberrancy (Table 2). A wide QRS tachycardia in an older patient with heart disease is usually ventricular tachycardia. Clinical symptoms depend on the severity of the underlying cardiac disease as well as the rate of the ventricular tachycardia. It may be associated with hemodynamic deterioration, or may accelerate and degenerate into ventricular defibrillation (Fig. 7).

Torsade de pointes is a polymorphic ventricular tachycardia associated with a long QT interval. This rhythm is difficult to treat,

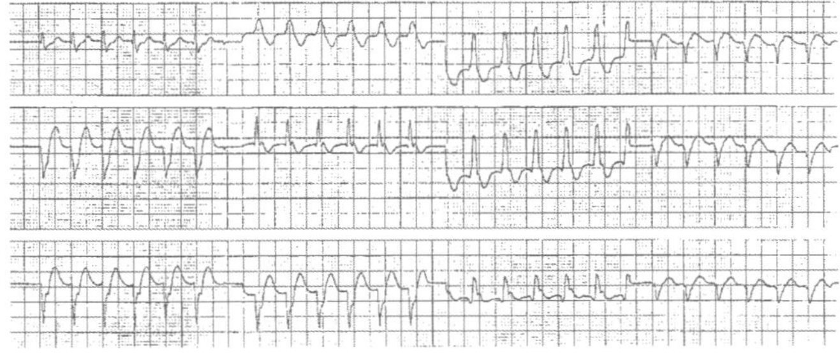

Fig. 7 Ventricular tachycardia. Criteria from Table 2 help with the diagnosis.

frequently degenerates into ventricular fibrillation, and tends to recur. It may be seen in patients with inherited long QT syndrome, but may also be caused by drugs, such as class IA, IC, and III antiarrhythmics and phenothiazines, electrolyte abnormalities such as hypocalcemia, hypokalemia, and hypomagnesemia, slow heart rates, and acute brain injury. Long-acting antihistamines and erythromycin have also been associated with this arrhythmia. Torsade may be associated with a bradycardia and triggered by an early ventricular premature complex which falls on a long bizarre T wave that occurs following a pause. It is due to delayed repolarization and tends to recur if the underlying cause is not or cannot be corrected.

Ventricular flutter is an extremely rapid form of ventricular tachycardia with rates of over 200 beats/min. It may also degenerate into ventricular fibrillation which is a totally disorganized ventricular activity caused by a random pattern of re-entry circuits. This arrhythmia may be the terminal event in cardiac death.

Bibliography

Andries, E.W., Brugada, J., Brugada, P., Mont, L., and Smeets, J. (1991). A new approach to the differential diagnosis of a regular tachycardia with a wide QRS complex. *Circulation*, **83**, 1649–59.

Beaudette, S.P. and Mittleman, R.S. (1996). Supraventricular tachycardias. In *Intensive care medicine* Vol. I (ed. J.M. Rippe, R.S. Irwin, M.P. Fink, and F.B. Cerra), pp. 422–34. Little, Brown, Boston, MA.

Davis, W.R. (1992). Cardiac dysrhythmias. In *Critical care* (2nd edn) (ed. J.M. Civetta, R.W. Taylor, and R.R. Kirby), pp. 1185–9. J.B. Lippincott, Philadelphia, PA.

Sayers, M.C. and Humphries, J. O. (1984). Tachyarrhythmias. In *Textbook of critical care* (ed. W.C. Shoemaker, W.L. Thompson, and P.R. Holbrook), pp. 445–59. W.B. Saunders, Philadelphia, PA.

Zipes, D.P. (1992). Specific arrhythmias. Diagnosis and treatment. In *Heart disease: a textbook of cardiovascular medicine* (4th edn) (ed. E. Braunwald), pp. 667–718. W.B. Saunders, Philadelphia, PA.

3.7.2 Therapeutic strategy

M. C. Sayers and C. M. Gasperetti

Key messages

- Always seek a correctable etiology.
- Treat the patient, not the monitor.
- Carotid sinus massage may assist diagnosis and treatment.

Introduction

The initial step in the management of tachyarrhythmia is to determine whether there is any hemodynamic compromise. The patient should be quickly evaluated for any change in mental status, blood pressure, and respiration, as well as for any signs of acute distress. Care must be taken to treat the patient and not the apparent dysrhythmia, as artifacts may be produced by patient movement or equipment problems. If the patient is hemodynamically unstable, whether due to a rapid supraventricular or a ventricular arrhythmia, electrical cardioversion is the treatment of choice, using sedation if the patient is alert. Evidence of hemodynamic compromise must be carefully documented, as mild hypotension may be due to a rapid rate and resolve quickly after intravenous drug administration as the arrhythmia is slowed or terminated. This may be true even if the medication utilized has potential vasodilatory and/or negative inotropic effects.

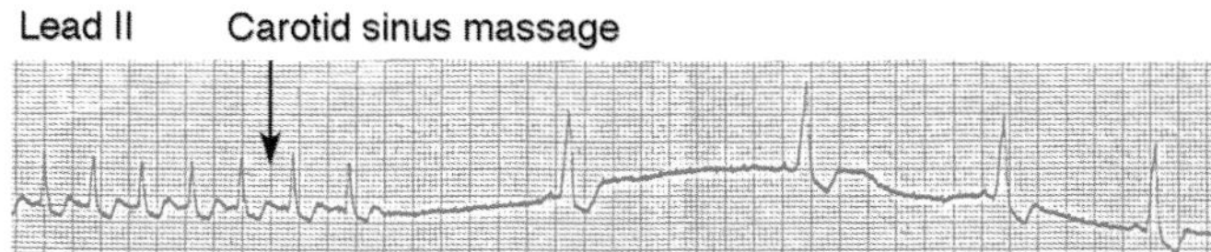

Fig. 1 Carotid sinus massage terminating a supraventricular tachycardia.

If the patient is hemodynamically stable, the diagnosis and etiology of the arrhythmia should be sought. Treatment is geared towards eliminating the underlying mechanism if possible and using appropriate medication.

In the case of supraventricular tachyarrhythmias, carotid sinus massage may not only help with the diagnosis but may also be therapeutic in certain supraventricular arrhythmias that utilize re-entry circuits (Fig. 1). These arrhythmias may be interrupted, provided that the sinoatrial or atrioventricular node is involved, since these are the only areas that are affected by increased vagal tone. Otherwise, carotid sinus massage or a Valsalva maneuver will increase atrioventricular block without interrupting the re-entry circuit.

Sinus tachycardia

The treatment is to eliminate the underlying etiology as this rhythm disturbance always occurs secondary to a precipitating factor. Efforts to control the tachycardia without determining the etiology could potentially harm the patient. Beta-blockers may be effective in slowing the sinus node. However, if the tachycardia is due to a low cardiac output, using a β-blocker may further decrease myocardial function. In contrast, a β-blocker may be very helpful in a patient who is thyrotoxic or hypertensive. Relieving the patient's pain or anxiety may also help to resolve the sinus tachycardia once it is established that there are no correctable causes. In patients who have suffered myocardial infarction, persistent sinus tachycardia may necessitate pulmonary artery catheter placement to aid evaluation of myocardial function and volume status.

Atrial premature contractions

Atrial premature contractions are not treated unless they are believed to precipitate atrial tachycardia, flutter, or fibrillation. Removing precipitating factors, including drugs such as caffeine and theophylline as well as metabolic abnormalities, may decrease atrial premature contractions. Congestive heart failure and infection may also precipitate atrial premature contractions that will resolve as the patient's condition improves. Occasionally, digitalis may be helpful, particularly if congestive heart failure is a factor. However, the most effective medication is an antiarrhythmic class IA drug such as quinidine or procainamide.

Sinoatrial and atrioventricular nodal re-entry tachycardias

Sinus node re-entry tachycardia, atrioventricular nodal re-entry tachycardia, atrioventricular reciprocating tachycardia, and paroxysmal junctional tachycardia are included in this category as all can be treated in a similar fashion. Because of the re-entry circuit, carotid sinus massage will either terminate the arrhythmia or have no effect. The next treatment of choice is adenosine which has a very short half-life and is as effective as verapamil. An initial dose of 6 mg is given by rapid intravenous bolus; if this is ineffective, a dose of 12 mg may be administered. Calcium-channel antagonists such as verapamil and diltiazem may be used; however, the patient must be closely monitored for hypotension when using these agents. Other medications that may be used acutely are β-blockers and digitalis. However, the onset of action of cardiac glycosides is delayed by about 30 min when given intravenously. Overdrive pacing has also been shown to be effective in terminating re-entry tachycardias, and this can be accomplished by atrial or esophageal pacing. If the patient is hemodynamically unstable, electrical cardioversion is indicated, with a voltage of 25 to 50 J usually being effective.

Any precipitating factors should be corrected and, if chronic therapy is required, verapamil, diltiazem, β-blockers, and digitalis, as well as class IA and IC antiarrhythmics, may be used. Patients with known atrioventricular reciprocating tachycardia with an accessory pathway should not be given verapamil or digitalis on a chronic basis since they may also exhibit atrial fibrillation and/or atrial flutter.

Intra-atrial re-entry tachycardia

This tachycardia is considered separately because the re-entry circuit is confined to the atrium and does not involve either sinoatrial or atrioventricular nodes. As a result, carotid sinus massage may cause an increase in atrioventricular block but will not terminate the tachycardia. However, termination can occur with atrial or esophageal pacing since it is a re-entry circuit. Acutely, an increase in atrioventricular block can also be obtained with calcium antagonists, digoxin, or β-blockers. Class IA or IC antiarrhythmics may be helpful on a more chronic basis.

Non-paroxysmal atrial tachycardia, multifocal atrial tachycardia, and automatic junctional tachycardia

These arrhythmias are due to an automatic mechanism and usually have precipitating factors. Carotid sinus massage does not affect the arrhythmia but may increase atrioventricular block, thereby slowing the ventricular response. Therapy is directed towards correcting the underlying mechanism, whether it be due to drug toxicity, electrolyte imbalance, pulmonary disease, or congestive heart failure. These arrhythmias can be very difficult to treat, particularly in an acutely ill patient. Class IA antiarrhythmics are occasionally as effective as those of classes IC and III. If these arrhythmias are not rapid and the patient is hemodynamically stable, simple observation may be indicated as the patient's acute illness is treated.

Atrial fibrillation and atrial flutter

Management depends on the clinical status of the patient and the duration of the arrhythmia. Even in a patient who is somewhat hypotensive, slowing the ventricular response with intravenous medication may increase the blood pressure. The best medication for slowing rapid atrial fibrillation and flutter is intravenous diltiazem. This can be used as a drip and titrated according to the patient's blood pressure and heart rate. Intravenous digoxin can also be given; however, the maximum effect of this medication is delayed an hour after intravenous administration. Its effectiveness is also limited in a patient who is acutely ill. Intravenous verapamil or a β-blocker may be alternative therapies for slowing the ventricular response. In the intensive care unit setting, the use of β-blockers may be limited in patients with chronic obstructive pulmonary disease.

Patients who have a bypass tract with a wide QRS complex may be difficult to differentiate from ventricular tachycardia, although the gross irregularity of atrial fibrillation is very helpful. However, medications that would be appropriate in either situation may be given. Intravenous lidocaine (lignocaine) or procainamide can be used since both may slow conduction in an accessory pathway and would also be appropriate treatment for ventricular tachycardia. Other class I drugs are also effective in decreasing the conduction down the bypass tract. In general, a wide complex tachycardia should not be treated with digoxin or verapamil. When a bypass tract is involved, these agents may increase conduction, resulting in acceleration of the ventricular response and possible degeneration into ventricular fibrillation. In the case of ventricular tachycardia, digoxin has no therapeutic effect and verapamil should not be used since it may increase the ventricular rate.

If the patient remains in atrial fibrillation or flutter, rate control with intrvenous medication is necessary. Intravenous diltiazem is usually the most effective medication as the patient is started on a class IA antiarrhythmic if necessary. Corvert® (ibutilide fumarate), a new medication with class III properties, may be effective (43–48 per cent) in converting atrial flutter or fibrillation to a normal sinus rhythm within 1 h of infusion. Patients who do not have an accessory bypass tract may be started on digitalis. This drug will help slow the ventricular response and may offer some assistance in maintaining a normal sinus rhythm.

In patients who have had atrial fibrillation for more than 2 days or present with this arrhythmia of unknown duration, electrical or chemical cardioversion should not be attempted until the patient has been anticoagulated for 2 to 3 weeks. Atrial thrombi begin to form within 48 h of the onset of atrial fibrillation, and so cardioversion at this point could result in an embolic event.

Anticoagulation for atrial flutter is more controversial; in most cases, this is a more unstable rhythm and less likely than atrial fibrillation to be chronic. A thromboembolic event is less likely with atrial flutter but is still a possibility. Currently, there are no specific guidelines to anticoagulate prior to converting atrial flutter to a normal

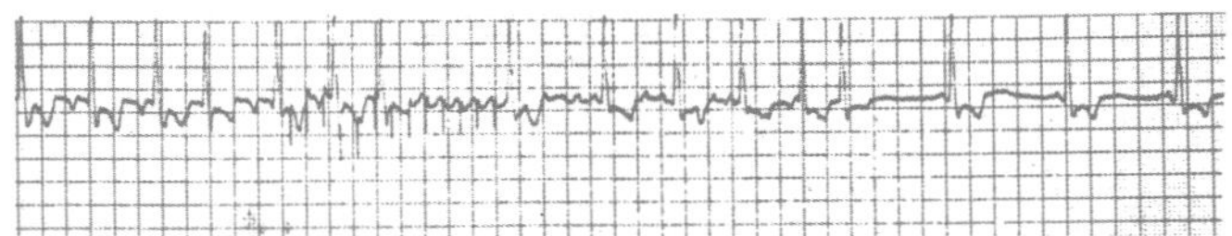

Fig. 2 Conversion of atrial flutter to a normal sinus rhythm by rapid atrial pacing.

sinus rhythm. In addition to medication, overdrive atrial pacing may be an effective mechanism for converting atrial flutter to atrial fibrillation or a normal sinus rhythm (Fig. 2). Atrial fibrillation cannot be converted by pacing because of the random depolarization that is occurring.

Ventricular tachyarrhythmias

Ventricular premature complexes are commonly seen in the critically ill and are not usually treated unless the patient develops ventricular tachycardia. In the setting of acute myocardial infarction, premature ventricular contractions are considered as warning arrhythmias and may progress to ventricular tachycardia. However, unless this develops, the current recommendation is not to treat isolated premature ventricular contractions as the toxic effects of antiarrhythmic drugs do not warrant the potential benefits. During the first 48 h following myocardial infarction, most episodes of ventricular tachycardia are non-sustained asymptomatic events that do not alter prognosis.

Occasionally, an accelerated idioventricular rhythm occurs with rates between 50 and 100 beats/min. More commonly seen after acute myocardial infarction, it is not associated with lethal ventricular arrhythmias and does not alter short- or long-term prognosis. It is frequently seen during episodes of sinus bradycardia, and the administration of 0.5 to 1 mg atropine will usually restore normal conduction by increasing the sinus rate. However, atropine possesses arrhythmogenic properties and may also result in sinus tachycardia which will increase myocardial oxygen consumption. Therefore this rhythm is not treated unless there is hemodynamic compromise.

Treatment of ventricular tachycardia depends on the condition of the patient. Frequently, short runs of non-sustained ventricular tachycardia are not treated if asymptomatic, and treatment of the underlying problem may be all that is necessary. Lidocaine is usually the drug of choice with symptomatic non-sustained or sustained ventricular tachycardia. However, if it is injected too quickly, transient toxic levels may cause atrioventricular block or central nervous system abnormalities. Amiodarone may be administered as a continuous infusion, giving additional 150-mg boluses over 10 min for breakthrough ventricular tachycardia/fibrillation to a total of 1000 mg in the first 24 h. Other medications that can be used include intravenous procainamide and bretylium. If the patient is stable, an alternative would be ventricular pacing to interrupt the re-entry circuit. Occasionally, however, artificial pacing may increase the ventricular rate and degenerate into ventricular fibrillation. If medications are ineffective in converting the ventricular tachycardia and ventricular pacing is not available, elective d.c. synchronized cardioversion, starting as low as 50 J, may be effective. The patient should be adequately sedated prior to this procedure. If the patient becomes unstable at any point, immediate synchronized d.c. cardioversion (100–360 J) is indicated.

Beta-blockers may be helpful in acute myocardial infarction, as ventricular tachycardia during this time period may be due to an automatic focus. Antiarrhythmic agents may be proarrhythmic when used in this setting and the patient must be observed closely. Non-pharmacological intervention, such as intra-aortic balloon counterpulsation, emergency coronary artery bypass surgery, or angioplasty, may be necessary to stabilize arrhythmias related to ischemia.

Patients with torsade de pointes respond poorly to lidocaine, procainamide, and bretylium, and actually may be made worse by class IA, IC, or III antiarrhythmics. Treatment requires prompt correction of any metabolic abnormalities, and particular attention should be given to potassium and magnesium replacement. Any potentially offending medications should be withheld, and intravenous catecholamines or rapid rate pacing may help prevent recurrence.

Ventricular fibrillation or flutter requires immediate electrical cardioversion at 360 J. If this is ineffective, administering a 100-mg lidocaine bolus followed by repeat d.c. cardioversion is indicated. Bretylium and procainamide are other agents that may be used. As with any arrhythmia, metabolic abnormalities such as acidosis, hypoxemia, hypokalemia, and hypomagnesemia should be corrected. Prophylaxis for recurrent ventricular fibrillation is similar to that for ventricular tachycardia, and care must be taken to search for any precipitating factors.

Bibliography

Andries, E.W., Brugada, J., Brugada, P., Mont, L., and Smeets, J. (1991). A new approach to the differential diagnosis of a regular tachycardia with a wide QRS complex. *Circulation*, 83, 1649–59.

Beaudette, S.P. and Mittleman, R.S. (1996). Supraventricular tachycardias. In *Intensive care medicine*, Vol. I (ed. J.M. Rippe, R.S. Irwin, M.P. Fink, and F.B. Cerra), pp. 422–34. Little, Brown, Boston, MA.

Davis, W.R. (1992). Cardiac dysrhythmias. In *Critical care* (2nd edn) (ed. J.M. Civetta, R.W. Taylor, and R.R. Kirby), pp. 1185–9. J.B. Lippincott, Philadelphia, PA.

Sayers, M.C. and Humphries, J. O. (1984). Tachyarrhythmias. In *Textbook of critical care* (ed. W.C. Shoemaker, W.L. Thompson, and P.R. Holbrook), pp. 445–59. W.B. Saunders, Philadelphia, PA.

Zipes, D.P. (1992). Specific arrhythmias. Diagnosis and treatment. In *Heart disease: a textbook of cardiovascular medicine* (4th edn) (ed. E. Braunwald), pp. 667–718. W.B. Saunders, Philadelphia, PA.

3.8 Bradyarrhythmias

3.8.1 Causes and diagnosis

C. M. Gasperetti and M. C. Sayers

Key messages

- Bradyarrhythmias are due to (i) primary cardiac disorders and (ii) acute and chronic non-cardiac medical disorders.
- Patients can present with bradyarrhythmias or with syncope.
- Causes include blocked atrial premature beats, sinus node dysfunction, and atrioventricular node conduction disorders, as well as metabolic and electrolyte impairment.

Sinus node dysfunction

Patients with sinus node dysfunction have either slow sinus beats or intermittently dropped or absent beats. They may be asymptomatic, although any symptoms must be related directly to the sinus dysfunction to assess whether pacing may be of benefit.

Blocked atrial premature beats

Blocked atrial premature beats are the most common cause of bradycardia in the critical care setting. They must be distinguished from sinus pauses by searching for P waves buried within the T wave (Fig. 1). The timing of cannon A waves has diagnostic significance. For blocked atrial premature beats, they occur before the first heart sound associated with the pause, and for complete heart block they are unrelated to heart sounds.

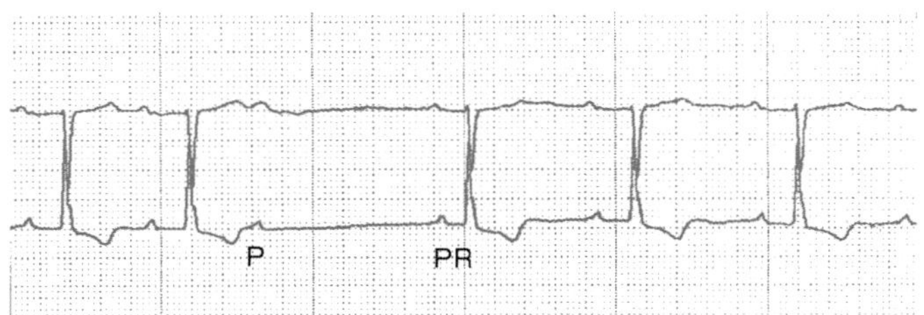

Fig. 1 Blocked atrial premature beat. Note the P wave buried in the T wave of the second sinus beat. The third conducted beat follows a shortened PR interval. This occurs because the interval between the atrial premature beat and the next P wave is longer, allowing the PR segment time to shorten.

Sinus bradycardia

Sinus bradycardia occurs when the sinus rate slows to below 60 beats/min. Although it may occur during sleep, in the young, or in athletes at rest, patients may experience fatigue, presyncope, or syncope. It may represent an exaggerated increase in vagal tone, for example during nausea, vomiting, or vasovagal episodes. It can occur in the settings of acute ischemia or infarction, acute hypertension, acute respiratory compromise, hypercapnia, acidemia, hypothermia, and hypothyroidism. It can contribute to increased ventricular arrhythmias. If it occurs in response to calcium-channel blockers, digoxin, or antiarrhythmic drug therapy, this may suggest an underlying conduction abnormality.

Sinoatrial exit block and sinus arrest

Sinoatrial exit block occurs when there is a block between the sinoatrial node and atrial tissue. It is characterized by absence of P waves with a pause that is a multiple of the sinus or PP interval. It occurs most often in elderly patients and usually suggests a chronic problem. Other causes include myocarditis, myocardial infarction, excess vagal stimulation, atrial fibrosis, quinidine, procainamide, and digoxin. Sinus arrest occurs when there is no sinus beat and the

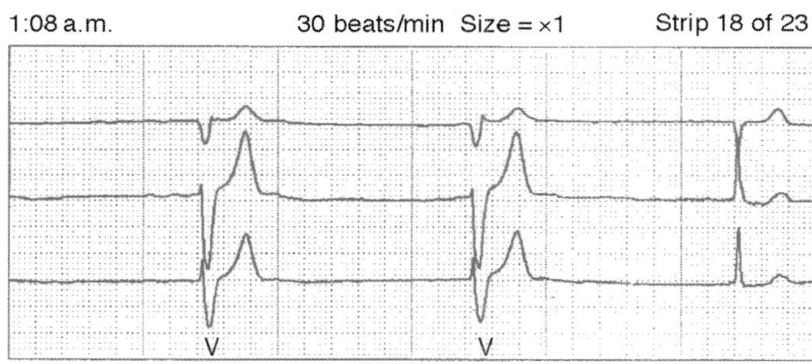

7:46 a.m. Start of VRun 130 beats/min Size = ×1 Strip 19 of 23

VT

Fig. 2 Ventricular escape complexes (V) following sinus arrest with resumption of conducted narrow QRS beat. The second rhythm strip shows an episode of ventricular tachycardia following an interval of bradycardia (not shown).

dropped P waves are not multiples of the PP interval (Fig. 2). If the pauses are lengthy, symptoms of syncope or presyncope may occur. Ischemia, conduction system disease, digoxin, β-blockers, and calcium-channel blockers can also cause sinus arrest.

Sick sinus syndrome

Sick sinus syndrome consists of periods of supraventricular tachycardia alternating with periods of bradycardia. The symptoms depend on the degree of tachycardia, the length of the pauses, and the presence of underlying cardiac dysfunction. Treating the tachycardia can result in worsening of the bradycardia and may require pacemaker implantation.

Bundle branch block

Patients with bifascicular block and symptoms of syncope or near syncope should have high-degree atrioventricular block excluded. If seen in association with atrioventricular block, this is evidence of infranodal conduction disease and often indicates a need for pacing. Left bundle branch block often signifies underlying cardiac disease, although it can result from electrolyte disturbance. In the presence of left bundle branch block, electrocardiography cannot exclude and may mask myocardial infarction. Right bundle branch block may be seen with acute pulmonary hypertension or embolus, or as a complication of right heart catheterization. Its significance during acute myocardial infarction is discussed below.

Atrioventricular conduction block

Impulse propagation from the atrioventricular node through the His–Purkinje system, bundle branches, and fascicles is blocked. The site of block should be determined as this can indicate the likelihood of a stable escape rhythm and the risk of progression to complete heart block (Table 1). Patients with atrioventricular conduction abnormalities may have abnormalities in autonomic tone, particularly heightened vagal stimulation. Atrioventricular conduction block may be due to myocardial ischemia, chronic hypertension, digoxin toxicity, β-blockers, calcium-channel blockers, myocarditis, acute rheumatic fever, myocardial calcification, aortic stenosis, mitral stenosis, and congenital atrioventricular block.

First-degree atrioventricular block

First-degree atrioventricular block occurs when the PR interval is prolonged (0.20 s or longer, or 0.18 s or longer if the heart rate exceeds 90 beats/min). If the QRS interval is normal, the delay in conduction is in the atrioventricular node. With a prolonged QRS, the conduction delay may be in the His–Purkinje system. It is usually asymptomatic, and may be due to drugs (e.g. digoxin, β-blockers, calcium-channel blockers), intrinsic conduction system disease, and hypertension. It can be ominous if it is due to endocarditis with abscess formation involving the conduction system.

Second-degree atrioventricular block

Type I (Wenckebach or Mobitz I) second-degree atrioventricular block occurs when there is intermittent failure of the atrial beats to reach the ventricle. It is characterized by progressive lengthening of the PR interval with subsequent shortening of the RR interval until a P wave is not conducted. The PR interval is not constant, and the longest interval (between the non-conducted and conducted beats) is less than twice the normal sinus interval. The periodic pauses create the impression of group beating (Fig. 3). Physical examination reveals regular A waves with progressive softening of the first heart sound. Wenckebach periodicity is usually asymptomatic. It is exacerbated by drugs (digoxin, β-blockers, calcium channel blockers) and heightened vagal tone, but in the setting of acute myocardial infarction, particularly inferior infarction, it often responds to atropine.

Type II (Mobitz II) second-degree atrioventricular block occurs when there is intermittent failure of the atrial beats to reach the ventricle and there is a sudden non-conducted P wave unassociated

Table 1 Types of atrioventricular conduction abnormalities

Type of AV block	ECG interval	Conduction abnormality
First degree	PR segment	SA node or AV node or bundle of His or bundle of His and bundle branches
Second degree	Type I	AV node
	Type II	Bundle of His or bundle of His and bundle branches
Third degree	Narrow QRS	AV node
Complete	Wide QRS	AV node or bundle of His or bundle of His and bundle branches

AV, atrioventricular; SA, sinoatrial.

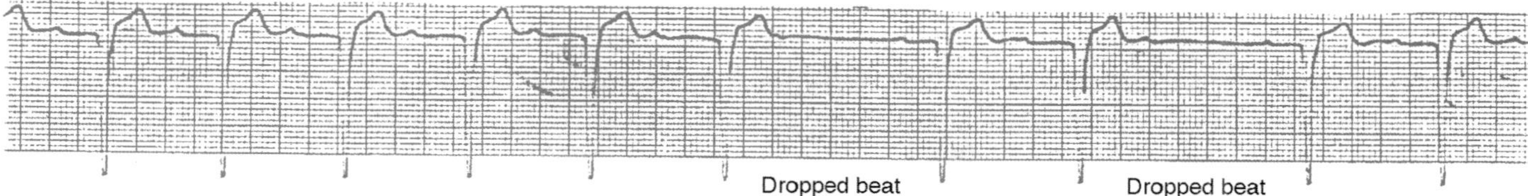

Fig. 3 Type I (Wenckebach) second-degree atrioventricular block. The rhythm strip shows progressive lengthening of the PR interval, with progressive shortening of the RR interval until a dropped beat occurs. The longest interval is less than twice the sinus interval and group beating, characteristic of the Wenckebach pattern, can be seen.

with any change in PR or RR intervals. Most often a sign of block in the bundle of His or lower in the conduction system, it is more likely than type I to progress to complete heart block and can be caused by ischemia. When it occurs with 2:1 periodicity, surface electrocardiography cannot distinguish it from type I and so invasive testing may be necessary. However, if the rate is fast and well tolerated, carotid sinus massage may slow it and lead to 1:1 conduction. Similarly, atropine or atrial pacing can change the atrial rate or the degree of atrioventricular node refractoriness and alter the periodicity in type I atrioventricular block. Although patients with Mobitz II block may be asymptomatic, it is a warning of potentially severe conduction system disease. In acute myocardial infarction a pacemaker is indicated for type II block because of the likelihood of progression to complete heart block.

Complete heart block

Complete heart block is diagnosed when there are independent atrial and ventricular rhythms, the atrial rate exceeds the ventricular rate, and the atrial beat does not conduct to the ventricles although there is clear opportunity for it to do so (Fig. 4). The first heart sound varies in intensity and atrial sounds may be heard during long diastoles. The arterial pulse pressure is wide. Complete heart block may occur in the atrioventricular node or at the level of the bundle of His (narrow QRS), or it may be below the bundle of His or in the Purkinje system (prolonged QRS). When the atrial and ventricular rates are similar, atrioventricular dissociation may be present without complete heart block. When the atrial and ventricular rates are almost identical but still independent, isorhythmic dissociation is present. Complete heart block can be caused by digoxin, β-blockers, antiarrhythmic drugs, calcium-channel blockers, infiltrate conduction system diseases, electrolyte disturbances, endocarditis, calcific aortic stenosis, and myxedema. Complete heart block can complicate acute inferior or anterior myocardial infarction.

Manifestations of digoxin toxicity include varying degrees of atrioventricular block, often in conjunction with atrial dysrhythmias. The pathognomonic conduction abnormality associated with digoxin toxicity is paroxysmal atrial tachycardia with varying block. Asystole is not uncommon in patients in the intensive care unit setting.

Non-cardiac causes of bradyarrhythmias

Frequent non-cardiac causes seen in the intensive care unit include severe electrolyte disturbances. These can be reflected in the ECG, as

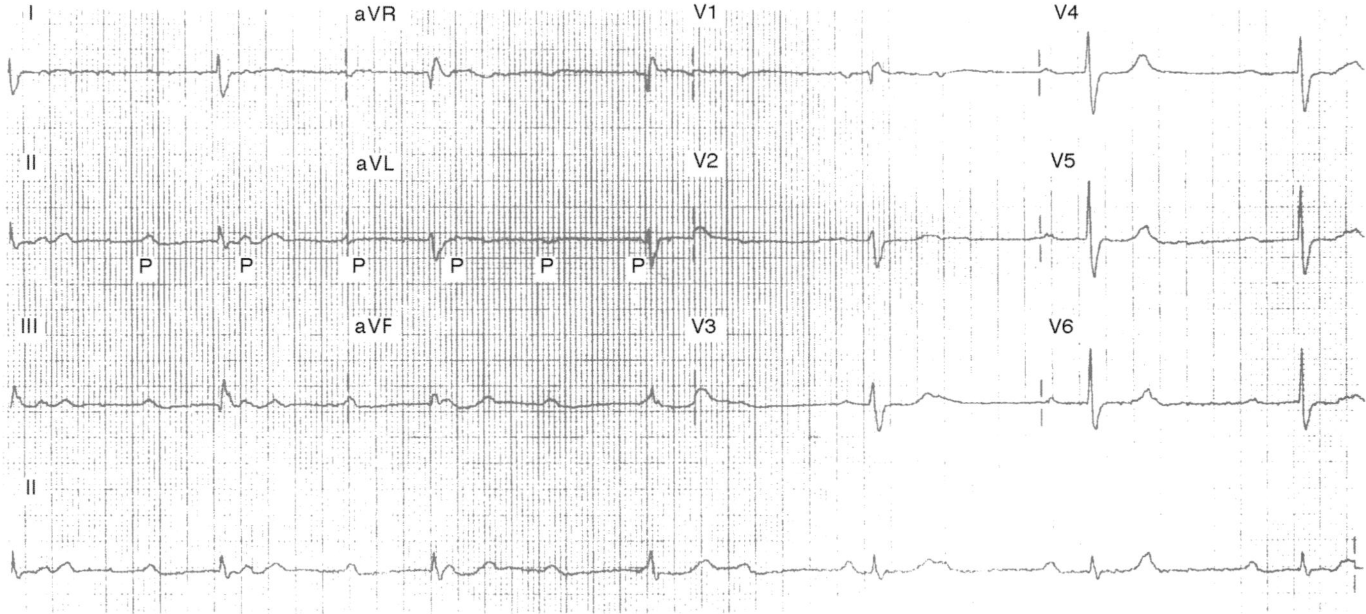

Fig. 4 Complete heart block in a patient with underlying right bundle branch block. Atrial rhythm and ventricular rhythm are independent, atrial rate is faster than ventricular rate, and the atrial beat does not conduct to the ventricle although clear opportunity for it to do so is noted.

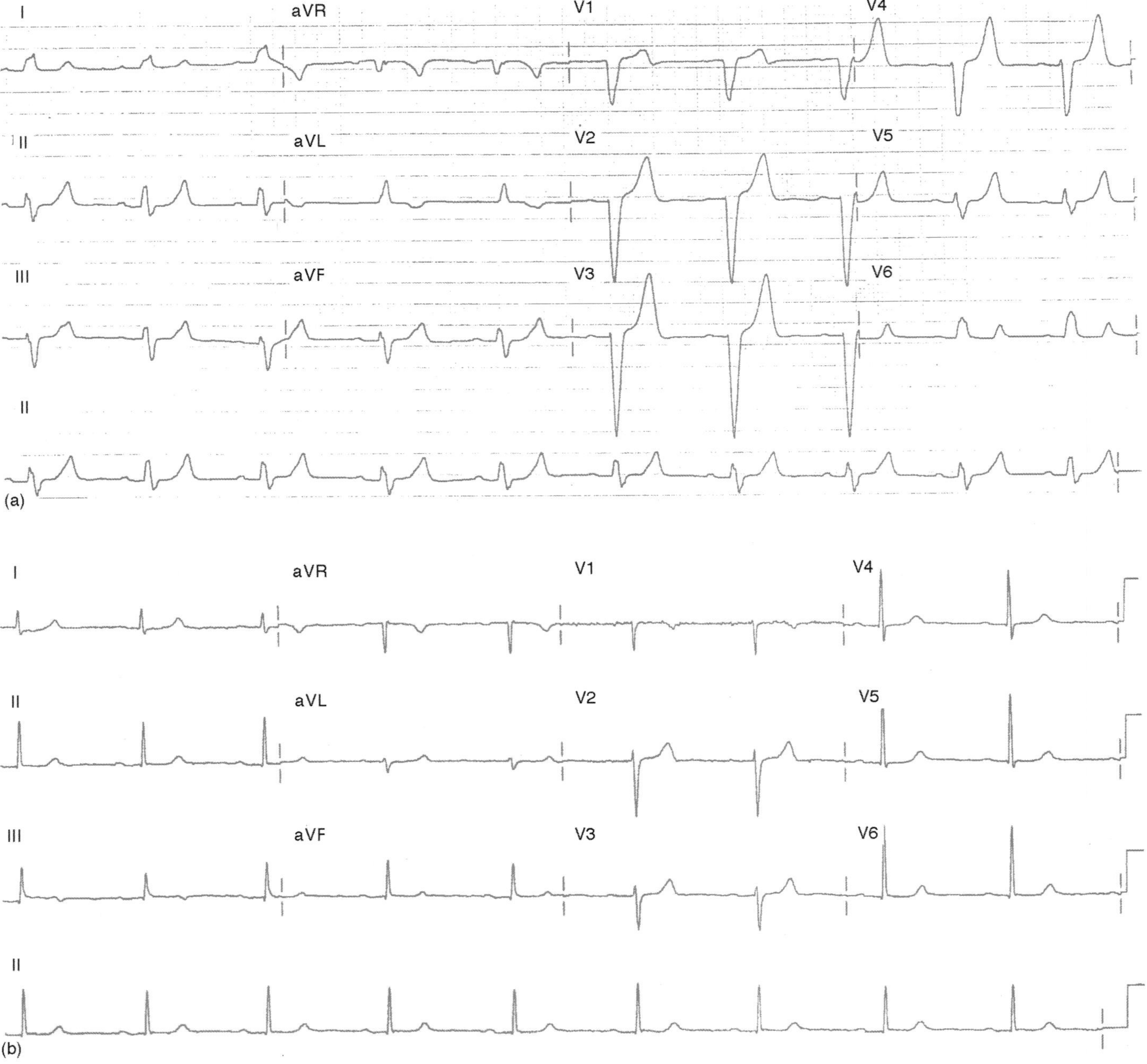

Fig. 5 Manifestations of hyperkalemia: (a) 12-lead ECG for a patient with potassium level of 7.2 mg/dl shows sinus rhythm with left bundle branch block, slurring of the QRS and ST segments, increased QT interval, and symmetrically peaked T waves; (b) with decrease in potassium to 5.2 mg/dl following treatment with calcium gluconate, bicarbonate, glucose, and insulin, there is narrowing of the QRS, resolution of ST segment shifts, decrease in QT interval, and resolution of T-wave changes.

can the response to therapy. Hyperkalemia is common and can be fatal. It is often due to severe renal failure and can also result from potassium salts of penicillin, digitalis toxicity, or acidosis. The action potential across the cell membrane depends upon the concentration of ions across the membrane. Hyperkalemia shortens the action potential duration and slows conduction. Severe hyperkalemia can have negative inotropic effects and can increase the threshold for capture of artificial pacemakers. Hyperkalemia can lead to non-conducted atrial beats, asystole, ventricular tachycardia, ventricular fibrillation, and conduction disturbances including both type I and type II second-degree atrioventricular block, accelerated idioventricular ventricular rhythm, and ventricular escape rhythms. ECG changes progress from a flattened PR interval to peaked T waves, prolonged QRS, and prolonged QT intervals (Fig. 5). Severe arrhythmias including ventricular fibrillation can occur without warning.

Hypokalemia can also cause severe arrhythmias, for example paroxysmal atrial tachycardia with block, atrioventricular block, atrioventricular dissociation, and ventricular fibrillation. It may increase the chance of digoxin toxicity and drug-induced long QT interval. ECG changes include a long QT interval, ST depression, decreased T waves, prominent U waves, and prolonged QRS duration. Hypokalemia can be worsened by coexisting hypocalcemia, which rarely causes rhythm abnormalities by itself but can lead to ST segment prolongation. Hypermagnesemia can prolong PR and QRS durations and lead to sinoatrial and atrioventricular blocks. Hypomagnesemia can worsen digoxin toxicity and facilitate ventricular arrhythmias.

Bibliography

Buxton, A.E. and Hurwitz, J.L. (1993). Disorders of cardiac rhythm and conduction in the medical intensive care unit. In *Principles and practice of medical intensive care* (ed. R.W. Carlson and M.A. Geheb), pp. 1049–88. W.B. Saunders, Philadelphia, PA.

Fowler, N.O. (1993). Bradyarrhythmias. In *Diagnosis of heart disease*, pp. 315–39. Springer-Verlag, New York.

Gunnar, R.M., *et al.* (1990). Guidelines for the early management of patients with acute myocardial infarction. ACC/AHA Task Force report. *Journal of the American College of Cardiology*, **16**, 257–8.

Homcy, C.J. (1993). Atrioventricular block. In *The practice of cardiology* (ed. K.A. Eagle, E. Haber, R.W. DeSanctis, and W.G. Austen), pp. 267–81. Little, Brown, Boston, MA.

Marriott, H.J.L. (1990). *Pearls and pitfalls in electrocardiography*. Lea & Febiger, Malvern, PA.

3.8.2 Therapeutic strategy

C. M. Gasperetti and M. C. Sayers

Key messages

- The hemodynamic consequences of the rhythm abnormality should be assessed initially.
- Possible reversible causes should be sought and corrected (hypoxemia, hypokalemia, acidemia, digoxin toxicity, myocardial ischemia or infarction).
- Complete heart block is associated with a two- to fourfold increase in mortality when it occurs in the setting of coexisting coronary heart disease or heart failure.
- Initial atropine doses below 0.5 mg should not be used because they can precipitate a paradoxical slowing of heart rate due to their vagomimetic effect.

Introduction

When bradycardias occur in the critical care setting, the first step should be to assess the hemodynamic consequences of the rhythm abnormality. Possible reversible causes should be sought and corrected (hypoxemia, hypokalemia, acidemia, digoxin toxicity, myocardial ischemia or infarction).

Sinus node dysfunction

Blocked atrial premature beats

Usually this requires no treatment other than that of the underlying cause.

Sinus bradycardia

This requires treatment only when hemodynamically significant. The rates at which sinus bradycardia need to be treated vary greatly. While rates up to 60 beats/min may be inappropriate for a patient in congestive heart failure, few tolerate rates below 30 beats/min. There may be increased ectopy in some patients, while ventricular premature beats may themselves induce a relative bradycardia if the effective heart rate is slowed. If symptomatic, the bradycardia should be treated with pacing. In the setting of acute myocardial infarction, sinus bradycardia should be treated if it is associated with hemodynamic compromise or is contributing to significant ectopy.

Atrioventricular conduction block

First-degree atrioventicular block does not have adverse prognostic implications and usually requires no treatment. However, it should prompt consideration of possible underlying heart disease, as it can be a manifestation of coronary and other structural heart disease.

Type I second-degree atrioventicular block is generally well tolerated and does not usually require treatment. However, it can also indicate coronary ischemia or infarction, and an etiological diagnosis must be established. It should be treated if it is symptomatic.

Type II second-degree atrioventicular block may require transvenous pacing. Determining the level of block is important in predicting whether an adequate escape focus will be present. Permanent ventricular pacing is indicated with high-grade atrioventicular block and symptoms of syncope or presyncope, whether the block is nodal or infranodal. If it is infranodal, pacing is recommended even if the patient is asymptomatic because of the high incidence of sudden death in these patients.

Complete heart block may be asymptomatic and well tolerated, particularly if it is congenital or has a high junctional rhythm. However, it is associated with a increase in mortality of a factor of two to four when it occurs in the setting of coexisting coronary heart disease or heart failure. Pacing can improve prognosis in these patients and may improve heart failure or mental status changes due to impaired perfusion.

An accelerated idioventricular rhythm can occur after myocardial infarction with rates usually ranging from 40 to 100 beats/min. Treatment is generally not necessary, although pacing may be necessary if a sustained accelerated idioventricular ventricular rhythm leads to symptoms.

Acute severe bradycardia

When acute severe bradycardia occurs in the critical care setting, the first step should be to administer 0.5 mg atropine intravenously as the initial bolus, followed by 0.5-mg increments up to a cumulative dose of 2 mg. Initial doses below 0.5 mg of atropine should not be used because they can precipitate a paradoxical slowing of heart rate owing to their vagomimetic effect. While atropine is being administered, the patient can be asked to cough several times. Particularly during episodes of vasovagally mediated bradycardia, cough may have a vagolytic effect resulting in a transient increase of heart rate. An intravenous saline bolus can also be given. If bradycardia has resulted in hypotension and vasodilation, the response to atropine can be delayed. Intravenous aminophylline may be tried in a patient who is not unstable. If not contraindicated by acute infarction or ongoing ischemia, isoproterenol (isoprenaline) 1.0 mg in 250 to

Table 1 Use of temporary pacing in acute myocardial infarction

Condition	Indication for temporary pacing
Asystole	Yes
First-degree AV block	No
Type I second-degree AV block	
Asymptomatic	No
Symptomatic	Yes*
Type II second-degree AV block	
Narrow QRS	Possibly*
Wide QRS	Yes
Third-degree AV block	
Narrow QRS	Possibly*
Wide QRS	Yes
Complete heart block	
Inferior MI	Possibly*
Anterior MI	Yes
LBBB	
New	Yes
Old	Possibly†
RBBB	
New	Possibly†
Old	No
RBBB with LAHB or LPHB	
New	Yes
Old	Possibly‡
Incessant VT, overdrive pacing	Yes

AV, atrioventricular; MI, myocardial infarction; LBBB, left bundle branch block; RBBB, right bundle branch block; LAHB, left anterior hemiblock; LPHB, left posterior hemiblock; VT, ventricular tachycardia.
*Symptomatic bradycardia associated with low blood pressure or ischemia that is not responsive to atropine.
†If associated with first-degree AV block.
‡Bifascicular block of unknown duration is a relative indication.

500 ml of 5 per cent gluose can also be administered at a rate which increases the pulse to 60 beats/min.

Indications for temporary pacing

If bradycardia persists or there are contraindications to isoproterenol, pacemaker placement is indicated. A transcutaneous pacemaker can be positioned on the chest wall before insertion of a temporary pacing lead. Indications for temporary pacing include symptomatic bradycardias occurring during sinus node dysfunction, carotid hypersensitivity, complete heart block, and after cardiac surgery. Pacing may be used prophylactically during second-degree atrioventicular block or with right heart catheterization in a patient with left bundle branch block.

Patients with symptomatic bradycardia (asystole, sinus bradycardia, sinus pauses, type I second-degree atrioventicular block) associated with hypotension and unresponsive to atropine are treated with transvenous pacing. Patients who demonstrate increased likelihood of progression to complete heart block by either involvement of the bundle branches (see below) or type II second-degree atrioventicular block also require transvenous pacing.

Indications for temporary pacemaker insertion during myocardial infarction are based on the patient's hemodynamic status and symptoms during bradycardia, and the likelihood of the development of complete heart block with an inadequate escape mechanism (Table 1). The type of heart block and its severity in acute myocardial infarction is dependent on the coronary artery involved. The dominant coronary artery (right coronary artery in 90 per cent of patients and left circumflex in 10 per cent) gives rise to the posterior descending artery which supplies the atrioventicular node, the proximal bundle of His and the proximal right bundle. The left anterior descending artery, through its septal perforators, supplies the left bundle branch and left anterior fascicle, and the right bundle branch beyond its origin. The left posterior fascicle receives a dual blood supply. During inferior myocardial infarction, atrioventicular nodal ischemia often leads to PR prolongation and type I second-degree atrioventicular block. If there is complete heart block, this occurs progressively and the escape complex is often narrow. Sinus bradycardia can also be caused by either sinoatrial ischemia or reflex bradycardia. These conduction abnormalities are often responsive to atropine or can lead to accelerated junctional rhythms. During anterior myocardial infarction, manifestation of bundle branch involvement is secondary to extensive involvement of one or both of the bundle branches. Complete heart block appears suddenly and is a manifestation of infranodal block. The most common preceding bundle branch involvement in this setting is right bundle branch block and left anterior hemiblock, followed by type II second-degree atrioventicular block and new bilateral bundle branch block. Isolated left anterior hemiblock or right bundle branch block is less likely to progress to complete heart block.

Bibliography

Buxton, A.E. and Hurwitz, J.L. (1993). Disorders of cardiac rhythm and conduction in the medical intensive care unit. In *Principles and practice of medical intensive care* (ed. R.W. Carlson and M.A. Geheb), pp. 1049–51. W.B. Saunders, Philadelphia, PA.

Fowler, N.O. (1993). Bradyarrhythmias. In *Diagnosis of heart disease*, pp. 315–39. Springer-Verlag, New York.

Gunnar, R.M., *et al.* (1990). Guidelines for the early management of patients with acute myocardial infarction. ACC/AHA Task Force report. *Journal of the American College of Cardiology*, **16**, 257–8.

Homcy, C.J. (1993). Atrioventricular block. In *The practice of cardiology* (ed. K.A. Eagle, E. Haber, R.W. DeSanctis, and W.G. Austen), pp. 267–81. Little, Brown, Boston, MA.

Marriott, H.J.L. (1990). *Pearls and pitfalls in electrocardiography.* Lea & Febiger, Malvern, PA.

3.9 Endocarditis

3.9.1 Pathophysiology and causes

A. P. R. Wilson

Key messages

- Endocarditis caused by *Staphylococcus aureus* follows a short aggressive course, unlike the more indolent streptococcal disease.
- Underlying cardiac lesions, most frequently mitral valve prolapse, are present in 73 per cent of patients.
- Prosthetic valve endocarditis accounts for a third of all cases.
- Development of the vegetation precedes bacterial colonization.
- Alpha-hemolytic streptococci are the most common pathogens in native valve endocarditis and coagulase-negative staphylococci are the most common in prosthetic valves.

Introduction

The incidence of endocarditis is between 0.3 and 1.3 episodes per 1000 admissions per year and has been affected by the reduction in rheumatic heart disease, prolonged survival of patients with congenital or degenerative heart disease, and the increase in nosocomial endocarditis. The mitral valve is the most commonly affected, followed by the aortic valve and then the mitral and aortic valves together.

Successful management of this disease requires close co-operation between physician, surgeon, and microbiologist. The intensive care unit (ICU) usually becomes involved in this management just prior to or following surgery to replace an infected valve or when the patient develops a severe complication, for example peripheral, pulmonary, or cerebral embolism or pulmonary edema (Fig. 1).

Diagnostic criteria

Invasion by organisms of the endothelium can involve the valve leaflets, the chordae tendineae, the endocardium, the valve rings, and the great vessels in aortic coarctation or patent ductus arteriosus. The clinical signs and symptoms are variable and often non-specific, and the vegetations are not always readily apparent on imaging. The most widely accepted diagnostic criteria were those of von Reyn *et al.* (1981), which included definite, probable, possible, and rejected categories (Table 1). Durack *et al.* (1994) observed that some patients with a clinical diagnosis of endocarditis failed to satisfy these criteria, particularly when the diagnosis was made by echocardiography or acute endocarditis occurred in the absence of previous cardiac disease. A revised set of criteria were produced following evaluation of 405 episodes of endocarditis (Table 2). In 69 cases of endocarditis which were proved pathologically, the sensitivity of the new criteria (excluding pathology) was 80 per cent compared with 51 per cent for the von Reyn criteria.

Acute endocarditis is caused by *Staphylococcus aureus*, *Streptococcus pyogenes*, and *Streptococcus pneumoniae.* It affects normal heart valves and follows a short aggressive course with high fever, leukocytosis, and death in a few weeks if untreated. 'Subacute' disease develops on already damaged valves, follows a prolonged indolent course with low-grade fever and night sweats, and is caused by α-hemolytic streptococci. Although responsible for a third of all cases of endocarditis, infection of valvular prostheses often fails to demonstrate the diagnostic clinical features. Division into early (operative) and late cases is arbitrary, and may be taken at 60 days, 4 months, or 1 year.

Blood cultures must be collected before starting antibiotic treatment; only three sets are necessary, taken 1 h apart preferably from different sites. Bacteremia is low grade but continuous, and blood cultures will be positive in 95 per cent of cases. In streptococcal disease, 96 per cent of the first blood cultures are positive, as are 98 per cent of one of the first two. For other causes, 86 per cent of first cultures are positive, as are almost all of one of the first two. If cultures are negative, repeated specimens are rarely helpful unless the patient was given antibiotic therapy in the previous 2 weeks (only

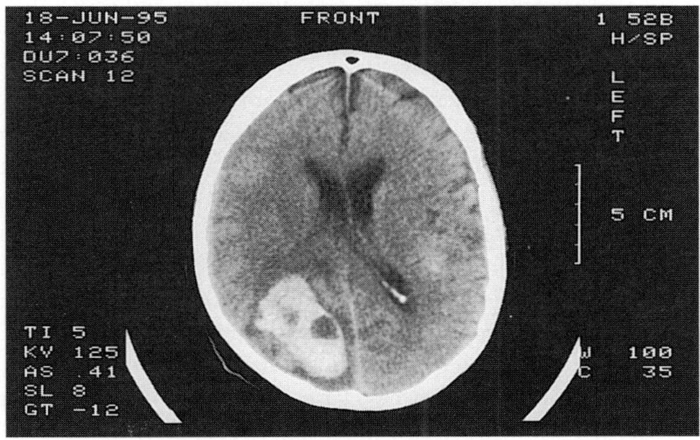

Fig. 1 CT scan of a patient with *Staph. aureus* endocarditis who was admitted with a stroke. A large cerebral abscess developed to which the patient succumbed.

Table 1 Criteria for diagnosis of infective endocarditis

Definite

Direct evidence of infective endocarditis based on histology from surgery or autopsy, or on bacteriology (Gram stain or culture) of valvular vegetation or peripheral embolus

Probable

Either persistently positive blood cultures (two of two positive, three of three positive, or ≥70% of cultures positive in four or more cultures taken) plus one of:
(1) New regurgitant murmur
(2) Predisposing heart disease (valvular or congenital heart disease or cardiac prosthesis) and vascular signs (petechias, splinter hemorrhages, Roth spots, Osler's nodes, Janeway lesions, aseptic meningitis, glomerulonephritis, and pulmonary, central nervous system, coronary, or peripheral emboli)

Or negative or intermittently positive blood cultures plus three of:
(1) Fever
(2) New regurgitant murmur
(3) Vascular signs

Possible

Either persistently positive blood cultures plus one of:
(1) Predisposing heart disease
(2) Vascular phenomena

Or negative or intermittently positive blood cultures with all three of the following:
(1) Fever
(2) Predisposing heart disease
(3) Vascular signs

Or if α-hemolytic streptococci, two or more positive blood cultures, without an extracardiac source, plus fever

Rejected

Either endocarditis unlikely as other diagnosis apparent

Or endocarditis likely and empirical antibiotic treatment warranted

Or culture-negative endocarditis diagnosed clinically but excluded by postmortem

Data from von Reyn *et al.* (1981).

65 per cent of first cultures are positive). At least half of cases of fungal endocarditis are culture negative. A normochromic normocytic anemia is usual, but a leukocytosis is common only in acute endocarditis. Both the erythrocyte sedimentation rate and the C-reactive protein are markedly elevated, and rheumatoid factor is detectable in half the cases of subacute disease. Urinalysis will often reveal microscopic hematuria and proteinuria.

Echocardiography is important in confirming the presence of vegetations, underlying cardiac lesions, and any complications, but can only confirm and not exclude the diagnosis. Planar vegetations on a prosthetic or myxomatous valve are easily missed and the skill of the operator is a major factor in its success (Fig. 2). Transthoracic echocardiography may fail to demonstrate ring abscesses. Although prolongation of the PR interval on the electrocardiogram can suggest the presence of an abscess in the conducting system, transesophageal echocardiography is often the most reliable means of diagnosis. Vegetations are demonstrable by transthoracic imaging in 70 per cent of clinically suspected cases compared with 80 to 100 per cent when the transesophageal approach is used. An increase in diagnostic accuracy from 46 to 92 per cent is obtained for abscesses. Diagnosis of vegetations on prosthetic valves is difficult by conventional echocardiography (20–30 per cent) but rises to 90 per cent accuracy when a transesophageal method is used. Obesity and emphysema can reduce the quality of a transthoracic image but not of a transesophageal image. Echocardiography can be used to monitor the size of vegetations and hence the outcome of treatment. Therefore it is important in establishing the need for surgery.

Risk factors

The average age of patients with endocarditis is 58 years and men outnumber women. A falling incidence of rheumatic heart disease, the aging population, recognition of diseases such as mitral valve prolapse, surgical correction of congenital heart disease, and placement of prosthetic heart valves have changed the population at risk. Between 1938 and 1967, rheumatic heart disease was implicated in 38 per cent of cases compared with 6 per cent between 1980 and 1984 (McKinsey *et al.* 1987). The proportion of endocarditis associated with the aortic and mitral valves is now similar, whereas mitral valve disease used to be twice as common.

Underlying cardiac lesions were present in 73 per cent (46 out of 63) of patients in one series, with mitral valve prolapse being the most usual, followed by degenerative disease of aortic or mitral valves, congenital heart disease, and rheumatic heart disease (McKinsey *et al.* 1987). Aortic regurgitation precedes endocarditis in a third of patients compared with a sixth of patients for aortic stenosis. Mitral

Table 2 Revised criteria for diagnosis of infective endocarditis

Definite

Pathological criteria: micro-organisms demonstrated by culture or histology in a vegetation or in a vegetation that has embolized or in an intracardiac abscess *or* vegetation or intracardiac abscess present, confirmed by histology showing active endocarditis

Clinical criteria

Two major criteria or one major and three minor criteria or five minor criteria

Major criteria

(1) Positive blood culture for infective endocarditis as shown by typical organism from two separate blood cultures (α-hemolytic streptococci, HACEK group, community-acquired *Staph. aureus* or enterococci without primary focus) or persistently positive blood culture, i.e. blood cultures drawn 12 h apart or all of three or a majority of four or more separate blood cultures with the first and last >1 h apart

(2) Evidence of endocardial involvement: positive echocardiogram for infective endocarditis (oscillating intracardiac mass on valve or supporting structures or in the path of regurgitant jets or on implanted material in the absence of an alternative anatomical explanation, or abscess or new partial dehiscence of prosthetic valve) or new valvular regurgitation (increase or change in pre-existing murmur not sufficient)

Minor criteria

(1) Predisposing heart condition or intravenous drug use

(2) Fever $\geqslant 38.0\,^{\circ}$C

(3) Vascular signs (major arterial emboli, septic pulmonary infarcts, mycotic aneurysm, intracranial hemorrhage, conjunctival hemorrhages, Janeway lesions)

(4) Immunological phenomena (glomerulonephritis, Osler's nodes, Roth spots, rheumatoid factor)

(5) Microbiological evidence (positive blood culture not meeting major criterion excluding single positive cultures for coagulase-negative staphylococci and non-endocarditis organisms) or serological evidence of active infection with organism consistent with endocarditis

(6) Echocardiogram consistent with infective endocarditis but not meeting major criterion

Data from Durack *et al.* (1994).

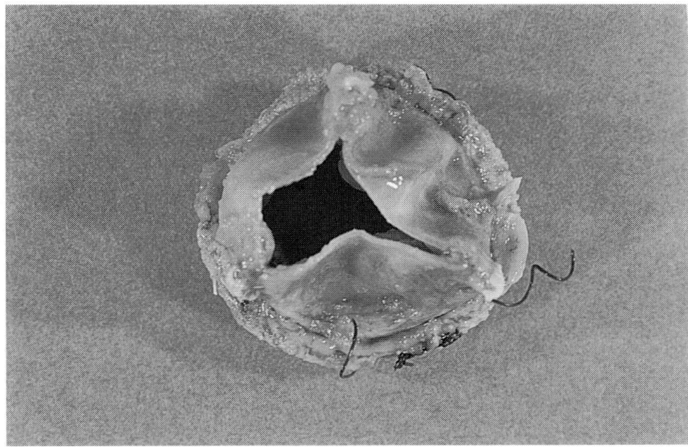

Fig. 2 A xenograft showing small vegetations that could easily be missed on echocardiography.

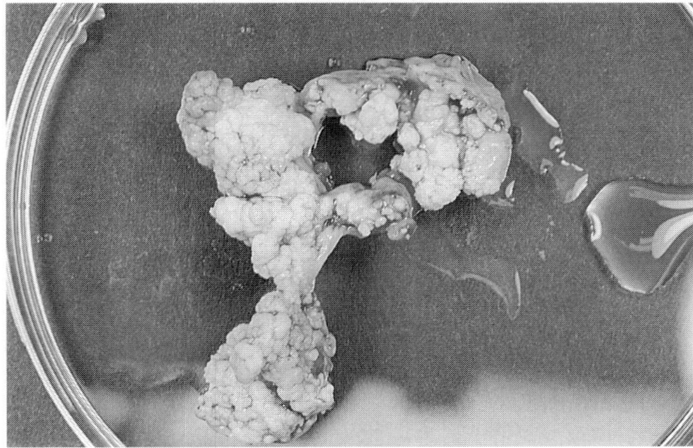

Fig. 3 A massive vegetation caused by a coagulase-negative staphylococcus originally seeded from an infected Hickman catheter in a leukemic patient. The patient had been treated with large doses of teicoplanin for 3 months, but the vegetation continued to enlarge. The patient died during removal of the valve.

regurgitation precedes another 21 to 33 per cent of cases (Michel and Acar 1995). Mitral valve prolapse is present in 5 per cent of the population, and the risk of endocarditis is increased in those with valvular thickening ($>$ 5 mm), occurring in up to 6 per cent of such patients. Tricuspid valve disease is rare.

Congenital heart disease is present in 10 to 20 per cent of patients with infective endocarditis (McKinsey *et al.* 1987). Patent ductus arteriosus, ventricular septal defect, tetralogy of Fallot, and a bicuspid aortic valve are the most common. Of 1347 patients with ventricular septal defect followed for 16 years, 32 developed endocarditis (Michel and Acar 1995). Ventricular septal defect is most likely to predispose to endocarditis if uncorrected and associated with aortic regurgitation. Partial correction of congenital heart disease is associated with a greater risk of endocarditis than if correction is complete. A patch or residual ventricular septal defect is the usual site of infection in tetralogy of Fallot. The risk of endocarditis in congenital aortic stenosis depends on the severity of the lesion. A bicuspid aortic valve is present in 1 to 2 per cent of the population and accounts for 20 per cent of cases of endocarditis. Patent ductus arteriosus and coarctation of the aorta are rarely associated with endocarditis because these malformations are now usually corrected in infancy.

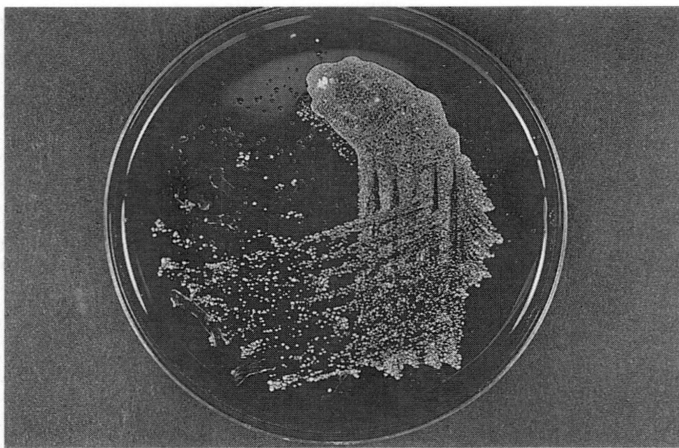

Fig. 4 Growth of the organism from the excised valve of the patient in Fig. 3.

Fig. 5 A Starr–Edwards valve showing a large vegetation.

Hypertrophic obstructive cardiomyopathy predisposes to endocarditis, particularly on the aortic or mitral valves or the subaortic endocardium. Vegetations of the aortic and pulmonary valves occur on the ventricular surface of the valve, and those of the mitral and tricuspid valves occur on the atrial surface. Approximately 5 per cent of patients may develop endocarditis, depending on the severity of obstruction.

Previous infective endocarditis is a frequent risk factor, particularly in patients with cardiac disease, continued drug abuse, or dental decay. Intravenous drug abuse is commonly associated with the development of endocarditis in the younger patient. Central venous and pulmonary artery catheters, feeding lines, and intracardiac pacemakers present other risk factors, usually for right-sided endocarditis, if not quickly removed following infection (Figs 3 and 4).

Up to a third of cases of infective endocarditis are due to infection of a prosthetic valve. In a review of 21 series, prosthetic valve endocarditis was found to occur in 0.3 to 1 patient per 100 patient-years (de Gevigney *et al.* 1995). Infection of the aortic valve may be more common than that of the mitral valve and is more likely when more than one prosthetic valve is present. Opinion is divided as to the relative risk of infection of bioprosthetic and mechanical valves, but age and length of cardiopulmonary bypass are important (Fig. 5). The risk of endocarditis rises from 3 per cent at 1 year to 5 per cent at 10 years after insertion. Endocarditis occurs in 0.1 to 7 per cent of patients with permanent pacemakers (de Gevigney *et al.* 1995).

Pathogenesis

Animal models show that the normal endothelium is not usually susceptible to bacterial infection, but when it is damaged platelets and fibrin are deposited, giving rise to non-bacterial thrombotic endocarditis. If bacteremia occurs, the organisms will adhere and colonize these lesions. Bacteria are then covered by further layers of platelets and fibrin, protecting them from the host defense. Organisms within vegetations divide at only a very slow rate, reducing the efficacy of some normally bactericidal antibiotics.

Endothelial damage is usually caused by a valvular lesion or prosthesis leading to turbulent blood flow. Vegetations are found on the ventricular surface of the aortic and pulmonary valves, on the atrial surface of the tricuspid and mitral valves, and on the surface to which flow is directed through a septal defect. The degree of turbulence and mechanical stress are important in the ease with which infection is established. Some virulent bacteria can establish infection by direct invasion of the endocardial surface. Bacteremia will follow mucosal trauma of the oropharynx, the gastrointestinal tract, or the genitourinary tract. Generally there are less than 10 organisms per milliliter of blood. It commonly occurs during dental surgery and less frequently during endoscopy, sigmoidoscopy, urethral catheterization, or cystoscopy. Intravenous catheters, pacemakers, wound infections, and urinary infections may be sources in the critically ill.

Production of an extracellular polysaccharide, dextran, by some organisms improves their binding to vegetations. *Streptococcus mutans*, *Streptococcus bovis*, dextran-positive *Streptococcus mitior*, and *Streptococcus sanguis* all produce dextran, and, in contrast with other streptococci, their presence in the blood is more likely to be associated with endocarditis than with other diseases. Bacterial strains, particularly *Strep. sanguis,* associated with endocarditis are highly active in promoting platelet aggregration. Once covered by further layers of platelets and fibrin, bacteria are protected from phagocyte cells and, despite low metabolic activity, a very high density can be achieved.

Vegetations vary widely in size, and tend to be larger and more easily disrupted in acute disease. They can form emboli to other organs or, in the case of the tricuspid valve, into the lung. As the vegetation enlarges, the valve leaflet or the chordae tendinae are perforated or ruptured, causing regurgitation. Staphylococcal and some streptococcal infections cause abscess formation in the cardiac wall at the root of insertion of the valve, resulting in heart block or (with artificial valves) a paraprosthetic leak. Myocardial and renal infarctions are found in half of patients at postmortem. Mycotic aneurysms occur particularly in streptococcal disease. Cerebral emboli are common and usually pass into the middle cerebral artery, causing infarction, abscesses, or hemorrhage (Fig. 1). Hyperplasia of the follicles within the spleen results in splenomegaly, but infarcts are also found in almost half of patients at postmortem. Petechias develop in a third of patients. Osler's nodes are less common and are the result of immune complex deposition in the dermal vessels. Janeway lesions are caused by septic emboli reaching the skin. Roth spots are formed by local swelling and hemorrhage in the retina.

Humoral and cellular immunity is stimulated, resulting in an elevated plasma γ-globulin and splenomegaly. Rheumatoid factor and hypocomplementemia are found in 25 to 50 per cent of patients.

Circulating immune complexes contribute to the formation of splinter hemorrhages, Osler's nodes, purpura, glomerulonephritis, and arthritis. Immunofluorescence has demonstrated immune complexes in blood vessels in purpura. Except in the acute stages of the illness, immune complexes and complement are deposited in the basement membrane of the kidney, forming a diffuse glomerulonephritis. Focal glomerulonephritis is common and is recognized on biopsy by segmental fibrinoid necrosis and crescent formation.

Pathogens

The α-hemolytic streptococci are the most common pathogens, particularly *Strep. sanguis*, *Strep. bovis*, *Strep. mutans*, and *Strep. mitis. Strep. mutans* is a prominent component of dental plaque. *Strep. bovis* is found in the gastrointestinal tract; when it is found to cause bacteremia, a search should be made for carcinoma or other lesions of the colon. Nutritionally deficient streptococci (*Streptococcus adjacens* and *Streptococcus defectivus*) require addition of cysteine to culture media to allow detection. Enterococci are responsible for approximately 10 per cent of cases and affect elderly men, usually after genitourinary procedures. The organisms characteristically have poor susceptibility to most agents. *Streptococcus pneumoniae* is responsible for only 1 per cent of cases, but causes an aggressive infection with abscess formation and death in half of those affected.

A third of cases of endocarditis are caused by staphylococci, usually *Staph. aureus* in native valve infections and *Staphylococcus epidermidis* in prosthetic valve endocarditis. *Staph. aureus* attacks intact valves and rapidly causes abscesses in the valve ring extending into the conduction system. Death occurs in up to half of cases. In drug abusers the disease is less severe but the organisms are often multiply resistant. *Staph. epidermidis* is the most common cause of early prosthetic valve endocarditis (occurring within 60 days of surgery), probably from intraoperative contamination. The rate of mortality and complications remains high despite aggressive surgical and antibiotic treatment.

Gram-negative bacilli are uncommon in endocarditis, with *Salmonella* species, *Escherichia coli*, and *Serratia marcescens* being the most frequent. Most patients have previously normal valves and mortality is high. *Pseudomonas* species cause endocarditis in intravenous drug abusers, predominantly affecting young men, who present with embolic complications and heart failure. Early surgery is needed to avoid a fatal outcome. *Hemophilus* species (*Hemophilus aphrophilus*, *Hemophilus parainfluenzae*) cause 1 per cent of cases of endocarditis, resulting in a chronic illness with common neurological complications. Endocarditis due to *Actinobacillus actinomycetemcomitans*, *Cardiobacterium hominis,* or *Kingella* species presents similarly but is rare.

Gram-positive bacilli rarely cause endocarditis. *Corynebacterium diphtheriae* (non-toxigenic) and *Listeria* species cause an aggressive disease. Lactobacilli are difficult to eradicate with medical treatment and infection follows dental procedures. Anaerobes (usually *Bacteroides fragilis, Fusobacterium necrophorum, Clostridium* species, *Fusobacterium nucleatum*, or *Propionibacterium acnes*) occasionally cause endocarditis (1 per cent of cases). The organisms are often mixed with other species and derive from the gastrointestinal tract or mouth. Intravenous drug abusers are susceptible to infection with *Staph. aureus*, often methicillin-resistant *Staph. aureus* (MRSA), streptococci, *Pseudomonas aeruginosa*, *Candida* species, and enterococci. Q-fever endocarditis is a complication of chronic infection with the rickettsial organism *Coxiella burnetii*, and diagnosis depends on serological methods rather than blood culture. Following dissemination in the blood, the organism can evade host defense and bactericidal activity of antibiotics by multiplying in the phagolysosomes of host cells.

Fungal endocarditis is a feature of patients who have a prolonged stay in the ICU, drug abusers, or following cardiac surgery. *Candida albicans* is the most common pathogen in patients subjected to long-term catheterization, whereas drug abusers tend to be infected with other species (*Candida parapsilosis*, *Candida tropicalis*). Aggressive surgical and medical treatment is necessary, but cure rates are low. The vegetations are large and poorly penetrated by amphotericin, spread occurs within the cardiac muscle, and emboli are frequent.

References

de Gevigney, G., Pop, C., and Delahaye, J.P. (1995). The risk of infective endocarditis after cardiac surgical and interventional procedures. *European Heart Journal*, **16** (Supplement B), 7–14.

Durack, D.T., Lukes, A.S., Bright, D.K., and Duke Endocarditis Service (1994). New criteria for diagnosis of infective endocarditis: utilization of specific echocardiographic findings. *American Journal of Medicine*, **96**, 200–9.

McKinsey, D.S., Ratts, T.E., and Bisno, A.L. (1987). Underlying cardiac lesions in adults with infective endocarditis: the changing spectrum. *American Journal of Medicine*, **82**, 681–8.

Michel, P.L. and Acar, J. (1995). Native cardiac disease predisposing to infective endocarditis. *European Heart Journal*, **16** (Supplement B), 2–6.

von Reyn, C.F., Levy, B.S., Arbeit, R.D., Friedand, G., and Crumpacker, C.S. (1981). Infective endocarditis: an analysis based on strict case definitions. *Annals of Internal Medicine*, **94**, 505–17.

3.9.2 Prevention

A. P. R. Wilson

Key messages

- There is little evidence that antibiotics can prevent endocarditis in susceptible individuals, but their administration is routine practice.
- Only a small proportion of cases are due to medical and surgical procedures.
- Antibiotics must be active against the likely pathogens and be administered before the procedure.
- Patients at risk in the intensive care unit will usually require a glycopeptide and an aminoglycoside.
- A cephalosporin or flucloxacillin plus an aminoglycoside are used in cardiac surgery, but courses need not be prolonged more than 24 h.

If a prosthetic valve is seeded during cardiac surgery, the patient will usually recover and leave hospital only to return up to a year later needing life-threatening surgery. Even if the patient recovers, life expectancy is reduced. Multiple antibiotic resistance in organisms causing endocarditis is making treatment more difficult. Therefore use of antibiotics to prevent endocarditis is considered essential, but there is little firm scientific evidence for present advice mainly because of the rarity of the disease. Furthermore, only 10 per cent of cases of endocarditis can be ascribed to bacteremia caused by invasive procedures.

Prevention of endocarditis in patients with abnormal heart valves can be achieved by many general measures, for example regular dental care (covered by antibiotic prophylaxis), reducing the duration of procedures in areas of resident bacterial flora, limiting the number and duration of intravascular catheterizations, and ensuring that bacteriuria has been treated before a urological procedure. Gingival inflammation can be reduced in patients at risk by encouraging brushing, use of chlorhexidine or fluoride mouthwash, and cleaning. Chlorhexidine applied to the gingiva or irrigated in the gingival sulcus prior to dental extraction will reduce bacteremia.

The rationale for the use of antibiotics in the prevention of endocarditis derives from animal models and clinical experience. Animal models depend on the presence of a plastic catheter and simulate prosthetic valve endocarditis rather than native valve disease. Dental extraction results in a bacteremia of about 100 CFU/ml after bacteria have been diluted and cleared by the reticuloendothelial system. However, there is no obvious relationship between the number of circulating bacteria and the likelihood of developing endocarditis. The ability to adhere to platelets and fibrin is more predictive in models.

Many individual cases of endocarditis following dental procedures have been reported, but the risk of developing endocarditis must be very low. Endocarditis following bacteremia associated with urological procedures has been recorded but the risk is similarly low. Underlying cardiac abnormalities (e.g. patent ductus arteriosus, prosthetic valves, aortic valve disease, or previous endocarditis) can greatly increase the risk of endocarditis. Mitral valve prolapse is common but merits antibiotic prophylaxis if it causes a murmur.

In humans, case–control studies suggest a wide range of efficacy of prophylaxis for native valves (van der Meer *et al.* 1992). Medical and surgical procedures cause only a small proportion of cases of endocarditis. van der Meer and colleagues estimated that, if prophylaxis had been given for all procedures in 275 patients with known lesions in the preceding 180 days, 47 (17 per cent) of the cases might have been prevented. If given only in the last 30 days, 23 (8 per cent) of the cases would potentially be prevented. Only eight of 48 patients having a procedure in the previous 6 months had actually been given antibiotic prophylaxis. In 52 cases of endocarditis preceded by a procedure that could induce bacteremia, mitral valve prolapse was present in a third and prosthetic valves in a fifth. However, most had received an antibiotic during the procedure.

Prophylaxis should be given at the time of the procedure in doses sufficient to ensure therapeutic concentrations. Antibiotics must be active against the likely pathogens (α-hemolytic streptococci and enterococci). Amoxicillin (amoxycillin) is recommended because of its good gastrointestinal absorption and sustained serum concentrations. However, a combination of a penicillin and an aminoglycoside is more likely to be bactericidal in high-risk cases. Antibiotics should be started just before the procedure (no more than 1–2 h before) and not continued for more than 6 to 8 h to avoid the emergence of resistant bacteria.

Procedures causing gingival bleeding should be covered by prophylaxis, but this does not include endotracheal intubation or the adjustment of orthodontic appliances. Tonsillectomy, adenoidectomy, and other surgical operations to the respiratory mucosa should be covered. If the urine is infected, instrumentation must be covered by an antibiotic active against the pathogen isolated. Flexible bronchoscopy with biopsy, cardiac catheterization, endoscopy with biopsy, liver biopsy, and urethral catheterization in the absence of infection do not need prophylaxis. Patients having colonoscopy or sigmoidoscopy probably do not need prophylaxis unless there is a prosthetic valve or previous endocarditis.

Patients in the intensive care unit (ICU) are likely to have a resistant bacterial flora and the standard high-risk regimens may not be adequate. At-risk patients who are having a tracheostomy inserted should be given vancomycin (1 g intravenously) or teicoplanin (12 mg/kg intravenously) with gentamicin before the procedure.

Table 1 European consensus for the prevention of endocarditis in at-risk patients*

Patient groups

High risk: prosthetic valves, cyanotic congenital heart disease, previous infective endocarditis

Low risk: valvular heart disease, congenital heart disease not causing cyanosis, hypertrophic obstructive cardiomyopathy

Procedures

Dental work, tonsillectomy, adenoidectomy, esophageal dilatation or surgery or endoscopic laser procedures, sclerosis of esophageal varices, abdominal surgery, instrumentation of ureter or kidney, surgery of prostate or urinary tract

Regimens

Minimal regimen to be used for low cardiac risk, dental work, single procedure, out-patients or local anesthesia

Amoxicillin 3 g orally or clindamycin 300–600 mg orally as a single dose

Maximal regimen to be used for high cardiac risk, gastrointestinal or urological procedures, multiple procedures, in-patients, or general anesthesia

Amoxicillin or ampicillin 2 g IV+gentamicin 1.5 mg/kg IV/IM then amoxicillin or ampicillin1–1.5 g at 6 h

or Vancomycin 1 g IV over >1 h as an infusion+gentamicin 1.5 mg/kg IM/IV then vancomycin 1 g IV over 1 h at 12 h

IV, intravenous; IM, intramuscular

*The clinician is allowed discretion in individual patients to form intermediate precautions between minimal and maximal regimens.

Data from Leport *et al.* (1995).

Table 2 Recommendations of the American Heart Association for the prevention of endocarditis

Patient groups

Prosthetic cardiac valves, previous endocarditis, congenital cardiac abnormalities, rheumatic and other valvular disease, hypertrophic cardiomyopathy, mitral valve prolapse with regurgitation

Procedures

Those causing gingival or mucosal bleeding, tonsillectomy, operations of intestinal or respiratory mucosa, rigid bronchoscopy, sclerotherapy for varices, esophageal dilatation, gallbladder surgery, cystoscopy, urethral dilatation, urethral catheterization if urine infected, prostatic surgery, drainage of infected tissue, vaginal hysterectomy, vaginal delivery in presence of infection

Regimens

Dental, oral, and upper respiratory tract procedures

Amoxicillin 3 g 1 h before procedure then 1.5 g at 6 h

or Erythromycin ethylsuccinate 800 mg or erythromycin stearate 1 g orally 2 h before procedure then half the dose at 6 h

or Clindamycin 300 mg orally 1 h before procedure then 150 mg at 6 h

If unable to take orally

Ampicillin IV/IM 2 g 30 min before procedure then 1 g at 6 h (or 1.5 g amoxicillin orally) *or* clindamycin 300 mg IV 30 min before procedure then 150 mg IV/orally at 6 h

High risk: ampicillin IV/IM 2 g+gentamicin 1.5 mg/kg 30 min before procedure then amoxicillin 1.5 g orally 6 h or IV 8 h *or* vancomycin IV 1 g over 1 h before procedure

Genitourinary and gastrointestinal tract procedures

Ampicillin+gentamicin then amoxicillin as above parenteral regimen

or Vancomycin IV/IM 1 g+gentamicin IV/IM 1.5 mg/kg 1 h before procedure

IV, intravenous; IM, intramuscular
Data from Dajani *et al.* (1990).

Insertion of a chest drain would require an appropriate agent if the effusion is likely to be infected, but the microbiologist should be consulted. Cannulation of peripheral veins and arteries does not need prophylaxis, nor does placement of a central venous catheter unless it is likely to be difficult or is a guidewire insertion into a previously infected site.

Recommendations for prophylactic regimens differ from country to country (Tables 1 and 2). Most countries agree that dental procedures (causing gingival bleeding) and oropharyngeal, gastro-intestinal, and urological procedures present a risk, but recommendations vary with respect to bronchoscopy, colonoscopy, and vaginal hysterectomy (Leport *et al.* 1995). A single large oral dose of amoxicillin (3 g), as used in Europe, results in a high ratio of serum level to likely minimum inhibitory concentration of the pathogens and in animals was effective against bacteremia that would produce infection in 90 per cent of controls. An additional dose can be used if the likely inoculum is high, and in the United States all patients are required to take a second dose. Clindamycin or erythromycin is the usual oral alternative. Ampicillin or amoxicillin can be given intravenously 1 h before a procedure, with an oral dose 6 h later. High-risk patients are given aminoglycosides in addition, and vancomycin or teicoplanin is used for patients allergic to penicillins. Parenteral prophylaxis is used for gastrointestinal and urological procedures, and aminoglycosides are included in the regimen to provide synergistic activity against enterococci.

Only a minority of patients receive adequate prophylaxis or are properly advised about its use. In a German study, 75 per cent of cardiologists issued written and oral instructions to patients but 60 per cent of procedures were not covered (Gohlke-Barwolf *et al.* 1995). Only 57 per cent of dentists administered antibiotics according to published recommendations. The need for prophylaxis was recognized by 53 per cent of patients at risk but, of patients recently having dental procedures, only 66 per cent (71/107) with prosthetic valves and 46 per cent (10/22) with abnormal native valves received antibiotic prophylaxis. It is a common error to administer antibiotic prophylaxis more than 30 to 60 min before the procedure, but this is not justified and risks the emergence of a resistant bacterial flora. Prolongation of prophylaxis presents a similar risk and increases the chance of adverse effects. The high oral dose of amoxicillin provides a sufficient serum concentration, but the parenteral route is preferable in most patients in the ICU.

Many α-hemolytic streptococci are tolerant to the action of amoxicillin, but prophylaxis is successful even below the minimum bactericidal concentration. Some benefit is gained even if antibiotics are administered after the bacteremia has occurred. Prevention of adherence to fibrin has been demonstrated at low antibiotic concentrations, rendering the bacteria susceptible to host defense. Low numbers of bacteria adhering to vegetations are eventually eradicated in the absence of antibiotics, and inhibition of adherence of greater numbers by antibiotics may allow this process to continue (Blatter and Francoli 1995). It is possible that bactericidal agents are produced by the platelets in the vegetation.

Patients having open-heart surgery are at high risk of bacteremia. Coagulase-negative staphylococci and coryneforms are commonly present in the blood during surgery and can be disseminated from central venous and peripheral catheters in the early postoperative period. During surgery, host defense is compromised by complement activation after contact with plastics in the pump, antibiotics are affected by protein binding and hemodilution, a large operative area is at risk, many staff are involved, and surgery is often prolonged. Blood accumulating in the chest cavity is returned to the circulation by suction, together with contaminating organisms. If prosthetic valve endocarditis develops soon after surgery, it is often fatal. Patients with dental disease should have treatment or extractions, under antibiotic cover, before surgery.

The use of antibiotic prophylaxis is routine during cardiac surgery. Flucloxacillin plus an aminoglycoside or a cephalosporin is a common choice in the United Kingdom, while cephalosporins are often used alone in the United States. In most Scandinavian countries, single-drug antistaphylococcal agents are used for less than 48 h. Animal models suggest that antibiotic prophylaxis is effective only up to 4 h after surgery, and in humans 2-day or even single-dose regimens are effective in preventing wound infection. Although the efficacy of prophylaxis has been demonstrated in the prevention of wound infection, the effect on endocarditis has not been unequivocally shown in a controlled trial. Improvements in surgical technique are probably responsible for most of the improvement in the rate of postoperative endocarditis, but a risk of early endocarditis remains in 0.5 per cent of patients. The coagulase-negative staphylococci associated with early prosthetic valve endocarditis are usually resistant to prophylactic antibiotics, suggesting that colonization or bacteremia has occurred in the operating room or the ICU. Therefore use of antibiotics beyond 24 h is to be discouraged.

Regimens including aminoglycosides would be expected to be more active than cephalosporins against the majority of coagulase-negative staphylococci, but both regimens work well in practice. The rarity of endocarditis makes direct comparisons difficult. If aminoglycosides are used, only 24-h dosing should be given. Some surgeons ask for 48-h administration to cover intravascular catheterization, in which case a single trough level should be taken the morning after surgery. If it is higher than 2 mg/l, the gentamicin should be stopped. Even a 2-day course of aminoglycoside is associated with a significantly higher serum creatinine after surgery than other antibiotics. Vancomycin has been used, particularly in units where methicillin-resistant *Staphylococcus aureus* (MRSA) is prevalent, but it is associated with an increased need for inotropes in the early postoperative period.

References

Blatter, M. and Francoli, P. (1995). Endocarditis prophylaxis: from experimental models to human recommendation. *European Heart Journal*, **16** (Supplement B), 107–9.

Dajani, A.S., *et al.* (1990). Prevention of bacterial endocarditis. Recommendations by the American Heart Association. *Journal of the American Medical Association*, **264**, 2919–22.

Gohlke-Barwolf, C., Rost, M., and Roskamm, H. (1995). Prevention of bacterial endocarditis in Germany. *European Heart Journal*, **16** (Supplement B), 110–13.

Leport, C., Horstkotte, D., Burckhardt, D., and The Group of Experts of the International Society for Chemotherapy (1995). Antibiotic prophylaxis for infective endocarditis from an international group of experts towards a European consensus. *European Heart Journal*, **16** (Supplement B), 126–31.

van der Meer, J.T.M., van Wijk, W., Thompson, J., Vandenbroucke, J.P., Valkenburg, H.A., and Michel, M.F. (1992). Efficacy of antibiotic prophylaxis for prevention of native valve endocarditis. *Lancet*, **339**, 135–9.

3.9.3 Treatment

A. P. R. Wilson

Key messages

- The presentation of endocarditis varies widely and fever in the presence of a cardiac lesion may be the only initial signs.
- Serum concentrations of β-lactams and glycopeptides should be kept well above the minimum inhibitory concentration, but high peaks are more important for aminoglycosides and quinolones.
- Serum assays of aminoglycosides must be performed every 1 to 2 days.
- Penicillin G (benzylpenicillin) is used to treat infections due to α-hemolytic streptococci but gentamicin is added if not fully susceptible.
- Both flucloxacillin and nafcillin with gentamicin are effective against *Staphylococcus aureus*, but the aminoglycoside is probably only needed for the first 5 days for native valves.
- Glycopeptides are reserved for methicillin-resistant staphylococci and some enterococci.

Introduction

Bacteria are present in large numbers in vegetations (10^9 organisms/g); they divide slowly and are protected from host defenses. Antibiotics, particularly the glycopeptides, may not penetrate well into the substance of a vegetation. Medical treatment of prosthetic valve endocarditis without replacement of the valve will often fail, even if the organism is apparently susceptible *in vitro*. A persistent fever without obvious cause, blood cultures positive for a typical organism, echocardiographic evidence of vegetations, abscess, or paraprosthetic leak, and a documented new regurgitant murmur are the common indications for treatment. Vascular and immunological signs (e.g. arterial emboli, microscopic hematuria, or subungual hemorrhages in a non-manual worker) can be helpful. However, patients rarely present in the same way, and in some cases fever in the presence of a cardiac lesion or prosthetic valve may be the only initial grounds for suspicion.

Treatment of infective endocarditis must be tailored to the species and antibiotic susceptibility of the pathogen, and the presentation of the disease and its complications. The minimum inhibitory concentration (MIC) and minimum bactericidal concentration can be used to guide choice of antibiotic. A bactericidal regimen is required, usually in the form of a synergistic combination of antibiotics which protect each other from the emergence of resistance. However, patients with large vegetations, intracardiac abscess, continued fever, emboli, or congestive cardiac failure require early surgery.

Penicillins and glycopeptides (vancomycin and teicoplanin) have a relatively slow bactericidal activity (Besnier and Choutet 1995). The extent of bacterial killing depends on the time during which serum concentrations exceed the MIC. Aminoglycosides and quinolones have a faster bactericidal effect which depends on concentration (e.g.

high peaks). If the concentration falls below the MIC, regrowth of the organism does not occur until after a delay; this is known as the post-antibiotic effect. Gram-positive bacteria may not grow for 2 to 3 h after removal of β-lactams or vancomycin. The slow rate of division of bacteria in vegetations reduces the bactericidal effect that β-lactams can exert, resulting in relapse if the treatment time is too short. Autoradiographic methods in animal models show that some antibiotics (e.g. teicoplanin) do not diffuse into a vegetation but are limited to the periphery. Others (e.g. penicillin) form a gradient to the center of the vegetation, and still others (aminoglycosides and quinolones) diffuse throughout the vegetation.

When using β-lactams or glycopeptides, the dose should be sufficient to maintain concentrations in the vegetation above the MIC. High serum concentrations may be necessary, particularly in the case of teicoplanin in staphylococcal endocarditis. High peaks are necessary for aminoglycosides and quinolones. The postantibiotic effect will maintain suppression of Gram-negative bacteria, but not enterococci, between doses. Parenteral therapy is normally necessary, although ciprofloxacin and rifampin (rifampicin) can be given by mouth. To maintain high trough concentrations, penicillins should be given every 4 h (i.e. three to four half-lives) or as a continuous infusion. Ceftriaxone and teicoplanin can be given once daily because of their long half-lives, but teicoplanin requires three loading doses of 6 mg/kg at 12-h intervals to achieve therapeutic concentrations.

Serum monitoring of aminoglycosides at intervals of 1 to 2 days is essential to avoid potentially toxic concentrations, particularly in the critically ill (trough, below 2 mg/l; peak, 5–10 mg/l), although low levels may be acceptable when used against susceptible streptococci (trough, below 1 mg/l; peak, 4 mg/l). Nephrotoxicity may be reduced by administration of gentamicin once daily (7 mg/kg/day), but experience in endocarditis is limited. Vancomycin concentrations should be 10 to 15 mg/l at trough to ensure efficacy without nephrotoxicity, and trough concentrations of teicoplanin in excess of 20 mg/l are associated with improved outcome in staphylococcal endocarditis if it is used alone.

Recurrence of fever can be due to uncontrolled infection, but it is sometimes a reaction to the treatment itself or is due to thrombophlebitis. Further blood cultures should be taken during treatment to ensure bacterial eradication. The serum bactericidal titer is sometimes determined; it is defined as the greatest dilution of serum, taken at trough and peak of antibiotic administration, which still inhibits the infecting pathogen *in vitro*. It is limited in its clinical use and a low titer does not necessarily predict failure of treatment. Persistent bacteremia and fever suggest a cardiac abscess or metastatic abscesses, and surgery should be considered.

Streptococci

The α-hemolytic streptococci are the most common pathogens isolated. Although frequently sensitive to penicillin, a proportion display tolerance; the minimum bactericidal concentration of penicillin exceeds the MIC by a factor of 32 or more. The American Heart Association does not recommend determination of the minimum bactericidal concentration (Wilson *et al.* 1995).

For streptococci (including *Streptococcus bovis*) which are highly susceptible to penicillin (MIC 0.1 mg/l), treatment with penicillin G is curative in almost all patients (Table 1). Penicillin can be given in six equal doses or as a continuous infusion. Ceftriaxone is effective and is more convenient if treatment is eventually completed as an outpatient. Penicillin or ceftriaxone alone is to be preferred in the elderly patient or those at high risk of ototoxicity or nephrotoxicity with aminoglycosides. Gentamicin shows synergism with penicillin in bactericidal activity, but addition of gentamicin has not been demonstrated to improve cure rates significantly. However, some recommend an initial 2-week course of gentamicin (Simmons *et al.* 1985). Serum concentrations of gentamicin need only be 3 to 4 mg/l at peak and less than 1 mg/l at trough. Infections of prosthetic valves should be treated for 6 weeks with penicillin combined with gentamicin for the first 2 weeks (Wilson *et al.* 1995). Patients allergic to β-lactam antibiotics can be treated with ceftriaxone if there is no history of immediate type hypersensitivity or, alternatively, vancomycin or teicoplanin for 4 weeks. Vancomycin must be infused over 1 h to avoid 'red man' syndrome and hypotension, and is usually assayed. Teicoplanin (not available in the United States) does not suffer these disadvantages but fever can be an adverse effect.

Combination treatment should be used for treating infections caused by α-hemolytic streptococci which are not fully susceptible to penicillin (MIC > 0.1 mg/l) and in patients with a mycotic aneurysm, with symptoms that have been present for more than 3 months, who have relapsed, or who are on long-term penicillin. If the MIC is less than 0.5 mg/l, gentamicin can be given for the first 2 weeks of treatment with penicillin. If the MIC is greater than 0.5 mg/l or the streptococci are nutritionally deficient variants, gentamicin should be given for 4 weeks at a dose just sufficient to keep peak levels of 3 mg/l and trough levels below 1 mg/l. Serum assays two or three times weekly are essential to minimize the risk of cumulative toxicity. Vancomycin or teicoplanin can be used as alternatives without an aminoglycoside.

The β-hemolytic streptococci and *Streptococcus pneumoniae* are infrequent causes of endocarditis. In some areas, pneumococci increasingly have a reduced susceptibility to penicillin (MIC > 0.1 mg/l) or are resistant (MIC > 2 mg/l). Treatment with penicillin G or a glycopeptide should be decided by laboratory results. Group A streptococci should be treated with penicillin G, but other β-hemolytic streptococci may require addition of gentamicin for the first 2 weeks if susceptibility is reduced.

Enterococci

Enterococci are resistant to low concentrations of penicillin (MIC = 2 mg/l) and vancomycin (MIC = 4 mg/l). They appear to be more susceptible to teicoplanin (MIC = 0.5 mg/l), but neither glycopeptide is bactericidal when used alone at 16 times the MIC. Cephalosporins are ineffective. A combination of penicillin, ampicillin, or a glycopeptide with an aminoglycoside is synergistic and bactericidal for strains with low-level gentamicin resistance. Combinations are more effective in sterilizing vegetations in animal models, and improved efficacy has been found in limited clinical studies.

Penicillin is less active *in vitro* against enterococci than ampicillin, but there have been no studies to suggest a difference in efficacy. Gentamicin will need to be administered for 4 weeks with penicillin G or ampicillin. The risk of ototoxicity can be reduced by ensuring peak levels of 3 mg/l and trough levels below 1 mg/l. A 6-week course may be necessary if symptoms have been present for more than 3 months before treatment. If toxicity prevents completion of aminoglycoside treatment, penicillin should be given for at least 6 weeks. Patients allergic to penicillin or infected by resistant organisms

Table 1 Medical treatment of infective endocarditis

Description	Probable pathogen	Treatment	Dose
Native valve and late prosthetic valve endocarditis of slow onset	α-Hemolytic streptococci penicillin MIC ⩽0.1 mg/l	Penicillin G 4 weeks ±gentamicin 2 weeks or ceftriaxone or vancomycin 4 weeks or teicoplanin 4 weeks	6–12 g/day 80 mg/12 h 2 g/24 h 15 mg/kg/12 h 6 mg/kg/24 h
	MIC >0.1 mg/l	Penicillin G 4 weeks +gentamicin 2–4 weeks or vancomycin or teicoplanin	6–12 g/day 80 mg/12 h
Native valve in older men	Enterococci	Penicillin G 4 weeks +gentamicin 4 weeks or ampicillin 4 weeks +gentamicin 4 weeks or teicoplanin+gentamicin or vancomycin+gentamicin	11–18 g/day 1 mg/kg/8 h 12 g/day 1 mg/kg/8 h 6 mg/kg/day 1 mg/kg/8 h 15 mg/kg/12 h 1 mg/kg/8 h
Rapid onset or drug abuser	*Staph. aureus*	Flucloxacillin/nafcillin 4–6 weeks or penicillin G (if sensitive) or vancomycin 4–6 weeks or teicoplanin 4–6 weeks *Any*+gentamicin 5 days or 2 weeks	2 g/4 h IV 9.6–12 g/day 15 mg/kg/12 h 6–12 mg/kg/day 1 mg/kg/8 h
Early prosthetic valve endocarditis	Coagulase-negative staphylococci	Vancomycin 4 weeks or teicoplanin 4 weeks or flucloxacillin/nafcillin *Any*+rifampin 4 weeks and/or gentamicin 2 weeks	15 mg/kg/12 h 6 mg/kg/day 2 g/4 h IV 300 mg orally every 8–12 h 1 mg/kg/8 h
Immune suppression long-term ICU patient	*Candida* or *Aspergillus* species	Amphotericin B 4 weeks	3 g total dose
Culture negative	—	Penicillin G 4 weeks+ gentamicin 2 weeks	6–12 g/day 80 mg/8 h

IV, intravenous.

should be treated with a glycopeptide, but vancomycin tends to increase the risk of nephrotoxicity with aminoglycosides. Combinations of teicoplanin and gentamicin are superior to teicoplanin alone in animal models and are recommended clinically. An adequate sustained concentration of teicoplanin may be more effective than intermittent high peaks.

Enterococci showing a high-level resistance to aminoglycosides are not subject to any synergism between penicillin and gentamicin. Up to a quarter of *Enterococcus fecalis* and half of *Enterococcus fecium* are in this category. If the organism is resistant to gentamicin and streptomycin, high doses of penicillin or ampicillin given over 8 weeks can be curative in some patients but others will require surgery. Intrinsic penicillin resistance and β-lactamase production are present in increasing numbers of isolates of enterococci, but vancomycin or teicoplanin plus an aminoglycoside can be used to provide synergistic killing. Vancomycin-resistant enterococci are now causing outbreaks of infection in the United States and Europe, often centered in intensive care units. Those of the *vanA* genotype are resistant to both vancomycin and teicoplanin, and treatment can only be decided in consultation with the microbiology laboratory. If the resistance to aminoglycoside is low level, a combination of vancomycin, penicillin, and aminoglycoside can be tried. Enterococci of the less common *vanB* genotype have lower-level resistance to vancomycin and remain susceptible to teicoplanin.

Staphylococcus aureus

Endocarditis on a native valve caused by *Staph. aureus* is usually a rapidly progressive disease with a high mortality (up to 40 per cent). Intravenous drug abusers develop infection on the tricuspid valve which is more amenable to treatment (mortality up to 15 per cent). Almost all isolates are resistant to penicillin, but penicillin G should be used if the organism is susceptible. Flucloxacillin, nafcillin, or oxacillin are effective in disease caused by methicillin-sensitive staphylococci.

Combinations of β-lactam antibiotics and aminoglycosides exert a more rapid bactericidal activity than either drug alone both *in vitro* and in vegetations in animal models. However, clinical trials using aminoglycosides for the first 2 weeks of treatment failed to show any marked improvement in duration of fever, mortality, or occurrence of complications (Bayer 1993). The American Heart Association recommends that gentamicin be used for 3 to 5 days to ensure rapid clearance of the organism and to minimize abscess formation, but then stopped (Wilson *et al.* 1995). A 2-week course of gentamicin is suggested for prosthetic valve endocarditis. United Kingdom recommendations are that gentamicin be given for no more than 2 weeks (Simmons *et al.* 1985).

Patients allergic to penicillin should be treated with vancomycin or teicoplanin for 4 to 6 weeks, preferably with aminoglycosides initially. Vancomycin serum concentrations should be 25 to 30 mg/l at peak and 10 mg/l at trough, and teicoplanin should be kept over 20 mg/l if used alone. Gentamicin may not improve outcome of treatment with vancomycin and risks increased nephrotoxicity. However, if the organism is susceptible, a course of at least 3 to 5 days is recommended, particularly if the patient experiences persistent bacteremia, neurological sequelae, or bone metastases (Bayer 1993). The bactericidal activity of glycopeptides is slower than that of penicillins, and they should not otherwise be preferred.

Infection with methicillin-resistant staphylococci must be treated with glycopeptides. Vancomycin (30 mg/kg/day) or teicoplanin (6–12 mg/kg/day) are given for 4 to 6 weeks. Vancomycin alone results in a slow response with up to 50 per cent of blood cultures remaining positive at 1 week. Rifampin (600 mg/day orally) has high activity *in vitro* and can be given concomitantly. Resistance can develop, but gentamicin can be added as a third drug for 2 weeks.

Coagulase-negative staphylococci

Early prosthetic valve endocarditis is commonly caused by coagulase-negative staphylococci which are characteristically resistant to methicillin (Wolff *et al.* 1995). Late endocarditis is usually caused by more susceptible organisms. Teicoplanin is less active than vancomycin against *Staphylococcus hemolyticus* and some strains of *Staphylococcus epidermidis*. Vancomycin or teicoplanin plus rifampin and/or gentamicin is suitable. For methicillin-susceptible strains, flucloxacillin, nafcillin, or oxacillin are used instead of the glycopeptide. United States recommendations are that three drugs be used to prevent emergence of resistance to rifampin (Wilson *et al.* 1995). Surgery is usually required for paraprosthetic leak or valvular root abscess before a response to treatment can be expected.

Other organisms

Endocarditis caused by *Hemophilus parainfluenzae*, *Hemophilus aphrophilus*, *Actinobacillus actinomycetemcomitans*, *Cardiobacterium hominis*, *Eikenella corrodens*, and *Kingella kingae* (the HACEK organisms) is often missed because the organisms grow slowly in blood culture. Strains resistant to ampicillin have been recognized, but susceptibility testing is difficult. Ampicillin, cefotaxime, or ceftriaxone are given for 4 or 6 weeks for a prosthetic valve. For Gram-negative bacteria (e.g. *Pseudomonas aeruginosa*), susceptibility to β-lactamase, aminoglycosides, and fluoroquinolones should be determined.

Fungal endocarditis

Fungal endocarditis has become more common with the increased use of long-term intravenous catheters and devices, parenteral nutrition, immune suppression, drug abuse, and prolonged antibiotic treatment. *Aspergillus* and *Candida* species are the usual causes, and almost half the patients in some series have recently had cardiac surgery. Amphotericin B alone can be used for patients who are not immune suppressed or in organ failure. Most patients require surgery in addition to antifungal agents. Survival rates of 25 per cent overall increase to 58 per cent in those having surgery. Surgery should be performed soon after diagnosis because the large vegetations are liable to embolize. Amphotericin B is associated with fever, chills, and renal failure, and high total doses (2.5–3 g) have been thought necessary to overcome the poor penetration of the drug. Liposomal and colloidal forms avoid the adverse effects even at higher doses but tend to be less potent and should be used only if the conventional form is not tolerated. Animal models suggest that fluconazole may be effective in endocarditis caused by *Candida albicans* but there are few clinical data.

Culture-negative endocarditis

Initial empirical treatment of endocarditis is penicillin plus gentamicin or, if there is an acute aggressive onset, flucloxacillin (or similar) plus gentamicin. Patients with prosthetic valves should be treated with a glycopeptide plus an aminoglycoside with or without rifampin. If blood cultures remain negative, it is probable that the patient has received previous antibiotics. The nutritionally deficient streptococci, HACEK organisms, anaerobes, *Brucella* species, and *Listeria* species may take a week or more to be recognized in culture. Rarely, failure to respond to penicillin and gentamicin may be the result of infection with *Coxiella*, *Bartonella*, or *Chlamydia* species, and serological tests are required. Fungal endocarditis may appear culture negative. Surgical management will be necessary if there is no response in the presence of clear vegetations on echocardiography. Collagen vascular diseases, sarcoidosis, malignancy, tuberculosis, and viral infections (including HIV) should be eliminated. Recurrence of fever after an initial response may be caused by the antibiotic itself and resolves after a change of treatment. *Coxiella* infection requires prolonged treatment with doxycycline, possibly for life, and surgery is often necessary.

Surgery

Surgery is commonly required before the end of antibiotic treatment if the valve is destroyed, leading to cardiac failure. Cardiac abscesses are more common in patients with prosthetic valves and those with staphylococcal infection. The onset of atrioventricular heart block is an ominous sign. Extension of the abscess into supporting structures makes surgical treatment difficult and can be avoided by early echocardiographic assessment and intervention. Vegetations of dimensions over 10 mm tend to carry a greater risk of embolization and are less likely to be cured by medical treatment alone.

The purpose of surgery is to remove infected tissue, obliterate abscess cavities, close any fistula or defect, and remove vegetations, usually by valve replacement. The timing of surgery can be difficult and depends on the presence of a prosthetic valve, the site, complications and organism, the delay in diagnosis, the efficacy of antibiotic treatment, and the progression of the disease. If the patient remains cardiovascularly compensated, surgery can be delayed until completion of antibiotic treatment, depending on the degree of dysfunction. Severe aortic regurgitation usually requires surgery because of the poor eventual prognosis. Severe mitral regurgitation (usually papillary rupture or valvular perforation) may open to a limited repair. If pulmonary edema develops or cardiac output falls and renal failure develops, surgery will usually have to be performed as an emergency procedure following ultrasound examination. Transient pulmonary edema may be caused by an arrhythmia or chordae tendinae rupture, but if only aortic valve disease is found, surgery should be considered after 3 weeks of antibiotic treatment.

Persistence of bacteremia during antibiotic treatment can be caused by resistance of the pathogen, intracardiac or metastatic abscesses, or mycotic aneurysms. Staphylococcal prosthetic valve endocarditis inevitably requires valve replacement, as do half of the cases of staphylococcal native valve endocarditis. Extension of an abscess into the interventricular septum causes atrioventricular block and indicates the need for urgent surgery. Aortic and prosthetic valve disease is most often complicated by abscess formation (up to 50 per cent) which makes surgical correction difficult and risky.

Surgery is needed when a new murmur appears or emboli are produced. Perivalvular leaks are common following valve replacement. Transesophageal echocardiography is the most sensitive means of confirming the presence of an abscess. Surgical intervention should not await sterilization by antibiotics because the prognosis is poor. Appearance of a pansystolic murmur in native aortic valve disease suggests septal perforation and, if confirmed by echocardiography, is another indication for surgery. Emboli occur in a third of cases of endocarditis and account for a quarter of the deaths, particularly when caused by staphylococci, fungi, and Gram-negative bacilli. More than two episodes of embolization should be an indication for surgery to remove the vegetation.

Infected valves can be replaced by mechanical or bioprosthetic valves. Homografts are often used because they do not need anticoagulation and facilitate repair of abscess cavities. Some cases of mitral valve disease can be adequately treated by excision of the vegetations. Tricuspid valve disease in drug abusers can be treated by excision of the valve or the vegetations. Insertion of bioprosthesis is not always necessary in these patients, and there is a risk of recurrence if abuse continues. Mortality in native valve disease has been reported in 5 to 18 per cent of cases, being higher in those with annular abscesses or aneurysms or multiple valve replacement. Surgery for prosthetic valve endocarditis results in death in 13 to 30 per cent of cases depending on the time since insertion of the valve.

References

Bayer, A.S. (1993). Infective endocarditis. *Clinical Infectious Diseases*, **17**, 313–22.

Besnier, J.M. and Choutet, P. (1995). Medical treatment of infective endocarditis: general principles. *European Heart Journal*, **16** (Supplement B), 72–4.

Simmons, N.A., *et al.* (Working Party of the British Society for Antimicrobial Chemotherapy) (1985). Antibiotic treatment of streptococcal and staphylococcal endocarditis. *Lancet*, **ii**, 815–17.

Wilson, W.R., *et al.* (1995). Antibiotic treatment of adults with infective endocarditis due to streptococci, enterococci, staphylococci and HACEK microorganisms. *Journal of the American Medical Association*, **274**, 1706–13.

Wolff, M., Witchitz, S., Chastang, C., Regnier, B., and Vachon, F. (1995). Prosthetic valve endocarditis in the ICU. *Chest*, **10**, 688–94.

3.10 Severe hypertension

3.10.1 Pathophysiology and causes

E. A. Panacek

Key messages

- If severe hypertension is allowed to persist for long enough, it will eventually cause vascular or end-organ damage. The organs most often affected include the heart, the kidneys, and the central nervous system.
- When the upper limit of cerebral autoregulation is exceeded for a period of time, hypertensive encephalopathy results, with cerebral edema, petechial hemorrhages, and microinfarcts. These patients typically present with nausea, vomiting, severe headaches, visual changes and, in extreme cases, mental obtundation or seizures.
- Increased systemic blood pressure, which increases afterload on the heart, also increases myocardial workload. Afterload is a major determinant of myocardial oxygen demand. The acute onset of pulmonary edema or myocardial ischemia is a common presentation of a hypertensive emergency.

Introduction

Hypertension is one of the most common medical problems in the United States, affecting 25 to 30 per cent of the adult population. The majority will have only mild (70 per cent) or moderate (20 per cent) hypertension. At some time, 10 per cent will develop severe hypertension, which is generally, although not absolutely, defined as a diastolic blood pressure above 110 mmHg. Improved screening and the availability of improved medications continue to reduce the number of patients who will ever develop severe hypertension (Joint National Committee 1993). However, despite advances, severe hypertension remains a relatively common problem in the acute or critical care setting. Traditionally, patients with severe hypertension have been categorized as being emergency, urgent, or non-urgent cases. These categories have been defined by the presence or absence of acute organ dysfunction (emergencies), the presence or absence of underlying at-risk medical conditions (urgencies), or the demonstration of acute histopathological changes as a result of the elevated pressure (emergencies) (Houston 1989; Panacek 1994). The United States Joint National Committee on the Detection, Evaluation and Treatment of High Blood Pressure continues to define hypertensive emergencies as 'those situations that require *immediate* blood pressure reduction to prevent or limit target organ disease' (Joint National Committee 1993). Hypertensive urgencies are those situations 'in which it is desirable to reduce blood pressure within 24 h', whereas non-urgent severe hypertension may be treated in a routine manner or referred for close follow-up. A description of the categories of hypertension is given in Tables 1 and 2. However, sharp distinctions between these categories and even the use of specific blood pressure ranges are not always practical. Of those who develop severe hyper-

Table 1 Classification of blood pressure in adults aged 18 years or older

Blood pressure range (mmHg)		Category*
Systolic	Diastolic	
<130	<85	Normal
130–139	85–89	High normal
140–159	90–99	Mild hypertension (stage 1)
160–179	100–109	Moderate hypertension (stage 2)
180–209	110–119	Severe hypertension (stage 3)
⩾210	⩾120	Very severe hypertension (stage 4)

*When systolic and diastolic pressures fall into different category ranges, the higher category is used to establish the patient's classification status.
Modified from Joint National Committee (1993).

Table 2 Classification of severe hypertension

Hypertensive emergencies

Definition: marked hypertension with resultant acute organ system impairment

Therapy required immediately, usually parenteral agents

Specific conditions

- Diastolic blood pressure >110 mmHg, with:
  - Intracranial hemorrhage
  - Cerebrovascular accident
  - Hypertensive encephalopathy
  - Eclampsia
  - Pheochromocytoma crisis
  - Papilledema or acute retinal hemorrhages
  - Acute renal insufficiency or failure
  - Acute pulmonary edema
  - Active hemorrhage
  - Intraoperative period
- Any persistent hypertension with:
  - Acute aortic dissection
  - Acute myocardial ischemia

Hypertensive urgencies

Definition: marked hypertension with increased risk of acute morbidity, but not yet acute organ impairment

Therapy required within several hours, usually oral agents

Specific conditions

- Diastolic blood pressure >110 mmHg with:
  - Underlying cardiac or renal impairment, or CNS disease
  - Underlying coronary artery disease
  - Epistaxis
- Perioperative hypertension

Non-urgent severe hypertension

Substantial hypertension without any of the above conditions

Therapy not required acutely in all cases, but recurrent evaluation is imperative

CNS, central nervous system.
*Modified from Joint National Committee (1993).

tension, less than 10 per cent will ever have a true 'hypertensive emergency'. More commonly, patients with severe hypertension will fit a hypertensive urgency or non-urgency category.

Systemic effects

The height of a blood pressure measurement, in and of itself, does not independently define a true hypertensive emergency, urgency, or a non-urgency, as the host response to hypertension varies tremendously. Underlying conditions often determine how well an individual tolerates a given degree of hypertension. Patients with chronic hypertension tolerate much higher elevations in blood pressure than previously normotensive persons. However, most emergencies and urgencies occur with a diastolic blood pressure above 110 mmHg. If severe hypertension is allowed to persist for long enough, it will eventually cause vascular or endorgan damage. At a microvascular level, there develops a recurrent cycle of increased vascular reactivity, increased release of vasoconstrictors, endothelial damage, platelet aggregation, fibrin deposition, myointimal proliferation, and narrowed arterioles. The organs most often affected by these changes include the heart, the kidneys, and the central nervous system. An important part of this process is a pressure diuresis, often resulting in hypovolemia despite clinical findings suggestive of congestive heart failure. Because of damage to the blood vessels, autoregulatory abilities are impaired and are less predictable.

In addition to severe hypertension, in some settings any degree of hypertension is poorly tolerated and can seriously impair the patient's condition and therefore constitutes a hypertensive emergency. Examples include any persistent elevations in mean arterial pressure coincident with an acute aortic dissection, an acutely failing or ischemic heart, or active hemorrhage.

Central nervous system effects

Autoregulation is highly prominent in vascular beds of the heart, kidneys, and lungs, but is most important in the cerebral circulation. In the normal state, cerebral autoregulation maintains a stable cerebral blood flow rate despite fluctuations in mean arterial pressure over a range of 60 to 150 mmHg. When the upper limit of cerebral autoregulation is exceeded for a period of time, hypertensive encephalopathy results, with cerebral edema, petechial hemorrhages, and microinfarcts. This is usually associated with very high pressures (e.g. diastolic pressure generally exceeding 140 mmHg). These patients typically present with nausea, vomiting, severe headaches, visual changes and, in extreme cases, mental obtundation or seizures (Powers 1993).

Individuals with normal cerebral autoregulation can generally tolerate relatively rapid reductions in their systemic blood pressure, but this may not be true for patients with hypertensive emergencies, particularly those with hypertensive encephalopathy. Chronic elevation in systemic pressure, even without underlying cerebral vascular disease, can impair normal autoregulatory capability such that a substantial decrease in cerebral blood flow occurs at a higher mean arterial pressure than would be true in normal individuals. In individuals with chronic untreated severe hypertension, the lower limit of cerebral autoregulation can be reset to occur at a higher mean arterial pressure, such as 110 mmHg, instead of the 60 to 70 mmHg found in normotensive individuals. These changes in cerebral autoregulation have direct clinical relevance and influence the target blood pressure in patients with hypertensive emergencies. In both normotensive and hypertensive individuals, the lower limit of cerebral autoregulation is generally about 25 per cent below the baseline mean arterial pressure. However, the baseline mean arterial pressure is usually not known, and so a target range of 110 to 120 mmHg is generally recommended during treatment (Powers 1993; Panacek 1994).

Cardiac effects

Increased systemic blood pressure, which increases afterload on the heart, also increases myocardial workload. This can have both acute and chronic adverse effects. Afterload is a major determinant of myocardial oxygen demand; thus sudden increases can induce or aggravate myocardial ischemia and worsen underlying heart failure. The acute onset of pulmonary edema or myocardial ischemia is a common presentation of a hypertensive emergency.

Longstanding increased afterload, even if not severe, eventually produces a failing heart. Although the mechanisms are not entirely clear, this has been shown in both clinical practice and animal models. Increased afterload can also cause isolated diastolic dysfunction. This results in a relatively stiff ventricle with increases in end-diastolic pressure and decreases in end-diastolic filling volume. The ventricle is then less tolerant of increased preload volumes and is prone to failure.

Differential diagnosis

Most common causes

Because blood pressure is an objective measurement, the diagnosis of severe hypertension is usually straightforward. The difficulty lies in properly categorizing the patient with severe hypertension into an emergency, urgency, or non-urgency category, as outlined in Table 2. Although diagnosis of a 'hypertensive emergency' or 'urgency' describes the relative severity of the hypertension, it does not establish the etiology. Table 3 lists the major etiologies in order of frequency. These can generally be categorized as aggravation of underlying hypertension, an induced hypersympathomimetic state, or 'other'. Important causes in the latter category include eclampsia and the perioperative period. Both these conditions involve systemic vasoconstriction and relative hypovolemia.

After measures to stabilize the patient's hypertension have been initiated, attention should be given to historical and other factors which may help elucidate the etiology of the severe hypertension. This is clinically pertinent as different therapeutic approaches are often indicated for different etiologies.

Table 3 Causes of severe hypertension

Causes
Most common
Antihypertensive withdrawal syndromes (medication non-compliance)
Accelerated rise in a known hypertensive patient
Intraoperative period in cardiac surgery
Common
Effects of other drugs
Sympathomimetics (e.g. cocaine, decongestants)
Steroids
Non-steroidal anti-inflammatory drugs
Eclampsia
Less common
Endocrinopathies (e.g. hyperthyroidism, Cushing's disease)
Acute glomerulonephritis
Ingestion of tyramine in patients on monoamine oxidase inhibitors
Pheochromocytoma
Acute vasculitis

Pseudomalignant hypertension

Elderly patients with diabetes or renal failure and diffuse vascular disease can occasionally develop a condition called 'pseudomalignant hypertension' (Messerii 1985). As a result of advanced atherosclerosis, their blood vessels become relatively non-compressible. This makes it very difficult, and sometimes impossible, to measure an accurate cuff blood pressure and can result in artefactually elevated blood pressure measurements. There are numerous case reports of such patients receiving unnecessarily aggressive antihypertensive agents in efforts to control their artefactual hypertension, which even result in shock or cardiac arrest of the patient. In such cases, intra-arterial catheter monitoring shows the true pressure to be much lower than that measured by sphygmomanometry. These patients can sometimes be detected by use of a procedure called Osler's maneuver (Messerii 1985) whereby a blood pressure cuff on the arm is inflated above the auscultated systolic pressure. The radial and brachial arteries are then directly palpated; if either remains palpable, the patient is said to have a 'positive test'. Such patients have actual intra-arterial pressures 20 to 60 mmHg lower than those determined by cuff. If a patient has a positive Osler maneuver, only direct intra-arterial monitored measurements are absolutely reliable. Obviously, automated non-invasive blood pressure devices cannot be considered reliable in such cases. This condition should be suspected if a patient continues to have severely elevated cuff blood pressures despite appropriate therapeutic measures, or if the patient's overall condition appears inconsistent with his or her measured blood pressure.

References

Houston, M.C. (1989). Pathophysiology, clinical aspects, and treatment of hypertensive crises. *Progress in Cardiovascular Diseases*, **32**, 99.

Joint National Committee on Detection, Evaluation, and Treatment of High Blood Pressure (1993). Fifth Report. *Archives of Internal Medicine*, **153**, 154–83.

Messerii, F.H. (1985). Osler's maneuver and pseudohypertension. *New England Journal of Medicine*, **312**, 1548.

Panacek, E.A. (1994). Hypertensive emergencies and urgencies. In *Emergency cardiac care* (ed. W.B. Gibbler and T.P. Aufderhide), pp. 528–48. Mosby Year Book, St Louis, MO.

Powers, W.J. (1993). Acute hypertension after stroke: the scientific basis for treatment decisions. *Neurology*, **43**, 461–7.

3.10.2 Management

E. A. Panacek

Key messages

- If evidence of acute organ impairment is confirmed, the patient is a true 'hypertensive emergency'; therapy should be initiated immediately without waiting for the results of further laboratory tests or other studies.
- Aggressive therapies should never be initiated based on a single blood pressure measurement.
- Hypertensive urgencies are much more common than emergencies. Therapy should be initiated within the initial few hours, at least within 24 h, and oral medications are usually adequate.
- Sodium nitroprusside is the gold-standard parenteral agent for hypertensive emergencies against which all others are compared. It is the most potent, consistently effective, and highly recommended agent for hypertensive emergencies.

Clinical evaluation

The initial evaluation of a patient with hypertension focuses on confirming an accurate blood pressure and, if it is severely elevated, categorizing the patient as emergent, urgent, or non-urgent. If evidence of acute organ impairment is confirmed, the patient is a true 'hypertensive emergency'; therapy should be initiated immediately without waiting for the results of further laboratory tests or other studies. Further information relating to history, physical findings, and laboratory values can be obtained after therapeutic interventions have been started (Table 1). Evidence of specific acute endorgan dysfunction is sought. Acute cardiac impairment would be evidenced by ischemic changes on the ECG, anginal symptoms, or evidence of acute congestive heart failure with pulmonary edema. Evidence of related renal impairment includes new onset of hematuria, refractory oliguria, or acute elevation in blood urea nitrogen and creatinine. As most patients with severe hypertension are relatively hypovolemic, a normal urine output cannot be expected. Central nervous system involvement is evidenced by changes in the level of alertness, focal neurological findings, new severe headaches, new seizures, or papilledema. Any evidence of an acute aortic dissection, in the setting of even mild elevations in blood pressure, also constitutes a hypertensive emergency. Although somewhat controversial, severe perioperative elevations (e.g. diastolic pressure above 110 mmHg), even without acute organ impairment, are generally considered emergencies or, at least, urgencies.

Without delaying the initial therapy, some patients require additional specialized studies, for example CT head scan for neurological concerns, radiological studies if aortic aneurysm or dissection is suspected, and cardiac isoenzymes if myocardial ischemia is suspected.

Blood pressure should be measured repeatedly, both initially and after initiating treatment. Aggressive therapies should never be initiated based on a single blood pressure measurement. For ongoing monitoring, an intra-arterial catheter has the advantages of accuracy and continuous measurement although, being invasive, it does carry potential complications. Non-invasive automatic cuff blood pressure monitoring is an alternative option although accuracy is more variable; measurements should initially be confirmed by manual cuff readings. Repeat measurements should be very frequent when therapy is first started. Concurrent with measures to stabilize the hypertension, an evaluation for potential etiologies is initiated.

Hypertensive emergencies constitute a heterogeneous population of patients and so therapeutic endpoints should not be generalized. Target blood pressure and the time frame in which this should be achieved will vary, depending upon setting and etiology (Table 2).

Table 1 Evaluation of the patient with severe hypertension

Confirm blood pressure measurement
Measure in both arms
Consider 'pseudomalignant' hypertension
History
Duration of hypertension
Length of current symptoms
Symptoms of angina, heart failure, dissecting aneurysm, or neurological impairment
Medication history
Illicit drug use
Physical examination
General mental status
Evidence of heart failure
Focal neurological findings
Optic fundus for papilledema, hemorrhages, etc.
Adequacy of peripheral pulses
Diagnostic studies (all patients)
Electrolytes, blood urea nitrogen, creatinine
Urinalysis
ECG
Potential other studies (selected patients)
Toxicology screening
Head CT scan
Aortic angiogram
Endocrine testing
Urine tests for pheochromocytoma
Pregnancy test
Chest radiography

Table 2 Initial therapeutic options in hypertensive emergencies

Condition	Therapy of choice	Alternative therapy
Rapid reduction of blood pressure to near-normal levels		
Acute myocardial ischemia	Nitroglycerin IV, β-blockers	Nitroprusside, Ca-channel blocker
Acute congestive heart failure	Nitroglycerin IV	Nitroprusside
Acute aortic dissection	Beta-blockade (esmolol) and nitroprusside	Trimetaphan, labetalol
Complicated hypercatecholamine state	Labetalol	Esmolol and phentolamine
Intraoperative	Nitroglycerin, nitroprusside	Ca-channel blocker, β-blocker
Reduce MAP by 20%–25% (or to 110–120 mmHg) in less than 3 h		
Hypertensive encephalopathy	Nitroprusside	Labetalol, nifedipine
Eclampsia	Hydralazine	Labetalol, nifedipine
*Reduce MAP by 20%–25% (or to 110–120 mmHg) gradually over 6–12 h**		
Intracranial hemorrhage	Nitroprusside	Labetalol
Ischemic cerebrovascular accident	Nitroprusside	Labetalol
Subarachnoid hemorrhage	Ca-channel blocker	Nitroprusside

MAP, mean arterial pressure, IV, intravenous.
*If patient's neurological condition worsens, discontinue therapy and allow blood pressure to rise.

Treatment of specific conditions

Hypertensive encephalopathy

This is a result of hyperperfusion of the brain when the upper limit of cerebral autoregulation has been exceeded. Patient symptoms commonly progress over a period of 2 to 3 days, as opposed to an intracranial hemorrhage which generally has a more abrupt onset. The therapeutic goal is a reduction in mean arterial pressure by 20 to 25 per cent or to 110 to 120 mmHg over a period of 2 to 3 h (Calhoun and Oparil 1990; Tietjen *et al.* 1996). The drug of choice is sodium nitroprusside by constant intravenous infusion. An alternative is labetalol by small pulse doses or constant intravenous infusion. If the patient does not exhibit clinical improvement as the blood pressure is reduced, other diagnoses should be considered. Normally, patients with hypertensive encephalopathy improve in direct proportion to the control achieved in their blood pressure.

Intracranial hemorrhage

There is no consensus regarding the degree to which elevated blood pressure should be reduced in patients with intracranial hemorrhage. In the setting of subarachnoid hemorrhage, there are concerns regarding vasospasm; it is feared that reductions in blood pressure can further impair perfusion to an already ischemic zone. However, the complication rate in patients with subarachnoid hemorrhage is increased when blood pressure is particularly high (e.g. systolic blood pressure above 160–180 mmHg) (Tietjen *et al.* 1996). In addition, trials utilizing calcium-channel blockers (specifically nimodipine) have shown a small benefit, although it is unclear whether it is due to their effect on the blood pressure (Calhoun and Oparil 1990). The recommendation is to decrease blood pressure gradually by 20 to 25 per cent, or to an mean arterial pressure of 110 to 120 mmHg, over a period of 6 to 12 h. Initial therapeutic agents include calcium-channel blockers (e.g. nimodipine or nifedipine), starting with low doses, or sodium nitroprusside in patients unable to take oral medication. If blood pressure reduction appears to worsen the patient's overall status, the antihypertensive agents should be discontinued immediately and blood pressure allowed to return to moderately elevated levels. Experiences with hypertensive intraparenchymal cerebral hemorrhage are limited and there no controlled trials. Currently, the recommendations are the same as those for subarachnoid hemorrhage, with similar goals for target blood pressure and time frame (Calhoun and Oparil 1990).

Acute ischemic stroke

This is the classic condition in which concerns about cerebral vascular autoregulation must be seriously considered. The therapeutic goal is a decrease in mean arterial pressure of 20 to 25 per cent, or a mean arterial pressure of 110 to 120 mmHg. Experience with calcium-channel blockers has been less encouraging in this setting. Sodium nitroprusside is the preferred agent of choice, although sometimes even topical nitropaste can be effective in milder cases (Panacek 1994).

Acute aortic dissection

Regardless of whether surgery is indicated, the initial focus of care for all acute dissections is blood pressure control. A unique aspect is the issue of 'shearing forces,' whereby a wide pulse pressure, separate from an elevated mean arterial pressure, increases the likelihood of tearing of the aortic wall. Therefore the goal is an arterial pulse waveform which is more blunted than usual, in addition to lowering mean blood pressure (Panacek 1994). Sodium nitroprusside is the most reliable agent for the careful reduction of blood pressure to normal levels; however, it may actually widen the pulse pressure. To prevent this, a β-blocker is added. Although propranolol or other β-blockers may be used, the most titratable agent is esmolol given by constant infusion. Initial blood pressure control in these patients should take priority over angiographic studies. This is one setting in which blood pressure may be cautiously reduced, to levels below normal if necessary, to help control pain symptoms.

Acute cardiac impairment

When acute cardiac impairment occurs with severe hypertension, immediate blood pressure reduction will decrease cardiac work and myocardial oxygen demand. If neurological complications are not an active concern, the pressure should be quickly reduced to near-normal levels (diastolic blood pressure 85–90 mmHg). With ongoing myocardial ischemia the agent of choice is intravenous nitroglycerin (glyceryl trinitrate). However, this agent is not a very effective antihypertensive, and so sodium nitroprusside may be needed. A loop diuretic may also be considered in appropriate patients.

Eclampsia

Pre-eclampsia is the onset of hypertension, edema, and proteinuria after week 20 of gestation. Eclampsia includes the presence of seizures. Both pre-eclampsia and eclampsia most often occur near term. Definitive therapy for the severe hypertension seen with eclampsia is delivery of the fetus, but this is not always possible. Because of the concerns regarding placental hypoperfusion, only severe hypertension (e.g. diastolic blood pressure above 110 mmHg) is treated, with the goal of rapid but not complete normalization of blood pressure. Based on extensive experience in this setting, the agent of choice is hydralazine given in intermittent intravenous doses. In addition, infusions of magnesium sulfate can help, particularly if convulsions occur. Importantly, a number of antihypertensive medications, including nitroprusside and angiotensin-converting enzyme inhibitors, are contraindicated in this setting. Diuretics are relatively contraindicated as they can increase complications including seizures.

Hypercatecholamine states

These conditions usually constitute hypertensive urgencies or non-urgencies and are not associated with specific organ system damage. They are usually treated with oral medications, or simply observed. However, if the hypertension is particularly severe (e.g. diastolic blood pressure of 140 mmHg), unresponsive to oral medications, or associated with specific complications, parenteral medications are required. Intravenous labetalol, given by intermittent pulse therapy or constant infusion, has been successful in this setting. Second-line choices are the combined use of β-blockade with a vasodilator. Diazepam alone can be very effective for the hyperadrenergic state of cocaine or amphetamine toxicity (Panacek 1994).

Perioperative hypertension

Intraoperative and immediate postoperative severe hypertension is usually associated with systemic vasoconstriction with relative intravascular hypovolemia. It is generally best treated with a vasodilator (Halpern 1995). In patients with ischemic heart disease, it is important to limit heart rate to avoid increased myocardial ischemia. Nitrovasodilators are generally first-line therapy for perioperative hypertension. Intravenous β-blockers can be cautiously used to control heart rate, recognizing that they can also depress myocardial contractility. Other options include dihydropyridine or calcium-channel blockers such as nicardipine or isradipine.

Hypertensive urgencies

Hypertensive urgencies are much more common than emergencies. Therapy should be initiated within the initial few hours, at least within 24 h, and oral medications are usually adequate (Panacek 1994). The most popular agent for urgencies is currently nifedipine. Clonidine hydrochloride is a common alternative. There is also increasing experience with oral angiotensin-converting enzyme inhibitors and with intravenous or oral labetalol. The hypertension should be reduced but need not be normalized. These patients require close follow-up but not usually hospital admission.

Uncomplicated severe hypertension

Many more patients are seen with transient asymptomatic hypertension than with a hypertensive emergency or urgency. The hypertension is generally related to other conditions such as pain, anxiety, withdrawal symptoms, or as a compensatory response to mild intravascular volume loss. If the patients are asymptomatic, not perioperative, and without active bleeding, the prognosis is stable and the underlying disorder is the focus of care. Such patients are at greater risk from excessive hypotension resulting from the unnecessary use of antihypertensive agents than from the hypertension itself (Fagan 1989). The approach is to treat the patient, not the blood pressure. Close follow-up is important, rather than the immediate initiation of antihypertensive therapy.

Antihypertensive agents

Tables 3 and 4 list by category agents most commonly used to treat severe hypertension. Table 2 gives medical conditions for which each agent is mainly used.

Sodium nitroprusside is the gold-standard parenteral agent for hypertensive emergencies against which all others are compared. It is the most potent, consistently effective, and highly recommended agent for hypertensive emergencies. Its kinetics are almost ideal. Its use is complicated by the fact that it is photosensitive and so must be shielded by foil or other cover. Concerns regarding excessive side-effects and complications have been raised, including the development of cyanide toxicity, a metabolic product of nitroprusside. The cyanide is normally further metabolized by the liver and cleared by the kidneys; however, impairment of either of these organs predisposes the patient to this complication. It is seen most commonly, but not only, in patients on high dosages for longer than 48 h. If cyanide toxicity is suspected, arterial and venous blood gases are evaluated, seeking metabolic (lactic) acidosis with a decreased arteriovenous O_2 difference. When this is present, the infusion

Table 3 Parenteral agents for severe hypertension

Agent	Dose	Onset	Duration	Concerns
Vasodilators				
Nitroprusside	0.3–10 mg/kg/min IV	Seconds	3–5 min	Cyanide toxicity, photosensitive
Nitroglycerine	5–100 mg/min IV	1–5 min	5–10 min	Venodilator, methemoglobin
Diazoxide	50–150 mg IV bolus or 15–30 mg/min infusion	1–5 min	4–24 h	Hypotension, tachycardia, hyperglycemia, fluid retention
Hydralazine	2.5–20 mg IV or IM	10–30 min	2–4 h	Tachycardia, fluid retention, lupus syndrome
Fenoldopam	0.05–1.5 mg/kg/min IV	4 min	8–10 min	Limited experience
Nicardipine	5–15 mg/h IV	1–5 min	3–6 h	Reflex tachycardia, cerebral vasodilation
Enalaprilat	0.625–1.25 mg IV	10–15 min	6 h	Angio-edema, renal dysfunction
Adrenergic inhibitors				
Trimetaphan	0.5–5 mg/min IV	1–5 min	10 min	Autonomic side-effects, bronchospasm, tachyphylaxis
Labetalol	20–80 mg IV or 1–2 mg/min infusion	5–10 min	4–8 h	Beta-blocker side-effects
Phentolamine	1–10 mg IV as needed	1–2 min	10–20 min	Dysrhythmias, tachycardia
Esmolol	25–300 mg/kg/min	2 min	10–20 min	Beta-blocker side-effects

IV, intravenous, IM, intramuscular.

Table 4 Oral agents for severe hypertension

Agent	Dose	Onset	Duration	Concerns
Nifedipine	10 mg/30 min (40 mg max.)	5–15 min	3–6 h	Hypotension, tachycardia
Clonidine	0.1–0.2 mg/h (0.8 mg max.)	30–60 min	6–8 h	Sedation, dry mouth
Captopril	6.25–25 mg	15–30 min	4–6 h	Acute renal failure, angio-edema
Labetalol	100–300 mg	30–60 min	6–12 h	Beta-blocker effects predominate

should be discontinued immediately. Cyanide antidote kits can be used for severe cases. Hydroxycobalamin also counteracts cyanide and can be given either prophylactically or as acute treatment in those countries where it is approved (not the United States). The other serious concern with nitroprusside is excessive hypotension, precipitating myocardial infarction or stroke. This complication can be seen with any agent used to treat severe hypertension, and may not be increased with nitroprusside.

Nicardipine hydrochloride is a relatively new intravenous agent in the dihydropyridine category of calcium-channel blockers. Its overall effects are similar to those seen with nifedipine, but with less reflex tachycardia and less negative inotropic properties. Its efficacy is generally similar to that of nitroprusside, but with a much longer half-life. It vasodilates cerebral vessels and can raise intracranial pressure. It is gaining acceptance for the treatment of acute perioperative hypertension (Halpern 1995).

Fenoldopam mesylate is another newly approved agent for use in severe hypertension. It is a pure dopaminergic agonist, causing vasodilation in the renal, mesenteric, and splanchnic beds. In clinical trials it appears to have an efficacy relatively equal to that of nitroprusside with similiar kinetics, but without concerns regarding cyanide toxicity or photosensitivity. Because it has positive inotropic properties and specifically increases renal perfusion, it may have particular application in subsets of patients. It could replace much of the use of nitroprusside in the future (Panacek 1994).

Potent diuretics generally do not have a primary role in the management of patients with severe hypertension. Many of these patients are intravascularly volume depleted, particularly those with eclampsia or during the perioperative period. Subsets of patients with active congestive heart failure will require diuresis in addition to antihypertensives.

Patient disposition

All patients who fit the diagnosis of a hypertensive emergency require admission to a monitored bed, preferably in an intensive care unit (Calhoun and Oparil 1990). Regardless of the initial response to therapy, they should be observed closely in a monitored setting for at least 24 h. No matter how complete the blood pressure response in the emergency department or other setting, no patient with a 'hypertensive emergency' should be discharged home. They are prone to wide fluctuations in blood pressure during the initial treatment period, owing to altered vasomotor autoregulatory responses.

Patients with a hypertensive urgency may be discharged home if they show response to initial therapy and diagnostic evaluations show no evidence of acute organ impairment. Such patients should be referred for follow-up within a week and be given close instructions to return immediately if their condition deteriorates. If an adequate blood pressure response cannot be achieved over several hours, admission is necessary. Patients with non-urgent asymptomatic severe hypertension do not require acute hospital admission or aggressive initial treatment, but should have close follow-up and re-evaluation.

References

Calhoun, D.A. and Oparil, S. (1990). Treatment of hypertensive crises. *New England Journal of Medicine*, **323**, 1177.

Fagan, T.C. (1989). Acute reduction of blood pressure in asymptomatic patients with severe hypertension: an idea whose time has come and gone. *Archives of Internal Medicine*, **149**, 2169.

Halpern, N.A. (1995). Today's strategies for treating postoperative hypertension. *Journal of Critical Illness*, **10**, 478–90.

Panacek, E.A. (1994). Hypertensive emergencies and urgencies. In *Emergency cardiac care* (ed. W.B. Gibbler and T.P. Aufderhide), pp. 528–48. Mosby Year Book, St Louis, MO.

Tietjen, C.S., Hurn, P.D., Ulatowski, J.A., and Kirsch, J.R. (1996). Treatment modalities for hypertensive patients with intracranial pathology: options and risks. *Critical Care Medicine*, **24**, 311–22.

3.11 Severe capillary leak

3.11.1 Pathophysiology

T. M. Moore and A. E. Taylor

Key messages

- Oxidation is the major factor producing endothelial barrier damage.
- Oxygen radicals are associated with activation, rolling adherence, and transmigration of leukocytes at endothelial barriers.
- Ischemia–reperfusion injury involves neutrophils and communication between these cells, the endothelium, and surrounding parenchymal cells.
- Endotoxin damage is a longer inflammatory response associated with cytokine release by macrophages and plasma cells and their subsequent effect on the endothelial barrier.
- Tumor necrosis factor and interleukin 1 are cytokines required to produce endothelial damage associated with sepsis and ischemia–reperfusion.
- Cell cAMP levels modify endothelial barrier integrity by controlling the tension in the actin–myosin light-chain complex inside the endothelial cell relative to their effects on intracellular junction size.
- Specific antibodies against various cytokines, leukocyte adherence and rolling factors, and intracellular mechanisms responsible for producing damage in plasma and endothelial cells will be important treatment tools in the critically ill.

Inflammation associated with ischemia–reperfusion

Oxygen radicals

Since the classic work of Granger (1988), numerous studies have elucidated mechanisms which produce microvascular barrier damage following ischemia–reperfusion. Oxygen radicals are the major mediators of this damage. Radicals such as O_2^- and H_2O_2 are released in response to both xanthine oxidase and leukocyte activation in postischemic tissues. In the presence of Fe^{3+}, O_2^- and H_2O_2 react to produce the hydroxyl radical ($OH^·$) which, in turn, can directly produce endothelial membrane damage (Fig. 1(a)). In addition, oxygen radicals can stimulate release of inflammatory mediators such as platelet activating factor and/or cyclo-oxygenase and lipoxygenase products. These compounds can promote leukocyte adhesion to the microvascular endothelium and directly activate endothelial cell receptors causing disruption of the endothelial barrier (Moore *et al.* 1996*a*).

Rolling and adherence factors

Leukocyte–endothelial cell interactions which occur after ischemia–reperfusion are mediated by adhesion molecules (Fig. 1(b)). Leukocytes can initially roll along the microvascular endothelium through an L-selectin (leukocyte-expressed) interaction with a counter-ligand derived from endothelial cells. In addition, endothelial-derived oxygen radicals can mobilize preformed P-selectin, another rolling factor, to the surface of the endothelial cell. Endothelial-derived radicals can also stimulate production of platelet activating factor, which in turn causes rolling leukocytes to adhere firmly to the endothelial cell. Platelet activating factor promotes firm adherence by facilitating leukocyte integrin binding to integrin ligands expressed by endothelial cells. The leukocyte integrins mediating firm adherence are CD11a/CD18 and CD11b/CD18; these are also known as LFA-1 and MAC-1 respectively. The adherence factors expressed by endothelial cells belong to the immunoglobulin superfamily of proteins and are known as intercellular adhesion molecules (ICAM-1 and ICAM-2).

The firm adhesion of leukocytes to endothelial cells activates leukocytes, causing release of oxygen radicals and proteases. Furthermore, firm adhesion and subsequent migration of leukocytes across the endothelial barrier stimulates active contraction of endothelial cells. When endothelial cells contract, the microvascular barrier becomes very leaky to even large plasma proteins. Interstitial edema will then develop since microvascular permeability to both fluid and protein is greatly increased (Taylor *et al.* 1994).

The process described above is an oversimplification, but studies in many organs have confirmed that rolling and adherence factors are necessary for the production of acute microvascular injury by ischemia–reperfusion, and that oxygen radicals formed from xanthine oxidase activation initiate this process.

Reversal of endothelial damage in the lung induced by ischemia–reperfusion

Blocking oxygen radical production or leukocyte adhesion in ischemic and reperfused tissues prevents the occurrence of endothelial damage. However, in the clinical setting, patients may present with established inflammation and so measures need to be taken to reverse pre-existing microvascular injury. Exciting data which may prove to be clinically important have been obtained from studies of lung ischemia–reperfusion in which the role of the intracellular

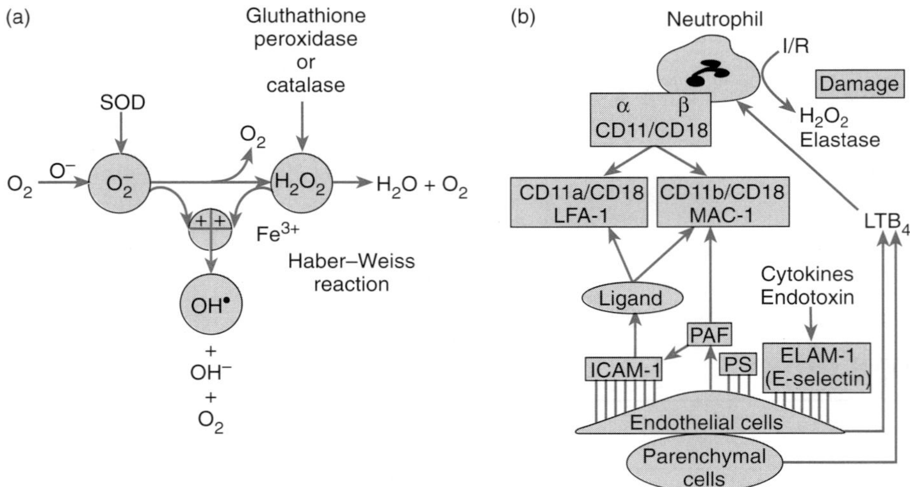

Fig. 1 (a) Mechanism of production of the hydroxyl radical (OH·) from the superoxide anion (O_2^-) and hydrogen peroxide (H_2O_2) in the presence of Fe^{3+} (O_2, molecular oxygen; SOD, superoxide dismutase). (b) Schematic representation of neutrophil–endothelial interactions during ischemia–reperfusion (I/R) (PS, P-selectin).

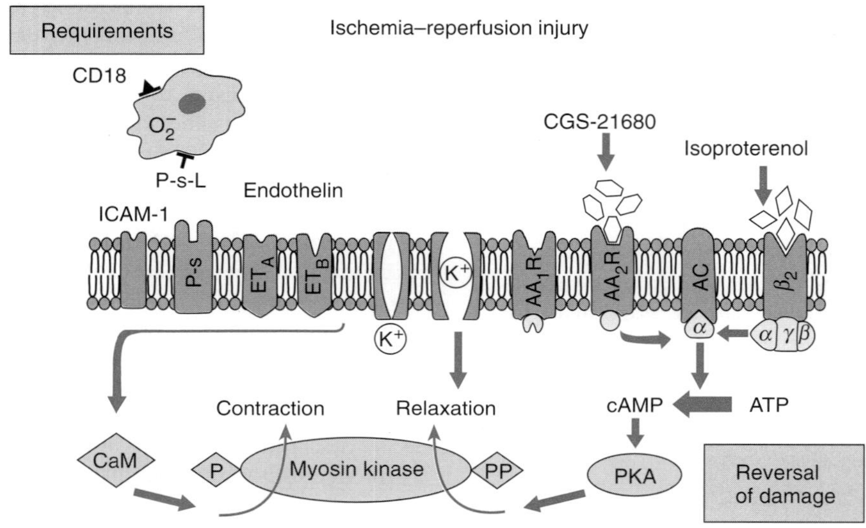

Fig. 2 Representation of factors that produce endothelial damage in lungs subjected to ischemia–reperfusion. Causative factors include integrins (CD18), P-selectin ligand (P-s-L), P-selectin (P-s), intercellular adhesion molecule 1 (ICAM-1), endothelin, activated calmodulin (CaM), and activated myosin light-chain kinase. On the right are shown factors that can reverse the damage: opening of ATP-sensitive K^+ channels, activating adenosine A_2 receptors (AA_2R), activating β-adrenergic receptors (β_2), and activation of protein kinase A.

second messenger cAMP has been evaluated. It has been shown that cAMP-elevating compounds can actually reverse endothelial barrier damage associated with the inflammatory response. It was established several years ago that increased microvascular permeability due to histamine exposure in skeletal muscle could be reversed with the cAMP-elevating compounds epinephrine (adrenaline) and isoproterenol (isoprenaline). Recent studies of endothelial damage induced by ischemia–reperfusion in isolated rat lungs have also shown that elevating cAMP by activation of β-adrenergic and adenosine A_2 receptors, direct stimulation of adenylyl cyclase, inhibition of cAMP phosphodiesterase, or addition of cAMP analogs can reverse increased microvascular permeability. Thus changes in cAMP levels during the inflammatory process appear to regulate endothelial permeability (Moore *et al.* 1996*a*).

It is not yet clear how cAMP prevents and reverses the increases in microvascular permeability that occur during inflammation. However, the reversal effect of cAMP was dependent on protein kinase A in an isolated rat lung model of ischemia–reperfusion. An antagonistic effect of increased intracellular Ca^{2+} levels on cAMP production has also been demonstrated. Ca^{2+} entering isolated pulmonary endothelial cells grown in culture inactivated a specific isoform of adenylyl cyclase, causing decreased cAMP production and increased monolayer permeability. Thus the inflammatory process induced by ischemia–reperfusion may produce microvascular injury by increasing endothelial cell Ca^{2+} levels which in turn causes decreased cAMP production.

In addition to a direct effect of Ca^{2+} on adenylyl cyclase, increased intracellular Ca^{2+} levels can stimulate endothelial contraction by a mechanism dependent on calmodulin and myosin light-chain kinase. Activation of myosin light-chain kinase is necessary for endothelial damage to occur in isolated rat lungs subjected to ischemia–reperfusion. Therefore the reversal effects of cAMP in lungs damaged by ischemia–reperfusion occur because the activity of myosin light-chain kinase is inhibited, thus promoting endothelial cell relaxation and decreased barrier permeability (Khimenko *et al.* 1996).

Regardless of the detailed mechanism by which cAMP acts, its usefulness as a therapeutic agent for preventing inflammatory-induced microvascular injury is well known. Analogs of cAMP are now included in some preservation solutions for storing lungs for transplantation. This appears to suppress the initial lung injury that occurs following reperfusion of lungs in the transplant recipient.

Figure 2 summarizes some of the known factors which both produce and reverse microvascular damage associated with lung ischemia–reperfusion. Xanthine-oxidase-derived oxygen radicals promote leukocyte (neutrophil) sequestration in pulmonary microvessels that is dependent on P-selectin, CD18, and ICAM-1. Subsequent activation of endothelial cell myosin light-chain kinase by increased intracellular Ca^{2+} causes endothelial cell contraction and the microvascular barrier becomes highly permeable to fluid and solute. This process is prevented by blocking leukocyte rolling and adherence factors and inhibiting xanthine oxidase. It can also be reversed by elevating intracellular cAMP levels (Moore *et al.* 1996*a*).

Inflammation associated with sepsis

When sepsis is present, endothelial damage occurs through at least two different pathways (Fig. 3). A rapid phase of damage occurs with activation of the serum complement system. This is followed by a slower phase of injury due to production of certain tissue cytokines. The rapid phase more closely resembles the acute inflammatory response occurring with ischemia–reperfusion because various components of the serum complement system activate neutrophils and other leukocytes causing generation of oxygen radicals, proteases, and arachidonic acid products. These factors cause acute damage to endothelial cells and increase microvascular pressures, both of which promote rapid tissue edema formation. The slower phase occurs because macrophages become activated and release cytokines, such as tumor necrosis factor-α and interleukin 1 (IL-1). These cytokines can cause up-regulation of neutrophil adherence factors, disruption of the endothelial cell cytoskeleton, and up-regulation of the endothelial-cell-expressed rolling factor E-selectin (ELAM-1). Leukocytes can then adhere to the endothelium and further disrupt the endothelial barrier, which in turn results in tissue edema formation over hours to days. The rapid and slow phases of endothelial damage have been discussed here as if they occur in series, but in practice it is difficult to separate the effects of the two systems in such a fashion.

Because the mechanisms producing endothelial injury during sepsis are so complex, many problems are created for the clinician who must treat sepsis-induced inflammation in patients. Blocking systems which are activated during inflammation have not yet been shown to be particularly beneficial in treating the critically ill patient. Therefore much more research must be conducted using experimental animal models before the roles of all factors involved in the inflammatory process can be elucidated.

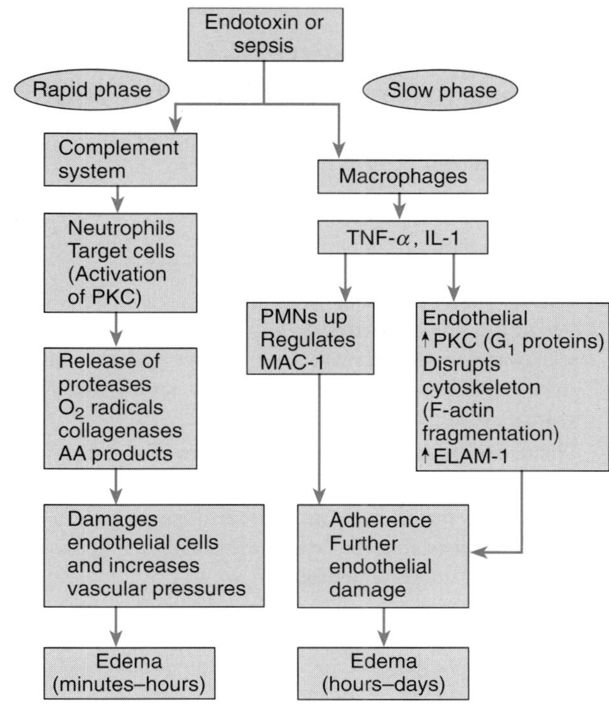

Fig. 3 Schematic diagram showing the ability of endotoxin to produce lung damage: TNF-α, tumor necrosis factor-α; PKC, protein kinase C; MAC-1, leukocyte integrin CD11b/CD18; ELAM-1 (endothelial–leukocyte adhesion molecule), E-selectin; AA, arachidonic acid.

Conclusions

Our knowledge of how fluid and solutes cross the endothelial barrier has now been expanded to incorporate cellular events, but the major findings from animal models of endothelial barrier dysfunction have not been adequately tested in the clinical setting. However, the recent finding that nitric oxide can protect the lung against ischemia–reperfusion damage in animal models (Moore *et al.* 1996*b*) may have important clinical implications, since this compound is already in use for treating pulmonary hypertension. It is quite possible that exciting new therapies for the successful treatment of endothelial barrier damage caused by the inflammatory process will be developed in the near future.

References

Granger, D.N. (1988). Role of xanthine oxidase and granulocytes in ischemia–reperfusion injury. *American Journal of Physiology*, **255**, H1269–75.

Khimenko, P.L., Moore, T.M., Wilson, P.S., and Taylor, A.E. (1996). Role of calmodulin and myosin light-chain kinase in lung ischemia/reperfusion injury. *American Journal of Physiology*, **271**, L121–4.

Moore, T.M., Khimenko, P.L., and Taylor, A.E. (1996*a*). Endothelial damage caused by ischemia and reperfusion and different ventilatory strategies in the lung. *Chinese Journal of Physiology*, **39**, 65–81.

Moore, T.M., Khimenko, P.L., Wilson, P.S., and Taylor, A.E. (1996*b*). Role of nitric oxide in lung ischemia and reperfusion injury. *American Journal of Physiology*, **271**, H1970–7.

Taylor, A.E., Khimenko, P.L., Barnard, J.W., Rizzo, A., and Moore, T. (1994). Protection and reversal in ischemia–reperfusion tissue injury. In *Reperfusion injuries and clinical capillary leak syndrome* (ed. B.A. Zikria, M.O. Oz, and R.W. Carlson), pp. 493–516. Futura, Armonk, NY.

3.11.2 Management of acute non-cardiogenic pulmonary edema

L. Oppenheimer

Key messages

- Until a direct methodology to diagnose capillary leak is developed, it is impossible to know whether patients with edema and low pulmonary artery wedge pressure have a quickly resolving capillary injury or slowly resolving hydrostatic edema.
- The use of positive end-expiratory pressure remains the most important therapeutic modality.
- Mortality remains high in non-cardiogenic pulmonary edema, emphasizing the fact that it is a mere reflection or symptom of multiorgan disease.

Introduction

In all tissues, fluid constantly crosses the vasculature into the interstitium, from which it is drained either by vascular reabsorption or lymphatic transport. For organs to maintain normal hydration, fluid filtration and drainage must balance. When filtration exceeds drainage, excessive fluid accumulates in the tissues and edema ensues.

Forces governing fluid exchange across vessels

At the end of the nineteenth century, Starling described the physical forces governing transvascular fluid filtration. His observations were simple but crucial. In his experiments, he observed that hemorrhage produced a decrease in hematocrit. He assumed that hypotension induced extravascular fluid reabsorption which resulted in hemodilution. However, for similar degrees of hypotension, reabsorption was significantly less when plasma rather than crystalloid was placed in the interstitium. He concluded that the forces governing transvascular fluid exchange included not only the difference in hydrostatic pressure across the vessels, but also the difference between intravascular and extravascular protein concentration.

Oncotic properties of proteins and other macromolecules

The thermodynamic principles describing how transvascular concentration differences of protein and other macromolecules (commonly known as oncotic gradients) induce fluid exchange are beyond the scope of this chapter (Oppenheimer 1990). The effect of oncotic gradients can be readily and directly measured in hydrostatic pressure units using clinical oncometers. A simple oncometer is shown in Fig. 1. Two compartments (A and B) are in communication but are separated by a semipermeable membrane which allows fluid exchange but completely prevents protein exchange. Compartment B contains pure crystalloid solution and compartment A contains plasma. The protein concentration difference (oncotic gradient) will draw fluid from B to A. This can be prevented by applying pressure to A. This applied pressure represents the oncotic effect measured in hydrostatic units (Fig. 1(a)). Alternatively, one could have allowed proteins to draw fluid into compartment A and measured the oncotic gradient as the difference in height between A and B (Fig. 1(b)). However, net accumulation of fluid in A dilutes the protein and decreases the oncotic effect.

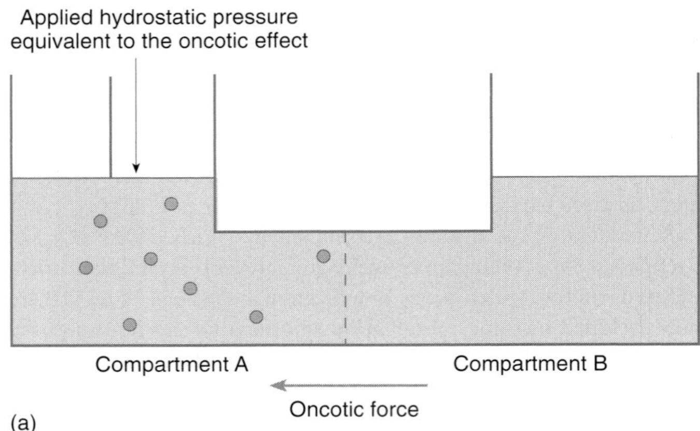

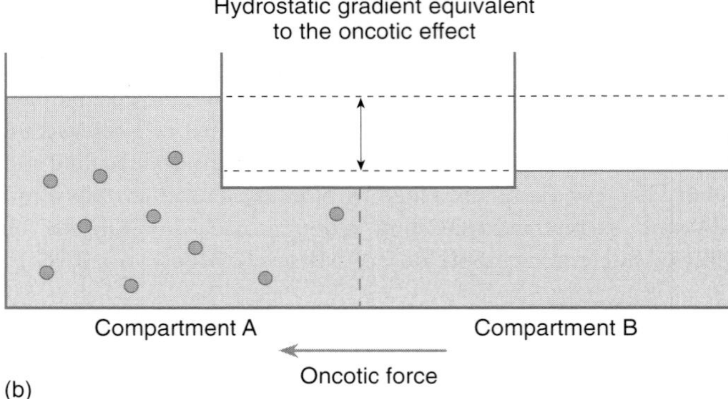

Fig. 1 (a) Direct hydrostatic pressure equivalent to the oncotic effect is applied. (b) Fluid is allowed into compartment A as the result of the oncotic effect. There is autodilution of proteins. The hydrostatic gradient so developed counteracts the oncotic effect.

Oncotic gradients offset hydrostatic gradients

Vascular membranes in fluid-exchanging vessels restrict, but often do not totally impede, protein escape from the vasculature to the interstitium. To the extent that they do, the resulting difference between vascular and interstitium protein concentrations results in an oncotic gradient $\Delta\pi$ which opposes fluid filtration driven by the intravascular–interstitial hydrostatic gradient ΔP. The net filtration gradient is obtained by subtracting $\Delta\pi$, expressed in hydrostatic units as shown above, from ΔP. For any net gradient, the fluid filtration rate Q_F is directly related to the membrane surface area and its permeability to fluid. Surface area and fluid permeability are combined and expressed as the filtration coefficient K_F. All the indices mentioned above are combined in an expression which has become known as the Starling equation:

$$Q_F = K_F(\Delta P - \Delta\pi). \quad (1)$$

Strictly speaking, every protein fraction exerts its own gradient. For simplicity, in the following discussion it will be assumed that the protein gradient reflects the transvascular albumin gradient, which is the most relevant component in humans.

The degree of restriction to protein escape varies. It is complete in the brain and is lowest in the lung (estimates of albumin retention in the vascular compartment vary between 60 and 80 per cent). Membrane permeability to protein results in a decreased oncotic effect for two reasons: protein accumulation in the interstitium reduces the protein concentration gradient across the exchanging vessels, and, for any given gradient, the oncotic effect is reduced by the degree of protein permeability (Hancock *et al.* 1989). Therefore eqn (1) needs to be corrected using what has become known as the reflection coefficient σ:

$$Q_F = K_F(\Delta P - \sigma\Delta\pi). \qquad (2)$$

The pulmonary capillary hydrostatic pressure is assumed to be halfway between the mean pulmonary and left atrial pressures, at approximately 15 mmHg above atmospheric pressure. Since the interstitial pressure is close to atmospheric pressure, $\Delta P = 15$ mmHg.

The oncotic pressure of normal plasma measured with a clinical oncometer is about 20 mmHg. Since about only 80 per cent of plasma albumin is retained by the vasculature, the transvascular protein gradient will be reduced accordingly. Because the reflection coefficient is 80 per cent of ideal (often expressed as $\sigma = 0.8$), the oncotic effect is reduced correspondingly. Consequently $\Delta\pi = 0.8 \times 0.8 \times 20$ mmHg $= 12.8$ mmHg which, when subtracted from ΔP, leaves a small imbalance favoring continuous fluid filtration. The filtrate must be drained by the mechanisms listed previously for the lungs to stay dry.

Pathophysiology of pulmonary edema

A force imbalance resulting in filtration rates which can no longer be accommodated by tissue fluid clearance mechanisms may be caused by the following.

1. An increased hydrostatic gradient secondary to the increased left atrial pressure of acute heart failure and fluid overload.

2. A decreased oncotic gradient which may be the result of protein dilution with crystalloid solutions or the result of increased vascular membrane permeability to protein—the 'capillary leak or injury'. In the lung, both the capillary and alveolar membrane may be injured.

These basic mechanisms are used on a daily basis to assist in the differential diagnosis, which remains one of exclusion. Pulmonary edema with normal or low pulmonary artery wedge pressure (a clinical measurement of left atrial pressure) points the diagnosis towards capillary leak. This approach can be misleading. Several hours may be required for edema to clear after the hydrostatic gradient has been normalized (Mayers *et al.* 1987). This is particularly likely in patients who have been acutely resuscitated with large volumes of crystalloid solutions. The hydrostatic gradients may be high due to the volume infusion. In addition, protein dilution decreases the gradient opposing edema formation and edema quickly ensues. By the time that patients are admitted to the intensive care unit for treatment of respiratory failure, the left atrial pressure may be normal but no capillary lesion may be present. Edema may resolve in 48 h. Until a direct methodology to diagnose capillary leak is developed, it is impossible to know whether such patients had a quickly resolving capillary injury or a slowly resolving hydrostatic edema.

Treatment of non-cardiogenic edema

The treatment of non-cardiogenic edema has been reviewed recently (Hudson 1995). Treatment modalities are typically considered as either supportive or directed against the injury.

Supportive modalities

The use of positive end-expiratory pressure remains the most important therapeutic modality and is described elsewhere in this book. The aim is to achieve adequate oxygenation with the lowest possible FiO_2 and the lowest possible mean airway pressures. A high FiO_2 and high ventilatory pressures and volumes may result in further lung injury. However, in some cases, despite careful manipulation of the ventilator settings, it is impossible to achieve adequate oxygenation. If that is the case, options include paralysis (to minimize desaturation associated with patient efforts), accepting lower oxygen saturation and decreasing the frequency of endotracheal tube suctioning (acute desaturation), or increasing the hemoglobin to improve the oxygen-carrying capacity. Ultimately, one may have to accept ventilation with pure oxygen and try other available ventilatory approaches.

None of the alternative ventilatory modes has been proved to be superior. Therefore it is necessary to try them individually on each patient. The most commonly used modes are inverse ratio ventilation and high frequency ventilation. Inverse ratio ventilation requires sedation and often paralysis. The inspiratory time is gradually prolonged as long as it is hemodynamically possible. Presumably, longer inspiratory times may recruit air spaces and improve gas exchange. There is some anecdotal evidence that high frequency ventilation has beneficial results at frequencies in the range of 300 breaths/min.

There are anecdotal reports of the use of intracorporeal and extracorporeal gas exchange devices in highly specialized centers (Hudson 1995). Liquid ventilation with perfluorocarbons which carry dissolved oxygen has been used. Very limited experience is available.

Pharmacological supportive therapies must still be considered experimental. They include vasoactive agents aimed at diverting blood flow away from edematous areas in an effort to minimize the intrapulmonary shunt (NO and almitrine). Surfactant replacement seems to be successful in children but has yet to be proved beneficial in the adult population (Hudson 1995).

Therapies aimed at the inflammatory injury

It is generally agreed that non-cardiogenic edema is the result of direct injury of the pulmonary capillary and, frequently, the alveolar membrane. This mechanism of injury has many characteristics of an inflammatory response. Many therapies against general or specific components of the inflammatory response have been developed. Generally, these are encouraging in the laboratory but are

Table 1 Pharmacological agents

General agents to inhibit inflammation
Corticosteroids (recently also used for the treatment of late fibrosis with anecdotal good results)
Prostaglandin E_1
Ketoconazole
Ibuprofen
Fibronectine
Specific agents against inflammation
Antiendotoxins
Anticytokines
IL-1-receptor antagonist
Antioxidants
Antiproteases
Anticomplement therapy

unsuccessful in clinical practice. A summary of pharmacological agents is given in Table 1.

Manipulation of oncotic forces

This approach remains experimental and a controlled study is presently under way. With experience, the principles and pitfalls of this approach are becoming clear. While we cannot yet claim success, a description is included for the benefit of readers. It is used when other support measures have failed.

Rationale

Injury to the pulmonary capillary and alveolar membranes allows increased escape of plasma proteins, particularly albumin, which has a molecular weight of 69 kDa. All molecules in every endogenous protein fraction have the same molecular weight and the membrane injury affects them all equally. Conversely, artificial plasma expanders such as dextrans and polystarch solutions contain a continuous range of fractions, some of which may be retained by the injured membrane. Thus it may be possible to generate an oncotic gradient based on the retained fractions. There may be concerns about the potential adverse effects due to the accumulation of smaller fractions in the interstitium. These concerns have not been substantiated. Using hydroxyethyl starch solutions we have been able to reduce edema accumulation in an animal model of lung injury despite the accumulation of smaller fractions in the extravascular space. In our clinical studies we have used Pentaspan (Dupont), a hydroxyethyl starch solution containing fractions with molecular weights ranging from 10 to 450 kDa. The larger fractions are broken down into smaller fragments in the first 2 h. Thereafter, the average half-life of the effect is about 18 h. Fractions with molecular weights below 30 kDa are excreted in the urine and can be ultrafiltered. The macromolecules are ultimately broken down and metabolized.

Successful treatment will induce fluid reabsorption not only from the lungs but from all tissues. Intravascular fluid reabsorption results in increased filtration pressures and autodilution of the oncotic agent. Therefore it is essential to maintain the intravascular volume as constant as possible by vigorous diuresis or by an alternative avenue for fluid removal, such as an ultrafiltration circuit.

Monitoring and treatment protocol

We use hematocrit determinations to monitor changes in the circulating blood volume (ΔCBV):

$$\Delta CBV = (Hct_{(t)} - Hct_{(initial)})/\ Hct_{(initial)} \tag{3}$$

where $Hct_{(t)}$ is the hematocrit at any time t. $Hct_{(t)}$ is referenced to the initial value. In ultrafiltration circuits, we use our own on-line monitor. We have found that $HCT_{(t)}$ is more useful than filling pressures for monitoring purposes. Increases in circulating blood volume frequently allow readjustment of inotropes and vasopressors, thus changing the filling pressures. Of course, hematocrit is only useful in the absence of bleeding. If frequent samples are taken, it is necessary to change the reference every few hours.

Loading phase

All fluid infusions are reduced to a minimum. After vigorous diuresis is induced (of the order of 500 ml/h) or comparable ultrafiltration rates are established, Pentaspan is infused to match fluid elimination (to maintain euvolemia). Hematocrit determinations (at least every 15 min) first remain steady and then start decreasing, suggesting an increase in circulating blood volume. As long as urine or ultrafiltra-

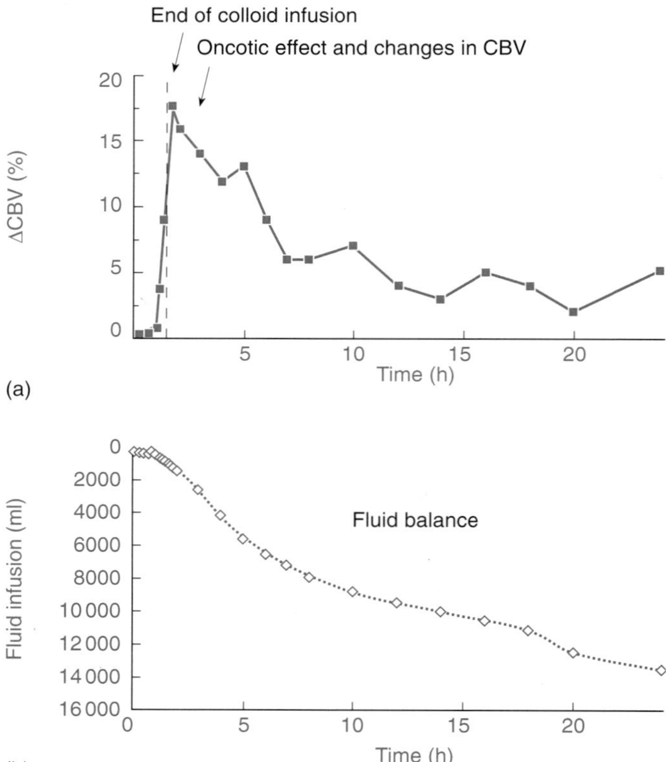

Fig. 2 (a) An example of changes in circulating blood volume achieved with colloid therapy. The colloid infusion is maintained until the circulating blood volume increases (loading phase). Thereafter, all fluid is restricted and fluid removal is continued until the circulating blood volume returns towards normal. Note how the circulating blood volume never becomes negative. (b) Summary of the fluid balance.

Table 2 Summary of clinical results

Patient population	Number of patients	Diuresis/ ultrafiltration	Negative balance (liters)	Resolved edema	Survival
Burns	4	2/2	12.5 (2.8)	4	2
Lung rejection	3	1/2	4.7 (1.8)	3	3
Lung contusion	4	4/0	6.2 (2.8)	3	2
Aspiration	3	2/1	5.3 (1.6)	3	3
Sepsis	15	9/6	10.3 (3.8)	12	7
Total	29	18/11		25	17

tion rates match the rate of infusion, the increase in circulating blood volume must reflect fluid mobilization. We generally allow for increases of the order of 20 per cent (Fig. 2), which usually permit a decrease in inotropic and vasopressor support. At this point the colloid infusion is stopped. Conversely, if 2 to 2.5 liters of Pentaspan have been infused without any change in blood volume, the injury is assumed to be non-amenable to this treatment.

Mobilization phase

After the colloid has been discontinued and fluid infusion is restricted, diuresis or ultrafiltration is continued as long as the circulating blood volume remains positive and the filling pressures are adequate. With this colloid load it is not unusual to induce negative balances of the order of 5 to 10 liters (Fig. 2) without hemodynamic consequences. These patients are often diffusely edematous and this approach probably mobilizes fluid from normal tissues first. This fluid must be mobilized before significant improvement in gas exchange can be expected. If reductions in blood volume occur before improvements in gas exchange take place, additional colloid may be required to increase the concentration of large fractions. This may lead to oncotic pressures above 30 mmHg which result in decreased glomerular filtration rates, subsequent oliguria, and a biochemical picture similar to renal failure. No further colloid should be infused unless an ultrafiltration circuit is established. If the renal function has not been compromised for other reasons, this picture subsides as the colloid is metabolized (1–2 days). Table 2 summarizes our results with this approach (Oppenheimer 1995, unpublished data). Again, the experimental nature of these results must be emphasized.

Does treatment influence outcome?

Despite the significant efforts to improve outcome, progress has been slow. Non-cardiogenic edema still has a high mortality. The worst prognosis can be expected in opportunistic and interstitial pneumonias (close to 90 per cent mortality). Trauma patients have the best prognosis (about 40 per cent mortality). Non-cardiogenic edema associated with sepsis still has a mortality of about 70 per cent (Artigas *et al.* 1991). These disappointing figures emphasize the fact that, in many instances, non-cardiogenic edema is a mere reflection or symptom of multiorgan disease. Successful resolution of non-cardiogenic edema may not affect the outcome of the underlying process. These considerations may explain our results (Table 2) where the mortality is higher than the degree of success in controlling low-pressure edema.

References

Artigas, A., Carlet, J., Le Gall, J.R., Chastang, C.L., Blanch, L., and Fernandez, R. (1991). Clinical presentation, prognostic factors, and outcome of ARDS in the European Collaborative Study (1985–1987). In *Adult respiratory distress syndrome* (ed. W.M. Zapol and F. Lemaire), pp. 37–63. Marcel Dekker, New York.

Hancock, B.J., Hoppensack, M., and Oppenheimer, L. (1989). Do transvascular forces in isolated lobe preparations equilibrate? *Journal of Applied Physiology*, **67**, 628–35.

Hudson, L.D. (1995). New therapies for ARDS. *Chest*, **108**, 79S–91S.

Mayers, I., Stimpson, R., and Oppenheimer, L. (1987). Delayed resolution of high pressure pulmonary edema or capillary leak. *Surgery*, **101**, 450–8.

Oppenheimer, L. (1990). Lung fluid movement and its relevance to the management of patients with increased lung water. *Clinical Intensive Care*, **1**, 103–10.

3.12 Pericardial tamponade

3.12.1 Pathophysiology and causes

W. C. Shoemaker

Key messages

- Cardiac tamponade is a low-flow state resulting from accumulation of blood or fluid in the pericardial sac that limits ventricular filling, stroke volume, and cardiac output.
- It may progress slowly with some degree of compensation, or develop rapidly into an acute tamponade syndrome with potentially lethal cardiogenic shock
- With slowly developing effusions, the pericardial sac is able to dilate and can accommodate 2 liters or more of fluid.
- Pulsus paradoxus is a useful sign in slowly developing pericardial effusions, but is not very useful in rapidly developing tamponade from penetrating injuries.
- Definitive diagnostic information can be provided by echocardiography, CT, and magnetic resonance imaging, and confirmed by pericardiocentesis.

Pathophysiology

Cardiac tamponade is a low-flow state due to increased pericardial pressure producing constriction of the cardiac chambers. It results from accumulation of blood or fluid in the pericardial sac that limits ventricular filling, stroke volume, and cardiac output. Hemopericardium and pericardial effusions are produced by a wide variety of traumatic and medical conditions (Table 1). The syndrome is uncommon, usually occurs insidiously, and carries a high mortality if unrecognized or inadequately treated. It may progress slowly with some degree of compensation, or develop rapidly into an acute tamponade syndrome with potentially lethal cardiogenic shock.

The pericardial fluid confined by the relatively unyielding pericardial sac takes up space and exerts pressure. The intrapericardial pressure is transmitted across the cardiac wall to increase intracardiac pressures; this has proportionally greater effects on atrial and end-diastolic pressures than on ventricular pressures. The reduced filling pressure and the volume encroachment by the pericardial fluid limits ventricular filling and stroke volume. Thus reduced stroke volume is the major circulatory defect.

With penetrating cardiac injuries, blood that leaks into the free pericardial space is quickly decompressed into the pleural space or mediastinum. Pericardial blood accumulates and produces tamponade when bleeding is sufficiently rapid that its egress from the pericardial cavity is blocked by clots that form faster than they are lysed. A rather high incidence of clots (60 per cent) have been reported in penetrating cardiac wounds explored soon after injury; non-clotting pericardial blood, which is diagnostic of tamponade, is present because this blood has already clotted and spontaneously lysed.

Secondary compensatory circulatory responses include increases in central venous pressure, heart rate, and peripheral vascular resistance. The rise in peripheral resistance maintains blood pressure when cardiac output begins to fall. The increased heart rate improves cardiac output when the latter is limited by stroke volume, The adrenal stress response, which increases arterial blood pressure and peripheral resistance, will further increase central venous pressure, thereby augmenting the limited cardiac filling.

Decompensation occurs with progression of the disorder. When compensatory circulatory mechanisms maintaining hemodynamic stability fail, the adrenal stress response is exhausted and acidosis with low cardiac output further impairs the coronary circulation, leading to hypotension, shock, and death.

Clinical appearance

Hemopericardium from penetrating injuries

Tamponade from hemopericardium after penetrating cardiac injury occurs when fluid accumulating in the pericardial sac is not decompressed into the pleural cavity via the pericardial rent. The incidence of tamponade is 2 per cent of stab or gunshot wounds to the chest,

Table 1 Causes of tamponade

Causes of tamponade
Penetrating trauma to the chest, back, or upper abdomen
Blunt trauma with fracture of left fourth, fifth, or sixth ribs
Thrombolytic therapy
Lymphomas and other malignancies
Tuberculosis and bacterial effusions
Rheumatic fever and other non-bacterial inflammatory effusions
Ischemic rupture of the left ventricular free wall after acute myocardial infarction
Postcardiac surgery
Perforation of the right atrium or ventricle during cardiac catheterization, biopsy of myocardial wall, introduction of pacemaker, or placement of central venous catheters

back, and upper abdomen; fatal tamponade has been reported after a one-inch stab wound in the right mid-axilla. After penetrating injury, the full clinical picture of tamponade may develop within minutes, or slowly over periods of days. This is because the knife or gunshot also tears the pericardial sac which allows the pericardial blood to decompress into the pleural cavity or mediastinum. Tamponade occurs when the egress of pericardial blood is obstructed by clots that form in the pericardial space shortly after laceration of a cardiac chamber; large clots were found in 60 per cent of patients operated on for tamponade. Although tamponade from penetrating cardiac wounds has a low incidence, disastrous consequences occur when the diagnosis is delayed or missed. The lack of specificity of routine monitoring means that a high degree of suspicion should be exercised in trauma patients.

Iatrogenic tamponade occurs with manipulation of instruments and catheters that inadvertently perforate the wall or the right atrium or ventricle during cardiac catheterization or biopsy, or during placement of pacemakers or central venous catheters. This occurs most commonly in patients with ischemic or severely diseased ventricles. In contrast with penetrating cardiac wounds, tamponade accidentally occurring after catheter manipulation may be fatal unless rapidly corrected by pericardiotomy and drainage because there is no route for spontaneous decompression of the pericardial fluid.

Pericardial effusion

At the other end of the temporal spectrum, effusions that occur in uremia, neoplasms, post-myocardial infarction (Dressler's syndrome), and bacterial and non-bacterial inflammatory conditions may produce tamponade slowly over periods of weeks or months. In slowly developing patterns, the pericardial sac is able to dilate and can accommodate 2 liters or more of pericardial fluid.

Postcardiac surgery

After cardiac surgical operations, accumulation of seromas can sometimes occur. The fluid is usually found in loculated areas either within or outside the pericardial space; when the loculations become large, whether inside or outside the confines of the pericardium, the tamponade syndrome may be produced.

Pulsus paradoxus

The so-called paradoxic pulse or pulsus paradoxus is an exaggerated pattern of change in arterial pressure with ventilation. By definition, it occurs when the systemic arterial pressure decreases by more than 15 mmHg with inspiration. Normal variation in arterial pressures is less than 10 mmHg and is not clearly evident by examination with stethoscope and sphygmomanometer. Pulsus paradoxus may be detectable by careful examination of the radial pulse, but is more easily detected with direct recording of arterial pressures and the ventilatory cycle. It is a useful sign in slowly developing pericardial effusions, but is not very useful in rapidly developing tamponade from penetrating injuries.

Investigations

Central venous pressure values of 15 mmHg or higher suggest tamponade, particularly if they are rising over a short observation period. Routine chest radiographs may reveal left or right hemothorax, the most common clinical condition associated with tamponade from penetrating injuries. Hemomediastinum occasionally occurs. The 'water-bottle' appearance of the heart is rare in acute tamponade from penetrating cardiac injuries, but may occur in slowly developing pericardial effusions from medical conditions. Pulmonary congestion is usually absent in patients with tamponade compared with the pulmonary congestion and cardiac dilatation seen in patients with congestive heart failure. Non-specific ECG changes include low voltage, depressed or altered ST segments, and evidence suggestive of subepicardial ischemia. ECG patterns are rarely diagnostic of tamponade, but are performed routinely to exclude other cardiac conditions.

Echocardiography provides definitive diagnostic information. Normally, a single echo is reflected from the posterior wall of the ventricles; two distinct echoes from this area indicate pericardial fluid, and the space between these echoes is a reflection of the volume of fluid.

CT and magnetic resonance imaging of the chest accurately show small chambers of the heart and the pericardial effusion. Limited time precludes these studies in rapidly evolving tamponade after penetrating chest injuries. Angiocardiography with radio-opaque dye can outline the right atrial lumen and the thickness of the atrial wall; a wall thickness greater than 6 mm indicates increased pericardial fluid.

The diagnosis is confirmed if non-clotting blood is obtained by pericardiocentesis, but a negative tap does not rule out the diagnosis. False-negative taps were seen in 19 per cent of hemopericardium from penetrating chest injuries because of clotted pericardial blood

Clinical diagnosis

Stage one

The presence of tachycardia, muffled heart sounds, increased area of cardiac dullness, and central venous pressure values of 15 mmHg and rising suggests early tamponade in patients with chest stab wounds.

Stage two

The addition of hypotension completes Beck's triad and indicates that physiological compensations are inadequate to maintain circulatory integrity. This actually represents an advanced stage of the disease in which arrest is likely.

Stage three

Fall in central venous pressure and severe hypotension suggests that circulatory deterioration is advanced and cardiac arrest is imminent. Mortality in tamponade from penetrating wounds increases from 25 per cent without arrest to 65 per cent after arrest. The fall in central venous pressure from levels of 15 to 20 mmHg or more, together with normotension and normal heart rate and rhythm, suggests circulatory improvement without significant tamponade

Differential diagnosis of elevated central venous pressure

Acute tamponade must be differentiated from other conditions with an acutely elevated central venous pressure, such as acute right ventricular failure, chronic obstructive lung disease, constrictive pericarditis, pulmonary embolism, and acute exacerbation of chronic bronchitis. It must also be differentiated from a high central venous pressure following excessive or too rapid fluid administration, abdominal distention from paralytic ileus or ascites, increased intrathoracic pressure from pneumothorax or hemothorax, airway obstruction, infusions of vasopressors, and a clotted or non-functional central venous pressure line.

Bibliography

Beall, A.C., Diethrich, E.B., Crawford, H.W., Cooley, D.A., and DeBakey, M.E. (1966). Surgical management of penetrating cardiac injuries. *American Journal of Surgery*, **112**, 686–92.

Naclerio, E.A. (1964). Penetrating wounds of the heart: experience with 249 patients. *Diseases of the Chest*, **46**, 1–12.

Shoemaker, W.C., *et al.* (1970). Hemodynamic alterations in acute pericardial tamponade after penetrating injuries of the heart. *Surgery*, **67**, 754–64.

Wilson, R.F. and Bassett, J.S. (1966). Penetrating wounds of the pericardium or its contents. *Journal of the American Medical Association*, **195**, 513–18.

Yao, S.T., Vanecko, R.M., Printen, K.J., and Shoemaker, W.C. (1968). Penetrating wounds of the heart. *Annals of Surgery*, **168**, 67–78.

3.12.2 Management

W. C. Shoemaker

Key messages

- Early recognition and treatment of acute tamponade, before cardiogenic shock and cardiac arrest occur, is crucial.
- Volume loading and, if necessary, vasoactive agents may be necessary to maintain blood pressure and flow while preparing for drainage.
- The pericardial collection may be drained by needle pericardiocentesis or surgically via a subxiphoid pericardial window.
- After the life-threatening situation has been resolved by drainage, the cause should be sought and corrected where possible, for example antibiotics for a bacterial etiology or surgical repair of a myocardial laceration.

Introduction

The major acute problem in tamponade from penetrating chest and upper abdominal wounds is early recognition and treatment before cardiogenic shock and cardiac arrest occur. This may entail maintenance of blood pressure and flow with vasoactive agents until the pericardial collection has been drained. With pericardial effusions arising from a non-surgical cause, the tamponade syndrome is more likely to progress slowly; this allows more time for diagnostic procedures and trials of conservative measures.

Therapeutic management

In suspected cases of tamponade, regular measurement of arterial pressure, heart rate, and central venous presure should be made at intervals of 5 to 15 min. Hematocrit, urine output, and arterial blood gases should also be measured, but at less frequent intervals. While the patient is being prepared for drainage, a trial of volume loading may improve the hemodynamic picture. This confirms that hypovolemia and myocardial insufficiency are not major problems. Volume loading is not contraindicated by high central venous pressures, as tamponade is produced by inadequate ventricular filling rather than ventricular failure.

Figure 1 shows a clinical algorithm designed for the immediate treatment of patients suspected of having acute pericardial tamponade from blunt or penetrating injury of the chest or upper abdomen, after thoracotomy, cardiac catheterization, or placement of central venous catheters, and for patients with pericarditis, malignancies, or granulomatous diseases of the chest who become symptomatic.

Pericardiocentesis

Large pericardial effusions that occur in medical conditions lend themselves to pericardiocentesis. However, hemopericardial tamponade from penetrating cardiac wounds should rarely have pericardiocentesis because in the very early stages the pericardial blood usually contains large clots which cannot be aspirated, and these lesions, which may rapidly be fatal, require immediate surgical drainage and repair of the cardiac laceration. Pericardiocentesis is rarely used for diagnostic purposes in acute penetrating trauma because of the high incidence of false-negative results occurring when the aspirating needle enters a clot.

A subxiphoid approach is commonly used, preferably under echocardiographic, fluoroscopic, or CT control. The drainage needle can be connected to the chest wall electrode (V lead) of the ECG as a characteristic injury pattern (ST-segment changes or multiple ventricular ectopics) is produced if the tip comes in contact with the myocardium. No fluid may be aspirated if the cannula tip is in the wrong location or there is a hemopericardium with organized blood clot. Thus a screening technique facilitates placement and reduces the risk of the procedure.

After placing the patient in a semisupine position, skin cleansing, placement of sterile towels, and installation of a local anesthetic into skin and underlying soft tissues, an 18 gauge cannula is advanced from beside the xiphisternum under the costal margin, aiming toward the left scapula, until fluid can be readily aspirated. Samples of the fluid should be taken for microbiological, histological, and biochemical analyses, as appropriate. The pericardial sac can be drained to dryness or, alternatively, a guide wire can be introduced through the cannula which is then removed, allowing placement over the guide wire of an indwelling catheter which is usually of the pigtail type. The catheter can be removed, after several days if necessary, when no further fluid is draining. Complications of pericardiocentesis include damage to the ventricle or a coronary artery, arrhythmias, and pneumothorax.

Surgical procedures

A subxiphoid pericardial window may be performed through a small incision at the left lateral side of the xiphoid cartilage. The diaphragm

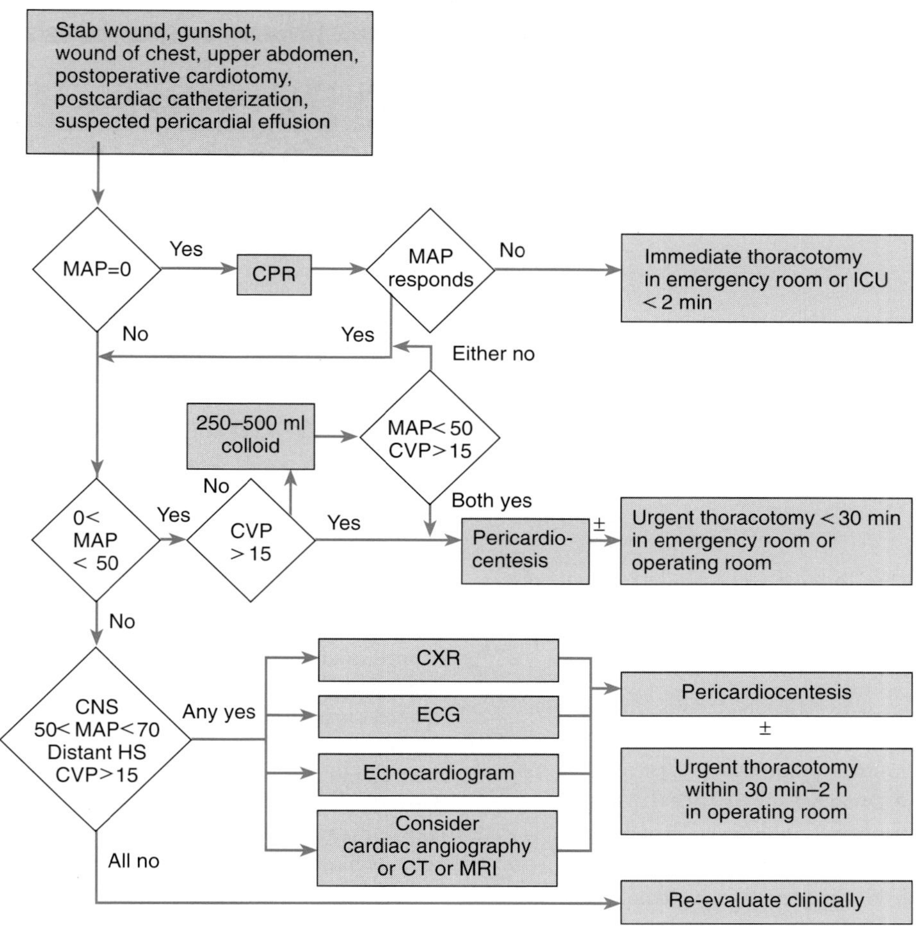

Fig. 1 Clinical management algorithm: CNS, central nervous system; CPR, cardiopulmonary resuscitation; CVP, central venous pressure; CXR, chest radiography; ICU, intensive care unit; MAP, mean arterial pressure; MRI, magnetic resonance imaging; HS, heart sounds.

is reflected inferiorly, the pericardium is exposed, and, if blood or fluid is present, a 4 cm × 4 cm window of pericardium is excised and the perforation of the cardiac wall repaired. The pericardial sac may be approached through the standard left anterolateral thorocotomy at the fourth or fifth left interspace. Inspection, palpation, and ballotment of the pericardium may reveal significant amounts of pericardial fluid under pressure, which may be drained by a pericardial window.

Treatment of the underlying cause

After the life-threatening situation has been resolved by catheter or surgical drainage of the collection, the cause should be sought and corrected where possible, for example appropriate antibiotic therapy for a bacterial etiology or surgical repair of a myocardial laceration.

Bibliography

Shoemaker, W.C. (1975). Algorithm for early recognition and management of cardiac tamponade. *Critical Care Medicine*, **3**, 59–63.

Shoemaker, W.C. (1995). Pericardial tamponade. In *Textbook of critical care* (3rd edn) (ed. S.M. Ayres, A. Grenvik, P. Holbrook, and W.C. Shoemaker), pp. 562–7. W.B. Saunders, Philadelphia, PA.

Yao, S.T., Vanecko, R.M., Printen, K., and Shoemaker, W.C. (1968). Penetrating wounds of the heart: a review of 80 cases. *Annals of Surgery*, **168**, 67–78.

3.13 Pulmonary hypertension

3.13.1 Pathophysiology and causes

D. Pappert and K. J. Falke

Key messages

- Pulmonary hypertension is often secondary to other cardiopulmonary diseases.
- Acute pulmonary hypertension is ameliorated by treatment of the underlying disease.
- Chronic pulmonary hypertension results in vascular remodeling.
- Recovery from chronic pulmonary hypertension is possible and depends on vascular endothelial integrity.
- The prognosis of primary pulmonary hypertension is generally fatal.
- Therapy is focused on a reduction in pulmonary artery pressures and the prevention of right ventricular dysfunction or failure.

Physiological considerations

Unlike the systemic circulation, the pulmonary circulation can be characterized as a 'high-flow–low-pressure system', tolerating up to three- to fourfold increases of cardiac output without significant increases in pulmonary artery pressures under physiological conditions. Vertical distribution of pulmonary blood flow is not equal but is influenced by gravitational forces as well as by the relationship between pulmonary artery pressure, pulmonary venous pressure, and alveolar pressure. Most of the blood flow is distributed to the dependent regions of the lung. Exercise redistributes flow to the more apical parts by recruiting vascular space. This homogenizes pulmonary blood flow across the lung, improving ventilation–perfusion distribution and thereby gas exchange.

The pulmonary vascular resistance (PVR) is mathematically defined by cardiac output (CO), pulmonary arterial pressure (PAP), and pulmonary artery wedge pressure (PAWP):

$$\mathrm{PVR} = (\mathrm{PAP}_{\mathrm{mean}} - \mathrm{PAWP})/\mathrm{CO}.$$

In healthy subjects, PVR decreases when pulmonary blood flow rises. Forty-six per cent of PVR lies within the alveolar capillary bed and arterioles. Although initially considered as mostly passive, the regulation of pulmonary vascular tone is by active vasoconstriction or relaxation of vascular smooth muscle. This is mediated by chemical stimuli (e.g. hypoxia) or by humoral or neuronal transmitters (catecholamines, nitric oxide, prostaglandins) (Murray 1994).

Etiology and classification of pulmonary hypertension

Pulmonary hypertension is the consequence of either acute or chronic pathological mechanisms resulting in a loss of the independence of pulmonary artery pressure and cardiac output. Although pulmonary hypertension can be characterized by more or less arbitrary pulmonary artery pressure limits under static conditions, it is better defined by an increasingly linear relationship between vascular resistance and cardiac output when blood flow is elevated (Table 1). Whereas in healthy subjects the cross-sectional area of the pulmonary vascular bed increases with elevated cardiac output owing to recruitment of apical vessels, a restriction of this mechanism causes a linear dependence of blood flow and pulmonary artery pressures which correlates with disease severity. The pulmonary circulation then behaves like the systemic circulation and no longer as a 'high-flow–low-pressure system' (Enson 1989).

Pulmonary hypertension can be differentiated by (a) its underlying pathophysiological mechanism (Table 2), (b) clinically by its pulmonary or extrapulmonary origin (Table 3), or (c) its anatomical location (i.e. pre- or postcapillary). Many pathological conditions can eventually result in pulmonary hypertension, but in the intensive care unit (ICU) most patients develop pulmonary hypertension in relation to the acuteness of the underlying disease. Other patients, in whom pulmonary hypertension is either latent or develops subacutely, are clinically asymptomatic unless severe structural changes in pulmonary vasculature result in progressive deterioration of pulmonary or cardiovascular function. Advanced stages of chronic

Table 1 Definition of pulmonary hypertension

Under resting conditions at age >2 weeks	
	$\mathrm{PAP}_{\mathrm{systolic}} > 30$ mmHg
	$\mathrm{PAP}_{\mathrm{mean}} > 15$ mmHg
	$\mathrm{PAP}_{\mathrm{diastolic}} > 10$ mmHg
Under exercise conditions (twofold increase in basal oxygen consumption)	
	$\mathrm{PAP}_{\mathrm{systolic}} > 30$ mmHg
	$\mathrm{PAP}_{\mathrm{mean}} > 20$ mmHg
	$\mathrm{PAP}_{\mathrm{diastolic}} > 14$ mmHg

Table 2 Causes of pulmonary hypertension

Pulmonary	Extrapulmonary
Amyloidosis	Mitral valve dysfunction
Primary pulmonary hypertension	Left ventricular failure
Hypoxic vasoconstriction	Patent ductus arteriosus
Vasculitis	Portal hypertension
High-altitude sickness	Pulmonary embolism
Pulmonary arteriovenous fistulas	Chronic thromboembolic disease
Ventricular septal defect	Sepsis
Pulmonary infection	Liver cirrhosis
Pulmonary edema	
Pulmonary fibrosis	
Emphysema	

Table 3 Pathophysiological mechanisms resulting in pulmonary hypertension

Increased pulmonary blood flow	Ventricular septum defect
	Atrial septum defect
	Patent ductus arteriosus
	Sepsis
Increased pulmonary venous pressure	Left heart failure
	Mitral valve disease
Vasoconstriction	Hypoxia
	Hypercarbia
	Chronic obstructive pulmonary disease
	Acute respiratory distress syndrome
Pulmonary vascular obstruction	Thromboembolic disease
	Pulmonary embolism
	Sickle cell disease
	Pulmonary edema
Pulmonary vascular obliteration	Pulmonary vasculitis
	Interstitial lung diseases
	Primary pulmonary hypertension

pulmonary hypertension represent a major therapeutic challenge because of vascular remodeling and compensatory transformation of the pulmonary circulation including the right ventricle.

Acute pulmonary hypertension

Acute pulmonary hypertension in the ICU often represents a clinical problem secondary to acute respiratory failure, left heart failure, or pulmonary embolism. The pathophysiological consequences mirror the severity of the disease and subside with successful therapy.

Acute respiratory failure and sepsis

In acute respiratory failure, several mechanisms promote the development of pulmonary hypertension. The inflammatory character of the disease leads to neutrophil accumulation within the lung and subsequent release of cytokines, free radicals, and other cytotoxic metabolites (Gossage and Christman 1994). Increased permeability of the pulmonary capillary bed leads to a subsequent fluid shift to the interstitium, eventually resulting in pulmonary permeability edema. The increase in interstitial pressure compromises capillary blood flow, thus promoting ventilation–perfusion mismatch with hypoxia and respiratory acidosis. Because a high shunt fraction and a patchy distribution of atelectasis is characteristic in acute respiratory failure, hypoxic pulmonary vasoconstriction is an important factor in redistribution of blood flow to ventilated parts of the lung in order to reduce shunt perfusion and prevent hypoxic vasoconstriction in non-injured parts of the lung. The high airway pressures produced by mechanical ventilation cause distortion and high shear forces in the capillary bed of the overinflated part of the diseased lung; this hampers blood flow owing to mechanical compression of the capillaries. Alterations in the vascular endothelium, increases in procoagulatory and fibrinolytic activity, and reductions or stasis of blood flow promote deposition of fibrin and microemboli (Zapol and Snider 1977). These pathological features are not unique to acute respiratory failure but occur in other inflammatory states, including sepsis, where respiratory failure is secondary to the disease process. The reduction of the cross-sectional area of the pulmonary vascular bed restricts pulmonary blood flow without further increases in pulmonary artery pressures. The need for an elevated cardiac output to augment oxygen delivery in sepsis further increases the degree of pulmonary hypertension. Acute pulmonary hypertension ameliorates the ongoing healing process, the restitution of endothelial integrity, normoxia, and normalized acid–base status.

Pulmonary edema

Permeability edema plays an important role in the development of pulmonary hypertension. It is often associated with abnormal left heart function, for example left ventricular failure and mitral valve disease. Blood volume and vascular resistance within the pulmonary circulation are directly dependent upon left ventricular end-diastolic pressure. After initial compensation and pulmonary congestion, which can be visualized radiographically and is a typical feature of left heart failure, fluid shift occurs across the capillary endothelium as determined by the imbalance between hydrostatic and colloid osmotic pressures across the capillary wall. Pulmonary hypertension results from mechanical compression of the vessels and, in severe cases, is augmented by hypoxia and respiratory acidosis when fluid shifts lead to alveolar flooding, thereby limiting gas exchange. In most cases pulmonary hypertension resolves upon reversibility of the underlying pathophysiological process, once a diagnosis has been made and appropriate therapy initiated.

Chronic pulmonary hypertension

This is consequent to a persistent increase in pulmonary vascular resistance resulting in remodeling of the pulmonary vascular bed. It may either follow a process of acute onset or result from chronic inflammatory, immunological, or toxic processes.

Primary pulmonary hypertension

This diagnosis is established when no evidence of a secondary cause can be established. The pulmonary artery wedge pressure should be normal (Table 1). Although etiology cannot be attributed to any single disorder, the predominant target group, young women of childbearing age, is rather homogenous with a mostly fatal outcome. The pulmonary arteriopathy with plexiform lesions, pronounced intimal proliferation, and medial hypertrophy are not exclusive to primary pulmonary hypertension but can also be found in chronic thromboembolism. Evidence exists that a reduced fibrinolytic activity promotes thrombosis leading to vascular obstruction. Deposits of immune complexes, fibrin deposits, and lymphocytic infiltration are demonstrable. In the etiology of primary pulmonary hypertension, a familial predisposition has been demonstrated in some cases, as well as an association with certain drugs (e.g. oral contraceptives, sulfonamides, and antiarrhythmic drugs). Primary pulmonary hypertension has been associated with HIV infection in some patients. In pulmonary veno-occlusive disease the injury and structural changes are localized in the pulmonary veins and venules. The primary site of an increased vascular resistance is the postcapillary vessels, resulting from intimal alterations with organized and partly recanalized thrombi, medial hypertrophy, and arterialization, the hallmark of this etiology. Alterations in pulmonary arterial vasculature appear to be secondary to these features.

There are a number of therapeutic approaches to the management of primary pulmonary hypertension, but therapy remains largely symptomatic and no definite cure is possible. The course of the disease is determined by the successful relief of pulmonary hypertension. Right heart failure is the cause of death in the majority of the patients (63 per cent) and survival rate 5 years after diagnosis is 22 to 38 per cent (Rubin *et al.* 1993).

Secondary causes of chronic pulmonary hypertension

Chronic pulmonary hypertension develops when the underlying disease is not diagnosed because the pathophysiological changes are mild and do not result in clinical symptomatology (e.g. congenital heart diseases, thromboembolic disease), or when the primary disease becomes chronic despite adequate therapy (e.g. inflammatory or infectious diseases). The pathophysiological alterations of the pulmonary circulation are progressive and are mainly determined by the time course of vascular remodeling. All entities except primary pulmonary hypertension result in pulmonary hypertension secondary to an underlying disease (Table 3).

Mechanisms of vascular remodeling

Persistent pulmonary hypertension, whether acute in onset or chronic and slowly progressive, induces structural changes to the pulmonary arterial tree. Remodeling of pulmonary arteries is a response to high flow and high pressures and/or any underlying inflammatory or toxic process. Shear stress to vascular endothelial cells and the subsequent restructuring of their surface properties and shape are considered important factors for remodeling. Pulmonary arteriosclerosis is the most common initial alteration, presenting as a thickening and hypertrophy of smooth muscle of the vascular medial layer. An increase in connective tissue can also be observed. The consequence is a narrowing of the arterial lumen, already the major site of pulmonary vascular resistance under physiological conditions. Furthermore, a reduced compliance of the vascular wall to changing blood flows enhances the development of pulmonary hypertension. The anatomical localization of the injury is of major importance for the disease process because a destruction of alveolar vessels, as seen in severe emphysema, does not raise pulmonary vascular resistance to the same extent as obliteration of large-caliber pulmonary arteries such as segmental arteries.

The late presentation of the patient with chronic pulmonary hypertension represents a major problem in the treatment of chronic pulmonary hypertension. At this stage of the disease vascular changes are advanced and therapy is often restricted to a symptomatic therapeutic approach.

Pathophysiological consequences of pulmonary hypertension

Exertional shortness of breath may represent the first discrete clinical symptom of pulmonary hypertension. Dyspnea at rest and pathologic heart sounds, particularly a loud secondary pulmonary sound, indicate advanced-stage pulmonary hypertension. Under experimental conditions, pulmonary artery pressures increase when more than two-thirds of the vascular bed is occluded. In patients with a thromboembolic episode, a 70 per cent obstruction of the vascular bed is still compatible with life.

When pulmonary hypertension results from increased pulmonary blood flow or pulmonary venous congestion, such as with a ventricular septal defect, the right heart slowly adapts to the slow rise in vascular resistance and responds by hypertrophy. When pulmonary artery and systemic pressures eventually equalize, the left-to-right direction of blood flow through the shunt reverses (Eisenmenger syndrome) and a further increase in vascular resistance due to hypoxemia results in right ventricular failure. Provided that the alveolar capillary bed is not affected by the disease process, oxygenation is primarily restricted not by a diffusion limitation but by the inadequate response of cardiac output to increased oxygen consumption. When the process is diffuse and the pulmonary capillaries are also involved, a variety of factors, not exclusively a limitation of diffusion capacity, reduce blood oxygenation.

In the late course of the chronic disease, impairment in right ventricular function owing to the increase in right ventricular afterload leads to a reduction in contractility, thus lowering cardiac output and oxygen delivery. Insufficiency of the tricuspid valve and regurgitation combined with high central venous pressures are typical clinical findings. The combination of reduced arterial oxygen content, critical tissue perfusion of right ventricular muscle, and a further decrease in coronary blood flow eventually induces right ventricular failure.

Prognosis and outcome

Remodeling of the vessel structure is potentially reversible when caused by mechanical forces. A reduction in blood flow and perfusion pressure also normalizes shear stresses to intact endothelial cells which are essential for reversal of vascular remodeling. The potential recovery of the pulmonary circulation has been demonstrated in many clinical entities, for example after correction of mitral valve disease, pulmonary artery banding in congenital pulmonary hypertension, or closure of cardiac septal defects. Depending on the duration of pulmonary hypertension, it may take years for the vessels to restructure and pulmonary artery pressures to normalize. Involvement of the right and left ventricles in the disease process must also be considered as this will influence or limit restoration of a normal pulmonary circulation.

In contrast with chronic pulmonary hypertension resulting from increased blood flow or high pressures, remodeling of pulmonary vasculature is almost irreversible in patients when the disease process results from chronic inflammatory processes, immunological alterations, or thrombogenic obliteration of the lumen. Interstitial fibrosis and destruction of the regular histological pattern, including loss of endothelial cell integrity, determines the extent of future deterioration.

Conclusions

The clinical course and prognosis of pulmonary hypertension is mainly determined by successful cessation of the disease process. Increased pulmonary blood flow, hypoxic vasoconstriction, elevated left ventricular preload, and increased pressure are factors which promote vascular remodeling in acute pulmonary hypertension. While acute pulmonary hypertension, for example during sepsis or acute respiratory failure, is relieved with the cure of the underlying disease, persistent pulmonary hypertension leads to intimal hypertrophy and endothelial alterations. Provided that endothelial integrity remains, recovery from pulmonary hypertension is possible. Recovery may take years and depends on hemodynamic alterations to the whole pulmonary circulation, including the right and left ventricles. If pulmonary hypertension is secondary to immunological or metabolic processes and obliteration or destruction of the vascular texture occurs, the disease process is rarely reversible. Outcome is eventually fatal, particularly in primary pulmonary hypertension, and therapy is mainly restricted to symptom relief. Right ventricular failure is the main cause of death in endstage pulmonary hypertension.

References

Enson, Y. (1989). Pulmonary hypertension and its consequences. In *Textbook of pulmonary diseases* (4th edn) (ed. G.L. Baum and E. Wolinsky), Chapter 55. Little, Brown, Boston, MA.

Gossage, J.R. and Christman, B.W. (1994). Mediators of acute pulmonary hypertension. *Seminars in Respiratory and Critical Care*, **15**, 190–8.

Murray, J.F. (1994). General principles and diagnostic approach. In *Textbook of respiratory medicine* (2nd edn) (ed. J.F. Murray and J.A. Nadel), Chapter 52. W.B. Saunders, Philadelphia, PA.

Rubin, L.J. (Chairman) (1993). ACCP consensus statement. Primary pulmonary hypertension. *Chest*, **104**, 236–50.

Zapol, W.M. and Snider, M.T. (1977). Pulmonary hypertension in severe acute respiratory failure. *New England Journal of Medicine*, **296**, 476–80.

3.13.2 Management

R. Rossaint and K. Falke

Key messages

- Chronic pulmonary hypertension results in remodeling of pulmonary vessels.
- Surgical interventions should be considered at an early stage to avoid or reverse remodeling of pulmonary vessels.
- Both intravenously infused and inhaled vasodilator have been shown to lower pulmonary artery pressure.
- Inhaled nitric oxide and aerosolized epoprostenol (prostaglandin I_2) may induce a selective vasodilation in ventilated lung regions.
- The inhalatory strategy may be advantageous in disease associated with increased pulmonary right-to-left shunting.

Introduction

Pulmonary hypertension is a feature common to many acute and chronic lung diseases. Acute pulmonary hypertension may be reversible if the provoking stimulus is removed, whereas chronic pulmonary hypertension is often irreversible owing to fixed structural changes in the pulmonary arterial wall (Fig. 1). Mechanical factors, including vasoconstriction due to hypoxemia or mediators, passive pulmonary venous hypertension, and luminal thrombosis, are primarily involved in acute pulmonary hypertension. In chronic

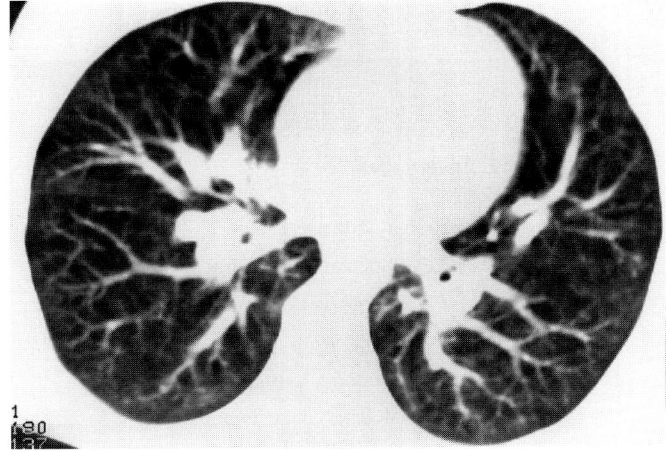

Fig. 1 CT scan of a patient with chronic pulmonary hypertension. Note the abrupt and extensive reduction in the diameters of the pulmonary arteries as a result of the remodeling of pulmonary vessels.

pulmonary hypertension, remodeling of pulmonary vessels occurs with medial hypertrophy, intimal proliferation, fibrosis, and more complex changes such as plexiform lesions resulting in narrowing of the vascular lumen and decreased distensibility of the vascular wall. Altered hemodynamics due to increased blood flow or pressure, and vascular injury, as a response to inflammation or direct toxic effects of drugs, are two important factors involved in vascular remodeling. This remodeling process is reversible if the hemodynamic stress is reduced in the presence of still intact endothelial cells. Studies in the 1960s demonstrated that correction of mitral valve stenosis or surgical banding of pulmonary arteries in congenital pulmonary hypertension was associated with a regression of medial hypertrophy.

Therapeutic considerations for congenital or acquired cardiac malformations associated with high pulmonary flow or pressure should include early surgical correction of these malformations. However, heart–lung or lung transplantation often represents the only therapeutic option in some of these patients, as well as in others suffering from severe idiopathic pulmonary hypertension. In the interim period, these patients have to be treated with vasodilators since right heart failure is the predominant cause of death in this pathophysiological constellation. Pharmacological interventions are also necessary in diseases associated with acute pulmonary hypertension, since right ventricular dysfunction due to increased afterload may complicate acute respiratory distress syndrome, cardiac surgery, and heart transplantation.

Under normal physiological conditions, the functional significance of the right ventricle appears minimal; it acts more or less as a passive conduit responsible for accepting venous blood and propelling it through the pulmonary system to the left ventricle. However, development of sudden pulmonary hypertension induces a decrease in right ventricular ejection fraction and increased right ventricular volumes. Provided that right ventricular contractile function is not depressed, the right ventricle may be able to pump sufficiently to maintain a normal cardiac output; in this situation a severely reduced right ventricular ejection fraction indicates right ventricular dysfunction rather than failure. However, if pulmonary hypertension increases further and/or right ventricular contractile function is depressed, the right ventricle may not be able to generate a sufficient output, resulting in right ventricular failure. Moreover, an increased right ventricular end-diastolic volume may cause a left shift of the ventricular septum which, in turn, may alter left ventricular performance. Thus right ventricular failure may influence the systemic circulation and may even limit survival. In addition to these secondary alterations in left ventricular performance, the leftward septal shift may result in an increased wedge pressure, possibly enhancing the development of interstitial pulmonary edema.

Intravenously administered vasodilators used to be the only pharmacological option to treat acute and chronic pulmonary hypertension. However, in the last few years, inhaled nitric oxide (NO) and aerosolized epoprostenol (prostaglandin I_2) have been demonstrated to be suitable alternatives, particularly in the treatment of acute pulmonary hypertension.

Intravenously administered vasodilators

As intravenous vasodilator infusions lower pulmonary artery and effective pulmonary capillary pressures, right ventricular function may be improved and interstitial edema formation reduced (Table 1). However, the intravenous use of conventional vasodilators such as nitroglycerin or epoprostenol is limited because of diffuse dilatation of the whole vasculature. Global vasodilatation of the pulmonary vasculature increases blood flow to areas of intrapulmonary shunt which further reduces the already compromised $P\text{ao}_2$, particularly in patients with acute respiratory distress syndrome. Moreover, concurrent dilatation of the systemic vasculature may result in a dose-dependent arterial hypotension which may possibly lead to ventricular ischemia and consequent heart failure. Nevertheless, vasodilator agents play a major role in the therapeutic strategy of primary pulmonary hypertension. Continuous epoprostenol infusion not only improves the well being of patients considered for heart–lung transplantation, but also doubles median survival time (Higenbottam *et al.* 1993).

Vasodilator therapy is often required to treat increased pulmonary artery pressures after heart transplantation. Currently used intravenous vasodilators include epoprostenol (2–30 ng/kg/min), prostaglandin E_1 (5–250 ng/kg/min), and sodium nitroprusside (0.3–2 g/kg/min). In patients without hemodynamic signs of acute right heart failure, as reflected by normal values of right ventricular

Table 1 Advantages and disadvantages of intravenous vasodilators

Advantages	Disadvantages
Pulmonary artery pressure ↓	Systemic arterial pressure ↓
Right ventricular ejection fraction ↑	Risk of ventricular ischemia ↑
Left atrial pressure ↓	Intrapulmonary shunt ↑
Pulmonary capillary pressure ↓	$P\text{ao}_2$ ↓
Interstitial edema formation ↓	
Left ventricular outflow impedance ↓	
Cardiac output ↑	
Survival in chronic pulmonary hypertension ↑	

end-diastolic volume, central venous pressure, stroke volume, and cardiac output, epoprostenol appears superior to NO inhalation and the best choice for intravenous vasodilator therapy after heart transplantation (Kieler-Jensen *et al.* 1995). Epoprostenol reduces both right and left ventricular outflow impedance, which is a desired effect in most transplanted patients (Kieler-Jensen *et al.* 1995).

Inhaled vasodilators

In contrast with intravenously delivered vasodilators, inhalation of low concentrations of NO (0.01–40 ppm) as well as aerosolization of epoprostenol (2–20 ng/kg/min) may induce a selective vasodilation of ventilated lung regions.

Inhaled NO

Inhaled NO relaxes smooth muscles in arteries and veins in a similar way to endogenous NO by activating soluble guanylate cyclase and increasing cyclic guanosine 3′,5′-monophosphate. Since NO is inactivated by hemoglobin as soon as it enters the blood stream, the dilatory effect of low concentrations of inhaled NO is restricted to the pulmonary vasculature of ventilated lung areas (Table 2). Thus elevated pulmonary artery pressure is selectively decreased in various diseases. In addition, the ratio of $Pa\mathrm{O_2}$ to $\mathrm{F_{IO_2}}$ may increase in patients with a high intrapulmonary shunt (Rossaint *et al.* 1993).

In patients with acute respiratory distress syndrome, the increase in ratio of $Pa\mathrm{O_2}$ to $\mathrm{F_{IO_2}}$ and the decrease in pulmonary artery pressure may be associated with increases in right ventricular ejection fraction and a decrease in right ventricular volume. In contrast with intravenous epoprostenol, which has a similar but greater effect, inhaled NO does not usually increase cardiac output in this group of patients. However, in some patients suffering from severe acute respiratory distress syndrome, inhaled NO may cause an increased right ventricular ejection fraction which will be followed by a rise in cardiac index.

In congenital heart disease, pulmonary hypertension may result from increased pulmonary blood flow or from pulmonary venous obstruction resulting in hypertrophy and hyperplasia of pulmonary artery smooth muscle and pulmonary vasoconstriction. After corrective cardiac surgery the pulmonary vascular bed may be unable to regress sufficiently to accommodate the postoperative hemodynamic changes and this may lead to right ventricular failure. This is a common cause of death after surgical correction of congenital heart diseases with a left to right shunt. In order to determine the vasodilatory capacity of the pulmonary vascular bed and thereby predict the postoperative risk of life-threatening suprasystemic pulmonary hypertension, NO inhalation has been used during preoperative cardiac catheterization. Inhaled NO exhibits a far greater vasodilatory effect than hyperoxic breathing. Moreover, adding NO inhalation to the routine testing protocol for pulmonary vascular reactivity during preoperative cardiac catheterization allows identification of patients with preserved vascular reactivity who did not respond to infused epoprostenol or inhaled pure oxygen but did respond to inhaled NO. If preoperative inhaled NO produces pulmonary vasodilatation, this gas may be used to relieve severe postoperative pulmonary hypertension. Inhaled NO has recently been used to treat moderate and severe pulmonary hypertension in children undergoing open-heart surgery. In some cases NO inhalation caused not only a reduction in pulmonary artery pressure, but also an increase in pulmonary blood flow resulting in a simultaneous decrease in $Pa\mathrm{CO_2}$, improvements in arterial oxygenation, progressive weaning from epinephrine (adrenaline), and reversal of oligoanuria.

In patients with pulmonary hypertension undergoing cardiac surgery, inhaled NO decreased pulmonary artery pressure and pulmonary vascular resistance but did not affect central venous pressure, cardiac output, or systemic hemodynamics. These observations are in agreement with those made in patients suffering from chronic pulmonary hypertension, demonstrating that inhalation of 40 ppm NO did not increase cardiac output despite a decrease in pulmonary vascular resistance and a constant central venous pressure. Inhaled NO does not, in general, appear to influence right ventricular function in patients with chronic pulmonary hypertension despite its effect on pulmonary vascular resistance. However, a small number will derive benefit. In patients receiving circulatory support with a ventricular assist device, 20 ppm NO decreased pulmonary artery pressure and pulmonary vascular resistance, as well as increasing $Pa\mathrm{O_2}$, arterial pressure, and mixed venous oxygen saturation. These alterations may reflect improved right ventricular function and increased cardiac output from the natural heart. Thus inhaled NO may improve systemic hemodynamics when right ventricular failure and an elevated pulmonary vascular resistance are limiting factors.

Inhaled NO has also been evaluated in the pre- and postoperative treatment of patients receiving orthotopic heart transplantation. Pulmonary hypertension is a common preoperative feature. An increased pulmonary vascular resistance may induce right ventricular

Table 2 Advantages and disadvantages of inhaled NO

Advantages	Disadvantages
Pulmonary artery pressure ↓	Sophisticated delivery technique
Right ventricular ejection fraction ↑	$\mathrm{NO_2}$ toxicity
Intrapulmonary shunt ↓	In left heart failure, left atrial pressure ↑
$Pa\mathrm{O_2}$ ↑	Bleeding tendency ↑
Pulmonary capillary pressure ↓	
Interstitial edema formation ↓	
In right heart failure, cardiac output ↑	

Table 3 Advantages and disadvantages of aerosolized epoprostenol

Advantages	Disadvantages
Pulmonary artery pressure ↓	Sophisticated titration necessary
Right ventricular ejection fraction ↑	'Spillover' with possible side-effects
Intrapulmonary shunt ↓	Bleeding tendency ↑
$Pa\text{O}_2$ ↑	
Pulmonary capillary pressure ↓	
Interstitial edema formation ↓	
In right heart failure, cardiac output ↑	

failure after implantation of the donor heart. Therefore the extent of the elevated pulmonary vascular resistance as well as the prospect of pharmacological treatment should be evaluated preoperatively in transplant candidates. If reversibility is shown, patients can be accepted for orthotopic heart transplantation; otherwise, heart–lung transplantation may be considered more suitable. NO inhalation has also been used as a bridge to heart–lung transplantation to manage right ventricular failure with concurrent systemic hypoperfusion. After heart transplantation, NO inhalation appears superior to epoprostenol infusion if pulmonary hypertension and right ventricular failure are associated with systemic hypotension.

Aerosolized epoprostenol

Aerosolized epoprostenol may induce a selective pulmonary vasodilatation despite its half-life of 2 to 3 min (Pappert *et al.* 1995). This selective vasodilatory effect is seen in patients with acute respiratory distress syndrome and after surgical repair of congenital heart defects, but not in patients with chronic pulmonary hypertension. Like inhaled NO, aerosolized epoprostenol may induce improvements in arterial oxygenation in both acute respiratory distress syndrome and postoperative pulmonary hypertension (Table 3). As with inhaled NO, a dose-dependent effect is seen with aerosolized epoprostenol. An increase in the dose of aerosolized epoprostenol may cause a 'spill-over' of the prostanoid into the systemic circulation, thereby reducing systemic vascular resistance. In contrast with NO inhalation, which allows exact measurement of the concentration administered by chemiluminescence or electrochemical fuel cells, it is not presently possible to define the dose of aerosolized epoprostenol acting upon the lung since there will be losses in the nebulizer chamber, ventilator tubing, the endotracheal tube, and the large airways. Moreover, the active aerosol fraction deposited in the alveolar spaces is dependent on physical and physiological factors which vary from patient to patient. It is estimated that the aerosol fraction deposited in the alveolar space during mechanical ventilation is below 10 per cent.

Conclusions

Congenital or acquired cardiac malformations associated with high pulmonary flow or pressure should be corrected at an early stage to reverse or avoid remodeling of pulmonary vessels. In some of these patients, as well as in others suffering from severe idiopathic pulmonary hypertension, heart–lung or lung transplantation often represents the only therapeutic option. In the interim period, these groups of patients should be treated with vasodilators since right heart failure is the predominant cause of death. Whereas intravenously infused vasodilators used to be the only pharmacological option available to treat chronic and acute pulmonary hypertension, inhaled NO and aerosolized epoprostenol may now be regarded as promising alternatives, particularly in the treatment of acute pulmonary hypertension.

References

Higenbottam, T.W., *et al.* (1993). Prostacyclin (epoprostenol) and heart–lung transplantation as treatments for severe pulmonary hypertension. *British Heart Journal*, **70**, 366–70.

Kieler-Jensen, N., Lundin, S., and Ricksten, S.E. (1995). Vasodilator therapy after heart transplantation: effects of inhaled nitric oxide and intravenous prostacyclin, prostaglandin E_1, and sodium nitroprusside. *Journal of Heart and Lung Transplantation*, **14**, 436–43.

Pappert, D., Busch, T., Gerlach, H., Lewandowski, K., Radermacher, O., and Rossaint, R. (1995). Aerosolized prostacyclin versus inhaled nitric oxide in children with severe acute respiratory distress syndrome. *Anesthesiology*, **82**, 1507–11.

Rossaint, R., Falke, K.J., Lopez, F., Slama, K., Pison, U., and Zapol, W.M. (1993). Inhaled nitric oxide in adult respiratory distress syndrome. *New England Journal of Medicine*, **328**, 399–405.

3.14 Pulmonary vascular obstruction

3.14.1 Capillary occlusion

B. Zwissler and K. Messmer

Key messages

- Acute pulmonary microvascular occlusion may be present in several diseases commonly encountered in critically ill patients and may involve arterioles, capillaries, and venules.
- Mechanisms include active vasoconstriction, microthrombosis, microembolism, endothelial edema, mechanical compression, and vascular remodeling.
- Hemodynamic and gas exchange deterioration may result (e.g. right heart failure, hypoxemia).
- Diagnosis requires analysis of underlying diseases, exclusion of macrovascular lesions, and careful evaluation of both pulmonary circulation and gas exchange.
- Management focuses on recruitment of occluded vessels (e.g. by optimizing lung volume, positioning, fluid restriction) and dilation of constricted vessels (e.g. by increasing $P\text{aO}_2$, decreasing $P\text{aCO}_2$, alkalinization, intravenous or inhaled vasodilators).

Introduction

Occlusion of pulmonary vessels may occur at both macro- and microcirculatory levels, and the disease process may be acute or chronic. Here we focus on the mechanisms, pathophysiology, diagnosis, and management of acute pulmonary microvascular occlusion involving arterioles, capillaries, and venules. This may be present in several diseases commonly encountered in intensive care patients, including acute respiratory insufficiency of different origins, sepsis, and disseminated intravascular coagulation (DIC). Microvascular occlusion due to chronic or idiopathic disease will not be discussed.

Mechanisms

Both functional and structural changes may contribute to pulmonary microvascular occlusion (Fig. 1). These mechanisms, although described separately, are actually linked together and may aggravate each other.

Vasoconstriction

Pulmonary capillaries contribute approximately 40 per cent, arteries 50 per cent, and veins 10 per cent to total pulmonary vascular resistance. Only arteries and veins actively regulate their diameters through smooth muscle cells. Physiologically, a balance exists between vasodilating agents (e.g. prostacyclin, nitric oxide) and vasoconstricting agents (e.g. thromboxane, endothelins, serotonin). In health the effects of vasodilators predominate, keeping pulmonary

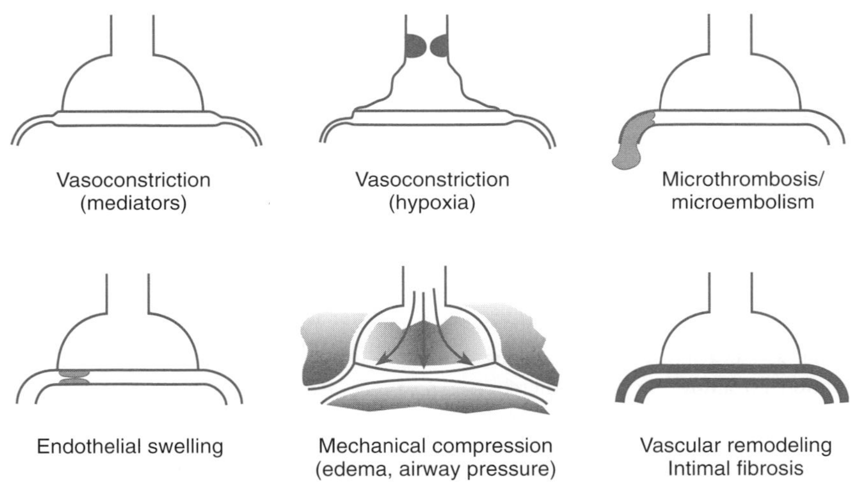

Fig. 1 Mechanisms of acute microvascular occlusion.

vascular tone low. Any increase in vasoconstrictor activity and/or decrease in vasodilator activity will cause pulmonary vasoconstriction.

Small pulmonary arteries, arterioles, and probably venules constrict as a response to alveolar hypoxia. (The problem is actually more complex. In pulmonary arteries of diameter greater than 500 µm, vascular tone is determined by arterial $P\text{O}_2$, with maximum hypoxic pulmonary vasoconstriction occurring at 40 to 50 mmHg (5.3–6.6 kPa). In contrast, vessels of diameter less than 500 µm mainly respond to local $P\text{O}_2$, i.e. the gradient between alveolar and mixed venous $P\text{O}_2$.) This mechanism is called hypoxic pulmonary vasoconstriction and its molecular basis remains unclear. Hypoxic pulmonary vasoconstriction prevents non-ventilated alveoli from being perfused, thereby reducing intrapulmonary shunt and maintaining $P\text{aO}_2$. It is particularly active in the presence of profound hypoventilation (e.g. respiratory drive impairment, sleep apnea, neuromuscular disorders) and/or ventilation–perfusion mismatch (e.g. acute respiratory distress syndrome, chronic airflow limitation).

Microthrombosis

Microthrombosis is a common cause of pulmonary microvascular occlusion. Several mechanisms may contribute to local platelet aggregation and entrapment:

(1) increase in systemic procoagulant activity (e.g. DIC);

(2) increase in local endothelial procoagulant activity (e.g. mediated by tumor necrosis factor);

(3) reduction in local anticoagulant activity (e.g. endothelial dysfunction);

(4) stasis secondary to microvascular hypoperfusion.

Microembolism

Microemboli may contribute to pulmonary vascular occlusion, particularly in trauma patients who present with a reduced capacity for phagocytosis. Consequently, clearing of platelets, fibrin, fat, and other particulate matter is decreased, allowing embolic occlusion.

Edema of endothelial cells

Edema of endothelial cells occurs subsequent to both direct (trauma) and indirect (mediators, O_2 radicals) endothelial damage. Swollen endothelial cells with rarified cytoplasm and enlarged endoplasmic reticulum may encroach upon the capillary lumen and obstruct local perfusion.

Mechanical compression

A few areas exist within the lung where alveolar and/or venous ('outflow') pressure exceeds pulmonary inflow pressure, thereby impeding effective blood flow. This phenomenon may be aggravated by alveolar edema. Furthermore, any increase in transpulmonary pressure (e.g. by positive end-expiratory pressure) will cause overdistension of more compliant alveoli with subsequent vascular hypoperfusion.

Vascular remodeling

In the late phase of acute diseases (e.g. acute respiratory distress syndrome), vascular remodeling may occur due to intimal fibrocellular proliferation and may contribute to considerable reductions in pulmonary vascular cross-sectional area.

Pathophysiology

Independent of the mechanism(s) involved, pulmonary microvascular occlusion adversely affects both cardiovascular and respiratory systems.

Cardiovascular effects

The increase in pulmonary outflow resistance is detrimental for two reasons.

1. Elevation of right ventricular afterload: if the mean pulmonary artery pressure in the non-adapted right ventricle exceeds 40 mmHg, this may lead to right ventricular failure and cardiogenic shock.
2. (Micro)vascular occlusion increases hydrostatic pressure in the lungs, thereby promoting pulmonary edema. This, in turn, compromises gas exchange and may further increase pulmonary artery pressure.

Respiratory effects

Pulmonary microvascular occlusion increases the fraction of lung tissue which is ventilated but not perfused (high $\dot{V}_a/Q$). Consequently, pulmonary dead-space ventilation rises, resulting in impaired CO_2 elimination and hypercapnia. Theoretically, microvascular occlusion should not impair oxygenation, as it increases rather than decreases $\dot{V}_a/Q$. Clinically, however, hypoxemia is often observed for several reasons.

1. Reduced pulmonary blood flow may alter the surface-active properties of the lung and cause postocclusion atelectasis. Reperfusion of these badly ventilated regions, for example after break-up of microthrombi, reduces $P\text{aO}_2$ because of local $\dot{V}_a/Q$ mismatch.
2. Pulmonary edema commonly associated with microvascular occlusion may impair alveolar ventilation, lower $\dot{V}_a/Q$, and increase shunt.
3. The release of vasoactive mediators will impair $\dot{V}_a/Q$ matching, even in areas unaffected by local vessel occlusion.
4. Pulmonary hypertension may cause paradoxic intracardiac right-to-left shunting through a patent foramen ovale.
5. Low mixed venous $P\text{O}_2$ due to a low cardiac output may amplify the negative impact of shunted blood on $P\text{aO}_2$.

It is not clear whether reduced pulmonary capillary transit time due to increased blood flow velocity in unobstructed vessels contributes to hypoxemia.

Diagnosis

In contrast with macrovascular events (e.g. pulmonary embolism), neither the site nor the severity of microvascular pulmonary occlusion can be adequately visualized using present technology. Hence diagnosis is based on underlying disease, exclusion of macrovascular lesions, and accurate cardiorespiratory evaluation.

Underlying disease

Acute pulmonary microvascular occlusion should always be suspected if one of the following conditions is present: hypoxemia/hypercapnia; pulmonary edema; signs of endothelial dysfunction (e.g. sepsis); coagulopathy (e.g. DIC); continuous positive-pressure breathing; suspected microembolism.

Exclusion of macrovascular lesions

Macrovascular pulmonary occlusion must be excluded using available technology (e.g. chest radiography, echocardiography, Doppler ultrasonography, computed tomography, ventilation–perfusion scintigraphy, angiography). The diagnosis of macrovascular occlusion does not exclude additional microvascular occlusion.

Cardiovascular evaluation

Pulmonary hypertension is the hallmark of pulmonary microvascular occlusion. Hence pulmonary artery pressure (PAP) provides a rough measure of disease severity. It is not ideal as it will not distinguish between micro- and macrovascular occlusion. Furthermore, PAP does not rise unless more than 50 per cent of pulmonary vessels are occluded and hence may fail to detect significant microvascular occlusion. In addition, it is affected passively by intrathoracic pressure, pulmonary blood flow, and pulmonary downstream pressure, all of which may change unpredictably.

These drawbacks are partially overcome by calculating pulmonary vascular resistance (PVR) as

$$\mathrm{PVR} = \frac{\mathrm{PAP} - \mathrm{PAOP}}{\mathrm{CO}}$$

where PAOP is the pulmonary artery occluded pressure and CO is cardiac output. Unfortunately, PVR overestimates the true vascular resistance and fails to reflect changes over time when the critical closing pressure of pulmonary vessels (which ideally is equal to PAOP) greatly exceeds PAOP ('vascular waterfall' phenomenon). However, this is precisely what happens in severe microvascular occlusion, thereby limiting the diagnostic value of PVR.

The problem can be solved by directly assessing the pulmonary vascular closing pressure (CP) by measuring PAP at different pulmonary blood flows (Fig. 2). CP is then obtained by extrapolating the pressure–flow curve to a pulmonary blood flow of zero. PVR is now calculated more correctly as

$$\mathrm{PVR} = \frac{\mathrm{PAP} - \mathrm{CP}}{\mathrm{CO}}.$$

The slope of the pressure–flow relationship reflects the incremental resistance upstream of vascular closure. Hence, any shift of the pressure–flow curve to the left reflects increased vascular resistance

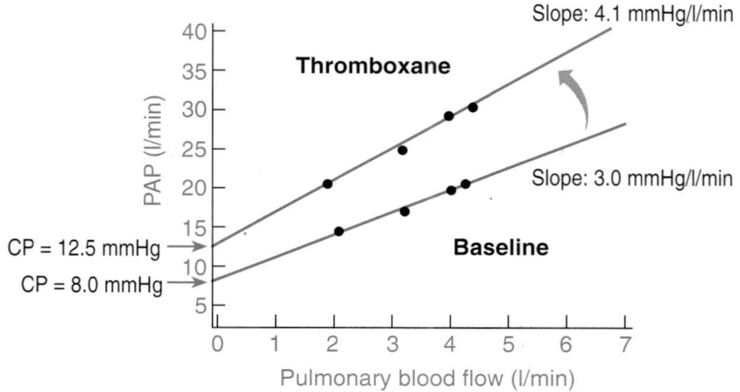

Fig. 2 Pressure–flow plots and critical closing pressure (CP) of the lung. Pressure–flow plots and CP values are shown both at baseline and after induction of experimental pulmonary hypertension by infusion of thromboxane in dogs. See text for further explanation.

(Fig. 2). Clearly, CP and pressure–flow plots are the optimal tools for evaluating pulmonary vascular pathophysiology, but unfortunately they are difficult to obtain in patients.

In contrast, the gradient between end-diastolic pulmonary artery pressure ($\mathrm{PA_{ed}}$) and PAOP can be assessed easily and may distinguish postcapillary from a (pre)capillary origin of pulmonary hypertension. Physiologically there should be no major gradient. If (pre)capillary resistance rises, $\mathrm{PA_{ed}}$ will markedly exceed PAOP. Conversely, if pulmonary hypertension is caused by left ventricular failure and pulmonary venous congestion, $\mathrm{PA_{ed}}$ and PAOP will both rise with no gradient.

Pulmonary capillary pressure (PCP) is the major determinant of transvascular filtration in the lung (Fig. 3). PCP provides information on the risk of pulmonary edema formation and the efficacy of vasodilator therapy, and it also allows identification of the relative contributions of arterial resistance ($\mathrm{PVR_{art}}$) and venous resistance ($\mathrm{PVR_{ven}}$) to total PVR:

$$\mathrm{PVR} = \frac{\mathrm{PAP} - \mathrm{PAOP}}{\mathrm{CO}}$$

$$\mathrm{PVR_{art}} = \frac{\mathrm{PAP} - \mathrm{PCP}}{\mathrm{CO}}$$

$$\mathrm{PVR_{ven}} = \frac{\mathrm{PCP} - \mathrm{PAOP}}{\mathrm{CO}}.$$

An example is shown in Figs 2 and 3. Microvascular occlusion was induced by infusion of the potent pulmonary vasoconstrictor thromboxane. Mean PAP rose from 16 to 27 mmHg while CO remained stable at 4 l/min. Based on PAOP values of 8 mmHg and 9 mmHg respectively, 'conventional' PVR can be calculated as 160 dyn s/cm^5 at baseline and 360 dyn s/cm^5 with thromboxane. The assumption of an increased PVR is confirmed by a steepening of the pressure–flow plot (Fig. 2) and an increase in the $\mathrm{PA_{ed}}$–PAOP gradient from 6 to 17 mmHg; CP increased from 8 to 12.5 mmHg. Using CP instead of PAOP for calculation of PVR does not change baseline resistance but results in a PVR value of 290 dyn s/cm^5 with thromboxane. This analysis demonstrates that 'conventional' PVR overestimates 'true' PVR by 24 per cent in this situation. PCP rose from 10 to 15 mmHg with thromboxane (Fig. 3). Baseline $\mathrm{PVR_{art}}$ and $\mathrm{PVR_{ven}}$ values were

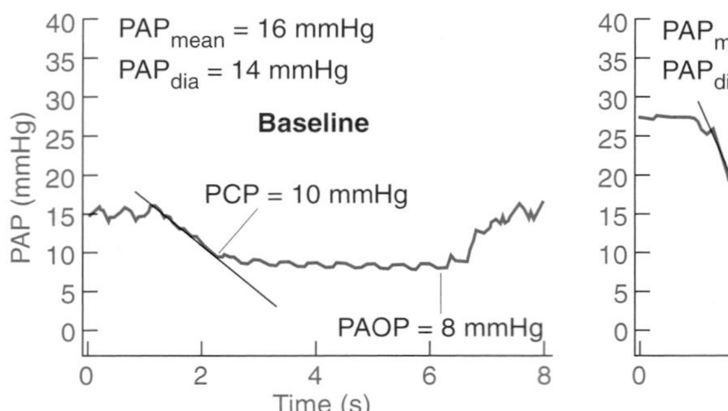

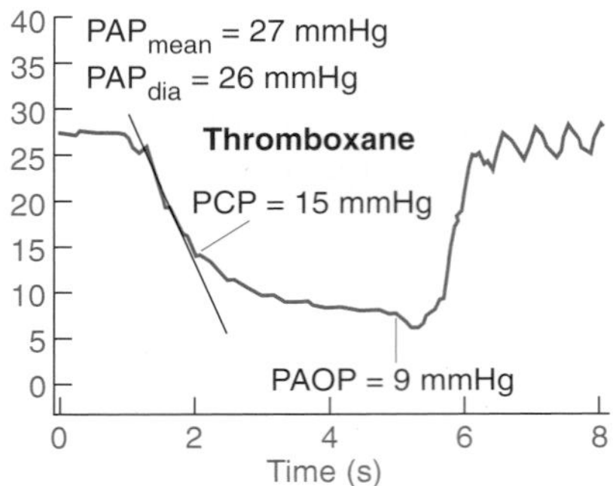

Fig. 3 Pulmonary capillary pressure (PCP). The vascular pressure transient after pulmonary artery occlusion is analyzed at end-expiration and the inflection point between the rapid and slow components is identified by visual inspection (arrow). See text for further explanation.

120 dyn s/cm^5 and 40 dyn s/cm^5 respectively, and 246 dyn s/cm^5 and 123 dyn s/cm^5 respectively with thromboxane. These data suggest that the relative venous contribution to total PVR was 25 per cent at baseline and 33 per cent during infusion of thromboxane.

Finally, the vasoconstrictor component of pulmonary hypertension can be quantified by testing pulmonary vascular reactivity. Inhalation of oxygen or nitric oxide and infusion of the prostaglandin I_2 analog epoprostenol have been used for this purpose. A reduction of pulmonary vascular resistance with 'therapy' suggests that vascular occlusion is partly due to active pulmonary vasoconstriction, responds to pharmacological therapy, has occurred recently (in chronic disease vascular reactivity is almost lost), and affects micro- rather than macrovessels.

Evaluation of gas exchange

Hypercapnia in the presence of normal alveolar ventilation is a hallmark of (micro)vascular occlusion. Therefore arterial and mixed venous $P\text{CO}_2$ should be analyzed and the physiological dead-space quantified:

$$\frac{V_D}{V_T} = \frac{P\text{aCO}_2 - P\bar{\text{e}}\text{CO}_2}{P\text{aCO}_2}$$

where $P\text{aCO}_2$ and $P\bar{\text{e}}\text{CO}_2$ are arterial and mean expiratory $P\text{CO}_2$, respectively. A V_D/V_T ratio of 0.3 to 0.45 is commonly observed in ventilated patients, while values above 0.5 to 0.6 indicate severe impairment of the $\dot{V}_a/\dot{Q}$ ratio. The diagnostic value of evaluating arterial oxygenation is limited as there is no close relationship between the degree of microvascular occlusion and the severity of hypoxemia. Also, hypoxemia is not specific as mechanisms other than microvascular occlusion may decrease $P\text{aO}_2$. Interestingly, the relative extent of impairment of oxygenation and CO_2 elimination may help identify the size of vessels involved in vascular occlusion. Occlusion of small arteries (<100 µm) mainly increases intrapulmonary shunt, causing severe hypoxemia, while occlusion of larger vessels (>50–1000 µm) predominantly increases V_D/V_T, causing severe hypercapnia.

Management

Pulmonary microvascular occlusion requires urgent intervention as it compromises cardiorespiratory function and may increase mortality. Treatment of the underlying disease and symptomatic therapy of right ventricular failure are essential. In addition, management includes recruitment of occluded vessels and/or dilatation of constricted vessels.

Recruitment of occluded vessels

Efforts should be made to optimize lung volume. Reinflation of atelectatic lung areas will improve arterial oxygenation and resolve hypoxic pulmonary vasoconstriction, thereby decreasing pulmonary vascular resistance. Mechanical ventilation is often required, but may be a two-edged sword as it affects pulmonary vessels differently depending on the actual lung volume and the level of positive airway pressure applied. If lung volume is high, any increase in airway pressure will further increase capillary resistance by direct mechanical compression. In addition, overinflation may aggravate alveolar epithelial and capillary injury and enhance pulmonary edema. In contrast, positive airway pressure reduces pulmonary vascular resistance in acute lung disease characterized by high extra-alveolar lung vessel resistance due to low lung volumes. The ventilatory pattern should be adjusted to achieve adequate arterial oxygenation and oxygen transport at the lowest airway pressure, while keeping alveoli open at end-expiration. Positioning (e.g. prone positioning, kinetic therapy) of the ventilated patient may also contribute to recruitment of non-ventilated lung in dependent lung regions. If significant pulmonary edema is present, fluid restriction and diuretics will alleviate vascular compression by reducing extravascular lung water. It is not clear whether patients with acute microvascular occlusion will benefit from systemic anticoagulation or antiplatelet agents.

Vasodilation of constricted vessels

Microvascular occlusion induced by active pulmonary vasoconstriction responds to vasodilatation. As a first step, $F\text{IO}_2$, $P\text{aO}_2$, and if possible $P\bar{\text{v}}\text{CO}_2$ should be increased to counteract hypoxic pulmonary vasoconstriction. Since acidosis (pH < 7.2) aggravates pulmonary vasoconstriction induced by hypoxia, correction of acidosis or even alkalinization by infusion of bicarbonate and/or hyperventilation may effectively lower pulmonary artery pressure. Intravenous vasodilators have also been shown to lower pulmonary artery pressure and pulmonary vascular resistance effectively in the presence of active pulmonary vasoconstriction (Table 1). Unfortunately, their use is limited in critically ill patients as systemic hypotension and/or

Table 1 Treatment of pulmonary hypertension by intravenous vasodilators

Drug	Dose for infusion*	$t_{½}$ (min)
Nitroprusside	0.5–0.8 μg/kg/min	3.5
Isoproterenol	0.01–0.5 μg/kg/min	3.8
Prostaglandin E_1	0.05–0.4 μg/kg/min	NA
Epoprostenol	1.0–20 ng/kg/min	3.0
Nitroglycerin	0.5–8.0 μg/kg/min	2.7

NA, no information available; $t_{½}$, plasma half-life.
*The dose should be titrated up or down to the most efficient value for lowering pulmonary artery pressure without major compromise of arterial pressure.

hypoxemia may occur. In contrast, inhaled vasodilators selectively dilate vessels within the lungs without causing hypotension or hypoxemia. Both inhaled nitric oxide (NO) and epoprostenol (prostaglandin I_2) aerosolization are used at present as investigational drugs in humans. They relax smooth muscle cells by stimulating the formation of intracellular cGMP and cAMP respectively. After inhalation they have been shown to reduce pulmonary artery pressure and pulmonary vascular resistance in patients with severe pulmonary endothelial injury (e.g. acute respiratory distress syndrome).

Experimental therapies

Experimental therapies aiming at reducing the sequelae of microthrombosis, the liberation of mediators, and/or ischemia–reperfusion injury include administration of plasminogen activators, pentoxifylline, antibodies to tumor necrosis factor or interleukin 1, cyclo-oxygenase inhibitors, surfactant, Ca^{2+} antagonists and inhibitors of neutrophil–endothelial interaction. Their clinical benefit remains to be established.

Bibliography

Hudson, L.D. (1995). New therapies for ARDS. *Chest*, **108** (Supplement), 79S–91S.

Leeman, M. (1991). The pulmonary circulation in acute lung injury: a review of some recent advances. *Intensive Care Medicine*, **17**, 254–60.

Marshall, B.E., Marshall, C., Frasch, F., and Hanson, C.W. (1994). Role of hypoxic pulmonary vasoconstriction in pulmonary gas exchange and blood flow distribution. *Intensive Care Medicine*, **20**, 291–7.

Pappert, D., Rossaint, R., Slama, K., Gruning, T., and Falke, K.J. (1994). Influence of positioning on ventilation–perfusion relationships in severe adult respiratory distress syndrome. *Chest*, **106**, 1511–16.

Zwissler, B., Welte, M., and Messmer, K. (1995). Inhaled vasodilatory drugs and gases. *Current Opinion in Anaesthesiology*, **8**, 557–64.

3.14.2 Pulmonary embolus

E. P. Wilson and R. A. White

Key messages

- Pulmonary embolism accounts for up to 150 000 deaths annually in the United States.
- Mortality is approximately 10 per cent in patients properly diagnosed and treated.
- Presenting features include dyspnea, chest pain, cough, and hemoptysis; with major emboli, there is cardiovascular collapse.
- Diagnostic modalities include 12-lead electrocardiography, chest radiography, arterial blood gas analysis, ventilation–perfusion scanning, CT, and, for definitive diagnosis, pulmonary angiography.
- Classification in terms of severity is useful for determining treatment strategies and prognostication.
- Treatment is supportive, with oxygen, fluid loading, and inotropes if necessary. Judicious use of heparinization, thrombolysis, and embolectomy, depending on disease severity, has been shown to reduce mortality and morbidity significantly.

Incidence

Pulmonary embolism is the most devastating complication of deep venous thrombosis, accounting for up to 150 000 deaths annually in the United States, with 285 000 to 600 000 cases recorded annually. Venous thrombosis is generally asymptomatic at the time of presentation with pulmonary embolism. Since death occurs within 30 min of embolization in two-thirds of fatal cases, clinical suspicion and efficient diagnosis and treatment are of paramount importance. In patients properly diagnosed and treated for pulmonary embolism, the mortality approaches 10 per cent.

Pathophysiology

The majority of lethal pulmonary emboli originate in the deep pelvic veins and the veins of the lower extremities, although a significant percentage embolize from the cavernous sinus, internal jugular vein, right heart, and deep venous system of the arm (5–15 per cent). Predisposing conditions include carcinomatosis, prostatic and pancreatic cancer, congestive heart failure, sepsis, obstructive pulmonary disease, and stroke.

Massive embolization to the main pulmonary arteries can lead to immediate cardiovascular collapse and death related to vasovagal shock, right heart failure, or acute pulmonary insufficiency. Multiple small emboli, more often than larger emboli, can cause diffuse bronchospasm and vasoconstriction through the release of systemic and locally active substances such as serotonin, bradykinin, and prostaglandins, although their role in death from pulmonary embolus has not been clearly shown. The occlusion of pulmonary artery branches leads to unperfused alveoli, resulting in a ventilation–perfusion mismatch. Ultimately, surfactant production is impaired, with edema and atelectasis leading to a congestive state that grossly resembles pulmonary infarction, although, because of the dual blood supply of the lung, true infarction occurs in less than 10 per cent of cases.

The clinical sequelae of pulmonary emboli as a consequence of this cascade include the following: hyperventilation, as evidenced by an increased minute ventilation, pulmonary restriction, indicated by a decreased vital capacity, bronchoconstriction, with its attendant decrease in forced expiratory volume per unit time (FEV_1), hypoxemia, and pulmonary hypertension.

Presentation

Patients presenting with dyspnea, chest pain, and/or hemoptysis in the presence of venous thrombosis should be evaluated for pulmonary embolus, although only a quarter will have all three symptoms. In a national co-operative study (Urokinase Pulmonary Embolism Trial 1973), dyspnea was found to occur in 81 per cent of patients, with chest pain in 72 per cent, cough in 53 per cent, and hemoptysis in 34 per cent. Associated findings of tachypnea (88 per cent) and tachycardia (30 per cent) were also noted, and additional symptoms, such as apprehension, can present in up to half of these patients (Table 1). In mechanically ventilated patients, an acute fall in oxygen saturation, elevation in mean pulmonary artery and right atrial pressures, or onset of otherwise unexplainable cardiovascular collapse should raise suspicions of pulmonary embolism (Table 2). Approximately one-third of patients with pulmonary embolism will also present with findings consistent with lower-extremity venous thrombosis. Other common physical findings include rales, accentuated P2, fever, S1 or S2 gallop, diaphoresis, cyanosis, and hypotension. Because of the non-specific nature of the signs and symptoms of pulmonary embolus, the differential diagnosis must also include myocardial infarction, pneumonia, pneumothorax, and atelectasis.

Table 1 Pulmonary embolism in spontaneously respiring patients

Symptoms	Signs
Dyspnea	Tachycardia
Chest pain	Tachypnea
Cough	Rales (crackles)
Apprehension	Pleural rub
Hemoptysis	Accentuated P2
	S1 or S2 gallop
	Diaphoresis
	Orthostasis/hypotension

Table 2 Pulmonary embolism in mechanically ventilated patients

Signs
Acute deterioration in oxygenation, hypercapnia, or acidosis
Acute rise in mean pulmonary artery pressure or central venous pressure
Unexplained tachypnea or tachycardia
Unexplained cardiovascular collapse

Diagnostic studies

Screening

Patients with unrecognized pulmonary emboli frequently suffer additional embolizations which may be fatal. Therefore efficient use of diagnostic modalities is warranted once pulmonary embolism is clinically suspected. Several easily obtainable tests should be performed early in the diagnostic work-up.

ECG findings in patients with pulmonary embolism include T-wave changes, T-wave inversion, right bundle branch block, arrhythmia, and P pulmonale. Although these findings are non-specific, they may help to alert to other conditions in the differential diagnosis, such as myocardial infarction.

Arterial blood gas analysis can suggest pulmonary embolism in the face of the above signs and symptoms. The characteristic findings of hypoxemia with a respiratory alkalosis is frequently seen, with a direct correlation between the size of the embolus and decreased oxygenation. Arterial oxygen ($Pa{O_2}$) is generally less than 80 mmHg (10.7 kPa) on room air; however, up to 12 per cent of patients with angiographically detected pulmonary embolism will have higher values. The alveolar–arterial (A–a) oxygen gradient is commonly elevated, suggesting a ventilation–perfusion discrepancy, and can easily be evaluated at the bedside using the following equation to determine expected normal values:

$$\text{A–a gradient} = 4 + (\text{age}/4).$$

Blood gas analysis may be less reliable in patients with chronic obstructive pulmonary disease.

Plain film radiographic studies are required in any patient presenting with the symptoms of pulmonary embolus. Recent posteroanterior chest films (within 6 h) are also required when further work-up determines that ventilation–perfusion scanning is necessary. The chest radiographic findings most commonly seen include pulmonary consolidation and hemidiaphragmatic elevation, with each being seen in 40 per cent of patients. As a screening device, this quickly obtainable study can suggest pneumonia, atelectasis, pleural effusion, or pneumothorax as alternative diagnoses. The most interesting finding in patients with pulmonary embolus may be a normal chest radiograph, which can be seen in nearly 40 per cent of cases.

Historically, several radiographic findings were believed to be specific for pulmonary embolus, although more recently these features have been determined to be neither sensitive nor specific. Hampton's hump (a round density at the pleural margin pointing towards the hilum) and Westermark's sign (pulmonary vascular congestion together with regional oligemia) have both been shown to be poor predictors of pulmonary embolus.

CT scanning can be a useful addition to the screening armamentarium. Contrast-enhanced CT scanning, particularly spiral CT with its fast acquisition times, which was initially found to discover mural thrombi in the pulmonary vasculature when scanning for other diagnoses, can delineate acute thrombi in second- to fourth-order pulmonary vessels. It can also discern oligemic lung parenchyma, a decrease in the number and caliber of vessels, and a decreased attenuation of the signal in this region. It is also sensitive for diagnosing chronic thromboembolic disease, as it shows both the central mural thrombi and occlusion of smaller arteries. Lung parencyhmal infarction can also be determined, although only 10 per cent of pulmonary emboli will actually result in infarction. A thorough knowledge of bronchovascular anatomy is necessary, as a diagnosis of pulmonary embolism may be misinterpreted by intersegmental lymph nodes or from partially opacified pulmonary veins and thrombosed arteries.

Confirmation

Armed with a strong clinical suspicion and screening tests which do not point to other obvious causes, the examiner must confirm the diagnosis of pulmonary embolus. At this time, if there are no contraindications, heparinization should be initiated to prevent further propagation of thrombus and worsening of the clinical condition. Once a baseline partial thromboplastin time is obtained, a heparin bolus of 100 to 150 units/kg is given, followed by continuous intravenous infusion of 10 units/kg/h. Provided that there are no identified active bleeding sites or disorders, heparinization is generally safe, even in the postoperative patient. A perfusion lung scan, or ventilation–perfusion scan, should then be obtained. This test is the most accurate non-invasive test available to make the diagnosis of pulmonary embolus (PIOPED Investigators 1990).

The perfusion lung scan determines regions of decreased pulmonary arterial perfusion, taking into consideration the fact that several pathological states, including pulmonary embolus, will produce this abnormality. An intravenous injectate of a radionuclide, such as technetium-99m, is delivered prior to nuclear imaging. Areas of capillary trapping of radioactive aggregates (albumin- or microsphere-bound radionuclide) are visualized and denote the distribution of pulmonary blood flow. This perfusion photoscan can be compared with images produced by the ventilation portion of the test. Xenon gas inhalation can be used to increase the sensitivity of a ventilation–perfusion scan, although, if the perfusion phase is performed first and is positive, a ventilation scan may be unnecessary. However, there are several instances where positive perfusion defects correlate with areas of radiographic consolidation or cases of subsegmental defects, when further confirmation by pulmonary angiography may be necessary.

In 1990, the Prospective Investigation of Pulmonary Embolism Diagnosis (PIOPED) results were published in an attempt to validate the role of perfusion lung scans in the diagnosis of pulmonary embolus. The study was a multicenter trial involving over 900 patients which compared ventilation–perfusion scan results with pulmonary angiography, long held as the gold standard in the diagnosis of pulmonary embolus. Scans were categorized as normal or as a high, intermediate, low, or very low probability of pulmonary embolism. When findings were correlated with pulmonary angiograms, perfusion lung scans were found to have an overall sensitivity of 98 per cent, although the specificity was only 10 per cent. There was also a 4 per cent false-negative rate. Patients with high-probability scans were found to have angiographic evidence of pulmonary embolus in 88 per cent of cases, while intermediate- and low-probability scans were positive for embolus in 33 per cent and 12 per cent of cases respectively. In this trial, clinical acumen was predictive in only 20 per cent of cases.

Table 3 Indications for pulmonary angiography

Inconclusive/negative $\dot{V}/Q$ scan with continued clinical suspicion
Inconclusive $\dot{V}/Q$ scan with contraindication to anticoagulation
Pre-existing pulmonary disease rendering $\dot{V}/Q$ scan useless
Planning thrombolytic therapy
Planning embolectomy
Recurrent pulmonary embolus
Rapid clinical deterioration

$\dot{V}/Q$ scan, ventilation–perfusion scan.

Non-invasive studies are frequently inconclusive in diagnosing pulmonary embolism (Table 3). Pulmonary angiography should be performed to confirm the diagnosis in cases where treatment may be urgently needed or anticoagulation is contraindicated. This invasive study carries a morbidity of 1 to 4 per cent and a mortality below 1 per cent. Findings on pulmonary angiography positive for pulmonary embolism include intravascular filling defects or complete vessel occlusion. Other less specific findings suggesting the diagnosis include delayed vessel filling, asymmetric filling, oligemia, or a prolonged arterial phase. Defects which do not correspond to perfusion scan findings may be artifacts on the angiogram. In patients with inconclusive non-invasive studies and in whom pulmonary angiography is deemed dangerous, as in patients with recent myocardial infarction or arrhythmias, the decision to treat empirically must be based on the clinical situation. Techniques such as digital subtraction and selective lobar or segmental injections may reduce the morbidity of the procedure.

Classification and treatment

The ability to classify pulmonary embolism in terms of severity is useful when determining treatment strategies and patient prognosis. The classification scheme in Table 4 utilizes symptomatology, blood gas derangements, and hemodynamic factors to place patients in one of five categories which correlate with a percentage of vascular occlusion. Class I patients are usually discovered on screening studies performed for other reasons. These patients have less than 20 per cent occlusion of the pulmonary arterial circulation and are asymptomatic. Patients with class II pulmonary embolism (20–30 per cent vascular occlusion) have the complex of findings associated with classic minor embolism: chest pain, hyperventilation, and anxiety, together with mild hemodynamic changes. Class III and IV ($>$ 30 per cent vascular occlusion) patients present with more significant hypoxia as well as hemodynamic instability and shock. In class V pulmonary embolism ($>$ 50 per cent vascular occlusion), shock and syncope are common features. Supportive measures are instituted based on clinical condition. Oxygen should be effectively delivered to provide a $Pa\text{O}_2$ above 60 mmHg (8 kPa). Pulmonary artery pressure should be monitored to help guide fluid resuscitation, particularly in patients in classes II to V where fluid is administered to help support hypotension and increase right ventricular stroke volume. Endpoints

Table 4 Classification of pulmonary embolism

Class	Symptoms	Arterial blood gas	Pulmonary vascular occlusion (%)	Hemodynamic monitoring
I	None	Normal	<20	Normal
II (minor)	Chest pain, anxiety, cough, hyperventilation	$Pa\text{O}_2$ 80 mmHg (10.7 kPa) $Pa\text{CO}_2$ <35 mmHg (4.7 kPa)	20–30	Tachycardia
III (major)	Dyspnea, orthostasis	$Pa\text{O}_2$ <65 mmHg (8.7 kPa) $Pa\text{CO}_2$ <30 mmHg (4 kPa)	30–50	CVP, mean PAP >20 mmHg
IV (massive)	Dyspnea, shock	$Pa\text{O}_2$ <50 mmHg (6.7 kPa) $Pa\text{CO}_2$ <30 mmHg (4 kPa)	>50	CVP, mean PAP >25 mmHg, BP <100 mmHg
V(chronic)	Syncope, dyspnea	$Pa\text{O}_2$ <50 mmHg (6.7 kPa)	>50	CVP, mean PAP >40 mmHg, BP <100 mmHg, CO low

BP, blood pressure; CVP, central venous pressure; mean PAP, mean pulmonary artery pressure; CO, cardiac output.
Adapted with permission from Greenfield (1993).

of therapy are an increase in overall cardiac output and urine production, an improvement in stroke volume, and stabilization of blood pressure. With continued cardiovascular compromise, the short-term use of norepinephrine (noradrenaline) has beneficial inotropic and pressor effects. However, the use of epinephrine (adrenaline), isoproterenol, dopamine, and dobutamine should all be considered and utilized as each individual clinical situation warrants. Hydralazine may also reduce pulmonary vascular resistance, decreasing right ventricular afterload.

Heparin therapy is instituted by administering a loading dose of 100 to 150 units/kg intravenously, followed by continuous infusion of 10 units/kg/h, and monitoring the partial thromboplastin time with a goal of 1.5 to 2 times control values. This management, together with supplemental oxygen therapy, may be all that is necessary for patients with class I and II pulmonary embolism. The heparin should be continued for 7 to 10 days to prevent propagation of clot, while the thrombus becomes adherent to the wall of the vessel during that time. The patient should be given oral warfarin (coumarin) while on heparin to allow several days of overlap while the prothrombin time becomes therapeutic (1.5 to 2 times control). This regimen prevents further embolization in over 90 per cent of these patients. Duration of therapy is controversial; however, at least 6 months of anticoagulation is recommended.

In patients with class III and IV pulmonary embolism, thrombolytic therapy should be initiated in an attempt to overcome hemodynamic compromise and improve clinical findings more rapidly than with heparinization. However, thrombolytic therapy has not been shown to improve overall mortality (National Heart, Lung, and Blood Institute 1974). The three agents most commonly used are streptokinase, urokinase, and tissue plasminogen activator. Streptokinase is effective systemically at a loading dose of 250 000 units, followed by infusion of 100 000 units/h intravenously for 12 to 24 h. Systemic administration of urokinase begins with a loading dose of 300 000 units and is followed by continuous infusion for 12 to 24 h at 300 000 units/h. Tissue plasminogen activator should be used at a dose of 0.5 mg/kg administered over 4 h. Regional (directed) infusion of these agents, particularly urokinase, can achieve faster clearance of thrombi (Goldhaber *et al.* 1987).

Contraindications to the use of thrombolytics include surgery within the previous 7 days, gastrointestinal bleeding in the previous 3 months, history of bleeding diathesis, hypertensive related cerebrovascular accident, or active intracranial process. If initiation of thrombolytic treatment is followed by a bleeding episode, the agent should be discontinued and 5 mg of aminocaproic acid administered to reverse the lytic state.

The short half-life of these drugs (streptokinase, 80 min; urokinase, 15 min; tissue plasminogen activator, 4 min) may be a consideration in their use. Heparin should be instituted following thrombolytic therapy as outlined previously.

Surgical management

Surgical thromboembolectomy must be considered in compromised patients (classes III, IV, and V) in whom there is a poor response to thrombolytic therapy or the clinical deterioration is rapid (Robison *et al.* 1986). The procedure is performed through a median sternotomy, with the patient on full cardiopulmonary bypass. An incision is made over the main pulmonary artery, with the thrombus being retrieved using forceps and a balloon thromboembolectomy catheter. This procedure carries a mortality approaching 50 per cent, with complications that include massive endobronchial hemorrhage and reperfusion pulmonary edema.

Suction catheter embolectomy has been advocated for acute embolic events (< 72 h). The catheter can be manipulated from a remote site, such as the femoral or jugular vein, and controlled using fluoroscopy. Repeated passes of the suction device are necessary for retrieval, with the fall of pulmonary artery pressure as an indicator of effectiveness.

Prophylaxis

Prevention of further embolization can best be accomplished with adequate anticoagulation with heparin. In patients with contraindications to anticoagulation or with recurrent pulmonary emboli while maintaining a therapeutic partial thromboplastin time, some form of vena caval interruption, with either the placement of an intraluminal filter or an extraluminally applied serrated clip, should be undertaken. Filter devices such as the Greenfield filter have rates of preventing significant recurrent emboli of 95 per cent and long-term patency of 97 per cent. Emboli as small as 3 mm can be successfully trapped. These devices can be placed from remote access sites via percutaneous techniques under local anesthesia with a low morbidity and mortality.

Conclusions

The diagnosis and treatment of pulmonary embolism, with its attendant 30 per cent mortality in untreated patients, is predicated on the physician's suspicion, although a paucity of signs and symptoms may be present. An algorithmic approach utilizing non-invasive screening tests and invasive pulmonary angiography when appropriate, together with the judicious use of supportive care, anticoagulation, thrombolytic therapy, and surgical thromboembolectomy, has been shown to produce significant reductions in morbidity and mortality. When indicated, selective use of caval filters is also effective prophylaxis for recurrent embolization. Future diagnostic studies may include CT in chronic embolizations and single-photon-emission CT (SPECT) scanning, although their role in supplanting angiography has not been fully elucidated.

References

Goldhaber, S.Z., *et al.* (1987). Thrombolytic therapy in acute pulmonary embolism: current status and future potential. *Journal of the American College of Cardiology*, **10** (Supplement B), 96B–104B.

Greenfield, L.J. (1993). *Surgery*, p.1773. J.B. Lippincott, Philadelphia, PA.

National Heart, Lung, and Blood Institute (1974). Urokinase–streptokinase embolism trial: phase 2 results. A cooperative study. *Journal of the American Medical Association*, **229**, 1606–13.

PIOPED Investigators (1990). Value of ventilation–perfusion scan in acute pulmonary embolism. Results of the Prospective Investigation of Pulmonary Embolism Diagnosis (PIOPED). *Journal of the American Medical Association*, **263**, 2753–9.

Robison, R.J., Fehrenbacher, J., Brown, J.W., Madura, J.A., and King, H. (1986). Emergent pulmonary embolectomy: the treatment for massive pulmonary embolism. *Annals of Thoracic Surgery*, **42**, 52–5.

Urokinase Pulmonary Embolism Trial (1973). A national cooperative study. *Circulation*, **47** (Supplement 2), 1–108.

4 The gastrointestinal and hepatic systems

4.1 Normal physiology

4.1.1 The gastrointestinal system

U. Haglund

Key messages

- The gastrointestinal tract secretes 6 to 9 liters of fluid every 24 h.
- Peristalsis is dependent on intact innervation.
- The splanchnic area receives 25 per cent of cardiac output.
- The small intestine regains postoperative motor function within hours.
- Inhibition of gastric acid secretion may cause bacterial overgrowth.
- Mucosal ischemia is an important pathogenetic factor for stress ulceration.
- Mucosal blood flow is less affected than that of the muscularis layer in critical illness.
- During critical illness villous blood flow may be maintained but oxygenation becomes impaired owing to increased demand and decreased utilization and extraction.
- The villous countercurrent exchanger causes a reduced villous oxygen delivery in hypotensive states.

Introduction

The function of the gastrointestinal tract is to serve as a reservoir for ingested food and fluid, to transport and digest ingested material so that it can be absorbed, and to transport non-digestible and non-absorbable material so that it can be removed from the body. The physiology of the gastrointestinal tract is well adapted to these functional needs, and the operation of this system includes co-ordination of several mechanisms such as motility, secretion, active and passive absorption, and blood flow. These separate mechanisms, together with very special structural arrangements, make the gut well adapted to absorptive needs. In this short overview we concentrate on those aspects of gastrointestinal physiology that are likely to have an impact on the care of the critically ill patient rather than attempting to cover all its many special features.

Motility

The gastrointestinal tract has an active motor function with spontaneous pacesetting activities in all areas. The spontaneous activity is influenced by nervous and humoral activities, which are both stimulating and inhibiting. The esophagus exhibits both peristaltic and non-peristaltic movements. Orderly peristalsis is dependent on intact innervation. With regard to motor function, the stomach can be divided into a proximal and a distal part. The former acts as a reservoir and the fundus has an active receptive relaxation which depends on intact vagal innervation. Gastric emptying depends on peristaltic movements which involve the entire stomach and bring small portions of the antral content through the relaxed pylorus. This function also depends on intact innervation. Retropulsion, caused by closure of the pylorus at the end of the antral contraction, generates mixing of the gastric contents. Increased intragastric volume increases the rate of gastric emptying. Gastric motility is stimulated by the amount and consistency of the intraluminal content. Fatty acids as well as acid in the duodenum are potent inhibitors of gastric emptying. Solids are retained by the antrum, and in normal states the majority of particles emptied through the pylorus have dimensions of less than 1 mm. After a meal fluids are emptied immediately while solids are retained in the fundus of the stomach to be reduced.

The small intestine has a basal electrical 'rhythm' (slow waves). The frequency of these slow waves is highest in the proximal part of the duodenum. These slow waves are the pacesetters, and some give rise to action potentials. Slow waves are propagated along the entire gut, and the mechanical activity is also propagated in this manner. During certain periods every slow wave gives rise to an action potential, and such complexes migrate along the entire small intestine. The electrical action also has a mechanical consequence. This activity has a periodicity of 60 to 90 cycles/min, so that when one cycle ends in the ileum a new one starts in the duodenum.

Colonic motility provides the prerequisite for a steady slow flow of the bowel content. This slow flow is interrupted periodically by mass contractions causing an urge to defecate.

Critical illness, intra-abdominal surgery, and intraperitoneal abscess usually cause paralytic ileus. This condition typically affects all levels of the gastrointestinal tract through one or more of the following mechanisms: increased activity of intrinsic or extrinsic inhibiting neurons, decreased activity of intrinsic or extrinsic stimulating neurons, or increased levels of inhibitory hormones. In peritonitis a complete cessation of motor function is likely; drainage and/or removal of the causative factor has to be performed before motor activity will recommence. Local paralysis affecting an intestinal loop close to a regional inflammatory process constitutes another possibility, for example duodenal paralysis in a patient with acute pancreatitis.

The influence of critical illness and intra-abdominal surgery on motility varies between levels of the gastrointestinal tract. The stomach, and particularly the colon, are characteristically slow in regaining adequate motor function, while the small intestine will usually have returned to adequate motor activity within hours. This is the background to early enteral feeding in critically ill and post-operative patients.

Secretion

The gastrointestinal tract with its associated glands normally secretes 6 to 9 liters of fluid every 24 h. Most of this volume is secreted by or delivered to the small intestine (e.g. by the pancreatic juice and with the bile), and most of it is reabsorbed there. In a healthy individual, slightly less than 1 l/day passes from the distal ileum into the colon and less than 0.5 liters leaves the body as stool. Increased secretion (as in cholera) and/or decreased fluid absorption in the small intestine (as in mechanical ileus) may rapidly cause significant fluid shifts. This may not necessarily be detected as increased losses or diarrhea since significant amounts could also be lost as stagnant fluid in the distended gut. Such a situation is always present to some extent during obstruction of the distal small intestine and/or colon.

The secreted fluid is not just important in terms of volume. The content of electrolytes or protein in specific regions, including gastric juice, bile, and the highly protein-rich pancreatic juice, warrants special mention.

Gastric acid and pepsin secretion

Gastric juice contains a very high concentration of hydrochloric acid (pH 1) and pepsin. The physiological role of gastric acid is probably to facilitate digestion by reducing the size of certain food particles and also to prevent gastric colonization by micro-organisms. Acid secretion takes place in the parietal cells located in the deep part of the gastric glands in the body and the fundus of the stomach. Pepsin is secreted by the chief cells located in the same area. Acid secretion is regulated by nervous (vagal) and humoral (gastrin) mechanisms (Olbe and Haglund 1986). Pepsin secretion is less well studied than acid secretion. However, both seem to be regulated by very similar mechanisms. Gastrin is produced by the G cells in the gastric antrum. There are gastrins of different sizes: G17 (containing 17 amino acids) is dominant in the gastric mucosa while G34 predominates in serum. Gastrin secretion is reduced when the antral mucosa becomes acidic and is stimulated by a neutral antral environment. The expectation and swallowing of food and the presence of food in the stomach constitute the strongest biological stimulating mechanisms for acid and pepsin secretion. There is a very complex interaction between stimulating and inhibiting nervous and humoral factors regulating acid secretion.

Peptic ulcer disease

Peptic ulcer disease is strongly associated with acid and pepsin secretion. However, ulcers are not the simple result of acid hypersecretion, but rather are due to an imbalance between aggressive factors (mainly acid and pepsin) and the mucosal defense capacity (Olbe and Haglund 1986). Patients with duodenal ulcer generally tend to secrete more acid than normal subjects, while patients with gastric ulcers tend to have a low acid secretion. Thus the mucosal defense mechanisms of gastric ulcer patients are likely to be impaired. Intramucosal infection by *Helicobacter pylorii* causes chronic gastritis; this seems to be an important causative factor, particularly in duodenal ulcer disease.

The duodenum receives the (often) acidic contents of the stomach and regulates the speed of gastric emptying by humoral mechanisms. Histamine appears to have a central role in the stimulation of acid secretion, since histamine-2 receptor antagonists may effectively block acid secretion regardless of the mode of stimulation (histamine, gastrin, or vagal nerve stimulation). In addition to histamine, the parietal cells also have receptors for gastrin and acetylcholine. The final step in acid secretion, which takes place against a very significant concentration difference, is regulated by the proton pump H^+,K^+-ATPase. Specific acid inhibitors directed against the proton pump are available. The critically ill patient is likely to secrete only small amounts of acid, although the acidity of the secreted volume may well be within the normal range.

The gastric mucosa has a unique capacity to defend itself against acid. Several factors co-operate, such as the thin lining of mucus on the surface epithelium, the secretion of bicarbonate by the superficial epithelial cells, a high mucosal blood flow, and the high content of bicarbonate in the blood of the gastric superficial mucosa (as a consequence of maintaining electroneutrality within the parietal cells during acid secretion, i.e. the alkaline tide). With the exception of the alkaline tide, most known defensive mechanisms depend on endogenous production of prostaglandin E_2 by the gastric mucosa.

The content of the stomach is normally effectively sterile. However, bacterial overgrowth is likely to occur following even short periods of inhibition of gastric secretion and/or gastric emptying. This may cause aspiration of contaminated gastric juice resulting in pneumonia, which is a common complication in critically ill patients. The simultaneously impaired motor function seen in the critically ill increases the risk of regurgitation of gastric contents. Acid inhibition has been shown to increase the risk of nosocomial pneumonia in the intensive care setting.

Acute mucosal ulceration

Peptic ulcers may recur during critical illness. However, other forms of acute ulceration of the gastroduodenal mucosa are more commonly encountered in critically ill patients without actual association with peptic ulcer disease. This mucosal injury is often referred to as a stress ulcer. However, the pathological anatomical definition of this lesion is an erosion (a superficial lesion not penetrating the mucosa). In contrast, an ulcer by definition penetrates at least down to the muscular layer.

Stress ulceration may be found in critically ill patients, usually as multiple superficial erosions on top of the gastric ridges in the acid-producing part of the stomach. The clinical manifestation of stress ulceration is acute gastrointestinal hemorrhage. This is less common today than 10 to 20 years ago, but an incidence of 1 to 3 per cent in critically ill patients has recently been reported. Bleeding is less likely to occur if the pH of the gastric contents can be kept normal or close to normal. Endoscopic examination of the stomach within the first 24 h of severe injury has revealed that almost all patients develop mucosal injury. Several risk factors for hemorrhage have been identified, including sepsis, shock, acidosis, and other coexisting components of the multiple organ failure syndrome. Impaired mucosal oxygenation has been demonstrated to be an important pathophysiological factor for the development of bleeding stress ulcers. The most important preventative factor is recognized to be optimal intensive care, providing adequate oxygenation of the peripheral tissues.

A special form of acute ulceration is the Curling ulcer seen in major burn injury. Burn patients often have single ulcers which may penetrate quite deep into the gastric or upper duodenal wall and, in contrast with stress ulcers, may even cause perforation. Cushing ulcers, which are seen in patients with neurological injury, fall between the two other acute ulcer forms. For completeness, a fourth type of acute mucosal ulceration caused by drugs, mainly non-steroidal anti-inflammatory drugs (NSAIDs), should be noted. There are often multiple erosions combined with deeper ulcers. Characteristically, these acute ulcerations are most often found in the gastric mucosa while peptic ulcers are most often seen in the duodenum. The pathogenesis of NSAID-induced ulcers is complex but is probably related to inhibition of prostaglandin synthesis by a non-selective inhibition of the enzyme cyclo-oxygenase.

Blood flow

The splanchnic area receives approximately 25 per cent of cardiac output. The blood supply is rich, not to satisfy the nutritional needs of the gastrointestinal tract but to allow transportation of absorbed nutrients. Therefore the oxygen content of the portal vein is normally quite high. Blood flow to the gastrointestinal tract can be characterized as several circuits coupled in parallel supplying the different parts of the wall. Each circuit contains sections coupled in series — the resistance vessels, precapillary sphincters, capillaries, postcapillary sphincters, and capacitance vessels. The last mentioned vessels contain the bulk of the regional blood volume, while the ratio of pre- to postcapillary sphincters regulates the mean hydraulic capillary pressure and hence the direction and amount of fluid movement across the huge gastrointestinal capillary surface area. The number of capillaries open to flow varies continuously and is determined by the activity of the precapillary sphincters. For example, intestinal capillaries have a much higher permeability than skeletal muscle capillaries. Of the different sections coupled in parallel, the mucosal sections are the most interesting from a functional and structural point of view.

Gastrointestinal blood flow is regulated by central nervous and humoral mechanisms as well as by local mechanisms often referred to as 'local metabolites' (Lundgren 1989). Stimulation of the sympathetic nerves to the gut causes a rapid reduction of local blood volume and an initial sharp reduction of blood flow which, after 1 to 2 min, starts to increase despite continuous nerve stimulation. Blood flow then levels off at an only slightly reduced value. This phenomenon is characteristic of the splanchnic vasculature and is referred to as local escape from central sympathetic vasoconstriction influence.

Sympathetic nerve stimulation does not cause any change of regional mean capillary pressure in contrast with the situation in skeletal muscle. Local control (mainly vasodilating in action) often takes precedence over central control (mainly vasoconstricting) in situations when the regional needs for oxygen or removal of metabolites are threatened. During cardiac tamponade and hemorrhage the reduction in gastric and intestinal blood flows tends to be significantly greater than the fall in cardiac output. The renin–angiotensin axis seems to play an important role in this disproportionate gastrointestinal vasoconstriction while the sympathetic nervous systems do not. In sepsis, however, changes in gastrointestinal blood flow tend to follow changes in cardiac output.

The vascular supply of the villous region of the small intestine has a very special arrangement (Lundgren 1989). The centrally located artery supplying each villus does not branch until it reaches the tip, and the blood flow from the tip is directed through a subepithelial network of capillaries and venules. The distance between the two sets of vessels is less than 20 μm, allowing diffusion for easily diffusible substances, i.e. a counter-current exchanger. Although this system probably developed to prevent rapid changes in the water volume of the body, it has significant functional consequences such as hindering blood transportation of absorbed lipid-soluble substances. Therefore large fatty acids are transported by the lymph. It also creates a mechanism for short-circuiting oxygen, causing low oxygen pressure at the tips of the villi. At low blood flow velocities (e.g. generalized mucosal vasodilation caused by reduced arterial blood pressure), more time is available to create a diffusion equilibrium between the contents of the ascending and descending limbs of the villous vascular loop. This situation may cause an almost anoxic situation at the tips of the villi despite fairly well maintained blood flow volumes.

Mucosal injury is not only caused by hypoxia but also by reperfusion. Indeed, reperfusion injury as a concept was first demonstrated in the small intestine. Reperfusion causes increased generation of oxygen radicals, initially mainly by the hypoxanthine system. Endothelial injury caused by these oxygen radicals activates polymorphonuclear leukocytes, exacerbating tissue injury through microcirculatory disturbances.

The gastrointestinal tract may compensate for reduced blood flow delivery by increased oxygen extraction. This can be effectively achieved down to a blood flow of about 50 per cent of normal (Bulkley *et al.* 1986). However, this takes place at the expense of liver oxygen delivery; the liver becomes hypoxic very rapidly if the hepatic arterial supply cannot compensate for the reduced portal oxygen delivery. Because of this compensatory mechanism, oxygen consumption of the gut remains normal over a wide range of blood flow. In sepsis, however, oxygen comsumption is increased by 50 to 70 per cent. Thus normal splanchnic blood flow does not guarantee adequate intestinal mucosal oxygenation (Antonsson *et al.* 1995). At the same time the small intestinal mucosa is quite susceptible to hypoxic injury, ranging from increased mucosal permeability to loss of the villous layer (Haglund 1993). As a consequence, the barrier function between intraluminal contents, which may be infected to some extent, and the supposedly sterile interior is lost. Small intestinal mucosal injury may also cause release of toxic materials such as cardiotoxic factors, aggravating the shock state.

Mechanical small intestinal obstruction causes increased intraluminal pressure and intestinal distension. Severe and prolonged intestinal obstruction increases the intraluminal pressure and causes reduction of intramural blood flow and a significant reduction in the number of perfused capillaries. This is particularly obvious during intestinal contractions when intraluminal pressure can be extremely high. Absorption of intraluminal fluid in the small intestine, which is of the order of 5 l/day or more, becomes significantly impaired by increased intraluminal pressures; intestinal absorption may eventually be converted to net secretion. Thus obstruction of the small intestine constitutes a significant threat to homeostasis.

Absorption

Most nutrients are normally digested, enzymatically split into small basic components, and absorbed in the proximal part of the small intestine. Absorption may be passive or active. The villous structure of the small intestine significantly increases the absorptive area. This area is further increased by the microvilli at the luminal end of the enterocytes. The enterocytes are formed by mitotic active cells at the bottom of the crypts and then move up to the tip of the villus before they are lost. The lifespan of a small intestinal enterocyte is 2 to 3 days.

In addition to adequate intestinal blood flow, the absorptive process depends on reasonably well maintained secretion of bile and pancreatic juice, as well as on secretion of fluids and enzymes by the gastrointestinal tract mucosa itself. The mucosa produces enzymes (e.g. proteases and disaccharidases) necessary for the final breakdown of proteins and carbohydrates to small peptides, amino acids, and monosaccharides. Long-chain fatty acids, bile acids, and vitamin B_{12} are examples of substances absorbed mainly or exclusively in the distal part of the small intestine (the distal ileum), while most molecules are absorbed in the jejunum. The function of the proximal part of the small intestine (jejunum) can easily be taken over by the distal part, but the reverse does not hold true. Absorption of fluid and basic nutrients, such as those used in various nutritional formulas, are maintained during critical illness and recover rapidly following intra-abdominal operations. Pathologically prolonged postoperative paralysis, which might be caused by an intraperitoneal postoperative abscess, may impair the absorptive capacity of the small intestine; more clinically evident is the prolonged inhibition of the motor function of the stomach and colon in these situations.

Immune function

The small intestine and its mesentery contain a very high proportion of the body's lymphocytes and also a very large number of macrophages. Normal function of these cells is essential for the defense mechanisms against intraluminal bacteria and bacterial products as well as other non-self agents that could be introduced by the alimentary route. The intestinal T lymphocytes are important antibody producers, while the macrophages are essential for initiating and regulating defense reactions in different pathological conditions through their secretion of cytokines, mainly tumor necrosis factor and interleukins such as IL-1, IL-2, IL-6, etc. Activation of macrophages to secrete cytokines may initiate all aspects of an inflammatory response reaction. Intestinal ischemia may, on the one hand, activate this high number of intestinal macrophages, but on the other hand hypoxia is likely to impair the immune cell function of the gut. However, the detailed mechanisms of activation, the different chain reactions involved, and how the control of the immune function is exerted are not yet fully understood for normal function. Similarly, the influence of ischemia on the normal immune repsonse is also unknown.

References

Antonsson, J.B., Engström, L., Rasmussen, I., Wollert, S., and Haglund, U.H. (1995). Changes in gut intramucosal pH and gut oxygen extraction ratio in a porcine model of peritonitis and hemorrhage. *Critical Care Medicine*, **23**, 1872–81.

Bulkley, G.B., Kvietys, P.R., Parks, D.A., Perry, M.A., and Granger, D.N. (1986). Relationship of blood flow and oxygen consumption to ischemic injury in the canine small intestine. *Gastroenterology*, **89**, 852–7.

Haglund, U. (1993). Hypoxic damage. In *Pathophysiology of shock, sepsis, and organ failure* (ed. G. Schlag and H. Redl), pp. 314–19. Springer-Verlag, Berlin.

Lundgren, O. (1989). Physiology of the intestinal circulation. In *Splanchnic ischemia and multiple organ failure* (ed. A. Marston, G.B. Bulkley, R.G. Fiddian-Green, and U.H. Haglund). Edward Arnold, London.

Olbe, L. and Haglund, U. (1986). Gastroduodenal physiology and pathophysiology. In *Surgery of the stomach and duodenum* (4th edn) (ed. L.M. Nyhus and C. Wastell), pp. 89–126. Little, Brown, Boston, MA.

4.1.2 The hepatic system

S. Beloucif and D. M. Payen

Key messages

- There are two inflow circuits: high-pressure high-resistance hepatic artery blood mixes with low-pressure low-resistance portal vein blood to form sinusoidal capillaries.
- The portal vein provides two-thirds of total hepatic blood flow and the hepatic artery provides the other third. However, because of differences between the arterial and venous oxygen contents, two-thirds of the hepatic oxygen supply is derived from the hepatic arterial circulation.
- The liver is the main site of protein synthesis, including albumin and coagulation factors. It is essential for carbohydrate metabolism by increasing glycogen synthesis when the serum glucose concentration is high, and degrading glycogen or converting amino acids to glucose by gluconeogenesis when the serum glucose concentration is low. The liver is also responsible for both chylomicron degradation and lipoprotein synthesis.
- An intrahepatic regulation of hepatic protein synthesis occurs during inflammation, with a shift towards enhanced production of acute phase proteins such as C-reactive protein, serum amyloid A, lipopolysaccharide-binding protein, and complement components.
- The liver protects the organism from the deleterious accumulation of useless metabolites and various toxic compounds of both endogenous and exogenous origin.

Introduction

A knowledge of the normal physiology of the hepatic system is relevant to the understanding of numerous pathophysiological conditions, including the development of multiple organ failure states, splanchnic hypoperfusion, and metabolism of endogenous substances and drugs. The particular arrangement of the liver's circulation between two inflow circuits (hepatic arterial and portal venous circulations) complicates the exploration of this regional circulation.

Anatomical considerations

High-pressure high-resistance hepatic artery blood flow mixes with low-pressure low-resistance portal vein blood to form sinusoid capillaries. These vessels are characterized by a thin fenestrated endothelium that allows easy transfer of blood constituents through the space of Disse and communication with hepatocyte and Kupffer cells. Within the liver, the triangular portal space is formed by ramifications of the portal vein, the hepatic artery, and the bile ducts. Between the portal spaces, centrilobular veins progressively mix to form the hepatic veins.

Several histological descriptions of this arrangement have been proposed:

(1) the hepatic lobule, with a centripetal circulation towards the centrilobular vein, and a centrifugal circulation of the bile ducts;

(2) the portal lobule, or portion of the hepatic parenchyma perfused by a portal space (with a centrifugal circulation and a centripetal biliary excretion);

(3) the hepatic acinus.

The acinus represents a portion of the hepatic parenchyma between two centrilobular veins and is the functional unit of the liver, centered by a terminal portal venule, a terminal arteriole, a bile duct, lymphatic vessels, and nerves. Blood flows through the capillary sinusoids and is drained at the periphery of the acinus by terminal hepatic venules. The sinusoids are perfused by a mixture of hepatic arterial and portal venous blood joining at capillary level. Three different zones can be distinguished according to the perfusion of the liver: a well-perfused central zone close to the smaller vascular ramifications of the portal space, an intermediate zone, and a less perfused area close to the centrilobular veins.

Because of the dual vascularization within the liver originating from hepatic artery and portal vein, shunts and sphincters have been described. Vascular shunts exist between the hepatic artery, the portal vein or sinusoids, and the hepatic veins. The existence of sphincters at hepatic arteriolar and venous presinusoidal levels has also been suggested (Richardson 1982). Although specific muscular formation has not been formally identified, it has been suggested that arterial and venous presinusoidal 'sphincters' contract independently and regularly. This may be due to a myogenic phenomenon. Therefore a given sinusoid capillary can be vascularized by blood with different oxygen contents. Considering global flows, the portal venous flow Q_{pv}, the outflow of the mesenteric circulation, is roughly 25 to 33 per cent of cardiac output, with Q_{pv} representing two-thirds of total hepatic blood flow Q_{hep} and hepatic arterial flow Q_{ha} representing the other third. However, because of the differences in arterial and venous oxygen contents, two-thirds of the hepatic oxygen supply depends on the hepatic arterial circulation.

Monitoring hepatic blood flow in humans is difficult; this is in part due to its dual circulation. Doppler techniques are promising, but are limited to patients undergoing surgery. In contrast, monitoring of hepatic venous oxygen saturation is relatively easily performed using a fiber-optic pulmonary artery catheter inserted into the hepatic vein, and is useful clinically because of the linear relationship between hepatic oxygen delivery and hepatic vein $P\text{O}_2$. In the intensive care unit, this monitoring can help diagnose hepatic hypoperfusion and assess the response to therapies aimed at increasing liver blood flow.

Hepatic pressure–flow relationships

Liver perfusion must be maintained in a wide variety of clinical conditions characterized by either low downstream right atrial pressures (e.g. during hypovolemic shock) or high downstream pressures (e.g. pulmonary hypertension). The existence of hepatic vascular waterfalls then represents a protective mechanism for the liver against high increases in central venous and hepatic venous pressures. By using an extracorporeal bypass from the right atrium to the pulmonary artery, flow can be controlled while right atrial pressure P_{ra} is varied through a Starling resistor. Inflating the Starling resistor increases P_{ra} and decreases flow, but P_{pv} begins to increase only after P_{ra} reaches a certain level, demonstrating the existence of a vascular waterfall phenomenon downstream from the portal vein.

By using an *in situ* isolated liver bypass preparation with independent control of Q_{pv} and Q_{ha}, during which hepatic venous pressure P_{hv} could be maintained constant using a Starling resistor, vascular waterfalls have been demonstrated in both the portal venous and hepatic arterial circulations (Beloucif *et al.* 1995). In both circuits, the analysis of pressure–flow relationships was performed at incremental levels of P_{hv}. In the portal venous circulation, the critical closing pressure was about 4 mmHg, with a classical waterfall physiology observed as P_{hv} was raised. In the hepatic arterial circuit, the critical closing pressure (about 8 mmHg) showed a constant positive pressure difference of 5 mmHg over P_{hv} as the latter was increased. Analysis of the slopes of the pressure–flow relationships showed that portal venous resistance decreased when P_{hv} was greater than the portal venous critical closing pressure, consistent with recruitment/distention phenomena in this circuit. In contrast, hepatic arterial resistance was unchanged as P_{hv} was increased (Brienza *et al.* 1995).

During mechanical ventilation, particularly when positive end-expiratory pressure is applied, liver blood flow can be reduced by a decrease in venous return/cardiac output, an increase in transhepatic vascular resistance due to diaphragmatic descent, a transmission to the sinusoids of the increase in P_{ra}/P_{hv}, or an increase in abdominal pressure acting on intrahepatic closing pressures. Analysis of the pressure–flow relationships in the portal vein during mechanical ventilation demonstrated the existence of a passive distensibility with classic critical closing pressure, whereas the hepatic arterial critical closing pressure always remained higher than the outflow pressure (Brienza *et al.* 1995).

Regulation of the hepatic circulation

Portal venous flow

Q_{hep} and its distribution between Q_{ha} and Q_{pv} is determined by the hepatic arterial resistance, the intra- and prehepatic portal resistance, and Q_{pv}. Q_{pv} is not intrinsically regulated *per se*. The portal venous resistance R_{pv} is low since about 25 per cent of cardiac output passes through a pressure gradient of 4 to 5 mmHg between the portal and hepatic veins. This low resistance is also about a tenth of the resistance through the splanchnic organs. Q_{pv} depends on the outflow of the extrahepatic splanchnic organs and not of the liver itself (Richardson 1982). Finally, Q_{pv} is not autoregulated (i.e. the relationship between pressure and flow is linear, and reactive hyperemia is absent). The principal intrinsically regulated variable is portal venous pressure (Richardson 1982), with R_{pv} modifying flow to maintain a relatively constant P_{pv}. P_{pv} depends primarily on the degree of constriction or dilatation of the mesenteric and splanchnic arterioles and on intrahepatic resistance. During low Q_{pv} states, high R_{pv} tends to maintain P_{pv} relatively constant whereas, in the presence of a high Q_{pv}, a decrease in R_{pv} would prevent the deleterious effect of an increase in P_{pv} on mesenteric venous drainage.

Hepatic arterial resistance and extrinsic control of Q_{hep}

Hepatic arterial resistance R_{ha} is high, with a large pressure drop from aorta to hepatic vein, and its changes will affect Q_{ha}. R_{ha} can be modified by hormones and nerves (extrinsic factors). The composition of portal blood and the gastrointestinal hormones seem to be the most important determinants of the extrinsic regulation of Q_{hep}. During postprandial hyperosmolarity, Q_{hep} is increased and a hepatic arterial vasodilation is observed with secretion of secretin, pentagastrin, cholecystokinin–pancreozymin, and vasoactive intestinal peptide. Glucagon also increases Q_{hep}, mainly by increasing Q_{pv} secondary to an intestinal vasodilatation.

Intrinsic control of Q_{ha}: autoregulation and the hepatic arterial buffer response

Q_{ha} can also be modified by two specific intrinsic controls: autoregulation of Q_{ha} and the hepatic arterial buffer response. The autoregulation of a given vascular bed is defined by its ability to modify its vasoactive status in response to changes in perfusion pressure so as to maintain blood flow relatively constant within given pressure limits. Beyond these limits, flow becomes linearly related to pressure. Such pressure–flow autoregulation has been described in the hepatic arterial bed, although it also depends on hepatic metabolic status to a relatively small extent. This intrinsic factor of hepatic artery circulation persists after α- or β-blockade, but it is influenced by hepatic metabolic status with a higher gain in fed compared with fasted animals.

Hepatic arterial buffer response is a unique intrinsic regulatory mechanism which serves to maintain total Q_{hep}, and therefore hepatic clearance of hormones and drugs, as steady as possible. It is dependent on a local adenosine washout mechanism. When Q_{pv} is reduced, less adenosine is washed away; the increased concentration of adenosine dilates the hepatic artery with an increase in flow. In contrast, a high Q_{pv} prevents accumulation of adenosine, thus reducing Q_{ha}. This phenomenon has been well characterized by Lautt (1985), who initially constructed a cumulative dose–response curve and demonstrated that increasing doses of intraportal adenosine were accompanied by an increase in Q_{ha}. After the adenosine infusion was stopped, the superior mesenteric artery was occluded, causing Q_{pv} to fall to zero. This was accompanied by an immediate increase in Q_{ha}. Finally, when the superior mesenteric arterial occlusion was released, there was a postocclusive hyperemia and a concomitant decrease in Q_{ha}.

Physiologically, nitric oxide (NO) can modulate hepatic arterial resistance but has no effects on portal venous resistance. In normal conditions, specific inhibition of NO synthase with N^{G}-nitro-L-arginine methyl ester (L-NAME) increased the slope of the P_{ha}–Q_{ha} relationship at an unchanged back-pressure (i.e. vasoconstriction). This effect was reversed by L-arginine. In contrast, modifications of the L-arginine–NO pathway had no effect on P_{pv}–Q_{pv} relationships (Ayuse *et al.* 1995). After inhibition of NO synthase, the hepatic arterial buffer response was still present in the hepatic artery, consistent with an NO-mediated attenuation of the gain of hepatic arterial buffer response in the normal state. This preparation was also used following endotoxemia, with suppression of hepatic arterial buffer response and hepatic arterial autoregulation independent of NO (Ayuse *et al.* 1995).

In isolated perfused rat livers, NO donors blocked the increase in portal venous pressure induced by phenylephrine; following endotoxinemia, this vasoconstrictive response, which was mediated by endogenous NO, was markedly reduced. However, after endotoxin injection *in vivo*, Q_{ha} and Q_{pv} were reduced by NO inhibition but maintained by NO donors or L-arginine. This questions the usefulness of NO inhibition during human septic shock. At the hepatic microcirculatory level, NO synthase inhibition during sepsis markedly increased leukocyte adherence to periportal and centrilobular regions, reducing the number of open sinusoids.

Metabolic function and detoxification

The liver receives the effluent of the splanchnic circulation and is located directly upstream from the pulmonary circulation and cardiac pump, an arrangement adapted to its involvement in metabolic and detoxification functions. The liver is the main site of protein synthesis, for example albumin, which is a key element in maintaining oncotic pressure, and a site of binding for numerous hormones and drugs. It also synthesizes coagulation factors, which may (e.g. factors II, VII, IX, X) or may not (factors V, XI, XII, XIII, and fibrinogen) be dependent on vitamin K. The liver is essential for carbohydrate metabolism by increasing glycogen synthesis when the serum glucose concentration is high, and by degrading glycogen or converting amino acids to glucose by gluconeogenesis when serum glucose concentration is low. With regard to fat metabolism, the liver is responsible for both chylomicron degradation via acetyl-coenzyme A and lipoprotein synthesis during which fatty acids are incorporated. An intrahepatic regulation of hepatic protein synthesis has been described during inflammation, with a shift towards enhanced production of acute phase proteins such as C-reactive protein, serum amyloid A, lipopolysaccharide-binding protein, and complement components. This condition is the result of cross-talk between the hepatocyte and the Kupffer cell (the hepatic resident macrophage), leading to an amplification of the induction of proinflammatory cytokines such as interleukins 1 and 6.

Finally, the liver is crucial in protecting the organism from the deleterious accumulation of useless metabolites and various toxic compounds of both endogenous and exogenous origin because of its high concentration of oxydoreductases, such as the dehydrogenases involved in alcohol oxidation and the mono-oxygenases involved in the biotransformation of many drugs. In conclusion, the special organization of the vascular and parenchymal structure of the liver is adapted to the crucial functions of this unique organ, i.e. regulation of systemic blood flow, nutrient metabolism, and xenobiotic epuration.

References

Ayuse, T., Brienza, N., Revelly, J.P., Boitnott, J.K., and Robotham, J.L. (1995) Role of nitric oxide in porcine liver circulation under normal and endotoxemic conditions. *Journal of Applied Physiology*, **78**, 1319–29.

Beloucif, S., *et al.* (1995) Distinct behavior of portal venous and arterial vascular waterfalls in porcine liver. *Journal of Critical Care*, **10**, 104–14.

Brienza, N., Revelly, J.P., Ayuse, T., and Robotham, J.L. (1995) Effects of PEEP on liver arterial and venous blood flow. *American Journal of Respiratory and Critical Care Medicine*, **152**, 504–10.

Lautt, W.W. (1985). Mechanism and role of intrinsic regulation of hepatic arterial blood flow: hepatic arterial buffer response. *American Journal of Physiology*, **249**, G549–56.

Richardson, P.D.I. (1982). Physiological regulation of the hepatic circulation. *Federation Proceedings*, **41**, 2111–16.

4.2 Gastrointestinal hemorrhage

4.2.1 Upper gastrointestinal hemorrhage

H. J. Dworken

Key messages

- Stabilize vital signs before performing diagnostic studies.
- Explain procedures to patients and family.
- Obtain surgical consultation early.

Introduction

Here, we deal primarily with massive losses of blood from the digestive tract and their causes. However, since small hemorrhages often enlarge spontaneously, it is implicit that urgent attention must be paid to indications of even slight bleeding episodes, i.e. those manifested by positive testing of stools for occult blood or a single melenic stool.

A truly melenic stool is usually shiny black and almost invariably represents bleeding from above the ileocecal valve. A dull black stool is often not true melena and is negative to occult blood testing. Such stool coloring is usually the result of ingested medications or foods, particularly iron-containing supplements or bismuth, or dark-green leafy vegetables like spinach. The greater the rate of bleeding, the looser and redder the stool becomes. The higher the level of bleeding in the digestive tract, the greater is the fetid odor of the feces. This is probably a result of partial digestion of blood proteins in the upper intestine. Grossly bloody stools, or actual blood clots, can come from any level of the bowel, with the more malodorous and the more darkly red coming from higher levels. Significant gastrointestinal bleeding from above the cecum is usually accompanied by elevation of the blood urea nitrogen (BUN) level. This is due to the large protein bolus in the lumen of the small intestine and the reduced renal circulation resulting from compromise of blood volume.

First steps

Certain actions must be taken quickly in patients with gastrointestinal blood loss, regardless of the location of the hemorrhage. Significant losses from any site in the digestive tract are usually manifested by signs and symptoms of decreased circulating volume, sometimes even before the bloody stools and hematemesis have appeared. Attention should be paid immediately to circulatory compromise, well before any efforts are made to diagnose the specific cause of blood loss. Thus tachycardia, air hunger, or postural changes require instant administration of intravenous fluids and the typing of at least 3 units (1.5 liters) of blood for possible transfusion. These urgent first steps will often avoid complications of stroke and cardiac or renal impairment, particularly in older patients. Endoscopic and radiological studies should not be undertaken until measures to correct losses of volume have been instituted. At this point, it is also wise to call for a surgical consultation since the need for extreme intervention cannot be predicted and the assistance of an experienced surgeon in case management is often very helpful.

At the time of the patient's arrival in the emergency area the hematocrit may not show a very large fall because reactive hemodilution has not yet occurred. The presence of microcytosis indicates that blood loss is either chronic and continuous or that a hemorrhage has occurred previously. The potential presence of a disorder of blood clotting should also be assessed by performance of coagulation studies and by enquiry regarding the use of anticoagulants. BUN and creatinine levels should also be obtained. Elevation of the former in the presence of a normal creatinine points to bleeding from above the colon, while elevation of both these modalities is more suggestive of renal disease.

Once the intravenous infusion is running, the passage of a nasogastric tube is indicated. The presence of fresh blood in the aspirate usually suggests active upper gastrointestinal bleeding, while coffee-ground-colored material indicates that bleeding has either discontinued or is not very active. Most authorities recommend that ice-water lavage through the nasogastric tube be commenced after the initial aspiration of bloody or coffee-ground contents, since cooling of the stomach may quickly reduce the rate of blood loss and lavage preceding endoscopy often makes that procedure more successful. Two to three liters of ice-water lavage should be employed, with aspiration of contents after each 500 ml. Using saline instead of water is not recommended in case the patient has disease of the liver or heart and possible borderline salt retention. Lavage is of doubtful value in those bleeding patients in whom gastric aspirates show no evidence of bleeding. Emergency endoscopy is probably not necessary in patients whose gastric aspirates are either clear or coffee-ground color in character (Lin *et al.* 1995).

When vital signs become stabilized, it is well to pause and explain to the patient and the hidden patient(s), i.e. the relatives or friends sitting or pacing nervously in the waiting room, just what lies ahead in diagnostic or therapeutic measures. The most co-operative patients and families are those who understand what is at stake and are appraised of the treatment program as early as possible. The need for endoscopy and possible therapeutic measures that might be indicated should be carefully explained, and any informed consent given as soon as circulatory stabilization has been achieved. If permission to use blood products is not given, the risks of this refusal must be emphasized.

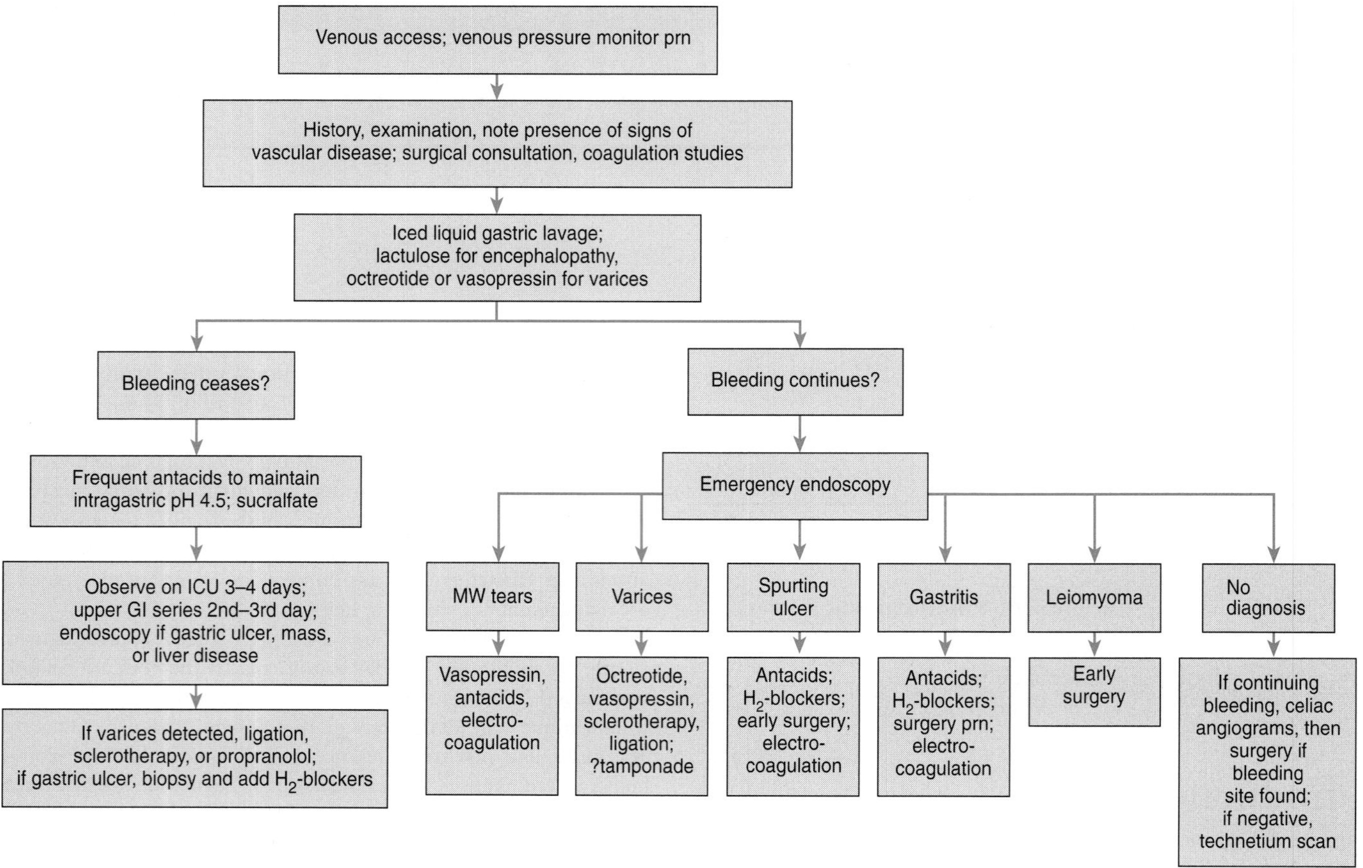

Fig. 1 Treatment program for upper gastrointestinal hemorrhage: ICU, intensive care unit; GI, gastrointestinal; MW, Mallory–Weiss; prn, whenever necessary.

Upper gastrointestinal hemorrhage (above the lower ileum)

The treatment program for upper gastrointestinal hemorrhage is shown in Fig. 1.

Esophageal sources other than varices

Hemorrhage from the esophagus, generally associated with hematemesis, is caused most commonly by mucosal tears at the gastroesophageal junction (Mallory–Weiss lesions). This disorder is usually precipitated by bouts of retching or vomiting of clear non-bloody gastric contents, indicating that the retching itself may have led to the subsequent bleeding. The syndrome is often preceded by an alcoholic debauch, or following ingestion or administration of nauseant drugs. The diagnosis is often suspected by a history suggesting that the contraction of abdominal and diaphragmatic muscles has led to stretch forces on the esophagogastric mucosa. It is confirmed by endoscopic visualization of the lesions themselves. Bleeding often ceases spontaneously, or can be controlled by endoscopic coagulation of the oozing areas with a bipolar (Bicap) electric probe. Patients with Mallory–Weiss syndrome should be examined for subcutaneous emphysema, which indicates the presence of tears through the esophageal wall into the mediastinum (Boerhaave syndrome) and requires early surgical repair.

Reflux esophagitis is another common cause of esophageal bleeding. This diagnosis is suspected when guaiac-positive stools or small amounts of melena are discovered in a patient experiencing frequent bouts of heartburn, particularly when in a recumbent position. The disorder results from repeated reflux and retention of acid-peptic gastric contents in the esophagus as a result of both reduced resting pressure at the level of the inferior esophageal sphincter and delayed esophageal emptying. While the inflammatory process is usually mild and bleeding is slight, actual mucosal dysplasia (Barrett epithelium) leading to ulcers and strictures may occur in severe chronic cases. In immunocompromised patients in particular, several types of mycoses (especially candidiasis) or viral infections may cause extensive esophagitis and dysphagia, usually with minor bleeding. These disorders are generally responsive to specific chemotherapeutic drugs. Cancer of the esophagus is unlikely to cause major bleeding unless the tumor penetrates into a large submucosal blood vessel. Rarely, a bronchogenic carcinoma may erode into the esophagus and cause heavy bleeding, usually preceded by a period of dysphagia.

Gastroduodenal sources

While melena and hematemesis alone may be caused by any lesion in the upper gastrointestinal tract, the appearance of coffee-ground vomitus or gastric aspirate indicates that blood has been in contact with acidic gastric contents. Erosive gastroduodenitis is suspected when a patient with upper gastrointestinal bleeding gives a history of having recently ingested non-steroidal anti-inflammatory drugs or significant amounts of alcohol. These patients may also give a history of upper abdominal discomfort, occurring irregularly for days or weeks prior to the bleeding episode, and may have noted that liquid antacids provided them with temporary relief. The alcoholic patient with erosive gastritis may be either a chronic or a binge drinker; bleeding from such lesions may occur at any age. Stress ulceration is a process similar to erosive gastritis, but occurs primarily in patients confined to intensive care units, suffering from severe injuries, infections, or cardiovascular problems, and not exposed to non-steroidal anti-inflammatory drugs. Their lesions appear to result from transient mucosal ischemia.

Lesions of erosive gastroduodenitis or stress ulcers are most often multiple and tend to occur mainly in the gastric antrum. On microscopic examination, the acute ulcerations demonstrate loss of epithelium and inflammatory cells in the mucosa. The erosions themselves do not usually penetrate into the submucosa and are not surrounded by fibrous tissue.

Hemorrhages from erosive gastroduodenitis or stress ulcers vary greatly in intensity, depending on the extent of the disease process. Upper gastrointestinal endoscopy is necessary to confirm the diagnosis and to estimate the rate of bleeding. Coagulation of the acute ulcers with the Bicap or the heater probe is effective therapy; the disorder is unlikely to recur once the causative agents are discontinued and the patient is removed from life-support systems.

Hemorrhage from stress ulcerations has become a large enough problem during intensive care that most units now advise some sort of prophylaxis in patients at risk. Antacids, H_2-blocking agents, and sucralfate have been used to prevent bleeding, and all have been beneficial. However, pneumonia has been a significant complication in ventilated patients on agents that raise gastric pH levels. Because of better protection from bleeding and lower levels of pneumonia, sucralfate presently appears to be the drug of choice in preventing stress ulcers (Tryba 1991). With regard to non-steroidal anti-inflammatory drugs and upper gastrointestinal hemorrhage, risk increases with age and with dosage, and the danger lasts for about a month after the medication is discontinued (MacDonald *et al.* 1995). Misoprostol (400–800 mg daily) reduces both the risk of hemorrhage from non-steroidal anti-inflammatory drugs and the histological gastritis that they cause.

Peptic ulcers may occur in the esophagus as a complication of reflux esophagitis, the stomach, or the duodenum. They may also be found in the jejunum in patients who have previously undergone a partial gastrectomy with gastrojejunostomy (Billroth 2 anastomosis), and are occasionally multiple (mainly duodenal and jejunal) in patients with the Zollinger–Ellison (hypergastrinemia) syndrome. Hemorrhage is a common complication of peptic ulcer at any site, and occurs in about 20 per cent of patients with untreated lesions. Most patients with peptic ulcer have been symptomatic prior to bleeding. Many report a previous diagnosis of ulcer and may have been treated in the past with sucralfate, H_2-blocking agents, or gastric acid pump inhibitors. Untreated patients often relate a history of epigastric pain relieved for several hours by meals or antacids, and occurring between meals and while asleep. Loss of weight and vomiting are uncommon symptoms, suggestive rather of obstruction of the gastric outlet. Patients with the Zollinger–Ellison syndrome frequently complain of diarrhea suggesting malabsorption (fatty foul-smelling stools) and loss of weight, and occasionally report lack of success of conventional therapy for peptic ulcer disease. The diarrhea in these patients is believed to result from inactivation of lipases in the highly acidic upper intestinal lumen, and may also be caused by excesses of vasoactive intestinal polypeptide from an associated neuroendocrine tumor.

Once cardiovascular stability has been achieved, ulcer-bleeding suspects should be treated with an intravenous H_2-blocking agent (e.g. ranitidine, 6.25 mg/h) and immediate endoscopy. This procedure will identify the site, depth, and size of the ulcer(s), and whether the ulcer is single or multiple and actively bleeding. A pumping artery in the ulcer base may be difficult to coagulate endoscopically, but surprisingly good results have been obtained using a combination of electrocoagulation of the ulcer base and multiple injections of about 1 ml of epinephrine (adrenaline) (0.005 per cent) in hypertonic (3.6–7.1 per cent) saline around the circumference of the ulcer. Caputi Iambrenghi *et al.* (1995) found that 25 of 86 patients (29 per cent) with recent ulcer hemorrhage re-bled on 'conservative' treatment, while only three of 34 patients (8.8 per cent) re-bled after receiving periulcer injections of epinephrine and a sclerosing agent.

In instances where the ulcer is not bleeding at the time of endoscopy but instead contains a large clot, electrocoagulation of the clot itself often prevents further bleeding. There is still disagreement among endoscopists as to whether the clot should be removed in these cases before attempting coagulation. Where a visible vessel is seen in the base of the ulcer, there is a slightly lower success rate with electrocoagulation; multiple treatments and, occasionally, surgery are indicated. One study (Grosso *et al.* 1995) has shown that re-bleeding from ulcers containing a visible vessel (19 per cent) is essentially the same if patients are treated with either perendoscopic injection of sclerosants or with intravenous omeprazole (40 mg twice daily). If multiple ulcers are seen in the lower stomach or duodenum, serum gastrin levels should be measured for the possible presence of a gastrinoma.

The symptoms of gastric and duodenal ulcers are indistinguishable. Upper gastrointestinal endoscopy and barium-contrast radiography are the only techniques available for a precise diagnosis. This distinction should be made as early as possible because of the possibility of dysplastic or malignant changes in gastric lesions. About 2 to 5 per cent of gastric ulcers appear in malignant tissue. Since changes of carcinoma are not always readily apparent on gross endoscopic examination, multiple biopsies should be taken from the circumference and base of gastric ulcers. In lesions that are actively bleeding, or have recently bled, it is probably best to wait several days before obtaining biopsies. The treatment of bleeding peptic ulcers should include the use of intravenous H_2-blocking agents for the first week or so, and a bismuth–antibiotic program if *Helicobacter pylori* is detected.

Other causes

Leiomyomata are uncommon, and usually asymptomatic, causes of upper gastrointestinal hemorrhage. They generally arise from the muscularis layer of the upper stomach and present as ruptures of the small arteries with which these tumors are amply supplied. The bleeding vessel may be seen to pulsate during endoscopy. Bleeding can usually be controlled by electrocoagulation. However, recurrence is very common, and definitive control can be achieved only by surgical removal of the tumor.

Small intestinal obstructive lesions can bleed extensively, usually late in their course. They have generally been symptomatic for weeks or months before hemorrhage occurs, causing recurrent attacks of upper gastrointestinal cramping or colic, usually after meals. With some types of small intestinal obstruction, patients may be asymptomatic for several days between attacks. Vomiting, fecal in nature at times, may terminate one of these attacks. Loss of weight is common. Patients may also recognize that shifting to a softer or more liquid diet will lessen the frequency and severity of symptoms. Some patients may have noted abdominal bloating, audible borborygmi, and irregularity of their bowel pattern. Particularly in Crohn's disease, diarrheal episodes may have preceded the onset of bleeding.

History of a previous aortic graft suggests the possibility of graft rupture into the small intestine. Intestinal bleeding in patients with a record of episodic cramping pain after meals is suggestive of mesenteric vascular insufficiency (abdominal angina). Gastrointestinal hemorrhage in these subjects occurs when necrotic ulcers appear in areas of ischemia. Physical signs in patients with small intestinal bleeding vary with the amount of blood lost and the resulting effects on hemodynamics. Abdominal examination may show distention, borborygmi and visible peristaltic rushes, or a palpable mass. Palpation of a pulsating mass in the upper mid-abdomen suggests the presence of a ruptured aortic aneurysm. Auscultation and/or palpation of a systolic bruit in the same area supports the diagnosis of aortic aneurysm or mesenteric arterial sclerosis. Gastric aspiration is usually clear. Plain films of the abdomen are very valuable, and can demonstrate free air in the case of bowel perforation, a closed-loop obstruction in cases of volvulus, distended small bowel loops from luminal obstruction, or abnormal air–fluid mucosal patterns in Crohn's disease. Gastrointestinal hemorrhage in patients with small intestinal obstruction represents either bowel strangulation or mucosal ulceration. The presence of fever and polymorphonuclear leukocytosis is a strong indication for taking blood cultures and consulting a surgeon, since these findings suggest probable bowel necrosis. The place of endoscopy in patients whose hemorrhage suggests a small intestinal site is uncertain. In the case of positive abdominal films, there is probably no need for endoscopy. In the symptomatic bleeding patient with no physical signs, a negative upper endoscopy and colonoscopy can have strong predictive value. The site of continued bleeding can occasionally be located with isotopic studies, using technetium-tagged red blood cells as markers, or by mesenteric angiography. It is probable that, in the near future, enteroscopy will be widely available to locate bleeding sites in the jejunum.

Hemobilia, i.e. bleeding into the biliary tree or pancreatic duct, presents most commonly as upper gastrointestinal hemorrhage associated with right upper-quadrant abdominal pain and jaundice. The pain is steady and severe. Since trauma is one of the most common causes of hemobilia, patients should be asked about previous episodes of blunt injury to the right lower chest, and procedures such as percutaneous biopsy of the liver, cholangiopancreatography, or Oddi sphincterotomy that they may have undergone in the previous 4 to 6 weeks. Some cases result from injuries to the common bile duct by gallstones or by rupture of a hepatic artery aneurysm into the ductal system. Endoscopy may show blood emerging from the ampulla of Vater, but angiography is probably most diagnostic and may also provide the best therapeutic vehicle by allowing fluoroscopic embolization of the bleeding vessel.

References

Caputi Iambrenghi, O., Lospalluti, M., Piccinni, G., and Lippolis, A. (1995) Is endoscopic injection hemostasis of bleeding gastrointestinal ulcers obsolete in 1995? *Surgical Endoscopy*, **9**, 1090–9.

Grosso, C., *et al.* (1995). Non-bleeding visible vessel treatment: perendoscopic injection therapy versus omeprazole infusion. *Scandinavian Journal of Gastroenterology*, **30**, 872–5.

Lin, H.J., Wang, K., Perng, C.L., Lee, C.H., and Lee, S.D. (1995). Emergency endoscopy is needed in patients with a bleeding peptic ulcer and bloody nasogastric aspirate—a prospective randomized study (Abstract). *Gastroenterology*, **108**, A150.

MacDonald, T.M., Morant, S.V., Robinson, G.C., Shield, M.J., Murray, F.E., and McDevitt, D.G. (1995). The risk of serious upper GI complications is constant with continuous NSAID therapy: results of a record linkage study in 52 382 exposed patients (Abstract). *Gastroenterology*, **108**, A156.

Tryba, M. (1991). Prophylaxis of stress ulcer bleeding: a meta-analysis. *Journal of Clinical Gastroenterology*, **13** (Supplement 2), S44–55.

4.2.2 Variceal bleeding

H. J. Dworken

Key messages

- Saline and large-volume lavage should be avoided to prevent water retention.
- Blood should be washed promptly from the intestine to avoid encephalopathy.
- Endoscopic ligation of varices appears to be the most effective non-surgical therapy.

Introduction

Varices in the upper intestinal tract result from sustained hypertension in the portal venous system. Hypertension arises from blockading lesions in the portal system, the liver, or the hepatic veins. Intrahepatic arteriovenous shunts that develop in the liver as a result of hepatocellular necrosis and hepatic sinusoidal damage also increase portal pressure. Although varices are located primarily in the submucosa of the lower esophagus, they are not uncommon in the upper wall of the stomach, and may even occur in the duodenum and jejunum. Portal hypertension is caused by disease proximal to the hepatic sinusoids (presinusoidal), in the sinusoidal area, or distal to the sinusoids (postsinusoidal). Extrahepatic presinusoidal obstruction is rarely seen, but may result from pylethrombosis or portal vein obstruction by tumor. Such lesions cause portal hypertension and varices, but not hepatocellular damage. Intrahepatic presinusoidal obstruction occurs in patients with liver disease that is confined to the bile ducts and portal areas, without significant damage to the hepatic parenchyma. Among these presinusoidal disorders are portal fibrosis from schistosomiasis, early lesions of primary biliary cirrhosis, granulomatous disorders such as sarcoidosis, and metastases of cancer to the liver. Liver function is usually well preserved in these patients, but varices and their complications may occur.

Sinusoidal obstruction of the portal tract is most commonly the result of cirrhosis of the liver. This develops from cellular necrosis followed by scarring, destruction of the reticulin framework of the liver, and regenerating nodules. These processes damage sinusoidal pathways and in many cases lead to intrahepatic arteriovenous shunts. Passage of blood at pre-, intra-, and postsinusoidal levels is slowed as portal hypertension increases. Extrahepatic postsinusoidal obstruction leading to portal hypertension and varices is infrequent but significant, and may be the result of hepatic vein thrombosis (Budd–Chiari syndrome) or, rarely, chronic congestive heart failure.

Portal hypertension leads to the opening of collateral passages for the flow of blood from mesenteric and retroperitoneal veins to the heart. This blood is partially oxygenated and may be deficient in coagulation factors. For example, prothrombin levels are often reduced because of decreased hepatocellular function, while platelet counts are diminished because of splenic enlargement and resultant increased platelet consumption.

The normal flow of blood through collateral vessels is reversed since alternative passages to the heart are opened by the high pressures in the portal vein and its tributaries. Collateral veins distend, causing varices. Where these thin-walled vessels are close to the lumen of the alimentary canal, there is a heightened tendency for them to rupture. Bleeding can be massive, particularly if blood clotting is also deficient. The most common sites for variceal rupture to occur are in the mucosa of the lower esophagus and the upper stomach.

Signs and symptoms

Rupture may commence without premonitory symptoms. In some cases, bleeding is preceded by an attack of clear vomiting due, for example, to an alcoholic debauch or the ingestion of non-steroidal anti-inflammatory drugs. Signs of circulatory collapse depend on the amount of blood lost. The patient may have no history of preceding liver disease. More commonly, however, earlier attacks of jaundice, ascites, hepatitis, or a long history of alcohol intake, raise suspicions of chronic disease of the liver with variceal rupture. If the bleeding patient becomes disoriented, tremulous, or somnolent with the hemorrhage, the onset of hepatic encephalopathy is suspected.

On physical examination, liver disease in a patient with upper gastrointestinal hemorrhage is suggested by the presence of jaundice, cutaneous lesions such as palmar erythema, spider angiomas, hematomas, dilated abdominal veins, or asterixis. The presence of an enlarged liver or spleen, or ascites, increases the likelihood of portal hypertension and variceal bleeding. Leukopenia, hypoprothrombinemia, and elevations in the levels of aminotransferases and bilirubin also support the diagnosis of liver disease and, by inference, the presence of variceal hemorrhage.

Diagnosis

Not everyone with liver disease and upper gastrointestinal hemorrhage is bleeding from varices. Nearly half of such patients bleed from mucosal lesions including peptic ulcers, gastric erosions, or esophageal tears. Any patient with chronic disease of the liver is at great risk of becoming encephalopathic from an upper gastrointestinal hemorrhage at any site as a result of both circulatory compromise and the presence of intestinal contents high in protein. Because of this, efforts to provide circulatory support and reduce the amount of blood in the intestine should be undertaken promptly. The latter goal is best achieved by the use of mineral laxatives and by gastric lavage with ice water. Saline lavage should be avoided because of the danger of fluid retention and the initiation or aggravation of ascites.

Definitive diagnosis of the bleeding site can be made only by upper gastrointestinal endoscopy, after lavage of the stomach, and following efforts to restore circulatory homeostasis. Prior to endoscopy, vitamin K (phytomenadione) should be administered if the prothrombin time is prolonged, even though response to this vitamin may be poor because of hepatic damage. Fresh frozen plasma and platelet transfusions should also be given where indicated. Before endoscopy, bleeding patients with signs and symptoms of hepatic encephalopathy should be sedated very lightly, if at all, in order to avoid worsening that state.

General and specific management

The endoscopist should be prepared to institute direct and indirect forms of therapy once the diagnosis of bleeding varices is made. Indirect methods consist of the use of vasopressin, octreotide, or propranolol to reduce pressure in the portal system. Vasopressin reduces portal flow by causing intense arterial spasm, and is best administered intravenously at the rate of 0.2 to 0.6 IU/min, with the smaller dosages being reserved for use in elderly patients whose course may be complicated by the onset of angina pectoris. The addition of nitroglycerin (glyceryl trinitrate) to the vasopressin seems to reduce the onset of complications while not reducing the benefits of this therapy.

Octreotide, a longer-lasting analog of somatostatin, has been found to inhibit portal blood flow, and may be more effective than vasopressin in stopping variceal hemorrhage without the danger of that drug's complications (Goff 1993). An initial dose of 250 μg intravenously is recommended, followed by an infusion of 250 μg/h. The β-adrenergic blocking agent propranolol has also been used to decrease portal pressure and has been shown to reduce variceal bleeding at times. The goal of this treatment is to reduce the resting heart rate to about 60 beats/min.

Since variceal hemorrhage recurs frequently and the mortality increases with successive bleeding episodes, most therapeutic efforts today are directed towards elimination of varices by endoscopic, radiological, or surgical techniques. Although none of these approaches is perfect, there is no doubt that they have distinct advantages over previous approaches.

Tamponade of varices, using an intraesophageal balloon mounted on a medium-sized rubber tube, was often applied in the past when direct endoscopic measures aimed at obliterating varices were not available. The goal of this treatment was to compress varices in the esophagus and upper stomach until clots formed at the bleeding site(s). The balloons were deflated after 1 or 2 days, and the tube left in place for another day to make certain that bleeding did not immediately recur. It was clear that recurrence of hemorrhage was very frequent and that tamponade was not definitive treatment in any way. This procedure may still be useful for buying time for a heavily bleeding patient when immediate access to a well-equipped endoscopy suite is not available, when the use of vasopressin or somatostatin does not appear to be reducing the bleeding, and when hepatic encephalopathy, worsening renal function, and progressive hepatic failure are imminent.

The triple-lumen Sengstaken–Blakemore tube is passed through the nose, and its position checked by fluoroscopy to see that the free distal tip of the tube is pointing towards the pylorus, that the gastric 'anchoring' balloon is located in the cardia, and that the esophageal balloon is properly distended. Traction is then applied to the external, or upper, end of the tube assembly, making sure that the tube is padded with soft plastic foam where the side presses against the nostril. In some centers, the esophageal balloon is not distended in the belief that pressure in esophageal varices is adequately reduced by compression of the more proximal gastric veins. Others feel that heavier intragastric traction is necessary if the esophagus is not compressed, and there is danger of causing greater pain and gastric and nasal damage by this approach. The middle open lumen of the tube is aspirated frequently to determine whether bleeding has been arrested. Some physicians recommend placing a second tube in the esophagus above the esophageal balloon to prevent aspiration of mucus and blood. Patients with a tamponade assembly in place must be under constant care because of the danger of aspiration and the frequent need for sedation.

Endoscopic obliteration of varices

There are several ways of doing this, including intravariceal injection of sclerosing agents, ligation of varices with rubber bands, clipping, and various combinations of these techniques. Each method requires several sessions, and the percentage of full obliteration is variable. Both sclerosing and ligation lead to esophageal ulcerations around the varix. These are usually smaller and are less likely to lead to esophageal stenosis after ligation than after sclerosis. Portal hypertensive gastropathy, i.e. engorgement with predisposition of gastric mucosa to bleed, may occur spontaneously in patients with portal hypertension, and is aggravated by courses of variceal sclerosis but not after variceal ligation. A meta-analysis of seven studies comparing sclerotherapy with ligation has shown that ligation is superior by leading to lower recurrences of bleeding, lower mortality, fewer complications, and fewer treatment sessions (Laine and Cook 1995). Another study has shown that ligation of varices compares favorably with variceal clipping (Koutsomanis *et al.* 1995).

Portosystemic shunting by interventional radiology

Successful shunting of blood from the portal to the systemic circulation has been shown to reduce or obliterate esophagogastric varices. In recent years it has become possible to establish such channels by a process called transjugular intrahepatic portosystemic shunting. Under radiological guidance, a catheter is passed into the liver through the jugular vein and advanced until a portal channel is ruptured. A stent is then passed to maintain the permanency of this traumatically established portosystemic shunt. Portal pressures drop from hypertensive to normal levels after establishment of the shunts, varices are reduced in number, and repeat variceal hemorrhages greatly reduced. The procedure appears to be much more effective than sclerotherapy of varices but requires very great technical skill (Skoens *et al.* 1995).

Surgical portosystemic shunting

When all else fails to stop recurrent variceal bleeding, and when hepatic function and nutrition are not extremely impaired, one can resort to intrabdominal portosystemic shunting operations. One study demonstrated that, when abstinence from alcohol is assured and protein intake is regulated sufficiently well to avoid episodes of hepatic encephalopathy, 67 per cent of 220 mainly alcoholic cirrhotics survived for at least 10 years after emergency shunting, 70 per cent had improved hepatic blood tests, and only 8 per cent had recurrent portosystemic encephalopathy (Orloff *et al.* 1994). These results were achieved by a very experienced group, and may not be applicable in hospitals with less experience.

References

Goff, J.S. (1993). Gastroesophageal varices: pathogenesis and therapy of acute bleeding. *Gastroenterology Clinics of North America*, **22**, 779–800.

Koutsomanis, D., Sebti, F., and Essaid, A. (1995). A randomized study of endoscopic hemostatic clipping versus endoscopic variceal ligation (Abstract). *Gastroenterology*, **108**, A1102.

Laine, L. and Cook, D. (1995). Endoscopic ligation compared with sclerotherapy for treatment of esophageal variceal bleeding: a meta-analysis. *Annals of Internal Medicine*, **123**, 280–7.

Orloff, M.J., Orloff, M.S., Rambotti, M., and Girard, M. (1994). Emergency portacaval shunt for bleeding esophageal varices in 400 unselected patients with cirrhosis (Abstract). *Gastroenterology*, **106**, A956.

Skoens, J., Semba, C., and Dale, M. (1995). Transjugular intrahepatic portosystemic shunts. *Annual Reviews of Medicine*, **46**, 95–102.

4.2.3 Lower gastrointestinal hemorrhage

H. J. Dworken

Key messages

- Perform stool cultures and microscopy in all cases of new bloody diarrhea.
- Inquire about exposure to persons with similar complaints.
- Do not attribute bright bleeding to hemorrhoids without demonstration of a bleeding point.

Introduction

Although many bleeding lesions in the lower gastrointestinal tract are either self-limited or intermittent, they often lead to considerable losses of blood, and earliest attention must be given to replenishing circulating volume. Diagnostic procedures should be undertaken only after adequate intravenous access has been established and vital signs stabilized. Patients with lower bowel hemorrhage rarely show true melena. Rather, intestinal contents show different shades of red—bright if from the lower third of the colon, and darker and more fetid if more proximal.

The program of diagnosis of lower gastrointestinal hemorrhage is summarized in Fig. 1.

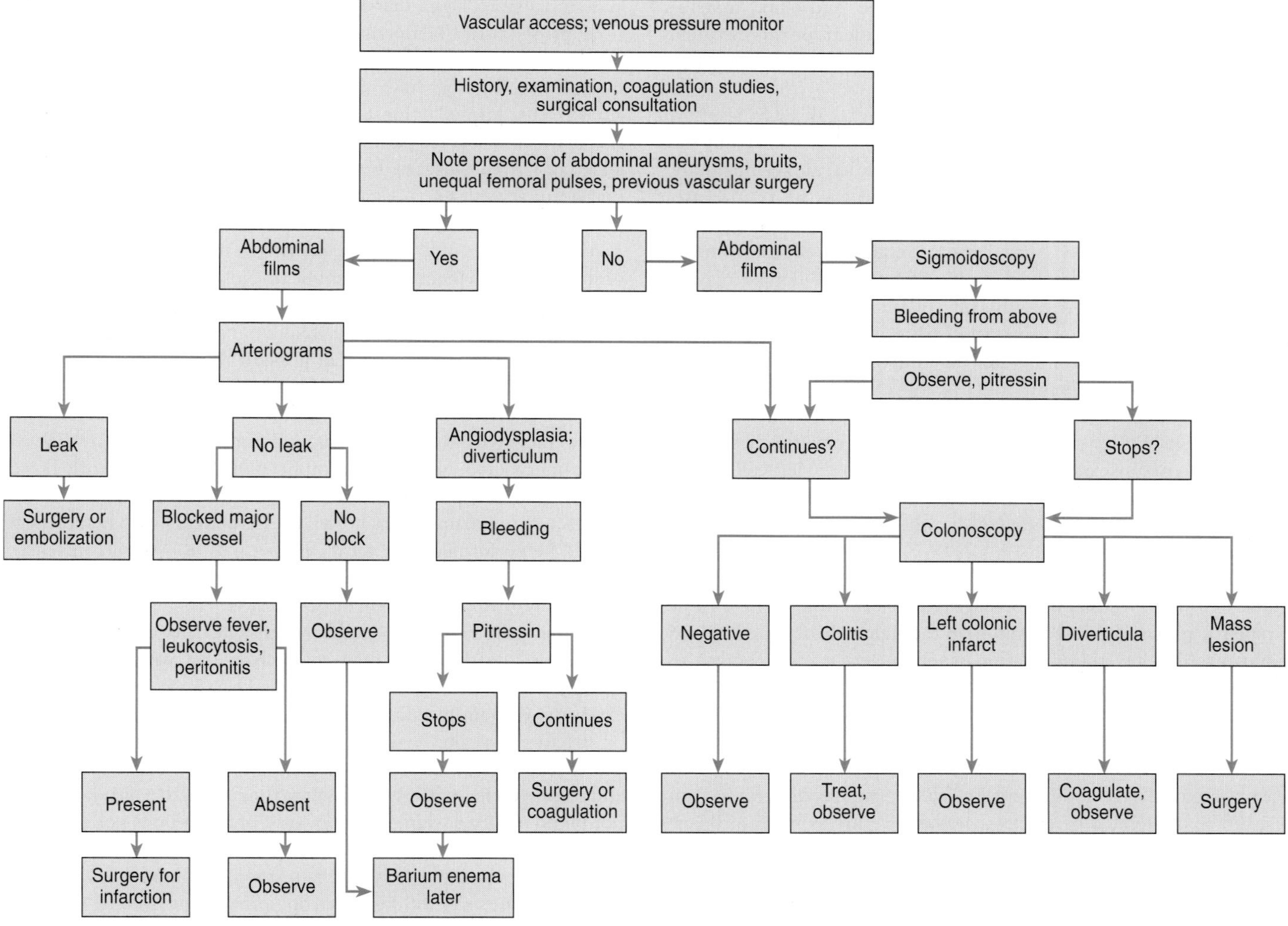

Fig. 1 Program of diagnosis of lower gastrointestinal hemorrhage.

History and findings

Premonitory symptoms are often absent, or may be merely a sudden urge to defecate. Patients with bleeding diverticulosis may offer a previous history of constipation and lower abdominal cramping, while those with angiodysplasia (vascular ectasias), a common cause of colonic blood loss, may have noted no previous bowel complaints. Hemorrhages from subjects with ulcerative colitis or Crohn's disease are often preceded by months or weeks of intermittent cramps, diarrhea, and occasional or frequent blood-tinged stools. At times, massive colonic bleeding can also arise from an intussuscepting large polyp or from cancer. These may also have been preceded by intermittent episodes of cramps.

A history of contemporaneous diarrhea, often with fever, among family members or personal contacts prior to onset of the hemorrhage suggests the possibility of acute colitis, such as infections with salmonellosis, *Campylobacter*, *Yersinia*, or typhoid organisms. There is often a recent history of foreign travel or ingestion of unpurified food or water. Sigmoidoscopy in these cases may show a diffusely erythematous mucosa, with numerous petechias and some friability. Ulcerations may be seen, particularly with amebiasis.

Pseudomembranous colitis due to overgrowth of *Clostridium difficile* is anticipated in patients with severe diarrhea and bleeding who give a history of concurrent or recent ingestion of broad-spectrum anbtibiotics. The normal bacterial flora have been altered by the antibiotics, causing overgrowth of the toxic Clostridia. Specific diagnosis depends on the detection of clostridial toxin in the stool, but the possibility of pseudomembranous colitis becomes very likely when a thick mucosal membrane is detected during proctosigmoidoscopy.

Blood loss can be a late and severe complication of any invasive diarrheal process, and patients are often dehydrated and moderately anemic when they present to the emergency room. In all invasive acute colitides, except amebiasis, microscopic examination of stool shows large numbers of white cells, while in amebiasis itself motile parasites may be detected. Specific diagnosis awaits the performance of stool cultures and definitive microscopy.

Radiation colitis should be suspected when a patient with bloody diarrhea has a record of radiation therapy to the abdomen. This therapy may have occurred months, or even several years, before the hemorrhage. Radiation leads to thinning of the exposed colonic mucosa and to considerable increases in vascularity. Sizable colonic hemorrhages may occur, similar to those seen among some patients with ulcerative colitis. Apart from rehydration, and blood transfusions when indicated, there is no specific treatment for this disorder. Glucocorticoids are of limited value with severe bleeding, and symptoms are usually chronic.

Infarctions of the left colon are most commonly caused by low flow states. Descending colonic infarcts are commonly announced by the sudden onset of steady and severe pain in the left upper quadrant. Bleeding arises from mucosal ulcerations and may be very severe. Examination often shows mild distention of the abdomen with reduced bowel sounds. Auscultation frequently demonstrates a prominent bruit. A plain radiographic film of the abdomen may be diagnostic in showing a gas-dilated colonic loop whose ulcerated margins have a typical 'thumb-print' pattern. Colonoscopy may be diagnostic by demonstrating normal mucosa in the rectum and dusky reddish-blue ischemic mucosa above. The rectum is spared from infarction because its blood supply arises from branches of the iliac artery, while lower colonic blood comes from the mesenteric vasculature. Because of collateral blood supply to the left colon from both the upper and lower mesenteric arteries, ischemic necrosis and ulceration often heal spontaneously when circulatory stability is restored.

Colonic mucosal angiodysplasia is frequently encountered as a cause of lower intestinal hemorrhage (Reinus and Brandt 1994). The dysplastic process is most common in the right side of the colon and appears to develop from dilatation of submucosal veins secondary to repeated episodic contraction of the muscularis layers of the bowel. Dilatation gradually extends retrograde into the capillary bed, weakens precapillary sphincters, and promotes the appearance of subepithelial arteriovenous communications (Boley *et al.* 1977). The lesions are readily seen during colonoscopy, and bleeding points can usually be obliterated by electrocoagulation. Similar types of telangiectatic lesions have been found in both the lower small intestine and the colon of patients with aortic stenosis, and also many who have been subjected to radiotherapy of the abdomen or pelvis. Patients with enterocolic bleeding from hereditary hemorrhagic telangiectasias (Osler–Weber–Rendu disease) can often be identified by the presence of oral spider-like vascular lesions. While subjects with one of the variants of von Willebrand's disease often bleed heavily from the alimentary tract and can be identified by the detection of a prolonged bleeding time, the hemorrhage is most often from a single site such as an ulcer or a neoplasm.

Colonic diverticulosis is by far the most frequent cause of left-sided lower bowel hemorrhage in Western societies. The incidence of diverticulosis increases with age, probably due to the relative weakness of collagen in older patients, along with increased intraluminal pressures arising from diets low in residue. Diverticula usually appear at sites in the colonic muscularis penetrated by mucosal end-arteries. Thus they are closely apposed to the walls of these vessels and may rupture into them easily. The resulting hemorrhage is often quite heavy, but can frequently be stopped by endoscopic cauterization. Angiography may well demonstrate the bleeding point if it is not seen on endoscopy, and continued bleeding may be arrested by the use of intra-arterial vasospastic agents or thrombi.

Inflammatory bowel disease, Crohn's disease, and chronic ulcerative colitis are well recognized, although fairly uncommon, causes of lower gastrointestinal hemorrhage. In Crohn's disease, bleeding arises from deep, and often multiple, mucosal ulcerations. In chronic ulcerative colitis, it is usually due to extensive mucosal inflammation and friability. Patients with Crohn's disease have usually suffered from abdominal cramps and diarrhea with weight loss for long periods of time prior to the hemorrhage, and children may show signs of delayed growth and development. Subjects suffering from chronic ulcerative colitis will have noted small amounts of blood in their stools, either formed or loose, for days or weeks before a large hemorrhage. It is unusual for patients aged over 40 to experience their first attack of Crohn's disease, but the onset of chronic ulcerative colitis may occur at any age.

Further history in patients with inflammatory bowel disease may also reveal attacks of fever in the present or previous episodes, or the occurrence of erythema nodosum or unexplained attacks of arthritis (colitic arthropathy). Hemorrhage from inflammatory bowel disease is rarely continuous and seldom requires blood transfusions. Since bleeding comes from the colon, or possibly the lower small intestine, blood urea nitrogen is almost always in the normal range. Plain radiographs of the abdomen may demonstrate abnormal mucosal patterns in bowel segments dilated with gas, and can also demonstrate massive 'toxic megacolon', a serious complication of inflammatory bowel disease that can presage a spontaneous perforation of the bowel. Endoscopy in inflammatory bowel disease patients with hemorrhage should be performed with great care because of probable increased bowel fragility. Bowel preparation should be minimal, or omitted completely, owing to the risk of perforation. Proctoscopy in chronic ulcerative colitis shows mucosal friability, usually extending to the anal verge, and bleeding may be intense. In Crohn's disease the rectum may be completely normal or else show patchy areas of erythema without bleeding. The presence of an anal fissure, fistula, or stricture is strong evidence in favor of Crohn's disease. In each disorder the disease process can be more severe in the sigmoid colon and upper reaches of the bowel than in the rectum. If a bleeding area is not found on proctosigmoidoscopy or if preparation is not adequate, it is often expedient to defer further endoscopy until bleeding has stopped and circulatory stability is restored. If blood loss continues and the diagnosis is still not clear, angiography or isotopic scanning can be very helpful in locating the bleeding area.

Meckel diverticula form in the lower ileum as a result of the embryonic failure of closure of the vitelline (omphalomesenteric) duct which connects the intestine with the fetal yolk sac, and occur in about 2 per cent of newborns. Almost half of these anomalies contain aberrant gastric glands and secrete acid and pepsinogen. In 5 to 10 per cent of patients with Meckel lesions, peptic ulcers form in the ileal wall opposite to the diverticular lumen. Lower bowel hemorrhage is the chief complication of these ulcers, and other symptoms such as pain are usually absent. Bleeding may be very heavy, but is usually self-limited, and generally occurs in patients below 20 years of age. Bloody stools are not melenic and are often quite loose. Endoscopy and barium contrast radiographs are usually normal, except in those rare incidents when barium may demonstrate a diverticulum arising from the lower ileum. Patients are quite comfortable between attacks of bleeding, and the diagnosis requires a high level of suspicion. Angiography while actively bleeding may demonstrate the actual area of extravasation, and scanning with sodium pertechnetate, an isotope with affinity for gastric parietal (oxyntic) cells, has great specificity. Unfortunately, this test is not very sensitive, so that the diagnosis might only be made at exploratory surgery if bleeding continues or recurs.

Anorectal disorders, internal hemorrhoids, fissures, and fistulas are probably the most common causes of gross bleeding at stool, although they are rarely, if ever, responsible for massive hemorrhage. There are two pitfalls to be encountered in dealing with patients with the single complaint of rectal bleeding. The first is to attribute the blood loss to internal hemorrhoids without actually viewing hemorrhoids and bleeding from them on anoscopy. The second is failing to remind oneself that over a period of time hemorrhoids may bleed quite extensively and lead to a debilitating blood-loss anemia. Beyond demonstrating hemorrhoids, an anoscopic examination without prior preparation may allow one to obtain a sample of stool from the vault of the rectum. If this specimen is negative for occult blood, one can assume that there is no current bleeding from above. However, the rule still holds: if a bleeding point is not demonstrated by anoscopy, do not attribute gross bleeding to hemorrhoids alone. Thus examination of stools over several days and flexible sigmoidoscopy may be necessary to provide a complete evaluation. Treatment with stool softeners and hemorrhoidal suppositories over a period of at least a fortnight generally leads to cessation of hemorrhoidal bleeding.

If a patient with rectal bleeding proves to have a microcytic hypochromic anemia, it must be assumed that a moderate to severe iron-deficiency state exists and has been present for some time. Even if such a patient has demonstrably bleeding anal veins, it is proper to be suspicious that another gastrointestinal site of blood loss may coexist, since it is reasonable to presume that a subject with bleeding hemorrhoids alone would have been compelled to see a physician long before a clinical iron-deficiency state supervened.

The diagnosis of an anal fissure or fistula in a young patient should also lead to a consideration of the possible coexistence of Crohn's disease elsewhere in the intestinal tract, particularly if there is also a history of concurrent abdominal cramps, diarrhea, or fever. Endoscopic or radiological evaluation should be undertaken in these subjects.

References

Boley, S.J., *et al.* (1977). On the nature and etiology of vascular ectasias of the colon. Degenerative lesions of aging. *Gastroenterology*, **72**, 650–68.

Reinus, J.F. and Brandt, L.J. (1994) Vascular ectasias and diverticulosis. Common causes of lower bowel bleeding. *Gastroenterology Clinics of North America*, **23**, 1–20.

4.3 The acute abdomen

4.3.1 Perforated viscus

J. W. M. Greve and G. Ramsay

Key messages

- Check for important physical signs.
- Select diagnostic tools carefully.
- Aim for aggressive and early intervention when a perforation is suspected.
- Be aware of the patient's history and recent interventions.

Introduction

Perforations can occur in all organs forming the digestive tract. Patients can present to the intensive care unit (ICU) with a severe systemic inflammatory response syndrome or sepsis as a result of a perforated viscus. Alternatively, a perforation can develop in the course of a severe illness during ICU treatment, resulting in a deterioration for which the cause can be very difficult to determine. Causes of perforation and appropriate treatment strategies are summarized in Table 1.

The most frequent cause of a perforated viscus is inflammation (large bowel, appendix, gallbladder) or erosion (peptic ulcer). However, perforations can be due to trauma, can be iatrogenic as a result of diagnostic interventions, or can be a postoperative complication due to anastomotic leakage (Table 2).

Diagnosis

As with every patient admitted to the ICU, early diagnosis is of vital importance. The history and clinical picture can provide important clues. Recent diagnostic and therapeutic interventions such as colonoscopy, endoscopic retrograde cholangiopancreatography, or abdominal surgery will clearly direct the search for the cause of the illness.

Signs on physical examination depend largely on the site of the perforation. Subcutaneous emphysema in the neck region can be the result of an esophageal perforation, but has also been reported after duodenal, gastric, or large bowel perforation. Absent liver dullness on abdominal percussion is a classical sign of intra-abdominal air. Absence of bowel sounds is indicative of peritonitis accompanying bowel perforation, but can also be the result of a general severe illness (multiple organ failure) or the use of sedatives and analgesic medication. Further signs of peritonitis are usually difficult to interpret in the typical ICU patient, who is usually too sick or sedated to be able to respond to percussion and palpation of the abdomen or is on artificial ventilation. Moreover, these patients are often on medications (corticosteroids, non-steroidal anti-inflammatory drugs, chemotherapy) that will suppress many of the important clinical symptoms.

Blood chemistry and hematology are of limited value in the diagnosis of a perforated viscus. However, a markedly elevated serum alkaline phosphatase may be the only discriminating finding indicating gallbladder perforation (Madl *et al.* 1992).

Valuable information can be gained from the use of radiological diagnostic tools. Simple radiography of the thorax can reveal a pneumomediastinum in the case of esophageal perforation with or without mediastinitis. Free abdominal air on a plain abdominal radiograph is often proof of a perforation of a peptic ulcer in either stomach or duodenum or of a perforated small or large bowel. However, free air cannot be visualized in 30 to 50 per cent of patients with a perforated peptic ulcer, in 70 per cent with a perforated small bowel, and in 63 per cent with a perforated large bowel (Winek *et al.* 1988). Conversely, in patients with blunt abdominal trauma, intra-abdominal air can be present without a perforated viscus (Hamilton *et al.* 1995). CT scanning is of value in detecting the cause of an intra-abdominal catastrophe, particularly when abscesses that can be treated by percutaneous drainage are suspected. However, the patient must be relatively stable to be able to withstand transportation to the radiology department. Suboptimal monitoring during CT scanning should also be taken into account. In postsurgical patients, oral (water-soluble) contrast or a rectal contrast enema may enhance the possibility of detecting an anastomotic leakage. A fluid collection surrounding a thick-walled gallbladder suggests perforation of this organ. The value of CT scanning in the detection of a perforated viscus in trauma patients is still under debate (Hamilton *et al.* 1995), but intraperitoneal fluid of unknown origin, retroperitoneal, mesenteric, or intramural air, or a pneumoperitoneum may provide important information.

Ultrasound is another valuable tool in the diagnosis of intra-abdominal pathology. Its advantage is the possibility of examining the patient in the ICU. Drawbacks are the investigator-dependent reliability and technical limitations (e.g. air interposition). Ultrasound is still the first choice for detection of intra-abdominal fluid collections or gallbladder abnormalities.

In certain circumstances more invasive techniques may be appropriate. Peritoneal lavage can be used to exclude bowel perforation in patients with blunt abdominal trauma (Wisner *et al.* 1990). Laparoscopy in experienced hands may detect a perforated viscus, with the advantage of potential laparoscopic repair (e.g. perforated peptic ulcer).

Treatment

General surgical principles apply to the treatment of a perforated viscus. As soon as the diagnosis has been made and the patient has

Table 1 Synopsis of the cause of perforation and treatment strategy per organ

Esophagus	
Cause	Foreign body, endoscopy (dilatation), Boerhaave syndrome, trauma (gunshot)
Diagnosis	Radiography of thorax, contrast swallow, CT scan
Treatment	Early: drainage (and suture), primary resection
	Late: excessive debridement and drainage, resection (specialized center)
Stomach	
Cause	Peptic ulcer, malignancy, salmonella enteritis, interventional endoscopy, trauma
Diagnosis	Clinical presentation, history (NSAIDs), abdominal radiography, laparoscopy
Treatment	Suture and omental patch (excision of the lesion for tissue diagnosis), in selected cases resection (tumor, outlet stenosis)
Gallbladder	
Cause	Acute cholecystitis (cholecystolithiasis, acalculous), porcelain gallbladder
Diagnosis	Blood chemistry (alkaline phosphatase), ultrasound (CT scan)
Treatment	Cholecytestectomy, partial resection and drainage
Duodenum	
Cause	Peptic ulcer, ERCP, blunt or sharp abdominal trauma, infection (CMV, tuberculosis), pseudomembranous enterocolitis, diverticula
Diagnosis	Blood chemistry (amylase), plain abdominal radiography, CT scan (retroperitoneal air)
Treatment	Suture and drainage, exteriorization with a rubber balloon catheter
Jejunum, ileum	
Cause	(Uncommon) diverticula, Meckel's diverticulum, inflammatory bowel disease, trauma (stab wound, gunshot), postsurgical, carcinoma
Diagnosis	Plain abdominal radiography, laparoscopy
Treatment	Resection and primary anastomosis. In severe peritonitis, exteriorization (stomata) and debridement
Appendix	
Cause	Acute appendicitis, carcinoid tumor, fecolith, foreign body
Diagnosis	Clinical presentation, ultrasound, laparoscopy
Treatment	Resection and debridement of the abdominal cavity, ileocecal resection
Large bowel	
Cause	Diverticulitis, carcinoma, interventional endoscopy, ischemic colitis, ulcerative colitis, pseudomembranous colitis, postsurgical (anastomotic leak)
Diagnosis	Plain abdominal radiography, water-soluble contrast enema
Treatment	Resection of the affected colon, exteriorization of the bowel with stomata (Hartmann procedure), extensive debridement and lavage of the abdominal cavity, second look (24–48 h)

NSAIDs, non-steroidal anti-inflammatory drugs; ERCP, endoscopic retrograde cholangiopancreatography; CMV, cytomegalovirus.

Table 2 Causes of perforation on ICU admission

Type	No. of patients
Colonic	
Diverticulosis	22
Carcinoma	20
Injury	12
Ischemia (low flow)	9
Anastomotic leak	4
Total	67
Non-colonic	
Duodenum	18
Stomach	9
Small intestine	9
Gallbladder	7
Appendix	4
Total	47

Reproduced with permission from Barie *et al.* (1996).

been resuscitated, no further delay should be allowed. A perforated viscus is always an emergency. Conservative treatment by means of gastric suction and resuscitation may be considered in patients with a perforated peptic ulcer. However, the patient's condition must improve within 6 to 12 h; if not, laparotomy should follow immediately. Treatment of intra-abdominal abscesses due to a perforated bowel with CT-guided percutaneous puncture and drainage should only be considered if the perforation has been treated appropriately.

In the compromised (e.g. septic) patient, surgery should be aimed at the most reliable treatment strategy. The first aim of the laparotomy is removal of all foreign material, drainage of all abscesses, and extensive peritoneal lavage. There is no place for primary anastomosis of the bowel in the severely contaminated abdomen.

Particularly in the large bowel, the affected part should be exteriorized, if necessary with multiple stomata. Usually, this also holds true for the small bowel; however, primary anastomosis is possible in selected cases. Resection of the perforation site should be performed when the perforation is caused by an inflammation or a tumor (perforated appendix, diverticulitis, perforated colonic carcinoma, etc.).

In the case of a perforated gallbladder, the affected organ should be removed. However, when a dense inflammatory mass prevents recognition of important anatomic structures, partial resection and placement of drains in the remaining gallbladder lumen is a treatment option.

A perforated peptic ulcer is sutured and the perforation is covered with an omental patch. Extensive surgery to reduce acid production (e.g. highly selective vagotomy or antrum resection and vagotomy) increases the operative risk and is not indicated, since modern medication (H_2 receptor antagonists, proton pump inhibitors) is equally effective in healing and preventing peptic ulcers. Exceptions to this rule are the large perforated duodenal ulcer that cannot be closed by suture and the severely deformed pyloric region with obvious outlet obstruction, in which case a resection is necessary.

After adequate surgical debridement there are three options: close the abdomen with a resorbable mesh to allow drainage of the abdominal cavity, the so-called 'open abdomen' where the abdominal wall is left open, or primary suture of the abdominal wall. In the latter case a thorough follow-up and a very low threshold for relaparotomy (within 24–48 h) are the secrets to successful treatment. In the case of an open abdomen, the use of a wound manager will facilitate proper care of the patient. Repeat exploration of the abdominal cavity in these patients can be performed in the ICU. When the abdomen is closed with a resorbable mesh, placement of a large self-adhesive plastic sheet prevents excessive fluid loss and dehydration of the underlying bowel loops. The strategy adopted depends largely on the experience and preference of the surgeon.

A very difficult category of patients are those with a perforation of the esophagus. The perforation should be treated aggressively and as soon as possible. Depending on the cause, the location, and the size of the perforation, different treatment strategies are possible. A small perforation as a result of an endoscopy can be treated conservatively with suction in the esophagus and antibiotics, if necessary combined with drainage of the thoracic cavity. A large distal perforation (e.g. after dilatation) can be treated by suture of the perforation and plication with the gastric fundus. In cases of a large perforation or in the presence of a tumor, primary resection may be necessary. This should only be performed by a surgeon with long experience in esophageal surgery and preferably in a referral hospital. In patients with severe sepsis (multiple organ failure) due to esophageal perforation, deviation of the esophagus to the neck and local debridement and drainage of the perforation may be a suitable treatment option to get the patient through the first stage of the therapy.

References

Barie, P.S., Hydo, L.J., and Fischer, E. (1996). Development of multiple organ dysfunction syndrome in critically ill patients with perforated viscus. *Archives of Surgery*, **131**, 37–43.

Hamilton, P., Rizoli, S., McLellan, B., and Murphy, J. (1995). Significance of intra-abdominal extraluminal air detected by CT scan in blunt abdominal trauma. *Journal of Trauma, Injury, Infection, and Critical Care*, **39**, 331–3.

Madl, C., *et al.* (1992). Diagnosis of gallbladder perforation in acute acalculous cholecystitis in critically ill patients. *Intensive Care Medicine*, **18**, 245–6.

Winek, T.G., Mosely, H.S., Grout, G., and Luallin, D. (1988). Pneumoperitoneum and its association with ruptured abdominal viscus. *Archives of Surgery*, **123**, 709–12.

Wisner, D.H., Chun, Y., and Blaisdell, F.W. (1990). Blunt intestinal injury. Key to diagnosis and management. *Archives of Surgery*, **125**, 1319–23.

4.3.2 Ischemic bowel

A. Marston

Key messages

- Causes include arterial embolism, arterial thrombosis, venous thrombosis, and non-occlusive infarction.
- Acute intestinal ischemia most commonly presents with severe abdominal pain out of proportion to objective physical findings.
- When the ischemia becomes transmural a peritonitic reaction occurs and the clinical state becomes that of a desperate illness with tachycardia, hypotension, fever, ileus, and rebound tenderness.
- Acute intestinal ischemia involves loss of body fluids, the effects of bacterial invasion, and the absorption of toxic factors into the portal and systemic circulations.
- The decision to operate will often depend upon clinical suspicion as there is no specific test (except angiography) for detecting ischemia.

Introduction

The gastrointestinal tract has an abundant collateral blood supply, and so vascular accidents occur less frequently in the intestine than in heart, brain, kidney, or limbs. However, in an ageing population, the incidence of acute intestinal ischemia is increasing. In a community of 250 000 an average of one case occurs each month; therefore it is a common surgical emergency for which the mortality has remained at 70 to 100 per cent since 1930 (Clavien *et al.* 1987). It was observed more than 50 years ago that the difficulty in arriving at an early diagnosis undoubtedly accounts for the high mortality. The delay in diagnosis is often unavoidable owing to the lack of definite symptoms, even though the damage to the intestinal tract is severe. Although it is easy to say that an aggressive therapeutic approach offers the only hope of reducing mortality, the reality is different as in the early stages of this disease the clinical features are vague and non-specific, and routine laboratory data and radiographs are not diagnostic.

Etiology

Causes include arterial embolism, arterial thrombosis, venous thrombosis, and non-occlusive infarction. Approximately 50 per cent of cases are secondary to superior mesenteric artery occlusion, 25 per cent to non-occlusive infarction, and 5 to15 per cent to venous thrombosis.

The gut mucosa represents an important barrier, impairment of which may lead to a cascade of events ending in multisystem organ failure. Tonometric measurement of intestinal pH (pHi) has been shown to correlate well with blood flow and has a strong predictive value in forecasting surgical outcome (Fiddian-Green and Baker 1987).

Embolism and thrombosis

Arterial emboli to the gastrointestinal tract most commonly lodge in a bifurcation or branch of the superior mesenteric artery. Approximately one-third of patients will have experienced a previous embolus to the brain or an extremity (Marston *et al.* 1991).

The classic clinical setting is acute onset severe mid-abdominal pain followed by vomiting, diarrhea, and blood-stained stools in a patient who has an arrhythmia or has recently sustained a myocardial infarction. Patients with arterial embolization are typically younger than those with acute ischemia secondary to a thrombosis.

Thrombosis of the superior mesenteric artery occurs most often within the proximal few centimeters of the origin (Clavien *et al.* 1987). This is often associated with advanced atherosclerosis affecting other branches of the aorta. A history of intestinal angina (food-related abdominal pain) is present in approximately 50 per cent of these patients.

Non-occlusive infarction

Non-occlusive intestinal infarction was first described in 1958 by Ende. It is recognized with increasing frequency and, at present, accounts for approximately 25 per cent of all cases of acute ischemia. Mucosal blood flow reduction occurs in a wide variety of clinical situations, including major surgery, sepsis, and cardiogenic and hemorrhagic shock (Marston *et al.* 1991). This may not prove crucial in a young healthy individual with a resilient splanchnic circulation, but in a patient whose vessels are less adaptable and in whom other organ systems may be impaired, a vicious circle may be initiated, leading to a catastrophic series of life-threatening events.

The patient usually shows evidence of low cardiac output and hypotension. Many are receiving digoxin. Unexpected development of abdominal problems in a patient admitted to the intensive care unit (ICU) with a myocardial infarction, arrhythmia, or valvular heart disease, or who has recently undergone cardiac surgery, should arouse suspicion of such an infarct. Other potential causes of non-occlusive intestinal ischemia are the Arthus or Schwartzmann phenomena or vasoconstrictive bacterial toxins such as clostridial exotoxin.

Clinical features

Acute intestinal ischemia most commonly presents with severe abdominal pain out of proportion to objective physical findings. This deceptive clinical picture may last for a few hours to several days. The pain is usually periumbilical but may sometimes occur in the right upper quadrant or right iliac fossa. With an embolus the pain usually occurs suddenly, while in arterial and venous thrombosis or non-occlusive ischemia the onset may be more insidious. Occasionally, particularly in non-occlusive infarction, pain may be absent altogether and the conspicuous clinical feature is silent abdominal distension. More often, this is preceded by intense peristaltic activity resulting in spontaneous evacuation of the bowel. Bloody diarrhea is not evident until several hours into the course of the disease when mucosal infarction has begun, but the stools may test positive for occult blood at an early stage.

Physical findings are few. Bowel sounds may be hyperactive or normal. Tenderness is mild or absent, and temperature and heart rate are normal. Later, the pain becomes more constant and localized. Heart rate rises and blood pressure falls. Bowel sounds decrease and abdominal distension occurs. The intravascular volume contracts as fluid is lost into the intestinal lumen, bowel wall, and peritoneal cavity. The temperature, if elevated, is usually below 38°C.

When the ischemia becomes transmural a peritonitic reaction occurs and the clinical state becomes that of a desperate illness with tachycardia, hypotension, fever, ileus, and rebound tenderness. Pain and distension interfere with respiratory movements, resulting in anxiety, restlessness, air hunger, and cyanosis. Urine output decreases as dehydration proceeds and eventually the patient succumbs to multiorgan failure.

Laboratory findings

No specific laboratory test is of use in the early stages of the disease. A leukocytosis of over 15 000/mm^3 associated with hemoconcentration (hematocrit > 45 per cent) is common. Changes in amylase, liver enzymes, and inorganic phosphate do not generally occur until the clinical state of the patient is so obviously grave that laparotomy is mandatory.

Radiographic changes

The plain radiograph is normal in the earliest stages. Its main use is to exclude perforation or obstruction. Later, a non-specific picture of distended small gut with air-fluid levels develops; this is difficult to distinguish from changes found in peritonitis or early mechanical obstruction. Thickening of the bowel wall with gas bubbles and loop separation have been reported but are not constant or reliable features. The presence of gas in the mesenteric or portal veins was previously regarded as a lethal event. However, there have been reports of successful revascularization in such patients. Ultrasound imaging, CT scan, and magnetic resonance imaging are not easy to use in the emergency situation and have not been evaluated.

Angiography

Clearly, not every patient with the rather indeterminate symptoms and signs described above should be submitted to angiography. If peritonitis is present, this will only delay a necessary laparotomy. Moreover, a normal arteriogram does not exclude ischemia as up to one-third of cases are non-occlusive in nature. Conversely, the radiographic finding of an occluded or stenosed superior mesenteric artery does not necessarily imply that this is the cause of the patient's pain. Such lesions are quite common, and are often asymptomatic and of no hemodynamic significance; if discovered on an arteriogram, they may distract attention from the true cause of the problem. Nevertheless, percutaneous arteriography is widely available and can be performed safely in most patients provided that they are not allergic to the contrast medium, are well hydrated, and have adequate renal function. If an embolus is present, the surgeon can embark on a laparotomy with a view to embolectomy and gut revascularization.

In addition to documenting an embolus, angiography may also identify vasoconstriction, as shown by spasm and narrowing of the branches of the superior mesenteric artery and decreased filling of the intramural vessels. Given these findings, some authors have advocated an 'aggressive' management, using selective injections into the superior mesenteric artery of vasodilators such as papaverine (30–60 mg/h), phenoxybenzamine, or prostaglandin E_1. Although major reductions in mortality from acute intestinal ischemia have been claimed for these regimens, it is difficult to know when exactly to use them and their role in the total context of unexplained abdominal pain remains unclear. Vasospasm is a more complex matter than an angiographic appearance, and its significance is difficult to assess in patients who recover without operative proof of the diagnosis.

Clots within the mesenteric system may be dissolved with streptokinase. A catheter is positioned in the superior mesenteric artery immediately proximal to the occlusion and an infusion begun. The arteriogram is repeated after 1 to 3 h and, if no resolution occurs, the concentration of streptokinase is increased. If the patient's abdominal condition deteriorates, a laparotomy is performed. Preliminary reports have been encouraging, but streptokinase therapy causes complications such as anaphylaxis, febrile reactions, and bleeding. Newer agents such as urokinase or recombinant tissue plasminogen activator may prove to have an application.

Management

Resuscitation and support

Acute intestinal ischemia involves loss of body fluids, the effects of bacterial invasion, and the absorption of toxic factors into the portal and systemic circulations.

Fluid loss initially consists of water, electrolytes, and protein. This results in hemoconcentration and a fall in circulating volume, leading to impaired flow to both the gut wall itself and the rest of the body. Replacement is guided by urine output, hematocrit, and central venous or pulmonary artery wedge pressure. Depending on the hemodynamic status, liberal quantities of balanced saline, colloid, or protein solutions are given until the patient is adequately resuscitated. Monitoring of cardiac output provides additional accuracy both before and during operation.

Bacterial invasion and toxemia clearly demand the use of antibiotics. In the necessary absence of laboratory control, the chosen regimen will often include an aminoglycoside, such as gentamicin, coupled with metronidazole to combat anaerobic organisms. Polymyxin B may have a specific protective effect against mucosal ischemia and endotoxemia. In fact, although antimicrobial therapy is logical, the acute and sporadic nature of these problems precludes controlled trials and there is no concrete evidence of any effect on mortality.

Metabolic acidosis is due to the combination of poor tissue perfusion, hemoconcentration, and absorption of the products of gut necrosis. To this is added a respiratory component, resulting from impaired respiratory movement and increased blood viscosity with intrapulmonary sludging. Measurements of $P\text{CO}_2$ and arterial pH will guide the amount of support required.

Additional measures

Laboratory evidence suggests that α-blocking agents (phenoxybenzamine), β-stimulators (isoproterenol), and inotropic agents (dopamine and glucagon) increase flow in the mesenteric circuit and may preserve viability. Clinical information is not available, and it is doubtful whether, in the presence of a blocked superior mesenteric artery or massive shutdown of the minute vessels, they penetrate the gut wall. Additionally, the hypotension that these agents (particularly phenoxybenzamine) produce may be difficult to control.

The gut mucosa is uniquely rich in xanthine dehydrogenase which, in the circumstances of ischemia, converts to xanthine oxidase resulting in the production of toxic reactive oxygen metabolites including superoxide (O_2^-), peroxide (H_2O_2), and the hydroxyl radical (OH^-). This oxidative tissue damage is particularly severe during reperfusion. The process can be mitigated by the use of superoxide dismutase which 'scavenges' oxygen, catalase which reduces H_2O_2, and dimethylsulfoxide which scavenges OH^-. Another agent which may be useful in the preservation of acute mesenteric ischemia is allopurinol which blocks xanthine oxidase. Laboratory results suggest that, to be effective, the therapy must be given before the ischemic injury occurs. Although no controlled information is to hand, it would seem a wise precaution to cover operations for elective reconstruction of the mesenteric arteries with superoxide dismutase and allopurinol.

Digitalis and other cardiac glycosides constrict the mesenteric vessels and can provoke intestinal necrosis (Bulkley and Meilahn 1988). Nonetheless, in the control of congestive heart failure and atrial fibrillation, their effect will be to raise cardiac output and improve mucosal flow.

Surgery

The decision to operate will often depend upon clinical suspicion as there is no specific test (except angiography) for detecting ischemia. The first step may be laparoscopy which will identify most causes of peritonitis. If the diagnosis is still in doubt, the abdomen is opened through a long midline incision. In early cases the small intestine may appear healthy. Closer inspection reveals a loss of the normal glistening appearance of the serosa and absent pulsation in the mesenteric vascular arcades. As ischemia progresses, the bowel becomes pale, blue, and edematous. All too often, however, a massive infarction is found, ruling out any procedure except resection.

If the small bowel appears viable, the superior mesenteric artery is exposed. In the case of an embolus, a pulse will be felt in the root of the mesentery between the duodenum and the inferior border of the pancreas, and the distal vessels will be soft and normal. The patient is heparinized and a longitudinal arteriotomy is made. Catheter embolectomy is performed, taking care to avoid overinflating the balloon which can result in arterial rupture or intimal injury. After proximal and distal embolectomy, the arterial bed is flushed with heparinized saline, and the arteriotomy is closed with a fine vascular suture.

Once adequate arterial flow has been established and, if possible, documented by completion arteriography, the bowel must be inspected carefully to determine the need for resection. A 20- to 30-min period of watchful waiting after revascularization will often reveal adequate perfusion of segments initially thought to be non-viable.

Intraoperative assessment of intestinal viability remains an unsolved problem. Many techniques designed to determine adequate perfusion of the gut wall, including electromyography, pH recordings, fluorescein injections, and Doppler laser ultrasound, have been advocated, tried, and abandoned for reasons of practicality and cost. Most surgeons fall back on clinical judgment in deciding whether to resect (Ballard *et al.* 1993). The criteria usually employed for viability are the return of color and pulsation and the resumption of peristalsis in the injured segment. The difficulty is often resolved by resecting short bowel loops of doubtful viability, if the patient's general condition is satisfactory, and exteriorizing the ends. Primary anastomosis is dangerous in these circumstances and should not be attempted.

Thrombosis of the superior mesenteric artery in the presence of viable bowel may justify a reconstructive vascular procedure. Extensive thromboendarterectomy of the midaortic region, including the origins of the visceral vessels, has been advocated, although most surgeons would now prefer a bypass graft from the infrarenal aorta to the superior mesenteric artery using autologous saphenous vein. For the surgeon inexperienced in reconstruction of the visceral arteries (surely, the majority), it is safer, in case of doubt, to resect the gut rather than to attempt to revascularize it.

Postoperative care

The patient returns to the ICU with a central venous catheter, possibly a pulmonary artery catheter, tracheal intubation, urethral catheter, gastrostomy, and appropriate drains in place. Fluid replacement, antibiotics, and inotropic agents are continued.

Classical teaching advocated that, regardless of clinical state, any patient who had undergone revascularization of the gut should be re-explored the following day because of the difficulty in assessing success at the first operation. Although this is true theoretically, it has now become apparent that, since surgeons have learned to avoid primary anastomoses, such operations are seldom required. An interesting modification is to introduce a wide-bore drain, through which a laparoscope can be introduced to inspect the bowel, at the time of the initial operation.

The total infarct

Sadly, it is exceptional to find recoverable bowel; the most common situation that the emergency surgeon encounters is a major infarction of the midgut loop. If the patient is otherwise recoverable, it is worthwhile resecting large lengths of ischemic bowel and, following correction of metabolic deficits, restoring intestinal continuity. The capacity of the mucosa to regenerate was quite unappreciated before the advent of total parenteral nutrition because, at that time, all the patients died. Provided that nutrition can be maintained for a few weeks, many people with only a few centimeters of bowel remaining will eventually become able to feed themselves, although some will require lifelong parenteral support. Clearly, the outlook for patients who have lost their bowel because of trauma or strangulation will be much better than for those with underlying vascular disease.

References

Ballard, J.M., Stone, W.M., Hallett, J.W., Pairolero, P.C., and Cherrey, K.J. (1993). A critical analysis of adjuvant techniques used to assess bowel viability. *American Surgeon*, **59**, 309–11.

Bulkley, G.B. and Meilahn, J.E. (1988). Vasoactive humoral mediators and the splanchnic circulation in shock. In *Perspectives in shock research* (ed. R.F. Bond), pp. 91–100. Alan R. Liss, New York.

Clavien, P.A., Muller, C., and Harder, F. (1987). Treatment of mesenteric infarction. *British Journal of Surgery*, **74**, 500–3.

Fiddian-Green, R.G. and Baker, S. (1987). The predictive value of pH measurements for complications of cardiac surgery. *Critical Care Medicine*, **8**, 153–6.

Marston, A., Bulkley, G.B., Fiddian-Green, R.G., and Haglund, U.H. (ed.) (1991). *Splanchnic ischemia and multiple organ failure*. Edward Arnold, London.

4.3.3 Intra-abdominal sepsis

H. van Goor and R. J. A. Goris

Key messages

- The conventional approach of managing intra-abdominal infection may fail in patients with severe peritonitis caused by colonic perforation or leakage of intestinal anastomoses.
- In critically ill patients peritoneal defense mechanisms fail to control infection and may have detrimental effects on the host.
- Diagnostic laparoscopy and peritoneal lavage are useful tools in the diagnosis of intra-abdominal sepsis.
- The ideal operative approach for patients with severe forms of intra-abdominal sepsis has not yet been established. Recovery may depend more upon the ability to restore host defense mechanisms than upon surgical technique.

Introduction

Infection of the abdominal cavity is a serious and potentially life-threatening condition. When untreated, about 90 per cent of patients will die from sepsis. The conventional management approach is elimination of the source of peritoneal contamination, removal of purulent material, parenteral antibiotic administration, and hemodynamic and respiratory support. This results in complete recovery in most patients. However, it may fail in patients with severe infection caused, for example, by perforation of the colon, leakage of intestinal anastomoses, and pancreatic necrosis. The mortality of these conditions remains high, varying from 20 to 60 per cent (McLauchlan *et al.* 1995). Many patients have persistent or recurrent intra-abdominal infection and eventually die from multiple organ dysfunction.

Pathophysiology

After bacterial contamination of the abdominal cavity, a complex series of local events is initiated by the host in order to eradicate invading bacteria. This local host response consists of mechanical clearance of fluid and bacteria via diaphragmatic lymphatics, phagocytosis and destruction of bacteria by phagocytic cells, and sequestration of bacteria.

After intra-abdominal inoculation, bacteria begin to disappear from the abdominal cavity immediately, even before the influx of phagocytic cells. Bacteria are recoverable from the right thoracic lymph duct within 6 min of intra-abdominal inoculation and from blood within 12 min. Bacteria are transported by abdominal fluid circulation via the paracolic gutters to the left and right subdiaphragmatic spaces (Autio 1964) (Fig. 1). They are absorbed into the diaphragmatic lymphatics via stomata between specialized peritoneal mesothelial cells, and are then carried to the bloodstream (Allen 1936) (Fig. 2).

The second line of defense is phagocytosis with killing of bacteria by phagocytic cells. Phagocyte influx into the abdominal cavity is mediated by bacterial cell products (e.g. lipopolysaccharide), complement cleavage products (specifically C5a), inflammatory proteins (kinins, histamine), and cytokines such as tumor necrosis factor and interleukins 1, 6, and 8. Lipopolysaccharide, derived from Gram-negative enteric bacteria, is a particularly strong stimulus for release of cytokines by peritoneal macrophages. These enteric bacteria are most commonly found in intra-abdominal infectious processes. After chemotaxis, invading organisms must be ingested and destroyed. In particular, encapsulated micro-organisms are able to resist ingestion by human phagocytes and so must first be opsonized with a sufficient amount of specific antibody. On contact of the opsonized bacteria with the phagocyte, fusion processes are initiated, followed by degranulation of antimicrobial products that eventually kill the bacteria.

The third line of defense is bacterial sequestration. Fibrin plays a crucial role in walling off infection by incorporating bacteria in its matrix and by creating a physical barrier against dissemination through formation of intra-abdominal adhesions. Fibrin also impairs clearance of bacteria from the abdominal cavity by occluding the diaphragmatic stomata. Local fibrin deposition appears to stimulate the inflammatory response by increasing vascular permeability and chemotaxis of neutrophils and monocytes. Through all these actions, fibrin protects the host against early systemic spread of bacteria.

In critically ill patients these peritoneal defense mechanisms may fail to control intra-abdominal infection. As a consequence, the peritoneal cavity is flooded with infectious material and the peritoneal defense mechanisms may now have detrimental effects on the host. At least five interacting mechanisms are responsible for the failure to control infection: fluid exudation into the abdominal cavity, impairment of immunological defense mechanisms, adjuvant substances, bacterial synergism, and fibrin.

1. The influx of large amounts of protein-rich exudate into the abdominal cavity produces massive third-space fluid shifts. These exudates impair bacterial opsonization by diluting opsonins and reduce the ability of phagocytic cells to reach and phagocytose bacteria. Absorption of these intraperitoneal exudates and bacteria is disturbed by occlusion of the diaphragmatic stomata.

2. Immunological defense mechanisms are impaired by dilution of chemotaxic factors and opsonins by adjuvant substances and fibrin. Adjuvant substances (e.g. bile salts, gastric mucin, feces, barium sulfate, necrotic tissue, blood components, and foreign bodies) inhibit migration, phagocytosis, and the killing capacity of neutrophils and result in premature release of oxygen radicals by these cells. Impairment of phagocytosis may be due to (i) physical entrapment of neutrophils preventing locomotion, (ii) reduced opsonization of bacteria in a fibrinous environment, and (iii) premature release of lysosomal enzymes in the presence of activated complement components along fibrin strands.

3. Adjuvant substances not only impair immunological defense mechanisms but promote bacterial growth. For example, hemoglobin enhances bacterial growth and increases bacterial virulence. *In vitro* addition of iron to a culture medium enhances growth of *Escherichia coli*.

4. Lethality and the formation of residual abscesses are enhanced

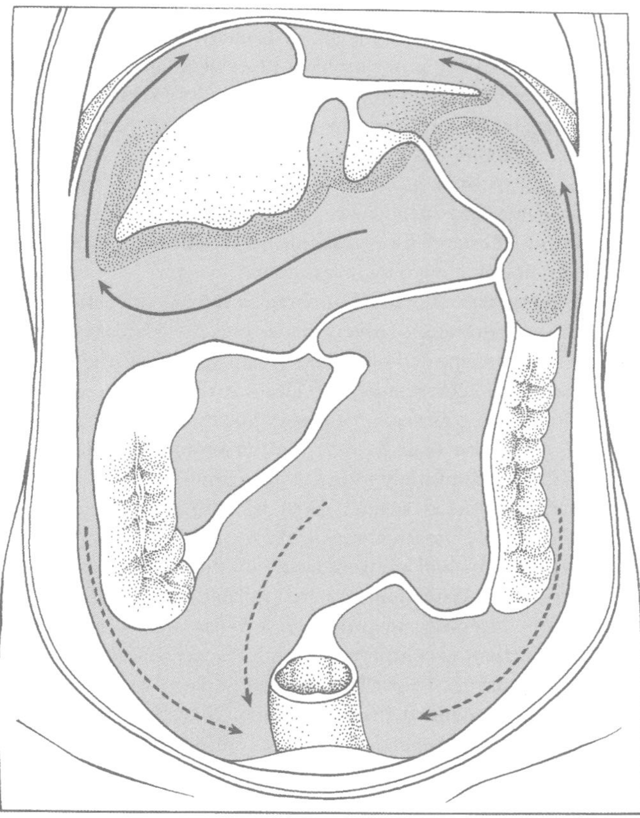

Fig. 1 Circulation of fluid and particles in the abdominal cavity: solid arrows indicate the direction of flow, generated by diaphragmatic movement and absorption of material from the diaphragmatic lymphatics; broken arrows demonstrate the effect of gravity in the upright position.

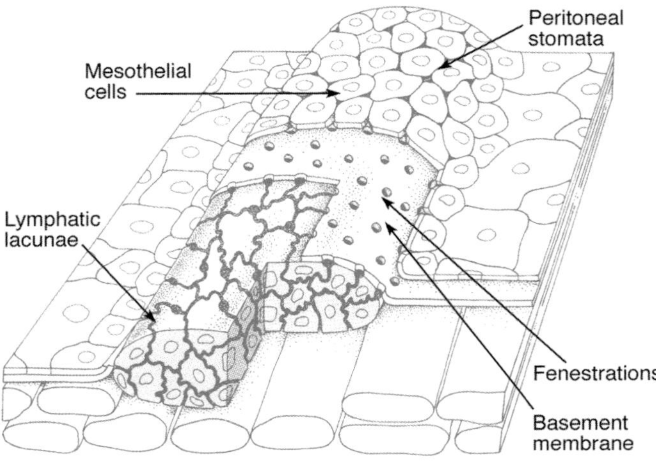

Fig. 2 Schematic view of the peritoneal stomata and underlying lymphatic tissue. Over the lymphatics, mesothelial cells are small, with stomata between the cells. Through these stomata and corresponding fenestrations of the basement membrane, the abdominal cavity communicates with lymphatic lacunas which are lined by lymphatic endothelium.

more by a polymicrobial intra-abdominal infection than by a monomicrobial infection. Synergism between anaerobic and aerobic bacteria plays an important role. In particular, the synergism between *Bacteroides fragilis* and *E. coli* has been studied both *in vitro* and *in vivo*. Possible mechanisms of synergy are (i) the ability of one species to provide growth nutrients for its bacterial partner, (ii) the ability of one species to impair host defenses, permitting its copathogen to survive and exert its intrinsic virulence, and (iii) the ability of the species to optimize local environment, thereby enhancing bacterial proliferation.

5. Fibrin entraps bacteria, providing a protected environment in which they can proliferate almost unaffected by neutrophils. Fibrin also impairs neutrophil function and blocks diaphragmatic stomata, thereby preventing absorption of intra-abdominal fluid and bacteria. Impairment of neutrophil function by bacteria, fibrin, and adjuvant substances is accompanied by premature extracellular release of neutrophil enzymes. These enzymes can damage viable tissue, resulting in necrosis. In addition to enzymes liberated by neutrophils, bacteria themselves produce a number of exoenzymes (e.g. hyaluronidase and lipase) that cause tissue destruction. In such an environment, characterized by hypoxia and a low pH, neutrophil function is further impaired and bacterial growth is stimulated. Tissue necrosis, fibrin, and bacteria augment the severity of the inflammatory process with continuous influx of neutrophils and macrophages. This process may readily become self-perpetuating, even in the absence of bacteria.

Diagnosis

Diagnosis is usually clinical; the predominant symptom is invariably intra-abdominal pain which is often steady, severe, and aggravated by movement. Anorexia and nausea are often accompanying symptoms. Most patients look unwell and in acute distress. Body temperature is usually elevated but, in cases of severe septic shock, hypothermia may be present. Tachycardia is common. Abdominal tenderness, particularly referred rebound tenderness, is the hallmark of intra-abdominal sepsis. The abdomen is distended and bowel sounds are heard occasionally or are even absent.

Leukocytosis with a shift to the left supports the clinical diagnosis of peritonitis; however, leukopenia may occur in overwhelming sepsis. An erect abdominal radiograph is particularly useful for demonstrating free air under the diaphragm, which is an indication of a perforated viscus, particularly a perforated duodenal ulcer. However, this diagnostic tool is of limited value in critically ill patients. Abdominal ultrasound and CT scan are indicated when an intra-abdominal abscess is suspected. These modalities may also be used in treatment. CT- or ultrasound-guided drainage has become the treatment of choice for unilocular abscesses located in subphrenic, paracolic, and presacral regions. However, multilocular abscesses, which are located in difficult areas for CT- or ultrasound-guided drainage, are often present in critically ill patients (Fig. 3).

When the diagnosis of intra-abdominal sepsis is uncertain, for example with altered consciousness as seen in sedated ventilated patients or with the use of immunosuppressive drugs, diagnostic peritoneal lavage or diagnostic laparoscopy are useful tools. Diagnostic peritoneal lavage appears to be a safe and accurate method of determining the presence of intra-abdominal sepsis necessitating

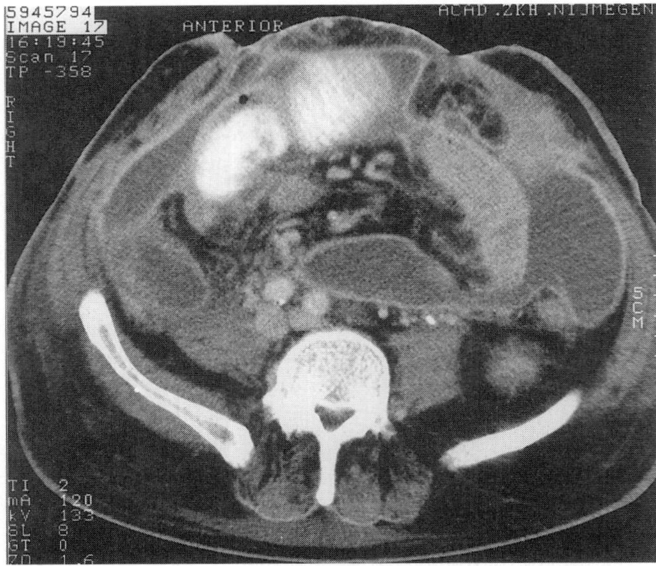

(a)

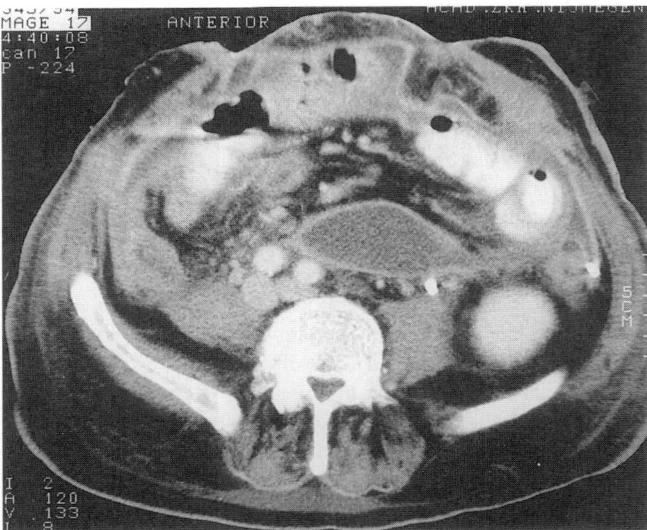

(b)

Fig. 3 (a) Multilocular intra-abdominal abscesses following treatment of diffuse peritonitis caused by perforation of the small bowel during laparoscopic hernia repair. (b) After percutaneous drainage: a residual abscess located ventrally from the aorta had to be drained surgically.

operation. The presence of more than 500 white blood cells/mm^3 after a 1-liter saline lavage correlates best with the presence of intra-abdominal sepsis. With diagnostic laparoscopy the error rate was reduced from 40 to 10 per cent in a group of patients with acute abdominal pain whose requirement for laparotomy was considered uncertain (Paterson-Brown *et al.* 1989).

Management

Important measures include fluid resuscitation, antibiotic therapy, and in most cases surgery.

Fluid resuscitation entails administration of fluids and monitoring of fluid status. In the critically ill, resuscitative measures must be directed by invasive monitoring techniques such as the pulmonary artery catheter. Controversy over the use of colloid versus crystalloid for resuscitation of the septic hypovolemic patient remains unanswered. Crystalloid resuscitation, which requires much larger volumes, appears to be as effective as colloid. Because crystalloid is readily available and inexpensive, it may well be the solution of choice. The use of starch for isovolemic hemodilution in these conditions is the subject of ongoing investigation.

Antibiotic therapy should be initiated as soon as the clinical diagnosis of intra-abdominal sepsis is made. The choice of agent should be based on the suspected offending organism and on the ability of the antibiotics to achieve adequate levels at the site where needed. Intra-abdominal sepsis is almost always polymicrobial. Gram-negative enteric bacteria (e.g. *E. coli*) and anaerobic bacteria (e.g. *B. fragilis*) are most commonly found. Most clinicians agree that antibiotic therapy directed against these bacteria is the mainstay of treatment of intra-abdominal sepsis. We use a combination of cefuroxime, gentamicin, and metronidazole as the initial treatment of diffuse peritonitis. When cultures are available, antibiotic treatment is adjusted if necessary. In principle, a 5-day antibiotic course is given. If the patient is compromised with respect to susceptibility to opportunistic pathogens, an antibiotic that is effective against enterococci should be included in the regimen. Moreover, enterococcal bacteremia or the recovery of enterococci from a residual or recurrent intra-abdominal infection is an indication for coverage by the appropriate antibiotic.

In some critically ill patients with evidence of sepsis and multiple organ dysfunction, a so-called tertiary peritonitis develops. This type of peritonitis is persistent and diffuse, responds poorly to drainage, and presents a difficult problem with respect to antibiotic management. The usual empirical therapy is no longer valid, as the infections are caused by *Staphylococcus epidermidis, Pseudomonas aeruginosa*, *Candida* species, and enterococci, rather than by *E. coli* and *B. fragilis*. Microbial cure is unusual in these situations despite efficacy of appropriate antibiotics for such infections. This failure in cure is probably caused by general failure of host defense mechanisms.

Operative management

Surgery is the primary treatment modality for managing intra-abdominal infection caused by perforation of a hollow viscus, transmural necrosis of the gastrointestinal tract, or pancreatic necrosis. The elimination or control of the infectious source and the reduction of peritoneal contamination by debridement and lavage are the cornerstones of primary operative treatment.

Source of intra-abdominal sepsis

In general, the source of intra-abdominal sepsis is eliminated by resection, closing the perforated viscus or resection of any necrosis. For pathology located in the large intestine, resection is usually performed followed by creation of a proximal enterostomy (Hartmann's procedure). In this location, primary anastomosis is avoided because of the high risk of anastomotic dehiscence. The risk associated with primary anastomosis of the small intestine following resection of a diseased segment is considered to be much lower. However, if peritoneal soiling is particularly extensive, resection plus proximal and distal enterostomy is advocated. Suturing a perforation is now mainly performed for perforated peptic ulcer. Additionally, it may be considered for a single small foreign body or traumatic perforation of the small intestine, if diagnosed early.

Debridement and lavage

Reduction of peritoneal contamination by debridement and intra-operative lavage is intended to prevent residual infection. As discussed, adjuvant substances such as foreign material, necrotic tissue, fibrin, bile, blood, or intestinal contents enhance the severity of the infectious process by stimulation of bacterial growth and impairment of neutrophil and macrophage function. Radical debridement—eliminating all adherent fibrinous deposits on peritoneal surfaces—was initially promising but later criticized because of the high risk of excessive bleeding from serosal surfaces and intestinal perforations. Intra-operative lavage has become a standard procedure, although the efficacy of intra-operative lavage has not been well documented in clinical studies. The addition of antibiotics or antiseptics in the lavage solution does not improve outcome in patients with intra-abdominal sepsis who are already receiving appropriate systemic antibiotics.

There is a great variance in surgical strategies to be followed after elimination of the source, debridement, and intra-operative lavage, particularly in patients with severe intra-abdominal sepsis. These strategies vary from a conservative 'wait and see' policy to the most aggressive method of 'planned relaparotomy'. Four strategies can be distinguished: a 'wait and see' policy with or without tube drainage, continuous postoperative peritoneal lavage, 'open' drainage (laparostomy), and 'planned relaparotomy'.

In the 'wait and see' policy, surgeons rely on host defenses to eradicate bacteria and adjuvant substances that remain intraperitoneally after the first operation. However, in severe forms of intra-abdominal infection, host defenses are unlikely to do this. This is illustrated by the high incidence of residual intra-abdominal infection in patients with diffuse peritonitis treated according to the 'wait and see' policy (McLauchlan *et al.* 1995). In this policy, surgeons also rely on clinical signs of residual or recurrent intra-abdominal infection in deciding whether or not to re-explore the abdominal cavity (i.e. 'on demand' relaparotomy). Clinical signs of residual intra-abdominal infection are often blunted in patients who are critically ill and cared for in intensive care units. Furthermore, diagnostic investigations such as ultrasound or CT for the detection of residual infection are often inconclusive or impossible to perform. Therefore, most 'on demand' relaparotomies are performed on the basis of otherwise unexplained progressive organ dysfunction or bacteremia. These relaparotomies are technically difficult, have a high morbidity, and often do not reverse organ dysfunction even when infectious foci are encountered and drained.

The role of tube drainage appears to be minor in diffuse intra-abdominal sepsis, mainly because of the rapidity with which drains are sealed off by fibrin deposition. Tubes placed intraperitoneally may cause visceral and vessel wall erosion with fistula formation and bleeding. Moreover, drains impair neutrophil function and potentiate infection. The only indications for drain insertion are evacuation of well-defined abscesses and provision of a preferential pathway for escape of visceral secretions such as bowel contents, pancreatic juice, and bile.

The concept of continuous postoperative peritoneal lavage has recently received more attention as a means of reducing residual intra-abdominal infection. At the time of initial surgery several peritoneal tubes are placed in 'strategic' positions to irrigate the abdominal cavity in the postoperative phase. Lavage with 10 to 20 liters of dialysis solution is advised to ensure dispersion of the solution throughout the abdominal cavity and to prevent fluid loculi. The mechanical action of continuous postoperative peritoneal lavage detaches necrosis, debris, and fibrin from surrounding tissues. Despite the potentially beneficial effects of continuous postoperative peritoneal lavage, reports on this method show conflicting results. Only for necrotizing pancreatitis has continuous postoperative peritoneal lavage been found a valuable adjunct to debridement of pancreatic necrosis.

The third strategy, leaving the abdomen open (laparostomy), was initially applied in France. The laparostomy should allow free drainage of exudate and simple access to the abdominal cavity after the first operation. In practice, free drainage becomes impossible within 1 to 2 days as a consequence of rapid adhesion formation by fibrin deposition. For the same reason, safe access to all parts of the abdomen will be difficult. The main advantage of this technique seems to be abdominal decompression. Leaving the abdomen open carries a risk of evisceration of abdominal contents, fistula formation, and abdominal wall hernia. In conjunction with nursing problems, these are considered the major limitations of this technique. These drawbacks have generated the application of alternative methods combining abdominal decompression with covering of intra-abdominal organs by, for example, Velcro analog (Sigma Medical, Apeldoorn, The Netherlands).

The concept of 'planned relaparotomy' has evolved from the technique of leaving the abdomen open. Re-exploration of the abdominal cavity is performed every 24 to 48 h to remove residual infectious material, such as necrosis, fibrin, and bacteria, and to prevent fluid loculi. The abdomen may be left open or temporarily closed by means of a zip or Velcro analog to relieve intra-abdominal pressure and facilitate re-exploration (Fig. 4).

The success of this approach has been attributed to improved elimination of the bacterial inoculum, necrotic material, and fibrin, and the possibility of early detection and correction of complications. Favorable results compared with more conventional surgical therapies in terms of mortality and residual abscess formation have been

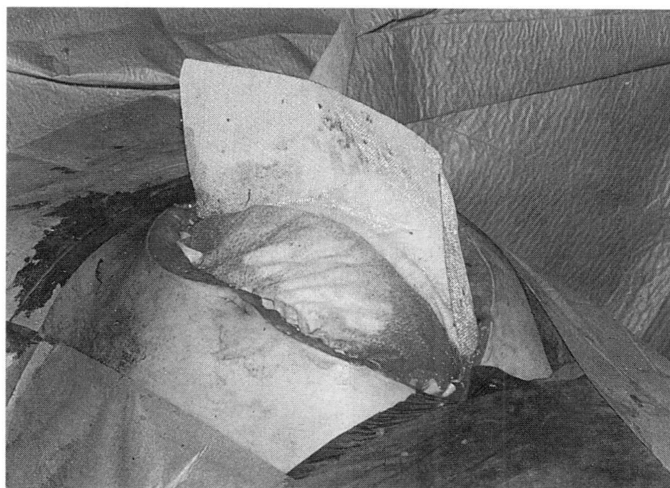

Fig. 4 The Velcro analog (artificial burr), consisting of two sheets of polypropylene–nylon material, sutured to the fascia in a patient with severe intra-abdominal sepsis.

reported, although prospective randomized studies are still lacking. More recently, disadvantages have been brought to light: unnecessary re-explorations, bleeding due to frequent manipulation of abdominal contents and removal of fibrin from serosal surfaces, 'spontaneous' fistula formation, interference with anastomotic healing by detaching adhesions and removing fibrin deposits, excessive fluid and protein losses, and large abdominal wall defects.

The ideal operative approach for severe intra-abdominal sepsis has yet to be established. More conventional approaches, such as 'wait and see' or continuous postoperative peritoneal lavage, do not seem to prevent residual or recurrent intra-abdominal infection and are associated with a high mortality. The method of 'planned relaparotomy' seems to decrease the rate of residual intra-abdominal infection. However, this method has a high complication rate and improvement in patient survival has yet to be clearly shown.

In some critically ill patients with the multiple organ dysfunction syndrome (MODS), signs of intra-abdominal sepsis persist in the absence of well-defined foci of residual infection or after successful surgical treatment of intra-abdominal infection. It has been hypothesized that bacteria in the gastrointestinal tract represent an undrained abscess that leads to the development of MODS. Interaction between these bacteria and cells of the monocyte–macrophage lineage releases cytokines and other cellular products which act as the effectors of end-organ dysfunction. The association between the disease processes leading to MODS and the egress of bacteria from the gastrointestinal tract (bacterial translocation) has been well established in animal studies (Wells *et al.* 1988). However, clinical studies do not unequivocally support the relation between bacterial translocation and MODS. It is obvious that administration of antibiotics or a reoperation without clear indication will not alter outcome in these patients. Other approaches such as reducing bacterial translocation, modulation of cytokine release, and alteration of the target-organ effects of cytokines are the subject of study at present. Mortality will depend on the ability to restore immunological host defense mechanisms.

References

Allen, L. (1936). The peritoneal stomata. *Anatomical Record*, **67**, 89–103.

Autio, V. (1964). The spread of intraperitoneal infection: studies with roentgen contrast medium. *Acta Chirurgica Scandinavica*, **123**, 1–31.

McLauchlan, G.J., Anderson, I.D., Grant, I.S., and Fearon, K.C.H. (1995). Outcome of patients with abdominal sepsis treated in an intensive care unit. *British Journal of Surgery*, **82**, 524–9.

Paterson-Brown, S., Vipond, M.N., Simms, K., Gatzen, C., Thompson, J.N., and Dudley, H.A.F. (1989). Clinical decision making and laparoscopy versus computer prediction in the management of the acute abdomen. *British Journal of Surgery*, **76**, 1011–13.

Wells, C.L., Maddeus, M.A., and Simmons, R.L. (1988). Proposed mechanisms for the translocation of intestinal bacteria. *Reviews of Infectious Diseases*, **10**, 958–79.

4.3.4 Acute acalculous cholecystitis

M. J. Shapiro

Key messages

- Acute acalculous cholecystitis is gallbladder inflammation without calculi.
- It may be due to ischemia, and risk factors include ventilation, dehydration, fasting, narcotics, transfusion, hyperalimentation, wounds, and renal failure.
- In addition to a right upper quadrant mass, tenderness, and fever, an ultrasonogram showing a thickened gallbladder wall, pericholecystic fluid, subserosal edema, intramural gas, or mucosal membrane sloughing is consistent with the diagnosis.
- Cholecystectomy rather than cholecystostomy is the preferred treatment.

Introduction

Acute acalculous cholecystitis is the abrupt onset of inflammation of the gallbladder without evidence of calculi; it accounts for 2 to 15 per cent of all cases of acute cholecystitis. The incidence is 0.4 to 4 per cent in burn patients, 0.5 to 1.6 per cent in postoperative and post-trauma patients, and up to 4 per cent in patients on hyperalimentation.

Pathophysiology

Although the pathophysiology of acute acalculous cholecystitis is not clear cut, many hypotheses have been proposed. Visceral hypoperfusion of the gallbladder may lead to ischemia in critically ill hypotensive patients. Narcotics such as morphine, which are used in the intensive care unit (ICU), may lead to biliary stasis and sphincter of Oddi spasm, causing gallbladder distention and possible ischemia. In the fasting state, the gallbladder may not contract, and when it does contract against a sludge-filled cystic duct, as may occur with initiation of oral feeding after a period of fasting, acute acalculous cholecystitis may occur secondary to the increase in intraluminal pressure. Lack of neurohumoral stimuli may occur in the biliary tract in the fasting state with a decrease in gastrointestinal secretion which may lead to an increase in bile viscosity and concentration, an increase in water absorption, or an increase in mucin secretion. This may lead to acute acalculous cholecystitis.

Risk factors

Acute acalculous cholecystitis occurs in patients with chronic debilitating disease, or with superimposed critical illness after major burn injury. It may be associated with a decrease in gastric motility and prolonged ileus. Risk factors for the development of acute acalculous cholecystitis include mechanical ventilation, hyperalimentation, dehydration, fasting, narcotic utilization, massive blood transfusions, open wounds and abscesses, and chronic renal failure. Up to 4 per cent of patients with an ileus and on hyperalimentation may develop

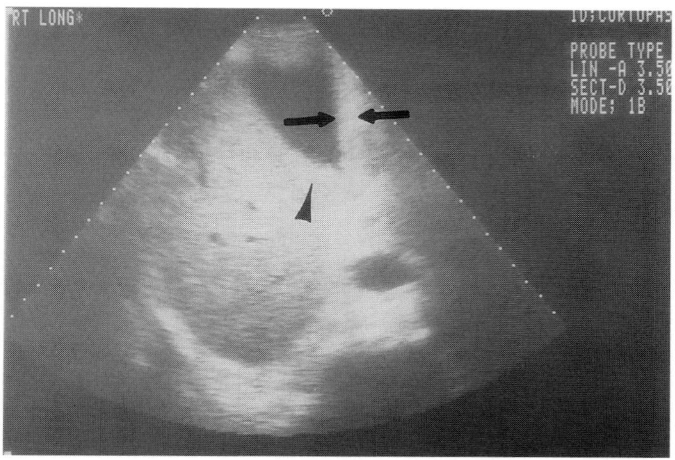

Fig. 1 Ultrasonogram showing a thickened gallbladder wall (between arrows) and sludge (single arrow) in a patient with operatively confirmed acute acalculous cholecystitis.

acute acalculous cholecystitis. It has been suggested that the use of positive end-expiratory pressure (PEEP), which produces an increase in bile and biliary stasis with a decrease in portal flow, may also be associated with acute acalculous cholecystitis.

Diagnosis

The signs and symptoms on clinical examination are not consistent, although the presence of a palpable right upper quadrant mass with tenderness and fever in an elderly male should raise suspicion of acute acalculous cholecystitis. In a study of 22 ICU patients, only plasma alkaline phosphatase and urea levels were found to be significantly elevated when comparing the presence of necrosis or gangrene versus inflammation of the gallbladder. 'Liver function' studies are non-specific, particularly in the ICU patient who may have vascular insufficiency leading to an elevation in serum chemistries for other reasons. Although technetium hepato-iminodiacetic acid (HIDA) scans can be perfomed in the ICU and have 100 per cent sensitivity and 90 per cent specificity for acute acalculous cholecystitis, false-negative results have been reported. The use of the synthetic cholecystokinin octapeptide, sincalide, for cholescintigraphy has a high incidence of false negativity. In addition, some of the ICU patients may not be fed enterally, and thus radionuclide studies will give false-positive results, accounting for a high sensitivity but low specificity.

Ultrasound is a portable and less expensive study which can be done at the bedside in the ICU (Fig. 1). Significant findings include a gallbladder wall thickness of 4 mm or more, the presence of pericholecystic fluid, subserosal edema in the absence of ascites, intramural gas, and sloughing of the mucosal membrane. The presence of biliary sludge is a soft sign, and in the presence of right upper quadrant pain coupled with normal liver function and amylase levels is only 75 per cent accurate. A sonographic Murphy's sign of gallbladder distension and a thickened wall on ultrasound with right upper quadrant pain induced with the ultrasound probe may be a useful diagnostic tool. CT scanning may also be helpful, with positive findings for acute acalculous cholecystitis being similar to those seen on ultrasound. However, CT scanning requires transportation of patients out of the ICU with its inherent risks.

Therapy

There is still some debate about the appropriate therapeutic approach to patients with acute acalculous cholecystitis. Although patients have been successfully observed with supportive therapy only, there have also been reports of failure of conservative therapy. Percutaneous non-transhepatic and transhepatic cholecystostomy under ultrasound guidance has been used with some success; however, a mortality of up to 10 per cent has been reported. Owing to the lack of correlation between symptomatology and pathology, and the high incidence (59 per cent) of necrosis or gangrene on pathological review of the gallbladder wall, cholecystectomy is the definitive procedure of choice. The use of laparoscopy in the ICU is inviting; however, it has not been critically analyzed. Nonetheless, owing to the presence of necrosis or gangrene in a significant number of patients and the lack of correlation of physical findings with the presence of acute acalculous cholecystitis, removal of the gallbladder, if feasible, is the safest option.

Bibliography

Barie, P.S. and Fischer, E. (1995). Acute acalculous cholecystitis. *Journal of the American College of Surgeons*, **180**, 232–44.

Jennings, W.C., Drabek, G.A., and Miller, K.A. (1992). Significance of sludge and thickened wall in ultrasound evaluation of the gallbladder. *Surgery, Gynecology and Obstetrics*, **174**, 394–8.

Kurzweil, S.M., Shapiro, M.J., Andrus, C.H., Wittgen, C.M., Herrmann, V.M., and Kaminski, D.L. (1994). Hyperbilirubinemia without common bile duct abnormalities and hyperamylasemia without pancreatitis in patients with gallbladder disease. *Archives of Surgery*, **129**, 829–33.

Shapiro, M.J., Luchtefeld, W.B., Kurzweil, S., Kaminski, D.L., Durham, R.M., and Mazuski, J.E. (1994). Acute acalculous cholecystitis in the critically ill. *American Surgeon*, **60**, 335–9.

4.4 Pancreatitis

4.4.1 Pathophysiology and diagnosis

T. Dugernier and M. S. Reynaert

Key messages

- Early differentiation between edematous (90 per cent) and necrotizing (10 per cent) pancreatitis is important for both prognostication and therapeutic strategy.
- Clear-cut identification of an etiological factor impacts on therapy, prognosis, and prevention of recurrences. Alcohol abuse and biliary stones or sludge account for more than 85 per cent of episodes.
- Abdominal pain, vomiting, and hyperamylasemia are diagnostic cornerstones, and CT is the imaging method of choice for diagnostic dilemmas.
- Both severity and prognosis depend on the magnitude of the inflammatory reaction initiated in the retroperitoneal space, the presence and extent of regional necrosis, and any subsequent bacterial contamination.
- Multisystem organ dysfunction syndrome typically complicates the early phases and is best ascribed to spillage of inflammatory mediators, activated enzymes, and vasoactive substances from the retroperitoneal areas into the systemic circulation.
- Mortality associated with severe acute pancreatitis still ranges between 10 and 20 per cent.

Introduction

Diagnosis of severe acute pancreatitis can be difficult; indeed, 30 to 40 per cent of fatal cases pass undetected until postmortem discovery. Outcome is highly unpredictable at the onset, as severe local, regional, and/or remote complications may arise subsequently. The pathogenesis is still poorly understood so that truly specific treatment is lacking; management strategies and their timing remain a matter of controversy.

Most cases (80–90 per cent) of acute pancreatitis are mild, self-limiting, and morphologically characterized by edematous interstitial inflammation that resolves rapidly following a period of bowel rest, pain control, and fluid/electrolyte replacement. The disease course is seldom complicated and overall mortality is below 3 per cent. However, 10 to 20 per cent progress to a devastating illness, which is classified as severe acute pancreatitis on the basis of a fatal attack or failure to settle on simple supportive therapy because of local and/or systemic complications. Morphological hallmarks are extensive pancreatic and retroperitoneal inflammation with superimposed patchy or generalized areas of necrosis and hemorrhage in the pancreas and surrounding tissues. The propensity and degree of severity of both local and remote complications are closely linked to the magnitude of the inflammatory reaction and the presence and extent of ensuing regional necrosis (Beger *et al.* 1988). Severe acute pancreatitis still carries a mortality between 10 and 20 per cent.

Etiology

Epidemiological studies performed using similar admission criteria and over different time periods are limited to a few geographical areas. All indicate that the incidence of acute pancreatitis is increasing. Biliary stones and alcohol abuse account for 70 per cent of cases; gallstone impaction is more common in females, and chronic alcoholism is more common in males. A minority (5–10 per cent) are caused by various other conditions (Table 1). Acute idiopathic pancreatitis occurs in 20 to 40 per cent, although biliary sludge can be

Table 1 Etiological factors in acute pancreatitis

Biliary stones
Alcohol abuse
Idiopathic (**biliary sludge** in 65%)
Postoperative (gastric, biliary, cardiopulmonary bypass)
Endoscopic retrograde cholangiopancreatography
Abdominal trauma (including iatrogenic)
Drug induced (e.g. metronidazole, valproate, azathioprine, tetracycline, steroids, diuretics)
Infections (e.g. hepatitis, cytomegalovirus, Coxsackie virus, mumps, ascaridiasis)
Hypertriglyceridemia/hyperchylomicronemia
Hypercalcemia
Systemic vasculitis
Tumors
Inherited or acquired abnormalities of pancreatic and/or biliary ducts and papilla

Causes associated with recurrent acute pancreatitis are indicated in bold type.

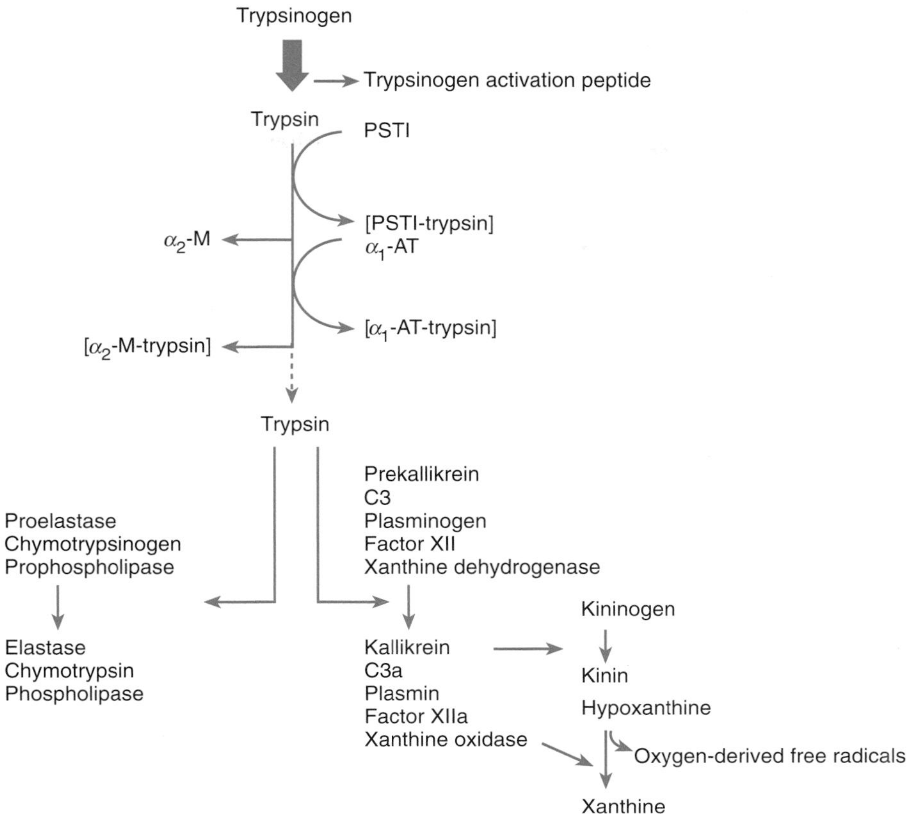

Fig. 1 Excess intra-acinar generation of trypsin results in protease–antiprotease imbalance, secondary glandular and periglandular activation of pancreatic proenzymes and factors of the cascade systems of proteases, and induction of an oxidant stress via the hypoxanthine–xanthine oxidase pathway: α_1 -AT, α_1-antitrypsin; α_2-M, α_2-macroglobulin; PSTI, pancreatic secretory trypsin inhibitor. (Modified with permission from Ohlsson (1988).)

demonstrated in approximately two-thirds of these. A causal association has been made between biliary sludge and acute pancreatitis.

Most etiological factors have a significant impact on investigation, prognosis, and therapeutic approach. Recurrent attacks should trigger appropriate investigation for biliary sludge or very small stones, hypertriglyceridemia, drug-induced pancreatitis, and occult abnormalities in the biliary and/or pancreatic duct.

Pathophysiology of regional necrosis

It is not clear how the inflammatory necrotizing process in the retroperitoneal space is induced. Proposed pathophysiological mechanisms are largely based on animal models. Three key factors, acting sequentially, may account for regional necrosis:

(1) intraglandular activation of pancreatic enzymes, which is probably responsible for pancreatic autodigestion;

(2) overstimulation of inflammatory effector cells (e.g. macrophages, polymorphonuclear leukocytes);

(3) vascular mechanisms (ischemia/reperfusion, hemorrhage).

The earliest morphologically detectable lesions in animal models involve acinar cells (necrosis), followed later by interstitial hemorrhage and inflammation (Steer 1992). Intra-acinar activation of proteolytic/lipolytic pancreatic enzymes, in particular trypsin, may be the initial mechanism triggering glandular autodigestion and terminating in regional necrosis (Steer 1992). As with ischemia, trypsin catalyzes proteolytic conversion of xanthine dehydrogenase to xanthine oxidase (Fig. 1). Depending upon the amount of active trypsin, secondary exhaustion of the antiprotease systems, and the extent of acinar cell injury, these cascade activations extend to the gland and the peripancreatic region. Transient impaction of biliary stones in the terminal common bile duct or spasm/edema of the sphincter of Oddi persisting after stone passage may conceivably lead to premature zymogen activation (Steer 1992). Alcohol may decrease activity of trypsin inhibitors, increase production of lysosomal hydrolases, and impair separation between the latter and zymogens in the Golgi apparatus.

Other factors which determine the severity of pancreatitis include local microcirculatory disturbances and massive glandular invasion by macrophages and polymorphonuclear neutrophils (Gross *et al.* 1993). The close interplay between the direct toxic effect of prematurely activated digestive enzymes, local microcirculatory impairment, and toxic substances released by overactivated neutrophils and macrophages is responsible for the severity of the inflammatory reaction and the ensuing necrosis (Fig. 2). Although premature intra-acinar activation of zymogens seems to be the initial step in this chain of events, the respective roles of these three mechanisms in the development of pancreatic necrosis remain controversial (Gross *et al.* 1993).

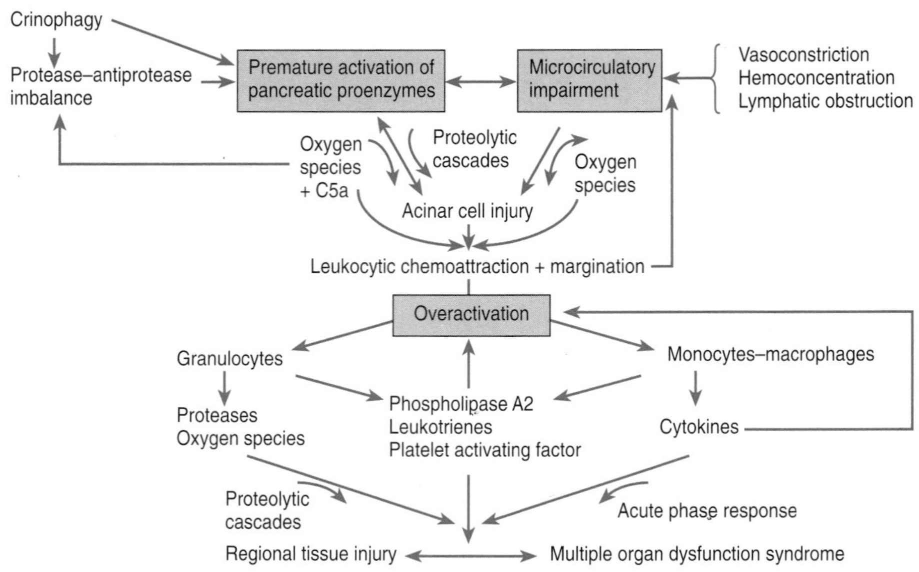

Fig. 2 Pathophysiology of pancreatic necrosis and subsequent multiple organ dysfunction syndrome: interrelation between premature activation of trypsin, microcirculatory impairment, and overstimulation of macrophages and polymorphonuclear neutrophils (PMNs).

Pathophysiology of remote organ dysfunction

The clinical course of severe acute pancreatitis can be divided into an early 'toxemic' phase (0–15 days), characterized by the emergence of distant organ damage, and a later 'necrotic' phase when local complications prevail (Table 2). However, the two phases are often superimposed, particularly when necrosis becomes infected early on.

Systemic complications characterizing the toxemic phase, in particular cardiocirculatory and respiratory failure, contribute significantly to morbidity and mortality. Most are mediated by spillage into the systemic circulation of pancreatic enzymes and toxic substances including proinflammatory cytokines generated retroperitoneally by the inflammatory reaction and the necrotizing process. Among the various routes of autointoxication of the organism (pancreatic veins, transperitoneal and transpleural absorption), retroperitoneal and peripancreatic lymphatics which enter the thoracic duct play a predominant role.

The hemodynamic profile of the early phase of severe acute pancreatitis is usually hyperdynamic, although severe myocardial depression occasionally occurs. Contributory factors include intravascular volume depletion secondary to increased vascular permeability, abdominal fluid sequestration and/or gastrointestinal and retroperitoneal hemorrhage, direct or proteolytic activation of the contact system, a myocardial depressant factor synthesized by acinar cells, and release of prostanoids (platelet-activating factor) and cytokines by activated leukocytes in the necrotic areas.

Metabolic sequelae include an increase in resting energy expenditure averaging 125 per cent of predicted values, elevated protein catabolism, unsuppressed hepatic gluconeogenesis, and peripheral insulin resistance. Hepatic and peripheral insulin resistance, β-cell dysfunction, and hyperglucagonemia result in hyperglycemia. Diabetic ketoacidosis or non-acidotic diabetic coma may be presenting features, particularly in patients with hyperlipidemia. Hypocalcemia and hypertriglyceridemia are also seen.

Respiratory complications commonly arise from pleural effusions and elevated/limited excursion of the diaphragm due to pain/ileus with secondary pulmonary atelectasis, hypoxemia, and pulmonary infection. All patients with severe acute pancreatitis develop pleural effusions, and 20 per cent suffer early acute respiratory distress.

Disseminated intravascular coagulation mostly results from the proteolytic effects of circulating free trypsin, although it is usually clinically silent. Oliguria and mild azotemia are usually attributed to intravascular volume depletion, although histological studies have shown deposits of fibrin and fibrinogen in the glomerular capillaries. Thus impaired renal perfusion and ischemia may be related to a trypsin-induced hypercoagulable state.

Central nervous system disorders include psychosis, confusion, coma, and seizures. Alcohol withdrawal, fat embolism, intravascular coagulation, hypoxemia, and hyperosmolarity have all been incriminated.

The multiple organ dysfunction syndrome seen with severe acute pancreatitis is indistinguishable from that seen in septic shock or following extensive tissue injury. It supports a final common pathway and reflects the capacity of the necrotizing inflammatory process to produce and release multiple mediators, enzymes, and vasoactive substances into the systemic circulation (Gross *et al.* 1993).

Pathophysiology of local complications of severe acute pancreatitis

Over 50 per cent of patients with severe acute pancreatitis proceed to a spontaneous and uneventful resolution.

Table 2 Clinical manifestations of severe acute pancreatitis—a model of multiple organ dysfunction syndrome

Remote organ damage			
Cardiac	Hyperdynamic circulatory status Myocardial depression Hypovolemia Low O_2 extraction ability	→Cardiocirculatory failure	
Pulmonary	Pleural effusion Atelectasis ARDS	→Respiratory failure	
Renal	Oliguria Tubular necrosis	→Acute renal failure	**Toxemic phase (0–15 days)**
Hematological	DIC	→Severe hemorrhage	↓
Metabolic	Hypocalcemia Hyperglycemia Hypertriglyceridemia Acidosis Hypermetabolism	→Malnutrition Ketoacidosis ARDS	↓
Central nervous system	Encephalopathy	→Pulmonary atelectasis and aspiration	↓
Loco-regional complications			↓
Infection	Infected pancreatic necrosis Pancreatic abscess Infected pseudocyst		↓
Hemorrhage	Retroperitoneal (pseudoaneurysm) Gastrointestinal Pseudocyst		**Necrotic phase (>15 days)**
Disruption of pancreatic duct	Pancreatic ascites Pleural/pericardial fistula Acute pseudocyst		
Adjacent organs	Thrombosis of splenic vessels Bowel infarction Gastric/duodenal necrosis Obstructive jaundice (biliary stricture)		

ARDS, acute respiratory distress syndrome; DIC, disseminated intravascular coagulation.

With necrotizing pancreatitis, 40 to 60 per cent develop infection of the necrosis. This is more common in biliary pancreatitis. The risk of pancreatic sepsis is maximum in the third week, although bacterial invasion has been documented in 30 to 50 per cent of patients within 14 days of the onset of symptoms. The earlier infection occurs, the higher is the mortality, as early mixing of bacteria with ongoing enzymatic and necrotizing processes appears to generate a highly toxemic course and amplifies distant tissue injuries. In these early infected cases, culture usually yields a single organism of a Gram-negative species (75 per cent) originating from the digestive tract by translocation, perforation of a hollow viscus, infected bile, or following retrograde cholangiopancreatography, or via hematogenous or lymphatic spread. Anaerobic organisms are present in only 10 per cent of cultures. Infected pancreatic necrosis, where necrotic tissue is the dominant element, differs from pancreatic abscess by the absence of a well-loculated collection of pus, an earlier presentation, and a much higher mortality (Beger *et al.* 1988).

Pancreatic abscess consists of a collection of pus, enclosed by inflammatory walls, resulting from liquefaction of necrotic areas inside or outside the pancreas that have become secondarily infected. Over 50 per cent are polymicrobial with a predominance of enteric bacteria. *Candida albicans* is often cultured in those previously treated with broad-spectrum antibiotics. In contrast with infected pancreatic necrosis, pancreatic abscess occurs when the active phase of pancreatitis is over and it tends to present with a more indolent process. Remote complications are less frequent, mortality is lower, and sometimes a state of relative well being interplays between the toxemic phase and the clinical emergence of abdominal sepsis.

Acute pseudocyst develops in 10 to 20 per cent of patients with severe acute pancreatitis. It is more common in alcoholic pancreatitis, but hemorrhage and infection are more likely to complicate cysts associated with gallstones. In contrast with pancreatic fluid, collections which represent a mixture of inflammatory exudate, necrotic tissue, and blood, an acute pseudocyst consists of an effusion of pancreatic juice rich in amylase that lacks an epithelial lining and has gradually become enclosed by fibrous walls over a period of 4 to 6 weeks. About 50 per cent disappear spontaneously within 1 to 2 months. Pancreaticopleural or pancreaticopericardial fistula or, more commonly, pancreatic ascites may ensue if rupture occurs. This peritoneal exudate is a major pool of pancreatic enzymes and a potential source of toxic substances. About 30 to 50 per cent become secondarily symptomatic via infection, adjacent extrinsic organ compression, or bleeding.

Serious gastrointestinal complications include bleeding, mechanical obstruction, colonic necrosis, and/or development of perforation or fistula. Persistent abdominal sepsis and fecal fistula are usual presenting features. Vascular thrombosis may manifest as gastrointestinal bleeding, perforation, fistulas, or late stricture. Obstructive jaundice is usually due to biliary stricture and fibrosis, and not to extrinsic compression by a pseudocyst.

Diagnosis

Acute pancreatitis

The gold standards are direct inspection at laparotomy and microscopic examination. However, routine diagnostic cornerstones remain abdominal pain, nausea, vomiting, and hyperamylasemia. Clinical signs, although suggestive, are notoriously non-specific. Remote organ failure and metabolic disturbances such as hyperglycemia and hypocalcemia, although rarely present at the onset, should alert the clinician.

Hyperamylasemia has a high sensitivity (> 95 per cent) within 24 h of symptom onset but a lower specificity (70 per cent). The magnitude of increase in serum amylase does not correlate with disease severity. Furthermore, normal amylase levels are encountered in cases of hyperlipidemic pancreatitis, acute exacerbations of chronic pancreatitis, and particularly with delayed measurement.

The diagnostic value of isoamylase determination (pancreatic versus salivary) remains controversial, although the lipase assay is as sensitive as but more specific than amylase. Other pancreatic enzymes, such as immunoreactive trypsin, elastase, and phospholipase A_2, are elevated in acute pancreatitis but, except in cases of delayed admission, add little to the diagnosis of suspected cases and a normal or only mildly raised amylase. Moreover, these tests are not performed routinely, they are expensive, and their prognostic significance remains doubtful. Rarely, elevation of amylase in pleural or ascitic fluid may coexist with a normal serum level and so is occasionally of some diagnostic utility.

Plain abdominal radiography may reveal adynamic ileus, increased gastrocolic separation, sentinel loops, and the colon cut-off sign at the splenic flexure. It aids diagnosis of other potentially confounding abdominal emergencies, such as hollow viscus perforation or mesenteric infarction. Peritoneal aspiration, smelling the fluid, and Gram staining are useful, since the presence of peritoneal infection on admission is unusual in acute pancreatitis and suggests other pathology that mandates urgent laparotomy.

Although convenient, non-invasive, inexpensive, and repeatable, ultrasound examination can be inconclusive and occasionally misleading. It is highly operator dependent and in 30 to 40 per cent the gland cannot be evaluated, mainly due to excessive overlying bowel gas. It is inaccurate for detection of pancreatic necrosis or regional infection and for delineating the extent of peripancreatic fluid collections.

Abdominal contrast-enhanced CT is the imaging method of choice for cases that pose a diagnostic dilemma. Satisfactory evaluation is obtained in almost all patients, and diagnostic abnormalities (pancreatic swelling, areas of attenuation in the gland, and peripancreatic fluid collections) persist after 1 week in 80 per cent. The specificity of an admission CT is nearly 100 per cent; although sensitivity is about 85 per cent, most episodes undetected by CT are mild. It may provide significant assistance where enzyme levels are normal or falsely elevated. Importantly, CT allows a reliable appraisal of the presence and extent of intra- and extrapancreatic necrosis by high-resolution CT and prolonged bolus scanning techniques (dynamic pancreatography). This provides obvious diagnostic and prognostic information and implications with regard to surgical intervention. Dynamic pancreatograms should usually be obtained within 48 h of admission for severe episodes. Focal areas of diminished enhancement after administration of contrast material allow differential diagnosis between edematous and necrotizing pancreatitis with an overall accuracy rate exceeding 90 per cent. The size of these defects correlates closely with the amount of devitalized pancreatic tissue assessed at operation and confirmed by histomicroangiographic studies. Ancillary CT signs of necrotizing pancreatitis include marked glandular enlargement, lowered pancreatic density values on precontrast CT, and extensive extrapancreatic extension of inflammation. Although the degree of pancreatic necrosis correlates with the extent of peripancreatic fluid collections, these are not mutually inclusive.

Failure to diagnose acute pancreatitis is generally ascribed to an atypical presentation, namely the absence of abdominal pain. This may occur during artificial ventilation with analgesia/sedation, or because shock, severe gastrointestinal bleeding, diabetic coma, or respiratory failure may be the presenting manifestations of a painless episode or conceal abdominal pain. In rare instances (up to 5 per cent) a definite diagnosis can only be made at laparotomy.

Diagnosis of the etiological factor: impact on recurrences

A causal factor should be sought (Table 1). Because of the immediate therapeutic implication of calculus impaction and total prevention of further attacks after removal of stones from the biliary tree, early detection is mandatory. Up to 20 per cent of apparently alcohol-related attacks occur in patients who also have gallstones. Female gender, age (> 50 years), and early increases (within 48 h of the onset of symptoms) in serum alanine aminotransferase above 100 IU/l, alkaline phosphatase above 300 IU/l, and bilirubin above 40 mmol/l (2.3 mg/dl) predict the presence of stones in 70 to 90 per cent of patients with gallstone-associated pancreatitis. Plain abdominal radiography reveals stones in up to 15 per cent of cases; however, when the gallbladder is outlined, stones are accurately detected by ultrasonography in over 90 per cent of cases. As overlying bowel gas precludes satisfactory evaluation in 20 to 30 per cent, the overall sensitivity for early ultrasound examination detecting gallstones is decreased to 70 per cent. Sandy stones without acoustic shadows and microlithiasis also produce false negatives. Common bile duct dilatation does not necessarily equate with stone impaction, since swelling of the pancreatic head may be responsible. Although specific, the CT scan has a sensitivity of only 30 to 50 per cent in identifying gallstones.

Common bile duct stones and impaction are best diagnosed by endoscopic retrograde cholangiopancreatography (ERCP), although this risks exacerbating pancreatitis and introducing infection into devitalized pancreatic areas. In 10 to 20 per cent technical failure occurs due to duodenal stenosis by pancreatic swelling, papillary edema, and previous gastrectomy. Because of these technical difficulties and the potential morbidity, ERCP should only be carried out in patients whose clinical course fails to improve despite full intensive care unit (ICU) support and in whom ampullary or common bile duct stone impaction is suspected by ultrasound and/or clinical/biochemical signs of acute cholangitis. Percutaneous transhepatic cholangiography avoids the risk of exacerbation of pancreatitis but, for calculus obstruction of the ampulla, does not allow the immediate therapeutic option of endoscopic papillotomy and stone retrieval.

Biliary sludge and microlithiasis can be diagnosed by ultrasonography, ERCP, and, with more accuracy, biliary drainage and subsequent microscopic analysis. There is a substantial initial false-negative rate. Recurrence of hypertriglyceridemia-related pancreatitis can be prevented with diet and lipid-lowering agents.

Accurate identification of the causal factor of the attack in the later phase is important, since recurrent episodes may be prevented in selected cases by appropriate endoscopic sphincterotomy or stenting, surgical sphincteroplasty, and other operative interventions.

Table 3 Diagnostic work-up for acute pancreatitis

The acute episode
Clinical signs/remote organ failures
Serum amylase/lipase assays
Abdominal plain radiography
Abdominal contrast-enhanced CT
Peritoneal aspiration
Laparotomy
The causal factor
Patient history (alcohol, drugs, trauma, surgery, etc.) and demographic data
Serum aminotransferases, alkaline phosphatase, bilirubin
Serum calcium
Serum triglycerides, lipoprotein electrophoresis
Ultrasonography and abdominal plain radiography
ERCP+bile sampling+sphincter of Oddi manometry
Percutaneous transhepatic cholangiography
Remote complications
Pulse rate, blood pressure, urinary output, temperature
Fluid balance, central venous pressure, pulmonary artery catheterization
SvO_2 monitoring, indirect calorimetry, gastric tonometry
Arterial blood gas analysis, chest radiography
Blood glucose, serum electrolytes, urea, creatinine, coagulation tests, hematocrit, platelet/white blood cell counts
Local complications
Abdominal contrast-enhanced CT
Ultrasonography
ERCP+pancreatic ductography
Abdominal angiography
Contrast studies (digestive tract, fistulous tracts)

Recognition of complications

Remote organ dysfunction typically supervenes early in the attack and justifies close ICU monitoring. Regular assessment of pulmonary gas exchange, acid–base status, fluid balance, central venous pressure determination, and, in selected cases, pulmonary artery catheterization, indirect calorimetry, and gastric tonometry, are valuable diagnostic and management tools which contribute to understanding of the pathophysiological mechanisms. Relevant biochemical work-up includes regular determinations of blood glucose, electrolytes, calcium, urea and creatinine, coagulation tests, hematocrit, and full blood count.

Local complications typically prevail after day 15, although severe arterial hemorrhage, intestinal or colonic perforation, or infection of pancreatic necrosis may occur earlier. Close monitoring of the regional inflammatory process and necrosis is mandatory throughout, particularly in those patients who are deteriorating or not responding to conservative management and in whom surgery or percutaneous drainage might be contemplated. Repeated abdominal CT scanning is preferable, particularly for diagnosis, accurate localization and follow-up (diameter, walls) of any acute pseudocyst, pseudoaneurysm formation, pancreatic hemorrhage, pancreatic necrosis, peripancreatic fluid collections, and, occasionally, fistulous tracts and portal, mesenteric, or splenic vein thrombosis.

Bacterial or fungal infection of devitalized areas inside and outside the pancreas is responsible for 80 per cent of deaths from severe acute pancreatitis. Early identification is crucial, since drainage and often surgical debridement are mandatory. The associated mortality is directly related to any delay in diagnosis and proper treatment. Persistent remote organ failure or unexpected clinical deterioration (epigastric pain, vomiting, abdominal distension with guarding or rebound) should raise the suspicion of regional infection and prompt appropriate investigation. Clinical and biochemical indicators of pancreatic infection include the combination of fever exceeding 38.5°C, base excess above 4 mmol/l, and hematocrit below 35 per cent. However, precise differentiation between inflammatory sterile and infected collections cannot be made reliably on clinical and laboratory grounds. Fever and leukocytosis are absent in 15 to 20 per cent with infected pancreatic necrosis or pancreatic abscess who may present with a disarmingly indolent course. Conversely, sterile necrosis may mimic abdominal sepsis. Serial CT evaluation is the best method for early definite diagnosis of infection, either by demonstrating extraintestinal gas in or outside the gland (15–20 per cent) or, more reliably, by enabling percutaneous fine-needle aspiration sampling of inflammatory collections with subsequent Gram staining and aerobic/anaerobic cultures. This technique is safe and

Table 4 Multiple prognostic factors in acute pancreatitis: variables, time of collection, advantages, and definition of severe acute pancreatitis

Ranson scoring system	Imrie scoring system	APACHE II score
At admission or diagnosis		**Advantages**
Age > 55 years		Easy to obtain on admission
WBC > 16 × 10^9/l (16 000/mm^3)		Better accuracy (sensitivity and specificity)
Blood glucose > 10 mmol/l (200 mg/dl)		Reassessment possible
LDH > 350 IU/l		Monitoring of pancreatic necrosis
SGOT > 250 Sigma Frankel units/l		Prediction of regional infection
Within 48 h		
Serum calcium < 2 mmol/l (8 mg/dl)	WBC > 15 × 10^9/l (15 000/mm^3)	
Hematocrit fall > 10%	Blood glucose > 10 mmol/l (200 mg/dl) (no diabetes)	
$Pa\text{O}_2$ > 8 kPa (60 mmHg)	BUN > 16 mmol/l (46 mg/dl) (no response to IV fluids)	
Base deficit > 4 mmol/l	LDH > 600 IU/l	
BUN rise > 1.8 mmol/l (5 mg/dl)	AST > 200 IU/l	
Fluid sequestration > 6 liters	Plasma albumin < 32 g/l (3.2 g/dl)	
	Plasma uncorrected calcium < 2 mmol/l (8 mg/dl)	
	$Pa\text{O}_2$ < 8 kPa (60 mmHg)	
SAP if more than two criteria are met during the first 48 h after admission		SAP if score greater than 9 on admission

AST, aspartate aminotransferase; BUN, blood urea nitrogen; LDH, lactate dehydrogenase; $Pa\text{O}_2$, arterial partial pressure of oxygen; SAP, severe acute pancreatitis; SGOT, serum glutamic oxaloacetic transaminase; WBC, white blood cell count.
Modified with permission from Reynaert *et al.* (1990).

accurate (>90 per cent correct) for early recognition of regional infection. It enables a more timely and rational basis for percutaneous or operative interventions and for optimal antibiotic treatment, avoiding the excessive mortality rate associated with delayed recognition. Importantly, since bacterial contamination is heightened by necrosis, needle aspiration should be preferentially directed to all non-perfused areas as identified by intravenous contrast. Similarly, all potentially septic foci such as pseudocysts should be sampled. With a protracted course and initial sterile sampling, serial percutaneous aspiration should be performed. Preoperative CT is helpful in localizing infected collections and areas of necrosis to be removed by the surgeon.

Ultrasonography is useful for delineating the number, size, and location of acute pseudocysts, pancreatic ascites, and portal or splenic thrombosis. As with CT, it may guide fine-needle aspiration of suspected areas of infection.

Angiography should be the preoperative diagnostic step for ruptured pseudoaneurysm and subsequent severe hemorrhage. Other indications for angiography include identification of splenic vein thrombosis with subsequent variceal hemorrhage.

Elevated fluid amylase and protein levels are diagnostic clues of pancreatic fistula. ERCP or percutaneous pancreatic ductography under CT or ultrasound guidance further aid diagnosis and management.

The varying diagnostic modalities for the acute episode, the causal factor, and each stage of the attack are outlined in Table 3.

Prognostic stratification

Early assessment and prediction of severity is important to avoid costly and invasive monitoring and treatment in the majority of patients who run a benign course. In severe forms, it is necessary to monitor the patient closely from admission through to an intensive care area, to anticipate early and late complications, to consider aggressive treatment directed at necrosis and its local and systemic consequences, and to judge response to therapy.

Prognostic systems

Ranson and Imrie used clinical and laboratory criteria to provide early classifications of pancreatic damage (Table 4). The high proportion of alcoholic patients in the original studies produced misleading results with regard to those with a gallstone etiology. This shortcoming led to slight modifications to extend accuracy; however, the poor specificity of Ranson scoring and the limited sensitivity of the Imrie system impair their practical value for early assessment of acute pancreatitis. In addition, laboratory data collection is often incomplete or takes 48 h to obtain, thereby introducing potential errors or delaying

Table 5 Single prognostic factors for early detection of severity and prediction of early/late complications

	Criteria for SAP	Advantages	Drawbacks
Peritoneal lavage	>20 ml of ascitic fluid or dark-colored ascitic fluid or peritoneal lavage fluid darker than a pale straw color	Speed Availability Correct misdiagnosis Selection for therapeutic lavage	Invasive Insensitive in gallstone pancreatitis Poor predictor of late complications
Hypoxemia	$Pa\text{O}_2 < 8$ kPa (60 mmHg)	Diagnostic tool (respiratory insufficiency)	Insensitive (reflects one aspect of SAP)
α_2-Macroglobulin	<1.5 g/l (150 mg/dl)	Indicator of necrosis	Limited availability Poor predictor of late complications
C-reactive protein	>300 mg/l on day 2 >120 mg/l on day 8	Speed Simplicity Probably correlates with necrosis Predicts early/late complications	Limited sensitivity 48 h delay
Granulocyte elastase Phospholipase A_2 Trypsinogen activation peptides		Early indicators of severity and necrosis Sound pathophysiological basis	Limited availability Further evaluation needed
Abdominal CT	Enhancement defects Peripancreatic fluid collections	Reliable detection of necrosis and its extent Diagnosis and prediction of early/late regional complications and outcome Complementary to physiological assessment	Expensive and complex Limited availability Limited specificity Not convenient for severely ill patients

SAP, severe acute pancreatitis.
Modified with permission from Reynaert *et al.* (1990).

effective management of ongoing complications. As these systems only include early data, they may be valuable in identifying those who will benefit most from early intensive care. However, acute pancreatitis is a dynamic process; although these systems correlate with the extent of necrosis and with the likelihood of abscess and death, they do not enable accurate prediction of late septic complications on an individual basis.

Other scores for quantifying severity of acute illness have proved more accurate in identifying severe attacks and in predicting pancreatic necrosis and regional infection. On admission, an APACHE II score of at best 13 correctly predicted outcome in 77 per cent of attacks and identified 63 per cent of severe attacks. After 48 h, outcome was correctly predicted in 88 per cent of attacks, compared with 69 per cent and 84 per cent for the Ranson and Imrie scores respectively. Sequential scoring may be most useful for monitoring pancreatic necrosis.

Prognostic markers

Prognostic markers are listed in Table 5.

Clinical signs

Even experienced clinicians failed to identify on hospital admission 60 to 70 per cent of cases classified as severe episodes. Clinical indicators of severity, irrespective of the bacteriological status of necrosis, are based on impending remote organ failures and on the extent of retroperitoneal inflammation (abdominal mass, guarding and rebound tenderness, Cullen's sign (periumbilical staining), and the Grey–Turner sign (flank staining)). After 48 h, diagnostic accuracy rises to 80 per cent and yields similar results to scoring systems of prognostic criteria.

As 'prune-juice'-colored ascitic fluid is present in over half the patients with necrotizing pancreatitis who undergo early surgery, peritoneal fluid obtained by abdominal paracentesis may be sampled. By considering the smell, volume, and color of free ascitic fluid or the color of lavage fluid using a standard color chart, 90 per cent of eventual fatalities and 72 per cent of severe cases can be identified within 5 h of admission. In addition to its greater speed and availability, additional benefits of diagnostic peritoneal lavage are the correction of an erroneous clinical diagnosis of pancreatitis and selection of the patient for therapeutic lavage. There is an 0.8 per cent risk of visceral puncture and a lower sensitivity when compared with multiple laboratory criteria, particularly in patients with a gallstone etiology. Since it is based on early data, monitoring of the course of pancreatitis is impossible and accurate prediction of late complications poor.

Biochemical markers

Hypoxemia ($Pa\text{O}_2 < 8$ kPa (60 mmHg)) is a marker of severe attack and therapeutic need. Other useful biochemical indicators of severity include antiproteases (α_2-macroglobulin and α_1-antiproteinase),

C-reactive protein, granulocyte elastase, phospholipase A_2, and trypsinogen activation peptides (Gross *et al.* 1993).

Imaging procedures

CT severity scoring systems correlate with the Ranson score, the length of hospital stay, the development of serious complications, and mortality (Balthazar *et al.* 1990). Although patients without regional necrosis had no mortality and a 6 per cent complication rate, mortality and morbidity rose to 29 per cent and 94 per cent respectively in those with more than 30 per cent necrosis. Similarly, patients with one or more peripancreatic fluid collections had a 14 per cent mortality and 54 per cent morbidity. Importantly, morbidity rose to 80 per cent in those with both necrosis and fluid collection visualized on initial CT. The likelihood that infected pancreatic necrosis or pancreatic abscess will develop is highest in patients with associated necrosis and fluid collections. As late development of necrosis may occasionally be demonstrated, this raises the question of optimal timing of the initial CT and justifies follow-up CT examinations in those with a protracted course. However, considerable variance between a mild clinical course and initial CT appearance exists in many patients, and so therapeutic implications are not straightforward.

Stratification

The major determinants of outcome are the extent of pancreatic necrosis and subsequent release of toxic substances, the presence of extrapancreatic necrosis, and secondary bacterial contamination of these devitalized areas. With advances in care, most patients survive the multiple and remote organ failures that characterize the early stage of the attack. Pancreatic infection is now the leading cause of death.

Patients with many signs of severity may have an uncomplicated course, and some with few signs may progress to necrosis, infection, and multiple organ failure. Nevertheless, these multiple prognostic criteria allow identification of most patients with severe episodes within a few hours of admission. Importantly, the timely value of each prognostic criterion relies on the successive phases of acute pancreatitis: inflammation, necrosis, and bacterial contamination. Early and sequential assessment of the severity of acute pancreatitis with complementary indicators adapted to each stage of the disease is a prerequisite to any improvement in therapy and survival. Whether early prediction of severity really influences treatment and outcome remains controversial.

References

Balthazar, E.J., Robinson, D.L., Megibow, A.J., and Ranson, J.H.C. (1990). Acute pancreatitis: value of CT in establishing prognosis. *Radiology*, **174**, 331–6.

Beger, H.G., Büchler, M., Bittner, R., Block, S., Nevalainen, T., and Roscher, R. (1988). Necrosectomy and postoperative local lavage in necrotizing pancreatitis. *British Journal of Surgery*, **75**, 202–12.

Gross, V., Leser, H.-G., Heinisch, A., and Schölmerich, J. (1993). Inflammatory mediators and cytokines — new aspects of the pathophysiology and assessment of severity of acute pancreatitis? *Hepato-Gastroenterology*, **40**, 522–30.

Ohlsson, K. (1988). Acute pancreatitis. Biochemical, pathophysiological and therapeutic aspects. *Acta Gastroenterologica Belgica*, **LI**, 3–12.

Reynaert, M.S., Dugernier, T., and Kestens, P.J. (1990). Current therapeutic strategies in severe acute pancreatitis. *Intensive Care Medicine*, **16**, 352–62.

Steer, M.L. (1992). How and where does acute pancreatitis begin? *Archives of Surgery*, **127**, 1350–3.

4.4.2 Conservative management

T. Dugernier and P. J. Kestens

Key messages

- Basic medical management includes intensive care support of failing organs, nasogastric suction, analgesia, antibiotic prophylaxis, and nutritional support. Bacterial translocation and subsequent pancreatic infection may justify early jejunal feeding and selective gut decontamination.
- The conservative therapeutic approach favors non-surgical removal of toxic substances released by the inflammatory necrotizing process, relying upon peritoneal lavage and drainage of thoracic lymph. Early endoscopic dislodgement is advocated for impacted ampullary stones and worsening remote organ failure or acute cholangitis. Surgery is delayed until massive hemorrhage, complicated pseudocyst, or bacterial contamination of necrosis are demonstrated. Careful serial monitoring of pancreatic necrosis is mandatory for early identification.
- In selected cases with local complications, particularly fully liquified collections (e.g. pancreatic abscess and infected pseudocyst), percutaneous drainage is a valid alternative to surgery.
- Most therapeutic modalities are still directed towards the complications. Future therapies should aim at early reduction of the volume of regional necrosis by blocking premature activation of pancreatic proenzymes, enhancing the glandular microcirculation, and modulating mediators of the inflammatory reaction.

Basic medical management

This is largely empirical and includes withholding of oral feeding, nasogastric suction for paralytic ileus, correction of hyperglycemia, fluid, electrolyte, and acid–base disorders, and vigorous volume replacement (Table 1).

Sedatives and non-morphine-based analgesics are often required for agitation and pain. Epidural analgesia has the advantage of increasing splanchnic and pancreatic blood flow, which is often rapidly impaired in the setting of severe acute pancreatitis. Theoretically, it may limit the progression toward pancreatic necrosis and prevent bacterial translocation from the digestive tract.

As morbidity caused by early remote organ dysfunction has been controlled by advances in intensive care, bacterial contamination of pancreatic necrosis has emerged as the major determinant of mortality. Early prophylactic antibiotics and selective digestive decontamination have been shown to reduce the incidence of pancreatic infection. It seems wise to consider both measures in view of the 40 per cent incidence of early bacterial contamination of necrosis, the mechanisms underlying bacterial invasion of necrotic areas, and the potential for further bacterial/endotoxin translocation. Considering their penetration into pancreatic tissues, the microbial spectrum most prevalent in infected necrosis, and the potential to

Table 1 Basic medical management of severe acute pancreatitis

Withholding of oral feeding plus nasogastric suction
Correction of hyperglycemia and fluid, electrolyte, and acid–base disorders
Sedation and analgesia (including epidural)
Prophylactic antibiotics and selective digestive decontamination
Nutritional support including early jejunal feeding
Volume replacement
Supportive care for organ failures (cardiocirculatory, pulmonary, renal, coagulation)

achieve therapeutic and antimicrobial activity in the pancreas, quinolones or imipenem may be used as first-line agents.

Aggressive nutritional support has become standard therapy, since constant features of severe acute pancreatitis include the inability of the gastrointestinal tract to function effectively for long periods and a concurrent increase in the basal metabolic rate caused by the inflammatory process. In addition, there is a high prevalence of premorbid nutritional depletion. The impact on the primary disease process by suppressing pancreatic secretion with total parenteral nutrition remains controversial and probably of little concern. Nutritional support should be initiated early in the course of the attack; this may be given parenterally as intravenous substrate infusions including fat emulsions do not stimulate exocrine pancreatic secretion. Lipids are usually well tolerated and contribute towards the caloric substrate needed to meet the increased energy needs of these glucose-intolerant patients. However, intravenous lipid solutions should not be given to either the rare patient with hypertriglyceridemia-induced acute pancreatitis or those with lipemia. Triglyceride levels usually fall quickly when oral intake is withheld and insulin administered. Plasma exchange should be undertaken in refractory cases with marked hypertriglyceridemia.

Both gastric/duodenal feeding and enteral fat stimulate exocrine pancreatic function which may be deleterious with respect to the pathophysiology of severe acute pancreatitis. Enzyme output appears to be proportional to the fat content in the diet and to duodenal inflow of nutrients. In patients requiring surgery for complications of severe acute pancreatitis, nutritional support can be given through a feeding jejunostomy with a semi-elemental diet incorporating medium-chain triglycerides. Early placement of a nasojejunal tube may be a valid, yet unsubstantiated, alternative to allow early enteral feeding and gut decontamination in these patients. Proper positioning of the tube, careful exclusion of jejunoduodenal reflux, and prevention of duodenal injury are prerequisites for the use of this technique. Nevertheless, early jejunal feeding improves nitrogen–caloric balance, helps to preserve or restore the integrity of the gut–blood barrier by providing luminal nutrients to the mucosa, and is of immunotherapeutic benefit in those patients at risk of developing bacterial translocation and multiple organ failure. The value of diets supplemented with enterocyte or colonocyte substrates such as arginine, glutamine, and short-chain fatty acids remains to be determined. Low-fat oral feeds should be reinstituted carefully after completion of the acute phase of the attack.

Acute pancreatitis should be regarded as the disease prototype that generates multisystem organ failure even in the absence of sepsis. Consequently, intensive care support of these patients shares many common therapeutic features with other acute illnesses requiring intensive care unit (ICU) admission. Circulatory shock, acute respiratory failure, renal dysfunction, and metabolic derangements involving glucose and calcium homeostasis, hepatic function, and the hematopoietic system all mandate appropriate therapy and ICU support.

Conservative management

Much controversy surrounds whether, when, and why patients with severe acute pancreatitis should be subjected to surgery. Whatever the stage of the attack, two extremes of therapeutic strategy have been advocated, although neither has yet been substantiated by a controlled trial.

The first strategy is early surgical removal of necrotic material (necrosectomy). Reported indications for surgery during either the early or the late phase of the attack are based on clinical, morphological, and/or microbiological criteria. Massive retroperitoneal hemorrhage, an acute abdomen, early infected pancreatic necrosis, extensive regional necrosis, and persistence of remote organ dysfunctions despite ICU therapy for 3 to 5 days usually dictate surgery during the toxemic phase. Pancreatic abscess, symptomatic acute pseudocyst, fistula, or compression/perforation of a hollow viscus also mandate operative procedures during the necrotic phase. Almost half of those who undergo surgery do so for persistent organ failures without pancreatic infection and on the basis of CT findings of large necrotic areas and laboratory/clinical signs of inflammation which cannot differentiate between sterile and infected necrosis. Morphological data such as extensive sterile pancreatic necrosis are used as criteria for surgery. Some even claim that prophylactic necrosectomy, irrespective of the patient's condition and the bacteriological status of the necrosis, may reduce the incidence of sepsis, and that waiting for proof of infection delays operative intervention and increases mortality in those with large necrotic areas. Regardless of the surgical procedure undertaken, early, aggressive, complete, and often repeated debridement of unviable tissues is performed. This is because necrosis persists due to oozing of activated pancreatic enzymes and toxic substances, thus demanding prolonged drainage and repeated operative interventions. When undertaken during the toxemic phase, this surgical drainage results not only in continuous removal of necrotic and/or infected tissues but also in diversion of toxic substances and active enzymes released in the pancreatic region. These act synergistically, lessening the inflammatory necrotizing process and reducing the mortality of severe acute pancreatitis to 7 or 15 per cent, depending on whether the necrosis is sterile or infected. Whatever surgical approach and method of drainage are used, about 20 per cent require reoperation because of hemorrhage, fistula, or, most commonly, recurrent or incompletely drained infected areas with ongoing sepsis. Additional morbidity includes mechanical ileus, stomach outlet stenosis, and incisional hernias.

The objectives of the step-care conservative approach are to prevent and/or treat any organ damage in the early phase and to allow spontaneous healing of regional necrosis during the later phase, with precautions against emergence of local complications, in particular bacterial invasion of necrotic areas. Aggressive surgical

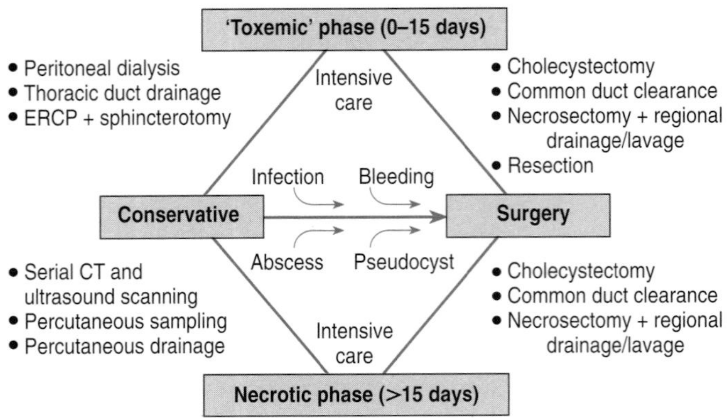

Fig. 1 Schematic representation of the conservative therapeutic strategy that may be considered in both phases of severe acute pancreatitis. In addition to basic medical therapy and intensive care support, non-surgical removal of toxic substances released by the inflammatory necrotizing process is carried out while serial monitoring of necrotic areas is undertaken to avoid missing the emergence of regional infection. Surgery is contemplated only if severe retroperitoneal hemorrhage, complicated pseudocyst, or infected pancreatic necrosis/abscess supervene. A shift from a conservative to a surgical approach remains controversial in the case of extensive sterile necrosis and worsening multisystem organ dysfunction syndrome.

debridement of necrosis, with its attendant morbidity, is avoided unless regional infection, severe hemorrhage, or symptomatic pseudocyst occur and fail to settle on conservative therapeutic options (Fig. 1) (Rau *et al.* 1995). The step-care conservative management of the early 'toxemic' phase of severe acute pancreatitis includes percutaneous peritoneal lavage, thoracic duct drainage, and endoscopic retrograde cholangiopancreatography (ERCP) with emergent papillotomy for acute gallstone pancreatitis and impacted ampullary stones.

The last procedure should be carried out without delay since early endoscopic dislodgement of impacted stones reduces morbidity when undertaken within 72 h of admission (Winslet *et al.* 1991). However, it remains largely unproven that relief of ductal hypertension by whatever means lessens the inflammatory process and impedes progression toward pancreatic necrosis. Although this procedure is achieved at no risk of exacerbating the disease process, hemorrhage, or pseudocyst formation, it still carries a substantial risk of introducing bacteria into necrotic areas. Thus it should not be considered routinely in severe gallstone pancreatitis, but only when signs of cholangitis develop or when suspicion of stone impaction arises on clinical, biochemical, and ultrasonic grounds in a patient not responding to ICU therapy. Alternatively, temporary relief from cholangitis may be obtained by percutaneous or nasobiliary drainage of the biliary tree. As the stone is impacted in the ampulla for some time in more than 60 per cent of patients during the first 48 to 72 h of the attack, some authors have advocated early (within 48 h) cholecystectomy and common duct exploration with subsequent choledocotomy. However, since 95 per cent of stones pass spontaneously into the intestine during the first week and early biliary surgery carries a prohibitive mortality in severe acute pancreatitis, it is common practice to delay cholecystectomy until signs of the inflammatory phase have subsided. Cholecystectomy should be performed during the same hospital admission. In poor surgical candidates a wide endoscopic sphincterotomy may obviate the need for a cholecystectomy.

Percutaneous peritoneal lavage (usually 1–2 l/h of dialysis fluid or normal saline) should be considered early whenever dark brown ascitic fluid is obtained. Lavage is usually discontinued when the returned fluid is clear. Striking reduction of early cardiorespiratory dysfunction and early mortality has been demonstrated, in particular for alcohol-induced pancreatitis. These benefits are ascribed to the early removal of toxic substances before they reach the systemic circulation. Peritoneal lavage is also useful in patients with coexisting renal failure and/or fluid overload. However, it seems unlikely that enough lavage fluid gains access to the lesser sac where the concentration of toxic agents must be greatest. This therapy neither slows progression of local tissue injury nor prevents the development of necrosis and infection, and thus fails to influence overall mortality. A dramatic reduction in the incidence and subsequent mortality from pancreatic infection has been obtained by using peritoneal lavage for an extended period ($>$ 7 days). Since evidence exists that bacteria can translocate transperitoneally to the pancreas as viable organisms within macrophages, the removal of these phagocytic cells containing gut-derived bacteria and chemotactic factors may account for these benefits. Hazards of this procedure include the risk of visceral injury when introducing the catheter into the peritoneum, increased respiratory distress due to abdominal distension by lavage fluid, fluid overload if the lavage fluid is not returned, hyperglycemia, and bacterial contamination of the peritoneal cavity.

Lymphatic pathways play a pivotal role in the systemic transfer of active enzymes and toxic products released by the pancreas. Diversion of these substances by drainage of the thoracic duct lymph should be considered in those patients with persistent or worsening remote organ dysfunctions (Dugernier *et al.* 1989). This is carried out surgically with a 7 Fr pulmonary artery catheter. This type of drainage has been found to be particularly useful for patients with persistent cardiocirculatory failure and/or impaired pulmonary gas exchange, including acute respiratory distress syndrome, during the toxemic phase of the disease. Thoracic duct drainage should be considered early in the course of failure of conventional conservative maximum supportive therapy. The beneficial effects of lymphatic drainage with regard to prevention and correction of life-threatening complications of the early phase are substantiated by the finding of substantial amounts of enzymatically active trypsin, neutrophil myeloperoxidase and elastase, and cytokines in the lymph. Like peritoneal lavage, lymphatic drainage fails to influence the incidence of locoregional complications of acute pancreatitis.

Although the role of surgery in severe hemorrhage, hollow viscus perforation, complicated pseudocyst, and pancreatic infection remains undisputed, the step-care conservative approach helps to manage the early remote organ dysfunctions. The question arises as to whether these patients with extended regional necrosis and systemic complications but without infection should be subjected to early surgery and its attendant morbidity. The success of early surgical debridement is partly attributed to the prolonged postoperative lavage of the peripancreatic region. This removal of toxic substances can be achieved almost non-invasively by the conservative approach. Nevertheless, the major challenge for those who adopt this strategy is early differentiation between sterile and infected necrosis. Patients with abdominal sepsis not treated by surgical or percutaneous drainage will be jeopardized, since this condition is almost

Table 2 Therapeutic interventions during the early phase of severe acute pancreatitis

	Indications
Supportive care	Severe acute pancreatitis (all cases)
Peritoneal lavage (+protease inhibitor?)	Alcohol-related pancreatitis and early distant organ failures
Thoracic duct drainage	Severe acute pancreatitis and early cardiorespiratory failure
Endoscopic sphincterotomy	Severe biliary pancreatitis
	+gallstone ampullary impaction
	+persistent organ failures
	+acute cholangitis
Laparotomy	Diagnostic uncertainty
Angiographic embolization	Massive retroperitoneal hemorrhage
Necrosectomy+vessel ligation	
Percutaneous drainage	
Necrosectomy+prolonged closed drainage/lavage *or*	Infected pancreatic necrosis Worsening multiple organ failure and extensive sterile necrosis
necrosectomy+open packing *or*	
necrosectomy+sump drainage *or*	
percutaneous drainage+lavage	

Modified with permission from Reynaert *et al.* (1990).

100 per cent lethal if left undrained. Precise criteria for moving from intensive conservative therapy to surgery are essential. Apart from severe arterial hemorrhage and an acute abdomen (usually by early perforation of a hollow viscus), a more rational basis for operative intervention is provided by a reliable and rapid differentiation between sterile and infected necrosis. This diagnostic work-up is best achieved by serial CT examination, repeated needle-aspiration sampling of necrotic areas, and monitoring of C-reactive protein or elastase.

As an increasing number of patients survive the early phase of severe acute pancreatitis without operative debridement of necrosis, many enter the necrotic stage with large areas of necrosis. Considering the rate of spontaneous and uneventful resolution of even extensive necrotic areas, a policy of conservative medical management remains sensible during this phase of the attack.

Percutaneous drainage

Selected cases of retroperitoneal hemorrhage without infection and prolonged circulatory embarrassment are successfully managed conservatively or with ultrasound- or CT-guided percutaneous drainage. However, in the majority, surgical vessel ligation (pseudoaneurysm) and thorough debridement are mandatory to control sepsis-associated hemorrhage and prevent recurrence of bleeding. Angiography and arteriographic embolization, when feasible, should precede the operative procedure.

Ultrasound- or CT-guided percutaneous drainage of infected collections using large-bore drainage tubes, apart from its therapeutic potential in selected cases, may buy time for severely ill patients until they can withstand formal surgical debridement and drainage (Lee *et al.* 1992). This therapeutic approach is highly demanding as it usually requires insertion of multiple catheters and a prolonged drainage/lavage period with close clinical scrutiny, regular abdominal CT scans, and catheter repositionings. The rather small caliber of the drains and the thick necrotic material often preclude complete evacuation of the collections, so that septic recurrences are common and dictate surgical operative intervention. Factors that adversely affect percutaneous drainage include complex and not fully liquified collections, as well as a central localization. Complications are common and include gastrointestinal and colocutaneous fistulas, bleeding, pneumothorax, and, in particular, the possibility of delaying necessary surgery.

Up to half of patients with pseudocyst remain free of symptoms and can be safely managed non-operatively with careful clinical and ultrasound follow-up studies. Some clinicians consider that, irrespective of symptoms, a diameter of more than 5 cm mandates drainage owing to the risk of complications. Ultrasonic- or CT-guided percutaneous aspiration and drainage has a role as a temporary measure in severely ill patients who are poor candidates for surgery. This approach increases the risk of secondary infection of a sterile pseudocyst, and the small caliber of the catheter often precludes complete aspiration of necrotic material. A 50 per cent recurrence rate has been reported, particularly if obstruction of the downstream part of the duct is outlined on pancreatography and if pancreatic ductal communication to the cyst persists. Percutaneous catheter cystogastrostomy and endoscopic cystogastrostomy or duodenostomy have been proposed. Internal drainage should be avoided in patients with pancreatic ascites unless the cyst leak can be incorporated into a Roux-en-Y cyst jejunostomy. In selected cases,

Table 3 Therapeutic interventions during the late phase of severe acute pancreatitis

	Indications
Conservative treatment	Sterile necrosis
Necrosectomy+closed drainage/lavage *or* +open packing *or* +sump drainage	Extensive sterile necrosis Pancreatic abscess
Ultrasound- or CT-guided percutaneous drainage	Acute pseudocyst
	Pancreatic abscess
	Enteric fistulas
Internal drainage	Acute pseudocyst (symptomatic or diameter > 5 cm)
External drainage	Acute pseudocyst (thin-walled/infected)
Formal pancreatic resection	Acute pseudocyst (multiple, hemorrhagic)
Cholecystectomy+common duct clearance	Gallstone pancreatitis
	'Idiopathic' pancreatitis+biliary sludge+crystals
ERCP+sphincterotomy	Gallstone pancreatitis (poor surgical candidate)
	'Idiopathic' pancreatitis+biliary sludge+crystals
	Abnormalities of the pancreatobiliary system
Biliary surgery (sphincteroplasty, pancreatojejunostomy, etc.)	Abnormalities of the pancreatobiliary system
Segmental resection	Gastric and colonic necrosis

Modified with permission from Reynaert *et al.* (1990).

total parenteral nutrition, percutaneous drainage, and somatostatin may lead to fistula closure in cases of recurrent pseudocyst or persistent pancreatic ascites.

Pancreaticopleural fistula should be managed conservatively with somatostatin, total parenteral nutrition, and tube thoracostomy for up to 2 to 3 weeks. Pleural effusion recurs in half of cases and requires operative intervention (excision or decompression of the leak) after visualization of the site of disruption of the pancreatic duct with a pancreatogram.

Early and late therapeutic interventions during severe acute pancreatitis are summarized in Tables 2 and 3 respectively.

Future avenues

Treatment is still directed mainly towards the complications. Despite constant advances in the understanding of the pathophysiological mechanisms underlying the progression towards regional necrosis and the emergence of local or remote organ damage, the basic triggering event still remains controversial. There is no specific drug that can block premature activation of pancreatic proenzymes. The close interrelation between pathophysiological mechanisms, the complexity of the enzyme cascade activations, and the multiplicity of mediators involved in the inflammatory necrotizing process allow little hope for the development of a single specific agent capable of halting or reversing this amplification once initiated. Nevertheless, potential therapies are being developed that may be useful, if given early, to limit the volume of regional necrosis, prevent early systemic organ damage, reduce the incidence of secondary infection of necrotic areas, and thus improve prognosis.

To date, attempts to neutralize proteases by antienzyme therapy or by replenishing antiproteases with fresh frozen plasma, and treatment aimed at resting the pancreas (glucagon, calcitonin, somatostatin) have failed to produce any consistent and substantial benefit in terms of morbidity and mortality. If the primary pathophysiological event is inhibition of digestive enzyme secretion with secondary localization of digestive zymogens with lysosomal hydrolases inside acinar cells, interventions that restore secretion should be of greater benefit than inhibition of pancreatic exocrine secretion.

Overactivation of leukocytes in the pancreatic area and subsequent release of proinflammatory mediators are increasingly recognized as major determinants of local and remote organ damage. In animal experiments, early inhibition of inflammatory mediators protects against both local and systemic complications. Potential targets and future therapies might include free-radical scavengers, phospholipase inhibitors, lipoxygenase pathway inhibitors, interleukin-1 receptor antagonist, anti-tumor necrosis factor polyclonal antibody or soluble tumor necrosis factor receptors, and interleukin 10 to modulate the local secretion of proinflammatory cytokines.

The other potential therapeutic avenue is the prevention/correction of pancreatic ischemia. Experimental data suggest an early and specific disturbance of the pancreatic microcirculation characterized by low capillary blood flow. This could convert an initially edematous condition into progressive necrosis. Hemodilution with dextran 60/70 appears far more effective than conventional fluid therapy in enhancing regional blood flow and preventing tissue necrosis during the early days of the attack.

References

Dugernier, T., *et al.* (1989). Prospective evaluation of thoracic-duct drainage in the treatment of respiratory failure complicating severe acute pancreatitis. *Intensive Care Medicine*, **15**, 372–8.

Lee, M.J., *et al.* (1992). Acute complicated pancreatitis: redefining the role of interventional radiology. *Radiology*, **183**, 171–4.

Rau, B., Pralle, U., Uhl, W., Schoenberg, M.H., and Beger, H.G. (1995). Management of sterile necrosis in instances of severe acute pancreatitis. *Journal of the American College of Surgeons*, **181**, 279–88.

Reynaert, M.S., Dugernier, Th., and Kestens, P.J. (1990). Current therapeutic strategies in severe acute pancreatitis. *Intensive Care Medicine*, **16**, 352–62.

Winslet, M.C., Imray, C., and Neoptolemos, J.P. (1991). Biliary acute pancreatitis. *Hepato-gastroenterology*, **38**, 120–3.

4.4.3 Surgical management

B. Rau and H. G. Beger

Key messages

- Acute pancreatitis can be classified into interstitial edematous pancreatitis, necrotizing pancreatitis, postacute pseudocyst, and pancreatic abscess.
- Discrimination between interstitial edematous and necrotizing pancreatitis is by contrast-enhanced CT and serum markers of necrosis.
- Initial management should be conservative in an intensive care environment.
- Indications for surgery include persisting pulmonary, renal, cardiocirculatory, or abdominal complications, sepsis, and proven infection of pancreatic necrosis. The timing for surgical management should be a preferably late intervention during the second week of the disease. The basic principle of surgical treatment is a combination of necrosectomy and an additional technique to manage the peripancreatic space.
- There are three different techniques of surgical management after initial necrosectomy: conventional treatment with sump drainage; open procedures with open abdominal management (open packing) or temporary abdominal closure with scheduled reoperations; closed procedures with continuous closed lavage of the lesser sac.

Introduction

Acute pancreatitis is an inflammatory disease ranging from mild self-limiting discomfort to a severe illness with fatal outcome. With respect to morphological, radiological, and bacteriological criteria the disease comprises four different entities: acute interstitial edematous pancreatitis, acute necrotizing pancreatitis, postacute pancreatic pseudocyst, and pancreatic abscess. These pathomorphological conditions closely correlate with the clinical severity and with local and systemic complications.

Classification of acute pancreatitis

Macroscopic and histological features of interstitial edema and of intra- and peripancreatic fluid collections are present in 70 to 80 per cent of cases. Peripancreatic fat necrosis may be present, but associated organ dysfunction is minimal. Conservative treatment usually leads to an uneventful recovery with morphological and functional *restitutio ad integrum*.

Necrotizing pancreatitis is the clinically severe form of acute pancreatitis and develops in 20 to 30 per cent of cases. Macroscopic characteristics are focal or diffuse areas of devitalized pancreatic parenchyma and fat necrosis involving peripancreatic and retroperitoneal areas. The pancreatic exudate contains various vasoactive toxic substances which are responsible for remote organ failure. The extent of necrosis usually correlates with clinical severity, the incidence of organ failure, and systemic complications. Infection arises in 30 to 70 per cent of patients with necrotizing pancreatitis and is caused predominantly by gut-derived Gram-negative organisms. Infection of pancreatic necrosis is currently the main determinant of mortality in acute pancreatitis.

Postacute pancreatic pseudocysts and abscesses are rare consequences of acute pancreatitis. They do not usually occur before the fourth week after onset of symptoms and are both characterized by an inflammatory non-epithelialized wall delineating the process. Pseudocysts contain pancreatic juice and liquefied necrotic tissue, are sterile in most cases, and carry a low morbidity and mortality. Peripancreatic fluid collections, usually arising in the early course of severe acute pancreatitis, often lack a defined wall and represent a different morphological entity. Abscess appears as a collection of pus with little or no necrosis; cultures reveal bacteria or fungi. The clinical picture is usually one of abdominal sepsis without signs of acute pancreatitis. Abscesses should not be confused with infected necrosis as they differ in clinical expression and associated mortality. The incidence of these entities of acute pancreatitis is shown in Table 1.

Table 1 Incidence of acute pancreatitis

	Patients	Percentage
Interstitial edematous pancreatitis	992	71
Necrotizing pancreatitis	286	21
Sterile	196	68
Infected	90	32
Pancreatic abscess	36	3
Postacute pseudocyst	82	6
Total	1396	100

Data from Department of General Surgery, University of Ulm, Germany, between May 1982 and December 1995.

Table 2 Discrimination between edematous and necrotizing acute pancreatitis

Biochemical markers of necrosis
C-reactive protein >120 mg/l
Lactate dehydrogenase >270 IU/l
Phospholipase A2 >3.5 IU/l
PMN-elastase >120 μg/l
Trypsinogen activation peptide >2.0 nmol/l
Morphological discrimination
Dynamic contrast-enhanced CT

Discrimination between edematous and necrotizing pancreatitis

Early discrimination between acute interstitial edematous pancreatitis and necrotizing pancreatitis is the most important diagnostic step in patient management. Serum markers and imaging procedures allow early and accurate discrimination (Table 2). Serum C-reactive protein and lactate dehydrogenase levels as well as contrast-enhanced CT are the clinical gold standards, with an accuracy above 90 per cent in detecting necrotizing pancreatitis. Moreover, helical dynamic CT scanning allows precise volumetric assessment of the amount of intra- and extrapancreatic necrosis or peripancreatic exudate. Guided-needle aspiration with Gram staining and culture of the aspirate should be considered when clinical and laboratory signs of sepsis develop in the course of proved necrotizing pancreatitis to confirm or exclude infection of necrosis. The use of multiple-factor scoring systems (e.g. Ranson or APACHE II scores) may provide additional information in terms of severity stratification of acute pancreatitis.

Conservative management

Since there is no specific pharmacotherapy for acute pancreatitis, initial treatment is based on the principles of intensive care unit (ICU) support with close monitoring of vital functions, adequate analgesia, and fluid replacement. Special attention must be paid to pulmonary, renal, and cardiocirculatory function.

The first step in management is transfer of the patient to the ICU. Acute interstitial edematous pancreatitis usually responds to basic

Table 3 Definitions of organ complications and therapeutic measures in necrotizing pancreatitis

Organ complication	Definition	Therapy
Pulmonary insufficiency	$Pa\text{O}_2 < 60$ mmHg (8 kPa)	Humidified O_2
	$Pa\text{O}_2 < 60$ mmHg (8 kPa) despite O_2	Mechanical ventilation
Renal insufficiency	Serum creatinine > 2.0 mg/dl (180 μmol/l)	Diuretics
	BUN > 30 mg/dl	Dopamine 3 μg/kg/min
	Anuria	Hemodialysis, hemofiltration
Shock	Syst BP < 80 mmHg > 15 min	Fluid replacement Catecholamines
Sepsis	T > 38.5°C, WBC > 16.0/mm^3, BE < − 4 mmol/l, platelets < 150/mm^3 positive blood culture/aspirate for > 48 h	Fine-needle aspiration Antibiotics
Metabolic disorders		
Hypocalcemia	Serum calcium < 8.5 mg/dl (2.13 mmol/l)	Calcium gluconate 10%
Hyperglycemia	Serum glucose > 200 mg/dl (11 mmol/l)	Insulin
Metabolic acidosis	BE < −2.5 mmol/l	$NaHCO_3$
DIC	Prothrombin time < 70%	Fresh frozen plasma
	Partial thromboplastin time > 45 s	
Biliary pancreatitis	Increased bilirubin, AST, alkaline phosphatase	Endoscopic papillotomy for impacted stones

AST, aspartate aminotransferase; BE, base excess; BUN, blood urea nitrogen; DIC, disseminated intravascular coagulation; syst BP, systolic blood pressure; *T*, temperature; WBC, white blood count.

Table 4 Indications for surgery in necrotizing pancreatitis

Clinical criteria	Persisting organ failure or systemic complications despite ICU treatment for more than 72 h
	Pulmonary insufficiency
	Renal insufficiency
	Shock
	Sepsis
	Abdominal complications
	Peritonitis
	Ileus
Bacteriological criteria	Positive fine-needle aspiration

treatment within days. Unless biliary tract stones are present there is no indication for surgery. However, necrotizing pancreatitis is associated with a high incidence of systemic and local complications, of which the most devastating is infection of intra- or extrapancreatic necrotic tissue. The prophylactic use of antibiotics able to penetrate into the pancreas is recommended for such patients. The supportive treatment of organ and systemic complications includes the whole spectrum of ICU measures (Table 3).

Conservative treatment should be applied over a minimum of 3 to 5 days unless a specific indication for surgery arises. There is ongoing controversy about the role of non-surgical management of clinically severe sterile necrotizing pancreatitis, unless there is documented infection with clinical signs of sepsis, as some of these patients can be successfully treated conservatively. In the absence of prospective randomized data, indications for surgical management of sterile necrosis require further investigation.

Indication and timing for surgical management

Surgical intervention in patients with severe necrotizing pancreatitis with local and systemic complications should be based on the patient's response to optimum ICU treatment as shown in Table 3. Several clinical and morphological criteria may be considered as indications for surgery if the patient fails to improve after 3 to 5 days. Clinical criteria are signs of acute abdomen with peritonitis and ileus, persisting or deteriorating organ failure such as pulmonary or renal insufficiency, and systemic complications such as persisting or progressing sepsis syndrome, metabolic derangement, coagulation disorders, and shock. With respect to morphological criteria, documented infection of pancreatic necrosis is a generally accepted indication for surgical intervention if clinical signs of sepsis are present (Table 4). It should be stressed that no controlled data exist

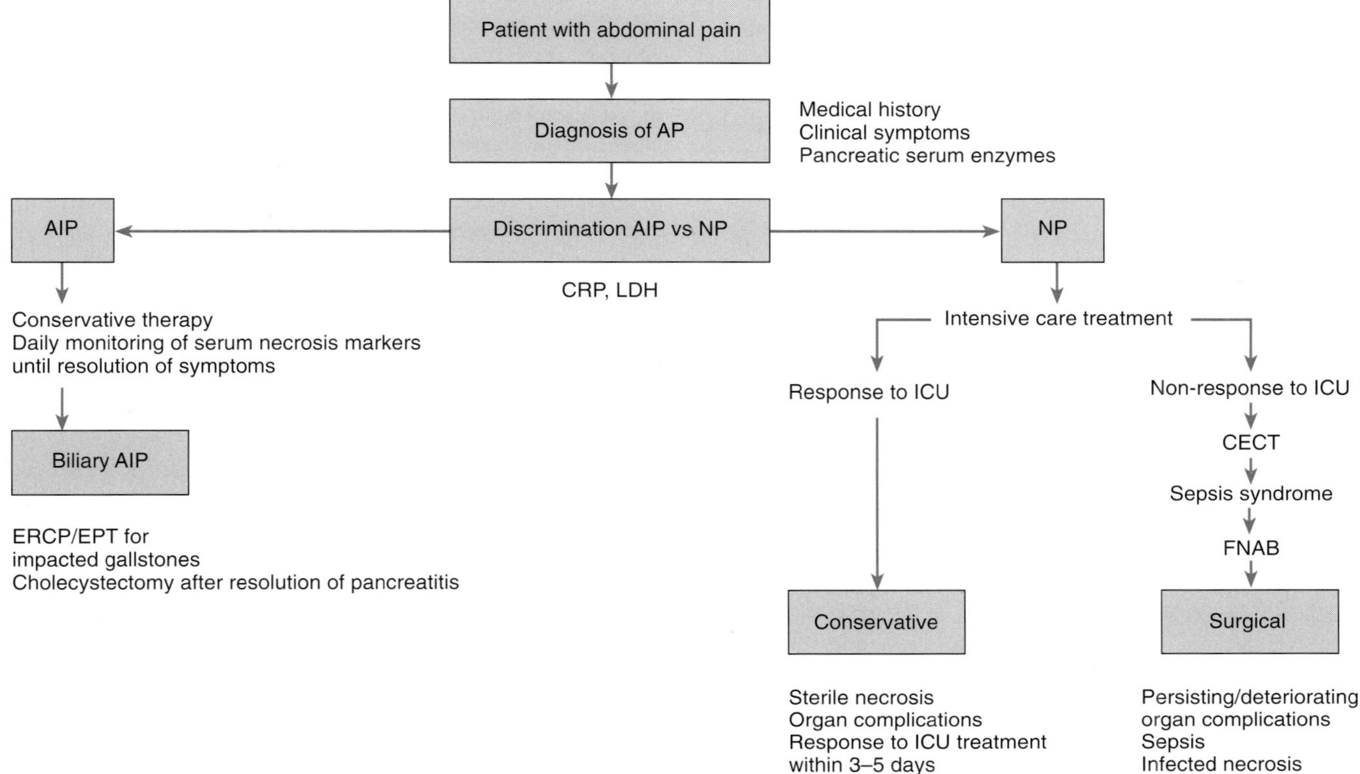

Fig. 1 Management of acute pancreatitis: AP, acute pancreatitis; AIP, interstitial edematous pancreatitis; NP, necrotizing pancreatitis; ERCP, endoscopic retrograde cholangiopancreatography; EPT, endoscopic papillotomy; CRP, C-reactive protein; LDH, lactate dehydrogenase; CECT, contrast-enhanced CT; FNAB, fine-needle aspiration biopsy.

Table 5 Current surgical treatment modalities in necrotizing pancreatitis

Conventional treatment
Resection or necrosectomy with drainage
Reoperations on demand
Open procedures
Resection or necrosectomy and scheduled relaparotomies
Open abdominal management (open packing)
Temporary abdominal closure (semi-open procedure)
Closed procedures
Necrosectomy and continuous closed local lavage
Reoperations on demand

regarding optimum treatment, either surgical or conservative, in patients with complicated sterile necrotizing pancreatitis.

Ongoing discussion persists regarding the optimum timing of surgery in necrotizing pancreatitis which can be performed early within the first 7 days or late during the postacute phase in the second week. Operative intervention in the second week of the disease is recommended for patients who do not respond to maximum ICU treatment The rationale for delayed surgery is to wait until demarcation of necrosis has occurred. Early intervention in the acute phase (days 1 to 7) is indicated only in those few patients with a fulminant course or with proven infection of necrosis (Fig. 1).

Techniques of surgical treatment

The surgical treatment of necrotizing pancreatitis has three major goals: (i) evacuation of necrotic tissue and exudate which releases toxic compounds into the circulation and accounts for remote organ failure, (ii) prevention of complications such as infection and pancreatic abscess by removing pancreatic necrosis, and (iii) preservation of pancreatic tissue which is still vital (this strongly influences the quality of long-term results concerning endo- and exocrine pancreatic function). A variety of different approaches have been advocated, ranging from conservative non-resecting methods to aggressive extensively resecting procedures.

As surgical treatment centers on debridement of necrotic tissue, simple peritoneal dialysis cannot be considered adequate as its effects are restricted to the abdominal cavity and do not address the necrotizing process in the retroperitoneal space. Controlled trials have shown no reduction in overall mortality of severe acute pancreatitis.

Pancreatic resection modalities—partial or total pancreaticoduodenectomy—aim at radical removal of the gland. Even in patients with macroscopically total pancreatic necrosis, pancreatectomy is rarely justified as in many cases only the superficial areas of the gland are necrotic. This type, known as 'surface pancreatitis', can easily be mistaken intraoperatively for total pancreatic necrosis. The removal of still viable pancreatic parenchyma and healthy duodenum, stomach, and biliary tracts poses additional stress on the critically ill patient. Therefore resecting surgical modalities carry the risk of overtreatment, are linked to high postoperative and increased late morbidity, and have been abandoned.

The most appropriate surgical procedure for necrotizing pancreatitis is the careful removal of necrosis in combination with an additional technique to provide further evacuation of infected peripancreatic exudates and to promote further debridement. This change in intraoperative management has decreased mortality from above 80 per cent to about 20 per cent. Three competitive techniques have been introduced to deal with the peripancreatic space after necrosectomy (Table 5).

Conventional drainage

The conventional surgical approach includes debridement, intraoperative lavage, and Penrose/sump drainage. Reoperations, usually for persistent intra-abdominal infection, are necessary in more than a third of these patients. As a consequence of technical improvements (e.g. more aggressive debridement, wider drainage, and insertion of multiple sump and rubber drains), a decrease in hospital mortality of necrotizing pancreatitis or pancreatic abscess to 24 per cent could be achieved. If this form of drainage is chosen, a low threshold for re-explorations is imperative.

Open and semi-open management

The open and semi-open techniques are based on primary surgical necrosectomy and frequent planned reoperations. After initial debridement, the involved necrotic area is packed with a ring of non-adherent gauze, the abdomen is left open, and the packing is changed every 24 to 48 h under sedation in the ICU. Other centers have introduced a semi-open procedure by inserting a synthetic mesh or a zipper in the abdominal fascia to allow easy re-entry for repeated lavage. Experienced centers have been able to reduce mortality rates to a favorable 15 per cent in patients with necrotizing pancreatitis. However, the schedule of repeated intra-abdominal manipulation results in a high incidence of bowel complications, pancreatic fistulas, and hemorrhage. Consequently, surgical treatment and the need for ICU therapy is often mandatory for several weeks.

Closed management

Since the early 1980s we have advocated surgical debridement (necrosectomy) supplemented by intra- and postoperative closed local lavage of the lesser sac for the treatment of necrotizing pancreatitis. Surgical debridement is performed either digitally or by the careful use of instruments to preserve still vital pancreatic parenchyma. Any tissue becoming necrotic after necrosectomy is rinsed out by the lavage fluid in the postoperative period. After surgical debridement an intraoperative lavage is performed using 6 to 12 liters of isotonic saline to clear the surface of the pancreatic bed. For postoperative continuous local lavage, large-bore single-lumen (Ch 24–28) and double-lumen (Ch 16–18) catheters are placed into the lesser sac. The gastrocolic and duodenocolic ligaments are sutured to create a closed compartment for a regionally restricted lavage. Initial postoperative continuous lavage is performed using 24 liters of commercial dialysis fluid per day. This is discontinued when the effluent is clear with no signs of active pancreatic enzymes or positive bacteriology. The overall mortality of closed management in necrotizing pancreatitis ranges from 9 to 23 per cent.

Late results

Complete morphological and functional recovery of the pancreas was previously assumed to be characteristic after resolution of even severe acute pancreatitis. However, recent evaluations have demonstrated that exocrine insufficiency was present in 80 per cent of patients, and endocrine dysfunction, including a subclinical or insulin-dependent diabetes mellitus, developed in 50 per cent within the first 4 months after surgical treatment. Functional and morphological abnormalities can persist for years, but have a tendency to ameliorate within 12 to 14 months following the acute attack. The degree of these derangements was found to parallel the extent of necrosis. Moreover, etiological factors obviously play an additional role in long-term outcome, for example pre-existing pancreatic alterations in alcoholic pancreatitis.

Bibliography

Beger, H.G. (1991). Surgery in acute pancreatitis. *Hepato-Gastroenterology*, **38**, 92–6.

Beger, H.G., Bittner, R., Block, S., and Büchler, M. (1986). Bacterial contamination of pancreatic necrosis. *Gastroenterology*, **49**, 433–8.

Block, S., Maier, W., Bittner, R., Büchler, M., Malfertheiner, P., and Beger, H.G. (1986). Identification of pancreatic necrosis in severe acute pancreatitis: imaging procedures versus clinical staging. *Gut*, **27**, 1035–42.

D'Egidio, A. and Schein, M. (1991). Surgical strategies in the treatment of pancreatic necrosis and infection. *British Journal of Surgery*, **78**, 133–7.

Uhl, W., Buchler, M., Malfertheiner, P., Martini, M., and Beger, H.G. (1991). PMN-elastase in comparison with CRP, antiproteases and LDH as indicators of necrosis in human acute pancreatitis. *Pancreas*, **6**, 253–9.

4.5 Disordered gastric motility

4.5.1 Vomiting

G. J. Dobb

Key messages

- Many different pathological and pharmacological stimuli cause vomiting.
- Identification of the central neurotransmitters involved in vomiting has provided the basis for current antiemetic treatment.
- Pulmonary aspiration is the major complication of vomiting in patients with impaired consciousness or bulbar dysfunction.
- Antiemetics are used to relieve distress and minimize loss of fluid and electrolytes.
- The 5-hydroxytryptamine$_3$ receptor antagonists are the most effective antiemetics for most causes of vomiting, but are expensive.

Table 1 Conditions associated with nausea and vomiting

Gastrointestinal
Gastroenteritis
Gastric irritation (poisons, alcohol, blood, etc.)
Obstruction
Pseudo-obstruction
Pancreatitis
Other
Migraine
Subarachnoid hemorrhage
Raised intracranial pressure (tumor, hematoma, etc.)
Metabolic disturbance (diabetic ketoacidosis, uremia, hyponatremia)
Severe pain (myocardial infarction)
Fear
Psychogenic vomiting
Bulimia
Drugs (opiates, cytotoxics, theophyllines, etc.)
Snake bite and other envenomations
Anesthesia
Acoustic neuroma
Ménière's disease
Sea and travel sickness
Mountain sickness
Weightlessness
Radiation
Pregnancy
HELLP syndrome

Introduction

Nausea and vomiting are common but debilitating symptoms of illness affecting not only the gastrointestinal tract but also other organs (Table 1). The diversity of associations suggests that a variety of stimuli may trigger a final pathway leading to the conscious appreciation of nausea and the active process of retching and vomiting. Vomiting must be distinguished from the passive process of reflux of gastric contents. Passive reflux of gastric contents occurs in patients with a hiatus hernia or an incompetent lower esophageal sphincter if abdominal pressure exceeds pharyngeal pressure. Maintaining a head-up posture usually eliminates reflux of gastric contents into the mouth and pharynx.

Pathophysiology

Although much of the experimental work has been in cats and ferrets, the process in humans is probably similar. Nausea, retching, and vomiting are separate but related processes. Nausea usually precedes vomiting, and this phase may be accompanied by gastric relaxation, retroperistalsis, salivation, and vagally mediated vasomotor changes. During retching the abdominal muscles, diaphragm, and intercostal muscles contract while the glottis remains closed. The pressure gradients make the gastric contents flow between stomach and esophagus. Strong abdominal muscle contraction and relaxation of the upper esophageal sphincter lead to vomiting with ejection of gastric contents. This is followed by lethargy, shivering, and a sensation of generalized weakness.

Many of the afferent stimuli from intra-abdominal organs and stimuli associated with cardiac pain are transmitted through vagal and sympathetic afferents to the 'vomiting center' in the parvicellular reticular formation in the lateral medulla. This is adjacent to the chemoreceptor trigger zone of the area postrema which has chemoreceptors sensitive to changes in both blood and cerebrospinal fluid. Afferent stimuli from the chemoreceptor trigger zone, the vestibular apparatus, and other parts of the brain also interact with the 'vomiting center' (Mitchelson 1992). Some experimental work suggests this center may be more a physiological than an anatomical structure. Nevertheless ablation of the area postrema prevents vomiting induced by many drugs, including opiates, cytotoxic drugs, and digoxin, in most species. The area postrema is rich in dopaminergic receptors, encephalins, which cause dopamine release, and 5-hydroxytryptamine (5-HT) receptors. Cholinergic receptors are found in the nucleus ambiguous and the lateral vestibular nucleus. Histamine H_1 receptors are involved in some of the central pathways. Definition of the neurotransmitters involved in vomiting has assisted understanding of the effects of antiemetic drugs.

Efferent stimuli during vomiting pass to the pharynx, larynx, upper gastrointestinal tract, respiratory and abdominal muscles, vasomotor center, and salivary and sweat glands.

Vomiting in the intensive care patient

The frequency of vomiting during intensive care stay is poorly documented. It will be greatly affected by patient case mix. As shown in Table 1, vomiting is frequently a symptom of underlying disease. Of particular concern in the intensive care unit (ICU) are postoperative vomiting, vomiting associated with neurological disease and surgery, and hyperemesis gravidarum.

Complications of severe or protracted vomiting include dehydration, electrolyte disturbance, and alkalosis. There is a risk of pulmonary aspiration of gastric content in all patients with impaired consciousness or bulbar dysfunction. Aspiration during obstetric procedures may cause Mendelson's syndrome.

Postoperative nausea and vomiting

Most studies do not distinguish between nausea and vomiting. The frequency varies with age, being greatest in the age group between 6 and 16 years (34–51 per cent) and less in children under 5 years (about 20 per cent) and adults (14–40 per cent). It is more common in women than men and after major abdominal, laparoscopic, and ear, nose, and throat surgery (up to 70 per cent). The use of older inhalational anesthetics and opiates is associated with a greater frequency of nausea and vomiting (Kenny 1994). All patients are at risk of pulmonary aspiration during recovery from anesthesia if they vomit or regurgitate stomach contents.

Vomiting and neurological disease

Neurological disease and surgery are commonly associated with vomiting. The frequency is up to 77 per cent in patients with subarachnoid hemorrhage and of this order or higher during migraine, in patients with raised intracranial pressure caused by tumor or hematoma, and after surgery for acoustic neuroma. Vomiting in the presence of impaired consciousness increases the risks of pulmonary aspiration. In a clinical trial (Pugh *et al.* 1996) metoclopramide 10 mg intravenously was more effective than ondansetron 8 mg intravenously in reducing nausea and vomiting after neurosurgery.

Hyperemesis gravidarum

Vomiting in early pregnancy, often after wakening, is very common (70–80 per cent) and usually self-limiting. The etiology is unknown. Rarely (approximately 2 per 1000), the symptoms are very severe and protracted with complications that may include Wernicke's encephalopathy and death. A desire to avoid or delay antiemetic treatment during early pregnancy may cause severe dehydration and ketoacidosis. Past reports show associations between hyperemesis gravidarum and fetal growth retardation, congenital abnormalities, and prematurity. In recent reports no effect has been found on the outcome of pregnancy, perhaps because of improved patient management.

The antihistamine antiemetics (e.g. promethazine) have been widely used in pregnancy. They appear to be safe and reasonably effective. Nasogastric or nasojejunal feeding is used to maintain nutrition and hydration (van de Ven 1997); only if these are unsuccessful should parenteral nutrition be used. Steroids have been used but are controversial.

Management

If possible, the underlying cause of vomiting is corrected by surgery or other means. However, this is not possible for many causes of vomiting, and symptomatic treatment with antiemetics is used to relieve distress and minimize further fluid and electrolyte loss. Prolonged vomiting may cause dehydration, metabolic alkalosis, and hypokalemia. Hypomagnesemia also occurs and may need correction before the plasma potassium concentration can be maintained. Infusion of 0.9 per cent saline with potassium and magnesium supplements is usually sufficient to correct dehydration.

Clinical trials can be used to guide antiemetic treatment. However, there appears to be considerable variation in the effects of different antiemetics in patients with similar causes of vomiting. Antiemetics include prokinetics, phenothiazines, butyrophenones, antihistamines, anticholinergics, and 5-HT_3 receptor antagonists.

Prokinetics

Metoclopramide and domperidone are dopamine-D_2 receptor antagonists with predominant action on the chemoreceptor trigger zone. Both are well absorbed from the gut. For vomiting in the ICU, 10 mg metoclopramide is usually given intravenously. A daily dose of 3 to 10 mg/kg has been used for vomiting during cancer chemotherapy. The most common side-effects of metoclopramide are extrapyramidal reactions. Renal failure prolongs the excretion half-life from approximately 4 to 14 h. Domperidone penetrates the blood–brain barrier poorly, and so central side-effects are rare; however, no intravenous form is available as this causes cardiac arrest.

Phenothiazines

Prochlorperazine is the most commonly used phenothiazine for nausea and vomiting. Others include perphenazine and chlorpromazine. They block dopamine-D_2 receptors in the chemoreceptor trigger zone and have variable anticholinergic affects. Intravenous doses of prochlorperazine for nausea or vomiting are 5 to 12.5 mg. Side-effects are frequent, and include hypotension, sedation, extrapyramidal reactions, and, more rarely, neuroleptic malignant syndrome. The phenothiazines are less effective than alternative drugs for vomiting during cancer chemotherapy, but are commonly used after anesthesia or to minimize opiate-induced nausea and vomiting.

Butyrophenones

Actions and adverse effects are similar to those of phenothiazines. Clinical trials show that droperidol is an effective postoperative antiemetic in doses of 0.25 to 5 mg intravenously. Smaller doses appear as effective as larger doses and are probably less likely to cause hypotension.

Antihistamines

Antihistamines act principally on vestibular efferents, and so can be effective for motion sickness (e.g. patient transport), Ménière's disease, or other vestibular disturbances. Promethazine 12.5 to 50 mg orally or intravenously is widely used for nausea and vomiting during pregnancy. It has pronounced sedative effects. Cyclizine is an antihistamine with additional anticholinergic effects used for postoperative nausea and vomiting. The usual dose is 50 mg intramuscularly.

Anticholinergics

Hyoscine 0.2 to 0.4 mg intramuscularly is sometimes used in conjunction with anesthesia. The antiemetic effect appears to be short lived, and it can cause drowsiness, mental confusion, and an unpleasantly dry mouth. Alternative drugs are preferred in the ICU.

5-HT$_3$ receptor antagonists

5-HT$_3$ receptor antagonists are expensive compared with other antiemetics but are often effective for otherwise unresponsive nausea and vomiting. They block vagal afferent pathways and 5-HT receptors in the area postrema. Ondansetron, tropisetron, and granisetron are much more active at the 5-HT$_3$ receptors than at other 5-HT receptors. Their efficacy is similar (Morrow *et al.* 1995). They are effective in nausea and vomiting induced by cancer chemotherapy and radiotherapy. For postoperative nausea and vomiting, ondansetron 8 mg is more effective than either metoclopramide 10 mg or droperidol 1.25 mg, all given intravenously. A combination of ondansetron and droperidol can be effective in vomiting after neurosurgery or removal of acoustic neuroma which is difficult to control. The 5-HT$_3$ receptor antagonists have few side-effects; the most common is headache.

Other care

Adequate analgesia and reassurance can alleviate some of the distress associated with vomiting. Conscious patients need a suitable receptacle. Washing and mouth care are essential.

Some studies suggest benefit from acupuncture at the P6 point, and corticosteroids and cannabinoids have been used to treat vomiting induced by cytotoxic drugs.

References

Kenny, G.N.C. (1994). Risk factors for post-operative nausea and vomiting. *Anaesthesia*, **49** (Supplement), 6–10.

Mitchelson, F. (1992). Pharmacological agents affecting emesis: a review (Part 1). *Drugs*, **42**, 295–315.

Morrow, G.R., Hickok, J.T., and Rosenthal, S.N. (1995). Progress in reducing nausea and emesis. Comparisons of ondansetron (Zofran), granisetron (Kytril) and tropisetron (Navoban). *Cancer*, **76**, 343–57.

Pugh, S.C., Jones, N.C., and Barsoum, L.Z. (1996). A comparison of prophylactic ondansetron and metoclopramide administration in patients undergoing major neurosurgical procedures. *Anaesthesia*, **51**, 1162–4.

van de Ven, C.J.M. (1997). Nasogastric enteral feeding in hyperemesis gravidarum. *Lancet*, **349**, 445–6.

4.5.2 Large nasogastric aspirates

G. J. Dobb

Key messages

- Large gastric aspirates inhibit enteral nutrition.
- The risk of regurgitation or vomiting is increased.
- Transpyloric enteral tube placement allows enteral feeding when gastric aspirates remain high.
- Acute gastric dilatation can complicate many causes of critical illness.

Introduction

Gastric motility and gastric emptying are impaired during critical illness, although small-bowel absorption and motility is often adequate to maintain enteral nutrition. Large gastric aspirates can delay or prevent intragastric tube feeding. Large nasogastric aspirates prevent enteral feeding in 17 per cent of patients and are the most common reason for decreasing or stopping it. Gastric residual volumes of 150 to 300 ml, or less than twice the hourly volume of feed being given, have been used to define the upper limit of 'normal'. Patients with residual volumes of less than 300 ml usually tolerate gastric feeding, but larger residual volumes require gastric tube decompression and transpyloric feeding.

Large gastric aspirates increase the risk of regurgitation, vomiting, and aspiration of stomach contents, particularly in patients who are sedated or whose consciousness or bulbar function is impaired.

Pathophysiology

The main components of gastric motility are as follows:

(1) relaxation to accommodate food with only a small increase in intraluminal pressure;

(2) contractions to mix food with gastric juice and reduce particle size;

(3) controlled gastric emptying.

Relaxation of the gastric fundus is mediated mainly through efferent vagal inhibitory neurons with some additional contribution from the action of cholecystokinin. Psychogenic stimuli, swallowing, and afferent impulses from mouth and esophagus initiate gastric relaxation. These factors are bypassed during intragastric tube feeding.

After relaxing to accommodate food, the proximal stomach has low-frequency tonic contractions which pass food towards the distal stomach where gastric smooth muscle cells have an innate electrical activity of approximately 3 cycles/min. Muscle contraction occurs when depolarization exceeds the threshold. Conduction of depolarization from the middle of the body of the stomach through the antrum to the pylorus causes the peristaltic waves that pass distally through the stomach. Contraction in the distal stomach is enhanced by vagal cholinergic nerves and inhibited by the sympathetic innervation. Gastrin increases the frequency of contractions.

Gastric emptying of liquids depends on the pressure gradient across the pylorus and is not regulated by pyloric activity. Solids empty more slowly. As contractions reach the distal antrum, the pylorus closes and solid material is squeezed and ground before being passed back into the body of the stomach. When digestible solids have been reduced to small particles, they pass from the stomach with the liquid material. The effects of different food contents on the rate of gastric emptying are summarized in Table 1. Receptors in the proximal small bowel are sensitive to pH, osmolality, and fatty acid content. Hormonal and local neural pathways then regulate gastric emptying. Motilin and neurotensin increase gastric emptying. Secretin, catecholamines, and β-endorphins may delay gastric emptying. During fasting, undigested solids are propelled from the stomach by the migrating motor complex—a wave of electromechanical activity that normally occurs approximately every 90 min and passes from the stomach distally to the ileum. Motilin is released from the proximal ileum and modulates gastric contraction during the migratory motor complex. Migrating motor complexes occur more frequently during sleep and are inhibited by stress.

During constant infusion of nasogastric feeds in healthy volunteers, gastric volume reaches a maximum 2.25 h after starting the feed and then decreases to a steady state volume after 6 h. There is little information on the usual pattern of gastric emptying in the critically ill. In a study of 10 patients with burns affecting more than 20 per cent of body surface area no significant difference was found compared with normal volunteers, although others have found gastroparesis in many critically ill patients with development of large nasogastric aspirates when fed into the stomach. Autonomic imbalance and reduced motilin secretion are probably responsible.

Table 1 The effect of feed content and properties on the comparative rate of gastric emptying

Fast	Intermediate	Slow
Liquid		Solid
Body temperature		Cold or hot
Carbohydrate	Protein	Fat
Neutral solution		Acid solution
Isotonic solution		Hyper- or hypotonic solution

Table 2 Conditions associated with acute gastric retention

Pain	Renal colic, biliary colic, recent surgery, trauma, etc.
Trauma	Retroperitoneal hematoma, intraperitoneal blood, head injury
Infection	Intraperitoneal sepsis, generalized septicemia
Inflammatory	Pancreatitis
Metabolic	Diabetic ketoacidosis, myxedema, acute porphyria, hepatic encephalopathy, hypoxia
Electrolyte	Hypokalemia, hypomagnesemia, hypocalcemia, hypercalcemia
Other	Immobilization (e.g. paraplegia), after abdominal surgery
Drug induced	Narcotics, sedatives, anticholinergics

Causes

Conditions associated with acute gastric retention are summarized in Table 2. Chronic problems with gastric emptying may become evident when nasogastric feeding and regular gastric aspiration are needed during critical illness (e.g. gastric outlet obstruction from scarring or tumor, diabetic gastroparesis, hypothyroidism, or previous truncal vagotomy). Intolerance to intragastric feeding is particularly common in patients with traumatic brain injury and is independent of the opiates that many receive.

Complications

Large gastric aspirates are not a benign problem. Complications include impaired nutrition, nausea, vomiting, pulmonary aspiration of gastric contents, loss of water and electrolytes, and acute gastric dilatation.

Acute gastric dilatation may complicate any condition associated with acute gastric retention. When it is severe, there is upper abdominal distension, hiccoughs are common, and patients who are not deeply sedated or unconscious usually have obvious discomfort and tachycardia. Copious vomiting or regurgitation is usually a late manifestation. The first indication of acute gastric dilatation in the critically ill is often a hugely distended stomach seen on a chest radiograph taken for other reasons.

Management

Correctable underlying causes of gastric retention (e.g. electrolyte abnormality, pain) should be treated, but gastric emptying is often slow to improve.

Acute gastric dilatation

Treatment is gastric aspiration, continuous gastric drainage, and replacement of fluid and electrolytes.

Promotion of gastric emptying

Gastric residual volumes can be affected by patient position; they are usually lowest when the patient is lying on the right side with the pylorus dependent, and greatest when lying on the left side. If residual volumes remain large after the patient is positioned right side down, prokinetic agents are given, for example intravenous metoclopramide 10 mg every 6 h, domperidone 10 to 20 mg orally or 30 to 60 mg rectally every 4 to 8 h, and cisapride 10 mg via the nasogastric tube every 6 h. Only cisapride has been shown to reduce gastric residual volumes significantly in critically ill patients. Erythromycin is prokinetic, but the high frequency of nausea often makes it unsuitable.

Transpyloric feeding

Transpyloric feeding and continuous gastric drainage should be attempted in all critically ill patients with large nasogastric aspirates. There are few absolute contraindications other than intestinal obstruction, pseudo-obstruction, or fistula. A small clinical trial showed benefit and safety when feeding was started early after bowel resection. Many techniques have been described for transpyloric tube placement and a variety of tubes are commercially available. A soft 8 Fr hygromer-coated self-lubricating tube at least 100 cm long with a metal stylet and non-bulbous weighted tip is suitable for most adult patients.

Blind placement

Spontaneous passage of fine-bore tubes into the duodenum and beyond occurs in 15 to 50 per cent of patients, but is much lower in critically ill mechanically ventilated patients (5–15 per cent). Metoclopramide and cisapride probably have no effect on transpyloric passage, but intravenous erythromycin 200 mg may be effective, although less so in patients with brain injury. Weighted-tip tubes are easier to pass through the esophagus into the stomach although they do not assist passage through the pylorus.

A 'blind' technique for inserting Entriflex (Biosearch Medical Products, Somerville, NJ) and Corsafe (Corpak Inc., Wheeling, IL) feeding tubes with a 92 per cent success rate in intensive care patients has been described. The technique is simple, but the high success rate has yet to be confirmed.

Placement assisted by pH measurement

Blind placement can be assisted by repeated aspiration and testing aspirate pH. The usual intragastric pH (1.5–5.0) increases to 6.0 in the small bowel. Tubes of 6 to 10 Fr with a pH sensor at the tip are commercially available: Accusite pH sensor (Zinetics, Salt Lake City, UT), Entek (Rusch, Atlanta, GA), Biosearch (Biosearch Medical Products, Somerville, NJ), and Corpak (Corpak Inc., Wheeling, IL). Success rates of 80 to 90 per cent for correct placement are reported.

Placement assisted by gastroscopy

The endoscope allows direct vision of tube insertion. Techniques described include those in which the feeding tube is 'piggy-backed' to the endoscope and released when the tip is in the duodenum, simple visualization of a feeding tube passing through the pylorus, and using the gastroscope to insert a 2-m guidewire under direct vision with the feeding tube being passed over the guide after removal of the endoscope. Disadvantages include the risks of upper gastrointestinal endoscopy and the need for skilled staff and equipment. Successful placement should be over 90 per cent.

Placement assisted by fluoroscopy

This technique, which is the author's preference, is easily performed in the ICU with a portable image intensifier provided that the beds are radiolucent. A standard 8 Fr soft feeding tube at least 110 cm long with a wire stylet (e.g. Entriflex, Biosearch Medical Products, Somerville, NJ), can usually be passed. If it is difficult to pass the tube into the duodenum, its passage is usually assisted by intravenous erythromycin 200 mg. If this fails, it is often possible to pass a floppy J-tipped guidewire into the duodenum and jejunum under fluoroscopy and insert a feeding tube over the guidewire (e.g. the McLean Ring tube and guidewire (Cook Inc., Bloomington, IN)). With experience, the successful placement rate is approximately 95 per cent.

Jejunostomy

Percutaneous jejunostomy may be considered during laparotomy for intra-abdominal trauma, major gastrointestinal surgery, etc. Alternative techniques are formal surgical jejunostomy and needle-catheter jejunostomy. The latter is advocated by some in the United States as a means of allowing early enteral feeding, but is less popular in Europe and Australasia. Complications, including leaks, wound infection, peritonitis, and bowel obstruction, are relatively common.

Other techniques

Double-lumen tubes with both gastric and jejunal orifices are available for insertion from the nose or through a gastrostomy.

Complications of fine-bore enteral feeding tubes

Complications associated with the use of fine-bore feeding tubes are related to tube misplacement, blockage, and unintentional removal. Misplacement complicates 0.3 to 4 per cent of nasoenteric tube insertions, and so the tube-tip position should always be confirmed radiographically before commencing feeding. If doubt remains, confirmation of the tip position is assisted by injecting a small amount of radiocontrast material. Aspiration through fine-bore tubes increases the risk of tube blockage. Medications given through 8 Fr or smaller tubes should be in solution or as elixirs to reduce the risk of blockage. Unscheduled tube removal by patients (up to 30 per cent) and staff (up to 10 per cent) limits enteral nutrition. A nasal 'bridle' fixation technique can reduce unscheduled removal.

Bibliography

Carr, C.S., Ling, K.D.E., Boulos, P., and Singer, M. (1996). Randomised trial of safety and efficacy of immediate postoperative enteral feeding in patients undergoing gastrointestinal resection. *British Medical Journal*, **312**, 869–71.

Heyland, D., Cook, D.J., Winder, B., Brylowski, L., Van de Mark, H., and Guyatt, G. (1995). Enteral nutrition in the critically ill patient: a prospective survey. *Critical Care Medicine*, **23**, 1055–60.

Spapen, H.D., Duinslaeger, L., Diltoer, M., Gillet, R., Bossuyt, A., and Huyghens, L.P. (1995). Gastric emptying in critically ill patients is accelerated by adding cisapride to a standard enteral feeding protocol: results of a prospective, randomized, controlled trial. *Critical Care Medicine*, **23**, 481–5.

Weekes, E. and Elia, M. (1996). Observations on the pattern of 24-hour energy expenditure changes in body composition and gastric emptying in head-injured patients receiving nasogastric tube feeding. *Journal of Parenteral and Enteral Nutrition*, **20**, 31–7.

Zaloga, G.P. (1991). Bedside method for placing small bowel feeding tubes in critically ill patients. *Chest*, **100**, 1643–6.

4.5.3 Ileus and obstruction

D. C. Evans and J. C. Marshall

Key messages

- Intestinal dysmotility in the intensive care patient is a common problem.
- The clinical challenge is to distinguish ileus from obstruction without delay.
- The management of intestinal obstruction is surgical. Swift involvement of a general surgeon is warranted as soon as the diagnosis is suspected.
- Bowel obstruction is usually diagnosed clinically and confirmed by surgical exploration.
- Treatment of ileus is expectant. Correctable underlying causes of ileus (e.g. drugs, abdominal sepsis, and electrolyte disorders) should be addressed.

Introduction

The gastrointestinal tract, although rarely the center of attention in the intensive care unit (ICU), will either facilitate or impede the return to good health of every critically ill patient. Aside from its primary function of digestion and absorption of nutrients, a healthy gut plays a central role in maintaining metabolic homeostasis, meeting the body's caloric needs, assuring fluid balance, providing a physiological and bacteriological barrier to infection, and supporting immunocompetence. The bacteriological load of the gut is formidable, and the integrity of the gut membrane barrier is essential to continued survival. Yet the absence of measurable indicators of gut dysfunction in critical illness makes intestinal health a low priority when more visible needs are being addressed. Indeed, it typically takes a crisis, be it hemorrhage, diarrhea, or perforation, to alert the intensivist to the absolute requirement for reasonable gut health to be achieved and maintained.

The first presentation of a gastrointestinal problem is often the radiographic demonstration of fluid-filled loops of bowel with or without abdominal distension. Immediate concerns of ileus or obstruction are raised. Rapid differentiation between these two entities is important although difficult, as they present with similar clinical pictures but signify vastly different pathologies (Fromm 1993). Because mechanical bowel obstruction, albeit rare in the ICU patient, portends disaster if missed or acted upon late, a systematic approach to care is required.

Physiology of the healthy gastrointestinal tract

The physiology of normal intestinal function is complex, and a grasp of the basic principles of intestinal motility is invaluable to an understanding of gut motor dysfunction.

There are many determinants of gut motility (Quigley 1996). An intricate and interrelated neural network invests all layers of the bowel wall and responds to stimuli originating locally, in the autonomic nervous system, and in the brain. Disturbances at all levels can upset motility. The intrinsic nervous system of the bowel comprises Meissner's submucosal plexus investing the muscularis mucosae, Auerbach's myenteric plexus running between the longitudinal and circular muscular fibers, and a fine subserosal plexus running with the external longitudinal muscle fibers. The external nervous system is divided into two autonomic components: sympathetic activity which is generally adrenergic and inhibitory, and parasympathetic activity which is cholinergic and excitatory. Responding to distension through mechanoreceptors, and to other disturbances through chemoreceptors, efferents arising in the thoracolumbar splanchnic nerves act to inhibit motility through release of norepinepherine (noradrenaline). Circulating catecholamines do not appear to impair gut motility.

All portions of the small bowel maintain an intrinsic electrical control activity capable of prompting peristalsis. Electrical control activity frequency declines gradually along the bowel length. As the duodenum maintains the fastest and therefore overriding rate, it serves as the pacemaker for peristalsis. The gut behaves differently in fed and fasted (interdigestive) states. Migrating motor complexes are phasic peristaltic contractions occurring in 90- to 120-min cycles to clear the fasted small bowel of a regular build-up of enteric residue, desquamated cells, and bacteria. Slower more irregular contractions emerge postprandially to facilitate mixing and absorption. Such contractions are an intrinsic phenomenon and occur independent of vagal regulation. Giant migrating complexes are large-amplitude long-duration contractions which move enteric contents distally *en masse* and propagate uninterrupted to the ileocolic junction. Erythromycin enhances this activity. These more intense contractions may stimulate nociceptors causing intermittent crampy pain, particularly where obstruction is present. Migrating clustered contractions lasting 1 to 3 min are thought to occur randomly over short segments (10–30 cm) of small bowel to assist with postprandial clearance. These contractions increase in the early phase of small bowel obstruction. Neurohormonal reflexes involving vagal innervation and hormones such as motilin co-ordinate postprandial activity with the rest of the gut. Emesis occurs when retrograde giant complexes originating in the mid small bowel return intestinal contents to the stomach for expulsion. The colon manifests broad peristaltic activity based largely on migrating motor complex and giant migrating complex contractions (Sarna 1993).

The colon is extremely efficient in absorbing the 1.5 liters of water, sodium, potassium, and bile salts entering the cecum daily. When excess bile salts are delivered to the colon, as may occur in short gut or dysfunctional bowel syndromes, a secretory diarrhea results.

Although less rigidly regulated than cerebral or renal blood flow, there is significant autoregulation of intestinal perfusion such that gut oxygen uptake is maintained despite pressure fluctuations as great as 50 mmHg. Significant hypotension or hypoxia provokes a fall in vascular resistance and capillary recruitment that helps protect the gut from ischemia.

Mechanical obstruction

Mechanical obstruction is an uncommon cause of intestinal distension in ICU patients. However, it is important to exclude this diagnosis through careful consideration of the common causes of small and large bowel occlusion. Figure 1 presents an algorithm for the investigation of intestinal distension with the focus on distinguishing obstruction from ileus.

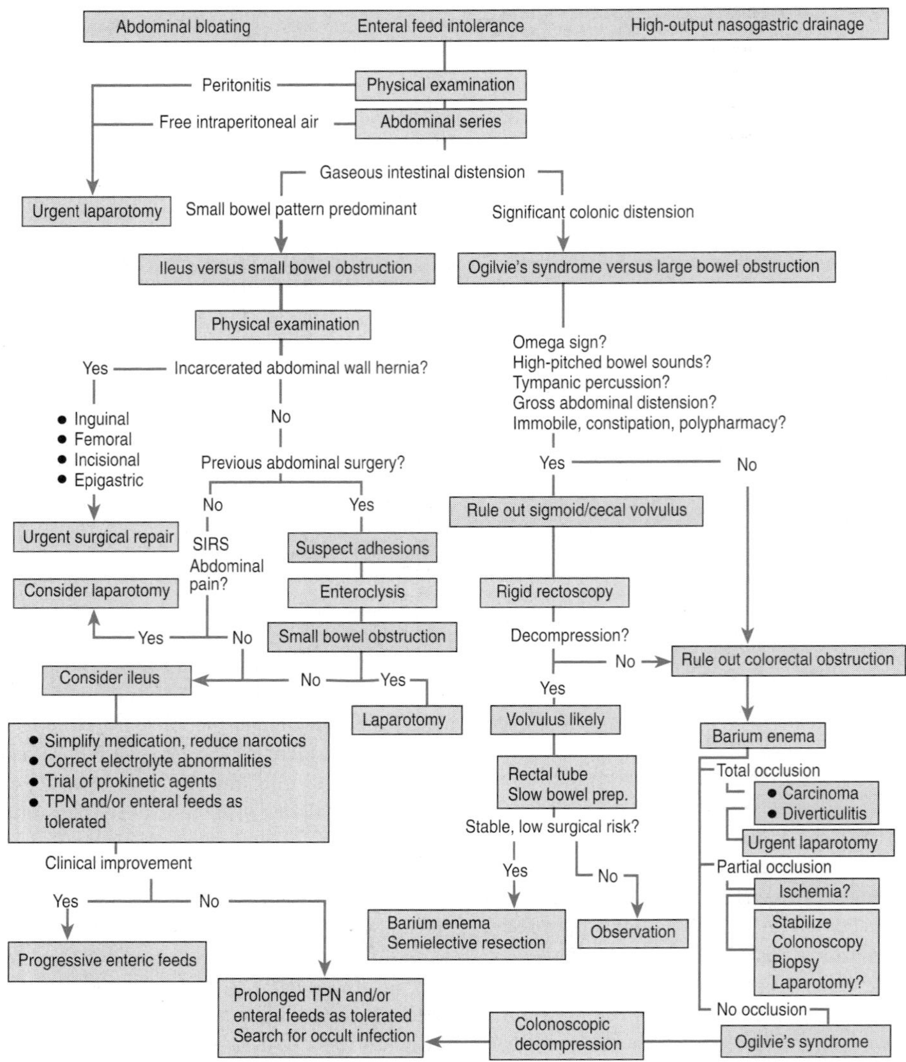

Fig. 1 Algorithm for the approach to ileus versus obstruction in the ICU patient: SIRS, systemic inflammatory response syndrome; TPN, total parenteral nutrition.

Patients experience crampy abdominal pain as peristaltic contractions continue against obstruction in the early phase. As ICU patients are often comatose, hyperactive bowel sounds, visible peristalsis, absence of stool, and compatible abdominal radiographs showing fixed-position air–fluid levels on serial films all promote the diagnosis of obstruction.

Before extensive investigation is undertaken for possible bowel obstruction, it is prudent to assess any intraintestinal or intraperitoneal tubes or drains present. Balloon catheters used as feeding or drainage conduits may cause intraluminal obstruction. Internal hernias and volvulus may occur around the point of fixation of jejunal feeding catheters, and extrafascial stomal prolapse at colostomy or ileostomy sites may also cause remediable obstruction. A digital rectal examination (or digital exploration of stomas), followed by gentle saline enemas, is an often neglected yet essential initial maneuver, which may relieve existing fecal impaction. Aside from the more common surgical causes of obstruction, one etiology seen occasionally in the coagulopathic patient is a submucosal intestinal hematoma which can arise spontaneously. The typical finding on an upper gastrointestinal contrast study is the 'picket-fence' appearance of the mucosa at the site of obstruction.

A closed-loop obstruction is a surgical emergency often caused by an internal small bowel hernia due to previous postoperative adhesions or distal colorectal occlusion (neoplasm or diverticulitis) with a competent ileocecal valve. Acute cecal distension to greater than 12 cm should prompt consideration of surgical decompression. A loop transverse colostomy, which may be performed under local anesthesia in an unstable patient, affords highly effective decompression and facilitates rapid stabilization of the patient in preparation for a delayed definitive procedure.

Luminal sequestration of fluids and electrolytes causing significant third-space losses occurs within 12 h of small bowel occlusion. When intraluminal pressure exceeds 20 cmH_2O, venous congestion of the bowel wall ensues and augmented capillary filtration enhances the accumulation of intraluminal and peritoneal free fluid. Bacterial overgrowth follows, and translocation of gut bacteria and endotoxin may occur. Overgrowth is fostered by gastric achlorhydria, diminished gut motility, obstruction, reflux, villus atrophy (as seen with prolonged total parenteral nutrition), altered gut flora (secondary to antibiotic use), intestinal distension, and sepsis.

The management of intestinal obstruction is surgical. Swift involvement of a general surgeon is warranted as soon as the diagnosis is suspected. Bowel obstruction is usually diagnosed clinically and confirmed by surgical exploration. Less commonly, radiological imaging is required to establish the diagnosis. If distal colonic obstruction is a possibility, a saline rectal lavage should be performed followed by a gastrograffin enema. If normal, the small bowel should be imaged using barium since water-soluble contrast media dilute readily in intestinal fluid, making interpretation difficult.

Ileus

Ileus is defined as a reversible decrease in gut motility occurring secondary to a physiological derangement and resulting in functional obstruction to the passage of enteric contents. In the ICU patient, ileus is far more frequent than obstruction and has a typically multifactorial etiology (Smith *et al.* 1977) (Table 1). Metabolic and pharmacological causes predominate; therefore correction of electrolyte and acid–base abnormalities is the first step in treatment. Medications that can alter gut motility should be avoided, particularly drugs which exhibit adrenergic and cholinergic activity.

Many ICU patients will have a history of recent abdominal surgery. Postoperative ileus, which is believed to be due to a somatic inhibitory reflex, is common and prolonged if there has been intestinal manipulation. Following laparotomy, ileus resolves in a predictable sequence that begins with the small bowel (5–10 h), followed by the stomach (up to 3 days), and finally the colon (up to 5 days). It is important to appreciate that bowel sounds arise only in the presence of fluid, air, and peristalsis. Their presence or absence is largely non-specific and offers little aid in reliably distinguishing ileus from obstruction.

The length of surgery or degree of manipulation cannot predict the extent of ileus. While adrenergic stimulation impairs gut motility, circulating catecholamines play no significant role. Adrenalectomy, adrenergic blockade, or epidural anesthesia to block efferent sympathetic fibers have negligible protective effects against postoperative ileus.

While small bowel resection and re-anastomosis interrupts propagation of migrating motor complexes for 30 to 40 days, the distal small bowel will still maintain its own intrinsic pacesetter rhythm. The postprandial peristaltic response and absorptive capacity of the bowel are unaffected by resection. This phenomenon accounts for the success of early postoperative feeding. Early feeding also prevents villus atrophy and helps to maintain normal gut physiology.

The diagnosis of ileus requires that mechanical obstruction be excluded. Serial abdominal films showing non-specific intestinal distension, with or without scattered air–fluid levels on upright or decubitus films, are the best adjunct to repeated clinical evaluation (Table 2).

After correction of metabolic abnormalities, medications should be reviewed and rationalized to reduce cumulative sympathomimetic effect. Opiates should be reduced, although not to the exclusion of adequate pain control as both pain and immobilization contribute to prolongation of ileus.

Treatment of ileus is expectant. Nasogastric decompression is not essential if there is no emesis. Intravenous hydration to provide adequate urine output is essential, and total parenteral nutrition should be initiated promptly if the conditions believed to be responsible for the ileus are unlikely to abate within 3 to 5 days. Parasympathomimetic agents (neostigmine, bethanechol) and sympatholytics have been used with some success. Prokinetic agents are frequently tried; metaclopramide enhances gastric and small bowel motility but with no reproducible capacity to relieve ileus. Cisapride, by augmenting acetylcholine release in the myenteric plexus, is of some benefit in the treatment of postoperative ileus (Verlinden *et al.* 1987). Even in the face of moderate ileus, it is often possible to initiate enteral feeds at a slow infusion rate (20–40 ml/h) and increase as tolerated.

Table 1 Common causes of ileus

Common causes of ileus
Metabolic
Hypokalemia
Hypocalcemia
Hypomagnesemia
Hypophosphatemia
Acidosis
Endocrinological
Diabetic gastroparesis
Hypothyroidism
Neurogenic
Pain (incisional, thoracolumbar)
Retroperitoneal hemorrhage
Pharmacological
Anesthetics
Analgesics (morphine)
Vasoactive agents (dopamine, nifedipine)
Psychotropic agents (tricyclics, phenothiazines)
Anticholinergics
Traumatic
Surgical manipulation (laparotomy)
Bowel resection
Infectious/inflammatory
Localized or diffuse peritonitis
Intra-abdominal abscess
Systemic inflammatory response syndrome
Viral enteritis (cytomegalovirus, herpes simplex virus)
Neoplastic
Mesenteric infiltration by tumor
Retroperitoneal malignancy

Conclusion

Intestinal dysmotility is a common problem in the ICU patient. The clinical challenge is to distinguish ileus from obstruction without

Table 2 Plain radiograph features of ileus versus obstruction

	Ileus	Obstruction
Small bowel		
Degree of gaseous distention	Mild to moderate	Moderate to severe
Gas distribution	Diffuse	Localized
Air–fluid levels	Diffuse; uniformity of width	Tend to be localized; varying widths
Evolution of gas pattern	May shift over time	Generally fixed
Site of occlusion	None seen	May be evident
Mucosal pattern	Unreliable	Unreliable
Colonic gas	Usually present	Diminished or absent
Large bowel		
Gas distribution	Excessive right colon dilatation (Ogilvie's sign)	Contiguous throughout colon
Sigmoid (omega) loop	Absent	Present in volvulus
Small bowel gas	Variable	Distal distension; may be absent if ileocecal valve competent
Air in rectum	May be present	Usually absent

delay and then to address correctable underlying causes of ileus, particularly abdominal sepsis. Ileus is no longer a contraindication to enteral feeding, and every effort should be made to use the gut early when the ICU stay is likely to be protracted.

References

Fromm, D.G. (1993). Ileus and obstruction. In *Surgery: scientific principles and practice* (ed. L.J. Greenfield), pp. 731–40. J.B. Lippincott, Philadelphia, PA.

Quigley, E.M.M. (1996). Gastric and small bowel intestinal motility in health and disease. *Gastroenterology Clinics of North America*, **25**, 113–45.

Sarna, S.K. (1993). Colonic motor activity. *Surgical Clinics of North America*, **73**, 1201–15.

Smith, J., Kelly, K.A., and Weinshilboum, R. M. (1977). Pathophysiology of postoperative ileus. *Archives of Surgery*, **112**, 203–6.

Verlinden, M., Michiels, G., Boghaert, A., de Coster, M., and Dehertog, P. (1987). Treatment of postoperative gastrointestinal atony. *British Journal of Surgery*, **74**, 614–17.

4.5.4 Diarrhea

G. J. Dobb

Key messages

- Severe diarrhea with dehydration can require intensive care treatment, particularly for infants, the elderly, or the debilitated.
- Prevention or correction of dehydration and electrolyte imbalance are the most important components of treatment.
- Complications of diarrhea caused by *Escherichia coli* include hemolytic–uremic syndrome and thrombotic thrombocytopenic purpura.
- *Clostridium difficile* colitis can cause diarrhea up to 6 weeks after antibiotic use.
- Altered intestinal function and side-effects of drug treatment are other common causes of diarrhea arising during intensive care.

Introduction

Diarrhea can be the predominant symptom of a debilitating illness requiring intensive care or it can arise as a complication of intensive care treatment. General management principles are similar but there are significant differences in investigation and treatment.

Diarrhea is an increase in the frequency, quantity, and liquidity of feces. Quantitative definitions (e.g. more than 200 g of feces daily) have been used, but it is hard to weigh or measure the quantity of stool in critically ill patients. Commonly, all three characteristics of the stool are changed in patients with diarrhea. Different types of diarrhea are summarized in Table 1.

Pathophysiology

Approximately 1500 ml of fluid normally passes into the cecum each day, although only 100 to 200 ml are lost in feces. The maximum colonic absorptive capacity is approximately 4500 ml/day. Diarrhea (increased fecal water loss) occurs when the volume of fluid entering the cecum exceeds the maximum absorptive capacity, if colonic water absorption is impaired, or if there is net loss of fluid into the colon. Colonic water absorption is linked to electrolyte absorption. Energy-dependent active transport of electrolytes across the colonic mucosa draws water through the mucosa into the intercellular space. Unlike the ileum, where absorbed electrolytes are accompanied by isosmotic quantities of water, the colon can transport water against an osmotic gradient of up to 50 mosmol/l. The permeability of the colon decreases from cecum to rectum, with the rate of water absorption being greater in the ascending colon than in the descending colon and rectum.

Mixing contractions, which correspond to the radiological haustra, retard flow and maximize absorptive function. Rapid transport through the proximal colon, for example in patients with carcinoid syndrome, is associated with diarrhea.

Secretory diarrhea, caused by toxins produced by *Vibrio cholerae*, enterotoxigenic *Escherichia coli*, *Staphylococcus aureus*, and *Clostridium*

Table 1 Types of diarrhea

Secretory diarrhea
Cholera
Enterotoxigenic *E. coli*
Excess gut hormone secretion (e.g. vipoma)
Osmotic diarrhea
Lactose intolerance
Laxative abuse
Sugar content of elixirs
Exudation
Excess mucus production
Increased mucosal permeability
Motility disorder
Amyloid
Scleroderma
Diabetes
Short bowel syndrome
After intestinal resection
Malabsorption

perfringens, is associated with increased salt and water loss into the small intestine. This is mediated by increased cyclic adenosine monophosphate (cAMP) concentrations generated in enterocytes in response to the bacterial toxins (Hansen and Skadhauge 1995). High intracellular cAMP concentrations phosphorylate enterocyte microvillus membrane proteins, including the protein of the chloride channel. This results in increased chloride secretion and inhibition of sodium chloride absorption, with the associated water loss causing diarrhea. Tumors secreting vasoactive intestinal peptide (vipomas) cause secretory diarrhea through the same mechanism. Other enterotoxins are believed to act by increasing enterocyte cytoplasmic free calcium concentrations (e.g. *Clostridium difficile*) or by depleting colonic mucosal ribosomal RNA, halting protein synthesis and killing the cell (e.g. *Shigella*).

Severe diarrhea

Admission to the intensive care unit (ICU) is rarely needed for a primary diarrheal illness, but may be indicated for severe dehydration, electrolyte imbalance, associated systemic symptoms, severe coexisting disease, or complications of the diarrheal illness (e.g. hemolytic–uremic syndrome). The most common cause of symptoms of such severity is severe infectious diarrhea.

Severe infectious diarrhea

It is estimated that in some countries between a quarter and a half of childhood mortality is related to diarrheal disease. The risk to adults from associated dehydration, electrolyte imbalance, and weight loss is less, but nevertheless can be life threatening. Intercurrent infection with pneumonia or septicemia is common. Opportunistic infections in patients with AIDS are an increasing cause of severe diarrhea but rarely require intensive care.

Food poisoning

Common causes of food poisoning in developed countries are *Salmonella, Shigella, Staph. aureus,* and *C. perfringens. Salmonella, Shigella, Yersina*, and *Campylobacter* are invasive and cause an inflammatory diarrhea which may lead to fever and abdominal pain. The stools can contain blood or mucus but more often are watery. Preformed enterotoxins cause diarrhea associated with *Staph. aureus*. It is usually a self-limiting disease with onset within hours of eating contaminated food. The enterotoxins are not absorbed, and, although the precise action is unclear, the result is increased net water and electrolyte secretion by the gut. *C. perfringens* type A causes a non-invasive non-inflammatory perfuse watery diarrhea, commonly accompanied by abdominal pain, starting 8 to 24 h after ingestion of contaminated food and lasting for 12 to 24 h. The *C. perfringens* type A enterotoxin is heat labile and impairs electrolyte transport and water absorption in the ileum. *C. perfringens* type C, which is usually acquired from undercooked pork and largely confined to New Guinea, Thailand, Nepal, and China, causes enteritis necroticans.

E. coli

At least five groups of *E. coli* causing diarrhea have been identified (Cantey 1993): enterotoxigenic, enteroinvasive, enteropathogenic, enterohemorrhagic, and enteroadherent. The enteroinvasive *E. coli* causes a similar syndrome to *Shigella*. The serotype 0157:H7 and other serotypes cause bloody diarrhea. The hemolytic–uremic syndrome complicates approximately 30 per cent of enterohemorrhagic infections. Thrombotic thrombocytopenic purpura is a less frequent complication.

V. cholerae

V. cholerae, which is an important cause of severe diarrhea in developing countries, may be acquired from water, rice, or seafood. The majority of people infected do not develop diarrhea; only 2 to 5 per cent develop the 'typical' severe diarrhea and a quarter develop mild to moderate diarrhea.

Investigation

Most infective diarrhea is self-limiting. Further investigations are indicated to identify the cause when diarrhea is severe, associated with systemic symptoms, or prolonged, or for public health reasons. Microscopy, culture, and serotyping identify parasites and pathogenic bacteria. Viruses can be identified by electron microscopy or immunoassay. Molecular methods using synthetic oligonucleotide probes and polymerase chain reaction amplification are now available for identifying many enteric pathogens. These have the potential to provide faster identification to guide treatment.

Treatment

Prevention and correction of dehydration and electrolyte imbalance are the most important components of treatment. Oral rehydration solutions are effective and help reduce ongoing water loss into the bowel lumen (Farthing 1988). Intravenous rehydration may be needed when dehydration is severe or if diarrhea is accompanied by vomiting. Severe diarrhea over 48 h can result in fluid loss of 4.5 to 8.5 liters and sodium depletion of 10 to 20 mmol/kg body weight. Replacement with isotonic saline and potassium supplements is usually adequate. After very severe diarrhea a metabolic acidosis can make part replacement with isotonic sodium bicarbonate more appropriate. Antidiarrheals provide symptomatic relief but should not be used when diarrhea is caused by invasive enteropathogens, i.e. when diarrhea is bloody or accompanied by fever. Cyclo-oxygenase inhibitors (indomethacin, aspirin) appear to be effective in experimental models of secretory diarrhea, but clinical information is limited. Chloride-channel-blocking agents are being investigated as treatment for secretory diarrhea.

The place of antibiotics is controversial. Antibiotic treatment may prolong carriage of *Salmonella* and antibiotics do not, for example, reduce the complications of enterohemorrhagic *E. coli* infection. Bacteremia and extraintestinal manifestations of infection are the usual indications for antibiotics, but these can be difficult to predict and so antibiotics are also given to neonates, the elderly, debilitated patients, and the immunosuppressed. Fluoroquinolones (norfloxacin, ciprofloxacin), trimethoprim–sulfamethoxazole (co-trimoxazole), tetracycline (not for children or pregnant women), and intravenous third-generation cephalosporins have been used for empirical treatment. Otherwise, treatment is guided by culture and sensitivity results.

C. difficile

C. difficile is a potential cause of diarrhea in any patient who has received antibiotics in the preceding 4 to 6 weeks. All antibiotics except parenteral aminoglycosides have been implicated. Infection can be asymptomatic but causes syndromes ranging from mild diarrhea to fulminant and fatal pseudomembranous colitis. Pseudomembranous colitis is more frequent in the elderly and the debilitated and in ICUs. Nosocomial acquisition of *C. difficile* can occur, and so patients with *C. difficile* present in stool cultures should be subject to enteric isolation procedures.

C. difficile produces toxins A and B which cause the colitis. These toxins can be detected by cytotoxic tissue culture assay or enzyme-linked immunoassays. About 80 per cent of patients with *C. difficile* toxin will be positive on the first stool sample tested, but up to three samples should be sent to exclude the diagnosis (Manabe *et al.* 1995).

Pseudomembranous colitis presents with diarrhea and abdominal pain or discomfort. Tenesmus, anorexia, nausea, or vomiting may also occur. Flexible procto/sigmoidoscopy may reveal the typical pseudomembranes, but their absence does not exclude the diagnosis; up to a third of patients have pseudomembranes only in the ascending colon. Toxic megacolon is a rare complication of pseudomembranous colitis. Increasing abdominal pain, abdominal distension, fever, and tachycardia in a patient with recent diarrhea are clues to the diagnosis. Half the patients with toxic megacolon present with shock or hypotension and all are sufficiently ill to need admission to an ICU. Toxic megacolon has a high morbidity; up to two-thirds of the patients may die despite active treatment.

Treatment includes cessation of antibiotics if possible, rehydration, and specific treatment to eradicate *C. difficile*. Metronidazole 250 mg or vancomycin 125 mg orally four times daily are alternatives for mild to moderate pseudomembranous colitis. Metronidazole is cheaper and, in most studies, equally effective. Severe colitis has been treated with increased doses of vancomycin (up to 500 mg orally every 6 h), with intravenous metronidazole 250 mg every 6 h added for toxic megacolon. Patients with toxic megacolon who deteriorate despite medical treatment need surgical review. Worsening abdominal pain and increasing colonic diameter are indications for subtotal colectomy.

Diarrhea developing during intensive care

Diarrhea is more common in intensive care than in a general hospital ward population. Reports suggest that 33 to 66 per cent of ICU patients have diarrhea at some time during their stay, with the frequency being partly definition dependent. Diarrhea developing during intensive care can be caused by reactivation of chronic disease, altered intestinal function during critical illness, or an adverse effect of treatment, including enteral nutrition.

Reactivation of chronic disease

Exacerbations of ulcerative colitis are occasionally seen during severe illness. Reactivation of *Strongyloides stercoralis* can be precipitated in immunosuppressed, malnourished, or debilitated patients. If the parasite disseminates throughout the body, mortality is high. Reactivation should be considered in the differential diagnosis of diarrhea in patients from the endemic regions of Eastern Europe, Africa, Southeast Asia, and the Caribbean.

Altered intestinal function in critical illness

Normal intestinal secretion, absorption, and motility is maintained by a complex interaction of over 30 hormones and hormone-like peptides and the influence of the autonomic nervous system. Inflammatory mediators released during critical illness, including prostaglandins, opioids, histamine, serotonin, cytokines, and nitric oxide, have an effect on intestinal function. Numerous mediators increase intestinal capillary permeability, possibly causing loss of protein and fluid into the intestinal lumen, or have a direct or indirect effect on enterocyte cAMP concentrations. Therefore it is not surprising that diarrhea is a fairly frequent non-specific accompaniment of severe illness, particularly in children. Shock and hypoxemia are independent variables predictive of diarrhea.

A number of factors may contribute to the diarrhea associated with altered intestinal function. Splanchnic vasoconstriction is a physiological response to hypotension. Ischemia reduces the supply of energy substrates to the gut and may limit, for example, energy-dependent reabsorption of sodium and water from the gut. Loss of enteral feeding deprives the gut mucosa of this contribution to mucosal nutrition. Colonic mucosa normally receives 60 to 70 per cent of its energy substrate needs from the bowel lumen. Without enteral nutrition, mucosal enzyme activity rapidly decreases and there is loss of villus height, inducing malabsorption through impaired digestion and reduced absorptive capacity. Changes to the intestinal flora are common in severe illness. Contributory factors are loss of enteral nutrition, colonization of the upper gastrointestinal tract by Gram-negative organisms or yeasts, the effects of drugs on intestinal secretion and motility, and bacterial selection in patients treated with antibiotics. An increase in bile acids reaching the colon promotes fluid and electrolyte secretion. Critical illness can reduce active reuptake of bile acids in the ileum, and an altered bowel flora will be less efficient at deconjugating bile acids so that they can be reabsorbed. Hypoalbuminemia may induce intestinal mucosal edema and impair absorption, but the association between hypoalbuminemia and diarrhea in critically ill patients is uncertain, with most studies showing no significant link. Constipation is common in patients confined to bed. Spurious diarrhea, i.e. fluid passing around impacted feces, must always be excluded by rectal examination as a potential cause of apparent diarrhea.

Table 2 Drugs associated with diarrhea

Laxatives
Antacids (e.g. magnesium trisilicate)
Antibiotics
Antiarrhythmics (e.g. quinidine, digoxin, procainamide)
Bronchodilators (e.g. theophylline)
Antihypertensives (e.g. angiotensin-converting enzyme inhibitors, α-methyldopa, β-blockers, hydralazine, guanethidine)
Cholesterol-lowering drugs
Cytotoxic drugs
Diuretics (e.g. thiazides, furosemide (frusemide))
Histamine H_2 receptor antagonists (e.g. cimetidine)
Non-steroidal anti-inflammatory drugs (e.g. indomethacin)
Prokinetics (e.g. metoclopramide, cisapride)
Prostaglandin analogs (e.g. misoprostol)

Diarrhea as a complication of treatment

Many different classes of drugs have diarrhea as a side-effect. The list in Table 2 is not exhaustive. Cumulative effects of laxatives given during hospital admission and antibiotics are amongst the most common drug-induced causes of diarrhea. When drugs are administered as elixirs, the amount of sorbitol given with an adult dose is often enough for it to act as an osmotic purgative (Edes *et al.* 1990).

Diarrhea occurring in patients receiving enteral tube feeding is usually, although often erroneously, blamed on the feed (Edes *et al.* 1990). Factors suggested as predisposing to diarrhea during enteral feeding are high feed osmolality, lactose intolerance, bacterial contamination, bolus (as opposed to continuous) feeding, lack of dietary fiber, and previous fasting. The role of fiber is uncertain, with one clinical trial showing a reduced frequency of diarrhea in enterally fed patients given fiber supplements and others showing no benefit. Prolonged fasting causes loss of small bowel mass and mucosal enzyme activity. When enteral feeding is reintroduced, malabsorption can cause diarrhea.

Management

Patients must be kept clean to maintain their dignity and comfort, to prevent excoriation of the skin, to minimize contamination of the surrounding area with potentially pathogenic organisms, and for a pleasant working environment. A soft rectal tube can reduce soiling when diarrhea is profuse, but care is needed to avoid mucosal ulceration or perforation. Enteric isolation precautions are needed if an infectious cause for diarrhea is suspected or proven.

Any fluid and electrolyte deficiency must be corrected and ongoing losses replaced. Fluids are usually given intravenously in the ICU, but oral rehydration solutions provide a potentially cheaper alternative. The main electrolyte losses are sodium, potassium, chloride, and bicarbonate, but losses of magnesium, zinc, and copper may also cause these ions to be depleted. When patients are receiving enteral feeds and develop diarrhea it is common practice to reduce the volume and/or dilute the feed. This is a logical response when the diarrhea occurs after prolonged fasting, but not under other circumstances.

The history taken should include recent travels, past history of bowel disease, and both current and recent medications. Physical examination may reveal dehydration, evidence of malabsorption or poor nutrition, or features of inflammatory bowel disease. A fever is so common in ICU patients as to be unhelpful in indicating an infective cause. Rectal examination is essential to exclude spurious diarrhea.

Laboratory investigations include plasma electrolyte and urea concentrations, and stool microscopy and culture. *C. difficile* toxin should be excluded in patients receiving current or recent antibiotics. A stool osmotic gap of more than 100 mosmol/kg (2 × stool $[Na^+ + K^+]$ − serum osmolality) demonstrates osmotic diarrhea from sorbitol-containing medications etc., enteral feeds, or malabsorption.

If an infective cause is excluded, treatment with an antidiarrheal agent produces rapid symptomatic relief in most patients. Loperamide 4 mg orally or enterally initially followed by 2 mg up to six times daily appears to be the most satisfactory agent.

References

Cantey, J.R. (1993). *Escherichia coli* diarrhea. *Gastroenterology Clinics of North America*, **22**, 609–22.

Edes, T.E., Walk, B.E., and Austin, J.L. (1990) Diarrhea in tube-fed patients: feeding formula not necessarily the cause. *American Journal of Medicine*, **88**, 91–3.

Farthing, M.J.G. (1988). History and rationale of oral rehydration and recent developments in formulating an optional solution. *Drugs*, **36** (Supplement 4), 80–90.

Hansen, M.B. and Skadhauge, E. (1995) New aspects of the pathophysiology and treatment of secretory diarrhoea. *Physiological Research*, **44**, 61–78.

Manabe, Y.C., Vinetz, J.M., Moore, R.D., Merz, C., Charache, P., and Bartlett, J.G. (1995). *Clostridium difficile* colitis: an efficient clinical approach to diagnosis. *Annals of Internal Medicine*, **123**, 835–40.

4.5.5 Constipation

G. J. Dobb

Key messages

- Constipation is common in patients needing intensive care.
- Drugs are frequently causative.
- Use a stepwise approach to treatment, starting with mild laxatives.
- Regular rectal examination is essential to detect a full rectum in unconscious patients.

Introduction

Constipation is the most common disorder of intestinal motility. Its community prevalence is estimated at 3 to 17 per cent. A period of constipation is very common in critically ill patients. Immobility, drugs, reduced food intake, abdominal surgery, and neurohumoral changes are contributory factors, but the frequency of constipation during intensive care is poorly documented.

'Normal' stool frequency varies between individuals; 'constipation' describes a variety of disturbances including infrequent defecation, excessive straining during defecation, stools that are too small or too hard, or a sense of incomplete defecation. When a frequency-based definition was used, only half of those describing themselves as constipated met the objective criteria.

Definition

Chronic constipation is defined as less than two bowel movements per week in the absence of mechanical obstruction, persisting after attempts with medical therapy and a high-fiber diet. Because constipation is common in the critically ill, intervention should be considered after 2 to 3 days or earlier if there has been no bowel movement for a period before admission to the intensive care unit (ICU).

Pathophysiology

Chronic constipation has three principal mechanisms: colonic inertia, pelvic floor disorders, and irritable bowel syndrome. Colonic inertia is the principal mechanism underlying acute constipation associated with critical illness.

Normal defecation is under a high measure of voluntary control. Rectal pressure rises during normal rectal emptying. There is an increase in intra-abdominal pressure accompanied by inhibition of the smooth and striated muscle of the rectal sphincter. During rectal emptying the pelvic floor descends, and this is followed by increased pelvic floor muscle activity at the end of defecation as it returns to the resting position. Colonic motor activity is usually localized to short segments and occurs in an uncoordinated way at irregular periods during the day. Several times a day, most often after morning wakening or in the late postprandial period, high-pressure contractions occur that propagate as a peristaltic wave over a long length of colon (mass movements) and are responsible for the transport of colonic content. These contractions are accompanied by an urge to defecate or precede defecation in conscious subjects. Colonic motor activity is inhibited during sleep and also by abdominal incision. Peristaltic contractions occur less frequently in patients with constipation.

Colonic muscle is innervated by the myenteric plexus which includes cholinergic and peptidergic neurons sensitive to vasoactive intestinal peptide, substance P, encephalins, somatostatin, and other mediators. It is also subject to stimuli mediated by the autonomic nervous system and other gastrointestinal hormones. The vagus provides the parasympathetic supply to the proximal colon and the distal colon is supplied through the pelvic nerves. The sympathetic supply is also divided; the proximal colon is supplied by splanchnic nerves arising from the superior mesenteric ganglia and the distal supply comes from lumbar nerves arising from the inferior mesenteric ganglia. Adrenergic stimulation inhibits colonic muscle. Loss of the pelvic parasympathetic supply interferes with the reflux control of defecation.

Causes

Table 1 summarizes causes of acute constipation affecting the critically ill. Many conscious patients find it difficult to use a bedpan or commode amongst the noise and relatively public setting of the ICU. Attention to privacy and the more normal background noise of radio music help to overcome psychogenic inhibition.

Most patients given narcotic analgesics during intensive care become constipated, as do most admitted after an overdose of tricyclic antidepressants.

Table 1 Factors contributing to acute constipation during intensive care

Exclude: Bowel obstruction
Pseudo-obstruction
Electrolyte abnormalities
High sympathetic tone
Drug induced
Anticholinergics
Opiates
Sympathomimetics
Calcium-channel blockers
Psychogenic
Neurological
Brain injury
Spinal injury
Abdominal or pelvic surgery
Dietary
Reduced food intake
Reduced fiber intake

The role of altered diet or reduced fiber intake remains somewhat controversial. In normal subjects bowel transit time is shorter, frequency of defecation is higher, and stool weight is greater in those with a higher fiber intake. However, there is no evidence that fiber-containing feeds prevent or control constipation in tube-fed patients.

Women are more prone to constipation, particularly during pregnancy. It has been suggested that constipation is more common in the elderly, although not all studies support this.

Management

Many different laxatives are available. A common first step is to increase the fiber content of diet or enteral feed. Enteral feeds were originally low residue, i.e. contained no fiber, but a variety of fiber-containing feeds are now available. The fiber is usually in a single form (e.g. methylcellulose). As noted above, there is no evidence that these prevent or control constipation. In some patients the addition of fiber increases wind, bloating, abdominal distension, and discomfort.

A management plan for bowel treatment in unconscious patients who are unable to swallow is shown in Fig. 1. Management of conscious patients is individualized depending on their usual bowel habit, eating, and symptoms. When patients have a daily bowel habit, are confined to bed, and have not defecated for 2 days, docusate sodium tablets (120–240 mg) are given as a fecal softener. More prolonged constipation is treated with a stepwise approach, moving to the next step if the previous treatment is ineffective (Table 2).

Cisapride promotes gastric emptying and absorption of nasogastric feeds. It also increases stool frequency in patients with chronic constipation. Diarrhea is a fairly common side-effect. Whether cisapride reduces the frequency of constipation in ICU patients has still to be confirmed.

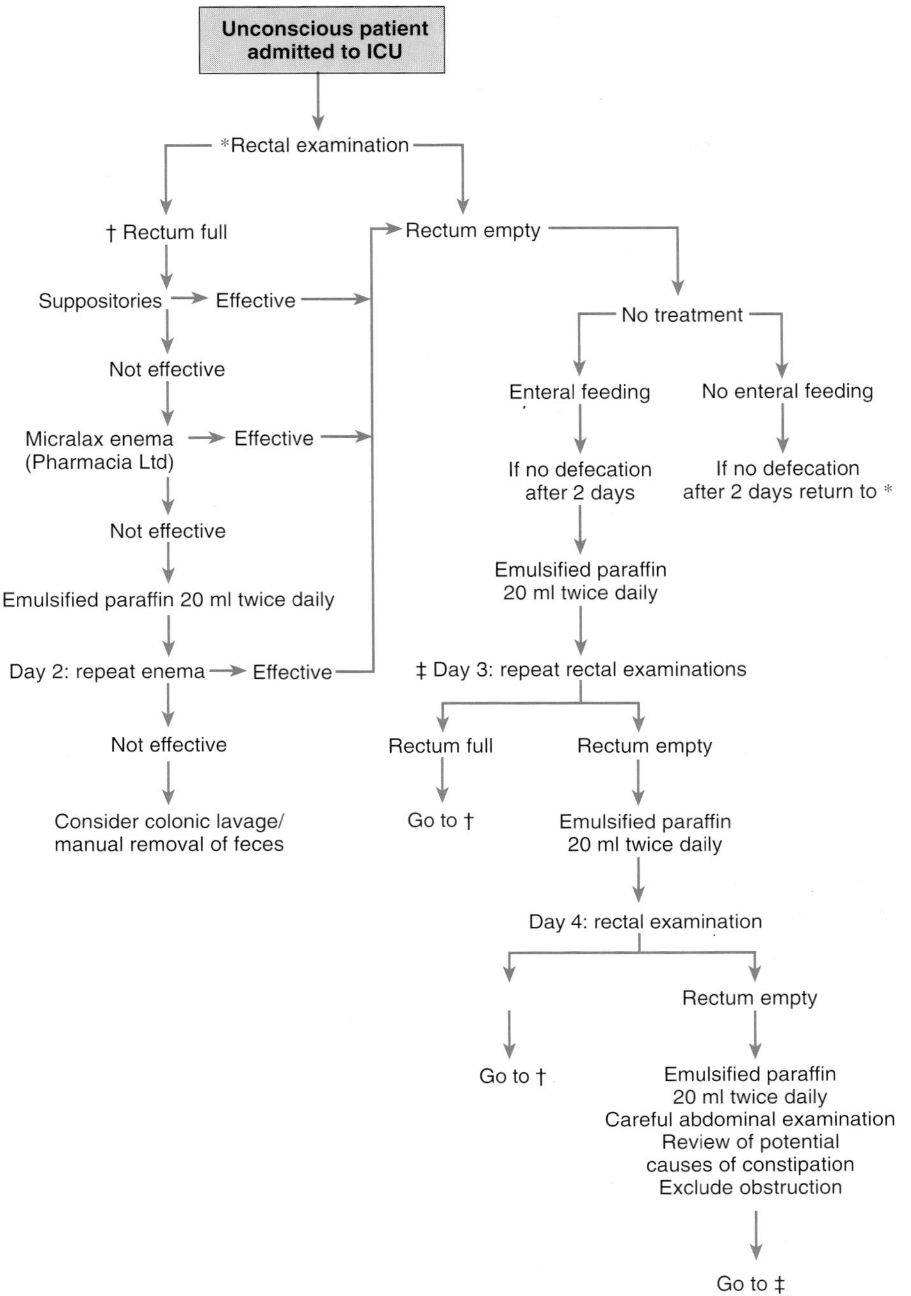

Fig. 1 Bowel management and treatment of constipation in unconscious patients needing intensive care.

Table 2 Progressive steps in treatment of constipation in conscious patients

1. Stool softener (e.g. docusate sodium)
2. Magnesium sulfate (Epsom salts)
3. Stimulants (e.g. Senokot)
4. Suppositories (e.g. glycerin or bisacodyl)
5. Enema (e.g. Micralax)
6. Colonic lavage±manual removal of feces

Bibliography

Ashraf, W., Park, F., Lof, J., and Quigley, E.M.M. (1996). An examination of the reliability of reported stool frequency in the diagnosis of idiopathic constipation. *American Journal of Gastroenterology*, **91**, 26–32.

Davies, G.J., Crowder, M., Reid, B., and Dickerson, J.W.T. (1986). Bowel function measurements of individuals with different eating patterns. *Gut*, **27**, 164–9.

Frankenfield, D.C. and Beyer, P.L. (1991). Dietary fiber and bowel function in tube-fed patients. *Journal of the American Dietetic Association*, **91**, 590–9.

Locke, G.R. (1996). The epidemiology of functional gastrointestinal disorders in North America. *Gastroenterology Clinics of North America*, **25**, 1–9.

Müller-Lissner, S.A. and the Bavarian Constipation Study Group (1987). Treatment of chronic constipation with cisapride and placebo. *Gut*, **28**, 1033–8.

4.6 Jaundice

4.6.1 Pathophysiology and causes

A. E. S. Gimson

Key messages

- Changes in hepatic intermediary metabolism and an acute phase protein response occur during sepsis. Hepatic dysfunction ranges from poor hepatic clearance of test drugs to frank hyperbilirubinemia and intrahepatic cholestasis.
- A wide range of causes of jaundice can occur during critical illness. Accurate diagnosis is essential.
- In the absence of direct involvement of the liver, there is an increasing prevalence of jaundice with increasing severity of illness.
- The prognostic significance of jaundice in a critically ill patient depends on the clinical context. It is of little significance during bacteremia, but denotes a poor prognosis during acute respiratory distress syndrome or multiple organ failure syndrome.
- In multiple organ failure syndrome the mechanism includes changes in sinusoidal blood flow due to plugging by activated leukocytes and platelets, vasoactive mediators including complement products, platelet activating factor, leukotrienes, cytokines, and nitric oxide, as well as direct metabolic effects such as those due to oxygen free radicals.

Introduction

The association of jaundice with sepsis has been noted for over 150 years, with early descriptions of 'pneumonia biliosa', but it is only more recently that the range of subtle disturbances of hepatic function in critically ill patients has become apparent. These later studies have demonstrated that the liver plays a crucial role in modulating the systemic response to sepsis, and the adequacy of this response is of profound prognostic importance.

Etiology

A wide range of causes of jaundice may occur in intensive care unit patients, and so accurate assessment of the etiology is crucial (Table 1). It is also important to exclude fulminant hepatic failure, the diagnosis and management of which is dealt with elsewhere. Exacerbations of underlying chronic liver disease due to surgical, septic, or traumatic stress may be the first presentation of the condition. Even in the absence of overt liver disease, alcoholics may be at greater risk of liver dysfunction during stress states owing to low hepatic glutathione levels. An increased bilirubin load on the liver during sepsis and reabsorption of hematomas are underlying factors in some cases but are unlikely to be the sole cause. Overt hemolysis, either drug-induced or following late transfusion reactions, may also occur. Hepatotoxic reactions to the numerous drugs used in the critically ill are important, and include reactions to anesthetic agents, antibiotics, and total parenteral nutrition. Early studies of patients receiving total parenteral nutrition, where predominantly glucose-based regimens were used, revealed hyperbilirubinemia in up to 26 per cent and abnormalities of alkaline phosphatase in 54 per cent by day 12. Excess use of carbohydrate has a widespread effect on hepatic function and can also alter drug metabolism. If solutions containing at least 40 per cent of calories in the form of lipid are used, the frequency is significantly lower.

Table 1 Causes of jaundice or abnormal liver blood tests during critical illness

Underlying chronic liver disease
Fulminant hepatic failure
Hemolysis
Drug hepatotoxicity
Liver dysfunction induced by total parenteral nutrition
Jaundice following cardiopulmonary bypass
Lobar pneumonia
Bacteremia
Systemic non-hepatic infection
Opportunistic infections in immunocompromised host
Multiple organ failure syndrome

Lobar pneumonia

Hyperbilirubinemia is well described in patients with lobar pneumonia. A frequency of 14 per cent was reported in a series from the United States, with the highest frequency in black males. Other series from Africa have reported higher frequencies. Although in some patients this jaundice may be related to glucose-6-phosphate dehydrogenase deficiency, this does not account for all cases.

Similarly, malnutrition or specific serological subtypes of streptococcus do not explain this high frequency. In many cases ultrastructural changes in the liver have shown hepatocyte swelling, dilatation of biliary canaliculas, and hepatocellular cholestasis, features which are more consistent with an interruption to bile transport across the biliary epithelium. Pneumonia related to *Legionella* and cryptogenic organizing pneumonia may also be associated with cholestasis.

Systemic non-hepatic infection and bacteremia

Hepatic involvement as just one component of a more generalized systemic infection occurs with some viruses (cytomegalovirus, Epstein–Barr virus), protozoa (*Coxiella*), parasites (malaria), and fungi (hepatosplenic candidiasis), particularly in immunosuppressed patients. Typhoid fever may cause an unusual hepatitis, but the jaundice in leptospirosis is due to hemolysis rather than hepatocellular injury.

Gram-negative sepsis in infants is often associated with hyperbilirubinemia, and may be due to lipopolysaccharide interference with bile-salt-independent bile flow. Bile salt pool and synthetic rates are low in infants, and impairment of bile-salt-independent flow may be more likely to result in a cholestatic syndrome. A large prospective study of 84 adult cases of bacteremia, 34 per cent of which were due to Gram-positive organisms, found liver blood test abnormalities in 65 per cent irrespective of the organism, the age or sex of the patient, the origin of the sepsis, or mean arterial pressure. Biochemical findings were of no prognostic significance, but in a series of cases with staphylococcal endocarditis mortality was five times higher in those presenting with jaundice. Abnormalities of liver function are also very common in the toxic shock syndrome. Jaundice due to a wide range of organisms, including *Mycobacterium avium intracellulare* complex and *Cryptosporidium*, may also develop in those with HIV infection.

Multiple systems organ failure

Jaundice as a distant manifestation of untreated intra-abdominal Gram-negative sepsis is well described, and occasionally can be sufficiently severe to mimic fulminant hepatic failure. Sequential failure of multiple organs, often as part of the sepsis syndrome or following severe trauma, occurs in up to 15 per cent of cases admitted to intensive care facilities. In a series of 92 cases with multiple systems organ failure, hepatic dysfunction occurred in over 80 per cent, commonly as the first organ failure. In a further series, cholestasis developed in 35 per cent following trauma and in 62 per cent of cases with intra-abdominal sepsis and multiple organ failure syndrome, and was a poor prognostic sign. In cases with acute respiratory distress syndrome, liver blood tests on the first day of diagnosis were predictive of survival, being worse in those who subsequently died. Histological studies demonstrated an intense canalicular cholestasis with little necrosis in the absence of significant splanchnic hypoperfusion. Acalculous cholecystitis with jaundice may develop late after surgery, trauma, or sepsis, and may run a rapid course with gallbladder perforation.

Although abnormalities of liver blood tests are the most obvious manifestation of multiple systems organ failure, there is increasing evidence of a wide range of more subtle changes in hepatocellular metabolic function that may be of profound prognostic importance. The initial response to sepsis is associated with increased gluconeogenesis and a reprioritization of hepatic protein synthesis. The former is driven in part by increased glucose counter-regulatory hormones (cortisol, glucagon, and epinephrine (adrenaline)), but increased uptake of alanine for gluconeogenesis is also important. In bacteremic burns patients, hepatic glucose output doubled and alanine uptake by the liver rose fivefold. Protein anabolism and catabolism both increase, with peripheral release of amino acids available for hepatic uptake and acute phase protein synthesis. This enhanced acute phase protein synthesis and concomitant reduction in albumin synthesis is mediated by the release from Kupffer cells of a wide range of mediators including the cytokines interleukin 1 (IL-1) and interleukin 2 (IL-2). While these changes are a normal physiological response to the stressed state, in some cases the adequacy of this response is impaired as liver failure develops. Thus in those bacteremic burns patients who developed multiple organ failure syndrome, glucose output and lactate and alanine uptake by the liver all fell dramatically compared with non-infected controls. Similarly, the central fractional clearance rate for amino acids (a measure of splanchnic amino acid clearance) is elevated during the septic state, but is lower in those cases who subsequently die compared with those who survive.

Mechanisms of hepatocellular dysfunction during multiple systems organ failure

Despite extensive investigation the causes of this hepatic dysfunction remain controversial. The two main hypotheses put forward, which are not mutually exclusive, are changes due to a reduction or heterogeneity of nutrient hepatic blood flow, and the metabolic consequences of the release of inflammatory mediators that are involved in the systemic response generating multiple systems organ failure. In experimental models using cecal ligation and puncture, an early reduction in nutrient blood flow was observed with a reduction in bioenergy status and oxygen consumption. These changes occur early and before other metabolic alterations. The liver may be particularly sensitive to this form of hypoxia as metabolic autoregulation of flow does not occur. Unfortunately, not all the evidence points to impaired hepatic perfusion, as both animal and human studies have found that there is an increased hepatic blood flow and oxygen consumption during the early hyperdynamic stages of sepsis. Furthermore, isolated hepatic mitochondria from endotoxic rats demonstrated depressed function even when placed in aerobic conditions. Nevertheless, it is still possible that any increase in global oxygen consumption may not be adequate to supply all metabolic demands during stressed states.

Even in the absence of global changes in hepatic blood flow, heterogeneity of sinusoidal perfusion may be an important mechanism resulting in marked reduction in nutrient delivery to some areas of the liver parenchyma. Such heterogeneity has been demonstrated in normal humans, and the fat-storing Ito cells within normal sinusoids have been shown to contract actively to a range of vasoactive mediators including endothelin-1. During stressed states, interruption to sinusoidal flow is exaggerated when Ito cells become activated and are transformed to myofibroblasts which exhibit increased responses to exogenous activators, or when sinusoids are blocked by activated neutrophils and platelets.

Inflammatory mediators and hepatocellular dysfunction

Hepatocellular dysfunction and multiple systems organ failure can occur after both sepsis and trauma without apparent infection; the unifying factor in all causes is an uncontrolled systemic inflammatory response. One hypothesis has been that the initiating event is activation or priming of polymorphonuclear cells by complement polypeptide C5a, gut-derived platelet activating factor, or lipopolysaccharide. Adhesion of these cells to hepatic endothelium occurs through selectins and intercellular adhesion molecules (ICAM-1), expression of which can be upregulated by other cytokines and platelet activating factor. Subsequent events may lead to generation of reactive oxygen species, cytokines, arachidonic acid derivatives, and nitric oxide (NO). Hepatic Kupffer cells may modulate this response and act as amplifiers, transmitting effects to sinusoidal lining cells and Ito cells controlling sinusoidal perfusion. Cytokines including IL-1, IL-6, and IL-8, which are liberated in large amounts during sepsis, have been implicated in many of the pathological processes occurring during endotoxemia, including alterations in hepatic protein synthesis. Tumor necrosis factor-α, released from neutrophils and endothelial cells by endotoxin, increases neutrophil activation, IL-1 production, and NO production by Kupffer cells. NO also depresses hepatic protein synthesis and, in turn, contributes to vascular hyporesponsiveness. The role of NO in modulating tissue injury is complex and may depend on the clinical context. Several studies have shown that it protects the liver from lipopolysaccharide-induced injury by acting as a vasodilator and partially reversing the effects of cytokines such as tumor necrosis factor-α. In contrast, during ischemia–reperfusion, NO production may contribute to hepatic injury by combining with oxygen free radicals to form peroxynitrite.

Bibliography

Sarfeh, I.J. and Balint, J.A. (1978). The clinical significance of hyperbilirubinemia following trauma. *Journal of Trauma*, **18**, 58–62.

Schwartz, D.B., Bone, R.C., Balk, R.A., and Szidon, J.P. (1989). Hepatic dysfunction in the adult respiratory distress syndrome. *Chest*, **95**, 871–5.

Sikuler, E., Guetta, V., Keynan, A., Neumann, L., and Schaeffer, F. (1989). Abnormalities in bilirubin and liver enzyme levels in adult patients with bacteremia. A prospective study. *Archives of Internal Medicine*, **149**, 2246–8.

te Boekhorst, T., Urlus, M., Doesburg, W., Yap, S.H., and Goris, R.J.A. (1988). Etiologic factors of jaundice in severely ill patients. *Journal of Hepatology*, **7**, 111–17.

Wilmore, D.W., Goodwin, C.W., Aulick, L.H., Powanda, M.C., Mason, A.D., Jr, and Pruitt, B.A. (1980). Effect of injury and infection on visceral metabolism and circulation. *Annals of Surgery*, **192**, 491–502.

4.6.2 Therapeutic strategy

A. E. S. Gimson

Key messages

- Therapy is determined by diagnosis; an accurate history and careful sequence of investigations are crucial.
- Once hemolysis is excluded, an ultrasound scan is the most important diagnostic modality.
- A therapeutic endoscopic retrograde cholangiopancreatography is indicated in selected cases with severe acute pancreatitis and/or cholangitis.
- Rigorous attempts must be made to identify occult sepsis.
- Attempts to improve splanchnic blood flow using dopexamine hydrochloride are logical, and may reduce morbidity and mortality in sick presurgical cases.
- Although dopexamine hydrochloride may improve splanchnic hemodynamics and hepatic oxidative capacity in patients with multiple systems organ failure, there is no evidence yet of survival benefit.
- Experimental studies inhibiting inflammatory mediators underlying hepatic dysfunction in multiple systems organ failure have not yet been translated into clinical trials.

Introduction

The most important part of a therapeutic strategy for managing jaundice in the critically ill is to make the correct diagnosis. A wide range of diagnostic tests and modalities are now available, not only to identify the site of interruption to bilirubin excretion but also to identify the etiological agent(s).

As in any patient with jaundice, an accurate history is crucial. The onset of the jaundice in relation to the critical illness, previous evidence of, or risk factors for, chronic liver disease, use of hepatotoxic drugs or anesthetics, and alcohol consumption will need to be recorded. Clinical examination may reveal stigmata of chronic liver disease (Table 1).

Investigations

A careful sequence of investigations is mandatory in patients with jaundice (Fig. 1). Interpretation of patterns of liver blood test abnormality is problematic as such tests carry low sensitivity and specificity. Nevertheless, they are of use in certain specific diagnoses.

A diagnostic strategy to exclude hemolysis is considered elsewhere but, in any patient with jaundice, the conjugated and unconjugated bilirubin fraction should be identified, in addition to a reticulocyte count and haptoglobin level. The latter is less discriminatory during critical illness because changes in haptoglobin synthesis may occur in this situation. Furthermore, the finding of an unconjugated hyperbilirubinemia does not confirm hemolysis as it may be a reflection of Gilbert's syndrome, a hereditary disorder due to abnormal gene expression of bilirubin-conjugating enzymes in which bilirubin levels may rise up to 70 μmol/l during stressed states.

On confirmation of a conjugated hyperbilirubinemia, the most important discriminatory test is an ultrasound scan of the liver, bile ducts, pancreas, and spleen, and Doppler interrogation of the hepatic vasculature. In experienced hands considerable additional information can be obtained from assessing liver texture.

Dilated bile ducts involving intra- or extrahepatic ducts require further delineation with abdominal CT scanning or endoscopic retrograde cholangiopancreatography (ERCP). In the presence of non-dilated ducts, evidence of portal hypertension suggests underlying chronic liver disease or Budd–Chiari syndrome, whereas

Table 1 Diagnostic strategy for jaundice in critically ill patients

History
Time of onset in relation to critical illness
Prior history of chronic liver disease, biliary/pancreatic disease
Risk factors for acquired liver disease (origin from high-risk geographic zone, alcohol consumption, blood/blood product transfusion, IV drug use, at-risk sexual activity, tattoos)
Hepatotoxic drugs/anesthetics/TPN
Investigations (Fig. 1)
Exclusion of hemolysis
Bilirubin, aminotransferases, alkaline phosphatase/γ-glutamyl peptidase, albumin, coagulation screen
Septic screen (blood, sputum, urine culture)
Serology for atypical pneumonia (*Mycoplasma*, *Legionella*)
Serology: IgM anti-HAV, IgM anti-HB core/HB surface antigen, anti-HCV, EBV, CMV
Radiology
Ultrasound of liver, biliary tree, spleen, pancreas
$\pm$ERCP/CT scan abdomen/nuclear medicine white cell scan
Others
Liver biopsy ($\pm$transjugular)
Metabolic function tests
Gastric pHi

IV, intravenous; TPN, total parenteral nutrition; HAV, hepatitis A virus; HB, hepatitis B; HCV, hepatitis C virus; EBV, Epstein–Barr virus; CMV, cytomegalovirus.

heterogeneity of liver texture with abdominal or other lymphadenopathy and splenomegaly raises the possibility of disseminated malignancy or lymphoma (Fig. 1). Occasionally, dramatic elevations of alkaline phosphatase and γ-glutamyl transpeptidase herald underlying malignant infiltrations or hepatosplenic candidiasis.

Ischemic hepatitis, resulting from a profound fall in hepatic oxygen delivery, is associated with a rapid rise in transaminases, returning towards normal within 2 to 3 days, and a prothrombin time that may not be significantly prolonged. Both viral hepatitis and drug hepatotoxicity may mimic this pattern, but a rapidly rising prothrombin time (INR) with evidence of hepatocellular damage always points towards acute liver failure.

Biochemical cholestasis can have a number of causes, including bacterial cholangitis, acute pancreatitis, drug hepatotoxicity, and parenteral nutrition, and it is also a common pattern with hepatic infiltrations and hepatosplenic candidiasis in the immunosuppressed. Similarly, in patients with multiple systems organ failure, the liver dysfunction is often just a conjugated hyperbilirubinemia with transaminases or alkaline phosphatase twice the upper limit of normal.

Supplementary radiology will depend on the clinical context, but ERCP may be necessary to confirm (and treat) cholelithiasis and nuclear medicine white cell scans may be important in identifying intra-abdominal or other septic foci. Intra-abdominal fluid collections, subphrenic abscesses, and pancreatic pathology can be detected by ultrasonography, but CT scanning is often a better radiological modality for the latter.

Occasionally, liver histology is important, as in the diagnosis of some causes of acute liver failure and of opportunistic infections in immunocompromised patients. In those with a significant coagulopathy this will need to be done by the transjugular route. In contrast, liver histology is usually not diagnostic in those cases with jaundice associated with sepsis, multiple systems organ failure, and cholestatic drug hepatotoxicity.

Management strategies

In many patients the identification of specific etiological factors will dictate the appropriate therapeutic strategy. Where possible, drugs, antibiotics, and anesthetic agents with a lower incidence of hepatotoxic side-effects should be used. The hepatic metabolism of many drugs is profoundly altered during critical illness and positive-pressure ventilation, with changes in volume of distribution, protein binding, and plasma clearance by the kidney or liver. Hepatic abnormalities due to total parenteral nutrition can be avoided by administering adequate protein and 20 to 50 per cent of non-protein

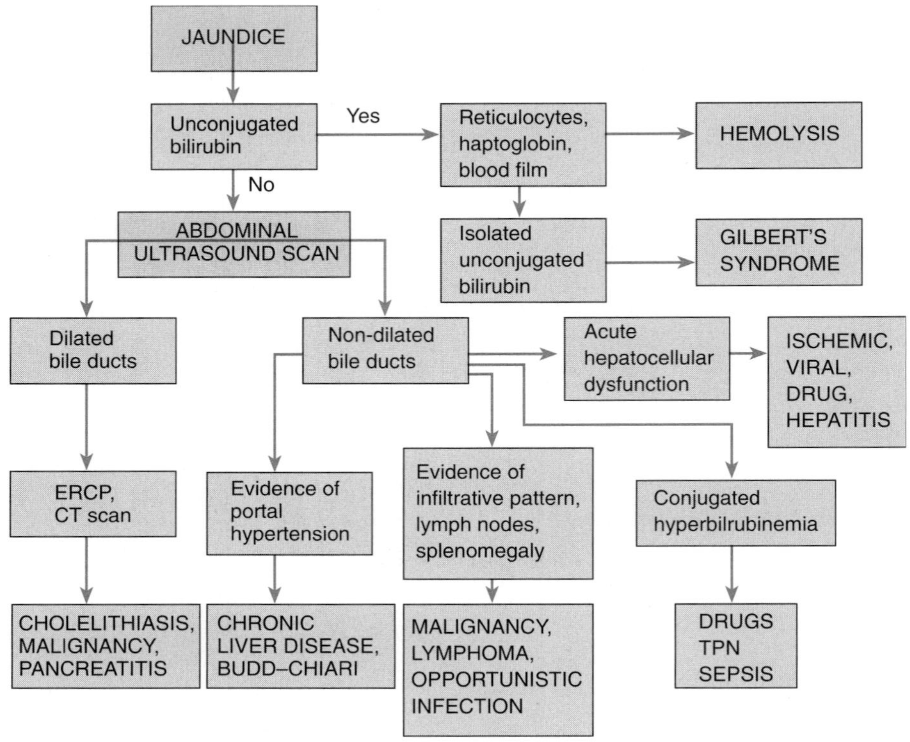

Fig. 1 Investigations of jaundice in the critically ill patient: TPN, total parenteral nutrition.

calories as fat. Cyclical administration of total parenteral nutrition and glucose may help to prevent these changes.

Management of sepsis and cholangitis

Distant sepsis is a common cause of liver dysfunction in critically ill patients owing to a range of inflammatory mediators which cause changes in hepatocellular micronutrient blood flow and/or metabolic dysfunction and cholestasis. The importance of aggressive treatment of septic foci cannot be overemphasized. In many cases the focus of infection is obvious, but unusual sources of sepsis in critically ill patients who develop jaundice during a prolonged stay in the intensive care unit (ICU) may include acalculous cholecystitis, purulent sinusitis, prosthetic vascular graft colonization, and endocarditis. Vascular catheter-related sepsis is also important, and consideration should be given to replacing all lines and assessing colony counts on all catheters in a jaundiced patient. In immunosuppressed patients, such as those who have just had organ transplantation or those with HIV carriage, special consideration of potential organisms is important. This necessitates not only the use of appropriate combinations of antibiotics and antifungal agents, dictated by presumed or actual sensitivity of the organism to the drugs, but also careful surgical drainage of abscesses or infected fluid collections. Although the identification of these is often difficult, particularly in trauma cases, the use of diagnostic blind laparotomy and second-look laparotomy is now less favored than previously. Rigorous use of radiological techniques, nuclear medicine white cell scans, serial cultures, and measurement of C-reactive protein are important in identifying significant collections. Early surgical intervention in trauma cases, with debridement of necrotic tissue and stabilization of fractures, is clearly better than a delayed or elective approach in this regard. Currently, there is no evidence that prophylaxis with selective bowel decontamination reduces the frequency of jaundice in the critically ill.

It has been suggested that early intervention with ERCP is important in cases of bacterial cholangitis due to choledocholithiasis. In a series of severe cases (according to the Glasgow criteria) of non-alcohol-related acute pancreatitis, ERCP within 72 h, including sphincterotomy if necessary, resulted in significant reductions in mortality (4 per cent versus 18 per cent) and rates of complication (24 per cent versus 61 per cent). Such emergency intervention is not necessary in mild pancreatitis. Similar results have been reported for the urgent management of severe acute cholangitis due to choledocholithiasis. In these studies ERCP was shown to be safe, but the procedure must be restricted to cases with a poor prognosis, i.e. those with septic shock, renal impairment or jaundice, fever, and rigors not resolving after antibiotics for 48 h.

Maintenance of hepatic blood flow and sinusoidal perfusion

Changes in hepatic blood flow and micronutrient sinusoidal flow have been described in animal models of sepsis and may mediate some of the changes in hepatic function seen in this situation.

Monitoring splanchnic and hepatic blood flow is complex and not routinely available. Indocyanine green clearance, a dye with a clearance that is dependent on hepatic blood flow, and the formation of monoethylglycinexy-lidide (MEGX) from lidocaine (lignocaine) are the most commonly used tests. Assessment of gastric mucosal pH (pHi) by tonometry shows poor correlation with splanchnic blood flow, but the presence of gastric mucosal acidosis may be of prognostic significance. Splanchnic oxygen consumption and its extraction ratio in hyperdynamic sepsis are elevated, despite increases in splanchnic blood flow, although imbalance between oxygen supply and demand may still occur. Although inflow into the splanchnic circulation is largely dependent upon mean arterial pressure, the usual relationship between perfusion pressure and organ flow does not always apply in the critically ill. Many drugs may modulate splanchnic blood flow. Enteral nutrition may increase it, while positive end-expiratory pressure ventilation (PEEP) may decrease it. PEEP may lower cardiac output, increasing the free hepatic venous pressure and affecting the relative ratio of portal venous to hepatic arterial blood flow. Such changes may have complex consequences, impairing the clearance of some inflammatory mediators and increasing that of others where increasing sinusoidal pressure increases trans-sinusoidal plasma flux. Intermittent positive-pressure ventilation and PEEP also reduced metabolic clearance of lidocaine and MEGX formation in an animal model of lung injury.

In animal models of sepsis, attempts to improve these parameters with volume loading, dopamine, positive inotropes (dobutamine), or vasoconstrictors have been unsuccessful. In contrast, dopexamine hydrochloride, a peripheral vasodilator which acts via adrenergic β_2 and dopaminergic DA_1 receptors but has no direct α-adrenoceptor activity, may have significant effects within the splanchnic circulation. In a porcine peritonitis model, dopexamine maintained cardiac output and portal venous flow better than volume loading alone and prevented hepatic ultrastructural changes. Furthermore, in studies where dopexamine and dobutamine were directly compared, despite better cardiac output, oxygen delivery, and arterial pressure with dobutamine, the hepatic histology was actually worse. These animal studies strongly supported the concept that measures to improve splanchnic perfusion with splanchnic vasodilators may prevent early deterioration in hepatic function with sepsis.

In a recent trial high-risk surgical patients were randomized to either best standard perioperative care or a protocol group whose oxygen delivery index was deliberately increased to above 600 ml/min/m^2 by using dopexamine. The protocol group achieved significantly higher oxygen delivery both pre- and postoperatively with a 75 per cent reduction in mortality (5.7 per cent versus 22.2 per cent, $p = 0.015$) and a halving of the mean number of complications per patient (0.68 ± 0.16 versus 1.35 ± 0.20, $p = 0.008$). It was not clear from this study whether the perioperative use of dopexamine prevents subsequent development of liver dysfunction due to sepsis, but in a small study in ICU patients low-dose dopexamine increased splanchnic blood flow as measured by gastric intramucosal pHi, the oxidative formation within the liver of the metabolite MEGX from lidocaine, and indocyanine green clearance. The lack of any change in systemic hemodynamic measurements suggests that these effects were the result of selective vasodilatation of the splanchnic vessels. Therefore dopexamine may be useful in the management of splanchnic ischemia in the critically ill.

Experimental therapeutic strategies

A wide range of therapies have been shown to improve splanchnic blood flow in animal models of ischemia–reperfusion. These strategies have centered on inhibiting or inactivating polymorphonuclear adhesion and activation, as well as inhibiting the downstream consequences of such polymorphonuclear activation. Xanthine oxidase inactivation, phospholipase A_2 inhibition, platelet activating factor antagonists, and P-selectin inhibition all inhibit polymorphonuclear sequestration in the splanchnic circulation after ischemia–reperfusion. Antibodies to IL-6 and IL-8 may also abrogate polymorphonuclear activation. The role of modulation of NO is unclear and may depend on ambient oxygen concentrations. While NO may protect the liver from lipopolysaccharide-induced injury, by acting as a vasodilator and partially reversing the effects of cytokines such as tumor necrosis factor-α, during ischemia–reperfusion NO production may contribute to hepatic injury by combining with oxygen free radicals forming peroxynitrite. The therapeutic use of such strategies will require careful clinical trials.

Bibliography

Biffl, W.L. and Moore, E.E. (1996). Splanchnic ischaemia/reperfusion and multiple organ failure. *British Journal of Anaesthesia*, **77**, 59–70.

Fan, S-T., Lai, E.C.S., Mok, F.P.T., Lo, C.-M., Zheng, S.-S., and Wong, J. (1993). Early treatment of acute biliary pancreatitis by endoscopic papillotomy. *New England Journal of Medicine*, **328**, 228–32.

Maynard, N.D., Bihari, D.J., Dalton, R.N., Smithies, M.N., and Mason, R.C. (1995). Increasing splanchnic blood flow in the critically ill. *Chest*, **108**, 1648–54.

4.7 Acute hepatic failure

4.7.1 Pathophysiology and causes

R. Williams

Key messages

- The new classification of acute liver failure (ALF) takes into account rapidity of onset and etiology as important indicators of likely progress.
- Microcirculatory dysfunction with covert tissue oxygen debt is the basis of multiple organ failure.
- Host defenses to bacterial infection are severely compromised early in the course of ALF.
- Reduction in cerebral blood flow and resultant cerebral ischemia are important in the development of cerebral edema.
- Acetaminophen (paracetamol) overdose is the most common cause of ALF in the United Kingdom. Antituberculosis drug therapy is responsible for an increasing number of cases.
- The frequency with which viral hepatitis A, B, C, D, E is the cause of ALF is related to the underlying prevalence of these infections in the country concerned. In many cases of presumed fulminant viral hepatitis, no specific agent is identifiable.

Introduction

Any consideration of the pathophysiology and etiology of acute liver failure (ALF) has to start with a description of the terminology that has been used in describing this very rare clinical syndrome. The first attempt at a formal definition described fulminant hepatic failure as 'a potentially reversible condition, the consequence of severe liver injury, in which the onset of hepatic encephalopathy was within eight weeks of the first symptoms of illness, and in the absence of pre-existing liver disease'. Even rarer are the cases described as late-onset hepatic failure in which hepatic encephalopathy develops between 8 and 24 weeks after the onset of jaundice. The latter is more readily identified than encephalopathy by both patients and clinicians, and allows more accurate timing of illness duration. Workers in France have proposed a different set of definitions, although also based on the interval from the onset of jaundice to the development of encephalopathy. Patients with encephalopathy developing within 2 weeks and between 2 and 12 weeks from the onset of jaundice are categorized as fulminant and subfulminant hepatic failure respectively (Benhamou 1991).

It has become clear that the rate of onset of illness is an important indicator of the likely prognosis. Paradoxically, it is the group of patients with the most rapid onset of encephalopathy who have the best chance of spontaneous recovery. To take account of this, we have recently proposed a new classification (Table 1) in which acute liver failure is taken as an umbrella term, with subgroups of hyperacute, acute, and subacute to reflect different clinical patterns of illness, etiology, and, most importantly, prognosis (O'Grady *et al.* 1993). In this classification acute liver failure becomes the core term, with 'hyper' and 'sub' being used as prefixes to separate the two cohorts at opposite ends of the clinical spectrum. Hyperacute liver failure is used to describe those patients who develop encephalopathy within 7 days of the onset of jaundice. The majority of cases in this group have acute acetaminophen (paracetamol) hepatotoxicity, but a proportion of those with viral hepatitis A or B may also present with cerebral edema. Survival in a group of 81 cases from this unit was 36 per cent. Acute and subacute hepatic failure includes those with a jaundice to encephalopathy time of 8 to 28 days. There was a similar high incidence of cerebral edema but the survival was lower (7 per cent of a cohort of 89 patients). In subacute hepatic failure, with a jaundice to encephalopathy time of 5 to 26 weeks, the outcome was also poor (14 per cent) although the frequency of cerebral edema was low (O'Grady *et al.* 1993).

In addition to the defining state, namely hepatic encephalopathy, severe coagulopathy and jaundice develop together with renal impairment and multiorgan failure. These clinical features may not all occur in every case and they are not specific to the etiology of ALF, although patterns of illness may be discerned that are to some extent indicative of the etiology. Thus those with acetaminophen overdose most often present with severe coagulopathy and signs of encephalopathy and the patient may not be jaundiced, whereas those with non-A non-B hepatitis are usually deeply jaundiced at presentation and are less likely to develop cerebral edema. Most of the damage to the liver has already been done by the time of presentation; histopathological examination of the liver shows collapsed hepatic parenchyma with few surviving hepatocytes and, depending on the time after onset, areas of regeneration.

Pathophysiology of ALF

The clinical signs and pattern of illness depend on the balance between three factors: the liver's ability to regenerate, the adverse metabolic consequences of the loss of a functioning liver, and the systemic effects of a wide range of toxic substances released into the serum from the necrotic liver. Although symptoms and signs of the different organ involvement are often considered separately, it is imperative to realize their interdependence (Fig. 1) and the central

Table 1 New classification of ALF based on time from first sign of jaundice to appearance of encephalopathy, with frequency of different etiologies in the three groups

	Frequency (%)		
	Hyperacute (0–7 days)	Acute (7–28 days)	Subacute (5–26 weeks)
Hepatitis A	55.2	31	13.8
Hepatitis B	62.5	29.9	8.3
Non-A non-B hepatitis	13.6	38.8	47.6
Idiosyncratic drug reactions	35.2	52.9	11.7

Data from O'Grady *et al.* (1993).

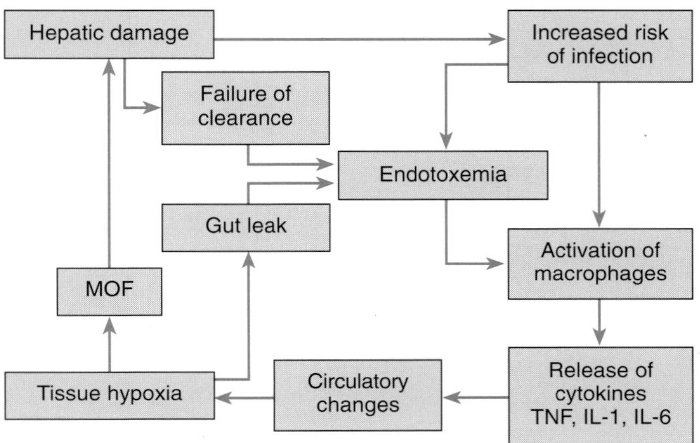

Fig. 1 Vicious cycle of events forming the basis of multiple organ failure: TNF, tumor necrosis factor; MOF, multiple organ failure.

driving force of the primary liver injury. As an early consequence of the acute liver injury, host defenses to infection are severely compromised and with secondary bacterial infection there is endotoxemia, activation of macrophages, and release of cytokines and tumor necrosis factor. The resulting clinical picture is similar to that of septic shock, with hypotension and other circulatory changes leading to tissue hypoxia and damage to a number of organs as well as the gut, and with yet further ischemic injury to the liver.

Microcirculatory dysfunction

Patients with ALF, as with other forms of critical illness, frequently demonstrate a covert tissue oxygen debt despite apparently adequate blood pressure and arterial oxygen saturation. Failure to maintain an adequate oxygen uptake to cells appears to be related to a combination of factors resulting in an inability to regulate delivery and extraction of oxygen at a cellular level. The basis of this microcirculatory dysfunction is poorly understood, but evidence is accumulating to suggest the importance of interactions between the endothelium, exogenous factors such as bacterial toxins, and cytokines, specifically tumor necrosis factor and interleukins 1 and 6. Endotoxin and other bacterial toxins stimulate the production of cytokines by activated macrophages which may be maintained in the circulation because of impaired Kupffer cell function. Activation and consumption of platelets with formation of microthrombi within various organs may lead to endothelial damage and release of further vasoactive compounds. This series of events, together with increased adhesion of activated leukocytes to endothelial cells, causes microcirculatory plugging with blood being shunted through non-nutritive arteriovenous channels. Nitric oxide and epoprostenol are important endothelial-derived factors whose role in the control of microcirculatory flow is being increasingly recognized. Patients with ALF have been demonstrated to have elevated levels of the end-products of nitric oxide (nitrite and nitrate) and of cGMP and citrulline, suggesting activation of the nitric oxide pathway. The ability of the endothelium to release prostaglandin I_2 or nitric oxide has an important bearing on the evolution of tissue hypoxia as endothelial interactions may limit release of these factors, thus further potentiating tissue hypoxia and end-organ damage.

In healthy individuals, physiological supply dependency of oxygen only occurs when Do_2 falls below a level of 330 ml/min/m^2. Any additional fall in delivery below this critical level will result in a fall in tissue oxygen uptake Vo_2, with the subsequent development of tissue hypoxia, anaerobic metabolism, and accumulation of lactate. Normal oxygen delivery is around 525 to 675 ml/min/m^2; at deliveries between 300 and 525 ml/min/m^2 oxygen consumption is maintained at normal levels by increasing extraction ratio in the periphery. In patients with ALF, as well as in those with severe sepsis and multiple trauma, 'pathological supply dependency' for oxygen is observed (Fig. 2). Oxygen consumption is then dependent upon delivery over a far greater range such that an increase in Do_2 will frequently result in an increase in Vo_2. An appropriate level for Vo_2 is difficult to estimate in patients with critical illness, but it is likely to be higher than that seen in the resting healthy state owing to the presence of fever, sepsis, inflammatory foci, and increased levels of circulating catecholamines.

The presence of pathological supply dependency for oxygen in ALF was first shown by Bihari *et al.* (1986) who noted that, following infusion of the microcirculatory vasodilator epoprostenol, there was a fall in systemic vascular resistance resulting in an increase in oxygen delivery and a significant increase in oxygen consumption. The patients who failed to survive had both a lower baseline Vo_2, indicative of a

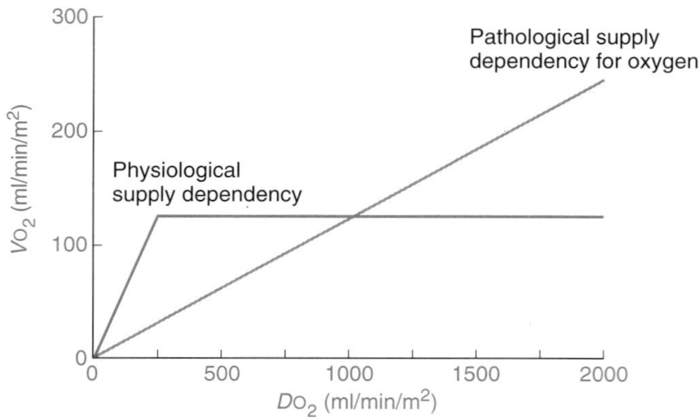

Fig. 2 Pathological supply dependency for oxygen.

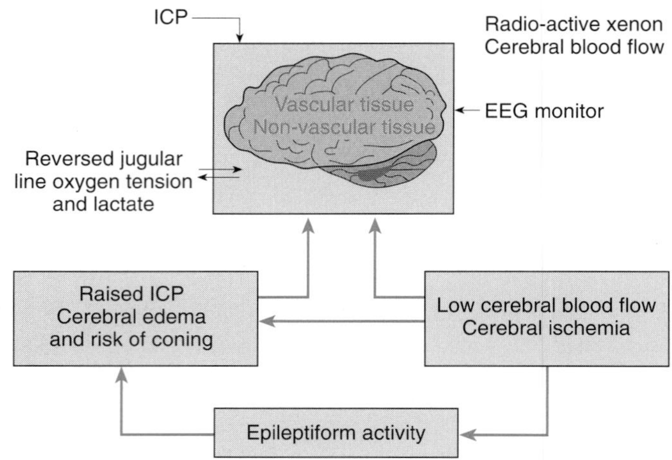

Fig. 3 Monitoring ischemia in cerebral edema: ICP, intracranial pressure.

large oxygen debt, and greater increases in $\dot{V}O_2$ following infusion of epoprostenol. In addition, the mixed venous lactate level (a measure of anaerobic metabolism) correlated inversely with systemic vascular resistance index, mean arterial pressure, and oxygen extraction ratio.

Intracranial hypertension and cerebral edema

Increases in intracranial pressure, as occur in the later stages of ALF, could be the result of increases in volume of any of the three main components within the skull, namely brain tissue, cerebrospinal fluid, and the cerebral vessels. The ventricular system is usually collapsed at CT scanning, and cerebral edema is reported to develop in up to 80 per cent of ALF patients progressing to grade IV encephalopathy.

The development of cerebral edema which underlies the rise in intracranial pressure probably depends on a variety of causative events. Increased permeability of the blood–brain barrier will result in leakage of protein-rich fluid into the extracellular space of the brain. The efficacy of mannitol in lowering intracranial pressure suggests that an intracellular cytotoxic mechanism is also important. The clinical signs of cerebral edema, namely systemic hypertension, decerebrate posturing, and abnormal pupillary reflexes, are generally attributed to brainstem compression. In an early series, cerebellar or uncal herniation was found at autopsy in 25 per cent of patients with cerebral edema. More recent experience shows that this is comparatively rare and that cerebral ischemia is the more likely underlying cause (Fig. 3).

A fall in cerebral blood flow as the level of coma deepens has been documented in a number of studies. In other studies observed values have been within the normal range (40–70 ml/min/100 g brain tissue), although investigators have questioned whether this level of cerebral blood flow in patients who are deeply comatose is surplus to demand and one group has reported that a cerebral hyperemia precedes or coincides with the presence of cerebral edema on CT scanning. Studies have shown that cerebral blood flow values for patients in grade IV encephalopathy varied widely between 12 and 50 ml/min/100 g (Fig. 4). In those patients who had developed clinical signs of cerebral edema, cerebral blood flow was higher and correlated significantly with the cerebral metabolic rate for oxygen

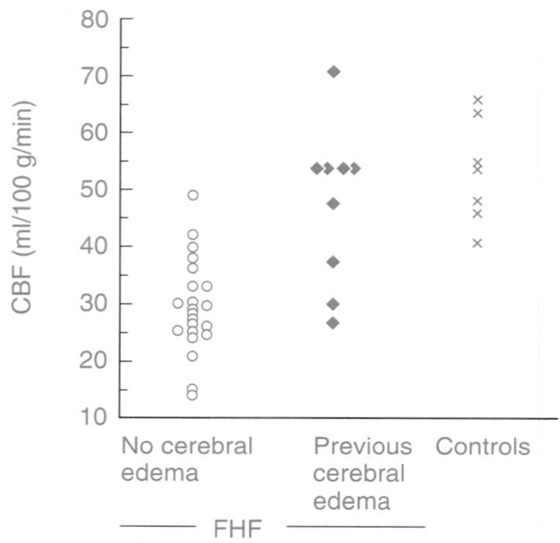

Fig. 4 Cerebral blood flow (CBF) in 30 patients with fulminant hepatic failure, nine of whom had previously had clinical signs of cerebral edema. (Reproduced with permission from Wendon *et al.* (1994).)

($CMRO_2$), which appears at variance with that seen in normal individuals where changes in cerebral blood flow are balanced by reciprocal changes in the proportion of oxygen extracted so that $CMRO_2$ remains unchanged (Wendon *et al.* 1994). The observation that many of these patients had cerebral lactate production is further evidence of an inadequate delivery of oxygen to the brain.

Etiology including geographical variation

Acetaminophen overdose, taken with deliberate suicidal intent, is the most common cause of ALF in the United Kingdom, accounting for 50 to 60 per cent of all cases referred (Spooner and Harvey 1993). In England and Wales, the number of deaths attributed to acetaminophen increased from seven in 1968 to 66 in 1973 and 144 in 1977, and by the late 1980s there were over 500 deaths annually. However, in a recent review of the 547 deaths recorded in 1990, there was evidence of acute hepatic damage in only 150. In the remaining cases, death could not be directly attributed to acetaminophen although it was a component of the fatal overdose (Spooner and Harvey 1993). The extensive press coverage afforded to fatal acetaminophen overdose cases in the United Kingdom coupled with the easy availability of the drug are the two most important factors underlying its use in suicide attempts. A recent study of overdose patients demonstrated that 63 per cent had chosen acetaminophen because it was easy to obtain and 53 per cent had obtained the drug specifically for the overdose; 36 per cent of the patient group knew that acetaminophen was dangerous and 78 per cent realized that it could be fatal.

Other countries have also recorded a substantial increase in the annual number of cases. The number of calls to poisons centers in the United States concerning acetaminophen overdoses increased from 60 000 in 1987 to 90 000 in 1993, although in the same year only 92 deaths were directly related to this cause. In France there are less than 109 deaths annually, which is thought to be the result of legally limiting the pack size to just 8 g. However, in one of the major liver centers in France the incidence of acetaminophen overdose as a cause of ALF has increased from less than 2 per cent of all cases before 1991 to 16 per cent currently.

Of considerable interest are the reports of severe hepatotoxicity following therapeutic doses of acetaminophen in patients who are concurrently taking enzyme-inducing drugs, such as phenytoin, or are chronic consumers of alcohol. In a series of 67 patients who developed hepatic injury after ingestion of acetaminophen taken with therapeutic intent, 64 per cent were considered to be 'alcoholic' or reported alcohol intakes above 80 g/day. The acetaminophen doses taken were within the recommended therapeutic range (< 4 g/day) in 40 per cent of cases. Almost 20 per cent of the patients died. Presumably, susceptibility is caused by induction of cytochrome P-4502EI by ethanol and depletion of glutathione by the effects of both alcohol and the frequently associated malnutrition.

The anesthetic agent halothane continues to feature among the drugs responsible for ALF, although much less frequently than previously. The exact mechanism of toxicity underlying this very rare idiosyncratic reaction is uncertain but appears to depend on both immunological and constitutional factors. A recent analysis of our case experience over the past 10 years revealed 18 cases of ALF due to halothane compared with 48 patients referred between 1965 and 1984. In many instances the recommended guidelines of the Committee on Safety of Medicines had been disregarded, with the patients having been re-exposed to halothane within a month of the last halothane anesthetic and despite having had documented reactions previously. Many other drugs, including monoamine oxidase inhibitors, non-steroidal anti-inflammatory drugs, gold, sodium valproate, cotrimoxazole, sulfonamides, disulfiram, ketoconazole, and phenytoin, may also rarely cause ALF. With the re-emergence of tuberculosis in many Western countries, an increasing number of cases of ALF from isoniazid–rifampicin–pyrazinamide hepatotoxicity are being encountered, usually as a result of a failure to monitor liver function properly after starting such therapy. A rare cause of acute hepatotoxicity in the United Kingdom, although it is relatively common in France, is ingestion of the poisonous mushroom *Amanita phalloides*.

During the past few years we have also seen cases of ALF following ingestion of ecstasy (3,4-methyldioxymethamphetamine). The spectrum of liver injury ranges from a heatstroke-type lesion to a more hepatitis-like injury.

Acute viral hepatitis accounts for up to 50 per cent of cases of ALF seen in mainland European centers and in North America. The increased incidence of viral hepatitis A as a cause reported from some countries may relate to a change in the time of exposure to the virus with infection occurring later in life. When ALF follows an acute infection with hepatitis B virus (HBV), viral replication has usually ceased by the time of presentation. Chronic HBsAg carriers who develop 'super-infection' with the delta virus (HDV) can develop ALF, and it is a rare development in carriers with a reactivation of HBV replication following immunosuppressive therapy for some other condition. The importance of mutant HBV strains in causing ALF, as demonstrated by molecular biology techniques, is the subject of conflicting reports. In many cases of presumed fulminant viral hepatitis, as referred to earlier, no specific viral hepatitis agent can be identified (i.e. hepatitis A, B, C, D, and E negative) despite the clinical presumption of viral hepatitis as the cause. In a recent study of 50 patients with seronegative non-A non-B fulminant hepatitis using the sensitive polymerase chain reaction analysis of hepatic DNA for detection of occult viral infection, negative results were obtained.

In general, the frequency with which the different viral hepatitis agents are responsible for ALF is related to the underlying prevalence of those infections in the country concerned. Thus hepatitis B is particularly common in the Far East, where carrier rates of HBV are high, but is rare in the United Kingdom where HBV infects less that 1 per cent of the population. Indeed, the proportion of ALF cases in the United Kingdom in which a viral etiology can be implicated is less than 30 per cent, with more than half of these having a seronegative hepatitis. In France, the proportion of viral-associated ALF is higher, with hepatitis B being the major viral agent (Benhamou 1991), and the figures are similar in the United States, with viral hepatitis accounting for 62 per cent of cases. In the Indian subcontinent a high proportion of cases fall within the non-A non-B category. Hepatitis E is more frequently encountered in subtropical areas and is associated with high mortality, particularly in pregnant women. However, a recent study showed that hepatitis E virus could be implicated in a small proportion of sporadic cases of non-A non-B hepatitis in which there was no history of travel abroad.

There is also considerable geographical variation with respect to hepatitis C as a cause of ALF. Instances of infection are rarely found in the West, whereas in Japan over half the patients from one series with acute or subacute liver failure had evidence of infection with HCV. In another series from Japan, dual infection with HCV RNA was observed in a number of patients with HBV ALF. This is similar to a report from France in which almost half of the HBsAg-positive patients had detectable HCV RNA in serum. In Taiwan, 40 to 50 per cent of cases of fulminant viral hepatitis are related to HCV infection.

The contribution of the newly described GB viral agent and of hepatitis G (which appear to be different serotypes of the same virus with some homology to the flaviviruses, particularly hepatitis C) to the non-A to E category of fulminant viral hepatitis is probably smaller in the West, although in a recent study from Japan three of six cases with fulminant hepatitis of unknown etiology were found to have positive signals for GBV-C genome in serum. Further viral infections which rarely result in ALF include cytomegalovirus, Epstein–Barr, herpes simplex, Coxsackie B, and echovirus. Other rare causes of ALF, which are important because of the possibilities of specific treatment, are acute fatty liver of pregnancy, lymphoma, ischemic hepatitis, and acute Budd–Chiari syndrome. Rarely, Wilson's disease has a fulminant presentation and can be identified from the associated hemolysis, splenomegaly, and Kaiser–Fleischer rings. Finally, an autoimmune hepatitis can also occasionally present as a fulminant illness.

References

Benhamou, J.-P. (1991). Fulminant and subfulminant hepatic failure; definition and causes. In *Acute liver failure: improved understanding and better therapy* (ed. R. Williams and R.D. Hughes), pp. 6–10. Mitre Press, London.

Bihari, D.J., Gimson, A.E., and Williams, R. (1986). Cardiovascular, pulmonary and renal complications of fulminant hepatic failure. *Seminars in Liver Disease*, **6**, 119–28.

O'Grady, J.G., Schalm, S., and Williams, R. (1993). Acute liver failure: redefining the syndromes. *Lancet*, **342**, 373–5.

Spooner, J.B. and Harvey, J.G. (1993). Paracetamol overdose — facts not misconceptions. *Pharmaceutical Journal*, **252**, 706–7.

Wendon, J., Harrison, P., Keays, R., and Williams, R. (1994). Cerebral blood flow and metabolism in fulminant liver failure. *Hepatology*, **19**, 1407–13.

4.7.2 Diagnosis and assessment

F. Hawker

Key messages

- There is no specific diagnostic test.
- Classical presentation is with jaundice, coagulopathy, and encephalopathy in the absence of previous liver disease.
- Liver function tests (apart from plasma bilirubin concentration) are of little value in the diagnosis or assessment of severity.
- Assessment involves consideration of the need to transfer the patient, determination of the cause, and assessment of prognosis.
- Both the King's College and the Clichy criteria have good positive predictive value but are unable to identify the subgroup at low risk of death.

Diagnosis of acute liver failure

Acute liver failure (ALF) is a clinical syndrome rather than a specific disease. It is diagnosed by the presence of a number of clinical and laboratory findings, although these may vary in severity from patient to patient. The classical triad is jaundice, coagulopathy, and encephalopathy in a patient in whom there is no previous evidence of liver disease. The subject has been reviewed elsewhere (Hawker 1993; Caraceni and Van Thiel 1995; Hoofnagle *et al.* 1995).

History

The history is dependent, to some extent, upon the cause (Table 1). There is usually a short prodromal influenza-like illness in an individual who was previously well. Nausea, vomiting, and right upper quadrant pain may be present. Jaundice, manifested by either discoloration of the skin or dark urine, is usually the first specific symptom. The full-blown syndrome of ALF may develop in days or weeks after first presentation. This variability has led to a relatively standardized nomenclature for subsets of ALF (O'Grady *et al.* 1993).

Clinical examination

On presentation, the patient with ALF is usually jaundiced. The liver is typically impalpable and there are no signs of chronic liver disease or portal hypertension, such as spider naevi or ascites. There may be active bleeding from skin puncture sites or after instrumentation (e.g. nasogastric tube insertion) reflecting the profound coagulopathy. Examination of the cardiovascular system usually reveals a normal or marginally decreased blood pressure, bounding pulses, and warm peripheries. A flow murmur may be present. Hemodynamic monitoring usually supports the presence of a high cardiac index, low systemic vascular resistance state. The major clinical signs are usually neurological and reflect the encephalopathy that is part of the ALF syndrome. Hepatic encephalopathy is classified into four grades (Table 2); the grade present at the time of neurological examination reflects the clinical findings. Agitated and combative behavior, often with delusional ideas and perseveration of speech, may be one of the earliest signs of progression of encephalopathy. At the same time there is a generalized increase in muscle tone which may progress to full decerebrate posturing. There is a parallel increase in the briskness of the deep tendon reflexes. An extensor plantar response and sustained clonus may be present as encephalopathy progresses. Lateralizing signs are almost never present and suggest an alternate or additional diagnosis such as intracerebral hemorrhage. Spontaneous hyperventilation is common and can result in significant respiratory alkalosis. Deep coma (unresponsive to painful stimuli) occurs with advanced encephalopathy (grade IV). The pupils may become dilated at this stage, particularly in response to noxious stimuli. This and other abnormalities of cranial nerve function usually suggest that cerebral edema is present. Hypertension is a common accompaniment of elevated intracranial pressure.

Laboratory tests

Liver function tests

A plasma bilirubin concentration above 300 μmol/l (18 mg/dl) always indicates severe disease (O'Grady *et al.* 1989). Lower levels do not necessarily preclude the diagnosis or indicate a good prognosis with acetaminophen (paracetamol) poisoning or other situations where the onset of liver failure is rapid. Plasma aminotransferase concentrations (alanine aminotransferase and aspartate aminotransferase) may be grossly elevated (up to 10 000 IU/l). However, increases are time dependent, and relatively normal plasma concentrations of these enzymes may be found if massive hepatic necrosis has occurred before presentation. Non-specific and minor elevations may occur in plasma alkaline phosphatase and γ-glutamyl transferase concentrations.

Blood glucose concentration

Hypoglycemia is a common but not universal manifestation of ALF. It is important because it is treatable with intravenous glucose. Signs of hypoglycemia may be masked by encephalopathy.

Coagulation studies

The international normalized ratio (INR) for prothrombin, previously the prothrombin time (PT), is always prolonged. It can be used as an index of disease severity (Table 3) provided that fresh

Table 1 Clinical and laboratory characteristics of the usual causes of ALF

Cause	History	Clinical examination	Laboratory tests
Viral hepatitis	Contact with virus	Non-specific	
HAV	Travel, epidemic		IgM anti-HAV
HBV	Contact with infected blood/body fluids		IgM anti-HBc, HbsAg, HbcAg
HCV	Contact with infected blood/body fluids		Anti-HCV, HCV RNA
HDV	HBV positive		Anti-ADV, HDAg in Serum
HEV	Pregnancy, travel		Anti-HEV
HGBV?	Contact with infected blood/body fluids		HGBV RNA
Acetaminophen poisoning	Paracetamol overdose Personal/financial problems Nausea and vomiting followed by apparent recovery	Non-specific Jaundice may be absent	Elevated serum acetaminophen relative to time from ingestion
Amanita mushroom poisoning	Mushroom ingestion Severe GI symptoms ?Multiple casualties	Non-specific Jaundice may be absent	Non-specific ↓ White cell count (bone marrow suppression) ↓ PO_4 (GI secretions)
Idiosyncratic drug reaction	Medication/anesthesia within 6 weeks	Fever, rash, joint pain	Non-specific Eosinophilia
Wilson's disease	Family history Neurological symptoms	Kayser–Fleischer rings ?Neurological signs	Ceruloplasmin <200 mg/l Liver copper >250 mg/g dry weight 24-h urine copper >1.6 μmol/day Hemolysis
Acute fatty liver of pregnancy	Third trimester of pregnancy Pre-eclampsia	Pregnant	Aminotransferase only moderately elevated, ↑ WCC, ↓ platelets, DIC
Budd–Chiari syndrome	Abdominal pain Thrombotic tendency	Hepatomegaly, ascites	LFTs non-specific Hepatic venous thrombosis

GI, gastrointestinal; WCC, white corpuscle count; DIC, disseminated intravascular coagulation; LFT, liver function test.

frozen plasma has not been given. Levels of all coagulation factors, except factor VIII, are reduced. Low-grade disseminated intravascular coagulation occurs in most patients, probably because of low levels of antithrombin III and protein C. Thrombocytopenia is common.

Plasma proteins

Plasma albumin and globulin concentrations may be normal on presentation. Plasma concentrations of proteins with a shorter half-life synthesized by the liver (e.g. prealbumin) are decreased.

Other laboratory tests

Many other laboratory tests are abnormal in ALF and, although not necessarily diagnostic, may guide management. Common abnormalities include respiratory alkalosis, metabolic alkalosis or acidosis, hypokalemia, hypomagnesemia, hyper- or hyponatremia, and renal failure.

Specific tests to determine the cause of acute liver failure

These are discussed below and summarized in Table 1.

Other tests

Liver biopsy is not indicated in ALF because sampling error may misrepresent the degree of necrosis and because of the potential for bleeding caused by the profound coagulopathy. However, imaging of the liver and its vasculature using either ultrasound or CT scanning techniques is useful for assessing liver size, excluding biliary obstruction or hepatic vein thrombosis, and assessing the liver vasculature if transplantation is contemplated. Other tests are used to assess cerebral function and structure in deeply encephalopathic patients. Advances in CT scanning technology make this test a more useful detector of cerebral edema than previous studies have suggested. Continuous monitoring of intracranial pressure is commonly used to detect episodes of cerebral edema despite the significant risk of

Table 2 Grades of acute hepatic encephalopathy

Grade	Description
I	Altered mood; impaired intellect, concentration, and psychomotor function, but arousable and coherent
II	Inappropriate behavior; increased drowsiness and confusion, but arousable and conversant
III	Stuporous but arousable, often agitated and aggressive
IV	Coma; unresponsive to painful stimuli

intracranial bleeding (Hoofnagle *et al.* 1995). Non-invasive methods of assessing cerebral function, such as sensory evoked potentials, may prove useful in the future.

Differential diagnosis

No other condition presents with deep jaundice, gross prolongation of INR, and encephalopathy in the absence of established liver disease. The major issue is to recognize the circumstance where a patient with chronic liver disease presents with an acute deterioration in liver function. This can usually be differentiated by the clinical history and the presence of signs of chronic liver disease such as spider naevi and ascites on clinical examination. Acute massive hepatic vein thrombosis or Budd–Chiari syndrome should be considered, particularly if the liver is of normal size or enlarged. In these circumstances ALF may be reversed by prompt treatment.

Assessment

Assessment of the patient with ALF firstly involves determination of the need for urgent life-saving intervention and resuscitation. Such measures may include endotracheal intubation and mechanical ventilation, treatment of shock, and administration of glucose if hypoglycemia is present. The next concern is assessment of the need and urgency for transfer of the patient to a liver unit. This will depend upon the expertise and facilities available at the place of presentation, the grade of encephalopathy and severity of illness at presentation (see adverse prognostic indicators below), and assessment of the rate of progression from the clinical history and examination. Regardless of these factors, the patient should be discussed with a specialist from a liver unit. If transportation is necessary, the patient should be accompanied by an experienced medical escort, and full monitoring and equipment to initiate therapy should be available. Endotracheal intubation and mechanical ventilation are usually indicated for transportation.

Assessment of cause

In general, the clinical features of ALF reflect the degree of hepatocellular damage and are independent of the cause. However, determination of the cause is important because specific therapy may be indicated in some instances, as determination of the etiology has implications for the spontaneous return of liver function and it may indicate the need for screening of family members (e.g. Wilson's disease) or other close contacts (hepatitis).

Important features to note in the history are shown in Table 1. The clinical examination is usually unhelpful in establishing the cause of ALF; however, if fulminant Wilson's disease is suspected, it is important to examine the eyes for Kayser–Fleischer rings. If present, these confirm the diagnosis, but their absence does not exclude it. Laboratory tests used to determine the cause of ALF are shown in Table 1.

Table 3 Criteria to identify patients with ALF who are unlikely to survive without liver transplantation

King's College criteria	Clichy criteria
PT > 100 s* or any three of the following:	Factor V < 20% (age < 30 years) Factor V < 30% (age > 30 years)
Age < 10 years or > 40 years Non-A, non-B hepatitis[†], halothane, or other drug reaction etiology Duration of jaundice before onset of encephalopathy > 2 days PT > 50 s[‡] Serum bilirubin > 300 μmol/l (18 mg/dl)	Coma or confusion
If induced by acetaminophen pH < 7.3 or PT > 100 s* and Creatinine > 300 μmol/l (3 mg/dl) Grade III or IV encephalopathy	

*PT > 100 s (UK) ≈ PT > 30 s (USA) ≈ INR > 9.7.
[†]Can substitute 'etiology unknown'.
[‡]PT > 50 s (UK) ≈ PT > 21 s (USA) ≈ INR > 4.2.

Assessment of prognosis and outcome

Overall, survival from ALF when all causes are considered is approximately 10 to 40 per cent with medical therapy alone, and 60 to 80 per cent with liver transplantation (Caraceni and Van Thiel 1995). Assessment of prognosis is the critical issue in decisions to perform urgent liver transplantation or other developing therapies. The most widely used criteria for assessment of prognosis (and therefore determination of the need for liver transplantation) were developed at King's College Hospital, London, using multivariate analysis of a series of 588 patients and were then applied prospectively to an additional group of 175 patients (O'Grady *et al.* 1989). The factors that were predictive in patients with acetaminophen poisoning were different from those in patients with other causes of ALF. These are shown in Table 3 and are frequently referred to as the 'King's College criteria'. Bernuau and coworkers from the Hôpital Beaujon, Clichy, France, have developed other criteria (the 'Clichy criteria'), also shown in Table 3. Both sets of criteria have good positive predictive value but a low negative predictive value, indicating that neither could identify a subgroup at low risk of death. The cause of ALF also has some importance in assessing outcome. In general, the prognosis for spontaneous recovery is relatively good for hepatitis A infection, acetaminophen poisoning, and acute fatty liver of pregnancy, intermediate for hepatitis B infection, and poor for idiosyncratic drug reactions, fulminant Wilson's disease, and unknown etiology.

References

Caraceni, P. and Van Thiel, D.H. (1995). Acute liver failure. *Lancet*, **345**, 163–9.

Hawker, F.H. (1993). *The liver: critical care management*, pp. 71–135. W.B. Saunders, London.

Hoofnagle, J.H., Carithers, R.L., Shapiro, C., and Ascher, N. (1995). Fulminant hepatic failure: summary of a workshop. *Hepatology*, **21**, 240–52.

O'Grady, J.G., Alexander, G.J.M., Hayllar, K.M., and Williams, R. (1989). Early indicators of prognosis in fulminant hepatic failure. *Gastroenterology*, **97**, 439–45.

O'Grady, J.G., Schaim, S.W., and Williams, R. (1993). Acute liver failure: redefining the syndromes. *Lancet*, **342**, 273–5.

4.7.3 Management

J. Wendon

Key messages

- Acute hepatic failure is seen in patients with acute hepatic necrosis whose clinical course is complicated by the development of encephalopathy and coagulopathy.
- Metabolic derangements, including hypoglycemia, hypophosphatemia, and metabolic acidosis, are frequently seen.
- There is a high incidence of renal failure and hemodynamic disturbance, usually with a hypotensive, hyperdynamic state.
- Management is supportive in the first instance. Most patients are significantly volume depleted and require aggressive volume resuscitation.
- The level of consciousness can change rapidly. Patients who deteriorate to grade III coma should be electively ventilated because of the risk of cerebral edema and to allow optimal treatment.
- Consideration should always be given to liver transplantation; in selected patients the mortality with medical treatment alone remains at 90 per cent.

Introduction

Acute hepatic failure (AHF) describes a constellation of symptoms associated with the sudden cessation of normal hepatic function and the subsequent development of encephalopathy and jaundice. This is often accompanied by coagulopathy, renal failure, sepsis, cardiovascular instability, metabolic acidosis, cerebral edema, and multiple organ failure. The speed of onset of complications may vary from a few days to a more prolonged process. The first attempt at a formal definition was made in 1970 when a potentially reversible condition, the consequence of severe liver injury, was described, with the onset of hepatic encephalopathy being within 8 weeks of the first symptoms and in the absence of pre-existing liver disease. Subsequently, patients with late-onset hepatic failure were defined as those where the onset of hepatic encephalopathy was within 8 to 24 weeks from the onset of jaundice. Similarly, liver failure in which encephalopathy developed within 2 weeks of jaundice has been described as fulminant and that in which the delay was 2 to 12 weeks has been described as subfulminant. More recently, the terms 'hyperacute' (encephalopathy within 8 days from the onset of jaundice), 'acute' (jaundice to encephalopathy, 8–28 days), and 'subacute' (jaundice to encephalopathy, 4–26 weeks) have been proposed. A high proportion of patients with hyperacute liver failure will develop grade IV coma and cerebral edema, although many will recover spontaneously. The subacute patients have a lower incidence of cerebral edema but more will require transplantation. As to etiology, acetaminophen (paracetamol) toxicity and mushroom poisoning predominate in the hyperacute group, while seronegative hepatitis predominates in subacute liver failure.

Causes of AHF are many and varied, and they will undoubtedly continue to increase as new viral agents are described and new drugs with potential hepatotoxic side-effects are prescribed. Alcohol-induced liver damage is not usually considered to be a cause of AHF as such patients often have chronic liver damage (cirrhosis and/or fibrosis). However, they often present as acute decompensations with ascites, jaundice, renal failure, encephalopathy, and a coagulopathy. The clinical diagnosis is usually easily reached and the management, as with many causes of AHF, is supportive, although some may also benefit from steroid therapy.

Investigation

Regardless of etiology, much of the liver damage has often occurred by the time of presentation. However, this should not result in a failure to define the cause since the treatment options may vary. For example, liver transplantation would be inappropriate for a patient with malignant infiltration, while chemotherapy is indicated as early as possible for lymphoma. Thus the work-up and management of all patients with acute hepatic necrosis should address etiology. Investigations and imaging will be dictated by the clinical scenario, but may include some or all of those shown in Table 1.

General management

Hypoglycemia is common. It arises as a consequence of impaired circulating insulin, impaired gluconeogenesis, and an inability to mobilize glycogen stores. Hypophosphatemia and hypomagnesemia are also frequent problems, particularly when urine output is maintained. A metabolic acidosis may be seen; this carries a poor prognosis and prompts consideration of transplantation.The etiology is multifactorial, relating to hepatic dysfunction, impaired lactate metabolism, and also tissue hypoxia in patients who are inadequately volume resuscitated. Patients with AHF are at increased risk of infections, both bacterial and fungal; this appears to be related to compromised immune function with impaired neutrophil and Kupffer cell function and deficiency of opsonins. Bacterial infection

Table 1 Investigation of acute liver failure

Studies	Investigation	Notes
Hematological	Full blood count and film	In any patient with hemolysis consider the diagnosis of Wilson's disease
	Hemolysis screen	Perform in patients with predominant unconjugated bilirubinemia
	INR and clotting profile	
	Procoagulant screen	Patients with possible Budd–Chiari syndrome or veno-occlusive disease
	Bone marrow aspirate and trephine	Where infiltration, lymphoma, or hemophagocytosis are being considered
Biochemical and immunological	Electrolyte, urea, and creatinine, liver function tests	
	Regular blood glucose	
	Urinalysis, urinary pH, creatinine clearance	
	Blood gases, lactate	If evidence of metabolic acidosis
	Toxicology screen	If indicated
	Blood urate	Elevated in fatty liver of pregnancy
	Copper studies (serum copper, ceruloplasmin, and urinary copper pre- and postpenicillamine challenge)	If Wilson's disease is being considered
	Immunoglobulins, autoantibodies	
Imaging	Liver ultrasound	Check patency of blood vessels, particularly hepatic veins in Budd–Chiari syndrome
	CT, hepatic venography, aortoporography, pressure studies	As indicated
Virology	Hepatitis A IgM	
	Hepatitis B (surface Ag, Ab and IgM core Ab initially; if positive, also e Ag and Ab)	
	Hepatitis C RNA	Hepatitis C is a very rare cause of AHF in North America and Europe
	Hepatitis E	IgG only available at present
Histology		Need guided by clinical situation: in critically ill patients with severe coagulopathy and likely to require transplantation, biopsy may be an inappropriate risk; in other patients, drug intervention may be modified by histology

INR, international normalized ratio.

rates of up to 80 per cent occur within a few days of admission. Some centers use prophylactic antibiotic regimens; however, a high index of suspicion should be maintained in all patients. All patients are given local oral antifungal treatment. The development of a coagulopathy is inherent to the disease process and is a good prognostic indicator. Therefore fresh frozen plasma is not given unless the patient is actively bleeding. Quantitative and qualitative defects in platelet function are observed; infusions are given as clinically required. Gastrointestinal bleeding is rare, but prophylaxis may be undertaken with agents such as sucralfate.

As in all critically ill patients, feeding should be instituted as soon as possible. In nearly all patients enteral nutrition is tolerated. In AHF a low-protein low-sodium diet is not appropriate; calories and protein should be given as the clinical situation dictates.

Derangement of the systemic circulation

The hemodynamic disturbances seen in AHF are similar to those of sepsis, i.e. elevated cardiac output and lowered systemic vascular resistance. Relative hypovolemia secondary to vasodilatation is frequent, and a pulmonary artery flotation catheter is often required to optimize fluid replacement. Colloid loading is almost inevitably required, usually to a greater degree than initially considered. Crystalloid is also necessary both to cover maintenance requirements and maintain euglycemia. The basis for microcirculatory dysfunction seen in liver failure is poorly understood, but evidence is accumulating to suggest the importance of interactions between the endothelium, exogenous factors such as bacterial toxins, and cytokines, particularly tumor necrosis factor and interleukins 1 and 6. Endotoxin and other bacterial toxins lead to cytokine production by activated macrophages and may be maintained in the circulation because of impaired Kupffer cell function and the presence of portosystemic shunts. Platelet activation and consumption, with formation of microthrombi within various organs, may lead to endothelial damage and release of further vasoactive compounds. Patients with AHF have elevated levels of citrulline, cyclic GMP, nitrite, and nitrate, suggesting activation of the nitric oxide pathway.

Hypotension is common in AHF and is characterized by both systolic and diastolic hypotension. Although initially responsive to fluid loading, vasopressors are frequently required, particularly in patients with concomitant cerebral edema. Agents such as dobutamine are not normally effective in increasing mean arterial pressure. The most efficacious agents are either epinephrine (adrenaline) or norepinephrine (noradrenaline), commencing at a dose of 0.1 μg/kg/min. Administration of these agents results in significant improvements in blood pressure, but the cardiac index and oxygen delivery are not always improved and there may be a fall in oxygen consumption. The role of agents to block the synthesis or effects of nitric oxide require further evaluation in clinical trials.

N-Acetylcysteine is given to many patients with AHF, regardless of etiology, at a loading dose of 150 mg/kg over 30 min followed by a maintenance dose of 150 mg/kg/day. The rationale for this has developed from the observation that patients receiving *N*-acetylcysteine following acetaminophen overdose after the period where it would act as an antidote have a better prognosis than those who do not receive it. There was a lower incidence of renal failure and less progression to deeper levels of coma despite similar levels of prolongation of prothrombin times. These findings were confirmed in a prospective study (Keays *et al.* 1991) in which treated patients displayed less hypotension and cerebral edema. More recent work has demonstrated that *N*-acetylcysteine results in increased cardiac output, oxygen delivery, oxygen extraction ratio, and hence oxygen consumption in patients with AHF regardless of etiology. The beneficial mechanism of the action of *N*-acetylcysteine is unclear, although it is a potent antioxidant and therefore may stabilize the effects of endothelium-derived relaxant factor and prevent endothelial damage by free radicals. It also repletes tissue sulfydryl groups either directly or by its action of increasing cysteine levels, and thus may restore full endothelium-derived relaxant factor activity by a similar mechanism to its reversal of tolerance to nitrates.

The role of agents that improve splanchnic flow requires further investigation. Another potentially beneficial therapy is high-volume plasmapheresis. Marked improvements in hemodynamic and neurological status have been reported (Larsen *et al.* 1994). The mode of action is presumably related to the removal of putative vasorelaxant compounds which accumulate or are produced in patients with AHF.

Cerebral management

Encephalopathy is characteristic of the syndrome of AHF. Patients may progress from being mildly confused through aggressive and unmanageable to deeply comatose over a matter of 24 h. Patients with AHF are also at risk of developing cerebral edema in association with deep levels of coma, a complication not seen in chronic liver failure. Increases in intracranial pressure may be caused by increases in volume of any of the three main components within the skull, namely brain tissue, cerebrospinal fluid, and cerebral vessels. An increase in the volume of brain tissue is reported to occur in up to 80 per cent of patients progressing to grade IV encephalopathy, although more recent work suggests that the incidence of cerebral edema in AHF has fallen to approximately 40 per cent.

The clinical signs of cerebral edema, i.e. systemic hypertension, decerebrate posturing, and abnormal pupillary reflexes, are generally attributed to brainstem compression. However, personal experience suggests that cerebellar or uncal herniation is comparatively rare with present-day treatment, being found at autopsy in only 25 per cent of patients with cerebral edema. This suggests that other factors are involved. The development of cerebral edema in any given setting probably depends on a variety of causative events. Increased permeability of the blood–brain barrier results in leakage of protein-rich fluid into the extracellular space of the brain. The efficacy of mannitol in the management of patients with raised intracranial pressure suggests that intracellular cytotoxic cerebral edema is also important. The role of hyponatremia remains to be clarified but appears to be important in the evolution of cerebral edema in some clinical settings.

Traditionally, cerebral blood flow was believed to be significantly increased in all patients with AHF. However, recent work has shown marked variations, with cerebral blood flow falling as the coma deepens. One study (Aggarwal *et al.* 1994) found that 25 per cent of patients studied had cerebral blood flow values above 50 ml/min/100 g, suggestive of a level greater than that required, while 52 per cent had a reduced cerebral blood flow. Cerebral hyperemia has also been reported to precede or coincide with the presence of cerebral edema on CT scanning. Using a radioactive xenon technique, we found that cerebral blood flow varied between 12 and 50 ml/min/100 g in patients in grade IV coma. Although such levels of cerebral blood flow have been considered adequate for the metabolic demands of patients in deep coma, the observation that many of these patients have cerebral lactate production may be suggestive of an inadequate delivery of substrate for metabolic demand in some clinical settings. Substantial increases in cerebral blood flow can be demonstrated following infusion of mannitol and *N*-acetylcysteine, and are associated with a rise in cerebral O_2 utilization and a fall in anaerobic metabolism. A fall in cerebral blood flow occurs following hyperventilation, which may thus have the potential to produce cerebral hypoxia. This may lead to cell membrane damage which, in association with Na^+,K^+-ATPase activity, could result in increased brain water and a further increase in cerebral edema.

Elective ventilation following development of grade III encephalopathy is required to control the airway and frequently to allow treatment in the aggressive agitated patient. Lactulose and non-absorbable antibacterials have no role in management. Patients should be nursed at 10° to 20° head-up tilt to maximize cerebral perfusion pressure. Patient stimulation should be kept to a minimum during this period. Early detection of cerebral edema is important in order to initiate appropriate management; many of the normal clinical signs are lost and treatment decisions may need to be made on the basis of systolic hypertension, tachycardia, or pupillary abnormalities. Routine use of intracranial pressure monitoring in all ventilated grade IV patients is frequently advocated, but the relative risks and benefits must be carefully considered in patients with a coagulopathy. A recent review (Blei *et al.* 1993) suggested that extradural systems were associated with the lowest incidence of complications. A reliable non-invasive method of monitoring cerebral edema would be of great assistance in clinical management. CT does not appear to be reliable or useful in clinical practice, except to exclude other intracranial pathology. We institute intracranial pressure monitoring in all patients who have demonstrated either pupillary abnormalities or unexplained systolic hypertension or who have developed grade III/IV coma and are proceeding to liver transplantation. Prior to insertion of a bolt, sufficient fresh frozen plasma and platelets should be administered to achieve an international normalized ratio (INR) of 2 or less and a platelet count above 50×10^9/l.

In patients at risk of cerebral edema, the preservation of an adequate cerebral perfusion pressure is critical. The minimum tolerable level of cerebral perfusion pressure to allow full neurological recovery remains controversial, with some claiming that levels below 40 mmHg for more than 24 h are still compatible with eventual full recovery (Davies *et al.* 1994). Jugular venous bulb blood sampling for oxygen saturation (Sjo_2) and arteriojugular lactate difference provide indirect markers of cerebral oxygen consumption and extraction. In addition to hemodynamic data, such information can aid therapeutic management. We assume cerebral blood flow and cerebral perfusion pressure to be adequate when Sjo_2 is between 55 and 75 per cent. If the cerebral perfusion pressure falls predominantly due to falls in mean arterial pressure and Sjo_2 is below 55 per cent, vasopressor therapy is instituted. When the cerebral perfusion pressure fall is predominantly due to a rise in intracranial pressure, mannitol 0.5 g/kg is the treatment of choice provided that plasma osmolarity does not exceed 320 mosm/l. Hyperventilation is not routinely advocated, although there may be a place for short periods of hyperventilation when Sjo_2 and intracranial pressure are raised, suggestive of hyperemia. All our patients receive *N*-acetylcysteine in view of the decreased incidence of cerebral edema and the rise in cerebral blood flow seen with this agent.

Cerebral function monitoring is of considerable value in detecting subclinical epileptiform activity, a highly detrimental occurrence resulting in increased metabolic demand in the face of potentially limited substrate delivery. The use of routine cerebral function monitoring has demonstrated a high incidence of epileptiform activity and explains the observed clinical situation of pupillary abnormalities with relatively low Sjo_2 and only a marginal increase in intracranial pressure.

Thiopental (thiopentone) (50-mg bolus followed by an infusion of 50 mg/h for up to 6 h) has proved successful in treating otherwise intractable cerebral hypertension. However, this drug has a tendency to lower blood pressure and cause hypothermia.

Renal management

Renal failure (urine output below 300 ml/day and serum creatinine above 300 μmol/l) despite adequate filling occurs in up to 70 per cent of patients with AHF. Hypovolemia and sepsis are important precipitating factors. In addition, altered levels of vasodilators and constrictors within the renal vasculature play an important role. Acetaminophen may cause direct renal toxicity as well as liver damage. The etiology of renal failure in AHF is usually multifactorial, comprising prerenal factors, hepatorenal failure, acute tubular necrosis, and direct nephrotoxicity.

Initial management consists of aggressive volume loading utilizing colloid and crystalloid as previously described. If oliguria persists, 'renal' dose dopamine ($\leqslant$ 2.5 μg/kg/min) or a furosemide (frusemide) infusion (10 mg/h) may be tried. Current opinion suggests a very limited role for dopamine. Renal replacement therapy will be needed if acidosis, hyperkalemia, fluid overload, or a rising creatinine develop. It should be instituted in any oliguric patient showing signs of cerebral edema in order to remove fluid following mannitol administration. Continuous hemodiafiltration is the best form of renal replacement therapy in AHF. High-volume hemofiltration may also be utilized, but care must be taken in the choice of replacement fluid since patients will be unable to metabolize many of the standard solutions in which lactate and acetate are the predominant buffers. A bicarbonate-buffered solution will be well tolerated. Intermittent hemodialysis should be avoided since the frequent complication of hypotension may result in significant falls in cerebral perfusion pressure. However, it is useful in patients who have undergone transplantation or in those following AHF who are hemodynamically stable and no longer at risk of cerebral edema. Anticoagulation can be difficult since platelets are large and sticky and circuits may clot rapidly despite a coagulopathy. Heparin is used unless active bleeding is present or an intracranial bolt is *in situ*. However, heparin resistance is common owing to concurrent antithrombin III deficiency. Epoprostenol 5 ng/kg/min is an alternative anticoagulant.

Transplantation and other treatment options

The basis of AHF management is supportive therapy in an intensive care setting, aiming to decrease the risk of development of other organ failures and to promote liver regeneration. The role of regenerative agents such as hepatocyte growth factor are still under investigation. Prevention and early treatment of infectious episodes appear to be of great importance.

Plasmapheresis is advocated by groups in Copenhagen and Japan, although controlled trials are needed. Similarly, extracorporeal liver-assist devices are exciting potential developments, although none have yet shown conclusive efficacy in the clinical setting with respect to outcome.

Transplantation remains the treatment of choice for a subgroup of patients with a poor prognosis. A variety of models have been developed, including assessment of liver volume (volumes below 700 cm^3 resulting in transplantation), assessment of percentage necrosis on biopsy (complicated by sampling error), and a variety of prognostic models. These need to be simple to apply in the clinical setting and to achieve a high level of sensitivity and specificity in discriminating survivors and non-survivors. Two models are commonly used in Europe. In the first, criteria for poor outcome in non-acetaminophen-induced liver failure have been established as the findings of coma or confusion in association with a factor V level below 20 per cent of normal in patients less than 30 years of age, or below 30 per cent of normal in those aged 30 years or above. In the second, the criteria separate acetaminophen and non-acetaminophen patients. For acetaminophen, the criteria are an arterial pH below 7.3 at greater than 24 h post-overdose and following volume resuscitation, or the concurrent findings of a prothrombin time longer than 100 s (INR > 6.5) plus a creatinine level above 300 μmol/l and grade III/IV encephalopathy. In non-acetaminophen patients the criteria are a prothrombin time longer than 100 s (irrespective of level of encephalopathy) or any three of the following: etiology of seronegative hepatitis or drug-induced liver failure; age less than 10 or more than 40 years; jaundice to encephalopathy time more than 7 days; prothrombin time more than 50 s; serum bilirubin above 300 μmol/l. Comparisons of these two models show them to be equally useful in the clinical setting. They are not applicable to pediatric patients, where different criteria exist, or to specific conditions such as Budd–Chiari syndrome, Wilson's disease, or fatty liver of pregnancy. Transfer to a center offering liver transplantation should ideally be undertaken prior to the development of criteria to allow optimal assessment and time to acquire a suitable graft. Outcome after transplantation is steadily improving; 8-year data from Europe (1988–1996) demonstrate a 52 per cent survival for a group of patients with a predicted 90 per cent mortality by medical management alone. Exciting opportunities are developing with the application of auxiliary liver transplantation which allows for the possibility of immunosuppression withdrawal in patients whose native liver regenerates. This and the progress in the fields of optimizing regeneration and liver support allow great scope for steady improvement in the prognosis of patients with AHF.

References

Aggarwal, S., Yonas, H., Obrist, W., Kang, Y., Martin, M., and Policare, R. (1994). Cerebral haemodynamic and metabolic changes in fulminant liver failure: a retrospective study *Hepatology*, **19**, 80–7.

Blei, A., Olafsson, S., Webster, S., and Levy, R. (1993). Complications of intracerebral pressure monitoring in fulminant hepatic failure. *Lancet*, **341**, 157–8.

Davies, M.H., Multimer, D., Lowes, J., Elias, E., and Neuberger, J. (1994). Recovery despite impaired cerebral perfusion in fulminant liver failure. *Lancet*, **343**, 1329–30.

Keays, R., Harrison, P.M., Wendon, J.A., Gimson, A.E.S., Alexander, G.J.M., and Williams, R. (1991). Intravenous acetylcysteine in paracetamol-induced fulminant hepatic failure. *British Medical Journal*, **303**, 1026–32.

Larsen, F.S., *et al.* (1994). Cerebral blood flow velocity during high volume plasmapheresis in fulminant hepatic failure. *International Journal of Artificial Organs*, **303**, 1026–32.

4.7.4 Management of complications

F. Hawker

Key messages

- The cerebral edema that accompanies or is part of fulminant hepatic encephalopathy is a common cause of death.
- Management of cerebral edema involves optimization of liver function, and monitoring of and measures to reduce the intracranial pressure.
- The most effective treatment of elevated intracranial pressure is mannitol infusion.
- Infection is common and is often fungal after the first week of management in the intensive care unit.
- Hypoglycemia must be excluded as a cause of deteriorating mental state.

Introduction

The definitive management of acute liver failure (ALF) involves measures to improve liver function. However, complications are often fatal before such measures can be undertaken or before liver regeneration occurs spontaneously. The severity and multiplicity of these complications are a testimony to the role of the healthy liver in maintaining normal physiology. Because of multiple organ involvement and conflicting therapeutic aims, the management of complications of ALF can be one of the most difficult challenges for the intensive care unit (ICU) (Lee 1993; Caraceni and Van Thiel 1995).

Hepatic encephalopathy and cerebral edema

The most lethal complication of ALF is the cerebral edema that accompanies or is part of fulminant hepatic encephalopathy. Over 50 per cent of patients with grades III and IV encephalopathy have episodes in which the intracranial pressure exceeds 30 mmHg (Caraceni and Van Thiel 1995). If intracranial pressure cannot be controlled and liver function cannot be improved (e.g. by liver transplantation), then this cerebral edema is usually fatal.

Management of cerebral edema

Cerebral edema should be presumed to be incipient or already present in all patients with advanced encephalopathy. Management generally involves the principles used in treating cerebral edema of any etiology. The airway should be secured if necessary (if compromised or to prevent aspiration), and mechanical ventilation instituted if there are abnormalities of respiratory drive or pulmonary gas exchange. Patients should be nursed with the head elevated to 20°; higher elevations may lead to a fall in cerebral perfusion pressure, probably because this maneuver results in a fall in mean arterial pressure due to peripheral vasodilatation. Venous return from the head should be unimpeded by malpositioning or tapes. Episodic agitation is common and may be associated with increased intracranial pressure (Munoz 1993). Although sedative drugs will obscure progression of neurological signs, judicious doses must be given in this situation to ensure patient safety. The most effective treatment of elevated intracranial pressure in ALF is mannitol infusion. In a controlled trial, infusion of 20 per cent mannitol (1 g/kg) reversed clinical signs of cerebral edema, reduced intracranial pressure by a mean of 22 mmHg, and increased survival. Subsequent studies have shown benefits with lower doses (0.5 g/kg), and have further demonstrated that mannitol infusion is associated with an increase in the cerebral metabolic rate for oxygen, and a decrease in cerebral lactate production (Wendon *et al.* 1994). Mannitol should be infused rapidly using a burette, but maximum reduction in intracranial pressure may not occur for 20 to 60 min. The dose can be repeated to control further episodes of intracranial hypertension, provided that the serum osmolality remains below 320 mOsmol/kg. Complications of mannitol therapy include dehydration and hyperosmolality if renal function is normal. In patients with renal failure, mannitol should only be used in conjunction with hemodialysis or hemofiltration, since fluid overload may exacerbate cerebral edema. Furosemide (frusemide) may be a useful adjunctive therapy to maintain the initial osmotic gradient established by mannitol.

Hyperventilation may be useful to decrease intracranial pressure during an acute elevation, but does not reduce the incidence or severity of cerebral edema in the longer term. This may be due to establishment of a new equilibrium between $P\text{aCO}_2$ and cerebral blood flow; alternatively, it may suggest that hyperventilation, with the consequent decrease in cerebral blood flow, may have adverse effects in ALF. In the study by Wendon *et al.* (1994), hyperventilation was associated with an expected decrease in cerebral blood flow but also a decrease in the cerebral metabolic rate for oxygen. However, mannitol and *N*-acetylcysteine infusion were associated with the potentially beneficial effect of an increase in cerebral metabolic rate for oxygen but also an increase in cerebral blood flow. These findings raise the possibility that cerebral ischemia may be a trigger to cerebral edema formation in ALF. This issue is speculative; nevertheless, current recommendations are that extreme hyperventilation should be avoided in patients with ALF and that $P\text{aCO}_2$ should be maintained in the range of 4.7 to 5.3 kPa (35–40 mmHg). Because of these findings, *N*-acetylcysteine requires further evaluation as a means of preventing cerebral edema in ALF. Barbiturate therapy (specifically thiopentone by infusion) has been used to control intracranial pressure in patients with ALF, but has not been subjected to a randomized controlled trial. Although reductions in intracranial pressure occur, the concomitant decrease in mean arterial pressure resulting from cardiovascular depression can result in no change or a decrease in cerebral perfusion pressure. This treatment may have a place in treating cerebral edema refractory to mannitol therapy when liver transplantation is imminent, but only when intracranial pressure and therefore cerebral perfusion pressure is monitored. There is no evidence that lactulose administration will delay or ameliorate the progression of encephalopathy or cerebral edema in ALF. Nevertheless, it is widely used in a dose of 30 ml thrice daily by gastric tube or enema, although bowel distension associated with its use may complicate liver transplantation. Use of neomycin is not recommended because of its nephrotoxicity and ototoxicity. Corticosteroids do not influence the incidence and severity of cerebral edema. The usefulness of benzodiazepine antagonists has not been established.

The role of intracranial pressure monitoring

Despite the well-described clinical signs in patients with ALF and cerebral edema, the correlation between the intracranial pressure level and specific clinical signs is poor. Consequently, intracranial pressure monitoring has become increasingly used in ALF to guide therapy of cerebral edema, to identify circumstances that increase intracranial pressure (e.g. endotracheal suction), to select patients for transplantation, and to assist in intraoperative management. Several uncontrolled studies suggest that intracranial pressure monitoring is of benefit, and slightly more than half of liver transplantation programs in the United States regularly use intracranial pressure monitoring for patients with ALF (Hoofnagle *et al.* 1995). Monitors in the extradural space are preferred since the complication rate (4 per cent) is lower than that associated with subdural (20 per cent) or parenchymal (22 per cent) monitors. Deaths caused directly by the monitor are due to intracranial bleeding or, more rarely, infection. The incidence is below 1 per cent for extradural monitors, but 4 to 5 per cent for other types. Most units using intracranial pressure monitoring place the monitors when grade IV encephalopathy develops. A cerebral CT scan is first obtained and the coagulopathy is reversed with infusions of fresh frozen plasma. In general, the goals of therapy are to maintain intracranial pressure below 20 mmHg and cerebral perfusion pressure above 50 mmHg. Recovery is unlikely when intracranial pressure is sustained above 40 mmHg, and a cerebral perfusion pressure below 40 mmHg for more than 2 h is considered to be a contraindication to proceeding with urgent liver transplantation. Although intracranial pressure monitors have been widely used for patients with ALF, the hazards must be considered. A death from intracranial hemorrhage in a patient who might otherwise have survived is a catastrophic event. The indications should be considered individually in each patient. Excellent results have been obtained without their use.

Coagulopathy

Despite the gross prolongation in the international normalized ratio (INR) for prothrombin with ALF, severe bleeding appears to have become less common in recent years. Sites of hemorrhage include the gastrointestinal tract, the nasopharynx, the respiratory tract, skin puncture sites, or the retroperitoneal space. Spontaneous intracerebral hemorrhage is unusual. Infusion of fresh frozen plasma is indicated before invasive procedures or if there is overt bleeding. Because the INR is a major consideration in the decision to undertake liver transplantation, fresh frozen plasma is not given routinely unless a decision to proceed with transplantation has been made. Vitamin K should be given parenterally on a daily basis. Blood losses should be replaced by blood transfusion, and platelet transfusions are indicated if the platelet count falls below $50\,000 \times 10^9$/l. Because of the risk of gastrointestinal bleeding, H_2 antagonists or omeprazole are given routinely.

Control of infection

Infections are common complications and significant causes of death in patients with ALF. Sepsis is secondary only to cerebral edema as a cause of death and typically occurs more than a week after ICU admission. Early infections are usually bacterial infections of the respiratory and urinary tracts. Infections occurring after the first week of ICU treatment are often fungal. Traditional signs of infection, such as elevations of temperature and peripheral white blood cell count, are often absent. Management involves regular culture of available secretions and early institution of appropriate antibiotic therapy. If empirical antibiotics are necessary, both Gram-positive and Gram-negative bacteria should be covered, and antifungal agents should be considered early. Aminoglycoside antibiotics should be avoided because of an increased risk of nephrotoxicity. Application of topical antifungal agents to the mouth and skin creases is a useful measure to reduce colonization and is particularly important before liver transplantation.

Pulmonary management

The airway may be compromised because of the decreased level of consciousness. Hyperventilation leading to respiratory alkalosis is a characteristic feature of hepatic encephalopathy. In the majority of patients who require endotracheal intubation and mechanical ventilation, the indication is airway management and treatment of cerebral edema. However, hypoxemia is the indication for ventilation in approximately one-third of patients. Complications of coma, such as bronchopneumonia, aspiration, or atelectasis, are the usual causes.

Hemodynamic management

Patients with ALF characteristically have a high cardiac index (frequently $>5\,l/min/m^2$) and a low systemic vascular resistance. The peripheral vasodilatation may result in hypotension but this is usually responsive to appropriate volume loading. If hypotension is not controlled by fluid replacement, inotrope infusions are indicated. However, hypertension may occur in patients with advanced encephalopathy, and is probably an appropriate response to maintain the cerebral perfusion pressure in response to an elevated intracranial pressure. In this circumstance, management of hypertension should involve assessment and treatment of the raised intracranial pressure. *N*-Acetylcysteine improves tissue perfusion, oxygen extraction ratio, and mean arterial pressure in patients with ALF from various causes. Despite these potentially beneficial effects, its effect on outcome has yet to be evaluated.

Renal failure

Renal failure occurs in approximately 10 per cent of patients, unless the cause is acetaminophen (paracetamol) poisoning when the incidence is very much higher, approaching 75 per cent in patients with grade IV encephalopathy. The mechanism in this circumstance is a direct nephrotoxic effect of acetaminophen. In patients without acetaminophen poisoning, many factors may contribute to the pathogenesis of renal failure. The usual form of renal failure is the hepatorenal syndrome, characterized by a progressive increase in plasma creatinine concentration, oliguria, low urine sodium concentration (usually $< 10\,mmol/l$), and histologically normal kidneys. Acute tubular necrosis may also occur as a result of hypotension, hypovolemia, or severe sepsis. It may be difficult to differentiate between these two syndromes, and often a diagnosis of hepatorenal syndrome can only be made after correction of the predisposing causes for acute tubular necrosis. Management includes monitoring of volume status, volume loading (although clearly this must be balanced against the risk of worsening cerebral edema), and optimization of hemodynamic status. Low-dose dopamine is commonly used to promote diuresis but, as in other conditions, its benefits are unproven. If renal failure develops, early dialysis is indicated. Continuous techniques of dialysis are preferred because of greater hemodynamic stability and consequent benefits to the intracranial pressure. Even if renal replacement therapy is required, survival can be as high as 30 to 50 per cent, depending upon the etiology.

Metabolic management

Hypoglycemia develops at some time in almost 50 per cent of patients with ALF (Munoz 1993). It should be prevented by a continuous infusion of 10 per cent glucose and must be considered during any deterioration in mental state. Both hyper- and hyponatremia may be present. The former is a consequence of treatment of cerebral edema with dehydration and is treated by judicious rehydration. Hyponatremia appears to be dilutional and is managed by volume restriction. Hypokalemia is present in approximately 50 per cent. It is closely related to the presence of metabolic alkalosis and may be both cause and result of this abnormality. It is treated with potassium supplementation, intravenously or via the nasogastric tube. Primary respiratory alkalosis is common in patients allowed to breathe spontaneously. It is a consequence of hepatic encephalopathy but the mechanism is unknown. No treatment is usually necessary. Metabolic alkalosis occurs in 25 to 50 per cent of patients; predisposing factors are hypokalemia and, perhaps, the liver's inability to synthesize urea. The treatment is correction of hypokalemia. Metabolic acidosis is common in patients with acetaminophen poisoning and is the most reliable indicator of a poor prognosis in this group. In other patients it is almost always associated with severe circulatory compromise. Management involves correction of the underlying cause, where possible. Nutritional therapy can be provided using conventional nutritional solutions, preferably by the enteral route. Caloric requirements are increased in ALF because of the markedly elevated basal energy expenditure. There is no evidence that protein restriction improves the outcome. Branched-chain amino acid solutions have not been shown to improve encephalopathy or overall mortality rate.

References

Caraceni, P. and Van Thiel, D.H. (1995). Acute liver failure. *Lancet*, **345**, 163–9.

Hoofnagle, J.H., Carithers, R.L., Shapiro, C., and Ascher, N. (1995). Fulminant hepatic failure: summary of a workshop. *Hepatology*, **21**, 240–52.

Lee, W.M. (1993). Acute liver failure. *New England Journal of Medicine*, **329**, 1862–72.

Munoz, S.J. (1993). Difficult management problems in fulminant hepatic failure. *Seminars in Liver Disease*, **13**, 395–413.

Wendon, J.A., Harrison, P.M., Keays, R., and Williams, R. (1994). Cerebral blood flow and metabolism in fulminant hepatic failure. *Hepatology*, **19**, 1407–13.

4.7.5 The effect of acute hepatic failure on drug handling

T. S. Leary and G. R. Park

Key messages

- Liver failure may cause unpredictable changes in the effect and elimination of drugs. Causes include changes in enzyme function and liver blood flow, in turn changing metabolism, deterioration in renal function, and alterations in receptor sensitivity and specificity.
- Caution is needed when prescribing drugs for these patients.

Introduction

Liver disease is common. It may be the reason for admission to the intensive care unit (ICU), for example after a paracetamol overdose. Alternatively, it may develop in the ICU, complicating another illness such as sepsis. Finally, patients may have a chronic disease complicated by another illness or an acute exacerbation.

There are several problems caused by hepatic failure which will influence the use of drugs (Fig. 1).

Reduced hepatic blood flow

Liver blood flow constitutes a quarter of cardiac output. As well as changes in total blood flow, in liver disease there may be shunting of

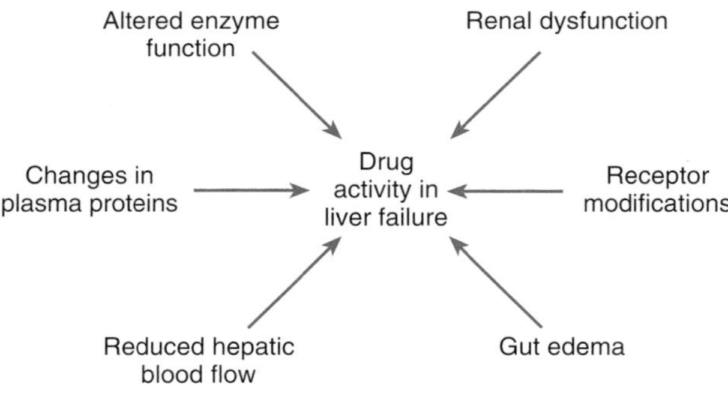

Fig. 1 Some of the many factors affecting pharmacodynamics and pharmacokinetics in liver failure.

Table 1 Some of the cytochrome P-450 enzymes and the drugs that they metabolize

Cytochrome P-450	Substrates (drugs)
2C11	Phenytoin
2D62	Tricyclic antidepressants
2E1	Volatile agents Acetaminophen
3A4	Benzodiazepines Local anesthetics Synthetic opioids

Where groups of drugs are given, not all drugs in this group will be metabolized by this enzyme. In addition other pathways may be involved.
Reproduced with permission from Park (1996).

blood so that it is not delivered to functioning hepatocytes. Shunts may be intrahepatic or extrahepatic (mesenteric or splenic) and may divert more than 50 per cent of blood flow in chronic hepatic disease. Moreover, inhalational and intravenous anesthetic agents, the hypotension associated with epidural blockade, sympathetic nervous system activity, and mechanical ventilation all reduce hepatic blood flow.

The amount of drug extracted from blood passing through the liver is dependent upon the metabolic activity of the liver, the amount of unbound drug, and liver blood flow. The hepatic extraction ratio (ER) is defined as follows:

$$\mathrm{ER} = \frac{(C_\mathrm{A} - C_\mathrm{V}) \times \mathrm{HBF}}{C_\mathrm{A} \times \mathrm{HBF}}$$

where C_A is the concentration of drug entering liver, C_V is the concentration of drug in the hepatic vein, and HBF is the hepatic blood flow. Thus, with high extraction drugs, clearance is dependent on the rate of hepatic blood flow.

Those drugs with a high extraction ratio, such as morphine (ER = 0.7) and lidocaine (lignocaine) (ER = 0.6), will have a reduced clearance with decreases in hepatic blood flow. When hepatic blood flow is reduced, the volume of distribution is usually unchanged and so the size of the first dose is unchanged. However, subsequent doses should be reduced in size or given less often.

The liver is particularly sensitive to hypoxia, with 70 per cent of its blood supply coming from the portal vein. Oxygen is needed by the cell for efficient drug metabolism to produce energy needed both for the reactions themselves and to make the enzymes. It is also needed as a substrate for drug oxidation, as a terminal electron acceptor, and for processes dependent upon the oxidation equilibrium (redox potential) of the cell.

The effects of altered enzyme function

Most drugs are metabolized from active fat-soluble compounds into inactive water-soluble compounds that can be excreted by the kidneys and bile. Usually this involves two phases of metabolism, so-called phase I and phase II metabolism.

Phase I metabolism usually involves the cytochrome P-450 superfamily of enzymes, performing reactions such as oxidation and hydroxylation. There are about 25 such enzymes, each identified by a number identifying the family, a letter identifying the subfamily, and a further number identifying the gene product. Each enzyme metabolizes multiple substrates; some examples relevant to critical care are shown in Table 1. The products of phase I metabolism may be highly active and, indeed, even toxic. For example, *N*-acetyl-*p*-benzoquinone, the phase I metabolite of acetaminophen (paracetamol), is hepatotoxic and causes the damage seen after an overdose of acetaminophen.

After phase I metabolism, further metabolism occurs involving the phase II enzymes. These commonly conjugate the phase I product with a group such as glutathione, glucuronic acid, or a sulfate.

The phase I enzymes are present in small amounts and are profoundly affected by disease. However, phase II enzymes are present in much larger amounts and are less affected by disease. Both acute and chronic liver disease reduce the amount of phase I enzymes, in turn reducing the ability of the liver to metabolize drugs dependent on this route of elimination. Those few drugs whose elimination is dependent only on phase II metabolism, such as propofol, morphine, and remifentanil, will have relatively normal metabolism in the presence of severe disease.

Benzodiazepines should be chosen with care. Those such as lorazepam and oxazepam which are metabolized mostly by glucuronidation may be preferred as their metabolism is not affected by liver disease. Diazepam and others dependent upon phase I pathways may have reduced elimination.

Enzymes responsible for the metabolism of most drugs are found in the greatest amounts in the liver, although concentrations may be higher in other tissues. The highest concentrations of the cytochrome P-450 enzymes are in the nose where they break down amines that cause smell, and in the adrenals where they make steroids.

Several studies performed during the anhepatic phase of liver transplantation have shown that many drugs are metabolized at these extrahepatic sites. The hepatic extraction ratio of propofol is greater than unity, showing that there is significant extrahepatic metabolism. However, the metabolic capacity of these sites is only about 10 per cent of that of the liver.

A third group of enzymes are becoming increasingly important for drug metabolism. These are the plasma and tissue esterases; for example, butyrl esterase is important in the metabolism of suxamethonium. Muscle relaxants such as mivacurium are being designed to avoid hepatic metabolism using this system of elimination. Because of their dependence on this one enzyme, they still risk genetic variability. Remifentanil, a new opioid, is metabolized by many esterases found both in the blood and elsewhere. Because it is metabolized by so many different esterases, its elimination is unlikely to be affected by disease or genetic variation.

Reduced plasma proteins

Hepatic failure is associated with a number of alterations which can influence the binding of drugs within the blood, including the following:

(1) decrease in serum albumin concentrations;

(2) appearance of altered or defective plasma proteins;

(3) accumulation of endogenous (e.g. bilirubin) and exogenous compounds which can displace drugs from albumin binding sites.

The influence of the consequent reduction in drug–protein binding is complex because it influences both the elimination and the distribution of the drug. With reduced binding there is an increase in the free fraction of the drug available for elimination; this helps to offset the reduction in drug metabolism caused by the liver disease. This effect is more significant with highly bound drugs; a 1 per cent reduction in the binding of a drug that is usually 99 per cent bound will double the free fraction of the drug. With an increase in proportion of active unbound drug, dosages may have to be reduced, despite normal total drug concentrations (bound and unbound), to avoid potential drug toxicity.

Reduced plasma drug binding will increase the volume of distribution by making more of the drug available for equilibration with extracellular tissues. This change will alter drug elimination half-life, irrespective of metabolism ($t_{0.5\ \mathrm{elim}} = \ln 2 \times V_\mathrm{d}/C_\mathrm{l}$ where $t_{0.5\ \mathrm{elim}}$ is the half-life, V_d is the volume of distribution, and C_l is clearance), thus making half-life a measure of little use in liver failure. The increase in the elimination half-life of lorazepam is purely a result of reduced binding.

Clearance of tolbutamide (based on total drug concentration) is increased in acute hepatic failure secondary to reduced protein binding. Although total drug levels may decrease, free drug levels are constant as intrinsic clearance (based on free drug levels) is unchanged. Thus no alteration in dose is required.

Disturbances of the enterohepatic circulation

As a result of physical obstruction or cholestasis of critical illness bile may not pass from the hepatocyte to the duodenum. Thus substances normally excreted by the biliary system (e.g. rifampicin and fusidic acid which are excreted unchanged in bile) may accumulate in the blood. Biliary obstruction may also induce some changes in hepatocellular function with a secondary effect on drug clearance. Accumulation of bile salts in the circulation will reduce drug–protein binding by competition. With reduced bile salts in the small bowel lumen, there will also be a reduction in absorption of fat-soluble drugs and nutrients. Edema of the small bowel mucosa, which is common in the critically ill, will reduce absorption of drugs from the gut and enterohepatic circulation. Portal hypertension will exacerbate this.

Impaired renal function

Progression of liver disease may result in hepatorenal syndrome. This is characterized by reduced glomerular filtration rate, hyponatremia with water intoxication, and almost complete sodium reabsorption. Since the main role of the kidney is excretion, the first dose of drugs used in renal failure is usually unchanged; again, it is subsequent doses that may need modification. One exception is suxamethonium which should be used with extreme caution in acute renal failure because of the risk of hyperkalemia. Other effects of renal disease may also change drug effects. These may include the following:

(1) disturbances of acid–base balance affecting drug activity and patient sensitivity;

(2) altered fluid balance affecting drug distribution;

(3) altering molecular states of metabolites (morphine-6-glucuronide becomes lipophilic);

(4) uremia displacing drugs from binding sites on plasma proteins, such as albumin globulins and lipoproteins.

Renal function can be assessed in many ways, and the estimates of the doses of drugs required based on the results. However, these methods are not without pitfalls and so clinical monitoring remains important. Measurement of serum urea is affected by both renal and hepatic function. Similarly, creatinine levels vary with age and muscle bulk. Creatinine clearance may give a good assessment of glomerular filtration rate, allowing prediction of the secretion of some drugs such as the aminoglycosides. However, other drugs, such as the penicillins, are secreted by tubular mechanisms. Their excretion may be preserved until the creatinine clearance is very low.

Altered receptor dynamics

The patient with liver disease may have alterations in the pharmacological effects of drugs in addition to, or independent of, changes in their pharmacokinetics, for example increased central nervous system sensitivity to opioids and other sedatives. These differences may result from increased numbers of GABA receptors. With chronic alcohol use, a tolerance to sedative agents may be explained by the reduction in mRNA that makes the GABA receptor (Mhatre and Ticku 1992).

While measures of the clinical effect of some drugs such as the catecholamines may be possible, for others it may not be so easy. Estimation of plasma concentrations in this latter group may also be of limited value because of altered pharmacodynamics.

Prescribing in liver failure

The use of drugs always carries some risks, but there are certain groups which should either be avoided or used with caution in liver disease. These include the following:

(1) drugs capable of causing hepatic damage (e.g. acetylsalicylic acid, methotrexate, chlorpromazine, methyldopa);

(2) drugs which can worsen liver function (e.g. contraceptive steroids, tetracycline, prednisone (in acute viral hepatitis));

(3) drugs which make complications of liver disease worse (e.g. cyclo-oxygenase inhibitors may increase fluid retention,

Table 2 Guidelines for adjusting the dose of drugs in liver disease

Extent of change in drug dose	Conditions or requirements
No or minor change	Mild liver disease Extensive renal excretion and normal renal function Metabolism by pathways spared by liver disease Drug metabolism is enzyme limited Flow-sensitive metabolism drug given IV No change in drug sensitivity
Decrease up to 25%	Elimination by liver less than 40% and normal renal function Flow-limited drug given IV with little change in protein binding Drug has large therapeutic ratio
Decrease more than 25%	Drug metabolism greatly affected by liver disease Narrow therapeutic range Significant change in protein binding Drug excreted by kidney and poor renal function Increased drug sensitivity

IV, intravenous.
Reproduced with permission from Wedlund and Branch (1983).

sedatives may worsen hepatic encephalopathy, and anticoagulants may worsen clotting).

There is wide variation in drug pharmacokinetics and pharmacodynamics in hepatic disease. Some suggestions for starting drugs are given in Table 2.

Although clinical experience and data on average clearances according to the degree of hepatic function derangement may provide an estimate for the first dose, subsequent adjustments in dose are often necessary. The best advice is as follows:

(1) use caution when giving a dose of a drug;

(2) use boluses rather than infusions whenever appropriate;

(3) follow blood concentration in the few drugs where it is useful;

(4) look for signs of drug toxicity.

Most drugs should be given intravenously, as this is both reliable and rapid. The oral route is best avoided because of the frequent ileus and reduced gut absorption that accompanies liver disease. Intramuscular and subcutaneous routes should be avoided because of unpredictable blood flow and the risks associated with coagulopathy. Other suitable routes include rectal (avoids first-pass metabolism) and into the airways. While giving bolus doses titrated to the patient's needs avoids accumulation, it may be associated with high and low levels of the drug, risking toxicity and inadequate treatment. Infusions of short-acting drugs may avoid these problems. However, many of these agents end their effects by redistribution (e.g. chlormethiazole), and once these sites are full, such drugs may change to become long-acting agents.

References

Mhatre, M.C. and Ticku, M.K. (1992). Chronic ethanol administration alters γ-aminobutyric $acid_A$ receptor gene expression. *Molecular Pharmacology*, **42**, 415–22.

Park, G.R. (1996). Molecular mechanisms of drug metabolism in the critically ill. *British Journal of Anaesthesia*, **77**, 32–49.

Wedlund, P.J. and Branch, R.A. (1983). Adjustments of medications in liver failure. In *The pharmacologic approach to the critically ill patient* (ed. B. Chernow), pp. 84–114. Williams & Wilkins, Baltimore, MD.

5 Nutrition

5.1 Normal physiology

5.1.1 Nutrition

K. N. Jeejeebhoy

Key messages

- Nutrients can be divided into macronutrients, such as proteins, carbohydrates, and fats, which are consumed in gram quantities, and micronutrients, which are consumed in microgram or milligram quantities but are essential for the metabolism of macronutrients.
- Nitrogen retention increases when either nitrogen or energy intake is increased.
- Energy is derived from all three macronutrients, namely proteins, carbohydrates, and fats.
- The metabolic rate is increased within an hour of eating or infusing nutrients.
- Trace elements and vitamins are essential for regulating metabolic processes as constituents of enzyme complexes required for the utilization of carbohydrates, proteins, and fats.

Introduction

Nutrition encompasses all processes which relate the intake of food to the structure and function of a living organism. It is concerned with the requirements of food and its metabolism. Maintenance of body composition and function depend upon adequate intake and metabolism of several nutrients. These nutrients interact and depend on each other for their action (Jeejeebhoy 1993; Shils *et al.* 1994). Nutrients can be divided into macronutrients, such as proteins, carbohydrates, and fats, which are consumed in gram quantities, and micronutrients, which are consumed in microgram or milligram quantities but are essential for the metabolism of macronutrients.

Macronutrients

Proteins

Proteins make up 16 per cent of body weight, half intracellular and half extracellular. Proteins are composed of 22 amino acids of which nine (histidine, threonine, lysine, phenylalanine, leucine, isoleucine, valine, methionine, and tryptophan) are essential. Normal protein requirement is 0.6 ± 0.15 g/kg/day (mean $\pm$ 2 SD) for adults.

Amino acid metabolism

Amino acids absorbed from the gastrointestinal tract enter the portal circulation and pass through the liver. Glutamine is taken up by the intestine where it is a source of energy. Most of the amino acids, except those with branched chains (leucine, isoleucine, and valine), are removed by the liver. The liver also receives alanine for gluconeogenesis from muscle and the intestine, where glutamine is transaminated to alanine.

The branched-chain amino acids entering the systemic circulation are taken up by muscles and other peripheral tissues for protein synthesis. Some nitrogen is released from muscle in the form of alanine and glutamine. Thus the muscle is a major reservoir of glutamine for immunocytes and the intestine.

Amino acids in excess of requirements are not stored but are used as metabolic fuel. The ultimate disposal of nitrogen from excess amino acids depends upon urea formation.

Obligatory nitrogen losses and nitrogen balance

Fasting nitrogen excretion rates are proportional to metabolic rate. Endogenous nitrogen excretion is about 2 mg/kcal/day. This value, which is called the obligatory nitrogen loss, varies from 41 to 69 mg/kg/day in humans.

Protein–energy interrelationship

In humans, increasing energy intake as either fat or carbohydrate increases nitrogen retention by 4 mg and 2 mg respectively per extra kilocalorie fed. Nitrogen retention increases when either nitrogen or energy intake are increased. The ability of proteins to modulate nitrogen retention independent of other factors is known as anabolic drive (Millward 1990).

Energy

Energy is derived from all three macronutrients, i.e. proteins, carbohydrates, and fats. Transformation to energy occurs in three stages. In the first, macromolecules are hydrolyzed to simpler molecules—proteins to amino acids, carbohydrates to monosaccharides, and fats to fatty acids and glycerol. In the second stage, these simpler molecules are degraded to two-carbon fragments and linked to coenzyme

Table 1 Energy requirements

	Age (years)	REE (kcal/kg/day)	Activity factor	TEE (kcal/kg/day)
Male	11–14	32	1.7	55
	15–25	26.7	1.67	45
	25–50	22.8	1.6	37
	51+	19.8	1.5	30
Female	11–14	28.5	1.67	47
	15–25	24.9	1.6	40
	25–50	21.9	1.55	36
	51+	19.7	1.5	30

REE, resting energy expenditure from the Harris–Benedict equation (see text).
TEE, total energy expenditure.

A as acetyl coenzyme A. In the final stage, acetyl coenzyme A enters the tricarboxylic acid (Krebs or citric acid) cycle and is oxidized to provide energy, which is stored as high-energy phosphates.

Essential fatty acids

Two unsaturated fatty acids, linoleic acid and linolenic acid, are called essential fatty acids because they cannot be synthesized *in vivo* from non-dietary sources. Both these acids have 18 carbon atoms, but the position of the first double bond, called the omega number, is between the sixth and seventh carbons for linoleic acid and between the third and fourth carbons for linolenic acid. Linoleic acid is converted by chain elongation to arachidonic acid, which is the precursor of prostaglandins of the E_2 series (e.g. prostaglandin E_2), while linolenic acid is chain elongated to eicosapentaenoic acid and decosahexanoic acid and is a precursor of prostaglandins of the E_3 series (e.g. prostaglandin E_3).

Energy requirements

Energy intake is dependent on output. Energy output is due to basal metabolism (defined as the energy expenditure 12 h postprandial and soon after awakening in the morning), to physical activity, to the thermogenic effect of eating, and to anabolic processes such as growth, pregnancy, and lactation. In turn these components depend on a number of factors which include the body surface area (derived from height and weight), age, and gender (Table 1).

Resting energy expenditure

Resting energy requirements for normal humans are usually predicted by the Harris–Benedict equation (Harris and Benedict 1919). The resting energy expenditure (REE) is given by

$$\text{males: REE} = 66.473 + (13.7516 \times \text{weight}) + (5.0033 \times \text{height}) - (6.7550 \times \text{age})$$

$$\text{females: REE} = 655.0955 + (9.5634 \times \text{weight}) + (1.8496 \times \text{height}) - (4.6756 \times \text{age})$$

where REE is in kilocalories per 24 h, weight is in kilograms, height is in centimeters, and age is in years.

Table 2 Activity factor and total energy expenditure

Activity factor		Duration	Integrated factor
Resting	1.0	9	9
Light	2.5	10	25
Heavy	7.0	5	35
Total			69

If REE=1500 kcal/day, then TEE=1500 × 69/24=4312 kcal/day

Activity factor

The activity factor depends upon the occupation and leisure activities of the person. It is expressed as a multiple of the REE as given in Table 2. The total energy expenditure per day (TEE) can be estimated by multiplying the activity factor by the number of hours during which it is applicable and summing the values for 24 h.

Thermic effect of feeding

The metabolic rate increases within an hour of eating or infusing nutrients. This effect is higher for glucose (6–8 per cent of intake) and lower for lipids (2–3 per cent of intake). In part, the thermic effect of feeding carbohydrate is due to increased catecholamine secretion which does not occur when lipids are infused. In patients with burns the thermic effect is increased to 23 per cent of the calories fed.

Pregnancy and lactation

Pregnancy increases energy requirements by 300 kcal/day during the second and third trimesters. Lactation increases energy requirements by 500 kcal/day.

Growth

The energy requirement for growth has been computed to be 5 kcal/g tissue. Since the weight gain in adolescence is about 5 kg/year with an increase in height of between 4 and 6 cm/year from the ages of 12.5 to 16.5 years, the excess energy requirement will be 25 000 kcal/year or 5000 kcal/cm. It is clear that any energy deficit will impair growth.

Electrolytes

The importance of fluid and electrolyte replacement for promoting tissue perfusion and ionic equilibrium is self-evident. In addition, it is necessary to give potassium, magnesium, phosphate, and zinc to ensure optimum nitrogen retention.

Sodium and chloride

Foods high in sodium are often high in chloride. Sodium and chloride are the most abundant extracellular ions and are responsible for maintaining the osmolality of the extracellular fluid. In a cold climate balance can be maintained with as little as 5 mmol/day, but this rises to 20 mmol/day in hot weather.

Potassium

Potassium, which is the main intracellular cation, is present in a variety of foods. Traditionally, orange juice and bananas are given as rich sources of potassium. The range of normal dietary intake is determined by the fact that potassium is not as well conserved as sodium and there are significant fecal and renal losses. To maintain body stores and normal plasma levels 40 to 50 mmol/day must be given.

Magnesium

Magnesium is the second most abundant intracellular cation. It is necessary for mitochondrial integrity and stability of nuclear DNA. It is also necessary for control of neuromuscular excitability. The amount of magnesium required to maintain balance in normal subjects has been estimated to vary from 0.125 to 0.187 mmol/kg/day. During pregnancy and lactation the requirements are increased to 0.25 mmol/kg/day.

Calcium

About 98 per cent of the total body calcium is in the skeleton. The intracellular concentration of calcium is extremely low as it is toxic to the cell. The concentration of extracellular calcium plays a very important part in nerve and muscle cell excitability. The calcium requirement is 60 mmol/day during adolescence up to the age of 24 and is reduced to 40 mmol/day for older persons. During pregnancy and lactation 60 mmol/day is considered a prudent intake.

Phosphorus

Phosphorus is the major intracellular anion. In the cell, among other functions, it is a part of buffer systems of energy-storing nucleotides (ATP), membranes (as phospholipids), and oxygen transfer systems in the form of erythrocyte 2,3-diphosphoglycerate. Phosphorus requirement is set at a 1:1 ratio with calcium. Therefore the phosphorus intake in normal subjects and orally fed patients should approximate the calcium intake.

Micronutrients

As discussed above, the major part of human dietary intake comprises water, proteins, carbohydrates, fats, and electrolytes. However, for the utilization of these nutrients it is essential to absorb other substances, called micronutrients, in relatively small and in some instances minute amounts. Trace elements are inorganic elements, whereas vitamins are complex organic compounds. Both are essential because they regulate metabolic processes as constituents of enzyme complexes required for the utilization of carbohydrates, proteins, and fats.

Trace elements

An essential trace element has the following characteristics:

(1) it is present in the healthy tissues of all living things;

(2) its tissue concentration is the same in all animals;

(3) withdrawal leads to a reproducible functional and/or structural abnormality;

(4) addition of the element prevents the abnormality;

(5) the abnormality is associated with a specific biochemical change;

(6) the biochemical change and the observed clinical abnormality are prevented and/or cured by giving the nutrient.

According to these criteria, iron, zinc, copper, chromium, selenium, and cobalt in vitamin B_{12} are essential.

Iron

Iron is an essential constituent of porphyrin-based compounds bound to protein such as hemoglobin and myoglobin. Normal males require about 1 mg of absorbed iron per day and menstruating females require 1.8 mg/day. Since only about 10 to 15 per cent of iron intake is absorbed, men require 10 mg/day and menstruating women 15 mg/day. In pregnancy the iron intake should be augmented by 15 mg/day. Lactation does not require increased iron intake.

Zinc

Zinc has been identified as a component of about 120 enzymes. It is an integral constituent of DNA polymerase, reverse transcriptase, RNA polymerase, tRNA synthetase, and the protein chain elongation factor. In volunteer studies, experimental mild zinc deficiency reduced thymulin levels and the ratio of CD4 to CD8 (Prasad *et al.* 1988). The recommended dietary zinc intake for adults is 15 mg/day. This should be increased by 7 mg/day during the first month of lactation and by 4 mg/day subsequently.

Copper

Copper is widely distributed in human tissues and is a component of enzymes such as cytochrome c oxidase, superoxide dismutase, dopamine β-hydroxylase, monoamine oxidase, and lysyl oxidase. In addition, 90 per cent of the plasma copper is in the form of ceruloplasmin. The normal diet supplies 2 to 4 mg/day, and on this intake deficiency has never been observed in adults.

Chromium

Chromium deficiency in animals has been found to cause a syndrome of glucose intolerance similar to that of clinical diabetes. The abnormalities found were corrected by giving chromium. This element is also important in promoting insulin action in peripheral tissues. Oral chromium requirements in the adult have not been determined.

Selenium

Selenium is interrelated with other antioxidants such as vitamin E. A deficiency of one can be partially corrected by giving the other. It is an integral part of the enzyme glutathione peroxidase which is associated with the control of free radicals. The dietary intake of selenium varies from a low of 0.23 to 0.33 μmol/day in New Zealand to a high of 1.9 to 2.8 μmol/day in Canada. The daily requirement is defined as 0.011 μmol/kg/day. This allowance should be increased by 0.12 μmol/day in pregnant women and by 0.24 μmol/day in lactating women.

Vitamins

Vitamins are essential nutrients which are active in minute quantities.

Fat-soluble vitamins

Vitamin A

Deficiency is caused by poor intake, malabsorption, or liver disease, and leads to night blindness, thickening of the conjunctiva (Bitot's spots), keratomalacia, and hyperkeratosis of the skin. The normal requirement is 1000 IU/day (3300 IU/day in intravenous feeding).

Vitamin D

Deficiency is caused by malabsorption and lack of exposure to sun, and leads to bone pain, fractures, osteomalacia, and rickets. The normal requirement is 200 IU/day.

Vitamin K

Deficiency is caused by severe malabsorption and cephalosporin antibiotics, and leads to bleeding diathesis. The normal requirement is 1 g/kg/day (10 mg/week in intravenous nutrition).

Vitamin E

Deficiency is caused by severe malabsorption and biliary obstruction, and leads to neuropathy and spinocerebellar degeneration. The normal requirement is 10 to 20 mg/day (10 mg/day in intravenous nutrition).

Water-soluble vitamins

Thiamine

Deficiency is caused by poor intake, malabsorption, alcoholism, and intravenous feeding with high glucose intake. Effects include neuropathy, cardiac failure, and severe acidosis without sepsis. The normal requirement is 0.5 to 1 mg/1000 kcal intake (3 mg/day in intravenous nutrition). Assessment of thiamine deficiency is by whole-blood transketolase activity with and without added thiamine pyrophosphate. In deficiency transketolase levels rise by more than 16 per cent with added thiamine pyrophosphate.

Riboflavin

Deficiency is caused by poor intake, malabsorption, and alcoholism. Effects include chelosis and conjuctival injection. The normal requirement is 0.6 mg/1000 kcal intake (3.6 mg/day in intravenous nutrition).

Niacin

Deficiency is caused by poor intake and malabsorption, and leads to the pellagra syndrome. The normal requirement is 6.6 mg of niacin equivalents per 1000 kcal intake (40 mg/day in intravenous nutrition).

Pyridoxine

Deficiency is caused by poor intake and malabsorption, and leads to stomatitis, glossitis, chelosis, depression, and confusion. The normal requirement is 1.6 to 2.0 mg/day or 0.038 mg/g protein intake (4 mg/day in intravenous nutrition).

Biotin

Deficiency is caused by poor intake and malabsorption, and leads to mental changes, myalgia, and dermatitis. The normal requirement is 30 to 100 g/day (4 mg/day in intravenous nutrition).

Vitamin B_{12} and folate

Deficiency is caused by poor intake, drugs, alcoholism (folate), and malabsorption. Vitamin B_{12} deficiency occurs with ileal and gastric pathology. Effects include megaloblastic anemia, glossitis, and stomatitis. Neuropathy or neuromyelopathy (subacute combined degeneration) occur with B_{12} deficiency. The normal requirements are 3 g/kg/day folate and 2 g/day vitamin B_{12}.

Ascorbic acid

Deficiency is caused by poor intake, abnormal diet, and possibly critical illness, and leads to the scurvy syndrome. The normal requirement is 60 to 100 mg/day (100 mg/day in intravenous nutrition).

References

Harris, J.A. and Benedict, F.G. (1919). Standard basal metabolism constants for physiologists and clinicians. In *A biometric study of basal metabolism in man*, pp. 223–50. Publication 279, Carnegie Institute of Washington. J.B. Lippincott, Philadelphia, PA.

Jeejeebhoy, K.N. (1993). Nutrient requirements and nutrient deficiencies in gastrointestinal diseases. In *Gastrointestinal disease* (5th edn) (ed. M.H. Sleisenger and J.S. Fordtran), pp. 2017–47. W.B. Saunders, Philadelphia, PA.

Millward, D.J. (1990). The hormonal control of protein turnover. *Clinical Nutrition*, **9**, 115–26.

Prasad, A.S., *et al.* (1988) Serum thymulin in human zinc deficiency. *Journal of Clinical Investigation*, **82**, 1202–10.

Shils, M.E., Olsen, J.A., and Shike, M. (ed.) (1994). *Modern nutrition in health and disease*. Lea & Febiger, Philadelphia, PA.

5.2 Nutritional failure

5.2.1 The metabolic and nutritional response to critical illness

G. Lavery

Key messages

- The nutritional and metabolic changes which occur in critical illness are driven by any combination of lack of nutrient intake (starvation), the endocrine/mediator response to the 'stress' of critical illness, and physiological insult to the gastrointestinal tract.
- The metabolic response to critical illness is typified by catabolism, protein loss, and immunosuppression.
- Nutritional support of the critically ill patient should supply raw material for protein anabolism, wound healing, and organ repair, supply energy and other substances needed to maintain cellular function, and, if possible, attenuate harmful aspects of the metabolic response.

Effects of starvation

The initial response to starvation is glycogenolysis to maintain blood glucose concentration. However, glycogen stores become exhausted within 2 to 3 days, and an alternative source of glucose is required to fuel tissues such as the brain which generates energy exclusively from glucose. Structural body protein is broken down to provide amino acids as a substrate for gluconeogenesis. This protein breakdown is halted by an adaptive process by which the brain uses ketones as its metabolic fuel. In addition, the body's basal energy requirement is reduced and is supplied mainly by oxidation of lipid stores (fat). Only when the fat stores are almost depleted does protein breakdown and gluconeogenesis again become significant.

Thus the result of prolonged starvation is a reduction in muscle mass (and strength) with a less marked reduction in visceral protein mass. The muscle weakness is accentuated by deficiencies in electrolytes and trace elements (e.g. calcium, magnesium, phosphate, and selenium). These factors have significant effects on diaphragmatic function, skeletal muscle function, and even myocardial performance.

The immune response is also generally impaired by malnutrition, although conventional markers (white cell count, immunoglobulin levels) are usually normal. There is atrophy of the spleen and lymph nodes, and an attenuation of the response to immunological challenge.

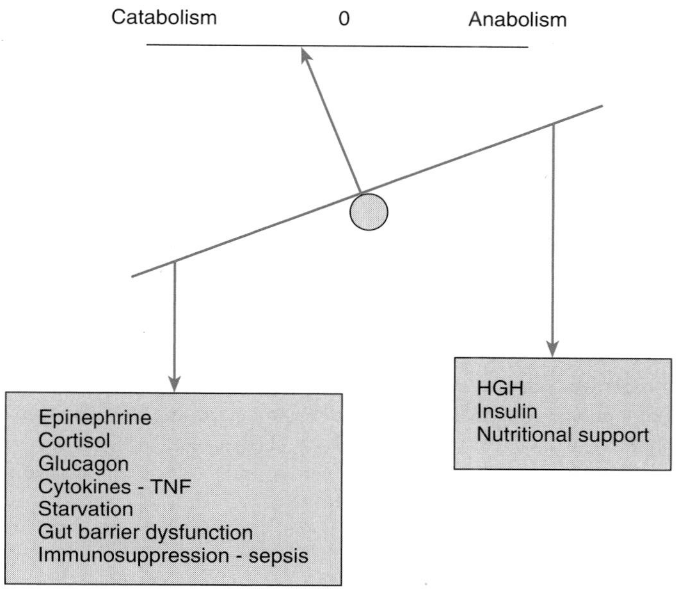

Fig. 1 Nutritional and metabolic imbalance in critical illness: TNF, tumor necrosis factor; HGH, human growth hormone.

Metabolic response to stress

Compared with the starving patient, the critically ill patient has a very different 'internal milieu' and exhibits greater and more prolonged negative nitrogen balance due to protein breakdown (gluconeogenesis). There is no adaptation to energy production from fat stores, and the basal metabolic rate is increased by 20 to 80 per cent. Metabolic and endocrine responses include stimulation of the sympathetic nervous system, release of epinephrine (adrenaline), glucocorticoids, and glucagon, suppression of prolactin, growth hormone, and thyroid hormone release, metabolic resistance to the effects of insulin, production of cytokine mediators, and release of eicosanoid precursors from membrane-bound phospholipids (Fig. 1).

The body's initial response to stress is primarily controlled by the locus ceruleus in the brainstem and the paraventricular nucleus of the hypothalamus (Michelson *et al.* 1994). When stimulated, the former releases norepinephrine (noradrenaline), stimulates the autonomic nervous system, and increases arousal. The latter produces corticotropin-releasing hormone which activates the hypothalamo–pituitary–adrenal axis. The locus ceruleus and corticotropin-releasing hormone systems exhibit mutual feedback in that stimulation of either system will lead to activation of both pathways.

Sympathetic nervous system activity and catecholamine release

Most physiological insults occurring in critical illness stimulate the sympathetic nervous system (e.g. pain, hypotension, hypoperfusion, hypoxia, hypercarbia, acidosis). This causes stimulation of the adrenal medulla by preganglionic sympathetic fibers and within minutes leads to increased plasma epinephrine levels. Norepinephrine levels are also increased since neuronal re-uptake does not keep pace with the increased release of this sympathetic neurotransmitter which therefore enters the plasma. This mechanism complements the release of norepinephrine and the sympathetic activity consequent on stimulation of the locus ceruleus.

The effects of the catecholamines, although complex, are often summarized as the promotion of the fight or flight response. These include stimulant effects on heart rate, cardiac contractility, and blood pressure. Regional blood flow is altered, with increased flow to skeletal muscle and decreased flow to skin and viscera (including the gut and kidneys). Sodium retention is promoted by a direct tubular effect and by the renin–angiotensin mechanism. This is a protective mechanism against actual or potential hemorrhage/hypovolemia. Epinephrine is a major catabolic hormone, promoting glycogenolysis, gluconeogenesis, and lipolysis. This causes an increase in blood glucose which is further compounded since epinephrine promotes glucagon secretion and inhibits the release of insulin.

Adrenal glucocorticoids in critical illness

Corticotropin-releasing hormone from the paraventricular nucleus is carried in the hypophyseal–portal circulation and stimulates release of ACTH from the anterior pituitary. The result is an increase in plasma cortisol levels with a loss of the normal circadian rhythm. During this period of sustained and increased cortisol synthesis there may be a reduction in the production of other adrenal steroid hormones. Conversely, adrenal insufficiency during critical illness is being reported more frequently, particularly in cases of systemic sepsis with shock.

Glucagon in critical illness

Glucagon is a peptide hormone produced in the α cells of the pancreatic islets in response to decreasing plasma glucose, exercise, and protein loading. It stimulates glycogenolysis, gluconeogenesis, and ketone production, and inhibits lipogenesis. Thus its major effect is to raise plasma glucose and promote catabolism.

In health, glucagon is inhibited by hyperglycemia, increased plasma insulin, and free fatty acids. In critical illness, however, glucagon levels are increased despite the presence of the above, and this is thought to be a response to increased catecholamines

Human growth hormone in critical illness

Human growth hormone is a 191-amino-acid protein produced by the anterior pituitary. Its effects can be summarized as a protein-sparing effect which protects lean body mass, a reduction in body fat (lipolysis), improvement in immunocompetence, sodium and water retention, hyperglycemia, and (potentially) increased incidence of neoplasia.

Human growth hormone is stimulated by the hypoglycemic state and also by exercise, sleep, high protein intake, and increased levels of leucine and arginine. Human growth hormone secretion is inhibited by long-term glucocorticoid administration and plasma free fatty acids (plasma free fatty acids are increased by epinephrine and glucagon). It is thought to exert its effects partially by direct action on tissues and partially through stimulation of the liver to produce insulin-like growth factor I, a 70-amino-acid chain.

Baseline human growth hormone is elevated and diurnal variation is reduced in many stressed individuals. However, continuing high levels of stress, as typified by critical illness, may be associated with reduced human growth hormone levels and insulin-like growth factor I is also usually decreased. This attenuates anabolic processes.

Thyroid function in critical illness

Laboratory measurements of thyroid function in critically ill patients are frequently abnormal and are always difficult to interpret. The most common abnormality is low T_3 syndrome. Most patients exhibiting this condition have a normal total thyroxine level and should be considered euthyroid. Low T_3 is probably due to deficient T_4 uptake by the tissues and thus reduced peripheral conversion of T_4 to T_3 (responsible for 80 per cent of total T_3 production). The enzyme which catalyzes the conversion (5′-monodeiodinase) is inhibited by glucocorticoids. In low T_4 and T_3 syndrome those exhibiting the lowest plasma T_4 levels have the highest mortality. Despite normal levels of free hormone, postmortem analysis has shown low tissue T_4 levels in brain, lung, liver, and kidney, but normal T_4 levels in myocardium and skeletal muscle. These abnormalities are commonly grouped together under the term 'sick euthyroid syndrome' (Rolih and Ober 1995).

The role of cytokines

These naturally occurring mediators are produced by monocytes, macrophages, and related cell lines. Cytokine production increases during critical illness and has effects which complement the counter-regulatory hormones (epinephrine, glucagon, and glucocorticoids). The presence of invading Gram-negative bacteria, circulating endotoxin, or widespread tissue injury are all potential triggers. The responses associated with cytokine action are pyrexia, loss of lean body mass, stimulation of acute phase protein synthesis, and changes in vascular tone and permeability.

Increased levels of the cytokine tumor necrosis factor have been found in patients with trauma, burns, and sepsis. It increases neutrophil and monocyte activation and lymphokine production. Fever and amino acid mobilization from skeletal muscle are also due, in part, to tumor necrosis factor. It also encourages vascular proliferation and collagen production and so enhances wound healing. When excessive production occurs, tumor necrosis factor causes excessive muscle wasting and nitrogen loss. It triggers the subsequent release of other cytokines (interleukins 1 and 6) as well as prostaglandins and the counter-regulatory hormones.

Eicosanoids (lipid metabolites)

Specific fatty acids derived from dietary lipids are incorporated into the structure of the phospholipid cell membranes. These phospholipids represent a reservoir of potential metabolic mediators, the nature of which depends on the fatty acids ingested (or infused). Under enzymatic action, the phospholipid membranes release fatty acid molecules which are converted to a number of eicosanoids. The intake of omega-6 fatty acids causes the release of arachidonic acid which produces thromboxane A_2 and prostaglandin E_2 via the cyclo-oxygenase pathway. The former is a potent vasoconstrictor while the latter is an immunosuppressant. The substitution of omega-3 for omega-6 fatty acids in the diet ultimately results in the production of eicosapentanoic acid, thromboxane A_3, and the 3 series prostanoids. Thromboxane A_3 is a less potent vasoconstrictor than A_2, and the 3 series prostanoids are less immunosuppressant than the 2 series.

Prolactin in critical illness

Prolactin, which is secreted by the anterior pituitary to promote lactation, also has an immunoregulatory role. Both T and B lymphocytes have prolactin receptors. The immunosuppressive effect of cyclosporin is probably due, in part, to the fact that it competes for the prolactin binding site on T cells.

Prolactin concentrations increase in acute stress but are normal or reduced in long-term stress such as that associated with critical illness. This may be explained by the prolonged increase in glucocorticoid production which is thought to inhibit prolactin release. This prolactin deficiency state would impair immune function, particularly cell-mediated immunity, although the physiological significance is unclear.

The effect of critical illness on the gut

The mucosal cells of the gastrointestinal tract have one of the highest turnover rates of any body tissue. Endothelial renewal depends on the division and migration of stem cells within the mucosal crypts. Therefore an intact gut mucosa depends on a balance between cell renewal and exfoliation. The intact mucosal layer, tight junctions between cells, lymphocytes, macrophages, and neutrophils in the submucosa and Peyer's patches, and gut-generated IgA all contribute to barrier function of the gut (Thompson 1995).

A fall in perfusion and tissue oxygenation, which occurs in many forms of critical illness, is a significant insult to the gut. Splanchnic hypoperfusion may persist after apparently adequate fluid resuscitation, and even short periods of circulatory compromise may result in prolonged gut ischemia/hypoxia. This may cause cell injury, necrosis, and loss of mucosal integrity, a state which may be exacerbated by coexisting malnutrition, bacterial overgrowth of the gut, and (possibly) reperfusion injury following periods of hypovolemia and hypotension.

Bacterial translocation (the migration of viable bacteria across the intestinal barrier to the liver, spleen, or mesenteric lymph nodes) and endotoxin translocation may occur following loss of barrier function. When such processes occur to a limited extent, which may be an everyday event, the Kupffer cells in the liver prevent spill-over into the systemic circulation. When there is a major deficiency in gut barrier function, systemic spill-over occurs because (i) the liver is overwhelmed by the amount of bacteria and/or endotoxin presented to it and (ii) often the cause of gut barrier failure (e.g. hypoperfusion) will also induce hepatic dysfunction, preventing efficient phagocytosis of bacteria and removal of endotoxin in portal blood.

Organisms which breech the mucosal element of the gut barrier may be phagocytosed by macrophages and transported to the mesenteric lymph nodes. Again, if the microbial load is excessive, these micro-organisms may reach the circulation via the lymph system. The presence of circulating endotoxin has itself been found to increase intestinal permeability in healthy humans.

Bacterial translocation has been implicated in the pathogenesis of multiple organ dysfunction in animal models of critical illness, but the clinical significance in humans is less clear. The overgrowth of pathogens in the upper gastrointestinal tract of critically ill patients may be followed by multiple organ failure in the majority. However, other work suggests no significant portal bacteremia after major trauma and no correlation between increased intestinal permeability and infective complications.

Implications for nutritional support in critical illness

Our current incomplete view of the response to critical illness is only of value if we use it constructively. When attempting to optimize nutritional support we should consider the following.

Structural changes in the gut might promote translocation

Whenever possible the enteral route should be used for nutritional support. Early enteral nutrition helps to preserve intestinal mucosal mass, barrier function, and (possibly) intestinal immune function.

Hypermetabolism, increased catabolism, and nitrogen loss

It would appear that, at least after trauma, optimally resuscitated patients have an increased metabolic rate within the first 24 h. The practice of supplying excess calories to avoid underfeeding has been shown to increase mortality in septic rats.

The tendency to hyperglycemia and glucose intolerance

High glucose intake is associated with hepatic lipogenesis and dysfunction. The high circulating levels of epinephrine, glucagon, and cortisol and the resistance to insulin mean that the critically ill patient is unlikely to tolerate a glucose load without significant hyperglycemia and its associated problems.

Use of nutrients for non-nutritional effects

Some substances have metabolic benefits which go beyond their nutritional value (Minard and Janu 1996). The amino acid glutamine is an important nutrient for enterocytes and prevents mucosal atrophy in experimental models of starvation. When administered enterally glutamine reduces bacterial translocation in rats, and when administered parenterally it appears to enhance gut immune function. Glutamine is also a natural building block for the reducing agent glutathione, and this may increase protection against cellular injury in ischemia–reperfusion.

Arginine is another conditionally essential amino acid involved in the process of protein synthesis, the production of nitric oxide (which has beneficial effects on gut microvascular responses to injury), and the secretion of human growth hormone and insulin. It stimulates immune function generally and has been associated with improved bactericidal activity and wound healing.

Nutritional benefit may also spring from manipulation of metabolic pathways. Utilization of omega-6 fatty acids (common in Western diet) results in the production of dienoic prostaglandins such as prostaglandin E_2 which has an immunosuppressive effect due to its effect on T cells. In contrast, if omega-3 fatty acids (found in abundant amounts in fish oil) are substituted, prostaglandin synthesis is switched from dienoic to trienoic pathways, thus reducing prostaglandin E_2 levels.

Nucleotides are the structural subunits for DNA and RNA synthesis. Dietary nucleotides are broken down in the small bowel lumen into nucleosides and purine or pyrimidine bases. These may be absorbed by the enterocytes. During metabolic stress, with rapid cell turnover, intestinal mucosal health may become dependent on, and limited by, the supply of nucleotides. Thus nucleotide supplementation may maintain or improve gut barrier function and therefore reduce septic complications.

Correction of specific nutrient and hormone deficiencies

In 1961, the administration of pituitary gland extract to a group of severely burned patients was found to reduce urinary nitrogen excretion and improve other nutritional parameters. By the 1980s, recombinant human growth hormone was available and had been shown to improve nitrogen balance, even at low caloric intakes in normal subjects. A similar improvement in nitrogen balance associated with the use of human growth hormone was demonstrated in patients receiving total parenteral nutrition after major gastrointestinal surgery. When administered to children with burns, human growth hormone reduces the healing time of skin donor sites and total time in hospital. Human growth hormone increases plasma levels of insulin-like growth factor I. Indeed, the benefit of human growth hormone administration is partly due to the action of insulin-like growth factor I which, given alone, has been shown to improve gut mucosal atrophy and reduce intestinal permeability in septic animals. However, questions about optimum dose, required duration of treatment, other effects, and indications for the use of human growth hormone and insulin-like growth factor I remain. They do not yet constitute therapies of proven benefit and so should not be used routinely.

Other factors may influence the metabolic and nutritional response. Since dopamine appears to depress pituitary function and thus inhibits prolactin and human growth hormone release (van den Berghe and de Zegher 1996), it may be prudent to avoid its use, at least in severe sepsis. Gut barrier function appears to be important in the overall response, and so maintenance of gastric acidity, aggressive cardiovascular resuscitation, and monitoring/optimization of splanchnic blood flow may be helpful. The presence of bile in the gut is important, since bile salts bind endotoxin in the gastrointestinal tract. Other strategies to prevent or attenuate the cytokine cascade have been less successful, although antibodies to endotoxin, tumor necrosis factor, and receptor antagonists have been (or are at present) under investigation.

References

Michelson, D.G., Gold, P.W., and Sternberg, E.M. (1994). The stress response in critical illness. *New Horizons*, **2**, 426–31.

Minard, G. and Janu, P. (1996). Status and clinical utility of pharmaconutrients. *Current Opinion in Critical Care*, **2**, 253–9.

Rolih, C.A. and Ober, K.P. (1995). The endocrine response to critical illness. *Medical Clinics of North America*, **79**, 211–24.

Thompson, J.S. (1995). The intestinal response to critical illness. *American Journal of Gastroenterology*, **90**, 190–200.

van den Berghe, G. and de Zegher, F. (1996). Anterior pituitary function during critical illness and dopamine treatment. *Critical Care Medicine*, **24**, 1580–90.

5.2.2 Pathophysiology of nutritional failure

S. M. Gabe and G. K. Grimble

Key messages

- After admission to the intensive care unit, feeding may be delayed for three common reasons: feed not prescribed, poor understanding of the urgency of nutritional support, and late prescription of total parenteral nutrition when enteral nutrition is poorly tolerated.
- Gastrointestinal motility is often impaired in critically ill patients receiving mechanical ventilation, and this is most marked in the stomach.
- True ileus is uncommon in the critically ill and is often stated to be present when there is simply gastric stasis.
- In addition to its effects on body composition, malnutrition is associated with impaired respiratory function, cardiac function, immune responsiveness, wound healing, and mobilization.

What is nutritional failure and can we recognize it?

Nutritional failure results when a patient's nutrient intake falls short of metabolic requirements. If prolonged, this will lead to malnutrition which is associated with an increased morbidity and mortality. Signs of nutritional failure are insidious, and the diagnosis is often only made when malnutrition has become established.

In malnutrition, as with any disorder, survival depends upon the maintenance of vital functions. Table 1 lists those which may be adversely affected by malnutrition.

Preventing the development of malnutrition during the course of critical illness is a major therapeutic goal in the intensive care unit (ICU). Whether this can be achieved by conventional nutritional therapies is still debatable as some studies suggest that the side-effects of current nutritional therapies may outweigh the possible benefits.

Table 1 Vital functions endangered in malnutrition

Vital function	Effect of malnutrition	Outcome
Vital organ fueling	Maintained (at the expense of muscle breakdown)	Muscle wasting and fat depletion
Respiratory function	1. Reduced vital capacity, respiratory rate, and minute volume	1. Increased risk of respiratory infection
	2. Difficulty in weaning from ventilator support	2. Prolonged ventilation
Cardiac function	Reduced cardiac output, myocardial contractility, and compliance	Decreased tissue perfusion
Tissue oxygenation	Decreased from anemia	Tissue hypoxia
Immune defense	Reduced host resistance	Immunodeficiency
Body temperature maintenance	Disturbed	Deficient thermoregulation
Water–electrolyte balance	Susceptibility to salt and water overload	Edema
Wound healing	Increased wound dehiscence, reduced tensile strength of skin	Breakdown anastomosis
Clearance of debris, free radicals, toxins	Decreased	Tissue damage, ? multiple organ failure

Mechanisms of nutritional failure in the critically ill

There are a number of factors which may limit the achievement of prescribed nutritional intake. This has been elegantly defined in an unusual study by Byers *et al.* (1995) who investigated the predictive factors for patients who were unable to take a regular oral diet for 5 days. Scoring systems, clinical criteria, and nutritional indices were all considered. The initial clinical decision to admit the patient to the ICU or ward had the highest impact on time taken to initiate normal oral feeding (17 days for ICU patients versus 3 days for ward patients). In descending order, the positive risk factors were the need for neurosurgery, endotracheal intubation, pelvic fracture, laparotomy, and emergency surgery. Surprisingly, body mass index had little effect. These results are obvious in one sense but do not give reasons why a patient requiring laparotomy, for example, should take 12 days to tolerate a normal diet. Clearly, the severity of the disease process is important in delaying normal feeding, but iatrogenic factors may also be significant.

Iatrogenic causes

Failure to initiate feeding

Before initiating nutritional support, the patient must be hemodynamically stable and adequately oxygenated. However, even after the patient has been stabilized, feeding may be delayed for a further 12 to 48 h for three common reasons:

(1) feed not prescribed;

(2) poor understanding of the urgency of nutritional support;

(3) late prescription of total parenteral nutrition when enteral nutrition is poorly tolerated.

Failure to feed patients in the first few days in the ICU will be most harmful to the already malnourished patient whose nutrition is of paramount importance. The consequences may be much less marked in the well-nourished septic patient.

Insufficient daily delivery of enteral nutrients

If fluid must be restricted, the first source of fluid input to be decreased or stopped is often the feed. Furthermore, enteral feed is often prescribed by volume (or calories), and regular cessation for drug delivery or routine tube aspiration, for example, results in underdelivery that is rarely corrected.

Drug-related causes

Opiates and sedatives cause gastric stasis by decreasing smooth muscle contraction and may further decrease absorption by causing both biliary spasm and reduced gastric secretions.

Disease-related causes

Gastric stasis

Gastrointestinal motility, particularly in the stomach, is often impaired in critically ill patients receiving mechanical ventilation (Dive *et al.* 1994). As many as 60 per cent of critically ill patients develop delayed or absent gastric motility as a result of drugs (particularly opioids and sedatives), sepsis, or their underlying disease.

Ileus

With a true ileus it is impossible to feed enterally. However, true ileus is uncommon in the critically ill and is often stated to be present when there is simply gastric stasis. An ileus will occur if there is peritonitis—an indication for surgery.

Metabolic causes

Exogenous glucose utilization is impaired in trauma, sepsis, and multiple organ failure, indicating that in these situations there is a derangement of cellular metabolism.

Summary

Thus there are several reasons why inadequate nutrient metabolism may occur in the critically ill patient, more than one of which may operate in a given patient. Iatrogenic causes are all too frequent and protocols should be established to prevent unnecessary delays in feeding.

Feeding the septic patient

Nutritional support is a cornerstone in the management of septic patients. However, well-controlled studies demonstrating this in this patient group are conspicuous by their absence. Despite this, there have been considerable efforts to determine what constitutes adequate and optimal nutritional support in septic patients.

Septic patients have been thought to have a resting energy expenditure of at least 1.5 times to twice their basal requirement in health. Protein and amino acid requirements should be 1 to 1.5 g/kg/day, corresponding to 15 to 20 per cent of the total caloric supply. The question of the best energy substrate in septic patients is controversial. However, in clinical practice both carbohydrates and lipids are usually administered. Glucose can be recommended as the main source of non-protein energy, while 20 to 40 per cent of the total energy supply should be provided as lipids.

The route of feeding is also controversial. Currently there is growing enthusiasm for early enteral feeding in all critically ill patients. This is endorsed by a reluctance to start total parenteral nutrition because of its potential complications and deleterious effects in sepsis. However, these views do not stand up to close scrutiny as there is still little evidence for the superiority of enteral over parenteral nutrition. Cerra *et al.* (1988) studied septic patients at risk of multiple organ failure and death, randomizing them to receive enteral nutrition or total parenteral nutrition. There was no difference in the incidence of multiple organ failure or mortality between the two groups.

There are several reasons for preferentially choosing the enteral over the parenteral route. The most important reason is to slow intestinal mucosal atrophy and help preserve the integrity of the mucosal barrier. The atrophy that accompanies prolonged bowel rest is associated with difficulty in refeeding and may be associated with bacterial translocation. Another factor favoring enteral feeding is that nutrition supplied enterally appears to be better utilized than equivalent substances provided by total parenteral nutrition (Brunetto *et al.* 1990). However, it is expected that enteral feeding will be poorly tolerated in 15 per cent of treated patients, necessitating the use of total parenteral nutrition.

Novel substrates — future or present?

Within the past 10 years there has been an increasing awareness that the response to trauma or sepsis can be changed by modifying the diet with specific nutrients. This strategy is additional to meeting macronutrient and micronutrient requirements which are raised in critical illness. These nutrients have been classed as 'novel substrates', or even 'nutraceuticals', and represent an attempt to achieve one of the the following:

(1) correct metabolic disturbances;

(2) modulate the immune system and the trauma response;

(3) preserve intestinal barrier function;

(4) maintain antioxidant defenses.

The underlying disturbance of trauma or sepsis is viewed as a metabolic 'set' which is driven by catabolic hormone and cytokine release. This results in tissue catabolism and mobilization of amino acids from skeletal muscle protein, supplying amino acids for acute-phase protein synthesis and gluconeogenisis (e.g. brain and kidney metabolism) and glutamine for the maintenance of the intestine and immune system functions. In addition, other substrates derived from glutamine (e.g. glutathione) are necessary for maintaining antioxidant defenses. The intestine is viewed as an 'infected wound' whose barrier function, once impaired, will lead to movement of bacteria across the mucosa and into the systemic circulation. Mobilization of lipid provides precursors for prostaglandin and leukotriene synthesis, thus mediating the inflammatory response, and the intensity of this may be modulated by varying the ratio of n-3 to n-6 polyunsaturated fatty acids in the lipid component of the diet.

Glutamine

Glutamine comprises over one-third of the amino acids present free in tissues and the severity of illness is associated with significant losses from muscle and blood. Cells with rapid turnover (e.g. immune system, mucosa) avidly metabolize glutamine. The gut converts it to CO_2, ammonia, and citrulline, which are precursors for liver urea synthesis, in proportion to the arterial concentration. The absence of glutamine in total parenteral nutrition solutions may lead to reduced cellularity in the intestine. There is also a growing suspicion that the ability of immune cells to respond to antigen challenge is impaired by low blood glutamine concentrations (e.g. 100–500 mol/l).

Ornithine-ketoglutarate

Ornithine-ketoglutarate will produce glutamine through metabolic conversion, and will also stimulate growth hormone secretion. It shares this property with glutamine and in addition will correct the underlying metabolic acidosis of trauma metabolism (Grimble 1993). Two trials have shown that the metabolic benefits of supplementation of total parenteral nutrition with ornithine-ketoglutarate or glutamine are equivalent (Grimble 1993).

Arginine

Arginine has many properties which play a central role in ammonia/urea metabolism and an ability to stimulate immune function. Its metabolite nitric oxide (NO) is able to control vascular tone and gut permeability.

Nucleotides

Nucleotides are absent from total parenteral nutrition regimens and enteral formulas even though the gut has limited ability to synthesize them for RNA/DNA synthesis *de novo*. Therefore it is possible that supplementation may help to maintain intestinal function (Grimble 1993).

The clinical consequences of nutritional failure

The obvious effects of malnutrition are reduction in weight, loss of fat, and progressive muscle wasting. In addition to its effects on body composition, malnutrition is associated with impaired respiratory function, cardiac function, immune responsiveness, wound healing, and mobilization (Table 1). Severely malnourished patients who require surgery are also at increased risk of postoperative complications and consequently have a longer hospital stay. Therefore one should not be too surprised that malnutrition can lead to delayed recovery from illness or injury in the hospital patient.

Infection

The relationship between nutrition and the immune system is complex since it is difficult to separate the effects of an illness from the immunodeficiency state. Often a vicious circle operates in the acute severely ill, in whom nutritional deficiency will impair the immune response, reducing protection against infection. The resultant infection can itself subsequently precipitate malnutrition. In general, severe protein deficiency has the most profound effect on cell-mediated immunity and a less pronounced effect on humoral immunity.

Cell-mediated immunity

In humans and animals severe protein deficiency impairs lymphocyte transformation in response to agents that stimulate T-cell proliferation *in vitro*, and the production of interferon by stimulated lymphocytes is reduced. Cutaneous delayed hypersensitivity reactions are consistently impaired in hospital patients with malnutrition. Although this has been used as a marker of protein energy malnutrition, suppression of skin tests is not specific as it occurs in other types of illness including acute infections. In general, the abnormalities of cell-mediated immune function patients are reversible with refeeding. Skin test responsiveness improves in malnourished children who receive a normal diet and circulating levels of lymphocyte subtypes rapidly return to normal.

Humoral immunity

Humoral immunity is less sensitive than cellular immunity to the effects of protein energy malnutrition. However, despite a normal or increased serum immunoglobulin concentration, the immunoglobulin turnover in response to an antigenic challenge may be impaired. Similarly, levels of secretory IgA in many external secretions are reduced in undernourished individuals despite a normal serum antibody response. These significant humoral abnormalities may contribute to the development of mucosal infections in malnourished patients.

Respiratory complications

The importance of nutrition for good respiratory function has been recognized since the early 1900s. Protein energy malnutrition leads to abnormalities in the control of ventilation and respiratory muscle function in addition to pulmonary structural changes and deterioration of pulmonary defense mechanisms. Respiratory muscles, like other skeletal muscles, are subject to catabolic changes during critical illness, causing a significant loss in muscle mass. Furthermore, the impairment in respiratory muscle function exceeds the loss of muscle mass.

In addition to impairment of respiratory muscle function protein energy malnutrition can affect the pulmonary structure. Malnutrition may also predispose to pulmonary infection as alveolar defence mechanisms are impaired due to reduced numbers of alveolar macrophages, with inhibited function and decreased IgA secretion.

Thus, in critically ill patients with lung disease, underfeeding may perpetuate respiratory muscle weakness and predispose to pulmonary infection. Conversely, overfeeding results in excessive CO_2 production, leading to an increase in the workload required to achieve steady state ventilation.

Cardiac complications

In the early part of this century it was believed that the heart was spared during starvation, a view which was widely held until the mid-1940s. It is now known that undernutrition is associated with decreases in heart rate, arterial and venous pressures, stroke volume, and cardiac output. Electrocardiographic changes included sinus bradycardia, increased Q–T interval, and diffusely decreased voltage.

Conversely, refeeding malnourished patients can precipitate congestive cardiac failure through several mechanisms:

(1) expansion of ventricular volume and increase in cardiac output may outpace the increase in ventricular mass;

(2) an increase in metabolic rate;

(3) excessive sodium retention;

(4) relative deficiencies in vitamins or minerals;

(5) residual depression of myocardial contractility.

Therefore caution and patience should guide the refeeding of severely undernourished patients and their cardiac status should be evaluated.

Morbidity and mortality

It has been repeatedly demonstrated that increased morbidity and mortality occur in surgical patients who are anergic to a delayed cutaneous hypersensitivity test.

Simple correction of undernutrition in surgical patients is certainly a very attractive concept, as the treatment is simple and basic. However, it has proved difficult to verify this even in the best of controlled studies because of other confounding factors. Although many such studies have been performed (over 50 to date), most can be criticized on the grounds of the trial design, the trial size, or the adequacy of the nutritional regimen used.

References

Brunetto, A.L., Pearson, A.D., Gibson, R., Bateman, D.N., Rashid, M.U., and Laker, M.F. (1990). The effect of pharmacological modification of gastric emptying and mouth-to-caecum transit time on the absorption of sugar probe marker molecules of intestinal permeability in normal man. *European Journal of Clinical Investigation*, **20**, 279–84.

Byers, P.M., *et al.* (1995). The need for aggressive nutritional intervention in the injured patient: the development of a predictive model. *Journal of Trauma*, **39**, 1103–9.

Cerra, F.B., McPherson, J.P., Konstantinides, F.N., Konstantinides, K.M., and Teasley, K.M. (1988). Enteral nutrition does not prevent multiple organ failure syndrome (MOFS) after sepsis. *Surgery*, **104**, 727–33.

Dive, A., Moulart, M., Jonard, P., Jamart, J., and Mahieu, P. (1994). Gastroduodenal motility in mechanically ventilated critically ill patients: a manometric study. *Critical Care Medicine*, **22**, 441–7.

Grimble, G.K. (1993). Essential and conditionally-essential nutrients in clinical nutrition. *Nutrition Research Reviews*, **6**, 97–119.

5.2.3 Assessing nutritional status

J.-M. Boles

Key messages

- All the traditional markers of malnutrition lose their specificity in the sick adult as each may be affected by a number of non-nutritional factors.
- Nutritional assessment is required for patients presenting with clinical evidence of malnutrition, patients with chronic diseases, such as cancer, alcoholism, liver disease, and kidney disease, patients with acute conditions accompanied by high catabolic rates, and geriatric patients.
- The initial nutritional status and the extent of the disease-related catabolic process are the main risk factors for nutritionally related complications. Subjective global assessment including careful recording of patient history, physical examination, and weight measures allow assessment and classification of nutritional status.
- Muscle function evaluated by hand-grip strength and serum albumin, which is the most useful biological test, provide an objective measure of risk assessment. Calculation of a nutritional index according to the patient's disease is helpful for determining a percentage risk of complications and thus the need for nutritional support.
- A high index of suspicion remains the best way of uncovering potentially harmful nutritional deficits.

Introduction

A normal nutritional status is a key element in the ability of a patient to overcome a critical illness. Normal body composition and function are maintained in adults by a daily diet providing nutrients meeting the needs of the individual. Malnutrition, the result of extended inadequate intake on body composition and function, is responsible *per se* for increased morbidity and mortality. Thus assessment of nutritional status must be performed in critically ill patients. All the traditional markers of malnutrition lose their specificity in the sick adult as each may be affected by a number of non-nutritional factors. Therefore interpretation of the results must be cautious.

Why assess nutritional status?

Nutrition and disease interact in several ways. One or combinations of at least two of the following processes are almost always present in critically ill patients: decreased nutrient intake, increased body requirements, and altered nutrient utilization. The frequency of malnutrition in hospital in-patients has been estimated to be between 30 and 50 per cent of both medical and surgical patients. There is an established relationship between initial nutritional status and in-hospital morbidity and mortality. Many complications are related to protein energy malnutrition: increased nosocomial infection rates due to diminished immune competence, delayed wound healing due to decreased ability to repair tissue, delayed weaning from mechanical ventilation due to altered vital functions, and frequent depression and psychological disturbances. Patients with protein energy malnutrition have two- to fourfold higher mortality. Moreover, adequate nutritional support has been shown to improve outcome and reduce the duration of hospital stay in different subsets of malnourished patients.

Therefore assessing nutritional status pursues several goals: determination of nutritional deficiencies and evaluation of risk factors of nutrition-related complications that could affect the outcome of the patient; evaluation of the need and potential value of nutritional support; monitoring the efficacy of nutritional support and therapeutic response.

Which patients should be assessed?

Obviously, patients with apparently normal physical build, normal diet intake, and no reason for significant increased nutrient requirements need no further investigation. However, the following types of patient require a more precise assessment.

1. Patients presenting with clinical evidence of malnutrition (marasmus or the hypoalbuminemic form of protein energy malnutrition or a mixed form).
2. Patients with chronic diseases such as cancer, alcoholism, liver disease, and kidney disease, particularly those undergoing aggressive treatment, such as chemotherapy, which impairs nutrient absorption and/or utilization.
3. Patients with acute conditions accompanied by high catabolic rates, such as severe sepsis, trauma, and emergency surgery.
4. Geriatric patients: undernutrition is more frequent in older persons because aging is associated with a physiological anorexia, as well as other causes such as poor dentition, economic problems, and chronic illness. Nutrition-mediated complications are more frequent in these patients.

How can nutritional status be assessed in current practice?

Nutritional assessment should include assessment of body composition, the presence and duration of inadequate nutrient intake, and the degree and duration of metabolic stress. The main markers of nutritional assessment in healthy adults are shown in Table 1. All the current criteria for objective evidence of malnutrition are non-specifically affected by many diseases and are subject to wide errors;

Table 1 Indices of nutritional assessment

Anthropometric measurements
Weight and related formulas: percentage of normal weight, percentage of usual weight
Mid-arm circumference (MAC)
Triceps skinfold thickness (TST)
Mid-arm muscle circumference: MAMC=MAC – (TST × 3.14)
Biological tests
Plasma proteins
Albumin (half-life, 18–21 days)
Transferrin (half-life, 8–10 days)
Thyroxin-binding pre-albumin (half-life, 2 days)
Retinol-binding protein (half-life, 12 h)
Urinary
Creatinine height index
3-Methyl-histidine 24-h excretion
Muscle function testing
Hand-grip strength
Body composition
Bioelectrical impedance method

also, disease and inactivity alone can result in the same effects as malnutrition. Sophisticated methods of measuring body composition have been developed, such as multiple isotope dilution methods, dual-photon absorption, and γ-neutron activation. Because of technical complexity, scientific limitations, and high cost, none of these methods is of clinical utility in routine critical care. In current practice, a comprehensive assessment of nutritional status relies on a step-by-step clinically based approach and cautious interpretation of measurements and results.

Clinical assessment

Recording the patient's history and physical examination, including weight measurement, must remain the first steps of standard nutritional assessment.

History

The history includes dietary habits, nutrient intake, and interference between nutrition and the disease process itself. The latter may be responsible for either inadequate intake (because of anorexia, nausea, or dysphagia for example) or excessive losses (because of vomiting, diarrhea, or malabsorption for example).

Physical examination

Signs of nutritional deficiency, such as muscle wasting, loss of subcutaneous fat, skin rashes, hair thinning, edema, ascites, fingernail abnormalities such as koilonychia, glossitis, and other mucosal lesions, are sought. Apathy and lethargy can also be attributed to malnutrition. Particular signs of specific nutrient deficiencies must also be noticed.

Estimation of weight loss

A loss of 10 per cent of the usual body weight over a 6-month period or 5 per cent over a 1-month period are indicative of a compromised nutritional status. However, the sensitivity of correctly predicting actual weight loss and the predictive power are both less than 75 per cent. Weight and weight variations do not reflect nutritional status or nutritional support efficacy when edema or dehydration, whatever the cause, are or were present.

Other anthropometric measurements

Anthropometric measurements sufficient to define the nutritional status in healthy individuals may be affected by non-nutritional factors. For example, bedridden patients will lose muscle mass without malnutrition. Therefore anthropometric measurements must be interpreted with care.

Measurements include mid-arm circumference and triceps skinfold thickness measured with a skin caliper. Mid-arm muscle circumference, which is calculated from the preceding two measures, reflects skeletal muscle. Triceps skinfold thickness reflects fat stores. High coefficients of variation between observers suggest that they should always be recorded by the same observer. Values obtained are compared with those of matched normal controls. Relating the values in any one person to a normal value in order to identify that person as malnourished or normal depends upon several factors, one of which is the range of the normal value. These measurements are of no value in cases of subcutaneous emphysema or generalized edema. Because of slow variations, these measures cannot be used to evaluate nutritional support efficacy.

Functional tests

Functional changes, such as a reduction in muscle power due to reduced nutrient intake, occur long before demonstrable anthropometric changes and are better predictors of complications than other anthropometric measurements. Therefore muscle function can be considered as a specific measure of the effect of withdrawing nutrients and refeeding. Two methods can be used in critically ill patients, according to their state of consciousness.

1. Assessment of hand-grip strength (of the non-dominant side) with a hand-grip dynamometer is reserved for co-operative patients; it has been shown to correlate with arm muscle circumference and to be the most sensitive test for predicting post-operative complications.

2. Measurement of the contraction of the adductor pollicis muscle in response to an electrical stimulus of the ulnar nerve at the wrist can be performed even in unconscious patients. The combination of an abnormal force–frequency curve and a slow relaxation rate is the most specific and sensitive predictor of nutritionally associated complications in preoperative patients.

Plasma proteins

Plasma proteins reflect the visceral protein mass. They include albumin, transferrin, thyroxin-binding pre-albumin, and, in patients with normal kidney function, retinol-binding protein.

Serum albumin level is the most widely used measure of plasma protein in nutritional assessment, although a fall in albumin level is usually the consequence of its metabolism and reflects more the severity and duration of the metabolic stress than the nutritional status itself. Several studies have demonstrated that a serum albumin level below 30 g/l correlates with complications, but careful studies have shown it to be less predictive than clinical judgment. Sensitivity to predicting complications is better when measurements of serum albumin and transferrin are combined.

Although it is dependent on the iron status, transferrin has a better response than albumin to nutritional repletion.

Thyroxin-binding pre-albumin and retinol-binding protein, which have short half-lives, are particularly useful for following the efficacy of nutritional support. Other acute-phase reactants are markers of the injury response and its intensity rather than markers of nutritional status.

Creatinine height index

The daily urinary creatinine excretion is closely correlated with the lean body mass as determined by potassium-40 counting. Averaged over 3 consecutive days, it is matched with normal controls for sex and height. The creatinine height index is a reliable index of muscle mass. It cannot be used in patients with renal failure or rhabdomyolysis.

Immune competence

Cellular immunity is the most sensitive component of malnutrition, but reduced immune competence is not specific to malnutrition, thus making it a poor predictor of such a state in sick patients. The lymphocyte subtype count, which correlates best with undernutrition in marasmus where metabolic stress is absent, cannot be used for that purpose in critically ill patients.

Many factors non-specifically alter delayed cutaneous hypersensitivity in the absence of malnutrition: viral, bacterial, and granulomatous infections; uremia, cirrhosis, hepatitis, trauma, burns, and haemorrhage; drugs such as steroids, immunosuppressants, and cimetidine; general anesthesia and surgery. Although anergy resulting from malnutrition can be reversed by nutritional support, delayed cutaneous hypersensitivity is valueless in assessing the state of nutrition as one or more of the above factors are generally present in critically ill patients. However, simply draining an abscess can reverse anergy. Hence, measuring delayed cutaneous hypersensitivity is not recommended in assessment of nutritional status.

Subjective global assessment

Subjective global assessment (SGA) scores the following six features only based on history and physical examination of the patient.

1. Weight change: loss in past 6 months (zero, < 5 per cent, > 10 per cent); change in past 2 weeks (increase, stable, or decrease (in the case of recent weight gain, previous loss is not considered)).
2. Dietary intake: no change or suboptimal intake, liquid diet, or hypocaloric fluids or starvation.
3. Gastrointestinal symptoms for more than 2 weeks (none, anorexia and nausea, vomiting, diarrhea).
4. Functional capacity (normal, suboptimal work, ambulatory, or bedridden).
5. Stress (none, minimal, or high).
6. Physical signs: loss of subcutaneous fat over the triceps and/or chest; muscle wasting of the deltoids, temporals, and/or quadriceps; fluid retention (edema, ascites); mucosal lesions (glossitis, skin rash) suggestive of deficiency.

The physician may weight one factor or another to accommodate for interactions between the findings. After compilation of data, the patient is classified into one of three classes.

1. Well nourished: no or minimal restriction of food intake and/or absorption with minimal change or improvement in function and body weight.
2. Moderate malnutrition: clear evidence of food restriction with functional changes but little evidence of any changes in body mass.
3. Severe malnutrition: changes in both intake and body mass with poor function.

Controlled studies have shown that SGA gives reproducible results that identify patients at risk of nutritionally mediated complications. In a large study, SGA increased the identification of the likelihood of complications by a factor of 7. But the exact sensitivity of the method varies depending upon the clinical material and the population studied.

Nutritional indices

Several nutritional indices have been developed using mathematical and statistical methods to identify patients at risk of nutritionally mediated complications. These indices, selecting significant nutritional parameters, were designed and generally validated in specific groups of patients, usually cancerous or surgical. Thus their utilization cannot be extended to other types of patient.

The most widely studied is the prognostic nutritional index (PNI) which is calculated from multiple variables:

$$\text{PNI} = 158 - 16.6(\text{albumin}) - 0.78(\text{TST}) - 0.2(\text{serum transferrin}) - 5.8(\text{DCH}).$$

This equation correctly predicts the percentage risk of postoperative complications, particularly in populations with a high incidence of malnutrition. Adequate nutritional support in patients with a high PNI has been shown to improve postoperative outcome.

Bioelectrical impedance analysis

This technique, which is based on the electrical impedance of body tissues, can be performed at the bedside. It provides a reliable estimate of total body water under most conditions. Fat-free mass and body fat can be estimated from equations using values derived from bioelectrical impedance analysis. It can be a useful technique for body composition analysis in healthy individuals and in many medical conditions except those involving major disturbances of water distribution. As such disturbances are frequent in critically ill patients, the use of this technique cannot be recommended in such a setting.

Bibliography

Bozzetti, F. (1987). Nutritional assessment from the perspective of a clinician. *Journal of Parenteral and Enteral Nutrition*, **11** (Supplement), 115S–21S.

Daley, B.J. and Bistrian, B.R. (1994). Nutritional assessment. In *Nutrition in critical care* (ed. G.P. Zaloga), pp. 9–33. Mosby Year Book, St Louis, MO.

Jeejeebhoy, K.N. (1990). Assessment of nutritional status. In *Clinical nutrition: enteral and tube feeding* (2nd edn) (ed. J.L. Rombeau and M.D. Caldwell), pp. 118–26. W.B. Saunders, Philadelphia, PA.

Jeejeebhoy, K.N., Detsky, A.S., and Baker, J.P. (1990). Assessment of nutritional status. *Journal of Parenteral and Enteral Nutrition*, **14** (Supplement), 193S.

Vellas, B.J., Guigoz, Y., Garry, P.J., and Albarede, J.L. (ed.). (1994). *The mininutritional assessment: facts and research in gerontology.* Springer, New York.

5.3 Nutritional support

5.3.1 Enteral nutrition

G. J. Dobb

Key messages

- Whenever possible, enteral nutrition should be used in preference to parenteral nutrition.
- Early enteral nutrition has been shown to improve outcomes in some groups of critically ill patients.
- Enteral feed administration sets must be incompatible with those used for intravenous infusion.
- Many patients can be fed through a transpyloric feeding tube when intragastric feeding has failed.
- The value of newer expensive feeds supplemented with glutamine, arginine, nucleotides, omega-3 fatty acids, and other nutrients has still to be proved.

Introduction

The place of nutritional support, both enteral and parenteral, is controversial. In many settings there is little evidence that patient outcome is improved, and the indications for enteral nutrition are not well defined. Nevertheless, it is clear that nutrients are needed for protein synthesis, for organ function, and to sustain life. Starvation causes delayed healing, impaired immunity, impairment of gut and muscle function with generalized weakness, and eventual death.

The need for intensive care often arises after a period of illness accompanied by poor nutrition and weight loss. Most patients needing intensive care are anorexic or unable to eat. They are also metabolically stressed by the severity of their illness, injury, or major surgery. Without nutritional support there is rapid loss of body weight and muscle mass, and this is not fully reversible even with full nutritional support.

Indications

Enteral nutrition is a routine part of intensive care in all patients who are unable to eat, unless there is either a generalized ileus or access to the gastrointestinal tract is lacking, usually because of previous trauma or surgery and absence of a gastrostomy or jejunostomy. Enteral nutrition is always preferred to parenteral nutrition. In patients with blunt or penetrating trauma, comparisons of enteral and parenteral nutrition show that enteral nutrition within 24 h of injury is well tolerated and associated with a lower frequency of infection. Early nasojejunal feeding is possible even after intestinal resection and primary anastomosis. The aims of nutritional support have been defined (American Society for Parenteral and Enteral Nutrition Board of Directors 1993) as including the following:

(1) detection and correction of pre-existing malnutrition;

(2) prevention of progressive protein energy malnutrition;

(3) optimizing the patient's metabolic state;

(4) reduction of morbidity and time to convalescence.

A dietary history and clinical examination indicate if there is pre-existing malnutrition. Surveys show that up to one-third of hospital in-patients have unrecognized malnutrition. Objective indices of poor nutritional status include a body mass index (body mass/(height)^2) below 19 kg/m^2, a weight loss of more than 10 per cent of body weight in the preceding 3 months, and mid-arm muscle circumference (mid-arm circumference (cm) − 3.14 × triceps skinfold thickness (cm)) below the fifth percentile, but subjective clinical assessment is no worse an indicator of nutritional status. Hypoalbuminemia and lymphopenia are too non-specific to be useful in the critically ill.

Methods

A team approach to nutrition has been recommended. A dietitian helps to provide a high-quality service sensitive to the needs of individual critically ill patients. The roles of team members must be clearly defined.

Nutritional requirements can be estimated from indirect calorimetry and nitrogen balance studies, approximated from formulas (e.g. the Harris–Benedict equation), or based on simple calculations (e.g. protein, 1.5–2.0 g/kg/day; non-protein energy, 25–35 kcal/kg/

Table 1 Electrolyte requirements

Electrolyte	Common requirement (mmol/day)
Sodium	70–120
Potassium	50–100
Magnesium	5–15
Calcium	5–10
Phosphate	20–40

day (105–145 kJ/kg/day)). There is no evidence that more sophisticated methods for estimating needs lead to better patient outcomes. Most commercially available enteral feeds provide 100 to 180 kcal/g (420–765 kJ/g) of protein nitrogen, although feeds providing as little as 90 kcal/g (380 kJ/g) of nitrogen are promoted for critically ill patients (e.g. after major trauma). Excessive energy, particularly as carbohydrate, increases fat synthesis and carbon dioxide production. The increased carbon dioxide production may delay weaning from ventilation of patients with limited ventilatory reserve. Neither hypercaloric nor isocaloric nutritional support prevent protein catabolism in medical patients with multiple organ failure, but supplying this much energy increases energy expenditure, urea production, and glucose and lactate concentrations. Hypocaloric nutrition in terms of usual energy estimates may be better in patients with early multiple organ failure. Enteral nutrition can also be constrained by restrictions on fluid intake, glucose intolerance, renal impairment, large gastric aspirates, and fasting for procedures. Intensive care patients commonly receive only about 75 per cent of estimated needs.

Enteral nutrition is frequently increased in steps to the estimated daily need by increasing feed volume and concentration, but this process is usually unnecessary unless enteral feeding starts after prolonged starvation.

Administration

Enteral feeds should be colored distinctively to minimize the risk of confusion with fluids intended for intravenous use. The administration set must be sterile and have connectors incompatible with intravenous infusions.

Care is needed to prevent bacterial colonization of enteral feeds. High colonization counts have been found in up to 24 per cent of enteral feed reservoirs at 24 h. It is generally accepted that enteral feeds should not be left hanging at the bedside for more than 24 h, and many units limit this to 12 h at room temperature. Bacteria can migrate retrogradely from the patient to the feed reservoir. Other factors influencing the frequency of bacterial contamination are the number of manipulations during feed preparation, poor preparation technique, and feed composition.

Enteral feeds are usually given continuously by gravity feed or pump-assisted infusion. Preliminary studies suggest that intermittent bolus feeds given every 4 h have a more positive affect on nitrogen balance than the same quantity of continuous feeds. However, although bolus feeds are associated with periods when intragastric pH is significantly lower, the frequency of gastric colonization, ventilator-associated pneumonia, and death is unchanged from continuous feeding (Bonten *et al.* 1996). Many units routinely use pump-assisted infusion of enteral nutrition solutions. Pump assistance may reduce the frequency of diarrhea.

Bowel sounds are a poor indicator of small bowel activity, particularly in patients subject to tracheal intubation and mechanical ventilation. Their absence should not delay a trial of enteral nutrition.

Routes

Patients tolerate nasal tubes better than oral tubes, but the nasal route is associated with more frequent bleeding during insertion, erosion of the nares, and sinusitis. Nasal intubation is relatively contraindicated in patients with a fractured base of skull because of the risk of intracranial penetration. Nasogastric feeding usually starts using a 12- to 14-FG tube to allow aspiration of gastric contents to check feed absorption and administration of viscous elixirs or crushed tablets. If the nasogastric aspirates are large (i.e. more than 200–300 ml), transpyloric tube placement should be considered. Many different fine-bore enteral feeding tubes (less than 12 FG and usually 8 FG) are available. The designs vary, but all require a wire stylet or guidewire to facilitate insertion. Most have a weighted tip, although this probably does not assist transpyloric placement. Less than 50 per cent of fine-bore tubes pass through the pylorus spontaneously within 24 h of insertion, and the proportion appears to be much less in patients needing intensive care. Prokinetic drugs such as cisapride (10 mg every 6 h), metoclopramide (10–20 mg intravenously every 6 h), or erythromycin (100–200 mg intravenously) are used to assist transpyloric placement. Fine-bore tubes can also be guided through the pylorus, using fluoroscopy or endoscopy, so that the tip lies in the duodenum or jejunum. Naso- or orogastric tubes are misplaced in 0.3 to 4 per cent of blind insertions, with the most serious complications following accidental tracheal intubation. Risks can be reduced by inserting the fine-bore tube through a larger esophageal introducer. The position of the tip of a feeding tube must always be confirmed by radiography before feeding starts.

Some centers favor creation of a feeding jejunostomy when critically ill patients undergo laparotomy. This allows early enteral nutrition in most patients, but complications are common compared with nasojejunal intubation. They include leaks, peritonitis, wound infection, and bowel obstruction.

The use and timing of percutaneous endoscopic gastrostomy varies between different intensive care units. The principle indication is for long-term enteral nutrition in patients with impaired consciousness or dysphagia. The tube is less likely to be accidentally removed or become blocked, medication can be given through the tube more easily, and the sinus and middle-ear problems associated with nasogastric intubation are avoided. Overall complication rates from percutaneous endoscopic gastrostomy are as high as 25 per cent. They include wound infection, leaks, and peritonitis. It is suggested that feeding by percutaneous endoscopic gastrostomy reduces the risk of pulmonary aspiration and chest infection, but this has not been confirmed.

Feeds

A wide range of manufactured enteral feeds are available in both liquid and powder forms. They vary in carbohydrate, fat and protein sources, amino acid distribution, osmolality, and electrolyte, mineral, and vitamin content. Some feeds are 'enhanced' by the addition of fiber, additional amino acids (branched-chain amino acids, essential amino acids, glutamine, arginine), specific fatty acids, nucleotides, and other nutrients. In practice, most patients will tolerate a lactose-free near-isotonic feed providing approximately two-thirds of the non-protein energy as carbohydrate and the remainder as fat.

Most standard formulations contain 1 kcal/ml (4.2 kJ/ml). Feeds containing 1.5 to 2 kcal/ml (6.3–8.4 kJ/ml) are available if fluid restriction is a priority. Electrolyte requirements (Table 1) vary widely in critically ill patients, depending principally on urinary and gastrointestinal losses. Electrolyte content will affect the choice of feed when sodium or potassium restriction is important; these electrolytes can always be added to a feed when supplementation is needed. Sodium restriction may be needed in patients with heart failure or edema, and care is needed with the potassium and magnesium content of feeds in patients with renal failure. Extra potassium, magnesium,

Table 2 Complications associated with enteral nutrition

Complications of feeding tube insertion
Trauma and bleeding
Perforation: retropharyngeal space, esophagus, stomach
Pneumomediastinum, pneumothorax
Complications of having a tube in place
Patient discomfort
Erosion of nares
Sinusitis
Aerophagy
Tube displacement or obstruction
Wound infection, leaks, peritonitis (with gastrostomy, jejunostomy)
Misplacement of feed
Pneumonitis
Pleural effusion or empyema
Complications during feeding
Nausea, abdominal distension
Regurgitation and vomiting
Pulmonary aspiration of feed
Intestinal pseudo-obstruction
Diarrhea
Interaction with medications (reduced absorption)
Complications related to feed content
Hyperglycemia
Hypercarbia
Electrolyte abnormality
Azotemia
Deficiency disorders after long-term use

and phosphate are often needed when feeding is started after a period of fasting.

The vitamin and trace element needs of critically ill patients are poorly defined. It is likely that there are important interactions with the pathophysiology of critical illness. For example, vitamins are involved in oxidative phosphorylation and interact with oxidant mediators, while trace elements have direct antioxidant activity and act as cofactors for antioxidant enzymes. Both vitamins and trace elements have effects on wound healing and immune function. The recommended daily allowances for vitamins are based on the needs of healthy young adults. Commercial feeds commonly contain the recommended daily allowance in a volume equivalent to about 2000 kcal (8.4 MJ). The needs of critically ill patients are likely to be higher than the recommended daily allowances for at least some vitamins and losses may be increased in continuous hemodiafiltration (e.g. the water-soluble vitamins including folic acid and vitamin C). Therefore it is usual to give vitamin supplements to critically ill patients even if they are receiving their full estimated needs for protein and energy.

Specific feeds with targeted protein-to-energy or carbohydrate-to-fat ratios have been developed and are promoted for specific groups of patients (e.g. patients after trauma, patients with respiratory failure). The value of a very high protein-to-energy ratio is unproven. Feeds with reduced carbohydrate and increased fat content (e.g. providing two-thirds of the non-protein energy as fat and one-third as carbohydrate) facilitate glycemic control in diabetics and can assist weaning from mechanical ventilation in patients with carbon dioxide retention by minimizing the respiratory quotient.

Elemental feeds contain protein as peptides or amino acids. They have been used after prolonged fasting, and in patients with short bowel syndrome, radiation enteritis, pancreatitis, and pancreatic insufficiency. Elemental feeds do not protect intestinal growth factors in the same way as standard protein sources and therefore do not prevent gut atrophy. A clinical trial in critically ill patients has shown no advantage over standard feeds (Mowatt-Larssen *et al.* 1992).

Specific nutrients

Some nutrients have specific biological effects other than their use as energy sources or substrates for protein synthesis.

Glutamine

Glutamine is a major metabolic fuel for the enterocytes of the gut mucosa, lymphocytes, and macrophages. It is the most abundant amino acid in plasma and is the principle interorgan transporter of nitrogen between sites of synthesis (muscle and liver) and sites of utilization (gut, lymphocytes, and lung). Glutamine is a precursor of nucleotides and glutathione. While not usually regarded as an 'essential' amino acid, it becomes conditionally essential during critical illness. In most enteral formulations glutamine is present in only low concentrations. When added to parenteral nutrition it improves nitrogen balance and gut barrier function, and reduces infection and duration of mechanical ventilation. Glutamine supplementation of enteral feeds has still to be shown to improve patient outcome (Souba 1997), although experimental studies suggest that it should be beneficial.

Branched-chain amino acids

The plasma concentrations of the branched-chain amino acids (valine, leucine, isoleucine), which can act as a source of glutamine, decrease during critical illness. Clinical trials of feed supplementation with branched-chain amino acids after surgery and during sepsis show no overall benefit. There have been extensive studies of the effect of supplementing feeds with these amino acids in patients with severe liver disease. In acute hepatic encephalopathy they do not affect outcome, but in chronic hepatic encephalopathy they may allow a greater total protein intake without causing deterioration.

Arginine

Arginine is a non-essential amino acid in adults, but it is also a potent stimulator of prolactin, growth hormone, glucagon, and insulin release, and a precursor of nitric oxide. Indirect measures of immune function have been improved by arginine supplementation in both experimental and clinical studies. However, clear benefit in critically ill patients has still to be shown, and it is of concern that large arginine supplements (6 per cent of total energy intake) decreased survival in guinea pigs with peritonitis.

Nucleotides

Nucleotides form part of the normal diet but are absent from standard enteral feeds. They can be synthesized from glutamine, but considerable upregulation of this pathway is needed if the body's needs are to be satisfied in the absence of dietary intake. *In vitro* experiments show that human enterocyte growth is stimulated by purines or pyrimidines, but only when the supply of glutamine is limited. Similar effects have been shown in lymphocytes. Experimental studies have demonstrated a link between dietary nucleotide restriction and mortality from infection. In humans, the bacteria dying in the bowel lumen would normally provide larger amounts of nucleotide than the content of 'nucleotide-supplemented' enteral feeds, and improved patient outcome has not yet been shown.

Fiber

Bacterial breakdown of fiber in the colon produces the short-chain fatty acids acetate, propionate, and butyrate. These are important substrates for the cells of the colonic mucosa and their uptake enhances absorption of water and electrolytes from the bowel lumen. Fiber also binds bile salts which would otherwise be irritants to the colonic mucosa, promotes glucose absorption, and provides substrate for the normal bowel flora. Nevertheless, published clinical trials do not provide firm evidence for the routine use of fiber-supplemented enteral feeds (Scheppach *et al.* 1990). No adverse effects have been reported from including fiber in enteral feeds, and most newer formulations contain a fiber source.

Omega-3 fatty acids

Inflammatory mediators including prostaglandins, leukotrienes, and platelet activating factor are products of omega-6 fatty acids such as linoleic acid and its product, arachidonic acid. Most standard enteral formulas contain omega-6 fatty acids as the principle fat source. Fish oils with omega-3 fatty acids have anti-inflammatory effects in animal models, probably because production of arachidonic acid metabolites is reduced. In human volunteers dietary supplements of omega-3 fatty acids reduce monocyte production of interleukin 1α, interleukin 1β, and tumor necrosis factor. Experimental studies show varying effects of omega-3 fatty acids on outcome in animal models of sepsis. A preliminary study shows a trend to reduced infection in patients given fish oil as their fat source after abdominal surgery.

'Immune-enhancing' enteral feeds

Feeds specifically supplemented with some of the nutrients described above have been formulated with the aim of improving immune function and so reducing the risk of infection in patients known to have impaired host resistance, i.e. the critically ill and after major surgery or trauma. Most clinical trials have been small and some have failed to provide equal energy and total protein to the new feed and the control feed patients. Multiple subgroup analysis complicates assessment of principle endpoints. The best studied feed is Impact (Sandoz Nutrition, Minneapolis, MN) which is supplemented with arginine, nucleotides, and omega-3 fatty acids. Other products include Immune-Aid (McGaw Inc., Irvine, CA), which is also supplemented with glutamine and branched-chain amino acids, and Replete (Clintec Nutrition, Deerfield, IL) with only glutamine and omega-3 fatty acid supplements. The effect of these feeds remains controversial and they are expensive compared with standard enteral feeds. Well-designed large clinical trials are still needed to show whether they improve outcome in critically ill patients.

Growth factors

Growth factors can now be manufactured using recombinant techniques. Growth hormone, insulin-like growth factor, and epidermal growth factor have been investigated. Growth hormone improves intestinal uptake of amino acids and nitrogen retention (Souba 1997). Side-effects include hyperglycemia and hypercalcemia. Insulin-like growth factor improves nitrogen balance in critically ill patients, but it can cause hypoglycemia. Growth factors appear promising as adjuncts to enteral feeding, but they are expensive and still have to be shown to improve patient outcome.

Complications

Complications of enteral nutrition include regurgitation and aspiration, hyperglycemia, electrolyte disturbance (especially low concentrations of potassium, phosphate, magnesium, and zinc), abdominal distension, and diarrhea. A more comprehensive list is shown in Table 2. The overall complication rate from enteral nutrition is low, and is certainly much less than that associated with parenteral nutrition. Complications can be reduced by regular observation and monitoring.

References

American Society for Parenteral and Enteral Nutrition Board of Directors (1993). Guidelines for the use of parenteral and enteral nutrition in adult and paediatric patients. *Journal of Parenteral and Enteral Nutrition*, **14**, 1SA–52SA.

Bonten, M.J., *et al.* (1996). Intermittent enteral feeding: the influence on respiratory and digestive tract colonization in mechanically ventilated intensive-care-unit patients. *American Journal of Respiratory and Critical Care Medicine*, **154**, 394–9.

Mowatt-Larssen, C.A., Brown, R.O., Wojtysiak, S.I., and Kudsk, K.A. (1992). Comparison of tolerance and nutritional outcome between a peptide and a standard enteral formula in critically ill hypoalbuminemic patients. *Journal of Parenteral and Enteral Nutrition*, **16**, 20–4.

Scheppach, W., Burghardt, W., Bartram, P., and Kasper, H. (1990). Addition of dietary fiber to liquid formula diets: the pros and cons. *Journal of Parenteral and Enteral Nutrition*, **14**, 204–9.

Souba, W.W. (1997). Nutritional support. *New England Journal of Medicine*, **336**, 41–8.

5.3.2 Parenteral nutrition

G. Iapichino and D. Radrizzani

Key messages

- Parenteral feeding of a well-nourished patient able to take enteral nutrition within 4 to 5 days or with a relatively minor injury is unnecessary.
- Parenteral feeding should be performed when enteral nutritional support is strictly contraindicated or enteral nutrition is unable to cope with energy and nutrient requirements.
- The nutrient intake must be within an appropriate range (20–35 kcal/kg and 0.1-0.3 g/kg/day nitrogen) to avoid over- or underfeeding (the first is considerably more dangerous).
- The absence of foodstuff in the gut lumen is a disadvantage of the parenteral feeding approach.
- Parenteral nutrition in the critically ill is usually via a multiple-lumen central venous catheter with one lumen dedicated exclusively to nutritional fluids.

Introduction

The administration of nutrients to critically ill patients is a supportive measure. Its importance relative to the specific therapy of the underlying disease is often underestimated. The maintenance of physiological balance is one of the foundations of intensive care, without which the outcome is often death (Nitenberg 1993).

Thus nutritional support, by supporting protein synthesis and controlling catabolism, should maintain the energy equilibrium and restore, or at least limit, the reduction of body energy stores and the wasting of lean tissue.

Indications for parenteral feeding

Patients can be classified according to nutritional (normal or malnourished) and metabolic status.

1. Well-nourished and moderately catabolic (10–15 g urea nitrogen lost daily) patients need therapy if unable to take oral nutrition within 10 days.

2. Well-nourished and severely catabolic (> 15 g urea nitrogen lost daily) patients who are unable to eat need immediate protein-sparing therapy, i.e. anticatabolic support.

3. Malnourished patients (> 10 per cent weight loss in the previous 3 months), particularly if catabolic and with oral intake impossible for more than 5 days, need immediate nutritional support.

Artificial feeding of a well-nourished patient, able to take enteral nutrition within 4 to 5 days or with a relatively minor injury is unnecessary. Parenteral feeding should be performed in the following circumstances.

1. Enteral nutritional support is strictly contraindicated (gut obstruction, high output (i.e. > 0.5 l/h), jejunal or ileal fistula, severe non-hypovolemic gut ischemia, gut failure due to extensive resection or absorption impairment).

2. Enteral nutrition is unable to cope with energy and nutrient requirements (mixed nutrition).

Energy needs

Energy requirement is considered to be equal to measured or assessed energy expenditure. By feeding a patient, we try to replace oxidized endogenous substrates by diet-derived substrates. Therefore it should be more appropriate to consider nutrient balance rather than energy balance.

The energy production rate of a patient in basal conditions (supine subject in a thermoneutral environment and after a 12-h fast) can be determined by indirect calorimetry, calculated using formulas, or estimated.

Measurement may be appropriate in more clinically unstable and catabolic patients, particularly those with respiratory failure, and in patients requiring prolonged intensive treatment. Basal energy production rate (kcal/day) is predicted by the Harris–Benedict equations:

$$\text{males: } 66.473 + (13.752 \times \text{weight}) + (5.003 \times \text{height}) - (6.755 \times \text{age})$$

$$\text{females: } 655.095 + (9.563 \times \text{weight}) + (1.850 \times \text{height}) - (4.676 \times \text{age})$$

where weight is in kilograms, height is in centimeters, and age is in years. These equations should only be corrected for stress (final value no more than 1.15 to 1.25 times the calculated value except for burned patients). Calculated or measured values, minus the amount of energy derived from protein oxidation (nitrogen loss (g) × 26.6), are utilized to plan the non-protein calorie intake. In overweight patients we suggest using the formula with the ideal body weight to avoid further overstimation.

A rough estimate of 20 to 35 kcal/kg is reasonably adequate in clinical practice.

Nitrogen requirement

Nitrogen requirement is evaluated by measuring daily nitrogen loss corrected for body urea pool variation according to the following equations:

$$\text{males: urea (g/24 h)} + [\Delta(\text{blood urea (g/l/day)}) \times \text{body weight (kg)} \times 0.50] + 3$$

$$\text{females: urea (g/24 h)} + [\Delta(\text{blood urea (g/l/day)}) \times \text{body weight (kg)} \times 0.60] + 3.$$

The determination of basal nitrogen loss must be performed during a 24-h fast, owing to the mandatory contribution of protein amino acid intake to the urea production rate (about 30–40 per cent of the supply (Iapichino *et al.* 1984)). The value of the nitrogen loss represents the need to maintain the actual lean body mass, i.e. the nitrogen balance, which is a reasonable goal for non-depleted injured patients in the acute stress phase.

In malnourished patients, nitrogen support could be increased to replace body protein stores.

Energy-nitrogen relationship

Although nitrogen excretion and retention are affected by calorie intake in both stressed well-nourished and non-stressed malnourished patients, nitrogen intake is the main determinant of nitrogen balance (Iapichino *et al.* 1984). In both classes of patients the effect of calorie intake is comparable and accounts for a nitrogen retention of about 1 mg/kg/kcal/day. Also, the effect of nitrogen intake is comparable and results in the retention of about 60 per cent of nitrogen supply.

It is easier to obtain nitrogen balance in malnutrition because malnourished patients show nitrogen losses that are consistently less than those of stressed patients. The simple replacement of fasting nitrogen losses leads to a negative nitrogen balance owing to incomplete utilization of the load. Depending on the patient's metabolic environment (stress or malnutrition), nitrogen clearance ability (renal and hepatic function), and planned nitrogen balance (replacement or maintenance) it may be necessary to add nitrogen to fasting losses. Even in the most favorable conditions (non-stressed depleted patient), it is unlikely that a nitrogen balance of more than 2 to 3 g/day will be obtained.

A rough estimate of 0.11 to 0.3 g/kg/day nitrogen supply is reasonably adequate in clinical practice.

Timing

Critically ill patients show daily negative cumulative energy and protein balance (1–2 per cent of body cell mass). The aim of treatment is to provide early support, preventing tissue wasting and nutritional deficit rather than correcting them later. After the start of treatment, cell energetics (sodium pump and calcium kinetics) rapidly recover, resulting in an immediate improvement in cell function (Wilmore and Carpentier 1993).

Obviously, it is mandatory to start treatment when hemodynamic, water, electrolyte, and acid–base balances reach equilibrium.

Planning the nutritional regimen

Energy and nitrogen

The approach with respect to the nutrient balance depends on either the amount of nutrients necessary for a particular aim (e.g. optimizing nitrogen utilization) or the possibility of safely utilizing body fat stores to satisfy energy requirements. In theory, patients with preserved fat stores can tolerate negative nutrient balance, i.e. a hypocaloric regimen, while depleted patients require promptly oxidizable nutrients to reach energy balance and a further intake to replace depleted stores (net positive nutrient balance).

Two concepts are important in planning the nutritional metabolic treatment in the critically ill.

1. The nutrient intake must be within an appropriate range (20–35 kcal/kg and 0.1–0.3 g/kg/day nitrogen) to avoid over- or underfeeding (the former is considerably more dangerous).

2. Any variation within this range is performed stepwise until the therapeutic goal or the range limit is reached.

The calorie figure is difficult to evaluate when there is no measure of energy production. The upper limit of supply (to avoid overfeeding in non-burn patients) is 35 kcal/kg, even if the estimated value is higher. If the estimated value is less than the top limit, this value must be considered to be the maximum calorie intake. A higher nitrogen intake load can be planned according to the therapeutic goal: a near nitrogen balance versus a positive balance ($\leqslant$ 3 g/day).

Insulin administration and adjuvant therapies

Insulin is useful for controlling and avoiding hyperglycemia. If it is added to the parenteral nutrition bag, the suggested intake ranges from 1 IU per 10 g glucose in malnourished minimally stressed patients to 1 IU per 4 to 5 g of glucose in severely catabolic patients. When resistance is severe, it is safer to administer insulin (1–3 IU/h) with a syringe pump. Several anabolic agents are effective in, or should improve the efficacy of, the conventional nutritional-metabolic approach to protein metabolism in stress conditions. Hormones like insulin, growth hormone, and insulin-like growth factors are particularly promising (Wilmore and Carpentier 1994).

Vitamins, minerals, and water

The recommended dietary allowances for parenteral administration of vitamins and trace elements have been reviewed elsewhere (Nitenberg 1993). The optimum dose and frequency of administration has not been defined in detail in critically ill patients. The need to provide increased amounts of zinc (large intestinal losses), magnesium, and vitamin E to prevent acute deficiencies and to maximize antioxidant potential has been stressed (Wilmore and Carpentier 1994).

Water supply must be planned strictly according to patient need. The safest way is to fill a bag with the planned nutrient load at the highest concentration. This results in the possibility of adjusting fluid intake without changing the nutrient program.

Fuel substrates

Nutrients providing calorie sources are carbohydrates and lipids. Carbohydrates are the main energy substrate for humans. Alternative substrates, such as xylitol, sorbitol, or fructose, have been used in parenteral nutrition, but offer no advantage over glucose. Lipids are available as emulsions containing medium-chain triglycerides. The choice between glucose and fat is a topic of much discussion in artificial nutrition. We can address the question by analyzing the key points useful for therapeutic decisions.

Anabolic action

The aim of nutritional support in catabolic patients is to attain the best anabolic drive. This is achieved by administering an adequate protein supply (Iapichino *et al.* 1984), and optimizing the hormonal control of the body metabolism that is deranged by stress or injury (i.e. reaching an 'optimal' plasma insulin level either spontaneously or with planned insulin supply to overcome resistance). There is an increasing body of evidence that the best anabolic drive is achieved by parenteral feeding with a predominantly glucose system (Wilmore and Carpentier 1993).

Gas exchange modifications and diet-induced thermogenesis

When planning the calorie load the following points must be remembered.

1. Diet-induced thermogenesis depends primarily on the overall energy intake as well as the nutrient used (30–40 per cent of intake for protein, 6–8 per cent for glucose, and 3 per cent for fat).
2. Continuous infusion of nutrients at a rate balancing the energy expenditure seems to suppress diet-induced thermogenesis (Wilmore and Carpentier 1993).

Gas exchange largely reflects the composition of the oxidized fuels. In particular, carbon dioxide (CO_2) production per kilocalorie is maximum for glucose oxidation (100 per cent), and decreases to 93 per cent for protein oxidation and to 76 per cent for fat oxidation. In contrast, oxygen consumption per kilocalorie is maximum for lipids and decreases to 92 per cent for glucose. A mixture of medium- and long-chain triglycerides increases minute ventilation by 12 per cent, CO_2 production by 15 per cent, and oxygen consumption by 20 per cent compared with long-chain triglycerides (Chassard *et al.* 1994). Moreover, the supply of any substrate at a rate exceeding the energy requirement results in storage as fat. Liposynthesis from glucose has a high energy cost (12 per cent of stored energy) and yields a large amount of CO_2. The effect of a substrate load given at a rate not exceeding the energy demand will be almost negligible.

Substrate oxidation

One reason given for the preferential use of fatty emulsions in critically ill patients is the belief that they are easily utilized for energy and therefore could spare the energy drawn from endogenous fat fuels during fasting. In hypermetabolic fasting patients, as in fasting volunteers, endogenous fats are extensively oxidized. Moreover, in normal and stressed patients, the rate of oxidation of exogenous glucose (given without fat) increases with the load and the period of infusion. The oxidation of endogenous lipid is progressively reduced even if endogenous fat continues to be oxidized when glucose intake meets energy needs (non-protein respiratory quotient $RQ_{np} \leqslant 1$).

When glucose intake exceeds energy expenditure, the patient produces more energy (up to 100 per cent ($RQ_{np} > 1$)) from the glucose. However, when both carbohydrates and lipids (long-chain fatty triglycerides) are given, RQ_{np} (i.e. the proportion of glucose and fat oxidized) seems to be determined by the carbohydrate load. The energy not drawn by the carbohydrate load to reach the energy balance is derived from fat. Indirect calorimetry does not separate exogenous from endogenous fat.

The exogenous lipid load neither induces further lipid oxidation nor results in a decrease in carbohydrate oxidation (lower RQ_{np}) compared with that oxidized during the supply of the same amount of carbohydrate without lipid. After a meal or during continuous parenteral nutrition, oxidized lipid is largely derived from endogenous stores; dietary fat is mainly stored and probably oxidized later.

Lipid versus lung function

Intravenous fat emulsion induces changes in the pulmonary function, generally consisting of a decrease in $Pa{O_2}$ and in the pulmonary diffusion capacity for carbon monoxide. In critically ill ventilated patients with normal lung conditions, only an increase in pulmonary vascular resistance or a minor increase in venous admixture is seen.

The magnitude of the related impairment of gas exchange may be greater, reaching clinical relevance, in the presence of acute lung injury with or without sepsis syndrome. The adverse effects consist of a decrease in the $Pa{O_2}/Fi{O_2}$ ratio and increases in the mean pulmonary pressure, the venous admixture, and pulmonary resistances. It is not clear whether slow infusion of fat emulsions is of any clinical relevance in acute lung injury. Fat emulsions composed of medium-chain triglycerides reduce the omega-6 polyunsaturated fatty acid load and seem to be safe in septic patients.

Lipids and the inflammatory response

In animal injury models, standard lipid emulsions may induce enhanced inflammatory reaction and have an adverse effect on immunological function, resulting in increased susceptibility to infection.

Intravenous emulsions enriched with omega-3 fatty acids (fish oil), whose metabolic products (eicosanoids) have less biological action in influencing the cell membrane composition and response to inflammatory stimuli, are currently under investigation.

The modulation of the immune response in patients with persistent hypermetabolism and organ failures following trauma, burns, surgery, or sepsis remains a fascinating field of research.

Amino acids

Bearing in mind the amounts necessary to obtain nitrogen balance in stressed and malnourished patients, the amino acid composition should predominantly consist of essential and semi-essential amino acids (histidine, arginine). Taurine, cystine, and tyrosine are the most important non-essential amino acids. Other constituents, up to the desired nitrogen intake, could include glutamine and glutamic and aspartic acids (the major participants in transamination reactions), serine, proline, and glycine.

Arginine, a conditionally dispensable amino acid, has a potential role in immunomodulation and a positive influence on post-traumatic nitrogen metabolism and the wound-healing process by enhancing collagen deposition. It has been shown to be a unique substrate for production of the biological effector molecule nitric oxide. Nevertheless, there is no evidence that exogenous arginine, even in large doses, modulates the rate of synthesis of nitric oxide.

Glutamine is an important metabolic fuel for the cells of the gut and the immune system. It is also involved in the regulation of muscle and liver protein balance, probably mediated by an increase in cellular hydration, a triggering signal, or protein anabolism.

Special problems and requirements

Acute respiratory disease syndrome

Nutritional metabolic treatment in patients with acute respiratory disease syndrome should be by the enteral route as soon as possible. This approach will reduce the supply of omega-6 polyunsaturated fatty acids and improve the supply of omega-3 polyunsaturated fatty acids. If parenteral feeding is necessary, procedures that increase oxygen consumption, CO_2 production, ventilation, and ventilation–perfusion maldistribution, impair body temperature control, or favor inflammatory reaction or immunodepression should not be used

(Wilmore and Carpentier 1993). However, to satisfy energy needs with rapidly available calories and to curtail protein loss, i.e. to maintain vital physiological balance in such critically ill patients, it may be advisable to administer 1000 to 1300 kcal preferentially as glucose for short-term treatment or 80 to 90 per cent glucose plus 10 to 20 per cent fat for treatment lasting for more than 5 to 6 days, together with amino acids and insulin.

Acute renal failure

The implementation of continuous renal replacement therapy in the intensive care unit setting has drastically reduced the requirement for a different nutritional approach compared with other critically ill patients. The estimated energy production rate could be increased if warming equipment is not available. It is easy to calculate the heat loss if the body and fluid temperatures and the volume of daily replacement fluids are known. Fat substrate (both long- and medium-chain) should be administered with caution because of reduced disposal; carbohydrates (with or without insulin supplementation) appear to be tolerated better.

Body nitrogen losses are assessed by measuring urea nitrogen plus dialysate urea and amino acid losses (5–15 g/day). A general amino acid solution is generally indicated for replacement, despite the reported particular need for histidine, arginine, taurine, and tyrosine. Folinic acid (1 mg) and vitamin C (500 mg) supplementation during hemofiltration is suggested, with no variation in trace elements and other vitamins.

Enteral nutrition should be used whenever possible.

Acute liver failure

Critically ill patients may present with acute liver failure or may develop it as part of multiple-organ failure. In both cases patients are generally catabolic, malnourished, and intolerant of water, sodium, and protein. The aim of treatment is to reduce catabolism, thus decreasing the plasma level of poorly tolerated aromatic and sulfated amino acids, and to support protein synthesis.

Extreme caution is needed when treating these patients. Calorie load (avoid overfeeding) and composition ($\geqslant$ 70 per cent glucose), as well as amino acid load (30–50 g/day) and composition (titrate infusion of branched-chain, aromatic, and sulfurated amino acids on plasma profile), must be carefully checked (Nitenberg 1993).

Adverse effects

The absence of foodstuff in the gut lumen is a disadvantage of parenteral feeding. Careful attention is required to avoid hypokalemia, hypophosphatemia and vitamin B deficiency.

Tolerance of substrate load

Substrate utilization or storage may result in increased oxygen consumption and/or CO_2 production, with resulting increases in cardiac output and ventilation. Monitoring of heart rate, respiratory rate, and blood gas analysis is very useful, and unexplained increases should be controlled.

Tolerance of nitrogen load

Nitrogen supply results in increased urea production. An increase of 30 per cent in plasma urea, above the normal range and not due to concentration, means poor nitrogen utilization.

An excess of nitrogen load over disposal may result in metabolic hyperchloremic acidosis (chlorinated amino acids (arginine, lysine)) and derangement of plasma amino acid levels.

Intravenous feeding catheters

Central venous catheters are generally inserted in the operating theater or emergency room for reasons primarily related to life support rather than nutrition. The site of vascular access is usually the internal jugular or the subclavian vein.

During critical illness, the catheter is multipurpose. It is used to deliver drugs and other infusions including blood products. Usually it is a multiple-lumen catheter, with one lumen dedicated exclusively to nutritional fluids. The catheter is not tunneled so that it can be changed over a guidewire.

If the only reason to maintain a central vascular catheter is artificial long-term nutrition, a silicon catheter, tunneled for prevention of infection, should be considered.

References

Chassard, D., Guiraud, M., Gauthier, J., Gelas, P., Berrada, K.R., and Bouletreau, P. (1994). Effects of intravenous medium-chain triglycerides on pulmonary gas exchanges in mechanically ventilated patients. *Critical Care Medicine*, **22**, 248–51.

Iapichino, G., *et al.* (1984). The main determinants of nitrogen balance during total parenteral nutrition in critically ill injured patients. *Intensive Care Medicine*, **10**, 251–4.

Nitenberg, G. (1993). Enteral and parenteral nutrition. In *Pathophysiologic foundations of critical care* (ed. M.R. Pinsky and J.F.A. Dhainaut), pp. 42–81. Williams & Wilkins, Baltimore, MD.

Wilmore, D.W. and Carpentier, Y.A. (ed.) (1993). *Metabolic support of the critically ill patient.* Springer-Verlag, Berlin.

Wilmore, D.W. and Carpentier, Y.A. (1994). Round Table Conference on Metabolic Support of the Critically Ill Patient. *Intensive Care Medicine*, **20**, 298–9.

6 The renal system

6.1 Normal physiology

6.1.1 The renal system

J. Reichman and M. Brezis

Key messages

- Although comprising only 0.5 per cent of total body weight, the kidneys receive between 20 and 25 per cent of the cardiac output.
- Renal blood flow is mostly directed to the cortex to optimize glomerular filtration and bulk solute reabsorption. In contrast, blood flow to the renal medulla is low to preserve osmotic gradients and to enhance urinary concentration.
- A decline in glomerular filtration rate diminishes sodium filtration and excretion, while catecholamines augment proximal tubule sodium reabsorption.
- Approximately 180 liters of water are filtered daily through the glomeruli, of which roughly 1 per cent (2 liters) is normally excreted.

Introduction

The primary function of the kidneys is to maintain body fluid and ionic homeostasis by tightly regulating excretion and reabsorption of these substances (Fig. 1). Although comprising only 0.5 per cent of total body weight, the kidneys receive between 20 and 25 per cent of the cardiac output. This constitutes a greater blood flow per unit weight of tissue than other vital organs and ensures efficient filtration of the blood. The volume of daily urinary excretion is the residuum of an ultrafiltration of roughly 180 liters of plasma per day and the reabsorption of more than 99 per cent of this ultrafiltrate. Despite wide fluctuations in renal arterial perfusion pressure (80–180 mmHg), renal plasma flow is strictly preserved, governed by autoregulation of the renal vascular resistance (afferent and efferent arterioles). Renin–angiotensin, prostaglandins, thromboxanes, and the sympathetic nervous system are the putative principal mediators of arteriole tone (Neild 1990). Approximately 20 per cent of the daily renal plasma flow (600 ml/min) is filtered via the glomeruli, resulting in a glomerular filtration rate of 120 ml/min. Glomerular filtration rate is contingent on the rate of plasma flow, the balance of hydrostatic and oncotic pressures traversing the capillary wall, and the permeability and total surface area of the filtering capillaries. Creatinine clearance and plasma creatinine are rough approximations of the glomerular filtration rate. The kidney also serves as a selective mechanical and electrostatic barrier to plasma protein excretion, effectively restricting the passage of molecules larger than inulin (5200 Da) and negatively charged proteins (albumin).

Intrarenal distribution of blood flow

Renal blood flow is mostly directed to the cortex to optimize glomerular filtration and bulk solute reabsorption. In contrast, blood flow to the renal medulla is low to preserve osmotic gradients and to enhance urinary concentration (Fig. 1). Within the medulla, tubules and vasa recta are disposed in a hairpin pattern to maximize urine concentration by countercurrent exchange. Oxygen diffuses from arterial to venous vasa recta, leaving the outer medulla deficient in oxygen. In this region, the medullary thick ascending limb is responsible for the generation of an osmotic gradient by active NaCl reabsorption, a process that has high oxygen requirements. Medullary hypoxia under normal conditions has been documented using oxygen

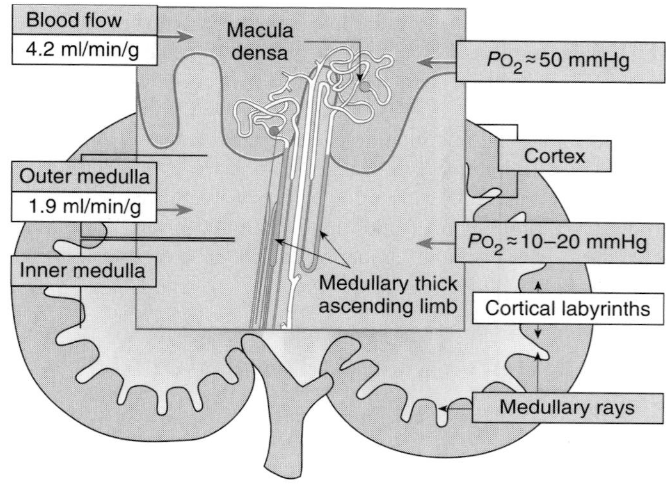

Fig. 1 Anatomical and physiological features of the kidney comparing cortex and medulla.

microelectrodes in several mammalian species including humans (Brezis and Rosen 1995). Renal medullary hypoxia is an obligatory price that the mammalian kidney pays for successful urinary concentration.

Tubular transport

The glomerular filtrate initially enters the proximal tubule of the nephron where an estimated two-thirds of it is isosmotically reabsorbed, coupled with the active transport of sodium. Proximal sodium reabsoption is chiefly an active process, relying on Na^+,K^+-ATPase pumps. The primary anions of the extracellular fluid, choride and bicarbonate, accompany sodium reabsorption via a Na^+/H^+ countertransport exchange system in order to maintain electrical neutrality. Bicarbonate, glucose, and amino acids (cotransported with sodium) are almost complete reabsorbed in the proximal tubule. This nephron segment, displaying a high permeability to water, also links volume and solute reabsorption. The rate of ionic and water proximal tubule reabsorption is invariably subject to the kidney's interpretation and regulation of the body's physiological requirements. Finally, the proximal tubule is a major site of creatinine and organic acid (uric acid) secretion. Physical factors modulating proximal tubule reabsorption include hydrostatic and oncotic peritubular capillary pressures. Owing to highly selective protein ultrafiltration, glomerular capillary oncotic pressure rises and subsequently transfers the augmented pressure to the peritubular capillaries. Several hormones, such as atrial natriuretic peptide, modulate proximal tubule reabsorption.

The thick ascending limb is impermeable to water while active transport occurs by a Na:K:2Cl cotransport mechanism, inhibited by furosemide (frusemide). Approximately 25 per cent of the filtered sodium and chloride and 15 per cent of the filtered water entering the loop are ultimately reabsorbed. Consequent to greater solute than water reabsorption, a hypotonic tubule fluid and hypertonic medullary interstitium are generated. The distal convoluted tubule, interposed between the loop of Henle and the collecting tubules, is virtually impermeable to water and reabsorbs roughly 10 per cent of the filtered sodium load. This reabsorption is load dependent and is inhibited by thiazide, amiloride, and spironolactone. The cortical and medullary collecting ducts constitute the final segments of the nephron, where water permeablility is negligible in the absence of ADH which enhances water reabsorption. Aldosterone augments sodium reabsorption and potassium/hydrogen secretion via stimulation of Na^+,K^+-ATPase pumps. The final urine volume and osmolality are precisely tuned in the collecting tubules. The kidney can produce minimum and maximum urinary concentrations of 50 mosmol/l and 1200 mosmol/l respectively. Since daily waste product excretion (e.g. urea) adds up to about 600 mosmol, minimum obligatory urine volume amounts to 500 ml/day (600 mosmol/1200 mosmol/l).

Regulation of sodium chloride

The kidneys dictate sodium excretion and reabsorption with precision, typically reabsorbing over 99 per cent of some 25 000 mmol of sodium filtered daily through the glomeruli. An increment in salt intake induces water retention and a subsequent transient gain in body weight until a compensatory increase in urinary sodium excretion ensues. Opposite effects are observed when sodium intake is reduced. Control of sodium excretion is mainly dependent on glomerular filtration rate and sodium reabsorption, modulated by various hormones. Extracellular fluid volume, not extracellular fluid sodium, is the critical factor determining sodium balance.

A decrease in absolute or effective blood volume and a subsequent decline in arterial blood pressure precipitate a decrease in renal perfusion pressure (renal plasma flow) and glomerular filtration rate. The decline in glomerular filtration rate is proportionally less than the drop in renal plasma flow, yielding an increase in the filtration fraction (ratio of glomerular filtration rate to renal plasma flow) and in the protein concentration in the efferent arterioles and peritubular capillaries. Sympathetic activity consequent upon a drop in blood pressure induces renal vasoconstriction. The decline in glomerular filtration rate diminishes sodium filtration and excretion, while catecholamines augment proximal tubule sodium reabsorption. Stimulation of angiotensin II during episodes of volume depletion vasoconstricts mostly the glomerular efferent arterioles, preserving glomerular filtration rate, but reduces peritubular capillary hydrostatic and interstitial pressure. This pressure reduction, combined with the increase in peritubular capillary protein concentration, facilitates sodium and water reabsorption in the proximal tubules during hypovolemia. Extracellular fluid volume expansion results in opposite compensatory effects (Vander 1991).

Endocrine effects on sodium regulation are illustrated in Table 1.

Water regulation

Approximately 180 liters of water are filtered through the glomeruli daily, of which roughly 1 per cent (2 liters) is normally excreted. Daily

Table 1 Endocrine effects on sodium regulation

Stimulators of Na reabsorption
Aldosterone (directly)
Angiotensin II (indirectly via aldosterone)
Potassium (indirectly via aldosterone)
ACTH (indirectly via aldosterone)
Renin (indirectly via angiotensin II)
Renal sympathetic nerves (indirectly via renin)
Insulin, cortisol, estrogen, growth and thyroid hormones
Inhibitors of Na reabsorption
Sodium intake
Increased renal perfusion pressure (indirectly via renin inhibition)
Increased renal NaCl delivery (indirectly via renin inhibition)
Atrial natriuretic peptide
Glucagon, progesterone, PTH
Prostaglandins, kinins, dopamine
ADH
Ouabain-like agents

urine volume may vary from 400 ml to more than 15 liters, depending on water requirements. The kidneys strictly preserve plasma volume and osmolality within a limited range. Two-thirds of filtered water is isosmotically reabsorbed in the proximal tubule, largely dependent on sodium reabsorption. The remaining water reabsorption along the nephron is contingent on the interaction of renal interstitium osmolality, tubule water permeability, and ADH levels. ADH, which is produced in the hypothalamus, plays a crucial role in regulating plasma osmolality by increasing water reabsorption along the distal convoluted tubule and collecting ducts. Hypothalamic osmoreceptors sense deviations in plasma tonicity (changes of 1 per cent), enhancing and inhibiting ADH output with increased and decreased urine osmolality respectively. Large disturbances in extracellular fluid volume (10 per cent) also modify ADH secretion via baroreceptor mediators, sometimes overruling the regulation by tonicity; for example, increased ADH complicates and exacerbates hyponatremia associated with volume depletion or cardiac failure. Catecholamines, angiotensin II, prostaglandins, and nicotine may stimulate ADH secretion via baroreceptor activation. Other non-osmotic stimuli of ADH release are nausea, pain, anxiety, drugs, hypothyroidism, hypopituitarism, and pulmonary and cerebral disorders. Isoproterenol (isoprenaline) increases ADH release (by stimulating β-adrenergic receptors), while it is indirectly suppressed by the vasopressor norepinephrine (noradrenaline), dopamine antagonists such as haloperidol and promethazine (which inhibit the emetic center), and ethanol. Abnormal ADH release may also occur in tumors; the syndrome of inappropriate release of ADH is a difficult diagnosis since it requires the exclusion of all the causes mentioned above for non-osmotic ADH release.

Potassium balance

The kidneys finely regulate potassium balance by both reabsorption and secretion. Two-thirds of filtered potassium is isosmotically reabsorbed in the proximal tubule, with a fixed fraction (80–90 per cent) being reabsorbed prior to the distal convoluted tubule. Distal nephron segments (distal convoluted tubule and collecting ducts), capable of both reabsorbing and secreting potassium, are the principal sites for homeostatic control of potassium excretion. Potassium concentration in renal distal tubule cells, the size of the cell and tubular lumen electrochemical gradients, and luminal flow rate all influence potassium secretion which is mediated by Na^+,K^+-ATPase pumps. Factors directly correlated with potassium excretion include increased dietary potassium, plasma potassium, aldosterone levels, tubular flow rate, sodium excretion, diuretics (furosemide, thiazide, acetazolamide), and alkalosis. Conversely, dietary potassium depletion, low plasma potassium or aldosterone levels, diminished tubular flow rate or sodium excretion, primary adrenal insufficiency, and acidosis reduce potassium excretion. Additionally, potassium-sparing diuretics, trimethoprim, inhibitors of renin release (β-blockers), and angiotensin-converting enzyme inhibitors diminish potassium excretion. Finally, the renal response to potassium overload is rapid and entails a concerted action of multiple regulatory mechanisms, while renal compensation for potassium depletion is slower and potentially paradoxical. If potassium intake is restricted in a patient undergoing surgery, stress-induced aldosterone secretion may augment potassium excretion. Furthermore, concurrent infusions of saline will enhance tubule flow rate and potassium excretion. Another example is combined volume and potassium depletion which stimulates the renin–angiotensin–aldosterone system, promoting potassium excretion (Velazquez and Wright 1993).

Acid–base balance

The kidney normally maintains plasma pH within a narrow range (7.36–7.44) by reabsorbing filtered bicarbonate and subsequently regenerating bicarbonate. The body produces daily roughly 50 to 100 mmol (1 mmol H^+/kg/day) of non-volatile or fixed acids (phosphoric and sulfuric acids derived from protein catabolism, lactic acid, ketone bodies). Bicarbonate functions as an essential buffer, minimizing disturbances in pH. Bicarbonate undergoes virtually complete reabsorption in the proximal tubules provided that a critical serum threshold level of 24 to 26 mmol/l is not exceeded. Plasma bicarbonate levels above this threshold cause the excess bicarbonate to be excreted, consequently returning plasma bicarbonate to threshold levels. Angiotensin II and parathyroid hormone (PTH) increase and decrease proximal tubule bicarbonate reabsorption respectively. Urinary hydrogen ions are buffered and excreted as titrable acid, combining with phosphorus or sulfur. This process accounts for one-third of acid excretion, but the primary mode of acid excretion is ammoniagenesis. Proximal tubule cells metabolize glutamine, forming ammonium (ammonia combined with H^+). Titration and ammoniagenesis regenerate bicarbonate, replacing the consumed buffered bicarbonate. Factors augmenting renal H^+ secretion (Na–H^+ and K–H^+ exchange) include volume depletion, acidosis, hypokalemia, and hyperaldosteronism, whereas the opposite effects diminish renal H^+ secretion. Anion gap metabolic acidosis due to an accumulation of non-volatile acids develops when the glomerular filtration rate falls below 50 per cent, reflecting a reduction in nephron mass and ammoniagenesis. Patients with stable chronic renal failure exhibit relatively mild acidosis. Bone serves as an additional alkaline buffer to combat persistent H^+ excess, contributing to renal osteodystrophy.

Vitamin D, calcium, and phosphorus homeostasis

Vitamin D is converted to its active form 1,25-dihydroxyvitamin D by hydroxylation in the kidney, stimulating bone, intestine, and kidney to raise calcium and phosphorus levels. Hypophosphatemia, elevated PTH output, and low 1,25-dihydroxyvitamin D levels increase vitamin D production, while the opposite effects decrease its synthesis. Progressive loss of renal tissue impairs the kidney's ability to generate activated vitamin D.

The kidney plays a major role in calcium homeostasis, housing the hydroxylase enzyme which activates vitamin D. Furthermore, PTH directly enhances distal nephron calcium reabsorption. Roughly 60 per cent of the plasma calcium is filterable; the remainder is bound to albumin. Approximately 98 per cent of filtered calcium is reabsorbed and parallels sodium reabsorption in nephron segments up to the distal convoluted tubule. Loop diuretics promote calciuria by inhibiting sodium and calcium transport in the thick ascending loop of Henle, whereas thiazide diuretics, acting at the distal convoluted tubule, induce natriuresis and hypocalciuria. Finally, acidosis and alkalosis decrease and increase calcium binding respectively. Hence the metabolic acidosis of renal failure may increase ionized calcium levels while total serum calcium remains unchanged.

Several factors, namely dietary phosphorus load and PTH, influence renal regulation of serum phosphorus. An estimated 90 to 95 per cent of phosphorus is filterable; the remainder is protein bound. Roughly 5 to 20 per cent of filtered phosphorus is normally excreted. Increases in PTH levels and phosphorus intake hinder phosphorus reabsorption and promote phosphorus excretion. Other factors, such as extracellular fluid volume expansion, calcitonin, glucocorticoids, metabolic acidosis or alkalosis, and glucagon, induce hyperphosphaturia, whereas 1,25-dihydroxyvitamin D, insulin, growth hormone, and respiratory acidosis generate hypophosphaturia. Phosphorus retention typically occurs when the glomerular filtration rate falls below 25 ml/min.

Magnesium regulation

About 1 per cent of total body magnesium is in the extracellular fluid, of which 80 per cent is filterable and the remainder is protein bound. Roughly 95 per cent of filtered magnesium is normally reabsorbed and the kidney's range of magnesium excretion extends from 0.5 to 70 per cent. Extracellular fluid volume expansion, increased magnesium intake, calcium, or glucose load, alcohol ingestion, loop diuretics, cisplatin, and aminoglycosides promote magnesium excretion. Magnesium depletion and elevated PTH levels enhance magnesium reabsorption.

Red blood cell mass regulation

Erythropoietin, a hormone produced mainly by the kidney, stimulates bone marrow differentiation and production of red blood cells. Decreased renal oxygen delivery secondary to hypoxia, anemia, or ischemia augments erythropoietin synthesis. Additional stimulants of erythropoietin production include vasoconstrictors such as norepinephrine, ADH, and angiotensin. Suppressants of erythropoietin synthesis include hyperoxia, increased red cell blood volume, and reduced renal mass. Furthermore, the adenosine antagonist theophylline inhibits erythropoietin production.

Renal metabolism

In addition to its higher blood flow per weight than other vital organs, the kidney exhibits a higher rate of oxygen consumption than all other organs except the heart. The kidney normally plays a major role in the catabolism of low-molecular-weight proteins (insulin, glucagon, PTH). Reductions in glomerular filtration rate reduce protein filtration, hence decreasing renal protein catabolism and elevating plasma protein levels. Renal degradation of larger proteins (albumin) is normally minimal, although increased protein excretion (nephrotic syndrome) significantly enhances intrarenal protein catabolism.

References

Brezis, M. and Rosen, S. (1995). Hypoxia of the renal medulla—its implications for disease. *New England Journal of Medicine*, **332**, 647–55.

Neild, G.H. (1990). Endothelial and mesangial cell dysfunction in acute renal failure. In *Acute renal failure in the intensive therapy unit* (ed. D. Bihari and G. Neild), pp. 77–89. Springer-Verlag, Berlin.

Vander, A.J. (1991). Control of sodium and water excretion. In *Renal physiology* (ed. A.J. Vander), pp. 112–38. McGraw-Hill, New York.

Velazquez, H. and Wright, F.S. (1993). Tubular potassium transport. In *Diseases of the kidney* (ed. R.W. Schrier and C.W. Gottschalk), pp. 181–206. Little, Brown, Boston, MA.

6.2 Oliguria

6.2.1 Diagnosis and assessment

C. Ronco and A. G. Fabris

Key messages

- Oliguria is a state in which the urine volume is less than 500 ml/day or less than 20 ml/h. Anuria applies to urine volumes below 50 to 100 ml/day.
- Prerenal azotemia should be suspected in all conditions characterized by hypovolemia.
- In elderly patients, whose kidneys are already nephrosclerotic, hypotension and markedly reduced renal perfusion may cause anuria for many hours.
- The absence of cellular elements in the urinary sediment or protein is most compatible with prerenal and postrenal azotemia.

Introduction

Oliguria and anuria are two clinical conditions frequently encountered in patients admitted to intensive care departments. By definition, oliguria is a state in which the urine volume is less than 500 ml/day or less than 20 ml/h. A normal individual on a standard diet excretes about 600 mosmol/24 h of solute to maintain adequate metabolic balance. Urea generally constitutes 40 to 50 per cent of the solute load, while sodium and potassium salts represent the bulk of the balance. Because the maximum urinary concentration is 1200 mosmol/kg H_2O, 500 ml/24 h represents the minimum amount of urine output required to excrete such a solute load.

Anuria is defined as a state in which there is a total absence of urine. However, there is general agreement that this term applies to urine volumes of less than 50 to 100 ml/day.

Diagnostic assessment

Clinical assessment

Two important questions arise in oliguria and anuria.

1. Are these conditions accompanied by renal failure?
2. If the answer is positive, is the renal failure acute or chronic?

Since oliguria and anuria are not always synonymous with renal impairment, the first question is very important because a correct diagnosis avoids unnecessary and expensive procedures. If an individual is on a low-protein diet or if a patient is unable to break down protein to urea, as in cirrhosis of the liver, both the amount of urea to be excreted and the amount of water required to excrete it will be diminished. Since urea constitutes half the normal solute load to be excreted, the urine output necessary to maintain metabolic balance will be reduced to the oliguric range. Oliguria may also represent the extreme physiological condition in which a normal kidney reduces fluid loss in dehydration.

Once the diagnosis of renal failure has been established by an increase in azotemia and creatininemia, the first step is the differentiation of acute from chronic renal failure. Long-standing and unknown renal impairment may be suspected from a history of hypertension, hematuria or proteinuria, polyuria and nocturia, pruritis, neuropathy, and symptoms related to chronic anemia, as well as by the presence of predisposing diseases such as diabetes or a family history of renal disorders such as polycystic kidneys. Objective findings, such as hypertensive retinopathy, band keratopathy, uremic pallor, and skin pigmentation, support the suspicion of chronic renal disease. Although the kidneys are usually small in chronic renal failure, it should be remembered that the kidney may be of normal size or enlarged in chronic uremia secondary to diabetes mellitus, amyloidosis, or polycystic disease.

In acute renal failure, conventionally divided into prerenal, renal, and postrenal forms, a meticulous anamnestic history, a full clinical examination, and an adequate laboratory and radiological assessment are the fundamental steps in establishing a correct diagnosis in a disease characterized by a broad spectrum of clinical presentations. Additionally, clinical acumen plays an important role in recognizing those forms, such as prerenal and postrenal, in which prompt therapy could rapidly reverse the factors responsible for renal impairment.

Prerenal azotemia should be suspected in all conditions characterized by hypovolemia such as hemorrhage, gastrointestinal, renal, or skin losses, and capillary leak. Clinical assessment includes symptoms of thirst and evidence of orthostatic hypotension and tachycardia, decreased skin turgor, dry mucous membranes, and reduced axillary sweating. Definitive diagnosis cames from prompt resolution of acute renal failure after expansion of extracellular fluid volume. Prerenal azotemia may also be encountered in clinical conditions such as nephrotic syndrome and hepatic or cardiac failure, characterized by decreased effective plasma volume; clinical examination may reveal signs of chronic liver disease, advanced cardiac failure, and nephrotic syndrome.

In intrinsic renal failure, a history of prolonged hypotension or surgery, recent administration of nephrotoxic medications or radiocontrast agents, previous excessive exercise, or alcohol or drug abuse, supportive of rhabdomyolysis, may indicate acute tubular necrosis. Recent ingestion of drugs associated with fever, rash, or arthralgias suggests allergic interstitial nephritis.

Table 1 Indices used for the differential diagnosis of prerenal and renal azotemia

Diagnostic index	Prerenal azotemia	Renal azotemia
FE_{Na} (%)	<1	>1
Urinary Na concentration (mmol/l)	<20	>20
Urinary Cr/plasma Cr	>40	<20
Urinary BUN/plasma BUN	>8	<3
Urine osmolality (mosmol/kg H_2O)	>500	<350
Plasma BUN/Cr	>20	<15
Renal failure index UNa/UCr/PCr	<1	>1
FE_{Ua}	<11	>20

FE_{Na}=(UNa × PCr)/(PNa × Ucr) × 100, fractional excretion of Na^+; UNa, urine sodium concentration; UCr, urine creatinine concentration; PNa, plasma sodium concentration; PCr, plasma creatinine concentration; FE_{Ua}, fractional excretion of uric acid.

A history of brown-colored urine following a throat or skin infection or a butterfly rash associated with arthritis, serositis, photosensitivity, Raynaud's phenomenon, or abdominal pain, fever, weight loss, and palpable purpura, sometimes associated with inflammation of the upper airways, may be suggestive of acute glomerulonephritis, systemic lupus erythematosus, and vasculitis.

Acute renal arterial occlusion may be secondary to atrial fibrillation or recent myocardial infarction as a source of emboli. Suggestive clinical features include flank pain, fever, nausea, or vomiting, while lactate deydrogenase, glutamate oxaloacetate transaminase, and alkaline phosphatase may be elevated.

Patients with postrenal azotemia may present suprapubic or flank pain or a history of oliguria alternating with polyuria. Prostatic disease should be suspected in patients with a history of nocturia, frequency, hesitancy, postvoid dribbling, and an enlarged prostate gland on rectal examination. A pelvic exploration may also reveal a pelvic mass in females. In patients with bladder-neck obstruction, bladder distension may be evident by dullness of the lower abdomen.

Urinalysis

One of the fundamental steps of laboratory diagnosis is the analysis of urine and its composition. Since the majority of cases of acute renal failure are accompanied by oliguria, the presence of anuria, particularly if it is abrupt, suggests either obstruction of the urinary tract or thrombosis or embolism of the renal arteries. It should be remembered that hypotension and markedly reduced renal perfusion may cause anuria for many hours in elderly patients whose kidneys are already nephrosclerotic. Acute cortical necrosis observed in pregnant women and rapidly progressive glomerulonephritis may also be accompanied by anuria, and this condition rarely occurs in acute tubular necrosis. Finally, anuria following a pelvic operation should lead to the suspicion that the ureters have been ligated during the surgical procedure. Often, the diagnosis of anuria may be delayed because previously formed urine is found on catheterization of the bladder, and a variable period of time is required to confirm that no more urine has been produced.

The absence of cellular elements in the urinary sediment or protein is most compatible with prerenal and postrenal azotemia, although transparent hyaline casts may be seen in the former and pyuria and bacteriuria in the latter if infection is present.

Heavy proteinuria may occur in prerenal azotemia if severe congestive heart failure is present. An active sediment with pigmented muddy-brown granular casts, tubular epithelial cell casts, and cellular debris are characteristic of ischemic or nephrotoxic acute tubular necrosis; however, casts may be absent in about 20 to 30 per cent of these patients (Levinsky *et al.* 1981). A few red and white blood cells per high power field and tubular proteinuria in the range of 1 g/day may also be present. Red blood cell casts almost invariably indicate a glomerular injury; they are accompanied by dysmorphic red blood cells and sometimes by white blood cells and by epithelial and granular casts. Glomerular proteinuria exceeding 1 g/day, often in the nephrotic range, is also present in glomerulonephritides where the diagnosis can be confirmed by the positivity of immunological markers.

Heavy proteinuria is also frequent in myeloma. White blood cells and granular casts suggest interstitial nephritis; eosinophiluria (between 1 and 50 per cent of urine leukocytes) is common in drug-induced interstitial nephritis. Heavy proteinuria is frequent in patients with allergic interstitial nephritis secondary to non-steroidal anti-inflammatory drugs, because in these patients glomerular lesions resembling those of minimal-change glomerulonephritis are associated with acute interstitial inflammation. A red, red-brown, or brown coloration of urine, positive for hemoglobin by dipstick, and a negative microscopic examination for red blood cells are suggestive of hemoglobinuria or myoglobinuria. The differentiation between hemolysis and rhabdomyolisis is carried out by inspection of spun serum which is pink in hemolysis and normally clear in rhabdomyolisis (Glassock 1989). Additionally, muscular lesions are characterized by tests confirming elevation of creatine phosphokinase, lactate dehydrogenase, transaminases, uric acid, aldolase, and myoglobin (Glassock 1989). Finally, with regard to the presence of crystals in the urine, an abundance of urate should raise the possibility of tumor lysis syndrome, while a high level of oxalate should suggest a diagnosis of ethylene glycol or methoxyfluorane toxicity.

The assessment of urine composition may give considerable help in distinguishing prerenal from intrinsic renal disease (mostly acute tubular necrosis). Since tubular function is well preserved in prerenal azotemia, tubules avidly reabsorb sodium and water at the proximal as well as the distal tubules. Thus the small amount of urine eliminated is almost free of sodium and is highly concentrated. The solutes that give the urine its high concentration are primarily urea and other waste products that are excreted in a small volume of water. Therefore in prerenal azotemia the urinary sodium concentration is less than 20 mmol/l, the urine osmolality is greater than 500 mosmol/kg H_2O, and the urine-to-plasma ratios of blood urea nitrogen (BUN) and creatinine are greater than 8 and 40 respectively (Brady *et al.* 1996). In contrast, in acute tubular necrosis the urine osmolality is less than 350 mosmol/kg H_2O, the urinary sodium concentration is greater than 20 mmol/l, and the urine-to-plasma ratios of BUN and creatinine are less than 3 and 20 respectively (Brady *et al.* 1996). The ratio of BUN to creatinine in plasma is of some value in evaluating prerenal azotemia. Since urea clearance is flow dependent, the decreased urine flow associated with intact tubular function is associated with reduced urea clearance in prerenal azotemia (Schrier and Conger 1980). Therefore the increase in BUN may be more rapid than the increase in serum concentration of creatinine, since creatinine clearance is not flow dependent (Schrier and Conger 1980). Consequently a BUN-to-creatinine ratio above 20 suggests prerenal failure, while in acute tubular necrosis this ratio is less than 15. Additional urinary values provide good clues to differentiating prerenal from acute renal failure. One of these, the fractional excretion of sodium, relates sodium clearance to creatinine clearance; it is less than 1.0 per cent in prerenal azotemia and greater than 1.0 per cent in renal failure (Espinel 1976).

Table 1 summarizes urinary indices utilized in oliguric renal failure. None of them has an absolute value in differential diagnosis because of overlapping factors. Prior treatment with osmotic agents such as mannitol or glucose gives a urine osmolality much higher than expected. Urinary sodium concentration may be affected by the use of diuretics or the presence of bicarbonaturia. In the latter condition urinary sodium concentration frequently exceeds 20 mmol/l, since bicarbonate, an anion, must be accompanied by increased cation losses (sodium and potassium) to maintain electrical neutrality. In this setting urinary chloride is a better indicator of prerenal failure than urinary sodium, and its concentration should be very low (< 20 mmol/l). Acute urinary tract obstruction for a few hours may be associated with renal ischemia and thus with a urinary picture similar to that in prerenal azotemia (Schrier and Conger 1980). Urine/plasma creatinine or urea ratios and serum urea nitrogen/creatinine ratio are of limited value in differential diagnosis, particularly for elderly people, in whom the concentrating mechanism is frequently impaired, while the mechanism of sodium reabsorption is preserved.

Additional evaluation

Additional evaluation may be of aid in the diagnosis of acute uremia. Severe hypocalcemia at the onset of oliguria occurs in rhabdomyolysis and pancreatitis. Very high levels of uric acid accompany tumor lysis, trauma, or muscle damage. Eosinophilia is frequent in acute interstitial nephritis. Hemolytic uremic syndrome and thrombotic thrombocytopenic purpura are characterized by hemolytic anemia, thrombocytopenia, and the presence of schistocytes, and disseminated intravascular coagulation is characterized by anemia, thrombocytopenia, and prolonged prothrombin, partial thromboplastin, and thrombin times associated with low fibrinogen titers and elevated levels of fibrin degradation products.

Ultrasonography can evaluate the number and the size of the kidneys, the thickness of the parenchyma and its echogenicity, and the presence of hydronephrosis (it should be remembered that its absence does not completely rule out obstruction), and Doppler tests provide information on renal arterial flow.

The only way to carry out a correct diagnosis in oligoanuric patients in whom prerenal and postrenal causes can be excluded and renal causes are not completely clear is a renal biopsy.

References

Brady, H.R., Brenner, B.M., and Lieberthal, W. (1996). Acute renal failure. In *The kidney* (5th edn) (ed. B.M. Brenner), pp. 1200–52. W.B. Saunders, Philadelphia, PA.

Espinel, C.H. (1976). The FE_{Na} test: use in the differential diagnosis of acute renal failure. *Journal of the American Medical Association*, **236**, 579–81.

Glassock, R.J. (1989). Hematuria and pigmenturia. In *Textbook of nephrology* (ed. S.G. Massry and R.J. Glassock), pp. 491–500. Williams & Wilkins, Baltimore, MD.

Levinsky, N.G., Alexander, E.A., and Venkatachalam, M.A. (1981). Acute renal failure. In *The kidney* (ed. B.M. Brenner and F.C. Rector Jr), pp. 1181–1236. W.B. Saunders, Philadelphia, PA.

Schrier, R.W. and Conger, J.D. (1980). Acute renal failure: pathogenesis, diagnosis, and management. In *Renal and electrolyte disorders* (ed. R.W. Schrier), pp. 375–407. Little, Brown, Boston, MA.

6.2.2 Management

J.J. Cordingley and M.G.A. Palazzo

Key messages

- Restoration of renal perfusion by optimization of blood volume and correction of low cardiac output and hypotension are the most important objectives for achieving improved renal function.
- Speed in initiating treatment is particularly important in preventing progression to established acute tubular necrosis.
- Dopamine and diuretics are frequently administered to improve urine output but their efficacy in reducing the need for acute renal replacement therapy has not been substantiated by prospective controlled clinical trials.
- The use of natriuretic peptides awaits the results of further clinical studies.

Introduction

In over 80 per cent of patients who develop oliguria and acute renal failure (ARF) the cause can be directly related to hypovolemia associated with a predisposing condition such as diabetes mellitus, hypertension, or use of nephrotoxic drugs. The aggression with which patients with impending renal failure are managed requires a judgment on whether generous fluid loading and inotropes in a patient with limited cardiorespiratory reserve would result in a better outcome than early use of hemofiltration.

Avoidance of risk factors

Patients at obvious risk of ARF are usually easy identified and include those with existing renal impairment or those who have had prolonged episodes of renal hypoperfusion or sepsis, or who have been exposed to large doses of nephrotoxic drugs. However, there are many patients with serum creatinine concentrations in the upper range of normal who have 'concealed' impairment of renal function and who are susceptible to minor renal insults. These include small patients (low muscle mass), diabetics, and hypertensives. The reason for their exquisite sensitivity is the non-linear relationship between renal function and serum creatinine concentration. Serum creatinine rises above the upper level of laboratory normal ranges only after some 50 per cent of glomeruli fail (Fig. 1).

There are a number of risk factors which are potentially avoidable or to which exposure can be substantially reduced.

Nephrotoxic agents

The use of drugs with known nephrotoxic effects should be avoided where possible. However, there may be situations in which an assessment of the relative risks is in favor of a potentially nephrotoxic agent, for example use of an aminoglycoside in the management of Gram-negative septicemia.

When nephrotoxic agents are used care should be taken to minimize adverse effects by monitoring drug levels. In addition, toxicity may be reduced by adopting an optimal dosing regimen. For example, nephrotoxicity of aminoglycosides may be reduced by once daily administration (Hatala *et al.* 1996) and close monitoring of trough concentrations. A different formulation, such as use of liposomal amphotericin B, may reduce nephrotoxicity.

Other risk factors

A number of preventable renal insults due to episodes of hemodynamic instability may be inadvertently repeated throughout a patient's stay in intensive care and result in renal impairment and oliguria. These insults may follow changes in patient position, interruption of inotrope infusions, chest physiotherapy, hypotension following an excessive dose of induction agent for emergency tracheal intubation, or episodes of atrial fibrillation precipitated by overlooked hypokalemia and hypomagnesemia.

It should be noted that positive pressure ventilation alone may result in activation of physiological mechanisms directed at sodium and water retention through reduced atrial natriuretic peptide release and stimulation of aldosterone and ADH. The reduction in urine output is frequently reversed by fluid loading.

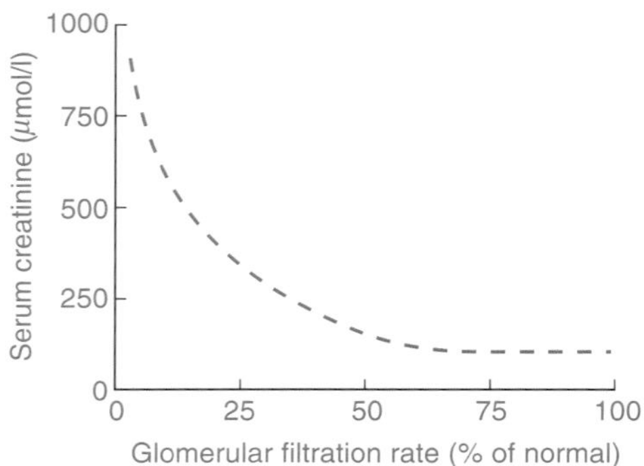

Fig. 1 The relationship between glomerular filtration rate and serum creatinine.

Practical management of oliguria and impending ARF

The general principles of the management of oliguria are to exclude urinary tract obstruction and restore normal intravascular volume, adequate cardiac output, and blood pressure. This treatment is supplemented by the use of low-dose dopamine and diuretics. The efficacy of such an approach is variable and is determined by premorbid renal function, the size and duration of the insult, and the period that has elapsed without restorative therapy.

It is important to note that, although urine output is the most commonly used bedside monitor of renal performance, an increase in urine output does not necessarily imply an improved glomerular filtration rate. For example, low-dose dobutamine has been demonstrated to increase glomerular filtration rate but not urine output, while low-dose dopamine increased urine output but not glomerular filtration rate in hemodynamically stable critically ill patients (Duke *et al.* 1994).

Exclude urinary tract obstruction

The initial management of the oliguric patient is to exclude urinary tract obstruction. Anuria is usually due to urinary catheter obstruction and is remedied by flushing. It should be noted that anuria is uncommon in patients with either impending or established oliguric acute tubular necrosis, who frequently continue passing urine volumes of up to 20 ml/h.

Obstruction of the urinary tract can be confirmed by ultrasound of the renocalyceal system which may demonstrate hydronephrosis. Patients with evidence of obstruction require urgent intervention to provide free drainage of urine.

Elevated intra-abdominal pressure can be an important factor in producing oliguria and may follow abdominal trauma or postoperative intra-abdominal bleeding. Occasionally it may be associated with ascites under tension. The mechanism of oliguria secondary to raised intra-abdominal pressure is not clearly established but is probably due to a combination of increased renal vascular resistance, reduction in cardiac output, and direct pressure on the renal pelvis.

A bedside estimate of intra-abdominal pressure can be conveniently obtained by transducing a urinary catheter (water manometer) or femoral venous line. A clinically significant elevated pressure is more than 30 mmHg above the pubis. An abdomen generating such pressures in the presence of oliguria should be considered for decompression.

Correction of hypovolemia

Correction of hypovolemia is the most important aspect of the management of oliguria. Administration of fluid optimizes cardiac performance and reverses the compensatory mechanisms which mediate fluid retention.

Overt hypovolemia is managed by rapid fluid resuscitation, preferably by warm colloids which may achieve normal intravascular volume status faster than crystalloids. Ideally, patients should have a pulmonary flotation catheter in place for assessment of left ventricular stroke volumes in response to intermittent fluid challenges.

In some patients aggressive fluid administration may lead to interstitial pulmonary edema due to increased pulmonary capillary permeability. Pulmonary interstitial edema can decrease pulmonary compliance, and such a trade-off may be acceptable. However, other patients may unpredictably develop alveolar edema in which oxygenation deteriorates rapidly. Unfortunately, in the oliguric patient this may not be easily reversible other than by venesection or hemofiltration. Diuretics and vasodilators may be useful in this situation but usually only temporarily. Judging how rapidly intravenous fluids should be given, particularly in the non-ventilated patient with a permeability defect, may become a matter of trial and error. The fear of alveolar edema, particularly in patients with poor oxygenation from a primary pulmonary insult, may be justification to refrain from aggressive fluid therapy, an acceptance of ARF, and early management with hemofiltration.

Maintenance of adequate cardiac output and blood pressure

Following optimization of intravascular volume, the passage of small volumes of concentrated urine, with a low sodium concentration and high osmolality, may be continuing evidence for inadequate renal blood flow. This may be secondary either to low cardiac output or systemic arterial hypotension.

Cardiac output

An increase in cardiac output generally increases renal blood flow and glomerular filtration provided that blood pressure is maintained. It is not possible to recommend a specific target for cardiac output to restore renal perfusion. The adequacy of cardiac output for an individual is that required to achieve acceptable tissue perfusion and is extremely variable between patients.

When cardiac output is thought to be insufficient after fluid loading it may be improved by a combination of vasodilators and inotropes. The choice depends on whether the clinician wishes to increase cardiac output through preload, afterload, contractility, or heart rate. Most drugs will alter at least two of these simultaneously.

Blood pressure

Renal blood flow and glomerular filtration rate are normally autoregulated over a range of mean arterial blood pressures from 60 to 160 mmHg. However, autoregulation may be impaired in ischemic acute tubular necrosis and sepsis, or shifted to the right in patients with chronic hypertension or renovascular disease. Thus renal blood flow may be more directly related to systemic arterial blood pressure than expected.

Elevation of mean arterial blood pressure with norepinephrine (noradrenaline) improves glomerular filtration rate and urine output in oliguric hemodynamically stable surgical patients and patients with hyperdynamic septic shock. It is important that vasoconstrictors are only used with full hemodynamic monitoring and regular patient assessment in order to avoid potential falls in tissue perfusion, which are more likely to occur in hypovolemic patients and those with low cardiac output.

The ideal blood pressure to aim for must again be subject to individual patient assessment which is mostly dependent on such factors as premorbid blood pressure and cardiac function.

Specific measures

Dopamine

Low-dose dopamine infusion (1–3 g/kg/min) has been used for more than 25 years as part of the management of oliguria. The renal actions of dopamine include renal arteriolar vasodilatation and a diuretic effect at the proximal tubule (Duke and Bersten 1992). However, dopamine has a significant inotropic effect even at so-called 'renal-dose' infusion rates, and it has not been possible to separate any specific renal increase in glomerular filtration rate from that due to an increase in cardiac output. In a number of studies dopamine has been shown to increase glomerular filtration rate, urine output, and induce natriuresis. However, there have been no prospective randomized controlled studies to determine whether administration of low-dose dopamine to oliguric patients reduces the requirement for renal support.

Furosemide (frusemide)

There is experimental evidence that furosemide has renoprotective properties by improving tubular oxygen balance. This is thought to be due to a reduction for the requirement of Na^+,K^+-ATPase and an increase in vasodilatory prostaglandin E_2 concentrations which combine to improve oxygen balance in the medullary tubule.

Furosemide has traditionally been used in large bolus doses (up to 1 g) in the management of oliguria when other measures have failed. This approach may increase urine output but has not been shown to reduce the time course or mortality of renal failure. Indeed, large doses of furosemide can be nephrotoxic with concurrent use of aminoglycosides or cephalosporins.

It is notable that intravenous infusion of furosemide has a greater diuretic effect than intravenous bolus administration (van Meyel *et al.* 1992).

Mannitol

Boluses of mannitol are also commonly used in oliguric patients and frequently result in increased urine output. Mannitol is an osmotic diuretic but is also an oxygen free-radical scavenger, increases renal tubular flow, causes a rise in renal prostaglandin E_2 concentrations, and reduces cell swelling following ischemia.

It is important that intravascular volume should be optimized prior to the use of any diuretic agents as further depletion may worsen renal perfusion.

Natriuretic peptides

Natriuretic peptides (atrial natriuretic peptide (ANP), ANP analogs, and urodilatin) increase glomerular filtration pressure by renal afferent arteriolar dilatation and efferent arteriolar constriction. They also increase glomerular permeability and enhance sodium and water excretion. ANP infusion reverses the antinatriuretic and antidiuretic effects of positive pressure ventilation in intensive care patients (Andrivet *et al.* 1991).

ANP infusion can result in systemic hypotension but this is less likely with urodilatin, a natriuretic peptide produced in the kidney.

Further studies in general intensive care patients are required to explore the potential of these agents to reduce the incidence of acute tubular necrosis. It is worth noting that aggressive intravascular fluid loading will also increase ANP production and may achieve similar effects. The potential advantage of ANP is that it may be given instead of volume and therefore may minimize other risks of volume loading.

Calcium antagonists

Although there have been no prophylactic studies with calcium antagonists among critical care patients, they have been found to reduce the incidence of post-transplant acute tubular necrosis.

Theophylline

Theophylline and related drugs have a number of potentially beneficial effects in maintaining renal function. These include an inotropic action, a proximal tubular diuretic effect, and antagonism of adenosine. In the kidney, adenosine production is increased under hypoxic conditions and reduces glomerular filtration rate as part of the tubuloglomerular feedback mechanism.

Animal models of ARF have demonstrated a potential protective effect of theophylline and other adenosine antagonists. In the clinical setting administration of theophylline prior to radiocontrast media has been shown to prevent a reduction in glomerular filtration rate.

Prophylaxis of ARF in specific conditions

Contrast media nephropathy

Renal dysfunction following radiocontrast media is rare in patients with normal renal function but more common in those with renal disease, particularly diabetic nephropathy.

The most important prophylactic measure against contrast nephropathy is maintenance of hydration. Risk of renal dysfunction is also reduced by using smaller volumes of contrast and avoiding repeated doses at short time intervals. Use of non-ionic contrast media is associated with less renal dysfunction in patients with pre-existing renal disease. Prophylaxis with mannitol, furosemide, dopamine, and ANP has had variable results.

Rhabdomyolysis

The primary therapy for restricting renal damage after rhabdomyolysis is early optimization of intravascular volume and induction of a diuresis. Alkalization of the urine has also been shown to be of benefit in animal models and has been recommended from clinical experience. Following establishment of a diuresis, sodium bicarbonate should be infused to maintain a urine pH above 6.5 but should be carefully monitored to avoid hypocalcemia.

References

Andrivet, P., *et al.* (1991). Hormonal interactions and renal function during mechanical ventilation and renal function and ANF infusion in humans. *Journal of Applied Physiology*, **70**, 287–92.

Duke, G. and Bersten, A. (1992). Dopamine and renal salvage in the critically ill patient. *Anaesthesia and Intensive Care*, **20**, 277–302.

Duke, G., Briedis, J., and Weaver, R. (1994). Renal support in the critically ill patients: low-dose dopamine or low-dose dobutamine? *Critical Care Medicine*, **22**, 1919–25.

Hatala, R., Dinh, T., and Cook, D. (1996). Once-daily aminoglycoside dosing in immunocompetent adults: a meta-analysis. *Annals of Internal Medicine*, **124**, 717–25.

van Meyel, J., Smits, P., Russel, F., Gerlag, P., Tan, Y., and Gribnau, F. (1992). Diuretic efficiency of furosemide during continuous administration versus bolus injection in healthy volunteers. *Clinical Pharmacology and Therapeutics*, **51**, 440–4.

6.3 Acute renal failure

6.3.1 Pathophysiology

J. Reichman and M. Brezis

Key messages

- Prerenal acute renal failure is consequent to renal hypoperfusion and denotes a functional impairment without renal parenchymal damage. The injury is reversible once perfusion is restored.
- Between 25 and 40 per cent of cases of acute renal failure are ascribed to intrinsic renal damage, predominantly induced by prolonged renal hypoperfusion (as for prerenal) or nephrotoxins (contrast, antibiotics, rhabdomyolysis), and most often by their combination.
- Renal ischemia induces tubular epithelial damage, impairing tubular reabsorptive properties and urinary concentration.
- Renal hypoperfusion combined with administration of nephrotoxic agents (gentamicin, radiocontrast, indomethacin) induces intrarenal injury, possibly in part by disrupting the balance between medullary oxygen demand and supply.

Introduction

Acute renal failure (ARF) is a clinical entity characterized by an abrupt (hours to days) decline in glomerular filtration rate, resulting in the retention of nitrogenous waste products and pertubation of water, electrolyte, and acid–base balance. It is typically associated with daily increments in serum urea and creatinine levels greater than 3.57 mmol/l and 44.2 mol/l respectively.

ARF is frequently encountered, occurring in approximately 5 per cent of all hospital patients and up to 30 per cent of intensive care unit admissions. Oliguria (urine output < 400 ml/day) is observed in 50 per cent of cases of ARF, with an associated mortality of 50 to 80 per cent. Non-oliguric ARF has a better prognosis with a mortality of 15 to 40 per cent.

As illustrated in Table 1, a significant decline in glomerular filtration rate (of more than 50 per cent) may derive from prerenal azotemia, acute intrinsic renal failure, or obstructive uropathy (postrenal azotemia).

Prerenal azotemia

Prerenal ARF is consequent to renal hypoperfusion and denotes a functional impairment without renal parenchymal damage. The injury is reversible once perfusion is restored.

Multiple disturbances precipitate the prerenal state (Table 1). In response to effective hypovolemia, several compensatory neurohumoral mechanisms are activated in concert to preserve arterial blood pressure, namely the sympathetic nervous system, the renin–angiotensin–aldosterone system, ADH release, and endothelin production. The enhanced sympathetic response to hypovolemia and cardiac failure serves to reduce renal blood flow, while glomerular filtration rate is initially maintained due to a disproportionate increase in efferent arteriolar tone (angiotensin-mediated). Moreover, renal production of vasodilatory prostaglandins attenuates afferent arteriolar vasoconstriction (Ballerman *et al.* 1991). Pharmacological agents such as angiotensin-converting enzyme inhibitors and nonsteroidal anti-inflammatory drugs are particularly hazardous in the setting of decreased effective blood volume, as they interfere with renal compensatory mechanisms and precipitate a decline in glomerular filtration rate.

Hepatic disease results in pooling of blood within the portal circulation and diminished serum albumin levels, hence lowering effective blood volume and renal perfusion. As liver failure progresses, renal cortical vasoconstriction and systemic vasodilatation further reduce glomerular filtration rate. This combined liver and kidney failure (called hepatorenal syndrome) is probably related to vasoactive substances. Among the many putative mediators are vasoconstrictors (e.g. catecholamines, thromboxane, endothelin) and systemic vasodilators (e.g. nitric oxide), and diminished intrarenal synthesis of prostaglandin and bradykinin may also play a role. Patients with hepatic failure are exquisitely sensitive to further hemodynamic insults (volume depletion from paracentesis or diuresis, hemorrhage, sepsis, cyclo-oxygenase inhibitor administration) which often induce ARF.

Renal impairment, characterized by oliguria, is often observed in sepsis. Despite apparent maintainance of arterial blood pressure in early sepsis, the kidney construes a volume-depleted state and responds by enhancing sodium reabsorption. Elevated renin, angiotensin, catecholamines, or nitric oxide (with peripheral vasodilation) may account for the stimulation of sodium retention. Endotoxins contribute to the initiation of ARF by many effects, including sympathetic vasoconstriction, increased production of cytokines, angiotensin, endothelin, or nitric oxide, volume contraction, and disseminated intravascular coagulation. Unfortunately, direct antagonists of lipopolysaccharides or cytokines have not been conclusively effective in alleviating the renal dysfunction of sepsis, which often degenerates to acute intrinsic renal failure.

Table 1 Causes of acute renal failure

Prerenal failure

Absolute decrease in blood volume
- Hemorrhage
- Skin losses (burns, sweating)
- Gastrointestinal losses (diarrhea, vomiting, surgical gastric drainage)
- Renal losses (diuretics, diabetes insipidus, salt-losing nephropathy, osmotic diuresis, adrenal insufficiency)
- Burns

Decrease in effective blood volume
- Cardiac dysfunction and/or low cardiac output (myocardial infarction, tamponade, valvular disease, arrhythmias, pulmonary embolism, cor pulmonale, positive-pressure mechanical ventilation)
- Liver disease with ascites
- Low plasma oncotic pressure (hypoalbuminemia)
- Fluid pooling (pancreatitis, peritonitis)

Peripheral vasodilation
- Sepsis
- Anaphylaxis
- Antihypertensive drugs
- Anesthesia

Renal vasoconstriction
- Hypercalcemia
- Norepinephrine (noradrenaline), epinephrine (adrenaline)
- Cyclosporine
- Amphotericin B
- Cirrhosis with ascites

Renal hypoperfusion with impairment of renal autoregulatory responses
- Cyclo-oxygenase inhibitors, angiotensin-converting enzyme inhibitors

Hyperviscosity syndrome
- Multiple myeloma, macroglobulinemia, polycythemia

Renal arterial occlusion
- Bilateral thromboembolism, renal artery stenosis, surgery/trauma, angioplasty, dissecting aneurysms, vasculitis

Intrinsic renal failure

Glomerular
- Primary glomerulonephritis (membranoproliferative)
- Secondary glomerulonephritis (hemolytic uremic syndrome, thrombotic thrombocytopenic purpura, disseminated intravascular coagulation, toxemia of pregnancy, postinfectious, accelerated hypertension, radiation nephritis, scleroderma, systemic lupus erythematosus, Goodpasture's syndrome, Wegener's granulomatosis)

Vascular
- Vasculitis, malignant hypertension, renal vein thrombosis, vasopressors, eclampsia
- Hyperviscosity syndrome, iodinated contrast agent, non-steroidal anti-inflammatory drugs

Acute tubular necrosis
- Ischemia (as for prerenal failure)
- Toxins (contrast, aminoglycosides, heavy metals, cyclosporine, amphotericin B, cisplatin, ethylene glycol, acetaminophen (paracetamol) intoxication)
- Pigments (hemoglobin, myoglobin)
- Proteins (in multiple myeloma)
- Crystals (uric acid, oxalate, sulfonamides)

Tubulointerstitial nephritis
- Allergic (β-lactams, sulfonamides, trimethoprim, rifampin (rifampicin), cyclo-oxygenase inhibitors, diuretics, captopril)
- Infectious (pyelonephritis, cytomegalovirus, HIV, candidiasis, tuberculosis)
- Infiltration (lymphoma, leukemia, sarcoidosis)
- Radiation

Renal allograft rejection

Postrenal failure

Ureteric
- Calculi, blood clot, papillary necrosis, neoplasm, retroperitoneal fibrosis or hemorrhage, fungus balls, ligation, inflammatory aortic aneurysm

Bladder outflow obstruction
- Prostatic hypertrophy, calculi, blood clot, neurogenic bladder, neoplasm, prostatitis

Urethra
- Phimosis, meatal stenosis, stricture, neoplasm, congenital valve

Acute intrinsic renal failure

Between 25 and 40 per cent of cases of ARF are ascribed to intrinsic renal damage, predominantly induced by prolonged renal hypoperfusion (as for prerenal failure) or nephrotoxins (contrast, antibiotics, rhabdomyolysis), and most often by their combination (Dinour and Brezis 1998). Acute intrinsic renal failure is synonymous with vasomotor nephropathy and acute tubular necrosis. Although acute intrinsic renal failure and the prerenal state represent two extremes of renal hypoperfusion, intrinsic failure implies renal parenchymal damage (tubular epithelium) that is not immediately reversible upon elimination of the precipitating factor and restoration of blood flow.

Putative mechanisms by which renal ischemia impairs glomerular filtration include reduction in glomerular perfusion, renal tubule cast or cellular debris obstruction, and backleak of glomerular filtrate through injured tubule epithelium. ARF is often multifactorial, involving more than one mechanism of injury.

Tubule effects

Tubular backleak plays a minor role in mild or non-oliguric forms of toxic or ischemic ARF. As the severity of renal ischemia increases, the contribution of tubular wall permeability to the suppression of glomerular filtration is enhanced. Immunoglobulin light-chain cast precipitation may complicate plasma cell dyscrasias and intraluminal pigments may play a role in rhabdomyolysis and hemoglobinuria, although the correlation between cast deposition and renal failure is weak. Tubular dilatation may be seen with intratubular precipitation of crystals (uric acid, oxalate, acyclovir, methotrexate, sulfonamides), as typically seen in the tumor lysis syndrome

Vascular and hypoxic effects

The initial phase of ischemic acute intrinsic renal failure is characterized by a reduction of more than 50 per cent in renal blood flow, a disproportionate lowering of glomerular filtration rate, and a loss of renal autoregulation. Despite receiving a greater blood flow per unit weight of tissue than most other vital organs, the kidney is highly vulnerable to oxygen deprivation owing to the countercurrent urinary concentrating system. The most vulnerable nephron segments are the straight portion of the proximal tubule (S_3 segment) and the medullary thick ascending loop of Henle because of their hypoxic milieu and high rates of oxygen consumption (active solute transport and reabsorption), while the inner medulla exhibits less metabolic demands (Brezis and Rosen 1995).

Tubuloglomerular feedback

Tubuloglomerular feedback or secondary intrarenal vasoconstriction is a regulatory mechanism which lowers glomerular filtration rate in response to increased solute delivery at the macula densa and may contribute to the decline in glomerular filtration rate in ARF. Renal ischemia induces tubular epithelial damage, impairing tubular reabsorptive properties and urinary concentration. Activation of tubuloglomerular feedback serves to conserve volume loss, precluding massive diuresis despite defective reabsorption, and to reduce the workload to the distal tubule as part of a co-ordinated system to preserve medullary oxygenation (Fig. 1).

Intrarenal balance of medullary oxygenation

The renal medulla works normally on the brink of anoxia. Regulation of oxygen supply and demand in this region is delicately regulated by multiple homeostatic systems (Fig. 1). Oxygen demand is reflected by the rate of tubular reabsorption, which is determined by the glomerular filtration rate, delivery of urine to the thick limbs, and control of solute transport by local mediators (Brezis and Rosen 1995). Nonsteroidal anti-inflammatory drugs inhibit prostaglandin synthesis,

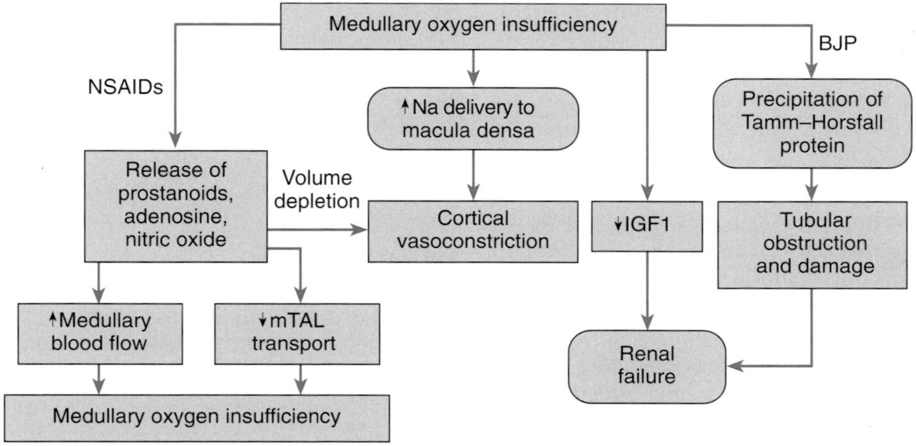

Fig. 1 Putative mechanisms leading from medullary cell hypoxia to renal failure. Physiological homeostatic signals that improve medullary oxygenation (by increased blood flow and decreased transport) and often contribute to reduced renal function are shown on the left. Some pathophysiological consequences of more advanced medullary hypoxia, such as tubular damage and reduced insulin-like growth factor 1 (IGF-1), are shown on the right. The potential adverse effects of some nephrotoxicants and volume depletion are shown. Non-steroidal anti-inflammatory drugs (NSAIDs) disable the beneficial prostanoid-mediated medullary vasodilatatory response to local hypoxia. Volume depletion enhances the tubuloglomerular feedback reflex decrease in glomerular filtration. In myeloma kidneys, Bence–Jones proteins (BJP) coprecipitate with the Tamm–Horsfall protein released by damage to medullary thick ascending limbs, increasing the likelihood of tubular obstruction and renal failure from other insults (such as NSAIDs, volume depletion, or radiographic contrast agent). Renal failure results from tubular obstruction (by casts), backleak of glomerular filtrate from the lumen to blood (through damaged epithelium), impaired intrarenal microcirculation such as activation of tubuloglomerular feedback (by increased distal delivery of solute to the macula densa), and failure of the local production of growth factors.

precipitating a decline in medullary blood flow and oxygenation. Adenosine, a by-product of ATP released during tissue hypoxia, may augment tubuloglomerular feedback and induce cortical vasoconstriction (reducing glomerular filtration rate) with medullary vasodilatation and tubular transport inhibition to limit medullary hypoxia (Brezis and Rosen 1995). Similarly, nitric oxide and endothelin induce medullary vasodilatation, and endothelin also causes cortical vasoconstriction. All these vasoactive mediators appear to be co-ordinated to preserve medullary oxygen balance in renal hypoperfusion. Deficient or excessive stimulation of these renal defensive mechanisms may generate medullary oxygen insufficiency and/or ARF (Fig. 1).

Nephrotoxins

Several drugs (non-steroidal anti-inflammatory drugs, mannitol, amphotericin, polyene antibiotics), radiocontrast agents, and endogenous toxins (myoglobin) can precipitate medullary hypoxia in rat kidney. Synergism among simultaneous insults, mediated by distinct mechanisms, is often observed in the pathogenesis of ARF. Renal hypoperfusion combined with administration of nephrotoxic agents (gentamicin, radiocontrast, indomethacin) induces intrarenal injury, possibly in part by disrupting the balance between medullary oxygen demand and supply. The kidney is subjected to both dimished oxygen supply and increased oxygen demand due to reduced perfusion and enhanced sodium reabsorption respectively. As nephrotoxins accumulate in zones prone to hypoxia (medulla), oxygen demand may be further increased by toxin-induced mitochondrial and cellular membrane injury. Additional processes observed in combined hypoxic and toxic insults include disruption of renal defensive mechanisms (non-steroidal anti-inflammatory drugs), enhanced tubuloglomerular feedback, and precipitation of crystals. Synergism between renal hypoperfusion and toxic insults in regions of poor oxygenation may constitute a major mechanism of ARF. Sufficient volume repletion and judicious use of nephrotoxic agents defends against combined renal insults and curbs the incidence of ARF (Brezis and Rosen 1995).

Aminoglycoside nephrotoxicity complicates between 0.5 and 30 per cent of patients treated. Suggested modes of toxicity include mitochondrial disruption, cell membrane damage, and immune mechanisms. Renal failure typically occurs roughly a week after treatment onset, with recovery occurring about a week after cessation of the drug. Important risk factors include age, prior renal or hepatic dysfunction, peak and/or trough levels, and duration of therapy (Linton 1990). Amphotericin causes dose-related ARF via direct tubular epithelial cell membrane toxicity. Volume depletion predisposes patients to gentamicin, cisplatin, and amphotericin nephrotoxicity, and aggressive volume repletion clearly obviates drug-induced ARF. In the setting of cardiac failure, nitroprusside reduces systemic vascular resistance more than renal vascular resistance. This may generate a steal phenomenon, diverting blood flow from the kidney and precipitating ARF (Linton 1990).

Contrast nephropathy more often afflicts those patients with combined risk factors, particularly prior renal dysfunction, diabetes mellitus, and volume depletion. Suggested pathogenetic mechanisms of contrast-induced ARF include direct tubular cell toxicity, intraluminal precipitation of proteinaceous casts or uric acid crystals, intrarenal vasoconstriction, alterations in red cell morphology, and immune complex injury (Heyman *et al.* 1994). Endothelial-derived vasorelaxors (nitric oxide and prostaglandins) are impaired in patients with diseases frequently associated with contrast nephropathy, namely diabetes mellitus, hypertension, and atherosclerosis. Hence impaired endothelial-derived vasorelaxation may play an important role in contrast nephropathy, inducing severe vasoconstriction and medullary hypoxia (Brezis and Rosen 1995).

Rhabdomyolysis and hemolysis

Rhabdomyolysis and hemolysis may cause acute intrinsic renal failure, but the precise mechanism is unclear. Neither hemoglobin nor myoglobin are nephrotoxic in animals. Postulated modes of toxicity include direct tubular epithelial cell toxicity, intraluminal cast precipitation, and nitric oxide inhibition. Volume depletion and acidosis are predisposing factors.

Other causes of acute intrinsic renal failure

Other less common etiologies of ARF include acute allergic interstitial nephritis, characterized by inflammatory cell infiltration and interstitial edema. It is typically induced by antibiotics (penicillins, cephalosporins, and sulfonamides), diuretics (hydrochlorothiazide), non-steroidal anti-inflammatory drugs, allopurinol, and angiotensin-converting enzyme inhibitors (captopril). Renal artery or vein occlusion and glomerulonephritis may also reduce glomerular filtration rate and generate ARF, the etiology of which is often suspected by the clinical setting.

Postrenal azotemia

Postrenal ARF indicates obstruction of the urethra or bladder neck, bilateral ureteric obstruction, or unilateral ureteric obstruction in patients with one functioning kidney or chronic renal failure. Common etiologies are illustrated in Table 1. Early obstruction is characterized by increased intraluminal pressure prior to the area of obstruction followed by a reduction in glomerular filtration rate. Arteriolar vasoconstriction ensues, thus further diminishing glomerular filtration rate. This type of ARF is readily recognized by ultrasonography of the kidneys and the bladder.

References

Ballerman, B.J., Zeidel, M.L., Gunning, M.E., and Brenner, B.M. (1991). Vasoactive peptides and the kidney. In *The kidney* (ed. B.M. Brenner and F.R. Rector Jr), pp. 510–83. W.B. Saunders, Philadelphia, PA.

Brezis, M. and Rosen, S. (1995). Hypoxia of the renal medulla — its implications for disease. *New England Journal of Medicine*, **332**, 647–55.

Dinour, D. and Brezis, M. (1998). Acute renal failure. In *Renal toxicology* (ed. R.S. Goldstein). Elsevier Science, Oxford, in press.

Heyman, S.N., Rosen, S., and Brezis, M. (1994). Radiocontrast nephropathy: a paradigm for the synergism between toxic and hypoxic insults in the kidney. *Experimental Nephrology*, **2**, 153–7.

Linton, A.L. (1990). Drug-induced acute renal failure. In *Acute renal failure in the intensive therapy unit* (ed. D. Bihari and G. Neild), pp. 157–66. Springer-Verlag, Berlin.

6.3.2 Diagnosis

R. Bellomo and C. Ronco

Key messages

- Acute renal failure is a syndrome, not a nosological diagnosis. Its presence demands that an etiological diagnosis be established. The best clinical approach to an etiological diagnosis is achieved by considering prerenal, parenchymal, and postrenal causes.
- Prerenal causes of renal failure are responsible for the vast majority of cases seen in the intensive care unit.
- An accurate history, a focused clinical examination, microscopic examination of urine, and a renal ultrasound examination are usually sufficient to establish the correct diagnosis in the vast majority of cases.

Acute renal failure

The term acute renal failure (ARF), like acute respiratory failure and congestive cardiac failure, simply and broadly defines a clinical syndrome. This syndrome is characterized by an abrupt decrease in glomerular filtration rate, a rapid deterioration in renal function, and the accumulation in blood of nitrogenous waste products.

Unfortunately, unlike other syndromes seen in the intensive care unit (ICU), such as the acute respiratory distress syndrome and the sepsis syndrome, there are no consensus criteria for the diagnosis of ARF let alone the assessment of its severity. Such lack of consensus results in great variability in epidemiology and in diagnostic and therapeutic information. This is because the diagnosis, incidence, causes, and effects of therapy of ARF will clearly depend on how this syndrome is defined (Bellomo and Ronco 1995).

Despite these very serious nosological concerns, ARF is currently diagnosed in practice when the urea and creatinine concentrations in the blood rise above normal in the setting of a suspected renal insult. Such a rise is frequently, although not always, associated with decreased urinary output. It also has a characteristic clinical course: the glomerular filtration rate remains fixed at low levels until recovery, the serum creatinine rises steadily at a rate of approximately 50 to 100 mol/l each day, and urinary output remains relatively fixed until recovery starts. Such recovery often takes place after 1 to 2 weeks, but will take longer if the patient remains critically ill. However, recovery may take a shorter time (3–5 days) if the insult has been moderate in severity and the patient has promptly recovered from the initial injury.

Such recovery is typically characterized by a plateau in the serum urea and creatinine levels, a daily increase in urinary output, and later a progressive decrease in nitrogenous waste products. This syndrome is frequent in the ICU (Groeneveld *et al.* 1991).

Despite the usual straightforward nature of the diagnosis of ARF, critical care physicians often have to deal with several clinical questions.

1. Does this patient presenting with an elevated plasma urea and serum creatinine concentration have acute, rapidly progressive, or chronic renal failure?
2. What is the etiology of this patient's acute renal failure?
3. What investigations should be performed to confirm the diagnosis and establish an etiology?

Acute, rapidly progressive, or chronic renal failure?

The differentiation of acute from rapidly progressive or chronic renal failure is important because it helps in the etiological diagnosis and can usually be made on history alone. In most cases seen in the ICU, one is dealing with patients with known near-normal or moderately impaired renal function who may have been in hospital for some time and experience a serious ischemic injury to the kidney secondary to hypotension, decreased cardiac output, severe blood loss, severe sepsis, or a combination of these factors. Such patients rapidly develop a rising serum creatinine and plasma urea concentration, commonly with, but occasionally without, concomitant oliguria.

In some cases, however, the clinical presentation does not provide clear information regarding the course and rapidity of onset of renal dysfunction. In these patients, history and physical examination may point to a subacute illness (particularly of a vasculitic type) which may have caused a rapidly progressive loss of renal function. Clues to a vasculitic disorder include weeks of recurrent fevers and sweats, myalgias, a syndrome consistent with sinusitis, hemoptysis, dyspnea, arthralgias, pleuritic chest pain, a rash, evidence of synovitis, and the like.

In other patients, the presentation may be one of severe hypertension with renal failure. In these patients, it may be unclear whether the hypertension has caused renal failure or whether it is a manifestation of an as yet undiagnosed chronic renal failure. A number of clinical findings may be helpful in the diagnosis (see below). Other patients may present with an acute illness and a possible mild to moderate ischemic injury to the kidney, and be found to have blood levels of urea and creatinine which are markedly higher than would be expected given the nature and severity of the insult. In such patients, one may be dealing with true ARF and a misassessment of the severity of the insult or with previously unrecognized chronic renal failure with a small acute component. A history of nocturia, difficulties with micturition, previous renal disease, diabetes, chronic analgesic ingestion, kidney stones, and a family history of kidney disease all suggest pre-existent chronic renal failure. Physical examination may display the sallow skin coloration of uremia, pruritic excoriations, palpably enlarged kidneys, a full bladder, and an enlarged prostate, which are all clues to chronic renal disease.

Very few laboratory studies are helpful in distinguishing chronic from acute renal failure. However, an unexplained normochromic normocytic anemia is suggestive of chronic renal failure. An abdominal radiograph or ultrasound may assist in the diagnosis by showing the presence of either small scarred kidneys or large polycystic kidneys. The presence of nephrocalcinosis or dilatation of the pelvicaliceal system are also useful diagnostic clues. Finally, radiographic studies of the outer clavicle or hands can document the presence of renal osteodystrophy or osteomalacia, findings not seen in acute renal failure.

Prerenal, renal, or postrenal kidney failure?

In general, several conditions affecting the kidneys typically present with ARF. Others do so only occasionally (Table 1). In the ICU, by far the most common cause of acute renal failure is diminished renal blood flow (so-called prerenal ARF). This type of ARF can be induced by a variety of insults ranging from hemorrhage to septic shock, and from severe diarrhea to cardiogenic shock. However, the final common pathway appears to be outer medullary ischemia and hypoxia (Heyman *et al.* 1995). The clinical circumstances and severity of the insult vary greatly and need to be profound if normal kidneys are to develop severe oliguria or anuria, but less profound if one is already dealing with a patient who has prior renal impairment, is elderly, or has significant renal artery atheroma. There may be functional failure of the kidney following mild to moderate renal ischemia. Cell death is absent or minimal. However, if the injury is severe enough, cell death will occur and the influx of inflammatory cells will establish and maintain ARF (Johnson and Rokaw 1995). In many of these patients, ARF is one of the manifestations of multi-organ failure. This type of ARF accounts for more than 90 per cent of cases seen in our ICU. Thus, while one should always keep an open mind, other causes of ARF are only considered if there are atypical clinical features as previously described.

In all cases of ARF it is advisable to perform several screening diagnostic investigations to exclude parenchymal (renal) or obstructive (postrenal) causes of ARF.

A very accurate drug-intake history should be taken to assess the possibility of nephrotoxic ARF. Attention should be paid to the use of angiotensin-converting enzyme inhibitors, non-steroidal anti-inflammatory drugs, and antibiotics. The finding of eosinophils in the urine suggests a drug-induced allergic nephropathy. A full blood examination, including a blood film, and standard biochemical tests for electrolytes and liver function are mandatory.

Table 1 Disorders causing ARF

Disorders which typically present with ARF
Acute tubular necrosis
Hypovolemia
Cardiovascular collapse
Nephrotoxins
Hypersensitivity nephropathy, hemolytic–uremic syndrome, acute bilateral pelviureteric obstruction, bilateral arterial occlusion, bilateral renal vein thrombosis, acute urate nephropathy, acute hypercalcemic nephropathy, malignant hypertension, cryoglobulinemia, postpartum nephropathy
Disorders which occasionally present with ARF
Rapidly progressive glomerulonephritis
Goodpasture's syndrome
Microscopic polyarteritis, multiple myeloma, membranous lupus nephropathy, Waldenström's macroglobulinemia, acute endocarditis
Disorders which rarely present with ARF
Diffuse proliferative lupus nephritis
IgA nephropathy
Wegener's granulomatosis
Allergic granulomatosis

The urine should always be tested by dipstick for blood and protein, and sent to the microbiology laboratory for culture to exclude urinary tract infection. Urine microscopy, looking for evidence of possible glomerular disease (red cell casts, white cell casts, fragmented red cells), is mandatory if blood is present in the urine.

In a number of centers, renal ultrasonography is also considered a mandatory component of the work-up in a patient with ARF. If the patient presents to hospital with suspected ARF, this approach is warranted as many patients will be found to have urinary tract obstruction. If the patient is known to produce normal amounts of urine and develops ARF in the hospital or in the ICU in association with a clear hypovolemic, hypotensive, or hypoxic insult, we do not routinely perform renal ultrasonography.

If obstruction is suspected (history and symptoms of prostatism, history and symptoms of urolithiasis, no clear hypovolemic or hypotensive insult) and demonstrated on ultrasonography, a number of investigations are available and may be needed to diagnose the site of obstruction. They include cystoscopy, retrograde pyelography, nephrostomy followed by instillation of radiocontrast into the renal pelvis, and radionuclide excretion studies.

It is important to realize that obstructive uropathy can occur without ultrasound evidence of hydronephrosis. Therefore, if clinical suspicion is strong, further investigations (as outlined above) are needed. Intravenous pyelography may occasionally become necessary, but it is generally not needed for diagnostic purposes in severe ARF of the critically ill and is generally contraindicated in these patients.

If there are clinical and laboratory features suggesting that ARF is due to parenchymal renal disease, several other investigations may become necessary. A markedly elevated sedimentation rate, an elevated or abnormally low white cell count in the absence of infection or the presence of eosinophilia, a blood film suggesting hemolysis, serological studies positive for antineutrophil cytoplasmic antibodies, the presence of anti-DNA autoantibodies, the detection of cryoglobulins, the detection of antiglomerular basement membrane antibodies, and so on are all important diagnostic clues to parenchymal renal failure in the appropriate clinical context. Uncommonly, the final diagnosis rests upon the performance of a renal biopsy which can safely be performed in mechanically ventilated patients under ultrasound guidance.

Rarely, ARF may be due to vascular disease (large-vessel vasculitis, severe bilateral atheroma, renal vein thrombosis, cholesterol embolism, etc.). In this situation, renal vein thrombosis may be detected by contrast CT of the abdomen or it may require venography.

Arterial disease can now be reliably detected by the peripheral intravenous infusion of radiocontrast and the use of spiral CT technology (Olbricht *et al.* 1995).

A working algorithm for the assessment of the etiology of ARF in the ICU is summarized in Fig. 1.

Prerenal azotemia or established acute renal failure?

In the context of a hypovolemic, hypotensive, or hypoxic insult, having excluded renal or postrenal cause of ARF, several authorities recommend the performance of biochemical investigations to help distinguish between so-called 'prerenal azotemia' (the kidney is underperfused and stressed but still functioning appropriately) and

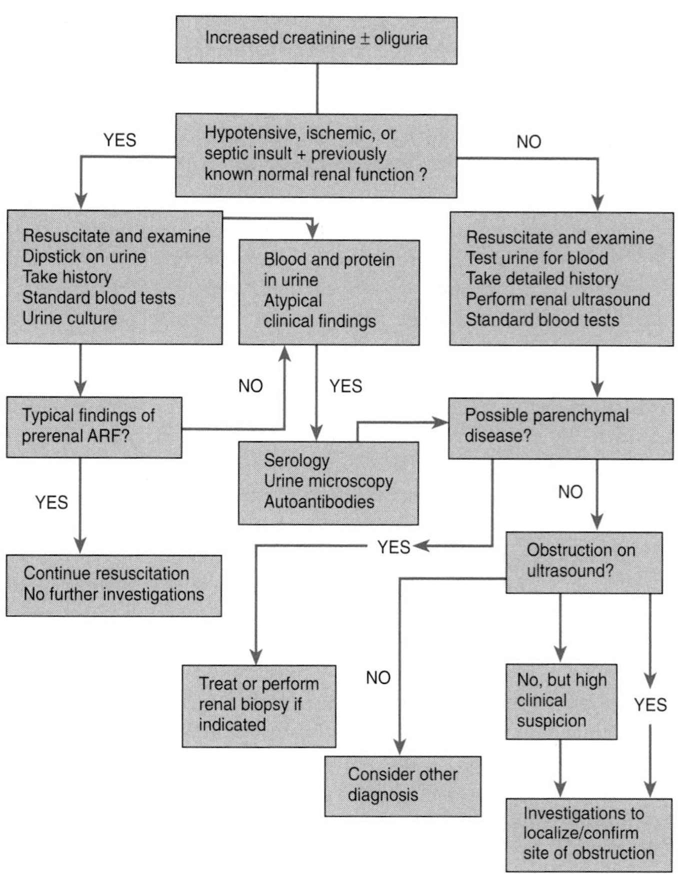

Fig. 1 Illustration of a recommended diagnostic approach in critically ill patients with suspected ARF.

'acute renal failure' (normal cell function has been lost because of the severity of the insult). These investigations typically include the following measurements:

(1) urinary sodium concentration, which is low (< 20 mmol/l) when a functioning kidney is seeking to retain it in an appropriate response to diminished renal perfusion;

(2) the fractional excretion of sodium, which is also low (< 1 per cent) for the same reasons;

(3) the urinary 'concentration', which is relatively preserved with a high specific gravity (> 1020) and a high urinary osmolality (> 500 mosmol/l), if the normal renal response to hypoperfusion is preserved.

Such measurements have several serious flaws in the ICU setting. They have not been validated in critically ill patients, their specificity and sensitivity are poor, and their implications for treatment are unclear. In the ICU, all resuscitation is typically conducted rapidly, is guided by invasive hemodynamic monitoring, and is aggressively directed toward the restoration of vital organ perfusion. Biochemical evidence pointing to 'prerenal azotemia' should not influence either the degree or the speed of resuscitation. Equally, biochemical evidence suggesting 'established ARF' should certainly not inhibit appropriate and speedy resuscitation and re-establishment or maintenance of an optimal circulating blood volume and cardiac output. In this regard, although the clinical assessment of the patient's volume status is a vital first step in the resuscitation process, it is often inexact. Assessment of skin turgor, moistness of mucous membranes, axillary moistness, orthostatic pressure drop, pulse rate, jugular venous pressure, and estimation of the patient's fluid balance are too imprecise to guide resuscitation in critically ill patients. Invasive hemodynamic monitoring (central venous pressure, arterial pressure, and, in patients with cardiac dysfunction or need for inotropic support, pulmonary artery catheterization) is needed to guide rational resuscitation and is strongly recommended in most patients with oliguric ARF.

Conclusions

The syndrome of ARF is frequent in the ICU and its presence is easily confirmed by simple biochemical tests and clinical findings. In the vast majority of cases, ARF is due to a severe prerenal insult. This typically requires no specific investigations but, rather, prompt resuscitation. Occasionally, postrenal or renal causes of ARF need to be considered and either confirmed or excluded by specific tests.

Rarely, ARF can be confused with rapidly progressive or chronic renal failure. Clinical clues and some relatively simple investigations are usually sufficent to clarify the diagnosis.

References

Bellomo, R. and Ronco, C. (1995). Acute renal failure in the intensive care unit: which treatment is best? In *Acute renal failure in the critically ill* (ed. R. Bellomo and C. Ronco), pp. 385–406. Springer-Verlag, Berlin.

Groeneveld, A.B.J., Tran, D.D., van der Meulen, J., Nauta, J.J.P., and Thijs, J.G. (1991). Acute renal failure in the medical intensive care unit: predisposing, complicating factors and outcome. *Nephron*, **59**, 602–10.

Heyman, S.N., Fuchs, S., and Brezis, M. (1995). The role of medullary ischemia in acute renal failure. *New Horizons*, **3**, 597–607.

Johnson, J.P. and Rokaw, M.D. (1995). Sepsis or ischemia in experimental acute renal failure: what have we learned? *New Horizons*, **3**, 608–14.

Olbricht, C.J., *et al.* (1995). Minimally invasive diagnosis of renal artery stenosis by spiral computed tomography angiography. *Kidney International*, **48**, 1332–7.

6.3.3 Management

R. Bellomo and C. Ronco

Key messages

- Maintenance of sufficient intravascular volume, normal or high normal cardiac output, and an adequate mean arterial pressure (typically > 80 mmHg) are important goals of hemodynamic resuscitation in patients at risk of acute renal failure.
- There is currently no convincing evidence that specific drugs, such as dopamine, mannitol, or loop diuretics, exert a clinically important level of renal protection in patients at risk of renal failure.
- Patients with acute renal failure should be treated like all other critically ill patients, and physiological and biochemical homeostasis, as well as adequate nutrition, should be diligently pursued.

- Despite the lack of randomized controlled trials to provide convincing and clear support for their use, continuous renal replacement therapies have major practical and physiological advantages over standard intermittent hemodialysis in critically ill patients and are probably the preferred approach to renal replacement therapy in the intensive care unit.
- Maintenance of good resuscitation, avoidance of hypotension, adequate nutrition, prompt treatment of sepsis, use of biocompatible dialysis membranes, and attention to physiological and biochemical details are important clinical means of preventing further renal injury in established renal failure.

Introduction

The management of acute renal failure (ARF) in critically ill patients can be divided into four major aspects:

(1) prevention of ARF;

(2) prevention of complications of ARF;

(3) treatment with renal replacement therapy;

(4) prevention of further injury to the kidney.

Each of these aspects requires separate detailed discussion.

The prevention of acute renal failure

Critically ill patients are at particular risk of developing ARF. They are likely to experience hypotension because of sepsis, low cardiac output, major fluid losses, and hemorrhage. They often require treatment with drugs (aminoglycosides, amphotericin, non-steroidal anti-inflammatory drugs, radiocontrast agents, etc.) which can induce nephrotoxicity and which are often given under pathophysiological conditions which increase such toxicity. In addition, such patients are often in a state of immunological derangement which is associated with the release of renal vasoconstrictive substances and the induction of parenchymal inflammation. For these reasons, the intensivist and nephrologist must be extremely diligent in the treatment of all the above pathophysiological states and cautious in the prescription of drugs which may damage the kidney.

In this context, adequate fluid resuscitation and maintenance of a normal cardiac output are vital goals of resuscitation in all patients. In the intensive care unit (ICU), such resuscitation must always be guided by invasive monitoring techniques (central venous pressure and arterial pressure measurements at a minimum, and pulmonary artery occlusion pressure and thermodilution measurement of cardiac output in many cases).

While fluid resucitation is generally prompt in the ICU and support of cardiac output is equally aggressive, an important therapeutic goal in the prevention of renal failure, i.e. the maintenance of an adequate renal perfusion pressure, is commonly underemphasized. In fact, much experimental and clinical evidence indicates that global renal blood flow and glomerular filtration rate are 'pressure dependent' once they operate at values below their physiological autoregulation limits (mean arterial pressure of approximately 70–75 mmHg for renal blood flow and approximately 80–85 mmHg for glomerular filtration rate) (Bersten and Holt 1995). Despite this well-established knowledge, lower levels of mean blood pressure are often tolerated unnecessarily in the ICU. Renal blood flow and glomerular filtration rate decrease precipitously once the mean arterial pressure falls below 75 to 80 mmHg. This decrease is not abrogated by the presence of a high cardiac output or the aggressive administration of intravenous fluids. More importantly, autoregulation of renal blood flow is lost once renal injury has occurred (Kelleher *et al.* 1987). Thus maintenance of an adequate renal perfusion pressure is even more important in the patient with early renal dysfunction (Fig. 1).

While an adequate renal perfusion pressure can often be achieved with fluid resuscitation alone, several patients who have received adequate fluid resuscitation and have a high-normal to above-normal cardiac output remain hypotensive due to systemic vasodilatation and sepsis. These patients are at high risk of renal failure, and their blood pressure should be sustained to near-normal levels (> 80 mmHg) with vasoactive agents such as norepinephrine (noradrenaline), epinephrine (adrenaline), or α-dose dopamine. In addition, it is important to note that urine output and glomerular filtration rate appear to be preserved in some patients despite hypotension or a low cardiac output. These patients have maximum activation of compensatory mechanisms (afferent arteriolar vasodilatation and efferent arteriolar vasoconstriction). The administration of drugs that interfere with such compensatory mechanisms (e.g. non-steroidal anti-inflammatory drugs) can precipitate ARF.

The above physiological considerations are supported by all available clinical experience. Such experience suggests that vasopressor support in these patients results in improved renal function and increases urinary output (Bersten and Holt 1995). Animal data are also strongly suggestive of a beneficial effect of vasopressor treatment on renal blood flow during sepsis.

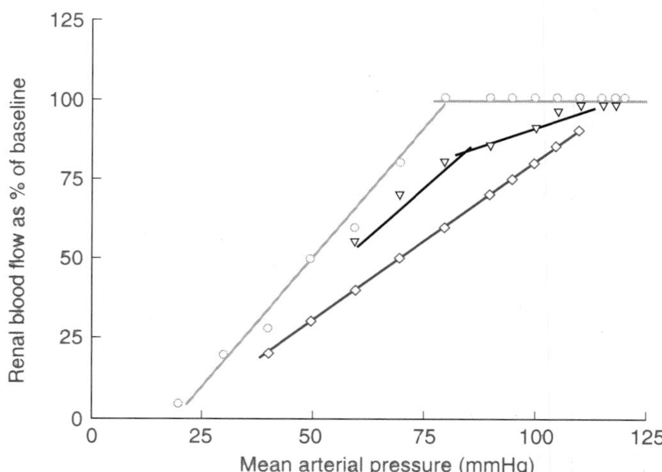

Fig. 1 The relationship between blood pressure and renal blood flow. The normal pressure–flow relationship (○) shows a steep increase in renal blood flow as the mean pressure rises to the autoregulation level. Once the autoregulation plateau is reached (mean arterial pressure of 75–80 mmHg), further increases in blood pressure do not increase blood flow to the kidney. If ARF develops, renal blood flow autoregulation fails and renal blood flow is even more dependent on adequate levels of perfusion pressure (□). As renal function recovers after ARF, the pressure–flow relationship begins to return to normal after 1 to 2 weeks (∇).

In all cases it is always important to consider whether increased intra-abdominal pressure may be responsible for inadequate renal perfusion pressure (renal perfusion pressure is the difference between mean arterial pressure and intra-abdominal pressure) despite adequate fluid and pressor support. Such pathophysiological increases in intra-abdominal pressure may be found in patients with intra-abdominal clot, severe unrelieved ileus, severe tense ascites, massive retroperitoneal hemorrhage, etc. Under these circumstances, renal failure will become inevitable unless such an increase in intra-abdominal pressure is relieved.

Once adequate resuscitation, maintenance of a normal cardiac output, and normalization of arterial blood pressure have been achieved, attention must be focused on making the correct diagnosis with regard to the etiology of the ARF.

If obstruction and parenchymal renal disease have been excluded, nephrotoxins have been removed, and so-called 'prerenal' renal failure has been diagnosed, attention is focused on whether additional therapeutic measures may benefit the kidney.

Such additional measures are typically thought to involve the administration of drugs with the intent of restoring renal blood flow to normal or near-normal levels or of maintaining glomerular filtration rate and urinary output, thus avoiding oliguric renal failure. Several drugs have been used and recommended for these purposes, including so-called low-dose dopamine, mannitol, furosemide (frusemide), and other loop diuretics. While there is controlled evidence that high-dose furosemide will increase urine output in patients with ARF, thus increasing the frequency of polyuric renal failure, there is no convincing evidence that any of these agents has a beneficial effect on the course of renal dysfunction.

A particular situation for the prevention of ARF is found in patients who already have renal dysfunction and are about to receive a radiocontrast agent. Radiocontrast agents are known to induce renal ischemia and medullary hypoxia, and are frequently needed for diagnostic purposes. A recent randomized controlled study (Solomon *et al.* 1994) clearly demonstrated that administration of saline to prevent intravascular volume depletion was superior to administration of saline plus mannitol or saline plus furosemide, adding further evidence against the use of so-called nephroprotective drugs.

The administration of potential nephrotoxins, such as non-steroidal anti-inflammatory drugs, angiotensin-converting enzyme inhibitors, aminoglycoside antibiotics, glycopeptide antibiotics, amphotericin, cyclosporin A, and platinum-containing antineoplastic agents, must be closely monitored in all critically ill patients, and must occur only in situations where alternative drugs are not available and the benefit to the patient outweighs the associated risks.

Finally, the prevention of ARF requires close monitoring of renal function. Hourly urinary output measurements, daily serum creatinine and urea measurements, and regular monitoring of serum electrolytes are a mandatory part of the care of critically ill patients.

Prevention of complications of acute renal failure

Once ARF develops, urinary output declines and nitrogenous waste products accumulate. The patient is at risk of serious and potentially lethal complications which must be prevented.

Particularly dangerous is the accumulation of potassium. In the presence of tissue injury or necrosis, such as in rhabdomyolysis or trauma, the serum potassium concentration can rapidly rise to levels which affect cardiac electrical activity, resulting in serious atrial and ventricular arrhythmia and possible cardiac arrest. Serum potassium must be monitored regularly and if it is rapidly rising above acceptable levels ($>$ 6 mmol/l), emergency treatment with glucose and insulin, bicarbonate, and inhaled salbutamol must be instituted according to its severity while renal replacement therapy is rapidly introduced.

Other electrolyte disturbances, including hypocalcemia and hyperphosphatemia, also occur in patients with ARF.

Although not of immediate danger to the patient, maintenance of calcium–phosphate homeostasis is advisable. This may require cautious calcium administration with or without cholecalciferol, but it is more often correctly treated by the introduction of renal replacement therapy.

Major cardiovascular complications of ARF include pulmonary edema, hypertension, and uremic pericarditis. Neurological complications include uremic encephalopathy and seizures. Uremia also causes platelet dysfunction and promotes gastrointestinal inflammation and ulceration. All these complications can be prevented by the early use of renal replacement therapy.

In many critically ill patients with ARF, appropriate nutritional support is unnecessarily withheld because of the fear of inducing fluid overload, hyperglycemia which is difficult to control, and rapidly deteriorating uremia. Because of such medical management and the common presence of hypercatabolism in these patients, malnutrition becomes another complication of ARF. Once again the early application of adequate renal replacement therapy is the best form of prevention.

Treatment with renal replacement therapy

In the majority of patients with severe ARF, residual renal function is so limited that renal replacement therapy becomes necessary to avoid complications and provide optimal levels of physiological support. Renal replacement therapy is now available in the form of continuous hemofiltration, intermittent hemodialysis, or peritoneal dialysis.

All these therapies have advantages and disadvantages. However, the use of peritoneal dialysis for the dialytic treatment of severe ARF in the ICU in developed countries is now uncommon. This is because standard peritoneal dialysis is associated with limited urea clearances, a high risk of peritonitis ($>$ 15 per cent), and limited ability to maintain volume control.

Intermittent hemodialysis remains the most common form of renal replacement therapy in American ICUs, but in Europe it has been partially superseded by continuous hemofiltration. In Australia, continuous hemofiltration has almost completely superseded intermittent hemodialysis.

Traditional indications for starting renal replacement therapy (urea level $>$ 35 mmol/l, fluid overload, pericarditis or other clinically important uremic complications, hyperkalemia, and severe metabolic acidosis) apply poorly to the intensive care environment where a more aggressive approach is advisable (Table 1). The emerging philosophy of renal replacement therapy aims at preventing metabolic and cardiovascular derangements (maintaining homeostasis) rather than having to treat them.

Table 1 Austin and Repatriation Medical Center criteria for the initiation of renal replacement therapy in critically ill patients

1. Oliguria (urine output <500 ml/day)
2. Anuria (urinary output <50 ml over 12 h)
3. Plasma urea concentration >30 mmol/l
4. Serum creatinine >400 μmol/l
5. Hyperkalemia (serum potassium concentration >6 mmol/l and rising)
6. Pulmonary edema not responsive to diuretics
7. Severe metabolic acidosis (pH <7.1 or base deficit >10 mmol/l)
8. Uremic encephalopathy
9. Uremic pericarditis
10. Uremic neuropathy

The presence of one of the above criteria is sufficient for the initiation of renal replacement therapy. The simultaneous presence of two of these criteria mandates the rapid initiation of renal replacement therapy.

The maintenance of fluid homeostasis in critically ill patients who receive significant amounts of nutritional support, blood, blood products, multiple medications, and variable amounts of resuscitation fluids becomes difficult even with the frequent use of hemodialysis. The rapid intravascular fluid and solute shifts associated with its application are poorly tolerated by such critically ill patients, particularly if cardiac dysfunction is present or the blood pressure requires support with vasopressor drugs. Under these circumstances, continuous hemofiltration techniques provide steady control of fluid balance, full and steady control of uremia, and easy maintenance of homeostasis. These features make them more suited to the intensive care environment even though their probable superiority is only supported by retrospective data (Bellomo and Mehta 1995).

Prevention of further injury to the kidney

The development of established ARF and the initiation of renal replacement therapy do not absolve the treating physician from continuing to take care of the damaged and, hopefully, recovering kidney. The recovering kidney is unable to maintain normal autoregulation of its blood flow and therefore is highly susceptible to hypotensive injury. Biopsy studies have confirmed the significance of this by showing areas of fresh ischemia in patients who are recovering from ARF and have sustained renewed septic or hypotensive insults. Therefore it is of great importance to continue to pay attention to all the factors mentioned in the previous discussion concerning the prevention of renal injury.

More recently, evidence has emerged (Hakim *et al.* 1995) that the type of membrane used during replacement therapy may make a significant difference to renal and patient recovery during ARF. While the issue of whether all patients with ARF should be treated with synthetic membranes remains controversial, cellulose-based membranes have certainly been shown to induce marked complement and leukotriene activation. This may be associated with delayed renal recovery as well as an increased incidence of sepsis when compared with synthetic (biocompatible) membranes. However, synthetic membranes appear capable of absorbing significant quantities of inflammatory molecules and thus may attenuate the inflammatory response associated with sepsis.

Conclusions

The management of ARF is multifaceted. It involves prevention of renal injury by adequate and closely monitored fluid resuscitation, maintenance of cardiac output and near-normal blood pressure, and avoidance of nephrotoxins. It requires prevention of all the metabolic complications associated with the uremic state and the early initiation of adequate biocompatible renal replacement therapy. Such therapy should preferably be continuous. Finally, in the presence of ARF it is mandatory to pay close attention to all hemodynamic, dyalitic, metabolic, and pharmacological details to ensure the recovery of the organ and the patient.

References

Bellomo, R. and Mehta, R. (1995). Acute renal replacement in the intensive care: now and tomorrow. *New Horizons*, **3**, 760–7.

Bersten, A.D. and Holt, A.W. (1995). Vasoactive drugs and the importance of renal perfusion pressure. *New Horizons*, **3**, 650–61.

Hakim, R.M., Wingard, R.L., and Parker, R. (1994). Effect of the dialysis membrane in the treatment of patients with acute renal failure. *New England Journal of Medicine*, **331**, 1338–42.

Kelleher, S.P., Robinette, J.B., and Conger, J.D. (1987). Sympathetic nervous system in the loss of autoregulation in acute renal failure. *American Journal of Physiology*, **246**, F379–86.

Solomon, R., Werner, C., Mann, D., D'Elia, J., and Silva, P. (1994). Effects of saline, mannitol, and furosemide on acute decreases in renal function induced by radiocontrast agents. *New England Journal of Medicine*, **331**, 1416–20.

6.4 Established renal failure

6.4.1 The effect of renal failure on drug handling

A. Olyaei and W. M. Bennett

Key messages

- Uremia affects every organ system and alters the pharmacokinetic behavior of drugs commonly used in intensive care units.
- Accumulation of the active moiety of drugs should be considered in patients with acute renal failure.
- Renal function should be evaluated before initiation of drug therapy.
- Drug dosing should be modified to enhance drug efficacy and reduce drug toxicity.

Introduction

Although significant advances have been made in critical care therapeutics over the last two decades, pharmacotherapy in these patients remains perplexing and difficult. Achieving the proper balance between ensuring efficacy and avoiding toxicity in critically ill patients remains a vital issue in patients with impaired renal function. The physiological abnormalities of uremia, in addition to ongoing pathological factors, alter the pharmacodynamics and pharmacokinetics of the majority of drugs used in the critical care setting. Drugs with a pharmacologically active moiety or active metabolite that is primarily excreted in the urine may accumulate to toxic concentrations secondary to decreased renal excretion. Therefore a knowledge and understanding of the pharmacokinetic and pharmacodynamic variability of medications used during renal failure is essential for critical care clinicians who must make appropriate dosage adjustments to ensure efficacy while minimizing toxicities (Rudy and Brater 1994).

Pharmacokinetic considerations in renal failure

The renal system is pivotal, directly or indirectly, in determining the pharmacokinetic properties of the majority of medications used in intensive care units. Pharmacokinetics is defined as the time course of drug behavior in the body. Thus it is the study of drug absorption, distribution, metabolism, and excretion. The use of patient-specific pharmacokinetic parameters allows the clinician to modify clinical response according to serum drug concentrations. Pharmacokinetic variability is greater in intensive care patients. The multiple pathological processes that occur in this patient group may affect many organs of the body which, in turn, influence absorption, distribution, metabolism, and elimination. Predominantly, the distribution and elimination phases (and to a lesser extent absorption and metabolism) are affected by renal dysfunction (Rudy and Brater 1994).

Absorption

Bioavailability is defined as the extent and rate of drug absorption. There is limited information on alteration of drug absorption in patients with renal dysfunction. Nausea, vomiting, impaired gastric motility, edema of the gastrointestinal tract, increased gastric pH secondary to elevated ammonia, and administration of aluminum-containing antacids can all diminish the bioavailability of oral medications. Little information is available regarding the effect of renal dysfunction on the absorption and bioavailability of medications given by non-oral routes such as rectal, intramuscular, and subcutaneous administration (Matzke and Milliken 1992).

Distribution

After absorption from the gastrointestinal tract, a drug enters the central compartment and is bound to plasma proteins. Any changes in protein binding may alter the volume of distribution, which is defined as the ratio of drug distribution in plasma to the concentration in the tissues at steady state. Agents that are highly protein bound and distribute mostly into the intravascular compartment have a small volume of distribution. Agents that are hydrophobic, distributing widely in adipose tissue, have a large volume of distribution. Protein binding and volume of distribution change in an unpredictable fashion in patients with azotemia. Any alteration of protein binding affects the volume of distribution. This is more common with acidic drugs such as penicillin, phenytoin, and valproic acid. Diminished protein binding secondary to hypoalbuminemia, accumulation of organic acid, or a change of binding configuration can cause an increase in the concentration of free active drug in plasma. Displacement from protein-binding sites can enhance the pharmacodynamics of most drugs. In patients with renal failure, it is best to monitor only free or 'unbound' drug concentrations when an assay is available and to make dosage adjustments according to these concentrations (Matzke and Milliken 1992; Bennett *et al.* 1994; Rudy and Brater 1994).

Excretion

Unchanged drug and active and inactive drug metabolites are mostly excreted in the urine through the processes of glomerular filtration and active tubular secretion. Tubular reabsorption can also modify the net drug excreted (Matzke and Milliken 1992; Bennett *et al.* 1994). Drugs bound to plasma proteins cannot be filtered because of the effective glomerular filtration barrier. Only free unbound drug is filtered at the glomerular membrane in the absence of glomerular injury. The degree of plasma protein binding, rate of filtration, drug concentration, and perfusion pressure correlate with the amount of drug that is filtered. Active tubular secretion is a carrier-mediated process that takes place primarily in the proximal tubules and is dependent on the plasma concentration, renal clearance, and filtration rate of the drug. In contrast, a passive reabsorption process takes place in both the distal and proximal tubules and depends mostly on the gradients between tubular fluid and plasma (Matzke and Milliken 1992; Bennett *et al.* 1994; Rudy and Brater 1994).

Dosage modification

It is best to follow a stepwise process for dosage modification in renal failure. The method outlined here will help clinicians achieve a desirable clinical response while minimizing the risk of adverse drug reactions and toxicity (Bennett *et al.* 1994). Start dosage modification by (1) assessing renal function by measuring or estimating creatinine clearance, (2) choosing a loading dose, (3) choosing a maintenance dose, and (4) performing therapeutic drug monitoring.

Assessing renal function

Glomerular filtration rate is a true index for measuring renal function. More commonly, however, glomerular filtration rate is estimated using creatinine clearance. Glomerular filtration rate in adults can be estimated from the patient's serum creatinine, weight, age, and gender by using the Cockroft and Gault equation (Cockroft and Gault 1976):

$$C_{cr} = \frac{(140 - \text{age}) \times \text{IBW}}{(72 \times S_{cr})} \times 0.85 \text{ (for women)}$$

where C_{cr} is the creatinine clearance (ml/min), IBW is the ideal body weight (kg), and S_{cr} is the serum creatinine (mg/dl = (μmol/l)/88.4). The ideal body weight is defined as follows:

$$\text{IBW(men)} = 50\text{ kg} + 2.3\text{ kg per inch over 5 feet}$$
$$(0.9\text{ kg per cm over 152 cm})$$

$$\text{IBW(women)} = 45.5\text{ kg} + 2.3\text{ kg per inch over 5 feet}$$
$$(0.9\text{ kg per cm over 152 cm}).$$

Serum creatinine is produced at a constant rate and has relevant characteristics that allow it to be used in the estimation of glomerular filtration rate. This equation should be used only when serum creatinine is stable and there are not continuing changes in renal function. The Cockroft and Gault equation is subject to significant error in patients with diminished muscle mass (the elderly, those with hepatic cirrhosis, and malnourished patients) (St. Peter 1993). In unstable patients, the estimated glomerular filtration rate is often an educated guess based on individual patient characteristics.

Choosing a loading dose

For a given drug, an average of four to five half-lives are required to reach therapeutic plasma concentrations. Loading doses are used routinely to provide rapid induction and achieve adequate therapeutic plasma concentrations in a short time, assuming that this is clinically feasible. The loading dose in patients with renal failure is usually no different from that in patients with normal renal function (Bennett *et al.* 1994).

Half-life

Half-life is defined as the time required for the plasma drug concentration to decrease to 50 per cent of its original concentration. Drug half-life is a simple concept and is most commonly utilized to estimate the time to achieve steady state plasma concentrations (Rudy and Brater 1994). However, it is frequently misunderstood. The half-life of a drug does not explain solely the drug excretion process; in contrast, it is a by-product of drug distribution and clearance (metabolism and excretion). In patients with renal failure, the half-life and the time to reach steady state are usually prolonged. If doses remain unchanged, the steady state plasma concentration in a patient with renal failure state will be higher than that in a patient with normal renal function. A loading dose is usually given, if feasible, to reduce the time required to achieve a therapeutic plasma concentration. Reduction of the total dose (by decreasing the dose or extending the dosing interval) will avoid higher steady state plasma concentrations (Rudy and Brater 1994).

Choosing a maintenance dose

Most drugs are given at regular intervals to maintain therapeutic plasma concentrations (Bennett *et al.* 1994). In the patient with renal failure, maintenance doses should be modified to avoid toxicity. The maintenance dose is normally adjusted by lowering the dose per dosing interval or by extending the dosing interval. For example, for an anticonvulsant to sustain plasma concentrations above a seizure threshold, a reduced dose given at a 'normal' dosing interval is more appropriate. Increasing the dosing interval is more appropriate for a drug with a long half-life. A combined method of lowering the dose and extending the dosing interval may be needed for the majority of drugs.

Maintenance doses, volume of distribution, percentage excreted unchanged, and plasma protein binding of the more common drugs used in intensive care are given in Table 1.

Performing therapeutic drug monitoring

Therapeutic drug monitoring is defined as applying clinical pharmacokinetics to achieve a desirable clinical response by measuring and interpreting plasma drug concentrations. Therapeutic drug monitoring is particularly important for drugs with a narrow therapeutic index, for drugs which are subject to considerable intra- and interpatient variability, for drugs which are used for chronic diseases such as seizures and arrhythmias, for evaluating the possibility of drug toxicity and suspicion of non-compliance, for evaluating possible drug–drug, drug–food, and drug–disease interactions, and for patients with unstable renal function. Therapeutic drug monitoring should be used with some caution in patients with renal dysfunction. Clinical response must be used as the endpoint of pharmacotherapy; simply achieving and maintaining

Table 1 Required dose intervals and maintenance doses for common intensive care drugs in renal failure

Drug	First dose	First interval	GFR > 50	GFR 10–50	GFR < 10
Aminoglycoside					
Amikacin	Individualize	Individualize	Individualize	Individualize	Individualize
Gentamicin	Individualize	Individualize	Individualize	Individualize	Individualize
Tobramycin	Individualize	Individualize	Individualize	Individualize	Individualize
Cephalosporines					
Cefamandole	0.5–1.0 g	Every 4–8 h	Every 6 h	Every 6–8 h	Every 12 h
Cefazolin	0.5–1.5 g	Every 6 h	Every 8 h	Every 12 h	Every 24–48 h
Cefmetazole	2.0 g	Every 6–12 h	Every 16 h	Every 24 h	Every 48 h
Cefoperazone	1–2 g	Every 12 h	100%	100%	100%
Cefotaxime	1.0 g	Every 6 h	Every 6 h	Every 8–12 h	Every 24 h
Cefotetan	1–2 g	Every 12 h	100%	50%	25%
Cefoxitin	1–2 g	Every 6–8 h	Every 8 h	Every 8–12 h	Every 24–48 h
Ceftazidime	1–2 g	Every 8 h	Every 8–12 h	Every 24–48 h	Every 48 h
Ceftriaxone	0.2–1.0 g	Every 12 h	100%	100%	100%
Cefuroxime sodium	0.75–1.5 g	Every 8 h	Every 8 h	Every 8–12 h	Every 12 h
Miscellaneous antibacterial					
Aztreonam	12.5 mg/kg	Every 8–12 h	100%	50%–75%	25%
Clindamycin	150–300 mg	Every 6 h	100%	100%	100%
Erythromycin	150–300 mg	Every 6 h	100%	100%	50%–75%
Imipenem	0.25–1.0 g	Every 6 h	100%	50%	25%
Metronidazole	7.5 mg/kg	Every 6 h	100%	100%	50%
Teicoplanin	6.0 mg/kg	Daily	Every 24 h	Every 48 h	Every 72 h
Vancomycin	500 mg	Every 6 h	Individualize	Individualize	Individualize
Penicillins					
Ampicillin	250 mg–2 g	Every 6 h	Every 6 h	Every 6–12 h	Every 12–24 h
Mezlocillin	1.5–4.0 g	Every 4–6 h	Every 4–6 h	Every 6–8 h	Every 8 h
Nafcillin	1–2 g	Every 4–6 h	100%	100%	100%
Penicillin G (benzylpenicillin)	0.5–4 million IU	Every 6 h	100%	75%	20%–50%
Piperacillin	3–4 g	Every 4 h	Every 4–6 h	Every 6–8 h	Every 8 h
Ticarcillin	3.0 g	Every 4 h	1–2 g every 4 h	1–2 g every 8 h	1–2 g every 12 h
Antifungal agents					
Amphotericin	20–40 mg	Daily	Every 24 h	Every 24 h	Every 24–36 h
Fluconazole	200–400 mg	Daily	100%	100%	100%
Antiviral agents					
Acyclovir	5.0 mg/kg	Every 8 h	5 mg/kg every 8 h	5 mg/kg every 12–24 h	2.5 mg/kg every 24 h
Foscarnet	60 mg/kg	Every 8 h induction doses	28 mg/kg	15 mg/kg	6 mg/kg
Ganciclovir	5.0 mg/kg	Every 12 h	Every 12 h	Every 24–48 h	Every 48–96 h

(*continued*)

Table 1 (*continued*)

Drug	First dose	First interval	GFR > 50	GFR 10–50	GFR < 10
Antihypertensive agents					
Clonidine	0.1–0.6 mg	Twice daily	100%	100%	100%
ACE inhibitors					
Captopril	25 mg	Every 8 h	100% every 8–12 h	75% every 12–18 h	50% every 24 h
Enalapril	5–10 mg	Every 12 h	100%	75%–100%	50%
Lisinopril	5–10 mg	Every 24 h	100%	50%–75%	25%–50%
Beta-blockers					
Atenolol	50–100 mg	Every 24 h	100% every 24 h	50% every 48 h	30%–50% every 96 h
Esmolol	50–150 μg/kg/min	By infusion	100%	100%	100%
Labetalol	200–600 mg	Twice daily	100%	100%	100%
Metoprolol	50–100 mg	Twice daily	100%	100%	100%
Propranolol	80–160 mg	Twice daily	100%	100%	100%
Sotalol	160 mg	Every 24 h	100%	30%	15%–30%
Diazoxide	150–300 mg bolus		100%	100%	100%
Hydralazine	25–50 mg	Three times daily	Every 8 h	Every 8 h	Every 8–16 h
Nitroprusside	0.25–8.0 μg/kg/min	By infusion	100%	100%	100%
Antiarrhythmic agents					
Adenosine	3–6 mg IV bolus		100%	100%	100%
Amiodarone	800–2000 mg	Load	100%	100%	100%
Bretylium	5–30 mg/kg	Load	100%	25%–50%	25%
Lidocaine (lignocaine)	50 mg over 2 min	Repeat every 5 min × 3	100%	100%	100%
Procainamide	350–400 mg	Every 3–4 h	Every 4 h	Every 6–12 h	Every 8–24 h
Propafenone	150–300 mg	Every 8 h	100%	100%	100%
Calcium-channel blockers					
Amlodipine	5.0 mg	Every 24 h	100%	100%	100%
Diltiazem	90 mg	Every 8 h	100%	100%	100%
Nimodipine	30 mg	Every 8 h	100%	100%	100%
Verapamil	80 mg	Every 8 h	100%	100%	100%
Cardiac glycosides					
Digitoxin	0.1–0.2 mg	Every 24 h	100%	100%	50%–75%
Digoxin	1–1.5 mg load	Load	100% every 24 h	25%–75% every 36 h	10%–25% every 48 h
Diuretics					
Acetazolamide	250 mg	Every 6–12 h	Every 6 h	Every 12 h	Avoid
Bumetanide	1–2 mg	Every 8–12 h	100%	100%	100%

(*continued*)

Table 1 (*continued*)

Drug	First dose	First interval	GFR > 50	GFR 10–50	GFR < 10
Ethacrynic acid	50–100 mg	Three times daily	Every 8–12 h	Every 8–12 h	Avoid
Furosemide (frusemide)	40–80 mg	Twice daily	100%	100%	100%
Metolazone	5–10 mg	Every 24 h	100%	100%	100%
Torsemide (torasemide)	5.0 mg	Twice daily	100%	100%	100%
Miscellaneous cardiac drugs					
Amrinone	5–10 μg/kg	Min	100%	100%	50%–75%
Dobutamine	2.5–15 μg/kg	Min	100%	100%	100%
Dopamine	ND	< 10 min	100%	100%	100%
Milrinone	15–75 μg/kg IV	Load	100%	100%	50%–75%
Nitroglycerine	Many routes and methods		100%	100%	100%
Nitroprusside	0	10 min	100%	100%	100%
Barbiturates					
Pentobarbital (pentobarbitone)	30 mg	Three to four times daily	100%	100%	100%
Phenobarbital (phenobarbitone)	50–100 mg	Two to three times daily	Every 8–12 h	Every 8–12 h	Every 12–16 h
Secobarbital (quinalbarbitone)	30–50 mg	Three to four times daily	100%	100%	100%
Benzodiazepines					
Chlordiazepoxide	5–100 mg	Every 24 h	100%	100%	50%
Diazepam	5–40 mg	Every 24 h	100%	100%	100%
Lorazepam	1–2 mg	Two to three times daily	100%	100%	100%
Midazolam	Individualized		100%	100%	50%
Narcotics and narcotic antagonists					
Alfentanil	Anesthetic induction		100%	100%	100%
Butorphanol	2.0 mg	Every 3–4 h	100%	75%	50%
Codeine	30–60 mg	Every 4–6 h	100%	75%	50%
Fentanyl	Anesthetic induction		100%	75%	50%
Meperidine (Demerol®)	50–100 mg	Every 3–4 h	100%	75%	50%
Methadone	2.5–10 mg	Every 6–8 h	100%	100%	50%–75%
Morphine	20–25 mg	Every 4 h	100%	75%	50%
Naloxone	2.0 mg IV		100%	100%	100%
Sufentanil	Anesthetic induction		100%	100%	100%

(*continued*)

Table 1 (*continued*)

Drug	First dose	First interval	GFR > 50	GFR 10–50	GFR < 10
Anticonvulsants					
Carbamazepine	200 mg	Twice daily	100%	100%	100%
Phenytoin	1000 mg	Load	100%	100%	100%
Primidone	250–500 mg	Four times daily	Every 8 h	Every 8–12 h	Every 12–24 h
Sodium valproate	15–60 mg/kg	Every 24 h	100%	100%	100%
Corticosteroids					
Betamethasone	0.5–9.0 mg	Daily	100%	100%	100%
Dexamethasone	0.75–9.0 mg	Daily	100%	100%	100%
Hydrocortisone	20–500 mg	Daily	100%	100%	100%
Methylprednisolone	4–48 mg	Daily	100%	100%	100%
Neuromuscular agents					
Atracurium	0.4–0.5 mg/kg	Load	100%	100%	100%
Doxacurium	0.025–0.05 mg/kg		100%	50%	50%
Etomidate	0.2–0.6 mg/kg		100%	100%	100%
Fentanyl	0.0002–0.05 mg/kg		100%	100%	100%
Gallamine	0.5–1.5 mg/kg		75%	Avoid	Avoid
Ketamine	1–4.5 mg/kg		100%	100%	100%
Metocurine	0.2–0.4 mg/kg		75%	50%	50%
Neostigmine	15–375 mg	Daily	100%	50%	25%
Pancuronium	0.04–0.1 mg/kg		100%	50%	Avoid
Pipecuronium	100 μg/kg		100%	50%	25%
Propofol	2–2.5 mg/kg		100%	100%	100%
Pyridostigmine	60–1500 mg	Daily	50%	35%	20%
Succinylcholine (suxamethonium)	0.3–1.1 mg/kg	Load	100%	100%	100%
Sufentanil	1–30 μg/kg		100%	100%	100%
Tubocurarine	0.1–0.2 mg/kg		75%	50%	Avoid
Vecuronium	0.08–0.1 mg/kg	Load	100%	100%	100%

GFR, glomerular filtration rate; ACE, angiotensin-converting enzyme; IV, intravenous.

a plasma concentration in the drug's therapeutic range is neither sufficient nor adequate (Rudy and Brater 1994).

Drugs in this chapter are listed by generic name and therapeutic use. It should be emphasized that Table 1 is intended to provide a rough guideline to assist the clinician in drug therapy for patients with renal dysfunction (with some limitations). This recommendation should only be used as a starting point. Dosages may vary or a drug should be completely avoided according to a patient's other pathological factors.

Renal replacement therapy, particularly continuous arteriovenous hemofiltration and continuous venovenous hemofiltration, is gaining popularity in the management of renal failure in hemodynamically unstable patients in the intensive care setting. Drug replacement and extent of drug removal by these extracorporeal processes should be considered for designing appropriate and effective drug therapy for patients with acute renal failure. The specific properties of the drug that influence removal of active moiety are as follows:

(1) molecular weight;

(2) water solubility;

(3) protein binding;

(4) volume of distribution;

(5) dialysis clearance compared with plasma clearance.

For example, as the molecular weight of the drug increases, removal during extracorporeal processes decreases. Once the molecular weight exceeds 500 Da, conventional dialysis is not very effective in drug removal. Lipophilic drugs and agents with a large volume of distribution should not be expected to dissolve in aqueous dialysate and be removed from the blood during dialysis. Drugs with a high degree of protein binding are not significantly removed by the extracorporeal process, with the exception of plasmapheresis, because of the formation of a large drug–protein complex. Therefore, to ensure efficacy and avoid toxicity, appropriate replacement should be considered during dialysis (Bennett *et al.* 1994).

References

Bennett, W.M., Aronoff, G.R., Golper, T.A., Morrison, G., Brater, D.C., and Singer, I. (1994). *Drug prescribing in renal failure: dosing guidelines for adults*. American College of Physicians, Philadelphia, PA.

Cockroft, D.W. and Gault, M.H. (1976). Prediction of creatinine clearance from serum creatinine. *Nephron*, **16**, 31–41.

Matzke, G.R. and Millikin, S.P. (1992). Influence of renal function and dialysis on drug disposition. In *Applied pharmacokinetics: principles of therapeutic drug monitoring* (ed. W.E. Evans, J.J. Schentag, W.J. Jusko, and M.V. Relling), pp. 8–42. Applied Therapeutics, Vancouver, WA.

Rudy, A.C. and Brater, D.C. (1994). Pharmacokinetics. In *Essentials of critical care pharmacology* (ed. B. Chernow), pp. 3–17. Williams & Wilkins, Baltimore, MD.

St. Peter, W.L. and Halstenson, C.E. (1994). Pharmacologic approach in patients with renal failure. In *Essentials of critical care pharmacology* (ed. B. Chernow), pp. 41–79. Williams & Wilkins, Baltimore, MD.

6.4.2 The effect of chronic renal failure on critical illness

R. G. Woolfson

Key messages

- Patients with chronic renal failure must be considered as having a multisystem disease which is likely to complicate their management following admission to the intensive care unit.
- The aims of management are twofold: firstly, conservation of renal function since any further loss is often irrecoverable; secondly, protection of the patient from underlying cardiovascular disease, coagulopathy, and predisposition to gastrointestinal hemorrhage.
- Other factors, such as malnutrition, immune depression, and endocrine and metabolic abnormalities, can complicate both diagnosis and management.

Management of the critically ill patient with chronic renal failure

The management of patients with chronic renal failure must include conservation of residual renal function, an understanding of both the multisystem complications of uremia, and dialysis or immunosuppressive therapy when necessary.

Underdiagnosis of chronic renal failure

An individual's plasma creatinine concentration is determined by its production from skeletal muscle (dependent on age, gender, and ethnicity) and its renal clearance due mainly to glomerular filtration. The inverse relationship between plasma creatinine and glomerular filtration rate means that small increments to a near normal plasma level correspond to much larger decrements in glomerular filtration rate. For example, a doubling of plasma creatinine within the normal range (from 60 to 120 μmol/l) represents a halving of glomerular filtration rate. This subtle biochemical relationship and the absence of referable symptoms and signs means that cases of chronic renal failure may go unrecognized on the intensive care unit as easily as elsewhere.

Conservation of residual renal function

The common causes of chronic renal failure are listed on Table 1, although the proportional distribution varies with age. With adequate medical management only a few patients will go on to develop endstage renal failure; however, it may be precipitated by critical illness since the chronically insufficient kidney is more sensitive to hypoperfusion injury, more likely to develop acute tubular necrosis, and less likely to recover from an acute insult. The development of acute on chronic renal failure commonly results from multiple small insults, such as a combination of hypovolemia and nephrotoxicity.

Glomerular and renovascular disease are commonly associated with retention of salt and water leading to edema and hypertension, whereas in tubulointerstitial disease due, for example, to chronic obstruction or interstitial nephritis, the kidney fails to conserve sodium and patients are normotensive. Patients with chronic renal failure cannot compensate for hypovolemia or hypotension and are at risk from a further acute reduction in glomerular filtration rate. The risk is greatest in those with contracted circulations, as found for example in patients with tubulointerstitial disease or nephrotic syndrome and in renal failure associated with liver disease. Therefore fluid status and cardiovascular indices must be assessed in the context of the underlying renal pathology and corrections made rapidly. Finally, the dosages of drugs excreted by the kidneys should be modified in patients with chronic renal failure and, where possible, nephrotoxic drugs should be avoided.

Table 1 Common causes of chronic renal failure

Chronic glomerulonephritis
Chronic pyelonephritis
Diabetic nephropathy
Hypertensive nephrosclerosis
Hereditary renal disease (e.g. polycystic kidney disease)
Obstructive uropathy

Cardiovascular disease

Vascular disease affecting coronary, cerebral, and peripheral circulations accounts for more than 50 per cent of overall morbidity and mortality in endstage renal failure. Hypertension, due to sodium retention, hyper-reninemia, and other factors, occurs early in renal failure and leads to the development of left ventricular hypertrophy. Hypertension (pressure overload), anemia, volume overload (including overdevelopment of arteriovenous fistula), and hyperparathyroidism have all been implicated in the development of myocardial fibrosis which results in progressive diastolic heart failure and an increased risk of arrhythmias (Massry and Smogorzewski 1996). Myocardial disease is compounded by accelerated atherogenesis secondary to an adverse lipid profile, hypertension, glucose intolerance, increased oxidative stress, deposition of calcium phosphate salts in the vascular wall, and other factors which underlie the increased prevalence of coronary artery disease responsible for silent ischemia, angina, and subendocardial and transmural infarction in patients with chronic renal failure. This range of cardiac pathology may influence the choice of inotropes and pressor agents.

Calcification of the aortic and mitral valves in patients with renal failure results from a persistently elevated calcium phosphate solubility product and may predispose the septicemic patient to endocarditis which can be difficult to diagnose by echocardiography.

Nutrition

The patient with chronic renal failure is commonly malnourished (Avram and Mittman 1994). Restriction of dietary protein ameliorates symptoms of uremia which occur when the glomerular filtration rate has fallen to 40 per cent of normal, and most patients spontaneously modify their protein intake to accommodate this. However, in the absence of expert dietetic input and in the context of an inadequate intake of essential amino acids, this may lead to a negative nitrogen balance in which protein degradation exceeds synthesis and there is marked loss of lean body mass. Energy intake is also important, with calorie malnutrition contributing to a negative nitrogen balance. Physical examination and calculation of the body mass index may suggest protein malnutrition, with hypoalbuminemia and reduced levels of prealbumin and transferrin providing biochemical evidence although these may be misleading in the context of massive proteinuria. Levels of iron, zinc, red cell folate, 1,25-dihydroxyvitamin D, and water-soluble vitamins may be low in chronic renal failure, with pyridoxine deficiency contributing to immune depression and anemia. Supplements should be provided.

Metabolic abnormalities

Acidosis

A glomerular filtration rate below 30 ml/min is associated with reduced ability to excrete H^+ ions due to impaired generation of ammonia and failure to excrete endogenously produced acid. This leads to a metabolic acidosis with increased anion gap as sulfates, phosphates, and organic anions accumulate. When severe, the acidosis may cause dyspnea and require correction by bicarbonate administration.

Potassium

Kaliuresis is impaired in chronic renal failure with normal plasma levels maintained by an adaptive increase in secretion by the remaining nephrons and by increased colonic excretion under the influence of aldosterone. Potassium-sparing diuretics, angiotensin-converting enzyme inhibitors, and non-steroidal anti-inflammatory drugs can precipitate hyperkalemia, and the composition of fluid replacement and feeding should be considered with care.

Calcium

Phosphate retention with reciprocal hypocalcemia and reduced dihydroxylation of 25-hydroxyvitamin D are triggers for the development of secondary hyperparathyroidism. Parathyroid hormone stimulates osteoclastic activity and liberates calcium from the bones in an effort to normalize serum calcium levels, and over time a minority of patients will develop hypercalcemia due to autonomous or tertiary hyperparathyroidism. The adverse neuromuscular effects of hypocalcemia are offset by the acidosis of renal failure which increases the ionized component of the total plasma calcium. Acute correction of the acidosis can precipitate tetany and fits during which bones weakened by renal osteodystrophy can fracture.

Gastrointestinal tract

Upper gastrointestinal tract

The incidence of upper gastrointestinal disorders, including gastritis, ulceration, and hemorrhage, is increased in chronic renal failure (Kang 1993) and may reflect increased gastrin levels in response to hypochlorhydria, reduced renal clearance, stress, and hyperparathyroidism. Biliary reflux due to gastrin-induced pyloric sphincter incompetence coupled with reduced basal bicarbonate excretion by the pancreas may be contributory, but there appears to be no predisposition to infection with *Helicobacter*. Clinically significant disease is often silent, and bleeding is very commonly precipitated by pro-ulcerogenic drugs such as aspirin, non-steroidal anti-inflammatory drugs, and steroids. Critically ill patients should receive enteral feeding and H_2 antagonists, as prophylaxis with treatment of established gastritis or ulcers is the same as for non-uremic patients, although drug dosages may require adjustment.

There is morphological change in the small bowel with reduction in villus height, increased crypt depth, mucosal edema, and inflammatory cell infiltrate into the lamina propria which is responsible for increased mucosal permeability. Abnormalities in bile acid secretion contribute to diarrhea, and uremic patients are susceptible to prolonged ileus.

Lower gastrointestinal tract

Colonic abnormalities include ulceration, diverticulosis, diverticulitis, spontaneous perforation, and increased incidence of angiodysplasia in patients undergoing hemodialysis.

Liver

Abnormalities in liver function tests are common in dialysis patients and reflect hepatic congestion, viral infection, or drug hepatotoxicity.

Pancreas

Levels of most pancreatic hormones are elevated in uremia owing to reduced clearance. Similarly, amylase levels are mildly raised, but an increase by a factor of 3 or more indicates pancreatitis.

Hematology

Anemia

The normochromic normocytic anemia of chronic renal failure results from erythropoietin deficiency and increased red cell destruction; fragmented (burr) cells and a slightly increased reticulocyte count are typical. Iron absorption is normal, but dietary intake may be decreased due to reduced meat consumption and anorexia. Folate supplementation is required by patients undergoing regular hemodialysis. Correction of the anemia is possible by the regular parenteral administration of recombinant erythropoietin, although the onset of action is slow and the hormone is ineffective in the context of chronic inflammation or infection (raised C-reactive protein) so that anemia will rapidly reappear in treated patients who develop acute inflammation or infection.

Neutrophils

Neutrophil production in renal failure is normal or slightly increased. Functional defects of chemotaxis, phagocytosis, degranulation, and oxidative metabolism and phagocytic activity have been reported.

Lymphocytes

Lymphocyte counts are low in renal failure and, although the distribution amongst subsets remains normal, patients are anergic with increased sensitivity to infective agents, particularly viruses and mycobacteria, and a poor response to active immunization. Abnormalities in the function of both T and B cells have been identified.

Coagulopathy

In renal failure, there are no consistent major alterations in coagulation factors or tests, no impairment of fibrinolysis, and no reduction in platelet production. However, prolongation of the bleeding time correlates with clinical bleeding, and impaired aggregation, adhesion, and interaction of platelets with the vessel wall underlie the primary hemostatic defect in uremia (Zachee *et al.* 1994). The pathogenesis of this platelet dysfunction is multifactorial, and defects in calcium homeostasis, purine nucleotide metabolism, prostaglandin production, and secondary hyperparathyroidism have been implicated. The prolonged bleeding time can be shortened by elevation of the hematocrit above 0.35, infusion of cryoprecipitate (enriched in factor VIII), or the administration of deamino-D-arginine vasopressin (0.3 μg/kg intravenously over 30 min or subcutaneously) which increases levels of circulating von Willebrand factor but may be relatively contraindicated in patients with critical coronary or cerebrovascular disease. Conjugated estrogens, which produce a longer-lasting reduction in bleeding time, are also used.

Infection

Urinary infection is common, particularly in patients with chronic renal failure who are immunosuppressed, have abnormal lower tracts, or are diabetic. In the presence of polyuria, lower bacterial counts should be accepted as being significant; the presence of pyuria is often more useful.

Endocrinology

Patients with chronic renal failure have glucose intolerance due to reduced tissue sensitivity to insulin and decreased sensitivity of pancreatic β cells to glucose (decreased secretion). This is of importance in insulin-dependent diabetics in whom the response to exogenous insulin is also diminished, although this is compensated by reduced renal clearance of insulin.

Goiter, or at least a degree of thyroid gland enlargement, is common in renal failure. There is an increased incidence of hypothyroidism with low levels of total T_3 and T_4 but normal levels of the free hormone (Hochstetler *et al.* 1994). The presence of normal levels of thyroid-stimulating hormone remains a valid indicator for differentiating between clinical and biochemical hypothyroidism.

The incidence of pituitary–adrenal axis disease is not increased in patients with chronic renal failure (who have not received steroid therapy), and the diurnal rhythm of cortisol secretion should be retained although absolute cortisol production and ACTH levels may be increased. Excess or inadequate production of cortisol can produce symptoms and signs similar to those present in uremia. If required, adrenal function can be tested by 24-h cortisol profiles and the ACTH test, and pituitary function can be assessed by the high-dose dexamethasone test or the corticotropin-releasing hormone test.

Raised levels of prolactin in renal failure result from reduced renal clearance and increased pituitary release due to reduced dopaminergic inhibition.

References

Avram, M.M. and Mittman, N. (1994). Malnutrition in uremia. *Seminars in Nephrology*, **14**, 238–44.

Hochstetler, L.A., Flanigan, M.J., and Lim, V.S. (1994). Abnormal endocrine tests in a hemodialysis patient. *Journal of the American Society of Nephrology*, **4**, 1754–9.

Kang, J.Y. (1993). The gastrointestinal tract in uremia. *Digestive Diseases and Sciences*, **38**, 257–68.

Massry, S.G. and Smogorzewski, M. (1996). The heart in uremia. *Seminars in Nephrology*, **16**, 214–21.

Zachee, P., Vermylen, J., and Boogaerts, M.A. (1994). Hematologic aspects of end-stage renal failure. *Annals of Hematology*, **69**, 33–40.

7 The neurological system

7.1 Normal physiology

7.1.1 The nervous system

M. W. Nicolle and G. B. Young

Key messages

- Diseases of the nervous system seen in the intensive care unit can occur as a result of involvement of either the central or the peripheral nervous system. Knowledge of the relevant neuroanatomy is crucial to localize the lesion.
- Consciousness comprises alertness and awareness. Alertness depends on the ascending reticular activating system. The cerebral cortex, with extensive intracortical connections and interactions with subcortical structures, governs awareness and cognition. For disorders of cognition, it is best to describe the clinical features and to localize the particular deficit.
- Although many central nervous system functions are localized in discrete regions, the latter have rich interconnections with other brain components. Control of ventilation is ultimately dependent on centers in the medulla, but limbic and forebrain motor areas have strong temporarily overriding influences. Clinical and electrophysiological evaluation can determine the sites in the central and peripheral nervous systems that produce disorders of ventilation. The hypothalamus is the principal center for homeostasis, but peripheral receptors and higher central nervous system structures have considerable influence.
- The motor system is organized in a heirarchical fashion, including executive planning centers and regions for postural control and fine movements. These are integrated with other regions, many subcortical, which co-ordinate and smooth the movements. Sensory feedback is an essential component of motor performances.
- A knowledge of how nerve action potentials are generated, and of the basic mechanism of neuromuscular transmission and excitation–contraction coupling of muscle, is important to comprehend the pathophysiological mechanisms behind diseases of the peripheral nervous system seen in the intensive care unit, and also to understand their treatment.

Introduction

Diseases of the nervous system are common in the intensive care unit (ICU), both as primary processes and as complications of multiple organ failure in the critically ill patient. Thus coma may be the result of a structural lesion of the central nervous system or of a metabolic encephalopathy. Similarly, the demyelinating polyneuropathy of the Guillain–Barré syndrome may require management in the ICU, whereas patients in the ICU for prolonged periods for other reasons may secondarily develop a critical illness polyneuropathy. An understanding of basic neurophysiology is important if these disorders are to be recognized and treated.

Central nervous system

Consciousness

Consciousness consists of two principal components, alertness and awareness. Alertness, or wakefulness, including electroencephalographically documented arousal, is dependent on the ascending reticular activating system (ARAS), located in the tegmentum of the upper brainstem and diencephalon and projecting through the midline and intralaminar nuclei of the thalamus to the cerebral cortex (Kinomura *et al.* 1996) (Fig. 1). Activation of the ARAS allows

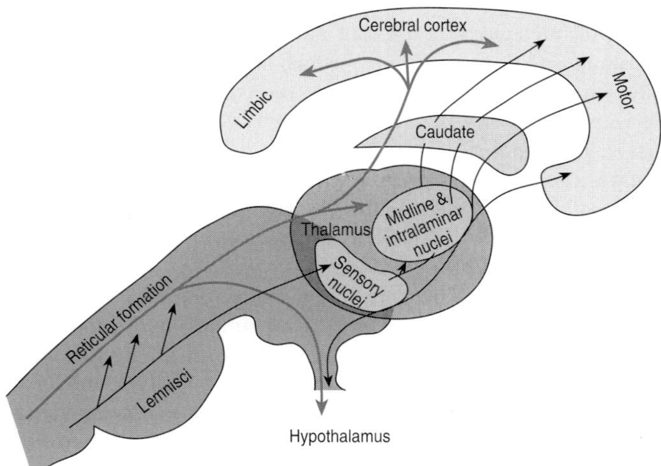

Fig. 1 The ascending reticular system and its projections. This diagram depicts the brainstem, containing the reticular formation and the ascending sensory pathways (lemnisci), the thalamus, the caudate nucleus, and higher structures (limbic system, motor areas, and cerebral cortex). Projections to the hypothalamus are also shown.

sensory information to be transmitted to the cerebral cortex for further processing. Awareness, or 'content', refers to higher functions including judgment, memory, emotions, and interpretation of sensory information. This depends on the integrated function of the cerebral cortex and the subcortical structures, particularly the thalamus.

Control of ventilation

Breathing requires the integrity of both the central and peripheral nervous systems at multiple levels (Berger *et al.* 1977). Peripheral receptors located in aortic and carotid bodies are activated by decreased arterial oxygen, pH, or blood pressure, and by increased carbon dioxide tension. This leads to increased central ventilatory drive. In addition, mechanical receptors along the respiratory tract can facilitate or inhibit breathing, or affect the breathing cycle in a reflex fashion (e.g. sneezing, coughing, sniffing, or brief apnea).

Centrally, ventilation is controlled by the cerebral cortex, brainstem ventilatory centers and their descending pathways, and spinal cord segments. Activation of limbic structures tends to inhibit ventilation, while stimulation of the motor and premotor areas of the frontal lobes may increase it. Such modulation of breathing allows for certain voluntary activities including speech.

The main automatic breathing centers lie in the brainstem. The pneumotaxic center in the pontine tegmentum switches off inspiration. The medullary center includes the dorsal respiratory group, the nucleus of the tractus solitarius, and the ventral respiratory group in the nuclei ambiguus and retroambiguus. The dorsal respiratory group contains inspiratory neurons which drive the ventral respiratory group. The ventral respiratory group contains both inspiratory and expiratory regions. Fibers from the ventral respiratory group descend in the lateral columns of the spinal cord to innervate spinal cord intercostal and phrenic neurons for inspiration and other intercostal and abdominal muscles for expiration.

Homeostasis and autonomic control

The nervous system plays the directing role in achieving homeostasis, or the maintenance of a constant internal environment. The hypothalamus and limbic systems, which are closely linked in this function, act to regulate endocrine secretion and the autonomic nervous system and to influence behavior through emotions and drives. The hypothalamus plays a central role in synthesizing information from various sources and in directing a response. It is the 'final pathway' for emotional expression and serves as the most important center for homeostasis, including autonomic and endocrine control. Its roles include the regulation of intravascular volume, maintenance of temperature, motivation for drinking and eating, and activities involved in pleasure or aversion. In addition, it serves as a pacemaker for cyclical activities, including the wake–sleep and menstrual cycles.

Endocrine function is regulated by the hypothalamus through its control of pituitary hormones by releasing and inhibiting hormones or factors (Table 1). For example, thyrotropin-releasing hormone produced in the hypothalamus flows through the portal system to the anterior pituitary gland where it stimulates the release of thyrotropin, which in turn stimulates the production and release of thyroid hormone from the thyroid gland.

The autonomic nervous system is under the control of the hypothalamus and the nucleus of the solitary tract in the medulla. There are important influences from the cerebral cortex and limbic system as well as from ascending sensory pathways. The autonomic nervous system, with its three major divisions (sympathetic, parasympathetic, and enteric), is largely an involuntary system. It allows the body to adjust to rapidly changing external events (the 'flight or fight' response of the sympathetic division) and to regulate internal activities (blood pressure, temperature, airway and breathing, urinary function, and digestion by the parasympathetic and enteric divisions). The autonomic nervous system also regulates pupil size and plays a major role in reproductive activity, at least in males.

Motor control

Movements can be divided into three main classes: voluntary activity, rhythmic motor patterns, and reflex responses. The highest-order activity is voluntary movement which allows for expression of the will and a purposeful response to the environment (e.g. reading, speaking, and performing calculations). Such activity is goal directed and largely learned, and improves with practice. In rhythmic motor patterns, the initiation and termination may be voluntary, but the rhythmic activity itself does not require conscious participation (e.g. chewing, walking, and running). Reflex responses are simple stereotyped responses that do not involve voluntary control (e.g. deep tendon reflexes or withdrawal of a limb from a hot flame).

Motor control is carried out in a hierarchical yet parallel fashion in the cerebral cortex, the brainstem, and the spinal cord. Modulating influences are provided by the basal ganglia and cerebellum through the thalamus.

Cerebral cortex

In the precentral gyrus, or primary motor cortex, components are represented in maps for the action of certain muscle groups on the opposite side of the body (homunculus) (Fig. 2). There is bilateral representation for central groups of muscles that work in unison (e.g. upper face, pharynx). Strong sensory input into the motor cortex and feedback from basal ganglia and cerebellar modulating centers allow for sensorimotor integration and modulation so that movements are smooth and accurate. Integrated postural movements are represented in the adjacent premotor and supplementary motor areas which also receive sensory feedback. Communication occurs between these cortical motor areas, and each sends descending pathways down the corticospinal tracts to the brainstem and spinal cord. Apart from the direct communications of the primary motor cortex with the cell bodies of lower motor neurons, these primary and supplementary centers also act indirectly via brainstem centers that have motor pathways to local spinal or brainstem segmental circuits, and eventually to the lower motor neuron cell bodies.

The basal ganglia and cerebellum

The basal ganglia, consisting of the caudate, putamen, globus pallidus, substantia nigra, subthalamic nucleus, and related nuclei in the brainstem, play an important role in movements, as evidenced by the hypokinetic/rigid and hyperkinetic disorders seen with lesions of various components. However, their role in the initiation and control of movement cannot be isolated from the motor activities of the cortex and brainstem centers discussed previously.

Table 1 Hypothalamic and pituitary factors

Hypothalamic substance	Affected anterior pituitary hormone	Principal function of pituitary hormone
Releasing		
Thyrotropin-releasing hormone	Thyrotropin	Stimulates production and release of thyroid hormone
	Prolactin	Lactation
Corticotropin-releasing hormone	Adrenocorticotropic hormone	Release of cortisol from adrenal cortex
	Beta-lipotropin	Prohormone for β-endorphin
Gonadotropin-releasing hormone	Luteinizing hormone and follicle-stimulating hormone	Responsible for menstrual activity, spermatogenesis, stimulation of sex hormones by gonads
Growth-hormone-releasing hormone	Growth hormone	Linear growth, anabolic activity, insulin antagonism
Prolactin-releasing factor	Prolactin	Milk production and secretion
Melanocyte-stimulating hormone-releasing factor	Melanocyte-stimulating hormone	Central role in sleep, stimulation of melanocytes
	Beta-endorphin	Opiate precursor
Inhibitory		
Prolactin release inhibiting hormone (dopamine)	Prolactin	
Growth hormone release inhibiting hormone (somatostatin)	Growth hormone, thyrotropin	
Melanocyte-stimulating hormone-releasing hormone release inhibiting factor	Melanocyte-stimulating hormone	
Hypothalamus–posterior pituitary hormone		
Antidiuretic hormone		Increases retention of free water by the kidney
Oxytocin		Involved in labor

The cerebellum plays a more obvious role in co-ordinating movements by providing feedback to the motor cortex, as well as by providing important influences on eye movements through brainstem connections and on postural activity through projections down the spinal cord.

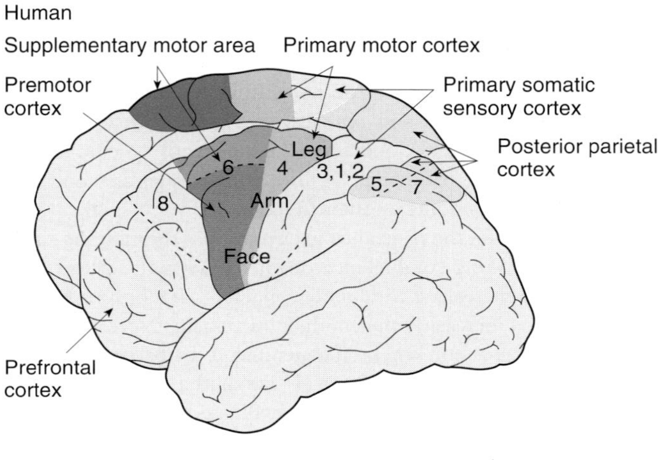

Fig. 2 The principal motor control areas of the cortex are shown on the convexity (in front) and the medial aspects (behind) of the cerebral hemispheres (Reproduced with permission from Kandel *et al.* (1991).)

Brainstem systems

Brainstem centers are organized into medial, lateral, and aminergic systems. Collectively, these integrate vestibular, visual, and somatosensory inputs for the control of eye movements and, through projections to the spinal cord, provide for postural adjustments. For example, these centers keep the images on matching regions of the retinas when the head moves by causing conjugate eye movements in the opposite direction to which the head is turned. This is the basis for the 'doll's eyes' test in clinical neurology in which the head is rapidly turned and the eyes move conjugately in the opposite direction, demonstrating the integrity of much of the brainstem.

Spinal segments

The anterior horn cell bodies that send out motor axons are organized in neuronal pools, medial for proximal and lateral for distal muscles. There are local circuits involving interneurons that allow for complex movements at local segmental levels. For example, in making a movement some muscles are prime movers, while others are brought into a supportive role and antagonist muscles are inhibited to allow the

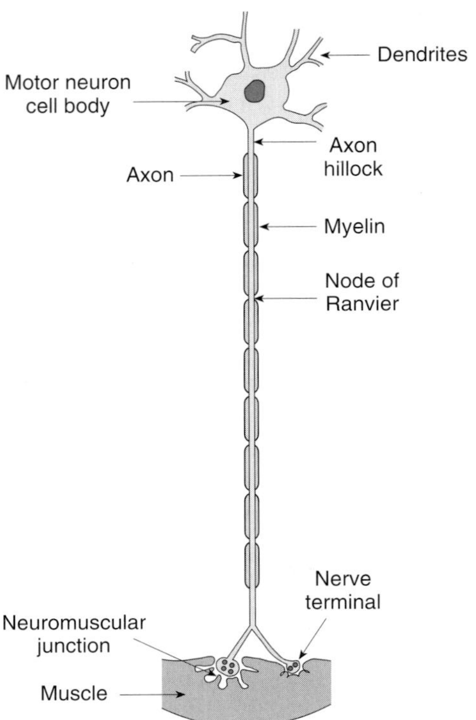

Fig. 3 Schematic diagram showing the motor neuron and its components, the neuromuscular junction, and muscle.

movement to proceed in a smooth fashion. Such activity is possible well below the conscious level.

Peripheral nervous system

The peripheral nervous system has both motor and sensory components. The former includes the motor neuron cell body in the anterior horn of the spinal cord and its peripheral axonal process traveling through the ventral root and eventually the peripheral nerve (Fig. 3). The motor nerve terminal, together with the muscle endplate and the synapse between the two, comprises the neuromuscular junction. The peripheral sensory axon, beginning at receptors in cutaneous and deep structures, as well as muscle and tendon receptors, travels back through peripheral nerves to its cell body located in the dorsal root ganglion. Its central process, traveling through the dorsal root, enters the spinal cord in the region of the dorsal horn.

The basic mechanisms for conveying motor and sensory information are similar, involving the generation and transmission of action potentials. In the ICU, the motor component is arguably more relevant, and the following discussion focuses on this.

Peripheral nerves

Generation of action potentials

Changes in the resting membrane potential determine whether action potentials are generated (Kandel *et al.* 1991). Motor neurons have a resting membrane potential of −70 mV (intracellular negative), with a relatively high intracellular concentration of K^+ and a low intracellular concentration of Na^+. This is maintained by Na^+,K^+-ATPase, which pumps Na^+ ions out of, and K^+ ions into, the axon cytoplasm. Depolarization, or a change to a less negative membrane potential, is excitatory. If a threshold level of depolarization is reached, an action potential is generated. Repolarization restores the membrane potential to its normal value after depolarization. Hyperpolarization generates a more negative membrane potential and decreases the likelihood of generating an action potential.

The motor neuron has an extensive dendritic tree (Fig. 3) which receives synaptic inputs from several sources, including the central processes of sensory axons, interneurons (largely inhibitory), and the descending pyramidal (corticospinal) motor pathway. These 'synaptic receptor potentials', produced by the interaction of neurotransmitters with receptors (chemically gated ion channels) on the motor neuron cell surface, produce local changes in the membrane potential (Kandel *et al.* 1991). Excitatory inputs, from Ia afferents mediating stretch reflexes and from descending corticospinal fibers, are probably mediated by the interaction of glutamate with receptors on dendritic spines, resulting in Na^+ influx and depolarization. Inhibitory inputs are mediated by the interaction of either glycine or GABA with receptors on the dendritic base or on the motor neuron cell body . This results in an influx of Cl^-, causing hyperpolarization. Reflecting the spatial and temporal summation of these excitatory and inhibitory inputs, an integrative potential is generated at the axon hillock, the most proximal part of the peripheral process.

Conduction and myelin

The action potential must somehow reach the nerve terminal. If the integrative potential produces sufficient depolarization to reach threshold, voltage-gated Na^+ channels in the axolemma open, and a depolarizing influx of Na^+ produces an all-or-none action potential and the opening of still more Na^+ channels. Simultaneously, depolarization induces the opening of voltage-gated K^+ channels. The resultant efflux of K^+ produces hyperpolarization. Although these two processes are initiated simultaneously, a delay in the opening of K^+ channels permits an initial depolarization before the repolarizing K^+ efflux restores the membrane potential. Eventually, the concentrations of intracellular Na^+ and K^+ are restored by the actions of Na^+,K^+-ATPase. Conduction velocity down the axon is increased in axons of larger diameter and in those insulated by myelin.

In the peripheral nervous system, Schwann cell processes form myelin. Voltage-gated Na^+ channels are concentrated at the nodes of Ranvier, which are gaps between Schwann cell processes (Fig. 3). Voltage-gated K^+ channels are located more diffusely in the axolemma under the myelin sheath of the paranode and internode. Additional ion channels in the axolemma modulate the balance between depolarization and repolarization (Waxman 1993). These include distinct Na^+, K^+, and Ca^{2+} channels with specific localizations at the nodal, paranodal, or internodal axolemma under the myelin sheath, depending on their function.

Myelin increases the resistance to current loss through the surface of the axon, making conductance down the axon the path of least resistance and increasing conduction velocity. Larger axons have less resistance, also increasing the conduction velocity. Nevertheless, the amplitude of the action potential generated at the bare membrane of the axon hillock decays as it travels down the axon. However, if sufficient depolarization reaches the voltage-gated Na^+ channels concentrated at the node of Ranvier, an influx of Na^+ further depolarizes the axon and regenerates the action potential (Kandel *et al.* 1991). Repeating this process at nodes every 1 to 2 mm further down the axon produces 'saltatory conduction' in myelinated fibers.

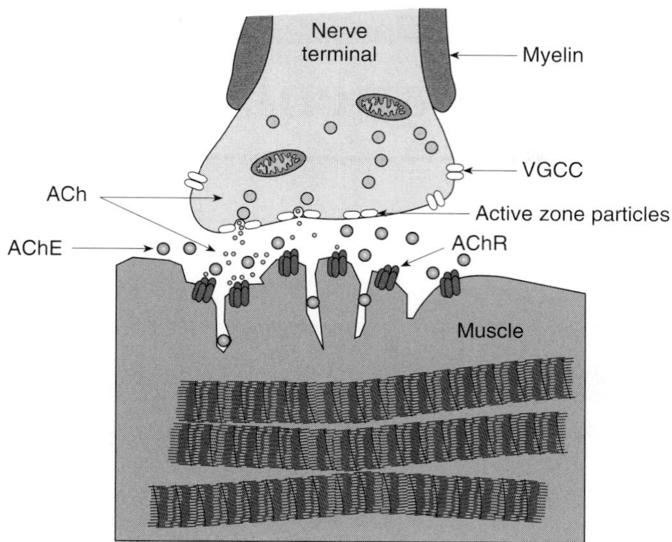

Fig. 4 The neuromuscular junction. The nerve terminal, synaptic cleft, and muscle endplate are shown: ACh, acetycholine; AChE, acetylcholinesterase; VGCC, voltage-gated Ca^{2+} channels; AChR, acetylcholine receptors.

Demyelination of peripheral nerves, as occurs in the Guillain–Barré syndrome, slows conduction and may result in conduction block, manifest clinically as weakness. In time, remyelination, or even redistribution of voltage-gated Na^+ channels over the full surface of the axolemma, may restore conduction. In some immune-mediated neuropathies, antibodies against myelin gangliosides may react with the extracellular portion of voltage-gated Na^+ channels, interfering with their function and impairing conduction.

Neuromuscular junction and neuromuscular transmission

The translation of an action potential in the motor nerve to muscle contraction begins at the neuromuscular junction (Fig. 3). Depolarization opens voltage-gated Ca^{2+} channels concentrated in the nerve terminal membrane (Fig. 4) (Keesey 1989). Ca^{2+} influx promotes the mobilization and fusion of preformed vesicles containing acetylcholine with active zone particles on the intracellular surface of the nerve terminal, and the exocytosis of their contents. Voltage-gated Ca^{2+} channels are blocked by high concentrations of extracellular magnesium, explaining the neuromuscular blockade seen in hypermagnesemia. In the Lambert–Eaton myasthenic syndrome, antibodies directed against voltage-gated Ca^{2+} channels prevent Ca^{2+} influx and therefore decrease acetylcholine release in response to depolarization (Keesey 1989).

After release, acetylcholine crosses the synaptic cleft and interacts with acetylcholine receptors on the crests of muscle postsynaptic junctional folds (Fig. 4). The opening of acetylcholine receptors allows the movement of Na^+ into the muscle fiber and the generation of an excitatory endplate potential. When of sufficient amplitude, nearby voltage-gated Na^+ channels are opened and an action potential is generated. Acetylcholine is hydrolyzed by acetylcholinesterase, terminating its action. In myasthenia gravis, anti-acetylcholine receptor antibodies interfere with neuromuscular transmission, resulting in muscle weakness (Keesey 1989).

Muscle

Excitation–contraction coupling is the process linking the generation of an action potential with muscle contraction (Kandel *et al.* 1991). Sodium influx triggers an action potential which propagates over the surface of the muscle fiber and into its interior via the T-tubule system. This spreading depolarization acts at voltage-sensitive channels located in terminal cisterns of the T tubules adjacent to the sarcoplasmic reticulum. As a result, Ca^{2+} is released from the sarcoplasmic reticulum into the intracellular space of the muscle fiber. Intracellular Ca^{2+} binds to troponin molecules located on actin polymers of the thin filaments. This induces a conformational change in actin, exposing a binding site for the myosin heads of thick filaments. The subsequent rotation of the myosin heads pulls the thin filaments over the thick filaments, shortening the muscle sarcomere, the basic unit of contraction in muscle. In a process dependent on ATP and Ca^{2+}, the myosin heads detach from actin binding sites and rotate back to the resting position, ready for reattachment to actin. When muscle depolarization is complete, Ca^{2+} is pumped back into the sarcoplasmic reticulum and new myosin–actin cross-bridges can no longer form, resulting in cessation of contraction. If ATP is present, the existing myosin–actin contacts detach and the muscle relaxes.

Conclusions

From the above discussion, it is apparent that a wide variety of disorders common in the ICU can affect the central and peripheral nervous systems. Even critical illness itself, with its complicated balance of organ failure, may affect the nervous system in many ways. Knowledge of the basic neurophysiological mechanisms involved is essential to the recognition, understanding, and treatment of these disorders.

References

Berger, A.J., Mitchell, R.A., and Severinghaus, J.W. (1977). Regulation of respiration. *New England Journal of Medicine*, **297**, 138–43.

Kandel, E.R., Schwartz, J.H., and Jessell, T.M. (ed.) (1991). *Principles of neural science*. Appleton & Lange, Norwalk, CT.

Keesey, J.C. (1989). Electrodiagnostic approach to defects of neuromuscular transmission. *Muscle and Nerve*, **12**, 613–26.

Kinomura, S., Larsson, J., Gulys, B., and Roland, P.E. (1996). Activation by attention of the human reticular formation and thalamic intralaminar nuclei. *Science*, **271**, 512–15.

Waxman, S.G. (1993). Molecular dissection of the myelinated axon. *Annals of Neurology*, **33**, 121–36.

7.2 The unconscious patient

7.2.1 Pathophysiology

E. Oschatz and A. N. Laggner

Key messages

- The principle pathophysiological alterations which may lead to impaired consciousness occur either in the bilateral cerebral cortex or in the brainstem reticular activating system.
- The changes at cellular level that may lead to impaired consciousness are cellular hypoxia, cellular substrate deficiency, and activation of cerebrodepressant receptors.

Introduction

In critical illness impaired consciousness is often the first sign of a severe pathological process. Consciousness is defined as an awareness of self and the environment, which requires both arousal and content. There are different types of depressed consciousness and even coma, the most severe form of absolute unconsciousness, is a heterogeneous state. Coma is a state of unresponsiveness from which the patient, who appears to be asleep, cannot be aroused by verbal and physical stimuli to produce any meaningful response; therefore the diagnosis of coma implies the absence of both arousal and content. Coma must be considered as a symptom with varied clinical accompaniments, numerous causes, different natural modes, and several management modes. Stupor is a state of unconsciousness from which the patient can be awakened to produce inadequate responses to verbal and physical stimuli. Somnolence is a state of unconsciousness from which the patient can be fully awakened (Tintinalli *et al.* 1992).

Topography of cerebral affection

Although there are many specific causes of unconsciousness, the sites of cerebral affection are either the bilateral cerebral cortex or the brainstem reticular activating system. The most common causes of bilateral cortical disease are deficiencies of oxygen, metabolic disorders, physical injury, toxins, postconvulsive coma, and infections (Table 1).

Table 1 Unconsciousness due to bilateral cortical disease

Cause of cortical impairment	Clinically relevant examples
Hypoxia/anoxia	Cardiac arrest Global cerebral ischemia Sinus thrombosis
Metabolic	Hypoglycemia Hepatic encephalopathy Septic encephalopathy Diabetic ketoacidosis Hyperosmolar diabetic coma Hypernatremia Hyponatremia Uremia Myxedema coma Thyroid storm
Physical injury	Hypothermia, hyperthermia Trauma
Toxin	Alcohols Barbiturates Antidepressants Benzodiazepines
Postconvulsive	Epilepsy
Infection	Meningitis

The reticular activating system maintains the state of wakefulness through continuous stimulation of the cortex. Any interruption may lead to unconsciousness. The reticular activating system can be affected in three principal ways: by supratentorial pressure, by infratentorial pressure, and by intrinsic brainstem lesions (Table 2).

Supratentorial lesions produce impaired consciousness by enlarging and displacing tissue. The skull is a limited area and when additional volume accumulates, pressure is directed to the point of least resistance. Thus the temporal lobes may be forced through the tentorial notch, compressing brainstem structures and cranial nerves. By pressing on the brainstem through this remote mechanism, the reticular activating system is also compressed. The brainstem lies infratentorial in the posterior fossa with the ventricular aqueduct of Sylvius, the fourth ventricle, and the cerebellum. An increase of infratentorial pressure may force the content of the posterior fossa upward through the tentorial notch or downward through the foramen magnum. This again causes compression of the reticular activating system and unconsciousness. Lesions which affect the brainstem itself damage the reticular activating system directly (Plum and Posner 1982; Tintinalli *et al.* 1992).

Table 2 Unconsciousness due to lesions of the reticular activating system

Cause of RAS impairment	Clinically relevant examples
Supratentorial pressure	Bleeding with perifocal edema (EDH, SDH, ICB) Ischemia with perifocal edema Infection with concomitant edema Tumor with perifocal edema Traumatic lesion with perifocal edema
Infratentorial pressure	Bleeding with perifocal edema (ICB) Ischemia with perifocal edema Infection with concomitant edema Tumor with perifocal edema Traumatic lesion with perifocal edema
Intrinsic brainstem lesion	Bleeding with perifocal edema (ICB) Ischemia with perifocal edema Infection with concomitant edema Tumor with perifocal edema Traumatic lesion with perifocal edema

RAS, reticular activating system; ICB, intracranial bleeding; EDH, epidural hemorrhage; SDH, subdural hemorrhage

Alterations at cellular level

Many of the enzymatic reactions of neurons, glial cells, and specialized cerebral capillary endothelium in the brain must be catalyzed by the energy-yielding hydrolysis of adenosine triphosphate (ATP) to adenosine diphosphate (ADP) and inorganic phosphate. Without a constant and generous supply of ATP, cellular synthesis slows or stops, neuronal functions decline or cease, and cell structures quickly fall apart. The brain depends entirely on the process of glycolysis and respiration within its own cells to provide its energy needs. The oxygen-requiring process of respiration is more efficient than glycolysis in generating the energy needs of the brain. The enzymes of the glycolytic pathway can increase their rates only about sixfold; glycolysis alone is unable to meet the energy needs of the human brain, even if the circulation and other support systems could sustain an increase in the delivery of glucose. This insufficiency in glycolytic enzymatic action explains the constant need for oxygen to ensure appropriate cerebral function. Even a short interruption of blood flow or oxygen supply threatens tissue vitality (Plum and Posner 1982).

During anoxia, the level of ATP rapidly decreases to near zero. This causes an increase in glutamate, an excitatory amino acid, which acts through the *N*-methyl-D-aspartate (NMDA) receptor and finally leads to high levels of free intracellular calcium $[Ca^{2+}]_i$. The addition of 2-deoxyglucose to cells, which acts as an ATP sink, causes a rapid increase in $[Ca^{2+}]_i$, activating phospholipase A_2 which breaks down membrane phospholipids into free fatty acids, particularly arachidonic acid. This causes increased activity of the cyclo-oxygenase pathway producing prostaglandins (including thromboxane A_2), the lipoxygenase pathway producing leukotrienes, or both. Furthermore, the hydrolysis of ATP via AMP leads to an accumulation of hypoxanthine. Increased $[Ca^{2+}]_i$ enhances the conversion of xanthine dehydrogenase to xanthine oxidase, priming the neuron for the production of the oxygen free radical O^{2-} intracellularly, once O_2 is reintroduced during reperfusion (Safar and Bircher 1988).

In hypoglycemia loss of substrate leads to increased metabolism of endogenous substrates in the brain, membrane depolarization, energy depletion, and increased levels of intracellular calcium. The neuronal injury and death induced by sustained glucose deprivation also result in part from the neurotoxic effects of glutamate, acting through NMDA receptors to stimulate the cellular uptake of calcium and to activate lipases, in a manner analogous to the mechanism of anoxic neuronal injury.

The development of coma in hepatic encephalopathy has been attributed to high blood levels of ammonia, alterations of the plasma amino acid profile (increased levels of branched-chain amino acids and decreased levels of aromatic amino acids), release of false neurotransmitters (octopamine and phenylethanolamine), and activation of the GABA system. Recent experimental studies of septic encephalopathy suggest also involvement of the GABA system.

In patients with diabetic ketoacidosis impaired consciousness is assumed to be due to increasing extracellular hydrogen ion concentration, followed by reduced oxygen consumption and decreased glycolysis in brain cells. In hyperosmolar non-ketotic diabetic coma, osmotic diuresis leads to exsiccation and neuronal dehydratation, thereby reducing cerebral cellular metabolism. Hypernatremia causes neuronal dehydratation with consecutive rupture of the vessels. When it develops slowly, this cellular dehydratation could be balanced by the synthesis of 'idiogenic' osmotic substances (amino acids, particularly taurine). Acute hyponatremia can cause brain swelling when water moves from plasma to brain, lowering brain osmolality to match that of the hypotonic plasma. Hyponatremic brain edema is normally prevented by the transport of osmotically active solutes out of the brain cells by processes involving the Na^+,K^+-ATPase pump, amino acids, and the calcium channels. This preventive mechanism may be altered in premenopausal women and hypoxia.

In uremia increased permeability of cell membranes allows circulating organic acids access to the brain and causes an alteration in mental status. In addition, the cerebral oxygen consumption declines, glycolysis and energy utilization are reduced, and the sodium and potassium flux is decreased (Plum and Posner 1982). Myxedema coma and thyroid storm can also cause alterations in consciousness by decreasing or increasing cerebral oxygen consumption and the synthesis of nucleoprotein and protein in neurons and synapses (Plum and Posner 1982).

Hyperthermia (body temperature above 40 °C) has a direct toxic effect on brain and consciousness by denaturing cellular enzymes and therefore leading to cell death, cerebral edema, and local hemorrhage. Hypothermia (body temperature below 30 °C) also has a severe impact on cerebral function by reducing the cellular metabolism until irreversible structural damage occurs (Rippe *et al.*

1991). In head injury an initial impact produces several degrees of mechanical neuronal and axonal damage. Secondary lesions occur as a consequence of elevated intracranial pressure due to cellular edema and hypoxic cell injury (Tintinalli *et al.* 1992).

Many drugs in common use must be considered as causes of unconsciousness. Alcohol intoxication may lead to impaired consciousness by depressing cerebral activity due to interference with ion transport at the cell membrane rather than the synapses. Barbiturates mediate their effect through the inhibitory GABA synapses of the brain and thus interfere with cerebral oxidative enzymes and depress cellular metabolism. Tricyclic antidepressants reduce the production of serotonin and norepinephrine (noradrenaline) in the cerebral fluid. They are also competitive antagonists of the muscarinic acetylcholine receptors and much of the central nervous system toxicity is an exaggeration of the central and peripheral side-effects. Benzodiazepines attach to the polysynaptic terminals where GABA is released, cause hyperpolarization, and therefore potentiate the GABA effect. Opiates (synthetic and natural) produce a number of clinical effects that are also induced by endogenous morphine-like substances (endorphins), as well as substances in related groups (encephalins and dynorphins), acting on specific neurons. In these neurons different receptor sites may be present, of which the m, k, and s receptors appear to be associated with the opioid action. The main effect of opiates is mediated by the m receptor, whose stimulation has a central depressant effect (Ellenhorn and Barceloux 1988).

Postconvulsive coma may occur after seizures and is mediated by neuronal damage due to both increases in neuronal metabolic demand and decreases in energy supply. The degree of hyperthermia during an episode of seizure seems to be correlated with the degree of neuronal damage (Rippe *et al.* 1991). In cerebral infection, coma may be caused by hypoxia due to vascular damage, increased cerebral pressure due to cell edema, and energy deficiency due to competition with micro-organisms for nutrients.

References

Ellenhorn, M.J. and Barceloux, D.G. (1988). *Medical toxicology; diagnosis and treatment of human poisoning.* Elsevier, New York.

Plum, F. and Posner, J.A.B. (1982). *The diagnosis of coma and stupor.* F.A. Davis, Philadelphia, PA.

Rippe, J.M., Irwin, R.S., Alpert, J.S., and Fink, M.P. (1991). *Intensive care medicine.* Little, Brown, Boston, MA.

Safar, P. and Bircher, N.G. (1988). *Cardiopulmonary cerebral resuscitation*, pp. 229–37. W.B. Saunders, Philadelphia, PA.

Tintinalli, J.E., Krome, R.L., and Ruiz, E. (1992). *Emergency medicine. A comprehensive study guide.* McGraw-Hill, London.

7.2.2 Causes and diagnosis

T. P. Bleck and A. R. Webb

Key messages

- Two major neuroanatomic structures are necessary for consciousness: the reticular activating system and the cerebral hemispheres.
- Evaluation must be rapid, since morbidity is often related to how quickly therapy is begun. Many of the diffuse metabolic or toxic causes of coma resolve without long-term damage to the central nervous system, whereas acute structural disease, particularly if causing herniation, may be rapidly fatal.
- Damage to the midbrain affects both the reticular activating system and pupil reactivity, whereas metabolic disease produces coma but usually leaves the light reflex intact.

Introduction

Unconsciousness is produced in one of three ways: diffuse bilateral involvement of the cerebral hemispheres, injury to the brainstem reticular activating system, or a combination of focal injuries to the cortex and brainstem. Herniation is a special case of insult to the reticular activating system caused by a cerebral space-occupying lesion which produces a mechanical shift of brainstem structures.

Consciousness has two necessary components: arousal and content. Specific structures in the central nervous system modulate arousal; lesions in these areas render the patient unable to respond despite intact sensory afferents. Unconsciousness is the condition in which the patient makes no appropriate responses to stimuli, either external (e.g. pain) or internal (e.g. thirst). This state does not exclude posturing and other reflex movements.

Causes of unconsciousness

There are a large number of causes of coma, and it is convenient to classify them as those producing diffuse bihemispheric dysfunction and those with structural lesions producing mass effect (Brock and Bleck 1995). Some of the major causes are listed in Table 1. Nonconvulsive status epilepticus may present as unresponsiveness. Subtle signs, such as eyelid fluttering, mild facial twitching, or nystagmus, may be the only evidence that the patient is having a seizure. Similarly, a postictal state may show prolonged unresponsiveness and include focal signs of hemiparesis (Todd's paralysis) or posturing.

Psychiatric conditions mimicking coma include catatonia, conversion reactions, and feigned coma. In these conditions, pupillary responses and ocular movements will be normal. The motor examination may demonstrate normal tone in supposedly paretic limbs, and the patient will have normal movements when asleep.

Table 1 Causes of coma

Presentation as diffuse bihemispheric process (usually symmetric, often with intact brainstem reflexes)

Drug intoxication
- Barbiturates
- Opiates
- Alcohol
- Antidepressants
- Benzodiazepines
- Sedatives/tranquilizers
- Drugs of abuse
- Antipsychotic agents
- Anticholinergics
- Salicylates

Metabolic
- Hypo/hypernatremia
- Hypoglycemia
- Non-ketotic hyperosmolar coma
- Diabetic ketoacidosis
- Hypothyroidism
- Uremia
- Hepatic encephalopathy
- Hypo/hyperthermia
- Hypercarbia
- Hypoxia

Other causes
- Toxins, particularly CO_2, methanol
- Meningitis/encephalitis (viral or bacterial)
- Sepsis
- Hypertensive encephalopathy
- Subarachnoid hemorrhage (low grade)
- Head trauma, mild
- Seizures/postictal

Presentation as brainstem process (often asymmetric, impaired brainstem function)

Brainstem infarction/hemorrhage
Posterior fossa tumor
Herniation from any cerebral space-occupying lesion (tumor, stroke, hemorrhage)
Subarachnoid hemorrhage (higher grades)
Trauma, severe

The anatomy of consciousness

Two major neuroanatomic structures are necessary for consciousness: the reticular activating system and the cerebral hemispheres. The reticular activating system is primarily responsible for arousal mechanisms, and the hemispheres influence the content of consciousness (Brock and Bleck 1995).

The reticular activating system receives input from all major afferent tracts, and projects widely to the thalamus, the basal forebrain, and the cerebral hemispheres. The crucial segment of the reticular activating system for arousal is between the rostral midbrain and the midpons. Isolated lesions to this portion of the reticular activating system produce coma, whereas lower lesions do not. Damage to the thalamus or hypothalamus can also alter consciousness, which is understandable given the large interconnection between these structures and the reticular activating system, but bilateral involvement is usually required.

Focal lesions in the cerebral cortex tend not to alter arousal but instead affect the content of consciousness. To affect arousal, large areas of both hemispheres need to be involved, on either a structural or a metabolic basis. Large focal processes such as tumor or infarction may alter the contralateral hemisphere by pressure effects, or by disrupting circulation or metabolism. Dominant hemisphere lesions may be more significant in affecting arousal than lesions in the non-dominant hemisphere.

Herniation

Herniation occurs when pressure of a mass lesion forces brain tissue to shift from one intracerebral compartment to another, and it can cause unconsciousness when pressure on the brainstem disrupts the reticular activating system.

The total volume contained in the cranial vault is limited, and a mass lesion must cause some shift of the intracerebral contents. Initially a mass lesion displaces cerebrospinal fluid, but eventually a limit is reached and intracranial pressure increases. Brain tissue is highly inelastic and so this pressure causes herniation.

Central herniation is presumed to be due to a pressure cone forcing the brain out towards its only exit, the foramen magnum. Uncal herniation occurs when the medial temporal lobe is forced over the tentorial edge into the space beside the lateral midbrain. This compresses the third cranial nerve and particularly the parasympathetic fibers to the pupil traveling around the outside of the nerve, causing unilateral dilation of the pupil on the side of the lesion, the 'blown pupil'.

Ropper (1986) has ascribed many of the signs of herniation to horizontal displacement of the brainstem. In this model, third-nerve paresis results from stretching of the nerve because of an increase in the distance from its origin on the anterior surface of the midbrain to the point at which it is tethered to the cavernous sinus. Acute pupillary dilation often corrects within minutes of initiating therapy for increased intracranial pressure, which is more consistent with lateral displacement than with irreversible uncal herniation.

Downward herniation can distract the basilar artery, which is tethered to the skull base, away from the brainstem and cause hemorrhage or infarction. Eventually, the downward movement increases subtentorial pressure and forces the cerebellar tonsils out through the foramen magnum. Tonsillar herniation can also occur if a lumbar puncture is performed on a patient with elevated intracranial pressure, thus generating a large transforamenal pressure gradient. This is often catastrophic, as acute pressure on the medulla causes sudden respiratory arrest. Upward herniation of brainstem structures through the tentorium is possible in the setting of large posterior fossa masses which increase subtentorial pressure.

Evaluation of unconsciousness

Evaluation must be rapid since morbidity is often related to how quickly therapy is begun. Many of the diffuse metabolic or toxic causes of coma resolve without long-term central nervous system damage, whereas acute structural disease may rapidly be fatal, particularly if causing herniation.

A key tenet of the care of the comatose patient is that initial treatment must proceed simultaneously with diagnosis (Brock and Bleck 1995). Acute coma (seconds to minutes) suggests cerebrovascular disease, either hemorrhagic or ischemic, or cardiac arrest. A recent history of head injury may indicate a subdural or epidural hematoma. A subacute course (many minutes to hours) may suggest intoxication or infection, while a more prolonged period of altered mental status might be caused by a central nervous system tumor or a systemic metabolic disturbance. Weakness or falling to one side suggests a focal lesion. A history of epilepsy may point to a postictal state. Witnesses should be carefully questioned about possible toxic ingestion.

Laboratory studies

Laboratory studies may not aid in acute management, but frequently help with assessment later in the patient's course. Serum biochemistry will identify major metabolic derangements. A complete blood count should be obtained, since either central nervous system or systemic infections can cause coma. Urinalysis should be routine, as urosepsis may present first as altered mental status and often occurs without fever in the elderly. Hypothyroidism can cause coma of unknown etiology, and so a thyroid-stimulating hormone level needs to be determined.

A urine toxicological screen for drugs of abuse is almost always needed, since drug overdose is one of the most common causes of coma of unknown etiology (Plum and Posner 1982). Drug abuse is frequently implicated in cases of trauma. Mixed intoxication with combinations of alcohol, barbiturates, opiates, and benzodiazepines may be seen, and these agents should be screened for routinely. Finally, unconsciousness with focal signs can be caused by cocaine intoxication, which can cause cerebral vasculitis and stroke.

Blood gas analysis is necessary to assess hypoxia, hypercarbia, or abnormal pH. Hypoxia may be due to drug intoxication. Particular acid–base abnormalities are often helpful in diagnosing metabolic encephalopathies. Respiratory acidosis, with hypoxia, occurs with hypoventilation from respiratory depressants. Metabolic acidosis may suggest particular toxic ingestions (e.g. salicylates or alcohols) or diabetic ketoacidosis, uremia, sepsis, or lactic acidosis. Compensatory respiratory alkalosis can be seen with many of the metabolic acidoses. Pure respiratory alkalosis, with hyperventilation, may suggest psychogenic coma.

Examination

A neurological examination is performed to assess the etiology of unconsciousness. The goal is to determine whether there is a bihemispheric process or a reticular activating system problem, and particular attention is paid to signs of herniation (Brock and Bleck 1995).

One neurological cause of pseudocoma that needs to be excluded immediately is the 'locked-in' state (Hawkes 1974). In this condition, which is usually due to pontine infarction or hemorrhage, all cortical control except that of vertical gaze is disconnected. Often the patient is only able to look upward, but may not even be capable of opening the eyelids. If, after the eyelids are opened, the patient can follow the command to look up, he or she is not comatose but locked in, and needs studies directed towards identifying a pontine lesion.

Pupillary responses

The pupils should be examined, preferably with a bright light in a darkened room. The pupillary light reflex requires intact sympathetic and parasympathetic systems to dilate and constrict respectively. The key paradigm to remember is that damage to the midbrain affects both the reticular activating system and pupil reactivity, whereas metabolic disease produces coma but usually leaves the light reflex intact (Plum and Posner 1982).

Specific structural lesions produce particular pupillary patterns. Hypothalamic lesions, either by direct involvement or secondary to increased pressure from above, interrupt the efferent sympathetic pathways, producing small reactive pupils. Unilateral diencephalic dysfunction may cause Horner's syndrome of unilateral pupillary constriction and ptosis, which may be a an early sign of herniation. Dorsal midbrain damage interrupts the parasympathetic efferents, and the pupils become slightly large and unreactive but may spontaneously fluctuate in size (hippus). Central midbrain lesions damage both sympathetic and parasympathetic tracts, producing fixed, often irregular, midposition pupils. This is most frequently seen in the setting of true transtentorial herniation and generally implies a poor outcome. Pontine lesions, usually hemorrhagic, interrupt the descending sympathetic fibers and irritate the parasympathetic fibers, producing pinpoint pupils. More caudal lesions affect only the sympathetic system, again causing Horner's syndrome. Finally, a unilateral fixed dilated pupil suggests third-nerve compression and herniation (Plum and Posner 1982).

Small reactive pupils are the hallmark of drug intoxication, particularly by opiates, and metabolic disease, but there are a few exceptions. The light reflex is usually resistant to metabolic disease, but it may be suppressed in the setting of severe drug overdose, particularly barbiturates. Severe opiate intoxication may mimic the pinpoint pupils of pontine hemorrhage. Anticholinergics may produce large unreactive pupils associated with altered mental status, as may glutethemide intoxication. Anoxia may cause fixed dilated pupils, which become reactive if cerebral oxygen delivery is restored in time.

Eye movements

If purposeful movements, such as visual tracking movements looking toward a loud noise, are absent, check for spontaneous roving eye movements. Roving movements are often seen with metabolic encephalopathies. All these findings imply intact cortical control of the brainstem. A fixed deviation of the eyes usually means that there is a hemispheral lesion on the side towards which the eyes deviate, often associated with a contralateral hemiparesis. Isolated pontine lesions affecting the conjugate gaze centers cause paresis of gaze toward the damaged side, often associated with ipsilateral facial paresis and contralateral weakness of the arm and leg. A fixed downward gaze is seen with midbrain compression from above. Inability of one eye to move medially is seen with upper brainstem lesions to the medial longitudinal fasciculus on the side of the abnormal eye, and is called an internuclear ophthalmoplegia.

If no spontaneous movement is found, the cervico-ocular reflex should be tested (doll's eyes maneuver); an important caveat is to ensure prior to testing that there is no possibility of a cervical cord or spine lesion. The reflex is tested by rapidly turning the head from midline to one side and observing the eye movements. In the intact brainstem this produces a contralateral conjugate eye movement, the net effect of which is to keep the eyes seemingly fixed on a point in space. After a few moments the eyes should return to midposition. The head should then be turned in the opposite direction to check for symmetry of the response. Failure of the reflex in either direction implies brainstem dysfunction. The reflex also works in the vertical plane and should be tested in a similar fashion. If this maneuver fails or is untestable because of neck injury, the vestibular–ocular reflex may be assessed by caloric testing. This is done by elevating the patient's head to 30°, if possible, and rapidly instilling about 50 ml of ice water with a syringe. It is important to check first that the ear canal is clear and that the tympanic membrane is not damaged, which is a relative contraindication to the test. Cold water instilled against the tympanic membrane produces cooling of the adjacent semicircular canal, increasing the local density of the endolymph, and creating a net flow towards the cooler side. This direction of flow mimics the head turning away from the stimulated side, and therefore causes reflex slow eye movement towards the stimulus. In an intact brain, the frontal eye fields attempt to override this brainstem-driven tonic eye deviation, producing rapid saccades away from the stimulus (nystagmus), but with cortical damage the eyes will maintain a fixed deviation. Cold caloric testing is a potent stimulus to the brainstem and may produce gaze deviation even when head turning fails to do so. The key observation is that whenever conjugate gaze occurs, regardless of the stimulus, it implies an intact brainstem in the region of the reticular activating system (Plum and Posner 1982).

Eye deviation from hemispheral lesions can usually be overcome by these maneuvers, whereas with pontine lesions the eyes will not cross midline. Total lack of response can be seen with severe brainstem dysfunction, drug ingestion (particularly barbiturates, narcotics, and phenytoin), neuromuscular blockade, or bilateral vestibular lesions (Plum and Posner 1982).

Motor responses

The patient is observed for spontaneous movement or, if none is present, response to stimulus. The type of motor response and its symmetry provide important clues to assessing the location and severity of focal deficits. Any asymmetry of the motor patterns suggests a contralateral focal cerebral lesion. Comatose patients may show purposeful movements, such as reaching for their endotracheal tube or localizing painful stimuli, which require an intact sensory system, an efferent motor system, and cortical processing. Patients who cannot localize pain may withdraw from focal stimuli, again requiring functioning afferent and efferent tracts.

Abnormal motor responses include decorticate and decerebrate posturing. Decorticate posturing consists of flexion of the arms and extension of the legs, while in decerebrate posturing both the arms and the legs extend. The important principle of localization is that both forms of abnormal posture can occur with hemispheral as well as brainstem lesions. Prognostically, decerebrate posturing is worse than decorticate posturing . Any comatose patient who develops either form of abnormal posturing needs intervention for acute worsening.

Extension of the arms with weak flexion of the legs or absent leg movement implies severe structural damage of the pontine tegmentum. This finding indicates severe brainstem dysfunction and carries a grave prognosis (Turazzi and Bricolo 1977). Total loss of tone does not necessarily mean upper brainstem damage and can be associated with spinal cord or medullary transection (spinal shock), peripheral nerve injury or disease, or neuromuscular blockade.

Finally, care must be taken not to confuse reflex activity with other responses. In particular, triple flexion of the lower extremity is a reflex signaling upper motor neuron dysfunction. It may look like spontaneous movement of the leg away from painful stimuli, but it is a reflex; the important finding is that the reflex response is very rapid and stereotyped.

Respiration

Respiration is controlled by brainstem structures with mediation by cortical influences, and specific respiratory patterns have localizing value (Brock and Bleck 1995). Unfortunately, these patterns are often not noticed in patients receiving mechanical ventilatory support. The most common abnormal respiratory pattern is Cheyne–Stokes respiration, in which there is a sequential waxing and waning of tidal volume, including periods of apnea. It can be seen in non-comatose patients with congestive heart failure, hypoxia, or occasionally during normal sleep, and is associated with bihemispheric dysfunction in unconscious patients. Respiratory centers in the brainstem increase or decrease the respiratory rate in response to elevated or lowered $P\text{aCO}_2$ levels respectively. However, there is frontal lobe control such that, even with very low $P\text{aCO}_2$ levels, respiration does not stop but only slows, with a reduced tidal volume, until the $P\text{aCO}_2$ normalizes. When bihemispheric dysfunction is present on a structural or metabolic basis, the modulating influence of the cortex is lost and Cheyne–Stokes respiration is seen.

Damage to the upper brainstem reticular formation is reported to cause sustained hyperventilation, called central neurogenic hyperventilation or central reflex hyperpnea. Tachypnea is often seen in comatose patients, but other causes, particularly hypoxia, neurogenic pulmonary edema, or metabolic disarray, are more likely. Diagnosis of true central neurogenic hyperventilation requires an increased $P\text{aO}_2$ and a decreased $P\text{aCO}_2$, without other metabolic changes or drug intoxication.

Apneustic, cluster, and ataxic breathing are patterns associated with lesions of the mid-lower pons, upper medulla, and caudal medulla respectively, and all provide inadequate ventilation so that mechanical support is needed. With apneusis, a patient has a prolonged inspiratory pause, or respiration may consist of cycles of quick inhalation–pause–exhalation–pause. Cluster breathing consists of several rapid shallow breaths followed by a pause, while ataxic breathing is irregular brief respirations of small random tidal volume. Finally, apnea is of poor localizing value and may be seen secondary to cardiac arrest, multifocal brain lesions, drug overdose, spinal cord transection, or primary pulmonary process.

CT scanning

Any unconscious patient with focal signs will need a CT scan, as many will have potentially treatable problems. The differential diagnosis of coma with focal signs includes tumor, increased intracranial pressure, intracranial hemorrhage, central nervous system infection, and stroke. Unfortunately, some structural lesions may not produce focal signs but only brainstem dysfunction. Because of this, most comatose patients with coma of unknown etiology will need early CT scanning. Any comatose patient may also have fallen, and be at risk of a traumatic subdural or epidural hematoma, in addition to the primary problem. Traumatic hematomas may appear after the initial CT, and so acute worsening of any patient is an indication for repeat scanning.

Lumbar puncture and antibiotics

Other than suspected bacterial meningitis, a lumbar puncture is needed in any patient for whom the cause of coma is still unknown after initial evaluation to look for non-bacterial meningitis or encephalitis, particularly viral or fungal, or occult subarachnoid hemorrhage.

References

Brock, D.G. and Bleck, T.P. (1995). Coma: a practical approach. In *Pulmonary and critical care medicine*, Vol. 3 (ed. R.C. Bone, D. Dantzker, R. George, R. Matthay, and H. Reynolds), pp. 1–11. Mosby Year Book, Chicago, IL.

Hawkes, C.H. (1974). 'Locked-in' syndrome: report of seven cases. *British Medical Journal*, **4**, 379–82.

Plum, F. and Posner, J.B. (1982) *The diagnosis of stupor and coma*. F.A. Davis, Philadelphia, PA.

Ropper, A.H. (1986). Lateral displacement of the brain and level of consciousness in patients with an acute hemispheral mass. *New England Journal of Medicine*, **314**, 953–8.

Turazzi, S. and Bricolo, A. (1977). Acute pontine syndromes following head injury. *Lancet*, **ii**, 62–4.

7.2.3 Management

T. P. Bleck and A. R. Webb

Key messages

- If intubation is required the stability of the cervical spine must be assessed.
- A key part of the immediate evaluation is the patient's vital signs, which may themselves help decide the cause of unconsciousness.
- Hypoglycemia can be catastrophic to the central nervous system, with the degree of injury determined by the length of time and level of the low blood glucose.
- Prior to giving glucose, thiamine 1 mg/kg must be administered intravenously to prevent precipitation of acute Wernicke's encephalopathy.
- Specific causes of unconsciousness such as elevated intracranial pressure, drug intoxication, seizures, meningitis, or metabolic encephalopathy must be sought and treated.

Initial management

Resuscitation

Acute care must always start with basic life support: a patent airway, ventilation, and circulation. An easily reversible cause of coma, if treated sufficiently rapidly, is hypoxia secondary to airway obstruction or pulmonary disease. However, central nervous system lesions can also cause abnormal respiratory patterns. Quickly observing the patient's respiratory pattern may help with localization as described above.

Comatose patients are frequently intubated for one of two reasons: ventilatory failure or airway protection. Airway protective reflexes of gagging and coughing may be lost in coma, increasing the risk of aspiration. Also, the tongue and oropharynx relax, increasing the chance of airway obstruction. The aspiration risk is increased if gastric lavage is used for suspected toxic ingestion without a cuffed endotracheal tube in place.

Prior to intubation, the stability of the cervical spine must be assessed, particularly in patients with trauma. Also, patients who lose consciousness acutely may fall and injure their cervical spine. Because patients with altered mental status may be unable to tell the examiner about neck pain, all comatose patients should be treated as if their neck is injured unless a reliable witness can attest to the absence of a fall or other potential for neck injury.

A patient in a hard cervical collar may be difficult to intubate orally as neck extension cannot and should not be attempted; therefore nasotracheal intubation may be preferred. The one exception to nasal intubation would be if there is a suspected basilar skull fracture. If raised intracranial pressure is at all possible, 100 mg lidocaine (lignocaine) or 300 mg thiopental (thiopentone) should be administered intravenously 1 min prior to intubation to blunt the rise in intracranial pressure normally associated with intubation.

A key part of the immediate evaluation is the patient's vital signs, which may themselves help to decide the cause of unconsciousness. Severe hypotension can be sufficient to cause symptomatic central nervous system hypoperfusion and should be corrected urgently with fluids and/or vasopressors. Conversely, a severely elevated blood pressure may cause hypertensive encephalopathy, which is a neurological emergency requiring rapid treatment.

Urgent corrective measures

Hypoglycemia can be catastrophic to the central nervous system, with the degree of injury determined by the length of time and level of the low blood glucose (Malouf and Brust 1985). Fingerstick determinations cannot be relied upon, as the glucometers can be less accurate at low values; a measured serum glucose may in reality be much lower than that estimated by fingerstick methods. However, a fingerstick glucose reading below 4.0 mmol/l should prompt urgent replacement of glucose. A dose of 50 ml of 50 per cent glucose in water should be given immediately to any patient with coma of unknown etiology. This will have no detrimental effect on other causes of coma (except Wernicke's encephalopathy which is discussed below). Even in the case of coma produced by hyperglycemic states, the marginal increase in total body glucose will not adversely affect treatment or generate central nervous system damage. In the case of hypoglycemia, glucose replacement may produce rapid reversal of unconsciousness.

Prior to giving glucose, thiamine 1 mg/kg must be administered intravenously to prevent precipitation of acute Wernicke's encephalopathy (confusion, ataxia, and ophthalmoplegia) with associated necrosis of the midline gray structures leading to permanent memory loss. Alcoholics are at particular risk because of poor general nutrition. Although most patients will not be thiamine deficient, the potentially terrible result if missed makes it imperative to administer thiamine before giving glucose.

Narcotic overdose is a common cause of coma in emergency room patients, as well as in hospital in-patients, particularly those in intensive care. The classic findings in narcotic intoxication are coma, small reactive pupils, shallow respiration, and hypotension. Unfortunately, not all patients display these findings. Pupillary responses in particular may be unreliable. Ingestion of other drugs (e.g. anticholinergics) may mask pupil findings. Hypoxia, brainstem lesions, or barbiturates may also blunt the response. Because the physical signs can be unreliable and narcotic overdose is common, any patient with coma of unknown etiology should be given naloxone 0.2 to 0.4 mg by slow intravenous injection.

Management of specific causes

Elevated intracranial pressure

If increased intracranial pressure is suspected, treatment needs to be initiated immediately, particularly if herniation is identified. The patient's head should be elevated to between 30° and 45°, and if possible the neck should be placed in a neutral position to facilitate venous return. Adequate ventilation and oxygenation are essential as hypoxia or hypercarbia will increase cerebral blood flow and intracranial pressure. Hyperventilation to lower the patient's $Pa\text{CO}_2$ to approximately 25 to 30 mmHg (3.5–4.0 kPa) will acutely reduce cerebral blood volume and intracranial pressure, but this effect is transient and other interventions to control intracranial pressure will need to follow.

For severe rises in intracranial pressure, mannitol should be administered at a dose of 0.25 to 0.5 g/kg every 4 to 6 h, taking care to correct serum electrolytes, osmolality, and volemic status. The mechanism of action of this agent is uncertain (extracellular space dehydration versus intravascular volume expansion) but its utility is not. Administration of steroids is somewhat controversial; it is useful for tumors and abscesses, probably efficacious in meningitis, and of no benefit in stroke or anoxia. If the cause of coma is unknown and the patient has evidence of increased intracranial pressure, an initial intravenous dose of 4 to 6 mg dexamethasone may be given empirically while diagnostic studies are undertaken.

Other factors that should be considered in the acute management of intracranial pressure include adequate sedation, with the addition of neuromuscular blockade if necessary to stop excess muscle activity, low ventilator pressure settings, which allow for good central venous return, and suppression of fever, which accelerates cerebral metabolism and thereby raises oxygen consumption, resulting in increased cerebral blood flow and higher intracranial pressure. Fever also accelerates neuronal damage from other causes. Finally, care should be taken not to lower blood pressure acutely (unless hypertension is very severe) to maintain cerebral perfusion pressure.

Investigation of the patient with raised intracranial pressure requires a CT scan to exclude a space-occupying lesion and neurosurgical support if this is found. Alternative considerations requiring specific treatment include cerebrovascular accident, subarachnoid hemorrhage, and metabolic encephalopathy.

Drug poisoning

When drug overdose or toxin ingestion is suspected, activated charcoal at a dose of 50 to 100 mg should be given to prevent system absorption. Gastric lavage may also be useful and is preferable to emesis. Use of both charcoal and lavage requires airway protection, and in patients with mental status changes this usually mandates intubation prior to treatment. For suspected benzodiazepine intoxication, flumazenil given in divided doses up to 1 mg in total may produce dramatic arousal, but is contraindicated in cases where tricyclic antidepressants may also have be consumed as there is an increased risk of seizures and status epilepticus. Flumazenil is also reported to improve mental function transiently in patients with hepatic coma, probably by reversing the GABAergic effect of accumulating endogenous benzodiazepine receptor ligands.

Seizures

Status epilepticus is another cause of coma requiring emergency action. Clinical generalized convulsive status epilepticus is readily recognized, but non-convulsive status epilepticus may be difficult to discern. Often the only indication of non-convulsive status epilepticus may be subtle twitching of individual muscles, particularly the face, rhythmic eye movements, or blinking. Status epilepticus of either type needs rapid treatment both to suppress the seizures and to prevent their recurrence. Lorazepam 2 mg every 2 to 3 min to a total of 0.1 mg/kg is often effective in terminating seizures. At the same time 20 mg/kg of phenytoin should be administered (at a rate no greater than 50 mg/min), with an additional 5 mg/kg to be given if seizures continue. If fosphenytoin is available, it is used in the same doses, but may be given at a rate up to 150 mg/min. Further therapy may include adding phenobarbital (phenobarbitone) if discrete seizures recur. If refractory status epilepticus develops, high-dose benzodiazepines, propofol, or pentobarbital (pentobarbitone) may be used. An emergency electroencephalogram is indicated for any patient with suspected non-convulsive status epilepticus or a patient being put in drug-induced coma to control seizures.

Meningitis

Emergency antibiotic therapy is indicated for any patient with suspected bacterial meningitis. In this life-threatening illness, treatment should not be delayed to obtain a CT or a lumbar puncture. Lumbar puncture results will not be significantly compromised if obtained immediately after antibiotic administration, and even if cultures are negative, bacterial antigens will remain positive.

Supportive care

Fluid therapy and vital organ support must be ongoing. The unconscious patient is totally dependent on nursing care, with particular attention paid to eye, mouth, and pressure area care. Nutritional requirements will usually be provided by nasogastric tube.

The unconscious patient may be capable of supporting respiration without mechanical ventilation. However, protection of the airway may not be possible after the cessation of respiratory support. For this reason, an early tracheotomy is useful to provide some airway protection from oropharyngeal secretions.

Monitoring unconsciousness

The Glasgow Coma Scale is a common method of monitoring the progress of coma, producing a score between 15 (normal) and 3 (deep coma) and taking into account eye opening, motor responses, and verbal responses. Frequent assessment of the Glasgow Coma Scale gives an objective measure of progress. Otherwise the prognosis deteriorates with the duration of coma; patients with postanoxic coma for 3 days rarely survive without severe disability.

Poor prognostic features are listed in Table 1. Outcome from coma is primarily dependent on the cause, assuming that appropriate steps are taken to avoid secondary injury from hypoxia, hypotension, etc. In cases of severe head injury, recent data give an overall mortality of 37 per cent, and good or moderate outcome at 43 per cent (Chesnut

Table 1 Features associated with a poor prognosis

Decerebrate posturing and rigidity for >24 h in non-trauma patients
Decerebrate posturing and rigidity for >2 weeks in trauma patients
Absent vestibulo-ocular reflex for >24 h
Absent pupillary reflexes for >24 h following anoxic brain injury
Absent pupillary reflexes for >3 days

et al. 1993). Coma from anoxia is associated with a poor prognosis, with 90 per cent mortality at 1 year if coma lasts for more than 6 h (Levy *et al.* 1985). Septic encephalopathy carries a mortality of 35 to 53 per cent depending on severity (Young *et al.* 1990). Hepatic coma is associated with good recovery in only 27 per cent of patients (Levy *et al.* 1981). Outcome from drug ingestion, if the patient survives until admission to hospital, is quite good.

References

Chesnut, R.M., *et al.* (1993) The role of secondary brain injury in determining outcome from severe head injury. *Journal of Trauma*, **34**, 216–22.

Levy, D.E., *et al.* (1981). Prognosis in nontraumatic coma. *Annals of Internal Medicine*, **94**, 293–301.

Levy, D.E., *et al.* (1985) Predicting outcome from hypoxic–ischemic coma. *Journal of the American Medical Association*, **253**, 1420–6.

Malouf, R. and Brust, J.C. (1985). Hypoglycemia: causes, neurological manifestations and outcome. *Annals of Neurology*, **17**, 421–30.

Young, G.B., *et al.* (1990). The encephalopathy associated with septic illness. *Clinical Investigations in Medicine*, **13**, 294–304.

7.3 Seizures

7.3.1 Pathophysiology and causes

T. P. Bleck

Key messages

- Seizures in critically ill patients are most commonly due to drug effects, metabolic or toxic disorders, or intracranial mass lesions (tumor, abscess, or hematoma).
- Conditions producing seizures tend either to increase neuronal excitation or to impair neuronal inhibition.
- The development of seizures as part of a metabolic disorder or drug effect do not mean that the patient will have epilepsy.
- A few generalized disorders (e.g. non-ketotic hyperglycemia) may produce partial (focal) seizures.

Introduction

Seizures result from imbalances between excitation and inhibition, and between neuronal synchrony and dyssynchrony. Current models of human seizures implicate the cerebral cortex in their genesis, although thalamic mechanisms (particularly the thalamic reticular formation) are involved in the synchronization of cortex neurons. These same systems are involved in the maintenance of consciousness and in sleep–wake cycling. The precipitants of a seizure at a particular time in the critical care setting are often pharmacological, even in the patient with a predisposition to seizures. Others may reflect normal physiological rhythms; for example, sleep stage transitions are typically times of increased cortical and thalamocortical neuronal synchrony.

A clear distinction should be drawn between seizures, which are events occurring to the patient, and epilepsy, which is a chronic condition characterized by the tendency to have repeated seizures as a consequence of a central nervous system disorder. The majority of patients experiencing seizures in the intensive care unit (ICU) do not have pre-existing epilepsy, and their chances of developing epilepsy in the future are usually more dependent on the cause than on the number or intensity of seizures that they experience. However, because of other deleterious neuronal and systemic effects of seizures, their rapid diagnosis and suppression during a period of critical illness is almost always necessary.

The major excitatory neurotransmitters in the cerebral cortex are excitatory amino acids, particularly glutamate. These transmitters control a variety of ion channels and second-messenger systems, some of which are linked to the production of nitric oxide. The ion channels gated during excitation allow the entry of sodium, which depolarizes the postsynaptic neuron, and calcium, which can trigger both intracellular metabolic processes and the transcription of a variety of gene products. During normal brain function, these processes are required for learning and memory functions, among others.

The line dividing normal neuronal physiology from the excessively rapid hypersynchronous depolarizations which characterize a seizure is quite narrow, and appears to depend upon several mechanisms which can be deranged by critical illnesses. Failure to remove glutamate and potassium from the extracellular space, functions performed predominantly by astrocytes, occurs in trauma, hypoxia, ischemia, and hypoglycemia. Loss of normal inhibition (which is provided in the cortex primarily by γ-aminobutyric acid (GABA) and, to a lesser extent, glycine) occurs during withdrawal from alcohol and other hypnosedative agents, or in the presence of GABA antagonists such as penicillin or imipenem. The deleterious effect of the ICU environment on sleep–wake cycling alters thalamocortical rhythmicity, but its contribution to seizures is only just beginning to be studied.

Many of these mechanisms are common to other conditions, such as stroke and trauma, for which the patient may require critical care. It is hoped that, in the future, a more direct approach to the correction of these mechanisms will improve the outcome of these patients. At present, the mainstay of therapy is pharmacological interference with the processes that trigger or maintain seizures (particularly by employing GABA agonists, such as benzodiazepines, or agents such as phenytoin that affect sustained high-frequency repetitive firing). Although the currently available agents are relatively effective in this role, they all have significant adverse effects and do not interfere with the process of epileptogenesis.

Many of the central problems regarding epileptogenesis remain to be solved, but the process involves several steps which are relevant to the intensive care specialist. Focal cerebral cortical disorders, whether or not accompanied by seizures in their acute stages, may lay the foundation for later epilepsy. Conditions such as trauma, hemorrhages, abscesses, and neoplasms all produce physical distortions of the adjacent neurons and astrocytes, and the extracellular space. Deposition of iron in the cortex from the breakdown of hemoglobin appears to be particularly epileptogenic. The timely diagnosis and management of these conditions may prevent later epilepsy or lessen its severity.

Etiology

In the critical care patient, seizures may either prompt the patient's admission (because of status epilepticus), or develop as a complication of another illness. The physician must determine rapidly if a treatable etiology is present. The patient suffering a single seizure, or a few discrete seizures, in the setting of a critical illness may not require antiseizure therapy, but this decision depends on the etiology of the seizure and the confounding effects of other illnesses. The patient presenting in status epilepticus requires aggressive treatment unless an underlying condition renders such intervention pointless.

Discrete seizures

Seizures occurring during another critical illness usually represent a central nervous system manifestation of a systemic disorder. In a prospective study of neurological complications of critical medical illnesses, seizures were the second most common problem encountered (and were almost as frequent as metabolic encephalopathy). The most frequent causes were cerebrovascular disease, central nervous system infections, metabolic encephalopathies, neoplasms, hypoglycemia, and osmolar disorders (including non-ketotic hyperglycemia). A retrospective analysis found that drug withdrawal was the most common etiology of seizures. Table 1 summarizes common causes of seizures in ICU patients.

Iatrogenic precipitant seizures should be rapidly detected and the cause removed, if possible, instead of treating the patient only with antiseizure agents. Seizures occurring because of the abrupt withdrawal of GABA agonists (e.g. hypnosedatives such as barbiturates and benzodiazepines) are managed by using a drug of the same class, followed by a very slowly decreasing dose to prevent subsequent withdrawal seizures. The actual agent selected may be different to that from which the patient is withdrawing; the choice depends on the other circumstances of the patient. Alcohol withdrawal seizures constitute a special type of hypnosedative withdrawal, because several neurochemical systems are involved in addition to GABA and other withdrawal phenomena may be anticipated once seizures begin. Although seizures due to withdrawal of alcohol (and other hypnosedative agents) are usually thought to be primarily generalized, a substantial minority of alcohol withdrawal seizures have focal components. These localized cortical disturbances probably reflect pre-existing areas of cortical hyperexcitability, as might follow remote head trauma, which are rendered more excitable by the loss of inhibition.

Acute metabolic disturbances commonly produce seizures in ICU patients, but one must consider that seizures may be symptoms of structural disorders which are made manifest by metabolic disorders. Generalized convulsions are typical of metabolic disturbances, but two common exceptions are seizures due to non-ketotic hyperglycemia or hypoglycemia which often have focal components. Acute hyposmolar states (i.e. those developing over hours) frequently present with generalized convulsions, as well as evidence of increased intracranial pressure. Hyposmolar states developing over days are rarely associated with seizures; these patients more commonly present with weakness or confusion without seizures or loss of consciousness. Hypocalcemia is frequently listed as a cause of seizures in children and adults, but it is rarely the sole cause of convulsions beyond the neonatal period and must not be used as an excuse to avoid investigation for other etiologies, particularly structural lesions. The same is true of hypomagnesemia.

Status epilepticus

When status epilepticus develops in a critically ill patient, the etiologies are the same as those of discrete seizures. Status epilepticus,

Table 1 Common etiologies of seizures in critically ill patients

- Anoxic encephalopathy (must be distinguished from myoclonic activity)
- Central nervous system infections
  - Brain abscess
  - Encephalitis
  - Meningitis
  - Subdural empyema
- Cerebrovascular disorders
  - Central nervous system vasculitis
  - Cortical vein thrombosis
  - Intracerebral hemorrhage
  - Ischemic stroke
  - Lupus cerebritis
  - Subarachnoid hemorrhage
  - Thombotic thrombocytopenic purpura
- Drug intoxication
  - Antibiotics with proconvulsant effects
    - High-dose penicillins in patients with renal failure
    - Imipenem–cilastatin in patients with renal failure or damaged blood–brain barrier function
  - Cocaine and other central stimulants
  - Dopamine antagonists
  - Meperidine metabolites (e.g. normeperidine)
  - Theophylline
  - Tramadol
  - Tricyclic antidepressants
- Drug withdrawal
  - Alcohol withdrawal
  - Hypnosedative drug withdrawal (primarily benzodiazepines and barbiturates)
  - Withdrawal of other anticonvulsants in patients with a history of seizures or a seizure diathesis
- Head trauma
- Metabolic encephalopathies
- Acute severe hyposmolality (e.g. water intoxication)
  - Hepatic failure (seizures are rare)
  - Renal failure
    - Acute renal failure
      - Drug or metabolite intoxication
      - Hypertensive encephalopathy
    - Chronic renal failure
      - Drug or metabolite intoxication
      - Dialysis dementia
- Neoplasms
  - Primary
  - Metastatic

particularly in patients with systemic disorders affecting the central nervous system, may not manifest as convulsions and should be considered in any case of prolonged change in mental status. In recent series, the most frequent causes for those patients admitted to ICUs primarily because of status epilepticus were withdrawal from alcohol or antiseizure drugs, anoxia, head trauma, central nervous system infection, acute metabolic disturbances, and tumors.

Bibliography

Alldredge, B.K. and Lowenstein, D.H. (1993). Status epilepticus related to alcohol abuse. *Epilepsia*, **34**, 1033–7.

Bleck, T.P., Smith, M.C., Pierre-Louis, J.C., Jares, J.J., Murray, J., and Hansen, C.A. (1993). Neurologic complications of critical medical illnesses. *Critical Care Medicine*, **21**, 98–103.

DeLorenzo, R.J., Pellock, J.M., Towne, A.R., and Boggs, J.G. (1995). Epidemiology of status epilepticus. *Journal of Clinical Neurophysiology*, **12**, 316–25.

Fountain, N.B. and Lothman, E.W. (1995). Pathophysiology of status epilepticus. *Journal of Clinical Neurophysiology*, **12**, 326–42.

Schuchardt, V. and Bourke, D.L. (1994). Alcoholic delirium and other withdrawal syndromes. In *Neurocritical care* (ed. W. Hacke, D. Hanley, K. Einhäupl, T.P. Bleck, and M. Diringer), pp. 835–9. Springer-Verlag, Berlin.

7.3.2 Assessment

T. P. Bleck

Key messages

- Recognizing seizures in critically ill patients may be difficult because of the effects of other diseases and their treatments.
- An attempt to make an etiological diagnosis should be made in all critically ill patients with new seizures.
- Since seizures are frequently the first sign of an otherwise unsuspected intracranial mass in critically ill patients, almost all intensive care patients with new seizures need a neuroimaging study.
- Cerebrospinal fluid analysis and electroencephalography are important investigations in these patients.

Introduction

Patients with seizures enter the intensive care environment either because of a complication of another critical illness or for the management of status epilepticus. The diagnostic studies required vary with the presentation and must often wait for the institution of emergent therapies, such as antibiotics in cases of suspected bacterial meningitis and antiseizure treatment for status epilepticus.

Table 1 Recognizing seizures in critically ill patients

1. Make a positive diagnosis of the etiology of altered awareness
2. In patients receiving neuromuscular junction blockade, evidence of sympathetic overactivity (e.g. pupillary dilatation, hypertension, tachycardia) may reflect seizure activity as well as pain or inadequate sedation. Only EEG monitoring can reliably distinguish these possibilities
3. In patients with intact neuromuscular transmission, observe for a prolonged postictal state after a recognized seizure: facial myoclonus, unexplained fluctuations in the level of consciousness, and nystagmoid eye movements
4. In most situations, one can transiently discontinue neuromuscular junction blockade to assess the patient's neurological status without harming the patient. If this cannot be done, one should consider intermittent (or continuous) EEG monitoring for patients receiving proconvulsant agents (e.g. imipenem–cilastatin) even in the absence of autonomic signs suggesting seizures
5. Hypnosedative agents may diminish the motor manifestations of seizures or status epilepticus without attenuating the electrical activity or preventing the consequent neuronal damage; thus, one should either lower the dose of these drugs occasionally to assess the patient's neurological status or perform intermittent (or continuous) EEG monitoring
6. Although elevations in serum prolactin and ACTH levels commonly accompany seizures, many common disorders and treatments in ICU patients may produce similar elevations (e.g. neuroleptic agents, which are also proconvulsant). Furthermore, this neuroendocrine response fatigues with frequent seizures. Thus studies of these substances are often unreliable markers of seizure activity in ICU patients
7. Decerebrate (extensor) posturing and opisthotonic posturing are occasionally seen in ICU patients, and may be confused with seizure activity. If they cannot be distinguished on clinical grounds, an EEG should be obtained. Short-acting neuromuscular junction blockade may be necessary (e.g. vecuronium 0.1 mg/kg with appropriate ventilatory and airway management). Succinylcholine (suxamethonium) should usually be avoided in this setting because of the possible precipitation of severe hyperkalemia

Discrete seizures

Recognition of a generalized seizure is straightforward in previously alert patients not receiving neuromuscular junction blocking agents; however, such patients are less common in intensive care practice than in other areas. Partial (focal) seizures that do not secondarily generalize are also difficult to detect in these patients. The intensive care unit (ICU) staff must maintain a high index of suspicion for the development of seizures in such patients, and maintain a correspondingly low threshold for neurological consultation and electroencephalographic studies. Table 1 presents some guidelines for the recognition of seizures in the ICU. One should strive to make a definitive diagnosis of the etiology of altered alertness, even if that diagnosis is metabolic encephalopathy, to avoid missing the possibility of such remediable disorders as seizures or status epilepticus.

Table 2 Classification of epileptic seizures

I. Partial seizures (seizures beginning locally)
- (A) Simple partial seizures (consciousness not impaired)
  1. With motor symptoms
  2. With somatosensory or special sensory symptoms
  3. With autonomic symptoms
  4. With psychic symptoms
- (B) Complex partial seizures (with impairment of consciousness)
  1. Beginning as simple partial seizures and progressing to impairment of consciousness
     - (a) Without automatisms
     - (b) With automatisms
  2. With impairment of consciousness at onset
     - (a) Without automatisms
     - (b) With automatisms
- (C) Partial seizures (simple or complex), secondarily generalized

II. Generalized seizures (bilaterally symmetric, without localized onset)
- (A) Absence seizures
  1. True absence (petit mal)
  2. Atypical absence
- (B) Myoclonic seizures
- (C) Clonic seizures
- (D) Tonic seizures
- (E) Tonic–clonic seizures (grand mal)
- (F) Atonic seizures

III. Unclassified seizures

When one suspects a seizure, an EEG should be obtained without delay. This is of particular importance when a patient does not return to baseline mental status, which suggests progression to non-convulsive status epilepticus. Most patients will need a brain imaging study in this setting, since a majority of critical care patients with new onset seizures have intracranial structural lesions. Although magnetic resonance imaging (MRI) is the method of choice in patients with new onset seizures, critically ill patients may not be able to undergo this procedure if they depend on infusion pumps for vasoactive drugs. MRI can be performed on mechanically ventilated patients using non-ferromagnetic ventilators. CT scanning remains a useful alternative; a scan is obtained before and after contrast enhancement, unless the patient has a serious contraindication to contrast material, because many causes of seizures in these patients are not apparent without enhancement. The need for other studies, such as lumbar puncture for cerebrospinal fluid analysis, depends on the likely diagnostic possibilities. If bacterial meningitis is suspected at any stage, appropriate antibiotic therapy should be instituted without waiting to obtain a lumbar puncture.

Classifying the seizure according to the International Classification of Epileptic Seizures aids in the etiological diagnosis and often has therapeutic implications (Table 2).

Status epilepticus

The usual definition of status epilepticus involves continuous or recurrent seizure activity without recovery, typically with a minimum duration of 30 min. However, this duration is arbitrary and, while important for epidemiological studies, is not useful at the bedside. The average duration of monitored single seizures is about 1 min, and since most seizures end spontaneously within 4 min, one should consider treatment for status epilepticus after about 5 min.

Status epilepticus is easily diagnosed when the patient suffers prolonged convulsions or has numerous partial seizures without an intervening recovery of normal alertness. More challenge arises when status epilepticus occurs without obvious motor manifestations, particularly if the patient has other reasons for unresponsiveness. In such circumstances, careful observation for small clonic movements (e.g. twitches of the corner of the eye) may be very rewarding. Other neurological symptoms such as aphasia may be signs of status epilepticus. The need for an EEG in this circumstance is indisputable. About 20 per cent of status epilepticus is non-convulsive. The EEG appearance of different forms and durations of status epilepticus is quite variable.

Bibliography

Bleck, T.P. (1993). Why isn't this patient awake? *Journal of Intensive Care Medicine*, 8, 155–6.

Bleck, T.P. (1995). Seizures in the critically ill. In *Critical care medicine. Principles of diagnosis and management* (ed. J.E. Parrillo and R.C. Bone), pp. 1271–83. Mosby Year Book, Chicago, IL.

Bleck, T.P. (1995). Seizures in the critically ill. In *Textbook of critical care* (3rd edn) (ed. S.M. Ayres, A. Grenvik, P.R. Holbrook, and W.C. Shoemaker), pp. 1545–55. W.B. Saunders, Philadelphia, PA.

Fagan, K.J. and Lee, S.I. (1990). Prolonged confusion following convulsions due to generalized nonconvulsive status epilepticus. *Neurology*, **40**, 1689–94.

7.3.3 Management

T. P. Bleck

Key messages

- The first priorities in managing seizure patients remain airway, breathing, and circulation, and the attempt to keep the patient from harm.
- Single seizures do not always require either pharmacological termination or prophylaxis against subsequent spells. These decisions must be made individually.
- Currently, the most effective drug for terminating generalized convulsive status epilepticus is lorazepam 0.1 mg/kg.
- Refractory status epilepticus should be treated aggressively in an attempt to prevent further brain damage.

Discrete seizures

The first concern in caring for any patient experiencing a seizure is to protect him or her from harm as far as possible. This is usually a straightforward task in the general hospital patient (e.g. preventing aspiration, removing sharp objects from the patient's vicinity), but assumes a new level of complexity in patients in intensive care who may be dependent on devices with connections of limited or no flexibility (e.g. intra-aortic balloon pumps, high-frequency oscillatory ventilators). In such circumstances, immediate neuromuscular junction blockade may be necessary. Following this, the patient should be observed for seizure classification and for possible intervention to maintain the airway and adequate ventilation. When emergency therapy is indicated, the physician should proceed as discussed below. In most cases, one should expeditiously attempt to make to an etiological diagnosis, and to institute appropriate treatment if possible.

Whether to administer antiseizure drugs after one seizure in a critically ill patient is often a complex decision, The causes of the patient's seizure may be unknown or difficult to reverse. Another seizure may pose substantial risk to a patient with tenuous cardiopulmonary function, which may be sufficiently great that medications are justified after one fit. In this circumstance, two treatment modalities are currently available. Our center employs a long-acting benzodiazepine (e.g. lorazepam 0.1 mg/kg, or clonazepam) unless contraindicated by its effect on the patient's level of alertness. An acceptable alternative is phenytoin (20 mg/kg) or fosphenytoin, as discussed below. The impending release of intravenous valproate may represent a useful alternative.

The choice of a maintenance antiseizure compound for those who remain at risk of seizures is beyond the scope of this chapter. However, the physicians caring for the patient in the intensive care unit should realize that physicians treating the patient subsequently usually continue the initial agent for months or years, and should communicate the bases for their decision for later reconsideration.

Status epilepticus

Attention to basic life-support issues (airway, breathing, and circulation) is central to the management of generalized convulsive status epilepticus. While the rapid termination of this condition is often the best way to deal with airway problems, emergent endotracheal intubation may be necessary for the patient unable to ventilate adequately or if treatment to terminate status epilepticus causes respiratory depression. Blind nasotracheal intubation is preferable for patients able to move air spontaneously, since neuromuscular junction blockade is not required. If laryngoscopic intubation is required in a patient with suspected intracranial hypertension, premedicate the patient with lidocaine (lignocaine) 1 mg/kg or thiopental (thiopentone) 3 mg/kg to blunt the increase in intracranial pressure associated with laryngoscopy.

During the first 30 to 60 min of status epilepticus most patients are hypertensive; after this, blood pressure usually declines to normal or hypotensive values. Many drugs used to terminate status epilepticus can induce hypotension, and so the clinician should always be ready to support the circulation with fluids and vasopressors.

Since hypoglycemia may be associated with status epilepticus, either as an etiology or as a consequence of prolonged seizure activity with autonomic failure, the blood glucose should be rapidly determined. Since the techniques employed for bedside determination lose accuracy outside the normal range, 'borderline' hypoglycemic values should be treated (along with thiamine administration). Non-ketotic hyperglycemia frequently presents with epilepsia partialis continua, which does not typically respond to antiseizure agents, but which usually remits with rehydration and control of the blood sugar.

Therapy for status epilepticus involves three principles: terminate the condition, prevent its recurrence, and manage its complications. The recently completed United States Department of Veterans Affairs study of status epilepticus compared four treatment regimens for generalized convulsive status epilepticus (lorazepam, diazepam followed by phenytoin, phenytoin alone, and phenobarbital (phenobarbitone) alone). This study showed that lorazepam 0.1 mg/kg terminated 63 per cent of generalized convulsive status epilepticus episodes and was marginally superior to the other treatment arms. In another important preliminary result, about 20 per cent of patients remained in electrical status epilepticus after their clinical seizure activity stopped. This reinforces the clinical suspicion of non-convulsive status epilepticus in patients who do not begin to awaken within 15 to 20 min after the apparent termination of their seizures.

Recommended lorazepam doses vary from 0.05 to 0.2 mg/kg (up to 8 mg in adults) administered at a rate of 0.04 mg/kg/min. Increasing the dose above 8 mg does not appear to improve the response rate. If this drug fails, phenytoin is frequently chosen as a second-line agent. However, the aggregate response rate to all second-line drugs in the Department of Veterans Affairs study was about 9 per cent, and the response to third-line drugs was 3 per cent. This suggests that additional conventional treatments have limited utility after the first choice fails. If phenytoin is used, it should be given in an initial dose of 20 mg/kg, at a rate of up to 1 mg/kg/min (maximum rate, 50 mg/min). This process generally takes over 30 min to complete, during which the majority of patients who failed lorazepam will continue in status epilepticus.

At this juncture, the patient is refractory to standard therapy and one might use one of the 'definitive' treatments. These more aggressive treatments should be given in intensive care units, since the patients require ventilatory support and frequently need hemodynamic support as well. EEG monitoring is also necessary, since these drugs typically abolish clinical evidence of seizure activity before achieving electrographic control.

We use either midazolam or propofol at this juncture. Midazolam therapy starts with a loading dose of 0.2 mg/kg and an initial maintenance dose of 0.1 mg/kg/h. The maintenance dose is titrated to produce clinical and electrographic suppression of seizures, rather than a burst-suppression EEG pattern. This agent is effective in over 90 per cent of cases of refractory status epilepticus pooled from three centers, but produces substantial tachyphylaxis. When the infusion rate reaches 2.0 mg/kg/h, we arbitrarily change to another drug, usually pentobarbital. This is often required in those situations in which the patient has a prolonged stimulus for epileptogenesis, such as encephalitis. Typical doses for propofol in status epilepticus range from 2 to 15 mg/kg/h; we also use a loading dose of 2 mg/kg. The costs of these doses for either drug are quite high, and often approximately equal. The caloric content of propofol, which is administered in a lipid solution, is about 1.1 kcal/ml.

Recommendations for pentobarbital loading doses vary from 5 to 12 mg/kg, with maintenance doses beginning at about 1 mg/kg/h. An EEG goal of burst suppression is traditional, but since there is no clear evidence to support this, we now choose seizure suppression instead. A pentobarbital dose adequate to produce burst suppression is often necessary to suppress all seizures, and there are several examples of seizures arising from a background of this pattern. Thus intermittent sampling of the EEG to insure that this pattem is present is not sufficient to be certain that no seizures are occurring. The extent to which occasional seizures produce further brain damage in this setting is unknown.

Adverse effects of pentobarbital include hypotension (from both venodilation and diminished myocardial contractility), immune suppression, poikilothermia, interference with the clearance of pulmonary secretions, and loss of gastrointestinal motility. Almost all adult patients will require pulmonary artery catheterization to provide the information necessary for the management of fluids and vasoactive drugs. Total parenteral nutrition may be required. Pulmonary embolism prophylaxis is necessary, and gastric mucosal protection (H_2 blockade or sucralfate) is advisable.

The optimal duration of these therapies is uncertain. To prevent recurrence of refractory status epilepticus, we typically institute a regimen of phenytoin and high-dose phenobarbital (aiming for a phenobarbital serum concentration of 50–100 µg/ml or higher) before withdrawing the agent being used for definitive control of status epilepticus. These choices are modified on the basis of the patient's history. This withdrawal of therapy is attempted once or twice daily, and requires both EEG and clinical absence of seizures to be considered successful.

Many other agents have been reported to be useful in terminating status epilepticus, but none have been demonstrated to be superior to those discussed herein.

Complications of status epilepticus include neuronal damage, rhabdomyolysis, hyperthermia, and cerebral edema. The only current approach to neuronal damage is to prevent it by rapid termination of status epilepticus, but excitatory amino acid antagonists are one of a number of promising neuroprotective treatments currently under study. Rhabdomyolysis from sustained muscular overactivity may be sufficient to produce renal damage; it may be preventable if treated early with volume expansion and alkalinization with sodium bicarbonate. Hyperthermia usually resolves after status epilepticus is controlled, but may contribute to neuronal damage prior to this point. External cooling is usually sufficient to control it once status epilepticus is terminated. Cerebral edema due to status epilepticus is vasogenic in experimental models, and may be managed with steroids and mannitol in the rare cases in which it produces symptoms. However, if marked edema is present, the likelihood that both status epilepticus and the edema reflect an underlying pathological process should be considered.

One change in drug therapy will most likely occur in the very near future; intravenous phenytoin will probably be replaced by fosphenytoin in most centers. This phenytoin prodrug is converted to phenytoin within about 20 min of administration. Its major advantage in status epilepticus will be the increased safety of its administration; in contrast with phenytoin, no adverse effects occur if it is extravasated. It can be administered more rapidly that phenytoin (up to 150 mg/min of phenytoin equivalent), and may be used either intravenously or subcutaneously. In studies to date, fosphenytoin does not appear to produce hypotension at the same frequency or severity as the intravenous formulation of phenytoin.

Bibliography

Bebin, M. and Bleck, T.P. (1994). New anticonvulsant drugs. *Drugs*, **48**, 153–71.

Chang, C.W.J. and Bleck, T.P. (1995). Status epilepticus. *Neurological Clinics*, **13**, 529–48.

Parent, J.M. and Lowenstein, D.H. (1994). Treatment of refractory generalized status epilepticus with continuous infusion of midazolam. *Neurology*, **44**, 1837–40.

Ramsay, R.E. and DeToledo, J. (1996). Intravenous administration of fosphenytoin: options for the management of seizures. *Neurology*, **46** (Supplement 1), S17–S19.

Treiman, D.M., *et al.* (1996). Treatment of generalized convulsive status epilepticus: a multicenter comparison of four drug regimens. *Neurology*, **46** (Supplement), A219.

7.4 Raised intracranial pressure and cerebral edema

7.4.1 Pathophysiology of generalized cerebral edema

L. F. Marshall

Key messages

- Increases in intracranial volume that are not caused by the tumor itself or by hemorrhage, for example, are generally described as cerebral edema. However, the term is also used for increases in intravascular volume within the intracranial space which are more correctly described as brain swelling.
- Vasogenic edema occurs as a result of increased capillary permeability. Such changes in permeability can be caused by the breakdown of capillary endothelial cells, by relative opening of tight junctions, or because of failure of certain transport mechanisms within endothelial cells.
- Cytotoxic edema refers to an increase in the brain-water content because of major disturbance in ionic homeostasis within the brain. Cerebral ischemia and trauma are the major causal factors.
- Interstitial or hydrocephalic edema is less frequent. Under these circumstances edema is a result of the movement of water from the ventricles through the brain because of obstruction of the outflow of cerebrospinal fluid.

Increases in intracranial volume as a result of a variety of processes within the brain are among the most dreaded consequences of neurological disease. The lethality of many disorders of the central nervous system, such as infection, neoplasms, and traumatic injury, stem in large part from the secondary reaction of the brain to insults. Increases of volume within the intracranial space are dangerous because the brain resides within the skull, which is a closed box with no substantial possibility of venting. While the displacement of cerebral spinal fluid out of the intracranial space is a mechanism which gives the system some capacitance, this is a relatively limited defense mechanism for compensating for increases in intracranial volume.

Increases in intracranial volume that are not caused by the tumor itself or by hemorrhage, for example, are generally described as cerebral edema. However, the term is also used for increases in intravascular volume within the intracranial space which are more correctly described as brain swelling. Because these processes often occur concurrently, they are discussed under the general rubric of brain swelling, with brain edema referring specifically to an increase in the water content of the brain.

In a series of classic papers, Klatzo divided brain edema into two types (Klatzo 1967). The first of these is vasogenic edema which occurs as a result of increased capillary permeability. Such changes in permeability can be caused by the breakdown of capillary endothelial cells, by relative opening of tight junctions, or because of failure of certain transport mechanisms within endothelial cells. The blood–brain barrier is extremely efficient in keeping out even small ions. In general, water entering the brain under normal conditions does so without solute and only because of hydrostatic forces. Therefore the osmotic gradient is directed at returning the water to the plasma. The term 'blood–brain barrier breakdown' has also been used to describe vasogenic edema; however, the term is somewhat imprecise. Increases in brain water content with vasogenic edema are often quite dramatic because the fluid which results from increased capillary permeability is usually rich in proteins and is similar to plasma. A number of factors influence the rapidity and severity of vasogenic edema. The first, obviously, is the degree of impaired capillary permeability. The second is the systemic arterial pressure and its relationship to local tissue pressure. The third is the influence of edema which is already present on the brain microcirculation. The continued deterioration of patients under these circumstances is often a result of the spread of edema, resulting in brain ischemia. This can lead to the second type of edema, cytotoxic edema, and to the progressive breakdown of both astrocytes and neurons. This breakdown generates both a significant osmotic load and the release of substances such as excitatory amino acids, catecholamines, and other biological products which may influence both vasogenic and cytotoxic edema.

Cytotoxic edema refers to an increase in the brain-water content because of major disturbance in ionic homeostasis within the brain. Although there are a number of causes of cytotoxic brain edema, clearly cerebral ischemia and trauma are the major causal factors. When the brain is ischemic, energy metabolism falls rapidly and thus the ability of the brain to pump ions actively also begins to deteriorate. Transport systems which normally move charged substances, glucose, and amino acids fail relatively quickly. Sodium enters the cell at an excessively rapid rate, pulling water with it and causing a subsequent temporary decrease in the extracellular space. Because the relative concentration of sodium in the extracellular space drops as it enters the astrocytes and neurons, sodium enters the extracellular space from the blood in an attempt to maintain osmotic equilibrium.

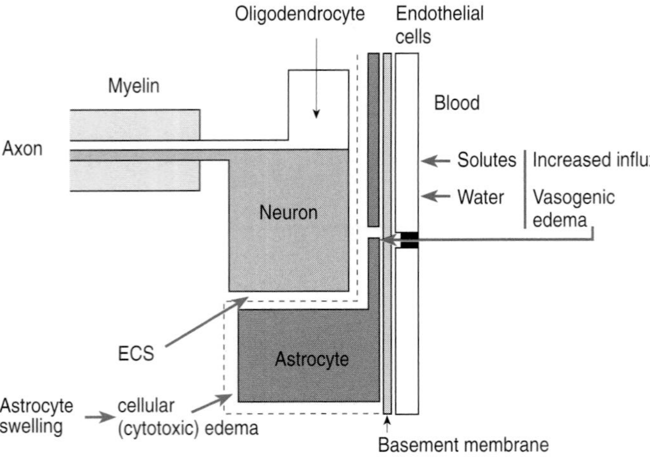

Fig. 1 Cytotoxic edema may release vasoactive compounds which increase capillary permeability and lead to vasogenic edema: ECS, extracellular space. (Reproduced with permission from Kimelberg (1995).)

The failure of active transport mechanisms in the brain, when coupled with cellular injury, results in the release of a number of vasoactive compounds which may further affect the blood–brain barrier or, in the case of glutamate for example, may produce a neurotoxic effect on the cell resulting in edema. Kimelberg (1995) has postulated that cytotoxic edema is frequently exacerbated by the release of excitatory amino acids from already swollen astrocytes which then produce neurotoxic injury. If this injury is of sufficient severity to cause damage to the cell, calcium influx and cell death will further perpetuate edema formation. Thus a vicious cycle may develop where initially cytotoxic edema is primary, but the release of these potent factors results in impaired capillary permeability with ensuing vasogenic edema (Fig. 1). This again results in ischemia because of impairment, from edema formation, of the microcirculation and a vicious cycle ensues which, if it is not adequately interrupted, can lead to coma and death.

Interstitial or hydrocephalic edema is less frequent. Under these circumstances edema is a result of the movement of water from the ventricles through the brain because of obstruction of the outflow of cerebrospinal fluid. Brain imaging will reveal a zone of edema directly adjacent to the ventricles, almost always bilaterally.

While the classification of various types of edema is useful in attempting to define specific treatment for certain disorders, it should also be recognized as somewhat arbitrary as cytotoxic and vasogenic edema frequently occur concurrently, although in varying proportions. In fact, each of these processes may cause the other.

Recent studies of closed head injury in laboratory models have demonstrated, using diffusion-weighted imaging, that the acute phase of brain injury is characterized primarily by cytotoxic edema (Ito *et al.* 1996). This contrasts with the previous view that vasogenic edema is a primary factor early in acute head injury.

Further evidence to support the fact that cytotoxic edema occurs early in head injury and that vasogenic edema is primarily a late phenomenon is provided by the data of Bullock *et al.* (1990), who used single-photon emission CT (SPECT) mapping of blood–brain barrier defects in humans, and by a study using gadolinium-diethylenetriaminepentaacetic acid (Gd-DTPA) enhanced magnetic imaging in human head injury (Lang *et al.* 1990). The latter study failed to show evidence of altered blood–brain barrier permeability to gadolinium within the first 4 days following head injury. Thus it can be concluded that therapeutic strategies in head injury need to take account of this new information regarding the time course of various types of edema.

Another example of a mixed cytotoxic edema is that associated with the treatment of hyperglycemia. If blood glucose is reduced too rapidly, a dramatic influx of water into the cerebral tissue can occur in an attempt to regain osmotic equilibrium and this will result in significant cytotoxic edema. This will lead to generalized cerebral ischemia, resulting in further cytotoxic edema because of ischemia.

The process of cytotoxic edema is quite complex because it has not yet been possible to identify all the osmols resulting from cellular injury which can draw water from the intravascular space, and because these osmols and other substances may differ quite substantially depending on the intracranial process responsible for their generation.

Brain edema in particular and increases in intracranial volume in general are particularly dangerous because of the limited capacitance of the brain to tolerate increases in volume. This relationship is described in the Monroe–Kelley doctrine, which states that the relationship between volume and pressure within the intracranial space is not linear and that increases in volume initially result in no increase in pressure, but as one moves to the right on the curve, small additional increments in intracranial volume result in dramatic rises in intracranial pressure. This relationship is made more complex by the fact that the intracranial space, which for many years was believed to be a single compartment, is made up of many compartments which eventually result in generalized increments in pressure but can act as semi-independent compartments producing focal compression. The most dramatic examples of this are separation of the supratentorial from the infratentorial compartment because of obstruction at the tentorium and the substantial pressure differential resulting from a balloon of equal size when inflated in the temporal versus the high parietal region. In the former, pressures within the brainstem are significantly higher than when the lesion is in the parietal lobe. This appears to be because there is a substantial dampening effect of the falx which reduces the transmission of pressure downward; however, when a lesion is located in the temporal lobe, the vector force is directed at the brainstem resulting in higher pressures and much more risk to the patient. This explains why, in the pre-CT era, patients with temporal lobe gliomas were often discovered only after acute deterioration which occasionally was fatal.

The influence of focal versus generalized edema is also quite important. As indicated earlier, the transmission of forces from lesions within the temporal lobe appears to be substantially greater than for disorders of the parietal region and for more generalized processes, where much larger increments in volume and much higher pressures are necessary to produce coma and/or death than when a focal lesion either compresses the brainstem directly or where there are significant amounts of midline shift (of the order of millimeters) resulting in distortion or kinking of the brain stem. Neurosurgeons generally use the degree of midline shift associated with focal lesions to determine whether or not emergency surgical intervention is required, particularly in patients with traumatic injuries, because shifts in excess of 5 mm usually result in compression of the brainstem.

Aggressive surgical treatment of focal lesions of the brain, particularly traumatic contusions and hematomas, and also tumors which result in edema formation and focal inflammatory processes such as brain abscesses, has developed because of a number of laboratory investigations which have clearly shown that these lesions are edematogenic, i.e. they serve to generate edema themselves and their removal will result in the cessation of that process. Some of the clearly demonstrated improvement in the outcome of acute severe head injury is due to early aggressive removal of intraparenchymal hemorrhages and brain contusion before brain edema becomes fully developed.

References

Bullock, R., Statham, P., Patterson, J., Wyper, D., Hadley, D., and Teasdale, E. (1990). The time course of vasogenic oedema after focal human head injury—evidence from SPECT mapping of blood brain barrier defects. *Acta Neurochirurgica*, **51** (Supplement), 286–8.

Ito, J., *et al.* (1996). Characterization of edema by diffusion-weighted imaging in experimental traumatic brain injury. *Journal of Neurosurgery*, **84**, 97–103.

Kimelberg, H.K. (1995). Current concepts of brain edema: review of laboratory investigations. *Journal of Neurosurgery*, **83**, 1051–9.

Klatzo, I. (1967). Presidential address. Neuropathological aspects of brain edema. *Journal of Neuropathology and Experimental Neurology*, **26**, 1–13.

Lang, D.A., Hadley, D.M., Teasdale, G.M., Macpherson, P., and Teasdale, E. (1990). Trauma and brain oedema (II): gadolinium-DTPA enhanced magnetic resonance imaging in human head injury. *Acta Neurochirurgica*, **51** (Supplement), 293–5.

7.4.2 Pathophysiology of space-occupying lesions

E. M. Trinidad and T. H. Milhorat

Key messages

- A pressure gradient between intracranial compartments is a much greater hazard than generalized intracranial hypertension because shift of intracranial structures caused by differences in compartmental pressures can lead to compression and infarction of herniated tissue.
- A sudden change in intracranial pressure, as is seen in acute intracranial hemhorrage, may lead to acute and fulminating neurological decline.
- Tumors may cause a gradual increase in intracranial pressure, which is tolerated with few symptoms but is predisposed to rapid deterioration on slight changes in intracranial pressure with a relatively innocuous stimulus such as a cough, sneeze, or Valsalva maneuver.
- Unilateral transtentorial herniation is caused by a supratentorial mass lesion and is initially associated with an ipsilateral dilated pupil followed by decline in mental status and contralateral hemiparesis.
- Central transtentorial herniation, commonly seen in the pediatric population with hydrocephalus, is initially associated with diencephalic dysfunction and paresis of upward gaze (Parinaud's syndrome) rather than dilated pupils.
- Transforaminal herniation which presents with neck tilt or nuchal rigidity must be distinguished from subarachnoid hemorrhage and/or meningitis because diagnostic lumbar puncture and resultant decreases in spinal cord pressure can lead to further herniation and infarction of tissues as the posterior inferior cerebellar arteries are compressed at the foramen magnum.
- Upward cerebellar herniation can compress the aqueduct of Sylvius with resultant hydrocephalus and exacerbation of the process.

Introduction

Increased intracranial pressure is a frequent complication of disorders that consume intracranial space such as brain tumors, hydrocephalus, intracranial hemorrhage, and brain swelling or edema. Although it is generally assumed that increased intacranial pressure is dangerous in its own right, extraordinary elevations of intracranial pressure may be tolerated surprisingly well by patients with pseudotumor cerebri, obstruction of the superior vena cava, and dural sinus thrombosis (Bell and McCormick 1972; Crockard *et al.* 1992; Cheek *et al.* 1994).

In contrast, the development of a pressure gradient between intracranial compartments is a much greater hazard than a generalized rise in intracranial tension. This gradient favors a shift of intracranial structures from the compartment of greater pressure toward the compartment of lesser pressure, and may lead to compression and distortion of structures as well as ischemic infarction of herniated tissue. If the shift occurs suddenly, symptoms will develop in an acute and fulminating manner. However, if the shift occurs gradually, there may be few outward signs of distress. Under such circumstances, neurological function may deteriorate at any time.

Herniation syndromes

The effects of space-occupying lesions that produce secondary shifts in intracranial contents can be described as herniation syndromes which characterize the direction and effects of the displacements.

Unilateral transtentorial herniation

This is perhaps the best known herniation syndrome and is caused by an elevation of pressure in one supratentorial compartment of sufficient magnitude to displace the temporal lobe on the affected side over the lateral edge of the tentorium (Fig. 1). The herniation may be more marked anteriorly (uncus) or posteriorly (hippocampal gyrus), and is usually accompanied by displacement of the ipsilateral cingulate gyrus under the falx. With the descent of the uncus through the incisural notch, the ipsilateral third nerve is compressed against the free edge of the tentorium or some other resistant structure such as the petroclinoid ligament or the superior cerebellar artery. This produces unilateral pupillary dilatation followed by other signs of oculomotor palsy.

With continuing descent of the uncus through the incisural notch, the midbrain is displaced laterally and downward. Consciousness is affected at this stage owing to compression of the ascending reticular formation, and an alteration in cardiorespiratory function may occur

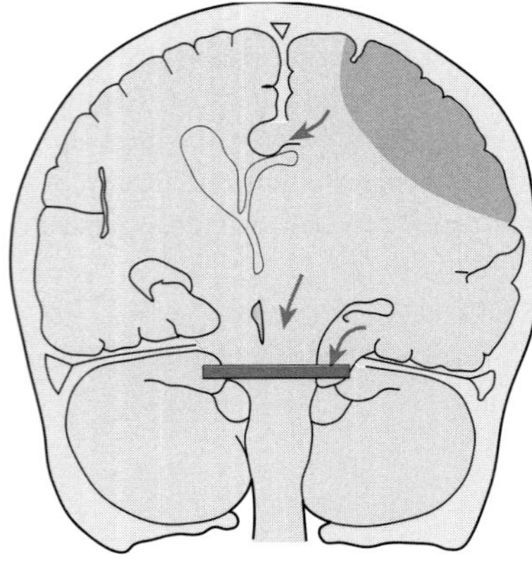

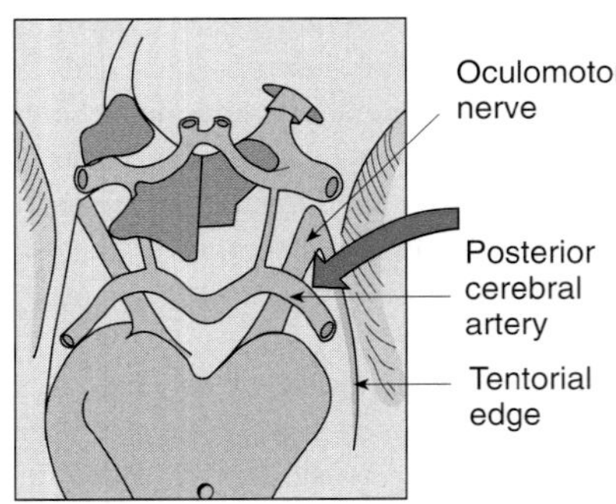

Fig. 1 Diagram of unilateral transtentorial herniation secondary to supratentorial mass lesion. The temporal lobe is herniated through the incisural notch and the midbrain is displaced laterally and downward. Note the associated herniation of the cingulate gyrus under the falx. The inset is a basal view of the brain at the level of the tentorial notch showing the relationship of the oculomotor nerve to the posterior cerebral artery and the free edge of the tentorium. (Reproduced with permission from Milhorat (1978).)

as a consequence of distortion of the basal diencephalon. The most common respiratory patterns associated with transtentorial herniation are Cheyne–Stokes respiration (alternating phases of hyperpnea and apnea) and central neurogenic hyperventilation. In the majority of cases the herniating temporal lobe will compress the ipsilateral cerebral peduncle, producing weakness and decorticate posturing on the opposite side of the body. However, in approximately 20 per cent of cases the contralateral cerebral peduncle is shifted laterally against the opposite edge of the tentorium, producing ipsilateral motor signs (false localizing signs) (Crockard *et al.* 1992). Occasionally, large uncal herniations will occlude the ipsilateral posterior cerebral artery with subsequent infarction and swelling of the occipital lobe. This can create a vicious cycle of hemicranial hypertension which further accentuates the herniation process.

Once signs of midbrain encroachment appear, rostral–caudal deterioration proceeds rapidly. This is characterized by deepening stupor and coma, decorticate or decerebrate posturing, and profound alterations in cardiorespiratory function. Occlusion of the aqueduct of Sylvius may accelerate this process by obstructing the flow of cerebrospinal fluid (Plum and Posner 1992; Cheek *et al.* 1994). Death usually results from brainstem hemorrhages (Duret hemorrhages) which characteristically develop in the central core area at any point from the diencephalon to the lower pons.

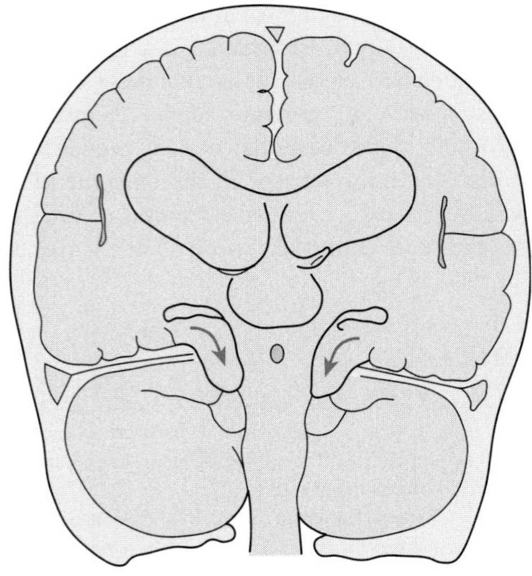

Fig. 2 Diagram of central transtentorial herniation secondary to hydrocephalus. The dorsal half of the brainstem is compressed from side to side, producing a pear-shaped deformity of the midbrain. (Reproduced with permission from Milhorat (1978).)

Central transtentorial herniation

This is a clinically distinct syndrome which occurs more frequently in infants and children. It is seen typically in patients with acute or progressive hydrocephalus, and reflects symmetric downward displacement of both cerebral hemispheres through the tentorial incisura (Fig. 2). The early signs are due to progressive impairment of diencephalic function rather than oculomotor nerve palsy (Crockard *et al.* 1992). Initially the pupils may be small or pinpoint and, if examined superficially, appear to have lost their reaction to light (Crockard *et al.* 1992; Plum and Posner 1992). With progressive central transtentorial herniation, the dorsal half of the brainstem is compressed from side to side, producing a pear-shaped elongation of the midbrain and upper pons (Bell and McCormick 1972; Crockard *et al.* 1992). The resulting distortion of the quadrigeminal plate region can lead to a paralysis of upward gaze (Parinaud's syndrome), and pituitary necrosis and diabetes insipidus may occur in exceptional cases. Although the syndromes of central and unilateral transtentorial herniation are clinically distinct in the early stages, both merge into a similar picture as rostral–caudal deterioration advances.

Transforaminal herniation

Displacement of one or both cerebellar tonsils through the foramen magnum (Fig. 3) is a common occurrence in patients with posterior fossa mass lesions and usually precedes upward cerebellar herniation. Although the condition may be tolerated for considerable periods of time without noticeable symptoms, neck tilt or nucal rigidity is often evidence on physical examination. Such findings must be carefully distinguished from neck stiffness resulting from meningitis or subarachnoid hemorrhage. Lumbar puncture in a patient with transforaminal tonsillar herniation can result in a sudden reduction in pressure within the spinal canal and further descent of the cerebellar tonsils through the foraminal notch. In questionable cases, CT or magnetic resonance imaging should be performed to exclude a posterior fossa mass.

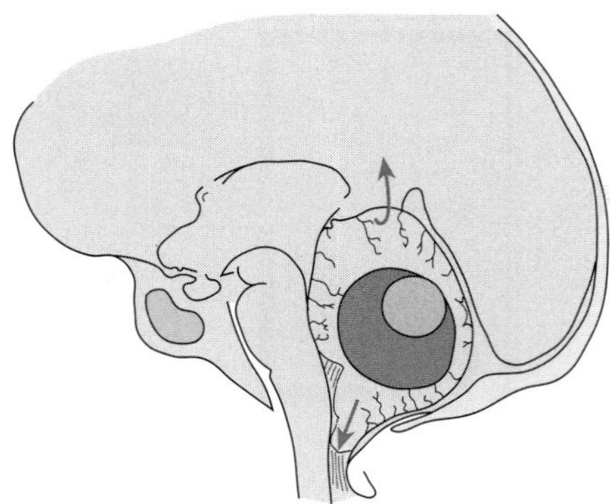

Fig. 3 Diagram of upward (transtentorial) and downward (transforaminal) cerebellar herniation secondary to infratentorial mass lesion. (Reproduced with permission from Milhorat (1978).)

A constant hazard of transforaminal cerebellar herniation is sudden respiratory arrest. This can occur at any time, occasionally in response to otherwise innocuous stimuli such as a cough or sneeze. In rapidly developing herniations, the cerebellar tonsils as well as the medulla and the upper cervical spinal cord can undergo ischemic infarction as a consequence of compression of one or both posterior inferior cerebellar arteries against the rim of the foramen magnum (Bell and McCormick 1972; Plum and Posner 1992).

Upward cerebellar herniation

This occurs with large or superior cerebellar masses and produces a clinical picture which is indistinguishable from the central herniation syndrome. With progressive crowding of the tentorial notch, the pineal gland and suprapineal recess are displaced superiorly and the superior cerebellar and posterior cerebral arteries may be compressed against the free margins of the tentorium. This can lead to ischemic infarction of the superior cerebellum and occipital lobes which further complicates matters. If the precipitating lesion has not already obstructed the flow of cerebrospinal fluid, upward herniation of the cerebellum will often do so at the level of the aqueduct of Sylvius. This produces a counterbalancing effect in which the tentorial notch is invaded from below by the herniating cerebellum and from above by the descending temporal lobes (Fig. 3). The combination is invariably fatal unless immediate decompression of both the supra- and infratentorial compartments is provided.

References

Bell, W.E. and McCormick, W.F. (1972). *Increased intracranial pressure in children*. W.B. Saunders, Philadelphia, PA.

Cheek, W.R., Marlin, A.E., McLone, D.G., Reigel, D.H., and Walker, M.L. (ed.) (1994). *Pediatric neurosurgery. Surgery of the developing nervous system* (3rd edn). W.B. Saunders, Philadelphia, PA.

Crockard, A., Hayward, R., and Hoff, J. (ed.) (1992). *Neurosurgery: the scientific basis of clinical practice* (2nd edn). Blackwell Scientific, Boston, MA.

Milhorat, T.H. (1978). *Pediatric neurosurgery*, pp. 34–7. F.A. Davis, Philadelphia, PA.

Plum, F. and Posner, J.B. (1992). *Diagnosis of stupor and coma* (3rd edn). F.A. Davis, Philadelphia, PA.

7.4.3 Diagnosis of raised intracranial pressure

A. Mallick and N. M. Dearden

Key messages

- Clinical diagnosis of raised intracranial pressure is made when pressure is above 20 mmHg, i.e. twice normal.
- Symptoms include headache, vomiting, dizziness, changes in mental function, and visual disturbances.
- Typical clinical signs are papilledema, cranial nerve palsies, neck stiffness, bradycardia, and hypertension (Cushing's response).
- Investigations include skull radiography, CT scan, magnetic resonance imaging, and direct measurement of intracranial pressure.

Introduction

The skull is a rigid box containing the brain parenchyma (80 per cent), blood (10 per cent), and cerebrospinal fluid (10 per cent). Intracranial pressure (ICP) in supine humans ranges from 5 to 13 mmHg (0.5–1.3 kPa) and is determined by the relationship between the volume allowed and the volume of the contents.

Raised ICP is diagnosed when pressure is persistently greater than 20 mmHg (2.6 kPa); it is the most common cause of death in patients with neurological conditions. Intracranial hypertension also complicates systemic conditions including post-hypoxic and metabolic encephalopathies. Since raised ICP may compromise cerebral blood flow, it is important to be familiar with its pathophysiology and diagnosis.

The most common cause of raised ICP is an expanding intracranial mass. When the mass volume exceeds the displaceable volume of venous blood or cerebrospinal fluid, ICP rises (Fig. 1); this is known as the Monroe–Kelley doctrine. Initially ICP rises slowly in the compensation phase. The compensatory mechanisms include displacement of cerebrospinal fluid and reduced cerebral blood volume. As compensatory mechanisms are exhausted, ICP increases as a function of the relationship between ICP and volume (compliance) and of the rate of expansion of the intracranial mass.

An expanding mass distorts the adjacent brain tissue and can generate pressure gradients in the cranium. The local deformation is reflected more distally in mechanical shifts and herniations resulting in compression and ischemia.

Four patterns of herniation can occur: subfalcine (cingulate gyrus), transtentorial (uncal or temporal), tonsillar (cerebellar or through the foramen magnum), and transcalvarial (traumatic) (Fig. 2). Subfalcine herniation may entrap the branches of the anterior cerebral artery. In transtentorial herniation the hippocampus and uncus of the temporal lobe herniate through the tentorial notch compressing the midbrain, and entrapping the occulomotor and abducens nerves and branches of the posterior cerebral artery. When the tonsils of the cerebellum

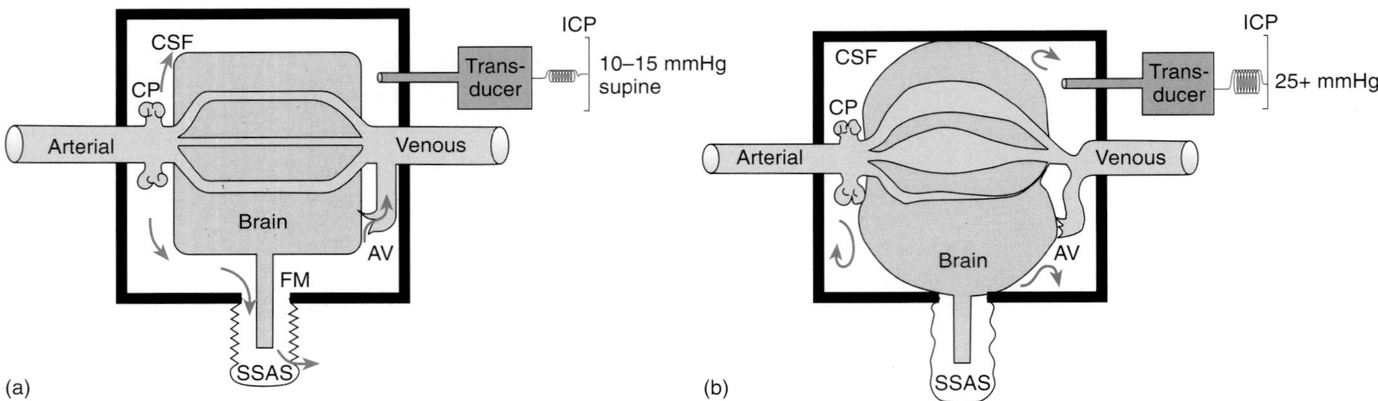

Fig. 1 (a) Schematic representation of skull and normal intracranial contents with arrows indicating the direction of cerebrospinal fluid (CSF) flow. (b) Schematic representation of intracranial contents in the decompensated stage of raised ICP; cerebral vessels are compressed and abnormally dilated in different regions of the brain, and cerebrospinal fluid pathways are blocked preventing translocation and absorption. SSAS, spinal subarachnoid space; FM, foramen magnum; CP, choroid plexus; AV, arachnoid villi.

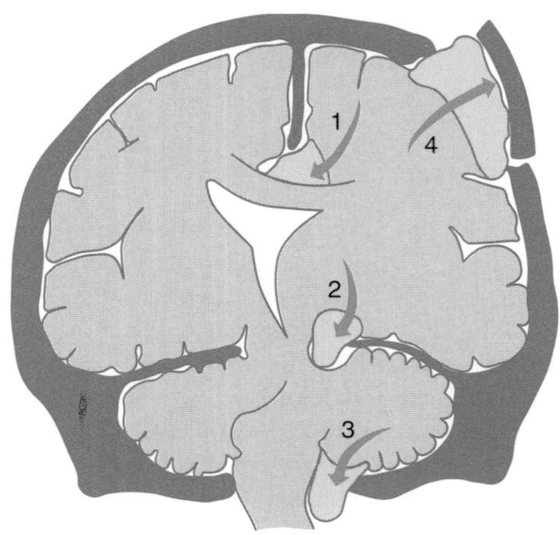

Fig. 2 Schematic representation of four different types of herniation of brain: (1) subfalcine (cingulate gyrus); (2) transtentorial (uncal or temporal); (3) tonsillar (cerebellar or through the foramen magnum); (4) transcalvarial (traumatic).

herniate through the foramen magnum (tonsillar herniation), the medulla is compressed.

Clinical features of raised intracranial pressure

There are no specific signs or symptoms of raised ICP. Most are produced indirectly as a result of distortion, shifts, and herniations of brain tissue. The classic triad of headache, papilledema, and vomiting is encountered in two-thirds of patients with space-occupying lesions, and at least two of these symptoms occur in the remainder. While more common causes account for each alone, their concurrence invokes suspicion of raised ICP. There is no consistent correlation between the height of ICP and severity of the symptoms.

Symptoms of raised intracranial pressure

Headache

The cardinal symptom of raised ICP is headache with an explosive onset, as in the case of intracranial hemorrhage, or slowly progressive and worse in the morning with intracranial tumor. Morning headache is attributed to an elevated ICP with recumbent position and an increase in $Pa\text{CO}_2$ during sleep, and is provoked by maneuvers which raise ICP such as coughing, straining, and sneezing. Dilatation, traction, and stretching of the blood vessels, together with compression of basal dura, are implicated.

Vomiting

Although commonly seen in patients with posterior fossa tumors, vomiting is a late feature of raised ICP from any cause and frequently accompanies morning headache. It is often projectile and unexpected without preceding nausea. Vomiting further raises the ICP and worsens the headache.

Dizziness

Symptoms of lightheadedness or dizziness are common in patients with raised ICP, probably attributable to pressure on the brainstem.

Changes in mental function

The majority of patients with raised ICP will show evidence of alteration of mental functions, from mild unusual irritability to specific memory defects and reduced ability to handle conceptual thoughts. Change in personality, a degree of drowsiness, fatigue, lethargy, alteration in sleep, and difficulty in pursuing normal daily occupation are described. Impairment of conscious level implies a serious rise in ICP.

Visual disturbances

The most frequent visual symptom is blurring with intermittent loss of vision (amblyopic attacks). Visual acuity may decline with hydrocephalus because of direct pressure on the optic nerve and chiasma. Diplopia usually results from abducens nerve palsy.

Signs of raised intrcranial pressure

Papilledema

Papilledema, the cardinal sign of raised ICP, may be unilateral or absent even in the presence of a large intracranial mass. Its incidence is higher in children but unusual at the extremes of life. Absence of papilledema does not exclude raised ICP. Since pupillary response is an important assessment of patients with raised ICP, mydriatics should not be used to perform fundoscopy.

Cranial nerve palsies

Abducens and occulomotor palsies predominate. The abducens, which is the longest intracranial nerve, is commonly compressed or distorted with raised ICP, and diplopia results as a false localizing sign that is rarely related to a specific anatomical lesion within the cranium. However, occulomotor nerve palsy is frequently localizing, usually occurring on the same side as a supratentorial mass, a characteristic feature of transtentorial herniation.

Neck stiffness

If raised ICP causes tension upon meninges in the posterior fossa, pain on neck movement and discomfort on attempted neck flexion may ensue.

Cushing's response

Bradycardia and hypertension, as described by Cushing, are usually late signs of raised ICP. In clinical practice they are often preceded by deteriorating conscious level and are believed to be due to brainstem distortion or ischemia.

Respiratory changes

Sudden rises in ICP often involve slow and deep respiratory movements and breathing may become irregular (e.g. Cheyne–Stokes breathing). Neurogenic pulmonary edema and acute respiratory distress syndrome are also associated with raised ICP.

Other signs

Pupillary dilatation, bilateral ptosis, impaired upward gaze, extension to pain, and respiratory irregularity are due to either tentorial and tonsillar herniation or direct involvement of the brainstem. In young children with open fontanelles, the signs and symptoms of raised ICP are delayed or even absent. They include bulging fontanelles, progressive frontal bone protrusion, intermittent headaches, irritability, and vomiting.

Investigations

Skull radiography

Erosion of dorsum sella, thinning of the skull vault, loss of convolutions over the frontal and parietal bones, and sutural diastasis in children occur when ICP is chronically raised. Skull radiography is not informative in acute elevation of ICP.

CT scan

Conditions associated with raised ICP, such as hematoma, tumor, abscess, hydrocephalus, and brain swelling, may be shown. Midline shift, obliteration of cerebrospinal fluid cisterns, dilatation of the contralateral ventricle, and effacement of the cortical sulci may be demonstrated (Fig. 3). Diffuse cerebral swelling is associated with loss of the third ventricle and basal cisterns.

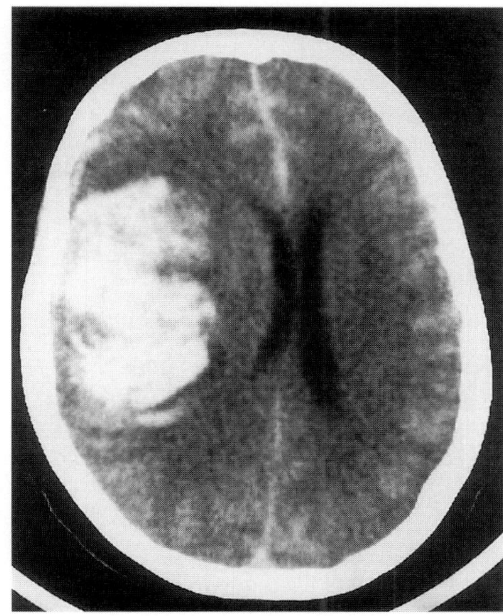

Fig. 3 CT scan shows intracerebral hematoma with features of raised ICP.

Magnetic resonance imaging

Magnetic resonance imaging shows the brain tissue water content, and more clearly demonstrates midline and posterior fossa structures.

Monitoring of intracranial pressure

The absolute level of ICP for diagnosis or monitoring the effects of treatment modalities can only be determined by direct and continuous monitoring. ICP monitoring relies on conversion of hydraulic pressure into electrical signals via a transducer. Transducers may be hydraulically linked to extradural, subdural, or ventricular spaces or directly implanted into the brain. Ventricular fluid pressure remains the most accurate and reliable for clinical use. ICP monitoring may be used to determine the need for shunting hydrocephalus, during care of neurosurgical patients following sustained head injury, subarachnoid hemorrhage, or posterior fossa surgery, and in management of patients with metabolic encephalopathies. It is also useful for guiding treatment modalities such as hyperventilation, mannitol infusion, and barbiturate administration during management of head injury, and in conjuction with measurement of mean arterial pressure allows calculation of cerebral perfusion pressure.

Conclusions

The pathophysiology of raised ICP and its effects on cerebral perfusion are increasingly well understood. The conditions raising ICP should be considered in differential diagnosis (Table 1). Since raised ICP is usually diagnosed at an earlier stage, fully developed clinical features are increasingly uncommon. Clinical signs of raised ICP are

Table 1 Causes of raised intracranial pressure

Non-vascular causes
- Hematoma/contusion
- Mass lesion (tumor)
- Cerebral edema (raised brain water)
- Increased CSF outflow resistance
- Hydrocephalus
- Benign intracranial hypertension

Vascular causes—increased cerebrovascular volume

Arterial
- Loss of autoregulation
- Blood pressure above autoregulation threshold
- Raised $Pa\text{CO}_2$
- Lowered $Pa\text{O}_2$

Venous
- Venodilatation
- Venous engorgement
  - Jugular vein compression
  - Raised intrathoracic pressure
  - Secondary to CSF outflow obstruction

CSF, cerebrospinal fluid.

unreliable, and monitoring and control of raised ICP is essential for effective care of some neurosurgical patients.

Bibliography

Bingaman, W.E. and Frank, J.I. (1995). Malignant cerebral edema and intracranial hypertension. *Neurologic Clinics*, **13**, 479–509.

Ghajar, J. (1995). Intracranial pressure monitoring techniques. *New Horizons*, **3**, 395–9.

Lang, E.W. and Chesnut, R.M. (1994). Intracranial pressure. Monitoring and management. *Neurosurgery Clinics of North America*, **5**, 573–605.

Lang, E.W. and Chesnut, R.M. (1995). Intracranial pressure and cerebral perfusion pressure in severe head injury. *New Horizons*, **3**, 400–9.

North, B. and Reilly, P. (1990). *Raised intracranial pressure: a clinical guide*. Heinemann Medical, Oxford.

7.4.4 Management of cerebral edema

H. Klein and N. M. Dearden

Key messages

- Brain edema can be classified into five types. In clinical practice several types may coexist but identification of the predominant type may guide therapy.
- In some conditions non-specific intracranial pressure reduction therapy can be used to improve cerebral perfusion pressure, reduce edema formation, and increase edema resolution.
- Edema associated with hydrocephalus or brain tumors may best be treated using cerebrospinal fluid drainage and steroids respectively.
- Increased understanding of the biochemical mediators of edema genesis is leading to successful experimental therapeutic intervention with novel agents. Their efficacy in the clinical situation remains a subject for investigation.

Introduction

Brain edema is an abnormal accumulation of fluid within the brain parenchyma which produces a volumetric enlargement of the brain tissue. This contrasts with brain swelling which is an increase in cerebral blood volume, cerebrospinal fluid volume, or brain tissue mass. In this chapter we classify cerebral edema, discuss its diagnosis, and describe therapeutic interventions.

Anatomical considerations

Volume homeostasis

The Monroe–Kelley doctrine states that the volume of the craniospinal compartment which comprises blood, cerebrospinal fluid, and brain parenchyma remains constant. Increased intracranial volume will initially lead to displacement of these three components and reduced production of cerebrospinal fluid (volume compensation). Initially, intracranial pressure (ICP) will rise slowly, but progressive exhaustion of the compensatory mechanisms causes an exponential rise in ICP. Normal recumbent ICP is between zero and 13 mmHg. A sustained ICP in excess of 20 mmHg is pathological. Elevation of ICP reduces the cerebral perfusion pressure, which is defined as the difference between mean arterial pressure and ICP. At cerebral perfusion pressures below 50 mmHg, autoregulation of cerebral blood flow is compromised. If cerebral blood flow falls below 18 ml/100 g/min, the metabolic requirements of brain cells are not met, the cells become ischemic, and energy-dependent membrane pumps fail (Fernandez and Landoldt 1996). A vicious cycle develops, with intracellular migration of water (hydrostatic edema) causing a further rise in ICP. Falling cerebral perfusion pressure may invoke a compensatory rise in mean arterial pressure in an attempt to maintain it (Cushing response).

Blood–brain barrier

Ionic equilibrium is maintained across the cerebral vascular endothelium preserving cerebral volume homeostasis. The major functions of the blood–brain barrier are as follows (Rissal 1994):

(1) protection of the brain from the blood milieu via tight junctions which prevent uncontrolled passage of molecules;

(2) specialized active transport of selected molecules;

(3) metabolism or modification of substances.

The ability of solute to cross the blood–brain barrier depends on both properties of the barrier and properties of the solute. Exchange across the barrier is easy for small lipid-soluble uncharged molecules and difficult for large polar substances. Most metabolic substrates (glucose and amino acids) are charged and polar, and so specialized carrier systems exist to enable their transport across the blood–brain barrier. Facilitated diffusion allows equilibrium down concentration gradients between plasma and brain, and energy-dependent transport mechanisms exist against electrical or concentration gradients.

Water passage follows osmotic forces across the intact blood–brain barrier since tight junctions resist hydrostatic forces. If the blood–brain barrier is disrupted around areas of tumor, infection, or trauma, this osmotic gradient is lost and hydrostatic egress of protein-rich fluid provokes vasogenic cerebral edema. Increased permeability to ions and larger molecules such as albumin forms the basis of tests examining the integrity of the blood–brain barrier.

Classification of brain edema

Brain edema was first described at autopsy by Reichardt in 1905. With the introduction of electron microscopy it became apparent that edema was present both intracellularly and extracellularly. Klatzo (1994) differentiated vasogenic edema, where fluid accumulates extracellularly in association with a dysfunctional blood–brain barrier, from cytotoxic edema where abnormal intracellular water uptake occurs. More recently, osmotic, hydrostatic, and interstitial edema have been described. Clinically, several of the five potential forms of edema may coexist (Dearden and Miller 1991).

Vasogenic edema

Breakdown of the blood–brain barrier surrounding brain tumors and following infection or trauma disrupts fluid homeostasis and hydrostatic efflux of protein-rich fluid from the intravascular into the extracellular space occurs. Vasogenic edema is generated in the gray matter (more blood vessels) but spreads to the white matter where there are fewer cells and brain compliance is greater. Edema accumulation increases with fever, hypercarbia, hypoxia, and hypertension, and falls with hyperventilation, hypotension, and raised ICP but at the risk of cerebral ischemia. Studies utilizing CT and magnetic resonance imaging (MRI) have demonstrated pathways of edema formation and spread (Fig. 1). The initial pressure required to open channels is high, but thereafter bulk flow allows edema to spread with little pressure gradient. Edema resolves mainly by drainage into the ventricles or subarachnoid cerebrospinal fluid, down a pressure gradient between edematous subependymal regions and the cerebrospinal fluid in ventricles, and from uptake by glial cells. A small amount is reabsorbed into capillaries. The resolution of vasogenic brain edema is inversely related to ICP.

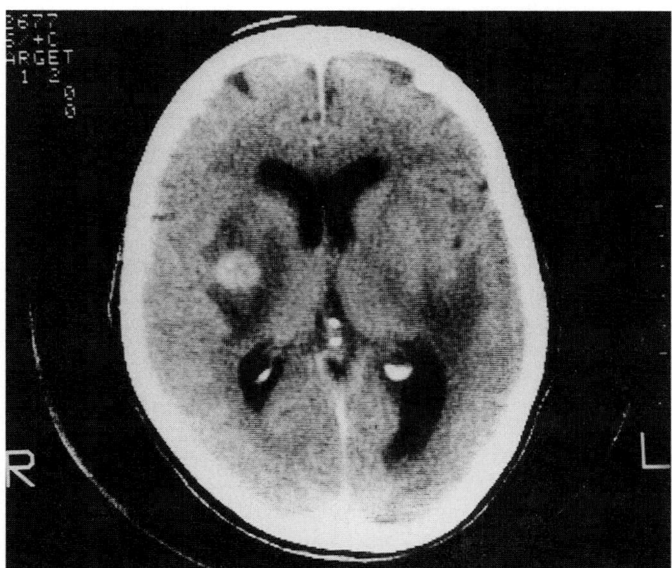

Fig. 1 Vasogenic edema in a patient with cerebral metastases. The CT scan shows a hypodense area around the tumor in the right hemisphere.

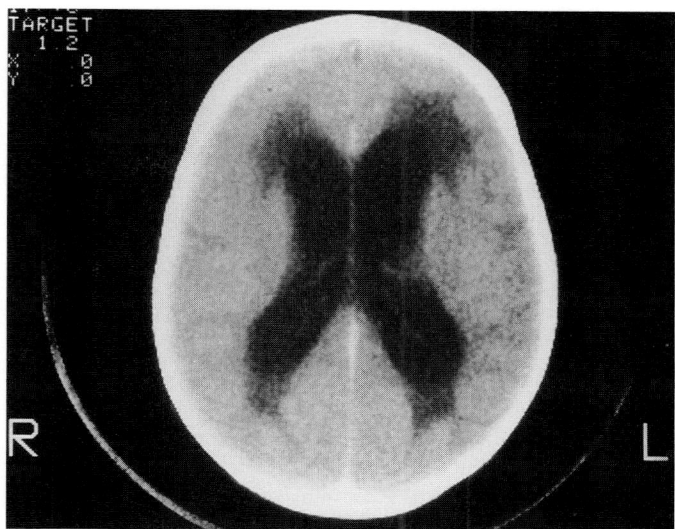

Fig. 2 CT scan showing acute hydrocephalus with interstitial edema around the acute angles of the ventricles.

Cytotoxic edema

Normal and glial homeostasis is maintained by energy-dependent ion pumps in the cell membrane which remove sodium from the inside of cells and concentrate potassium intracellularly. Any situation that interferes with the energy processes, either by lack of substrates and oxygen or by toxicity, will lead to failure of these pump mechanisms. Water and sodium influx into the cell leads to intracellular cytotoxic edema. Reduced cerebral perfusion pressure, vascular distortion or spasm, and vascular obstruction may give rise to ischemic cytotoxic edema if cerebral blood flow falls below 20 ml/100 g/min. Reduction of cerebral blood flow below 18 ml/100 g/min is associated with the release of excitatory amino acids. At this stage the situation is potentially reversible by adequate reperfusion. Further reduction in cerebral blood flow to around 10 ml/100 g/min leads to failure of the potassium and calcium membrane pumps and irreversible damage. Therefore development of ischemic edema depends on both the severity and the duration of reduction of cerebral blood flow and may be global or regional depending on the underlying pathology.

Interstitial edema (hydrocephalic edema)

Interstitial edema is most commonly associated with obstructive hydrocephalus and predominantly accumulates at the acute angles of the ventricular system (Fig. 2). Continued production of cerebrospinal fluid results in ventricular dilatation, and raised cerebrospinal fluid pressure forces a protein-poor transudate into the adjacent extracellular spaces while the blood–brain barrier remains intact.

Hydrostatic edema

Hydrostatic edema is due to a sudden rise of hydrostatic pressure in the cerebral capillaries. Egress of filtrate across the capillary membrane exceeds colloid osmotic reabsorption. The blood–brain barrier remains intact and the edema fluid, which has a low protein content, accumulates largely extracellularly. Clinically, this can occur after removal of an intracranial hematoma when the sudden rise in cerebral perfusion pressure causes regional increased transcapillary filtration.

'Osmotic' edema (oncotic edema)

This intracellular edema occurs without failure of the sodium potassium pump in acute dilutional states, particularly hyponatremia, and therefore water migrates intracellularly. The edema shows a generalized distribution. The oncotic gradient favors fluid movement into the cell. Causes include iatrogenic infusion of large amounts of hypotonic fluids (e.g. 5 per cent glucose) and inappropriate ADH secretion.

Clinical examples of cerebral edema

Cerebral edema is associated with a variety of clinical conditions (tumor, infection, trauma, hydrocephalus), although more than one form is usually present. However, the potential to identify the main mechanism of edema formation may be helpful in allowing more appropriate use of interventional therapy since different forms of edema are likely to respond to different forms of treatment.

Traumatic edema

Until recently it was believed that post-traumatic brain swelling was due to increased cerebral blood volume and a breakdown of the blood–brain barrier (vasogenic edema). However, evidence from MRI studies shows that, while the 'brain swelling' in a diffuse head injury is due to cerebral edema, the blood–brain barrier remains intact and overall cerebral blood volume is reduced. Mechanisms of post-traumatic edema appear predominantly ischemic and cytotoxic.

Edema after ischemic stroke

This description arises from vascular occlusion models in animals. The initial component after vascular occlusion is cytotoxic, owing to the energy depletion of membrane ion pumps which results in intracellular swelling, and is potentially reversible. After reperfusion vasogenic edema occurs due to biphasic opening of the blood–brain barrier. Autoregulation is lost during ischemia and reperfusion causes a short-lived breakdown of the blood–brain barrier sufficient to promote influx of vasoactive substances which subsequently lead to a permanent breakdown of the barrier (Betz and Crockard 1992).

Confirmation of brain edema

The diagnosis of cerebral edema is usually made after clinical presentation of symptoms or signs relating to raised ICP. The differential diagnosis will include mass lesions. Diagnosis is usually made by CT or MRI appearance. On CT scanning, edema is characterized by a low-density area with a shift of surrounding tissues due to the dilution of lipid structures with water. CT has limitations in detecting edema in the white matter and tends to underestimate its severity. Edema associated with abscesses is well visualized, but edema after acute cerebral hemorrhage is rarely visible. Abnormal blood–brain barrier characteristics can be demonstrated by timed measurements of contrast or radionuclide uptake, but interpretation depends on the size of the markers. MRI is a very sensitive but less specific imaging technique. It is particularly useful for detecting edema in the white matter. Magnetic resonance relaxation time, mapping, and marker-enhanced imaging can be used to interrogate blood–brain barrier integrity. Extravasation of larger molecules into the surrounding tissue can be visualized using marker substances.

Current therapy

Therapy can be classified as treatment for 'brain swelling' (raised ICP) or as specific therapy to reduce edema (Dearden and Miller 1991). ICP therapy aims to place the cranial contents on a more favorable part of the intracranial compliance curve. Reduction of ICP resulting in improved cerebral perfusion pressure may reduce brain ischemia and help to resolve edema. Cerebral blood volume may be reduced by head-up tilt, hyperventilation, and hypnotic agents. Cerebrospinal fluid volume may be reduced by using intraventricular drainage and will decrease hydrocephalic edema. If ICP reduction improves cerebral perfusion pressure, compromised cerebral blood flow may also be restored.

Diuretics such as furosemide (frusemide) and acetazolamide are not widely employed to reduce the production of cerebrospinal fluid because of unfavorable effects on electrolytes and acid–base balance.

Methods of ICP/edema reduction include osmotherapy, steroids, and manipulation of osmotic and oncotic pressure (Dearden and Miller 1991).

Osmotherapy

Hyperosmotic agents (e.g. mannitol) exert a dehydrating effect on the brain parenchyma. Several mechanisms of action have been proposed.

Hyperosmotic agents build up an osmotic gradient since they are small non-diffusible molecules and therefore water is withdrawn where the blood–brain barrier is intact. In theory they will not work across an impaired blood–brain barrier. However, there is evidence that mannitol works in areas with impaired blood–brain barrier with a reduced effect.

Hyperosmotic agents decrease plasma viscosity and improve the rheological properties of blood and plasma, thereby improving cerebral blood flow. They improve energetic metabolism and acidosis and reduce the spread of edema.

Steroids

Steroids have been shown to be useful in chronic perifocal edema around some cerebral tumors or brain abscesses (vasogenic brain edema) where the symptoms improve in 24 h and the ICP reaches baseline values after 48 to 72 h. There is no compelling evidence of efficacy on raised ICP or outcome after severe head injury. The consensus is that pure glucocorticoids are preferable to agents with mineralocorticoid effects.

Manipulation of colloid osmotic pressure

The theory behind using high colloid osmotic pressure for edema reduction is that osmotic diuretics only transiently reduce ICP and a 'rebound' occurs, whereas albumin is able to maintain an osmotic gradient for up to 2 weeks. It has been used successfully in contusional edema and cerebral edema surrounding hemorrhage.

Future therapeutic possibilities

A variety of mediators and biochemical factors have been implicated in the genesis of edema and the swelling of nerve cells. They form the basis of potential future pharmacological intervention.

References

Betz, A.L. and Crockard, A. (1992). Brain edema and blood–brain barrier. In *Neurosurgery: the scientific basis of clinical practice* (2nd edn) (ed. A. Crockard, R. Hayward, and J. Hoff), pp. 353–71. Blackwell Scientific, Boston, MA.

Dearden, N.M. and Miller, J.D. (1991). Management of brain edema in head injury. In *Advances in brain resuscitation* (ed. Takeshita, Siesjo, and Miller), pp. 221–32. Springer, Tokyo.

Fernandez, C. and Landoldt, H. (1996). Blood–brain barrier and cerebral oedema. In *Manual of neurosurgery — Neuro 96* (ed. J.D. Palmer), pp. 377–84. Churchill Livingstone, Edinburgh.

Klatzo, I. (1994). Evolution of brain edema concepts. *Acta Neurochirurgica*, **60** (Supplement), 3–6.

Rissal, W. (1994). Molecular biology of blood brain barrier ontogenesis and FCT. *Acta Neurochirurgica*, **60** (Supplement), 109–12.

7.5 Non-traumatic intracranial hemorrhage

7.5.1 Assessment

H. M. Fernandes and A. D. Mendelow

Key messages

- Suspect intracranial hemorrhage in any patient who presents with an acute change in conscious level.
- Intracranial hemorrhage causes over 30 per cent of all strokes, and the younger the patient the greater its probability. Mortality and morbidity remain high.
- Primary brain damage from the hemorrhage will be complete at time of presentation. Secondary brain damage can be prevented or ameliorated by prompt assessment and optimal management.
- Correction of hypoxia and hypotension and identification and treatment of raised intracranial pressure are absolutely mandatory in the prevention of secondary brain damage.
- Do not treat reactive hypertension as low blood pressure may reduce cerebral perfusion pressure.

Etiology and pathology

Non-traumatic intracranial hemorrhage may occur within the subarachnoid space (subarachnoid hemorrhage), within the brain parenchyma (intracerebral hemorrhage), or into the ventricular system itself (intraventricular hemorrhage). Often a subarachnoid hemorrhage may present with an associated intracerebral hematoma. Likewise an intracerebral hematoma, depending upon its anatomical position, may burst into the subarachnoid space. Intraventricular hemorrhage is commonly associated with either of the former. Rarely, spontaneous hemorrhage can occur into the subdural or epidural spaces. Intracranial hemorrhage may be supratentorial or originate in the posterior fossa.

Beyond the third decade subarachnoid hemorrhage is most commonly due to rupture of a cerebral aneurysm, usually situated at the origin of the branches of the major cerebral arteries. A further 10 to 20 per cent of cases of subarachnoid hemorrhage are due to arteriovenous malformations and tend to occur in the younger patient. However, in 20 per cent of all subarachnoid hemorrhage no cause is detected. Intracerebral hemorrhage is more common in the older patient (60 years and above), in whom it is secondary to hypertension and located in the basal ganglia. Lobar hemorrhages constitute a further 40 per cent and are often related to underlying amyloid angiopathy. Coagulation abnormalities, intracranial tumors, and drug abuse make up some of the rare causes of bleeding.

The final outcome of these patients is related to the amount and location of neuronal loss that occurs both at the time of the hemorrhage (primary brain damage) and in the subsequent minutes, hours, and days (secondary brain damage). Primary brain damage is caused by the hemorrhagic destruction of brain tissue itself and is unalterable at the time of presentation. Further neuronal death secondary to cerebral ischemia around the hematoma is the final common cause of secondary brain damage (Mendelow *et al.* 1984). This is potentially preventable and forms the basis of patient management. Despite the differing etiologies and locations of intracranial hemorrhage, these patients tend to present with a similar spectrum of symptoms and signs, varying from headache only to deep coma, with or without a lateralizing neurological deficit. Assessment of these patients occurs in two parts: first a rapid appraisal of the vital signs and conscious level of the patient, with simultaneous correction of any reversible factors that may be contributing to the processes of secondary brain damage, and second, after stabilization, a more detailed examination to aid in the differential diagnosis and prognosis. A schematic approach to assessment is described.

Cerebral ischemia and secondary brain damage

Regulation of cerebral blood flow

Cerebral ischemia after intracranial hemorrhage has several causes. Normally, cerebral blood flow is kept fairly constant, despite changes in cerebral perfusion pressure over a mean arterial pressure range of approximately 50 to 150 mmHg, by a process termed autoregulation. Cerebral perfusion pressure is defined as the difference between mean arterial pressure and intracranial pressure. Cerebral blood flow may be markedly increased by a rise in $P\text{aCO}_2$, secondary to vasodilatation. Autoregulation is lost during hypercapnia.

Global ischemia

Following intracranial hemorrhage, the autoregulation reflex is also lost and cerebral blood flow becomes passively dependent upon cerebral perfusion pressure. Thus systemic hypotension and respiratory difficulties can affect both cerebral perfusion pressure and cerebral blood flow. Therefore provision of an adequate airway, mean arterial pressure, and gas exchange are mandatory for optimal management.

Focal ischemia

Around the hematoma there is a zone of ischemic tissue, which itself is surrounded by a penumbra of functionally impaired but recoverable neuronal tissue. Treatment directed at prevention of global ischemia will also lessen this focal ischemia and thus minimize neuronal death.

Intracranial pressure

Intracranial pressure is commonly increased after intracranial hemorrhage. It is elevated by both the actual space-occupying effect of an intracerebral hematoma and cerebral edema. Cerebral edema, defined as an abnormal collection of fluid within the brain substance, develops secondary to cerebral ischemia. Hydrocephalus, a dilatation of the ventricular system, will also increase intracranial pressure. Intraventricular hemorrhage into the third or fourth ventricle with obstruction of cerebrospinal fluid outflow is usually responsible, but a cerebellar hemorrhage may cause extrinsic compression of the fourth ventricle in the posterior fossa. Identification and prompt treatment of raised intracranial pressure are again important in preservation of cerebral perfusion pressure.

Cerebral vasospasm

Peculiar to subarachnoid hemorrhage is a phenomenon termed cerebral vasospasm, i.e. narrowing of the cerebral arteries in response to the presence of subarachnoid blood and its breakdown products. It begins immediately, but manifests its clinical effects on day 3 following subarachnoid hemorrhage, peaks at days 6 to 8, and then gradually subsides after the second week. It causes cerebral ischemia and continuing or new neurological deficit. The diagnosis is clinical, but changes can be detected on angiography and serial blood velocity studies using transcranial Doppler ultrasonography. Treatment is prophylactic with use of the calcium antagonist nimodipine. Established vasospasm can be treated by raising cerebral perfusion pressure with blood volume expansion and pressor agents.

Table 1 The Glasgow Coma Scale

Parameter	Response	Numerical value
Eye opening	Spontaneous	4
	To speech	3
	To pain	2
	None	1
		Subtotal 4
Best verbal response	Orientated	5
	Confused	4
	Inappropriate	3
	Incomprehensible sounds	2
	None	1
		Subtotal 5
Best motor response	Obeys commands	6
	Localizes pain	5
	Flexion to pain	4
	Abnormal flexion to pain	3
	Extension to pain	2
	None	1
		Subtotal 6
		Total 15

Patient assessment

Immediate evaluation

History

Although not often possible, attempts should be made to differentiate trauma from spontaneous hemorrhage.

Airway, breathing, circulation (ABC)

A rapid assessment of the airway, respiration, blood pressure, and pulse, with intervention if necessary, forms as important a part of the evaluation of patients following intracranial hemorrhage as with any other critically ill patient. If in doubt about the airway and respiration of a comatose patient (not localizing pain), it is probably better to secure the airway before further deterioration. Care must be taken in all patients to ensure that $Paco_2$ levels are normal or low to prevent cerebral vasodilatation, which acts to increase intracranial pressure. Hypoxia and hypotension must also be corrected. However, great care must be taken in the treatment of hypertension, as this may be a physiological (Cushing) response to raised intracranial pressure. Treatment of this, in the face of disordered autoregulation, would reduce cerebral perfusion pressure and thus induce cerebral ischemia.

Conscious level

A brief assessment of the conscious level will have taken place as part of examination of the airway. However, a more detailed appraisal is useful at this stage as a guide to the severity of the hemorrhage and the presence of raised intracranial pressure.

The Glasgow Coma Scale (GCS) (Teasdale and Jennett 1974), although developed for use after head injury, provides a useful numerical value for the level of consciousness in any patient. It is based on the best motor response, verbal response, and level of eye opening. A normal conscious level will score 15 (Table 1). Aphasia and hemiplegia are common findings in patients following intracranial hemorrhage and may reflect local destructive damage. Thus the best motor response (of the non-plegic limbs) is the most accurate assessor of the conscious level. This examination must be repeated at regular intervals and particular attention paid to the deteriorating patient. A falling GCS usually indicates rising intracranial pressure, and appropriate intervention should be taken (hyperventilation, osmotic agents, surgical intervention).

Pupillary size and reaction

Pupillary size and reaction reflect a balance between the parasympathetic (constrictor) and sympathetic (dilatory) pathways in the afferent optic nerve, the efferent oculomotor nerve, and the ciliary ganglion. Although hemorrhagic lesions in the thalamus and brainstem produce abnormalities, raised intracranial pressure, transtentorial herniation, and compression of the oculomotor nerve will produce ipsilateral pupillary dilatation and eventually an unreactive large pupil. Thus an immediate assessment of the pupils can provide information about the location of any hemorrhage and act as a guide to tentorial herniation from raised intracranial pressure. Repeated examination will detect any increasing mass lesion, particularly in the paralysed patient.

Table 2 The WFNS grading system for assessment of subarachnoid hemorrhage patients

WFNS grade	GCS score	Motor deficit
I	15	Absent
II	13–14	Absent
III	13–14	Present
IV	7–12	Present/absent
V	6	Present/absent

Further examination

Neurological examination following intracranial hemorrhage is useful in assessing the level of focal damage to the patient, with serial examinations acting as a monitor of further damage. Grading systems exist for the classification of patients after subarachnoid hemorrhage (see below), but none are specific to intracerebral hematoma. Assessment of these patients includes general, cardiorespiratory, and abdominal examination.

Brainstem function tests

In an unconscious patient requiring respiratory support a limited examination of the brainstem reflexes (corneal, oculocephalic, pupillary reaction) may provide information about the severity and location of hemorrhage and guide early prognosis.

Further neurological examination

Cranial nerve examination is often difficult in these patients, but important diagnostic information can be gained from the presence of a gaze paresis or abnormal eye movements such as nystagmus. Papilledema will not occur with acutely raised intracranial pressure, but acute subhyaloid hemorrhage is present in up to 20 per cent of cases and is pathognomic of subarachnoid hemorrhage.

Increased reflexes, clonus, and extensor plantar responses are common after intracranial hemorrhage with or without raised intracranial pressure; if asymmetrical, they may provide information about the laterality of the lesion. Motor and other long-tract signs are usually contralateral to the side of hemorrhage. Neck stiffness can be detected in an unconscious patient and, although more common in subarachnoid hemorrhage, is not diagnostic.

Grading systems for subarachnoid hemorrhage

There are many grading systems for the further assessment of subarachnoid hemorrhage patients. These aid rational management and prognosis. The World Federation of Neurological Surgeons (WFNS) have produced an internationally recognized system (Drake 1988) in which the GCS of the patient is combined with the presence or absence of neurological deficit to provide a five-point scale (Table 2).

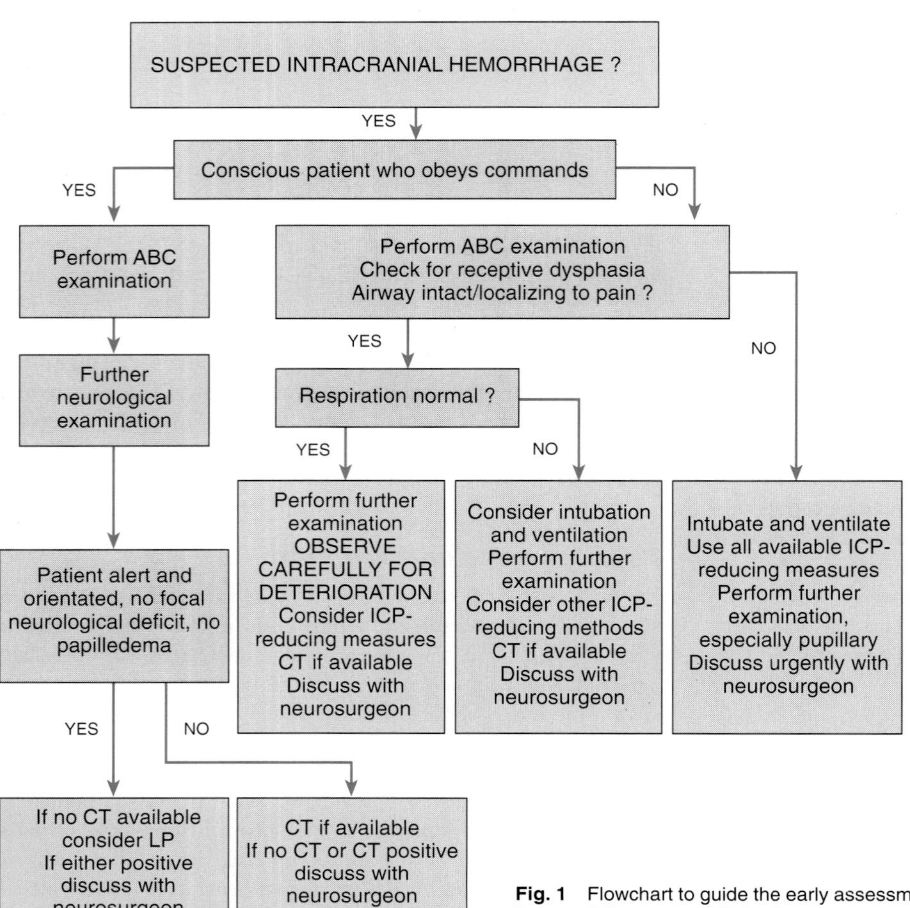

Fig. 1 Flowchart to guide the early assessment of patients following intracranial hemorrhage: ICP, intracranial pressure; LP, lumbar puncture.

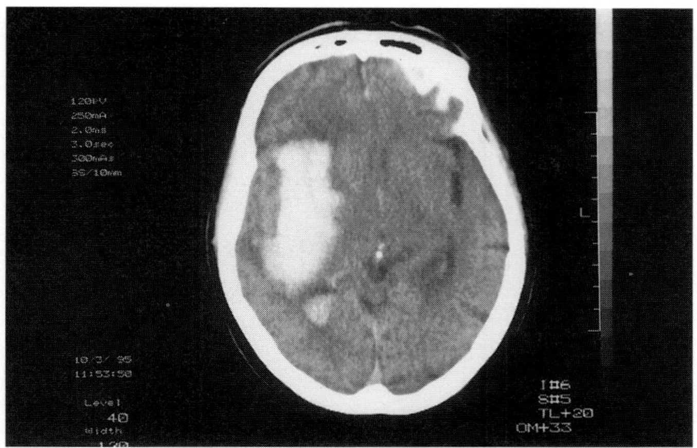

Fig. 2 CT scan showing a large right putaminal intracranial hemorrhage, with associated intraventricular hemorrhage, causing mass effect and midline shift.

Investigation

A flowchart summarizing the early assessment and investigation of intracranial hemorrhage patients is shown in Fig. 1.

Laboratory indices

Samples for urea, electrolytes, and blood count should be taken in all patients. Arterial blood gas measurement is recommended in patients requiring any respiratory support and should be repeated at regular intervals. Clotting studies should also be performed.

Computed tomography

CT has revolutionized the assessment and treatment of intracranial disease. It provides accurate information about the location, extent, and sequelae of any intracranial hemorrhage (Fig. 2). Often the underlying cause can be detected and decisions about immediate management taken. CT frequently forms part of the very early assessment of any patient with an altered conscious level but it should not precede patient stabilization.

Cerebral angiography

Details of the cerebral vasculature are essential in the management of all subarachnoid hemorrhage patients in whom surgical intervention is considered. It is now commonly performed via femoral artery puncture under local anesthetic. An adequate study should reveal any underlying aneurysmal or arteriovenous malformation. In pure intracerebral hematoma, angiography is performed depending upon the clinical suspicion of an underlying vascular cause. In an unconscious or rapidly deteriorating patient with an intracerebral hematoma strongly suspicious of an underlying aneurysm, a limited angiographic study may be performed prior to emergency evacuation of the hematoma. Magnetic resonance angiography and CT angiography are being used with increasing frequency to diagnose an underlying vascular cause.

Other investigations

In any conscious patient with a normal head CT, lumbar puncture is mandatory. Even in the acute setting, CT will be negative in 10 per cent of all subarachnoid hemorrhages, and meningitis should be excluded in every patient with an altered conscious level. Single-photon-emission CT (SPECT) scanning will display the extent of ischemia surrounding the hematoma. Intracranial pressure monitoring may serve as a guide to cerebral perfusion pressure and mean arterial pressure management.

References

Drake, C.G. (1988). Report of World Federation of Neurological Surgeons Committee on a universal subarachnoid hemorrhage grading scale. *Journal of Neurosurgery*, **68**, 985–6.

Mendelow, A.D., Bullock, R., Teasdale, G.M., Graham, D.I., and McCulloch, J. (1984). Intracranial hemorrhage induced at arterial pressure in the rat. Part 2: Short term changes in local cerebral blood flow measured by autoradiography. *Neurological Research*, **6**, 189–93.

Teasdale, G. and Jennett, B. (1974). Assessment of coma and impaired consciousness. A practical scale. *Lancet*, **ii**, 81–4.

7.5.2 Medical management

M. J. G. Harrison

Key messages

- The medical management of cerebral hemorrhage includes care of the unconscious patient, management of elevated arterial blood pressure, reversal of therapeutic anticoagulation, treatment of seizures, and control of raised intracranial pressure.
- Prevention of recurrence depends principally on long-term treatment of hypertension.

Causes

Although only a minority of strokes are due to spontaneous hemorrhage (approximately 15 per cent), their importance lies in their severity, high mortality, and different management. The hemorrhage usually occurs as a result of the rupture of a Charcot–Bouchard aneurysm, a microscopic aneurysm that develops in small intracerebral vessels usually as a response to sustained arterial hypertension and the formation of microatheroma (Russell 1963). Some hemorrhages are due to rupture of penetrating arteries themselves. In elderly patients, particularly those with the features of Alzheimer's disease and/or vascular dementia, amyloid angiopathy may be the cause of single or multiple superficial hematomas. Rarely, arteriovenous malformations, mycotic aneurysms, or anticoagulant drugs are responsible. Abuse of amphetamines or cocaine is another cause. Endocarditis also needs to be excluded in the intravenous drug user (Warlow 1991) (Table 1).

Pathophysiology

The rupture of a vessel or microaneurysm causes the sudden development of a hematoma as the blood emerges at arterial pressure. Bleeding may be temporarily arrested but then continues, leading to a slowly evolving clinical deficit. The development of a mass effect (Fig. 1) due to the volume of the hematoma and associated edema causes both a generalized increase in intracranial pressure and the

Table 1 Causes of cerebral hemorrhage

Causes of cerebral hemorrhage
Hypertension/atheroma
Anticoagulant medication
Antithrombotic medication (e.g. aspirin)
Amyloid angiopathy
Sickle cell disease/trait
Arteritis
Cerebral venous thrombosis
Vascular malformation
Moya moya syndrome
Disseminated intravascular coagulation
Mycotic aneurysm
Cerebral tumors
Hemophilia and other inherited coagulation/platelet diseases
Thrombocytopenia
Snake venom
Asparaginase

Reproduced with permission from Warlow (1991).

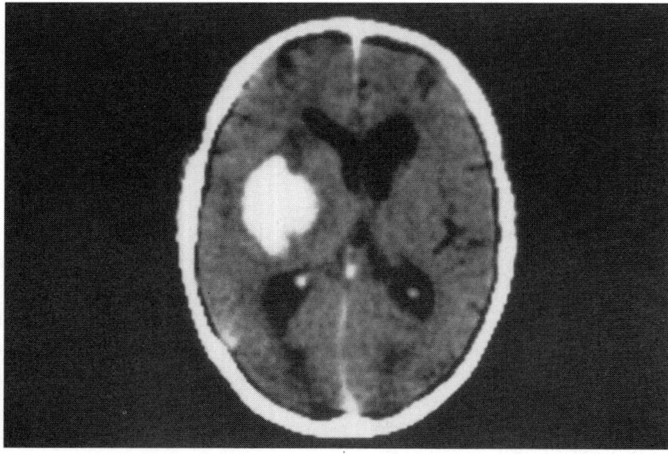

Fig. 1 CT showing cerebral hemorrhage with mass effect.

development of pressure gradients which cause shift across the midline or between compartments. If intracranial pressure rises rapidly, the patient may complain of severe headache and lapse into immediate coma. In others developing a mass effect less acutely the clinical features include declining alertness, the appearance of a third-nerve palsy, or the development of an extensor plantar response ipsilateral to a supratentorial hematoma or bilateral extensor plantar responses with a cerebellar bleed. The hematoma disrupts some tissue, but the neurological deficit is also due to the acute displacement of fiber tracts and may not be permanent. Blood may escape into the cerebrospinal fluid, causing blood staining and later xanthochromia of cerebrospinal fluid, but lumbar puncture is contraindicated because of the risk of herniation if pressure gradients are exaggerated by removal of fluid. The hemorrhage may also breach the ventricular wall with blood appearing in the ventricles (Fig. 2). The blood in a hematoma resolves over 10 days to 3 weeks. Hematomas are generally found in the basal ganglia, particularly in the putamen and thalamus, the lobar white matter, and the cerebellum or pons (Ojemann and Mohr 1975). Rarely, a hematoma in the region of the external capsule slowly splits white matter fibers, extending itself fore and aft, and undercutting the gray matter of the cortex. If a hematoma is lobar, the chances of finding an arteriovenous malformation, tumor, or amyloid, or the patient being on anticoagulants, are all higher. Angiography is often indicated in normotensive patients with lobar hematomas, although it may still miss a small angioma and amyloid angiopathy is not detectable.

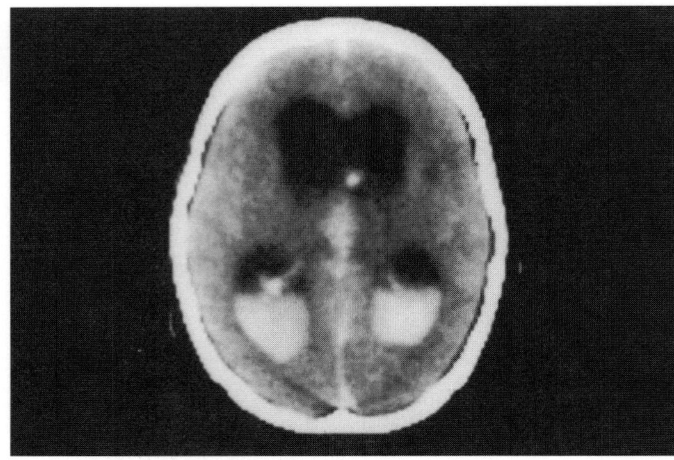

Fig. 2 CT scan showing blood in the ventricular system. (Reproduced with permission from Aminoff (1989).)

Management

Management is summarized in Table 2.

Table 2 Management of cerebral hemorrhage

1. Protect airway
   Maintain oxygenation
   Check fluid balance
2. Treat raised intracranial pressure
3. Control blood pressure
4. Reverse coagulopathy
5. Prevent complications: seizures
6. Prevent complications: deep vein thrombosis, pulmonary embolism
7. Good nursing care (skin problems, nutrition, hygiene)
8. Prevent recurrences
9. Consider prognosis

Immediate care

Careful medical management of the patient with cerebral hemorrhage can be lifesaving and is important even in those who are going to receive surgical treatment. Many patients will be in a coma when found or become unconscious within a few hours. These will need the standard care of the unconscious patient. The priority lies with the preservation of an airway and maintenance of oxygenation. Intubation will be followed by the need for a tracheostomy if coma persists for more than a few days. The patient should be nursed in the 20° to 30° head-up position. The bladder will usually need to be catheterized, and fluid balance is critical with good intravenous access.

Raised intracranial pressure

Raised intracranial pressure is accompanied by drowsiness and respiratory depression; $P\text{aO}_2$ is likely to fall, and $P\text{aCO}_2$ to rise. The latter leads to intracranial vasodilatation and further increase in intracranial pressure. This vicious cycle needs addressing, with attention to the level of $P\text{aCO}_2$. There is little evidence that corticosteroids are of benefit to the raised pressure of cerebral hemorrhage, but pressure is reduced by hyperventilation and/or mannitol. Mannitol is particularly relevant in the short term, for example when preparing a deteriorating patient for transfer to the operating room.

Blood pressure control

With increased intracranial pressure there is an increase in cerebral venous pressure and a reduction in perfusion pressure. Therefore it is hazardous to lower arterial pressure (Lavin 1986), despite the fact that recurrent or continued bleeding is more likely at elevated blood pressure. Hypertension is also believed to exacerbate vasogenic edema around the hematoma. Autoregulation of cerebral blood flow is lost after cerebral hemorrhage, as after cerebral infarction, thus adding to the risk of inducing cerebral ischemia if systemic blood pressure falls. In practice it may be necessary to reduce high arterial pressure if it is causing other problems such as left ventricular failure, or if it is very high with attendant fears of rebleeding (e.g. 180/120). The choice of pharmacological regimen should be for agents with a smooth slow effect (e.g. intravenous β-blockers). Pressures as 'normal' as 120/80 may be hazardous; therefore no more than a modest reduction (e.g. 25 per cent) should be attempted.

Coagulopathy

Most cerebral hemorrhages occur in patients with arterial disease risk factors (hypertension etc.), but about 10 per cent accompany the use of anticoagulants or other treatments affecting fibrinolysis or thrombosis. If the patient is on heparin when a bleed occurs, it should be reversed with protamine, and if on warfarin with fresh frozen plasma and vitamin K (Feldman 1993). Hemorrhage on warfarin is unusual unless the international normalized ratio (INR) is over 5. If the patient has been on thrombolytics, for example for coronary occlusion, protamine and ε-aminocaproic acid may be needed after advice from a hematologist. Such bleeds are usually lobar in distribution. Factor VIII will be needed in a hemophiliac. If the patient is thrombocytopenic or bleeding persists, platelet transfusions may be necessary. Hemorrhage into a tumor is particularly likely in the context of high-grade gliomas, metastases, or melanoma.

Seizures

Seizures may complicate the development of a cerebral hemorrhage in up to 25 per cent of patients. They often occur in the first 48 h and should be treated promptly. Prophylactic antiepileptic drugs are appropriate with superficial hematomas. They are probably not indicated for deeply situated hemorrhages.

Deep vein thrombosis and pulmonary embolism

The prevention of deep vein thrombosis and pulmonary embolism in an unconscious or severely immobilized patient, for example with a dense hemiplegia, would normally entail the use of subcutaneous heparin. A single trial has shown that this regimen can be employed safely even in patients with cerebral hemorrhage but, because of the theoretical risk, compression stockings are usually preferred.

Nursing care

After a few days calories should be provided. Prophylaxis against deep vein thrombosis and pulmonary embolism needs to be considered (see above). Protection of pressure areas becomes important if immobility and a depressed conscious level persist. Physiotherapy, speech therapy, and early mobilization may all be appropriate, depending on the patient's progress.

Prevention of recurrence

Prevention of recurrence depends on control of blood pressure, review of the need for anticoagulation, and surgical or other management of arteriovenous malformations (Fig. 3) or underlying neoplasms. Special problems arise when the hemorrhage is due to an underlying vasculitis. Cocaine use (either chronic or reuse after a period of withdrawal) may trigger hemorrhage. In some the bleed is

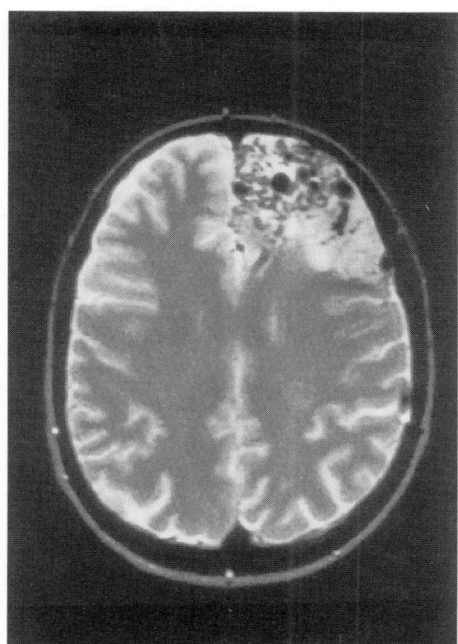

Fig. 3 Magnetic resonance image showing arteriovenous malformation in the frontal pole.

due to hypertension-provoking hemorrhage from an arteriovenous malformation rather than a vasculitis, but the drug must be stopped in either case. In systemic vasculitis steroids and immunosuppression may be relevant.

Prognosis

The best predictor of outcome is the size of the hematoma. Massive cerebral hemorrhage is rapidly fatal and no treatment is indicated. Small hematomas recover like minor ischemic strokes, and management rests on control of risk factors, particularly hypertension, to prevent recurrence and avoidance of aspirin. It is the medium-size hematomas that demand urgent and vigorous treatment since these are life threatening through the development of mass effect. Neurosurgical advice should always be sought in these cases. The surgical management is discussed in detail elsewhere. Elderly patients and those with multiple bleeds (e.g. with amyloid angiopathy) fare badly. Functional recovery can be surprisingly good in younger subjects, and recurrence is rare and usually due to poor control of hypertension.

References

Aminoff, M.J. (1989). *Neurology and general medicine*, p. 120. Churchill Livingstone, New York.

Feldmann, E. (1993). Intracerebral hemorrhage. In *Clinical atlas of cerebrovascular disorders* (ed. M. Fisher), pp. 11.1–11.16. Wolfe, London.

Lavin, P. (1986). Management of hypertension in acute stroke. *Archives of Internal Medicine*, **146**, 66–8.

Ojemann, R.G. and Mohr, J.P. (1975). Hypertensive brain hemorrhage. *Clinics in Neurosurgery*, **23**, 220–44.

Russell, R.W.R. (1963). Observations on intracerebral aneurysms. *Brain*, **86**, 425–42.

Warlow, C.P. (1991). *Handbook of neurology*. Blackwell Scientific, Oxford.

7.5.3 The role of surgical management

H. M. Fernandes and A. D. Mendelow

Key messages

- Surgical evacuation of an intracerebral hematoma may be lifesaving.
- Further surgical intervention is usually aimed at the prevention of rebleeding.
- Medical measures to reduce intracranial pressure can reduce secondary brain damage while surgical intervention is awaited.
- The surgical management of hypertensive intracerebral hemorrhage remains very controversial and clinical trials to evaluate the role of surgery are in progress.
- Do not treat reactive hypertension as low systemic blood pressure may reduce cerebral perfusion pressure.

Introduction

Non-traumatic intracranial hemorrhage may present with bleeding into the subarachnoid space (subarachnoid hemorrhage) or bleeding into brain parenchyma itself (intracerebral hematoma), with or without intraventicular extension (intraventricular hemorrhage). Neurological damage caused directly by the hemorrhage (primary brain damage) is usually unalterable at the time of presentation, and thus the surgical management of these patients concentrates on the prevention of secondary brain damage. This secondary damage may be caused by a further hemorrhage or the effects of an intracerebral hematoma. Therefore operative intervention may be necessary to prevent rebleeding, to evacuate clot, or both.

An intracerebral hematoma may be present after intracranial hemorrhage secondary to an intracranial aneurysm or an arteriovenous malformation or following hypertension. Surgical evacuation of the hematoma may be lifesaving and carried out urgently. However, surgery to prevent rebleeding differs depending on the cause. If appropriate and possible, it may be carried out at the same time as clot evacuation, but more often it is performed some days or weeks after the initial hemorrhage.

Evacuation of hematoma

The presence of an intracerebral hematoma may be responsible for the comatose state of a patient or for continuing neurological deterioration.

In aneurysmal intracerebral hematoma surgical treatment of both the aneurysm and the hematoma has been shown to be far superior to conservative management (Heinsanken *et al.* 1988), but the value of surgical evacuation of intracerebral hematoma from other causes (particularly hypertensive type hemorrhage) remains controversial. There is experimental evidence from models of intracerebral hematoma that early surgical evacuation reduces ultimate infarction volumes (Nehls *et al.* 1990; Hoff 1997). However, surgical practice remains largely individual to both patient and surgeon. Factors such as the size and site of the intracerebral hematoma, patient age and condition, and time since ictus have a varying influence, but as a rule of thumb surgical evacuation of hematoma is often carried out in those patients who remain initially well after their hemorrhage but subsequently deteriorate. The variability of views about surgery amongst British neurosurgeons is wide (Fernandes and Mendelow 1997).

Further neurological damage following the initial hemorrhage may be due to global cerebral ischemia following a reduction in cerebral perfusion pressure because of increasing intracranial pressure, or may be focal due to the direct toxic effects of the hematoma products on the surrounding brain. There also appears to be a penumbra of functionally impaired but potentially recoverable tissue around the ischemic core surrounding the hematoma, and this may be preserved with clot removal (Nehls *et al.* 1990). Surgical evacuation of the clot will not affect neurological damage caused by the primary destructive effect of the hemorrhage in the ischemic tissue immediately surrounding the intracerebral hematoma; indeed, the surgical approach may add to it. However, it may prevent secondary damage by removing the mass effect of the hematoma, therefore reducing intracranial pressure and allowing cerebral perfusion pressure to rise. This in turn may reperfuse tissue in the penumbra. Therefore preparation of these patients for surgery needs to be rapid.

Assessment of airway, respiration, and blood pressure both initially and at continuing intervals is crucial. Continuing assessment of the conscious level of the patient if possible and the pupillary size and response is also important. Urea, electrolyte, blood count, and arterial blood gas measurements should be performed urgently. Four units of blood should be cross-matched. Clotting studies should also be performed and, if abnormal, the patient should be rapidly treated with vitamin K and fresh frozen plasma. During the time taken to perform these procedures, medical options should be considered to reduce intracranial pressure. These include hyperventilation, to maintain $P\text{aCO}_2$ between 3.5 and 4.5 kPa (26 and 34 mmHg), and administration of osmotic agents such as mannitol and furosemide (frusemide). Reactive hypertension should not be treated, as low systemic blood pressure may reduce cerebral perfusion pressure.

It may also be necessary, depending on the nature and site of the hemorrhage and the preference of the surgeon, to perform a cerebral angiogram prior to operation. This is usually limited, because of time, to a single carotid injection ipsilateral to the hematoma if an aneurysm or arteriovenous malformation is suspected.

Surgical evacuation of intracerebral hematoma is usually performed via craniotomy, particularly if secondary to a ruptured aneurysm or arteriovenous malformation. However, some hypertensive type hemorrhages may be evacuated stereotactically or endoscopically through a burr hole or trephine.

Further surgical management

Intracranial aneurysm

Over half of non-traumatic intracranial hemorrhage is secondary to the rupture of an intracranial aneurysm. About 20 per cent will have an associated intracerebral hematoma as illustrated in Fig. 1. Aneurysmal rebleeding carries a high mortality and morbidity, with the peak incidence in the first 24 to 48 h, but with a continuing risk reaching 50 per cent over a month.

The best method of prevention of rebleeding, despite many attempts in the past to control 'transmural pressure' and the use of antifibrinolytic agents, remains obliterating the aneurysm from the intracranial circulation. This can be done surgically with a clip at craniotomy, as illustrated in Fig. 2, or by the endovascular route, filling the aneurysm sac with coils (embolization). Endovascular techniques, although much less invasive, have not been evaluated in the long term and do not have the advantages of open operation, i.e. drainage of cerebrospinal fluid, removal of intracerebral hematoma or cisternal clot (believed to be helpful in the prevention of vasospasm), and control of proximal vessels in the event of an intraoperative aneurysmal rupture.

The timing of surgical clipping or embolization remains another neurosurgical controversy, but there is a growing trend to perform this in the first few days after hemorrhage to prevent rebleeding and allow the maximal treatment of vasospasm, particularly in the patient with a good grade. In patients with WFNS (World Federation Neurological Surgeons) grades 3 to 5, treatment of the aneurysm is usually delayed pending neurological improvement.

Preoperative preparation of such patients should follow on from their medical management at ictus. An assessment and any necessary treatment of general medical problems should be carried out, with particular attention to the respiratory system. All patients should receive adequate hydration (2–3 l/24 h), and the calcium-channel blocker nimodipine should be given to prevent deterioration secondary to vasospasm. If nimodipine cannot be adminstered orally (60 mg every 4 h), it should be given via the nasogastric or intravenous route. The dose should be decreased if it is thought to have a deleterious effect on systemic blood pressure. A recent full blood count, urea, electrolytes, and clotting profile should be available, and four units of blood should be cross-matched. Hyponatremia is common secondary to cerebral salt wasting and hemodilution, and may require correction. Fluid restriction should be avoided because of the adverse effect on cerebral perfusion pressure.

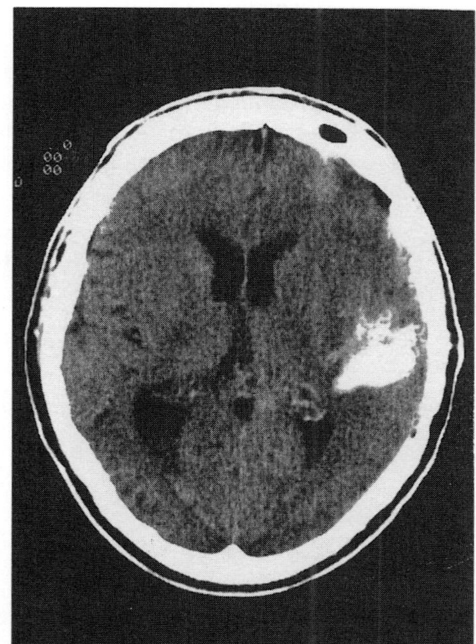
(a)

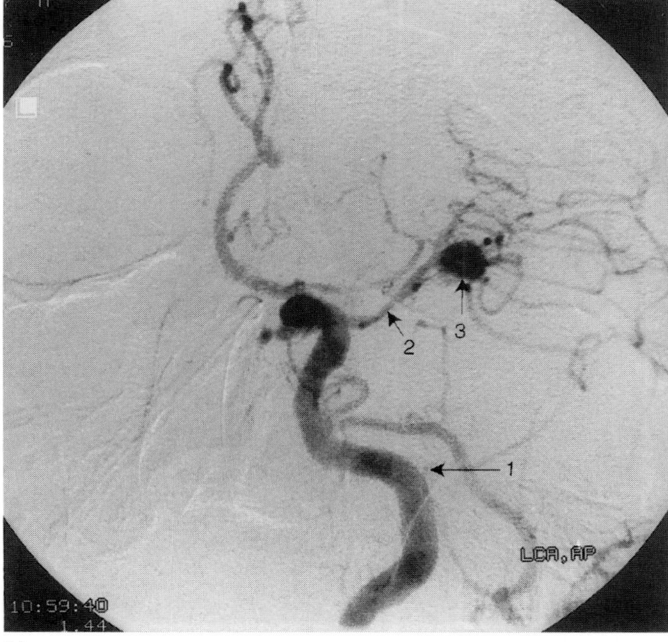

(b)

Fig. 1 (a) CT brain scan showing hemorrhage into the left temporal lobe. (b) This was later confirmed, after cerebral angiography, to be secondary to rupture of a left middle cerebral artery aneurysm: 1, left internal carotid artery; 2, left middle cerebral artery; 3, left middle cerebral artery aneurysm.

Arteriovenous malformation

Arteriovenous malformations (Fig. 3) are congenital vascular anomalies in which an abnormal communication exists between an artery and a vein. Hemorrhage associated with arteriovenous malformations is usually more benign than that associated with aneurysms, and the risk of rebleeding is much less (6 per cent in the first year and 2–3 per cent per year thereafter) (Wilkins 1985). However, as they usually present in younger patients (even children), the lifetime risk of rehemorrhage can be substantial.

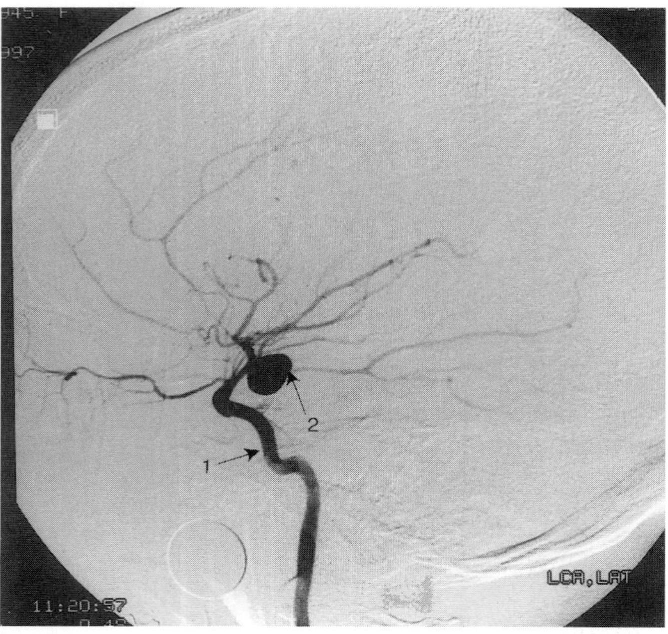

(a)

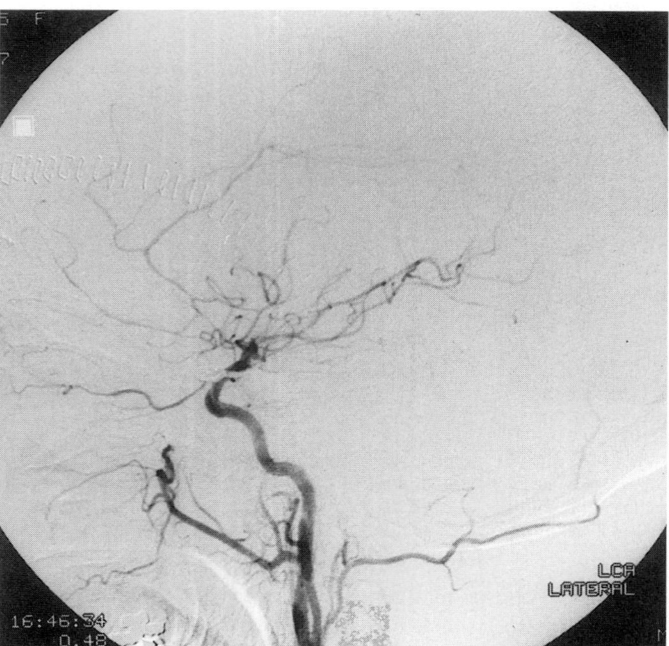

(b)

Fig. 2 (a) Left lateral carotid angiogram performed after subarachnoid hemorrhage revealing a large posterior communicating artery aneurysm and distal narrowing of cerebral vessels secondary to vasospasm: 1, left internal carotid artery; 2, left posterior communicating artery aneurysm. (b) A repeat postoperative angiogram shows successful surgical clipping of the aneurysm.

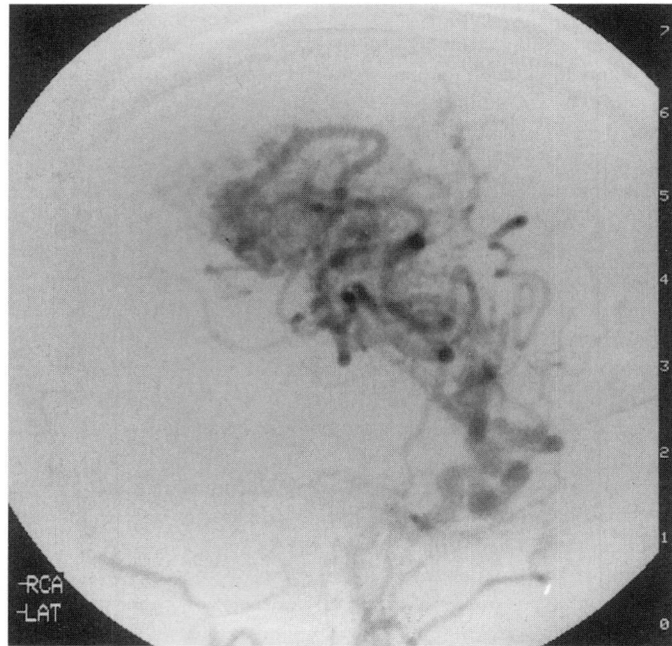

Fig. 3 Cerebral angiogram showing a large arteriovenous malformation.

The type and timing of surgical intervention for arteriovenous malformations is dictated more by their size, position, and nature than by the clinical condition of the patient. Complete excision may be possible by craniotomy alone, or complete obliteration by embolization alone. Often, however, a combination of the above techniques is necessary. With many large lesions treatment can be carried out in stages. In those arteriovenous malformations that are unsuitable for the above approach (too deep-seated, in an eloquent area of brain, feeding artery, also end artery) treatment with stereotactic radiosurgery may be possible.

Preoperative preparation of these patients may be in the acute posthemorrhage period or some time later. Again, an assessment of the general medical condition of the patient and routine laboratory investigations are necessary. Nimodipine has been used following hemorrhage, and in some cases antiepileptic treatment and steroids are prescribed prophylatically.

Primary intracerebral hemorrhage

Intracerebral hemorrhage as described may be secondary to an underlying lesion such as an aneurysm or arteriovenous malformation, but 'spontaneous' or 'primary' intracerebral hemorrhage forms a further subgroup. It most commonly occurs in the basal ganglia area (Fig. 4), often secondary to hypertension. Lobar hemorrhages are also common and may be secondary to amyloid angiopathy. Coagulation disorders, tumors, and drug abuse are rarer causes of 'primary intracerebral hemorrhage'.

Apart from the first few hours following ictus, rebleeding is not a common problem and so surgical intervention generally concerns evacuation of the clot only. Herein lies controversy. Probably apart from the rapidly deteriorating patient, there is little agreement

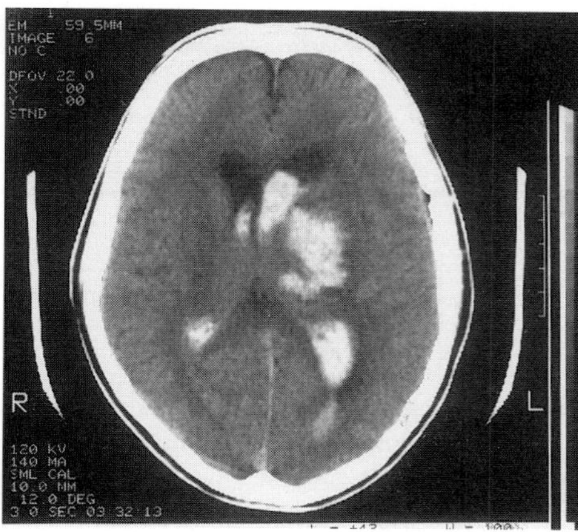

Fig. 4 CT brain scan showing 'typical' hypertensive intracerebral hemorrhage in the left putaminal area. The hemorrhage is causing significant mass effect, with midline shift and has also ruptured into the lateral ventricular system.

about if and when surgical evacuation should be performed (Fernandes and Mendelow 1997).

As this type of hemorrhage is more common in the older patient, assessment of the general medical condition is an important part of preoperative preparation. The recognition that many hematomas are secondary to anticoagulation or antithrombotic treatment indicates that care should be taken to perform clotting studies, particularly if no history is available from the unconscious or dysphasic patient.

Hydrocephalus

Hydrocephalus can commonly present in these patients. This may be due to obstruction of the cerebrospinal fluid pathways by intraventricular or subarachnoid hemorrhage, or to extrinsic compression of the fourth ventricle by a posterior fossa hematoma. Hydrocephalus acts to increase intracranial pressure, and is surgically treated by drainage of cerebrospinal fluid. In the acute stages this is usually by placement of an external ventricular drain through a frontal burr hole. Although some types of communicating hydrocephalus can be treated by lumbar drainage, this is not possible with an intracerebral hematoma because of the danger of increasing brain shifts and herniation.

In the moribund patient an external ventricular drain can be inserted under local anesthesia, but in any case preoperative preparation follows the general pattern described elsewhere, with the use meanwhile, if appropriate, of medical measures to reduce intracranial pressure.

Cerebellar hematomas

Ten per cent of all primary intracerebral hematomas occur in the cerebellum, usually related to hypertension. There may be associated hydrocephalus. Surgical intervention is often urgently required, and is most often carried out in moribund or deteriorating patients. Evacuation is by suboccipital craniectomy and can be combined with placement of an external ventricular drain. Preoperative preparation of the patient is as described above, with emphasis placed on rapidity and medical management of raised intracranial pressure meanwhile.

References

Fernandes, H.M. and Mendelow, A.D. (1997). Spontaneous intracerebral haemorrhage: a snapshot of U.K. neurosurgical management. *Journal of Neurology, Neurosurgery and Psychiatry*, in press.

Heinsanken, O., *et al.* (1988). *Acta Neurochirurgica*, **90**, 81–3.

Hoff, J. T. (1997). The rationale for a more aggressive surgical approach to some spontaneous intracerebral hemorrhages. *Journal of Neurology, Neurosurgery and Psychiatry*, in press.

Nehls, D.G., Mendelow, A.D., Graham, D.I., and Teasdale, G.M. (1990). *Neurosurgery*, **27**, 674–82.

Wilkins, R.H. (1985). Natural history of intracranial vascular malformations: a review. *Neurosurgery*, **16**, 421–30.

7.6 Meningitis

7.6.1 Pathophysiology and causes

M. Levin and S. Nadel

Key messages

- Bacterial meningitis is the most common form of serious intracranial infection.
- Over 75 per cent of cases occur in children under 15 years of age.
- Pathogenic organisms colonize the nasopharyngeal mucosa of a susceptible host.
- The cerebrospinal fluid inflammatory response is responsible for most of the deleterious effects of the infection.
- Understanding this response has allowed a more rational approach to therapy.

Introduction

Meningitis is the most common form of intracranial infection. Acute bacterial meningitis remains a significant cause of morbidity and mortality throughout the world. The current morbidity and mortality depend on the causative organism and the duration of symptoms prior to therapy.

Here, we review the epidemiology, pathophysiology, clinical presentation, diagnosis, and approach to therapy of meningitis. We necessarily concentrate on acute bacterial meningitis, as patients with this disease will tend to be more severely ill and are more likely to require intensive care management.

Epidemiology of acute bacterial meningitis

The annual incidence of bacterial meningitis in the developed world is about 3 to 5 cases per 100 000. It has been estimated that over 75 per cent of all cases of bacterial meningitis occur in children under 15 years of age, and it is one of the most common life-threatening infections encountered by the medical profession.

The causative organisms will vary according to the population studied and the age of the study group, and may rise considerably, for example during epidemics of meningococcal infection. Currently, *Neisseria meningitidis* and *Streptococcus pneumoniae* are the most common causes of bacterial meningitis outside the neonatal age group.

The microbial epidemiology of meningitis is changing in older children and adults, in whom nosocomial meningitis accounts for an increasing proportion of infections. Many of these are associated with recent neurosurgical intervention or trauma. In such cases, *Pseudomonas aeruginosa*, enterococci, *Staphylococcus aureus*, and the coagulase-negative staphylococci are the most common causative organisms. Table 1 lists the common causes of bacterial meningitis associated with different underlying conditions.

Pathophysiology

In order to cause meningitis, bacteria must colonize and penetrate through the nasopharyngeal or oropharyngeal mucosa, survive

Table 1 Organisms associated with the development of meningitis according to different underlying conditions

Underlying condition	Organism
Basilar skull fracture	*Strep. pneumoniae* *N. meningitidis* *H. influenzae* type b *Staph. aureus* *Strep. pyogenes* (group A) Gram-negative bacilli
Post-neurosurgical procedure	Coagulase-negative staphylococci *Staph. aureus* Gram-negative bacilli
With cerebrospinal fluid shunt	Coagulase-negative staphylococci *Staph. aureus* Gram-negative bacilli *Corynebacteria* (diphtheroids) *Propionibacteria acnes* *Bacillus* species
Nosocomial (without neurosurgery)	*Staph. aureus* Gram-negative bacilli Enterococci *Candida* species
Immune deficiencies (including hyposplenism)	*Strep. pneumoniae* *N. meningitidis* *H. influenzae* type b *L. monocytogenes* *Salmonella* species *Cryptococcus neoformans* *Nocardia* species Gram-negative bacilli Enteric bacilli *Ps. aeruginosa*

and multiply in the blood stream, evading antibacterial immunological mechanisms, and then invade through the blood–brain barrier.

Colonization and penetration of the nasopharyngeal mucosa

Infection cannot occur until colonization of the host has taken place, usually in the upper respiratory tract. Pathogenic bacteria such as meningococci are usually transferred by droplet spread. Crowding facilitates transmission, and is considered to be an important factor in the high attack rates of meningococcal infection in military recruits. There is a suggestion that invasive meningococcal infection develops shortly after colonization of a new host. Most individuals are asymptomatic following colonization. There is some evidence that invasion of pathogenic organisms is more common following upper respiratory infection with influenza A, although the evidence is less clear for other respiratory pathogens, inhalation of dry or dusty air, and passive smoking among children.

Studies in animal and human subjects have substantiated that secretory IgA is superior to other immunoglobulin isotypes in protection from viral and bacterial mucosal pathogens. IgA1 is the predominant IgA subclass expressed on the mucosa of the nasopharynx (approximately 95 per cent). All the main pathogenic bacteria which colonize the nasopharyngeal mucosa produce specific IgA1 proteases. These inactivate secretory IgA1, while IgA2 is not cleaved. The biological significance of these enzymes is unclear but some non-pathogenic colonizers of the nasopharynx (e.g. *Neisseria lactamica*) do not produce IgA proteases.

Attachment of bacteria to nasopharyngeal epithelial cells depends on the presence of bacterial virulence factors such as fimbriae and pili. The presence of these filament-like bacterial appendages has been shown to mediate epithelial adhesion of *Hemophilus influenzae* and *N. meningitidis*. Non-piliated meningococci bind inefficiently to mucosal surfaces. *Strep. pneumoniae* contains carbohydrate adhesins which may mediate its adhesion to epithelial surfaces.

The organisms which most commonly cause meningitis possess a polysaccharide capsule which is a potent virulence factor. Presence of the capsule causes inhibition of complement-mediated polymorphonuclear phagocytosis and inhibits immune lysis of the organism. Non-encapsulated strains of the common meningitis-causing organisms are of low virulence and rarely cause invasive disease in the immunocompetent host. The alternative complement pathway plays an important accessory role in prevention of infection by encapsulated organisms, particularly *Strep. pneumoniae*. The spleen plays a vital role in removing non-opsonized material from the circulation, particularly in non-immune individuals who do not have specific antibodies. Therefore individuals who lack a functional spleen for whatever reason (i.e. sickle cell disease, congenital asplenia, etc.) are at high risk of serious invasive infection, including meningitis, from the encapsulated organisms.

Thus mucosal invasion and penetration occurs in those individuals where local mucosal immunity fails to contain bacterial proliferation. The polysaccharide capsule partially protects the invading organism by inhibiting the complement-mediated phagocytic capacity of local leukocytes, and some evidence suggests that locally produced IgA may inhibit the bactericidal effect of IgG and IgM (Berkowitz *et al.* 1996).

Penetration of the blood–brain barrier

Once in the bloodstream, the invading organism may either be neutralized by the combined action of circulating antibacterial factors, including antibodies, complement proteins, and phagocytic cells, or, for reasons which are not yet fully understood, may survive and proliferate in the bloodstream to initiate the bacteremic stage of the illness.

Most episodes of meningitis arise by hematogenous spread of the causative organism, with a primary bacteremia and secondary focal infection of the meninges by seeding. The mechanisms by which circulating organisms penetrate the blood–brain barrier and seed the cerebrospinal fluid remain unclear. What does appear clear, however, is that a high-grade bacteremia is necessary, with more than 10^2 organisms/mm^3 of blood to enable meningeal seeding. The organisms which cause meningitis appear to exhibit tropism for the meninges. This may be partly due to bacterial adhesive structures such as fimbriae which bind the circulating bacteria to receptors on the choroid plexus and cerebral capillary endothelial cells, thus facilitating bacterial translocation into the cerebrospinal fluid.

Other organisms, particularly those which may proliferate in the intracellular environment, such as the mycobacteria (including *Mycobacterium tuberculosis*) and salmonellae, may enter the cerebrospinal fluid despite having been phagocytosed by circulating monocytes or polymorphs. It is thought that these organisms may cause meningitis by entering the cerebrospinal fluid by normal white cell trafficking.

After penetrating the blood–brain barrier and entering the subarachnoid space, bacterial multiplication may proceed rapidly as the cerebrospinal fluid is a relatively poor immune environment. The cerebrospinal fluid contains low concentrations of specific antibody and complement components, and thus has a poor capacity for opsonization and bactericidal activity. In addition, bacterial capsular polysaccharide is antiphagocytic.

The cerebrospinal fluid inflammatory response

Once the organism has invaded the cerebrospinal fluid, a number of bacterial components, particularly lipopolysaccharide or lipo-oligosaccharide of Gram-negative organisms and peptidoglycan are the major determinants of meningeal inflammation. The techoic acid of Gram-positive organisms and the peptidoglycan components of both Gram-positive and Gram-negative organisms have been shown to be potent inducers of inflammation in the cerebrospinal fluid and to impair blood–brain barrier function on direct intracisternal inoculation in experimental animals. Similarly, direct inoculation of lipopolysaccharide from *H. influenzae* or the lipo-oligosaccharide of *N. meningitidis* into the cerebrospinal fluid of experimental animals causes an intense inflammatory reaction with influx of leukocytes, increase in protein and lactate, and a decline in cerebrospinal fluid glucose concentration. These studies have provided overwhelming evidence that bacterial endotoxins and other bacterial cell wall constituents are important in the initiation of the inflammatory changes within the cerebrospinal fluid and disturbance of function of the blood–brain barrier (Quagliarello and Scheld 1992).

The inflammatory changes in the cerebrospinal fluid occur several hours after inoculation of bacteria or cell wall constituents into the meningeal space of experimental animals. This has led to the hypothesis that elaboration and release of host mediators are instrumental in the development of the inflammatory changes. Animal models have established that tumor necrosis factor-α and interleukin 1β (IL-1β) are important mediators of the initial meningeal inflammation. Levels of tumor necrosis factor-α and IL-1β, together with interleukin 6 (IL-6), increase in the cerebrospinal fluid of animals following intracisternal inoculation of meningococcal lipo-oligosaccharide, and this rise in cytokine levels precedes cellular influx and protein exudation. These cytokine mediators have been shown to stimulate the release of other factors in the inflammatory cascade, including platelet activating factor, interleukin 8 (IL-8), and interferon-γ. Release of these proinflammatory mediators causes upregulation of cellular adhesion molecules, which include molecules such as the integrins, selectins, and the IgG superfamily, on the surface of peripheral blood leukocytes and vascular endothelial cells of the blood–brain barrier. This results in attraction, attachment, and migration of leukocytes into the cerebrospinal fluid. Once present in the cerebrospinal fluid, polymorphonuclear neutrophils are activated and undergo degranulation and release of proteolytic enzymes, cationic proteins, and reactive oxygen species. These products further alter the integrity of the blood–brain barrier, thus interrupting its primary functions, i.e. active transport and facilitated diffusion of nutrients (including glucose and other metabolites) and secretion of cerebrospinal fluid. Increased permeability of the blood–brain barrier results in leakage of albumin and other macromolecules into the cerebrospinal fluid, causing vasogenic edema. The presence of the anaphylotoxins (C3a, C5a) in the cerebrospinal fluid due to the protein leak further encourages passage of polymorphonuclear neutrophils into the cerebrospinal fluid, thus accentuating the inflammatory process. Toxic products of neutrophil activation and other inflammatory cells cause cytotoxic edema and damage to surrounding cells.

Recently, nitric oxide (NO) has been implicated as a potential neurotoxic factor. NO is produced by a variety of cell types and is involved in processes such as smooth muscle relaxation, regulation of cell-mediated cytotoxicity, inhibition of platelet aggregation, and neuronal signaling. Cytokines, such as those described above, together with lipopolysaccharide induce the formation of large amounts of NO via the enzyme inducible NO synthase (iNOS) in microglia and astrocytes. This implies a possible role for NO in the host response to bacterial meningitis. A recent study has shown increased levels of cerebrospinal fluid nitrite and nitrate (degradation products of NO) in patients with bacterial meningitis, which correlated with the level of tumor necrosis factor-α in the cerebrospinal fluid.

The importance of the cytokines has been confirmed in studies where direct inoculation of tumor necrosis factor-α, IL-1β, and interferon-γ into the cerebrospinal fluid of experimental animals induces inflammatory changes which closely resemble those which are seen with lipopolysaccharide inoculation. In addition, the inflammatory response to lipopolysaccharide in the cerebrospinal fluid can be reduced by the simultaneous inoculation of antibodies to tumor necrosis factor-α, IL-1β, or both (Saez-Llorens *et al.* 1990). However, the use of antibodies against these cytokines does not completely abrogate the inflammatory response, suggesting that other mediators also play an important role.

Apart from the proinflammatory cytokines described above, anti-inflammmatory cytokines, such as interleukin 10 (IL-10) have recently been shown to be present in the cerebrospinal fluid of children with bacterial meningitis. IL-10 inhibits the production of chemokines by polymorphonuclear neutrophils, and downregulates production of intercellular adhesion molecule 1 (ICAM1) as well as chemoattractant proteins by endothelial cells. The significance of IL-10 and other host-produced anti-inflammatory agents in bacterial meningitis has not yet fully been elucidated.

Cerebral edema and thrombosis

In addition to the development of vasogenic and cytotoxic edema as described above, interstitial edema may be caused by impaired reabsorption of cerebrospinal fluid by the arachnoid villi, a phenomenon which has been demonstrated in experimental models of meningitis. The consequences of increased secretion and diminished reabsorption of cerebrospinal fluid, and breakdown of the blood–brain barrier, are an increase in brain water content and cerebrospinal fluid volume, with the development of severe brain edema, relative hydrocephalus, and therefore increased intracranial pressure. If this increase in intracranial pressure is severe, it will lead to a reduction in cerebral blood flow. In experimental models of meningitis, cerebral blood flow first increases due to local vasodilatation, probably induced by oxygen-free radicals and local NO production from leukocytes, vascular smooth muscle, vascular endothelium, and glial cells. This suggests that cerebral blood volume is increased, at least early in the disease, and this probably contributes to raised intracranial pressure. These changes are paralleled by an increase in cerebrospinal fluid lactate levels, possibly indicating tissue hypoxia. However, the origin of the increased cerebrospinal fluid lactate is unclear. Increased lactate may contribute to the increase in cerebral blood flow, as acidosis is a vasodilator. Following this initial hyperemic stage, cerebral blood flow then decreases, presumably as a consequence of raised intracranial pressure, but again the reduction in cerebral blood flow is likely to be multifactorial. Reduction in cerebral blood flow has been correlated with levels of tumor necrosis factor-α and cerebrospinal fluid endothelin. Changes in cerebral blood flow may also be a consequence of a loss of cerebrovascular autoregulation, which has been demonstrated to occur in bacterial meningitis. Cerebral blood flow is normally maintained at constant levels irrespective of systemic arterial pressure. Once autoregulation has been lost, cerebral blood flow is totally dependent on systemic pressure. Blood flow may become inadequate if systemic hypotension occurs. Thus the rise in intracranial pressure, together with systemic hypotension which is common in bacterial sepsis, may result in cerebral hypoperfusion.

All patients with bacterial meningitis are likely to have raised intracranial pressure as part of their disease process. In a recent study the mean opening pressure at the time of lumbar puncture was $180 \pm 70\,\text{mmH}_2\text{O}$.

Intracranial pressure is an important determinant of the cerebral perfusion pressure, and its level appears to be correlated with outcome of bacterial meningitis. Several studies have demonstrated that morbidity and mortality are highest in those individuals with a cerebral perfusion pressure below 30 mmHg. It is likely that the reduction of cerebral perfusion pressure is more a consequence of increased intracranial pressure than of systemic hypotension.

Vasculitis, vascular spasm, and thrombosis have been demonstrated to occur in large intracerebral arteries and veins in children and adults with bacterial meningitis. As these blood vessels lie over the surface of the brain and are bathed in infected cerebrospinal fluid containing cytokines and other mediators, bacterial products, and leukocytes, they are particularly susceptible to vascular damage. These findings are often associated with the presence of neurological deficits on clinical examination, secondary to cerebral infarction.

Another aspect of central nervous system disturbance is alteration in cerebral metabolism. The increased cerebrospinal fluid lactate has already been referred to. Another universal finding in bacterial meningitis is the reduction in cerebrospinal fluid glucose concentration. The etiology of this is unclear, but is believed to involve increased metabolism of glucose by leukocytes and bacteria, and disturbance of glucose transport into the cerebrospinal fluid (Pfister *et al.* 1994).

Excitatory amino acids, such as glutamate, which interact with neuronal *N*-methyl-D-aspartate receptors are believed to play a significant role in other models of neuronal injury, and several reports have documented a rise in glutamate concentrations in brain interstitium and cerebrospinal fluid of experimental animals with pneumococcal and *Escherichia coli* meningitis. Animal models of bacterial meningitis treated with the glutamate antagonist kynurenic acid showed reduced neuronal injury compared with untreated control animals.

All these recent advances in our understanding of the pathophysiology and molecular mechanisms leading to the clinical consequences of bacterial meningitis will improve the delivery of rational therapies for this life-threatening infection.

References

Berkowitz, I.D., Berkowitz, F.E., Newton, C., Willoughby, R., and Ackerman, A.D. (1996). Meningitis, infectious encephalopathies and other central nervous system infections. In *Textbook of pediatric intensive care* (ed. M.C. Rogers), pp. 1039–91. Williams & Wilkins, Baltimore, MD.

Pfister, H.W., Fontana, A., Tauber, M.G., Tomasz, A., and Scheld, M.W. (1994). Mechanisms of brain injury in bacterial meningitis: workshop summary. *Clinics in Infectious Diseases*, **19**, 463–79.

Quagliarello, V. and Scheld, W.M. (1992). Bacterial meningitis: pathogenesis, pathophysiology and progress. *New England Journal of Medicine*, **327**, 864–72.

Saez-Llorens, X., Ramilo, O., Mustafa, M.M., Mertsola, J., and McCracken, G.H., Jr (1990). Molecular pathophysiology of bacterial meningitis: current concepts and therapeutic implications. *Journal of Pediatrics*, **116**, 671–84.

7.6.2 Therapeutic approach

M. Levin and S. Nadel

Key messages

- Patients may present with non-specific features, particularly neonates, infants, postneurosurgical patients, and the elderly.
- Any patient with mental status changes and fever should be assumed to have meningitis and treated as such until the diagnosis is excluded.
- Lumbar puncture may be hazardous in patients with meningococcal septicemia or features of raised intracranial pressure.
- Antibiotic resistance amongst the most common organisms causing meningitis is becoming more common and antibiotic therapy should be adjusted accordingly.
- Adjunctive therapy with anti-inflammatory agents such as steroids has a rational basis for use. Clinical trials have suggested that steroids may be beneficial in community-acquired bacterial meningitis if given before or with the first dose of antibiotics.
- Vaccine research is progressing, so that effective vaccines against all the common causes of bacterial meningitis should be available shortly.

Clinical features

Classically, patients with bacterial meningitis present with headache, fever, photophobia, vomiting, neck stiffness, and alteration of mental status. These are present at some stage in 85 per cent of patients. Cranial nerve palsy (particularly nerves III, IV, VI, and VII), focal neurological signs (10–20 per cent of cases), and seizures (up to 30 per cent of cases) may also occur. Papilledema is rarely seen (about 1 per cent of cases). With disease progression, signs of raised intracranial pressure (ICP), including coma, hypertension, bradycardia, and altered respiratory status, become more likely.

The presentation of meningitis may be non-specific, particularly in young children and the elderly. The elderly may be afebrile and confused or obtunded, without classical signs of meningitis. Post-neurosurgical meningitis may present with very subtle alterations in mental status, without any other signs. A high index of suspicion is necessary in all groups of patients, particularly febrile patients with altered mental status (Segretti and Harris 1996).

There appear to be two main patterns of clinical presentation of community-acquired meningitis. The first is a prolonged history of non-specific symptoms, including lethargy, fever, and myalgias, progressing over several days. It may be impossible to be accurate about the time of onset of meningitis. The second presentation is an acute fulminating course, with manifestations of both sepsis and meningitis. This is frequently associated with raised ICP leading to brain herniation.

Cutaneous manifestations may be seen depending on the etiology of the meningitis. Approximately 50 per cent of patients with meningococcal meningitis and most patients with meningococcal septicemia, with or without meningitis, will have a petechial or purpuric rash which may start as maculopapular.

Seizures during the early stages occur in 20 to 30 per cent of patients with bacterial meningitis. Generalized seizures are not associated with a worse outcome. Focal seizures are more likely to be associated with persistent neurological sequelae, as are focal neurological signs and cranial nerve palsies.

Diagnosis

A diagnosis of acute bacterial meningitis is made by obtaining cerebrospinal fluid by lumbar puncture. Cerebrospinal fluid should be obtained whenever meningitis is suspected.

Lumbar puncture is generally a safe procedure, but there are recognized contraindications: signs of raised ICP, infection of the skin at the lumbar puncture site, the presence of bleeding diatheses, and cardiovascular compromise.

In considering the risks of lumbar puncture in patients with bacterial meningitis, particularly those with suspected meningococcal disease, additional factors should be considered.

Patients with meningococcal disease are often in compensated septic shock. Performing a lumbar puncture may cause an increase in respiratory and cardiovascular workload due to stress and patient positioning, and may result in acute clinical deterioration. In addition, the airway may become obstructed and respiratory function may be compromised by limiting chest expansion. The patient position may also interfere with venous return and thus limit cardiac output. Patients with meningococcal disease are likely to have a coagulopathy, which would be another relative contraindication to lumbar puncture. They may have raised ICP as a consequence of meningitis and reduced cerebral perfusion due to shock. Lumbar puncture in these circumstances may result in acute brainstem herniation (Cartwright 1995).

We recommend that lumbar puncture is contraindicated in patients with suspected meningococcal disease until acute cardiovascular compromise has been adequately treated. This would also be true for any patient with signs of shock.

If any features of raised ICP are present, the risk of cerebral herniation following lumbar puncture is considerable and the procedure should be deferred.

In bacterial meningitis the cerebrospinal fluid white cell count is usually over 1000/μl, with a neutrophil predominance. About 10 per cent of patients will have lymphocytic cerebrospinal fluid, particularly in listerial meningitis.

Reduced cerebrospinal fluid glucose concentration (less than 60 per cent of serum glucose measured simultaneously) is found in 70 per cent of cases. The cerebrospinal fluid protein concentration is elevated in almost all cases of bacterial meningitis.

Nearly all cases of bacterial meningitis as compared with viral meningitis have a combination of cerebrospinal fluid glucose below 2 mmol/l, a ratio of cerebrospinal fluid to serum glucose below 0.23, cerebrospinal fluid protein above 2.2 g/l, and cerebrospinal fluid white cell count above 2000/μl or 1180 neutrophils/μl. However, any one of these should raise the suspicion of bacterial meningitis, even if other cerebrospinal fluid results are atypical.

The cerebrospinal fluid Gram stain is positive in 60 to 90 per cent of cases and is extremely specific. However, the sensitivity is only 40 to 60 per cent and depends on the observer, the time spent inspecting the specimen, and whether the patient has had recent antibiotic therapy.

Cerebrospinal fluid culture is positive in up to 85 per cent of patients, but this falls to below 50 per cent in those who have received prior antibiotic therapy.

In patients with cerebrospinal fluid findings typical of bacterial meningitis, but where the Gram stain is negative, latex agglutination testing of the cerebrospinal fluid for bacterial antigen improves identification of the causal organism. Kits are available for the detection of *Hemophilus influenzae* type b (Hib), *Streptococcus pneumoniae*, *Neisseria meningitidis*, *Escherichia coli* K1, and group B streptococci. The sensitivity of the test varies from 50 to 100 per cent depending on the organism, but the specificity is high.

The polymerase chain reaction has recently been introduced for diagnosis of meningococcal meningitis and will become increasingly useful once it comes into widespread clinical usage (Tunkel and Scheld 1995).

Differential diagnosis

The main differential diagnosis is acute viral meningitis. Although in its early stages viral meningitis may be associated with neutrophil pleocytosis in the cerebrospinal fluid, lymphocyte predominance develops within 12 to 24 h. Other causes of cerebrospinal fluid pleocytosis include adverse drug reactions (particularly to non-steroidal anti-inflammatory agents), chemical meningitis (e.g. due to intrathecal administration of chemotherapeutic agents), carcinomatous meningitis, and parameningeal foci of infection. In all these, the cerebrospinal fluid white cell count is commonly lower than in acute bacterial meningitis. However, patients should be assumed to have bacterial meningitis and treated appropriately until it is definitely excluded. In patients where the diagnosis is unclear, the lumbar puncture should be repeated 12 to 24 h later to assess whether the cerebrospinal fluid findings have changed. In bacterial meningitis the findings will be similar despite initiation of appropriate therapy, whereas in viral meningitis the pleocytosis should have altered.

Patients who present with focal neurological signs or papilledema will usually not have acute bacterial meningitis. In these, brain abscess, intracranial hemorrhage, or tumor form part of the differential diagnosis. Lumbar puncture should not be performed in these patients until a space-occupying lesion has been excluded by CT scan or magnetic resonance imaging. If there is any suspicion of bacterial meningitis, antibiotic therapy should be commenced following blood culture, and the patient should be reviewed by a neurosurgeon to decide the most appropriate management and whether to obtain cerebrospinal fluid by ventricular drainage.

Occasionally these patients will have chronic meningitis, for example due to tuberculosis or crytocococcus.

Management

Recent years have brought major advances in our understanding of pathophysiological events occurring in the central nervous system in bacterial meningitis. Much of this knowledge has come from studies of animal models which have been extended into clinical studies. As a result there have been a number of important changes in clinical management, the most important being the introduction of anti-inflammatory treatment.

Antimicrobial therapy

Prompt diagnosis of bacterial meningitis is essential. The longer the delay prior to starting appropriate antimicrobial therapy, the more likely it is that the disease will cause sequelae or be fatal. Tunkel and Scheld (1995) recommend that appropriate antimicrobial therapy for bacterial meningitis should be commenced within 30 min of presentation for medical care, even if delay prior to obtaining cerebrospinal fluid is unavoidable. In this situation, choice of antimicrobial is empirical and is based on the most likely causative organisms for the patient's age. Antimicrobial therapy should not be delayed whilst awaiting organization or results of imaging studies in the case of patients with focal neurological signs or papilledema. Even if lumbar puncture is delayed, the cerebrospinal fluid findings will still be helpful and a bacteriological diagnosis may be obtained by latex agglutination or polymerase chain reaction.

In patients who are immunosuppressed or postsurgical, or where cerebrospinal fluid leak is present, antimicrobial therapy should be broadened to include other Gram-negative organisms, staphylococci, or any possible suspected opportunistic organisms.

Recommendations on empirical antimicrobial therapy in bacterial meningitis have recently been adapted to take account of changing susceptibility patterns of the commonly isolated organisms.

Despite reports of penicillin-resistant meningococci in certain countries, these have not proved to be a problem in the United Kingdom or the United States. Even if they have been isolated, patients have been cured with standard doses of penicillin. Other organisms have recently developed much more worrying patterns of antimicrobial resistance. Beta-lactamase-producing Hib accounts for about 30 per cent of all isolates of this organism, and chloramphenicol-resistant Hib accounts for up to 50 per cent of isolates in some parts of the world. Fortunately, the incidence of bacterial meningitis due to Hib has dramatically decreased since the introduction of Hib vaccine.

The recent development of most concern is the growing incidence of high-level penicillin resistance in *Strep. pneumoniae*, with countries in Eastern Europe reporting almost 60 per cent of isolates as penicillin resistant. In the United Kingdom approximately 3 per cent of isolates are highly penicillin resistant. Alarmingly, there are now reports of high-level cephalosporin resistance, conferring resistance to the third-generation cephalosporins. Factors increasing the likelihood of infection with a resistant strain include the patient's age (under 10 or over 50 years), immunosuppression, prolonged hospital stay, infection by serotypes 14 and 23, and frequent or prophylactic use of antibiotics.

Most authorities in Europe and North America now recommend a third-generation cephalosporin such as cefotaxime or ceftriaxone as first-line treatment for bacterial meningitis. These agents are active against all common meningeal pathogens (except *Listeria monocytogenes*), the common organisms are usually sensitive (except rarely *Strep. pneumoniae*), they penetrate excellently into cerebrospinal fluid, and they are single agents which can be given once to three times daily.

If *Strep. pneumoniae* meningitis is strongly suspected, there is a case for including vancomycin in empirical therapy until the antimicrobial susceptibility pattern of the isolate is determined. Vancomycin penetrates cerebrospinal fluid adequately in the presence of meningeal inflammation and its combination with a third-generation cephalosporin may be synergistic for meningitis caused by a highly penicillin-resistant organism. In addition, rifampin (rifampicin) may have a role in patients who fail to respond to this combination.

If *L. monocytogenes* is suspected, addition of penicillin or ampicillin is indicated. Cephalosporins have no activity against this organism. Vancomycin is active and may be adequate initial therapy whilst awaiting culture results.

In the postneurosurgical patient, initial therapy must cover Gram-negative organisms including coliforms, *Pseudomonas aeruginosa*, and skin flora, as well as community-acquired organisms. Broad antimicrobial coverage is essential in these patients. One possible combination is vancomycin, ceftazidime, and an aminoglycoside.

At present there are no data suggesting that quinolones or extended-spectrum macrolides are useful for empirical therapy of bacterial meningitis. However, early studies with the carbapenem meropenem look promising as single-agent therapy for a wide range of pathogens, including those causing bacterial meningitis.

The duration of antimicrobial therapy depends on the age and immune status of the patient, the etiological agent, and the clinical course or development of complications. There is no universally accepted standard. As little as 7 days of therapy or less would be appropriate for meningococcal meningitis. In Hib meningitis 10 days is the conventional wisdom, while for pneumococcal meningitis it is 14 days. Meningitis due to *L. monocytogenes* should be treated for 14 days, extending to 21 days in the immunocompromised host.

Duration of therapy may need to be prolonged because of complications such as development of brain abscess or subdural empyema, prolonged fever, or development of nosocomial superinfection.

Anti-inflammatory treatment

Unfortunately, even with appropriate antimicrobial therapy, the risk of morbidity and mortality from bacterial meningitis remains high. The evidence that the damaging process within the brain is mediated by activation of host inflammatory pathways, which is triggered by release of endotoxin and other bacterial constituents, often accentuated by use of powerful bactericidal antibiotics, has led to the hypothesis that injury to the brain may be reduced by anti-inflammatory treatment. In experimental animals, the inflammatory process can be reduced using antiendotoxins (e.g. polymyxin B), blockers of endotoxin binding to macrophages (e.g. bactericidal permeability increasing factor or anti-CD14 antibodies), antibodies against cytokine mediators (e.g. interleukin 1β or tumor necrosis factor-α), antibodies against neutrophil adhesion molecules, or agents which inhibit neutrophil and macrophage activation (e.g. steroids, pentoxifylline, or non-steroidal anti-inflammatory agents) (Feigin 1992).

Many anti-inflammatory agents have proved beneficial in animal experiments, and convincing evidence for a beneficial effect from steroids in humans has now emerged. Initial studies conducted in the United States demonstrated clear benefit from the use of dexamethasone in reducing the severity of neurological sequelae, particularly nerve deafness. Subsequent studies in Costa Rica, Egypt, and Switzerland have all reported similar results. Most patients in these studies have suffered from Hib or *Strep. pneumoniae* meningitis. Firm conclusions as to the efficacy of dexamethasone in reducing neurological damage in meningococcal meningitis cannot be drawn from these studies. However, the pathophysiological events occurring in meningococcal meningitis are unlikely to differ significantly from those seen in other forms of bacterial meningitis.

The benefit of dexamethasone appears greatest if it is administered early in the course of the illness, preferably prior to antibiotic administration. Current recommendation is to administer dexamethasone 0.15 mg/kg every 6 h for 4 days, starting prior to, simultaneously with, or within 4 h of the first dose of antibiotics (Tunkel and Scheld 1995).

Few side-effects have been documented in patients receiving dexamethasone. In particular there are no reports of delayed cerebrospinal fluid sterilization or treatment failure, although gastrointestinal bleeding is seen in a small proportion of patients.

The early reports of a beneficial effect of dexamethasone should be interpreted with caution. Steroids may downregulate important host anti-inflammatory mediators such as interleukin 10, and this may be potentially harmful. Other studies of patients with bacterial meningitis where appropriate therapy is delayed suggest a worse outcome in patients treated with dexamethasone.

Many other anti-inflammatory agents are likely to be beneficial in reducing central nervous system injury, but there have been no clinical studies on which to base recommendations for their routine use.

Supportive care

There has been increasing recognition that antibiotic administration is only one component of management of patients with meningitis. Neurological derangement often coexists with circulatory insufficiency, impaired respiration, metabolic derangement, and convulsions. Measures to detect and correct any coexisting physiological derangement are important in improving prognosis.

Management of raised intracranial pressure

Elevated ICP is almost universal in bacterial meningitis. Signs of raised ICP include altered level of consciousness, altered pupillary responses, hyper- or hypotension, reduction in resting pulse rate, and altered respiratory pattern. Papilledema is a late sign of raised ICP. Raised ICP should be suspected in any patient with a severely depressed level of consciousness and measures should be instituted in order to prevent brainstem compression and herniation.

Simple measures to reduce ICP include nursing the patient in a quiet environment in a head-up position at 20° to 30° from horizontal. Other interventions include the use of osmotic agents, fluid restriction, and control of cerebrovascular tone through manipulation of arterial CO_2 concentration.

Patients with bacterial meningitis may have been vomiting or have had reduced fluid intake in the days preceding admission. Further fluid restriction may impair circulating volume and further reduce cardiac output. In the past, emphasis on the possibility of inappropriate ADH secretion (SIADH) has led to the practice of fluid restriction in all patients with meningitis, even in the face of severe hypovolemia. Recent studies indicate that increased levels of ADH, which are seen in individuals with meningitis, represent an appropriate response to dehydration. With adequate rehydration the level returns to normal.

Correction of hypovolemia improves cardiac output and may have a beneficial effect on cerebral blood flow. In patients with incipient shock, the use of inotropic agents as well as colloid infusions may be important in optimizing cerebral perfusion.

Sedation and control of convulsions

A patient with a significantly reduced level of consciousness due to bacterial meningitis should not be sedated, even if extremely irritable or combative. Irritability or combativeness may indicate hypoxia due to disturbed respiratory drive. The addition of hypnotic or tranquilizing agents may precipitate respiratory failure or respiratory arrest and a further rise in ICP. Simple analgesics or antipyretic agents alone should be used in those patients who are not critically ill. Patients who require endotracheal intubation and mechanical ventilation should receive a combination of drugs to provide analgesia, amnesia, and sedation, occasionally with a muscle relaxant.

Barbiturates have been used to treat refractory raised ICP. However, large doses of barbiturates may impair cardiac output and should only be used in patients with cardiovascular stability who should be carefully monitored. Thiopental (thiopentone) is particularly useful for induction of anesthesia prior to endotracheal intubation in patients with raised ICP.

Seizures occur within 48 h of presentation in 20 to 30 per cent of patients with bacterial meningitis. Seizures are particularly dangerous in patients with raised ICP as they cause extreme metabolic demands and increased cerebral blood flow, and may precipitate a further rise in ICP. Convulsions may be difficult to detect in patients who are pharmacologically paralysed for mechanical ventilation. In such patients cerebral function monitoring should be used to detect seizure activity. The use of anticonvulsant treatment in non-ventilated patients may precipitate respiratory arrest; therefore careful observation of ventilation should be undertaken during the treatment of seizures. Short-acting anticonvulsants (e.g. diazepam or paraldehyde) can be used to control acute seizures, and barbiturates or phenytoin are generally used for longer-term control (Berkowitz *et al.* 1996).

Complications

Complications of bacterial meningitis vary according to the etiological organism, the duration of symptoms prior to initiation of appropriate therapy, and the age and immune status of the patient.

Temporary problems include development of hemodynamic instability and disseminated intravascular coagulopathy, particularly in meningococcal infection, SIADH or other dysregulation of the hypothalamic–pituitary axis (e.g. diabetes insipidus), and an acute increase in ICP.

Focal neurological signs may develop in the early stages of meningitis, but are more common later. Although the mechanisms described above, including vasculitis and thrombosis, may explain these clinical findings, awareness of conditions that may require neurosurgical intervention is necessary. These include development of subdural empyema, brain abscess, and acute hydrocephalus. Subdural effusions are more common after Hib meningitis, but can occur with any organism. They usually resolve spontaneously, but the presence of significant and persistent neurological symptoms, including seizures, paresis, raised ICP, and development of empyema, are indications for drainage.

Cerebral abscess must also be considered in any child who deteriorates neurologically, usually following the acute phase of bacterial meningitis, and is often accompanied by persistent fever. Other causes of focal neurology include the development of ischemic areas or infarction, caused by vasculitis, vascular spasm, or venous thrombosis.

Duration of fever varies. Fever persisting beyond day 10 is considered prolonged, whereas a new fever following defervescence for 24 h is considered to be secondary or recurrent. Duration of fever depends on the etiological agent. After 5 days of appropriate therapy, more than 85 per cent of patients with pneumococcal or meningococcal meningitis will be afebrile, whereas only 68 per cent will be afebrile with Hib meningitis. In addition, about 80 per cent of the latter will develop secondary fever. Most commonly, recurrent fever is due to a nosocomial infection such as thrombophlebitis, but subdural empyema, disseminated sepsis such as osteomyelitis, pericarditis, and endocarditis must be considered. It is unusual for fever to be caused by persistence of the organism within the meninges. However, with the emergence of drug-resistant organisms it cannot be assumed that persistent or recurrent fever is not due to continued presence of live bacteria within the cerebrospinal fluid.

Prevention

Chemoprophylaxis is used to prevent transmission and development of invasive meningococcal and Hib infection in susceptible individuals who are close contacts of the index case. Fifty per cent of the increased risk of meningococcal infection in household contacts occurs within 7 days of the presentation of the index case. Half of this occurs in the first 48 h. Therefore decisions regarding institution of chemoprophylaxis should not be delayed by waiting for microbiological confirmation.

Fear and anxiety within the family, school, or community frequently follows the occurrence of a case of meningococcal disease. Close collaboration between physicians caring for the affected patient and those responsible for community and public health advice is always essential. Rifampin (rifampicin) should be administered to all household contacts as soon as possible after the index case has been diagnosed. Ciprofloxacin or ceftriaxone are alternative, although less widely used, prophylactic agents.

Conclusions

Recent years have seen enormous advances in our understanding of the pathophysiology of bacterial meningitis. This has lead to improvements in therapy, both antibacterial and supportive. However, new challenges, such as the emergence of multidrug-resistant bacteria, will inevitably mean that the intensive care clinician will continue to be faced with the management of this potentially life-threatening infection.

References

Berkowitz, I.D., Berkowitz, F.E., Newton, C., Willoughby, R., and Ackerman, A.D. (1996). Meningitis, infectious encephalopathies and other central nervous system infections. In *Textbook of pediatric intensive care* (ed. M.C. Rogers). Williams & Wilkins, Baltimore, MD.

Cartwright, K. (ed.) (1995). *Meningococcal disease*. Wiley, Chichester.

Feigin, R.D. (1992). Bacterial meningitis beyond the neonatal period. In *Textbook of pediatric infectious diseases* (3rd edn) (ed. R.D. Feigin and J.D. Cherry), pp. 421–9. W.B. Saunders, Philadelphia, PA.

Segretti, J. and Harris, A.A. (1996). Acute bacterial meningitis. *Infectious Disease Clinics of North America*, **10**, 797–809.

Tunkel, A.R. and Scheld, W.M. (1995). Acute bacterial meningitis. *Lancet*, **346**, 1675–80.

7.7 Encephalitis

7.7.1 Pathophysiology and causes

R. P. Paczynski and D. B. Clifford

Key messages

- The acute encephalitis syndrome features a triad of fever, head pain, and acute neurological dysfunction.
- Herpes simplex encephalitis is the most common form of lethal non-epidemic encephalitis. Morbidity and mortality are high. Pharmacotherapy is available, but is only effective if begun promptly.
- Parasitic infections remain important treatable causes of acute encephalitis in the developing world. Central nervous system toxoplasmosis is common in the immunosuppressed.
- Epidemic arboviral and rabies encephalitis remain important causes of morbidity and mortality worldwide. Management is limited to supportive care.
- Mechanisms of secondary brain injury superimpose hyoxia–ischemia on the primary infectious–inflammatory process.

Introduction

Encephalitis occupies a position of increasing importance owing to several developments:

(1) wider availability of effective pharmacotherapy against several important pathogens;

(2) the increasing prevalence of encephalitic disease in the setting of immunosuppressive syndromes;

(3) greater recognition of the epidemiological importance of encephalitis in the developing world.

The 'acute encephalitis syndrome' features a triad of fever, head pain, and neurological dysfunction.

General pathophysiology

The term encephalitis encompasses a diverse spectrum of infectious and non-infectious illnesses that range in clinical severity from the benign to the rapidly lethal. Inflammation of neural parenchyma is the cardinal pathological feature, and demonstration of pleocytosis on examination of cerebrospinal fluid is an essential step in confirming the diagnosis in the appropriate clinical setting. When encephalitis occurs in concert with inflammation of the coverings of the brain or its myelinated neural tracts, the terms meningoencephalitis and encephalomyelitis respectively are appropriately applied. The neurological damage resulting from encephalitis may result from three general mechanisms: direct cytotoxic effects of specific pathogens, damage resulting from the host's immunological response, and non-specific secondary injury mechanisms such as cerebral edema and vascular obstruction. In large part secondary injury mechanisms exacerbate neurological damage by superimposing hypoxia–ischemia. Most encephalitides are infectious illnesses. As with infectious disease in general, the spectrum of causal agents will depend on the socio-economic and geographic milieu. For example, in the developing world, primary central nervous system infections or systemic infections with prominent central nervous system involvement dominate neurological illness in general, particularly among adolescents and young adults. Parainfectious (or postvaccinial) encephalitis and paraneoplastic encephalitis are rare conditions, characterized by diffuse neurological damage, that are believed to result from unusual manifestations of the immunological response to foreign antigens or other related mechanisms.

Viral encephalitis

A wide variety of viruses cause encephalitis (Table 1). Viral pathogens produce damage or destruction of neurons, glia, and to a lesser extent cerebrovascular elements by establishing lytic infections of individual cells. After appropriating the host cell's metabolic machinery, viruses replicate rapidly and are released into the extracellular space through membrane rupture or the less catastrophic process of persistent budding from host cell membranes over variable periods of time. Viral antigens and the spilled contents of the host cell are potent chemoattractants which initiate an inflammatory response. Many viral pathogens establish latency in extra-axial neural tissues such as peripheral nerves (e.g. varicella zoster) or cranial ganglia (e.g. herpes simplex). In addition to lytic infection and establishment of extra-axial latency, some viruses are capable of producing a persistent infection within the central nervous system that leads to more gradual destruction of neurons or compromise of optimal cell function.

Aseptic meningitis

Aseptic meningitis is the term used to describe the common acute or subacute syndrome which occurs in association with a variety of viral upper respiratory infections or episodes of gastroenteritis. Typical clinical features are fever, mild nuchal rigidity, malaise, and mild clouding of sensorium. Cerebrospinal fluid most commonly bears a mild to moderate lymphocyte-predominant pleocytosis. Frequent causes are enterovirus, adenovirus, the paramyxovirus which causes mumps, and, increasingly, HIV. Rarely, a true encephalitis (more severe alterations of sensorium, seizures, focal neurological deficits) or an encephalomyelitis may occur in association with these usually benign infections. Management in the intensive care setting is rarely indicated, and treatment is symptomatic.

Herpes simplex encephalitis

The herpes family (herpes simplex I and II, cytomegalovirus, varicella zoster, Epstein–Barr) deserves special emphasis because of both the high rates of morbidity and mortality associated with these infections and the availability of effective pharmacotherapy for some. All herpes

Table 1 Viral encephalitides

Disease entity	Comment
DNA-containing viruses	
Herpes group	
Herpes simplex I	Most common form of non-epidemic encephalitis in adults; high morbidity and mortality
Herpes simplex II	A major cause of neonatal ecephalitis; also causes encephalitis in immunodepressed adults; may also cause GBS or myelitis
Cytomegalovirus	May cause encephalitis in association with HIV infection
Varicella zoster	Rare cause of encephalitis in association with shingles
Epstein–Barr	Infectious mononucleosis complicated by CNS involvement (<1%); may cause GBS, cranial neuropathies, myelitis
Papovavirus	Causes progressive multifocal leukoencephalopathy in immunodepressed patients
Poxvirus	Vaccinia (virtually eradicated) infrequently complicated by CNS inflammation
Adenovirus	Multiple adenovirus serotypes infrequently associated with encephalitis; more typically cause benign aseptic meningitis syndrome
RNA-containing viruses	
Arbovirus (togavirus)	
Equine encephalitis (eastern, western, Venezuelan) St Louis encephalitis (flavivirus) Japanese encephalitis (flavivirus) Tick-borne encephalitis Rubella	Eastern equine encephalitis associated with highest morbidity and mortality; St Louis and Japanese encephalitis similar serologically and are most common causes of epidemic encephalitis in USA and Asia respectively; most CNS pathology occurs with congenital rubella syndrome
Arbovirus (bunyavirus)	
California encephalitis	Serogroup includes the LaCrosse virus; low morbidity and mortality; common cause of encephalitis in midwest USA
Rhabdovirus (lyssavirus)	
Rabies	May manifest as an acute 'limbic' encephalitis or more rarely as an acute paralytic syndrome; frequently fatal; still common in Asia and southeastern Europe
Picornavirus ('small' (pico) RNA viruses)	
Poliovirus	Acute poliomyelitis may also manifest encephalitis
Coxsackie virus Echovirus Enterovirus 70, 71, 72	Coxsackie B and various echovirus serotypes are the most common causes of the rare encephalitic forms of aseptic mengingitis; several enterovirus serotypes are rare causes of polio-like syndromes or encephalitis
Arenavirus	Lymphocytic choriomeningitis; virus carried by mice; low morbidity; other arenaviruses cause hemorrhagic fevers
Orthomyxovirus and paramyxovirus	Causes of measles and mumps respectively; infection infrequently complicated by acute meningoencephalitis syndrome or persistent infection (subacute sclerosing panencephalitis)
Retrovirus	HIV important cause of chronic CNS infection (AIDS dementia syndrome); acute infection with HIV may cause aseptic meningitis

CNS, central nervous system; GBS, Guillain–Barré syndrome.

group viruses contain double-stranded DNA, and drugs of proven efficacy specifically target the viral encoded thymidine kinase enzyme that catalyzes viral DNA replication.

Herpes simplex encephalitis is the most common form of lethal non-epidemic encephalitis in the West, with an annual incidence between 1 and 2 per 500 000 persons. Without therapy, mortality exceeds 70 per cent and only about 10 per cent of survivors will regain normal neurological function. In children aged less than 6 months, adolescents, and adults, herpes simplex encephalitis is a focal brain infection, typically localized to one or both temporal lobes or somewhat less frequently the orbitofrontal and mesiofrontal (limbic) regions. In contrast with the neonatal form, which is commonly caused by herpes simplex type II, 'adult' herpes simplex encephalitis is almost always caused by herpes simplex type I and may be associated with primary infections or reactivations. In adults, herpes simplex virus type II central nervous system infections are generally seen in immunodepressed individuals and present as a viral meningitis or, less commonly, as an encephalomyelitis.

The pathogenesis of herpes simplex encephalitis remains illusive. It is generally agreed that the route of viral entry into the central nervous system is via neural structures, in particular the trigeminal ganglia and the olfactory bulbs after viral penetration of the nasal mucosa. However, encephalitis generally occurs on reactivation of the virus at a later time (approximately two-thirds of cases). Once established, the progression of herpes simplex encephalitis is usually rapid and necrotizing. Cerebrospinal fluid is frequently hemorrhagic and the pleocytosis is mild to moderate with a predominance of mononuclear forms. Recent experimental work with herpes simplex encephalitis has emphasized the capacity of herpes simplex I to suppress facets of T-cell activation essential for effective immune surveillance without interfering with the generalized inflammatory response to tissue necrosis. As the infection spreads, the parenchymal inflammation and associated brain swelling are generally severe. Brain swelling reflects a combination of cytotoxic and vasogenic edema and cerebral hyperemia (vascular engorgement).

Cytomegalovirus encephalitis

Encephalitis associated with cytomegalovirus is rare in individuals with normal immune systems. In patients with iatrogenic immunosuppression (e.g. organ transplant recipients), cytomegalovirus may cause an acute and frequently fatal meningoencephalitis with a generalized inflammation of the brain. In patients with AIDS, cytomegalovirus may cause a more subtle encephalitic illness, pathologically focal or multifocal, with uncertain clinical correlates and natural history. A more aggressive subacute cytomegalovirus encephalitis seen in as many as 10 per cent of AIDS patients typically strikes the subependymal lining, causing a non-focal encephalitis associated with prominent mental status changes and cranial nerve abnormalities. Without intervention, progression to death in a matter of a few months is frequent. This syndrome may be mistaken for AIDS dementia complex, but it is more aggressive and is associated with high levels of cytomegalovirus DNA in the cerebrospinal fluid. Prognosis is poor and reflects the advanced state of AIDS-associated immunodepression typically present at the time of diagnosis. Therapeutic trials involving the nucleoside analog ganciclovir and the pyrophosphate analog foscarnet are under way.

Epidemic or arboviral encephalitis

A large number of small RNA viruses exist in the wild which alternate between blood-sucking insect vectors and a vertebrate host, most frequently birds or small mammals. Humans and horses are incidental hosts. Entry into the central nervous system occurs via hematological spread after a period of primary replication in subcutaneous tissues or muscle. Acute encephalitides associated with the arthropod-borne viruses (arboviruses) are highly variable in the extent to which they produce serious neurological damage. Higher morbidity and mortality have been consistently reported in association with eastern equine encephalitis (mortality rates over 50 per cent in some series), Japanese encephalitis, and St Louis encephalitis, in descending order of severity. Control of insect populations and containment of epizootics are the central focus of management. Reflecting its contemporary epidemiological importance, trials of vaccination against Japanese encephalitis are under way in several regions throughout Asia. Therapy for arboviral encephalitis is supportive, since no specific antiviral therapy is available for Japanese encephalitis. During the 1933 epidemic in St Louis, Missouri, St Louis encephalitis was associated with a mortality of 20 per cent. Mortality has been 2 to 12 per cent in subsequent epidemics, probably reflecting improvements in supportive care.

Rabies

Now rare in North America and Western Europe, rabies remains comparatively common in Asia and some portions of southeastern Europe. Over 50 000 cases of rabies are reported to the World Health Organization annually. Worldwide, infected canines remain the major vector of human rabies. It is caused by a large single-stranded RNA containing a rod-shaped (rhabdo) lyssa group virus that is transmitted to humans from the saliva through bites or skin abrasions. Passage to the brain occurs via axonal transport through sensory and motor nerve fibers without an intervening viremia. The incubation period may be only a few days in the case of severe bites, particularly on the scalp and face, and is typically lengthy when the site of entry is in the lower extremities. As with herpes simplex encephalitis, rabies may initially manifest as an infection of the limbic regions of the brain. Characteristic symptoms and signs are dramatic and include agitated delirium, extremely high fevers, hypersalivation, and violent bouts of hydro- and aerophobia (phobic spasms). Less commonly, rabies may manifest as a Guillain–Barré syndrome (paralytic or dumb rabies) with relative preservation of sensorium. Survival after establishment of rabies within the brain and the onset of symptoms is exceptional. Presently, the focus of management is immediate provision of both passive and active immunization to individuals exposed or likely to have been exposed to a rabid animal (Table 2). Research continues to focus on bolstering the characteristically deficient immune response to infection and the possible role of 'excitotoxicity' (hyperstimulation of central nervous system glutamate receptors) in the clinical manifestations of encephalitic rabies. Management remains essentially supportive and comfort oriented.

Parasitic encephalitis

The acute encephalitis syndrome may be associated with a wide variety of parasitic infections. With the exception of cerebral toxoplasmosis, parasitic encephalitis is rare in the United States and Western Europe. Encephalitides caused by tapeworms, nematodes,

Table 2 Rabies postexposure prophylaxis schedule*

Not previously vaccinated	Local wound cleansing	Immediate thorough cleansing with soap and water
	HRIG[†]	20 IU/kg body weight; up to half infiltrated around wound(s) and the rest given IM in gluteal region; do not exceed recommended dose (excess may suppress active production of antibody in response to vaccine)
	Vaccine	HDCV or RVA[‡] 1.0 ml IM in deltoid region (avoid gluteal region or overlap with HRIG injection site), once each on days 0, 3, 7, 4, and 28
Previously vaccinated	Local wound care	As above
	HRIG	Not to be administered
	Vaccine	HDCV or RVA 1.0 ml IM (deltoid) once on days 0 and 3

HDCV, human diploid cell vaccine; HRIG, human rabies immune globulin; IM, intramuscularly; RVA, rabies vaccine, adsorbed.

*Recommendations apply to USA.

[†]Efficacious regimens involving purified antirabies sera of equine origin have been used where HRIG is not available.

[‡]Purified VERO cell rabies vaccine (Pasteur–Merieux) 0.1 ml IM at two different sites on days 0, 3, and 7, followed by one 0.1-ml booster on days 30 and 90 may also be used.

Data from the Centers for Disease Control (1991).

Table 3 Parasitic infections causing acute encephalitis syndrome

Disease entity	Comment
Protozoan	
Toxoplasmosis	Common cause of acute or subacute encephalitis in immunodepressed patients (particularly with AIDS); presentation as mass lesion common; sulfadiazine and pyrimethamine treatment of choice
Amebic encephalitis	Rare cause of acute encephalitis; free-living amoebae (*Naegleria* and *Acanthamoeba*) and *Entamoeba histolytica*; often rapidly lethal; effective treatment with amphotericin B or metronidazole has been described
Trypanosomiasis	'Sleeping sickness' and Chagas disease; CNS symptoms typically slowly progressive; acute encephalitic episodes may punctuate chronic course
Cerebral malaria	Acute or subacute onset of cerebral symptoms; however, CNS pathophysiology distinct from encephalitis: capillary occlusion, vasculitis, and microhemorrhages
Helminthic	
Cysticercosis	*Taenia solium* (pork tapeworm) is a common cause of acute or subacute encephalitis in developing world; infection via oral–fecal spread of ova; multifocal encystment of adult worm within CNS typical; may present with hydrocephalus, stroke-like syndrome, or status epilepticus; albendazole and praziquantel effective against cysts
Schistosomiasis	*Schistosoma japonicum* a rare (3%–5% of schistosomiasis cases) cause of fulminating meningoencephalitis; focal pathology includes granulomatous tumors; praziquantel drug of choice
Trichinosis	Infection with the roundworm *Trichinella spiralis* complicated by acute focal or multifocal CNS involvement in about 10%–15% of cases; corticosteroids and thiabendazole moderately effective
Echinococcus	Infection with tapeworm parasite *Echinococcus granulosus* causes focal CNS lesions (hydatid cysts) and encephalitis in about 2% of cases; albendazole or praziquantel may shrink cysts; corticosteroids useful in suppressing inflammation provoked by degenerating tapeworms
Paragonimiasis	Infection with lung flukes *Paragonimus westermani* and *Paragonimus mexicanus* are complicated by acute meningoencephalitis or chronic granulomatous infection in 10%–15% of cases; 'soap bubble' calcifications on CT scan of brain; praziquantel and corticosteroids may be useful
Strongyloidiasis	Intestinal infection with the nematode *Strongyloides stercoralis* may occasionally cause a disseminated infection with encephalitis in the immunodepressed patient; concurrent Gram-negative bacteremia common; thiabendazole effective against the organism
Angiostrongylus	*Angiostrongylus cantonensis* or rat lungworm not uncommonly produces hemorrhagic encephalitis or radiculomyelitis in Asia, most often in children; peripheral eosinophilia is common

CNS, central nervous system.

and schistosomes remain important causes of morbidity in populations throughout the developing world (Table 3). In most of these diseases, however, the burden of parasitic involvement is outside the central nervous system. Because effective pharmacotherapy is available against many parasitic infections, it is always important to consider this possibility in patients presenting with an acute encephalitis syndrome who may have risk factors for parasitic infestation.

Toxoplasmosis is caused by the protozoan *Toxoplasma gondii*, and is one of the most frequent opportunistic central nervous system infections in patients with AIDS. After ingestion of raw meats or soil contaminated with protozoan oocysts (typically from feline fecal material), organisms are disseminated hematogenously, invade cerebral and retinal blood vessels, and form miliary granulomas. Most cases of central nervous system toxoplasmosis are believed to represent reactivation of a latent infection. In the immunocompromised host, recrudescent parasites replicate unchecked. The most common clinical presentation is that of an awake patient, with or without fever or evidence of systemic illness, presenting with signs of a focal mass lesion (single or multiple toxoplasmic abscesses). Early initiation of specific pharmacotherapy can result in dramatic improvement in the neurological picture. Up to 80 per cent will respond to a combination of pyrimethamine and sulfadiazine. Chronic pharmacological suppression of the parasitic infection is required.

Neurocysticercosis is caused by encystment of larvae of the pork tapeworm *Taenia solium* in brain tissue. It is a very common central nervous system infection throughout India, Southeast Asia, China, Central and South America, Mexico, and to a lesser extent the southwestern United States. Central nervous system involvement occurs in the majority of cases, and most commonly presents with focal seizures or a circumscribed neurological deficit. A more dramatic acute presentation featuring stupor or coma may result from the development of obstructive hydrocephalus or rupture of a cyst, provoking an intense inflammatory response.

Acute disseminated encephalomyelitis

Acute disseminated encephalomyelitis is an often devastating acute illness that occurs infrequently during the course of childhood viral exanthems, following vaccinations, or less commonly in the aftermath of benign viral infections during adulthood. It is believed to reflect the consequences of an aberrant immunological response to infection, rather than the direct effects of pathogens on central nervous system tissue. An immune response mounted against autoantigens (e.g. myelin basic protein associated with damaged central nervous system white matter) with structural similarities to foreign antigens may explain many cases of acute disseminated encephalomyelitis. Unlike the viral and parasitic acute encephalitides described above, pathological involvement is focused in white matter regions deep within the brain. In fulminant forms, intense inflammation and diffuse microhemorrhages may occur (acute hemorrhagic leukoencephalitis). Worldwide, measles virus appears to be the most consistent precipitant of acute disseminated encephalomyelitis, with an incidence of approximately one case per thousand infections. However, postvaccinial acute disseminated encephalomyelitis is very rare, occurring at a rate of less than one case per million recipients. These facts alone provide adequate justification for the continuation of vigorous vaccination programs against measles. In Western populations, where measles is now comparatively rare, acute disseminated encephalomyelitis is observed sporadically in association with varicella and non-specific upper respiratory infections. Mortality from acute disseminated encephalomyelitis is generally high (20–30 per cent), with many of the deaths resulting from intercurrent illness during the typically lengthy period of recovery, but long-term survivors may make remarkably full functional recoveries.

Bibliography

Arribas, J.R., Storch, G.A., Clifford, D.B., and Tselis, A.C. (1996). Cytomegalovirus encephalitis. *Annals of Internal Medicine*, **125**, 577–87.

Ellner, J.J. (1985). Central nervous system infections in the intensive care unit. In *Handbook of critical care neurology and neurosurgery* (ed. R.J. Henning and D.L. Jackson), pp. 241–59. Praeger, New York.

Esiri, M.M. and Kennedy, P.G.E. (1992). Virus diseases. In *Greenfield's textbook of neuropathology* (5th edn) (ed. J. Hume Adams and L.W. Duchen), pp. 335–48. Oxford University Press, New York.

Harrison, M.J.G. and McArthur, J.C. (1995). *AIDS and neurology*. Churchill Livingstone, Edinburgh.

Jubelt, B. and Miller, J.R. (1989). Viral infections. In *Merritt's textbook of neurology* (8th edn) (ed. L.P. Rowland), pp. 96–119. Lea & Febiger, Philadelphia, PA.

7.7.2 Therapeutic approach

R. P. Paczynski and D. B. Clifford

Key messages

- Elevated intracranial pressure (ICP) forebodes a poor prognosis. Maintenance of optimal cerebral perfusion is the central goal of mangement in patients with brain swelling and elevated ICP.
- Sustained hyperventilation should be avoided in the management of elevated ICP.

Introduction

Herpes simplex encephalitis is used as a model to define the general approach to the evaluation and management of life-threatening acute encephalitis. Aspects of medical management particular to specific forms of viral and parasitic infection are presented in Tables 1 and 2.

The importance of early diagnosis

Institution of antiviral therapy with acyclovir (aciclovir) can dramatically reduce the mortality and morbidity associated with herpes simplex encephalitis, but only if started early. The prognosis is poor once the patient with this typically rapidly progressing infection reaches the stage of stupor and coma. Therefore there is an abundant literature focusing on the early diagnosis of herpes simplex encephalitis.

Typical clinical presentation includes signs and symptoms of focal neurological impairment or personality change in the setting of an acute or subacute febrile illness. However, focal neurological abnormalities such as partial seizures or hemiparesis do not occur in a majority of patients, and lateralized signs may be subtle and/or

Table 1 Pharmacotherapy for herpes group encephalitis

Treatment	Comments
Herpes simplex encephalitis	
Acyclovir 10 mg/kg every 8 h intravenously for 10–14 days (infuse at constant rate over about 1 h)	Oral acyclovir does not penetrate into CNS well Base dosing on lean body weight Adequate hydration essential in avoiding crystallization of drug in urine/renal toxicity If diagnosis uncertain or response to empirical therapy poor, consider brain biopsy
CMV encephalitis	
Ganciclovir 5 mg/kg twice daily induction Ganciclovir 5 mg/kg daily maintenance (7.5 mg/kg for relapses) or Foscarnet 90 mg/kg twice daily induction (infuse no longer than 1 mg/kg/min) Foscarnet 90 mg/kg daily maintenance (adjust for renal function)	Both effective against CMV retinitis; clinical trials for encephalitis under way Drug precipitation and hypocalcemia may occur with faster rates

CMV, cytomegalovirus; CNS, central nervous system.

transitory. Fever, head or neck pain, and altered mentation are almost invariably present; however, these features are non-specific, and very early in its course herpes simplex encephalitis may be indistinguishable from common viral infections with features of aseptic meningitis. Herpes simplex encephalitis presentation also overlaps a variety of conditions that can cause focal neurological impairment and signs of systemic illness, such as bacterial abscess, stroke from septic emboli, central nervous system vasculitis, cerebral venous thrombosis, or a parameningeal source of bacterial or fungal infection. Thus the major challenge facing the clinician encountering symptoms and signs generally suggestive of herpes simplex encephalitis is to maintain a high index of suspicion and a low threshold for following through with investigations that will confirm or refute the diagnosis.

Characteristic paraclinical features include periodic lateralized epileptiform discharges (PLEDs) on electroencephalography, which are virtually pathognomonic if bilateral (BiPLEDs). Magnetic resonance imaging (MRI), particularly using the T_2-weighted spin echo technique, may reveal the typical mesial temporal and frontal localization of herpes simplex encephalitis very early in the clinical course. High-resolution CT and nuclear medicine scans are less sensitive to lesions and have considerably less spatial resolution than MRI. Clinical signs of peripheral infection with herpes simplex infrequently accompany herpes simplex encephalitis, and only a minority (10–15 per cent) of patients will have recoverable virus shed from mucous membranes. Polymerase chain reaction assays for viral DNA in cerebrospinal fluid have superseded serologic analyses. Where available, a polymerase-chain-reaction-based assay of cerebrospinal fluid is the diagnostic procedure of choice. When judged by the gold standard of brain biopsy, large contemporary series indicate sensitivity in excess of 95 per cent if cerebrospinal fluid specimens are processed promptly. Specificity appears to be 90 per cent or greater.

Acyclovir and brain biopsy

Since acyclovir is generally a safe drug, empirical administration in cases of suspected herpes simplex encephalitis is justified on clinical grounds or when polymerase chain reaction diagnostic tests are equivocal or unavailable. The role of brain biopsy in establishing the diagnosis of herpes simplex encephalitis remains controversial. Advocates of routine biopsy for patients with suspected herpes simplex encephalitis point out the lack of pathognomonic clinical features in early herpes simplex encephalitis, which frequently create

Table 2 Therapy for parasitic encephalitis

Treatment	Comments
Cerebral toxoplasmosis	
Pyrimethamine 200 mg loading dose, then 50–75 mg/day orally plus sulfadiazine 1–1.5 g every 6 h plus leucovorin supplement 10 mg/day or Clindamycin 600 mg three times daily orally + pyrimethamine	Chronic drug therapy required Single-agent therapy less effective Leucovorin required to prevent vitamin deficiency
Neurocysticercosis	
Parenchymal neurocysticercosis Albendazole 15 mg/kg/day per 1 month or Praziquantel 50 mg/kg/day for 2 weeks Subarachnoid neurocysticercosis Steroids and/or immunosuppressants then albendazole 15 mg/kg/day for 1 month then use niclosamide 2 g orally once if concurrent intestinal infection suspected	Concomitant treatment with corticosteroids recommended for control of inflammation if pericyst swelling severe Praziquantel more expensive than albendazole, but 2 weeks of therapy must be adequate Anticonvulsant regimen may require modification during drug therapy If accompanied by hydrocephalus, ventricular shunt should be placed before initiation of drug therapy Repeat course of antiparasitics if response is inadequate Children should receive 1–1.5 g oral niclosamide, dose gauged to body weight

diagnostic confusion, and that acyclovir therapy is not completely without risk, particularly with regard to acute renal toxicity. However, the increasing availability of polymerase chain reaction assays for viral DNA makes these arguments less persuasive. Acyclovir therapy can be suspended if the subsequent clinical course suggests an alternative diagnosis or polymerase chain reaction analysis of cerebrospinal fluid is negative. Brain biopsy is indicated when compelling clinical or radiographic evidence suggests an alternative diagnosis, or the patient continues to deteriorate despite appropriate acyclovir therapy.

Supportive management

Although early delivery of acyclovir results in improved outcome from herpes simplex encephalitis, morbidity and mortality remain high. Patients who present for medical attention with stupor or coma have a mortality rate of approximately 25 per cent and a very poor prognosis for good functional recovery whether or not acyclovir is administered. A substantial number of these individuals succumb to secondary brain injury. Factors likely to produce additional morbidity by secondary injury mechanisms include severe cerebral edema with attendant elevation of intracranial pressure (ICP), incarceration of major vessels (e.g. posterior cerebral artery at the tentorial hiatus), and compromised oxygen delivery to the brain secondary to hemodynamic and/or pulmonary dysfunction. A consistent program of medical management aimed at reducing secondary injury may improve an individual's chances of good functional recovery. In addition to the foundations of supportive care necessary for any critically ill patient, a number of specific points deserve additional emphasis.

Airway and ventilatory management

Severely lethargic or stuporous/comatose patients should be electively intubated and mechanically ventilated to protect the airways from aspiration, assure optimal oxygenation, and prevent retention of carbon dioxide. Hypercarbia can result in cerebral hyperemia, aggravation of elevated ICP, and abrupt clinical deterioration in the marginally compensated patient. Hasty intubation under crisis conditions is a common but often preventable contributor to morbidity. Use of short-acting sedative premedication, oronasal topical anesthesia, adequate preoxygenation, and gentle laryngoscopic technique may prevent surges in ICP during the process of intubation. Nasotracheal intubation may be the route of choice in some cases.

Agitation and disordered ventilation secondary to bilateral cerebral or brainstem involvement (particularly Cheyne–Stokes respiration, episodic hyperventilation) are common in viral encephalitis, particularly herpes simplex encephalitis and rabies. However, pharmacological suppression of abnormal breathing patterns or behavior should not generally not be practised unless there is compromised gas exchange or increased work of breathing represents a significant physiological stress in an individual patient. Too liberal use of sedatives will obviously result in sacrifice of the neurological examination and should be undertaken only with appropriate justification, and then with caution. In non-intubated patents with severe agitation that threatens patient safety or the provision of care, short-term use of haloperidol, administered intravenously, is a particularly useful strategy because this agent generally does not suppress respiratory drive and has little adverse effect on hemodynamics.

Fluid and intracranial pressure management

The documented association of increased ICP with poor outcome from herpes simplex encephalitis may simply indicate that elevated ICP is a marker of disease severity rather than a specific mechanism of secondary injury requiring active management. Nevertheless, simple measures to reduce or prevent further elevations of ICP and prevent complications (brain ischemia) are central to supportive management. Recommended practices include the following.

1. Elevate the patient's head by 20° to 30° to facilitate venous return.
2. Maintain a mild fluid restriction (1.5–2 l/day for the average-sized adult) with thoughtful allowance for increased insensible losses associated with the febrile state.
3. Avoid hypotonic replacement fluids or feeding solutions, which may facilitate brain edema by osmotic mechanisms.
4. Arterial blood pressure management should center on maintenance of an adequate cerebral perfusion pressure, recognizing that normal autoregulation of cerebral perfusion may be impaired. Cerebral perfusion pressure is approximately equal to the difference between mean arterial pressure and ICP, and a level of 70 to 90 mmHg is a practical general goal in this setting.

Bolus administration of osmotic diuretics such as 25 per cent mannitol (0.25–1.5 g/kg intravenously) or 10 per cent glycerol (200–250 ml via slow intravenous infusion) are extremely effective in acutely reducing elevated ICP. In the case of mannitol, onset of effect is usually observed within a few minutes of administration, although peak effects may not occur for 20 to 30 min. Repeated administration of osmotic diuretics should be practised only with caution, and cerebral perfusion pressure management should always take precedence over treatment of isolated ICP values when the latter is monitored. Serial use of osmotic diuretics produces an insidious intravascular volume depletion, wasting of several key plasma constituents (particularly potassium, phosphate, and magnesium), and potential iatrogenic worsening of ICP secondary to a number of mechanisms.

Although hyperventilation is a reliable means of acutely lowering ICP, the effect is transient. Induced respiratory alkalosis rapidly results in reduction of cerebral blood volume through cerebral vasoconstriction; however, 'pH compensation' occurs in the brain within hours. Further, severe hyperventilation may produce cerebral ischemia secondary to marked vasoconstriction. Sustained hyperventilation offers no clear benefit and is not without risk (e.g. barotrauma, compromised venous return to the heart with reduction of cardiac output). For all these reasons, it is prudent to gauge ventilatory parameters to achieve normocarbia in most cases. Acute hyperventilation is then an option available to the clinician in the event of an 'ICP crisis'. In patients who have been hyperventilated, return to normocarbia should be accomplished gradually.

In cases where ICP is refractory to medical management, neurosurgical consultation should be considered. Acute obstructive hydrocephalus can complicate some forms of encephalitis, and ventriculostomy with metered drainage of cerebrospinal fluid can provide definitive control over ICP. Attempts to control refractory elevations of ICP with aggressive pharmacological measures such as 'barbiturate coma' are generally ineffective and can introduce new complications into the course of management. In selected cases, surgical decompression of the temporal lobe may be a lifesaving procedure that provides some hope for functional recovery. Serial monitoring of blood flow velocities in vulnerable vascular territories with transcranial Doppler ultrasonography may provide information to help with the fine tuning of fluid and hemodynamic management and with decisions about neurosurgical intervention. A trend of increasing velocities in a threatened vascular territory often indicates progressive vascular narrowing from compression.

Seizure control

Seizure activity frequently complicates encephalitis. Recurrent seizures can lead to permanent damage to vulnerable neurons, and generalized convulsive activity may exacerbate problems with ICP and produce systemic complications. The focus of initial management is prompt establishment of adequate levels of a long half-life parenteral anticonvulsant such as phenytoin, which should be given as a slow intravenous load (e.g. 18–20 mg/kg by slow intravenous infusion at a rate not exceeding 50 mg/min) followed by a maintenance dose of 3 to 5 mg/kg/day. Phenytoin is a particularly useful agent in the neurological intensive care setting because it is a relatively non-sedating anticonvulsant with minimal adverse effects on hemodynamics when administered properly. Seizure activity inadequately suppressed with phenytoin monotherapy can usually be managed by adding a second agent with complementary anticonvulsant properties such as a barbiturate or benzodiazepine. In addition to their general anesthetic and anticonvulsant properties, all barbiturates are myocardial depressants and vasodilators. Adverse hemodynamic sequelae are not uncommon. Because of these and other shortcomings of barbiturates, there has been more interest in recent years in the use of benzodiazepine infusions (e.g. midazolam or lorazepam) for the control of status epilepticus. Continuous EEG monitoring may be warranted in refractory cases in which more subtle (non-convulsive) seizure activity may be difficult to identify on clinical grounds.

Corticosteroids

When encephalitis becomes a life-threatening condition complicated by cerebral swelling and raised ICP, pharmacological doses of corticosteroids are a reasonable intervention. Steroid agents are particularly effective in retarding the formation of vasogenic edema, which is a probable source of pathological brain bulk in many forms of severe encephalitis. It should be noted that initiation of antiparasitic medication may cause an intensification of cerebral swelling as material from dying organisms provokes an inflammatory response. Dexamethasone (10 mg intravenously initially followed by 4–6 mg every 6 h intravenously or the same dose enterally) is the agent of choice. In cases of toxoplasma encephalitis, when a diagnostic/therapeutic trial of treatment is undertaken rather than a brain biopsy, it is important not to confuse interpretation of the therapeutic response by using corticosteroids.

Bibliography

Barnett, G.H., Ropper, A.H., and Romeo, J. (1988). Intracranial pressure and outcome in adult encephalitis. *Journal of Neurosurgery*, **68**, 585–8.

Bergen, D.C. (1996). The world-wide burden of neurologic disease. *Neurology*, **47**, 21–5.

Paczynski, R.P. (1997). Osmotherapy: basic concepts and controversies. *Critical Care Clinics of North America*, **13**, 105–29.

Whitley, R.J. and Lakeman, F. (1995). Herpes simplex virus infections of the central nervous system: therapeutic and diagnostic considerations. *Clinical Infectious Diseases*, **20**, 414–20.

Whitley, R.J., Cobbs, C.G., and Alford, C.A. (1989). Diseases that mimic herpes simplex encephalitis: diagnosis, presentation, and outcome. *Journal of the American Medical Association*, **262**, 234–9.

7.8 Acute weakness

7.8.1 Pathophysiology and causes of neuropathy

C. F. Bolton

Key messages

- Clinical assessment in the intensive care unit is often difficult and misleading.
- Neurological examination and electrophysiological testing of the peripheral nervous system are essential.
- Phrenic nerve conduction and needle electromyography are often valuable.
- Magnetic resonance imaging of the cervical spinal cord, measurement of serum creatine phosphokinase, and muscle biopsy may further establish the diagnosis.
- Conditions such as spinal cord compression and Guillain–Barré syndrome require specific treatment.
- Adequate investigation, as described above, invariably provides important information on prognosis.

Introduction

Neuropathies severe enough to require admission to an intensive care unit (ICU) may present in two main ways (Table 1). Firstly, limb and respiratory muscle weakness may develop rapidly over days in a sufficiently severe fashion to require intubation, ventilation, and management in an ICU (Zifko and Chen 1996). The main aim in management will initially be to stabilize the patient, but the cause of the weakness must be investigated while the patient is still in the ICU. Secondly, the patient may develop difficulty in weaning from the ventilator and variable limb weakness during management in the ICU after admission for other causes, i.e. severe systemic illness or trauma.

In both instances emergency evaluation is required, and the differential diagnosis should be approached by considering disorders of the spinal cord, peripheral nerves, neuromuscular junction, and muscle (Table 1). The investigations may involve magnetic resonance imaging to exclude neoplastic infectious or traumatic conditions involving the high cervical spinal cord, electrophysiological studies to investigate disorders of the peripheral nervous system, and measurements of creatine phosphokinase and possibly muscle biopsy to investigate for a primary myopathy. Accurate diagnosis is important because specific treatments may be instituted, such as immunosuppression for Guillain–Barré syndrome or myasthenia gravis. It is important in rendering a prognosis. Patients with amyotrophic lateral sclerosis will have a poor prognosis, but it is much better for patients with critical illness polyneuropathy if the sepsis and multiple organ failure can be brought under control.

Polyneuropathies presenting before admission to the ICU

The conditions listed in Table 1 are traditional, and are often given in standard texts. Guillain–Barré syndrome has recently been reviewed by Hahn (1996). Other polyneuropathies (Table 2), particularly amyotrophic lateral sclerosis, may present as acute respiratory failure (Zifko and Chen 1996).

Here, we emphasize common and important, but less well-recognized, neuropathies which develop after admission to the ICU (Tables 2 and 3).

Polyneuropathies presenting after admission to the ICU

These polyneuropathies usually present after days or weeks when patients have been managed in the ICU for a variety of primary illnesses or injuries. It will be noted that there is difficulty in weaning from the ventilator. After cardiac and lung conditions have been excluded, a neuromuscular condition should be suspected if, on attempted discontinuation of ventilation, respiration becomes rapid and there is a rise in blood CO_2. There may be variable signs of abdominal paradox (inward movement of the abdominal wall during inspiration) or respiratory alternans (alternation of rib cage and abdominal movement). Measurements of vital capacity, maximum inspiratory pressure, breathing frequency, and tidal volume may give inconclusive results. Unilateral damage to the phrenic nerve due to operative trauma is often undiagnosed, despite chest radiography and fluoroscopy.

Electrophysiological testing, measurements of creatine phosphokinase, and sometimes muscle biopsy are again necessary to investigate the problem thoroughly. Complications of the sepsis and multiple organ failure syndrome, now termed the systemic inflammatory response syndrome, will be the underlying cause for the neuromuscular condition in most instances (Table 3). This is an important phenomenon, since the incidence of this syndrome in ICUs may be as high as 50 per cent.

Table 1 Differential diagnosis in the ICU of the syndrome of rapidly developing weakness of limb and respiratory muscles

	Before admission to the ICU	After admission to the ICU
Disorders of the spinal cord	Traumatic myelopathy Acute epidural compression due to neoplasm infection Acute transverse myelitis Acute ischemia	
Acute polyneuropathies	Motor neuron disease Guillain–Barré syndrome Axonal forms of Guillain–Barré syndrome	Critical illness polyneuropathy
Chronic polyneuropathies	Chronic inflammatory demyelinating polyneuropathy Diabetic polyneuropathy	Chronic polyneuropathies plus sepsis
Neuromuscular transmission defects	Myasthenia gravis Lambert-Eaton myasthenic syndrome	Neuromuscular blockers
Myopathy	Muscular dystrophy (Duchenne, myotonic, etc.) Acute necrotizing myopathy (myoglobinuria)	Cachectic myopathy Necrotizing myopathy of intensive care Thick-filament myopathy

Critical illness polyneuropathy

This polyneuropathy (Zochodne *et al.* 1987; Leijten and de Weerd 1994; Bolton 1996) occurs in 70 per cent of patients who have the systemic inflammatory response syndrome. However, clinical signs are often absent and it remains an occult condition in many ICUs worldwide. Nonetheless, it is important to establish the diagnosis since it is an important cause of difficulty in weaning from the ventilator and problems in rehabilitation after the acute illness has been treated in the ICU.

Only in more severe critical illness polyneuropathy will there be obvious limb weakness and reduced deep tendon reflexes. Signs of sensory impairment are difficult to test in the ICU. Electrophysiological studies are essential to establish the diagnosis. These will reveal a primary axonal degeneration of first motor and then sensory fibers. Phrenic nerve conduction and needle electromyography of the diaphragm will disclose that the difficulty in weaning from the ventilator is due to involvement of the nerves and muscles of respiration (Zifko and Chen 1996). Creatine phosphokinase is either normal or mildly elevated. Muscle biopsy will reveal the presence of both acute and chronic denervation of muscle. Often the muscle is involved directly by the sepsis to varying degrees; this causes muscle necrosis, the severity of which is reflected in the degree of elevation of creatine phosphokinase.

If the sepsis and multiple organ failure can be successfully treated (the mortality remains as high as 50 per cent), recovery from the critical illness polyneuropathy can be expected. This will occur in a matter of weeks in mild cases and in months in more severe cases. Electrophysiological studies will provide valuable information as to the time it may take for successful weaning from the ventilator.

Axonal motor neuropathy associated with the use of neuromuscular blocking agents

A number of reports (Bolton 1996) have associated the use of neuromuscular blocking agents, particularly vecuronium or pancuronium bromide, with the development of polyneuropathy. These drugs are used to ease mechanical ventilation. The clinical features are similar to critical illness polyneuropathy. In our opinion, this is simply a variant of that condition, with sepsis as an important underlying factor. The practical implications are that these neuromuscular blocking agents should be used as sparingly as possible, and should be promptly discontinued if there are signs of a polyneuropathy. Like critical illness polyneuropathy, the prognosis for recovery is good.

The pathophysiology of the neuromuscular complications of systemic inflammatory response syndrome (sepsis)

Retrospective and prospective studies (Bolton 1996) have failed to incriminate a variety of potential causes of critical illness polyneuropathy, including types of primary illness or injury, Guillain–Barré syndrome, medications including aminoglycoside antibiotics and neuromuscular blocking agents, and specific nutritional deficiencies. We have speculated that sepsis is the cause (Bolton 1996). The severity of the polyneuropathy can be quantified from electrophysiological data. It tends to be more severe with time in the ICU and with increasing blood glucose and decreasing serum albumin concentrations. All these factors are recognized manifestations of the sepsis and multiple organ failure syndrome.

The microcirculation is disturbed in sepsis (Bolton 1996). Blood vessels supplying peripheral nerves lack autoregulation, rendering them particularly susceptible to such disturbances. Moreover, cytokines that are secreted in sepsis have histamine-like properties which may increase microvascular permeability (Zochodne *et al.* 1987). The resulting endoneural edema could induce hypoxia by increase in intercapillary distance and other mechanisms. Severe energy deficits would result and induce a primary axonal degeneration, most likely distally, if highly energy-dependent systems

Table 2 The main features of polyneuropathies that may be associated with ICU care

Type of polyneuropathy	Precipitating factors	Signs and symptoms	CSF	Microbiology and immunology	Electrophysiology	Morphology	Treatment	Course
Critical illness polyneuropathy	SIRS	Difficulty weaning from ventilator	Near normal	Non-specific	Evidence of axonal motor and sensory polyneuropathy	Primarily axonal degeneration of distal peripheral nerves	Treat SIRS	Recovery
Acute inflammatory demyelinating polyneuropathy, (ALDP)	*C. jejunii* Preceding diarrheal illness EBV, CMV, HIV Vaccinia Trauma	Symmetric progressive (days to 4 weeks) motor weakness and mild sensory loss, DTRS↓±autonomic features	Albumino-cytological dissociation may be normal early on	Serology IgA, IgM to C. jejuni anti-GM_1, anti-GD1B, stool culture	Demyelinating with conduction block prolonged distal latencies/F waves MU↓, FR↑±spontaneous activity	Primarily demyelinating process, secondarily axonal process	Plasmapheresis IVIg	75% complete recovery, 30% require ICU admission
Acute motor sensory axonal neuropathy (AMSAN)	*C. jejuni*	All features of AIDP+unusually severe course	As above	Serology as above	Unresponsive nerves, abundant spontaneous activity	Motor and sensory axonal	Plasmapheresis IVIg Efficacy	Worse than AIDP
Acute motor axonal neuropathy (AMAN)	*C. jejuni*	Features of AIDP motor only, early and severe respiratory problems; often children and young adults	As above	Serology as above	Motor CMAP↓, conduction block, MU↓, FR↑, SNAP normal	Motor axonal	Plasmapheresis IVIg ? Efficacy	Somewhat worse than AIDP

CMAP, compound muscle action potentials; CMV, cytomegalovirus; DTR, deep tendon reflexes; EBV, Epstein–Barr virus; FR, firing rate; IVIg, intravenous immunoglobulin; MU, motor units; SIRS, systemic inflammatory response syndrome; SNAP, sensory nerve action potentials.

Table 3 Neuromuscular conditions associated with the systemic inflammatory response syndrome in the ICU

Conditions	Incidence	Clinical features	Electromyography	Creatine phosphokinase	Muscle biopsy
Polyneuropathy					
Critical illness polyneuropathy	Common	Flaccid limbs and respiratory weakness	Axonal degeneration of motor and sensory fibers	Near normal	Denervation atrophy
Motor neuropathy	Common with neuromuscular blocking agents	Flaccid limbs and respiratory weakness	Axonal degeneration of motor fibers	Near normal	Denervation atrophy
Neuromuscular transmission defect					
Transient neuromuscular blockade	Common with neuromuscular blocking agents	Flaccid limbs and respiratory weakness	Abnormal repetitive nerve stimulation studies	Normal	Normal
Myopathy					
Thick-filament myopathy	Common with steroids, neuromuscular blocking agents, and asthma	Flaccid limbs and respiratory weakness	Abnormal spontaneous activity	Elevated	Central loss of thick filaments
Disuse (cachectic myopathy)	Common (?)	Muscle wasting	Normal	Normal	Normal or type 2 fiber atrophy
Necrotizing myopathy of intensive care	Rare	Flaccid weakness, myoglobinuria	Abnormal spontaneous activity in muscle	Markedly elevated	Panfascicular muscle fiber necrosis

Reproduced with permission from Bolton (1996).

involving axonal transport of structural proteins are involved. The predominantly distal involvement may explain why recovery time in some patients may be surprisingly short, conforming to the short length of nerve through which axonal regeneration takes place. It is also possible that cytokines themselves may have a direct toxic effect on peripheral nerve. To our knowledge, this as not been demonstrated in either humans or experimental animals. However, tumor necrosis factor decreases the resting transmembrane potential of skeletal muscle fibers *in vitro* and also induces muscle proteolysis in animals.

Disturbances of the microcirculation to nerve and muscle may also explain the effects of neuromuscular blocking agents and steroids. Neuromuscular blocking agents, notably vecuronium or its metabolite 3-desacetyl-vecuronium, could have a direct toxic effect on peripheral nerve axons through increased capillary permeability induced by the sepsis. These drugs may also cause functional denervation through their prolonged neuromuscular blocking action. The result would be denervation atrophy of muscle and a relatively pure motor neuropathy.

We have always been concerned that antibiotics, particularly aminoglycosides with their known neural toxicity, might cause critical illness polyneuropathy. They might gain access to the peripheral nerves as a result of increased capillary permeability. However, there has been no statistical proof that antibiotics cause peripheral nerve dysfunction in sepsis (Bolton 1996). Nonetheless, this possibility should be explored by fundamental experiments.

We suspect that many asthmatic patients who develop acute myopathy when they are treated with neuromuscular blocking agents and steroids suffer from systemic inflammatory response syndrome, since infection is often a precipitating event in acute severe asthma. Animal experiments (Bolton 1996) have shown that a thick-filament myopathy similar to that seen in humans can be induced if the muscle is first denervated by nerve transection and then steroids are given. Thus, in humans, critical illness polyneuropathy and the additional effects of neuromuscular blocking agents would denervate muscle and then steroids would induce the typical myopathic changes. The rapidly evolving myopathy characterized by destruction of thick filaments throughout the muscle fibers and the acute necrotizing myopathy of intensive care may simply represent further stages of this process (Bolton 1996).

References

Bolton, C.F. (1996). Sepsis and the systemic inflammatory response syndrome (SIRS): neuromuscular manifestations. *Critical Care Medicine*, **24**, 1408–16.

Hahn, A.F. (1996). The management of inflammatory neuropathies. In *Baillière's clinical neurology: critical care in neurology* (ed. C.F. Bolton and G.B. Young), pp. 627–44. Baillière Tindall, London.

Leijten, F.S.S. and de Weerd, A.W. (1994). Critical illness polyneuropathy: a review of the literature, definition and pathophysiology. *Clinical Neurology and Neurosurgery*, **96**, 10–19.

Zifko, U. and Chen, R. (1996). The respiratory system. In *Bailliere's clinical neurology: critical care in neurology* (ed. C.F. Bolton and G.B. Young), pp. 477–95. Baillière Tindall, London.

Zochodne, D.W., *et al.* (1987). Polyneuropathy associated with critical illness: a complication of sepsis and multiple organ failure. *Brain*, **110**, 819–42.

7.8.2 Pathophysiology and causes of myopathy

C. F. Bolton

Key messages

- As with neuropathy, clinical assessment may be misleading.
- Neurological consultation, magnetic resonance imaging of the cervical spinal cord, electrophysiological testing, and, in particular, measurements of creatine phosphokinase and muscle biopsy usually establish the diagnosis.
- Neuromuscular blocking agents and steroids, together with sepsis, may be important causes of the myopathy in critical illness.
- The results of investigations are important, largely for prognostic reasons.

Introduction

As with neuropathies, patients with a defect in neuromuscular transmission or myopathies may present either before or after admission to the intensive care unit (ICU), usually depending upon the underlying cause. However, in both instances the approach is similar. There should be systematic investigation for disorders of the high cervical spinal cord, peripheral nerve, neuromuscular junction, and muscle.

Traditional texts have listed only conditions such as myasthenia gravis, Lambert-Eaton myasthenic syndrome, botulinium toxicity, acute myoglobinuria, and muscular dystrophy as myopathies to be considered in the ICU. Here, we describe the less well-known myopathies that present after admission to the ICU.

Patients presenting with neuromuscular transmission defects after admission to the ICU

Non-competitive neuromuscular blocking agents such as succinylcholine (suxamethonium) are rarely used in the ICU setting because they may induce prolonged blockade, circulatory collapse, and hyperkalemia.

Sepsis and multiple organ failure in themselves will not cause a defect in neuromuscular transmission (Bolton 1996). However, whenever competitive neuromuscular blocking agents are used to ease mechanical ventilation, there is a risk of muscle weakness as a complication. This may occur through the induction of a pure axonal motor neuropathy.

A second mechanism is a defect in neuromuscular transmission which occurs when neuromuscular blocking agents such as vecuronium or pancuronium bromide are given in the presence of renal failure. The action of these drugs is prolonged beyond hours to a number of days after they have been discontinued. Nerve stimulation studies show a typical postsynaptic defect. There is a decrement of the compound muscle action potential at slow rates of stimulation. The prognosis for recovery of muscle strength is quite good, although an accompanying critical illness (polyneuropathy) is often present and may further prolong recovery.

The third situation results in the induction of a thick-filament myopathy (see below).

Myopathies presenting after admission to the ICU

Myopathies presenting in this situation are frequently associated with sepsis and share theoretical pathophysiological mechanisms with neuropathy.

Thick-filament myopathy

A distinctive syndrome occurs in children or adults in the setting of sudden severe asthma or at the post-transplant state (Lacomis *et al.* 1996). Endotracheal intubation and placement on a ventilator is necessary. High-dose steroids and neuromuscular blocking agents to ease ventilation are given, often for a number of days. On attempted weaning from the ventilator, it will be noted that the patient has severe neuromuscular respiratory insufficiency and limb weakness. Ophthalmoplegia may be present. Creatine kinase levels are often considerably elevated. Repetitive nerve stimulation studies are usually normal. Sensory conduction is normal, as is motor conduction, except for a low-amplitude compound muscle action potential. On needle electromyography, motor unit potentials tend to be polyphasic and of low amplitude and short duration, indicating a primary myopathy. The muscle may be electrically unexcitable on direct stimulation, suggesting inactivation of sodium, chloride, or potassium channels. Muscle biopsy shows a loss of structure centrally in muscle fibers. This has been shown by electron microscopy to be due to destruction of the thick myosin filaments (Danon and Carpenter 1991). Denervation of muscle, secondary to either critical illness polyneuropathy or the neuromuscular blocking agent, probably predisposes to this distinctive pathological change (Bolton 1996).

Recovery is quite rapid. The clinical and electrophysiological features of this syndrome are usually so distinctive that muscle biopsy is often unnecessary; this is a worthwhile consideration in children because of the disfiguring scar.

Although the subject is still controversial, the author believes that neuromuscular blocking agents should be used to ease ventilation in asthmatics only when there are clear-cut indications, and at as low a dosage and for as short a period as possible. The use of steroids should also be limited as much as possible.

Cachectic myopathy

Cachectic myopathy, disuse atrophy, and catabolic myopathy are often cited as complications of critical illness. However, even though they cause muscle weakness and wasting, all are ill defined in clinical terms (Bolton 1996). Motor and sensory nerve conduction studies, needle electromyography of muscle, and creatine kinase levels are all normal. Muscle biopsy may be normal or show type 2 muscle fiber atrophy, which is a non-specific finding.

Acute necrotizing myopathy of intensive care

Rarer, but better defined, is 'acute necrotizing myopathy of intensive care' (Zochodne *et al.* 1994). It may be precipitated by a wide variety of infective, chemical, and other insults, basically involving the differential diagnosis of acute myoglobinuria. It would be expected to occur with increased frequency in ICUs, in which there is a high incidence of trauma, infection, and use of various medications. Thus Zochodne *et al.* (1994) reported 11 cases, in which there was severe weakness with high levels of creatine phosphokinase and often myoglobinuria, in ICUs. Electrophysiological studies were consistent with a severe myopathy, and muscle biopsy showed widespread necrosis of muscle fibers. Rapid and spontaneous recovery is expected in milder cases, but in more severe cases the prognosis may be poor.

We have observed mild elevations of creatine kinase and scattered necrosis of muscle fibers on muscle biopsy in some critically ill patients, suggesting primary involvement of muscle as well as denervation atrophy. This may be due to a reduction in bioenergetic reserves as measured by phosphorus-31 magnetic resonance spectroscopy, since in two of our patients the ratio of phosphocreatine to inorganic phosphate was much lower than would be expected from denervation of muscle alone (Bolton *et al.* 1994). These abnormalities returned to normal as the patients recovered from the critical illness and the polyneuropathy. Nonetheless, biopsies performed in 11 of our ICU patients, because of uncertainties about the nature of the neuromuscular condition, were dominated by denervation atrophy, presumably secondary to a critical illness polyneuropathy.

References

Bolton, C.F. (1996). Sepsis and the systemic inflammatory response syndrome (SIRS): neuromuscular manifestations. *Critical Care Medicine*, **24**, 1408–16.

Bolton, C.F., Young, G.B., and Zochodne, D.W. (1994). Neurological changes during severe sepsis. *Current Topics in Intensive Care*, **1**, 180–217.

Danon, M.J. and Carpenter, S. (1991). Myopathy and thick filament (myosin) loss following prolonged paralysis with vecuronium during steroid treatment. *Muscle and Nerve*, **14**, 1131–9.

Lacomis, D., Giuliani, M.J., Van Cot, A., and Kramer, D.J. (1996). Acute myopathy of intensive care: clinical, electromyographic, and pathological aspects. *Annals of Neurology*, **40**, 645–54.

Zochodne, D.W., Ramsay, D.A., Saly, V., Shelley, S., and Moffatt, S. (1994). Acute necrotizing myopathy of intensive care: electrophysiological studies. *Muscle and Nerve*, **17**, 285–92.

7.8.3 Assessment

J. Coakley

Key messages

- Weakness is not always due to neurological disorders and may occur in a large variety of systemic disorders.
- Weakness may complicate critical illness in addition to being a cause of admission to the intensive care unit.
- The assessment of weakness may have to follow urgent intervention (e.g. mechanical ventilation) if it is severe and life threatening.

Introduction

Weakness is a symptom of a huge variety of acute medical and surgical problems. Acute weakness may be a condition which either precipitates intensive care admission, or manifests itself during or after intensive care treatment. Clearly the most life-threatening complications are failure to protect the airway or maintain adequate ventilation function, and these should be assessed urgently prior to a more leisurely examination of the rest of the patient. Support should be initiated prior to diagnosis if necessary.

A list of some common causes of acute weakness is shown in Table 1. This is not an exhaustive list; for instance, there is a two-volume textbook devoted to disorders of the peripheral nerves alone!

General considerations

Any acute illness may present with weakness as a major symptom, although weakness is typically considered a neurological dysfunction. Therefore it is important to exclude clinically significant systemic disorders as well as primary disorders of the nervous system. It is worth bearing in mind that symptoms are often atypical in the elderly and the seriously ill.

In extreme cases, intubation of the trachea and mechanical ventilation may be required pending further diagnostic tests. However, assuming that the patient is not moribund, attention should be paid to the various systems of the body. Respiratory disorders, particularly exacerbation of chronic lung disease, asthma, and pneumonia, may all cause weakness. These should be excluded on clinical examination and chest radiographs. Cardiac failure and cardiogenic shock may also render patients weak, and clinical signs may be deceptive. Septic shock or the systemic inflammatory response syndrome may be associated with global weakness. Advanced renal and hepatic failure may also lead to weakness. Metabolic disorders such as hypokalemia and hypocalcemia should be excluded, as should anemia. Endocrine disorders, including thyrotoxicosis, myxedema, Addison's disease, Cushing's syndrome, and disorders of the pituitary gland, may lead to muscle involvement. In all the above scenarios, treatment must be directed at the underlying cause of the weakness, as well as supporting the patient's general condition.

Table 1 Causes of weakness

Brain
Trauma
Intracerebral hemorrhage
Subarachnoid hemorrhage
Cerebral infarction
Cerebral infections
Drugs (legal or illicit)
Demyelinating disorders
Status epilepticus
Spinal cord
Trauma
Infection
Demyelinating disorders
Syringomyelia
Peripheral nerves
Congenital peripheral neuropathies
Acquired peripheral neuropathies
Demyelinating disorders
Porphyria
Critical illness neuropathy
Neuromuscular junction
Myasthenia gravis
Neuromuscular blocking agents
Muscle
Congenital myopathies
Muscle dystrophies
Inflammatory myopathies
Endocrine myopathies
Critical illness myopathy
Systemic disorders
Numerous

Clinical neurological assessment

Cranial nerves

These nerves can be assessed fairly rapidly. The olfactory nerve is not commonly assessed formally unless the patient complains of a recent alteration in sense of smell. The integrity of cranial nerve II should be checked by assessing the visual fields. Bitemporal hemianopia (tunnel vision) suggests a lesion at the optic chiasm, commonly a pituitary tumor. Homonymous hemianopia indicates a post-chiasmal lesion opposite the side of the visual defect. An upper and outer defect suggests a lesion of the opposite cortex in the temporal region, while central visual loss indicates a retrobulbar problem, usually multiple sclerosis. Sudden total visual loss, usually in one eye, may occur in retinal detachment, retinal venous or arterial occlusion, glaucoma, or vitreous hemorrhage.

The pupillary reflexes are controlled by cranial nerves II and III. The pupils should constrict in response to light and convergent gaze. If the direct reflex is absent but the convergent reflex is not, a lesion of the brainstem or ciliary ganglion should be suspected. The opposite suggests visual cortical blindness. The pupils may be small in narcotic poisoning, pontine hemorrhage, or Horner's syndrome. Large pupils are associated with mydriatic agents, lesions of nerve III, or the Holmes–Adie syndrome. It should be remembered that many disorders of eye movement and pupil size are congenital, and the patient or relatives should be asked if any abnormalities noted are long-standing. The patient's eyes should be examined for ptosis. When complete, this indicates palsy of nerve III, which is also associated with a downward and outward gaze and a dilated pupil. Partial ptosis may occur in myasthenia gravis, myotonic dystrophy, facio scapulohumeral muscular dystrophy, and syphilis. If ptosis occurs in the presence of a small pupil, absence of sweating, and anophthalmos on the affected side, the cause is a disorder of the sympathetic nervous supply to the eye (Horner's syndrome). The disorder may be located anywhere in the sympathetic chain from the brainstem, the cervical and upper thoracic cord, the sympathetic chain, and the stellate ganglion.

Nerves III, IV, and V are assessed together. The range of eye movements should be noted. Failure of lateral gaze implies a nerve VI palsy, diplopia on downward gaze indicates a nerve IV palsy, and the consequences of nerve III palsy have been mentioned above.

The three divisions of the trigeminal nerve should be assessed (including the corneal reflex) by touching with cotton wool. The motor function is tested by clenching the jaw and opening the mouth against resistance.

Nerve VII is assessed by asking the patient to close the eyes tightly and bare the teeth. Unilateral weakness is always due to a lower motor neuron lesion. Nerve VIII is tested by assessing hearing and also by the caloric response in cases of suspected brainstem death. Cranial nerves IX and X can be tested by documenting gag and cough reflexes and assessing the ability to swallow. Nerves XI and XII are tested by asking the patient to shrug the shoulders and protrude the tongue respectively.

Lesions of upper motor neurons

These may occur as a consequence of disorders of brain or spinal cord. Upper motor neuron weakness of the arms begins distally and spreads proximally. Flexor muscles are usually relatively spared, and therefore the arm assumes a flexor posture. The flexor muscles of the leg are affected to a greater extent than the extensors, and proximal muscles are affected early. Thus the leg assumes an extensor posture. Eventually both limbs develop 'clasp knife' rigidity and clonus may be demonstrated. Reflexes are brisk and the plantar response is typically extensor. Wasting of the muscles is not usually a prominent early feature.

Lesions of lower motor neurons

Anterior horn cell

Disease affecting the anterior horn cell may be congenital (e.g. spinal muscular atrophy) or acquired (e.g. motor neuron disease, poliomyelitis). Weakness and wasting may be variable, patchy, and asymmetrical, and muscle fasciculations are common, indicating chronic denervation with spontaneous contraction of enlarged motor units. Reflexes are reduced or absent.

Nerve roots

Weakness and wasting occur only in the distribution of the nerve root involved. Pain and sensory loss also occur. Lesions of the nerve root commonly arise as a consequence of disk disease, mainly in cervical and lumbosacral regions.

Peripheral nerves

In generalized disorders of the peripheral nerves, the distribution is usually symmetrical and predominantly distal, with proximal spread occurring later. Where lesions are due to entrapment or diabetes there is often asymmetrical and patchy distribution. In Guillain–Barré syndrome, proximal weakness is the norm. Reflexes are usually absent and the tone is floppy. The degree of wasting depends upon the duration of the neuropathy.

Neuromuscular junction

This is involved in myasthenia gravis, myasthenic syndromes, and following administration of depolarizing and non-depolarizing muscle relaxants.

Muscle

In primary muscle diseases weakness is usually proximal and symmetrical. Inflammatory myopathies (polymyositis, dermatomyositis) and some metabolic disorders (McArdle's disease, carnitine palmityl transferase deficiency) are usually accompanied by muscle pain and weakness, but many other primary disorders of muscle are painless. The myopathy associated with myxedema often causes muscle pain, and muscle tenderness is common in Guillain–Barré syndrome. The reflexes are usually preserved in myopathies unless wasting is severe, as in advanced Duchenne muscular dystrophy.

Investigation of weakness

Investigation must be guided by as precise an anatomical clinical diagnosis as is feasible given the condition of the patient.

Radiological investigations

Plain radiography of the skull and spine is particularly valuable in trauma victims. Such radiographs may also be appropriate for the investigation of systemic disease such as disseminated malignancy. Their use as detectors of structural change within the central nervous system has diminished greatly since the advent of CT and magnetic resonance imaging (MRI). Radionuclide investigations have also been superseded by these techniques. If a central nervous system lesion is suspected on clinical assessment, a CT scan of the brain before and after intravenous contrast injection should be obtained. CT scanning of the spine may be useful in cases of trauma, particularly when plain radiographs are inadequate. Myelography has largely been replaced by MRI. Cerebral angiography may be required in cases of subarachnoid or other intracranial hemorrhage to delineate intracerebral aneurysms or angiomata.

Electroencephalography

An electroencephalogram is a non-invasive, harmless, and inexpensive investigational technique. It is mainly used to show epileptic activity. It may also help to differentiate some causes of impaired consciousness and coma. It can help to distinguish the effects of metabolic and toxic changes from cerebral hypoxia, encephalitis, and in some cases subclinical epilepsy. It also reveals characteristic features in conditions such as herpes encephalitis. It may be used to give prognostic information in suspected ischemic damage or head injury. It may also help to differentiate locked-in syndrome from persistent vegetative state.

Electromyography

This may be helpful in differentiating weakness primarily due to nerve involvement from that due to muscle involvement. Disorders of the nerves may be demyelinating or axonal. In demyelinating disorders loss of the myelin insulating sheath results in a reduced nerve conduction velocity, whereas axonal loss leads to a reduction in the action potential obtained as a result of stimulating the nerve. Gross loss of muscle mass may also lead to reduction in the action potential detected in the muscle following stimulation of the nerve supplying it. Thus it is possible to differentiate predominantly axonal neuropathies from those due to demyelination and, by examining a wide variety of nerves neurophysiologically, it may be possible to discern whether the neuropathy is generalized or localized. Localized neuropathies are often due to entrapment syndromes such as carpal tunnel syndrome or root lesions. Electrical recordings of muscle activity may also be made. These may be taken from either the surface of the muscle or from needles inserted within the muscle. In myasthenia gravis and muscle relaxant toxicity there is usually a decrement of the response of muscle to repetitive nerve stimulation. In the Eaton–Lambert syndrome associated with bronchial carcinoma there is usually an increment in the muscle response to repetitive stimulation. There are also characteristic features differentiating neuropathic causes of muscle weakness and myopathic causes. A normal muscle is electrically silent, but, following denervation, affected muscle fibers may develop spontaneous contractions known as fibrillations. Surviving axons attempt to reinervate denervated muscle; thus during voluntary muscle movement the motor unit potentials are larger than normal but fewer in number. In myopathic disorders the loss of muscle fibers leads to small and shorter motor unit potentials. Occasionally, myopathic disorders may also be associated with fibrillations which occur particularly in the inflammatory myopathies.

Cerebrospinal fluid

When performed in the investigation of chronic neurological disorders, such as multiple sclerosis or acute disorders such as Guillain–Barré syndrome, lumbar puncture is reasonably safe. However, if a patient presenting with weakness is too drowsy to give a history, lumbar puncture is contraindicated prior to CT scanning to exclude raised intracranial pressure. Any suggestion of raised intracranial pressure means that a lumbar puncture is contraindicated. The absence of papilledema does not preclude raised intracranial pressure.

Muscle biopsy

Muscle biopsy is indicated in the diagnosis of myopathies and neuropathies. A variety of percutaneous and open techniques have been described. Muscle biopsy specimens may be examined histologically, histochemically, and biochemically. Muscle biopsy appearances can be used to differentiate between primarily myopathic disorders, in which a variety of myofibrillar abnormalities may be seen, and chronic neuropathic disorders, in which changes due to reinervation (principally fiber-type grouping) may be seen.

Spirometry

The most obvious life-threatening complication of weakness is the involvement of muscle associated with coughing and swallowing. It is important to note that arterial blood gas analysis is practically useless in the assessment of impending ventilatory failure. Patients with respiratory muscle weakness will characteristically develop abnormal blood gases shortly after respiratory arrest. In the early stages of respiratory muscle weakness the clinical signs are subtle, but there may be dyspnea of effort and the use of accessory muscles of respiration. Paradoxical abdominal movement may suggest diaphragmatic weakness. Difficulty in swallowing, coughing, and speaking suggests impending respiratory failure and may also expose the patient to the risk of aspiration pneumonia. Patients with respiratory muscle weakness and a consequent inability to increase tidal volumes compensate by increasing the respiratory rate. Tachypnea with a respiratory rate of more than 30 breaths/min is an ominous sign. Therefore the most useful bedside test is measurement of tidal volume and vital capacity. A vital capacity of less than 1 liter or a fall of 50 per cent from normal should lead to intensive care referral. Vital capacity should be measured at least twice daily, and more often if the disease is progressing rapidly. If the patient develops severe difficulty in swallowing or coughing or develops an aspiration pneumonia, the airway should be secured and mechanical ventilatory support provided.

Neuromuscular complications of critical illness

There has recently been considerable growth in the literature describing these complications. The muscle may be weak as a result of atrophy, which can progress at an alarming rate in the catabolic critically ill patient. This atrophy may occasionally be complicated by a severe necrotizing myopathy. The neuromuscular junction may be blocked by the injudicious use of muscle relaxants, particularly when administered by infusion to patients with renal or hepatic failure. Finally, an acute axonal neuropathy (the critical illness neuropathy) may complicate severe illness, often when there is concomitant sepsis and multiple organ failure. The assessment and management of these is no different from any of the other causes of weakness. The long-term prognosis is usually good.

Conclusions

Weakness can usually be assessed clinically. It may be a manifestation of a systemic illness or of a primarily neurological disease. Investigation should be guided by the clinical signs. It is important that life-saving measures, principally intubation of the trachea and provision of mechanical ventilatory support, should never be delayed pending a diagnosis.

Bibliography

Bannister, R. (1992). *Brain and Bannister's clinical neurology* (7th edn). Oxford University Press.

Bleck, T.P. (1996). The expanding spectrum of critical illness neuropathy. *Critical Care Medicine*, **24**, 1282–3.

Bolton, C.F. (1996). Sepsis and the systemic inflammation response syndrome: neuromuscular manifestations. *Critical Care Medicine*, **24**, 1408–16.

7.8.4 Management

J. Coakley

Key messages

- Most aspects of the management of acute weakness are symptomatic or supportive (e.g. prevention of thromboembolism, nutritional support, psychological support).
- Management may of necessity precede detailed assessment and diagnosis.
- Patients with acute weakness are at greatly increased risk of respiratory failure and arrest, which may occur before blood gases become grossly abnormal.

Introduction

The management of the acutely weak patient can be divided into two parts. The first is the specific management of the cause precipitating the weakness, and the second is general supportive care which is necessary to keep the patient alive until the cause of the weakness has resolved.

Specific therapy

In cases where weakness is due to systemic disease the underlying process should be corrected. Very few of the primary neurological conditions leading to weakness are amenable to specific therapy. Patients with polymyositis and other inflammatory myopathies may respond to treatment with corticosteroids and immunosuppressive agents. Surgical stabilization or treatment of the traumatized patient should be prompt. Infective causes of weakness should be treated with appropriate antibiotics.

General therapy

General management of severe weakness is summarized in Table 1.

Table 1 Management of severe weakness
Airway
Respiration
Nutrition
Pressure areas
Mobility
Prophylaxis against thromboembolism
Pain control
Psychological problems

Respiratory support

In patients presenting with acute weakness the respiratory muscles should be assumed to be involved until proven otherwise. Clinical assessment and simple bedside measures are the best methods of determining the requirement for mechanical ventilation. A vital capacity of less than 1 liter or less than 50 per cent of the predicted value, a respiratory rate of more than 30 breaths/min, a poor cough, or a clinical impression of impending ventilatory failure should lead to an urgent referral to the intensive care unit (ICU). A vital capacity of less than 30 ml/kg leads to reduced capacity to cough, and less than 10 ml/kg causes ventilatory failure. The airway should be secured and mechanical ventilatory support should be instituted if the respiratory function falls below the limits given above. When a decision has been made to institute mechanical ventilation, this should be explained to the patient. It is important to note that succinylcholine (suxamethonium) should never be used where there is muscle damage or denervation. This is because denervated or traumatized muscle releases large amounts of potassium into the circulation when depolarizing agents are used, and this may prove fatal.

Prior to intubation of the trachea, an attempt should be made to ensure that the stomach is empty when respiratory muscle weakness has developed slowly and is not likely to progress with great rapidity. Therefore patients should remain nil by mouth for at least 4 h. In an emergency the stomach may be emptied by inserting a nasogastric tube. The patient should be preoxygenated for approximately 5 min and then anesthesia induced with a suitable induction agent (e.g. propofol, etomidate, thiopental (thiopentone), or midazolam). Muscle relaxation should be ensured with a non-depolarizing muscle relaxant (e.g. atracurium or vecuronium). A cuffed endotracheal tube should be inserted and the cuff inflated while an assistant maintains cricoid pressure, which should only be released once the airway is secured adequately and the endotracheal cuff inflated.

The precise mode of mechanical ventilation depends on the state of the patient. Initially synchronized intermittent mandatory ventilation may be appropriate, and infusions of sedative and analgesic agents may be required to keep the patient comfortable. Initial ventilator settings should include a respiratory rate of between 10 and 12 breaths/min with a tidal volume of about 10 ml/kg. These settings may be titrated in the light of subsequent blood gas analysis. It is usually acceptable to maintain arterial carbon dioxide tension within the normal range with oxygen saturation of over 95 per cent. If the patient is capable of making respiratory effort, pressure support ventilation may be appropriate. For patients with severe weakness which is likely to run a prolonged course (e.g. Guillain– Barré syndrome) it is prudent to perform an early tracheostomy so that the prolonged use of sedative agents may be avoided.

Nutritional support

In the vast majority of patients presenting with acute weakness the gastrointestinal system remains functional. Therefore parenteral nutrition is usually unnecessary. A nasogastric tube should be inserted, or alternatively a percutaneous endoscopic gastrostomy tube may be used. There is little evidence that any particular enteral feed is superior in terms of outcome. However, it is important to ensure that sufficient quantities of nitrogen and calories are administered to overcome the negative balances of each associated with critical illness and immobility. If a tracheostomy tube has been inserted, it is usually possible to allow the patient to eat normally during the day, if necessary supplementing the oral diet with enteral feeding overnight. Severely weak patients are likely to be unable to feed themselves without help. Nursing staff and relatives should be encouraged to assist in the feeding of the patient.

Stress ulceration

Stress ulceration is a potentially life-threatening complication of critical illness. It is unlikely to occur in patients who are receiving enteral nutrition via a nasogastric tube or eating normally. Therefore, for most patients presenting with acute weakness, it is probably only necessary to consider stress ulcer prophylaxis in the very early stages of the acute illness. Antacids may be used to maintain a gastric pH above 3.5. This is effective in reducing the incidence of gastrointestinal haemorrhage but, depending on the agent used, may result in either diarrhea or constipation. In view of the immobility of most of these patients, neither of these complications is desirable. H_2 receptor antagonists may be administered either orally or intravenously to maintain gastric pH above 3.5. There is some evidence that use of these agents, by raising gastric pH, may allow Gram-negative organisms to flourish in the stomach with transmission up into the oropharynx and down into the lungs, causing a secondary pneumonia. Sulcralfate may be administered either orally or nasogastrically and forms a protective coating over inflamed areas of mucosa. It is probably associated with a lower incidence of secondary pneumonias than is the case with agents which alter gastric pH.

Prevention of thromboembolism

Patients with weakness are immobile and therefore at high risk of deep venous thrombosis or pulmonary embolism. In most patients the combination of low-dose heparin (5000 IU subcutaneously two to three times daily) coupled with early mobilization reduces the incidence of both complications. Elasticated antiembolism stockings should be used routinely. If the patient is too weak to move, passive leg exercises should be encouraged.

Psychological problems

Patients with profound weakness present a significant challenge to intensive care clinicians. They are unlikely to require profound sedation and analgesia to facilitate mechanical ventilation, and are frequently acutely aware of their total dependence on ventilatory support. They are likely to suffer from sleep deprivation, partly due to the increased noise levels in the critical care environment. Their relatively long stay in the ICU coupled with the often very slow rate of recovery may produce psychological problems. Therefore these patients are prone to a number of disorders, including delirium, psychosis, anxiety, and depression. These problems are best averted by the attention of an experienced and compassionate team of nurses, doctors, and physiotherapists who keep the patient fully informed about progress at all stages. Despite this, some patients develop severe anxiety states and in such cases an anxiolytic such as diazepam may be helpful. The development of other complications such as hallucinations or delusions should be treated by the addition of haloperidol or chlorpromazine. Severe depression may require treatment with antidepressant agents.

Pain

Pain is a frequent complication of Guillain–Barré syndrome and may also occur in patients who are immobile for other reasons. It is characteristically exacerbated by movement and touch. The cause of this pain is not clear. It should initially be treated with simple analgesics such as acetaminophen (paracetamol), although this often proves ineffective. In such cases tricyclic antidepressants, anticonvulsants, or narcotic analgesics may be required. In exceptionally severe cases epidural infusion of opiates may be of use. In resistant cases it may help the patient if anesthetists with an interest in pain control are involved in the management.

Pressure sores

Patients who are unable to move are at high risk of developing pressure sores. These may be prevented by careful attention to nutritional support coupled with frequent turning of the patient. Pressure-relieving mattresses may also be helpful in reducing the incidence of pressure sores, but established sores in very immobile patients usually require the provision of an air fluidized bed.

Physiotherapy

Patients with acute weakness require the services of the physiotherapy department to maintain adequate respiratory function and to preserve limb function. In the early stages, when cough is poor and respiratory excursion minimal, the patient should be assessed by the physiotherapist. Assisted coughing and breathing exercises should be carried out, and it may be necessary for the nursing staff to perform tracheal suction on a more frequent basis. The aims of treatment are to maintain a clear airway and prevent collapse of the basal sections of the lung.

The limbs should be subjected to frequent passive movements in an attempt to prevent the development of contractures. Splinting of the joints may sometimes be necessary. As the patient recovers strength, physiotherapy advice should be taken to ensure adequate positioning of the patient and to facilitate early mobilization.

Bowel care

Immobile patients are particularly prone to develop constipation. This may be ameliorated to some extent by the inclusion of fiber supplements in the enteral feed, but sometimes requires the addition of laxative drugs.

Weaning from mechanical ventilation

Patients with respiratory muscle weakness should never be forced into rapid weaning maneuvers or confidence in the attendant staff will be lost. Extreme patience is often required and it must be understood that the patient will breathe when he or she is ready and not before. Patients who are recovering from respiratory muscle weakness are often acutely aware of subtle changes in ventilator settings such as respiratory rate, tidal volume, and pressure support. Therefore it is important to inform the patient prior to altering ventilator settings. If the patient is totally dependent upon ventilator-generated breaths, it is prudent to reduce these gradually while simultaneously delivering a degree of pressure support to spontaneously generated breaths sufficient to allow an acceptable tidal volume. Once mandatory ventilator breaths have been removed and the patient is breathing solely with pressure support, the level of pressure support may be gradually reduced. However, it is often necessary to increase the pressure support delivered to the patient overnight to allow adequate sleep. A patient who has been totally removed from ventilatory support should be observed breathing unaided for approximately 24 to 48 h prior to discharge to a general ward. In some cases it is prudent to observe the patient on a high-dependency unit, although many hospitals in the United Kingdom do not have such a facility. The tracheostomy should be removed when the patient has fully recovered the ability to swallow, breathe, and cough adequately.

7.8.5 Specific treatment for Guillain–Barré syndrome

D. Hunt and D. J. Cook

Key messages

- Guillain–Barré syndrome is an uncommon condition associated with peripheral nerve demyelination which leads to both weakness and sensory and autonomic involvement.
- The overall prognosis for patients with Guillain–Barré syndrome is relatively good.
- Ten to fifteen per cent of patients will have long-term disabilities.
- The condition is associated with a mortality of 3 per cent, and 10 to 20 per cent of patients became ventilator dependent at some point in the course of their illness.
- Supportive care, including careful monitoring for respiratory compromise, is essential.
- Plasma exchange and intravenous immunoglobulin hasten recovery and help to prevent respiratory failure in patients who are unable to walk without assistance.

Introduction

Guillain–Barré syndrome, the most common acute polyradiculopathy, is generally a benign, spontaneously resolving condition. However, its course may be complicated by ventilatory failure and arrhythmias. Additionally, 10 to 15 per cent of patients are left with severe neurological deficits.

Epidemiology

Guillain–Barré syndrome has an annual incidence of about 0.4 to 1.7 per 100 000. People of all ages can be affected, but the frequency tends to increase slightly in older age groups.

About 70 per cent of patients with Guillain–Barré syndrome have a recognized preceding prodromal illness or other antecedent event. Such events include viral infections due to cytomegalovirus, Epstein-Barr virus, and HIV (Parry 1993). Bacterial infections preceding some Guillain–Barré syndrome cases include *Campylobacter jejuni*, *Mycoplasma pneumoniae*, and *Borrelia burgdorferi*. Recent interest has focused on *C. jejuni* as this has been associated with several severe outbreaks of Guillain–Barré syndrome (Hughes and Rees 1994). Vaccinations are also well recognized antecedent events. For example, in the swine influenza vaccine of 1976 the incidence of Guillain–Barré syndrome amongst those who were immunized was almost 20 times the expected rate. Likewise, vaccination with oral polio vaccine and with older preparations of rabies vaccine (prepared from neural tissue) have both been associated with Guillain–Barré syndrome. Newer preparations of the rabies vaccine have not been associated with this syndrome. There has also been an association with a number of systemic illnesses, including certain malignancies and some connective tissue disorders.

Pathogenesis

The underlying pathogenesis of Guillain–Barré syndrome is still not well understood. It is believed to be an inflammatory process involving an autoimmune response to Schwann cells or peripheral nerve myelin antigens. Pathologically, it is characterized by discrete foci of inflammation with segmental demyelination of peripheral nerves.

The two main hypotheses which have been proposed suggest that these inflammatory foci are secondary to either a cellular immunity process or a humoral factor. Early studies in fatal cases supported a humoral factor. Subsequently, however, cellular infiltrates with lymphocytes and macrophages were detected in cases soon after the onset of the illness. The hypothesis that Guillain–Barré syndrome primarily represents a cell-mediated process gained further acceptance after the development of an animal model, namely experimental allergic neuritis. Injection of homogenized neural tissue led to the development of an acute polyneuropathy in these animals. Pathologically, multiple foci of demyelination were noted which were very similar to those found in human cases of Guillain–Barré syndrome. The cellular hypothesis is also supported by recent studies of activation markers on peripheral T cells, and by the finding of soluble interleukin-2 receptors in the serum of patients with Guillain–Barré syndrome (Hughes and Rees 1994).

Clinical features

The hallmark of Guillain–Barré syndrome is acute paralysis. However, there are many other signs and symptoms that may be associated with the syndrome, as well as atypical variants of the classical ascending paralysis which may complicate diagnosis of the disorder.

The most common and most widely recognized feature of Guillain–Barré syndrome is acute motor weakness. This usually begins in the lower limbs and then progresses to a variable degree to involve the upper limbs, cranial nerves, and thoracic musculature. Involvement is generally relatively symmetrical, although some asymmetry is not inconsistent with the diagnosis. Limb involvement may be more proximal, more distal, or about equal.

Cranial nerve involvement is common, with signs of nerve VII weakness being the most frequent. Laryngeal and pharyngeal weakness is far rarer, but can lead to severe complications such as aspiration pneumonia if undetected.

Thoracic and diaphragmatic muscular weakness leading to respiratory compromise occurs in about a third of cases. Classically, this was the most ominous sign, and in the era before intensive care units (ICUs) accounted for the majority of the fatalities associated with Guillain–Barré syndrome.

Examination of the reflexes shows them to be absent or severely reduced. This is true for both weakened muscle groups and muscles without marked clinical involvement.

Symptoms of distal parasthesia and deep aching pain are quite common, being present in about two-thirds of cases. Objective sensory loss is less frequent and will most often involve proprioception and vibration, although pain and temperature modalities may also be affected.

Autonomic nervous system dysfunction, which may take the form of cardiac rhythm disturbances, labile blood pressure, or altered gastrointestinal or genitourinary function, is another prominent aspect of Guillain–Barré syndrome. This aspect of the disease (particularly cardiovascular involvement) accounts for the majority of fatalities from Guillain–Barré syndrome now that ventilatory support is available. The most common cardiac arrhythmia is persistent sinus tachycardia. Other dysrhythmias, such as bradycardias, atrioventricular blocks, atrial fibrillation, and ventricular tachycardia, may also occur. Blood pressure abnormalities may include marked postural hypotension as well as swings from hypertension to hypotension, and vice versa, which are difficult to manage. Finally, both the syndrome of inappropriate ADH secretion and diabetes insipidus have been described in Guillain–Barré syndrome and associated with the autonomic involvement.

The first symptoms of weakness generally appear about 2 to 3 weeks after the prodromal illness, and will progress over the course of the next 1 to 3 weeks. Symptoms will remain stable for a similar period of time, after which slow recovery will generally occur over a period of months.

One of the less typical presentations should also be mentioned. The Miller–Fisher syndrome is a triad of ataxia, areflexia, and ophthalmoplegia. It is generally believed to be a more benign form of Guillain–Barré syndrome, but still has the potential for rapid respiratory compromise as well as autonomic dysfunction (Parry 1993).

Diagnosis

No specific test is available to diagnose Guillain–Barré syndrome definitively. Rather, it is necessary to consider the signs and symptoms and their time course, as well as cerebrospinal fluid findings and electromyography results. Electromyography is important both for diagnosis of the condition and as a tool to assist in predicting the long-term prognosis. The classical findings are reduced nerve conduction velocities and conduction blocks. Prolonged distal latencies may be seen, as well as absent F and H responses. Responses consistent with axonal degeneration may also be present. Patients with mean distal motor amplitudes less than 20 per cent of normal have a poor prognosis overall (McKhann 1988). However, it is still not possible to make strong predictions on an individual patient basis.

The diagnostic process is complicated by the wide range of possible clinical presentations, and in view of this the National Institute of Neurological and Communicative Disorders and Stroke established a set of diagnostic criteria (Table 1) to act as a general guide (Ashbury *et al.* 1978).

Differential diagnosis of Guillain–Barré syndrome includes several infectious processes, toxic exposures, and porphyria. Infections include the now uncommon disorder of poliomyelitis, which is a

Table 1 Diagnostic criteria for the Guillain–Barré syndrome

Features required for diagnosis
Progressive motor weakness involving more that one limb; weakness may be mild, or severe with total paralysis of all extremities and the trunk, bulbar and facial paralysis, and external ophthalmoplegia
Areflexia

Features strongly supportive of the diagnosis
Clinical features
- Progression to maximal involvement by 2–4 weeks
- Relative symmetry
- Mild sensory symptoms or signs
- Cranial nerve involvement
- Recovery usually complete and beginning 2–4 weeks after progression stops
- Autonomic dysfunction
- Absence of fever at onset of neuropathic symptoms

CSF features
- Elevated CSF protein level
- CSF cell count of 10 or fewer mononuclear leukocytes per ml
- Electrodiagnostic features strongly supporting the diagnosis
- Evidence of nerve conduction slowing or block at some point in the illness; distal latencies may be increased

Features casting doubt on the diagnosis
Marked persistent asymmetry of weakness
Persistent bladder or bowel dysfunction
Bladder or bowel dysfunction at onset
More than 50 mononuclear leukocytes per ml in CSF
Sharp sensory level

Features that rule out the diagnosis
Current history of hexacarbon abuse
Diagnosis of acute intermittent porphyria
History or finding of recent diphtheritic infection
Clinical and laboratory features of lead neuropathy
Purely sensory syndrome
Definite diagnosis of poliomyelitis, botulism, functional paralysis, or toxic neuropathy

CSF, cerebrospinal fluid.
Reproduced with permission from Ashbury *et al.* (1978).

febrile illness associated with an asymmetrical pattern of weakness as well as meningeal irritation and a marked cerebrospinal fluid pleocytosis. Diphtheritic neuropathy is very similar to Guillain–Barré syndrome. A history of pharyngitis and a marked delay between this episode and the onset of weakness (often about 3 months) help to distinguish the two conditions. Other toxic conditions include botulism, hexane inhalation, arsenic ingestion, tick paralysis, and ingestion of marine animal toxins. Hepatic porphyria also needs to be considered; these patients will also have a history of abdominal pain, and the weakness will generally develop after exposure to barbiturates or other drugs.

Treatment

Supportive measures

Supportive care involves careful monitoring for respiratory compromise and autonomic instability. It also involves preventing and treating the complications of immobility and altered pharyngeal and thoracic muscle function. Thus patients need to be assessed frequently for progression of thoracic muscle weakness. Measurement of vital capacity several times daily is a useful guide. It is important to remember that arterial blood gases are not very useful markers as hypoxia and hypercarbia are late signs of respiratory failure. Monitoring patients closely for impending respiratory failure is crucial even if they have minimal weakness, since thoracic muscle involvement may develop and progress very suddenly. Sinus tachycardia due to autonomic instability is frequent. Any other arrhythmia or blood pressure instability should be another indication for ICU admission. Thromboembolic disease prophylaxis and aggressive treatment of any thrombotic complications are important. Aspiration pneumonia is common, and needs appropriate preventive and treatment strategies.

Immune-modulation therapies

Plasma exchange

Plasma exchange has been investigated extensively in Guillain–Barré syndrome. It involves exchanging approximately 250 ml/kg of plasma over a period of 10 to 14 days. The principle problems with this approach are its cost and potential for cardiovascular compromise. Intensive monitoring is generally required in view of the latter concern.

We have recently completed a systematic overview of the trials evaluating the role of plasma exchange in Guillain–Barré syndrome (Lang *et al.* 1994). Five clinical trials (two large randomized trials with over 200 patients in each, and three smaller trials with about 30 patients) were subjected to a methodological critique. One of the smaller trials was randomized, while the other two used alternate allocation. None of the studies was blinded. Patients, whose ages averaged between 44 and 54 years, were enrolled 1 to 2 weeks after the onset of symptoms. The majority of subjects were unable to walk unassisted on entry into the trials. Total plasma exchange volumes of approximately 200 ml/kg and eight plasma volumes were used in two of the studies; slightly smaller volumes were used in the other three trials. The exchanges were carried out over 1 to 2 weeks. Patient follow-up was for 6 months in one trial and for 1 year in the other four.

Overall, these studies indicated that plasmapheresis hastens recovery and helps to prevent respiratory failure. The median time to walking independently was shortened from 85 to 53 days in the Guillain–Barré Syndrome Study Group trial, and from 111 to 70 days in the French Co-operative trial. Muscle strength was regained more rapidly. For example, the Guillain–Barré Syndrome Study Group found that 59 per cent of treated patients had improved by one grade at 1 month compared with only 39 per cent in the control group ($p < 0.01$); the motor grading scale used is shown in Table 2. Pooling the trials gave a common odds ratio of 2.49 (95 per cent confidence interval 1.5–4.1, $p < 0.0002$) for improvement by one grade at 1 month.

The French Co-operative trial reported that only 21.4 per cent of their treated patients required ventilatory support if they were randomized prior to developing respiratory failure. This contrasts with the 42.6 per cent that they noted in their control group.

Mortality was unaffected (3 per cent in the Guillain–Barré Syndrome Study Group trial and 6 per cent in the French Co-operative trial). Likewise, long-term disability does not appear to be altered by plasmapheresis. The French Co-operative study found a rate of severe disability of 11 per cent at 1 year. Interestingly, the same trial

Table 2 Motor grading scale

Grade 0	Healthy
Grade 1	Minor signs or symptoms
Grade 2	Able to walk 5 m without a walker or support
Grade 3	Able to walk 5 m with a walker or support
Grade 4	Bed or chair bound
Grade 5	Requiring assisted ventilation
Grade 6	Dead

noted that complete recovery differed significantly between the two arms of the trial at 1 year (60 per cent in the treated group and only 39 per cent in the control patients). Relapse rates did not differ significantly in the two largest trials. With regard to complications, the only significant difference noted was that the patients receiving plasmapheresis in the French Co-operative trial had a higher rate of septicemia (12.8 versus 4.5 per cent).

The French Co-operative trial also compared two different exchange fluids, namely albumin and fresh frozen plasma. No significant differences were noted with respect to the neurological endpoints. However, several patients receiving fresh frozen plasma developed viral hepatitis, and the frequency of incidents (including fevers and skin rashes) was higher in this group. The authors concluded that fresh frozen plasma should not be used as a replacement fluid in view of these associated risks and the lack of any demonstrated benefit over albumin.

Intravenous immunoglobulin

Since plasma exchange requires relatively intense monitoring and specialized equipment and is not universally available, and intravenous immunoglobulin has been found to be beneficial in chronic inflammatory demyelinating polyneuropathy, interest has shifted to this alternative immune-modulation therapy.

Only one trial to date has assessed this treatment in a randomized fashion; 147 patients who were unable to walk 10 m unassisted and whose symptoms had appeared within the previous 2 weeks were randomized to plasma exchange (200–250 ml/kg over 7–14 days) or intravenous immunoglobulin (0.4 g/kg daily for 5 days) in a non-blinded fashion. Patients with a history of a severe allergic reaction to properly matched blood or with a known history of selective IgA deficiency were excluded. The patients were then followed for 6 months. More immunoglobulin-treated patients had improved by at least one motor grade at 1 month than patients who had received plasmapheresis (53 versus 34 per cent; 19 per cent difference with a 95 per cent confidence interval of 3 to 34 per cent). They also noted that at 2 weeks fewer patients were ventilated in the immunoglobulin group (27 per cent versus 42 per cent, $p < 0.05$). Relapses were similar in the two groups during the follow-up period.

More recently, several small case series have noted an increased relapse rate with intravenous immunoglobulin. One reported a relapse in five out of seven patients, while another noted deterioration in almost half of the 15 patients treated. Further randomized trials comparing intravenous immunoglobulin with plasma exchange are in progress.

Corticosteroids

For many years, corticosteroids were believed to be beneficial in Guillain–Barré syndrome. Numerous cases series suggested that they hastened recovery and improved overall outcome. However, two randomized, controlled trials assessed this question and arrived at a different conclusion. The first trial randomized patients with an acute neuropathy that was not secondary to a metabolic cause to either high-dose oral prednisolone or placebo for 1 week. Throughout the follow-up period of 1 year, a trend towards greater improvement was found in the placebo group, and this reached statistical significance at 3 months for patients who entered the trial within 1 week of the onset of weakness (2.5 grades of improvement in the placebo group versus 0.9 grades in the treated patients). The second trial was larger, randomizing 242 patients with Guillain–Barré syndrome within the first 2 weeks of the illness to either 500 mg of solumedrol (methylprednisolone) intravenously for 5 days or placebo. No significant differences in outcome were noted.

Conclusions

Guillain–Barré syndrome is an uncommon condition associated with peripheral nerve demyelination which leads to both weakness and sensory and autonomic symptoms. The condition generally begins several weeks after a prodromal illness, and usually progresses over the next few weeks. It generally has a good prognosis, reflected in an overall recovery rate of 80 to 85 per cent by 12 to 18 months. However, about 10 to 15 per cent of patients are left with severe long-term disabilities, such as an inability to perform self-care activities or walk. Between 10 and 20 per cent of patients will become ventilator dependent at some time during the course of their illness, and about 3 to 4 per cent will die, usually because of aspiration pneumonia or cardiovascular instability. Treatment and careful observation of these patients are important to prevent the complications of immobility and to provide timely respiratory support should it become necessary. Recently, two immune-modulation therapies (plasmapheresis and intravenous immunoglobulin) have been shown to have a positive influence on the course of Guillain–Barré syndrome.

References

Ashbury, A.K., Amason, B.G., Karp, H.R., and McFarlin, D.E. (1978). Criteria for diagnosis of Guillain–Barré syndrome. *Annals of Neurology*, **3**, 565–6.

Hughes, R.A.C. and Rees, J.H. (1994). Guillain–Barré syndrome. *Current Opinions in Neurology*, **7**, 386–92.

Lang, J.D., Cook, D.J., and Hunt, D.L. (1994). Plasma exchange for Guillain–Barré syndrome: a systematic overview. *Journal of General Internal Medicine*, **9** (Supplement 2), 32.

McKhann, G.M., *et al.* (1988). Plasmapheresis and Guillain–Barré syndrome: analysis of prognostic factors and the effect of plasmapheresis. *Annals of Neurology*, **23**, 347–53.

Parry, J.P. (1993). *Guillain–Barré syndrome*. Thieme, New York.

7.9 Acute myasthenia

7.9.1 Pathophysiology

L. Loh

Key messages

- Myasthenia gravis is an autoimmune disease which causes muscle weakness.
- The anti-acetylcholine receptor antibody causes a marked reduction of acetylcholine receptor density in the post-junctional membrane.
- There is a reduction in the safety margin for neuromuscular transmission which accounts for the fatiguable muscle weakness.
- Symptomatic improvement in muscle power can occur with anticholinesterase medication.
- Removal of the thymus gland often produces a reduction in anti-acetylcholine receptor antibody and an improvement in symptoms.

Introduction

Myasthenia gravis is a disorder of neuromuscular transmission. The disease is brought about through the action of a circulating antibody which dramatically reduces the density of acetylcholine receptors on the postjunctional membrane of the neuromuscular junction (Newsom-Davis 1992; Drachman 1994).

The safety margin

Figure 1 shows a schematic diagram of the structure of the neuromuscular junction (Dreyer 1982). Vesicles containing quanta of acetylcholine molecules are generated and stored in the presynaptic nerve terminal, and are released from active zones in the presynaptic membrane. The active zones are sited opposite the synaptic folds of the postsynaptic muscle membrane where the acetylcholine receptors are most densely concentrated. When the nerve terminal is at rest, quanta of acetylcholine molecules are occasionally liberated spontaneously from vesicles. Some acetylcholine molecules will be destroyed by the enzyme cholinesterase in the synaptic cleft, but some molecules attach to acetylcholine receptors and cause a small change in membrane potential called the miniature endplate potential. This change of about 0.8 mV is small compared with the resting membrane potential of -70 mV and is not sufficient to stimulate a muscle contraction. However, when a nerve impulse arrives at the nerve terminal, a large number of acetylcholine quanta are released.

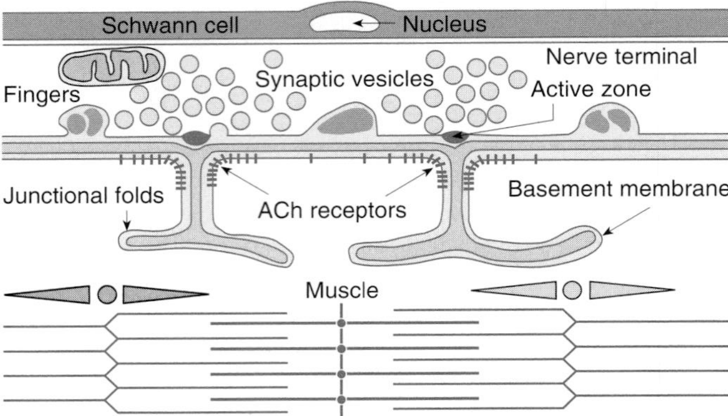

Fig. 1 Schematic diagram of the neuromuscular junction showing active zones on the presynaptic membrane and synaptic vesicles located opposite the junctional folds of the postsynaptic membrane. Acetylcholine receptors are located in high density at the crests of the junctional folds. (Adapted with permission from Dreyer (1982).)

The change in membrane potential greatly exceeds a critical firing threshold and muscle contraction takes place. The safety factor for neuromuscular transmission is the margin by which the change in membrane potential exceeds the critical firing threshold. In myasthenia gravis both structural changes in the architecture of the neuromuscular junction and dynamic alterations in the turnover of acetylcholine receptors erode the safety margin and efficiency of neuromuscular transmission.

The effects of the anti-acetylcholine receptor antibody

Eighty to eighty-five per cent of patients with myasthenia gravis have an identifiable and quantifiable antibody found in the IgG fraction of plasma which is responsible for the changes which take place at the neuromuscular junction. This is called the anti-acetylcholine receptor antibody and has the following effects.

1. The antibody attaches to the acetylcholine receptor and increases the rate of breakdown and internalization of the receptor.
2. The antibody may also block the receptor to the action of acetylcholine.
3. The antibody causes destruction of the postjunctional folds through complement-mediated lysis. This hinders the insertion of new acetylcholine receptor.
4. The result is a marked reduction in the density of acetylcholine receptor.

5. There is a widening of the synaptic cleft and the receptor area is also increased. This increases the destruction of acetylcholine by cholinesterase and decreases the efficiency of acetylcholine targeting a receptor.
6. The result of all the above is a progressive erosion of the safety margin. Therefore successful neuromuscular transmission is markedly affected by small and subtle changes in acetylcholine release and other factors, and this gives rise to the decrement in transmission with repetitive stimulation and the characteristic fatiguable muscle weakness.

Anticholinesterase therapy

Anticholinesterase drugs inhibit the action of cholinesterase and increase the survival of acetylcholine molecules in the synaptic cleft. This improves the probability of successful neuromuscular transmission. Some anticholinesterase drugs are clinically useful in improving muscle power in myasthenia gravis. However, if too much anticholinesterase is present, persistent depolarization of the muscle endplate occurs and this cholinergic state causes failure of neuromuscular transmission.

The role of the thymus gland

The removal of the thymus gland frequently improves the clinical state of patients with myasthenia gravis. The reason for this is not clear. Cells in the thymus gland have been shown to produce anti-acetylcholine receptor antibody. However, the amount of antibody produced is small, and most is produced in the periphery outside the thymus gland. Acetylcholine-receptor-like tissue has been identified in thymic tissue and may act as an antigenic stimulus for antibody production. The thymus gland may be a source of lymphocytes which stimulate antibody production elsewhere in the body. The removal of the thymus gland is followed by a decrease in antibody level in those patients with a thymus which appears histologically active.

The precise trigger for the onset of myasthenia gravis is not known.

References

Drachman, D.B. (1994). Myasthenia gravis. *New England Journal of Medicine*, **330**, 1797–1810.

Dreyer, F. (1982). Acetylcholine receptor. *British Journal of Anaesthesiology*, **54**, 115–30.

Newsom-Davis, J. (1992). Myasthenia gravis. *Medicine International*, **100**, 4168–71.

7.9.2 Assessment

L. Loh

Key messages

- Myasthenia gravis occurs in all age groups and races with a predominance in younger females.
- Thymoma occurs in 10 per cent of patients.
- The disease affects several muscle groups with differing severity: ocular muscles, limb muscles, bulbar muscles, and respiratory muscles.
- Diagnosis is made on the history of fatiguable muscle weakness and confirmed by electromyography studies and anti-acetylcholine receptor antibody assay.
- Eighteen per cent of patients are antibody negative.

Epidemiology

The incidence of myasthenia gravis in the general population is 1 in 20 000. The disease occurs in all countries. There is no familial tendency but some HLA types are common. Frequently, other members of the family may suffer from other autoimmune diseases such as rheumatoid arthritis, thyrotoxicosis, and pernicious anemia. The female-to-male ratio is 3:2. The onset of the disease tends to be earlier in females than in males, with a peak incidence in the second decade in females and in the fourth decade in males. The age distribution is wide in both sexes, with the disease occurring in all age groups. Ten per cent of myasthenia gravis patients have an associated thymoma. This tumor is usually benign, but may occasionally become locally invasive. Thymomas should generally be excised to prevent local damage to important structures such as the phrenic nerves and the great vessels in the mediastinum.

Presentation

The characteristic presentation of myasthenia gravis is fatiguable muscle weakness which is improved by rest and usually improved by the administration of an anticholinesterase drug. The muscle weakness can affect all voluntary muscle groups but often one muscle group is more affected than others; as the disease progresses, more muscle groups become involved.

Ocular muscles

Frequently the presenting symptoms in myasthenia gravis are diplopia and ptosis. Sometimes these are the only symptoms, and if there is no progression to other muscles within 2 years of onset, it is likely that the disease will remain as ocular myasthenia gravis.

Generalized muscle weakness

The proximal muscles tend to be mostly affected. Weakness of the shoulders and hips impairs mobility and normal activity. There may be difficulty in supporting the head upright.

Bulbar weakness

Some patients complain of difficulty with chewing and speech (phonation due to vocal cord weakness and articulation due to weakness of the tongue and lips). In more severe cases, difficulty with swallowing and protection of the airway, causing choking and aspiration of oral secretions, can give rise to major problems. A characteristic nasal speech and inability to smile properly allows easy recognition of this condition. A typical stance of a severe myasthenia gravis patient is with arms folded across the chest and one fist held below the chin keeping the jaw closed and the head up to enable the patient to see ahead from below the drooping eyelids.

Respiratory muscle weakness

The inability to take a deep breath and to cough efficiently is often associated with bulbar weakness and is a dangerous combination. Inability to swallow oral secretions and to clear secretions aspirated

Table 1 Classification of myasthenia gravis

I	Ocular signs and symptoms only
IIA	Mild generalized muscle weakness with slow progression; no crises; drug responsive
IIB	Moderate generalized muscle weakness; skeletal and bulbar muscle involvement but no crises; drug response less than satisfactory
III	Acute fulminating presentation; rapid progression of severe symptoms with respiratory crises and poor drug response; high incidence of thymoma
IV	Late severe myasthenia; respiratory crises and poor drug response as in III but with slower progression over 2 years or more from class I or II

into the trachea is a crisis situation which can lead to acute respiratory failure.

Clinical grading of disability

Several methods of staging disability in myasthenia gravis have been used in the past. The best known is that of Osserman and Genkins (1971) which divides myasthenics into four classes (Table 1). More useful clinical scoring systems have been devised based on the patient's ability to perform a few simple tasks (Besinger *et al.* 1983) (Table 2). These are clinically relevant as they reflect the response to treatment.

Diagnosis of myasthenia gravis

History

The diagnosis is usually made on a history of fluctuating fatiguable muscle weakness, made worse by exercise of the muscle group involved and improved by rest, which tends to be worse towards the end of the day. Examination should confirm the motor weakness with no sensory changes.

Response to edrophonium

Edrophonium (Tensilon®) is a short-acting anticholinesterase which can be administered intravenously. In myasthenia gravis a transient improvement in muscle strength occurs within 30 s of the administration of 4 to 8 mg of edrophonium. The effect only lasts for about 60 s. A positive response may not be seen in a myasthenic patient treated with anticholinesterase drugs if the patient is in a cholinergic state. In fact such a patient can be made considerably worse by the test. A patient on anticholinesterase therapy should be pretreated with intravenous atropine first to prevent a bradycardia, and resuscitation facilities should be available when performing the test.

Table 2 Clinical scores for patients with myasthenia gravis

Weakness grades	None 0	Mild 1	Moderate 2	Severe 3
Muscles of limbs and trunk				
Arms outstretched (90°, standing) (s)	>240	90–240	10–90	<10
Leg outstretched (45°, supine) (s)	>100	30–100	0–30	0
Head lifted (45°, supine) (s)	>120	30–120	0–30	0
Grip strength (decrement after 10 maximal closures) (%)	<15	15–30	30–75	>75
Vital capacity (liters)				
Male	3.5	2.5–3.5	1.5–2.5	<1.5
Female	2.5	1.2–2.5	1.2–1.8	<1.2
Oropharyngeal muscles				
Facial muscles	Normal	Mild weakness on lid closure, snarl	Incomplete lid closure	No mimic expressions
Chewing	Normal	Fatigue on chewing solid foods	Only soft foods	Gastric tube
Swallowing	Normal	Fatigue on normal foods	Incomplete palatal closure, nasal speech	

The table shows the clinical score. For calculation, grades of each item are added and divided by the number of items tested. All patients were tested under optimal anticholinesterase treatment. The upper test limit was taken from patients and healthy controls without neuromuscular weakness.

Electromyography

Fade is seen on repetitive stimulation at about 3 Hz in myasthenia gravis, and fade and post-tetanic facilitation are seen when using tetanic stimulation. Jitter is also seen because of variable conduction at the nerve terminal.

Anti-acetylcholine receptor antibody

An immunoprecipitation assay is available for the anti-acetylcholine receptor antibody. An assay value greater than 2×10^{-10} mmol/l is diagnostic of myasthenia gravis. There is no close relationship between antibody level and severity of illness in a population of individuals. However, in a particular individual there is a direct relationship between antibody level and clinical weakness. Fifteen to twenty per cent of patients with myasthenia gravis have no detectable anti-acetylcholine receptor antibody and yet respond to treatments designed to lower antibody levels (Soliven *et al.* 1988). It is assumed that in these individuals an antibody is present but is as yet unidentified.

Thymic scan

It is wise to scan the anterior mediastinum with CT or magnetic resonance imaging for evidence of thymoma, as this will affect management.

Other myasthenic conditions

Congenital myasthenia gravis

Some patients have a congenital abnormality in the type of acetylcholine receptor on the postjunctional membrane (Vincent *et al.* 1997). There are several variants which are manifest at an early age and produce fatiguable weakness, but do not respond to thymectomy or immunosuppression.

Lambert–Eaton myasthenic syndrome

This is a disorder of neuromuscular transmission where the primary defect is an autoimmune interference of presynaptic voltage-gated calcium channels which control acetylcholine release from the nerve terminal (Lennon *et al.* 1995). There is a deficiency of acetylcholine release giving rise to muscular weakness but, rather than fatiguing, the muscle power improves with activity. An antibody has not been identified as yet, but the condition is improved with plasma exchange and immunosuppression. About 50 per cent of patients with this condition have an associated neoplasm, usually an oat cell carcinoma of the lung.

References

Besinger, U.A., Toyka, K.V., Hömberg, M., Heininger, K., Hohlfeld, R., and Fateh-Moghadam, A. (1983). Myasthenia gravis: long-term correlation of binding and bungarotoxin blocking antibodies against acetylcholine receptors with changes in disease severity. *Neurology*, **33**, 1316–21.

Lennon, V.A., *et al.* (1995). Calcium-channel antibodies in the Lambert-Eaton syndrome and other paraneoplastic syndromes. *New England Journal of Medicine*, **332**, 1467–74.

Osserman, K.E. and Genkins, G. (1971). Studies on myasthenia gravis. Review of a twenty-year experience in over 1200 patients. *Mount Sinai Journal of Medicine*, **38**, 497–573.

Soliven, B.C., *et al.* (1988). Seronegative myasthenia gravis. *Neurology*, **38**, 514–17.

Vincent, A., Newland, C., Croxen, R., and Beeson, D. (1997). Genes at the junction—candidates for congenital myasthenic syndromes. *Trends in Neuroscience*, **20**, 15–22.

7.9.3 Treatment

L. Loh

Key messages

- Anticholinesterase drugs are useful in producing a temporary improvement in neurotransmission but are of no benefit to the underlying pathology.
- Anticholinesterase drugs can produce a cholinergic weakness.
- Thymectomy is the treatment of first choice in those in the younger age group.
- Longer-term immunosuppression using steroids or azathioprine alone or in combination is beneficial.
- Temporary reduction of anti-acetylcholine receptor antibody activity can be obtained with plasma exchange, IgG immunoadsorption, or intravenous immunoglobulin.

Introduction

Several forms of treatment are available for myasthenia gravis. It is through the clarification of the immune disturbance in this disease that management has improved and mortality and morbidity have been dramatically reduced in the last 20 years.

Anticholinesterase drugs

Pyridostigmine is probably the anticholinesterase of choice in the majority of patients with myasthenia gravis. The half-life of the drug is reasonably long (90 min) but it does not accumulate easily. It can be given orally as a tablet or elixir and has relatively mild cholinergic side-effects. It is frequently used to improve muscle strength, but only produces symptomatic improvement and does not correct the basic pathophysiological problem. If given in too large a dosage, anticholinesterase drugs produce cholinergic neuromuscular blockade and increase muscle weakness. In the longer term, these drugs can damage the neuromuscular junction itself. Anticholinesterase drugs can produce cholinergic side-effects such as excessive salivation, abdominal cramps and diarrhea, sweating, and bradycardia. An average dose of pyridostigmine would be 30 to 60 mg four or five times daily.

Thymectomy

Removal of the thymus gland in the younger age group (under 50 years old) is likely to produce remission (return to normal activity without medication) in 25 per cent of cases and partial remission (return to normal activity with the addition of medication) in a further 50 per cent, but 6 months or more may be required to achieve remission. Thymectomy is the first treatment of choice in women of child-bearing age since the operation may prevent the need for immunosuppressant drugs. Patients over the age of 50 respond less well to thymectomy because the gland is more atrophic in middle age. Patients who have a thymoma should have a thymectomy in order to decrease complications from the tumor. However, as thymomas tend to occur in the older age group, the myasthenic symptoms are less likely to benefit from thymectomy alone.

Thymectomy can be performed through a trans-sternal incision. This approach exposes the thymus gland well and makes total removal more certain. It is the method most favored currently and probably the safest when excising large thymomas. A supraclavicular transcervical incision can be used if there is no thymoma present, but total thymectomy may not be achieved. More recently, minimally invasive thoracoscopic thymectomies have been performed with success and with minimum complications and postoperative morbidity (Yim *et al.* 1995). Further studies of this novel technique are required. The thymus gland can be irradiated and thymomas are relatively radiosensitive. Radiotherapy following surgery may be useful in locally invasive tumors.

Steroids

Immunosuppression with steroids such as prednisolone is effective in decreasing anti-acetylcholine receptor antibody levels and improving clinical state in the majority of patients with myasthenia gravis. An alternate-day steroid regimen is used starting with a low dose of prednisolone (e.g. 5 mg) and then increasing the alternate-day dose by 5 mg each week until a maintenance dose of 1 mg/kg on alternate days is reached. This dose is then maintained until the patient has been in remission for several months. The steroids are then reduced slowly over months until a minimum effective dose is achieved. Starting with a high dose of steroid or increasing the dose too fast may cause a deterioration of muscle power before improvement begins (Pascuzzi *et al.* 1984). However, remission is achieved faster. If a high-dose regimen is contemplated, the patient is best managed in an intensive care environment where respiratory support is rapidly available.

Azathioprine

This form of immunosuppression is also successful in producing remission in myasthenia gravis. A dose of 2.5 mg/kg/day is used. However, 6 months or more may be required to achieve remission. The side-effects of this drug are less pronounced than those of steroids. Liver function must be checked regularly. Some patients cannot tolerate the drug. Because of the lengthy period required to achieve remission with azathioprine, a combined regimen of steroids and azathioprine is commonly used. This strategy has the benefit of more rapid steroid remission and better long-term maintenance control, and also a higher overall success rate than either regimen alone (Myasthenia Gravis Clinical Study Group 1993).

Plasma exchange

The anti-acetylcholine receptor antibody is in the IgG fraction of the plasma. The level of the antibody can be rapidly reduced by plasma exchange. A common course would be five daily sessions with 2 to 3 liters of plasma being removed at each exchange and replaced by a plasma substitute. Improvement in the clinical state may be seen within 3 days of the start of plasma exchange and improvement is optimal by about day 10. The benefit may be maintained for a month or more but is not sustained. Repeated courses are often needed to bolster improvement while waiting for other forms of immunosuppression to have an effect. The technique is often useful to achieve an optimal clinical state prior to thymectomy.

IgG immunoadsorption

It is now possible to pass the separated plasma through a gel column of tryptophan-linked polyvinylalcohol resin. The IgG fraction is absorbed onto the surface of the gel, but a substantial amount of albumin can be returned to the patient and protein replacement is unnecessary. The clinical benefits are otherwise similar to plasma exchange (Grob *et al.* 1995).

Intravenous immunoglobulin

Several studies have shown a temporary clinical improvement in myasthenia gravis following the administration of intravenous immunoglobulin. The results are comparable to those of plasma exchange but the administration of the immunoglobulin is considerably easier (Gajdos 1994). The precise mechanism of action of immunoglobulin in myasthenia gravis is yet to be determined.

References

Gajdos, P. (1994). Intravenous immune globulin in myasthenia gravis. *Clinical and Experimental Immunology*, **97** (Supplement 1), 49–51.

Grob, D., *et al.* (1995). Treatment of myasthenia gravis by immunoadsorption of plasma. *Neurology*, **45**, 338–44.

Myasthenia Gravis Clinical Study Group (1993). A randomised clinical trial comparing prednisone and azathioprine in myasthenia gravis. Results of the second interim analysis. *Journal of Neurology, Neurosurgery and Psychiatry*, **56**, 1157–63.

Pascuzzi, R.M., Coslett, H.B., and Johns, T.R. (1984). Long-term corticosteroid treatment of myasthenia gravis: report of 116 patients. *Annals of Neurology*, **15**, 291–8.

Yim, A.P.C., Kay, R.L.C., and Ho, J.K.S. (1995). Video-assisted thoracoscopic thymectomy for myasthenia gravis. *Chest*, **108**, 1440–3.

7.9.4 Crisis in myasthenia gravis

L. Loh

Key messages

- The combination of bulbar and respiratory muscle weakness can lead to acute respiratory failure in myasthenia gravis. This is called a crisis.
- The three main causes of crisis in myasthenia gravis are myasthenic crisis, cholinergic crisis, and steroid crisis.
- The crisis is managed by first relieving the respiratory distress by intubation and ventilation, and then carefully assessing and treating the cause and reviewing treatment.

Introduction

Myasthenic weakness of the ocular or generalized type is distressing and inconvenient. Bulbar weakness and respiratory muscle weakness giving rise to respiratory failure and requiring control of the airway and artificial ventilation is life threatening. This situation is termed a crisis. There are three main types of crises in myasthenia gravis: myasthenic crisis, cholinergic crisis, and steroid crisis.

Myasthenic crisis

Several situations can precipitate a myasthenic crisis.

1. A sudden worsening of the myasthenic weakness may be the result of an inability to take or absorb the prescribed anticholinesterase therapy (e.g. persistent vomiting, coma, perioperative starvation).
2. Infection appears to worsen the myasthenic state. The weakness could be the result of a sudden rise in antibody production in general.
3. Mental or physical stress also seems to influence and exacerbate myasthenic weakness, although the cause is not understood.
4. The natural progression of the untreated disease can lead to bulbar and respiratory weakness and a crisis state.

Management

Although the response to edrophonium may be strongly positive in this situation, it would be unwise to rely on adjustment of medication alone to salvage the situation in the first instance. If the patient is in a distressed state with breathing difficulties, it is essential to control the airway first and to provide satisfactory ventilation and oxygenation. This means rapid and efficient intubation. Once ventilation is under control, the situation can be reassessed. An edrophonium test can be performed and suitable anticholinesterase medication prescribed. Pyridostigmine can be administered as an elixir via a nasogastric tube.

Any precipitating factors should be corrected (e.g. treatment of infection). Further treatment of the myasthenia should be reviewed. Plasma exchange, intravenous immunoglobulin, or IgG immunoadsorption should be considered with the aim of producing a rapid improvement in clinical state. Long-term management with steroids and azathioprine should also be assessed.

If extubation is unlikely to be achieved within a few days, tracheostomy should be considered as a means of providing a safe and more comfortable weaning from ventilation.

Cholinergic crisis

The most common cause of a cholinergic crisis is anticholinesterase overdose in a self-medicating patient. Initially, the patient, who is already marginally cholinergic and feeling slightly weak, may increase the dose of anticholinesterase to achieve optimal strength. On finding the medication is less effective and not appreciating the symptoms and dangers of the cholinergic state, the patient increases the dose even further. If the bulbar and respiratory muscles become weak, the excessive salivation of the cholinergic state can precipitate acute respiratory failure.

Medically prescribed overdose of anticholinesterase therapy can occur. Not infrequently, some muscle groups are more cholinergic than others. If an edrophonium test is performed and it is observed that one muscle group improves with edrophonium, the patient may be thought to be myasthenically weak and the anticholinesterase is increased. Failure to examine or take note of the lack of improvement or worsening in power in other important muscle groups, such as the bulbar and respiratory muscles, may lead the physician to misinterpret the edrophonium test and precipitate the patient into a cholinergic crisis. It is wiser to maintain the patient slightly on the myasthenic side rather than to attempt to achieve optimal strength and run the risk of making the patient cholinergic.

Management

The edrophonium test in the cholinergic patient should be performed with care. It is possible to precipitate a crisis with edrophonium. Signs of the cholinergic state are excessive salivation, abdominal cramps and diarrhea, small pupils, and sweating. As with the myasthenic crisis, if there is respiratory distress it is prudent to control the airway with intubation and ventilate in the first instance. Nutrition and medication should be administered through a nasogastric tube.

Anticholinesterase therapy should be stopped and only restarted when the patient is definitely myasthenic. Meanwhile treatment should be reviewed, and plasma exchange and other means of improving the situation should be considered in the same way as with a myasthenic crisis.

Steroid crisis

The steroid crisis is the result of too rapid an increase in steroid dosage at the start of steroid therapy (Seybold and Drachman 1974; Johns 1987). The patient can rapidly become weak and unresponsive to anticholinesterase therapy. The safe procedure is to manage the patient in the intensive care unit. If indicated, the patient should be intubated and ventilated until the situation improves. Usually the crisis has resolved and remission started within 7 to 10 days.

Drug interactions

Muscle relaxants

Non-depolarizing muscle relaxants act by blocking the acetylcholine receptors on the postjunctional membrane. In myasthenia gravis the number of acetylcholine receptors is so dramatically reduced that muscle paralysis occurs with very low concentrations of relaxant drug (Baraka 1992). Thus myasthenic patients are extremely sensitive to non-depolarizing muscle relaxants, and profound and prolonged paralysis may result from the administration of normal doses. Atracurium is the non-depolarizing muscle relaxant of choice since its breakdown and elimination is reasonably predictable. However, neuromuscular transmission should be monitored continuously when using non-depolarizing muscle relaxants.

The following drugs have been shown to exacerbate myasthenic weakness: aminoglycoside antibiotics; ciprofloxacin; quinine, quinidine, and procainamide; pencillamine (Kuncl *et al.* 1986).

References

Baraka, A. (1992). Anaesthesia and myasthenia gravis. *Canadian Journal of Anaesthetics*, **39**, 475–86.

Johns, T.R. (1987). Long-term corticosteroid treatment of myasthenia gravis. *Annals of the New York Academy of Sciences*, **505**, 568–83.

Kuncl, R.W., Pestronk, A., Drachman, D.B., and Rechthand, E. (1986). The pathophysiology of penicillamine-induced myasthenia gravis. *Annals of Neurology*, **20**, 740–4.

Seybold, M.E. and Drachman, D.B. (1974). Gradually increasing doses of prednisone in myasthenia gravis. *New England Journal of Medicine*, **290**, 81–4.

7.10 Tetanus

7.10.1 Pathophysiology

M. Hopley and J. Lipman

Key messages

- Tetanospasmin, a toxin produced by *Clostridium tetani*, is responsible for all the manifestations of tetanus.
- Hematogenously spread to neural tissue, tetanospasmin inhibits presynaptic release of neurotransmitters.
- Inhibition of inhibitory transmitter release results in muscular spasms, rigidity, and autonomic dysfunction.

Tetanus

Tetanus has affected man throughout recorded history. It is an acute, often fatal, disease caused by exotoxins produced by *Clostridium tetani*. Tetanus is characterized by generalized muscle rigidity, spasms, and autonomic instability.

Pathogenesis

The obligate anaerobe *C. tetani* is a spore-bearing Gram-positive bacillus. Spores are found in animal and human feces and exist ubiquitously in soil. *C. tetani* cannot breach intact skin or mucous membranes; the portals of entry include puncture wounds, lacerations, surgical wounds, burns, gangrene, chronic ulcers, dog bites, injection sites, dental infections, otitis media, and the genital tract following abortion or childbirth. The injury itself may be trivial, and in up to 20 per cent of cases there is no history or evidence of a wound. Clostridial spores germinate in oxygen-poor environments (e.g. necrotic tissue) and toxins are liberated by vegetative forms. *C. tetani* infections remain localized, but the toxin is distributed widely via the bloodstream, taken up into endings of motor nerves, and transported into the nervous system. Motor neuron endplates in skeletal muscle are affected by decreasing release of acetylcholine. Decreased neurotransmitter release by spinal neurons impairs polysynaptic reflexes and causes autonomic dysfunction.

The pathogenesis of tetanus is summarized in Table 1.

Toxins

The vegetative forms of *C. tetani* release two exotoxins: tetanolysin, which is of little clinical importance, and tetanospasmin, also known as tetanus toxin, which is responsible for all the manifestations of tetanus.

The genetic information for tetanospasmin is encoded on a single plasmid. Infections with strains of *C. tetani* lacking this plasmid do not result in tetanus. It is synthesized as a single polypeptide chain, and post-translational changes (disulfide bridge formation and strand cleavage) produce one heavy chain (100 kDa) and one light chain (50 kDa). Hematogenous spread is followed by uptake at neuromuscular junctions mediated by the heavy chain of tetanospasmin. Retrograde transportation to the cell body and across the synaptic cleft into the presynaptic neuron then occurs. Once within the presynaptic neuron, the light chain of tetanospasmin mediates inhibition of transmitter release.

The release of neurotransmitter in response to neuronal depolarization requires docking of synaptic vesicles with the presynaptic membrane. Calcium-mediated fusion of vesicle and membrane then occurs, releasing neurotransmitter into the synaptic cleft. Because the light chain of tetanospasmin is a zinc-dependent endoprotease, it is thought to cleave one of the docking proteins, specifically vesicle-associated membrane protein, also known as synaptobrevin (Fried 1995). This explains the finding that 'intoxicated' presynaptic terminals contain normal amounts of neurotransmitter but release subnormal quantities when stimulated. Angiotensin-converting enzyme is also a zinc-dependent endoprotease. Tetanospasmin-mediated production of angiotensin II may contribute to hypertension in tetanus. Treatment of tetanospasmin-inhibited neurons with angiotensin-converting enzyme inhibitors or zinc chelators has restored neurotransmitter release (Dayanithi *et al.* 1994). The importance of these findings still has to be tested clinically.

Physiological consequences

Lower neuron

Inhibition of acetylcholine release results in muscular weakness, but other features of tetanus tend to overshadow this.

Spinal neurons

Trans-synaptic transport of tetanospasmin and consequent inhibition of neurotransmitter release in spinal neurons is responsible for the major physiological manifestations of tetanus. Inhibitory neurons are more affected than excitatory neurons. Inhibitory neurons are responsible for modulation of excitatory signals from higher centers and local reflex arcs. Loss of this inhibition leads to both a higher firing rate in the lower motor neuron and a loss of reflex relaxation of antagonistic muscle groups during movement. The effects on the α motor neuron are an increased basal tone, manifesting clinically as rigidity. Clinically, spasms are the result of exaggerated uncoordinated reflex responses to minimal stimuli.

Table 1 Pathogenesis of tetanus

	Site of injury	Blood	Lower motor neuron	Spinal neuron
Pathological	*C. tetani*	Tetanospasmin spread hematogenously	Heavy chain Trans-synaptic uptake and retrograde transport Light chain Disruption of neurotransmitter release	
Physiological	Little/no inflammation	Protective antibodies 0.01 IU	Neuromuscular blockade	Decreased inhibition Increased firing rate Uncoordinated reflexes
Clinical	20% not apparent		Weakness	Muscular rigidity and spasm Autonomic crisis
Therapeutic	Surgery Metronidazole	Immunization Passive Active		GABA agonists Neuromuscular blockade Sympatholysis Morphine α and β blockade Magnesium Epidural

Autonomic nervous system

The increased basal sympathetic tone may manifest as tachycardia and alteration of bladder function and bowel motility. Episodes of marked sympathetic overactivity occur, with increased vascular resistance, increased central venous pressure, increased cardiac output, hypertension, pallor, cyanosis of the digits, pyrexia, and sweating. These sympathetic crises are thought to be mediated by loss of spinal inhibition of the adrenal glands. Therefore epidural anesthesia may attenuate or eliminate them.

The role of the parasympathetic nervous system in the pathogenesis of tetanus is uncertain. Episodes of bradycardia, loss of peripheral vascular resistance, low central venous pressure, and a profound fall in blood pressure are seen, and are frequently preterminal. Sudden and repeated cardiac arrests occur, particularly in intravenous drug abusers with tetanus. Most authors have attributed these events to total absence of sympathetic tone, since they do not respond to atropine. However, they may be caused by catecholamine-induced myocardial damage or direct damage to the brainstem. Whatever the mechanism, patients afflicted with the autonomic dysfunction of tetanus are at risk of sudden death (Bleck and Brauner 1997).

Tetanospasmin could theoretically spread to higher neurological centers. The possible clinical consequences of this are uncertain.

Clinical syndromes

The incubation period (time from injury to onset of symptoms) varies from 2 to 60 days, as does the period of onset (from first symptom to first spasm). Ninety per cent of cases present within 15 days of infection. The incubation period and the period of onset are of prognostic importance, with shorter times signifying more severe disease. Variations in the site of toxin action result in four clinical syndromes (Oh and Lipman 1996).

Local tetanus

Rigidity confined to a limb or muscle group associated with the site of injury is known as local tetanus. Non-progressive local tetanus is uncommon and has a low mortality (1 per cent). Non-immune patients progress to generalized tetanus.

Cephalic tetanus

Cephalic tetanus is a form of local tetanus. It results from head and neck injuries, infections of the eye, oral infections, and otitis media. The cranial nerves, particularly cranial nerve VII, are frequently involved and the prognosis is poor. Cephalic tetanus has been described in fully immune patients.

Generalized tetanus

Generalized tetanus progresses in a descending fashion. Presenting symptoms are pain and stiffness of the head, neck and shoulders. Rigidity of the masseter muscle, trismus, or lockjaw is the presenting sign in 75 per cent of cases with generalized tetanus. Risus sardonicus, a characteristic clenched-teeth expression, results from involvement of the facial muscles. Abdominal muscle rigidity may be present. Spasms, with flexion and adduction of the arms, extension of the legs, and opisthotonos, follow. Initially precipitated by external stimuli, spasms increase in intensity and duration. Fractures and tendon separation have been described. Involvement of the larynx and diaphragm are life threatening. Autonomic dysfunction occurs a few days after the development of spasms.

Recovery

Recovery from the effects of tetanus requires removal of the source of tetanospasmin. Neuronal recovery involves the synthesis of new presynaptic proteins, and in the case of α motor neurons sprouting of the terminal end to produce new synapses. This process may take as long as 4 weeks.

References

Bleck, T.P. and Brauner, J.S. (1997). Tetanus. In *Infections of the central nervous system* (2nd edn) (ed. W.M. Scheld, R.J. Whitley, and D.T. Durack), pp. 629–53. Raven Press, New York.

Dayanithi, G., *et al.* (1994). Exploring the functional domain and the target of the tetanus toxin light chain in neurohypophysial terminals. *Neuroscience*, **58**, 423–31.

Fried, G. (1995). Synaptic vesicles and release of transmitters: new insights at the molecular level. *Acta Physiologica Scandinavica*, **154**, 1–15.

Oh, T.E. and Lipman, J. (1996). Tetanus. In *Intensive care manual* (4th edn) (ed. T.E. Oh), Chapter 47. Butterworth Heinemann, Oxford.

7.10.2 Prevention and management

H. Hon and J. Lipman

Key messages

- Active immunization with tetanus toxoid is the cornerstone of prevention.
- The diagnosis is based entirely on clinical findings.
- The aim of therapy is to eliminate the source of toxin, eliminate unbound toxin, prevent muscle spasms, and provide supportive care.
- Natural infection does not confer immunity. All patients recovering from tetanus need to be actively immunized.

Prevention

Tetanus is an entirely preventable disease. Unfortunately, cases continue to occur almost exclusively in inadequately vaccinated or unvaccinated individuals.

Naturally acquired immunity does not occur, nor is it conferred by infection. Therefore all patients recovering from tetanus must be actively immunized. Tetanus can both relapse and recur if this is not ensured.

Active immunization

Aluminum phosphate adsorbed tetanus toxoid should be used for active immunization against tetanus as it is more immunogenic than other preparations.

In the first year of life, diphtheria and tetanus toxoids and pertussis vaccine (DTP), are recommended at ages 2, 4, 6, and 12 months. Diphtheria and tetanus toxoid (Td) are recommended at age 11 to 12 years, but may be given from ages 14 to 16 years (Centers for Disease Control 1995).

A primary series of tetanus immunizations for susceptible adults consists of three doses of combined Td toxoid, with the first two doses given at least 4 weeks apart and the third dose given 6 to 12 months after the second. An incomplete primary series of diphtheria and tetanus toxoids should be completed with Td.

Unvaccinated pregnant women should receive two doses of Td 4 to 8 weeks apart before delivery, preferably during the last two trimesters. Pregnant women who have not had a complete vaccination series should complete the three-dose series.

Boosters every 5 to 10 years are necessary to ensure continued protection in all age groups. Side-effects of tetanus toxoid are mild but uncommon.

Routine wound care

All wounds require thorough cleaning and surgical debridement to remove foreign and necrotic material. Although chemoprophylaxis is not always feasible or effective, antibiotics should be prescribed for septic or contaminated wounds. Penicillin or metronidazole both have good activity against *Clostridium tetani*. The need for tetanus toxoid, with or without human tetanus immunoglobulin, depends on both the condition of the wound and the patient's vaccination history (Table 1). The Centers for Disease Control recommend the administration of human tetanus immunoglobulin to all patients with tetanus-prone wounds who are unvaccinated or whose vaccination history is unknown (Centers for Disease Control 1991). The cost of such a practice in third-world countries is prohibitive and requires re-evaluation. When tetanus toxoid and human tetanus immunoglobulin are given concurrently, separate syringes, separate sites, and adsorbed toxoid only should be used (Table 1).

Management

The major objectives of management are to assess and stabilize the patient, to eliminate the source of toxin, to neutralize circulating unbound toxin, to minimize the effect of toxin already bound in the nervous system, and lastly, but no less importantly, to provide general supportive care.

The diagnosis is clinical. There are no laboratory tests specific to tetanus. Wound cultures are non-specific and only positive in less than 30 per cent of cases. The differential diagnosis includes dystonic drug reactions, hypocalcemia, local temporomandibular disease, peritonsillar infections, dental abscesses, various other local oral diseases, meningitis, encephalitis or intracranial hemorrhage, rabies, strychnine poisoning, peritonitis, spinal injury, convulsive status epilepticus, and psychiatric disorders.

Assessment

Several tetanus severity assessment scores which are useful in management decision-making and prognosis have been developed. Neonates and patients with generalized spasms, cephalic tetanus, an incubation period of 7 days or less, an onset time to generalized spasm of less than 48 h, and more than 12 reflex spasms per 24 h generally have severe tetanus and consequently a poorer prognosis (Bleck and Brauner 1996). Intubation and prolonged ventilatory support are essential components of the management of severe tetanus.

Stabilization

In the early stages of tetanus, the patient's life is most at risk from laryngeal and respiratory muscle spasm. The airway must be secured promptly by endotracheal intubation or tracheostomy. Dysphagia is an important symptom suggesting impending laryngeal spasm.

Table 1 Guidelines for tetanus prophylaxis in routine wound management

Vaccination (doses)	Time since last dose (years)	Tetanus-prone wound		Minor wound	
		Td*	HTIG	Td*	HTIG
⩾3	<5	No	No	No	No
⩾3	5–10	Yes	No	No	No
⩾3	>10	Yes	No	Yes	No
<3/unknown	—	Yes	Yes	Yes	No

Tetanus-prone wound is defined as follows: wound age >6 h; wound depth >1 cm; contamination with dirt, feces, or saliva; stellate configuration, avulsion or crush injury, devitalized tissue, burns, frostbite, and missile wounds.
Td, tetanus and diphtheria toxoid 0.5 ml intramuscularly; HTIG, human tetanus immunoglobin 250 IU intramuscularly.
*DTP if age <7 years; Td if pertussis vaccine is contraindicated.
Adapted from Centers for Disease Control (1991).

Eradication of the organism

Wound care

Once human tetanus immunoglobulin has been given, the infected site (if found) should be thoroughly cleaned and all necrotic tissue should be extensively debrided.

Antibiotics

Although questionable, antibiotics are administered to eradicate the vegetative toxin-producing forms of *C. tetani*. Recommended antibiotic regimens include the following:

(1) metronidazole 500 mg intravenously every 6 h for 10 days;

(2) benzylpenicillin (penicillin G) 0.6 to 1.8 g intravenously every 6 h for 10 days (NB penicillin, a GABA antagonist, may aggravate the spasms);

(3) erythromycin or clindamycin are alternatives.

Toxin neutralization

Human tetanus immunoglobulin will at best neutralize only circulating toxin, and does not affect toxins already fixed in the central nervous system. Symptoms already present are not ameliorated.

Present recommendations for human tetanus immunoglobulin are 3000 to 6000 IU intramuscularly, although a single 500-IU dose may be as effective. Equine antitetanus serum 1500 to 10 000 IU can be used after testing and desensitization if human tetanus immunoglobulin is not available.

Intrathecal administration of tetanus antitoxin is still controversial. Meta-analysis of available studies suggests no benefit.

Side-effects of human tetanus immunoglobulin include fever, shivering, chest or back pains, tachycardia, and hypotension.

Suppression of toxin effects

Muscle spasms

Tetanus of mild severity can be managed with minimal stimulation, careful monitoring, and sedation sufficient to ablate reflex spasm.

Neuromuscular blockade in addition to adequate sedation and analgesia may be required in severe tetanus, where spasms usually preclude effective ventilation. Pancuronium bromide can be used safely as a neuromuscular blocker in tetanus. The level of neuromuscular blockade being sustained should be monitored by either direct observation of respiratory effort or peripheral nerve stimulation. An attempt to taper neuromuscular blockade is usually made after 14 days. However, it may be necessary to continue for longer periods in patients with severe tetanus.

Heavy sedation (predominantly morphine and benzodiazepines), used alone, may prevent muscle spasms and improve autonomic dysfunction. Barbiturates, neuroleptic agents, and propofol have also been used. Patients can be weaned slowly from sedation once neuromuscular blockers are no longer necessary.

Debatable therapies include dantrolene and continuous or intermittent intrathecal baclofen.

Autonomic dysfunction

In severe tetanus, autonomic dysfunction characterized by extreme variations in blood pressure, heart rate, and cardiac rhythm remains one of the major causes of death. Optimal therapy has not been defined, but includes adrenergic blockade, heavy sedation, magnesium sulfate, clonidine, and epidural blockade.

Adrenergic blockade

Traditionally, a combination of peripherally acting α- and β-receptor blockers have been used. Phenoxybenzamine, phentolamine, bethanidine, and chlorpromazine can block α receptors. Chlorpromazine is probably the best because of its additional central sedative effects. Propranolol (pure β-receptor blocker) and labetalol (α- and β-receptor blocker) have had limited success. We do not advise unopposed β-blockade as deaths from acute congestive cardiac failure have occurred. Removal of β-mediated vasodilatation in the muscle causes a rise in systemic vascular resistance, and the β-blocked myocardium may not be able to maintain cardiac output. The short-acting intravenous β-blocker esmolol is useful for controlling the tachycardia and hypertension of the crises; however, catecholamine levels remain raised (Lipman and Oh 1996). This is worrying because excessive catecholamine secretion is associated with myocardial damage.

Heavy sedation

It would appear more logical to decrease catecholamine output. This can be done by means of centrally acting sedative drugs. Benzodiazepines and morphine are the preferred drugs.

Morphine acts centrally to minimize the effects of tetanospasmin by the following mechanisms: replacing deficient endogenous opiates, blocking sympathetically mediated peripheral venoconstriction, and reducing sympathetic α tone by decreasing central nervous system sympathetic efferent discharge. In our unit up to 2500 mg/day of morphine has been needed to ablate autonomic dysfunction. No adverse effects have been documented when using such doses, and enteral feeding is usually well tolerated.

Benzodiazepines increase the affinity and efficacy of GABA at its receptor. They are currently the best agents available for the relief of spasms and rigidity. Lorazepam is suggested for long-term sedation. Very large doses (diazepam 3400 mg/day) may be required and are well tolerated, with cardiac output falling minimally or even increasing.

Magnesium

A loading dose of 70 mg/kg magnesium sulfate followed by an infusion of 1 to 4 g/h to keep serum levels between 2.5 and 4 mmol/l reduces systemic vascular resistance, pulse rate, and to a small extent cardiac output. The release of epinephrine (adrenaline) from the adrenal gland and norepinephrine (noradrenaline) from peripheral nerve terminals is inhibited and the sensitivity of receptors to these neurotransmitters is reduced. Magnesium also has a marked neuromuscular blocking effect, and may reduce the intensity of muscle spasms. However, it cannot be used without sedation and serum levels are difficult to maintain. Serum calcium needs to be monitored and occasionally requires supplementation.

Clonidine

Clonidine (Sutton *et al.* 1990) and epidural blockade are less well documented therapies which warrant further investigation.

Supportive treatment

In addition to the therapy aimed specifically at tetanus, the patient also needs standard intensive care. Fluids, electrolytes, and nutrition should be provided, and steps should be taken to prevent problems of prolonged immobilization. Tracheostomy is recommended in patients requiring prolonged intubation. The patient (and, in the case of a neonate, the mother also) must be actively immunized. Where possible, supportive psychotherapy should be offered to both the patient and his or her family.

Complications

Muscle spasms disappear after 1 to 3 weeks but residual stiffness may persist. Most survivors recover completely by 6 weeks. Cardiovascular complications, including cardiac failure, arrhythmias, pulmonary edema, and hypertensive crises, can occur and may be fatal. Twenty per cent of autopsies show no obvious cause of death. Other complications include those associated with hypoxia, mechanical ventilation, rhabdomyolysis, sepsis, fluid and electrolyte problems, bed sores, bone fractures, contractures, deep vein thrombosis, embolic phenomena, and particularly pneumonia.

Outcome

Recovery from tetanus is thought to be complete; an average mortality of 10 per cent would seem to be reasonable for most intensive care units. Mortality is highest in neonates and geriatric patients. Neurological problems such as seizures, myoclonic jerks, EEG changes, intellectual or emotional impairment, sleep disturbance, and decreased libido can occur.

References

Bleck, T.P. and Brauner, J.S. (1997). Tetanus. In *Infections of the central nervous system* (2nd edn) (ed. W.M. Scheld, R.J. Whitley, and D.T. Durack), pp. 629–53. Raven Press, New York.

Centers for Disease Control (1991). Diphtheria, tetanus, and pertussis: recommendations for vaccine use and other preventive measures. Recommendations of the Immunization Practices Advisory Committee (ACIP). *Morbidity and Mortality Weekly Report*, **40**, 1–25.

Centers for Disease Control (1995). Recommended childhood immunization schedule—United States, January 1995. *Morbidity and Mortality Weekly Report*, **43**, 959–60.

Lipman, J. and Oh, T.E. (1996). Tetanus. In *Intensive care manual* (4th edn) (ed. T.E. Oh), Chapter 47. Butterworth Heinemann, Oxford.

Sutton, D.N., Tremlett, M.R., Woodcock, T.E., and Nielsen, M.S. (1990). Management of autonomic dysfunction in severe tetanus: the use of magnesium sulphate and clonidine. *Intensive Care Medicine*, **16**, 75–80.

7.11 Brainstem death and management of the potential organ donor

7.11.1 The diagnosis of brainstem death

C. J. Hinds

Key messages

- The diagnosis of death has evolved from recognition of the traditional signs (absence of heart sounds, pulse, and breathing) to the diagnosis of total brain death and, in some countries, to establishing death of the brainstem.
- Death can be conceived as a state in which there is irreversible loss of the capacity for consciousness combined with irreversible loss of the capacity to breathe (and hence maintain a heart beat). Neither would be sufficient alone.
- The diagnosis of brain death can be based on etiological preconditions (certain diagnosis of irremediable structural brain damage due to a disorder that can cause brain death), exclusion of certain specified conditions which might contribute to or cause the coma, and clinical testing.
- Once brain death has been confirmed, mechanical ventilation should be discontinued as soon as possible. This should not be viewed as withdrawing support to allow a patient to die, but rather as ceasing a futile intervention in a patient who is already dead.

Introduction

Death is best viewed as a process during which the function of various organs deteriorates progressively; the precise sequence and time course of the organ failures depends on the nature of the underlying disease and treatment. Therefore a confident diagnosis of death is dependent on defining, with certainty, the point at which this process becomes irreversible. This decision is not always straightforward. The traditional signs of death (unconscious, not breathing, absent pulse) are easily detectable major events, following which the process of dying normally rapidly accelerates and passes the point of no return. However, the institution of cardiopulmonary resuscitation can interrupt, and sometimes reverse, this process so that cessation of heart beat and breathing do not necessarily indicate irreversibility. Therefore the context in which cardiorespiratory arrest occurs must be taken into account when deciding whether cessation of breathing and heart beat should be considered to represent death. Mechanical ventilation can prolong the process of dying in patients whose respiratory center has ceased to function as a result of irreversible brain injury. A person should not be pronounced dead unless the brain is dead; cardiorespiratory arrest is indicative of death only when it persists long enough for the brain to die.

History of brain death

Brain death was first described by two French physicians (Mollaret and Goulon 1959) and termed *coma dépassé* (a state beyond coma). They differentiated *coma dépasse* from *coma prolongé*, the latter being the condition which is now termed 'persistent vegetative state'. In 1968 the Ad Hoc Committee of the Harvard Medical School defined brain death as irreversible coma, with the patient being totally unreceptive and unresponsive, with absent reflexes and no spontaneous respiratory effort during a 3-min period of disconnection from the ventilator. The report unambiguously proposed that this clinical state should be accepted as death. A few years later Mohandas and Chou (1971) suggested that in patients with known but irrepairable intracranial lesions irreversible damage to the brainstem was 'the point of no return' and that the diagnosis 'could be based on clinical judgment', thereby introducing the important concepts of etiological preconditions and a purely clinical diagnosis. Another important contribution was the memorandum issued by the Conference of Medical Royal Colleges (1976) which has underpinned clinical practice in the United Kingdom and many other countries ever since. This emphasized that 'permanent functional death of the brainstem constitutes brain death' and that this should only be diagnosed in the context of irremediable structural brain damage, after exclusion of certain specified conditions which might contribute to or cause the coma. This report also described the use of simple clinical tests to establish the permanent loss of brainstem function. A second memorandum in 1979 equated brainstem death with death itself. In the United States, guidelines issued by a governmental commission state that an individual who has sustained either irreversible cessation of circulatory and respiratory functions or irreversible cessation of all functions of the entire brain, including the brainstem, is dead (Table 1).

The diagnosis of death has evolved from recognition of the traditional signs (coma, absence of heart sounds, pulse, and breathing) to the diagnosis of total brain death and, in some countries, to establishing death of the brainstem. It has been suggested that death should now be conceived as a state in which there is irreversible loss of the capacity for consciousness combined with irreversible loss of the capacity to breath (and hence maintain a heartbeat) (Pallis and Harley 1996). Neither would be sufficient alone. Both are essentially

Table 1 Criteria used for the diagnosis of death in the United States

EITHER

(A) An individual with irreversible cessation of circulatory and respiratory function is dead

1. Cessation is recognized by an appropriate clinical examination
2. Irreversibility is recognized by persistent cessation of functions during an appropriate period of observation and/or trial of therapy

OR

(B) An individual with irreversible cessation of all functions of the entire brain, including the brainstem, is dead

1. Cessation is recognized when evaluation discloses findings of (a) and (b):
   (a) cerebral functions are absent
   (b) brainstem functions are absent
2. Irreversibility is recognized when evaluation discloses findings of (a) and (b) and (c):
   (a) the cause of coma is established and is sufficient to account for the loss of brain functions
   (b) the possibility of recovery of any brain functions is excluded
   (c) the cessation of all brain functions persists for an appropriate period of observation and/or trial of therapy

brainstem functions, and therefore death of the brainstem is considered by many to be equivalent to death of the individual.

Ethical considerations

Although they have never been tested in law, the recommendations of the Conference of Royal Colleges (1976), or similar guidelines elsewhere in the world, are believed by most practising clinicians to be a correct and ethical means of diagnosing brain death. Therefore death can be declared once death of the brainstem has been confirmed, and most would argue that mechanical ventilation should then be discontinued as soon as possible. This should not be viewed as withdrawing support to allow a patient to die, but rather as ceasing a futile intervention in a patient who is already dead. Therefore it is clear that, even if transplantation therapy did not exist, the ability to diagnose brain death with confidence contributes to the humane practice of intensive care, and most clinicians find the decision to discontinue ventilating a brain-dead patient relatively straightforward. The only situation in which it might be considered inappropriate to discontinue ventilating a brainstem-dead patient is when she is pregnant. In some such cases it has proved possible to support the mother for up to 3 months to permit growth of the fetus until delivery by Cesarean section can be performed safely.

Diagnosis of brainstem death

The diagnosis of brainstem death (Table 2) must only be considered when the patient is in apneic coma as a result of irremediable structural brain damage caused by a disorder which is known to be capable of producing brain death.

The possibility that the patient's condition can be attributed to the effects of drugs, hypothermia, or metabolic or endocrine disorders must have been excluded before proceeding to the assessment of brainstem function. It is recognized that some metabolic and endocrine disturbances are a likely accompaniment of brainstem death (e.g. hypernatremia, diabetes insipidus), but these are the effect rather than the cause of the patient's condition and do not necessarily preclude the diagnosis of brainstem death.

Seizure activity and posturing, whether decorticate or decerebrate, are incompatible with a diagnosis of brain death. However, it is important to recognize that reflex movements of the limbs and trunk may occur after brainstem death. The physician must be able to explain the significance of such movements to the patient's relatives and friends, including the fact that they are reflex rather than voluntary.

A variety of acquired neuropathies and myopathies may complicate critical illness. Extreme neuromuscular weakness should not be confused with true apnea. It does not affect pupillary, corneal, oculocephalic or vestibulo-ocular reflexes, and therefore is usually easily distinguished from brainstem death. However, in all such cases brainstem death criteria are not satisfied because the diagnosis of a disorder that can lead to brain death has not been established, and there is no evidence of irremediable structural brain damage.

In the United Kingdom it is recommended that the examination of brainstem function should be performed by two physicians, either the consultant in charge of the case and one other (who should be clinically independent of the first and registered for more than 5 years) or the consultant's deputy (provided that he or she has been registered at least 5 years and has adequate appropriate experience) and one other. Neither should be a member of the transplant team. The tests should be performed on two separate occasions, with the interval between the two being agreed by all the staff involved. The first examination should normally be performed a minimum of 6 h after the onset of coma. However, when a patient has suffered from cardiac arrest, hypoxia, or severe circulatory insufficiency with an indefinite period of cerebral hypoxia, or is suspected of having cerebral air or fat embolism, it may take longer to establish the diagnosis. In such cases it is recommended that the first examination should be delayed for at least 24 h after restoration of an adequate circulation. In some patients the primary pathology may be unclear, and a confident diagnosis may only be reached by continued clinical observation and further investigations. A neurologist should be consulted if the diagnosis is in doubt.

In the United States various authoritative groups and institutions have issued guidelines for the diagnosis of brain death. These guidelines have legal implications and some of the recommendations vary between states, for example in the suggested time period over which the diagnosis is established and in recommendations for the use of confirmatory tests. In general the preconditions and exclusions are similar to those recommended in the United Kingdom. The cause of coma must be known and that cause must be adequate to explain the coma. The physical examination is also similar (Table 3), although the intention is to demonstrate absent cerebral function as well as loss of all brainstem reflexes and the recommendations for apnea testing differ in detail from those in the United Kingdom. Most authors in the United States recommend that the findings should be confirmed on at least two occasions separated by about 12 h, although some suggest 24 h or more.

Table 2 Criteria used for the diagnosis of brainstem death in the United Kingdom

Preconditions

1. The patient is in apneic coma (i.e. unresponsive and on a ventilator, with no spontaneous respiratory efforts).
2. Irremediable structural brain damage due to a disorder that can cause brainstem death has been diagnosed with certainty (e.g. head injury, intracranial hemorrhage).

Exclusions

1. The possibility that unresponsive apnea is the result of poisons, sedative drugs, or neuromuscular blocking agents must be excluded. The drug history should be carefully reviewed and adequate time allowed for the persistence of drug effects to be excluded. A nerve stimulator may be used to ensure that the patient is not paralysed. Blood and urine should be tested for the presence of drugs if there is any doubt.
2. Hypothermia must be excluded as a cause of coma. It is recommended that the central body temperature should be more than 35°C.
3. There must be no significant metabolic or endocrine disturbance that could produce or contribute to coma or cause it to persist. There should be no profound abnormality of the plasma electrolytes, acid–base balance, or blood glucose levels.

Assessment of brainstem function

1. Before proceeding to a formal assessment of brainstem function it is recommended that the absence of oculocephalic reflexes should be confirmed (i.e. when the head is rotated from side to side, the eyes move with the head and therefore remain stationary relative to the orbit). In a comatose patient whose brainstem is intact, the eyes will rotate relative to the orbit (i.e. doll's eye movements are present). It is then necessary to establish that all brainstem reflexes are absent.
2. The pupils should be fixed and unresponsive to bright light. Both direct and consensual light reflexes should be examined. (Difficulty in interpretation may be experienced when there has been direct trauma to the eye and/or optic nerve, and if topical mydriatics have been used.) Pupil size is irrelevant, although most often they will be dilated.
3. Corneal reflexes should be absent. (Again, there may be difficulty when the eyelids are bruised and edematous due to trauma.)
4. Vestibulo-ocular reflexes should be absent ('caloric testing'). If 20 ml of ice-cold water is slowly instilled into the external auditory meatus, it will stimulate the tympanic membrane. If the patient's brainstem is intact, reflex eye movements will occur within 20–30 s, provided that access to the tympanic membrane is not prevented (e.g. by blood or wax in the meatus). This should be checked by direct inspection. It should be remembered that gentamicin can cause end-organ poisoning, and central pathways may be impaired by drugs. Severe trauma may prevent caloric testing on one side or the other.
5. There should be no motor responses within the cranial nerve territory to painful stimuli applied centrally or peripherally. Spinal reflexes may be present.
6. There must be no gag or cough reflexes in response to pharyngeal, laryngeal, or tracheal stimulation.
7. Spontaneous respiration should be absent. This test is crucial to the diagnosis of brainstem death. One commonly used technique is to ventilate the patient with 5% carbon dioxide in 95% oxygen for 10 min and then disconnect the ventilator for a further 10 min. Oxygenation is maintained by insufflation with 100% oxygen at high flows via a catheter placed in the endotracheal tube. The patient is observed for any signs of spontaneous respiratory efforts. A blood gas sample should be obtained during this period to ensure that $Paco_2$ is sufficiently high to stimulate spontaneous respiration ($>$6.7 kPa, 50 mmHg). It must be remembered that a few patients with severe chronic obstructive pulmonary disease are dependent on a hypoxic drive to respiration. Confirmatory tests such as electroencephalography and assessment of cerebral blood flow (e.g. carotid angiography) may be misleading (see text) and should be reserved for those cases in which there is uncertainty regarding the cause and reversibility of coma.

Confirmatory tests

A recent review of the criteria for diagnosing brainstem death reiterated that confirmatory tests (e.g. imaging, electroencephalography, or evoked potentials) should not presently form part of the diagnostic requirements for brainstem death (Working Group convened by the Royal College of Physicians 1995). Similarly, in New York and most other states of the United States, clinical criteria are considered to be sufficient to diagnose brain death. Confirmatory tests are reserved for those cases in which there is uncertainty regarding the cause and the reversibility of the coma.

Electroencephalography

Some still consider it to be important to confirm absence of cerebral electrical activity before concluding that a patient is brain dead. Others recommend the use of evoked potentials to assist in the diagnosis since these will be preserved when electroencephalographic silence is attributable to drugs. In the United Kingdom most argue that electroencephalography would only be relevant if the objective was to establish death of the whole brain, not just the brainstem, and that, in any case, an isoelectric electroencephalogram cannot exclude activity in deeper areas of the brain and may have other causes, such as

Table 3 United States recommendations for clinical testing for brain death

1. Conditions which may mimic brain death (e.g. hypothermia, drug overdose) must be excluded
2. Spinal reflexes should not be tested and may be present
3. Decorticate and decerebrate posturing must be absent, but can be difficult to distinguish on occasion from complex spinal reflexes
4. Pupils must be unresponsive to light but need not be dilated or symmetrical
5. Corneal reflex must be absent
6. Oculovestibular reflexes must be absent
7. Apnea must be confirmed
8. Ciliospinal, jaw, and snout reflexes, as well as plantar responses, may continue in brain death and should not be tested

barbiturate coma. Patients have been reported in whom the electroencephalogram was isoelectric but brainstem reflexes were preserved, although this is extremely unusual.

Demonstrating absence of cerebral blood flow

Absence of cerebral blood flow may also be used to confirm death of the whole brain. Four-vessel cerebral angiography was used in the past, but radionuclide cerebral scanning, which cannot document absent flow in the vertebrobasilar circulation, has been introduced more recently. Color flow duplex scanning may be used to complement radionuclide cerebral scanning. Although total absence of cerebral blood flow provides compelling evidence of brain death, findings can be equivocal in the presence of unequivocal clinical signs of brain death.

Validity of brainstem death criteria

A small minority of clinicians continue to dispute the validity of brainstem death procedures. They claim that it is not legally necessary to diagnose brainstem death in order to discontinue mechanical ventilation and hence that the concept is only required to facilitate organ transplantation. They further suggest that the clinical brainstem tests in routine use are not exhaustive, that they cannot ensure the permanent absence of all brainstem function, and that their satisfaction does not ensure that the brain has been destroyed or that higher brain function has ceased irreversibly. A few have also claimed that there has been no satisfactory 'controlled trial' to assess the prognostic value of the clinical signs of brainstem death. However, a total of over 1000 cases where the patient has been supported after brainstem death has been established clinically have been reported in the world literature. The largest single series was from three neurosurgical centers in the United Kingdom. Asystole was the invariable outcome, usually within a few days.

References

Conference of Medical Royal Colleges and their Faculties in the United Kingdom (1976). Diagnosis of death. *British Medical Journal*, **ii**, 1187–8.

Mohandas, A. and Chou, S.N. (1971). Brain death—a clinical and pathological study. *Journal of Neurosurgery*, **35**, 211–18.

Mollaret, P. and Goulon, M. (1959). Le coma dépassé (mémoire préliminaire). *Revue Neurologique*, **101**, 3–15.

Pallis, C. and Harley, D.H. (1996). *ABC of brainstem death*. BMJ, London.

Working Group convened by the Royal College of Physicians and endorsed by the Conference of Medical Royal Colleges and their Faculties in the United Kingdom (1995). Criteria for the diagnosis of brainstem death. *Journal of the Royal College of Physicians of London*, **29**, 381–2.

7.11.2 Managing the potential organ donor

C. J. Hinds

Key messages

- The increase in organ donors has failed to keep pace with the accumulation of patients with endstage organ failure who might benefit from transplantation.
- Retrieval of multiple organs should now be the objective in all brain-dead potential donors.
- Assiduous supportive treatment is essential to prevent deterioration of initially suitable donors. This will both increase donation rates and improve graft survival and function.
- The incidence of complications increases progressively after brain death and, although adequate time must be allowed to confirm the diagnosis, unnecessary delays must be avoided.
- Aggressive supportive treatment of the donor should be continued intraoperatively.

Introduction

As a result of advances in immunosuppression, surgical techniques, and intensive care, transplantation of cadaveric organs is now a life-saving and life-enhancing treatment for patients with endstage organ failure.

Improving organ retrieval rates from individuals

Medical contraindications to donation

General medical contraindications, such as age, infection, and malignancy, may prevent as many as 18 per cent of those confirmed as brainstem dead from becoming organ donors. Moreover, on occasions there may be important disagreements between transplant co-ordinators, intensive care clinicians, and transplant surgeons regarding general and organ-specific contraindications to donation. This emphasizes the importance of early discussion with the transplant team or co-ordinator about every brain-dead potential donor.

Elective ventilation

Although the number of donors can be substantially increased by transferring patients with lethal cerebrovascular accidents and incipient brainstem death to the intensive care unit for elective mechanical ventilation until brainstem death is confirmed (Feest *et al.* 1990), this practice raises a number of legal and ethical issues and is associated with a variety of practical difficulties. Some have suggested that the acceptance of elective ventilation for organ donation, a procedure from which the patient cannot benefit, represents the first step along a particularly dangerous road. The legality of elective ventilation is also questionable, and currently this practice is proscribed in the United Kingdom. Many have also been concerned that elective ventilation would place unacceptable demands on limited facilities and might affect staff morale adversely. Perhaps more worrying is the suggestion that the procedure could produce significant numbers of patients in a persistent vegetative state, although experience to date suggests that such an outcome is unlikely provided that a strict protocol is used.

Securing permission for organ retrieval

This involves attempting to determine the previously stated wishes of the deceased. If this is not possible, then permission for, or lack of objection to, organ retrieval is sought from the next of kin. In some instances there may be difficulty in deciding who this is, in which case the most senior, the closest, or at least the most concerned relative should be identified and asked to canvass views within the family and to act as spokesperson. In the event of disagreement amongst the relatives it might be considered reasonable to accede to the wishes of the nearest or most interested, or, in a situation where there are several close relatives, the majority view. If there is a serious unresolvable dispute, it may be wiser not to proceed to organ donation.

Difficulties may also occur when the next of kin attempt to override the clearly stated wish of the deceased to donate organs in the event of his or her death. This may arise, for example, when the patient was carrying a donor card or had registered his or her wishes on a computer database. Currently the generally accepted legal view in the United Kingdom is that the person lawfully in possession of the body is the hospital management, and therefore the relatives cannot veto the patient's wishes. In practice, however, once the issues have been fully discussed, it is almost always advisable to concur with the wishes of the relatives.

All potential donors must be tested for HIV, and it is recommended that permission for this should be obtained from the relatives. A positive test must be revealed to those at risk of contracting the disease from the deceased. This requires sensitivity and tact, combined with the provision of appropriate counseling and follow-up services.

It is important that discussions with the relatives should specifically address the issue of multiple rather than limited organ retrieval.

Multiple organ donation

Retrieval of multiple organs, which can be performed without jeopardizing the function of individual organs, should now be the objective in all brain-dead potential donors. Assiduous supportive treatment is essential to prevent deterioration of initially suitable donors as this will both increase donation rates and improve graft survival and function. Complications related to the profound physiological disturbances consequent on brainstem death are common (Table 1) and may jeopardize organ function. The incidence of complications increases progressively after brainstem death and, although adequate time must be allowed to confirm the diagnosis, unnecessary delays must be avoided. Meanwhile meticulous intensive care is required to sustain organ perfusion and to improve graft survival and function. In particular a high level of nursing case, with a bedside nurse-to-patient ratio of 1:1 or sometimes 2:1, is required. The introduction of a comprehensive standardized donor management regimen, including pulmonary artery catheterization and hormone replacement, has been shown to increase the proportion of hearts which are suitable for transplantation (Wheeldon *et al.* 1995).

Table 1 Consequences of brain death
Hypotension
Requirement for multiple transfusions
Arrhythmias
Cardiac arrest requiring cardiopulmonary resuscitation
Disseminated intravascular coagulation
Pulmonary edema
Cardiovascular collapse
Hypoxia
Acidosis
Bacteremia
Diabetes insipidus

Criteria for organ donation

In all cases brain death must have been established and the consent of the next of kin and, where appropriate, the coroner must have been obtained. Blood samples should be obtained for screening for hepatitis and HIV, as well as blood group determination and tissue typing. The organs must be removed by a surgeon who has examined the donor and is satisfied with the diagnosis of brainstem death. The procedure must be performed under full operating conditions in theater. If there is any doubt, the local transplant co-ordinator should be contacted early for advice on donor/individual organ suitability.

Kidney

Kidney donors must be brainstem dead, 2 to 70 years old, adequately perfused and hydrated, and artificially ventilated. They should be free from hepatitis antigen and HIV, malignant disease (except primary brain tumor), systemic infection, chronic urinary tract infection (acute urinary tract infection related to catheterization is not a contraindication), and renal disease. The following are not necessarily contraindications: diabetes, pneumonia, hypertension, or hypotension.

Heart or heart–lung

Donors should be brainstem dead, preferably under 50 years old, adequately perfused and hydrated, and artificially ventilated. They should be free from hepatitis antigen and HIV, malignant disease (except primary brain tumor), systemic infection, diabetes mellitus, ischemic heart disease, and cardiac murmurs. There should be no family history of heart disease and no history of heavy smoking.

A chest radiograph and 12-lead ECG should be obtained. Myocardial dysfunction associated with brain death may be reversible and is not necessarily a contraindication to donation. Rarely, coronary angiograms may be requested by the transplant team in male donors over 35 years old and female donors over 40 years old.

Liver

Most suitable kidney donors are also suitable liver donors. There should be no history of chronic drug or alcohol abuse.

Cornea

All adults are suitable donors provided that there is no history of eye disease, intraocular surgery, syphilis, hepatitis, HIV infection, Creutzfeldt–Jakob disease (CJD), including new variant CJD, or postinfectious polyneuritis. The eyes can be removed up to 24 h after cardiorespiratory arrest, but preferably within 1 h.

Supporting the organ donor

Mangement of the organ donor has been reviewed by Timmins and Hinds (1991).

Cardiovascular system

Pathophysiology

When brainstem death occurs because of raised intracranial pressure, the sudden onset of brainstem ischemia is associated with an agonal period of intense autonomic activity and large increases in circulating catecholamine levels. This may precipitate vasoconstriction, hypertension, myocardial ischemia, bradycardia, and various arrhythmias, including ventricular ectopics, supraventricular tachycardia, and heart block. Left atrial pressure may increase sufficiently to cause pulmonary edema. These changes may be associated with significant myocardial damage which can be severe and irreversible, particularly when the rise in intracranial pressure is sudden. Moreover, the myocardial cells are depleted of high-energy phosphates and glycogen; it has been suggested that this defect in oxidative metabolism might be caused by mitochondrial inhibition mediated by endocrine disturbances, in particular reductions in tri-iodothyronine (T_3) (see below). These changes, combined with desensitization of β receptors and marked alterations in β-adrenergic signal transduction, are associated with impaired right and left ventricular function, which could contribute to early post-transplantation cardiac failure. Administration of high-dose catecholamines may exacerbate myocardial dysfunction by decreasing β-adrenergic receptor numbers. The period of intense autonomic activity is followed by a fall in catecholamine levels, loss of sympathetic tone, and vasodilatation, sometimes with bradycardia and a nodal rhythm unresponsive to atropine.

Monitoring

Assessment of the cardiovascular system is particularly difficult in brainstem-dead patients. Central venous and intra-arterial pressure monitoring, as well as hourly measurements of urine volume, are essential and the core/peripheral temperature gradient should be monitored routinely, but these may be misleading as indicators of tissue perfusion. Therefore pulmonary artery catheterization has been recommended for those in whom blood pressure and urine output fail to improve despite achieving an apparently adequate central venous pressure, as well as for patients with valvular heart disease, hypertrophic cardiomyopathy, pulmonary edema, or left ventricular dysfunction secondary to direct cardiac trauma or cardiac tamponade. Pulmonary artery catheterization is also justified to guide the selection of the most appropriate pressor agent and the dose to be used.

Management

Hypotension may precipitate acute tubular necrosis and decrease graft survival. Initially hypotension should be treated with aggressive fluid resuscitation aiming to maintain the hematocrit at approximately 0.3. Intravascular volume expansion can be achieved using normal saline or lactated Ringer's solution, although many prefer to use colloidal solutions. There should be no hesitation in administering inotropes or vasopressors to those who fail to respond to volume expansion, although the use of high-dose inotropes may depress myocardial β-adrenergic receptors and adversely affect tissue blood flow. Most recommend dopamine as the first choice because at low infusion rates it causes vasodilation of renal and mesenteric vessels and therefore might help to preserve intra-abdominal organs. However, the direct tubular effect of dopamine could exacerbate polyuria and, at higher doses, vasoconstriction might impair graft survival. Because the release of endogenous catecholamines is believed to be an important factor in the development of myocardial damage after brain death, the use of dopamine, which can release norepinephrine (noradrenaline) from adrenergic nerve endings, might theoretically exacerbate myocardial injury. Dobutamine can be used in those who fail to respond to a moderate dose of dopamine and when specifically indicated (e.g. in those with myocardial contusion), although β-mediated peripheral vasodilation may exacerbate hypotension in patients with a low systemic vascular resistance. Isoproterenol (isoprenaline) has been used in conjunction with dopamine in pediatric donors, but otherwise its use is limited by tachycardia and peripheral vasodilation. Norepinephrine has been recommended as a short-term measure in potential organ donors with profound vasodilation in order to maintain systemic blood pressure during fluid resuscitation. Epinephrine (adrenaline) may be useful in those with refractory hypotension. In the unlikely event of donor surgery involving only the heart, it has been suggested that a pure α-agonist, which would maintain coronary perfusion pressure and limit tachycardia, might be the ideal choice.

Endocrine changes

Diabetes inspidus

Deficiency of ADH (diabetes insipidus) leads to an inappropriate diuresis, hypovolemia, hyperosmolality, and hypernatremia. Treatment initially involves replacing measured urine volume and electrolyte losses with appropriate fluids. Hypernatremia must be avoided by using hypotonic solutions such as 0.45 per cent saline or 5 per cent dextrose with added electrolytes, guided by plasma levels and urinary losses. It should be remembered that most colloidal solutions contain considerable quantities of sodium.

Treatment with vasopressin has been controversial. It causes a dose-dependent vasoconstriction that could be associated with a higher incidence of acute tubular necrosis and a lower rate of graft survival. Administration of vasopressin can be avoided by replacing the large volumes of dilute urine with an electrolyte soluton; this regime may be associated with hemodynamic improvement and a decrease in plasma renin activity. However, administration of a low-dose arginine vasopressin infusion may decrease plasma hyperosmolality, increase blood pressure, and reduce inotrope requirements. The current consensus favors the use of vasopressin, often given as a continuous intravenous infusion in the lowest dose (0.5–6 IU/h) required to reduce urine output to 1.5 to 3 ml/kg/h. However, there is considerable disagreement about the volume of urine above which vasopressin should be administered. Compared with vasopressin, desmopressin has a longer duration of action coupled with a greater antidiuretic effect relative to its pressor activity and therefore has been recommended as an alternative to vasopressin.

Thyroid hormones

Circulating levels of thyroxine (T_4) and T_3 fall in brain-dead experimental animals, while levels of thyroid-stimulating hormone remain unchanged. Similar reductions in T_3 have been reported in human donors. This reduction in circulating levels of thyroid hormones could be responsible for the inhibition of mitochondrial function, reduced aerobic metabolism, and deterioration in myocardial function seen after brainstem death. In support of this hypothesis, administration of T_3 can reverse donor myocardial dysfunction. However, the fall in T_3 levels in brainstem-dead patients is similar to that found in most severe illnesses (sick euthyroid syndrome) and replacement is of no proven benefit in such patients. Therefore T_3 supplementation may be unnecessary, and it is possible that the fall in T_3 is a protective mechanism.

Circulating levels of ACTH and cortisol fall after brain death in experimental animals, and lesser falls have also been recorded in organ donors. However, reductions in serum cortisol levels following brainstem death do not seem to be related to the incidence of hypotension. Glucagon and insulin levels also fall. The administration of supplementary T_3, cortisol, and insulin to organ donors may improve hemodynamics, reduce lactate levels, and reverse acidosis, thereby increasing retrieval rates of transplantable organs and improving graft function.

Hyperglycemia

Hyperglycemia is a frequent complication of brain death because of the administration of glucose-containing fluids to replace large urinary losses, reduced insulin secretion, and increased levels of catecholamines. The consequent increase in extracellular osmolality may lead to intracellular dehydration, an osmotic diuresis, and electrolyte disturbances; hyperglycemia may be associated with decreased pancreatic allograft survival. Therefore it is important to control blood glucose levels with intravenous insulin infusions.

Renal support

Episodes of hypotension can precipitate acute tubular necrosis and reduce graft survival. A donor urine output of less than 100 ml/h in the hour preceding nephrectomy has been shown to be associated with an increased incidence of acute tubular necrosis in the transplanted kidney. Some suggest that if urine output is low, despite an adequate circulating volume and perfusion pressure, a low-dose dopamine infusion should be administered. If this fails to produce a diuresis, small intravenous doses of furosemide (frusemide) (10 mg) or a furosemide infusion may be given. Administration of mannitol (0.5 g/kg) 1 h before harvest has been recommended.

Respiratory support

The aim of respiratory support is to optimize oxygen delivery to vital organs by ensuring adequate oxygenation with minimal cardiovascular depression. The risk of pulmonary infection should be minimized by aseptic tracheal suction and postural drainage, as there are no constraints on these procedures once the diagnosis of brainstem death has been established. $P\text{aCO}_2$ should be allowed to return to normal levels in order to avoid the vasoconstriction associated with hypocarbia. Since carbon dioxide production falls after brainstem death, this may require a considerable reduction in minute volume or the addition of dead-space. The inspired oxygen concentration should be adjusted to achieve adequate saturation of hemoglobin with oxygen but, particularly if the lungs are to be transplanted, the fractional inspired oxygen concentration should, if possible, be maintained below 0.5 to 0.6 to avoid oxygen toxicity. High levels of positive end-expiratory pressure (PEEP), which can reduce cardiac output and exacerbate ventilator-induced lung injury, should be avoided, although if the lungs are to be transplanted, a low level of PEEP (5 cmH_2O) is recommended to prevent alveolar collapse. Other recommendations for potential lung donors include the avoidance of barotrauma by keeping peak inspiratory pressure below 30 cm H_2O and the use of colloids rather than crystalloids to maintain the circulating volume.

Coagulation

Tissue fibrinolytic agents and plasminogen activators are released from areas of ischemic or necrotic brain. Fibrin deposits and free hemoglobin may jeopardize the function of transplanted organs. To minimize these problems, support with blood products and early procurement of organs is advised. Epsilon-aminocaproic acid should not be given because of the risk of microvascular thrombosis in donor organs.

Hypothermia

Temperature regulation is markedly impaired after brainstem death because of a reduction in heat production secondary to the fall in metabolic rate, loss of muscular activity, and peripheral vasodilation. Although mild hypothermia may protect vital organs, temperatures below 32 °C can result in decreased cardiac performance and coagulopathy. Moreover, brainstem death cannot be diagnosed in a hypothermic patient (core temperature below 35 °C). Core temperature should be maintained above 35 °C by increased insulation and active warming of the patient.

Intraoperative management

The principles of donor management described above should be continued into the intraoperative period. To avoid reflex neuromuscular activity a muscle relaxant that does not depress the cardiovascular system (e.g. pancuronium bromide) should be used. Nitroglycerin, nitroprusside, and isoflurane have all been recommended to control reflex hypertension. Hypotension may be precipitated by manipulation of the heart and great vessels or flushing donor organs with cold preservation solution. Administration of pressor agents, particularly in high doses, may jeopardize organ perfusion, and some recommend pulmonary artery catheterization during retrieval surgery in multiple organ donors to minimize the use of inotropes and vasopressor drugs. If cardiac arrest occurs, cardiopulmonary resuscitation should be instituted while procurement of liver and kidneys proceeds rapidly with cross-clamping of the aorta at diaphragm level and infusion of cold preservation solution into the distal aorta and portal vein.

References

Feest, T.G., Riad, H.N., Collins, C.H., Golby, M.G.S., Nicholls, A.J., and Hamad, S.N. (1990). Protocol for increasing organ donation after cerebrovascular deaths in a district general hospital. *Lancet*, **335**, 1133–5.

Timmins, A.C. and Hinds, C.J. (1991). Management of the multiple-organ donor. *Current Opinion in Anaesthesiology*, **4**, 287–92.

Wheeldon, D.R., Potter, C.D.O., Oduro, A., Wallwork, J., and Large, S.R (1995). Transforming the 'unacceptable' donor: outcomes from the adoption of a standardized donor management technique. *Journal of Heart and Lung Transplantation*, **14**, 734–42.

7.11.3 Management of the relatives of a potential organ donor

B. Watson and V. Morgan

Key messages

- The family must be given time to accept the inevitability of death and to understand the concept of brainstem testing.
- Staff must understand and accept the validity of brainstem testing.
- Where there are no medical contraindications, the possibility of organ donation should be offered to the family.
- Whatever the outcome, the family need to be supported throughout and made aware of the time-scale of events.
- Organ donation gives a positive outcome for both the family and intensive care unit staff.

Introduction

Death is not uncommon in the intensive care unit (ICU); however, situations requiring the diagnosis of brainstem death are relatively rare (Gore *et al.* 1989). A significant aspect of the diagnosis of brainstem death is that it is often carried out following an acute event and therefore relatives have usually had only a short time in which to prepare themselves. This is an important consideration when deciding to approach relatives for organ donation, and time must be allowed for the family to understand fully the implications of the tests before such a request is made.

Unless there are medical contraindications, all patients who have been diagnosed as brainstem dead should be considered for organ donation. It is important to establish suitability prior to approaching the family in order to spare them the added stress of addressing this issue when it may be to no avail (Crombie and Watson 1987).

Where there is an established program of organ donation and transplantation, there is usually widespread public awareness and many families may broach the subject spontaneously. Even if the family do not mention organ donation, this is no indication that they would not agree to donate if asked. In either case ICU staff have a duty to carry out the wishes of the relatives and the donor whenever possible.

Brainstem-dead patients

It is essential that all staff understand the brainstem tests and accept their validity, as confusion may compromise their support in helping the family to come to terms with the death. If staff are able to cope personally with death and grief, it is easier for them to communicate and establish a rapport with grieving relatives without becoming detached or overwhelmed (Morgan 1995).

The family should have been kept informed and supported throughout the patient's time in the ICU. They need to know when the brainstem tests are to be done so that they can begin their adjustment to the finality of the situation. Lack of communication can have a detrimental affect on the family and may influence their long-term memories of events surrounding the death. Once death has been declared, it is important that staff use the word 'dead' in order to reinforce this finality. Other aspects of care which could be misleading, such as talking to the patient after death, must also be carefully considered (Morgan 1995).

When the cause of death is straightforward and there is no legal requirement for the coroner to be notified, the death certificate can be issued. It is widely accepted that relatives may not assimilate and retain facts given at this acute stage, and it might be advisable to provide information setting out the subsequent course of events in the form of a booklet.

Making the request

In countries which have transplant programs utilizing cadaveric organs there will also be a national/state system, which is legal, professional, or both, facilitating organ retrieval. Some countries, many in Europe, have presumed consent (opting out) laws which allow the removal of organs and tissue for transplantation without the need to obtain consent from the next of kin unless the deceased has previously stated that he or she would not wish this to happen. In practice, however, most countries with presumed consent still offer the family the right of veto even though this is not necessarily a legal requirement. The alternative system is one of informed consent (opting in). Here, either the deceased has previously expressed a wish to donate, which the family then honors, or the next of kin make the decision on behalf of the deceased.

In the United States many states have introduced the legislation of required request which makes it mandatory for staff to approach the family about organ donation if the patient is suitable.

Whatever system is in place, religious and cultural beliefs must be respected. Worldwide, the beliefs of major religions pose no difficulties for organ donation as the gift of life is considered to be the ultimate gift. However, the personal feelings of the individual must be respected, as cultural beliefs can have a very strong influence on attitudes to death and the integrity of the body.

Each ICU should have a protocol for organ donation and all grades of staff should be made aware of the steps to follow in the event of identification of a potential donor. The responsibility for actually approaching the family may always be taken by the same people, and it is important that other members of staff know who they are (Stein *et al*. 1995). If there is a transplant co-ordinator, there will already be an established procedure with a service provided for advice and referrals.

Allowing the family to make their decision based on what they know the patient would have wanted is their right. In donating, they have the chance of gaining comfort from the knowledge that some good will have come from the unexpected death (Watson and Morgan 1995). Donation, in helping others to live, is a positive outcome for family and staff alike, and staff can also gain comfort from knowing that by offering the choice, whatever the outcome, they have done all that they can to support the family.

The referral process

Once brainstem death has been confirmed, the referral process can begin if the family have no objection to organ donation. At this point it needs to be established which organs and tissues are suitable for donation and which the family wish to donate, as it is their prerogative to make the choice. Even if seemingly blanket permission is given, it is essential to obtain consent for each organ/tissue to be removed and to ascertain whether it should be used for clinical purposes only or can be used for research.

The family must never be put under pressure to donate and should be reassured that they can change their minds at any time. Their welfare is paramount, and if they wish to be with the patient when ventilation is withdrawn then this must take priority over organ donation. It is far better for the relatives to leave the hospital knowing that they have been supported throughout the decision-making process rather than feeling that they have been coerced into doing something with which they are not, and will never be, comfortable.

The referral process can take some hours and, again, the family need to be aware of the time-scale. If they wish to stay until the patient is transferred to the operating room, then of course they must be allowed to do so. However, experience has shown that most families say their last goodbyes before then. They should also be given the opportunity to view the body afterwards, as for some families this is a necessary part of the transition through the grieving process.

Follow-up

All families should be given the opportunity to receive written follow-up information on the outcome of the donation, and this is frequently provided by the transplant co-ordinator. It should be made clear beforehand that both donor and recipient anonymity will be maintained and that no personal details will be released. Even if the family does not wish to be given recipient information, it is considered good practice to send a letter offering condolences and acknowledging their thoughtfulness in agreeing to donation. This also establishes a point of contact if they should wish to make enquiries at a later date.

References

Crombie, A. and Watson, B. (1987). Donors welcome. *Nursing Times*, 18 February, 29.

Gore, S.M., Hinds, C.J., and Rutherford, A.J. (1989). Organ donation from intensive care units in England. *British Medical Journal*, **299**, 1193–7.

Morgan, V. (1995). A discussion of the legal and ethical issues surrounding brainstem death testing and consent for cadaveric organ donation. *Care of the Critically Ill*, **11**, 20–2.

Stein, A., Hope, T., and Baum, J.D. (1995). Organ transplantation: approaching the donor's family. *British Medical Journal*, **310**, 1149–50.

Watson, B. and Morgan, V. (1995). Brainstem death, organ donation and the role of the transplant coordinator. *TECHNIC*, **145**, 8–10.

7.11.4 The organ donation process

I. J. A. Davidson and D. A. Payne

Key messages

- Organ donation should be increased by educating the public and family, emergency, and intensive care physicians.
- Strict adherence to ethical principles, national and international co-ordination, and medical priorities for organ attribution are essential.

Introduction

Organ donation is very personal and intensely emotional. For socio-economic, cultural, political, and other reasons, organ procurement and transplantation vary greatly between countries and societies. The organ shortage is striking and tragic in view of the large number of existing potential donors. Recent studies have estimated the annual identified donor pool in the United States to be approximately 7000–11 000, or 29 to 44 donors per million population. Since the actual number of recovered cadaveric donors is approximately 5400, this translates into a procurement efficiency of only 40 to 65 per cent depending on the donor criteria used. The number of patients waiting for solid organs has increased in recent years (55 994 in November 1997 in the United States). If we consider kidney donation alone, there are currently 37 859 patients on the waiting list for a kidney transplant in the United States (Table 1). In 1995, 68 870 new renal failure patients began dialysis. Only about 12 000 of the roughly 214 000 patients on dialysis treatment receive a transplant from 7710 cadaver donors and 3389 living donors. Each year, more than 2000 dialysis patients on the transplant waiting list die because of a lack of transplantable organs.

In the United States, the overall solid organ donation rate ranges from 15 to 38 donors per million population, with the national

Table 1 Solid organ waiting list (15 November 1997) and transplants performed in 1996

	Waiting list	Transplants performed in 1996
Kidney	37 859	12 039
Liver	9323	4062
Pancreas	356	172
Kidney+ pancreas	1596	850
Intestine	94	45
Heart	3869	2343
Heart+ lung	233	39
Lung	2588	805
Total	55 918	20 355

Data from United Network for Organ Sharing Registry.

average at approximately 20 donors per million and an average of 3.5 organs recovered from each donor. If the donor rate was increased by just 5 per million, over 4500 additional lives could be saved each year. However, only about 1 per cent of all in-patient hospital mortalities will satisfy the existing criteria for brain death determination.

Each organ donor may potentially improve or save the lives of seven or more individuals by providing two kidneys, one heart, two lungs, one liver, and one pancreas. Additionally, a liver may be split to benefit two recipients. Also, organs traditionally not investigated for transplantation, such as the small intestine, are now being utilized with increasing frequency.

Some strategies for increasing organ donation are listed in Table 2. To compensate for lack of available cadaveric organs, many centers have extended their living donor program to include unrelated donors such as spouses and friends. The outcome of such unrelated donation is actually better than cadaver donation, probably reflecting a shorter ischemia time and better-quality organs. Living related partial liver and pancreas (tail) donation are also used in a few selected centers. Research efforts should be directed towards determining the causes of denial of consent for organ donation. By identifying the salient characteristics of donor versus non-donor families, and regional and cultural differences in the acceptance of organ donation, effective measures to improve procurement efficiency can be instituted.

Table 2 Suggested strategies to increase organ donation

Use of living donors (related and unrelated)
Government interaction by supportive legislation
Education of the public and medical professionals
Use of marginal donors
Changes in hospital policies regarding organ donation
Routine referral of all deaths to an organ procurement organization

Most Western societies have national organizations which promote, direct, and regulate organ procurement and transplantation. In some countries, including Belgium and Austria, consent for organ donation is presumed unless the citizen signs a document negating donation. However, in the majority of Western countries, organ donation continues to rely on consent from the next of kin, regardless of the expressed wishes of the potential donor. Thus the very emotional and personal decision to donate organs must be made at the time of death. There has been no demonstrable tendency toward increased organ procurement, despite numerous federal and state programs designed to heighten donor awareness and opportunity.

Clearly, education of health care personnel is a key factor. The process of brain death determination and the medical management of the potential donor have been shown to be unclear and misunderstood. The paradox of attempting to maintain life while establishing brain death creates confusion, and may lead to inappropriate requests for donation and to emotional discomfort among those who provide care in intensive care units (ICUs) and operating rooms. Many donor families express distrust, which may be exacerbated by staff uncertainty. In the future, as expansion of the potential organ donor pool is contemplated to include severely brain-damaged patients who are clearly alive by today's legal and medical standards, the confusion and distrust may be further compounded. Health professionals should play a leading role in clarifying these issues before policies and laws are changed and formed. To do so effectively, health professionals must first clarify their own thinking, since the ability to obtain organs from suitable donors depends largely on their attitude and commitment (Youngner *et al.* 1989).

Another proposed strategy for increasing the potential donor pool is procurement from marginal donors. While the inclusion of non-heart-beating cadaveric donors would certainly increase the supply of available organs, these donors have not been widely used because of the detrimental effects of anoxia and warm ischemic time. Previous advances in pulsatile perfusion have enabled kidneys to be successfully procured from this population, but attempts to resuscitate extrarenal organs have not generally been undertaken. Experimental use of new fluid solutions, cooling techniques, and cardiopulmonary bypass have demonstrated limited early success in animal models, leading to the speculation that multiorgan procurement from this donor population may be a viable adjunct to current utilization of heart-beating brain-dead donors (Matsuno *et al.* 1993).

As a means to the end of establishing the highest possible ethical standards with regard to the organ donation and transplantation process, the International Transplant Society has given general guidelines for cadaver organ distribution. The initial caveat is that no transplant surgeon or team should be involved directly or indirectly in the buying or selling of organs or tissues, or in any transplant activity involving commercial gain to himself or to an associated hospital or institution. Other proposed guidelines include the following.

1. The best possible use must be made of the donor gift.
2. Organs should be transplanted to the most appropriate recipient, judged solely on the basis of medical and immunological criteria.

3. Usable organs should never be wasted. In the majority of the cases, an organ will be used within an established network, and only if it cannot be placed should it be offered to other recognized networks.
4. Sharing of organs should only be arranged via national and/or regional organ-sharing networks.
5. Priorities in the assignment of organs cannot be influenced by political considerations, gifts, special payments, or favoritism to particular groups.
6. Transplant surgeons and physicians should not advertise regionally, nationally, or internationally.

Also, perhaps in response to practices in some countries, the International Transplant Society has summarized the following guidelines for donation by unrelated living donors.

1. Living unrelated donors should be used only when a satisfactory cadaver or living related donor cannot be found.
2. It must be established by the patient and transplant team alike that the motives of the donor are in the best interest of the recipient and are not self-serving or for profit. The motivation and medical suitability of the donor should be evaluated by physicians independent of the potential recipient, the recipient's physicians, and the transplant team. An independent donor advocate should be assigned to the unrelated donor to ensure that informed consent is made without pressure, to enhance personal attention throughout the donation period, to ensure official expressions of gratitude, and to aid with subsequent problems. In the exceptional case where the emotionally related donor is not a spouse or second-degree relative, the donor advocate should attempt to ensure further that the donation was one of true altruism and was not self-serving or for profit.
3. Active solicitation of living unrelated donors for profit is unacceptable.
4. Living unrelated donors must be of legal age.
5. The living unrelated donor must satisfy the same ethical, medical, and psychiatric criteria used in the selection of living related donors.
6. It should be clearly understood that no payment to the donor by the recipient, the recipient's family, or any other supporting organization can be allowed. However, reimbursement for loss of earnings and any other expenses directly related to the donation is acceptable.

The diagnostic and operative procedures on the donor and recipient must be performed only in recognized institutions, with staff experienced in living related and cadaveric transplantation. It would be expected that the donor advocate should be a member of the same institution, but not a member of the transplant team.

The ICU and organ donation

The role of the ICU is critical to the organ donation process and goes beyond issues of medical management. Initially, there must be a supportive attitude to organ donation and transplantation on the part of the ICU medical leadership, including ongoing staff education. Furthermore, there should be distinct protocols clearly identifying the referral process for potential donors, as well as who will manage the potential donor until brain death declaration and through the time of operative excision of organs. These protocols must be updated to reflect changing medical and technological developments. In large part, the success of a donor referral program depends on supportive enforcement and monitoring of protocols and constructive evaluation in cases of system failure.

Interaction with the donor family is of crucial importance in obtaining consent for donation. Appropriate staff should be available to assist in family support and education, but ideally a trained organ procurement co-ordinator should make the request for donation. Analysis of the data of many organ procurement organizations has shown that trained personnel have the highest consent rates followed by the hospital nursing staff and then the physician. Involvement of an organ procurement organization co-ordinator allows the family to separate caregivers from the transplant team, thereby eliminating the perception of conflict of interest. An additional important aspect of the consent process is to allow family members the time to deal with the death of their loved one before the option of organ donation is brought to their attention. This process, which seeks to separate the notification of death from the request for donation, is often referred to as 'decoupling' (Nathan *et al.* 1991). Development of such an active organ recovery program at all hospitals would substantially increase the number of organ donors (Dawidson *et al.* 1993).

The role of the transplant team

There is an obvious conflict of interest between potential donors, attending physicians, and the transplantation team. It should be made quite clear that the potential donor's physician has no role in the transplantation procedure. Similarly, a member of the transplant team should never directly participate in the diagnosis of brain death. However, once the patient is legally declared brain dead, the transplant team may and should be involved in the management of the donor up to and through the surgical organ recovery. The transplant team members should also participate in educational efforts and create positive feedback and recognition among the personnel involved in the donation process.

The role of organ procurement personnel

The majority of actual organ donors die in an ICU within 36 h of admission. A wide range of services for the community hospital should be provided, including professional education, evaluation of potential donors, and the co-ordination of organ recovery when suitable donors have been identified. Initial notification by hospital staff of the potential for donation is the first and most crucial step. If the patient information suggests potential for donation, an organ procurement co-ordinator is dispatched to the referring hospital. Once at the site, the organ procurement co-ordinator further evaluates the patient's past medical history, current hospital course, and the laboratory findings indicative of patient status and organ function. If donation is deemed appropriate, the co-ordinator must then orchestrate the process of brain death as it relates to the consent process.

Hospitals and organ donation

Depending on tradition, government regulations, the medical community, and the local culture, the organ donor process may vary widely between countries. It is estimated that in the United States 5600 additional lives could be saved annually from transplant organs by improved hospital practices related to donation requests. This could be achieved through active prospective identification of available donors, the majority of whom will be in an ICU setting or come through emergency room triage. According to a recent study, 27 per cent of potential donors were lost because health professionals did not identify them or did not make a proper effort to ask about donation or refer them to the appropriate organ procurement agency; there was no difference between large and small hospitals in handling donation (Gortmaker *et al.* 1996).

In the future, animals may be the main source of donor organs. For this to be realized, new effective and safe combinations of immunosuppressive drugs and protocols are needed. Another avenue may be to insert human genetic material into animal organs to induce tolerance without the need for immunosuppressive drugs. Much further research is needed in this exciting and promising area of xenotransplantation.

References

Dawidson, I., *et al.* (1993). Increasing organ donation at a large inner city public hospital. *Transplantation Proceedings*, **25**, 3139.

Gortmaker, S.L., *et al.* (1996). Organ donor potential and performance: size and nature of the organ donor shortfall. *Critical Care Medicine*, **24**, 432–9.

Matsuno, N., Sakuri, E., Uchiyama, M., Kozaki, K., Tamaki, I., and Kozaki, M. (1991). Use of *in situ* cooling and machine perfusion preservation for non-heart beating donors. *Transplantation Proceedings*, **25**, 3095–6.

Nathan, H.M., *et al.* (1991). Estimation and characterization of the potential renal organ donor pool in Pennsylvania. *Transplantation*, **51**, 142–9.

Youngner, S.J., Landerfeld, C.S., Coulton, C.J., Juknialis, B.W., and Leary, M. (1989). 'Brain death' and organ retrieval. A cross-sectional survey of knowledge and concepts among health professionals. *Journal of the American Medical Association*, **261**, 2205–10.

7.12 Sleep disturbance

7.12.1 Pathophysiology

N. P. Hirsch and R. S. Howard

Key messages

- Normal sleep produces physiological changes which may be poorly tolerated in critically ill patients.
- Many disease states and therapeutic interventions disrupt normal sleep and this may result in serious physiological consequences.
- Sleep deprivation is common in patients nursed in the intensive care unit; sequelae of this disruption include increased catabolism, increased difficulty in weaning patients from mechanical ventilation, and possibly decreased immune integrity.

Introduction

Sleep produces major physiological changes which, while well tolerated in health, may produce profound detrimental effects in critically ill patients. Furthermore, conditions existing in intensive care units (ICUs) may result in disruption of normal sleep architecture, and this may contribute to increased morbidity and mortality.

Physiology of normal sleep

Observation and electrophysiological (EEG) examination of sleep have led to the recognition of two distinct phases of sleep: slow-wave or non-rapid eye movement (NREM) sleep (approximately 80 per cent of total sleep time), and rapid eye movement (REM) sleep (20 per cent of total sleep time). Following a period of 10 to 20 min of sleep latency, individuals enter NREM sleep which is divided into four stages, each characterized by its EEG pattern. Stage I NREM, which accounts for 10 per cent of NREM sleep, is identified by a decrease in α (awake) activity and the appearance of both rhythms (4–7 Hz) and β rhythms (above 13 Hz). Stage II takes up about 50 per cent of NREM sleep and is characterized by the appearance of K spindles and sleep spindles (12–14 Hz) with less than 20 per cent of activity (below 4 Hz). Stages III and IV are marked by progressive increases of activity and represent deeper and more restful sleep. REM sleep occurs 60 to 90 min after the onset of sleep and recurs at 90-min intervals throughout the night. It is more frequent and of greater duration towards the waking hours. EEG findings include desynchronization with low-voltage fast-wave activity and variable amounts of activity and 'sawtooth' waves. Two phases of REM sleep are recognized. The tonic phase, which lasts throughout REM sleep, is marked by hypotonia of the neck, facial, and antigravity muscles. Superimposed on the tonic phase are phasic events during which rapid eye movements occur and myoclonic twitching of the facial muscles is seen.

Physiological changes during sleep

Although sleep results in changes to most physiological systems, attention will be directed to those affecting the respiratory and cardiovascular systems.

Respiratory system

Although respiratory rate tends to increase during NREM sleep (and becomes irregular during REM sleep), minute ventilation falls by 0.5 to 1.5 l/min mainly as a result of a decrease in tidal volume. This decrease is most pronounced during REM sleep. The resultant alveolar hypoventilation produces a mean fall of 3 to 10 mmHg (0.4–1.3 kPa) in $Pa\text{O}_2$ and an increase of 2 to 8 mmHg (0.2–1.1 kPa) in $Pa\text{CO}_2$. Owing to the shape of the oxyhemoglobin dissociation curve, oxygen saturation is well maintained if the individual has a normal awake $Pa\text{O}_2$. The most important factor contributing to the decrease in minute ventilation appears to be the decrease in resting tone of the upper airway musculature, which leads to increased airway resistance and, particularly in the supine position, predisposes to intermittent upper airway obstruction. In addition, intercostal muscle tone is progressively lost as sleep deepens and virtually disappears in REM sleep. During these stages, minute ventilation is provided entirely by the diaphragm. Other factors involved include a decreased chemoreceptor sensitivity to hypercarbia and hypoxemia, the absence of voluntary control of respiration present during wakefulness, and a reduction of metabolic rate during sleep. Furthermore, the supine position during sleep is associated with a decreased functional residual capacity. Therefore airway closure may occur at resting lung volumes, resulting in ventilation–perfusion mismatch. Laryngeal reflexes, cough, and ciliary activity are also decreased during both phases of sleep, resulting in an increased tendency to pulmonary aspiration of saliva and gastric contents.

Cardiovascular system

Changes in cardiovascular variables during sleep reflect alterations in autonomic system activity. During NREM sleep there is an increase in parasympathetic tone which, in REM sleep, is accompanied by an inhibition of sympathetic activity. Therefore there is a reduction in resting heart rate during NREM sleep, which decreases further during REM sleep. However, sympathetic tone varies during the latter and intermittent tachyarrhythmias are commonly seen. Similarly, arterial blood pressure decreases during NREM sleep and is variable during REM sleep. Peripheral resistance falls during both phases of sleep, and thus cardiac output decreases. This is particularly marked during early morning REM sleep and may explain the increased incidence of sudden death in the early morning of patients with cardiorespiratory disease. Pulmonary artery pressure increases during sleep and reflects the changes in $P\text{aO}_2$. Cerebral blood flow increases during NREM and REM sleep, and this may reduce central chemoreceptor activity.

Sleep problems in the ICU

Polysomnography has demonstrated severe sleep abnormalities in patients nursed in the ICU. They include increased sleep latencies, sleep fragmentation (rapid changes in sleep stage), increases in stage I sleep, and decreased REM sleep. A number of often coexisting factors contribute to the resultant sleep deprivation (Table 1).

Disease and sleep

Respiratory disease

Patients with chronic obstructive airways disease and restrictive pulmonary disease are commonly admitted to the ICU with exacerbation of their underlying condition. Typically these patients show increased sleep latencies, and decreased total sleep time and REM sleep. Furthermore, frequent arousals related to hypoxemia, hypercarbia, coughing, and treatment with methylxanthines decrease the time spent in NREM sleep. Obstructive sleep apnea, an under-recognized condition characterized by cyclical upper airway obstruction resulting in hypoxemia and hypercarbia, produces fragmented sleep with frequent arousals and sleep stage changes. Its estimated prevalence is 2 to 4 per cent of the adult male population, and therefore its presence in the ICU, either as the primary disorder or complicating another condition, is common.

Cardiovascular disease

The hypoxemia that accompanies sleep may exacerbate myocardial ischemia in patients with coronary artery disease. Nocturnal angina is well recognized and is often related to REM sleep. It may be aggravated by an increase in ventricular ectopy and the autonomic changes associated with sleep. Following myocardial infarction, sleep efficiency may be disrupted for up to 9 days. Patients with congestive heart failure, particularly when associated with Cheyne–Stokes respiration, show increases in stage I sleep with marked decreases in NREM sleep. Arousals are common and are often related to the hyperventilation phase of Cheyne–Stokes respiration.

Table 1 Etiology of sleep disturbance in the ICU

- Pre-existing disease states
- Effects of medication
- Effects of surgery and anesthesia
- Environmental effects
  - Noise
  - Lighting
  - Nursing and medical staff intervention
- Psychological effects

Neurological disease

Sleep disturbance may occur as a consequence of neurological disorders affecting central sleep mechanisms (e.g. in myotonic dystrophy and progressive supranuclear palsy). However, sleep disruption due to neurological disease is most commonly associated with conditions which affect the respiratory pathways, leading to nocturnal hypoventilation and sleep disruption as a consequence of microarousals. Neuromuscular disorders and primary muscle disease may present with respiratory insufficiency and sleep disorders either primarily or in the postoperative period. Examples of such disorders include motor neuron disease, multiple sclerosis, hereditary sensorimotor neuropathy, inflammatory muscle disease, and acid maltase deficiency.

Renal disease

Sleep disturbance is commonly reported in patients with chronic renal failure and consists of a decrease in total sleep time and NREM sleep. Obstructive sleep apnea is common in patients receiving hemodialysis, and the periodic limb movements common in patients with chronic renal failure may contribute to the sleep disruption.

Effects of medication on sleep

A variety of drugs commonly used in the ICU may disrupt sleep. Benzodiazepines reduce (or abolish) REM sleep and stage III and IV sleep and increase stage II sleep. After treatment is discontinued 'REM rebound' occurs, and patients may have an increased number of vivid and frightening dreams. Opioid drugs tend to suppress REM activity; in addition, morphine increases the frequency of arousals and stage I sleep. Tricyclic antidepressant drugs depress REM sleep. Theophylline, in higher doses, decreases total sleep time and increases sleep fragmentation. Intravenous catecholamines may increase wakefulness.

Effects of anesthesia and surgery on sleep

Profound sleep disturbance is common for at least 6 days following anesthesia and surgery. The magnitude of the disturbance appears to be more related to the magnitude of surgery, and anesthesia seems to be of less importance. In general, NREM sleep is reduced for 4 nights postoperatively and REM sleep is abolished for the first (and sometimes the second) postoperative night. There is a significant REM rebound thereafter. This may explain the increased incidence of hypoxemia during this period. Etiological factors involved in the sleep disturbance include the surgical stress response, fever, pain, opioid administration, starvation, age, psychological factors, preoperative sleep architecture, and environmental factors.

Environmental factors and sleep

Environmental stimuli remain the most potent causes of sleep fragmentation in the intensive care setting. Patients report noise as being the most disturbing factor disrupting sleep; levels are often greater than internationally recommended levels. Common sources include mechanical devices such as ventilators and alarms, ambient noise (often consistently above 80 dB), and conversations within the unit. Noise not only prevents a patient from falling asleep, but also decreases NREM sleep and causes frequent arousals. Constant bright lighting also disrupts sleep and interferes with normal circadian rhythms. High room temperature disturbs sleep, and nursing and medical observation and intervention inevitably fragment sleep.

Sequelae of sleep disruption

Sleep deprivation has profound effects on the mental health of the patient. It appears to be the primary cause of the reversible confusional state which characterizes the 'ICU syndrome'. Rapid resolution of symptoms often occurs following resumption of a normal sleep pattern.

Less evident is the effect of sleep deprivation on metabolism, respiratory mechanics, and the immune system. Evidence suggests that sleep (particularly NREM sleep) promotes increased protein synthesis, amino acid uptake by cells, and therefore anabolism. In contrast, sleep deprivation appears to favor catabolism, which may adversely effect wound healing. The effects of sleep disruption on respiratory function are more immediately tangible. Upper airway muscle dysfunction, decreased respiratory muscle strength and stamina, and decreased ventilatory responses to hypoxemia and hypercapnia have been reported in sleep-deprived patients; these obviously have profound implications for successful weaning of patients from mechanical ventilation. The effect of sleep on the immune response has been less well studied. However, initial studies of blood lymphocyte DNA synthesis, interferon production by lymphocytes, and leukocyte phagocytosis in sleep-deprived individuals suggest that healthy sleep aids tissue repair and increases resistance to bacterial and viral infection.

Conclusion

Sleep disruption occurs commonly in patients nursed on the ICU, and may affect morbidity and mortality in these critically ill individuals. Many etiological factors contribute to the sleep disorder, and they need to be considered before effective treatment can be provided.

7.12.2 Therapeutic strategy

N. P. Hirsch and R. S. Howard

Key messages

- There should be a rational approach to the treatment of sleep disruption in the intensive care unit and this should be directed towards the underlying cause.
- Measures include treatment of underlying and coexisting disease, treatment of pain and anxiety, and sensible modification of the intensive care unit environment.
- If the above measures prove ineffective, a rational choice of drug therapy should be instituted.

Introduction

Although sleep disturbances and their sequelae are becoming increasingly recognized in the critically ill patient, studies of the effectiveness of treatment rarely appear. However, it is possible to propose a number of strategies that will restore a normal sleep architecture in this group of patients.

Treatment of underlying and coexisting disease

Patients with intrinsic lung disease exhibit sleep disruption, often associated with hypoxemia and other factors. Supplemental oxygen (at 2 l/min), delivered via nasal cannula, has been shown to decrease sleep latency and increase all stages of sleep, including non-rapid eye movement (NREM) and rapid eye movement (REM) sleep. This improvement is believed to be related to a decrease in the frequency and severity of episodic nocturnal oxyhemoglobin desaturation.

Patients suffering from obstructive sleep apnea are effectively treated with nasal continuous positive airway pressure (CPAP) which, by acting as a 'pneumatic splint', results in a patent upper airway during sleep. Tracheostomy should be considered in cases of severe obstructive sleep apnea that is unresponsive to nasal CPAP.

A number of neurological, neuromuscular, and primary muscle disorders result in nocturnal alveolar hypoventilation with consequent sleep disruption. Although success in treating nocturnal desaturation has been reported using negative pressure ventilation, the equipment required is rather cumbersome, particularly for the disabled patient. Furthermore, this type of ventilation is commonly associated with upper airway obstruction during sleep. Until recently, the only alternative was conventional positive-pressure ventilation via a tracheostomy. However, non-invasive positive-pressure ventilation via a nasal or face mask has now become widely accepted as the method of choice in these patients. Reversal of nocturnal hypoventilation often rapidly restores normal sleep patterns.

Postoperative sleep disturbances and treatment of pain and psychological factors

Pain, anxiety, and depression are potent causes of sleep disruption in critically ill patients. Adequate and effective analgesia is essential. Unfortunately, many of the major analgesic drugs contribute to sleep disruption. Therefore a balanced appraisal of their use is necessary. A reduction in the magnitude and duration of the surgical procedure may be achieved by minimally invasive techniques, and regional analgesic techniques may be more appropriate in some patients. Noise, nocturnal nursing procedures, starvation, and increased room temperature are sleep-disturbing factors which could be modified in the postoperative period.

Anxiety and depression may be eased by careful explanation of treatments, but antidepressant therapy may be necessary. Antidepressant drugs suppress REM sleep and the sedative effects of most cyclic antidepressants parallel their anticholinergic properties, so the tertiary amines (amitriptyline, trimipramine, doxepin, imipramine, and clomipramine) are the most soporific and anticholinergic.

Modification of the intensive care unit environment

The modern intensive care unit (ICU) promotes poor sleep architecture. A number of measures can be implemented to attempt to make the overall environment more conducive to effective sleep.

Control of noise

A number of simple measures can be taken to decrease the noise level in an ICU. Their effectiveness should be monitored regularly. Patient and utility areas should have effective sound insulation. Ideally, the unit should be composed of single rooms. Speech is particularly disruptive to sleep, and conversations near the bedside should be limited to those necessary for patient care and should be directed away from the patient. Equipment such as ventilator and infusion pumps should not be placed at the head of the bed. Furthermore, the use of alarms remote from the patient should be considered. The volume of telephones and pagers should be reduced to a minimum, and noisy equipment (e.g. squeaky chairs, beds) should be repaired rapidly.

Lighting

Measures should be taken to restore day–night cycles in the unit in an attempt to restore circadian rhythms. During the day curtains should be open and the patient nursed as near to a window as possible. A large clock should be visible to the patient. If conscious, the patient should be reminded that it is daytime. At night, lights should be dimmed or extinguished.

Control of medical and nursing intervention

It should be the aim in the ICU to maximize the patient's period of uninterrupted sleep during the night. Only essential monitoring should be carried out and unnecessary observation should be omitted. Intervention from physicians, physiotherapists, radiographers, and phlebotomists should not start too early in the morning.

Pharmacological treatment of sleep disturbance

Despite the promotion of appropriate conditions for sleep, many patients suffer from insomnia in the ICU setting. In this situation, drug treatment will be necessary in the short term. Benzodiazepines, despite their tendency to disrupt normal physiological sleep, are the most widely used agents. Short-acting agents, such as temazepam or midazolam, should be used in order to decrease the occurrence of residual sedation during the following day. Benzodiazepines should be avoided in patients with nocturnal alveolar hypoventilation as they may exacerbate hypoxemia and hypercarbia. In addition, by increasing pharyngeal hypotonia, this group of drugs may worsen obstructive sleep apnea. In these clinical situations, the newly introduced drugs zolpidem and zopiclone should be considered. They decrease sleep latency and increase total sleep time while preserving a normal sleep architecture with maintenance of NREM and REM sleep. Tolerance and dependence are less likely to occur, and daytime impairment also appears to be reduced. They may prove to be the hypnotic drugs of choice for use in the ICU.

Education of medical and nursing staff

Education of medical and nursing staff in the mechanisms and importance of sleep disturbance is essential. Personnel should be taught to keep unnecessary interventions to a minimum, to ask specific questions about the quality of their patients' sleep, and to recognize the signs and symptoms of sleep deprivation.

Conclusion

The restoration and maintenance of normal sleep architecture in the ICU requires the treatment of the medical conditions that disrupt sleep, modification of the ICU environment, and a rational choice of hypnotic drugs.

Bibliography

Krachman, S.L., D'Alonzo, G.E., and Criner, G.J. (1995). Sleep in the intensive care unit. *Chest*, **107**, 1713–20.

Rosenberg-Adamsen, S., Kehlet, H., Dodds, C., and Rosenberg, J. (1996). Postoperative sleep disturbances: mechanisms and clinical implications. *British Journal of Anaesthesia*, **76**, 552–9.

Schwab, R.J. (1994). Disturbances of sleep in the intensive care unit. *Critical Care Clinics*, **10**, 681–94.

7.13 Agitation and confusion

7.13.1 Causes

S. Heckers and T. Stern

Key messages

- Delirium is a syndrome with multiple etiologies comparable with conditions such as heart failure or renal insufficiency.
- Delirium can be conceptualized as an acute and often reversible alteration of consciousness resulting from temporary or permanent damage of cortical and subcortical structures as well as from perturbations of several neurotransmitter systems.
- Many prescription drugs and several illicit drugs cause delirium. Of particular interest are those used commonly in intensive care settings.
- Manic patients or patients with agitated/psychotic depression might present as agitated and confused. Since mood disturbances often coexist with delirium, it is difficult to exclude a primary mood disorder in a delirious patient.
- Seizures, particularly complex partial seizures, can produce a behavioral syndrome that mimics delirium.

Differential diagnosis of delirium

The clinician should organize the etiological search to allow for an efficient, yet thorough, evaluation and treatment. The following outline divides this search into three major steps: exclude potentially life-threatening causes; exclude harmful effects of prescription or illicit drugs; be aware of the complete list of conditions, illnesses, and medications that can lead to delirium (Fig. 1 and Tables 1, 2, 3, and 4).

Exclude potentially life-threatening causes

The mnemonic WWHHHHIMP can guide the clinician during the rapid assessment of serious and potentially life-threatening causes of delirium (Fig. 1).

Wernicke's encephalopathy is strongly suggested by the triad of confusion, ataxia, and ophthalmoplegia. Additional support for this diagnosis can be derived from a history of alcohol dependence, by abnormal liver function tests, and by an elevated red blood cell mean corpuscular volume.

Withdrawal states occur after discontinuation of illicit or prescription drug use, even when small dosages are used for short periods of time. Full serum and urine toxicological screens, as well as obtaining an extensive drug-use history, are helpful.

Hypertensive encephalopathy, Hypoxia, and Hypoglycemia can be assessed with vital signs, results of arterial blood gases, and a serum glucose level. Again, it is helpful to know details of the medical history (hypertension, pulmonary disease, diabetes mellitus).

Intracranial bleed, Meningitis, and encephalitis almost always produce focal neurological signs as well as headache, loss of consciousness, and seizures. Fever and elevated white blood cell count should prompt further studies, including a lumbar puncture.

Poisons that may lead to delirium include pesticides, solvents, or heavy metals.

Exclude the impact of drugs

Many prescription drugs and several illicit drugs cause delirium (Tables 3 and 4). Of particular interest are those used commonly in intensive care settings (Table 4). A thorough review of the medications given, i.e. study of the nurse's records rather than the doctor's order book, is essential. The clinician must also be aware of drug–drug interactions that can lead to significant elevations of one or more drug levels even when used at deceivingly low dosages.

Drug overdose, either accidental or as a result of a suicide attempt, is a frequent cause of delirium in the medical intensive care unit (ICU). For example, at the Massachusetts General Hospital, 5 per cent of all ICU admissions during a 5-year period were the direct result of a drug overdose; most of these individuals were delirious at some point during their ICU stay.

Continue the search

When life-threatening causes of delirium and the impact of several specific drugs have been excluded, the clinician will often proceed to empirical treatment of the delirious patient. However, it is important to continue the search for an underlying cause, which can be found in more than 85 per cent of cases. Table 1 offers the mnemonic I WATCH DEATH to organize the differential diagnosis of delirium. In view of the mortality associated with acute confusional states, the mnemonic is appropriate.

Pathophysiology of delirium

Delirium is a syndrome with multiple etiologies comparable with conditions such as heart failure or renal insufficiency. Attempts to subdivide delirium based on different clinical presentations and to correlate them with objective measures have not been successful (Trzepacz 1994). Electroencephalography, studies of cerebral blood

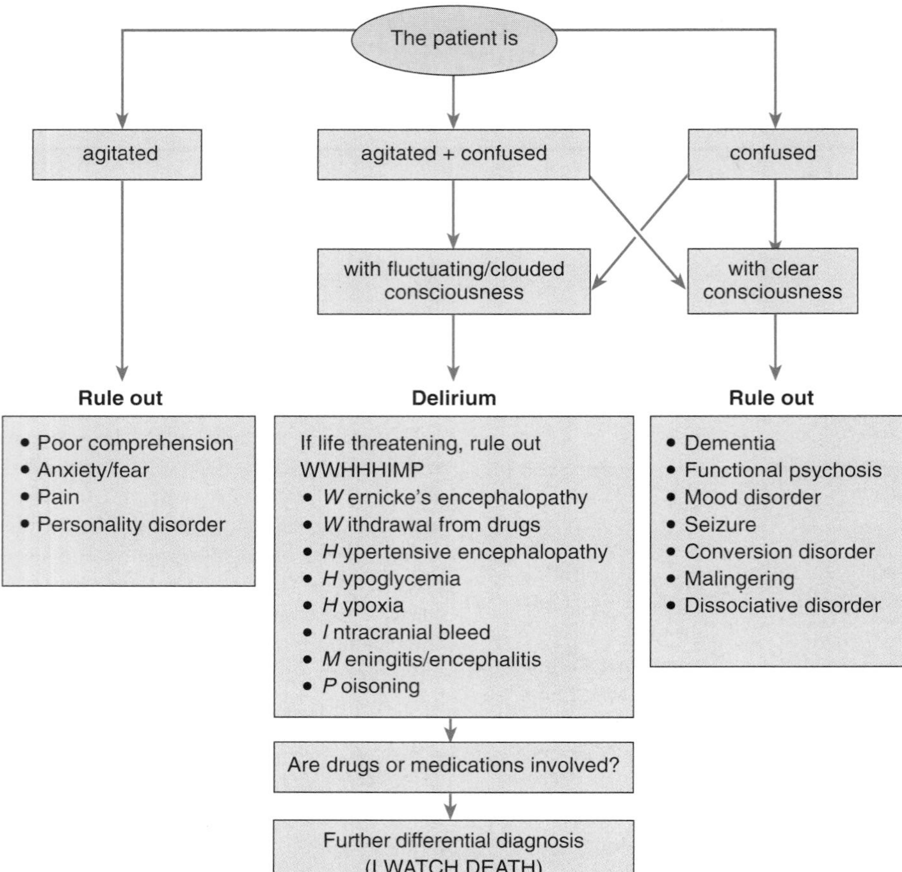

Fig. 1 The agitated/confused patient.

Table 1 Differential diagnosis of delirium (I WATCH DEATH)

Infectious	Encephalitis, meningitis, syphilis
Withdrawal	Alcohol, barbiturates, sedative–hypnotics
Acute metabolic	Acidosis, alkalosis, electrolyte disturbance, hepatic failure, renal failure
Trauma	Heat stroke, severe burn, postoperative state
CNS pathology	Abscess, hemorrhage, hydrocephalus, seizure, stroke, tumor, vasculitis
Hypoxia	Anemia, CO poisoning, hypotension, pulmonary/cardiac failure
Deficiencies	Vitamin B_{12}, niacin, thiamine
Endocrinopathies	Hyper- or hypoadrenocortisolism, hyper- or hypoglycemia
Acute vascular	Hypertensive encephalopathy, shock
Toxins or drugs	Medications (see Tables 3 and 4)
Heavy metals	Lead, manganese, mercury

Adapted with permission from Wise and Gray (1994).

flow, modulation of neurotransmitter levels, and the study of hemispheric differences in the development of delirium are the principal methods used to understand the pathophysiology of delirium (Trzepacz 1994).

Electroencephalograph abnormalities are seen in virtually all delirious patients. Generalized slowing is typically seen in hypoactive states and in the majority of hyperactive states. Some hyperactive patients (e.g. those with delirium tremens) show low-voltage fast activity. Electroencephalograms can be read as normal if the patient demonstrates a baseline voltage above normal and only drops into the normal range during the delirious state.

Single-photon-emission CT and positron-emission tomography studies have shown globally increased cerebral blood flow in delirium tremens. Subclinical hepatic encephalopathy is associated with reduced cortical cerebral blood flow and with both increased and decreased subcortical cerebral blood flow.

Neurotransmitters have rarely been measured directly in delirious patients, but their involvement has been implied based on the effect of several drugs associated with delirium. Impaired cholinergic neurotransmission, induced by anticholinergic drugs or seen in elderly patients, can result in or facilitate delirium; physostigmine is effective in treating anticholinergic delirium. Reduction of dopaminergic neurotransmission at the dopamine D_2 receptor via typical neuroleptics is highly effective in the treatment of delirium. GABAergic neurotransmission has been found to be increased in hepatic encephalopathy and decreased during withdrawal from sedative agents. Benzodiazepines are effective in treating types of

Table 2 Work-up of delirium

Physical examination	Vital signs (current and over the last 24 h) Cardiac and pulmonary examination; skin and extremities Neurological examination
Mental status examination	Consciousness Psychomotor activity Mood/affect Thinking Perception Insight/judgment Test for memory, attention, reasoning, apraxia, dysgraphia, dysnomia
Primary tests	Electrolytes, blood urea nitrogen, creatinine, glucose, calcium, phosphate, liver function tests Hematocrit, differential white cell count, mean corpuscular volume, sedimentation rate Toxicological screen (blood, urine) Urinalysis Pulse oximetry/arterial blood gas ECG Chest radiography
Further tests	Electroencephalogram Head CT scan Lumbar puncture Additional blood tests (VDRL or FTA-ABS, thiamine, folate, thyroid function tests, heavy metal screen)

VDRL, Venereal Disease Research Laboratory; FTA-ABS, fluorescent treponemal antibody absorption.

delirium. Increased serotonergic activity, as seen in the 'serotonin syndrome', leads to delirium.

Cerebrovascular accidents of the right hemisphere, particularly those involving the parietal and frontal lobes, are more likely to produce delirium than are insults to the left hemisphere. Recent anoxia or strokes of the basal ganglia and thalamus (again with a predilection for the right hemisphere) can also result in delirium.

Delirium can be conceptualized as an acute and often reversible alteration of consciousness (affecting primarily attention, orientation, and memory) resulting from temporary or permanent damage of (mainly right-sided) cortical and subcortical structures as well as from perturbations of several neurotransmitter systems.

Differential diagnosis of agitation and confusion (other than delirium)

Agitation

Poor comprehension due to a language barrier or intellectual impairment should always be considered and excluded. Sensory impairment, either pre-existing or recently developed in the ICU, puts the patient at greater risk of development of hallucinations and agitation (e.g. black-patch delirium after ocular surgery).

Anxiety is a prominent and expected reaction to conditions leading to intensive care (e.g. myocardial infarction). Most patients develop anxiety which lasts for only a few days. However, some do not return to a normal level of concern and psychomotor activity. A thorough review of the patient's medical history, with emphasis on previous panic attacks, agoraphobia, or generalized anxiety, is then crucial. Weaning from chronic ventilatory support often generates considerable anxiety which can be minimized with skillful weaning techniques.

Pain is often undetected when communication with the patient is impaired. Abnormal vital signs and agitation may be its only indicators. If the patient reports pain and becomes agitated, it is important to use analgesics.

If all other sources of agitation are excluded, consider that the patient might have a personality structure that makes it difficult to cope with the stresses imposed by critical care. The patient might have suffered serious injuries, lost family members, or lost perspective on life after hospital discharge. Patients with rigid, obsessive, and controlling styles are particularly prone to agitation in the ICU. Others may develop extreme anger towards the treatment team; this anger can set off a defensive counterattack that further worsens the relationship. Early recognition of such personality traits can lead to appropriate psychiatric intervention and can help the patient to avoid being restrained or signing out against medical advice.

Confusion

Confusion (i.e. impaired attention, memory, orientation, and several other deficits) is not specific for delirium. If the cognitive deficits precede hospital admission, do not have an acute onset, and do not fluctuate during the day, several other conditions should be considered (Fig. 1).

Elderly patients with pre-existing dementia often become delirious. It is important to identify those individuals who have a stable course of dementia and who are not acutely confused to spare them extensive assessment and treatment with neuroleptics or benzodiazepines. A thorough neuropsychiatric evaluation of the patient performed before the diagnosis of delirium was made can serve as baseline for repeated assessments. A more global cognitive decline and absence of lucid intervals makes a diagnosis of dementia more likely.

Patients with a functional psychosis (e.g. schizophrenia or mania) often present with telltale signs: a long psychiatric history, normal laboratory test results, a normal level of consciousness, and well-systemized delusions. However, severely disorganized psychotic patients, when agitated and confused, can resemble the patient with delirium.

Manic patients or patients with agitated/psychotic depression might present as agitated and confused. Since mood disturbances often coexist with delirium, it is difficult to exclude a primary mood disorder in a delirious patient. However, it is important to exclude secondary mood disorders, such as those due to drugs, infection, neoplasms, or seizures (particularly of the right temporal lobe).

Seizures, particularly complex partial seizures, can produce a behavioral syndrome that mimics delirium. An abnormal electroencephalogram and focal neurological deficits help to make the

Table 3 Drugs that can cause delirium (extensive)

Antibiotic	**Anti-inflammatory**	**Sedative–hypnotic**
Acyclovir (aciclovir)	ACTH	Barbiturates
Amphotericin	Corticosteroids	Glutethimide
Cephalexin	Ibuprofen	Benzodiazepines
Chloroquine	Indomethacin	
Isoniazid	Naproxen	**Sympathomimetic**
Rifampin (rifampicin)	Phenylbutazone	Amphetamine
		Phenylephrine
Anticholinergic	**Antineoplastic**	Phenylpropanolamine
Antihistamines	5-Fluorouracil	
Chlorpheniramine		**Miscellaneous**
Benztropine	**Antiparkinson**	Aminophylline
Biperiden	Amantadine	Bromides
Antispasmodics	Carbidopa	Chlorpropamide
Atropine	Levodopa	Cimetidine
Belladonna alkaloids		Disulfiram
Diphenhydramine	**Analgesics**	Drug withdrawal
Phenothiazines	Opiates	Alcohol
Promethazine	Salicylates	Barbiturates
Scopolamine (hyoscine)	Synthetic narcotics	Benzodiazepines
Tricyclic antidepressants		Lithium
Trihexyphenidyl (benzhexol)	**Cardiac**	Metrizamide
	Beta-blockers	Metronidazole
Anticonvulsant	Propranolol	Podophyllin
Phenobarbital (phenobarbitone)	Clonidine	Propylthiouracil
Phenytoin	Digitalis	Quinacrine
Valproic acid	Disopyramide	Theophylline
	Lidocaine (lignocaine)	Timolol ophthalmic
	Mexiletine	
	Methyldopa	
	Quinidine	
	Procainamide	

Adapted with permission from Wise and Gray (1994).

Table 4 Common drugs in the ICU that can cause delirium

Drug group	Agent
Antiarrhythmics	Lidocaine, mexiletine, procainamide hydrochloride, quinidine sulfate
Antibiotics	Penicillin, rifampin
Anticholinergics	Atropine sulfate
Antihistamines	Non-selective: diphenhydramine hydrochloride, promethazine hydrochloride H_2-blockers: cimetidine, ranitidine
Beta-blockers	Propranolol hydrochloride
Narcotic analgesics	Meperidine hydrochloride, morphine sulfate, pentazocine

Adapted with permission from Tesar and Stern (1986).

diagnosis. Neuroleptics, a primary treatment for most cases of delirium, should be avoided in those with ongoing seizures since these agents can lower the seizure threshold.

Conversion reaction, malingering, and dissociative disorder, although rarely seen in the ICU, can present with confusion and agitation. The cognitive deficits associated with these conditions are often inconsistent and may include disorientation to person (a feature that is rarely seen in delirium).

References

Tesar, G.E. and Stern, T.A. (1986). Evaluation and treatment of agitation in the intensive care unit. *Journal of Intensive Care Medicine*, **1**, 137–48.

Trzepacz, P.T. (1994). The neuropathogenesis of delirium. *Psychosomatics*, **35**, 374–91.

Wise, M.G. and Gray, K.F. (1994). Delirium, dementia, and amnestic disorders. In *Textbook of psychiatry* (ed. R.E. Hales, S.C. Yudofsky, and J.A. Talbott), pp. 311–53. American Psychiatric Press, Washington, DC.

7.13.2 Assessment

S. Heckers and T. Stern

Key messages

- Agitation is defined as a state of increased psychomotor activity, often accompanied by increased volume and rate of speech, as well as labile and inappropriate affect.
- The hallmark of delirium is fluctuation or clouding of consciousness, i.e. a reduced clarity of awareness of the environment.
- The differential diagnosis and treatment of the purely agitated state differ from that of confusion. A confusional state may change several times over a period of hours, requiring repeated assessments.
- Delirium can be diagnosed if the patient shows several features during parts of the evaluation process. Three domains (i.e. attention, orientation, and memory) are typically impaired in delirious patients.

Introduction

Agitation and confusion are the most common mental status changes seen in the critical care setting. They are associated with a wide variety of pathological conditions that affect normal brain function and might indicate (a) significant or even life-threatening changes in the patient's medical condition, (b) pre-existing brain dysfunction, or (c) maladaptive behavior in response to stress in the intensive care unit (ICU). Prompt assessment and therapeutic intervention are essential for adequate management of both agitation and confusion.

Definition

Agitation is defined as a state of increased psychomotor activity, often accompanied by increased volume and rate of speech, as well as labile and inappropriate affect.

Confusion is a layman's term; it is not a well-defined mental state. Webster's Dictionary (3rd edition, 1986) defines the 'state of being confused' as one that has a 'lack of certainty, orderly thought, or power to distinguish, choose, or act decisively.' Confusion with an acute onset has been called delirium, acute confusional state, acute brain syndrome, acute cerebral insufficiency, or ICU psychosis. For the purpose of this chapter, we prefer the terms delirium and confusional state. The hallmark of delirium is fluctuation or clouding of consciousness, i.e. a reduced clarity of awareness of the environment. Delirium can present with agitation (hyperactive subtype), with decreased psychomotor activity or somnolence (hypoactive subtype), or with a mixture of hyper- and hypoactivity. Alterations of three cognitive domains (attention, orientation, and memory) contribute primarily to the clinical presentation of delirium.

Incidence and risk factors

The prevalence of delirum in medically and surgically ill patients is 11 to 16 per cent and the incidence varies between 4 and 31 per cent (Wise and Gray 1994). The highest incidence is found in the surgical ICU, followed by the coronary care unit and medical and surgical wards (Lipowski 1990). Several risk factors for the development of delirium have been identified.

1. Elderly patients tend to develop delirium more frequently than younger patients, most likely as a result of pre-existing medical conditions (including impaired brain function and the frequent administration of numerous medications).
2. A history of brain damage (e.g. stroke, dementia), drug dependency, or AIDS increases the risk of delirium.
3. Postcardiotomy and burn patients are particularly prone to delirium. The prevalence of delirium after cardiac surgery is about 32 per cent and seems to be correlated with several factors (e.g. time on bypass, decreased cardiac output, embolic events, and poor nutritional status as indicated by low levels of albumin) (Smith and Dimsdale 1989). The incidence of delirium in burn patients is approximately 18 to 30 per cent; it increases with age and with the severity of burns.

Psychosocial and environmental factors, particularly sleep deprivation, have not been closely linked to an increased risk of delirium.

Morbidity and mortality

Delirium is often defined as a reversible and relatively short-lasting impairment of brain function. In fact, the majority of delirious patients recover without observable sequelae, but the exact percentage is unknown (Lipowski 1990). Patients in drug withdrawal states may develop seizures. Some patients fail to recover from the acute confusional state, progress to stupor or coma, and die.

Delirium and agitation increase the risk of complications (e.g. decubiti and aspiration pneumonia) and tend to prolong the length of hospital stay. A substantial number of elderly patients who develop delirium as in-patients (22 to 76 per cent according to different studies) die during their stay in hospital; about 25 per cent of delirious hospitalized patients die within 6 months of discharge (Wise and Gray 1994).

Initial assessment

Unfortunately, no pathognomonic diagnostic test is available; the assessment relies on a thorough physical and mental status examination, review of the history, and routine laboratory tests.

As a first step, the clinician should determine whether the patient is agitated, confused, or both. The differential diagnosis and treatment of the purely agitated state differ from that of confusion. A confusional state may change several times over a period of hours, requiring repeated assessments. Confusion can have an acute onset (in the hospital), may develop over weeks to years, and/or may be complicated by agitation.

Clinical features of confusional states

Confused patients need to be assessed repeatedly. Delirium can be diagnosed if the patient shows several features during parts of the evaluation process. The mental activities altered in confusional states are listed in Table 1. Three domains (i.e. attention, orientation, and memory) are typically impaired in delirious patients; other features are more variable.

Table 1 Mental activities altered in delirium (confusional state)

Attention	Decreased
Orientation	Impaired (time > place > > person)
Memory	Impaired
Thinking	Illogical, poor insight and judgment, delusional
Perception	Illusions, hallucinations (visual > auditory > tactile)
Arousal	Decreased or increased
Sleep–wake cycle	Altered, may be reversed
Affect	Labile, irritable
Neurological deficits	Tremor, myoclonus, asterixis, constructional deficit, dysnomia, dysgraphia

Attention deficits can be regarded as the *sine qua non* of confusional states. The patient is easily distracted and cannot maintain his or her focus of attention during a task. Inattention and distractability are often easily observed during the interview. In less obvious cases, inattention can be assessed at the bedside by asking the patient to name the months of the year in reverse order or to repeat several numbers in sequence.

Memory and orientation are typically impaired, at least in part, because of attention deficits. Memory deficits are readily apparent when the patient is asked to encode new information, for example to recall three words, either immediately or after a delay of 5 min. The patient may also show deficits in the recall of previously encoded information. Some patients will be partially or completely amnestic for the delirious episode.

Impaired orientation is the third main feature of confusional states. Except for lucid intervals, delirious patients are typically disoriented with regard to time and to place, but rarely to person.

Thinking is often altered in acute confusional states. Such patients are disorganized, irrational, and manifest impaired reasoning. Thinking can be grossly impaired; delusions and paranoia may develop. Poor insight and judgment about medical treatment often make decision-making impossible. Simple tests of reasoning can be quite helpful. For example, the patient is asked to explain the following story: 'I have a friend named Frank Jones. His feet are so big that he has to put his pants on over his head. Can Mr. Jones do that?' The patient may explain that the story is impossible (i.e. cortical function intact), respond with humor (i.e. limbic understanding of the absurdity of the story), or show no reaction. The delirious patient often responds with a smile but lacks a clear understanding of the story.

Altered perception in the form of illusions or, less often, hallucinations is seen in some confusional states. Visual hallucinations are more common than auditory or tactile hallucinations. Confusional states secondary to sedative–hypnotic withdrawal are more likely to present with hallucinations.

Several additional features are seen in some confused patients. For example, the hyperactive confused patient with delirium tremens displays increased arousal, labile affect, and decreased sleep, whereas the hypoactive confused patient with hepatic encephalopathy shows decreased arousal, flat affect, and increased sleep (to the degree of somnolence or coma).

Neurological signs such as tremor, myoclonus, or asterixis are seen in some types of delirium (e.g. hepatic encephalopathy). Other deficits such as impaired constructional ability, word-finding difficulties (dysnomia), or writing disturbances (dysgraphia) are more generally found. The clock-drawing test often provides a rapid screen for the presence and degree of delirium. Dysgraphia, which is easily tested by asking the patient to write a sentence, is a very sensitive, albeit non-specific, test for delirium.

Several rating scales are now available for the efficient and reliable assessment of acute confusional states (Wise and Gray 1994). These are particularly helpful when repeated assessments of a patient are performed by different observers.

If the patient displays the clinical features seen in acute confusional state, i.e. delirium, it is important to investigate the duration, rapidity of onset, and course over time. The differential diagnosis of acute confusional states includes chronic conditions of impaired brain function that lead to similar cognitive deficits. For example, functional psychosis, dementia, and depression show many of the features seen in delirium, but do not show the fluctuation and clouding of consciousness.

References

Lipowski, Z.J. (1990). *Delirium: acute confusional states.* Oxford University Press, New York.

Smith, L.W. and Dimsdale, J.E. (1989). Postcardiotomy delirium: conclusions after 25 years? *American Journal of Psychiatry*, **146**, 452–8.

Wise, M.G. and Gray, K.F. (1994). Delirium, dementia, and amnestic disorders. In *Textbook of psychiatry* (ed. R.E. Hales, S.C. Yudofsky, and J.A. Talbott), pp. 311–53. American Psychiatric Press, Washington, DC.

7.13.3 Therapeutic management

S. Heckers and T. Stern

Key messages

- Narcotics and drugs with anticholinergic properties account for a large proportion of drug-induced confusional states. The first step in treating drug-induced delirium is to stop the offending drug and to find an alternative compound without similar adverse effects.
- The successful treatment of withdrawal-induced agitation and confusion requires the replacement of the same substance or a cross-reacting compound with less adverse effects and a better pharmacokinetic profile.
- When a specific cause of agitation and confusion cannot be identified or corrected, it is still important to stabilize the patient and to ensure his or her safety.
- Haloperidol, a high-potency neuroleptic of the butyrophenone class that blocks postsynaptic D_2 dopamine

receptors, has an extensive record of safety and efficacy in critically ill patients and is most often used to treat the agitated delirious patient.

- Agitation due to panic attacks, generalized anxiety, or fear of being in the ICU should be treated primarily with a benzodiazepine.

Introduction

The clinician treating the agitated or confused patient should always consider two avenues of patient management: starting specific treatment if the etiology of the altered mental state is reasonably clear, and treating empirically to avoid immediate danger while awaiting clarification of the specific etiology.

Specific treatment of agitation and confusion

A thorough examination of the patient and a limited set of laboratory tests can often make the diagnosis of several life-threatening causes (think WWHHHIMP). Maintenance of normal blood pressure, circulation, and blood oxygenation, correction of metabolic derangements, and treatment of local or systemic infections are essential to good medical care. Such treatment averts many mental status changes. Drug toxicity and withdrawal, anxiety, pain, and inadequate coping skills also contribute to agitation and/or confusion. Specific treatment strategies for each of these problems are presented below.

Eliminate drug toxicity and treat drug withdrawal

The regular use of prescription drugs, the abuse of illicit drugs, and the withdrawal from either type of drug are all common causes of agitation and confusion.

Narcotics and drugs with anticholinergic properties account for a large proportion of drug-induced confusional states. The first step in treating drug-induced delirium is to stop the offending drug and to find an alternative compound without similar adverse effects. If necessary, antidotes can be administered.

1. Naloxone hydrochloride (0.4 mg subcutaneously or intravenously) is often used to reverse acute confusional states due to the use of narcotics. Repeated doses can be given if long-acting narcotics have been responsible for delirium. Meperidine toxicity, leading to hallucinations, irritability, myoclonus, and seizures, is caused by stimulation of the central nervous system by the metabolite normeperidine and may require additional aggressive treatment with barbiturates or benzodiazepines.

2. Physostigmine (1–2 mg), infused slowly intravenously as a one-time dose or given as a continuous drip, can effectively treat anticholinergic delirium.

3. Intravenous verapamil has been used for the treatment of phencyclidine intoxication.

4. The benzodiazepine antagonist flumazenil has been used to reverse the effects of benzodiazepine excess.

Intense agitation and visual hallucinations are often seen in states of drug withdrawal. Such symptoms should raise the clinician's suspicion of withdrawal syndromes. Fever, tremulousness, diaphoresis, and tachycardia are further corroborating signs. The successful treatment of withdrawal-induced agitation and confusion requires the replacement of the same substance or a cross-reacting compound with less adverse effects and a better pharmacokinetic profile.

Treat anxiety, pain, and inadequate coping skills

Anxiety is a common reaction to physical or psychological trauma and to treatment in intensive care units (ICUs). It can be seen as a transient reaction in a previously healthy individual or as a manifestation of a pre-existing anxiety disorder. The treatment team should make every effort to put the patient at ease to avoid potentially harmful reactions (e.g. pulling out intravenous lines, refusing treatment, and leaving against medical advice). Direct discussion of the patient's medical condition and a compassionate stance by the clinician are crucial in guiding the anxious patient. A psychiatrist, a calm family member, benzodiazepines, and neuroleptics can each be helpful in reducing anxiety; their uses are discussed below.

Pain, particularly if not reported by the patient because of limited communication in the intensive care setting, often leads to agitation and jeopardizes the patient's recovery. Insufficient amounts of analgesics may have been administered, for fear that the patient might become addicted. Addiction rarely develops unless the patient has a history of drug-seeking behavior or of chronic pain refractory to numerous interventions.

Patients with rigid, obsessive, and controlling styles or with poor regulation of affect (e.g. with irritability, anger outbursts, and manipulative or hostile attitudes) often decompensate in the critical care setting. Furthermore, they are uncooperative and may induce a rageful counterattack by members of the treatment team. A non-judgmental but firm stance, which gives the patient as much control as possible while ensuring his or her safety, is essential for the survival of both the patient and the caregiver. Pharmacological or mechanical restraints should be avoided if possible.

Non-specific treatment of agitation and confusion

When a specific cause of agitation and confusion cannot be identified or corrected, it is still important to stabilize the patient and to ensure his or her safety. Pharmacological intervention, mechanical restraints, and supportive measures (to reorient and calm the patient) are central to treatment.

Pharmacological treatment

Neuroleptics and benzodiazepines are the primary drugs used to manage the agitated and/or confused patient. If the patient's agitation cannot be controlled by these two classes of medications, narcotics and paralyzing agents can be used as a last resort (Table 1).

Neuroleptics

Haloperidol, a high-potency neuroleptic of the butyrophenone class that blocks postsynaptic D_2 dopamine receptors, has an extensive record of safety and efficacy in critically ill patients and is most often used to treat the agitated delirious patient. Other typical neuroleptics

Table 1 Drugs used to treat delirium and agitation

Drug	Route	Onset (min)	Peak effect (min)	Starting dose	Important adverse effects
Neuroleptics					
Haloperidol	IV, IM Oral	5–20 30–60	15–45 120–240	Mild agitation: 0.5–2 mg Moderate agitation: 5–10 mg Severe agitation: >10 mg	Cardiac arrhythmia (QTc increased, torsades de pointes) Extrapyramidal side-effects
Droperidol	IV, IM	3–10	15–45	2.5–10 mg	Hypotension
Chlorpromazine	IV, IM	5–40	10–30	25 mg	
Benzodiazepines					
Diazepam	IV Oral	2–5 10–60	5–30 30–180	2–5 mg	Respiratory depression May aggravate delirium
Lorazepam	IV, IM SL Oral	2–20 2–20 20–60	60–120 20–60 20–120	1–2 mg 0.5–1 mg 0.5–1 mg	
Midazolam	IV, IM	1–2	30–40	0.05–0.15 mg/kg	
Narcotics					
Morphine sulfate	IV, IM	1–2	20	4–10 mg	Resiratory depression May aggravate delirium
Paralytics					
Metocurine iodide	IV	1–4	2–10	0.2 – 0.4 mg/kg	Hypotension
Pancuronium bromide	IV	0.5–1	5	0.04 – 0.1 mg/kg	Tachycardia, hypertension

IV, intravenous; IM, intramuscular; SL, sublingual.

(e.g. chlorpromazine, droperidol, thiothixene, and trifluoperazine) can also be used. High-potency neuroleptics cause less hypotension, sedation, and anticholinergic symptoms than low-potency neuroleptics, but cause more extrapyramidal side-effects (i.e. acute dystonia, akathisia, parkinsonism). Haloperidol is approved by the United States Food and Drug Administration for intramuscular and oral use only, but it can be given intravenously.

Typically, the agitated patient will be given haloperidol at a starting dose of 0.5 to 5 mg. The elderly and those with known central nervous system dysfunction (e.g. dementia, stroke) require doses as low as 0.5 mg two to three times daily; doses of 2 to 5 mg two to four times daily are not uncommon.

Parenteral administration, at a dose of approximately half the oral dose, is indicated when the oral route is not available or when rapid onset of effect is desired. Intramuscular administration is effective, but repeated injections are painful, cause elevation of creatine phosphokinase levels, and are time consuming. Intravenous administration is preferred when frequent dosing is required, i.e. in a severely agitated patient where systematic escalation of the dosage will probably be required. Intravenous injection of haloperidol requires that the intravenous line be flushed first with saline, because haloperidol precipitates with both heparin and phenytoin. The rate of infusion is not critical, but should be extended over 5 min if the patient is hypotensive. When a given dose of intravenous haloperidol fails to calm the agitated patient, doubling the previous dose after 20 min is recommended. If the patient becomes calm, the effective dose should be repeated at the next dosing interval. Although injection of boluses of more than 50 mg is rarely necessary, individual boluses of 150 mg and as much as 1200 mg total daily have been reported. Continuous infusion of intravenous haloperidol has also been used safely and effectively to control severe agitation.

All neuroleptics predispose to extrapyramidal side-effects. Acute effects, i.e. acute dystonia, akathisia, and parkinsonism, are more likely to develop in the ICU than are tardive dyskinesia and neuroleptic malignant syndrome, which are both severe complications of long-term neuroleptic use. Oral and intramuscular administration of haloperidol is much more likely to cause extrapyramidal side-effects than is intravenous use. Acute dystonia, particularly the life-threatening laryngeal dystonia, should be treated promptly with benztropine mesylate (1–2 mg intravenously) or diphenhydramine (25–50 mg intravenously). Akathisia responds well to reduction of the neuroleptic dose and concomitant administration of a β-blocker (e.g. propranolol 10–20 mg two to three times daily) or a benzodiazepine (e.g. diazepam or lorazepam 0.5–1 mg two to three times daily). Parkinsonian side-effects occur more often in older individuals and respond to treatment with an anticholinergic agent, which unfortunately can aggravate the acute confusional state.

Low-potency neuroleptics tend to cause hypotension, whereas high-potency neuroleptics have little effect on cardiovascular function. However, high doses of haloperidol have been associated with prolongation of cardiac conduction, leading to QTc increases. In some patients, particularly those with pre-existing dilated ventricles and a history of alcohol abuse, torsades de pointes arrhythmia has developed. Monitoring of the ECG, the magnesium level, electrolytes, and medications which have an effect on cardiac conduction is recommended; discontinuation of intravenous haloperidol may be necessary if the QTc increases by more than 25 per cent from the baseline value.

Benzodiazepines

Agitation due to panic attacks, generalized anxiety, or fear of being in the ICU should be treated primarily with a benzodiazepine. Diazepam at a starting dose of 2 to 5 mg, and lorazepam or clonazepam at a starting dose of 0.5 to 1 mg are effective in calming the anxious and agitated patient. In the case of panic attacks, maintenance treatment (e.g. clonazepam 0.5 mg three times daily) is often recommended. Psychotic episodes and manic presentations have also been managed exclusively with clonazepam.

Midazolam, a water-soluble imidazolebenzodiazepine with a rapid onset of action and an elimination half-life of 1 to 4 h, has been used successfully in the treatment of agitation. Intramuscular administration of 2 to 3 mg of midazolam calms the patient within 5 to 10 min. Continuous infusion of midazolam, with a mean infusion rate of approximately 0.6 to 6 g/kg/min, is an effective treatment for severely agitated patients.

Many agitated patients benefit from the combined use of a neuroleptic and a benzodiazepine. Coadministration of these two classes of drugs allows for lower doses of neuroleptics, which reduces the risk of extrapyramidal side-effects. Benzodiazepines are also effective in treating neuroleptic-induced akathisia. Lorazepam has been used most frequently in conjunction with neuroleptics. In contrast with diazepam, lorazepam offers the advantage of fewer cardiopulmonary side-effects, a shorter elimination half-life, and less accumulation in patients with hepatic dysfunction.

Benzodiazepines can cause respiratory depression if large doses are given. Elderly patients and those with chronic obstructive pulmonary disease are particularly prone to developing apnea or respiratory arrest. All benzodiazepines can lead to clouded consciousness, mimicking the mental status changes seen in acute confusional states.

Narcotics and non-depolarizing muscle relaxants

Morphine sulfate can be used for pain control and to sedate the agitated patient in the ICU. Parenteral administration leads to a prompt effect that lasts for 4 to 5 h. Respiratory depression and aggravation of confusion are potential adverse effects.

Intubation, sedation, and paralysis using metocurine iodide or pancuronium bromide are the final treatment options when other measures fail to control severe agitation. Cardiovascular effects of metocurine iodide and pancuronium bromide, pulmonary complications associated with intubation, and traction injuries that develop during paralysis are potential adverse effects of this treatment strategy.

Mechanical restraints

The severely agitated patient may attempt to get out of bed or to pull out intravenous cannulas, an endotracheal tube, or an intra-aortic balloon pump. These actions are potentially lethal and require time-consuming supervision by nursing staff. Even when optimally medicated with a neuroleptic and/or a benzodiazepine, the calm but confused patient might still perform dangerous maneuvers. In such situations it is recommended that the patient be placed in mechanical restraints. The indication should be clearly documented in the medical records and the need for restraints assessed continuously. Adverse effects, such as strangulation and brachial plexus injury, can be avoided by use of proper techniques and frequent changes in the position of the restraints. To avoid further exacerbation of the agitated behavior, patients with obsessive, overly controlling, or affectively unstable personalities should be placed in restraints only if absolutely necessary.

Supportive measures

The confused patient benefits from frequent reorientation to time and place. Adequate lighting can reduce the likelihood of illusionary misperceptions. The presence of a calm and reassuring family member and nursing staff is helpful for the patient. A 'sitter' can also help reduce anxiety. Education of the staff, the patient's family, and the patient about the nature of confusion and/or agitation is calming and therapeutic for all the parties involved.

Bibliography

Stoudemire, A., Moran, M.G., and Fogel, B.S. (1990). Psychotropic drug use in the medically ill: Part I. *Psychosomatics*, **31**, 377–91.

Stoudemire, A., Moran, M.G., and Fogel, B.S. (1991). Psychotropic drug use in the medically ill: Part II. *Psychosomatics*, **32**, 34–46.

7.14 Pain

7.14.1 Pathophysiology

M. Schäfer and C. Stein

Key messages

- Pain initiates a wide range of physiological and psychological alterations that initially help the body to maintain homeostasis. However, these effects may become harmful if pain remains untreated or undertreated and may cause morbidity and mortality.
- Pain from surgical trauma consistently elicits increased sympathetic nervous system activity and adrenal catecholamine secretion. This leads to enhanced myocardial oxygen consumption which can be detrimental in patients with underlying cardiac disease.
- The neuroendocrine stress response after painful trauma is characterized by an increased secretion of catabolic hormones and a decreased secretion of anabolic hormones. Hyperglycemia, increased protein breakdown, and a negative nitrogen balance are the consequences.
- The transition from acute to chronic pain may critically depend on disturbances of intrinsic control mechanisms as well as on effective therapeutic measures initiated at the earliest possible time.

Introduction

Research in the field of pain has expanded considerably in the past 25 years. However, despite a better understanding of pain mechanisms, acute pain is not effectively treated. More than 50 per cent of medical and surgical in-patients experience 'excruciating pain' during their stay in hospital, and this is also a major problem in intensive care units (ICUs) (Murray and Plevak 1994). While non-intubated patients can tell health care providers about their discomfort, intubated patients cannot effectively communicate and are at higher risk of inadequate analgesia.

Pain is 'an unpleasant sensory and emotional experience associated with actual or potential tissue damage or described in terms of such damage'. The biological function of pain is to prevent further tissue injury and to support the healing process. However, the body benefits only briefly from this effect, and prolonged pain can result in an adverse outcome. In addition to the surgical trauma, pain itself initiates a wide range of physiological and psychological alterations that initially help the body to maintain homeostasis. However, these effects may become harmful if pain remains untreated or undertreated and may cause morbidity and mortality (Kehlet 1993).

Respiratory effects

Pain from abdominal or thoracic surgery commonly results in a deterioration of the patient's pulmonary function. The painful stimulus triggers spinal reflex responses which result in spasms of the chest and/or abdominal wall muscles. Such 'muscle splinting' also affects the diaphragm. As a result of these reflexes, the patient is unable to breathe or cough sufficiently and voluntarily reduces inspiratory or exspiratory efforts. These changes lead to decreases in tidal volume, forced vital capacity, functional residual capacity, and forced expiratory volume. Subsequently, atelectasis occurs which worsens the ventilation–perfusion ratio and finally results in hypoxemia and an increased risk of pulmonary infection. Adequate pain treatment prevents these adverse effects. For instance, in patients undergoing abdominal aortic aneurysm repair, significant decreases in time of intubation, time in the ICU, and cardiac and pulmonary complications have been achieved by agressive pain treatment (Kehlet 1993).

Cardiovascular effects

Pain from surgical trauma consistently elicits increased sympathetic nervous system activity and adrenal catecholamine secretion. As a consequence, blood pressure, heart rate, and cardiac output increase. This leads to enhanced myocardial oxygen consumption which can be detrimental in patients with underlying cardiac disease. Inadequate treatment of pain in such patients increases the risk of perioperative myocardial ischemia. Additionally, an increased sympathetic tone may lower the heart's fibrillation threshold and render it more susceptible to arrhythmic events. The incidence of cardiovascular complications is reduced by effective pain relief (Kehlet 1993).

In the peripheral circulation, increased sympathetic stimulation may alter regional blood flow. Decreased blood flow will reduce oxygen delivery, resulting in local acidosis and impaired wound healing. Together with changes in coagulability, this may also lead to venous thrombosis and pulmonary embolism. Epidural analgesia can prevent these sequelae and improve outcome (Kehlet 1993).

Neuroendocrine and metabolic effects

The neuroendocrine stress response after painful trauma is characterized by an increased secretion of catabolic hormones (ACTH, cortisol, glucagon, catecholamines) and a decreased secretion of anabolic hormones (insulin, testosterone, growth hormone). This results in hypermetabolism and release of substrates (glucose, amino acids, fat) from peripheral stores (muscle, fat tissue, liver). Hyperglycemia, increased protein breakdown, and a negative nitrogen balance are the consequences. In addition, the release of aldosterone and vasopressin is enhanced and results in sodium and water retention. Nociceptive impulses from the injured area contribute significantly to the initiation and maintainance of these effects, since regional anesthesia can potently reduce the stress response (Kehlet 1993). Other factors such as cell breakdown products or inflammatory mediators also play a role. Stress-induced activation of the hypothalamopituitary axis may cause suppression of immune functions and promote impaired wound healing and sepsis. Finally, the neuroendocrine stress response can cause molecular changes; for example, it can induce expression of heat shock proteins in the adrenal cortex, aorta, and vena cava (Kehlet 1993).

Psychological effects

Psychological distress can easily derive from unrelieved pain and plays a major role in the initiation and maintenance of the stress response. Pain can promote anxiety and fear. Anxiety stimulates the fight or flight response that focuses an individual's attention on avoiding further injury. Anxiety by itself can initiate the neuroendocrine stress response. Both anxiety and pain can lead to insomnia. Pain, anxiety, and sleep deprivation increase the perception of each other, resulting in a vicious cycle of mutual reinforcement.

Effects on the central and peripheral nervous systems

Pain initiates multiple plastic changes within the nervous system at peripheral, spinal, and supraspinal levels (Coderre *et al.* 1993; Schaible and Grubb 1993). Tissue injury increases sensitivity to stimuli that would not normally produce pain (allodynia), exaggerates responsiveness to noxious stimuli in the injured area (primary hyperalgesia), and leads to a spread of pain and hypersensitivity to uninjured adjacent areas (secondary hyperalgesia). These changes in sensitivity are caused by a reduction in nociceptive thresholds of large Aβ sensory fibers (allodynia, secondary hyperalgesia) and thinly myelinated A and unmyelinated C fibers (primary and secondary hyperalgesia). Continuous nociceptive input from C and A fibers can trigger changes in the excitability of neurons in the spinal cord, which can persist even after the noxious stimulus has gone ('central sensitization') (Schaible and Grubb 1993). This sensitization of spinal cord neurons is reflected by increased spontaneous activity, increased responsiveness to afferent inputs ('wind up'), prolonged afterdischarges to repeated stimulation (persistent pain), and the expansion of the peripheral receptive fields of individual neurons ('central neuroplasticity').

The most distressing feature of these pathological processes is that they are associated with alterations in neuronal gene expression which outlast the primary nociceptive input and therefore may predispose to chronic pain (Coderre *et al.* 1993). A possible approach to preventing central sensitization after surgical trauma is to start adequate treatment of acute pain as early as possible, i.e. pre- or intraoperatively. This has led to the concept of 'pre-emptive analgesia'. The basic hypothesis is that analgesic treatment before the onset of a surgical trauma will be more effective than after surgery, when central sensitization has already been established. The practical value of this concept awaits further clinical studies (Coderre *et al.* 1993).

The pathological changes in the excitability of spinal neurons are initiated by the transmission of nociceptive stimuli in the dorsal horn of the spinal cord via release of neuropeptides or excitatory amino acids from primary afferent neurons. Neuropeptides, such as substance P, calcitonin gene-related peptide (CGRP), and neurokinin A are clearly involved, and the synthesis and release of these substances is upregulated in the spinal cord after peripheral injury. Excitatory amino acids (glutamate, aspartate) are also released in the dorsal horn upon noxious stimulation; they activate postsynaptic *N*-methyl-D-aspartate (NMDA) and/or non-NMDA receptors and thereby facilitate the onset of central sensitization. Neuropeptides and excitatory amino acids trigger these alterations in membrane excitability through interactions with second-messenger systems such as intracellular calcium and protein kinases. Specific antagonists at neurokinin or NMDA receptors can effectively inhibit pain in animals (Coderre *et al.* 1993; Schaible and Grubb 1993).

Inhibitory mechanisms within the spinal cord are being activated to counteract hyperalgesia and pain. Animal models of inflammatory pain have demonstrated an increase in the production and release of endogenous morphine-like substances (enkephalins) in the spinal cord. These opioid peptides can activate either postsynaptic opioid receptors on spinal cord neurons or presynaptic opioid receptors on primary afferent neurons. Both these actions result in decreased excitabilities of the respective neurons and, in the latter case, a diminished release of excitatory neuropeptides or amino acids. Eventually, these actions lead to the inhibition of pain (Coderre *et al.* 1993; Schaible and Grubb 1993).

Neuroimmune interactions

Multiple pathophysiological changes occur at the site of an injury. These result from mutual neuroimmune interactions and contribute to plasticity at the peripheral terminals of sensory neurons (Stein 1995). In response to trauma, an inflammatory process develops quickly, characterized by the migration of polymorphonuclear leukocytes into the affected area in the early phase followed immediately afterwards by lymphocytes, macrophages, and monocytes. Products from cell breakdown (prostaglandins, protones), plasma leakage (bradykinin), and inflammatory cells (cytokines, histamine, serotonin) directly stimulate (bradykinin, protones) and/or indirectly sensitize (prostaglandins) primary afferent neurons. This results in a reduction of nociceptive thresholds of C and A primary afferent nerve endings (hyperalgesia). A subpopulation of these nociceptors ('silent nociceptors') which are unresponsive under normal conditions may be activated in inflammation, leading to an expansion of their peripheral receptive fields (Schaible and Grubb 1993).

The primary afferent nerve ending itself contributes to these changes. It contains various neuropeptides such as substance P and CGRP. Under inflammatory conditions, the neuronal content of these peptides is upregulated and their axonal transport towards the peripheral nerve endings is increased. This upregulation is partially caused by nerve growth factor which is abundant in inflamed tissue. Substance P and CGRP are released into the injured tissue where they can modulate the secretion of mediators and cytokines from resident immune cells and may contribute to an increased inflammatory response. This closes the vicious cycle of mutually reinforcing effects between immune cells and primary afferent nerve terminals resulting in enhanced hyperalgesia and pain (Schaible and Grubb 1993).

Concurrently, however, counteractive endogenous mechanisms are being established to inhibit inflammatory pain at the site of the injury. These mechanisms are also based on an interaction between the immune and nervous systems. Opioid receptors are synthesized in dorsal root ganglia and present on peripheral sensory nerve fibers. Owing to an enhanced axonal transport towards peripheral nerve terminals, their number is upregulated during inflammation. When activated by exogenous opiates (morphine), these receptors can mediate potent analgesia, similar to the activation of opioid receptors within the central nervous system but without central side-effects (Stein 1995). In addition, injured tissue contains elevated concentrations of endogenous opioid peptides. These peptides (β-endorphin, enkephalins, dynorphin) are synthesized within resident lymphocytes, monocytes, and macrophages, and can be released by stressful stimuli. This release can be triggered by a paracrine action of endogenous corticotropin-releasing factor, which is abundantly expressed in inflamed tissue, or by exogenous corticotropin-releasing factor and interleukin 1β, resulting in potent inhibition of inflammatory pain (Stein 1995).

Conclusions

Pain can have detrimental effects on almost every organ system. The most readily apparent effects are those on the respiratory and cardiovascular systems. More prolonged effects develop via the expression of genes for neuropeptides, cytokines, growth factors, and other compounds. Interactions between the immune and nervous systems are the basis for the development and also the control of inflammatory pain. Immediately after tissue injury, a cascade of local events is initiated that leads to increased sensitivity. In parallel, intrinsic counteractive mechanisms, which are capable of controlling pain by central and peripheral opioid actions, can come into play. The transition from acute to chronic pain may critically depend on disturbances of intrinsic control mechanisms as well as on effective therapeutic measures initiated at the earliest possible time.

References

Coderre, T.J., Katz, J., Vaccarino, A.L., and Melzack, R. (1993). Contribution of central neuroplasticity to pathological pain: review of clinical and experimental evidence. *Pain*, **52**, 259–85.

Kehlet, H. (1993). General vs. regional anesthesia. *Principles and practice of anesthesiology*, pp. 1218–34. Mosby Year Book, St Louis, MO.

Murray, M.J. and Plevak, D.J. (1994). Analgesia in the critically ill patient. *New Horizons*, **2**, 56–63.

Schaible, H.G. and Grubb, B.D. (1993). Afferent and spinal mechanisms of joint pain. *Pain*, **55**, 5–54.

Stein, C. (1995). Mechanisms of disease: the control of pain in peripheral tissue by opioids. *New England Journal of Medicine*, **332**, 1685–90.

7.14.2 Assessment

A. R. Molloy and M. J. Cousins

Key messages

- Pain in the critical care setting is predominantly associated with nociception.
- Neuropathic pain must always be considered as it responds poorly to drugs used for nociceptive pain and requires different treatments.
- Pain should be assessed and recorded along with other routine measurements to provide an overview of the 'pain state'.
- The pain management team should be involved early in patients with pain which is responding poorly to treatment.

Introduction

Practitioners are expected to be skilled in contemporary acute and chronic pain assessment and treatment, but evidence suggests that their training is often inadequate and that pain continues to be undertreated in all areas including critical care (Dracup and Bryan-Brown 1995). Improvements in knowledge and patient care have been the goals of the International Association for the Study of Pain which was founded in 1974. The Agency for Health Care Policy and Research was established by the United States Department of Health and Human Services to develop practice guidelines for assessment and management of pain (Agency for Health Care Policy and Research 1992). These practice guidelines, amongst others, are relevant to critical care management.

Concepts of pain

Pain can be defined as 'an unpleasant sensory and emotional experience associated with actual or potential tissue damage, or described in terms of such damage' (Merskey and Bogduk 1994). Thus pain should be conceptualized as more than a hard-wired sensory system. It is a complex perception which is multidimensional with emotional and behavioral components (i.e. suffering). As it is a subjective experience, it cannot be objectively quantified.

The perception of acute pain is one of a number of bodily responses triggered by nociception. In addition, physiological responses include secretion of stress hormones that promote catabolism, hyperglycemia, and sodium and water retention, increased sympathetic output resulting in tachycardia, hypertension, and ischemia, and increased vagal tone (e.g. slowing of peristalsis). In the critical care setting the focus is on the somatic (sensory) components of pain intensity, site, and temporal characteristics. The term 'low pain threshold' should not be used. Individuals have a uniform sensation threshold which is independent of their cultural background. However, there are differences in pain perception threshold and pain tolerance levels which appear to be influenced by cultural factors.

Assessment of pain

Assessment involves the measurement of both physiological responses to nociception and the patient's verbal and behavioral communications (pain behavior). Acute pain is viewed as a biologically advantageous and time-limited experience or state (Chapman *et al.* 1992). Sympathetic responses may indicate the presence of acute pain, but over time these responses habituate and their absence does not exclude pain. In the communicative patient, the location of pain should be consistent with the injury or the surgical procedure. If not, another process should be considered.

Numeric pain intensity scale
Draw a vertical line to indicate your current level of pain

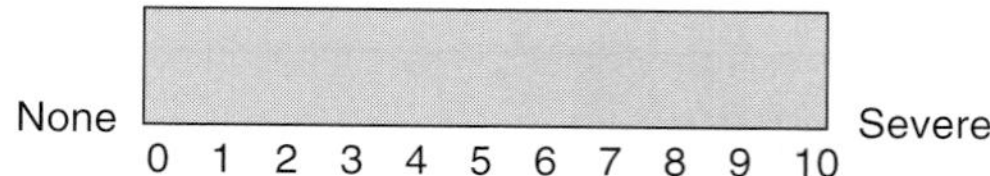

Box intensity scale
Place an x on the scale below to indicate your current level of pain

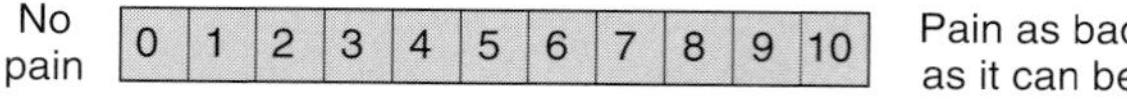

Fig. 1 Examples of pain intensity rating scales.

Pain may be rated as 'mild', 'moderate', or 'severe'. Alternatively, a visual or a verbal analog scale from zero (no pain) to 10 (the worst pain that can be imagined) may be used (Fig. 1).

Although such scores condense a multidimensional experience into a single rating, they are useful management tools if recorded at regular intervals (e.g. every second waking hour) together with other parameters to provide a profile of the pain state. Pain on movement should also be recorded as 'mild', 'moderate', or 'severe' in order to trigger appropriate interventions. It cannot be assumed that patients will progress at the same rate, and in more complex cases measurements should be taken more regularly. The power of such data may be increased by assessing other components. Thus the ratings may be changed so that zero indicates no unpleasantness or distress and 10 indicates severe unpleasantness or distress. A satisfaction scale may similarly be constructed. Intubated patients may experience difficulties in communicating about their pain. Possible solutions are to indicate painful areas on a body chart and words to describe their pain from a descriptive list. Patients experience procedural and activity-related pain which should be treated proactively with regular assessments of outcome. Incorporating the information into a flowchart will standardize documentation and help to improve clinical practice. More comprehensive tools exist, such as the McGill Pain Questionnaire, which are part of a comprehensive multidisciplinary assessment.

The components of the pain experience are shown in Fig. 2.

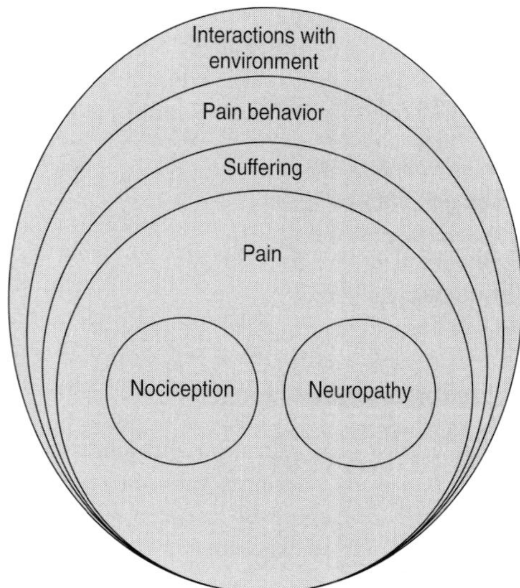

Fig. 2 Diagram adapted from Loeser's formulation identifying the different components of the pain experience. This emphasizes that nociception, neuropathy, pain, and suffering are personal and private events whose existence can only be inferred. Only pain behavior can be observed and quantified by independent observers.

Nociceptive pain

A noxious stimulus such as surgery produces afferent discharges in A + C fibers and the release of mediators in the periphery termed the 'inflammatory soup'. Sensitization of high-threshold nociceptors occurs, and this process is termed primary hyperalgesia (i.e. increased pain intensity within the injury in response to a painful stimulus). Afferent transmission to the dorsal horn results in the phenomenon of 'wind-up' and central sensitization through an action on *N*-methyl-D-aspartate (NMDA) receptors leading to calcium influx and production of nitric oxide. This central process is reflected in the periphery by secondary hyperalgesia, which is an area of increased sensitivity to stimuli in the undamaged area surrounding the injured tissue. Sensitization is associated with 'clinical' pain and manifested as spontaneous pain, hyperalgesia, and allodynia.

The central nervous system receives and processes afferent information if nociception occurs. The pathways are not 'hard-wired' and display plasticity. There is no relationship between the input (afferent stimuli) and output of dorsal horn neurons partly due to descending influences from higher centers. Thus two individuals may have a different pain experience from the same noxious stimulus.

Neuropathic pain

Neuropathic pain syndromes are associated with trauma, disease, or damage to the peripheral or central nervous system. The mechanisms are poorly understood but appear to be predominantly associated with central mechanisms such as 'wind-up' and reorganization of spinal cord connectivity. Initial mechanisms may produce other changes or evolve into other processes that alter the nature of the pain over time. Neuropathic pain occurs commonly in critical care. Early diagnosis allows for patient reassurance and appropriate early intervention to improve outcome. Assessment of neuropathic pain can be difficult. Firstly, a thorough examination may be painful and distressing for the patient because of the presence of hyperalgesia and allodynia. With patience and care the essential aspects can be assessed. Secondly, patients find it hard to describe some of the pain that they are experiencing because of differences in language, experience, and frames of reference (Turk and Melzack 1992). Features which help to distinguish neuropathic from nociceptive pain are listed in Table 1.

Table 1 Features of neuropathic pain
Pain in the absence of ongoing tissue damage
Pain in an area of sensory loss
Paroxysmal or spontaneous pain
Allodynia (pain in response to a non-painful stimulus)
Hyperalgesia (increased response to a painful stimulus)
Dysesthesias (unpleasant abnormal sensations such as burning or electric shocks)
A delayed onset after injury is characteristic

Burning pain with hyperalgesia or allodynia, sweating, vasomotor instability, edema, and trophic changes suggest the diagnosis of complex regional pain syndrome type 1 (reflex sympathetic dystrophy) or type 2 (causalgia) (Merskey and Bogduk 1994).

Suffering

Suffering is invariably associated with pain, but suffering and pain are distinct and may occur independently. Suffering includes disruption of mood, thoughts, plans, and sense of well being. There may be economic distress, family/marital problems, and various losses resulting from the impact of pain on the patient's life. Disruption of the sleeping cycle, anger, anxiety, helplessness, and the side-effects of medication may add to demoralization and increased suffering. To some extent the degree of suffering is related to the adequacy of the patient's coping skills. Patients who 'catastrophize' about their pain tend to suffer more than those who do not.

Pain behavior and interactions with the environment

Acute pain may last for minutes (e.g. cannulation) or weeks to months following a burn. In a small but significant proportion of patients, pain persists and becomes chronic or there are recurring episodes of acute pain.

In this situation management involves assessment of the 'whole picture' and not just the pain. Interactions between psychological, environmental, and somatic factors have been shown to influence the nature, intensity, and persistence of pain and disability (Turk and Melzack 1992). Patients' reaction to their pain (known as pain behavior), their moods, their beliefs, their coping strategies, 'internal conversations', and interactions with their family and environment, in particular the ward, also need consideration. Pain behavior refers to speech, posture, and facial expressions (e.g. grimacing) which would lead one to infer that the individual has pain. Pain behavior has been shown to be influenced by a variety of factors other than an identifiable pathological process. This is a highly specialized area, and its thorough evaluation and treatment require the skills of a multidisciplinary pain management team including a clinical psychologist or psychiatrist with expertise in pain assessment and management. Treatment may be required to address psychological factors including 'maladaptive' behavior and to encourage 'well' behavior.

References

Agency for Health Care Policy and Research (1992). *Acute pain management: operative or medical procedures and trauma*. Publication 92–0032, US Department of Health and Human Services, Rockville, MD.

Chapman, R.C., Donaldson, G.W., and Jacobson, R.C. (1992). Measurement of acute pain states. In *Handbook of pain assessment* (ed. D.C. Turk and R. Melzack), pp. 332–43. Guildford Press, New York.

Dracup, K. and Bryan-Brown, C.W. (1995). Pain in the ICU: fact or fiction? *American Journal of Critical Care*, 4, 337–9.

Merskey, H. and Bogduk, N. (ed.) (1994). *Classification of chronic pain: descriptions of chronic pain syndromes and definition of pain terms*. IASP Press, Seattle, WA.

Turk, D.C. and Melzack, R. (1992). The measurement of pain and the assessment of people experiencing pain. In *Handbook of pain assessment* (ed. D.C. Turk and R. Melzack), pp. 3–12. Guildford Press, New York.

7.14.3 Management

A. R. Molloy and M. J. Cousins

Key messages

- Pain continues to be undertreated in critical care.
- Assessment and the application of available therapies will result in excellent analgesia in the majority of patients.
- Some patients will require more sophisticated techniques and this should trigger a 'pain consultation'.
- Evidence is accumulating that effective pain management is associated with improved outcomes.

Introduction

Effective acute pain management is an essential component of contemporary critical care. This is for compelling humanitarian reasons combined with the need to minimize physiological and neurobiological changes. Increasing basic scientific and clinical evidence support pre-emptive and aggressive provision of analgesia for acute pain to minimize suffering, decrease morbidity, reduce costs, and improve outcome. Fears that pain therapies may compromise the critically ill are unfounded provided that analgesia is administered with due care and expertise.

There are many novel and sophisticated analgesic techniques that may be appropriate in difficult cases, but practitioners should not adopt approaches that are outside their experience or ability to monitor appropriately. Fortunately, careful administration of standard drugs and the application of non-drug approaches will provide excellent results in the majority of patients.

The essential component of effective pain management is constant re-evaluation with regular assessment of pain intensity and side-effects. Defining outcomes as 'normalization' of physiological variables in the unconscious patient or a score of less than 4 out of 10 on a visual or verbal analog scale in the conscious and communicating patient will assist in appropriate adjustments to treatment. Failure to achieve pain relief within the skills of the practitioner should prompt a referral to the pain management team as is normal practice with other medical problems.

An algorithm for pain management in critical care is shown in Fig. 1.

Pre-emptive analgesia

Pre-emptive analgesia refers to interventions aimed at preventing the development of central sensitization, resulting in reduced requirement for analgesia and a lower incidence of chronic pain syndromes. This is of clinical relevance, as approximately 70 per cent of patients attending our pain center have chronic pain arising from an acute noxious event. Acute and chronic pain models reveal a greater effect from local or systemic analgesia administered before rather than after a noxious stimulus, but the clinical evidence is not so clear cut. Methodological reasons may account for this.

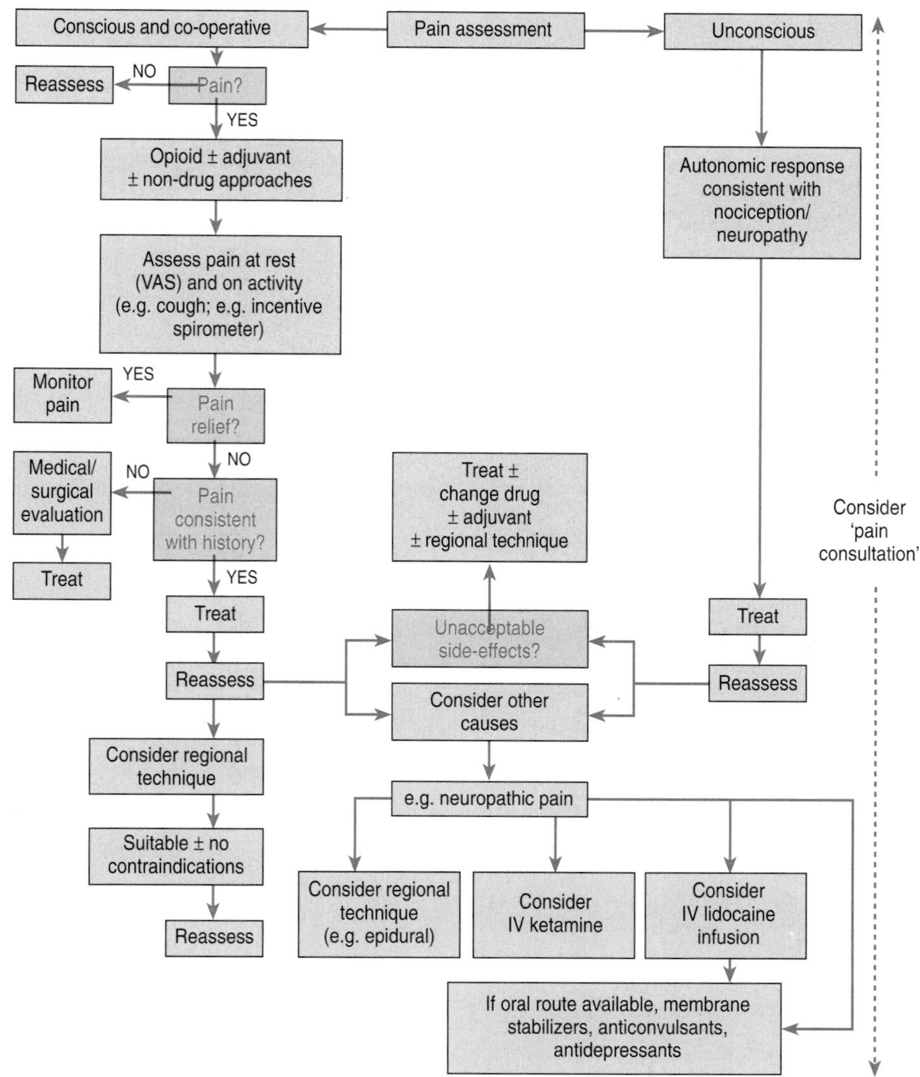

Fig. 1 Algorithm for pain management in critical care: VAS, visual analog scale; IV, intravenous.

Non-drug techniques

Causes of agitation and sympathetic response, such as acid–base or blood gas disturbance, distended bladder, or undetected pathology, should be considered and treated. Anxiety will be helped by communication, reassurance, and careful preparation for procedures. For mild pain, initially consider therapies such as heat, cold, transcutaneous electrical nerve stimulation (TENS), relaxation, and distraction. Psychological treatments are important particularly if pain becomes chronic or out of keeping with identified pathology.

Drug techniques

Significant changes in the pharmacological management of pain are occurring due to a combination of improved understanding of the neurobiological basis of pain, advances in pharmacokinetics and pharmacodynamics, the clinical application of new drugs, and the use of new analgesic technologies. The current analgesic drug options consist of the following:

(1) opioids;

(2) non-steroidal anti-inflammatory agents;

(3) local anesthetics;

(4) adjuvants.

The choice of analgesia should be based on the location and type of pain (nociceptive or neuropathic), with allowance made for individual patient factors. Intravenous analgesia is used in unconscious patients, those with pain at multiple sites, and in the presence of contraindications to regional techniques. Intramuscular and subcutaneous routes are avoided. Regional techniques confer significant advantages in selected patients.

Opioids: advances in pharmacokinetics and pharmacodynamics

Drugs are often chosen on the basis of differing elimination half-lives. Opioids with long elimination half-lives (e.g. morphine) may be chosen for long procedures and infusions, whereas opioids such as alfentanil with a shorter elimination half-life are used for brief procedures (Hughes *et al.* 1992). This is simplistic as all intravenous drugs are described by multicompartmental pharmacokinetics. Rational selection of an opioid for use in critical care should utilize new pharmacokinetic concepts such as biophase (i.e. effect site) and context-sensitive half-life (the time required for the central compartment drug concentration to decrease by 50 per cent). Pharmacodynamic variables are also important. The principal pharmacodynamic differences between the opioids are in the half-time of equilibration between the biophase and plasma (e.g. 5–6 min for fentanyl and 1.5 min for alfentanil) and their potency.

Morphine

Morphine is widely used in critical care. Glucuronidation of morphine produces two metabolites, morphine-3-glucuronide and morphine-6-glucuronide.

The concentrations of morphine-3-glucuronide and morphine-6-glucuronide may be up to 10 times and 100 times respectively the concentration of morphine in the plasma. Morphine-3-glucuronide tends to antagonize morphine, and morphine-6-glucuronide is up to 10 times more effective than morphine in tests of analgesia. The metabolites depend on glomerular filtration and accumulate in renal failure. This is a significant drawback in critical care patients as it may result in prolonged weaning, increased morbidity, and increased costs. Multicompartmental pharmacokinetic data are not available for morphine. Following titration (front-loading), the elimination half-life (approximately 3 h) is used to calculate the approximate infusion rate as the dose of morphine required in a 3-h period is half the front-loading dose. Suboptimal analgesia requires retitration as adjustment of the infusion alone requires four to five half-lives to steady state.

Fentanyl, alfentanil, sufentanil, and remifentanil

Research has focused on the semisynthetic opioids fentanyl, alfentanil, sufentanil, and remifentanil which are more expensive than morphine; further studies are needed as their use may be associated with a reduction in overall costs and improved outcomes. Electroencephographic analysis and computer simulation are used to characterize these drugs, although there is no relationship with the level of analgesia. These correlate well with standard clinical doses, but studies have not taken into account the altered pharmacokinetics and dynamics in the critical care setting. Multicompartmental modeling suggests that the time to peak biophase concentration following an intravenous bolus is a function of both the rate of decline of plasma concentration and the rate of equilibration of plasma concentration with the site of the drug effect. Infusion regimes require an exponentially decreasing series of infusion rates before steady state is reached. These concepts are extremely relevant to the infusion of analgesics and sedative agents; for example, a fall of 80 to 90 per cent in opioid concentration is required for adequate ventilation after an opioid–oxygen anesthetic.

Reference to Fig. 2 shows that if a rapid decrease in opioid concentration at the effect site is required, alfentanil is the drug of choice for operations lasting longer than 6 to 8 h and sufentanil is preferable for operations lasting up to 6 to 8 h. Fentanyl accumulates and thus may be associated with prolonged recovery if used for more than 1 h.

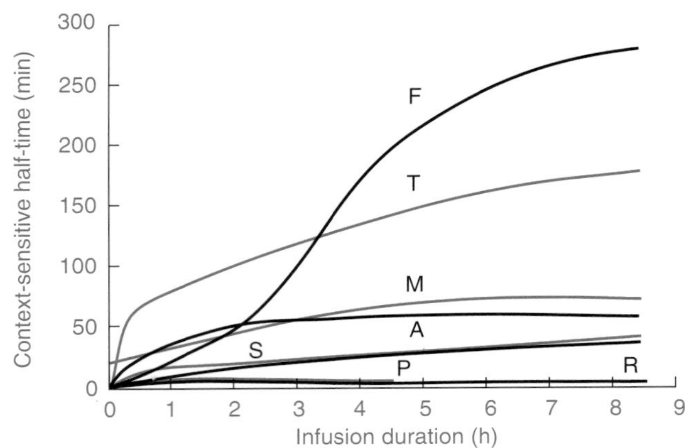

Fig. 2 Context-sensitive half-lives as a function of infusion duration for each of the pharmacokinetic models simulated: F, fentanyl; A, alfentanil; S, sufentanil; R, remifentanil. Thiopental (thiopentone) (T), midazolam (M), and propofol (P) are included for comparison.

Remifentanil is an exciting new opioid which is inactivated by blood and tissue esterases. The context-sensitive half-life is 3 to 4 min; it is constant over a 15-fold variation in effect-site concentration and is independent of duration of infusion. This allows for rapid titration of effect with ideal characteristics for critical care if a rapid reversal of effect is required. All these drugs appear to provide significant advantages over morphine, but further studies are required.

Patient-controlled analgesia

Patient-controlled analgesia minimizes interpatient pharmacokinetic and pharmacodynamic variability to allow suitable patients to titrate a predetermined bolus of opioid against their pain (Ballantyne *et al.* 1993). The dose, lockout period, and bolus infusion period are set by the practitioner. Most patients experience an individually determined satisfactory level of analgesia. Usage is significantly influenced by factors such as mood, anxiety, and expectation of recovery. Patient-controlled analgesia should not be used if the negative feedback loop is not intact. A maximum limit of half the predicted hourly rate may be used if an infusion combined with patient-controlled analgesia is desired. Meperidine (pethidine) may be used for periods of 2 to 3 days, but prolonged use is associated with dysphoria and accumulation of normeperidine (norpethidine) with a risk of convulsions.

Codeine and tramadol

Codeine is a prodrug metabolized by *ortho*-demethylation to morphine. Approximately 8 per cent of the Caucasian population are poor metabolizers, and this percentage is increased by medications such as fluoxetine. Absorption by the intramuscular route is variable, with a fivefold difference in peak blood concentration and a ninefold variation in minimum effective analgesia concentration. Codeine is not a suitable analgesic for the critical care patient.

Tramadol is an atypical opioid which is an agonist for μ receptors and inhibits uptake of serotonin and norepinephrine (noradrenaline), producing less sedation and respiratory depression.

New technologies

Target-controlled infusion, patient demand target-controlled infusion, and computer-assisted continuous infusion are some of the analgesic technologies that are undergoing development. They facilitate the maintenance of a selected effect-site concentration through an infusion pump linked to a pharmacokinetically modeled computer program. Such technologies allow the use of opioids such as alfentanil which has a low tissue solubility, ensuring a rapid onset and short latency to peak effect and short duration of action.

Non-steroidal anti-inflammatory drugs

Non-steroidal anti-inflammatory drugs such as ketorolac trometamol (ketorolac tromethamine) are used as components of 'balanced' analgesia because of their opioid-sparing effects in acute pain management. Potential adverse effects, such as gastric ulceration, renal impairment, and increased bleeding, limit their use in critical care. The synthesis of non-steroidal anti-inflammatory drugs selective for the isomer of cyclo-oxygenase responsible for inflammatory effects may improve the side-effect profile. Acetaminophen (paracetamol), traditionally available for oral or rectal administration, has been released in an injectable form. Dipyrone is a pyrazolone non-opioid analgesic that is much more potent than acetaminophen and probably has a 'central' analgesic action. It has minimal effects on the kidneys, gut, and platelet function. However, it has excellent antispasmodic activity and is valuable when both pain and visceral spasm are present. A rare occurrence of agranulocytosis has limited its use in some countries, but it is widely used in Europe. It is available orally, intramuscularly, and for intravenous infusion. Currently, its place in acute pain relief is being re-evaluated.

Regional analgesia

Peripheral neural blockade

Percutaneous conduction blockade can be performed on most major nerves using anatomical landmarks. Techniques are described in many texts (Cousins and Bridenbaugh 1988). Single-shot techniques (e.g. brachial plexus, femoral nerve, and intrapleural space) have been employed successfully. Potential problems are difficulties with placement, neural damage, hematoma formation, and infection. These techniques provide excellent analgesia compared with parenteral opioid administration, but the potential complications need to be justified on clinical grounds.

Epidural and intrathecal administration

Epidural analgesia has been shown to improve outcome in high-risk patient populations (Lui *et al.* 1995).

Epidural and subarachnoid administration of local anesthetics and opioids may be as single agents or in combination. Subarachnoid administration with local anesthetics is used for many surgical procedures but has a short duration of action. Continuous infusion may be used but requires close monitoring. Clinically significant analgesia is achieved with subarachnoid administration of opioids, for example morphine 0.5 tol mg with a duration of action of 8 to 24 h. Bupivacaine has been the local anesthetic of choice for epidural infusion. A low concentration (e.g. 0.125 per cent) with opioid minimizes motor block. Fentanyl is highly lipid soluble and thus needs to be administered close to the appropriate spinal cord segments. Morphine is less lipid soluble and thus can be administered in the lumbar region, from which it will spread cranially to treat thoracic, cervical, or head pain.

Ropivacaine is a new enantioselective local anesthetic with greater sensory–motor separation, less cardiotoxicity, and higher clearance than bupivacaine, thus offering distinct advantages over bupivacaine.

Management of neuropathic pain

Neuropathic pain may be associated with a variety of complex processes and interactions at many levels of the nervous system. Current recommendations are for early intervention to attenuate the cellular processes that may lead to persistent pain. For example, *c-fos* is expressed within minutes and dorsal horn sprouting occurs within days. More than one pain-generating mechanism is likely to be operative in the individual patient, and treatment aims to target these with a variety of pharmacologically distinct drugs. Despite this, many patients continue to experience pain despite optimal use of available therapies.

Therapies are limited in the critical care setting. Neuropathic pain tends to be poorly responsive to parenteral opioids (Siddall and Cousins 1995). For regional pain, epidural opioids are combined with bupivacaine (usually more than 0.125 per cent), and clonidine, an α_2 agonist, may often be administered. Diffuse pain, such as that occurring after a spinal cord injury, may respond to the membrane-stabilizing effects of subcutaneous or intravenous lignocaine (lidocaine) (loading dose followed by 50–100 mg/h) or ketamine (10–30 mg/h), which is an *N*-methyl-D-aspartate antagonist and may be of benefit in suitable patients. Oral agents such as mexiletine, which has a membrane-stabilizing action, or anticonvulsants (e.g. sodium valproate, carbamazepine, and gabapentin) may be tried. Complex regional pain syndrome type 1 (reflex sympathetic dystrophy) or type 2 (causalgia) may respond to sympathetic nerve blocks or regional sympathetic blockade with guanethidine or bretylium. If these interventions fail, treatments are similar to those for neuropathic pain. Marked lower-limb edema may be a component of complex regional pain syndrome and act as a peripheral 'generator'. In severe cases intense spinal analgesia with bupivacaine is used with monitoring in a critical care unit to reverse proposed postcapillary sphincter spasm. Early involvement of the chronic pain service is recommended.

Outcomes

Currently, there are no data for outcomes associated with pain management in critical care. Studies are urgently needed. Inferences may be drawn from the postoperative setting which support the pre-emptive and aggressive provision of analgesia.

References

Ballantyne, J.C., Carr, D.B., Chalmers, T.C., Dear, K.B.G., Angelillo, I.F., and Mosterller, F. (1993). Postoperative patient-controlled analgesia: meta-analyses of initial randomised control trials. *Journal of Clinical Anesthesia*, **5**, 182–93.

Cousins, M.J. and Bridenbaugh, P.O. (ed.) (1988). *Neural blockade in clinical anesthesia and management of pain* (2nd edn). J.B. Lippincott, Philadelphia, PA.

Hughes, M.A., Glass, P.S., and Jacobs, J.R. (1992). Context-sensitive half-time in multicompartment pharmacokinetic models for intravenous anesthetic drugs. *Anesthesiology*, **76**, 334–41.

Lui, S., Carpenter, R.L., and Neal, J.M. (1995). Epidural anesthesia and analgesia. Their role in postoperative outcome. *Anesthesiology*, **82**, 1474–1506.

Siddall, P.J. and Cousins, M.J. (1995). Recent advances in pain management. *Australia and New Zealand Journal of Surgery*, **65**, 674–85.

8 The metabolic and endocrine systems

8.1 Normal physiology

8.1.1 The endocrine system

D. Watson

Key messages

- The main function of the endocrine system and the stress response is to maintain cardiovascular and metabolic homeostasis in an integrated fashion.
- The best known neuroendocrine response to injury is the increase in ACTH secretion and the resultant increase in adrenocortical secretion of glucocorticoid hormones.
- Exercise, pyrogens, and some psychological conditions increase the secretion of growth hormone. Glucagon and growth hormone help maintain a nutrient supply to metabolizing tissues.
- The cytokines may attenuate the magnitude of the post-injury metabolic response by direct influences upon hypothalamic and pituitary hormone production.

The endocrine system in critical illness

The circulating levels of ACTH–cortisol, thyroid-stimulating hormone–thyroxine, epinephrine (adrenaline), norepinephrine (noradrenaline), vasopressin, growth hormone, prolactin, insulin, and glucagon all increase in severe illness. These changes represent an exaggeration of normal host responses. There then develops a reduced thyroxine response to thyroid-stimulating hormone and a reduced synthesis of insulin-like growth factors in response to growth hormone. It is not clear at present of the significance of this tissue resistance or of the benefit of treating these derangements.

Hypothalamic–pituitary–adrenal axis

The central part of the endocrine system is the hypothalamic–pituitary–adrenal axis. This includes the paraventricular nucleus, which secretes corticotropin-releasing hormone (CRH) and arginine vasopressin, and the noradrenergic neurones clustered within the locus ceruleus. CRH stimulates secretion of pro-opiomelanocortin, simultaneously releasing ACTH from the anterior pituitary gland. The CRH and noradrenergic systems are both stimulated by serotoninergic and cholinergic systems and are inhibited by the γ-aminobutyric acid (GABA) and opioid–peptide systems.

The best known neuroendocrine response to injury is undoubtedly the increase in ACTH secretion and the resultant increase in adrenocortical secretion of glucocorticoid hormones. It is obvious that this increase is of great importance. In its absence, tolerance to stress is reduced. In critically ill patients cortisol levels are positively correlated with APACHE, TISS (therapeutic intervention scoring system), and ISS (injury severity score). In patients suffering myocardial infarction cortisol levels reflect both infarct size and mortality. There are also significant associations between burn size and cortisol levels.

Recently the hypothalamic–pituitary–adrenal axis was found to undergo a biphasic change during critical illness. In the first phase, lasting a few days, the high cortisol concentrations appear to be induced by augmented ACTH release. In the second phase there is a discrepancy between low ACTH and high cortisol concentrations, suggesting that cortisol release is stimulated through an alternative pathway.

Sympathetic activation also results in the liberation of norepinephrine from postganglionic nerve endings and the adrenal medulla as well as the release of epinephrine from the adrenal medulla (Fig. 1). In addition to cardiovascular disturbances, other contributory factors to the secretion of catecholamines cannot be excluded and pulmonary stimuli may be particularly important. Furthermore, many other factors are known to influence catecholamine secretion, including trauma, hypoxia, hypercarbia, hypoglycemia, body temperature, and acid–base status.

Renin secretion from the juxtaglomerular cells is stimulated by catecholamines from the adrenal medulla and the postganglionic renal sympathetic nerves that surround the juxtaglomerular cells (Fig. 2). Any decline in blood pressure produced by injury also contributes to the stimulation of renin secretion. The angiotensin generated by the renin secretion helps to maintain tissue perfusion pressure by maintaining blood pressure, increasing aldosterone secretion, increasing vasopressin secretion, and stimulating sodium and water retention.

Various other stressful stimuli also increase vasopressin secretion. Decreased extracellular fluid volume is a particularly strong stimulus since vasopressin contributes to the maintenance of extracellular fluid volume by retaining water and causing vascular smooth muscle to contract. Therefore vasopressin may be seen to have a significant role in maintaining blood pressure when blood volume is reduced.

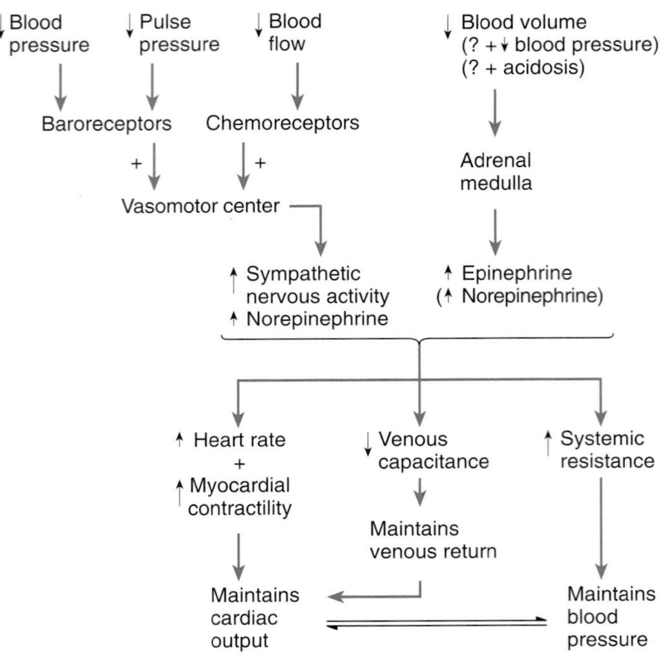

Fig. 1 The sympathoadrenal response to shock. (Reproduced with permission from Hinds and Watson (1996).)

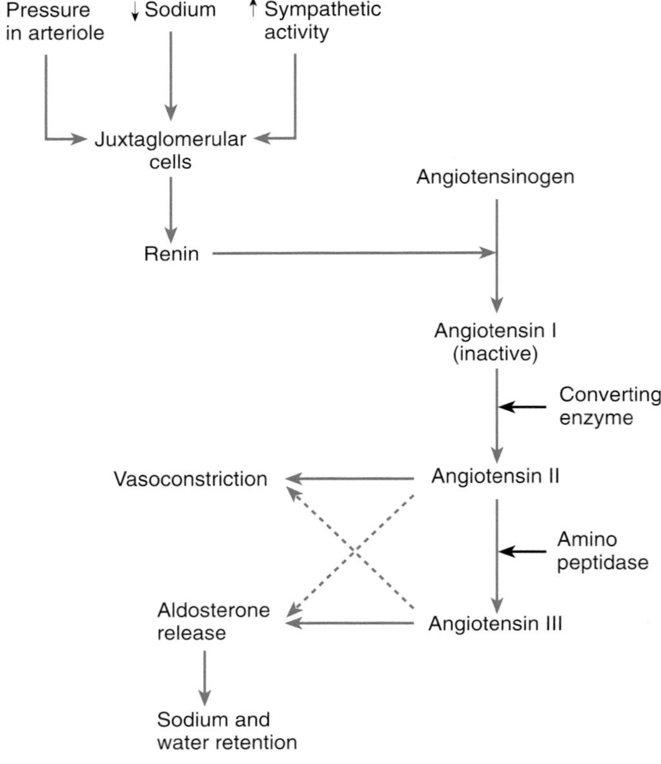

Fig. 2 The renin–angiotensin system in shock. (Reproduced with permission from Hinds and Watson (1996).)

Pituitary–thyroid axis

Thyroid-stimulating hormone stimulates the preferential release of thyroxine (T_4) from the thyroid gland. T_4 is then peripherally deiodinated into its active metabolite tri-iodothyronine (T_3) or into reverse T_3 (rT_3) which is thought to be biologically inactive. Both T_4 and T_3 exert feedback inhibition at the pituitary and hypothalamic level. Critical illness is characterized by a low T_3 or 'sick euthyroid syndrome'. In mild and severe illness T_3 production is rapidly decreased by inhibited conversion of T_4 to T_3. There is then a reciprocal increase in the inactive thyroid metabolite rT_3. With more sustained illness T_4 levels may also fall because of reduced T_4 binding to its carrier proteins leading to accelerated metabolism. More importantly, thyroid-stimulating hormone secretion is also suppressed; for example, the normal night-time surge in thyroid-stimulating hormone is not present in patients with sick euthyroid syndrome and the response of thyroid-stimulating hormone to thyrotropin-releasing hormone is frequently blunted. The degree of T_3 suppression with concomitantly low thyroid-stimulating hormone concentrations correlates positively with disease severity and negatively with outcome. The changes in T_3 and T_4 concentrations also correlate inversely with changes in circulating catecholamine levels.

The pathogenesis of sick euthyroid syndrome remains unclear, although recent observations have indicated that interleukin 1 (IL-1) and tumor necrosis factor induce the changes observed in this syndrome. It remains questionable whether these alterations in tests of thyroid function reflect deficiency of thyroid hormone, although an increase in thyroid-stimulating hormone concentrations is viewed as a marker of recovery from life-threatening illness.

Gonadal responses to critical illness

Luteinizing hormone is secreted in a pulsatile fashion by the pituitary gland and stimulates testicular secretion of testosterone in men. Testosterone levels fall promptly within 24 h of the onset of severe illness, but there is no compensatory rise in gonadotropins. Instead there is a fall in levels which persists for the duration of illness. In postmenopausal women, severe illness is also associated with relatively low concentrations of gonadotropins which correlate directly with outcome. These findings are compatible with an illness-induced hypogonadal syndrome.

In men, testosterone is the most important of the endogenous anabolic steroids and various catabolic states result in low testosterone concentrations. It is interesting to speculate that the changes in gonadal steroids, unlike those observed with the thyroid hormones, reflect actual deficiency states and that the lack of testosterone/estrogen contributes to the catabolism of critical illness.

Growth hormone and insulin growth factor: effects on metabolism

Stressful stimuli increase glucagon via stimulation of the sympathetic nerves to the pancreas and the β-adrenergic receptors in the pancreatic islets. Exercise, pyrogens, and some psychological conditions also increase the secretion of growth hormone. Glucagon and growth hormone help to maintain a nutrient supply to metabolizing tissues.

Growth hormone is secreted from the anterior pituitary gland in a pulsatile fashion following stimulation from hypothalamic growth hormone releasing hormone. Release of growth hormone is inhibited by somatostatin. Growth hormone binds to its own specific receptor and regulates the expression of insulin growth factor 1 (IGF-1) in many tissues, particularly the liver. The metabolic actions of IGF-1 are modulated by at least six different circulating binding proteins. Growth hormone causes lipolysis and antagonizes insulin effects, as well as having anabolic actions mediated by IGF-1. In fasting and serious illness there is a reduction in the indirect anabolic actions mediated by IGF-1 and an increase in lipolysis and insulin antagonism, probably associated with increased basal concentrations of growth hormone. Such changes might benefit fasting sick patients by providing circulating metabolic substrate.

Injury metabolism

Much of the observed metabolic profile of the injured patient cannot be explained solely by excessive activity of one or more components of the classical stress hormone response. Not least, the catabolic effects of one hormone (e.g. cortisol) may be counterbalanced by the elevated levels of another (e.g. growth hormone). It now appears that additional mediators, many of which are generated acutely in response to injury or infection, also appear capable of influencing systemic and cellular metabolism in a manner reminiscent of injury. Of these, the cytokine class of polypeptides, such as tumor necrosis factor, are most readily identifiable as altering systemic and tissue substrate handling in a fashion characteristic of severe injury or infection.

Cytokines and the classical stress hormone system

Current evidence points to a significant interdependence of the neuroendocrine axis and the cytokine mediator systems. For example, the exogenous administration of tumor necrosis factor elicits increased secretion of ACTH, cortisol, catecholamines, and growth hormone. These interdependent mediator systems also exhibit negative feedback. For example, glucocorticoids reduce tumor necrosis factor and IL-1 transcription. The cytokines may also serve to attenuate the magnitude of the postinjury metabolic response by direct influences upon hypothalamic and pituitary hormone production. This is suggested by the thyroid hormonal profile commonly observed in critically ill patients which may serve to limit the degree of catabolic processes in the seriously ill. If this challenge is of a modest reparable degree and of limited duration, the initial counter-regulatory hormone and cytokine responses promote the mobilization of energy stores and the maintenance of substrate availability for critical organ function. In contrast, the hormonal response to prolonged insult erodes metabolic reserves and potentially limits recovery from life-threatening illness.

Influence of anesthesia on the endocrine-metabolic response to surgery

Stress-induced changes in pituitary hormones are blocked or inhibited by regional anesthesia with local anesthetics. Thus the normal increase in plasma ACTH, β-endorphin, vasopressin, prolactin, and growth hormone is not observed, while the minor decrease in follicle-stimulating hormone and luteinizing hormone in female patients is accelerated. Thyroid-stimulating hormone changes appear to be unaffected. The cortisol response to procedures on the lower part of the body is blocked by regional anesthesia, and correspondingly changes in aldosterone and renin are generally diminished. The increase in epinephrine and norepinephrine is blunted.

The hyperglycemic response to surgery, considered to be due to a combination of increased glycogenolysis and insulin resistance, is reduced or blocked by regional analgesia. This is not mediated by inhibition of a single hormonal response to trauma but probably through the simultaneous inhibition of the cortisol, catecholamine, glucagon, and growth hormone responses. Amelioration of these endocrine changes may be advantageous in those patients with jeopardized cardiovascular reserves or coexisting metabolic disorders (e.g. diabetes).

Bibliography

Goldstein, B. and Woolf, P.D. (1996). The autonomic and neurohumoral response to critical illness. *Clinical Intensive Care*, 7, 297–307.

Hinds, C.J. and Watson, D. (1996). *Intensive care: a concise textbook* (2nd edn), pp. 66–7. W.B. Saunders, London.

Reichlin, S. (1993). Neuroendocrine–immune interactions. *New England Journal of Medicine*, **329**, 1246–52.

Van den Berghe, G. and de Zegher, F. (1996). Anterior pituitary function during critical illness and dopamine treatment. *Critical Care Medicine*, **24**, 1580–90.

8.2 Electrolyte disturbance

8.2.1 Hypernatremia

A. I. Arieff

Key messages

- Hypernatremia occurs primarily when water ingestion is inadequate to correct for fluid losses.
- Hypernatremia affects those at the extremes of age and is frequently iatrogenic.
- The mainstay of treatment is free-water administration, usually 77 mmol/l NaCl, with correction over 48 h, as overly rapid correction may be complicated by cerebral edema.
- The mortality from hypernatremia in adults exceeds 50 per cent, and it has not been demonstrated to be improved by therapy.

Introduction

Hypernatremia is one of the most commonly encountered electrolyte abnormalities in a hospital population, with an incidence in adults of 0.3 to 4.1 per cent (Long *et al.* 1991; Palevsky *et al.* 1996). Most patients have had severe concomitant diseases, making the attribution of particular signs and symptoms to hypernatremia difficult (Long *et al.* 1991; Palevsky *et al.* 1996). There is a paucity of experimental data, in either humans or laboratory animals, on the appropriate therapy for hypernatremia (Ayus *et al.* 1996).

Causes of hypernatremia

The many important causes of hypernatremia (Fig. 1) include nasogastric hyperalimentation, diabetic coma, acute renal failure, improper mixture of dialysis fluid, dehydration secondary to either fever or elevated ambient temperature, and diabetes insipidus (Snyder *et al.* 1987; Palevsky *et al.* 1996). In the elderly, hypernatremia is often associated with infirmity and inability to obtain water, leading to gradual desiccation (Snyder *et al.* 1987). Diabetes insipidus is associated with hypernatremia only when the patient has impaired thirst or is unable to obtain water. A subgroup of patients often develop hypernatremia secondary to excessive administration of hypertonic NaCl or $NaHCO_3$. Such patients are generally critically ill, and intravenous bicarbonate leads to a dangerously elevated plasma osmolality with infrequent survival. A similar outcome may be observed after inadvertent intravenous hypertonic NaCl for therapeutic abortion.

Clinical manifestations of hypernatremia

In elderly patients, pre-existing mental status abnormalities may make it difficult to detect new neurological findings. In addition, since hypernatremia frequently occurs in the setting of a coexistent pathological process, it may be difficult to ascribe any particular symptoms to hypertonicity. Gastroenteritis is frequent, and nausea and vomiting are often present without diarrhea. Seizures are frequent and roughly correlate with the degree of hypernatremia. Even when asymptomatic, untreated hypernatremia is a wasting condition associated with a high mortality (over 40 per cent) (Snyder *et al.* 1987; Palevsky *et al.* 1996). Associated laboratory abnormalities may include metabolic acidosis, azotemia, hypophosphatemia, hyperglycemia, and rhabdomyolysis (Singhal *et al.* 1990).

Among elderly patients hypernatremic at the time of hospital admission, the majority were women (Long *et al.* 1991). Although nearly half of the hypernatremic patients had a febrile illness, other associated conditions were frequent. The most important of these included the postoperative state, physical handicap, diuretic administration, excessive intravenous solute administration (NaCl, $NaHCO_3$, and nutritional supplements), and diabetes mellitus (Snyder *et al.* 1987; Palevsky *et al.* 1996). Depression of sensorium is frequent and highly correlated with the degree of hypernatremia, and is an independent predictor of mortality (Snyder *et al.* 1987; Palevsky *et al.* 1996).

Morbidity and mortality of hypernatremia

Hypernatremia is associated with substantial long-term morbidity and mortality. Snyder *et al.* (1987) reported a mortality of 42 per cent in elderly patients with hypernatremia, a finding recently confirmed with mortalities of 41 to 54 per cent reported in subsequent studies (Long *et al.* 1991; Palevsky *et al.* 1996). Neurological morbidity, as assessed by changes in level of care, was present in 38 per cent of surviving patients (Snyder *et al.* 1987).

Treatment of hyperosmolar states

The therapeutic goal in hypernatremia is reduction of plasma osmolality towards normal by the administration of an excess of free water. Removal of solute is necessary only after inadvertant administration of NaCl; it is not usually a part of therapy and is accomplished by dialysis or diuretic administration. When free-water administration is planned, the major therapeutic questions are the type of fluid to be given and its rate of administration. Most agree that patients with hypernatremic dehydration should be treated with fluid which

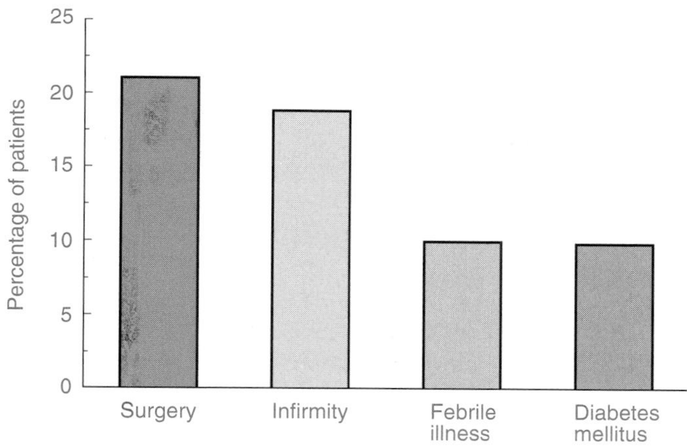

Fig. 1 The major causes of hypernatremia in a geriatric hospital population. About two-thirds of all cases are due to surgery, infirmity, febrile illness, or diabetes mellitus. (Data from Snyder *et al.* (1987).)

provides free water in excess of electrolytes. Fluid therapy is usually calculated to be administered over about 48 h (Snyder *et al.* 1987). Despite such recommendations, few human or experimental animal data are available for the ideal rate of fluid administration (Ayus *et al.* 1996). The pathophysiology of cerebral edema complicating rapid correction of experimental hypernatremia in rabbits has recently been described (Ayus *et al.* 1996), highlighting the dangers of overly rapid correction (less than 24 h) of hypernatremia. In adults with hypernatremia, 280 mmol/l glucose in water (5 per cent dextrose in water) has commonly been administered, but recent information suggests that therapy of hypernatremia with glucose-containing solutions (280 mmol/l glucose in water) may lead to cerebral intracellular lactic acidosis with increased mortality. Based on current knowledge, recommendations for treatment of chronic hypernatremia in adults are as follows.

1. If there is evidence of circulatory collapse, the patient should receive initial resuscitation with colloid which is sufficiently rapid to correct shock and stabilize the circulation.
2. Fluid deficit should be estimated on the basis of serum sodium and estimated total body water. The deficit should be given over a 48-h period, aiming for a decrement in serum osmolality of approximately 1 to 2 mOsm/kg/h. Maintenance fluids, which include replacement of urine volume with hypotonic fluid, are given in addition to the deficit.
3. Hypotonic fluid should be administered. The usual replacement fluid will be 77 mmol/l NaCl, and solutions containing glucose should be avoided if possible.
4. Plasma electrolytes should be monitored at frequent intervals, usually every 2 to 4 h. Many adult patients with hypernatremia have serious underlying systemic illness, and many such patients appear to die from their underlying illness rather than from hypernatremia. Close attention should be given to the treatment of associated medical conditions.

References

Ayus, J.C., Armstrong, D.L., and Arieff, A.I. (1996). Effects of hypernatraemia in the central nervous system and its therapy in rats and rabbits. *Journal of Physiology*, **492**(1), 243–55.

Long, C.A., Marin, P., Bayer, A.J., Shetty, H.G.M., and Pathy, M.S.J. (1991). Hypernatraemia in an adult in-patient population. *Postgraduate Medical Journal*, **67**, 643–5.

Palevsky, P.M., Bhagrath, R., and Greenberg, A. (1996). Hypernatremia in hospitalized patients. *Annals of Internal Medicine*, **124**, 197–203.

Singhal, P.C., Abramovici, M., and Venkatesan, J. (1990). Rhabdomyolysis in the hyperosmolar state. *American Journal of Medicine*, **88**, 9–12.

Snyder, N.A., Feigal, D.W., and Arieff, A.I. (1987). Hypernatremia in elderly patients. A heterogeneous, morbid, and iatrogenic entity. *Annals of Internal Medicine*, **107**, 309–19.

8.2.2 Hyponatremia

A. I. Arieff

Key messages

- Hyponatremia can result in permanent brain damage, including cerebral infarction, blindness, a persistent vegetative state, or death.
- Brain damage from hyponatremia can occur in anyone, but is most frequent in menstruant women and prepubertal children.
- The most frequent symptoms of hyponatremia are headache, nausea, and emesis.
- The indication for active therapy of hyponatremia is central nervous system symptomatology.
- Brain damage associated with hyponatremia is usually due to hypoxia. Although improper therapy can lead to brain damge, this occurs in less than 4 per cent of cases.

Introduction

The kidney is important in the pathogenesis of hyponatremia, but the target organ for sequelae which produce morbidity and mortality is the brain. Brain edema associated with hyponatremia can result in several devastating clinical entities, such as pulmonary edema, central diabetes insipidus, cerebral infarction, cortical blindness, and a persistent vegetative state. Seizures, coma, and death in patients with hyponatremia have been described for over 60 years, as has recovery following active treatment with hypertonic NaCl. Although it was often perceived that hyponatremic deaths were due to associated underlying medical conditions, it has become clear over the last decade that hyponatremic encephalopathy can progress to death or permanent brain damage in otherwise healthy individuals (Fraser and Arieff 1990; Ayus *et al.* 1992).

Clinical settings associated with brain damage from hyponatremia

Hyponatremia occurs with a similar frequency among men and women, but brain damage occurs predominantly in young (menstruant) females and prepubertal individuals (Fraser and Arieff 1990; Ayus *et al.* 1992). There are at least two distinct causes of hyponatremic brain damage: hyponatremic encephalopathy, and improper therapy of symptomatic hyponatremia. The vast majority of brain damage from hyponatremia occurs in three clinical settings: the postoperative state; psychogenic polydipsia; patients taking diuretics.

Postoperative hyponatremia

The incidence of postoperative hyponatremia in the United States is about 250 000 cases per year, with an overall morbidity of approximately 5 per cent (Ayus and Arieff 1996). In the majority of cases, patients tolerated the surgery without complications and the symptoms of encephalopathy developed between 1 and 2 days postoperatively. Initial symptoms usually include progressive headache, nausea, emesis, and weakness, which are often mistakenly believed to be routine postoperative symptoms. If the symptoms represent hyponatremic encephalopathy and are not treated, the patient may progress to seizures, obtundation, respiratory arrest, coma, and brain damage (Fraser and Arieff 1990; Ayus and Arieff 1995). Thus symptomatic hyponatremia in postoperative patients is particularly dangerous, whatever the level of serum sodium, and should be treated promptly. Premenopausal women with hyponatremic encephalopathy are at particular risk of developing brain damage. Men and postmenopausal women are far less likely to develop encephalopathy from hyponatremia (Ayus *et al.* 1992). Respiratory arrest and seizures occur at a significantly higher ($p < 0.001$) plasma sodium in women (95 per cent confidence limits, 95–130 mmol/l) than in men (95 per cent confidence limits, 92–115 mmol/l), and at a significantly higher ($p < 0.01$) plasma sodium in menstruant women than in postmenopausal women. Among patients with postoperative hyponatremia, recent studies suggest that the frequency of permanent brain damage in those with encephalopathy is about 20 per cent (Ayus and Arieff 1996).

Psychogenic polydipsia

Another common clinical setting for symptomatic hyponatremia is psychogenic polydipsia. If renal diluting capability is intact, a normal individual can theoretically excrete up to 20 l/day, so that development of hyponatremia without elevated plasma ADH levels requires an intake of over 20 l/day in a 60-kg adult. In practice, most patients with psychogenic polydipsia who develop hyponatremia have ingested substantially less water; rather, they have an abnormal urinary diluting capacity with elevated plasma ADH levels, and many are schizophrenic. Beer potomania is a variation of psychogenic polydipsia, where the hyponatremia is associated with massive ingestion of beer, poor nutrition, impaired urinary dilution, and probable low plasma ADH levels.

Pharmacological agents

A number of pharmacological agents may interfere with the ability of the kidney to excrete free water. These include large numbers of sedatives, hypnotics, oral hypoglycemic agents, tranquilizers, narcotics, antidepressants, and diuretics. In most such instances, there is excessive net retention of ingested or infused free water. The diuretics most commonly asociated with hyponatremia are thiazides and loop diuretics. In patients with thiazide-associated hyponatremia, there is often an idiosyncratic reaction to the drug in addition to polydipsia and a substantial loss of sodium and potassium in the urine.

Clinical manifestations of hyponatremia

The signs and symptoms of hyponatremia are directly related to the development of cerebral edema, increased intracranial pressure, and cerebral hypoxia. The early symptoms of hyponatremia (from any cause) may include apathy, weakness, muscular cramps, nausea, vomiting, and headache. More advanced clinical manifestations include impaired response to stimuli, myoclonus, urinary incontinence, and hallucinations. More severe clinical manifestations of hyponatremia may include decorticate posturing, seizures, and respiratory arrest.

Hypoxia and hyponatremic encephalopathy

Hypoxemia is a major factor contributing to brain damage in patients with hyponatremia (Ayus and Arieff 1995; Ayus and Arieff 1996). From a clinical standpoint, a hypoxic episode (such as respiratory arrest) frequently complicates hyponatremic encephalopathy and is the critical event which determines the likelihood of survival without permanent brain damage. If the patient has intact cognitive responses to verbal stimuli and does not have respiratory insufficiency, complete recovery is possible following appropriate active therapy (Ayus *et al.* 1992). Hypoxia impairs several pathways for homeostatic brain ion transport, decreasing the effectiveness of the brain's compensatory adaptation to hyponatremia. Respiratory arrest often occurs abruptly in patients with hyponatremic encephalopathy, and such patients infrequently survive without permanent brain damage (Ayus and Arieff 1995). The most frequent prelude to respiratory arrest is delayed initiation of therapy in a symptomatic patient. Since the possibility of hypoxia complicating symptomatic hyponatremia far exceeds that of brain injury due to inappropriate therapy (Fig. 1), there is essentially no modern medical rationale for failure to treat patients with symptomatic hyponatremia actively.

Management of hyponatremia

Asymptomatic hyponatremia

Active therapy with hypertonic NaCl is not indicated in patients with asymptomatic hyponatremia. If the patient is volume depleted, isotonic (154 mmol/l) NaCl is usually the fluid of choice, and if there is a hormone deficiency (cortisol, aldosterone, thyroid hormone) appropriate replacement is indicated. If the patient has received a drug which may interfere with renal handling of sodium or water, it should be discontinued whenever possible. Water restriction is of theoretical benefit in selected disorders, but practical considerations limit its usefulness. With successful fluid restriction (less than 800 ml/day), correction of plasma sodium by more than 1.5 mmol/l daily cannot be expected. Thus water restriction is only appropriate in a patient whose hyponatremia is asymptomatic.

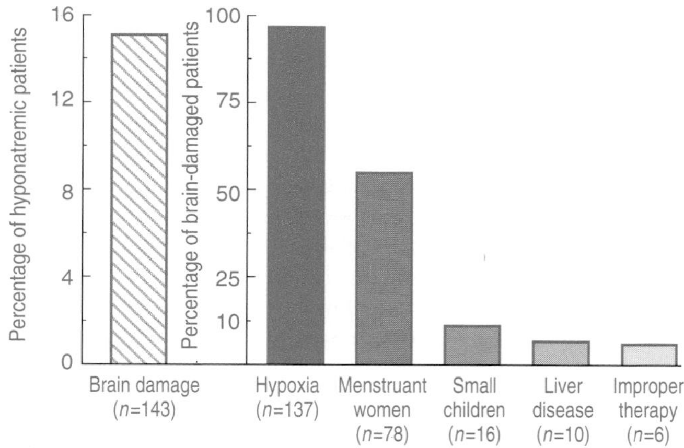

Fig. 1 The major risk factors associated with permanent brain damage among 958 hospital in-patients with hyponatremia (serum sodium below 128 mmol/l). Most patients (96 per cent) suffered a hypoxic episode because of failure to initiate active therapy in a timely manner. In only 4 per cent of patients suffering permanent brain damage could improper therapy for hyponatremia be implicated in the outcome. The incidence of hyponatremic encephalopathy in 11 published series from our laboratory comprising the 958 hospital in-patients with hyponatremia was 23 per cent (220/958). The overall morbidity among patients with hyponatremic encephalopathy was 15 per cent.

Several medical regimens are available for patients with stable asymptomatic hyponatremia. Demeclocycline in doses above 600 mg/day can produce a state of nephrogenic diabetes insipidus and has been used to treat patients with the syndrome of inappropriate ADH secretion. Other drug regimens of potential benefit in the treatment of chronic hyponatremia (urea, ADH inhibitors) are still experimental.

Symptomatic hyponatremia

When a patient with hyponatremic encephalopathy experiences respiratory arrest, therapy is unlikely to result in an outcome without brain damage (Ayus *et al.* 1992). Among the 1 per cent of postoperative patients who become hyponatremic, more than 15 per cent develop encephalopathy (Ayus and Arieff 1996). Every postoperative patient is at risk of developing hyponatremia, and prophylactic measures are indicated. The most important measure includes the avoidance of intravenous hypotonic fluid to postoperative patients unless the patient is hypernatremic. The use of hypotonic fluid in postoperative patients has no place in modern medical practice as it can potentially lead to permanent brain damage or death; isotonic (154 mmol/l) NaCl is virtually always preferable.

Symptomatic hyponatremia is a medical emergency, with a morbidity in excess of 15 per cent (Ayus and Arieff 1996). In patients with hyponatremic encephalopathy, the morbidity and mortality following correction by water restriction alone are unacceptable. Patients with hyponatremic encephalopathy should be constantly monitored, preferably in an intensive therapy unit. The first step in management of such patients is to ascertain that there is a secure airway, with assisted ventilation if required. Therapy should be initiated with intravenous hypertonic sodium chloride (514 mmol/l), using an infusion pump, with administration of a loop diuretic such as furosemide (frusemide). The infusion is designed to increase plasma sodium at a rate of about 1 mmol/l/h. If the patient has active seizures, the rate of hypertonic fluid administration should be adjusted such that the rise in plasma sodium is about 8 to 10 mmol/l/h over the initial 4 h, or until cessation of seizure activity. Therapy with hypertonic NaCl should be discontinued when either the patient becomes asymptomatic or the plasma sodium reaches a value in the range of 124 to 132 mmol/l. The plasma sodium should not be increased by more than 20 mmol/l during the initial 48 h of therapy and should not be acutely elevated to hyper- or normonatremic levels.

These guidelines require modification if patients are symptomatic at higher levels of plasma sodium (124–131 mmol/l) (Fraser and Arieff 1990; Ayus *et al.* 1992). During the period that active correction of symptomatic hyponatremia is being carried out, plasma electrolytes should be monitored every 2 h until the patient has become neurologically stable. This regimen may require modification in patients with renal or cardiac disease. An estimate of the patient's total body water should be made at the start of therapy of symptomatic hyponatremia. Correction of the plasma sodium should initially be planned using intravenous 514-mmol/l NaCl, often combined with a loop diuretic (furosemide). The technique is as follows: infuse 514-mmol/l NaCl at a rate calculated to increase plasma sodium at a rate of 1 mmol/l/h for 24 h. For a 54-kg woman (assuming 26 liters of total body water) who is symptomatic and whose plasma sodium is 108 mmol/l, the aim is to raise the plasma sodium to about 128 mmol/l in 24 to 48 h. This is accomplished by infusing (using an infusion pump) 514-mmol/l NaCl at a rate calculated to increase plasma sodium at 1 mmol/l/h, i.e. (26 liters $\times$ 20 mmol/l) = 520 mmol NaCl in 48 h. Using 514-mmol/l NaCl, this will be

$$\frac{(520\ \text{mmol})/(514\ \text{mmol/l})}{48\ \text{h}} = 21\ \text{ml/h}.$$

Individuals most susceptible to hyponatremic brain damage

Although previous medical opinion suggested that major factors which might lead to permanent brain damage were related to the magnitude and duration of the hyponatremia, recent studies have demonstrated that neither were important factors associated with the development of brain damage (Ayus *et al.* 1992). Rather, the age, gender, and reproductive (hormonal) status of the patient, as well as the presence of encephalopathy, were the most important predictive factors. The most susceptible groups were menstruant women and prepubescent children (Fraser and Arieff 1990; Ayus *et al.* 1992). Suggestions that patients with 'chronic' hyponatremia are more likely than those with 'acute' hyponatremia to develop brain damage as a complication of therapy with hypertonic NaCl are not supported by clinical evidence. Improper therapy can lead to permanent brain damage in hyponatremic patients, but the vast majority who develop complications have had essentially no therapy for their hyponatremia (Fraser and Arieff 1990). Improper therapy may be a factor in at most about 4 per cent of brain damage associated with hyponatremia (Ayus and Arieff 1996) (Fig. 1). Timely treatment of symptomatic hyponatremia with respiratory support and carefully administered hypertonic NaCl is usually associated with recovery (Ayus *et al.* 1992).

Possible complications of therapy

In the past, there has been some controversy about the rate at which hyponatremia should be corrected. Some authors have suggested that development of the rare neurological syndrome central pontine myelinolysis might be related to overly 'rapid' correction of hyponatremia. However, a number of studies have shown that cerebral demyelinating lesions develop only when patients with hyponatremia are inadvertently made normonatremic or hypernatremic during treatment, have an absolute increase in plasma sodium exceeding 25 mmol/l in the first 24 to 48 h of therapy, suffer a hypoxic event, or have hepatic insufficiency. Modern diagnostic criteria show that virtually all patients with central pontine myelinolysis have had severe medical conditions, including alcoholism, advanced liver disease, burns, sepsis, or various malignancies (Tien *et al.* 1992). Cerebral demyelinating lesions may result from improper therapy of hyponatremia in laboratory animals, but a human analog is uncommon (Tien *et al.* 1992). Extrapontine cerebral demyelinating lesions in hyponatremic patients are sometimes mistakenly diagnosed as central pontine myelinolysis, even though the pons is rarely involved, thus negating the diagnosis. Such lesions are sometimes mistakenly described as central pontine myelinolysis, but are characteristic of hypoxia.

References

Ayus, J.C. and Arieff, A.I. (1995). Pulmonary complications of hyponatremic encephalopathy. Noncardiogenic pulmonary edema and hypercapnic respiratory failure. *Chest*, **107**, 517–21.

Ayus, J.C. and Arieff, A.I. (1996). Brain damage and postoperative hyponatremia: the role of gender. *Neurology*, **46**, 323–8.

Ayus, J.C., Wheeler, J.M., and Arieff, A.I. (1992). Postoperative hyponatremic encephalopathy in menstruant women. *Annals of Internal Medicine*, **117**, 891–7.

Fraser, C.L. and Arieff, A.I. (1990). Fatal central diabetes mellitus and insipidus resulting from untreated hyponatremia: a new syndrome. *Annals of Internal Medicine*, **112**, 113–19.

Tien, R., Arieff, A.I., Kucharczyk, W., Wasik, A., and Kucharczyk, J. (1992). Hyponatremic encephalopathy: is central pontine myelinolysis a component? *American Journal of Medicine*, **92**, 513–22.

8.2.3 Hyperkalemia

J. Salmon

Key messages

- Because the kidney is responsible for 90 per cent of potassium excretion, most true hyperkalemia results from renal insufficiency
- Glucose and insulin reduce plasma potassium; insulin promotes potassium entry into cells by mechanisms separate from glucose entry, and glucose is required to prevent hypoglycemia
- Sodium bicarbonate promotes entry of potassium into cells by reducing hydrogen–potassium exchange
- Promotion of a diuresis with loop diuretics or thiazides will increase potassium excretion.
- Calcium salts should be reserved for patients with severe hyperkalemia or life-threatening dysrhythmias.

Physiology of potassium balance

Plasma potassium levels are determined by the balance between absorption and excretion and by internal shifts between the extracellular fluid and cells. Normal daily intake is 50 to 150 mmol, of which 5 to 15 mmol appears in the stools and sweat. Potassium excesses are usually excreted rapidly; however, if this is impossible, plasma potassium rises sharply, with each increase of 1 mmol/l above 4 mmol/l corresponding to an increase in total body potassium of 50 to 100 mmol. Plasma potassium reflects total body potassium only in stable states at normal pH. In acutely ill patients, particularly those with acid–base disturbances, plasma and total body potassium may vary in opposite directions, with plasma potassium rising in response to acidosis. These changes are more marked with metabolic than with respiratory disturbances.

There is no control of gastrointestinal absorption of potassium; 85 to 90 per cent is absorbed and 10 to 15 per cent is excreted in the feces. Short-term control of potassium balance is exerted by changes in renal excretion and by shifts of potassium into and out of cells. Beta-blockers and α-adrenergic agonists impair cellular potassium uptake. Long-term control is effected mainly by the rate of renal excretion, controlled by intrinsic renal mechanisms and aldosterone. The range of renal excretion is normally 3 to 300 mmol/day. In subjects with a chronically high potassium intake, renal adaptation allows the excretory capacity to reach 500 mmol/day. Renal responses to excessive intake are relatively rapid; approximately half an acute load appears in the urine within 12 h.

Hyperkalemia is defined as a plasma potassium level greater than 5.0 mmol/l. This may be associated with low, normal, or high total body potassium. True hyperkalemia should be distinguished from pseudohyperkalemia, which may follow *in vitro* hemolysis and lysis of white blood cells (usually only at levels above $200 \times 10^9/1$) or platelets (usually only when the platelet count is above $750 \times 10^9/1$).

A more detailed discussion of hyperkalemia is given by Tannen (1992).

Adaptations to a raised potassium intake

A chronically high intake of potassium produces an increase in the distal tubular excretory capacity for potassium; this capacity may increase by a factor of 10 to 20. The main mechanisms are an increase in the surface area and the Na^+,K^+-ATPase concentration of the basolateral membranes in the cortical collecting ducts. Similar increases in Na^+,K^+-ATPase are seen in the colon. Raised potassium concentrations stimulate aldosterone release; this not only stimulates Na^+,K^+-ATPase but also increases potassium conductance in the luminal membrane of the principal cells in the collecting ducts.

Other mechanisms that may be relevant include the following: a reduction in potassium absorption in the thick ascending limb contributes to kaliuresis, and a high medullary potassium concentration inhibits sodium reabsorption in the thick ascending limb, thereby increasing sodium delivery to the distal tubule where it is exchanged for potassium. In the distal tubule, potassium moves into the urine down electrical and chemical gradients determined by the luminal concentration of sodium and its accompanying anion. In the cortical collecting ducts, active transport of potassium into the urine takes place. This is further stimulated by ADH.

Causes of hyperkalemia

Because the kidney is responsible for 90 per cent of potassium excretion, most true hyperkalemia results from renal insufficiency.

Hypertrophy of renal tubules usually maintains the ability to balance potassium until the glomerular filtration rate falls below 10 ml/min, provided that dietary intake is normal. An increased potassium intake or tissue release may lead to hyperkalemia at much higher glomerular filtration rates, as may the effects of drugs that modify renal handling of potassium. Hyperosmolar infusions may also produce hyperkalemia (Conte *et al.* 1990) (Table 1).

Pathophysiology of hyperkalemia

The effects of hyperkalemia are seen primarily at the membranes of excitable tissues. Signs and symptoms may be absent; treatment depends upon suspicion and regular monitoring of potassium levels. This is particularly important in conditions where potassium may be released in large quantities in association with developing renal dysfunction, for example rhabdomyolysis, tumor lysis syndrome, reperfusion of ischemic tissue, severe burns, succinylcholine (suxamethonium), etc. Rapid metabolic changes in such circumstances may produce lethal hyperkalemia in a short period of time.

Treatment of hyperkalemia

Irrespective of the cause, the primary goals are reduction of plasma potassium towards the safe range and myocardial protection if required. Secondary goals are the correction of coexisting metabolic disturbances and treatment of the underlying cause. It is usual to start trying to correct plasma potassium at levels above 5.5 mmol/l. Treatment should include stopping drugs that promote potassium retention and reviewing potassium intake in feeds and infusions. At levels above 6.5 mmol/l, aggressive measures should be adopted to reduce potassium whilst calcium salts should be given if there are cardiac arrhythmias or broadening of the QRS complex. Temporary short-term control is achieved most effectively by promoting potassium shifts into cells (Allon 1993); these methods all require 15 to 30 min to take effect (Lens 1989). Longer-term control involves promoting increased potassium excretion.

Glucose–insulin infusions

Insulin promotes potassium entry into cells by mechanisms separate from glucose entry; glucose is also required to prevent hypoglycemia. There is no general agreement on the dose to be used. Commonly used prescriptions include 15 units of soluble insulin in 50 ml of 50 per cent glucose given centrally over 20 to 30 min, and 15 to 20 units of insulin in 500 ml of 10 or 20 per cent glucose over 30 to 60 min. All methods will reduce plasma potassium by approximately 1 mmol/l over a period of 30 to 60 min and will usually maintain potassium at the lower level for 3 to 6 h, after which the treatment may be repeated. This may be sufficient to maintain plasma potassium levels at acceptable levels, but is usually inadequate in hypercatabolic patients and those with established renal failure; in such cases it will buy time for definitive treatment.

Table 1 Causes of hyperkalemia

Pseudohyperkalemia
Faulty venesection technique
In vitro hemolysis
Extreme leukocytosis
Thrombocytosis
Transcellular shift
Acidosis
Diabetic ketoacidosis
Hyperosmolality
Beta-blockade
Drugs
Succinylcholine
Digoxin
Rhabdomyolysis
Tumor lysis syndrome
Hemolysis
Malignant hyperthermia
Exercise
Hyperkalemic periodic paralysis
Renal
Severe renal failure
Excess potassium load with renal dysfunction
Addison's disease
Aldosterone deficiency
Hyporeninemic hypoaldosteronism
Drugs
Non-steroidal anti-inflammatory drugs
Angiotensin-converting enzyme inhibitors
Cyclosporine
K^+-sparing diuretics
Heparin
Enzymatic defects
Tubular disorders

Correction of circulatory disturbances

Where possible, hypovolemia should be corrected before bicarbonate is given. Metabolic acidoses secondary to hypovolemia will usually correct with appropriate volume replacement.

Sodium bicarbonate

Sodium bicarbonate promotes entry of potassium into cells by reducing hydrogen–potassium exchange. The usual dose is 44 to 132 mmol given rapidly. It is valuable in extreme cases of hyperkalemia, but should be used cautiously in hypovolemic acidotic patients where volume replacement is a more appropriate treatment for acidosis. It should not be given with calcium salts.

Beta-2 agonists

Nebulized β_2 agonists may be valuable where intravenous access is a problem and have been shown to be effective in children, but their place in the treatment of adults is not yet established (Lens *et al.* 1989; McClure *et al.* 1993). Anxiety has been expressed about the use of drugs with positive chronotropic and arrhythmogenic effects in severely ill patients, particularly those with coexistent myocardial disease. The usual adult doses are 0.5 mg intravenously over 10 to 15 min or 10 to 20 mg by nebulizer over 10 min.

Diuretics

Promotion of a diuresis with loop diuretics or thiazides will increase potassium excretion. This is aided by increasing sodium delivery to the distal tubule. In patients with prerenal failure, volume loading and diuretics will promote potassium excretion. In cell lysis syndromes, an osmotic diuretic is more appropriate.

Hemofiltration and/or hemodialysis

This is usually unnecessary in patients with intact renal function. Patients with predictably transient hyperkalemia can usually be maintained at acceptable potassium levels by using a combination of glucose–insulin, sodium bicarbonate, and forced diuresis, possibly supplemented with β_2 agonists. In patients where these methods fail and in those with established acute or acute-on-chronic renal failure, hyperkalemia usually reflects a generalized metabolic disturbance and is an indication to start renal replacement therapy. *In extremis*, peritoneal dialysis has been used successfully in the management of hyperkalemic cardiac arrest pending institution of more conventional treatments.

Calcium

Calcium salts should be reserved for patients with severe hyperkalemia or life-threatening dysrhythmias. The usual dose is 10 ml of 10 per cent calcium gluconate given rapidly, repeated where necessary up to 30 ml.

Cation-exchange resins

These can be administered orally or by retention enema, with the latter being more rapidly effective. Although these resins have a place in the management of isolated renal failure, gut function in critically ill patients is not dependable and dialysis or hemofiltration is usually a better choice.

Special cases

Rhabdomyolysis and cell lysis syndromes

In rhabdomyolysis, as in tumor lysis syndrome and massive hemolysis, plasma potassium levels may rise very rapidly. Provided that rapid action is taken, it is usually possible to prevent lethal hyperkalemia. The key is to promote and maintain a large diuresis with alkalinization to promote solubility of, for example, myoglobin or urate. Urine outputs of up to 12 l/day have been advocated but 4 to 6 l/day is usually adequate. Following resolution of the acute phase of these syndromes, total body potassium will be low and will require replenishment.

Diabetic ketoacidosis

In diabetic ketoacidosis, hyperkalemia may develop as a result of metabolic acidosis, insulin deficiency, and reduced renal excretion. Hyperkalemia usually accompanies a large whole-body potassium deficit. It is unusual for extreme methods of potassium control to be needed, and potassium levels will usually return to normal with replacement of the volume and insulin deficits. Very occasionally it is necessary to give calcium chloride or gluconate for myocardial protection, and a period of hemofiltration or dialysis may be required in patients with coexistent oliguric renal failure. In most patients it is necessary to start potassium replacement as plasma potassium levels fall into the upper part of the normal range; thereafter plasma potassium levels should be checked frequently (every 1–2 h).

References

Allon, M. (1993). Treatment and prevention of hyperkalaemia in end-stage renal disease. *Kidney International*, **43**, 1197–1209.

Conte, G., *et al.* (1990). Acute increase in plasma osmolality as a cause of hyperkalaemia in patients with renal failure. *Kidney International*, **38**, 301–7.

Lens, X.M., Montoliu, J., Cases, A., Campistol, J.M., and Revert, L. (1989). Treatment of hyperkalaemia in renal failure: salbutamol v. insulin. *Nephrology Dialysis Transplantation*, **4**, 228–32.

McClure, R.J., Prasad, V.K., and Brocklebank, J.T. (1993). Treatment of hyperkalaemia using intravenous and nebulised salbutamol. *Archives of Disease in Childhood*, **70**, 126–8.

Tannen, R.L. (1992). Hypo-hyperkalaemia. In *Oxford textbook of clinical nephrology* (ed. S. Cameron, A.M. Davison, J.-P. Grunfeld, D. Kerr, and E. Ritz), Vol. 2, Chapter 7.2, pp. 895–917. Oxford University Press.

8.2.4 Hypokalemia

J. Salmon

Key messages

- Plasma potassium may bear no relation to total body potassium; hypokalemia may be present in patients with low, normal, or high total body potassium.
- Diuretics increase potassium excretion by increasing delivery of sodium to the distal tubule.
- Severe vomiting and the accompanying hypochloremic alkalosis leads to a shift of potassium into cells and also increased potassium loss from the kidney.
- Acidosis favors potassium shifts from the cell to the extracellular fluid in exchange for hydrogen ions.
- ECG changes of hypokalemia consist of ST depression, flattening of the T wave, and prominent U waves. Subsequently there may be widening of the QRS complex and atrioventricular block.

Physiology of potassium balance

In patients with chronic alterations in potassium balance, each fall in plasma potassium of 1 mmol/l in the range 2 to 4 mmol/l reflects a total body deficit of 200 to 300 mmol. In acutely ill patients, particularly those with acid–base disturbances, plasma and total body potassium may vary in opposite directions, with plasma potassium falling in alkalosis. These changes are more marked with metabolic than with respiratory disturbances.

Short-term control of potassium balance is exerted by changes in renal excretion and by shifts of potassium into and out of cells (Dawson 1987). Aldosterone, insulin, and epinephrine (adrenaline) (via β_2 receptors) promote potassium uptake by cells. Long-term control is effected mainly by the rate of renal excretion, controlled by intrinsic renal mechanisms and aldosterone. Renal response to dietary potassium depletion is slow; excretion falls gradually over 7 to 14 days so that total body deficits of 200 mmol or more may develop.

The major physiological effects of changes in potassium levels are seen in excitable tissues. As extracellular potassium falls, so membranes become hyperpolarized and more excitable.

A more detailed discussion of hypokalemia is given by Tannen (1992).

Causes of hypokalemia

Hypokalemia and potassium depletion are common in critically ill patients (Table 1). Serum potassium may bear no relation to total body potassium; hypokalemia may be present in patients with low, normal, or high total body potassium.

Table 1 Causes of hypokalemia

Potassium depletion
Non-renal (urine K^+ <20 mmol/day)
Inadequate intake
Excessive sweating
Gastrointestinal losses
Renal losses (urine K^+ >20 mmol/day)
Renal tubular acidosis
Diabetic ketoacidosis
Chloride depletion
Mineralo- and glucocorticoid excess
Magnesium depletion
Antibiotic therapy
Leukemia
Bartter's syndrome
Osmotic diuresis
Transcellular shift
Alkalosis
Insulin excess
Beta-2 adrenergic stimulation
Poisoning
Theophyllines
Barium
Toluene
Hypokalemic periodic paralysis

Renal handling of potassium

Renal excretion is independent of glomerular filtration rate and the amount filtered. Active secretion of potassium by Na^+,K^+-ATPase is stimulated by aldosterone, corticosteroids, and potassium. Potassium excretion is increased by high tubular flow and by increased delivery of sodium to the distal tubule.

Diuretics

Diuretics (except for triamterene, amiloride, and spironolactone) increase potassium excretion by increasing delivery of sodium to the distal tubule. Carbonic anhydrase inhibitors increase potassium loss by increasing the tubular concentration of bicarbonate which acts as a non-reabsorbable anion.

Mineralocorticoids and hyperaldosteronism

Secondary hyperaldosteronism due to Bartter's syndrome, renin-secreting tumors, and malignant hypertension may produce severe potassium depletion.

Bartter's syndrome may be mimicked by bulimia nervosa, laxative and diuretic abuse, and primary renal tubular disorders such as cystinosis. In Bartter's syndrome there is failure to retain chloride on a high-chloride diet, whereas in bulimia and following diuretic abuse chloride is retained avidly.

Vomiting

Because of the low concentration of potassium in upper gastrointestinal secretions, even severe vomiting is insufficient on its own to cause profound potassium depletion. The accompanying hypochloremic alkalosis leads to a shift of potassium into cells and also increased potassium loss from the kidney.

Diarrhea

Stool contains 50 to100 mmol/l of potassium. Diarrhea is usually associated with significant losses of bicarbonate and hence a metabolic acidosis. This may be further increased by the effects of hypovolemia; the consequent shift of potassium from cells may mask the true extent of potassium depletion. In certain conditions, such as villous adenomas of the rectum and non-insulin-secreting islet cell tumors, diarrhea may contain high chloride concentrations. In such cases potassium loss occurs without an accompanying acid–base disturbance; hence hypokalemia tends to be more profound for a given whole-body potassium deficit. Laxative abuse similarly produces a profound hypokalemia without alterations in acid–base status. Patients with ureterosigmoidostomies, where urine is allowed to stagnate in the colon, may develop a profound hypokalemic hypochloremic acidosis. Potassium losses are increased in conditions of rapid intestinal transit, small bowel fistulas, intestinal drains, malabsorption, and small bowel bypass.

Extracellular–intracellular shifts

Potassium movement into cells is enhanced by insulin, β_2 agonists, mineralocorticoids, and theophyllines. Acidosis favors potassium shifts from the cell to the extracellular fluid in exchange for hydrogen ions; alkalosis has the opposite effect. These changes are more pronounced with metabolic than with respiratory disturbances of acid–base balance. Vitamin B_{12} favors potassium uptake and may contribute to acute profound hypokalemia during the treatment of megaloblastic anemias. Acute shifts may be seen with hypokalemic periodic paralysis.

Effects of hypokalemia

The most important effects of hypokalemia clinically are on myocardial cells and to a lesser extent on smooth and striated muscle. As the major intracellular cation, potassium is important in the maintenance of cell volume.

Effects on cardiac muscle

ECG changes of hypokalemia consist of ST depression, flattening of the T wave, and prominent U waves. Subsequently there may be widening of the QRS complex and atrioventricular block. More significant are arrhythmias, ranging from single or multiple ectopics to atrial or ventricular tachycardias and fibrillation. Membrane hyperpolarization and prolongation of the refractory period lead to prolongation of the action potential and an increased tendency to re-entry tachycardias.

ECG changes are rare at serum potassium levels above 3 mmol/l, and arrhythmias are uncommon in healthy subjects at these levels. In critically ill patients, particularly those with myocardial disease, even small reductions in plasma potassium may be associated with dysrhythmias, and it is common intensive care practice to maintain plasma potassium levels at the upper end of the normal range.

Effects on other muscles

Profound potassium depletion may lead to impaired skeletal muscle function with weakness, hyporeflexia, and, in extreme cases, paralysis which may prevent weaning from mechanical ventilation. Muscle weakness may be exacerbated by accompanying hypophosphatemia as potassium depletion leads to renal tubular phosphate wasting. Gut motility may be reduced, leading to paralytic ileus. Vascular smooth muscle contractility may be reduced, leading to reduced pressor responses to catecholamines and angiotensin.

Renal effects

Hypokalemia may produce a state resembling diabetes insipidus as a result of reduced medullary solute concentrations. There is an association between potassium depletion and metabolic alkalosis, particularly in the hypovolemic patient. This may be partly due to shifts of hydrogen ions into cells but probably also reflects increased renal retention of sodium and bicarbonate.

Chronic potassium depletion produces morphological changes in the kidney with the development of cytoplasmic vacuoles in proximal and distal cortical tubular cells.

Metabolic effects

These include reduced protein synthesis, a negative nitrogen balance, glucose intolerance, a metabolic alkalosis, and reduced carbohydrate synthesis.

Treatment of hypokalemia

Underlying abnormalities and causes should be corrected where possible. Potassium should be replaced in conjunction with correction of other metabolic and electrolyte abnormalities. If there are symptoms or ECG signs, or [K^+] is below 2 mmol/l, this should be replaced rapidly (up to 40 mmol/h). At higher potassium levels and in asymptomatic patients, intravenous correction at rates of 10 to 20 mmol/h are usually adequate. Potassium replacement is usually performed using potassium chloride, although potassium phosphate can be used if there is concomitant phosphate deficiency. *In extremis* (lethal arrhythmias in the presence of severe hypokalemia), potassium can be given by rapid infusion of 50 to 100 ml of 40 mmol/l KCl over 2 to 5 min. In patients with coexistent metabolic acidosis and hypokalemia, potassium should be replaced before correction of the acidosis in order to prevent further severe falls in serum potassium.

After the extracellular deficit has been corrected, there will usually also be a whole-body intracellular deficit to correct. There is no single satisfactory way of judging the adequacy of potassium repletion, but looking for an increase in urinary potassium, a tendency for plasma potassium to stay elevated after potassium infusion, or the correction of a metabolic alkalosis may all give useful clues. Continuing replacement of magnesium should accompany potassium replacement. Extracellular fluid potassium may be restored in the magnesium-deficient patient, but magnesium is needed for potassium to enter cells and prevent continuing renal potassium wasting (Solomon 1987). Hypokalemia in hyperaldosteronism may be minimized by reducing sodium intake. Spironolactone may also be useful for reducing renal potassium wasting.

References

Dawson, D.C. (1987). Cellular mechanisms for K transport across epithelial cell layers. *Seminars in Nephrology*, 7(3), 185–92.

Solomon, R. (1987). The relationship between disorders of K^+ and Mg^{++} homeostasis. *Seminars in Nephrology*, 7(3), 253–62.

Tannen, R.L. (1992). Hypo-hyperkalaemia. In *Oxford textbook of clinical nephrology* (ed. S. Cameron, A.M. Davison, J.-P. Grunfeld, D. Kerr, and E. Ritz), Vol. 2, Chapter 7.2, pp. 895–917. Oxford University Press.

8.2.5 Hypomagnesemia

H. G. Wakeling and M. G. Mythen

Key messages

- Hypomagnesemia is very common in the critically ill and is associated with higher mortality and longer hospital stay.
- It has multiple causes including increased magnesium demand during illness, increased magnesium losses, and iatrogenic magnesium wasting and redistribution.
- Experimentally, magnesium-deficient animals have a higher mortality following endotoxin challenge which is reduced significantly by magnesium replacement.
- 1 mmol = 24 mg = 2 mEq of magnesium; 1 g $MgSO_4$ = 4 mmol ≈ 98 mg elemental magnesium.
- Daily requirements: 0.4 mEq/kg/day orally; 0.1 to 0.2 mEq/kg/day parenterally.

Introduction

Magnesium is the second most abundant intracellular cation. It is an essential cofactor for many cellular enzyme and membrane systems. The unbound intracellular ion plays a critical role in enzymatic processes concerned with the transfer, storage, and utilization of high-energy phosphates. Therefore it is essential for protein, fat, nucleic acid, and carbohydrate metabolism, and for energy-dependent cellular functions including membrane stability and ion permeability, action potential generation, neurochemical transmission, and cellular second-messenger systems. Potentially, magnesium deficiency can severely disrupt normal cellular function. More detailed discussions are given by Cronin (1990), Zagola and Roberts (1990), and Salem *et al.* (1991).

The total serum magnesium is 0.8 to 1.2 mmol/l, representing only 1 per cent of the total body content, with approximately 31 per cent being intracellular and 67 per cent in bone. The serum magnesium is 55 per cent ionized with 15 per cent chelated to other salts. Only 30 per cent is protein bound; therefore the fluctuation of plasma concentration with changes in plasma proteins is reduced compared with the more highly bound calcium ion. As with other predominantly intracellular ions, the plasma concentration does not neccessarily correlate with cellular stores or the active ionized fraction. However, numerous studies suggest that the serum level is clinically important. Normal adult requirements are 5 mg/kg/day or 0.4 mEq/kg/day orally, or 0.1 to 0.2 mEq/kg/day parenterally; during pregnancy and critical illness these requirements are increased to 10 to 15 mg/kg/day. Magnesium is absorbed in the jejunum and ileum. Usually, 30 to 40 per cent of the oral intake is absorbed, but this may increase to 70 per cent in magnesium deficiency.

Magnesium balance is regulated primarily by renal excretion. It is filtered by the glomeruli and is reabsorbed in the proximal convoluted tubule and the ascending limb of the loop of Henle. Reabsorption of magnesium is increased by parathyroid hormone, vitamin D, hypocalcemia, hypothyroidism, and volume depletion. However, metabolic acidosis, diuretics, alcohol, mineralocorticoids, volume expansion, saline, and catecholamines all reduce reabsorption. The renal threshold T_m for magnesium is very close to the normal plasma concentration. Supplementing for deficiency must continue for several days, as a proportion of each dose will be renally excreted if the plasma concentration exceeds T_m.

Magnesium metabolism is intimately involved with that of other ions. It is essential for normal potassium metabolism. Since magnesium is a cofactor for the Na^+,K^+-ATPase pump, deficiency causes intracellular potassium depletion and renal potassium wasting. Therefore hypokalemia is a common finding in hypomagnesemia. Refractory hypokalemia is often resolved by correcting the accompanying magnesium deficit. Calcium metabolism is also linked to magnesium in a complex way. Magnesium deficiency impairs parathyroid hormone secretion and also causes end-organ resistance to both parathyroid hormone and vitamin D, resulting in hypocalcemia. Beta-adrenoceptor stimulation drives magnesium intracellularly, reducing serum concentration; exercise may cause a temporary rise in concentration with hemoconcentration and cellular leakage.

Etiology and diagnosis

The causes of magnesium deficiency are presented in Table 1. Magnesium deficiency is often multifactorial, and many therapeutic measures taken during critical illness may exacerbate it further. For example, catecholamine infusions, diuretics, and aminoglycoside antibiotics all worsen hypomagnesemia.

As serum magnesium accounts for only 1 per cent of the total body magnesium, the serum concentration may be normal in patients with magnesium deficiency. However, most hypomagnesemic patients will have some degree of magnesium deficiency. The normal serum magnesium is 0.8 to 1.2 mmol/l depending on the laboratory reference range. This is affected by many factors including the release of magnesium from cells by acidosis and alterations in protein binding. If spectrophoretic methods are used, hemolysis and raised bilirubin will cause overestimation by absorbing light of similar wavelengths. If both the serum magnesium concentration and 24-h urinary magnesium are both normal, it is unlikely that significant deficiency exists. A magnesium loading test may be performed aimed at detecting increased magnesium retention in deficiency, but, as in the 24-h urinary measurement, it requires good renal function for reliable results. Therefore these tests are of very limited value in critical care.

Intracellular measurement of magnesium has been used experimentally for muscle, bone, and heart, but these tests are not used clinically. Some workers have tried using the magnesium content of erythrocytes and circulating lymphocytes as indices of tissue magnesium content, but these do not correlate well with the magnesium loading test.

Symptoms and signs

Symptoms of pure magnesium deficiency are non-specific and very variable. Many may be due to coexisting hypocalcemia. They are difficult to separate from those commonly present in critical illness. They include lethargy, apathy, generalized weakness, anorexia, nausea, and vomiting. Signs and metabolic features of magnesium deficiency are summarized in Table 2.

Table 1 Causes of magnesium deficiency

Decreased intake
Decreased oral intake
Prolonged magnesium-free parenteral fluids
Inadequate nutritional support during illness
Chronic ethanol abuse
Increased demand
Childhood pregnancy
Critical illness
Redistribution
Endocrine
Catecholamines
Insulin, glucose, and amino acid administration
Hyperthyroidism
Hungry bones postparathyroidectomy
Hemodilution
Chelation after transfusion with citrated blood
Cardiopulmonary bypass surgery
Recovery from hypothermia
Sepsis syndrome
Alkalosis
Renal losses
Renal parenchymal disease (acute tubular necrosis and interstitial nephritis)
Hypercalcemia
Hyperaldosteronism
Syndrome of inappropriate ADH secretion
Hyperthyroidism
Drugs
Diuretics (loop, thiazide, osmotic)
Antimicrobials (aminoglycosides, amphotericin B, pentamidine)
Digoxin
Cyclosporine
Chemotherapy (cisplatin, methotrexate)
Acidosis and diabetic ketoacidosis reduce tubular reabsorption
Hypophosphatemia and hypokalemia
Alcohol
Magnesium-deficient dialysate fluids in chronic renal failure
Gastrointestinal losses
Malabsorption syndromes: short loop, sprue, and celiac disease
Prolonged diarrhea, steatorrhea, and laxative abuse
Primary hypomagnesemia
Pancreatitis
Prolonged nasogastric drainage, chronic bowel and biliary fistulas
Skin losses
Burns

Clinical relevance of hypomagnesemia in critical care

Cardiovascular system

After acute myocardial infarction, serum potassium and magnesium fall. Hypomagnesemia may have an arrhythmogenic role as the incidence of serious arrhythmias is higher in these patients. A decrease in the available magnesium causes impairment of the magnesium-dependent Na^+,K^+-ATPase pump, resulting in an increase in intracellular sodium concentration and a decrease in intracellular potassium concentration. This in turn stimulates the exchange of sodium and calcium across the sarcolemma, raising cytosolic calcium concentrations which is a major factor in myocardial injury. Reduction in available magnesium also affects the cAMP second-messenger system, reducing adenyl cyclase activity. This alters the response of myocardial and vascular smooth muscle to catecholamines and other vasoactive substances. Incubation of blood vessels in magnesium-deficient buffer solution causes an increase in tension and enhanced contractile responses to catecholamines and angiotensin. Hypomagnesemia is associated with increased vasomotor tone, hypertension, and coronary artery spasm. It is implicated in sudden-death ischemic heart disease. This is more common in areas with magnesium-deficient water supplies, and myocardial magnesium content is lower in these patients. In addition, arrhythmias and coronary artery spasm may be caused by magnesium deficiency.

Magnesium replacement has been shown to prevent arrhythmias and improve survival after myocardial infarction. Even in normomagnesemia, magnesium has been useful in treating supraventricular and ventricular arrythmias, particularly torsade des points and arrhythmias associated with digitalis toxicity. Hypomagnesemia is common following cardiopulmonary bypass surgery, and magnesium replacement reduces the frequency of ventricular arrhythmias.

Chronic ethanol abuse

Alcoholic patients in the intensive care unit are particularly prone to magnesium deficiency and hypomagnesemia. The reasons include dietary insufficiency, diarrhea, steatorrhea, and increased renal losses induced by alcohol. Although the serum magnesium may be normal on admission, it falls markedly on alcohol withdrawal and will be missed unless looked for daily. This fall is multifactorial; causes include ATP resysnthesis as alcohol levels decline (which promotes cellular uptake of magnesium) and refeeding which allows phosphorylation. Catecholamine levels are high, which drives magnesium intracellularly and may further exacerbate hypomagnesemia by promoting lipolysis and allowing precipitation of magnesium with free fatty acids. The signs of hypomagnesemia listed in Table 2 are found in withdrawing alcoholic patients who may also exhibit delirium and psychosis. They are strikingly similar to those seen in delirium tremens. However, no study has been able to correlate hypomagnesemia and the onset of delirium tremens, and patients survive without magnesium replacement. Nevertheless, replacement therapy may prevent aggravation of withdrawal symptoms by hypomagnesemia and is justified for all alcoholic patients during withdrawal.

Renal failure

In chronic renal failure magnesium balance is often very positive owing to the inability of the kidneys to excrete the ion. Significant retention occurs when the glomerular filtration rate falls below 30 per

Table 2 Signs and metabolic features of hypomagnesemia

Neuromuscular
Trousseau's sign
Chvostek's sign
Muscle fasciculations
Tremor
Spasticity
Hyporeflexia
Confusion, disorientation, ataxia, nystagmus
Depression, irritability, psychosis
Parasthesias
Seizures
Cardiovascular
Arrhythmias
Hypocontractility
Increased systemic vascular resistance
Hypertension
Vasospasm
Increased sensitivity to digoxin
Metabolic
Hypokalemia
Hypocalcemia
Hypophosphatemia
Electrocardiography
Prolongation of the PR and QT intervals
ST depression and T-wave inversion
Electromyography
Myopathic-like configurations

cent of normal. However, some patients with predominantly tubulo-interstitial disease have magnesium wasting. Magnesium deficiency may be seen in patients on renal replacement therapy if the dialysates for hemodialysis and ambulatory peritoneal dialysis are magnesium free and where non-magnesium phosphate binders are used.

Diabetes mellitus

In diabetes mellitus patients are often found to be borderline magnesium deficient and have a low-normal plasma magnesium. This may be a risk factor for some of the long-term complications of diabetes. In diabetic ketoacidosis, the high blood glucose level acts as an osmotic diuretic which increases renal magnesium loss and is exacerbated further by a reduction in tubular reabsorption caused by the acidemia. Patients with ketoacidosis may have an initially high plasma magnesium as acidemia results in magnesium moving extracellularly in a similar way to potassium. However, once treatment has commenced with fluid and insulin, there may be a precipitous fall in plasma magnesium as it moves with potassium intracellularly. This may result in symptomatic hypomagnesemia.

Table 3 Treatment of hypomagnesemia

General measures
Stop magnesium-lowering medications
Treat underlying cause
Treat other electrolyte disturbances
Symptomatic deficiency
600 mg elemental magnesium intravenously over 3 h
Follow with 600–900 mg intravenously over 24 h
Life-threatening arrhythmias
1–2 g magnesium sulfate bolus intravenously over 3–5 min
Follow with maintenance infusion of 1–2 g/h
Prolonged magnesium deficiency
600–1200 mg/day of elemental magnesium orally for several days (the elemental content depends on the preparation)

Systemic inflammatory response syndrome

Magnesium may have an important role in the inflammatory response. Hypomagnesemia has been shown to be an adverse factor in the survival of critically ill patients. Patients admitted to a medical ward or medical intensive care unit who were hypomagnesemic had approximately twice the mortality rate of similar patients who had normal plasma magnesium levels (Rubeiz *et al.* 1993). The development of the systemic inflammatory response syndrome following an insult is complex. There has been much animal work to show that hypomagnesemia leads to a proinflammatory state with increased levels of inflammatory cytokines including interleukins 1 and 6, and tumor necrosis factor-α. Magnesium deficiency also has a membrane oxidant effect, with protein oxidation products being detectable in rat brain and kidney after 3 weeks on a magnesium-free diet. The response to an oxidant stress results in greater membrane lipid peroxidation and reduction in endothelial cell viability in the presence of magnesium deficiency. The effect of magnesium deficiency on the mortality following endotoxin challenge has been evaluated in rats (Salem *et al.* 1995). A significant increase in mortality was found in magnesium-deficient rats compared with controls. Magnesium replacement in rats immediately prior to endotoxin improved the survival over other deficient animals. Even when the magnesium was given after the endotoxin challenge, there was a significant improvement in survival. Plasma magnesium concentration rose during sepsis which correlated with cell damage. This rise may explain the lack of association between hypomagnesemia and the sepsis syndrome. However, there is no proof that magnesium replacement improves the survival of patients with the systemic inflammatory response syndrome.

Treatment of magnesium deficiency

Prevention of magnesium deficiency requires consideration of several factors: first the normal magnesium requirements, second the increased requirements and increased losses due to the disease state, and finally iatrogenic magnesium depletion and redistribution.

The treatment of magnesium deficiency is summarized in Table 3.

Care should be taken when using magnesium therapy, particularly with large intravenous doses or when there is coexisting renal failure. It is advisable to monitor vital signs, respiratory rate, and the electrocardiogram during acute magnesium replacement, and to measure plasma levels of magnesium, calcium, and potassium. Magnesium may cause some reduction in vasomotor tone and myocardial contraction, but overall there is generally little detrimental cardiovascular effect. Signs of hypermagnesemia should be sought and are usually apparent when the serum magnesium exceeds 3 mmol/l. They include loss of deep tendon reflexes, lethargy, muscle weakness leading eventually to muscle paralysis, and respiratory failure. Cardiovascular effects include hypotension from reduced vascular resistance and cardiac conduction changes including prolonged sinoatrial conduction time, atrioventricular refractory period, and PR and AV intervals, which explain the antiarrhythmic effects. If the above signs develop, administration of magnesium should cease and the serum level should be measured.

References

Cronin, R.E. (1990). Magnesium disorders. In *Fluids and electrolytes* (2nd edn) (ed. J.P. Kokko and R.L.Tannen), pp. 631– 49. W.B.Saunders, Philadelphia, PA.

Rubeiz, G.J., Thill-Baharozian, M., Hardie, D., and Carlson, R. (1993). Association of hypomagnesemia and mortality in acutely ill medical patients. *Critical Care Medicine*, **21**, 203–9.

Salem, M., Munoz, R., and Chernow, B. (1991). Hypomagnesemia in critical illness: a common and critically important problem. *Critical Care Clinics*, **7**, 225–52.

Salem, M., Kasinski, N., Munoz, R., and Chernow, B. (1995) . Progressive magnesium deficiency increases mortality from endotoxin challenge: protective effects of acute magnesium replacement therapy. *Critical Care Medicine*, **23**, 108–18.

Zagola, G.P. and Roberts, J.E. (1990). Magnesium disorders. *Problems in Critical Care*, **4**, 425–36.

8.2.6 Hypercalcemia

N. B. Eastwood and M. G. Mythen

Key messages

- Mild hypercalcemia is common.
- Various symptoms and signs, including nausea, vomiting, altered conscious level, and hypertension, may be seen in moderate or severe hypercalcemia.
- Rehydration is the first step in treatment, whatever the etiology of the hypercalcemia.

Table 1 Symptoms and signs of hypercalcemia

Central nervous system	Depression, impaired memory, lethargy, psychosis, ataxia, stupor, confusion, coma, neuromuscular weakness, proximal myopathy, hypotonia, reduced deep tendon reflexes
Cardiovascular	Hypertension, bradycardia (even asystole), short QT interval on ECG
Renal	Polyuria, renal stones, hyperchloremic acidosis, nephrocalcinosis
Gastrointestinal	Anorexia, nausea, vomiting, constipation, peptic ulcer, pancreatitis
Other	Band keratopathy, tissue calcification (e.g. heart valves), bone disease

Introduction

Hypercalcemia, defined as a total serum calcium concentration greater than 2.5 mmol/l (10.2 mg/dl), or a serum ionized calcium concentration greater than 1.3 mmol/l (5.3 mg/dl), results from an imbalance between calcium influx (from bone resorption and gastrointestinal absorption) and efflux (by renal excretion). Calcium has widespread metabolic and electrophysiological roles; hence there are a variety of clinical presentations of hypercalcemia ranging from asymptomatic individuals to those in the coma of hypercalcemic crisis (Table 1). Most of the symptoms and signs are non-specific and are present with all etiologies of hypercalcemia. Relatively more intense symptoms are described with higher serum concentrations of calcium and with a more acute onset of hypercalcemia. The incidence of hypercalcemia is approximately 400 new cases per million population per year, but may exceed 3 per cent of the hospital population. Most of these are mild episodes, and the incidence of severe hypercalcemia, arbitrarily defined as serum calcium greater than 3.5 mmol/l or 14 mg/dl (Kaye 1995), is not known. More detailed discussions are given by Payne *et al.* (1973), Benabe and Martínez-Maldonado (1994), and Strewler and Rosenblatt (1995).

In blood, calcium has three forms: 45 per cent is present as the free divalent cation, 45 per cent is bound to serum proteins (principally albumin), and 10 per cent is complexed with anions such as citrate or phosphate. The concentration of free calcium ions in plasma is near saturation, as defined by the solubility product constant for calcium phosphate; hence tissue calcification may be seen in chronic hypercalcemia.

Variation in serum albumin concentration causes a corresponding change in total serum calcium concentration. By adjustment of the measured total serum calcium concentration one can normalize this result to allow interpretation based on the usual laboratory reference range. Empirical formulas to achieve this correction are available:

$$\text{corrected } [Ca^{2+} (\text{mmol/l})] = \text{measured } [Ca^{2+} (\text{mmol/l})] - 0.25[\text{albumin (g/dl)}] + 1.0 \qquad (1)$$

$$\text{corrected } [Ca^{2+} (\text{mg/dl})] = \text{measured } [Ca^{2+} (\text{mg/dl})] - [\text{albumin (g/l)}] + 4.0. \qquad (2)$$

Critically ill patients frequently have reduced serum albumin concentrations and consequently may have low total serum calcium concentrations. Indeed, ionized hypercalcemia may be overlooked in the presence of a low total serum calcium concentration associated with a reduced serum albumin concentration.

Measurement of ionized serum calcium concentration may be carried out using an ion selective electrode, and the result needs no correction based on the serum albumin concentration. The normal range for serum ionized calcium concentration is 1.1 to 1.3 mmol/l (4.6–5.3 mg/dl).

Acute alkalemia increases calcium binding to serum albumin and reduces the concentration of free ionized calcium; such alkalemia may be seen, for example, in iatrogenic hyperventilation in the management of the patient with raised intracranial pressure. Acute acidemia has the opposite effect. Usually, pH-induced changes in serum ionized calcium concentration are small.

It should be noted that hypercalcemia does not present with coagulopathy, even though calcium has a vital role in blood coagulation.

Etiology and diagnosis

The causes of hypercalcemia are listed in Table 2. Hyperparathyroidism and malignancy together constitute approximately 80 per cent of the hospital caseload.

Hypercalcemia occurs in approximately 10 per cent of cancer patients, most commonly those with lung, breast, kidney, prostate, or hematological malignancies (including multiple myeloma). The source of the calcium is the skeleton, secondary to either osteolytic bony metastases or humoral factors (paraneoplastic hypercalcemia) such as parathyroid hormone (PTH), PTH-like substance, prostaglandins, 1,25-dihydroxyvitamin D, or cytokines (including interleukins 1α and 6, and tumor necrosis factor). The malignancy is only occasionally occult once hypercalcemia is evident, and prognosis is poor with only 30 per cent surviving for 1 year. An exception to this rule is multiple myeloma which presents with hypercalcemia, where there may be a more chronic course.

Primary hyperparathyroidism increases calcium concentration secondary to an increase in bone resorption and in absorption of calcium from the gastrointestinal tract and renal tubular fluid. The patient may be hypophosphatemic, since PTH enhances renal phosphate excretion; PTH-induced bicarbonate wasting may also lead to a hyperchloremic acidosis. Hyperparathyroidism is usually due to a single adenoma (80 per cent); 1 to 3 per cent present with carcinoma, and the rest have multiglandular benign disease. Rarely, hyperparathyroidism may be part of the multiple endocrine neoplasia syndrome. Bone disease is now a rare feature of hyperparathyroidism; indeed the majority of patients are asymptomatic.

Chronic immobilization, which is a common characteristic of the critically ill patient, may lead to marked skeletal calcium liberation. The degree of hypercalcemia may be particularly dramatic where high rates of bone turnover are present, such as in Paget's disease or thyrotoxicosis.

Hypercalcemia is seen in up to a third of patients in the diuretic phase of acute renal failure, particularly acute renal failure secondary to rhabdomyolysis. Following successful renal transplantation for chronic renal failure, mild hypercalcemia may be seen in up to a third of patients; the condition resolves spontaneously within 2 years.

A variety of other critically ill patients become hypercalcemic. The condition seems to be multifactorial and has been reported in 15 per cent of surgical critical care patients (Forster *et al.* 1985); associated factors include systemic inflammatory response syndrome, multiple transfusions, parenteral hyperalimentation, renal failure, and dialysis. These patients have elevated serum PTH concentrations, and it is suggested that parathyroid hypertrophy follows repeated hypocalcemic episodes and the gland displays autonomous PTH secretion at a later stage.

Table 2 Etiology of hypercalcemia

Malignancy	10% of all cancer patients
Hyperparathyroidism	
Immobilization	
Calcium therapy	Including milk alkali syndrome
Renal	Renal failure or following renal transplant
Granulomatous disease	Sarcoidosis
Following hypocalcemia	
Hypocalciuric hypercalcemia	Familial, benign
AIDS related	
Drug induced	Thiazides, lithium, estrogens, tamoxifen, theophylline
Endocrine	Hyperthyroidism, multiple endocrine neoplasia syndrome (1, 2a), acromegaly, Addison's disease, pheochromocytoma
Other	Hyperalimentation, idiopathic, chronic liver disease, vitamin A or D overdose

The cause of hypercalcemia in a particular instance is often clear. A concise history and full physical examination is essential. Particular attention should be given to drug and vitamin intake, calcium supplements, any family history of hypercalcemia, and signs or symptoms attributable to malignancy. Investigations should include a chest radiograph, a 12-lead electrocardiogram, serum PTH concentration, serum electrolyte and cortisol concentrations, and thyroid function tests. Serum and urine electrophoresis is carried out to identify multiple myeloma, the most common occult malignancy to cause hypercalcemia. A bone scan will identify osteolytic metastases in malignant hypercalcemia. This strategy will elucidate the etiology in the large majority of patients. A physician with experience in this area should be involved early in the patient's course.

Treatment

Mild asymptomatic hypercalcemia requires no specific treatment. The definitive treatment of moderate or severe hypercalcemia depends on the specific diagnosis, but the initial treatment and resuscitation is the same in all cases. Patients with potentially life-threatening manifestations (altered conscious level, bradycardia, short QT interval on electrocardiogram) should be treated immediately as there is a risk of coma and death in these patients.

The mainstay of treatment is rehydration, which should be commenced with 0.9 per cent saline solution intravenously. Intravenous therapy will correct the inevitable dehydration, and the added sodium load reduces calcium reabsorption in the proximal renal tubule. Once the patient is adequately hydrated, forced saline diuresis is initiated, supplemented with intravenous furosemide (frusemide) as required. Furosemide inhibits calcium reabsorption in the distal renal tubules. The aim of therapy is to induce a diuresis of approximately 5 liters over each 24-h period. This therapy must be individualized, and, in patients with cardiovascular or renal failure, fluid management should be guided by invasive monitoring such as a central venous or pulmonary artery flotation catheter. In renal failure, hemodialysis using a low-calcium dialysate is an effective alternative to forced diuresis. In all cases, it is important to monitor serum calcium, electrolyte, magnesium, and phosphate concentrations carefully and to correct any abnormalities. Hypophosphatemia exacerbates hypercalcemia and should be addressed promptly. Other monitoring should include continuous electrocardiography, blood pressure, and neurological observations.

Glucocorticosteroids are useful in hypercalcemia attributable to excess calcium absorption from the gastrointestinal tract, since they antagonize vitamin D. They may be effective in myeloma-associated cytokine release and lymphoma, but high doses may be required (prednisolone 60–100 mg/day). High-dose corticosteroid therapy may be safely stopped abruptly at the end of a course of up to 1 week, but if longer courses are given it is necessary to taper the dose before discontinuation. In the presence of high-dose corticosteroid, it may be necessary to increase the dose of furosemide administered, as salt and water retention is a common effect of high-dose corticosteroids.

Calcitonin inhibits osteoclast function and is effective within 2 h of subcutaneous injection (4 IU/kg every 12 h); a reduction in serum total calcium concentration of 2 mg/dl is typical, although tachyphylaxis develops within days. Calcitonin should be commenced early, and is safe during the rehydration phase of treatment. Calcitonin is not contraindicated in renal failure. Side-effects are benign; flushing, transient nausea, and local hypersensitivity at the injection site may be seen. Calcitonin is ineffective in approximately 25 per cent of patients.

Several other drugs are available which prevent osteoclast resorption of bone and will reduce serum calcium over the short term. They each have a delayed action; serum calcium concentration falls after 2 to 3 days of therapy. These agents include biphosphonates (diphosphonates), plicamycin (mithramycin), and gallium nitrate. Use of these drugs is secondary to rehydration, and they are best prescribed by a physician experienced in their use since some of the agents (particularly plicamycin and gallium nitrate) may induce nephrotoxicity.

Phosphates were previously given routinely to control hypercalcemia, and intravenous preparations may still be given in life-threatening hypercalcemia. Phosphates are not indicated if hyperphosphatemia coexists with hypercalcemia.

Verapamil, a calcium-channel blocker, has been used successfully to control severe cardiovascular manifestations of hypercalcemia (arrhythmia and hypertension) before measures to control serum calcium concentration have taken effect.

Whichever medical therapy is used, the effect is transient. Definitive correction of the underlying pathology should be undertaken once the patient's condition is stabilized.

References

Benabe, J.E. and Martínez-Maldonado, M. (1994). Disorders of calcium metabolism. In *Maxwell and Kleeman's clinical disorders of fluid and electrolyte metabolism* (5th edn) (ed. R.G. Narins), Chapter 32. McGraw-Hill, New York.

Forster, J., Querusio, L., Burchard, K.W., and Gann, D.S. (1985). Hypercalcemia in critically ill surgical patients. *Annals of Surgery*, **202**, 512–18.

Kaye, T.B. (1995). Hypercalcemia: how to pinpoint the cause and customize treatment. *Postgraduate Medicine*, **97**, 153–60.

Payne, R.B., Little, A.J., Williams, R.B., and Milner, J.R. (1973). Interpretation of serum calcium in patients with abnormal serum proteins. *British Medical Journal*, **4**, 643–6.

Strewler, G.J. and Rosenblatt, M. (1995). Mineral metabolism. In *Endocrinology and metabolism* (3rd edn) (ed. P. Felig, J.D. Baxter, and L.A. Frohman), Chapter 23. McGraw-Hill, New York.

8.2.7 Hypocalcemia

T. J. Gan

Key messages

- Hypocalcemia is a common disorder in patients in the intensive care unit.
- Measurement of ionized calcium is necessary for the accurate diagnosis of hypocalcemia.
- Definitive treatment depends on recognition and correction of the underlying disorders.
- Acute hypocalcemia needs to be treated urgently to bring the serum calcium back within the normal range. This may then be followed by addressing the primary disorders.

Introduction

Calcium is an essential cation required for many biological processes, including excitation–contraction coupling in muscles, hormonal secretion, neurotransmission, cell division, blood coagulation, and bone structure. It is also a major intracellular messenger, is needed for cellular processes that require movement, and is required by many enzymes for normal activity (Tohme and Bilezikian 1993). Deficiency of calcium is common in patients with critical illness. Total serum calcium concentrations are low in 70 to 90 per cent of intensive care unit (ICU) patients. However, the occurrence rate of ionized hypocalcemia varies from 15 to 50 per cent, depending on the type of patient treated.

Physiology of calcium homeostasis

The calcium content of the adult human body is approximately 25 mg/kg lean body weight, more than 99 per cent of which is found within the skeleton. The remainder, termed the miscible pool, is divided into three fractions. Approximately 40 per cent is bound to serum proteins, primarily albumin, 10 per cent is complexed with serum anions (chelated form), and 50 per cent is in free ionized form. The free component is the major physiologically active form of plasma calcium. In the adult, the ionized calcium content is normally between 1.00 and 1.15 mmol/l, although the range is technique specific and may vary between laboratories. Proper handling of the blood specimen with minimal delay prior to measurement using anaerobic precautions similar to those for blood gas analysis is recommended.

Many attempts have been made to correct the total serum calcium concentration mathematically for alterations in circulating albumin levels and arterial pH. However, total serum calcium and calculated ionized calcium levels have been poor predictors of the physiologically active ionized calcium fraction. Despite high sensitivity, total serum and calculated ionized calcium values lack specificity in predicting ionized hypocalcemia. Serum ionized calcium level remains the best measure of calcium delivery in critically ill patients.

Circulating calcium is closely regulated by parathyroid hormone (PTH) and vitamin D through their effects on bone, kidney, and gut. A drop in circulating calcium elicits an increase in PTH which stimulates osteoclastic mediated mobilization of calcium from bone as well as renal reabsorption of calcium. PTH also stimulates 1-α-hydroxylation of 25-hydroxycalciferol by the kidney, leading to the formation of the active vitamin D metabolite 1,25-dihydroxycalciferol. This metabolite stimulates absorption of dietary calcium from the duodenum and colon. The 1,25-dihydroxy form of vitamin D is also necessary for the acute mobilization of calcium from bone in response to PTH, which may be important for the acute response to a fall in circulating calcium. Vitamin D is a fat-soluble vitamin and is stored within the body. However, the active metabolite 1,25 dihydroxycalciferol has a half-life of less than 24 h. Therefore, if 1-α-hydroxylase activity is impaired, functional vitamin D deficiency may develop rapidly despite adequate total body stores of vitamin D. In addition, phosphate levels in humans also modulate this enzyme, and hyperphosphatemia can directly suppress vitamin D conversion to 1,25-dihydroxycalciferol. Other hormones, such as growth hormone and estrogens, are also thought to influence 1-α-hydroxylase activity, but the clinical significance of these effects has not been clarified. Although vitamin D deficiency may impair the acute mobilization of calcium from bone, the chronic mobilization in response to increased PTH does not depend on vitamin D. Therefore hypocalcemia may result from an inadequate PTH response to hypocalcemia, vitamin D deficiency, or skeletal resistance to either hormone (Zaloga 1992).

Causes of hypocalcemia

The causes of hypocalcemia in critically ill patients (Table 1) include primary hypoparathyroidism (rarely seen) and secondary hypoparathyroidism following neck surgery (common). These patients usually remain asymptomatic; however, patients may occasionally develop parasthesias, laryngeal spasm, or tetany. The hypocalcemia occurring in patients with sepsis has multiple etiologies. Some patients have dietary vitamin D deficiency, while others have acquired hypoparathyroidism, renal 1-hydroxylase deficiency, or peripheral resistance to calcitriol. Renal 1-hydroxylase deficiency usually occurs when renal insufficiency accompanies sepsis. A common finding in these patients is their impaired ability to mobilize skeletal calcium owing to PTH or vitamin D deficiency (Zaloga and Chernow 1986).

Vitamin D deficiency is being increasingly recognized as an important cause of hypocalcemia in patients in the ICU. Many of these patients are chronically ill, malnourished, and have minimal sunlight exposure. Although these vitamin-D-deficient patients usually have a normal serum calcium as out-patients, they are unable to maintain normal levels during critical illness. Hypocalcemia with hypophosphatemia are commonly associated with vitamin D deficiency.

Hyperphosphatemia may be a cause of hypocalcemia as a result of calcium precipitation and suppression of renal 1-hydroxylation of 25-hydroxyvitamin D. The most common causes of this syndrome include phosphorus administration (e.g. during treatment for diabetic ketoacidosis), rhabdomyolysis, and the tumor lysis syndrome following chemotherapy. This results from the massive and sudden

Table 1 Causes of hypocalcemia

- Increased protein binding affinity
  - Alkalosis
  - Elevated free fatty acids
  - Complexing anions, exogenous phosphate, phosphorus burn, renal failure, tumor lysis syndrome, rhabdomyolysis
- Citrate
  - Blood transfusion
  - Radiographic contrast
- Edetate
  - Radiographic contrast
- Bicarbonate
- Pancreatitis
- PTH deficiency congenital: parathyroid aplasia, DiGeorge syndrome
- Surgical
  - Transient: postparathyroid adenoma resection
  - Prolonged or permanent: post-total parathyroidectomy, disruption of vascular supply
- Infiltration: sarcoidosis, Wilson's disease, hemochromatosis, metastatic carcinoma
- Suppression: hypomagnesemia: pancreatitis, aminoglycosides, pentamidine, loop diuretics, amphotericin B
- Hypermagnesemia
- Drugs: aluminum, asparaginase, doxorubicin, cytosine arabinoside, cimetidine
- Critical illness
- Idiopathic
  - Early onset: autoimmune polyglandular syndrome type 1
  - Late onset: sporadic, Kenny syndrome
- PTH resistance: pseudohypoparathyroidism types Ia, Ib, Ic, II
- Vitamin D deficiency
  - Lack of absorption: nutritional, small bowel syndrome, pancreatic insufficiency, cholestyramine
  - Lack of metabolism: hepatic or renal disease, vitamin-D-dependent rickets
  - Vitamin D resistance: vitamin-D-resistant rickets
- Increased calcitonin: toxic shock syndrome
- Inhibited bone resorption: estrogen, calcitonin, bisphosphonates, plicamycin
- Rapid bone formation: hungry bone syndrome, hungry tumor syndrome

release of phosphate from injured cells. Hypocalcemia secondary to hyperphosphatemia from an exogenous source is rare. It may be caused by excessive phosphate taken orally as a laxative or by a high-phosphate enema. Phosphorus burns may also cause hyperphosphatemia and hypocalcemia.

Hypocalcemia is occasionally seen in trauma patients with fat embolism. The hypocalcemia probably results from a combination of calcium chelation and increases in protein binding of calcium. Bone blood flow decreases to 10 to 20 per cent of control values during hypotension. Toxic shock syndrome may be associated with hypocalcemia. Although calcitonin levels are elevated, the exact pathogenesis of the hypocalcemia remains poorly defined.

Diagnosis of the various causes of hypocalcemia depends on measurements of the ionized calcium value. Patients with these diseases frequently have alterations in serum protein levels, acid–base status, free fatty acid concentrations, and received blood and albumin infusions. Hence estimation of ionized calcium based on serum total calcium and albumin levels is unreliable.

Clinical manifestations of hypocalcemia

Hypocalcemia may present with a variety of clinical signs and symptoms (Table 2) that relate to increased neuronal irritability. Cardiovascular manifestations are the most commonly encountered features in critically ill hypocalcemic patients. Patients may develop hypotension, cardiac insufficiency, arrhythmias (e.g. bradycardia, prolonged QT interval, and ventricular tachycardia and fibrillation), and failure to respond to drugs that act through calcium-related mechanisms (e.g. digoxin, norepinephrine (noradrenaline), and dopamine). Hypocalcemia should always be considered in patients with hypotension that responds poorly to fluids or pressor agents. Hypocalcemia may also present as laryngospasm in critically ill patients after neck surgery. Patients who had been doing well and were extubated after neck surgery have required reintubation as a result of laryngospasm. Tetany with muscle spasm is rarely seen in the ICU. Chovestek's and Trousseau's signs are indicators of peripheral nerve irritability during hypocalcemia (Lebowitz and Moses 1992).

Hypocalcemic symptoms are notoriously difficult to elicit in critically ill patients. Anticonvulsant drugs, sedation, and paralysis may mask signs of neuronal irritability. Intubated, sedated, and paralyzed patients may be unable to express symptoms such as anxiety and parasthesias. Psychiatric manifestations of hypocalcemia (such as depression and psychosis) are non-specific and are frequently seen in critically ill patients as a result of their illness. Hence it is important to monitor calcium levels routinely in all patients.

Table 2 Clinical features of hypocalcemia

Neuromuscular
Tetany
Chvostek's and Trousseau's signs
Muscle weakness
Muscle spasm
Paresthesias
Hyperactive reflexes
Seizures
Cardiovascular
Bradycardia
Hypotension
Impaired contractility
Digitalis insensitivity
QT and ST prolongation
Cardiac arrest
Respiratory
Bronchospasm
Laryngeal spasm
Psychiatric
Anxiety
Confusion
Depression
Irritability
Dementia
Psychosis

Adapted from Zaloga (1992).

Evaluation of hypocalcemia

When there is clinical suspicion of hypocalcemia, an ionized calcium level is mandatory. In the presence of acute life-threatening symptoms, intravenous calcium chloride is recommended. After normalization of ionized calcium, a search for the underlying causes should begin. History and physical examination should alert the physician to various medications and associated medical illnesses that may be responsible. In the event that this is unrevealing, measurements of serum magnesium, PTH, and phosphorus may be helpful in evaluating the causes of hypocalcemia.

Treatment of hypocalcemia

Emergency treatment

Acute symptomatic hypocalcemia is a medical emergency that requires intravenous calcium therapy. Initial treatment consists of the administration of a calcium bolus (100–200 mg of elemental calcium over 10 min) followed by a maintenance calcium infusion of 1 to 2 mg/kg/h. Calcium chloride (10 per cent) contains 27.2 mg Ca/ml and calcium gluconate (10 per cent) contains 9 mg Ca/ml.

Subsequent rate adjustments should be based on serial calcium measurements obtained every 2 to 4 h. Measurement of serum magnesium is important in patients with ionized hypocalcemia because hypo- and hypermagnesemic hypocalcemia respond poorly to calcium therapy and magnesium replacement may be all that is needed to correct the hypocalcemia (Reber and Heath 1995). Potassium deficiency protects patients from hypocalcemic tetany, and correction of hypokalemia without correction of hypocalcemia may induce tetany. Administration of drugs that may aggravate hypocalcemia (e.g. furosemide (frusemide)) should be discontinued or another drug substituted, if possible. Calcium must be administered cautiously to patients receiving digitalis, since hypercalcemia predisposes to digitalis toxicity (Zaloga and Chernow 1986). Optimal therapy requires frequent monitoring of serum calcium, magnesium, phosphorus, potassium, and creatinine levels, as well as electrographic and hemodynamic status. Intravenous calcium salt is very irritating to the veins and tissue injury may result from extravasation.

Long-term treatment

Long-term therapy, consisting of oral calcium and vitamin D, is indicated in most cases of hypocalcemia. The goal of long-term treatment is maintenance of serum calcium in the low-normal range. This alleviates hypocalcemic symptoms, while avoiding hypercalciuria, and provides a cushion against treatment-induced hypercalcemia. Oral elemental calcium 1 to 3 g/day in divided doses is usually sufficient. Several pharmacological preparations of calcium are available. Calcium citrate is more soluble than calcium carbonate, particularly in patients taking H_2 antagonists or proton-pump inhibitors and those with achlorhydria.

The principal effect of vitamin D is to increase gut calcium absorption, although it also has an effect on bone resorption. Various forms of vitamin D may be used depending on the specific abnormality of vitamin D synthesis. Vitamin D_2 (ergocalciferol — the parent compound) requires hepatic and renal transformation for full activation. For most cases of hypocalcemia, usually hypoparathyroidism, vitamin D_3 (cholecalciferol) is sufficient. 25-Hydroxyvitamin D (calcifediol — the product formed by hepatic conversion of vitamin D) may be used in severe liver disease, and 1,25-dihydroxyvitamin D (calcitriol — the active metabolite formed in the kidney) is most appropriate in vitamin D resistance, severe renal impairment, and patients with hypoparathyroidism unresponsive to other forms of vitamin D therapy. Vitamin D requirements may vary based on disease activity, interacting drugs, and dietary calcium intake. Hypercalciuria may occur during treatment. In this situation, a thiazide diuretic can be helpful in reducing urinary calcium excretion.

References

Lebowitz, M.R. and Moses, A.M. (1992). Hypocalcemia. *Seminars in Nephrology*, **12**, 146–58.

Reber, P.M. and Heath, H. (1995). Hypocalcemic emergencies. *Medical Clinics of North America*, **79**, 93–106.

Tohme, J.F. and Bilezikian, J.P. (1993). Hypocalcemic emergencies. *Endocrinology and Metabolism Clinics of North America*, **22**, 363–75.

Zaloga, G.P. (1992). Hypocalcemia in critically ill patients. *Critical Care Medicine*, **20**, 251–62.

Zaloga, G.P. and Chernow, B. (1986). Hypocalcemia in critical illness. *Journal of the American Medical Association*, **256**, 1924–9.

8.2.8 Hyperphosphatemia

J. V. Booth and M. G. Mythen

Key messages

- Severe hyperphosphatemia is rare.
- The most common causes of hyperphosphatemia are renal failure and transcellular shift from intracellular to extracellular fluid.
- Clinical effects are often due to concommitant hypocalcemia.
- Treatment commences with elucidating etiology.
- Further treatment consists of decreasing phosphate load, increasing renal excretion, or causing transcellular shifts from extracellular to intracellular fluid.

Introduction

A fasting plasma phosphate concentration above 1.5 mmol/l (47 mg/l) in adults indicates hyperphosphatemia. Further classification defines mild to moderate hyperphosphatemia as up to 2.6 mmol/l (80 mg/l) and severe hyperphosphatemia as greater than 3.2 mmol/l (100 mg/l). The most common cause of hyperphosphatemia is renal failure. In analyzing plasma phosphate levels care must be taken to exclude pseudohyperphosphatemia. Spurious levels are produced by paraproteinemias, hyperlipidemia, hemolysis, and hyperbilirubinemia (Larner 1995). Clinically unexplained increases in plasma phosphate levels should initiate a search for potential causes of pseudohyperphosphatemia. Life-threatening hyperphosphatemia is much less common than similarly severe hypophosphatemia, and when it does occur it is usually the result of liberation of intracellular stores (Lau 1986).

Etiology

A logical approach to hyperphosphatemia is to divide the etiologies into increased load, decreased renal excretion, or transcellular shifts from the intracellular to the extracellular space (Table 1). The most common cause of hyperphosphatemia in the critically ill patient is endstage renal failure, and hyperphosphatemia is rare in the presence of a normal or only moderately impaired renal function. Other causes pertinent to the critically ill patient include transcellular shifts caused by acidosis, rhabdomyolysis, and malignant hyperpyrexia, as well as exogenous phosphate from laxatives and enemas.

Clinical effects

The clinical sequelae of hyperphosphatemia are related to its effects on the serum calcium concentration. When the solubility product of phosphate and calcium is exceeded, calcium phosphate precipitation occurs. This results in ectopic calcification (Table 2). Mild hyperphosphatemia such as that seen in hyperthyroidism has no obvious clinical effects. However, when a substantial increase in phosphate occurs with a parallel precipitous fall in serum calcium, tetany,

Table 1 Systematic etiology of hyperphosphatemia

Decreased renal excretion	
Decreased GFR	Chronic renal failure
Increased tubular reabsorption	Hypoparathyroidism Hypomagnesemia Acromegaly Hyperthyroidism Diphosphonate therapy
Transcellular shift from ICF or ECF	Metabolic or respiratory acidosis Tumor lysis syndrome Sepsis Malignant hyperpyrexia Rhabdomyolysis Acute hemolysis
Increased phosphate intake	Laxatives or enemas Injudicious IV phosphate therapy

GFR, glomerular filtration rate; ICF, intracellular fluid; ECF, extracellular fluid; IV, intravenous.

Table 2 Clinical sequelae of hyperphosphatemia

Hypocalcemia	Skeletal muscle tetany Seizure activity Prolongation of QT interval on ECG Arrhythymias Hypotension
Metastatic calcification	Vascular occlusion from arterial/venous calcification Periarticular calcification Band keratopathy Crystal deposits in heart, lungs, kidneys, and corneas
Others	Secondary hyperparathyroidism Progression of chronic renal failure

hypotension, and even convulsions can occur. Although hypocalcemia commonly parallels hyperphosphatemia, chronic increases in serum phosphate levels may not show these changes (Bourke and Yanagawa 1993).

Management

Mild hyperphosphatemia does not require treatment. However, the cause requires elucidation and may itself require treatment. Severe acute hyperphosphatemia, most often caused by transcellular shifts or exogenous phosphate therapy, is often adequately treated by removing the source of phosphate load and accelerating renal excretion with intravenous saline and diuretic therapy. Transcellular shifts can be treated by removing the cause of the shift (e.g. acidosis) or administering glucose and insulin infusions to induce a rapid transcellular shift from extracellular to intracellular fluid. Where severe or symptomatic hypocalcemia is coexistent, intravenous calcium administration is required.

In the presence of renal failure other strategies are required. Avoidance of excess dietary intake is the initial starting point, followed by gastrointestinal chelating agents to decrease intestinal absorption. Aluminum gels were used originally, but it has been demonstrated that these agents result in substantial side-effects over time, particularly in patients with poor renal function. As a result calcium salts are now the preferred agents for limiting absorption of phosphate in the intestinal tract. However, care must be taken to prevent hypercalcemia as a result of using these agents. Where hypercalcemia may be a problem, occasional supplementation with aluminum binders may be useful (Bourke and Yanagawa 1993). This dual therapy may limit complications from use of one or other agent alone. Hemodialysis is rarely required, but is effective in lowering serum phosphate levels particularly when accompanied by renal failure.

References

Bourke, E. and Yanagawa, N. (1993). Assessment of hyperphosphatemia and hypophosphatemia. *Clinics in Laboratory Medicine*, **13**, 183–207.

Larner, A. (1995). Pseudohyperphosphatemia. *Clinical Biochemistry*, **28**, 391–3.

Lau, K. (1986). Phosphate disorders. In *Fluids and electrolytes* (ed. J.P. Kokko and R.L. Tannen), pp. 398–416. W.B. Saunders, Philadelphia, PA.

8.2.9 Hypophosphatemia

J. V. Booth and M. G. Mythen

Key messages

- The incidence of hypophysphatemia in patients in critical care units is 25 to 30 per cent.
- Only severe phosphate depletion has significant clinical sequelae.
- The most common etiologies of hypophosphatemia include glucose infusions and refeeding syndrome.
- Hypophosphatemia is a contributing cause to respiratory muscle weakness.
- A graduated dosing scheme is effective for replacement therapy.

Introduction

Phosphate is the major intracellular anion in the body and is required for the metabolism of fat, carbohydrate, and protein in oxidative phosphorylation. Low-plasma phosphate levels may not necessarily equate with total body phosphorus deficiency and may be due to a transcellular shift from the extracellular to the intracellular space. Spuriously low phosphate levels may also be seen in patients receiving mannitol therapy and in patients with multiple myeloma, Waldenstrom's macroglobulinemia, and monoclonal gammopathies. Plasma levels may also demonstrate diurnal variations, with the highest levels measured between 2 a.m. and 6 a.m. and the lowest between 8 a.m. and noon, although this may not be evident in patients in the critical care setting. However, clinical features of phosphorus

Table 1 Plasma phosphate levels

	mmol/l	mg/l
Normal	0.8–1.5	25–47
Low	0.5–0.7	15–24
Severe	<0.5	<15

Table 2 Common causes of severe hypophosphatemia

Iatrogenic*	65%
Iatrogenic+fasting	15%
Septicemia	9%
Iatrogenic+septicemia	8%
Iatrogenic+gastrointestinal losses	6%
Iatrogenic+hypokalemia	4%

*Iatrogenic includes glucose infusions, refeeding syndrome, and drugs.

depletion have only been described in the presence of low plasma inorganic phosphate levels, and therefore plasma levels are crucial in diagnostic decision making. Plasma levels should be checked twice a week in critically ill patients (Table 1).

Incidence

Studies indicate that the incidence of hypophosphatemia is about 3 per cent of the general hospital in-patient population and the incidence of severe hypophosphatemia is about 0.2 per cent (Halevy and Bulvik 1988). However, evidence is accumulating that the incidence is much higher in the critical care population, with one study demonstrating an incidence of 28 per cent (Kruse *et al.* 1992).

Etiology

Several factors may contribute to hypophosphatemia in critically ill patients. The most common factors in these patients include infusions of glucose and insulin as well as the refeeding syndrome (Table 2). Insulin production stimulated by glucose infusions or high caloric intake after a period of inadequate feeding results in transcellular shifts in phosphate from the extracellular to the intracellular space.

The condition can also be ascribed to inadequate input, transcellular shifts, or excessive losses (Table 3). If the cause is not clear, a spot urinary phosphate level is helpful. If the concentration is less than 1.2 mmol/l (40 mg/l) despite significant hypophosphatemia, excessive renal losses can be excluded. If the urinary level is elevated, excessive renal losses, such as diuretic drug administration or the diuretic phase of acute renal failure, are the most likely cause (Bourke and Yanagawa 1993).

Clinical effects

Mild to moderate hypophosphatemia has not been conclusively linked with significant clinical problems such as severity of illness, concomitant electrolyte abnormalities, length of stay, or outcome (Kruse *et al.* 1992). Major clinical sequelae occur only as a result of severe depletion. Acute problems of phosphate deficiency are related to either decreased production of 2,3-diphosphoglycerate (2,3-DPG) in red blood cells with a concomitant shift in the oxygen dissociation curve to the left, or the reduction in the formation of high-energy phosphate bonds seen in ATP etc. A list of these clinical effects is given in Table 4. Of particular relevance to the critically ill population is the decreased oxygen delivery as a result of both decreased 2,3-DPG and decreased cardiac output from decreased contractility and

Table 3 A systematic etiology of hypophosphatemia

Inadequate input	Gastrointestinal phosphate binders, antacids, laxatives, oral iron Starvation Vomiting or nasogastric suctioning Malabsorption syndrome including ileus Relative or absolute vitamin D deficiency
Transcellular shifts	Alkalosis Carbohydrate load Drugs (e.g. insulin, epinephrine (adrenaline), steroids, β_2 agonists) Strenuous exercise Recovery from hypothermia Alcohol withdrawal
Excessive losses	Phosphaturic agents Acute volume expansion Massive diuresis Diuretic phase of acute renal failure Primary or secondary hyperparathyroidism Dialysis Primary reabsorptive defect Postrenal transplant Osteomalacia Increased filtered load from cellular leakage

Table 4 Clinical manifestations of hypophosphatemia

Cardiac	Decreased contractility Possible cardiomyopathy
Respiratory	Respiratory failure Failure to wean
Skeletal muscle/ bone	Osteomalacia Myalgia Weakness Rhabdomyolysis
Hematological	Granulocyte dysfunction Leukocyte dysfunction Platelet dysfunction
Central nervous system	Weakness Numbness Seizures
Renal	Acute tubular necrosis

Table 5 Phosphate replacement strategies

Phosphate concentration	Dose
>0.7 mmol/l	0.16 mmol/l/kg over 4–6 h
0.5–0.7 mmol/l	0.32 mmol/l/kg over 4–6 h
<0.5 mmol/l	0.64 mmol/l/kg over 8–12 h

Doses should be in addition to normal phosphate intake via supported nutrition. Doses should be adjusted in circumstances where complications are more likely to occur (see text for details).

occasionally a cardiomyopathy. Hypophosphatemia is also implicated as a reversible cause of failure to wean from ventilatory support, although it is unlikely that this is often a significant factor on its own. Although there are no studies relating the independent predictive value of severe hypophosphatemia to outcomes, treatment is relatively safe and effective and therefore should be implemented in all cases.

Management

Phosphate replacement strategies must take into account the etiologies of the deficiency, the level of deficiency, and the side-effects of phosphate replacement. A strategy is outlined in Table 5. Complications of replacement therapy include hypocalcemia, hyperphosphatemia, hypernatremia, dehydration, renal failure, hypotension, phlebitis, and metastatic calcification. Therefore care must be taken when using replacement therapy in patients prone to these side-effects. Other electrolytes should be monitored closely during replacement therapy.

References

Bourke, E. and Yanagawa, N. (1993). Assessment of hyperphosphatemia and hypophosphatemia. *Clinics in Laboratory Medicine*, **13**, 183–207.

Halevy, J. and Bulvik, S. (1988). Severe hypophosphatemia in hospitalized patients. *Archives of Internal Medicine*, **148**, 153–5.

Kruse, J., Al-Douahji, M., and Carlson, R.W. (1992). Hypophosphatemia in critically ill patients: incidence and associations. *Critical Care Medicine*, **20**, S104.

8.3 Metabolic acidosis

8.3.1 Pathophysiology and causes

D. M. Forrest and J. A. Russell

Key messages

- Acid–base balance is maintained by buffers, the most important of which is the bicarbonate–carbon dioxide system.
- Metabolic acidosis is due to loss of bicarbonate or gain of hydrogen ions.
- Acid is secreted and bicarbonate is reclaimed and regenerated by the kidneys; ammoniagenesis and secretion by the kidneys is crucial.
- Acidemia (lowering of systemic pH) can produce intracellular acidosis which causes widespread organ dysfunction, particularly in the cardiovascular system.
- Metabolic acidosis is classified on the basis of the anion gap (the difference between measured cations and anions) as normal or increased anion gap acidosis.

Introduction

Metabolic acidosis is a frequent and potentially serious aspect of critical illness. Acidosis increases hydrogen ion concentration $[H^+]$, which compromises cellular protein function (Schmidt 1992). Clinical assessment of acid–base disorders is based on measurable alterations in the extracellular compartment; inferred changes in intracellular acid–base balance may be inaccurate (Schmidt 1992; Tønnessen 1995). Acidosis and acidemia are not synonymous. Acidemia refers to an abnormally low blood pH (e.g. pH < 7.36), while acidosis refers to the underlying process which causes acidemia (Barnett and Schmidt 1992; Schmidt 1992). Understanding the pathophysiology of metabolic acidosis is fundamental to management.

Physiological response to metabolic acidosis

Metabolic acidosis is characterized by a decline in bicarbonate concentration $[HCO_3^-]$ through either loss of HCO_3^- or gain of H^+ (Barnett and Schmidt 1992). Accumulated hydrogen ions can be removed effectively by buffering, net excretion, and/or metabolism of fixed acids.

Buffering

Weak acids and their conjugate bases readily take up or release H^+ throughout the physiological pH range in order to mitigate changes of systemic pH. Cellular and plasma proteins and phosphates are important buffers extra- and intracellularly, but bone provides the largest repository of base and is an important buffer in chronic acidosis (Emmett *et al.* 1992; Schmidt 1992). However, the bicarbonate–carbon dioxide (CO_2) buffering system is the most significant because the sum of the acid (CO_2) and its conjugate base (HCO_3^-) can vary in a regulated manner, i.e. CO_2 can be removed readily by respiration and excretion of HCO_3^- is regulated by the kidney (Emmett *et al.* 1992).

CO_2 is controlled by regulation of respiration through central chemoreceptors in the medulla (not the cerebrospinal fluid) and by carotid body receptors, cortical effects, and sensory input from the chest wall (Schmidt 1992). Respiratory compensation for metabolic acidosis is usually developed fully within 12 to 24 h and results in a fall in $Pa\text{CO}_2$ of approximately 0.134 kPa (1 mmHg) for each 1-mmol/l fall in plasma HCO_3^-. In response to a primary change of $Pa\text{CO}_2$, HCO_3^- is regulated at the renal tubule by reabsorption of filtered HCO_3^- (reclamation) and by HCO_3^- regeneration. The latter process reconstitutes HCO_3^- which has been consumed as a result of introduction of fixed acids or has been lost (Emmett *et al.* 1992). Renal HCO_3^- reclamation and regeneration is developed fully in 48 to 72 h.

Intracellular buffering is critical because the intracellular volume is twice that of the extracellular fluid volume and the intracellular compartment has much higher buffering capacity per unit volume (Tønnessen 1995). Hemoglobin (Hb) in red blood cells is a major intracellular buffer because of its many histidine residues, particularly in the deoxygenated state (Haldane effect).

CO_2 is a major source of acid and has a relatively complex system of excretion and metabolism. In tissues, CO_2 is produced by cellular metabolism, diffuses into the plasma compartment, and then diffuses across the red cell membrane. In the red cell, carbonic anhydrase (CA) catalyzes the formation of carbonic acid from CO_2 and water (H_2O) (reaction (1) below). Carbonic acid dissociates, releasing H^+, which is buffered by hemoglobin, and HCO_3^-, which diffuses out of the cells into the bloodstream in exchange for chloride (Cl^-), thus raising plasma HCO_3^- (reaction (2)):

$$\begin{array}{l} \qquad\quad \text{CA} \qquad\qquad\quad \text{CA} \\ CO_2 + H_2O \overset{(1)}{\leftrightarrow} H_2CO_3 \overset{(2)}{\leftrightarrow} H^+ + HCO_3^- \rightarrow H^+ + Hb \rightarrow H^+\text{–}Hb \\ \qquad\qquad\qquad\qquad\qquad\quad \rightarrow HCO_3^- \rightarrow \text{plasma} \rightarrow [HCO_3^-] \Uparrow \end{array}$$

In the lungs, CO_2 escapes from the circulation and the process is reversed. These mechanisms allow much larger quantities of CO_2 to be transported as HCO_3^- than would be possible if it were dissolved,

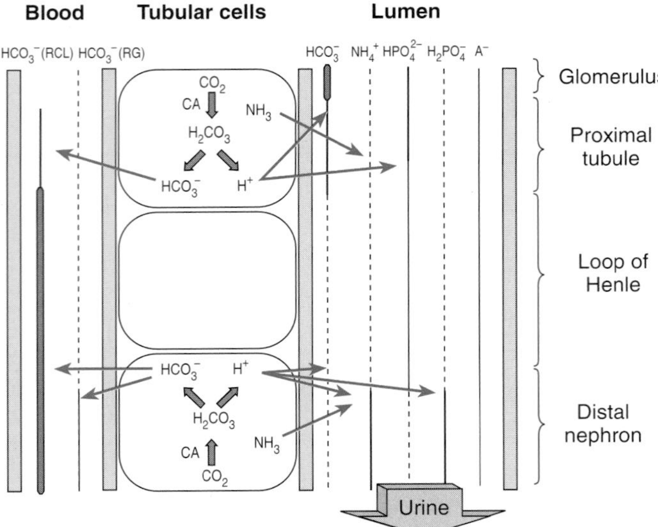

Fig. 1 Renal acid–base homeostasis. Filtered bicarbonate (HCO_3^-) is reclaimed and titrated bicarbonate is regenerated. The diagrams from left to right represent the renal microcirculation, then renal tubular cells, and finally the tubular lumen. Relative ion concentrations are indicated by the width of the line under each heading. Secretion of hydrogen ion (H^+) in the proximal tubule is largely responsible for reclamation (RCL) of equimolar amounts of HCO_3^-. Further H^+ secretion is used to titrate ammonia (and inorganic phosphate), particularly in the distal nephron. This regenerates (RG) HCO_3^-. (CA, carbonic anhydrase; A^-, anions.) (Reproduced with permission from Schmidt (1992).)

and therefore facilitate highly effective buffering of acid (Emmett *et al.* 1992; Schmidt 1992; Tønnessen 1995).

Net acid excretion

Bicarbonate regulation is intimately linked to net acid secretion throughout the kidney tubule. In the proximal tubule, where 85 to 90 per cent of filtered HCO_3^- is resorbed, reclamation of HCO_3^- is mediated by secretion of H^+ (Fig. 1); reabsorption of HCO_3^- is enhanced by increased CO_2 tension, volume contraction, and hypokalemia (Emmett *et al.* 1992). H^+ secretion in the distal tubule and collecting ducts is associated with HCO_3^- generation intracellularly and its subsequent transport into the blood (Emmett *et al.* 1992).

Urinary ammonium excretion is an important mechanism of net acid excretion. Ammonia (NH_3) is produced in the proximal renal tubular cells, enters the lumen, and is then reabsorbed in the loop of Henle. In the distal nephron, NH_3 diffuses back into the lumen where it is trapped by acidification to produce ammonium (NH_4^+). The net result of this process is secretion of acid as NH_4^+ and generation of an equimolar quantity of HCO_3^- (Emmett *et al.* 1992). Renal ammonia synthesis and thus handling of net acid load is augmented by acidosis (positive feedback) and hypokalemia, and is inhibited by low glomerular filtration rate, low levels of glutamine (due to malnutrition), high levels of alternative renal fuel sources (e.g. fat in parenteral nutrition), and hyperkalemia (Emmett *et al.* 1992; Schmidt 1992).

Metabolism of fixed acids

Metabolic acidosis may also be relieved by metabolism of fixed acids to neutral products (such as glucose) or to H_2O and CO_2 (which can be removed by respiration). Metabolism of acid removes both the organic anion and H^+, and may produce HCO_3^- (Emmett *et al.* 1992).

Effects of metabolic acidosis

The regulation of intracellular pH is mediated by activation or inhibition of carrier proteins which facilitate movement of charged particles across cell membranes (Tønnessen 1995). Carrier proteins can be modified by intracellular and extracellular pH, hormones (such as parathyroid hormone, epinephrine (adrenaline), and angiotensin II), and drugs (such as non-steroidal agents, amiloride, and furosemide (frusemide)) (Tønnessen 1995).

Intracellular acidosis has detrimental effects on important cellular functions because alteration of protein charge impairs ATP production. Furthermore, acidemia facilitates the detrimental effects of oxygen free radicals, Ca^{2+}, and other toxic products. Paradoxically, intracellular acidosis may also protect cells against ischemic injury by slowing the activity of most intracellular enzymes. Rapid normalization of intracellular pH during reperfusion can change cellular metabolism suddenly, generating free radicals and other toxins and inducing cell injury (Tønnessen 1995). The effects of acidemia on organ function are summarized in Table 1.

Pathogenesis of different forms of metabolic acidosis

Metabolic acidosis is conveniently classified on the basis of changes in the anion gap. The anion gap is the difference between measured cations (the most important of which quantitatively is sodium (Na^+)) and measured anions (Cl^- and HCO_3^-) and is normally 12 ± 2 mmol/l. Unmeasured anions, such as albumin, inorganic phosphate, sulfate, and organic anions, account for the anion gap. Normal anion gap acidosis is caused by HCO_3^- loss with replacement by Cl^-. Increased anion gap acidosis is caused by accumulation of an organic acid (such as lactic acid), which generates unmeasured anions (such as lactate) when HCO_3^- is consumed by buffering of H^+ (Barnett and Schmidt 1992; Emmett *et al.* 1992).

Increased anion gap metabolic acidoses

The major causes of increased anion gap acidosis are lactic acidosis, ketoacidosis, uremia, salicylate overdose, and poisoning by methanol, ethylene glycol, and other toxins. Lactic acidosis is an important cause of increased anion gap acidosis in the critically ill. The pathway of lactate metabolism is shown in Fig. 2. Under normal conditions, lactate is produced by a variety of tissues and is metabolized in the liver and kidneys, where it may be converted back to glucose. Hyperlactatemia occurs if lactate production exceeds excretion. Hyperlactatemia in the presence of acidemia defines lactic acidosis.

Lactic acidosis is caused by hypoxic and non-hypoxic mechanisms (Table 2) (Barnett and Schmidt 1992; Emmett *et al.* 1992; Schmidt 1992; Stacpoole 1993). During tissue hypoxia, the ratio of NADH to NAD^+ (redox ratio) rises and/or pyruvate levels increase, so that conversion of pyruvate to lactate is stimulated and lactate accumulates (see Fig. 2). Anaerobic glycolysis increases pyruvate because it is the only mechanism of energy production during hypoxia.

Tissue hypoxia can be classified as stagnant (low flow), anemic, or hypoxic hypoxia. Global tissue hypoxia due to circulatory shock is the most common etiology of hypoxic hyperlactatemia in the intensive care unit (ICU). In addition, it is important to consider regional hypoperfusion, particularly of the gastrointestinal tract, and severe exertion (including acute severe asthma, heat stroke, and seizures) as causes of hyperlactatemia. Uncoupling of oxidative phosphorylation is also important and may occur with salicylate, cyanide (a product of nitroprusside metabolism), methanol, isoniazid, and carbon

Table 1 Potential effects and mechanisms of injury on vital organ function in acidemia

Organ	Effect	Mechanism
Heart	Decreased contractility despite sympathoadrenal stimulation Tachy- and bradydysrhythmias	Intracellular acidosis Inhibition of Ca^{2+} exchange Reduced actin–myosin interaction Reduced catecholamine effect in presence of hypoxia
Vasculature	Systemic arterial vasodilatation Venoconstriction Pulmonary vasoconstriction	Intracellular acidosis Reduced catecholamine effect
Respiratory system	Depressed diaphragmatic function	Intracellular acidosis
Brain	Cerebral edema	Altered action of membrane ionic carriers
Liver	Reduced lactate clearance	Intracellular acidosis and impaired gluconeogenesis
Hemoglobin	Reduced affinity for oxygen	Low pH shifts oxyhemoglobin dissociation curve to right (Bohr effect)
	Increased affinity for oxygen	Persistent low pH reduces red cell 2,3-diphosphoglycerate
Other	Metabolic bond disease Protein catabolism	Bone buffering of chronic acidemia Increased glutamine synthesis

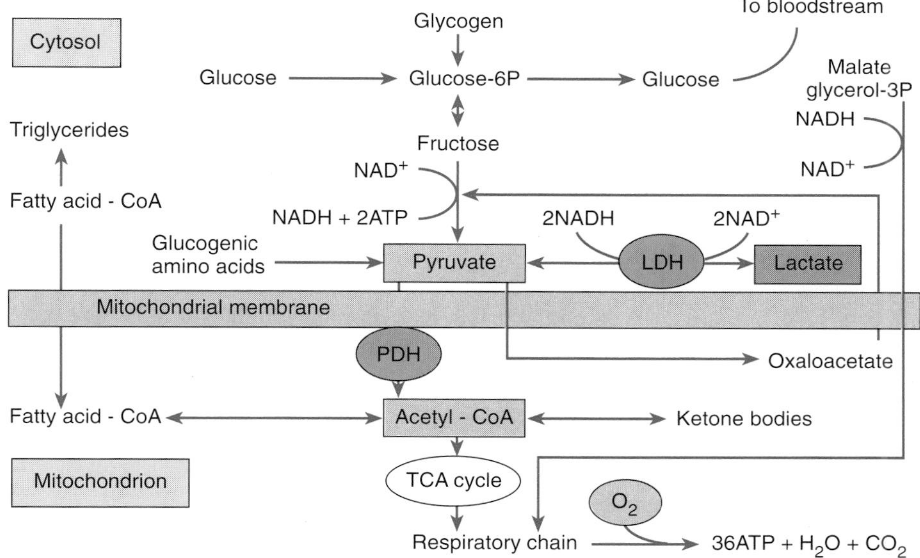

Fig. 2 Metabolism of lactate. Increased glycolysis or accumulation of pyruvate or cytosolic NADH favors metabolism of pyruvate to lactate, particularly if pyruvate cannot be metabolized in the mitochondrion, as in anaerobic conditions. Cytosolic NAD^+ is regenerated by transportation of malate and glycerol-3-phosphate to the mitochondrion, where they are oxidized through the respiratory chain. Anaerobic metabolism of 1 mol of glucose to lactate yields only 2 mol of ATP, whereas aerobic metabolism produces 38 mol of ATP. H^+ is produced by hydrolysis of ATP. (LDH, lactate dehydrogenase; PDH, pyruvate dehydrogenase; TCA cycle, tricarboxylic acid cycle.)

monoxide (CO) intoxication. Rarely, hyperlactatemia is caused by severe anemia, malignancies, and CO poisoning.

There are several potential non-hypoxic mechanisms of lactic acidosis (Table 2) in critically ill patients. Hyperlactatemia can also be caused by decreased metabolism by liver and kidney. Therefore mild hyperlactatemia can occur in patients who have hepatic failure, cirrhosis, or renal failure. Non-hypoxic lactic acidosis is a more important mechanism in the ICU than is generally appreciated.

The primary pathophysiological abnormality of ketoacidosis is low insulin concentration and high levels of counter-regulatory hormones. In the absence of insulin, fatty acids are oxidized in the liver to keto-acids, primarily acetoacetate and β-hydroxybutyrate. This response is normal in the fasting state. Thus starvation induces modest keto-acidosis which is self-limiting. Diabetic ketoacidosis is caused by severe insulinopenia and hyperglucagonemia. Furthermore, keto-acidosis is exacerbated by associated volume depletion.

Alcoholic ketoacidosis can be serious and is due to the combined effects of starvation (as described above) and alcohol withdrawal (which induces a marked elevation of counter-regulatory hormones). Alcoholic ketoacidosis is frequently accompanied by hypoglycemia

Table 2 Hypoxic and non-hypoxic mechanisms of lactic acidosis

Mechanism	Example
Hypoxic	
Global ischemia	Shock states Severe anemia Severe arterial hypoxemia CO poisoning
Focal ischemia	Gastrointestinal ischemia
Non-hypoxic	
Delayed clearance of lactate	Hepatic and/or renal dysfunction
Arteriovenous or physiological peripheral shunts	True anatomical peripheral shunts ?Physiological peripheral shunts in sepsis due to pathological vasodilation
Dysfunction of pyruvate dehydrogenase	Sepsis, endotoxemia Thiamine deficiency High catecholamine levels (e.g. stress, pheochromocytoma)
Uncoupling of oxidative phosphorylation	Cyanide, salicylate, biguanides, methanol metabolites Severe phosphate depletion Congenital enzyme defects
Protein degradation: amino acid conversion to pyruvate and lactate	Severe catabolism
Accelerated aerobic glycolysis	Sepsis Seizures Salicylate Large fructose loads (in parenteral nutrition) Malignancies

and lactic acidosis (due to volume depletion and ethanol-induced alteration in the redox ratio).

Renal failure initially reduces ammoniagenesis, which may cause a hyperchloremic normal anion gap acidosis. The increased anion gap acidosis of renal failure is caused by accumulation of unmeasured anions, such as phosphate and sulfate, but particularly the anions of non-volatile endogenous organic acids (such as peptides and amino acid derivatives). There is concomitant reduction in HCO_3^- owing to buffering in the face of diminished HCO_3^- reclamation and regeneration. In chronic renal failure, HCO_3^- is stabilized by the enormous buffering capacity of bone (Emmett *et al.* 1992; Schmidt 1992).

Poisoning with salicylate causes complex acid–base disturbances (Barnett and Schmidt 1992; Emmett *et al.* 1992). Salicylate itself is an organic acid, but it also stimulates respiratory centers, resulting in hyperventilation and respiratory alkalosis. The latter induces glycolysis, further accelerated by hypoglycemia (due to salicylate-stimulated insulin secretion). There are several causes of increased anion gap acidosis in salicylate poisoning. Salicylate uncouples oxidative phosphorylation, thus inhibiting normal aerobic metabolism and causing a mild lactic acidosis. In addition, ketoacidosis may develop. Thus the increased anion gap of salicylate poisoning is caused by accumulation of salicylate, lactate, ketoacids, and other organic acids.

The toxicity of alcohol intoxications is related to accumulation of their metabolites (Barnett and Schmidt 1992; Emmett *et al.* 1992; Schmidt 1992). Metabolism of methanol produces formaldehyde and formic acid, which causes an increased anion gap acidosis. Formate also inhibits cytochrome oxidase and induces lactic acidosis. Ethylene glycol is oxidized to glycolic and oxalic acids.

Paraldehyde is a rare cause of increased anion gap acidosis. It is metabolized to acetaldehyde and subsequently acetic acid, but this does not seem to account for the acidosis (Emmett *et al.* 1992).

Normal anion gap metabolic acidosis

Normal anion gap acidosis is also quite common in critically ill patients and may coexist with increased anion gap acidosis. It is characterized by hyperchloremia and is due to HCO_3^- loss and Cl^- gain. Thus it is also called hyperchloremic acidosis. Normal anion gap acidosis can be classified by the presence or absence of normal urinary NH_4^+ excretion, which can be detected by calculating the urinary anion gap as follows:

$$\text{urinary anion gap} = ([Na^+] + [K^+]) - [Cl^-].$$

This value is negative in the presence of normal (unmeasured) NH_4^+ excretion, as in diarrhea with gastrointestinal HCO_3^- loss. It is positive or neutral in the absence of renal acidification and NH_4^+ excretion. For example, this occurs with a distal renal tubular dysfunction due to mineralocorticoid defect. Thus calculation of the urinary anion gap is occasionally useful in assessing mixed or complex metabolic acidosis in the critically ill.

Proximal renal tubular acidosis is a rare cause of normal anion gap acidosis in the critically ill and is due to a defect in proximal tubule secretion of H^+ and reclamation of HCO_3^- (type II renal tubular acidosis) (Emmett *et al.* 1992; Schmidt 1992). Bicarbonate loss is offset by normal ammoniagenesis and NH_4^+ excretion. Proximal renal tubular acidosis may be due to HCO_3^- wasting in isolation or with other tubular resorption abnormalities (Fanconi syndrome), inhibition of carbonic anhydrase by acetazolamide, other disorders of amino acid and carbohydrate metabolism, Wilson's disease, myeloma (due to the toxic effects of Bence-Jones and other proteins on tubular cells), hyperparathyroidism, vitamin D deficiency, defects or inhibition by acetazolamide or Sulfamylon® of tubular carbonic anhydrase, and interstitial renal diseases (such as Sjögren's syndrome, amyloidosis, renal transplant rejection, and renal vein thrombosis).

Excessive loss of HCO_3^- in the gastrointestinal tract also causes normal anion gap acidosis. Bicarbonate is lost due to diarrhea, pancreatic or biliary fistulas, ureteric diversions to the gastrointestinal tract, particularly ureterosigmoidostomy, and vomiting in the presence of achlorhydria (Emmett *et al.* 1992; Schmidt 1992).

Less common causes of hyperchloremic acidosis with normal NH_4^+ excretion include recovery from chronic hypocapnia (associated with a HCO_3^- diuresis), administration of HCl and HCl precursors, such as NH_4Cl and Cl^--containing amino acids in parenteral nutrition, and administration of large volumes of HCO_3^--depleted solutions (such as normal saline) in the setting of volume contraction (Emmett *et al.* 1992).

Distal renal tubular acidosis is another uncommon cause of normal anion gap acidosis in the critically ill. It may be primary or secondary, due to familial hypercalciuria and nephrocalcinosis, hyperparathyroidism, hypervitaminosis D, hyperthyroidism, and tubular epithelial defects caused by drugs (such as amphotericin B, lithium, toluene, analgesics, particularly phenacetin and acetaminophen (paracetamol), and cyclamates). Tubulointerstitial defects, such as those occurring with polyclonal gammopathies, Sjögren's syndrome, chronic renal transplant rejection, chronic obstructive uropathy, and chronic active hepatitis, are important causes.

Hyperkalemic renal tubular acidosis is caused by a mineralocorticoid defect resulting from impaired adrenal gland aldosterone synthesis (i.e. adrenal insufficiency), reduced angiotensin II stimulation of aldosterone (as occurs in hyporeninemia in diabetes or pharmacological prostaglandin or angiotensin-converting enzyme inhibition), and renal resistance to mineralocorticoid (from receptor abnormalities, tubular destruction, or pharmacological interference by spironolactone, amiloride, and cyclosporin). In hyperkalemic renal tubular acidosis, reduced mineralocorticoid activity decreases Na^+ resorption and K^+ and H^+ secretion, which therefore causes impaired acid excretion, reduced tubular trapping of NH_3 as NH_4^+, volume depletion, and hyperkalemia.

References

Barnett, V. and Schmidt, G. (1992). Acid–base disorders. In *Principles of critical care* (ed. J.B. Hall, G.A. Schmidt, and L.D.H. Wood), pp. 1954–65. McGraw-Hill, New York.

Emmett, M., Alpern, R.J., and Seldin, D.W. (1992). Metabolic acidosis. In *The kidney: physiology and pathophysiology* (2nd edn) (ed. D.W. Seldin and G. Giebisch), pp. 2759–836. Raven Press, New York.

Schmidt, G.A. (1992). Acid–base and electrolyte homeostasis. In *Principles of critical care* (ed. J.B. Hall, G.A. Schmidt, and L.D.H. Wood), pp. 52–83. McGraw-Hill, New York.

Stacpoole, P.W. (1993). Lactic acidosis. *Endocrinology and Metabolism Clinics of North America*, **22**, 221–45.

Tønnessen, T.I. (1995). Intracellular pH and electrolyte regulation. In *Textbook of critical care* (ed. S. Ayres, A. Grenvik, P. Holbrook, and W. Shoemaker), pp. 172–87. W.B. Saunders, Philadelphia, PA.

8.3.2 Therapeutic strategy

D. M. Forrest and J. A. Russell

Key messages

- Metabolic acidosis is classified as increased or normal anion gap acidosis.
- Increased anion gap acidosis is common in the intensive care unit, is often serious, and has a short differential diagnosis in the critically ill.
- The cornerstone of management of metabolic acidosis is correction of the underlying pathophysiological cause.
- Sodium bicarbonate is not necessary in the management of most patients who have metabolic acidosis. Exceptions include profound metabolic acidosis and some specific causes of metabolic acidosis

Diagnostic approach to metabolic acidosis

The approach to the management of metabolic acidosis is based on correcting underlying pathogenic processes.

Metabolic acidosis is the consequence of bicarbonate (HCO_3^-) loss or hydrogen ion (H^+) gain. The normal respiratory compensation decreases carbon dioxide (CO_2) tension in the arterial blood ($Pa\text{CO}_2$), such that it falls by 0.134 kPa (1 mmHg) for every 1 mmol/l decrease in HCO_3^- concentration. If $Pa\text{CO}_2$ is not appropriate for the HCO_3^- concentration, another primary respiratory acidosis or alkalosis is diagnosed. Inadequate respiratory compensation (so that $Pa\text{CO}_2$ is higher than expected) indicates respiratory acidosis, which is usually indicative of patient fatigue or interference with ventilatory drive. Because of the critical importance of respiratory compensation in the mitigation of changes to systemic pH, impairment of ventilation heralds marked worsening of acidemia, may indicate impending respiratory collapse, and requires prompt correction.

Metabolic acidoses can be classified as increased or normal anion gap (Table 1). The anion gap ($[Na^+]-\{[Cl^-]+[HCO_3^-]\}$) should always be determined to direct appropriate management. The normal anion gap is 12 ± 2 mmol/l. However, the normal anion gap may be lower in critically ill patients because of hypoalbuminemia. The normal range of the anion gap is reduced by 3 mmol/l for every 10 g/l reduction in albumin. Other causes of a lowered normal anion gap (e.g. increased unmeasured cations with myeloma, lithium poisoning, or severe hypermagnesemia) are unusual in the critically ill (Emmett *et al.* 1992).

Approach to increased anion gap metabolic acidosis

An approach to the assessment of increased anion gap acidosis is summarized in Table 2.

Ketoacidosis can be diagnosed by measurement of ketones in serum and urine; blood glucose should also be measured. Ketoacidosis generates both acetone and β-hydroxybutyrate. However, only acetone is detected when measuring urine and serum ketones. If both acetone and β-hydroxybutyrate are present, increased ketone levels are detected. If only β-hydroxybutyrate is present, ketone levels are normal. In conditions of tissue hypoxia, the major ketone is β-hydroxybutyrate, not acetone. Therefore, in a patient who has both diabetic ketoacidosis and tissue hypoxia (e.g. in the setting of concomitant shock), serum ketones may be normal and the diagnosis of diabetic ketoacidosis may be missed. Lactic acidosis may occur in conjunction with diabetic ketoacidosis.

Uremia is detected by elevated urea and creatinine. However, severe protein depletion suppresses these levels, hepatic disease impairs urea formation, and blood in the gastrointestinal tract raises urea concentration. Salicylate poisoning can be recognized by increased serum salicylate levels.

Alcohol poisoning which causes increased anion gap acidosis (methanol and ethylene glycol) also increases the osmole gap. With increased anion gap acidosis, the osmole gap (Halperin and Goldstein 1988; Emmett *et al.* 1992) should always be assessed as follows:

$$\text{osmole gap} = (\text{measured osmolarity}) - (2[Na^+] + [\text{urea}] + [\text{glucose}] + [\text{ethanol}])$$

where the osmole gap is expressed in milliosmoles per liter and $[Na^+]$, [urea], [glucose], and [ethanol] are in millimoles per liter.

Table 1 Different forms of metabolic acidosis

Increased anion gap (mnemonic, KUSMAL)	Normal anion gap (mnemonic, If NoGO P, you're DeHD)
Ketoacidosis	**No**rmal NH_4^+ excretion
Uremia	**P**roximal RTA
Salicylate poisoning	**G**astrointestinal HCO_3^- loss
Methanol	**O**ther (postchronic hypocapnia, HCl, HCl precursors, and Cl^- administration)
Other **A**lcohols and toxins	**De**creased NH_4^+ excretion
Lactate	**H**yperkalemic RTA
	Distal RTA

RTA, renal tubular acidosis.

Table 2 Diagnosis of increased anion gap metabolic acidosis

Etiology	Diagnostic test
Ketoacidosis	Ketones (urine and serum) Blood glucose
Uremia	Urea, creatinine
Salicylate poisoning	Salicylate (+acetaminophen (paracetamol))
Methanol and other alcohols	Serum osmolar gap Urinalysis, fundoscopy
Lactate	Serum lactate
Ancillary tests	Electrolytes, albumin, ethanol Ionized Ca^{2+}, Mg^{2+}, PO_4^{2-}

The normal osmole gap is less than 10 mosmol/l. Methanol and ethylene glycol poisoning (but not ethanol intoxication) increase the osmole gap. Other clues to the specific form of alcohol poisoning include the presence of dilated pupils with decreased light reflex (methanol), hyperemia of the optic disk and retinal edema (methanol), and oxalate crystalluria (ethylene glycol).

Approach to normal anion gap metabolic acidosis

The approach to metabolic acidosis with a normal anion gap is shown in Fig. 1. Note that it is first important to ensure that the anion gap is indeed normal and not spurious (i.e. not greater than normal when corrected for albumin). Note also that urinary electrolyte determinations in the intensive care unit (ICU) are often confounded by concomitant administration of medications, particularly diuretics.

Management of specific forms of metabolic acidosis

The four steps of management of metabolic acidosis are primary resuscitation, assuring adequate ventilation, correction of the underlying cause of acidosis, and finally consideration of alkali therapy. Primary resuscitation (assurance of airway, breathing, circulation, etc.) is fundamental to managing critically ill patients who have metabolic acidosis. Then, assurance of adequacy of ventilation is paramount. Once the correct diagnosis has been established, management is directed to correcting the underlying cause of the acidosis. Finally, exogenous HCO_3^- has a role in some forms of acidosis. However, in others, bicarbonate is not beneficial and may be potentially harmful.

Increased anion gap metabolic acidosis

Correction of underlying pathophysiology is perhaps most critical in lactic acidosis (Arieff 1991; Emmett *et al.* 1992; Stacpoole 1993). In the critically ill, pathogenesis is often related to hypoxic and non-hypoxic mechanisms. Oxygen delivery and tissue perfusion must be improved with blood and/or volume infusion, inotropes, and pressors, and oxygen consumption must be limited. Therapy should be instituted to correct non-hypoxic causes, including treatment of sepsis, administration of thiamine, treatment of concomitant salicylate, methanol, ethylene glycol, cyanide or carbon monoxide toxicity, insulin administration in diabetic ketoacidosis, and elimination of sorbitol and fructose in parenteral alimentation solutions.

The use of bicarbonate for treatment of lactic acidosis is controversial because bicarbonate does not improve cardiovascular performance and has important complications, including hypokalemia, hypocalcemia, increased CO_2 production, and sodium and volume overload. In a prospective randomized double-blind trial by Cooper *et al.* (1990) comparing sodium bicarbonate with equimolar sodium chloride in critically ill patients who had lactic acidosis, there were no differences between sodium bicarbonate and sodium chloride in hemodynamic effects. However, sodium bicarbonate significantly increased $P\text{aCO}_2$ and significantly decreased ionized calcium levels. The lack of benefit of sodium bicarbonate occurred even in patients who had profound lactic acidosis (i.e. arterial pH < 7.1) (Stacpoole 1993). Therefore we recommend that bicarbonate is rarely, if ever, useful in the treatment of lactic acidosis.

Exogenous bicarbonate is converted to CO_2 which diffuses freely across cell membranes and causes intracellular acidosis. Hence CO_2 levels can increase after bicarbonate administration, particularly if CO_2 elimination is impaired by decreased cardiac function and

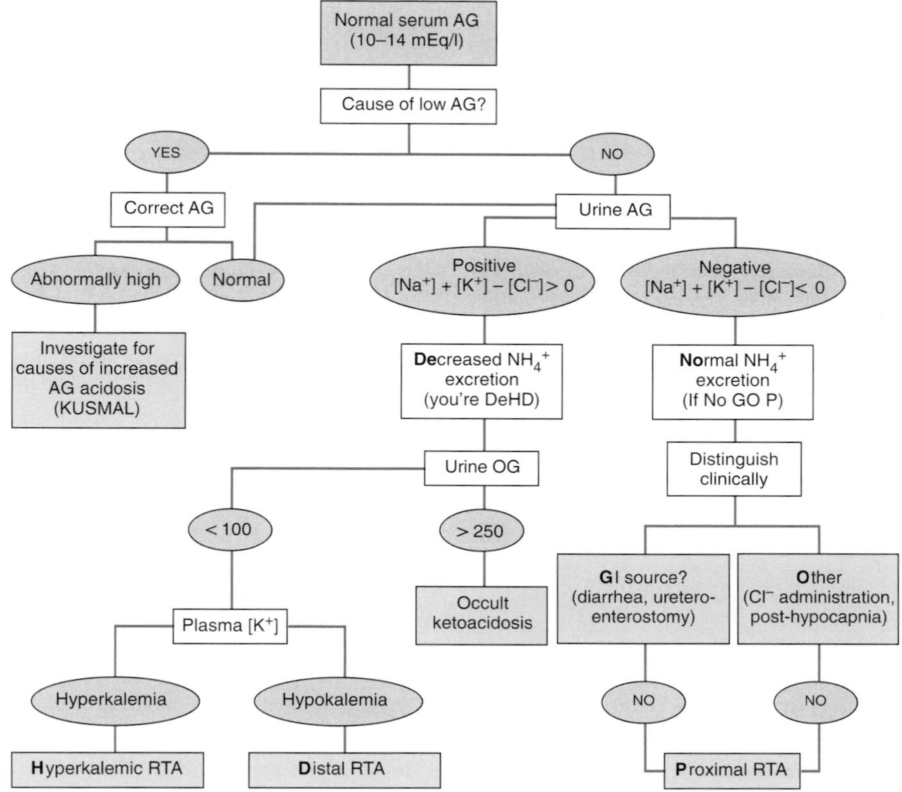

Fig. 1 Approach to normal anion gap metabolic acidosis. The mnemonic for normal anion gap (AG) acidosis is 'if No GO P, you're DeHD'. If the serum AG is normal and there are no features suggesting a cause of a low serum AG, the urinary AG should be calculated from urinary electrolytes. If the urine AG is positive, it is the result of either diminishing NH_4^+ secretion or the presence of osmotically active ketoacids. This can be distinguished by determining the urinary osmolar gap (OG): OG = (measured urinary osmolality) – {2([Na^+]$_u$ + [K^+]$_u$) + (urinary urea) + (urinary glucose)}. The clinical situation and/or plasma [K^+] determine probable final diagnosis, but special tests may be necessary to confirm this or to differentiate further various forms of renal tubular acidosis (RTA).

diminished ventilatory capacity. Further, arterial alkalinization does not necessarily effect alkalinization in ischemic tissues, where pH may be influenced more by CO_2 production than by arterial bicarbonate level (Arieff 1991; Emmett *et al.* 1992; Stacpoole 1993).

Alternative means of alkalinization which do not have the same side-effects as bicarbonate have been investigated (Arieff 1991; Emmett *et al.* 1992; Stacpoole 1993). Agents which do not generate CO_2, or at least not to the same extent as bicarbonate, have theoretical advantages over bicarbonate. These agents include sodium carbonate, Carbicarb® (a combination of equimolar amounts of sodium carbonate and bicarbonate), THAM (tris(hydroxymethyl)-aminomethane), and Tribonate® (a combination of sodium bicarbonate, THAM, phosphate, and acetate). None has been evaluated adequately to recommend its clinical use in the critically ill. Dichloroacetate can also be used to correct lactic acidosis. Dichloroacetate increases pyruvate dehydrogenase activity, lowers lactate levels, and corrects lactic acidosis. However, in a prospective randomized clinical trial, there was no difference in mortality of patients with lactic acidosis (lactate > 5 mmol/l) who were treated with dichloroacetate compared with those treated with bicarbonate (Stacpoole 1993).

The fundamental principles of management of diabetic ketoacidosis are correction of volume contraction by infusion of normal saline, correction of hyperglycemia and ketoacidosis by infusion of insulin, and frequent clinical assessment and monitoring of glucose, sodium, and potassium to guide glucose, sodium, and potassium therapy. One of the hallmarks of diabetic ketoacidosis is volume contraction (Halperin and Goldstein 1988; Emmett *et al.* 1992), which can precipitate concomitant circulatory collapse and lactic acidosis. Hence adequate volume resuscitation is essential. Since the process is the consequence of relative insulinopenia with elevated counter-regulatory hormones, administration of insulin is a critical component of therapy to correct hyperglycemia and to halt ketone production. Interruption of insulin administration should be avoided. When blood glucose is 15 mmol/l or less, glucose-containing intravenous solutions should be substituted for normal saline.

Ketoacids are converted to HCO_3^- during insulin infusion, which is one reason why exogenous bicarbonate is unnecessary in the management of diabetic ketoacidosis. Because of the potentially deleterious effects of acidemia, some authors recommend exogenous bicarbonate if the pH is markedly depressed (e.g. pH < 7.1), particularly if there is hemodynamic compromise. However, exogenous HCO_3^- also causes side-effects and complications, as discussed under lactic acidosis above. In diabetic ketoacidosis, in particular, exogenous bicarbonate can precipitate dangerous hypokalemia, hypocalcemia, paradoxical intracellular acidosis in the myocardium and brain, hypernatremia, and volume overload (Arieff 1991; Stacpoole 1993).

Overcorrection of acidemia may produce alkalemia (particularly as metabolism of ketoacids and lactate converts them to further HCO_3^-), which causes the oxyhemoglobin dissociation curve to shift to the left, diminishes 2,3-diphosphoglycerate production, and promotes its breakdown, inhibiting oxygen unloading to tissues. This may worsen tissue hypoxia. Alkalemia also favors glycolysis, which increases pyruvate and hence lactate formation. An increase in pH may also cause hypokalemia and hypocalcemia, potentially complicated by dysrhythmias. Finally, if administered as a bolus, sodium bicarbonate is hypertonic and may induce circulatory overload with pulmonary edema, hypernatremia, and hyperosmolarity with cerebral damage. Indeed, in clinical studies in ketoacidosis, bicarbonate use has not been beneficial and has been associated with hypokalemia. While some authors advocate its judicious use in cases of extreme acidemia, we advocate not using it at all, preferring optimization of corrective therapies and supportive measures. We recommend treatment using volume infusion and insulin, and recommend bicarbonate only for profound metabolic acidosis. Other alkalinizing agents (see above) have not been studied.

The foundation of management of alcoholic ketoacidosis is glucose infusion, volume repletion, thiamine, potassium, phosphate, and magnesium infusion, and, less often than in diabetic ketoacidosis, insulin infusion. Insulin is usually unnecessary and may be dangerous, since glycogen depletion predisposes to hypoglycemia (Emmett *et al.* 1992). Glucose (to interrupt ketogenesis and replete glycogen stores) and volume repletion are essential. Potassium deficits must be anticipated and are often worsened as a result of simultaneous magnesium depletion (which may not be reflected by serum levels) due to malnutrition and renal wasting. Hypophosphatemia is also common. Thiamine deficiency is frequent and can exaggerate lactate production (which may be induced by an increased redox ratio due to metabolism of ethanol).

Treatment of the metabolic acidosis of uremia focuses on general non-dialytic and dialytic management of renal failure. Uremic metabolic acidosis is most efficiently treated by dialysis (Halperin and Goldstein 1988; Emmett *et al.* 1992), particularly in oliguric renal failure. Metabolic acidosis is usually less severe in non-oliguric renal failure and can often be managed without exogenous bicarbonate. Bicarbonate may be indicated for severe acidemia (e.g. pH $\leqslant$ 7.1) if there is hemodynamic compromise and if there is profound hyperkalemia with electrocardiographic abnormalities. Sodium bicarbonate should be used cautiously, because of the increased risk of volume overload with impaired renal function.

The basic principles of the treatment of the metabolic acidosis of salicylate poisoning are interruption of absorption of salicylate, cardiopulmonary support, infusion of potassium and sodium bicarbonate, thiamine supplementation, and dialysis. The approach to salicylate poisoning can be remembered by the mnemonic GASP'DT (**G**lucose, **A**lkali, **S**aline, **P**otassium, **D**ialysis, **T**hiamine). Hyponatremia may worsen the lactic acidosis because of cellular enzymatic impairment by salicylate and respiratory alkalosis. Hence saline volume resuscitation is essential. A saline diuresis also increases salicylate excretion, which is further enhanced by urinary alkalinization.

Sodium bicarbonate administration is indicated in salicylate toxicity because alkalinization of blood prevents absorption of salicylate into the brain. A low pH favors the non-ionized form of salicylate, which diffuses readily across cell membranes. Thus alkalinization of the blood bars salicylate entry to the central nervous system (where it exerts serious toxicity). Secondly, maintenance of high urine pH traps salicylate in the renal tubule, facilitating its removal. Finally, hemodialysis effectively removes salicylate and should be considered early in patients who have severe salicylate toxicity (serum salicylate > 100 mmol/l), oliguric renal failure, and profound metabolic acidosis (because this prevents urinary clearance of salicylate).

Glucose administration is particularly important in salicylate toxicity because of possible cerebral hypoglycorrhachia, which aggravates salicylate toxicity. Potassium and magnesium deficits should be repleted. Thiamine should be administered to prevent worsening of hyperlactatemia should the patient be deficient.

Therapy of methanol and ethylene glycol intoxication is based on preventing formation of toxic products and on their removal. In patients who have an increased anion gap acidosis and an increased osmole gap, exogenous ethanol infusion should be administered early before final confirmation of methanol or ethylene glycol toxicity. Methanol and ethylene glycol cause 'drunkenness' similar to that caused by ethanol (hence their consumption), but have no other direct toxicity. Infusion of exogenous ethanol inhibits metabolism of methanol and ethylene glycol by alcohol dehydrogenase because ethanol is oxidized preferentially by alcohol dehydrogenase. Methanol, ethylene glycol, and their toxic metabolites are removed by dialysis (Halperin and Goldstein 1988; Emmett *et al.* 1992). The alcohol dehydrogenase inhibitor 4-methylpyrazole may be of benefit, but is not yet generally available. Folate supplementation is necessary in methanol poisoning to facilitate formate breakdown, and pyridoxine and thiamine have similar effects on glyoxalate metabolism in ethylene glycol overdose. Thiamine should be considered in all patients with such overdose to counteract the impairment by alcohol metabolites of cellular machinery, which promotes lactic acidosis.

Exogenous bicarbonate may be of use in preventing toxicity of methanol and ethylene glycol by trapping ionized metabolites in the blood (as in salicylate poisoning), but dialysis removes these toxins more effectively. Therefore dialysis should never be delayed in preference for alkalinization. In severe ethylene glycol poisoning, alkalemia may also precipitate hypocalcemia.

Normal anion gap metabolic acidosis

These conditions are less common and less life-threatening in the ICU setting. In general, they are best treated by reversal of underlying processes. Where this is not possible, and particularly with severe acidemia, bicarbonate is often indicated, especially for renal tubular acidosis (Emmett *et al.* 1992). Correction of hyperkalemia is more important in hyperkalemic distal renal tubular acidosis. Exogenous bicarbonate may also be indicated in hyperchloremic acidosis associated with gastrointestinal HCO_3^- loss, particularly that due to pancreaticobiliary fistulas and ureteroenterostomies (Emmett *et al.* 1992). Acetazolamide, nicotinic acid, and citrate may also be of use.

References

Arieff, A.I. (1991). Indications for use of bicarbonate in patients with metabolic acidosis. *British Journal of Anaesthesia*, **67**, 165–77.

Cooper, D.J., Walley, K.R., Wiggs, B.R., and Russell, J.A. (1990). Bicarbonate does not improve hemodynamics in critically ill patients who have lactic acidosis. *Annals of Internal Medicine*, **112**, 492–8.

Emmett, M., Alpern, R.J., and Seldin, D.W. (1992). Metabolic acidosis. *The kidney: physiology and pathophysiology* (2nd edn) (ed. D.W. Seldin and G. Giebisch), pp. 2759–836. Raven Press, New York.

Halperin, M.L. and Goldstein, M.B. (1988). Metabolic acidosis. In *Fluid, electrolyte and acid–base emergencies*, pp. 40–96. W.B. Saunders, Philadelphia, PA.

Stacpoole, P. (1993). Lactic acidosis. *Endocrinology and Metabolism Clinics of North America*, **22**, 221–45.

8.4 Metabolic alkalosis

8.4.1 Pathophysiology and causes

S. Brimioulle and R. J. Kahn

Key messages

- Metabolic alkalosis is the most common acid–base disorder in hospital patients.
- It is associated with significant respiratory, cardiovascular, and neurological complications.
- Metabolic alkalosis is generally initiated by alkali administration, digestive losses, and diuretic therapy.
- Subsequently it is maintained by blood volume depletion, potassium depletion, and chloride depletion.

Introduction

Metabolic alkalosis is the most frequently observed acid–base abnormality in hospital patients and in postoperative patients. In a study of 13 430 acid–base analyses, it has been identified in 40 per cent of all samples and in 51 per cent of abnormal samples. Metabolic alkalosis causes significant clinical and metabolic complications, and thus deserves early recognition and appropriate management. More detailed discussions are given by Harrington and Kassirer (1982), Jacobson and Seldin (1983), Sabatini and Kurtzman (1994), and Galla (1995).

Definitions

The term alkalosis refers to the abnormal processes tending to decrease the amount of acid or to increase the amount of alkali in the organism, and thus to increase blood pH. Respiratory alkalosis results from a primary decrease in $Pa\text{CO}_2$, and metabolic alkalosis from a primary increase in bicarbonate concentration. Alkalemia refers to a blood pH value above 7.45, regardless of its mode of generation.

In the normal subject, acidosis and alkalosis elicit an adaptative response with tends to bring the pH back to its normal value (Table 1). Metabolic alkalosis causes a decrease in ventilation and an increase in $Pa\text{CO}_2$. Reference to this secondary change as respiratory acidosis should be avoided because it is a normal compensatory response and not an abnormal process due to respiratory dysfunction.

Compensatory responses are essentially proportional to the primary acid–base disturbance and are effective in limiting the change in pH. However, two points should be kept in mind. First, respiratory responses are rapid and reach their maximum within hours, whereas renal responses are complete after only 3 to 5 days. For example, if $Pa\text{CO}_2$ rises to 70 mmHg, arterial blood pH first decreases to 7.20 (acute condition) and then progressively increases to reach 7.32 (chronic condition). Therefore the recent history must be taken into account for the correct assessment of compensation. Second, although compensatory responses limit the change in pH, they do not bring the pH back to normal. A pH of 7.40 in a patient with chronic hypercapnia of 70 mmHg should not be interpreted as an appropriate response because of the so-called normal pH, but as a superimposed metabolic alkalosis. Similarly, 'overcompensation' does not occur and must be interpreted as indicating the presence of another acid–base disturbance.

Table 1 Normal compensatory responses to primary acid–base disturbances

Acid–base disturbance	Expected secondary response
Metabolic acidosis	$\Delta Pa\text{CO}_2 = 1.20\ (0.16) \times \Delta\text{HCO}_3^-$
Metabolic alkalosis	$\Delta Pa\text{CO}_2 = 0.70\ (0.09) \times \Delta\text{HCO}_3^-$
Respiratory acidosis, acute	$\Delta\text{HCO}_3^- = 0.10\ (0.75) \times \Delta Pa\text{CO}_2$
Respiratory acidosis, chronic	$\Delta\text{HCO}_3^- = 0.35\ (2.60) \times \Delta Pa\text{CO}_2$
Respiratory alkalosis, acute	$\Delta\text{HCO}_3^- = 0.20\ (1.50) \times \Delta Pa\text{CO}_2$
Respiratory alkalosis, chronic	$\Delta\text{HCO}_3^- = 0.50\ (3.75) \times \Delta Pa\text{CO}_2$

Data from *in vivo* studies in humans or intact animals submitted to an isolated primary acid–base disturbance. The term 'chronic' refers to the delay of 3 to 5 days required to reach a new steady state. Bicarbonate concentration (HCO_3^-) is expressed in mmol/l and $Pa\text{CO}_2$ in mmHg or kPa (value in parentheses).

Generation and maintenance

Metabolic alkalosis is primarily due to either a gain in alkali or a loss of non-respiratory acid. The most common causes are listed in Table 2.

In normal conditions, kidneys are able to excrete large amounts of bicarbonate and rapidly reverse the alkalosis. Therefore alkalosis only persists if, for any reason, the kidneys contribute to maintaining the bicarbonate excess. The principal maintenance factors are chloride

Table 2 Generation of metabolic alkalosis

Mechanism	Example
Alkali intake or administration	
Bicarbonate	Treatment of metabolic acidosis
Citrate	Blood transfusion, plasmapheresis
Carbonate	Antacids, milk alkali syndrome
Acetate	Dialysis, parenteral nutrition
Acid losses, digestive	
Gastric	Vomiting, gastric suction
Fecal	Chloride-losing diarrhea
Acid losses, urinary	
Increased tubular sodium content	Diuretics
Hypercapnia	Post-hypercapnic alkalosis
Hyperaldosteronism	Hypovolemia, Cushing's syndrome
Non-reabsorbable anions	Penicillin, carbenicillin

depletion, decreased effective blood volume, potassium depletion, and mineralocorticoid stimulation. Chloride depletion is the most common and most important factor, and its prominent role has been emphasized repeatedly. Decreased glomerular filtration rate is usually observed but is not mandatory, at least in some models. The exact mechanisms by which these factors contribute to maintaining the alkalosis and their relative importance in different conditions has yet to be clarified. In many cases, and particularly in critically ill patients, several maintenance factors are present and contribute to the typical picture of hypochloremic hypokalemic metabolic alkalosis. Identification of maintenance factors is necessary because they can maintain the alkalosis even when the generating factors have disappeared.

Alkali administration

Isolated alkali ingestion or administration only causes persistent metabolic alkalosis when the absorption is massive and prolonged. Much more commonly, metabolic alkalosis occurs after alkali administration in patients who present or develop oliguric renal failure. Typical situations in critically ill patients are abundant transfusions of citrated blood for hemorrhagic shock, or sodium bicarbonate therapy for metabolic acidosis due to circulatory arrest. Considerable alkalosis may develop when the excess lactate is metabolized and bicarbonate cannot be eliminated. In patients with renal failure, metabolic alkalosis can also result from the administration of lactate in the hemofiltration replacement fluid.

Digestive loss of acid

Gastric alkalosis occurs when substantial amounts of acid gastric juice are eliminated from the organism by vomiting or gastric suction. Chloride is lost together with hydrogen ions. Bicarbonate generated during acid secretion is not neutralized by more distal buffering in the gut, and blood bicarbonate concentration increases above the renal reabsorption ability. Bicarbonaturia is associated with natriuria, kaliuria, and fluid excretion aggravating the gastric electrolyte and fluid losses. Whether, how, and how much the chloride, sodium, potassium, and fluid deficits individually contribute to the maintenance of the alkalosis has been the subject of much debate. The question is of considerable pathophysiological interest, but of less clinical importance because all the deficits can be corrected concomitantly to reverse the alkalosis.

Diarrhea due to metabolic defects and villous adenomas can cause metabolic alkalosis owing to failure of the the colonic mucosa to reabsorb chloride adequately. Diarrhea with normal chloride concentration can also cause metabolic alkalosis when large fluid, sodium, and potassium losses result in sodium and water retention at the expense of hydrogen ion and potassium excretion by the kidney.

Urinary loss of acid

Most diuretics generate metabolic alkalosis by directly causing losses of chloride, sodium, potassium, and fluid to the urine. Volume contraction, aldosterone secretion, chloride depletion, and potassium depletion all contribute in some way to enhancement of proximal bicarbonate reabsorption and/or distal proton excretion.

Post-hypercapnic alkalosis is a two-step process commonly observed in critical care units. First, sustained respiratory acidosis determines a compensatory retention of bicarbonate and a concomitant excretion of chloride by the kidney. If chloride is not available when the hypercapnia is reversed, the bicarbonate excess cannot be eliminated. Such a post-hypercapnic alkalosis commonly occurs when patients with congestive heart failure due to respiratory disease are submitted to sodium chloride restriction or diuretic therapy.

Mineralocorticoid excess alone causes mild metabolic alkalosis, but the alkalosis is much more serious when potassium depletion is also present. The mechanisms by which the two factors combine to generate the alkalosis remain to be fully clarified. Mineralocorticoid excess, potassium depletion, and chloride depletion all contribute to the alkalosis in Bartter's syndrome.

It has been reported that severe isolated potassium depletion generates metabolic alkalosis by causing renal chloride wasting. Penicillin and related dugs can generate mild metabolic alkalosis, possibly as non-reabsorbable anions. Hypercalcemia may be associated with metabolic alkalosis or acidosis.

Clinical manifestations

The compensatory response to metabolic alkalosis is a decrease in ventilation and an increase in $P\text{aCO}_2$. Multiple cases of severe hypercapnia that are clearly due to metabolic alkalosis have been reported. Whether the response is or is not always present, and whether its magnitude depends on the cause of the alkalosis or on associated factors, has been questioned in numerous studies. It is now accepted that the increase in $P\text{aCO}_2$ is proportional to the increase in bicarbonate concentration and is independent of the mode of generation of the alkalosis. $Pa\text{CO}_2$ normally increases by about 0.7 mmHg for each increase of 1 mmol/l in bicarbonate concentration and can exceed 80 mmHg (10.7 kPa) in severe cases. Hypoxemia is almost always present, and has generally been attributed to the decreased ventilation and the resulting atelectasis. Metabolic alkalosis also inhibits hypoxic pulmonary vasoconstriction, which contributes to the deterioration of pulmonary gas exchange and to increasing hypoxemia in patients with respiratory failure (Brimioulle *et al.* 1991).

Cardiovascular complications mainly consist of various cardiac arrhythmias: supraventricular and ventricular premature beats, supraventricular tachycardia, atrioventricular block, ventricular tachycardia, and fibrillation. The arrhythmias seem to result from the associated potassium depletion rather than from the increase in pH, and are enhanced by myocardial ischemia and digitalis therapy. Alkalosis has a small positive inotropic effect, at least in animal studies.

Neurological complications generally consist of altered orientation and comportment, decreased consciousness, and neuromuscular hyperexcitability. Coma and grand mal seizures may occur, but are generally precipitated by concomitant hypovolemia, hypercapnia, hypoxemia, or other metabolic disturbances (e.g. hyperammonemia in hepatic failure). Muscle weakness may occur in the presence of severe potassium depletion.

Alkalosis shifts the oxyhemoglobin dissociation curve to the left, thereby increasing oxygen uptake in the lung and decreasing oxygen release to peripheral tissues. Alkalosis also increases red cell 2,3-diphosphoglycerate concentration, which has an opposite effect and returns the dissociation curve to its original position after 6 to 8 h. Alkalosis increases the oxygen consumption, and increases the blood lactate concentration by a pH-induced enzymatic activation. Potassium depletion is almost invariably present in metabolic alkalosis and contributes to its clinical manifestations. Calcium concentration is typically normal, and the hyperexcitability results from decreased calcium ionization or an associated magnesium depletion. Alkalosis enhances the conversion of the ammonium ion to ammonia.

Mixed acid–base disorders are common in critically ill patients and should be actively sought. Particular attention should be paid to the combination of respiratory acidosis and metabolic alkalosis, which is very common but frequently misdiagnosed when the resulting pH remains close to 7.40. Spontaneous correction is unlikely, since each disturbance contributes to enhance the other. Metabolic alkalosis may cause serious alkalemia when the patient is mechanically ventilated, and serious hypercapnia and hypoxemia when weaning is attempted.

Diagnosis

Determination of urinary electrolyte concentrations is helpful for identifying causal and maintenance factors in metabolic alkalosis, and for monitoring the treatment efficiency. Chloride concentration allows patients to be separated into two major groups and to predict their response to therapy. Urinary chloride is generally lower than 10 mmol/l in patients with digestive losses, post-hypercapnic state, and after diuretic therapy. It is typically higher than 50 mmol/l (if chloride intake is preserved) in the case of alkali intake, urinary acid losses, and during diuretic therapy. Metabolic alkalosis is responsive to chloride in the former group, but is chloride resistant in the latter. Urinary pH is of little diagnostic value in metabolic alkalosis, particularly in critically ill patients.

References

Brimioulle, S., Vachiéry, J.L., Lejeune, P., Leeman, M., Mélot, C., and Naeije, R. (1991). Acid–base status affects gas exchange in canine oleic acid pulmonary edema. *American Journal of Physiology*, **260**, H1080–6.

Galla, J.H. (1995). Metabolic alkalosis. In *Fluid, electrolyte and acid–base disorders* (ed. A.I. Arieff and R.A. DeFronzo), Chapter 6, pp. 199–222. Churchill Livingstone, New York.

Harrington, J.T. and Kassirer, J.P. (1982). Metabolic alkalosis. In *Acid–base* (ed. J.J. Cohen and J.P. Kassirer), Chapter 9, pp. 227–306. Little, Brown, Boston, MA.

Jacobson, H.R. and Seldin, D.W. (1983). On the generation, maintenance, and correction of metabolic alkalosis. *American Journal of Physiology*, **245**, F425–32.

Sabatini, S. and Kurtzman, N.A. (1994). Metabolic alkalosis. In *Maxwell and Kleeman's clinical disorders of fluid and electrolyte metabolism* (5th edn) (ed. R.G. Narins), Chapter 29, pp. 933–56. McGraw-Hill, New York.

8.4.2 Therapeutic strategy

S. Brimioulle and R. J. Kahn

Key messages

- Prevent metabolic alkalosis: restrict alkali and diuretic therapy.
- Treat underlying disease and suppress generating factors.
- Identify and reverse all maintaining factors.
- If there are clinical complications or alkalosis is severe, give intravenous HCl.

Introduction

Metabolic alkalosis can result in serious clinical respiratory, cardiovascular, and neurological complications. Severe alkalemia is associated with increased mortality rate in critically ill patients. Thus effective prevention or therapy of metabolic alkalosis is warranted. More detailed discussions are given by Harrington and Kassirer (1982), Sabatini and Kurtzman (1994), and Galla (1995).

Prevention

Metabolic alkalosis commonly occurs in settings where it could have been anticipated. Therefore prevention of the generating factors is frequently possible, for example by limiting bicarbonate administration in acute metabolic acidosis, preventing acid losses due to gastric drainage (e.g. H_2 receptor antagonists), restricting diuretic administration, or providing chloride on reversal of sustained hypercapnia. Prevention of the development of maintenance factors may also be possible by providing sufficient amounts of fluids, sodium, potassium, and chloride.

Table 1 Clinical experience of HCl infusion in the treatment of metabolic alkalosis: administration modalities and complications

N	Concentration (mmol/l)	Rate (mmol/h)	Amount (mmol)	Complications
8	150–200	8–25	150–1200	*
20	100–200	?	8–1200	*
17	100	17–25	100–400	*
12	150	18	60–475	*
21	???–200†	4–15	60–2100	—
24	1000	60	Mean 90	—
15	250	25	150–300	—
15	200	8	200–450	—
15	250	25	75–300	—
8	1000	65	250–500	—

*Venous and/or cutaneous necrosis.
†HCl diluted in a solution of amino acids acting as buffer.

Generating factors

When metabolic alkalosis has developed, the initial treatment, whenever possible, is the correction of the underlying disease and elimination of the generating factors. Suppression of further bicarbonate generation or reabsorption may include the use of H_2 receptor antagonists or proton pump inhibitors in gastric alkalosis, the use of prostaglandin synthetase inhibitors or angiotensin-converting enzyme inhibitors in Bartter's syndrome, the use of antiemetic and antidiarrheic drugs in digestive losses, reduction or discontinuation of diuretic therapy, etc. Treatment of the underlying disease and elimination of generating factors should always be followed by the identification and reversal of maintenance factors, which can maintain the alkalosis even when the generating factors have disappeared.

Maintenance factors

Metabolic alkalosis is always characterized by a chloride deficit, except in the possible case of mild hypochloremia associated with fluid excess. Chloride depletion is typically associated with blood volume and potassium depletion, so that chloride is best administered as sodium chloride and potassium chloride solutions. The blood volume deficit is variable, and large amounts of sodium chloride should be given until blood volume and natremia are corrected (as shown by an increase in urinary sodium excretion if renal function is normal). The potassium deficit can reach 1000 mmol, and potassium urinary wasting will persist until the alkalosis is corrected. Potassium chloride should be infused at a rate of 150 to 300 mmol/day to restore and maintain the serum potassium concentration above 4 mmol/l. Rectal administration of calcium chloride has been proposed, owing to the chloride reabsorption ability of the colonic mucosa, but this route only allows limited amounts of fluids. When given in sufficient amounts, sodium chloride and/or potassium chloride solutions generally reverse the alkalosis within a few days. Although mineralocorticoid excess should be identified as a maintenance factor when present, its correction is less important in critically ill patients because it causes only mild metabolic alkalosis by itself. If reversal of maintenance factors is ineffective or is not possible (e.g. in renal or cardiac failure), or if severe clinical complications require rapid correction, more specific approaches can be considered.

Hemodialysis and hemofiltration

Hemodialysis and peritoneal dialysis, when required because of renal failure, are effective in reversing metabolic alkalosis if bicarbonate concentration is decreased and chloride concentration is increased in the dialysate. Similarly, continuous hemofiltration allows treatment of metabolic alkalosis by adapting the composition of solutions used to compensate fluid losses through the membrane. Metabolic alkalosis can be treated by other methods, so that these techniques are not required to correct the alkalosis in the absence of renal failure.

Acetazolamide

Acetazolamide, a carbonic anhydrase inhibitor, is a diuretic drug which enhances renal excretion of water, sodium, potassium, and bicarbonate. Although increased excretion of water, sodium, and potassium are generally unwelcome in metabolic alkalosis, acetazolamide has been proposed as a treatment for alkalosis in patients with cardiac and respiratory failure. Its diuretic effect may be useful in edematous cardiac failure, provided that potassium losses can be matched effectively by an appropriate intake so that the potassium deficit can actually be reversed. It is more hazardous in respiratory failure, and particularly in hypercapnic patients. On the one hand, the effects of acetazolamide persist for 24 to 48 h, which may be a problem when the drug is found to decrease the pH and not the $P\text{aCO}_2$. On the other hand, acetazolamide inhibits carbonic anhydrase and thus interferes with tissue CO_2 elimination. In patients with chronic obstructive pulmonary disease, it has been shown to cause a sustained and significant increase in the difference between tissue CO_2 tension and $P\text{aCO}_2$. Thus the decrease in $P\text{aCO}_2$ reported during acetazolamide therapy reflects, at least in part, altered tissue removal rather than improved pulmonary elimination of CO_2. Acetazolamide has also been reported to cause hematuria and renal failure. It should be avoided in renal and hepatic failure, and is ineffective in anuric patients.

Acidifying agents

Treatment of metabolic alkalosis with chloride-containing acids has been suggested, but most of these substances have been associated with deleterious side-effects. Lysine chloride and arginine chloride can cause hyperglycemia and hyperkalemia, and worsen pre-existing renal failure. Ammonium chloride can result in cardiac, respiratory, and neurological complications, particularly in cases of rapid administration and in patients with renal or hepatic failure. Thus these substances are best avoided in the treatment of metabolic alkalosis.

Hydrochloric acid

Hydrochloric acid (HCl) is proposed as the most efficient way of rapidly and selectively correcting the deficits in hydrogen and chloride ions which characterize metabolic alkalosis. Suggested indications for HCl therapy are (Brimioulle *et al.* 1989; Brimioulle and Kahn 1990):

(1) serious clinical complications due to the alkalosis (coma, arrhythmias);

(2) contraindications to or ineffectiveness of other treatments;

(3) severe alkalemia (arterial pH $>$ 7.55);

(4) respiratory failure with hypercapnia and/or hypoxemia.

The efficiency of HCl in reversing the alkalosis has been reported in several clinical studies (Table 1).

One case of pulmonary edema due to HCl therapy, which was attributed to the amount of fluid required to correct the alkalosis, has been reported. Hemolysis has been reported in animal experiments but not in clinical series. Vascular and perivascular necrosis have been observed in earlier series, after HCl infusion into a peripheral vein. Extensive tissue necrosis has only been reported after leakage or infusion to extravascular sites.

HCl can be infused into the superior vena cava without complications, provided that adequate precautions are respected. HCl is prepared in a glass bottle as a 250- or 500-mmol/l solution (25 or 50 ml of 36 per cent HCl in 1 liter of sterile water). After very careful verification of the correct position of the catheter tip on a chest radiograph, the solution is infused at a rate of 1 ml/kg/h without being mixed with non-electrolytic solutions. Arterial blood gases are measured every 2 h (500 mmol/l) or 4 h (250 mmol/l), and the infusion is discontinued when the alkalosis is reversed (e.g. base excess below zero). When rapid correction is required, a 1000-mmol/l solution may be infused at the same rate while blood gases are measured hourly (Brimioulle and Kahn 1990). Animal experiments have shown that there are no complications when HCl is administered at a rate of 2 mmol/kg/h, but a syndrome similar to adult respiratory distress syndrome commonly occurs at a rate of 3 mmol/kg/h. Metabolic alkalosis may relapse after achievement of the HCl infusion if generating or maintenance factors have not been corrected.

References

Brimioulle, S. and Kahn, R.J. (1990). Effects of metabolic alkalosis on pulmonary gas exchange. *American Review of Respiratory Diseases*, **141**, 1185–9.

Brimioulle, S., Berré, J., Dufaye, P., Vincent, J.L., Degaute, J.P., and Kahn, R.J. (1989). Hydrochloric acid infusion for treatment of metabolic alkalosis associated with respiratory acidosis. *Critical Care Medicine*, **17**, 232–6.

Galla, J.H. (1995). Metabolic alkalosis. In *Fluid, electrolyte and acid–base disorders* (ed. A.I. Arieff and R.A. DeFronzo), Chapter 6, pp. 199–222. Churchill Livingstone, New York.

Harrington, J.T. and Kassirer, J.P. (1982). Metabolic alkalosis. In *Acid–base* (ed. J.J. Cohen and J.P. Kassirer), Chapter 9, pp. 227–306. Little, Brown, Boston, MA.

Sabatini, S. and Kurtzman, N.A. (1994). Metabolic alkalosis. In *Maxwell and Kleeman's clinical disorders of fluid and electrolyte metabolism* (ed. R.G. Narins), Chapter 29, pp. 933–56. McGraw-Hill, New York.

8.5 Hyperglycemia

8.5.1 Pathophysiology

G. Bishop and K. Hillman

Key messages

- Blood sugar is normally controlled by the interplay between insulin and glucagon.
- Diabetic emergencies have varying degrees of hyperglycemia, acidosis, ketosis, and hypovolemia.
- Hyperglycemia is a common accompaniment of the stress response to critical illness.

Introduction

Glucose is the most important energy source for almost all body tissues (Walmsley and White 1994). The blood sugar is usually tightly controlled by the interplay between the hormones, insulin, and glucagon. Both are secreted by the islet cells of the pancreas in response to the nutrient composition of the blood perfusing the pancreas. Insulin transports glucose into cells, storing energy in the form of glycogen and fat. Glucagon, in contrast, mobilizes energy via glycogen breakdown, lipolysis, and gluconeogenesis. In the day-to-day control of blood glucose, insulin inhibiting the secretion of glucagon is the dominant effect.

Hyperglycemia in the critically ill

Hyperglycemia can be either the primary problem, as in diabetic emergencies, or a result of other processes, such as hyperglycemia secondary to sepsis.

Diabetic emergencies

Diabetic ketoacidosis

Diabetic ketoacidosis is the most common form of diabetic emergency. It is part of the spectrum of diabetic emergencies (Table 1) which have varying degrees of hyperglycemia, acidosis, and ketosis.

Diabetic ketoacidosis typically occurs in type I or insulin-dependent diabetics. The main problem is an absolute or relative lack of insulin, leading to unrestrained oversecretion of glucagon (Fig. 1). This process will not be turned off until insulin inhibits the secretion of glucagon. Most tissues do not starve as they are able to use fuel other than glucose.

As the blood glucose increases, changes occur in body fluid compartments and in the renal handling of sodium and water (Foster and McGarry 1989). Initially, extracellular glucose causes water to move out of cells into the extracellular fluid to maintain osmotic equilibrium. The renal threshold for glucose is exceeded, and glucose spills into the urine causing an osmotic diuresis. Typically, more water than sodium is lost, with varying potassium, phosphate, and magnesium losses.

The intracellular fluid volume becomes depleted because water moves into the extracellular fluid, which in turn becomes depleted as a result of the osmotic diuresis. Eventually this process can lead to severe hypovolemia and shock. Organ perfusion may be compromised, resulting in impaired function such as oliguria and decreased level of consciousness. Poor tissue perfusion and oxygen delivery may cause lactic acidosis.

The level of dehydration, hypovolemia, and electrolyte loss is also influenced by oral fluid intake, the presence of vomiting, diarrhea, fever, and pre-existing renal function.

Glucagon stimulates the production of ketone bodies which are utilized for energy by the brain and by skeletal and cardiac muscle. The hydrogen ion produced with each ketone body soon overwhelms the acid–base buffering capacity and leads to a high-anion-gap metabolic acidosis. Ketone body production far outweighs the peripheral utilization of ketone bodies as fuel.

The pathophysiology results in the clinical picture of hyperglycemia, acidosis, dehydration, polyuria, polydipsia, and electrolyte disturbance.

Hyperosmolar non-ketotic coma

Patients with hyperosmolar non-ketotic coma are hyperglycemic, hyperosmolar, and severely dehydrated by the same mechanisms as patients with diabetic ketoacidosis.

The major difference is that, unlike in patients with ketoacidosis, ketone bodies are either not produced or produced in small amounts. Classically, patients with hyperosmolar non-ketotic coma are undiagnosed type II or non-insulin-dependent diabetics and, as such, may have enough residual insulin to inhibit lipolysis, and hence ketone body production, but not enough to prevent hyperglycemia (Cefalu 1991). Hyperosmolarity may also inhibit lipolysis. Patients with hyperosmolar non-ketotic coma are often profoundly dehydrated because of prolonged glycosuria and polyuria. Their ability to take oral fluids to correct the dehydration may be limited because of their age and coexisting dementia or the diabetic process itself, producing confusion and coma. Patients with hyperosmolar non-ketotic coma often have coexisting illness such as myocardial ischemia or decreased renal function. When hypovolemia supervenes on this background, decreased organ perfusion, ischemia, and lactic acidosis may result.

Table 1 A comparison of clinical and biochemical indices in the spectrum of diabetic emergencies

	Diabetic ketoacidosis	Hyperglycemic non-ketotic coma	Euglycemic ketoacidosis
Diabetic history	Known insulin-dependent or new diabetic	Mild or undiagnosed type II diabetic	Insulin-dependent diabetic
Age	Young	Old	Young
Onset	Days	Days to weeks	Hours to days
Change in level of consciousness	0/+	++	0
Acidosis	+++	0/+	+++
Fluid loss	++	+++	0/+
Blood sugar	++	+++	0/+

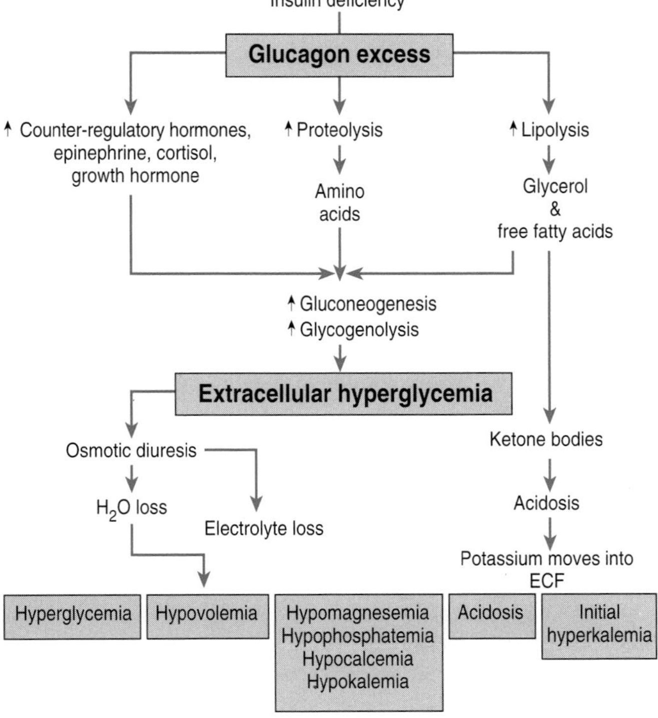

Fig. 1 Pathophysiology of diabetic ketoacidosis: ECF, extracellular fluid. Insulin deficiency leads to a glucagon excess which mobilizes energy substrates to produce glucose and ketone bodies.

Patients with hyperosmolar non-ketotic coma are often very hyperglycemic. The osmotic diuresis produces severe hypovolemia and urine output falls, secondary to decreased renal blood flow. Glucose is no longer lost in the urine at the same rate, resulting in a sharp increase in the serum levels.

Euglycemic ketoacidosis

These patients have ketoacidosis with only a slightly elevated blood sugar. They are typically young insulin-dependent diabetics who have a concurrent illness and have administered enough insulin to keep their blood sugar only mildly elevated but not enough to prevent ketone body production.

Secondary hyperglycemia

Stress response

The metabolic response to critical illness is to mobilize energy substrates. The catabolic hormones epinephrine (adrenaline), cortisol, growth hormone, and glucagon all increase blood glucose by gluconeogenesis and glycogenolysis. Insulin secretion and its peripheral action are inhibited by cortisol and glucagon. Insulin levels are usually high despite a high blood sugar. Inflammatory mediators such as cytokines and interleukins can also result in mobilization of energy stores (Kinney 1995).

Other causes of hyperglycemia

1. Intravenous glucose infusions (e.g. intravenous nutrition).
2. Lack of insulin owing to pancreatic disorders (e.g. pancreatitis and pancreatectomy).
3. Endocrine disorders involving the catabolic hormones usually produce only moderate glucose intolerance. Examples include Cushing's syndrome, pheochromocytoma, acromegaly, and thyrotoxicosis.
4. Impaired glucose tolerance (e.g. pregnancy or secondary to stress). Critical illness in a known diabetic often produces hyperglycemia.
5. Some drugs (e.g. salicylates, thiazide diuretics, steroids, and β-agonists such as epinephrine).

Effects of hyperglycemia in the critically ill

Glycosuria is common in the critically ill; the renal threshold for glucose is usually exceeded at blood sugar levels of more than 10 mmol/l. The subsequent osmotic diuresis will result in losses of water and electrolyte, particularly sodium, potassium, magnesium, and phosphate. Failure to prevent glycosuria will result in dehydration. A blood glucose level of more than 12 mmol/l inhibits white cell function and may predispose to infection.

References

Cefalu, W.T. (1991). Diabetic ketoacidosis. *Critical Care Clinics*, 7, 89–108.

Foster, D.W. and McGarry, J.D. (1989). Acute complications of diabetes: ketoacidosis, hyperosmolar coma, tactic acidosis. In *Endocrinology* (ed. L.J. DeGroot *et al.*), Chapter 87. W.B. Saunders, Philadelphia, PA.

Kinney, J.M. (1995). Metabolic responses of the critically ill patient. *Critical Care Clinics*, 11, 569–85.

Walmsley, R.N. and White, G.H. (1994). *A guide to diagnostic clinical chemistry*, pp. 222–68. Blackwell Scientific, Melbourne.

8.5.2 Assessment

G. Bishop and K. Hillman

Key messages

- Mild hyperglycemia results in an osmotic diuresis, fluid shifts, and impairment of white cell function.
- Hyperglycemia accompanied by acidosis, ketosis, hypovolemia, and electrolyte abnormalities is a medical emergency and needs rapid assessment and stabilization of airway, breathing, and circulation.

Introduction

Hyperglycemia in the hospital setting is usually a result of primary diabetic emergencies, such as diabetic ketoacidosis, or secondary to other factors such as serious illness. Mild hyperglycemia (up to 12 mmol/l) is often detected on a routine biochemistry screen in the intensive care unit. The cause can usually be elucidated by a careful history and a review of interventions, intravenous therapy, and drug treatment. Possible causes include excessive intravenous glucose, a lack of insulin (e.g. with diabetes or pancreatitis), or peripheral resistance to the action of insulin, as occurs in the critically ill. Increased amounts of catabolic hormones, such as epinephrine (adrenaline), growth hormone, or cortisol, either given as a drug or increased secondary to an endocrine disorder or stress can also cause hyperglycemia. If hyperglycemia occurs in a previously stable patient, a new stress such as underlying sepsis or a change in drug therapy should be considered.

Mild hyperglycemia can be associated with an osmotic diuresis and fluid shifts between the intracellular, extracellular, and intravenous compartments. Even relatively low glucose concentrations (above 12 mmol/l) have a clinically significant detrimental effect on phagocyte function. Hyperglycemia also predisposes to infection, particularly by *Candida albicans*.

Severe hyperglycemia accompanied by acidosis, ketosis, hypovolemia, and electrolyte abnormalities is a medical emergency requiring rapid assessment and treatment.

Initial assessment

Immediate airway–breathing–circulation

If the patient is conscious and able to speak, airway patency, breathing, and circulation can be assessed in a systematic fashion. If the patient cannot speak, check his or her level of consciousness and pupils, and look for localizing signs. If the Glasgow Coma Scale is less than 9, the airway needs protecting urgently with an endotracheal tube. A meticulous assessment of breathing and circulation should follow. Commence monitoring of continuous oxygen saturation, ECG, and non-invasive blood pressure measurement.

History

A history should elicit what may have precipitated this diabetic crisis. The most common causes are infection, upper respiratory tract infection and gastroenteritis, or inadequate insulin therapy (Berger and Keller 1992). Check the patient's recent oral food and fluid intake, as well as his or her insulin and oral hypoglycemic drug use. Ask about other drugs, particularly illicit drugs and alcohol. Discuss the patient's normal diabetic control and diabetic complications such as nephropathy, neuropathy, and retinopathy. A history may be available from relatives or ambulance officers if the patient is unconscious.

Neurological

Less than 10 per cent of patients with diabetic ketoacidosis have a Glasgow Coma Scale of less than 9 (Lebovitz 1995). However, the majority of patients with hyperosmolar non-ketotic coma are confused or comatose. The higher the osmolality or corrected serum sodium, the more likely are patients with hyperosmolar non-ketotic coma to be comatose on presentation. These patients may also have pre-existing or new neurological abnormalities, such as a stroke.

Respiratory

Minute volume is always increased in patients with diabetic ketoacidosis unless they are unconscious. This is a compensatory ventilatory response to the acidosis. Ketoacids are produced as a direct result of the lack of insulin and lactate due to generalized ischemia. The gas exchange needs to be assessed with arterial blood gases and underlying respiratory pathology excluded.

A chest radiograph may show hyperinflated lung fields consistent with hyperventilation. It may also point to precipitating illness, such as congestive cardiac failure or a pneumonia.

Cardiovascular

Patients have varying degrees of dehydration and hypovolemia. Hyperglycemia and osmotic diuresis leads to intracellular and interstitial depletion, with sunken eyes, dry mucous membranes, and loss of skin turgor. Hypovolemia is common, and is manifested by hypotension, tachycardia, oliguria, poor peripheral perfusion, and lactic acidosis.

Patients with hyperosmolar non-ketotic coma have usually suffered greater fluid depletion than patients with diabetic ketoacidosis because they have been hyperglycemic for longer and consequently have had greater water and electrolyte loss from osmotic diuresis. These patients also often have restricted access to oral fluids.

Renal

Remember that urine output does not reflect intravascular volume and is driven by glycosuria. Oliguria is a late sign and reflects decreased renal blood flow, secondary to a decrease in intravascular volume.

Test the urine for glucose and ketones. Estimate acetoacetate with dipsticks. The ratio of acetoacetate to β-hydroxybutyrate in diabetic ketoacidosis is usually 1 to 3. Therefore ketones may seem inappropriately low. Serum measurement is more accurate.

Gastrointestinal

Abdominal pain is common with diabetic ketoacidosis and may be related to autonomic dysfunction or pancreatitis as a result of high levels of triglycerides. Careful abdominal examination is essential in order to exclude other pathology.

Peripheral

Look for evidence of skin lesions such as abscesses, cellulitis, ulcers, and necrosis. Diabetics often have a peripheral neuropathy which may impair their sensation. Therefore a full neurological examination should be performed when the patient is able to co-operate.

Investigations

Electrocardiography

An ECG is essential as myocardial ischemia is a common precipitant of diabetic emergencies.

Hematology

The hemoglobin and white cell count should be checked. Neutrophilia with degranulation (left shift) suggests bacterial infection. A white cell count between 10×10^9 and 20×10^9/l may be secondary to dehydration.

Biochemistry

The biochemistry in diabetic ketoacidosis and hyperosmolar non-ketotic coma is often complex and difficult to interpret.

1. Sodium concentrations can be normal, low, or, high. The sodium concentration in patients with hyperosmolar non-ketotic coma is usually high, and reflects severe intracellular dehydration and total body water loss. The serum sodium depends on the following:
   (a) the diluting effect of movement of intracellular water to the extracellular space, particularly in the early stages of the illness;
   (b) loss through osmotic diuresis;
   (c) underlying renal function;
   (d) vomiting of oral fluid intake and diarrhea.

2. Hyperglycemia interferes with the measurement of sodium as the increased osmolarity in the extracellular compartment draws water from the intracellular compartment and dilutes the sodium:

$$\text{corrected sodium} = \text{measured sodium} + \frac{\text{glucose}}{4}.$$

3. Free fatty acid and triglyceride levels are high and will interfere with sodium measurement, making it falsely low.

4. Hyperkalemia is common initially, despite a total body deficit of potassium as a result of the diuresis. Hyperkalemia is mainly due to release of intracellular potassium because of the increased tissue catabolism driven by the insulin deficiency. Acidemia causing cellular potassium–hydrogen exchange plays a minor role.

5. A high-anion-gap metabolic acidosis is present in diabetic ketoacidosis with varying degrees of respiratory compensation. If patients with hyperosmolar non-ketotic coma have a metabolic acidosis, it is usually secondary to high lactate levels as a result of hypovolemia or due to precipitating factors such as septicemia or acute myocardial infarction. Urea and creatinine are often elevated secondary to dehydration. Creatinine and amylase may be falsely raised secondary to interference with the assay by ketone bodies.

Magnesium and phosphate levels are often low owing to excessive loss in the urine.

Microbiology

Both diabetic ketoacidosis and hyperosmolar non-ketotic coma may be precipitated by infection, either bacterial or viral. It is important to look particularly for gastroenteritis, urinary tract infection, and pneumonia. Blood cultures should be taken, particularly if the patient is febrile.

Drugs

Check the patient's usual medications, both diabetic and non-diabetic. A recent change of medication may affect his or her diabetic control.

References

Berger, W. and Keller, U. (1992). Treatment of diabetic ketoacidosis and non-ketotic hyperosmolar diabetic coma. *Baillière's Clinical Endocrinology and Metabolism*, **6**, 1–22.

Lebovitz, H.E. (1995). Diabetic ketoacidosis. *Lancet*, **346**, 767–72.

8.5.3 Management of hyperglycemia

P. Bristow and K. Hillman

Key messages

- Restore circulation.
- Exclude reversible causes.
- Examine the patient's need for drugs which may cause hyperglycemia.
- If the patient is receiving intravenous nutrition, do not infuse more than 80 ml/h 25 per cent glucose in total.
- Consider lipid infusion as a source of calories if the patient is on intravenous nutrition.
- Feed enterally if possible.
- Insulin infusion should be titrated to effect.

Introduction

The management of hyperglycemia is related to the cause, for example control of the underlying diabetes, control of the illness predisposing to hyperglycemia, or critical care treatments which predispose to elevated blood glucose.

General management

The initial management of hyperglycemia involves assessment and resuscitation of the airway, breathing, and circulation. Hyperglycemia may have resulted in an osmotic diuresis, dehydration, and decreased circulating fluid volume. Even minor degrees of hypovolemia should be rapidly corrected. An adequate circulation is essential for insulin to be delivered to cells.

Modifying precipitating factors

The response to critical illness

In the critical care setting, hyperglycemia is often related to neuroendocrine changes in the patient. These changes are a response to the severe illness. Improving the patient's underlying condition is an important step in correcting hyperglycemia.

Intravenous nutrition

Intravenous nutrition usually includes hyperosmolar glucose solutions to supply calories. The seriously ill patient often has obligatory catabolism and fails to utilize carbohydrate calories normally. Elevation of the blood sugar level is the common sequela. Glucose solutions should not be infused at a rate that will cause blood glucose levels to exceed 12 mmol/l. Intravenous nutrition should be commenced at 20 ml/h of 25 per cent glucose solution for at least 4 h. If the blood sugar level remains below 12 mmol/l, the rate should be increased to 40 ml/h. Depending on the blood sugar, the rate can be increased to 60 and then 80 ml/h. The latter rate supplies 2000 kcal/day. Insulin by infusion may be needed if hyperglycemia develops when the rate of glucose infusion is less than 4 mg/kg/min (approximately 80 ml/h of a 25 per cent glucose solution for a lean body mass of 80 kg). Glucose infusions should not exceed 4 mg/kg/min (American Society for Parenteral and Enteral Nutrition 1993).

If hyperglycemia develops, fat is a useful alternative source of up to 40 per cent of the non-protein calories (Cerra 1987). Approximately 2000 kcal will be delivered by a rate of 50 ml/h of 25 per cent glucose and 16 ml/h of 20 per cent triglyceride emulsion. The use of intravenous fat emulsions to decrease glucose administration is routine in many countries.

Potassium and trace minerals (zinc, chromium) are important for glucose homeostatasis. The level of potassium should be corrected, and trace element supplements should be routinely administered weekly.

The indications for intravenous nutrition should be continuously reviewed. Enteral feeding may be associated with better patient outcomes and is tolerated by most seriously ill patients.

If the patient needs insulin while receiving intravenous nutrition, it should initially be given as an insulin infusion. Once the patient is stable for more than 24 h, the insulin requirements for the previous 24 h can be added to the nutrition bottles and run continuously over 24 h. This is most appropriate when the patient is transferred to a bed on the ward. The blood sugar level will still need regular monitoring every 4 h. The addition of insulin directly to the nutrition flasks has the advantage that if intravenous nutrition stops temporarily, so does the insulin. The disadvantage is that the dose cannot be easily adjusted.

Drugs

Steroids were previously widely used in the critically ill. Their use should be limited to where benefit is likely, and the dose should be reduced if possible.

The internal milieu of the critically ill patient is dominated by the so-called counter-regulatory hormones glucagon, cortisol, and epinephrine (adrenaline). Administration of catecholamines exacerbates this effect and should be reduced if possible. This will obviously depend on improvement in the underlying pathophysiolgy.

Furosemide (frusemide) and thiazides have the potential to exacerbate hyperglycemia.

Specific measures

Insulin

Unfortunately, despite the maneuvers described above, the seriously ill are often hyperglycemic. Blood glucose levels above 12 mmol/l are associated with impaired neutrophil function and polyuria. In the critically ill, blood sugar levels should be kept below 12 mmol/l with insulin. More meticulous glucose control to less than 7 mmol/l may be indicated in the pregnant patient.

In areas with adequate staffing, such as the intensive care unit, insulin by intravenous infusion is the best method of delivery (Sajbel *et al.* 1984). Intravenous boluses are less effective because the hyperglycemia may be an ongoing process and the half-life of intravenous insulin is only about 4 min.

A useful regimen is to dilute 50 units of human Actrapid® insulin in 50 ml of normal saline. The infusion tubing should be adequately purged with 5 ml of saline before commencing the infusion to overcome the possibility of insulin binding to the plastic. The infusion should be run by a syringe driver through a dedicated lumen to prevent bolus administration when other fluids are given.

Analogous to an inotrope, the insulin infusion needs to be titrated to effect. The rate should be adjusted on the basis of the patient's fingerprick capillary (or arterial cannula if present) glucose concentration measured by a glucose reflectance meter (i.e. a bedside glucose analyzer or a blood gas analyzer). Measurements should be made hourly until the glucose stabilizes between 5 and 12 mmol/l, when measurement every 2 h is adequate. Formal blood glucose should be measured at least twice daily to ensure that the reflectance meter is correctly calibrated. If the reflectance meter gives a blood glucose above 20 mmol/l, a formal blood sugar measurement should be performed as the meter may be less accurate above this value.

If the blood glucose level is between 12 and 15 mmol/l, an appropriate rate to start the insulin infusion is 1 unit/h. If the glucose level is between 15 and 20 mmol/l, the insulin infusion should commence at 2 units/h. If the glucose level exceeds 20 mmol/l, the insulin infusion should be started at 3 units/h (Hongsermeier and Bistrian 1993).

In the insulin-requiring diabetic the insulin infusion could be started at approximately two-thirds the total daily administered insulin divided by 24. For example, if the total daily subcutaneous insulin is 54 units, commence at 36 units/day (two-thirds of 54) which is 1.5 units/h.

These are only guidelines to begin therapy. The important principle is to titrate the infusion according to response. If the glucose level is falling by more than 5 mmol/l/h, the insulin infusion should be slowed. If the glucose level is rising, the infusion rate should be increased until the glucose is under control.

Sliding-scale subcutaneous insulin

Subcutaneous insulin has only a gradual effect on the blood sugar level as the depot of subcutaneous insulin is mobilized. A sliding-scale subcutaneous regimen should only be used when the circulation is adequate, as absorption may vary depending on perfusion. It should be used where fluctuations in glucose level are not likely to be great (Knapke *et al.* 1989).

References

American Society for Parenteral and Enteral Nutrition Board of Directors (1993). Guidelines for the use of parenteral and enteral nutrition in adult and pediatric patients. Section IV: Nutrition support for adults with specific diseases and conditions. *Journal of Parenteral and Enteral Nutrition*, **17**, 12SA–26SA.

Cerra, F.B. (1987). The hypermetabolism organ failure complex. *World Journal of Surgery*, **11**, 173–81.

Hongsermeier,T. and Bistrian, B.R. (1993). Evaluation of a practical technique for determining insulin requirements in diabetic patients receiving total parenteral nutrition. *Journal of Parenteral and Enteral Nutrition*, **17**, 16–19.

Knapke, C.M., Ponting Owens, J., and Mirtallo, J.M. (1989). Management of glucose abnormalities in patients receiving total parenteral nutrition. *Clinical Pharmacy*, **8**, 136–44.

Sajbel, T.A., Dutro, M.P., and Radway, P.R. (1987). Use of separate insulin infusions with total parenteral nutrition. *Journal of Parenteral and Enteral Nutrition*, **11**, 97–9.

8.5.4 Management of hyperglycemic diabetic emergencies

P. Bristow and K. Hillman

Key messages

- The initial management of diabetic emergencies is aimed at airway, breathing, and circulation, and not at insulin therapy.
- Rapidly correct hypovolemia and slowly reduce blood sugar and replace water losses.
- Regularly measure the electrolytes potassium, phosphate, and magnesium, and replace as necessary.

Introduction

Hyperglycemic diabetic emergencies fall into two broad categories: diabetic ketoacidosis and hyperosmolar non-ketotic coma. Both emergencies have a relatively high mortality: approximately 5 per cent for diabetic ketoacidosis and 50 per cent for hyperosmolar non-ketotic coma (Hillman 1991; Hammond and Wallis 1992; Lebovitz 1995). The high mortality is probably related as much to the underlying chronic organ impairment and precipitating factors as to the acute physiological disturbance. For all these reasons, patients with diabetic emergencies are at high risk of further pathophysiological insults, such as hypoxia and ischemia, and they require rapid and competent resuscitation. It is recommended that these high-risk patients are managed in an intensive care environment until they are fully resuscitated and stable.

Patients with diabetic ketoacidosis have decreased or absent insulin levels, whereas patients with hyperosmolar non-ketotic coma have more of a peripheral resistance to insulin. Hyperglycemia, which leads to glycosuria and an osmotic diuresis, occurs in both types of patients. The osmotic diuresis can result in large fluid and electrolyte losses. Lipolysis, free fatty acid release, and ketone formation also occur in patients with diabetic ketoacidosis.

Precipitating factors in both conditions include infection, myocardial infarction, thromboembolic disorders, and poor patient compliance with medication.

The principles of management for the two conditions are similar. Common aspects of management will be considered below.

Initial assessment

Most current textbooks and reviews begin by recommending insulin replacement as first-line treatment for diabetic emergencies. Although this seems logical, it is not the highest priority. Neither the lack of insulin nor the high blood glucose are in themselves an immediate threat to the patient's life. It is the pathophysiological sequelae of the high blood glucose such as hypovolemia and hypoxia which demand immediate attention. Patients with diabetic emergencies require the same standardized approach as any other life-threatening emergency—airway, breathing, and circulation (Bratton and Krane 1992).

Control and protection of the airway is the first priority. Approximately 10 per cent of patients with diabetic ketoacidosis are initially unconscious, and the majority of patients with hyperosmolar non-ketotic coma are comatose. Control of the airway may simply require positioning or an oral airway. However, there should be a low threshold for intubating an unconscious patient. Patients with a Glasgow Coma Scale of 9 or less require intubation.

Attention to breathing is the next priority. Patients with diabetic ketoacidosis are normally hyperventilating as a result of the metabolic acidosis. The lung fields on chest radiography are typically over-expanded. Comatose patients are at risk of aspiration. Patients with hyperosmolar non-ketotic coma are usually older and more obese than those with diabetic ketoacidosis. This, combined with a decrease in the level of consciousness, can predispose to basal collapse and pneumonia. Oxygen requirements need to be assessed with pulse oximetry and arterial blood gases. For a conscious patient, continuous positive airway pressure via a mask may be beneficial for prevention and treatment of basal collapse. Occasionally, intubation and artificial ventilation are also necessary. Once the airway and breathing have been assessed and initial treatment undertaken, hypovolemia should be corrected.

Fluid requirements

Fluid loss is almost inevitable in diabetic emergencies. The average loss for diabetic ketoacidosis is between 5 and 10 liters, and it is even higher for hyperosmolar non-ketotic coma. The loss is due to an osmotic diuresis caused by glycosuria. The resulting urine is isotonic with plasma, and therefore urine osmolality is not a guide to the state of dehydration.

The urinary fluid losses come from all body fluid compartments in relatively equal proportions, resulting in generalized dehydration. However, the first priority in any patient with fluid loss is to correct the hypovolemia or shock. The most efficient fluids for resuscitating the intravascular space are colloids. They also result in less expansion of the interstitial space compared with crystalloid solutions such as isotonic saline. Although isotonic saline is still recommended in many texts as the first-line fluid for resuscitation in diabetic ketoacidosis, it neither replaces the total body water adequately nor corrects hypovolemia efficiently. Moreover, crystalloid solutions are associated with a higher incidence of pulmonary and peripheral edema than colloid solutions. The hypovolemia of diabetic emergencies is no less life threatening than the hypovolemia of blood loss. In fact, because of the relatively high incidence of chronic impairment of organs, such as heart and kidney, in diabetic patients, as well as a higher incidence of peripheral cardiovascular disease, it could be argued that it is even more critical to correct hypovolemia rapidly in a diabetic patient than in, for example, a young hypovolemic trauma patient.

Possibly even more important than identifying the ideal fluid is the rapidity with which normal perfusion is re-established (Hillman *et al.* 1996). Even minor degrees of ischemia can result in organ damage and increased morbidity and mortality. This may result from ischemia of the gastrointestinal mucosa, leading to bacterial translocation and eventually multiorgan failure.

Many standard textbooks still recommend inflexible fluid regimens based on absolute fluid amounts over certain time periods. In diabetic emergencies it is important to infuse the right amount of fluid. This will vary from patient to patient. Some patients have minimal fluid losses and may not require any intravascular replacement. Others are severely shocked, and require large amounts of fluid replacement.

Fluid requirements should initially be guided by simple vital sign measurements such as pulse rate, systemic blood pressure, and peripheral perfusion. Urine output is a misleading guide to intravascular volume because of the glycosuria and polyuria. Fluid challenges should be given until the circulation is fully restored. Extracellular water will move into the intracellular space when insulin is commenced, exacerbating hypovolemia. It is important to monitor the patient's circulatory state closely and increase the rate of fluid resuscitation if necessary. Fine tuning of intravascular fluid requirements may eventually be achieved with more complex measurements such as central venous pressure, pulmonary artery wedge pressure, and cardiac output. Overaggressive unmonitored resuscitation may also lead to problems, for example in those with coronary artery disease. In summary, inflexible fluid regimens should be avoided and hypovolemia corrected within minutes not hours.

The polyuria in diabetic ketoacidosis and hyperosmolar non-ketotic coma also results in fluid losses from the interstitial space and intracellular space. However, these losses are not immediately life threatening and can be corrected with water in the form of 5 per cent glucose at a rate between 100 and 200 ml/h, depending on the degree of dehydration. There is some evidence that rapid replacement of water losses may predispose to cerebral edema. There is a minimal amount of extra glucose in 5 per cent glucose solution and this is readily controlled, if necessary, by increasing the insulin infusion. The sodium deficits are more than adequately replaced by the sodium content of the infused colloid solution.

The sodium concentration of urine as a result of the osmotic diuresis is between 50 and 70 mmol/l. Therefore it could be argued that a fluid such as half-normal saline would be an even more appropriate fluid replacement, as it approximates what is lost. While this makes sense, it does not address the immediate problem of hypovolemia. Hypotonic saline would correct shock even more slowly than isotonic saline because it would be distributed not only to the interstitial space but also to the intracellular space. However, in the case of minimum fluid losses with no hypovolemia, a solution such as half-normal saline would be an ideal replacement fluid.

It is important to encourage the patient's own system of controlling fluid requirements as soon as possible. Oral fluids should be commenced as soon as the patient is resuscitated, stable, co-operative, and with a functioning gastrointestinal tract.

In summary, correct hypovolemia rapidly with colloid using simple vital signs as a guide. Avoid inflexible fluid recipes. Simultaneously replace the body's water losses with a 5 per cent glucose infusion at a rate of 100 to 200 ml/h. When there is no hypovolemia, use a solution similar to what is being lost (e.g. half-normal saline). Encourage oral fluid intake as soon as possible. Patients with hyperosmolar non-ketotic coma usually have greater fluid losses than those with diabetic ketoacidosis.

Insulin

Hyperglycemia is an obvious metabolic accompaniment of diabetic emergencies and therefore it is tempting to correct it to normal values rapidly. However, hyperglycemia itself is not life threatening, although the metabolic sequelae and fluid losses can be. The high blood glucose levels may have been reached over days and sometimes weeks. The body's osmolar and metabolic homeostasis has slowly adjusted to this new biochemical environment. When insulin is given, the glucose is encouraged to move into the cell, causing water, potassium, magnesium, phosphate, and other compounds to enter the cell also. If this occurs too rapidly, it can cause cell swelling, hypokalemia, hypomagnesemia, and hypophosphatemia, and exacerbate hypovolemia (Fig. 1). For this reason it is not necessary, and may in fact be dangerous, to give an initial insulin bolus.

Insulin should be commenced as an intravenous infusion at a low level (1 unit/h). It is then easy to measure blood glucose levels regularly and adjust the rate accordingly, as with any other drug infusion. Although the initial blood glucose levels of patients with hyperosmolar non-ketotic coma are higher than those with diabetic ketoacidosis, the former are more sensitive to insulin than the latter.

As with fluid regimens, there are many inflexible recommendations about the amount of insulin required for diabetic emergencies. Some patients may only ever require a low infusion rate (e.g. 1 unit/h), whereas others may require higher infusion rates (e.g. 15 units/h). Insulin needs should be tailored to individual patients. An average dose is misleading and inappropriate. Blood glucose should be performed initially and then at least hourly until a trend becomes obvious. Remember that the initial decrease in blood glucose is related mainly to the dilutional effect of fluid resuscitation. Measurement of blood glucose with bedside reflectance devices is inaccurate above levels of approximately 20 mmol/l. Blood must be sent to the laboratory for measurement until levels return to values below 20 mmol/l. Glucose can also be measured on some blood gas analyzers.

Aim to achieve a slow and smooth reduction in blood glucose over at least 24 to 72 h. So-called insulin resistance may be related to inadequate resuscitation and failure to transport the insulin to the peripheral cells.

Insulin may adsorb to glass and plastic, prompting the use of carrier solutions to deliver it. This appears to be more a theoretical

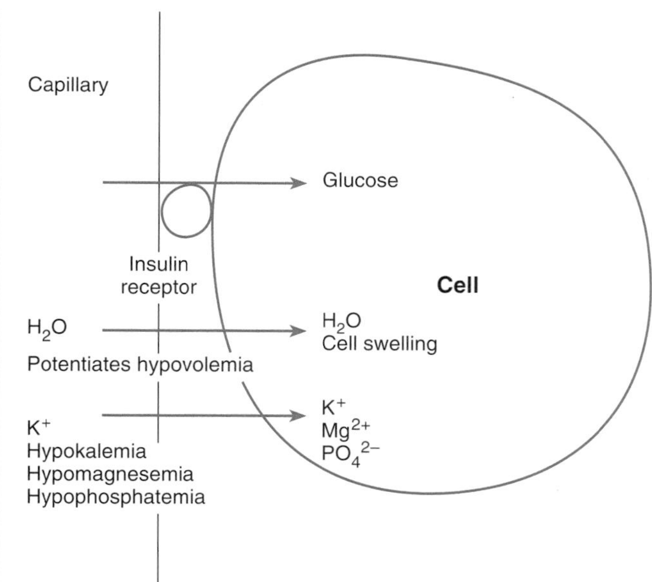

Fig. 1 Effects of insulin therapy: insulin not only reduces the blood glucose levels, but also causes water and electrolytes to move into cells, with potential adverse effects.

than a practical concern and ignores the aim of reducing blood glucose levels by titrating the rate of insulin delivery rather than achieving absolute insulin levels.

Potassium

Potassium losses are often greater than sodium losses. While patients are often initially hyperkalemic, hypokalemia commonly occurs after treatment has commenced in diabetic ketoacidosis and is related to absolute losses, hemodilation, the movement of glucose back into the cell, and correction of acidosis. It is the most common reversible cause of death during the management of diabetic ketoacidosis.

As with fluid and insulin requirements, there is no place for inflexibility in the recommended potassium replacement requirements. It is important to measure the serum potassium initially and approximately 1 h after the commencement of insulin in the resuscitation phase. This will give an indication of the trend. There may also be chronic or acute renal impairment, interfering with potassium excretion. The potassium should be measured at least every 2 h in the early stages of resuscitation and then less regularly as the levels achieve a more predictable pattern.

A rate of between 5 and 30 mmol/h of potassium is usually required during resuscitation of diabetic emergencies. Because of the relatively high concentrations, it should be delivered via a large vein or central venous catheter. Continuous ECG monitoring is necessary with such high fluxes of extra- and intracellular potassium levels.

Phosphate

Phosphate deficiency is common in diabetic emergencies and is mainly a result of phosphaturia. Moreover, serum levels fall during treatment. There is increasing awareness of the dangers of hypophosphatemia, particularly in the critically ill. Therefore, as with potassium, it would seem logical to measure phosphate regularly and replace it as necessary to normal levels. As phosphate is usually given in a potassium-containing solution, simultaneous infusions of potassium may have to be slowed or stopped.

Magnesium

As with potassium and phosphate, hypomagnesemia is common in diabetic emergencies, sometimes resulting in serious complications. Serum magnesium levels should be measured initially and then at least daily, and then corrected as necessary.

Bicarbonate

The use of sodium bicarbonate has become less common in the management of diabetic emergencies. Administration of sodium bicarbonate may result in paradoxical intracellular acidosis, hyperosmolality, hypernatremia, and hypokalemia, as well as delayed ketone body and lactate metabolism. Moreover, giving sodium bicarbonate does not improve the rate of metabolic recovery, even in the presence of severe acidosis. Oxidation of ketone bodies in itself produces bicarbonate.

General measures

Depending on the severity of presentation, continuous monitoring of systemic blood pressure, pulse oximetry, and ECG should be considered. Initial investigations include blood glucose, serum electrolytes, urea and creatinine levels, ketones, hemoglobin and full blood count, triglycerides, and arterial blood gas analysis.

Underlying cause

Intercurrent diseases such as infection, cerebrovascular disease, and myocardial infarction should be excluded as part of initial management.

Although the cause of cerebral edema during resuscitation of patients with diabetic emergencies is not known, overzealous correction of blood sugar and excessive fluid replacement have been implicated. If the level of consciousness does not improve, the patient becomes comatose during treatment, or there are focal neurological signs, a CT scan should be performed.

The incidence of thromboembolic episodes may be increased in diabetic emergencies as a result of an increased hematocrit, underlying vascular disease, and a hypercoagulable state. Prophylactic heparin should be used unless there is a contraindication.

Chest physiotherapy, particularly in elderly, obtunded, or obese patients, may reduce the incidence of complications.

Antibiotics should only be commenced when there is definite evidence of infection.

Education

All patients need educating about their diabetes, with the aim of giving them as much autonomy as possible over the control of their disease.

References

Bratton, S.L. and Krane, E.J. (1992). Diabetic ketoacidosis: pathophysiology, management and complications. *Journal of Intensive Care Medicine*, **7**, 199–211.

Hammond, P. and Wallis, S. (1992). Cerebral oedema in diabetic ketoacidosis. *British Medical Journal*, **305**, 203–4.

Hillman, K. (1991). The management of acute diabetic emergencies. *Clinical Intensive Care*, **2**, 154–62.

Hillman, K., Bishop, G., and Bristow, P. (1996). Fluid resuscitation. *Current Anaesthesia and Critical Care*, **4**, in press.

Lebovitz, H.E. (1995). Diabetic ketoacidosis. *Lancet*, **345**, 767–72.

8.6 Hypoglycemia

8.6.1 Pathophysiology

R. B. Noone and M. G. Mythen

Key messages

- Hypoglycemia is most frequently due to glucose production that is inadequate to meet metabolic demands.
- The central nervous system is dependent on serum glucose for fuel needs and the body normally has an intricate system for maintaining serum glucose within a tight range.
- The most common reason for loss of glucose control is the effects of drugs including insulin, other hypoglycemic drugs, and ethanol.
- Organ failure is another cause of hypoglycemia in hospital patients. This is usually due to massive failure of the liver.

Introduction

Hypoglycemia can lead to significant morbidity and mortality and is common among hospital patients, particularly diabetics and the critically ill. The glycogen stores in the brain are only sufficient for a few minutes, and, unlike other body tissues, the brain cannot acutely metabolize circulating free fatty acids. Thus plasma glucose is the primary source of metabolic fuel for the brain and maintenance of glucose levels is crucial for survival. The definition of hypoglycemia is clinical: a low serum glucose level combined with symptoms. Symptoms are classified as adrenergic (anxiety, palpitations, diaphoresis, irritability, and tremor, usually beginning at glucose levels of about 3.2 mmol/l) and neuroglycopenic (hunger, dizziness, tingling, blurred vision, difficulty in thinking, and faintness, beginning at about 2.8 mmol/l, and deterioration of mental function which begins at about 2.7 mmol/l). Hypoglycemia occurs when utilization of glucose exceeds production and release to the plasma.

Glucose homeostasis

An elaborate system maintains serum glucose levels between meals. Glucose utilization is about 2 mg/kg/min and the brain uses half of this. About 4 h after a meal the liver switches from net glucose storage to net release to maintain serum levels. About 70 to 80 per cent of this release comes from liver glycogenolysis and the rest comes from gluconeogenesis.

The adult liver stores about 40 to 50 g of glucose as glycogen and this is depleted after 24 to 36 h of fasting. Gluconeogenesis from lactate, pyruvate, glycerol, and amino acids accounts for about 25 per cent of glucose production between meals and increases as glycogen stores are depleted. During prolonged fasting, amino acids provide almost all glucose and muscle proteins are degraded. Glucose utilization by all tissues is reduced. Ketone levels rise and begin to provide a major fuel source for the brain.

Glucagon, catecholamines, cortisol, and growth hormone prevent diversion of serum glucose away from the brain during insulin excess and mobilize glucose and alternative fuels during fasting. Glucagon affects the liver directly and is a potent stimulator of glycogenolysis and gluconeogenesis.

Epinephrine (adrenaline) has a more complex effect: α-adrenergic stimulation inhibits insulin secretion, and β-adrenergic stimulation causes glucagon release as well as direct limitation of tissue glucose utilization. Epinephrine stimulates liver glucose production directly, primarily through β_2-receptor activation. Growth hormone initially has an insulin-like effect, reducing serum glucose, but then limits glucose utilization and stimulates production after several hours. Cortisol also causes glucose production and inhibits use after 2 to 3 h. Physiological maintenance of glucose levels is primarily due to the effect of glucagon, followed by the catecholamines to a lesser extent. The role of the catecholamines seems to be more important in states of relative glucagon insensitivity, for example in patients with insulin-dependent diabetes.

Clinical hypoglycemia

Although low serum glucose can be caused by excessive glucose loss or use, hypoglycemia is most commonly the result of liver glucose production that is inadequate for the rate of glucose utilization. This may be due to regulatory, enzyme, or substrate effects. The clinical syndromes are commonly divided into postabsorptive (fasting) and postprandial (reactive) hypoglycemia. Reactive hypoglycemia is uncommon and may follow gastric surgery. Rarely, it is due to inborn enzyme defects. Idiopathic or functional hypoglycemia is a controversial syndrome. If it exists, it is certainly rare. Most patients encountered clinically have postabsorptive hypoglycemia. Drugs or organ failure are the most common causes in the hospital or emergency room patient.

Drugs

The majority of hypoglycemic episodes are due to drugs, most commonly insulin, sulfonylureas, and alcohol. Patients with insulin-dependent diabetes on conventional insulin therapy will have one symptomatic episode of hypoglycemia per week on average, and those on intensive insulin therapy will have two per week. An estimated 4 per cent of deaths of insulin-dependent diabetes patients are due to hypoglycemia. The majority of patients with insulin-dependent diabetes have deficient glucagon response to low plasma glucose. Many patients have diminished adrenergic response. Some patients, particularly those with long-standing diabetes, lose the classic symptoms of hypoglycemia. This syndrome of hypoglycemia unawareness makes complications much more frequent and dangerous. Recent studies of patients with insulinomas suggest that the mechanism of unawareness may be the hypoglycemia itself and that intensive insulin therapy may worsen this problem.

Sulfonylureas are also prone to cause hypoglycemia. Such episodes may be long lasting, particularly with first-generation drugs. High insulin and low C peptide concentrations characterize hypoglycemias from insulin injection, and inappropriately high levels of both substances are common with sulfonylureas. These drugs can be detected in the urine or plasma.

Hepatic oxidation of ethanol ultimately to acetate leads to the accumulation of NADH and diminished NAD^+ which is needed as a cofactor for several steps of gluconeogenesis. Hypoglycemia tends to occur in food-deprived individuals within 6 to 24 h after moderate intake of ethanol. Mortality may be as high as 10 per cent.

Other drugs are known to decrease serum glucose. Salicylates in large doses can cause hypoglycemia in children and, rarely, in adults. Pentamidine, a β-cell toxin, causes hypoglycemia in a dose-dependent manner by stimulating insulin release. It may lead to late development of diabetes.

Organ failure

Extensive liver disease is required to cause hypoglycemia. This can happen with fulminant hepatitis, fatty liver from starvation or alcohol ingestion, and cholangitis with biliary obstruction. Hypoglycemia is unusual with common forms of cirrhosis or hepatitis, although glucose metabolism may be altered.

Severe cardiac failure may lead to hypoglycemia; the pathogenesis is not known, but is thought to be from hepatic congestion, cellular hypoxia, or substrate limitation. Some patients with renal failure develop hypoglycemia, and this is not well understood either. It may be due to cachexia with limited substrate for glucose production or to inhibition of gluconeogenesis itself.

Sepsis commonly leads to hypoglycemia. A combination of increased glucose utilization and failure of glucose production seem to be involved.

Tumors

Insulinomas have an incidence of approximately one in one million people. They account for 66 to 85 per cent of patients with endogenous hyperinsulinism. They usually have dimensions of 1 to 2 cm and 99 per cent of them occur within the mass of the pancreas; 5 to 10 per cent are malignant. Typical symptoms of hypoglycemia tend to occur in the late afternoon or some hours after a meal. Diagnosis requires demonstration of Whipple's triad: low serum glucose, spontaneous symptoms during documented hypoglycemia, and resolution of symptoms with glucose administration. Rare non-β-cell tumors can cause related syndromes. Most are mesenchymal, but occasional epithelial neoplasms such as hepatomas and adrenocortical and carcinoid tumors can cause hypoglycemia.

8.6.2 Therapeutic strategy

I. C. Sanderson and M. G. Mythen

Key messages

- Given the wide range of 'normal' blood glucose values, diagnosis of pathological hypoglycemia depends on the clinical recovery from neuroglycopenic symptoms after an intravenous infusion of glucose.
- Glucose 25 g given intravenously (50 ml of 50 per cent glucose) usually reawakens a comatose hypoglycemic patient in 5 to 10 min.
- Intramuscular glucagon (1 mg) is an effective and portable alternative to intravenous glucose for the majority of hypoglycemic episodes in insulin-dependent diabetics.
- In the malnourished, thiamine deficiency will be accelerated if circulating glucose substrate is rapidly restored, causing acute and potentially fatal Wernicke's encephalopathy.
- Prolonged hypoglycemia, particularly in association with hypoxemia and hypotension, leads to cerebral edema. Recovery may be delayed despite the attainment of normoglycemia.

Introduction

Hypoglycemia is a clinical diagnosis with the finding of a low blood glucose in association with symptoms due to an inadequate supply of glucose to the brain, a condition termed neuroglycopenia (Andreani *et al.* 1987). Hypoglycemia can cause inconsistent and multiple symptom complexes at a variety of blood glucose thresholds according to the pathological state. Healthy subjects made hypoglycemic by exogenous insulin may experience symptoms at 3.0 mmol/l or lower, whereas prolonged pathological hyperinsulinism causes insulin resistance and symptoms may not be seen until levels of 2.2 mmol/l are reached. The elderly are less tolerant, with a higher neuroglycopenic threshold at 3.5 mmol/l.

Hypoglycemic neuronal distress provokes a reflex increase in sympathetic outflow as well as complex central effects. The symptoms of acute neuroglycopenia, usually caused by insulin overdose, include hunger, anxiety, disorientation, and restlessness. Accompanying signs are tremor, tachycardia, profuse sweating, facial flushing, ataxia, and reduced consciousness. Catecholamine and glucagon-mediated counter-regulatory mechanisms, which act to mobilize glucose from the liver, are triggered. Death is more likely when counter-regulation is impaired, as in the case of the inhibition of gluconeogenesis by alcohol, or if hypotension or hypoxemia are added to the insult. Uncomplicated acute hypoglycemic coma rarely leads to permanent brain dysfunction provided that it is rapidly corrected. However, there is evidence that diabetics who experience repeated hypoglycemic episodes suffer from memory loss.

Experimental and clinical evidence suggests that hypoglycemia elicits a dominant centrally mediated response rather than a direct effect on the adrenal medulla. The signs of neuroglycopenia (apart from sweating) can be abolished by β-blockade but persist after adrenalectomy. In some diabetics increased sweating may be the only warning of impending coma.

Classic neuroglycopenic symptoms can occasionally be seen in diabetics at normoglycemic blood levels during rapid correction of hyperglycemia. In patients with diseased coronary arteries, angina is a common presenting symptom brought on by the increase in cardiac work due to sympathetic activation. In the intensive care unit, a heavily supported patient may show only the signs of tachycardia and sweating.

Subacute neuroglycopenia, such as that caused by an insulin-secreting tumor, causes mainly central symptoms without peripheral sympathetic effects. Behavioral manifestations, amnesia, focal neurological signs, and decreased consciousness are typical. Chronic neuroglycopenia may only manifest with personality changes or psychosis.

Diagnosis and clinical presentation

A blood glucose measurement is part of the initial investigation of any patient presenting with disorientation or coma. Diagnosis is made by the measurement of blood glucose using arterialized capillary blood from a fingerprick. Whole-blood glucose measured in this way will be 10 to 15 per cent lower than corresponding plasma samples. If hypoglycemia is found, blood should be drawn for plasma glucose in a fluoride or oxalate tube, together with plasma insulin, C peptide, and cortisol. To assist future diagnosis, samples should be taken before treatment is initiated. A low ratio of C peptide to insulin suggests exogenous insulin administration, whereas a normal ratio suggests an endogenous source of insulin.

Given the wide range of 'normal' blood glucose values, diagnosis of pathological hypoglycemia depends on the clinical recovery from neuroglycopenic symptoms after an intravenous infusion of glucose.

Patients with hypoglycemia will usually present to critical care in one of the following groups.

1. Brittle diabetics who have missed a meal or taken an inappropriate dose of insulin; often these younger patients will have been found in coma, and may have already been treated *in situ* with intramuscular glucagon or a glucose infusion.

2. An older group or their relatives with access to oral hypoglycemic drugs who present in coma having intentionally or otherwise taken an overdose, sometimes in combination with additional prescribed agents and alcohol.

3. Alcohol-induced hypoglycemia; most, but not all, of these patients are chronic alcoholics. Hypoglycemia is an incidental but dangerous finding as these patients are often assumed to be in coma of alcoholic etiology. A significant proportion of custodial deaths are attributed to this misunderstanding. Paradoxically, the most severe cirrhotics rarely develop alcohol-induced hypoglycemia as their liver function is so poor that they cannot generate sufficient NADH to inhibit gluconeogenesis.

4. The hospital in-patient with an insulin-secreting tumor who is being prepared for surgery. Rarely, hypoglycemia is part of the initial presentation of this disease.

5. As part of the initial presentation of Addison's disease, where hypoglycemia is an additional finding in a patient with glucocorticoid deficiency.

6. As part of the symptomatology of a severe illness, such as fulminant liver disease, acute and chronic pancreatitis, malaria, or sepsis.

7. As an endstage in the symptoms of starvation, particularly in the edematous variety of kwashiorkor. Paradoxically, starvation is rarely associated with hypoglycemia in its early stages, despite exhaustion of glycogen stores, as ketone bodies can act as energy substrates for the brain. Hypoglycemia is occasionally seen and is a poor prognostic sign in anorexia nervosa.

Management

Once hypoglycemia is diagnosed, the aim of treatment is to correct it rapidly and maintain a blood glucose level that prevents neuroglycopenia. Uncomplicated hypoglycemia requires a level of 1.1 mmol/l or less to trigger the events that lead to brain damage. Mitochondrial disruption leads to neuronal death with cerebral edema, gliosis, and permanent sequelae as a result. However, hypoxemia, hypotension, and epileptic activity all act synergistically with hypoglycemia to reduce the substrates necessary for neuronal respiration, and will increase the chance of an unfavorable outcome.

Brain damage from hypoglycemia tends to occur in the 'watershed' areas of critical cerebral perfusion. Slowness to recover from the cerebral symptoms of neuroglycopenia is a measure of the degree of the hypoglycemic insult and may reflect cerebral edema.

Therefore it is essential to resuscitate the patient and his or her cellular milieu according to clinical priorities, restoring oxygenation and the circulation before or simultaneously with the correction of the blood glucose. Bedside monitoring should include frequent estimations of blood glucose to prevent a relapse into dangerous hypoglycemia. Treatment may be necessary for several days, particularly in drug-induced hypoglycemia or severe liver disease. Overdose of the long-acting sulfonylurea chlorpropamide ($t_{1/2} = 32$ h) commonly requires 5 to 7 days of glycemic support. The endpoint of therapy is when the hypoglycemic stimulus diminishes and/or iatrogenic hyperglycemia begins to occur, which is an indication to withdraw treatment over 24 h.

Glucagon

Endogenous glucose production can sometimes be increased by stimulating the counter-regulatory mechanisms of glycogenolysis and gluconeogenesis. Since a glucose infusion requires venous access, intramuscular glucagon (1 mg) is an effective and portable alternative to intravenous glucose for the majority of hypoglycemic episodes in insulin-dependent diabetics. It will often awaken a comatose patient in 10 to 15 min to the stage where an oral glucose load can be safely ingested. A second dose may be needed, but further doses should be resisted as failure to respond implies glycogen exhaustion, inadequate gluconeogenesis, or the onset of cerebral edema, for which alternative therapy should be instituted.

In other circumstances, glucagon may increase morbidity. Where residual β-cell activity exists, glucagon stimulates rebound insulin secretion which acts to reduce its glycemic effect. This response is exaggerated and can lead to rebound hypoglycemia in conditions with hyperinsulinism and coincident insulin resistance, such as insulinoma, obesity, acromegaly, Cushing's syndrome, and dystrophia myotonica. Glucagon will have little effect in those who have exhausted their glycogen reserve or in whom gluconeogenesis is impaired, and so should be used with caution in alcohol-induced hypoglycemia or severe liver disease.

Glucose infusions

Given an average glucose pool of 15 to 20 g in the adult and glucose consumption between 120 and 180 mg/min, the standard therapy is 25 g of glucose given intravenously (50 ml of 50 per cent glucose), which usually reawakens a comatose patient in 5 to 10 min. With a plasma half-life of 60 to 80 min, the blood glucose level will fall after this initial bolus depending on the hypoglycemic stimulus and the metabolic state, potentially leading to a relapse into neuroglycopenic symptoms.

If the awakened patient has sufficient protection of the airway (i.e. positive swallowing, gag and cough reflexes), an oral glucose load should be given as soon as possible to maintain blood glucose levels, followed by a high-carbohydrate meal. If an oral load is contraindicated, it is imperative to add an infusion of glucose to initial parenteral therapy (e.g. 200 ml/h of 10 per cent glucose infusion in an adult), with the rate adjusted as necessary to obtain normoglycemia.

Glucose solutions of 10 per cent or stronger are acidic, hyperosmotic, and highly irritant to tissues (the pH of 10 per cent glucose is 5.6, and the osmolarity is 523 mosmol/l). They will thrombose any venous access and should be followed with a flush of 0.9 per cent saline.

Other pharmacological options

In some circumstances the hypoglycemic stimulus is so powerful that an infusion of glucose alone is insufficient to restore normoglycemia. This can occur with insulin-secreting tumors. Additional therapeutic measures in these circumstances are aimed at reducing endogenous insulin production and promoting gluconeogenesis.

Diazoxide is related to the thiazide diuretics, emphasizing the inhibition of insulin secretion by β-cells but interestingly causing salt and water retention. It has powerful vasodilator activity, causing reflex tachycardia. An infusion of 3 to 8 mg/kg/day in 5 per cent glucose or a bolus of 30 to 300 mg intravenously are suggested dosages.

Somatostatin and its longer-acting analog octreotide are often used for resistant hypoglycemia caused by insulinoma. However, these drugs have variable effects on gut hormone release and can cause a fall in blood glucose in some patients. Their use should be closely monitored with serial blood glucose measurements.

Glucocorticoids bind to intracellular receptors and alter protein and receptor synthesis to exert their multiple pharmacodynamic effects, of which stimulation of gluconeogenesis, muscle catabolism, and lipolysis all act to increase blood glucose. Intravenous administration of hydrocortisone 100 mg every 6 h is a useful adjunct that may help in resistant hypoglycemia.

Thiamine and Wernicke's encephalopathy

Caution should be exercised in the correction of hypoglycemia in the malnourished. Thiamine deficiency will be accelerated if circulating glucose substrate is rapidly restored, causing acute and potentially fatal Wernicke's encephalopathy. This devastating syndrome is often misdiagnosed as brainstem stroke and results from cellular necrosis in structures surrounding the fourth ventricle, with the sudden onset of symptoms which include ocular palsies, ataxia, and mental confusion. Thiamine will dramatically reverse the early symptoms of Wernicke's encephalopathy, often within a few hours, but the probability of permanent dysfunction increases with the delay in administration. For this reason, supplementary thiamine 50 mg intramuscularly is usually given prior to glucagon or glucose administration if there is any doubt concerning the nutritional status of the patient. Some emergency protocols mandate the use of thiamine as part of the initial management of coma, given its lack of side-effects.

Post-hypoglycemic coma

Prolonged hypoglycemia, particularly in association with hypoxemia and hypotension, leads to cerebral edema. Recovery may be delayed despite the attainment of normoglycemia, a condition historically termed 'irreversible hypoglycemic coma' but better described as 'post-hypoglycemic coma'. The delay probably reflects the severity of underlying brain damage. In addition to the measures outlined above to restore normoglycemia, additional interventions to reduce cerebral swelling are required, such as head elevation, hyperventilation, high-dose steroids (dexamethasone 8 mg every 6 h), diuretics, and mannitol (0.5–1 g/kg). The prognosis for patients with this condition is varied, with outcomes ranging from complete recovery to persistent vegetative state and death (7–16 per cent). The only prognostic factor is the duration of coma before attempts at correction.

Reference

Andreani, D., Marks, V., and Lefebvre, P.J. (1987). *Hypoglycemia*. Serono Symposia Publications, Vol. 38. Raven Press, New York.

8.7 Hypoadrenalism

8.7.1 Pathophysiology

L. G. Thijs

Key messages

- Glucocorticoids act on numerous tissues via the intracellular cytosol–nuclear receptor mechanism.
- Glucocorticoids are important for cardiovascular function with permissive effects on the vasopressor actions of norepinephrine (noradrenaline) and angiotensin II.
- Failure to produce aldosterone results in renal sodium wasting with concomitant water loss and potassium retention.
- In acute hypoadrenalism (adrenal crisis), hemodynamic instability and shock, which is often unresponsive to volume loading and vasopressor agents, are prominent features.
- The concept of relative adrenal insufficiency in critically ill patients, representing an inadequate adrenal response to severe stress, has emerged.

Introduction

Hypoadrenalism is an uncommon condition in the intensive care unit (ICU), but it has important consequences and is potentially life-threatening when not recognized in time. Hypoadrenalism is either primary, as the result of adrenal pathology, or secondary as a consequence of hypothalamic–pituitary insufficiency (deficient secretion of pituitary ACTH or hypothalamic corticotrophin-releasing hormone (CRH)). Iatrogenic hypoadrenalism caused by suppression of the hypothalamic–pituitary–adrenal axis by glucocorticoid administration is a relatively common condition.

Causes

The causes of primary hypoadrenalism are listed in Table 1. Tuberculosis, once a major cause of primary hypoadrenalism, has been largely replaced by autoimmune (Addison's) disease (either isolated or as part of a polyglandular autoimmune syndrome) as the leading cause of hypoadrenalism in the industrialized countries. Addison's disease has a marked female predilection and usually occurs in the third to fifth decades of life. Recently, AIDS has increasingly been recognized as a cause of primary adrenal dysfunction due to necrotizing adrenalitis.

Table 1 Causes of primary hypoadrenalism

Autoimmune (Addison's) disease
Infections
Tuberculosis
HIV-associated
Histoplasmosis
Blastomycosis
Coccidoidomycosis
Other
Malignancy
Primary
Metastatic disease (lung, breast, lymphoma, gastrointestinal, melanoma)
Medications
Ketoconazole
Etomidate
Suramin
Rifampin
Phenytoin
Other
Infiltrative disease
Sarcoidosis
Amyloidosis
Hemochromatosis
Adrenal hemorrhage
Sepsis
Anticoagulant therapy
Infarction
Other
Adrenoleukodystrophy
Congenital adrenal hypoplasia

Disseminated fungal infections, particularly histoplasmosis, can cause adrenal insufficiency in endemic areas. Metastatic infiltration of the adrenal glands is common, particularly in lung and breast cancer, but clinically recognized adrenal insufficiency is uncommon. Some drugs (e.g. ketoconazole, etomidate, suramin) inhibit cortisol biosynthesis, whereas others (e.g. rifampin (rifampicin), phenytoin) accelerate the metabolism of cortisol and these can induce adrenal insufficiency in patients with a limited hypothalamic–pituitary–adrenal axis reserve.

Pathophysiology

In Addison's disease a combined autoantibody and cell-mediated immune reaction against adrenocortical cells ultimately results in fibrous cortical replacement with relative sparing of the medulla. Clinical signs and symptoms do not become manifest until at least 90 per cent of the adrenocortical tissue is destroyed. In other conditions (e.g. tuberculosis, adrenal hemorrhage) the medulla is also affected. However, there is no distinct clinical syndrome of adrenal medullary insufficiency. This could be explained by the fact that catecholamines are produced by other tissues such as the sympathetic nerves. In primary hypoadrenalism both glucocorticoid and mineralocorticoid secretion are deficient. In secondary/tertiary hypoadrenalism primarily glucocorticoid deficiency occurs. In both conditions adrenal androgen secretion is also affected which has some clinical implications in women (e.g. loss of pubic and axillary hair).

Glucocorticosteroids

Glucocorticoid production (cortisol is the principal hormone) in the adrenal zona fasciculata and zona reticularis is controlled primarily by ACTH released from the anterior pituitary. Glucocorticoids inhibit the release of CRH and ACTH as part of a negative feedback mechanism. Glucocorticoids act on numerous tissues via the intracellular cytosol–nuclear receptor mechanism. They affect the sense of well-being, appetite, and mood. They promote hepatic gluconeogenesis and glycogen storage, promote protein catabolism, and decrease peripheral utilization of glucose in tissues. These concerted actions result in an increase in blood glucose concentration, which subsequently increases circulating insulin levels. Glucocorticoids are important for cardiovascular function. Glucosteroids have permissive effects on the vasopressor actions of norepinephrine and angiotensin II. These hormones increase synthesis of β-adrenergic receptors, reverse β-adrenergic dysfunction, and increase coupling between cell membrane β-adrenergic receptors and second-messenger systems.

Glucosteroids increase free water excretion by the kidneys and are also important for potassium homeostasis. They have mineralocorticoid-like renal effects, specifically in the absence of aldosterone. Glucocorticoids affect gastrointestinal motility. These hormones increase circulating neutrophils but impair their recruitment into an inflamed area and decrease the number of circulating lymphocytes, monocytes, eosinophils, and basophils. Effects on calcium homeostasis are inhibition of intestinal calcium absorption, increase of renal calcium excretion, and inhibition of osteoblast function. These various activities of glucocorticoids largely explain the clinical manifestations of glucocorticoid deficiency.

Mineralocorticoids

The synthesis and secretion of the principal mineralocorticoid aldosterone in the adrenal zona glomerulosa are stimulated by potassium ions, ACTH, serotonin, and β-endorphins, and are inhibited by dopamine, somatostatin, and atrial natriuretic peptide. However, the primary regulatory system of aldosterone is the renin–angiotensin system. Renin release from the kidney juxtaglomerular apparatus is stimulated by the integrated effects of a number of mechanisms, including a decrease in renal blood flow (through volume-sensitive stretch receptors in the renal afferent arterioles), β-adrenergic activity, and a low sodium concentration of the tubular fluid sensed by the macula densa. Renin converts angiotensinogen to angiotensin I, which in turn is cleaved by the angiotensin-converting enzyme to form angiotensin II which is a strong vasoconstrictor and stimulator of aldosterone secretion. Aldosterone is much less under the control of ACTH than cortisol. The principle actions of aldosterone are promotion of sodium reabsorption linked to excretion of potassium, and to a lesser extent hydrogen ions, in the distal renal tubules and collecting ducts. Aldosterone also affects sodium excretion via the gut and the salivary and sweat glands.

Failure to produce aldosterone results in renal sodium wasting with concomitant water loss and potassium retention. Partial compensation occurs with stimulation of the renin–angiotensin system and ADH secretion.

Hemodynamic alterations

Hemodynamic instability and shock, which is often unresponsive to volume loading and vasopressor agents, are prominent features of acute hypoadrenalism (adrenal crisis). Several mechanisms contribute to cardiovascular failure. Mineralocorticosteroid deficiency resulting in sodium and water loss and subsequent hypovolemia, which may be aggravated by vomiting and diarrhea, is the major pathogenetic factor.

However, glucocorticoids are also of prime importance in maintaining adequate cardiovascular function as shock may occur in glucocorticoid deficiency. Glucocorticoid deficiency is associated with reduced myocardial contractility which responds to corticosteroid treatment. Withdrawal of glucocorticosteroid treatment in patients with chronic hypoadrenalism results in a pronounced fall in lymphocyte β_2-adrenergic receptor density, which may reflect changes in cardiovascular β-receptors, and a reduced cAMP response to isoprenaline. This is associated with impaired left ventricular relaxation and diastolic filling which may contribute to cardiovascular failure in adrenal crisis. Patients with substituted chronic hypoadrenalism have significantly lowered levels of plasma epinephrine (with normal norepinephrine levels), indicating that high local adrenocortical glucocorticoid concentrations are necessary for normal epinephrine production via adrenal medullary enzyme induction. This may have implications for patients in adrenal crisis. Glucocorticoid deficiency is associated with a reduced peripheral vascular adrenergic tone, which may be due to a reduced sensitivity to pressor effects of angiotensin II and norepinephrine, decreased synthesis of renin substrate, and loss of glucocorticoid-induced inhibition of vasodilating prostaglandins. Such mechanisms may explain the hyperdynamic profile with a high cardiac output and a low peripheral vascular resistance observed in patients with hypoadrenalism after correction of hypovolemia.

Impaired renal water excretion

Mineralocorticoid deficiency impairs water excretion by mechanisms activated by a decrease in effective circulating volume. These include non-osmotic stimulation of ADH secretion and impaired renal free water generation owing to reduced distal fluid delivery in the kidneys. However, glucocorticoid deficiency is usually of primary importance, as cortisol replacement rapidly increases the rate of water excretion and raises the plasma sodium concentration to normal. Plasma levels of ADH are usually inappropriately elevated in glucosteroid-deficient patients and return to normal upon glucocorticoid replacement. Persistent ADH secretion in patients with glucocorticoid deficiency is probably due to a loss of hypotonic suppression of ADH release owing to glucocorticoid deficiency *per se*, which is aggravated secondarily by multiple non-osmotic stimuli such as nausea and hypoglycemia. Both sodium loss by mineralocorticoid deficiency and ADH-related retention of free water contribute to the development of hyponatremia.

Adrenal dysfunction in critical illness

Normal adrenal function is vital to survival from severe (septic) insults. Plasma cortisol concentrations increase during critical illness and this rise under conditions of stress appears to be a protective mechanism. In severe illness there is a relative shift in the metabolism of adrenal pregnenolone away from mineralocorticoid and adrenal androgen pathways and toward the glucosteroid pathway.

Increased plasma concentrations of cortisol in critical illness reflect a balance between an increased cortisol secretory rate and a reduced (hepatic) degradation. Septic patients with hepatic disease may have higher cortisol levels than septic patients without liver disease (Bouachour *et al.* 1995). A decreased level of corticosteroid-binding globulin — a specific transport protein for glucocorticoids — further increases free cortisol concentrations. Marked elevated cortisol concentrations have been found in patients with medical or surgical illness, severe bacterial infections, or septic shock (Rothwell *et al.* 1991; Moran *et al.* 1994; Bouachour *et al.* 1995; Soni *et al.* 1995) and in patients in a general ICU (Jurney *et al.* 1987). However, there is a wide range in cortisol levels with (low) normal to extremely high values (Rothwell *et al.* 1991; Moran *et al.* 1994). Also, in these conditions the adrenal cortex is usually able to augment further cortisol secretion upon ACTH stimulation (Bouachour *et al.* 1995).

Although ACTH levels are usually elevated also, this is not observed in all patients (Bouachour *et al.* 1995; Soni *et al.* 1995). There is probably an altered regulation of the hypothalamic–pituitary–adrenal axis. Several explanations have been suggested, including diminished sensitivity of the pituitary to glucosteroids, synergism of other hormones such as vasopressin with CRH, and stimulation of the activity of cytokines on the hypothalamic–pituitary–adrenal axis.

Although plasma cortisol concentrations are usually elevated compared with normal values in these patients, they may be insufficient to meet the tissue requirements. The concept of relative adrenal insufficiency in critically ill patients, representing an inadequate adrenal response to severe stress, has emerged (Rothwell *et al.* 1991). However, it is difficult to define an inadequate level of plasma cortisol in these conditions and it may vary with the severity of the insult. A basal level of above 500 nmol/l has been considered adequate, but it has not been demonstrated that somewhat lower levels are insufficient (Jurney *et al.* 1987; Moran *et al.* 1994) and in some cases higher levels may be required. Mortality usually increases with increases of plasma cortisol concentrations, probably reflecting the severity of illness (Jurney *et al.* 1987; Moran *et al.* 1994).

Adrenocortical insufficiency has been documented in critically ill patients with severe infections, particularly disseminated meningococcemia and other Gram-negative sepsis. This is usually related to hemorrhagic necrosis of the adrenal gland, which has a poor outcome when not adequately treated. However, absolute hypoadrenalism is a rarity in severe sepsis (Rothwell *et al.* 1991; Bouachour *et al.* 1995).

Several studies have demonstrated that an inadequate adrenocortical function or reserve, assessed as a suboptimal response to synthetic ACTH (tetracosactrin), occurs during severe illness. Various criteria have been used to assess inadequate adrenal response: less than 500 nmol/l as a maximum plasma cortisol concentration following stimulation (Jurney *et al.* 1987; Bouachour *et al.* 1995; Soni *et al.* 1995) and/or a rise of less than 200 nmol/l (Moran *et al.* 1994; Bouachour *et al.* 1995) or 250 nmol/l (Rothwell *et al.* 1991; Bouachour *et al.* 1995) 30 or 60 min after stimulation. Relative adrenal insufficiency defined in this way appears to be common, particularly in septic shock (Rothwell *et al.* 1991; Moran *et al.* 1994; Bouachour *et al.* 1995; Soni *et al.* 1995), but the exact incidence may vary widely with the criteria used (Bouachour *et al.* 1995). Basal levels of cortisol do not differ between patients who respond adequately or inadequately to ACTH stimulation, supporting the importance of this test in unmasking relative hypoadrenalism (Rothwell *et al.* 1991; Bouachour *et al.* 1995).

Critically ill patients with a poor response to ACTH usually have a higher mortality rate than those who respond adequately (Rothwell *et al.* 1991; Soni *et al.* 1995), but this has not been documented in all studies (Bouachour *et al.* 1995). Cortisol replacement in these patients may improve hemodynamics and sometimes (short-term) survival (Soni *et al.* 1995), but controlled studies are lacking at present. It has been suggested that this relative hypoadrenalism occurs *de novo* and is reversible, but pre-existing subclinical deficient adrenal function cannot usually be ruled out.

Several mechanisms could be involved in the pathogenesis of relative adrenal insufficiency in critical illness, but at present they are largely hypothetical. Inadequate perfusion of areas of the hypothalamic–pituitary–adrenal axis, the presence of circulating substances which inhibit adrenal function or ACTH action, and diminished pituitary function are possible mechanisms (Soni *et al.* 1995). In critical illness the hypothalamic–pituitary–adrenal axis is activated by, among others, mediators of the systemic inflammatory response. There is extensive bidirectional communication between the hypothalamic–pituitary–adrenal axis and the immune system. Several cytokines (interleukin 1, interleukin 2, and tumor necrosis factor) can activate the hypothalamic–pituitary–adrenal axis. They seem to act primarily at the level of CRH stimulation, but also have stimulatory or inhibitory actions on the pituitary and on adrenal cortical cells. In contrast, glucocorticoids modulate the inflammatory response and decrease release of proinflammatory cytokines, creating a negative feedback loop. Sepsis may affect adrenocortical function by various mechanisms: decreased binding of corticotropin to the membrane receptor, inhibition of cAMP production, and inhibited conversion of cholesterol to pregnenolone. Transforming growth factor-β, another immune modulator, can inhibit the actions of ACTH on adrenal cortical cells, and certain macrophage and neutrophil secretagogues may inhibit steroidogenesis. Corticostatins, a family of peptides induced by various cells including granulocytes, are able to compete with ACTH for its binding receptor on the adrenal cell and thus can inhibit adrenal cortical steroidogenesis.

Stimulated human mononuclear cells can produce ACTH-like peptides which can be postulated to interfere with ACTH binding to its receptor sites and possibly alter the adrenal response to ACTH (Soni *et al.* 1995). These various mechanisms could be responsible for a reduced effect of ACTH on the adrenal cortex in critically ill patients. Insufficient secretion of ACTH by the pituitary gland in response to stress might be another mechanism responsible for a relatively low plasma cortisol concentration (Soni *et al.* 1995). The complex interactions of these various mechanisms may be responsible for the large variations in plasma cortisol levels in critically ill patients. In some patients this may result in an inadequate adrenal response which could possibly be considered as part of the multiple organ dysfunction syndrome which may complicate severe illness.

References

Bouachour, G., Tirot, P., Gouello, J.P., Mathieu, E., Vincent, J.F., and Alquier, P. (1995). Adrenocortical function during septic shock. *Intensive Care Medicine*, **21**, 57–62.

Jurney, T.H., Cockrell, J.L., Jr, Lindberg, J.S., Lamiell, J.M., and Wade, C.E. (1987). Spectrum of serum cortisol response to ACTH in ICU patients. Correlation with degree of illness and mortality. *Chest*, **92**, 292–5.

Moran, J.L., Chapman, M.J., O'Fathartaigh, M.S., Peisach, A.R., Pannall, P.R., and Leppard, P. (1994). Hypocortisolaemia and adrenocortical responsiveness at onset of septic shock. *Intensive Care Medicine*, **20**, 489–95.

Rothwell, P.M., Udwadia, Z.F., and Lawler, P.G. (1991). Cortisol response to corticotrophin and survival in septic shock. *Lancet*, **337**, 582–3.

Soni, A., *et al.* (1995). Adrenal insufficiency occurring during septic shock: incidence, outcome, and relationship to peripheral cytokine levels. *American Journal of Medicine*, **98**, 266–71.

8.7.2 Therapeutic strategy

L. G. Thijs

Key messages

- In gradually developing primary hypoadrenalism, increasing weakness, fatigue, lassitude, anorexia, myalgia, and weight loss are the first symptoms.
- In acute adrenal failure symptoms are generally non-specific and may include vomiting, diarrhea, abdominal pain, weakness, myalgias, joint pains, lethargy, confusion, delirium, or even coma.
- A basal cortisol concentration below 275 mmol/l ($<10\,\mu$g/dl) strongly suggests hypoadrenalism, and a level of 550 nmol/l (20 μg/dl) or above argues strongly against it. A short ACTH (tetracosactrin) stimulation test should be performed in virtually all patients in whom the diagnosis is considered.
- Dexamethasone (4 mg) or hydrocortisone (100 mg) are the glucocorticoid preparations of choice in the emergency situation and should be immediately injected intravenously.
- Correction of hypovolemia requires a large volume (usually at least 3 liters) of 0.9 per cent saline which also improves hyponatremia.

Introduction

Acute adrenal hypofunction may develop during acute injury or disease in patients with known hypoadrenalism or may be the first presentation. The stress of injury unmasks a pre-existing insufficient hypothalamic–pituitary–adrenal axis in these patients. Acute hypoadrenalism can also develop with adrenal hemorrhage. In critically ill patients the adrenal response to stress may be insufficient to meet the metabolic demands, and this condition is referred to as relative *de novo* hypoadrenalism. Deficient aldosterone production can occur in conditions other than Addison's disease.

Clinical manifestation

The clinical presentation of adrenal insufficiency is variable, depending on its pathogenesis, and ranges from vague non-specific symptoms to a sudden cardiovascular collapse (adrenal crisis).

Chronic hypoadrenalism

The onset of clinical manifestation is usually insidious with gradual development of signs and symptoms, each of which is non-specific alone. Increasing weakness, fatigue, lassitude, anorexia, myalgia, and weight loss are the first symptoms in gradually developing primary hypoadrenalism. Weight loss, which is mostly due to anorexia but can also be caused by dehydration, may be substantial. Vomiting and constipation, apathy, depression, and even psychosis may also occur. Vomiting and abdominal pain often herald an adrenal crisis. In some instances a history of craving for salt or salty foods may be present. Blood pressure is usually low and signs of dehydration are common. Hyperpigmentation, particularly of the lips, buccal mucosa, gingiva, palmar and plantar creases, areolas, and previous scars, is characteristic of primary hypoadrenalism but may be absent in hypoadrenalism of recent onset. Hyperpigmentation may be generalized, creating a darkish cast to the skin resembling a sun tan. The ACTH precursor pro-opiomelanocortin, which is produced by the pituitary, is also the precursor of β-melanocyte-stimulating hormone which also increases. This explains the hyperpigmentation observed in (long-standing) Addison's disease which is absent in secondary/tertiary hypoadrenalism. In many instances patients with chronic hypoadrenalism may be able to survive in the basal state but develop an adrenal crisis following trauma or infection. The main symptoms and signs of chronic hypoadrenalism are summarized in Table 1.

Acute adrenal insufficiency (adrenal crisis)

Acute adrenal insufficiency is a true endocrinological emergency. It usually presents as shock in a previously undiagnosed patient with (primary) hypoadrenalism who has been subjected to major stress or in a patient with established hypoadrenalism who does not or is not able to increase glucocorticoid replacement during an intercurrent illness or injury. A febrile illness, bacterial infection, trauma, surgery, dehydration, or any other intercurrent illness in these patients can trigger acute adrenal failure as the diseased adrenal cortex is unable to respond adequately to this type of stress. Acute hypoadrenalism can also result from bilateral adrenal hemorrhage, embolus or thrombosis, and, rarely, adrenal vein thrombosis (Rao *et al.* 1989; Orth *et al.* 1992). In the intensive care setting bilateral hemorrhage is most commonly seen in overwhelming sepsis, classically in meningococcemia but also in other Gram-negative

Table 1 Clinical manifestations of chronic hypoadrenalism

Symptoms
General: fatigue, tiredness, weight loss, salt craving
Gastrointestinal: anorexia, nausea, vomiting, constipation, abdominal pain, diarrhea
Cardiovascular: postural dizziness, syncope
Skeletomuscular: myalgia, arthralgia, weakness
Central nervous: confusion, lethargy, apathy, psychosis
Reproductive: menstrual abnormalities, decreased libido
Signs
Weight loss
Hyperpigmentation
Dehydration
(Orthostatic) hypotension
Vitiligo
Auricular calcification
Pubic and axillary hair loss
Laboratory findings
Hyponatremia
Hyperkalemia
Hypercalcemia
Hypoglycemia
Azotemia
Eosinophilia
Lymphocytosis
Anemia
Mild metabolic acidosis

sepsis (e.g. *Pseudomonas aeruginosa*), use of anticoagulants, and HIV infection.

Symptoms are generally non-specific and may include vomiting, diarrhea, abdominal pain, weakness, myalgias, joint pains, lethargy, confusion, delirium, or even coma. Most patients have fever due to the precipitating infection, but also without infection and often out of proportion to any minor infection. Fever and abdominal pain and tenderness can mimic an acute surgical abdomen. Acute flank, lower chest, or back pain can be present in patients with acute adrenal hemorrhagic necrosis in addition to other signs and symptoms of adrenal crisis (Rao *et al.* 1989). Abdominal rigidity or rebound may be observed (Rao *et al.* 1989). Shock in adrenal crisis usually presents as hypovolemic shock due to a marked loss of plasma volume, but after correction of hypovolemia a hyperdynamic profile is often present, mimicking septic shock (Bouachour *et al.* 1994). Shock due to adrenal crisis may be masked by or contribute to circulatory shock associated with other concomitant critical illness, particularly septic shock.

Diagnosis

The most important factor in diagnosing hypoadrenalism in the intensive care unit (ICU) setting is a high index of clinical suspicion as its clinical presentation is usually non-specific. Previous or present use of corticosteroids, use of anticoagulants, a history of tuberculosis, metastatic malignancy, and (ongoing) sepsis are among the conditions that may be associated with acute hypoadrenalism and should rouse this suspicion. Adrenal crisis should be considered in (septic) shock that is unresponsive to volume expansion and vasopressor agents and in (postoperative) patients with unexplained abdominal complaints, fever, and/or hypotension. Clinical features of chronic hypoadrenalism when present (e.g. hyperpigmentation) and laboratory findings are helpful. However, in acute cases laboratory abnormalities are often less pronounced than in chronic hypoadrenalism. Hyponatremia with a high urinary sodium concentration, hyperkalemia, and hypoglycemia are important clues to the diagnosis of hypoadrenalism. The sodium-to-potassium ratio will almost always be below 30. However, hyperkalemia is not a feature of secondary hypoadrenalism. Eosinophilia, leukocytosis, increased blood urea and calcium concentrations, and a mild metabolic acidosis support the diagnosis of hypoadrenalism, but these abnormalities (with the exception of eosinophilia) are common in critically ill patients. The basis of the diagnosis of adrenal insufficiency depends on the demonstration of inadequate cortisol production. A basal cortisol concentration below 275 mmol/l ($< 10\,\mu g/dl$) strongly suggests the diagnosis, and a level of 550 nmol/l ($20\,\mu g/dl$) or above argues strongly against it. However, in virtually all patients in whom the diagnosis is considered, a short ACTH (tetracosactrin) stimulation test should be performed because of the variable and pulsatile nature of the response of the hypothalamic–pituitary–adrenal axis to stress. The test involves the intravenous administration of 0.25 mg tetracosactrin and measurement of plasma cortisol concentrations before stimulation and 30 and/or 60 min afterwards. A rise in plasma cortisol concentration between 30 and 60 min to a peak of at least 600 mmol/l or more virtually excludes primary hypoadrenalism (Patel *et al.* 1991).

This is an accurate and safe diagnostic test in the ICU setting. However, an adequate response does not eliminate secondary/tertiary adrenal insufficiency of recent onset. Measurement of ACTH levels prior to the ACTH test will show low-normal or decreased concentrations in these cases. Available provocative tests, which centrally stimulate the hypothalamic–pituitary–adrenal axis, are the insulin-induced hypoglycemia and metyrapone tests (Grinspoon and Biller 1994). They can help to discriminate between primary and secondary/tertiary hypoadrenalism but are not practical in the ICU. A corticotropin-releasing hormone stimulation test can test the ability to secrete sufficient ACTH, but an exaggerated and prolonged ACTH response is observed in primary hypothalamic disorders. Prolonged ACTH stimulation tests can also be helpful in distinguishing primary from secondary/tertiary adrenal insufficiency, but their use in critically ill patients is rather cumbersome.

In addition to laboratory investigations, diagnostic imaging modalities such as computer tomography and magnetic resonance imaging are useful in the diagnosis of adrenal insufficiency. Both techniques are able to identify anatomical lesions in the area of the hypothalamic–pituitary–adrenal axis.

Treatment

Acute adrenal failure is a life-threatening emergency that requires immediate and appropriate treatment. Prompt recognition and treatment is critical for survival. The mainstay of treatment is administration of glucocorticoids and rapid correction of hypovolemia. When there is a strong clinical suspicion, therapy should not be delayed to perform diagnostic tests or await laboratory results. Blood samples should be taken for plasma cortisol measurement and other laboratory tests (in any case electrolytes and glucose) and treatment initiated immediately. Dexamethasone (4 mg) or hydrocortisone (100 mg) are the glucocorticoid preparations of choice in the emergency situation and should be immediately injected intravenously. When there is no time for a short ACTH stimulation test, dexamethasone is preferred as it does not significantly interfere with the plasma cortisol assay so that this test can be performed at a later stage. The effects of dexamethasone last for 12 to 24 h. When the ACTH test has been performed, hydrocortisone, which has adequate mineralocorticoid activity when given in large doses, can be substituted for dexamethasone. It is given intravenously at least every 6 h during the first 24 h, starting immediately after the ACTH test. In patients with concomitant hypothyroidism (in secondary/tertiary hypoadrenalism), glucosteroids should always be given prior to thyroid hormone replacement. Thyroid hormone increases glucocorticoid metabolism and therefore may worsen the hypoadrenal state. If relative adrenal insufficiency is suspected in critically ill patients, hydrocortisone 100 mg given intravenously every 6 to 8 h should be considered after a short ACTH stimulation test. Therapy can be adapted when the results of this test are available. Mineralocorticoid adminstration in the case of suspected primary hypoadrenalism is not useful acutely as it takes several days for the sodium-retaining effects to become manifest and adequate sodium replacement can be achieved by intravenous saline infusions (Orth *et al.* 1992).

Correction of hypovolemia requires a large volume (usually at least 3 liters) of 0.9 per cent saline which also improves hyponatremia. Additional glucose is necessary to combat hypoglycemia. Invasive hemodynamic monitoring could be helpful in guiding fluid therapy, but is not usually needed to manage acute hypoadrenalism unless required for adequate treatment of precipitative conditions such as septic shock. Hyperkalemia rarely exceeds 7.0 mmol/l and usually responds to correction of hypovolemia, glucose administration, and glucocorticoids. However, monitoring of potassium levels is necessary as total body potassium may be decreased due to vomiting and diarrhea, and potassium supplementation may be needed at a later stage. After this emergency treatment the precipitating cause of adrenal crises should be sought and treated appropriately. The probability of a bacterial infection and sepsis should always be taken into consideration.

Once the patient has stabilized, the hydrocortisone dose can generally (depending on the course of the underlying disease) be gradually tapered (after the first 24 h) until a maintenance dose of 20 to 30 mg/day is achieved. Most patients who present with adrenal crisis have primary hypoadrenalism and therefore also require (lifelong) mineralocorticosteroid replacement. Fludrocortisone can be started when the patient is able to take oral medication and the glucocorticosteroid therapy is near or at its normal maintenance dose. The usual initial dose is 0.1 mg/day and can be adjusted to achieve a normal blood pressure without excessive sodium retention or hypokalemia.

Isolated mineralocorticoid deficiency

Deficient aldosterone production can occur in patients who have a normal cortisol production (Orth *et al.* 1992). The most common form is hyporeninemic hypoaldosteronism which is often associated with diabetes mellitus but also with other disease states. Several mechanisms have been proposed to explain deficient renin production (Orth *et al.* 1992). Isolated hypoaldosteronism also occurs in congenital hypoaldosteronism which is due to an inherited isolated defect in aldosterone biosynthesis. Several other defects in the renin–angiotensin–aldosterone system, with resulting functional hypoaldosteronism, have been described. Heparin suppresses aldosterone biosynthesis with a subsequent rise in plasma renin activity. Dopamine also has a direct inhibitory effect on aldosterone secretion. Both these drugs are frequently used in the ICU. The diagnosis of isolated hypoaldosteronism must be considered in patients with unexplained (chronic) hyperkalemia which is the most important (clinical) manifestation. This may be asymptomatic, but may cause muscle weakness and cardiac arrhythmias in a minority of patients (Orth *et al.* 1992). Sodium depletion and significant renal sodium loss are not usually clinically significant. A mild renal tubular acidosis may be present. The diagnosis of hyporeninemic hypoaldosteronism can be established by a low plasma renin activity and a low plasma aldosterone concentration. In many patients with mild isolated hypoaldosteronism no therapy is required. Drugs that tend to promote hyperkalemia should be avoided and coexisting diseases, such as hypertension, congestive heart failure, and diabetes mellitus, should be adequately treated. When mineralocorticoid replacement therapy is considered necessary to control hyperkalemia, fludrocortisone with a starting dose of 0.1 mg/day can be given.

References

Bouachour, G., Tirot, P., Varache, N., Gouello, J.P., Harry, P., and Alquier, P. (1994). Hemodynamic changes in acute adrenal insufficiency. *Intensive Care Medicine*, **20**, 138–41.

Grinspoon, S.K. and Biller, B.M.K. (1994). Clinical review. Laboratory assessment of adrenal insufficiency. *Journal of Clinical Endocrinology and Metabolism*, **79**, 923–31.

Orth, D.N., Kovacs, W.J., and Debold, C.R. (1992). The adrenal cortex. In *Williams textbook of endocrinology* (8th edn) (ed. J.D. Wilson and D.W. Foster), pp. 489–536. W. B. Saunders, Philadelphia, PA.

Patel, S.R., Selby, C., Dornan, T.L., and Jeffcoate, W.J. (1991). The short Synacthen test in acute hospital admissions. *Clinical Endocrinology*, **35**, 259–61.

Rao, R.H., Vagnucci, A.H., and Amico, J.A. (1989). Bilateral massive adrenal hemorrhage: early recognition and treatment. *Annals of Internal Medicine*, **110**, 227–35.

8.8 Hypopituitarism

8.8.1 Pathophysiology

G. Van den Berghe

Key messages

- Prolactin subserves several physiological functions including effects on the immune system. Hypoprolactinemia is associated with immune suppression.
- Growth hormone deficiency induces protein hypercatabolism, deficient lipolysis, and altered carbohydrate metabolism, a metabolic dysfunction associated with reduced lean body mass and bone mineral density as well as impaired peripheral muscle and myocardial function.
- Acute corticotropin deficiency induces an Addisonian-like crisis characterized by electrolyte abnormalities, which are usually mild, and shock. The latter is unresponsive to vasopressor therapy until glucocorticoids are replaced.
- Thyrotropin deficiency induces central hypothyroidism, characterized by adverse effects on all tissues and organ systems including abnormal systolic and diastolic myocardial function, renal and central nervous system dysfunction, disturbances of water and sodium handling and of lipid and glucose metabolism, malabsorption syndrome, and biliary dysfunction.
- Gonadotropin deficiency induces secondary hypogonadism, characterized in men by reduced circulating androgen levels associated with protein hypercatabolism.

Introduction

Hypopituitarism is the clinical syndrome which results from failure of the anterior pituitary gland to secrete adequately one or more of its hormones. In contrast with the posterior pituitary, the anterior lobe is an endocrine gland. Prolactin, growth hormone, corticotropin, thyrotropin, and the gonadotropins (luteinizing hormone and follicle-stimulating hormone) are all primarily under control of regulatory hormones secreted by the hypothalamus and released into the hypothalamic–pituitary portal system. The somatotropic, thyrotropic, and gonadotropic axes are key determinants of normal growth and anabolism. Prolactin, the pituitary–adrenal axis (cortisol and dehydroepiandrosterone sulphate (DHEAS)), and growth hormone play important roles in immunological homeostasis.

Hypopituitarism can result from intrinsic hypothalamic lesions, intrinsic pituitary lesions, or extrinsic extrasellar disease which impinges on, infiltrates, or destroys the hypothalamus, the pituitary stalk, or the pituitary gland. All disorders can have an 'organic' or 'functional' origin; the latter has the potential for reversibility (Table 1).

A critical care specialist may have to deal with hypopituitarism as a recognized pre-existing disease that requires appropriately continued therapy during intercurrent illness, an acute exacerbation of an undiagnosed insufficiency evoked by surgery or disease, the sequela of recent neurosurgery, cerebral trauma, infection, or shock, or a temporary dysfunction induced by the stress of the critical condition itself and/or its therapy.

The clinical impact of acquired hypopituitarism in critical care conditions depends on which axes are involved in the lesion and on the severity and duration of the pituitary failure.

Prolactin deficiency

Prolactin is secreted by the pituitary lactotropes under hypothalamic control by prolactin-inhibiting factors (predominantly dopamine)

Table 1 Etiology of hypopituitarism

Hypothalamic
Hypothalamic or central nervous system disease
Organic
Developmental
Traumatic
Inflammatory
Neoplastic
Idiopathic
Functional
Stress
Anorexia nervosa
Renal failure, hepatic failure, uncontrolled diabetes mellitus
Prolonged critical illness
Drugs (dopamine, somatostatin, glucocorticoids, etc.)
Stalk trauma or compression
Pituitary
Organic
Developmental aplasia or hypoplasia
Traumatic or postneurosurgical
Inflammatory or infiltrative (infectious or systemic disease)
Ischemic (Sheehan postpartum pituitary necrosis, diabetes mellitus, arteritis, eclampsia)
Pituitary apoplexia
Neoplastic, miscellaneous
Functional
Drugs (dopamine, somatostatin)
Extrasellar disease

and to a lesser extent by prolactin-releasing factors (including thyrotropin-releasing hormone) (Thorner *et al.* 1992). In addition to its role in the induction of maternal lactogenesis, prolactin has other physiological functions including effects on the immune system (Murphy *et al.* 1995).

Prolactin receptors are present on human T and B lymphocytes, and T lymphocytes depend on prolactin for maintenance of immune competence. In mice, inhibition of prolactin release results in impaired lymphocyte function, depressed lymphokine-dependent macrophage activation, and death from a normally non-lethal exposure to bacteria. Cyclosporin is known to compete with prolactin for a common binding site on T cells, and its immunosuppressive effect is thought to be exerted in part by blocking the prolactin receptor. Suppression of prolactin secretion by the dopamine agonist bromocriptine is emerging as an adjuvant immunosuppressive strategy in autoimmune disease and after organ transplantation.

Serum prolactin concentrations are assumed to rise in response to acute stress, possibly as part of a natural defense against infections. However, in chronic stress conditions prolactin levels are no longer elevated. In critically ill subjects, dopamine infusion induces severe hypoprolactinemia that has been linked with impaired T-lymphocyte proliferative response (Devins *et al.* 1992; Van den Berghe and de Zegher 1996). The lack of prolactin secretion during dopamine administration or prolonged stress may be a codeterminant of immune dysfunction (anergy) in the critically ill.

Growth hormone deficiency

Growth hormone is a polypeptide with anabolic effects on protein metabolism and with lipolytic and immunostimulating properties (Thorner *et al.* 1992). It is secreted by pituitary somatotropes in a pulsatile fashion that is believed to result principally from a dynamic interaction between two hypothalamic neuropeptides, the stimulatory growth-hormone-releasing hormone and the inhibitory somatostatin. Growth hormone has direct and indirect anabolic actions; the principal mediator of the latter is insulin-like growth factor I (IGF-I). To a certain extent, growth hormone action is reflected in serum concentrations of IGF-I, which is mainly generated by the liver and bound to six specific binding proteins, predominantly to IGF-binding protein 3. In critically ill patients, serum IGF-I correlates well with conventional nutritional indices such as nitrogen balance. The immunostimulating properties of growth hormone are manifested in direct action on T lymphocytes and indirect effects through IGF-I on neutrophils (Murphy *et al.* 1995).

Growth hormone deficiency is probably the most common endocrine expression of hypopituitarism. Acquired growth hormone deficiency in adulthood results in reduced muscle strength and exercise capacity, reduced thermoregulation and sweating ability, subnormal kidney function, decreased lean body mass and bone mineral density, abnormal thyroid hormone, lipid, and carbohydrate metabolism, myocardial dysfunction, and impaired social well being, and it leads to increased mortality due to cardiovascular disease (De Boer *et al.* 1995). Growth hormone replacement therapy in growth-hormone-deficient adults improves these symptoms.

During prolonged critical illness, growth hormone secretion is characterized by reduced pulse amplitude and sometimes by elevated interpulse levels and is associated with low circulating IGF-I and IGF-binding protein 3 levels (Van den Berghe and de Zegher 1996). Dopamine infusion further attenuates pulsatile growth hormone secretion and lowers serum IGF-I. Through this endocrine side-effect, prolonged dopamine administration in the critically ill presumably contributes to the maintenance of fat depots and the failure to induce protein anabolism despite optimal feeding. In catabolic conditions requiring new protein synthesis, prolonged suppression of growth hormone secretion may contribute to impaired wound healing, delayed recovery, or atrophy of the enteral mucosa and muscle weakness. Likewise, exogenous and endogenous glucocorticoids and somatostatin may have a deleterious effect on the somatotrope axis, and hence anabolism, in severe illness.

Corticotropin deficiency

Pituitary corticotropin normally stimulates the adrenal cortex to secrete a cascade of steroids including cortisol, and its own secretion is principally stimulated by hypothalamic corticotropin-releasing hormone and arginine vasopressin (Thorner *et al.* 1992). Cortisol exerts negative feedback inhibition at the basophil cell level.

An intact pituitary adrenocortical axis is of vital importance, particularly in critical illness (Van den Berghe and de Zegher 1996). The stress of critical illness normally induces elevated serum cortisol concentrations with loss of circadian rhythm. This hypothalamic–pituitary–adrenal response to prolonged stress induced by trauma or sepsis was recently found to be biphasic. In the first phase, lasting for a few days, the high cortisol levels appear to be induced by augmented corticotropin release, which in turn is presumably driven by cytokines and the noradrenergic system. In the second phase, there is a discrepancy between low corticotropin and high cortisol levels, suggesting that cortisol release is stimulated through an alternative pathway. The hypercortisolism elicited by disease or trauma can be interpreted as an attempt of the organism to mute its own inflammatory cascade, thus protecting itself against possible endogenous over-responses as virtually all the components of the immune response are inhibited by cortisol. Moreover, the acute cortisol-induced shifts in carbohydrate and protein metabolism result in instantly available energy and postpone anabolism. Therefore inability to activate the hypothalamic–pituitary–adrenocortical axis acutely in response to illness or stress hampers survival if left untreated.

DHEAS, the most abundant steroid secreted by the zona reticularis of the adrenal cortex after adrenarche, is under pituitary control, possibly through corticotropin and/or prolactin (Thorner *et al.* 1992). Pan-hypopituitarism is always associated with low levels of DHEAS. It circulates as a prohormone and is peripherally converted into the active dehydroepiandrosterone (DHEA), which is a weak androgen and an anabolic agent, as well as a modulator of the immune response. DHEA stimulates the Th1-cell function directly through a specific intracellular receptor and enhances T-lymphocyte function indirectly through a glucocorticoid-antagonizing effect. Low levels of DHEAS, which occur gradually with advancing age and acutely in severe illness, may contribute to T-lymphocyte dysfunction.

Corticotropin deficiency is almost always accompanied by deficiency of other pituitary hormones. The usual presentations are similar to those of primary adrenocortical failure, except for two characteristics.

1. Because of corticotropin and related pro-opiomelanocortin-derived peptide deficiency, the characteristic hyperpigmentation

of Addison's disease is absent and patients have pallid skin and diminished tanning after exposure to sunlight.

2. Because of the primary dependence of aldosterone secretion on the renin–angiotensin system, the clinical features of mineralocorticoid deficiency are usually absent in corticotropin deficiency in the unstressed state.

Dehydration, volume depletion, and electrolyte abnormalities are not usually seen. Hypotension is less severe except in acute presentations. Although corticotropin deficiency is usually partial, a full-blown Addisonian-like crisis with shock may occur, with the latter unresponsive to vasopressor therapy until glucocorticoids are replaced. Such an acute presentation can occur in the undiagnosed corticotropin-deficient patient during acute illness or surgery, in the glucocorticoid-treated patient whose glucocorticoid dosage was not increased in such circumstances, and in pituitary apoplexia.

Dopamine infused as part of intensive care suppresses serum concentrations of DHEAS whereas the increased cortisol levels appear to be dopamine independent, suggesting a differential regulation of adrenal androgen and cortisol metabolism in critical illness (Van den Berghe and de Zegher 1996). This DHEAS suppression may be mediated in part by the concomitant hypoprolactinemia. Low serum levels of prolactin, growth hormone, and DHEAS in the presence of high cortisol concentrations may contribute to the anergy-type immune dysfuction, the protein hypercatabolism, and the myopathy distinctively associated with prolonged critical illness.

Thyrotropin deficiency

Thyrotropin originates in the pituitary thyrotropes and stimulates the thyroid gland to release thyroxine (T_4) preferentially (Thorner *et al.* 1992). This prohormone is peripherally de-iodinated into either its active metabolite tri-iodothyronine (T_3) or into reverse T_3, which is believed to be biologically inactive. T_3 exerts hormone action through nuclear and non-nuclear mechanisms. Both T_4 and T_3 exert feedback inhibition at the pituitary and hypothalamic levels. T_3 is an important endogenous inotropic, lusitropic, and afterload-reducing factor. Normal levels of T_3 are also required for growth hormone secretion and responsiveness, protein synthesis, and fuel utilization by muscle. Plasma T_3 appears to be the sole source of tissue T_3 for the heart, liver, and kidney, Other organs or tissues, such as the brain, the pituitary, and brown fat, are capable of generating T_3 locally.

As with most other pituitary hormone deficiencies, thyrotropin deficiency can be congenital or acquired, familial or sporadic, partial or complete, organic or functional. It may occur as an isolated deficiency or as part of pan-hypopituitarism. The ensuing hypothyroidism is labeled 'central' ('secondary' in the case of pituitary origin and 'tertiary' in the case of hypothalamic origin).

Primary and central hypothyroidism are in general clinically indistinguishable from each other except for the severity of the manifestations. No tissue or organ system escapes the adverse effects of thyroid hormone insufficiency and the clinical manifestations include decreased contractility and relaxation of the heart often in the face of elevated peripheral vascular resistance, pericardial and pleural effusions, renal insufficiency, fluid retention and hyponatremia, malabsorption syndrome, biliary sludge and dyskinesia, ileus, atrophy of the intestinal mucosa and the tegumentum, anemia, muscle and joint pains, stiffness and paresthesias, impaired lipid and glucose metabolism, and loss of cognitive abilities and thermoregulatory capacity. In central hypothyroidism, there is however less likelihood of periorbital edema, hoarseness, weight gain, and hypercholesterolemia.

In prolonged critical illness, thyrotropin levels are low in the face of low circulating levels of free T_4 and T_3, an entity labeled 'low-T_3 syndrome'. Although it is still a matter of controversy, recent data increasingly support the concept that, in addition to the observed changes in peripheral thyroid hormone metabolism, there is a certain degree of central hypothyroidism induced by the prolonged critical condition. Tumor necrosis factor and interleukins 2 and 6 could be involved in the pathogenesis. Moreover, low-dose dopamine infusion induces or aggravates the low-T_3 syndrome in critical illness through a direct inhibition of thyrotropin release and effects on thyroid hormone conversion, with the duration of dopamine treatment correlating with the severity of the T_3 deficiency (Van den Berghe and de Zegher 1996). Other medications, such as glucocorticoids and somatostatin, may also suppress pituitary thyrotropin release, and glucocorticoids may inhibit T_4 to T_3 conversion. Iatrogenic central hypothyroidism may perpetuate the failing anabolism as well as other problems often encountered during prolonged critical illness, such as diminished cognitive status with lethargy, somnolence, or depression, ileus and gallbladder dysfunction, pleural and pericardial effusions, glucose intolerance and insulin resistance, hyponatremia, normocytic normochromic anemia, and deficient clearance of triglycerides.

Gonadotropin deficiency

Gonadotropin deficiency may be part of the clinical syndrome of pan-hypopituitarism or it may be isolated (Thorner *et al.* 1992). Impaired ovarian function in women and suppressed testicular function in men influence fertility and metabolism. Testosterone deficiency in men, a result of deficient luteinizing hormone secretion, is relevant to critical care medicine as testosterone is the most important of the endogenous anabolic steroids, and a decrease of its availability in men results in a negative nitrogen balance that can be restored by testosterone supplementation.

In prolonged critical illness, low luteinizing hormone levels and a profound hypoandrogenemia are consistently present. Infusion of dopamine, and to a lesser extent of opioids also, tends to blunt the secretion of luteinizing hormone further. Iatrogenic suppression of the male gonadal axis may delay recovery by blocking one of the endogenous pathways to anabolism.

References

De Boer, H., Blok, G.J., and Van der Veen, E.A. (1995). Clinical aspects of growth hormone deficiency in adults. *Endocrine Reviews*, **16**, 63–86.

Devins, S.S., Miller, A., Herndon, B.L., O'Toole, L., and Reisz, G. (1992). Effects of dopamine on T-lymphocyte proliferative responses and serum prolactin concentrations in critically ill patients. *Critical Care Medicine*, **20**, 1644–9.

Murphy, W.J., Rui, H., and Longo, D.L. (1995). Effects of growth hormone and prolactin: immune development and function. *Life Sciences*, **57**, 1–14.

Thorner, M.O., Vance, M.L., Horvath, E., and Kovacs, K. (1992). The anterior pituitary gland. In *Williams textbook of endocrinology* (8th edn) (ed. J.D. Wilson and D.W. Foster), pp. 221–310. W.B. Saunders, Philadelphia, PA.

Van den Berghe, G. and de Zegher, F. (1996). Anterior pituitary function in critical illness and dopamine treatment. *Critical Care Medicine*, **24**, 1580–90.

8.8.2 Therapeutic strategy

G. Van den Berghe

Key messages

- Growth hormone therapy improves the metabolic and functional abnormalities associated with growth hormone deficiency in adults. Growth hormone as a therapeutic strategy to improve the catabolic state of critical illness is currently under intense investigation and should still be considered experimental.
- Treatment of corticotropin deficiency is lifesaving and therefore takes precedence over that of all other anterior pituitary hormones. The goal is to restore the eucortisol state, which is achieved by hydrocortisone administration, intravenously in critical care conditions.
- Treatment of thyrotropin deficiency is achieved by administration of thyroxine associated with the more active metabolite tri-iodothyronine in critical care conditions and in case myxedema-like coma is present.
- Treatment of gonadotropin deficiency in men is partially achieved by replacing the gonadal steroid testosterone which has been shown to improve the negative nitrogen balance associated with male hypogonadism.
- Substitution therapy of hyopituitarism with glucocorticoids, growth hormone, thyroid hormones, and eventually androgens prevents evolution to a life-threatening condition and optimizes the potential for recovery from intercurrent illnesses.

Introduction

Treatment of hypopituitarism is based on recognition and removal of the cause where possible and on substitution of the deficient hormones. The three potential options for hormone substitution are as follows.

1. Administer the hormone of the affected target gland or one of its synthetic analogs. This is the most practical and cheapest way.
2. Administer the anterior pituitary hormone (e.g. growth hormone).
3. Administer the hypothalamic regulatory hormone (e.g. gonadotropin-releasing hormone and growth hormone-releasing hormone in secondary hypopituitarism).

Treatment of prolactin deficiency

Prolactin is currently not available for therapy. Future studies are needed to evaluate the therapeutic potential of prolactin release induced by thyrotropin-releasing hormone for optimizing immune function during critical illness.

Treatment of growth hormone deficiency

Administration of human recombinant growth hormone to adults with growth hormone deficiency, who are already receiving appropriate thyroid, adrenal, and gonadal hormone replacement, increases lean body mass, muscle-to-fat ratio, exercise capacity, and muscle strength, augments cardiac output and lowers cholesterol and low-density lipoprotein concentrations, normalizes renal blood flow and glomerular filtration rate, stimulates the impaired peripheral conversion of T_4 to T_3, increases bone turnover, and improves thermoregulation and psychological well being. However, it is difficult to define criteria for diagnosis and treatment of adult-onset growth hormone deficiency accurately, as production of growth hormone and insulin-like growth factor I also declines in otherwise healthy elderly subjects in association with decreased muscle mass and increased adiposity (Ho and Hoffman 1995). Moreover, administration of growth hormone in healthy aging individuals also evokes beneficial effects comparable with those obtained in growth-hormone-deficient adults.

A rough guideline for growth hormone treatment in adulthood is 0.012 mg/kg/day (0.025 IU/kg/day). Side-effects include transient water and sodium retention, arthralgias and myalgias, increased fasting plasma glucose and insulin levels, and a small increase in systolic blood pressure.

The therapeutic potential of the new peptide and non-peptide growth hormone secretagogues, which are also orally active, is currently being investigated.

Treatment in catabolic states of critical illness

The finding that pulsatile growth hormone secretion is suppressed in critically ill subjects, together with the fact that growth hormone is a potent anabolic agent, provides a basis for investigating the potential benefit of growth hormone in critical illness (Wilmore 1991). Preclinical data on growth hormone treatment in the catabolic state show that growth hormone promotes wound healing in perioperative, fasted, and tumor-bearing rats, and enhances postoperative immune function. Clinical studies on growth hormone treatment have demonstrated decreased carbohydrate and increased fat oxidation, improved (albeit not restored) nitrogen balance, and beneficial effects on wound healing and grip strength as well as on immune function (Wilmore 1991; Voerman *et al.* 1995). However, positive results are not consistently found, and a selective target group of critically ill patients who will predictably benefit from growth hormone therapy remains to be defined.

Doses used for this indication tend to be much higher than those employed for replacement therapy in growth hormone deficiency and the treatment period is much shorter (16–24 IU/day subcutaneous for 2 to 3 weeks). Side-effects include fluid and salt retention and glucose intolerance, often requiring diuretics and exogenous insulin.

The use of growth hormone in the critically ill should still be considered to be experimental. More adequately randomized controlled clinical trials investigating its effect on outcome parameters are required. Whether natural or synthetic growth hormone secretagogues have therapeutic potential in critical illness remains to be studied.

Treatment of corticotropin deficiency

As corticotropin deficiency is potentially a life-threatening disease, its treatment takes precedence over that of all other anterior pituitary hormones. The goal of treatment in corticotropin deficiency is to restore the eucortisol state. This is achieved by administration of cortisol or one of its glucocorticoid analogs in a dose equivalent to daily normal cortisol production and in a way that mimics normal diurnic rhythm. The dose is increased during periods of acute stress to mimic augmented cortisol production physiologically occurring in such conditions.

Maintenance glucocorticoid therapy can be given as hydrocortisone 20 to 30 mg/day, cortisone acetate 25 to 37.5 mg/day, or prednisone 5 to 7.5 mg/day. Two-thirds should be given in the morning and the remaining third in the afternoon. High-potency long-acting glucocorticoids such as dexamethasone are not recommended for glucocorticoid replacement. Under conditions of acute medical or surgical illness, the dose should be increased to two to three times the usual maintenance dose, and injectable steroids (e.g. 100 mg hydrocortisone hemisuccinate intramuscularly) should be used whenever inadequate enteral absorption is suspected.

If corticotropin deficiency is partial, the patient may be able to secrete enough cortisol to take care of day-to-day needs, and will only need treatment in episodes of stress or illness. These are the patients who present with acute corticotropin deficiency in the critical care unit and for whom prompt recognition and treatment is vital. A blood sample is drawn for serum cortisol and corticotropin before intravenous administration of a soluble glucocorticoid: hydrocortisone 100 mg followed by 50 to 100 mg every 6 h on the first day, 50 mg every 6 h on the second day, and 25 mg every 6 h on the third day, tapering to a maintenance dose by the fourth to fifth day.

To prevent acute corticotropin deficiency in the perioperative phase, 100 mg of hydrocortisone hemisuccinate is given intramuscularly or intravenously on call to the operating room, 50 mg is given in the recovery room, and 50 mg is given every 6 h for the first 24 h. If the patient's progress is satisfactory, the dose can be reduced to 25 mg every 6 h for 24 h and then gradually tapered to maintenance dose by the third to fifth day. However, if fever, hypotension, or any other complication occurs, the dose of hydrocortisone is increased to 200 to 400 mg/day in divided doses every 6 h followed by gradual tapering to maintenance dose. In prolonged critical conditions, the maintenance dose should be kept two to three times the basal need.

Special attention should be given to patients with concomitant diabetes insipidus, as lack of cortisol may prevent polyuria because cortisol is needed for free-water clearance. Glucocorticoid therapy may induce or aggravate diabetes insipidus in these patients. Another specific condition is the post-hypophysectomy phase of Cushing's disease, characterized by a high vulnerability to Addisionian-like crisis.

Drugs such as phenytoin, barbiturates, and rifampin (rifampicin) can accelerate glucocorticoid metabolism by induction of microsomal enzyme activity and can increase the glucocorticoid replacement dose requirements. If this increased requirement is not met, adrenal crisis may occur.

Treatment of thyrotropin deficiency

The goal of treatment is to restore the euthyroid state and is achieved by the administration of thyroid hormones. Synthetic L-thyroxine (T_4) and L-tri-iodothyronine (T_3) are preferred to animal thyroid extracts because of their uniform potency. In non-critical care conditions, T_4 is the agent of choice because endogenous regulation of conversion to T_3 prevents abrupt fluctuations in the active thyroid hormone concentration and activity. The usual adult replacement dose of L-thyroxine in basal conditions is between 100 and 150 µg/day. A reduction of 20 to 40 per cent is required in the elderly (Larsen and Ingbar 1992). To establish the adequacy of therapy, reliance is placed on clinical assessment accompanied by determinations of serum T_4, T_3, and thyroid-hormone-binding globulin. Since thyrotropin levels are either normal or low in central hypothyroidism, they cannot be used as indicators of attainment of a euthyroid state.

In the presence of hypothyroid coma or precoma, thyroid hormone replacement is as urgent as in true myxedema coma, with the latter still having a high mortality (over 80 per cent) if not treated promptly. It is well known that the 5′-deiodinase enzyme responsible for the conversion of T_4 to T_3 is less active in critically ill patients. It has also been shown that the conversion always lags by a few days. Therefore the efficacy of substituting only T_4 in this emergency situation could be questioned. The continuous infusion of the active thyroid hormone T_3 together with the prohormone T_4 seems a better alternative within the safety of a well-staffed intensive care unit, but this option is made difficult because there is no commercially available intravenous form of T_3. The disadvantages of oral T_3 preparations in intensive care conditions are obvious. Within these limitations, it has been recommended that intravenous T_4 (200–300-µg bolus followed by a second 100 µg after 24 h) should be used together with enteral T_3 (25 µg every 12 h) until the patient is conscious and taking oral thyroxine.

When the critical condition is induced not only by central hypothyroidism but also by the intercurrent illness, the addition of intravenous T_3 (if available) is a superior strategy because of the near absence of peripheral conversion of T_4. Our therapeutic regimen in the acute phase of critical illness in adults is 200 µg/day of T_4 intravenously and 0.5 to 0.6 µg/kg/day of T_3 infused over 24 h, tapered to maintenance-dose oral thyroxine in the recovery phase (Van den Berghe, unpublished observations). Twice weekly control of serum thyroid hormone levels seems to be justified in the critical phase to avoid overdosage.

Drugs such as rifampin and phenytoin augment T_4 and T_3 clearance and therefore require higher doses. Interference with absorption of thyroid hormones by biliary and intestinal disorders, as well as by dietary fibers and binding resins such as cholestyramine, necessitates dose adaptation.

Administration of thyroid hormone to treat thyrotropin deficiency with unrecognized corticotropin deficiency may precipitate an Addisonian crisis because thyroid hormone accelerates cortisol metabolism and increases the glucocorticoid requirements. Therefore evaluation of the corticotropin–cortisol axis is critical in any patient with thyrotropin deficiency.

The thyrotropin deficiency as part of the low-T_3 syndrome of critical illness has recently been shown to be partially overcome by administration of thyrotropin-releasing hormone. Whether this finding will lead to therapeutic applications remains to be clarified.

Treatment of gonadotropin deficiency

The aim of treatment is to restore gonadal function. This is accomplished by replacing gonadal steroids and, when indicated, restoring fertility potential by the administration of gonadotropins or gonadotropin-releasing hormone. The latter is not a priority in critical care conditions. In contrast, the catabolic effects of testosterone deficiency in men may contribute to the loss of protein from vital organs and tissues, and, in particular, aggravate muscle wasting. The critical condition in itself is associated with profound hypoandrogenemia, and treatment with testosterone or anabolic testosterone analogs has been shown to improve nitrogen balance without having the potential to normalize it.

Administration of testosterone is preferred in the parenteral form (testosterone enanthate or cypionate 100–300 mg intramuscularly every 2–4 weeks). The effectiveness can be monitored by the patient's clinical condition and by measuring serum levels. Overdosage may lead to salt and fluid retention, edema, excessive sexual stimulation, priapism, gynecomastia, aggressive behavior, polycythemia, and worsening of benign prostatic hypertrophy in middle-aged and elderly males. Synthetic testosterone analogs have been designed to weaken the androgen properties while selecting the anabolic effect on protein synthesis.

Conclusion

Hypopituitarism has adverse effects on metabolic and immunological homeostasis. Precise diagnosis and substitution therapy with glucocorticoids, growth hormone, thyroid hormones, and eventually androgens prevents evolution to a life-threatening condition and optimizes the potential for recovery from intercurrent illnesses.

Prolonged critical illness without pre-existing hypothalamic pituitary disease is also associated with metabolic and immunological malfunctions which determine morbidity and mortality. The management and prevention of the protein hypercatabolism and the anergy-type immune suppression remain primary objectives of critical care medicine and subjects of intense investigation. It is conceivable that iatrogenic suppression of pituitary function in critically ill patients should be avoided in order not to impair further their potential for recovery.

Whether or not the endogenously occurring hypopituitarism in the presence of elevated cortisol levels is to be considered beneficial, adaptation within current standards of intensive care remains at present controversial. Studies of the effects of substituting the different suppressed anterior pituitary axes during critical illness are warranted and should provide an answer to this question in the near future.

References

Ho, K.Y. and Hoffman, D.M. (1995). Defining growth hormone deficiency in adults. *Metabolism*, **44**, 91–6.

Larsen, P.R. and Ingbar, S.H. (1992). The thyroid gland. In *Williams textbook of endocrinology* (8th edn) (ed. J.D. Wilson and D.W. Foster), pp. 357–487. W.B. Saunders, Philadelphia, PA.

Voerman, B.J., Strack van Schijndel, R.J.M., Groeneveld, A.B.J., de Boer, H., Nauta, J.P., and Thijs, L.G. (1995). Effects of growth hormone in critically ill non-septic patients: results from a prospective, randomized, placebo-controlled trial. *Critical Care Medicine*, **23**, 665–73.

Wilmore, D.W. (1991). Catabolic illness: strategies for enhancing recovery. *New England Journal of Medicine*, **325**, 695–702.

8.9 Hyperthyroidism

8.9.1 Pathophysiology

D. Watson

Key messages

- Autonomous hypersecretion and deliberate overdose of thyroid hormones are the most common causes of severe hyperthyrodism.
- Thyrotoxicosis may be manifest by adverse changes in every organ system.
- Pyrexia and altered mental status are the hallmarks of thyrotoxic crisis, which is a rare but life-threatening complication of severe hyperthyroidism.

Introduction

Hyperthyroidism is a clinical syndrome caused by exposure of the body tissues to excess circulating levels of free thyroxine and/or tri-iodothyronine. In over 90 per cent of patients hyperthyroidism is caused by one of the following:

(1) Graves' disease;

(2) toxic multinodular goiter;

(3) toxic solitary nodule;

(4) deliberate overdose of thyroid hormones.

Excess pituitary secretion of thyroid-stimulating hormone, intrinsic thyroid stimulating activity of human chorionic gonadotrophin in hydatidiform mole and choriocarcinoma, ovarian teratoma containing thyroid tissue, and metastatic differentiated carcinoma of the thyroid are unusual causes of hyperthyroidism.

Graves' disease

Graves' disease is the immunologically mediated form of hyperthyroidism which is associated with diffuse thyroid enlargement, ophthalmopathy, and pretibial myxedema. It results from the presence of IgG antibodies, known as thyroid-stimulating immunoglobulins, directed against the thyroid-stimulating hormone receptor of the thyroid follicular cell. Binding of these antibodies to the thyroid-stimulating hormone receptor stimulates thyroid hormone production via the adenylate cyclase system. Other immunoglobulins, termed thyroid-growth immunoglobulins, may stimulate growth of the thyroid without affecting hormone production. The pathogenesis of the opthalmopathy and dermopathy is poorly understood, although both are believed to be immunologically mediated. There is a rise in pressure within the orbit caused by an increased interstitial fluid content and chronic inflammatory cell infiltrate affecting extraocular muscles and fat. This results in forward movement of the eye (proptosis) and impairment of extraocular muscle function.

Toxic multinodular goiter

It is likely that in some patients the natural history of a simple diffuse goiter is a progression over decades to a multinodular goiter with areas of autonomous function. The stimulus to this progression may be thyroid-growth immunoglobulins. As the amount of autonomous tissue increases, hyperthyroidism develops in middle or old age.

Toxic solitary nodule

The nodule is a follicular adenoma which autonomously secretes excess thyroid hormones and inhibits endogenous thyroid-stimulating hormone secretion with subsequent atrophy of the rest of the thyroid gland.

Special situations

Hyperthyroidism in pregnancy

The coexistence of pregnancy and hyperthyroidism is unusual as anovulatory cycles are common in thyrotoxic patients and autoimmune disease tends to remit during pregnancy. Hyperthyroidism in pregnancy is almost always due to Graves' disease.

Factitious hyperthyroidism

This situation occurs when an emotionally disturbed person, often with a medical background, abuses a thyroid hormone preparation, most frequently thyroxine.

Iodide-induced hyperthyroidism

Iodide, in the form of dietary supplements or medication (e.g. antitussive agents, amiodorone, or radio-angiographic contrast agents) can induce hyperthyroidism, particularly in patients who are relatively iodide deficient.

Thyrotoxic crisis

Thyrotoxic crisis is a life-threatening increase in the severity of the clinical features of hyperthyroidism. It may be the presenting form of the condition or develop after subtotal thyroidectomy in an inadequately prepared patient or within a few days of iodine-131 therapy. It is most commonly precipitated by infection in a patient who is inadequately controlled by antithyroid drugs. Patients present with a constellation of symptoms of exaggerated thyrotoxicosis such as fever, severe agitation, often with a psychotic component, and tachyarrhythmia, sometimes with cardiac failure. Gastrointestinal symptoms may include diarrhea, nausea, vomiting, and jaundice.

Pathophysiological consequences of hyperthyroidism

Thyroid hormone exerts its effects directly via specific nuclear receptors and through central activation of the sympathoadrenal system. The increased basal metabolic rate, often exacerbated by pyrexia due to impaired thermoregulation, is associated with an increase in oxygen consumption and carbon dioxide production and a hyperdynamic circulation. Consequently, patients are at risk of developing arrhythmias and myocardial ischemia, while respiratory muscle weakness may lead to ventilatory failure.

Thyroid hormones influence the concentration and distribution of various neurotransmitters. Neurological complications of severe hyperthyroidism include irritability and tremor, delirium, stupor, and seizures. The myopathy of hyperthyroidism may cause weakness of the striated muscles of the pharynx, resulting in dysphagia, as well as affecting the eye, resulting in ophthalmoplegia. Proximal muscle weakness is common and may be the dominant manifestation in some individuals.

Hypercalcemia is the most common life-threatening electrolyte abnormality seen in hyperthyroidism. Hematological manifestations include anemia, thrombocytopenia, and eosinophilia with lymphocytosis.

Patients with hyperthyroidism have a high incidence of gastrointestinal hypermotility with malabsorption associated with gastritis. Rarely, fulminant hepatic necrosis can occur.

Bibliography

Reasner, C.A. and Isley, W.L. (1991). Thyrotoxicosis in the critically ill. *Critical Care Clinics*, 7, 57–74.

8.9.2 Therapeutic strategy

D. Watson

Key messages

- Sometimes treatment for presumed thyroid disease will have to be initiated before the results of diagnostic tests are available.
- The determination of the rapidity with which the patient needs to be treated is of utmost importance.
- Therapy should reduce both serum thyroid hormone levels and the action of thyroid hormones on peripheral tissues as well as treat any precipitating event.
- A combination of antithyroid drugs and iodine may decrease serum tri-iodothyronine levels in days although the metabolic response lags behind.

Introduction

If the patient is extremely unwell with a clinical diagnosis of thyrotoxicosis, therapy should precede laboratory confirmation of the diagnosis. Nevertheless, blood should be taken for determination of circulatory levels of total or free thyroxine, thyroid-stimulating hormone, and random cortisol before starting therapy. Specific therapeutic strategies to reduce acutely the synthesis and secretion of thyroid hormone and measures to control the effects of circulating thyroid hormone (e.g. the management of cardiac arrhythmias) must be accompanied by treatment of any precipitating cause (Reasner and Isley 1991).

General measures

General measures should include the provision of supplemental oxygen and intravenous fluids to maintain circulating volume, cardiac output, blood pressure, and tissue oxygen delivery. Salicylates should be avoided since they displace thyroid hormone from binding proteins and thereby exacerbate the hypermetabolism. In cases of known or suspected overdose of thyroid hormones gastric lavage with charcoal is indicated, perhaps combined with the administration of cholestyramine to increase fecal elimination of the hormone. Treatment of overdosage is otherwise symptomatic and supportive, employing measures to counteract effects of excess thyroid hormone.

The principle differential diagnosis for thyrotoxicosis is sepsis (high temperature and leukocytosis), and when there is any doubt broad-spectrum antibiotics should be commenced.

Specific antithyroid treatments

Propylthiouracil impedes formation of thyroid hormone and prevents conversion of thyroxine to the metabolically active hormone tri-iodothyronine. The recommended dose is propylthiouracil 200 mg every 6 h by mouth or nasogastric tube if the patient is unable to take oral medication. An alternative is carbimazole 15 mg orally every 6 h. In very severe cases potassium iodide or Lugol's iodine may be required, but their administration should be delayed for approximately 1 h following the first dose of propylthiouracil. This is to minimize the possibility of massive hormone release following iodination. Dosing schedules vary, but potassium iodide 60 mg orally three times daily, or potassium iodide saturated solution (eight drops every 6 h orally), or sodium iodide 1 g intravenously twice daily are appropriate. Alternatively, Lugol's iodine can be administered as ten drops of a solution containing 130 mg/ml iodine diluted in milk or water twice daily. Lithium carbonate may be useful in patients with iodine sensitivity. The initial dose of 300 mg every 6 h should be adjusted to maintain serum lithium levels of approximately 1 mmol/l. A more detailed discussion is given by Vanderpump *et al.* (1996).

Recently, radiographic contrast dyes containing iodine have been suggested as alternative preparations in the event that antithyroid drugs cannot be used because of a previous history of reactions such as agranulocytosis or hepatotoxicity. The cholecystographic contrast agent sodium ipodate (Oragrafin) is metabolized to yield iodide, and a daily dose of 1 g orally is recommended (Wu *et al.* 1982).

Partial thyroidectomy is effective in relieving hyperthyroidism, but patients should be made euthyroid before surgery. This can be achieved by the use of carbimazole for several months before planned elective operation. The administration of radio-iodine is simple, but in patients with severe hyperthyroidism thyrotoxic crisis has been reported following radio-iodine. Glandular stores of thyroid hormones should also be depleted by administration of antithyroid drugs for several weeks before radio-iodine therapy.

Measures to control the effects of hyperthyroidism

Propranolol (80 mg orally or 1–2 mg intravenously as required) blocks the β-adrenoreceptor effects of thyroxine and is particularly useful for reducing sinus tachycardia. It also impairs conversion of thyroxine to tri-iodothyronine and inhibits release of thyroid hormone. Propranolol is contraindicated in patients with cardiac insufficiency or air-flow limitation, although newer more specific β-blockers such as atenolol can be used in patients with congestive cardiac failure, bronchospasm, or peripheral vascular disease. Rapid atrial fibrillation should be managed by standard protocols involving digoxin administration (up to 1.5 mg as a loading dose) following correction of hypokalemia and anticoagulation. Thyrotoxicosis accelerates the metabolism of digoxin. Conversely, patients with thyrotoxicosis are more sensitive to warfarin because vitamin-K-dependent clotting factors are metabolized more rapidly. Verapamil may also be useful in this setting to reduce the ventricular rate. Steroids should also be given (e.g. hydrocortisone 100 mg every 8 h intramuscularly) since relative hypoadrenalism is believed to occur in severe hyperthyroidism because of the accelerated metabolism of cortisol. Glucocorticoids may also reduce the conversion of thyroxine to tri-iodothyronine. Cooling may follow the administration of chlorpromazine (50–100 mg intramuscularly).

The response to treatment of acute hyperthyroidism should be monitored by recording clinical vital signs such as pulse and temperature as well as regular serum tri-iodothyronine estimations. Usually the hyperthyroidism is brought under control after 10 days and the iodide preparations and propranolol are withdrawn. The patient is then maintained on carbimazole 15 mg every 6 h (Table 1).

Table 1 Recommended treatment of severe hyperthyroidism

General measures
- O_2 administration
- Adequate hydration
- Fever reduction

Reduce blood hormone levels
- Inhibit hormone synthesis: propylthiouracil, carbimazole
- Inhibit hormone release: iodide preparations
- Inhibit peripheral tri-iodothyronine formation: propylthiouracil, glucocorticoids, propranolol

Adrenergic blockade of thyroid hormone effects: propanolol, atenolol

Treatment of congestive cardiac failure and arrhythmias
- According to standard protocols

Diagnosis and treatment of precipitating event (e.g. Graves' disease) by alternative therapies, i.e. radio-iodine, subtotal thyroidectomy. Patients should be rendered biochemically euthyroid before surgery. Use of radio-iodine should only be authorized by a clinician experienced in the administration of radio-active substances

Hyperthyroidism in pregnancy

Thyroid function should be assessed in women with hyperemesis gravidarum. Women known to have Graves' disease may decide on definitive treatment of their condition prior to becoming pregnant. If Graves' disease is diagnosed during pregnancy, it is important to use the smallest dose of antithyroid drugs by reviewing regularly to maintain maternal concentrations of free thyroid hormone and thyroid-stimulating hormone within their respective normal ranges. This is to avoid fetal hypothyroidism and goiter. Most recommend that carbimazole should be discontinued 4 weeks before the expected date of delivery to avoid any possibility of fetal hypothyroidism at the time of maximum brain development. If subtotal thyroidectomy is necessary because of poor drug compliance or hypersensitivity, it is most safely performed in the middle trimester. Radio-iodine is contraindicated because it invariably induces fetal hypothyroidism. If antithyroid agents are given postdelivery, breast feeding is not precluded. As propylthiouracil is excreted less in breast milk, it is preferable to use this agent at the lowest effective dose. Patients already receiving carbimazole need not normally change provided that the daily dose is 20 mg or less. In dire circumstances when patients are not responding to conventional therapy plasmapharesis should be considered (Derkson *et al.* 1984).

References

Derkson, R.H., van de Wiel, A., Poortman, J., der Kinderen, P.J., and Kater, L. (1984). Plasma exchange in the treatment of severe thyrotoxicosis in pregnancy (case reports). *European Journal of Obstetrics, Gynecology, and Reproductive Biology*, **18**, 139–48.

Reasner, C.A. and Isley, W.L. (1991). Thyrotoxicosis in the critically ill. *Critical Care Clinics*, **7**, 57–74.

Vanderpump, M.P.J., Ahlquist, J.A.O., Franklyn, J.A., and Clayton, R.N. (1996). Consensus statement for good practice and audit measures in the management of hypothyroidism and hyperthyroidism. *British Medical Journal*, **313**, 539–44.

Wu, S.-Y., Shyh, T.-P., Chopra, I.J., Solomon, D.H., Huang, H.-W., and Chu, P.-C. (1982). Comparison of sodium ipodate (Orografin) and propylthiouracil in early treatment of hyperthyroidism. *Journal of Clinical Endocrinology and Metabolism*, **54**, 630–4.

8.10 Hypothyroidism

8.10.1 Pathophysiology

S. Arlot and A. Fournier

Key messages

- Thyroid hormones act on most tissues in the organism via nuclear T_3 receptors.
- Biological actions vary depending on the target cell. The main actions are influence on cell growth and maturation (central nervous system, bone) and modulation of metabolism (carbohydrates, lipids, proteins, drugs).
- Thyroid hormones are thermogenic agents, stimulating respiratory oxygen consumption and heat production. The latter is important in hypothyroidism, mainly in myxedema coma.

Calorigenic deficit

Thyroid hormones regulate basal metabolism, i.e. basal heat production. They stimulate energy production during rest and muscular exercise. If thyroid hormones are lacking, cold-response synthesis of proteins involved in calorigenesis is reduced, and heat production and oxygen consumption diminish, leading to cardiovascular, respiratory, and neurological adjustments (Nicoloff and LoPresti 1993). Body core temperature falls, even at ambient room temperatures (Mariotti *et al.* 1995). Furthermore, the thermogenic response to epinephrine (adrenaline) is reduced in hypothyroidism, impairing the ability to maintain a normal core temperature (Nicoloff and LoPresti 1993).

Cardiovascular changes

Most cardiovascular changes are reversible on treatment with L-thyroxine.

Bradycardia

Reduction of pulse rate parallels reduction of basal metabolism. In addition, thyroid hormones have a direct effect on the atrial and sinoatrial node cells: the duration of the action potential is prolonged and diastolic repolarization is slowed down (Bürgi and König 1988). Although serum catecholamine levels are normal, there is reduced sensitivity to these hormones via a reduction of β-receptor expression. Several causes are suspected: reduction of the number of β-receptors, altered regulation of G proteins (modification of ratio of Gs to Gi subunits), or enhanced phosphodiesterase activity. Production of cyclic AMP and β-adrenergic effects are reduced, but α-adrenergic response is preserved (Nicoloff and LoPresti 1993).

Low cardiac output

T_4 stimulates skeletal muscle and myocardial growth, while acting on myosine metabolism. In hypothyroidism, there is a reduction of myosine ATPase and reduced myocardial contractility, leading to a reduction of stroke volume (Bürgi and König 1988). Furthermore, hypothyroidism *per se* can induce congestive heart failure, even if concomitant heart disease is not already present.

Cardiomegaly

Cardiomegaly, which is observed on chest radiographs, can be explained by heart dilatation aggravated by pericardial effusion, which itself is due to increased capillary permeability. It also contributes to the low voltage observed on ECG (Bürgi and König 1988).

Conduction abnormalities seen on ECG

The following cardiac changes are obvious on ECG (Tachman and Guthrie 1984):

(1) low-voltage complexes;

(2) sinus bradycardia;

(3) long QT interval;

(4) flattened or inverse T waves.

Diastolic hypertension

Peripheral vasoconstriction is always found in hypothyroidism and causes diastolic hypertension. In order to maintain a normal body core temperature, chronic peripheral vasoconstriction redirects blood distribution from the peripheral to the central organs, lowering cutaneous blood flow and leading to a cold pale skin. The lack of diastolic hypertension is unusual and suggests hypothyroid decompensation and myxedema coma (Nicoloff and LoPresti 1993).

Coronary atherosclerosis

Several vascular risk factors are present in hypothyroidism. Hyperlipidemia and hypertension play a role in atherosclerosis, mainly on the coronary bed. Silent ischemia is usual and may become clinically manifest during thyroid hormone replacement therapy, which increases O_2 requirements (Tachman and Guthrie 1984; Bürgi and König 1988).

Respiratory changes

In hypothyroidism, there are alterations in pulmonary function. In myxedema coma, pulmonary infections frequently precipitate respiratory dysfunction.

Sleep apnea syndrome is of mixed type (Bürgi and König 1988):

(1) central, via reduction of cerebral ventilatory response to hypercapnia;

(2) obstuctive, via reduction of airway (tongue enlargement).

Central bradypnea and reduction in ventilatory muscle performance lead to alterations of alveolocapillary gas exchange. These abnormalities may explain impaired gas exchanges, with CO_2 retention leading to severe hypercapnia and hypoxemia (Nicoloff and LoPresti 1993).

Cerebral and neurological changes

Thyroid hormones act on protein and nucleic acid synthesis, cerebral lipids, myelinization, and numerous enzymatic activities. In particular, they stimulate synthesis of synaptosome proteins and enzymes associated with neural endings (Nicoloff and LoPresti 1993). Progressive mental changes, which are not merely due to cerebral atherosclerosis, occur in hypothyroidism. Somnolence, lethargy, and coma, sometimes with seizures, are common features of myxedema coma (Tachman and Guthrie 1984). These neurological disturbances are made worse by concurrent hyponatremia (Mariotti *et al.* 1995).

Gastrointestinal changes

Constipation is usual in moderate hypothyroidism. Intestinal pseudo-obstruction and paralytic ileus can occur in myxedema coma (Tachman and Guthrie 1984). Lack of thyroid hormones induces a slowing in gastric emptying and in the motor activity of the small intestine and colon. Amplitude and frequency of colon muscular contractions are decreased because of neuropathy (altered impulse transmission at the myoneural junction) and/or intestinal ischemia (Bürgi and König 1988). Intestinal absorption of drugs, even oral L-thyroxine, is usually preserved.

Hematological changes

Normocytic anemia is always present, and hematocrit is around 30 to 35 per cent (Bürgi and König 1988). Decreased erythropoietin secretion and erythrocyte production lead to low erythrocyte mass which is sometimes associated with relatively preserved hematocrit because of concomitant hypovolemia. If anemia is of the macrocytic type, pernicious anemia must be suspected (Mariotti *et al.* 1995). If anemia is severe (hematocrit < 30 per cent), additional factors should be looked for such as blood loss by gastrointestinal bleeding, folic acid deficiency, and iron deficiency (Mariotti *et al.* 1995).

Thyroid hormone therapy increases volemia before erythrocyte production. Therefore hematocrit falls by 5 to 6 per cent during the first days of treatment because of 'dilution anemia' (Bürgi and König 1988).

The total leukocyte count decreases during hypothyroidism and rarely exceeds 10 000 even if infection is the precipitating event. If band cells are present, suspicion of sepsis is high (Nicoloff and LoPresti 1993).

Blood coagulation is often altered in hypothyroidism. Fibrinolytic activity is increased, with high plasminogen and reduced activation of the plasminogen inhibitor (Bürgi and König 1988). These abnormalities may play a dramatic role in hemorrhages associated with severe hypothyroidism. Acquired von Willebrand's coagulopathy via non-specific action of thyroid hormones on protein synthesis has been described, but it is reversible after L-thyroxine treatment (Tachman and Guthrie 1984).

Metabolic changes

Hyponatremia is a constant feature of myxedema coma. It is a dilution hyponatremia, with high radiosodium space, which is reversible after restriction of water intake (Bürgi and König 1988). Indeed, the ability of hypothyroid kidneys to excrete water is more impaired than their ability to excrete sodium. This is mainly explained by the reduction in glomerular filtration rate, leading to a decreased delivery of sodium to the thick ascending limb of Henle where free water is formed (Bürgi and König 1988). Furthermore, inappropriately high plasma levels of vasopressin are usually present, leading to water retention. Reduction of glomerular filtration is explained by a decreased renal blood flow secondary to reduction of cardiac output and increased vascular resistance favored by thickening of the basement membrane of the vessels (Bürgi and König 1988). Acid–base metabolism is disturbed in myxedema coma with respiratory acidosis.

Muscular changes

Muscle contraction and relaxation are slowed in hypothyroidism. Intramuscular metabolism is altered because of modification of Ca^{2+}-ATPase activity and fluctuations of cytoplasmic Ca^{2+}. The activities of several enzymes involved in muscle glycogen synthesis and catabolism are variable. Some muscular enzymes (transaminases, creatine phosphokinase, lactic dehydrogenase, and their isoforms) are secreted in large amounts in the bloodstream, confusing an ECG diagnosis of myocardial infarction (Bürgi and König 1988).

Skeletal changes

Bone metabolism is low when thyroid hormones are lacking. Levels of parathyroid hormone are often increased, but its biological effects on kidney or bone are reduced. 1,25-OH vitamin D concentrations are increased, leading to decreased calciuria, which may explain the hypercalcemia occasionally seen in severe hypothyroid patients (Bürgi and König 1988).

Dermatological changes

Cutaneous changes have been the classic feature of hypothyroidism ever since the first description of myxedema by Ord in 1878. Peripheral vasoconstriction and decreased cutaneous metabolism are expressed as a cold, dry, and pale skin. The accumulation of mucopolysaccharides and proteins in the interstitial spaces of the skin induces thinning of the epidermis and deep wrinkles (Nicoloff and LoPresti 1993). Hair is scarce, with slow growth. Only 40 per cent of hypothyroid patients have typical cutaneous infiltration.

Changes in lipid metabolism

Plasma lipids are often altered in hypothyroidism (Tachman and Guthrie 1984). Hyperglyceridemia is due to increased hepatic output of very-low-density lipoproteins and decreased catabolism of these lipids by low plasma post-heparin lipolytic activity. Hypercholesterolemia is explained by reduced clearance of low-density lipoproteins, reduced hepatic synthesis of cholesterol, and slowing down of secretion into bile (Bürgi and König 1988).

Lipolytic and lipogenic enzyme activities are altered. These abnormalities may play a role in coronary atherosclerosis.

Changes in carbohydrate metabolism

The capacity to counter-regulate hypoglycemia is impaired in hypothyroidism. The risk of severe and prolonged hypoglycemia is high. Insulin clearance is reduced, and sensitivity to endogenous or exogenous insulin is increased (Bürgi and König 1988). Gluconeogenesis is decreased due to reduced availability of amino acids and glycerol. The impaired responses to epinephrine and glucagon prolong hypoglycemia and aggravate the mental status (Nicoloff and LoPresti 1993).

Changes in drug metabolism

In hypothyroidism, there is a general slowing down of most metabolic processes, particularly those of drugs, whose tolerance is therefore reduced (Nicoloff and LoPresti 1993).

1. Digoxin: the half-life is longer because of a slowing down of the rate of peripheral metabolism and reduced glomerular filtration. A high risk of digoxin intoxication is present in myxedema coma.
2. Diuretics: overdosage is frequent, aggravating hypovolemia in myxedema coma.
3. Tranquilizing agents, sedatives, narcotics, and morphine: these drugs must be prescribed with caution because of the risk of overdose or severe respiratory depression. Sometimes these drugs convert moderate hypothyroidism into myxedema coma.
4. Cortisol: basal plasma cortisol levels may be normal since there is reduction of the secretion rate, endogenous metabolism may be reduced by 50 per cent, and stress response may be inadequate. Cortisol treatment is commonly used in myxedema coma, together with L-thyroxine treatment (Bürgi and König 1988).

Conclusion

Thyroid hormones act on a large variety of tissues and metabolic processes. The clinical signs of mild or moderate hypothyroidism are well known to physicians, even if the disease is oligosymptomatic.

Myxedema coma is a rare condition with a mortality of around 50 per cent. The diagnosis is sometimes easy if treatment with L-thyroxine has been stopped for a long time. In most cases, diagnosis is more difficult and three main clinical features need to be present for the condition to be suspected: altered mental status, defective thermoregulation, and a precipitating event, often in an old patient.

References

Bürgi, U. and König, M.P. (1988). Clinical pathophysiology and metabolic effects of hypothyroidism. *Clinical Endocrinology and Metabolism*, **2**, 567–89.

Mariotti, S., Franceschi, C., Cossariza, A., and Pinchera, A. (1995). The aging thyroid. *Endocrine Reviews*, **16**, 686–715.

Nicoloff, J.T. and LoPresti, J.S. (1993). Myxedema coma. A form of decompensated hypothyroidism. *Endocrinology and Metabolism Clinics of North America*, **22**, 279–90.

Tachman, M.L. and Guthrie, G.P., Jr (1984). Hypothyroidism: diversity of presentation. *Endocrine Reviews*, **5**, 456–65.

8.10.2 Therapeutic strategy for hypothyroidism

S. Arlot and A. Fournier

Key messages

- Successful management requires early diagnosis and rapid appropriate therapeutic measures in an intensive care unit.
- Raising of the body temperature must be progressive at a speed inversely proportional to the degree of hypothermia, particularly when the temperature is below 30 °C.
- Concurrent hypovolemia aggravates tolerance to blood loss. Transfusion with whole blood is the fastest and most effective treatment for hypotension and anemia, particularly when hematocrit is below 30 per cent.
- Preferred treatment is L-thyroxine 500 µg intravenously on day 1, and then 100 µg/day.

Introduction

Myxedema coma is a rare life-threatening medical emergency. Successful management requires early diagnosis and quick appropriate therapeutic measures in an intensive care unit. Most clinical signs are quickly reversible with hormonal treatment by thyroid hormone supplementation. Hormonal treatment is usually begun before a definite diagnosis has been made. Indeed, the fact that immediate treatment is the key to its success has led to the aphorism 'When in doubt, treat' (Nicoloff and LoPresti 1993). The patient's condition should improve in 24 to 48 h.

Diagnosis of myxedema coma

Diagnosis in a patient with known hypothyroidism, who stopped treatment a long time ago and shows typical clinical features, is easy.

It is more difficult when hypothyroidism has not yet been diagnosed. The association of altered mental status, hypothermia, and a precipitating event should lead to suspicion, particularly in an elderly patient (Nicoloff and LoPresti 1993). Hyponatremia, hypoxia, hypercapnia, and anemia are always present in this condition (Table 1) (Mariotti *et al.* 1995). Specific treatment should be started after blood samples have been obtained for assays of thyroid-

Table 1 Biological signs in myxedema coma

Constant	Frequent	Rare
Anemia	Hypoglycemia	Hypercalcemia
Hyponatremia	Elevated creatinemia	
Hypoxia		
Hypercapnia		
Respiratory acidosis		
Elevated CPK		
Elevated LDH		

CPK, creatine phosphokinase; LDH, lactate dehydrogenase.

stimulating hormone, free tri-iodothyronine, free thyroxine, and cortisol. The diagnosis is positive when the results are in agreement with primary hypothyroidism, i.e. elevated plasma thyroid-stimulating hormone and dramatic reduction of both free tri-iodothyronine and free thyroxine, thus excluding a 'euthyroid sick syndrome' or low tri-iodothyronine syndrome. The latter syndrome is observed in critically ill patients with normal thyroid function and low plasma free tri-iodothyronine (Roti *et al.* 1993). In more severe illnesses low plasma free thyroxine may also be present, but plasma thyroid-stimulating hormone remains normal. At present, the results of thyroid hormone assays can be obtained within a short time, for example within an hour by chemiluminescence immunoassay.

Symptomatic treatment

Patients must be treated in an intensive care unit and appropriate supportive measures started. Oxygen, central venous catheter insertion, nasogastric tube insertion, and bed-sore prevention are the first non-specific measures to be taken.

Hypothermia

Raising the body temperature must be progressive at a speed inversely proportional to the degree of hypothermia, particularly when the temperature is below 30 °C. The patient should be maintained at room temperature (20–22 °C) and covered with blankets. This external warming allows gradual vascular expansion in a patient with hypovolemia and peripheral vasoconstriction (Nicoloff and LoPresti 1993).

If aggressive external or internal warming is used, peripheral vasodilatation occurs too rapidly and precipitates cardiovascular collapse.

Respiratory assistance

Upper airways must be kept free to ensure proper ventilation. When the patient's condition is worrying (e.g. severe hypoxia or hypercapnia), respiratory assistance should be implemented without delay by endotracheal intubation and mechanical ventilation in that order (Nicoloff and LoPresti 1993).

Table 2 Safety of methods of cardiovascular assistance

Safe	With caution	Dangerous
Blood transfusions	Digitalis	Vasoactive drugs
Colloid solutions	Diuretics	Alpha-adrenergic agents
	Afterload-reducing agents	Transvenous ventricular pacing

Table 3 Factors indicating unfavorable prognosis

Hypotension	High plasma T_3 during treatment
Hypothermia (< 35 °C)	Depth of coma
Collapse	Bradycardia (< 50 beats/min)
Hyponatremia (< 120 mmol/l)	Pathological repolarization on ECG
Old age	Respiratory acidosis

Tissue hypoxia is often aggravated because of anemia. Ventilatory function can be made worse by a pulmonary infection which acts as a frequent precipitating event.

Cardiovascular assistance

Continuous cardiac monitoring is vital (Table 2).

Hypotension indicates an unfavorable prognosis (Table 3) and is sometimes a sign of an irreversible phase of the illness, since hypothyroid patients usually have mild diastolic hypertension (Hylander and Rosenqvist 1985; Roti *et al.* 1993). Several other factors, such as blood loss by gastrointestinal bleeding (which should be sought and treated), aggressive warming, and overuse of diuretics, may aggravate hypotension. Concurrent hypovolemia aggravates tolerance to blood loss. Transfusion with whole blood is the fastest and most effective treatment for hypotension and anemia, particularly when hematocrit is below 30 per cent. Careful monitoring of central venous pressure is required during the transfusion. In the absence of severe anemia, colloid solutions can be used instead of blood transfusions.

Hypotension can occur as a complication of a silent myocardial infarction. Diagnosis may be difficult because of high levels of muscular enzymes and abnormalities on ECG due to severe hypothyroidism itself (Nicoloff and LoPresti 1993). Echocardiography may help to make a definite diagnosis.

Caution is needed in treating congestive heart failure. Digitalis, if used, should be given at low dose, since drug metabolism is slowed during hypothyroidism (Nicoloff and LoPresti 1993). The risk of overdose is high for digitalis, diuretics, and afterload-reducing agents. These drugs should be prescribed with caution and at low dosage in such patients.

Vasoactive drugs such as dopamine are ineffective and may be dangerous, precipitating ventricular arrhythmias; α-adrenergic agents are not indicated as vasoconstriction already exists and may induce tachyarrhythmias (Nicoloff and LoPresti 1993). Transvenous ventricular pacing may dramatically reduce cardiac output, which is already reduced. If pacing is necessary, an atrioventricular system should be used.

Fluids and electrolytes

Fluid status should be monitored closely. Since dilutional hyponatremia is always present water restriction should be instituted first.

If hypoglycemia appears, 10 per cent hypertonic glucose-containing solutions should be administered. If the glucose osmotic diuresis is insufficient to correct hyponatremia, intravenous loop diuretics at progressive doses may be used to decrease the negative free-water clearance (Nicoloff and LoPresti 1993). If severe hyponatremia and seizures are present, hypertonic saline infusion must be considered: hypertonic sodium chloride (514 mmol/l) should be delivered by a constant infusion pump at a rate set to increase the sodium concentration by about 1 mmol/l/h until neurological signs have disappeared.

Infections and antibiotic therapy

Diagnosis of concomitant infection is difficult because the leukocyte count rarely exceeds 10 000/mm^3, even if sepsis is present. Any band forms should be interpreted as signifying bacterial infection (Nicoloff and LoPresti 1993). Furthermore, these patients are unable to generate a fever in response to such infections, and bacterial infection is the most common precipitating illness for myxedema coma.

Bacterial samples must be taken in sputum, urine, and blood, and broad-spectrum intravenous antibiotic coverage must be prescribed until the culture results are reported. Antibiotic therapy can be stopped if the culture results are negative and no signs of infection appear. If the culture results are positive, antibiotic coverage should be adapted to the infective agent found.

Hormonal therapy

Cortisone therapy

Glucocorticoids are used empirically because of the risk of coexistent adrenal insufficiency. Hydrocortisone hemisuccinate is administered intravenously (300 mg/day) until cortisolemia results are reported (Nicoloff and LoPresti 1993). If cortisol levels are appropriate to the stress levels, hydrocortisone can be safely discontinued. If cortisol levels are low or normal, a coexistent adrenal insufficiency (Schmidt's syndrome) is likely. Hydrocortisone should be continued for 7 to 10 days, and then reduced to a physiological dose of oral hydrocortisone (30 mg/day).

Thyroid hormone therapy

Choice of thyroid hormone (thyroxine or tri-iodothyronine), its optimal dose, and the route of administration remain controversial. To exert its full biological effect, thyroxine must be converted into tri-iodothyronine by an enzymatic step involving 5′-deiodinase.

In the past, large doses of L-tri-iodothyronine were used with a high and unacceptable mortality of 50 to 80 per cent (Mariotti *et al.* 1995). Tri-iodothyronine induced a too rapid increase in oxygen consumption in elderly patients, adversely affecting their cardiovascular and respiratory functions (Mariotti *et al.* 1995). More recently, low doses of L-tri-iodothyronine have been used, either orally or intravenously, with better therapeutic results. However, intravenous L-tri-iodothyronine is no longer available. The use of oral L-tri-iodothyronine is justified when the activity of 5′-deiodinase is low, preventing generation of tri-iodothyronine from thyroxine (Mariotti *et al.* 1995). In critically ill patients or patients receiving drugs such as amiodarone which affect 5′-deiodinase activity, L-tri-iodothyronine at low dosage is preferable to L-thyroxine because high levels of tri-iodothyronine during treatment are a risk factor for fatal outcome (Hylander and Rosenqvist 1985). A dose of 25 μg orally every 6 h gives a good clinical response (Roti *et al.* 1993).

The clinical response with L-thyroxine is more progressive but still sufficiently rapid: metabolic effects can be observed as soon as 6 h after oral or intravenous administration, and the cerebral state improves within 24 to 48 h (Arlot *et al.* 1991). L-Thyroxine enters tissues very slowly, and allows gradual saturation of target cells and a progressive increase in oxygen consumption. On day 1, a single intravenous bolus of 300 to 500 μg L-thyroxine may be given but the use of a continuous infusion of the same dose is preferred. If intravenous L-thyroxine is not available, 500 μg of L-thyroxine should be administered orally via the nasogastric tube (Arlot *et al.* 1991). On subsequent days, a dose of 100 μg/day of L-thyroxine is given by either the intravenous or the oral route. Monitoring of serum concentrations of thyroid-stimulating hormone, free tri-iodothyronine, and free thyroxine is not clinically useful during intensive care treatment. Nevertheless, if it is done it could show a rapid increase of plasma free thyroxine and a slower increase of free tri-iodothyronine, whereas thyroid-stimulating hormone levels decrease only after a few weeks. The goal of the maintenance treatment is to normalize the thyroid-stimulating hormone plasma levels which takes several months (Mariotti *et al.* 1995).

Conclusion

Myxedema is rare, but often occurs in elderly patients with poor cardiovascular condition. Old age, delayed diagnosis, and severe coexistent illness make its prognosis worse. Thyroid hormone treatment reverses most clinical and biological signs, even if cardiovascular or respiratory assistance is needed. Currently, use of L-thyroxine is preferred even though its dosage and route of administration remain empirical.

References

Arlot, S., *et al.* (1991). Myxoedema coma: response of thyroid hormones with oral and intravenous high-dose L-thyroxine treatment. *Intensive Care Medicine*, **17**, 16–18.

Hylander, B. and Rosenqvist, U. (1985). Treatment of myxoedema coma — factors associated with fatal outcome. *Acta Endocrinologica*, **108**, 65–71.

Mariotti, S., Franceschi, C., Cossariza, A., and Pinchera, A. (1995). The aging thyroid. *Endocrine Reviews*, **16**, 686–715.

Nicoloff, J.T. and LoPresti, J.S. (1993). Myxedema coma. A form of decompensated hypothyroidism. *Endocrinology and Metabolism Clinics of North America*, **22**, 279–90.

Roti, E., Minelli, R., Gardini, E., and Braverman, L.E. (1993). The use and misuse of thyroid hormone. *Endocrine Reviews*, **14**, 401–23.

9 Poisoning

9.1 Principles of management

9.1.1 Role of toxicology assessment

A. Jaeger

Key messages

- An accurate and useful toxicological analysis needs good collaboration between the physician and the analytical toxicologist.
- Indications for toxicological analysis include assessment of diagnosis, evaluation of severity and prognosis, indication and evaluation of treatments, and medicolegal implications.
- Interpretation must take into account kinetic and toxicodynamic variation, type of poisoning, and factors such as age and underlying diseases.

Introduction

Diagnosis of acute poisoning is based on history, symptoms, biomedical investigations, toxicological analyses, and sometimes therapeutic tests (Jaeger *et al.* 1991). Toxicological analytical methods are now widely available, but extensive and specific quantitative analyses are expensive. Therefore the indications should be carefully considered by the physician according to the following objectives: assessment of diagnosis, evaluation of severity and prognosis, indication and evaluation of treatments, and medicolegal implications. The interpretation needs to take several factors into account depending on the poison and the patient.

Methods

Before a toxicological analysis is performed, the physician and the analytical toxicologist should discuss the following points (Lheureux *et al.* 1995).

1. Which substance (the parent compound and/or the metabolite(s)) should be analyzed and in what biological sample(s)?
2. What type of analytical method (qualitative or quantitative) and what specificity are needed?
3. Is the analysis useful for the management of the patient?
4. Within what time limit should the results be available for the physician?

A qualitative or semiquantitative analysis of the parent compound (or the active or non-active metabolites) may be adequate for diagnostic assessment. A quantitative analysis of the parent compound is mandatory for kinetic studies. Analysis of the parent compound and the active metabolite(s) is needed for toxicodynamic assessment (symptom–concentration relationship) (Jaeger *et al.* 1994; Lheureux *et al.* 1995). For instance, in ethylene glycol poisoning, analysis of ethylene glycol concentrations is useful for the diagnosis, but glycolate concentrations are more relevant for the evaluation of the severity and prognosis. Therefore the analytical toxicologist should be precisely informed about the indications and the objective of the analysis.

Indications

Assessment of diagnosis

The usefulness of systematic toxicological screening in poisoned patients has not been established. In several reports, the concordance between the drug(s) suspected clinically and the drugs detected by toxicological screens ranged from 26 to 96 per cent and was dependent on the physician, the age of the patient, and the drug ingested (Lheureux *et al.* 1995). The results rarely contributed to the management of the patient. Toxicology screens give qualitative or semiquantitative results for drugs which are frequently involved in poisonings, but they do not usually include toxins which induce severe poisoning and for which analysis is essential for the prognosis or the treatment (e.g. theophylline, digoxin, lithium, carbon monoxide, methanol, ethylene glycol, etc.) (Jaeger *et al.* 1994; Lheureux *et al.* 1995).

In practice, four different situations may be observed.

1. Poisoning is definite, the toxin(s) is known according to the history, and the symptoms are related to the toxin(s) and dose. A toxicological analysis is not absolutely necessary if it has no prognostic, therapeutic, or medicolegal implications.
2. Poisoning is definite and the toxin(s) is known, but the symptoms are not related to the suspected toxin(s) or to the dose. A toxicological analysis is indicated in order to detect other toxins which may have been ingested.

Table 1 Examples of clinical findings in specific poisonings

Clinical finding	Toxin
Cardiovascular disturbances with intraventricular block	Tricyclic antidepressants, class I antiarrhythmics, chloroquine, some β-blockers
Atrioventricular block	Digoxin, digitoxin, β-blockers
Anion gap	Methanol, ethylene glycol, acetone
Osmolal gap	Ethanol, methanol, ethylene glycol, acetone
Hypoglycemia	Insulin, oral antidiabetics
Hypokalemia	Chloroquine, theophylline
Hyperkalemia	Digitalis
Pseudohyperchloremia	Bromine and bromide
Decreased prothrombin level	Oral anticoagulants, rodenticides
Methemoglobinemia	Methemoglobin-forming agents
Decreased plasma or erythrocyte cholinesterase level	Organophosphates
EEG disturbances	Sedatives and psychotropic drugs
Oxalate crystals in urine	Ethylene glycol
Gastric opacities on radiographs	Metals, halogenated hydrocarbons

3. Poisoning is suspected because of symptomatology (toxic symptoms or syndromes), but the toxin(s) is unknown. Only a toxicological analysis can confirm or refute poisoning.

4. Poisoning must be excluded by toxicological analysis in patients presenting with disturbances of the central nervous system (trauma, brain death, elderly patients), cardiovascular symptoms, or convulsions.

According to the history and the symptoms, the analysis should be directed towards specific drugs or groups of drugs.

Biomedical abnormalities may provide diagnostic clues for toxins which are not usually included in routine screens. They reflect the toxic effects and are sometimes more useful than the measurement of plasma drug concentrations in the management of the patient (Jaeger *et al.* 1991; Lheureux *et al.* 1995). Some examples of disturbances suggesting specific poisonings are given in Table 1.

Therapeutic tests, such as naloxone in opiate and flumazenil in benzodiazepine poisoning, may also confirm the diagnosis (Jaeger *et al.* 1991).

Evaluation of severity and prognosis

In order to establish the relationship between the severity and the blood/plasma concentrations, the analysis must be specific and quantitative and sometimes include the active metabolites. The relationship is dependent on the mechanism of toxicity (Jaeger *et al.* 1994).

Functional toxins (e.g. barbiturates, benzodiazepines, meprobamate, cardiotropic drugs, lithium, theophylline, etc.) impair the function of one or more organs. Patients recover without sequelae if no complications occur during the poisoning. Their toxicity is directly related to the concentration at the target organ or receptor; symptoms appear when the plasma concentration exceeds a threshold level, and the severity increases with the concentration (Fig. 1). The duration of the toxicity is dependent on the plasma half-life and the decrease of the concentration at the target organ. For instance, in barbiturate, meprobamate, or ethanol poisoning, the severity of disturbances of the central nervous system and coma is closely related to the plasma concentration. In acute theophylline poisoning, toxicity is minor at concentrations between 20 and 40 mg/l, moderate at concentrations between 40 and 100 mg/l, and severe at concentrations above 100 mg/l. If the parent compound is metabolized into active metabolites which have not been analyzed, there is not usually a relation between plasma parent drug concentrations and symptoms (Jaeger *et al.* 1994).

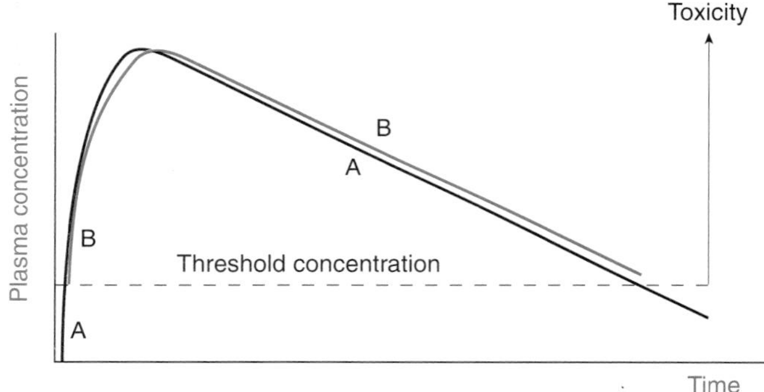

Fig. 1 Toxicokinetic–toxicodynamic relationship for a functional poison (monocompartment kinetics): curve A, concentration; curve B, toxic effect.

Lesional toxins (paraquat, acetaminophen (paracetamol), colchicine, amatoxins, heavy metals, etc.) induce cellular or organ damage. The severity depends on the maximum concentration which has been (or will be) reached at the target organ. If cellular damage has occurred, symptoms may not improve even though the toxin has been eliminated from the target organ. The interpretation has to

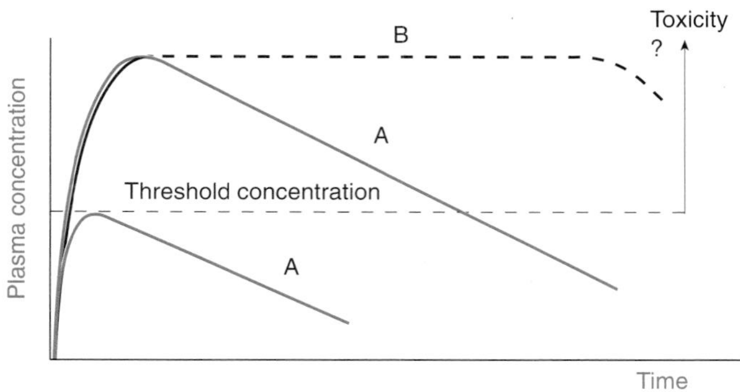

Fig. 2 Toxicokinetic–toxicodynamic relationship for a lesional poison: curve A, concentration; curve B, toxic effect.

take into account the plasma concentration and the time at which this concentration has been measured. Depending on the delay following ingestion, the same plasma concentration may be non-toxic, toxic, or lethal (Fig. 2). In these poisonings, plasma concentrations have a prognostic value: risk of lethal outcome in paraquat poisoning (Proudfoot *et al.* 1979), and risk of hepatitis in acetaminophen poisoning (Rumack *et al.* 1981).

Some toxins act by both mechanisms: the parent compound is a functional toxin, but after a delay cellular damage due to prolonged cellular hypoxia (carbon monoxide, cyanide) or to the accumulation of cytotoxic metabolites (methanol, ethylene glycol) may occur. The interpretation is based on the kinetic data (plasma concentrations of the parent compound and metabolites) and on the time after ingestion or the duration of the exposure (Jaeger *et al.* 1994). In acute short exposure to carbon monoxide, the symptoms correlate well with carboxyhemoglobinemia levels. In prolonged exposure, the severity depends not only on the carboxyhemoglobin level but also on the duration of the cerebral hypoxia. The potential toxicity of methanol is related to the methanol concentration measured in the early phase of the poisoning. The real toxicity depends on the concentrations of the toxic metabolites. If the patient is seen in a later phase, severe symptoms may be present despite low methanol concentrations.

Indication and evaluation of treatments

The management of the poisoned patient is mostly supportive and based on anamnestic, clinical, and biological data. Toxicological quantitative analyses are mandatory for some treatments (Jaeger *et al.* 1994). Depending on the analytical results and the clinical data, the physician will estimate the indication of the following:

(1) alkaline diuresis in salicylate poisoning;

(2) repeated dose of oral activated charcoal in phenobarbital or theophylline poisoning;

(3) hemodialysis in lithium, methanol, ethylene glycol, and salicylate poisoning, and hemoperfusion in carbamate or theophylline poisoning;

(4) ethanol or 4-methylpyrazole in ethylene glycol or methanol poisoning, *N*-acetylcysteine in acetaminophen poisoning;

(5) chelating agents in metal poisoning or deferoxamine (desferrioxamine) in iron poisoning.

In digitalis poisoning, the indication for Fab fragments is essentially based on clinical and biochemical (hyperkalemia) criteria, but previous confirmation of the diagnosis by measurement of digitalis plasma concentration is recommended.

Evaluation of the methods used for decontamination or enhancing elimination must be based not only on clinical improvement but also on precise kinetic data which vary depending on the technique used.

Medicolegal implications

Quantitative and specific analyses are indicated if the poisoning may have medicolegal consequences (e.g. occupational or criminal poisoning).

Interpretation

Apart from the mechanism of toxicity, other factors must be taken into account in the interpretation of analytical and kinetic data (Jaeger *et al.* 1994; Lheureux *et al.* 1995).

Type of poisoning

Similar plasma concentrations may be associated with different severity depending on the type of poisoning: acute, acute on chronic, or chronic. Toxic symptoms appear at lower plasma concentrations in chronic theophylline poisoning than in acute poisoning; convulsions and severe dysrhythmias may appear at concentrations between 40 and 70 mg/l, and the probability of developing convulsions is 50 per cent when the peak concentration exceeds 40 mg/l whereas in acute poisoning the same probability is only observed if the peak concentration is above 120 mg/l. In chronic lithium poisoning, severe disturbances of the central nervous system may appear at supratherapeutic plasma concentrations ($>$ 1.2 mmol/l), whereas in acute poisoning no toxicity has been reported at concentrations ranging up to 8 mmol/l (Jaeger *et al.* 1994). Similar severity of digoxin poisoning is observed at lower plasma digoxin concentrations in chronic overdose than in acute poisoning.

With some toxins (barbiturates, ethanol), a tolerance may be observed in patients treated or poisoned chronically. In acute ethanol poisoning with similar blood ethanol levels, symptoms are less severe in chronic alcoholics than in non-tolerant individuals. Patients treated chronically with barbiturates are more tolerant to acute barbiturate toxicity and the duration of coma is often shorter because of an increase in hepatic elimination by enzyme induction.

Age

In chronic theophylline overdoses with the same plasma concentrations symptoms and prognosis are more severe in elderly patients than in young adults. For a given plasma concentration, the cardiotoxic effects of digoxin are more severe in adults than in children.

Underlying diseases and toxic symptoms

An underlying disease or toxic symptoms, such as hypoxemia and shock, may strongly modify the toxicodynamics (Jaeger *et al.* 1994). In theophylline poisoning, the risk of toxicity and the plasma half-life are increased in patients with congestive heart failure because of impaired elimination by hepatic metabolism. Patients with epilepsy are at higher risk of developing convulsions in poisoning with drugs which may induce convulsions. In poisoning with cardiotropic drugs, the toxicity is increased in patients with chronic heart diseases. In acute meprobamate overdoses, the plasma half-life is increased in patients with shock.

Concurrent ingestion of other drugs

The ingestion of drugs with anticholinergic effects may prolong the gastrointestinal absorption of other drugs. In poisoning with several cardiotoxic drugs with synergistic effects severe symptoms may appear even if the plasma concentration of each drug individually is at a therapeutic level.

The dose ingested

Dose-dependent kinetics, with a change from first-order to zero-order kinetics, have been reported in massive theophylline and salicylate poisoning.

References

Jaeger, A., Kopferschmitt, J., Sauder, Ph., Flesch, F., and Tournoud, C. (1991). Diagnosis in clinical toxicology. *Archives of Toxicology*, Supplement 15, 29–39.

Jaeger, A., Berton, C., and Kempf, J. (1994). Basis of kinetics in clinical toxicology. In *Yearbook of intensive care and emergency medicine* (ed. J.L. Vincent), pp. 707–15. Springer-Verlag, Berlin.

Lheureux, P., Askenasi, R., and Maes, V. (1995). Du bon usage du laboratoire en toxicologie. 2e partie: utilité clinique et interprétation des résultats. *Réanimation Urgences*, 5, 341–52.

Proudfoot, A.T., Stewart, S.M., Levitt, T., and Widdop, B. (1979). Paraquat poisoning: significance of plasma paraquat concentrations. *Lancet*, **ii**, 330–2.

Rumack, B.H., Peterson, R.C., Koch, G.G., and Amara, I.A. (1981). Acetaminophen overdose: 662 cases with evaluation of oral acetylcysteine treatment. *Archives of Internal Medicine*, **141**, 380–5.

9.1.2 Elimination techniques

S. W. Borron and C. Bismuth

Key messages

- Elimination techniques should be evaluated on a case by case basis and never viewed as routine.
- Gastric emptying is associated with significant complications and should be utilized in a selected group of patients.
- Activated charcoal is a useful adjunct with rare complications, but is not efficacious in some poisonings, particularly those involving metals.
- Enhanced elimination (diuresis) is applicable to a small group of toxins. While alkaline diuresis is generally safe, acid diuresis is generally not indicated because of the potential for complications.
- Extracorporeal elimination methods (dialysis and hemoperfusion) are effective for rare toxins with small apparent volumes of distribution, limited protein binding, and toxicity which correlates with plasma concentrations.
- Certain antidotes work by modifying the toxicokinetics of the associated toxin, thus providing the opportunity to affect the toxin's elimination favorably.

Introduction

The mainstay of acute poisoning management is intensive, although not necessarily invasive, supportive care. Toxin elimination is most efficiently accomplished first by a functioning liver and secondly by the kidney and lungs. Elimination techniques must be viewed as complementary gestures which cannot replace thoughtful attention to the vital functions of these body systems of self-defense. Elimination methods are rarely well studied, often inefficient, and at times harmful. Nonetheless, an understanding of available elimination techniques is necessary for the overall effective management of poisoning.

Gastrointestinal decontamination

Gastric emptying

Gastric emptying, although logical in theory, is often associated with an insignificant reduction in toxin body burden and a significant potential morbidity (Table 1). There is a trend away from gastric emptying and towards the use of effective adsorbents (Lheureux and Askenasi 1991).

Gastric lavage, where indicated, should be performed with a large-bore evacuation tube designed for this purpose. Lavage through a Salem sump or nasogastric tube is counterproductive. Lavage is contraindicated after the ingestion of caustics and should be employed after ingestion of petroleum solvents only if associated highly toxic products (e.g. organophosphate pesticides) are present. Attention to the airway during gastric emptying maneuvers cannot be overemphasized. Endotracheal intubation, although it is only partially protective, should be performed prior to lavage if there is any question regarding the patient's capacity for self-protection of the airway. Correct positioning of the patient in a left-decubitus

Table 1 Reported complications of gastric lavage
Aspiration pneumonia
Inadvertent tracheal intubation
Hypothermia
Cardiac dysrhythmias
Hypoxemia
Epistaxis and injury to nasal structures
Esophageal perforation
Esophageal spasm
Mallory–Weiss hemorrhage
Pneumothorax
Empyema
Pneumoperitoneum
Gastric rupture
Hypernatremia
Water intoxication
Pulmonary edema

Trendelenberg posture is optimal, even in the intubated patient. Lavage with warm water (45 °C) is preferable and should be administered in 4-ml/kg aliquots until the return is clear.

Gastric lavage remains a viable option under the following conditions.

1. Overdose is of massive quantity or of highly toxic potential (e.g. colchicine, chloroquine, cyanide, lithium) and treatment is early (less than 1–2 h).
2. There is a possibility of concretion formation (Table 2).
3. The ingested toxins are capable of lesion-inducing (as opposed to functional) toxicity (Table 3).
4. Adsorbents are ineffective.

Concretions should be considered when toxicity is more persistent than expected and/or plasma concentrations remain elevated relative to the known plasma half-life, and may be identified by plain or contrast radiography (depending on the radio-opacity of the product), ultrasound, or endoscopy. Removal of concretions by endoscopy should be considered in the presence of persistent toxicity. Gastrotomy should be reserved for life-threatening intoxication where a less invasive alternative is not feasible.

Emetics have no place in the treatment of intensive care patients.

Adsorbents

Activated charcoal

Activated charcoal is an effective adsorbent of many medications and of certain active metabolites excreted in the bile. Certain agents are not well adsorbed by charcoal (Table 4). Charcoal administration is rarely associated with side-effects, although pulmonary aspiration may be quite serious. It should not be administered in the presence of ileus. Rather than relying on a fixed dose of activated charcoal, it is recommended that the dose exceed the mass of the drug by a factor of 10. In practice, however, the initial dose given is of the order of 1 to 2 g/kg body weight.

Repeated doses of activated charcoal have been proposed for many drugs (Table 5), but are perhaps most useful for those which have important enterohepatic recirculation (phenobarbital (phenobarbitone), phenytoin, colchicine, theophylline). This pulsed-dose method of charcoal administration has been shown to be effective in reducing the elimination half-lives of some drugs (phenobarbital) and the toxicity of others (theophylline), although positive results are not uniform.

It is extremely important in pulsed dosing to use activated charcoal without associated laxatives following the first daily dose. Multiple doses of cathartics may result in severe diarrhea and dehydration as well as electrolyte disorders.

Table 2 Drugs which may form concretions (bezoars)

Aspirin
Barbiturates
Ethchlorvynol
Iron
Glutethimide
Meprobamate
Sustained-release formulations

Table 3 Examples of toxins inducing specific lesions

Toxin	Lesion
Acetaminophen (paracetamol)	Hepatocellular destruction
Cadmium salts	Proximal renal tubular destruction
Mushrooms, cyclopeptide	Hepatocellular destruction
Paraquat	Pulmonary fibrosis
Pyriminyl (Vacor®)	Pancreatic β-cell destruction

Table 4 Some toxins for which activated charcoal is ineffective

Alcohols and glycols
Mineral acids
Alkalis
Heavy metals
Iron
Lithium
Petroleum solvents

Table 5 Drugs for which multiple doses of activated charcoal may be appropriate

Carbamazepine
Colchicine
Digitoxin
Glutethimide
Meprobamate
Organochlorine pesticides (chlordecone, aldrin, dieldrin, DDT, DDE)
Phenobarbital
Theophylline
Valproic acid

Other adsorbents

Fuller's earth has been suggested in paraquat poisoning, although activated charcoal is effective in its absence. Immediate adsorbent therapy is of utmost importance in this intoxication and should be instituted with a readily available adsorbent. Prussian blue (potassium ferric hexacyanoferrate) 250 mg/kg/day in four divided oral doses is effective for thallium and cesium poisoning, although pharmaceutical preparations may be impossible

Purgatives

Cathartics

Cathartics, such as magnesium citrate, magnesium sulfate, or sorbitol, have unproven efficacy in removing toxins (Lheureux and Askenasi 1991), although in theory they may help to reduce the risk of charcoal-induced obstruction. However, the intemperate administration of magnesium salts in overdose has resulted in severe toxicity.

Whole-bowel irrigation

Whole-bowel irrigation, employing polyethylene glycol bowel preparation solutions, has shown promise in serious intoxications for which activated charcoal is ineffective (iron) (Tenenbein 1985) and in the acceleration of elimination of intact 'body-packer' packets of heroin and cocaine. Its efficacy in other intoxications is presumptive, although it should be considered in ingestions of other compounds which are poorly adsorbed by charcoal, particularly those with long half-lives such as certain heavy metals (lead and others) and radionuclides (cesium, plutonium, etc.).

Enhanced elimination

Forced diuresis

Forced diuresis is the administration of high volumes of intravenous fluids alone, or in combination with diuretics, with the goal of enhancing urinary excretion. It is fraught with the potential for complications, including pulmonary and/or cerebral edema, hypokalemia, hypernatremia, and hyponatremia (Garrettson and Geller 1990). Furthermore, its efficacy has been demonstrated for only a few toxins (Table 6). A urine output of 2 to 4 ml/kg/h is a reasonable goal of therapy.

Acid diuresis

Acid diuresis, using ammonium chloride or other acidifying agents, can increase the elimination of weak bases such as amphetamines and phencyclidine. However, its use is not recommended because of the risk of inducing metabolic acidosis which may aggravate the rhabdomyolysis frequently associated with these toxins and result in renal failure.

Alkaline diuresis

The administration of alkalinizing agents is of interest in the treatment of intoxication by certain weak acids (Table 6) including salicylates and barbiturates. If the urine is alkalinized, more of the excreted drug becomes ionized, making it unavailable for tubular reabsorption (ion trapping). Addition of 50 to 100 mmol of sodium bicarbonate per liter of intravenous fluids, administered at a rate sufficient to maintain a urine flow of 2 to 4 ml/kg/h is a reasonable starting point. Monitoring of urine pH (ideally at least 8.0), arterial blood gases (maintain below 7.5), and plasma electrolytes is recommended. Potassium supplementation may be required.

Table 6 Agents susceptible to enhanced elimination by forced diuresis

Neutral diuresis*	Alkaline diuresis	Acid diuresis†
Halides (bromide, fluoride, iodine, iodide)	Aspirin	Amphetamines
Pentobarbital	Chlorphenoxy herbicides	Phencyclidine
Secobarbital (quinalbarbitone)	Phenobarbital	Strychnine
Thallium		

*Neutral diuresis refers to administration of normal saline with or without diuretics to promote supranormal urine output. It should be employed with extreme caution because of the risk of pulmonary and cerebral edema and electrolytemias.
†Acid diuresis is not recommended because of the risk of inducing acidemia and renal failure in the presence of myoglobinuria.

Extracorporeal elimination

Hemoperfusion, hemodialysis, and peritoneal dialysis

Extracorporeal elimination techniques have been suggested as a means of reducing body burdens of a variety of toxins. While of unquestionable utility in the treatment of drug-induced renal failure and of demonstrated benefit in certain intoxications (e.g. methanol and ethylene glycol), extracorporeal methods are often recommended in the literature in the absence of adequate evidence for their efficacy. Clinical evaluation of these techniques is inadequate (Bismuth 1990).

Indications for extracorporeal elimination and choice of technique

In order for a patient to benefit from extracorporeal methods of elimination, the toxin must satisfy several conditions.

1. The apparent volume of distribution of the toxin should be limited (ideally no more than 1 l/kg body weight). Toxins with apparent volumes of distribution above 60 l/kg cannot be effectively removed by these methods, and those with intermediate values require prolonged therapy.
2. Protein binding should be limited.
3. There should be a positive relationship between plasma concentration and toxicity.
4. Elimination should follow a single-compartment model.

The choice of technique is based on equipment availability, patient characteristics (hemodynamic stability and renal function), and relative efficiency of the method. The mnemonic I STUMBLE (Table 7) can be used to recall some agents for which extracorporeal elimination is of potential benefit. In the case of attendant renal failure, the choice is clearly in favor of hemodialysis. Hemoperfusion, using a

Table 7 A useful mnemonic for drugs susceptible to elimination by extracorporeal methods

I	Isopropanol/isoniazide
S	Salicylates
T	Theophylline
U	Uremia induced by toxins
M	Methanol
B	Barbituric acid
L	Lithium
E	Ethylene glycol

coated charcoal filter, is superior to dialysis for phenobarbital and theophylline intoxication.

In assessing hemoperfusion and/or hemodialysis from a biological standpoint, plasma or blood concentrations are often used as an indication of the efficiency of the process. Like clinical parameters, these may be misleading, as plasma and blood concentrations are frequently poorly representative of the total body burden of the toxin. Furthermore, the simple observation of evolving plasma concentrations fails to consider an often large (up to 80 per cent) concomitant hepatic clearance of the compound and the role of continuing absorption of the compound from the gastrointestinal tract.

Figure 1 illustrates how surveillance of blood or plasma concentrations may be misleading. Likewise, clearance data may be misleading. The meaning of plasma clearance in a specific intoxication is dependent on the apparent volume of distribution of the toxin, the protein binding, and alternative routes of elimination (other than kidney). For example, in intoxication by cyclic antidepressants, even though efficient plasma clearance may be obtained, the net result is minimal owing to their large apparent volumes of distribution which indicate a wide diffusion of the substance in the body. In addition, even with identical clearance over two equal periods of time, the portion of toxin cleared by extracorporeal elimination may vary tremendously depending on the contribution of extrarenal elimination (e.g. hepatic metabolism) of the product.

A better approach to evaluating extracorporeal methods is measurement of the toxic agent in the dialysate (hemodialysis), or measurements of the concentration on the arterial and venous sides of the hemoperfusion filter multiplied by the blood flow across the system. In this way, a ratio of eliminated toxin to ingested dose can be calculated to assess efficiency. If the blood flow through the extracorporeal system is unknown, the toxicokinetics of the product can be compared with its spontaneous pharmacokinetics.

Hepatic clearance of drugs for which plasma and urine concentrations are available and reliable may be calculated, based on the formula for apparent volume of distribution (Fig. 2).

Toxicokinetic antidotes

Antidotes work by two different global mechanisms, i.e. by affecting either the toxicodynamic properties of the intoxicant (hemodynamics, binding at tissue sites) or the toxicokinetic properties (absorption, distribution, elimination) (Baud *et al.* 1995). Because

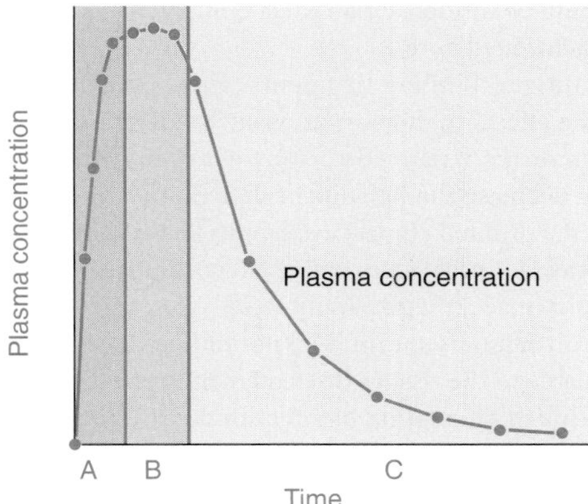

Fig. 1 Schematic representation of plasma drug concentrations after ingestion. The plasma concentrations of an ingested toxin are the sum of many factors, including absorption, distribution, metabolism, and excretion. Thus evaluation of efficacy of elimination methods by plasma concentrations is impractical. In time period A of this schematic curve, the concentration is strongly influenced by continued gastrointestinal absorption. Results obtained during this period may be erroneously interpreted as inadequacy of extracorporeal elimination. In time period B, the combination of ongoing absorption, distribution into various tissue compartments, and elimination results in a plateau effect. Results obtained during this period may be erroneously interpreted in either direction. In time period C, the elimination and metabolism of the drug predominate and plasma concentrations fall at a faster rate. Results obtained during this period may be erroneously attributed to efficacy of extracorporeal elimination.

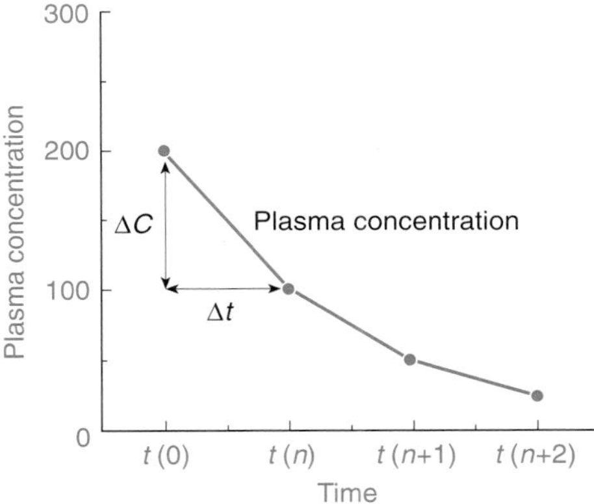

Fig. 2 Calculation of hepatic clearance. The relationship of drug quantity in the body to to its blood concentration is

$$Q\,(\text{mg}) = V_d\,(\text{l/kg}) \times C\,(\text{mg/l}) \times W\,(\text{kg})$$

where Q is the amount of drug, V_d is the apparent volume of distribution of the drug, C is the concentration of the drug, and W is the body weight. Thus the amount of drug eliminated can be calculated by subtracting the concentration at time t_n (blood concentration at a given time n after presentation) from the concentration at time t_o (initial blood concentration) and multiplying by the body weight and apparent volume of distribution. The amount of drug excreted in the urine is then subtracted from this total. In the absence of significant lung elimination, the remaining drug elimination is accounted for by hepatic clearance (metabolism and biliary excretion).

the latter group of antidotes may affect elimination, several examples are briefly mentioned here.

Digitalis antigen-binding fragments (Fabs) have very rapid pharmacodynamic effects (reduction of tissue binding and improvement in hemodynamics with reduction in dysrhythmias) and also dramatically decrease the half-life of digoxin in overdose. Likewise, colchicine Fabs (limited clinical availability) have shown promise not only in decreasing pharmacodynamic effects, but also in acting favorably on the pharmacokinetic parameters.

Hydroxocobalamin (vitamin B_{12a}) not only reduces the amount of cyanide available to the tissues (toxicodynamic effect) but diminishes the 1-h half-life of circulating blood cyanide by its rapid irreversible binding of the toxin.

4-Methylpyrazole is virtually devoid of toxicodynamic effects; rather, it acts as a pure toxicokinetic antidote for ethylene glycol and methanol. However, because in this instance it is the metabolic products of these toxins which are responsible for their danger, 4-methylpyrazole achieves its benefit by blocking metabolism, thus prolonging the half-lives of these compounds and allowing the kidneys to eliminate the unchanged product. If renal failure is present, the toxins can then be removed by dialysis.

Conclusion

Elimination techniques, while potentially useful in clinical toxicology, must be carefully evaluated for their indications in each specific poisoning. While they may be life saving in certain instances, these techniques are almost never as efficient as the intact human organs of elimination, and thus primary attention should be directed to the preservation of vital functioning of the liver, kidney, and lungs through intensive supportive care.

References

Baud, F.J., Borron, S.W., and Bismuth, C. (1995). Modifying toxicokinetics with antidotes. *Toxicology Letters*, **82/83**, 785–93.

Bismuth, C. (1990). Biological valuation of extra-corporeal techniques in acute poisoning. *Acta Clinica Belgica*, **45**, 20–8.

Garrettson, L.K and Geller, R.J. (1990). Acid and alkaline diuresis. When are they of value in the treatment of poisoning? *Drug Safety*, **5**, 220–32.

Lheureux, P. and Askenasi, R. (1992). Prévention de l'absorption digestive des toxiques: une réévaluation critique. *Réanimation Urgences*, **1**, 751–9, 915–26.

Tenenbein, M. (1985). Whole bowel irrigation for toxic ingestions. *Journal of Toxicology. Clinical Toxicology*, **23**, 177–84.

9.2 Management of specific poisons

9.2.1 Salicylate

J. A. Henry

Key messages

- The fatal adult dose of aspirin is 500 mg/kg; symptoms of toxicity may be minimal even in severe cases.
- Common early features of salicylate overdose are irritability, tinnitus, hyperventilation, nausea, vomiting, and abdominal pain.
- Respiratory alkalosis, metabolic acidosis, ketosis, hypoglycemia, and hyperglycemia may all occur.
- Gastric lavage may be performed up to 12 h after ingestion in large aspirin overdoses because, owing to the insolubility of salicylate in gastric acid, concretions may form within the stomach.
- Hemodialysis is indicated if there are severe metabolic changes, or if the salicylate level is over 800 mg/l (5.76 mmol/l) after rehydration.
- If the salicylate level is over 600 mg/l (4.32 mmol/l) urinary alkalinization (intravenous sodium bicarbonate 8.4 per cent) should be used to maintain a urine pH over 7.5.

Introduction

Salicylates can cause severe poisoning with relatively few symptoms, so that every case of ingestion must be taken seriously. Deaths occur because of a lack of understanding of the complexity and severity of salicylate overdose. Salicylates are found in aspirin (acetylsalicylic acid), benorylate, methyl salicylate (oil of wintergreen) which is used in liniments, choline salicylate which is included in some teething gels, and salicylic acid which has keratolytic properties. Teething gels rarely cause severe poisoning because the amount of salicylate contained is small. Aspirin comes as tablets or capsules (75–500 mg), sometimes in combination with other analgesics or decongestants.

Toxicity and mechanisms

The fatal adult dose of aspirin is 500 mg/kg (Temple 1981), and toxicity may be life threatening at plasma levels of over 750 mg/l (5.40 mmol/l) in adults.

Salicylates have a wide variety of actions. In overdose, this leads to a number of physiological and metabolic derangements. The mechanisms which lead to the clinical effects are outlined in Fig. 1. Nausea and vomiting are due to prostaglandin effects in the stomach. The most apparent metabolic effect in the early stages is due to stimulation of the respiratory center, causing hyperventilation which leads to a respiratory alkalosis. Later, a raised anion gap metabolic acidosis supervenes due to increased lactate levels, stimulation of lipid metabolism with ketone body production, and inhibition of aminotransferase. The body compensates by excreting bicarbonate, sodium and potassium ions, and water, reducing the body's buffering capacity, and resulting in an anion gap metabolic acidosis which enhances transfer of the salicylate ion across the blood–brain barrier. Salicylates also uncouple oxidative phosphorylation, decreasing ATP production, increasing oxygen utilization and carbon dioxide and lactate production, and contributing to the metabolic acidosis. The energy which should be used to produce ATP is diverted to mitochondrial heat production, leading to compensatory sweating and flushing, with the body temperature usually remaining normal in adults although pyrexia may occur in children. Vomiting, hyperventilation, and sweating all lead to volume depletion which may further complicate electrolyte disturbances. Consciousness is not usually lost, but in severe cases central nervous system toxicity gradually manifests itself as confusion, coma, and convulsions. Indeed, salicylate lethality has been related to central nervous system dysfunction when a 'critical brain salicylate level' is reached (Hill 1973).

Aspirin is rapidly converted to salicylic acid which is further metabolized to five main metabolites: salicyluric acid, salicyl phenolic glucuronide, gentisic acid, acyl glucuronide, and gentisuric acid. The pathways for the first two of these involve saturable hepatic enzymes, so that when a large amount of salicylate is ingested these routes quickly become saturated. This results in a change from first-order kinetics (where elimination is proportional to the plasma concentration) to zero-order kinetics (where only a certain amount is eliminated irrespective of the plasma concentration). Thus salicylate may accumulate following mild therapeutic overdoses, particularly in children, and prior therapeutic use of the drug may increase the toxicity of an acute overdose.

Under zero-order kinetics the amount of salicylate excreted unchanged in the urine increases. With very high plasma concentrations, up to 80 per cent of salicylate can be excreted unchanged in the urine. This pathway is sensitive to changes in urine pH; as the pH rises (i.e. the urine becomes more alkaline), excretion of salicylate is enhanced. This is the basis of urinary alkalinization in salicylate overdose.

Clinical features

Common early features of salicylate overdose are irritability, tinnitus, hyperventilation, nausea, vomiting, and abdominal pain. Further symptoms include sweating, flushing, deafness, tremor, hypokalemia,

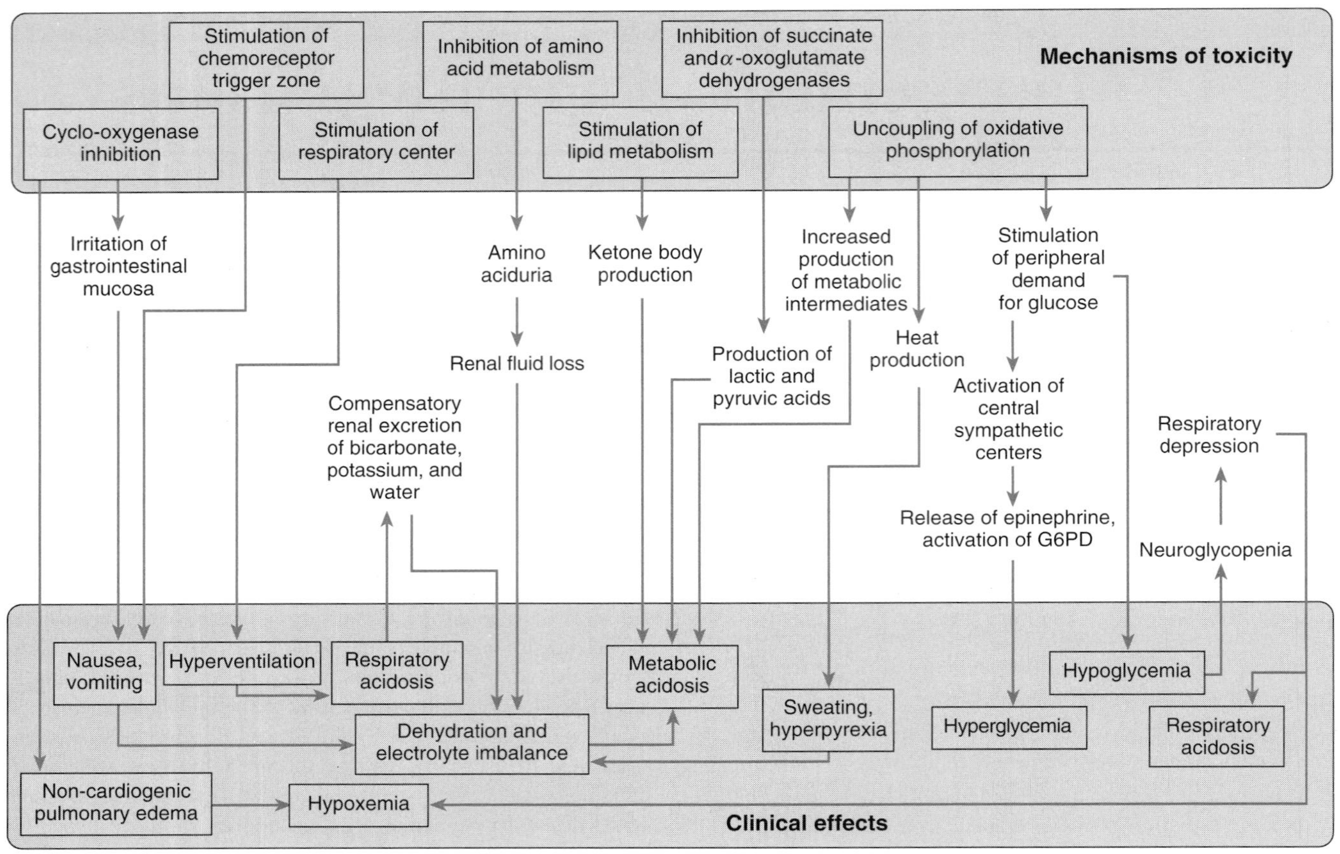

Fig. 1 Mechanisms of toxicity responsible for the clinical effects seen in salicylate poisoning: G6PD, glucose-6-phosphate dehydrogenase.

and hypernatremia. The combination of vomiting, hyperventilation, and sweating may lead to severe volume depletion. Other findings may include hypoprothrombinemia, pyrexia (usually in children), confusion, drowsiness, delirium, coma, and convulsions (more common in children).

Respiratory alkalosis, metabolic acidosis, ketosis, hypoglycemia, and hyperglycemia may all occur. The biochemical pattern of a respiratory alkalosis together with metabolic acidosis is characteristic of salicylate poisoning. The patient usually presents with a combined respiratory alkalosis and metabolic acidosis, with a blood pH in the range 7.40 to 7.46. Later, as metabolic compensation fails, the arterial pH may fall below 7.40. The most important clinical signs of serious toxicity are a falling plasma pH, hypoxemia, and the development of pulmonary edema. Confusion and depressed consciousness are serious signs, usually indicating that salicylate has entered the central nervous system. These symptoms may be improved by correction of metabolic acidosis. The likelihood of respiratory alkalosis increases with age until 12 years, when the adult picture of respiratory alkalosis followed by metabolic acidosis occurs. The underlying mechanisms for these age-dependent differences in acid–base balance are poorly understood. Since acidosis enhances transfer of the salicylate ion across the blood–brain barrier, it is necessary to employ more active therapy at lower salicylate concentrations in children.

Death is usually due to cardiorespiratory arrest; attempts at resuscitation are unlikely to be successful.

Management

Assessment of the severity of salicylate intoxication involves clinical assessment, metabolic assessment (e.g. a plasma pH of 7.45 bodes well, but a pH of 7.30 indicates severe toxicity), and measurement of the plasma salicylate level.

Every patient suspected of salicylate toxicity should be detained until the severity has been assessed biochemically. The plasma salicylate level should be measured on presentation and again after 3 h in case it is rising. Severe volume depletion commonly causes an artificially elevated blood salicylate concentration. Absorption may continue for many hours, leading to a delay of up to 24 h before the peak salicylate concentration is reached. Therefore it is advisable to monitor the salicylate concentration every 3 to 4 h after the initial measurement to ensure that the peak level has been reached. At the same time, arterial blood gases and plasma biochemistry should be checked.

Gastric decontamination should be performed. Gastric lavage may be useful up to 12 h after ingestion in large aspirin overdoses because salicylate is insoluble in gastric acid, so that it may form concretions within the stomach and induce pylorospasm. Repeated doses of activated charcoal should be given because salicylate may be retained in the stomach for many hours after ingestion. This treatment has been shown to shorten the plasma half-life of salicylate. Activated charcoal adsorbs salicylate but desorption may occur (Filippone *et al.* 1987). This can be minimized by giving repeated doses of activated charcoal every 4 h. This regimen also appears to enhance passage of the salicylate ion from the blood back into the gut ('gastrointestinal dialysis') (Hillman and Prescott 1985). As this procedure is effective and non-invasive, it is recommended for mild to moderately severe salicylate toxicity.

Central venous pressure should be measured early, and the patient should be adequately rehydrated with intravenous fluids and any metabolic acidosis or hypoglycemia corrected. Hemodialysis is recommended for severe salicylate intoxication. This includes any patient with markedly raised salicylate concentration, coma, convulsions, acidemia or hypoxemia, or mild central nervous system effects not responding to the correction of acidosis.

If there are severe metabolic changes or the salicylate level remains over 800 mg/l (5.76 mmol/l) after rehydration, hemodialysis is indicated and the patient should be referred urgently for treatment. Hemodialysis effectively eliminates salicylate and corrects any metabolic disturbances. If the level is over 600 mg/l (4.32 mmol/l), urinary alkalinization (intravenous sodium bicarbonate 8.4 per cent) should be commenced. The aim is to produce a urinary pH of over 7.5 as this enhances renal salicylate elimination by trapping the salicylate anion in the renal tubule. Renal elimination of salicylate is enhanced 10- to 20-fold when urine pH increases from 5 to 8 (Prescott *et al.* 1982). In the past forced alkaline diuresis (i.e. fluid plus alkali) has led to non-cardiogenic pulmonary edema, particularly in smokers and the elderly. Alkalinization of the urine without increasing fluid output is the recommended regime for moderate salicylate toxicity. Sodium bicarbonate should be administered intravenously to produce a urine pH between 7.5 and 8.5; the arterial pH should not rise above 7.6. It is difficult to produce alkaline urine if the patient is hypokalemic, and so the serum potassium must be corrected.

References

Filippone, G.A., Fish, S.S., Lacouture, P.G., Scavone, J.M., and Lovejoy, G.H., Jr (1987). Reversible adsorption (desorption) of aspirin from activated charcoal. *Archives of Internal Medicine*, **147**, 1390–2.

Hill, J.B. (1973). Salicylate intoxication. *New England Journal of Medicine*, **288**, 1110–13.

Hillman, R.J. and Prescott, L.F. (1985). Treatment of salicylate poisoning with repeated oral charcoal. *British Medical Journal,* **291**, 1472.

Prescott, L.F., Balali-Mood, M., Critchley, J.A., Johnstone, A.F., and Proudfoot, A.T. (1982). Diuresis or urinary alkalinisation for salicylate poisoning? *British Medical Journal*, **285**, 1383–6 .

Temple, A.R. (1981). Acute and chronic effects of aspirin toxicity and their treatment. *Archives of Internal Medicine*, **141**, 364–9.

9.2.2 Acetaminophen (paracetamol)

A. Rhodes and J. Wendon

Key messages

- Acetaminophen is predominantly metabolized in the liver; 90 per cent of the drug undergoes glucuronidation or sulfation to a non-toxic inactive compound, which is excreted in the urine, and 5 per cent is oxidized to *N*-acetyl-*p*-benzoquine imine which is rendered non-toxic by glutathiones.
- In acetaminophen poisoning there is an increase in the formation of the toxic *N*-acetyl-*p*-benzoquine imine, which saturates the glutathione defense mechanism and causes hepatic necrosis.
- The presence of symptoms or signs in acetaminophen poisoning should be treated as an indicator of hepatotoxicity. Liver function tests start to become deranged from 12 h with the serum aspartate aminotransferase peaking at 72 h.
- Any patient with a plasma acetaminophen level after 4 h which is greater than the standard treatment line or greater than half the standard line in a high-risk patient should be treated with the antidote *N*-acetylcysteine. The antidote should also be given if the timing is suspect or inaccurate.
- The management of fulminant hepatic failure resulting from acetaminophen overdose is based on resuscitation and preparation for urgent and safe transfer to a specialist liver failure unit.

Introduction

Acetaminophen (paracetamol, *N*-acetyl-*p*-aminophenol) is the most common analgesic used in the United Kingdom and is responsible for 150 deaths annually in England and Wales. This represents 3 per cent of all overdose deaths. Acetaminophen overdose is the leading cause of acute liver failure in the United Kingdom, being attributable for over 50 per cent of all acute referrals to a liver failure unit. Survival rates for acetaminophen-induced fulminant hepatic failure have improved in recent years, probably as a consequence of improvements in intensive care, although the mortality still remains high (at around 20 per cent of all admissions) (Makin *et al.* 1995).

Hepatotoxicity of acetaminophen

In therapeutic doses acetaminophen is predominantly metabolized in the liver, where 90 per cent of the drug undergoes either glucuronidation or sulfation to a non-toxic inactive compound which is excreted in the urine. The remainder of the drug is dealt with in two ways: 5 per cent is excreted unchanged in the urine, and the rest undergoes oxidation by the hepatic mixed-function oxidase system (cytochrome P-450 2E1) to the unstable toxic compound *N*-acetyl-*p*-benzoquine imine (NAPQI). NAPQI is rendered non-toxic by conjugation with hepatic intracellular glutathiones and is then excreted via the biliary system.

In acetaminophen poisoning there is saturation of the glucuronidation and sulfation pathways, which leads to an increase in the formation of the toxic NAPQI. The glutathione defense mechanism becomes depleted with even moderate doses of acetaminophen, and when it falls to less than 30 per cent of the normal level NAPQI is able to accumulate and bind covalently to the liver in a centrilobular fashion, causing hepatic necrosis via arylation and oxidation of key calcium-regulating proteins.

The minimum dose of acetaminophen believed to cause hepatic damage is 125 mg/kg (which is equivalent to 15 standard 500-mg tablets for a 60-kg person). Acetaminophen doses above 250 mg/kg produce hepatotoxicity in 60 per cent of cases, and severe liver damage (serum aspartate aminotransferase (AST) above 1000 IU/l) is inevitable when doses in excess of 350 mg/kg are absorbed. The most common method of predicting the potential risk of an overdose is through plasma levels and time from ingestion of tablets. These are plotted on a semilogarithmic graph (Prescott *et al.* 1979); 60 per cent of patients who lie above a line joining a plasma level of 200 mg/l at 4 h and 30 mg/l at 15 h will develop severe hepatic damage if untreated. A similar graph is used in the United States, except that the line is modified to be 25 per cent below the standard United Kingdom plot (Rumack *et al.* 1981).

High-risk groups

Several patient groups are believed to represent a higher risk than normal for severe hepatic damage following acetaminophen ingestion. Any state that leads to either induction of the cytochrome P-450 pathway or depletion of the glutathione system will result in the development of higher NAPQI levels and hepatic damage. Induction of the cytochrome pathway and depletion of glutathione can be caused by malnutrition, fasting, and chronic alcohol abuse. A variety of drugs (phenytoin, carbamazepine, primidone, phenobarbital (phenobarbitone), rifampin (rifampicin), and isoniazid) induce the cytochrome enzymes, while severe glutathione depletion may occur in sequential overdoses over several days. Many patients are unable to give an accurate history, which makes the timing of ingestion unreliable and the plasma levels meaningless. These groups include psychiatric patients, alcoholics, and patients who have taken an accidental staggered overdose. Any patient in these high-risk groups should be offered treatment at a plasma level of greater than half the standard line; treatment should also be offered if there is any evidence of severe hepatic dysfunction (raised AST), coagulopathy, renal failure, or encephalopathy.

Clinical course of acetaminophen overdose

The majority of overdoses are of small quantities of acetaminophen and the patients are not unwell. Thus the presence of symptoms or signs should be treated as an indicator of hepatotoxicity. Nausea and vomiting occur within a few hours of ingestion, soon followed by diffuse abdominal pain and hepatic tenderness. Liver function tests start to become deranged from 12 h with the AST peaking at 72 h. Jaundice becomes biochemically apparent at 24 h and rapidly deepens. A coagulopathy can be demonstrated only a few hours after overdose, with the prothrombin time being prolonged and continuing to rise until day 3. Oliguric or polyuric renal failure occurs in 11 per cent of severe overdoses, with or without the presence of acute liver failure, and is best monitored with the plasma creatinine as urea production is impaired in hepatic dysfunction. The renal failure can result from prerenal hypovolemia as well as a direct nephrotoxic effect of the acetaminophen. Once acute liver failure has developed, renal failure will be present in 70 per cent of cases. From 72 h many patients develop encephalopathy which worsens until 5 days after ingestion. This may occur earlier in the case of severe overdoses and persist for many days, and can be complicated by the development of cerebral edema.

There is often a profound metabolic disturbance with hypophosphatemia, hypoglycemia, and a metabolic acidosis. Hypoglycemia can occur within the first 24 h after overdose as a result of decreased gluconeogenesis and glycogen mobilization as well as an increased insulin concentration. The metabolic acidosis can occur early within the first 12 h owing to a derangement in hepatic lactate metabolism. Over subsequent hours, this worsens because of a further reduction in hepatic lactate clearance and an increased peripheral lactate generation due to tissue hypoxia. Hypovolemia is often profound, secondary to vomiting and vasodilatation, and worsens the tissue hypoxia and peripheral lactate generation.

Initial management

Absorption of acetaminophen from the stomach is not complete for at least 4 h after ingestion, and may take much longer with slow-release preparations. Therefore plasma levels before this time are misleading, and attempts should be made to limit further absorption of the drug with either activated charcoal or syrup of ipecacuanha. Special care should be taken with protection of the airway, particularly in mixed overdoses where sedation and central nervous system depression can be an early feature. Further management is based around a plasma acetaminophen level related to the time of ingestion, and a serum creatinine and prothrombin time if the patient presents more than 24 h after ingestion, the overdose is greater than 12 g of acetaminophen, or the patient is in one of the high-risk groups. An abnormality in either the creatinine or the prothrombin time can be an early indication of severe hepatic damage and the only sign of a serious overdose in high-risk patients. Any patient with a plasma acetaminophen level after 4 h which is greater than the standard treatment line, or greater than half the standard line in a high-risk patient should be treated with the antidote *N*-acetylcysteine (NAC). NAC should also be administered if the timing is suspect or inaccurate.

NAC acts by a number of mechanisms to prevent hepatotoxicity. The main action is to increase hepatic glutathione production by acting as a source of cysteine, which replenishes intracellular glutathione and allows an increase in conjugation of NAPQI. It can also act as both a sulfur donor and a direct substitute for glutathione, which causes an increase in both sulfation of acetaminophen to its non-toxic metabolite and conversion of NAPQI back to acetaminophen. NAC is most effective when given within 8 h of the overdose; however, it remains a safe and effective treatment even after 15 h. NAC is administered as an intravenous infusion, starting with a loading dose of 150 mg/kg over 15 min followed by 50 mg/kg for 4 h and then 100 mg/kg for 16 h. Following the initial NAC regimen, the presence of coagulopathy, renal dysfunction, or symptoms and signs attributable to the acetaminophen overdose should be treated with further infusions of NAC at 150 mg/kg every 24 h and daily monitoring of the international normalized ratio (INR) and plasma creatinine concentration. Allergic reactions to NAC are rare and are usually only seen with the initial loading dose. In cases where true anaphylaxis has occurred, oral NAC or methionine are available as substitutes.

Prognostic markers

A variety of markers are related to outcome and are used clinically to direct treatment strategies. Both serum creatinine levels and prothrombin time have been shown to correlate with survival. A serum creatinine concentration of less than 100 μmol/l (1.13 mg/dl) is associated with a survival rate of 65 per cent, but this falls to just 23 per cent when the serum creatinine is greater than 300 μmol/l

(3.39 mg/dl). A peak prothrombin time of less than 90 s correlates with 80 per cent survival, but a peak prothrombin time of greater than 180 s is associated with a survival of less than 10 per cent. The presence of metabolic acidosis which does not correct with fluid resuscitation is a particularly poor prognostic sign, with a survival rate of less than 10 per cent. The degree of encephalopathy is also related to outcome, with a grade I or II encephalopathy associated with 95 per cent survival, but this falls to less than 22 per cent if the patient develops grade IV encephalopathy or cerebral edema. Patients who become hypotensive and require vasopressors are also at very high risk, with less than 10 per cent survival. Referral to a specialist liver unit for discussion and subsequent transfer should be made if any of the complications in Table 1 develop (Makin and Williams 1996).

Treatment of fulminant hepatic failure

The management of fulminant hepatic failure resulting from acetaminophen overdose is based on good intensive care. Optimum management should be directed towards resuscitation and preparation for urgent and safe transfer. Control and protection of the airway is fundamental, particularly as these patients can rapidly become deeply encephalopathic. Sedatives are contraindicated, and patients who cannot be managed safely should be ventilated. Intravenous fluid resuscitation is always necessary and is best directed by either central venous or pulmonary artery pressures. Hypotension may require intravenous vasopressors as well as intravenous fluid. Volume expansion can often correct a mild metabolic acidosis, but bicarbonate infusions should be avoided as the trend in serum bicarbonate is crucial with respect to prognosis (Table 2). These patients often experience profound metabolic disturbances, particularly hypoglycemia and hypophosphatemia, which need to be corrected with infusions of 10 per cent glucose and potassium phosphate. The development of coagulopathy and thrombocytopenia can make insertion of venous catheters difficult; however, fresh frozen plasma should be avoided. The prothrombin time is the main prognostic factor with regard to transplantation (Table 2) (O'Grady *et al.* 1989).

All patients should be treated with NAC, which should be restarted with the initial loading regimen if there is doubt as to the dose received and then continued at 150 mg/kg/day until the INR has returned to less than 2. Patients should also be given a replacement dose of vitamin K as well as prophylactic broad-spectrum antibiotics and antifungals (oral fluconazole 50 mg once daily) as sepsis in this group of patients is common and is associated with a particularly poor outcome. If the patient achieves transplant criteria (Table 2) or develops renal failure or grade III/IV encephalopathy, intravenous amphotericin B should be started. Renal failure should be treated with continuous hemofiltration since intermittent dialysis can precipitate cerebral edema.

Forty-five per cent of patients with acetaminophen-induced fulminant hepatic failure and grade III/IV encephalopathy develop cerebral edema (Makin *et al.* 1995). The main aim in the management of these patients is to prevent further surges in intracranial pressure. They should be deeply sedated and paralysed with neuromuscular blocking agents. Stimulation, movement, and physiotherapy should be kept to a minimum. Seizures can be difficult to assess, and so prophylactic anticonvulsants should be started. Cerebral perfusion can be assessed from jugular venous bulb oximetry and intracranial pressure monitoring. Traditionally, a cerebral perfusion pressure of 50 mmHg should be obtained. Hyperventilation should be avoided and a close control on plasma osmolality maintained. Surges in intracranial pressure should be treated with intravenous mannitol (0.5–1.0 mg/kg) but the serum osmolality should not be allowed to exceed 320 mosmol/l.

Patients who develop the transplantation criteria should be offered an orthotopic liver transplant. With a transplant these patients have a survival rate of over 80 per cent, whereas patients developing these criteria but not receiving a transplant have a survival rate of only 40 per cent. Patients who do not develop the transplant criteria have a survival of 90 per cent.

Table 1 Guidelines for referral to a specialized liver unit

Progressive coagulopathy
INR >2 at 24 h or >4 at 48 h or >6 at 72 h
Renal impairment with plasma creatinine $>200\,\mu$mol/l (2.26 mg/dl)
Metabolic acidosis unresponsive to fluid resuscitation (pH <7.35, HCO_3^- <18 mmol/l)
Hypotension
Encephalopathy
Hypoglycemia

Table 2 King's College transplantation criteria

pH <7.3 which fails to correct with fluid
or
A normal pH with all three of the following
(a) Prothrombin time >100 s
(b) Creatinine $>300\,\mu$mol/l (3.39 mg/dl)
(c) Grade III or grade IV encephalopathy

References

Makin, A.J. and Williams, R. (1996). Acetaminophen overdose and acute liver failure: modern management. In *Yearbook of intensive care and emergency medicine* (ed. J.-L. Vincent), pp. 659–71. Springer Verlag, Berlin.

Makin, A.J., Wendon, J., and Williams, R. (1995). A 7-year experience of severe acetaminophen-induced hepatotoxicity (1987–1993). *Gastroenterology*, **109**, 1907–16.

O'Grady, J.G., Alexander, G.J.M., Hayllar, K.M., and Williams, R. (1989). Early indicators of prognosis in fulminant hepatic failure. *Gastroenterology*, **97**, 439–45.

Prescott, L.F., Illingworth, R.N., Critchley, J.A.J.H., Stewart, M.J., Adam, R.D., and Proudfoot, A.T. (1979). Intravenous *N*-acetylcysteine: the treatment of choice for acetaminophen poisoning. *British Medical Journal*, **2**, 1097–1100.

Rumack, B.H., Peterson, R.C., Koch, G.G., and Amara, I.A. (1981). Acetaminophen overdose: 662 cases with evaluation of oral acetylcysteine treatment. *Archives of Internal Medicine*, **141**, 380–5.

9.2.3 Opiates

G. Conti and A. Gasparetto

Key messages

- Opiate poisoning is a common emergency.
- Patients present with coma, cyanosis, hypoventilation or respiratory arrest, and pinpoint miosis.
- Naloxone reverses overdose of all opioids.
- Common complications are acute pulmonary edema, rhabdomyolysis, and acute renal failure.

Introduction

Opiate overdose may occur in a large number of clinical situations, such as accidental or intentional overdose of drugs prescribed for therapeutic use or, more commonly, in the context of drug abuse. Opiates (mainly heroin) are among the most commonly used narcotics and are responsible for the majority ($>$ 80 per cent) of deaths directly related to drug abuse. Therefore heroin and morphine addicts are often referred to emergency care departments or intensive care units for the treatment of life-threatening complications of drug abuse.

Diagnosis of opiate poisoning

Opiate overdose is generally reported after intravenous injection, although it can also occur after subcutaneous injection, oral intake, or intranasal heroin use ('snuffing'). It is usually observed after accidental overdose or voluntary increase of the standard dose in addicts.

The clinical presentation is characterized by coma, cyanosis, hypoventilation or respiratory arrest, miosis, and bradycardia; hypertension, muscle spasm, and itching are also often associated. All these symptoms promptly regress after intravenous administration of naloxone (*N*-allil-noroxymorphone), a pure opioid antagonist (Evans *et al.* 1973).

Treatment with naloxone

Naloxone reverses overdose from all opioids, including heroin, morphine, methadone, pentazocine, propoxyphene, dextromethorphan, nalline, and diphenoxylate, without inducing respiratory depression even in the absence of opiates. Therefore it can be used in the mixed or unknown overdose where it can be both diagnostic and therapeutic.

Naloxone administration is indicated when the patient presents signs of central nervous system depression and hypoventilation or respiratory arrest, or as a diagnostic tool when the etiological agent for overdose is unclear. When there is no sign of respiratory depression and the neurological status is only slightly compromised, the administration of supportive therapy with careful clinical observation may be sufficient.

The initial dose of naloxone most frequently recommended for opiate intoxication is an intravenous bolus of 0.4 to 2 mg. This dose can be also given intralingually or intratracheally in the absence of vascular access. Larger doses may be necessary to reverse opiate effects completely; however, if a cumulative dose of 10 mg has been administrated without response, the diagnosis of opiate overdose must be questioned.

The effective naloxone dose may have to be readministered every 30 to 60 min because of the much longer half-life of most opiates.

In the treatment of an overdose of long-acting opioids, a continuous infusion of naloxone should be used after the initial bolus, with two-thirds of the initial dose administered on an hourly basis. As naloxone requirements can vary during the infusion period, close monitoring of the patient is always required.

Although naloxone-induced pulmonary edema has been reported in healthy young subjects after general anesthesia, few adverse effects have been reported for the treatment of opiate overdose apart from agitation and acute withdrawal crisis for large doses of naloxone.

Table 1 Complications of opiate overdose

Acute pulmonary edema
Aspiration pneumonia
Cerebral vascular accidents
Brain edema
Seizures
Rhabdomyolysis
Acute renal failure

Elimination of opiates

Emesis is not recommended in cases of oral exposure to an opiate overdose because of the risk of central nervous system depression and seizures. However, a single oral dose of activated charcoal appears to be able to enhance the elimination of opiates; whole-bowel irrigation may provide an efficient and safe treatment for asymptomatic 'body-packers'.

When promptly treated and not complicated by acute pulmonary edema, aspiration pneumonia, or synergism with other drugs, opiate intoxication is generally easy to manage and the overall mortality rate is low. However, patients are prone to several life-threatening complications which are listed in Table 1.

Rhabdomyolysis and acute renal failure

Acute renal failure is often observed in heroin addicts as a major complication of rhabdomyolysis. Rhabdomyolysis is generally produced by a mechanism of 'pressure necrosis' due to profound sedation; this mechanism is aggravated by hypoxia, acidosis, and hypovolemia, which often coexist. The diagnosis is easy in the presence of a dramatic increase of the muscular isoenzymes aspartate transaminase, alanine transaminase, creatine phosphokinase, and lactic dehydrogenase. If urine analysis is performed during the early phase, large amounts of heme and myoglobin may be found (Conti *et al.* 1990).

Although difficult to prove, a direct nephrotoxic effect of opiates can be considered. The clinical picture is similar to that of non-traumatic rhabdomyolysis and the treatment is non-specific (alkaline diuresis or hemodyalisis in case of oligoanuria).

Non-cardiogenic pulmonary edema

Non-cardiogenic pulmonary edema is a recognized complication after opiate overdose. It is generally abrupt in onset (immediate to 2 h) after intravenous heroin overdose, although delayed onset (4–24 h) has been described following 'snuffing' or oral ingestion of methadone.

Although there is considerable evidence of the non-cardiogenic origin of opiate-induced pulmonary edema (Frand *et al.* 1972; Katz *et al.* 1972), many pathophysiological aspects remain obscure. The major functional factors that have been considered are profound hypoxia, a particular hypersensitivity to the drug or a contaminating excipient, and a direct toxic action of the drug.

In severe cases the clinical picture is similar to acute respiratory distress syndrome; the patients are severely hypoxic and hypercapnic, and their chest radiograph shows diffuse bilateral pulmonary infiltrates. Despite the similarities to acute respiratory distress syndrome, opiate-induced pulmonary edema is generally characterized by a rapid response to mechanical ventilation, with positive end-expiratory pressure, and a favourable outcome.

When hemodynamic data are analyzed in the first hour after the onset of pulmonary edema, the patients seem to be relatively hypovolemic with low right atrial pressure and reduced pulmonary artery occlusion pressure, confirming the absence of cardiogenic mechanisms in the genesis of pulmonary edema.

References

Conti, G., Teboul, J.L., and Gasparetto, A. (1990). Acute heroin intoxication. In *Update in intensive care and emergency medicine* (ed. J.L. Vincent), pp. 478–81. Springer-Verlag, Berlin.

Evans, L.E., Swainson, C.F., Roscoe, P., and Prescott, L.F. (1973). Treatment of drug overdosage with naloxone, a specific narcotic antagonist. *Lancet*, i, 721–4.

Frand, U.I., Shim, C.S., and Williams, M.H. (1972) Heroin-induced pulmonary edema. *Annals of Internal Medicine*, **77**, 29–35.

Katz, S., Aberman, A., Frand, U.I., Stein, I.M., and Fulop, M. (1972). Heroin pulmonary edema: evidence for increased pulmonary capillary permeability. *American Review of Respiratory Diseases*, **106**, 472–4.

9.2.4 Benzodiazepines

P. Lheureux and M. Vranckx

Key messages

- Benzodiazepines are the drugs most frequently involved in acute poisoning.
- Benzodiazepine overdose usually has a good prognosis:
  most patients do well with only careful observation and prevention of complications;
  supportive care including intubation, ventilation, and fluid administration may be required in some cases.
- Care should be taken with the following:
  elderly people, patients with chronic obstructive pulmonary disease, or liver disease;
  short-acting compounds;
  association with other central nervous system depressants.
- Flumazenil is a useful diagnostic tool and may help to improve alertness and prevent the need for respiratory support in some patients.

Introduction

Benzodiazepines have a wide pharmacological profile (namely hypnosedative, anxiolytic, muscle relaxant, amnestic, and anticonvulsant properties) and a high therapeutic index. They have replaced most older hypnosedatives, particularly barbiturates, in clinical use. Abuse potential has been recognized more recently.

Mechanism of action

The classical benzodiazepines (agonists) enhance GABAergic transmission, the main inhibitory system in the central nervous system. Their effects are mediated through binding to specific sites that are part of a supramolecular complex in association with the GABA receptors and the chloride channel (Möhler and Richards 1988). Another group of ligands of the benzodiazepine receptors are able to induce opposite effects compared with classical benzodiazepines (anxiogenic, central nervous system stimulants, proconvulsants, etc.). The physiological significance of these 'inverse agonists' is still poorly understood. More recently, new sleep-inducers with different chemical structures, but able to bind to the same receptors (imidazopyridines such as zolpidem, cyclopyrrolones such as zopiclone), have been introduced.

The pharmacokinetics of benzodiazepines are heterogeneous. Onset of action is highly variable; higher lipid solubility is associated with faster digestive absorption and distribution in the central nervous system. Duration of action is related to the elimination half-life of the parent compound and its active metabolites. Biotransformation may be impaired in elderly patients and those with liver disease.

Assessment

Clinical evaluation

Pure benzodiazepine overdose is rarely severe, particularly in previously healthy people. It is characterized by progressive central nervous system depression, ranging from drowsiness, somnolence, and confusion to stupor and flaccid coma. Paradoxical agitation, disinhibition, and aggressiveness may be observed. Anterograde amnesia is always present. Respiratory depression, hypotension, or hypothermia may be observed when other central nervous system depressants (particularly alcohol) are associated or in people with increased susceptibility, such as the elderly, in whom muscle relaxant effects may be particularly prolonged. Central nervous system depression may last in patients with liver disease, while the risk of respiratory depression is increased in those with chronic obstructive pulmonary disease. Other risk factors include massive ingestion and short-acting or highly sedative compounds (midazolam, triazolam, temazepam, flunitrazepam).

Morbidity and mortality associated with benzodiazepine overdosage are not negligible, particularly in elderly people, but are usually related to indirect complications, mainly respiratory problems (Höjer *et al.* 1989).

Withdrawal syndrome may develop after acute poisoning in chronically treated patients or chronic abusers.

Analytical toxicology

Qualitative screening may help to confirm the presence of benzodiazepines and to detect associated substances. Some derivatives (e.g. alprazolam, clonazepam, temazepam, or triazolam) are not easily detected. Determinations of serum levels (parent drugs and metabolites) may occasionally help to differentiate benzodiazepine-induced central nervous system depression from other etiologies. However, a diagnostic test with flumazenil is likely to provide adequate information more rapidly. Routine toxicological testing does not help to assess the severity or the prognosis, or to guide the management, as there is no strong correlation between blood levels and clinical presentation. In particular, very high levels with minimal impairment may be observed in tolerant people.

Other investigations

When multiple drug overdose is suspected or in coma of unknown etiology, an ECG should always be obtained before administration of flumazenil. Usual chemistry values, EEG, CT scan, and analysis of cerebral spinal fluid may be useful for excluding trauma, cerebrovascular accident, infection, or metabolic disturbances. In prolonged coma, aspiration pneumonia and rhabdomyolysis must be excluded.

Management

General supportive care

Treatment of benzodiazepine overdose is mainly supportive. Close neurological and respiratory observation is needed; intubation and ventilatory support may be required to prevent aspiration and atelectasis. Hypotension usually responds to volume expansion alone. The usual measures to prevent ocular and pressure-induced complications during coma must be taken.

Prevention of digestive absorption

Administration of activated charcoal either orally or through a gastric tube must be considered if the patient is admitted soon after drug ingestion. More invasive techniques such as induction of emesis or gastric lavage are not indicated if only benzodiazepines have been ingested, since the risks associated with these procedures, particularly in stuporous patients, outweigh the potential clinical benefit (Lheureux and Askenasi 1992). In mixed overdoses, the need for gastric aspiration and lavage before charcoal administration must be carefully evaluated in each case.

Antidotes

The risks of non-specific antagonists (physostigmine, doxapram, aminophylline) outweigh the toxicity of benzodiazepines. Naloxone has proved to be inactive.

Flumazenil is an antagonist of the benzodiazepine receptors with a high affinity for these sites. it has no significant intrinsic effects, and rapidly antagonizes the properties of both agonists and inverse agonists.

Flumazenil rapidly reverses central nervous system and respiratory depression. Arousal usually occurs 30 to 60 s after intravenous administration and may last for 1 to 2 h. The rate of gradual relapse of sedation depends on the substances involved and their metabolites. Slow injection is recommended because of the frequent side-effects associated with sudden arousal, including anxiety, palpitations, nausea, and vomiting. Measures to prevent aspiration (lateral position, suction equipment) must be taken. The usual dose consists of 0.1 to 0.2 mg/min every 30 s until sufficient alertness and adequate respiration and reflexes are obtained; full recovery should not be the goal. Continuous infusion of 0.5 to 2 mg/h may be needed to maintain the effect (Lheureux *et al.* 1991) since flumazenil has a much shorter half-life (about 1 h) than most agonists.

Flumazenil can be used as a diagnostic tool (Sprenger *et al.* 1994). However, it must be administered with caution in patients with history of chronic benzodiazepine abuse because of the risk of acute withdrawal syndrome associated with agitation, delirium, and seizures. Flumazenil should not be used in patients who are also intoxicated with proconvulsant or arrhythmogenic compounds

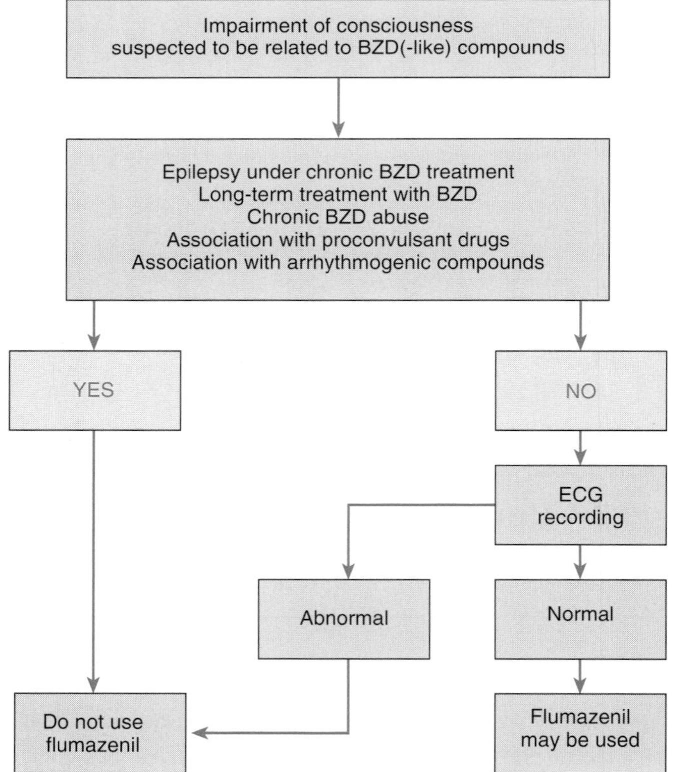

Fig. 1 Protocol for the rational use of flumazenil in patients suspected of benzodiazepine (BZD) intoxication.

(antidepressants, theophylline, antiarrhythmics, chloral hydrate, halogenated hydrocarbons, etc.) because of the risk of unmasking the toxic effects of these substances when the protective effects of benzodiazepines are antagonized. It should not be used in patients who present ventricular arrhythmias or an enlarged QRS complex (> 100 ms) because simultaneous intoxication with cardiotoxic drugs may be suspected. Administration of flumazenil is contraindicated in patients with epilepsy, particularly if they are on long-term treatment with benzodiazepines (Lheureux *et al.* 1991) (Fig. 1).

References

Höjer, J., Baehrendtz, S., and Gustafsson, L.L. (1989). Benzodiazepine poisoning: experience of 702 admissions to an intensive care unit during a 14-year period. *Journal of Internal Medicine*, **226**, 117–22.

Lheureux, P. and Askenasi, R. (1992). Place of activated charcoal and gastric emptying in acute toxic ingestions: a critical reappraisal. In *Update in intensive care and emergency medicine* (ed. J.L. Vincent), pp. 656–60. Springer-Verlag, Berlin.

Lheureux, P., Vranckx, M., and Askenasi, R. (1991) Use and misuse of flumazenil in clinical toxicology. In *Update in intensive care and emergency medicine* (ed. J.L. Vincent), pp. 482–90. Springer-Verlag, Berlin.

Möhler, H. and Richards, J.G. (1988). The benzodiazepine receptor: a pharmacological control element of brain function. *European Journal of Anaesthesiology*, Supplement 2, 15–24.

Sprenger, H., Sharpe, M.D., and McLachlan, R.S. (1994). Flumazenil as a diagnostic tool in the differential diagnosis of coma in a critically ill patient. *Canadian Journal of Anaesthesiology*, **41**, 52–5.

9.2.5 Tricyclic antidepressants

J. A. Henry and M. C. Colbridge

Key messages

- Tricyclic antidepressants can be fatal in overdose owing to cardiac arrhythmias resulting from their proarrhythmic effect or profound hypotension resulting from their negative inotropic effect.
- Because tricyclic drugs delay gastric emptying, peak plasma levels occur within 2 to 8 h of a therapeutic dose, and peak overdose toxicity is usually seen within 6 h of ingestion.
- In cases seen within 2 h of ingestion, gastric lavage followed by 50 g of activated charcoal is indicated.
- All patients should be observed with ECG monitoring for at least 6 h after ingestion.
- If the patient is showing signs of worsening toxicity, has cardiovascular instablity or a wide QRS complex on the ECG, or is acidotic, sodium bicarbonate should be administered to maintain the arterial blood pH between 7.45 and 7.55.

Introduction

Tricyclic antidepressants are likely to be taken in overdose by patients under treatment for depression, and are the most common cause of intensive care unit (ICU) admission for poisoning in most Western countries. The older tricyclic drugs, such as amitriptyline and dothiepin, are particularly toxic in overdose. Although the mechanism of tricyclic toxicity is now well understood, there have been few practical therapeutic advances in its management and good supportive care remains the mainstay. Non-tricyclic drugs are generally less toxic in overdose, although the monoamine oxidase inhibitors may cause a hyperthermic syndrome.

Mode of action

Tricyclic antidepressants are believed to act by centrally blocking the reuptake of norepinephrine (noradrenaline) and/or 5-hydroxytryptamine (serotonin, 5-HT). In addition, they block the parasympathetic nervous system (anticholinergic action) and the peripheral reuptake of norepinephrine. Their effect on the heart is similar to that of the class Ia antiarrhythmic agents (membrane-stabilizing or quinidine-like effect) at low doses, which in overdose is highly toxic to the heart and is the main mechanism responsible for a fatal outcome. Deaths from tricyclic antidepressant overdose are due to the negative inotropic effect resulting in profound hypotension and electromechanical dissociation, or the proarrhythmic effect, causing intraventricular conduction delays leading to ventricular arrhythmias (Pentel and Benowitz 1986) (Fig. 1).

Atypical tricyclic drugs

Lofepramine and amoxapine are atypical of the group. Lofepramine appears to be safer in overdose, and is considerably less cardiotoxic than the older tricyclics. Cardiac arrhythmias have been reported with amoxapine, and there is also a high incidence of fatal convulsions, rhabdomyolysis, and acute renal failure. Amoxapine is the

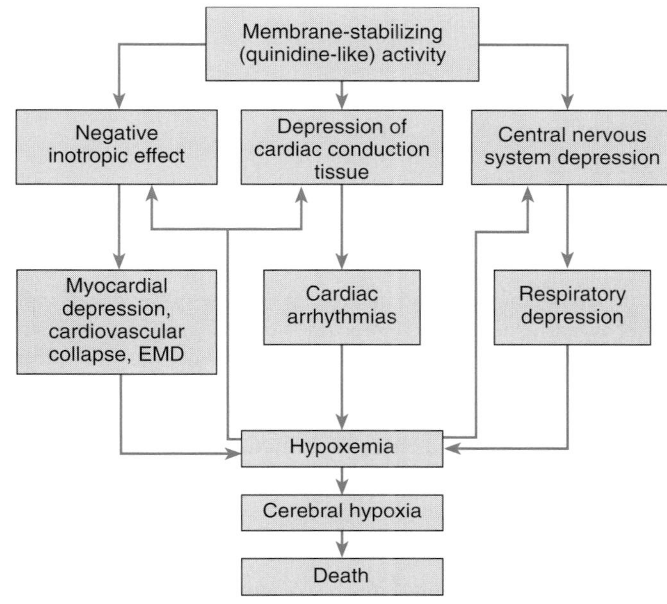

Fig. 1 Mechanisms of the cardiac toxicity of tricyclic antidepressants: EMD, electromechanical dissociation.

most toxic of the antidepressants in terms of deaths per million prescriptions (Henry *et al.* 1995).

Onset of effects

Because tricyclic drugs delay gastric emptying, peak plasma levels usually occur within 2 to 8 h of a therapeutic dose. After an overdose, peak levels may occur even later than this. Although the individual tricyclic drugs have differences in side-effects and pharmacokinetics, most behave similarly in acute overdose. Life-threatening signs will usually develop within 6 h of ingestion or not at all (Dziukas and Vohra 1991).

Clinical effects

Mild to moderate clinical effects of tricyclic antidepressant overdose are mainly anticholinergic and include dilated pupils, hallucinations, dry mouth, tachycardia, and urinary retention, as well as drowsiness, ataxia, and agitation. Metabolic acidosis and hypokalemia can occur. Severe poisoning may result in deep coma, hyper-reflexia, convulsions, hypotension, cardiac arrhythmias, respiratory depression, and pulmonary edema.

The patient may deteriorate over several hours following ingestion, with agitation, hallucinations, or confusion, giving way to deepening coma. During this phase, muscle tone is often increased with brisk tendon reflexes and extensor plantar reflexes, which may cause diagnostic problems unless this feature of tricyclic toxicity is known. Moderately dilated pupils and tachycardia are caused by the anticholinergic effects of the drug. However, although these features are useful in diagnosis, the anticholinergic effects are not responsible for serious toxicity.

The most dangerous problem is cardiac toxicity, which may be associated with widened QRS complexes (Boehnert and Lovejoy 1985), ventricular arrhythmias, severe hypotension, metabolic acidosis, heart block, and asystole. Respiratory depression is also common in severe cases. Death is mainly due to hypotension and cardiac arrhythmias (Pentel and Benowitz 1986).

Rare complications following tricyclic antidepressant overdose include rhabdomyolysis, myoglobinuria, hyperthermia, acute renal failure, and disseminated intravascular coagulation, usually secondary to intractable convulsions.

Management

Gastric decontamination

Induction of emesis should be avoided because of the risk of rapid onset of central nervous system depression and convulsions. Gastric lavage (with the airway protected if there are any but the mildest features of toxicity) should be followed by 50 g of activated charcoal within 2 h of ingestion. Because of gastric stasis it may be worth administering activated charcoal up to 6 h after large ingestions (> 1 g).

Enhanced elimination

Because tricyclic antidepressants have a large volume of distribution and enterohepatic recirculation accounts for only 15 per cent of their elimination, repeated doses of charcoal are unlikely to have a significant effect. A further dose of charcoal (50 g in adults) after 2 to 4 h should be considered only in large overdoses or with sustained release preparations. Diuresis, hemodialysis, and hemoperfusion are ineffective and are not recommended for elimination of tricyclic antidepressants.

Continuing management

All patients should be observed with ECG monitoring for at least 6 h after ingestion. Patients who are symptomatic after this time should be admitted and monitored until asymptomatic; those who are asymptomatic can be discharged. Electrolytes and arterial blood gas status should be measured and corrected, and adequate oxygenation and ventilation maintained.

Intravenous fluids should be used to maintain the blood pressure if necessary. Inotropic agents usually have little effect on tricyclic-induced hypotension, but glucagon may be effective. Once oxygenation has been attended to, diazepam is the treatment of choice for convulsions, taking care not to cause respiratory depression.

If the patient is acidotic, is showing definite signs of toxicity, or has cardiovascular instablity or a wide QRS complex on the ECG, sodium bicarbonate should be administered to maintain the arterial blood pH between 7.45 and 7.55. This often produces a marked improvement. The initial dose is 1 to 2 mmol/kg by slow intravenous infusion. The plasma potassium concentration should also be maintained at the upper end of normal.

Drug therapy should be avoided where possible, but if arrhythmias are unresponsive to sodium bicarbonate, phenytoin may be effective (18 mg/kg intravenously over 20 min). Intravenous esmolol or atenolol may also be given for arrhythmias and repeated as necessary. Lidocaine (lignocaine) has been used to treat ventricular dysrhythmias, but may precipitate convulsions.

Glucagon has been successfully used to treat a massive imipramine overdose (Sener *et al.* 1995). In this case, 10 mg intravenously immediately raised the blood pressure from 71/30 to 110/70 mmHg and the QRS interval shortened from 129 to 89 ms. It should be considered in hypotensive cases which fail to respond to volume expansion and sodium bicarbonate administration. Magnesium sulfate has been successful in reversing tricyclic-induced ventricular tachycardia in experimental animals, but human data are limited.

Class Ia antiarrhythmic drugs (quinidine, procainamide, and disopyramide) are contraindicated because their effects on myocardial contractility are similar to those of tricyclics and they will exacerbate the cardiotoxic effects. Use of physostigmine is no longer recommended to treat tricyclic poisoning as it has caused convulsions, dysrhythmias, and cardiac arrest. Flumazenil should not be used to reverse a concomitant benzodiazepine overdose. It has produced convulsions and ventricular dysrhythmias in the presence of tricyclic drugs.

If there is evidence of pulmonary edema or if respiratory depression is causing hypoxemia, assisted ventilation should be commenced. If there is severe unresponsive hypotension or a cardiac arrest occurs, the circulation should be supported by external chest compression for as long as necessary. This should be done in the case of asystole and should be employed even when electromechanical dissociation is apparent on the ECG. This has been carried out for 3 to 5 h with full recovery (Orr and Bramble 1981). In an ICU setting an aortic balloon pump or cardiopulmonary bypass may be necessary. Measures such as these are justified in severe tricyclic antidepressant overdose because this type of poisoning is essentially reversible, provided that hypoxic cerebral damage can be prevented.

References

Boehnert, M.T and Lovejoy, F.H. (1985). Value of the QRS duration versus the serum drug level in predicting seizures and ventricular arrhythmias after an acute overdose of tricyclic antidepressants. *New England Journal of Medicine*, **313**, 474–9.

Dziukas, L.J. and Vohra, J. (1991). Tricyclic antidepressant poisoning. *Medical Journal of Australia*, **154**, 344–50.

Henry, J.A., Alexander, C.A., and Sener, E.K. (1995). Relative mortality from overdose of antidepressants. *British Medical Journal*, **310**, 221–4.

Orr, D.A. and Bramble, M.G. (1981). Tricyclic antidepressant poisoning and prolonged external cardiac massage during asystole. *British Medical Journal*, **283**, 1107–8.

Pentel, P.R. and Benowitz, N.L. (1986). Tricyclic antidepressant poisoning. Management of arrhythmias. *Medical Toxicology*, **1**, 101–21.

Sener, E.K., Gabe, S., and Henry, J.A. (1995). Response to glucagon in imipramine overdose. *Clinical Toxicology*, **33**, 51–3.

9.2.6 Amphetamines and ecstasy

T. K. Hartung and J. A. Henry

Key messages

- The sympathomimetic effects of the amphetamines cause tachycardia, potentially fatal cardiac arrhythmias, myocardial infarction, cardiomyopathy, and hypertension which may lead to cerebrovascular accidents.
- The central stimulant properties can lead to euphoria, confusion, agitation, anorexia, hallucinations, psychosis, and convulsions.
- Hyperthermia may be associated with amphetamines, particularly when 3,4-methylenedioxymethamphetamine (MDMA, ecstasy) is used as a 'dance drug', with possible complications of hypotensive collapse, disseminated intravascular coagulation, rhabdomyolysis, and acute renal failure.
- In some cases excessive water drinking has led to acute hyponatremia cerebral edema which, rarely, may be fatal.
- There is no specific antidote for amphetamine toxicity. Management should be symptomatic and supportive.

Introduction

Amphetamine (β-phenylisopropylamine) and other phenethylamine derivatives have been widely abused for many years; this contrasts with their relatively limited value as therapeutic agents. Amphetamine was first marketed as an over-the-counter nasal decongestant in the United States in 1932, but following reports of abuse and toxicity it later became a controlled drug.

Street amphetamine sulfate is an off-white powder with an average purity of usually no more than 5 per cent. The rest may be made up of other less potent stimulants (such as caffeine or ephedrine) or inert chemicals (such as glucose, vitamin C, chalk, or talc). Formulations are sold as powder or tablets, and may be ingested, snorted, smoked, or injected intravenously; the usual daily adult dose of amphetamine is 5 to 15 mg. Commonly used terms are whiz, uppers, speed, sulfate, and sulf. A crystalline smokable form of methamphetamine ('ice'), an amphetamine derivative with marked central stimulant effects, is rapidly absorbed from the lungs, resulting in almost immediate clinical effects which might last for up to 15 h.

3,4-Methylenedioxymethamphetamine (MDMA, ecstasy) and other mood-enhancing amphetamine derivatives (3,4-methylenedioxyethamphetamine (MDEA, eve) and 3,4-methylenedioxyamphetamine (MDA, adam)) were widely abused in the United States in the mid-1980s. Their pharmacological profile can explain their widespread use as 'dance drugs' across Europe and Australia in the early 1990s; it has been suggested that as many as half a million people abuse MDMA every week in the United Kingdom. The stimulant and mood-enhancing properties of MDMA make it an ideal adjunct to so-called 'rave' dancing, where it facilitates heavy and prolonged exertion. Ecstasy is usually taken orally. It is generally sold in tablet form, or less commonly enclosed in capsules of varying colors; the usual street doses are 30 to 150 mg.

Pharmacology

Amphetamines and other indirectly acting sympathomimetic agents are characterized by their ability to stimulate the central nervous system in addition to their peripheral α- and β-adrenergic actions.

Sympathomimetic agents are derived from the phenylethylamine molecule and increase the norepinephrine (noradrenaline) concentration in the synaptic cleft and presynaptic nerve terminal by blocking the reuptake of norepinephrine into the presynaptic nerve terminal. A second mode of action is by inhibiting the active uptake of norepinephrine into storage vesicles. Thirdly, they inhibit mitochondrial monoamine oxidase, an enzyme responsible for norepinephrine degradation. Some amphetamines with a hydroxyl substitution in the β position (e.g. ephedrine) are able to stimulate sympathetic receptors directly. The central excitatory effects are mediated by indirect noradrenergic mechanisms in the cortex, the reticular activating system, and the medullary respiratory center. A phenomenon unique to the amphetamines is 'aggregation toxicity'. Animal studies show that the behavioral and toxic effects of amphetamines are massively enhanced if mice given the drug are grouped together as opposed to being alone. In addition to 'aggregation toxicity', acute toxicity in solitary mice is increased by loud noise, high ambient temperature, and dehydration. These factors may have human relevance. MDMA, MDA, and MDEA are phenethylamine derivatives with unique psychoactive properties causing heightened emotional and sensory experiences, without producing marked visual or sensory distortions. They promote the release of serotonin and block its reuptake, and cause serotonin depletion; they are neurotoxic to serotonin terminals in every animal species studied (Green *et al*. 1995).

Amphetamine is readily absorbed across mucosal surfaces. Peak plasma concentration is seen within minutes of intravenous injection, within 30 min of intramuscular or topical administration, and within 3 h of ingestion. Plasma concentrations are low (ng/ml) because of extensive tissue distribution (3–6 l/kg). It is 20 per cent bound to plasma proteins, and cerebrospinal fluid levels are 80 per cent of those found in plasma. Following typical recreational doses, the duration of effects is usually around 4 to 8 h, but large doses and alkaline urine can prolong the effects to 24 h or more. Renal excretion of amphetamine, a weak base, largely depends on urine pH because of trapping of ions in the renal tubule. In patients with a urine pH below 6.6 the half-life ranges from 7 to 14 h; with a urine pH above 6.7 the range is 18 to 34 h. Following MDMA doses of 75 to 150 mg, effects occur within 1 h and last for 4 to 12 h.

Clinical presentation

Acute toxicity

The individual response varies greatly and toxicity correlates poorly with the amount of drug taken. Typical effects following amphetamine ingestion include wakefulness, appetite suppression, talkativeness, hyperactivity, and euphoria. The patient has a dry mouth and dilated pupils and may be sweating. Tendon reflexes are brisk. Cardiovascular effects such as tachycardia and hypertension are common. Signs of moderate toxicity following MDMA use include nausea, increased muscle tone, hyper-reflexia, muscle pain, trismus (jaw-clenching), blurred vision, agitation, anxiety, and a slight rise in body temperature. Vomiting, abdominal pain and diarrhea may also occur. Signs of severe toxicity include delirium, coma, and convulsions.

Acute complications

Cardiac complications

Cardiac arrhythmias are common — usually a marked sinus tachycardia, but ventricular tachycardia may occur. Deaths which occur soon after ingestion are usually due to cardiac arrhythmias such as ventricular fibrillation. Other cardiac problems induced by amphetamines include myocardial ischemia and infarction, left ventricular dysfunction, and acute cardiomyopathy.

Severe hyperthermia

Hyperthermia may result from muscular hyperactivity, prolonged convulsions, or a serotonin syndrome. Several deaths have occurred in the context of MDMA abuse as a dance drug at 'rave' parties (Henry *et al.* 1992). Once the organism has become overheated and fluid requirements are not met, a vicious cycle is set in motion giving rise to disseminated intravascular coagulation (DIC), rhabdomyolysis, and acute renal failure (O'Connor 1994).

Hyponatremic encephalopathy

Excessive fluid consumption in the absence of significant losses through prolonged exertion compounded by the presence of the syndrome of inappropriate ADH secretion (SIADH) has led to acute symptomatic hyponatremia coagulation following MDMA abuse in a few cases (Hartung *et al.* 1998). Symptoms include nausea, vomiting, and a mute state with disorientation and convulsions; death due to cerebral edema is a rare complication.

Cerebrovascular accidents

Amphetamines are widely recognized as a cause of cerebrovascular accidents secondary to their potent sympathomimetic effects or to vasculitis. Most of these patients present with severe headache and nausea; they are usually conscious but confused and disorientated. Death may occur from cerebral infarction, cerebral hemorrhage, or subarachnoid hemorrhage.

Renal complications

Rarely, renal ischemia and infarction may occur due to vasculitis. More commonly, renal failure follows non-traumatic rhabdomyolysis.

Hepatitis

Although isolated liver toxicity due to toxic hepatitis is uncommon, MDMA abuse should enter the differential diagnosis in young patients presenting with unexplained jaundice or hepatomegaly.

Intra-arterial injection

Injection of amphetamine into an artery usually occurs accidentally and may cause vasospasm and subsequent ischemia of the affected extremity.

Chronic toxicity

Paranoid psychosis with visual, tactile, or olfactory hallucinations is a well-known manifestation of chronic amphetamine abuse. Dyskinesias, compulsive or repetitive behavior, and impaired performance are also recognized chronic sequelae. Recovery is usually rapid after withdrawal of the drug, but occasionally the condition becomes chronic. Poor dietary habits lead to nausea, vomiting, diarrhea, malnutrition, weight loss, dermatological changes, and various infectious complications. Paranoid reactions, anxiety and panic attacks, and depression can occur following MDMA abuse. Of even greater concern, its neurotoxic potential with destruction of serotoninergic neurons could cause mental illness after use of the drug has ceased.

Although tolerance occurs, withdrawal symptoms are not as severe as those with other drugs. The peak effect, with apathy, depression, lethargy, anxiety, sleep disturbances, myalgia, abdominal pain, and increased appetite, are seen 2 to 3 days after the last dose and can last for up to 6 days.

Laboratory findings

Arterial blood gas analysis

The most common finding is a respiratory alkalosis which may be combined with a metabolic acidosis. A respiratory acidosis may occur in the patient with seizures or cardiovascular collapse.

Electrolytes, urea, creatinine, osmolality, glucose, and thyroxine

Volume depletion, as indicated by elevated urea, creatinine, and sodium, is frequently present. Adrenergic overstimulation can lead to hypokalemia, while hyperkalemia together with hypocalcemia can be found secondary to rhabdomyolysis when raised serum urea and creatinine may indicate renal failure. Hyponatremia may result from excessive water drinking together with SIADH; a low plasma osmolality is accompanied by concentrated urine with a high osmolality. Sympathetic overdrive will often result in hyperglycemia. Transient hyperthyroxinemia may result from heavy amphetamine abuse.

Creatine phosphokinase and transaminases

These may be markedly elevated due to rhabdomyolysis associated with hyperthermia, seizures, or excessive physical activity. The creatine kinase MB fraction may be used to confirm suspected myocardial damage.

Hematological investigations

Leukocytosis is common with marked clinical toxicity. DIC may produce a characteristic coagulation profile.

Urine testing

Routine urinalysis may reveal glycosuria and give a false-positive test for hemoglobin because of the presence of myoglobin. A specific test for myoglobin should be used to confirm this.

Toxicology

Blood and urine tests may be positive for amphetamines and may help to exclude other possible diagnoses. However, the diagnosis is made clinically in the first place, and, even when confirmed, plasma concentrations correlate poorly with clinical effects.

Diagnosis

Although rapid laboratory facilities may be available in some centers, clinical findings are of utmost importance in establishing the diagnosis and assessing its severity. The differential diagnosis of acute amphetamine intoxication is extensive. Organic conditions such as hypertensive crisis, pheochromocytoma, thyrotoxicosis, and withdrawal states, as well as acute psychotic states, may mimic the clinical features of amphetamine toxicity. It may be clinically difficult to distinguish amphetamine toxicity from that due to drugs and chemicals causing similar symptoms (cocaine, anticholinergic agents, hallucinogens, lithium, nicotine, and organic compounds such as camphor, solvents, salicylates, and methylxanthines).

Treatment

Treatment is mainly symptomatic and supportive. However, complications must be anticipated and treated speedily. All symptomatic cases should be observed for at least 6 h and can be discharged once they have become asymptomatic. Patients with evidence of moderate or severe toxicity or unstable vital signs must be admitted and may require intensive monitoring (Linden *et al.* 1985).

After stabilizing the patient, gastrointestinal decontamination should be considered. Emesis is contraindicated as convulsions can set in rapidly. Although vagal stimulation from gastric lavage may precipitate seizures or cardiac arrhythmias, lavage should be performed in the stable patient within 2 h of ingestion of more than 2 mg/kg body weight. Activated charcoal avidly binds amphetamine and 50 g (10–15 g for a child) should be given orally or by narrow-gauge tube.

Agitation and seizures

Diazepam is the treatment of choice for agitated, aggressive, hyperactive, or convulsing patients. Neuroleptic drugs should be avoided as they may cause hypotension and lower the seizure threshold. Intractable seizures can be treated with phenytoin or barbiturates, and paralysis and ventilation may be necessary. Patients with a suspected cerebrovascular lesion will require a diagnostic CT scan.

Cardiovascular complications

Tachycardia and hypertension should be treated with β-blockers and/or α-blockers. The use of β-blockers may result in unopposed α action with severe hypertension which can be effectively treated with sodium nitroprusside.

Lidocaine (lignocaine) is the drug of choice for ventricular arrhythmias; phenytoin and propranolol are alternatives. Adenosine or verapamil may be used for supraventricular dysrhythmias.

Hyperthermia

In severe cases management should be prompt and aggressive since hyperthermia can be rapidly fatal. Immediate vigorous rehydration is essential to allow thermoregulation. Dantrolene should be administered if the patient has a temperature over 39 °C. Paralysis and mechanical ventilation will be required for the patient with continuing muscle stiffness or a serotonin syndrome. Rhabdomyolysis, renal failure, and DIC require symptomatic treatment.

Hyponatremia

Patients with acute hyponatremic encephalopathy usually recover spontaneously over 24 h and require fluid restriction only. The use of intravenous furosemide (frusemide) and mannitol may be considered if there is evidence of cerebral edema. If raised intracranial pressure is suspected, administration of hypertonic saline may prevent cerebral damage.

Accidental intra-arterial injection usually requires angiography of the affected limb to exclude a thrombotic occlusion or a surgically correctable lesion. Treatment is supportive with pain relief, heparinization, and intravenous or intra-arterial nitroprusside. Sympathetic nerve block, aspirin, corticosteroids, dextran-40, epoprostenol, papaverine, phentolamine, tolazoline, and phenoxybenzamine have been tried with varying effectiveness.

The clinical efficacy of acid diuresis remains unproven. The risks of cardiovascular complications, renal failure, and adverse metabolic consequences outweigh the risks of enhanced elimination. The amount removed by means of peritoneal dialysis or hemodialysis is small and does not justify their routine use.

References

Green, A.R., Cross, A.J., and Goodwin, G.M. (1995). Review of the pharmacology and clinical pharmacology of 3,4-methylenedioxymethamphtamine (MDMA or 'ecstasy'). *Psychopharmacology*, **119**, 247–60.

Hartung, T.K., Schofield, E., Short, A.I., Parr, M.J.A., and Henry, J.A. (1998). Hyponatremic states following MDMA ingestion (unpublished).

Henry, J.A., Jeffreys, K.J., and Dawling, S. (1992). Toxicity and deaths from 3,4-methylenedioxymethamphetamine. *Lancet*, **340**, 384–7.

Linden, C.H., Kulig, K.W., and Rumack, B.H. (1985). Amphetamines. *Trends in Emergency Medicine*, **7**, 18–32.

O'Connor, B. (1994). Hazards associated with the recreational drug 'ecstasy'. *British Journal of Hospital Medicine*, **52**, 507–14.

9.2.7 Digoxin

E. M. Antman and T. W. Smith

Key messages

- Cardiac glycoside administration should be discontinued, and the use of catecholamines should be avoided if possible.
- Serious rhythm disturbances require admission to an intensive care unit. All arrhythmias require continuous ECG monitoring for treatment.
- Potassium repletion should be considered unless:
  - serum potassium is below 5 mmol/l when the patient is first seen;
  - renal insufficiency is present;
  - atrioventricular block is present or conduction is prolonged (PR interval >0.26 s);
  - the patient has taken a large overdose of digitalis (serum potassium may rise to dangerously high levels).
- Bradyarrhythmias causing hypotension or a significant reduction in cardiac output may be treated initially with intravenous atropine (typically 0.5–1 mg in adults).
- Cardiac arrhythmias due to enhanced automaticity that are not overtly life threatening may require suppression with conventional antiarrhythmic therapy (intravenous lidocaine (lignocaine) or phenytoin) in addition to potassium supplementation.
- Cases involving large accidental or suicidal cardiac glycoside ingestions and those with potentially life-threatening arrhythmias, hyperkalemia, or both should be treated with digoxin-specific antibody (Fab) therapy.

Introduction

Digitalis glycosides have a narrow therapeutic index. Therefore clinicians must determine which individual patients have a favorable risk–benefit ratio for digitalis use and promptly identify signs and symptoms of digitalis toxicity.

Assessment

No specific serum digoxin level can be relied on to differentiate clearly between toxic and non-toxic states. The probability of overt toxicity rises sharply at levels of 2.0 to 3.0 ng/ml or above. However, toxicity cannot be categorically excluded at lower serum levels, particularly in the presence of cardiac disease. Electrocardiographic manifestations of digitalis toxicity are characterized by disturbances of impulse formation or conduction (at atrial, atrioventricular junctional, or ventricular levels), or a combination of both types of phenomenon (Antman and Smith 1995). Clinical symptoms of digitalis toxicity typically include anorexia, nausea and vomiting, and visual symptoms, as well as a variety of non-specific complaints such as weakness, fatigue, headache, dizziness, and psychiatric disturbances ('digitalis delirium').

Prompt identification of digitalis-toxic arrhythmias is essential for successful clinical management. The following general therapeutic measures apply to patients with clinically evident digitalis toxicity.

Management of digoxin toxicity

Potassium repletion

Potassium repletion should only be undertaken under closely monitored conditions because of the risk of provoking more troublesome arrhythmias, marked hyperkalemia, and even death (Antman and Smith 1995). Either the intravenous or the oral route of administration may be used. We prefer the latter when the rhythm disturbance is not immediately life threatening. The rate of intravenous infusion of potassium should be limited to less than 0.5 to 1 mmol/min. Normal saline rather than 5 per cent glucose should be used in cases of potassium depletion (< 3.5 mmol/l).

Bradyarrhythmias

Bradyarrhythmias causing hypotension or a significant reduction in cardiac output may be treated initially with intravenous atropine (typically 0.5–1 mg in adults). We insert a temporary demand ventricular pacemaker if atropine fails to resolve the problem in less than 5 min. Infusion of β-adrenergic agonists such as isoproterenol (isoprenaline) is best avoided in view of the potential for provoking more serious arrhythmias.

Conventional antiarrhythmic agents

Lidocaine (lignocaine) is of value for the management of digitalis-related arrhythmias, and we employ it in cases of tachyarrhythmias of less than life-threatening severity (e.g. paroxysms of non-sustained ventricular tachycardia) (Antman and Smith 1995). Lidocaine is

Table 1 Calculation of the equimolar dose of digoxin-specific Fab fragments

Calculation of body load of digoxin

Ingested amount (mg) × bioavailability of digoxin tablets

$$= \text{mg} \times 0.8 \quad \frac{\text{serum digoxin concentration (ng/ml)} \times 5.6^{*} \times \text{weight (kg)}}{1000}$$

Calculation of Fab fragment dose

$$\frac{\text{MW Fab} = 50\,000}{\text{MW digoxin} = 781} = 64 \times \text{body load (mg)} = \text{Fab dose (mg)}$$

$$\frac{\text{body load of digoxin (mg)}}{0.6\,\text{mg neutralized/40-mg vial}} = \text{number of vials of Fab fragments}$$

*Volume of distribution of digoxin in average adult (l/kg). For digitoxin, use 0.56 rather than 5.6.
MW, molecular weight.
Modified with permission from Antman *et al.* (1990).

Table 2 Examples of calculation of the equimolar dose of digoxin-specific Fab fragments

Case 1

A 75-year-old man (weight 70 kg) with chronic coronary heart disease and atrial fibrillation receiving maintenance therapy with digoxin 0.25 mg daily. He becomes confused and takes two of his digoxin tablets daily for 2 weeks, and presents with complaints of weakness and palpitations. ECG shows sustained ventricular tachycardia at 150 beats/min. SDC=3.0 ng/ml

$$\text{Body load of digoxin} = \frac{\text{SDC} \times 5.6 \times 70}{1000} = \frac{3.0 \times 5.6 \times 70}{1000} = 1.176\ \text{mg} \approx 1.2\ \text{mg}$$

$$\text{Dose of Fab fragments} = \frac{1.2\ \text{mg}}{0.6\ \text{mg neutralized/40-mg vial}} = 2\ \text{vials}$$

Case 2

A 40-year-old woman (weight 70 kg) with no history of heart disease ingests 100 tablets of digoxin 0.25 mg in a suicide attempt. She presents 8 h later with nausea, vomiting, hypotension, complete heart block, and an idioventricular escape rhythm at 35 beats/min. SDC result not yet returned from laboratory (and not needed for Fab dose determination). Serum potassium=6.1 mEq/l.

$$\text{Body load} = (\text{ingested amount} \times 0.8) = 25\ \text{mg} \times 0.8 = 20\ \text{mg}$$

$$\text{Dose of Fab fragments} = \frac{20\ \text{mg}}{0.6\ \text{mg neutralized/40-mg vial}} \approx 34\ \text{vials}$$

Case 3

A 60-year-old woman (weight 65 kg) with chronic rheumatic heart disease and mitral regurgitation maintained on digoxin 0.25 mg daily for control of the ventricular rate in atrial fibrillation. SDC = 2 ng/ml on maintenance therapy. She ingests 75 tablets of digoxin 0.25 mg during a period of depression and presents 36 h later with fascicular tachycardia, an unusual form of ventricular tachycardia originating in or near the left anterior fascicle. Serum potassium = 5.9 mEq/l.

$$\text{Ingested amount} \approx 18\ \text{mg digoxin (tablets)}$$

$$\text{Body load} = (\text{ingested amount} \times 0.8) + \frac{\text{SDC} \times 5.6 \times 65}{1000} = 14.4 + 0.728 \approx 15\ \text{mg}$$

$$\text{Dose of Fab fragments} = \frac{15}{0.6\ \text{neutralized/40-mg vial}} = 25\ \text{vials}$$

SDC, serum digoxin concentration.

administered as serial intravenous 100-mg boluses every 3 to 5 min (to a total dose of 300 mg) until either a therapeutic effect or lidocaine toxicity develops. This may then be followed by continuous infusion of 15 to 50 μg/kg/min if further suppression of the arrhythmia is needed. Slow intravenous infusion of phenytoin (100 mg every 5 min, not to exceed a total dose of 1000 mg) is also useful for digitalis-toxic arrhythmias (e.g. ectopic automatic atrial tachycardia). Clinical experience with β-blockers, quinidine, and procainamide has been less favorable than that with lidocaine.

Cardioversion

The view that d.c. cardioversion can be hazardous in individuals receiving cardiac glycosides is based on studies reporting that the shock provoked serious ventricular arrhythmias (refractory ventricular tachycardia or fibrillation). Indeed, toxic levels of digitalis can lower the threshold for postshock ventricular arrhythmias. Usual therapeutic digoxin levels do not increase the risk of serious postshock ventricular arrhythmias. The following approach is recommended when considering d.c. cardioversion in digitalized individuals.

1. Electrolyte imbalances should be corrected, fever suppressed, and hypoxia and anxiety treated before d.c. cardioversion.
2. When there is overt digitalis toxicity, elective d.c. cardioversion is not performed.
3. Under all circumstances, the smallest energy that is likely to be effective is used. We start with 25 to 50 J, with subsequent increments of 25 to 50 J as needed.

Digoxin-specific antibody (Fab fragments) for life-threatening toxicity

Because of the widespread tissue binding of digoxin, hemodialysis or hemoperfusion is of limited value for prompt reversal of life-threatening toxicity. Advanced digitalis intoxication should be treated with purified digoxin-specific polyclonal antibody fragments (Fab fragments) obtained from sheep immunized with digoxin. The following protocol for administration of Fab fragments is recommended (Antman *et al.* 1990; Hickey *et al.* 1991).

1. The dose of Fab fragments is calculated to be equal on a mole-for-mole basis to the amount of digoxin or digitoxin in the patient's body, estimated from the medical history, determinations of serum digoxin or digitoxin concentrations, or both (Table 1). Examples of the calculation of the body load of digoxin to be neutralized are shown in Table 2 (Antman *et al.* 1990).
2. The Fab fragments should be administered intravenously through a 0.22-μm membrane filter over 15 to 30 min, unless the gravity of the clinical situation demands more rapid infusion.
3. Serum potassium should be monitored at least every hour for the first 4 to 6 h after Fab treatment.
4. Typically, the total extracellular digoxin concentration rises dramatically, but such digoxin is pharmacologically inactive because only the unbound form can associate with the inhibitory site on Na^+,K^+-ATPase. For this reason, as well as because of technical problems imposed on assay systems by the presence of high-affinity Fab fragments, measurements of serum digoxin concentration are not reliable indicators of the state of digitalization for about 1 to 2 weeks after Fab administration.

References

Antman, E.M. and Smith, T.W. (1995). Digitalis toxicity: an overview of mechanisms, manifestations, and management. In *Cardiac arrhythmias* (ed. W. Mandel), pp. 1051–73. J.B. Lippincott, Philadelphia, PA.

Antman, E.M., Wenger, T.L., Butler, V.P., Jr, Haber, E., and Smith, T.W. (1990). Treatment of 150 cases of life-threatening digitalis intoxication with digoxin-specific Fab antibody fragments. Final report of a multicenter study. *Circulation*, **81**, 1744–52.

Hickey, A.R., *et al.* (1991). Digoxin immune Fab therapy in the management of digitalis intoxication: safety and efficacy results of an observational surveillance study. *Journal of the American College of Cardiology*, **17**, 590–8.

9.2.8 Cocaine

M. E. Palmer and L. R. Goldfrank

Key messages

- The mainstays of therapy for cocaine toxicity are sedation with benzodiazepines and rapid cooling; the use of phenothiazines or butyrophenones for sedation should be avoided.
- Chest pain should be treated with oxygen, nitroglycerin (glyceryl trinitrate), and aspirin as usual, but benzodiazepines with their proven efficacy to treat sympathetic excess should substitute for β-adrenergic antagonists which increase myocardial oxygen demand and decrease myocardial perfusion.
- Refractory chest pain and hypertension may be treated with phentolamine to reduce coronary vasospasm and afterload.
- Calcium-channel blockers should be avoided.
- Sodium bicarbonate is the drug of choice for treating wide-complex dysrhythmias and ongoing rhabdomyolysis.

Introduction

Cocaine is a highly addictive drug which has become readily affordable; its abuse has increased 10- to 20-fold in the United States over a recent decade. By 1990, over 40 per cent of all drug-related fatalities were due to cocaine (National Institute on Drug Abuse 1990). Diagnosis, assessment, and management of acute toxicity, particularly agitated delirium and ischemic events, have become increasingly important to the critical care physician.

Diagnosis

Although cocaine is a local anesthetic that blocks sodium reuptake and depolarization at peripheral nerves, its protean systemic effects are more pertinent to the critical care setting. In the central nervous system, cocaine blocks the reuptake of norepinephrine (noradrenaline), serotonin, and dopamine at the presynaptic junction. It also promotes synthesis and release of norepinephrine and excitatory amino acids. These central effects lead to excessive sympathetic discharge and increased sensitivity to catecholamines at peripheral receptors, causing a hypermetabolic condition; hypertension, tachycardia, tachypnea, and hyperthermia result. The patient may be agitated with diaphoresis, dilated reactive pupils, active bowel sounds, and hyperactive reflexes. Such adrenergic excess is similar to that caused by amphetamines or salicylate intoxication, or as a manifestation of thyrotoxicosis, pheochromocytoma, heatstroke, central nervous system infection or hemorrhage, or ethanol or sedative–hypnotic withdrawal. Cocaine intoxication is usually relatively short-lived, and this may help focus the differential diagnosis. History of abuse can often be elicited with pointed questions and reassurance about confidentiality. Paraphernalia found at the scene, such as pipes, lines of powder, or crack-laced cigarettes, can help. On physical examination, nasal mucosal erythema or mucosal ulceration and track marks on the skin should be sought since cocaine can be taken intranasally, injected, or smoked.

The route of administration determines the onset and duration of effects. Leaves of the *Erythroxylon coca* plant are extracted to form a white powder, a hydrochloride salt, which is 'snorted' intranasally or injected. Intranasal absorption is self-limited because of cocaine's vasoconstrictive action. The peak effect usually occurs within 30 min and lasts about 1 h. Pure alkaloid cocaine crystals that can withstand pyrolysis are known as 'crack'. Alternatively, when cocaine is dissolved in alkaline solution and extracted with ether, the term 'free-base' is applied. Intravenous and inhalational use result in a more intense and immediate effect (seconds to minutes) which typically terminates within half an hour. Abstinence or withdrawal results in feelings of depression, anhedonia, and drug-seeking behavior. 'Mules' or 'body-packers' smuggle large quantities of cocaine in their gastrointestinal tract or vagina as part of an organized criminal effort. These vials, rubber bags, or condoms may not always be detected radiographically. Life-threatening intoxication can occur if a single bag ruptures. 'Body-stuffers' typically ingest less well-packaged drugs, without prior preparation, in a rushed attempt to avoid apprehension by legal authorities. Exposure to a massive dose of cocaine should be suspected in any patient with a sudden onset of a sympathetic toxic syndrome if he or she is an international passenger or is in police custody. Absorption from the gut can take up to 2 h and lasts for several hours, but direct toxic vasospastic effects can lead to critical complications such as bowel ischemia.

The diagnosis can be confused by the presence of adulterants. The combination of heroin and cocaine (known as a 'speedball') results in mixed opioid and sympathetic findings, with the predominant symptoms dependent on dosing ratios. Some patients ingest organophosphates with the belief that they can prolong cocaine's effects by preventing its metabolism by serum cholinesterase. Diverse other agents, such as talc, quinine, arsenic, or strychnine, have also been added to cocaine.

Laboratory diagnosis can help to define active or recent use. Higher urinary drug concentrations make urine screens for cocaine metabolites more effective than blood sampling. Cocaine has a half-life of under 2 h and is metabolized within 24 h by liver and plasma esterases (forming ecgonine methyl ester), *N*-demethylation (forming norcocaine), and non-enzymatic hydrolysis (forming benzoylecgonine). Benzoylecgonine can subsequently be detected in the urine for at least 2 to 3 days. In the presence of alcohol, cocaine is transesterified to form the toxic metabolite cocaethylene.

Assessment and management of life-threatening manifestations

The essential approach to an agitated patient necessitates physical control. Restraining the patient with multiple personnel, soft cloth restraints, and benzodiazepines may be lifesaving and necessary before continuing evaluation of the airway, breathing, and circulation. Is the airway protected? Is the respiration depressed or shallow? Is the patient in circulatory collapse? High-flow oxygen, cardiac monitoring, and intravenous access must be established while obtaining vital signs, including a core temperature. Determination of bedside glucose is a priority. Blood should be sent for a complete blood count, electrolytes, blood urea nitrogen, creatinine, and arterial blood gas. Liver function studies (serum aspartate aminotransferase and serum alanine aminotransferase) should be performed, and creatinine phosphokinase levels should be determined. An ECG should be obtained to evaluate myocardial ischemia or infarction, atrial or ventricular dysrhythmias, and conduction abnormalities. Urine analysis should also be performed at the bedside. If the dipstick is positive, microscopy is required to differentiate between hematuria and myoglobinuria. Normal saline fluids should be started to replete volume, and sodium bicarbonate should be added to the infusion if there is evidence of rhabdomyolysis. Both these agents (sodium chloride and hypertonic sodium bicarbonate) may assist in the treatment of cocaine-associated type Ic antidysrhythmic sodium-channel blockade. If the patient is agitated, intravenous benzodiazepines should be given and rapid cooling measures instituted. Activated charcoal 1 g/kg should be given empirically to begin gastrointestinal decontamination if cocaine or other agents have been ingested.

The potential effects of cocaine (Goldfrank and Hoffman 1991) are diverse (Table 1). Here, we focus on the effects most commonly encountered in a critical care setting.

Agitated delirium and seizures

Agitated delirium, with or without seizures, is life threatening. The clinical manifestations are comparable to delirium tremens associated with ethanol or sedative–hypnotic withdrawal. Sedation and cooling must be achieved immediately. Since cocaine's effects begin centrally, the most direct approach is to provide centrally acting sedatives. Benzodiazepines are the drug of choice. An agent such as diazepam is ideal because of its rapid onset of action and the ability to give repetitive doses (10 mg in 5-min intervals depending on clinical status) in a situation where minutes may make a difference. Phenothiazines and butyrophenones are contraindicated since they may precipitate a dystonic reaction, impair heat dissipation, lower the seizure threshold, or precipitate torsades des pointes.

Hyperthermia and rhabdomyolysis

Hyperthermia and rhabdomyolysis are caused by psychomotor agitation, seizures, and concomitant vasoconstriction. Rapid cooling with water and an ice bath coupled with a convection system as well as sedation help treat the potentially lethal increase in temperature. Cooling and sedation also prevent acidemia which may induce dysrhythmias from sodium-channel blockade and increase the risk of acute tubular necrosis by causing precipitation of urinary myoglobin. Normal saline with sodium bicarbonate is appropriate in these situations not only to compete with sodium-channel blockade, but also to alkalinize the blood (to pH 7.40–7.55) and urine (to pH 7.50–8.00).

Hypertension

Cocaine-induced hypertension results predominantly from α- and β-adrenergic-induced tachycardia, inotropy, and afterload. It has been implicated in aortic dissection and cerebrovascular accidents, as well as in myocardial ischemia and infarction, because of increased myocardial oxygen demand and concomitant vasoconstriction. These life-threatening complications must be considered in those who use cocaine.

Benzodiazepines are the first line of treatment to decrease sympathetic outflow. If adequate sedation alone does not control blood pressure, the persistence of the α-adrenergic effects of cocaine may

Table 1 Multisystem effects of cocaine

Venous
Vasculitis, superficial and deep venous thrombosis and thrombophlebitis
Metabolic
Hyperthermia, lactic acidosis, weight loss
Eyes
Mydriasis (reactive), vertical and horizontal nystagmus, pseudoexopthalmos, lid lag, keratopathy, blindness
Nose and throat
Epistaxis, atrophic mucosa, ulcerations, septal perforations (snorting), sinusitis, epiglottitis
Cardiovascular
Tachycardia (SVT, VT) or bradycardia, hypertension, dysrhythmias (VF), vasomotor collapse, acute heart failure (high-output), endocarditis, rupture or dissecting aneurysm, coronary vasospasm, angina, myocardial infarction, cardiomyopathy, myocarditis
Respiratory
Rapid and irregular breathing, diffusion abnormalities, pneumonia, pulmonary edema, infarction. Secondary to smoking: pneumothorax, pneumomediastinum, pneumopericardium, hoarseness, bronchitis, bronchiolitis obliterans
Gastrointestinal
Anorexia, nausea, vomiting, diarrhea, abdominal pain, gangrenous or ischemic bowel, perforation
Skin
Self-induced sores or linear excoriations, 'coke burns', phlebitis secondary to intravenous cocaine, pallor and infarction, vasculitis
Psychiatric
Decreased REM sleep, agitation, hallucinations, anxiety or depression, psychosis, paranoia, suicidal ideation, altered tactile sensation ('cocaine bugs')
Neurological
Tonic–clonic seizures, tremor, headache, restlessness, confusion, subarachnoid hemorrhage, vasculitis, intracerebral hemorrhage, cerebral infarction, brain abscess, movement disorders
Musculoskeletal
Muscle infarction, rhabdomyolysis, myoglobinuria, wound botulism
Genitourinary
Renal vasculitis, acute tubular necrosis, renal infarction, priapism (urethral use), paraphimosis (urethral use), testicular infarction
Obstetric
Spontaneous abortion, abruptio placentae, lower birth weight, prematurity, congenital abnormalities, growth retardation
Great vessels
Aortic dissection and rupture, hypertension

Reproduced with permission from Lewin *et al.* (1994).

be responsible. An α-adrenergic antagonist such as phentolamine may decrease the vasoconstriction, or direct vasodilators such as nitroprusside or nitroglycerin (glyceryl trinitrate) may be employed.

Beta-adrenergic antagonists are contraindicated. Even β_1-specific agents have some activity at β_2-receptors that can counteract existing vasodilation. Peripheral α-adrenergic effects would then be unopposed, with a resultant worsening of vasoconstriction and afterload. Even agents with both α- and β-adrenergic antagonism (such as labetalol) have relatively greater β effects (a β/α potency ratio of at least 7:1). In addition, β-adrenergic antagonists such as propanolol have been shown to worsen coronary vasoconstriction and to increase coronary vascular resistance and myocardial oxygen demand in patients during coronary catheterization (Lange *et al.* 1990). Calcium-channel blockers should also be avoided since animal studies have shown morbidity with decreased time to seizure and ventricular fibrillation.

Myocardial ischemia

Those patients who have chest pain following cocaine use have a 6 per cent incidence of myocardial infarction which is comparable with that of other groups (Hollander *et al.* 1994). About 40 per cent of patients using cocaine complain of chest pain. Pleuritic or sharp chest pain is usually deemed 'non-cardiac'. Approximately one-third of cocaine abusers who have myocardial infarctions report the pain as pleuritic or sharp. Infarctions that may be silent, occurring in young patients (typically 19–40 years) with or without ECG changes or underlying atherosclerotic disease, makes the assessment more complex than that of traditional chest pain patients.

Although giving clinically atypical presentations, cocaine has a classic pathophysiology for myocardial ischemia: increased myocardial demand in the face of diminished oxygen supply. Cocaine abusers have increased oxygen demand with high sympathetic stimulation and development of left ventricular hypertrophy. Oxygen supply is diminished through coronary artery vasospasm, increased platelet aggregation, and premature atherosclerosis. Ischemia with ST segment elevations may occur for 2 to 6 weeks after abuse. Cocaine administration to humans undergoing angiography causes vasoconstriction that is more pronounced in diseased segments. Cocaine enhances α_1-mediated platelet aggregation which may contribute to thrombogenesis and is also believed to promote development of premature atherosclerosis.

Since as many as 50 per cent of patients abusing cocaine may have elevated serum creatinine phosphokinase, detection of myocardial infarction by cardiac enzymes can be confounded because the MB isoenzyme can also be elevated due to skeletal muscle damage. Immunoassay for cardiac troponin I may be a more specific test, given that it has no cross-reactivity with human skeletal muscle; however, this test is not readily available. Rapidly falling creatinine phosphokinase levels may be more suggestive of rhabdomyolysis than of myocardial infarction. At present, however, exclusion or diagnosis of myocardial infarction by enzymes in the setting of rhabdomyolysis may be impossible.

The treatment of cocaine-induced myocardial ischemia differs from that of traditional ischemic heart disease (Nelson and Hoffman 1995). First, sedation with benzodiazepines is essential to decrease myocardial oxygen demand. Second, the use of β-adrenergic antagonists is contraindicated since these may worsen coronary vasospasm. The third difference is that phentolamine, an α-adrenergic antagonist, may be useful for persistent ischemic pain and associated hypertension. It has been shown to reduce cocaine-induced coronary vasospasm. Finally, as discussed below, sodium bicarbonate is the treatment of choice for wide-complex dysrhythmias induced by cocaine.

The similarities of treatment of cardiac ischemia/infarction include the use of oxygen, nitroglycerin, and aspirin to maximize oxygen delivery and further combat damage from platelet aggregation. Nitroglycerin has been demonstrated to reduce cocaine-induced coronary vasospasm. Calcium-channel blockers have been inadequately studied in humans, and animal data are equivocal. Their use should be considered only after sedation has been achieved. Verapamil may reverse coronary artery vasoconstriction and therefore may have a role in treating refractory ischemic pain.

Although thrombolytic agents have been used safely in patients with documented cocaine-induced myocardial infarction, no efficacy data exist. Caution is advised in using thrombolytics or other anticoagulants in those patients with hypertension and its attendant complications, particularly since most patients with cocaine-induced myocardial infarction will generally have good outcomes independent of therapy. Selection of appropriate patients for thrombolytic therapy is further complicated by a high incidence of abnormal ECGs in this population. Coronary catheterization and angioplasty may be the preferred approach in patients who do not improve after treatment with oxygen, benzodiazepines, nitrates, aspirin, and possibly phentolamine.

Dysrhythmias

Cocaine has a transient central vagotonic effect producing bradycardia that quickly resolves and may not be clinically appreciated. More commonly, cocaine produces tachydysrhythmias, such as sinus tachycardia and atrial fibrillation, as well as ventricular tachycardia and ventricular fibrillation. Both cocaine and lidocaine (lignocaine) have type I antidysrhythmic effects and possess proconvulsant activity associated with sodium-channel blockade. Hypertonic sodium bicarbonate is the agent of choice for wide-complex tachydysrhythmias (Nelson and Hoffman 1995).

Pulmonary complications

In addition to aortic dissection and cardiac ischemia/infarction, cocaine-related chest pain may be attributed to pneumothorax, pneumomediastinum, and pneumopericardium resulting from the barotrauma of inhalational use. Asthma and pulmonary infarction are also reported.

Other acute ischemic events

Mesenteric ischemia and infarction are surgical emergencies. They should be suspected in patients using cocaine who present with abdominal pain. In addition, spontaneous abortion and placenta previa are cocaine-induced emergencies that require immediate obstetric intervention. Cerebral infarction may also result from cocaine abuse.

References

Goldfrank, L.R. and Hoffman, R.S. (1991). The cardiovascular effects of cocaine. *Annals of Emergency Medicine*, **20**, 165–75.

Hollander, J.E., *et al.* (1994). Prospective multicenter evaluation of cocaine associated chest pain. *Academic Emergency Medicine*, **1**, 330–9.

Lange, R.A., *et al.* (1990). Potentiation of cocaine-induced coronary vasoconstriction by beta-adrenergic blockade. *Annals of Internal Medicine*, **112**, 897–903.

Lewin, N.A., Goldfrank, L.R., and Hoffman, R.S. (1994). Cocaine. In *Goldfrank's toxicologic emergencies* (5th edn), p. 849. Appleton & Lange, Norwalk, CT.

National Institute on Drug Abuse (1990). *Annual Medical Examiner Data. Data from the Drug Abuse Warning Network.* Statistical Series I, Number 10-B. US Department of Health and Human Services, Rockville, MD.

Nelson, L. and Hoffman, R.S. (1995). How to manage acute MI when cocaine is the cause. *Journal of Critical Illness*, **10**(1), 39–43.

9.2.9 Beta-blockers

S. W. Metcalf

Key messages

- Death in β-blocker poisoning is usually due to pump failure rather than cardiac dysrhythmias.
- The degree of hypotension reflects the severity of toxicity; bradycardia is not useful in assessing toxicity as it occurs at therapeutic levels.

- Pulmonary artery wedge pressure and cardiac output should be measured in the seriously ill patient and in patients with pre-existing cardiac disease.
- In hypotension the treatment of choice is glucagon and a massive dose may be needed.

Introduction

Beta-adrenergic receptor blockers are one of the most widely prescribed classes of drug in clinical medicine. Cardiovascular instability presents considerable therapeutic problems in cases of serious poisoning; β_1-receptor cardioselectivity is by no means absolute. Receptor blockade leads to a reduction in intracellular cAMP concentration. In the heart, decreased intracellular cAMP inhibits sodium and calcium influx during phase zero of the action potential.

Death in β-blocker poisoning is usually due to pump failure, an effect caused by 'quinidine-like' membrane stabilization, rather than cardiac dysrhythmias. Cardioselectivity is lost in overdose and membrane stabilization becomes a feature common to most β-blockers. Individual response to β-blockers varies considerably and, with the exception of sotalol, blood levels do not correlate well with toxicity; therefore diagnosis should be on clinical grounds. Sotalol levels correlate with the degree of QTc prolongation and the risk of arrhythmias (Love 1994*a*). In addition, toxicity may persist even after the β-blocker is undetectable in the blood; the mechanism underlying this is unknown.

Table 1 Clinical effects of β-blocker poisoning

Nausea and vomiting
Sinus bradycardia
Hypotension
Pulmonary edema
Bronchospasm
First-degree atrioventricular block
Complete atrioventricular dissociation
Intraventricular conduction delays
Disappearance of P waves
Asystole
Drowsiness or even coma
Mydriasis
Hallucinations
Convulsions
Respiratory depression
Renal failure
Hypoglycemia
Hyperkalemia

Diagnosis and assessment

Beta-blockers are well absorbed and toxic effects usually begin within 30 min of ingestion, peaking within 1 to 2 h. With the exception of sotalol, toxicity is almost always apparent within 6 h (Love 1994*b*). Toxic effects are listed in Table 1. Patients can deteriorate very rapidly and should be carefully monitored. The most consistent findings are bradycardia and hypotension. The degree of hypotension reflects the severity of toxicity; bradycardia is not useful in assessing toxicity as it occurs at therapeutic levels. Tachycardia has been reported with partial agonists such as pindolol. Progressive pulmonary edema and bronchospasm are uncommon unless there is pre-existing cardiac or respiratory disease. Effects on the central nervous system are unusual with hydrophilic agents such as atenolol. When present, coma and respiratory depression are poor prognostic features and are always accompanied by severe cardiovascular depression. Coma without cardiovascular depression suggests that other substances have been ingested. Although convulsions are seen as part of the central effect, they may also be caused by hypoglycemia.

Management

Initial resuscitation begins with immediate establishment of a patent airway. The induction of emesis is contraindicated because of the risk of rapid patient deterioration and seizures. Gastric lavage is preferred and should be undertaken if a sufficient dose has been ingested. For atenolol and popranolol this is in excess of 200 mg or double the daily dose if the patient is already on therapy. Gastric lavage should be preceded by atropine (0.6 mg intravenously) to prevent unopposed vagal stimulation which can lead to cardiovascular collapse. Activated charcoal (50 g) should be given.

Maintaining the patient's fluid volume and estimating inotrope infusion rates may be difficult because of the combination of peripheral dilatation and myocardial depression. Therefore pulmonary artery wedge pressure and cardiac output should be measured in the seriously ill patient and in patients with pre-existing cardiac disease.

The atropine dose can be repeated to a maximum of 3 mg. Atropine may elevate heart rate but have little effect on increasing blood pressure since the fall in cardiac output is predominantly due to a reduction in cardiac contractility. If systolic blood pressure is above 90 mmHg (or if there are broad complexes on the ECG or there is coma), the treatment of choice is glucagon (10 mg intravenously followed by infusion of 1–10 mg/h) which activates adenyl cyclase independently of β-receptors. It elevates both heart rate and force of contraction, and reverses hypotension more effectively than isoproterenol (isoprenaline) or epinephrine (adrenaline). Since large doses may need to be given, phenol-free diluents must be used to prevent phenol toxicity. Isoproterenol (5–20 g/min infusion) should be used if glucagon is ineffective, but it has the disadvantage of causing vasodilation leading to a fall in diastolic pressure, particularly if used with cardioselective β-blockers. Glucagon may cause nausea and vomiting. Other inotropes such as dopamine, dobutamine, and amrinone have been used with variable success. Combined therapy using various combinations of glucagon and inotropes may be successful where monotherapy is ineffective. Massive doses of glucagon and inotropes may be needed to maintain cardiac output (up to 10 times the normal therapeutic dose). Since toxicity may be prolonged even after clearance of the drug from the body, inotropes may need to be continued for several days. Epinephrine and norepinephrine (noradrenaline) should not be used because of inappropriate peripheral vasoconstriction. Combined therapy with several inotropes may be successful where single-agent therapy is not. Intravenous calcium has been used successfully to reverse hypotension, and intra-aortic balloon counterpulsation may be necessary.

Sotalol differs from other β-blockers in that, in addition to causing bradycardia and hypotension, it may cause severe ventricular tachyarrhythmias including ventricular tachycardia and ventricular fibrillation. Deterioration may be sudden and without warning. Lidocaine (lignocaine) and intravenous magnesium are the antidysrhythmics of choice. Class IA agents and other Class III agents such as amiodorone and bretylium should be avoided (Muse and Linden 1996). Overdrive electrical pacing may be needed to control the rate in sotalol-induced tachyarrhythmias. Some authors recommend that patients with a heart rate of less than 50 beats/min despite initial inotrope therapy should also be electrically paced (Heath 1984). However, the heart may be refractory to normal pacing potentials.

Convulsions should be initially controlled with diazepam and phenytoin if necessary. Hypoglycemia should always be considered as a possible cause of fits. Recovery is usually complete within 48 h.

References

Heath, A. (1984). β-Adrenoceptor blocker toxicity: clinical features and therapy. *American Journal of Emergency Medicine*, **2**, 518–25.

Love, J.N. (1994*a*). Beta blocker toxicity after overdose: when do symptoms develop in adults? *Journal of Emergency Medicine*, **12**, 799–802.

Love, J.N. (1994*b*). β-Blocker toxicity: a clinical diagnosis. *American Journal of Emergency Medicine*, **12**, 356–7.

Muse, D.A. and Linden, C.H. (1996). Beta-blocker poisoning. In *The clinical practice of emergency medicine* (2nd edn) (ed. A.L. Harwood-Nuss, C.H. Linden, R.C. Luten, S.M. Shepherd, and A.B. Wolfson), pp. 1271–4. Lippincott–Raven, Philadelphia, PA.

9.2.10 Cyanide

D. G. Penney and W. E. Porter

Key messages

- Although there are relatively few deaths attributed to cyanide each year, it is important to be able to make a rapid diagnosis in order to provide treatment for cure.
- As in any comatose patient, 100 per cent oxygen, glucose, thiamine, and naloxone should be given.
- When severe cyanide poisoning (coma, shock, ischemic ECG changes, or arrhythmias) is diagnosed, sodium thiosulfate should be administered followed by sodium nitrite if CO poisoning can be excluded.
- Other management approaches include artificial respiration, hyperbaric oxygen, vascular volume expansion, bicarbonate infusion, anticonvulsant drugs, and cardiovascular support.
- Long-term follow-up should occur in order to watch for signs of neurological deterioration and sequelae such as parkinsonian syndrome, myocardial damage, and goiter.

Introduction

Cyanide is a deadly poison which animals and humans encounter from a number of sources. Cyanide salts and hydrocyanic acid are ubiquitous in industrial and chemical manufacturing such as electroplating, mineral refining, pest control, and tanning of hides. It is well known that cyanide is present naturally in a variety of plants, including apricot, peach, plum, and apple seeds.

Reports indicate that overzealous use of nitroprusside, a potent vasodilator, can lead to cyanide toxicity and death from long-term use. Cyanide is also released in the duodenum after ingestion of the cytotoxin amygdalin.

There is often a combination of carbon monoxide (CO) and cyanide intoxication in victims of fire and smoke inhalation. Cyanide is released in the thermal breakdown of natural fibers (wool and silk), nylon, polyurethane foam, and asphalt (Salkowski and Penney 1994). Although CO is the more common toxic agent, cyanide has caused death in fire victims without CO intoxication. Since cyanide inhalation may cause diminished strength and co-ordination, a victim's ability to leave the source of the toxin may be impaired. Cyanide poisoning in fire victims may be indicated by the presence of plasma lactate concentrations above 10 mmol/l (Baud *et al.* 1991).

Although cyanide is a diffuse substance in the environment, deaths are uncommon. Many factors may predispose people to developing cyanide toxicity, including poor nutrition status, with a concomitant decrease in the sulfur and albumin pool, and smoking. Vegans, persons who suffer pernicious anemia secondary to vitamin B_{12} deficiency, or persons with renal or hepatic insufficiency resulting in decreased elimination are also at risk.

Pathophysiology

The effects of cyanide have a rapid onset and are extremely toxic because it is easily absorbed through the skin, lungs, eyes, and gastrointestinal tract. Cyanide produces histotoxic cellular hypoxia by initially binding to the protein portion of cytochrome oxidase and then to the ferric form of iron. This reaction forms a relatively stable but reversible complex. Over 40 enzyme systems in the human body are potentially inactivated by cyanide. The most important of these are the cytochrome oxidases. This intramitochondrial system consists of the cytochrome a–a_3 complex of the electron transport chain. By binding to this enzyme complex, cyanide inhibits electron transfer to molecular oxygen, thus blocking ATP generation. This results in the reduced cellular utilization of oxygen and increased venous $P\text{O}_2$. The impairment of cellular oxygen utilization reduces aerobic respiration with a decrease in pyruvate conversion in the Krebs citrate cycle; lactic acid formation then increases, producing metabolic acidosis (Salkowski and Penney 1994).

Elimination of cyanide is primarily through the enzymes rhodanese and mercaptopyruvate (80 per cent), which catalyze the transfer of sulfur from the sulfur donor to cyanide. This forms thiocyanate which is subsequently excreted by the kidneys. The half-life of thiocyanate is 2.7 days. Renal insufficiency may lead to thiocyanate toxicity. The endogenous sulfur pool is small and the reaction is too slow to prevent acute cyanide toxicity. The reaction can be accelerated (as much as 30-fold) by adding sulfur in the form of sodium thiosulfate (Sylvester *et al.* 1983). The remaining cyanide can react with hydroxocobalamin (vitamin B_{12a}) to form cyanocobalamin (vitamin B_{12}) or oxidize to CO_2 and formate with small amounts of HCN.

Clinical signs and symptoms

Cyanide poisoning is difficult to diagnose because the symptoms are non-specific. It has diffuse negative effects on the cardiovascular, respiratory, central nervous, and endocrine systems. If available, a good patient history is an important part of the diagnosis. The well-known bitter almond smell from ingestion is detectable in only 60 per cent of the population. Onset of symptoms varies greatly, ranging from 35 min to 14 days (Salkowski and Penney 1994). Inhalation of cyanide may lead to very rapid demise, while difficulties may not appear for several hours after the ingestion of amygdalin. Cutaneous absorption of cyanide to toxic levels requires a large surface area and long exposure.

Arteriolization of venous blood is due to the inability of tissues to use oxygen and has been used as a significant symptom in cyanide poisoning.

The symptoms of cyanide poisoning begin with central nervous system stimulation followed by depression (faintness, headache, anxiety, and giddiness). These symptoms may mimic those of hypoxia; however, cyanosis will not be present unless respiratory failure has occurred. Eventually nausea, ataxia, palpitations, convulsions, and unconsciousness develop (Table 1). A wide array of cardiac symptoms may develop such as bradycardia, delayed arteriovenous conduction leading to arteriovenous block, ventricular complex prolongation, ischemic changes, ST elevation or depression, shortened ST segment with fusion of the T wave into the QRS complex, and supraventricular or ventricular arrhythmias (Salkowski and Penney 1994). Non-cardiogenic pulmonary edema may also develop after inhalation or ingestion.

Cyanide affects discrete areas of the central nervous system, i.e. the globus pallidus, putamen, basal ganglia, and cerebral cortex (Salkowski and Penney 1994). Spongy degeneration and focal demyelination of the cerebellar white matter are indicative of cyanide toxicity. These changes may explain the parkinsonian-like effects in the long-term sequelae of cyanide poisoning. Acute cyanide toxicity may cause whole-brain calcium elevation, thus disrupting neurotransmitter release with development of cyanide-induced seizures and tremors.

Table 1 Signs and symptoms of cyanide poisoning

Low levels		
Anxiety	Weakness	Headache
Confusion	Vertigo	Ataxia
Giddiness	Bitter almond odor	Nausea
Moderate levels		
Hyperpnea	Dyspnea	Tachycardia
Tachypnea	Irregular respiration	
High levels		
Unconsciousness	Blood arterialized	Lactic acidosis
Coma		Opisthotonos
Arrhythmias	Convulsions	Asystole
Hypotension	Bradycardia	Paralysis
Dilated pupils	Sweating	Death
	Unreactive pupils	

Reproduced with permission from Salkowski and Penney (1994).

Laboratory screening

Whole-blood cyanide levels

Cyanide levels have a direct correlation with the symptoms produced. However, a period of 4 to 6 h is usually needed to measure levels, and thus they are useful only for confirmation of poisoning. A new test paper (Cyantesmo, Gallard-Schlesinger, New York) allows rapid semiquantitative screening for cyanide. A level of zero to 30 μg/dl is normal, 0.05 to 0.5 μg/dl is associated with smoking, 0.5 to 1.0 μg/dl is associated with hypertension and tachycardia, 1.0 to 3.0 μg/dl is associated with decreased mental state and may be fatal, and a level above 3.0 μg/dl is usually fatal (Salkowski and Penney 1994).

Arterial blood gases

These show metabolic acidosis (with an anion gap) and normal Po_2. Calculated oxygen saturation will be normal, but measured oxygen saturation will be slightly decreased. This is secondary to some cyanide binding to hemoglobin, thus decreasing oxygen-binding capacity and the corresponding saturation.

Plasma thiocyanate level

Normal levels can be up to 10 μg/ml, while lethal levels range from 50 to 200 μg/ml.

Management and treatment of cyanide poisoning

Once cyanide poisoning is diagnosed, the main goals of treatment are preventing cellular anoxia as well as inactivating and removing the cyanide. This is accomplished by aggressive cardiorespiratory support with 100 per cent oxygen and, if symptoms are severe, administration of antidote (e.g. Eli Lilly Cyanide Antidote Kit). Health care providers should not inhale the victim's expired air nor perform mouth to mouth resuscitation. The victim's clothes and skin can be contaminated; thus proper clothes removal and washing of the skin must occur.

Mechanical ventilation may be necessary for initial management. Oxygen increases the rate of conversion of reduced cytochrome oxidase to the oxidized form, allowing the electron transport chain to function again (Salkowski and Penney 1994). The use of hyperbaric oxygen is controversial, but may be beneficial (Gossel and Bricker 1990), particularly with concomitant CO poisoning. Sodium bicarbonate should be administered if the pH is below 7.16. Anticonvulsants are used to treat any seizures that may develop. In addition, antiarrhythmics and vasopressors may be necessary for corresponding arrhythmias or hypotension. Gastric lavage with activated charcoal may be beneficial if administered shortly after oral ingestion of cyanide. Emesis induction is contraindicated because of the possibility of rapidly induced unconsciousness, coma, or aspiration.

Cyanide antidotes (Table 2) include amyl nitrite, sodium nitrite, and sodium sulfate. Nitrites act by inducing the formation of methemoglobin, which has a stronger affinity for cyanide than cytochrome oxidase. This reaction forms cyanomethemoglobin which protects from further cyanide toxicity. It is believed that nitrites

Table 2 Administration of the Eli Lilly Cyanide Antidote Kit

1. Amyl nitrite pearls are crushed and inhaled for 15–30 s every 2–3 min
2. When IV access is available, amyl or sodium nitrite 300 mg IV (10 ml of 3% solution) is given over 2–4 min; this will produce methemoglobinemia at the desirable 27% level
3. Sodium sulfate, a sulfur donor, is administered as 12.5 g IV
4. If in 2 h symptoms reappear or persist, steps 2 and 3 are repeated

IV, intravenous
Reproduced with permission from Rindone and Sloane (1992).

may have an additional protective mechanism. It is suggested that methemoglobin be kept between 25 and 40 per cent (Rindone and Sloane 1992) to prevent adverse effects, such as cyanosis. These levels should be followed serially to prevent toxicity. If severe methemoglobin toxicity develops, the treatment of choice is 1 to 2 mg/kg of methylene blue. It must be used with extreme caution because of its potential to release cyanide. A serious adverse effect associated with nitrite usage is the production of severe hypotension with rapid infusion and subsequent cardiovascular collapse.

Sodium thiosulfate donates sulfur which enhances rhodanase activity by increasing the rate of conversion of cyanide to sodium thiocyanate (Rindone and Sloane 1992). The coadministration of chlorpromazine or colbaltous chloride potentiates the effectiveness of sodium thiosulfate (Salkowski and Penney 1994). Sodium thiocyanate toxicity resembles cyanide toxicity of the central nervous system. Symptoms include muscle weakness, nausea, vomiting, hallucinations, twitching, coma, and death. Sodium thiocyanate is mildly neurotoxic (e.g. tinnitus, miosis, hyper-reflexia) at 1 mmol/l and may be life threatening at levels three to four times higher. Prolonged exposure may produce reversible hypothyroidism because of competition with iodine.

Other cyanide poisoning treatments are available. With hydroxocobalamin (vitamin B_{12a}), cyanide replaces the hydroxy group to form cyanocobalamin (vitamin B_{12}) which is subsequently eliminated via the kidney. At therapeutic doses (20–70 mg/kg), vitamin B_{12a} does not change mean blood pressure, left ventricular end-diastolic pressure, or the PR or QT intervals of the ECG. Hydroxocobalamin is a safe, rapid, and powerful antidote for cyanide poisoning which is recommended during the prehospital stage; however, its routine use is impractical and expensive, particularly for long-term use for nitroprusside prophylaxis (Rindone and Sloane 1992). Hydroxocobalamin is not yet available in the United States (Salkowski and Penney 1994).

A variety of drugs remain experimental or are not approved for use in the United States. In Germany, 4-dimethylaminophenol hydrochloride (4-DMAP) is used; its action is similar to that of nitrites. Although the production of methemoglobin is faster than with sodium nitrite, efficacy is no better than with nitrites. 4-DMAP does not produce the same profound hypotension as nitrites, but has been associated with renal failure in animals. Hydroxylamine is also a rapid methemoglobin former, but it is less toxic than 4-DMAP and provides anticonvulsive action by augmenting γ-aminobutyric acids in the brain, which may be decreased in cyanide poisoning.

Cobalt EDTA, a very effective treatment used in the United Kingdom, can cause severe side-effects, particularly when cyanide poisoning does not exist. Metabolic acidosis, hypotension, massive edema, and ventricular tachycardia have been found in animal and human studies. Because of cobalt's side-effects (diaphoresis, angina, and vomiting), cobalt EDTA should be reserved only for severe cyanide poisoning.

Administration of α-ketoglutarate in two doses of 50 mg/kg 20 min apart has been demonstrated to increase the 50 per cent lethal dose (LD_{50}) for nitroprusside 1.7 times and to decrease cyanide by 30 per cent. Sodium thiosulfate 1 g/kg alone increased LD_{50} 5.5 times with an 88 per cent reduction in cyanide. The coadministration of α-ketoglutarate and sodium thiosulfate increased the LD_{50} by 6.9, i.e. an overall cyanide reduction of 98 per cent. Cyanide and α-ketoglutarate form α-ketoglutaric-cyanhydrin, which is subsequently excreted through the kidneys. Pyruvate slightly potentiates the antidotal effect of thiosulfate, but not that of nitrites.

References

Baud, F.J., *et al.* (1991). Elevated blood cyanide concentrations in victims of smoke inhalation. *New England Journal of Medicine*, **325**, 1761–6.

Gossel, T.A. and Bricker, J.D. (1990). *Principles of clinical toxicology* (2nd edn). Raven Press, New York.

Rindone, J.P. and Sloane, E.P. (1992). Cyanide toxicity from sodium nitroprusside: risks and management. *Annals of Pharmacotherapy*, **26**, 15–19.

Salkowski, A.A. and Penney, D.G. (1994). Cyanide poisoning in animals and humans: a review. *Veterinary and Human Toxicology*, **36**, 455–66.

Sylvester, D.M., Hayton, W.L., Morgan, R.L., and Way, J.L. (1983). Effect of thiosulfate on cyanide pharmacokinetics in dogs. *Toxicology and Applied Pharmacology*, **69**, 265–71.

9.2.11 Alcohols

D. Jacobsen

Key messages

- Methanol and ethylene glycol poisoning share many characteristics, often presenting with metabolic acidosis of unknown origin.
- A common failure is delayed diagnosis of poisoning with fatal consequences, although effective treatment is available.
- Other alcohol poisoning is less severe and presents without pronounced metabolic acidosis.

Ethanol

Diagnosis

The most important diagnostic procedure in the patient with alcohol on the breath is to exclude other causes for the coma, such as head injuries, hypoglycemia, or coingestion of drugs or chemicals. Ethanol is easily determined by gas chromatography, which also allows the determination of methanol, isopropanol, and acetone. Ethanol elevates serum osmolality by 24 mosmol/kgH_2O per 1 g/l (100 mg/dl).

Assessment

The lethal ethanol dose for adults is about 3 to 5 g/kg in adults, corresponding to a lethal serum concentration of 110 to 180 mmol/l (5–8 g/l), and is lowest among chronic alcoholics with poor nutrition. Fatalities are often related to alcohol-induced organ complications such as cardiomyopathy, malnutrition, and secondary infections.

Typical features of ethanol intoxication are central nervous system depression, hypotension, hypothermia, and respiratory problems, including aspiration of vomit into the lungs.

As ethanol replaces normal calories in chronic alcohol abusers, ketoacidosis (alcoholic acidosis) may develop when the blood ethanol level is zero or close to zero. If pronounced ketoacidosis occurs, methanol poisoning can be excluded by a normal ophthalmoscopy and ethylene glycol poisoning by lack of renal failure and characteristic urine sediment (see later).

Management

If the patient is seen within 1 to 2 h of imbibing a substantial quantity of ethanol, gastric aspiration and lavage should be considered. Activated charcoal is of no value in pure ethanol intoxication, but as many patients have coingested other drugs which potentiate the toxic effects of ethanol, it should be considered in these cases.

Symptomatic and supportive treatment should be given as needed. Hemodialysis effectively removes ethanol but is rarely indicated as the clinical response to supportive measures is usually good. The indication for hemodialysis should be based on a clinical evaluation and a documented high blood ethanol concentration (> 110 mmol/l or 5 g/l).

Isopropanol

Diagnosis

The clinical diagnosis of isopropanol intoxication can be made from central nervous system depression and acetone breath. The specificity of Ketostix (sensitive only for acetoacetate) is far too low to be of practical value.

Both isopropanol (17 mmol/l or 1 g/l; osmolal contribution, 18 mosmol/kgH_2O) and acetone will increase the osmolal gap. The anion gap is usually normal but may be slightly increased due to ketosis (and some lactate) in the alcoholic and lactate in the hypotensive patient.

Isopropanol is best determined by a standard gas chromatographic method where ethanol, methanol, and acetone are also determined.

Assessment

The lethal dose is quite variably given as 1 to 4 ml/kg, i.e. about twice as toxic as ethanol.

Blood isopropanol levels in the range of 34 to 68 mmol/l (2–4 g/l) are often seen without complications other than coma and slight respiratory depression. Others have survived with levels as high as 95 mmol/l (5.6 g/l) although hemodialysis was neccessary. Elimination is mainly due to hepatic metabolism to acetone; the serum half-life of isopropanol is 6 to 7 h.

The typical patient will present with central nervous system depression, depressed respiration, gastritis with abdominal pain, and slight hypothermia. Aspiration pneumonia may be present, particularly if the patient is found outdoors. The very young and the most severe alcoholics may be hypoglycemic.

Management

If the patient is seen within 1 to 2 h, gastic aspiration and lavage should be considered. Activated charcoal is of no value.

The symptomatic treatment should follow the established principles of intensive care when needed. Mechanical ventilation should be liberally employed if respiratory depression is accompanied by pneumonia.

Hemodialysis will effectively remove isopropanol and should be considered at blood ispropanol levels above 68 mmol/l (4 g/l) and on clinical criteria if coma and sustained hypotension are present.

Methanol

Diagnosis

Methanol is determined by gas chromatography. The clinical diagnosis of methanol poisoning is difficult in the absence of a history of ingestion. The objective signs of ocular toxicity of methanol include dilated pupils, which are partially reactive or non-reactive to light, and optic disk hyperemia with blurring of the disk margins.

If a specific analytical method for methanol is not available, the anion and osmolal gaps should be calculated. The anion gap (AG) can be calculated using the formula

$$AG = ([Na^+] + [K^+]) - ([Cl^-] + [HCO_3^-]).$$

It normally varies from 10 to 16 mmol/l. In the absence of circulatory failure, diabetes, alcoholism, and uremia, an increased anion gap clearly indicates poisoning with salicylate, methanol, or ethylene glycol.

The osmolal gap (OG) is the difference between the measured osmolality O_m and the calculated osmolality O_c in serum:

$$OG = O_m - O_c.$$

Normally, the osmolality of serum is determined by sodium (and its accompanying anions), glucose, and urea according to the formula (SI units)

$$O_c = (1.86[Na] + [urea] + [glucose])/0.93.$$

The reference range for the osmolal gap is 5 ± 7 mosmol/kgH_2O.

The intoxicants best able to increase the osmolal gap are those which have a low molecular weight and are present in high mass units, i.e. high molar concentrations. The lower alcohols and glycols are such substances. A methanol concentration of 32 mmol/l (1 g/l) increases the osmolal gap by 32/0.93 = 34 mosmol/kgH_2O. The osmolal contribution of methanol is so significant that interference from other causes will only occur at methanol levels below 16 mmol/l (0.5 g/l).

Only methanol and ethylene glycol regularly cause severe metabolic acidosis and elevation of both the anion and osmolal gaps. However, if ethanol is coingested with methanol or ethylene glycol, there will be no metabolic acidosis until most of the ethanol is metabolized owing to its antidotal effect. In such circumstances calculation of the gaps must be repeated.

In the late stages of methanol or ethylene glycol poisoning, most of the alcohol or glycol may be metabolized to its acidic metabolite. In this situation there is a pronounced metabolic acidosis with a high anion gap. Owing to low alcohol/glycol levels, the osmolal gap may be close to normal values, particularly in ethylene glycol poisoning because of its higher molecular weight and thus smaller molar contribution. In this situation, a small or normal osmolal gap does not eliminate the possibility of toxic alcohol ingestion.

Assessment

The lethal dose of methanol is about 1 g/kg (1.2 ml/kg). The minimum dose causing permanent visual defects is unknown.

Methanol itself is not very toxic, and there is usually a latent period of about 12 to 24 h between ingestion and the occurrence of symptoms. This is the time required for sufficient amounts of formic acid to accumulate. Since ethanol inhibits methanol metabolism, concomitant intake of ethanol may lengthen the latent period.

Clinical features are non-specific and include weakness, anorexia, headache, and nausea accompanied by increasing dyspnea (hyperventilation) as the metabolic acidosis develops. Visual symptoms (blurred vision) may appear first or together with the features described above. A few cases may present with acute pancreatitis or even methemoglobinemia.

There is usually little improvement of impaired visual acuity after the acute stage. The patient should always be evaluated by an ophthalmologist as the incident may have legal and insurance implications. The most severely poisoned patients may suffer from a Parkinson-like syndrome due to methanol-induced symmetrical lesions in the putamen.

The possibility of other methanol victims in addition to the treated patient should always be considered.

Management

The treatment should follow the established principles of intensive care when needed. If the patient is seen within 1–2 h, gastric lavage should be performed. Activated charcoal is of no value.

The specific treatment of methanol poisoning includes administration of sodium bicarbonate to combat the metabolic acidosis, ethanol to inhibit metabolism of methanol to formate, and hemodialysis to remove methanol and formate.

Aggressive bicarbonate treatment must be given to correct the metabolic acidosis, to counteract the continuous production of organic acids, and to decrease the amount of undissociated formic acid and its access to the central nervous system. The aim should be full correction of the metabolic acidosis, i.e. as much as 400 to 600 mmol bicarbonate during the first few hours.

The therapeutic blood ethanol concentration of 22 mmol/l (1 g/l) may be attained by giving a bolus dose of 0.6 mg/kg of absolute ethanol, followed by 70 to 150 mg/kg/h intravenously (or orally) in dextrose–saline, with the highest maintenance dose for drinkers (1 g ethanol = 1.2 ml). Ethanol solutions above 20 per cent should be given via a central catheter. Monitoring of the blood ethanol level is important, particularly during hemodialysis which removes it. As a rule of thumb, the maintenance dose of ethanol should be doubled during hemodialysis.

Hemodialysis is indicated for any degree of visual impairment, provided that methanol or formate metabolic acidosis is still present, severe metabolic acidosis (base deficit above 15–20 mmol/l), blood methanol levels above 20 mmol/l (60 mg/dl), and if more than 40 ml has been ingested (adults). Hemodialysis should be performed for at least 8 h, as the blood half-life of methanol is 40 to 50 h (low pulmonary and renal excretion), with effective ethanol treatment.

Folinic acid (leucovorin) 50 to 70 mg (adult dose) should be given intravenously every 4 h for 24 h in order to enhance formate metabolism.

Ethylene glycol

Diagnosis

Ethylene glycol in biological fluids can be determined by gas chromatography. If specific analysis is not available, the use of the anion and osmolal gaps may also point to the diagnosis (see above). An ethylene glycol concentration of 16 mmol/l (1 g/l) increases the osmolality by 16/0.93 = 17 mosmol/kgH_2O. Urine microscopy will reveal needle-shaped (most frequent) or envelope-shaped calcium oxalate crystals among different types of cast. These findings may be delayed and therefore microscopy should be repeated if the initial result is negative.

Assessment

The lethal dose of ethylene glycol is 1 to 2 ml/kg. Like methanol, the toxicity of ethylene glycol is mediated through its metabolites of which glycolate and oxalate are the most important. However, unlike methanol, ethylene glycol causes central nervous system depression and inebriation in a similar manner to ethanol. The toxicity of ethylene glycol results from a combination of the severe metabolic acidosis caused by glycolic acid and the precipitation of calcium oxalate crystals, with impaired organ function as a result.

Ethylene glycol poisoning is characterized by an initial central nervous system depression phase with inebriation progressing to coma. Following a short latent period of 4 to 12 h, the signs and symptoms due to the metabolites start to appear. The increasing accumulation of glycolic acid leads to severe metabolic acidosis and increasing hyperventilation. For unknown reasons, elevated blood pressure and tachycardia are usually prominent features.

Without adequate treatment, the patient will deteriorate rapidly. In addition to central nervous system depression (cerebral edema occurs), convulsions, oliguric renal failure, and respiratory problems usually develop. Pulmonary infiltrations may be observed radiologically but these changes are thought to be non-infective in origin. In severe cases acute respiratory distress syndrome may develop.

Management

If the patient is seen within 1 to 2 h of ingestion, gastric aspiration and lavage should be considered. Activated charcoal is of no value.

Bicarbonate should be given as in methanol poisoning. Rapid correction of the acidosis by bicarbonate in these patients may provoke tetanic signs, particularly when hypocalcemia is already present (see later). Ethanol or 4-methylpyrazole (see below) is given to inhibit ethylene glycol metabolism (ethanol dose as for methanol poisoning). Hemodialysis removes ethylene glycol and glycolate and helps to correct metabolic acidosis. If blood ethylene glycol values are not available, hemodialysis should be performed for 8 h. Acute renal failure is a strong indication for hemodialysis as renal excretion is the dominant elimination pathway for ethylene glycol when its metabolism is blocked by ethanol.

Tetany and seizures should be treated with calcium intravenously. However, calcium should not be given for hypocalcemia *per se* as this may increase precipitation of calcium oxalate crystals in the tissues. If calcium gluconate or chloride is not effective, convulsions should be treated conventionally.

4-Methylpyrazole (fomepizole, Antizol) will most probably be accepted by the United States Food and Drug Administration (1997/8) as an alternative antidote to ethanol in ethylene glycol poisoning. The advantages of Antizol are better inhibition of alcohol dehydrogenase, lack of central nervous system depression, and slower elimination, particularly during hemodialysis. The dosage regimen is provided by the manufacturer.

Diethylene glycol

The few clinical and experimental data available for diethylene glycol poisoning show similarities with ethylene glycol toxicity. Therefore diethylene glycol poisoning should probably be treated as recommended for ethylene glycol poisoning.

Bibliography

Aabakken, L., Johansen, K.S., Rydningen, E.B., Bredesen, J.E., Øvrebø, S., and Jacobsen, D. (1994). Osmolal and anion gaps in patients admitted to an emergency medical department. *Human and Experimental Toxicology*, **13**, 131–4.

Baud, F.J. (1988). Treatment of ethylene glycol poisoning with intravenous 4-methylpyrazole. *New England Journal of Medicine*, **319**, 97–100.

Jacobsen, D. and McMartin, K.E. (1986). Methanol and ethylene glycol poisonings. Mechanism of toxicity, clinical course, diagnosis and management. *Medical Toxicology*, **1**, 309–34.

Jacobsen, D. and McMartin, K.E. (1997). Antidotes for methanol and ethylene glycol poisoning. *Clinical Toxicology*, **35**, 127–43.

Jacobsen, D., Hewlett, T.P., Webb, R., Brown, S., Ordinario, A.T., and McMartin, K.E. (1988). Ethylene glycol intoxication: evaluation of kinetics and crystalluria. *American Journal of Medicine*, **84**, 145–82.

9.2.12 Carbon monoxide

S. G. Simonson and C. A. Piantadosi

Key messages

- Carbon monoxide (CO) poisoning commonly occurs during exposure to products of incomplete combustion resulting from fires and faulty gas furnaces.
- Carboxyhemoglobin (COHb) is relatively stable and does not carry oxygen; hence CO poisoning causes tissue hypoxia, and CO can enter the body tissues and bind to cellular hemoproteins.
- The correlation between COHb and clinical severity of poisoning is poor.
- Neurological and cardiac symptoms are common during CO poisoning, and 3 to 30 per cent of poisoning victims develop a delayed neurological syndrome.
- Oxygen is the treatment of choice for mild poisoning, and in more severe cases hyperbaric oxygen is useful to decrease the body store of CO and prevent the delayed neurological syndrome.

Introduction

Carbon monoxide (CO) is one of the most common causes of accidental poisoning because it is a ubiquitous by-product of incomplete combustion. It is a non-irritating, colorless, tasteless, and odorless gas that is difficult to identify when it is present in the environment. It is produced during combustion of carbon compounds such as propane, gasoline, natural gas, wood, coal, and tobacco. The routine use of such fuels in automobiles, furnaces, space heaters, charcoal grills, and fireplaces creates numerous potential exposures in routine daily activities. Fortunately, with proper operation and adequate ventilation, these sources rarely result in toxic CO exposure. However, firefighters and fire victims are at risk of serious exposure. CO is a major contributor to morbidity and mortality in smoke inhalation and fire victims; hence these patients require evaluation for CO poisoning. A less common source of CO is the metabolism of methylene chloride in products such as paint thinner.

CO is produced naturally in the body as a result of enzymatic degradation of heme by heme oxygenase. Heme metabolism results in a very low basal level (1 per cent) of carboxyhemoglobin (COHb) unless there is hemolysis to increase it. COHb levels are increased during pregnancy and by some drugs such as diphenylhydantion. Smokers and urban dwellers exposed to automobile exhaust fumes commonly have COHb levels higher than normal, but these rarely exceed 10 per cent.

Pathophysiology

The acute physiological effects of CO poisoning result from a reduction in oxygen delivery to tissues because of CO binding to Hb. The CO dose received by the body is related to the partial pressure of CO inhaled, the duration of the exposure, and the ventilation of the patient during the exposure. CO inhaled into the lungs is rapidly and reversibly bound to Hb. Hb binds CO with an affinity greater than 200 times that for oxygen. This relationship is described by the equation

$$[\text{HbCO}]/[\text{HbO}_2] = M(P_{\text{CO}}/P_{\text{O}_2})$$

where the equilibrium constant M is 200 to 230 for human Hb. This relationship shows how a relatively small partial pressure of CO can form significant amounts of COHb. In addition to decreasing the Hb binding sites available for oxygen, CO also shifts the oxyhemoglobin dissociation curve to the left, increasing the affinity of Hb for any oxygen molecules bound to it. This effect further decreases oxygen delivery to the tissues.

Clinical experience and experiments in animals indicate that the binding of CO to Hb and the resulting tissue hypoxia do not completely explain all the toxicity seen after CO poisoning (Coburn 1979). Indeed, the correlation between COHb level and the severity of poisoning is poor. Thus additional mechanisms have been proposed to explain the toxic effects of CO. CO binding to intracellular proteins containing iron (hemoproteins) plays a role in some cases of CO poisoning. Myoglobin, cytochrome a–a_3, and guanylate cyclase are candidates for the cellular effects of CO. CO uptake by binding to intracellular proteins is increased with hypoxia. The role of CO binding to cellular hemoproteins in clinical CO poisoning is unclear, but up to half the total body store of CO after an exposure may be in the tissues. Leukocyte-mediated vascular (endothelial) injury may also occur after CO poisoning (Thom 1993).

Clinical features

The clinical presentation of CO poisoning can be quite variable depending on the intensity of the exposure. Short-term exposure to a high concentration of CO can result in loss of consciousness with few other symptoms, whereas longer exposures to lower concentrations may present with multiple symptoms. An 'influenza-like' presentation can occur with CO poisoning in milder forms. Headache is common when exposure results in COHb levels near 15 per cent, followed by malaise, nausea, dizziness, ataxia, syncope, seizures, and coma at higher exposure intensities. Other neurological manifestations can include visual disturbances, confusion and mental status changes, and peripheral neuropathy. Tachypnea, tachycardia, and hypotension may be present. Myocardial ischemia can occur, and in patients with coronary artery disease the exertional threshold required to produce angina is decreased. The brain and heart are the primary sites of clinical toxicity because of their high metabolic rate for oxygen; however, other systems can be affected. Rhabdomyolysis due to muscle necrosis may occur in severe poisoning. Renal dysfunction can occur with rhabdomyolysis. Primary damage to the lung is not common as a result of CO; however, patients with pre-existing pulmonary disease may experience worsening shortness of breath. Aspiration may result from vomiting or loss of consciousness and inability to protect the airway. Pulmonary edema can accompany CO poisoning, but this is more common in the setting of smoke inhalation. Bullous skin lesions have been reported but are not common. Additionally, the 'cherry-red' appearance frequently ascribed to CO victims is rare and should not be expected.

Assessment

The diagnosis of CO poisoning is confirmed by a history of exposure to a source of CO (usually a combustion source), symptoms, and an elevated COHb level measured in the blood. In clinical practice all criteria may not be present in a given patient. COHb levels above 50 per cent are frequently associated with loss of consciousness; however, loss of consciousness can be seen with lower COHb levels. Death is common at levels over 70 per cent, but otherwise the COHb level correlates poorly with symptoms and severity of poisoning. This is because patients have usually been removed from the source of CO and have frequently had a period of oxygen breathing before COHb levels have been measured. Perhaps more importantly, COHb levels do not reflect the extravascular load of CO. It is not uncommon for patients who have had a significant exposure to CO to continue to have symptoms even though the COHb level in the blood has decreased to normal. Thus COHb serves more to confirm exposure and suggest the diagnosis of CO poisoning than to predict severity of poisoning.

In most clinical laboratories COHb level is routinely measured using a co-oximeter. Arterial blood gas analyzers do not measure COHb but measure gas partial pressures ($P\text{O}_2$, $P\text{CO}_2$) and pH, and calculate Hb saturation based on $P\text{O}_2$. This distinction is important because the arterial $P\text{O}_2$ is frequently normal in CO poisoning; hence estimates of Hb saturation based on $P\text{O}_2$ nomograms will also be normal, giving no indication of CO poisoning. The co-oximeter directly measures Hb saturation and COHb. When a discrepancy exists between the measured Hb saturation and the measured $P\text{O}_2$, CO should be considered as a cause. COHb levels are usually measured on arterial blood samples to allow simultaneous evaluation of oxygen content and pH, but venous samples can also be used for COHb. COHb is stable and, if a co-oximeter is not readily available, it can be measured on an anaerobically sealed anticoagulated blood specimen many hours later without significant loss of accuracy.

Estimations of oxyhemoglobin saturation by pulse oximetry are not valid in the presence of COHb, i.e. pulse oximetry cannot be used to exclude CO poisoning. In fact, the error of the pulse oximetry measurement approximates the level of COHb.

CO poisoning may cause other laboratory abnormalities. Creatine kinase may be elevated from either skeletal or cardiac muscle. Serum creatinine may rise with rhabdomyolysis-induced renal failure. Metabolic acidosis from decreased oxygen delivery can lower serum bicarbonate concentration and create an elevated anion gap. An extremely high lactate in a fire victim should raise suspicion of concomitant cyanide poisoning as a result of combustion of plastics.

The electrocardiogram may be normal in CO poisoning or demonstrate tachycardia, ventricular arrhythmias, non-specific ST-T wave changes, or changes consistent with myocardial infarction. The chest radiograph is usually normal but may show edema or evidence of aspiration pneumonia. Magnetic resonance or CT imaging of the brain may be normal early, which may be helpful in ruling out other central nervous system pathology (stroke, hematoma). Cerebral edema and characteristic lesions of CO toxicity occur if the insult is severe. The most common findings on brain CT in CO poisoning are low-density lesions in the cerebral white matter, particularly the centrum semiovale, and the globus pallidus. Cortical atrophy and ventricular dilation can also be seen. The presence of abnormal findings on CT scanning, most notably white matter lesions, may have prognostic value; however, anatomic localization of lesions correlates poorly with clinical neurological findings.

Management

Management of CO poisoning begins by removing the victim from the CO-contaminated environment, followed by administration of 100 per cent oxygen which should be continued until symptoms have resolved and the COHb is less than 5 per cent. Altered mental status may necessitate endotracheal intubation to maintain an adequate airway. In severe cases, hypotension will require fluid administration and vasopressor support. Specific organ toxicity should be evaluated with ECG, cardiac enzymes, urinalysis, and serum creatinine. A drug or toxin screen may be indicated in some patients.

Symptoms related to CO poisoning may persist despite treatment in severe cases, and new or recurrent symptoms attributable to the CO exposure may occur days to weeks after the exposure (Smith and Brandon 1973). This syndrome, which is referred to as the delayed neurological syndrome, occurs in 3 to 30 per cent of patients after

CO poisoning. Delayed neurological syndrome is characterized by persistent headache, confusion and difficulty in concentrating, memory and personality changes, depression, incontinence, or parkinsonian features. The mechanism is unknown. Risk factors for delayed neurological syndrome include age over 40, loss of consciousness, brain CT abnormalities, and protracted severe CO exposures. Delayed neurological syndrome tends to improve with time, and the majority of those developing it will have resolution within 2 years.

Oxygen can be administered in higher doses in a hyperbaric chamber than is possible with normobaric oxygen. Generally accepted indications for hyperbaric oxygen are loss of consciousness, any degree of neurological impairment or persistent symptoms regardless of the COHb level, anginal pain or ischemic ECG changes, or a prolonged exposure and a COHb level above 20 to 25 per cent. Hyperbaric oxygen can be of benefit for several reasons. First, the relationship between oxygen and CO for Hb binding sites is competitive, and so higher oxygen concentrations enhance the kinetics of CO removal. The elimination half-life of COHb in patients breathing air is 4 to 6 h. This is reduced to 1 h for 100 per cent normobaric oxygen and 23 min for oxygen at 3 atm absolute. Second, hyperbaric oxygen can provide enough oxygen dissolved in the plasma to supply the oxygen requirements of the body independently of Hb. At 3 atm absolute the amount of oxygen dissolved in plasma is approximately 5 to 6 ml/dl. Third, extravascular tissue loads of CO are removed more effectively by hyperbaric oxygen (Brown and Piantadosi 1992). Fourth, hyperbaric oxygen can reduce cerebral edema in the injured brain. Finally, clinical evidence indicates that hyperbaric oxygen can prevent or at least decrease the incidence of delayed neurological syndrome in patients surviving CO poisoning (Thom *et al.* 1995).

Batteries of tests have been developed to aid in the diagnosis and quantitation of neuropsychiatric abnormalities associated with CO poisoning. Some tests are too cumbersome to be given prior to treatment, but others can be done prior to treatment or even in an emergency department to screen for subtle deficits that otherwise would be missed. The tests can be repeated to verify improvement after treatment (acute effect) or to discover decrements in performance consistent with delayed neurological syndrome (subacute or chronic effect).

The pregnant patient/fetus presents special circumstances in CO poisoning. Analogous to adult Hb, fetal Hb has a higher affinity for CO than for oxygen ($M = 180$). Fetal Hb has a slightly higher affinity for CO than adult Hb; thus the fetus can behave as a 'sink' for CO uptake. This effect can also result in slower CO removal. For this reason and perhaps because of lower oxygen tension in the fetus, COHb levels at steady state tend to be higher than maternal levels. The fetal nervous system may be more susceptible to the effects of CO and hypoxia, resulting in fetal loss or developmental delays after birth as a result of CO poisoning. This has been used to justify administration of hyperbaric oxygen to any pregnant patient with a significant CO exposure.

The outcome after CO poisoning depends on the intensity and duration of the exposure and the timely administration of oxygen. Mild poisoning in younger patients without loss of consciousness tends to have a good neurological outcome. It is difficult to predict the outcome in other cases. In patients with loss of consciousness, hyperbaric oxygen administered within 6 to 8 h offers the best chance of improving the clinical outcome after poisoning.

References

Brown, S.D. and Piantadosi, C.A. (1992). Recovery of energy metabolism in rat brain after carbon monoxide hypoxia. *Journal of Clinical Investigation*, **89**, 666–72.

Coburn, R.F. (1979). Mechanisms of carbon monoxide toxicity. *Preventative Medicine*, **8**, 310–22.

Smith, J.S. and Brandon, S. (1973). Morbidity from acute carbon monoxide poisoning at three-year follow-up. *British Medical Journal*, **i**, 318–21.

Thom, S.R. (1993). Leukocytes in carbon monoxide-mediated brain oxidative injury. *Toxicology and Applied Pharmacology*, **123**, 234–47.

Thom, S.R., Taber, R.L., Mendiguren, I.I., Clark, J.M., Hardy, K.R., and Fisher, A.B. (1995). Delayed neuropsychologic sequelae after carbon monoxide poisoning: prevention by treatment with hyperbaric oxygen. *Annals of Emergency Medicine*, **25**, 474–80.

9.2.13 Corrosives

S. W. Metcalf

Key messages

- Corrosive ingestion causes a spectrum of clinical effects ranging from insignificant oral erythema to full-thickness esophageal burns with perforation and rapid death.
- Development of esophageal strictures is seen in up to 70 per cent of patients following deep esophageal ulceration.
- Most corrosive incidents involve acids and alkalis which destroy tissues by various mechanisms including ionization, oxidation, reduction, and denaturation of cellular components.
- Early endoscopy is mandatory in all cases where significant ingestion may have occurred.
- Corticosteroids may reduce the incidence of strictures in patients with second- and third-degree burns.

Introduction

A corrosive or caustic substance destroys tissues, causing chemical burns. Acutely, corrosive ingestion causes a spectrum of clinical effects ranging from insignificant oral erythema to full-thickness esophageal burns with perforation and rapid death. The complications of corrosive poisoning are shown in Table 1. Recovery from the acute insult is frequently complicated by the development of esophageal strictures which are seen in up to 70 per cent of patients following deep esophageal ulceration. Corrosive injury is also carcinogenic, with the risk of esophageal carcinoma increased at least 1000-fold.

Corrosive poisoning may be deliberate by the suicidal or accidental. Purposeful poisoning often involves the ingestion of large quantities of strong corrosives, leading to severe and often fatal injury.

Pathophysiology

Most corrosive incidents involve acids and alkalis which destroy tissues by various mechanisms including ionization, oxidation,

Table 1 Complications of corrosive poisoning

Death
Esophageal and gastric perforation
Mediastinitis
Pneumothorax
Pneumonitis/acute respiratory distress syndrome
Upper airway edema
Esophageal stricture
Leather bottle stomach
Tracheo-esophageal fistula
Esophago-aortofistula
Gastrocolic fistula
Secondary infection and septicemia
Disseminated intravascular coagulation
Carcinoma esophagus

reduction, and denaturation of cellular components. Alkalis are associated with more severe tissue damage and a higher incidence of esophageal injury than acids. Acids coagulate proteins, resulting in a burn covered by eschar which then acts as a protective barrier limiting deeper penetration by acid. Alkalis dissolve proteins and saponify fats, leading to liquefactive necrosis and tissue softening which rapidly sloughs allowing deeper tissue penetration. In addition to deeper tissue penetration by alkali, the esophageal squamous epithelium with its slightly alkaline pH has more natural resistance to acids. Typically, esophageal injury is reported in 6 to 20 per cent of cases of acid ingestion. Acids are more toxic to the stomach where there is limited buffering capacity against them.

Strong alkalis are much more commonly used in poisoning in the West. In one reported series from India, where strong acids are widely available, acid ingestion led to esophageal injury in 87.8 per cent of patients, gastric injury in 85.4 per cent, and duodenal injury in 34.1 per cent; acute complications occurred in 39.1 per cent, with death in 12.2 per cent (Zargar 1989).

The corrosive potential of any given solution is dependent on its pK_a (a measure of its ease of dissociation), its pH (acids with $pH < 2$ or alkalis with $pH > 12.5$ will cause ulceration of the esophagus), its concentration, and the volume of solution. The corrosiveness of an acid or base is not solely dependent on its ability to donate or to accept protons respectively. For example, hydrofluoric acid is weaker than nitric acid in terms of pK_a but is much more corrosive, since it forms highly reactive species such as the fluoride ion in aqueous solution. The degree of tissue damage is further modified by variables such as the duration of contact between the corrosive substance and tissue and whether any food has been ingested. Severe tissue damage can occur with the briefest of contacts. Exposure of cat esophagus to 30 per cent sodium hydroxide solution for only 1 s causes transmural necrosis.

Contact between corrosive and tissue leads to hyperemia and edema, followed by necrosis and ulceration of the mucosa. Deeper injury involving the muscle layer may be exacerbated by local vessel thrombosis. The inflammatory phase is associated with fibroblast infiltration and collagen deposition with granulation tissue evident by 10 days. The early phase following corrosive damage is characterized by tissue with a low tensile strength and susceptibility to perforation. Repair is usually complete within 6 weeks, but scar formation and retraction can persist for months and may result in stricture formation or shortening of the segment of gut involved. Sites most commonly affected by corrosive ingestion are the oropharynx, the esophagus, and the stomach, with the duodenum and the jejunum less frequently involved. The esophagus is most vulnerable at its areas of natural anatomical narrowing: in the cricopharyngeal area, at the indentations made by the aortic arch and the left main bronchus, and at the diaphragmatic hiatus. The stomach is most severely affected in the antrum, where corrosive pools and antral damage may be aggravated by associated pylorospasm. Esophageal strictures are the principal complication occurring in up to 30 per cent of patients following esophageal injury. Strictures may be single, multiple, focal, or circumferential, but they will only develop if there is damage to the muscular layer. Stricture formation is usually early, with 58 per cent of patients developing stricture within 1 month, 80 per cent within 2 months, and almost all within 8 months. Occasionally a stricture may develop many years after injury.

Diagnosis and assessment

Burning and blistering of the mouth suggests that there may be serious injury to the esophagus and stomach. In most cases there is severe pain in the mouth, anterior chest, or abdomen. The patient may have dysphagia with drooling of saliva. There may be vomiting with or without hematemesis. Excruciating retrosternal pain usually follows esophageal perforation, which is usually accompanied by subcutaneous emphysema of the neck and muscle rigidity of the left chest wall. On auscultation a pleuropericardial friction rub may be heard (Hamman's sign) and there may be signs of a left-sided hydropneumothorax. Peritonitis indicates gastric perforation.

Any respiratory distress is ominous and should alert one to possible upper airway compromise. Stridor is indicative of supraglottic edema, but hoarseness and even muteness are suggestive of laryngeal injury. Respiratory embarrassment may also be due to chemical pneumonitis or even pulmonary edema. As respiratory tract edema typically evolves over hours, patients must be observed carefully. Fever and tachycardia frequently accompany even moderate poisonings. Although certain signs such as stridor may be predictive of severe esophageal injury, these should not be relied upon to gauge severity as the absence of oropharyngeal burns does not exclude severe esophageal damage. Early endoscopy is mandatory in all cases where significant ingestion may have occurred.

An erect chest radiograph may show mediastinal air in esophageal perforation; there may be an associated hydropneumothorax or pleural effusions which are often small and usually left-sided. Rarely, contrast studies using water-soluble dyes may be needed to demonstrate esophageal perforation. Pneumonitis, which may be due to aspiration or inhalation of volatile vapors, may be evident. Free air under the diaphragm confirms gastric perforation. An abdominal film is less helpful but may show ileus.

Arterial blood gas measurement and continuous pulse oximetry are essential for any case where respiratory signs are evident or significant corrosive ingestion may have occurred. There may be metabolic acidosis which may reflect acid absorption or result from extensive

Grade	Appearance
0	Normal examination
1	Hyperemia and edema of the mucosa
2(a)	Superficial ulcerations, erosions, friability, blisters, exudates; hemorrhages with white exudates
2(b)	As for 2(a) but with circumferential ulcerations
3(a)	Transmural necrosis, with multiple and deep ulcerations and areas of necrosis; mucosa is dusky or black, and the entire lumen may be obliterated
3(b)	As for 3(a) plus extensive necrosis

Table 2 Endoscopic grading of corrosive injury

tissue damage. Although most corrosive acids or alkalis are neutralized on contact with tissue, significant quantities may be absorbed, leading to systemic toxicity. For example, ammonia or phenol ingestion may lead to significant methemoglobinemia. Hemolysis and leukocytosis are occasionally seen. Disseminated intravascular coagulation may be seen in severe cases, but fibrinolysis is more common. Since corrosive ingestion may be an act of suicide, other substances may have been taken, and plasma acetaminophen (paracetamol) and salicylate levels should be determined.

The mainstay of investigation is endoscopy which should be performed as soon as is practicable after patient stabilization, ideally between 6 and 24 h after ingestion. Endoscopy should be performed by an experienced operator using a small-diameter flexible endoscope. The endoscope should be passed through damaged areas unless there is severe necrotizing damage. Reported series describing the use of early endoscopy show negligible risk of perforation. Burns are classified according to the criteria shown in Table 2. Assessment of burn severity and extent may be difficult as there are often concomitant burns of different stages present. The distinction between grade 2(b) and grade 3 burns may be particularly difficult.

Management

Resuscitation begins with control of the airway. Stridor and hoarseness are suggestive of upper airway obstruction and early endotracheal intubation is essential. Since upper airway edema may progress over the first 24 h, all cases involving potentially significant corrosive ingestion and any patients with respiratory embarrassment should undergo fiber-optic laryngoscopy. Prophylactic endotracheal intubation should be undertaken if laryngeal edema is seen. Tracheobronchial involvement is seen in up to 10 per cent of cases of severe corrosive burns. Intravenous fluids should be administered and the patient kept nil by mouth. Patients with severe injury will lose large volumes of body fluids and require generous fluid resuscitation. Central venous pressure measurement and urinary catheterization will help guide replacement therapy.

The type of corrosive, the amount ingested, and the time of ingestion should be determined where possible, and the Regional Poisons Unit should be consulted. Any corrosive solids or crystals should be wiped from the mouth which may then be gently irrigated with water. Emetics are contraindicated as the friable esophagus may be re-exposed to the corrosive substance and there is a risk of pulmonary aspiration. Indeed, antiemetics should be administered if the patient is vomiting. Activated charcoal has no role and will coat the gut lining, making endoscopic assessment of tissue damage extremely difficult. Gastric lavage has traditionally been avoided but may be useful in diluting corrosives; however, only small aliquots of water (not more than 5 mg/kg) should be used to avoid inducing emesis (Dowsett and Linden 1996). The pH of the stomach contents should be measured. An alkaline aspirate is indicative of severe gastric injury. Water must be the only diluent used, and attempts to neutralize corrosives are liable to cause even further tissue damage due to the vigorous exothermic reaction that may ensue.

Antibiotics should be administered early. Necrotic burns are rapidly invaded by pathogenic organisms, particularly *Escherichia coli* and *Klebsiella* species, and sepsis is a major cause of death. In addition to anticipating sepsis and possibly reducing the risk of perforation, there is evidence that antibiotic administration may decrease the incidence of stricture formation.

Patients with grade 1 burns can usually be discharged after observation for 24 h. No specific treatment is needed, but patients should be reviewed at 4 weeks and undergo a barium study to exclude stricture formation, as initial endoscopy may have underestimated the grade of burn. Patients with grade 2 burns are at risk of perforation and careful hospital observation is needed. Most patients with grade 2a burns recover quickly and can be discharged within 2 weeks. Again, careful follow-up is needed as strictures may develop. Patients with grade 2b and grade 3 burns will require a prolonged stay in hospital; many will perforate or develop strictures that are resistant to dilatation. Surgery is essential for patients with perforation or extensive grade 3 burns, but the mortality is high and the surgery is technically difficult, involving dissection of necrotic and inflamed tissues. Surgery may also allow the placement of a gastrotomy tube or the fashioning of a feeding jejunostomy since the institution of early nutrition improves the overall prognosis.

Dilatation therapy should not be carried out before 4 weeks as it confers no benefit and is associated with perforation (Shaffer *et al.* 1994). Severe and extensive stricturing should be treated surgically, where esophagectomy and colonic or jejunal interposition yield good results. Burns of the lips and mouth can lead to severe chronic disability, and the oral surgeon should be consulted at an early stage. Customized mouth protectors and extraoral splints may help to limit the extent of scar contracture.

The most controversial issue is the use of corticosteroids to prevent stricture formation. Patients with grade 1 burns do not develop strictures and so corticosteroids are not indicated. Corticosteroids may reduce the incidence of strictures in patients with second- and third-degree burns, although previous reports indicate that strictures develop in most patients with third-degree burns regardless of any therapy (Howell *et al.* 1992). Although evidence available at present is insufficient to recommend corticosteroid use, there appears to be no increase in morbidity or mortality from these drugs. Corticosteroids are absolutely contraindicated where there is perforation or hemorrhage.

Novel pharmacological approaches that have been used with good result to prevent stricture formation in animal models include D-penicillamine and β-aminoproprionitirile. More recently, the sequential use of epidermal growth factor and interferon-γ has been shown to reduce the frequency of residual stenosis significantly in corrosive esophageal burns of rats (Berthet *et al.* 1994). The applicability of such treatments to humans remains to be demonstrated.

References

Berthet, B., di Costanzo, J., Arnaud, C., Choux, R., and Assadourian, R. (1994). Influence of epidermal growth factor and interferon γ on healing of esophageal corrosive burns in the rat. *British Journal of Surgery*, **81**, 395–8.

Dowsett, R.P. and Linden, C.H. (1996). Corrosive ingestions. In *The clinical practice of emergency medicine* (2nd edn) (ed. A.L. Harwood-Nuss, C.H. Linden, R.C. Luten, S.M. Shepherd, and A.B. Wolfson). Lippincott–Raven, Philadelphia, PA.

Howell, J.M., Dalsey, W.C., Hartsell, F.W., and Butzin, C.A. (1992). Steroids for the treatment of corrosive esophageal injury: a statistical analysis of past studies. *American Journal of Emergency Medicine*, **10**, 421–5.

Shaffer, R.T., Carrougher, J.G., Kadakia, S.C., and Levine, S.M. (1994). Update on caustic ingestions: how therapy has changed. *Journal of Critical Illness*, **9**, 161–74.

Zargar, S.A., Kochhar, R., Nagi, B., Mehta, S., and Mehta, S.K. (1989). Ingestion of corrosive acids; spectrum of injury to upper intestinal tract and natural history. *American Journal of Gastroenterology*, **97**, 702–7.

9.2.14 Pesticides and agricultural chemicals

K. Koyama

Key messages

- Organophosphates cause acute respiratory and circulatory failure due to cholinesterase inhibition. Atropine sulfate is effective for muscarinic signs due to all types of organophosphates. Pralidoxime is effective for most of the highly toxic organophosphates, but is less effective for those with moderate toxicity.
- Paraquat is highly toxic with delayed and refractory pulmonary fibrosis. No known therapies are significantly effective to improve prognosis. Blood paraquat concentration 4 to 24 h after ingestion has reliable prognostic value.
- Glufosinate causes central nervous system signs and circulatory failure after a latent period of 8 to 38 h. Respiratory support must be provided for all patients, particularly those who have ingested 100 ml or more.

Organophosphates

Organophosphates, which were widely used as insecticides after the Second World War, have poisoned many by intentional or accidental ingestion and by dermal and respiratory exposure. They cause acute respiratory and circulatory failure due to cholinesterase inhibition. Lethal organophosphates (tabun, sarin, and VX) have been used as chemical weapons, in regional wars, and recently by terrorists.

Pathophysiology and diagnosis

The toxicities of individual compounds differ widely: highly toxic nerve agents include sarin (GB), tabun (GA), soman (GD), and VX; highly toxic insecticides include tetraethyl pyrophosphate (TEPP), disulfoton (Di-Syston), parathion, methyl parathion, and EPN; moderately toxic insecticides include dichlorvos (DDVP), trichlorfon (DEP), fenitrothion (MEP), and malathion. Systemic poisoning can occur as a result of dermal exposure.

Organophosphates inhibit acetylcholinesterase and cause accumulation of acetylcholine resulting in overstimulation of nicotinic and muscarinic receptors. Decreased activity of serum and red blood cell cholinesterase and cholinergic signs, such as miosis, perspiration, and fasciculation, have diagnostic value. The clinical signs usually develop very rapidly.

Muscarinic effects include miosis, perspiration, salivation, lacrimation, vomiting, diarrhea, increased bronchial secretion, pulmonary edema, bradycardia, and hypotension.

Nicotinic effects include muscular twitching, fasciculation, and muscular weakness or paralysis (including respiratory muscles). Tachycardia, hypertension, and hyperglycemia frequently develop and are suspected to be caused by stimulation of sympathetic ganglia.

Central nervous system effects include disturbance of consciousness and depression of the respiratory and circulatory centers.

There may be intermediate paralytic syndrome, i.e. muscular weakness involving the proximal limb and respiratory muscles may occur 24 to 96 h after exposure. Delayed motor-dominant peripheral neuropathies with Wallerian degeneration of myelinated axons may also occur 1 to 2 weeks after exposure. However, the mechanism and treatment of these delayed toxicities have not yet been established.

Management

Mouth-to-mouth ventilation should be avoided in highly toxic organophosphate poisoning and lethal gas inhalation, because of the likelihood of secondary poisoning of medical staff. Bag-and-mask ventilation, various types of airway, and active compression–decompression cardiopulmonary resuscitation are recommended instead. Intratracheal intubation with oxygen and antidote administration should be performed at the scene of poisoning or in the ambulance.

Atropine sulfate is effective for muscarinic signs due to all types of organophosphates. Intravenous infusions of 1 mg for mild cases, 1 to 2 mg for moderate cases, and 2 to 5 mg for severe cases are recommended. Additional doses should be administered until profuse bronchial secretion and bradycardia disappear. Pupil size is not an appropriate index because miosis is not the target of treatments in systemic poisoning.

Pralidoxime (PAM), a cholinesterase reactivator, is effective for most of the highly toxic organophosphates but less effective for most of the moderately toxic compounds except dichlorvos and diazinon. Intravenous slow infusions of 1 g (over 15 min) are recommended for symptomatic patients, with an additional 1 g for severe cases. Continuous infusion of 0.5 g/h for 24 h after ingestion could be used if necessary. Pralidoxime becomes ineffective 36 to 48 h after ingestion because of irreversible binding between the organophosphate and cholinesterase.

When organophosphates are ingested, gastric lavage and/or activated charcoal and cathartics are essential. The effects of diuresis and hemopurification are obscure.

Sarin and VX terrorism in Japan

The nerve gas sarin, used by terrorists from a religious organization, injured over 500 people (seven fatalities) in Matsumoto in June 1994, and over 5000 people (11 fatalities) in Tokyo in March 1995. In slight and moderate cases, rhinorrhea, dim vision with prominent miosis, headache, or ocular pain occurred. In severe cases, coma, foamy bronchial secretion, respiratory weakness, fasciculation, convulsion, tachycardia, and hypertension were recognized. Atropine and/or pralidoxime improved consciousness and respiration. Pralidoxime produced rapid recovery in serum and red cell cholinesterase. In surviving patients, extubation could be conducted on the first or second day. Complete recovery was obtained when patients received advanced life support before cardiac arrest developed. The same terrorist group also used VX in three serial incidents between 1994 and 1995; three people were injured, of whom one died. Dermal exposure or intramuscular injections were suspected.

Carbamate

Insecticide carbamates cause very rapid inhibition of cholinesterase. The clinical signs are similar to those of organophosphate poisoning. In general, spontaneous recovery of the enzyme activity is rapid, and the clinical signs disappear within 24 h. Therefore there is no indication for pralidoxime administration. Atropine should be used as in organophosphate poisoning.

Paraquat

Paraquat, a widely used herbicide introduced in 1962, has caused many fatalities worldwide (Bismuth and Hall 1995). The most common liquid formulation contains 20 per cent by weight of paraquat ion, a dark blue-green pigment, and an emetic. One mouthful of the 20 per cent concentrate can produce fatal respiratory failure due to pulmonary fibrosis. The effects of treatments are not satisfactory.

In Japan, the 24 per cent by weight formulation was changed in 1986 to the present formulation containing 5 per cent paraquat dichloride and 7 per cent diquat dibromide. However, the prognosis of patients has not significantly improved.

Pathophysiology and diagnosis

In human oral paraquat poisoning, full recovery can be expected if ingestion is less than 20 mg/kg, death may occur on ingestion of 20 to 40 mg/kg, and death usually occurs if more than 40 mg/kg is ingested (Vale *et al.* 1987).

Paraquat actively accumulates in the lung. The mechanisms of toxicity are suspected to be lipid peroxidation in biological membranes caused by superoxide free radicals generated from the oxidoreduction of paraquat.

Immediately after ingestion, vomiting and diarrhea with the blue-green pigment of the formulation occur. Painful erosions of oral and pharyngeal mucosa and the gastrointestinal tract become evident within 1 to 2 days. Moderate renal and hepatic dysfunction may also appear. Hypoxemia due to pulmonary edema and alveolitis occurs within a few days of ingestion, and develops into intractable pulmonary fibrosis resulting in death within a period of weeks. The central nervous system disturbances do not develop until the severe hypoxemia progresses. If ingestion is massive, patients die of cardiac failure within a few days.

Adding a grain of sodium hydroxide and the smallest spatula blade full of sodium dithionite to 10 ml of urine or gastric aspirate produces a blue coloration if 1 ppm or more of paraquat is present. The results of this test on urine on admission roughly predict the severity of poisoning.

Management

Gastric lavage, activated charcoal and cathartics, diuresis, and charcoal hemoperfusion are used, but do not significantly improve the prognosis. Oxygen administration should be restricted as far as possible to avoid excessive superoxide formation. Administration of free-radical scavengers, radical-catabolizing enzymes or their activators such as liposomal superoxide dismutase, vitamin E, vitamin C, niacin, riboflavin, selenium, *N*-acetylcysteine, desferrioxamine, clofibrate, and chlorpromazine have had some effects in animal experiments. However, the effects of these drugs on human paraquat poisoning are unclear. The results of treatment might appear in the patients in the border area of the survival curve presented by Proudfoot *et al.* (1979).

Prognosis

Blood paraquat concentration in relation to the time from ingestion (4–24 h) is valuable for predicting survival or death (Proudfoot *et al.* 1979). Most patients whose plasma paraquat concentrations do not exceed 2.0, 0.6, 0.3, 0.16, and 0.1 ppm at 4, 6, 10, 16, and 24 h after ingestion will survive.

Glufosinate

Poisoning caused by the suicidal ingestion of Basta®, a herbicide containing 20 per cent glufosinate ammonium and 33.3 per cent surfactant, has been reported only in Japan, but may develop worldwide because it is now available in major European, American, and Asian countries. Severe central nervous system disorders develop after a long latency period (8–38 h), when the patient appears asymptomatic and general treatments for poisoning have been completed.

Pathophysiology and diagnosis

The acute human oral toxic dose of Basta® that causes delayed central nervous system disorders has been estimated at 1.6 to 1.8 ml/kg, i.e. 320 to 360 mg/kg of glufosinate ammonium (Koyama *et al.* 1995*a*). In reported cases, 85 per cent of patients who ingested 100 ml or more of Basta®, and 30 per cent of patients who ingested less than 100 ml, developed respiratory failure requiring mandatory ventilation.

Glufosinate is a potent inhibitor of glutamine synthetase which catalyzes the formation of glutamine from glutamate and ammonia. However, the pathophysiology of human poisoning has not been clarified. The surfactant component is suspected to cause gastrointestinal irritation and circulatory failure.

Immediately after ingestion, patients complain of nausea and vomiting, but these symptoms subside within a few hours. However, 8 to 38 h after ingestion, patients develop disturbance of consciousness and respiratory arrest. Other delayed signs are agitation, tremor, seizures, nystagmus, ataxic gait, pyrexia, etc. Retrograde partial amnesia is also recognized on recovery. In severe cases, patients have died of refractory circulatory failure which is believed to be due to the surfactant component.

The diagnosis can be obtained by analyzing the patient's urine using paper chromatography.

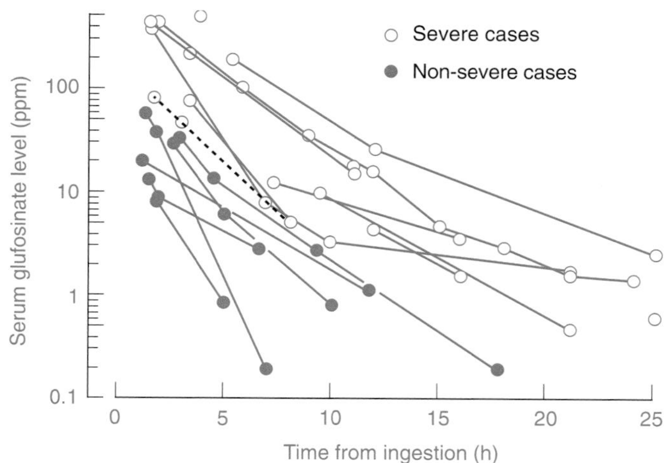

Fig. 1 Changes in serum glufosinate level in 20 reported cases of acute oral Basta® poisoning: ○ level in patients with severe poisoning ($n = 14$) who developed coma, respiratory arrest, or generalized convulsion; • level in patients with non-severe poisoning ($n = 6$) who did not develop these signs. Serial data from the same patient are connected by solid lines. The dotted line connecting 70 ppm at 2 h and 5 ppm at 8 h appears to discriminate between severe and non-severe patients. (Reproduced with permission from Koyama *et al.* (1997).)

The serum glufosinate level related to time from ingestion in several reported cases is shown in Fig. 1. Central nervous system signs develop after the serum glufosinate level decreases, and persist for several days. It may be possible to predict the development of delayed signs from the serum glufosinate level.

Management

Adsorption of glufosinate to activated charcoal is only 3.4 mg/g (Hoechst Japan Ltd, personal communication). The surfactant, which has a hydrophobic long carbon chain in its structure, is expected to be adsorbed by activated charcoal. Diuresis is believed to be essential, because most of the absorbed glufosinate is eliminated in the urine. Hemoperfusion and hemodialysis are performed in series to eliminate surfactant and glufosinate respectively (Koyama *et al.* 1995*b*). However, the clinical effects of hemopurification are obscure.

It is essential to observe vital signs closely in hospital for at least 2 days regardless of the amount ingested, to provide respiratory support in all patients, particularly those who have ingested 100 ml or more, and to consider the development of delayed central nervous system signs even if the primary treatment is complete.

Prognosis

Mortality is 19.4 per cent, from all reported cases. The circulatory shock suspected to be due to the surfactant and the failure of management of unexpected respiratory arrest are the main causes of death.

References

Bismuth, C. and Hall, A.H. (ed.) (1995). *Paraquat poisoning*. Marcel Dekker, New York.

Koyama, K., Matsuo, H., Saruki, K., and Andou, Y. (1995*a*). The acute oral toxic dose of a herbicide containing glufosinate and a surfactant (abstract). *Clinical Toxicology*, **33**, 519.

Koyama, K., Yamashita, M., and Andou, Y. (1995*b*). Hemodialysis is effective for glufosinate elimination (abstract). *Clinical Toxicology*, **33**, 546.

Koyama, K., *et al.* (1997). Relationship between serum glufosinate level and development of serious poisoning caused by the ingestion of Basta®, a herbicide containing glufosinate. *Journal of the Japanese Association for Acute Medicine*, **8**, 617–18.

Proudfoot, A.T., Stewart, M.S., Levitt, T., and Widdop, B. (1979). Paraquat poisoning: significance of plasma-paraquat concentrations. *Lancet*, **ii**, 330–2.

Vale, J.A., Meredith, T.J., and Buckley, B.M. (1987). Paraquat poisoning: clinical features and immediate general management. *Human Toxicology*, **6**, 41–7.

10 The hematological system

10.1 Normal physiology

10.1.1 The blood cells

K. K. Hampton and C. Wardrop

Key messages

- Quantitative deficiencies of hematopoietic cells and hemostatic proteins are common in the critical care setting.
- Deficiencies are frequently multiple and acquired rather than congenital.
- Deficiency states can be due to a combination of decreased production, loss from the intravascular space, increased consumption, and premature destruction.
- Localized bleeding usually responds to treatment, but generalized bleeding from two or more sites simultaneously often reflects a breakdown of hemostatic homeostasis and indicates disseminated intravascular coagulation.
- Hemostasis is usually adequate if the hematocrit is above 30 per cent, the platelet count is above 80×10^9/l, the fibrinogen is above 1 g/l, and the activated partial thromboplastin time and partial thromboplastin time are no more than 3 to 5 s longer than control values.

Table 1 Normal hematological indices

Hemoglobin (g/dl)	M F	13.0–16.5 11.5–15.5
Hematocrit	M F	0.40–0.50 0.37–0.47
Red blood cells ($\times 10^{12}$/l)	M F	3.8–5.6 3.4–5.2
Mean corpuscular volume (fl)		84–99
Reticulocytes ($\times 10^9$/l)		20–70 (0.5%–1.6%)
Red cell mass (ml/kg)	M F	25–35 20–30
Plasma volume (ml/kg)		35–45
White blood cells ($\times 10^9$/l)	Total Neutrophils Lymphocytes Monocytes Eosinophils Basophils	4.0–10.5 2.0–8.0 1.0–4.0 0.1–0.8 0.1–0.5 0.01–0.1
Platelets ($\times 10^9$/l)		150–400

Blood consists of cellular and fluid plasma components. Red cells, white cells, and platelets originate in the bone marrow, while plasma proteins are synthesized primarily in the liver. In common with most physiological systems, cell numbers and plasma protein concentrations are regulated within narrow limits under normal conditions (Table 1). Frequently recognized hematological abnormalities in the critically ill are predominantly quantitative; qualitative defects are rare. Likewise, most hematological abnormalities are secondary to non-hematological single- or multiple-organ pathology. Acquired defects are much more common than congenital defects, and multiple abnormalities of both cellular and hemostatic components are more frequent than the single defects usually identified in routine hematological practice.

Anatomically, the cellular components comprise cells in circulating blood, bone marrow, lymph nodes, and spleen. Red cells are primarily concerned with oxygen carriage, white cells with immunity and control of infection, and platelets with maintaining vascular integrity. The hemostatic system is responsible for maintaining blood fluidity and, at the same time, prevents blood loss by initiating rapid, localized, and appropriate blood clotting at sites of vascular damage. This system is complex, comprising both cellular and plasma elements, i.e. platelets, coagulation and fibrinolytic cascades, the natural anticoagulation pathway, and endothelial cells.

Hematopoiesis, the production of the cellular elements, occurs in adults within the marrow spaces of the bones of the axial skeleton. Hematopoiesis originates from the pluripotent stem cell which has the capacity for both replication, to maintain stem cell numbers, and differentiation, by which it gives rise to cells of the erythroid, myeloid, and lymphoid series as well as megakaryoctes, which are the precursors of platelets (Fig. 1). Development and differentiation of hemopoietic cells is under the control of glycoprotein cytokines which operate in a paracrine or endocrine fashion. These act at early,

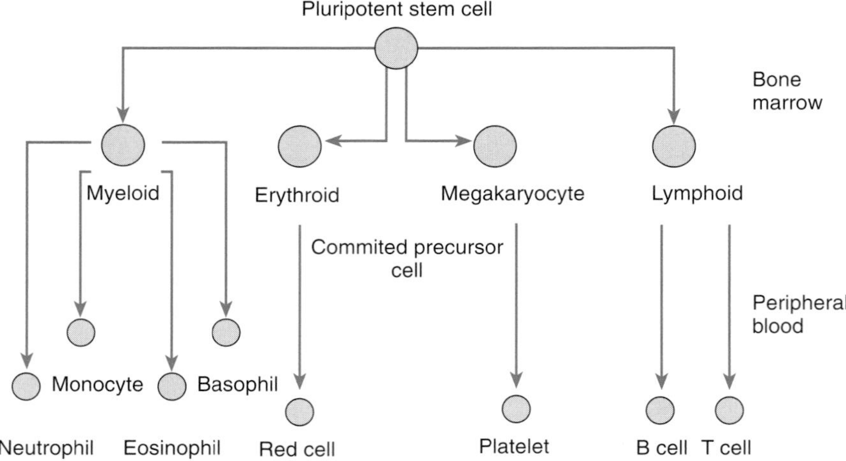

Fig. 1 Outline of hematopoietic development.

intermediate, and late stages of cell differentiation; target cell responsiveness depends upon both cytokine production and expression of specific receptors on the target cells. Early-acting cytokines include stem cell factor and interleukin 3, while cytokines acting on more differentiated cells include granulocyte colony-stimulating factor, granulocyte-macrophage colony-stimulating factor, thrombopoietin, the recently described platelet growth factor, and erythropoietin, produced in the kidney and controlling erythropoiesis. Availability of human recombinant forms of these proteins has resulted in their usage in the treatment of anemic and neutropenic states under some circumstances. There are also inhibitory cytokines (e.g. tumor necrosis factor, transforming growth factor, and, probably, interleukin 2) which act in a negative regulatory manner to inhibit hematopoiesis. These are implicated in pathological conditions resulting in the suppression of hematopoiesis, in particular the anemia of chronic disorder.

Erythrocytes are the most common circulating hemopoetic cells. The precursor forms within the marrow are nucleated but, when released into the peripheral blood, they no longer contain a nucleus. Red cells have a finite lifespan of around 120 days, at which time they are destroyed primarily in the reticuloendothelial system. Red cells recently released from the bone marrow contain trace amounts of RNA that can be stained with supravital stains and recognized as reticulocytes. The reticulocyte count can give a vital insight into the synthetic function of the marrow. Red cell numbers can be decreased by loss from the body (hemorrhage), failure of synthesis due to primary marrow disease or deficiency of iron, vitamin B_{12}, and folate, depressed erythropoiesis in the anemia of chronic disease, and premature red cell destruction (hemolysis). Measurement of hemoglobin concentration depends not only on the number of red cells present and their individual hemoglobin content, but also on the plasma volume in which they are distributed. A diminished plasma volume will spuriously elevate hemoglobin, while depletion of red cells and hemoglobin will be masked if there is concomitant loss of plasma volume. Consequently, measurement of plasma volume and red cell mass using isotopic techniques can sometimes be highly informative.

White cells divide into five groups, namely neutrophils, lymphocytes, monocytes, eosinophils, and basophils. Neutrophils, which form the majority, function primarily as a defense mechanism against pathogenic bacteria and have a relatively short half-life of about 6 h. A significant proportion are in a marginating rather than a circulating pool which can be rapidly mobilized, leading to the rapid development of a neutrophil leukocytosis following physiological stress, infection, trauma, burns, etc. Lymphocytes can be divided into B, T, and natural killer cells. B cells are particularly important in the humoral immune response, while T cells are primarily involved in the cellular immune response. It is becoming increasingly clear that B and T cells must act together for efficient functioning of the body's immune system.

Platelets are highly organized and structured particles of megakaryocyte cytoplasm. They contain no nuclei but express specific surface glycoproteins allowing them to bind to adhesive endothelial, subendothelial, and plasma proteins. They can undergo adhesion, aggregation, and subsequently a release reaction which liberates the contents of granules contained within their cytoplasm, resulting in augmentation of the aggregation reaction. Activated platelets also provide a negatively charged phospholipid surface upon which proteins of the coagulation cascade can assemble. Although they are individually very small, aggregated platelets form a platelet plug which can block vascular defects and precipitate a stabilizing fibrin mesh over the top. To perform this task efficiently, platelets need to be present in sufficient numbers and have normal function. Normal hemostasis, assuming normal platelet function, is maintained with platelet counts above 80×10^9/l; diminution below this level results in a progressively prolonged bleeding time. At levels below 20×10^9/l there is a much heightened risk of spontaneous hemorrhage.

The requirement for blood to remain fluid while confined to the intravascular compartment, but to clot rapidly when the vessel is damaged or breached, and the need for this response to be rapid, localized, and self-limiting has resulted in the evolution of a complex hemostatic mechanism encompassing the coagulation and fibrinolytic cascades, platelets, the natural anticoagulant pathway, and endothelial cells themselves (Table 2). The coagulation cascade consists of a stepwise series of reactions in which circulating inactive zymogen serine proteases are sequentially activated by proteolytic cleavage, resulting in the formation of active enzymes (Fig. 2). These multiple steps allow both biological amplification and multiple points of control. Although the coagulation cascade was historically divided

Table 2 Normal coagulation indices

APTT	27.4–40.3 s
Prothrombin time	12.3–16.1 s
Thrombin time	Control±3 s
Fibrinogen	1.7–3.1 g/l
D-dimer	<0.25 µg/l
Bleeding time	<8 min
Coagulation factors II, V, VII, VIII, IX, X, XI, XII	50%–150%
Antithrombin activity	80%–120%
Protein C activity	73%–121%
Protein S	
Total	55%–125%
Free	21%–53%
Activated protein C resistance ratio	>2.0
von Willebrand factor antigen	50%–150%
von Willebrand factor ristocetin cofactor activity	50%–200%

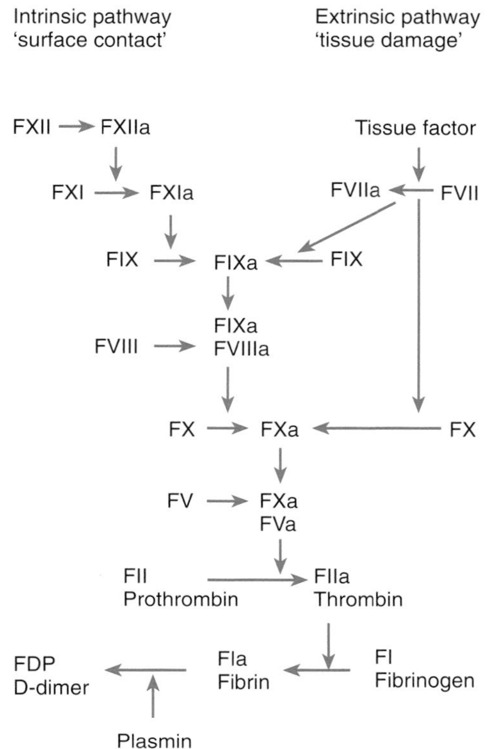

Fig. 2 The coagulation cascade.

into intrinsic and extrinsic pathways, it is clear that, under the vast majority of physiological circumstances, coagulation is activated by tissue factor exposure and hence proceeds down the extrinsic pathway. However, the tissue factor–factor VIIa complex can activate both factor X and factor IX, which has resulted in an extension of the final common pathway. The physiological role of the intrinsic pathway is unclear; in addition to initiating coagulation, activation of factor XI and factor XII also results in activation of the fibrinolytic cascade and the complement system, and the generation of vasoactive inflammatory proteins such as bradykinin. The purpose of the coagulation cascade is to generate the enzyme thrombin which cleaves soluble fibrinogen to form insoluble fibrin. Thrombin is also able to initiate a potent positive feedback pathway whereby thrombin activates factors V and VIII into their activated form; these increase the catalytic efficiency of the generation of factors Xa and IIa generation by over 1000-fold. Thrombin is also a potent initiator of platelet aggregation. Uncontrolled generation of thrombin is the hallmark of disseminated intravascular coagulation; thus thrombin generation is tightly regulated by the natural anticoagulant pathway (Fig. 3). This consists of antithrombin, a potent and rapid inhibitor of free thrombin which acts as a suicide substrate, rapidly neutralizing thrombin by forming thrombin–antithrombin complexes in which thrombin is devoid of proteolytic activity. In addition, free thrombin is bound by the protein thrombomodulin, which is expressed on the endothelial cell surface and alters the substrate specificity of thrombin. As a result, it is no longer able to clear fibrinogen or activate factors V or VIII but is now able to activate protein C. Activated protein C, with its cofactor protein S, degrades activated factors Va and VIIIa. Consequently, thrombin bound to thrombomodulin acts as an anticoagulant protein and initiates a feedback pathway to limit the extent of coagulation. Defects in the natural anticoagulant pathway, in the form of deficiencies of antithrombin, protein C, and protein S, are associated with a predisposition to excessive venous thrombosis known as thrombophilia. Acquired deficiencies of these proteins are extremely common in the critical care setting.

The fibrinolytic cascade is analogous to the coagulation cascade but generates the enzyme plasmin rather than thrombin. Plasmin degrades rather than generates fibrin. The fibrinolytic system is also controlled by specific inhibitors as well as feedforward and feedback pathways. Although congenital abnormalities of the fibrinolytic pathway proteins are uncommon, fibrinolytic therapy for the dissolution of thrombi in coronary and cerebral arteries and within the deep veins of the legs and pulmonary vasculature has proved effective.

Historically, endothelial cells were considered biologically neutral, lining blood vessels without apparent effect on hemostasis. It is now clear that endothelial cells are, under normal circumstances, actively anticoagulant. They produce nitric oxide and prostacyclin, resulting in vasodilatation and inhibition of platelet aggregation, express thrombomodulin to initiate the natural anticoagulant pathway on its surface, express glycosaminoglycans and heparinoids to inhibit the coagulation cascade, and produce tissue plasminogen activator to initiate fibrinolysis. However, endothelial cells which have become activated, for example by a bacterial endotoxin, no longer produce nitric oxide and prostacyclin, express tissue factor on their surface to initiate the coagulation cascade, downregulate thrombomodulin expression, and produce inhibitors of the fibrinolytic cascade rather than activators. Adhesion molecules such as von Willebrand factor and selectins are also released or expressed, resulting in an actively procoagulant endothelial cell surface.

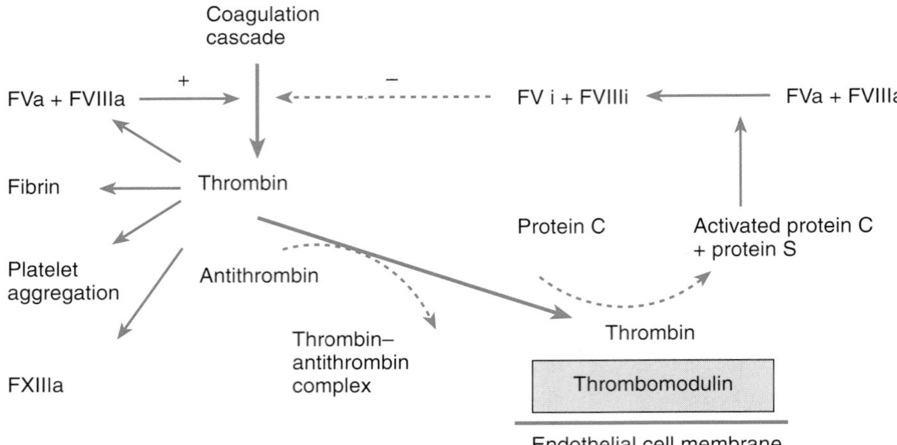

Fig. 3 The natural anticoagulant pathway.

The anticoagulant heparin acts by potentiating the thrombin-inhibiting activity of antithrombin. Unfractionated heparin prolongs the thrombin time and the activated partial thromboplastin time (APTT), by which it is monitored, but recently introduced low-molecular-weight heparins do not significantly affect APTT as they have more activity against factor Xa than thrombin. Consequently, monitoring, if necessary, requires an anti-factor Xa assay. The orally acting anticoagulant warfarin inhibits post-translational carboxylation of glutamic acid residues on coagulation factors II, VII, IX, and X, and also on proteins C and S, resulting in the synthesis of inactive forms of these proteins. Warfarin has a long half-life and complex pharmacokinetics with multiple drug interactions. A reversal of the warfarin effect may require use of fresh frozen plasma, parenteral vitamin K, or, in life-threatening situations, specific coagulation factor concentrates.

Thromboprophylaxis in patients immobilized for prolonged periods can be achieved with low doses of both unfractionated and low-molecular-weight heparins. Neither require monitoring at these doses. Thrombosis is more common in the presence of low levels of antithrombin, protein C, and protein S which can be either congenital or acquired. Acquired deficiency is extremely common in the critical care situation, where these proteins are rapidly depleted as a response to increased coagulation activation and synthesis is frequently impaired due to compromised hepatic function.

Although it may not be possible to normalize hemostatic abnormalities completely in the critically ill, bleeding will be reduced if hematocrit can be maintained above 30 per cent, platelet count above 80×10^9/l, fibrinogen above 1 g/l, and APTT and prothrombin time no more than 4 to 5 s above the upper limit of normal. Under these circumstances it is usually safe to perform invasive procedures including placement of central venous and arterial lines. Excessive bleeding under these conditions is usually due to either failure of local hemostasis (e.g. postoperatively) or the development of disseminated hemostatic failure.

Bibliography

Baglin, T. (1996). Disseminated intravascular coagulation: diagnosis and treatment. *British Medical Journal*, **312**, 683–7.

Chong, B.H. (1995). Heparin-induced thrombocytopenia. *British Journal of Haematology*, **89**, 431–87.

Hillarp, A., Dahlback, B., and Zoller, B. (1995). Activated protein C resistance: from phenotype to genotype and clinical practice. *Blood Reviews*, **9**, 201–12.

Meade, T.W. (ed.) (1994). Thrombophilia. *Baillière's Clinical Haematology*, **7**, 441–756.

Warkentin, T.E. and Kelton, J.G. (1994). Management of thrombocytopenia. In *Hemostasis and thrombosis: basic principles and clinical practice* (3rd edn) (ed. R.W. Colman, J. Hirsh, V.J. Mader, and E.W. Salzman), pp. 469–83. J.B. Lippincott, Philadelphia, PA.

10.2 Disordered coagulation

10.2.1 Pathophysiology

A. Nathwani and S. J. Machin

Key messages

- The hemostatic system is carefully balanced and controlled so that inappropriate bleeding or thrombosis does not occur.
- Normal hemostasis is an interaction between circulating platelets, coagulation factors, the vascular endothelial cell, the fibrinolytic system, and various inhibitory pathways.
- Acute disseminated intravascular coagulation is a major breakdown in the hemostatic process mediated mainly by excessive tissue factor expression and cytokine generation.
- Rapid assessment of the hemostatic process and appropriate therapeutic intervention is dependent on a thorough understanding of these various pathways.

Introduction

Hemostasis is a remarkable achievement of evolution which reconciles two conflicting requirements, namely that blood should clot extremely rapidly in order to prevent exsanguination and yet preserve liquidity in the circulation. It requires a highly complex, and intricate balance between the vascular endothelium, the coagulation cascade, anticoagulant mechanisms, the fibrinolytic system, platelets, leukocytes, cytokines, and blood flow dynamics.

In the critically ill patient, abnormalities of hemostasis are common and relate to the deregulation of host defense homeostatic mechanisms by trauma, infection, surgery, drugs, cytokines, and instability of the cardiorespiratory axis. Therefore the critically ill patient can present with either bleeding or thrombosis and in some cases with both.

The normal hemostatic pathway

When a blood vessel is injured, an immediate vasoconstriction of the injured vessel and adjacent small arteries and arterioles is responsible for an initial slowing of blood flow to the areas of injury. A series of interactions then results in the formation of a platelet plug which is rapidly stabilized by cross-linking fibrin strands. These procoagulant processes are limited and localized by a number of control mechanisms. It is convenient to divide them into five functional components: platelet function, coagulation cascade, natural anticoagulants, fibrinolysis, and endothelium.

Platelet function

Following trauma, a break in the endothelial lining exposes platelets to the subendothelial connective tissue. This leads to a series of reactions which include the following:

(1) adherence of platelets to the exposed collagen, a process that is potentiated by circulating von Willebrand factor;

(2) release of platelet granules which contain ADP, serotonin, fibrinogen, lysosomal enzymes, and platelet factor 4;

(3) stimulation of prostaglandin synthesis leading to the formation of thromboxane A_2 (TXA_2).

The released enzymes, ADP, and TXA_2 all promote platelet aggregation and the formation of a primary platelet plug (primary phase). Apart from temporarily sealing the breach in the endothelial lining, the platelet plug provides a surface on which procoagulant reactions can take place. Quantitative and qualitative platelet abnormalities are common in the critically ill (Table 1) and are manifest by symptoms of bleeding or bruising. Bleeding times are often used to assess platelet function.

Coagulation cascade

The coagulation cascade is responsible for the formation of a stable fibrin clot. It consists of an integrated cascade of enzyme reactions which amplifies a relatively small stimulus to produce a large response. It is now widely accepted that initiation of the cascade occurs when circulating coagulation factor VII is exposed to tissue factor expressed by the subendothelial tissues at the site of injury. This results in the formation of a tissue factor–factor VIIa complex which, through its actions on factors IX and X, leads to the generation of thrombin (Fig. 1). Thrombin cleaves fibrinogen to fibrin, and activates factor XIII and platelets. As well as enhancing clot formation, thrombin also limits coagulation by activating protein C through its interaction with thrombomodulin. Screening tests, such as the prothombin time (PT), activated partial thromboplastin time, and thrombin time, can be used to assess the function of the extrinsic, intrinsic, and thrombin–fibrinogen pathways respectively. Congenital or acquired defects of the major coagulation factors can result in hemorrhage in the brain, gastrointestinal tract, muscles, joints, and soft tissues.

Table 1 Hemostatic abnormalities in critically ill patients

Disease	Hemostasis
Liver disease	All clotting factors decreased except factor VIII and vWF Decreased ATIII, PC, PS Decreased platelet count Platelet dysfunction Dysfibrinogenemia Increased fibrinolysis (rare) DIC (in patients with acute hepatic failure)
Renal failure	Major abnormality is platelet dysfunction Decreased clotting factors rare in the absence of uremic enteritis, liver disease, or DIC Decreased ATIII deficiency in nephrotic syndrome
Cardiopulmonary bypass surgery	Platelet function defect (the most common and relates to duration of surgery) Decreased fibrinogen Decreased factors FII, V, VII, XI, X DIC Overheparinization is common; this causes increased APTT but bleeding is rare
Head injury	Increased FDPs DIC
Massive transfusion	Decreased factors V and VIII DIC
Warfarin	Decreased factors II, VII, IX, X Decreased PS, PC
Heparin	Inhibits action of Xa Thrombocytopenia (rare)
Thrombolytic agents	Increased plasmin, FDPs Decreased fibrinogen
DIC	Decreased factors V, VIII, IX, XI Decreased fibrinogen Decreased PC, PS, ATIII Decreased platelets Increased FDPs

vWF, von Willebrand factor; ATIII, antithrombin III; PC, protein C; PS, protein S; APTT, activated partial thromboplastin time; FDP, fibrin degradation product.

Natural anticoagulation mechanisms

There are two major anticoagulation pathways based around antithrombin III, a serine protease, and the vitamin-K-dependent protein C–protein S pathway. Antithrombin III is a potent inhibitor of thrombin and factor Xa, and can also inactivate factors IXa, XIa, and XIIa. It action is greatly potentiated by heparin or heparan sulfate and represents the mechanism by which heparin exerts its anticoagulant effect. Thrombin activates protein C through its interaction with thrombomodulin, a receptor on the vascular endothelium. Activated protein C in conjunction with protein S exerts its anticoagulant activity by enzymatic cleavage of factors Va and VIIIa which are important cofactors in the procoagulant pathway. Protein C also enhances the fibrinolytic system by protecting tissue plasminogen activator from inhibitors, resulting in enhanced conversion of plasminogen to plasmin. Specific antigenic and functional assays have been designed to detect abnormalities of these anticoagulant proteins. Protein C may play an important role in the pathogenesis of disseminated intravascular coagulation (DIC). Quantitative and functional defects of antithrombin III, protein C, and protein S are associated with a high incidence of thrombosis.

Fibrinolytic system

Fibrinolysis (like coagulation) is a normal hemostatic response to vascular injury. Plasminogen, a β-globulin proenzyme in blood and tissue fluid, is converted to plasmin by tissue plasminogen activator released from endothelium by stimuli such as thrombin, trauma, exercise, and venous stasis. Once released, tissue plasminogen activator binds to fibrin and converts thrombus-bound plasminogen into plasmin, thus localizing fibrinolysis to the blood clot. Plasmin splits both fibrin and fibrinogen, producing a variety of fibrin degradation products. The activity of plasmin is balanced by several physiological inhibitors, the principal one being α_2-antiplasmin. Two common therapeutic inhibitors of plasmin are ε-aminocaproic acid and tranexamic acid. In contrast, the fibrinolytic agent streptokinase forms complexes with plasminogen and enhances plasmin production.

Endothelium

Under normal conditions endothelial cells form a metabolically active antithrombotic interface between blood and tissue by synthesizing numerous anticoagulant molecules such as glycosaminoglycans, thrombomodulin, nitric oxide, and prostacyclin. Following a hemostatic challenge, endothelial cells can substantially modify their behavior. Acute responses, epitomized by reactions to thrombin formation, include the rapid synthesis and secretion of mediators designed to assist in localizing and limiting the formation of a hemostatic plug (e.g. nitric oxide, prostacyclin). A series of slower-onset alterations in endothelial function take place, particularly on exposure to cytokines or bacterial endotoxins. These changes, which include expression of tissue factor, tend to render the endothelial surface more procoagulant. Such changes may be valuable when they occur locally, but could be detrimental if they take place systematically. This has led to the concept that therapy directed at antagonizing the effects of cytokines on the endothelium could be valuable in the critically ill patient.

Abnormalities of hemostasis in the critically ill patient

The most common abnormality of hemostasis in the critically ill patient is DIC. This is an acquired condition which is commonly associated with sepsis, extensive surgery, hemorrhage, acute cardiovascular insult, malignancy, and obstetric complications. The clinical consequences of DIC are a potentially devastating consumptive thrombohemorrhagic disorder. Recent advances suggest a key role for cytokines such as tumor necrosis factor and interleukin 1 (IL-1) in the pathogenesis of DIC. These cytokines are expressed

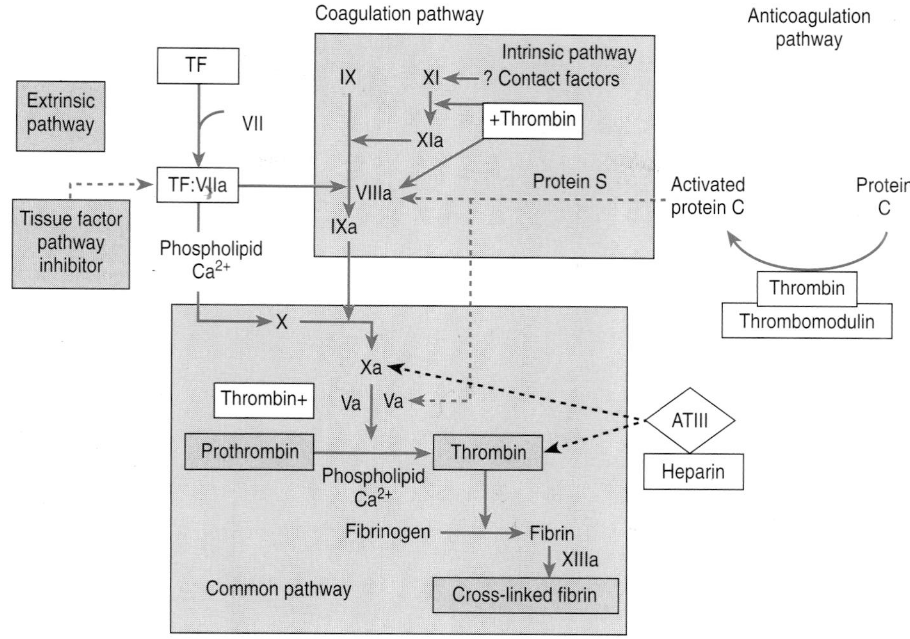

Fig. 1 The coagulation and anticoagulation pathways: TF, tissue factor; TFPI, tissue factor pathway inhibitor. Coagulation is initiated when factors VII or VIIa in plasma come into contact with tissue factor at the site of vascular injury. Limited quantities of factors IXa and Xa are generated before there is feedback inhibition of tissue factor–factor VIIa complex by tissue factor pathway inhibitor in concert with factor Xa. The generation of factor Xa is then amplified through the action of factor VIIIa on factor IXa; the latter is initially produced by tissue factor–factor VIIa complex and supplemented by factor XIa. The mechanism(s) for *in vivo* factor XI activation remain to be established. The broken lines indicate inhibitory pathways.

very early following an acute insult and have the ability to influence a variety of cellular processes. Two hemostatic pathways appear to be particularly prone to modulation by these cytokines. One is the tissue-factor-mediated pathway of coagulation. Because tissue factor is a potent initiator of coagulation, its expression is highly regulated. It is constitutively expressed in cells corresponding to biological boundary layers such as the skin epidermis, adventitia around blood vessels, mucosal epithelium lining the gut, and capsules surrounding organs. Thus tissue factor forms a 'hemostatic envelope' encasing blood vessels, organ structures, and the entire organism. Cells which are normally in direct contact with blood (e.g. leukocytes) do not express tissue factor. However, *in vitro* studies show that monocytes and endothelial cells can express tissue factor following exposure to bacterial endotoxins, tumor necrosis factor, IL-1, activated complement, and immune complexes. These inflammatory mediators further exacerbate the procoagulant state by downregulating thrombomodulin expression on the endothelium, thereby inactivating the protein C anticoagulant pathway. Such a procoagulant state is likely to overwhelm other control mechanisms and is believed to be the basis of DIC. There are *in vivo* data to support this theory. Monocytes recovered from the blood and other anatomical sites (e.g. peritoneum) of rabbits treated with endotoxin express strong tissue factor activity. In baboons, injection of endotoxins results in fatal DIC which can be ameliorated by prior infusion of an anti-tissue-factor antibody. Specific blockers of tumor necrosis factor and IL-1 also diminish mortality in experimental endotoxic shock models. In humans, elevation of monocyte tissue factor expression is associated with septicemia, malignancies, and obstructive jaundice; these are all conditions with a high incidence of DIC. Patients suffering from meningococcal septicemia have an acquired protein C deficiency. The prognosis in these patients is generally poor, but infusion of protein C concentrates has led to a rapid perceived improvement in the patient's condition. Another mechanism which downregulates the protein C pathway is related to the fact that protein S forms a reversible inactive complex with C4b binding protein. C4b binding protein is an acute phase reactant whose levels increase during inflammation resulting in a decrease in active protein S.

The laboratory abnormalities found in DIC and other disorders commonly seen in the intensive care unit are summarized in Table 1.

Bibliography

Mannucci, P.M. and Giangrande, P.L.F. (1994). Acquired disorders of coagulation. In *Haemostasis and thrombosis*, Vol. 2 (ed. A.L. Bloom, C.D. Forbes, D.P. Thomas, and E.G.D. Tuddenham), pp. 949–68. Churchill Livingstone, Edinburgh.

Staudinger, T., Locker, G.J., and Frass, M. (1996). Management of acquired coagulation disorders in emergency and intensive care medicine. *Seminars in Thrombosis and Hemostasis*, **22**, 93–104.

10.2.2 Disseminated intravascular coagulation

D. Cummins

Key messages

- Most cases of disseminated intravascular coagulation (DIC) in the intensive care setting are due to sepsis.
- DIC typically manifests as hemorrhage, thrombosis, and/or a typical pattern of abnormalities in laboratory coagulation tests.
- DIC may need to be differentiated from liver disease, vitamin K deficiency, thrombotic thrombocytopenic purpura, and certain postoperative states.
- The essentials of management include the identification and treatment of the underlying cause, the correction of hypoxia, and the maintenance of adequate tissue perfusion.
- Replacement of platelets and/or coagulation factors should be reserved for patients who are bleeding and those who are to undergo surgical intervention.

Introduction

Disseminated intravascular coagulation (DIC) is a pathological process involving global activation of hemostasis and the formation of soluble or insoluble fibrin within the circulation. It has many causes, but sepsis probably underlies at least half of all cases seen in the critical care setting. DIC may manifest as hemorrhage (90 per cent), widespread microvascular thrombosis with multiorgan dysfunction, or an incidental finding in laboratory coagulation tests. Its management is controversial, in part because of the large number of conditions with which it is associated (Table 1) and a dearth of prospective clinical trials. A rational approach to therapy requires some understanding of its pathogenesis (Fig. 1).

Table 1 Some clinical disorders that may be complicated by DIC

Infections	Bacterial Viral Other (fungal, rickettsial)
Obstetric conditions	Abruptio placentae Retained dead fetus syndrome Amniotic fluid embolism Fulminating pre-eclampsia
Trauma	Surgery Burns Heat stroke Head trauma Crush injuries Fat embolism
Shock	Hypovolemia from any cause
Malignancy	Metastatic tumors Mucin-secreting adenocarcinomas
Liver disease	Cirrhosis Acute hepatic necrosis Cholestasis
Immunological	ABO-incompatible blood transfusion
Important world-wide causes	Falciparum malaria Snake venoms

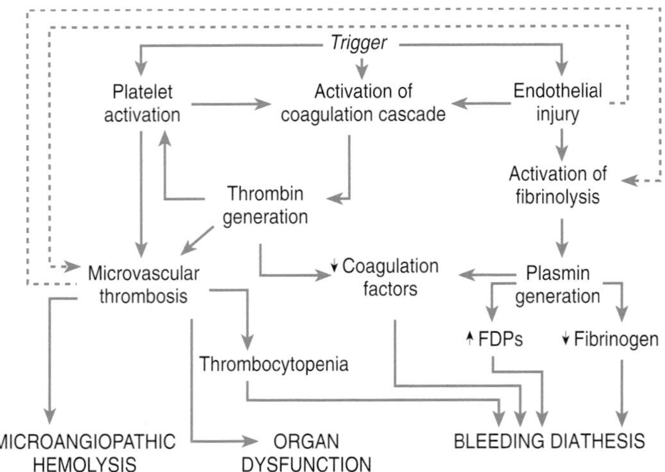

Fig. 1 The pathophysiology of DIC.

Pathogenesis

DIC involves the activation of hemostasis by a powerful or persistent 'trigger' and the subsequent interaction of several physiological and pathophysiological processes. Possible triggers include bacterial endotoxin, tissue factor, shock, burns, immune complexes, snake venoms, and substances released by malignant cells. The most important mode of hemostatic activation is believed to be via the extrinsic coagulation pathway by tissue factor released from damaged endothelium and/or circulating leukocytes. Hemostasis may also be activated via the intrinsic pathway, via a part of the coagulation cascade that bypasses the normal activation mechanisms, or via a direct effect on platelets or endothelium (Fig. 1). These effects in turn may be potentiated by other factors, particularly complement activation, the release of cytokines such as tumor necrosis factor and interleukin 1, and reticuloendothelial blockade.

The result of these processes is the release of free thrombin which leads to widespread deposition of fibrin within the microvasculature and, as part of a secondary fibrinolytic response, the generation of plasmin and other proteolytic enzymes. Despite its compensatory purpose, plasmin contributes adversely to hemostatic integrity by degrading coagulation factors and by cleaving fibrinogen and fibrin to fibrin(ogen) degradation products, substances which directly inhibit fibrin clot formation and platelet function. Obstruction of the microcirculation by platelet aggregates and fibrin strands leads to thrombocytopenia and, in about 50 per cent of cases, microangiopathic hemolysis.

Conceptually, acute DIC can be regarded as having two phases: an initial or 'compensated' phase, and a subsequent phase of decompensation. In at least some forms of DIC, consumption of the physiological inhibitors antithrombin III and protein C may underlie transition from the first to the second phase. This may have implications for therapy. Thus the overall picture is characterized by bleeding from loss of coagulation factors, thrombocytopenia, and platelet dysfunction, and a clotting tendency manifest by widespread microvascular thrombosis. The underlying disease, the strength and nature of the trigger stimulus, and the relative contributions of coagulation, fibrinolysis, and their potentiators all play a role in determining the clinical presentation and eventual outcome.

Etiologies

Most cases of acute DIC in the intensive care unit setting are due to sepsis, particularly involving Gram-negative organisms. One of the earliest and most severe forms described occurs with meningococcal septicemia, in which widespread bleeding and shock are associated with bilateral adrenal hemorrhage and a grave prognosis (Waterhouse–Friederichsen syndrome). DIC has since been linked with a wide variety of Gram-positive and Gram-negative infections. Viruses, particularly varicella, cytomegalovirus, the hepatitis viruses, and even HIV, occasionally cause DIC, possibly via immune-complex generation.

DIC may complicate a variety of obstetric conditions (Table 1) in which it develops against the background of the prothrombotic state of pregnancy. In amniotic fluid embolism and placental abruption, release of placental thromboplastins into the maternal circulation is believed to be important in the pathogenesis. Amniotic fluid embolism is manifest by an acute onset of respiratory failure and hypotension followed by coagulopathy and hemorrhagic shock; although rare (about 1 in 50 000 deliveries), the fatality rate is about 80 per cent. In abruptio placentae, the hemostatic disturbance is related to both the shock state and the degree of placental separation; in cases involving large amounts of concealed hemorrhage and fetal death, DIC is invariable and often severe. In contrast, DIC complicating the retained fetus syndrome is typically of a slow progressive type; bleeding is a late feature and is preceded by declining renal function.

DIC may follow various types of trauma. Specific examples include burns, heat stroke, open head wounds, and crush injuries, although hypovolemia of any cause may trigger its development. Acidosis, dehydration, and hypoxia are important contributory factors.

DIC is a recognized complication of hepatic cirrhosis, acute massive hepatic necrosis, and conditions associated with protracted (more than 5 days) intrahepatic or extrahepatic cholestasis. The pathogenesis may involve release of tissue thromboplastins and impaired hepatic clearance of fibrin degradation products and activated coagulation factors.

Other disorders, usually seen outside the intensive care setting, are also associated with DIC. Cancer, particularly involving metastatic and/or mucus-secreting tumors is an important cause; the DIC in such cases is typically of a chronic compensated type. DIC was long believed to be a common complication of acute promyelocytic leukemia; recent work suggests that bleeding is more often due to primary fibrinolysis. Worldwide, important causes of DIC include falciparum malaria and snake venoms. DIC complicating the latter often entails direct activation of specific coagulation factors and the generation of non-coagulable blood.

Diagnosis

DIC should be suspected in patients who develop hemorrhage, thrombosis, and/or a typical pattern of abnormalities in laboratory coagulation tests, particularly if they have one of the disorders listed in Table 1. The most frequent overt manifestation of DIC is bleeding, which may take the form of petechiae, mucosal oozing, wound hematomas, spontaneous soft tissue bruising, and bleeding from venepuncture sites and intra-arterial lines. DIC should always be suspected when hemorrhage occurs at multiple sites. Thrombosis usually involves small vessels and manifests principally as renal, pulmonary, and/or neurological dysfunction. However, large vessels also may be affected, particularly in patients with circulatory failure. Thrombosis may occur in central venous catheters.

The blood film contains red cell fragments, microspherocytes, and other signs of microangiopathic hemolysis. Laboratory evaluation of DIC should also include a platelet count, prothrombin time, partial thromboplastin time, and fibrinogen concentration. Measurement of serum fibrin degradation products is sometimes useful. In most cases there is prolongation of the prothrombin and partial thromboplastin times, hypofibrinogenemia, thrombocytopenia, and elevated fibrin degradation products; however, the pattern of changes is extremely variable and none of these tests, either alone or in combination, is specific. In recent years, an assay has become available for D-dimer, a neo-antigen formed when plasmin digests cross-linked fibrin. The assay allows distinction between fibrinolysis and fibrinogenolysis and thus may be a more reliable indicator of DIC than the traditional assay. Its main advantage is that it can be performed on routine plasma samples rather than on specially prepared serum samples.

Other more sophisticated tests which may have a role include assays of the physiological inhibitors antithrombin III and protein C, markers of platelet activation, such as platelet factor 4 and β-thromboglobulin, and measures of increased hemostatic activity such as thrombin–antithrombin III and plasmin–α_2-antiplasmin complexes. Assays of antithrombin III and/or protein C may be of value if their specific replacement is to be considered. The other tests are not widely available and do not predict the need for therapy; however, they may help to identify subtle degrees of hemostatic activation in patients with subclinical DIC.

Differential diagnosis

DIC needs to be differentiated from various other medical conditions. Prolongation of the prothrombin and partial thromboplastin times, thrombocytopenia, hypofibrinogenemia, and a raised level of fibrin degradation products all may occur in liver disease; in difficult cases, assay of factor VIIIc may be helpful (low in DIC but normal or raised in liver disease). Critically ill patients may develop a coagulopathy due to vitamin K deficiency; a therapeutic trial of the vitamin may clarify whether such a deficiency is present. Widespread thromboses, multiorgan dysfunction, thrombocytopenia, and microangiopathic hemolysis all occur in thrombotic thrombocytopenic purpura; however, major coagulation abnormalities are uncommon in this disorder. Finally, a raised level of fibrin degradation products may be found after major thrombosis, thrombolytic therapy, and major surgery, although this seldom creates major diagnostic difficulty.

Management

The aim in managing patients with DIC is to gain time until the underlying disease process can be brought under control. The precipitating cause should be identified and appropriate therapy instigated without delay. Intravenous broad-spectrum antibiotics should be given in septicemia and intravascular volume replaced in oligemic shock. In patients with abruptio placentae or the retained fetus syndrome, prompt evacuation of the uterus may be lifesaving. Irrespective of the underlying cause, general supportive measures are of prime importance, particularly correction of hypoxia and the maintenance of adequate tissue perfusion.

Replacement therapy should be given to patients who are bleeding or who are to undergo a surgical procedure. Patients with significant prolongation of prothrombin or partial thromboplastin times (more than 1.5 times control) should receive fresh frozen plasma (15 ml/kg), which contains all the major coagulation proteins. Clinically important hypofibrinogenemia (< 0.8 g/l) warrants fibrinogen replacement with either cryoprecipitate or a fibrinogen concentrate. Use of cryoprecipitate (0.2 bags/kg) exposes the patient to multiple donors and carries a small but significant risk of viral transmission; fibrinogen concentrates are safer but expensive. Significant thrombocytopenia ($< 50 \times 10^9$/l) represents an indication for platelet transfusions (1 IU/10 kg body weight). The response should be monitored by repeating the platelet count and coagulation profile immediately after the replacement therapy and then every 4 to 6 h.

There is increasing interest in the potential role of antithrombin III and protein C concentrates. Antithrombin III is rapidly consumed in DIC, and the magnitude of the initial decrease in its level has prognostic significance. Experimental and clinical studies suggest that antithrombin III concentrate may have a role in selected cases, although confirmation will require a large multicenter trial. Protein C concentrate has been shown to prevent DIC in an animal model, but its role in humans remains to be established.

Use of heparin in DIC is controversial. It may have a role in cases resulting from direct activation of the coagulation cascade, such as amniotic fluid embolism and ABO-incompatible blood transfusion, and in patients who develop large-vessel thrombosis. A low-dose continuous infusion (500–1000 IU/h) has been advocated. In patients with major hemorrhage, heparin therapy is potentially hazardous and should generally be considered only when adequate replacement therapy has failed to control the bleeding.

Bibliography

Anonymous (1991). Transfusion for massive blood loss. In *Standard haematology practice (The British Committee for Standards in Haematology)* (ed. B. Roberts), pp. 198–206. Blackwell Scientific, Edinburgh.

Anonymous (1994). Disseminated intravascular coagulation. Objective laboratory diagnostic criteria and guidelines for management. *Clinics in Laboratory Medicine*, **14**, 729–68.

Bloom, A.L. and Thomas, D.P. (ed.) (1987). *Haemostasis and thrombosis* (2nd edn). Churchill Livingstone, London.

Fourrier, F., Chopin, C., Huart, J.-J., Runge, I., Caron, C., and Goudemand, J. (1993). Double-blind, placebo-controlled trial of antithrombin III concentrates in septic shock with disseminated intravascular coagulation. *Chest*, **104**, 882–8.

Gilbert, J.A. and Scalzi, R.P. (1993). Disseminated intravascular coagulation. *Emergency Medicine Clinics of North America*, **11**, 465–80.

10.2.3 Thrombosis

H. Cohen and C. Ozanne

Key messages

- Diagnosis of venous thromboembolism in the intensive care unit requires a high index of clinical suspicion.
- Anticoagulate initially with closely monitored intravenous heparin by continuous infusion.
- Monitor the platelet count of patients receiving any form of heparin for early evidence of heparin-induced thrombocytopenia.
- Consider thrombolysis and caval filters.
- Consider underlying medical and hematological causes of thromboembolic disease (including rare disorders).
- Treat thrombotic thrombocytopenic purpura early with intensive plasma exchange and fresh frozen plasma replacement.

Introduction

Thromboembolic disease in the intensive care unit (ICU) patient presents a complex diagnostic and therapeutic challenge. Many of these severely ill patients have a degree of hemostatic failure, usually secondary to hepatic and/or renal dysfunction, which complicates the management of the underlying thrombotic problem. Optimal management depends upon careful clinical assessment and appropriate laboratory testing, and requires close liaison between the ICU and the hematology department.

Venous thromboembolism

Thromboprophylaxis for the ICU patient

Critically ill patients frequently have several coexistent risk factors for thromboembolic disease, in particular prolonged immobility, surgery, and indwelling vascular access catheters. Therefore they may benefit from thromboprophylaxis. However, they may also be at increased risk of hemorrhage due to recent surgery, hemostatic dysfunction secondary to renal failure (which also reduces heparin clearance), severe hepatic disease, or thrombocytopenia. These factors are relative contraindications to anticoagulation. As these patients form a highly heterogeneous group, the risk–benefit ratio will vary from patient to patient. There is a paucity of data on the risks and benefits of thromboprophylaxis or anticoagulant regimens in this group of patients.

Subcutaneous heparin, either unfractionated or low molecular weight, is the most common thromboprophylaxis used. Compared with unfractionated heparin at standard dosage, low-molecular-weight heparin has been shown to have increased efficacy in the high-risk situation of orthopedic surgery. Its disadvantages in critically ill patients are the need for subcutaneous absorption, which may be impaired due to poor peripheral perfusion, and the lack of easy reversibility due to the long plasma half-lives of these drugs. Low-molecular-weight heparin can be monitored to allow more precise control using an anti-activated factor X (anti-factor Xa) plasma level. However, this would not be feasible on a frequent basis in most institutions. A logical approach may be the administration of continuous low-dose intravenous unfractionated heparin at an initial dose of 500 IU/h. This has several advantages in the ICU setting; it is easily monitored using routine laboratory tests and, as it has a short plasma half-life, it can be stopped shortly before surgical procedures or the insertion of indwelling catheters. This regimen should be monitored, aiming to ensure that the activated partial thromboplastin time ratio remains less than 1.5.

Management of venous thromboembolism in the ICU

In the ICU patient both deep vein thrombosis and pulmonary embolism are likely to be silent and may therefore go unrecognized. Clinical suspicion of venous thromboembolism merits urgent consideration of therapeutic anticoagulation with intravenous heparin. Objective testing for thromboembolic disease, although necessary, may be logistically problematic and therefore restricted, limiting precise definition of the incidence of thromboembolic events. Prior to the commencement of anticoagulation, a careful risk assessment with consideration of the relative contraindications for full therapeutic anticoagulation must be undertaken.

Heparin acts by binding to and potentiating the action of antithrombin, which inhibits activated factor X and the action of thrombin. A standard regimen is an initial 5000-IU intravenous bolus of unfractionated heparin followed by a continuous intravenous infusion of 25 000 to 35 000 units over 24 h to ensure rapid achievement of full therapeutic anticoagulation. The activated partial thromboplastin time is measured at 4 h, aiming to maintain a ratio of 1.5 to 2.5. Dosage adjustments are followed by monitoring activated partial thromboplastin time every 4 h until stable anticoagulation is achieved, and then every 12 to 24 h. Omission of the bolus dose and reduction of the initial infusion rate, with the aim of maintaining the activated partial thromboplastin time ratio at 1.5, is advisable in patients with platelet counts of less than $50 \times 10^9/l$, within 24 h of surgery, or with a recent cerebrovascular event or gastrointestinal bleed. Bleeding associated with heparin over-anticoagulation can be reversed by protamine sulfate. The platelet count should be monitored daily to allow early detection of heparin-induced thrombocytopenia. Intravenous unfractionated heparin is the mainstay of treatment, but thrombolysis, surgical intervention, and inferior vena caval filters may be required. Oral anticoagulation with warfarin should be deferred until the patient is stable and able to take oral medication, usually after discharge from the ICU. Warfarin should be started using a standard protocol which has been agreed locally. The international normalized ratio (INR) is measured daily, aiming for a therapeutic range of 2.0 to 3.0. The heparin infusion is stopped once the INR exceeds 2.0. The coumarins are vitamin K antagonists which act by inhibiting γ-carboxylation of factors II, VII, IX, and X. The duration of oral anticoagulant therapy is usually 3 to 6 months for a deep vein thrombosis or pulmonary embolus. Longer-term anticoagulation is considered for recurrent events and in patients with an ongoing risk factor.

Thrombolysis

Thrombolytic therapy should be considered in patients with massive pulmonary embolism, and in young patients with extensive thrombosis compromising the vascular function of the affected limb. Thrombolysis improves radiographic resolution, right ventricular function, and pulmonary perfusion in small randomized trials. In addition, it may reduce recurrence rates and subsequent pulmonary hypertension. The currently available agents for thrombolysis are streptokinase, anisolated plasminogen streptokinase activator complex, recombinant tissue plasminogen activator, and urokinase. These agents have been widely used in the treatment of myocardial infarction. Recombinant tissue plasminogen activator is given as 100 mg over 2 h (10 mg over 1 to 2 min followed by 90 mg over 2 h). A continuous intravenous infusion of unfractionated heparin is subsequently administered when the activated partial thromboplastin time ratio is less than 2.0, and the heparin dose is then titrated to maintain an activated partial thromboplastin time ratio of 1.5 to 2.5. Streptokinase is given as a 30-min infusion of 600 000 units followed by 100 000 IU/h for 24 to 72 h. The thrombin time is monitored to maintain a value two to five times that of the control. The quoted risk of major bleeding varies from 3 to 5 per cent, depending on patient selection and the thrombolytic regimen used. The mortality rate is approximately 1.6 per cent. Bleeding associated with thrombolytic therapy requires treatment with fresh frozen plasma and cryoprecipitate. Consideration should be given to the use of aprotinin and other antifibrinolytic agents. Contraindications for thrombolysis include recent major surgery or trauma, organ biopsy, risk of gastrointestinal or genitourinary bleeding, pregnancy, hemostatic disorders, aortic dissection, pericarditis, severe hypertension, recent intracranial events or surgery, puncture of a non-compressible vessel, and cardiopulmonary resuscitation for over 10 min.

Indications for inferior vena caval filters

The insertion of percutaneous inferior vena caval filters should be considered in patients with thromboembolic disease in whom full anticoagulation is contraindicated, when thrombus extension or further pulmonary emboli occur despite adequate therapeutic anticoagulation, or in the presence of large free-floating thrombi of the iliac vein or inferior vena cava. Filters are effective in preventing fatal pulmonary embolus and are associated with few serious side-effects. Permanent filters have high long-term patency rates of up to 96 per cent at 12 years.

Massive pulmonary embolism

This condition is likely to be seen either as an indication for admission to the ICU or occurring in an existing patient. Sudden changes in hemodynamic variables, such as a rise in pulmonary artery pressure and reduction in cardiac index, may indicate the occurrence of a large pulmonary embolus. Severe right heart failure, hypoxemia, and death may occur quickly, with an estimated mortality rate of 43 to 80 per cent in the first 2 h. Diagnostic measures may necessarily be limited. Pulmonary angiography is the gold standard test, but echocardiography may be useful and more readily performed. Oxygen, inotropic agents, thrombolysis, surgical embolectomy, and ventilatory support may all be required. The placement of an inferior vena caval filter should be considered once the patient is stable.

Predisposition to thromboembolic disease

Consideration should be given to the identification of potential underlying conditions that may lead to venous thromboembolism. These include medical disorders (Table 1) which may require specific treatment. In individuals who develop thromboembolism at a relatively young age, i.e. under 45 years, an underlying hemostatic defect (thrombophilia) should be sought (Table 2). This may be particularly relevant where there is extension of venous thromboembolism despite adequate treatment, as replacement therapy with specific factor concentrates may be indicated. The prevalence of thrombophilic defects in individuals with a history of venous thromboembolism depends on selection criteria; the overall prevalence of antithrombin, protein C deficiency, and protein S deficiency varies between 6.5 and 26 per cent. Activated protein C resistance is by far the most common heritable thrombophilic defect with a reported prevalence of between 21 and 64 per cent in patients with deep vein thrombosis. The newly recognized prothrombin gene mutation, with a G to A transition at position 20210, occurs in 18 per cent of selected cases (Poort *et al.* 1996). Congenital dysfibrinogenemia is very rare. However, it is important to be aware of these defects in the ICU setting as they may present with venous thromboses in unusual sites or arterial thrombosis.

Unusual causes of thromboembolic disease

Antiphospholipid antibodies

Antiphospholipid antibodies (lupus anticoagulant and/or anticardiolipin antibodies) are a heterogeneous group of immunoglobulins directed against various protein–phospholipid complexes. Antiphospholipid antibodies occur in systemic lupus erythematosus or alone (the primary antiphospholipid syndrome). They are associated with recurrent miscarriage, arterial or venous thrombosis which is frequently recurrent, thrombocytopenia, and neurological and dermatological manifestations.

Catastrophic antiphospholipid syndrome

Patients with this rare condition often present critically ill with multiple vascular occlusions; death, usually from cardiopulmonary arrest, frequently results (Asherson 1992). There is widespread organ damage with renal dysfunction, hypertension, and respiratory distress with chest radiograph appearances typical of adult respiratory distress syndrome. Thrombotic skin manifestations are common (digital gangrene or ulceration, acrocyanosis, and livido reticularis), as are central nervous system symptoms and signs. Small- and large-vessel thrombosis is a hallmark of the disease, both arterial and venous. There may be a history of systemic lupus erythematosus or primary antiphospholipid syndrome. Laboratory features in addition to the presence of antiphospholipid antibodies include thrombocytopenia, leukocytosis, and an elevated erythrocyte sedimentation rate. Creatinine is often elevated and liver function tests may be abnormal. Treatment must be started early, although it is often delayed owing to the difficulty of diagnosing this rare condition. Anecdotal reports suggest that the most effective treatment modalities are full-dose anticoagulation, possibly with the addition of a thrombolytic agent, and plasma exchange. Ancrod or immunosuppressive agents such as cyclophosphamide may also be useful, but steroid therapy does not appear to be effective.

Acquired deficiencies of the naturally occurring anticoagulants

Acquired deficiencies of the naturally occurring anticoagulants, proteins C and S, and antithrombin III occur in liver disease, in disseminated intravascular coagulation (DIC), in acute thrombosis and in association with some chemotheraputic agents. In addition, antithrombin deficiency may also be seen in heparin therapy, estrogen administration, major surgery, protein-losing enteropathy, and nephrotic syndrome. Protein C and S deficiencies occur in patients receiving warfarin, with antiphospholipid syndrome, and with systemic lupus erythematosus. Protein C deficiency also complicates cardiopulmonary bypass and hemodialysis, while protein S deficiency is associated with pregnancy. Efficacy of specific factor concentrate replacement has not been proven in these situations.

Table 1 Medical conditions predisposing to venous thrombosis*

Myeloproliferative conditions
Primary proliferative polycythemia
Essential thrombocythemia
Systemic lupus erythematosus
Nephrotic syndrome†
Protein-losing enteropathy†
Heparin-induced thrombocytopenia
Paroxysmal nocturnal hemoglobinuria
Behçet's disease
Bone marrow transplantation‡
Liver transplantation
Splenectomy§
Malignancy

*The majority of these disorders may be associated with arterial as well as venous thromboembolism.
†Antithrombin deficiency.
‡Veno-occlusive disease of the liver.
§In the presence of ongoing hemolysis.

Table 2 Hemostatic defects predisposing to venous thrombosis

Established familial thrombophilic defects
Antithrombin deficiency
Protein C deficiency
Protein S deficiency
Activated protein C resistance/factor V Leiden mutation
Prothrombin gene mutation (20210)
Dysfibrinogenemia
Homocystinuria
Acquired thrombophilia
Antiphospholipid antibodies
Acquired deficiencies of the naturally occurring anticoagulants
Heparin-induced thrombocytopenia

Heparin-induced thrombocytopenia

Patients receiving heparin treatment who develop a new thrombosis, either arterial or venous, or become thrombocytopenic should have this diagnosis considered. Heparin-induced thrombocytopenia occurs in 1 to 3 per cent of patients receiving full-dose intravenous unfractionated heparin (Chong 1995), with a lower incidence in prophylactic regimens and with low-molecular-weight heparins.

There are two types of heparin-induced thrombocytopenia: type I, in which the platelet count often falls within the normal range, usually remaining above 100 × 10^9/l, and recovers without alteration in management, and type II which constitutes a severe complication of therapy. Platelet counts often fall below 60 × 10^9/l and are associated with life- and limb-threatening thromboses (myocardial infarction, cerebrovascular accidents, pulmonary emboli, large-vessel thromboses). Early recognition and prompt treatment has been shown to improve mortality from 23 to 12 per cent. Heparin-induced thrombocytopenia is an autoimmune phenomenon, usually occurring within 4 to 14 days of the initiation of heparin. Platelet counts rise on withdrawal of the drug but rapidly fall if the patient is re-exposed. The autoantibody is directed against a complex of platelet factor 4 and heparin, and is bound by platelet Fcγ receptors. *In vitro* testing with platelet aggregometry or microplate enzyme-linked immunosorbent assay (ELISA) for the platelet factor 4–heparin antibody are available. The mainstay of treatment is immediate withdrawal of heparin and institution of an alternative agent such as Orgaran, which has a cross-reactivity of approximately 10 per cent compared with 80 per cent for the low-molecular-weight heparins.

Table 3 Medical disorders predisposing to microvascular thrombosis

DIC
Thrombotic thrombocytopenic purpura/hemolytic uremic syndrome
Myeloproliferative diseases
Diabetes mellitus
Primary antiphospholipid syndrome
Systemic lupus erythematosus and related vasculitic disorders
Pre-eclampsia
Coumarin-induced skin necrosis
Purpura fulminans
Homozygous protein C or S deficiency
Associated with bacterial infection
Postinfectious (usually viral)*

*Acquired protein S deficiency due to an autoantibody.

Microvascular thrombosis

Microvascular thrombosis is a characteristic of a variety of conditions (Table 3), including DIC, thrombotic thrombocytopenic purpura, and the purpura fulminans syndrome. These conditions may also be associated with large-vessel thrombotic manifestations, both arterial and venous. In DIC clinical attention is often directed towards the bleeding manifestations; microvasular thrombosis causing endorgan damage may go unrecognized unless it is overt. The microangiopathic process in thrombotic thrombocytopenic purpura causes widespread small-vessel platelet thrombi with neurological, renal, and other endorgan damage, as well as hemolysis, red cell fragmentation, and thrombocytopenia. Thrombotic thrombocytopenic purpura differs from the hemolytic uremic syndrome in which the microvascular thrombosis is restricted to the kidney. The pathogenesis of thrombotic thrombocytopenic purpura is unknown, but both high-molecular-weight multimers of von Willebrand factor and calpain, a calcium-dependent cysteine protease, have been implicated. The mainstay of treatment is early and intensive plasma exchange with fresh frozen plasma replacement (Rock *et al.* 1991), to which about 70 per cent of patients respond. Refractory patients may benefit from the use of either cryosupernatant or solvent-detergent-treated plasma, which lack high-molecular-weight multimers of von Willebrand factor. Antiplatelet agents, steroids, and other immunosuppressive agents may also have a role in the management of these patients. Thrombotic thrombocytopenic purpura is associated with infections, pregnancy, drugs, HIV, malignancy, organ transplants, and collagen vascular disorders.

Purpura fulminans includes a group of heterogeneous conditions associated with rapidly progressive widespread thrombosis in the subcutaneous microcirculation, leading to necrosis of the dermis and subcutaneous fat and peripheral gangrene. It may occur in neonates as a result of homozygous or compound heterozygous deficiencies of proteins C and S. The severity of this potentially lethal condition is proportional to the functional protein C or S levels, which may be very low or undetectable. Acquired purpura fulminans generally occurs as a complication of infection. It is a cardinal feature of meningococcal sepsis, where there is usually laboratory evidence of consumptive coagulopathy and reductions in levels of all the naturally occurring anticoagulants, with the latter appearing to correlate with disease severity. In an attempt to improve the prognosis of this often fatal disorder, intensive plasma replacement and the recently available protein C concentrates have been administered to a few severely affected children. Although plasma protein C levels can be normalized using this approach, the impact on morbidity and mortality awaits prospective controlled clinical trials. Purpura fulminans also occurs in other bacterial infections including group A and B hemolytic streptococci, *Streptococcus pneumoniae*, *Staphylococcus aureus*, *Hemophilus influenzae*, and *Hemophilus egyptius*.

In a third group of patients, purpura fulminans occurs a few days to weeks after the onset of a febrile illness. The illness may be complicated by impaired perfusion of limbs or digits, peripheral gangrene, and major endorgan dysfunction due to thromboembolic phenomena affecting the lungs, heart, or kidneys. This condition has been shown to occur in chidren in association with primary herpes zoster infection where it is caused by an acquired protein S deficiency due to autoantibodies directed against protein S. Levin *et al.* (1995) reported five such cases; in four of these the acquired protein S deficiency was precipitated by varicella, and in the fifth after a non-specific viral-like illness. Immediate heparinization, aggressive plasma replacement therapy, and, in cases complicated by major vessel thrombosis, tissue plasminogen activator have been used to limit thromboembolic complications.

Peripheral arterial thrombosis

The patient will generally present with pain and dysesthesia in the affected limb which will lack pulses and may be pale and cooler than the opposite normal limb. Arterial thrombosis often occurs in an atherosclerotic vessel but may occur in a normal vessel. Care should be taken to exclude an embolic event, which is the main differential diagnosis and requires a search for the underlying cause. Doppler studies and angiography should be performed to confirm the diagnosis. Treatment options are surgery, balloon angioplasty, or intra-arterial thrombolysis, and should be considered in consultation with vascular surgeons and hematologists.

Local arterial thrombolysis

The best figures for primary success rates for local thrombolysis range from 69 to 85 per cent. Early recurrence occurs in approximately 10 per cent. Eight per cent of patients require amputation following failure of therapy. After successful therapy 80 per cent will be patent at 2 years. Recombinant tissue plasminogen activator is more effective than streptokinase and achieves earlier lysis. The thrombolytic agent is infused through an arterial catheter with its tip embedded in the proximal end of the thrombus. Recombinant tissue plasminogen activator is infused at a rate of 0.5 mg/h and progress is monitored angiographically. The average time to achieve reperfusion is 22 h. Vascular surgical back-up must be available. Thrombolytic therapy must not be given if there is evidence of irreversible ischemic damage.

References

Asherson, R.A. (1992). The catastrophic antiphospholipid syndrome. *Journal of Rheumatology*, **19**, 508–12.

Chong, B.H. (1995). Heparin-induced thrombocytopenia. *British Journal of Haematology*, **89**, 431–9.

Levin, M., Eley, B.S., Louis, J., Cohen, H., Young, L., and Heyderman, R.S. (1995). Postinfectious purpura fulminans caused by an autoantibody directed against protein S. *Journal of Paediatrics*, **127**, 355–63.

Poort, S.R., Rosendaal, F.R., Reitsma, P.H., and Bertina, R.M. (1996). A common genetic mutation in the 3-untranslated region of the prothrombin gene is associated with elevated plasma prothrombin levels and an increase in venous thrombosis. *Blood*, **88**, 3698–703.

Rock, G.A., *et al.* (1991) Comparison of plasma exchange with plasma infusion in the treatment of thrombotic thrombocytopenic purpura. *New England Journal of Medicine*, **325**, 393–7.

10.2.4 Thrombocytopenia

C. N. Harrison and S. J. Machin

Key messages

- A low platelet count should be confirmed by a repeat sample. A blood film should be requested and a screen performed for clotting abnormalities and fibrin degradation products (or equivalent).
- The patient's history, including the drug history (particularly heparin), should be checked. Any medications causally implicated or worsening platelet function (e.g. aspirin) should be discontinued.
- Hemorrhage and any features of disease should be sought and further tests performed as appropriate (e.g. autoimmune screen, bone marrow).
- Advice should be sought from a hematologist.
- Platelet concentrates should be used in life-threatening hemorrhage. Otherwise, they may not be appropriate (e.g. in thrombotic thrombocytopenic purpura).
- The platelet count should be maintained above 15×10^9/l. A higher threshold should be set if the patient is unstable. To cover invasive procedures the platelet count should be elevated to 30 to 50×10^9/l.

Introduction

Thrombocytopenia is both common and multifactorial in the critically ill (Table 1). The risk of bleeding cannot be directly related to platelet count, but is unlikely until this falls below 80×10^9/l and is increased below 10 to 20×10^9/l. Platelet function (modified by drugs such as aspirin), the integrity of the vessel wall, and the components of the coagulation cascade all contribute to hemorrhagic risk.

The pattern of bleeding due to thrombocytopenia is purpuric and is found on the shins, flexor areas, or pressure areas; small ecchymoses or mucous membrane bleeding are also seen. Subconjunctival and retinal hemorrhage are commonly found in those profoundly thrombocytopenic patients in whom life-threatening intracranial

Table 1 Pathological classification of causes of thrombocytopenia

Defective platelet production	
Bone marrow aplasia	Hereditary (e.g. Fanconi's anemia) Acquired (e.g. chemotherapy, drugs)
Metabolic disorders	Uremia Megaloblastic anemia (e.g. folate) Drugs (e.g. methotrexate, alcohol)
Marrow infiltration	Leukemia, lymphoma, carcinoma
Megakaryocyte abnormalities	Rare (e.g. May–Hegglin anomaly)
Diminished platelet survival	
Immune	
Alloantibodies	Post-transfusion purpura
Autoantibodies	Idiopathic thrombocytopenic purpura
Other	Drugs (e.g. heparin, penicillin) Autoimmune disorders Infections (e.g. AIDS, malaria, Epstein–Barr virus) Lymphoproliferative disorders
Non-immune	DIC Sepsis Microangiopathic processes Thrombotic thrombocytopenic purpura Extracorporeal circulation (e.g. dialysis)
Loss from circulation	Massive or exchange transfusion Splenomegaly (e.g. Felty's syndrome)
Artifactual	EDTA-induced platelet clumping

and gastrointestinal hemorrhages are a major risk. Hematomas or hemarthroses are rare unless there is a coexisting hemostatic disorder.

Assessment

Knowledge of the patient's medical and drug history, previous blood counts, and examination for features of hemorrhage and underlying disorder can aid diagnosis and management. Initially, a full blood count and film will help to determine if other lineages are involved, and may be of use in identifying the etiology (e.g. platelet clumping producing an artefactually low count, malarial parasites, or features of a microangiopathic process). A sample should also be sent to determine whether a coagulopathy is also present. Further tests should be chosen in clinical context. Discussion with a hematologist regarding likely etiology, further investigations (e.g. bone marrow, specific antibody tests), and management is then advised.

Management

Life-threatening hemorrhage is treated with platelet concentrates and correction of coagulopathy and, if possible, other risk factors. Platelet concentrates do not always result in reproducible increments. Thus it is important to measure platelet counts post-transfusion and to give further doses accordingly. If the patient is well, without sepsis or bleeding, a count of $15 \times 10^9/l$ is acceptable. If there is instability or invasive procedures are planned, a count of 30 to $50 \times 10^9/l$ is desirable. Once the situation is stable or in the absence of life-threatening hemorrhage, platelet concentrates may paradoxically worsen thrombocytopenia and risk the development of refractoriness to future transfusions.

Metabolic causes of platelet production failure, particularly acute folate deficiency (dietary or due to consumption) or B_{12} deficiency, should be treated individually either by replacement or by withdrawal of any toxic substance, such as alcohol or nitrous oxide.

Antibodies, either autoimmune or alloimmune in origin, cause thrombocytopenia by interaction between platelets and the reticuloendothelial system. Autologous antibodies may arise due to an underlying condition. Drugs may cause thrombocytopenia by this mechanism. After bleeding, heparin-induced thrombocytopenia is the most important side-effect of heparin. It is divided into two types.

Type I is experienced by as many as 10 per cent of those receiving intravenous heparin. It occurs early in therapy and is due to direct heparin–platelet interaction. The platelet count falls slightly and transiently.

Type II occurs after 4 to 20 days of therapy in 1 to 5 per cent of patients. An unexplained thrombosis and a 50 per cent reduction in platelet count suggests this diagnosis. Bleeding is infrequent, but thrombosis, either arterial or venous, contributes to the 20 to 30 per cent mortality. A recent theory of the mechanism of type II heparin-induced thrombocytopenia is interaction of platelet factor 4, a platelet granule protein, with heparin to form platelet factor 4–heparin complexes which stimulate antibody production. Immune complexes activate platelets and endothelial cells; this dual activation is the likely explanation for the combination of thrombocytopenia and thrombosis. A test for the platelet factor 4–heparin antibody is available using enzyme-linked immunosorbent assay (ELISA). Rapid diagnosis is important, as mortality increases with the delay between the first event and stopping heparin. Thrombosis should be managed with anticoagulants. There is a risk of cross-reactivity with low-molecular-weight heparins, but traditional anticoagulants or drugs such as Org 10172 or hirudin can be used. Heparin must be avoided in the future.

Non-immune platelet consumption occurs frequently in the critically ill. The most common cause is disseminated intravascular coagulation (DIC). Sepsis may result in thrombocytopenia by mechanisms other than or in addition to DIC. These include marrow failure as a direct consequence of sepsis or of drugs used in its treatment. Other mechanisms include hemophagocytosis (probably immune related), platelet deposition, and platelet consumption. A rare cause of platelet consumption is thrombotic thrombocytopenic purpura or the related hemolytic uremic syndrome. Fever, fluctuating neurological signs, renal impairment, thrombocytopenia, and microvascular hemolysis form the diagnostic pentad of hemolytic uremic syndrome. Associated causes, including malignancy, cyclosporin A, autoimmune disorders, and pregnancy, may be identified. Clinical suspicion, fragmented red cells on blood film, and normal coagulation aid the diagnosis. Treatment by large-volume plasma exchange and replacement with fresh frozen plasma should be instituted urgently. It is important not to give platelets unless there is life-threatening hemorrhage, as they may fuel the thrombotic process and result in disability or death.

HELLP (**h**emolysis, **el**evated **l**iver enzymes, **l**ow **p**latelets) syndrome occuring in pregnancy or early postpartum may be difficult to distinguish from thrombotic thrombocytopenic purpura. It is likely to be a variant of pre-eclampsia, and is treated supportively and by delivery.

Extracorporeal circulations cause thrombocytopenia and reversible loss of platelet function by virtue of platelet activation. Replacement therapy should be used for extreme thrombocytopenia and/or bleeding. Heparin-induced thrombocytopenia should be considered where heparin has been used as the anticoagulant.

Massive transfusion causes hemodilution resulting in a low platelet count which may manifest as microvascular ooze. Full blood count and coagulation should be monitored as a coagulopathy can occur. Platelet concentrates should be given to maintain the count above $50 \times 10/^9l$.

Bibliography

Chong, B.H. (1995). Heparin induced thrombocytopenia. *British Journal of Haematology*, **89**, 431–9.

Hewitt, P.E. and Machin, S.J. (1990). ABC of transfusion. Massive blood transfusion. *British Medical Journal*, **300**, 107–9.

Machin, S.J., Kelsey, H., Seghatchian, M.J., Warwick, R., and Mackie, I.J. (1995). Platelet transfusion. *Thrombosis and Haemostasis*, **74**, 246–52.

Rock, G., *et al.* (1991). Comparison of plasma exchange with plasma infusion in the treatment of thrombotic thrombocytopenic purpura. *New England Journal of Medicine*, **325**, 393–7.

10.3 The neutropenic patient

10.3.1 Pathophysiology

A. C. Nathwani and A. H. Goldstone

Key messages

- Neutropenia is defined as a reduction of the absolute neutrophil count below the normal range. This is influenced by gender, age, physiological status, and ethnic origin.
- An unexpected neutropenia should always be confirmed on a blood film, which may also provide clues to the likely etiology. Bone marrow examination may be necessary when history or blood film examination do not reveal an underlying cause.
- Neutropenia occurs when neutrophil production by the marrow is outpaced by peripheral utilization or, conversely, when neutrophil survival in the peripheral circulation is decreased due to infection or immune destruction.
- The majority of acquired neutropenic states are either due to viral infections or are iatrogenic as a result of drugs. Severe bacterial sepsis can also cause neutropenia.
- Numerous genetic disorders are associated with neutropenia but represent a minority of cases seen in clinical practice.

Introduction

Neutrophils account for 40 to 75 per cent of the total white cell count in the peripheral blood and are essential components of the host defense system. They are characterized by a dense nucleus consisting of two to five lobes and pale cytoplasm packed with granules whose contents are essential for killing and degrading micro-organisms. Neutrophils are derived from pluripotent stem cells located in the bone marrow under the influence of growth factors, interleukins 1 and 3 (IL-1 and IL-3), granulocyte colony-stimulating factor, and granulocyte macrophage colony-stimulating factor over a period of 6 to 10 days. In the normal state, the bone marrow storage compartment contains 10 to 15 times the number of neutrophils found in peripheral blood. Following their release from the marrow, they spend 6 to 10 h in the circulation before moving in response to endotoxins, cytokines, and chemokines into tissues where their lifespan is of the order of 4 to 5 days.

Through their cell-surface receptors, neutrophils can respond to chemotactic stimuli that direct them to the site of infection. These stimuli include breakdown products of bacterial walls and the complement cascade (e.g. C5a). Following stimulation, the polarized neutrophil adheres to the endothelium, passes through it by diapedesis, and migrates towards the site of infection. Once there, it attacks bacteria by phagocytosis and kills them by multiple mechanisms, the most important of which are release of bactericidal enzymes from the granules and the respiratory burst pathway which produces bactericidal free radicals from molecular oxygen.

Apart from their major role in the defense against bacterial infections, neutrophils also contribute to defense against fungal and viral infections. Consequently, quantitative or qualitative abnormalities of neutrophils are associated with severe and occasionally fatal infections with these pathogens.

Diagnosis of neutropenia

Neutropenia is defined as a reduction of the absolute neutrophil count below the normal range. However, the normal concentration is influenced by gender, age, physiological status, and ethnic origin. To avoid a misdiagnosis of neutropenia, it is imperative that any definition of the normal range takes account of these factors. For example, the normal neutrophil count for an adult Caucasian male is usually between 1.5×10^9 and 6.5×10^9/l, and for females it is between 1.8×10^9 and 7.4×10^9/l. In contrast, the neutrophil count for a normal baby during the first day of life is much higher at 4.8×10^9 to 17.1×10^9/l. Many individuals of African, Afro-Caribbean, and Middle Eastern origin have average neutrophil counts that are significantly lower than those of adult Caucasians but can still mount a normal response to infection.

Automated full blood count analyzers have made a great impact on the definition of the normal neutrophil range because of their speed and accuracy in assessing neutrophil numbers. However, in certain circumstances they produce spuriously low counts, for example following excessive clumping of neutrophils in the presence of a paraprotein, and disintegration of neutrophils due to excessive delay between blood withdrawal and analysis. An unexpected neutropenia should always be confirmed on a blood film, which may also provide clues to the likely cause. Bone marrow examination may become necessary in circumstances where history or blood-film examination do not reveal an underlying cause.

General mechanisms of neutropenia

Under normal physiological conditions, a stable equilibrium exits between marrow neutrophil production and peripheral utilization. Neutropenia occurs when neutrophil production by the marrow is

Table 1 Acquired neutropenic states

Drug-induced neutropenia	**Immune neutropenia**
Direct drug toxicity	*Alloimmune neutropenia*
Chlorambucil	Isoimmune neonatal
Vinblastine	
Adriamycin	*Autoimmune neutropenia*
Hydroxyurea	Idiopathic
	Rheumatoid arthritis
Idiosyncratic	Systemic lupus erythematosus
Thiazides	Sjögren's syndrome
Carbimazole	Felty's syndrome
Chloramphenicol	Hodgkin's disease
	Crohn's disease
Immune	Autoimmune hepatitis
Aminopyrine	Thymoma
Chlorpropamide	Pure white cell aplasia
Phenytoin	
Neutropenia with infectious diseases	**Nutritional deficiency**
Acute bacterial	Vitamin B_{12} and folate
Gram-negative sepsis	Copper
Typhoid fever	Anorexia nervosa
Chronic bacterial	
Tuberculosis	**Marrow infiltration**
Brucellosis	Malignant
Viral	Myelofibrosis
Influenza	Osteopetrosis
Infectious mononucleosis	
Infectious hepatitis	**Other causes**
Kawasaki disease	Lymphomas
HIV	Glycogen storage disorder
	Leukopheresis
	Hemofiltration

outpaced by peripheral utilization. For example, abnormalities of hemopoietic stem cell development and decreased proliferation of neutrophil precursors in the marrow can reduce neutrophil production. Conversely, decreased neutrophil survival in the peripheral circulation owing to infection or immune destruction can also give rise to neutropenia. Many authors have classified neutropenia into broad categories of decreased production and increased elimination, akin to the classification of anemia. However, since specific pathophysiological mechanisms remain poorly defined for most neutropenias, these attempts have met with limited success. Here, acquired neutropenias, which account for the majority of clinical cases, will be considered separately from inherited neutropenias. Disorders of neutrophil function will also be discussed briefly.

Acquired neutropenias

The majority of acquired neutropenic states are either due to viral infections or are iatrogenic as a result of drugs (Table 1).

Infections

A mild self-limiting reduction in absolute neutrophil count is most commonly seen following an acute influenza infection. Neutropenia develops during the acute viremic phase of the infection and can persist for up to a week without serious sequelae. Concomitant diminution of other cell lines is unusual in this setting, and diagnosis is aided by a history of an influenza-like illness and reactive lymphocytes with toxic granulation of the neutrophils on the blood film. Protracted neutropenia can occasionally occur with viral hepatitis, Kawasaki disease, and infectious mononucleosis. The mechanism of neutropenia is unclear but probably results from direct suppression of myelopoiesis and increased utilization of neutrophils.

Severe bacterial sepsis can also cause neutropenia, probably as a result of direct damage of myeloid precursors by bacterial endotoxin and increased adherence of neutrophils to the vascular endothelium. Bacterial-induced neutropenia tends to be more common in patients with reduced marrow neutrophil reserve, due to either a primary hematological disease or previous treatment with myelosuppressive drugs or irradiation. Some patients are malnourished or have a history of alcohol abuse. Chronic infections causing splenomegaly, such as tuberculosis, brucellosis, kala-azar, and malaria, also cause neutropenia, but this is probably secondary to splenic sequestration.

Drugs

Drug-related neutropenia generally falls into two categories: direct drug toxicity and immunological. Drugs causing neutropenia by direct toxicity can be further subdivided into those that have a predictable dose-dependent effect on the myeloid precursors and those with an idiosyncratic and dose-independent response.

Dose-dependent mechanism

If given in sufficient dosage, certain drugs will consistently cause neutropenia and also suppress other hemopoietic elements. They act mainly, if not entirely, by interfering with cell division in the following manner:

(1) inhibition of purine or pyrimidine biosynthesis or incorporation into DNA (e.g. mercaptopurine, 6-thioguanine, hydroxyurea);

(2) cross-linkage of DNA, thereby impeding RNA formation (e.g. nitrogen mustard, chlorambucil);

(3) disruption of the microtubules of the mitotic spindle (vinblastine).

For the most part these agents are used in the treatment of malignant diseases. The degree and reversibility of neutropenia is determined by the dose and type of drug.

Idiosyncratic mechanism

An ever-growing list of drugs cause neutropenia unpredictably and in only a few patients exposed to them. These include chloramphenicol, carbimazole, and phenothiazines. Typically, neutropenia becomes evident 1 to 2 weeks following initial exposure. The basis of the drug reaction is unclear, but it is probably due to a genetic aberration in enzymatic responsiveness. Treatment consists of immediate withdrawal of the implicated drugs and aggressive management of any underlying infection. Hemopoietic recovery usually follows within 1 to 2 weeks.

Immunogenic mechanism

Although rarely used, aminopyrine remains historically important as the prototype for immunogenic neutropenia. Other drugs recently implicated include phenytoin, chlorpropamide, levamisole, and clozapine. The mechanisms involve development of an antibody to a complex between the drug and specific neutrophil proteins. Administration of the drug to a sensitized individual results in massive leukoagglutination and removal of neutrophils from the circulation. Neutropenia is reversible within 1 to 2 weeks upon withdrawal of the drug.

Other acquired causes of neutropenia

Neutropenia due to bone marrow infiltration or obliteration

A number of conditions cause extensive infiltration of the bone marrow compartment resulting in moderate to severe neutropenia associated with a reduction in the platelet count and hemoglobin level. These include neoplastic infiltration with metastatic carcinoma, acute or chronic leukemia, myeloma, and sarcoma. Irradiation, osteosclerosis, and myelofibrosis cause neutropenia as a result of damage to the marrow stroma resulting in fibrosis or obliteration.

Neutropenia due to nutritional deficiency

Cachexia and debilitated states associated with protein undernutrition, vitamin deficiency (vitamin B_{12} or folate), or trace element deficiency (copper) are occasionally accompanied by neutropenia. Neutropenia is usually mild and may be associated with anemia and thrombocytopenia.

Autoimmune neutropenia

Neutropenia due to neutrophil autoantibodies can be either idiopathic or associated with other autoimmune disorders such as rheumatoid arthritis, systemic lupus erythematosus, or Felty's syndrome. Patients usually have a selective neutropenia, although autoimmune hemolytic anemia and thrombocytopenia may also occur. Demonstration of neutrophil antibodies is required to differentiate the idiopathic variant from benign chronic idiopathic neutropenia. The bone marrow usually shows an excess of early myeloid precursors with very few mature neutrophils. Spontaneous remission sometimes occurs, but treatment with intravenous immunoglobulin or steroids is often required.

Neutropenia associated with splenomegaly

Various diseases, including sarcoidosis, lymphoma, and Gaucher's disease, cause splenomegaly and are usually associated with thrombocytopenia and anemia. The neutropenia is probably due to sequestration, but some investigators have suggested an immune mechanism. It is usually clinically insignificant and splenectomy is rarely indicated.

Hemodialysis neutropenia

A striking but transient neutropenia has been observed during hemodialysis and leukopheresis. It is probably due to margination and sequestration of neutrophils within the pulmonary circulation.

Inherited neutropenic syndromes

Numerous genetic disorders are associated with neutropenia, but they represent a minority of cases seen in clinical practice (Table 2). Some of the more common disorders are discussed below.

Table 2 Some inherited disorders causing neutropenia

Kostmann's syndrome
Cyclical neutropenia
Reticular dysgenesis
Familial benign neutropenia
Familial severe neutropenia
Chediak–Higashi syndrome
Transcobalamin II deficiency
Schwachman–Diamond syndrome
Pearson's syndrome
Myelokathexis
Lazy leukocyte syndrome

Kostmann's syndrome

Kostmann's syndrome is an autosomal recessive disease characterized by severe neutropenia (often $< 0.5 \times 10^9$/l) and frequent infection during the first month of life. In the absence of bone marrow transplantation, most children die during the first year of life. Treatment with supraphysiological doses of granulocyte colony-stimulating factor causes neutrophil differentiation and a reduction in the incidence of infection and mortality.

Cyclical neutropenia

This rare congenital disorder is characterized by regular oscillations of blood neutrophil count, usually with a periodicity of 14 to 21 days. Most cases occur sporadically, although a third have an autosomal dominant pattern of inheritance. The onset of neutropenia is usually in childhood when patients present with recurrent infections. Diagnosis may be difficult unless serial blood counts are performed. Treatment with granulocyte colony-stimulating factor shortens the period of neutropenia sufficiently to avoid infection.

Functional neutrophil disorders

An increased susceptibility to infection is well recognized in conditions associated with defects in phagocytic function. These defects occur in many systemic conditions, including diabetes mellitus, uremia, malnourishment, and alcohol ingestion, and following corticosteroid therapy. There are also well-recognized but exceedingly rare inherited defects of phagocyte function, the most striking being chronic granulomatous disease in which neutrophils are unable to kill catalase-positive micro-organisms such as staphylococci, *Serratia*, *Candida*, and *Aspergillus*. These organisms survive intracellularly, resulting in chronic local granulomatous inflammation. The defect in chronic granulomatous disease involves an enzyme in the respiratory burst of oxidative metabolism that accompanies phagocytosis. Unless the disease has previously been recognized in the family, the diagnosis is usually made when the clinical picture of recurrent infection prompts investigation of neutrophil and immune function.

Bibliography

Dale, D. (1995). Neutropenia. In *Williams hematology* (ed. E.Beutler, M.A. Lichtman, B.S. Coller, and T.J. Kipps), pp. 815–29. McGraw-Hill, New York.

Nimer, S.D. and Golde, D.W. (1995). Molecular, cellular, and clinical biology of phagocytic cells. In *Blood. Principles and practice of hematology* (ed. R.I. Handin, S.E. Lux, and T.P. Stossel), pp. 513–43. J.B. Lippincott, Philadelphia, PA.

10.3.2 Management of neutropenia and its consequences

A. C. Nathwani and A. H. Goldstone

Key messages

- Neutropenia can be 'mild' (1.0–1.8 × 10^9/l), 'moderate' (0.5–1.0 × 10^9/l), or 'severe' (< 0.5 × 10^9/l).
- Patients with severe neutropenia, particularly when levels are below 0.2 × 10^9/l, are at significant risk of infection, usually by bacterial invasion arising from the mouth and intestinal tract.
- The spectrum of bacteria isolated during febrile neutropenia has changed in the last decade from predominantly Gram-negative to Gram-positive organisms.
- Other factors have important influences on the risk and type of infection: the duration of neutropenia, defects in other immune system components, nutritional status, previous splenectomy, skin and mucous membrane integrity, and the use of blood components.
- Severely neutropenic patients are unable to mount a normal inflammatory response. Normal signs and symptoms of infection may be absent or muted, and the patient can deteriorate rapidly. It is imperative to monitor closely and treat immediately if infection is suspected.
- In severe neutropenia it is essential to start treatment promptly (before results of cultures) with an empirical antibiotic regimen that is bactericidal and has a broad spectrum of activity, particularly against virulent Gram-negative bacteria such as *Pseudomonas aeruginosa*.

Introduction

Neutrophils play a vital role in the immunological defense system of the body, particularly against bacteria but also against fungi and viruses. Consequently, impairment in neutrophil function, either qualitative or quantitative, results in an increased susceptibility to infection. These infections can disseminate rapidly and, if not treated promptly or appropriately, can be fatal. With the widespread use of myelosuppressive agents for the treatment of malignancies, and in a variety of autoimmune conditions, the incidence of neutropenia and associated infection is on the increase (Beutler 1995). Therefore it is essential that clinicians are familiar with both the unusual infections that occur in these patients and the manner in which the underlying disease and/or its therapy can modify the clinical presentation. This chapter focuses on the infective complications of neutropenia.

The risk of infection and predominant pathogens

When a patient is found to be neutropenic, the peripheral blood neutrophil count serves as a rough guide to the relative seriousness of the disorder. This degree of neutropenia can be 'mild' (1.0–1.8 × 10^9/l), 'moderate' (0.5–1.0 × 10^9/l), or 'severe' (< 0.5 × 10^9/l). Patients with severe neutropenia, particularly those with neutrophil levels exceeding 0.2 × 10^9/l, are at significant risk of infection, usually by bacterial invasion arising from the mouth and intestinal tract, such as the Gram-negative bacilli *Escherichia coli*, *Klebsiella* species, and *Pseudomonas aeruginosa* (Table 1). These organisms are responsible for considerable morbidity and mortality as they can lead to overwhelming septic shock and disseminated intravascular coagulation (DIC).

With the widespread use of indwelling catheters and the alteration of normal gut flora by prophylactic antibiotics, the spectrum of bacteria isolated during febrile neutropenia has changed in the last decade from predominantly Gram-negative to Gram-positive organisms such as *Staphylococcus epidermidis* and α-hemolytic streptococci. Although Gram-positive organisms have also been associated with the septic shock syndrome and DIC, they are generally regarded as less virulent than Gram-negative bacilli. Further changes in the microbiological flora have occurred because of the extensive use of third-generation cephalosporins, resulting in the emergence of multiresistant *Enterobacter* species and unusual bacteria (e.g. *Stenotrophomonas maltophilia*, *Acinetobacter* species, and other pseudomonads) as frequent pathogens in neutropenic patients.

Apart from the severity of neutropenia, other factors which have an important influence on the risk and type of infection include the duration of neutropenia, defects in other components of the immune system (due to either the underlying disease or treatment), nutritional status, previous splenectomy, the integrity of skin and mucous membranes, and the use of blood components because of the risk of blood-borne infection. For example, fungal infections are very common in patients with prolonged neutropenia and in those treated with broad-spectrum antibiotics or corticosteroids. Particularly vulnerable are bone marrow transplant patients, in whom the incidence of deep mycoses reaches 30 per cent. Similarly, the risk of opportunistic infections (e.g. *Pneumocystis carinii*) is much higher in neutropenic patients with human immunodeficiency virus (HIV) infection. In contrast, a defect of humoral immunity, as in chronic lymphocytic leukemia and multiple myeloma, will predispose neutropenic patients to infection from encapsulated organisms such as *Streptococcus pneumoniae* or *Hemophilus influenzae* because of lack of appropriate opsonization.

Clinical evaluation

Vital clues to the source of infection may be obtained from a carefully taken history and a detailed physical examination. However, it must be appreciated that severely neutropenic patients are unable to mount a normal inflammatory response; thus the normal signs and symptoms of infection may be absent or muted. In addition, the clinical state of neutropenic patients can deteriorate rapidly with resultant septic shock, DIC, or acute respiratory distress syndrome. Therefore it is imperative that all severely neutropenic patients are closely monitored and treatment commenced as soon as infection is suspected.

History

Recognition that the patient is neutropenic by virtue of the underlying disease (e.g. neoplasm) or some form of treatment (e.g. chemotherapy) should be followed by a careful review of the following factors:

Table 1 Immunological defects and associated pathogens in neutropenic patients

Type of immune defect	Common associated pathogens
Isolated neutropenia Post-chemotherapy Acute leukemia Myelodysplasia	Gram-negative bacilli: *E. coli*, *Klebsiella* species, *Ps. aeruginosa* Gram-positive cocci: *Staph. epidermidis*, *Staph. aureus* Gram-positive bacilli: *Bacillus* species, *Corynebacterium* species, *Clostridium* species Fungi: *Aspergillus fumigatus*, *Aspergillus flavus*, *Candida* species Viruses: herpes simplex, varicella zoster
Associated T-cell dysfunction HIV Hodgkin's lymphoma Non-Hodgkin's lymphoma	Gram-negative bacilli: *Salmonellae*, *Legionella pneumophila* Gram-positive bacilli: *Listeria monocytogenes*, *Nocardia* species Mycobacteria: *Mycobacterium tuberculosis* and other atypical mycobacteria Fungi: *Cryptococcus neoformans*, *Candida* species, *Histoplasma capsulatum*, *Coccidium* Viruses: herpes simplex, varicella zoster Protozoa: *P. carinii*, *Toxoplasma gondii*, *Cryptosporidium* Helminth: *Strongyloides stercoralis*
Associated B-cell dysfunction Multiple myeloma Chronic lymphocytic leukemia Hypogammaglobulinemia	Gram-negative bacilli: *H. influenzae* Gram-positive cocci: *Strep. pneumoniae*, *Staph. aureus* Gram-negative cocci: *Neisseria meningitides*

(1) duration of the neutropenia and symptoms of infection (the history should be broadened to determine whether fever and evidence of infection developed in the community or in the hospital setting);

(2) treatment for underlying disorders such as radiation, drug treatment (e.g. steroids), and transfusion of blood products;

(3) history of previous infections (e.g. tuberculosis, HIV) and antibiotic treatment.

Physical examination

Fever is usually the first and often the only sign of infection in a neutropenic host. Particular areas of concern during physical examination should be the head and neck for evidence of central nervous system infection, the oropharynx for evidence of bacterial or fungal pharyngitis, and the lungs, abdomen, perirectal area, and integument for evidence of any infectious process therein. Cardiovascular status has an important bearing on outcome.

Investigations

The severity of neutropenia and the degree of anemia or thrombocytopenia can be established with a full blood count. CD4 lymphocyte counts may be useful in assessing the risk of opportunistic infections. Blood cultures and cultures from presumed sites of infection (e.g. urine, stool, and sputum) are essential before the commencement of antimicrobial therapy. It may be necessary to repeat these cultures to improve the likelihood of recovering fastidious organisms. Chest radiography should be performed routinely and sinus radiography may be useful. CT may be of value in the assessment of infection of the lungs and the brain. If the patient has any central nervous system signs and symptoms, a lumbar puncture should be performed urgently provided that there is no evidence of raised intracranial pressure.

Treatment

In any severely neutropenic patient it is essential to start treatment empirically with an antibiotic regimen that is bactericidal and has a broad spectrum of activity, particularly against the more virulent Gram-negative bacteria such as *Ps. aeruginosa*. There is no time to wait for the results of cultures as delay in the commencement of antibiotics is associated with an extremely high mortality. The initial clinical manifestation of any infectious process is usually fever. Fever suggestive of infection is defined as an elevation of core body temperature in excess of 38 °C on two occasions in 24 h or 38.5 °C on one occasion in the absence of any obvious non-infectious cause of fever (e.g. blood transfusion). Such a definition is designed to prompt immediate clinical evaluation and early empirical antimicrobial therapy, but the absence of a fever or the presence of hypothermia does not exclude an infective process.

It is not possible to discern on clinical grounds alone the difference between infectious fever and fever secondary to a neoplasm, drug reaction, or hypersensitivity reaction (including reactions to blood products). The basic tenet that underlies all major initial therapeutic choices should be the clinical status of the patient. If the patient's immune status is severely impaired, as manifested by a circulating normal neutrophil count of less than 0.5×10^9/l, or the clinical state appears to be deteriorating, it is prudent to obtain the cultures mentioned above and begin empirical antimicrobial treatment. If the neutrophil count is greater than 0.5×10^9/l and there is a reasonable likelihood that the fever is due to a hypersensitivity reaction or underlying disease, then diagnostic measures including cultures should be obtained and the patient observed very carefully over the ensuing hours.

Choice of empirical therapy

Any empirical antibiotic regimen should provide broad-spectrum cover, have additive or synergistic effect against the more virulent Gram-negative pathogens, and decrease the emergence of resistant organisms. Knowledge of local prevalence and antimicrobial

Table 2 Some empirical antimicrobial regimens for neutropenic patients

Regimen	Dose schedules	Toxicity
Piperacillin	250–300 mg/kg/day in 4 divided doses	Penicillin allergy
or		Hypokalemia
Ceftazidime	6 g/day in 3 divided doses	Skin rashes, diarrhea
±		
Gentamicin	4–5 mg/kg/day in 2 or 3 divided doses (adjust on basis of blood levels)	Nephro- and ototoxicity
Ciprofloxacin	200–400 mg IV every 12 h	Nausea, not for use in children
+		
Vancomycin	15 mg/kg/day in 2 doses (adjust on basis of blood levels)	Nephrotoxicity, red man syndrome
Imipenem	0.5–1 g every 6 h	Nausea, seizures

IV, intravenous.

resistance of pathogens is crucial in the selection of the most appropriate empirical antibiotic regimen. Traditional choices (Table 2) have included an aminoglycoside such as gentamicin paired with a broad-spectrum agent of another class that provides additional cover against *Ps. aeruginosa* (e.g. a β-lactam compound such as piperacillin or a cephalosporin such as ceftazidime). Nephrotoxicity has been the major problem with such combinations, particularly when used simultaneously with other nephrotoxic agents such as vancomycin or amphotericin B. Therefore combinations of two β-lactams have emerged in an effort to overcome aminoglycoside nephrotoxicity. While such combinations are equally efficacious, their major disadvantage has been the selection of unusual bacteria or resistant organisms such as multiresistant *Enterobacter* species. Some centers have advocated monotherapy with an agent such as ceftazidime or imipenem. These agents may be adequate to sustain a neutropenic patient for the first 48 h until the results of cultures are available, when specific antimicrobial therapy can be instituted.

Because of the increasing incidence of Gram-positive bacteremia, many investigators have advocated the empirical use of an antibiotic with potent Gram-positive activity such as vancomycin or teicoplanin. Vancomycin causes significant toxicity (renal and skin), and there is no evidence that delay in instigating treatment for Gram-positive infection is associated with increased mortality. Therefore empirical Gram-positive therapy should be reserved for severely neutropenic patients with the following:

(1) septic shock;

(2) infected intravenous catheters (most catheter-associated bacteremias usually settle with appropriate antibiotics without catheter removal but, in the setting of a persistent bacteremia, catheter removal may become necessary);

(3) streptococcal sepsis which is a frequent complication of febrile neutropenia.

Duration and modification of empirical antibiotics

Antibiotics may be discontinued after 3 to 5 afebrile days provided that the patient is very carefully observed. Some clinicians prefer to continue antibiotics until the neutrophil count rises above $0.5 \times 10^9/l$, but this is associated with an increase in fungal infections, emergence of resistant bacteria, and an increased risk of antibiotic-related toxicity. In patients with persistent culture-negative fever 72 h after the commencement of empirical antimicrobial therapy, re-evaluation including a full medical examination, repeated cultures, and chest radiography should be performed. Metronidazole should be added in cases of severe mucositis, necrotizing gingivitis, perianal tenderness, and acute abdominal pain. However, in the absence of any clues to the etiology of persistent fever, antifungal treatment should be commenced (Fig. 1).

Use of hemopoietic growth factors

Many published studies suggest that hemopoietic growth factors such as granulocyte colony-stimulating factor and granulocyte macrophage colony-stimulating factor shorten the duration of neutropenia when given prophylactically to patients with solid tumors (Kim and Demetri 1996). The incidence of fever also appears to be curtailed, with an associated reduction in the use of systemic antibiotics and in-patient stay. However, there is no evidence to date that the routine use of these growth factors improves survival.

Use of granulocyte transfusion

Although granulocyte transfusion has been available for many years, its use has waned over the past decade, partly because of advances in antimicrobial therapy and the availability of hemopoietic growth factors. In addition, early studies showed that no clinical benefit was demonstrated, primarily because the collection method prevented sufficient numbers of granulocytes from being collected, thus requiring frequent transfusions, and the function of the cells collected was defective. There have also been concerns regarding the risk of cytomegalovirus infections, transfusion-associated graft versus host disease, and the development of acute respiratory distress syndrome following transfusion of granulocytes. Efforts to improve granulocyte collection following granulocyte colony-stimulating factor administration to donors is now being considered. The use of granulocyte colony-stimulating factor offers many potential advantages. Its ability to increase granulocyte numbers would allow collection of

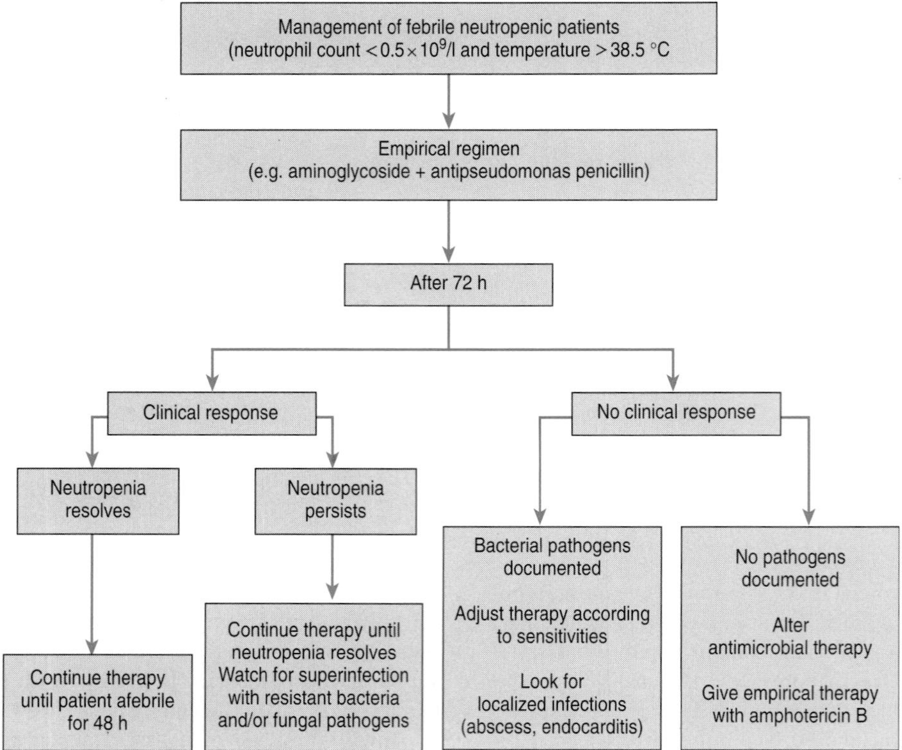

Fig. 1 Empirical approach in febrile neutropenic patients.

enough cells to reflect physiological production. In addition, it offers the advantage of granulocyte collection with potentially enhanced functional capability and increased survival secondary to delayed apoptosis. In a recent study, a high yield of granulocytes was obtained from family members or volunteer donors following granulocyte colony-stimulating factor administration. Post-transfusion, granulocyte recovery and survival were excellent and sustained, with normal *in vitro* function (Wuest 1996). Further studies are required to evaluate efficacy and to monitor long-term toxicity of granulocyte colony-stimulating factor in donors and granulocytes in recipients.

Treatment of non-bacterial infections

Fungal infections

Up to a third of febrile neutropenic patients unresponsive to empirical antibiotics have systemic infections with either *Candida* or *Aspergillus*. Serious fungal infections also occur in patients undergoing bone marrow transplantation and organ transplantation, and in those who have received high-dose corticosteroid therapy. The mortalities from these infections are in excess of 35 per cent and 80 per cent respectively. This provides the rationale for administering empirical systemic antifungal drugs in patients with a persistent culture-negative fever 72 to 96 h after the commencement of empirical antimicrobial therapy or in the presence of pulmonary infiltrates.

The newer imidazoles, such as fluconazole and itraconazole, have been extremely useful in the treatment of *Candida*, *Cryptococcus*, and *Coccidiomycosis* but have little activity against *Aspergillus*. Amphotericin B still remains the gold standard for the treatment of most fungal infections in immunocompromised hosts. The problems with amphotericin B are that it must be given intravenously, preferably through central venous access, and it is associated with severe side-effects which include fever, rigor, hypotension, and renal toxicity. Concomitant use of antibiotics such as gentamicin and vancomycin may exacerbate renal toxicity in patients receiving amphotericin. Lipid-based amphotericin compounds have been developed in an effort to reduce toxicity and improve efficacy.

Viral infections

Viral infections such as herpes simplex, varicella zoster, cytomegalovirus, and adenoviruses frequently occur in neutropenic patients, particularly those with impaired cell-mediated immunity. These infections can be primary, for example when a renal transplant patient who is seronegative for cytomegalovirus is grafted with a kidney from a seropositive donor. More often they result from reactivation of latent virus, as when a patient with Hodgkin's disease develops varicella zoster following irradiation or chemotherapy.

Herpes simplex virus infection usually presents as a mucocutaneous infection which is easy to diagnose except in patients with mucocutaneous bleeding. Acyclovir (aciclovir) is the treatment of choice and is relatively free from toxicity as it can only be converted into its active form by herpes viruses and not by the host cells. It is excreted by the kidneys and the dose should be reduced according to creatinine clearance. Because of poor bioavailability of orally administered acyclovir, the intravenous route is recommended for neutropenic patients with severe infection.

Varicella infection is not uncommon in children with malignancy (e.g. acute lymphocytic leukemia) who are receiving chemotherapy. As a preventive measure, immunocompromised children with no immunity against varicella should receive varicella zoster immune globulin (VZIG) as soon as possible if exposed to either chickenpox or varicella zoster. Those with active infection should be treated with intravenous acyclovir. Foscarnet is usually effective in patients with acyclovir resistance.

Cytomegalovirus is an important opportunistic pathogen following organ transplantation. It is associated with a variety of clinical syndromes including pneumonitis, colitis, hepatitis, mononucleosis, retinitis, and encephalitis. In bone marrow transplant patients, the risk of cytomegalovirus pneumonitis is further increased in patients with graft versus host disease. Mortality from cytomegalovirus infections is high, particularly in patients with pneumonitis. Two agents, ganciclovir and foscarnet, are currently available for the treatment of cytomegalovirus. Major side-effects of these drugs include bone marrow and renal toxicity.

Mycobacterial infection

In the neutropenic host, infection with *Mycobacterium tuberculosis* usually results from reactivation of latent infection and is more common in conditions such as myelodysplasia, chronic lymphocytic leukemia, and hairy cell leukemia. These patients are prone to atypical presentations, rapid progression, and extrapulmonary and disseminated mycobacteriosis. Infections with non-tuberculous mycobacteria (e.g. *Mycobacterium avium* complex and *Mycobacterium kansasii*) are more common in patients with HIV infection. Treatment of tuberculosis should be with multiple drugs (three or four drugs), and should be individualized on the basis of sensitivity testing, degree of immunodeficiency, and adverse effects. The non-tuberculous mycobacteria have heterogeneous *in vitro* antibiotic susceptibility patterns, and treatment regimens must be designed on a species-specific basis.

Parasitic infections

P. carinii is a ubiquitous extracellular organism that causes asymptomatic latent infection in normal hosts but rarely causes disease. However, 85 per cent of patients with AIDS develop *P. carinii* pneumonia, which may also be seen in bone marrow and solid organ transplant patients as well as in patients with leukemia, lymphoma, and primary immune deficiency. Drugs such as fludarabine are increasingly being used in the treatment of low-grade lymphomas. By their action on adenosine deaminase they cause a reduction in CD4 T-lymphocyte count and therefore predispose patients treated with these agents to infection with *P. carinii*.

The standard treatment of *P. carinii* pneumonia is with trimethoprim plus sulfamethoxazole (co-trimoxazole) or parenteral pentamidine. Other drugs that have been used include dapsone, atovaquone, clindamycin, and primaquine. Other life-threatening parasitic infections such as malaria and babesiosis, and infections with *Strongyloides* species, can also occur in the neutropenic patient. These infections can sometimes emerge decades after a forgotten exposure in an endemic area.

Prevention of infections

More than 80 per cent of infections in the neutropenic patient can be traced to host endogenous flora. Surveillance cultures in these patients are useful, although they may have a low predictive value. They provide hints of the possible causative micro-organism and may influence therapeutic decisions. They are usually limited to potential sites of infection such as cultures from the nares, throat, perianal area, and stool for bacteria, fungi, and viruses. Routine sterilization with non-absorbable oral antimicrobials consisting of colistin, neomycin, nystatin, and amphotericin has been advocated. Some of these antimicrobials are often poorly tolerated. Systemic antibiotics such as ciprofloxacin have been successfully used to prevent Gram-negative infections in neutropenic patients. Unfortunately, increasing and indiscriminate use in the community has diminished their value as resistant bacteria have developed. Another problem with this approach is superinfection with Gram-positive organisms. Ancillary measures such as handwashing, special diets, and reverse isolation are all valuable preventive measures.

Conclusion

The past decade has produced tremendous advances in the support of neutropenic patients undergoing chemotherapy or transplant procedures in the form of potent broad-spectrum antimicrobials, hemopoietic growth factors, and stem cell transfusions. Hemopoietic growth factors and stem cell therapy have reduced the period of neutropenia following chemotherapy and have allowed dose intensification of chemotherapy with a possibility of better cure rates for some malignancies. Because of their potent broad-spectrum activity, the newer antimicrobials have made the neutropenic phase relatively safer with a significant reduction in morbidity and mortality. Many patients with congenital neutropenia (e.g. Kostmann's syndrome and cyclical neutropenia) have also benefited from these advances. Unfortunately, these supportive measures are all very expensive and have important implications for the health care budget. Therefore it is imperative that the potential benefits of these measures are properly validated. Clear guidelines for their use in clinical practice are essential.

References

Beutler, S.M. (1995). Treatment of infection in the immunocompromised host. In *Williams hematology* (ed. E. Beutler, M.A. Lichtman, B.S. Coller, and T.J. Kipps), pp. 166–72. McGraw-Hill, New York.

Kim, S.K. and Demetri, G.D. (1996). Chemotherapy and neutropenia. *Hematology/Oncology Clinics of North America*, **10**, 377–95.

Wuest, D.L. (1996). Transfusion and stem cell support in cancer treatment. *Hematology/Oncology Clinics of North America*, **10**, 397–429.

10.4 Anemia

10.4.1 Pathophysiology

E. C. M. van Gorp and H. ten Cate

Key messages

- Anemia is caused by decreased production of red cells or by an increase in red cell or whole blood loss.
- Bone marrow disease or hematinic deficiencies lead to underproduction of red cells.

Introduction

Anemia is defined as a reduction in red cell mass, resulting in a lowered hemoglobin and hematocrit level, and leading to a decrease in the oxygen-carrying capacity of the blood. The main causes, diagnostic methods, and effects of anemia in the critically ill patient will be discussed.

Etiology

Anemia is caused by a decreased production of red cells or by an increase in red cell or whole blood loss (Feingold 1986; Bunn 1994). Loss of blood may be caused by hemorrhage, while loss of red cells is due to a reduced red cell lifespan resulting from enhanced elimination from the circulation. A decreased production of red cells may be the result of either hematinic deficiency or a primary bone marrow disease. In both cases, there is a decreased production capacity.

Clinical presentation

Anemia may be the first presenting symptom of a serious underlying disease such as malignancy or ulceration in the gastrointestinal tract. The clinical picture may vary considerably depending on the severity as well as the rapidity of onset of the anemia. The underlying physical condition will also influence the clinical picture. A severe anemia developing gradually over a long period of time may present with only slight complaints, whereas the patient with anemia of similar magnitude but due to acute bleeding may develop signs of myocardial infarction, stroke, transient cerebral ischemic attacks, and cardiac failure, which eventually may lead to (cold) shock.

In general, complaints of the anemic patient in the intensive care unit will include fatigue, dyspnea, palpitations, angina pectoris, dizziness, headache, syncope, tinnitus, vertigo, gastrointestinal symptoms (e.g. anorexia, nausea), abnormal menstruation, and (cold) shock.

On physical examination special attention should be focused on findings such as pallor, compensatory mechanisms such as tachycardia, wide pulse pressure, hyperdynamic precordium, a systolic precordial ejection murmur, icterus, and splenomegaly, with the last of these being suggestive of hemolysis.

Increased blood and red cell loss

Blood loss

The development of anemia due to bleeding follows a pattern determined by the origin and severity (acute or chronic) of hemorrhage. In the acute phase, with loss of blood volume, there may be no significant changes in hemoglobin level or hematocrit. Only after hemodilution occurs as a compensatory mechanism to restore blood volume will a reduced hemoglobin concentration and hematocrit be detected. For this reason, a single hemoglobin determination is not sufficient to determine or exclude blood loss in the acute situation and therefore should be regularly repeated.

Anemia resulting from an episode of acute bleeding may be normochromic and normocytic until the reticulocyte response occurs. This response, which normally occurs after 24 to 48 h, may be delayed if some underlying bone marrow disease is present. After the increase in reticulocytes, which may be accompanied by a mild leukocytosis and/or thrombocytosis, the anemia will have a macrocytic aspect because of the predominance of younger red cells which have a larger volume than old erythrocytes.

Chronic blood loss due to hemorrhage from a discrete lesion may remain clinically undetected and eventually result in a microcytic anemia due to depletion of the body's iron reserves. This phase will be accompanied by a normal or low reticulocyte count.

Reduced cell life

A reduced red cell life may be caused by a group of hereditary or acquired disorders, with intracorpuscular or extracorpuscular abnormalities leading to hemolysis (Table 1).

The hallmark of all hemolytic anemias is a significant increase in reticulocyte count. However, there is a 48-h gap between the fall in hematocrit and a noticeable rise in newly formed red blood cells. In addition, the appearance of the blood smear and the findings of elevated serum levels of lactate dehydrogenase and bilirubin, and a decrease in haptoglobin, may help to confirm the diagnosis of hemolytic anemia.

Table 1 Classification of anemias and additional laboratory tests

Classification	Etiology	Blood smear	Specific laboratory test
Increased loss	Blood loss	Normochromic normocytic	High reticulocytes
	Reduced red cell life (hemolysis)	Normochromic normocytic	High reticulocytes, elevated LDH and unconjugated bilirubin, low/undetectable haptoglobin
	Red cell interior defect		
	Enzyme defect (G6PD deficiency)		G6PD
	Hemoglobinopathies (thalassemia)	Target cells	
	Sickle cell anemia	Sickle cells	Hb electrophoresis
	Red cell membrane abnormality	Normochromic normocytic	
	Hereditary spherocytosis	Spherocytes	
	Paroxysmal nocturnal hemoglobinuria		Acidified Ham test
	Spur cell anemia	Spur cells	
	Extravascular factors	Normochromic normocytic	
	Splenomegaly		
	(Auto)antibody	Spherocytes	Coombs test
	Drugs		
	Intravascular factors		
	Trauma in the circulation (e.g. prosthetic heart valve)	Helmet cells/schistocytes	
	Microangiopathic hemolytic anemia		
	HUS	Fragmented red blood cells	
	TTP		
	DIC	Fragmented/helmet cells	Low platelets/decreased fibrinogen, thrombin time prolonged, PTT/PT prolonged FDP/D-dimers elevated
Reduced red cell production	Deficiency		
	Iron deficiency	Microcytic, hypochromic	High/normal reticulocytes, low iron
	Vitamin B_{12}, folic acid deficiency	Macrocytic, hypochromic	Low vitamin B_{12}, folic acid
	Bone marrow disease/failure	Normochromic normocytic	Low reticulocyte count
	Systemic (chronic) disease	Normochromic normocytic	Cold agglutinin/IgM antibody

LDH, lactate dehydrogenase; G6PD, glucose-6-phosphate dehydrogenase; Hb, hemoglobin; HUS, hemolytic uremic syndrome; TTP, thrombotic thombocytopenic purpura; DIC, disseminated intravascular coagulation; PTT, partial thromboplastin time; PT, prothrombin time; FDP, fibrin degradation products.

Decreased red blood cell production

Deficiencies of iron, vitamin B_{12}, or folic acid on the one hand and/or the presence of primary bone marrow disease on the other may result in a decreased production of red blood cells (Table 1). Underlying chronic diseases (i.e. malignancies, systemic diseases, metabolic disturbances) or (chronic) infections may also result in a decreased red cell production. Anemias in the latter group of disorders are characterized by an underproduction of red blood cells owing to a decreased availability of iron. Depletion of the body iron stores during chronic infection probably has an immunological etiology and is modulated by the production of cytokines. This mechanism is well demonstrated in some recent animal studies, suggesting that iron depletion has a protective effect in the critically ill septic patient (Kent *et al.* 1994).

Diagnostic approach

The initial test is the determination of hemoglobin level and hematocrit. In the case of acute bleeding, hemoglobin determination should be repeated at regular intervals as hemodilution and equilibration take 24 to 48 h to be completed and will be affected by (rapid) intravenous fluid replacement. Furthermore, the expected variability in hemoglobin testing is an important factor in the interpretation of the laboratory results.

In the work-up of the patient with anemia, the readily obtainable results of red cell smear and red cell indices will aid classification of the anemia. When the reticulocyte count is high, red cell production is sufficient. If the reticulocyte count is low, a bone marrow examination is indicated to diagnose a primary bone marrow failure.

Particularly in cases of microcytic hypochromic anemia, regular testing for occult blood loss in the feces may be helpful in identifying the site of blood loss. However, a hypochromic anemia with a negative occult feces test and no other explanation for blood loss must lead to further examination of the gut by endoscopic or radiographic examination.

Calculation of red cell indices and the reticulocyte count are also particularly helpful in the diagnosis and classification of anemias due to underproduction of red cells.

Microcytic anemia

This represents a group with decreased availability or synthesis of one of the major constituents of the hemoglobin molecule (iron, porphyrin, or globin). Anemia during chronic inflammation or in malignancy may be slightly microcytic due to a defect in the availability of iron, although these disorders are more often associated with normocytic anemia. As mentioned earlier, in the case of microcytic anemia a source of (chronic) bleeding should be regarded as the most important factor to exclude first.

Macrocytic anemia

A deficiency of either vitamin B_{12} or folic acid results in the impairment of the replication of DNA, leading to large macrocytic cells in both the bone marrow and peripheral blood. Liver disease, alcoholism, aplastic and hemolytic anemia, and acute blood loss may also be associated with a macrocytic anemia.

Normocytic anemia

Normocytic normochromic anemia often accompanies chronic disorders such as rheumatoid arthritis, (chronic) infections, diabetes mellitus, hypothyroidism, or renal failure. These disease states may result in a decreased production of red cells which is probably due to production of toxins and cytokines (Kent *et al.* 1994). Finally, diseases associated with intrinsic bone marrow pathology will lead to underproduction of red cells and result in a normocytic anemia.

Compensatory mechanisms

Anemia will lead to a decreased capacity of oxygen transport, eventually resulting in tissue hypoxia and even necrosis when compensatory mechanisms are insufficient. In response to tissue hypoxia, 2,3-diphosphoglycerate in the red cells will increase, leading to increased oxygen dissociation at any given $P\text{O}_2$. Cardiac compensatory mechanisms, particularly an increase in heart rate, result in an increase in cardiac output. Blood loss followed by compensatory fluid replacement reduces blood viscosity, resulting in an increased tissue perfusion and a decreased cardiac workload.

In the patient with cardiopulmonary compromise, these compensatory mechanisms may not be sufficient and cardiac decompensation, angina pectoris, and even myocardial infarction may result.

There are no accurate means of predicting failure of the compensatory mechanisms to anemia. Thus it is essential to obtain information regarding the patient's medical history. In particular, any history of heart and/or vascular disease should influence the decision as to whether or not the anemia should be corrected. However, criteria for transfusion are arbitrary and vary considerably.

References

Bunn, H.F. (1994). Anemia. In *Harrison's principles of internal medicine* (13th edn) (ed. K.J. Isselbacher *et al.*), pp. 313–17. McGraw-Hill, New York.

Feingold, A.O. (1986). Disorders of the red blood cells. In *Principles and practice of emergency medicine* (2nd edn), (ed. G.R. Schwartz *et al.*), pp. 1132–42. W.B. Saunders, Philadelphia, PA.

Kent, S., Weinberg, E.D., and Stuart-Macadam, P. (1994). The etiology of the anemia of chronic disease and infection. *Journal of Clinical Epidemiology*, **47**, 23–33.

10.4.2 Hemolysis

J. Isbister

Key messages

- The diagnosis of hemolytic anemia in the critical care setting can be challenging due to the multifactorial nature of patients' pathophysiology and multiple therapies which may be complicated by hemolytic anemia.
- A problem-oriented approach to the cause of hemolysis centers around examination of the blood film, which requires close liaison with the hematology laboratory.
- In the critical care setting the hemolysis may have an acute onset and the marrow may be slow in responding due to suppression from shock or sepsis.

Introduction

The diagnosis and investigation of hemolytic anemia in the critical care medicine setting can be challenging because of the multifactorial nature of the patient's pathophysiology, confounding factors, and multiple therapies which may be complicated by hemolytic anemia. As hemolysis is an important diagnosis to establish and may present in a plethora of clinical settings and in many deceptive ways, it is important to have a good problem-solving approach to the initial assessment and definitive diagnosis of hemolysis. In the critical care setting hemolysis is usually of acute onset as a feature of the primary disease or a complication of therapy.

When should acute hemolysis be suspected?

Presentation may be as follows.

1. Acute anemia and its pathophysiological effects
2. Jaundice
3. Acute intravascular hemolysis
4. Activation of proteolytic systems and the consequences thereof:
   - complement and kinin systems → shock, anaphylactic/anaphyloid reactions
   - cytokine activation and release → cytokine storm, systemic inflammatory response syndrome
5. Dark urine: hemoglobinuria, bilirubinuria

6. Laboratory presentation: hemoglobinemia, hyperbilirubinemia, high lactate dehydrogenase
7. Oliguria ± renal failure
8. Cold agglutinins → microvascular occlusion

The primary disease may be one of acute hemolysis presenting as a medical emergency.

1. Hereditary red cell disorders and hemoglobinopathies
   - (a) Hereditary spherocytosis: with acute infection ± aplastic crisis
   - (b) Sickle cell disease: with acute infection, hypoxia, shock ± aplastic crisis
   - (c) Glucose-6-phosphate dehydrogenase deficiency: with infection, favism, drugs
   - (d) Unstable hemoglobinopathies: with infection, drugs
2. Acquired primary acute hemolytic diseases
   - (a) Acute autoimmune hemolytic anemia:
     acute cold agglutinin syndrome (e.g. mycoplasma infection, cytomegalovirus)
     acute warm hemolytic anemia (idiopathic, systemic lupus erythematosus)
     paroxysmal cold hemoglobinuria
     paroxysmal nocturnal hemoglobinuria

The primary disease may have hemolysis as a feature or complication.

1. Microangiopathic syndromes
   - (a) Disseminated intravascular coagulation
   - (b) Thrombotic thromocytopenic purpura
   - (c) Hemolytic uremic syndrome
   - (d) HELLP syndrome
2. Infection
   - (a) Clostridial septicemia
   - (b) Malaria and babesiosis
   - (c) Viral associated hemophagocytic syndrome
3. Toxic or traumatic
   - (a) Burns
   - (b) Mechanical: extracorporeal procedures (cardiopulmonary bypass, blood salvage)
   - (c) Snake bite
   - (d) Near drowning
   - (e) Large intravenous hypotonic solutions or sterile water

Acute hemolysis may be related to therapy.

1. Blood transfusion
   - (a) Immediate acute incompatible transfusion reaction (usually ABO errors)
   - (b) Delayed hemolytic transfusion reaction
   - (c) Stored blood hemolysis secondary to incorrect storage or mechanical hemolysis.
2. Drug induced
   - (a) Immune-mediated, particularly antibotics, non-steroidal anti-inflammatory drugs, quinine, quinidine
   - (b) Metabolic oxidative hemolysis (Heinz body hemolysis), particularly in glucose-6-phosphate dehydrogenase deficiency and unstable hemoglobinopathy

Questions to ask in the setting of suspected hemolysis

What clinical features are suggestive or supportive of the diagnosis of hemolysis?

1. Family history
2. Past history (e.g. jaundice, gallstones, leg ulcers)
3. Recent history (e.g. jaundice, dark urine, pains, Raynaud's phenomenon, shivers and sweats, pyrexia, blood transfusion, drug administration, toxin exposure, envenomation, burns)

Are there any laboratory features of hemolysis?

Figure 1 summarizes the pathways of hemoglobin degradation and the biochemical abnormalities found in hemolysis.

Naked eye examination

Red or brown plasma in the presence of dark urine is highly suggestive of hemolysis. Clear plasma in the presence of dark urine suggests myoglobinuria. Jaundice with clear urine suggests hemolysis, but with dark urine (frothy on shaking) suggests conjugated hyperbilirubinemia.

Urinalysis

Urobilinogen in the urine indicates increase bilirubin turnover. Bilirubinuria indicates conjugated hyperbilirubinemia. Positive testing for 'blood' should be further examined. Frank bright red hematuria or 'smoky' urine indicates the presence of red cells. Dark 'muddy' urine suggests the presence of oxidized hemoglobin or myoglobin. It is desirable to examine the urine further, both microscopically and biochemically. The differentiation of hemoglobinuria from myoglobinuria needs expert knowledge and careful interpretation.

Blood film examination

Hemolysis may be suspected on blood film examination. The presence of anemia with polychromasia raises this possibility. Specific red cell changes may be highly suggestive or diagnostic of hemolysis and so the blood film should be reviewed.

Hyperbilirubinemia

If liver and biliary function are normal, an unconjugated hyperbilirubinemia occurs in hemolysis. The only common differential diagnosis is that of Gilbert's syndrome which is usually a diagnosis of exclusion. Routine techniques for measuring conjugated bilirubin are indirect and not truly accurate. A normal person does not have conjugated bilirubin present, but up to one-third of the bilirubin may appear conjugated by these methods. There are several clinical situations in which there may be a combination of hemolysis in conjunction with impaired hepatobiliary handling of conjugated bilirubin. The bilirubin transport system may be stressed by hemolysis, with hepatic

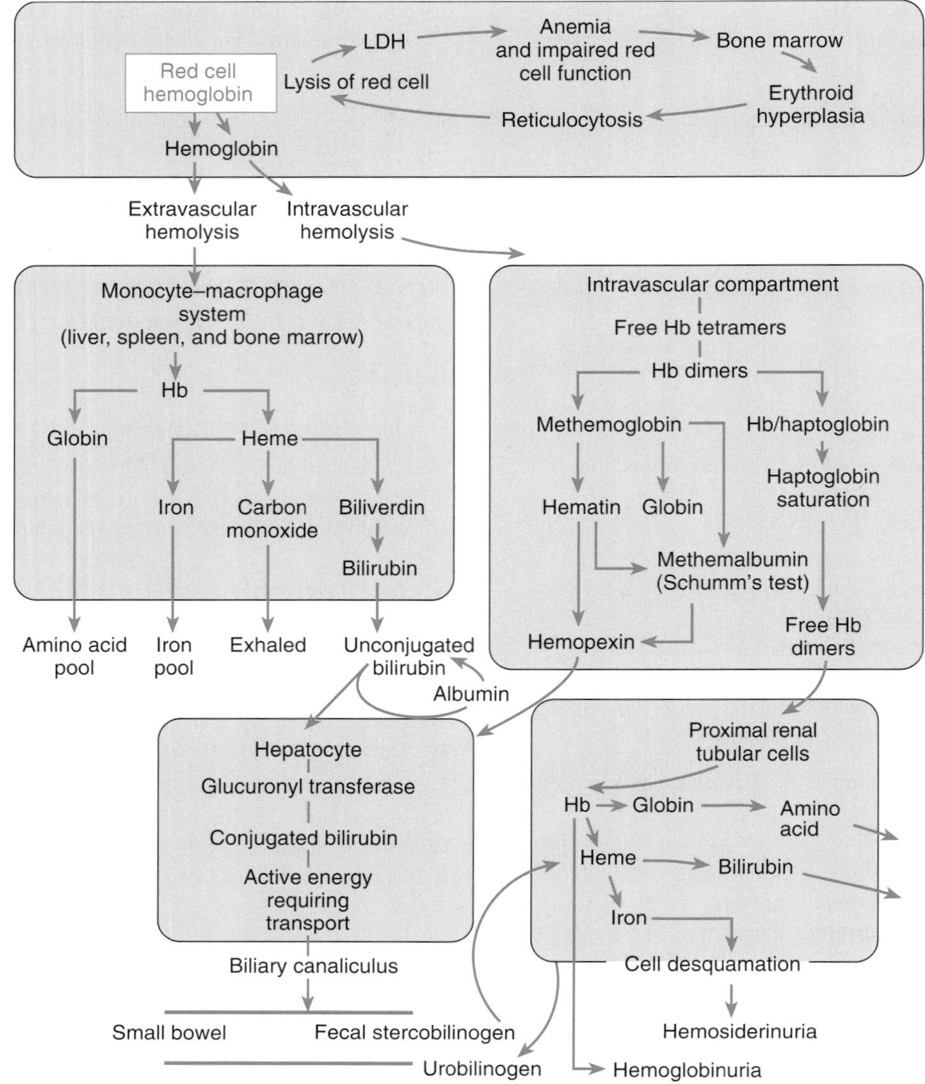

Fig. 1 Pathways of hemoglobin degradation and the biochemical abnormalities found in hemolysis.

conjugation proceeding normally but with a build-up occurring at the excretory level. Excluding obvious biliary obstruction, the usual defect seen in seriously ill patients (who are high-risk candidates for excess bilirubin production due to hematoma resorption or hemolysis secondary to blood transfusion, drugs, infections, etc.) is the active transport of conjugated bilirubin from the hepatocyte to the biliary canaliculus. Critically ill patients with acute hemolysis are likely to have impaired bilirubin transport owing to the effects of shock or sepsis. Under these circumstances bilirubin from hemolysis will be rapidly conjugated but excretion will be delayed; this is manifest as conjugated hyperbilirubinemia. Any sudden rise in the bilirubin level (conjugated or unconjugated), particularly if other liver function tests have not risen, is highly suggestive of hemolysis. Resorption of hematoma is a common trap in the diagnosis of hemolysis.

Reticulocytosis

In the critical care setting the hemolysis may have an acute onset and the marrow may be slow in responding because of marrow suppression from shock or sepsis. Bone marrow failure due to associated or unrelated causes (e.g. aplastic crisis, malignant infiltration, folate deficiency) will prevent reticulocytosis and can result in profound and rapidly fatal anemia.

Elevated lactate dehydrogenase

An isolated marked elevation of lactate dehydrogenase, with other 'profile' enzymes (aspartate aminotransferase (AST), alanine aminotransferase (ALT), creatinine kinase, γ-glutamyl transferase, and alkaline phosphatase) remaining normal, is highly suggestive of red cell destruction (hemolysis or ineffective erythropoiesis). Red cells lack mitochondria, which explains why the other 'standard' enzymes for tissue damage (e.g. ALT, AST) are normal or only marginally elevated. Lactate dehydrogenase has five isoenzymes, with LD1 being present in red cells. Muscle damage will markedly elevate all enzymes, particularly creatine kinase, thus helping to differentiate hemolysis from rhabdomyolysis.

Haptoglobins

Reduced haptoglobins are not a reliable indicator of hemolysis. Haptoglobins are acute phase reactants and may be elevated in infec-

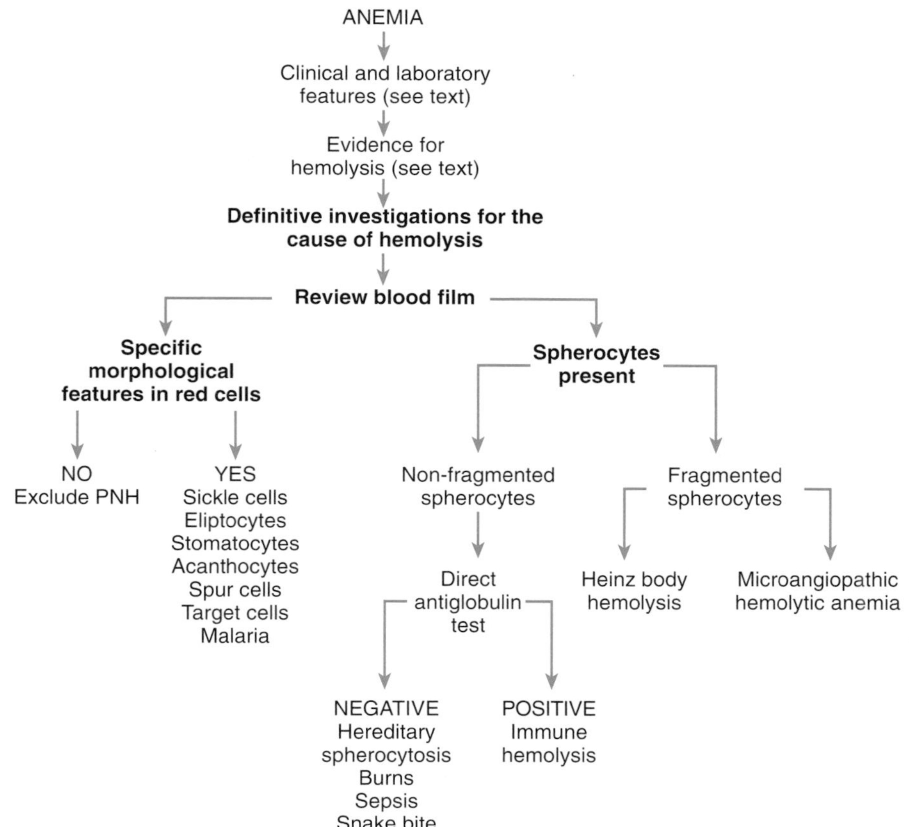

Fig. 2 Algorithm for the investigation of hemolytic anemia.

tion and inflammation but, in contrast, may be reduced by blood transfusion as a result of non-surviving stored red cells. If they are present or increased, significant hemolysis can be excluded, but their reduction or absence must be interpreted with caution.

Plasma hemoglobin

Plasma hemoglobin levels are susceptible to several measurement and interpretative difficulties. Blood may be hemolysed during collection and small elevations are probably not of clinical significance.

Schumm's test

In severe intravascular hemolysis free hemoglobin is released into the circulation, initially binding to haptoglobins which are rapidly cleared from the circulation by the liver. This clearance mechanism is limited. In its tetrameric form free hemoglobin is not filtered by the kidney and accumulates in the blood, splitting into dimers which can be filtered as hemoglobinuria. Hemoglobin is oxidized to methemoglobin and the globin chains are separated from the heme molecule. This metheme binds to hemopexin which is responsible for its clearance from the circulation and conservation of the iron, but when hemopexin is overloaded the metheme binds to albumin, forming methemalbumin (positive Schumm's test).

Hemosiderinuria

In chronic intravascular hemolysis or a week after an acute episode, examination of the urine for iron may confirm the presence of hemosiderinuria.

What steps should be taken to establish the cause of hemolysis?

There are over 50 causes of hemolysis and investigation can be rather daunting. The mechanistic classifications of hemolytic anemia into intracorpuscular, membrane, and extracorpuscular types have served well in extending our understanding of the causes, but can be a hindrance in attempting to find a way through the diagnostic maze of complex investigations. As a result of these mechanistic approaches, the clinician has been tempted to request investigations in a 'poker machine pathology' fashion, in the hope of hitting the jackpot. To make matters worse, the hemolytic episode may have been acute and self-limiting, and investigations are carried out during the 'aftermath'. A problem-oriented approach centers around examination of the blood film, which requires close liaison with the hematology laboratory.

What does the blood film show?

Figure 2 shows an algorithm for the investigation of hemolytic anemia. Whatever the cause of hemolysis, the end result is rupture and destruction of the red cell. Rupture involves membrane damage which need not be complete from the outset (except in fulminant intravascular hemolytic reactions), and so degrees of development of red cell damage may be observed in the circulating red cells. If membrane is lost evenly from the red cell, as seen in hereditary spherocytosis and autoimmune hemolysis, smooth regular hyperchromic spherocytes will be seen in the blood film. If the spherocytes are irregular fragmented cells, there may be specific morphological changes, for example Heinz body

Table 1 The management of acute hemolytic anemia

Cause	Diagnosis	Specific therapy
Microangiopathic hemolysis	Feature of underlying disease	Non-specific therapy for hemolysis and for underlying disease
Infection and toxin related	Feature of underlying disease	Non-specific therapy for hemolysis and for underlying disease Rarely, red cell exchange in malaria and babesiosis
Blood transfusion reaction	Awareness and early recognition Serological test checking blood groups and cross-matching Direct antiglobulin test and antibody screen	Cease transfusion and resuscitate+non-specific therapy for hemolysis and for underlying disease
Drug-induced hemolysis	Awareness and early recognition Checking for likely drugs Follow-up tests for confirmation	Cease drug and future drug alert
Sickle cell disease	History, sickling test Hemoglobin electropheresis	Hydration, treat infection, analgesia, consultation
Glucose-6-phosphate dehydrogenase deficiency	History, males, specific tests	Non-specific therapy for hemolysis, exclude drug induction

spherocytes and microangiopathic hemolytic anemia. In some hemolytic anemias the red cell morphology itself may be diagnostic (e.g. sickle cells, malarial parasites, autoagglutination).

What specific investigations should be performed to confirm the final diagnosis?

When investigations lead to an understanding of the mechanisms of hemolysis, it is then possible to pursue more specific investigations. Some of these tests are complex and require expertise. With failure of the definitive test to confirm the diagnosis, clinicians are sometimes surprised at how difficult it can sometimes be to establish firmly the cause of hemolysis. In some hemolytic diseases it is essential that the tests be performed at the time of hemolysis; in other cases the tests will only be diagnostic when the patient is not hemolysing.

Causes of hemolytic anemia

Table 1 outlines the causes, diagnosis, and management of red cell hemolysis.

General management measures for hemolytic anemias

During acute hemolysis there may be a marked systemic reaction with activation of several proteolytic systems in the plasma, such as the kinin–kininogen, complement, and coagulation systems, and also the release of cytokines. A shock state may occur, with the development of prerenal or renal failure. Acute apprehension and pain in the abdomen, chest, or back may be severe and lead to the suspicion of other diagnoses. Attention to circulatory perfusion and urinary output is essential, and an alkaline diuresis should be established. The general principles for the management of acute hemolysis are as follows.

1. Early diagnosis:
   - awareness of possible causes and clinical settings in which hemolysis may occur
2. Early therapy:
   - removal of immediate cause and/or treatment of underlying condition
   - resuscitation if necessary
   - establish and maintenance of an alkaline diuresis
3. Recognition of complicating factors:
   - coagulopathy, renal failure, poor marrow function, liver impairment
   - prevention of venous thromboembolism
   - associated respiratory failure (e.g. mycoplasma, thrombotic thrombocytopenic purpura, systemic lupus erythematosus)
4. Documentation:
   - avoid recurrence in future

Bibliography

Gottschall, J.L., Neahring, B., McFarland, J.G., Wu, G.-G., Weitekamp, L.A., and Aster, R.H. (1994). Quinine-induced immune thrombocytopenia with hemolytic uremic syndrome: clinical and serological findings in nine patients and review of literature. *American Journal of Hematology*, **47**, 283–9.

Isbister, J.P. (1990). Molecular disorders of the red cell: haemoglobinopathies, enzymopathies and membrane defects. In *Scientific foundations of anaesthesia* (4th edn) (ed. C. Scurr, S. Feldman, and N. Soni), Chapter 40. Heinemann, Oxford.

Isbister, J.P. (1994). Hemopoietic system. In *Laboratory medicine: the selection and interpretation of clinical laboratory studies* (ed. D.A. Noe and R.C. Rock), Chapter 13. Williams & Wilkins, Baltimore, MD.

Petz, L.D. (1993). Drug-induced autoimmune hemolytic anemia. *Transfusion Medicine Reviews*, **7**, 242–54.

Serjeant, G.R. (1993). The clinical features of sickle cell disease. *Baillière's Clinical Haematology*, **6**, 93–115.

10.4.3 Blood transfusion

P. Hewitt and F. Regan

Key messages

- Transfusion of wrong blood does occur in the intensive care unit.
- Irradiate blood for severely immunodeficient patients or blood from a first-degree relative.
- Transfuse with caution in conditions of hyperviscosity.
- If O RhD-negative blood is used in an emergency, switch to ABO- and RhD-compatible blood as soon as possible.
- In massive transfusion, give fresh frozen plasma and platelets according to clinical and laboratory indices, not fixed prophylactic replacement formulas.

Introduction

The aim of treatment of anemia is to ensure an adequate oxygen supply to the tissues. Oxygen delivery depends on cardiac output and arterial oxygen content, which in turn depends on hemoglobin concentration and oxygen carried per gram of hemoglobin (Mollison *et al.* 1994).

Physiological compensatory mechanisms for anemia

In the presence of anemia, cardiac output increases to maintain oxygen delivery. This occurs whether the anemia arises from acute blood loss or is of gradual onset. Less immediately, the red cell content of 2,3-diphosphoglycerate rises, shifting the oxygen dissociation curve to the right so that, at any given level of hemoglobin oxygen saturation, oxygen is given up more readily to the tissues. Provided that normal blood volume is maintained, oxygen delivery is enhanced by a moderate degree of anemia. This is due to a reduction in blood viscosity and increased cardiac output, leading to increased extraction of oxygen from blood perfusing the tissues (Wilkerson *et al.* 1987). However, ischemia may result if severe acute anemia occurs in the presence of coronary or cerebral atherosclerosis.

Factors contributing to anemia

In critically ill patients, anemia is often multifactorial. Causes include hemorrhage, drugs, nutritional deficiencies, repeated blood sampling, renal failure, anemia of chronic disease, or delayed hemolytic transfusion reactions. If possible, the underlying cause(s) should be established and treated. However, transfusion of red cells is still often required.

Indications for transfusion

In general, red cells are used to replace blood lost in hemorrhage or to treat symptomatic anemia where hematinics (iron, vitamin B_{12}, or folate) alone are inappropriate or insufficient. In the absence of acute hemorrhage, the decision to transfuse a critically ill anemic patient depends on many factors but cannot be made universally on a fixed threshold level of hemoglobin or hematocrit. The potential benefits must be weighed against the risks.

Table 1 Risks of transfusion

Infections
Wrong blood
Immune
Red cell alloimmunization
HLA alloimmunization
Transfusion-associated graft versus host disease
Transfusion-related acute lung injury
Metabolic
Circulatory

Table 2 Estimated rates of transfusion-transmitted infection in the United States and the United Kingdom

Infection	Estimated rates in USA	Estimated rates in UK
HIV	1:420 000	1:3 000 000
Hepatitis B	1:37 000	1:20 000
Hepatitis C	1:62 000	1:200 000

Risks of transfusion

The risks of transfusion are summarized in Table 1.

Infection

The risks of transfusion-transmitted infection remain a concern to many patients, although they are now lower than ever (Contreras and Barbara 1993; Busch 1995). Transmission of HIV, hepatitis B, and hepatitis C is rare (Table 2); however, other agents, such as human T-cell lymphotrophic virus (HTLV), malaria, and trypanosomiasis (Chagas' disease), are important in certain geographical areas. In an attempt to avoid such infections, the use of directed donations (from relatives or friends) may be requested, but studies in the United States have shown that these are no safer than allogeneic blood donations. Bacterial contamination of blood components still occurs occasionally, and all red cell units should be examined for hemolysis before infusion.

Incorrect blood

A more common hazard is infusion of wrong blood with potentially fatal consequences of ABO incompatibility. Transfusion of wrong blood is usually due to an error in blood sampling for compatibility testing or in identifying the correct unit or patient for transfusion; less frequently, incorrect testing within the laboratory is the cause. Such errors have occurred in the intensive care unit setting (Baele *et al*. 1994). When these errors occur, they should be discussed with transfusion staff as advice can be given on patient management; treatment of intravascular hemolysis may be required, depending on the nature of the ABO incompatibility. If RhD-positive blood is given to an RhD-negative recipient, exchange transfusion may be appropriate, depending on the age, gender, and underlying condition of the patient. Secondly, the exact incidence of errors is not accurately defined, as reporting has been erratic. A central registry for confidential reporting of such errors in the United Kingdom is now established.

Immune complications

Red cell alloimmunization

Red cell transfusion leads to red cell antibody formation in some patients, particularly certain populations such as those receiving multiple transfusions or with sickle cell disease. Phenotyped red cells are then required. Occasionally, delayed hemolytic transfusion reactions occur when antibodies in the recipient's serum are below the threshold for detection.

Human leukocyte antigen (HLA) alloimmunization

Patients may develop HLA antibodies to antigens on white cells present as contaminants in red cell preparations. HLA antibodies can cause non-hemolytic febrile transfusion reactions or reduced survival of transplanted organs, or the patient may become refractory to platelet transfusions.

Allergic reactions

Rarely, hypersensitivity to plasma proteins causes urticaria or even anaphylaxis; this is less likely when using red cell concentrates rather than whole blood as most of the donor plasma has been removed.

Transfusion-associated graft versus host disease

This may occur in severely immunodeficient patients or in patients receiving blood from a first-degree relative. It can be prevented by irradiation of red cells (or platelets) to destroy donor leukocytes.

Transfusion-related acute lung injury

If leukocyte antibodies are present in the donor unit of red cells (or platelets or fresh frozen plasma), they can interact with leukocyte antigens present on the recipient's white cells. A severe and potentially fatal reaction follows, with fever, dyspnea, and hypotension. A chest radiograph typically shows numerous perihilar nodules and infiltration of the lower lung fields without cardiac enlargement. Leukodepletion is not effective in preventing transfusion-related acute lung injury as it cannot remove antibodies present in the donor's blood. Implicated donors should be removed from the donor panel.

Immune modulation

Currently, there is much debate about possible immune-modulating effects of blood transfusion which may have additional effects on the recovery of a critically ill patient.

Metabolic disturbances

These are mainly relevant in massive transfusions (see below).

Circulatory overload

Circulatory overload occurs particularly in patients with underlying cardiovascular disease or conditions predisposing to hyperviscosity (e.g. acute leukemia, multiple myeloma). Transfusion may precipitate complications of hyperviscosity, such as endorgan hemorrhage, including retinal and cerebrovascular bleeds. Transfusion should be undertaken with caution; leukopheresis may be indicated in acute leukemia before red cells are given.

Consent for transfusion

In some countries the informed consent of the patient or the relatives is obtained before an elective transfusion is performed. This is not routine practice in the United Kingdom, although it is still under debate.

Management of red cell transfusion

Different red cell components are indicated under different circumstances. Red cell components available include the following.

Whole blood

Storage at 4 °C is optimal for red cell viability, but leaves whole blood devoid of lasting functional platelets and certain coagulation factors (factor VIII levels decrease to 50 per cent after 1 day). Therefore whole blood is usually separated within hours of donation so that each component can be stored under optimal conditions. Red cell concentrate is stored at 4 °C, plasma is frozen quickly for storage as fresh frozen plasma or prior to fractionation for factors VIII and IX, and platelets are stored at 22 °C, with continuous agitation to maintain platelet function, for a maximum of 5 days.

The concept of fresh whole blood, only a few hours old and containing functional platelets and coagulation factors, is now obsolete, as testing to exclude viral infections takes 24 h. Consequently, it is preferable to use red cell concentrates and supplement with fresh frozen plasma or platelets for hemostatic problems, or with crystalloid or colloid solutions if volume expansion is required. Therefore whole blood is rarely indicated, except in massive transfusions.

Supplemented red cell concentrate

After blood has been donated, most of the plasma is removed and 100 ml of nutrient solution (e.g. saline adenine glucose mannitol (SAGM)) is added. The final hematocrit is approximately 60 per cent.

Standard red cell concentrate

Red cells are concentrated and suspended in about 80 ml of a 4:1 mixture of donor plasma and an anticoagulant–nutrient solution, giving a hematocrit of 65 to 75 per cent. Supplemented red cell concentrate is now replacing standard concentrate.

Special requirements

Phenotyped red cells

Some patients develop red cell alloantibodies when exposed to foreign red cell antigens during transfusion, or during pregnancy where fetal hemorrhage at delivery sensitizes the mother. Before transfusion, all patients are screened for red cell alloantibodies. The antibody is defined and ABO- and RhD-compatible red cells lacking the corresponding antigen are selected to prevent a hemolytic transfusion reaction. Identification of the antibody (if not previously known) may take several hours. In an emergency, if there is a high risk of the patient dying immediately through lack of red cell transfusion, it is preferable to give red cells which are simply ABO and RhD compatible, and to treat any subsequent hemolysis later. This should be discussed with transfusion medical staff.

Leukocyte-depleted (filtered) red cells

Leukodepletion may be performed within a closed system in the laboratory or using a bedside leukocyte filter. Non-hemolytic febrile transfusion reactions are mostly due to the development of HLA antibodies in the recipient which react with HLA antigens on donor leukocytes. Therefore leukocyte-depleted components may be used prophylactically to prevent HLA sensitization in patients who are chronically transfusion dependent, or who are potential recipients of organ transplants. Similarly, leukodepleted red cells may be used to lessen reactions in patients with a history of severe non-hemolytic febrile transfusion reactions. Leukocyte depletion of red cells can also be used to prevent transmission of cytomegalovirus infection if cytomegalovirus seronegative red cells are not available.

Irradiated red cells

Red cell irradiation can prevent graft versus host disease in severely immunodeficient patients (e.g. following bone marrow transplantion, intrauterine (fetal) transfusions, and those with cellular immune deficiencies). Irradiated red cells are also indicated in patients receiving red cells from a first-degree relative, where donor leukocytes may not be recognized as foreign by recipient leukocytes as many HLA antigens are common to both. Following irradiation, the potassium content rises; therefore, once irradiated, red cells should be transfused as soon as possible.

Cytomegalovirus-negative red cells

These are indicated for cytomegalovirus-seronegative patients in whom cytomegalovirus infection could be severe, for example bone marrow transplant recipients who have received marrow from a cytomegalovirus-seronegative donor, cytomegalovirus-seronegative HIV-positive patients, and neonates weighing less than 1.5 kg. All patients who may in future require a bone marrow transplant and who are cytomegalovirus seronegative, or whose status is unknown, require cytomegalovirus-negative blood components.

Administration of red cells

Transfusion of 4 ml/kg of red cell concentrate should raise hemoglobin levels by approximately 10 g/l (1 g/dl).

Timing

Red cells are stored at 4 $\pm$ 2 °C. Any blood left at higher temperatures for more than 30 min before transfusion begins should be returned to the blood bank and discarded. Transfusion of a unit of red cells should be completed within 4 h.

Apparatus

Red cells are given via a blood-giving set containing an in-line 170-μm filter to remove occasional debris. A corresponding platelet-giving set is indicated for platelet infusions. Microaggregate filters (40 μm) have been used in large-volume transfusions to prevent pulmonary complications but, on the basis of current evidence, their benefit is unproven. Microaggregate filters are used in the arterial line of an extracorporeal circuit, where a decrease in postoperative morbidity and cerebrovascular emboli has been shown. Blood warmers are indicated when a large volume of red cells is given in a short time or where the patient is known to have clinically significant cold agglutinins. In addition, they are appropriate in critically ill patients who would otherwise use extra energy in warming the infused blood.

Additives

No drugs should be added to red cells as these may have potentially harmful effects on the stored red cells. Also, many drugs are prepared in solutions of an optimal pH which may not prevail in stored blood. Similarly, calcium-containing solutions should not be infused with blood, as anticoagulants used in stored blood depend on calcium depletion.

Management of massive transfusion

Massive transfusion is defined as a transfusion of blood equivalent to the patient's normal blood volume given within 24 h. Severe hemorrhage causes acute hypovolemic shock leading to underperfusion of vital organs and tissues. Mortality increases with the duration and severity of shock. Urgent resuscitation is needed to restore the circulating blood volume and oxygen-carrying capacity of blood and to maintain hemostasis and biochemical balance.

Following a massive transfusion, mortality and morbidity are high; this is mainly due to the underlying condition causing the blood loss rather than the massive transfusion itself. Some conditions predispose to abnormalities of hemostasis. Prolonged hypotension, extensive tissue damage, and obstetric complications can cause disseminated intravascular coagulation (DIC), and established liver or renal disease may impair coagulation and exacerbate metabolic complications associated with massive transfusion.

Restoration of blood volume

Initially, crystalloid and colloid solutions may be used, but when replacement of blood lost reaches 40 per cent of the patient's blood volume, transfusion of red cells is required. Red cell concentrates may be used initially, but whole blood is preferable after the first 4 units of an anticipated massive transfusion, as some coagulation factors and proteins which may contribute towards normal hemostasis and colloid osmotic pressure are present in the plasma of whole blood.

Once the need for a massive transfusion is established, blood samples should be taken for compatibility testing (cross-matching) and also for baseline hemoglobin, platelet, coagulation, and biochemical measurements (Table 3). Normally, the patient's ABO and RhD group are determined, a red cell antibody screen is carried out, and the patient's blood is tested for compatibility with the donor's red cells. If blood is needed more urgently and the patient's ABO and RhD group are already known, ABO- and RhD-compatible blood can be issued without waiting for compatibility testing. If the patient's ABO and RhD group and antibody status are not known, then O RhD-negative blood may be given until the patient's group is determined. Switching to ABO- and RhD-compatible blood should be done as soon as possible for two reasons: first to avoid unnecessary use of O RhD-negative red cells, and second to minimize the volume of group O plasma in whole blood transfused to a group A, B, or AB patient as group O plasma contains anti-A and anti-B antibodies. Up to 10 units of group O whole blood can be given before the anti-A and anti-B antibodies present will cause significant problems with subsequent red cells of the patient's own ABO group.

In an emergency, giving group O RhD-negative red cells to patients of unknown ABO and RhD groups avoids the major problem of ABO- and RhD-incompatible transfusion reactions, but does not prevent reactions due to antibodies of other red cell antigen systems. Many of these can cause subsequent hemolytic transfusion reactions, although they are usually less severe than ABO-incompatible transfusion reactions. These can be treated as they occur.

Table 3 Minimum blood tests at the beginning of massive transfusion

Compatibility testing ('cross-match')
Hemoglobin or hematocrit, platelet count
Prothrombin time, partial thromboplastin time, thrombin time
Biochemistry: urea and electrolytes, creatinine

Complications of massive transfusion

Hemostasis

Hemostasis may already be abnormal due to the underlying condition. During massive transfusion of red cells, platelet numbers fall as few functioning platelets exist in whole blood that has been stored for more than 48 h. Factors V and VIII are reduced after storage for a few days, and remaining levels may be diluted if large volumes of intravenous crystalloid or colloid are given. In addition, DIC may be provoked by release of thromboplastin-like material from platelets, white cells, and red cells broken down during storage of red cells and because of partial activation of coagulation factors. The extent of hemostatic derangement varies widely and is not predictable according to the volume of red cells transfused. Therefore prophylactic replacement formulas (e.g. platelets and fresh frozen plasma after every eight units of red cells) are not recommended. It is preferable to monitor hemostasis and use fresh frozen plasma when abnormalities of the coagulation system exist. Cryoprecipitate is indicated in DIC when the fibrinogen level is below 0.8 g/l. Thrombocytopenia below 50×10^9/l contributes to microvascular bleeding from mucosal surfaces (e.g. mouth, gastrointestinal tract, epistaxis), wounds, and puncture sites. A standard adult dose of platelets is usually 6 to 8 units of platelets (individually or pooled) which gives approximately 1 unit of platelets per 10 kg body weight. One unit of platelets contains at least 0.55×10^9 platelets in 50 ml of plasma. The patient's platelet count may be monitored to guide further requirements if bleeding continues.

Metabolic complications

Oxygen-carrying capacity

Anticoagulant–preservative solutions (e.g. citrate, phosphate, dextrose, adenine) in all red cell components help to maintain the 2,3-diphosphoglycerate levels and therefore the oxygen-carrying capacity of red cells for up to 14 days during storage. It is reasonable to use blood less than 14 days old for massive transfusion.

Hyperkalemia

A raised potassium level from stored red cells is usually transient and is not often a problem by itself, although with hypocalcemia and hypothermia, cardiac arrhythmias may result. Also, hypokalemia may develop within 24 h of transfusion of stored red cells as potassium returns to the red cells following transfusion.

Hypocalcemia

This may arise following a massive transfusion, as stored blood contains anticoagulants which cause calcium depletion. Clinical manifestations are rare, and prophylactic use of intravenous calcium to correct the hypocalcemia may cause more problems than it prevents.

Hypothermia

Transfusion of large volumes of red cells stored at 4 °C may cause hypothermia; therefore blood should be given via a blood warmer.

Citrate toxicity

Citrate is present in the anticoagulant–preservative solution of red cells. Toxicity is rare, as citrate is metabolized rapidly by the liver and, to exceed this metabolism, red cells would need to be given at a rate of 1 unit every 5 min.

pH balance

Stored red cells are acidotic, but metabolism of citrate produces an alkalosis. Therefore pH imbalance due to red cell transfusion itself is rare, but hypovolemic shock and the underlying conditions may cause an acidosis.

Plasma colloid osmotic pressure

Plasma colloid osmotic pressure falls as large volumes of crystalloids and red cells without plasma are given. Infusion of colloid may be required to counteract this.

It should be noted that previous standard regimens for managing massive transfusions included fresh frozen plasma, platelets, alkalizing agents, and calcium in fixed amounts according to the volume of red cells transfused. These are often unnecessary and therefore wasteful of limited resources, and may even be harmful. Current recommendations are to give supplements according to each patient's needs based on clinical and laboratory indices.

Future developments

Alternatives to the transfusion of red cells are being sought in order to overcome some of the hazards associated with transfusion. An appropriate substitute must have an adequate oxygen-carrying capacity, release oxygen readily to the tissues, and be non-toxic. Modified hemoglobin solutions are currently undergoing trials, but difficulties remain with the stability of hemoglobin compounds as some are too rapidly metabolized to be useful.

References

Baele, P.L., *et al.* (1994). Bedside transfusion errors: a prospective survey by the Belgium SAnGUIS group. *Vox Sanguinis*, **66**, 117–21.

Busch, M.P. (1995). Incidence of infectious disease markers in blood donors: implications for residual risk of viral transmission by transfusion (abstract). In *Program and Abstracts of National Institutes of Health Consensus Development Conference on Infectious Disease Testing for Blood Transfusions, Bethesda, MD, 9–11 January 1995*, pp. 29–30. National Institutes of Health, Washington, DC.

Contreras, M. and Barbara, J. (1993). Transfusion-transmitted infections and their impact on virology. *Current Medical Literature, Virology*, **4**, 67–72.

Mollison, P.L., Engelfriet, C.P., and Contreras, M. (1994). Blood transfusion. In *Clinical medicine*, pp. 55–6. Blackwell Scientific, Oxford.

Wilkerson, D.K., Rosen, A.L., Gould, S.A., Sehgal, L.R., Sehgal, H.L., and Moss, G.S. (1987). Oxygen extraction ratio: a valid indicator of myocardial metabolism in anemia. *Journal of Surgical Research*, **42**, 629–34.

10.5 Polycythemia

10.5.1 Pathophysiology

C. A. J. Wardrop

Key messages

- Because of almost invariable falls in intravascular red cell and plasma volumes, polycythemias are uncommon as problems in intensive care patients, but do occur sporadically.
- Based upon measured red cell and blood volume, polycythemias are classified as primary (polycythemia rubra vera), secondary (e.g. to chronic or inappropriate erythropoietin secretion), or apparent (marked by contracted plasma volume).
- Primary polycythemia, a myeloproliferative syndrome, poses risks of vascular occlusive complications.
- Polycythemia secondary to chronic hypoxia enhances systemic oxygen transport and organ perfusion, unless plasma volume contraction occurs, with inordinate increases in blood viscosity.
- Contracted plasma volume syndromes ('apparent polycythemia') may be associated with increased risk of stroke and myocardial infarction.

Introduction

The ability of the blood to function for oxygen delivery and organ perfusion is governed by hematocrit and blood volume (Jones *et al.* 1990). Blood volume is the sum of total red cells and plasma in circulation.

In critical illness, the patient's total circulating red cells — the red cell mass or volume — tends to fall because of inevitable losses of blood plus inhibition of red cell production by any septic or inflammatory disease, renal failure, and/or any myelotoxic drug. Thus red cell production falls, or ceases, in patients in the intensive care unit (ICU) setting. Because of damage to vascular endothelial integrity, the intravascular plasma volume also tends to fall. Therefore the hematocrit appears 'falsely high' and poorly reflects the red cell lack. These invariable influences on the ICU patient reduce the chance that polycythemia will be a problem, although this will occur sporadically.

In health, normal erythropoiesis keeps the peripheral blood hematocrit between 0.37 and 0.47 for women and between 0.40 and 0.50 for men. The red cell volume is normally 25 ± 5 ml/kg for women and 30 ± 5 ml/kg for men. The scatter reflects the problems of expressing red cell volume per kilogram body weight; the higher the body's proportion of (relatively avascular) fatty tissue, the lower is the red cell volume when expressed in relation to body weight. Official guidelines call for expression of blood volume relative to body surface area, which is derived from height and weight.

Polycythemia is marked by an excess of red cells and a supranormal hematocrit value. Measurements of red cell and plasma volume are needed for diagnostic categorization. Standard hematology techniques based on the dilution principle use radio-chromium or radiotechnetium for labeling of autologous red cells followed by reinjection and quantitation of the dilution; the higher the dilution, the larger is the pool of red cells in the circulation. The plasma volume can be estimated using radio-iodine-labeled albumin. 'Older' labels of albumin as indicators of plasma volume have included indocyanine green and Evans blue, but, like radio-iodinated albumin, these albumin labels lead to an overestimate of plasma volume in critical illness because of the continuity of intravascular plasma albumin with extravasated albumin, water, and salt. Therefore blood volume, based on red cell volume divided by hematocrit, is a more reliable indicator of intravascular volume than is the sum of red cell volume and the volume of distribution of albumin in such patients.

Whenever polycythemia is suspected, the hematological diagnosis is not complete without knowledge of the measured red cell volume and calculated blood volume, as elevated hematocrit values may reflect increases in red cells and/or reductions of plasma volume in the circulation.

The main categories of polcythemia include primary polycythemia (polycythemia rubra vera), secondary (or compensatory) polycythemia, and contracted plasma volume syndromes ('apparent' polycythemia, 'stress' polycythemia, 'relative' polycythemia, or 'pseudo'-polycythemia)

Primary polycythemia (polycythemia rubra vera)

This is a myeloproliferative syndrome of unknown cause leading to chronic proliferation of reticuloendothelial tissues in the liver, spleen, and bone marrow with excessive blood cell production. It most commonly affects middle-aged and elderly patients. The proliferative emphasis in hemopoiesis may involve red cells, white cells, and platelets, in differing balances, but the combination of abnormal red cell and platelet proliferation, particularly thrombocythemia, with platelet counts above 450×10^9/l is a very important cause of vascular occlusive complications in hospital patients. Typically, the patient has a ruddy complexion, generalized pruritus, particularly after taking a warm bath, hepatosplenomegaly, and digital and/or coronary and/or cerebral ischemia, and is at risk of intra-abdominal, intrcranial, coronary, and limb vascular occlusions. The blood findings show supranormal red cell, white cell, and platelet counts. The blood picture commonly shows evidence of iron-deficient erythropoiesis, with biochemical evidence of increased purine turnover in hyperuricemia. The bone marrow is hyperplastic with diagnostically useful megakaryocytic hyperplasia and pleomorphism, which are very conspicuous on histological examination of the bone marrow and marrow smears. Commonly, such patients present with vascular, neurological, or cardiological complications as a consequence of blood hyperviscosity and vascular occlusion affecting the digits, heart, and/or brain. If the importance of platelet counts above $450 \times 10^9/l$ and/or red cell counts of $6 \times 10^{12}/l$ or more is overlooked, underlying myeloproliferation may be missed. Inappropriate surgical and/or medical treatment, which fails to deal with the hematological disease upon which the clinical syndromes are based, may then be given. These patients are, of course, a minority of those presenting with vascular insufficiency and ischemia, but, because of the remediable nature of the myeloproliferation, it is particularly important to make the diagnosis, preferably at first presentation. Such patients may have no further vascular occlusive or ischemic organ complications once the myeloproliferation is treated. Hematological examinations upon which diagnosis depend include blood count, measured blood volume indices, and marrow examination — the last of these is particularly important when the red cell volume is raised.

Secondary (or compensatory) polycythemia

This involves red cell proliferation, with measurable increases in red cell volume, and is almost always secondary to arterial hypoxemia. Cyanotic heart disease, chronic bronchitis, emphysema, and other pulmonary causes of arterial hypoxemia are particularly prominent. In these disorders the polycythemia represents a beneficial natural reaction to the hypoxemia leading to a hypervolemic erythrocytosis. This, in association with hematocrit values of 0.55 to 0.65 and a red cell volume of approximately 55 to 65 ml/kg, promotes oxygen carriage and transport to the vital organs in patients with cyanotic heart and/or lung diseases (Erslev and Caro 1984). Comparable hematological findings are made in high-altitude dwellers where the erythropoietic response is essentially beneficial to oxygen transport. It is notable that when these patients suffer falls in hemoglobin concentration, for example from 19 to 14 g/dl after blood loss, the clinical signs and symptoms of inadequate oxygen transport become much worse. If the hematocrit exceeds 0.70, a state of 'decompensated erythrocytosis' may arise, leading to impairment of systemic oxygen transport and regional perfusion (Rosove *et al.* 1986). This may result from increases in blood viscosity, particularly when the total blood volume is not elevated, such as when the plasma volume falls. In the individual patient with decompensated erythrocytosis, it is not clear whether inordinate erythrocytosis or transient falls in plasma volume have occurred. This uncertainty makes the therapy of the 'excessive' polcythemia difficult and not based on rational theory.

Smoking and drinking to excess also tend to elevate red cell mass, reduce plasma volume, and raise hematocrit, thereby impairing oxygen transport. Also, carbon monoxide will bind to hemoglobin, inactivating the function of some of the circulating hemoglobin and impairing the release of oxygen by inducing a 'left shift' in the oxygen dissociation curve. Measurement of carboxyhemoglobin levels in smoking patients, or in those possibly exposed to carbon monoxide, is relevant not only in diagnosis but also in assessing the ability of the blood to transport and release oxygen (Moore-Gillon and Pearson 1986).

Occasionally, the basis for secondary polcythemia may be renal, hepatic, uterine, or even intracranial tumors; hemoglobinopathy is another etiology. These conditions are conspicuous by their rarity.

Contracted plasma volume syndromes ('apparent' polycythemia, 'stress' polycythemia, 'relative' polycythemia, or 'pseudo'-polycythemia)

This is seen particularly in patients with hypertension and a lifestyle involving excessive smoking and drinking (Pearson 1991). Some of these patients are on diuretic therapy for hypertension, which further contracts plasma volume. Again, hematological diagnosis by estimation of blood volume is needed for certainty and rational management. High-dose diuretic therapy, for example furosemide (frusemide) given in a dose of 400 mg or more daily, will contract the plasma volume; when the diuretics are discontinued, the hematocrit may fall within hours from, say, 0.65 to 0.50. It is not really known whether the increased frequency of vascular occlusive diseases in such patients is the result of their lifestyle, their excess weight, or their excessively viscous blood, but interaction between these factors clearly multiplies the risks of vascular occlusive complications.

References

Erslev, A.J. and Caro, J. (1984). Secondary polycythaemia: a boon or a burden? *Blood Cells*, **10**, 177–91.

Jones, J.G., Holland, B.M., Hudson, I.R.B., and Wardrop, C.A.J. (1990). Total circulating red cells versus haematocrit as the primary descriptor of oxygen transport by the blood. *British Journal of Haematology*, **76**, 288–94.

Moore-Gillon, J. and Pearson, T.C. (1986). Smoking, drinking, and polycythaemia. *British Medical Journal*, **292**, 1617–18.

Pearson, T.C. (1991). Apparent polycythaemia. *Blood Reviews*, **5**, 205–13.

Rosove, M.H., Perloff, J.K., Hocking, W.G., Child, J.S., Canobbio, M.M., and Skorton, D.J. (1986). Chronic hypoxaemia and decompensated erythrocytosis in cyanotic congenital heart disease. *Lancet*, **ii**, 313–15.

10.5.2 Therapeutic strategy

C. A. J. Wardrop

Key messages

- Primary polycythemia, posing risks of vascular occlusive disease, demands urgent treatment with venesection, cytotoxic agents, and aspirin.
- Although secondary polycythemia benefits oxygen transport in hypoxemic patients, venesection may be necessary to reduce hematocrit values in excess of 0.65.
- 'Dilution-venesection' should be considered for patients with apparent ('stress') polycythemia and hematocrit values of more than 0.55.

Introduction

When polycythemia has been diagnosed and categorized as primary, secondary, or due to a contracted plasma volume, the therapeutic decision regarding clinical management has to be taken.

Primary polycythemia

Optimal management includes the following.

1. Venesection, with isovolemic replacement using crystalloids or colloids to reduce the hematocrit to less than 0.50, and preferably to about 0.45, is used to correct the blood hyperviscosity associated with the erythrocytosis of untreated polycythemia vera.

2. At the same time as the venesection is started, hydroxyurea is given in a dose ranging from 500 mg twice daily to 500 mg five times daily to control hemopoiesis over the first 10 to 14 days of treatment.

3. If more rapid effects are needed to reduce the thrombocythemia when the patient has presented with organ ischemia associated with platelet counts in excess of 750×10^9/l, cytarabine 0.5 g can be given intravenously on one or two occasions for acute curtailment of platelet production during the following 2 to 7 days. Reduction of circulating platelet counts by platelet pheresis is ineffective and is not worthwhile.

4. Platelet function can also be reduced using aspirin in a dose of 150 mg on alternate days.

5. Once the diagnosis is confirmed as primary polycythemia in an adult over 60 years of age, radio-phosphorus (^{32}P) 185 mBq (6 mCi) is given intravenously for longer-term control, which lasts for months to 2 years in most patients. Radio-phosphorus is the definitive treatment for such patients. Most patients presenting with polycythemia rubra vera or thrombocythemia may have a normal lifespan because of the success of this treatment in maintaining remission for many years.

6. In younger patients, control of the myeloproliferation is possible using hydroxyurea in a maintenance dose of 1 to 2 g daily, with out-patient clinic review every few weeks or months depending on the individual patient's response. The long-term risks of this treatment seem to be less than those associated with alkylating agents (e.g. bulsulphan), but the latter may be useful in an initial dose of 2 to 6 mg daily if hydroxyurea is not well tolerated.

Secondary polycythemia

In this condition the hypervolemic erythrocytosis, with a hematocrit of up to 0.65, benefits systemic oxygen transport and regional perfusion (Erslev and Caro 1984; Jones *et al.* 1990).

Hypoxemic patients, for example those with cyanotic congenital heart disease or chronic bronchitis with respiratory failure, may benefit from correction, by venesection, of hyperviscosity due to 'decompensated' erythrocytosis associated with a hematocrit above 0.65. This 'decompensation' process is poorly understood and so rational management is difficult, but in practice a hematocrit above 0.65 should be reduced to between 0.60 and 0.65 (Rosove *et al.* 1986). High-dose diuretics reduce blood volume with transient increases in hematocrit which do not reflect plasma volume in the steady state. It is better to use angiotensin-converting enzyme inhibitors than diuretics for heart failure in patients with secondary polycythemia. When doubt arises regarding management of secondary polycythemia, blood volume measurement clarifies the basis of the increases in hematocrit and hemoglobin concentration, allowing rational therapeutic strategies to be used.

Of course, the basic treatment of secondary polycythemia is effective management of the hypoxemia due to cardiopulmonary disease or, very occasionally, the tumor of kidney, uterus, liver, or brain which leads to excessive erythrocytosis through inappropriate erythropoietin production.

Severe warnings should be given to patients discharged from the intensive care unit who may resume smoking. Heavy alcohol intake may also hazard the circulation through its effect of contracting the plasma volume. Excess alcohol intake has also been blamed for inadequate ventilation or overnight apnea, leading to rises in red cell volume, hematocrit, and blood viscosity (Moore-Gillon and Pearson 1986).

Apparent ('stress') polycythemia

Apparent ('stress') polycythemia (Pearson 1991) is poorly understood, but the contracted plasma volume may be improved by occasional venesection if the hematocrit exceeds 0.55. Although not proven, it is thought that a change of lifestyle for patients with a contracted plasma volume may be beneficial. Therefore such patients should be firmly encouraged to stop both smoking and excessive alcohol intake. Their doctors should avoid prescribing them diuretics, using vasodilators and angiotensin-coverting enzyme inhibitors for antihypertensive and heart failure therapy. The link between stress polycythemia and vascular occlusive diseases is not well understood, and the patient's lifestyle, for example smoking habit, may be more important than a contracted plasma volume and high hematocrit in predisposing to vascular occlusive complications. It is possible that reducing high hematocrit values, for example from 0.55–0.60 to 0.50 or less, may help blood flow and organ perfusion and oxygenation.

References

Erslev, A.J and Caro, J. (1984). Secondary polycythaemia: a boon or a burden? *Blood Cells*, **10**, 177–91.

Jones, J.G., Holland, B.M., Hudson, I.R.B., and Wardrop, C.A.J. (1990). Total circulating red cells versus haematocrit as the primary descriptor of oxygen transport by the blood. *British Journal of Haematology*, **76**, 288–94.

Moore-Gillon, J. and Pearson, T.C. (1986). Smoking, drinking, and polycythaemia. *British Medical Journal*, **292**, 1617–18.

Pearson, T.C. (1991). Apparent polycythaemia. *Blood Reviews*, **5**, 205–13.

Rosove, M.H., Perloff, J.K., Hocking, W.G., Child, J.S., Canobbio, M.M., and Skorton, D.J. (1986). Chronic hypoxaemia and decompensated erythrocytosis in cyanotic congenital heart disease. *Lancet*, **ii**, 313–15.

10.6 Sickle crisis

10.6.1 Pathophysiology

G. R. Serjeant

Key messages

- Sickle cell trait is not a disease, but problems may develop under hypoxic conditions.
- The sickle cell gene is not confined to people of African origin.
- Sickle hemoglobin is an efficient carrier of oxygen, and transfusion of patients at their steady state hemoglobin level is not justified to increase oxygen carriage.
- Different genotypes of sickle cell disease manifest highly variable clinical courses.
- Most patients are well between complications, many of which have precipitating factors which can sometimes be avoided.

Nomenclature

Sickle cell disease includes a group of disorders characterized by a pathological process consequent on red cell abnormalities associated with high levels of sickle hemoglobin (HbS).

Sickle cell trait

Inheritance of the sickle cell gene from one parent and the gene for normal adult hemoglobin (HbA) from the other results in a benign condition called sickle cell trait in which HbS levels (20–45 per cent) are not high enough to cause the complications of sickle cell disease under normal physiological conditions. Under unusual conditions, such as respiratory depression induced by anesthesia or drugs, environmental conditions of high-altitude climbing or in unpressurized aircraft, cyanotic heart disease, or severe pulmonary pathology, hypoxemia sufficient to induce sickling-related pathology may occur.

Sickle cell disease

Several genotypes fall within the definition of sickle cell disease. Inheritance of the sickle cell gene from both parents results in homozygous sickle cell disease. Inheritance of the sickle cell gene from one parent and another abnormal gene, such as that for hemoglobin C (HbC), β^{+}-thalassemia, or β°- thalassemia, from the other results in sickle cell–HbC disease, sickle cell–β^{+}-thalassemia, or sickle cell–β°-thalassemia. These conditions manifest HbS levels of 50 to 98 per cent, which are sufficient to cause problems under normal physiological conditions. Sickle cell–HbC disease with 50 to 55 per cent HbS and sickle cell–β^{+}-thalassemia with 65 to 75 per cent HbS generally have mild manifestations, whereas homozygous sickle cell disease and sickle cell–β°-thalassemia with 80 to 98 per cent HbS are variable but generally severe. The relative frequencies of these major genotypes at birth in Jamaica are shown in Table 1.

Table 1 Relative frequencies of the major genotypes of sickle cell disease at birth in Jamaica

Homozygous sickle cell disease	1 in 300 births
Sickle cell–hemoglobin C disease	1 in 500 births
Sickle cell–β^{+}-thalassemia	1 in 3000 births
Sickle cell–β°-thalassemia	1 in 7000 births

Racial distribution

It is commonly and incorrectly assumed that the sickle cell gene is confined to people of African origin. Most of the disease encountered in North and South America, the Caribbean, the United Kingdom, and France is of African origin, but the gene also occurs around the Mediterranean, particularly in Sicily and southern Italy, northern Greece, southern Turkey, the Arabian peninsula, and central India. Patients of Turkish origin are commonly seen in Germany, and Greek patients are widely scattered in North and South America, the United Kingdom, and Australia. In India, the gene largely affects people in the interior of Orissa, Madhya Pradesh, and Maharastra; although large numbers are affected, people from these populations are not commonly seen elsewhere.

Pathophysiology of sickling

High intracellular concentrations of HbS polymerize on deoxygenating, raising the intracellular viscosity, causing membrane changes, and deforming some of the red cells into the characteristic sickle shape. Such abnormal red cells have difficulty in negotiating the capillary beds and are prematurely destroyed, leading to a hemolytic anemia and flow obstruction within blood vessels causing death or damage of the tissues supplied.

Sickle cell disease presents a wide spectrum of pathology. Hemolytic aspects include the problems of rapid red cell destruction, anemia, jaundice, pigment gallstones and those associated with the compensatory bone marrow response such as aplastic crisis, increased requirements of folic acid, and an increased metabolic rate. The consequences of vaso-occlusion are generally more serious and depend upon the vessels affected, but may include stroke, retinal problems, splenic abnormalities, including acute and chronic sequestration and the loss of normal splenic function rendering patients prone to overwhelming septicemia, pulmonary problems including infection, infarction, and pulmonary sequestration, and damage to the bone marrow resulting in the painful crisis, localized avascular necrosis of bones, and involvement of the femoral head.

Hemoglobin levels and transfusion

During steady state, when patients are clinically well, hemoglobin levels vary within a narrow range characteristic of the individual patient and the genotype. Most patients with homozygous sickle cell disease have steady state levels of 6 to 9 g/dl, although levels may be as low as 5 g/dl or as high as 12 g/dl. HbS within the red cell manifests a low oxygen affinity, with more oxygen being unloaded per gram of hemoglobin than is the case with HbA. The oxygen affinity varies between patients, but those with the lowest steady state hemoglobin levels have the most marked shifts in their oxygen dissociation curve. A patient with a steady state hemoglobin level of 6 g/dl has a low oxygen affinity and maintains that level because oxygen delivery is near normal; thus a higher level is unnecessary. Transfusing a patient up from his or her steady state level does not necessarily improve oxygen delivery since, although hematocrit increases, the oxygen dissociation curve returns towards normal. Therefore a knowledge of the patient's steady state hemoglobin level is important for management, and transfusion up from that level achieves little in terms of oxygen delivery.

Genotype variation

All genotypes of sickle cell disease have similar manifestations, although the frequency and severity are usually greater in homozygous sickle cell disease and sickle cell–β°-thalassemia than in sickle cell–hemoglobin C disease and sickle cell–β^{+}-thalassemia. The only exception to this pattern is retinal disease which is more severe in the latter two conditions. The clinical complications described below apply principally to homozygous sickle cell disease and to a lesser extent to other genotypes.

Hemolysis and low levels of vaso-occlusion are continuous features of sickle cell disease even when the patient is clinically well (steady state), but the disease is also punctuated by a variety of acute events commonly called crises. Crisis is an overused term in sickle cell disease and should never be used without an adjective describing its nature.

Common acute clinical problems encountered in the management of sickle cell disease include the aplastic crisis, acute splenic sequestration, septicemia, stroke, painful crisis, acute chest syndrome, pregnancy, and postoperative management.

Aplastic crisis

The shortened red cell survival is compensated by a marked increase in bone marrow activity achieved by both an expansion of bone marrow volume and an increased rate of turnover. Bone marrow activity is approximately 40 times normal; this is reflected by a mean reticulocytosis of 10 to 12 per cent.

Pathogenesis

Human parvovirus infection destroys red cell precursors, causing reticulocytes to disappear from the peripheral blood and the hemoglobin level to fall at a rate determined by the red cell survival. The period of aplasia is normally 7 to 10 days, but since average red cell survival in homozygous sickle cell disease is 10 to 12 days, hemoglobin may fall to life-threatening levels unless maintained by transfusion.

Acute splenic sequestration

Acute splenic sequestration is one of the earliest life-threatening complications. It is most common in the first 3 years of life but is occasionally seen in young adults.

Pathogenesis

The spleen undergoes acute enlargement, trapping circulating red cells and leading to an acute anemia. The mechanism is believed to be jamming of poorly deformable red cells in the interendothelial slits of the spleen as cells move from the red pulp back into the splenic sinuses when returning to the circulation. The fact that a low fetal hemoglobin (HbF) is a risk factor supports this etiology. Obstruction of this pathway leads to an acute accumulation of red cells in the pulp, splenic enlargement, and a rapidly developing anemia.

Septicemia

The early loss of splenic function is associated with an increased risk of overwhelming septicemia, which is most apparent with *Streptococcus pneumoniae* and *Hemophilus influenzae* type b. Effective prophylaxis against *Strep. pneumoniae* has been associated with increased prominence of *H. influenzae* type b and *Salmonella* as causes of septicemia. Although the association of *Salmonella* osteomyelitis with sickle cell disease is well established, *Salmonella* septicemia may be overlooked in the absence of bone involvement; in a Jamaican series the mortality was 22 per cent.

Pathogenesis

The mechanism is believed to be loss of the normal filtering function of the spleen because of the damage caused by poorly deformable red cells. The risk commences as early as 4 months and is greatest in the first 3 years of life.

Stroke

Stroke is predominantly a problem of childhood in homozygous sickle cell disease with a median age of 6 years and an incidence of 8 per cent by the age of 14 years, but it may also occur in adolescence and later adult life.

Pathogenesis

The etiology of stroke and its high prevalence in childhood is not understood, although hemiplegias are usually associated with blockage of major cerebral vessels such as the carotid, vertebral, anterior, or middle cerebral arteries. The mechanism whereby an essentially small-vessel disease gives rise to major arterial occlusion is not known. Once a stroke has occurred, the risk of recurrence is 50 to 70 per cent within 3 years of the first event. Subarachnoid hemorrhage, usually associated with aneurysms, is a common cause of cerebrovascular disease in adult homozygous sickle cell patients.

Painful crisis

Painful crisis characteristically causes severe pain and, occasionally, swelling of the juxta-articular areas of the long bones, sternum, ribs, thoracic and lumbar spine, and pelvis. These sites have in common the presence of active bone marrow and the common pathology is bone marrow necrosis, as evidenced by the aspiration of dead bone marrow from painful sites and the loss of normal activity on bone scans. The mechanism leading to avascular necrosis is controversial, and several lines of evidence conflict with the usual assumption of localized vaso-occlusion.

Pathogenesis

The most common precipitating factor for painful crisis in Jamaica is skin cooling, which is not known to promote sickling but is well established as a cause of reflex shunting of blood away from the skin. Genotypes of sickle cell disease associated with a reduction in intravascular sickling such as homozygous sickle cell disease with homozygous α^{+}-thalassemia and sickle cell–β°-thalassemia result in more frequent painful crises. The distribution of bone pains shows a significant excess of a bilateral symmetrical pattern, with pain often involving both knees simultaneously or developing within hours. These observations are difficult to explain by a mechanism of random vaso-occlusion. A new theory more consistent with these findings envisages the painful crisis as resulting from a centrally mediated reflex shunting of blood away from the bone marrow so that the nutrient supply is no longer adequate to supply the high metabolic demands, resulting in marrow necrosis. It is believed that the avascular necrosis of bone marrow does not induce acute pain but is associated with a variety of sensations, reported as numbness, heaviness, and paresthesia, representing a prodrome of the painful crisis. The subsequent inflammatory response to dead bone marrow with increased vascularity and outpouring of inflammatory exudate serve to increase intramedullary pressure and cause the severe pain characteristic of the painful crisis.

Precipitating factors

Patients with sickle cell disease do not have continual painful crises but events tend to follow well-defined conditions. In addition to skin cooling (winter, rain, swimming, cold showers, washing cars or clothes), other precipitating events include physical exertion, dehydration, emotional stress, and infections. The physiological paths whereby these events cause bone marrow necrosis are usually unclear, although skin cooling has been shown to invoke shunting of blood away from the skin. It is presumed that deep shunts accommodating the diverted blood flow may offer less resistance than the bone marrow and so also compromise bone marrow blood flow.

Risk factors

In addition to precipitating factors, risk factors rendering patients more prone to painful crises include certain genotypes such as homozygous sickle cell disease with homozygous α^{+}-thalassemia and sickle cell–β°-thalassemia, male sex, age 15 to 29 years, high total hemoglobin and low HbF levels, and pregnancy and the immediate postpartum period. Common to several of these factors is the high total hemoglobin level, so that a logical approach to prevention might be venesection to lower the hemoglobin level.

Acute chest syndrome

Acute chest syndrome applies to a spectrum of pulmonary events having in common pulmonary symptoms (cough, shortness of breath, pleuritic pain), clinical signs of pulmonary consolidation, or radiological evidence of new pulmonary infiltrates.

Pathogenesis

There are components of infection, infarction, and pulmonary sequestration. A slow and only partial response to antibiotics or anticoagulation compared with the treatment of pneumonia or pulmonary infarction in otherwise normal people suggests a more complex pathology; it is likely that primary infections are complicated by areas of secondary infarction and vice versa. A third pathological process is sequestration in the pulmonary capillaries which may lead to a rapid clinical deterioration. Once started, sequestration impairs oxygen absorption from the alveoli, leading to a vicious circle of red cell deoxygenation, further sickling, and further sequestration. Its reversal may be complete and rapid following exchange transfusion, implying a vascular phenomenon without evidence of either infection or infarction. Infarction of the sternum or ribs, which limits chest movement and ventilation, may be associated with secondary areas of lung collapse or consolidation, the incidence of which can be markedly reduced by incentive spirometry. Fat embolism has also been recognized recently as a common pathology.

Pregnancy

Pregnancy in the mother with homozygous sickle cell disease holds risks for both mother and child. Fetal loss is increased at every stage of pregnancy and full-term babies usually have low birth weights. For the mother the risk of painful crisis and of the acute chest syndrome is increased, particularly in the third trimester and immediate postpartum period. Maternal mortality is approximately 1 per cent.

Pathogenesis

The mechanism of these changes and of the low birth weight are unknown. It is assumed that placental infarction compromises fetal growth, but there is no direct evidence for this and regular transfusion in the mother from the middle trimester does not improve fetal outcome or increase birth weight.

Bibliography

Balkaran, B., Char, G., Morris, J.S., Serjeant, B.E., and Serjeant, G.R. (1992). Stroke in a cohort study of patients with homozygous sickle cell disease. *Journal of Pediatrics*, **120**, 360–6.

Koshy, M., Burd, L., Wallace, D., Moawad, A., and Baron, J. (1988). Prophylactic red-cell transfusions in pregnant patients with sickle cell disease. A randomized cooperative study. *New England Journal of Medicine*, **319**, 1447–52.

Platt, O.S., *et al.* (1991). Pain in sickle cell disease. Rates and risk factors. *New England Journal of Medicine*, **325**, 11–16.

Serjeant, G.R., *et al.* (1981) Outbreak of aplastic crises in sickle cell anaemia associated with parvovirus-like agent. *Lancet*, **ii**, 595–7.

Serjeant, G.R., De Ceulaer, C., Lethbridge, R., Morris, J.S., Singhal, A., and Thomas, P.W. (1994). The painful crisis of homozygous sickle cell disease—clinical features. *British Journal of Haematology*, **87**, 586–91.

10.6.2 Prevention and management of complications

G. R. Serjeant and M. Singer

Key messages

- The general principles of treatment are oxygenation, hydration, and maintenance of oxygen delivery.
- Acute chest syndrome remains a major cause of mortality.
- Exchange transfusion may be life saving in acute pulmonary sequestration.
- Serious complications may accompany pregnancy or the postoperative period.

General principles

The most serious life-threatening complications occur predominantly in early childhood, although all are seen, albeit much less frequently, in adult life. The principal events in adult life requiring intensive care include the acute chest syndrome and complications occurring postoperatively or during pregnancy.

General principles of treatment include oxygenation, hydration, and maintenance of oxygen delivery. There are no risks with oxygen therapy, but there are no controlled observations confirming benefit. Hydration is important but again lacks experimental support, although the increased blood viscosity is likely to be deleterious and there is *in vitro* evidence that a hypertonic medium promotes sickling. High skin losses in a tropical setting, particularly with fever, inability to concentrate the urine, and reluctance or inability to drink, imply that dehydration may develop rapidly. There is sometimes reluctance to give intravenous fluid because of the belief that cardiac and renal damage diminish the patient's ability to handle a fluid load, but the dangers of dehydration are far greater. Most patients can handle fluid loads well, and intravenous fluid intakes should be at least 2 liters plus urinary output, monitored by a central venous line if there are concerns over the degree of dehydration or of overhydration. Blood transfusion may be indicated for two reasons, namely maintenance of oxygen delivery and dilution or replacement of red cells containing sickle hemoglobin (HbS). A knowledge of the patient's steady state hemoglobin level is valuable in assessing oxygen delivery. If the hemoglobin falls more than 2 g/dl below steady state levels, the underlying cause should be investigated and, if not treatable directly, may require 'top-up' transfusion. Since transfused blood of an AA genotype has a higher oxygen affinity than HbS, the aim should be to transfuse to at least 1 g/dl above the steady state hemoglobin level. If the aim is to replace HbS-containing cells, this is best achieved by exchange tranfusion which is most simply performed by removing 500 ml of blood followed by the transfusion of 2 units and repeating the procedure after 4 h.

Aplastic crisis

Clinical features

Aplastic crises predominantly affect patients under 15 years of age, although they may occur at older ages. They occur in epidemics with 3 to 4 years between peaks, and cluster in families with a 50 per cent risk of a susceptible sibling being affected within 3 weeks. The patient presents with tiredness, marked pallor, and shortage of breath. Clinical diagnosis is based on the combination of a markedly lowered hemoglobin level and absence of reticulocytes from the peripheral blood or, if reticulocytes are present, a daily marked increase consistent with the recovery phase of aplasia. Laboratory diagnosis depends on the demonstration of parvovirus-specific DNA, IgM, or IgG.

Management

Treatment requires a single transfusion. In most cases the clinical course is so predictable and uncomplicated that transfusion may be performed as an out-patient with follow-up to ensure that the anticipated reticulocytosis occurs. Immunity is lifelong, and a recurrence of aplastic crisis induced by human parvovirus has never been reported. This augurs well for a vaccine which has recently been developed in the United States and is currently undergoing clinical trials.

Acute splenic sequestration

Clinical features

The principal clinical features are splenomegaly of 3 to 6 cm below the costal margin and pallor. In extreme cases the child may present moribund in peripheral circulatory failure. The cumulative incidence in the Jamaican Cohort Study increased from 0.255 by 2 years to 0.265 by 3 years and 0.297 by 5 years. Events recurred in 50 per cent, with further recurrences occurring at shorter intervals. Outcome is less severe in adults, who usually present with symptoms of increasing anemia.

Management

Immediate treatment is transfusion. Prophylactic splenectomy is recommended after two attacks. Parental education has had a marked impact on outcome, with the mortality falling by 90 per cent as earlier presentation allowed time for transfusion.

Septicemia

Clinical features

Most present with high fever (above 40 °C) in a seriously sick patient, although pneumococcal septicemia occasionally presents without high fever. The risk of pneumococcal septicemia is greatest in the first 3 years of life, although it continues, albeit at a lesser level, throughout life. There are no reliable incidence figures in adults for *Streptococcus pneumoniae*, *Hemophilus influenzae* type b, or *Salmonella*, although there is a clinical impression that septicemias tend to be due to either *Pneumococcus* or *Salmonella*.

Management

Treatment requires intravenous antibiotics and supportive therapy. Initial antibiotic coverage, pending the results of blood culture, should be against the organisms mentioned above.

Prevention of pneumococcal infection

Penicillin provides effective prophylaxis and may be given either orally twice daily (erythromycin in the case of penicillin sensitivity) or by monthly injections of long-acting depot preparations. The pneumococcal vaccine does not confer protection against all serotypes, frequently failing to protect against serotypes 6, 9, and 23, although there is evidence of increasing antigenicity with age. The current Jamaican protocol uses intramuscular injections of penicillin from 4 months to 4 years and then the 23 valent vaccine with the last injection of penicillin. There is little naturally acquired immunity during penicillin prophylaxis and the child is highly susceptible when penicillin is stopped; therefore it is vital to give the vaccine before the protective effect of penicillin ceases. Studies in the United States are currently comparing protocols stopping penicillin at the age of 5 years or continuing indefinitely. Two recent developments may cause this protocol to be revised, namely the increasing emergence of penicillin-insensitive pneumococci and the availability of a conjugated pneumococcal vaccine which may be effective as early as 2 to 4 months. This latter development is particularly important in view of penicillin-resistant strains, but its effectiveness in children with homozygous sickle cell disease awaits confirmation in clinical trials. In adults, the currently available pneumococcal vaccine appears to generate protective antibody levels.

Stroke

Clinical features

There is a wide spectrum of clinical features, although the most common pattern is of a hemiparesis or hemiplegia of usually sudden onset, which may resolve partially or completely on follow-up. Recurrent episodes are common. Strokes in adults commonly manifest the clinical signs of a subarachnoid hemorrhage.

Management

Since the cause of the initial episode is unknown, prevention is not possible and treatment is currently confined to prevention of recurrence. This is most commonly attempted by a chronic transfusion program aiming to maintain HbS levels below 30 per cent and preferably below 20 per cent. This is initiated by exchange transfusion followed by regular top-up transfusions. Prevention of accumulation of iron requires regular chelation therapy with desferrioxamine, and difficulties with venous access may require installation of a permanent port. Despite the use of white cell and platelet-poor blood, minor incompatibilities are common and severe transfusion reactions may occur. The required duration of treatment is unknown, and high recurrence rates after stopping transfusion programs (even after 10 years) have led to the conclusion that treatment may be needed for life. There is a need to understand the risk factors for the initial stroke so that this may be avoided if possible. Investigation and management of subarachnoid hemorrhage should be the same as for this complication in other conditions, although the use of brain-shrinking agents such as mannitol or dexamethasone, and the use of hypotension and of hypothermia, are controversial issues on which there are insufficient data to draw conclusions.

Painful crisis

Clinical features

Pain of variable severity affects the juxta-articular areas of the long bones, particularly adjacent to the knees, the elbows, the shoulders and ankles, the sternum and ribs, the spine, particularly the lumbar region, and the pelvis. Pain may be localized to one or two sites, although multiple sites are frequently involved. Occasionally pains are described as 'all over'. Severity varies from transient pain barely limiting function to a severe debilitating pain requiring hospital admission and parenteral pain relief for days. Painful areas may be tender to palpation, although this is not always the case. Fever is common, and dehydration may occur if the patient has been reluctant to drink. The urine is commonly reported as dark or red and contains increased porphyrin levels at the time of the painful crisis.

Management

Recognized precipitating factors should be avoided, and the patient should be taught to seek out possible antecedent factors since avoiding these conditions may reduce the frequency of painful crises. The ability to cope with pain is influenced by many social and cultural factors, but reassurance and detailed explanation may do much to alleviate the sense of impending doom in the painful crisis.

Patients should be examined for infections and other precipitating factors which may need treatment. In addition to rest and reassurance, patients may be dehydrated and require fluids either by mouth or intravenously. Pain relief should be tailored to the patient's needs. Simple oral medications such as acetaminophen (paracetamol) are normally taken at home before attending but, if not, may contribute to pain relief. Stronger analgesics include codeine, pentazocine, pethidine, and morphine. These may be given by either intermittent parenteral routes or patient-controlled analgesia, which has the advantage of more sustained blood levels, better pain control at a lower overall consumption of drugs, and the psychological benefit of patient control. Analgesic requirements vary markedly between patients and in different cultures. In Jamaica, adequate pain relief may be achieved by modest levels of drugs, particularly in the context of other supportive measures.

The painful crisis is rarely a cause of admission to intensive care, although it may complicate an acute chest syndrome, pregnancy, and the postoperative period. It occasionally presents with features of disseminated intravascular coagulation.

Other treatments

There are no controlled data on the benefits of oxygen, although this is commonly used. Hydroxyurea, which has been shown to ameliorate painful crisis frequency in a selected group of severely affected patients, may be useful in selected cases although it has potentially serious side-effects and should not be used without careful monitoring. It is also a long-term prophylactic measure and should not be used for short-term acute treatment. Transfusion is commonly employed without good reason, and some patients are placed on chronic transfusion regimens to prevent painful crises. It is likely that the long-term complications of such treatment far outweigh the short-term benefits.

Natural history

Painful crises become less frequent and less severe after the age of 30. They cease in most patients aged over 40 years. This is useful in reassuring patients and helping them to cope with painful episodes.

Acute chest syndrome

Clinical features

There is usually cough, fever, dyspnea, pleuritic pain, and signs of consolidation. There is usually radiological evidence of pulmonary infiltrates, although occasionally a dramatic clinical picture may be associated with minimal radiological change. Some patients exhibit sudden clinical deterioration with rapidly increasing shortage of breath, falling oxygen saturations, and progressive radiological opacity of the lungs. This is usually the picture of acute pulmonary sequestration which has a high mortality without prompt intervention.

Management

Pathogens are rarely isolated from homozygous sickle cell adults with the acute chest syndrome, although it is usual practice to give antibiotics (e.g. cefuroxime plus erythromycin) which may at least prevent secondary infection. Oxygen is given and physiotherapy is used to encourage ventilation and to drain infected areas. Pulse oximetry gives reliable information in pigmented subjects and values do not appear to be offset by HbS. However, desaturation is common in the steady state and a quarter of steady state homozygous sickle cell adolescents in the Jamaican Cohort Study had values below 90 per cent. Therefore changes in oxygen saturation are more important than actual levels and continuous monitoring is vital in patients with the acute chest syndrome. Patients demonstrating rapid clinical deterioration and progressive falls in oxygen saturation may indicate acute pulmonary sequestration. A $Pa\text{O}_2$ below 75 mmHg (10 kPa) has been suggested as an indicator for both intensive care admission and immediate exchange transfusion. This is facilitated by central venous cannulation which also enables central venous pressure monitoring. The aim of exchange transfusion is to reduce levels of HbS; some recommend partial exchange to lower HbS levels below 20 to 30 per cent, whereas others advocate as complete a replacement with adult hemoglobin as possible. The optimal degree of exchange transfusion has yet to be verified by a controlled study, but sufficient evidence exists to confirm that transfusion *per se* is beneficial. HbS levels should be monitored every 2 to 3 days with further top-up or exchange transfusions as indicated by departmental policy.

Pulmonary artery catheterization is occasionally warranted if hemodynamic compromise is marked or if doubt remains about the adequacy of volume resuscitation. It is important to maintain normovolemia as well as normoxemia. Invasive or non-invasive mechanical ventilation is indicated when there is inability to maintain adequate gas exchange and the patient is fatiguing. It is important to maintain adequate oxygen saturations throughout, although no benefit has been shown in administering oxygen to achieve complete hemoglobin saturation and supranormal arterial oxygen tensions.

Pregnancy

Management

Patients should be monitored carefully in antenatal clinics and delivered in hospital. Iron and folic acid supplementation is standard practice since the requirements of these are increased by pregnancy. Chronic transfusion is commonly performed from the middle trimester; although there is no evidence of fetal benefit, the risks of painful crises and acute chest syndrome in the mother may be reduced. The obstetric experience of centers using transfusion therapy does not differ markedly from those which do not. The management of the painful crisis and acute chest syndrome is similar to that in non-pregnant patients, although delivery may be induced if the acute chest syndrome appears life threatening. Delivery is normally by the vaginal route unless there are obstetric indications for operative delivery.

Postoperative management

Patients with homozygous sickle cell disease are prone to a variety of complications requiring surgery such as hypersplenism, gallbladder disease, avascular necrosis of the femoral head, and osteomyelitis. Perioperative transfusion has become common practice without evidence of a beneficial effect over patients managed without transfusion. A recent study of preoperative transfusion showed no difference other than in transfusion-related complications between a group of patients aggressively managed by lowering the HbS level to below 30 per cent and a more conservative approach raising the hemoglobin level to 10 g/dl.

General anesthesia is readily induced in patients at their steady state hemoglobin level, and blood should be available for replacement of blood lost at surgery. The greatest risk is in the immediate postoperative period with anesthesia-induced respiratory depression and limited chest movement, particularly with upper abdominal surgery such as the gallbladder and spleen. Preoxygenation for 20 min before induction, replacement of blood lost at surgery if necessary, maintenance of oxygen levels during the postoperative period, adequate hydration, pain relief, and physiotherapy, as practised in Jamaica, achieve postoperative outcomes similar to those obtained using preoperative transfusion elsewhere.

Bibliography

Davies, S., Luce, P.J., Win, A.A., Riordan, J.F., and Brozovic, M. (1984). Acute chest syndrome in sickle cell disease. *Lancet*, **i**, 36–8.

Emond, A.M., Collis, R., Darvill, D., Higgs, D.R., Maude, G.H., and Serjeant, G.R. (1985). Acute splenic sequestration in homozygous sickle cell disease: natural history and management. *Journal of Pediatrics*, **107**, 201–6.

Gaston, M.H., *et al.* (1986). Prophylaxis with oral penicillin in children with sickle cell anemia. *New England Journal of Medicine*, **314**, 1593–9.

Lanzkowsky, P., Shende, A., Karayalcin, G., Kim, Y.J., and Aballi, A.J. (1978). Partial exchange transfusion in sickle cell anemia. *American Journal of Diseases of Children*, **132**, 1206–8.

Serjeant, G.R. (1992). *Sickle cell disease* (2nd edn). Oxford University Press.

Vichinsky, E.P., *et al.* (1995). A comparison of conservative and aggressive transfusion regimens in the perioperative management of sickle cell disease. The Preoperative Transfusion in Sickle Cell Disease Study Group. *New England Journal of Medicine*, **333**, 206–13.

11 Trauma, burns, and physical disorders

11.1 Multiple trauma

11.1.1 Pathophysiology of chest trauma

P. A. Carli and M. Lejay

Key messages

- The mechanism of injury (blunt or penetrating, fall or crash) will factor into the extent of injury.
- The pathophysiology of respiratory distress is multi-factorial based on air exchange, pleural pressures, and ventilatory requirements.
- Shock can have many thoracic etiologies such as tension pneumothorax, cardiac tamponade, or hemorrhage.

Introduction

The mechanism by which the trauma occurs is an important factor in its severity. In cases of penetrating injuries the type of the weapon used and the history assumes a great importance in the occurrence of intrathoracic injuries. For example, major differences in injury will be observed with low- or high-energy bullets, and with the direction and distance of the gunfire. Similarly, stab wound injuries differ according to the type of knife used. The pathophysiology for blunt trauma is more complex and involves various forces. Chest trauma is frequently observed in motor vehicle accidents and falls. Speed, direction of impact, and the type of seatbelts and airbags present influence injuries by direct impact. Ejected and rollover victims are particularly at risk of deceleration injuries. In falls, the height, the site of primary impact, and the type of surface on which the patient lands modify the pattern of injury and the outcome. In high-speed impact or explosions, blast injuries of the lung may also be observed.

Previous significant medical history may effect the outcome. A history of smoking with chronic obstructive pulmonary disease or cardiac problems may contribute to respiratory failure (LoCicero and Mattox 1989).

From a pathophysiological point of view severe chest trauma leads to two major consequences, respiratory distress and shock.

Respiratory distress

Several factors may contribute to acute respiratory distress and failure.

Pneumothoraces can occur secondary to penetrating injury, rib fractures, intra-alveolar overpressure (barotrauma), or bronchial disruption. A tension pneumothorax produces the risk of immediate respiratory insufficiency, with increasing intrapleural pressure shifting the mediastinum towards the non-injured side. This can occur with a mechanically ventilated patient, where a small well-tolerated pneumothorax in a spontaneously breathing patient may suddenly be converted to a large tension pneumothorax after positive airway pressure is instituted. If it is not relieved quickly, this condition may be fatal.

Multiple rib fractures and flail chest (at least three contiguous ribs, each fractured in two places) are commonly associated with visceral injury. Flail chest can lead to paradoxical ventilation, where the flail segment moves inwards during expiration and outwards during inspiration. This paradoxical movement modifies intrapleural pressures and generates inefficient ventilation by the *pendelluft* phenomenon (Maloney *et al.* 1960) where, with each inspiration, respiratory gases pass from the flail side (which appears to be collapsing) into the controlateral normal lung. During expiration the process reverses, with gas passing from the normal lung into the flail side (which appears to expand). The consequence of *pendelluft* is progressive respiratory distress with hypoxemia and hypercapnia. Treatment for this form of respiratory failure is based on systematic early mechanical ventilation (pneumatic internal stabilization) or, rarely, surgical stabilization of the chest wall.

However, the *pendelluft* phenomenon may not be responsible for respiratory failure in flail chest. Rather, it may be due to chest wall instability which increases the work of breathing and pain, compromising cough and ventilation and leading to atelectasis and pneumonitis. Thus not all patients require intubation but rather alleviation of pain, so that effective tidal volumes and vital capacities can be generated (Bolliger and Van Eeden 1990; Cicala *et al.* 1990).

Pulmonary contusion is common with blunt chest injury and is an important factor leading to respiratory failure. Pulmonary contusion occurs as a result of disruption of the alveolocapillary integrity and may also be seen with high-velocity bullet impact or blast injury. Initially focal, the pulmonary contusion may progress adversely, affecting gas exchange with a reduction of compliance and increased pulmonary shunting. Pulmonary contusion, and consequently hypoxia, can be made worse by shock and volume loading and may progress to acute respiratory distress syndrome (Fulton and Peter 1973). The treatment of pulmonary contusion is supportive in that hypoxemia justifies mechanical ventilation with positive end-expiratory pressure.

Shock

There are many causes of severe hypotension in chest injured patients. A hemothorax of volume 2 liters or more caused by bleeding from parietal vessels (intercostal, internal mammary, great vessels) or the pulmonary vasculature or parenchyma can quickly lead to shock. Bleeding from the aorta, the hilar vessels, and the heart is massive and rapidly fatal, even with immediate surgical treatment.

Aortic rupture may occur in blunt trauma patients subjected to rapid deceleration. The intact adventitia may contain the hemorrhage temporarily, allowing time for diagnosis and surgical intervention. However, any delay may be fatal. Penetrating cardiac or great vessel trauma may cause pericardial tamponade which sometimes saves the patient from exsanguination, particularly if the pericardium is torn, allowing decompression into the pleural space. Cardiac contusion rarely causes shock and is generally self-limiting.

Air embolism is a rare but possible cause of intractable shock and cardiac arrest in the chest trauma patient. When airway and vascular injury coexist, as in penetrating trauma, embolization is possible with rapid and dramatic consequences.

References

Bollinger, C.T. and Van Eeden, S.F. (1990). Treatment of multiple rib fractures: randomized controlled trial comparing ventilatory with nonventilatory management. *Chest*, **97**, 943–8.

Cicala, R.S., Voeller, G.R., Fox, T., Fabian, T.C., Kudsk, K., and Mangiante, E.C. (1990). Epidural analgesia in thoracic trauma: effects of lumbar morphine and thoracic bupivacaine on pulmonary function. *Critical Care Medicine*, **18**, 229–31.

Fulton, R.L. and Peter, E.T. (1973). Physiologic effect of fluid therapy after pulmonary contusion. *American Journal of Surgery*, **126**, 773–7.

LoCicero, J. and Mattox, K.L. (1989). Epidemiology of chest trauma. *Surgical Clinics of North America*, **69**, 15–19.

Maloney, J.V., Schmutzer, K.J., and Raschke, E. (1960). Paradoxical respiration and *pendelluft*. *Journal of Thoracic and Cardiovascular Surgery*, **41**, 291–8.

11.1.2 Pathophysiology of abdominal trauma

J. Biert and R. J. A. Goris

Key messages

- Gunshot wounds produce more hollow viscus injury and subsequent infectious complications than stab wounds. This is due to the mechanism of stabbing; victims are usually stabbed in the anterior abdominal wall.
- Blunt injury usually results in more widespread damage, even at some distance from the point of impact; essentially, there are crush lesions and lesions due to sudden high pressure.
- High intra-abdominal pressure may decrease normal splanchnic blood flow, creating an 'abdominal compartment syndrome' with a low flow state to the bowels, leading to ischemia and necrosis, or possible reperfusion syndrome after reduction of intra-abdominal pressure.
- Pelvic fractures are complicated by severe hemorrhage and shock in up to 35 per cent of patients. The urethra is injured in 3 to 21 per cent of pelvic fracture in males, usually at the prostatic urethra.
- Severe tissue injury results in the systemic release of inflammatory mediators, which may cause an inflammatory reaction in remote organs, such as the lung, resulting in acute respiratory distress, systemic inflammatory response, and multiple organ failure.

Introduction

Patients with abdominal trauma can present with a wide variety of clinical signs and symptoms. Initial investigation can be misleading, and injury is only suspected in a later phase when clinical signs occur. Medical attention may be misdirected to other more obvious injuries of the head, spine, thorax, or extremities. Delay in establishing the correct diagnosis and treatment may lead to increased morbidity and mortality. Injury to the abdomen can be divided into the following groups:

(1) interrupted tissue continuity (wounds, fractures, parenchymal injury);

(2) perforation and leakage of intestines;

(3) hemorrhage from vessel rupture or parenchymal injury;

(4) necrosis and ischemia;

(5) lesions resulting from the propagation of high pressures through tissues (i.e. ruptured diaphragm with pelvic compression injury).

Knowledge of underlying pathophysiological processes is required to understand the events taking place after abdominal trauma and to prevent later sequelae.

Trauma mechanism

In the early phase after injury, the pathophysiology of blunt and penetrating injury may be very different. Stab wounds lead to a well-defined limited injury. Bullets produce more extensive tissue damage, depending on the type and kinetic energy of the missile (high versus low velocity). Gunshot wounds produce more hollow viscus injury and subsequent infectious complications than stab wounds. This is due to the mechanism of stabbing; victims are usually stabbed in the anterior abdominal wall.

Predicting the trajectory of a bullet or a knife wound after the skin has been entered is difficult; the line of damage may follow a ricochet pathway or it can be altered due to postural changes.

Blunt injury usually results in more widespread damage, even at some distance from the point of impact; essentially, there are crush lesions and lesions due to sudden high pressure. Both blunt and penetrating trauma may lead to hemorrhage, tissue necrosis, and organ damage. Late complications of the two types of injury are similar.

Local effects

Gastric dilatation

Acute gastric dilatation is a common occurrence after multiple trauma. Gastric dilatation is the result of increased aerophagia due to agitation or pain, reflex gastric ileus by visceral and somatic nerve stimulation, or intubation or ventilatory problems. It may lead to vomiting, aspiration with subsequent pneumonia, and shock. Physical examination, ultrasound investigation, and diagnostic peritoneal lavage are hindered by gastric dilatation, and should be postponed until a nasogastric tube is inserted. The use of muscle relaxants at intubation can provoke esophageal reflux; this should be foreseen, and the cervical esophagus should be compressed manually (Sellick maneuver).

Ileus

Generalized ileus is common in trauma patients. The presence of a retroperitoneal hematoma due to fractures of spine or pelvis, or other lesions, causes a reflex inhibition of peristalsis and bowel wall muscle tone. This results in generalized dilatation, particularly of the small bowel, and lack of propulsion of bowel contents. Vomiting is stimulated by vagal and sympathetic afferent nerves to the medulla oblongata. Large quantities of fluid and electrolytes continue to be secreted into the lumen of the small bowel. Protein is lost into the bowel lumen and the bowel wall, which becomes edematous. Distension of the small bowel stimulates secretion but does not enhance absorption. The contents of the large intestine may reflux through an incompetent ileocecal valve, resulting in bacterial overgrowth of the small bowel. The bowel contents produce gas which, if not discharged, leads to increased bowel dilatation. Flatus and defecation stop, leading to fecal impaction in the terminal colon and rectum.

Bacterial translocation

The clinical importance of bacterial translocation in trauma patients is still a matter of controversy (Nieuwenhuijzen and Goris 1996).

Bacterial translocation is promoted by one or more of the following three factors: bacterial overgrowth, impaired host immune defense, and physical disruption of the mucosal barrier of the gut. In patients suffering from abdominal trauma, these conditions arise owing to intestinal injury, hemorrhagic shock, administration of antibiotics, and parenteral feeding.

Increased intestinal permeability after major trauma is a well-known phenomenon (Roumen *et al.* 1993); however, bacterial translocation is not observed after trauma (Nieuwenhuijzen and Goris 1996).

The incidence of bacterial translocation in non-trauma patients shows a distinct variability. If the mesenteric lymph nodes are sampled, 6 to 81 per cent contain bacteria depending on the bacterial determination technique (cultures versus electron microscopy). The relationship between increased permeability and bacterial translocation with subsequent infectious complications has not been clearly established.

Experimentally, elevated serum endotoxin results in splanchnic ischemia and increased intestinal permeability. However, in patients suffering from severe trauma or hemorrhagic shock, endotoxin levels, if elevated, are independent of increased intestinal permeability. It has been suggested that increased intestinal permeability, bacterial translocation, endotoxemia, and infectious complications are essentially independent phenomena which occur frequently in seriously injured patients.

Intra-abdominal pressure

In dogs, it has been shown that increasing the intra-abdominal pressure by instillation of saline into the peritoneal cavity leads to acute renal failure and anuria. An intra-abdominal pressure of 15 mmHg may decrease urinary output, and 30 mmHg induces anuria. The same mechanism has been described in patients suffering from serious hemoperitoneum. The mechanism of this phenomenon remains unclear. Arterial renal flow has been reported to be normal in some patients, with normal renograms. It has been suggested that renal cortical pressure is elevated in association with increased intra-abdominal pressure, and as a result acute tubular necrosis may develop (Harman *et al.* 1983).

It has been shown experimentally that, with increased intra-abdominal pressure, the serum level of ADH is increased, leading to oliguria and anuria. There may also be increased pressure in the renal vein and the inferior vena cava (Richards *et al.* 1984). Increased external and internal abdominal pressure in human volunteers with a measured inferior vena cava pressure of 20 mmHg resulted in decreased renal plasma flow, glomerular filtration rate, and tubular glucose reabsorption. This suggests intrarenal shunting.

Apart from effects on renal function, increased intra-abdominal pressure may lead to occlusion of the inferior vena cava. It can also lead to difficulties in ventilation due to diaphragmatic excursion. High intra-abdominal pressure may also decrease normal splanchnic blood flow, creating an 'abdominal compartment syndrome' with a low flow state to the bowels, leading to ischemia and necrosis, or possible reperfusion syndrome after reduction of intra-abdominal pressure.

Effects on pulmonary function

Abdominal trauma influences pulmonary function. Diaphragmatic tears and abdominal wall injuries have a direct effect on muscles used for inhalation. Rapid superficial breathing due to abdominal pain leads to hypoventilation, with atelectasis and pneumonia. Expectoration is decreased after a laparotomy. Abdominal distension leads to increased pressure on the diaphragm, with subsequent decrease of vital capacity followed by atelectasis.

Hypoxia, hemorrhagic shock, sepsis, or pulmonary embolism with release of acute phase mediators into the pulmonary circulation have indirect effects, resulting in macrophage and neutrophil activation, pulmonary microvascular alterations, and the development of acute respiratory distress syndrome and multiple organ failure.

Trauma and pregnancy

In pregnant women sustaining severe abdominal injuries, the life of both mother and child are at risk. The duration of pregnancy is an important factor, since a small pregnant uterus is well protected against external influences by the bony pelvis. Fetal loss can be expected in 20 to 34 per cent of seriously injured patients; this is the result of placental separation with hemorrhage, uterine laceration, fetal head injury, or umbilical cord compression. Maternal shock and hemorrhage induce catecholamine-mediated uterine artery vasoconstriction and decreased uteroplacental perfusion. Maternal circulation is compensated at the expense of fetal circulation. Maternal mortality is not changed compared with non-pregnant patients. Fetal outcome has been related to injury severity scores, serum lactate, and serum bicarbonate levels, which correspond to placental hypoperfusion and hypoxia. Wearing seat belts does not increase the rate of fetal loss.

Specific effects of trauma on organs and organ function

Diaphragm

Blunt traumatic diaphragmatic rupture is due to an increase in intrathoracic or intra-abdominal pressure. Usually, the left hemidiaphragm is involved. The diagnosis of a right-sided or bilateral rupture is relatively rare, although reported figures are increasing (right-sided in 20 per cent and bilateral in 6 per cent of patients). A right-sided tear is seen more frequently in victims on the passenger side of the car, implying a vector-related mechanism. The tear is usually located at the posterolateral side. A left-sided tear can lead to visceral herniation and strangulation of stomach, bowel, or spleen. However, strangulation is a rare phenomenon, and morbidity and mortality are related to other injuries. There is no difference in postoperative course between left- and right-sided tears. Penetrating trauma leads to small diaphragmatic holes, which are repaired during laparotomy.

Stomach, duodenum, and small intestine

Different mechanisms of blunt gastrointestinal trauma lead to diverse patterns of injury. Mortality can range from 10 to 30 per cent. Punctate or slit-like perforations usually occur at the antimesenteric border, and are caused by a sudden increase in intraluminal pressure while closed loops are formed. Alternatively, the bowel can be crushed against the vertebral column. In this case, the bowel wall is locally contused and lacerated. Sometimes the mesentery is injured as well. Complete avulsion of the mesentery root, although uncommon, may lead to early death from hemorrhage (Dauterive *et al.* 1985). Hemorrhagic shock is associated with acute circulatory and functional changes of the gastrointestinal mucosa (Haglund 1986).

Duodenal lesions are particularly troublesome. Although the duodenum is reasonably well protected in the retroperitoneum, its relatively fixed position makes it vulnerable to injury. Combinations with pancreatic injury are not uncommon, particularly in penetrating trauma. Owing to its location, duodenal injury may be missed initially; this leads to ongoing leakage or hemorrhage in the retroperitoneal space, with subsequent infection. At the time that the infection is discovered, sepsis may already have set in.

Spleen

Although the spleen is protected by the left lower ribs and is held in place by peritoneal ligamentous attachments, it is the most common solid organ injured in blunt trama. Its has a spongy structure, with little collagenous tissue, and the capsule is rather thin. Thus the spleen is easily injured, leading to considerable hemorrhage in a short time. The capsule may remain intact while the parenchyma is injured, although this occurs in only 2 per cent of cases. This leads to a subcapsular hematoma with absence of signs in the immediate post-traumatic period. Delayed rupture may occur weeks later, leading to hemorrhagic shock.

The spleen is the principal organ for clearing the circulation of non-opsonized bacteria. Post-traumatic splenectomy can be complicated years later by an overwhelming post-splenectomy infection (OPSI), with an annual risk of 0.5 per cent.

These considerations have led to the understanding that splenic salvage in trauma is important. Techniques for splenorrhaphy include absorbable mesh-wrap, partial splenectomy, or conservative treatment. Results of splenic autotransplantation are still not convincing.

If splenectomy is inevitable, postoperative pneumococcal vaccine should be administered. As the effectiveness of this vaccination is still uncertain, patients should be informed about the risk of OPSI. Further prevention of OPSI includes early evaluation by a physician in the case of fever above 38.5 °C and a generous use of prophylactic antibiotics.

Kidney

The kidneys are protected in part by the lower ribs, and injury should be suspected in patients with fractures of ribs 8 to 12. It is diagnosed by the presence of micro- or macroscopic hematuria, ultrasonography, or CT scan. In severe cases, renal artery or venous avulsion may lead to hemorrhagic shock and possible organ loss.

Renal function itself is also endangered by hemorrhagic shock, and the result may vary from reversible tubular necrosis to total irreversible renal failure.

Large vessels

Injury to the inferior vena cava leads to massive hemorrhage and is associated with mortalities of 21 to 66 per cent in patients reaching the hospital alive. Retrohepatic caval injuries are difficult to control, leading to extensive blood loss, hypothermia, coagulation disorders, and subsequent organ failure, if not death. Temporary vascular isolation of the liver, using a Pringle maneuver for liver hilus occlusion together with an atrial–caval shunt, permits the surgeon to repair the caval injury or perform ligation of hepatic veins with partial liver resection.

Major aortic or large arterial injury is usually less difficult to control. The aorta can be clamped supradiaphragmatically through a left thoracotomy, infradiaphragmatically, or infrarenally.

Female reproductive tract

The female reproductive organs may be injured in penetrating trauma or as a complication of pelvic fractures. The normal menstrual cycle is often disturbed by trauma, with long periods of amenorrhoea.

Pelvic fractures

Pelvic fractures are complicated by severe hemorrhage and shock in up to 35 per cent of patients. The urethra is injured in 3 to 21 per cent of pelvic fracture in males, usually at the prostatic urethra. Bladder injury is reported in 10 per cent of patients with a pelvic fracture. Rupture is extraperitoneal in 60 per cent of patients, usually at the anterior wall or bladder neck. A full bladder (e.g. after alcohol consumption) usually perforates the dome intraperitoneally.

Another associated lesion is a hollow viscus tear which is reported in 1 to 3 per cent of patients. Rectal tears need to be addressed agressively with a proximal colostomy diversion to prevent pelvic abscesses (8–46 per cent), sepsis (8–15 per cent), and rectal fistulas (8–23 per cent). Injury to the non-pregnant uterus, vagina, ovaries, or Fallopian tubes may also be associated with pelvic fractures.

Owing to immobility, patients with pelvic fractures are at higher risk of thromboembolic disease than other trauma patients. This may be due to venous thrombosis in the lesser pelvis. Deep venous thrombosis is found in 6 per cent of patients, and pulmonary emboli are reported in 0.5 to 8.3 per cent of these.

Abdominal wall

Traumatic abdominal wall hernias may result from both penetrating and blunt trauma. In blunt trauma, the injury pattern is usually a shearing force over the abdomen in different directions, distributed over underlying bony prominences such as the pelvic rim. The use of a lap-belt restraint in an obese person may contribute to this finding. The result may be a rectus hematoma, a full-thickness tear of the anterior or lateral abdominal wall, or a total disruption of the abdominal wall with evisceration of the intestines. CT scanning may help to identify clinically unrecognized cases.

Liver

Hepatic trauma can result in massive hemorrhage and shock, or, if the capsule is intact (in blunt trauma), it can lead to subcapsular or intrahepatic hematoma. In a later phase, continuous expansion of this hematoma may cause sudden massive bleeding. Other sequelae are hemobilia and intra- or perihepatic abscesses caused by infection of devitalized tissue.

Furthermore, shock is associated with acute circulatory and metabolic disorders of the liver. The demand on liver function (e.g. for the metabolism of lactic acid) is increased during or following shock. Liver ischemia is associated with changes in glucose metabolism and decreased protein synthesis, leading to a decreased production of coagulation factors with subsequent coagulation disorders. In severe conditions, liver failure is seen as part of multiple organ failure, with hyperbilirubinemia and elevated liver enzymes. This is associated with a high mortality (Haglund 1986).

Pancreas

The pancreas is located retroperitoneally, and is protected by the vertebral column and muscles in the back and by the viscera in front. Therefore pancreatic injury is often associated with injury of other organs.

Blunt trauma can cause pancreatic injury to any extent; in severe cases, the pancreatic gland is fractured over the spinal column. This can be seen in children who have fallen against bicycle handlebars and in victims of motor vehicle accidents.

Penetrating trauma, particularly bullet wounds, can lead to extensive necrosis and ductal injury. While an intact duct may lead to few or no problems, unrecognized ductal injury is followed by early or late hemorrhage, fistulas, pancreatitis, intra-abdominal sepsis, and pseudocysts.

Late exocrine or endocrine dysfunction is rare and is related to extensive surgical resection.

Colon and rectum

Injuries to the colon and rectum may lead to abdominal sepsis and subsequent multiple organ failure following massive fecal contamination, and are associated with a high mortality rate. The introduction of antibiotics has improved outcome significantly. Antibiotic treatment should be initiated as soon as possible and directed against both aerobic and anaerobic bacteria.

Blunt injury can result in both devascularizing lesions, most often located in the right colon, and perforating injuries, which are often seen in the sigmoid colon as a result of the formation of closed loops (Dauterive *et al.* 1985). Diversion is usually necessary.

Remote effects

Inflammatory and immune effects

Severe tissue injury results in the systemic release of inflammatory mediators, which may cause an inflammatory reaction in remote organs, such as the lung, resulting in acute respiratory distress, systemic inflammatory response, and multiple organ failure.

Early after injury, elevated blood levels of C3a, C5a, C3d, C5b–C9, elastase, tumor necrosis factor-α, and interleukin 6 (IL-6) have been shown to correlate with the severity of injury as well as with the onset of acute respiratory distress syndrome. In addition, plasma cytokine levels (IL-6 and C-reactive protein) are significantly higher with 'classical' than with laparoscopic cholecystectomy, indicating that the extent of operative trauma also correlates with the systemic inflammatory response.

In general, the release of inflammatory mediators in trauma causes an immediate, marked,and transient vasodilatation. The microcirculation of the traumatized tissue is isolated from neurohumeral or metabolic control.

Organ dysfunction

The systemic inflammatory response, which is activated by proinflammatory mediators released by local phagocytes in the injured area, is characterized by a progressive and sequential dysfunction of multiple remote organs, such as the lung, kidney, brain, and clotting system, and of intraperitoneal organs such as the liver and the gut (multiple organ failure). The presence and severity of these organ dysfunctions can be quantified with the multiple organ failure score (Table 1) (Goris *et al.* 1985). Patients with multiple organ failure are regularly characterized by fever, generalized permeability edema, and increased susceptibility to hospital-aquired infections, particularly bronchopneumonia and urinary tract infections. Despite intensive support of failing organ systems, the mortality of full-blown multiple organ failure is still about 60 per cent (Roumen *et al.* 1993).

Effects of hypothermia

Hypothermia is commonly observed in trauma patients. Immobility leads to heat loss by conduction and convection. Frequently alcohol is involved, leading to peripheral vasodilatation. Medical treatment can intensify heat loss by exposing the body. The decrease in central body temperature is correlated with the Injury Severity Score. In severely injured patients, particularly those with head injuries, shivering is not observed, indicating altered thermoregulation.

Table 1 Multiple organ failure score

Organ system	Grade 1	Grade 2
Pulmonary	Mechanical ventilation	Mechanical ventilation
	PEEP $\leqslant$ 10 cmH_2O	PEEP $>$ 10 cmH_2O
	$Fio_2 \leqslant 0.4$	$Fio_2 > 0.4$
Cardiac	Hypotension $>$ 100 mmHg	Hypotension $<$ 100 mmHg
	Dopamine $\leqslant$ 10 μg/kg/min	Dopamine $>$ 10 μg/kg/min
	Nitroglycerine $\leqslant$ 20 μg/kg/min	Nitroglycerine $>$ 20 μg/kg/min
Renal	Serum creatinine $\geqslant$ 2 mg/dl (175 μmol/l)	Dialysis
Hepatic	Serum bilirubin $\geqslant$ 2 mg/dl (33 mmol/l) or SGOT $\geqslant$ 25 U/l	Serum bilirubin $\geqslant$ 6 mg/dl (100 mmol/l) or SGOT $\geqslant$ 50 U/l
Hematological	Platelets $< 50 \times 10^9$/l and/or WBC $\geqslant 30 \times 10^9$/l	DIC WBC $< 2.5 \times 10^9$/l or $\geqslant 60 \times 10^9$/l
Gastrointestinal	Acalculous cholecystitis Stress ulcer	Perforated gallbladder Bleeding from ulcer $>$ 2 units blood/24 h Necrotizing enterocolitis Pancreatitis
Central nervous	Diminished responsiveness	Severely disturbed responsiveness and/or diffuse neuropathy

PEEP, positive end-expiratory pressure; SGOT, serum glutamic oxaloacetic transaminase; WBC, white blood cells.
Data from Goris *et al.* (1985).

Hypothermia poses a severe threat to the patient. A core temperature below 35 °C alters the coagulability of the blood by retarding the function of coagulation enzymes. The platelet count is decreased due to sequestration of platelets in the liver and spleen. Platelet aggregation is decreased and fibrinolytic activity is increased. Disseminated intravascular coagulation (DIC) is promoted by a number of factors, all contributing to further hemorrhagic shock. Platelet dysfunction appears to be the most important of these factors.

A temperature below 32 °C in trauma patients hardly ever leads to survival without rewarming using rapid warming fluid infusers, peritoneal lavage, hemodialysis, or cardiovascular bypass. The combination of severe hypothermia and hemorrhagic shock is considered detrimental. Therefore some prefer to keep operative procedures in the initial phase as limited as possible with emphasis on control of surgical bleeding (and resection of injured bowel), allowing for a rewarming of the patient and thus obtaining control over the coagulation physiology before restorative surgery is performed 24 to 48 h later.

References

Dauterive, A.H., Flancbaum, L., and Cox, E.F. (1985). Blunt intestinal trauma. A modern-day review. *Annals of Surgery*, **201**, 198–203.

Goris, R.J.A., te Boekhorst, T.P.A., Nuytink, J.K.S., and Gimbrere, J.S.F. (1985). Multiple organ failure: generalised autodestructive inflammation? *Archives of Surgery*, **120**, 1109–15.

Haglund, U. (1986). Gastro-intestinal, hepatic and renal complications of shock and trauma. In *The scientific basis for the care of the critically ill* (ed. R.A. Little and K.N. Frayn), pp. 293–306. Manchester University Press.

Harman, P.K., Kron, I.L., McLachlan, H.D., Freedlender, A.E., and Nolan, S.P. (1983). Elevated intra-abdominal pressure and renal function. *Annals of Surgery*, **196**, 594–7.

Nieuwenhuijzen, G.A.P., Deitch, E.A., and Goris, R.J.A. (1996). MODS: uncontrolled infection and sepsis? *European Journal of Surgery*, **162**, 259–73.

Roumen, R.M.H., *et al.* (1993). Cytokine patterns in patients after major vascular surgery, hemorrhagic shock and severe blunt trauma: relation with subsequent ARDS and MOF. *Annals of Surgery*, **218**, 769–76.

11.1.3 Assessment

R. F. Wilson and J. G. Tyburski

Key messages

- The systematic approach to assessment includes (1) triage, (2) primary survey (ABCs), (3) resuscitation, and (4) secondary survey (head to toe).
- During the initial evaluation, the highest priority is the patency of the airway. With blunt trauma, the cervical spine should be stabilized, particularly during securing of the airway.
- Hemorrhage is the major preventable cause of trauma deaths in the hospital. Hypotension following injury should be considered to be due to hypovolemia until proven otherwise.

- The adequacy of resuscitation is assessed by continued monitoring of blood pressure, pulse pressure, pulse rate, ventilatory rate, urine output, arterial blood gases, and core temperature.
- The secondary survey does not begin until the primary survey is completed and adequate resuscitation has been initiated. All areas of the body are completely examined. Indicated radiography and special procedures, such as peritoneal lavage, can be performed during this time.

Introduction

Optimal initial treatment of seriously injured patients requires rapid complete assessment of the patient and simultaneous institution of life-preserving therapy. The systematic approach to assessment endorsed by the Committee on Trauma of the American College of Surgeons (1993) includes (1) triage, (2) primary survey (ABCs), (3) resuscitation, and (4) secondary survey (head to toe).

Triage

Two types of triage situation may exist. In the usual situation, the number of patients and the severity of their injuries do not exceed the ability of the facility to render care, and the patients with the most life-threatening problems are treated first. However, if the number of patients and/or the severity of their injuries exceed the capability of the facility and staff, the patients with the greatest chance of survival with the least expenditure of time, equipment, supplies, and personnel, are managed first (American College of Surgeons 1993).

Primary survey

During the primary survey, the patients are rapidly assessed and resuscitation of vital functions is begun based on the ABCs of trauma care:

A Airway maintenance with cervical spine control

B Breathing and ventilation

C Circulation with hemorrhage control

D Disability: neurological status

E Exposure/Environmental control: completely undress the patient but prevent hypothermia

Airway assessment with cervical spine control

During the initial evaluation, the highest priority is the patency of the airway. This is assessed by inspection of the upper airway for foreign bodies, blood, or vomitus, and identification of facial, mandibular, or tracheal/laryngeal injuries that may result in airway obstruction. The presence of inspiratory stridor persisting after the tongue is pulled forward is an important sign of subtotal upper airway obstruction. With blunt trauma, the cervical spine should be stabilized, particularly during securing of the airway.

Breathing

Once the airway has been assured, the patient's chest is exposed to assess ventilation. The symmetry and extent of chest wall expansion can be checked by inspection. The adequacy of exchange in the lungs is initially evaluated by auscultation. Percussion may reveal the presence of air or blood in the chest. Palpation may reveal tender areas or crepitation, indicating fractures. A pulse oximeter should be applied rapidly. Decreased breath sounds on one side which persist even though checks have ensured that the endotracheal tube is not inserted too far, the presence of hypotension, neck vein distension, and deviation of the trachea suggest the presence of a tension pneumothorax. If there is hypotension with no neck vein distension, the trachea is midline, and the chest is dull to percussion, a significant hemothorax is likely to be present.

Circulation with hemorrhage control

Blood volume and cardiac output

Hemorrhage is the major preventable cause of trauma deaths in the hospital. Hypotension following injury should be considered to be due to hypovolemia until proven otherwise. Important information on the cardiovascular system can be obtained by evaluating the level of consciousness, the skin, the pulse, and neck vein distension.

Level of consciousness

When circulating blood volume is reduced, the brain is one of the first organs to be impaired, resulting in an altered level of consciousness, restlessness, or combativeness.

Skin

If the skin over the chest or abdomen is cool, clammy, and pale and/or mottled, the patient is probably hypovolemic. A patient with warm pink skin, particularly in the extremities, is rarely significantly hypovolemic. Brisk capillary refill on the forehead, the nailbed, or the thenar eminence is also generally a sign of adequate tissue perfusion.

Pulse

A rapid thready pulse is characteristic of hypovolemia. An irregular pulse can be a warning of cardiac impairment. Absent femoral and carotid pulses indicate a need for immediate resuscitative action to restore a severely depleted blood volume. Full regular peripheral pulses at a normal rate are usually signs of a relatively normovolemic patient.

Neck vein distension

Hypotension without neck vein distension suggests hypovolemia. Hypotension with neck vein distension suggests the presence of a tension pneumothorax or pericardial tamponade. Beck's triad (hypotension, neck vein distension, and muffled heart tones) is very unreliable for diagnosing pericardial tamponade (Wilson and Bassett 1966); muffled heart tones are often missed, and neck vein distension may not occur while the patient is hypovolemic.

Bleeding

External blood loss is usually obvious, except perhaps from the scalp, and is managed by direct manual pressure on the wound. However, internal blood loss may be much greater and is easily missed. Blood losses seen with various fractures are usually at least 1000 to 2000 ml for a fractured pelvis, 500 to 1000 ml for a fractured midshaft femur, and 250 to 500 ml for other long-bone fractures. Hematomas the size of a fist usually represent at least 500 ml of blood. An increase of 2.5 cm in circumference in the thigh usually represents at least 500 ml of blood, and an increase of 2.5 cm in the circumference of the abdomen represents about 1000 ml of blood. Any decrease in breath sounds on one side of the chest usually represents at least 500 ml of blood, and complete loss of breath sounds probably represents at least 1500 to 2000 ml of blood.

Disability (neurological evaluation)

The initial rapid neurological evaluation includes pupillary size and reaction and the patient's response to verbal and/or painful stimuli. The Glasgow Coma Scale (GCS) is based on three behavioral responses: eye opening, best verbal response, and best motor response (Table 1). As a general rule, the head injury is mild with a GCS of 13 to 15, moderate with a GCS of 9 to 12, and severe with a GCS of 3 to 8.

An altered level of consciousness indicates the need for immediate re-evaluation of the patient's oxygenation, ventilation, and perfusion. Alcohol and/or other drugs may also alter the level of consciousness. However, if hypoxia and hypovolemia have been ruled out, changes in the level of consciousness should be considered to be due to trauma of the central nervous system until proven otherwise.

Table 1 Glasgow Coma Scale

	Points
A. Eye opening	
Spontaneous	4
To voice	3
To pain	2
None	1
B. Verbal response	
Oriented	5
Confused	4
Inappropriate words	3
Incomprehensible sounds	2
None	1
C. Motor response	
Obeys commands	6
Localizes pain	5
Withdraw (pain)	4
Flexion (pain)	3
Extension (pain)	2
None	1
Total GCS points (A+B+C)	

Exposure/environmental control

Patients with possible multiple trauma should be completely undressed, usually by cutting off the garments, to facilitate a thorough examination and assessment. However, it is imperative to cover them after the examination to prevent hypothermia. If an inflated pneumatic antishock garment is in place, it should not be removed until each compartment can be deflated slowly without a drop in systolic blood pressure of more than 5 mmHg. If inflating the abdominal compartment causes respiratory difficulty, a diaphragm injury should be suspected.

Resuscitation

Airway

The airway should be protected in all patients and secured in those patients whose ventilation is not adequate. In most patients with severe trauma or any airway problem, endotracheal intubation is preferred. If endotracheal intubation is not possible, a prompt coniotomy (cricothyroidotomy) may be required.

After an endotracheal tube is inserted, care must be taken to ensure that it is not in the esophagus. This can be checked by listening over the chest and epigastrium. If the breath sounds are better over the epigastrium, the tube is probably in the esophagus. If the tube is in the esophagus, it should be kept there as a guide until the endotracheal tube has been properly placed. Esophageal intubation is also confirmed if the end-tidal $P\text{CO}_2$ during expiration is less than 10 mmHg (1.3 kPa).

Breathing, ventilation, and oxygenation

Absent or decreased breath sounds on one side of the chest suggest the presence of a hemothorax or pneumothorax. Although a hemothorax should be dull on percussion and a pneumothorax should be tympanitic, it is often difficult to differentiate between the two without a chest radiograph. The patient should not be hyperventilated vigorously after an endotracheal tube is inserted because this decreases venous return and increases the chances of a pneumothorax or air embolism, particularly if the lung is injured.

If an arrhythmia, hypotension, and/or changes in the central nervous system occur during or immediately after endotracheal intubation, the presence of an increasing pneumothorax or air embolism should be suspected.

Circulation

A minimum of two large-caliber intravenous catheters should be established. Upper extremity peripheral lines are preferred in most cases. Central venous lines should be established if no peripheral access is possible and/or if central venous pressure monitoring is needed. When the first intravenous line is inserted, blood should be drawn for full blood count, type and cross-match, and baseline chemistry studies. A pregnancy test should be performed in females of childbearing age.

A balanced electrolyte, such as Ringer's lactate, is preferred as the initial resuscitation fluid, and 2 to 3 liters of solution may be required in the first 5 to 15 min to achieve an appropriate response in severely hypovolemic patients. If the patient is still hypotensive after rapid intravenous administration of 2 to 3 liters of crystalloids, this is often evidence of continued severe bleeding that generally needs to be controlled in the operating room.

Hypothermia can develop quickly in the emergency department if the patient is left uncovered and if the resuscitation fluids have not been warmed. A rectal or esophageal probe or thermistor built into a special bladder catheter can be used to monitor core temperature. The use of a high-flow fluid warmer to heat fluids and blood to 39 to 42 °C is recommended. The mortality and incidence of coagulopathy rise if the core temperature falls below 35 °C. If the core temperature falls below 32 °C, most patients become acidotic and coagulopathic and have greatly increased morbidity and mortality.

ECG monitoring of all patients with severe trauma is recommended. Dysrhythmias, including unexplained tachycardia, atrial fibrillation, premature ventricular contractions, and ST segment changes, may indicate cardiac contusion. A prolonged QT interval suggests hypocalcemia, and peaked T waves suggest the presence of hyperkalemia. Electromechanical dissociation may indicate cardiac tamponade, tension pneumothorax, and/or profound hypovolemia. If bradycardia, aberrant conduction, or premature beats are present, hypoxia, hypoperfusion, and/or hypothermia may be present.

Urinary and gastric catheters

Urinary catheters

If the patient is severely injured or is stable but cannot urinate, a catheter should be inserted into the bladder after ascertaining that there is no evidence of a urethral injury (such as blood at the urethral meatus, penile or scrotal hematomas, or a displaced prostate on rectal examination).

Gastric tubes

A gastric tube is inserted to reduce gastric distension and the chances of aspiration. Blood in the gastric aspirate may represent oropharyngeal (swallowed) blood, traumatic insertion, or actual injury to the esophagus or stomach. Deviation of a nasogastric tube to the right in the middle or upper chest on a chest radiograph suggests a traumatic rupture of the aorta. If the end of the tube enters the abdomen and then goes up into the chest, it suggests a diaphragmatic rupture.

Monitoring

The adequacy of resuscitation is assessed by continued monitoring of blood pressure, pulse pressure, pulse rate, ventilatory rate, urine output, arterial blood gases, and core temperature. These values need to be recorded every 5 to 15 min in critically injured patients.

1. The systolic blood pressure is often a poor measure of actual tissue perfusion; however, changes in pulse pressure often correlate well with changes in stroke volume. Class III hemorrhage (30–40 per cent of the blood volume loss), as reflected by a decreased systolic blood pressure and pulse pressure, generally indicates a need for blood as part of the fluid resuscitation.
2. Heart rates above 90 beats/min and rising are usually a sign of increasing hypovolemia or hypoxemia. A heart rate above 120 to 140 beats/min implies a class III or class IV hemorrhage.
3. In the absence of diuretics or osmotic substances in the urine, such as with glycosuria, the urine output serves as a guide to the general adequacy of tissue perfusion. A urine output below 0.5 ml/kg/h usually indicates hypovolemia.
4. Ventilatory rate and arterial blood gases should be used to monitor the patient's ventilation and oxygenation. A respiratory rate above 30 to 40 breaths/min and/or evidence of increased efforts to breathe are strong indicators of a need for endotracheal intubation and ventilatory assistance.
5. Pulse oximetry is a valuable adjunct for continuously monitoring the adequacy of tissue oxygenation; however, the $S\text{aO}_2$ from the arterial blood gas may be 2 to 3 per cent less than the saturation shown on the pulse oximeter.
6. End-tidal carbon dioxide monitoring showing an end-tidal $P\text{CO}_2$ of 10 mmHg (1.3 kPa) or more can confirm that an endotracheal tube is properly positioned. In addition, an end-tidal $P\text{CO}_2$ below 25 to 30 mmHg (33–40 kPa) in a patient with a $P\text{aCO}_2$ of 40 mmHg (5.3 kPa) or more implies that the patient is inadequately resuscitated (Domsky *et al.* 1995).

Laboratory evaluation

Serial hematocrit measurements may provide an estimate of the amount of ongoing blood loss, as long as hemodynamic stability is maintained. Persistently elevated amylase levels suggest injury of the pancreas. An initially elevated serum amylase level is of little diagnostic help, particularly if there are facial injuries with salivary gland injury. Fractionation of serum amylase, if available, may be of some assistance in differentiating pancreatic injury. A urinalysis showing 50 or more red cells per high-power field usually indicates a need for radiological evaluation of the urinary tract.

In patients with head or chest injuries, frequent arterial blood gas analyses are mandatory to guide respiratory therapy. Measurements of serum osmolarity may be required every 6 h in patients who are receiving mannitol therapy for raised intracranial pressure. Blood glucose and electrolyte values should also be evaluated every 6 to 12 h in such patients.

Radiography

The three radiographs that should generally be obtained for patients with multiple trauma are lateral cervical spine, anteroposterior chest, and anteroposterior pelvis. These films can be taken in the resuscitation area with a portable X-ray unit, but should not interrupt the resuscitation process. After all life-threatening injuries have been identified and treated, complete cervical spine, thoracic, and lumbar spine films can be obtained as needed. In the patient with penetrating truncal injuries, films pertinent to the site(s) of wounding should be obtained, if patient condition allows.

Secondary survey

The secondary survey does not begin until the primary survey is completed and adequate resuscitation has been initiated. All areas of the body are completely examined. Indicated radiography and special procedures, such as peritoneal lavage, can be performed during this time. Complete evaluation of the patient also requires repeated examination of the patient's mental status, chest, abdomen, and vital signs.

History

The history should include as much detail as possible on the mechanism of trauma, and many injuries can be predicted from the direction and amount of the trauma. Other items of particular importance are current medication, previous surgery, allergies, use of alcohol and/or recreational drugs, tetanus prophylaxis, and the time of the last meal.

Physical examination

Head

The entire scalp and head are examined for lacerations, contusions, and evidence of fractures. Although edema around the eyes may later preclude an in-depth examination, the eyes should be evaluated for injury, visual acuity, and pupillary size. If there is any suspicion that the patient may have a significant intracranial lesion requiring surgery, a prompt head CT should be performed; however, in general a patient should not be sent for a CT scan while hemodynamically unstable.

Maxillofacial examination

Many facial fractures can be detected more readily by physical examination than by radiography. If the teeth do not fit together normally, a maxillary or mandibular fracture should be suspected and Panorex films should be taken. Radiographs or CT scans of the face are not usually necessary in the emergency department if the patient is going to be admitted.

Cervical spine and neck

Patients with head, facial, or neck trauma should be assumed to have an unstable cervical spine injury, and the neck should be immobilized until all aspects of the cervical spine have been adequately studied and an injury excluded. However, the absence of neck pain, neck tenderness, and distracting injuries in an awake and alert adult who has not taken mind-altering drugs is usually sufficient to exclude a cervical spine injury.

If a cervical collar is in place, it can be removed to examine the neck while another person stabilizes the head. Cervical spine tenderness, subcutaneous emphysema, tracheal deviation, and laryngeal fracture may be noted on palpation if the patient is not intubated. Any change in voice should raise suspicion of a laryngeal injury. The carotid arteries should be auscultated for bruits.

Patients with a high spinal cord injury are vasodilated and tend to have a systolic blood pressure of only 80 to 90 mmHg. In addition, the pulse rate tends to be normal and the skin is relatively warm and dry. If a patient with a high spinal cord injury has tachycardia or a systolic blood pressure less than 80 mmHg, he or she is probably hypovolemic but may also have cardiac injury.

With penetrating neck trauma, wounds that extend through the platysma should not be explored manually in the emergency department. This type of injury requires evaluation in the operating room or with angiography, endoscopy, and/or contrast studies to rule out vascular and aerodigestive tract injuries.

Chest

Inspection of the chest, both anterior and posterior, may identify conditions such as sucking chest wounds and flail segments. When the back is to be examined, the patient must be logrolled to avoid aggravating any spinal injury that may be present. The entire chest cage must be carefully palpated. Contusions and hematomas of the chest wall should alert the physician to the possibility of internal injury. However, up to 30 to 40 per cent of patients with severe intrathoracic injuries, particularly younger patients, have little or no external evidence of trauma.

Echocardiography is a non-invasive and relatively inexpensive means of evaluating patients with a possible myocardial contusion or tamponade. It can also detect acute valvular injury, pericardial effusions, or intracavitary clots.

Chest radiography can detect many pneumothoraces that are not apparent clinically; however, up to 1000 ml of hemothorax may not be apparent on a supine chest radiograph. A certain number of 'occult' pneumothoraces are seen on CT scans of the chest or abdomen but not on chest radiographs. An 'occult pneumothorax' can be observed if the patient is not on a positive pressure ventilation. Rib fractures may be present, but up to 50 per cent of these are not visualized on the initial radiographs, and radiographs with rib detail add nothing to management.

Although complete absence of a mediastinal hematoma on CT scan may be helpful for excluding traumatic rupture of the aorta, an aortogram should be performed in suspicious cases. Transesophageal echocardiography can be very useful for evaluating a possible traumatic rupture of the aorta, particularly in patients too unstable to go to the angiography suite, but it should not replace aortography (Shapiro *et al.* 1991; Kearney *et al.* 1993).

Abdomen

A normal initial examination of the abdomen does not exclude a significant intra-abdominal injury because some injuries may not cause pain or tenderness for up to 12 to 24 h. In addition, alcohol, illicit drug use, or a head or spinal cord injury may preclude a reliable examination. Close observation and frequent re-evaluation of the abdomen, preferably by the same observer, is important for detecting such injuries early. Fractures of the pelvis or the lower rib cage may also hinder adequate examination of the abdomen, because pain from these areas may be elicited when palpating the abdomen.

Patients with equivocal physical findings, unexplained hypotension, or an impaired sensorium secondary to trauma, alcohol, or other drugs should be considered as candidates for a peritoneal lavage. CT scans and ultrasound can also be used to evaluate the abdomen. The CT scan is particularly useful for evaluation of retroperitoneal structures, such as the kidneys and pancreas, but can easily miss bowel injuries.

Ultrasound is being used increasingly as a screening tool for hemoperitoneum in blunt trauma. In a collected series of 4000 trauma patients, ultrasound was used to detect hemoperitoneum and/or pericardial effusion in injured patients with a 93.4 per cent sensitivity and 98.4 per cent specificity (Rozycki and Shackford 1996).

Perineum, rectum, and vagina

The perineum should be examined carefully for contusions, hematomas, lacerations, and urethral bleeding. On rectal examination the physician should assess for blood, a high-riding prostate, the presence of pelvic fractures, the integrity of the rectal wall, and the quality of the sphincter tone. In females, a vaginal examination is essential to assess for the presence of blood or injuries in the vaginal vault. A pregnancy test should be performed on all females of childbearing age.

Musculoskeletal

The extremities should be carefully examined for tenderness, crepitation, or abnormal motion. The stability of any pelvic fractures should also be evaluated as a possible source of major occult blood loss. Impaired sensation and/or loss of voluntary muscle strength may be due to nerve or vessel injury or compartment syndrome.

Perfusion of injured extremities should be monitored closely. An arteriogram is the gold standard for evaluating arterial injuries, but Doppler ultrasound examination is a good screening modality. If compartment syndrome is suspected, compartment pressure can easily be determined at the bedside with a hand-held monitor or using an arterial line set-up. Normal compartment pressures range from zero to 8 mmHg, and decompression should be considered for pressures greater than 30 mmHg.

Neurological examination

A comprehensive neurological examination includes not only motor and sensory evaluation of the extremities, but also continued re-evaluation of the patient's level of consciousness and pupillary size and response. The intracranial pressure should be monitored in any patient with suspected intracranial mass lesion. A decrease of 2 or more in the Glasgow Coma Scale should prompt urgent neurosurgical evaluation.

Re-evaluation

The trauma patient must be re-evaluated constantly to assure that new findings are not overlooked and to discover any deterioration early. As initial life-threatening injuries are managed, other life-threatening problems, severe injuries, and underlying medical problems may become evident. Suspicion must be particularly high in patients who have altered consciousness due to head injury or the effects of drugs or alcohol.

References

American College of Surgeons (1993). *Advanced trauma life support program for physicians*. American College of Surgeons, Chicago, IL.

Domsky, M., Wilson, R.F., and Heins, J. (1995). Intraoperative end-tidal carbon dioxide values and derived calculations correlated with outcome: prognosis and capnography. *Critical Care Medicine*, **23**, 1497–1503.

Kearney, P.A., *et al.* (1993). Use of transesophageal echocardiography in the evaluation of traumatic aortic injury. *Journal of Trauma*, **34**, 696–702.

Rozycki, G.S. and Shackford, S.R. (1996). Ultrasound. What every trauma surgeon should know. *Journal of Trauma*, **40**, 1–4.

Shapiro, M.J., *et al.* (1991). Cardiovascular evaluation in blunt thoracic trauma using transesophageal echocardiography (TEE). *Journal of Trauma*, **31**, 835–40.

Wilson, R.F. and Bassett, J.S. (1966). Penetrating wounds of the pericardium or its contents. *Journal of the American Medical Association*, **195**, 513–18.

11.1.4 Management of the chest injury

J. E. Mazuski and R. M. Durham

Key messages

- Most patients with chest injuries are treated non-operatively, with only 10 to 15 per cent requiring surgery.
- Initial therapy includes ensuring an adequate airway and ventilation, controlling sites of bleeding, and restoring adequate tissue perfusion.
- Closed-tube thoracostomy is the primary therapy for most patients with pneumothorax or hemothorax as a result of either penetrating or blunt trauma.
- Injuries to the heart and the aorta should be suspected in patients who have suffered significant blunt injuries to the thorax, particularly those due to deceleration forces.

Introduction

Chest injuries are a significant cause of morbidity and mortality; they are directly responsible for 20 to 25 per cent of deaths due to trauma, and contribute to another 25 per cent of those deaths. Both blunt and penetrating traumas produce these injuries, but the general principles of management are similar regardless of the mechanism. However, the actual diagnostic and treatment modalities employed depend on the specific thoracic structures which have been damaged. The vast majority of patients with chest injuries can be treated non-operatively, with only 10 to 15 per cent requiring operative intervention (McSwain 1992).

General management

Initial assessment

In general, the diagnostic evaluation of patients with chest injuries is similar to that of other trauma patients. The initial emphasis is on ensuring that there is an adequate airway and ventilation, controlling sites of bleeding, and restoring adequate tissue perfusion (Kshettry and Bolman 1994). Patients with significant chest injuries may present with severe respiratory distress or frank respiratory failure and require immediate institution of assisted ventilation.

Tension pneumothorax is a relatively common cause of respiratory distress and shock in the patient with chest trauma. Tracheal deviation and decreased breath sounds contralateral to the side of the deviation strongly suggest this diagnosis and call for immediate needle decompression and/or tube thoracostomy. Early tube thoracostomy, without waiting for a chest radiograph, is warranted in patients with penetrating chest wounds who present with significant respiratory distress or hypotension.

Patients with significant hemorrhage due to chest injuries require fluid resuscitation to restore adequate circulation and prevent ongoing deficits in tissue perfusion. This is usually accomplished with crystalloid infusions initially, followed by packed red blood cells if the patient remains refractory to the administration of 2 liters of isotonic fluid. The use of delayed resuscitation for patients with penetrating truncal trauma remains controversial, and there is no information about its applicability to patients with blunt chest trauma.

Although the usual cause of hypotension in trauma patients is hemorrhagic shock, patients with chest trauma may also have cardiogenic shock. Blunt injuries to the chest may produce significant myocardial contusion with resultant myocardial dysfunction. Acute cardiac tamponade due to bleeding within the pericardial sac may be produced by both penetrating and blunt mechanisms. This should be suspected in patients who have refractory shock and evidence of elevated central venous pressures. Decompression of the pericardium may be life saving in this latter group of patients.

A chest radiograph should be obtained during the initial resuscitative efforts. If this study shows a significant pneumothorax or hemothorax, closed-tube thoracostomy should be performed. The chest radiograph should also be evaluated for bony and soft tissue injuries, pulmonary parenchymal infiltrates, deviation of the tracheobronchial tree, lesions of the mediastinum, particularly widening of the mediastinal silhouette, and abnormal diaphragmatic shadows. These findings will help decide what additional diagnostic or therapeutic interventions need to be undertaken.

Emergency room thoracotomy is primarily used in patients with penetrating chest trauma who sustain a cardiac arrest in the emergency department or shortly prior to arrival. A small but significant number of these patients, particularly those who have suffered stab injuries to the heart, may be salvaged. In general, emergency room thoracotomy is not warranted for patients with blunt chest trauma, since their survival with this procedure is negligible. The procedure should only be performed by those who can repair the injuries.

Indications for operative intervention

A modest number of patients with chest trauma require early operative intervention. The most common indication for thoracotomy is significant blood loss into the chest cavity. Guidelines for operation are an initial loss of at least 1500 cm^3 of blood upon placement of the chest tube, or continued bleeding of 200 to 300 cm^3/h. However, patients with ongoing hemodynamic instability require operative intervention in order to obtain hemostasis even if these arbitrary threshold values are not exceeded. Thoracotomy is also indicated if a large hemothorax cannot be adequately evacuated from the chest by tube thoracostomy, although the timing of this procedure is controversial. Other indications for early thoracotomy or median sternotomy are injuries to the heart, aorta, or other great vessels, tracheal or major bronchial injuries, and esophageal injuries. Diaphragmatic injuries are usually approached through the abdomen because of the common association with other intra-abdominal injuries.

Subsequent management

The subsequent management of patients with chest injuries is quite similar to that of other critically ill patients. These patients generally require monitoring and support of their respiratory and cardiovascular functions.

Injuries to the chest commonly result in decreased compliance of the chest wall and the pulmonary parenchyma. Chest wall compliance may be altered directly by injury to the ribs and associated structures, or by attendant pain and muscle spasm. Compliance of the lung may be reduced as a result of pulmonary contusion or aspiration pneumonitis. In addition, the presence of increased extravascular lung water, produced indirectly as a result of shock and resuscitation, may also contribute to decreased pulmonary compliance.

These changes in compliance result in a decrease in the functional residual capacity, which ultimately results in increased work of breathing. This predisposes the patient to respiratory dysfunction. Another factor that compromises respiratory reserve is the increased metabolic need of the injured patient. If the patient is unable to meet these respiratory demands, respiratory failure ensues.

Management of these respiratory problems is primarily supportive. The indications for intubation and institution of mechanical ventilation in patients with chest trauma are similar to those for other critically ill patients. In general, patients who have excessive work of breathing or are developing refractory hypoxia or hypercarbia require ventilatory assistance. Many patients with severe chest wall injuries, particularly those with a flail chest, have paradoxical motion during spontaneous breathing. However, a flail chest alone is not an indication for mechanical ventilation if the patient has adequate gas exchange.

The use of non-invasive ventilatory assistance has proved beneficial in avoiding intubation and mechanical ventilation in some patients with imminent respiratory failure due to exacerbation of chronic pulmonary disease. The effectiveness of these modalities in patients with chest trauma has not yet been ascertained. Nonetheless, it may be reasonable to attempt these non-invasive techniques in selected patients with chest trauma. However, there may be logistical difficulties in using a ventilatory mask in patients who have associated maxillofacial injuries.

Patients with chest injuries require excellent pain control in order to facilitate adequate pulmonary physiotherapy. Adequate pain control alone may improve functional residual capacity and thereby enhance pulmonary function. Dosing of narcotic analgesics on an as-needed basis is a much less effective method of controlling pain than other modalities, such as patient-controlled analgesia. Other methods of pain control, including intrapleural and epidural analgesia, are efficacious in patients with chest injury and avoid some of the systemic effects of narcotics. Non-steroidal anti-inflammatory agents may also be used to supplement the effects of narcotics and allow a lower dose of opiates to be administered.

Cardiovascular complications related to chest injury are relatively common. Myocardial dysfunction may result from myocardial contusion, prolonged hypoperfusion, or pre-existing cardiac disease. Intravenous fluids should be used to maintain an adequate intravascular volume. Patients with an increased afterload should be treated with vasodilators to enhance myocardial performance. If these maneuvers are unsuccessful, patients with poor tissue perfusion should receive inotropic support, at least until their cardiac function improves.

Fluid resuscitation is best guided by the physiological performance of the myocardium. Excessive volume beyond that needed to optimize myocardial function is neither necessary nor desirable. However, there is no support for the notion that patients with chest injuries should be kept hypovolemic ('dry') in order to prevent further fluid extravasation into damaged lungs. Patients who are inadequately resuscitated are likely to suffer the deleterious consequences of prolonged shock.

Many patients with chest injuries will require the placement of a pulmonary artery catheter in order to assess cardiac performance accurately. Myocardial compliance may be significantly altered in these patients. Thus the measured pulmonary capillary wedge pressure may not be as good an indicator of intravascular volume as it is in other critically ill patients. It may be useful to monitor the cardiac output in response to a volume challenge in order to estimate the adequacy of volume replacement. Another useful tool is the calculated right ventricular end-diastolic volume, which can be obtained using pulmonary artery catheters capable of determining right ventricular ejection fraction. The calculated end-diastolic volume has been shown to be a better predictor of preload than the wedge pressure, and has been successfully used to guide fluid therapy in patients with altered myocardial compliance.

Chest and lung injuries

Chest wall injuries

Management of chest wall injuries is primarily supportive in nature. Patients with large open wounds to the chest should have operative debridement and closure in conjunction with drainage of the pleural space. Patients with flail chest do not require specific therapy directed toward the flail segment; rather, their need for mechanical ventilation is related to the underlying pulmonary contusion or other parenchymal injury. Injuries of the first and second ribs are generally due to a substantial transmission of force, and are associated with injuries to the great vessels of the thoracic outlet. However, in the absence of suggestive clinical signs or evidence of mediastinal hemorrhage on the initial chest radiograph, these injuries alone are not an indication for invasive vascular studies. Sternal fractures suggest the possibility of an underlying myocardial contusion, but are of little consequence by themselves.

Pneumothorax

Traumatic pneumothoraces are common with both blunt and penetrating mechanisms, and can be caused by direct penetration of a missile or other object through the chest wall, a closed laceration of the lung parenchyma from a fractured rib, or a sudden increase in intrathoracic pressure. During the initial assessment, clinical evidence of a tension pneumothorax calls for immediate needle decompression or tube thoracostomy of the pleural space involved. However, most pneumothoraces are diagnosed by chest radiography.

In most cases, traumatic pneumothoraces are treated with tube thoracostomy. A posterior chest tube is usually placed because an associated hemothorax is common, even if it is not obvious on the initial chest radiograph. Under selected circumstances, patients with small pneumothoraces may be observed. This mandates careful re-examination and repeat chest radiography to ensure that there is no progression of the pneumothorax. Small pneumothoraces which are not evident on the initial chest radiograph may be identified in trauma patients undergoing chest or abdominal CT scans. In general, most of these 'occult pneumothoraces' can be safely observed according to these same principles. However, observation is not recommended for patients with small or occult pneumothoraces who are on positive-pressure ventilation or who will need to undergo operative procedures under general anesthesia.

Once a chest tube has been inserted, it is placed on suction. A follow-up chest radiograph is obtained to ensure that the lung has been fully re-expanded. If there is no evidence of an air leak, the tube can usually be removed after 48 to 72 h. Some clinicians prefer to leave the chest tube on water seal for 6 to 24 h prior to removal to ensure that a recurrent pneumothorax will not develop. In any case, a follow-up chest radiograph should be taken following the removal of the chest tube to ensure that a pneumothorax has not redeveloped.

If the patient has a continued air leak, a repeat chest radiograph should be checked for a persistent pneumothorax. If this is found, an additional chest tube is placed, usually directed anteriorly and superiorly. Nearly all air leaks will eventually seal if the lung is fully re-expanded and the chest tube is kept on suction. Generally, an air leak must persist for at least 14 days before operative intervention is considered. The primary exception to this rule is the patient with a major bronchial injury, which is suggested by a very large air leak and an inability to re-expand the lung with closed-tube thoracostomy.

Hemothorax

Hemothorax is generally due to bleeding from the pulmonary parenchyma or the chest wall. In most cases bleeding will stop spontaneously once the lung has been re-expanded using tube thoracostomy. However, as indicated previously, ongoing bleeding is an indication for thoracotomy.

If the initial chest tube does not adequately drain the hemothorax, additional tubes should be placed in an effort to clear the blood completely. Patients who have a retained hemothorax are at risk of the development of empyema or a fibrothorax with lung entrapment. If the hemothorax persists despite efforts to obtain complete drainage, it is likely that the blood in the pleural space has clotted. Some have advocated early operative intervention to remove the clotted hemothorax. If an operation is undertaken within the first few days after an injury, the blood clot is usually not very adherent to the pleura and lung, and can be removed quite easily. Indeed, it may be possible to evacuate the clot using thoracoscopy and thereby avoid a thoracotomy incision. If early intervention to remove a retained clot is not undertaken, it may be best to wait for up to 6 weeks before undertaking thoracotomy. The operation is usually easier to perform once the initial inflammatory reaction has subsided, and in some cases the patient will resorb the clot during this period and not require an operation.

Another common reason for delayed operation in patients with chest injuries is an empyema. This may occur following either blunt or penetrating injuries, although it is somewhat more common after the latter. Thoracentesis should be performed if a patient has a new or retained pleural effusion and systemic signs of infection. The pleural fluid is tested for pH, protein, and lactate dehydrogenase levels, and sent for culture. Appropriate antibiotics should be administered if the fluid is infected. Initially, an attempt should be made to drain the infected pleural fluid using closed-tube thoracostomy. Non-operative management has a lower likelihood of success when there is a significant amount of retained clot, if the fluid is quite viscous, or if it has a low pH. Operatively, a localized posterior empyema may be approached with a posterior rib resection and open drainage. A larger empyema generally requires thoracotomy with decortication and drainage, particularly if an extensive pleural peel has developed.

The use of antibiotics at the time of chest tube insertion has been advocated as a means of decreasing the risk of empyema and other infections. Several prospective studies have failed to demonstrate that prophylactic antibiotics prevent subsequent infections. In contrast, a recent meta-analysis of selected studies suggested that there was a statistically significant decrease in the incidence of empyema among patients receiving antibiotic prophylaxis prior to chest tube insertion. Thus the role of antibiotics remains controversial.

Pulmonary parenchymal injuries

Parenchymal infiltrates occurring after trauma may have several causes. Pulmonary contusions and hemorrhages are the direct result of forces to the chest wall or penetration of the lung parenchyma. Aspiration is also common in traumatically injured patients, and may not be distinguishable from pulmonary contusion. Multiply injured patients are also at risk of the development of acute lung injury. It is not uncommon for several of these mechanisms to contribute to the early development of pulmonary dysfunction in trauma patients.

Pulmonary contusions and other parenchymal injuries are almost always treated non-operatively. The mainstay of therapy is supportive pulmonary care, including mechanical ventilation as needed. Rarely, patients with large devitalized segments of pulmonary parenchyma may require thoracotomy for formal operative debridement. However, most patients with this degree of injury will have been operated on early because of hemodynamic instability or ongoing hemorrhage.

Tracheobronchial injuries

In general, injuries to the trachea or major bronchi require operative repair. As indicated previously, a major bronchial injury is suggested by a pneumothorax which cannot be adequately decompressed with conventional tube thoracostomy or by an overwhelming air leak. The diagnosis is usually confirmed by bronchoscopy. Patients with smaller bronchial injuries can be treated by close observation if the lung can be completely re-expanded with chest tubes and there is no other reason for urgent thoracotomy.

Mediastinal injuries

Heart and pericardium

Injuries to the heart and pericardium occur as a result of both blunt and penetrating trauma. Lacerations to the heart are generally due to a penetrating injury, although occasionally they may be produced by a blunt mechanism. Most patients with cardiac lacerations die at the scene or soon after the injury. Those who survive generally have relatively small contained injuries.

Patients with cardiac lacerations may present with shock due to cardiac tamponade. However, the classic signs of cardiac tamponade, including hypotension, muffled heart tones, and jugular venous distention, may not be present or may be difficult to appreciate owing to associated injuries or accompanying hypovolemia. In general, the diagnosis should be considered in any patient who has a penetrating injury to the central portion of the chest, or who has evidence of continuing hypoperfusion in the face of an elevated central venous pressure.

Many patients with cardiac lacerations will require emergency department thoracotomy because of severe hemodynamic compromise. Optimally, however, the patient should be transported to the operating room for definitive therapy. Patients with a suspected cardiac injury who show any signs of hemodynamic instability require immediate operative intervention. Needle pericardiocentesis may be attempted as an emergency maneuver for the temporary relief of tamponade, but this does not eliminate the need for surgery. Depending on the suspected site of injury, the heart is usually approached through a median sternotomy or left thoracotomy. The tamponade is relieved by opening the pericardium and the perforation is then directly sutured. It is almost never necessary to place the patient on cardiac bypass for these injuries.

In many patients with penetrating chest trauma, a cardiac injury is suspected because of proximity of a wound to the heart, even though the patient does not manifest any clinical signs or symptoms of such an injury. If the patient is to undergo an operative procedure for some other reason, subxiphoid pericardiotomy may be performed to determine whether blood is present within the pericardial sac. If it is not, the patient can be observed. Alternatively, transthoracic or transesophageal echocardiography can be used in stable patients who do not require surgery. However, there have been occasional reports of cardiac injuries which were not detected by echocardiography. Thus careful clinical follow-up of these patients is mandatory even if the echocardiogram is negative.

The most common cardiac injury associated with blunt trauma is myocardial contusion. This injury usually results from application of a compressive force to the anterior part of the chest. Thus the portion of the heart usually affected is the right ventricle. The major complications of this injury are hemodynamic compromise and arrhythmias.

The best approach for diagnosing myocardial contusion is controversial. An electrocardiogram is generally obtained for suspected myocardial contusion. Occasionally, a right ventricular injury pattern is detected. The sensitivity of the electrocardiogram can be improved with the use of right-sided leads, although this is not done routinely. Most patients with myocardial contusion show only non-specific ECG changes, such as sinus tachycardia or minor ST–T wave alterations. Thus the ECG results are relatively unreliable as a definitive test for this injury. The same applies to determinations of creatine kinase MB fractions. These have not been found not to be very specific for this injury, and their routine measurement in patients with blunt chest trauma is questionable (Christensen and Sutton 1993). Radionuclide imaging of the heart may allow the diagnosis to be made, but it is usually not immediately available. The best test currently available is the echocardiogram, which can detect significant areas of abnormal myocardial contractility. The importance of screening for myocardial contusion is debatable because patients who do not develop either significant arrhythmias or hemodynamic compromise within a few hours after blunt chest trauma are unlikely to develop any sequelae, even if a myocardial contusion is definitively diagnosed by echocardiography.

Aortic and great vessel injury

Penetrating injuries to the aorta and great vessels require operative management. This needs to be performed immediately if the patient is hemodynamically unstable. However, if the patient is hemodynamically stable, diagnostic arteriography is used to determine whether there is a significant vascular injury. Patients with penetrating injuries within the area bordered by the sternal notch, the xiphoid process, and the nipples laterally, or who have evidence of transmediastinal missile passage, are at risk of these injuries. The arteriography findings will delineate the optimal operative approach.

In blunt trauma, the aorta is most commonly injured at the level of the ligamentum arteriosum, close to the take-off of the left subclavian artery. The possibility of blunt aortic injury should be considered in patients who have had a major deceleration injury, such as occurs with a head-on motor vehicle crash at a high rate of speed or a fall from a long distance. Aortography is the definitive diagnostic test for blunt aortic injury, and should be obtained in patients who have a highly suspicious mechanism or a suggestive chest radiograph. The most sensitive radiographic sign of an aortic laceration is mediastinal widening of more than 8 cm at the level of the aortic knob. However, as many as 28 per cent of patients with a proven aortic injury may have a normal chest radiograph, and 90 per cent of patients with mediastinal widening will have a negative aortogram (Rosenthal and Ellis 1995). Alternative screening techniques, such as the chest CT, have been advocated, but these have not yet been proven to be sufficiently sensitive to eliminate the need for arteriography (Durham *et al.* 1994).

Transesophageal echocardiography is being successfully performed instead of aortography in selected patients. This technique appears to be quite sensitive for detecting injuries of the descending aorta. Transesophageal echocardiography may be particularly useful for patients requiring ongoing resuscitation or other intensive care management, since it does not require transport to an angiographic suite. However, transesophageal echocardiography does not adequately visualize the upper ascending aorta or aortic arch. While not nearly as common as injuries at the aortic isthmus, injuries to the root of the aorta or to the innominate artery can be missed by transesophageal echocardiography.

Most aortic injuries should be surgically repaired. However, non-operative management, similar to that used in patients with aortic dissections, has been successfully utilized in carefully selected patients in whom only an intimal injury is identified. Some details of the operative procedure, such as the use of a graft versus direct repair of the aorta and the use of a shunt while the aorta is cross-clamped, are still controversial.

Esophagus

Penetrating injuries to the esophagus are relatively uncommon, and blunt injuries to the esophagus are rare. In patients not undergoing surgical exploration of a penetrating wound, the diagnosis of esophageal injury can be made using radiographic studies or endoscopy. A combination of esophagography and rigid esophagoscopy appears to be the most sensitive technique for identifying these injuries. However, frequently only one of these studies is performed unless there is a very high clinical suspicion of an esophageal injury.

In general, esophageal injuries are managed operatively. Most of these injuries can be directly repaired, particularly if they are diagnosed and treated within 24 h. If the injury is more than 24 h old, mediastinal inflammation may make it impossible to close the perforation. In those circumstances, drainage, with or without diversion, may be the only option.

There have been a few case reports in which esophageal injuries have been successfully managed non-operatively in a manner similar to that employed for certain iatrogenic esophageal perforations. The general guidelines for non-operative management of iatrogenic esophageal injuries are that the perforation is small, the area of extravasation is contained and has free communication back to the esophagus, and the patient exhibits no clinical signs and symptoms of ongoing infection. Whether or not these guidelines can be safely extended to allow non-operative management of patients with traumatic esophageal injuries awaits a more definitive study.

References

Christensen, M.A. and Sutton, K.R. (1993). Myocardial contusion: new concepts in diagnosis and management. *American Journal of Critical Care*, **2**, 28–34.

Durham, R.M., *et al.* (1994). Computed tomography as a screening exam in patients with suspected blunt aortic injury. *Annals of Surgery*, **220**, 699–704.

Kshettry V.R. and Bolman, R.M., III (1994). Chest trauma: assessment, diagnosis, and management. *Clinics in Chest Medicine*, **15**, 137–46.

McSwain, N.E. (1992). Blunt and penetrating chest injuries. *World Journal of Surgery*, **16**, 924–9.

Rosenthal, M.A. and Ellis, J.I. (1995). Cardiac and mediastinal trauma. *Emergency Medicine Clinics of North America*, **13**, 887–902.

11.1.5 Management of abdominal injury

M. Poeze and G. Ramsay

Key messages

- The approach to abdominal injuries is determined by the mechanism, the location and whether the injury is pentrating or blunt.
- Diagnostic peritoneal lavage, CT, ultrasound, and laparoscopy each have their indications for diagnostic evaluation.
- In general, intra-abdominal penetration requires early laparotomy.
- Blunt injuries to the liver and spleen may be observed in the intensive care unit if the patients are hemodynamically stable, although invasive monitoring may be necessary.

Introduction

Trauma is the leading cause of death among young people in the West. Over 150 000 deaths due to trauma occur annually in the United States, and abdominal injuries account for a small but significant number of these. Unrecognized abdominal injuries are a frequent cause of death, which often occurs in the intensive care unit (ICU). Peritoneal signs are often subtle, and are underestimated because of pain from associated extra-abdominal trauma or masked by head injury or intoxicants. Moreover, the peritoneal cavity is a potential reservoir for major occult blood loss, resulting in late therapy. The primary factor in assessing and managing abdominal trauma is not the accurate diagnosis of a specific type of abdominal injury, but rather recognizing that an intra-abdominal injury exists and that operative intervention is required (Romenofsky *et al.* 1993).

A flowchart for the management of abdominal injury is shown in Fig. 1.

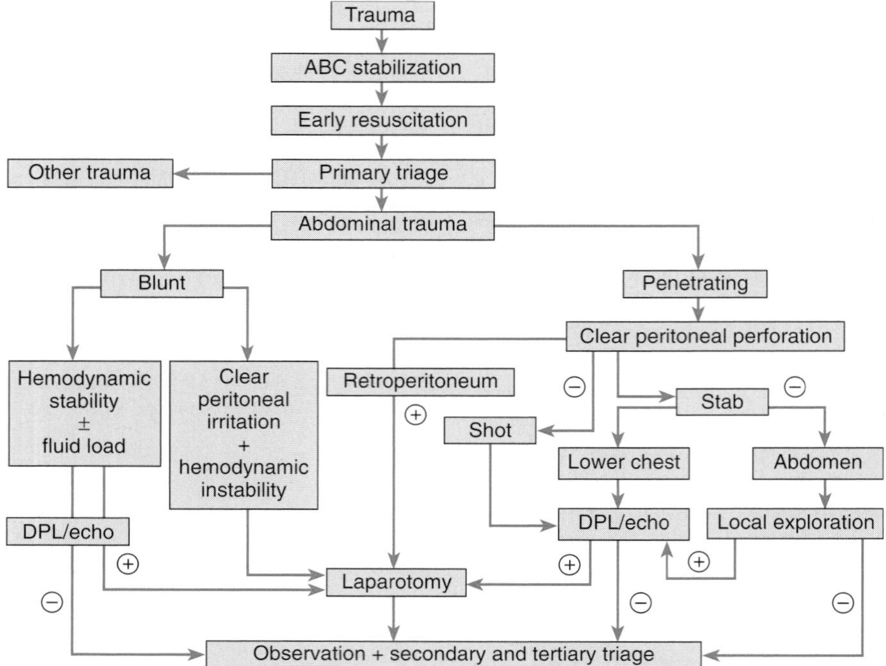

Fig. 1 Flowchart for management of abdominal injury: DPL, diagnostic peritoneal lavage. Note that there are exceptions to each category.

Assessment of abdominal injury

Traumatic abdominal injuries are classified as penetrating or blunt injuries. Penetrating injuries are caused when an object, such as a bullet, crushes and tears the tissues it traverses. The energy transfer from the penetrating object to the tissues determines the degree of injury, and the wounding track determines the pattern of injury. In penetrating wounds to the upper abdomen the attending physician must be aware that abdominal injury can occur between the fifth intercostal space and the pelvis. This also means that one must suspect abdominal injuries in a patient with penetrating chest injuries.

The decision for operative intervention in patients with penetrating abdominal trauma is relatively simple compared with that in patients with blunt abdominal trauma. Overt signs of peritoneal irritation, obvious traverse of the abdominal cavity, or acute blood loss are indications for immediate surgery. However, physical examination can be misleading; patients with penetrating trauma in the back or flank (retroperitoneum) can have massive bleeding or perforation of intestines without obvious signs but with a risk of visceral perforation of up to 20 per cent.

There is a lower risk of serious visceral injury with stab wounds than with gunshot wounds. Less than 50 per cent require an emergency laparotomy. In the stable patient, local wound exploration can identify peritoneal perforation. If there is no violation of the peritoneum, discharge can be considered. Abdominal stab wounds with hypotension and abdominal distension, peritonitis, significant evisceration, or other clear signs of visceral injury require prompt laparotomy. Gunshot wounds are difficult to explore locally, and exploratory laparotomy is usually indicated.

Blunt trauma, which is usually due to vehicle collisions or falls, produces injury by transferring energy through acceleration and deceleration forces. The pattern and severity of injury are determined by the magnitude and orientation of the acceleration–deceleration change to the victim. The organs predominantly involved in blunt abdominal trauma are the liver, spleen, and kidney, although hollow visceral perforation can occur with incorrect seatbelt use. Correct seatbelt use can decrease the complication rate of blunt abdominal trauma during pregnancy. Hypotension, abdominal distention, peritonitis, or other signs of visceral injury due to blunt abdominal trauma are indications for early intervention.

Diagnostic peritoneal lavage, ultrasonography, and CT scanning are modalities used in patients with no clear-cut indication for laparotomy. The specific condition, indications, contraindications, and practical problems will dictate the type of study performed (Feliciano 1991).

Diagnostic peritoneal lavage has a high sensitivity (96–98 per cent) and a low false-negative rate (0.4 per cent) in detecting the presence of significant intra-abdominal hemorrhage. Indications for diagnostic peritoneal lavage include intra-abdominal bleeding with hypotension in blunt trauma and stable patients with anterior stab wounds. Contraindications are a clear indication for laparotomy, previous abdominal surgery, significant obesity, and pregnancy. The last three are relative contraindications. Diagnostic peritoneal lavage is simple and cheap with a low complication rate. Its disadvantages are that it misses diaphragmatic, retroperitoneal, and thoracic injuries, it requires an invasive procedure, and a positive result does not identify the exact visceral injury. Since many less serious splenic and hepatic injuries are observed, an unnecessary laparotomy may be performed based on the results of diagnostic peritoneal lavage.

Ultrasonography can be used in blunt trauma as an alternative to diagnostic peritoneal lavage to evaluate intra-abdominal bleeding in the hypotensive patient, and is preferred in combined thoracic–abdominal injuries. Clear indications for emergency laparotomy, significant obesity, and subcutaneous emphysema are contraindications for the use of ultrasonography. The advantages of ultrasonography are that the thorax and retroperitoneum can be examined at the same time, and that it is cheap and simple to use. It may also define the injured organ and has good sensitivity (84 per cent). Its limitations are the need for at least 70 ml of intraperitoneal fluid (compared with 20 ml in diagnostic peritoneal lavage), 25 per cent failure in the detection of injuries of the liver or spleen, and reliance on skilled staff.

CT is sometimes used instead of diagnostic peritoneal lavage to evaluate blunt abdominal trauma. However, it is time consuming and requires specialized personnel. The accuracy rate is high (92 per cent) and it is able to define the location and magnitude of intra-abdominal injuries in hemodynamically stable patients with blunt or penetrating trauma to the back, buttocks, or flank. It may be useful in cases of significant hematuria, in evaluating the genitourinary system when intravenous contrast is used, and in examining the extent of pelvic fractures. One limitation is the failure to detect injuries to the intestines.

Laparoscopy is less sensitive and less specific as a diagnostic tool, has higher costs, and may require a general anesthestic, and at present has little role in the work-up of abdominal injuries. Its use may be limited to confirming intra-abdominal penetration in penetrating trauma.

Management of specific injuries

Priorities for laparotomy are to secure hemostasis, to control contamination, to identify injuries, and to repair defects (Weigelt 1990).

Intestinal injuries

An aggressive approach to penetrating adominal trauma will detect the majority of intestinal injuries. A high index of suspicion in patients with abdominal trauma, combined with diagnostic peritoneal lavage or contrast studies, can indicate the diagnosis, but if in doubt it is better to operate as early as possible. Most blunt and penetrating injuries of the intestines can be managed by debridement and simple closure, although devastating transection or devascularization of the hollow organs requires resection. In cases of penetrating colonic injuries, the trend is towards safe primary repair. Full-thickness penetrating colonic injuries without severe tissue loss, devascularization, or heavy contamination in low-risk patients with a high trauma score, low penetrating abdominal trauma index, and few associated intra-abdominal injuries can usually be treated by primary closure. Otherwise, it may be best to perform a diverting colostomy.

Pancreatic trauma

Management of the patient with a pancreatic injury depends on its extent and complexity. Procedures range from external drainage to total pancreaticoduodenectomy depending on the intraoperative findings. Because of mortality in excess of 50 per cent for the most serious injuries, patients should initially be monitored in an ICU, observing for the occurrence of pancreatic necrosis with secondary infection, abcess formation, pseudocyst, or glucose abnormalities.

Genitourinary tract injury

The presence of major renal injury does not, in itself, constitute an indication for mandatory exploration. The main indication for surgical intervention in renal trauma is an injury with major hemorrhage, such that the patient remains hemodynamically unstable despite adequate non-operative measures. Patients who are not treated operatively and those with major injuries should be monitored closely for clinical deterioration in an ICU setting. Repeat renal imaging using CT, ultrasound, or radionuclide studies may be helpful for following the injury.

Pelvic fractures and associated injuries

Pelvic fractures are the third most frequent injuries in victims of motor vehicle accidents and are associated with abdominal injuries. The overall mortality has been reported to be greater than 15 per cent, rising to 50 per cent in patients with open pelvic fractures; hemorrhage remains the principal cause of death. Since pelvic fractures are often combined with other injuries and significant non-obvious bleeding, patients with severe pelvic fractures should initially be treated in the ICU. A cystogram should also be considered to rule out bladder perforation. Therapy depends on the presence of an open pelvic fracture, whether a stable or unstable fracture pattern exists, or whether there is a hemodynamic instability. Early external or internal reduction and fixation of many types of pelvic fractures may be necessary for mobilization of the patient. Major pelvic fracture bleeding is usually venous and can be managed by angiographic embolization. External fixation and application of military antishock trousers (MAST) can be helpful in control of bleeding from the pelvic venous plexus. Patients who do not respond to non-operative measurements must be taken to the operation room for laparotomy.

Non-operative management

There has been a marked change towards a more conservative approach in the treatment of blunt splenic and hepatic injuries. This approach is generally confined to patients with minor liver injuries without signs of hemodynamic instability and other intra-abdominal injuries. However, patients treated non-operatively must be monitored closely and continously re-evaluated in an ICU setting.

Traumatized patients and intensive care

Mortality for abdominally traumatized patients can be divided into three phases with time. The first phase occurs with major injuries such as disruption of major abdominal vessels. These patients often die at the scene of the accident and, other than accident prevention, little can be done. The second peak in mortality occurs within hours ('the golden hour'), and the third peak is related to complications from the injury and may be delayed by days or weeks. Many important procedures determining survival, such as aortography, operation, reduction of fractures, or transfusions, should have been applied before the patient arrives in the ICU. Trauma care is a continuum addressing the three phases with everyone working as a team, which should include an intensivist for serious injuries.

The primary and secondary surveys should be repeated in the ICU. The vital signs should be maintained and monitored closely using invasive monitoring if necessary. Prophylactic application of goal-directed therapy has been shown to be beneficial in some cases. These findings contrast with the findings of goal-directed therapy in septic patients (Apolone *et al.* 1996).

The secondary survey is a detailed examination from head to toe, including radiological examination. It also implies a critical review of the primary survey. The tertiary survey or routine in-hospital follow-up must be repeated regularly to identify and treat those factors that can cause a second or third insult and to detect missed injuries.

The clinician should have a high index of suspicion for the development of intra-abdominal hypertension. The increase in intra-abdominal pressure may be due to hemoperitoneum, hemo-retroperitoneum, or extensive postoperative bowel edema. The patient will have a drop in cardiac output, a rise in peak airway pressure, and a significant decrease in lung volumes. Renal failure may develop if early intervention or prevention is not undertaken. Abdominal pressure can be measured easily, and a high pressure (25–40 mmHg) is a strong consideration for surgery after the patient has been stabilized by rewarming, fluid, and coagulation factor supplementation. Intra-abdominal hypertension can be prevented by leaving the abdominal wall open after surgery in patients with extensive bowel edema.

Hemorrhage is the most common complication in the first 24 h after operation. Patients with large transfusion requirements and hypotension at the end of the operation are most prone to this complication. Assessment of hemostasis may identify a non-surgical cause. In a small number of severely injured patients bleeding cannot be stopped entirely at the end of the procedure and packing will be required (damage control). When packing is necessary, patients are taken to the ICU where hypothermia and coagulation disturbances are corrected. A second look is planned 24 to 48 h later, when the packing is removed and definitive hemostasis is obtained. In cases where a splenectomy is performed, the patient should be given pneumococcus prophylaxis, usually just before discharge to allow an appropriate immunological response.

Blunt diaphragmatic injuries are difficult to diagnose and may not be found in penetrating abdominal injuries during abdominal exploration. These patients may develop a hemothorax from abdominal blood entering through the defect. A high index of suspicion is necessary to recognize such an injury. Tears of the mesenteric blood supply due to blunt trauma may lead to intestinal infarction and present several days later with peritonitis. Thus serial examinations are important but may be limited by head injury or by chemical or physiological paralysis. Injuries to the liver and spleen can initially bleed very slowly during the hypotensive period, but bleeding can become significant after reperfusion, leading to hypotension and the need for rescanning, diagnostic peritoneal lavage, and/or operation.

Advances in the early management of critically injured patients have led to improved survival, but the multiple injuries expose the patient to the risk of developing systemic inflammatory response syndrome, septic shock, and multiple organ dysfunction syndrome (McGuire and Pearl 1990). Various inflammatory factors and sepsis have been implicated in the development of multiple organ dysfunction syndrome. Early diagnosis and management are essential for a reduction of mortality from abdominal sepsis. Aggressive hemodynamic support just prior to surgical intervention may be helpful (Bickell *et al.* 1994). The occurrence of multiple organ dysfunction syndrome is a late sequel in the third phase of abdominal trauma, and mortality increases exponentially as an increasing number of organs fail. Therefore multiple injuries must be treated immediately and simultaneously if possible. A fractured femur must be managed with the other injuries; repair must not be delayed for several days. Failure of trauma patients to progress or the development of organ failure with no obvious cause is reason to consider a diagnosis of postoperative intra-abdominal sepsis, which should be searched for.

References

Apolone, G., *et al.* (1996). Comment mettre en évidence, corriger et prévenir l'hypoxie tissulaire chez les malades de réanimation. *Réanimation Urgences*, **5**, 163–71.

Bickell, W.H., *et al.* (1994). Immediate versus delayed fluid resuscitation for hypotensive patients with penetrating torso injuries. *New England Journal of Medicine*, **331**, 1105–9.

Feliciano, D.V. (1991). Diagnostic modalities in abdominal trauma. *Surgical Clinics of North America*, **71**, 241–56.

McGuire, G.P. and Pearl, R.G. (1990). Sepsis and the trauma patient. *Critical Care Clinics*, **6**, 121–46.

Romenofsy, M.C., *et al.* (1993). *Advanced trauma life support*. American College of Surgeons, Chicago, IL.

Weigelt, J.A. (1990). Duodenal injuries. *Surgical Clinics of North America*, **70**, 529–39.

11.1.6 Rhabdomyolysis and compartment syndrome

J.-M. Boles

Key messages

- Traumatic rhabdomyolysis due to crush, electrocution, or other cause can have a very rapid onset.
- Aggressive resuscitation, alkalinization, and decompression of elevated compartment pressures are critical for therapy.

Introduction

Rhabdomyolysis is a clinical and biological syndrome resulting from skeletal muscle injury with release into the plasma of myoglobin, creatine phosphokinase, and other muscle cell contents. Rhabdomyolysis, myoglobinuria, and renal failure were initially described in victims trapped during the bombing of London in 1940 and 1941. Muscle damage, either extensive or localized, may occur in all traumatic circumstances and/or may be the consequence of a wide array of etiological factors and underlying conditions. The complications of rhabdomyolysis include hemodynamic and metabolic disturbances, acute renal failure, and compartment syndrome. Therefore prevention of acute renal failure and early recognition of the compartment syndrome are major clinical goals.

Clinical presentation

The clinician may be confronted with clinical evidence or suspicion of acute traumatic rhabdomyolysis in different clinical settings. The crush syndrome occurs when people are trapped in vehicles or under rubble after catastrophes such as explosions or earthquakes. These patients often have flaccid paralysis and sensory loss that are unrelated to the distribution of nerves in the affected limbs. Victims

Table 1 Etiological and other factors contributing to rhabdomyolysis

Traumatic causes
Crush syndrome
Shock
Burns
Electrocution, lightning stroke
Muscle hyperactivity (strenuous physical exercise, seizures, delirium tremens, repeated defibrillation)
Non-traumatic causes
Muscle compression (postural syndrome, coma)
Muscular ischemia (acute arterial occlusion, gas embolism, CO)
Temperature abnormalities (malignant hyperthermia, hypothermia)
Systemic infection (viruses including HIV, bacterial infections, etc.)
Toxins (heroin, methadone, LSD, amphetamines, cocaine, marijuana, ecstasy, alcohol, toluene, ethylene glycol, organophosphates, strychnine, snake venom, scorpion bites)
Drugs (amphotericin B, isoniazid, fenfluramine, phenformin, clofibrate, glutethimide, salicylates, phenylcyclidine, succinylcholine (suxamethonium), monoamine oxidase inhibitors, etc.)
Metabolic disorders (hypocalcemia, hypokalemia, hypophosphatemia, hyperosmolar state, severe hyponatremia, hypothyroidism, hyperthyroidism)
Inflammatory muscle disorders (polymyositis and dermatopolymyositis)
Hereditary enzymopathies
Carbohydrate metabolism abnormalities
Lipid metabolism abnormalities
Other

immobilized under fallen masonry appear to be initially relatively protected from the systemic consequences of the muscle injury. Decompression of the limbs during extrication releases muscle cell components into the circulation, which in turn accelerates the development of shock, hemoconcentration, hyperkalemia, and acute renal failure. These potentially lethal consequences must be prevented.

Muscle destruction may or may not be obvious in patients subjected to constant direct electrical shock. Extensive burns, particularly if circular, can cause a tourniquet-like compression which leads to muscle cell lysis. In both these situations rhabdomyolysis may be severe and fatal.

The possibility of acute traumatic rhabdomyolysis must be raised under several circumstances in which muscle damage may not be clinically apparent. Muscle compression secondary to coma, post-trauma or postseizure, or postural syndrome is sufficient to generate acute rhabdomyolysis. Electrocution or thundershock may also cause muscle necrosis. Postexertional rhabdomyolysis may occur after strenuous physical exercise, particularly in unprepared military recruits or occasional sportsmen. Other etiological and contributing factors include hypothermia, heat stroke, metabolic derangements, drug or cocaine abuse, acute or chronic alcoholism, and hereditary or acquired myopathies (Table 1).

Myalgias are the most common symptom. Muscle stiffness, edema visible with compression, paresthesias, and skin lesions are infrequent findings. Dark brown urine may be the only visible sign; an orthotoluidine reactive-paper test may be falsely positive for blood and protein.

Diagnosis

The diagnosis of rhabdomyolysis made by the measurement of serum creatine phosphokinase and serum and urinary myoglobin is usually systematic. Serum creatine phosphokinase should be at least six times the upper normal value (100 IU/l) for diagnosis. Serum creatine phosphokinase begins to rise 2 to 12 h after the injury and reaches its peak value, ranging from several hundred IU per liter in mild muscle injury to over 200 000 IU/l in the full-blown crush syndrome, 1 to 3 days after injury; the serum level slowly declines at a rate of 50 per cent every 48 h. Therefore serum creatine phosphokinase should be measured daily for at least 3 days to follow the extent of muscle destruction and its regression. If the serum creatine phosphokinase level remains elevated, ongoing muscle injury, necrosis, and/or compartment syndrome should be sought. Creatine phosphokinase isoenzymes should be assayed only in the case of concomitant heart injury, in which case the elevated MB fraction has a cardiac etiology and the MM fraction is indicative of rhabdomyolysis. The measurement of other serum muscle enzymes such as lactate dehydrogenase or aldolases, which are also elevated in acute traumatic rhabdomyolysis, is of no practical use and is not measured routinely. Heavy-chain fragments of serum myosin and cardiac troponin T have recently been shown to reach long-lasting elevations in rhabdomyolysis, but their diagnostic and prognostic usefulness has not yet been evaluated. Therefore their routine measurement cannot be recommended.

The normal concentration of serum myoglobin ranges up to 85 ng/ml, of which half is bound to an α_2 globulin. Myoglobin is an oxygen-binding respiratory protein found in muscle at a concentration of 2.5 g per 100 g of skeletal muscle. It is released from injured muscle, and it increases before creatine phosphokinase and decreases more rapidly owing to its clearance by renal excretion and its metabolism to bilirubin. A close correlation has been demonstrated between the percentage and depth of burned surface and the concentration of serum myoglobin and creatine phosphokinase. There is no correlation between serum and urinary myoglobin. Myoglobinuria appears when serum concentrations exceed 1500 ng/ml, suggesting that at least 100 to 200 g of muscle have been injured. The presence or absence of visible myoglobinuria does not reflect the extent of muscle damage. Furthermore, owing to its metabolites, a normal myoglobin serum level does not exclude the diagnosis of acute traumatic rhabdomyolysis.

Evaluation of the metabolic consequences of muscle destruction includes the measurement of serum and urinary electrolytes, including calcium and phosphorus, and renal function indices. These measurements should be repeated once daily until they are normal to monitor treatment-induced modifications.

Complications

Acute traumatic rhabdomyolysis can cause three types of complication: acute renal failure, the compartment syndrome, and calcium and phosphate derangement.

Table 2 Flow of solutes and water across skeletal muscle cell membrane in rhabdomyolysis

Compound	Consequences
Influx from extracellular compartment into muscle cells	
Water, sodium, chloride, calcium	Hypovolemia and hemodynamic shock, progressing to acute renal failure, hypocalcemia, increased cytosolic calcium, activation of cytotoxic proteases
Efflux from injured muscle cells	
Potassium	Hyperkalemia
Purines from disintegrating cell nuclei	Hyperuricemia, nephrotoxicity
Phosphate	Hyperphosphatemia, aggravation of hypocalcemia, metastatic calcification
Lactic and other organic acids	Metabolic acidosis and aciduria
Myoglobin	Nephrotoxicity, aggravated in case of coexisting oliguria, aciduria, and uricosuria
Thromboplastin	DIC with activation
Creatine kinase	Elevation of serum level
Creatinine	Increased serum creatinine-to-urea ratio

Adapted from Better and Stein (1990).

The liberation of large amounts of myoglobin and other muscle cell contents in the blood leads to hyperkalemia, hyperphosphatemia, hypocalcemia, and metabolic acidosis (Table 2). These abnormalities begin when serum urea and creatinine concentrations are still within normal limits. Hyperkalemia may cause life-threatening ventricular arrhythmias, justifying continuous electrocardiographic monitoring.

Myoglobinuria associated with intravascular volume depletion, which is seen in crush syndrome, leads to acute tubular necrosis. The latter is the consequence of renal hypoperfusion, acid urine pH, tubular obstruction caused by precipitation of heme proteins, and possibly tubular cell anoxia. Delayed volume replacement, even if adequate in quantity, results in the development of acute renal failure within a day.

Patients with acute renal failure in this setting have a high-anion-gap metabolic acidosis, hyperuricemia, and hyperphosphatemia. Hemoconcentration and occasionally thrombocytopenia may suggest the onset of disseminated intravascular coagulation (DIC).

Factors predicting the development of acute renal failure have been identified as peak creatine phosphokinase elevation, degree of serum potassium elevation on admission, degree of phosphate elevation on admission, decrease in serum albumin, and clinical evidence of dehydration. Acute renal failure has been seen in patients with a mean creatine phosphokinase level of 40 500 IU/l accompanied by significant hypovolemia, or in patients with urinary myoglobin concentrations in excess of 1000 ng/ml. None of these factors have been confirmed prospectively.

Within minutes of trauma, intramuscular pressure in injured muscle may exceed arterial blood pressure. Ischemic interference with the regulation of muscle cell volume causes cell swelling, which results from the accumulation of intracellular solutes, an increase in the leakiness of the cell membrane, and a reduction in active ionic extrusion. Intracompartmental pressure rises rapidly in muscle groups confined in tight fibrous sheaths with low compliance, such as the calf or forearm. The compartment syndrome appears when intracompartmental pressure exceeds arteriolar perfusion pressure, obliterating the circulation to the affected region and causing muscle and myoneuronal ischemic damage within hours. Stretching of the muscle increases the leakiness of the skeletal muscle cells and nerve cells to calcium ions, which may explain the vulnerability and sensitivity of skeletal muscle to mechanical pressure of short duration. The intramuscular pressure in patients with compressed or wedged limb injuries may increase to more than 240 mmHg, causing rhabdomyolysis independent of ischemia. Full-blown rhabdomyolysis and anterior tibial compartment syndrome may occur in such patients despite normal arterial pedal pulses and warm skin.

Many symptoms may mimic spinal injury. However, the neurological examination reveals normal anal sphincter and urinary bladder function, which may help to exclude the presence of an acute spinal cord lesion in chemically or organically paralyzed patients with rhabdomyolysis. Diagnosis relies upon repeated physical examination and measurement of the intramuscular compartment pressure. Severe worsening of myalgias, development of paresthesias, and tense muscle compartments with myoedema should provoke the measurement of the intramuscular pressure by means of manometry using a probe and a transducer. Values in excess of 30 to 40 mm Hg confirm the diagnosis and prompt the need for decompression (e.g. fasciotomy, escherotomy).

Rhabdomyolysis is often associated with hypocalcemia due to a shift of extracellular calcium into the injured muscle. Sarcolemmic Na^+,K^+-ATPase activity is impaired in damaged muscle. Attenuating the activity of this ion pump would diminish the extrusion of sodium from the sarcoplasm and interfere indirectly with the efflux of calcium from the cell (Table 2). Thus an increase in the cytosolic free-calcium level would activate neutral proteases which, in turn, disrupt myofibrils and lead to muscle damage with rhabdomyolysis. Serum monitoring may initially show hypocalcemia. A ^{99m}Tc-diphosphate scan would confirm the calcium shift. Release of muscle-fixed calcium leads to secondary hypercalcemia with elimination of calcium excess by the kidneys as the syndrome resolves. Secondary hypercalcemia of less than 3 mmol/l is clinically uneventful and may be secondary to abnormal control of 1,25-$(OH)_2$-vitamin D_3 metabolism. Therefore monitoring of serum (and urinary) calcium levels at regular intervals during the course of acute traumatic rhabdomyolysis is recommended.

Management

Preventing the systemic and renal complications of the crush syndrome requires early removal of the offending agent and vigorous aggressive intravenous volume replacement. During the extrication of an injured person from a collapsed building or a wrecked automobile, isotonic saline solution should be infused initially at a rate of 1.5 l/h as soon as one of the limbs has been freed and until the person is in hospital. Once the systemic circulation has been stabilized and the presence of adequate urine flow has been confirmed, forced mannitol–alkaline diuresis therapy for prophylaxis against hyperkalemia and acute renal failure should be undertaken. Fluids should be given so as to provide a forced diuresis of several liters daily and to maintain urinary pH above 6.5 until myoglobinuria disappears (usually by the third day). Serum chemistries and pH need to be followed. Bladder catheterization is needed to monitor ouput. The infusion of bicarbonate may be gradually discontinued after 36 h. The use of loop diuretic agents, which acidify the urine, is not recommended. Mannitol–alkaline diuresis therapy has the two beneficiary actions of a salutary hypokalemic action and solubility of myoglobin and urate. Unless there is a danger of hyperkalemic-induced arrhythmias, calcium infusion is not necessary. The benefits of such a regimen to control hyperkalemia and acidosis and to prevent myoglobinuric acute renal failure have been seen after extensive traumatic rhabdomyolysis due to entrapment in a collapsed building for a mean period of approximately 12 h. Intravenous saline treatment was initiated before complete extrication. Within 2 h of extrication, patients were evacuated to a hospital where they received forced solute–alkaline diuresis therapy. Despite clear-cut clinical and laboratory evidence of extensive rhabdomyolysis, none developed either acute renal failure or even transitory azotemia. The large positive fluid balance observed was due to the sequestration of fluid into the injured muscles, and the administration of large fluid volumes appears to be well tolerated in previously healthy young adults. The treatment of elderly patients requires a more cautious administration of intravenous fluids.

Treatment of acute renal failure includes forced diuresis and renal replacement therapies. Forced diuresis must not be undertaken in a hypovolemic patient. Continuous arterio- or venovenous hemofiltration is an efficient method of dialyzing myoglobin from the plasma and may speed the recovery of renal function. It is preferable to conventional hemodialysis, particularly in cases of hemodynamic instability. Most patients recover normal renal function within 3 to 4 weeks. The morbidity is linked to either inadequate management of metabolic complications or the underlying health status in elderly patients. Adequate therapy of hyperkalemia and hyperphosphatemia should be instituted following serum chemistries.

The management of the local injury of the crushed limb depends on its clinical condition. The skin has an extraordinary capacity to withstand pressure, and serves as a barrier against infection even when bruised over crushed limbs. Therefore surgical exploration of limbs with acute traumatic rhabdomyolysis should not be performed unless there is an overriding reason to do so, for example re-establishing arterial inflow.

Repeated physical examination and measurement of compartment pressures will guide therapy. A Doppler scan provides information about arterial inflow but not about compartment pressures. Fasciotomies should be performed when intracompartmental pressure exceeds 30 to 40 mmHg. Intense bleeding when a fasciotomy is performed to relieve intracompartmental pressure is no proof of the viability of muscle, as dead muscle may also bleed profusely. Therefore dead and live muscle should be distinguished by the response to mechanical or electric stimulation.

Early amputation may be necessary in some cases of multiple and complex limb lesions with worsening of the clinical condition and continuous elevation of the serum creatine phosphokinase and/or myoglobin despite adequate treatment.

Thus traumatic rhabdomyolysis is a common disorder which requires both intervention as early as possible, preferably before hospital admission, and constant visual monitoring to detect the occurrence of a compartment syndrome.

Bibliography

Berns, J.S., Cohen, R.M., and Rudnick, M.R. (1991). Removal of myoglobin by CAVH-D in traumatic rhabdomyolysis. *American Journal of Nephrology*, **11**, 73.

Better, O.S. and Stein, J.H. (1990). Early management of shock and prophylaxis of acute renal failure in traumatic rhabdomyolysis. *New England Journal of Medicine*, **322**, 825–9.

Better, O.S., Rubinstein, I., and Winaver, J. (1992). Recent insights into the pathogenesis and early management of the crush syndrome. *Seminars in Nephrology*, **12**, 217–22.

Boles, J.-M., Renault, A., Tchoua, R., and Garo, B. (1994). Diagnostic, étiologies et prise en charge des rhabdomyolyses aiguës. In *Actualités en réanimation et urgences*, pp. 263–79. Arnette, Paris.

Lofberg, M., Tahtela, R., Harkonen, M., and Somer, H. (1995). Myosin heavy-chain fragments and cardiac troponins in the serum in rhabdomyolysis. *Archives of Neurology*, **52**, 1210–14.

Ward, M. (1988). Factors predictive of acute renal failure in rhabdomyolysis. *Archives of Internal Medicine*, **14**, 1553–7.

11.1.7 Fat embolism

J. M. Ryan and S. Metcalf

Key messages

- Fat embolism syndrome is the most common cause of death secondary to long-bone and pelvic fractures.
- The classic triad of respiratory distress, petechial rash, and neurological function is diagnostic of fat embolism syndrome.
- Young males with long-bone fractures are most at risk, and such patients demand awareness and a high index of suspicion.
- Management is largely supportive and is aimed at maintaining an adequate Po_2, if necessary by advanced techniques such as positive end-expiratory syndrome.

Introduction

Fat embolism was first described by Zenker (1862). More than a century later there is still confusion and controversy surrounding its definition, incidence, pathogenesis, assessment, and management (Peltier 1969; Lindeque *et al.* 1987).

Fat embolism is distinguished from fat embolism syndrome in that the former refers to the presence, usually asymptomatic, of fat droplets in the lung parenchyma and systemic circulation following major trauma, particularly closed long-bone or pelvic fractures. The incidence of fat embolism is unknown, but it is probably present in the majority of patients suffering from long-bone or pelvic fracture. Most remain asymptomatic or develop mild self-limiting hypoxia.

Fat embolism syndrome is clinically significant and may be a separate pathological entity or a more severe manifestation of fat embolism. It occurs less frequently and is the most common cause of death secondary to long-bone and pelvic fractures (Acosta and Peltier 1992).

Pathophysiology

Droplets of fat are released primarily from fat depots in the marrow of long bones, but also from the soft tissues and organs such as liver, enter the venous circulation, and reach the lungs, where they become lodged in the pulmonary capillaries. This behavior is similar to pulmonary embolism.

Fat embolism syndrome is characterized by pulmonary insufficiency following trauma and is a specific variant of acute lung injury. It appears to be one of a number of triggering mechanisms leading to lung pathology indistinguishable from adult respiratory distress syndrome.

The mediated inflammatory response associated with fat embolism syndrome and resulting in acute lung injury is probably the same for all variants of adult respiratory distress syndrome irrespective of etiology. The initiating event in fat embolism syndrome appears to be the release of injurious free fatty acids by pulmonary lipase. The free fatty acids result in 'injury' leading to an inflammatory cascade. Thereafter events follow a common pathway. Important initial mediators include activated complement and neutrophils. These mediate the release of oxygen free radicals and other cellular toxins such as proteases, arachidonic acid, and platelet activating factor which, in turn, directly damage vascular and respiratory endothelium, resulting in the further release of activated neutrophils, thus perpetuating the injury cycle. The final pathway is a pathological diffusion barrier and ultimately breakdown of the capillary alveolar membrane barrier, leading to leakage of protein-rich fluid through the alveolar cell wall into the alveolar sacs. This is further compounded by impaired surfactant production leading to alveolar collapse.

There are other systemic features distinguishing fat embolism syndrome. A common presenting feature is an altered mental state. The etiology may be complex. In multiple system injury it may be difficult to distinguish neurological features related to traumatic primary brain injury from hypoxia secondary to acute lung injury or hypovolemia. Where no head injury is present, neurological symptoms may present dramatically and are easily related to hypoxia from the acute lung injury or hypovolemia. Severe neurological disturbance may result from fat embolization to the brain without lung involvement. Here, arterial blood gas analysis is normal and cerebral signs are due to edema, microinfarction, and demyelination (Scopa *et al.* 1994).

Hemorrhagic shock may be one factor responsible for the entry of fat droplets into the circulation. Other factors include disseminated intravascular coagulation (DIC), sepsis, and site of injury. The incidence, morbidity, and mortality are greater with proximal than with more distal injury. Other features which characterize fat embolism syndrome as a discrete variant of adult respiratory distress syndrome include petechial hemorrhages of skin, and renal, retinal, subconjuctival, and myocardial involvement by droplet emboli.

Incidence

The true incidence of fat embolism syndrome is unknown but is under-reported. It is most closely associated with fractures of the pelvis and long bones of the lower extremity, with a reported incidence of 3.4 per cent for tibial fractures and 9 per cent for femoral fractures, rising to 20 per cent for fractures of both tibia and femur (Peltier 1969). Although injury is the main triggering factor leading to fat embolism syndrome, orthopedic procedures such as hip arthroplasty and intramedullary nailing for lower limb fractures may lead to the release of marrow fat into the circulation.

Table 1 Signs and symptoms of fat embolism syndrome

Diagnostic
$P\text{O}_2 < 8$ kPa (60 mmHg)
Petechial rash
Unexpected neurological signs
Supportive
Associated hypovolemia and tachycardia
Hypothermia
Fever
Sudden drop in hemoglobin
Sudden thrombocytopenia
Increased erythrocyte sedimentation rate
Fat globules in urine/sputum
Retinal changes

There is no consistent relationship between the amount of fat released and the probability of developing fat embolism syndrome. The condition has been reported in association with isolated fractures of bones such as the clavicle and vertebrae where the amount of fat released is small. Moreover, fat embolism syndrome may develop in any condition where there is potential for fat release into the circulation, such as muscle injury and burns.

Of those patients who develop clinically evident fat embolism syndrome, 20 per cent show a fulminant course with a mortality approaching 50 per cent. Although the condition may occur at any age, it is most commonly seen in those most at risk of serious trauma, i.e. young males.

Clinical diagnosis

The classic triad of respiratory distress, petechial rash, and neurological dysfunction is notable (Table 1). The onset of these symptoms and signs usually occurs 12 to 72 h after injury. The key to making the diagnosis is to have a high index of suspicion. In centers where arterial blood gas analysis is routine for patients suffering from long-bone fractures, 30 per cent show evidence of hypoxia which is usually self-limiting but may progress to respiratory failure.

Respiratory dysfunction characterized by a low $P\text{O}_2$ (< 8.0 kPa (60 mmHg)) in a patient at risk should arouse suspicion that fat embolism syndrome may be present. Initially, compensatory mechanisms lead to hypocapnia and a mild respiratory alkalosis. If the condition progresses, a mixed metabolic and respiratory acidosis supervenes. In patients with multiple injuries, respiratory distress and hypoxia may be compounded by associated chest trauma and underlying lung injury. Further diagnostic difficulties may arise as multiple rib fractures alone may precipitate fat embolism syndrome.

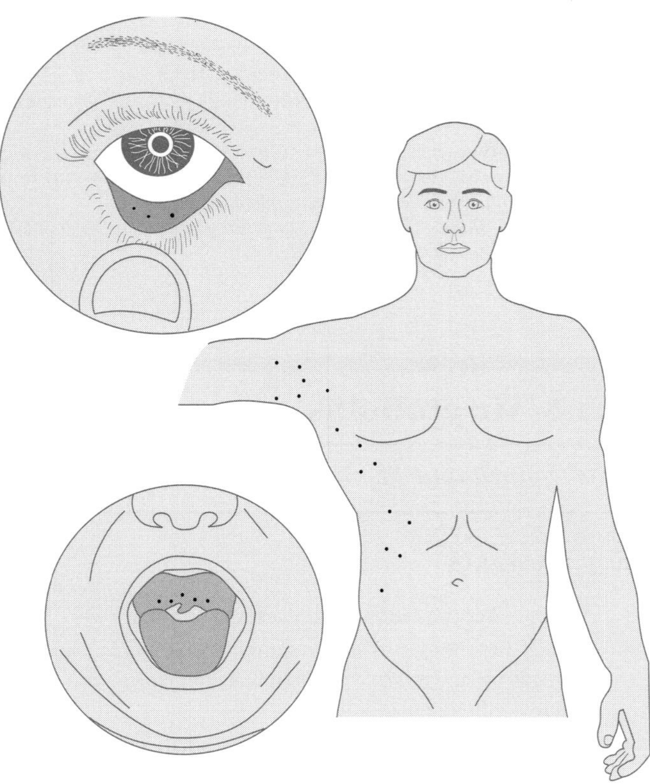

Fig. 1 Typical distribution of the petechial rash. (Reproduced with permission from Peltier (1970).)

Although fat embolism syndrome may develop at any time from 12 h to 7 days after injury, the risk is greatest in the first 48 h. Therefore any significantly injured patient should be repeatedly assessed, bearing fat embolism syndrome in mind. Pulse oximetry is useful in a well-perfused patient.

Neurological dysfunction ranges from mild restlessness and agitation to convulsions and coma. The signs may be related to hypoxia of pulmonary origin, fat embolization of the brain, or a combination of both. Further confusion may arise if there is an altered mental state related to head injury or inadequate resuscitation in the face of multiple system injury. Careful systemic assessment and investigation are required.

The hallmark sign in a patient at risk is the petechial rash (Fig. 1). It should be sought for specifically and is most commonly found on the trunk, conjunctiva, sclera, and buccal mucosa. Petechias appear in crops but are of short duration, usually fading after 48 h. They are associated with capillary fragility, not direct skin embolization. Profound thrombocytopenia and massive blood transfusion can also give rise to petechias and should be clinically considered.

Fat embolism syndrome is usually accompanied by fever ($> 38\,^{\circ}C$) and tachycardia (100–120 beats/min). In the absence of sepsis, fat embolism syndrome should be strongly considered. Fundoscopy should be performed in all cases, as fat may be seen in the retinal arterioles. Fat embolization in the eye may also provoke non-specific retinal exudate and hemorrhage.

In conclusion, a high index of suspicion is needed for patients at risk. Early and repeated arterial blood gas analysis is the most important single investigation.

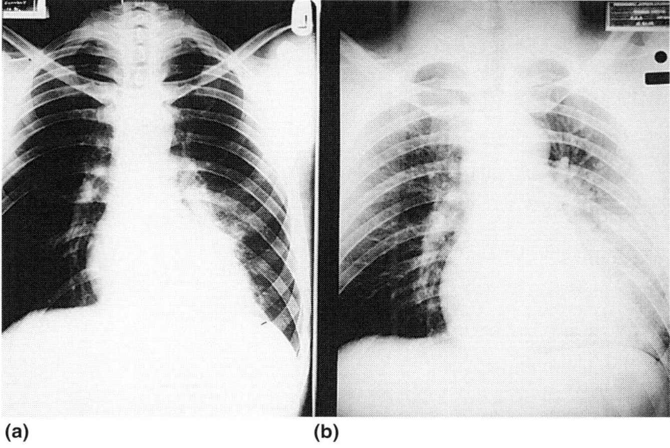

Fig. 2 Two chest radiographs of a case of fat embolism syndrome: (a) a normal film taken after admission; (b) a film taken 18 h later showing diffuse pulmonary infiltrates consistent with a diagnosis of fat embolism syndrome.

Additional diagnostic aids

Chest radiography may show rib fractures or pulmonary contusion. However, approximately a third of patients who develop fat embolism syndrome have additional radiographic changes most often characterized by diffuse parenchymal infiltrates (Fig. 2). Unfortunately, these signs bear little relationship to disease severity or even blood gas parameters. Although not diagnostic, serial chest radiographs are advised and, combined with other investigations, provide useful guidance for diagnosis and the effects of injury.

Since neurological features take many forms, there should be early recourse to CT. This may allow exclusion of direct intracranial injury as a confounding factor. CT may also reveal cerebral edema which, while not specific for fat embolism syndrome, is supportive evidence of the diagnosis and demands appropriate treatment. Magnetic resonance imaging may be more specific for fat embolization of the brain and often shows demyelinating lesions (Scopa *et al.* 1994). The prognosis for neurological recovery remains good and appears to be unrelated to the nature of the original neurological signs.

The following, while not diagnostic, may provide supportive evidence.

1. Thrombocytopenia ($< 150\,000/mm^3$), which may be accompanied by a drop in hemoglobin, may occur.
2. Coagulation derangements range from minor abnormalities to DIC.
3. Lipase may be elevated in association with fat embolism and fat embolism syndrome, but is non-specific. The same is true of lipuria and fat droplets in the sputum.
4. Electrocardiographic changes, usually of right heart strain, occur in up to 75 per cent of patients with fat embolism syndrome, but are non-specific.
5. Staining of tracheal aspirate for fat and, more recently, staining of bronchoalveolar lavage fluid for fat-laden cells are recommended, but there is lack of specificity with these techniques.

Table 2 Summary of management of fat embolism syndrome

High index of suspicion
High-flow oxygen therapy
Airway and ventilation
PEEP
Aggressive shock management
Hemodynamic monitoring and support
Definitive fixation of fractures
Nutritional support

Management

All trauma victims require high-flow supplemental oxygen utilizing a rebreathing bag or intubation. Ventilation should be assisted if appropriate. Serial arterial blood gas analyses and pulse oximetry should be used to identify and prevent hypoxia (Table 2).

If fat embolism syndrome presents unexpectedly after an interval, or is initially unrecognized, more aggressive measures may be required such as endotracheal intubation and mechanical ventilation, often using high positive end-expiratory pressure (PEEP) to open and recruit collapsed alveoli. Such patients must be managed in an intensive care setting where invasive monitoring is possible.

Mortality appears to be related to the degree and duration of hypovolemic shock. Initial management should involve early recognition and vigorous resuscitation, initially with warm Ringer's lactate followed by blood. Hypothermia is a further aggravating factor which may be exacerbated by the administration of cold resuscitation fluids.

Injured tissues and fractures must be handled gently, avoiding unnecessary manipulations, with early splint application. All lower limb and pelvic fractures require early reduction and immobilization, with external or internal fixation where appropriate. This particularly applies to pelvic and femoral shaft fractures.

Controversial measures for treating fat embolism syndrome include the use of corticosteriods. Although improvement in clinical outcome has not been consistently shown, studies indicate that their early use may reduce the overall incidence of fat embolism syndrome in particular subsets of patients (Lindeque *et al.* 1987). However, their routine use cannot be recommended at present. There have been trials of many other agents, including aspirin, dextran, and alcohol, but to no avail.

In summary, management is largely supportive if not preventive. Supplemental oxygen must be delivered to maintain a $P\text{O}_2$ of greater than 8 kPa (60 mmHg). There should be early recourse to intubation and ventilation using PEEP if adequate oxygenation is not achievable by simpler measures. The key is early consideration and intervention.

References

Acosta, J.A. and Peltier, L.F. (1992). Fat embolism syndrome. In *Principles and practice of emergency medicine* (ed. G.R. Schwartz, C.G. Cayten, M.A. Mangelsen, T.A. Mayer, and B.K. Hanke), pp. 1210–13. Lea & Febiger, Philadelphia, PA.

Lindeque, B.G.P., Schoeman, H.S., Dommisse, G.F., Boeyens, M.C., and Vlok, A.L. (1987). The fat embolism syndrome. A double-blind therapeutic trial. *Journal of Bone and Joint Surgery (British Volume)*, **69B**, 128–31.

Peltier, L.F. (1969). Fat embolism. A current concept. *Clinical Orthopedics and Related Research*, **66**, 241–53.

Peltier, L.F. (1970). Fat embolism. *Orthopedic Clinics of North America*, **1**, 13.

Scopa, M., Magatti, M., and Rossitto, P. (1994). Neurologic symptoms in fat embolism syndrome: case report. *Journal of Trauma*, **36**, 906–7.

Zenker, F.A. (1862). *Beiträge zur normalen und pathologischen Anatomie der Lunge*. Braunsdorf, Dresden (in German).

11.1.8 Air embolus

C. O'Donnell and M. Gavalas

Key messages

- Air can enter the venous or the arterial circuit.
- Venous air embolus usually requires a large bolus, causes aggregation and release of chemical mediators, and can lead to cardiovascular collapse.
- Arterial air embolism occurs with a small bolus which usually goes to the cerebral and coronary circulation and leads to cardiovascular collapse.
- Early use of a hyperbaric chamber may be lifesaving.

Introduction

Air embolism refers to the obstruction of a blood vessel by the impaction of an air bubble. Both the pulmonary arterial circulation (venous embolism) and the systemic arterial circulation (arterial embolism) can be involved. Although the etiology, pathophysiology, assessment, and management of air emboli are mainly dependent upon which part of the circulation has been involved, there is considerable overlap.

Etiology

Air can enter the venous system in a number of situations (Table 1). The best known example is where air enters through a catheter into a peripheral or central vein. This can occur during insertion or removal, or when the catheter is *in situ* if venous access is left open to the atmosphere. Venous air embolism may be caused by either blunt or penetrating chest trauma. Many surgical procedures, either involving the lung parenchyma directly or extrathoracic in origin, may precipitate venous emboli formed from air or other gases. Barotrauma may be iatrogenic (positive-pressure ventilation) or due to a blast injury when significant venous air embolism can occur if extra-alveolar air is driven along a pressure gradient into the systemic veins. Ingestion of hydrogen peroxide is a rare cause of venous air embolism.

Arterial air emboli (Table 2) can occur as a result of any of the causes discussed above. Air enters the venous circulation and can pass into the systemic circulation via a functioning right to left shunt (usually a patent foramen ovale). Even in the absence of a clinically apparent shunt, pulmonary arteriolar to pulmonary venous air spill-over

Table 1 Causes of venous air embolism

Intravenous cannulas
Insertion
Removal
In situ
Chest trauma
Blunt
Penetrating
Surgical procedures
Cardiothoracic
Neurosurgical
Otolaryngoscopic
Obstetric
Laparoscopic
Arthroplastic
Barotrauma
Intermittent positive-pressure ventilation
Blast injuries
Miscellaneous
Hydrogen peroxide

occurs once the finite capacity of the lungs to filter out air bubbles is exceeded.

Direct entry of air into the arterial circulation can also occur via indwelling arterial catheters, penetrating injuries, and during surgical procedures. However, self-contained underwater breathing apparatus (scuba) diving is the most common cause of patients presenting to emergency departments with arterial air embolism (dysbaric arterial air embolism).

Dysbaric arterial air embolism is caused by gas bubbles entering the systemic circulation from ruptured alveoli and pulmonary veins. According to Boyle's law, the volume of a fixed mass of gas is inversely proportional to pressure, and therefore compressed air can expand into surrounding tissues. During ascent after diving, delicate alveoli and pulmonary veins are damaged by increasing volumes of gas as pressure decreases. This is marked when high intrapulmonary pressures are reduced when divers surface and remove their masks, and is exacerbated if the divers inadvertently fail to exhale fully on ascent so that gas can pass rapidly into the circulation.

Table 2 Causes of arterial air embolism

As for venous air embolism (Table 1)
Indwelling arterial catheters
Surgical and traumatic arteriotomies
Dysbaric air embolism

Pathophysiology

The effects of air in the venous system on cardiovascular and pulmonary physiology are complex. The basic cardiovascular dysfunction in venous air embolism is obstruction to right ventricular ejection. This affects either the pulmonary artery when a large bolus is implicated, or the pulmonary arterioles when microbubbles are involved. The pathophysiological cascade of venous air embolism is shown in Fig. 1.

Macrobubble pathophysiology causes circulatory standstill, leading to complete cardiorespiratory arrest. The fatal dose of air embolism in adults has not been scientifically established, but estimates from animal studies are of the order of 500 ml.

Microbubbles in the pulmonary circulation cause histopathological changes. The bubbles obstruct blood flow, and subsequently platelets, neutrophils, erythrocytes, fibrin, and fat aggregate on the bubbles and local endothelial cells. Neutrophil-released mediators and platelet activating factor play important roles. Platelet activating factor may contribute to cardiac deterioration from air embolism through its effects on pulmonary arteriolar vascular tone. The superoxide anion, hydrogen peroxide, leukotrienes, and thromboxanes, which are neutrophil derived, are implicated in pulmonary injury.

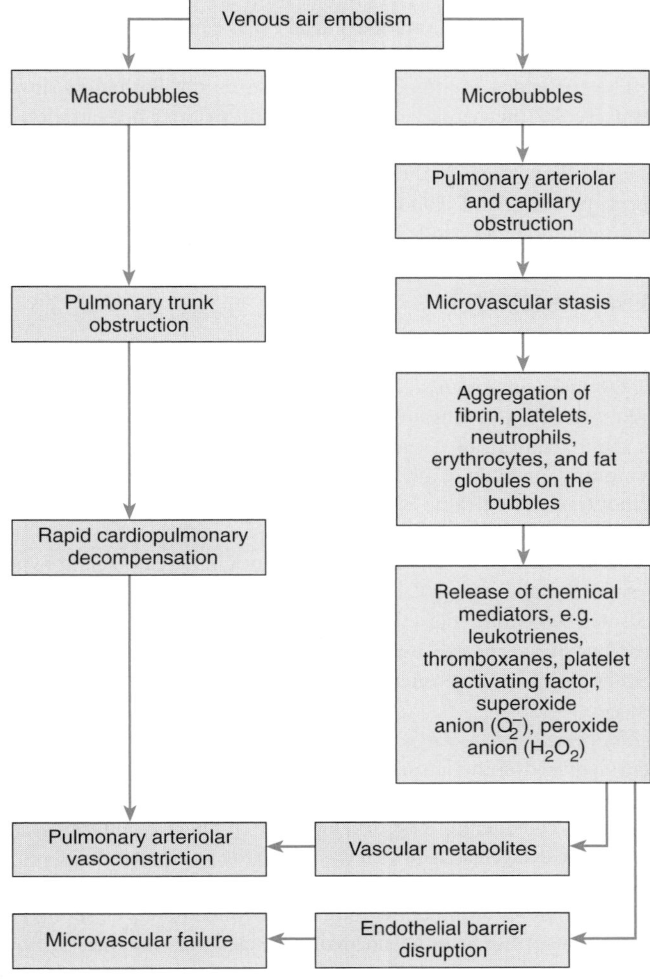

Fig. 1 Pathogenesis of venous air embolism.

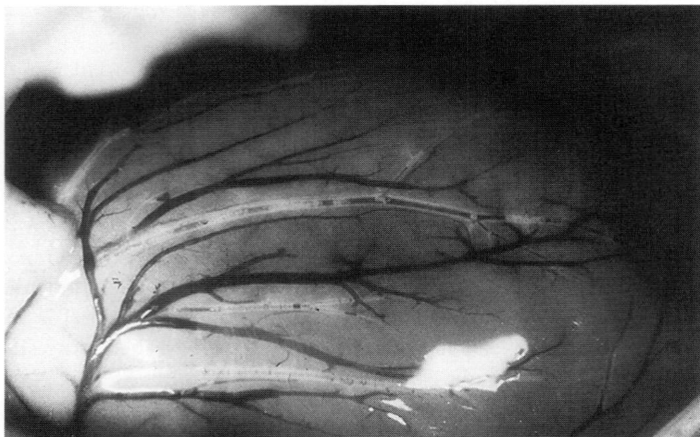

Fig. 2 Air in the coronary arteries.

Administration of superoxide dismutase, catalase, and methylprednisolone can negate superoxide anion, hydrogen peroxide, and mediators derived from arachidonic acid and ameliorate the deleterious effects of venous air embolism in dogs.

Disturbances of cardiopulmonary physiology due to venous air embolism confined to the right heart circulation are profound, but the consequences of arterial embolization are even more dramatic. Less than 1 ml of air in a coronary artery may cause a fatal myocardial infarction (Fig. 2).

The pathology of arterial gas embolism is not limited to simple intravascular mechanical obstruction. The delayed presentation of the condition is a consequence of a wide range of pathophysiological changes. As with gas in the venous system, local accumulation of fat, fibrin, platelets, and blood cells exacerbates local ischemia and can activate complement and clotting cascades.

Assessment

Venous air embolism

The sequelae of venous air emboli are variable. Air transported to the right ventricle and lungs may be completely asymptomatic owing to its rapid removal, or it can rapidly lead to cardiovascular collapse. Symptoms include cough, dyspnea, wheezing, chest pain, and hemoptysis. Often there is a feeling of impending death, confusion, and collapse as cardiovascular dysfunction worsens.

Significant emboli can result in tachypnea, tachycardia, and hypotension. A right ventricular heave may also be present. The classical mill-wheel (systolic–diastolic) murmur is rare and depends upon the presence of a very large amount of air in the right ventricle. A harsh systolic murmur or absence of a murmur is a more frequent occurrence.

Neurological sequelae with a depressed Glasgow Coma Scale and neurological dysfunction are probably the result of hypoxia and hypotension, although the possibility of paradoxical embolism should always be considered. The development of chest crepitations and wheeze are delayed signs which are due to the accumulation of extravascular fluid in the lungs.

Abnormalities found on investigation depend upon the severity of the embolism but may be normal. The electrocardiogram is nonspecific with sinus tachycardia and evidence of right heart strain.

Chest radiography may demonstrate non-cardiogenic pulmonary edema or, more commonly, evidence of barotrauma (pneumothorax, pneumopericardium, interstitial emphysema, pneumomediastinum, or subcutaneous emphysema).

Arterial blood gases reveal hypoxemia, metabolic acidosis due to dependence upon anaerobic respiration, and a low, normal, or high CO_2 level depending on the severity of the insult.

Monitoring by oximetry and capnography can detect early falls in both oxygen saturation and end-tidal CO_2 concentration, and are reliable indicators of evolving venous gas embolism.

Invasive hemodynamic measurements and computed indices are diagnostic adjuncts in air embolism. Pulmonary arterial pressure can be normal or raised depending on whether the embolus is trapped low down in the pulmonary outflow tract or high in the pulmonary trunk. Central venous pressure may be elevated with a normal or low wedge pressure. The right ventricular stroke work is initially elevated in an attempt to compensate for the profound increase in afterload. Invasive lines can serve diagnostic and therapeutic purposes, as conduits for air aspiration. Ultrasonography, particularly in the form of transesophageal echocardiography, is of value in the diagnosis of venous air embolism and has the added advantage of detecting any paradoxical emboli (Cucchiara *et al.* 1984). Transesophageal echocardiography should be considered in any patient with cardiovascular instability and a history suspicious of air embolism.

Venous air embolism has a broad spectrum of presentation. The differential diagnosis often includes the numerous causes of acute cardiorespiratory collapse and neurological dysfunction. Pulmonary angiography and CT scan of the brain are used to exclude alternatives such as pulmonary thromboembolism and cerebrovascular accidents.

Arterial air embolism

Neurological symptoms and signs occurring within 10 min of surfacing from a scuba dive are usually due to arterial gas embolism — 'the ten minute rule' (Kizer 1987).

The clinical features of arterial gas embolism are specific to the body region embolized and the cause of the embolus (e.g. thoracic trauma). The most common sites of emboli are the cerebral and coronary circulations.

The symptoms relate to the artery embolized, for example chest pain and dyspnea (coronary artery) and alterations in consciousness and subtle mental changes (cerebral artery). Alveolar rupture, which is often associated with arterial gas embolism, manifests as hemoptysis and pleuritic chest pain.

The signs are primarily of major organ dysfunction. Associated clinical findings include livedo reticularis, Liebermeister's sign (a well-defined area of glossal pallor), and air in the retinal vessels.

Air in the systemic circulation may be difficult to detect. Its absence does not exclude clinically significant arterial gas embolism, as minute amounts of gas can cause significant symptoms. Diagnostic investigations are aimed at either directly demonstrating air in the arterial system or locating the possible source of arterial gas emboli. Transesophageal echocardiography can be used as a sensitive noninvasive procedure to diagnose intrathoracic air embolism on both sides of the circulation (Saada *et al.* 1995). Cerebral air embolism seen on CT scanning is diagnostic.

Transcranial Doppler ultrasonography is a non-invasive tool used to monitor middle cerebral artery blood flow during carotid endarterectomy. Air emboli of diameter as small as 30 μm are detected as short-duration high-intensity 'chirps' on the ultrasound flow waveform.

Biochemical and hematological evidence of arterial gas embolism is non-specific. It includes a rise in creatinine kinase and a fall in hematocrit.

Standard radiographs often provide an accessible method of supporting a diagnosis of arterial air embolus, such as looking for barotrauma. Subtle findings of extrathoracic ectopic air on skull and abdominal radiographs can help to confirm the clinical diagnosis.

It should be emphasized that the diagnosis of arterial air embolus is often clinically based. A high index of suspicion is vital for the successful resuscitation of patients with a paucity of abnormal diagnostic investigations.

Management

Airway, breathing, circulation, and disability (of neurological function) (ABCDs) defines a prioritized sequence of evaluation and treatment which provides a safe and effective method of managing any patient with a suspected air embolus in its early stage.

Speed of diagnosis and management is essential for all types of gas embolism. Failure of a patient to make a dramatic response to these first-line measures indicates the need to consider rapid institution of hyperbaric therapy.

All patients who have gas embolism have the potential to be hypoxemic. Thus immediate administration of 100 per cent oxygen is necessary. In an unconscious patient any obstruction in the oropharynx must be cleared under direct vision. If protective airway reflexes are intact, as demonstrated by a normal gag reflex, a tight-fitting oxygen mask with a reservoir should be used to provide flow rates of more than 12 l/min. If reflexes are depressed or absent, the airway must be secured at the earliest opportunity by intubation. Administration of 100 per cent oxygen optimizes tissue oxygen supply and attenuates bubble size by causing 'nitrogen washout'. It is essential to avoid exposure to nitrous oxide.

Circulatory arrest is a clinical diagnosis. Basic life support followed by advanced life support should be instituted. Any potentially lethal arrhythmias should be treated (Chamberlain *et al.* 1994).

Closed chest compressions may have the additional benefit of dissipating any embolic gas into the pulmonary arterial tree. The circulation should be supported with warmed intravenous crystalloid solutions. Other therapeutic maneuvers that rapidly treat circulatory compromise include the prevention of further embolization (e.g. decompression of pneumothorax or occluding an open central venous catheter), attempted aspiration of air via indwelling central venous catheters, and placing the patient in the left lateral decubitus and/or Trendelenberg position. The last of these maneuvers prevents obstruction of the pulmonary outflow tract.

Patients requiring positive-pressure ventilation need strategies to keep their peak airway pressures below 50 cmH_2O. Lowering positive end-expiratory levels, reducing tidal volume, and permissive hypercarbia may significantly decrease systemic air embolism in the presence of barotrauma (Saada *et al.* 1995).

Delayed circulatory embarrassment may be negated by adminstration of 8 mg dexamethasone followed by a maintenance dose depending on the patient's clinical course.

An abnormal neurological status that does not respond rapidly to the above measures is an absolute indication for hyperbaric therapy. Seizure prophylaxis using standard anticonvulsant therapy should be considered in any patient with even the slightest alteration of the sensorium.

Hyperbaric centers in the United Kingdom and the Royal College of Physicians of London stress the significance of early compression in a hyperbaric chamber when first-aid measures fail to achieve complete resolution (James 1993).

Any acute symptoms or signs caused by air embolism can be ameliorated by hyperbaric oxygen therapy. Increased pressure has several effects on a fixed mass of gas. The volume of free gas is reduced (Boyle's law) and gas enters solution (Henry's law). Increasing the partial pressure of oxygen dissolved in the plasma permits supplemental oxygen delivery to the tissues. Increasing the partial pressure of oxygen in the alveoli favors nitrogen diffusion from bubbles in the pulmonary circulation and the alveoli, facilitating elimination.

The principles of hyperbaric therapy are rapid compression to several atmospheres followed by intermittent 100 per cent oxygen and 'air breaks' to prevent the symptoms of oxygen toxicity. Treatment periods are of the order of 4 to 6 h. The compression is carried out slowly to permit the gas dissolved in the plasma to equilibrate with the alveolar gas. It must be emphasized that rapid transfer of the patient to a hyperbaric chamber greatly improves the prognosis. Treatment instituted within 4 h is usually successful.

References

Chamberlain, D., *et al.* (1994). Management of peri-arrest arrythmias. A statement for the Advanced Cardiac Life Support Committee of the European Resuscitation Council, 1994. *Resuscitation*, **28**, 151–9.

Cucchiara, R.F., Nugent, M., Seward, J.B., and Messick, J.M. (1984). Air embolism in upright neurosurgical patients: detection and localisation by two-dimensional transesophageal echocardiography. *Anesthiosology*, **60**, 353–5.

James, P.B. (1993). Dysbarism: the medical problems from high and low atmospheric pressure. *Journal of the Royal College of Physicians of London*, **27**, 367–74.

Kizer, K.W. (1987). Dysbaric cerebral air embolism in Hawaii. *Annals of Emergency Medicine*, **16**, 535–41.

Saada, M., *et al.* (1995). Systemic gas embolism complicating pulmonary contusion. Diagnosis and management using transesophageal echocardiography. *American Journal of Respiratory and Critical Care Medicine*, **152**, 812–15.

11.1.9 The role of surgical management of fractures

R. J. A. Goris and J. Biert

Key messages

- Early osteosynthesis of major fractures in traumatized patients leads to a reduction in morbidity and mortality.
- Patients with severe brain or thoracic injuries benefit from early fracture stabilization.

Introduction

Various controversies surround the surgical management of extremity injuries in multiple traumatized patients. Such injuries may induce life-threatening systemic complications, and local complications, possibly occurring in the intensive care unit (ICU), may result in severe late incapacitation. The timing of the operative stabilization of major long-bone and pelvic fractures, attempted limb salvage, and treatment of compartment syndromes must be co-ordinated with the patient's overall condition.

Timing of osteosynthesis

Performing operative stabilization of major fractures (pelvis, femur) on the day of injury is an attractive option. The young and active trauma patient is still in optimal nutritional and immunological condition. This condition will deteriorate during the following days or weeks, with colonization by nosocomial bacteria, infectious complications, and wound infection increasing the risks of secondary surgery. The presence of one or more 'floating' major fractures makes nursing difficult and the patient uncomfortable. Stabilizing fractures allows optimal nursing, as the patient can be moved and turned, preventing decubitus, and mobilization can be started. Since fewer analgesics are needed, the patient is more alert. Thus active mobilization should start at an early stage, as a prerequisite for the prevention of thromboembolic complications and for attaining optimal functional outcome of the fractured extremities.

In severely traumatized patients, 90 per cent of late deaths are caused by the acute respiratory distress syndrome and the multiple organ dysfunction syndrome. An important factor, correlating with late death, is the presence of one or more major fractures which have not been stabilized by external or internal fixation. Early osteosynthesis in polytraumatized patients results in a reduction in morbidity and mortality, because of a lower incidence of acute respiratory distress syndrome and sepsis, and fewer ventilator and ICU days (Biert and Goris 1995).

Early osteosynthesis in patients with severe brain injury

Immobilization in a supine position, which is often mandatory with skeletal traction for major fractures, increases intracranial pressure. Positive-pressure ventilation, which is required for acute respiratory distress syndrome, increases intracranial pressure. Forced immobilization may induce or aggravate agitation and restlessness, increasing intracranial pressure. Operative fracture stabilization allows the brain-injured patient to be placed in a semirecumbent position at an early stage, which may help to decrease intracranial pressure. Early fixation also offers the best chance of functional recovery of extremity injuries, particularly if a permanent neurological deficit results.

Several clinical studies have shown a positive effect of early fracture stabilization in preventing secondary brain damage, reducing mortality, and facilitating nursing (Biert and Goris 1995). Also, the outcome of brain injury was better in patients with early fracture fixation than in a group receiving conservative treatment (Hofman and Goris 1991).

Early osteosynthesis in patients with severe thoracic injury

Patients with long-bone fractures and blunt chest trauma with pulmonary contusion have a higher incidence of pulmonary morbidity and death than patients with the same injury severity score but without long-bone fractures. Concern has been expressed about the potential detrimental effects on pulmonary function of early osteosynthesis in these patients, particularly with intramedullary reaming and nailing. Increased stress at a time when inflammatory mediators are maximally activated may indeed result in increased pulmonary inflammation (Waydhas *et al.* 1994).

However, in a retrospective study of patients stratified according to the severity of pulmonary contusion, the ratio of $P\text{aO}_2$ to FiO_2 showed no significant difference between those with and without early osteosynthesis of major fractures. Neither was there a difference in the incidence of acute respiratory distress syndrome, the duration of artificial ventilation, or the length of ICU stay between the two groups (van Os *et al.* 1994).

Temporary deterioration of lung function with a rise in pulmonary artery pressure, as seen during reaming and nailing of femur fractures, can be largely prevented by the use of unreamed nails. Secondary pulmonary damage can also be circumvented by using a two-stage procedure, with an external fixator applied first, followed later by internal fixation.

Prerequisites for early osteosynthesis in polytraumatized patients

Prolonged operations for extremity injuries in the polytraumatized patient require diligent preparation and monitoring, taking the ICU to the operation theater. The prerequisites are as follows.

1. The patient should have stable vital functions, a $P\text{aO}_2/\text{FiO}_2$ ratio above 280 mmHg (37 kPa), and a stable normal arterial blood pressure.
2. All important traumatic lesions should have been identified and, where appropriate, treated.
3. The patient should not be hypothermic (core temperature above 35 °C).
4. In the operating room normothemia should be maintained by using warming devices for the patient and for infusion fluids.
5. Blood coagulation should be within normal limits. Platelet counts should be above 75×10^9/l.
6. Monitoring of vital functions should be performed to the same level as in the ICU.
7. If blood loss is expected, cross-matched blood should be available and cell-saver use should be considered.
8. Patients with a Glasgow Coma Scale score of 8 or below should have a recent brain CT scan. If possible, an intracranial pressure measurement device should be implanted before starting the operation.
9. Osteosynthesis should be performed by a team experienced in all forms of osteosynthesis techniques.
10. Intramedullary reaming is controversial.

Which fractures should be operatively fixed early?

Dislocated joints should be repositioned as soon as possible, particularly if there is a pulse deficit. Fractures or fracture/dislocations of the spine, with progressive neurological deterioration, and depressed skull fractures take priority over other fractures. Fractures and joint injuries associated with vascular injuries should have a vascular supply provided within 6 h of injury. Open fractures and joint injuries should be treated within 6 h of injury because of the increasing risk of infection after this period. Operative stabilization of major joint reconstructions (i.e. of acetabular fractures and closed fractures of the forearm), bone grafting, and closure of large skin defects with skin grafts need not be carried out in the primary period.

Limb salvage

The development of microsurgical techniques has increased the chances of salvaging amputated or semiamputated limbs. In the polytraumatized patient, however, salvaging such extremities with severe soft tissue injury increases the systemic inflammatory response, increases the risk of local septic complications, and requires a prolonged operation and recovery in the primary period. Therefore salvage or replantation is not recommended in patients with severe head injury or other injuries requiring intensive treatment. A two-stage replantation procedure with initial osteosynthesis and revascularization, followed 3 days later by final debridement, nerve and tendon suturing, and soft tissue coverage, may help to reduce the length of the primary operation. Amputation in the upper extremity leads to major disability, but the prospects of a successful replantation are much better than for the lower extremity.

Preventive decompression of muscle compartments

After severe soft tissue injury, excessive swelling may be expected in the affected muscle compartments. This is due to local hemorrhage as well as increased capillary permeability. Since the fascial structures surrounding each compartment are rigid, an increase in compartmental volume leads to increased pressure, which eventually results in shutdown of capillary perfusion with subsequent necrosis. In the polytraumatized patient, this development is enhanced when arterial pressure is decreased or (local) venous pressure is increased. In addition, while elevation of the affected extremity decreases edema, traction on the affected extremity may increase compartmental pressure. Early clinical signs and symptoms of a developing compartment syndrome include paresthesias, although they may go unnoticed in the polytraumatized patient who is unable to communicate.

Compartment pressures can be measured using a specialized instrument or an arterial line set-up. Pressures above 20 mmHg require close observation, and pressures over 30 mmHg require fasciotomy. A knowledge of the various compartments is necessary; for example, all four compartments of the lower leg must be measured and surgically addressed if necessary.

Special considerations in polytrauma patients

Life-threatening injuries are treated before limb-threatening injuries. To reduce the time spent in the operating room, external fixation without anatomical reposition may be preferred over anatomical reposition and internal osteosyntheses. A definitive operation is performed at a later stage to achieve the optimal result.

The patient with an isolated extremity injury benefits from active exercise, muscle training, or early weight bearing to gain functional recovery of the injured limb at an early stage. Thus the occurrence of complications such as muscle wasting, joint stiffness, and thromboembolic disease is reduced. Polytraumatized patients require early and intensive physiotherapy, including early active and/or passive movement of the injured and uninjured limbs. Continuous passive movement devices are useful, particularly for joint injuries. Passive exercises, together with the application of removable splints or external fixators, should prevent joint contractures. The patient should undergo an intensive rehabilitation program, if necessary in a rehabilitation center.

Conclusion

Adequate stabilization of major fractures has become an integrated part of the care of the multiple traumatized patient. Early osteosynthesis leads to a lower rate of pulmonary insufficiency, lower mortality, and earlier mobilization of the patient. In severe pelvic fractures, operative stabilization also contributes to hemostasis. Other essential steps for an optimal functional outcome include adequate fasciotomy, splinting joints to prevent contractures, and early active and/or passive exercises. Salvage or replantation of a severely injured extremity should be considered with great care, as this may substantially lengthen the time in the operating theater and may contribute to later sepsis.

References

Biert, J. and Goris, R.J.A. (1995). *The integrated approach to trauma care. The first 24 hours*, pp. 219–32. Springer-Verlag, Heidelberg.

Hofman, P.A.M. and Goris, R.J.A. (1991). Timing of osteosynthesis of major fractures in patients with severe brain injury. *Journal of Trauma*, **31**, 261–3.

van Os, J.P., Roumen, R.M.H., Schoots, F.J., Heystraten, F.M.J., and Goris, R.J.A. (1994). Is early osteosynthesis safe in multiple trauma patients with severe thoracic trauma and pulmonary contusion? *Journal of Trauma*, **36**, 495–8.

Waydhas, C., *et al.* (1994). Operationsplanung von sekundären Eingriffen nach Polytrauma. *Unfallchirurgie*, **97**, 244–9.

11.2 Ballistic trauma

11.2.1 Pathophysiology

D. V. Shatz and O. C. Kirton

Key messages

- The science of ballistics involves internal ballistics (what happens to the bullet inside the weapon), external ballistics (flight characteristics), and terminal ballistics. When the target is living tissue, terminal ballistics is also referred to as wound ballistics.
- The number attached to any given bullet is its diameter in inches (e.g. .45 caliber = 0.45 inches) or in millimeters (e.g. 9 mm).
- The tissue damage caused by a bullet is that which is crushed in its path (permanent cavitation) and, in the case of non-elastic tissue, includes that stretched beyond its limits by temporary cavitation.
- The effective diameter of a bullet, and therefore the amount of tissue encountered by the bullet, is expanded beyond the original diameter for hollow-point and soft-point bullets.
- Muzzle velocities are markedly different between the typical handgun bullets and those fired from rifles. This difference is due in large part to the size of the cartridge, and therefore the amount of gunpowder exploded during discharge of the weapon.
- Yaw refers to the end-over-end motion of the projectile through its target.

Introduction

Most domestic and urban firearm injuries are due to handguns (Schwab 1993), but injuries from bullets fired from hunting or military assault rifles may be seen in civilian emergency rooms or military field hospitals. Although the mainstay of the acute management of missile injuries remains control of hemorrhage and repair of the structures injured, an understanding of ballistics and associated wounding patterns may assist the clinician in detecting subtle injuries and/or anticipating physiological changes attributable to the missile.

Ballistics — the science

Bullets

The science of ballistics is the study of the physical and flight characteristics of bullets. Internal ballistics refers to those characteristics immediately following the firing of the bullet while it remains within the gunbarrel. Once the projectile exits the weapon, a new set of variables are encountered (e.g. wind drag) and impart yet another array of properties upon the bullet. How the bullet reacts to these variables is known as external ballistics. The injury patterns caused by bullets upon striking tissue are referred to as terminal ballistics and will be discussed later.

What is commonly referred to as the 'bullet' is more properly called a cartridge, and is composed of several components (Fig. 1). Impact on the firing pin, located on the base of the cartridge, ignites a small primer, which leads to an explosive burn of the gunpowder within the case. The size of this case, and therefore the amount of gunpowder loaded, at least partially determines the velocity of the bullet. For instance, a .22 long rifle and a .223 Remington (United States military bullet fired from the M-16 rifle) round are similar in size and weight (Table 1), but have markedly different muzzle velocities (the speed of the bullet as it exits the muzzle of the gun). The muzzle velocity of the .22 long rifle is 383 m/s, while that of the .223 Remington is over 988 m/s. This pronounced difference in velocity is due primarily to the amount of gunpowder contained within the cartridge.

Although intimidating to the novice, bullet nomenclature is actually quite simple; the numbers represent the projectile diameter in either inches or millimeters. Therefore a 9 mm bullet is 9 mm in diameter and a .45 caliber bullet is 0.45 inches in diameter. The .308 Winchester bullet is the civilian version of the military 7.62 NATO cartridge (0.308 inches = 7.62 mm), and the common civilian .22 caliber is only 0.003 inches smaller than the 5.56 mm, one of the most common military rifle cartridges in the world.

The weight of the bullet and its tendency to tumble (end-over-end cartwheeling) and yaw (the degree of divergence from the straight-line path of the bullet) contribute to tissue injury and therefore are important considerations in the design of ammunition. Bullets range in weight from about 2 g for a .22 short to 15.6 g for a .44 caliber. A typical 9 mm bullet weighs about 8 g, while a .38 caliber generally weighs 10 g. The in-flight stability imparted to the bullet by the weapon's rifling deteriorates once it encounters the increased density of tissue. Tumble and yaw may be significant as the bullet travels through tissue. The cross-sectional area is increased significantly, increasing both the kinetic energy loss to the tissue and temporary and permanent cavitation.

Bullet design plays a key role in flight characteristics and wounding patterns. Bullets fired from early weapons were simply round balls of lead. Modern bullets have a variety of designs, ranging from simple

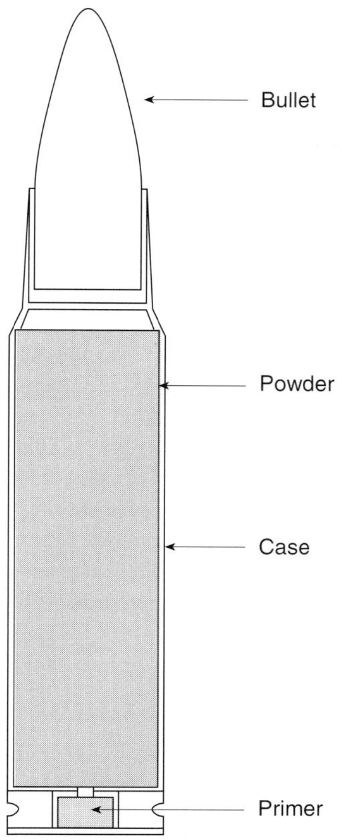

Fig. 1 A cartridge and its components.

round-nose lead bullets to jacketed bullets which begin to expand on contact with tissue (Fig. 2). The round-nose lead bullet is the cheapest. These bullets are frequently used for target practice, but are also often found in tissue wounds because of their availability. Since lead is irregular, these bullets have a higher friction coefficient as they travel down the barrel of the gun and during flight. Being relatively soft, lead bullets also tend to deform more on contact with tissue. Because of the addition of a thin outer layer of copper, full metal jacket bullets have a lower drag coefficient and do not usually deform or expand. However, expansion of the bullet beyond its original diameter is important when designing the missile with the intent to injure. Jacketed hollow-point bullets combine the advantages of a copper jacket with the expansion associated with a hollow-point bullet. These bullets expand by as much as two to three times their original diameter upon contact with tissue. However, the forward edge of an expanded hollow-point bullet is made of lead and is smooth and rounded. When it strikes a large blood vessel within inches of the end of its path, the vessel may be pushed aside so that injury is avoided. Bullets such as the Winchester Black Talon take this concept one step further, with a jacket whose leaves are exposed upon impact and remain attached to the expanding bullet (Fig. 3). These leaves act as six tiny knives lacerating tissue as the bullet travels through the body. Near the end of its course, these 'talons' continue their cutting action rather than pushing tissue aside. The 7.62 mm NATO SP (soft-point) is designed to fragment, sending multiple smaller missiles through the tissues. Similarly, the .223 caliber (5.56 mm) round bullet fired from the M-16 fractures at the cannelure (a circumferential groove around the bullet), again converting a single missile into several smaller ones with more diverse tissue injury.

The barrels of most firearms contain a set of grooves, known as rifling, that impart a rotational spin to the bullet as it travels down the barrel. This spin stabilizes the bullet as it exits the gun and travels through air. The resistance encountered during flight causes some degree of angulation off the in-line trajectory of the bullet, i.e. yaw (Fig. 4). Yaw angles are usually small, varying from 1.5° to almost 10°, and are known exactly only for military weapons.

Muzzle velocities are also an integral part of external and terminal ballistics. Velocities typically range from 240 m/s in handgun bullets

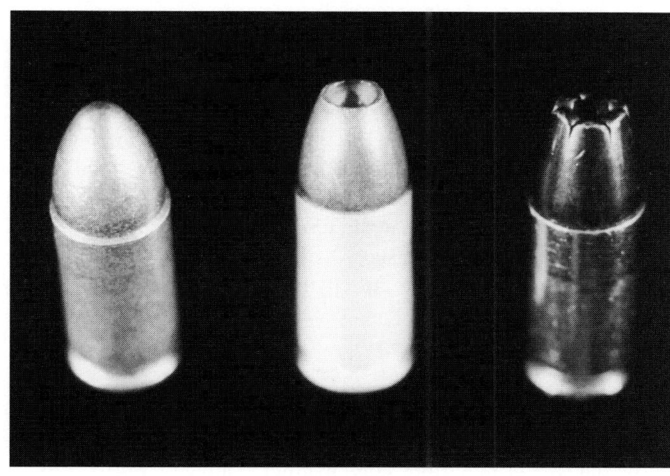

Fig. 2 Left to right, lead round-nose (LRN), jacketed hollow-point (JHP), and Black Talon bullets.

Table 1 Weight, muzzle velocity, and kinetic energy of some representative bullets

Caliber	Bullet weight (g)	Muzzle velocity (m/s)	Kinetic energy (J)
.32	4.6	278	177
.38 special	10.3	265	363
.45	16.3	262	562
.22 long rifle	2.6	383	192
.223 (M-16)	3.6	988	1753
.30 (7.62 mm, AK-47)	7.9	702	1999

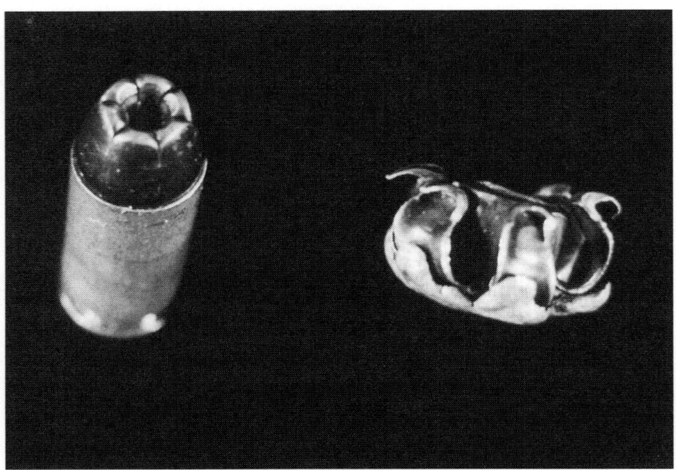

Fig. 3 .40 caliber Winchester Black Talon before and after firing. The leaves of the bullet's jacket are exposed upon impact and tear tissue in its path.

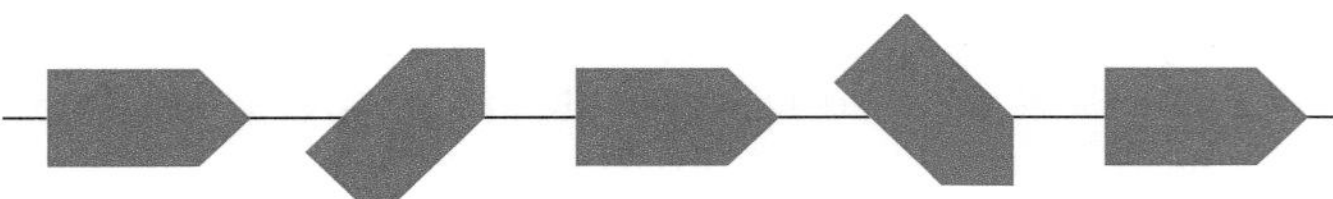

Fig. 4 During flight through air and tissue, bullets tend to yaw; this is measured as the divergence of the long axis of the bullet from the straight flight path.

to almost 1220 m/s in some military assault weapons. The use of the term 'high velocity' should be avoided for several reasons. First, wound damage can actually be less than that seen with lower-velocity missiles. The velocity of the bullet, even if it is known, will rarely dictate the care of the wound. Finally, the actual definition of 'high velocity' varies so much that the term is too imprecise to be used scientifically (British usage starts at 335 m/s, while American usage varies from 610 to 915 m/s). Specifying the actual velocity provides an unequivocal definition.

Bombs

In addition to the concussive and burn effects of a bomb blast, injuries are caused by propelled bomb fragments. These projectiles produce tissue damage in much the same manner as bullets, but in a much less structured fashion. The fragments are not aerodynamically sound, resulting in unpredictable flight patterns. Unlike bullets, tumble and yaw are prominent in bomb shrapnel. The surface area of the shrapnel striking the human body will vary depending on its position at the time of contact and the size of the fragment. Composition of the bomb casing (e.g. steel, plastic) will dictate the degree of deformation and fragmentation that the shrapnel undergoes upon contact with tissue. Lead fragments will deform much more easily than steel, and plastic will deform in yet another fashion. The size of the temporary and permanent cavities and the amount of kinetic energy transferred will differ not only by these variables but also by the distance between the victim and the bomb. Since these fragments are not aerodynamically designed, their velocity will dissipate more quickly than that of bullets. All these elements usually prevent fragments from penetrating deep into tissues. Fragmenting antipersonnel weapons release a large number of fragments, usually of small size (200–500 mg). The small fragment size and low velocity results in a kinetic energy of only 25 to 30 J. A small temporary cavity is produced, skin damage is limited, and necrotic tissue within missile tracks is nominal, allowing for spontaneous healing without surgical intervention in most cases (Bowyer *et al.* 1996). The degree of injury is related to the size of the explosive device, the distance between the device and the victim, and the surrounding medium (air or water). When combined with even a remote penetrating injury, the lungs seem to be the most susceptible to damage, presumably because of substances released from the necrotic wound track (Huang *et al.* 1996).

Ballistics — the wound

The quest for the perfect bullet by law enforcement agencies and bullet designers centers on the concept of 'stopping power'. This concept is a fallacy because no bullet is capable of total incapacitation in and of itself. Aim and the organ struck will determine the degree to which an assailant is incapacitated. Only a bullet striking the brain or the high cervical spinal cord will prevent an assailant from any further action. Hemorrhage to the point of cerebral hypoxia and hypoperfusion will render incapacitation when 25 per cent or more of the blood volume is lost. A male of average size, able to compensate fully for progressive hypotension, could maintain adequate mentation for at least 4.5 s, even with a freely bleeding transected aorta. This is sufficient time for the entire magazine of a semi-automatic weapon to be emptied before the assailant is neutralized (Newgard 1992).

The mechanics by which bullets inflict injury is frequently ascribed to the transfer of kinetic energy from the bullet to the tissue. In contrast with stored or potential energy, kinetic energy (KE) is the energy of an object in motion, and is given by the formula

$$\mathrm{KE} = mv^2/2$$

where m is the mass of the object and v is its velocity. Conceptually, the larger and faster the bullet, the more damage it causes. However, energy transfer is a key concept in the kinetic energy description of terminal ballistics. Theoretically, a faster lighter bullet could do

Table 2 Energy transfer of a 9-mm bullet with full metal jacket versus that of a jacketed hollow-point

Projectile	Striking velocity (m/s)	Exit velocity (m/s)	Energy transfer (J)
9 mm 7.5 g FMJ	345	283	146
9 mm 7.5 g JHP	343	95	409

FMJ, full metal jacket; JHP, jacketed hollow-point.

more damage than a slower heavier bullet because it has a higher velocity and therefore more kinetic energy (Table 1). However, the smaller bullet may actually travel through tissue without significant energy transfer and thus will be less effective in imparting damage.

Alteration in bullet design can profoundly change the amount of energy transferred. For example, a non-deforming bullet (e.g. full metal jacket) is more likely to travel easily through tissue, while a bullet that expands and deforms (e.g. hollow-point or soft-point) will be slowed during its travel. The difference between entry and exit velocity will determine the amount of energy transferred to the tissues (Table 2). Similarly, as a higher-velocity missile encounters tissue and is slowed by its yaw, more kinetic energy will be transferred to the tissue, theoretically inflicting more damage.

The results of studies of the actual physical effects of projectiles using ordnance gelatin, which closely simulates human skeletal muscle (Post and Johnson 1995), and high-speed videography correlate with injury patterns seen clinically. As the bullet travels through tissue, a temporary cavity (a pathway which collapses) and a permanent cavity are created.

References

Bowyer, G.W., Cooper, G.J., and Rice, P. (1996). Small fragment wounds: biophysics and pathophysiology. *Journal of Trauma*, **40** (Supplement), S159–64.

Huang, J.Z., Yang, Z., Wang, Z., and Leng, H. (1996). Study on characteristics of blast–fragment combined injury in dogs. *Journal of Trauma*, **40** (Supplement), S63–7.

Newgard, K. (1992). The physiological effects of handgun bullets. The mechanisms of wounding and incapacitation. *Wound Ballistics Review*, **1**, 12–17.

Post, S.M. and Johnson, T.D. (1995). A survey and evaluation of variables in the preparation of ballistic gelatin. *Wound Ballistics Review*, **2**, 9–20.

Schwab, C.W. (1993). Violence: America's uncivil war—Presidential Address, Sixth Scientific Assembly of the Eastern Association for the Surgery of Trauma. *Journal of Trauma*, **35**, 657–65.

11.2.2 Assessment

K. A. Kralovich and H. M. Horst

Key messages

- Unstable patients require operative exploration and a minimum of diagnostic studies.
- The velocity of the missile and the anatomical path of the missile can be used to predict injuries.
- Diagnostic studies are appropriate for selective management of neck, chest, and extremity injuries.

General assessment

The assessment of patients injured by ballistic trauma includes resuscitation, medical history, physical examination, radiographs, and other diagnostic tests. Advanced Trauma Life Support protocols are followed as in blunt trauma, but several additional concerns must be addressed. The time from injury, the type of weapon, the distance between the weapon and victim, and the amount of blood at the scene are important facts to attain. Low-velocity bullets cause organ injury within their anatomical path. High-velocity missiles, shotgun blasts, and bombs create additional injury via blast injury and cavitation. Complete patient exposure allows identification of all wounds and predicts the number of body cavities violated. Pulsatile hemorrhage is immediately controlled with pressure from a gloved finger or hand. Two large-bore (16 gage or greater) peripheral intravenous cannulas are placed in the upper extremity in patients with abdominal or lower extremity wounds, and in the lower extremities in patients with missile injuries to the upper extremities or subclavicular areas. Blood is obtained for type- and cross-matching. Tetanus prophylaxis and appropriate antibiotics are administered. Blood found in endotracheal tubes, nasogastric tubes, bladder catheters, the rectum, or the vagina imply internal organ damage. Extremity fractures are immobilized and peripheral pulses are noted. Ambient temperatures are elevated, the patient is covered, and intravenous fluids are warmed in order to avert hypothermia. Frequent monitoring of vital signs gages the effectiveness of the resuscitation. In general, patients with unstable vital signs require immediate surgical exploration, while patients with stable vital signs may benefit from additional diagnostic maneuvers prior to operation.

Neck injuries

Assessment of penetrating neck trauma requires special attention to the airway, breathing, and neurological status. The patient is questioned about the mechanism of injury, pain, respiratory distress, dysphagia, and voice changes. He or she is assessed for active bleeding, hematoma formation, subcutaneous emphysema, bruits, cranial nerve function, and hemispheric or upper-extremity neurological deficits. Victims with gunshot wounds in proximity to the cervical spine require immobilization until the vertebrae are assessed with radiographs. Probing of the wound or coughing induced by placement of nasogastric tubes may dislodge thrombus and result in significant hemorrhage, and should be done only in the operating room. Diagnostic studies are contraindicated in patients with uncontrolled bleeding. Hemorrhage is controlled digitally until the patient is prepared and draped for operation. The patient with pulsatile hemorrhage or expanding hematoma must have the airway controlled by early intubation, as delay may lead to tracheal deviation or compression necessitating surgical airway placement. Awake intubation is preferred because the patient maintains spontaneous respiration. A fiber-optic endoscope can improve visualization in difficult cases. Upper airway or laryngeal injuries require a surgical airway; intubation is contraindicated. Radiographs of the neck are useful in identifying bullet fragments, vertebral injury, tracheal compression, and subcutaneous air, and in helping to reconstruct missile trajectories.

In the stable patient, assessment varies with the anatomical location of the injury. The neck is divided into three zones (Fig. 1). Violation of the platysma mandates a thorough evaluation for vascular, airway, and esophageal injury. Physical examination has a sensitivity of approximately 80 per cent and a specificity of only 61 per cent for predicting vascular injuries. Arteriography is recommended, with a four-vessel study having an accuracy of around 95 per cent. Duplex scanning and intravenous digital subtraction angiography have not been investigated extensively in trauma. Zone I and zone III patients require angiography for diagnosis, possible therapeutic intervention, and planning of operative approach. Operative exposure of zone I injuries may require thoracotomy, sternotomy, clavicular excision,

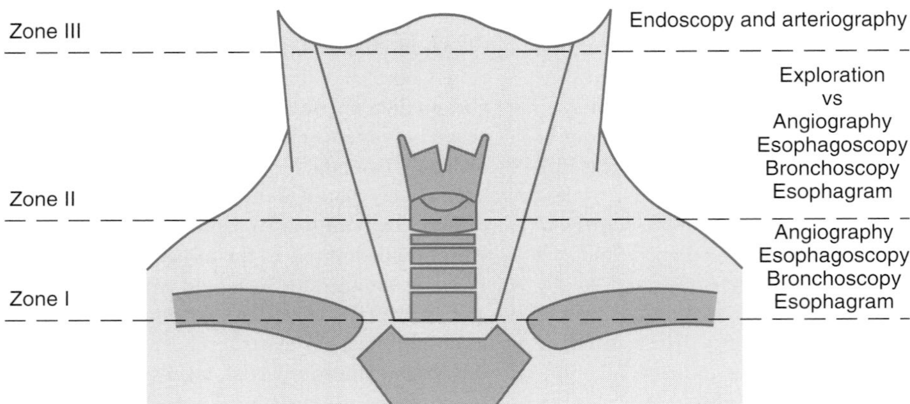

Fig. 1 Zone I extends from the sternal notch to the level of the cricoid cartilage and includes the thoracic outlet. Zone II extends from the cricoid to the angle of the mandible. Zone III lies above the angle of the mandible. The recommended diagnostic evaluation for each zone is shown.

and/or trap-door incisions. Zone III injuries are not easily accessible at exploration and require rotational osteotomy or subluxation of the temporomandibular joint to expose the injury. Angiography in zone III injuries allows for embolization of vertebral or external carotid injuries. Zone I and zone III injuries also require laryngoscopy and/or bronchoscopy to evaluate the airway; physical examination is not sufficient to evaluate esophageal injury. Rigid esophagoscopy and contrast esophagraphy have a sensitivity of 80 per cent individually, but when they are combined the sensitivity approaches 100 per cent. Flexible endoscopy may miss up to 50 per cent of esophageal injuries. Asymptomatic zone II injuries are effectively managed either by mandatory exploration protocols or selective plans which use double endoscopy, arteriography, and esophagraphy. Symptomatic patients are taken to the operating room. A subclass of zone II injuries are transcervical (injuries that cross the midline) and result in visceral injury in 83 per cent of cases; mandatory exploration is advocated. The need for mandatory arteriography in all selective protocols is being questioned at present.

Thoracic injuries

In the evaluation of patients sustaining ballistic chest trauma, examination includes attention to jugular venous distension, expanding hematoma, decreased breath sounds, subcutaneous emphysema, air leak, and heart sounds. Witnessed cardiac arrest from penetrating thoracic trauma is a clear indication for emergency department (resuscitative) thoracotomy. Unstable patients should have a 36 to 40 French chest tube placed on the side of injury. Upright chest radiographs (supine if vertebral injury is suspected) are obtained to assess the tract of the missile, hemopneumothorax, pneumomediastinum, enlarged cardiac silhouette, and underlying pulmonary contusion.

Hemopneumothorax is the most common injury and 85 per cent of patients can be treated with tube thoracostomy alone. If more than 1500 ml of blood is immediately returned, thoracotomy is usually indicated. Cardiac, pulmonary, great vessel, or intercostal vascular injury is frequently found. If the patient continues to bleed at more than 300 ml/h, exploratory thoracotomy must be considered. The hemothorax must be completely drained by placing additional chest tubes or using thoracoscopy.

A fully expanded lung may tamponade bleeding and reduce the risk of empyema and resultant fibrothorax. If full lung expansion is not obtained after placement of two functioning tubes, tracheobronchial injury should be suspected and bronchoscopy performed. Occasionally the chest tube will recover a patient's tidal volume and severe dyspnea will ensue, suggesting major airway or pulmonary injury for which further investigation such as bronchoscopy or thoracotomy is indicated. Deterioration after positive-pressure ventilation may be due to conversion of a simple to a tension pneumothorax or an air embolus secondary to traumatic bronchovenous fistula. Severe hypoxia requiring intubation may result from pulmonary contusion, pneumothorax, hemothorax, or aspiration of blood. If there are no contraindications, a nasogastric tube is placed and the return of blood should raise suspicions for esophageal or gastric injury. The standard assessment of transmediastinal injuries consists of either exploratory thoracotomy or a combination of arteriogram, esophagoscopy, bronchoscopy, and echocardiography (Mattox *et al.* 1996). Currently, support for exploratory thoracoscopy in this patient population is growing. However, the ability of this new modality to detect occult cardiac or esophageal injury awaits conformation from prospective trials (Simon and Ivatury 1995). In patients with tangential or peripheral chest wounds and a normal initial chest radiograph, the study must be repeated after 6 h in order to detect delayed pneumothorax, which has an incidence approaching 10 per cent. If the second study is normal, the patient is discharged.

Penetration of the anterior mediastinal rectangle (bounded by the sternal notch, the xiphoid, and the nipples) places the patient at risk of cardiac injury and tamponade. Beck's triad of distended neck veins, hypotension, and muffled heart sounds is present in only 10 per cent of cases; therefore all injuries must be fully assessed. Measurement of central venous pressure can be useful in differentiating hemorrhagic shock from tamponade. Chest radiographs may reveal an enlarged cardiac silhouette or pneumopericardium. Echocardiography, either transthoracic or transesophageal, may reveal the presence of fluid in the pericardium, necessitating exploration via either the subxiphoid window or thoracoscopy. Pericardiocentesis is of limited use in the trauma setting.

Thoracoabdominal injuries

Bullets which cross the anatomical boundary of the diaphragm (fourth intercostal space anteriorly, the sixth space in the mid-axillary line, or the eighth space posteriorly) mandate abdominal evaluation and

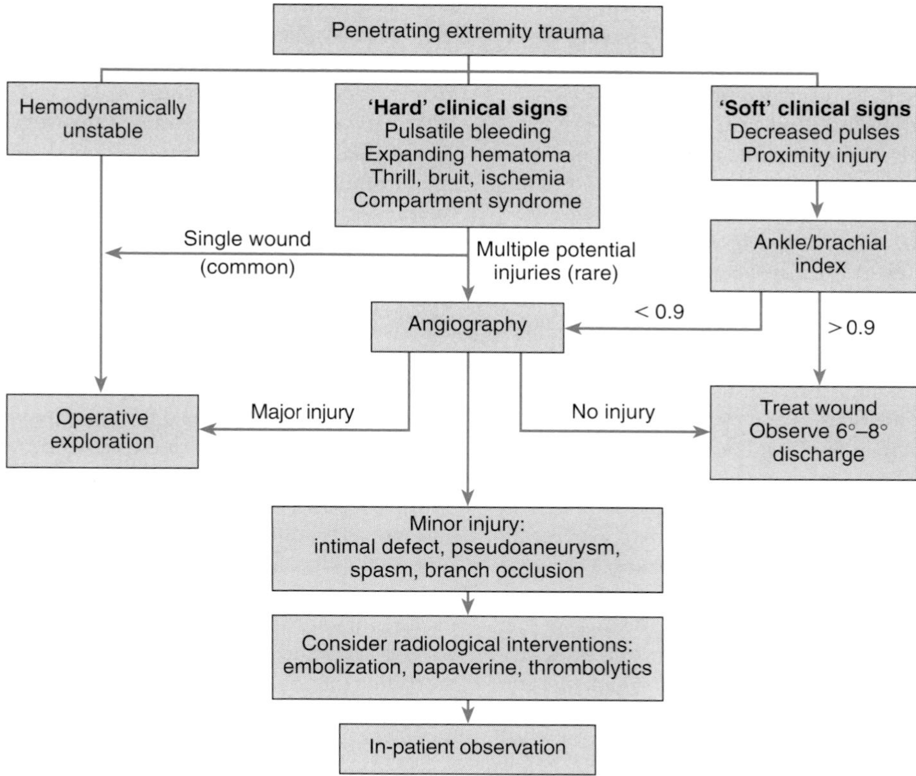

Fig. 2 An algorithm for extremity arterial injury emphasizes early operative exploration for obvious injuries. Arterial pressure indices determine which patients with equivocal examinations or proximity injuries require angiography. Arteriogram or operation remains the standard for defining the injury.

probably celiotomy. Physical examination is unreliable in the presence of thoracic trauma. Assessment requires surgical exploration, diagnostic peritoneal lavage, laparoscopy, ultrasound, and/or thoracoscopy. Abdominal exploration is appropriate for patients with hemodynamic instability without an obvious thoracic source of hemorrhage, if investigations or examination suggest injury. Opening a second body cavity adds to surgical stress, hypothermia, and anesthesia time (Hirshberg *et al.* 1995). Diagnostic peritoneal lavage is the most commonly performed procedure in thoracoabdominal injuries but has a limited ability to detect diaphragmatic and retroperitoneal injuries. In penetrating trauma, the red blood cell count criterion for a positive diagnostic peritoneal lavage may be reduced to as low as 1000/mm^3. Laparoscopy is useful in identifying peritoneal violation but remains unproven in assessing visceral injury. Ultrasound may detect intra-abdominal fluid. A 30° thoracoscope placed through the chest tube site can effectively evaluate the diaphragm for penetration without adding major surgical incisions.

Abdominal injuries

Abdominal gunshot wounds require laparotomy and few if any diagnostic tests are indicated. Anteroposterior and lateral films of the abdomen will enable the surgeon to reconstruct the path of the missile and anticipate the internal injuries. A single-shot intravenous pyelogram is advocated when renal injury is suspected in order to verify the presence of two kidneys, should nephrectomy be required. Shotgun wounds are difficult to evaluate. Distribution of the pellets may correlate with the depth of penetration since wide scatter implies a greater distance between gun and victim with lower-speed lower-energy pellets. Injuries covering 25 cm or less demand exploration, whereas scatter over more than 25 cm suggests that the patient can be evaluated with serial physical examinations (Glezer *et al.* 1993). Superficial bullet tracts that do not violate the peritoneum may have significant tenderness upon palpation. Lateral radiographs, tractotomy, and diagnostic peritoneal lavage are useful adjuncts in assessing these questionable cases.

Flank and back injuries are difficult to assess. Unstable patients receive a laparotomy. In stable patients, abdominal radiographs including a lateral view aid in assessing the tract of the projectile. Formal wound exploration, followed by diagnostic peritoneal lavage if the fascia is violated, determines the need for operation. A CT scan with triple contrast (intravenous, oral, and rectal) can evaluate the retroperitoneum, but hollow viscus injuries may still be missed. Suprapubic penetration may lead to pelvic visceral and extraperitoneal vascular injury without hypotension or peritoneal signs. With pelvic or perineal injury, rigid sigmoidoscopy may reveal rectal injuries below the peritoneal reflection, leading to repair and/or diversion.

Extremity injuries

Arteries, veins, nerves, bones, and soft tissues are at risk with missile injuries to the extremities. If the physical examination identifies pulsatile bleeding, expanding hematoma, thrills, bruits, or distal ischemia, exploration is required. These 'hard' physical signs are 99 per cent accurate for arterial injury. Physical examination is less accurate when pulses are palpable, as up to one-third of patients with arterial injury have a normal vascular examination. Diminished pulses or proximity to a major vessel mandates determination of arterial pressure indices using Doppler flow ultrasonography. A standard sphygmomanometer cuff is placed just above the ankle or wrist on the injured extremity and inflated until flow is occluded. The cuff is slowly deflated until a hand-held Doppler instrument detects flow in the vessels distal to the cuff, and this initial pressure is recorded. Measurements should be obtained from the posterior tibial and dorsal pedal arteries of a patient with an injured lower extremity and from the radial and ulnar arteries if the upper extremity has been wounded. A ratio is then calculated between the pressures in the suspect extremity and a similarly obtained pressure in an uninvolved brachial artery. An index of below 0.9 suggests arterial injury, and an arteriogram or exploration should be considered. Duplex scanning is accurate in assessing injury, but its availability limits its usefulness in emergency situations. The arteriogram remains the standard for diagnosing arterial injury but is rarely needed in penetrating extremity trauma. An arteriogram's main utility is to determine the level of injury in patients with multiple potential lesions (shotgun wounds) or a long tract. Arteriography may reveal minor injuries including small pseudoaneurysms, intimal defects, minor branch occlusions, and spasm which may be treated by interventional techniques, obviating the need for surgery. Because of the possibility of bullet emboli, the entire extremity must be assessed. Chest radiographs are obtained in cases of suspicious vascular injury to evaluate for embolization to the lungs, or in rare cases to the heart or arterial system via a patent foramen ovale. An algorithm for the assessment of extremity arterial injury is shown in Fig. 2.

Traumatized extremities are at risk of compartment syndrome. The first signs are a tense calf and pain on passive stretch. The initial symptoms are usually paresthesias. Diagnosis should be made before the 'six Ps' of paresthesia, pain, pallor, paralysis, poikilothermia, and pulselessness are evident. Measurement of extremity compartment pressures using a modified arterial line or self-contained instrument suggests the diagnosis if values are in excess of 30 mmHg. Early fasciotomy prevents further injury secondary to pressure-induced ischemia. When a missile injury results in compartment syndrome without associated fracture, arterial injury is assumed and exploration performed.

Massive tissue destruction by shotgun injuries, blast injuries, or high-velocity weapons requires operative debridement and removal of the wadding of the shotgun shell. Fractures are immobilized and evaluated with appropriate radiographs. Open fractures are contaminated and must be treated with irrigation, debridement, and antibiotics. Nerve injuries can be diagnosed by physical examination in conscious patients. Acute nerve injury secondary to ballistic injury is not addressed immediately. Six weeks is given to allow the neuropraxia to resolve before repair is attempted. Massive extremity injury with arterial, venous, neural, and bony injuries are deemed nonviable, and primary amputation should be performed as the mangled extremity predisposes to the systemic inflammatory response.

Bomb injuries

Bomb injuries result in blunt, penetrating, and thermal injury. Initial evaluation and resuscitation should progress as with other trauma. The physical examination may disclose dyspnea, chest pain, decreased breath sounds, hemoptysis, abdominal wall disruption, peritoneal signs, rectal bleeding, neurological deficits, auditory complaints, and extremity damage. Tympanic membrane rupture suggests exposure to pressures high enough to cause internal injury. Therefore tympanic membrane perforation, tinnitus, otalgia, vertigo, and hearing loss are markers for associated trauma and mandate assessment of the entire body.

Primary blast injury is caused by the impact of a wave of compressed air. Damage can be compounded by shock waves reflected off surrounding walls. Gas pockets in the alveoli are rapidly compressed and then re-expand, causing miniature secondary implosions. The blast also forces blood from the pulmonary vessels into areas of lesser density such as the alveoli. The resultant alveolar–venous disruption predisposes to air embolism. Retinal examination and CT scans may demonstrate embolized air. Positive-pressure ventilation and air transport are relatively contraindicated. Hyperbaric chambers may prove useful to treat hypoxia and air embolism. Gastrointestinal injuries typically involve rupture of hollow organs, particularly the colon owing to its larger air pockets. Laparotomy may be necessary as diagnostic peritoneal lavage is less sensitive for early small bowel perforation and CT scan may miss an unacceptable number of hollow viscus injuries. Delayed rupture of contused and ischemic bowel has been reported, and thus serial evaluation is advocated. Orthopedic injuries, usually diaphyseal fractures of long bones, result from the intense pressure of the shock wave. These injuries may be converted to traumatic amputations by the flailing motion of the limb in the trailing blast wind.

Secondary trauma results from flying debris. Explosive devices are frequently designed to propel metallic shards of shrapnel into the intended victims. Lacerations and soft tissue avulsions on the extremities are the most common injuries, but any body area may be traumatized. The majority of patients have between two and 20 injuries concentrated in one or two body areas. Wounds must be assessed for associated fractures, tissue loss of less than 2 cm, and absence of gross soil contamination or infection, as these patients are candidates for medical treatment instead of the traditional mandatory debridement (Bowyer 1996). Traumatic amputations can result from the cutting action of larger fragments. Assessment should proceed as previously outlined for each body cavity penetrated.

Tertiary injuries are due to acceleration of the body itself. Injuries are caused by impacting surrounding objects followed by rapid deceleration. A chest radiograph should be obtained to exclude a wide mediastinum, indistinct aortic knob, or other signs of mediastinal hematoma which may indicate an aortic transection.

Other associated problems include burns and inhalation injuries. The thermal injury may result in major fluid and electrolyte imbalances, protein losses, immunosuppression, infection, and eschars on the chest and abdomen which may lead to hypoventilation and require incision. Inhalation injuries may require bronchoscopy for evaluation and intubation for treatment.

References

Bowyer, G.W. (1996). Management of small fragment wounds: experience from the Afghan border. *Journal of Trauma*, **40** (Supplement), S170–2.

Glezer, J.A., Minard, G., Croce, M.A., Fabian, T.C., and Kudsk, K.A. (1993). Shotgun wounds to the abdomen. *American Surgeon*, **59**, 129–32.

Hirshberg, A., Wall, M.J., Allen, M.K., and Mattox, K.L. (1995). Double jeopardy: thoracoabdominal injuries requiring surgical intervention in both chest and abdomen. *Journal of Trauma*, **39**, 225–9.

Mattox, K.L., Wall, M.J., and Pickard, L.R. (1996). Thoracic trauma: general considerations and indications for thoracotomy. In *Trauma* (3rd edn) (ed. D.V. Feliciano, E.E. Moore, and K.L. Mattox), pp. 345–53. Appleton & Lange, Stamford, CT.

Simon, R.J. and Ivatury, R.R. (1995). Current concepts in the use of cavitary endoscopy in the evaluation and treatment of blunt and penetrating truncal injuries. *Surgical Clinics of North America*, **75**, 157–74.

11.2.3 Management

R. C. Karmy-Jones and J. W. Wagner

Key messages

- The degree of injury is related to the velocity and size of the bullet.
- Some patients with multiple injuries are best managed by abbreviated approaches to allow resuscitation and reversal of hypothermia prior to definitive treatment.

General assessment

The ABCs (airway, breathing, circulation) of Advanced Trauma Life Support remain the cornerstone in the treatment of ballistic trauma. After ensuring adequate ventilation, resuscitation with fluid and blood should be initiated early and in most instances continuously and aggressively. However, uncontrolled hypertension may aggravate bleeding, and therefore in thoracoabdominal injuries, where operative control of the bleeding site is needed, 'controlled hypotension' (systolic pressure between 80 and 90 mmHg) may be preferable until the patient is in the operating room. Hypothermia, which can result in rapid and significant metabolic and coagulation defects, should be prevented by heating the resuscitation and operating rooms, warming intravenous fluids and blood products, and performing intraoperative irrigation with warmed fluids. Active rewarming may be achieved by a number of techniques including heating ventilator circuits, rapid infusion devices, continuous arteriovenous rewarming devices, and, occasionally, various forms of bypass. Coagulation defects may also require supplemental platelet and plasma infusions.

All ballistic injuries can result in such severe anatomical and physiological disruption that the goal of surgery is simple ligation of non-life-essential vessels, closure of enteric injuries, packing of bleeding sites, and closure with towel clips or other temporary means. The patient should be transferred to the intensive care unit and aggressively rewarmed and resuscitated. Re-exploration and definitive repair is performed when the patient's condition has improved, usually within 24 to 48 h. Threatened limb loss may have to be accepted, or can be managed by extra-anatomical or retroperitoneal revascularization.

Neck injuries

The approach to neck injuries is based on the zone affected. Carotid injures in zone III (above the angle of the mandible), unless close enough to be approached with jaw dislocation, may require a combination of neurosurgical, maxillofacial, and radiological interventions. Traumatic pseudoaneurysms and arteriovenous fistulas (including carotid-cavernous sinus fistula) can often be embolized if they are at the level of the siphon or higher, although retrograde bleeding may still occur. Rarely, ligation of the internal carotid is necessary, and if neurological sequelae develop, a bypass from the external carotid to the middle cerebral artery may be considered. If ligation is needed to control massive bleeding, anticoagulation, although not universally practiced, should be considered (unless contraindicated by other major injuries) because of the risk of propagation of thrombus into the cerebral circulation. More proximal injuries in zones I or II, affecting the common or internal carotid arteries, may be managed by primary repair, interposition graft, or carotid–subclavian bypass graft. The external carotid can be ligated or repaired, depending on the extent of injury. Specific concerns exist in the patient who presents with a neurological deficit. Most of these result from ischemia due to interruption of cerebral perfusion and therefore are not hemorrhagic. Thus, revascularization rarely worsens the defect, and may improve a deficit or prevent it from occurring.

The vertebral artery is difficult to approach surgically and the extensive cranial circulation provides a source of retrograde flow that often complicates management. Options include ligation at both ends in the cervical spine (where it is more accessible), radiographic embolization, or placement of a Fogarty balloon catheter to promote thrombosis.

Vascular injuries in the thoracic inlet are approached through a variety of operative procedures, and occasionally cardiopulmonary bypass is needed. As many as a third of patients will have associated venous injuries.

Tracheal injury may be seen acutely following penetrating injuries, burn inhalation, blast trauma, or foreign body ingestion, but more often may be a sequela of prolonged intubation. Initial management must always involve ensuring a patent airway and adequate ventilation, usually by intubation. Flexible or rigid bronchoscopy may be needed for intubation, ventilation, and/or to assess cord function. Complete tracheal transection may require the distal trachea to be grasped with clamps in the thoracic outlet and pulled into the field. The trachea usually can be primarily repaired unless the defect is greater than 5 cm. Surgical practices include the use of absorbable sutures, limiting dissection to within 1 cm of either side of the proposed site (to prevent laryngeal nerve injury and devascularization) and avoiding tension (neck flexion, laryngeal or hypoglossal release, mobilization of the anterior trachea, or carinal release). If tracheostomy is necessary, it should be 2 to 3 cm away from the resection line. Viable tissue should be interposed between the trachea, esophagus, and carotid sheath to avoid fistula formation. Vocal cord function always needs to be assessed. Primary repair of transected recurrent nerves can be attempted, but more commonly unilateral paralysis (with unilateral cord abduction) is managed by injection to allow effective cough.

Thoracic injuries

Patients arriving in cardiac arrest may be candidates for rescusitative thoracotomy (Table 1). This procedure should generally be performed by individuals who are capable of handling intrathoracic injuries. Technical aspects include anterolateral thoracotomy at the nipple line (males) or the inframammary crease (females), opening the pericardium anterior to the phrenic nerve, open-heart massage, rarely atrial cannulation, and packing the thorax to control great vessel bleeding. Cross-clamping the thoracic aorta has been advocated as a routine part of resuscitative thoracotomy to improve cerebral and coronary perfusion, but the exposure may be difficult, and prolonged

Table 1 Outcome of resuscitative thoracotomy: predictive factors in patients who arrive in cardiac arrest

Negative predictors	Positive predictors
Absent vital signs at scene	Tamponade
Blunt trauma	Stab wound
No intubation	Isolated chest trauma
Extrathoracic injury	Vital signs at scene
Asystole	Signs of life on arrival
Dilated pupils	Cardiac activity by ECG

occlusion may result in both increased ventricular strain and splanchnic ischemia. Aortic occlusion may be better performed by temporary hand compression. Patients who have had emergency thoracotomy and cardiopulmonary resuscitation should be monitored for postpericardiectomy syndrome requiring non-steroidal anti-inflammatory agents. If acute neurological or ECG changes occur, or if there is arrest after intubation, consideration should be given to the occurrence of air embolism, necessitating resuscitative thoracotomy and hilar cross-clamping.

After thoracotomy or sternotomy, initial control of cardiac wounds can be achieved with digital pressure, skin staples, or placement of a Foley catheter. Definitive repair is obtained by using sutures buttressed with Teflon or pericardial pledgets. Distal coronary artery injuries can be managed by ligation, but more proximal injuries (usually the left anterior descending artery) may require cardiopulmonary bypass for repair. Traumatic penetrating septal defects can be managed expectantly, but left-sided valvular injuries will almost invariably require formal repair or valve replacement. Postoperative support may include an intra-aortic balloon pump or a partial bypass using heparin-bonded tubing.

Most thoracic injuries require only a chest tube (36–40 French) to allow complete drainage of blood and air. Open wounds can be temporarily covered with a gauze taped along three sides to create a 'flutter valve' prior to tube placement. The critical technical factors include preparation of the chest wall, placing the tube in the mid or anterior axillary line at the nipple line, using a finger to probe the thorax before placing the tube, directing the dissection over the top of a rib and avoiding the neurovascular structures that lie inferior to the ribs, ensuring that the 'last hole' is within the thorax (this can be done rapidly by measuring the tube from the entry site to the suprasternal notch and trimming if needed), and ideally placing the tube to lie posteriorly and apically. In patients who are unstable, time should not be wasted in trying to create a 'tunnel'; rather, one should go directly into the interspace below the skin incision. A thoracotomy is required if there is a massive air leak, an initial hemothorax greater than 2000 ml, or an initial hemothorax greater than 1000 ml followed by outputs of 300 ml/h for 2 to 3 h. Autotransfusion devices using a 40-μm filter should be considered. Patients requiring urgent exploration may be best managed with single-lumen intubation which avoids the risk of extending a bronchial tear (Wagner *et al.* 1996). If necessary, the tube can be advanced down the opposite bronchus. In stable patients with unilateral thoracic injuries without evidence suggestive of major tracheobronchial injury, double-lumen intubation and independent-lung ventilation is helpful.

Parenchymal injuries caused by low-velocity bullets do not usually need operative intervention, but pneumorrohaphy and individual ligation of the vessels are favored if persistent bleeding occurs. High-velocity missile and blast injuries, including shotgun injuries, may require complex debridement or lobectomy. Major hilar injuries should be managed by immediate clamping to avoid bleeding and air embolism, with early pneumonectomy if the lung cannot be salvaged. Postoperative interventions after pneumonectomy may include fluid restriction, diuretics, treatment of right ventricular failure with pulmonary vasodilating agents, and possibly selective intubation of the opposite bronchus to avoid barotrauma. Interval rethoracotomy to 'irrigate' the thorax and buttress the bronchial stump with a tissue pedicle may decrease the risk of empyema and bronchopleural fistula.

Patients with severe unilateral lung injury and poor compliance can be managed with placement of the 'good' lung upwards, independent lung ventilation, or selective intubation of the good lung. Severe post-traumatic respiratory distress may require diverse strategies such as the use of high-frequency positive-pressure ventilation, pressure control–inverse ratio ventilation, permissive hypercapnia, or occasionally extracorporeal membrane oxygenation. Rotating the patient (or bed) at intervals of 2 h may also be helpful. Pulmonary hypertension may require vasodilating agents such as dobutamine, isoproterenol (isoprenaline), nitroglycerin, or inhaled nitric oxide, although, in the setting of unilateral injury, pulmonary vascular dilation may increase shunt. Persistent hemothorax can be managed early by thoracoscopic drainage and irrigation, particularly if there is a question of secondary infection, but may require formal thoracotomy. Moderate hemothoraces (less than a third of the thorax) may resolve without significant fibrothorax if left alone. Traumatic psuedocysts are managed expectantly unless they become infected, in which case percutaneous drainage may suffice.

Hemoptysis may occur as a consequence of pulmonary contusion or major vascular injury. In the former situation, the hemoptysis is not usually 'exsanguinating'. If hemoptysis is persistent, the patient should be placed in the reverse Trendelenburg position, have supplemental oxygen administered as well as systemic antibiotics, and undergo bronchoscopy. With severe hemoptysis, the injured side should be placed dependent to prevent the uninjured lung from filling, and in the operating room a bronchial blocker should be placed to isolate the affected side. If lung isolation controls the bleeding, the patient should remain intubated and chemically paralyzed. Serial chest radiographs should be obtained to rule out hidden hemothorax. If there is evidence of continued bleeding after 24 h, or if the initial bleed was significant, pulmonary angiography with embolization should be considered. Persistent bleeding and/or inability to perform embolization is an indication for thoracotomy.

Esophageal injuries, whether cervical or intrathoracic, generally require urgent surgical intervention. Small shotgun pellet injuries without documented leak may be managed non-operatively (Fig. 1). If the injury is small, primary repair is sufficient. Larger injuries should be treated with resection, diversion, and/or reconstruction. Within 24 h, or if contamination is minimal and the patient is stable, primary repair with tissue buttressing (Grillo flap) or resection with primary repair or reconstruction is performed. Injuries initially operated on more than 24 h postinjury often require esophagectomy or exclusion with reconstruction later.

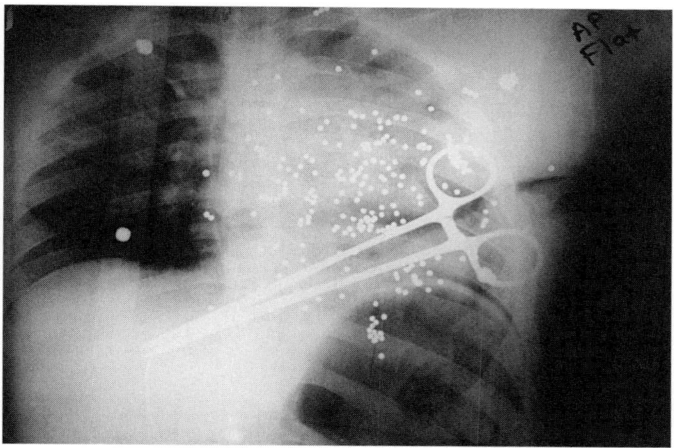

Fig. 1 This patient sustained a shotgun injury to the left chest. During laparotomy, esophagoscopy revealed a pellet 35 cm from the incisors. No leak was demonstrated by insufflation or injection of methylene blue. Postoperative Gastrograffin and thin barium studies also failed to demonstrate a leak. This injury was managed successfully non-operatively.

Abdominal injuries

Low-velocity injuries to both large and small bowel can usually be managed by primary repair, or by resection and anastomosis. Most civilian penetrating injuries do not require colostomy, although high-velocity missile injuries require more extensive debridement. AK47 injuries are characterized by multiple perforations, while M-16 injuries result in large contiguous areas of devascularization mandating stoma creation (Bellamy and Zajtchuk 1991). If extensive contamination has occurred, a second-look laparotomy after 24 to 48 h should be considered.

Most low-velocity missile injuries to the liver can be managed by drainage alone, hepatorrhaphy with ligation of vessels, or omental packing. High-velocity injuries result in more extensive damage, requiring radical debridement, possible segmentectomy or hepatectomy, or 'damage control' (packing with plans for a second-look operation). Transection of the extrahepatic biliary ducts can be managed by primary repair or biliary bypass. Postoperatively the patient should be monitored for possible abscess formation, bile leak, or cholangitis. Pancreatic injuries can be usually managed by resection and/or drainage. Complex pancreaticoduodenal injuries may require pancreaticoduodenectomy or duodenal exclusion. Octreotide to decrease pancreatic secretions may reduce the risk of pancreatic fistula formation, and has been suggested as an unproven adjunct.

Intra-abdominal vascular injuries are best managed by primary repair. Although contamination is generally a contraindication to prosthetic grafts, they have been used successfully when primary repair or the use of autogenous material is not possible. Infrarenal caval injuries can be ligated, repaired, or patched with saphenous vein. Injuries at the level of or above the renal veins require repair and carry a high mortality. Temporary vascular control may be achieved using shunts or a caval–axillary vein bypass with heparin-bonded tubing. As a final tool, circulatory arrest may be necessary.

Injuries to the extremities

Bleeding from the extremities is best controlled by direct pressure or intermittent application of a tourniquet. Prolonged use of a tourniquet can aggravate ischemia, nerve injury, and tissue death, and should be avoided if at all possible. Arterial injuries can be primarily repaired, resected and grafted, ligated, or bypassed, depending on the condition of the patient and the nature of the wound. Low-velocity injuries require minimal debridement and resection; primary anastomosis is frequently possible. High-velocity injuries result in much longer segments of intimal damage and grafting is usually needed. Options for bypass include autologous vein (e.g. saphenous) or prosthetic material; the latter has the disadvantage of decreased patency as well as increased risk of infection (Fig. 2). All arterial repairs should be covered. If the graft fails, or if temporary ligation is required because of the patient's overall condition, a bypass may be performed in a more controlled setting if there is the possibility of salvaging an ischemic limb. Postoperatively, patients should receive antiplatelet medication. When ligation is necessary, amputation rates depend on the site of injury and the available collateral supply. Limb loss occurs in 80 per cent of cases when both the superficial and deep femoral arteries are ligated, while ligation of the femoral artery alone results in limb loss rates of 25 to 50 per cent. If major venous injury requires ligation, a significant proportion of patients will experience limb swelling and edema which may progress to compartment syndrome. Management may include elevation, possible fasciotomy, and consideration of anticoagulation therapy (Shackford and Rich 1991).

High-velocity gunshot wounds or blast injuries resulting in complex orthopedic and arterial injuries are technically more challenging, and an attempt should be made to keep the ischemic time below 6 to 8 h. Approaches include stabilizing with external fixation devices followed by vascular repair, using a shunt during bone stabilization, and vascular repair followed by bone fixation.

Simple extremity swelling may be managed conservatively with elevation and ice packs. Compartment syndrome requires complete fasciotomy. A 'four-compartment' fasciotomy, using two incisions or one incision with fibulectomy, is required in the leg below the knee. The muscle should be viable and bulge freely after the fasciotomy. The wound is closed electively (primary and/or skin grafting).

Injuries caused by bombs and other explosives

Most bomb injuries involve combinations of thermal and orthopedic injuries. If possible, definitive surgery should be delayed for at least 24 h if there is suspicion of primary blast lung injury (Bellamy and Zajtchuk 1991; Stapczynski 1992; Karmy-Jones *et al.* 1994). If laparotomy is required, and the abdomen has suffered a thermal injury, the midline approach is best. The wound can be closed with stainless steel retention wires, which can be adjusted to allow for intra-abdominal swelling and, hopefully, prevent intra-abdominal hypertension. If a colostomy is required, it should not be matured until abdominal distension begins to diminish (Fig. 3). If a pulmonary blast injury is suspected, high-pressure ventilation and high-altitude aeromedical transport should be avoided if possible to prevent air embolism. The treatment of air embolism is supportive, with hyperbaric oxygen providing some benefit.

Land mines often result in traumatic amputations. Aggressive debridement and close monitoring of the wound are necessary.

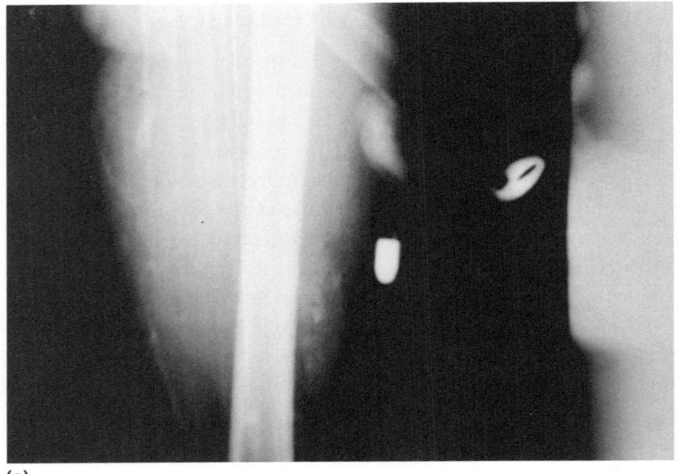

(a)

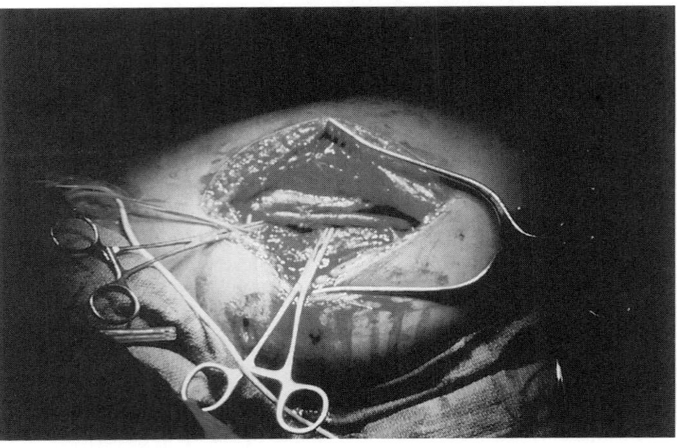

(b)

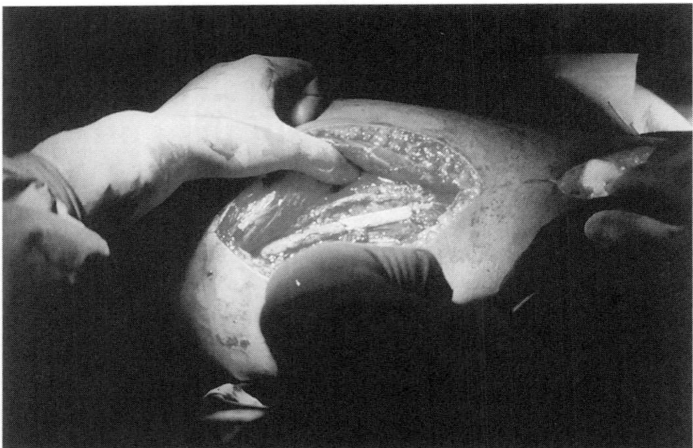

(c)

Fig. 2 (a) Emergency department arteriogram of a patient who sustained a gunshot wound that traversed the right and left thighs. The bullet is seen lying laterally in the left thigh and the study demonstrates an injury to the superficial femoral artery. (b) Operative findings demonstrating a transected superficial femoral vein and contused superficial femoral artery. (c) The vein has been ligated and the artery has been grafted with Gortex after debridement.

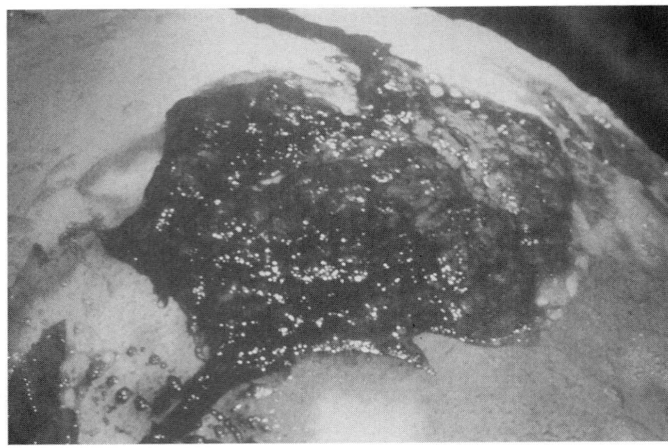

(a)

(b)

Fig. 3 (a) A patient who sustained second- and third-degree burns to his abdomen and chest from a parcel device. He also received penetrating injuries from metal fragments and presented with omental evisceration. (b) The abdomen was explored through a midline approach. The wound was closed with stainless steel retention sutures (another option would be to use a Silastic bag). The ostomy was not matured initially.

Grenade injuries can usually be managed similarly to shotgun injuries as debridement is not always required.

Complicated issues of wound management

Debridement is dictated by the magnitude of tissue destruction and contamination, the presence of foreign material (e.g. shotgun wadding) and/or tissue necrosis, and the time interval between injury and definitive treatment (Fig. 4). Low-velocity wounds usually only require local debridement of the entrance and exit sites. Wounds secondary to high-velocity missiles and bombs, and multi-fragment wounds (shotgun, grenade), require operative debridement more frequently.

High-velocity injuries alone are not an indication for debridement as many can be managed by drainage, antibiotics, and dressing

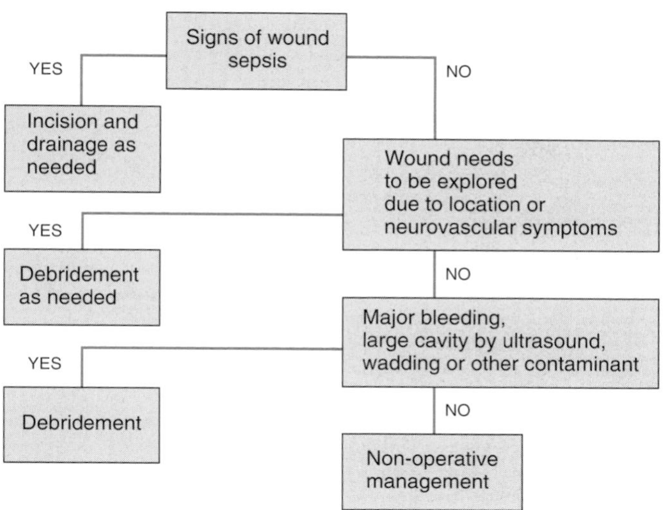

Fig. 4 Management of soft tissue wounds.

changes (Bellamy and Zajtchuk 1991). During debridement, devitalized tissue, determined by loss of muscle contractility, cyanosis, and/or lack of bleeding, should be carefully resected. Entrance and exit sites should be opened longitudinally, and the wound possibly opened to the deepest point. Fasciotomy and a second-look procedure should be considered for extreme injuries. If tissue destruction and/or contamination is too extensive, 'guillotine' amputation may be required, with the amputated limb being a potential source for donor skin. Defects in the chest and abdomen can be closed using muscle flaps, prosthetic mesh, and/or methylmethacrylate sandwiches. Tetanus prophylaxis should be given early. Systemic antibiotics appear to be effective in controlling initial contamination but do not prevent subsequent colonization of an open wound. Topical antibiotics and burn cream, such as silver sulfadiazine, are useful in small superficial wounds, while frequent saline dressing changes and/or whirlpool treatments can keep larger wounds clean. If necrotizing infections or wound sepsis develop radical debridement, systemic antibiotics and aggressive resuscitation directed by invasive monitoring may be required. Hyperbaric oxygen therapy is still unproven. High-velocity injuries to the gluteal area present a particular risk for massive tissue infection and a diverting colostomy may be required (Bellamy and Zajtchuk 1991). Ultimately, all wounds can be closed by secondary intention, skin grafts, omental grafts, or musculocutaneous flaps (pedicle, advancement, and/or free).

Myoglobinuria can be a complication of both compartment syndrome and massive soft tissue destruction. Prevention includes early fasciotomy and amputation rather than attempts at limb salvage. The aim is to prevent renal failure and to maintain a urine flow rate above 100 ml/h with aggressive hydration. Mannitol may be useful, acting as an osmotic diuretic as well as an O_2 radical scavenger. The use of sodium bicarbonate and furosemide (frusemide) remains controversial. Diuresis should be maintained until serum creatine phosphokinase levels return to near normal and the urine myoglobin clears (Fig. 5). Electrolytes, particularly potassium and calcium, should be monitored.

Retained missiles and foreign objects may present a particular problem. 'Dum-dum', hollow-point, and other missiles become deformed and lose up to 30 per cent of their mass after impact, resulting in complex injury tracts. 'Exploding' bullets may be triggered by ultrasound or cautery. If such a bullet is imbedded in an organ, resection of the organ rather than direct removal is the safest method of retrieval (Sykes *et al.* 1988). Lead poisoning is rarely seen with modern bullets, but the risk is increased if the bullet is retained in a synovial joint or is imbedded in the bones of a patient with osteoclastic lesions. Phosphorus grenades and bombs filled with soap may leave residue in the wound, so that debridement in a water bath is required. Until definitive surgery, the area should be soaked in saline gauze.

Missile embolization is a rare event that may occur following shotgun or small-bullet injury. In the Vietnam War, missile embolization occurred in 22 of 7500 vascular trauma patients with 80 per cent arterial embolization and 90 per cent migrating antegrade (i.e. towards the lung) (Bellamy and Zajtchuk 1991; Shackford and Rich 1991). If the missiles are large and/or are associated with pulmonary symptoms, treatment involves extraction by interventional radiology or operatively. In those rare patients with a patent foramen ovale, the

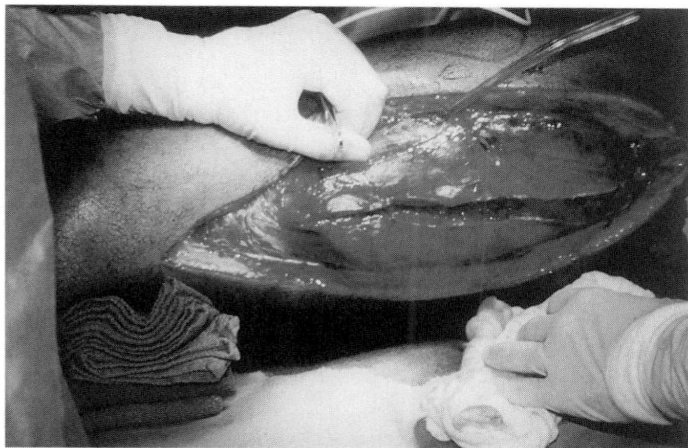

(a)

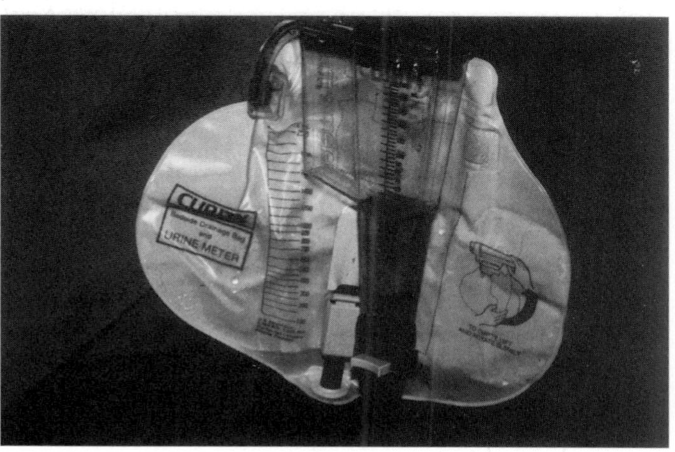

(b)

Fig. 5 (a) This patient required fasciotomies of the calf and thigh compartments following an electrical injury that resulted in compartment syndrome. (b) Dark urine characteristic of myoglobinuria in the patient in Fig. 3(a).

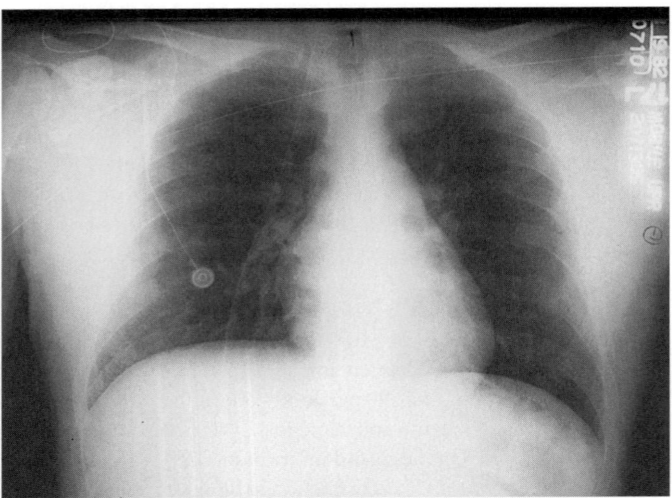

Fig. 6 A patient who had sustained a shotgun wound to the right thigh without obvious vascular trauma developed chest pain 24 h later. ECG suggested acute ischemia in the anterolateral area (left anterior descending distribution). This chest radiograph demonstrates pellets in the left chest. Coronary angiography confirmed coronary artery embolization, presumably through a patent foramen ovale. Operative removal was successful.

missiles may cross to the systemic circulation, subsequently embolizing to coronary, cerebral, or other vessels. Small pellets entering the coronary circulation can result in myocardial ischemia that may be severe enough to warrant operative extraction and coronary bypass (Fig. 6).

References

Bellamy, R.F. and Zajtchuk, R. (1991). The management of ballistic wounds of soft tissue. In *Textbook of military medicine*. Part I, Vol. 5, *Conventional warfare: ballistic, blast, and burn injuries* (ed. R.F. Bellamy and R. Zajtchuk), p. 163. Office of the Surgeon General, Washington, DC.

Karmy-Jones, R., Kissinger, D., Golocovsky, M., Jordan, M., and Champion, H.R. (1994). Bomb-related injuries. *Military Medicine*, **159**, 536–9.

Shackford, S.R. and Rich, N.H. (1991). Peripheral vascular injury. In *Trauma* (2nd edn) (ed. E.E. Moore, K.L. Mattox, and D.V. Feliciano), p. 639. Appleton & Lange, Norwalk, CT.

Stapczynski, J.S. (1992). Blast injuries. *Annals of Emergency Medicine*, **11**, 687–94.

Sykes, L.N., Champion, H.R., and Fouty, W.J. (1988). Dum-dums, hollow points, and devastators: techniques designed to increase wounding potential of bullets. *Journal of Trauma*, **28**, 618–23.

Wagner, J.W., Obeid, F.N., Karmy-Jones, R., Casey, G.D., Sorenson, V.J., and Horst, H.M. (1996). Trauma pneumonectomy revisited: the role of simultaneously stapled pneumonectomy. *Journal of Trauma*, **40**, 590–4.

11.3 Head injury

11.3.1 Pathophysiology

J. Jallo and R. K. Narayan

Key messages

- Brain injury is classified as focal (intracerebral hemorrhage, subdural or epidural hematoma) or diffuse (diffuse axonal injury). Focal injuries are more amenable to surgical intervention.
- Primary injury occurs at the moment of injury and includes contusion, laceration, diffuse axonal injury, and epidural and subdural hemorrhage.
- Secondary injuries are due to secondary processes, such as those caused by excitatory amino acids, free radicals, and nitric oxide, or secondary insults, such as hypotension, hypoxia, hyponatremia, and hypothermia.
- Maintenance of cerebral perfusion pressure is important in maximizing the probability of a favorable outcome following traumatic brain injury.

Introduction

Traumatic brain injury is a leading cause of disability and death in developed countries and increasingly in developing countries. The annual death rate from traumatic brain injury ranges from 14 to 30 per 100 000 population. Persons at the highest risk of brain injury are aged between 15 and 24 years, with the incidence in males being two to three times greater than that in females. The cost to society is great given the lost potential productivity of these usually young individuals, in addition to the need for prolonged support by family and community for those with persistent disability.

A number of clinical classifications of brain injury have been developed to help understand the pathology and predict outcome. The most widely used is the Glasgow Coma Scale (GCS). Although the postresuscitation GCS score correlates well with outcome, it does not provide the clinician with a morphological picture of the underlying brain injury.

Morphologically, brain injury may be broadly classified as either focal or diffuse, although these terms are not mutually exclusive. Focal injury implies intracerebral hemorrhage, subdural hematoma, or epidural hematoma; diffuse injury generally refers to clinical and surrogate radiographic manifestations of diffuse axonal injury. Diffuse axonal injury classically consists of focal lesions in the dorsolateral quadrants of the midbrain or the corpus callosum, and microscopic damage to axons. The distinction between focal and diffuse injury is clinically important in that focal injuries are potentially amenable to surgical intervention.

Over the past two decades, the concept of primary and secondary brain injury has acquired greater significance. Primary brain injury refers to damage that occurs at the moment of injury and includes brain contusion or laceration, diffuse axonal injury, epidural hemorrhage, and subdural hemorrhage. Secondary injury is caused by events that follow the initial brain insult and adversely affect outcome. Secondary injury can be due to secondary insults or to secondary processes that are set in motion by the initial trauma. Secondary insults include hypotension leading to reduced cerebral perfusion, brain edema, delayed hematomas, elevated intracranial pressure, hyperthermia, seizures, hyperglycemia, infections, and iatrogenic misadventures. These can be avoided or attenuated by careful attention to detail. These secondary insults should be distinguished from secondary processes that occur at a biochemical level in the brain in the minutes, hours, and days following injury.

Primary and secondary injury

Primary injury

The events that occur at the moment of injury, although possibly preventable, are not reversible. However, brain contusions and hemorrhages are potentially avoidable causes of death and disability if evacuated promptly. Intracranial hemorrhage is classified based on location as being intracerebral (within the brain parenchyma), subdural (beneath the dura mater covering the brain surface), or epidural (outside the dura mater). Combinations of these hemorrhages are not uncommon. Epidural hematomas are classically associated with hemorrhage from a lacerated middle meningeal artery with formation of a lenticular hematoma in the temporal fossa (Fig. 1(a)). Other causes of epidural hemorrhage include diploic bleeding from a skull fracture or injury to a venous sinus. Clinically, epidural hematomas classically present with a lucid interval where the patient may be alert initially but deteriorates to coma because of the rapidly increasing epidural mass. Prompt evacuation of a rapidly enlarging epidural hematoma is not only lifesaving, but may allow good recovery. Patients with epidural hematomas have the greatest potential for good recovery from prompt hematoma evacuation because the underlying brain is usually normal.

Acute subdural hematomas (Fig. 1(b)) are more common than epidural hematomas and are generally associated with a poorer prognosis. Hemorrhage in the subdural space frequently results from injury to bridging veins that drain blood from the cortical surface to the dural sinuses, or from lacerations of the brain surface. Prompt

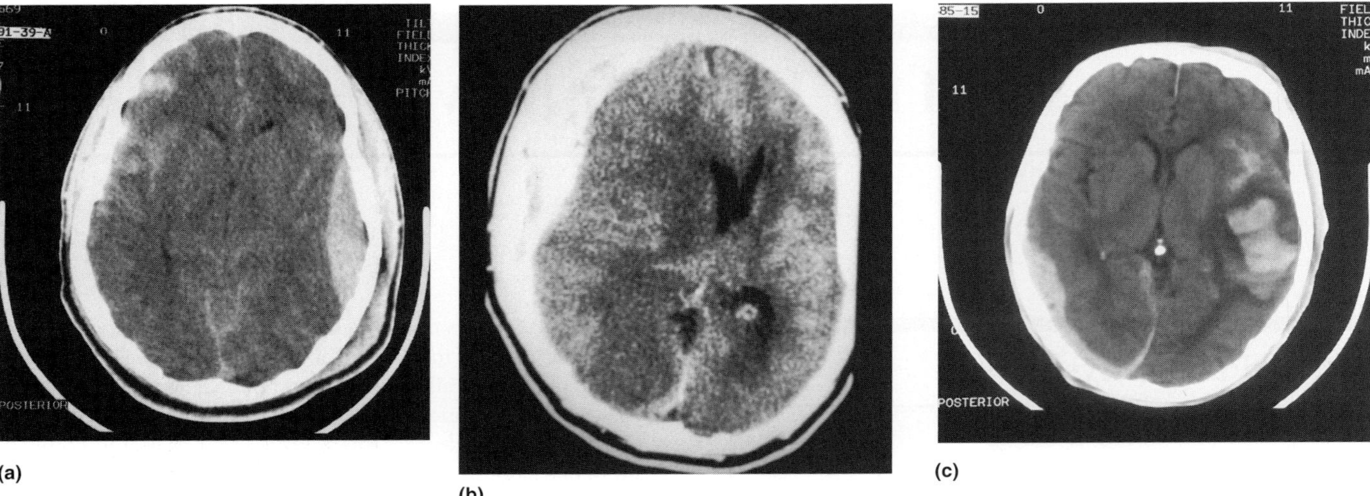

Fig. 1 Axial CT scans of (a) a left temporal acute epidural hematoma with compressed cisterns and contralateral contusions, (b) a right hemispheric subdural hematoma with a shift in midline, and (c) a left frontal and temporal intracerebral contusion and hematoma with an associated right subdural hematoma.

evacuation of an acute subdural hematoma may help to reduce the high mortality associated with these lesions.

Contusions or intracerebral hematomas (Fig. 1(c)) are common after traumatic brain injury and are frequently seen in the frontal and temporal lobes. They often coexist with subdural hematomas and develop over time; hence it is important to perform a follow-up CT scan 4 to 8 h after the initial CT scan in patients with demonstrated brain injury.

Secondary injury

Secondary neuronal injury after traumatic brain injury has recently received much attention, with new mechanisms being elucidated and previously described mechanisms being understood better. At a macroscopic level, secondary phenomena include edema, ischemia, necrosis, elevated intracranial pressure, and inadequate cerebral perfusion. At a cellular level, energy failure occurs with a cascade of events that contribute to secondary injury, including elevated levels of intracellular calcium, release of excitatory amino acids, generation of free radicals, and breakdown of the cellular cytoskeleton and membrane with vascular dysfunction eventually leading to cell death (Fig. 2).

Excitatory amino acids

The initial event in the sequence is believed to be the release of excitatory amino acids. Glutamate is the most studied of the excitatory amino acids, and the mechanism of injury has been termed excitotoxic injury. Following brain injury, excitatory amino acids are released from injured cells. Glutamate acts postsynaptically at five receptor subtypes which are described on the basis of the agent that specifically activates them. The *N*-methyl-D-aspartate (NMDA) receptor complex is an ion channel that allows passage of calcium and sodium ions. When activated by glutamate, the NMDA receptor allows calcium ions to enter the cell. Large influxes of calcium ions into neurons stimulate calcium-dependent enzymes with activation of proteases, kinases, phospholipases, and nitric oxide synthase. If unchecked, these processes can eventually lead to cell death with breakdown of the cytoskeleton, free-radical formation, alterations in gene expression and protein synthesis, and membrane dysfunction. Blocking the NMDA receptor improves neuronal survival *in vitro* and in animal models of neuronal injury. NMDA blockers are currently undergoing clinical trial.

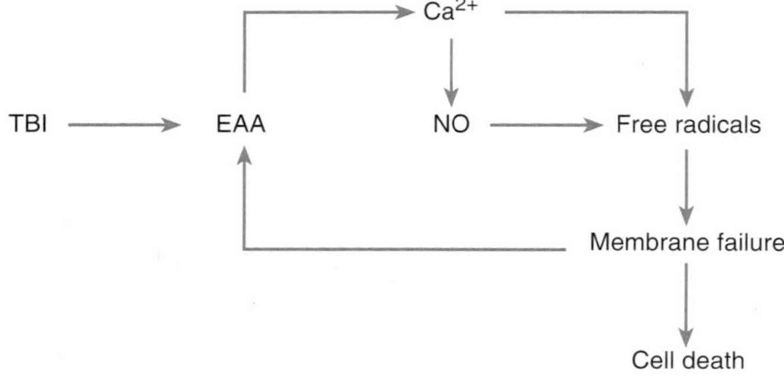

Fig. 2 Mechanisms of secondary injury: TBI, traumatic brain injury; EAA, excitatory amino acids; NO, nitric oxide.

Free radicals

Free-radical formation is an integral aspect of many mechanisms of secondary injury. The most common free radicals studied are superoxide (O_2^-), hydrogen peroxide (H_2O_2), hydroxyl (OH^-), and nitric oxide (NO). Free radicals are atoms or molecules with an unpaired electron in the outer orbit, making them highly reactive. Free iron is an important catalyst of free-radical-mediated injury and is readily available in injured and contused brain tissue. Free radicals damage endothelial cells and injure the brain parenchyma. In so doing, they disrupt the blood–brain barrier and are partly responsible for both vasogenic and cytotoxic edema. Once initiated, free-radical injury is a self-perpetuating process with increasing damage which generates more free radicals. Cells attempt to limit the injury caused by free radicals by binding them with free-radical scavengers such as vitamin E, ascorbic acid, superoxide dismutase, etc. However, these coping mechanisms may be overwhelmed in areas of severe brain injury or ischemia. Pharmacological agents that scavenge free radicals have been effective in reducing neuronal damage in animal models of brain injury. Unfortunately, a human trial of the free-radical scavenger PEG-SOD (polyethylene glycol superoxide dismutase) in severe head injury did not demonstrate an improved outcome.

Nitric oxide

Nitric oxide, a ubiquitous second messenger, has received much attention recently and has been implicated in a variety of cellular processes including vascular relaxation, neurotransmitter effects, and cytotoxicity. It is formed via the conversion of L-arginine to L-citrulline and nitric oxide catalyzed by nitric oxide synthase. The mechanism of nitric oxide cytoxicity is unclear, but it is closely associated with NMDA, calcium, and free-radical-mediated injury. Inhibitors of nitric oxide synthase have been found to reduce neuronal injury in animal models of brain injury. Many other mediators of secondary brain injury are under investigation including catecholamines, adenosine, cytokines, opioid peptides, and thyrotropin-releasing hormone.

Physiological response to brain injury

Brain edema

The cellular mechanisms of secondary brain injury described above contribute to brain edema, and subsequently to increased intracranial pressure, decreased cerebral perfusion, and cerebral ischemia. Brain edema has classically been described as either vasogenic or cytotoxic. Vasogenic edema results from disruption of the blood–brain barrier which is maintained by tight junctions between the endothelial cells that line the vessels of the brain. Injury to these cells allows extravasation of fluid and proteins into the interstitial space of the brain parenchyma. Disruption of endothelial cells may be primary, resulting from the initial impact or subsequent hemorrhage, or secondary, resulting from free-radical generation, cytokines, and other secondary mechanisms of injury.

Cytotoxic or cellular edema is edema of the cells themselves, whereas vasogenic edema is edema of the interstitial space. Cytotoxic edema results from failure of cellular ion homeostasis and membrane function. Secondary injury at a cellular level results in loss of ion homeostasis and membrane dysfunction. The time course of brain edema is variable. However, it is believed that vasogenic edema occurs early after injury and cytotoxic edema occurs in a more delayed fashion. Brain edema is often maximal 24 to 48 h after traumatic brain injury.

Brain edema is significant as a marker for injury and also as a cause of secondary injury. This is because the brain is in a closed container—the skull. Intracranial pressure is determined by the volumes of three elements contained in the skull: the brain parenchyma, the blood, and the cerebrospinal fluid. An increase in the volume of any of these elements will result in increased intracranial pressure after compensatory mechanisms are exhausted. As intracranial pressure rises, there is an initial passive release of cerebrospinal fluid into the spinal canal. There is also an innate compliance of the craniospinal axis that allows for an increase in intracranial contents without an initial increase in intracranial pressure. However, once the compensatory reserve is exhausted, there is a precipitous rise in intracranial pressure.

Intracranial pressure and cerebral blood flow

Intracranial pressure in an adult is normally below 15 mmHg. Sustained elevations in intracranial pressure above 20 mmHg are not well tolerated by the injured brain and have been associated with increased mortality. Sustained elevation in intracranial pressure may result in cerebral ischemia if cerebral perfusion is hampered and, if severe, can result in brain herniation. Brain herniation may occur across the falx and involve the cingulate gyrus, across the tentorial hiatus and involve the uncus, or through the foramen magnum and involve the cerebellar tonsils.

Cerebral ischemia results from inadequate cerebral perfusion. The ideal degree of cerebral perfusion in the injured brain is unknown. In the uninjured brain, cerebral blood flow is tightly regulated by myogenic, humoral, and neural mechanisms to maintain a constant flow despite changes in systemic arterial pressure and intracranial pressure. This autoregulation functions over a broad range of systemic arterial pressures from a lower limit of 65 mmHg to an upper limit of 140 mmHg and maintains cerebral blood flow at approximately 50 ml/100 g/min. If blood flow falls below 25 ml/100 g/min, electrical activity is lost and the electroencephalogram is flat, below 12 ml/100 g/min brainstem-evoked responses are lost, and below 10 ml/100 g/min brain failure results with loss of ion homeostasis and eventual cell death.

Cerebral perfusion pressure

It is of paramount importance to maintain an adequate cerebral blood flow to the injured brain. Unfortunately, there is no simple technique that allows reliable continuous bedside measurement of cerebral blood flow. As a result, the concept of cerebral perfusion pressure is invoked when discussing cerebral blood flow to the injured brain (Fig. 3). Cerebral perfusion pressure (CPP) is defined as the difference between mean arterial pressure (MAP) and intracranial pressure (ICP):

$$\text{CPP} = \text{MAP} - \text{ICP}.$$

The concept of autoregulation can be extended to discussions of cerebral perfusion pressure in the uninjured brain. Autoregulation in terms of perfusion pressure has a lower limit of 50 mmHg and an upper limit of approximately 140 mmHg. With a perfusion pressure below 50 mmHg, cerebral arterioles are maximally dilated and perfusion cannot match the brain's metabolic needs. If this state persists, brain ischemia follows. As previously noted, the brain tolerates poorly ischemic conditions which aggravate secondary injury. If the perfusion pressure is above 140 mmHg, cerebral arterioles are maximally constricted and blood flow increases passively with perfusion pressure.

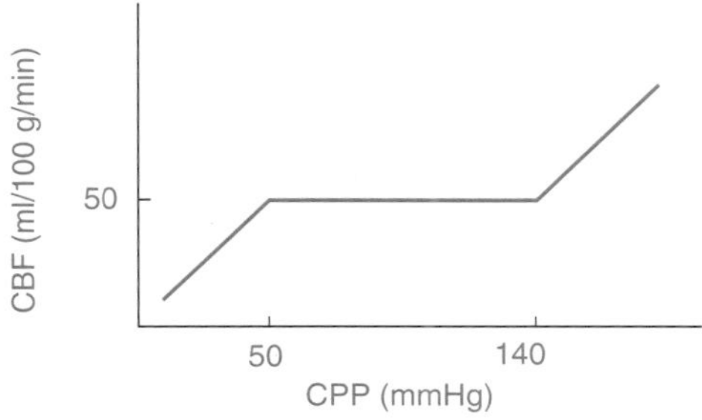

Fig. 3 Cerebral autoregulation curve: CBF, cerebral blood flow; CPP, cerebral perfusion pressure.

Discussions of cerebral autoregulation are generally predicated on the brain's physiological responses to changes in systemic arterial pressure. However, in these discussions it is assumed that the physiological responses occurring in an uninjured brain are also in place after brain injury. This has been a topic of debate and study. Up to half of severely head-injured patients will have some autoregulatory impairment. The autoregulatory curve appears to be shifted to the right with an increase in the lower limit at which autoregulation is effective. Patients with severe head injury probably require a cerebral perfusion pressure of 60 to 70 mmHg to maintain autoregulation of cerebral blood flow and prevent ischemic complications. Although no randomized trials assessing outcome based on cerebral perfusion pressure have been completed, a number of observations can be made from retrospective and historically controlled studies in severely head-injured individuals.

1. Outcome is significantly worse in patients with a cerebral perfusion pressure below 60 mmHg.
2. A significantly better outcome and survival rate are achieved if the cerebral perfusion pressure is above 80 mmHg in the first 48 h after head injury.
3. A more favorable outcome has also been anticipated when cerebral perfusion pressure is maintained above 70 mmHg.

Despite the absence of a prospective randomized study, the available data suggest that it is best to maintain cerebral perfusion pressure at 60 to 70 mmHg.

Clinical management and cerebral perfusion pressure

The brain tolerates ischemic conditions poorly. Frequently, the need to maintain cerebral perfusion pressure necessitates the use of vasopressors directed by the patient's cardiac condition. Elevations in intracranial pressure may be treated directly with mannitol and mild hyperventilation if refractory to ventricular drainage. Hypocapnia is a potent vasoconstrictor and controls elevations in intracranial pressure by limiting cerebral blood flow. Unfortunately, at $Pa\text{CO}_2$ levels below 25 mmHg (3.5 kPa), it may promote ischemic injury. Thus, although it is effective in reducing intracranial pressure, hyperventilation should not be allowed to reduce $Pa\text{CO}_2$ below 25 mmHg (3.5 kPa). Mannitol is also effective in reducing intracranial pressure. Its immediate effect on intracranial pressure has been ascribed to a rheological property. It also functions as a volume expander to improve perfusion pressure and as an osmotic diuretic which removes extravascular water from the brain. Bolus doses of 0.25 g/kg are effective in reducing intracranial pressure and may be repeated as needed. Care needs to be taken to monitor serum osmolality and to prevent levels from increasing above 320 mosmol. It is important to maintain a euvolemic state with the use of mannitol. Dehydration may compromise cerebral perfusion and could have other deleterious consequences. Fluid management is directed towards maintaining an adequate intravascular volume and avoiding a hyposmolar state, i.e. hypotension and hyponatremia must be avoided.

Conclusion

The ultimate outcome following a head injury is a function of the severity of the primary injury and the subsequent secondary insults and processes. Mortality from severe head injury has declined over the past two decades primarily as a consequence of improved critical care.

Bibliography

Cormio, M., Robertson, C.S., and Narayan, R.K. (1997). Secondary insults to the injured brain. *Journal of Clinical Neuroscience*, **4**, 132–48.

Lynch, D.R. and Dawson, T.M. (1994). Secondary mechanisms in neuronal trauma. *Current Opinion in Neurology*, **7**, 510–16.

Marmarou, A. (1994). Traumatic brain edema: an overview. *Acta Neurochirurgia*, **60** (Supplement), 421–4.

Miller, J.D., Piper, I.R., and Jones, P.A. (1996). Pathophysiology of head injury. In *Neurotrauma* (ed. R.K. Narayan, J.E. Wilberger, and J.T. Povlishock), pp. 61–70. McGraw-Hill, New York.

Siesjo, B.K. (1993). Basic mechanisms of traumatic brain damage. *Annals of Emergency Medicine*, **22**, 959–69.

11.3.2 Assessment

J. Jallo and R. K. Narayan

Key messages

- The cornerstone of assessment is the history and physical examination, notably the Glasgow Coma Scale score. Ancillary studies include CT scanning and, when indicated, monitoring of the intracranial pressure.
- CT scanning is the primary radiographic modality used for acute head injury, although magnetic resonance imaging may play a role.
- Intracranial pressure monitoring is a useful diagnostic and therapeutic guide.

Introduction

A system for evaluating the brain-injured patient is important for establishing the severity of injury, determining the need for surgery, monitoring progress, and assessing prognosis. A proper classification of the type of injury sustained allows reliable communication between medical personnel in the same and different centers regarding patients' conditions and outcomes. The clinical examination and the cranial CT scan are the two most important means of assessing these patients. The neurological examination should be brief and focused, and should be followed quickly by the CT scan. Laboratory tests such as complete blood count, serum electrolytes, blood urea nitrogen (BUN), glucose, toxicology screen, and arterial blood gases are usually performed in severely injured individuals. Cerebral angiography is occasionally used if a vascular injury is suspected. Near-infrared spectroscopy has been reported to be a good screening device for the presence of an intracranial hematoma. Intracranial pressure monitoring forms the basis for guiding subsequent therapies, and jugular venous oxygen saturation can yield important supplemental information on cerebral oxygenation. Other technologies, including positron emission tomography, single-photon emission CT, and microdialysis, are currently primarily of research interest.

Neurological examination

The Glasgow Coma Scale (GCS) (Table 1) is the cornerstone of the neurological evaluation of the brain-injured patient. It is the best studied of the numerous assessment scales that are available. Although it is fraught with significant interobserver variability, the GCS score has been shown in numerous studies to correlate strongly with outcome. The highest possible total score is 15 and the lowest is 3. Patients with a GCS score of 15 or 14 are often classified as having mild head injury, 13 to 9 is moderate, and 8 to 3 is severe. Patients with a severe head injury (GCS score 3–8) are unable to follow simple commands. The GCS score should be supplemented by an assessment of pupillary response to light and brainstem reflexes, i.e. oculocephalic and oculovestibular responses.

The importance of sequential neurological examinations cannot be overemphasized. When neurological assessment is clouded by sedation or paralysis, intracranial pressure monitoring and repeat CT scanning are invaluable.

Radiographic imaging

CT scanning

CT scanning has revolutionized the management of the head-injured patient. It has virtually replaced skull radiography because it provides images of the brain and other soft tissues. CT scanners have become widespread in the developed world and allow for rapid low-cost non-invasive imaging. A CT scanner is now considered essential in the optimal care of patients with brain injuries.

Fractures of the skull and skull base can be readily diagnosed. When skull base fractures are noted, cerebrospinal fluid leaks may be anticipated and injury to the facial nerve should be sought. Depressed skull fractures and those that are exposed due to a scalp laceration require surgical repair. Extracranial hematomas, both epidural and subdural, are readily identified on CT. If they are associated with a shift in midline of approximately 5 mm or more, surgical intervention is generally required. Parenchymal injuries such as contusions and intracerebral hematomas are also diagnosed by CT scan. Contusions often evolve in a delayed fashion, sometimes with associated clinical deterioration hours or even days after the initial injury.

Sequential follow-up studies are recommended after the initial CT scan, particularly when the patient's condition deteriorates. With continued improvement in emergency systems, many patients reach trauma centers within 1 h of the injury. Therefore the initial CT scan may not show the maximum pathology. Consequently, some neurosurgeons advocate routine scanning of patients with moderate and severe head injury 4 to 8 h after presentation in order to identify expanding hematomas and contusions. Other features to look for on a cranial CT scan include evidence of brain swelling such as the compression or absence of the basal cisterns, a shift in the midline, or herniation.

Table 1 Glasgow Coma Scale

Function	Score*
Eye opening	
Spontaneous	4
To voice	3
To pain	2
None	1
Verbal response	
Oriented	5
Confused	4
Inappropriate words	3
Incomprehensible sounds	2
None	1
Best motor response	
Follows commands	6
Localizes pain	5
Withdraws from pain	4
Abnormal flexion	3
Abnormal extension	2
None	1

*Lowest score, 3; highest score, 15.

Magnetic resonance imaging

Magnetic resonance imaging (MRI) is a more sensitive imaging modality than CT but is also more expensive, more technically demanding, and more time consuming. MRI takes approximately 30 to 40 min and requires a co-operative or sedated patient. These constraints generally make MRI impractical in the acute phase of injury. Patients with moderate to severe head injury are frequently uncooperative and require intubation and ventilatory support, making it impossible to obtain MRI scans without the availability of MRI-compatible ventilators. Despite these limitations, MRI is indicated if there is a focal neurological deficit or a prolonged period of unconsciousness not explained by CT. In addition to the conventional T1W and T2W images, other magnetic resonance techniques such as spin echo and gradient echo imaging are available for imaging brain injury. Other techniques of imaging brain injury are under investigation.

MRI is ideal for imaging diffuse axonal injury, which is characterized by multiple small focal areas of damage with increased signal intensity on T2W images. Lesions may be found in the lobar white matter, the corpus callosum, the dorsolateral brainstem, and the internal capsule. MRI is also more sensitive than CT in identifying hemorrhagic and non-hemorrhagic cortical contusions. Another advantage of MRI is its ability to image soft tissue at the skull base and posterior fossa. Imaging brainstem injury is particularly important in patients with prolonged loss of consciousness not explained by CT. In these cases MRI frequently demonstrates shear injury.

Laboratory tests

Consciousness may also be altered by electrolyte derangements, metabolic disturbances, or drug ingestion. Important laboratory studies include a complete blood count, serum electrolytes, glucose, BUN, creatinine, arterial blood gases, prothrombin and partial thromboplastin times, urinalysis, and toxicology screen to check for alcohol, narcotics, benzodiazepines, amphetamines, cocaine, and phencyclidine. Electrolytes are determined to ensure that the patient does not develop hyponatremia which may promote edema in injured brain tissue. It is also necessary to be aware of the syndromes of inappropriate ADH secretion (SIADH) and cerebral salt wasting, and to correct sodium disturbances if they develop. Hypoglycemia must be excluded in comatose patients, and hyperglycemia has been shown to be detrimental in animal models of brain injury. Measurement of arterial blood gases ensures adequate oxygenation and an appropriate response to hyperventilation. Hyperventilation should be mild to avoid excessive cerebrovascular constriction that may hamper cerebral blood flow. Coagulation disorders are frequent after severe head injury and may need to be corrected to prevent further hemorrhage.

Intracranial pressure monitoring

Intracranial pressure (ICP) data form the basis for guiding the management of comatose head injury patients whose clinical examination is usually compromised by sedation and/or chemoparalysis. The Guidelines for the Management of Severe Head Injury published by the American Association of Neurological Surgeons recommend that ICP monitoring be undertaken in patients with a head injury and an abnormal CT scan on admission. ICP monitoring may also be undertaken in patients with severe head injury and a normal CT scan if they have any two of three adverse characteristics including age over 40 years, unilateral or bilateral motor posturing, or systolic blood pressure less than 90 mmHg on admission, and are therefore at a high risk of developing raised ICP.

Once ICP monitoring is deemed necessary, a monitoring device is placed by the neurosurgeon. A variety of such devices are available. The oldest and most commonly used technique is a fluid-coupled catheter placed in the ventricular cavity (ventriculostomy). Newer devices that permit reliable pressure recording when placed in the brain parenchyma or lateral ventricle include fiber-optic catheters and strain gauges. In both these solid state systems, the transducer is at the catheter tip, eliminating the need for a fluid column to transmit pressure from the intracranial cavity to the transducer. Although convenient, these systems cannot be recalibrated once placed intracranially and are prone to measurement drift over time (days). Complications of ICP monitoring include malfunction, obstruction, malposition, infection, and hemorrhage. Bacterial colonization of the ICP monitoring device seems to increase after 5 days of placement. Frequent irrigation of fluid-coupled systems increases the risk of colonization. Although the risk of colonization increases over time, clinically significant intracranial infections are relatively uncommon. Infections are best treated by removing the infected catheter and treating the patient with intravenous antibiotics. Routine removal or changing of catheters at a fixed day after insertion has been debated, with no conclusive evidence as to the value of this policy. Measures to minimize infection include tunneling the catheter subcutaneously away from its insertion site and minimizing the invasion of fluid-coupled systems.

The upper limit of normal ICP is generally considered to be around 15 to 20 mmHg in adults. ICP is generally treated if it is above 20 to 25 mmHg. Treatment of elevated ICP is performed in a stepwise manner and continued until the pressure normalizes. ICP monitoring is continued as long as there is intracranial hypertension or ongoing therapy for intracranial hypertension. Post-traumatic edema is typically maximum at 48 to 72 h. However, it is not unusual for the ICP to remain elevated for days. Monitoring may generally be safely discontinued in patients who maintain a normal ICP for 24 h without requiring specific therapy.

Other monitoring techniques

Recent developments in neurocritical care monitoring are promising. New technology and application of state-of-the-art techniques promise to expand our understanding of the mechanisms of secondary injury and minimize the events that lead to poor outcome. The goal of neurocritical care is simple—to provide adequate nutrition and oxygen to injured brain tissue, and to prevent secondary insults to potentially viable brain. Cerebral blood flow is an indirect measurement of adequate tissue oxygenation and nutrition. Therfore a number of methods of assessing cerebral blood flow are available. These include transcranial Doppler ultrasonography, xenon CT, single-photon emission CT, and positron emission tomography. However, none of these techniques allows a reliable continuous bedside assessment of cerebral blood flow but provide snapshots of an immensely fluid process. Continuous bedside techniques include monitoring jugular venous oxygen saturation ($Sjvo_2$) and thermal diffusion local cerebral blood flow probes. Jugular venous oxygen monitoring provides an indirect measure of the balance between cerebral blood flow and metabolism and is sensitive at detecting episodes of global cerebral ischemia. Local cerebral blood flow probes based on thermal diffusion have been found to be quite sensitive in tracking cerebral blood flow, but are limited by artifacts and by the small region that they monitor. Near-infrared spectroscopy is a continuous bedside technique for monitoring changes in cerebral oxygenation. It is also sensitive in detecting extra-axial blood accumulation. Microdialysis can be performed in a continuous fashion at the bedside and may aid in guiding therapy in the future. Despite the promise of these new techniques, the cornerstone for assessment in neurocritical care units remains frequent neurological examination by the nursing and physician staff supplemented by ICP and periodic CT scan data.

Bibliography

American Association of Neurological Surgeons and the Brain Trauma Foundation—Joint Section on Neurotrauma and Critical Care (1995). *Guidelines for the management of severe head injury*, pp. 61–102, 119–62, 413–28. Brain Trauma Foundation, New York.

Diaz-Marchan, P.J., Hayman, L.A., Carrier, D.A., and Feldman, D.J. (1996). Computed tomography of closed head injury. In *Neurotrauma* (ed. R.K. Narayan, J.E. Wilberger, and J.T. Povlishock), pp. 137–50. McGraw-Hill, New York.

Hankins, L., Taber, K.H., Yeakley, J., and Hayman, L.A. (1996). Magnetic resonance imaging in head injury. In *Neurotrauma* (ed. R.K. Narayan, J.E. Wilberger, and J.T. Povlishock), pp. 151–62. McGraw-Hill, New York.

Marmarou, A. (1996). Pathophysiology of intracranial pressure. In *Neurotrauma* (ed. R.K. Narayan, J.E. Wilberger, and J. T. Povlishock), pp. 413–28. McGraw-Hill, New York.

Valadka, A.B. and Narayan, R.K. (1996). Emergency room management of the head-injured patient. In *Neurotrauma* (ed. R.K. Narayan, J.E. Wilberger, and J.T. Povlishock), pp. 117–36. McGraw-Hill, New York.

11.3.3 Medical management

A. Main and N. M. Dearden

Key messages

- Advanced Trauma Life Support protocols should be applied to the head-injured patient.
- Hypoxia must be avoided.
- Comatose patients, postoperative neurosurgical patients, patients with a changing neurological examination, intubated patients, and unstable patients should be admitted to the intensive care unit.
- Monitoring the intracranial pressure can guide therapy.
- Therapy may include mild hyperventilation, drainage of cerebrospinal fluid, osmotherapy, diuretics, hypnotics, hypothermia, and other medications.

Initial management

Care begins with resuscitation and diagnosis at the scene of the accident. The clinician should follow the ABCDE (airway, breathing, circulation, disability, exposure) rule of trauma management. If neurosurgical services are unavailable at the primary hospital, secondary referral and transport to a neurosurgical center may be indicated.

Initial periods of hypoxemia ($Pa\text{O}_2 < 8.5$ kPa (62.5 mmHg)) and hypotension (systolic blood pressure below 90 mmHg) in head injury double mortality compared with a similar normoxemic and normotensive group (Chesnut *et al.* 1993). Rapid diagnosis and therapy, with endotracheal intubation and ventilation and treatment of extracranial causes of hypotension, have led to a significant improvement in mortality and morbidity over the past two decades.

The ABCDE system

Airway and breathing

Hypoxia, which is common in head injury patients (30 per cent), must be avoided to prevent secondary brain injury (Chesnut *et al.* 1993). A crash induction technique, such as rapid-sequence intubation, maintains an immobile cervical spine and should be employed. The patient is ventilated to normoxia and mild hypocapnia ($Pa\text{O}_2 = 13$ kPa (97.5 mmHg), $Pa\text{CO}_2 = 4$–4.5 kPa (30–34 mmHg)), while maintaining sedation, analgesia, and paralysis as necessary. A thorough neurological examination should be completed only if it can be done safely.

Circulation

Hypotension (systolic blood pressure of 90 mmHg or less) occurs in about 30 per cent of patients with severe head injuries. Resuscitation must both treat the causes and correct the blood pressure. A high spinal cord transection can cause hypotension associated with bradycardia (Cushing's reflex).

Physiological saline, Hartmann's solution, or colloid solutions can all be administered initially via peripheral venous access. Blood should be given if hemorrhage is severe and persistent. Free water or 5 per cent glucose solution is contraindicated as it predisposes to cerebral edema and cerebral acidosis in ischemic conditions. Coagulopathies can occur and clotting profiles should guide appropriate replacement therapy.

In patients with severe head injury the early use of invasive arterial and central venous monitoring, as well as insertion of a urinary catheter and an orogastric tube, is mandatory to assess adequacy of resuscitation and prevent aspiration.

Disability

Early identification and evacuation of surgically correctable intracranial hematomas is mandatory and in part is guided by the Glasgow Coma Scale (GCS) and CT scanning (Tables 1 and 2).

A GCS sum score of 8 or less without eye opening is coma. In patients with a history of brain injury, an impaired conscious level should not be attributed to metabolic disturbance (hypoglycemia), intoxication (alcohol, drug overdose), or hypothermia. Repetitive assessment of the GCS allows the clinician to evaluate the efficacy of management and to chart deterioration. The GCS sum score and in particular the GCS motor score have prognostic value (Table 3).

In sedated or paralyzed patients pupillary responses to light need to be assessed frequently. In the absence of a history of ocular or local trauma, a single fixed and dilated pupil indicates temporal lobe herniation. Bilateral fixed and dilated pupils are a poor prognostic sign when they are due to transtentorial brain herniation affecting the brainstem. These signs are urgent indications for an intravenous bolus of 1 g/kg of 20 per cent mannitol over 15 to 20 min.

Periodic unprovoked cardiovascular instability in the chemically paralyzed and ventilated head-injured patient may indicate seizure activity. Seizures need to be controlled early since an increased oxygen demand raises cerebral blood flow and blood volume and increases intracranial pressure (ICP) if compliance is reduced. Prolonged seizures (status epilepticus) can lead to ischemia. Phenytoin or benzodiazepines are recommended but can cause cardiovascular depression and alter conscious level. Early neurosurgical consultation is also necessary.

Exposure

Full front and back exposure is indicated to reveal any further injury. Neck immobilization with log-rolling is necessary for posterior evaluation.

Transfer to a tertiary care center

Patient transfer can be fraught with danger if it takes place before resuscitation and stabilization are maintained, adequate monitoring and emergency equipment are unavailable, or communication with the accepting facility is inadequate. Comprehensive guidelines for the resuscitation and transfer of these patients are available (Gentleman *et al.* 1993).

Indications for the admission of brain-injured patients to intensive care are given in Table 4. Patients who are admitted postoperatively following major neurosurgical procedures may be ventilated

Table 1 Glasgow Coma Scale

Score	Eye opening (E)	Best verbal response (V)	Best motor response (M)
1	No eye opening to any stimulus*	No verbal response to any stimulus*	No motor response to any stimulus*
2	To pain	Sounds only*	Extending to painful stimuli*
3	On command	Words (inappropriate)	Abnormal flexion to painful stimuli*
4	Spontaneously	Confused sentences	Flexing to painful stimuli*
5	—	Orientated	Localizing to pain*
6	—	—	Follows commands

*Maximum scores for patients in coma.

Table 2 Indications for CT scanning of the resuscitated head-injured patient

1. Coma (GCS $\leqslant 8$) (40% of comatose head injuries have an intracranial hematoma)
2. Neurological signs (e.g. seizures)
3. Multiple injuries, including to head
4. Patients with skull fracture with or without improved consciousness
5. Patients with mildly impaired consciousness lasting 24 h

overnight and assessed within the next 24 h, but those with significant intracerebral trauma leading to unstable ICP or cerebral perfusion pressure or with multiple trauma require longer mechanical ventilation. Patients thought to have suffered severe brain trauma resulting in brainstem death are ventilated without sedation, analgesia, or paralysis until brainstem death testing can be performed and possible organ donation accomplished.

Intensive care management

Ventilation

Patients are ventilated to achieve mild hypocapnia ($P\text{aCO}_2 = 4$–4.5 kPa (30–34 mmHg), $P\text{aO}_2 > 13$ kPa (97.5 mmHg)) using sedation and analgesia but generally without muscle paralysis unless control of ICP or blood oxygenation dictates otherwise. Continuous pulse oximetry, indwelling arterial monitoring for blood pressure and arterial blood gas determination, and capnography aid cardiorespiratory management. If there are no vertebral injuries, head elevation reduces venous engorgement, as does securing the endotracheal tube with adhesive tape rather than a circumferential bandage. Attending to the patient at 30° to the horizontal minimizes any deleterious effects due to positive end-expiratory pressure (PEEP) which may be required to correct hypoxemia. Excessive PEEP or head elevation may adversely alter both mean blood pressure and ICP, worsening cerebral perfusion pressure. Better venous drainage by excessive head elevation may improve ICP (Feldman *et al.* 1993). A balance must be achieved. Use of a rigid cervical collar may also raise ICP by altering the cerebrospinal fluid dynamics, but neck stability is essential. Muscle relaxation to facilitate intubation is mandatory, but prolonged use in the intensive care setting makes neurological assessment difficult although it may be necessary to ensure adequate ventilation.

An ICP below 20 mmHg with an acceptable cerebral perfusion pressure and oxygenation may allow weaning to begin. Monitoring of ICP and hemodynamics should guide weaning, with sedatives used sparingly.

Some centers perform early tracheostomy in order to facilitate weaning and reduce sedation requirements, particularly in the presence of coma and cervical paralysis. Pulmonary complications are common in the head-injured patient with multiple trauma, and scrupulous attention must be paid to all forms of respiratory physiotherapy. The presence of spontaneous bilateral extensor

Table 3 Mortality based on GCS scores

GCS score	Mortality (%)
$\leqslant 8$	30–40
9–12	4
13–15	0.4

Table 4 Indications for admission to intensive care

1. Comatose patients with diffuse injury not surgically treatable
2. Postoperative patients who have undergone neurosurgery
3. Patients with an abnormal neurological state requiring intubation
4. Multiply injured patients with cardiovascular or respiratory instability
5. Patients thought to be brainstem dead (some debate)

posturing during weaning of a head-injured patient from a ventilator is usually considered as an indication to resume artificial ventilation.

Sedation and analgesia

The head-injured patient may be hypertensive or hyperadrenergic even in the face of hypovolemia. Protection from stress and the effects of coughing and straining against the ventilator is important with regard to ICP control; however, sedative-induced hypotension must be avoided. Judicious intravenous combinations of opioids (to blunt the hyperadrenergic state) and sedatives, which are also anticonvulsive, may be optimal.

Fluids and nutrition

Early enteral nutritional support should be attempted. Owing to gastric stasis, oro- or nasojejunal feeding may facilitate absorption and decrease aspiration. Enteral feeding is less invasive and carries a lower risk of infection than parenteral nutrition. Hyperglycemia, which worsens cerebral acidosis under ischemic conditions, is reduced when patients are fed enterally.

Protein loss should be prevented by caloric analysis and supplementation when deficient. Crystalloid solutions can be given to prevent prolonged cellular dehydration and maintain a normal serum sodium level, thereby ensuring a normal osmotic gradient across the blood–brain barrier.

Brainstem dysfunction can cause diabetes insipidus, which is characterized biochemically by hypernatremic dehydration in these patients when there is inadequate fluid replacement of urinary losses. Up to 4 µg of deamino-D-arginine vasopressin (DDAVP) intravenously can be used in the patient who has no other obvious cause of diuresis and is producing abundant dilute urine (specific gravity below 1005).

Seizure management

Patients at high risk of early post-traumatic seizure include those with intracerebral hematoma and penetrating cerebral injuries. Phenytoin and carbamazepine may be used proplylactially to increase the seizure threshold.

Sepsis

Pyrexia increases the cerebral metabolic requirement for oxygen and may also exacerbate raised ICP. High core temperatures in seriously injured trauma patients have been shown to be associated with a poor prognosis. A high temperature should be assumed to be associated with infection, and appropriate endotracheal, urine, and blood cultures should be sent. In patients with ventricular access for ICP monitoring, the cerebrospinal fluid should be sampled daily. Core temperatures above 38 °C should be actively treated using combinations of antipyretics, such as acetaminophen (paracetamol), and physical methods such as tepid sponging, fanning, and cooling blankets.

Elevated ICP

ICP monitoring should be considered for coma patients with a non-operative intracranial hemotoma, loss of ventricles and basal cisterns, diffuse brain swelling, multiple injuries, and profound motor deficits.

ICP monitoring with a fluid-filled intraventricular catheter can lead to bleeding and infection. Intraparenchymal monitoring using a fiber-optic transducer may not always correlate with ventricular devices, and therapeutic drainage of cerebrospinal fluid is not feasible. ICP is monitored contralateral to any focal pathology so that measurements obtained reflect pressures in undamaged brain. Subdural, subarachnoid, and epidural devices are even less accurate.

Cerebral perfusion pressure

Management of cerebral perfusion pressure aims to prevent ischemia by maintaining cerebral blood flow. Patients dying from head injury are known to suffer ischemic damage, and poorly controlled ICP and cerebral perfusion pressure are associated with increased mortality. Recent studies have reported reduced morbidity and mortality when cerebral perfusion pressure was maintained above 70 mmHg (Rosner and Daughton 1990). Falls in cerebral perfusion pressure during the first day after trauma are generally due to hypotension; subsequently they are increasingly attributable to raised ICP.

About 40 per cent of patients suffering severe brain injury have regional hypoperfusion associated with vasospasm induced by traumatic subarachnoid hemorrhage, and higher cerebral perfusion pressure levels may be needed to avoid ischemia. Studies evaluating the use of induced hypertension or calcium antagonist agents in this subgroup of patients are currently in progress. Systemic hypertension is a reflex consequence of brainstem ischemia and may temporarily improve cerebral perfusion. Artificial induction of a hypertensive state with vasocontrictors such as phenylephrine has been shown not to worsen ICP and, with altered autoregulation, may even improve it. Conversely, systemic hypotension may induce reflex cerebral arterial dilatation, causing a rise in ICP (Rosner *et al.* 1995).

Adequacy of cerebral perfusion pressure control therapy can be judged from indices of cerebral oxygenation and blood flow. Continuous monitoring of cerebral venous oxygen saturation ($Sjvo_2$) and brain tissue Po_2, and transcranial Doppler ultrasonography can guide therapy.

Treatment of raised ICP

The treatment of raised ICP allows adequate cerebral perfusion pressure and can prevent cerebral herniation. The aim of treating ICP and cerebral perfusion pressure is to minimize secondary insults and reduce the risk of cerebral ischemia.

A sustained ICP above 15 mmHg on the first day after injury demands investigation for cause. The accuracy of the transducer system should be verified, adequate cerebral venous drainage should be assured, hypoxemia and hypercapnia should be corrected, and adequate sedation to allow the patient to synchronize with the ventilator should be maintained. Ventilatory indices such as airway pressure and inspiratory-to-expiratory ratio should be reasonable, seizure activity must be excluded, core temperature must be in the range 35 to 37 °C, and serum biochemistry must be maintained (e.g. serum sodium concentration above 135 mmol/l). Once the ICP has risen to 20 to 25 mmHg treatment is mandatory, although after 48 to 72 h a level of 30 mmHg may be tolerated before treatment is necessary.

Failure of treatment or the onset of new neurological signs dictates an urgent CT scan to exclude newly developed lesions which may be amenable to surgery. Treatments for elevated ICP include drainage of cerebrospinal fluid, hyperventilation, osmotherapy, diuretics, hypnotics, and hypothermia.

Drainage of cerebrospinal fluid is continuous or intermittent via ventricular catheters against a back pressure of about 10 mmHg to avoid ventricular collapse, brain shifts, and loss of ICP monitoring capability. The Brain Trauma Foundation recommends this as a first-line maneuver in ICP control in those patients with ventricular catheters *in situ*. Potential problems include ventricular collapse, causing midline shift and obstructive hydrocephalus in the contralateral ventricle, and increased cerebral blood flow with formation of edema.

Hyperventilation is based on the sensitivity of the cerebral resistance vessels to CO_2. Hypocapnia to a $P\text{aCO}_2$ of 3.5 kPa (26 mmHg) reduces the cerebral blood flow and volume by 30 per cent, thus reducing ICP. If ventilation is to be effective, CO_2 sensitivity must be retained. In normal volunteers the effect of hyperventilation on cerebral blood flow lasts for about 4 h, but in head-injured patients it may well last longer.

Hyperventilation may work best during the early post-resuscitation phase of head injury if there is cerebrovascular engorgement secondary to absolute hyperemia where reduction in cerebral blood volume and flow should not be associated with ischemia. Cerebral blood flows measured in the ischemic range of 15 to 20 ml/dg/min (normal flow is 50 ml/dg/min) may be produced by aggressive ventilation. Hypocapnia can cause autoregulation failure and ICP may even rise, and, if blood pressure falls with CO_2 elimination, cerebral perfusion pressure may be further jeopardized, particularly during the early hypoperfusion phase of severe head injury during resuscitation. A worse outcome may be seen after prolonged aggressive hyperventilation to a $P\text{aCO}_2$ of about 2 kPa (15 mmHg) when compared with a normocapnic group. The use of the acid–base buffer THAM (tromethamine) reduces the harmful ischemia-related effects of hypocapnia, presumably by correcting cerebral acidosis. Jugular venous oxygen saturation monitoring in the intensive care unit (ICU) has demonstrated episodes of severe jugular venous desaturation during excessive hyperventilation. Jugular desaturation episodes have been shown to affect head injury outcome adversely.

The present recommendations are that prophylactic hyperventilation should be avoided, although acute neurological deterioration or failure of other therapies are an indication for hyperventilation. Cerebral venous oxygen saturation monitoring or the more robust brain tissue $P\text{O}_2$ measurement can be used to alert the clinician to impending cerebral ischemia ($S\text{jvO}_2 < 50$ per cent and brain $P\text{O}_2 < 20$ mmHg (2.7 kPa) respectively).

Intravenous 20 per cent mannitol solution works by two major mechanisms. Firstly, mannitol improves blood rheology, with its hemodilution leading to an increased cerebral blood flow and oxygen delivery. It works most rapidly at lower cerebral blood flows when given as a bolus of 0.5 to 1.0 g/kg over 20 min, repeated up to four times. Improved oxygen delivery in the presence of preserved cerebral autoregulation causes vasoconstriction, reduces cerebral blood volume, and leads to a fall in ICP. Secondly, there is delayed effect due to the osmotic gradient between plasma and cells which takes up to 30 min for onset and lasts for at least 8 h. Thus mannitol may be the first-line agent when raised ICP is associated with ischemic edema. Mannitol is also a free-radical scavenger and probably reduces production of cerebrospinal fluid.

Complications associated with mannitol infusions include the development of acute renal tubular necrosis, particularly in dehydrated hypotensive patients. The risk of serum hyperosmolality ($>$ 320 mosmol/kg) can occur with doses above 3 g/kg/day; thus serum osmolalities should be monitored. To keep dosages to a minimum, it is recommended that mannitol be given as a bolus. Its use prior to ICP monitoring is restricted to patients with acute neurological deterioration.

Another hypertonic agent, 7.5 per cent saline, has been used as an alternative to mannitol to correct hypovolemia and improve ICP control. Loop diuretics may control ICP by reducing brain water and ion content irrespective of blood–brain barrier integrity, slowing the formation of cerebrospinal fluid and reducing central venous pressure. Some centers use diuretics and mannitol together and replace losses with colloid solutions, thus maintaining normovolemia but promoting intracellular dehydration.

Analgesics and sedatives decrease cerebral metabolism, which leads to vasoconstriction, reduces cerebral blood volume, and lowers ICP. Their effect can be monitored by recording cerebral electric activity. As a further prerequisite to their therapeutic efficacy, CO_2 vasoreactivity should be preserved.

Barbiturates may be long acting and remain in the body for several days. They can cause cardiovascular collapse in hypovolemic patients, thus worsening cerebral perfusion pressure. Monitoring cerebral oxygenation and systemic filling pressures may augment their safe use. Utilization is associated with reduced immune function and increased infection risk in diffuse head injury. The use of barbiturates may be associated with a higher mortality than that occurring with mannitol. Owing to their toxicity, the barbiturate drugs should probably be restricted to selected patients who retain electrical activity and CO_2 reactivity after severe head injury and who fail to respond to other therapies.

Hypothermia, or lowering brain temperature, reduces both the metabolic component affected by analgesics and sedatives and the basal cerebral metabolism. A controlled trial of the effect of induced hypothermia on outcome following severe head injury is currently in progress in the United States.

Other medications under evaluation

Calcium-induced injury is central to neurocellular damage. Nimodipine, a calcium-channel antagonist, has been shown to have no overall efficacy, although it may be effective in a subgroup of patients with traumatic subarachnoid hemorrhage.

Steroids in high doses have not been shown to be effective and their administration is not recommended since they contribute to morbidity.

21-Amino steroids are much more potent antioxidants without any hormonal effects. They inhibit iron-dependent lipid peroxidation, thus preventing free-radical production. Tirilizad mesylate is undergoing phase III prospective evaluation.

Polyethylene glycol superoxide dismutase is an oxygen free-radical scavenger. Phase II clinical trials have shown safety and efficacy, although a recent unpublished phase III trial in the United States failed to confirm initial findings.

In situations where conventional ICP control has failed, reducing transcapillary pressure with dihydroergotamine as a precapillary vasoconstrictor and with systemic hypotensive agents while maintaining colloid oncotic pressure with albumin has led to improved mortality compared with a matched historical control group.

Conclusions

Brain injury is a common occurrence following trauma and consumes vast economic and manpower resources. Current medical management includes efficient resuscitation with prevention of hypotension and hypoxemia. Early scanning with medical intervention can be life-saving. Management in the ICU is focused on preventing secondary insults. By maintaining cerebral perfusion through manipulation of systemic blood pressure and/or ICP, it may be possible to minimize cerebral ischemia.

References

Chesnut, R.M., *et al.* (1993). The role of secondary brain injury in determining outcome from severe head injury. *Journal of Trauma*, **34**, 216–22.

Feldman, Z., *et al.* (1992). Effect of head elevation on intracranial pressure, cerebral perfusion pressure and cerebral blood flow in head-injured patients. *Journal of Neurosurgery*, **76**, 207–11.

Gentleman, D., Dearden, M., Midgley, S., and Maclean, D. (1993). Guidelines for resuscitation and transfer of patients with serious head injury. *British Medical Journal*, **307**, 547–52.

Rosner, M.J. and Daughton, S. (1990). Cerebral perfusion pressure management in head injury. *Journal of Trauma*, **30**, 933–41.

Rosner, M.J., Rosner, S.D., and Johnson, A.H. (1995). Cerebral perfusion pressure management: protocol and clinical results. *Journal of Neurosurgery*, **83**, 949–62.

11.3.4 The role of surgery

R. P. Boyer and J. J. Oro

Key messages

- Surgery is indicated for hematomas with significant mass effect and declining neurological status.
- Elevation of depressed skull fractures neither improves neurological status nor reduces the incidence of late epilepsy.
- Intracranial pressure monitoring may be indicated for a score of 8 or less on the Glasgow Coma Scale.

Introduction

Surgery for head injury has a varied role depending on the type of lesion, the imaging findings, the patient's neurological status, and the other associated injuries.

Extra-axial hematoma

Surgery is usually indicated for subdural or epidural hematomas 5 mm thick or more with a midline shift of more than 5 mm. In addition, all patients with an intracranial hematoma causing significant mass effect who demonstrate a declining level of consciousness or other focal neurological findings, or who have effacement of basilar cisterns on CT scanning, require craniotomy. A patient with a very thin subdural or epidural hematoma associated with midline shift from cerebral edema out of proportion to the mass of the hematoma may not benefit from surgical intervention.

Current operative management is assisted by rapid CT scanning. However, if CT scanning is not available or will be unduly delayed, emergent exploratory burr holes can be made.

Depressed skull fractures

A depressed skull fracture with the outer table of the skull lying at or below the level of the inner table is considered significant. These fractures are divided into closed (or simple) and open (or complex). They may tear the dura and lacerate the underlying brain, and subdural, epidural, or intracerebral hematomas may develop. An open fracture raises the concern that contaminated material may enter the cranial cavity.

Closed depressed fractures

Traditionally, elevation of a depressed skull fracture was performed to help improve focal neurological deficits. However, it appears that the cortical damage causing these deficits occurs at the time of impact and is not due to the subsequent pressure of the depressed fragments; therefore elevation of depressed fragments would not be expected to improve neurological recovery. Studies bear this out, as no difference can be shown between patients treated surgically and non-surgically (Jennett *et al.* 1974).

Post-traumatic epilepsy occurs after depressed skull fracture in 9.5 per cent of cases. However, the incidence of late epilepsy in patients with depressed skull fractures with no intracranial blood is not changed by elevating the fragment (Jennett *et al.* 1974). Therefore the only firm indication for elevating a closed depressed skull fracture is correction of a cosmetic deformity.

Open depressed skull fractures

Open fractures allow exposure of the brain to contaminated material; therefore the traditional practice is to explore, elevate, and debride these fractures as soon as possible, generally within 8 to 12 h. The concept of operative exploration has been challenged by van den Heever and van der Merwe (1989). They showed improved outcome in selected patients with relatively uncontaminated wounds who were treated by local debridement, scalp closure, and antibiotic coverage. Thus, while the vast majority of open depressed skull fractures should be treated in the operating room, there may be a role for conservative management in selected patients.

Frontal sinus fracture

Closed non-depressed linear fractures of either the anterior or posterior wall of the frontal sinus do not require surgical intervention, although some authors believe that all posterior fractures should be explored. Depressed anterior wall fractures should be explored and elevated, with obliteration of the sinus as needed. Depressed posterior wall fractures may tear the dura and lead to a leak of cerebrospinal fluid. Therefore they should be explored and the sinus obliterated.

Intra-axial contusions and hematomas

Contusions are areas of small hemorrhages surrounded by necrotic brain, usually at the gyral peaks in the frontal and temporal lobes where the brain is surrounded by uneven skull which it strikes during trauma. A hematoma is a mass of blood, free of intervening brain, which pushes the surrounding brain aside. Contusions may evolve and become larger, leading to increasing edema, necrosis, and extension of the contusion.

Small contusions may simply be observed, but large contusions, particularly those occurring in the temporal lobes, where a small mass can cause significant neurological deterioration, should have early surgery. If necessary, temporal or frontal lobectomies, including resection of up to 4 cm of the dominant temporal lobe or 6 cm of the non-dominant temporal lobe, may be performed. Contusions in eloquent brain may be observed and treated with aggressive management of increased intracranial pressure. Multiple contusions are treated in the same way. However, if one of these evolves and becomes a dominant intracranial mass, it may require removal. The majority of contusions can initially be treated non-surgically.

Post-traumatic intracerebral hematomas are uncommon. As with contusions, temporal hematomas are much more likely to cause neurological deficit than frontal or occipital hematomas. In a study of patients with intracerebral hematoma involving a single lobe, no patient with either frontal or parieto-occipital hematomas developed signs of transtentorial herniation, whereas 17 per cent with temporal lobe hematomas developed herniation syndrome. No patient with temporal hematomas less than 30 ml showed herniation findings, whereas 17 per cent of patients with hematomas larger than 30 ml had herniation findings (Andrews *et al.* 1988).

The surgical treatment of hematomas is similar to that of contusions. We recommend that all patients with a Glasgow Coma Scale (GCS) score of 8 or less receive intracranial pressure monitoring. Hematomas are evacuated in patients who are deteriorating neurologically, have midline shifts greater than 1 cm, medication-resistant intracranial pressure elevations, and temporal hematomas greater than 30 ml.

Delayed traumatic intracerebral hematomas commonly occur within the first 48 to 72 h following injury. They usually occur in areas of brain with small or unseen trauma and/or hypoxia. Significant hypotension following trauma appears to be an important contributor to the formation of these hematomas. Indications for surgical treatment of delayed hematomas are similar to those for primary traumatic hematomas.

Gunshot wounds to the head

Although regional differences exist, the incidence of civilian gunshot wounds, including gunshot wounds to the head, is increasing in most urban areas. The role of surgery in selected cases is to decrease secondary brain injury and thereby increase the quality of outcome. Debridement and removal of accessible fragments and watertight closure for patients with GCS scores of 6 to 15 is appropriate, as is evacuation of large hematomas causing significant mass effect. A difference of opinion exists in how to manage patients with GCS scores of 3 to 5. Levy *et al.* (1994) showed that these patients are not likely to benefit from surgical intervention, although a subgroup of patients who have reactive pupils and no evidence of subarachnoid hemorrhage, ventricular involvement, or fragmentation deserve close follow-up. Stone *et al.* (1995) treated patients with GCS scores of 3 with surgery if they had reactive pupils, and patients with GCS scores of 4 to 7 if they had some motor response, even with non-reactive pupils. Two patients with GCS scores of 3 had good outcomes.

Cranial stab wounds are much less common, but have the highest incidence of vascular lesions including traumatic aneurysms. Therefore angiography should be performed for all patients with cranial stab wounds. The weapon should remain in the head until it can be taken out in the operating room where any adverse consequences can be adequately controlled. Patients should receive broad-spectrum antibiotics for 10 days, and supratentorial stab wounds should receive prophylactic anticonvulsants.

References

Andrews, B.T., Chiles, B.W., Olsen, W.L., and Pitts, L.H. (1988). The effect of intracerebral hematoma location on the risk of brain-stem compression and on clinical outcome. *Journal of Neurosurgery*, **69**, 518–22.

Jennett, B., Miller, J.D., and Braakman, R. (1974). Epilepsy after nonmissile depressed skull fracture. *Journal of Neurosurgery*, **41**, 208–16.

Levy, M.L., Masri, L.S., Lavine, S., and Apuzzo, M.L.J. (1994). Outcome prediction after penetrating craniocerebral injury in a civilian population: aggressive surgical management in patients with admission Glasgow coma scale scores of 3, 4, or 5. *Neurosurgery*, **35**, 77–85.

Stone, J.L., Lichtor, T., and Fitzgerald, L.F. (1995). Gunshot wounds to the head in civilian practice. *Neurosurgery*, **37**, 1104–12.

van den Heever, C.M. and van der Merwe, D.J. (1989). Management of depressed skull fractures. Selective conservative management of nonmissile injuries. *Journal of Neurosurgery*, **71**, 186–90.

11.4 Spinal injury

11.4.1 Pathophysiology

C. Werner

Key messages

- Primary spinal cord insults are due to the immediate injury and occur prior to medical treatment.
- Secondary spinal cord insults are related to edema of neurons and glia, continuous mechanical compression of the cord, hypoxia, ischemia, and release of vasoactive and excitatory substances.
- Prevention of primary insults to the spinal cord may prevent secondary cord insults with preservation of vascular and cellular integrity.

Introduction

Complete spinal cord injury is a catastrophic event with a substantial impact on the patient and the supporting infrastructure (e.g. family, health services). Unless adequately treated, complete spinal cord injury is associated with lifelong disability (chronic bedridden state), pulmonary infections and possibly ventilator dependence, incontinence, infected decubiti, osteomyelitis, renal failure, and eventually death. The etiological factors for spinal injury include motor vehicle accidents (40–50 per cent), falls (20 per cent), sports injuries (10–15 per cent), gunshot wounds, and industrial or agricultural injuries. The majority of victims are males aged between 15 and 40 years.

There is great variation in the mechanism of injury, mechanical damage, and clinical symptoms of spinal cord injury. The forces of the trauma may induce injury ranging from mild cord contusion with minimal neurological deficits to severe spinal cord contusion with transverse functional dissection to complete anatomical spinal cord transection. Spinal cord injury can be divided into the primary insult and secondary insults. The primary insult represents the immediate impact and its consequences on the vertebral column and the central nervous system structures between the time of injury and the initial medical treatment. Primary spinal cord injury can only be reduced using preventive strategies such as automatic protection devices (e.g. air bags, seatbelts, helmets), laws (alcohol and drug tests), identification of high-risk recreational activities, and education about risk-taking. Secondary insults occur in response to the initial (primary) impact. This type of injury is related to a combination of edema of neurons and glia, continuous mechanical compression of the traumatized spinal cord segment (bone fragments, hematoma, edema), hypoxia, ischemia, and the release of vasoactive and excitatory substances which further promote catabolic processes and neuronal necrosis. In contrast with the primary insult, secondary insults are subject to several therapeutic interventions in order to minimize neuronal damage and improve neurological outcome. The basis for understanding the prevention and treatment of secondary insults is knowledge of the anatomy and physiology of the vertebral column and spinal cord, the regulation of spinal cord blood flow, biomechanical aspects, and the pathophysiological cascade that is initiated following spinal cord injury.

The anatomy of the vertebral column and spinal cord

The vertebral column consists of 30 vertebrae: seven cervical, 12 thoracic, five lumbar, five sacral (fused), and the coccygeal bone. With the exception of the first two cervical and the sacral vertebrae, all the bony elements of the spinal column articulate by intervertebral disks and posterolateral joints. Capsules and strong, yet elastic, ligaments with various points of origin and termination stabilize the spine in a fashion that provides both stability and flexibility. The vertebral column serves as a protective cavity for the spinal cord, which represents the caudal continuation of the brainstem. The spinal cord is suspended by means of a series of nerve roots, dentate ligaments, and three meninges to cover the cord and form the dural cavity which is filled with cerebrospinal fluid. The dura mater is an inelastic fibrous membrane which forms the dural sac. The epidural space (i.e. the space between the vertebral column and the spinal dura) contains fat, ligaments, small arteries, and venous plexus which supply and drain spinal cord tissue. The arachnoid membrane follows the inner layer of the dura and forms the subarachnoid space within which is the cerebrospinal fluid. The pia mater is directly applied to the surface of the cord. The spinal roots are divided into motor (anterior) and sensory (posterior) sections. They form the spinal nerves and leave the dura and the intervertebral foramina in a segmental fashion.

The physiology of spinal cord blood flow

The spinal cord is perfused by a single anterior spinal cord artery, which runs along the median fissure, and two posterior spinal cord arteries which are located posterolaterally to the cord. The spinal cord arteries run along the anterior and posterior longitudinal axes of the cord between the cervical and sacral segments. The anterior spinal cord artery originates at the level of the medulla from the vertebral arteries and supplies 75 per cent of the cord. Six to eight anterior radicular arteries form anastomoses with the anterior spinal cord artery every three to five segments. The most important radicular vessel is the artery of Adamkiewicz which contributes to the perfusion of the thoracolumbar segments. The relatively large distance between the radicular vessels results in the existence of watershed areas which have predelections for ischemic insults. The posterior spinal cord arteries originate from the posterior branch of the vertebral artery and supply 25 per cent of the spinal cord. Global spinal cord blood flow is lower than cerebral blood flow and shows a regionally specific pattern, with higher flow to the cervical and lumbar segments and lower flow to the thoracic segments. The regulation of spinal cord blood flow is similar to the regulation of cerebral blood flow. As spinal cord perfusion pressure (equal to the difference between mean arterial blood pressure and cerebrospinal fluid pressure) decreases or increases, the resistance vessels dilate or constrict to maintain the spinal cord blood flow constant. Spinal cord perfusion is also sensitive to changes in arterial carbon dioxide. During hypercapnia, spinal cord blood flow increases, whereas hypocapnia (e.g. during hyperventilation) decreases spinal cord blood flow. The similarity between cerebral and spinal cord blood flow regulation emphasizes the need to apply the therapeutic concepts and considerations of cerebral perfusion pressure management in head-injured patients to the management of spinal cord perfusion pressure following spinal injury.

Biomechanics

Primary insults are caused by external loading of the spine such as flexion, lateral flexion, extension, extension–rotation, tension, vertical compression, and shear stress. As a consequence, hemorrhage and rupture of ligaments and dislocation or fracture of vertebrae occurs, as well as contusion, laceration, and transection of neuronal or glial tissue. The most common cause of spinal cord injury is flexion, followed by compression and then extension. Spinal cord injuries most commonly involve the lower cervical spine and the thoracolumbar junction.

Pathophysiology

The force of the primary insult produces its initial damage in the central gray matter of the cord. The area of the injury is directly related to the force of the impact, and the damage may spread centrifugally into the white matter as the force of the impact increases. Delayed neuronal necrosis in the white matter may also occur in situations where the central but not the peripheral sections of the cord were injured initially. The mechanisms of both immediate and delayed neuronal necrosis are related to the induction of spinal cord ischemia and damage to the blood–cerebrospinal fluid barrier. The impact produces substantial disruption of the tissue with damage to interconnecting and vascular structures. The consequence of this is vasospasm, hemorrhage, disruption of the endothelial cell layer and its tight junctions, and in turn disruption of the blood–cerebrospinal fluid barrier with filtration of larger molecules and water into the extracellular and interstitial space. These alterations stimulate the synthesis of intercellular adhesion molecules which trigger leukocyte adhesion to the damaged endothelium. The immediate vascular injury, the damage of the blood–cerebrospinal fluid barrier, the interstitial edema, and the leukocyte adhesion reduce the microcirculation and induce spinal cord ischemia. This process is rapid in tissues with higher metabolic rates and flows, such as the central gray matter, compared with tissues with lower metabolic rates and flows, such as the peripheral white matter. Autoregulation of spinal cord blood flow may be impaired in patients with severe lesions or substantial edema formation. Hyperventilation may further decrease spinal cord blood flow if the reactivity to changes in arterial carbon dioxide is maintained.

The ischemic insult evokes a cascade of pathophysiological processes which may subsequently result in neuronal necrosis. The first level of the ischemic cascade is the accumulation of lactic acid due to anaerobic glycolysis. Decreases in pH raise the permeability of membranes, specifically those of the endothelial and glial cells. The subsequent formation of interstitial and intracellular edema further promotes decreases of spinal cord blood flow as the mechanical pressure on the cord is enhanced. Membrane potentials and synaptic transmission are based on effective ion pumps within cellular membranes, the metabolism of proteins, lipids, and carbohydrates, and the intracellular transport of molecules. All these cellular processes require the generation, consumption, and conservation of energy. In general, the central nervous system converts glucose and oxygen into usable forms of energy such as ATP. However, during ischemia the anaerobic metabolism is insufficient to maintain cellular energy states, the ATP stores deplete, and the membrane ion pumps fail. The depletion of the ATP stores is paralleled and enhanced by a terminal membrane depolarization with massive release of excitatory neurotransmitters, such as glutamate and aspartate, from stores within the presynaptic nerve terminals. The release of glutamate and aspartate activates *N*-methyl-D-aspartate (NMDA), non-NMDA, and voltage-dependent Ca^{2+} and Na^{+} channels. The subsequent pathological Ca^{2+} and Na^{+} influx induces activation of several catabolic intracellular processes. Ca^{2+} activates lipid peroxidases, proteases, and phospholipases which in turn increase the intracellular concentration of free fatty acids and free radicals. Subsequently, membrane degeneration of vascular and cellular structures leads to irreversible cellular death.

Systemic effects of spinal cord injury

The traumatic insult causing the spinal cord injury is associated with an immediate stimulation of central and peripheral sympathetic tone. Initially, the elevated sympathetic activity increases systemic arterial blood pressure and induces cardiac arrhythmias. At the stage of spinal shock with loss of neuronal conduction, the sympathetic excitation is closely followed by decreases in systemic vascular resistance, arterial hypotension, and venous pooling. Lesions above the level of T5 additionally present with severe bradycardia and cardiac dysfunction. The decreases in cardiac output combined with systemic hypotension further aggravate spinal cord ischemia in tissues with defective autoregulation.

Spinal cord injury may produce respiratory failure. The extent of respiratory complications is related to the level of the injured segments. Injuries above the level of C4 to C5 produce complete paralysis of the diaphragm with substantial decreases in tidal volume and consecutive hypoxia. With lesions below C6, the function of the diaphragm is maintained and there is incomplete respiratory failure due to paralyzed intercostal and abdominal musculature. As a consequence, arterial hypoxia and hypercapnia occur, both of which promote neuronal and glial acidosis, edema, and neuroexcitation. In summary, primary spinal cord injury creates a cascade of pathophysiological events which eventually induce secondary insults and determine outcome. Spinal cord ischemia and disruption of the blood–cerebrospinal fluid barrier are the most important complications following the initial impact. Acute systemic hypotension and respiratory failure with spinal shock further reduce spinal cord perfusion and oxygen delivery. The mechanical stress of the primary impact and the reduced tissue perfusion lead to accumulation of lactic acid, terminal membrane depolarization, ATP depletion, and release of excitatory neurotransmitters. The subsequent postsynaptic receptor stimulation triggers Ca^{2+} and Na^{+} influx, which in turn activates catabolic enzyme reactions and finally leads to cellular necrosis.

Bibliography

Albin, M.S. (1994). Spinal cord injury. In *Anesthesia and neurosurgery* (3rd edn) (ed. J.E. Cottrell and D.S. Smith), pp. 713–43. Mosby Year Book, St Louis, MO.

Hagberg, C., Welch, W.C., and Bowman-Howard, M. (1996). Anesthesia and surgery for spine and spinal cord procedures. In *Textbook of neuroanesthesia* (ed. M.S. Albin), pp. 1039–81. McGraw-Hill, New York.

Lam, A.M. (1992). Spinal cord injury and management. *Current Opinion in Anaesthesiology*, 5, 632–9.

MacKenzie, C.F. and Geisler, F.H. (1996). Management of acute cervical spinal cord injury. In *Textbook of neuroanesthesia* (ed. M.S. Albin), pp. 1083–1136. McGraw-Hill, New York.

Piek, J., Bock, W.J., and Vollmer, D.G. (1994). Spinal trauma. In *Neurocritical care* (ed. W. Hacke), pp. 708–24. Springer-Verlag, Berlin.

11.4.2 Management of cervical cord injury

C. Werner

Key messages

- Preservation of the airway, immobilization, oxygenation, ventilation when necessary, and resuscitation are necessary in managing patients with cervical cord injury.
- Administration of corticosteroids early and avoidance of hyperglycemia are part of management.

Introduction

Spinal cord ischemia and disruption of the blood–cerebrospinal fluid barrier are the most important complications following the primary impact. Acute systemic hypotension and respiratory failure with spinal shock further reduce spinal cord perfusion and oxygen delivery. The mechanical stress of the initial impact and the reduced tissue perfusion induce accumulation of lactic acid, terminal membrane depolarization, ATP depletion, and release of excitatory neurotransmitters. The subsequent postsynaptic receptor stimulation triggers Ca^{2+} and Na^{+} influx which in turn activates catabolic enzyme reactions, finally leading to cellular necrosis. Therefore treatment of acute cervical spinal cord injury is divided into assessing the injury and immobilizing the spine, initial medical treatment, diagnostics and indications for surgical treatment, anesthesia, and peritraumatic and perioperative concepts of neuroprotection.

Assessing the injury and immobilizing the spine

Spinal cord injury must be suspected in patients with neck pain, sensory and motor deficits, unconsciousness, intoxication, spondylitis or rheumatoid arthritis, head injury, and facial fractures. If spinal cord injury is suspected or cannot be excluded, the patient must be placed on a spine board with the head and neck immobilized in a neutral position using a rigid collar to reduce the risk of neurological deterioration from repeated mechanical insults. Total neck immobilization should not interfere with maintenance of the airway, and inadequate respiratory function must be avoided.

Initiation of medical treatment

Initial medical treatment aims at the decompression of the spinal cord and the reversal of neurogenic shock and respiratory failure. Spinal shock is associated with decreases in systemic vascular resistance, arterial hypotension, venous pooling, severe bradycardia, and decreased myocardial contractility. Consequently, treatment of neurogenic shock includes fluid replacement using crystalloid or colloid solutions to maintain arterial blood pressure, circulatory volume, renal function, and tissue oxygenation. Infusion of free water must be avoided, as this decreases plasma osmolarity and promotes spinal cord edema. Atropine may be administered to reverse bradycardia and increase cardiac output. Administration of vasopressors (e.g. norepinephrine (noradrenaline), phenylephrine, dopamine) prior to correction of the intravascular volume status may increase systemic vascular resistance (left ventricular afterload) and further impair myocardial contractility. Therefore volume replacement is the first step and administration of vasopressors the second step in the treatment of arterial hypotension and low cardiac output following acute cervical spinal cord injury.

The major early cause of death in patients with acute cervical spinal cord injury is respiratory failure. The extent of respiratory complications is related to the level of the injured segments. With lesions below C6, the function of the diaphragm is maintained and incomplete respiratory failure occurs related to paralyzed intercostal and abdominal muscles. Spinal cord injury above the level of C4–C5 produces complete paralysis of the diaphragm and the intercostal and abdominal muscle sections. As a consequence, patients are unable to cough or maintain adequate tidal volumes, which leads to retention of secretions, atelectasis, arterial hypoxia, and hypercapnia. Since adequate respiratory function is related to the level and degree of spinal cord injury and other trauma sustained at the time of injury, insufflation of oxygen, anterior displacement of the mandible, and minimal head extension at the occipital C1 level in conscious patients will avoid hypercapnia and hypoxia. Tracheal intubation may be indicated in unconscious patients, during shock, in patients with other major associated injuries, and during cardiovascular and respiratory distress. It is also indicated in conscious patients presenting with the following criteria: maximum expiratory force below $+20\,cmH_2O$, maximum inspiratory force below $-20\,cmH_2O$, vital capacity below 1000 ml, and presence of atelectasis, contusion, and infiltrate. In the emergency setting, cricothyroidotomy, tracheotomy, or jet ventilation via the cricothyroid membrane may be considered.

Diagnostics and indications for surgical treatment

Following the initial assessment of the patient, detailed diagnostic radiography defines the bony damage and compression of the spinal cord. First, lateral, anteroposterior, odontoid, and possibly oblique cervical spine radiographs are obtained. If there is no evidence of injury, flexion and/or extension views may be considered. If any of these radiographs suggest cervical spine abnormalities, specific radiological procedures such as cervical myelography, high-resolution CT scan, or magnetic resonance imaging will identify fractures, dislocation of bony fragments, and spinal cord contusion. In patients with a dislocated cervical fracture, decompression and anatomical bony realignment may be achieved with traction forces applied manually or with Halo or Gardner–Wells systems under radiological control. If the anatomical bony alignment procedures and traction forces fail to decompress the cord, surgical intervention to remove the lesion is required. The timing of surgical intervention remains controversial. While urgent surgical decompression or internal stabilization should be performed in all patients with deteriorating neurological status, some centers tend to defer surgical treatment in patients with spinal cord injury but stable neurological deficit.

Anesthetic management

Anesthetic management in patients with acute cervical spinal cord injury may require the following monitoring and instrumentation: electrocardiography; arterial catheter to measure arterial blood pressure; pulmonary artery catheter to measure cardiac output, pulmonary artery pressure, and pulmonary artery wedge pressure; sophisticated mechanical ventilation devices; end-tidal CO_2; pulse oximetry; blood gas analysis; somatosensory evoked potentials; temperature devices; nasogastric tube to reduce abdominal distension and risk of aspiration, and to measure gastric pH; urinary catheter; warming devices. The optimal technique of endotracheal intubation in patients with cervical spinal injury is related to the condition of the patient and his or her level of co-operation, the neurological deficit, and the training and experience of the individual performing the intubation. In conscious normoxic normocapnic patients without head injury, awake elective nasotracheal intubation may be performed following topical anesthesia of the nose, pharynx, and larynx using the blind nasal technique or fiber-optic devices. In patients with unstable injuries of the spine, neurological function frequently becomes worse as flexion of the neck occurs and thus should be avoided.

Consequently, collar devices and/or manual in-line traction should be used to immobilize the neck and support decompression of the spine during tracheal intubation. Manual in-line traction is performed by applying a traction force of less than 10 lb in the cephalad direction with the mastoid processes in line with the axis of the spine. In unconscious hypoxic hypercapnic patients with or without head injury, emergency oral intubation should be performed using rapid-sequence induction techniques and manual in-line traction. Thiopental (thiopentone) and succinylcholine (suxamethonium) may be used for rapid-sequence induction, since hyperkalemia from denervation sensitivity will not occur until 48 h after injury.

The concept of general anesthesia and sedation in patients with spinal cord injury is based on the maintenance of coupling between metabolism and spinal cord blood flow while achieving hypnosis, analgesia, and a 'relaxed cord'. This concept includes maintenance of normal to high systemic perfusion pressures, normoxia, and normocapnia. No particular anesthetic technique has been demonstrated to be superior, and total intravenous anesthesia using hypnotics (propofol, methohexital) in combination with narcotics (fentanyl, sufentanil, remifentanil) is as appropriate as any balanced technique using a combination of volatile anesthetics (halothane, isoflurane, sevoflurane) and narcotics. Although unproven, nitrous oxide should be avoided since it may produce neuroexcitation in ischemic tissues.

Peritraumatic and perioperative concepts of neuroprotection

Studies in laboratory animals and humans have shown that hyperglycemia is associated with worsened outcome following stroke or neurotrauma. The mechanisms by which normoglycemia may protect neuronal tissue include decreases in intracellular lactic acidosis and in membrane permeability with reduced cellular edema. Therefore plasma glucose concentration should be monitored closely and adjusted within the range of 100 to 150 g/dl (5.5–8.5 mmol/l).

The term 'lazeroids' is frequently used to describe compounds such as lipid peroxidase inhibitors, antioxidants, and free-radical scavengers. The proposed mechanisms by which these drugs reduce neuronal injury include improved spinal cord blood flow, increased order of lipid bilayers, free-radical scavenging and prevention of free fatty acid accumulation by inhibition of lipid peroxidation. The National Acute Spinal Cord Injury Study (NASCIS II) was a multicenter randomized placebo-controlled trial investigating the effects of methylprednisolone (a 30-mg/kg bolus followed by 5.4 mg/kg/h for 23 h) on motor function and sensory deficit at 6 weeks, 6 months, and 1 year after injury. Only patients receiving high-dose methylprednisolone within the first 8 h of injury showed a significant reduction in motor deficit and improved function of sensory tracts. The free-radical scavenger superoxide dismutase has been shown to be protective in animal models of cerebral ischemia and spinal cord injury. Despite the promising results from a phase II study, superoxide dismutase failed to be protective when tested in a multicenter randomized placebo-controlled trial in head-injured patients. The antioxidant tirilazad mesylate (U 74006 F) was protective during transient focal but not permanent focal cerebral and spinal cord ischemia in laboratory animals and humans. Tirilazad is currently under clinical investigation in combination with high-dose methylprednisolone in patients with spinal cord injury (NASCIS III).

In laboratory animals, deep systemic or local (brain, spinal cord) hypothermia protects neurons against both focal and global ischemia. Increasing interest has developed concerning the cerebral effects of mild hypothermia as a result of observations in laboratory animals and humans showing brain protection with small reductions in brain temperature during increased intracranial pressure and cerebral ischemia. It has been suggested that hypothermic protection is related to suppression of major biochemical processes such as decreases in cerebral metabolism, reduction of excitatory neurotransmitter release, and inhibition of accumulation of lipid peroxidation products and free-radical generation. Other studies indicate that small changes in temperature economize cerebral blood flow and prevent postischemic hyper- and hypoperfusion and formation of brain edema. The potential beneficial effects of mild hypothermia following spinal cord injury have yet to be investigated, but it is certainly important to maintain normothermia and to treat hyperthermia aggressively.

Glutamate and aspartate are known as excitatory neurotransmitters which stimulate *N*-methyl-D-aspartate (NMDA) and non-NMDA receptors. Since the activation of these receptors initiates catabolic intracellular processes, blockade of NMDA and non-NMDA receptors may protect cerebral tissue. NMDA receptor antagonists such as MK 801 or ketamine may provide some protection in animal models of focal ischemia, head injury, or spinal cord injury.

The proposed mechanisms of neuronal protection by Ca^{2+}-channel blockers include cerebral vasodilation, prevention of vasospasm, reduced Ca^{2+} influx, and modulation of free fatty acid metabolism. Unfortunately, the results in animal models are contradictory. While several studies have found decreases in neuronal injury and improved outcome following focal cerebral ischemia and spinal cord compression, others have failed to produce protection with Ca^{2+}-channel blockers. The mortality was higher and the neurological deficit was unchanged in patients with acute stroke treated with nimodipine within 6 h of the insult compared with placebo-treated patients. However, nimodipine may be beneficial in patients with symptomatic vasospasm following subarachnoid or traumatic hemorrhage provided that arterial blood pressure is maintained.

Conclusion

In patients with cervical spine injury, the medical treatment aims at decompression of the spinal cord and reversal of neurogenic shock (maintain spinal cord perfusion) and respiratory failure (maintain normoxia, avoid hyperventilation). Collar devices and/or manual in-line traction will immobilize the neck and support decompression of the spine. Treatment of neurogenic shock includes fluid replacement using crystalloid or colloid solutions to maintain arterial blood pressure, circulatory volume, renal function, and tissue oxygenation. Atropine will reverse bradycardia and increase cardiac output. Administration of vasopressors is appropriate only after restoring intravascular volume. Respiratory function is related to the level and degree of spinal cord injury and other trauma sustained at the time of injury. Tracheal intubation is indicated in unconscious patients, during shock, in patients with associated injuries, cardiovascular, or respiratory distress, and in conscious patients with vital capacity below 1000 ml. Plasma glucose concentration should be managed within the range of 100 to 150 g/dl, and normothermia should be maintained until the neuroprotective effects of mild hypothermia are confirmed clinically. Infusion of methylprednisolone (a 30-mg/kg bolus followed by 5.4 mg/kg/h for 23 h) is indicated within the first 8 h of injury.

Bibliography

Albin, M.S. (1994). Spinal cord injury. In *Anesthesia and neurosurgery* (3rd edn) (ed. J.E. Cottrell and D.S. Smith), pp. 713–43. Mosby Year Book, St Louis, MO.

Hagberg, C., Welch, W.C., and Bowman-Howard, M. (1996). Anesthesia and surgery for spine and spinal cord procedures. In *Textbook of neuroanesthesia* (ed. M.S. Albin), pp. 1039–81. McGraw-Hill, New York.

Lam, A.M. (1992). Spinal cord injury and management. *Current Opinion in Anaesthesiology*, **5**, 632–9.

MacKenzie, C.F. and Geisler, F.H. (1996). Management of acute cervical spinal cord injury. In *Textbook of neuroanesthesia* (ed. M.S. Albin), pp. 1083–1136. McGraw-Hill, New York.

Piek, J., Bock, W.J., and Vollmer, D.G. (1994). Spinal trauma. In *Neurocritical care* (ed. W. Hacke), pp. 708–24. Springer-Verlag, Berlin.

11.4.3 Management of thoracolumbar spinal cord injury

R. S. R. Fox

Key messages

- The initial aim is to prevent further injury and ameliorate symptoms.
- Autonomic dysfunction can lead to hypotension, ileus, and urinary retention.
- Depending on the level of injury, severe respiratory compromise can occur.
- Nutrition should begin early, and psychological concerns should be addressed.

Prevention of secondary and tertiary injury

Spinal injuries broadly encompass two distinct clinical entities, firstly those patients with spinal column injury but no neurological abnormality, and secondly those patients with varying degrees of neurological abnormality as a result of spinal column injury, vascular accident, or other cord pathology.

In the critical care setting those patients with injuries of the spinal column but no neurological abnormalities must be treated with the utmost caution in order to avoid inadvertently producing a spinal cord lesion. The patients most at risk are those who have suffered multiple trauma where imaging and examination of the spine has been cursory due to other clinically obvious and life-threatening injuries. All patients in these circumstances must be treated as though they have a spinal injury until proven otherwise (Burney *et al.* 1993).

The management of spinal column injury is based on straightforward orthopedic principles. If the injury is unstable, it should be immobilized and, if and when the patient is fit for surgery, internal fixation should be considered. The advantages of early fixation are that patient handling becomes much easier and that delayed fixation (2–6 weeks after injury) is technically harder and carries significantly greater morbidity.

If the spinal column injury is shown to be causing spinal cord compression or there is deteriorating neurological function, a decompressing procedure should be considered, even in the presence of complete neurological loss.

Immobilization requires keeping the spine straight in both the coronal and sagittal planes. There are well-recognized handling techniques designed to minimize any movement of the spine whilst allowing the patient to be moved or turned. However, these must be performed by staff trained in these techniques to minimize any risk to the patient.

There are also several specialized beds available for the management of spinal-injured patients. These beds are designed to make turning safer and easier in order to avoid the development of pressure sores. Low-air-loss beds and fluidized beds should not be used unless the spine has been stabilized surgically or externally.

The pathology of an injured spinal cord involves a complex of processes which includes edema formation, intracord hemorrhage, and release of inflammatory mediators, with alterations in vascular autoregulation in the area of the injury. A significant aspect of the care of spinal-cord-injured patients with other trauma or critical illness should be the avoidance of physiological states where blood flow and oxygen delivery to the cord are compromised, potentially aggravating the injury and increasing the degree of neurological deficit. Invasive cardiovascular monitoring or esophageal Doppler monitoring may be needed to maximize oxygen delivery.

Many techniques have been tried based on pathological and biochemical findings in the spinal cord lesion, but, even though several have been found to be successful experimentally, few have translated into clinical practice.

The National Acute Spinal Cord Injury Study II was one of the first to show an improved outcome in terms of neurological function at 6 months with the use of methylprednisolone given within 8 h of injury. The dose regimen is 30 mg/kg body weight as a bolus followed by an infusion of 5.4 mg/kg body weight per hour for 23 h (Bracken *et al.* 1990). A recent study by the same group has shown that patients started on methylprednisolone 3 to 8 h after injury should be maintained at 5.4 mg/kg body weight for 47 h (Bracken *et al.* 1997). However, there is a significant increase in the incidence of pneumonia.

The management of problems specific to spinal cord injuries

Many of the problems faced by spinal-cord-injured patients are related to autonomic dysfunction as well as muscle paresis.

Cardiovascular problems

The sympathetic innervation of the heart arises from T1 to T4 or T5, and that of the peripheral vasculature from T1 down to L1. Therefore any complete neurological injury to the cord from T5 upwards may lead not only to loss of peripheral vasomotor control but also loss of sympathetic chronotropic reflex ability along with unopposed vagal activity and intact but altered baroreceptor reflexes. Such a patient will present with extreme hypotension and bradycardia, occasionally leading to asystole, which will mask any associated pathological process due to trauma or illness. Although the bradyarrhythmias and hypotension will resolve in 2 to 6 weeks, immediate action may be required with the use of either temporary pacing and/or inotropes.

Neurogenic pulmonary edema is a recognized condition associated with spinal cord injury, possibly due to massive sympathetic outflow at the time of injury which causes systemic and pulmonary hypertension, leading to pulmonary capillary disruption. Therefore fluid replacement must be judicious.

Injuries below T5 will present with a varying degree of hypotension which is amenable to judicious fluid management. Central venous pressure monitoring alone may be an unreliable guide to fluid management in patients with high thoracic and cervical cord injuries, and so hemodynamic management should be guided by pulmonary artery catheterization or transesophageal Doppler monitoring.

Arrhythmias are commonly seen, primarly ventricular and supraventricular tachyarrhythmias, atrial fibrillation, and atrioventricular block. The possibility of myocardial contusion from a transthoracic injury should be considered, particularly if there are acute electrocardiographic changes with isoenzyme confirmation.

During the acute areflexic phase of a spinal cord injury, the peripheral vasculature will respond as expected to the effects of inotropes. At some point between 4 weeks and 3 months reflex activity reappears in the cord distal to the site of injury. Patients with lesions above T5 may be susceptible to autonomic hyper-reflexia, where a seemingly minor autonomic stimulus can lead to a massive sympathetic outflow below the level of the lesion, leading to severe hypertension, bradycardia or tachycardia, and sweating above the level of the lesion. In chronic high-level quadriplegics this can be a major cause of morbidity and mortality from cerebral haemorrhage, myocardial ischemia, or other vascular accidents.

Patients with chronic high lesions have increased sensitivity to the vasopressor effects of inotropes.

Respiratory problems

All patients with complete cord injuries from cervical to low thoracic will have a degree of respiratory compromise.

Continuous pulse oximetry, serial blood gas analysis, and vital capacity measurements should be performed. If the vital capacity is below 10 to 15 ml/kg in an adult, ventilatory support is likely to be needed; if the vital capacity approaches the expected tidal volume, ventilatory support is mandatory.

Caution should be exercised in patients with cervical or upper thoracic cord injuries who require ventilation. Reduced vasomotor control will lead to hypotension on initiation of positive-pressure ventilation and/or positive end-expiratory pressure. Absent sympathetic drive may result in higher resting bronchoconstrictor tone, and vagal stimulation from endobronchial suctioning will lead to bradyarrhythmias or asystole.

Patients with thoracic lesions have the use of intercostal and abdominal muscles, particularly as the level of the lesion descends. However, thoracic spinal injuries have a higher degree of associated injuries than cervical injuries, including fractured ribs, pulmonary contusion, and/or myocardial contusion, which will aggravate an already compromised respiratory musculature.

Gastrointestinal complications

Ileus, due to autonomic disruption, other intra-abdominal pathology, or retroperitoneal hematoma from vertebral fractures, is common in acute spinal cord injury. Nasogastric tubes should be placed and enteral feeding cautiously initiated, preferably beyond the ligament of Trietz.

Stress gastritis can occur and appropriate prophylaxis should be commenced. Coincidental intra-abdominal pathology is a diagnostic dilemma since symptoms are altered or absent because of paralysis. Some visceral sensation may be present as a result of vagal afferents, but any pain will be poorly localized and guarding will be absent. Any patient with pyrexia, leukocytosis, cardiovascular instability, nausea, and intolerance of enteral feeding should be investigated for intra-abdominal pathology, as morbidity and mortality are significantly increased if diagnosis is delayed.

Genitourinary problems

Acute urinary retention due to spinal shock will occur with all spinal cord injuries above S2. Catheterization under sterile conditions should be performed on admission and before retention develops in all patients.

Metabolism and nutrition

Nutritional requirements in the acute phase of a spinal cord injury are not well understood, although it is common practice to provide patients with an increased protein and calorie regime. Patients may manifest biochemical markers of malnutrition despite apparently adequate intake. Urinary nitrogen excretion and creatinine formation increase significantly due to muscle breakdown distal to the cord injury. When there are other major injuries associated with sepsis, patients should be treated as having a hypercatabolic response, requiring stress formulas on occasion.

Alteration of thermoregulation depends on the level of the cord injury. Distal to the lesion, thermoreception is lost, sweating and vasoconstriction are abolished, and the ability to increase muscle tone or shiver is lost. Patients with cervical cord injuries become poikilothermic, which results in frequent episodes of hypothermia unless appropriate measures are taken. The pyrexial response to infection may be masked, and sepsis may present as hypothermia.

Skin

Patients with spinal cord injuries are very susceptible to the formation of pressure sores. The combination of cutaneous anesthesia, immobility, poor nutrition, hypotension, and impaired vasomotor reflexes can rapidly lead to the environment necessary for the development of pressure sores. Once established, they can markedly prolong in-patient care. Early use of special rotating beds can be beneficial.

Pharmacological aspects

In general, pharmacological responses to drugs commonly used in critical care are unchanged in acute spinal cord injury. The responses to inotropes have been discusssed previously. The hyperkalemic response to the use of succinylcholine (suxamethonium) is well known, and the drug should be avoided for 24 h after injury. Muscle relaxants should be used with caution in the presence of an unstable spinal fracture as, theoretically, they increase the instability.

Muscle spasms may develop distal to the injury from the second to sixth week owing to enhanced spinal reflexes. If these cause difficulty with patient management, treatment with baclofen or dantrolene may be helpful, bearing in mind interactions with benzodiazepines and muscle relaxants.

Psychological problems

Patients with spinal cord injuries are prone to a number of psychological problems similar to those seen in other intensive care unit patients, in addition to dealing with their paralysis. Denial of the injury or its implications is common, and patients may appear unusually cheerful in relation to their degree of dysfunction. At some point during the early stages of the injury a reactive depression appears, which will respond to antidepressants if treatment is felt necessary. Sensory deprivation will occur in proportion to the level of neurological deficit, which itself can lead to confusion, visual hallucinations, and sensorimotor hallucinations. Family and familiarity with the environment can be helpful.

Prognosis

The life expectancy of patients with spinal cord injury has increased dramatically over the last few decades owing to the introduction of appropriate management of long-term morbidity. No firm prognosis can be made, as life expectancy depends more on the health care system in the area in which the patient resides than on other factors, although age and level of spinal cord injury do play a significant part.

References

Bracken, M.B., *et al.* (1990). A randomized, controlled trial of methyl prednisolone or naloxone in the treatment of acute spinal-cord injury. *New England Journal of Medicine*, **322**, 1405–11.

Bracken, M.B., *et al.* (1997). Administration of methylprednisolone for 24 or 48 hours or tirilazad mesylate for 48 hours in the treatment of acute spinal cord injury. *Journal of the American Medical Association*, **277**, 1597–1604.

Burney, R.E., Maro, R.F., Maynard, F., and Karonas, R. (1993). Incidence, characterics, and outcome of spinal cord injury at trauma centers in North America. *Archives of Surgery*, **128**, 596–9.

11.5 Burns

11.5.1 Pathophysiology

J. Hammond

Key messages

- Early burn wound edema is due to loss of microvascular integrity, impairment of cell membrane function, and hypoproteinemia.
- Loss of the sodium–potassium pump, the stimulation of inflammatory mediators, and release of oxygen free radicals may all lead to oxygen debt in a hypermetabolic state.
- Inhalation injury significantly increases morbidity and mortality.

Introduction

Burn injury, due to subthermal, suprathermal, chemical, or electrical causes, is a complex physiological phenomenon. It represents a formidable public health problem in both industrialized and emerging nations. In the United States, burn injury is the third leading cause of injury death and the incidence of serious burn injury is the highest of all industrialized countries. In addition to the almost 10 000 burn victims who die each year in the United States, over 100 000 require prolonged hospital care or rehabilitation. The cost to society of the long-term morbidity from loss of physical function or associated psychological and cosmetic impairment is incalculable.

The complexity of burn injury requires an understanding of cardiovascular, pulmonary, metabolic, and immunological derangements. Moreover, burn injury is a dynamic rather than static process, and as such the physiological and biochemical changes in the early and later stages may differ.

Local effects

The initiation of burn injury begins with damage to or loss of the victim's skin, which acts as an interface with the physical environment. The skin, which is the largest organ in the body, is not just a passive envelope, but a physiologically and immunologically active organ. In the immediate zone of injury, microcirculatory thrombosis and protein coagulation result in tissue death. A zone of ischemia exists between dead and living tissue in the early postburn period (Demling 1990). Impaired blood flow can persist in this zone of ischemia even after restoration of systemic hemodynamic parameters. The hallmark of tissue reaction in this zone is edema formation.

Local edema formation is a result of three pathophysiological vascular changes: loss of microvascular integrity resulting in fluid flux into the interstitium, impairment of cell membrane function resulting in cell swelling as a result of loss of cell membrane potential, and hypoproteinemia coupled with increased osmotic pressure in burned tissue. Derangement of cell membrane potential and sodium–potassium pump function also occurs systematically and may contribute to remote organ dysfunction, causing edema in non-burn tissue (Griswold 1994). The purpose of fluid resuscitation in the early postburn period is re-expansion of plasma volume within the extracellular space. This hinges on delivery of adequate amounts of sodium ion into the extracellular space so that cellular trans-membrane potentials return to normal.

While restoration of the sodium–potassium pump is central to control of intracellular swelling, edema is also exacerbated by the hypoproteinemia-based reduction in colloid oncotic pressure. Burn wound edema begins almost immediately, and non-burn tissue edema starts after 4 to 8 h. Human and animal studies demonstrate that the renin–angiotensin–aldosterone axis is appropriately responsive to postburn changes in plasma volume (Gore *et al.* 1996). Therefore, after restoration of plasma volume, sympathetic activity may play a predominant role in regulating renal blood flow.

The role of inflammatory mediators

The observation of progressive changes in the burn wound suggest the possibility that inflammatory mediators may play a role. Vasoactive mediators such as histamine, prostaglandins, and bradykinin can produce edema by direct action on endothelial cell permeability or indirectly by increasing microvascular hydrostatic pressure. While the role and significance of vasoactive substances and mediators is unclear, the concept of mediator-induced injury takes on clinical importance because of the theoretical possibility of injury modulation by mediator inhibition (Demling 1990). Histamine, released either as a direct result of heat or by stimulation of mast cells, is found in increased concentrations after thermal injury. Prevention of histamine release by H_2-blocking agents has been demonstrated to decrease the edema-producing effects of vasodilation and increased microvascular permeability.

The complement system is activated by an undefined mechanism. C5a appears to be the initiating factor in the inflammatory response, activating intravascular neutrophil secretion of histamine. C5a also catalyzes conversion of xanthine dehydrogenase into xanthine oxidase by interacting with endothelial cells leading to formation of oxygen radicals. Full-thickness burns exceeding 20 per cent of body surface area are associated with increased systemic oxidant activity, as measured by increased liver lipid peroxidation. Animal studies have shown that alterations in ATP production can be reversed by treatment with water-soluble antioxidants, suggesting a cause-and-effect relationship (Lalonde *et al.* 1996). C5a may also play a role as a cofactor in the production of tumor necrosis factor-α.

Oxygen free radicals may exert an indirect effect through the release of arachidonic acid metabolites. Prostaglandins released from burn tissue may contribute to edema formation. Moreover, burn injury is associated with increased plasma levels of nitric oxide, a potent endogenous vasodilator (Preiser *et al.* 1996). This appears to be due to increased release rather than decreased renal clearance. Nitric oxide may also participate in burn-related hemodynamic and immunological alterations.

Inhalation injury and respiratory burns

A concomitant inhalation injury, compounding a cutaneous burn, has a greater effect on mortality than either the age of the patient or the total body surface area of the burn. Inhalation injuries are associated with a reported mortality of 45 to 78 per cent, and mortality has been shown to be 20 to 40 per cent greater in patients with a combined cutaneous burn and inhalation injury than in patients with similar burns but without an inhalation injury (Herndon *et al.* 1985).

True respiratory burns of the lower pulmonary tree are rare, generally occurring only with inhalation of superheated steam. Moist air has 4000 times the heat-carrying capacity of dry air and interferes with the ability of the upper respiratory tract mucosa to dissipate heat efficiently. What is commonly thought of as an inhalation injury is generally smoke inhalation or carbon monoxide poisoning or both. These result in hypoxemia associated with bronchospasm and bronchorrhea.

Incomplete products of combustion, such as aldehydes, nitrogen dioxide, and hydrochloric acid, can cause direct parenchymal lung damage. Plastics and petroleum-based products such as polyvinyl chloride can release extremely toxic agents (e.g. benzene, phosgene, and isocyanates). These small smoke particles may cause damage either as direct irritants or as a mechanism for transmission of heat.

Carbon monoxide, which has an affinity for hemoglobin 200 times that of oxygen, will seriously impair oxygen delivery to the tissues. Carboxyhemoglobin levels of less than 15 per cent do not generally alter clinical outcome significantly; however, levels greater than 40 per cent are often associated with neurological or mental status changes. Treatment consists of the administration of high-flow oxygen, with or without intubation and hyperbaric oxygen. The concentration of carboxyhemoglobin is reduced by 50 per cent for each 40-min period of treatment with high-flow oxygen.

Pulmonary injury resulting from smoke inhalation results in histological and physiological changes in pulmonary function. The most prominent is an early marked increase in extravascular lung water. The severity and duration, usually from 24 h to 5 days, depends on the nature of the inhalation injury and the presence or absence of an associated cutaneous burn injury. An inflammatory exudate, containing high concentrations of thromboxane A_2 and glucuronidase, rapidly forms in the tracheobronchial tree. The end result of this cascade is progressive airway obstruction. Neutrophils present in the lung at the time of injury are trapped; additional neutrophils are recruited to the lung by release of chemotactic factors produced by stimulated pulmonary macrophages. Release of the oxygen free radicals and proteolytic enzymes results in fibronectin degradation and interstitial matrix disruption.

Systemic effects

A decrease in oxygen delivery in the immediate postburn period coupled with increased cellular demands for oxygen can result in an oxygen debt. Simultaneously, a massive catecholamine release from both the adrenal gland and the autonomic nervous system contributes to profound metabolic sequelae. This catecholamine discharge, coupled with resuscitation-associated edema in the bowel mesentery, results in a generalized ileus associated with burns covering more than 20 per cent of the body surface area.

The hypercatabolic state in patients with greater than 20 per cent body surface area burns is characterized by increased oxygen consumption, increased nitrogen excretion, and weight loss. The increased metabolic rate is proportional to the size of the burn up to 50 per cent total body surface area, where a plateau is reached at near-maximal levels of heat production. The metabolic rate may exceed normal levels by a factor of 2 to 2.5. The hypermetabolic state is characterized by the preferential use of lean body mass for gluconeogenesis.

The 'reset' of the hypothalamic temperature center, resulting in body temperature rises of 1 to 2 °C, is probably initiated by the release of pyrogenic substances such as interleukin 1 or wound-generated prostaglandins. Since evaporation of 1 ml of water requires 0.58 kcal, some portion of the hypercatabolism is associated with heat loss due to vaporization of water. Control of the external environment, by modulating temperature and humidity, has had only partial success in ameliorating the catabolic response. The regulatory hormones glucagon, cortisol, and epinephrine (adrenaline) play an important role in fueling the hypermetabolic state. The hypermetabolic response may also result in part from contamination of the burn wound with bacteria. Translocation of gut bacteria and release of endotoxin may also play a role. Hyperthyroidism does not appear to play a role. The end result of this stimulation is rapid loss of body nitrogen. The metabolic rate returns to normal once the wound is closed or grafted and inflammation is resolved.

References

Demling, R.H. (1990). Physiologic changes in burn patients. In *Scientific American Surgery* (ed. D. Wilmore *et al.*), pp. 1–8. Scientific American, New York.

Gore, D.C., Dalton, J.M., and Gehr, T.W. (1996). Colloid infusions reduce glomerular filtration in resuscitated burn victims. *Journal of Trauma*, **40**, 356–60.

Griswold, J.A. (1994). Sodium concentrations and burn shock resuscitation. In *Problems in general surgery: burns* (ed. C. Scott-Connor), pp. 566–76. J.P. Lippincott, Philadelphia, PA.

Herndon, D.N., Thompson, P.B., and Traber, D.L. (1985). Pulmonary injury in burned patients. *Critical Care Clinics*, **1**, 79–96.

LaLonde, C., Nayak, U., Hennigan, J., and Demling, R. (1996). Antioxidants prevent the cellular deficit produced in response to burn injury. *Journal of Burn Care and Rehabilitation*, **17**, 379–83.

Preiser, J.C., *et al.* (1996). Nitric oxide production is increased in patients after burn injury. *Journal of Trauma*, **40**, 368–71.

11.5.2 Assessment

J. B. Cone and R. D. Robertson

Key messages

- Intensive care units where serious burn injuries are treated require appropriate personnel and space.
- Size, depth, location, type of burn, inhalation, age, pre-existing conditions, and associated injuries are all important in predicting prognosis.
- Burn wounds must be observed constantly, treated, and covered at the most opportune time.

Introduction

Most severely burned patients will need the facilities of an intensive care unit (ICU), at least initially, for the management of hypovolemia, smoke inhalation, associated injuries, and pre-existing conditions. Ideally, such patients should be in a burn ICU; however, many ICUs will receive the occasional burned patient and the decision must be made either to keep and treat the patient or to transfer him or her to a specialized facility. Such decision making requires a knowledge of the support that an individual patient may need as well as the probable outcome. Approximately 10 per cent of burned patients will require prolonged intensive care, with an average stay of about 25 days (Lingnau *et al.* 1996).

The American Burn Association has produced detailed and specific recommendations for burn care facilities (American Burn Association 1990). However, a well-functioning burn ICU requires both an appropriate facility and a team of burn care specialists, including surgeons, intensivists, nurses, therapists, and many others. Burn ICUs must have space available for all the usual ICU equipment, such as ventilators and portable X-ray machines, but also need additional space for performing wound care. The potential for cross-infection is increased in an overcrowded unit. Individual rooms with isolation facilities and separate air-handling systems are strongly recommended for burns greater than 50 to 60 per cent. The unit must also have effective individual temperature control because of the excessive heat loss associated with burn injury. ICUs caring for burned patients have all the standard requirements, including adequate electrical and gas lines, and space for medications and charting, but the storage needs are generally much greater because of the large quantities of wound care supplies often needed daily for each patient. Typically, a unit in the United States will be directed by either a general surgeon or a plastic surgeon with additional experience in burn care and a head nurse/manager with both intensive care and burn experience. A number of medical specialists must be available, including additional intensivists, anesthesiologists, internists, psychiatrists, etc. The nursing care of seriously burned patients requires at least a one-to-one ratio, and this must often be supplemented by technicians and aides to assist with wound care and other activities. An array of other specialists is required to provide optimum care, including respiratory therapists, occupational and physical therapists, and social workers. The rehabilitation process is often deferred until the patient is past the critical stage of his or her injury. Unfortunately, such delays often result in permanent impairment or unnecessarily prolonged rehabilitation.

Initial assessment

The prognosis for patients with burn injuries is determined by the size, depth, location, and type of burn as well as related factors such as age, pre-existing conditions, and associated injuries (Dimick 1994).

The size of a burn injury is usually expressed in terms of the percentage of the body surface area involved. This measurement is utilized to predict resuscitation fluid volumes and to assess prognosis. Unfortunately, this number is not readily apparent on inspection of the patient. The simplest and most widely used guide for estimating burn wound size is the rule of nines (Fig. 1), which equates each major body part (e.g. a limb) to 9 per cent, or a multiple of 9 per cent, of the body surface area. This handy rule of thumb is sufficiently reliable for most purposes in adult burn patients, but introduces significant error for young children owing to their disproportionately larger heads. If burn size is being estimated in very young children or if precise estimates are necessary, the age-corrected diagrams shown in Fig. 2 are more appropriate. It is often difficult to apply either of these systems to small scattered burns, and the rule of palms may be helpful in such cases. The patient's palmar surface (not the examiner's) is approximately equal to 1 per cent of the body surface area.

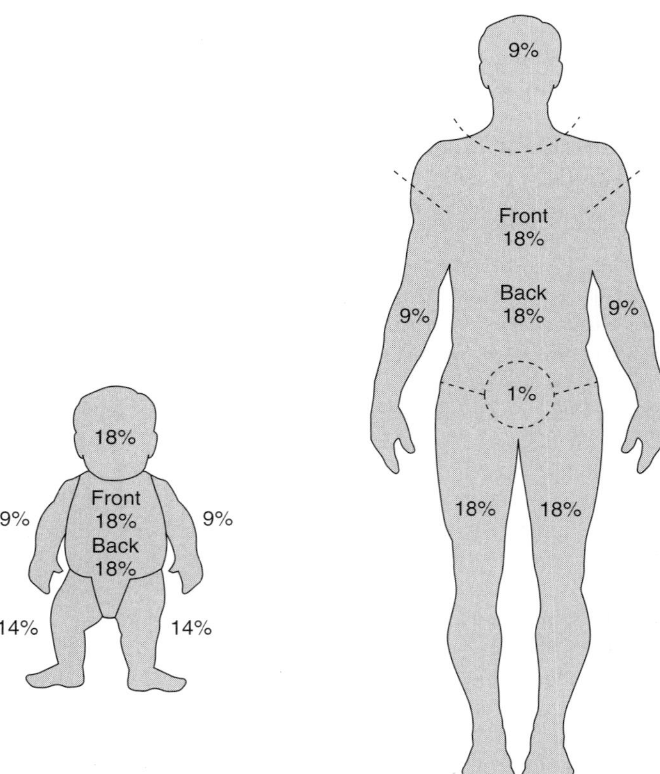

Fig. 1 The rule of nines. This is a simple rule of thumb for the estimation of burn size. Each major body part is either 9 per cent or a multiple of 9 per cent of the body surface area. This rule works well enough for routine purposes in adults but is much less reliable for young children owing to the disproportionately larger head.

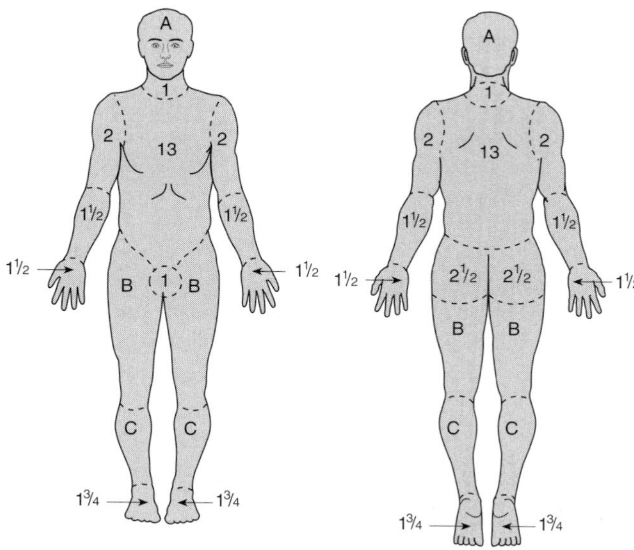

Relative percentage of areas affected by growth

	Age (years)					
	0	1	5	10	15	Adult
A-$\frac{1}{2}$ of head	$9\frac{1}{2}$	$8\frac{1}{2}$	$6\frac{1}{2}$	$5\frac{1}{2}$	$4\frac{1}{2}$	$3\frac{1}{2}$
B-$\frac{1}{2}$ of one thigh	$2\frac{3}{4}$	$3\frac{1}{4}$	4	$4\frac{1}{4}$	$4\frac{1}{2}$	$4\frac{3}{4}$
C-$\frac{1}{2}$ of one leg	$2\frac{1}{2}$	$2\frac{1}{2}$	$2\frac{3}{4}$	3	$3\frac{1}{4}$	$3\frac{1}{2}$

Total per cent burned — 2° + — 3° = ——

Fig. 2 The Lund–Browder chart. This chart allows more precise estimation of the body surface area injured. The area of burn is drawn on the figure and then estimated either visually or by planimetry. This chart also allows easy age-specific correction.

Burn wound depth is measured not in millimeters but in terms of which epidermal structures are destroyed. Thus a burn of depth 2 mm on the dorsum of an elderly hand is much 'deeper' than a burn of depth 2 mm on an adolescent's back. Traditionally, burn depth has been classified as first, second, or third degree. First-degree burns, similar to a sunburn, involve only the epidermis and are characterized by erythema and mild discomfort. Such injuries heal in a few days and produce essentially no systemic effects. The area involved in such epidermal burns should not be included in estimates of burn size since they do not contribute to fluid requirements or outcome determinations. Second-degree burns, more descriptively known as partial-thickness burns, represent destruction through the epidermis and to varying depths in the dermis. Such burns have the capacity to regenerate an epidermal layer from the epidermal cells lining skin appendages such as hair follicles and sweat glands which are embedded in the dermis. Many of these burns will heal within 2 to 3 weeks with an acceptable functional and cosmetic result. As the coagulation necrosis extends deeper into the dermis, fewer appendages remain and a longer period of time is required for re-epithelization of the wound. Healing in this fashion is often associated with excessive scarring and contractures which compromise mobility. Third-degree or full-thickness burns have no surviving epidermis or dermis. Such wounds can heal only by epithelial ingrowth from the margins and contracture. If these injuries are more than a few centimeters in size, they require skin grafting to heal.

Since many burn clinicians believe that patients with full-thickness and even deep partial-thickness burns benefit from early aggressive excision and skin grafting, it is crucial to be able to distinguish burn depth early in the patient's course. Superficial partial-thickness injuries characteristically form watery blisters, although these may not appear for several hours after the injury. When the blister is removed, the underlying tissue is moist, pink, and hypersensitive to touch. Because of the increased dermal blood flow, these wounds blanch when pressure is applied. In contrast, full-thickness wounds are firm, dry, and leathery. They have diminished sensitivity to light touch or needle prick, and may show coagulated dermal vessels. Classifying either of these wound types is relatively simple. Unfortunately, many patients have burns which do not fall clearly into either category. Such wounds may blister, but the underlying surface is mottled and does not blanch with pressure. Some sensation is preserved, but they are less sensitive than surrounding skin. Even experienced burn clinicians do not achieve an accuracy of better than 50 to 60 per cent in evaluating such wounds. In addition to physical examination, the mechanism of injury often provides valuable clues to burn depth. For example, scald burns without prolonged contact, such as a splash of hot liquid, often produce superficial partial-thickness wounds, while burning clothing almost always produces either deep dermal or full-thickness injury.

The area of the body involved in burn injury has little effect on resuscitation requirements or survival, but has a major impact on functional and cosmetic outcome. Traditional high-risk areas include the face, hands, feet, and perineum. Facial burns from flame should suggest the possibility of smoke inhalation. The eyes should be examined both directly and with the use of fluorescein and ultraviolet light to detect corneal or conjunctival injury. If burns involve the periorbital area, the patient should be warned to anticipate swelling which may prohibit opening the eyes and often lasts for 48 h. Advance warning will often avoid needless anxiety. Hand burns impair the patient's ability to feed and care for him- or herself. This impairment may last for only a few days as the result of pain and swelling; however, if the burn is deeper multiple surgical procedures and months of therapy may be required to produce acceptable function. Burns of the feet impair the patient's mobility and may lead to toe contractures and long-term problems with gait. Perineal burns present problems in controlling fecal contamination and the potential for scarring and contracture compromising genitourinary function.

Age is probably the most important of the non-burn-specific factors in determining outcome. Patients at either extreme of age have significantly poorer prognosis than children and young adults. There is still significant debate regarding the optimal approach to burn care in the elderly. Should they be treated with early aggressive excision or with a slower and more conservative strategy? Neither approach produces a significant chance of survival of the elderly with burns larger than 35 to 40 per cent (Anous and Heimbach 1986). Associated injuries such as fractures may complicate both the management of the burn wound and the patient's rehabilitation. The stress of a major burn injury may aggravate pre-existing medical conditions, leading to delays in necessary surgery or increases in morbidity and mortality.

Numerous authors have combined these factors to produce mathematical models of burn patient survival. Like all statistical methods they must be applied cautiously to the individual patient, but they do provide some generalizations regarding outcome.

Subsequent assessment

The primary goal of burn care is early effective wound closure. However, the burn wound is not a static entity. In the first few days following injury the wound may appear to become deeper. Later in the course areas heal and new wounds are created (i.e. donor sites for skin grafts). Initially, the burn consists of a central area of coagulation necrosis with a surrounding zone of stasis and poor perfusion due to microvascular injury. This zone of stasis frequently progresses to tissue necrosis as the cells die of ischemia, so that a burn may 'convert' from second to third degree over the first 48 to 72 h post-injury. After this initial period, further progression of the wound usually indicates infection which must be controlled aggressively. Frequent reassessment is essential since the remaining open wound is the primary determinant of patient management and prognosis.

Today, with improved wound care and earlier surgical excision, sepsis originating in the burn wound has been replaced by pulmonary sepsis associated with smoke inhalation as the leading cause of death. However, burn wound sepsis is still a major source of morbidity, including prolonged hospitalization and skin graft loss with the need for regrafting. Infection arising in the burn wound during the first 72 h postinjury usually appears as cellulitis surrounding the margins of the burn. The offending organisms are routinely either streptococci or staphylococci. Such infections may initially be difficult to distinguish from adjacent areas of first-degree burn with erythema, heat, and pain. However, serial examinations, often over only a few hours, will demonstrate progression. Such infections occur in about 30 per cent of admitted burn patients. The use of prophylactic antibiotics for 48 to 72 h remains a disputed issue. The cellulitis usually responds promptly once it is recognized and therapy is instituted.

Of much graver concern is the issue of invasive wound infection. It is unusual for such infections to occur prior to the end of the second postburn week. The organisms involved are predominantly Gram-negative enterics and staphylococci arising from the patient's own gastrointestinal tract in at least 80 per cent of cases. The precise organisms involved tend to be unit and institution specific, having previously colonized the patient's aerodigestive tract. Bacterial colonization of the burn wound is impossible to prevent, but tissue damage can usually be minimized if the number of organisms can be kept low. When the number of organisms reaches or exceeds 100 000 per gram of tissue, local defense mechanisms are often overwhelmed and invasion into underlying healthy tissue results. The eschar, if still present, will rapidly separate, demonstrating hemorrhagic discoloration of the underlying fat and violaceous edematous wound margins. If unchecked at this level, systemic sepsis may result. The mechanism by which the wound should be monitored to detect this progression in its earliest stages remains controversial. Visual inspection may demonstrate nothing as the bacterial burden is increasing. The quantity and character of wound drainage are often obscured by the topical antimicrobial applied. The earliest indication of local infection is often a failure to show the expected progress in wound healing. The superficial partial-thickness wound should demonstrate islands of epithelium growing to confluence at a rate sufficient to produce closure in 14 to 21 days, while wounds grafted with meshed skin grafts should show adherent dry grafts at 5 days and by 7 to 10 days most of the interstices of the mesh should be filled with epithelium. Unfortunately, failure of this progression has more than one possible cause. Inadequate nutrition, poor blood supply in the wound bed, and failure to immobilize grafts adequately can all produce similar findings. For these reasons, many burn units routinely monitor the burn wound with surveillance cultures. Such surveillance techniques range from simple swab cultures to quantitative cultures to histology performed on tissue biopsies. The pros and cons of these techniques have been extensively reviewed elsewhere (Heggers and Robson 1991). Early recognition that control of the wound has been lost offers the best chance of minimizing the resulting damage. Regardless of the technique chosen, changes must be made in the wound management protocol when the number or depth of invasion of organisms increases.

Recognition of systemic infection in the burned patient

Identification of systemic infection in most patients is based on the presence of fever, leukocytosis, erythema, and a hyperdynamic state. However, these signs are commonly present in burn patients in the absence of infection. Temperature elevations to 39 °C are common in the early postburn period. Hypothermia or a change in an established temperature pattern is a better indication of infection than simply the presence of a temperature elevation. Tachycardia and tachypnea result from the metabolic demands imposed by the injury. The best indicators of early sepsis in the burned patient are changes in established patterns. The patient who has an unexplained alteration in mental status, new glucose intolerance, and recent onset of a new ileus is septic until proven otherwise.

Conclusion

The burn wound is the dominant feature of the burned patient. It differentiates burned patients from other critically ill patients, accounts for the majority of the work involved in caring for the patient, and is responsible for most of the physiological derangements which are associated with critical burn injury. The central theme of modern burn care is support of the patient while achieving the most rapid wound closure that he or she can tolerate. As wound closure is accomplished, priorities shift from survival to function and appearance.

References

American Burn Association (1990). Hospital and prehospital resources for optimal care of patients with burn injury: guidelines for development and operation of burn centers. *Journal of Burn Care and Rehabilitation*, **11**, 98–104.

Anous, M.M. and Heimbach, D.M. (1986). Causes of death and predictors in burned patients more than 60 years of age. *Journal of Trauma*, **26**, 135–9.

Dimick, A.R. (1994). Initial management of thermal injuries using the advanced burn life support protocol. *Problems in General Surgery*, **11**, 559–65.

Heggers, J.P. and Robson, M.C. (ed.) (1991). *Quantitative bacteriology: its role in the armamentarium of the surgeon*. CRC Press, Boca Raton, FL.

Lingnau, W.W., Nguyen, T.T., Woodson, L.C., Herndon, D.N., and Prough, D.S. (1996). Critical care of burn complications. In *Total burn care* (ed. D.N. Herndon), pp. 319–45. W.B. Saunders, London.

11.5.3 Medical management

J. B. Cone and R. D. Robertson

Key messages

- Hypermetabolism necessitates early nutritional support.
- Early aggressive fluid resuscitation with correction of electrolyte abnormalities is crucial.
- Pain control is critical, although drug clearance is altered in the burned patient.
- Topical burn care and surgery should be complementary and initiated early.

Introduction

Burn injury, like most trauma, is a predominantly surgical disease, requiring the removal of devitalized tissue and wound closure. Many burn patients present with medical problems which complicate or delay wound closure if they are not addressed in addition to such acquired conditions as respiratory failure from smoke inhalation and renal failure due to inadequate resuscitation. Hypermetabolism and nutritional support, fluid and electrolyte management, altered pharmacokinetics, topical antimicrobials, and pain control are some of the issues arising with burned patients.

Hypermetabolism and nutritional support

The hypermetabolic response to burn injury surpasses, in both magnitude and duration, that associated with any other disease process. It can be thought of as consisting of two distinct components. First is the increased heat loss from the destruction of the skin's barrier function. Homeostasis in the face of such heat loss demands increased heat production and hence hypermetabolism. This component of demand can be reduced by decreasing the heat loss with occlusive dressings or by raising the ambient temperature. The second component appears to be mediated by humoral factors such as catecholamines, prostanoids, and cytokines from the burn wound, the gut, or elsewhere. Efforts to reduce this component of the demand have focused on agents such as propranolol and ibuprofen (Cone *et al.* 1994).

The hypermetabolic response is characterized by fever, tachycardia, tachypnea, and increased caloric consumption and protein catabolism. The increased metabolic rate appears approximately 24 h after the injury but does not peak until 7 to 10 days later. Thereafter, the metabolic rate decreases in proportion to the degree of wound closure which has been achieved by either spontaneous healing or successful skin grafting.

The initial nutritional assessment should be performed as soon as possible after injury and should include information on height, weight, age, size and depth of burn, and any pre-existing nutritional problems. Conditions such as alcohol abuse, excessive dieting, or gastrointestinal problems which have produced a deficit should be sought. With this information, an initial estimate of caloric and protein needs should be prepared. There is no agreement on how these data should be utilized to set nutritional goals. The widely utilized Curreri formula illustrates the high end of such estimates, often predicting a need for more than 5000 kcal/day. The Curreri formula was developed from studies of patients in a cool environment without dressings (Hildreth and Gottschlich 1996). It is important that each burn unit select a nutritional prediction system which was created under the same ambient conditions routinely utilized in that unit.

Portable metabolic carts allow the performance of indirect calorimetry at the bedside and individualization of metabolic calculations. Such measurements are not without problems. A single measurement over 20 to 60 min cannot be taken in isolation. Changes in temperature, pain from dressing changes, and activity of physical therapy all produce major alterations in metabolic rate.

Burned patients also suffer a marked protein catabolism. This wasting peaks in the second week postburn, but the losses remain supranormal throughout healing. Protein needs are difficult to measure with conventional nitrogen balance techniques because of wound losses, but measurement of urinary nitrogen often provides valuable information regarding protein supplementation. Protein needs are often estimated in relation to caloric needs such that 20 per cent of the kilocalories are administered as protein with a non-protein calorie-to-nitrogen ratio of 100:1.

Another approach to estimating the protein need is to use the formula

$$\text{protein need} = \text{g/kg} + 3\,\text{gm/\%burn day}.$$

Whether the needs are estimated from a formula or measured using a metabolic cart, once these goals have been met, the patient's nutritional status should be reassessed on a regular basis, noting weight change, wound healing, and visceral protein levels (i.e. prealbumin).

Patients with small to moderate burns (below 20 per cent) are often able to meet their needs orally, although this may require aggressive efforts on the part of the staff to select appealing foods and to motivate and feed patients. Larger burns (above 30 per cent), or those complicated by smoke inhalation, injuries, or medical conditions, commonly exceed the patient's ability to take oral nutrients and hence require nutritional support. Parenteral nutrition offers no advantage over enteral nutrition, is associated with a higher rate of complications, and is more expensive. The increased complication rate probably results not from the intravenous nutrients but from disuse of the gut. The enteral route is clearly preferred for nutritional support in burned patients, with parenteral nutrition being reserved for those who cannot be fed enterally.

The larger the burn, the more critical it is that nutrition is initiated early. Placement of enteral feeding tubes is often accomplished immediately after the airway, breathing, and circulation (ABCs) have been secured. There is nothing to be gained by waiting, and feedings are often better tolerated if started within the first few hours following injury. Specific additives, such as arginine, glutamine, ω_3 fatty acids, nucleic acids, etc., are being tested to enhance specific dysfunctions in the immune system, cell growth, etc. Initial results appear promising, but these new feeding formulas are significantly more expensive and in some cases may not be as well tolerated. Their final role is still unknown. Burned patients require increased intake of virtually all vitamins and minerals which have been studied. The requirements for supplemental phosphorus, zinc, and vitamin C are of particular importance.

Fluid and electrolyte management

Major thermal injury causes massive fluid shift from the vascular space to the interstitial and intracellular compartments, resulting in soft tissue edema and, if not replaced, hypovolemic shock. The management principles for burn shock are no different than those for any form of hypovolemic shock. However, since it is relatively simple to quantitate the severity of a burn injury and difficult to assess the intravascular volume loss visually, a number of formulas have been developed to predict the volume required to restore or maintain intravascular volume. These equations call for 2 to 4 ml of isotonic crystalloid per kilogram of body weight for each per cent of the body surface area burned in the first 24 h; half is given within the first 8 h following injury. The formulas provide a reasonable guide to initial resuscitation but should not be followed rigorously. Subsequent fluid replacement should be based on patient response. Patients who require significantly more than the predicted volume are often found to have either very deep burns or inhalation injury.

Fluid and protein may continue to leak from the vascular compartment for several days after injury, but the rate of loss slows after the first 24 to 48 h when soft tissue edema reaches maximal levels. Further fluid requirements will depend upon the type of resuscitation used as well as current wound care and renal function. If severe hypoproteinemia is present, spontaneous mobilization of the extravascular water will be delayed. Some patients require large quantities of free water to excrete the solute load.

Burned skin no longer forms an effective barrier to water loss into dressings or by evaporation. Such losses can be crudely estimated from the formula

$$\text{water loss(ml/h)} = 25 + \text{body surface area burned} \times \text{surface area (m}^2).$$

It is important to recognize that traditional indicators of volume status, such as urine output, may be unreliable in the first 10 to 14 days postburn, when large quantities of osmotically active breakdown products from injured tissue are released into the circulation and may induce an osmotic diuresis. This period also corresponds to the increase in nutrient delivery. If urine output is used to assess intravascular volume, it must be coupled with specific gravity, osmolality, etc.

Insensible water loss can be calculated from the difference between fluid intake and output and the change in weight, and can be replaced by low-sodium fluids since the most prevalent electrolyte abnormality during this period is hypernatremia from the loss of free water from the wound. When fluid balance is critical, fluid requirements can be calculated as the insensible loss from the preceding 24 h plus the desired urine output for the next 24 h.

Other electrolyte abnormalities, including hypokalemia, hypophosphatemia, hypocalcemia, and hypomagnesemia, are common, particularly in the first 10 to 14 days postinjury. Frequent monitoring may be required with more severe burns which need more aggressive replacement. When the patient is tolerating enteral nutrition, fluid and electrolyte replacement can be administered via feeding tube and the intravenous catheters can be removed.

Pharmacokinetics in burned patients

Burned patients experience a hyperdynamic circulation, fluid shifts, altered serum protein levels, and unusual fluid losses which often complicate the dosing of medications including antibiotics, analgesics, etc. Such alterations in clearance, distribution, and protein binding may not be critical for drugs with wide margins between therapeutic and toxic levels; however, where the margin is narrow, these effects must be taken into account and combined with frequent monitoring.

Higher than expected drug levels during the hypodynamic or resuscitation phase may result from decreased volume of distribution or because clearance is prolonged by decreased hepatic and renal blood flow. Decreased blood flow to organs and tissues may result in decreased absorption from the gastrintestinal tract or from subcutaneous or intramuscular sites. Therefore the safest approach is frequent small intravenous doses with regular monitoring of their effects.

A supranormal cardiac output, with increased blood flow to the splanchnic and renal beds, elevated temperature, hypoproteinemia, and major fluid shifts are common during the hypermetabolic phase of burn injury. Renal blood flow is significantly increased, but renal tubular function is depressed. Drugs cleared primarily by filtration may show accelerated elimination, while those secreted by the tubules may accumulate. Hepatic drug metabolism is more complex, but some generalizations are possible. Drug elimination by oxidation or reduction, hydroxylation, etc. will often be impaired, while conjugation reactions are preserved. Plasma protein levels are significantly altered during this period. Even when nutrition is well maintained, albumin synthesis is depressed as a component of the acute phase response, while synthesis of α_1-acid glycoprotein is increased. Drugs with significant albumin binding usually show an increase in the free drug, but the free fraction of drugs binding to α_1-acid glycoprotein will be reduced. Such changes in drug binding also lead to changes in renal clearance, since only the free component is readily filtered at the glomerulus. Drugs with a narrow therapeutic window should not be used in burned patients unless no other agent will achieve similar therapeutic goals, in which case the frequency of monitoring of both drug levels and effects must be increased.

Pain management

Pain management for burned patients is often unsatisfactory because of misconceptions about the sources of pain and the optimal use of various analgesic agents in burned patients, and because of such pragmatic factors as unit staffing. A common misconception is that deep burns, which are insensate to touch, are not painful. They frequently produce a severe dull pressure-like pain. Although the presence of pain can usually be recognized from physical signs such as pupillary dilation, sweating, tachycardia, tachypnea, and increased blood pressure, these findings are relatively poor indicators of its severity. Quantitation of pain severity should rely on the patient's own assessment via systems such as the visual analog scale or numerical and graphical rating scales (Kremer *et al.* 1981).

In the early postburn period, pain management relies on intravenous narcotics administered as periodic boluses or continuous infusion. Morphine is the most widely used agent, but fentanyl is popular because of its rapid onset, short duration of action, and fewer adverse effects such as histamine release and peripheral vasodilation. The use of patient-controlled intravenous narcotics has been very effective for older children and adults, as this not only provides better steady state pain relief but also gives the patient a sense of control. Regardless of the technique, it is important to realize that there will be pain breakthroughs during dressing changes or therapy, and supplemental medication should be given. Unfortunately, any episodes of poorly controlled pain may lead to fear and anxiety prior to the next procedure. The combination of an anxiolytic agent, such as a benzodiazepine, with the analgesic may reduce the narcotic requirement.

Intravenous catheters are a major risk of infection in the burned patient and should be removed as soon as possible. Although this is not usually possible during the intensive care phase, it should be recognized that if an intravenous catheter is not required for other reasons, a variety of effective non-intravenous accesses are available, including transdermal and sublingual fentanyl, intranasal butorphanol and midazolam, and oral agents such as sustained-release or immediate-release morphine (Siddall and Cousins 1995).

Topical antimicrobial agents

Burn injury not only produces thermal necrosis of skin, but also causes microvascular thrombosis, leaving the burn avascular and inaccessible to systemic antimicrobials. Topical antimicrobials are designed for direct application to the wound, with the goal of slowing colonization by bacteria and delaying invasive infection. No topical agent is capable of sterilizing the wound. The ideal topical agent should have activity against organisms likely to be encountered on the wound, should not interfere with wound healing, should not be painful or expensive, should have little or no systemic toxicity, and should be easy to apply and remove. Although topical agents may delay the onset of invasive sepsis, they cannot prevent it indefinitely. Surgical wound closure must be obtained as soon as possible.

Topical agents may also play a role in the therapy of invasive wound infections. Therapeutic agents require good penetration of avascular eschar to achieve therapeutic levels at the junction between eschar and viable tissue. Treatment for invasive infection often requires combination therapy with an effective topical agent to control bacterial growth on the wound plus systemic antibiotics to kill organisms invading viable tissue. Surgical debridement is often required to remove infected devitalized burn eschar which may remain on the wound.

Compared with the number and variety of systemic antibiotics available, there are relatively few suitable topical antimicrobials in widespread use.

Silver sulfadiazine, the most widely used topical agent, is bactericidal against a wide range of Gram-positive and Gram-negative enteric organisms. Application is daily and relatively painless, but penetration of the eschar is limited. Although microbial resistance is not common, it does occur and should be considered when colonization persists or the incidence of infection within the intensive care unit (ICU) is excessive.

Silver sulfadiazine has minimal toxicity but is a sulfonamide and should be used with caution in patients who report hypersensitivity to sulfa drugs. Minor skin rashes at the wound margins are common but rarely require a change in therapy. A significant leukopenia occurs in 10 to 15 per cent of patients, usually 2 to 5 days after initiating therapy. Neutrophil counts of less than 1000/mm^3 are common but are not associated with an increase in infection and usually disappear within 3 to 5 days.

Mafenide acetate is a sulfonamide derivative with excellent penetration of eschar and has the broadest antimicrobial spectrum of any available topical agent. Unfortunately, mafenide is painful to apply, particularly to partial-thickness areas. Better penetration of eschar is an advantage, but rapid absorption may leave the surface with subtherapeutic levels within hours so that more frequent dressing changes are required. Mafenide reaching the systemic circulation is metabolized to an effective carbonic anhydrase inhibitor. When applied to large surface areas for extended time periods or in the face of renal insufficiency the agent and its metabolite produce metabolic acidosis. Respiratory compensation can drastically increase the work of breathing, leading to respiratory failure. Bacterial resistance is infrequent, but does occur with some methicillin-resistant staphylococci. Mafenide is not usually a first-line agent but is held in reserve for established infection, particularly when needed to penetrate tissue. It is often used in conjunction with silver sulfadiazine; each agent is applied for 12 h in an effort to reduce both toxicity and the development of resistant strains.

Silver nitrate (0.5 per cent) solution in water is bacteriostatic for the organisms found in burn wounds. Bulky layers of dressing material are soaked in silver nitrate solution and applied to the wound, with additional solution applied to keep the wound moist. This constant application will leach electrolytes, particularly chloride, from the circulation. Silver nitrate is messy to use, as it stains everything with which it makes contact. However, the aqueous solution is useful over fresh grafts which will not tolerate the cleaning required to remove the vehicle of cream-based agents.

Although not essential for small burns, topical antimicrobial agents can preserve control of the burn wound for an extended although not indefinite period. They should not be used as a substitute for excision and grafting as soon as donor skin is available.

Conclusion

The management of critically burned patients strives to achieve the same goals as for all ICU patients: support of oxygen delivery, matching nutritional support with needs, treating or preventing infection, and relieving pain. However, pursuit of these goals must take account of the factor making the burned patient unique—the burn wound. The wound not only causes pain, but alters metabolism, provides a route for infection to bypass natural defenses, and alters pharmacokinetics.

References

Cone, J.B., Wallace, B.H., and Caldwell, F.T. (1994). The hypermetabolic response to burn injury. *Problems in General Surgery*, **11**, 577–89.

Hildreth, M. and Gottschlich, M. (1996). Nutritional support of the burned patient. In *Total burn care* (ed. D.N. Herndon), pp. 237–50. W.B. Saunders, London.

Kremer, E., Atkinson, J.H., and Ignelzi, R.J. (1981). Measurement of pain: patient preference does not confound pain measurement. *Pain*, **10**, 241–8.

Siddall, P.J. and Cousins, M.J. (1995). Recent advances in pain management. *Australian and New Zealand Journal of Surgery*, **65**, 674–85.

11.5.4 Role of surgical management

J. R. Saffle

Key messages

- In general, early excision and burn wound coverage is ideal.
- Tangential rather than fascial excision is preferred if possible.
- The timing of surgery must coincide with ongoing resuscitation, blood replacement, temperature control, and safe transport.

Introduction

The period following the Second World War has seen remarkable advances in burn treatment. Today, patients routinely survive injuries of 80 per cent total body surface area (TBSA) or more. Although advances in critical care, infection control, nutrition, and other areas have improved the support of patients with major burns, only the ability to remove large burns safely, and to obtain rapid coverage with skin or skin substitutes, has made it possible for burns to be closed and to heal. Burn surgery techniques are described by Heimbach and Engrav (1984) and Muller *et al.* (1996). Figure 1 gives an outline of the role of surgical therapy.

Types of surgery

A cornerstone of modern burn treatment is early excision, in which eschar is removed surgically before it can separate. Early excision and skin grafting of burn wounds reduces time in hospital and costs, improves functional and cosmetic outcomes, promotes early rehabilitation, relieves suffering, and almost certainly improves survival. Excision removes the eschar which causes much of the inflammatory response to burn injury and serves as a source of burn wound infection. The incidence of burn wound sepsis has declined dramatically since the widespread adoption of early excision.

Two common techniques are used for excision (Saffle and Schnebly 1994). Full-thickness, or fascial, excision of the skin and subcutaneous tissue to the level of muscle fascia is easy to perform, produces relatively little bleeding, and facilitates graft 'take'. However, removal of subcutaneous fat is disfiguring, and loss of padding around joints results in prolonged stiffness and discomfort. Fascial excision is frequently used for very deep extensive burns, particularly of the trunk; it can also be helpful in treating elderly patients, whose subcutaneous tissues may not support skin grafts. In tangential or layered excision, thin slices of eschar are removed with a dermatome until punctate capillary bleeding indicates viable tissue. Tangential excision requires experience and judgment to perform successfully. Bleeding from exposed capillary beds can be profuse and hard to control. Tangential excisions are ideal for deep partial-thickness injuries, in which small amounts of deep dermis can be salvaged. Such wounds accept grafts exceptionally well and scar relatively little. This technique is preferred for wounds of indeterminate depth, and for any areas (hands, face, etc.) where appearance and function must be optimized.

Once excised, burn wounds must be covered quickly or desiccation will produce a new layer of eschar. Split-thickness autograft is the cover of choice, but a growing variety of temporary coverings are available, including fresh and frozen cadaver allograft skin, acellular allodermis, porcine heterograft, collagen-impregnated silastic, and a variety of synthetic dermis-like coverings, some of which can be left in place permanently for eventual coverage with thin autografts or cultured epidermis.

Timing of surgery

Patient stability and the extent and depth of burn injury should determine the timing of burn surgery among the priorities of patient care. Small burns (< 10 per cent TBSA) are rarely life threatening and risk of infection is usually low; therefore surgery should be delayed until it is clearly indicated and can be performed with minimal risk. The surgeon may elect to follow small burns of indeterminate depth for 10 to 14 days to permit separation of superficial eschar, which will either confirm or obviate the need for surgery. Such observation is justified provided that the patient is stable and ambulatory, wound care and pain control are adequate, and there is no evidence of infection. In most cases, wounds that have not separated after 14 days will benefit from excision and skin grafting.

With increasing extent of injury, obtaining rapid wound coverage becomes progressively more urgent as a means of preventing infection. Deep burns of 10 to 20 per cent TBSA should be excised within 1 to 2 weeks of injury; larger wounds should be removed as soon as the patient's medical status will allow. In treating truly major injuries (⩾ 40 per cent TBSA), all but obviously superficial burns must be excised as the importance of removing eschar precludes observation of indeterminate wounds. Herndon (see Muller *et al.* 1996) has advocated complete removal of massive burns within 24 to 48 h of injury, but most experts limit each excision to 20 to 25 per cent TBSA; larger wounds are removed in a series of procedures over a period of 7 days. Even this more conservative approach means that patients are subjected to major operations within a day or two of injury, often before resuscitation and hemodynamic stability have been achieved. Careful planning and attention to perioperative care are critical for such procedures to be performed safely.

Perioperative care

Medical management of burn victims has been considered elsewhere, but some aspects require further discussion because of their importance in determining the success of surgical treatment. These issues include fluid balance, transfusion requirements, and body temperature control (Demling and LaLonde 1989).

Traditional burn shock resuscitation, performed with the primary goal of maintaining urine output, is known to 'run patients dry' and may fail to maintain intravascular volume. Often, blood pressure and cardiac output are supported partially, and sometimes largely, by elevated circulating catecholamine levels, even after resuscitation is complete. With general anesthesia, sudden hypotension can occur as underlying volume deficits are revealed. Even when fluid administration is adequate, stabilization of blood volume and oxygen delivery may require a considerably longer period than that of formal resuscitation. Capillary 'leakage' may persist and, coupled with increased

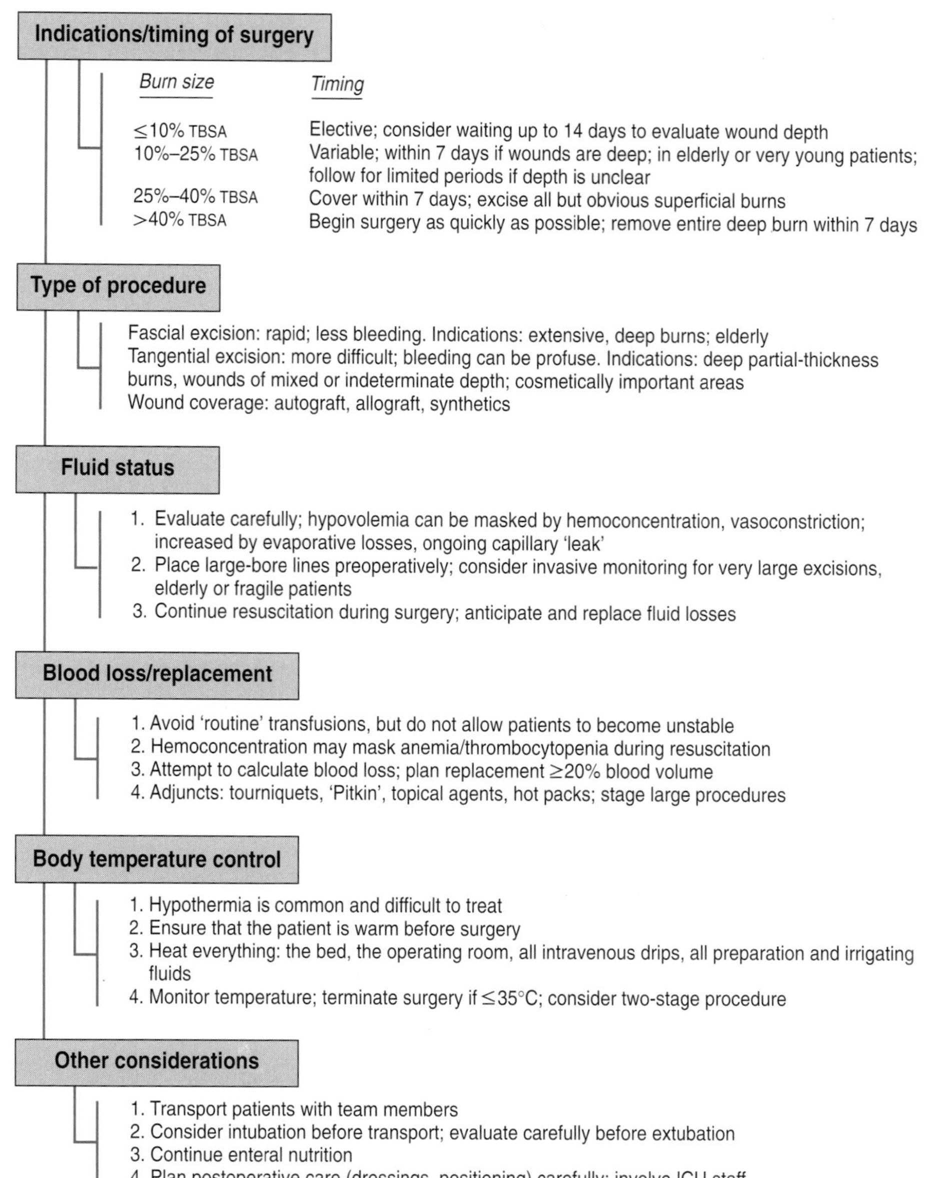

Fig. 1 Approach to the surgical therapy of the burn wound.

evaporative losses, continue the potential for hypovolemia days after injury. Bacteremia associated with excision of infected burn wounds can also produce hypotension.

The surgeon should secure large-bore intravenous lines prior to operation and consider arterial line placement for large procedures, particularly if surgery will limit access to extremities. To improve perfusion and stability in the early postburn period, some burn centers are resuscitating selected patients with the 'physiological' goals of maintaining ventricular filling pressures and normal or supranormal oxygen delivery. Monitoring pulmonary artery pressure and cardiac output can clarify fluid status, and can also help to prevent under- or over-resuscitation during surgery. Such monitoring should be considered when the patient is frail or elderly, or an unusually large excision is planned. Fluid administration should be continued during surgery at least as aggressively as preoperatively. The surgical team should anticipate blood loss (see below), and be prepared to infuse additional fluids to compensate for evaporative losses as they occur.

Heightened risks of disease transmission have caused a radical revision of transfusion practices in many burn centers. Surgeons should avoid 'routine' transfusions and may allow stable patients to develop significant anemia before red cells are replaced, particularly if aggressive infusion of crystalloids is performed before and during surgery. However, the surgeon should never permit patients to become or remain unstable because of reluctance to transfuse blood products. Over- and undertransfusion can be avoided by following blood counts regularly and by anticipating surgical blood loss. Hemoconcentration is common despite fluid resuscitation, and the finding of an elevated (or normal) hematocrit and/or platelet count may belie diminished red cell mass from hemolysis and the consumption of platelets and coagulation factors. As resuscitation is completed, a sudden fall in hematocrit and platelet count should be expected. The blood loss which accompanies major excisions can be predicted (Warden *et al.* 1982; Muller *et al.* 1996); if it exceeds 20 per cent of the circulating blood volume, blood replacement should be planned before instability develops.

A number of techniques can help to reduce blood loss and the need for transfusion. The use of tourniquets for excision of burned extremities provides improved visibility as well as reduced hemorrhage. Many centers infiltrate wounds prior to excision with crystalloid solution containing epinephrine (adrenaline) or phenylephrine — the so-called 'Pitkin' procedure. This technique is also valuable in reducing the blood loss from donor site harvesting. The punctate capillary bleeding produced by large tangential excisions can be difficult to control with cautery. Application of topical thrombin and epinephrine and use of hot (40 °C) packs are helpful in controlling this type of bleeding. For very large tangential excisions, some surgeons prefer a 'two-stage' procedure, in which excised wounds are wrapped snugly with moist bulky gauze dressings overnight and grafting is performed the following day. The two-stage technique helps to reduce bleeding and, by dividing the operation, permits a period of equilibration which limits the blood and heat loss from each half of the procedure, resulting in less overall stress for the patient.

Patients undergoing major burn excision are at high risk of developing hypothermia. The suppressed metabolism characteristic of the 'ebb' phase of injury is worsened by general anesthesia, and patient exposure, peripheral vasodilatation, and use of wet packs for hemostasis increase evaporative and conductive heat loss. Burn patients are often maintaining near-maximum energy expenditure and can be rewarmed only with great difficulty. The resulting hypothermia produces additional metabolic and cardiovascular stress, contributes to coagulation deficits, and impairs peripheral perfusion and skin graft 'take'. To prevent hypothermia, several routines should be adopted. Patients should go to surgery warm; dressing changes or hydrotherapy should not be performed for several hours preoperatively. The operating environment should be made as warm as possible; the patient should be surrounded by heating pads or blankets, radiant heaters, and heat lamps. All fluids should be infused through warmers, and fluids used for preparation for surgery and for packs should be warmed to 40 °C. Perhaps most importantly, the duration of surgical procedures should be limited to 2 to 3 h, and terminated if body temperature drops significantly. Performing major excision and grafting procedures in two stages, and limiting the extent of each excision to 20 to 25 per cent TBSA, also reduces the risks of hypothermia.

Successful surgery requires a team of experienced professionals. Transporting a severely burned patient to and from the operating room can be the most hazardous part of the operation and should be performed by team members. Surgery itself does not usually aggravate underlying pulmonary problems; excision of truncal eschar can improve ventilation, as can administration of some anesthetics. However, the addition of surgery to other stresses can precipitate or aggrevate pulmonary failure. The team may elect to intubate patients prior to transport, and each patient should be assessed carefully before extubation, with consideration given to the need for immobilization postoperatively and timing to next surgery. Enteral nutrition need not be withheld for surgery, and can be continued even in the operating room if necessary. Involvement of the intensive care unit staff can be helpful in co-ordinating postoperative care and in planning dressing changes and positioning, which can be complicated.

References

Demling, R.H. and LaLonde, C. (1989). Early wound excision and grafting. In *Burn trauma*, pp. 118–45. Thieme, New York.

Heimbach, D.M. and Engrav, L.H. (1984). *Surgical management of the burn wound.* Raven Press, New York.

Muller, M.J., Nicolai, M., Wiggins, R., MacGill, K., and Herndon, D.N. (1996). Modern treatment of a burn wound. In *Total burn care* (ed. D.N. Herndon), pp. 136–47. W.B. Saunders, Philadelphia, PA.

Saffle, J.R. and Schnebly, W.A. (1994). Burn wound care. In *Burn care and rehabilitation: principles and practice* (ed. R.L. Richard and M.J. Staley), pp. 119–45. F.A. Davis, Philadelphia, PA.

Warden, G.D., Saffle, J.R., and Kravitz, M. (1982). A two-stage technique for excision and grafting of burn wounds. *Journal of Trauma*, **22**, 98–103.

11.6 Drowning

11.6.1 Pathophysiology

P. G. Boysen and J. E. Szalados

Key messages

- Profound hypoxemia leads to morbidity and mortality from drowning.
- Freshwater drowning leads to alveolar collapse by interfering with surfactant production.
- Saltwater drowning leads to transudative pulmonary edema.
- Although significant electrolyte changes do not commonly occur, the cardiac and cerebral effects of fluid shifts can be lethal.

Epidemiology

Drowning is defined as asphyxia by submersion. It is the third leading cause of accidental death in the United States, accounting for approximately 8000 deaths each year. Globally, 140 000 deaths attributable to drowning occur annually. The incidence of drowning is three times higher in males than in females, and twice as high in the black compared with the white population (Ellis and Trent 1995). Near-drownings, defined as survival at least to the point of extraction from the water, occur three to five times as frequently as drownings.

Pulmonary pathophysiology

Profound hypoxemia is the final common pathway leading to cerebral anoxic brain injury, late morbidity, and death. Hypoxia occurs as a result of voluntary hyperventilation, aspiration-induced pulmonary edema, laryngospasm, or vagally mediated cardiac arrest. In those who survive the initial injury, the systemic inflammatory response syndrome and multiorgan dysfunction occur as a result of systemic perfusion deficit.

Typically, apnea can be maintained until a breakpoint, determined by arterial $Pa\text{CO}_2$ and $Pa\text{O}_2$, is reached when the drowning individual takes an involuntary breath. Although voluntary hyperventilation may raise the $P\text{AO}_2$ and $Pa\text{O}_2$ before submersion, prolonging the apneic time is secondary to the decrease in $Pa\text{CO}_2$. The breakpoint is determined by central and peripheral chemoreceptor activity. Following voluntary hyperventilation at rest, the breakpoint can be prolonged to an average of 146 s ($Pa\text{CO}_2$ = 46 mmHg (6.1 kPa); $Pa\text{O}_2$ = 49 mmHg (6.5 kPa)) or 85 s with exercise ($Pa\text{CO}_2$ = 49 mmHg (6.5 kPa); $Pa\text{O}_2$ = 43 mmHg (5.7 kPa)) (Levin *et al.* 1993). Although $Pa\text{CO}_2$ remains below the apneic threshold, owing to hyperventilation, victims lose consciousness from hypoxia, initiate an involuntary breath, and then aspirate. Ninety per cent of drowning victims aspirate, but most survivors do not aspirate more than 22 ml/kg of water. Hyperventilation also lowers the seizure threshold in susceptible individuals. Swallowed water may be regurgitated and aspirated, increasing the severity of bronchospasm and direct pulmonary injury (Mendelson's syndrome). Aspirated fluid and debris incite pulmonary edema, alveolar hemorrhage, and loss of pulmonary surfactant, resulting in a disruption of the alveolar capillary membrane, ventilation–perfusion mismatch, increased $P(\text{A}-\text{a})\text{O}_2$ gradient, decreased pulmonary compliance, and further hypoxia (Modell 1971). Pneumothorax is particularly likely to follow the aspiration of debris. Seventy per cent of drowning victims aspirate mud, sand, vomitus, or detritus.

Since seawater contains 2.9 per cent sodium chloride by weight, the hypertonic fluid causes a transudative pulmonary edema (alveolar flooding) as a result of the osmotic effect of salt within the alveoli, with preservation of the alveolar epithelial barrier (Cohen *et al.* 1992). Loss of fluid through the alveolar capillary membrane can cause hypovolemia, with intravascular volume contraction to as low as 65 per cent of normal. In contrast, hypotonic freshwater is rapidly absorbed from the alveoli, resulting in acute intravascular volume overload. Pulmonary edema following freshwater aspiration is a result of either acute heart failure from hypervolemia (transudate) or bronchorrhea from airway irritation (exudate).

Aspiration of freshwater interferes with surfactant production by type II pneumocytes and surfactant function, thereby increasing alveolar surface tension and promoting alveolar collapse. Saltwater does not affect surfactant and therefore alveolar collapse is not a prominent feature of saltwater aspiration. Approximately 50 per cent of near-drowning victims develop acute respiratory distress syndrome, most within 6 h.

Pulmonary edema, inactivation of surfactant, and alveolar collapse leading to alveolar hypoxia result in hypoxic pulmonary vasoconstriction which is an adaptive mechanism minimizing ventilation–perfusion mismatch. However, hypoxic pulmonary vasoconstriction also results in pulmonary hypertension and increased right ventricular afterload.

Ten per cent of drowning victims do not aspirate. In these patients, hypoxemia can follow either alveolar hypoxia from cessation of air movement or negative pressure pulmonary edema. Vigorous respiratory efforts against laryngospasm can generate intrathoracic pressures of $-90\ \text{cmH}_2\text{O}$, decreasing alveolar pressures and promoting fluid transudation across the alveolar capillary membrane.

Electrolytes and blood

Significant electrolyte abnormalities are uncommon unless large volumes are aspirated or the water aspirated contains high concentrations of dissolved salts as is the case in the Dead Sea, the Great Salt Lake, mineral pools, or spas. Similarly, hemolysis and rhabdomyolysis have been reported but are rare. Hemoglobinuria, myoglobinuria, and albuminuria are more likely to be due to ischemia than a direct result of absorbed water and electrolytes. Hemodilution and hemoconcentration occur with freshwater and saltwater aspiration respectively. Disseminated intravascular coagulation complicating freshwater near-drowning has been reported, but it is rare and the pathophysiological mechanism is not clear.

Cardiovascular pathophysiology

Pathological findings consistent with a massive outpouring of catecholamines have been demonstrated in the heart and pulmonary vascular endothelium. Contraction banding within the media of coronary arteries and ventricular myocytes and foci of myocardial necrosis may occur as a result of massive catecholamine release or decreased myocardial perfusion. Cardiac arrhythmias occur as a result of hypoxemia, respiratory and metabolic acidosis, and catecholamine surge. Vagally mediated cardiac arrhythmias occur as a result of peripheral alveolar flooding or the mammalian diving reflex. Ventricular fibrillation may follow myocardial ischemia and infarction or result from hypothermia.

Central nervous system pathophysiology

In all cases of near-drowning, the degree of central nervous system insult determines the prognosis. Hypoxemia and cardiopulmonary arrest result in generalized tissue hypoxia. Anoxic brain damage occurs within 4 to 6 min of cessation of effective cerebral blood flow. Global cerebral ischemia results in a diffuse axonal injury pattern characterized by generalized intracellular swelling. Histological changes consist of mitochondrial swelling, particularly in the cerebral cortex, hippocampus, and cerebellum. Intracranial pressure may be elevated secondary to the increase in brain volume, without mass effect, unless there is associated trauma. Aspiration of hypotonic freshwater results in a relative hypervolemia and decreased plasma oncotic pressure, further aggravating the cerebral edema initiated by hypoxia and loss of blood–brain barrier integrity. Cerebral edema may then increase intracranial pressure and further compromise cerebral blood flow during resuscitation.

Remarkably, some near-drowning victims survive submersion for periods of 1 h or more, usually in cold water. Survival in these situations appears to be the result of a specific temporal relationship between hypothermia and the diving reflex which together result in a state of protective hypometabolism (Gooden 1992). Hypothermia has cerebroprotective effects only at temperatures below 30 °C. The diving response, which is profound in 15 per cent of people, prevents aspiration of water, redistributes oxygen supply to the heart and brain, and decreases cardiac oxygen consumption. The pathophysiological importance of the diving reflex is controversial.

References

Cohen, D.S., Matthay, M.A., Cogan, M.G., and Murray, J.F. (1992). Pulmonary edema associated with salt water near-drowning: new insights. *American Review of Respiratory Disease*, **146**, 794–6.

Ellis, A.A. and Trent, R.B. (1995). Hospitalizations for near-drowning in California: incidence and costs. *American Journal of Public Health*, **85**, 1115–18.

Gooden, B.A. (1992). Why some people do not drown: hypothermia versus the diving response. *Medical Journal of Australia*, **157**, 629–32.

Levin, D.L., Morriss, F.C., Toro, L.O., Brink, L.W., and Turner, G.R. (1993). Drowning and near-drowning. *Pediatric Clinics of North America*, **40**, 321–36.

Modell, J.H. (1971). *The pathophysiology and treatment of drowning and near-drowning*. C.C. Thomas, Springfield, IL.

11.6.2 Management

P. G. Boysen and J. E. Szalados

Key messages

- Management begins with the ABC of basic resuscitation.
- Restoration of oxygen delivery is important.
- Cerebral protection is a high priority.

Introduction

Effective initial resuscitative treatment of the drowning and near-drowning victim has an impact on the extent of functional recovery. Since the prediction of outcome is extremely difficult, maximal resuscitation efforts are indicated unless there is obvious physical evidence of death. The correction of hypoxemia and the restoration of adequate tissue perfusion are the first priorities of emergency management. Standard American Heart Association protocols (airway, breathing, circulation (ABC)) for basic life support (BLS) and advanced cardiac life support (ACLS) for the management of cardiopulmonary arrest, including adequate cardiopulmonary resuscitation, are key determinants of survival and neurological preservation. Profound bradycardia, hypotension, hypothermia, or vasoconstriction make determination of absence of pulse difficult and this should not delay cardiopulmonary resuscitation. Rescue breathing and external cardiac compressions should be started in the water if feasible; however, extraction should not be delayed. The intensity and level of care should escalate as facilities and personnel become available. In the presence of hypothermia and in children, conventional time limits to resuscitation are not applicable. The presence of associated injuries (Table 1) should be considered. Stabilization of the cervical spine is important, particularly in diving accidents, to minimize the potential for devastating iatrogenic injury. Foreign matter such as gastric contents or debris in the oropharynx should be removed using standard techniques for clearing obstructed airways. The Heimlich maneuver is used only if difficulty in ventilation is encountered, since it delays cardiopulmonary resuscitation and may promote regurgitation, and there are better ways to clear the airway in an unconscious victim. Freshwater is rapidly absorbed from the airways and does not directly interfere with ventilation; drainage of saltwater is debatable. In most situations, neither emergency cardiopulmonary

Table 1 Injuries associated with drowning

Injuries
Pre-existing
Seizures
Myocardial ischemia/infarction
Cerebrovascular accident
Intoxication/drug overdose
Coexisting
Decompression sickness
Air embolism
Nitrogen, carbon dioxide, carbon monoxide poisoning
Envenomation
Anaphylaxis
Hypothermia
Pneumothorax
Traumatic
Cervical spine fracture/dislocation
Blunt head injury
Blunt chest/extremity injuries or fractures
Traumatic bleeding/exsanguination

bypass nor extracorporeal membrane oxygenation is possible or practical for the drowning victim, although they may facilitate warming while increasing oxygenation and perfusion.

Critical care of the near-drowning and drowning victim must address the following complications: acute lung injury, pulmonary edema, acute respiratory distress syndrome, shock (tissue perfusion deficit), hypothermia, cerebral hypoxic sequelae, infection, and associated injuries. Baseline physical and neurological examinations, electrocardiography, and laboratory and radiological studies as indicated will help guide resuscitation and intensive therapy.

Restoration of oxygen delivery and tissue perfusion

The restoration of adequate oxygen delivery ($D\text{O}_2$) to tissues presupposes adequate arterial oxygen content and cardiac output (CO):

$$D\text{O}_2 = [(\text{Hb} \times S\text{aO}_2 \times 1.34) + (0.0031 \times P\text{aO}_2)] \times \text{CO}.$$

It is necessary to secure the airway in patients with neurological compromise. Positive-pressure mechanical ventilation is indicated in the presence of pulmonary edema, acute lung injury or acute respiratory distress syndrome, severe bronchospasm, or therapeutic hyperventilation. Ventilatory support is titrated to minimize the work of breathing and ensure adequate arterial oxygen saturation ($S\text{aO}_2$) and $P\text{aCO}_2$. In spontaneously breathing patients continuous positive airway pressure with inspiratory pressure support will decrease the work of breathing and assist ventilation. Inspired oxygen concentration ($F\text{iO}_2$), positive end-expiratory pressure, mean airway pressure, inspiratory-to-expiratory ratio, and flow characteristics of ventilated breaths can be manipulated in order to recruit atelectatic alveoli, increase ventilation–perfusion matching, and increase blood oxygenation. Pulmonary management is largely empirical. Care must be used to ensure that positive airway pressure does not depress cardiac output, cause barotrauma, or increase intracranial pressure by impeding venous return. Ideally, the lowest inspired concentration of oxygen should be used to minimize hyperoxia and reperfusion injury (Rosenthal *et al.* 1995). Although controversial, hyperbaric oxygen therapy (2–3 atm at minimally acceptable $F\text{iO}_2$) may prevent neutrophil accumulation and lipid peroxidation associated with hyperoxia, oxygen free-radical toxicity, and reperfusion injury (Haywood *et al.* 1996).

Steroids have not been shown to be effective in acute lung injury or acute respiratory distress syndrome due to drowning. Likewise, antibiotic prophylaxis for pneumonia due to water-borne pathogens or for aspiration pneumonitis is not indicated. Instead, directed antibiotic therapy based on clinical signs of infection and culture results is more appropriate. The administration of artificial surfactant has been described in animal drowning models and in children with acute respiratory distress syndrome (Perez-Benavides *et al.* 1995), but has not been shown to improve survival in either group.

The management of the tissue perfusion deficit and pulmonary edema may necessitate invasive monitoring with either a pulmonary artery catheter or transesophageal echocardiography (Fig. 1). Lactic acidosis following drowning can be prolonged, and ensuring optimal oxygen delivery can aid in the management of metabolic acidosis. Although transesophageal echocardiography visualizes cardiac function and ventricular volumes better, a pulmonary artery catheter can be used to measure mixed venous oxygen saturation and calculate the oxygen extraction ratio. Although defibrillation is often necessary during the initial phases of resuscitation, the continued use of

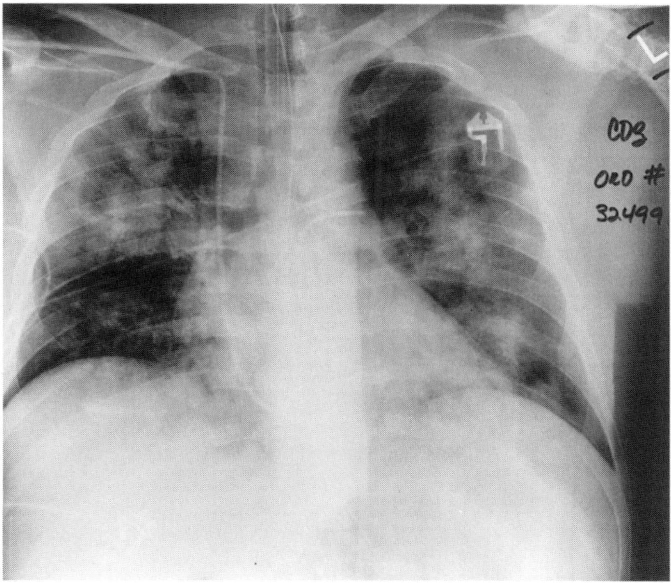

Fig. 1 Chest radiograph of a 36-year-old victim of drowning in freshwater. Bilateral infiltrates correlate clinically with the acute respiratory distress syndrome. A pulmonary artery catheter can be seen in the left main pulmonary artery and was used to optimize oxygen delivery.

inotropic therapy in drowning victims is associated with a poor prognosis since most drowning victims had good pre-existing cardiac function. The management of oliguria in the face of pulmonary edema is another indication for the invasive monitoring of volume status.

Hypothermia increases the incidence of cardiac arrhythmias, coagulopathy, and metabolic acidosis due to peripheral vasoconstriction and hypoperfusion. In general, forced-air convective heating is the most practical treatment of hypothermia. However, sudden vasodilation can precipitate hypotension and acid–base and electrolyte abnormalities from the washout of metabolic products following peripheral reperfusion. Shivering increases oxygen consumption and must be avoided in situations of tenuous oxygen delivery.

Cerebral protection

Neurological complications of cerebral anoxia secondary to drowning can be devastating and costly. The extent of neurological injury is the single most important predictor of morbidity and mortality, and correlates closely with the duration of immersion, water temperature, and the efficacy of early resuscitative efforts. No therapeutic measure has been shown consistently to prevent or reverse the sequelae of cerebral ischemia. Furthermore, neurological outcome cannot be reliably predicted (Oakes *et al.* 1982). The mainstay of neurological protection in the face of global cerebral ischemia is the control of intracranial pressure, prevention and control of seizures, and the maintenance of cerebral perfusion pressure and cerebral blood flow. Monitoring the intracranial pressure is not generally regarded as useful. Declining neurological status, as quantified by the Glasgow Coma Scale or increasing papilledema, correlates well with increasing intracranial pressure. Increases in intracranial pressure probably reflect the degree of neurological injury, and control of intracranial pressure alone does not imply functional neurological recovery. Increased intracranial pressure is inversely related to prognosis for functonal recovery.

Moderate hyperventilation to a $P\text{aCO}_2$ of 25 to 30 mmHg (3.3–4 kPa) results in cerebral vasoconstriction and can therapeutically decrease intracranial pressure for a period of 18 to 24 h. Osmotic diuresis with mannitol and forced diuresis with furosemide (frusemide) or torsemide can decrease cerebral edema. Increasing venous drainage will also help to decrease intracranial pressure. Coughing and straining, particularly on the endotracheal tube with suctioning, has been shown to result in prolonged elevations of intracranial pressure ('plateau waves') and should be minimized by sedation or neuromuscular blockade. Seizures markedly increase the cerebral metabolic rate and predispose to further intracellular acidosis and edema. Seizures may be latent and require a high index of suspicion and electroencephalographic diagnosis. Immediate pharmacological control of seizures is indicated. Cerebral perfusion pressure (CPP) is maximized by decreasing intracranial pressure (ICP), decreasing central venous pressure (CVP), and maximizing mean arterial pressure (MAP):

$$\text{CPP} = \text{MAP} - (\text{ICP or CVP}).$$

Although cerebral blood flow is a more reliable indicator of cerebral perfusion, special techniques are required for its measurement and it does not correlate well with functional neurological recovery. Maintenance of cardiac output and blood pressure by optimizing preload and contractility may increase cerebral blood flow, and moderate hemodilution may optimize the rheological properties of blood supplying oxygen and substrate. Since functional neurological recovery may best correlate with the brain's ability to utilize oxygen (difference between arterial and mixed venous oxygen content), jugular venous bulb cannulation may aid in prognosis but not in therapy (Connors *et al.* 1992).

Hyperglycemia has been shown to worsen neurological outcome in global cerebral ischemia and should be avoided. Hypothermia and the diving reflex have been hypothesized to be cerebroprotective at the time of immersion; however, therapeutic hypothermia confers no advantage.

The use of intravenous barbiturates in doses sufficient to produce an isoelectric EEG with the goal of decreasing cerebral metabolic rate and the critical level of oxygen extraction in the face of cerebral global ischemia, although an attractive hypothesis, has not been proven to be effective. Barbiturates may also decrease intracranial pressure. New strategies aimed at minimizing reperfusion injury following global cerebral ischemia include use of L and N calcium-channel blockers, deferoxamine (desferrioxamine), pyruvate dehydrogenase inducers such as dichloroacetate, and stimulators of aerobic metabolism distal to pyruvate dehydrogenase such as acetyl L-carnitine. However, although experimentally promising, they are clinically unproven.

In summary, strategies for the preservation of neurological function after global anoxia are aimed at the prevention of secondary neurological injury and are, in effect, empirical and supportive.

References

Connors, R., *et al.* (1992). Relationship of cross-brain oxygen content difference, cerebral blood flow, and metabolic rate to neurologic outcome after near-drowning. *Journal of Pediatrics*, **121**, 839–44.

Haywood, Y., Fiskum, G., Silbergleit, R., and Rosenthal, R. (1996). Neurologic effects of hyperbaric oxygen treatment after resuscitation from experimental cardiac arrest: a preliminary study. *Annals of Emergency Medicine*, **27**, 117.

Oakes, D.D., Sherck, J.P., Maloney, J.R., and Charters, A.C., III (1982). Prognosis and management of victims of near-drowning. *Journal of Trauma*, **22**, 544–9.

Perez-Benavides, F., Riff, E., and Franks, C. (1995). Adult respiratory distress syndrome and artificial surfactant replacement in the pediatric patient. *Pediatric Emergency Care*, **11**, 153–5.

Rosenthal, R.E., Miljkovic-Lolic, M., Haywood, Y., and Fiskum, G. (1995). Cerebral ischemia/reperfusion: neurologic effects of hyperoxic resuscitation from experimental cardiac arrest in dogs. *Annals of Emergency Medicine*, **25**, 137–8.

11.7 Electrocution

11.7.1 Pathophysiology

R. F. Armstrong

Key messages

- Electrical injury is produced by the flow of current.
- The current developed depends on the resistances encountered.
- Injuries include burns, cardiac disorders, muscle, and organ damage.

Introduction

In the United States electrical trauma accounts for approximately 1000 deaths per year and 5 per cent of admissions to burns units. In the United Kingdom there are an estimated 100 deaths and 2000 injuries annually (Cox 1992). Over 90 per cent are industrial accidents in young working males. In a study of 1231 electrocution accidents observed over a 10-year period at Electricité de France, Cabanes (1992) found a 2.4 per cent incidence of immediate death, probably from ventricular fibrillation, loss of consciousness in 7.2 per cent, trauma in 6 per cent, and burns in 93 per cent. Of the burn injuries, 77 per cent were arc burns, 15 per cent were electrothermal, and 6 per cent were mixed. The overall mortality was 3.3 per cent.

Electrical power

Alternating current is generated at power stations at a frequency of 60 Hz in the United States (50 Hz in the United Kingdom) and a voltage of 10 to 25 kV. This is fed into a grid of power lines and transmitted to the consumer via pylons or underground cables. The voltage is stepped up by transformers to 69 to 500 kV for transmission over long distances and then stepped down at the consumer end. The power lines are usually bare wires and are the cause of high-voltage accidents when touched by earthed ladders or cranes. Factories are supplied at voltages of 11 to 33 kV. Homes are supplied at 110 to 240 V.

Electrical injury

Electrical injury is determined by the flow of current. For a given voltage difference, the amount of current flowing (in amperes) depends on the resistance or impedance of the body. Traditionally, the body has been considered to have a resistance of about 500 to 1000, not including skin resistance. However, this figure varies markedly as it depends on the applied voltage, the duration of the shock, and the body current pathway. Skin resistance varies with the site, the condition of the skin, and the duration and area of contact (Bernstein 1991). Where the epidermis is thicker (i.e. soles of the feet and palms of the hands), the resistance may be much higher. Once skin burning has taken place, the skin resistance rapidly falls and current flow increases rapidly. Above 200 V, body impedance becomes very low.

Current effects

Following an electrical injury, current flow for any given voltage is difficult to estimate because the skin resistance, area of contact, and duration of contact vary and are usually unknown. Reported thresholds for the effects of hand-to-hand current flow are summarized in Table 1.

Injury occurs by several mechanisms, including skin burn at the point of contact, joule heating by flow of current to ground through the body, cell wall damage by current flow (electroporation), cardiac arrhythmias, thermal injury by arcing, lightning injury, and trauma due to falls.

Burns and electroporation

With high-voltage injuries there is charring and loss of tissue at the contact points; this is due to the current encountering high skin resistance (heat = amperage^2 × resistance × time). A low current contact may leave no evidence on the skin. Once penetration of the skin has occurred, current will flow along unpredictable routes through the body according to the resistances encountered. Any organ in the

Table 1 Current thresholds

0.5–5.0 mA	Startle reaction; victim may fall
>5.0 mA	(Cannot) 'Let go phenomenon'
20–30 mA	Respiratory arrest
50 mA for >2.0 s	Ventricular fibrillation
500 mA for <0.1 s	Ventricular fibrillation
>1.0 A	Asystole
>1.5 A	Muscle damage

Data from Bernstein (1994).

current path may be severely damaged as joule heating causes intolerable rises in temperature and tissue necrosis. Nerves, blood vessels, and muscle have the least resistance to current flow and thus have to endure the highest current density. Muscle injury predominates, but peripheral nerves are extremely vulnerable and persistent neuropathies are common. Thrombosis in damaged blood vessels results in devascularization and tissue necrosis. The high resistance and high heat capacity of bone leads to early heating and slow cooling, which is a possible factor in the extensive damage seen to periosseous muscle. So-called exit wound(s) may show worse damage than the contact site and have been described as blow-out or explosive injuries. Not all damage is heat induced. Electroporation, a process of electrical damage to cell membranes with subsequent lysis, may account for loss of muscle and nerve function without macroscopic alteration of tissue appearance.

Cardiac injuries

The cardiac effects of electrical shock are due to current flowing through the myocardium during the vulnerable period of the cardiac cycle (T wave). They range from arrhythmias and ventricular fibrillation at 50 mA to cardiac asystole at current levels between 1 and 10 A. Ventricular fibrillation is the main cause of death. Thoracic muscle and diaphragmatic tetany may lead to asphyxia and cardiac arrest. Patchy areas of necrosis scattered diffusely throughout the myocardium with evidence of reperfusion injury, possibly due to coronary artery spasm and ischemia, may be seen. Cardiac malfunction after electrical injury has been confirmed by echocardiographic evidence of left ventricular hypokinesia and reduced ejection fraction. Persistence of biventricular malfunction rather than rapid recovery of cardiac function may occur.

Arcing

Arcing occurs when a high-voltage current jumps an air gap. Because of the high resistance offered by air, very high temperatures (5000 °C) are reached, causing thermal burns to nearby victims which are often made worse by clothing catching fire. These arc burns are caused without passage of current through the body. They form a high proportion of industrial electrical injuries, are often small, and commonly affect the eye (Goubiere *et al*. 1994).

Lightning

Lightning kills about 150 to 200 people annually in the United States. It is caused by a build-up of electrical charge in the clouds which discharges to the ground. Voltages of 20 to 200 000 000 V have been reported. The current is unidirectional and achieves levels of 30 to 270 kA (Bernstein 1991). One flash is composed of a series of 10 to 15 strokes lasting for milliseconds. A direct hit is probably not survivable. Most victims are injured by sideflash from the high voltages developed when a nearby tree or other object is struck. Because of the extremely short exposure to the flash there is too little time for the skin to be penetrated by the current. The energy arcs over the outside of the person, producing the so-called 'flashover effect' (Cox 1992). As a result deep burns are infrequent.

Blunt trauma

Workmen receiving even minor shocks may fall as a result of a startle response. This may cause fractures, particularly to the vertebrae, as well as blunt trauma to abdominal organs. Higher voltages can cause violent muscle contractions, resulting in dislocations.

References

Bernstein, T. (1991). Electrical systems, terminology and components — relationship to electrical and lighting accidents and fires. In *Electrical hazards and accidents: their cause and prevention* (ed. E.K. Greenwald), pp. 1–27. Van Nostrand–Reinhold, Princeton, NJ.

Bernstein, T. (1994). Electrical injury: electrical engineers' perspective and a historical review. *Annals of the New York Academy of Sciences*, **720**.

Cabanes, J. (1992). Industrial electrical accidents and their complications observed by Electricité de France. In *Electrical trauma: the pathophysiology, manifestations and clinical management* (ed. R.C. Lee, E.G. Cravalho, and J.F. Burke), pp. 132–52. Cambridge University Press.

Cox, R.A.F. (1992). Lightning and electrical injury. *Journal of the Royal Society of Medicine*, **85**, 591–3.

Goubiere, E., Corbut, J.-P., and Bazin, Y. (1994). Functional consequences of electrical injury. *Annals of the New York Academy of Sciences*, **720**, 259–71.

11.7.2 Management

R. F. Armstrong

Key messages

- Following electrothermal injury or lightning strike:
  - cardiopulmonary resuscitation should be prolonged;
  - blunt trauma should be excluded;
  - underlying damage may be much worse than visible damage;
  - fluid requirements may be very high;
  - early escharotomy and fasciotomies prevent amputations.

Immediate resuscitation

The initial management of the electrically injured patient follows standard resuscitative measures. If cardiopulmonary arrest has occurred, basic and advanced life support techniques should be applied persistently and aggressively, given the young and often previously healthy state of the victim's myocardium. While resuscitation proceeds, efforts should be made to identify the voltage and the duration and site of contact, as well as the presence and condition of protective clothing. A history from the patient or witnesses of a fall should direct attention to the possibility of blunt trauma and bleeding from damaged internal organs (e.g. spleen). Examination should identify the presence of fractures and dislocations, particularly of vertebrae, and the area and depth of any cutaneous burn, and in particular should include a search for electrical contact points.

Once immediate resuscitation is completed, admission to a multidisciplinary intensive care unit should be expedited so that the

Table 1 Basic clinical management principles

Field response
Immobilize cervical spine
Support vital organ systems (may include CPR)
Fluid volume resuscitation
Transport to trauma center
Initial hospital treatment
Assess other injuries (fractures, hemorrhage, etc.)
Correct pH and electrolyte imbalance
Control cardiac arrythmias
Dilute and alkalinize myoglobin in urine
Diagnostic imaging (radiography, MRI, radionucleotide)
Tetanus prophylaxis
Early treatment
Transport to intensive care unit center
Decompress muscle and nerve compartments (<6 h)
Surgical debridement
Look for cardiac muscle damage
Second-look procedure (48 h): biological dressing or primary wound closure
Intermediate treatment
Wound closure
Begin surgical reconstruction
Nutritional support
Musculoskeletal splinting
Neurophysiological evaluation
Psychiatric consulting
Coworker evaluation
Late treatment
Rehabilitation (physical, psychological, occupational)
Additional reconstructive procedures as needed
Coworkers and supervisors

CPR, cardiopulmonary resuscitation; MRI, magnetic resonance imaging. Reproduced with permission from Lee (1994).

administration of oxygen, respiratory support, intravenous fluid, and pain relief can be continued with the benefit of comprehensive ECG, invasive, and blood gas monitoring. Lee (1994) has described a treatment strategy used in the care of the electrically injured patient (Table 1).

Fluid therapy

Large amounts of intravenous fluid need to be given, usually much greater than expected from the size of the cutaneous burn. Peripheral veins may be thrombosed. Ringer's lactate, Hartmann's solution, or saline are recommended resuscitation fluids. Luce and Gottlieb (1984) suggest 7 ml/kg per percentage of body surface area burned every 24 h. However, many centers prefer to resuscitate initially with colloid solutions and administer fluid guided by filling pressures, best stroke volume, and evidence of adequate tissue perfusion rather than according to formula. Common targets are pulse rate below 120 beats/min, mean arterial pressure above 60 mmHg, $Sa\text{O}_2 > 95$ per cent, urine output between 0.5 and 1.0 ml/kg/h, base deficit less than 3 mmol/l, a difference of less than 5 °C between core and peripheral temperature, and mixed venous oxygen saturation above 60 per cent. Frequent measurement of electrolytes, hemoglobin, coagulation status, creatinine phosphokinase, and arterial blood gases is essential and standard in the critical care environment.

Skin injury

Depending on the type of electrical injury, there may be only small marks on the skin or extensive charring at contact sites with or without adjacent cutaneous thermal damage. Thermal injury is often the result of clothing catching fire or of arcing from a high-voltage source causing high temperatures in the immediate surroundings of the patient. It is important to realize that the extent of the cutaneous injury bears no relation to underlying damage, which may be much worse. Escharotomy should be considered if there is evidence of any soft tissue compression, and debridement of necrotic tissue, tangential excision of the burn wound, and grafting should be performed according to normal surgical practice in burn management.

Underlying structures

The low resistance offered to current flow by nerve, muscle, and blood vessels accounts for the severity of damage to these structures. Muscle, in particular periosseous muscle, may be non-viable although surrounding tissues, fascia, and tendons are undamaged. Swelling and edema are worst in the first 24 h after the burn. These cause compression, particularly in the forearm and calf, and decompression should be performed by fasciotomies either on clinical grounds (pain, paresthesia, anesthesia, loss of pulse, tenseness on palpation) or by demonstration of raised compartment pressures (within 30 mmHg of diastolic pressure). Swelling may be accentuated by the administration of crystalloid, a dilemma familiar to those working with burned patients. At operation non-viable tissue should be removed and amputation may be necessary. In this situation the surgeon may be in the unenviable position of not knowing from visual examination whether tissues are dead or alive. Additional exploration is often necessary, and good results have been reported after repeated exploration of deep muscle compartments in the forearm and hand.

Abdominal organs

Although internal organ damage by current is rare, there are reports of paralytic ileus and necrosis of the gallbladder, colon, pancreas, and liver. These visceral injuries are often associated with burns of the abdominal wall.

Cardiac injury

The main cause of death following electric shock is cardiac arrest, usually as a result of ventricular fibrillation (Fontanarosa 1993). Both low and high voltages can produce life-threatening arrhythmias, with some developing several hours after the accident. High current flows (> 10 A) tend to cause asystole. Although myocardial infarction is rare, areas of necrosis in heart muscle and conducting tissue have been reported after electrocution. Diagnosis of cardiac injuries after high-voltage contact is difficult. ECG abnormalities (usually non-specific ST–T changes) may be transient or even absent, presumably due to the diffuse nature of the myocardial damage. Elevation of creatine kinase MB isoenzyme suggests cardiac injury, although under conditions of electrical stimulus it can be released from skeletal muscle. Patients at high risk of myocardial injury can be identified by extensive body surface burns and the presence of upper and lower body exit and entrance wounds. Clearly, if cardiac malfunction is suspected, echocardiographic and cardiac output studies are indicated. Myocardial injury can also result from low-voltage shocks. However, Australian experience suggests that after domestic 240-V shocks, asymptomatic patients with normal ECGs can be safely discharged home (Cunningham 1991).

Renal function

In the presence of extensive striated muscle damage there is release of myoglobin into the blood. Myoglobin is a low-molecular-weight protein which is filtered by the glomerulus and appears in the urine where it usually, but not always, produces a brown coloration. In the presence of myoglobin, urine will test positively for blood although there may be no microscopic evidence of red cells. In these circumstances renal failure may be averted by increasing the intravenous infusion rate to improve urine flow, using mannitol, and alkalinizing the urine (up to 500 ml/h of 1.26 per cent sodium bicarbonate) to a pH above 6. Other evidence of rhabdomyolysis is a plasma creatinine that is disproportionately high compared with urea. In these circumstances it is important to identify the source of the myoglobin and to perform appropriate fasciotomies or excise dead muscle.

Neurological function

Neurological disorders are seen more often after high-voltage injury than after low-voltage injury. Unconsciousness, seizures, and coma may result, with persistence of neurological deficit and cerebral edema in some victims. Grube and Heimbach (1992) described no sequelae in 87 per cent of patients with only transient loss of consciousness but a generally bad outcome after prolonged coma. Immediate or delayed peripheral neuropathy occurred in a significant number of patients.

References

Cunningham, P.A. (1991). The need for cardiac monitoring after electrical injury. *Medical Journal of Australia*, **154**, 765–6.

Fontanarosa, P.B. (1993). Electrical shock and lightning strike. *Annals of Emergency Medicine*, **22**, 378–87.

Grube, B.J. and Heimbach, D.M. (1992). Acute and delayed neurological sequelae of electrical injury. In *Electrical trauma: the pathophysiology, manifestations and clinical management* (ed. R.C. Lee, E.G. Cravalho, and J.F. Burke), pp. 133–52. Cambridge University Press.

Lee, R.C. (1994). Tissue injury from exposure to power frequency electrical fields. In *Advances in electromagnetic fields in living systems*, Vol. 1 (ed. J.C. Lin), p. 113. Plenum Press, New York.

Luce, E.A. and Gottlieb, S.E. (1984). 'True' high tension electrical injuries. *Annals of Plastic Surgery*, **12**, 321–6.

11.8 Noise

11.8.1 Pathophysiology

D. Balogh and A. Benzer

Key messages

- Constant exposure to noise causes irritation, cardiovascular changes, and altered sleeping patterns.
- Unexpected high-level noise (about 60 dB(A) in the intensive care unit (ICU)), such as alarms (approximately 80dB(A)), provokes an orienting reflex and a startle response. This may contribute to the ICU syndrome of ICU patients.
- Irritation due to noise influences the performance of personnel, and chronic exposure to noise may have health consequences for staff.

Introduction

Environmental noise is ubiquitous in the daily life of all industrialized nations. Noise is probably the most widespread single stressor in working conditions. There are strict guidelines setting upper limits for industrial workers, which are under regular control and observance. However, acoustic pollution and the internal sound level in a hospital are completely neglected. Noise annoys and irritates patients and personnel, particularly in intensive care units (ICUs) where an increasing number of technical devices produce noise and are monitored by acoustic alarms. Usually no effort is made to reduce noise in the ICU. As many patients are under continuous sedation or are clinically paralyzed, and are often in very critical life-threatening conditions, attention is concentrated on stabilizing vital functions.

Noise has many effects which have been investigated in a large number of laboratory studies (Abel 1990; Kjellberg 1990). There have been many studies of noise-induced hearing loss, but this problem only occurs in a noise environment exceeding 85 to 90 dB(A), which is not the case in an ICU. Non-auditory effects of noise, such as cardiovascular changes, irritation, and altered sleeping patterns, have been examined in industrial working populations and in residential populations living in areas exposed to noise (near an airport or highway). These effects may prevent out-patients from resting. Noise can also create psychological problems for hospital personnel, which reduce performance levels and affect the behavior of the working team.

Measurement and subjective perception

It should be remembered that objective measurements of sound do not represent the subjective perception of sound by the human ear.

Noise is usually measured as sound pressure level (SPL). The unit of noise is the decibel (dB(A)), which is proportional to the logarithm of sound intensity (sound energy per square meter per second). A consequence of this logarithmic scale is that an increase in sound pressure level of 10 dB(A) corresponds to a doubling of subjective loudness perception:

$$\mathrm{dB(A)} = 2\log_{10}\left(\frac{\text{A-weighted SPL}}{0.0002\ \text{dyn/cm}^2}\right).$$

The A-weighted sound pressure level measured in dB(A) allows for the fact that the human ear is most sensitive in the range 500 to 5000 Hz and is rather insensitive to very low and very high frequencies. Low-frequency noise is underestimated by the dB(A) scale, so that a 100-Hz tone raised by 10 dB(A) is experienced as four to five times louder (e.g. ventilation noise).

It is often believed that people become accustomed to noise, but in fact the annoyance increases with exposure time. The subjective response to noise depends not only on measurable facts, but also on many additional psychological factors (Kjellberg 1990).

Effects of noise on patients

In laboratory testing it is possible to demonstrate that a resting person exposed to a sudden unexpected high-level noise reacts with an 'orienting reflex', which is a direction of attention towards the noise source and an attempt to identify and judge the noise. This orienting reflex probably occurs in ICU patients who are not deeply sedated when audible alarms ring.

Resting persons primarily exhibit peripheral vasoconstriction and an increase in diastolic blood pressure after a sound pressure level of 70 dB(A), whereas systolic pressure is barely affected and the heart rate is lowered. No constant change in catecholamine or cortisol levels can be documented. Alarms exceeding 90 dB(A) provoke a startle response.

The EEG shows an arousal reaction with α-rhythm disruption in resting persons exposed to a sudden noise (hand clapping). This arousal reaction is still present in sedated patients, so that it must be assumed that even sedated and clinically paralyzed patients can perceive some noise (i.e. alarms).

Another important effect of noise is sleep disturbance. A low constant sound pressure level does not disturb sleep, but unexpected and unpredictable noise, such as an alarm, has a major effect on ICU patients and contributes to ICU psychosis. This is particularly likely

Table 1 Sound pressure levels of instrument alarms

Instrument	Alarm type	SPL (dB(A))				
		L_{eq}	L_1	L_{10}	L_{50}	L_{90}
Servo ventilator 900 C	I	78	82	80	77	73
Evita–Dräger	I	64	69	68	63	54
Sirecust 032 D	C	67	68	68	67	66
Hellige 610 (weak)	V	56	61	60	55	53
Hellige 610 (strong)	V	74	77	75	74	73
Diginfusa	C	62	64	62	61	61
Motor syringe	C	58	59	59	58	57
Volume-infusion pump	S	57	64	63	49	42
Hemodialysis	C	56	60	57	56	56

I, intermittent; C, continuous; S, single tone; V, variable.

to be seen in long-stay ICU patients such as those with severe burns and multiple injuries. Annoyance and constant irritation from noise may also contribute to this ICU syndrome, although this is difficult to quantitate.

Patients with a hearing impairment, which may occur after prolonged aminoglycoside treatment with concomitant administration of furosemide (frusemide), are even more irritated by noise, since it masks speech and makes understanding and communication more difficult (Kjellberg 1990).

Effect of noise on staff

ICU staff are constantly exposed to environmental noise. Noise in industrial workplaces is evident and therefore is regularly monitored. The acoustic environment of the ICU is neglected, although the growth of instrumentation and monitoring has caused a rapid change in workplace conditions in the ICU.

Annoyance and irritation due to noise can have behavioral consequences on personnel. It has been demonstrated in various test settings that performance level can be maintained during exposure to noise only at the price of increased stress, i.e. higher catecholamine levels. Annoyance also depends on the task being performed. A combination of difficult working conditions, such as night duty, exhaustion, and noise, alter the subjective response to noise. It has been shown that people tend to be less co-operative in such situations, which may cause problems in the ICU (Lercher *et al.* 1993).

The long-term health hazards for ICU staff resulting from exposure to noise have never been examined. Various epidemiological studies in occupational and residential groups chronically exposed to noise have shown an increase in blood pressure and a larger number of cardiovascular problems. Thus noise may contribute to the burn-out syndrome experienced by ICU staff (van Dijk 1986; Lercher *et al.* 1993).

Noise pollution in the ICU

Measurements show a constant background noise of 60 dB(A) in the ICU during both day and night. Ventilators with their pneumatic and mechanical noise, ECG signals, drainage suction, special pneumatic beds, the balloon pump, etc. exert a constant cacophony that can mask speech and interfere with communication.

Sound level peaks of 70 to 80 dB(A), caused by the triggering of various alarms, occur at irregular intervals. Table 1 shows records of various alarms; L_1 (sound pressure level for 1 per cent of time) indicates the short and very loud events that are the most disturbing (Balogh *et al.* 1993). It is essential to remember that short-duration events of elevated intensity are both frightening and annoying for conscious patients, particularly because it is difficult to distinguish between the noise from their own bed and that from a neighboring bed. The reaction of sedated patients has never been systematically examined.

On average 2.1 alarms/patient/h have been recorded in our ICU, but up to 42 alarms/h may be present for patients in very unstable conditions (Balogh *et al.* 1993). Other studies have registered more than 100 alarms/h in an open ICU ward.

Thus noise caused by personnel and machines should be minimized in the design of an ICU to avoid physiological stress to patients and staff.

References

Abel, S. (1990). The extra-auditory effects of noise and annoyance: an overview of research. *Journal of Otolaryngology*, **1** (Supplement), 1–13.

Balogh, D., Kittinger, E., Benzer, A., and Hackl, J.M. (1993). Noise in the ICU. *Intensive Care Medicine*, **19**, 343–6.

Kjellberg, A. (1990) Subjective, behavioral and psychological effect of noise. *Scandinavian Journal of Work and Environmental Health*, **16** (Supplement), 29–38.

Lercher, P., Hörtnagl, J., and Kofler, W. (1993). Work noise annoyance and blood pressure: combined effects with stressful working conditions. *International Archives of Occupational and Environmental Health*, **65**, 23–8.

van Dijk, F.J.H. (1986). Non-auditory effects of noise in industry. II. A review of the literature. *International Archives of Occupational and Environmental Health*, **58**, 325–32.

11.8.2 Management

D. Balogh and A. Benzer

Key messages

- The acoustic environment of the intensive care unit (ICU) is neglected and the noise problem is not recognized.
- In the ICU, alarms should be graded from life-threatening events to insignificant disturbances.
- Alert patients should be separated from patients requiring invasive treatment.
- Sound, conversation, discussion, and preparation of instruments should be separated from the bedside.

Introduction

The acoustic environment of the intensive care unit (ICU) is frequently neglected despite the increasing number of technical devices and instrumentation and their audible alarms. Noise seems to be a problem recognized by nursing staff (Topf 1989). Once the problem is recognized, strategies must be developed to ameliorate sound conditions (Lercher 1996).

There are various sources of ICU noise. In addition to the technical devices, the features of the building itself and the organization of daily work contribute to the sound pressure level in the ICU.

Equipment noise

Both constant background and intermittent noise occur in the ICU, both of which indicate undisturbed functioning of devices. There are also a number of audible alarm signals indicating equipment malfunction and limit overload.

The constant background noise of ventilators, suction devices, and intra-aortic balloon pumps, the periodic beep of the ECG, etc. are calming for staff and the conscious patient as they confirm that all machines in use are functioning normally. Therefore it is unnecessary to develop silent ventilators, balloon pumps, etc. because medical personnel have become accustomed to their particular sounds which are part of the controlling environment and are at low decibel (dB(A)) levels.

Most equipment used in the ICU incorporates alarms which consist of a visual and an audible component. Unfortunately, there is no uniform classification of alarms. They can be divided into two main categories: the first type are warnings which indicate that a life-support machine is not working in the desired manner but that the patient's condition has not yet deteriorated; the second type indicate emergencies, i.e. they monitor physiological function and show that a deleterious change in the patient has already occurred (Kerr and Hayes 1983).

As an illustration, Table 1 lists the 24 alarm variables in a mechanically ventilated patient on hemofiltration. When four patients are treated in one room, which is possible in an open ICU ward, up to 100 different alarms must be distinguished. An inexperienced nurse may easily become confused and unable to distinguish several concomitant alarms, which may have disastrous consequences (Cropp *et al.* 1994). The only way to cope with this problem is to switch off unimportant alarms (e.g. a motor syringe with no life-supporting medication) and to set the alarm range for a large scale. This change in alarm strategy must be made with the agreement of all responsible staff (doctors and nurses). As all device defects and changes in the patient's condition are indicated only after serious deterioration, the time available for solving problems is very short.

The design of new equipment demands the co-operation of industry, ICU personnel, and psychoacoustic experts in order to plan the noise profile of a device, i.e. the background noise and alarms. The increasing number of technical appliances changes the sound pressure level in the ICU; different categories of alarms could be used to distinguish life-threatening events from insignificant disturbances.

Table 1 Variables monitored by alarms on a ventilated and hemofiltrated ICU patient

	No. of variables monitored
Technical devices	
Ventilator + humidifier (volume pressure, O_2, temperature)	4
Hemofiltration (two pressure devices, airtrap)	3
Equiline (volume)	1
Six motor syringes	6
One infusion pump	1
Bed (temperature)	1
Patient	
ECG, MAP, PAP, CVP, end expiratory CO_2, $S\text{vo}_2$, pulse oximeter, temperature	8
Total for one bed	24

MAP, mean arterial pressure; PAP, pulmonary arterial pressure; CVP, central venous pressure; $S\text{vo}_2$, venous oxygen saturation.

Noise due to building characteristics

There is a difference between the noise exposure in an open ICU, where all rooms communicate with a central base station, and that in an ICU consisting of separate rooms. Nursing staff usually prefer an open ward because they can work together and there is always another nurse nearby to help in emergencies. However, the large number of technical devices operating in close proximity produces a terrible cacophony, which can barely be tolerated by conscious patients. Open wards may be suitable for heavily sedated patients requiring highly invasive instrumentation. Conscious patients are much more comfortable in separate rooms, each containing one or two patients.

The ceiling, floor, and furniture in the ICU should be designed such that reverberation and unnecessary noise are minimized. Acoustic insulation of scrub-up areas, sterilization, etc. should be optimized.

Noise caused during daily work

Personnel are often unaware of the noise that they create, including private background conversation, laughter, therapy discussion, teaching rounds with students, pagers, and telephones. The patient's circadian rhythms should be respected. A moderate conversation during daytime can be calming and agreeable to the patient. Therapeutic discussions should not be held at the bedside, and teaching rounds with residents and students should be kept brief and preferably out of earshot of the patient. Any new findings can be discussed later in a conference room.

Instrument preparation and equipment checks for new patients can be noisy, as can incoming trolleys carrying supplies of materials (Hodge and Thompson 1990). All work in the ICU must be analyzed with the aim of reducing noise pollution. If it is too noisy, efforts should be made to minimize the sound (Lercher 1996).

Management of noise pollution

The following measures can reduce noise pollution and noise effects.

1. Alert patients should be separated from unconscious patients requiring invasive therapy such as mechanical ventilation, hemofiltration, dialysis, extracorporeal membrane oxygenation, etc.
2. Headphones with selected music can screen patients from moderate external noise.
3. The patient's circadian rhythm should be respected; some alarms may be muted, conversation restricted, and activities which may cause additional noise, such as cleaning instruments, should not be performed at the bedside.
4. Sound insulation between ICU rooms is essential.
5. The choice of new equipment should include an assessment of its potential noise.
6. A graded system of alarms is mandatory. New equipment should co-ordinate several signals, so that the number of alarms can be reduced.
7. A patient who is under constant supervision needs only alarms that indicate life-threatening events.

References

Cropp, A.J., Woods, L.A., Raney, D., and Bredle, D.L. (1994). Name that tone. The proliferation of alarms in the intensive care unit. *Chest*, **105**, 1217–20.

Hodge, B. and Thompson, J.F. (1990). Noise pollution in the operating theatre. *Lancet*, **335**, 891–4.

Kerr, J.H. and Hayes, B. (1983). An 'alarming' situation in the intensive therapy unit. *Intensive Care Medicine*, **9**, 103–4.

Lercher, P. (1996). Environmental noise and health: an integrated research perspective. *Environment International*, **22**, 117–29.

Topf, M. (1989). Sensitivity to noise, personality, hardiness and noise-induced stress in critical care nurses. *Environment and Behavior*, **21**, 717–33.

11.9 Altitude- and depth-related critical illness

11.9.1 Pathophysiology related to altitude

W. Schobersberger and N. Mutz

Key messages

- Acute mountain sickness is associated with relative hypoventilation in relation to high-altitude hypoxia, water and sodium retention, and endothelial cell activation with increased capillary permeability.
- The severity of symptoms correlates with the degree of hypoxemia.
- Possible pathogenetic factors for the development of high-altitude cerebral edema include a depletion reduction in the active outward sodium transport, hydrostatic edema due to the increase in cerebral blood flow, a decrease in intracranial compliance, and hypoxia-induced angiogenesis and osmotic swelling.
- High-altitude pulmonary edema may be due to overperfusion, high permeability, or mechanical damage to pulmonary endothelial and epithelial cells.

Introduction

The altitude-related disorder seen in the intensive care unit is acute mountain sickness with its associated severe complications of high-altitude pulmonary edema and high-altitude cerebral edema. The relationship between acute mountain sickness and high-altitude pulmonary edema is not fully understood. Although most cases of high-altitude pulmonary edema are preceded by symptoms of acute mountain sickness, it can sometimes occur without evidence of these symptoms. In contrast, high-altitude cerebral edema appears to be a severe complication of acute mountain sickness.

Acute mountain sickness, high-altitude pulmonary edema, and high-altitude cerebral edema are disorders which typically can develop in healthy subjects with no pre-existing pulmonary, cardiac, or cerebral disease within hours to days at altitudes over 2500 to 3000 m above sea level. The prevalence depends mainly on the absolute altitude, the rapidity of ascent, the degree of high-altitude acclimatization, and individual susceptibility. Strenuous physical exercise and severe cold favor the occurrence of acute mountain sickness.

Acute mountain sickness

The complete pathophysiological sequence of acute mountain sickness from the benign self-limiting form to the severe life-threatening illness with clinical signs of cerebral edema is unknown at present. It has been established that the severity of the symptoms of acute mountain sickness correlate with the degree of hypoxemia and that the symptoms resolve on application of pure oxygen. Three pathophysiological aspects have been identified as contributing to the development of acute mountain sickness.

Relative hypoventilation in relation to high-altitude hypoxia

Hypoxia is a potent stimulator of resting ventilation by activating peripheral chemoreceptors. This hypoxic ventilatory response is not the same in all subjects. In several cases mountaineers suffering from acute mountain sickness have had a lower arterial partial pressure of oxygen and a lower arterial oxygen saturation than healthy persons at high altitude. Despite this fact, literature data are controversial since there is generally no correlation between the ventilatory drive in hypoxia and acute mountain sickness. Thus a low hypoxic ventilatory response does not appear to be the dominant pathophysiological factor predisposing to the development of acute mountain sickness.

Water and sodium retention

In healthy mountaineers with a balanced fluid intake hypoxia induces diuresis and natriuresis, whereas subjects developing acute mountain sickness have lower urinary water and sodium output. This 'antidiuresis' in subjects susceptible to acute mountain sickness is associated with no change in body weight, or even weight gain, in contrast with healthy subjects who tend to lose weight at high altitude. Possible explanations for sodium and water retention in acute mountain sickness are higher plasma concentrations of aldosterone at rest and a higher increase in aldosterone and ADH during physical exercise. Aldosterone and ADH override the diuretic and natriuretic effect of atrial natriuretic factor, which is often elevated in acute mountain sickness. Since atrial natriuretic factor increases capillary permeability, it may play a role in the formation of edema during hypoxia.

Endothelial cell activation with increased capillary permeability

The etiology of capillary leak leading to peripheral edema and proteinuria associated with acute mountain sickness is elusive. Although some inflammatory mediators (e.g. plasma eicosanoids, serum phospholipid bound arachidonic acid, urinary leukotriene E4 levels) are found to be elevated in patients with acute mountain sickness but without high-altitude pulmonary edema, others (e.g. IL-1-β, IL-2 and its soluble receptors, IL-6, IL-8, tumor necrosis factor-α and its soluble receptor) are not. Thus overall endothelial cell activation seems to be minor, although regional differences cannot be excluded.

High-altitude cerebral edema

High-altitude cerebral edema is seen less frequently than high-altitude pulmonary edema but, owing to its higher mortality (up to 100 per cent without treatment), it is the most dangerous form of altitude-related disease. After ascent to high altitudes, cerebral blood flow increases more in subjects with acute mountain sickness than in healthy persons. In addition, elevated cerebral spinal fluid pressure has been reported in several cases of high-altitude cerebral edema. Magnetic resonance imaging and CT brain scans indicate the predominant involvement of the white matter in high-altitude cerebral edema. Possible pathogenetic factors for the development of high-altitude cerebral edema include (1) a depletion of ATP stores with concomitant reduction in the active outward sodium transport leading to cellular edema, (2) hydrostatic edema due to the increase in cerebral blood flow which causes an alteration in the Starling forces, favoring fluid extravasation, (3) a decrease in intracranial compliance, and (4) focal capillary compression by hypoxia-induced angiogenesis and osmotic swelling.

High-altitude pulmonary edema

Most pathophysiological elements relevant to the development of acute mountain sickness are important in the pathogenesis of high-altitude pulmonary edema. It is hemodynamically characterized as non-cardiogenic pulmonary edema with pulmonary hypertension and normal or low wedge pressure. It is typically rapidly reversed after descent, particularly when oxygen is applied. The increased pulmonary arterial pressure seems to be the focus of the pathophysiological mechanism of high-altitude pulmonary edema. Hypoxia *per se* induces pulmonary vasoconstriction which tends to be more pronounced in subjects who are susceptible to high-altitude pulmonary edema than in healthy controls. The following hypotheses of pathogenesis have been proposed.

Overperfusion hypothesis

Since pulmonary hypertension due to arteriolar vasoconstriction is not uniform, alveolar edema is explained as overperfusion of unobstructed parts of the lung with no elevated pulmonary arterial pressure. The high blood flow could increase capillary filtration pressure, thus favoring the production of alveolar edema. Since inhaled nitric oxide improves arterial oxygenation in patients with high-altitude pulmonary edema by reducing pulmonary arterial pressure, a defect in nitric oxide production may be at least partially responsible for the pathogenesis of high-altitude pulmonary edema.

High-permeability hypothesis

High-altitude pulmonary edema is frequently associated with peripheral localized edema in the face, wrists, and ankles which may be due to a generalized permeability of central and peripheral capillaries. Possible mediators include metabolites of arachidonic acid (leukotrienes, prostaglandins, thromboxane), inflammatory cytokines, endothelin, free oxygen radicals, and adhesion molecules on the endothelial surface.

Stress failure hypothesis

The alveolar fluid of high-altitude pulmonary edema patients has a high content of large proteins and contains numerous erythrocytes, suggesting a permeability leak in pulmonary capillaries. The stress failure hypothesis incorporates hydrostatic and permeability components. Hydrostatic stress applied to pulmonary vessels induces mechanical damage to pulmonary endothelial and epithelial cells. In addition, breaks in the capillary endothelial layer expose the basement membrane with activation of neutrophils and thrombocytes. Thrombin formation and the presence of inflammatory mediators (e.g. leukotriene B4 and complement fragment C5a) in the bronchoalveolar lavage fluid are possible consequences favoring both interstitial and alveolar edema.

Conclusions

Many pathophysiological factors have been reported to be responsible for the genesis of acute mountain sickness, high-altitude pulmonary edema, and high-altitude cerebral edema. The precise mechanism by which hypoxia triggers the development of altitude-related disorders remains controversial and needs to be addressed in future investigations.

Bibliography

Bärtsch, P., *et al.* (1991). Enhanced exercise-induced rise of aldosterone and vasopressin preceding mountain sickness. *Journal of Applied Physiology*, **71**, 136–43.

Richalet, J.-P. (1995). High altitude pulmonary oedema: still a place for controversy? *Thorax*, **50**, 923–9.

Scherrer, U., *et al.* (1996). Inhaled nitric oxide for high-altitude pulmonary edema. *New England Journal of Medicine*, **334**, 624–9.

Schoene, R.B., *et al.* (1988). The lung at high altitude: bronchoalveolar lavage in acute mountain sickness and pulmonary edema. *Journal of Applied Physiology*, **64**, 2605–13.

Sutton, J.R., Houston, C.S., and Coates, G. (ed.) (1995). *Hypoxia and the brain.* Queen City Printers, Burlington, VT.

11.9.2 Pathophysiology related to depth

W. Hasibeder and N. Mutz

Key messages

- Drowning or near-drowning are the most frequent severe diving accidents, followed by decompression disorders.
- Symptoms of decompression sickness are classified as type I, which includes musculoskeletal pain, cutaneous

manifestations, and constitutional symptoms (fatigue, anorexia, malaise), or type II which include central nervous system and cardiorespiratory involvement.

- Chest pain, cough, dyspnea, tachypnea, cyanosis, and cardiovascular collapse, collectively referred to as 'the chokes', are caused by significant pulmonary gas embolism.
- Pulmonary barotrauma with air embolism is the second major cause of death in the recreational diving community.

Introduction

Diving accidents are rare and therefore do not present a frequent problem to the critical care practitioner. In the United States, which has a scuba diving population of 10 to 15 million people, approximately 100 diving fatalities are reported annually, with the majority due to human error from inexperience or overconfidence. Drowning or near-drowning are the most frequent severe diving accidents, followed by decompression disorders (Table 1). Among decompression disorders decompression sickness and pulmonary barotrauma may present as life-threatening illnesses with a very unique pathophysiology necessitating immediate recompression and specialized intensive care treatment.

Decompression sickness

Decompression sickness may occur if a sufficient volume of inert gas has dissolved in the tissues during diving and the release of ambient pressure during ascent is so rapid that tissue gas tension exceeds the local tissue pressure sufficiently to allow bubbles to form.

Table 1 Traditional classification of decompression disorders

Decompression sickness
Type I
Musculoskeletal pain
Cutaneous manifestations
Lymphatic obstruction
Constitutional symptoms (fatigue, malaise)
Arterial gas embolism ('blow-up')
Type II
Neurological
Cardiorespiratory ('chokes')
Vestibular/auditory
Barotrauma
Lung
Sinus
Ear
Dental
Gastrointestinal

Table 2 Frequency of the most common signs and symptoms of decompression sickness (935 cases)

Signs and symptoms	Occurrence (%)
Localized pain ('bends')	91.8
Numbness, paresthesia	21.2
Muscular weakness	20.6
Skin rash	14.9
Dizziness or vertigo	8.5
Visual disturbances	6.8
Paralysis	6.1
Unconsciousness	2.7
Dyspnea ('chokes')	2.0

Data from Rivera (1964).

The occurrence of symptoms is related to the number of bubbles formed, their location, and the presence of predisposing factors. Bubbles mechanically injure tissues by distortion, tearing, or the production of ischemia as a result of vessel compression. Bubbles within the circulation activate the coagulation, kininogen, and complement systems, leukocytes, and thrombocytes, leading to non-mechanical tissue injury. On rare occasions uncontrolled bubble formation leads to the fatal consequences of diffuse massive arterial air embolization known as the 'blow-up' syndrome. Symptoms of decompression sickness are classified as type I, which includes musculoskeletal pain, cutaneous manifestations, and constitutional symptoms (fatigue, anorexia, malaise), or type II, which includes central nervous system and cardiorespiratory involvement (Rivera 1964; Elliott and Moon 1993) (Table 2). Type I symptoms may rapidly progress to type II illness which is more serious, particularly when symptoms occur close to surfacing.

Type I decompression sickness

'Joint bends' (a deep aching or intense throbbing periarticular pain) most often affect the upper limbs, particularly the shoulder, asymmetrically in recreational divers. In caisson workers bends in the lower limbs, particularly the knee, predominate. The pain is probably produced by bubbles formed in well-innervated and non-compliant tissues like tendons and tendinous capsules. Tendons may have low and even intermittent blood flow, limiting the rate of gas clearance during ascent and thus favoring bubble formation.

'Skin bends', manifested as pruritus and urticaria affecting the hands, wrists, and trunk, are caused by subcutaneous microbubbles probably combined with a localized inflammatory reaction. Sometimes skin bends are accompanied by marbling or mottling of the affected skin area (cutis marmorata). Localized soft tissue swelling is believed to be due to obstruction of lymphatic vessels by microbubbles.

Type II decompression sickness

Chest pain, cough, dyspnea, tachypnea, cyanosis, and cardiovascular collapse, collectively referred to as 'the chokes', are caused by

Table 3 Differential diagnosis of decompression sickness and acute pulmonary barotrauma

	Decompression sickness	Acute pulmonary barotrauma
Onset of symptoms	>5–10 min after surfacing	Within 5 min of surfacing
Diving profile	Long deep dive	Independent of diving profile
Type I symptoms	Usually present	Joint and skin bends usually absent
Neurological	Spinal cord most often affected	Brain most often affected

significant pulmonary gas embolism. Intravascular bubbles may produce the adult respiratory distress syndrome by the activation of humoral and cellular blood elements (James 1993). If the lung filter is overwhelmed or acute pulmonary hypertension reopens a functionally closed foramen ovale, microbubbles may pass into the systemic circulation, leading to cerebral embolization and acute myocardial infarction. Although neurological symptoms may be present, symmetrical spinal cord lesions predominate in recreational diving accidents. These symptoms have been attributed to compromised venous spinal cord drainage due to intravascular blood–bubble interactions and external vessel compression. Untreated spinal decompression sickness leads to cord demyelination with relative preservation of axons (James 1993).

Pulmonary barotrauma

Pulmonary barotrauma with air embolism is the second major cause of death in the recreational diving community (McAniff 1980). Excessive transpulmonary pressures exceeding 98 Pa (100 cmH_2O) lead to rupture of the intra-alveolar septa and escape of gas into the interstitial space. Voluntary breath-holding, laryngospasm, bronchitis, bronchospasm, viscid pulmonary secretions, and pre-existing pulmonary disease may all prevent the escape of expanding alveolar gas during ascent. Interstitial gas may travel along perivascular sheaths and cause pneumothorax, mediastinal emphysema, pneumopericardium, pneumoperitoneum, subcutaneous emphysema, and even laryngeal nerve palsy. Gas entering pulmonary veins reaches the left heart and cause arterial gas embolization. Air embolism into coronary arteries produces myocardial infarction which often progresses into cardiac arrest (Bradley 1990). Hemoptysis (27.2 per cent), subcutaneous emphysema (22.7 per cent), and chest pain have been reported to be the most frequent thoracic pulmonary symptoms of acute pulmonary barotrauma.

Severe air embolism into the cerebral circulation leads to immediate occlusion of small arteries with distal ischemia, a temporary loss of the blood–brain barrier, and vasogenic and cytotoxic brain edema accompanied by systemic hypertension and sometimes severe arrhythmias. The last two are caused by excessive increases in plasma catecholamines (Bradley 1990). Neurological symptoms include, in descending order of frequency, coma, stupor, confusion, unilateral motor changes, convulsions, vertigo, visual disturbances, unilateral sensory changes, collapse, headache, and bilateral spinal neurological symptoms.

Since severe decompression sickness and acute pulmonary barotrauma may present with almost identical symptoms, a history and physical examination may provide the only clues for proper diagnosis and thus treatment of these disorders (Table 3).

References

Bradley, M.E. (1990). Pulmonary barotrauma. In *Diving medicine* (ed. A. Bove and J. Davis), pp. 188–91. W.B. Saunders, Philadelphia, PA.

Elliott, D.H. and Moon, R.E. (1993). Manifestations of the decompression disorders. In *The physiology and medicine of diving* (ed. P. Bennett and D. Elliott), pp. 481–505. W.B. Saunders, Philadelphia, PA.

James, P.B. (1993). Dysbarism: the medical problems from high and low atmospheric pressure. *Journal of the Royal College of Physicians*, **27**, 367–74.

McAniff, J.J. (1980). *U.S. underwater diving fatality statistics 1970–78*. Report URI-SSR-80-13, National Oceanic and Atmospheric Agency.

Rivera, J.C. (1964). Decompression sickness among divers: an analysis of 935 cases. *Military Medicine*, **129**, 314–34.

11.9.3 Management

M. Maggiorini

Key messages

- The symptoms of mild to moderate acute mountain sickness can be reduced by graded acclimatization or acetazolamide 500 mg slow release and ameliorated by bed rest and acetazolamide.
- Severe acute mountain sickness and/or high-altitude cerebral edema is treated with oxygen (4–6 l/min) and 4 mg dexamethasone every 6 h with urgent evacuation.
- High-altitude pulmonary edema is treated with oxygen and nifedipine 20 mg every 6 h with urgent evacuation. In the most severe cases dexamethasone 4 mg every 6 h and acetazolamide 250 mg twice daily should be added to nifedipine treatment.
- Repetitive Valsalva maneuvers are performed for barotrauma prophylaxis.
- Emergency treatment of decompression sickness consists of 100 per cent oxygen, fluid replacement, and urgent referral to a hyperbaric chamber.

The management of high-altitude illness

High-altitude illnesses include various grades of acute mountain sickness which must be distinguished from high-altitude pulmonary edema (Table 1). Mild to moderate acute mountain sickness is usually a self-limiting disease which improves with progressive acclimatization. In subjects who fail to acclimatize, acute mountain sickness or high-altitude pulmonary edema will progress into its most severe

Table 1 Severity of high-altitude illness and its associated symptoms and signs

Illness type	Severity	Symptoms	Signs
AMS	Mild	Mild headache, anorexia, nausea, sleeplessness	
	Moderate	Moderate headache (relieved by analgesics), anorexia, nausea, mild to moderate lightheadedness/dizziness, fatigue/weakness	
	Severe	Unrelieved headache, nausea/vomiting, severe lightheadedness/dizziness, severe fatigue/weakness	Body temperature >37.4 °C
HACE		Unrelieved headache, nausea/vomiting, severe lightheadedness/dizziness, severe fatigue/weakness	Confused, disoriented, falls asleep, ataxia (cannot hold the line in the heel-to-toe walking test), stupor, coma, body temperature > 37.4 °C
HAPE		Dyspnea at rest, orthopnea, dry cough, chest tightness or congestion, weakness or decreased exercise performance	Tachypnea (>25 breaths/min), tachycardia (>100 beats/min), body temperature >37.4 °C, peripheral cyanosis, pulmonary rales

AMS, acute mountain sickness; HACE, high-altitude cerebral edema; HAPE, high-altitude pulmonary edema.

form which, if unrecognized or untreated, leads to death (Ward *et al.* 1995).

Treatment of acute mountain sickness

Early recognition of the symptoms and signs is essential for the management of acute mountain sickness. If the symptoms and signs suggest the presence of mild to moderate acute mountain sickness, it is contraindicated to proceed to a higher sleeping altitude. A rest of at least 24 h should be taken during which the symptoms must be closely monitored. The aim of the treatment of severe acute mountain sickness and high-altitude cerebral edema is to improve the symptoms to the extent that a safe descent to low altitude (below 2500 m) is possible.

Treatment of mild to moderate acute mountain sickness

Mild to moderate acute mountain sickness can be treated by bed rest (Bärtsch *et al.* 1993) and/or acetazolamide. Acetazolamide is an inhibitor of carbonic anhydrase, which reduces the reabsorption of bicarbonate and sodium at the level of the renal tubules, leading to a metabolic acidosis within an hour of ingestion. An oral dose of acetazolamide 250 mg twice daily can be used for mild acute mountain sickness. Side-effects include peripheral paresthesias, polyuria, nausea, drowsiness, impotence, and myopia.

Treatment of severe acute mountain sickness and high-altitude cerebral edema

Severe acute mountain sickness and/or high-altitude cerebral edema should be treated with dexamethasone and bottled oxygen or a portable hyperbaric chamber if available. Dexamethasone is given at an initial dosage of 8 mg, followed by 4 mg every 6 h, and may provide longer-term clinical improvement than hyperbaric chamber treatment and more effective relief of the symptoms and signs of acute mountain sickness than acetazolamide (Hackett and Roach 1995).

Treatment of acute mountain sickness by simulated descent

The portable hyperbaric chamber or Gamow bag has become popular in the treatment of acute mountain sickness (Hackett and Roach 1995; Ward *et al.* 1995). It is a lightweight (approximately 7 kg) fabric bag which is inflated by a manual air pump. At an altitude of 4559 m the treatment pressure of 193 mbar is equivalent to a descent of 2250 m, making the chamber an inexhaustible reserve of oxygen. However, studies have failed to show efficacy, and thus optimal treatment is simulated descent combined with dexamethasone.

Prophylaxis of acute mountain sickness

Graded ascent is essential for the prophylaxis of acute mountain sickness. The current recommendations for graded high-altitude acclimatization are to spend 2 to 3 nights below 3000 m before going higher and thereafter not to increase sleeping altitude by more than 600 m/day (Hackett and Roach 1995). If a climber is still susceptible to acute mountain sickness, drug prophylaxis is recommended. The drug of choice is indisputably acetazolamide. Many studies have shown that acetazolamide 250 mg twice daily started 24 h before ascent is effective in preventing the symptoms of acute mountain sickness (Ward *et al.* 1995). A 500-mg capsule of the slow-release preparation may be as effective as a 250-mg dose twice daily. It is recommended that the prophylactic treatment is continued for the first day or two at high altitude. Although dexamethasone has been proved to be as effective as acetazolamide in the prophylaxis of acute mountain sickness, it cannot be recommended because of its well-known side-effects (Hackett and Roach 1995).

Treatment of high-altitude pulmonary edema

High-altitude pulmonary edema, which is the most common cause of death at high altitude, can be reversed if recognized early and treated properly. For many years the administration of bottled oxygen, furosemide (frusemide), and morphine was the only available medical treatment (Hackett and Roach 1995; Ward *et al.* 1995). However, the central respiratory depression and hypovolemia which may follow the administration of furosemide and morphine can be detrimental, and recent results with vasodilators have made their use obsolete. Since Oelz *et al.* (1989) reported that the symptoms and signs of high-altitude pulmonary edema were significantly improved by sublingual administration of 10 mg nifedipine initially, followed by slow release of 20 mg every 6 h, this drug has become the treatment of first choice. (The current recommendations for the treatment of high-altitude pulmonary edema are 20 mg of nifedipine every 6 h combined with oxygen (4–6 l/min) or hyperbaric therapy until the patient's condition improves (Hackett and Roach 1995; Ward *et al.* 1995).) Nifedipine treatment must be continued until descent is completed.

Severe high-altitude pulmonary edema is often combined with the symptoms and signs of severe acute mountain sickness and/or high-altitude cerebral edema. Addition of dexamethasone and acetazolamide may be of benefit for very sick climbers. The aim of treatment is to restore the patient's condition to the point where rapid and safe evacuation is possible. After evacuation to low altitude, hospital admission is needed only for the most severe cases of high-altitude pulmonary edema. Treatment consists of bed rest and oxygen given at a flow rate which is sufficient to keep arterial oxygen saturation above 90 per cent (Hackett and Roach 1995). Pulmonary infiltrates on chest radiographs and hypocapnic alkalosis may persist for a few days. If the condition of the patient and the radiographic findings do not improve within 48 h, other causes of pulmonary edema must be considered. Fever, which may be present in patients with high-altitude pulmonary edema, makes the differentiation between high-altitude pulmonary edema and pneumonia difficult (Ward *et al.* 1995). Bronchoalveolar lavage and right heart catheterization may help to distinguish high-altitude pulmonary edema from pulmonary infection or cardiogenic pulmonary edema.

Prophylaxis of high-altitude pulmonary edema

All climbers who have developed high-altitude pulmonary edema once should be informed that the relapse rate is up to 60 per cent. Thus, as for acute mountain sickness, adequate acclimatization is essential. Climbers who are susceptible to high-altitude pulmonary edema should be thoroughly instructed in the use of nifedipine as prophylaxis and treatment in future ascents. Intake of 20 mg of nifedipine three times daily can prevent high-altitude pulmonary edema effectively in susceptible climbers. An episode of high-altitude pulmonary edema is not a contraindication for future high-altitude exposures if susceptible climbers have been informed properly about preventive measures.

Management of disorders associated with depth

Dysbarism

Dysbarism is primarily a disorder of divers and compressed air workers exposed to increased atmospheric pressure; however, it may be seen in aviators and astronauts moving abruptly to altitudes above 5500 m. All pathological states caused by altered environmental pressure are subsumed in the term 'dysbarism'.

Treatment of barotrauma

Preventive measures are essential in order to avoid barotrauma of the external and middle ear, the labyrinth, and the sinus. Repetitive Valsalva and/or Frenzel maneuvers help to keep open the Eustachian tube which equalizes middle-ear and ambient pressure. Individuals suffering from upper respiratory infections or allergic rhinitis, which cause dysfunction of the Eustachian tube and produce edema of the mucosa around the sinus orifices, should avoid scuba diving. Treatment involves the use of analgesics, antihistamines, and vasoconstrictors. Antibiotics should be used if the tympanic membrane is ruptured (Kizer 1995).

The most serious barotrauma is pulmonary barotrauma of ascent, which results from expansion of air trapped in the lung, causing rupture of alveoli. Rupture of the alveoli leads to the pulmonary overpressurization syndrome, which includes mediastinal emphysema, pneumothorax, and air embolism. Treatment of mediastinal emphysema is usually conservative. If a pneumothorax is present, it must be drained before recompression treatment since it may convert into a tension pneumothorax (Kizer 1995).

Treatment of arterial air embolism

All cases of suspected arterial gas embolism must be referred to recompression treatment with hyperbaric oxygen therapy following the same strategy as used in decompression sickness (Kizer 1995; Tibbles and Edelsberg 1996) (see below). Emergency treatment using Trendelenburg positioning was promoted in the past, but it is no longer recommended (Kizer 1995). Hyperbaric oxygen therapy can improve the symptoms of air gas embolism up to 24 h after the onset of neurological manifestations.

Treatment of decompression sickness

All patients with suspected decompression sickness must be referred to hyperbaric oxygen therapy as quickly as possible, since the reduction of bubble size and alleviation of tissue hypoxia is essential treatment for this condition (Kizer 1995; Tibbles and Edelsberg 1996). The beneficial action of hyperbaric oxygen therapy may rely more on its favorable effect on biochemical action at the blood–gas interface, which leads to endothelial damage, activation of leukocytes, and alteration of hemostasis, than on the reduction of the bubble size (Tibbles and Edelsberg 1996). The physiological effect of hyperbaric oxygen therapy is based on Boyle's law, which states that the volume of a gas in an enclosed space is inversely proportional to the pressure exerted on it. At a pressure of 3 atm bubble volume is reduced by 67 per cent and the dissolved oxygen content is 6 ml/dl, which is

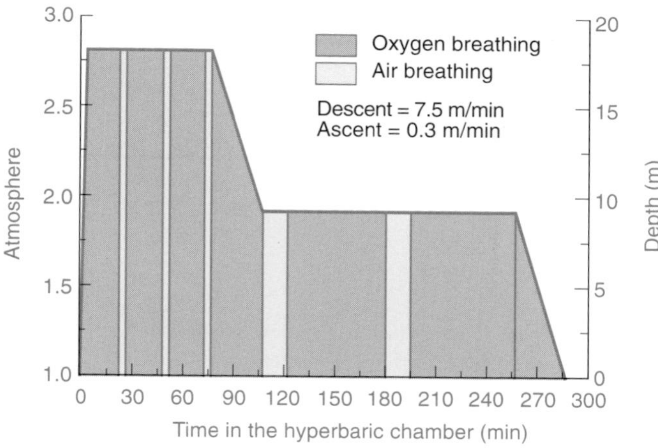

Fig. 1 Example of a hyperbaric oxygen therapy profile. The pressure in the hyperbaric chamber is rapidly (7.5 m/min) increased to 2.8 atm. The rate of descent between pressure steps is 0.3 m/min. To avoid oxygen toxicity, there are air-breathing breaks of 5 min and 15 min at 2.8 atm and 1.9 atm respectively. The total duration of the hyperbaric oxygen treatment depends on the severity of the decompression sickness.

sufficient to meet resting cellular requirements. The administration of 100 per cent oxygen at sea level increases the amount of oxygen dissolved in the blood from 0.3 to 1.5 ml/dl (Kizer 1995; Tibbles and Edelsberg 1996).

As emergency treatment, all patients suspected of having decompression sickness or arterial air embolism should receive 100 per cent oxygen and 1000 to 2000 ml of isotonic solution initially, followed by continuous infusion of isotonic solution at a flow rate which maintains urine output at 1 to 2 ml/kg/h to provide a favorable gradient for nitrogen washout and to support capillary perfusion (Kizer 1995). The administration of steroids is currently controversial. Acetylsalicylic acid (aspirin) up to 500 mg intravenously is sometimes recommended in severe decompression sickness because of its potentially beneficial effect on changes in homeostasis associated with the blood–gas interface (Kizer 1995). Benzodiazepines can be effective in improving vertigo-associated labyrinthine decompression sickness. The nearest center with a recompression chamber and experienced staff should be contacted. If a non-pressurized aircraft (helicopter) is used for patient transportation, the flight altitude should not exceed 300 m above sea level; it is better to use aircraft capable of maintaining sea-level cabin pressurization during flight (Kizer 1995). An example of a protocol for hyperbaric oxygen treatment is shown in Fig. 1. Hyperbaric oxygen therapy sessions sometimes have to be repeated, depending on symptoms, and may be efficacious several days after the incident.

Treatment of central nervous oxygen toxicity

Breathing 100 per cent oxygen for a prolonged period is toxic to the lung and brain. To avoid oxygen toxicity during hyperbaric oxygen therapy regimens, there should be intermittent breaks for air breathing (Fig. 1). Symptoms of central nervous oxygen toxicity, such as a grand mal seizure, are treated by immediate removal of the oxygen mask. In the case of a seizure the chamber pressure should be kept constant until seizure activity ceases in order to avoid a pulmonary overpressure accident. Anticonvulsant therapy is usually unnecessary because termination of oxygen breathing stops seizure activity. Recurrent seizures are rare and patients usually recover without sequelae (Kizer 1995).

References

Bärtsch, P., Merki, B., Hofstetter, D., Maggiorini, M., Kayser, B., and Oelz, O. (1993). Treatment of acute mountain sickness by simulated descent: a randomised controlled trial. *British Medical Journal*, **306**, 1098–1101.

Hackett, P.H. and Roach, R.C. (1995). High altitude medicine. In *Wilderness medicine* (3rd edn) (ed. P.S. Auerbach), pp. 1–37. Mosby, St Louis, MO.

Kizer, K.W. (1995). Scuba diving and dysbarism. In *Wilderness medicine* (3rd edn) (ed. P.S. Auerbach), pp. 1176–1208. Mosby, St Louis, MO.

Oelz, O., *et al.* (1989). Nifedipine for high altitude pulmonary oedema. *Lancet*, **ii**, 1241–4.

Tibbles, P.M. and Edelsberg, J.S. (1996). Hyperbaric-oxygen therapy. *New England Journal of Medicine*, **334**, 1642–8.

Ward, M.P., Milledge, J.S., and West, J.B. (1995). *High altitude medicine and physiology*. Chapman & Hall Medical, London.

11.10 Hyperthermia and pyrexia

11.10.1 Pathophysiology of fever

D. C. Macallan

Key messages

- Fever is due to a resetting of the thermoregulatory set-point and activation of heat-conserving mechanisms.
- Cytokines appear to be the endogeneous pyrogen, with interleukin 6 and prostaglandin E_2 being a common final pathway.

Introduction

Fever has been recognized as a characteristic sign of ill health for centuries. More recently it has been recognized as a cardinal sign of either infection or tissue injury. Body temperature is normally very tightly regulated by a feedback mechanism whereby the thermoregulatory center in the preoptic hypothalamus controls a variety of 'effector' mechanisms to maintain normothermia. In the critical care setting, fever is not usually the consequence of dysregulation of temperature control but rather of intact regulation with resetting of the thermoregulatory set-point at a higher level than 'normal'.

Determinants of body temperature

Body temperature itself is determined by a number of physical and physiological factors. At a purely physical level, it is a fundamental tenet that for any body, be it animate or inanimate, heat production must equal heat loss plus heat retention. Thus at steady state, i.e. zero heat retention, heat production and heat loss must be equal. If heat production exceeds loss, body temperature will rise. This may be the consequence of either increased heat production or reduced heat loss. Such a rise in body temperature will normally result in increased heat loss, by either active or passive mechanisms, and one of two consequences will follow: either normothermia will be restored or the balance between production and loss may only be regained at a new elevated temperature.

These thermodynamic considerations have two important implications. Firstly, rising temperature is not necessarily the consequence of increased heat production but rather of an imbalance between production and loss. Secondly, in the febrile state, if body temperature is elevated but stable, i.e unchanging, heat production and heat loss must once again be equal and are not necessarily abnormal. What is abnormal is the new set-point. Fever does not necessarily imply increased heat production.

The thermoregulatory center is located in the preoptic region of the hypothalamus near the floor of the third ventricle. This area contains heat-sensitive neurons as well as receiving neural inputs from other thermoreceptors. Humoral signals from circulating factors act primarily via the organum vasculosum of the lamina terminalis, an area of fenestrated capillaries which permits access for larger molecules, such as cytokines, to neuronal receptors. Also found in this area are fixed reticuloendothelial cells analogous to the Kupffer cells in liver, and these may have an important role to play in local cytokine synthesis and signaling (Saper and Breder 1994).

Perturbations in thermodynamics in the critical care setting

Although the primary cause of fever is resetting of the thermoregulatory set-point, both heat production and heat loss are likely to be perturbed in the critical care setting. Heat production is equal to the metabolic rate or total energy expenditure. The inflammatory response tends to increase energy expenditure, and this effect appears to be particularly marked with burn injuries. Fever itself results in increased energy expenditure by increasing the rate of metabolic reactions. Despite these factors, most intensive care unit (ICU) patients are not hypermetabolic in terms of total energy expenditure when compared with normally active individuals. Abolishing activity reduces energy expenditure by up to 50 per cent, and sedation and paralysis with mechanical ventilation reduce energy expenditure by at least a further 20 per cent (Manthous *et al.* 1995).

The dominant effector mechanism of thermoregulation is normally regulation of heat loss. Many features of the ICU environment perturb heat-loss regulation. Cutaneous blood flow may be altered by disease or drugs, particularly adrenergic agonists used as pressor agents, which cause peripheral vasoconstriction. Body surfaces may be exposed to the air and body fluids may be exposed to ambient temperature during procedures such as hemodialysis. Infusates may be administered without warming. In addition, behavioral mechanisms may be abolished by sedation; the patient may not be able to communicate feelings of warmth or cold. Finally, the shivering response, a thermoregulatory effector on the heat-production side of the equation, may be abolished by muscle relaxants.

On balance, most such features tend to reduce body temperature. This may be beneficial in terms of maintaining normothermia, but underlying 'fever' (resetting of the thermoregulatory set-point) may be masked or underestimated as a consequence and thus some of the value of fever as an indicator of pathology may be lost or underestimated, particularly during hemodialysis. Such effects should be

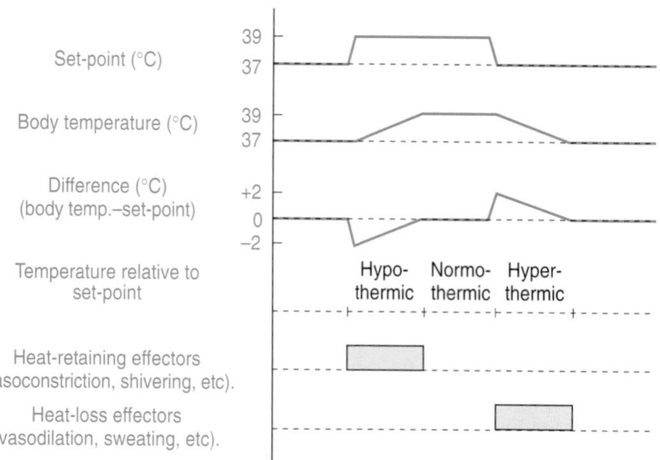

Fig. 1 Changes in actual and set-point temperatures during a febrile episode.

considered when assessing the clinical significance of an elevated temperature.

Fever as altered thermoregulatory set-point

Fever is characterized by activation of effector mechanisms to achieve and maintain a temperature higher than normal. Thus it should be distinguished from hyperthermia, in which the set-point is normal but effector mechanisms have failed to keep body temperature at the set-point. In fever, if the body temperature is below the elevated set-point, heat-conserving mechanisms will be activated as shown in Fig. 1. Thus, for example, a patient with a fever may have a set-point of 39 °C and a body temperature of 38 °C; such a patient will tend to feel cold (behavioral thermoregulation), have peripheral vasoconstriction, and may shiver. Therefore body temperature will tend to rise towards the new elevated set-point. In fever the set-point frequently varies in a cyclical fashion; thus as it falls, the actual body temperature may exceed the set-point and the patient will tend to feel hot, vasodilate, and sweat in order to lose heat.

Endogenous pyrogen

Early investigation of the nature of fever suggested the presence of a host-derived factor that resulted in elevation of body temperature. This was termed 'the endogenous pyrogen'. Initial studies demonstrated that interleukin 1 (IL-1) fulfilled many of the criteria of the endogenous pyrogen, but it has become apparent that several other cytokines also induce fever, in particular tumor necrosis factor, interferon-α, interferon-γ, and interleukin 6 (IL-6). Therefore it would appear to be oversimplistic to consider a single cytokine as 'the endogenous pyrogen'.

Many animal studies have investigated the possible roles of different cytokines in initiating fever. Difficulties in interpreting such data arise at several points: considerable species differences exist, several cytokines have differing actions when administered systemically or locally in the central nervous system, activation of cytokines tends to occur as a cascade and thus one cytokine rarely acts in isolation, and most cytokines are pleiotropic with a great deal

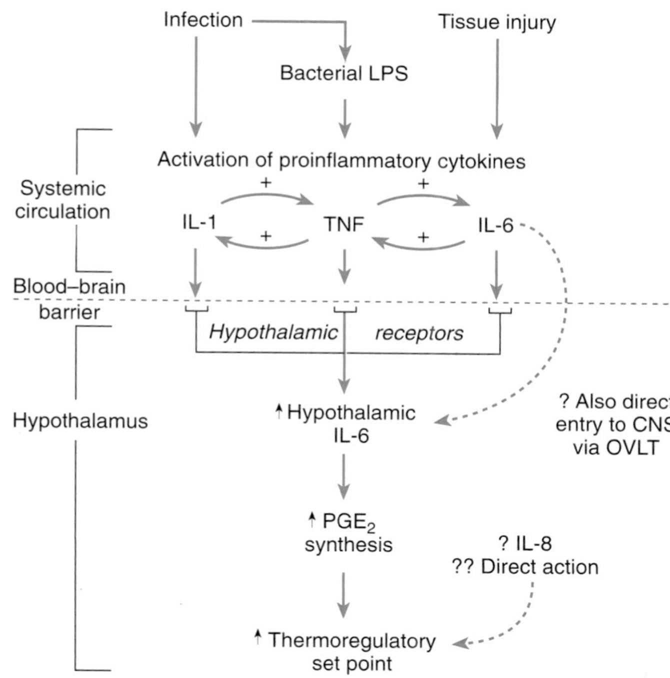

Fig. 2 Illustration of the probable sequence of events in the initiation of fever: OVLT, organum vasculosum of the lamina terminalis; CNS, central nervous system; TNF, tumor necrosis factor; LPS, lipopolysaccharide.

of functional overlap. In addition, it is not clear to what extent the presence of cytokines within the central nervous system is due to transport from the circulation across the blood–brain barrier, particularly at the organum vasculosum of the lamina terminalis, as opposed to local synthesis within the central nervous system.

Despite these difficulties, some consistent features can be observed. Cytokines act primarily on hypothalamic receptors; not infrequently, the 'neurotransmitter' released may be the same peptide as the cytokine that stimulated its release. Neurons producing mRNA for IL-1, IL-6, and tumor necrosis factor have been identified within the hypothalamus. Increased synthesis of IL-6 in the hypothalamus appears to be pivotal to the febrile response, and IL-6 receptors can be demonstrated in the preoptic hypothalamus. IL-6 probably represents a final common pathway for several cytokines. Thus IL-6-deficient mice fail to demonstrate a fever in response to lipopolysaccharide or IL-1. A schematic representation of the probable sequence of events is shown in Fig. 2. Although derived primarily from studies in small mammals, it is likely that a similar chain of cytokine signals prevails in humans (Kluger 1996).

Prostaglandins

Cytokine activity in the hypothalamus is thought to be mediated by increased prostaglandin production, primarily of the prostaglandin E_2 (PGE_2) class (Fig. 2). Thus microinjections of PGE_2 in the organum vasculosum of the lamina terminalis produce fever, and cyclo-oxygenase inhibitors such as aspirin and indomethacin, which block prostaglandin synthesis, inhibit fever following inflammatory stimuli. Prostaglandin F_2 may also have a role. Final cellular pathways are thought to involve cyclic AMP.

Endogenous cryogens

Cytokines do not always elevate body temperature. Several peptides reduce the thermoregulatory set-point; these are known as endogenous cryogens. Putative endogenous cryogens include α-melanocyte-stimulating hormone, AVP, and several inhibitors of proinflammatory cytokines. In addition, tumor necrosis factor, normally considered a 'proinflammatory' cytokine, may also act as an endogenous cryogen. Such mediators may be functionally significant; for example, there is evidence that some of the action of salicylates, in addition to their inhibition of cyclo-oxygenase, may be partially mediated by release of endogenous AVP.

Conclusions

Although heat production and heat loss are perturbed in the critical care setting, fever is the consequence of resetting of the thermoregulatory set-point and activation of heat-conserving mechanisms. Fever can be initiated by a number of cytokines which tend to trigger the whole cytokine cascade. IL-6 synthesis and action in the hypothalamus with induction of PGE_2 synthesis appear to be the final common pathway for most cytokines.

References

Kluger, M.J. (1996). Fever: role of pyrogens and cryogens. *Physiological Reviews*, **71**, 93–127.

Manthous, C.A., *et al.* (1995). Effect of cooling on oxygen consumption in febrile critically ill patients. *American Journal of Respiratory and Critical Care Medicine*, **151**, 10–14.

Saper, C.B. and Breder, C.D. (1994). The neurologic basis of fever. *New England Journal of Medicine*, **330**, 1880–6.

11.10.2 Assessment of fever

D. C. Macallan

Key messages

- The characteristics of fever, including peak, length, and recurrence, should be examined in addition to correlating onset with procedures or initiation of new medications.
- Careful examination should be performed prior to ordering tests which may have a low yield.
- Non-infectious conditions such as thyrotoxicosis or Addison's disease may also cause fever.

Introduction

The febrile patient in the intensive care unit (ICU) covers a broad range of possible pathologies with two broad clinical categories: those whose fever was present at the time of admission, or who had a history of fever prior to admission, and those who develop fever during their stay in the ICU. The second of these categories will be discussed here. Assessment of the patient with fever as part of the presenting illness will be determined by the other features of presentation. If the diagnosis is unclear, extensive assessment as for pyrexia of unknown origin may be indicated.

Infection is the most likely cause of fever developing during an ICU admission, although any pathology that results in tissue injury or necrosis can also result in fever. The reason for this is that tissue injury initiates the same cascade of cytokine activation as does infection. For example, it is well recognized that myocardial infarction is often followed by an acute-phase response with leukocytosis and fever. Similarly, tissue necrosis at pressure areas will elicit an inflammatory response, as does the extensive tissue necrosis that may follow meningococcal septicemia. Fever is unlikely to resolve until healing or debridement of such necrotic tissue has occurred.

Some conditions are characterized by the activation of inflammatory mediators without tissue damage or infection. Examples are immune-mediated transfusion reactions, some forms of drug reaction, and postinfective inflammatory conditions such as the reactive arthritis syndrome.

Recrudescence of old or occult infection is unusual but should be considered in assessment of the febrile patient. Reactivation of herpes simplex infection is well recognized in association with any febrile condition, and may itself promote or prolong fever. Reactivation of tuberculosis should also be considered in a febrile patient, particularly if from an endemic area.

Approach to assessment of development of fever in the ICU patient

In assessing a patient with fever the following questions must be considered.

1. What kind of fever is it?
2. When did it start?
3. Did it commence abruptly at a particular clinical event?
4. How high is it?
5. Is it constant or intermittent?

The chronology of a febrile episode is particularly important. If the patient had been in hospital for some time before admission to the ICU, temperature charts from that time period or any recent previous admissions should be reviewed and collated with the present charts. The pattern of fever may be helpful: traditionally, a high 'spiking' fever is associated with loculated pus and chronic low-grade fever may be associated with the presence of 'low-grade' pathogens such as coagulase-negative staphylococci on intravenous cannulas or prosthetic devices.

The recent in-patient history should be reviewed. Vital clues may be found in reviewing the true sequence of events, not just from clinical records but also from the patient if possible, the relatives, and other staff. This, in turn, should be related to the temperature chart. Attention should be given to the timing of transfusion with blood or blood products. Transfusion reactions are commonly associated with fever but such febrile episodes may be delayed.

Careful examination may reveal oral or nasal trauma or sepsis, lymphadenopathy, added pulmonary sounds, the appearance of a new murmur, abdominal mass or tenderness, or a swollen joint. Particular attention should be paid to the skin around line sites for signs of inflammation or pus, to pressure areas for necrosis of underlying tissues, and to limbs for evidence of deep venous thrombosis.

Fever is a common component of adverse drug reactions. It is often accompanied by a skin rash. The timing of the fever relative to changes in therapy is of particular note, and careful assessment of any changes in therapy should be related to the temperature chart.

When presented with a febrile patient it is tempting to initiate a batch of investigations as a 'fever screen'. Such an approach is not always the most appropriate. Firstly, fever is such a common phenomenon in ICU patients that it may not always represent infection. Secondly, non-targeted diagnostic tests tend to have a very low positive yield. In one study of 464 patients following major abdominal surgery, culture-proven infection was the cause of fever in 27 per cent. Of these infections, 74 per cent could be diagnosed from clinical findings and a single appropriate test. The yields from 'routine' blood, sputum, and urine cultures were particularly low, being 5 per cent, 9 per cent, and 10 per cent respectively. Thus diagnostic tests should be targeted according to clinical suspicion and therapeutic opportunity.

Most ICU-related infections are acquired from the hosts themselves. Microbiological screening is practiced in some units, and any information on colonization patterns of the individual patient should be reviewed. Infection may also be acquired from the ICU environment. Information regarding local colonization patterns and sensitivities, particularly for organisms such as *Pseudomonas* species, should be collected regularly and reviewed in the context of the individual patient.

Urinary tract infection accounts for about 40 per cent of nosocomial infections in ICU patients. It is easy to forget simple tests such as dipstix urinalysis. Catheter-related urinary infection is unlikely before 48 h but thereafter becomes progressively more likely.

Sputum Gram staining is rapid and may be useful, but it may be difficult to distinguish upper respiratory flora from pathogens until culture results are available. Less contaminated specimens may be achieved by bronchoscopic lavage. Colonization of the upper airways tends to occur after 48 h of intubation.

Blood cultures should be drawn during the febrile episode and wound swabs should be taken where appropriate. Central venous catheters are a potential site of infection. Although colonization is common, central venous catheters are unlikely to represent the source of fever if they have been in place for less than 48 h unless placement is known to be contaminated. Infection becomes more common with length of placement and length of stay in the ICU; in one study, 75 per cent of catheter-related infections occurred after the sixth day in the ICU. 'Through-line' blood cultures or the use of intravenous catheter microbiological 'brushes' may indicate the presence of a colonized catheter, and swabs from around an inflamed catheter entry site may reveal the pathogen. The cessation of fever after guidewire exchange of central venous catheters is perhaps the best diagnostic test and therapeutic maneuver simultaneously. Arterial cannulas appear to be less susceptible to infection. In a recent study no positive cultures were found until more than 96 h from the time of placement. Central line tips may be cultured but, if this is requested, care should be taken to ensure skin cleansing around the insertion site before the catheter is withdrawn to reduce contamination.

Candida infection of catheters is particularly hazardous because of the high mortality associated with candidal septicemia and should be considered, particularly if there is candidal colonization elsewhere such as in the oral cavity.

Chest radiography should be routine in ICU patients with fever and should normally be performed if they have been intubated.

Ultrasound examination is particularly useful for investigation of the abdomen, particularly in the right and left upper quadrants and the pelvis, and has the advantage of portability. Postsurgical collections may represent abscess or sterile fluid collections, and diagnostic guided aspiration may be performed for microbiological cultures and to determine if formal drainage is indicated. CT scanning may be able to identify collections not seen on ultrasound but is logistically more difficult.

Intracranial infection is best sought by CT scanning, and this should certainly be performed if lumbar puncture for cerebrospinal fluid examination is proposed because of the risks associated with the presence of cerebral abscess.

Scintigraphy can be used to identify localized sepsis not apparent on other investigations. Gallium citrate demonstrates collections of degranulating leukocytes by binding to released lactoferrin. Leukocytes can be labeled with indium-111 or technetium-99, and labeled antibody techniques are being developed. Scintigraphy has a low diagnostic yield when used as a 'blind' investigation but may disclose occult infection such as osteomyelitis. It is of value in determining whether structural lesions seen with other imaging represent inflammation, for example whether or not a collection seen on CT represents an abscess (Davis and Fink-Bennett 1994).

Measurement of activation of cytokines is not sufficiently specific to distinguish infective from non-infective causes of fever, but this may become a possibility in the future.

Several metabolic conditions result in fever, including Addison's disease (acute adrenal insufficiency), thyrotoxicosis, pheochromocytoma, and malignant neuroleptic syndrome. Salicylate poisoning and anticholinergic overdose are both associated with fever, as are venous thrombosis and pulmonary embolism.

Intracranial lesions such as emboli, tumors, head injury, or neurosurgery can cause 'neurogenic hyperthermia'. This occurs in the absence of elevated systemic cytokines and is thought to be due to direct effects on the hypothalamus. It tends to be intense and short lived, although a low-grade elevation of temperature may persist for several days. This may represent a true elevation of the set-point and be mediated by prostaglandin synthesis; it is suppressed by treatment with indomethacin (Moltz 1993).

Conclusions

Careful clinical assessment and review of the temperature charts are the starting points for assessment of the febrile ICU patient. Fever is such a common phenomenon that each episode must be assessed carefully; instigating a 'batch' of tests in response to fever is rarely a productive approach.

References

Davis, L.P. and Fink-Bennett, D. (1994). Nuclear medicine in the acutely ill patient. *Critical Care Clinics*, **10**, 383–400.

Moltz, H. (1993). Fever: causes and consequences. *Neuroscience and Biobehavioral Reviews*, **17**, 237–69.

11.10.3 Management of fever

D. C. Macallan

Key messages

- The occurrence of fever should initiate an investigation for an underlying cause which, once found, should be treated.
- Very high fevers are dangerous, whereas mild pyrexia may be protective.
- Antipyretics and antimicrobial agents can be helpful.
- Neutropenia, AIDS, immunosuppression, and infection in prosthetic heart valves and during pregnancy require special consideration.

Introduction

The development of elevated body temperature in response to infection may be a protective adaptive response. Blockade of the febrile response with salicylates in rabbits with bacterial or viral infection has been shown to reduce survival (Moltz 1993). In humans, several retrospective clinical trials have shown a correlation between the development of 'moderate' fever and survival from bacterial peritonitis and bacteremia.

Similarly, in sepsis syndrome, those patients with hypothermia have a poorer outcome than those who develop fever. However, it is unclear whether hypothermia in such a setting is itself the mediator of reduced survival or a marker of more severe underlying pathology.

Several possible mechanistic explanations for the beneficial effects of fever have been proposed. There is evidence that neutrophil chemotaxis and phagocytosis are enhanced.

Very high temperatures ($> 40\,°C$) cause neurological dysfunction, neuronal damage, and cardiac arrhythmias, and should always be treated. However, it is less clear whether moderate elevation of body temperature is detrimental. The major argument for treating fever is related to the increased demands that pyrexia makes upon oxygen consumption and delivery. Cooling febrile intensive care unit (ICU) patients by 2.4 °C reduces $\dot{V}O_2$ and energy expenditure by about 20 per cent. This is associated with a reduction in cardiac output from 8.4 to 6.5 l/min (Manthous *et al.* 1995). Thus, when oxygen delivery is limited or hypoxemic respiratory failure is a potential problem, cooling the febrile patient may result in a considerable reduction in the load upon the cardiorespiratory system. Such reduced requirements are likely to be beneficial. In addition, the symptomatic distress caused by high fever in patients who are not sedated should not be forgotten.

Approaches to the treatment of fever

Perhaps the most important aspect of pyrexia in the critical care setting is its 'indicator' function to alert the clinician to an underlying problem or new complication. Therefore fever should always be seen as a warning sign rather than a problem to be solved. However, fever is such a common phenomenon that it may not always have sinister significance, and a single elevated temperature reading should not be overinvestigated.

Clearly, the strategy of choice in managing fever is to treat the underlying disease, but direct measures to reduce body temperature may also be necessary. Antipyretics reduce fever by reducing the thermoregulatory set-point. Acetaminophen (paracetamol) is generally the drug of choice. Although a poor inhibitor of prostaglandin synthesis in peripheral tissues, and thus a poor anti-inflammatory agent when compared with salicylates or non-steroidal anti-inflammatory drugs, it appears to be a more potent inhibitor of prostaglandin synthesis in the central nervous system. It acts by reducing prostaglandin levels in the hypothalamus, thus blocking the effects of endogenous pyrogens. It may be given by the oral, nasal, or rectal route. Aspirin and non-steroidal anti-inflammatory drugs have a similar antipyretic effect via prostaglandin synthesis but are generally avoided because of potential effects on the gastrointestinal tract. Salicylates should be avoided in children because of the risk of Reye's syndrome.

Glucocorticoids reduce pyrexia but are only indicated when other considerations demand their use or in the context of allergic-type febrile transfusion or drug reactions. The antipyretic action of glucocorticoids is mediated by several different mechanisms. Although it is clear that cytokines are the primary mediators of fever, cytokine blockade does not appear to have a role to play in the management of fever *per se*, although such therapy may be of value in the sepsis syndrome.

Heat loss may be promoted by the use of tepid sponging (not cold sponging which would cause cutaneous vasoconstriction) or cooling blankets. Such measures are often very effective, but it should be remembered that, if the set-point is still elevated, one may end up 'competing' with the thermoregulatory effector mechanisms which may be attempting to retain heat. For example, cooling a patient to 38 °C when the set-point is still at 39 °C may result in shivering and vasoconstriction, with increased energy expenditure and discomfort, if the patient is conscious (Manthous *et al.* 1995). Therefore, where possible, taking over the effector mechanisms for regulation of heat loss should be combined with an attempt to reduce the set-point.

Strategies for management of the febrile patient

Although infection is the most common cause of fever in critically ill patients, there are many other potential causes of fever and anti-infective therapy may not always be appropriate. Despite this, many clinicians would utilize a broad-spectrum antibiotic in such a patient if fever was sustained or severe ($> 39\,°C$), once appropriate microbiological cultures had been taken. The choice of antibiotic should be guided by the nature of the underlying primary condition or any subsequent interventions and by knowledge of local antibiotic sensitivity patterns of common pathogens. A hierarchical approach may be appropriate, changing antibiotics after 48 to 72 h if there has been no response.

Central venous catheter infections become more common with increase of duration of use, and therefore some units have a policy of routine changes after a fixed period of time regardless of the presence or absence of signs of colonization or infection. If this is not the case, signs of sepsis in a patient whose catheters have been present for 72 h or more should prompt complete replacement or guide-wire exchange. Defervescence following such a maneuver suggests that the catheters were to blame.

It is not uncommon to encounter patients with persistent fever after several courses of broad-spectrum antibiotics in the absence of clear positive cultures. In this situation it may be appropriate, depending on the severity of the underlying condition, to discontinue antibiotics and repeat cultures after a 'diagnostic window' of 24 to 48 h. Pathogens partially susceptible to the prevailing antibiotic therapy may be impossible to culture in the presence of such treatment but may become apparent on cessation, or indeed the fever may remit as antibiotic drugs themselves are a not infrequent cause of fever as part of an allergic-type response. Similarly, other drugs may cause febrile reactions and it may be helpful to withdraw all drugs which may not be absolutely necessary during such a 'diagnostic window', particularly those known to be associated with febrile or allergic reactions.

Special cases

Neutropenia

Infection is more common and likely to be more severe in patients with neutropenia. Therefore broad-spectrum antibiotics should be started early and antifungal agents such as amphotericin added if fungal infection is suspected or if fever persists. Most units dealing with neutropenic patients have a predetermined hierarchy of agents used in an escalating fashion according to response. Fever often does not completely remit until neutrophil recovery begins to occur.

Immunosuppression and AIDS

Similar arguments for early 'blind' anti-infective therapy apply in patients immunosuppressed for other reasons, such as transplant recipients or those receiving steroid therapy. Transplant recipients often have impaired cell-mediated immunity and may be more susceptible to fungal, protozoal, or viral infection in addition to bacterial sepsis. A variety of opportunistic infections, such as *Pneumocystis carinii* pneumonia, cytomegalovirus infection, or *Mycobacterium intracellulare avium* infection, may cause fever in patients with AIDS, and these require investigation and treatment.

Prosthetic heart valves

Fever in patients with prosthetic heart valves should always alert the clinician to the possibility of infection on the prosthetic valve, particularly if bacteremia has been documented. If seeding onto the valve is suspected, it may be appropriate to continue antibiotic therapy for a period of up to 6 weeks, as if for endocarditis, even if fever and clinical signs have resolved.

Fever in pregnancy

Fever in pregnancy may be hazardous to the continuation of the pregnancy. In addition, high fevers in early pregnancy have been shown to be teratogenic in some animal models.

Conclusions

Fever is a common occurrence in ICU patients. Its major significance is to alert the clinician to possible infective or other complications. Mild fever may have beneficial effects in terms of survival from infection, but fever increases oxygen requirements and may compromise tissue oxygen delivery in severely ill patients. High fever is hazardous and should always be treated.

References

Manthous, C.A., *et al.* (1995). Effect of cooling on oxygen consumption in febrile critically ill patients. *American Journal of Respiratory and Critical Care Medicine*, **151**, 10–14.

Moltz, H. (1993). Fever: causes and consequences. *Neuroscience and Biobehavioral Reviews*, **17**, 237–69.

11.10.4 Pathophysiology of malignant hyperthermia

P. J. Halsall and F. R. Ellis

Key messages

- Malignant hyperthermia is thought to be caused by deregulation of Ca^{2+} homeostasis in skeletal muscle on exposure to triggering anesthetic agents, involving the mechanism of excitation–contraction coupling.
- The calcium channels of the triadic junction, the dihydropyridine and ryanodine receptors, are implicated in the etiology.
- Only 50 per cent of malignant hyperthermia families show genetic linkage to *RYR*1 (the ryanodine receptor).
- Marked genetic heterogeneity makes DNA testing for malignant hyperthermia dangerously unreliable at the present time.

Introduction

Malignant hyperthermia, an autosomal dominantly inherited condition, was first described in 1960. It was so named because it caused death associated with a very high body core temperature which developed during specific types of anesthesia. It is now recognized to be triggered by all the inhalational anesthetic agents and depolarizing muscle relaxants, for example succinylcholine (suxamethonium).

Description

With the marked improvement in monitoring during anesthesia, early diagnosis of a malignant hyperthermia crisis now occurs with a dramatic fall in mortality. The signs of malignant hyperthermia can be divided into two groups: metabolic stimulation and muscle abnormality. Malignant hyperthermia is often described as a metabolic storm with increased CO_2 production, falling SaO_2, increased heart rate, often with dysrhythmias, increased core temperature, and metabolic acidosis. The signs associated with muscle involvement include masseter muscle spasm, generalized rigidity, marked rise in creatine kinase, myoglobinuria, and a rise in serum K^+.

Mechanism

It is now recognized that skeletal muscle is the principle tissue involved in a malignant hyperthermia reaction. The demonstration by Kalow (1970) that malignant-hyperthermia-susceptible muscle developed spasm at lower concentrations of caffeine, which enhances the release of Ca^{2+} from the sarcoplasmic reticulum, than muscle that was not susceptible to malignant hyperthermia laid a basis for the pathophysiology of malignant hyperthermia as well as a screening test for this condition in humans. It was proposed that the deregulation of Ca^{2+} within the muscle cell could account for many of the signs in malignant hyperthermia. However, calcium is ubiquitous within the muscle cell and the exact site of the defect remains debatable. An increase in intracellular Ca^{2+} could come about by increased release from or decreased uptake by the sarcoplasmic reticulum, increased release from the mitochondria, or an increase in the entry of extracellular Ca^{2+} through the sarcolemma.

Whatever the source, the increase in Ca^{2+} in the sarcoplasm during a malignant hyperthermia reaction results in severe metabolic disturbances due to activation of the contractile proteins causing the contracture seen in malignant hyperthermia.

Membranes

A defect in the sarcolemma has been proposed as a cause of malignant hyperthermia. Halothane has been shown to cause depolarization of the sarcolemma in pigs susceptible to malignant hyperthermia and may enhance its permeability to calcium more than in normal muscle. Dantrolene, which is the only specific therapy for treating malignant hyperthermia but whose site of action is unknown, has been shown to reduce significantly the muscle contractures caused by K^+ which result in progressive depolarization of the sarcolemma and subsequent influx of extracellular Ca^{2+}. It has little effect on contractures caused by caffeine, which acts on the sarcoplasmic reticulum, or by dinitrophenol, which causes uncoupling of oxidative phosphorylation in the mitochondria.

Sarcoplasmic reticulum

As the sarcoplasmic reticulum is the main storage site of Ca^{2+}, and as caffeine has been shown to produce contractures in malignant-hyperthermia-susceptible muscle at much lower concentrations than in normal muscle, most experimental work has focused on the role of the sarcoplasmic reticulum in the pathophysiology of malignant hyperthermia.

Calcium is stored with calsequestrin in the terminal cisternas of the sarcoplasmic reticulum; it is released on depolarization and taken up again by means of an ATP-dependent Ca^{2+} pump. It has been shown that an increase in sarcoplasmic Ca^{2+} can itself cause release of further Ca^{2+}, a process known as Ca^{2+}-induced Ca^{2+} release. Although the role of this process in normal physiology is questionable, it has been postulated as a mechanism for malignant hyperthermia. Other studies of the sarcoplasmic reticulum suggest that a defect in the control of Ca^{2+} efflux from the sarcoplasmic reticulum may cause malignant hyperthermia.

Excitation–contraction coupling

The sarcoplasmic reticulum has recently assumed a more prominent role with detailed knowledge of the structural process involved in excitation–contraction coupling, the mechanism by which nervous stimulation is immediately converted into a muscle contraction in a tightly controlled co-ordinated manner.

The sarcoplasmic invaginations (T tubules) are aligned closely with the terminal cisternas of the sarcoplasmic reticulum. The junctional gap between the two has only recently been described on a molecular basis. There appears to be a direct link, known as the triadic junction, composed of the dihydropyridine receptor on the T tubule and the ryanodine receptor on the sarcoplasmic reticulum. The dihydropyridine receptor consists of five subunits and is believed to act as a voltage sensor or slow Ca^{2+} channel. The ryanodine receptor, so called because it binds with high affinity with the plant alkaloid ryanodine, exists as a tetramer of identical subunits and is similar in structure to the nicotinic acetylcholine receptor. It consists of a 'foot structure' lying within the cytoplasm, with the remainder lying in the sarcoplasmic reticulum. It acts as a Ca^{2+} channel, with Ca^{2+} and ATP acting synergistically to open the channel while Mg^{2+} and calmodulin inhibit opening.

It is believed that the altered kinetics of the ryanodine receptor Ca^{2+} channel are due to both increased release of Ca^{2+} by small increases in the cytoplasmic Ca^{2+} concentration and a reduction in the inhibitory effects of high Ca^{2+} concentrations.

Although the site of action of volatile anesthetic agents is uncertain, halothane can alter the voltage-dependent Ca^{2+} channels and Ca^{2+} release from the sarcoplasmic reticulum and can affect the Ca^{2+}-binding protein calmodulin which causes activation of many enzymes. It has recently been shown that propofol, which does not trigger malignant hyperthermia, does not stimulate ryanodine receptor activity but affects the dihydropyridine receptor and Ca^{2+}-ATPase in a similar manner to the inhalational anesthetics, supporting the hypothesis that the latter act on the ryanodine receptor (Fruen *et al.* 1995).

It is likely that excitation–contraction coupling occurs through a direct structural link rather than by a 'transmitter' effect using 1,4,5-trisphosphate. It is envisaged that the dihydropyridine receptor acts as a voltage sensor for changes in potential differences across the T-tubule membrane which in turn produce conformational changes in the ryanodine receptor, causing the Ca^{2+} channel to open. Although the precise mechanism of excitation–contraction coupling has yet to be verified, it is clear that the dihydropyridine and ryanodine receptors play crucial roles.

The role of the second messenger 1,4,5-trisphosphate in the pathophysiology of malignant hyperthermia is unclear; it has been postulated as a modulator of excitation–contraction coupling. As 1,4,5-trisphosphate levels are not altered after exposure to halothane, it seems unlikely that it has an important primary role in the onset and maintenance of malignant hyperthermia. A recent study suggests that malignant-hyperthermia-susceptible muscle is more responsive to 1,4,5-trisphosphate -induced Ca^{2+} release, which is prevented by dantrolene, than is normal muscle (Lopez *et al.* 1995).

Intracellular events

Whatever the mechanism underlying the pathophysiology of malignant hyperthermia, it is generally accepted that an uncontrolled influx of Ca^{2+} into the cytoplasm results in a severe metabolic disturbance within the cell and rapid depletion of ATP. In normal circumstances, muscle contraction, glycolysis, and mitochondrial function are regulated by intracellular Ca^{2+} concentration. Calcium is

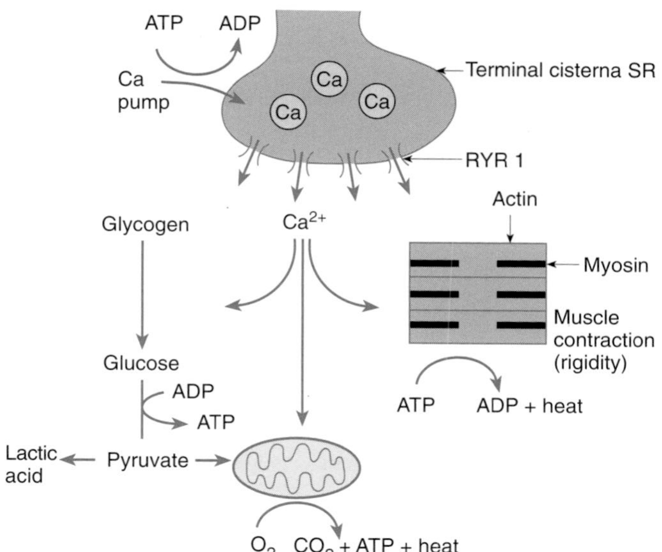

Fig. 1 Diagram showing the intracellular kinetics of calcium: SR, sarcoplasmic reticulum.

pumped back into stores by Ca^{2+}-ATPase to initiate muscle relaxation, while glycolysis and aerobic metabolism proceed just rapidly enough to maintain energy balance. In a malignant hyperthermia crisis Ca^{2+} is released into the cytoplasm at enhanced rates, overpowering the Ca^{2+}-ATPase pump. Glycolysis, aerobic metabolism, and anaerobic metabolism are activated in an attempt to replenish ATP levels. Phospholipase A_2 is also activated by Ca^{2+}, causing liberation of long-chain fatty acids and an increase in sarcoplasmic and mitochondrial membrane permeability, with further loss of Ca^{2+} regulation.

Damage to the sarcoplasmic membrane leads to loss of Mg^{2+} and K^+. If it is severe, larger molecules, namely creatine kinase and myoglobin, are also lost. As muscle is a major source of body heat, increased activity results in an increased core temperature. Thus a disruption of the normal control of Ca^{2+} regulation within the cell results in the severe metabolic disturbances seen in malignant hyperthermia and accounts for the observed clinical signs and symptoms (Fig. 1) (MacLennan and Phillips 1992).

The genetic model

Advances in DNA technology have supported the concept for a central role of the ryanodine receptor gene *RYR*1 in the pathophysiology of malignant hyperthermia. The identification of a series of markers around the region of *RYR*1 has enabled a detailed genetic analysis to be performed for families susceptible to malignant hyperthermia. Initially it was hoped that genetic studies would supersede the *in vitro* contracture test used to diagnose malignant hyperthermia and provide insight into the actual underlying defect. However, only 50 per cent of all malignant-hyperthermia-susceptible families studied worldwide segregated with *RYR*1, and furthermore *RYR*1 linkage could be firmly excluded in some families. Genetic heterogeneity has been described in both European and North American families. Searches have been performed in other potential candidate areas of the genome, in particular the regions for the protein subunits of the dihydropyridine receptor.

Eight different mutations in *RYR*1 have been identified, and approximately 20 per cent of families definitely linked to *RYR*1 have a mutation. Furthermore, the mutation does not always segregate with the *in vitro* contracture test result within a family. For example, an individual may be 'genetically' susceptible to malignant hyperthermia but malignant hyperthermia negative by the *in vitro* contracture test, or vice versa. The genetics of malignant hyperthermia is quite complex, and further elucidation is needed before DNA studies on malignant hyperthermia families can be considered as a possible alternative to the *in vitro* contracture test as a screening test (Hopkins *et al.* 1994).

Although malignant hyperthermia is rarely associated with central core disease either clinically or histologically, central core disease is commonly but not invariably associated with malignant hyperthermia (Halsall *et al.* 1996). Central core disease has been shown to be linked to *RYR*1 and several *RYR*1 mutations have been described in families with this disease.

Conclusions

It is possible that the clinical picture of malignant hyperthermia is the final common pathway exhibited by any number of defects, either alone or in combination, in the complex and interrelated kinetics of excitation–contraction coupling. The *in vitro* contracture test is able to detect this phenotype. This concept may explain the marked heterogeneity seen to date in DNA studies of malignant hyperthermia families.

References

Fruen, B.R., Mickelson, J.R., Roghair, T.J., Litterer, L.A., and Louis, C.F. (1995). Effects of propofol on Ca^{2+} regulation by malignant hyperthermia-susceptible muscle membranes. *Anesthesiology*, **82**, 1274–82.

Halsall, P.J., Bridges, L.R., Ellis, F.R., and Hopkins, P.M. (1996). Should patients with central core disease be screened for malignant hyperthermia? *Journal of Neurology, Neurosurgery and Psychiatry*, **61**, 119–21.

Hopkins, P.M., Halsall, P.J., and Ellis, F.R. (1994). Diagnosing malignant hyperthermia susceptibility. *Anaesthesia*, **49**, 373–5.

Kalow, W., Britt, B.A., Terrau, M.E., and Haist, C. (1970). Metabolic error of muscle metabolism after recovery from malignant hyperthermia. *Lancet*, **ii**, 895–8.

Lopez, J.R., Perez, C., Linares, N., Allen, P., and Terzic, A. (1995). Hypersensitive response of malignant hyperthermia-susceptible skeletal muscle to inosital 1,4,5-trisphosphate induced release of calicum. *Nauny-Schmiedeberg's Archives of Pharmocology*, **352**, 442–6.

MacLennan, D.H. and Phillips, M.S. (1992). Malignant hyperthermia. *Science*, **256**, 789–94.

11.10.5 Management of malignant hyperthermia

P. J. Halsall and F. R. Ellis

Key messages

- Successful management of malignant hyperthermia depends on prompt diagnosis.
- Prompt diagnosis depends on awareness of malignant hyperthermia by the anesthetist and modern non-invasive monitoring, particularly capnography.

- The specific therapy, dantrolene, should be available in every theater location.
- Previous uneventful anesthetic does not preclude malignant hyperthermia.
- Malignant hyperthermia patients should not be denied essential surgery.

Introduction

When malignant hyperthermia was first described the mortality was 70 to 80 per cent, but this has now fallen dramatically to less than 10 per cent worldwide; in the United Kingdom there have been only three deaths in the last 5 years. This vast improvement has been brought about by three factors:

(1) increased awareness and understanding of malignant hyperthermia by anesthetists;

(2) better intraoperative monitoring facilities, particularly capnography and pulse oximetry;

(3) the availability of an intravenous formulation of dantrolene.

The first two factors lead to the early diagnosis of malignant hyperthermia at a stage when prompt treatment is usually successful. Late recognition of malignant hyperthermia will probably result in the death of the patient despite treatment with dantrolene.

Clinical presentation of malignant hyperthermia

It is important to be aware of the different modes of presentation of malignant hyperthermia in order to be alert to the potential diagnosis. An attempt has been made to score the various modalities so as to give the likelihood of malignant hyperthermia (Larach *et al.* 1994). However, this method is very limited, particularly when data are missing, and no allowance is made for grading either the severity or the timing of the clinical signs.

In a previous study (Ellis *et al.* 1990), the various modes of clinical presentation of malignant hyperthermia were divided into eight groups ranging from the full-blown life-threatening fulminant case to a simple postoperative pyrexia. The results of *in vitro* contracture tests showed a good correlation with the eight categories such that, given a clinical scenario, the likelihood of malignant hyperthermia could be predicted (Table 1). The drawback of this method is that it depends entirely on the diligence and skill of the anesthetist interpreting the clinical situation and requires full documentation of all the clinical events.

Table 1 The likelihood of malignant hypothermia according to type of clinical presentation

Presentation	Incidence
Severe/fulminant	1.00
Moderate	0.88
Mild	0.14
Masseter spasm and muscle signs	0.76
Masseter spasm and metabolic signs	0.57
Masseter spasm alone	0.28
Unexplained anesthetic death/cardiac arrest	0.66
Others (postoperative pyrexias/rhabdomyolysis)	0.07

Management

The management of malignant hyperthermia should be considered in three stages: preoperative, intraoperative, and postoperative.

Preoperative management

A preoperative assessment of the patient must elicit a personal and family anesthetic history. However, a previous uneventful anesthesia does not preclude malignant hyperthermia as many malignant-hyperthermia-susceptible probands sustain an untoward reaction during subsequent administration of anesthesia. A history of unexplained cardiac arrest during general anesthesia in the family may be significant, and attempts should be made to establish the exact etiology. If malignant hyperthermia cannot be eliminated, the patient should be managed as potentially susceptible to malignant hyperthermia. A previous history of unexpected postoperative renal failure also suggests malignant hyperthermia.

A history of postoperative pyrexia without an eventful intraoperative course is unlikely to be associated with susceptibility to malignant hyperthermia. The only disorder clearly associated with malignant hyperthermia is central core disease. As this association is variable (Halsall *et al.* 1996), such patients should be treated as potentially susceptible to malignant hyperthermia and offered screening.

A flowchart for preoperative management is shown in Fig. 1.

Intraoperative management

Modern monitoring has greatly aided the early diagnosis of malignant hyperthermia. With the introduction of capnography it soon became apparent that the earliest sign of an impending malignant hyperthermia crisis was not an increase in temperature but an

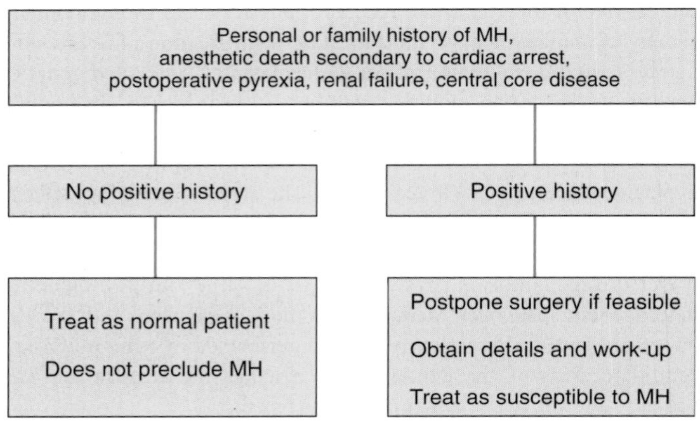

Fig. 1 Preoperative management of malignant hyperthermia (MH).

unexplained increase in end-tidal CO_2 concomitant with an unexplained tachycardia. Once other diagnoses have been excluded (e.g. problems with the anesthetic circuitry etc.), these two signs should alert the anesthetist to the possibility of malignant hyperthermia. At this stage core temperature recording should be established and monitored continuously.

The onset of a malignant hyperthermia reaction varies considerably; it may be sudden and rapidly life threatening as in fulminant crisis, or it may develop insidiously. The latter often occurs with low concentrations of the anesthetic vapors currently used to prevent 'awareness' and malignant hyperthermia may take 1 to 2 h to develop.

Once the diagnosis of malignant hyperthermia has been suspected, the following treatment plan should be instituted and modified based on signs and severity.

1. Withdraw all trigger agents (this includes all vapors), flush the anesthetic machine, and substitute a clean circuit. Ventilation may need to be increased to two to three times the minute volume to aid the removal of CO_2.
2. If feasible, abandon the procedure as soon as possible or continue with a 'malignant-hyperthermia-safe' technique (see later).
3. Give specific therapy: intravenous dantrolene 1 mg/kg. This can be repeated up to 10 mg/kg depending on the regression of signs. Frequently the response to dantrolene is rapid and obvious unless it is given too late. Dantrolene is expensive and rarely used, and so it is recommended that each operating theater should keep enough dantrolene to provide initial therapy while further amounts are obtained from a central store.
4. Start to cool the patient but avoid causing vasoconstriction, which impedes heat loss. Simple measures such as tepid sponging, fanning, and cool intravenous fluids are usually sufficient, although more aggressive therapy (e.g. cold gastric, peritoneal, or bladder lavage) may be needed.
5. Treat arrhythmias in the usual way. With a severe reaction a cardiac arrest may quickly ensue; thus metabolic acidosis and hyperkalemia should be corrected early.
6. Blood samples should be taken for serum K^+, arterial blood gases, and an initial creatine kinase. Hyperkalemia and acidosis can be corrected in the usual way.
7. The urine output should be observed for signs of impending renal failure, and catheterization may be required. A urine sample should be sent for myoglobin evaluation. If the urine is stained red-brown, suggesting the presence of myoglobin, diuresis should be promoted with fluids and perhaps mannitol, remembering that one 20-mg vial of dantrolene contains 3 mg of mannitol. If renal failure occurs, the patient will need to be dialyzed temporarily until the kidneys recover.
8. Although disseminated intravascular coagulation rarely develops, the clotting status should be checked.
9. Core temperature should be monitored for 24 to 48 h.
10. A repeat creatine kinase measurement should be performed after 24 h as it is slow to rise compared with myoglobinuria, which occurs soon after the initial event.
11. Consider the differential diagnosis of, for example, thyrotoxic storm, pheochromocytoma, septicemia, or respiratory or other infection. The appropriate investigations need to be performed, and include white blood count, chest radiography, thyroid function studies, and urine vanilmendalic acid.
12. Consider the possibility of a myopathic disorder, particularly the myotonic diseases if muscular dysfunction is prominent. An electromyogram (EMG) and/or a neurological opinion may be required.
13. After recovery the patient and/or the family should be counselled as to the potential implications of malignant hyperthermia and the necessity of confirming or refuting the clinical diagnosis by an *in vitro* contracture test.
14. Refer the patient to a malignant hyperthermia screening center if one is available.

Masseter muscle spasm

Masseter muscle spasm occurring after the administration of succinylcholine (suxamethonium) is generally regarded as a warning sign of malignant hyperthermia. Much of the controversy surrounding the significance of masseter muscle spasm stems from the difficulty in defining the spasm itself and that succinylcholine has been shown to increase jaw tension in normal patients. Unexpected difficulty with jaw opening after succinylcholine which hinders intubation by an experienced anesthetist should be considered as masseter muscle spasm, and the patient should be presumed to be potentially at risk from malignant hyperthermia. If feasible, the procedure should be abandoned, blood should be taken for initial and 24-h creatine kinase, and the first voided specimen of urine sent for myoglobin estimation. It is unlikely that the patient will exhibit any other signs of malignant hyperthermia, but these should be sought and treated appropriately. If the procedure is urgent, anesthesia can be continued with a 'malignant-hyperthermia-safe' technique once the patient is stable. The incidence of malignant hyperthermia susceptibility (from *in vitro* contracture tests) in patients presenting with masseter muscle spasm as the sole sign is 25 to 30 per cent, but this includes many patients who were not fully investigated so that masseter muscle spasm appeared to be the only apparent sign.

Patients with additional metabolic signs (e.g. tachycardia, pyrexia) have a 50 per cent incidence of being shown to be susceptible to malignant hyperthermia after *in vitro* contracture tests, and those with muscle signs (e.g. very high creatine kinase and myoglobinuria) have the highest incidence of 75 to 80 per cent (Ellis *et al.* 1990).

Postoperative management

The most frequent postoperative presentations are severe prolonged muscle stiffness (the patient can be immobile for 1 to 2 days, with significantly more severe pain than is usually associated with 'scoline pains'), myoglobinuria with or without renal failure, and a pyrexia developing postoperatively. These signs are usually missed by the anesthetist at the time, and are frequently noticed at a subsequent anesthesia. If the prominent signs are of severe rhabdomyolysis, an EMG and a neurological opinion are needed to exclude a myopathic condition.

The known or suspected patient

Malignant-hyperthermia-susceptible patients and their families should not be denied essential surgery on the grounds of malignant hyperthermia alone.

In the majority of situations a malignant-hyperthermia-safe technique will offer little increased risk except in certain rare circumstances when the avoidance of succinylcholine and anesthetic vapors is a potential hazard (e.g. crash induction for full stomach, difficult airway, etc.).

Family screening is recommended to avoid exposing other family members, which may include a large number of individuals, to a less than optimum anesthetic technique. Although family screening may seem a costly, time-consuming, and invasive procedure, only a small proportion of family members need to be screened to identify all malignant-hyperthermia-susceptible individuals (Ellis *et al.* 1986). There is no undisputed record of a patient who has been shown to be malignant hyperthermia negative by the *in vitro* contracture test having a subsequent malignant hyperthermia reaction during anesthesia.

Problem areas for malignant-hyperthermia-susceptible patients are as follows:

(1) full stomach and rapid-sequence induction;

(2) the difficult airway and inability to use inhalational induction;

(3) prevention of awareness when using inhalational drugs;

(4) provision of a vapor-free anesthetic machine, circuit, and ventilator.

A wide range of anesthetic agents are safe for malignant-hyperthermia-susceptible patients (Table 2). The anesthetist should have a knowledge and understanding of malignant hyperthermia, and anesthesia should be conducted in a well-equipped hospital operating theater. Adequate basic monitoring (i.e. ECG, non-invasive blood pressure, core temperature, capnography, and pulse oximetry) is essential for all procedures.

Ideally; a vapor-free anesthetic machine should be used. However, if this is not available, the vaporizers should be removed from the machine, which should then be flushed with oxygen for about 20 min to remove traces of vapor. A fresh clean breathing circuit should be used. Ventilators should also be flushed with oxygen before use.

Table 2 Agents which can be used in malignant hyperthermia patients

Thiopental (thiopentone), propofol
All analgesics
Droperidol, metoclopramide
Neostigmine+atropine or glycopyrrolate
Nitrous oxide
Benzodiazepines
All local anesthetic agents including lidocaine (lignocaine) and bupivacaine

There is still some uncertainty about the use of agents such as the phenothiazines following a reported malignant hyperthermia reaction many years ago. In view of their wide use it is unlikely that the phenothiazines play a significant part in triggering a malignant hyperthermia reaction. They are not contraindicated but should be used cautiously. However, they can cause neuroleptic malignant syndrome which clearly is a different condition (Krivosic-Horber and Adnet 1989).

The role of both intrinsic and extrinsic catecholamines remains unclear. A study of the effect of ephedrine on *in vitro* muscle contractures caused by halothane did show an increased response but at considerably greater concentrations than those used clinically. Drugs having a sympathomimetic effect should be used cautiously in malignant-hyperthermia-susceptible individuals.

References

Ellis, F.R., Halsall, P.J., and Harriman, D.G.F. (1986). The work of the Leeds Malignant Hyperthermia Investigation Unit, 1971–84. *Anaesthesia*, **41**, 806–15.

Ellis, F.R., Halsall, P.J., and Christian, A.S. (1990). Clinical presentation of suspected malignant hyperthermia during anaesthesia in 402 probands. *Anaesthesia*, **45**, 838–41.

Halsall, P.J., Bridges, L.R., Ellis, F.R., and Hopkins, P.M. (1996). Should patients with central core disease be screened for malignant hyperthermia? *Journal of Neurology, Neurosurgery and Psychiatry*, **61**, 119–21.

Krivosic-Horber, R. and Adnet, P.J. (1989). MH and neuroleptic malignant syndrome. *Beitrage zur Anaesthesiologie und Intensivmedizin*, **27**, 108–13.

Larach, M.G., *et al.* (1994). A clinical grading scale to predict malignant hyperthermia susceptibility *Anesthesiology*, **80**, 771–9.

11.10.6 Pathophysiology of heatstroke

A. M. S. Kashmeery

Key messages

- ATP depletion and Na^+,K^+-ATPase deficiency contribute to the pathophysiology of heatstroke.
- Oxygen delivery and consumption are impaired and correction is crucial for survival.

Introduction

Under normal conditions, the skin receives 8.6 per cent of the total circulating blood flow. Under conditions of stress, this rate may quadruple. When the blood temperature is elevated as little as 0.01 °C above 37 °C, heat dissipation is required to maintain this core temperature. Thermal changes are detected by two routes: skin sensory thermal nerve endings and blood temperature. Both detection mechanisms are integrated within the hypothalamus, which then sends impulses to several structures including the peripheral blood vessels and the sweat glands. The vasoconstrictory sympathetic nervous system impulses to the smooth muscles of the tunica media of the skin arterioles are inhibited, leading to vasodilation and sweat gland secretion. This allows efficient thermal exchange provided that environmental conditions, such as ambient temperature, humidity, and wind speed, are appropriate. External body cooling, which causes peripheral vasoconstriction, may actually be of little value by not allowing for heat dissipation (Ash and Kashmeery 1995)

Cellular events

Dense granules seen in the mitochondria following heat exposure appear to be heat shock proteins which enhance thermotolerance (Welch 1993).

The resemblance between the effects of antimuscarinic drugs (e.g. atropine), which produce a central anticholinergic syndrome, and heat illness is impressive. Heat-producing reactions, which occur at neuromuscular junctions (voluntary contractions, shivering), and sweat glands, which are both muscarinic, are under cholinergic nerve control.

Heat fatigue may be a manifestation of ATP depletion, in part secondary to sodium influx into the cells. This occurs when a high temperature increases membrane permeability to sodium, thereby dramatically stimulating the energy-consuming Na^+K^+-ATPase enzyme activity. The propagation of nerve impulses normally requires a Na^+ and K^+ gradient, which is constantly maintained by Na^+,K^+-ATPase. This enzyme is also responsible for maintaining cell volume by osmotically drawing water out of the cell in the direction of the net positive charges. When ATP is depleted, the activity of Na^+,K^+-ATPase is curtailed. This results in tissue edema and neural membrane depolarization, leading to the convulsions seen with heatstroke. In addition to the heat-induced energy depletion, high temperature is known to cause loss of cellular ATP synthesis. Thus, in the search for a primary lesion of heat injury, energy seems to be a focal point. Membrane destruction and protein denaturation have also been proposed, but with fewer advocates.

Deviation and compensation

With heat stress and exertion, homeostatic mechanisms become deranged and feedback mechanisms are activated (Fig. 1). This scheme is based on the assumption that heat load associated with excessive sweating may lead to hypovolemia and subsequent heatstroke shock. Early signs of heat illness do not seem to be associated with hypovolemia. However, the interactions of pertinent vasoactive forces seem to be more complicated. The skin of heatstroke patients is usually hot and dry. When sweating no longer occurs, this may indicate an ineffective or defective sympathetic response. Experimentally, sympathectomy will increase blood supply through vasodilation. Renin, which stimulates the formation of the potent vasoconstrictor angiotensin, has been found in high concentrations in the plasma of heat exhaustion patients (Kashmeery 1995*a*). A well-established lag in rectal temperature (Ash *et al.* 1992) and frequent occurrence of acute renal failure strongly suggest dramatic splanchnic vasoconstriction, a function controlled by the sympathetic nervous system. In addition, individuals with prolonged exposure to elevated environmental temperatures and physical stress have been shown to have an increase in neutrophils, which secrete vasodilators such as bradykinin which may serve to redistribute blood flow to the periphery. Thus the possibility that vasodilatory mechanisms overcome elevated sympathetic nervous system response in heatstroke patients cannot be excluded.

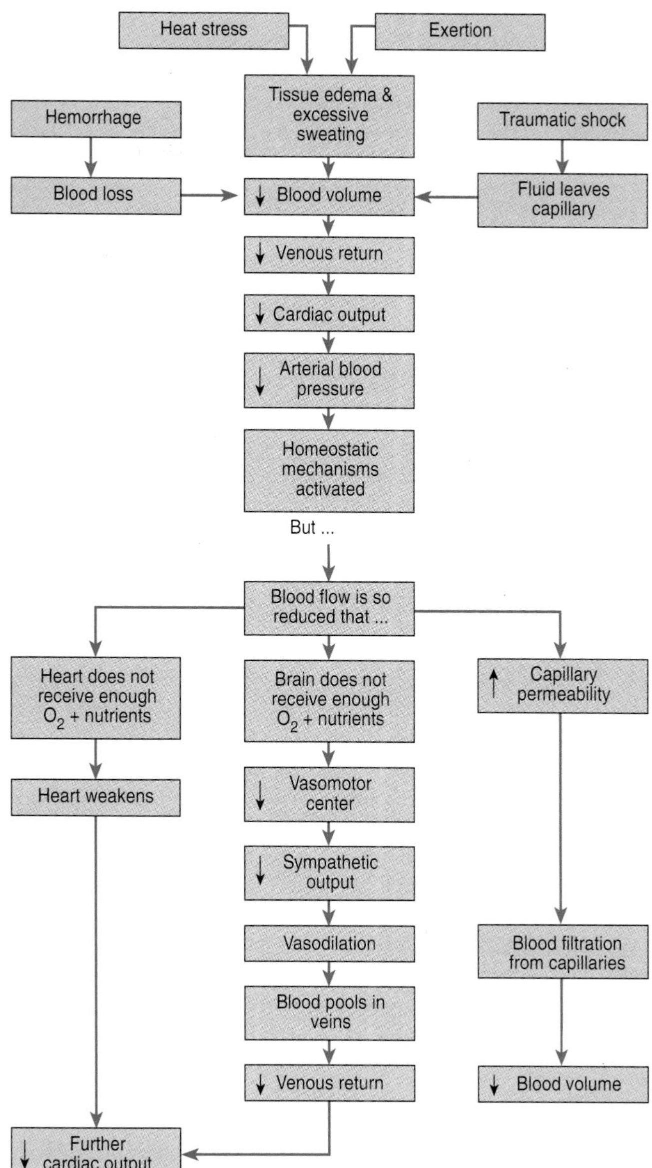

Fig. 1 In progressive circulatory shock, homeostatic mechanisms become inadequate when positive feedback mechanisms are triggered and shock is deepened.

Hypoxia, normoxia, and defective tissue oxygen delivery

Early heatstroke leads to inappropriate behavior and disorientation. In addition, elevated partial pressure of venous oxygen levels is evident (Kashmeery 1995*a*). This may be secondary to a reduced oxygen unloading capacity. In septic shock patients, this phenomenon has been attributed to arteriovenous shunting, which is unlikely in heatstroke patients who present with hot erythematous skin indicative of adequate blood perfusion. The delicate balance of mitochondrial energy production is susceptible to hypoxemia. Under hypoxic or normoxic conditions, a decrease in transmembrane potential, possibly due to a decrease in mitochondrial calcium, may result in inhibition of ATP synthesis. Experimentally, supplemental oxygen is unable to restore this uncoupling in mitochondria isolated from tissue previously exposed to hyperthermia or to restore the normal rate of substrate oxidation, even in the presence of calcium.

Oxygen delivery and oxygen consumption

The oxygen extraction ratio, which is given by oxygen consumption divided by oxygen delivery, is reflected by a large arteriovenous oxygen difference. Thus the elevated partial pressure of venous oxygen in heatstroke victims leads to a depressed oxygen extraction ratio. In a state of energy depletion, as is the case in heatstroke, the oxygen extraction ratio is expected to be high. The association of heatstroke with increased glucose utilization, and marked production of cytokine (Bouchama *et al.* 1993) and lactate, supports the assumption of impaired oxygen extraction in these patients. Victims of heatstroke also appear to have decreased hemoglobin concentrations (Kashmeery 1995*b*). Strenuous physical stress can, in combination with anemia and impaired oxygen extraction, lead to type A lactic acidosis, which is characterized by an increased load of hydrogen ions. Elevated urine pH in heat exhaustion patients might reflect impaired hydrogen ion secretion and/or impaired bicarbonate reabsorption in the kidney tubules.

The secretion of hydrogen ions, which takes place in the distal convoluted tubule of the nephron, is an active process requiring a constant energy supply. The source of this supply is oxidative phosphorylation, which requires adequate oxygen extraction. Minimizing hydrogen ion production is essential in managing type A lactic acidosis. This is achieved by improving tissue oxygen extraction. In both cases oxygen is a common factor in the manifestation of the disorder. However, it is not conclusively determined whether it is an aberration of oxygen delivery or oxygen consumption that is causing this manifestation.

Reversal of heatstroke pathophysiology

In surgical patients with low tissue oxygen tension, intravenous infusion of a small volume of normal saline results in a prompt rise in tissue oxygen tension readings. The administration of a cold intravenous infusion of less than 1 liter over less than 1 h has led to a significant reduction in blood temperature and general clinical improvement in heatstroke patients (Kashmeery 1995*b*). A significant reduction in the abnormally elevated venous oxygen content, implying an improved oxygen extraction ratio, is also seen in these patients. Owing to its established lag, reliance on rectal temperature should be minimized. Mild external body cooling eliminates shivering, which usually exacerbates the problem by accumulating further heat to the body core.

In hypovolemic shock the kidneys autoregulate. In heat exhaustion patients, who manifest minimal signs of hypovolemia, urine flow is sharply diminished. It is generally accepted that tissues like intestine and skin are peripheral and therefore dispensable in the hierarchy of blood-flow-demanding tissues under stressful conditions. Although the kidneys are as heavily innervated by the autonomic nervous system as these other tissues, they maintain adequate blood flow even under severe hypovolemia. Thus diminished urine flow associated with heat illnesses is not likely to be due to blood flow.

Oxygen extraction appears to be one of the important factors in the pathogenesis of heatstroke. Thus improvement in maximizing oxygen extraction and utilization, which is reflected in a large arteriovenous oxygen difference, can be crucial and life saving. Unless corrected early, the pathophysiology continues with significant morbidity and mortality.

References

Ash, C.J. and Kashmeery, A.M.S. (1995). Heatstroke: marathons to Mecca. In *Yearbook of intensive care and emergency medicine* (ed. J.L. Vincent), pp. 971–81. Springer-Verlag, New York.

Ash, C.J., Cook, J.R., McMurry, T.A., and Auner, C.R. (1992). The use of rectal temperature to monitor heat stroke. *Missouri Medicine*, **89**, 283–8.

Bouchama, A., Al-Sedairy, S., Siddiqui, S., Shail, E., and Rezeig, M. (1993). Elevated pyrogenic cytokines in heatstroke. *Chest*, **104**, 1498–1502.

Kashmeery, A.M.S. (1995*a*). Physiological studies on heat exhaustion victims among Mecca pilgrims: response of relevant hormones, and effect of cold IV infusion on recovery. *Acta Medica Austriaca*, **22**, 16–22.

Kashmeery, A.M.S. (1995*b*). Cold IV infusion vs. body cooling unit in heatstroke treatment. *International Journal of Health Sciences*, **6**, 195–200.

Welch, W.J. (1993). How cells respond to stress. *Scientific American*, **268**, 56–64.

11.10.7 Management of heatstroke

C. J. Ash and A. M. S. Kashmeery

Key messages

- Discontinue exposure to heat source.
- Use cooling methods, including shade, evaporative fanning (Mecca Body Cooling Unit, helicopters, open-air ambulance), air conditioning, ice bags, and cold intravenous infusions.
- Monitor clinical features including pulse rate, level of consciousness, and subjective sensation of coolness. If consciousness returns and shivering develops, cooling should probably be discontinued regardless of rectal temperature measurements. Vasoconstriction should be avoided.
- Maintain basic life support of airway, oxygenation, ventilation, intravenous fluids, and transfusion if necessary.
- Monitor fluid balance, electrolytes, hepatic enzymes, urine, blood gases, and lactate frequently, since initially patients may present in shock, repiratory alkalosis, metabolic acidosis, hypo- or hyperkalemia, and hyponatremia. Some degree of acidosis may be beneficial owing to a shift in the oxyhemoglobin curve. Fluid replacement must be assessed on an individual basis.
- Treat multisystem injuries on an individual basis.

Epidemiology

Heatstroke (classical or exertional) can be isolated or endemic, with the duration of heat exposure after collapse probably the most critical factor in determining mortality. Heatstroke occurring during military basic training involves short exposure time and rapid treatment (Ash and Kashmeery 1995), whereas under other conditions (e.g. military exercises, Mexican aliens locked in trucks, mine accidents, prisons, and stockades) exposure is severe and prolonged. Urban cases are usually sporadic but can be endemic during heat waves.

Examples of isolated cases of heatstroke include the following: occupational exertion, children in locked cars, exposure associated with mental illness or retardation, drug abuse (amphetamines, LSD, cocaine), alcohol abuse, injury, athletic training and competition, neuroleptic medication (haloperidol), and malignant hyperthermia (anesthesia).

Moderately high internal temperatures (42–43 °C) for very brief periods (5–10 min) produce little damage and probably only require removal of the heat source (Sminia *et al.* 1994). The heat syncope seen in Denver, Colorado, during the papal visit in 1993 resulted in 18 000 symptomatic victims with no mortality since immediate relief was available. However, heat waves in Chicago, Illinois, in 1995 and Adelaide, South Australia, in February 1993, with prolonged exposure, resulted in the full-blown picture of multisystem damage. The delayed discovery of comatose elderly victims was responsible for the increased mortality (Ash and Kashmeery 1995).

Air-conditioned ambulances are available at the Mecca pilgrimage (Hajj) but their arrival at the scene is often delayed due to severe traffic congestion. Owing to lack of shade and evaporative cooling, there is little relief from prolonged environmental exposure to heat, and so heatstroke is a serious problem.

Monitoring temperature

The pathophysiology and precise details of the cause of heatstroke have remained elusive. It has been suggested that methods of measuring temperature might be partially responsible (Ash and Kashmeery 1995). Recent experimental investigations of whole-body hyperthermia (Sminia *et al.* 1994) using esophageal temperature probes have demonstrated that there is a direct time–temperature relationship for thermal brain injury in humans and animals (Fig. 1) which is similar to the Arrhenius relation for chemical reactions and the well-established time–temperature relationship for surface burns.

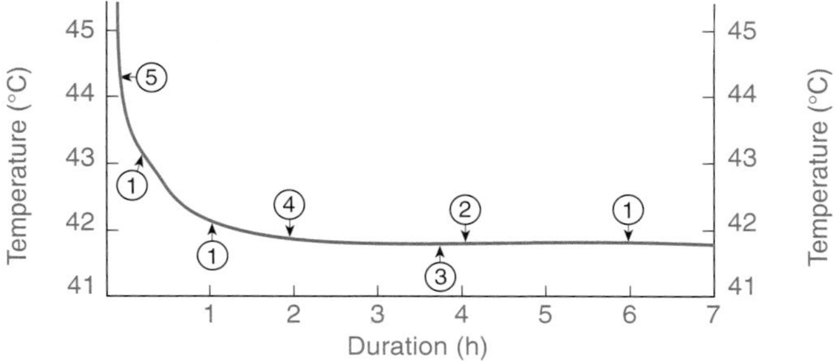

Fig. 1 Time–temperature relation for cellular injury to the brain: (1) Sminia *et al.* (1994); (2) Bull (1979) (cited by Sminia *et al.* (1994)); (3) Goffnet *et al.* (1977) (cited by Sminia *et al.* (1994)); (4) Eshel *et al.* (1990); (5) heart–lung preparation fails in 10 min at 45 °C.

Most previous monitoring of human heatstroke victims and animal heatstroke experiments has relied on measurements of rectal temperature (Ash and Kashmeery 1995). However, rectal temperature has shown little correlation with severity of symptoms. Fatal heatstroke has been described with a rectal temperature of 39.5 °C and survivors have registered up to 47 °C. Rapid changes in internal body temperatures may occur, and pronounced rectal temperature lag has been well documented during such rapid perturbation (Brengelmann 1987) even though many still consider it to be a reliable method of measurement.

Most emergency texts include methods of minimizing vasoconstriction and shivering, although this may indicate excessive vascular and brain cooling with lagging elevated rectal temperature. The time required to cool the brain and vascular tree below critical temperatures is probably much shorter than previously thought (Figs 2 and 3). Here, the rectal temperature lag is obvious. Esophageal, oral, and tympanic membrane temperatures follow pulmonary artery temperature closely but are subject to environmental error, particularly during treatment of heatstroke with ice water and evaporation. Tympanic membrane temperatures may vary by up to 0.5 °C in hot and cold environments. Most infrared ear thermometers are calibrated to read the midcanal temperature, and this region is more susceptible to error than the tympanic membrane. Esophageal sensor placement at the heart level is necessary for more accurate measurement. Innacurate recordings may be influenced by respiratory movement or visceral temperature. Swallowing associated with mouth breathing can also produce error. Oral temperatures require consciousness, adequate time, and close attention to proper placement without mouth breathing. Bladder temperature tends to follow visceral rather than vascular temperature (Martin *et al.* 1994). The most accurate site is the pulmonary artery and a Swan–Ganz catheter should be used (Brengelmann 1987).

Experimental heatstroke studies in animals using routes other than rectal monitoring have shed considerable light on the problem. Experiments in dogs using tympanic rather than rectal monitoring have revealed a startlingly different chain of events and results following exposure to temperatures of 43 °C (Ash and Kashmeery 1995). Using monkeys with epidural and esophageal monitoring during whole-body hyperthermia, Eshel *et al.* (1994) revealed a precise time–temperature relationship for brain injury with marked rectal lag.

Advanced heatstroke

Advanced heatstroke is a multisystem disease and treatment must be tailored to the individual. Coma, convulsions, severe dehydration, early respiratory alkalosis, compensated and uncompensated metabolic acidosis, hypovolemic shock, high-output cardiac failure, abnormal electrolytes, renal failure, severe generalized hemorrhage, disseminated intravascular coagulation, pulmonary edema, and liver necrosis can all occur.

With prolonged heat exposure, renal failure is most likely secondary to rhabdomyolysis but may be reversible. Myoglobinuria, acute tubular necrosis, renal congestion, and hemorrhage may also be present. The high cardiac output which results can lead to cardiac failure and subsequent shock. The workload imposed by hyperthermia is immense, particularly in exertional heatstroke where muscle function and heat loss requirements are combined.

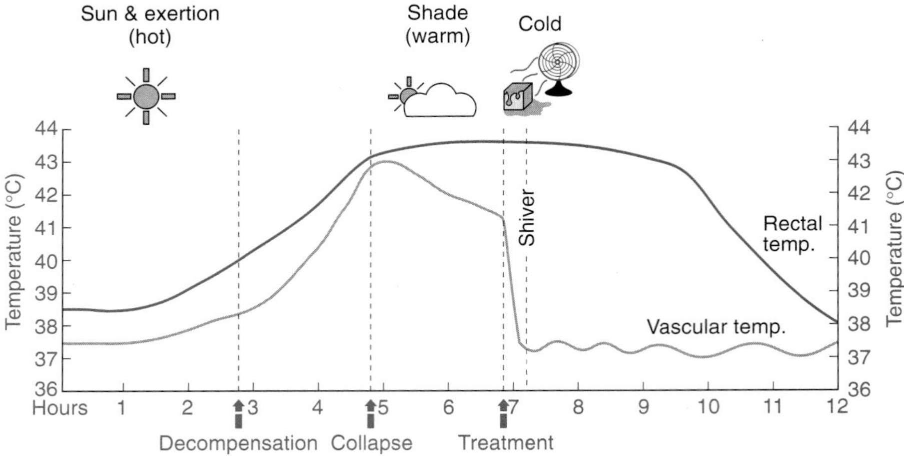

Fig. 2 Heatstroke: theoretical temperature changes in Mecca pilgrims.

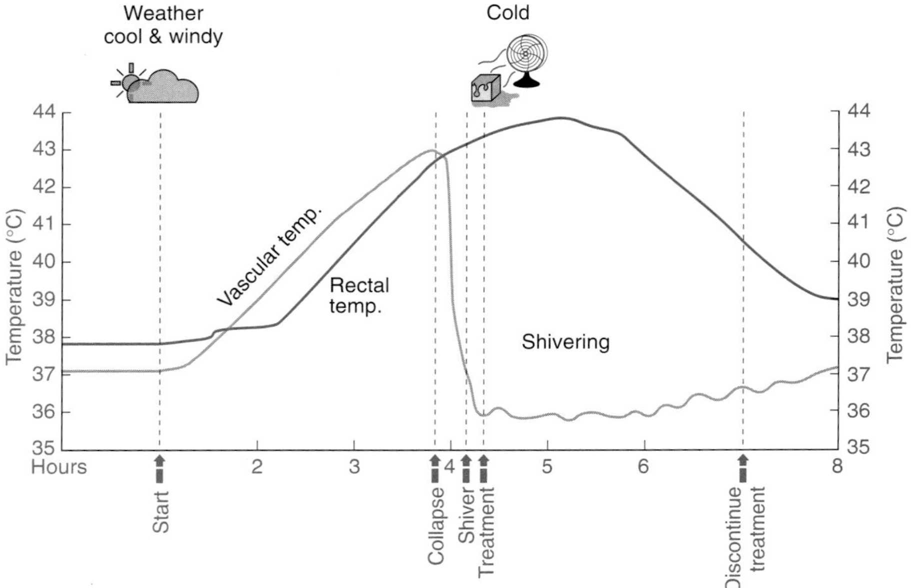

Fig. 3 Heat exhaustion: theoretical temperature changes in marathon runners.

Even in the absence of exertion, high cardiac output associated with dehydration may produce a pulse rate of 160 to 200 beats/min and a doubling of cellular metabolism. The body maintains central arterial pressure and metabolic needs at the expense of heat loss and workload. Arrythmias, such as ventricular fibrillation, are commonplace. Acute left heart failure is a common cause of death.

Hyperventilation occurs initially with respiratory rates of up to 60 breaths/min. Respiratory alkalosis and tetany can develop. Extensive pulmonary hemorrhage may occur with refractive pulmonary edema and infarction leading to death.

In advanced cases, hemorrhage from heatstroke can be severe and generalized. Fibrinolysis, depletion of clotting factors, prothrombin, and platelets, and capillary fragility are associated with a diffuse intravascular coagulopathy. Replacement of clotting factors, red cells, and platelets is necessary and may be accomplished by fresh whole-blood transfusions. Some authors have suggested heparin if disseminated intravascular coagulation is suspected, although its use is still questionable.

Diarrhea and vomiting commonly occur. Massive hematemesis and melena may be seen as part of a generalized hemorrhagic diathesis and carry a poor prognosis. Jaundice is not unexpected. Gram-negative endotoxemia and septic encephalitis may also occur.

Heatstroke victims may be in deep coma. Prolonged coma for over 1 h is an ominous sign. Seizures and cerebellar dysfunction may occur. Psychosis, combativeness, and confusion may be present, although the hypothalamus is usually spared. Cerebral edema, congestion, and widespread petechias may also accompany heatstroke.

Whole-body hyperthermia

The critical temperature of 42.2 °C for cellular heat injury in living animals and man has been established only recently as a direct result of the application of whole-body hyperthermia in the treatment of cancer. Patients have withstood up to 4 h exposure to internal temperatures of 41.8 °C, monitored by esophageal probes, without harmful effects. Cellular injury is irreversible at 45 °C and varies directly with the duration of exposure and exponentially with the temperature elevation. Cellular injury from secondary and tertiary protein denaturation occurs in 50 min at 42.2 °C, 20 min at 43 °C, and 10 min at 45 °C (Sminia *et al.* 1994).

Metabolism increases rapidly with increasing temperature until 42.2 °C is reached. At about this temperature, the cell's capacity to balance anabolism and catabolism is altered, phosphorylation is uncoupled, cell membranes become disrupted, the sodium pump fails, and cellular proteins are denatured. This involves all enzymes in living plants and animals, effectively halting cellular respiration and possibly being responsible for the more extensive injury and higher fatality seen with exertional heatstroke. Heat shock proteins, which are also found in sepsis, are manufactured by normal cells with sublethal heat exposure (40–45 °C) and account for the cardiovascular collapse.

Cases of heatstroke survivors with rectal temperatures of 47 °C are unexplained (Khogali 1994). At this temperature all living plant and animal cells (eukaryocytes) and most bacteria and algae (prokaryocytes) are rapidly destroyed, except for a special class of bacteria and fungi called thermophiles. Thermophiles belong to a new superkingdom, and are normally present in soil and feces; they may be responsible for the tolerance to high temperatures (47 °C) that would otherwise be incompatible with life.

Hypoxia has been suggested as one cause of cardiovascular collapse, although direct evidence is lacking. The lack of cellular enzymatic function associated with high temperatures is associated with normal tissue oxygen, high venous oxygen saturation, and low accumulation of lactate levels (Ash and Kashmeery 1995), suggesting, as discussed above, that oxygen is being delivered but not utilized.

Rebound temperature elevation has frequently been reported following aggressive external cooling. This is similar to the delayed elevation of temperature following prolonged whole-body hyperthermia. It is possible that this 'rebound' is not the result of inadequate cooling but is an expected response to the cellular injury which has occurred.

Implications for heatstroke

Rapid cooling of victims to below 42 °C is necessary as cells sustain injury above this temperature. Since cellular metabolism increases exponentially with temperature elevation and certain cellular changes are reversible, mere removal from the heat source and provision of a cool environment rapidly reduces body temperature below this critical value (Kashmeery and Ash 1995). Intravascular monitoring should be used to avoid the lag seen with rectal probes.

Prevention of vasoconstriction by avoidance of overcooling is important. Vasoconstriction and shivering are inhibited by sweating and hyperthermia, and it has recently been suggested that paradoxical shivering with elevated brain and vascular temperature does not occur (Ash and Kashmeery 1995). Shivering is commonplace after marathon races in cool weather.

The time-honored treatment of aggressive cooling has been modified significantly in recent years. The empirical use of immersion in an ice water bath has given way to various modalities of cooling by evaporation and fanning. Tapwater and fanning are as efficient as ice water (Ash and Kashmeery 1995). At Mecca, muslin gauze and evaporative cooling are as effective as the Mecca Body-Cooling Unit. To avoid vasoconstriction, internal cooling with cold intravenous infusions (up to 1000 ml at 12 °C for heat exhaustion and at 5 °C for heatstroke) has been used successfully at the Hajj (Kashmeery and Ash 1995).

References

Ash, C.J. and Kashmeery, A. (1995). Heatstroke: marathons to Mecca. In *Yearbook of intensive care and emergency medicine* (ed. J.L. Vincent), pp. 971–81. Springer-Verlag, New York.

Brengelmann, G.L. (1987). Dilemma of body temperature measurement. In *Man in stressful environments: thermal and work physiology* (ed. K. Shiraki and M.K. Yousef), pp. 75–122. Charles C. Thomas, Springfield, IL.

Eshel, G.M., Safar, P., and Stezoski, W. (1990). Evaporative cooling as an adjunct to ice bag use after resuscitation from heat-induced arrest in a primate model. *Pediatric Research*, **27**, 264–7.

Khogali, M. (1994). Heat-related illnesses. *Middle East Journal of Anesthesiology*, **12**, 531–72.

Martin, P.A., Robins, J.I., and Dennis, W.H. (1994). Monitoring body site temperatures during systemic hyperthermia. *Critical Care Medicine*, **15**, 163–4.

Sminia, P., van der Zee, J., Wondergem, J., and Haveman, J. (1994). Effect of hyperthermia on the central nervous system: a review. *International Journal of Hyperthermia*, **10**, 1–30.

11.11 Hypothermia

11.11.1 Pathophysiology

M.-D. Schaller and P. Eckert

Key messages

- Hypothermia is defined as a core temperature below 35 °C.
- When hypothermia develops unintentionally it is known as accidental hypothermia.
- In cold surroundings, the thermoregulatory mechanisms are set in gear to prevent heat loss and to increase heat production; immediate peripheral vasoconstriction and intense muscle shivering take place, and thyroid gland activation supervenes later.
- Accidental hypothermia can develop either in healthy individuals faced with extreme climatic conditions or in patients with abnormal thermoregulatory mechanisms.
- During hypothermia, enzymatic and biochemical reactions slow down, oxygen consumption falls, and the metabolic rate is very low.
- Hypothermia depresses all organ and system functions, resulting in decreased heart rate, blood pressure, and cardiac output, shock, bradypnea, confusion, muscle hypotony and rigidity, decreased gastrointestinal motility, and oliguria.
- Extremely severe hypothermia can mimic death with cardiorespiratory arrest, muscle rigidity, coma, and areactive dilated pupils, with no recordable activity on the electrocardiogram and electroencephelogram.
- An apparently dead hypothermic patient may survive after rewarming.
- Hypokalemia is expected during hypothermia. Extreme hyperkalemia in an apparently dead hypothermic patient may be a biological marker of death.

Definition

Hypothermia is defined as a core temperature (rectal, esophageal, tympanic, or central venous) below 35 °C. It can be induced for therapeutic purposes (cardiovascular surgery, brain protection). However, it may develop unintentionally, when it is known as accidental hypothermia. It is usual to classify accidental hypothermia as mild (35–32 °C), moderate (< 32–28 °C), or severe (< 28 °C). If the temperature is above 32°C, morbidity is not due to hypothermia but to the underlying disease. In contrast, hypothermia below 32 °C may itself be fatal with a mortality of up to 80 per cent.

Thermoregulation

Humans are homeotherms, i.e. they maintain their central body temperature constant despite variations in environmental temperatures. In warm surroundings heat is dissipated, whereas in a cold environment heat loss is prevented and heat production increases.

Complex mechanisms are involved in thermoregulation. Thermoreceptors are located throughout the skin, although predominantly in the face and hands. In addition, cold thermosensors have been identified in the upper gastrointestinal tract, the tongue, and the respiratory system. In the presence of cold, skin receptors are stimulated and impulses are sent through the dorsal nerve roots to the lateral spinothalamic tracts, the hypothalamus, and the cortex. Cortical stimulation results in a conscious sensation of cold. The hypothalamus receives information from the entire body, senses the arterial blood temperature directly, and co-ordinates mechanisms to maintain temperature. The posterior part of the hypothalamus plays a major role in controlling peripheral vasoconstriction, shivering, and metabolic rate. It also inhibits the anterior part of the hypothalamus, which is sensitive to warmth and is responsible for heat dissipation.

Thus, when humans are in cold surroundings, a conscious perception of this external threat prompts them to put on more clothes, find shelter, increase physical activity, and move elsewhere (behavioral adaptation).

Immediate vasoconstriction due to the activation of both the sympathetic nervous system and local cutaneous reflexes shunts peripheral venous blood to the inner part of the body, thus decreasing heat loss. This central pooling of venous blood beneath warm arterial blood circulating in the opposite direction forms a countercurrent exchange system, thus warming blood returning to the heart.

Heat is lost predominantly in the outer layers of the body through the four physical processes of conduction, convection, radiation, and evaporation. Conduction (transfer of heat by direct contact) is predominant during immersion, since the thermal conductivity of the body in water is 32 times higher than that in air. Convection (transfer of air heated by the body surface) is enhanced by wind, water currents, or large temperature gradients. Radiation (heat transfer by non-particulate means) occurs from unprotected skin surfaces in the presence of large temperature gradients. For example, about 50 per cent of total heat production can be lost from the uncovered head. Evaporation of water occurs primarily on skin and airway surfaces. It may account for a heat loss of 0.6 kcal/g of evaporated water.

Causes of hypothermia

Accidental hypothermia can develop in healthy individuals in extreme climatic conditions where cold defense mechanisms are overwhelmed or in patients with altered thermoregulatory mechanisms. The first situation is mainly encountered in young well-developed individuals while playing sports or working outdoors, and includes events such as snow avalanche accidents, immersion in cold water, or prolonged exposure to cold after falling into a crevasse or being trapped in a cave.

Hypothermia can be precipitated by alteration of thermoregulatory mechanisms; for example, alcoholic intoxication induces peripheral vasodilatation, decreases shivering capacity, depresses central thermoregulatory mechanisms, and impairs judgment. Alcoholics may stay outside for many hours wearing few clothes, either asleep or unconscious, in parks, on river banks, or in snow. Many drugs, such as tricyclic antidepressants, phenothiazines, or barbiturates, can lead to hypothermia either when used at toxic levels for self-poisoning or when administered in normal dose range for therapeutic purpose. Endocrine dysfunction, such as hypothyroidism, hypopituitarism, hypoglycemia, or diabetic ketoacidosis, may also result in hypothermia. Additional risk factors include central nervous system lesions (stroke, hemorrhage, trauma, tumor), spinal cord injuries, Parkinson's disease, Wernicke's encephalopathy, Alzheimer's disease, schizophrenia, anorexia nervosa, agenesis of the corpus callosum, or spontaneous periodic hypothermia.

The ability to conserve heat is lost when large areas of skin are injured, as in extensive burns. Newborn babies have a relatively large surface to body mass ratio, and premature infants have decreased subcutaneous tissue fat, leading to heat loss.

Decreased cold sensation, decreased muscle mass and shivering capacity, low physical activity, malnutrition, and impaired sympathetic nervous system activity and cardiac function make the elderly susceptible to hypothermia. Debilitating conditions, such as hepatic, renal, or cardiac failure, septic states, or malnutrition, may also contribute to hypothermia.

Pathophysiology and clinical findings

As the body temperature drops there is a progressive slowing in enzymatic and biochemical reactions. Oxygen consumption falls, reaching 50 per cent of normal at 30 °C and 10 to 20 per cent at 20 °C. This low metabolic rate allows organ preservation and survival with very low oxygen requirement. However, early in the development of accidental hypothermia, there is an increase in metabolic rate due to shivering, voluntary muscle activity, and involuntary muscle rigidity. Organ function is preserved provided that oxygen requirements are met.

The cardiovascular system

In the early phase of hypothermia, heart rate, cardiac output, central venous pressure, blood pressure, and systemic vascular resistance increase. This reflects an increased metabolic demand associated with shivering and enhanced sympathetic nervous activity and catecholamine release, elicited by cold skin, fear, and fight against the cold. As the body temperature decreases further, the heart rate slows and atrial fibrillation and ventricular ectopy occurs. Cardiac output decreases progressively, approaching 50 per cent at 28 °C. This is due to a fall in heart rate and stroke volume as a result of decreased enzymatic and cell function. Blood pressure is maintained by catecholamine release and sympathetic nervous activity, increased blood viscosity, and the direct effect of cold on the vessel walls. Below 28 °C, blood pressure falls and shock progressively develops. Systole increases, resulting in a flat pulse wave which is sometimes discordant with the almost normal blood pressure values measured invasively.

ECG modifications

Mild hypothermia depresses conductivity and pacemaker activity. As the body temperature falls, there is a progressive prolongation of the PR, QRS, and QT intervals, and non-specific changes in the ST segment and T waves. Broadening of the QRS complex may be due to the appearance of J waves. The Osborn J wave or camel-hump sign is typical of hypothermia, and is found consistently when the temperature is below 25 °C (Fig. 1). It is a secondary wave following the S wave. This extra deflection between the QRS complex and the ST segment is upright and more prominent in AVL and left chest leads. As the body temperature falls, its amplitude increases and T waves become inverted in the same leads. There are several hypotheses for the origin of the J wave: injury current, delayed ventricular depolarization, or early repolarization before delayed depolarization has occurred. It is so characteristic that its presence should prompt the search for hypothermia; differential diagnosis includes early repolarization, cerebral injury, and vasospastic angina.

Atrial fibrillation secondary to atrial distension, atrial flutter, atrioventricular junctional rhythm, and ventricular premature beats commonly appear below 30 °C. Ventricular fibrillation can occur below 28 °C, and asystole may appear at and below 20 °C.

Oscillation on the ECG baseline is due to muscle tremors and may be seen even in the absence of clinically obvious shivering.

Central nervous system and neuromuscular abnormalities

The metabolic rate of the brain declines in parallel with the fall in temperature, improving the chances of survival even after prolonged periods of circulatory arrest. Blood flow declines secondary to a fall in cardiac output, increased vascular resistance, and blood viscosity. Enzymatic and biological reactions slow down and cerebral function is depressed. Although the clinical expression may be variable, exhaustion, with tremulous speech and ataxia, may be noticed at 32 °C. Dysarthria, stupor, hallucinations, and dysphoria may appear with moderate hypothermia. Paradoxical undressing and aggressive behavior have been reported. At 30 °C, the patient is able to answer simple questions, but responds very slowly. Coma, hypotony, absence of deep tendon reflexes, and fixed and dilated pupils are present below 28 °C. Electrical activity progressively decreases below 28 °C and ceases completely around 15 to 20 °C with a silent electroencephalogram.

Muscle shivering reaches its peak at 32 to 33 °C and disappears completely at 30 °C. In severe hypothermia, muscle hypotony followed by rigidity occurs.

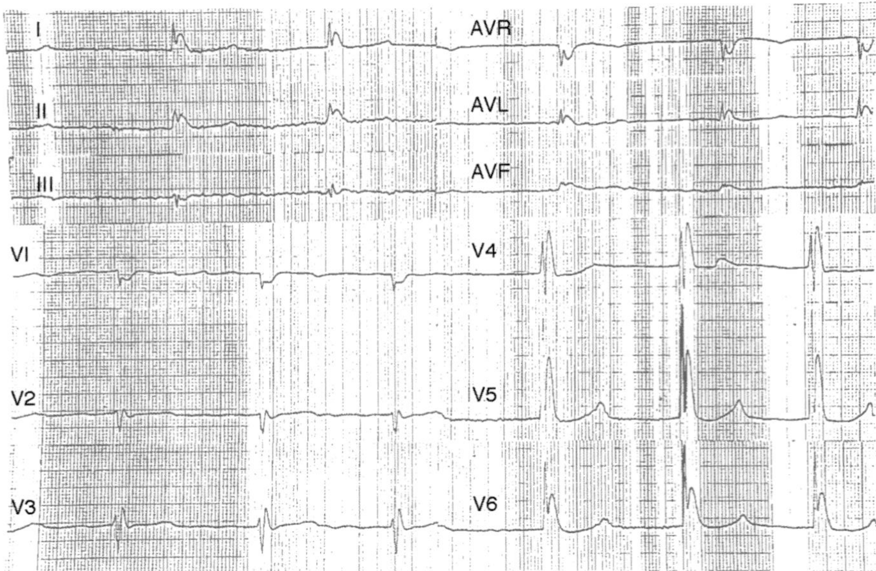

Fig. 1 ECG of an 87-year-old woman found with rectal temperature of 24.5 °C. The heart rate is irregular between less than 32 and 38 beats/min with no P wave. There is marked prolongation of the QT interval and oscillation of the baseline, possibly due to muscle tremors. The typical Osborn J waves can be seen in almost all leads.

Respiratory function

With mild hypothermia there may be an initial increase in respiratory rate; however, as hypothermia becomes more profound, respiratory rate, vital capacity, and minute ventilation all decrease. The cough reflex disappears, bronchial secretions increase, and mucous plugging, atelectasis, and pulmonary edema may develop. Bradypnea and apnea may occur in severe hypothermia.

Renal function

Hypothermia decreases sodium, chloride, glucose, and hydrogen ion transport, as well as sensitivity to ADH. This results first in increased diuresis, natriuresis, glucosuria, and metabolic acidosis. As a result of hypotension, low cardiac output, and increased blood viscosity, renal blood flow is reduced. The relative hypovolemia decreases the glomerular filtration rate, which accounts for the oliguria and anuria at lower temperatures.

Gastrointestinal function

The motility of the gastrointestinal tract decreases, with development of adynamic ileus and acute dilatation of the stomach and colon. Gastric hemorrhage and ulcerations are frequent; the incidence of Wischnevsky lesions found at autopsy approaches 80 per cent. These ulcerations develop more frequently in young healthy individuals and during slow-onset hypothermia. Cold-induced stress is attenuated in older debilitated patients; when the temperature falls rapidly, its effect is blunted.

Acute pancreatitis is common, but is sometimes limited to chemical expression only. Occasionally, a severe edematous or hemorrhagic and necrotic form may complicate and worsen outcome.

Hepatic function may be depressed, with impaired glucose utilization and lactate metabolism. Drug metabolism may be altered.

Endocrine function

There is an initial increase in catecholamine and cortisol release. Pituitary hormone release decreases with the exception of thyroid-stimulating hormone. Serum epinephrine (adrenaline) and norepinephrine (noradrenaline) levels rise immediately after exposure to cold, but thyroxine levels increase later. Below 30 °C, hormone and catecholamine secretion is blunted. Insulin secretion is also blunted which, in association with a peripheral glucose utilization impairment, leads to hyperglycemia.

Hematological findings

Hemoconcentration develops secondary to hypovolemia and fluid compartment shifts. Blood viscosity increases with decreasing temperature (by 2 per cent/°C). White blood cells and platelets may decrease at temperatures below 28 °C. Platelet dysfunction and disseminated intravascular coagulopathy are common in severe hypothermia.

Electrolytes and acid–base findings

Sodium, chloride, calcium, magnesium, and phosphate concentrations may be normal, high, or low. Owing to a shift of potassium into cells, hypokalemia is seen during hypothermia.

An initial respiratory alkalosis is followed by respiratory or mixed acidosis. Hypothermia modifies blood gas analysis values and shifts the oxyhemoglobin curve. Blood pH values are higher and $Pa\text{O}_2$ and $Pa\text{CO}_2$ values are lower than those determined at 37 °C. Since no normal physiological values have been established for hypothermia, it is more convenient for clinical use to perform blood gas analysis at 37 °C without correction for temperature.

Low temperature shifts the oxyhemoglobin curve to the left, decreasing peripheral oxygen delivery. Simultaneously, concomitant acidosis may shift the curve to the right, confusing the picture.

Table 1 Clinical signs in accidental hypothermia

Mild (<35–32 °C)	Moderate (<32–28 °C)	Severe (<28 °C)	20 °C
Heart rate ↑	Heart rate ↓ Cardiac output ↓ ECG: QRS-QT ↑ Atrial fibrillation	Heart rate ↓ Osborn J wave Ventricular fibrillation	Cardiac arrest Asystole
Blood pressure ↑	Blood pressure ↓	Blood pressure ↓ Shock	No pulse
Respiratory rate ↑		Respiratory rate ↓	Respiratory arrest
Shivering ++	Shivering ↓ Exhaustion Confusion Hallucinations Dysarthria	Muscle hypotony Coma Fixed dilated pupils	Muscle rigidity Absence of deep tendon reflexes Coma Fixed dilated pupils Flat EEG
Urine output ↑		Oliguria–anuria	

EEG, electroencephalogram.

Oxygen delivery may be further impaired by increased blood viscosity, low cardiac output, and respiratory failure.

Conclusions

It is important to understand and know the effects of hypothermia on organ function. Different clinical descriptions can be given according to the severity of hypothermia (Table 1). These signs may be modified by the causes of hypothermia and by concomitant disease.

In extremely severe hypothermia, patients appear to be clinically dead as they are in cardiorespiratory arrest. Attempts at rewarming are essential before death is declared.

Bibliography

Coniam, S.W. (1979). Accidental hypothermia. *Anaesthesia*, **34**, 250–6.

Jolly, B.T. and Ghezzi, K.T. (1992). Accidental hypothermia. *Emergency Medicine Clinics of North America*, **10**, 311–26.

MacKenzie, M.A., *et al.* (1992). Electrocardiographic changes during steady mild hypothermia and normothermia in patients with poikilothermia. *Clinical Science*, **82**, 39–45.

Paton, B.C. (1983). Accidental hypothermia. *Pharmacology and Therapeutics*, **22**, 331–77.

Reuler, J.B. (1978). Hypothermia: pathophysiology, clinical settings, and management. *Annals of Internal Medicine*, **89**, 519–27.

11.11.2 Management

M.-D. Schaller and P. Eckert

Key messages

- The diagnosis should be made and an accurate thermometer used to record very low core temperatures.
- Considering the high risk of ventricular fibrillation at 28 °C, hypothermic patients should be moved with caution.
- Any drug administered during hypothermia can reach toxic levels after rewarming, because of the modified pharmacokinetics associated with hypothermia.
- Hypotension, arrhythmias (e.g. atrial flutter or fibrillation, premature ventricular beats), and oliguria disappear with rewarming.
- Once the diagnosis of hypothermia is made, further heat loss must be prevented and rewarming started.
- Passive rewarming takes place via the patient's own heat production and ability to preserve heat loss.
- External active rewarming is achieved by applying a heat source to the body surface.
- Internal active rewarming includes administering warmed humidified air via an endotracheal tube, gastric, peritoneal, and pleural lavage, extracorporeal rewarming, and cardiopulmonary bypass.
- In mild hypothermia, most rewarming can be passive together with hot drinks and warmed humidified oxygen.

- In a moderately hypothermic patient, in coma or respiratory failure, active internal rewarming is recommended using warmed humidified oxygen delivered by mechanical ventilation and gastric lavage. In the presence of cardiovascular instability, peritoneal dialysis or warming by extracorporeal shunt may be required.
- In severe hypothermia with cardiorespiratory arrest, cardiopulmonary bypass is the best method of rewarming.

Introduction

The first step in the management of hypothermia is its diagnosis, which may be obvious if the patient has been immersed in cold water or spent the night outside in freezing conditions. Diagnosis is more difficult in those found unconscious at home after drug intoxication, endocrine dysfunction, or cerebrovascular accident. The second step is to prevent further heat loss, to monitor and avoid complications, and to sustain cardiorespiratory function. Finally, rewarming should be undertaken expeditiously. Management should be individualized, taking into account the underlying disease and cause of hypothermia, the clinical presentation, and the resources available.

Prognosis

Poor prognostic factors include advanced age, comorbid states, the circumstances and severity of hypothermia, and cardiorespiratory arrest. A patient who has been trapped under a snow avalanche with other injuries may have a worse outcome than a drug-intoxicated patient with hypothermia. Hypothermia associated with septic shock carries a high mortality.

The severity of hypothermia, the duration of exposure to cold, cardiac arrest, or a silent electroencephalogram cannot be used to predict the chance of survival in a clinically dead hypothermic patient. Such patients have been successfully rewarmed and resuscitated without any neurological sequelae after hours of cardiorespiratory arrest. The lowest recorded temperature in a survivor of accidental hypothermia is 14.2 °C. Thus clinically moribund hypothermic patients should be rewarmed before being declared dead. Extreme hyperkalemia (12–14 mmol/l) in hypothermic patients with cardiorespiratory arrest and non-reactive pupils is ominous and may imply that death has probably occurred. Hypothermia produces hypokalemia, unless it is associated with rhabdomyolysis secondary to trauma or acute heroin overdose.

Management before hospital admission

The patient should be removed from the cold environment as rapidly as possible, dressed in dry clothes, and covered with insulating blankets or sleeping bags. Frozen or potentially frozen extremities should not be rubbed to avoid secondary injuries and necrosis, but should be protected from further trauma. The diagnosis of hypothermia can be confirmed using a tympanic thermometer.

Warm drinks, but not alcohol, can be given to the conscious patient with stable hemodynamic parameters. Humidified and warmed oxygen should be delivered by face mask. Endotracheal intubation for apnea or low respiratory rate (2–4 breaths/min) should be performed without delaying hospital admission. Insertion of a catheter into a peripheral vein is difficult due to the intense venoconstriction, and central access may be necessary. Hypoglycemia should be suspected in a comatose person in whom hypothermia is obviously not due to an accident, and glucose should be administered intravenously.

The patient should be moved and transported gently since at 28 °C there is danger of inducing ventricular fibrillation simply by movement. In severe hypothermia, the pulse may be very weak and difficult to sense and the respiratory rate slow; cardiocirculatory arrest may develop at any time. Cardiopulmonary resuscitation should be initiated in all situations even if there is a risk of transforming a slow cardiac rhythm or ventricular tachycardia into ventricular fibrillation. It should be continued until hospital admission. Patients with moderate and severe hypothermia should be transferred directly to a hospital with appropriate rewarming facilities.

Hospital management

Physical examination

On admission, the diagnosis of hypothermia should be confirmed using a thermometer suitable for measuring very low temperatures. Rectal, esophageal, central venous, or tympanic probes are used to measure central or core temperature. The rectal temperature is usually 1 °C below the esophageal temperature because of the proximity of the great veins draining cold blood from the legs. This difference is reduced if the probe is inserted sufficiently deeply (10–15 cm) into the rectum. This is the most common site of measurement. A probe can be inserted into the esophagus, at the level of the right atrium, but may be influenced by warmed inhaled air, particularly during mechanical ventilation. The tympanic temperature should reflect central nervous system temperature; however, not all parts of the brain are of uniform temperature, with the scalp exposed and the external ear sometimes filled with snow. The temperature can also be measured in the urinary bladder using a Foley catheter or in central venous blood. It is recommended that the temperature is recorded simultaneously at two different sites and continuously monitored in order to follow therapy.

A 12-lead ECG should be recorded, followed by continuous monitoring. Vital signs including pulse, respiratory rate, blood pressure, pupil size and reflexes, consciousness, and neurological deficits should be documented. Unilateral or focal neurological deficits should raise suspicion of a subdural hematoma, cerebral hemorrhage, or stroke. Hypoglycemia, sepsis, and drug or alcohol intoxication may also lead to accidental hypothermia and should be investigated.

Laboratory tests

A complete blood cell count, coagulation and serum chemistries (creatinine, liver and pancreatic enzymes, lactate dehydrogenase, lactate, creatine phosphokinase, glucose, and electrolytes) should be drawn. Rapid serum potassium and glucose determination is crucial, as is blood gas analysis. The arterial blood gas should be analysed without temperature correction. Drug and/or alcohol screening and thyroid or adrenal gland function tests are obtained as indicated. Blood, urine, and sputum cultures should be obtained if sepsis is suspected. Radiography, including CT scanning, is performed as indicated. A baseline ECG is obtained.

General measures

To prevent fatal arrhythmias, minimal movement is recommended until the patient is warmed. Because of vasoconstriction, a central catheter is usually required with care taken to avoid myocardial stimulation. Because of the risk of inducing arrhythmia, a pulmonary catheter should not be used until normothermia has been re-established. Urinary bladder and gastric decompression catheters should be placed. In severe hypothermia, an arterial catheter should be inserted to monitor blood pressure and to obtain arterial blood gases.

The increased metabolic demand during rewarming requires oxygen administration via either a facial mask or an endotracheal tube. Endotracheal intubation is indicated for respiratory failure, airway protection, and rewarming. Ventricular fibrillation secondary to intubation has not been demonstrated.

Hypotension should be corrected by aggressive warm volume expansion. Vasopressors can induce dangerous arrhythmias. Dopamine may have a thermogenic effect.

Owing to unknown pharmacokinetics (altered cardiocirculatory conditions, decreased enzymatic activity and efficacy, augmented toxic effects), drugs should initially be used sparingly. Toxic levels may develop early as delivered medication is not metabolized, and marked toxicity may occur during rewarming, leading to life-threatening events. Thus expected hyperglycemia should not be treated. In decompensated diabetic patients, extremely high serum glucose levels may be an indication for a low dosage of insulin after the central temperature has reached 30 °C.

Although the risk of thromboembolism is increased, prophylactic heparin and leg compression should not be initiated in the early phase of the treatment of hypothermia. Arrhythmias are common at low temperatures, but atrial fibrillation, flutter, and ventricular premature beats should not be treated with antiarrhythmic agents as they are ineffective in restoring sinus rhythm and may actually induce other troublesome rhythms. Rewarming is the best therapy. Cardioversion of ventricular fibrillation has little chance of success at body temperatures below 30 °C. However, in view of case reports of successful defibrillation, up to three electrical countershocks should be given and cardiopulmonary resuscitation continued as the patient is rewarmed. Ventricular fibrillation may disappear spontaneously with rewarming.

Lidocaine (lignocaine) and procainamide have little efficacy in the treatment of ventricular tachycardia. However, bretylium has been successfully used to prevent and treat ventricular fibrillation in hypothermic dogs and is the preferred agent. Pacing, whether internal or external, is ineffective and may actually increase the risk of ventricular fibrillation. Bradycardia or atrioventricular block due to hypothermia resolves with return of normothermia.

Rewarming methods

Rewarming may be passive (endogenous heat production by the patient) or active (application of heat externally or internally). In active external rewarming, a heat source is applied to the skin surface, whereas in active internal rewarming heat is delivered to the core of the patient.

In passive rewarming the patient is first removed from the cold wet environment. The success of this method depends on the integrity of the thermoregulatory mechanisms. The temperature increases slowly (0.5–1 °C/h), depending on the physical conditions, comorbidities, and the degree of hypothermia.

External active rewarming involves applying a heat source on or over the body surface, which induces vasodilatation and improves circulation to superficial areas so that cold blood returns to the heart. The central core temperature may actually fall. However, the fall in temperature observed after removal from the cold environment may simply be due to the temperature gradient between the surface and internal tissue masses. This further drop in temperature may induce arrhythmias, further depress myocardial function, and lead to hemodynamic instability, shock, and death.

Heating pads, hot water bottles, circulating-water mattresses, and heating blankets should be used cautiously as they may cause burns. Immersion in a hot bath places the patient in an uncomfortable position which is inaccessible for physicians and nurses. Cardiopulmonary resuscitation and defibrillation are impracticable in water.

Warm air (heated to 43 °C) can be pumped through slits in a blanket. This method has been shown to increase body temperature by more than 2 to 3 °C/h. It should not be used for patients with cardiac arrest, hemodynamic instability, or hypothalamic dysfunction. Because of the vasodilatation that occurs when this method is used, rigorous and precise fluid resuscitation must be performed using warmed intravenous solutions. The volume of fluid administered is governed by hypotension and oliguria provided that no pulmonary edema has developed. The therapy should be guided by continuous monitoring.

Techniques for internal active rewarming include airway rewarming, gastrointestinal, peritoneal, or pleural cavity lavage, extracorporeal rewarming, and cardiopulmonary bypass.

A heated and humidified oxygen–air mixture may be administered through a face mask or endotracheal tube; the latter is more efficient because less heat is dissipated into the nasopharynx. The temperature increases by only 0.2 to 1.5 °C/h, although heat loss by respiration is prevented. Other advantages of this simple and readily available method include rewarming the heart and lungs first, restoring ciliary motility, and improving mobilization of bronchial secretions. Inspired air should be warmed to 40 °C in conscious patients and to 46 °C in intubated unconscious hypothermic victims. This procedure is relatively risk free.

Fluids (glucose 5 per cent, NaCl 0.9 per cent, or blood) should be warmed prior to intravenous infusion. Glucose or saline in plastic bags can be heated to 40 °C by microwave; however, this method usually cannot be used alone for rewarming, as the volume required is too large. A rapid fluid warmer–infuser can be used sucessfully.

Warmed solutions can also be administered through a nasogastric tube and aspirated after about 15 min. Bronchoaspiration of gastric contents is prevented by performing an endotracheal intubation in unconscious patients. Awake patients may partake of hot drinks. This will help to deliver heat to the heart and liver, thus promoting their functional recovery. Because of the small exchange surface of the stomach, this has a limited efficacy (0.5–1 °C/h).

Colonic lavage has been proposed for rapid rewarming of the liver and heart. However, this technique is not commonly used. Complications of colonic or gastric tube placement include perforation. Recent gastrointestinal tract trauma or surgery is a relative contraindication to these procedures.

Two peritoneal dialysis catheters can be placed and used to infuse warm dialysate (40–42 °C) which is then recovered from the peritoneal cavity. Because of the large surface exchange area, the temperature can rise by up to 5 °C/h. However, insertion of the catheters may cause intra-abdominal injuries. Recent abdominal trauma and intra-abdominal free air are contraindications and usually indicate laparotomy, during which warmed irrigation can be used.

Pleural lavage can be performed using two large-bore thoracostomy tubes. One tube is placed in the second or third anterior intercostal space at the mid-clavicular line, and the other is placed in the fifth intercostal space in the posterior axillary line. Warmed saline (39–42 °C) can be infused through the anterior tube and recovered in the posterior tube. Peritoneal or pleural lavages appear equally effective in raising temperature (5 °C/h). Complications include the risks of thoracostomy (lung perforation, vascular or cardiac injuries, infection) and the risk of increased intrathoracic pressure and mediastinal deviation due to accumulation of fluids, leading to pulmonary and myocardial dysfunction.

Hemodialysis and continuous arteriovenous or venovenous rewarming techniques are useful. Blood flows from a catheterized femoral artery or vein through a warming device back to a central vein. In the warming device, hot water (40 °C) separated from the blood by highly conductive layers is pumped in a countercurrent direction. These methods are effective, raising the temperature by 5 to 8 °C/h, and are technically feasible in most intensive care units. If anticoagulation is contraindicated, heparin-bonded tubing can be used.

Cardiopulmonary bypass provides the highest rewarming rate (up to 10 °C/h) under strict control. Fluid resuscitation and oxygen delivery are guaranteed, with optimal hemodynamic support being provided. All these advantages make this technique the best choice in the presence of cardiac arrest. Risks include hemolysis, hemorrhage, vascular perforation, and embolism. Heparin-bonded tubing can be used in trauma patients at high risk of bleeding, thus rendering systemic heparinization unnecessary.

Specific measures for rewarming

First, further heat loss should be prevented. Nurses and physicians should be aware that the patient's body temperature may drop during transport or when dressings are being changed. An insulating

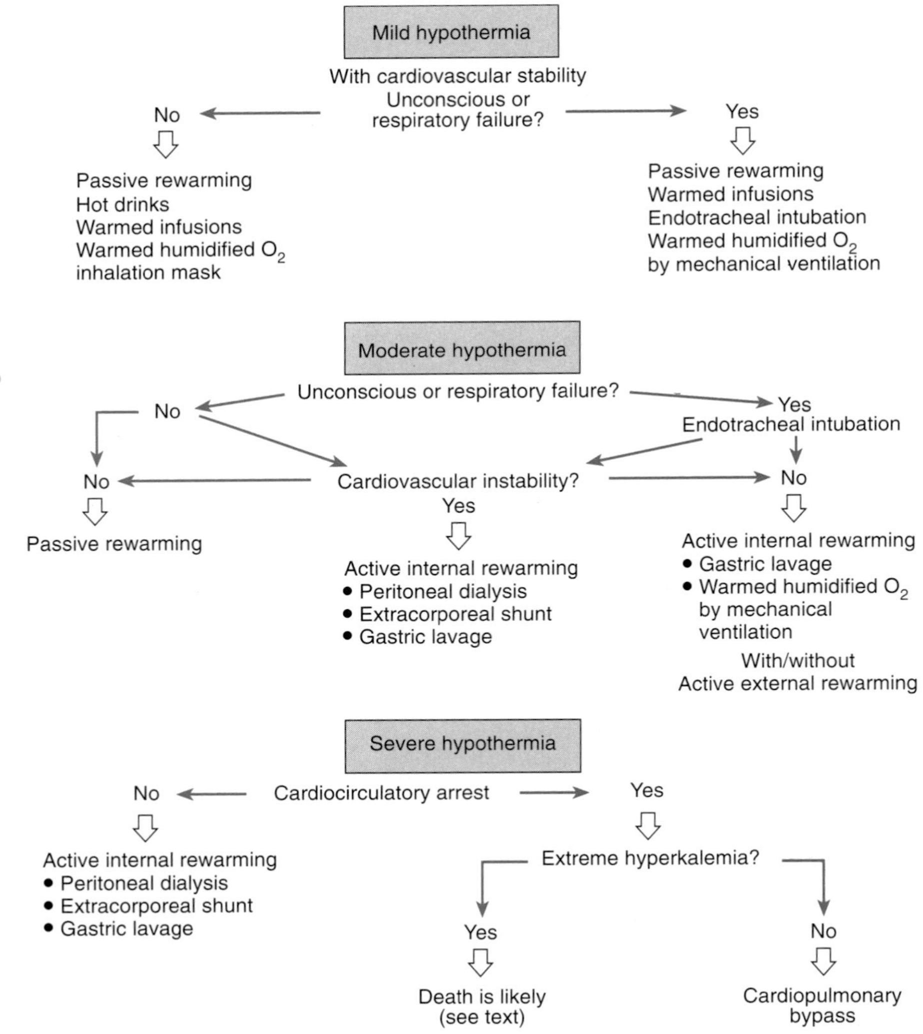

Fig. 1 Rewarming protocol for accidental hypothermia. In addition to this rewarming protocol, the diagnosis should be confirmed, the temperature measured, aggravating factors should be eliminated, cardiovascular and respiratory functions should be supported, and the temperature, heart rate, blood pressure, and respiratory rate should be monitored. If the rewarming process fails (i.e. the temperature rises by less than 0.5 °C/h or decreases), one should go to the next step, i.e. active rewarming.

aluminum blanket placed over the trunk can help to retain heat. Care should be taken to avoid prolonged exposure during radiography, catheter placement, or patient evaluation.

A protocol for rewarming a patient with accidental hypothermia is shown in Fig. 1.

Passive rewarming is usually all that is required to treat mild hypothermia. These patients are conscious with no hemodynamic instability and no respiratory failure. Treatment is simple and includes wrapping in an insulating blanket, inhalation of a heated and humidified oxygen–air mixture through a face mask, hot drinks, and, if indicated, warmed infusions. In athletic patients, rewarming occurs within 1 h. In patients with comorbid conditions (cardiomyopathy, respiratory failure, malnutrition, etc.), achievement of normothermia may take longer, requiring adjunctive measures. The same is true when thermoregulation is impaired. In unconscious patients, the warmed humidified oxygen–air mixture can be delivered through an endotracheal tube. If this method fails, i.e. the temperature rises by less than 0.5 to 1 °C/h or decreases, more invasive methods are required. Gastric lavage may be initiated. External active rewarming using forced air may be tried, provided that there is no hemodynamic instability and fluid resuscitation can be performed aggressively.

In cases of moderate hypothermia, particularly if the body temperature is below 30 °C, the shivering capacity disappears and rewarming should be active. In unconscious victims with no hemodynamic instability, a warmed humidified oxygen–air mixture is given via an endotracheal tube, warmed solutions are infused intravenously, and gastric lavage is attempted. If there is no increase in temperature, peritoneal irrigation, pleural lavage, or extracorporeal warming should be instituted, possibly with external active rewarming.

In cases of moderate hypothermia with hemodynamic instability, body cavity lavage, hemodialysis, or continuous arteriovenous or venovenous warming should be started sooner, particularly when the body temperature is 28 °C. In cases where dialysis is required, such as renal failure, hyperkalemia, or removal of other toxins, peritoneal irrigation or hemodialysis should be used.

Active internal core-rewarming methods are mandatory with severe hypothermia. Peritoneal irrigation, pleural lavage, and/or extracorporeal rewarming should be commenced in addition to administration of a warmed humidified oxygen–air mixture through an endotracheal tube, warmed irrigation, and intravenous fluids and gastric lavage. The goal is to raise the temperature rapidly to about 30 °C, thus minimizing the chance of life-threatening arrhythmias. The ideal rewarming method for hypothermic patients with cardiac arrest is cardiopulmonary bypass.

When the temperature falls below 20 °C, the patient is in cardiocirculatory arrest with non-reactive pupils and rigid trunk and extremities. Aggressive treatment should be undertaken. Endotracheal intubation may be difficult due to jaw rigidity and a surgical airway is sometimes required. Mechanical ventilation with warmed humidified gas is administered at a low rate (2–4 breaths/min). Cardiac massage is probably inefficient, and should be reserved for those in asystole. Intravenous access requires a central route. Immediate cardiopulmonary bypass may successfully rewarm and resuscitate patients after prolonged cardiac arrest. Extreme hyperkalemia may be used as an indicator of futility, although precise upper limits incompatible with life have not been established.

Bibliography

Britt, L.D., Dascombe, W.H., and Rodriguez, A. (1991). New horizons in management of hypothermia and frostbite injury. *Surgical Clinics of North America*, **71**, 345–70.

Danz, D.F. and Pozos, R.S. (1994). Accidental hypothermia. *New England Journal of Medicine*, **331**, 1756–60.

Larach , M.G. (1995). Accidental hypothermia. *Lancet*, **345**, 493–8.

Steinman, A.M. (1986). Cardiopulmonary resuscitation and hypothermia. *Circulation*, **74** (Supplement IV), IV-29–32.

Vretenar, D.F., *et al.* (1994). Cardiopulmonary bypass resuscitation for accidental hypothermia. *Annals of Thoracic Surgery*, **58**, 895–8.

11.12 Transport of the critically ill patient

11.12.1 Pathophysiology related to movement of the critically ill

B. J. Tortella and M. J. Shapiro

Key messages

- The transport of critically ill patients always involves some risk.
- The appropriate transport modality (ground versus air) and appropriate team composition are central to successful transport.
- Special considerations of altitude come into play when patients are transported by aircraft.
- Dalton's, Boyle's, and Charles' laws reveal that hypoxia occurs at altitude, gas expands at altitude, and gas expands as the temperature rises.
- Hypoxia, hypemia, and stagnant or histotoxic hypoxia can occur at altitude.
- Tachycardia occurs early with increase in altitude.
- Central nervous system signs are seen with altitude-induced hypoxia.
- Gas expands in the gastrointestinal tract and may lead to aspiration if gastric decompression is not performed.
- Barotitis media and barosinusitis should preclude crew from flying.

Introduction

Transportation of injured and critically ill patients is a relatively new development in history. Apart from the Spartan mothers, who admonished their warrior sons to 'Come back home with your shield, or on it', the first organized transport of patients did not occur until the French Revolutionary War (1792–1802). Here, Doctor Baron Dominique Jean Larey, who was the surgeon to Napoleon himself, popularized the idea of using wheeled carts (which he called flying ambulances) to transport the wounded to hospitals. Of course, this was not done until the battle was decided; thus the wounded lay on the battlefield for days. The first air medical transport in the modern era was when the British Royal Air Force organized a Medical Branch and formulated an organized plan for casualty evacuation which was first implemented in 1920 in Somaliland. Helicopters came into use shortly after their development by Sikorsky (1945) and Young (1946) during the Korean War and have flourished since.

Ground ambulance development occurred later in the United States, and up until the mid-1970s the most common ground vehicle that could transport a patient in a stretcher was a mortician's hearse. The development of emergency medical service systems had to await the concepts of rapid evacuation, learned in Vietnam, coupled with the ground-breaking work of field defibrillation by Pantridge in Ireland (Pantridge and Geddes 1967). The United States rapidly expanded the emergency medical service through government grants from the national Emergency Medical Services Act 1973 with the development of national emergency medical technician and paramedic curricula and the vehicles that we now recognize as the modern ground ambulance.

Indications for transport

The most common reason for domestic interfacility transport of a critically ill patient is the unavailability of appropriate medical staff or facilities in the patient's location. This involves transport of a patient to a higher and more complex echelon of care. In developed countries, complex cardiac problems comprise the largest percentage of medical patients requiring transport. An efficient interfacility transport system is essential in jurisdictions where invasive cardiology and cardiac surgery services are centralized, since the initial admitting hospital may not have capabilities commensurate with the patient's needs. Patients with complex vascular conditions (leaking abdmoninal aortic aneurysms and thoracic aortic aneurysms, both traumatic and atherosclerotic) frequently require interfacility transport. This is due not only to the need for advanced surgical procedures but also because of the requirement for an advanced critical care unit postoperatively.

The lack of adequate medical facilities or staff in some countries necessitates international transport of patients. International patient transports usually revolve around repatriation issues. Patients are usually more stable but are unable to travel on a commercial conveyance. Of United States citizens repatriated by air, 44 per cent had sustained trauma (45 per cent in motor vehicle crashes, 21 per cent in falls, and 8 per cent in recreational accidents). The next largest group were patients with cardiovascular diseases (28 per cent), with most having sustained a myocardial infarction, angina, or cardiac arrest. General medical conditions comprise 22 per cent, including gastrointestinal bleeding (17 per cent), respiratory diseases (13 per cent), and abdominal conditions (11 per cent).

Special considerations of altitude in air medical transport

Several special considerations must be taken into account when caring for patients being transported at altitude. Dalton's law states that the total pressure of a gas mixture is the sum of the individual partial pressures of all the gases in the mixture:

$$P_{\text{total}} = P_{\text{gas 1}} + P_{\text{gas 2}} + \ldots.$$

Breathing air contains 21 per cent oxygen. At sea level, where the barometric pressure is 760 mmHg, the partial pressure of oxygen is 160 mmHg (21 per cent of 760 mmHg is 160 mmHg). It is even lower (150 mmHg) if the partial pressure of water vapor (47 mmHg) is considered. As altitude increases, the barometric pressure decreases while the percentage of oxygen in breathing air remains the same. Thus at an altitude of 10 000 ft (3000 m), where the barometric pressure is 523 mmHg, the partial pressure of oxygen is 21 per cent of 523 mmHg or 110 mmHg. Simply put, the higher the altitude, the lower is the partial pressure of oxygen (Table 1). Although most helicopters operate below 8000 ft (2400 m), all critically ill patients being transported must receive supplemental oxygen.

Higher levels of ventilatory support (positive end-expiratory pressure etc.) may be required at altitude with pulse oximetry monitoring and ideally arterial blood gas analysis in flight if the transport is over long distances. The pilot can also be requested to make a slow ascent and descent to minimize the rate of pressure changes. A request can also be made to pressurize the cabin to sea level. Although this reduces the fixed-wing aircraft's range, it may be necessary in a patient with poor oxygenation. Altitude can also potentiate the effects of drugs, particularly sedatives.

The following simple equation can be used to determine the amount of supplemental oxygen required by a patient at a given altitude to maintain the level of supplemental oxygen received in the referring hospital:

$$\frac{\text{Fio}_2\text{ (at referring hospital)}\times\text{barometric pressure (at referring hospital)}}{\text{barometric pressure (inflight)}} = \text{Fio}_2\text{(during flight)}.$$

Thus a patient who was on 40 per cent oxygen in a referring hospital at sea level being transported at an altitude of 8000 ft (2400 m) would need at least 50 per cent oxygen ((0.40 × 760)/609 = 0.50) during transport (Blumen *et al.* 1992).

Boyle's law states that the pressure of a gas varies inversely with its volume ($P \propto 1/V$). Since atmospheric pressure decreases as altitude increases, gas contained within a fixed confine expands as altitude increases. In the clinical setting, Boyle's law can affect any body cavity or piece of equipment that contains an enclosed air space. Helicopters rarely ascend to very high altitudes, and so the volume changes are small. Fixed-wing aircraft should be pressurized at 5000 to 8000 ft (1500–2400 m). The endotracheal tube cuff will expand with increasing altitude, placing the patient at risk of tracheal pressure necrosis. Endotracheal tube cuff rupture may also occur which will result in loss of tidal volume and inability to ventilate or maintain positive pressure. The risk of aspiration also increases when the cuff is ruptured. The air medical crew must release gas from the cuff manually as altitude increases, or it can be filled with water rather than gas. Only a modest increase in volume (a factor of 1.2) is experienced in helicopters operating at altitudes of less than 5000 ft (1500 m). Volumes will double in fixed-wing aircraft operating at altitudes above 18 000 ft (5400 m). Similar attention should be given to Foley catheters if the balloon has been inflated with gas rather than water.

Charles' law states that when pressure is constant, the volume of a gas is nearly proportional to its absolute temperature. As the temperature increases, gas molecules will move faster and the volume of gas will increase:

$$V_1/V_2 = T_1/T_2$$

where V_1 is the initial volume, V_2 is the final volume, T_1 is the initial absolute temperature, and T_2 is the final absolute temperature (absolute temperature is defined as temperature in degrees celsius plus 273). Thus it is important that patients remain warm for hemodynamic stability and to maintain perfusion, and that they are not exposed to extreme temperature variations during transport. However, hyperthermia can lead to gas expansion.

Most of the marked pathophysiological changes occur in flight at an altitude over 8000 ft (2400 m). Sudden loss of cabin pressurization from a large defect results in rapid movement of air towards the hole. There is often a rapid drop in temperature, flying debris, noise, and fog formation due to condensation. Hypoxia rapidly ensues, and so supplemental oxygen should be applied immediately as the pilot attempts to bring the aircraft below 10 000 ft (3000 m). All catheters, chest tubes, and nasogastric tubes should be unclamped to combat the rapid gas expansion that may occur (Rodenberg and Blumen 1994).

Table 1 Dalton's law and hypoxia at altitudes

Altitude (ft) (m)	Barometric pressure (mmHg)	Pao_2 (mmHg)	Sao_2 (%)
0	760	159	98
5000 (1500)	639	130	97
8000 (2400)	609	128	93
10 000 (30050)	523	110	87
15 000 (4500)	429	90	84

Adapted with permission from Blumen *et al.* (1992).

Patients with pneumothoraces should be kept on suction drainage if possible (either continuous or intermittent). The tube thoracostomy must never be clamped during transport. If a suction system is not available, a simple one-way flap valve (Heimlich device) is useful. Patients with significant pneumocephaly from sinus fractures or a basilar skull fracture require special attention. Neurosurgical consultation should be obtained if possible. Consideration should be given to intracranial pressure monitoring or ventriculostomy if the patient has significant intracranial gas or will ascend to an altitude where gas expansion will occur. When possible, the patient should be placed in the conveyance head first and at a 10° to 15° head-up tilt since this permits better toleration of sudden decelerations.

Altitude effects on the respiratory system

Prevention of hypoxia during air medical transport is critical. Physical examination, pulse oximetry, and end-tidal CO_2 detection can assess adequacy of oxygenation and ventilation. As seen in Table 1, hypoxemia ensues with altitude and requires supplemental oxygen. If a pneumothorax is documented at altitude due to expansion of gas, this must be decompressed quickly. Needle thoracostomy followed by chest tube thoracostomy can be performed while the patient is being monitored. Any tube placed for decompression should be on continuous low suction. There are four categories of hypoxia which are exacerbated by flight and altitude.

1. Hypoxic hypoxia (altitude hypoxia) is due to compromised gas exchange at the alveolar capillary membrane. This is the most common etiology for hypoxia and may become apparent above 10 000 ft (3000 m). It may be caused by a ventilation–perfusion defect, airway obstruction, or inadequate oxygen delivery.
2. Hyperemic hypoxia (anemic hypoxia) results from a decrease in the oxygen-carrying capacity of blood due to anemia, blood loss, CO poisoning, drugs, or excessive smoking.
3. Stagnant hypoxia is due to poor circulation where the cardiac output and oxygen delivery are inadequate for oxygen consumption. This is caused by venous pooling, arterial spasm, vascular occlusion, or prolonged periods of positive-pressure ventilation.
4. Histotoxic hypoxia results from the inability of tissues to utilize available oxygen, secondary to such causes as CO poisoning, cyanide poisoning, ethanol ingestion, or narcotics.

The clinical manifestations of hypoxia can vary from those representative of hypoxia in general, mimicking mild alcohol intoxication with tachypnea, tachycardia, agitation, restlessness, and euphoria. Treatment includes eliminating the source and providing supplemental oxygen. Patients may require blood transfusion or correction of pre-existing conditions such as pneumonia, asthma, pneumothorax, cardiac disease, shock, or blood loss.

Altitude effects on the cardiovascular system

The cardiac system is more resistant to hypoxia than the respiratory and central nervous systems. Tachycardia may occur at approximately 4000 ft (1200 m), achieving a maximal rate at approximately 22 000 ft (6600 m). Initially, an increase in cardiac output will occur due to the tachycardia and selective vasoconstriction. However, this increased myocardial oxygen consumption in an already exisitng ischemic hypoxic myocardium can lead to bradycardia, hypotension, arrhythmias, and cardiac arrest.

Altitude effects on the central nervous system

When $Pa\text{O}_2$ decreases below 50 mmHg (6.7 kPa) cerebral hypoxia may ensue. An increase in cerebral blood flow may actually occur due to the vasodilatory effects of hypoxia, despite the tachypnea and vasoconstriction related to the decrease in $Pa\text{CO}_2$. Initially, patients may present with excitation, loquaciousness, euphoria, hyperactivity, and restlessness, followed by decreased attention, amnesia, deterioration of visual fields or depth perception, depression, impaired judgment, and confusion. In patients with recent ocular injury, ocular surgery, or progressive retinal disease, there is a high oxygen requirement because of the sensitivity of the retina.

The effective performance time of an experienced crew member exposed to an altitude of 18 000 ft (5400 m) without oxygen or pressurization is 20 to 30 min; at 25 000 ft (7500 m) it is 3 to 5 min and at 40 000 ft (12 000 m) it is 15 to 30 s. Arterial line monitoring and pulmonary artery catheter monitoring for volume status, blood pressure, oxygenation, and ventilation may be of value in prolonged flights with an unstable patient.

Altitude effects on the gastrointestinal tract

Efforts must be made to eliminate gas in the intestines prior to flight. Carbonated beverages, chewing gum, large meals, and pre-existing gastrointestinal disease can all lead to air in the stomach which can expand on ascent. This may lead not only to abdominal discomfort, but also to nausea, vomiting, shortness of breath, and hyperventilation. Distension of the abdominal contents may result in venous pooling, leading to syncope with subsequent tachycardia and hypotension. Thus patients with bowel obstruction or recent abdominal surgery must have a patent and unclamped nasal or oral gastric tube placed prior to transport. High-altitude fixed-wing transport in the early postoperative period is not without risk due to ileus and gas formation, leading to aspiration.

Altitude effects on obesity

Owing to the high concentration of nitrogen in adipose tissue, which can be released into the bloodstream in large quantities, leading to hypoxemia, the bariatric patient must be transported with great care. In addition, large concentrations of lipids can be released, leading to fat embolism. Severe dyspnea, chest pain, and petechias in the shoulders, neck, and axilla, pallor, and tachycardia in the obese patient should raise concern. Many of these effects can be minimized by utilizing 100 per cent oxygen 15 min prior to transport, which will help decrease the nitrogen levels in these patients.

Altitude in pregnancy

The fetal $P\text{O}_2$ at sea level is 32 mmHg (6.3 kPa), although the maternal $P\text{O}_2$ is 100 mmHg (13.3 kPa). Thus the change in $P\text{O}_2$ in the fetus at altitude is less than that in the mother; owing to the oxygen dissociation curve for fetal hemoglobin, it is more saturated at a lower Po_2 than the maternal hemoglobin. Thus attention should focus on hypoxia in the mother which, when corrected, should relieve fetal hypoxia, barring any mechanical etiology such as an amniotic fluid embolus. When the cervix is dilated more than 6 cm, an obstetrician should be part of the medical crew if possible (Rodenberg and Blumen 1994; Holleran 1996).

Barotitis media and barosinusitis

Barotitis media is a condition that occurs when pressure equalization between the middle ear and the atmosphere is impaired. Normally, gas escapes through the Eustachian tube about every 500 to 1000 ft (150–300 m) of ascent, i.e. when the pressure differential is about 15 mmHg. If a crew member has upper respiratory congestion, flight duty is not recommended, since barotitis media can cause severe pain, vertigo, nausea, and eardrum perforation. Crew members must often make a conscious effort to equalize pressures by opening the Eustachian tube with yawning or swallowing maneuvers. Similarly, barosinusitis is caused by congestion of the sinuses where swelling of the mucous membranes impairs pressure equalization between the sinus and the atmosphere. No flight duty should be assigned while these conditions are symptomatic. Oral decongestants can be prescribed.

Conclusions

The attending physician and crew must assess whether compelling reasons for transport exist and that these outweigh the risks of transport. Indications for transport vary but most revolve around bringing the patient to a higher and more sophisticated echelon of care. With the development of regional trauma systems, trauma patients comprise most of the critically ill patients transported. Ground and air medical helicopter ambulances, using a crew comprising nurses and paramedics, provide the majority of these transports. Specialized equipment used during transport must be self-contained, rugged, battery powered, and able to operate without interfering with the helicopter's electronic instrumentation. Transport distances over 150 miles require fixed-wing transport at an altitude which entails attention to changes in atmospheric oxygen tension which can lead to hypoxia, decreases in barometric pressure which can lead to dysbarism, and changes in gas volume which can lead to difficulties with endotracheal tubes and Foley catheters and exacerbation of a pneumothorax.

References

Blumen, I.J., Abernethy, M.K., and Dunne, M.J. (1992). Flight physiology: clinical considerations. *Critical Care Clinics*, 8, 597–618.

Holleran, R.S. (ed.) (1996). *Flight nursing. Principles and practice* (2nd edn). Mosby Year Book, St Louis, MO.

Pantridge, J.F. and Geddes, J.S. (1967). A mobile intensive care unit in the management of myocardial infarction. *Lancet*, **ii**, 271–3.

Rodenberg, H. and Blumen, I.J. (1994). *Air medical physician handbook.* Air Medical Physicians' Association, Salt Lake City, UT.

11.12.2 Minimizing risk during transport

J. E. Gilligan

Key messages

- The aim of any transport is to provide all the pre-existing intensive care unit in-patient services.
- The selection and training of staff, the equipment both inside and outside the vehicle, and preparation for transport requires much organization prior to flight.
- Altitude affects equipment and physiological spaces (e.g. pneumothorax), and an understanding of these changes is required.

Introduction

Adequate care of the critically ill during transportation is a commonly neglected area. Within hospitals, procedures lasting a few minutes (e.g. CT) may require extended preparation. Rationalization of health services may require the regionalization of specialized units, but most patients can be safely moved, even considerable distances, if effective measures are taken to minimize risk. Stabilization of circulation, ventilation, biochemistry, and thermoregulation before movement, using experienced staff, facilitates safe transport. Equipment, although similar to that in the intensive care unit (ICU), should be compact (Table 1). A general principle is that

Table 1 Ideal requirements for a transport modality

Adequate safety of operation
No abrupt movements in any axis
Sufficient room for at least one critically ill patient and supporting team, including an attendant at the head end
Adequate supply of energy and gases for life-support systems
Adequate lighting and internal climate control, including cabin pressurization
Tolerable noise and vibration levels
Adequate restraint devices for stretchers, occupants, and equipment
Easy embarkation and disembarkation of patient
A stretcher suitable for all phases of transport, obviating the need for changing stretchers between vehicles
Minimal secondary transport (i.e. road transport in air evacuation)
Adequate speed
Effective communication systems

Reproduced with permission from Oh (1997).

patients in transit should have at least the level of care that was provided in their initial location (ANZCA 1992).

Minimizing risk to patients and staff involves ensuring that the following are adequate:

(1) transport categories;

(2) selection and training of staff;

(3) communications;

(4) preparation for transport;

(5) care in transit;

(6) mode of transport.

Transport categories

Intramural transit occurs between departments (e.g. emergency department, ICU, radiology department) in a single institution. Extramural transport involves primary (prehospital) or secondary (interhospital) missions.

Primary mission teams may be called to scenes such as explosions and major road accidents, where they may be working with prehospital support personnel (ambulance officers, paramedics, police). Entrapped patients may require on-site blood and other infusion, anesthesia (local and general), intubation, controlled ventilation, pleural drainage, or limb amputation. The environment may limit access to patients and minimum intervention compatible with safe extrication should be performed, particularly if transit time is under 20 min.

Secondary missions involve interhospital movement, usually to a major medical center for more advanced diagnostic or therapeutic measures. Examples include the following.

1. Intensive care transport from rural communities for conditions such as severe asthma, poisoning, major trauma, septic shock, and multiple organ failure. Neonates require specialized pediatric transport teams.

2. Mobile surgical or obstetric teams capable of performing procedures such as evacuation of extradural hematoma or Cesarean section for obstetric bleeding are useful for patients in remote areas.

3. Long distance repatriation of the critically ill may include medical evacuation of injured tourists and military casualties.

Selection and training of staff

Ground or air services use dedicated personnel, or a medical center may use its own medical and nursing staff, usually from intensive care, anesthesiology, or emergency departments. These staff should be able to perform intubation, pleural drainage, cricothyrotomy, and resuscitation with intravenous access. Experience in Advanced Cardiac Life Support (ACLS) and Advanced Trauma Life Support (ATLS) is necessary.

Staff should be familiar with transport modalities, including loading, unloading, safety procedures, communications systems, and other characteristics. Safety uniforms, helmets, and identification are necessary. Staff should have normal fitness and not be prone to maladies, such as motion sickness or obstruction of the Eustachian tubes, which may interfere with their ability to provide care or their own well being.

Communications

Effective communication is essential. In the case of intramural movements (e.g. to radiology), waiting should be minimized and the transport party should be familiar with the route. In the case of extramural operations, a physician should be available at all times at the major referral center for urgent consultation by remote hospitals. Other consultative services (e.g. neurosurgery) should also be easily accessible.

At the remote hospital, the transportation team assumes management after examination of the patient and consultation with the staff and relatives. They direct urgent investigations and modifications to therapy before traveling, and obtain copies of relevant data and radiographs.

Contact between the transportation team, the transport authority, and hospital staff at the final destination is essential. A radio system, linking aircraft or road ambulance and recipient hospitals, should be used to alert the center of immediate needs, such as preparation for CT scanning or assembly of a trauma team. On arrival, the patient is handed over to those assuming care.

Preparation for transport

Following stabilization, the patient is packaged for travel. Unnecessary infusions and medications are discontinued, essential intavenous drugs and infusions are prepared, checked, and labeled, and infusion pumps are activated as required. Safety requirements include Luer-lock junctions and needle-free injection sites for bolus and intravenous additives to reduce the risk of disconnection and needlestick injury. Gloves, masks, eye protection, and sharps disposal bins must be available.

Typed and cross-matched blood likely to be used in transit is checked, reserving 'universal donor' O-negative blood for life-threatening situations. Central venous catheters needed for some infusions (inotropes, total parenteral nutrition in extended missions) are placed. If parenteral nutrition is stopped, rebound hypoglycemia should be prevented by intravenous glucose 10 per cent.

Arterial blood pressure cannulas are inserted if necessary, and all catheters and drains and the endotracheal tube are secured. In suspected cervical blunt injury, a rigid collar is applied, and such patients should be 'log-rolled' when being turned. Effective immobilization of fracture sites is achieved using an extension splint (e.g. the Hare design); correction of angulation may restore limb blood flow. An extremity with vascular occlusion and incipient compartment syndrome needs fasciotomy within 6 to 8 h. During long (e.g. international) trips, immobile limbs may swell, particularly if they are dependent; therefore plaster casts should be split before departure and limbs elevated. Plaster shears should be available. Finally, supplies such as oxygen, suction, and batteries are checked.

Using one stretcher only, which is compatible with transport vehicles, from the referring hospital bed to the receiving hospital bed enhances patient comfort and safety. In extramural missions, the patient, equipment (Table 2), and bedding are secured to the stretcher by safety harness. A reflecting 'space' blanket or plastic sheet protects the patient from adverse weather or helicopter rotor wash. Stretcher

Table 2 Requirements of mobile intensive care equipment (monitors, ventilators, and pumps)

Robust
Compact/light for use outside aircraft
Adequate range and number of parameters displayed
Operation in any orientation (e.g. on side)
Adequate performance with vibration, cabin temperature, and altitude changes
Adequate battery endurance and/or a.c. supply
No interference problems with other electronic equipment
Screen visibility in daylight and other light conditions
No mercury-containing equipment (corrosive and toxic)
Adequate anchorage in transit
Attachment to stretcher if portable
Stretcher compatibility with other transport

Reproduced with permission from Gilligan *et al.* (1996).

bridges for equipment are useful but are sometimes unwieldy. Eye protection is essential in bright sunlight, which also renders medical procedures such as laryngoscopy difficult.

Care in transit

Stabilized patients usually require only observation and adjustment of medication and fluids unless there is an an emergency requiring intervention (e.g. extubation). Ventilator-dependent patients need effective sedation and possibly neuromuscular blockade. Continuous positive airway pressure (CPAP) and BiPAP™ circuits may be bulky and use large volumes of gas. In non-intubated patients, loss of face seal may result in rapid desaturation. Thus some patients on CPAP may be more easily transported with controlled ventilation and a gastric decompression tube in place.

Respiratory rate, tidal volume, disconnection, overpressure, and fraction of inspired oxygen (FiO_2) should be monitored in patients on ventilators. End-tidal $P\text{CO}_2$ approximates $P\text{aCO}_2$, but some devices give variable readings at common cabin altitudes. Pulse oximetry displays pulse rate and oxygen saturation, and hypoxia can indicate disconnection. However, vasoconstriction may produce faulty readings and smaller devices appear to be more susceptible to artifacts. Transcutaneous oxygen measurement, which is used in neonates with thin skin, is not useful for adults.

Detection of ventilator or infusion disconnection requires close observation and reliable alarms, but noise and daylight may render alarms and monitors inaudible or unreadable, particularly during loading. Once the patient is aboard the aircraft, monitor alarms can be relayed through the medical crew's headsets.

Pleural drains should be attached to 'flutter' valves (e.g. Heimlich® valves) connected to collection bags for blood and air. Underwater seal drainage systems are awkward to transport; the reservoir, if not on suction, has to be prevented from tilting and must be located at least 30 cm below the patient to prevent aspiration of air and drainage fluid into the pleural cavity. However, these systems allow visualization of air leaks and are more feasible in large aircraft. Needle thoracostomy for suspected tension pneumothorax can be safely used only to gain a few minutes, pending formal thoracostomy tube placement.

Blood pressure is usually measured by aneroid gage and systolic palpation, as auscultation is usually difficult during road and air transport. Effective pneumatic cuff blood pressure transducers (piezocrystal microphone, Doppler generator, or oscillotonometric transducers) are bulky, and accuracy is limited in the hypotensive patient (Rutten *et al.* 1986). However, measurement by arterial catheter with a transducer gives continuous blood pressure surveillance. A simple alternative incorporates a heparin-filled line connected via a membrane isolator to a sphygmomanometer gage. Intermittent manual flushing is simpler than continuous flushing. Defibrillation procedures established in hospitals are followed in aircraft, ensuring that contact between patient and metal is avoided.

With transvenous pacing, capture should be assured before departure as external transthoracic pacing may be ineffective during transport. Additional space, and possibly a.c. electrical power, may be needed for intra-aortic balloon pumping.

Pressor infusion sleeves for emergency volume infusion need suspension from a convenient site, although a pack can be wedged under the patient's body if the delivery set is free of air. A dry heat warmer is safer than a water bath for transfusing cold blood during major blood loss. Adding warmed saline to packs may ameliorate hypothermia.

Urine catheter drainage systems establish urine volume, facilitate disposal, and obviate the discomfort of bladder distension. Wound loss should be similarly contained. Nasogastric intubation for ileus or gastric distension minimizes gas expansion at altitude and pulmonary aspiration.

Use of a rectal or esophageal temperature probe enables continuous display, but tympanic membrane measurement is a convenient alternative as oral and axillary temperature may be unreliable. Hyperthermic patients may require ice packs plus specific medications (e.g. dantrolene for malignant hyperpyrexia). Motion sickness medication and nasal spray for Eustachian obstruction should be available.

If there is renal failure, hemodialysis should be performed prior to transport but beware of temporary volume depletion. Peritoneal dialysis may impede respiration because of abdominal distension during inflow of fluid, and so intermittent positive-pressure ventilation or temporary hemodialysis may be required. During long trips hemofiltration can be performed using available compact systems. Compact equipment is also available for electrolyte and glucose measurements using fingertip or venous blood. Regular insulin should be used for sliding-scale therapy.

Patients with decompression sickness or arterial gas embolism are intolerant of even low altitudes (100–200 m) because bubble expansion exacerbates clinical effects (Edmonds *et al.* 1992). Sea-level cabin pressure or low-level helicopter flight is required. Transporting hyperbaric chambers to remote sites for initiation of treatment, with subsequent transfer to a larger chamber without pressure loss, is cumbersome, requires sizeable fixed-wing or helicopter transport, and is not recommended.

Documentation is continued in transit. If time zones are crossed, using the destination time zone provides a consistent time-scale.

Alternative methods, including a self-inflating bag for emergency manual ventilation, spare suction devices, exhalation, and positive end-expiratory pressure valves, and a reserve ventilator and regulators should be considered with long flights in case there is equipment malfunction. The provision of reserve intravenous pumps avoids unsatisfactory reversion to gravity infusion for critical infusions. Circulatory surveillance (blood pressure, $Sa\text{O}_2$) can be achieved using an aneroid sphygmomanometer and a 'pocket' pulse oximeter.

Mode of transport

Hospital design should allow intramural transfer between departments, preferably with orderlies to move beds or trolleys, and nursing and medical staff to maintain care in transit. Manual lifts onto trolleys have the potential of injuring patients and staff, and aids such as sliding devices are advocated. Some protocols recommend using up to five people for major shifts if no physical aids are available. Supporting equipment should be attached to the bed, as trailing devices are hazardous although sometimes unavoidable (e.g. balloon pumps).

Extramural safety requirements include a vehicle with adequate space, electrical power, and oxygen reserve (Table 1). Defensive parking at major incidents protects crews against other traffic. Aviation regulations require stretchers to tolerate the deceleration likely in an emergency, and securing patients and equipment by safety harness is easiest in the head-first posture. If the patient is stable, there is little clinical difference between the head-first and the feet-first position, despite the nose-up ascent posture of fixed-wing craft. Transverse positioning produces the least inertial change in body fluids, but few aircraft have the necessary cabin width.

During short trips, and particularly during 'hot' helicopter loading/unloading (i.e. with engines running), it is safer to close gravity-fed intravenous lines and place them on the litter than to elevate them on poles. When flow is restarted, entrained bubbles must be expelled.

Loading aircraft may be difficult because of the weight and size of the stretcher, patient, or equipment, small cabins, narrow aircraft doors and corridors, and the height of the fuselage from the ground.

The reduction of the partial pressure of oxygen in air at altitude affects tissue oxygenation (Henry's law) (Fig. 1). Also, reduction in total ambient pressure may allow entrapped gas in distensible cystic cavities to expand (Boyle's law). Denitrogenation (breathing 100 per cent oxygen) before and during exposure to altitude may help to reduce the amount of entrapped gas (e.g. intraocular air, pneumothorax, pneumoperitoneum, or entrapped intracranial air) (Hyldegaard and Madsen 1994). Nitrous oxide analgesia may be hazardous in patients with decompression sickness or with any entrapped gas, as a counterdiffusion effect may enhance bubble size.

With entrapped intraocular air, a sudden elevation of intraocular pressure (e.g. vomiting, coughing, straining, or hypoxia) may cause loss of vitreous and other ocular contents. Discharge of globe contents at high altitude has occurred. Experimentally, intraocular pressure is increased by about 40 mmHg at a cabin altitude of 8000 ft. These effects may be reduced by eye binding, antiemetics, sitting the patient upright, and breathing 100 per cent oxygen.

Trapped air in an enclosed rigid cavity (e.g. an obstructed nasal sinus) will maintain its pressure, even if the ambient air pressure is decreased. Compression of venous drainage and, later, arterial vascular supply in the cavity may produce pain, bleeding, or ischemic effects.

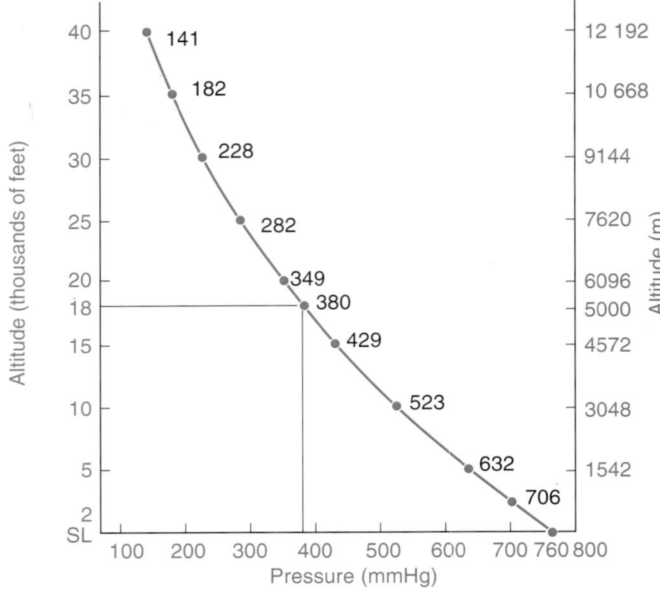

Fig. 1 Boyle's law and Henry's law: the relation between altitude and atmospheric pressure, which may affect oxygenation.

Endotracheal tube cuffs expand at altitude. If the cuff pressure on the tracheal mucosa becomes excessive, damage may occur. Thus cuff volume may require adjustment on ascent. Conversely, cuffs may shrink on descent and require reinflation. Cuff volume is usually not adjusted for short flights. Inflation with water may have a significant compressive effect on the tracheal mucosa and is not advised. Similarly, the inflation chambers of pneumatic antishock garments or military antishock trousers may expand at altitude. Possible compression of underlying tissues requires scrutiny to avert compartment syndrome or respiratory embarrassment. When intra-aortic balloon pumping is used, the effect of altitude on the balloon gas must be considered.

Similar changes occur in surface transport over mountainous terrain.

Cabin pressures in commercial airliners are usually maintained at about 550 mmHg (75 kPa), or an altitude of approximately 2000 to 2500 m (6000–8000 ft) at an actual altitude of 12 000 m (approximately 40 000 ft). (In aviation, altitude is commonly measured in feet and speed in nautical miles per hour where 1 nautical mile = 1.8 km.) Most patients travel well at conventional cabin pressures, provided that measures are taken to ensure adequate alveolar oxygen ($PA\text{O}_2$).

Air ambulances, executive charter aircraft, and military transport aircraft may maintain sea-level pressure by flying at lower less efficient altitudes where there is a greater likelihood of turbulence. Low-frequency (< 20 Hz) high-energy (large-amplitude) vibration is potentially damaging to various organs and may be encountered in both road and air transport, although in practice it is not usually a problem.

Aircraft cabin temperature may affect the well being of patient and crew and the resilience of any electromedical equipment, particularly when the aircraft is standing and the cabin temperature may range from 50 °C to below zero. Climate control of most aircraft requires the engines to be running, unless an auxiliary power unit is available.

External temperature reduction is approximately 2 °C per 1000 ft; this is sometimes a consideration in helicopter missions since the external temperature at 10 000 ft is about 0 °C.

The choice of fixed-wing aircraft or rotorcraft depends on the equipment carried, size, distance, and availability. Particular needs (e.g. altitude) should be discussed with the pilot before departure. Instrument flight rules (IFR), navigation instrumentation, weather radar, low-frequency radio navigation (e.g. Omega™), and global positioning system (GPS) navigation capability for both types of aircraft enhance safety in unexpected marginal weather. Any mission is at the sole discretion of the pilot and should be aborted if there is significant risk to the aircraft and crew.

Fixed-wing aircraft are faster than helicopters, but they are slower to be mobilized and require intermediary road transport. Helicopters conveniently operate up to a radius of 200 km (111 nautical miles), depending on range and fuel considerations. They can land near a location or deliver resources to inaccessible sites by winching or rappelling techniques. Noise is greater and altitude is lower than for fixed-wing operations. The trend to twin-engine helicopters is principally due to payload and safety. Larger helicopters require a crew member, separate from the medical team, to oversee safety and assist with navigation.

Fixed-wing aircraft are usually turbo-prop twin-engine design. Missions are commonly over 200 km. Other types of fixed-wing aircraft used include executive jets for urgent long-distance transport from an area not well served by commercial airlines, and commercial airliners for long flights (> 1000 km). However, not all carriers permit stretchers. A complex case, with a medical team, a stretcher, and equipment, may occupy 15 seats which is quite costly. Teams need to be self-sufficient in equipment, oxygen, and electrical power. A preflight assessment of the aircraft and utilization of hydraulic food or baggage lifts for loading is prudent. Military medical transport aircraft (e.g. Hercules C 130) are equipped with educated dedicated personel and medical equipment. They may have an intensive care unit configuration for long-distance transport (Gilligan *et al.* 1996)

If intercontinental transport is necessary, resolution of a disease or injury should generally be well established before evacuation, but occasionally, because of lack of local resources for example, it may be necessary to move a critically ill patient. Absence of more than 24 h from the home base may be involved, so that enough staff for conventional shifts (up to 12 h) to be worked and a route minimizing airport delays and aircraft changes will be required. Limited availability of supplies *en route*, language barriers, varied road transport, and uncertain legal status of the team on foreign soil with respect to emergency drugs and blood carried need to be considered.

References

ANZCA (Australian and New Zealand College of Anaesthetists and Australian College for Emergency Medicine) (1992). *Joint policy document. Minimum standards for transport of the critically ill.* Publication P23, ANZCA, Melbourne.

Edmonds, C., Lowry, C., and Pennefather, J. (1992). Transport of the diving injury. Aviation medevac. In *Diving and subaquatic medicine* (4th edn), p. 436. Butterworth-Heinemann, Oxford.

Gilligan, J.E., Goon, P., Maughan, G., Griggs, W., Haslam, R., and Scholten, A. (1996). An airborne intensive care facility (fixed wing). *Anaesthesia and Intensive Care*, **24**, 245–53.

Hyldegaard, O. and Madsen, J. (1994). Effect of air, heliox and oxygen breathing on air bubbles in aqueous tissues in the rat. *Undersea and Hyperbaric Medicine*, **21**, 413–24.

Oh, T. (ed.) (1997). *Intensive care manual* (3rd edn). Butterworth-Heinemann, Oxford.

Rutten, A.J., Ilsley, A.H., Skowronski, G.A., and Runciman, W.B. (1986). A comparative study of the measurement of mean arterial blood pressure using automatic oscillometers, arterial cannulation and auscultation. *Anaesthesia and Intensive Care*, **14**, 58–65.

11.12.3 Equipment required for in-hospital transfer

S. A. Ridley

Key messages

- Equipment used for in-hospital transfers should aim to recreate the intensive care environment outside the formal intensive care unit setting.
- The equipment should meet appropriate design standards, and several options exist for combinations of equipment to form a complete in-hospital system.
- The overall aim should be to produce a safe efficient system with simple, compact, and durable equipment.

Introduction

In-hospital transfers are required for movement of critically ill patients around the hospital; numerous therapeutic and diagnostic aids now help in the management of critical illness but many of these cannot be brought to the intensive care unit (ICU). Examples include surgical interventions in the operating theater, non-mobile diagnostic aids such as CT scanning, magnetic resonance imaging (MRI), cardiac catheterization, and angiography. Occasionally, journeys outside the ICU must be undertaken for non-operative therapy such as radiotherapy for relief of neoplastic compression of the airway or suppression of new bone formation following repair of acetabular fractures.

The aim of in-hospital transfer is to move the intensive care environment with the patient so that the transfer can be considered as an extension of the critical care continuum in a different part of the hospital.

The equipment required for in-hospital transfers is similar to that needed for interhospital transfer. Indeed, to avoid duplication of equipment and maximize staff familiarity with one standard equipment inventory, institutions may elect to use their interhospital mobile ICU for in-hospital transfers. Transfer equipment for in-hospital use only need not have such long battery life as wall-mounted a.c. supplies should be available at the destination. The in-hospital system may need to carry less emergency equipment, such as a d.c. defibrillator, as these should be available throughout the hospital. As the patient remains in hospital, there is the option of using bed-mounted systems which do not satisfy the size and weight restrictions imposed in interhospital transport by land or air.

General considerations

An in-hospital transfer should consist of all the elements which make up the usual ICU environment. This will include the attendant staff (medical, nursing, and technical) and the usual categories of ICU equipment. The equipment can be divided into four broad categories (Table 1). The first is the equipment needed for basic care such as the ICU bed, laundry, and sources of warmth and comfort. The second category covers life-support equipment such as the ventilator and other types of respiratory support, including an oxygen supply. The third and fourth categories comprise the monitoring equipment required to detect changes in physiological variables and the equipment required to treat the patient. The treatment equipment can be subdivided into that needed for continuing intensive therapy, such as infusion pumps for inotropic support, and that needed in an emergency which may arise because of physiological deterioration or equipment failure (e.g. a blocked endotracheal tube, loss of oxygen, power failure, avulsed intravenous infusion catheters).

Equipment standards

Critical care equipment is manufactured worldwide, but not all devices will satisfy all national standards. For example, a hand-operated suction device made by a well-known American manufacturer failed to attain one of the British Standards for such devices (Arnstein 1996). Equipment which is safe to use in one setting may not be safe in the rest of the hospital. For example, the use of mobile telephones in the ICU is discouraged because of the potential for interference with ventilators and infusion pumps (Richardson and Pickstone 1995), but their use in the rest of the hospital, particularly the corridors, is more difficult to control. Devices used for monitoring in MRI scanners must satisfy rigorous standards concerning non-ferrous components; for example, only a few pulse oximeters are safe for use near MRI scanners (e.g. Nonin 8600F0 with fiber-optic coupled sensor 8000FC (Cardiokinetics Ltd, Salford, UK)).

Mobility

The use of unmodified ICU or theater monitoring equipment is probably not realistic as its size and weight precludes a mobile role. Portable equipment measuring individual variables is available, as are combined units measuring a wide range of functions. The advantages and disadvantages of such equipment revolve around the combined size, total weight, and complexity of individual units against potential loss of all monitoring when a single main unit malfunctions. The battery life of the transport equipment need not be long (2–3 h) provided that the power cable is of sufficient length to reach the wall where the transfer is occurring. Multiple-plug extension boards should be avoided as there is a potential for current leakage between medical monitoring equipment directly connected to the patient (Hull 1993). For in-hospital transfers the patient's bed will usually be sufficiently mobile and so the extra equipment needed may be attached.

Table 1 Categories of equipment for in-hospital transfer

1	Basic care	ICU bed, warmth, comfort
2	Life support	Equipment to maintain: Airway Breathing Circulation
3	Monitoring equipment	
4	Treatment	Continuing ICU management Emergency treatment needed during transfer

Monitoring

A wide range of monitoring equipment which can be used for patient transfer is commercially available. The flexibility of a modular monitoring system may confer benefits in that the level of monitoring can be tailored to the patient's severity of illness with fewer indices needing measurement in the less ill patient. However, as the severity of illness increases, the connection and reconnection of monitoring lines and cables becomes more burdensome. Modules which retain the monitor set-up and can remain attached to the patient may be safer and more efficient. As with any form of patient transfer, the individual monitor used is not as important as its availability to measure the physiological variables deemed important.

Specific transfer problems

A number of special measures should be considered prior to the final promulgation of an equipment list. Color-coded intravenous lines will help to sort out the inevitable 'spaghetti' which occurs when patients are moved any distance from the ICU bed space (Cook and Seavell 1996). Special racks or cradles may need to be individually manufactured for carrying heavy objects such as oxygen cylinders. A special bag or pocketed carrier is needed so that drugs and fluids can be carried in an easily retrievable display allowing quick access and preparation. A paper record of the patient's physiological variables should be kept so that the information can be incorporated into the ICU daily chart upon return. Occasionally the transfer will be complicated by a special requirement such as extracorporeal membrane oxygenation or high-frequency jet ventilation.

Equipment required

The equipment chosen must be able to maintain a free-standing ICU environment for a few hours and to provide adequate monitoring to detect and, where necessary, treat abnormal physiological changes. Mishaps, such as disconnection of ECG leads and power failure (singly or in combination), occur frequently and require urgent correction. Equipment is also needed to maintain and safeguard the airway, augment or support breathing, and maintain stability of the circulation. An equipment list suitable for in-hospital transfer is shown in Table 2. Monitoring equipment should be added to this list; as mentioned earlier, the level of monitoring should be tailored to the patient's severity of illness. A checklist of equipment is also advisable so that the equipment is known to be complete and deficiencies are replenished.

Table 2 Equipment required for in-hospital transfer

Airway	
Laryngoscopes, spare batteries, and bulbs Sterile cuffed tracheotomy tubes (various sizes) Catheter mounts Disposable humidifiers Fixed-performance oxygen masks and piping Lubricating jelly	Sterile cuffed endotracheal tubes (various sizes) Endotracheal tube introducers (metal and gum elastic) Tape and adhesive dressing Magill forceps Airways, oral and nasal (various sizes) Suction catheters and suction device
Breathing	
Portable ventilators, ideally capable of volume-controlled ventilation, positive end-expiratory pressure generation, variable inspired oxygen fraction; they should be light-weight, simple, gas driven, robust, and reliable Disposable (or autoclavable) ventilator tubing	Oxygen cylinder with appropriate pressure-reducing valves for ventilator and fresh gas flow to face-mask Self-inflating resuscitation bag with anesthetic masks (various sizes)
Circulation	
Intravenous cannulas (various sizes) Cannulas appropriate for intra-arterial pressure monitoring Intravenous fluids and administration sets Three-way Luer lock taps and intravenous line extensions	Central venous catheters (various lengths and diameters) Sutures and dressings Infusion pumps with appropriate syringes
Monitoring devices (selected and appropriate to the patient's severity of illness) may include	
Central venous and pulmonary occlusion pressures Arterial oxygen saturation Temperature	Blood pressure, direct intra-arterial and/or non-invasive ECG End-tidal CO_2 tension
Emergency drugs (in addition to those required as part of the ICU therapy)	
Miscellaneous	
Nasogastric tubes (various sizes) Syringes/needles Gloves and aprons	Scissors Sterile plain and alcohol-soaked swabs Documentation for transfer and the patient's medical notes
Optional extras (which should be easily available from elsewhere in the hospital)	
Chest drains with Heimlich valves or underwater seals Disposable pressure transducers and pressure infusion bags should be connected to the patient prior to transfer	DC defibrillator Aneroid sphygmomanometer Blood filters

In-hospital transfer systems

The equipment used for in-hospital transfers has been combined to form complete transfer systems. These are generally not commercially available as hospitals have developed their own systems to suit their particular needs. Descriptions have been published and the systems appear to fall into three general types.

1. The same mobile ICU is used for in-hospital as well as interhospital transfers. This system has been used successfully by the Clinical Shock Study Group in Glasgow for many years but it involves an extra transfer of the patient and therefore may be inherently more dangerous (Runcie *et al.* 1991). Other mobile ICU suitable for interhospital transfers have also been described.
2. Carts and trolleys which clip or attach to the bed can be developed but may become redundant if the ICU obtains new incompatible beds.
3. The additional equipment may be attached to or carried on the bed. This option may be the most cost effective as specialist engineering work is minimized and purpose-made tables offered by the bed manufacturers can be used.

The most important considerations are ease of transfer, patient comfort, portability, set-up time, and ease of cleaning.

Drugs, fluid, and nutrition

The pharmacological support offered to the patient should continue during transfer. Continuous inotropic or vasodilator therapy may be mandatory and so must be continued irrespective of the complexity that this adds to the transfer. Emergency drugs (Table 3) and a checklist must be carried in case of an acute cardiorespiratory decompensation.

Table 3 List of drugs required for in-hospital transfer

Drugs for continued intensive care

Sedatives
- Midazolam
- Diazepam
- Etomidate
- Propofol

Neuromuscular blocking agents
- Succinylcholine (suxamethonium)
- Atracurium
- Vecuronium

Cardiovascular drugs
- Dopamine
- Dobutamine
- Dopexamine
- Norepinephrine (noradrenaline)
- Epoprostenol (prostacyclin)
- Nitrates
- Methoxamine
- Hydralazine
- Labetalol
- Verapamil
- Furosemide (frusemide)

Bronchodilators
- Aminophylline
- Albuterol (salbutamol)

Opioids
- Fentanyl
- Alfentanil
- Morphine
- Omnopon

Emergency resuscitation drugs

Epinephrine (adrenaline)
Isoproterenol (isoprenaline)
Lidocaine (lignocaine)
Atropine
Calcium
Bicarbonate

Conclusions

Equipment used for in-hospital transfers should be similar to that used in interhospital transport in that it should aim to recreate the ICU environment outside the formal ICU setting. The equipment should meet appropriate design standards; several options exist for combinations of equipment to form a complete in-hospital system. The overall aim should be to achieve a safe efficient system with simple, compact, and durable equipment.

References

Arnstein, F.E. (1996). A practical evaluation of four human-powered portable airway aspirators. *Anaesthesia*, **51**, 63–8.

Cook, T.M. and Seavell, C.R. (1996). Patient transfer; what to do about 'spaghetti'. *Anaesthesia*, **51**, 90–1.

Hull, C.J. (1993). Electrical equipment. In *Hazards and complications of anaesthesia* (2nd edn) (ed. T.H. Taylor and E. Major), p. 685. Churchill Livingstone, Edinburgh.

Richardson, N.H. and Pickstone, M. (1995). Interference from mobile phones in the ICU. *British Journal of Intensive Care*, **5**, 233–5.

Runcie, C.J., Reeve, W., Reidy, J., and Wallace, P.G.M. (1991). Secondary transport of the critically ill adult. *Clinical Intensive Care*, **2**, 217–25.

11.12.4 Equipment required for between-hospital transfer

A. G. Morrison and P. G. M. Wallace

Key messages

- The goal is to make the conditions during transport similar to those in the intensive care unit.
- Transport equipment, the mode of transport, and types of respiratory and circulatory support should be assessed for each patient.
- Owing to space constraints, monitoring may need to be adjusted, but patient safety should not be compromised.
- A wide range of medications are required, providing coverage for conditions similar to those seen in the intensive care unit.

Introduction

Many critically ill patients are transferred between hospitals, primarily for specialist therapy such as neurosurgery, admission to a staffed intensive care unit (ICU), or admission to specialized investigative facilities. The fundamental standards of care and equipment for between-hospital transfer are similar to those for intrahospital transfer, aiming to move ICU care with the patient (Runcie *et al.* 1991). The essential differences are that a transport vehicle is required and that staff accompanying the patient during transfer are isolated with no back-up of additional personnel or equipment. Everything that might be required must be transported with the patient. Equipment must be robust and suitable for use in difficult environments.

Equipment

Organization and planning

Every hospital should have a senior doctor responsible for the organization of transfers, including the maintenance and availability of suitable equipment. Dedicated transfer equipment ready for use should be stored in designated areas familiar to staff and close to transit locations. Staff involved in transfers should receive training to ensure that they select the appropriate equipment for each transfer and that they can operate it safely and correctly. Ideally, transfer equipment should be lightweight, robust, reliable, and familiar to the transfer team. Electrically powered equipment should be able to run on batteries which last for at least 4 h and can be recharged rapidly. The transfer team should check all equipment; checklists

Table 1 List of suggested transfer equipment

Respiratory	O_2 supply: cylinders and cylinder keys (2)
	Reducing valves (2) to Schraeder connector, spare Bodok seals
	Flowmeters (2)
	O_2 tubing, range of O_2 face-masks, nasal cannulas
	Range of oro- and nasopharyngeal airways
	Laryngoscopes (2), spare batteries and bulbs
	Range of cuffed endotracheal and tracheostomy tubes
	Tube introducer and gum elastic bougie, Magill forceps
	Ties for tube and lubricating jelly
	Disposable catheter mounts and humidifying filters
	Range of laryngeal mask airways
	Minitracheostomy kit
	Self-inflating resuscitation bag with reservoir and face-masks
	Mechanical ventilator
	Chest drain insertion kits, flutter valves and drainage bags
	Suction device and catheters
Circulation	Range of intravenous cannulas
	Range of arterial cannulas
	Multilumen central venous lines, pulmonary artery catheters and introducers
	Fluids: crystalloids and colloids
	Large-bore intravenous fluid administration sets
	Pressure bags
	50-ml Luer-lock syringes and extension sets
	Syringe drivers (2-3)
	Range of syringes and needles, three-way taps
	Adhesive tape, dressings
Monitoring	ECG, pulse oximetry, NIBP, two pressure channels, temperature, CO_2
	ECG electrodes
	Disposable pressure transducers
	Pressure bag of heparinized saline
Others	Sterile gloves, protective aprons
	Skin prep, razors
	Artery forceps, scalpels, scissors, sutures
	Nasogastric tubes and drainage bags
	Blankets, heat-reflective blankets
	Vacuum mattress
	Defibrillator

NIBP, non-invasive blood pressure.

provide a useful reminder of what to take for a transfer, what to bring back, and what needs to be restocked. A record of patient observations, drug therapy, and any interventions required should be made for every transfer, and should be kept to audit the transfer process, the outcome, and any problems associated with the transfer.

A full list of suggested equipment is shown in Table 1. Obviously the personnel involved (e.g. registered nurse, emergency medical technician, physician) and the type of patient will affect what is used.

Patient trolley

A trolley will be required to accommodate the patient for transfer both within and between hospitals. It must be safe and comfortable for the patient, and allow adequate access to accompanying staff for monitoring and resuscitation. Inevitably many pieces of equipment, such as monitors, a ventilator, pumps, and an oxygen cylinder, will have to be carried safely. An ordinary ambulance trolley is not designed for this purpose, and therefore most equipment will rest on the patient, be carried by accompanying staff, or be placed on the floor of the ambulance. This distracts the transfer team from monitoring the patient, increases the risks of accidental disconnections or extubations, may damage the equipment, and obscures observation of the monitors.

For these reasons, various modifications have been developed to improve standard trolleys with variable success. Removable shelves have been designed to be placed over the trolley above the patient. These allow good visibility and access to the equipment, but access to the patient is reduced and they can be cumbersome. Mounting brackets and poles allow the attachment of a limited amount of equipment. A shelf below the trolley provides a larger area to store equipment securely and does not impede access to the patient or equipment, but equipment may be difficult to see. Some modified trolleys require specialized mountings to secure them to the vehicle, thus precluding their more general use. Each hospital must develop a trolley system which permits efficient carriage of their chosen equipment, has the confidence of staff, and will securely fit in the vehicles usually employed for transfers.

The weight of a fully laden trolley, particularly with an obese patient, can be considerable and makes loading into an ambulance difficult. Ramps and hydraulic lifts ease this problem and reduce the risk of back injury to the transfer team.

Moving a patient from a land to an air ambulance is usually difficult. A semi-rigid vacuum mattress may assist by acting as a protective cocoon around the patient and allow easier handling. Some air ambulances use a sled system to move the patient into the aircraft and onto its special mountings.

Transfer vehicle

The type of vehicle used will depend on the distance to be traveled, the weather conditions, geographical considerations, and availability.

Air ambulances will generally be suitable for longer journeys and when road access is poor. Helicopters are recommended for journeys of 50 to 150 miles and for sites without landing strips; fixed-wing aircraft are used for journeys of over 150 miles.

Land ambulances are more suitable for shorter journeys of less than about 50 miles distance or 2 h duration. They provide good lighting and heating, have constant two-way communication, and carry a defibrillator, suction apparatus, and oxygen. All equipment must be checked routinely.

The availability of specialized ambulances or mobile ICUs varies from country to country. In the United Kingdom they are usually only available to dedicated transfer teams, but are more common in the United States and continental Europe. Designs vary widely, but usually include increased stores of oxygen and transfer equipment, more complex monitors, and mountings to secure the equipment.

The position of the patient in the ambulance varies according to design. The trolley must be well secured, but adequate access to the patient, particularly the head, is essential. Theoretically the effects of acceleration and deceleration will be diminished by fixing the trolley from side to side. In practice, a standard position with the head to the front of the ground vehicle poses few problems. With fixed-wing aircraft, owing to the effects of acceleration on take-off, patients with a propensity for an elevated intracranial pressure should be placed head forward.

Respiratory support

Sufficient oxygen must always be carried. The amount needed for a transfer should be calculated including the patient's requirements and gas required to drive ventilators and suction machines. An allowance of several hours of extra oxygen supply is necessary for safety. Oxygen is usually supplied in cylinders. Those of a useful size, i.e. containing over 600 liters, are bulky and heavy, but are readily available and familiar to staff. Liquid oxygen supplies are smaller and lighter; they are used by some transfer teams but require costly special containers and filling equipment.

A self-inflating manual resuscitation bag and a portable mechanical ventilator must always be available. Hand ventilation may cause hypo- or hyperventilation, and therefore a mechanical ventilator should be used routinely (Hurst *et al.* 1989). The portable ventilator must be robust, reliable, and versatile. It should offer a wide range of tidal volumes, frequencies, and inspiratory-to-expiratory ratios, and be able to deliver positive end-expiratory pressure. Most portable ventilators are gas driven and deliver either 100 per cent oxygen or an air mix of 45 to 65 per cent. Recent models offer continuous positive airway pressure and synchronized ventilation which some patients will tolerate, thus requiring less sedation.

There should be a variable-pressure limitation valve and a display of the airway pressure. A disconnection alarm is essential, but often is not included in the ventilator itself. Capnography is desirable.

Other equipment required to maintain and secure the airway is listed in Table 1.

Circulatory support

Many critically ill patients require large volumes of intravenous fluid and pharmacological support of their circulation. Most volumetric pumps for the delivery of fluid or drugs are too bulky and heavy for easy use, although portable devices are available. Fluid boluses may be delivered using a pressurized intravenous system. Drug infusions should be delivered by battery-powered syringe drivers. A minimum of two syringe drivers should be available and can also be used for sedative infusions. A generous amount of crystalloid and colloid solutions must be available. Blood products may be required and should be carried in a suitable container.

Monitoring

Technological improvements now allow sophisticated monitoring to continue during transfers. Continuous ECG, pulse oximetry, and arterial pressure measurement should always be used, although pulse oximetry may be difficult in the vasoconstricted patient. It should be possible to measure the blood pressure both non-invasively and directly (intra-arterial blood pressure).

Multimodal monitors, which are small, portable, and battery powered, are strongly recommended. Many can now offer ECG, pulse oximetry, temperature measurement, non-invasive blood pressure measurement, two pressure channels, and capnography. They minimize the number of items of equipment that must be transported with the patient and allow easier identification of alarms and changes in the patient's condition.

Ideally, the monitor should have multiple pressure channels. Where possible intra-arterial blood pressure should be measured directly. This is more accurate than non-invasive methods, particularly at the extremes of measurement (Runcie *et al.* 1990), and automated non-invasive blood pressure monitoring significantly reduces battery life.

Other channels can be used for central venous, pulmonary artery, or intracranial pressure monitoring. Many critically ill patients benefit from measurement of both continuous central venous pressure and pulmonary artery occlusion pressure measurement in the ICU. Most commonly only two pressure channels will be available during transport, and in these circumstances most clinicians will choose continuous monitoring of intra-arterial blood pressure and pulmonary artery pressure, and intermittent measurement of pulmonary artery occlusion pressure to assess left ventricular filling pressures. When a pulmonary artery catheter is *in situ*, a continuous display of its waveform is recommended to recognize inadvertent pulmonary artery occlusion.

Carbon dioxide (CO_2) monitoring offers considerable advantages in the care of the critically ill and should be used more widely in transfers. Portable battery-powered CO_2 monitors have recently become available (Donnelly *et al.* 1995). They provide a guide to the adequacy of ventilation and an apnea alarm which alerts the transfer team to disconnection. The expired CO_2 concentration should be compared with arterial blood gas analysis prior to departure.

Where battery power may be limited, static inverters can provide a 'mains voltage' during transfer. They are already used in many specialized transfer vehicles to power equipment without the need for internal batteries (e.g. continuous cardiac output monitors). Static inverters are also a useful back-up for battery failure.

Drugs

It is necessary to carry a wide range of drugs in adequate quantities for between-hospital transfers. A list of suggested drugs is shown in Table 2. They should be kept in a suitably robust container (e.g. a foam-lined box) ready for transfer. A checklist of drugs and quantities should be kept with the container, and the medical staff should confirm this before departure. Some drugs will require special storage arrangements; for example, succinylcholine (suxamethonium) and atracurium should be refrigerated, and opioids should be kept in a secure place.

Many drugs in the transfer box will rarely be used, but are still required for specific indications; therefore expiry dates need to be checked regularly. The help of a hospital pharmacist is useful for this and for the restocking of used items.

Resuscitation drugs should be available in the form used in the local hospital (e.g. Min-I-Jets) to aid rapid and accurate treatment of the patient.

Table 2 List of suggested drugs for between-hospital transfers

Sedation/ analgesia	Etomidate
	Midazolam
	Propofol
	Thiopental (thiopentone)
	Opioid of choice
Muscle relaxants	Succinylcholine (suxamethonium)
	Atracurium
	Vecuronium
Cardiovascular	Adenosine
	Epinephrine (adrenaline)
	Amiodarone
	Atropine
	Calcium
	Digoxin
	Dobutamine
	Dopamine
	Nitroglycerin (glyceryl trinitrate)
	Hydralazine
	Isoproterenol (isoprenaline)
	Labetalol
	Lidocaine (lignocaine)
	Methoxamine
	Norepinephrine (noradrenaline)
	Verapamil
Bronchodilators	Aminophylline
	Ipratroprium
	Albuterol (salbutamol)
Anticonvulsants	Diazepam
	Phenytoin
Others	Bicarbonate
	Glucose 50%
	Furosemide (frusemide)
	Heparin
	Hydrocortisone
	Magnesium
	Mannitol
	Metoclopramide
	Naloxone
	Potassium

Conclusions

Careful planning and organization of equipment is required to ensure safe transfers. High-quality equipment for transfers should be readily available in every hospital, although evidence suggests that it is not always used appropriately even when present (Hicks *et al.* 1994). Whatever equipment is available, accompanying staff require training and experience to gain the best advantage from its proper use.

References

Donnelly, J.A., Smith, E.A., Hope, A.T., and Alexander, R.J. (1995). An assessment of portable carbon dioxide monitors during interhospital transfer. *Anaesthesia*, **50**, 703–5.

Hicks, I.R., Hedley, R.M., and Razis, P. (1994). Audit of transfer of head-injured patients to a stand-alone neurosurgical unit. *Injury*, **25**, 545–9.

Hurst, J.M., Davis, K., Branson, R.D., and Johannigman, J.A. (1989). Comparison of blood gases during transport using two methods of ventilatory support. *Journal of Trauma*, **29**, 1637–40.

Runcie, C.J., Reeve, W.G., Reidy, J., and Dougall, J.R. (1990). Blood pressure measurement during transport. *Anaesthesia*, **45**, 659–65.

Runcie, C.J., Reeve, W., Reidy, J., and Wallace, P.G. (1991). Secondary transport of the critically ill adult. *Clinical Intensive Care*, **2**, 217–25.

11.12.5 Personnel required for transport

A. Bristow and M. J. Shapiro

Key messages

- The personnel necessary for transport depends not only on aeromedical and ground transport needs, but also on the type of patient.
- Training consists of developing a formal knowledge base relevant to the types of patients being transported and the mode of transport.
- Quality assurance and continuous quality improvement co-ordinated by the medical director can help to ensure that the most efficient and complete delivery of care is provided.

Introduction

The transport of a critically ill patient by ambulance or aircraft introduces the additional problems of acceleration, altitude, vibration, and thermal insult, which may worsen physiological instability and must be managed within a restricted space with limited lighting and equipment. The quality of the personnel transporting the patient is paramount for successful outcome. The qualifications of the transporting team should be at least equal to those of the personnel who would normally look after the patient within the hospital.

The minimal guidelines for aeromedical personnel in the United Kingdom specify that accompanying physicians should be from the specialty relevant to the patient's disease (Anonymous 1991). More recent recommendations for the transfer of head injury patients state that accompanying physicians should have at least 2 years experience in this specialty (Neuroanaesthesia Society 1996). In some situations a nurse may be replaced by a technician or a paramedic to provide optimal support. In the United States, most aeromedical and almost all ground ambulance personnel are non-physicians (nurses, paramedics, or emergency medical technicians). Unlike much of Europe, the United Kingdom, and Russia, in North America training and resources are such that physicians generally do not attend scene responses and rarely participate in interhospital transports. The replacement of various team members by other physicians, nurses, paramedics, or emergency medical technicians depends on the particular circumstances. In most cases, the placement of two physicians as part of a transport team duplicates resources and is not economically feasible. In addition, in the United States most aeromedical and group transport systems are privately owned and operated rather than government run; however, there are federal and state regulations which govern the operation of these services.

Flight programs are variable with respect to the types of mission that they undertake. The medical director must have an in-depth knowledge of the diagnosis and treatment of the types of patients being transported with protocols available to provide uniformity of care. Successful operation of an aeromedical transport program requires smooth integration of existing prehospital emergency medical service systems with referring hospitals. Thus a knowledge of prehospital care, the operation of emergency medical service systems, and regional referral patterns is required. The medical director should be licensed to practice medicine in the region in which the aeromedical transport service is based, should be able to recognize and stabilize critical patients with a wide range of medical and surgical conditions, and should have a knowledge of the emergency medical service and the extensive hospital referral systems including dispatch, communications, law, and regulations. The medical director should actively participate in the early prehospital management of the acutely ill or injured patient, should have management experience including quality assurance, and should have expertise in teaching medical and paramedical personnel (Thomas and Carrubba 1994; Mayfield and Lindstrom 1996).

Training

Training consists of three elements.

1. Professionals involved in transporting critically ill patients should have a knowledge base of training relevant to the disease states of the individual being transported. The patient may be injured or may have a cardiac history, and so the transport personnel must be trained in many disciplines.
2. The second element is formal training, which includes the principles listed in Table 1. The Royal College of Nursing of the United Kingdom has developed an in-flight nursing course, and more recently the British Trauma Society has recommended a syllabus for training individuals involved in intrahospital transport.
3. The third element includes education in transport medicine, which involves altitude medicine. Familiarity with different transport vehicles is necessary for those involved in aeromedical transport training. Personnel are supervised prior to being independently incorporated into the transport team.

Table 1 Syllabus for training transport personnel

1. Resuscitation of adults and children
2. Management of multiple injuries
3. Mechanical ventilation
4. Physical factors in the transport environment
5. Facilities and equipment for transferring patients
6. The role of patient escorts and other support staff
7. Communication methods
8. Safety procedures and regulations
9. Understanding and utilizing hemodynamic and non-invasive monitors

Educational models consist of a mixture of didactic, clinical, and procedural experiences. Procedural education includes ensuring that a flight team is familiar with indications, with the techniques for every procedure utilized in the aeromedical environment, and with complications. The flight team's knowledge, judgment, and performance can be evaluated by written test, oral examination, and field observation. The process of quality assurance and continuous quality improvement may help to maintain standards and improve patient care.

Initial flight crew training should include flight physiology and discussion of the various concepts of the pressure, volume, and temperature relationships of gases, including Boyle's, Henry's and Charles' laws. The composition, pressure, and temperature profiles of the earth's atmosphere and the physical and physiological effects of noise, vibration, and acceleration must be discussed. The principles of emergency medical services and mass casualty response, as well as search, rescue, and survival techniques, are also part of the curriculum. The crew should be familiar with hazardous materials and what to do when involved with them. Patient care training should include the disorders of neurological disease, respiratory disease, cardiovascular disease, trauma, high-risk obstetric and neonatal disease, and pediatric disease. The transport personnel should be familiar with aviation and aircraft safety and with equipment such as the battery master switch and fuel shut-off valve on helicopters, as well as the medical equipment which includes monitors, oximeters, mobilization equipment, and defibrillators. Infection control procedures should be included as part of the education.

The practice of transport

Transport personnel must be available at short notice. In the United States the independent transport companies are established in large part as for-profit organizations. In the United Kingdom, dedicated transport services are an exception, and it is often difficult to release physicians and nurses from other duties and patients. Transports may be time consuming, and when personnel are removed from in-hospital patient care, their absence must be covered. In the United Kingdom, some pediatric referral centers have developed their own

Table 2 Minimum patient information

Patient information (name, age, gender)
Clinical information (history, examination, investigations)
Indications for transfer
Times
Pick-up and disposal points
Treatment and procedures in transit
Complications
Follow-up
Severity and outcome score (e.g. sickness score in adults, PRISM in children)

retrieval teams. However, this is not a common practice in the United States.

Transport teams must be adequately clothed and equipped. Weather conditions fluctuate and environmental conditions may in part be influenced by the mode of transportation. Coveralls for flying, as well as helmets with headsets for communication and protection, are used. In the United Kingdom, the National Health Service pension and accident provisions are inadequate and usually hospitals must take out specific insurance policies. North American transport systems are usually self-insured or insured through their provider.

Transfer agreements should be established beforehand so that the administrative aspects of transporting patients do not delay the actual transport. The responsible consultant should be notified of each transfer before it takes place so that he or she has the opportunity to discuss the patient's requirements with the transport team. Methods such as flight following and electronic location of ground services have allowed agencies to attempt to cover large geographical areas in the shortest possible response time.

While a patient is in transit, comprehensive medical records should be kept much as if he or she was an in-patient. A minimum data set such as that shown in Table 2 should be recorded.

A hard copy of the transport data should be included in the patient's record at the receiving hospital. Quality improvement and continued quality assurance should be performed in order to ascertain that the best mode and safest method of transport were used. Morbidity and mortality meetings and audit meetings should be part of quality assurance for transport systems.

Conclusions

Considerable planning and training is necessary to develop transport teams and systems. A co-ordinated effort between institutions and health agencies is necessary. Without adequate resources, quality service cannot be provided. Rapid transport performed in a safe environment will fulfill the requirements necessary for transporting critically ill patients to tertiary care facilities.

References

Anonymous (1991). Medical helicopter systems — recommended minimum standards for patient management. *Journal of the Royal Society of Medicine*, **84**, 242–4.

Mayfield, T. and Lindstrom, A. (1996). Medical crew survey. *Air Medicine*, **73**, 21–5.

Neuroanaesthesia Society of Great Britain and Ireland, and the Association of Anaesthetists of Great Britain and Ireland (1996). *Recommendations for the transfer of emergency head injured patients to neurosurgical centres*. Neuroanaesthesia Society of Great Britain and Ireland, and the Association of Anaesthetists of Great Britain and Ireland, London.

Thomas, F. and Carrubba, C. (1994). *Air medical physician handbook*. Air Medical Physician Association, Salt Lake City, UT.

12 Obstetric problems

12.1 Normal physiology

12.1.1 Obstetrics

C. Nelson-Piercy and M. de Swiet

Key messages

- Peripheral vasodilatation occurs early in pregnancy, probably mediated by endothelium-dependent factors. These include nitric oxide synthesis upregulated by estradiol and possibly vasodilatory prostaglandins. To compensate, cardiac output increases by 40 per cent during pregnancy.
- Labor is associated with a further increase in cardiac output. Following delivery there is an immediate rise in cardiac output owing to the relief of inferior vena caval obstruction and contraction of the uterus which empties blood into the systemic circulation.
- Oxygen consumption is increased by about 20 per cent and the maternal metabolic rate by 15 per cent.
- The plasma volume increases progressively throughout normal pregnancy. Most of the 50 per cent increase occurs by 34 weeks and is positively correlated with the birth weight of the baby.
- There is physiological sodium (and water) retention during pregnancy. Eighty per cent of pregnant women develop some edema, particularly toward term, and it is not always a pathological sign.

Introduction

An understanding of the normal adaptive mechanisms in pregnancy is vital to enable accurate diagnosis, and to allow appropriate management of the critically ill obstetric patient. These remarkable changes begin soon after conception and continue with advancing gestation. Certain physical signs (e.g. resting sinus tachycardia, an ejection systolic murmur, or a third heart sound) considered abnormal in the non-pregnant woman may be normal findings in pregnancy. Conversely, what appears normal may be abnormal; some pathological conditions (e.g. pre-eclampsia) arising during pregnancy are, at least in part, a result of failure to adapt adequately to the pregnant state. Therefore a comprehensive understanding of the fundamental changes occurring within the different systems of the body during normal and abnormal pregnancy is required.

Cardiovascular adaptation

Cardiovascular changes during pregnancy are summarized in Table 1.

Peripheral vasodilatation occurs early in pregnancy, probably mediated by endothelium-dependent factors. These include nitric oxide synthesis upregulated by estradiol and possibly vasodilatory prostaglandins (prostaglandin I_2). To compensate for the consequent fall in systemic vascular resistance, cardiac output increases by 40 per cent during pregnancy. This is achieved via an increase in both stroke volume and heart rate, and begins early in gestation (Robson *et al.* 1989), so that by 8 weeks the cardiac output has already increased by 20 per cent. The maximum cardiac output is found at about 20 to 28 weeks. There is a minimal fall at term. An increase in stroke volume is possible because of the early increase in ventricular wall muscle mass and end-diastolic volume (but not end-diastolic pressure) seen in pregnancy. The heart is physiologically dilated and myocardial contractility is increased. Although stroke volume declines towards term, the increase in maternal heart rate (15–20 beats/min) is maintained, thus preserving the increased cardiac output. As well as the absolute increase in cardiac output, there is also significant redistribution with a marked increase in uteroplacental flow. Differences in the effects of pregnancy hormones on endothelium-derived mechanisms and vascular reactivity may underlie this redistribution (Weiner *et al.* 1995). Standardization of cardiac output by correcting for body surface area and using a cardiac index is not justified since the correlation between cardiac output and body surface area is poor in pregnancy (van Oppen *et al.* 1995).

Towards term, maternal position has a profound effect upon the hemodynamic profile of both the mother and the fetus. In the supine position, pressure of the gravid uterus on the inferior vena cava causes a reduction in venous return to the heart and a consequent fall in stroke volume and cardiac output. Turning from the lateral to the supine position may result in a 25 per cent reduction in cardiac output. Therefore pregnant women should be nursed in the left or right lateral position wherever possible. If the mother has to be kept on her back, the pelvis should be rotated so that the uterus drops forward and both cardiac output and uteroplacental blood flow are optimized.

Blood pressure is directly proportional to systemic vascular resistance and cardiac output. In early pregnancy, before the increase in cardiac output can adequately compensate for the fall in systemic

Table 1 Cardiovascular and respiratory adaptation to pregnancy

	Direction of change	Normal range for pregnancy
Cardiac output	↑	40%
Stroke volume	↑	
Heart rate	↑	15–20 beats/min
Blood pressure	↓ (first and second trimester)	
	→ (third trimester)	
CVP	→	
PAWP	→	
SVR and PVR	↓	25%–30%
Serum colloid osmotic pressure	↓	10%–15%
Oxygen consumption	↑	20%
Metabolic rate	↑	15%
Resting minute ventilation	↑	40%–50%
Tidal volume	↑	
Respiratory rate	→	
FRC	↓ (third trimester)	
VC	→	
FEV_1 and PEFR	→	
$Pa\text{O}_2$	→	
$Pa\text{CO}_2$	↓ (4.0 kPa (30 mmHg))	
Arterial pH	↑ (7.44)	

CVP, central venous pressure; PAWP, pulmonary artery wedge pressure; SVR, systemic vascular resistance; PVR, pulmonary vascular resistance; FRC, functional residual capacity; VC, vital capacity; FEV_1, forced expiratory volume at 1 s; PEFR, peak expiratory flow rate.

vascular resistance, blood pressure begins to decrease. Indeed, it is likely that vasodilation is the primary change in the circulation in pregnancy. Blood pressure continues to decrease in the first and second trimesters of normal pregnancy until the nadir in systolic and diastolic blood pressure is reached by about 24 to 28 weeks' gestation. From then on there is a steady rise to prepregnant levels until term. Because phase V (disappearance) of Korotkoff sounds may be audible at zero cuff pressure in pregnancy, it has been traditionally recommended in the United Kingdom that phase IV (muffling) should be taken as the diastolic reading. In fact phase V is more reproducible (Shennan *et al.* 1996) and correlates better with intra-arterial measurements of diastolic blood pressure. Since a blood pressure measured using phase V as the diastolic is lower than one measured using phase IV, hypertension defined using phase V will be more severe and therefore is more closely related to outcome. In the third trimester the blood pressure should be taken with the woman sitting or lying on her side with a 30° tilt. The upper arm (when using a cuff) should be at the same level as the heart.

Although both blood volume and stroke volume increase in pregnancy, pulmonary artery wedge pressure and central venous pressure do not increase significantly. Pulmonary vascular resistance, like systemic vascular resistance, decreases significantly in normal pregnancy. This is probably why women with fixed pulmonary vascular resistance and pulmonary hypertension (e.g. Eisenmenger's syndrome) tolerate pregnancy so poorly. Although there is no increase in the pulmonary artery wedge pressure, serum colloid osmotic pressure is reduced, making pregnant women particularly susceptible to pulmonary edema.

Intrapartum and postpartum hemodynamic changes

Labor is associated with a further increase in cardiac output (15 per cent in the first stage and 50 per cent in the second stage). Following delivery there is an immediate rise in cardiac output owing to the relief of inferior vena caval obstruction and contraction of the uterus which empties blood into the systemic circulation. Venous return and stroke volume are increased further by transfer of fluid from the extravascular space. Women with cardiovascular compromise are most at risk of pulmonary edema during the second stage of labor and the immediate postpartum period. Cardiac output has nearly returned to normal 2 weeks after delivery, although some pathological changes (e.g. hypertension in pre-eclampsia) may take much longer (see below).

Table 2 Hematological adaptation to pregnancy

	Direction of change	Normal range for pregnancy
Plasma volume	↑	50%
Hemoglobin	↓	10.5–14 g/dl
Leukocytes	↑	$(6–16) \times 10^9$/l
Platelets	↓ (some women)	
Clotting factors	↑ (particularly fibrinogen and factors VIII, IX, and X)	
Coagulation tests (activated partial thromboplastin time, thrombin time, prothrombin time)	Unchanged	

Blood pressure usually falls immediately after delivery, although it tends to rise subsequently reaching a peak 3 to 4 days postpartum. A considerable number of previously normotensive women may become transiently hypertensive following delivery. The explanation for this phenomenon may be related to return of normal vascular tone and a period of vasomotor instability while normal non-pregnant vasoregulation is re-established. This is of particular relevance for women who develop hypertension or pre-eclampsia since, although delivery theoretically 'cures' the problem, there may be a period of postpartum deterioration and some manifestations may take several weeks to resolve. Women with hypertension in pregnancy often require antihypertensive drugs postpartum.

Changes in respiratory function

Changes in respiratory function during pregnancy are summarized in Table 1.

Oxygen consumption is increased by about 20 per cent and the maternal metabolic rate by 15 per cent. This increase in demand is met by a 40 to 50 per cent increase in resting minute ventilation, resulting mainly from an increase in tidal volume rather than from an increased respiratory rate. This change in ventilation may be due to the respiratory stimulant effect of progesterone. Later in pregnancy, the elevation of the diaphragm caused by the enlarging uterus leads to a decrease in functional residual capacity, but diaphragm excursion is unaffected and so vital capacity is unchanged. A reduced functional residual capacity means that tidal ventilation occurs at lower lung volumes and may encroach on the closing volume. This may lead to pulmonary shunting in conditions where the closing volume is increased (e.g. obesity and the supine position).

Provided that the patient is placed in the semirecumbent left-lateral tilt position, there is no increase in the physiological pulmonary shunt and $Pa\text{O}_2$ remains greater than 13.3 kPa (100 mmHg). Maternal hyperventilation leads to a reduction in $Pa\text{CO}_2$ to 4.0 kPa (30 mmHg), and there is a compensatory fall in serum bicarbonate to 18 to 22 mmol/l. Therefore a mild respiratory alkalosis, with an arterial pH of 7.44, is normal in pregnancy.

The magnitude of the increase in ventilation occurring in pregnancy is similar to the increase in cardiac output, but the lungs have comparatively more reserve than the heart. Thus patients with respiratory insufficiency cope better with the increased demands of pregnancy than those with cardiac insufficiency.

Up to three-quarters of women experience a subjective feeling of breathlessness at some time during pregnancy, possibly due to an increased awareness of the physiological hyperventilation. This 'dyspnea of pregnancy' is most common in the third trimester and may lead to diagnostic confusion.

Hematological adaptation

Hematological adaptation to pregnancy is summarized in Table 2.

The plasma volume increases progressively throughout normal pregnancy. Most of the 50 per cent increase occurs by 34 weeks and is positively correlated with the birth weight of the baby. Because the expansion in plasma volume is greater than the increase in red cell mass, there is a fall in hemoglobin concentration, hematocrit, and red cell count. Despite this hemodilution, there is usually no change in mean corpuscular volume or mean corpuscular hemoglobin concentration.

The platelet count tends to fall progressively during normal pregnancy, although it usually remains within normal limits. In some women, the count will reach thrombocytopenic levels by term. Burrows and Kelton (1990) found that 7.6 per cent of 6715 pregnant women had thrombocytopenia ($(100–150) \times 10^9$/l) at term, but at least three-quarters of these women had 'pregnancy-associated thrombocytopenia'. This benign condition requires no intervention. Differentiation between idiopathic thrombocytopenic purpura and 'essential thrombocytopenia of pregnancy' is possible in specialist laboratories with antiplatelet antibody determination, but the assay is not readily available. Thrombocytopenia documented in the first half of pregnancy is less likely to be due to the pregnancy itself, and should alert the clinician to a possible diagnosis of idiopathic thrombocytopenic purpura.

Pregnancy causes a two- to threefold increase in the requirement for iron, and a 10- to 20-fold increase in folate requirements. These increased demands are often not met by diet alone, a fact which provides the rationale for routine supplementation with oral hematinics. All women planning a pregnancy are now advised to take 0.4 mg folate daily periconceptually as a prophylactic agent against neural

Table 3 Principal biochemical changes in pregnancy

	Direction of change	Normal range for pregnancy
Renal plasma flow	↑	60%–80%
Glomerular filtration rate	↑	50%
Creatinine clearance	↑	120–160 ml/min
Protein excretion	↑	<300 mg/24 h
Urea	↓	2.0–4.2 mmol/l (12–25 mg/100 ml)
Creatinine	↓	24–68 μmol/l (0.29–0.77 mg/100 ml)
Bicarbonate	↓	18–22 mmol/l
Albumin	↓	25–38 g/l
Alkaline phosphatase (total)	↑	133–418 IU/l (third trimester)
Uric acid	↓	100–270 μmol/l (1.7–4.5 mg/100 ml)

tube defects and other fetal abnormalities. In addition, women who have had a previous fetus with a neural tube defect and women taking anticonvulsants are advised to take 5 mg folate daily periconceptually.

Changes in the coagulation system during pregnancy produce a physiological hypercoagulable state (presumably to stop bleeding following delivery). The concentrations of certain clotting factors, particularly factors VIII, IX, and X, are increased. Fibrinogen levels rise significantly by up to 50 per cent. Fibrinolytic activity is decreased and the concentration of endogenous anticoagulants such as protein S falls. Thus pregnancy alters the balance within the coagulation system in favor of clotting, predisposing the pregnant and postpartum woman to venous thrombosis. This additional risk is present for at least 6 weeks following delivery. The *in vitro* tests of clotting remain normal in the absence of anticoagulants or a coagulopathy.

Renal adaptation

Renal adaptation to pregnancy is summarized in Table 3.

There is dramatic dilatation of the urinary collecting system during pregnancy. This may be the result of ureteral smooth muscle relaxation induced by progesterone or due to compression of the ureters by the enlarging uterus. Calyceal and ureteral dilatation is more pronounced on the right.

Renal plasma flow is increased by 60 to 80 per cent by the second trimester of pregnancy. Maternal position affects renal plasma flow in the same way as cardiac output; higher values are obtained in the left lateral position. There is a progressive fall in renal plasma flow throughout the third trimester until levels 50 per cent greater than prepregnancy values are reached at term. Glomerular filtration rate also increases significantly, and creatinine clearance rises by about 50 per cent. This results in a fall in the serum urea and creatinine. Protein excretion is increased and the upper limit of normal in pregnancy is taken as 300 mg/24 h.

There is physiological sodium (and water) retention during pregnancy. Eighty per cent of pregnant women develop some edema, particularly toward term, and it is not always a pathological sign. The pregnant woman has a decreased ability to excrete a sodium and water load, and this is most marked near term.

Changes in gastrointestinal and liver function

Changes in gastrointestinal motility during pregnancy include decreased lower esophageal pressure, decreased gastric peristalsis, and delayed gastric emptying. Gastrointestinal motility is generally inhibited during pregnancy, with reduced small- and large-bowel transit times. These changes may be partially responsible for the common symptoms of constipation, nausea, and vomiting in early pregnancy.

Despite the cardiovascular changes occurring in pregnancy, hepatic blood flow is unaltered. Liver metabolism is increased in pregnancy. Nearly all proteins synthesized in the liver are affected by pregnancy in a similar fashion to the effects of exogenous estrogens. The total serum protein concentration decreases, largely because of the fall of 20 to 40 per cent in serum albumin concentration. Some of this decrease may be explained by dilution due to the increase in total blood volume. In addition to the increase in fibrinogen levels, there are increases in the concentrations of ceruloplasmin, transferrin, and many of the specific binding proteins such as thyroid-binding globulin and corticosteroid-binding globulin. There is no significant change in bilirubin concentration during normal pregnancy, but the alkaline phosphatase concentration increases dramatically (two- to fourfold). This is largely due to placental production. The concentrations of other liver enzymes are not substantially altered, although the normal ranges for both alanine transaminase and aspartate transaminase are lower in all trimesters compared with ranges outside pregnancy.

Endocrine changes

The placenta is an endocrine organ, secreting both steroid and polypeptide hormones. Some of the hormonal changes occurring during pregnancy have a specific function relating to preservation of the corpus luteum, placentation, protection and growth of the fetus, and parturition. Others are the consequences of physiological adaptive processes in different systems. Although total levels of both thyroxine and cortisol are increased secondary to rises in hepatic synthesis of thyroid-binding globulin and corticosteroid-binding globulin respectively, levels of free thyroxine are not substantially altered by pregnancy. However, levels of free urinary and serum cortisol are raised up to threefold since cortisol production rate is increased. Therefore it is possible that a cortisol level falling within the normal non-pregnant range is indicative of hypoadrenalism in pregnancy. The placenta secretes both corticotrophin-releasing hormone and ACTH.

Serum concentrations of thyroid-stimulating hormone fall in the first trimester as concentrations of human chorionic gonadotrophin, to which it is structurally similar, increase. Hyperemesis gravidarum may be associated with a biochemical thyrotoxicosis with high levels of free thryoxine and suppressed thyroid-stimulating hormone. In the third trimester thyroid-stimulating hormone levels increase, and so the upper limit of the reference range is raised compared with the non-pregnant state. Similarly, the normal ranges for free thyroxine and T_3 are reduced.

The renin–aldosterone axis is profoundly altered by pregnancy, and levels of both angiotensin II and aldosterone are significantly increased. Urinary catecholamines are not affected by pregnancy.

Serum prolactin levels increase. Although basal pituitary growth hormone secretion is unchanged, human placental lactogen has very similar effects. Human placental lactogen, glucagon, and cortisol secretion from the placenta all contribute to the physiological state of relative insulin resistance seen in all pregnant women, particularly in the third trimester. Glucose handling is significantly altered in pregnancy; fasting levels of glucose are decreased, and serum levels following a meal or glucose load are increased compared with the non-pregnant state. Those women at one end of the spectrum for pregnancy will be classified as gestational diabetics depending on where the arbitrary line between normality and abnormality is drawn.

Pharmacological considerations and drug handling in pregnancy

The pharmacokinetics of drugs in pregnancy are influenced by the following physiological changes:

(1) delayed gastric emptying;

(2) increased plasma volume;

(3) increase in total body water;

(4) decrease in plasma protein concentration;

(5) increased drug metabolism by the liver;

(6) increased renal plasma flow.

Therefore some drugs undergo changes in distribution and elimination which influence the dose requirements in pregnancy. Although liver metabolism increases in pregnancy, liver blood flow does not. Drugs with a rate of elimination dependent on liver enzyme activity (e.g. phenytoin and theophylline) will be cleared more rapidly in pregnancy, but those whose elimination depends on liver blood flow (e.g. propranolol) show no change in clearance during pregnancy. Further caution is needed when interpreting plasma concentrations of drugs for therapeutic drug monitoring in pregnancy. There may be an artefactual fall in the measured levels of drugs which are highly protein bound (e.g. phenytoin) because of the fall in albumin concentration and consequent decrease in inactive protein-bound drug.

Consideration must also be given to the potential effects on the fetus of drugs administered to the mother. Lipid-soluble non-ionized drugs of low molecular weight cross the placenta more rapidly than polar drugs. However, most drugs will eventually cross to the fetus, and it is safer to assume passage across the placenta than to assume the converse. However, it must be stressed that the welfare of the mother is the greatest priority and, in the situation of obstetric emergencies and a critically ill pregnant woman, potentially life-saving drugs must not be withheld because of concern over possible effects on the fetus.

References

Burrows, R.F. and Kelton, J.G. (1990). Thrombocytopenia at delivery: a prospective survey of 6715 deliveries. *American Journal of Obstetrics and Gynecology*, **162**, 731–4.

Robson, S.C., Dunlop, W., Boys, R., and Hunter, S. (1989). Serial study of factors influencing changes in cardiac output during human pregnancy. *American Journal of Physiology*, **256**, H1060–5.

Shennan, A.J., Gupta, M., Halligan, A., Taylor, D.J., and de Swiet, M. (1996). Lack of reproducibility in pregnancy of Korotkoff phase IV as measured by mercury sphygmomanometry. *Lancet*, **347**, 139–42.

van Oppen, A.C., van der Tweel, I., Duvekot, J.J., and Bruinse, H.W. (1995). Use of cardiac index in pregnancy: is it justified? *American Journal of Obstetrics and Gynecology*, **173**, 923–8.

Weiner, C.P., Thompson, L.P., and Van Voorhis, B.J. (1995). The role of nitric oxide in female reproduction. *Fetal and Maternal Medicine Review*, **7**, 175–205.

12.2 Pre-eclampsia and eclampsia

12.2.1 Pathophysiology

D. O. C. Anumba and S. C. Robson

Key messages

- Pre-eclampsia, a multisystem disorder characterized by systemic hypertension and proteinuria, is a major cause of maternal death.
- Abnormal placentation is central to the pathogenesis of pre-eclampsia; there is failure of endovascular trophoblast invasion of the myometrial segments of spiral arteries.
- Endothelial dysfunction precedes clinical hypertension and accounts for the pathological changes seen in various organs.
- Uncontrolled lipid peroxidation and deportation of syncytiotrophoblast basement membrane elements from the placenta may mediate endothelial dysfunction.
- Neutrophil and platelet activation occurs, although this is probably secondary to endothelial dysfunction.

Introduction

Pre-eclampsia is a complex syndrome of human pregnancy characterized by systemic hypertension and proteinuria. Hypertension is most widely defined by an absolute systolic blood pressure of at least 140 mmHg and a diastolic blood pressure of at least 90 mmHg in the second half of pregnancy. However, the following definition suggested by Redman and Jefferies (1988) may be superior in excluding women with chronic hypertension: an initial diastolic blood pressure below 90 mmHg in the first half of pregnancy, a maximum diastolic blood pressure of 90 mmHg or more, and an increase in diastolic blood pressure of 25 mmHg or more. Blood pressure elevation should be ascertained on at least two consecutive occasions 4 to 6 h apart. Classically, the disorder is associated with significant proteinuria (greater than 300 mg/24 h). Pre-eclampsia occurs in 5 per cent of all pregnancies and affects predominantly primigravidas during the third trimester. Women at the extremes of reproductive age are more at risk. Pre-eclampsia may be associated with one or more of the following conditions: multifetal pregnancy, hydrops fetalis, chronic hypertension, coexisting renal disease, diabetes mellitus, thalassemia, rhesus incompatibility, family history of pre-eclampsia, or hydatidiform mole.

Twenty-five per cent of maternal deaths worldwide are related to hypertension in pregnancy. In the United Kingdom, pre-eclampsia and eclampsia accounted for 11 per cent of maternal deaths during the triennium 1988 to 1990, with a related perinatal mortality of 133 per 1000. Maternal death from the condition is most commonly from pulmonary complications and cerebral hemorrhage. Fetal demise is usually associated with fetal growth retardation and prematurity.

There is increasing support for the concept that hypertension is not a central pathogenic feature of the disease. Such changes as increased sensitivity to pressors, reduced plasma volume, altered proximal renal tubular function, and activation of the coagulation cascade antedate overt hypertension. Approximately 20 per cent of eclamptic patients, and 20 per cent of patients with the HELLP syndrome (a severe form of pre-eclampsia marked by hemolysis, abnormal liver function tests, and thrombocytopenia), are normotensive. Despite its impact on maternal and fetal health, the pathophysiological mechanisms of pre-eclampsia are not clearly established.

Hemodynamic findings

Normal pregnancy is characterized by a rise in cardiac output and a fall in systemic vascular resistance. Mean blood pressure gradually falls by 10 to 20 per cent, reaching a nadir by 16 to 22 weeks, and then rises to reach levels by term similar to, or even above, prepregnancy values. Pulmonary vascular resistance, central venous pressure, and pulmonary artery wedge pressure are unchanged.

The cardiovascular hemodynamic profile in pre-eclampsia has been reported to vary from a high-output low-resistance state to a low-output high-resistance state. This may reflect disease duration and severity, therapy, and measurement methods. Severe untreated disease appears to be characterized by intravascular volume depletion with low to normal cardiac output, low pulmonary artery wedge pressure, and high systemic vascular resistance. In pre-eclampsia, central venous pressure and pulmonary artery wedge pressure are poorly correlated, possibly reflecting differences in the preload of the right and left sides of the heart (Cotton *et al.* 1988). Low plasma colloid osmotic pressure, increased pulmonary capillary permeability, and iatrogenic fluid overload predispose women with pre-eclampsia to pulmonary edema. Left ventricular dysfunction may contribute in women with pre-existing chronic hypertension.

Pathophysiology

There is evidence that genetic factors, abnormal placentation, endothelial dysfunction, and neutrophil and platelet activation are involved in the pathophysiology of pre-eclampsia (Fig. 1).

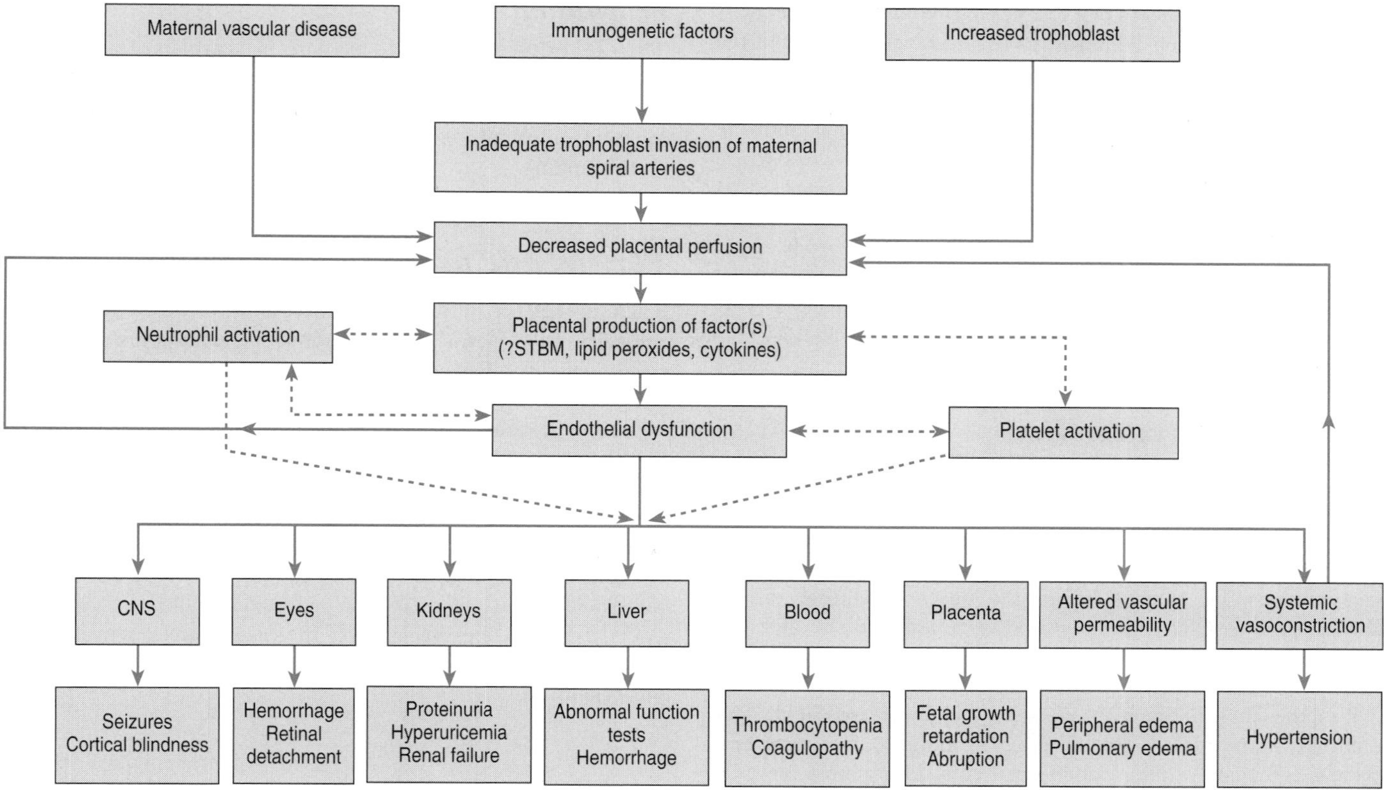

Fig. 1 Proposed model of the pathophysiology of pre-eclampsia: STBM, syncytiotrophoblast basement membrane; CNS, central nervous system.

Alterations in the renin–angiotensin–aldosterone system may also contribute.

Genetics

Numerous studies suggest a genetic susceptibility to pre-eclampsia. The genetic defect may result from dysfunction of the normal maternal immunological adaptation to the presence of a foreign antigenic load (the fetus) during pregnancy. Insufficient maternal production of 'blocking antibodies' or the production of autoantibodies may initiate abnormal placentation. The incidence of pre-eclampsia is five times higher in blood relatives of women affected by eclampsia than in a control population of their non-blood relatives. Two kinds of genetic models have been suggested: a simple recessive model with genes acting in the mother, or a dominant model with 50 per cent penetrance. The recessive model appears to fit best but is questioned by the reported lack of concordance in monozygous twins. Further evidence of a genetic basis in pre-eclampsia is suggested by the following observations:

(1) pre-eclampsia is more common with hydatidiform moles, where chromosomes are paternally derived;

(2) multiparous women with no previous history of pre-eclampsia may develop the condition when pregnant to a new partner;

(3) the incidence of eclampsia is higher in women born of eclamptic pregnancies than in their sisters born of non-eclamptic pregnancies;

(4) there are ethnic differences in the incidence of pre-eclampsia;

(5) pre-eclampsia is associated with some HLA types;

(6) Caucasian and Japanese women with the T235 molecular variant of the estrogen-related angiotensinogen gene have a higher incidence of pre-eclampsia.

Abnormal trophoblast invasion and placental pathology

The placenta is central to the pathogenesis of pre-eclampsia. The disease occurs in the absence of a fetus in molar pregnancy. Its incidence correlates with placental mass, being higher in twin pregnancy, molar pregnancy, and hydrops fetalis. The condition is cured by placental removal. Reduced uteroplacental perfusion has been demonstrated using radio-isotope and Doppler techniques. Ultrastructural studies of biopsy specimens from the placental bed have shown that, during normal placental development, endovascular trophoblast invades the decidual portion of the spiral artery at about 4 to 6 weeks of pregnancy and migrates along the lumen of the vessel wall, reaching the deciduomyometrial junction at 8 weeks. By 15 to 18 weeks, endovascular trophoblast can be found in the deeper parts of the myometrial segments of the spiral arteries. During this wave of migration, trophoblast penetrates the endothelium and replaces the intima and media by a fibrinoid material. The destruction of the tunica media results in distension of the spiral artery.

There is evidence that in pre-eclampsia, endovascular trophoblast migration and invasion is limited to the decidual segments of the spiral arteries (Pijnenborg *et al.* 1991). Absence of physiological invasion of the myometrial segments results in their remaining muscular, undilated, and responsive to vasomotor influences. This failure to convert the placental vasculature into a high-volume low-resistance system at least partially accounts for the decreased uteroplacental blood flow in pre-eclampsia. An additional lesion seen in some cases of pre-eclampsia is 'acute atherosis' in which there is fatty change in intimal cells, necrosis of the vessel wall, and luminal occlusion by aggregates of fibrin, platelets, and lipid-laden macrophages. This lesion may also contribute to reduced uteroplacental blood flow.

Endothelial dysfunction

The endothelium is important in the modulation of vascular tone and prevention of intravascular coagulation. Serum markers of endothelial injury and dysfunction such as fibronectin, von Willebrand factor, thrombomodulin, and type 1 plasminogen activator are increased several weeks or months before the onset of clinical disease (Ballageer *et al.* 1989). The increased disappearance rate of Evans blue dye from the vascular compartment in pre-eclampsia suggests altered endothelial cell permeability.

Endothelial dysfunction accounts for the morphological changes reported in the placental spiral arteries, kidneys, liver, umbilical vessels, and brain of pre-eclamptic patients. This almost certainly explains the multiorgan symptomatology of the condition (Fig. 1).

Endothelial dysfunction in pre-eclampsia is likely to lead to hypertension by altering the balance between vasodilators such as prostacyclin and nitric oxide (NO), and vasoconstrictors such as thromboxane and endothelin.

Endothelin

Endothelin (ET) is a small peptide with potent vasoconstrictor activity. Its major isoform in human plasma, endothelin-1 (ET-1), is produced from an intermediate (big ET-1). Plasma levels of ET-1 are reduced in normotensive pregnancy but increased in pre-eclampsia. The conversion of big ET-1 to ET is reduced in normal pregnancy and increased in pre-eclampsia. Injection of ET into rabbits causes vasospasm and hemolysis, elevated liver enzymes, and low platelet counts. These features are similar to those seen in the HELLP syndrome.

Nitric oxide

NO is produced in vascular endothelium and regulates basal vasodilator tone in humans. Impaired NO production may mediate the increase in systemic vascular resistance in pre-eclampsia. Chronic NO inhibition in pregnant rats produces a syndrome of hypertension, fetal growth retardation, and proteinuria similar to pre-eclampsia. Isolated arteries from women with pre-eclampsia have blunted responses to acetylcholine and bradykinin. The vasodilator response to these substances is known to be partly mediated by endothelial-derived NO. However, serum levels of NO metabolites (cyclic GMP, nitrites, and nitrates) in pre-eclampsia have been conflicting. Direct evidence of NO activity in pre-eclampsia is the subject of much research.

Prostanoids

There is evidence of increased placental biosynthesis of the vasoconstrictor eicosanoid thromboxane A_2 relative to prostacyclin in pre-eclampsia. The use of low-dose aspirin (a prostaglandin synthetase inhibitor) to prevent or alleviate early onset pre-eclampsia is based on this finding.

Platelet and neutrophil activation

There is evidence of increased platelet and neutrophil activation in pre-eclampsia. Plasma concentrations of neutrophil elastase (a specific marker for activation *in vivo*) are elevated, and correlate with those of von Willebrand factor, a marker of endothelial damage. Neutrophil elastase can cause endothelial dysfunction and promote further neutrophil activation and adherence. The mechanism underlying neutrophil activation is unknown. Plasma cytokine levels are increased in pre-eclampsia, and cytokines (particularly tumor necrosis factor-α and some interleukins) are known to trigger both endothelial cell and neutrophil activation. They may also mediate the increase in vascular cell adhesion molecule expression (VCAM-1) seen in pre-eclampsia. Increased platelet activation and adhesiveness may explain some features of the coagulopathy seen in pre-eclampsia.

Link between abnormal placentation and endothelial dysfunction

The link between abnormal trophoblast invasion and endothelial cell dysfunction in the pathogenesis of pre-eclampsia is unclear. It is postulated that factor(s) elaborated by the hypoperfused placenta are disseminated via the circulation to various vascular beds resulting in endothelial dysfunction and neutrophil activation. A number of substances have been proposed, but the current focus is on the possible role of 'oxidative stress' and deported trophoblast elements.

'Oxidative stress'

Lipid peroxidation normally occurs at low levels in all cells and tissues under the regulation of a variety of antioxidant mechanisms. Placental production of lipid peroxide has been localized to the trophoblast cells. Serum lipid peroxidation products are elevated in normal pregnancy and further augmented in women with pre-eclampsia (Hubel *et al.* 1996). Serum antioxidant activity is increased in normal pregnancy and reduced in pre-eclampsia, particularly in late gestation. Increased peroxidative activity may alter the structural and functional integrity of biological membranes and cause endothelial dysfunction.

Deported syncytiotrophoblast basement membrane

In pre-eclampsia, the microvilli of the syncytiotrophoblast are abnormally shaped, with focal areas of syncytiotrophoblast necrosis and loss. Syncytiotrophoblast microvillous membranes are shed into the maternal circulation in increased amounts. *In vitro* studies have demonstrated that syncytiotrophoblast basement membrane interferes with endothelial cell growth and may mediate endothelial dysfunction in pre-eclampsia.

The renin–angiotensin–aldosterone system

Pressor response to angiotensin II and other vasoconstrictors is enhanced in pre-eclampsia. The renin–angiotensin–aldosterone system is activated in normal pregnancy but suppressed in pre-eclampsia. Low plasma renin activity may contribute to the reduced intravascular volume in pre-eclampsia. Reports of serum angiotensin II levels in pre-eclampsia compared with those in normal pregnancy are conflicting. Angiotensin receptors are upregulated in pre-eclampsia, and high circulating angiotensin II levels may mediate the marked increase in blood pressure in at least some women with this condition.

Conclusion

Pre-eclampsia is a major cause of maternal mortality and morbidity. Defective trophoblast development leads to failure of the physiological dilatation of the spiral arteries and relative placental ischemia. Release of an as yet unidentified factor from the ischemic placenta causes endothelial dysfunction in a variety of vascular beds and organs. The vicious cycle of neutrophil and platelet activation, endothelial dysfunction, and vasoconstriction leads to the progressive organ damage seen in pre-eclampsia.

References

Ballegeer, V., Spitz, B., Kieckens, L., Moreau, H., Van Assche, A., and Collen, D. (1989). Predictive value of increased plasma levels of fibronectin in gestational hypertension. *American Journal of Obstetrics and Gynecology*, **161**, 432–6.

Cotton, D.B., Lee, W., Huhta, J.C., and Dorman, K.F. (1988). Hemodynamic profile of severe pregnancy-induced hypertension. *American Journal of Obstetrics and Gynecology*, **158**, 523–9.

Hubel, C.A., McLaughlin, M.K., Evans, R.W., Hauth, B.A., Sims, C.J., and Roberts, J.M. (1996). Fasting serum triglycerides, free fatty acids, and malondialdehyde are increased in pre-eclampsia, are positively correlated, and decrease within 48 hours post partum. *American Journal of Obstetrics and Gynecology*, **174**, 975–82.

Pijnenborg, R., *et al.* (1991) Placental bed spiral arteries in the hypertensive disorders of pregnancy. *British Journal of Obstetrics and Gynaecology*, **98**, 648–55.

Redman, C.W.G. and Jefferies, M. (1988). Revised definition of pre-eclampsia. *Lancet*, **i**, 809–12.

12.2.2 Management of pre-eclampsia

D. O. C. Anumba and S. C. Robson

Key messages

- Women with severe pre-eclampsia should be managed by a multidisciplinary team including experienced anesthetists, obstetricians, and, when appropriate, intensive care specialists.
- Treatment of severe hypertension is vital in order to avoid cerebral sequelae. The aim should be to keep mean arterial pressure below 125 mmHg or diastolic blood pressure below 110 mmHg.
- Optimal fluid management requires selective use of hemodynamic monitoring and plasma volume expansion. The use of diuretics should be limited to women with pulmonary edema.
- The value of routine seizure prophylaxis in all women with severe pre-eclampsia is unproven. If prophylaxis is deemed necessary, magnesium sulfate is the drug of choice.
- Epidural blockade is the optimal method for intrapartum analgesia and operative anesthesia. In order to avoid difficult or failed intubation, and the hypertensive response to intubation and extubation, general anesthesia should be avoided unless there is a coagulopathy.

Introduction

Pre-eclampsia is a multisystem disorder which classically presents with systemic hypertension and proteinuria. Various criteria have been used to define severe pre-eclampsia; those shown in Table 1 select a group of women at greatest risk of maternal and fetal complications. The last four United Kingdom Confidential Enquiries into Maternal Deaths have recommended that women with severe disease should be managed in appropriately staffed units with intensive care facilities. These women should be cared for by multidisciplinary teams, each comprising an experienced obstetrician, an anesthetist, and, when appropriate, an intensive care specialist. One of the aims of such a team is to set guidelines for management in consultation with the regional center. Such guidelines would assist trainee obstetricians and anesthetists to attain a consistent and safe level of care.

Principles of management

The only effective treatment of pre-eclampsia is delivery, and the principle of care is to support the mother and fetus until this has been achieved. Management dilemmas arise with early onset disease where the risk of maternal complications has to be weighed against the risks to the fetus of premature delivery. One small trial has compared pregnancy outcome in women with severe disease at less than 34 weeks gestation randomized to delivery or expectant management (with close maternal and fetal monitoring). There was a trend towards an improvement in neonatal outcome with no apparent increase in maternal complications (Odendaal *et al.* 1990). However, larger randomized trials are necessary to confirm the benefit of conservative management in this group of patients.

Patient selection

Protocols of care for women with pre-eclampsia should stipulate criteria for selecting patients whose severity of disease warrants high-dependency care (Table 1).

Screening for maternal and fetal complications

Blood pressure should be monitored by sphygmomanometry. Renal function should be assessed by measuring hourly urine output, serum creatinine, and potassium. Elevated serum alanine transaminase levels indicate hepatic dysfunction, while sustained ankle

Table 1 Patient criteria for high-dependency care

Hypertension (blood pressure ⩾140/90 mmHg) with proteinuria (⩾0.5 g/day or ⩾2+ on urinalysis) and one of the following
Epigastric pain, or headache, or visual disturbance
Clonus (⩾3 beats)
Platelet count <100 × 10^9/l, alanine transaminase >50 IU/l
Severe sustained hypertension (systolic blood pressure ⩾170 mmHg or diastolic pressure ⩾110 mmHg)
Eclampsia

clonus suggests cerebral involvement. Fundoscopy should be undertaken to exclude retinal hemorrhages, and hemostatic function should be screened by measurement of thrombin time and serum D-dimer levels. Hemoglobin estimation and a peripheral blood film may show evidence of hemolysis. Lung auscultation and pulse oximetry may indicate pulmonary edema. Placental dysfunction and fetal growth retardation should be excluded by cardiotocography and ultrasonography with Doppler assessment of the uteroplacental and fetal circulations.

Seizure prophylaxis

It is not possible to predict reliably which women with pre-eclampsia will develop eclampsia. Therefore the decision to administer seizure prophylaxis in women with pre-eclampsia is a balance between the small risk of eclampsia and the risks of anticonvulsant treatment. One study comparing magnesium and phenytoin in women with pre-eclampsia suggested that fewer women treated with magnesium sulfate developed eclamptic seizures (Lucas *et al.* 1995). However, the value of routine seizure prophylaxis has been questioned by numerous studies which have shown very low seizure rates in women with severe pre-eclampsia managed expectantly. No randomized trial has compared seizure rates in women managed with and without anticonvulsant therapy. Therefore, if seizure prophylaxis is deemed appropriate, magnesium sulfate is probably the drug of choice.

Blood pressure control

Strict blood pressure control is vital to prevent cerebral sequelae. Most units now measure blood pressure using automated oscillometric sphygmomanometers, although some of these devices may underestimate blood pressure. Mean differences of 5 mmHg (systolic) and 15 mmHg (diastolic, K4) have been reported for the Dynamap device, with individual observations showing discrepancies of up to 30 mmHg compared with standard mercury devices. Thus, if automated pressure measuring devices are being used as a guide to therapy, values should be validated against a standard mercury sphygmomanometer.

There is a loss of autoregulatory control within the cerebral circulation above a mean arterial pressure of 150 mmHg. Thus, to allow a safety margin, treatment of blood pressure should be commenced below this level. Blood pressure management should be guided by the following principles: keep mean arterial pressure below 125 mmHg or diastolic blood pressure below 110 mm Hg, avoid rapid precipitous decrements in blood pressure, and confirm blood pressure reduction by standard mercury sphygmomanometry.

Several agents are available for the acute treatment of severe hypertension. No evidence exists that any drug is superior; therefore the choice of drug therapy should depend on clinical experience and availability.

Hydralazine

Hydralazine (5 mg intravenously every 15–20 min) is the most widely used parenteral antihypertensive drug in pregnancy. It is a vasodilator which acts directly on the arterioles. It has the advantage that it increases cardiac output, and thus may improve placental and renal blood flow. Vasodilatation with hydralazine has been associated with fetal distress. This can be prevented by prior infusion of 400 to 500 ml of colloid (Paterson-Brown *et al.* 1994). The pharmacokinetics of hydralazine (maximal effect 15–20 min, duration of action 6–8 h) suggest that the drug is best given by intermittent intravenous bolus injections rather than by continuous infusion. Guidelines for the use of hydralazine are shown in Fig. 1. A single bolus of 5 mg reduces mean arterial pressure by a mean of 12 mmHg (95 per cent confidence limits, 10–14 mmHg). This effect is increased by up to two further boluses, although there is a tendency to tachyphylaxis. A sustained fall in mean arterial pressure to values below 125 mmHg can be achieved in 89 per cent of women by using up to three boluses of hydralazine (Paterson-Brown *et al.* 1994). Side-effects include tachycardia, flushing, nasal congestion, headache, tremors, and nausea. Tachycardia may warrant treatment with a β-adrenergic blocking agent. If repeated bolus hydralazine therapy fails to reduce mean arterial pressure below 125 mmHg, labetalol or nifedipine may be used.

Labetalol

Labetalol (20 mg intravenously every 15–20 min, increasing to 80 mg) is a combined α_1- and α_2-adrenergic blocking agent with some β-blocking activity. It causes arteriolar vasodilatation without significant tachycardia. It may exert beneficial effects on fetal lung maturation, and appears not to have any deleterious effects on uteroplacental blood flow at the above doses. However, higher doses have been associated with increasing fetal compromise, as suggested by a deterioration in umbilical artery Doppler waveform indices.

Nifedipine

Nifedipine (10 mg orally) acts primarily by inhibiting extracellular calcium influx into cells through slow calcium channels, leading to vasodilatation and decreased total peripheral resistance. It can be administered orally or sublingually, although the latter route may lead to precipitous falls in blood pressure with consequent fetal distress.

Fluid management

The hemodynamic derangement in severe pre-eclampsia (reduced plasma volume and low to normal cardiac output) increases the likelihood of oliguria and fetal distress, particularly following vasodilatation. Endothelial damage, low colloid osmotic pressure, and excessive fluid administration increase the risk of pulmonary edema. The principles of fluid management in severe pre-eclampsia are maintenance of strict fluid balance, selective hemodynamic moni-

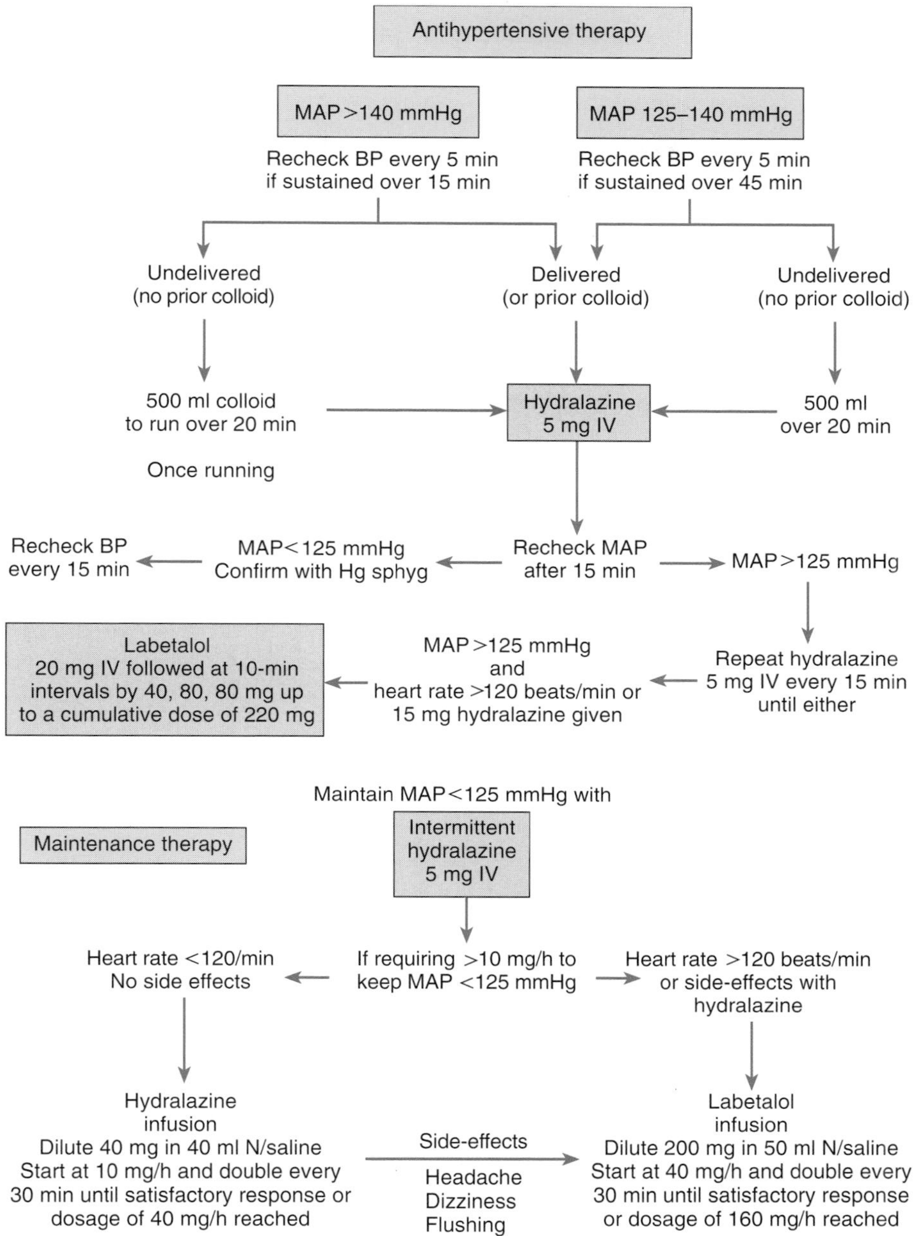

Fig. 1 Acute drug treatment of hypertension: MAP, mean arterial pressure.

toring, and selective plasma colloid expansion. Diuretic therapy should be limited to women with pulmonary edema.

Hemodynamic monitoring

The place of central venous and pulmonary artery wedge pressure monitoring in the management of women with severe pre-eclampsia is controversial. These procedures, particularly pulmonary artery catheterization, are not without risks, especially with concurrent coagulopathy. Therefore they should be instituted only when the benefits are thought to outweigh the hazards. Central venous pressure monitoring should be reserved for women with oliguria (defined as a urine output of less than 100 ml over 4 h) or hemorrhage (blood loss of more than 500 ml) and those having a Cesarean section. All women should receive maintenance crystalloid fluid, with subsequent management determined by urine output (Fig. 2). Oliguria with a low central venous pressure (< 4 mmHg) is an indicator of hypovolemia, and a reasonable indication for colloid (400–500 ml of 5 per cent albumin solution or an appropriate alternative). The role of colloid in women with oliguria, a normal central venous pressure, and normal serum creatinine is unproven, and conservative management appears justified. Women who are hypovolemic secondary to hemorrhage require transfusion with the aim of keeping the hemoglobin level above 10 g/dl. Dopamine (1–5 μg/kg/min) has been shown to increase urine output in euvolemic women.

Pulmonary artery catheterization should be considered in the following circumstances (Clark and Cotton 1988).

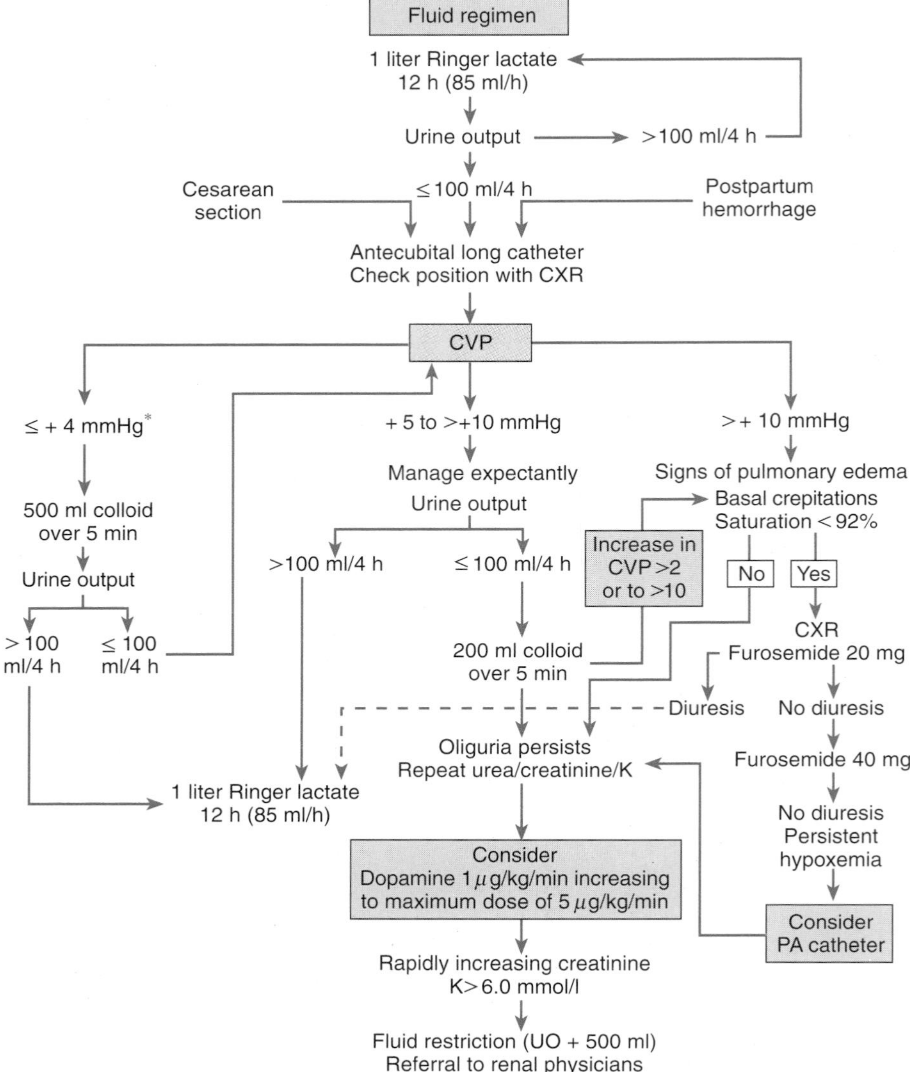

Fig. 2 Fluid regimen for management of severe pre-eclampsia: CVP, central venous pressure; CXR, chest radiography.

Hypertension unresponsive to standard vasodilator therapy

Some hypertensive patients who are unresponsive to standard doses of arterial vasodilators have a very high cardiac output. The use of β-blockers or a venodilator may be a more appropriate method of reducing blood pressure in such patients.

Pulmonary edema

Most cases of pulmonary edema are due to an abnormal colloid osmotic pressure–pulmonary artery wedge pressure (COP–PAWP) gradient. Such cases usually respond to oxygen and diuretic therapy. Although some authorities advocate pulmonary artery catheterization in all pre-eclamptic women with pulmonary edema, its use is probably best limited to patients with persistent hypoxemia. Measurement of cardiac output, COP–PAWP gradient, and left ventricular stroke work index allows clarification of the pathophysiological mechanism. Women with cardiogenic pulmonary edema associated with left ventricular failure may benefit from afterload reduction.

Persistent oliguria

In women with persisting oliguria, measurement of the pulmonary artery wedge pressure allows the most precise assessment of cardiac preload. Women with persisting hypovolemia are likely to benefit from further colloid. However, with normal renal function, expectant management without pulmonary artery catheterization appears to be justified.

Delivery and anesthetic management

Depending on maternal and fetal condition, there is usually justification for an attempt at vaginal delivery. However, this may be unsuccessful in primigravidas less than 34 weeks with an unfavorable cervix.

Continuous lumbar epidural analgesia has been shown to be safe and effective for women with pre-eclampsia provided that they do not have a coagulopathy. In addition to providing optimal pain relief, it is also helpful in the control of hypertension. Graham and Goldstein (1980) showed no fall in cardiac output in severely pre-eclamptic women who had epidural analgesia. Preloading the circulation with 500 to 750 ml of crystalloid minimizes the risk of hypotension and fetal compromise. Because the blood pressure rises during the active second stage of labor, women with severe hypertension should have forceps-assisted delivery or ventouse extraction. Intravenous syntocinon rather than preparations containing ergometrine should be used for the third stage of labor.

Epidural anaesthesia is also preferred for Cesarean section. The risks of failed or difficult intubation are avoided, and no single treatment consistently prevents the hypertensive response to intubation and extubation. Spinal anesthesia should be avoided because of the rapid onset of sympathetic block, potentially resulting in sudden and severe hypotension. Preloading the circulation with 500 to 750 ml of crystalloid and injecting bupivacaine (or a similar agent) in an incremental regimen appears to reduce the incidence of hypotension to about 10 per cent. To reduce this risk even further, ephedrine may be added to the second 500-ml bag of crystalloid which is infused during the establishment of the block. If general anesthesia is required, alfentanil and labetalol should be given 10 min prior to induction to minimize the hypertensive response to intubation.

Intensive care management

Indications for transfer of patients with severe disease to the intensive care unit include multiorgan failure, compromised myocardial function, persistent oliguria, persistent pulmonary edema, recurrent seizures, and resistant hypertension.

Conclusion

Management of severe pre-eclampsia should be co-ordinated by a multidisciplinary team of obstetricians, anesthetists, and, when appropriate, an intensive care specialist. Control of hypertension, fluid management, and safe expeditious delivery, at an optimum time to minimize maternal and fetal mortality and morbidity, are the main goals of therapy. High-dependency care should be reserved for the few critically ill patients who do not respond to standard therapy. There still remains a need for multicenter clinical trials to define the value of available hemodynamic interventions.

References

Clark, S.L. and Cotton, D.B. (1988) Clinical indications for pulmonary artery catheterization in the patient with severe pre-eclampsia. *American Journal of Obstetrics and Gynecology*, **158**, 453–8.

Graham, C. and Goldstein, A. (1980). Epidural analgesia and cardiac output in severe pre-eclamptics. *Anaesthesia*, **35**, 709–12.

Lucas, M.J., Leveno, K.J., and Cunningham, F.G. (1995). A comparison of magnesium sulfate with phenytoin for the prevention of eclampsia. *New England Journal of Medicine*, **333**, 201–5.

Odendaal, H.J., Pattinson, R.C., Bam, R., Grove, D., and Kotze, T.J.V.W. (1990). Aggressive or expectant management for patients with severe preeclampsia between 28–34 weeks' gestation: a randomized controlled trial. *Obstetrics and Gynecology*, **76**, 1070–5.

Paterson-Brown, S., Robson, S.C., Redfern, N., Walkinshaw, S.A., and de Swiet, M. (1994). Hydralazine boluses for the treatment of severe hypertension in pre-eclampsia. *British Journal of Obstetrics and Gynaecology*, **101**, 409–13.

12.2.3 Management of eclampsia

S. C. Robson

Key messages

- Eclamptic convulsions are due to localized cerebral vasoconstriction and hypoxia.
- The maternal case fatality rate from eclampsia in the United Kingdom is 1.8 per cent; 30 per cent will have major morbidity.
- Magnesium sulfate is the optimal drug to treat and prevent seizures.
- The ultimate treatment of eclampsia is delivery, but this should only be undertaken once seizures are controlled, severe hypertension is treated, and hypoxia is corrected.
- Eclamptic patients are at increased risk of developing hypertension and eclampsia in a subsequent pregnancy.

Introduction

Eclampsia is the occurrence of convulsions in association with the syndrome of pre-eclampsia. The incidence of eclampsia in the United Kingdom is 4.9 per 10 000 maternities (Douglas and Redman 1994), and comparable rates have been reported from the United States. There is little evidence in the United Kingdom that the incidence of the disease has changed substantially over the last 30 years (Douglas and Redman 1994).

Diagnosis

Eclamptic convulsions classically occur in the second half of pregnancy and up to 10 days after delivery. Postpartum convulsions occur in 30 to 40 per cent and, within this group, the first convulsion occurs more than 48 h after delivery in 30 per cent of cases (Sibai 1990). Eclamptic convulsions are epileptiform in type; a tonic phase of muscular contraction is followed by clonic movements that may last a minute or longer, after which the woman passes into a brief period of coma (postictal state). Although most convulsions occur without a preceding aura, many women will manifest excitability or hyperreflexia prior to the onset of the seizure.

The diagnosis of eclampsia is straightforward when convulsions occur in a woman admitted with pre-eclampsia. However, 38 per cent do not have established proteinuria and hypertension before the first fit, and less than 60 per cent have antecedent symptoms such as headache, visual disturbance, and epigastric pain (Douglas and Redman 1994). The diagnosis is more difficult if a woman is found unconscious. If she is hypertensive or has proteinuria, the working diagnosis is eclampsia. Observation over 30 min may confirm that she is postictal, but persistent coma or the development of localizing signs will raise the possibility of a cerebral accident. Other pathologies, including encephalitis, meningitis, drug withdrawal, and metabolic disturbances, together with iatrogenic causes (lidocaine (lignocaine) toxicity or water intoxication from inappropriate fluid and oxytocin use) should be considered if there are prolonged convulsions.

Emergency management

Every effort should be made to prevent the woman from injuring herself during the tonic–clonic phase of the convulsion. During the postictal phase it is important to maintain the integrity of the airway, administer oxygen, and avoid supine hypotension.

Most convulsions are self-limiting, but anticonvulsant therapy is indicated to prevent recurrent seizures. Magnesium sulfate is the preferred therapy, and 4 g should be given intravenously over 5 to 10 min (Eclampsia Trial Collaborative Group 1995).

Hypoxia and uterine hypertonus during the convulsion generally lead to fetal bradycardia. Under no circumstances should an emergency Cesarean section be undertaken for fetal distress, as the consequences of operating on an unstable mother may be catastrophic. The fetal heart rate usually recovers quickly once maternal hypoxia and acidemia are corrected.

Seizure prophylaxis

The results of the Collaborative Eclampsia Trial provide compelling evidence that magnesium sulfate is the anticonvulsant of choice in eclampsia. Women allocated magnesium had a 52 per cent lower risk of recurrent convulsions than those allocated diazepam (13.2 per cent versus 27.9 per cent) and a 67 per cent lower risk than those allocated phenytoin (5.7 per cent versus 17.1 per cent) (Eclampsia Trial Collaborative Group 1995). Although there was no difference in maternal mortality between the groups, women treated with phenytoin were more likely to be ventilated, to develop pneumonia, and to be admitted to intensive care facilities (Eclampsia Trial Collaborative Group 1995). Magnesium appears to act primarily by relieving cerebral vasospasm, although the drug has several other beneficial maternal effects including the production of the endothelial vasodilator prostacyclin. Recent evidence has also suggested that exposure to magnesium sulfate *in utero* reduces the risk of cerebral palsy in infants born weighing less than 1500 g.

Maintenance treatment with magnesium sulfate should be started immediately after the initial intravenous bolus. Intramuscular injections are painful and associated with abscess formation in 0.5 per cent (Eclampsia Trial Collaborative Group 1995). Therefore the intravenous route is preferred. Some authorities recommend a maintenance dose of 2 g/h on the basis that this dose consistently produces serum levels within the therapeutic range (2–4 mmol/l) (Sibai *et al.* 1981). However, a dose of 1 g/h was used in the Collaborative Eclampsia Trial and until a definite therapeutic benefit has been demonstrated using the higher dose, it is probably appropriate to use the lower dose. Maintenance anticonvulsant therapy should be continued for at least 24 h after the last convulsion.

The first sign of magnesium toxicity is the loss of deep tendon reflexes, which occurs when serum levels exceed 5 mmol/l. This is followed by respiratory depression and ultimately respiratory arrest. In most cases therapy can be monitored safely by hourly measurement of the patellar reflex and oxygen saturation without the need for serum levels. If deep tendon reflexes are absent, further doses of magnesium sulfate should be withheld until reflexes return. Significant respiratory depression should be treated with 1 g calcium gluconate given intravenously. Magnesium is excreted by the kidney, and regular monitoring of serum levels should be considered if the patient is oliguric (urine output below 100 ml in 4 h). If serum levels are not available, consideration should be given to reducing the maintenance dose to 0.5 g/h.

Recurrent convulsions will occur despite administration of magnesium sulfate in 5 to 15 per cent of cases (Eclampsia Trial Collaborative Group 1995). In a significant proportion of such cases magnesium levels will be subtherapeutic (Sibai *et al.* 1981) and therefore measurement of serum magnesium is advisable. Convulsions should be treated with a further bolus of 2 g of magnesium sulfate. Subsequent management will depend on the serum level and the clinical course. If the level is less than 2 mmol/l, the maintenance infusion should be increased and serum magnesium levels checked every 4 to 6 h. If further seizures occur despite therapeutic levels of magnesium, options include diazepam 10 mg intravenously or thiopental (thiopentone) 50 to 100 mg intravenously. Intubation may become necessary in such women to protect the airway and ensure adequate oxygenation. Further seizures should be managed by intermittent positive-pressure ventilation and muscle relaxation.

Treatment of hypertension

Reduction of severe hypertension (blood pressure of 160/110 mmHg or above, or mean arterial pressure above 125 mmHg) is mandatory to reduce the risk of cerebrovascular accident. Treatment may also reduce the risk of further seizures.

Delivery

The definitive treatment of eclampsia is delivery. Attempts to prolong pregnancy in order to improve fetal maturity are unlikely to be of value and expose the mother to the risk of further convulsions, hypertension, and ongoing organ dysfunction, while the fetus may die *in utero* from placental abruption. However, it is inappropriate to deliver an unstable mother even if there is fetal distress. Once seizures are controlled, severe hypertension is treated, and hypoxia is corrected, delivery can be expedited.

Vaginal delivery is preferable and this can be achieved safely in over 50 per cent of cases (Sibai 1990; Douglas and Redman 1994). Assisted instrumental delivery is advisable to avoid further increases in blood pressure during bearing down. Oxytocin, rather than syntometrine, should be given for the third stage of labor. In the absence of a coagulopathy, epidural analgesia is the preferred method of pain relief. Cesarean section is likely if there is fetal growth retardation and in preterm cases where the cervix is unfavorable. The optimal mode of anesthesia is controversial. General anesthesia is preferred by many anesthetists for reasons of speed and potential or real coagulopathy. However, this exposes the patient to the risk of difficult or failed intubation due to laryngeal edema and reflex hypertension during intubation and extubation. The hypertension response may be reduced by alfentanil (2 mg) and labetalol (20 mg) or magnesium sulfate (40 mg/kg), although no regimen consistently obtunds the pressor response. General anesthesia is preferred if the patient is comatose, has a coagulopathy, or has had recurrent seizures. Ventilation may be electively continued postoperatively to minimize the risks of hypoxemia, inhalation, and further seizures.

Management of associated complications

Major maternal complications occur in over 30 per cent of women with eclampsia (Fig. 1). In most cases these reflect the severity of the underlying pre-eclamptic process.

Renal complications

Acute renal failure occurs in approximately 5 per cent of women with eclampsia (Sibai 1990; Douglas and Redman 1994). Regular

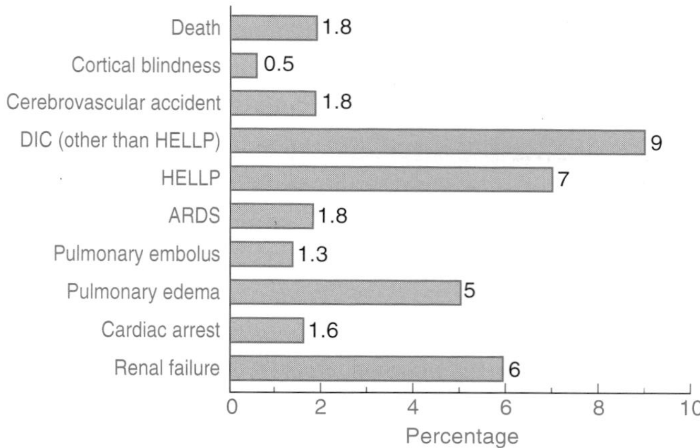

Fig. 1 Major maternal complications in eclampsia: DIC, disseminated intravascular coagulation; ARDS, acute respiratory distress syndrome. (Reproduced with permission from Douglas and Redman (1994).)

monitoring of urea, creatinine, and potassium is necessary, particularly if the patient is oliguric (urine output below 100 ml in 4 h). Optimizing ventricular preload is vital in women with oliguria. In most cases this can be achieved without recourse to pulmonary artery catheterization, but pulmonary artery wedge pressure monitoring may be particularly helpful if there is coexistent pulmonary edema. Low-dose dopamine (1–5 g/kg/min) may improve urine output in euvolemic pre-eclamptic women with oliguria. Once renal failure is established, fluid input should be restricted (urine output in preceding hour plus 30 ml/h). The majority of women have reversible acute tubular necrosis (Sibai *et al.* 1990). Dialysis is required in about 50 per cent of cases, with the average number of dialysis procedures being 1.9 and the average duration of oliguria–anuria being 4.6 days (Sibai 1990; Sibai *et al.* 1990).

Cardiopulmonary complications

Cardiac arrest occurs in 1.6 to 3.1 per cent of cases. During cardiopulmonary resuscitation it is vital that the uterus is displaced to the left. This is most readily achieved by manual displacement or by positioning a wedge under the right hip. If the fetus is viable, Cesarean section is advocated within 5 min of cardiopulmonary arrest to save both infant and mother. Because of the risk of pulmonary edema and acute respiratory distress syndrome, oxygen saturation should be monitored continuously. Pulmonary edema should be treated with oxygen, intravenous vasodilators, and furosemide (frusemide) .

Disseminated intravascular coagulation and hemolysis, elevated liver enzymes, and low-platelet (HELLP) syndrome

Liver transaminases, full blood count, and clotting studies should be monitored regularly. Elevated liver transaminases (> 50 IU/l) rarely require specific therapy. Hepatic hematoma is a rare complication and should be excluded by ultrasound in women with persistent epigastric pain and deranged liver function. Hemolysis may be suggested by a falling hemoglobin concentration and can be confirmed on a peripheral blood smear. Clotting abnormalities will be detected in around 15 per cent of women with eclampsia (Sibai 1990; Douglas and Redman 1994). Isolated thrombocytopenia rarely requires treatment unless the platelet count is less than 50×10^9/l and a Cesarean section is planned. Fresh frozen plasma will be required in uncompensated disseminated intravascular coagulation, particularly to cover delivery. The aim of therapy is to normalize the prothrombin time and partial thromboplastin time and to keep the fibrinogen level above 1 g/l.

Neurological complications

A variety of neurological complications can be associated with eclampsia. Transient deficits, including cortical blindness, occur in up to 3 per cent of cases (Sibai 1990; Douglas and Redman 1994). Cerebrovascular hemorrhage is less common but may lead to prolonged coma or death (Douglas and Redman 1994). Magnetic resonance imaging (MRI) is abnormal in under 50 per cent of cases, with the most common findings being cortical and subcortical areas of high signal intensity on T_2-weighted images in the posterior parietal and occipital lobes, particularly in the vascular watershed areas. These appearances are consistent with hypoxia-induced edema. Cerebral imaging (MRI or CT) is not indicated in patients with uncomplicated eclampsia. However, imaging is necessary to exclude hemorrhage and other serious abnormalities in women with focal neurological deficits or prolonged coma.

Outcome

The maternal case fatality rate from eclampsia in developed countries is 0.4 to 1.8 per cent (Sibai 1990; Douglas and Redman 1994). Perinatal mortality rates are between 5 and 8 per cent if deliveries under 24 weeks gestation are excluded (Sibai 1990; Douglas and Redman 1994). Approximately half the deaths occur in the neonatal period, with the remainder being stillbirths. Maternal recovery is usually complete, although hypertension may persist for several months (Douglas and Redman 1994). Eclamptic patients are at increased risk of developing hypertension and eclampsia in a subsequent pregnancy. They may also be at increased risk of chronic hypertension and diabetes mellitus.

Conclusion

Eclampsia is a rare but life-threatening complication of pregnancy. Few junior medical staff have experience of the condition. Therefore it is vital that all obstetric and intensive care units have clear and readily accessible management guidelines devised by a small team of local 'experts'. This team, which should include obstetricians, obstetric anesthetists, and intensivists, can provide advice in complicated cases.

References

Douglas, K.A. and Redman, C.W.G. (1994). Eclampsia in the United Kingdom. *British Medical Journal*, **309**, 1395–1400.

Eclampsia Trial Collaborative Group (1995). Which anticonvulsant for women with eclampsia? Evidence from the Collaborative Eclampsia Trial. *Lancet*, **345**, 1455–63.

Sibai, B.M. (1990). Eclampsia. VI. Maternal–perinatal outcome in 254 consecutive cases. *American Journal of Obstetrics and Gynecology*, **163**, 1049–55.

Sibai, B.M., Lipshitz, J., Anderson, G.D., and Dilts, P.V. (1981). Reassessment of intravenous $MgSO_4$ therapy in preeclampsia–eclampsia. *Obstetrics and Gynecology*, **57**, 199–202.

Sibai, B.M., Villar, M.A., and Mabie, B.C. (1990). Acute renal failure in hypertensive disorders of pregnancy. *American Journal of Obstetrics and Gynecology*, **162**, 777–83.

12.3 HELLP syndrome

12.3.1 Pathophysiology

P. F. Laterre and M. S. Reynaert

Key messages

- The **h**emolysis, **e**levated **l**iver enzymes, and **l**ow **p**latelets (HELLP) syndrome can be considered an additional criterion of pre-eclampsia.
- The etiology is unknown and is often observed in the absence of pre-eclampsia. An endothelial cell abnormality is thought to lead to fibrin deposition, secondary consumptive thrombocytopenia, and microangiopathic hemolysis.
- The platelet count separates the patients into two classes of severity: class 1 ($< 50 \times 10^9$/l) and class 2 ((50–100) $\times 10^9$/l).
- Clinical presentation is non-specific and includes malaise, right upper quadrant or epigastric pain, nausea, edema, headaches, and hypertension.
- Liver intraparenchymal and subcapsular hematoma are the most feared complications because of the risk of hepatic rupture.
- Prompt recognition by laboratory investigations and exclusion of hepatic complications are mandatory in order to reduce maternal morbidity.

Introduction

The entity of **h**emolysis, **e**levated **l**iver enzymes and **l**ow **p**latelets (HELLP) was first reported in three eclamptic patients by Pritchard in 1954, but its complete recognition and the 'HELLP acronym' were not described until 1982 (Weinstein 1982).

Definition

HELLP was initially described as a unique group of pre-eclamptic/eclamptic patients with the following diagnostic criteria (Weinstein 1982):

(1) elevated values of serum glutamic-oxaloacetic transaminase and/or serum glutamate-pyruvate transaminase;

(2) hemolysis with an abnormal peripheral blood smear containing burr cells and/or schizocytes plus polychromasia, consistent with microangiopathic hemolytic anemia.

Table 1 Diagnostic criteria of the HELLP syndrome

Hemolysis	Abnormal blood smear: burr cells/schizocytes/polychromasia Lactate dehydrogenase (>2 SD) Increased indirect serum bilirubin
Elevated liver enzymes	Raised SGOT and/or SGPT (>2 SD) Lactate dehydrogenase (>2 SD)
Thrombocytopenia	Class 1, $< 50 \times 10^9$/ml Class 2, (50–100) $\times 10^9$/ml

SD, standard deviation; SGOT, serum glutamic-oxaloacetic transaminase; SGPT, serum glutamate-pyruvate transaminase.

(3) thrombocytopenia defined as a platelet count below 100×10^9/ml.

Since this definition, diagnostic criteria have varied (Table 1). Martin *et al.* (1990) proposed a classification based on platelet count which correlated with both severity and time to recovery.

Pathophysiology

The exact cause is unknown. Considered a complication of severe pre-eclampsia, the HELLP syndrome was initially proposed as a sixth criterion of pre-eclampsia (Weinstein 1982), but it has since been observed in the absence of toxemia. An endothelial abnormality is thought to lead to fibrin deposition, tissue hypoperfusion, secondary microangiopathic hemolysis, and consumptive thrombocytopenia (Sibai *et al.* 1986).

The coagulopathy observed in HELLP is not disseminated intravascular coagulation (DIC), as fibrinogen levels and prothrombin, activated thromboplastin, and partial thromboplastin times are often within the normal range (Weinstein 1982). When defined as a plasma fibrinogen below 300 mg/dl, fibrin split products above 40 mg/l, and thrombocytopenia, DIC was observed in up to 38 per cent of HELLP cases (Sibai *et al.* 1986) and correlated with the degree of organ dysfunction and the number of maternal complications (Van Dam *et al.* 1989). The observation of a large number of newborn infants of mothers with HELLP syndrome demonstrating an abnormal blood smear, leukopenia, and thrombocytopenia (Weinstein 1982) is noteworthy. Immunological components may cross the placenta and cause these abnormalities. A circulating plasma immunological substance could explain the benefits of plasma exchange described in some patients with severe HELLP syndrome.

Table 2 Clinical symptoms of the HELLP syndrome

Symptoms	Percentage
General malaise	90–100
Right upper quadrant or epigastric pain	65–86
Nausea or vomiting	36–84
Edema	19–67
Headaches	31
Hypertension	10–27
Ascites	10–65
Visual disturbances	10
Bleeding	9
Jaundice	5

Organ dysfunction

Thrombocytopenia

Thrombocytopenia is secondary to increased consumption which is likely to occur by adherence to collagen exposed on damaged endothelium. Martin *et al.* (1990) suggested three categories based on the platelet count: class 1 when the platelet count is below 50×10^9/ml, class 2 when it is between 50 and 100×10^9/ml, and class 3 when it is between 100 and 150×10^9/ml. Class 1 patients had a higher mean value of lactate dehydrogenase, serum glutamic-oxaloacetic transaminase, serum glutamate-pyruvate transaminase, and indirect bilirubin than class 2 patients and a longer interval between delivery and recovery time. The degree of hypertension was not different between the groups. This classification may aid identification of more severe patients and facilitate earlier aggressive therapy.

Hemolysis

Microangiopathic hemolytic anemia is always present and an abnormal blood smear is a major diagnostic criterion. Red blood cell fragmentation is secondary to passage through blood vessels with endothelial damage and fibrin deposition. Because of the common feature of microangiopathic hemolytic anemia, some authors consider HELLP to be a form of thrombotic thrombocytopenic purpura or hemolytic uremic syndrome.

Liver

The classic hepatic lesions associated with HELLP include periportal hemorrhage and focal parenchymal necrosis, with hyaline deposits of fibrin and fibrin microthrombi in the sinusoids. These histological findings explain the right upper quadrant or epigastric pain and the elevation in liver enzymes. These enzyme abnormalities do not reflect the severity of histopathological change seen and could justify the need for rapid and aggressive delivery (Barton *et al.* 1992).

Blood flow obstruction in the sinusoids is the likely cause of liver distension, subcapsular hemorrhage, and hepatic rupture, the most feared complication of HELLP (Weinstein 1982). Increased hepatic artery resistance to blood flow has been demonstrated in pre-eclampsia with or without HELLP and is not implicated in the development of the syndrome. Some liver biopsy specimens have shown fat within hepatocytes; an overlap with acute fatty liver has been described, but clinical and histological findings can distinguish this syndrome from HELLP (Barton *et al.* 1992).

Clinical presentation

As with pre-eclampsia and acute fatty liver of pregnancy, the HELLP syndrome may present with non-specific symptoms such as malaise, nausea, vomiting, and epigastric pain (Weinstein 1982). Right upper quadrant pain is observed more frequently than in toxemia. Hypertension and edema may be present, particularly when HELLP accompanies pre-eclampsia, but this is not a constant feature and is often milder (Weinstein 1982; Sibai *et al.* 1986). Headaches and visual disturbances can be present. Over 10 per cent have ascites, and this correlates with increased liver dysfunction. Because HELLP may occur in the absence of the usual clinical findings of pre-eclampsia, a non-obstetric diagnosis such as hepatitis, gastroenteritis, or gall-bladder disease is often made.

The incidence of HELLP in patients with pre-eclampsia varies from 4 to 12 per cent according to the criteria used and the characteristics of the reporting center (Sibai *et al.* 1986). The clinical symptoms and signs encountered are listed in Table 2. Typically, patients with HELLP are slightly older than pre-eclamptic women, have a mean gestational age of 32 weeks (Weinstein 1982), and are more often primigravid and white (Weinstein 1982; Martin 1990). HELLP develops antepartum in 70 per cent and within the first postpartum week in the remaining 30 per cent. It is unlikely to recur with subsequent pregnancies.

Laboratory investigations should include a blood smear, lactate dehydrogenase and indirect bilirubin (to confirm microangiopathic hemolysis), liver enzymes, and a platelet count. Markers of DIC (e.g. fibrin split products, serum fibrinogen, and antithrombin III) may complete the picture of severity. Liver ultrasound with or without

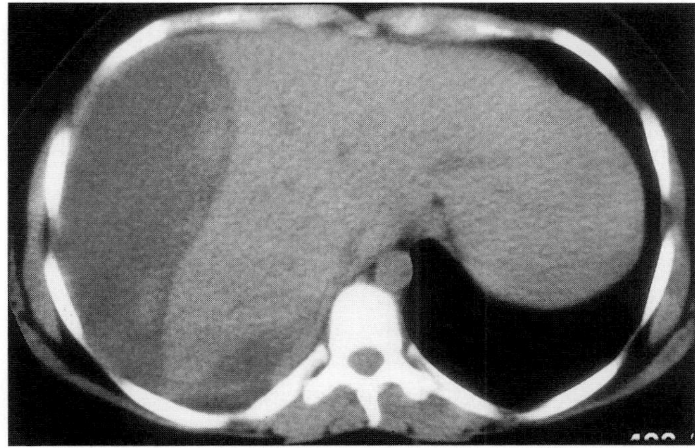

Fig. 1 Large liver subcapsular hematoma in a patient with the HELLP syndrome at 29 weeks of pregnancy.

Fig. 2 Liver intraparenchymal hematoma in a patient with eclampsia and HELLP.

CT should be performed to exclude subcapsular hematoma and to indicate the need for prompt and aggressive treatment (Figs 1 and 2).

References

Barton, J.R., Riely, C.A., Adamec, T.A., Shanklin, D.R., Khoury, A.D., and Sibai, B.M. (1992). Hepatic histopathologic condition does not correlate with laboratory abnormalities in HELLP syndrome (hemolysis, elevated liver enzymes, and low platelet count). *American Journal of Obstetrics and Gynecology*, **167**, 1538–43.

Martin, J.N., Blake, P.G., Lowry, S.L., Perry, K.G., Files, J.C., and Morrison, J.C. (1990). Pregnancy complicated by preeclampsia–eclampsia with the syndrome of hemolysis, elevated liver enzymes, and low platelet count: how rapid is postpartum recovery? *Obstetrics and Gynecology*, **76**, 737–41.

Sibai, B.M., Taslimi, M.M., El-Nazer, A., Amon, E., Mabie, B.C., and Ryan, G.M. (1986). Maternal–perinatal outcome associated with the syndrome of hemolysis, elevated liver enzymes, and low platelets in severe preeclampsia–eclampsia. *American Journal of Obstetrics and Gynecology*, **155**, 501–9.

Van Dam, P.A., Renier, M., Baekelandt, M., Buytaert, P., and Uyttenbroeck, F. (1989). Disseminated intravascular coagulation and the syndrome of hemolysis, elevated liver enzymes, and low platelets in severe preeclampsia. *Obstretrics and Gynecology*, **73**, 97–102.

Weinstein, L. (1982). Syndrome of hemolysis, elevated liver enzymes, and low platelet count: a severe consequence of hypertension in pregnancy. *American Journal of Obstetrics and Gynecology*, **142**, 159–67.

12.3.2 Therapeutic strategy

P.F. Laterre and T. Dugernier

Key messages

- High maternal and fetal morbidity and mortality characterize the syndrome.
- Termination of pregnancy and expeditious delivery are standard treatments.
- Conservative treatment to allow fetal maturity has been evaluated and appears promising in well-monitored patients fulfilling certain criteria. This hemodynamically guided therapy can control and sometimes even reverse the syndrome.
- Antepartum and postpartum steroid therapy is a possible adjunctive treatment.
- Plasma exchange with fresh frozen plasma has been proposed for persisting or worsening HELLP syndrome. However, delivery remains the most effective therapy and is an absolute indication in the presence of eclampsia, abruptio placentae, disseminated intravascular coagulation, fetal distress, or liver hematoma. Bleeding disorders, subcapsular liver hematoma, and hepatic rupture remain the most feared complications and must be promptly excluded.

Introduction

Termination of pregnancy by expeditious delivery was considered the therapy of choice in pre-eclamptic patients with the HELLP syndrome, an attitude justified by the high maternal and fetal morbidity and mortality (Sibai *et al.* 1993). Conservative treatment, including bed rest, magnesium sulfate, and antihypertensive medication, may be considered occasionally to allow prolongation of pregnancy when the syndrome occurs early. However, this may be of little benefit for the fetus since growth retardation is often present and will not be greatly modified by a few additional days of gestation.

Maternal and fetal outcome

The most feared, but fortunately rare, complications of HELLP are liver subcapsular hematoma and rupture with shock. Other maternal complications include aggravation of thrombocytopenia with coagulation disorders or disseminated intravascular coagulation (DIC). Depending on definition, DIC occurs in up to 21 per cent of cases, but seems mainly related to the presence of abruptio placentae which represents the second most common complication (Table 1). Acute renal failure, congestive heart failure, and pulmonary edema are frequent and occur more often with postpartum onset. Severe ascites occurs in up to 10 per cent and correlates with the development of pleural effusions and pulmonary edema (Sibai *et al.* 1993). Intra-abdominal hematoma or uterine bleeding are observed in the postpartum period following Cesarean section (Visser and Wallenburg 1995). Neurological complications can be expected when HELLP is associated with eclampsia. Maternal mortality is approximately 1 per cent in a tertiary care center but as high as 24 per cent in other centers (Sibai *et al.* 1993).

Conservative treatment

Perinatal morbidity and mortality are high (14–20 per cent) and are related to gestational age (Sibai *et al.* 1993; Visser and Wallenburg 1995). HELLP often occurs before week 32 of pregnancy (Sibai 1993), and so prolongation of gestation in such patients would be expected to reduce fetal morbidity and mortality. Conservative hemodynamic treatment was evaluated in patients with severe pre-eclampsia with ($n = 128$) and without ($n = 128$) the HELLP syndrome (Visser and Wallenburg 1995). Using pulmonary artery catheter monitoring, vasodilators were given in combination with plasma expanders to maintain the cardiac index between 3.5 and 4.6 l/min/m^2, the pulmonary artery wedge pressure between 10 and 12 mmHg, and the diastolic arterial blood pressure below 100 mmHg. After

Table 1 Complications of the HELLP syndrome

Complications	Percentage
Disseminated intravascular coagulation	2–21
Abruptio placentae	5–16
Ascites	8
Renal failure	8
Pleural effusions	6
Eclampsia	7–10
Liver hematoma	1
Acute respiratory distress syndrome	1

stabilization for 1 to 3 days, the pulmonary artery catheter was removed and the therapeutic regimen continued until delivery. Termination of pregnancy within 48 h of initiation of therapy occurred in 17 per cent because of fetal distress. In the remaining cases, pregnancy could be prolonged by a mean of 15 days. Maternal mortality did not occur and postpartum morbidity was low. Similar results were observed in pre-eclamptic patients without HELLP. Interestingly, complete antepartum resolution of the HELLP syndrome was observed in 43 per cent; the platelet count rose above 100 000/mm^3 before delivery in 61 per cent of cases, and liver rupture and renal failure did not occur. Thus conservative hemodynamically guided treatment may be used in selected patients in whom HELLP appears early in pregnancy and is associated with major maternal or fetal risks. The severity of the syndrome should be moderate (class 2), an initial stay in intensive care is mandatory, and the lack of any clinical or laboratory improvement should prompt expeditious delivery.

Treatment of HELLP associated with pre-eclampsia consists of hypertension control with hydralazine given in increasing dosage in combination with magnesium sulfate to reach a serum level of 2 to 3 mmol/l (Table 2). Nifedipine or labetalol are alternative treatments for hypertension. Sodium nitroprusside may be given for severe crisis and just before delivery, but should only be given for short periods (30 min) because of the risk of fetal cyanide toxicity. Intravascular volume should be closely monitored to avoid congestive heart failure or renal failure secondary to tubular necrosis (Sibai *et al.* 1993). Platelet transfusions are not indicated because of rapid consumption; these should be limited to the delivery period and for any postpartum complication.

Steroids

Antepartum steroids have been evaluated, although only one study has been prospective and randomized. Dexamethasone 10 mg twice daily was given to 12 patients with HELLP syndrome occurring between 24 and 37 weeks gestation (Magann *et al.* 1994*a*). The platelet count increased significantly in the steroid group compared with control patients. Concurrently, lactate dehydrogenase and serum glutamic oxaloacetic transaminase decreased and urinary output increased in the treated group. Time from study entry to delivery was prolonged in the steroid group; however, as this only averaged 1 day longer, the clinical relevance is doubtful. In addition, a postpartum deterioration in HELLP severity was observed, suggesting that disease stabilization by steroids is only a temporary phenomenon.

Respiratory distress syndrome, intraventricular hemorrhage, necrotizing enterocolitis, and death were less frequent in neonates whose mothers had been treated with steroids, and the postpartum infection rate was not increased. Thus antepartum steroids can be considered for selected patients with HELLP occurring early in pregnancy and of slow evolution. As fetal mortality is primarily related to gestational age, significant prolongation of pregnancy might improve neonatal survival. Accelerated fetal lung maturity appears to be the major benefit accruing from steroids given for antepartum HELLP (Magann *et al.* 1994*a*).

Postpartum steroids have also been proposed to accelerate recovery (Magann *et al.* 1994*b*). Dexamethasone given for 36 h was evaluated in a prospective randomized study; platelet count increased significantly by 24 h, and lactate dehydrogenase and serum glutamic oxaloacetic transaminase decreased significantly by 36 h. Thus shorter intensive care stays and reduction in health care costs could result from the use of postpartum corticosteroids.

Delivery

Termination of pregnancy and timing of delivery remain a matter of debate (Visser and Wallenburg 1995). Fear of maternal complications has driven the common rule of expeditious delivery (Sibai *et al.* 1993). Concurrent eclampsia, fetal distress, abruptio placentae, DIC, subcapsular liver hematoma, or progression to the class 1 syndrome (platelet count below 50 000/mm^3) should remain absolute indications for delivery (Table 2). Vaginal delivery, although preferred, is less common than Cesarean section because of fetal distress and an immature cervix. Cesarean section was performed in 42 to 79 per cent of cases reported by Sibai *et al.* (1993) and Visser and Wallenburg (1995). Incision should be made in the midline to enable better evacuation of intra-abdominal or liver subcapsular hematomas.

Platelet transfusions and fresh frozen plasma may be given just before incision to correct coagulation disorders and severe thrombocytopenia ($<$ 30 000/mm^3). Blood products and platelet transfusions can be continued during delivery and in the postpartum period because of diffuse oozing or intra-abdominal hematoma. Adequate control of hypertension should complete the treatment.

Plasma exchange

The HELLP syndrome shares certain similarities with thrombotic thrombocytopenic purpura and the hemolytic–uremic syndrome. HELLP syndrome persisting or worsening for more than 72 h after delivery has been treated by plasma exchange with fresh frozen plasma (Martin *et al.* 1995). One or two sequential plasma exchanges were able to arrest or reverse the syndrome in a limited number of patients, with a rapid increase in platelet count and reduction in transaminases. When HELLP was associated with other organ dysfunction, plasma exchange was of little benefit. Therefore plasma exchange has not been considered a routine therapy for HELLP because of its potential morbidity and limited benefit.

Table 2 Antepartum strategy for HELLP syndrome

	Hemodynamic treatment	Magnesium sulfate	Steroids (optional)	Delivery if
Class 2 Platelets 50 000–100 000/mm³	Fluid balance monitoring Hypertension control (SBP < 160 mmHg, DBP < 100 mmHg) Hydralazine Alternatives: nifedipine or labetalol (sodium nitroprusside only for crises and just before delivery) ICU stay if severe hypertension Consider invasive hemodynamic monitoring	Initial bolus of 4 g over 15 min followed by continuous IV infusion of 1–2 g/h (serum therapeutic range 2–3 mmol/l)	Consider fetal maturation	Subcapsular liver hematoma Abruptio placentae DIC Fetal distress Eclampsia Progression to class 1 Platelets and fresh frozen plasma transfusions if severe coagulation disorders or DIC and just before delivery
Class 1 Platelets < 50 000/mm³	Identical with class 2 plus delivery			

SBP, systolic blood pressure; DBP, diastolic blood pressure; IV, intravenous.

Table 3 Postpartum strategy for the HELLP syndrome

1. Preventive treatment of renal failure and congestive heart failure	2. Platelets and blood product transfusions	3. Prevent postpartum infection by systemic antimicrobials	4. Consider laparotomy eventually combined with arteriography
Fluid balance monitoring Adjust intravascular effective volume (consider central venous monitoring) Hypertension control Magnesium sulfate if antepartum pre-eclampsia/eclampsia	If Severe coagulation disorders DIC Diffuse oozing during laparotomy Intra-abdominal hematoma	If risk of Wound infection Endometritis Intra-abdominal infection	If Persistent intra-abdominal bleeding or hematoma Large liver subcapsular hematoma

Postpartum HELLP

Postpartum HELLP syndrome occurs in up to 30 per cent of patients; in the vast majority it appears within the first 48 h postpartum, although it can develop after a week (Sibai *et al.* 1993). Since delivery has been effected, treatment of maternal complications is the major priority (Table 3). Pulmonary edema and congestive heart failure are more frequent in postpartum-onset HELLP (Sibai *et al.* 1993), indicating the need for close monitoring of fluid balance after delivery and careful hypertension control. Steroids may facilitate blood pressure control and increase urinary output in the postpartum period (Magann *et al.* 1994*b*). Plasma exchange with fresh frozen plasma to accelerate the postpartum recovery has to be evaluated prospectively, and definitive indications should be determined. Platelets and blood product transfusions should be continued if severe coagulation disorders and bleeding persist. Postdelivery infection occurs in up to 31 per cent of patients (Weinstein 1982). Relaparotomy for bleeding is mandatory in about 5 per cent of those undergoing a Cesarean section (Sibai *et al.* 1993). Delayed closure of the section wound has been proposed when diffuse oozing is present; this can be achieved by 96 h postdelivery (Weinstein 1982).

Hepatic hemorrhage

Although occurring in only 2 per cent of patients, intrahepatic hemorrhage can be catastrophic and fatal (Sibai *et al.* 1993). Liver histopathology does not correlate with either biological data or liver enzymes. Patients with HELLP syndrome, in particular those with epigastric and right upper quadrant pain, should be evaluated by ultrasound to exclude an intrahepatic hematoma. CT may complete the diagnostic picture and help to differentiate an intact hematoma from a rupture of Glisson's capsule with secondary hemoperitoneum. A liver hematoma should be treated conservatively but does indicate expeditious delivery. Vaginal delivery is not the preferred method because of the fear of hepatic rupture. Cesarean section should be performed by a midline incision to allow good intra-abdominal hemostasis and eventual evacuation of the subcapsular liver hematoma if it is excessively large with a major risk of rupture. Evacuation can lead to bleeding and oozing. Supportive therapy includes platelet and blood product transfusions in combination with good hypertension control. If bleeding control cannot be achieved by vessel suture and ligation, packing of the liver is indicated; packs can be removed 24 to 48 h later.

Laparotomy is indicated for an acute abdomen with shock secondary to hepatic rupture, and liver rupture should be treated as for traumatic rupture. When shock can easily be controlled by massive transfusion, arteriography with a view to embolization should precede surgery.

Conclusion

Treatment of the HELLP syndrome remains a clinical challenge. Maternal morbidity has often indicated expeditious delivery, but fetal prematurity and mortality force clinicians to seek means of prolonging pregnancy. Conservative treatment has to be properly evaluated. Rapid recognition of the severity of the syndrome, a search for hepatic hemorrhage, and prompt therapy are clinical priorities.

References

Magann, E.F., Bass, D., Chauhan, S.P., Sullivan, D.L., Martin, R.W., and Martin, J.N. (1994*a*). Antepartum corticosteroids: disease stabilization in patients with the syndrome of hemolysis, elevated liver enzymes, and low platelets (HELLP). *American Journal of Obstetrics and Gynecology*, **171**, 1148–53.

Magann, E.F., Perry, K.G., Meydrech, E.F., Harris, R.L., Chauhan, S.P., and Martin, J.N. (1994*b*). Postpartum corticosteroids: accelerated recovery from the syndrome of hemolysis, elevated liver enzymes, and low platelets (HELLP). *American Journal of Obstetrics and Gynecology*, **171**, 1154–8.

Martin, J.N., Files, J.C., Blake, P.G., Perry, K.G., Morrison, J.C., and Norman, P.H. (1995). Postpartum plasma exchange for atypical preeclampsia–eclampsia as HELLP (hemolysis, elevated liver enzymes, and low platelets) syndrome. *American Journal of Obstetrics and Gynecology*, **172**, 1107–27.

Sibai, B.M., Ramadan, M.K., Usta, I., Salama, M., Mercer, B.M., and Friedman, S.A. (1993). Maternal morbidity and mortality in 442 pregnancies with hemolysis, elevated liver enzymes, and low platelets (HELLP syndrome). *American Journal of Obstetrics and Gynecology*, **169**, 1000–6.

Visser, W. and Wallenburg, H.C.S. (1995). Temporising management of severe preeclampsia with and without the HELLP syndrome. *British Journal of Obstetrics and Gynecology*, **102**, 111–17.

12.4 Postpartum hemorrhage

12.4.1 Pathophysiology and therapeutic strategy

H. El-Refaey and J. Kingdom

Key messages

- Uterine atony is the most common cause of postpartum hemorrhage.
- A patient with postpartum hemorrhage should initially receive uterotonic agents (Syntocinon, Syntometrine, ergometrine, and/or prostaglandins).
- Those who do not respond should be examined in theater under anesthesia to exclude other causes.

Introduction

The third stage of labor is potentially the most hazardous for the mother because of the risk of postpartum hemorrhage. Defined by the World Health Organization as postpartum blood loss in excess of 500 ml, it is a clinical diagnosis that encompasses excessive blood loss after delivery of the baby from a variety of sites (uterus, cervix, vagina, and perineum). Blood loss during the first 24 h after delivery is known as primary postpartum hemorrhage, while blood loss from 24 h up to 6 weeks after delivery is termed late or secondary postpartum hemorrhage.

This definition implies that any loss smaller than this is within normal limits and therefore can be tolerated without risk. This is certainly not the case in countries where severe anemia is common and where blood loss of as little as 250 ml may constitute a clinical problem. The incidence of postpartum hemorrhage ranges between 5 and 8 per cent where active management policies of labor are practiced, but may be as high as 18 per cent when a 'physiological' approach is the norm. The mainstay strategies for reducing postpartum bleeding include the use of oxytocic drugs, such as oxytocin and Syntometrine, or early clamping of the cord and delivery of the placenta by cord traction. These are collectively termed active management of the third stage.

Postpartum hemorrhage is one of the leading causes of maternal mortality, particularly in less developed countries. It is implicated in one-sixth of maternal deaths worldwide and is the most common reason for blood transfusion after delivery. Within the United Kingdom there are approximately five deaths per year from hemorrhage, representing 6.4 deaths per million maternities. In those parts of the world where blood replacement is not possible due to lack of resources, persistent severe hypotension postpartum leads to considerable morbidity including acute renal failure, partial or total necrosis of the anterior pituitary gland, and other organ system injury such as pancreatitis and acute respiratory distress syndrome.

Etiology of primary postpartum hemorrhage

Physiological control of postpartum bleeding occurs by contraction and retraction of the interlacing myometrial fibers surrounding maternal spiral arteries of the placental bed. Myometrial contraction compresses the spiral arteries and veins, thereby obliterating their lumens; hemostasis following placental separation is initially a mechanical process that is not primarily dependent upon an intact coagulation system. Primary postpartum hemorrhage due to uterine atony occurs when the relaxed myometrium fails to constrict these blood vessels, thereby allowing hemorrhage. Since up to a fifth of maternal cardiac output, or 1 l/min, enters the uteroplacental circulation at term, primary postpartum hemorrhage is capable of exsanguinating the mother within 5 min.

The placental bed

Uterine atony is responsible for the majority of primary postpartum hemorrhages originating from the placental bed. The main predisposing factors are prolonged labor followed by operative vaginal or abdominal delivery, uterine overdistension from polyhydramnios or multifetal pregnancy, and delivery following antepartum hemorrhage from either placenta praevia or abruption. Failure of the uterus to contract may also be due to retained placental fragments, either as disrupted portions or, more rarely, a succenturiate lobe. Fibroids may promote primary postpartum hemorrhage in a similar fashion. These factors should be considered if the uterus relaxes following normal uterotonic measures.

Disseminated intravascular coagulation (DIC) can aggravate any of the above etiological mechanisms. It is usually triggered by the degree of blood loss, coexisting abruption, and/or pre-eclampsia–eclampsia, and rarely by amniotic fluid embolism. When postpartum hemorrhage occurs following an antepartum hemorrhage, the scenario is particularly difficult since there have been two episodes of blood loss. The rare but serious complication of abruption is due to extravasation of blood into the myometrium, known as a Couvelaire uterus, which impairs the physiological uterine contraction–retraction hemostatic process. Persistent hemorrhage and coagulopathy in these circumstances can be fatal unless there is appropriate and rapid hematological support. DIC is a medical problem requiring rapid access to red cell concentrate, fresh frozen plasma, and platelets. With correct replacement and avoidance of hypotension, hypothermia, and volume overload, the situation should resolve.

Unless there is a traumatic injury to the genital tract necessitating a surgical approach to arrest hemorrhage, attempts to arrest bleeding via a laparotomy may do more harm than good since hysterectomy and/or ligation of the internal iliac arteries incur further hemostatic insults to the patient.

Extraplacental causes of postpartum hemorrhage

The surgical causes of primary postpartum hemorrhage include 'obstetric' lacerations of the cervix, vagina, paravaginal spaces, and perineum, and episiotomy. As a group these comprise about 20 per cent of all primary postpartum hemorrhage. Bleeding from uterine rupture may be revealed vaginally, but may be entirely concealed within the abdomen; this is a rare but extremely treacherous cause of postpartum hemorrhage and should be considered where shock is out of proportion to vaginal blood loss, particularly with vaginal delivery following a previous Cesarean section. Persistent bright red vaginal bleeding associated with a firm well-contracted uterus suggests one of these causes.

Laceration of blood vessels underneath the vaginal or vulval epithelium results in hematoma formation. In these circumstances the bleeding is concealed and may be particularly dangerous since, in inexperienced hands, it may go unrecognized for several hours until the patient becomes shocked. In the case of episiotomy and perineal tears, postpartum hemorrhage may result if the cut involves arteries or large varicosities, if it is extended and difficult to repair, if there is a delay between episiotomy and delivery, or if there is a delay between delivery and repair of the episiotomy.

Management of primary postpartum hemorrhage

The secondary complications of primary postpartum hemorrhage, namely unnecessary additional blood loss leading to shock and DIC, can usually be prevented by decisive obstetric action. This condition tests not only clinicial acumen but also an ability to mobilize and organize rapidly all available labor ward staff to deal with the problem. A properly managed case of severe postpartum hemorrhage indicates that the staff are well trained to deal with such disastrous situations, that they are familiar with carefully set guidelines, and that back-up services such as blood bank products and emergency hematological advice function quickly and to high standards.

Examination under anesthesia: the first operative procedure

The potential exists for mismanagement if the pathophysiology is not appreciated. If the initial management is incorrect, a cascade of problems may ensue. Most obstetric textbooks highlight the importance of ensuring initial intravenous access, the use of uterotonic drugs, and bimanual compression of the uterus. While these are reasonable 'first-aid' measures, this approach fails to address causes other than atonic postpartum hemorrhage.

The most crucial decision in the management of a case of postpartum hemorrhage is whether the bleeding is atonic (placental bed) or traumatic (extraplacental causes). While the former merely requires the use of uterotonic drugs, bimanual compression, and/or uterine packing, patients in the latter category will continue to deteriorate until dealt with surgically. The erroneous 'default' assumption of atony wastes both time and precious circulating blood volume—sadly until hypovolemic shock ensues, thereby precipitating a general anesthetic under the worst circumstances. Secondary DIC and systemic hypotension are the pathway towards maternal mortality, a postoperative situation that need not arise if prompt initial surgical action is taken.

A clinician faced with a woman with postpartum hemorrhage has to assume temporarily that the cause is uterine atony, provided that the uterus is soft and enlarged, becoming smaller and firm with the expulsion of clots and administration of a uterotonic agent (Syntometrine and/or an oxytocin infusion). Therefore this is a diagnostic/clinical test pending cross-matching of blood. Other uterotonic options include oral or rectal administration of inexpensive prostaglandins such as misoprostol. If the patient continues to bleed despite the use of these medications, the clinician needs to exclude genital tract trauma or retained placental tissue reliably by performing an adequate examination under anesthesia, bearing in mind that any final diagnosis of atonic uterus is by careful exclusion of surgical causes. It is worth noting that blood accumulating in the vaginal vault (from a surgical cause) can elevate the well-contracted uterus into the abdomen, giving the impression that the uterine cavity is full of blood.

Rather than simply examining the patient in the lithotomy position in the delivery room, it is better to examine the patient in the operating room under general anesthesia to ensure that an accurate diagnosis is made. This should be performed or attended by the most experienced clinician available. With the patient in the lithotomy position, under good light, and with at least one assistant, the entire genital tract should be examined. During the procedure it is likely that previous stitches used to repair an episiotomy or tear will be disrupted. This should not be of concern, since good access is needed to inspect the fornices fully. The uterus should be examined digitally to exclude uterine rupture or retained placental tissue. Having ensured that the uterus is intact, the obstetrician should then examine the cervix and vagina and repair any tears. Perineal repair sutures should be undone to ensure a complete examination. If no obvious cause is found, a diagnosis of intractable atonic postpartum hemorrhage can be made with certainty.

If vaginal bleeding continues following exclusion of trauma within the lower genital tract, it must be ensured that the uterus is firmly contracted. A second-line uterotonic agent (intramyometrial carboprost or rectal misoprostol) should be given if there is any doubt. If these measures fail to control the bleeding, the lower uterine segment should be rechecked using bimanual examination to confirm that it is intact. In instances where the bleeding is persistent, but not heavy, packing the uterus can be considered. This procedure is controversial since it may distract the obstetrician from hysterectomy. Some believe that packing will interfere with uterine contractility and that this procedure will only conceal further bleeding. We believe that it has a place as a temporary measure provided that it is done properly. It is most important to pack the fundus of the uterus thoroughly; a common mistake is to introduce a 'half-hearted' pack into the lower part of the uterus only, leaving a large cavity within which blood will accumulate. If inserted for too long, such a pack can interrupt the blood supply to the pelvic organs, in particular the base of the bladder, leading to fistula formation. The pack is also a focus for infection, and for this reason some obstetricians soak the pack in Betadine. It should be removed within 24 h as an operative procedure; significant postpartum hemorrhage can occur at this point and one should be in a position to proceed rapidly to hysterectomy to secure hemostasis.

The second operative procedure

More often than not atonic postpartum hemorrhage tends to stop when the measures described above are applied, although, when this is not possible, the obstetrician needs to consider selective vascular embolization or abdominal surgery in the form of internal iliac artery ligation or abdominal hysterectomy. Maternal age, parity, and number of surviving children all influence the choice to some extent, although the more severe the situation, the more rapidly one has to resort to hysterectomy. If the expertise to perform selective pelvic arterial embolization is available, this may be preferable to internal iliac artery ligation, but such a radiological approach demands both correction of coagulation and a reasonable degree of clinical stability as it may take up to 2 h to be accomplished successfully. For this reason, laparotomy with a view to hysterectomy and/or ligation of the internal iliac arteries is still the usual approach.

Secondary postpartum hemorrhage

Secondary postpartum hemorrhage usually results from an infective process within the decidua and myometrium, typically due to retained placental tissue fragments. The latter is suspected from subinvolution of the uterus and an open cervix through which tissue fragments may be found. Initial management is the taking of bacterial cultures, intravenous fluids, and blood, and antibiotics (e.g. Augmentin, or a cephalosporin and metronidazole) proceeding to exploration of the uterus. Ultrasound examination of the uterus is helpful to reduce the proportion of women undergoing exploration, but this should be done by an obstetrician with experience who is actively involved in the clinical management. Hemorrhage occurs from vessels in the infected myometrium; therefore the uterotonic measures outlined above for primary postpartum hemorrhage may be ineffective. A substantial secondary postpartum hemorrhage is more likely to result in hysterectomy than is a primary postpartum hemorrhage.

Conclusions

The successful management of postpartum hemorrhage requires, first and foremost, a clear understanding of the possible causes of hemorrhage and the consequences of failing to make the correct initial diagnosis. Management of the patient following postpartum hemorrhage, including volume replacement and correction of coagulopathy, follows general intensive care unit guidelines.

Bibliography

Benedetti, T.J. (1996). Obstetric hemorrhage. In *Obstetrics: normal and problem pregnancies* (3rd edn) (ed. S.G. Gabbe, J.R. Niebyl, and J.L. Simpson). Churchill Livingstone, London.

Donald, I. (1979). Postpartum haemorrhage. In *Practical obstetric problems* (5th edn), pp. 748–94. Lloyd-Luke, London.

El-Refaey, H., O'Brien, P., Morafa, W., Walder, J., and Rodeck, C. (1996). Misoprostol for third stage of labour. *Lancet*, **347**, 1257.

Hibbard, B.M., *et al.* (1996). Antepartum and postpartum haemorrhage. In *Report on confidential enquiries into maternal deaths in the United Kingdom 1991–1993*, Chapter 3. HMSO, London.

Prendiville, W.J., Harding, J.E., Elbourne, D.R., and Stirrat, G.M. (1988) The Bristol third stage trial: active versus physiological management of third stage of labour. *British Medical Journal*, **297**, 1295–1300.

12.5 Obstetric emboli

12.5.1 Pathophysiology

T. Dugernier and M.S. Reynaert

Key messages

- Venous and amniotic fluid thromboemboli are the leading cause of maternal mortality in Western countries.
- The increased frequency of venous thromboembolism during pregnancy should be ascribed to the combination of venous stasis with the physiological hypercoagulable state associated with pregnancy. Factors promoting thrombosis are maximally operative near term.
- The amniotic fluid embolism syndrome evolves typically in two successive phases. Presenting features include consistently acute cor pulmonale and, at times, hemorrhage, seizures, and fetal distress, which typically occur during labor and delivery or shortly thereafter. Acute respiratory distress syndrome and left ventricular failure, consumptive coagulopathy, and postanoxic encephalopathy complicate the course in survivors. This syndrome is accounted for by the introduction of abnormal amniotic fluid into the maternal circulation.

Introduction

Pulmonary embolism constitutes one of the most dreaded hazards of pregnancy. Although fat or air embolism may complicate the course of pregnancy in particular settings such as pulmonary laceration or pelvic/long-bone injury, venous and amniotic fluid thromboemboli deserve further consideration because of their specificity and prevalence during pregnancy and the puerperium, and their relative importance as a cause of maternal mortality.

Venous thromboembolism

Venous thromboembolism occurs in 2 to 5 per 1000 deliveries. As well as pregnancy itself, additional specific and non-obstetric risk factors include advanced maternal age, multiparity, obesity, Cesarean section, instrument-assisted delivery and other surgical procedures, prolonged bed rest, hemorrhage, trauma to the lower limbs, and inherited deficiencies in natural inhibitors of coagulation (Rutherford and Phelan 1991; Toglia and Weg 1996). The role of gestational age remains controversial. Traditionally, the risk of thrombosis was considered to be higher during the third trimester and immediately postpartum. Whereas more than 50 per cent of pulmonary emboli occur postpartum, 75 per cent of deep-vein thromboses present antepartum (Toglia and Weg 1996).

The risk of venous thromboembolism is five times higher in a pregnant woman than in a non-pregnant woman of similar age. This increased frequency is due to both the physiological hypercoagulable state associated with pregnancy and the venous stasis mediated by estrogens and local pressure effects of the gravid uterus on the inferior vena cava (Table 1). The levels of all clotting factors, except factors XI and XIII, increase during pregnancy. Plasma fibrinolytic activity also decreases as a result of placental inhibitors and, when the placenta separates, tissue thromboplastin is released into the circulation, further promoting thrombosis. These changes are prominent at term and in the immediate postpartum period, accounting for the maximum risk of venous thromboembolism around delivery. This hypercoagulable state is the result of stimulation of synthesis by the low-grade disseminated intravascular coagulation that takes place within the uteroplacental vasculature and that replaces the internal elastic lamina and smooth muscle in the spiral arteries with fibrin to enhance blood flow to the fetoplacental unit (Rutherford and Phelan 1991). Eventually, these alterations in the coagulation mechanisms that promote thrombosis help the mother to cope with the increased demand that arises on these components at the time of placental separation. By the third to fourth week following delivery these components of the coagulation and fibrinolytic systems have returned to their non-pregnant levels.

If left untreated, deep venous thrombosis will progress to pulmonary embolism in 25 per cent of patients, with a mortality of approximately 15 per cent.

Amniotic fluid embolism

Amniotic fluid embolism, although uncommon, is perhaps the most devastating critical illness specifically related to the pregnant state. The incidence is quoted as between 1 in 8000 and 1 in 80 000 live births with a fatality rate as high as 86 per cent (Peterson and Taylor 1970; Clark 1991). Together with pulmonary thromboembolism, this obstetric complication is the leading cause of maternal mortality in Western countries.

Most episodes of amniotic fluid embolism occur during labor and delivery. However, this condition has been reported as early as during first- and second-trimester abortion, as late as 48 h postpartum, and following abdominal trauma or any uterine manipulation such as Cesarean section or even amniocentesis. Increased maternal age, multiparity, premature placental separation, meconium in the amniotic fluid, and intrauterine fetal death are predisposing factors, but there is no firm association with vigorous labor and hypertonic uterine contractions (Peterson and Taylor 1970; Clark 1991).

Table 1 Pathophysiology of venous thromboembolism during pregnancy

Factors promoting thrombosis	Additional risk factors
Venous stasis (hormones, gravid uterus)	Advanced maternal age
Increased clotting factors	Multiparity
Decreased fibrinolysis	Cesarean section/instrument-assisted delivery
Tissue thromboplastin (placenta)	Hemorrhage
	Obesity
	Prolonged bed rest
	Trauma (lower limbs/endothelium)
	Inherited deficiencies in natural anticoagulants

Table 2 Symptoms of amniotic fluid embolism

- Phase 1
  - Presenting features: acute cor pulmonale
    - Respiratory failure
    - Circulatory collapse
    - Convulsions/altered consciousness
    - Hemorrhage
    - Fetal distress
- Phase 2
  - Secondary signs: left ventricular failure + (?)pulmonary capillary injury
  - Pulmonary edema
  - Consumptive coagulopathy + uterine atony
  - Postanoxic encephalopathy

Typical presenting features usually include sudden onset of respiratory failure and circulatory collapse, accompanied by seizures in 15 per cent of cases. Occasionally, fetal distress or hemorrhage due to disseminated intravascular coagulation herald this condition (15 per cent). Up to half the patients die within the first hour. Those who survive this initial cardiorespiratory insult face hypoxic brain damage, acute respiratory distress syndrome (up to 70 per cent), and uterine hemorrhage ascribed to consumptive coagulopathy and uterine atony (40 per cent) (Table 2).

The pathogenesis of this syndrome remains controversial. Entrance of abnormal amniotic fluid into the vascular circulation through uterine tears or endocervical veins seems to be the primary event. Transient trapping of amniotic debris in the uterine veins by uterine tone may explain the occasional delay between delivery and onset of the syndrome. Before week 12 of gestation the volume of amniotic fluid is probably insufficient to trigger this syndrome, although experimentally there is no close correlation between its severity and the volume of injected fetal debris. It is the introduction of abnormal amniotic fluid into the maternal circulation rather than the infusion of amniotic fluid *per se* that accounts for the syndrome (Clark 1991).

Uterine atony and consumptive coagulopathy result from a myometrial depressant effect of amniotic fluid and from procoagulant substances in amniotic fluid respectively. Early hemodynamic data are scarce in this catastrophic condition. Transient and intense pulmonary hypertension, secondary right ventricular insufficiency, and profound hypoxemia probably account for the initial phase of cardiorespiratory failure and for the bulk of maternal death. This pulmonary hypertension may be mediated by mechanical obstruction by amniotic debris or by pulmonary arterial spasm triggered by hypoxemia and vasoactive substances such as prostaglandin F_2 or mucin. Acute left ventricular dysfunction of obscure pathogenesis rapidly ensues in most surviving cases and accounts for the pulmonary edema that characterizes the second stage of amniotic fluid embolism (Clark *et al.* 1985). It has not been established whether it is the particulate contents, such as mucin from meconium, or humoral factors in the amniotic fluid, such as arachidonic acid metabolites, that are responsible for these hemodynamic effects.

References

Clark, S.L. (1991). Amniotic fluid embolism. In *Critical care obstetrics* (ed. S.L. Clark, D.B. Cotton, G.D.V. Hankins, and J.P. Phelan), pp. 393–410. Blackwell Scientific, Boston, MA.

Clark, S.L., Montz, F.J., and Phelan, J.P. (1985). Hemodynamic alterations associated with amniotic fluid embolism: a reappraisal. *American Journal of Obstetrics and Gynecology*, **151**, 617–21.

Peterson, E.P. and Taylor, H.B. (1970). Amniotic fluid embolism. An analysis of 40 cases. *Obstetrics and Gynecology*, **35**, 787–93.

Rutherford, S.E. and Phelan, J.P. (1991). Deep venous thrombosis and pulmonary embolus. In *Critical care obstetrics* (ed. S.L. Clark, D.B. Cotton, G.D.V. Hankins, and J.P. Phelan), pp. 150–79. Blackwell Scientific, Boston, MA.

Toglia, M.R. and Weg, J.G. (1996). Venous thromboembolism during pregnancy. *New England Journal of Medicine,* **335**, 108–14.

12.5.2 Therapeutic strategy

T. Dugernier and P. F. Laterre

Key messages

- Respiratory failure is usually the cardinal presenting feature of obstetric emboli, whatever their origin. Numerous other conditions, whether specifically related to pregnancy or not, may produce the same symptoms and must be carefully excluded.
- The diagnostic work-up and management of venous thromboembolism during pregnancy is basically similar to that for the non-pregnant patient. A diagnosis should be aggressively sought because of the therapeutic implications for the mother. Prompt attention to both maternal oxygenation and hemodynamic status is essential for fetal well being. Promotion of thrombus resolution and prevention of thrombus extension and dislodgement are further therapeutic priorities. These goals are best achieved with heparin and, at times, thrombolytic agents. The use of Coumadin (warfarin) should be restricted to the puerperium.
- The diagnosis of amniotic fluid embolism should be based on a typical clinical picture, i.e. the triad of hypotension, hypoxemia, and coagulopathy, that usually occurs around delivery. Treatment is mainly supportive and should be primarily directed at maternal oxygenation. Owing to inconsistent hemodynamic findings, cardiorespiratory management should be guided by pulmonary catheterization in those patients affected by successive right and left ventricular failure and potential alveolar-capillary membrane injury.

Diagnosis

Acute respiratory failure is usually the cardinal presenting feature of pulmonary embolism that requires critical care management. Apart from obstetric emboli originating from either amniotic fluid or deep vein thrombosis, a number of other conditions which may or may not be specifically related to the pregnant state can lead to acute respiratory failure (Table 1). These should be considered in every pregnant patient with respiratory distress, since they may be influenced significantly by the physiological, hormonal, or anatomical effects of pregnancy, resulting in an increase in incidence, a deterioration in symptomatology, or an alteration in management.

Table 1 Causes of acute respiratory failure in pregnancy and the puerperium

Specific to pregnancy
Amniotic fluid embolism
Sepsis (chorio-amniotitis)
Pulmonary edema
Severe toxemia
Tocolytics
Peripartum cardiomyopathy
Acute coronary artery dissection
Not strictly related to pregnancy
Venous, fat, or air embolism
Aspiration of gastric contents
Status asthmaticus
Pneumonia
Sepsis (pyelonephritis)
Necrotizing pancreatitis
Pulmonary edema
Valvular and congenital heart disease
Severe hypertension
Acute myocarditis

Venous thromboembolism

Investigation of suspected pulmonary embolism in pregnancy does not differ from investigation in the non-pregnant patient. The originating thrombosis is most frequently located in the veins of the calf or in the ileofemoral segment of the deep venous system. Signs and symptoms indicative of deep venous thrombosis or pulmonary embolism are unreliable during pregnancy, since swelling of the legs, dyspnea, and tachypnea are common, particularly near term.

Non-invasive studies, including impedance plethysmography and duplex Doppler ultrasonography, should be the initial diagnostic step for deep vein thrombosis. However, the use of these techniques has not been extensively validated in pregnant women and their sensitivity is far less satisfactory in detecting thrombi below the knee. If performed in late pregnancy, these procedures should be carried out in a lateral position to avoid false-positive results due to venous compression by the gravid uterus. Magnetic resonance imaging and venography, which is held to be the gold standard for diagnosis, should be restricted to clinically suggestive patients with equivocal non-invasive studies.

The ventilation–perfusion scan is the primary screening tool for the diagnosis of pulmonary embolism. It is self-supporting for clinical decisions when the results are interpreted as normal or as indicating a high probability of embolism. Electrocardiography, echocardiography, chest radiography, and arterial blood gas analysis may support the diagnosis or suggest other conditions (Table 1). A moderate hypoxemia is common in the supine position during late pregnancy owing to the decrease in functional residual capacity mediated by the mechanical effects of the gravid uterus on the diaphragm.

Equivocal cases in whom suspicion of embolism arises and a ventilation–perfusion scan shows a low or intermediate probability of pulmonary embolism should undergo pulmonary angiography. In these patients, negative non-invasive studies for detecting deep vein thrombosis should not obviate the need for angiography since they are likely to miss emboli that originate in the pelvic veins. False-negative rates of up to 57 per cent have been reported in this population.

Table 2 Differential diagnosis of amniotic fluid embolism

Dominant feature at onset	Potential diagnoses
Respiratory (50%)	See Table 1
Cardiocirculatory (30%)	Acute cor pulmonale (air, fat, or venous thromboembolism)
	Septic shock
	Hypovolemia (uterine rupture, atony, or laceration; placental abruption; retained tissues)
	Cardiogenic shock (myocardial infarction; congenital, valvular, or peripartum cardiomyopathy)
Hemorrhage–DIC (10%)	Placental abruption
	Toxemia
	Dead fetus syndrome/saline abortion
	Sepsis
	Intravascular hemolysis (post-transfusion etc.)
Seizures (10%)	Eclampsia
	Cerebral hemorrhage

DIC, disseminated intravascular coagulation.

The diagnosis of venous thromboembolism should be aggressively investigated in suspected cases because of its major immediate and future therapeutic implications. Moreover, the effects of radiation on fetal development should be of little concern, since the estimated exposure of the fetus to radiation during the combination of the aforementioned procedures is less than 5000 Gy, far less than the lowest dose associated with a teratogenic effect on the human fetus.

Amniotic fluid embolism

The mainstay for diagnosis of amniotic fluid embolism should be a typical clinical picture such as cardiorespiratory failure of abrupt onset that occurs during the third trimester of pregnancy or around delivery, and which is followed by the development of pulmonary edema and coagulopathy.

Demonstration of trophoblastic and squamous cells in the pulmonary artery vasculature is by no means pathognomonic of this condition. Similar cells have been recovered in the peripheral blood of normal pregnant women and in the pulmonary arterial circulation of patients undergoing pulmonary artery catheterization for a variety of medical indications other than amniotic fluid embolism. Any venepuncture may result in the introduction of adult squamous cells into the venous circulation; unfortunately, there are no reliable histological methods for differentiating adult epidermal from fetal squamous cells derived from the infant's skin and present in amniotic fluid. Other debris of presumed fetal or amniotic origin such as lanugo hair, mucin from meconium, and fat from the vernix caseosa may be demonstrated in the buffy coat of pulmonary arterial blood or in a wedged pulmonary capillary aspirate of patients with amniotic fluid embolism. Special stains are usually required to detect these elements whose diagnostic value remains controversial. In the future, antimucin monoclonal antibody and maternal zinc coproporphyrin I assay may reveal specific markers of meconium entrance into the maternal circulation.

As no single clinical or laboratory finding is pathognomonic of this syndrome, a careful exclusion of other critical illnesses is mandatory (Tables 1 and 2).

Treatment

Venous thromboembolism

The primary goals of therapy should be directed toward maintenance of oxygenation and cardiac output for both mother and fetus. This is achieved by traditional critical care support. The promotion of thrombus resolution and the prevention of thrombus extension and dislodgement are further therapeutic priorities, pending a complete diagnostic work-up.

Heparin is the safest anticoagulant in pregnancy because, unlike Coumadin (warfarin), its high molecular weight prevents it from crossing the placenta. Coumadin should be avoided throughout pregnancy as a characteristic embryopathy has been associated with its use in the first trimester. Central nervous system and ophthalmological abnormalities may result from exposure to these drugs in any trimester. In addition, fetal and neonatal hemorrhage and placental abruption are potential side-effects. Because of immature liver enzyme systems, the effect of oral anticoagulants on the fetus is much greater and lasts for much longer than the effect on the mother.

Women with venous thromboembolism presenting during pregnancy or the puerperium should initially be treated with a continuous intravenous infusion of heparin for 5 to 10 days according to current protocols for non-pregnant patients. If the thromboembolic event occurs antepartum, subcutaneous concentrated heparin

(20 000 IU/ml) should be given into the lateral abdominal wall every 12 h until delivery. The regimen should be dose adjusted to prolong the activated partial thromboplastin time measured 6 h after the injection to a value 1.5 times greater than control. To minimize the risk of recurrence, Coumadin is then administrated postpartum for at least 6 weeks (3 months if thromboembolism complicates the puerperium). Neither heparin nor Coumadin therapy is a contraindication to breast feeding. Postpartum suppression of lactation with estrogen markedly increases the incidence of thromboembolism and is contraindicated.

A separate entity from deep vein thrombosis is septic pelvic and puerperal ovarian vein thrombophlebitis. This is encountered in patients with endometritis who do not respond to antibiotics. A continuous intravenous infusion of heparin should be added to the antibiotic regimen for 10 to 20 days and then stopped.

Low-molecular-weight heparin does not cross the placenta. Limited experience with this agent has been gained during pregnancy, but its use is attractive because its longer half-life allows a once-daily regimen and because it lowers the risk of heparin-induced osteoporosis and thrombocytopenia.

Anticoagulant therapy usually does not affect vaginal delivery significantly. The risk of maternal hemorrhage is minimal if protamine sulfate is administered to patients with markedly prolonged activated partial thromboplastin times and to those subjected to Cesarean section. Regional anesthesia is usually contraindicated because of the risk of epidural hematoma.

Recurrence of pulmonary embolism in adequately anticoagulated patients, any absolute contraindication to anticoagulant therapy, and any serious complication of anticoagulation such as heparin-induced thrombocytopenia are potential indications for transvenous insertion of a permanent inferior vena cava filter in women with venous thromboembolism. These filters have the ability to trap emboli without significantly reducing flow in the inferior vena cava. Recurrence of pulmonary embolism after filter insertion is less than 5 per cent. If possible, anticoagulation should be continued for 3 months after filter insertion to allow lysis of initial thrombosis or any clot trapped by the filter. Apart from leg swelling, no serious fetal and maternal side-effects (e.g. perforation of the inferior vena cava wall) have been reported. However, there is a risk of dislodgement due to the dilated venous system and pressure effects during labor. Radiographic follow-up is mandatory to check for migration, angulation of the device, or perforation of the vein wall.

The use of thrombolytic agents should be restricted to life-threatening situations, such as impending or established acute cor pulmonale, because of the risk of maternal bleeding at the time of delivery and immediately postpartum.

Surgical or transvenous catheter embolectomy/fragmentation may occasionally be lifesaving in the expeditious management of massive pulmonary embolism with cardiovascular collapse in patients who are severely hemodynamically compromised and/or in whom thrombolytic therapy is inadvisable.

The therapeutic strategy for obstetric emboli should also encompass antepartum prophylaxis for women with a known hypercoagulable state, with any condition normally requiring long-term anticoagulant therapy, or with a history of previous venous thromboembolism related or not to pregnancy. The incidence of thromboembolism during pregnancy ranges from 5 to 70 per cent in these patients. Adjusted (1.5 times control value) rather than fixed-dose subcutaneous heparin should be started as early as possible in pregnancy and continued up to delivery. Coumadin should be given immediately postpartum for 6 weeks.

An adjusted-dose regimen is essential for effective prophylaxis or treatment during pregnancy to offset the physiological increases in plasma volume, renal clearance, and plasma levels of clotting factors.

Amniotic fluid embolism

Treatment of amniotic fluid embolism is only supportive and at times may begin with cardiopulmonary resuscitation. In this setting, Cesarean delivery results in greater maternal survival if the mother does not respond to standard cardiopulmonary resuscitation and the fetus is still undelivered. Given the consistent and early finding of profound hypoxemia with secondary potentially irreversible maternal brain damage, prompt attention should be paid to oxygenation. In view of inconsistent hemodynamic findings, with an initial presumptive phase of right ventricular failure due to severe pulmonary hypertension followed by left ventricular failure and pulmonary edema, central hemodynamic monitoring is recommended to guide inotropic support and fluid therapy in those patients who require an optimal cardiac preload despite alveolar-capillary membrane injury.

Disseminated intravascular coagulation and its attendant uterine bleeding require replacement therapy with clotting factors. Locoregional hemostasis is best achieved with uterine packing and with oxytocic agents or prostaglandin F_2 to control uterine atony. Arterial embolization and even hysterectomy are occasionally needed.

Amniotic fluid embolism is often, if not always, associated with, and at times may be heralded by, fetal distress. Therefore the fetal heart rate should be monitored if gestational age and maternal condition allow operative intervention for fetal distress.

Bibliography

Clark, S.L. (1991). Amniotic fluid embolism. In *Critical care obstetrics* (ed. S.L. Clark, D.B. Cotton, G.D.V. Hankins, and J.P. Phelan), pp. 393–410. Blackwell Scientific, Boston, Ma.

Clark, S.L., Montz, F.J., and Phelan, J.P. (1985). Hemodynamic alterations associated with amniotic fluid embolism: a reappraisal. *American Journal of Obstetrics and Gynecology,* **151**, 617–21.

Lee, W.L., Ginsburg, K.A., Cotton, D.B., and Kaufman, R.H. (1986). Squamous and trophoblastic cells in the maternal pulmonary circulation identified by invasive hemodynamic monitoring during the peripartum period. *American Journal of Obstetrics and Gynecology*, **155**, 999–1001.

Rutherford, S.E. and Phelan, J.P. (1991). Deep venous thrombosis and pulmonary embolus. In *Critical care obstetrics* (ed. S.L. Clark, D.B. Cotton, G.D.V. Hankins, and J.P. Phelan), pp. 150–79. Blackwell Scientific, Boston, MA.

Toglia, M.R. and Weg, J.G. (1996). Venous thromboembolism during pregnancy. *New England Journal of Medicine*, **335**, 108–14.

13 Infection and systemic inflammation

13.1 Normal physiology

13.1.1 Immunology

P. G. Medd and B. M. Chain

Key messages

- The immune system functions to distinguish self and harmless non-self from harmful non-self and having identified the latter deploys effector mechanisms to eliminate it.
- Immune responses can be both antigen specific and antigen non-specific. Non-specific responses are usually associated with the induction of an antigen-specific response.
- Antigen-specific responses control the adaptive immune response; these are mainly mediated by B and T lymphocytes. The B-lymphocyte antigen receptor is immunoglobulin (antibody) which can be secreted to recruit effector mechanisms. T lymphocytes recognize antigen in the context of major histocompatibility complex (MHC) molecules using the T-cell receptor.
- Fragments of antigen recognized by T cells are produced by antigen processing. MHC class I molecules present intracellular peptides, including viral and tumor peptides, to CD8 T cells, resulting in a cytotoxic response. MHC class II molecules present extracellular antigens to CD4 T cells, resulting in B-cell antibody production or macrophage activation.
- Interaction between T and B cells allows T cells to activate B cells bearing antibody specific for the antigen recognized by the T cell.
- Tolerance to self is maintained through T cells. Self-reactive T cells are eliminated in the thymus. In the periphery two signals from a professional antigen-presenting cell are required to activate T cells; these signals are controlled to avoid activating self-reactive T cells.
- Antibody effector mechanisms include neutralizing free antigens, facilitating pathogen phagocytosis, and the activation of complement.
- Cellular effector mechanisms can destroy infected cells (cytotoxic T cells), engulf pathogens, and act against parasites (eosinophils).
- The normal functioning of the immune system is remarkable for its invisible nature.

Introduction

The immune system has evolved to respond to molecules that are 'non-self' (or foreign) and distinguish them from those that are part of 'self' (or host). The system mounts a destructive response against the former, while being tolerant towards the latter. A contemporary variation of this hypothesis introduces an additional constraint and proposes that the immune system responds only to antigen (most commonly infection) associated with pathology, or damage, while remaining silent in the absence of damage-associated non-specific signals.

The major components of an immune response

The process of immunity can be compartmentalized as follows:

(1) starting a response;

(2) recognition of antigen and regulation of an antigen-specific response;

(3) producing an effector response to control infection.

Starting the immune response

Antigen-specific versus antigen-non-specific responses

An antigen is a molecular structure (often protein, but also nucleic acid, sugar, or lipid) which induces an 'adaptive' immune response specific for its unique molecular nature (sequence or shape). Traditionally, the measure of this response was the production of a soluble plasma protein (antibody) which specifically binds the antigen. As yet, this measure of immune response is the only one in routine clinical practice. We now know that the specific response can also be cellular, leading to cellular cytotoxicity or cytokine-dependent macrophage activation.

The cells of the immune system, particularly the tissue macrophage, can also mount less specific generalized responses against potentially dangerous micro-organisms. Most specific immune responses probably need to be associated with non-specific responses in order to get started. These non-specific responses are usually asymptomatic, and we know little of their molecular nature. However, when the response is excessive, immunopathology results in the process of inflammation which has been the subject of intense study.

Inflammation

Inflammation is a non-specific response of the microcirculation which follows tissue damage. After injury or lytic infections the local microvasculature dilates. Vessel wall permeability increases and this, coupled with the increased blood flow, leads to localized tissue edema. The chemical mediators of inflammation are diverse, and an individual mediator will usually be responsible for some, but not all, of the signs of inflammation. For example, interleukin 1 (IL-1), derived from macrophages and other sources, has systemic proinflammatory effects, causing the febrile response, hepatic acute phase protein release, and further leukocyte release from the bone marrow. Locally it increases vascular permeability and activates synthesis of matrix-destructive enzymes, thus amplifying inflammation.

The collective effect of all the inflammatory mediators is to generate the cardinal signs of inflammation: heat (*calor*), swelling (*tumor*), pain (*dolor*), redness (*rubor*), and loss of function (*lasae functio*) which results from the swelling and pain. Inflammation promotes the induction of an immune response in three ways.

The passage of soluble components of the immune system within blood (particularly antibody and complement) is increased. Both antibody and complement contribute to further release of inflammatory mediators, and hence amplify the ongoing response.

The passage through tissue of the principal effector cells of the immune system, which recirculate between blood and lymphoid tissue, is increased. In inflammation, the endothelium of the post-capillary venule expresses a range of adhesion molecules which allow the attachment of leukocytes and their movement through the vessel wall. Once within the tissues, these cells are further attracted to the site of inflammation by chemotaxis. Receptors on the cells are sensitive to some of the mediators released at the inflammatory site. The cellular infiltrate associated with acute inflammation initially consists principally of phagocytic neutrophils (or eosinophils in regions where helminth infections are common). Monocytes, which are the circulating precursors of macrophages, a principal phagocytic cell of the immune system, arrive at the inflammatory site 12 to 24 h after the onset of inflammation. Monocytes differentiate, activating as a phenotype with macrophage effector mechanisms. Activated macrophages, in turn, are major producers of inflammatory mediators and can play a central role in amplifying ongoing responses to pathological levels. In some cases, the site of infection is also infiltrated by T lymphocytes, which play a key role in macrophage activation.

Migration of 'antigen-presenting' cells out of damaged infected tissues into lymphoid organs is induced. Every tissue contains a population of hemopoetic 'antigen-presenting' cells (sometimes collectively referred to as the dendritic lineage) which function as 'sentries' of the immune system. In response to some local inflammatory mediators (e.g. cytokines IL-1 or tumor necrosis factor-α) these cells activate and migrate to the draining lymphoid tissue via the afferent lymphatics, carrying antigen from the local site. Once at their target tissue, they activate antigen-specific T lymphocytes and hence an antigen-specific response.

Recognition of antigen and regulation of an antigen-specific response

T and B lymphocytes are the principal components controlling the adaptive immune response. Each lymphocyte expresses a receptor which is specific for only one antigen. The total repertoire of all lymphocyte receptors is enormous (estimates range from 10^8 to 10^{10}), but all the receptors on any one cell will be identical. When a lymphocyte recognizes antigen, which occurs when the antigen-specific receptor on the lymphocyte is 'cross-linked' by antigen (with help from other cells if necessary), it will become activated and divide. All its progeny will express the same specificity of receptor, thus forming an amplified clonal population. This clonal population, by virtue of its increased size and its enhanced 'activation' state following antigen recognition, will respond more rapidly and strongly in response to a secondary exposure to the original antigen, but not to a new antigen. This model of immune function, which is described as the clonal selection hypothesis since antigen 'selects' a specific lymphocyte set from within a large repertoire, underlies the adaptive nature of both T and B response.

B lymphocytes

B lymphocytes are derived directly from bone marrow, and their antigen-specific receptor is an immunoglobulin (Ig) or antibody. The Ig structure is now known and can be found in any standard textbook of immunology. Functionally, the molecule can be divided into a variable region, which mediates antigen recognition, and a constant region, which determines isotype and mediates effector function. The constant region of Ig heavy chain can be one of five classes M, G, D, E, and A, each of which has further numbered subclasses. The Ig molecule is named after its heavy-chain class; for instance, a molecule with a G1 heavy chain is known as IgG1. Ig has a dual function, acting as an antigen-specific receptor mediating cellular activation when displayed on the surface of a B cell, but also acting as an effector molecule in its own right. Following recognition of antigen, B cells can differentiate into plasma cells and secrete soluble 'effector' Ig into the circulation.

The variable region forms an extended antigen-binding surface at the outermost tip of the molecule, consisting of three complementary-determining loop regions (CDRs). In general, antibody and antigen interact over a large (10–20 amino acid) area, and the side-chains of the amino acids, which make up the CDRs, form multiple low-energy interactions with antigen. The combination of these interactions gives rise to a very specific, and potentially high-affinity, binding between the three-dimensional surfaces of antigen and antibody. Destruction of antigen conformation usually results in loss of recognition by antibody.

The large repertoire of the antibody response resides in the extreme variability of the amino acids which make up the three

CDRs. This variability is achieved by a unique molecular mechanism which involves complex DNA rearrangements between families of multiple alternative exons, combined with a still mysterious process of selective somatic hypermutation. The variable region can be coupled to any of the heavy-chain constant-region genes, enabling an Ig molecule to maintain specificity while changing class, a process described as class-switching.

T lymphocytes

T cells recognize antigen via the T-cell receptor, a two-chain receptor which is homologous to Ig and has constant and variable regions. Unlike antibody, the T-cell receptor serves only antigen recognition and has no intrinsic effector function; thus its constant region is much simpler than that of Ig. The basic construction and genetics of the T-cell receptor are very similar to Ig; it too has three CDRs, the most variable of which is CDR3 which probably comes into contact with the antigen molecule.

However, there is a fundamental difference between T and B antigen recognition, since T cells interact only with 'processed' antigen bound to a major histocompatibility complex (MHC) molecule on a second cell. Therefore T-cell activation requires 'antigen presentation'; consequently T cells predominantly recognize the primary sequence of antigenic peptides and are insensitive to native conformation.

There are two principal types of antigen-presenting MHC molecule: class I and class II. Each MHC molecule contains a peptide-binding site which is capable of binding a large number of peptides, each predominantly 10 to 20 amino acids long. The antigen-binding region of the T-cell receptor interacts with the exposed surface of the peptide bound within the MHC binding site, and also with the sides of the binding site themselves.

Antigen-processing pathways

Class I MHC (HLA-A, HLA-B, and HLA-C in the human) contain mainly peptides derived from cytoplasmic proteins synthesized within the antigen-presenting cell itself. The peptides are produced within the cytoplasm by proteolytic degradation and transported into the endoplasmic reticulum via a specialized transporter of antigen peptide, where they bind within a nascent MHC chain during folding. The repertoire of peptides loaded is derived predominantly from normal cell proteins and in consequence is not seen by the immune system which is tolerant to self. However, in a virally infected or tumor cell, non-self or abnormal proteins may be synthesized by the cell's own machinery and these will also be presented on class I MHC and lead to a T-cell response, which is normally cytotoxic.

Class II MHC (HLA-DR, HLA-DP, and HLA-DQ in the human) contains predominantly peptides derived from membrane proteins of the antigen-presenting cell itself. Other peptides are derived from proteins taken up by endocytosis or phagocytosis. Therefore class II MHC will present any proteins derived from extracellular infectious agents (bacteria, parasites, or extracellular stages of viruses) which are taken up by antigen-presenting cells in the vicinity of infection.

MHC polymorphism

Both class I and class II MHC are polygenic and polymorphic. This presumably confers benefit since each isotype and allotype will bind a slightly different repertoire of peptides and hence increase the ability of the species to mount an effective immune response. However, the enormous polymorphism of the MHC is also the driving force for alloresponsiveness and hence the major barrier to transplantation.

MHC and T-cell subsets

MHC molecules interact with both the T-cell receptor and the coreceptor molecules CD4 and CD8 on the T-cell surface. Since CD8 interacts only with class I MHC, and CD4 only with class II MHC, and furthermore mature T cells express only CD4 or only CD8, this results in the creation of two classes of T cells recognizing different processed sets of antigen. Class-I-specific CD8 T cells are the major component of cellular cytotoxicity, while CD4 cells are the key cells regulating the immune response.

Cellular interactions as a hallmark of immunity

Few immunological responses are mediated by one cell type alone: two examples of some important cellular interactions regulating immunological responsiveness follow. Others are discussed later in the context of effector mechanisms.

T-cell help

B cells can produce antibody independently of T cells when many of their cell-surface Ig receptors are simultaneously bound to antigens which contain repeated identical recognition sites. These are mainly lipopolysaccharide or polysaccharide components of bacterial cell wall (T-independent antigens). However, most antigens also need to stimulate a CD4 T-cell response, which in turn 'helps' the B cells to differentiate into antibody-producing cells. Such antigens are bound by B cells, processed, and then presented to T cells via class II MHC. Because B cells only effectively present antigen bound by their own surface Ig, the epitope recognized by a B cell and the epitope recognized by a T cell helping it must be physically linked on the same molecule or within the same molecular complex. This process helps to prevent the production of autoreactive antibodies.

Once B and T cells interact via antigen, the most important signal mediating help is that given by the CD40 ligand on the surface of the T cell binding CD40 on the B-cell surface, which rescues B cells from apoptosis and stimulates them to divide. The T cell can also secrete many different cytokines (mostly belonging to the interleukin family) which regulate B-cell activation, differentiation, and proliferation. One major regulatory step in antibody formation is class-switching. IgM is expressed by all mature naive B cells, but B cells can switch their constant-chain usage to synthesize antibody of all five classes. The class switched to is determined by the cytokines secreted by the helper T cell. One very important example of this is that the cytokine IL-4 is absolutely essential for IgE production.

Tolerance to self

Because the process which generates antigen receptor diversity is random, mechanisms exist to avoid the initiation of a response to self-antigens. The acquisition of tolerance is a good example of the high degree of complexity of cellular interactions in immune function.

Thymic tolerance

B lymphocytes expressing self-reactive antigen-specific receptors are rarely tolerant and rely on the absence of T helper cells to avoid activation. T-cell self-tolerance, in contrast, occurs at many levels. The primary selection of the T-cell repertoire occurs within the thymus. T-cell precursors enter the thymus from the bone marrow via the blood stream. In the thymus they express an antigen-specific receptor and undergo two processes of selection. Self-antigens, derived from thymus and blood stream, are processed and presented. Those T cells which recognize any self-peptide with high affinity are induced to undergo programmed cell death (negative selection). Those T cells which recognize a self-peptide with intermediate affinity proliferate and differentiate into mature T cells (positive selection). Those T cells which do not recognize any peptide also undergo programmed cell death. Thus thymic education has the paradoxical outcome of selecting T cells which respond best to those non-self-peptides which are closely related (i.e. agonist analogs) to a self-antigen. This paradox may be of great significance in understanding autoimmune disease.

Peripheral tolerance

Although many self-reactive clones are deleted in the thymus, not all self-antigens will be found there. Antigen-presenting cells can exert a second level of tolerance induction. In order for a T cell to become activated, it requires two fundamentally different types of stimuli: one is antigen specific and is given by interaction of the T-cell receptor with antigen–MHC complex, and the other is antigen non-specific and depends on the engagement of secondary activation/adhesion molecules on the T-cell surface by appropriate counter-receptors on the stimulating antigen-presenting cell. Without correct co-stimulation T cells can become unreactive (anergic). The full extent of the molecular signals transmitting these 'co-stimulation' signals remains unknown, although interaction between the CD80(B7)–CD28 family of molecules is very important. The conditions regulating the expression of co-stimulatory molecules is also unknown, but it is clear that their expression is limited to a very few cell types and is regulated by the microenvironment surrounding the antigen-presenting cell, particularly by the extent of an inflammatory response. Thus responsiveness can be limited to appropriate sites of the body (e.g. lymphoid tissue) and to situations where antigen is found in the context of tissue damage.

Effector mechanisms

Humoral effectors

Antibodies are primarily effective in controlling organisms with a prolonged extracellular stage in their lifecycle, such as most common bacteria, some viruses, and some eukaryotic parasites. Their principle protective activities are as follows.

1. Neutralizing antibodies bind free extracellular particles, such as toxins and some virions, and prevent their actions on host cells.
2. Antibodies opsonize pathogens for phagocytosis. By opsonization, which literally means 'to spread with butter', antibody bound to a pathogen's surface can engage Fc receptors on the surface of phagocytes, activating their engulfment mechanisms.
3. IgM and IgG1, IgG2, and IgG3 isotypes can activate the 'classical' complement pathway. Complement is primarily an amplifying cascade of plasma proteinases and associated regulatory proteins. Binding of the first complement component Clq to an antibody–antigen complex activates the cascade and leads to the deposition of many molecules of C3b on the target antigen. C3b opsonizes for phagocytosis and activates the assembly of the lytic 'membrane attack complex', which then punches holes in the surface of the target cell.

Complement can also be activated without antibody by certain common bacterial and fungal cell-wall components ('alternative pathway'). The alternative complement pathway can also act to amplify the classical route. Complement by-products activate mast cells and act as chemoattractants for neutrophils, monocytes, etc.

Cell-mediated immunity

Cytotoxic T cells

Upon recognizing antigen–MHC complexes on a target cell, cytotoxic T cells are capable of destroying these cells. Cytotoxicity occurs either by secretion of pore-forming proteins, which help to lyse the cell and allow the entry of granzyme, a proteinase which activates the apoptotic pathway in the target cell, or by the expression of the surface protein Fas which engages Fas ligand on the surface of the target cell, leading to apoptosis of the target.

Activation of macrophages

T cells secreting the cytokines interferon-γ, IL-2, and tumor necrosis factor-β can activate cellular effectors, particularly macrophages, rather than antibody production. The subset of CD4 cells with this cytokine profile have been designated Th1 cells, in contrast with the Th2 subset characterized by secretion of IL-4. For example *Mycobacterium tuberculosis* survives in macrophage endosomes until Th1 cells activate them, enabling the macrophages to fuse their lysosomes with the endosomes in question. Activated macrophages can release their granule contents in 'frustrated phagocytosis' if they cannot engulf the opsonized material.

Eosinophils and mast cells

Eosinophils develop under the influence of IL-5 and act against parasites at inflammatory sites, in contrast with neutrophil activity against bacteria. They are activated by antibody-binding surface Fc receptors, some cytokines, and possibly intrinsic receptors for non-self. Like neutrophils and macrophages, they undergo a respiratory burst to produce reactive oxygen free radicals which kill effectively. They also release proteins damaging to helminths.

Mast cells contain large amounts of the vasoactive amines, such as histamine, which cause vasodilatation and increase vascular permeability, leading to urticaria. They also release other mediators of inflammation, particularly those derived from arachidonic acid. They bind IgE very tightly to their surface, and cross-linking of this leads to mast cell degranulation. Mast cells are potent cellular mediators of inflammation, whose physiological role is probably in defense against parasite infection.

How well does the immune system work?

There is no better demonstration of the importance of normal immunological function than the plethora of pathologies associated with acquired and many inherited immunodeficiencies. However, what these diseases so strikingly reveal is that most of the work of the normal immune system is invisible; it creates no symptoms, little systemic activity, and no pathology.

Bibliography

Ibrahim, M.A.A., Chain, B.M., and Katz, D.R. (1995). The injured cell: the role of the dendritic cell system as a sentinel receptor pathway. *Immunology Today*, **16**, 181–6.

Janeway, C.A. and Travers, P. (1996*a*). *Immunobiology*, pp. 5.1–5.29. Current Biology, London.

Janeway, C.A. and Travers, P. (1996*b*). *Immunobiology*, pp. 7.1–7.41. Current Biology, London.

Kuby, J. (1994). *Immunology* (2nd edn), pp. 369–90. W.H. Freeman, New York.

Springer, T.A. (1994). Traffic signals for lymphocyte recirculation and leukocyte emigration: the multistep paradigm. *Cell*, **76**, 301–14.

13.2 Infection control issues

13.2.1 Hygiene

G. L. Ridgway

Key messages

- By far the most important vehicle of transmission remains the hands of attendants.
- The antibiotic policy of the intensive care unit must be kept under continuous review so that there is no indiscriminant use of antibiotics.
- A clinical handwash using a proprietary skin disinfectant, must be performed at the beginning and end of each clinical session.
- Staff should not wear wrist watches, bracelets, rings other than plain bands, or ear or nose ornaments other than studs while carrying out clinical duties.
- Patients with unexplained diarrhea, infected eczema, or communicable diseases, or who are known or suspected carriers of multiple antibacterial-resistant bacteria, should be admitted to source isolation until cleared of transmission risk.

Introduction

Almost every item of equipment encountered in the intensive care unit (ICU), ranging from ventilators and dialysis machines, to intravenous and urinary catheters, to such mundane items as ECG electrodes, has been shown to be a source of nosocomial infection. By far the most important vehicle of transmission remains the hands of attendants.

The risk elements can be considered under four headings: the environment, the equipment, the attendant, and the technique.

The environment

The ICU will develop its own resident flora of bacteria, which will reflect the predominant case mix and antibiotic usage within the unit. The organisms causing nosocomial infection are constantly changing; hence surveillance noting both global and local trends is important. The staphylococcal infections of the late 1950s and 1960s yield first to coliforms and then *Pseudomonas*, only to return to the staphylococci, both coagulase-negative staphylococci and methicillin-resistant *Staphylococcus aureus* (MRSA). In addition, there has been the emergence of an *Enterobacter* species able to express extended β-lactamases, the inherent penem resistance of *Stenotrophomonas maltophilia*, and the spread of vancomycin-resistant (glycopeptide-resistant) enterococci. The effects of local practice are superimposed on these global problems. Thus heavy use of third-generation cephalosporins is likely to encourage the emergence of vancomycin-resistant enterococci, and the use of imipenem or meropenem may lead to problems with *S. maltophilia*. The antibiotic policy of the ICU must be kept under continuous review so that there is no indiscriminant use of antibiotics.

There are no requirements for special ventilation or air filtration systems other than those required for specific isolation purposes (Bauer *et al.* 1990). However, the environment must be kept clean. Special cleaning procedures are not required, other than to ensure that frequencies are optimal and that staff are trained in the technique of damp-dusting using detergent-containing water that is changed frequently and clean mops. All surfaces must be damp-wiped at least daily. Cleaning of clinical equipment, particularly that attached to patients, should be assigned to a specific attendant. The routine use of disinfectants for cleaning is unnecessary. Cleaning of isolation rooms when vacated may require disinfection with a general-purpose phenolic as recommended by the local infection control officer.

The equipment

All clinical equipment coming into contact with the patient must be clean or sterile. Care must be taken to ensure that no equipment becomes a reservoir of potentially pathogenic organisms. The reliability of bench-top sterilization is questionable unless the user is conversant with the time and temperature requirements and the load characteristics, and is sure that the necessary checks and maintenance are being carried out. Bench-top downward-displacement sterilizers (or gravity sterilizers) as set up in the United Kingdom are unsuitable for wrapped instruments, and must not be used for this purpose. The time and temperature relationships for steam sterilizers are different in the United States, the United Kingdom, and continental Europe. The recommended temperatures for such an autoclave in the United Kingdom are a minimum of 134 °C for a holding time of 3 min and a maximum of 137 °C. The cycle time for such a process is about 20 min. In contrast, the United States equivalent for a gravity-feed ('bench-top') autoclave is either 121 °C for a holding time of 20 min (standard gravity) or a 132 °C for a holding time of 10 min (high-speed or flash process). Machines are generally designed to conform with national guidelines. Therefore it is vital that the user checks that the recommended sterilization cycle for an article can be achieved by the sterilizers available. Further advice concerning United Kingdom practice is given in *Health Technical Memorandum 2010* (NHS Estates 1994).

The role of disinfectants must be carefully assessed. The hospital should have a formal disinfection policy which lays down the broad guidelines for supply and use of these agents. It is important that the legal aspects of the use of these agents is appreciated (such as complying with the Control of Substances Hazardous to Health (COSHH) legislation in the United Kingdom). Most disinfectants will support the growth of bacteria, and therefore can become an inadvertent source of infection. This is particularly true for chloroxylenols, diguanides (e.g. chlorhexidine), and quaternary ammonium compounds (e.g. cetrimide), all of which may become contaminated with environmental Gram-negative bacilli.

In general, surfaces should be kept clean with a detergent wash, and if necessary wiped over with a proprietary alcohol-impregnated cloth which is then discarded. Items must not be left to soak in containers of disinfectant unless a high-level disinfection process is being used for equipment that cannot be sterilized by any other technique (e.g. fiberoptic endoscopes). In such circumstances, a mechanical washer should be used to ensure optimal cleaning and disinfection. The use of chemical agents to sterilize equipment is not reliable.

Equipment labeled 'for single use only' must be discarded after single use. Attempts to reprocess such items may result in failure to sterilize or degradation of the material. If an adverse incident occurred in a patient, the user could be liable for damages in law. Some products may be for single-patient use or be recommended for a limited number of uses. It is important that the manufacturer's recommendations are followed.

The attendant

There is no doubt that the most important factor in the transfer of infection is the attendant's hands. Thus adherence to a strict policy of hand hygiene becomes the cheapest and most effective measure in the control of nosocomial infection. However, it remains the most difficult measure to enforce (Jarvis 1994; Sproat and Inglis 1994).

Handwashing is crucial for the protection not only of the patient but also of the attendant. A clinical handwash (lower forearm downward), using a proprietary skin disinfectant, must be performed at the beginning and end of each clinical session (Doebbeling *et al.* 1992). Between patients, either an alcohol rub (e.g. Hibisol®) should be used or the hands should be washed briefly but thoroughly in a surgical scrub detergent. Where a sterile procedure is being performed, a full surgical handwash (elbows to finger tips), together with the use of surgical-quality sterile gloves, is mandatory. Personnel must ensure that special attention is paid to the tips of fingers, particularly when using an alcohol rub. It is impossible to carry out a clinical handwash while wearing a wrist watch. Staff should not wear wrist watches, bracelets, rings other than plain bands, or ear or nose ornaments other than studs while carrying out clinical duties.

There is much confusion concerning latex gloves, which are used for two overlapping purposes: sterile gloves protect patients from contamination, and clean non-sterile examination gloves are used to protect the attendant from possible contamination with microorganisms from the patient. Before commencing a procedure, it is important to assess whether sterility is required so that the correct gloves are used. Examination gloves are presented in such a way that removal from the box involves contaminating the outer surface, underlining the non-sterile nature of this product. Therefore these gloves should be regarded as an extension of the operator's own skin. Thus they must be removed after contact with the patient and before writing notes or answering the telephone. Sterile gloves should be worn for aseptic procedures and discarded immediately on conclusion. Watches must not be worn under gloves (this suggests that a proper clinical handwash has not been performed). Under no circumstances should staff leave an isolation room wearing gloves of either sort. Suggestions that the use of examination gloves obviates the need for antiseptic handwashing seems to be a concession to slack practice, rather than an improvement.

A handwashing standard for ICU use is shown in Fig. 1, and an example of an over-basin display is shown in Fig. 2.

The technique

Invasive procedures are a major risk factor in nosocomial infection. Techniques require continual audit and reassessment to ensure that the chance of infection is not increased by inappropriate procedures or lack of appreciation of the risk. Particular areas of concern include intravenous and intra-arterial catheters, urinary catheters, chest and abdominal drains, hemofiltration, and dialysis. Because of their frequency of use, these techniques may lead to a laxity in basic hygiene. Where possible, the number of in-line taps should be kept to a minimum by rationalizing fluids and additives, and using welded connections with external closures. Closed drainage systems should be used.

ICU Handwashing Standard

At each WHB, there will be Hibiscrub™ and liquid soap in proper dispensers.
At each patient, there will be Hibisol™ in a proper dispenser.
No wrist watches, sleeves rolled to the elbow.

1. At the beginning of a session (i.e. on starting work or after a break), proper surgical **WASH IN HIBISCRUB™ for 2 minutes.***
   *There will then be residual disinfectant activity
2. Wear a new pair of GLOVES **before** breaking a drip line or attending to a drip site or touching a wound. Wear gloves during suction and when you are likely to touch blood or body fluids. Wash hands in liquid soap or use Hibisol™ *after removing gloves.*
3. Do as many procedures as possible using a non-touch technique.
4. Wash hands in Hibiscrub™ **after accidental soiling** with any body secretions (saliva, tracheal secretions, urine, feces, etc.)
5. Use Hibisol™ (or wash with liquid soap) if moving from one bed to another, or if moving to the patient from the telephone.

If a container of Hibiscrub™/Hibisol™ or soap runs out when you use it, *you* must ensure it is replaced.

Wipe pens, stethoscope head, and controls of all machines with alcohol at the beginning of each shift. Do not suck pencils/pens. Wipe them with an alcohol wipe if they fall on the floor.

University College London Hospitals Infection Control Team

Fig. 1 A handwashing standard poster for ICUs. NB This poster should be on a **yellow** background, indicating infection.

HANDWASHING

Use a skin disinfectant
Preferably chlorhexidine gluconate (e.g. Hibiscrub™)
Alternatively, povidone–iodine (e.g. Betadine™)

- Before an aseptic technique or invasive procedure
- Before contact with a susceptible patient or site, e.g. inspection of IVI site or wound or emptying urine drainage bag
- After contact with body secretions or excretions
- After handling contaminated equipment or laundry
- Before and after caring for patients on isolation precautions

Use liquid soap

- Before commencing drug rounds
- Before serving meals or drinks
- After cleaning spillages
- After using the toilet
- Before meal or beverage breaks

Please do not eat on the wards

- At the end of a span of duty

Use alcoholic chlorhexidine (e.g. Hibisol™)
ONLY AFTER HANDWASHING

- After dealing with patients in Source Isolation
- In special circumstances when dealing with several cases in one unit or during a procedure (consult Infection Control Nurse)
- Alcoholic chlorhexidine must ***not*** be used as a substitute for handwashing

University College London Hospitals Infection Control Team

Fig. 2 A handwashing poster. NB This poster should be on a **yellow** background, indicating infection.

Admissions policy

While admission to the ICU will depend on clinical need, it is important that other patients and staff are not exposed to unnecessary infection risk. Patients with unexplained diarrhea, infected eczema, or communicable diseases, or who are known or suspected carriers of multiple antibacterial-resistant bacteria, should be admitted to source isolation until cleared of transmission risk.

Maintenance of good hygienic practice is vital to ensure a smooth-running ICU that is a safe environment for both patients and staff. Co-operation between the infection control team and ICU staff is essential.

References

Bauer, T.M., Ofner, E., Just, H.M., Just, H., and Daschner, F.D. (1990). An epidemiological study assessing the relative importance of airborne and direct contact transmission of microorganisms in a medical intensive care unit. *Journal of Hospital Infection*, **15**, 301–9.

Doebbeling, B. N., *et al.* (1992). Comparative efficacy of alternative hand-washing agents in reducing nosocomial infections in intensive care units. *New England Journal of Medicine*, **327**, 88–93.

Jarvis, W.R. (1994). Handwashing — the Semmelweiss lesson forgotten? *Lancet*, **344**, 1311–12.

NHS Estates (1994). *Health Technical Memorandum 2010*. Part 1: *Management policy. Sterilization*. HMSO, London.

Sproat, L. J. and Inglis, T. J. (1994). A multicentre survey of hand hygiene practice in intensive care units. *Journal of Hospital Infection*, **26**, 137–48.

13.2.2 Isolation

G. L. Ridgway

Key messages

- Patients may require source isolation or protective isolation.
- Engineering considerations include source isolation, protective isolation, mechanics of intensive care unit isolation, and room utilities.
- Nursing considerations include dress, gloves, and masks.
- When the room is vacated, all surfaces must be thoroughly cleaned with warm water containing detergent and then dried.

Introduction

Two categories of patients admitted to the intensive care unit (ICU) need care delivered in isolation from other patients: immunocompromised patients who are susceptible to all kinds of infecting agents, and patients infected with communicable diseases or colonized or infected with an organism that may have important consequences for other susceptible patients in the ICU. In both cases the objective is to insert a barrier between these patients and all other patients in the ICU, hence the old term 'barrier nursing'. However, this term does not convey the important difference between the two groups. The susceptible patient is at risk from the environment and the people he or she comes into contact with and 'protective isolation' is the goal, whereas the infected patient may be a risk to people he or she is in contact with and other patients and 'source isolation' is the goal. Although the principles of protective and source isolation are very different, the two groups may overlap. For example, the susceptible burned patient may become colonized or infected with a multiple antibiotic-resistant organism.

When designing and planning an ICU, provision for isolation needs to be assured to avoid closure due to a problem of susceptibility or infection. Provision for both protective isolation and source isolation will be necessary in an ICU serving a diverse patient population.

Engineering considerations

Source isolation

It is essential to ensure that air is removed from the patient's immediate environment directly to the outside. Therefore the net outflow from the patient's room must be maintained at negative pressure, with the air ducted away from any other patient care area including areas outside the ICU. Air flow should be unidirectional across door openings, and it is important to ensure that there is no flow reversal between the top and bottom of the door in the absence of an airlock vestibule. Ideally, air should be ducted high up out of the building, but in such a way that there is no possibility of discarded air being drawn into the hospital's air intakes which should also be situated high on the building. Obviously, expert input is required at the design stage, taking into account climatic factors and the geography of nearby buildings which may affect wind patterns. The principles of design are given in *Health Technical Memorandum 2025* (NHS Estates 1994).

Protective isolation

In this case the concern is the source of the air, which should not be from the ICU environment. The room should be pressurized with respect to all other areas. The air-flow and unidirectional requirements are the same as for source isolation. Unless the ICU is dedicated to treating severely immunocompromised patients, it is not necessary to filter incoming air to operating room standards or higher, provided that it has been drawn from the top of the building with the design considerations discussed above.

Mechanics of ICU isolation

Provision of both protective and source isolation facilities adds cost and leads to difficulties with utilization of these specialized rooms. Two solutions are available.

1. The isolation unit can be a single room without an antechamber but with a reversible air flow. This will allow the room to be used for either protective or source isolation. The problems are (a) that patients requiring simultaneous source and protective isolation will have to be nursed under negative pressure, and (b) that great care is necessary to ensure that the air flow has been switched in the correct direction for the particular clinical need. A firm operating policy must be available, otherwise inevitably a source-isolated patient will be positive-pressure ventilated with possible dire consequences for the rest of the ICU.
2. The preferred solution is to have a single room separated from the rest of the ICU by an antechamber that acts as an air lock. Pressurized air can be delivered directly to the room. Exhaust ventilation from the lobby will allow simultaneous protective and source isolation to be carried out without further adjustment (NHS Estates 1993).

Room utilities

Carpeting should be avoided. If carpets are used, they must be of non-absorbent material with a close pile. Vinyl or similar covering with sealed joints and coving to the wall is preferred for ease of cleaning and maintenance, and to minimize the possibility of harboring organisms.

Clinical handwashing facilities are required in each room and in an isolation antechamber. A standard overflow-free basin without a plug and with an elbow-operated bi-flow tap should be supplied. Coat hooks should be available inside and outside the room; outer garments should be hung outside, but gowns used for source-isolated patients should be kept in the room, one to a hook, and hung inside out. Alternatively, disposable plastic aprons may be used. Aprons used in source isolation should be discarded before leaving the room. Equipment in the room should be kept to the necessary minimum.

Nursing considerations

Isolation should be done properly or not at all, and procedures should apply to all persons entering the room, including relatives and medical, paramedical, and nursing staff. Convenience is not a reason for isolation, and the psychological effects on the patient must also be considered. Knowledge of the epidemiology of a particular pathogen may allow some flexibility in management. The reason for isolation must be explained to the patient's relatives and, where possible, to the patient. A record of the need for isolation should be in the patient's nursing record. The infection control team must be informed and can assist with education if required.

The most important route of transmission is contact. The need for meticulous attention to handwashing cannot be overemphasized. Respiratory transmission is generally less of a problem, unless the attendant is susceptible to the infection (e.g. chickenpox or measles). Vaccination of staff against readily transmissible organisms is essential, and a record of vaccinations and previously acquired immunity should be kept and updated.

Laminated instructions, stating the main points of the isolation, should be displayed on the entry door. Examples are shown in Figs 1 and 2.

Dress

Outdoor clothing, jackets, and white coats are removed prior to entering the room. If the patient or any object in the room is to be contacted, sleeves are rolled above the mid-forearm and wristwatches removed. Hands are washed and dried before a protective apron is put on. Color-coded aprons will act as a prompt with regard to procedure; for example, yellow aprons should be used for source isolation, and white aprons for protective isolation.

SOURCE ISOLATION

TO CONTAIN INFECTION TRANSMITTED BY CONTACT, RESPIRATORY, FECAL–ORAL, AND INOCULATION ROUTES

ALL VISITORS must report to nurse in charge BEFORE entering room

REMOVE outdoor clothing or white coat
WASH HANDS before putting on protective clothing
Put on a new APRON before entering room
WEAR GLOVES for direct contact with patient, bed linen, body fluids, dressings, etc.

BEFORE leaving the room, DISCARD protective clothing into yellow bag and WASH HANDS

Keep the door CLOSED

WASH HANDS or use alcoholic chlorhexidine rub after leaving the room

EQUIPMENT must be DECONTAMINATED before it leaves this room

For details of precautions see the INFECTION CONTROL MANUAL

University College London Hospitals Infection Control Team

Fig. 1 Door card for source isolation. NB This poster should be on a **yellow** background, indicating infection.

PROTECTIVE ISOLATION

FOR THE PROTECTION OF IMMUNOCOMPROMISED PATIENTS AGAINST EXOGENOUS INFECTION

ALL VISITORS must report to nurse in charge BEFORE entering room

REMOVE outdoor clothing or white coat

WASH HANDS before putting on protective clothing

Put on a new APRON before entering room

WEAR GLOVES for direct contact with patient

AFTER leaving the room, DISCARD protective clothing into black bag and WASH HANDS

Keep the door CLOSED

For details of precautions see the INFECTION CONTROL MANUAL

University College London Hospitals Infection Control Team

Fig. 2 Door card for protective isolation. NB This poster should be on a **white** background, indicating infection free.

Gloves

Sterile gloves should be used for all aseptic procedures and discarded immediately after use. Non-sterile examination gloves are used for other contact with the patient, including body fluids. They must either be washed and dried after contact, or removed and discarded before handling notes, pens, or other communal items in the room.

Masks

Conventional masks serve no useful purpose for source isolation. If protection from respiratory organisms is necessary, a close-fitting mask or respirator is required. Masks worn for aseptic technique or for the severely immunocompromised should be of the high-efficiency type. They are effective for up to an hour or until damp. Cotton or paper masks are useless. Head coverings and overshoes are not required.

The end of the procedure

At the end of the procedure, protective clothing used for source isolation is either discarded in the room or, if the aprons are to be reused, hung inside out on single hooks in the room. In the latter case, the inside of the apron must be clearly marked. Protective isolation aprons are removed after leaving the room. All discarded materials from source isolation should be treated as infected waste and discarded in accordance with the local waste disposal policy. Persons leaving source isolation should wash their hands before leaving the room, and an alcohol rub can also be applied as a further precaution against transmission.

Room cleaning

Normal room cleaning and dusting, including equipment attached to the patient such as cardiac monitors and hemofiltration consoles, should continue during occupancy. Units must agree on a cleaning policy. Thoughtful explanation of why and how the isolation procedure is being carried out will allay unwarranted fears and ensure efficient cleaning.

Any equipment taken out of the room should be decontaminated according to the local disinfection policy. If equipment is sent for maintenance, proper labeling will be required to ensure safe handling by others.

When the room is vacated, all surfaces must be thoroughly cleaned with warm water containing detergent and then dried. Disinfectants are generally unnecessary, unless a highly infectious disease is involved or multiple-resistant bacteria (e.g. methicillin-resistant *Staphylococcus aureus* (MRSA)) are being contained. In this case, use of a phenolic disinfectant as a final surface wipe is recommended. All bed linen and curtains from source isolation should be sent for laundry, bagged as 'infected' in accord with local policy. Equipment in contact with patients should be wiped with water containing detergent followed by either a phenolic or alcoholic surface disinfectant. All materials are discarded as infected waste. The mop head is detached, bagged, and sent for laundry. Cleaning gloves are washed while still being worn, dried, and then placed with the cleaned mop bucket. Hands are thoroughly washed and dried.

References

NHS Estates (1993). *Nucleus. Isolation facilities study pack*. HMSO, London.

NHS Estates (1994). *Health Technical Memorandum 2025. Ventilation in health care premises*. HMSO, London.

13.2.3 Antimicrobial selection policies

A. P. R. Wilson

Key messages

- An antibiotic policy encourages rational, effective, and safe prescribing while reducing the threat of selection of antibiotic resistance.
- Antibiotics should be classed as unrestricted, available after discussion with the microbiologist or infectious disease physician, or unavailable.
- Meningitis should be treated with cefotaxime or ceftriaxone.
- Ceftazidime or piptazobactam are appropriate for treatment of ventilator-associated pneumonia more than 2 days after intubation.

Introduction

At least 80 per cent of patients receive antibiotic treatment during a stay in the intensive care unit (ICU), and the choice of antibiotic for empirical use can have a dramatic effect on the species and susceptibilities of the pathogens causing nosocomial infections. Ampicillin and imipenem are effective inducers of β-lactamase in Gram-negative bacteria. Substitution of ampicillin with cefuroxime results in a dramatic fall in the ampicillin resistance of Gram-negative isolates and a reduction in *Klebsiella* infections. However, the widespread use of cephalosporins has in turn been linked to the emergence of enterococci as major pathogens because of their general resistance to these agents.

Antibiotic treatment must aim to be rational, effective, and safe while posing the minimum threat of selection of bacterial resistance and remaining economical (Geddes 1988). The majority of infections in the ICU are hospital acquired and many are treated empirically. A policy should be decided for both treatment and prophylaxis of infection based on the common pathogens and their current antibiotic sensitivity patterns. The policy should be flexible and revised whenever indicated by a change in the properties of organisms causing bacteremia or serious infections. Therefore close liaison with the microbiologist and/or infectious disease physician is necessary.

Policies

Antibiotic policies usually place antibiotics in one of three categories: unrestricted, those that can be prescribed after discussion with a microbiologist or infectious disease physician, and those that are unavailable. Penicillin G (benzylpenicillin), amoxicillin (amoxycillin), nafcillin (or flucloxacillin), trimethoprim, cefuroxime (or cefotaxime or ceftriaxone), and metronidazole are usually unrestricted. Ceftazidime, ciprofloxacin, glycopeptides, amikacin, imipenem (or meropenem), and piptazobactam (or piperacillin) may fall into the second category of a restricted formulary. These antibiotics, which are commonly prescribed in the ICU, can be discussed during daily ward rounds with the microbiologist or infectious disease physician. Other cephalosporins, aminoglycosides, ureidopenicillins (depending on the hospital), and sulfonamides may not be available for use.

New antibiotics should be used if they have been demonstrated to be more effective than existing agents or if the pathogens are resistant to standard regimens. In this way their useful life can be prolonged. The dose needs to be adjusted to the severity of infection, the size of the patient, and renal or hepatic function. Intravenous administration should usually be stipulated for the critically ill, although the oral route can be just as efficacious for some agents. Intramuscular administration produces unreliable serum concentrations because of poor tissue perfusion in these patients. The duration of treatment should be tailored to the type of infection; treatment generally lasts for 5 to 7 days (or even longer for cardiac valve infection). Prolonged courses run the risk of superinfection or adverse reactions.

Empirical treatment

Antibiotic treatment is usually administered after cultures have been taken but before any results are available. The choice should be reviewed when culture results have been obtained, and an agent with the narrowest effective spectrum should be selected to reduce the pressure for emergence of resistance (Table 1). Unfortunately, the pathogen causing the infection may not be demonstrated or the cultures may be mixed or negative. The treatment should be continuously reviewed to continue, stop, or modify coverage for the most likely pathogens.

Table 1 Recommended antibiotic treatment for patients in the ICU

Infection	Treatment
Bacteremia	Cefuroxime
Abdominal sepsis	Cefuroxime/metronidazole
Febrile neutropenic episodes	Ceftazidime, ciprofloxacin with glycopeptide/amphotericin
Meningitis	Cefotaxime, ceftriaxone
Streptococcus group A	Penicillin G, debridement
Enterococcus species	Ampicillin or piperacillin
Methicillin-resistant *Staphylococcus aureus*	Teicoplanin or vancomycin
Lobar pneumonia	Penicillin G
Segmental pneumonia	Erythromycin
Nosocomial pneumonia	<3 days cefuroxime
	>3 days ceftazidime
Aspiration pneumonia	Clindamycin
Staphylococcal LRTI	Clindamycin
Urinary infection	Ceftazidime
Diarrhea	Withdraw antibiotics

LRTI, lower respiratory tract infection.

The presence of pulmonary consolidation or an abdominal collection of purulent material will often direct treatment of bacteremia. Without an obvious focus, a parenteral cephalosporin (cefuroxime, cefotaxime, or ceftriaxone) will be effective against the majority of likely causes unless the patient has had a prolonged hospital stay or previous antibiotics. Alternatively, amoxicillin–clavulanate (co-amoxiclav) or penicillin plus an aminoglycoside could be used. If the patient is hypotensive, the addition of an aminoglycoside, usually gentamicin, will improve the bactericidal activity of therapy but will not alter the antibacterial spectrum. If anaerobes are suspected, metronidazole or clindamycin should be added. Neither cephalosporins nor aminoglycosides are active against streptococci or enterococci, and a glycopeptide may be required in hospitals where methicillin-resistant *Staph. aureus* is common. Each microbiology laboratory should maintain lists of blood culture isolates so that the most likely species and recent sensitivity patterns are available to guide treatment (Fig. 1).

Patients with sepsis following abdominal or pelvic surgery may suffer a polymicrobial bacteremia and will require a broad-spectrum regimen such as cefuroxime and metronidazole either with or without gentamicin. An alternative is clindamycin plus gentamicin.

Patients developing a febrile episode during neutropenia or immune suppression are likely to have pseudomonal infection and

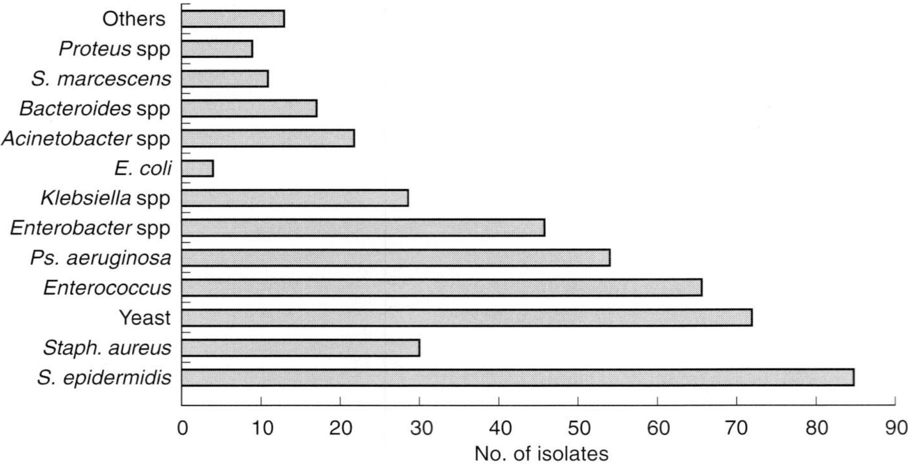

Fig. 1 Causative organisms in 300 surgical patients with septic shock. (Data from Shoemaker *et al.* (1993).)

should be treated without awaiting cultures. Ceftazidime, ciprofloxacin, or piptazobactam are common choices. Imipenem or meropenem have a wide spectrum of activity and are better reserved as second-line therapy. Gram-positive bacteremia is common, and is usually caused by coagulase-negative staphylococci seeded from long-term central venous catheters. These organisms are usually susceptible only to glycopeptides, and some argue that vancomycin or teicoplanin should form part of the initial regimen. However, Gram-positive bacteremia in these patients is rarely fatal, and outcome is similar if a glycopeptide is not started for 48 h when the effects of the initial antibiotic should be apparent. If there are signs of inflammation at the exit site, glycopeptides should be given immediately. Failure to respond to initial therapy, particularly in the presence of infiltrates on the chest radiograph, should prompt the addition of amphotericin B because of the likelihood of infection with *Aspergillus* or *Candida* species. Conventional formulations should be used unless the patient has previously experienced adverse effects with the agent or is developing renal failure.

Specific infections

Immediate treatment of meningitis is important in reducing mortality and is of greatest benefit when a petechial or purpuric rash, suggestive of meningococcal disease, is present. Penicillin G will often have been administered intramuscularly before the patient is admitted. Use of penicillin should override concerns of penicillin allergy unless there is a proven history of immediate-type hypersensitivity, in which case chloramphenicol should be given intravenously. Otherwise, empirical treatment of meningitis should be cefotaxime or ceftriaxone because of the increasing resistance of pneumococci and *Hemophilus influenzae* to penicillins or chloramphenicol. Penicillin G remains the drug of first choice for meningococcal disease. Treatment should be given for 10 to 14 days for pneumococcal meningitis and 7 to 10 days for other types.

Endocarditis should be treated in consultation with infectious disease physicians, microbiologists, and surgeons. The choice of antibiotic depends on the presence of a prosthetic valve, the bacterial species, and its susceptibility. Patients with staphylococcal or prosthetic valve endocarditis are most likely to require intensive care.

Infection caused by β-hemolytic streptococci (e.g. *Streptococcus* group A) should be treated with high-dose penicillin G (12–18 million units/day). Surgical debridement is also often required.

Enterococcal infection is treated with amoxicillin or a ureidopenicillin (azlocillin, piperacillin) combined with an aminoglycoside if Gram-negative infection is likely to be present. Enterococci resistant to penicillins must be treated with vancomycin or teicoplanin. Outbreaks of vancomycin-resistant enterococci are increasing and present a difficult therapeutic problem. Some may be susceptible to high doses of amoxicillin, teicoplanin, or chloramphenicol, but mortality from septicemia is high.

Staph. aureus bacteremia requires high-dose flucloxacillin (1–2 g every 4 h intravenously) or nafcillin given intravenously, usually in combination with an aminoglycoside. The risk of formation of metastatic abscesses is high but may not be clinically obvious for some weeks. Cellulitis can be treated with flucloxacillin alone in a dose of 1 g every 6 h to inhibit both staphylococci and streptococci. Staphylococcal pneumonia can be treated with clindamycin, which penetrates well into the lungs, or flucloxacillin plus fusidic acid. Infections with methicillin-resistant staphylococci should be treated with teicoplanin or vancomycin in combination with gentamicin or rifampin (rifampicin) (if suspectible) if the sepsis is deep seated.

Patients with a well-defined community-acquired lobar pneumonia and hemoptysis will probably have *Streptococcus pneumoniae* and should be treated with penicillin G. High pulmonary concentrations can be achieved using penicillin, but cephalosporins are preferred in areas where there is a high prevalence of penicillin-resistant pneumococci. A bilateral or segmental pneumonia suggests atypical infection and erythromycin can be effective. A common compromise is cefuroxime and erythromycin for severe community-acquired pneumonia. Patients developing pneumonia more than 72 h after admission (i.e. nosocomial pneumonia) can be treated with cefuroxime, cefotaxime, or ceftriaxone if ventilated for less than 2 to 3 days, as the usual pathogens are *Strep. pneumoniae*, *H. influenzae* or coliforms. Patients requiring long-term ventilation will usually be infected with Gram-negative aerobic flora including *Pseudomonas* species. Ceftazidime, ciprofloxacin, or piptazobactam are effective, but treatment should be reviewed after 3 to 5 days.

Aspiration pneumonia is a common event before or during an ICU admission, and the combination of anaerobes and upper respiratory flora can be treated with clindamycin or cefuroxime plus metronidazole. However, many episodes are chemical in nature and the low-pH gastric aspirate will not harbor bacteria. If this is the case, antibiotics serve little purpose and should not be started.

Legionnaire's disease may present as a severe pneumonia with organ failure unresponsive to cefuroxime. Patients respond to erythromycin (up to 4 g/day) and rifampin is often given in addition. Treatment should be continued for at least 2 weeks to prevent relapse.

Pneumonia due to *Pneumocystis carinii* is the most frequent presentation of AIDS and is a common reason for admission and ventilation. High-dose trimethoprim–sulfamethoxazole (co-trimoxazole) is the treatment of first choice, but adverse effects are common and alternatives include pentamidine or clindamycin plus primaquine.

Urinary tract infections in the critically ill are usually associated with Foley catheters and should not be a reason to commence antibiotic treatment unless the patient is symptomatic. *Escherichia coli*, *Pseudomonas aeruginosa*, *Klebsiella* species, and *Enterobacter* species are common pathogens and should be treated with ceftazidime or ciprofloxacin. Community-acquired infections or pyelonephritis can be treated with cefuroxime or another cephalosporin plus gentamicin if the patient is in shock.

Diarrhea is commonly caused by high osmolar nasogastric feeds, lactose intolerance, or ischemic damage rather than infection. However, the use of antibiotics is often associated with pseudomembranous colitis caused by *Clostridium difficile*. The correct response is to withdraw antibiotics, thus removing the selective environment for the organism. Most cases will then resolve. If disease is severe, antibiotics cannot be stopped, or there is dilatation of the colon, oral vancomycin or metronidazole should be given.

Severe soft tissue infections, such as necrotizing fasciitis and gas gangrene, require rapid surgical debridement, and this should take precedence over all other management. High doses of penicillin G and metronidazole are effective, but only as an adjunct to surgery and not as the sole treatment.

Adults suffering a primary infection with varicella can develop chickenpox pneumonia, which in its most severe form requires ventilatory support. Acyclovir (aciclovir) can be beneficial if given early in the disease, but prevention of spread, particularly among non-immune staff, is of prime importance. Viral encephalitis is treated with acyclovir and prolonged supportive therapy. Cytomegalovirus infection usually occurs in AIDS, transplant, or other immunosuppressed patients, and may present as retinitis or pneumonitis. Ganciclovir is the usual treatment but must be started early in the disease.

Prophylaxis

For the prevention of surgical wound infection, the antibiotic must be present in the tissues at the time of incision in sufficient concentrations to kill the bacteria most likely to contaminate the wound. It should not be of a broader spectrum than necessary. Antibiotics administered after surgery confer no additional benefit and pose the risk of the induction of bacterial resistance or adverse side-effects (Classen *et al.* 1992). A short course (24 h) of a bactericidal antibiotic should be started at induction of anesthesia. For operations where a site with normal flora is entered, a second-generation cephalosporin either with or without metronidazole is a common choice. In cardiac surgery, flucloxacillin plus an aminoglycoside can be used, although cephalosporins are effective despite the resistance of most coagulase-negative staphylococci. The glycopeptides are active against these organisms, but vancomycin can cause hypotension during surgery and teicoplanin may not achieve sufficiently high tissue levels after a single preoperative dose.

Selective decontamination

Nosocomial pneumonia is commonly preceded by colonization of the oropharynx by Gram-negative aerobes. Selective digestive decontamination (SDD) may use an oral antifungal agent (e.g. amphotericin B) and two non-absorbable antibiotics, such as tobramycin and polymyxin B, given both as an oral paste and by mouth. Rapid reduction in the Gram-negative flora is ensured by administration of a systemic antibiotic, usually cefotaxime, for the first 4 to 5 days. Although some trials have suggested that SDD prevents both nosocomial pneumonia and mortality, few were placebo controlled or blinded. Double-blind placebo-controlled trials show reductions in the number of episodes of pneumonia, urinary infection, and sinusitis, but usually not in mortality or costs (Silver and Bone 1993). Surprisingly, emergence of resistant bacteria during SDD is either rare or not easily demonstrated.

References

Classen, D.C., Evans, R.S., Pestotnik, S.L., Horn, S.D., Menlove, R.L., and Burke, J.P. (1992). The timing of prophylactic administration of antibiotics and the risk of surgical-wound infection. *New England Journal of Medicine*, **326**, 281–6.

Geddes, A.M. (1988). Antibiotic therapy—a resumé. *Lancet*, **i**, 286–9.

Shoemaker, W.C., Appel, P.L., Kram, H.B., Bishop, M.H., and Abraham, E. (1993). Sequence of physiologic patterns in surgical septic shock. *Critical Care Medicine*, **21**, 1876–89.

Silver, M.R. and Bone, R.C. (1993). Selective digestive decontamination in critically ill patients. *Critical Care Medicine*, **21**, 1418–20.

13.2.4 Selective decontamination

C. P. Stoutenbeek and H. K. F van Saene

Key messages

- Selective decontamination of the digestive tract is one of the best evaluated infection prevention techniques in intensive care.
- Selective decontamination reduces the risk of pneumonia in mechanically ventilated intensive care patients by 60 to 70 per cent.
- When the full regimen is applied mortality may be reduced by 20 per cent.
- The risk of emergence of resistance is low.

Introduction

Selective decontamination of the digestive tract (SDD) is an infection prevention technique used in granulocytopenic patients, transplantation surgery, and critically ill intensive care patients. It is based on two observations: first, most infections are endogenous, i.e. they are caused by aerobic Gram-negative bacteria, *Staphylococcus aureus*, and yeast species carried in the oropharyngeal or intestinal flora; second, the indigenous, mostly anaerobic, flora has important physiological functions and is seldom the cause of infection, and therefore should be preserved as much as possible.

Primary and secondary endogenous infections

Endogenous infections can be classified as primary or secondary. Primary endogenous infections are caused by potentially pathogenic micro-organisms already present in the throat or gastrointestinal tract on admission to the intensive care unit (ICU), whereas secondary endogenous infections are caused by potentially pathogenic micro-organisms acquired during the patient's stay in the ICU which secondarily colonize the throat and intestinal tract. Although both exogenous infections and secondary endogenous infections are caused by 'exogenous' micro-organisms, in the former the multiplication phase of potentially pathogenic micro-organisms is outside the body whereas in the latter it is inside the oropharynx or gastrointestinal tract. This difference is important from the point of view of prophylaxis, because the multiplication of pathogens inside the body can be prevented by SDD.

Normal flora

The normal indigenous flora consists of both anaerobic and aerobic micro-organisms. A healthy individual carries high concentrations of anaerobes (*Veillonella*, 10^8 CFU/ml saliva; *Bacteroides* species, 10^{12} CFU/g feces) and aerobic Gram-positive cocci (enterococci, 10^6 – 10^8 CFU/g feces; viridans streptococci, 10^6 –10^8 CFU/ml saliva) in the oropharynx and colon (CFU, colony-forming unit). *Escherichia coli* can be found in the feces of healthy individuals in a concentration of 10^5 CFU/g. Other aerobic Gram-negative bacilli are found in low concentrations and transiently. Healthy individuals frequently carry *Streptococcus pneumoniae*, *Hemophilus influenzae*, *Moraxella catarrhalis*, *Staph. aureus*, and *Candida* species in the oropharynx.

The normal intestinal flora is a very stable ecosystem. The digestive tract has powerful mechanisms (carriage defense) to control exogenous micro-organisms ingested with food and beverages. These include the following.

1. The integrity of the anatomy and physiology of the mucosa, which are covered by a protective blanket of mucus, prevents the adherence of micro-organisms.

2. Motility, including swallowing, chewing, deglutition, gastric emptying, and peristalsis, promotes the clearance of potentially pathogenic micro-organisms present in the daily food intake.

3. Gastric acidity.

4. Secretions such as saliva, gastric and pancreatic juice, bile, and mucus contain bactericidal products (e.g. lysozyme, peroxidase, S-IgA, and enzymes).

5. Mucosal cell renewal contributes to the removal of potentially pathogenic micro-organisms adhering to the mucosal cells.

6. The indigenous anaerobic flora inhibits exogenous bacteria by competing for food and producing bacterial toxins and volatile fatty acids (colonization resistance).

Impaired carriage defense and the abnormal carrier state

Carriage defense is the first line of defense to fail in critically ill patients. The adherence of Gram-negative bacilli to mucosal cells is greatly increased. Stasis, an increase in gastric pH, and mucosal atrophy all lead to an explosive overgrowth of aerobic Gram-negative bacilli. The intravenous administration of broad-spectrum antibiotics excreted via saliva, bile, and mucus may eradicate both the indigenous flora and the sensitive 'community' potentially pathogenic micro-organisms. Consequently, resistant Gram-negative bacilli, *Staph. aureus,* and *Candida* species are selected, followed by overgrowth in the oropharynx and gastrointestinal tract.

From a microbiological point of view, the impairment of carriage defense is associated with an abnormal carrier state. The incidence of abnormal carriage in mechanically ventilated patients is 60 to 90 per cent. High concentrations of mostly Gram-negative bacilli ($>10^8$/ml), but also *Staph. aureus* and *Candida* species, can be found in saliva, stomach, and gut as soon as 48 h after admission to the ICU.

Abnormal carriage predisposes to (secondary endogenous) infections and emergence of resistance, and is associated with multiple organ failure. The pathophysiological mechanisms by which infections and multiple organ failure may develop are (micro)aspiration, continuous spreading of potentially pathogenic micro-organisms to neighboring organ systems, translocation, and absorption of endotoxin.

SDD

SDD is a technique for preventing and treating abnormal carriage (van Saene *et al.* 1992). It has three essential elements:

(1) the application of topical non-absorbable antibiotics to both the oropharyngeal cavity and gastrointestinal tract for as long as the patient is at risk of infection;

(2) a short period of systemic antibiotic prophylaxis in the first few days after initiation of therapy;

(3) intensive bacteriological monitoring of the oropharynx and gastrointestinal tract throughout the treatment.

Topical non-absorbable antibiotics

The aim of the application of topical antibiotics is the selective elimination of the aerobic potentially pathogenic micro-organisms from the oropharyngeal cavity and gastrointestinal tract, thereby preventing endogenous infections and the emergence of resistance. The classical Groningen SDD regimen comprises polymyxin, tobramycin, and amphotericin B (PTA regimen) (Table 1). It covers all aerobic Gram-negative potentially pathogenic micro-organisms, *Staph. aureus*, and yeast, but has relatively little effect on the indigenous flora. Polymyxin is the cornerstone of SDD because it has a broad Gram-negative spectrum and there is no acquired resistance to this drug. However, the gap in the spectrum of polymyxin includes *Proteus* and *Providencia* species, and therefore it should be combined with another agent (e.g. an aminoglycoside or a fluoroquinolone). Tobramycin is the drug of choice because it has a potentiating bactericidal effect on *Pseudomonas* and *Acinetobacter* species.

The correct choice and dosage of the topical antibiotics is crucial for the success of SDD. Agents like neomycin, nalidixic acid, and

Table 1 PTA regimen

	Oropharynx	Gastrointestinal tract	Dosing
Polymyxin	2% in Orabase®	100 mg	Four times daily
Tobramycin	2% in Orabase®	80 mg	Four times daily
Amphotericin B	2% in Orabase®	500 mg	Four times daily

erythromycin which are used in the preoperative preparation for colorectal surgery or in granulocytopenic patients are unsuitable for decontamination because they are almost completely inactivated by feces.

The oropharynx is the most important reservoir of potentially pathogenic micro-organisms causing respiratory tract infections. Therefore it is essential to decontaminate both the oropharynx and the gastrointestinal tract.

In an effectively decontaminated patient only anaerobes, enterococci, viridans streptococci, and coagulase-negative staphylococci are found in throat or rectal cultures, since these micro-organisms are not sensitive to the topical antibiotics used. This is the aim of the technique and should not be regarded as overgrowth because the absolute numbers do not increase. Most controlled clinical trials show that the isolation of enterococci and *Staphylococcus epidermidis* in the oropharynx and rectum has no clinical sequelae.

Systemic antibiotic prophylaxis

A short-term systemic antibiotic prophylaxis is generally given during the first 3 to 4 days of SDD treatment with the following objectives:

(1) elimination from the oropharynx of Gram-positive potentially pathogenic micro-organisms which are insensitive to the topical antibiotics (e.g. pneumococci, hemolytic streptococci);

(2) early treatment or prophylaxis of primary endogenous infections in the first few days when the decontamination is not yet complete.

Early-onset infections in mechanically ventilated intensive care patients are invariably caused by potentially pathogenic micro-organisms carried by the patient in the throat or gastrointestinal tract and aspirated before or during intubation (i.e. primary endogenous infections). Primary endogenous infections, occurring mainly within the first 4 days in the ICU, constitute about 50 per cent of all respiratory infections developing in intensive care when no systemic antibiotic prophylaxis is given. Primary endogenous infections in patients who have not previously been in hospital (e.g. multiple trauma patients) are mostly caused by *Strep. pneumoniae, H. influenzae, Staph. aureus* and *E. coli,* whereas species from the *Klebsiella–Enterobacter* group are more frequently found in patients who have been in hospital prior to admission to the ICU.

Microbiological surveillance

Bacteriological monitoring is another essential element of SDD. It has the following objectives:

(1) to control the effectiveness of the decontamination procedure;

(2) to adjust the antibiotic regimen in case of colonization by resistant strains such as methicillin-resistant *Staph. aureus* (MRSA);

(3) to recognize exogenous infections and take appropriate measures.

Systematic reviews of SDD trials

To date there have been 46 trials of SDD and three meta-analyses have recently been performed (SDD Trialists' Collaborative Group 1993; Heyland *et al.* 1994; Kollef 1994). These meta-analyses differ not only in the number of trials and patients analysed, but also in the methodology. Kollef (1994) analysed only published trials including 2270 patients in 21 trials. Heyland *et al.* (1994) analysed 3395 patients in 25 published and unpublished randomized trials. The SDD Trialists' Collaborative Group (1993) analysed 4142 patients from 22 randomized trials, both published and unpublished; in order to perform an intent-to-treat analysis they collected all patient data through direct contact with the investigators.

All meta-analyses showed a reduction of 60 to 70 per cent in the relative risk of respiratory infections in patients receiving intensive care. The SDD Trialists' Collaborative Group found that SDD significantly reduced respiratory tract infections (odds ratio (OR), 0.37; 95 per cent confidence interval (CI), 0.31–0.43). The effect of SDD on pneumonia rate was independent of both the criteria used for diagnosis of pneumonia and the blinding of the study (SDD Trialists' Collaborative Group 1993; Heyland *et al.* 1994). Although the infection rate was lower in studies diagnosing pneumonia by the protected brush technique or by bronchoalveolar lavage than in studies using clinical criteria, it was demonstrated that the reduction of the odds ratio for pneumonia was similar in both subgroups of studies.

The effect of SDD on mortality is less clear. In the meta-analysis by the SDD Trialists' Collaborative Group the value of the common odds ratio for total mortality (OR, 0.90; CI, 0.79–1.04) suggested a moderate treatment effect, reaching statistical significance only when the subgroup of trials using both topical and systemic treatment was considered separately (OR, 0.80; CI, 0.67–0.97). This was confirmed by the meta-analysis by Heyland *et al.* (1994). In a subgroup analysis the largest absolute difference was found in the comparison of studies using regimens including intravenous antibiotics (risk ratio (RR), 0.80; CI, 0.68–0.95) and those that excluded systemic antibiotics (RR, 0.99; CI, 0.83–1.19).

The subgroup of studies using a systemic component also differs from those without a systemic component with respect to the decontamination regimen used: in the majority of the former both oral and intestinal decontamination with the PTA regimen were used, whereas in the latter many different topical antibiotic regimens were used and were applied only in the gastrointestinal tract without oropharyngeal decontamination. Thus the mortality difference between the two groups may be explained not only by the use of the systemic component but also by the effectiveness of the topical regimen. The discrepancy between the dramatic reduction in infection rate and the modest reduction in mortality clearly demonstrates that the underlying disease is a more important determinant of mortality than acquired infections. SDD might be particularly beneficial in patients with reversible underlying diseases (e.g. surgical patients)

The clinical relevance of the reduction in mortality can be assessed by the number of patients to be treated to prevent one death. Based on the results from trials in which combined topical and systemic treatment was used, 23 patients (range 13–139) would need to be treated with SDD to prevent one death (SDD Trialists' Collaborative Group 1993).

Inherent limitations of SDD

As a technique for infection prevention SDD has a number of limitations.

1. MRSA is frequently resistant to the tobramycin in the PTA regimen and may be selected under SDD. Therefore this technique should not be used in an environment where MRSA is endemic unless appropriate measures are taken, for example adding vancomycin to the regimen.
2. SDD has no effect on exogenous infections. If infections occur despite effective oral and intestinal decontamination, the origin of the infection is probably exogenous and appropriate measures should be taken. This may explain the apparent lack of effect of SDD in some studies.
3. Oropharyngeal decontamination using Orabase® containing 2 per cent PTA and decontamination of the stomach and upper gastrointestinal tract takes effect within 24 h. However, decontamination of the lower part of the gastrointestinal tract, the cecum, and the colon may take more than a week owing to the presence of a paralytic ileus. The risk of infection is high until the bowel is effectively decontaminated. Enemas or prokinetic drugs may accelerate the onset of decontamination.

Indications for SDD

Other indications have been studied in addition to the general indication in intensive care to prevent ventilator-associated pneumonia in patients requiring prolonged mechanical ventilation.

Preoperative SDD in high-risk surgery

Trials with preoperative SDD have shown that it is highly effective in reducing the risk of perioperative infection in high-risk surgery such as esophageal resections, liver transplantation, and small-bowel transplantation. The advantage of preoperative SDD is that intestinal decontamination can be achieved within 48 h before the operation when the bowel is still functioning, whereas when it is started postoperatively 7 to 10 days may be required before decontamination is complete.

SDD in acute pancreatitis

It is generally accepted that the initially sterile pancreatic necrosis secondarily infects with micro-organisms translocating from the gut. In a recent clinical study (Luiten *et al.* 1995) it has been shown that SDD reduced the infection-related morbidity (35 per cent) and mortality (22 per cent) in severe acute pancreatitis. There was no difference in the early mortality from multiple organ failure associated with the acute pancreatitis, but late infection-related multiple organ failure and death was reduced by SDD. The number of patients with infected necrosis and the number of laparotomies per patient were significantly lower in the group treated with SDD group than in the control group.

SDD in fulminant liver failure and burns

SDD has been shown to be very effective in preventing infectious complications by intestinal *Enterobacteriaceae* and *Pseudomonadaceae* in patients with acute fulminant liver failure waiting for transplantation and in burn patients. No controlled clinical SDD trials with mortality as the endpoint have been performed in burn patients to date.

Control of outbreaks

Outbreaks of infection with multiresistant potentially pathogenic micro-organisms in ICUs are generally caused by patients carrying these potentially pathogenic micro-organisms in the throat or gastrointestinal tract and thus acting as a reservoir for other patients. Prevention of transmission between patients is often extremely difficult owing to the lack of isolation facilities and the low nurse-to-patient ratio. In addition, only a few bacteria need be transmitted to the oropharynx of another patient to colonize the oral cavity and gastrointestinal tract and establish another reservoir. This chain of events can be stopped by decontaminating all long-stay patients in the ICU until the outbreak is over.

Emergence of resistance

Emergence of resistance under SDD includes resistance to the systemic antibiotic used in SDD, resistance to topical antibiotics, and the selection of indigenous aerobic Gram-positive cocci.

In general, resistance to systemic antibiotics develops in the oropharynx and gut, where high concentrations of potentially pathogenic micro-organisms are present and subinhibitory concentrations of antibiotics secreted via saliva or bile are reached. Superinfections are mostly secondary endogenous infections with resistant potentially pathogenic micro-organisms selected in the oral or intestinal flora. During SDD the oral cavity and gut are decontaminated with very high concentrations of topical antibiotics, preventing selection of resistance against the systemic antibiotics.

Acquired resistance of Gram-negative potentially pathogenic micro-organisms to polymyxin has never been described, although some strains (e.g. *Proteus* species) are intrinsically insensitive to polymyxin. Resistance to topical tobramycin should be related to the levels of tobramycin in the gut, being more than 100 times higher than the maximum attainable blood levels. Therefore strains considered to be resistant to tobramycin according to the standard criteria can often be eliminated by topical tobramycin. Emergence of resistance should not be analyzed by comparing the percentages of resistant isolates. It is important to exclude the copy strains (i.e. the same micro-organism from the same site from the same patient cultured more than once) and to compare the incidence of colonization/infection by resistant strains. Eliminating the sensitive Gram-negative potentially pathogenic micro-organisms with SDD results in a higher percentage of resistant isolates than is found without SDD, where the resistant isolates are diluted by the large number of sensitive potentially pathogenic micro-organisms. This may lead to the erroneous conclusion that emergence of resistance may be promoted by SDD. In most controlled SDD trials no evidence for emergence of resistance has been found. In many ICUs where the same antibiotic regimen for SDD has been used for more than 10 years there is still no evidence of emergence of resistance.

It has been suggested that SDD may lead to a Gram-positive selection. During successful decontamination, enterococci and *Staph. epidermidis*, which belong to the normal indigenous flora and are intrinsically insensitive to the topical antibiotics, are the only micro-organisms isolated. Controlled trials that specifically studied enterococci and *Staph. epidermidis* did not show increased colonization or infection with these micro-organisms during SDD.

MRSA is an inherent limitation of the PTA regimen. In an environment where MRSA is endemic, SDD may select this micro-organism. In this situation SDD should be reconsidered or vancomycin should be added to the PTA regimen.

References

Selective Decontamination of the Digestive Tract Trialists' Collaborative Group (1993). Meta-analysis of randomised controlled trials of SDD of the digestive tract. *British Medical Journal*, **307**, 525–32.

Heyland, D.K., Cook, D.J., Jaeschke, R., Griffith, L., Lee, H.N., and Guyatt, G.H. (1994). SDD of the digestive tract: an overview. *Chest*, **105**, 1221–9.

Kollef, M.H. (1994). The role of selective digestive tract decontamination on mortality and respiratory tract infections. A meta-analysis. *Chest*, **105**, 1101–8.

Luiten, E.J.T., Hop, W.C.J., Lange, J.F., and Bruining, H.A. (1995). Controlled clinical trial of SDD for the treatment of severe acute pancreatitis. *Annals of Surgery*, **222**, 57–65.

van Saene, H.K.F., Stoutenbeek, C.C., and Stoller, J.K. (1992). SDD of the digestive tract in the intensive care unit: current status and future prospects. *Critical Care Medicine*, **20**, 691–703.

13.3 Nosocomial infection

13.3.1 Pathophysiology and management

P. S. Barie and J. T. Cooke

Key messages

- At any given time, about 20 per cent of critically ill patients will have an infection acquired in the intensive care unit.
- The most important pathogens are enteric Gram-negative bacilli, *Pseudomonas*, and staphylococci. Enterococci and yeast can infect severely immunocompromised hosts.
- The patient, or another nearby, is often the source of the microbes. Health care workers, including physicians, are the common mode of transmission.
- Careful handwashing with microbicidal soap is the most effective means of prophylaxis.

Introduction

The critically ill patient, who is often elderly, immunosuppressed owing to trauma, malignant disease, or following solid-organ transplantation, already infected, or being administered broad-spectrum antibiotics (as treatment or prophylaxis), is a paradigm for the patient at risk of nosocomial infections. Therapeutic invasion of a normally sterile body cavity by virtue of surgery, invasive hemodynamic monitoring, or endotracheal intubation for anesthesia or prolonged ventilatory support also confers risk. The incidence of certain nosocomial infections, notably fungemia and Gram-positive bacteremia, has increased markedly within the past decade (Jarvis *et al.* 1991; Barie 1993). The type of intensive care unit (ICU) being studied may influence the observed prevalence of infection. For example, pneumonia has now supplanted surgical wound infection as the most common nosocomial infection in surgical patients in the United States (Jarvis *et al.* 1991), but catheter-related bacteremia is more common in both medical and pediatric units than in surgical ICUs. The cost of nosocomial infection is enormous in terms of morbidity and expenditure (Pittet *et al.* 1994) for both therapy and the necessary surveillance programs.

An extensive 1-day point-prevalence study (Vincent *et al.* 1995) was conducted to determine the prevalence of and risk factors for ICU-acquired infections, identify the predominant infecting organisms, and evaluate the relationship between ICU-acquired infection and mortality. A total of 1417 ICUs provided 10 038 patient case reports, of whom 4501 (44.8 per cent) were infected; nearly half of these (2064 or 20.6 per cent of the total population) had an ICU-acquired infection. Pneumonia accounted for nearly 50 per cent of the ICU-acquired infections, and bloodstream infections accounted for 12 per cent. The most frequently reported micro-organisms were enteric Gram-negative bacilli (34 per cent), *Staphylococcus aureus* (30 per cent, but 60 per cent of those strains were methicillin resistant compared with a prevalence of about 25 per cent overall), *Pseudomonas aeruginosa* (29 per cent), coagulase-negative staphylococci (19 per cent, usually *Staphylococcus epidermidis*), and fungi (17 per cent). Independent risk factors for ICU-acquired infection were an ICU stay of more than 48 h, mechanical ventilation, trauma, central venous catheterization, and stress ulcer prophylaxis (although the method was unstated).

The impact of nosocomial infection on mortality in a university hospital ICU was assessed by a prospective cohort study using standard definitions for nosocomial infections (Bueno-Cavanillas *et al.* 1994). The crude mortality risk was 2.48 times higher in patients with a nosocomial infection than in non-infected patients. By stepwise logistic regression analysis, the severity-adjusted risk of death in infected patients was 2.1 times greater than in patients without such an infection. Thus nosocomial infections increase the risk of death by a factor of 3 to 3.5.

Patients in the ICU often have symptoms and signs of systemic infection, but have no identifiable source of infection. This constellation is now referred to as the systemic inflammatory response syndrome. The distinction between infection and systemic inflammatory response syndrome, which is the clinical manifestation of the host response to infection or inflammation, can be very difficult to make at the bedside. Identification of bacteria in a specimen may reflect mucosal colonization rather than infection; this is a particular problem in the diagnosis of pneumonia. Moreover, many clinical conditions, including burns, pancreatitis, and aspiration of gastric contents, provoke a major inflammatory response in which the tissue is usually sterile initially but may become infected later. The acceptance of any positive cultures as prima facie evidence of infection may lead to the overuse of antibiotics, which is both costly and leads to morbidity such as allergy, visceral organ dysfunction (e.g. aminoglycoside nephrotoxicity), development of antibiotic-resistant bacterial infections, and disruption of endogenous flora with emergence of opportunistic bacterial and fungal pathogens (e.g. *Candida, Enterococcus, Clostridium difficile,* and *Pseudomonas*).

Common sense, cost-consciousness, and concern for the microbial ecology of patients and the facilities in which they are cared for mandates that broad-spectrum antibiotics be reserved for treatment of serious infections. Increasing bacterial resistance to currently available antibiotics is cause for alarm. As in the case of the enterococci, the gene that confers vancomycin resistance has been transferred experimentally to staphylococci. Confirmation of the reported emergence of clinical strains of vancomycin-resistant staphylococci, which many authorities now consider to be purely a matter of time, could render these common and dangerous infections untreatable.

Selected aspects of pathogenesis

Bacterial translocation

In addition to digestion and absorption of nutrients, a major function of the gastrointestinal tract is to serve as a barrier to microbial invasion of the host and to prevent systemic dissemination of bacterial toxins. During critical illness the gut barrier fails, allowing toxins and viable bacteria to escape the intestine and gain access to the circulation, primarily via mesenteric lymphatics and the portal circulation, leading to bacterial translocation. A number of components, including a normal enteric microflora and intestinal pH, enteral feeding, intraluminal secretion of immunoglobulin A, and clearance of escaping organisms by the reticuloendothelial system, contribute to the function of the gut as a barrier against intestinal microbial invasion of the host. A variety of direct and indirect insults to the gastrointestinal tract may lead to loss of effective barrier function. Increased intestinal permeability and the translocation of bacteria, endotoxin, or other toxins may then ensue.

Evidence for translocation in humans is largely circumstantial. Trauma patients in shock have bacteremia and endotoxemia in the absence of injury to a hollow viscus. Clinical reports suggest increased intestinal permeability after burns and sepsis, and increased permeability has been reported after endotoxemia in volunteers. Concern about nosocomial pneumonia in ICU patients has led to experiments to determine whether suppression of intestinal bacterial overgrowth will diminish this risk. Selective antibiotic decontamination of the digestive tract has been advocated to accomplish this goal; however, it lacks a defined role.

Nosocomial pneumonia

Hospital-wide nosocomial pneumonia has become more common than surgical wound infection, and is numerically second only to urinary tract infection. The majority of nosocomial pneumonias develop in ICU settings. The overall mortality approximates 35 per cent and may be as high as 50 per cent in tertiary care centers. The mortality of nosocomial pneumonia is related to the type of organism and the concomitant presence of bacteremia (the overall incidence is about 10 per cent, but mortality increases threefold in cases complicated by bacteremia). Mortality from pneumonia caused by Gram-negative organisms (most commonly *Pseudomonas, Enterobacter*, or *Klebsiella*) may be at least twice as high as that from pneumonia caused by Gram-positive organisms (usually *Staph. aureus*). Mortality is particularly common in *Pseudomonas* pneumonia, with as many as 70 per cent of affected patients dying.

Critically ill patients are at particular risk of the development of nosocomial pneumonia (Table 1). Both short-term endotracheal intubation and long-term mechanical ventilation increase the risk because protective reflexes are lost. Since the nasopharyngeal 'filter' is bypassed, tracheobronchial mucociliary clearance of particulates becomes essential, but it is suppressed by the presence of the tube. Other risk factors include increasing age and immunosuppression. Pre-existing chronic pulmonary disease may pose a risk independent of prolonged ventilator dependency. Although endogenous flora and lapses in disinfection protocols as caregivers move from patient to patient can explain the higher incidence of pneumonia due to enteric Gram-negative bacilli, other mechanisms may be responsible for the high incidence of pneumonia from *Pseudomonas*, which is not normally present in the human gastrointestinal tract.

Table 1 Intensive care interventions associated with nosocomial pneumonia

Use of H_2 receptor antagonists
Increased gastric pH
Enteric bacterial overgrowth
Bacterial translocation
Intracranial pressure monitoring
Prolonged mechanical ventilation
Loss of protective airway reflexes
Aspiration of gastric contents

The colonization of airway surfaces with Gram-negative bacilli is central to the pathogenesis of nosocomial pneumonia. Although nosocomial pneumonia may follow bacteremia, the majority of these pneumonias appear to be due to aspiration of pathogens that have colonized the upper airway. The oropharyngeal flora of hospital inpatients changes rapidly to reflect a predominance of Gram-negative organisms, even if the patient does not receive parenteral antibiotics or spend much time in the ICU.

Hospital admission alone leads to a 23 to 45 per cent incidence of bacterial colonization of the pharynx with aerobic Gram-negative bacilli (compared with an isolation rate of less than 2 per cent from healthy individuals), and 45 to 60 per cent of patients admitted to the ICU are colonized within 1 week of admission. Nosocomial pneumonia subsequently develops in about 25 per cent of colonized patients compared with only 3 per cent of patients whose oropharyngeal flora remains normal.

The diagnosis of nosocomial pneumonia poses both clinical and microbiological challenges, particularly in the intubated patient. The classical findings of fever, purulent sputum, leukocytosis, and a new pulmonary infiltrate actually represent infection in only about 45 per cent of bacterial pneumonias in ventilator-dependent patients. Clearly, better methods of detecting nosocomial pneumonia are needed. This issue is fundamental because it can be very difficult to distinguish colonization from infection, or bronchitis from pneumonia, in patients who are infected. The analysis of tracheal sputum aspirates collected via a conventional suction catheter is notoriously unreliable. Such specimens are often contaminated with upper airway flora and may not be representative of flora deep within the lung. Numerous invasive techniques have been described to circumvent these shortcomings. The advent of fiber-optic bronchoscopy has led to the development of additional techniques. The simple aspiration of sputum after introduction of the bronchoscope via the endotracheal tube or pharynx harbors the same potential for the confounding identification of colonizing organisms. The protected specimen brush circumvents many of these difficulties, with sensitivity rates for the diagnosis of pneumonia reported to be between 70 and 90 per cent in patient studies. Newer directional catheters may obviate the need for bronchoscopic introduction of the brush. Accuracy has been enhanced by using quantitative bacteriology to distinguish colonization from invasive infection. Finding at least 10^3 colony-forming units/ml by quantitative microbiology is a reasonable indicator of invasive infection by protected specimen brush. However, the very small sample size (estimated to be only 0.01 ml) is a potential shortcoming.

Bronchoalveolar lavage has increased the overall accuracy of bronchoscopic diagnostic techniques, particularly when combined with quantitative cultures. Instillation of 10 ml of saline into a pulmonary segment distal to a 'wedged' bronchoscope may sample as many as 10^6 alveoli. Such a comparatively large sample appears to be responsible for enhanced overall accuracy. The technique has been reported to be 88 per cent sensitive for the diagnosis of bacterial infection of the lower respiratory tract when bacterial counts exceeded 10^4 colony-forming units/ml, and no more than 1 per cent squamous epithelial cells were present in a differential cell count of the fluid.

Aspiration pneumonitis

The swallowing mechanism prevents aspiration of gastric contents almost completely in normal subjects, and the cough reflex usually clears the airway before injury is produced when aspiration does occur. However, the aspiration of foreign material into the lung carries a substantial morbidity and mortality. It is the leading mechanism in the pathogenesis of nosocomial pneumonia, and is a very common event in the ICU (Table 2). The average hospital stay is increased by 21 days in surgical patients who aspirate, and the mortality of aspiration exceeds 50 per cent in almost all reported series.

Aspiration of gastric contents produces a severe chemical pneumonitis, although the injury is attenuated if the pH of the aspirate is 2.4 or greater. Secondary bacterial infection does not invariably occur following aspiration. All forms of aspiration may eventually result in pneumonia, although community-acquired aspiration pneumonias are usually anaerobic whereas nosocomial events are mixed infections. The most common anaerobic organisms recovered following aspiration are *Bacteroides*, *Fusobacterium*, and *Peptostreptococcus*. Aerobic and facultative organisms are most commonly *Staph. aureus*, *Streptococcus pneumoniae*, enteric Gram-negative bacilli including *Escherichia coli* and *Klebsiella*, and *Pseudomonas*. However, because bacterial infection is not a certainty following aspiration, the common practice of presumptive administration of antibiotics should be discontinued. The presence of fever is not an indication for antibiotic therapy after aspiration, because chemical pneumonitis commonly produces fever. Specific therapy should be directed against a Gram stain of an adequate sputum specimen or a specific bacterial isolate.

Prevention of aspiration is undoubtedly the most important aspect of management. Meticulous airway care is essential. The patency of all indwelling nasogastric tubes must be continually confirmed. Careful monitoring of nasogastric tube feeding is extremely important. A bolus of feeding solution should never be given in an attempt to 'catch up' with an infusion that is behind schedule. The use of soft weighted-tip silastic catheters, positioned with the tip in the duodenum, is preferable.

Table 2 Factors predisposing to aspiration

General anesthesia
Endotracheal intubation
Narcotic analgesia
Drug and alcohol overdose
Head trauma or other central nervous system injury (e.g. stroke)
Gastric and intestinal motility disorders
Cardiopulmonary resuscitation
Tracheostomy
Nasogastric and nasojejunal tubes
Esophageal balloon tamponade of variceal hemorrhage

Selected nosocomial pathogens

Enterococcus

The incidence of enterococcal infections is increasing with the emergence of vancomycin-resistant strains. Enterococci are collectively the second most prevalent nosocomial isolate in the United States, representing more than 9 per cent. Enterococci were reported in 8 per cent of positive blood cultures taken because of suspected infection from a series of general and thoracic surgical patients, and represent an increasingly important cause of invasive burn wound infection.

Four enterococcal species can cause disease in humans. *Enterococcus fecalis* accounts for 85 per cent of isolates, whereas *Enterococcus fecium* accounts for 5 to 15 per cent of strains and *Enterococcus casseliflavius* and *Enterococcus durans* are rarely pathogens. Enterococcal infections may involve intra-abdominal or surgical wound foci, be complicated by bacteremia, or follow manipulation of the urinary tract, placement of intravascular catheters, or antibiotic therapy or prophylaxis with agents that are ineffective against enterococci (notably cephalosporin therapy).

The source of the enterococci in these infections is usually endogenous. Enterococci are part of the normal flora of colon, oropharynx, gallbladder, urethra, and the external female genitalia. The hospital environment may render endogenous flora more invasive. Antibiotic therapy may result in enterococcal overgrowth in fecal flora, with the potential for transmission by the fecal–oral route or by bacterial translocation.

However, there is evidence that extrinsically acquired strains of enterococci which may be resistant to aminoglycosides may cause disease, particularly in epidemic infections. Beta-lactamase production by gentamicin-resistant enterococci, the gene for which is in the same plasmid as that for gentamicin resistance, may make treatment of serious enterococcal infections with penicillins more difficult in the future.

Vancomycin resistance is troublesome. Approximately 15 per cent of enterococcal isolates in the United States are now resistant to vancomycin, with the vast majority being *E. fecium*. Alteration of the microbial ecology of the hospital by widespread use of vancomycin and third-generation cephalosporins has been implicated as an etiological factor. Patient-to-patient transfer by contaminated medical equipment has also been documented. Vancomycin-resistant strains are occasionally susceptible to ciprofloxacin or doxycycline, but the drugs are not bactericidal in clinically achievable concentrations in serum, and many strains now require chloramphenicol therapy. Vancomycin-resistant strains of enterococci are now widespread in hospitals, and are clearly present in the community in that they have been isolated from sewage and farm animals, including human foodstuffs.

Surgical drainage or removal of infected vascular catheters or prostheses is the cornerstone of therapy for enterococcal infections involving a discrete focus, such as an abscess. Empirical therapy for *Enterococcus* in intra-abdominal, non-biliary, or non-urinary infections does not appear necessary in good-risk patients, nor is therapy necessary if *Enterococcus* is isolated from a good-risk patient who is untreated but doing well. However, specific antibacterial therapy against enterococci is warranted in many of the circumstances typified by the seriously ill patient. If the patient has had enterococci cultured previously during hospitalization or if the infection is nosocomial, particularly with previous gastrointestinal or genitourinary surgery or previous antibiotic therapy, a drug effective against enterococci should be included in the regimen for both proved and presumptive abdominal or pelvic infections, or for burn or surgical wound infections. This is particularly true in patients with valvular heart disease or a prosthetic heart valve. Enterococcal bacteremia due to a sensitive strain requires parenteral antibiotic therapy with ampicillin, an acylureidopenicillin, or vancomycin (only for penicillin-allergic patients). As enterococcal bacteremia is often polymicrobial, appropriate therapy should be directed against all isolates. Appropriate therapy clearly reduces mortality. All patients must be monitored closely for the development of endocarditis.

Invasive fungal infections

Fungal infections have dramatically emerged as important nosocomial infections in the last decade. Between 1986 and 1990, *Candida* was the fourth most commonly cultured organism from all sites in the ICU setting in the United States. The increase in *Candida* bloodstream infections has been explosive, particularly in teaching hospitals. Fungal infections now account for nearly 8 per cent of nosocomial infections hospital-wide, and *Candida* is responsible for 80 per cent of these. Invasive fungal infections must be included in the differential diagnosis of hospital-acquired infection in every patient in the ICU.

Candida species are resident flora of the gastrointestinal tract and oropharynx in 25 per cent of normal individuals. The gallbladder is also a common site of colonization. The colonization rate doubles in hospital in-patients. Virtually all the previously mentioned factors that impair host defenses (e.g. extremes of age, diabetes, disease- or therapy-related immunosuppression) are factors in the colonization of the host by *Candida* (Table 3). Broad-spectrum antibiotic therapy is the most important risk factor owing to suppression of normal intestinal flora.

Patients colonized with *Candida* may suffer life-threatening disease if the infection becomes disseminated. Dissemination occurs only after mucosal or skin barriers have been breached, and either bloodstream infection or locally invasive disease is the result. Fungal infection is less common than either bacterial or viral infections after solid organ transplantation, although liver and bone marrow transplant patients are at substantial risk. Most critically ill patients have more than one major risk factor (Table 3).

The majority of infections in humans are caused by *Candida albicans*, although a few other species are increasingly isolated as the overall incidence of serious fungal infections continues to increase. *Candida tropicalis* is the second most common isolate, and accounts for as many as 25 per cent of cases. *Candida krusei* is common in patients with malignant hematological disease, whereas *Candida parapsilosis* is common in non-tumor-bearing patients and may primarily infect patients with lesser degrees of immunosuppression, such as those who develop catheter-related infections. An uncommon but dangerous pathogen is *Candida glabrata* (formerly *Torulopsis glabrata*), which also infects less immunosuppressed patients but carries a high mortality when disseminated.

Table 3 Risk factors for disseminated fungal infection*

Acute renal failure
Advanced age
Broad-spectrum antibiotics
Burns
Corticosteroid therapy
Diabetes mellitus
Gastrointestinal surgery, particularly multistage
Gram-negative sepsis
Head injury
Malignant disease
Multiple trauma
Organ transplantation
Parenteral nutrition
Peritonitis or intra-abdominal abscess

*Risk increases with increasing number of risk factors.

The mechanism by which *Candida* organisms express virulence is unknown. There is some evidence to suggest that *Candida* antigen may promote host immunosuppression by downregulating T-cell function, thereby initiating the process of tissue invasion; cell wall proteins may promote some of the serious manifestations of *Candida*-related disease.

Life-threatening *Candida* infection may present as either generalized sepsis or localized organ involvement. In the former circumstance, fungal sepsis can be indistinguishable from bacterial infection. Since fungal superinfection is commonplace during or after broad-spectrum antibiotic therapy, an indolent fungal infection may be mistaken for persistence of the bacterial infection or even for bacterial superinfection. Disseminated candidiasis may produce focal disease or microabscesses in many organs. Endophthalmitis may be present in as many as 10 per cent of disseminated infections, particularly when *C. albicans* is the pathogen. When present, the 'cotton-wool' exudates visible by fundoscopy are diagnostic and are strongly suggestive of systemic disease. Careful fundoscopy is mandatory when disseminated disease is suspected.

Hematogenous dissemination to skin can occur in critically ill patients. Lesions present as maculopapular eruptions 0.5 to 1.0 cm in diameter. *Candida* arthritis, myositis, and osteomyelitis are unusual but should not be discounted. Hepatosplenic candidiasis is common, particularly in patients with hematological tumors and neutropenia. This manifestation tends to be indolent, with fever, abdominal pain, and non-specific biochemical abnormalities of liver function. Although indolent in its presentation and relatively uncommon, candidal involvement of liver and spleen can be very difficult to eradicate because of the often profound immunosuppression typical in these patients.

Candida may affect the heart in many ways. Fungal pericarditis has been reported after burns and cardiac surgery. Myocardial involvement manifests with electrocardiographic abnormalities or arrhythmias, and may occur in up to 60 per cent of cases of disseminated disease. Fungal endocarditis may complicate cardiac surgery or intravenous drug abuse. *Candida* is an important pathogen in prosthetic valve endocarditis.

Contamination of the peritoneal cavity with fungi may follow perforation of a hollow viscus or dehiscence of an anastomosis, or perforation of an acutely inflamed gallbladder. Fungal peritonitis is also seen in patients undergoing chronic ambulatory peritoneal dialysis. Peritonitis rivals only catheter-associated infection as a cause of disseminated fungal infection in critically ill general surgery patients (Dean and Burchard 1996).

Since catheter-related infections frequently occur in critically ill patients, the presence of an indwelling central venous catheter should be suspected as the primary focus of infection in any patient with fungemia. However, *Candida* is also recognized as a cause of suppurative phlebitis of peripheral veins, a dangerous infection often characterized by a fulminant course after the onset of marked hyperpyrexia, a palpable cord with overlying cellulitis or induration, and the presence of intravascular pus.

Definitive diagnosis of invasive fungal infection is difficult, and is often made relatively late in the course of disease. Accurate timely diagnosis is of great importance because invasive infection accounts for substantial morbidity and mortality, and lack of antifungal therapy has been correlated with increased mortality. Therefore the decision to treat a patient with systemic antifungal therapy must often be made before the diagnosis has been established. In particular, a lack of positive blood cultures should not be counted as evidence against the presence of invasive candidiasis.

Other than the rare signs of cutaneous or ocular manifestations of invasive fungal infection, the clinician is reliant on the laboratory for the diagnosis. Unfortunately, the laboratory diagnosis of invasive fungal infection is often difficult, and sometimes the interpretation of even positive results from the laboratory is problematic. One of the major difficulties is that blood cultures are negative even in autopsy-proven disease up to 50 per cent of the time. False-negative blood cultures are particularly common in neutropenic patients, but can also occur in less immunocompromised patients. Moreover, a positive culture for *Candida* does not imply invasive infection in every circumstance, although the prevalence of the disease makes it prudent to consider every positive blood culture of importance.

Biopsy of infected tissue may have the highest diagnostic yield, but culturing of tissue samples may be as disappointing as blood cultures. However, the identification of fungal elements in tissue sections is extremely accurate. There are obvious difficulties in obtaining tissue for histology in critically ill patients, but skin lesions and percutaneous liver biopsies (particularly in cancer patients) are often diagnostic.

The difficulty with establishing the diagnosis of invasive fungal infection compounds the difficulty of the decision to treat the patient. Because it is hard to establish the presence of infection and mortality is high despite appropriate therapy, there are no definitive guidelines available to support the decision to treat. Currently, most authorities believe that most, if not all, patients with *Candida* fungemia should receive at least a short course of therapy even if fungemia is transient (e.g. a single positive culture with clinical improvement after removal of a colonized venous catheter), as untreated transient fungemia can lead to an invasive infection. Any positive blood culture requires a full course of therapy in a patient who is immunosuppressed or neutropenic, in whom eye lesions are present or *Candida* has been identified in tissue, or where there is a strong clinical suspicion of invasive infection. Anyone with a positive blood culture for *Candida* who is not treated requires very careful follow-up, including immediate removal and culturing of all central venous catheters, careful sequential eye examinations, and reculturing of multiple sites in 72 h.

In the surgical ICU, a decision to treat a patient with negative blood cultures must often be made; clinical suspicion is often the key factor in the decision-making process (Dean and Burchard 1996). Clinical suspicion of systemic disease with positive serology, tissue isolation, eye lesions, or *Candida* isolated from two sites (e.g. urine and sputum) justifies therapy; the last factor is commonplace in surgical illness. Because *Candida* peritonitis is being diagnosed with increasing frequency in critically ill surgical patients, an isolated finding of *Candida* in peritoneal fluid may sometimes warrant treatment.

Amphotericin B remains the treatment of choice for invasive fungal infections in critically ill patients. The drug acts by binding to sterols in the fungal cell membrane and increasing membrane permeability, and thus it can be fungicidal in high concentrations. After intravenous administration, the highest drug levels are achieved in lung, kidney, and spleen. Amphotericin B penetrates poorly into body cavities such as the subdural space or peritoneal cavity; thus direct intracavitary administration is sometimes necessary. Irrigation of the urinary bladder (50 mg in 1.0 liters of 5 per cent glucose at 50 ml/h) may also be employed for fungal cystitis. Dosage regimens for amphotericin B have been largely empirical, but accumulating evidence suggests that shorter courses of therapy may be appropriate, even for some serious *Candida* infections. A test dose (1.0 mg given over 4 h) is no longer considered mandatory, although anaphylaxis and ventricular fibrillation can occur. The usual dose is about 0.7 mg/kg/day. Some investigators suggest that total daily doses of up to 1.5 mg/kg/day may be given with safety, but it is not known whether these higher dosages confer a therapeutic advantage. A common regimen for amphotericin B administration is to treat patients other than those with endocarditis or invasive infections with 500 mg over 1 week. The decision to treat for another week is made on re-evaluation of the patient for signs of both efficacy and toxicity. Success with this type of regimen has been reported in treating surgical and trauma patients, and also for empirical therapy in immunosuppressed patients. However, a total dose of 1 to 2 g is still recommended for endocarditis, fungemia with dissemination or tissue invasion, and fungi other than *Candida*.

In addition to ototoxicity, amphotericin B may lead to nephrotoxicity in as many as 80 per cent of patients. Irreversible nephrotoxicity is unusual, particularly with smaller total doses and normal baseline renal function, but renal function may not return to baseline for months after a course of therapy. Renal damage may be attenuated by sodium loading or alternate-day therapy. Other important side-effects include renal tubular acidosis, hypokalemia, hypomagnesemia, cardiomyopathy, anemia, and thrombophlebitis. Liposomal amphotericin B has been approved in the United States. The total dose of active medication is probably the same, although toxicity is reduced. The cost of therapy may approach $500/day for a typical regimen of 5 mg/kg/day, although 1 g of liposomal amphotericin B can be administered to a 70-kg patient in only 3 days.

Fluconazole acts by inhibiting ergosterol biosynthesis and is relatively non-toxic. Adverse reactions may occur in 20 per cent of patients; they most commonly include nausea, headache, and skin rash. The recommended dose of fluconazole for serious infections is 400 to 800 mg on the first day followed by 200 to 400 mg/day for a minimum of 4 weeks, or 2 weeks after the resolution of symptoms. Randomized prospective studies of candidemia indicate that the effectiveness of fluconazole is comparable to that of amphotericin B (cure rates are about 80 per cent). The drug may be given orally or intravenously in the same dosage; the oral form is much less expensive. Infections of lesser magnitude (e.g. esophagitis) can be treated with 200 mg initially followed by 100 mg/day for 2 to 3 weeks (Dean and Burchard 1996).

Clostridium difficile

The Gram-positive bacillus *C. difficile* is the most common identifiable cause of antibiotic-associated diarrhea and pseudomembranous colitis. Although clindamycin, ampicillin, and the cephalosporins are most commonly implicated, virtually any antibiotic can cause the disorder. Usually, *C. difficile* related disease develops during or soon after antibiotic treatment, but it has been identified after a single dose of a cephalosporin and may develop as long as 6 weeks after antibiotics. Colitis may follow oral, intramuscular, intravenous, or topical antibiotic administration. The relative risk of induction of disease for various antimicrobials is not known with precision. Rare cases not related to antibiotic use have been associated with antineoplastic agents.

C. difficile is clearly a nosocomial pathogen. It is part of the nosocomial colonic flora of only 3 per cent of healthy adults, although colonization rates of 15 to 36 per cent of hospital patients have been reported. The epidemiology of *C. difficile* disease is complex. Two hypotheses exist for the origin of the diarrhea associated with *C. difficile* superinfection: endogenous activation of asymptomatically carried pathogens, and exogenous acquisition of organisms with resultant disease in hosts predisposed to superinfection by previous antimicrobial therapy.

Environmental contamination by *C. difficile* spores is well known, particularly in proximity to toilet facilities. Colonized and infected patients probably serve as a reservoir for spread of the infection to other patients via the hands of hospital personnel or such mundane instruments as rectal thermometers. Hand carriage of *C. difficile* by hospital workers has been documented, and the wearing of gloves has demonstrated a reduction in the incidence of disease. Other standard measures, such as enteric precautions and cohorting of patients, may minimize spread.

The diagnosis of *C. difficile* disease is sometimes difficult because standard diagnostic tests are fallible and there is a wide range of clinical disease (from asymptomatic to life-threatening manifestations of toxic megacolon). A latex agglutination test for detection of cytotoxin in fresh stool specimens is widely available, but false-positive and occasional false-negative results are possible. Because of the wide and rapid availability of these tests, they are increasingly supplanting sigmoidoscopy as the diagnostic test. Although sigmoidoscopy is required to diagnose pseudomembranous colitis *per se*, it is insensitive as a diagnostic test for *C. difficile* disease as only 25 to 50 per cent of culture- or cytotoxin-positive patients have pseudomembranes. However, endoscopy can be valuable when a rapid diagnosis must be made in a seriously ill patient, particularly if exploratory laparotomy is being contemplated. The pseudomembranes are most prevalent in the rectosigmoid colon, but may occur throughout the colon and occasionally in the terminal ileum.

The diarrhea often improves spontaneously after discontinuation of the offending antibiotic; thus not every patient who contracts the disease requires therapy. *C. difficile* is very sensitive to bacitracin, metronidazole, and vancomycin. Some strains of *C. difficile* are resistant to metronidazole, but controlled trials have shown efficacy equivalent to that of vancomycin. Metronidazole is the treatment of choice because of cost considerations and the need to reduce vancomycin usage for epidemiological reasons. Most patients become afebrile within 48 h, with resolution of diarrhea within 7 days even though laboratory diagnostic tests may remain positive for weeks. If the patient still requires antibiotic therapy for another condition, discontinuation of the inducing antibiotic, although desirable, is probably not necessary if specific therapy for *C. difficile* is given. Antidiarrheal (antimotility) agents are contraindicated because retention of the toxin may exacerbate mucosal damage.

Metronidazole may also be given intravenously in the case of ileus or other conditions that interfere with oral therapy. Patients who fail to respond to metronidazole therapy should receive vancomycin, which may also be given by enema or a stoma. Failure to respond to oral vancomycin is rare in patients with a functioning gastrointestinal tract if the diagnosis is correct. However, relapses that require retreatment are common (up to 20 per cent of cases). Nearly all patients respond to another course of metronidazole or vancomycin.

Pseudomonas aeruginosa

Ps. aeruginosa is a motile aerobic Gram-negative bacillus that may appear singly, in pairs, or in short chains. It can be isolated from soil, water, plants, and animals, including humans. Since it thrives in a moist environment, human colonization often occurs in sites such as the perineum and external ear. Normal human colonization rates are as follows: skin, zero to 2 per cent; nasal mucosa, zero to 3 per cent; throat, zero to 7 per cent; stool, 3 to 24 per cent. Potential hospital reservoirs include respiratory equipment, disinfectant solutions, sinks, mops, and diagnostic instruments such as endoscopes.

Between 1985 and 1991, *Ps. aeruginosa* was the fourth most common nosocomial isolate in the United States, accounting for 10 per cent of nosocomial infections. Hospital-wide, it is the leading cause of nosocomial pneumonia (17 per cent), the third most common cause of urinary tract infections (12 per cent), and the fifth most common cause of surgical wound infection (8 per cent). It is the most common isolate in the ICU (13 per cent), where it is the leading cause of pneumonia (18 per cent), the fourth most common cause of urinary tract infections (12 per cent), and the third most common cause of wound infections (11 per cent).

Ps. aeruginosa rarely causes disease in healthy patients, but it preys upon the immunologically defenseless. It is both invasive and toxinogenic. There is a clear relationship between oropharyngeal and respiratory colonization and bacterial adherence to epithelium, but adherence requires loss of fibronectin, sputum proteases (particularly in patients with cystic fibrosis), and cellular injury (e.g. viral infection or mechanical injury from an endotracheal tube). Under certain

Table 4 Relative importance of Gram-positive bacteria as nosocomial pathogens

Pathogen	Rank (Percentage)			
	Pneumonia	Bloodstream	Wound	Urine
Enterococci	9 (2)	3 (8)	2 (13)	2 (16)
Staph. aureus	2 (16)	2 (16)	1 (17)	9 (2)
Coagulase-negative staphylococcus (*Staph. epidermidis*)	8 (4)	1 (27)	3 (12)	10 (2)
Streptococci	11 (1)	9 (4)	9 (3)	13 (0)

Reproduced with permission from Jarvis *et al.* (1991).

conditions (particularly in cystic fibrosis), *Ps. aeruginosa* produces a polysaccharide glycocalyx biofilm that anchors it to surrounding cells and bacteria, and protects the bacteria from phagocytosis.

Invasiveness is regulated by proteases; elastase and alkaline protease are most closely associated with virulence. Tissue destruction of skin, lung, and cornea may occur, with cleavage of collagen, IgA, IgG, complement factors, and antiproteases. In addition to promoting dissemination, autolyzed tissue is a likely source of nutrients to perpetuate the process. A cytotoxin (leukocidin) is produced that is toxic to most eukaryotic cells by an action upon the cell membrane, mobilizing arachidonic acid to generate leukotrienes. Two hemolysins, phospholipase C and a heat-stable rhamnolipid, are produced and act synergistically to effect lipolysis and thereby promote tissue invasion. In pulmonary infections, phospholipase C is known to degrade phospholipids in surfactant, whereas the rhamnolipid inhibits mucociliary function in respiratory tract epithelium. In addition, exotoxin A is also produced by most clinical isolates. Exotoxin A is a potent inhibitor of protein synthesis, mediating systemic illness similar to the lipid A moiety of *Pseudomonas* endotoxin, in addition to having local effects on skin, cornea, and lung.

Ps. aeruginosa causes native tissue endocarditis in intravenous drug users, and can infect prosthetic valves. Depending upon the infected valve, consequences may include septic pulmonary or systemic emboli, conduction disturbances, intractable congestive heart failure, brain abscess, cerebritis, and metastatic skin lesions. Lower respiratory tract infections are common in patients with abnormal local respiratory or systemic host defenses. Pneumonia due to *Ps. aeruginosa* is fulminant and usually fatal (70 per cent). Radiographs reveal a diffuse bronchopneumonia that is typically bilateral and characterized by distinctive nodular infiltrates, sometimes with small areas of radiolucency. Such a pattern can be characteristic of pneumonia caused by *Staph. aureus*, but is unusual with other Gram-negative pneumonias. Bacteremic pneumonia is usually seen in the neutropenic cancer patient, but can occur in children or AIDS patients.

Bacteremia in the absence of pneumonia is common. Predisposition includes malignant disease, neutropenia, diabetes mellitus, organ transplantation, severe burns or trauma, AIDS, steroid or antibiotic therapy, intravascular catheters, and instrumentation of the urinary tract. Mortality is similar to that of *Ps. aeruginosa* pneumonia. Factors associated with death include persistent neutropenia, shock, inappropriate antibiotic therapy, lung or skin/soft tissue sources, inability to find the primary source, and renal failure.

Other ICU infections caused by *Ps. aeruginosa* include meningitis, brain abscess, eye infection, hematogenous bone and joint infection (osteomyelitis), urinary tract infection, and gastrointestinal infection. Central nervous infections can arise from direct extension (e.g. paranasal sinusitis, where *Ps. aeruginosa* is the most common nosocomial isolate), inoculation (surgery, head trauma), or hematogenous seeding. *Ps. aeruginosa* is a frequent and devastating ocular pathogen, causing rapidly progressive destructive infections that are difficult to treat because antibiotics penetrate poorly. It is a genuine medical emergency, as preservation of vision is rare after infection has progressed to endophthalmitis. Gastroenteritis due to *Ps. aeruginosa* usually affects infants or neutropenic cancer patients.

Antibiotic therapy of *Ps. aeruginosa* infections is often difficult. Pneumonia, bacteremia, and bone and soft tissue infections always require two-drug therapy such as an aminoglycoside and an anti-pseudomonal penicillin (e.g. piperacillin or ticarcillin with or without a β-lactamase inhibitor) or a cephalosporin (e.g. cefoperazone, ceftazidime, or cefepime). Imipenem–cilastatin or a fluoroquinolone (e.g. ciprofloxacin) also have excellent anti-pseudomonal activity and can be used with the aminoglycoside. The choice of agent, which often must be empirical because of the fulminant nature of these infections, must be based on knowledge of local susceptibility patterns. Oral ciprofloxacin alone may be sufficient to treat uncomplicated urinary tract infections. Ocular infections require topical aminoglycoside drops and sometimes intraocular injections combined with multidrug parenteral therapy. Ceftazidime is probably the best agent for central nervous system infections; it is not clear whether the addition of either intravenous or intrathecal aminoglycosides improves outcome in meningitis.

Methicillin-resistant staphylococci

Staph. aureus is a well-recognized pathogen capable of causing infections virtually anywhere in the body, and *Staph. epidermidis* is also emerging as a pathogen of critical importance (Barie 1993). Potential sites of nosocomial Gram-positive infections include the urinary tract, surgical wounds, intravascular loci (primary bacteremia, catheter-related infections, suppurative phlebitis), the lung and pleural space, the facial sinuses or parotid gland, and the peritoneal cavity. Infections of prosthetic devices are a particularly important subgroup of wound infections from staphylococci. Infections may be sporadic, or they may contribute to major nosocomial outbreaks.

Staphylococci possess many properties that enhance their virulence. Peptidoglycans in the bacterial cell wall activate complement and cause other immune responses, and the teichoic acid moiety promotes attachment to epithelial cells. Staphylococci also produce numerous cytotoxins, including α- and β-toxins and leukocidin. Capsular antigens, such as protein A, and a glycocalyx or 'slime' (particularly for *Staph. epidermidis*) decrease opsonization and antigen presentation. *Staph. aureus* is well established as the second most prevalent nosocomial Gram-positive pathogen (Table 4) (Jarvis *et al.* 1991). Coagulase-negative staphylococci, of which *Staph. epidermidis* is most commonly isolated, are less prevalent as nosocomial pathogens than *Staph. aureus,* but are recognized as formidable pathogens in surgical wounds, prosthetic devices such as vascular grafts, prosthetic joints, and bacteremia originating from intravascular catheters. Staphylococcal resistance to methicillin is an endemic problem in many hospitals; ongoing epidemiological surveillance is necessary to avert episodic epidemics. Approximately 25 per cent of *Staph. aureus* isolates are now resistant to methicillin (MRSA). Hope that newer antibiotics, such as the quinolones, would be effective against MRSA has largely dissipated; vancomycin remains the treatment of choice, and should be used even if *in vitro* susceptibility testing suggests an alternative. Of equal concern is the emergence of methicillin-resistant *Staph. epidermidis* (MRSE). Between 1980 and 1989, the incidence of MRSE in the United States increased from approximately 20 per cent to more than 60 per cent, reflecting the emergence and virulence of this pathogen. It is no longer true that a single positive blood culture for *Staph. epidermidis* can be dismissed as a contaminant.

References

Barie, P.S. (1993). Emerging problems in Gram-positive infections in the post-operative patient. *Surgery, Gynecology, and Obstetrics*, **177**, S55–64.

Bueno-Cavanillas, A., *et al.* (1994). Influence of nosocomial infection on mortality rate in an intensive care unit. *Critical Care Medicine*, **22**, 55–60.

Dean, D.A. and Burchard, K.W. (1996). Fungal infection in surgical patients. *American Journal of Surgery*, **171**, 374–82.

Jarvis, W.R., *et al.* (1991). Nosocomial infection rates in adult and pediatric intensive care units in the United States. *American Journal of Medicine*, **91** (Supplement 3B), 185S–191S.

Pittet, D., Tarara, D., and Wenzel, R.P. (1994). Nosocomial bloodstream infection in critically ill patients: excess length of stay, extra costs, and attributable mortality. *Journal of the American Medical Association*, **271**, 1598–1601.

Vincent, J.-L., *et al.* (1995). The prevalence of nosocomial infection in intensive care units in Europe: results of the European Prevalence of Infection in Intensive Care (EPIC) Study. *Journal of the American Medical Association*, **274**, 639–44.

13.3.2 Prevention

S. J. Harbarth and D. Pittet

Key messages

- The intensive care unit is a high-risk area for nosocomial infections because of both severe underlying patient disease and exposure to multiple invasive procedures and manipulations.
- Surveillance, consisting of collecting, tabulating, analyzing, and disseminating information on the occurrence of nosocomial infection, is an essential component of the prevention and control of infection.
- Surveillance of nosocomial infection may help to define and detect common or unusual sources of cross-infection or failures in management of patient care.

Introduction

The intensive care unit (ICU) is a high-risk area for nosocomial infections because of both severe underlying patient disease and patient exposure to life-saving invasive procedures and interventions. Nosocomial infection in the ICU increases patient morbidity, length of stay, and hospital costs. Ventilator-associated pneumonia and bloodstream infections in particular increase mortality rates. Many infections in the ICU are due to breaches in infection control and are avoidable if appropriate policies are in place (Pittet and Harbarth 1997).

Surveillance

Surveillance, consisting of the routine collection, tabulation, analysis, and dissemination of information on the occurrence of nosocomial infection, is an essential component of the prevention and control of infection. Surveillance of nosocomial infection may help to define and detect common or unusual sources of cross-infection or failures in management of patient care. The Study on the Efficacy of Nosocomial Infection Control (SENIC) Project, showed that surveillance of nosocomial infection was best conducted prospectively by a sufficient number of trained and independent infection control staff (Jarvis 1991). This team usually collaborates closely with the ICU team and reports the data collected to those responsible for patient care (Martin 1993).

Surveillance of nosocomial infection in the ICU can be classified as follows.

1. Site-specific surveillance focuses on particular types of infection.
2. Laboratory-oriented surveillance primarily collects data and resistance patterns of microbiological isolates. This approach has a low sensitivity since it does not detect non-microbiologically proven infections and a low specificity because of the high prevalence of colonized, rather than infected, patients in the ICU.
3. Outbreak surveillance deals only with the identification and control of outbreaks of infection.
4. Total surveillance considers all types of infection and attempts to correct problems as they arise.
5. In surveillance by objectives the infection control team identifies and prioritizes specific objectives to be met by surveillance. These might include reduction of the incidence of central venous catheter-associated infection, control of the spread of methicillin-resistant *Staphylococcus aureus*, or decreasing the incidence of ventilator-associated pneumonia (Haley 1995).

Total prospective surveillance by a full-time infection-control nurse is preferable, as it is the most effective method of collecting and interpreting data on nosocomial infection and subsequently reducing the impact of infection in the ICU. However, total surveillance of ICU-acquired infections is labor intensive and may be difficult and ineffective for many infection-control teams.

United States standards for hospital-wide surveillance specify one infection control nurse per 250 hospital beds. However, a single infection control nurse will have difficulty in surveying more than 30 to 35 ICU beds consistenly. In addition, continuous surveillance mandates daily ward visits with collection and interpretation of data.

If resources, staff, and time are limited, targeted surveillance may be all that can be done. Specific procedures, such as tracheotomies, or central venous catheters, may be targeted and/or specific infections, such as pneumonia or bacteremia, may be analyzed to identify possible infection control problems.

Whatever surveillance system is used, objective and reproducible data allowing meaningful conclusions should be generated. The infection control staff should collect data on infections not only in patients in the ICU, but also in discharged patients since a significant proportion of ICU-acquired infections are diagnosed 48 to 72 h after transfer to other hospital wards. Data on ICU-specific denominators should also be collected so that site-specific infection rates can be calculated based on the number of patients at risk, patient-days at risk, and number of days of indwelling urinary catheterization, central vascular cannulation, or ventilator support.

Host factors

Infection control measures in ICUs are traditionally directed at limiting person-to-person spread of infection and spread from other exogenous sources. These measures often fail because they may have little effect on the endogenous flora, which is an important source of infection in critically ill patients. Monitoring of antibiotic use can control the selection of resistant and sometimes more virulent micro-organisms from a patient's endogenous flora. The emergence of resistant endogenous strains is a particular problem with Gram-negative bacilli (e.g. *Pseudomonas* and *Enterobacter* species).

The normal host defense mechanisms of critically ill patients are impaired by underlying diseases or as a result of medical interventions (e.g. intravascular catheters and endotracheal tubes). Routine replacement of central venous catheters is recommended to prevent infection based on the observation that the risk increases exponentially after 3 to 5 days of catheterization. However, the duration of placement is still under debate, as is the use of guidewire exchange. If semiquantitative tip cultures are positive, a new catheter must be inserted at a new puncture site. Strict aseptic techniques (with sterile gowns, mask, and gloves) for catheter insertion should be applied. Written policies concerning the use of invasive devices should be implemented, and catheter insertion techniques and care should be standardized.

The gut flora serve as a main reservoir for endogenous nosocomial infection. The normally low pH of the stomach is often neutralized by H_2 blockers or antacids, promoting overgrowth of enteric micro-organisms associated with nosocomial infections like ventilator-associated pneumonia. Admission surveillance cultures of the gastrointestinal tract to identify the 'carrier state' of patients colonized with endogenous flora have been suggested. This may help to prevent infection by stopping the pathway from colonization to infection. Selective decontamination of the digestive tract (SDD) using topical antimicrobials to prevent infections from emerging from the endogenous gut flora is controversial, and should not be used routinely in the ICU setting.

Gut translocation may occur in ischemic gut, in obstruction, or during periods of hypothermia. Thus maintenance of adequate perfusion may help to maintain intact barrier functions.

Staffing

The extent and severity of illnesses affecting ICU patients dictate a high level of nursing care, with strict application of rigid barrier nursing techniques to prevent outbreak of nosocomial infection. Breakdown of these techniques during periods of understaffing or overcrowding has been associated with outbreaks of nosocomial infection. Therefore an ideal nurse-to-patient staffing ratio of 1:1 has been recommended to reduce breakdown of techniques leading to person-to-person transmission of pathogens within ICUs. However, this may not be cost effective.

Infection control guidelines for the ICU

Infection control guidelines for the ICU are directed against the spread of pathogens and include an emphasis on handwashing and isolating or cohorting infected or colonized patients. Aseptic techniques and early removal of indwelling devices are also essential

Table 1 Infection control measures in the ICU

Identify reservoirs
Colonized and infected patients
Environmental contamination, common sources
Halt transmission among patients
Improve handwashing practices (in particular hand disinfection)
Use aseptic techniques
Barrier precautions (gloves, gown) for colonized and infected patients
Eliminate any common source, disinfect environment
Separate susceptible patients
Close unit to new admissions if necessary
Halt progression from colonization to infection
Discontinue compromising factors when possible
Discuss SDD in specific cases
Modify host risk
Treat underlying disease and complications
Implement and control the use of appropriate antibiotic guidelines

Modified with permission from Weinstein (1991).

infection control measures. ICU infection control measures are listed in Table 1. Routine barrier precautions can significantly reduce endemic nosocomial infection rates.

Handwashing remains a cornerstone of the prevention of nosocomial infection in the ICU. Unfortunately, compliance rates of ICU handwashing practices average 30 to 40 per cent, and are even lower for physicians. Hand disinfection (using antiseptic solutions or alcoholic preparations) should be promoted; it is faster, easily achievable at the bedside in the busy routine of patient care, and more efficacous than handwashing with soap and water. The higher the compliance with handwashing recommendations, the lower is the need for isolation and barrier precautions. Unfortunately, there is so little confidence in handwashing practices that hospital isolation policies now assume non-compliance.

The use of gloves has been advocated. However, there is no evidence that the addition of gloves in routine intensive care practices has any advantage over adequate handwashing in controlling infections. Major arguments against the routine use of gloves in the ICU are that health care workers frequently do not remove gloves when moving from patient to patient and forget to wash their hands after glove removal.

Conclusion

ICU patients are at greatest risk of all hospital in-patients for both endemic and epidemic nosocomial infection. Surveillance of nosocomial infection and improvements in the reduction of endogenous and exogenous nosocomial infection have been shown to be efficacious in prevention. Handwashing, in particular hand disinfection, remains the cornerstone of infection control in the ICU.

References

Haley, R.W. (1995). The scientific basis for using surveillance and risk factor data to reduce nosocomial infection rates. *Journal of Hospital Infection*, **30** (Supplement), 3–14.

Jarvis, W.R., *et al.* (1991). Nosocomial infection rates in adult and pediatric intensive care units in the United States. National Nosocomial Infections Surveillance System. *American Journal of Medicine*, **91**, 185S–91S.

Martin, M.A. (1993). Nosocomial infections in intensive care units: an overview of their epidemiology, outcome, and prevention. *New Horizons*, **1**, 162–71.

Pittet, D. and Harbarth, S.J. (1997). The intensive care unit. In *Hospital infections* (4th edn) (ed. J.V. Bennett and P.S. Brachman). Little, Brown, Boston, MA.

Weinstein, R.A. (1991). Epidemiology and control of nosocomial infection in adult intensive care units. *American Journal of Medicine*, **91**, 179S–184S.

13.4 Infection in the immunocompromised

13.4.1 Drug-induced depression of immunity

M. Chapman and S. Peake

Key messages

- There exists a non-specific and a specific immune response against a foreign substance.
- Drugs influence the immune process by affecting neutrophils, killer cells, lymphocytes, and cytokine production and function.
- Immunosuppressive drugs include steroids, cyclosporine (cyclosporin), antilymphocyte antibodies, and chemotherapeutic agents.
- Treatment of the immunocompromised patient in intensive care involves careful prophylactic and empirical therapeutic use of antibiotics as well as immunomodulating substances such as granulocyte colony-stimulating factor and immunoglobulins.
- A number of drugs commonly used in the intensive care unit can affect immune function.

Introduction

Drug-induced immune suppression may be used intentionally to inhibit the immune response to allow organ transplant to be effective or for treatment of an autoimmune disease. However, immune suppression can also be an inadvertent complication of treatment. It occurs when cytotoxic agents are used in the treatment of malignancies where normal dividing cells are affected by the same mechanism that inhibits malignant cells. The most common result of drug-induced immune suppression is leukopenia.

The immune process

The immune response involves recognition of a substance as foreign and the steps taken to remove it. Two processes are involved, a rapid non-specific response and a response specific to the foreign substance. The non-specific natural response is mediated by natural killer and phagocytic cells (Dale *et al.* 1994; McBride *et al.* 1996).

The specific response is the adaptive response mediated by lymphocytes. B lymphocytes release a specific antibody which binds to the foreign substance. T lymphocytes are made up of a number of subsets including helper-1, helper-2, and cytotoxic T cells: helper-1 T cells interact with phagocytic cells to enhance microbial killing, helper-2 T cells are involved in B-cell development and antibody production, and cytotoxic T cells recognize and destroy virus-infected cells.

T cells do not recognize free antigen. The antigen is taken up, processed, and presented in the antigen-binding groove of one of the major histocompatibility complex (MHC) molecules on the surface of antigen-presenting cells (monocytes or lymphocytes). The T lymphocytes recognize these complexes.

Circulating cytokines are the mediators integral to the initiation, maintenance, and control of the inflammatory response. Cytokines are under elaborate neurohumoral control mediated by opioid, adrenergic, and cholinergic receptors expressed on leukocytes. Pro- and anti-inflammatory cytokines are kept in a delicate balance and control the inflammatory response. Systemic levels of cytokines are not indicative of local level or effect. Helper-2 T lymphocytes play a crucial role in the regulation of cytokine balance.

One of the consequences of cytokine activity is neutrophil adhesion to endothelium as a precursor of neutrophil migration into the tissue. Adhesion is mediated by adhesion molecules expressed on endothelial and neutrophil membranes.

Drugs may affect the immune process at any number of sites:

(1) neutrophil number or function;

(2) natural killer cell number or function;

(3) lymphocyte number (total population and subpopulations) or function;

(4) leukocyte adhesion;

(5) increased or decreased expression of MHC molecules;

(6) production of pro- or anti-inflammatory cytokines.

Immune suppression in the treatment of disease

Immune function may be suppressed to prevent organ transplant rejection which is mediated by T-cell activity. This involves the use of multiple drugs, each directed at different points in the T-cell activation cascade. They prevent T-cell activation, T-cell proliferation, and/or cytokine production. The use of combinations of agents allows the dosage of individual drugs to be reduced, thus minimizing side-effects. The targeted action of these drugs is becoming progressively more refined and specific, which also reduces side-effects.

Prevention of organ transplant rejection can be divided into three phases (Kahan and Ghobrial 1994).

1. Induction of prophylaxis, which often involves the use of cyclosporine (cyclosporin), steroids, an antimitotic such as azathioprine, and an antilymphocyte antibody.
2. Maintenance of prophylaxis, which involves the use of cyclosporine, steroids, and an antimitotic (at lower dose).
3. Treatment of acute rejection, which involves high-dose steroids and an antilymphocyte antibody.

Immunosuppression is now a common component of management of a number of other conditions including the following.

1. Asthma: steroids and methotrexate.
2. Inflammatory bowel disease: steroids, 5-aminosalicylic acid, 6-mercaptopurine, and cyclosporine.
3. Autoimmune connective tissue diseases: methotrexate is widely used for rheumatoid arthritis.
4. Certain skin disorders: pemphigus, pemphigoid, and pyoderma gangrenosa may all be treated with steroids in combination with other drugs such as azathioprine, cyclophosphamide, methotrexate, or cyclosporine.
5. Neurological diseases: multiple sclerosis may be treated with steroids and cyclophosphamide.

Immune side-effects of cytotoxic drugs

Combinations of immunomodulating drugs are used for their cytotoxic activity against malignant cells. Frequently used cytotoxic agents are listed in Table 1. As the effect is not specific against the malignant cells, immunosuppression is frequent.

Table 1 Commonly used cytotoxic agents

Alkylating agents
Oxazophosphorines (cyclophosphamide, ifosfamide)
Melphalan
Busulfan
Nitrosoureas (lomustine (CCNU), carmustine (BCNU))
Platinum compounds (cisplatin, carboplatin)
Anthracyclines
Doxorubicin
Daunorubicin
Antimetabolites
Purine analogs (6-mercaptopurine, azathioprine, fludarabine, 2-chlorodeoxyadenosine, 2-deoxycoformycin)
Pyrimidine analogs (5-fluorouracil, cytosine arabinoside)
Antifolates (methotrexate, edatrexate)
Vinca alkaloids
Vincristine
Vinblastine
Vindesine
Topoisomerase inhibitors
Etoposide
Teniposide
Antitumor antibiotics
Bleomycin
Dactinomycin
Distamycin
Mithramycin

Modified with permission from *Oxford textbook of medicine* (1996), p. 248.

Drugs used for immune suppression or chemotherapy

Steroids

Steroids lyse lymphocytes and thus decrease their total number. This is a dose-dependent response; a high steroid dose will reduce the total number of lymphocytes by a maximum of 50 to 75 per cent. Steroids also block the production and release of a number of inflammatory cytokines including prostaglandins and leukotrienes. They inhibit the non-specific immune response, modulate (downregulate) expression of human MHC class II (HLA-DR), and increase the number of neutrophils.

Lymphokine synthesis inhibitors (cyclosporine, tacrolimus)

The introduction of cyclosporine dramatically changed the management of organ transplant because its specific activity results in reduced immunosuppression. However, its use is limited by nephrotoxocity. Cyclosporine and its more recent derivative tacrolimus bind to separate but related proteins in the plasma, which are known as immunophilins. These complexes interact with receptors in the T-cell cytoplasm. The T cells fail to respond to stimulation because they fail to produce interleukin 2 (IL-2) which is necessary for proliferation to occur. Cyclosporine also suppresses the development of cytotoxic T cells and interferes with the functions of antigen-presenting cells and B cells. Both cyclosporine and tacrolimus act primarily on resting T cells. Other drugs have been developed that block lymphocyte signal transduction and inhibit nucleoside synthesis.

Cytokine signal transduction inhibitors (rapamycin, leflunomide)

Rapamycin is a macrolide antibiotic similar to cyclosporine and tacrolimus in that it binds to immunophilin (although on a different site) and blocks intracellular signal transduction at a different point. It inhibits activated T and B cells and does not block IL-2 production. Tacrolimus and rapamycin can be used synergistically because they act at separate sites.

Antilymphocyte antibodies (polyclonal antithymocyte globulin, antilymphocyte globulin, or monoclonal antilymphocyte antibody)

Antilymphocyte (thymocyte) globulin (ALG/ATG) is produced by injecting human lymphocytes into animals and raising antibodies against them. T lymphocytes from the thymus are often used as these are responsible for graft rejection. The active immunoglobulin is the IgG fraction and ALG/ATG is a complex biological product that has inherent composition variability between batches. After injection the ALG/ATG attaches to circulating lymphocytes. Owing to its molecular size it is confined to the vascular space so that only circulating lymphocytes are exposed. Lymphocytes are lysed by a complement-dependent process. The antibodies may also bind to the antigen receptor sites, thus inhibiting the effect of the lymphocytes.

Monoclonal antilymphocyte antibody (OKT3) is a monoclonal mouse antibody directed against the CD3 protein complex, a cell marker found only on the cell membrane of human T cells. It binds to CD3 and probably causes complement lysis of the cell or opsonizes the cell leading to its destruction.

Antimetabolites

Purine analogs (azathioprine, 6-mercaptopurine) block synthesis of DNA/RNA and other essential cell products by interfering with purine metabolism. They prevent proliferation of lymphocytes in response to an antigenic challenge and are only active as an immunosuppressant when given after the antigen. Azathioprine is converted to 6-mercaptopurine.

When used to treat inflammatory bowel disease, 6-mercaptopurine causes bone marrow depression in 3 per cent of patients and infection in 7 per cent. Azathioprine causes profound lymphopenia in 5 per cent which can occur at any time between 2 weeks and 11 years. A marked decrease in CD4 cells may persist for several years.

Antifolate medications include methotrexate, sulfasalazine, and aminopterin. Methotrexate decreases DNA/RNA and protein synthesis (thus inhibiting cell replication), inhibits production of leukotriene B_4 (a potent chemotactic agent), decreases neutrophil and monocyte chemotactic response to leukotriene B_4, decreases production of IL-1 and C5a, inhibits neutrophil adherence (by increasing adenosine production), and lowers the production of rheumatoid factor in patients with rheumatoid arthritis

When high doses are used, drug levels may be monitored and folinic acid may be given to allow bone marrow recovery. The toxicity (but not the efficacy) of methotrexate is mediated via the dihydrofolate reductase pathways.

5-Aminosalicylic acid

5-Aminosalicylic acid inhibits the production of arachidonic acid pathway inflammatory mediators, inhibits oxygen radical production, and scavenges oxygen radicals.

Gold

Gold usually lowers serum IgM and IgG levels, and occasionally causes severe IgA deficiency or hypogammaglobulinemia.

Management of patients with immune suppression

Overuse of antibiotics should be avoided as these predispose to infections with fungi and resistant organisms. Appropriate use of antibiotics is important, and protocols for prophylactic antibiotics should be followed. Prophylaxis against fungal infection with oral and nasogastric nystatin or low-dose intravenous amphotericin B (10 mg/day) may be used. Prophylaxis against *Pneumocystis carinii* and cytomegalovirus may also be used.

Infective complications can be difficult to diagnose in the immunocompromised patient. The temperature may not rise in the patient on steroids or the neutropenic patient, or the leukocytes may already be elevated by steroid use. Infection must be suspected if the patient's condition deteriorates, although there are often many possible causes for this in the postoperative organ recipient. Any deterioration suspected to be due to bacterial infection should be rapidly treated with broad-spectrum antibiotics after appropriate cultures have been taken, as infection can be rapidly progressive. Opportunistic organisms must be sought, including cytomegalovirus, *P. carinii*, *Legionella pneumophila*, and Epstein–Barr virus when clinically relevant.

More than 80 per cent of transplant recipients develop at least one infection after transplantation, and 40 per cent of deaths are due to infectious complications. About 55 per cent of post-transplant infections are caused by bacteria, 30 per cent are viral, and 15 per cent are fungal. Certain infections tend to occur at particular times after transplantation. The most common viral infection is cytomegalovirus, but the other herpes viruses (Epstein–Barr, herpes simplex, and varicella zoster) also occur. Symptomatic cytomegalovirus infection is associated with intense immune suppression, and particularly with the use of antithymocyte globulin. *Candida albicans* is the most common fungal infection. *P. carinii* pneumonia has become less of a problem since the routine use of prophylaxis with trimethoprim–sulfamethoxasole.

Granulocyte colony-stimulating factor (G-CSF) is a hematopoietic factor that plays a central role in precursor cell proliferation and differentiation into neutrophils (Demetri 1995). As well as increasing neutrophil numbers, it increases activity including chemotaxis, superoxide generation, bactericidal function, and phagocytic activity. The most extensively studied clinical application of G-CSF has been in chemotherapy-induced myelosuppression during the treatment of non-myeloid tumors. Its prophylactic use has been shown to reduce the duration of severe neutropenia, the incidence of febrile neutropenic episodes, the overall duration of intravenous antibiotic therapy, and the length of hospital stay. G-CSF is well tolerated; the most frequently reported adverse side-effect is mild to moderate bone pain.

Immunoglobulin administration may be of use for acquired hypogammaglobulinemia. Human cytomegalovirus immunoglobulin reduces the rate of severe cytomegalovirus-associated disease in patients undergoing orthotopic liver transplantation. However, it is not effective when the donor is cytomegalovirus positive and the recipient is cytomegalovirus negative, which suggests a need for additional prophylaxis in this high-risk group.

Immune side-effects of drugs commonly used in intensive care

The immune side-effects of drugs commonly used in intensive care are summarized in Table 2.

The importance of the immune side-effects of drugs used regularly in intensive care was highlighted by a retrospective study of 428 trauma patients (Watt and Ledingham 1984). It showed an increase in mortality from around 25 per cent to 47 per cent associated with the introduction of etomidate for sedation. When etomidate was discontinued, the mortality of this group of patients returned to 25 per cent. Subsequently, etomidate was found to inhibit basal cortisol production and abolish the stress response. This report has salutary lessons: first, drugs that are used routinely may significantly affect outcome by immune modulation; second, ongoing auditing is imperative so that changing responses to treatment in individual units can be monitored.

Most drugs, apart from those used specifically to affect the immune response, have not been extensively studied for their immune effects.

Table 2 Immune side-effects of drugs commonly used in intensive care

Drugs known to cause neutropenia by causing folate deficiency
Pyrimethamine
Phenytoin sodium (diphenylhydantoin)
Primidone
Barbiturates
Trimethoprim
Nitrofurantoin
Triamterene
Proguanil
Pentamidine

Drugs known to cause idiosyncratic neutropenia
Analgesics
  Indomethacin*
  Gold salts
  Pentazocine
  Para-aminophenol derivatives (acetaminophen (paracetamol), phenacetin)*
  Pyrazolon derivatives (aminopyrine, dipyrone, oxyphenbutazone, phenylphenbutazone)*
Antibiotics
  Cephalosporins
  Chloramphenicol*
  Clindamycin
  Gentamicin
  Isoniazid
  Para-aminosalicylic acid
  Penicillins and semisynthetic penicillins*
  Rifampin (rifampicin)
  Streptomycin
  Sulfonamides*
  Tetracyclines
  Vancomycin
Anticonvulsants
  Carbamazepine
  Phenytoin
Antidepressants
  Amitriptyline
  Amoxapine
  Desipramine
  Doxepin
  Imipramine
Antihistamines (H_2 blockers)
  Cimetidine
  Ranitidine
Antimalarials
  Amodiaquine
  Chloroquine
  Dapsone
  Pyrimethamine
  Quinine
Cardiovascular drugs
  Captopril
  Disopyramide
  Hydralazine
  Methyldopa
  Procainamide
  Propranolol
  Quinidine
  Tocainide

(continued)

Table 2 *(continued)*

Drugs known to cause idiosyncratic neutropenia *(continued)*
Diuretics
  Acetazolamide
  Chlorthalidone
  Chlorothiazine
  Ethacrynic acid
  Hydrochlorothiazide
Hypoglycemic agents
  Chlorpropamide
  Tolbutamide
Hypnotics and sedatives
  Chlordiazepoxide
  Benzodiazepines
  Meprobamate
Phenothiazines
  Chlorpromazine*
Antithyroid drugs
  Methimazole*
  Propylthiouracil*
  Carbimazole*
Other drugs
  Allopurinol
  Clozapine
  Levamisole
  Penicillamine
  Ticlopidine
  Ganciclovir (gancyclovir)

Drugs known to cause hypogammaglobulinemia or immunoglobulin deficiency
Gold
Penicillamine
Sulfasalazine
Phenytoin

*Most frequently reported.
Modified from Campbell and Smith (1996).

Drugs may have several contradictory effects at different points in the immune process.

The effects of modifying the stress response are confusing. While chronic blockade of cortisol release proved disadvantageous to outcome in the trauma patients given etomidate, there is believed to be some benefit in blocking the stress response during surgery. In fact, the postoperative rise in plasma cortisol concentrations has a destructive effect on some immune cells. Natural killer cell activity may be depressed following surgery under neuroleptanesthesia but not under epidural anesthesia. This may be due to the absence of a stress response in the epidural group.

Sedative drugs are used continuously for a prolonged period of time and so any depressant effect on immune function may have a significant effect on patient outcome. Morphine and midazolam are two of the sedative drugs most commonly used in the intensive care unit.

Midazolam appears to have a beneficial effect on immune function. It attenuates the rise in ACTH seen after surgery in experimental animals, inhibits the resulting fall in T-cell populations, and has no effect on neutrophil polarization.

Opioids inhibit the activity of natural killer cells in animals and humans, downregulate expression of human MHC class II (HLA-

DR), increase levels of tumor necrosis factor-α and IL-8, and then increase anti-inflammatory cytokines (IL-10, IL-lra, and tumor necrosis factor soluble receptors).

Propofol increases the production of tumor necrosis factor-α, IL-1α, and interferon-δ, reduces neutrophil polarization and chemotaxis, inhibits neutrophil respiratory burst activity, inhibits proliferation of B lymphocytes in the critically ill, decreases the total number of T lymphocytes, memory T lymphocytes and B lymphocytes in healthy patients, and increases the number of helper T cells.

Thiopental (thiopentone) depresses neutrophil activity. Long-term high-dose thiopental infusion used for head injury and status epilepticus may be associated with an increased susceptibility to bacterial pneumonia.

Epinephrine (adrenaline) increases natural killer cell numbers, alters T-cell population ratios, and inhibits mitogen-induced proliferation of lymphocytes. *In vitro*, it inhibits human monocyte adhesion; this effect can be blocked by propranolol and is believed to be mediated by cAMP since it can be mimicked by cAMP analogs.

Dobutamine increases natural killer cell number and activity and increases lymphocyte numbers.

Isoprotenerol (isoprenaline) inhibits the production of tumor necrosis factor-α *in vitro* but has no effect on IL-1β.

Dopamine *in vitro* produces an immediate and transitory decrease in T-cell proliferative response. This also occurs following hypophysectomy and with the use of other dopamine agonists such as bromocryptine. Animal studies have shown that this affects outcome. Dopamine can be blocked with dopamine antagonists such as metoclopramide or haloperidol. Dopamine antagonists also reverse the immunosuppressive effects of cyclosporine and steroids.

Phosphodiesterase inhibitors include methylxanthines, theophylline, milrinone, amrinone, pentoxifylline, and dipyridamole. Each has different effects on the immune response; however, all increase intracellular cAMP, which inhibits the release and action of pro-inflammatory cytokines. Pentoxifylline and milrinone decrease the production of tumor necrosis factor-α. Pentoxifylline also inhibits neutrophil adhesion.

Diltiazem restores macrophage antigen presentation function and IL-1 expression in hemorrhaged animals.

Separation of the direct effects of alcohol on immune function from the effects of poor nutrition and liver dysfunction is difficult. However, acute and chronic alcohol consumption are both associated with decreased phagocytic activity, decreased opsonization, and depressed cellular immunity. Impairment in cellular immunity is aggravated by malnourishment. With cirrhosis, further defects in the immune system occur. Some of the direct effects of alcohol are *in vitro* depression of monocyte function, including antigen presentation and cytokine production, increasing activity and number of natural killer cells, which are implicated in hepatic damage, and decreasing number of T cells.

H_2 blockers enhance immune function, inhibit suppressor T-cell function via H_2 receptors, and increase IL-2 production, cell-mediated cytotoxicity, and lymphocyte viability.

Heparin inhibits neutrophil adhesion.

Nitric oxide (NO) is a central participant in both the non-specific and specific immune response. Together with cytokines, NO is produced by stimulated macrophages. It kills many microbes and inhibits their replication, and increases lymphocyte proliferation. The significance of these effects with respect to attempts to inhibit NO in sepsis are as yet unknown.

Various antibiotics have been tested for their suppression of lymphocyte numbers at therapeutic concentrations. Minocycline and oxytetracycline are the most suppressive, followed by chloramphenicol, clindamycin, and tetracycline. Penicillin, carbenicillin, and cephalothin have no effect. Gentamicin stimulates the response.

Conclusion

It is important to maintain immunocompetence. Immunosuppression, related to either disease or treatment, is associated with a poor outcome. The immune effects of drugs used routinely are less obvious but may have just as significant an effect on outcome. The possibility of using drugs such as diltiazem, dopamine antagonists, or H_2 blockers to improve immune function are interesting concepts that need further exploration.

References

Campbell, E.W. and Smith, M.R. (1996). Hematology for primary care physicians. *Disease-a-Month*, **42**.

Dale, M.M., Foreman, J.C., and Fan, T-P.D. (1994). Introduction to the immunology and pathology of host defence mechanisms. In *Textbook of immunopharmacology* (ed. M.M. Dale, J.C. Foreman, and T.-P.D. Fan), pp. 1–17. Blackwell Scientific, Oxford.

Demetri, G.D. (1995) Hematopoietic growth factors. Defining the appropriate clinical role in multimodality cancer therapy. *Chest*, **107**, 255S–60S.

Kahan, B.D. and Ghobrial, R. (1994). Immunosuppressive agents. *Surgical Clinics of North America*, **74**, 1029–54.

McBride, W.T., Armstrong, M.A., and McBride, S.J. (1996). Immunomodulation: an important concept in modern anaesthesia. *Anaesthesia*, **51**, 465–73.

Watt, I. and Ledingham, I.McA. (1984) Mortality amongst multiple trauma patients admitted to an intensive therapy unit. *Anaesthesia*, **39**, 973–81.

13.4.2 Immune dysfunction associated with critical illness

G. J. Bellingan

Key messages

- The majority of deaths in patients who have been critically ill for more than 7 days are associated with sepsis.
- Critical illness is associated with multiple defects in immune function which reduce the capacity to fight infection.
- Immune dysfunction can be related both to excess production of proinflammatory mediators and, at other times, to overproduction of endogenous anti-inflammatory agents.
- Cytokines are not the only soluble factors implicated in the generation of immune dysfunction; eicosanoids, reactive oxygen intermediates, heat shock proteins, and acute phase reactants are also important.
- Hormonal, nutritional, and genetic factors, as well as associated organ failures (renal — uremia; gut — endotoxin/translocation; respiratory — hypoxia) and drugs, all contribute to impairment of normal immune function in critically ill patients.

Normal immune system

Immunity has two broad facets: natural immunity and the phylogenetically more advanced adaptive immunity which is divided into humoral and cellular components.

Innate or natural immunity

The body's most powerful barriers against invasion by micro-organisms are the skin and the mucosal membranes. Micro-organisms penetrating these defenses are usually rapidly phagocytosed; bacteria activate the alternative complement pathway which coats, opsonizes, and lyses them and generates chemotactic agents. Macrophages and neutrophils congregate in response to chemotactic signals and ingest the opsonized micro-organisms. Ingestion triggers a respiratory burst wherein reactive oxygen intermediates are formed which kill bacteria, while phagolysosome fusion exposes the bacteria to oxygen-independent methods of killing, including cathepsins and lysozyme.

Adaptive immunity

Humoral immunity

Humoral immunity provides specific adaptive immunity. Antigen-presenting cells, mainly dendritic cells and macrophages, recognize, ingest, and process foreign antigen. These antigens are presented to B cells in conjunction with class II and class I major histocompatability complex (MHC) molecules on the antigen-presenting cells. This process, which usually requires T cells, activates B cells to produce a specific immunoglobulin which opsonizes micro-organisms and activates complement, hence promoting their removal.

Cellular immunity

Cellular immunity also provides an adaptive immune response in which cytotoxic T cells are generated against target cells, usually virally infected cells, tumor cells, or allogenic graft cells. This requires helper T cells and appropriate interleukin 2 (IL-2) secretion, and allows distinction of self from non-self. Activated T cells also influence other immune cells, including macrophages, increasing their phagocytic and bacterial killing capacity.

Critical illness causes immune dysfunction

One of the earliest pieces of evidence that there was an immune defect in critically ill patients was provided by the delayed rejection of allografts or xenografts by severely burned patients. It is now well known that patients who do not succumb early as a direct result of critical illness, be it trauma, burn, or hemorrhage, have a high late mortality from sepsis, and this is attributed to defective immune function amongst other factors (Mannick 1993). Studies have demonstrated that shocked animals have a reduced ability to clear a bacterial challenge compared with normal animals, and this lasts for many days after trauma or hemorrhage. This immune defect extends beyond that of resistance to infection, as animals subjected to major trauma also exhibit reduced tumoricidal activity.

Anergy

Anergy is defined as a depression in the delayed hypersensitivity response. This impaired immune function can be demonstrated using simple skin tests against several common recall antigens including mumps, *Candida*, and purified protein derivative. Intradermal injections are given in the forearm and the reaction is measured 48 h later, with 10 mm of induration taken as positive and no response to any antigen considered anergic. Eight per cent of a random population are anergic, but this is significantly increased in postoperative patients with anergy related to the extent of the surgical procedure. The test is easy, inexpensive, and widely used, and some studies suggest that it can predict postoperative septic complications. However, delayed hypersensitivity skin tests do not correlate well with other tests of lymphocyte function, the results are altered by age, and drugs and responsiveness can be impaired by repeat testing. Furthermore, it is not clear which cells are being tested as the response requires antigen processing/presentation, activation of memory and helper T cells, and soluble mediator release.

Dysfunction of specific parts of the immune response

Critical illness can affect the protective immune mechanisms of the entire body, although it is generally observed that cellular immunity is depressed more than humoral immunity. Many of these changes can be related to the extent of injury. Patients who suffer major burns, severe trauma, major surgery, or severe hemorrhage have evidence of dysfunction of many of the immune system cell lines.

Lymphocytes

There is conflicting evidence about the effect of critical illness on circulating lymphocyte numbers, with reports of both a decrease in total lymphocytes (or T-cell and B-cell subpopulations) and a lymphocytosis after major surgery. Recently, a decrease in the ratio of helper to suppressor T cells, independent of any HIV effect, has been shown in critically ill patients. The mechanisms underlying these changes are multifactorial, although it is likely that cell sequestration to sites of endothelial activation accounts for most fluctuations in circulating counts. As lymphocytes are actively recirculating cells, passing from blood to inflamed site and back through the lymph, with other lymphocytes entering lymph nodes directly through specialized venule endothelium, changes in circulating lymphocyte counts may be unrelated to the total lymphocyte number.

Specific dysfunction of T cells has been shown after major surgery, with reduced T-cell proliferative responses to commonly used stimulating agents such as phytohemagglutinin; this can be correlated with the extent of surgery. Suppression is also seen in trauma patients and is mediated by a soluble factor, as normal lymphocytes exposed to serum from patients with severe trauma also exhibit reduced proliferative responses which are not seen when cells are exposed to serum from normal volunteers. This suppressive effect can be detected for up to 2 weeks and can be mimicked in animal studies, where 30 per cent hemorrhage reduces mitogen-induced proliferation of T lymphocytes for several days and is temporarily associated with a reduction in IL-2 production (Mannick 1993). Likewise, splenocytes have an impaired response to concanavalin A and are associated with depressed production of IL-2, IL-3, IL-6, and interferon-γ. Natural killer cell function is also reduced within hours of a burn injury, with function remaining impaired for days; again, this is mediated by a serum factor.

Evidence of specific dysfunction of B cells has not been so clear. With severe injury, total immunoglobulin levels are little altered and may even rise, although this is often due to a non-specific polyclonal increase that may obscure specific functional changes. Some patients exhibit a reduction in primary responses to antigen, while an impaired secondary response to common antigens has been reported in others. Responses to common bacterial antigens tend to remain intact (Mannick 1993).

Phagocytes

Macrophages secrete a complex array of inflammatory mediators that modulate inflammatory cell recruitment and the survival of leukocytes at the inflamed site. They kill bacteria, clear apoptotic cells, present antigen, activate resting T cells, and are central to appropriate wound healing. Despite their eclectic roles in inflammation and its resolution, macrophages have been implicated in the depressed immune function of critical illness. Because of their elaboration of prostaglandin E_2 and nitric oxide, which in high concentrations are both powerful immunodepressive agents, activated macrophages have been associated with impaired immunity. Critically ill patients have clear evidence of increased nitric oxide production with elevated nitrite and nitrate levels (the end-products of nitric oxide synthesis), and in animal models this is shown to be due to increased expression of inducible nitric oxide synthase (Albina and Reichner 1995).

Macrophages not only elaborate immunosuppressive factors but they also exhibit impaired function in response to critical illness. Antigen presentation by macrophages is depressed immediately after hemorrhage and remains so for up to 5 days. Macrophage capacity for release of proinflammatory cytokines is not demonstrably impaired.

Neutrophils are active phagocytes which can secrete tumor necrosis factor-α, IL-1β, and other important proinflammatory cytokines. Certain neutrophil functions are impaired in critical illness, and there is an association between the degree of neutrophil dysfunction and patient morbidity. Neutrophils exhibit reduced chemotaxis after burn injury, and some researchers have reported a reduction in bactericidal activity, the mechanisms of which are complex but are related to cellular hyperactivation, impaired neutrophil oxygen consumption, hydrogen peroxide production, lysosomal enzyme defects, and alterations in phagolysosome acidification (Mannick 1993). Furthermore, serum from critically ill patients can have suppressive effects, depressing neutrophil chemotaxis and bactericidal function. Patients with major burns or trauma have neutrophil subgroups with markedly reduced capacity for phagocytosis and production of hydrogen peroxide. These cells can be identified by their relative lack of expression of CD16, loss of which is also associated with neutrophil apoptosis.

Mechanisms underlying immune dysfunction in critical illness

No single mechanism can explain the many (often subtle) changes seen in the immune function of critically ill patients. A number of general factors need to be considered, including the patient's nutritional state, failure of other organs, and the influence of drugs. In addition, the inflammatory response itself, and particularly the balance of pro- and anti-inflammatory responses, is an important determinant of immune function (Bone 1996).

Malnutrition is the most common cause of secondary immunodeficiency worldwide. Malnourished patients are predisposed to infection, exhibit poor wound healing, and have a worse prognosis than well-nourished patients. Not only can poor nutrition contribute to the risk of critical illness, but critical illness can predispose to a poor nutritional state. The catabolic state associated with burns and major trauma is well recognized and is often exacerbated by difficulties in providing appropriate caloric intake. Starvation can reduce T-cell numbers and cell-mediated immune responses and can depress primary antibody responses, although total immunoglobulin levels usually remain relatively unaffected. A number of specific deficiencies are recognized in the long-term critically ill, including zinc deficiency which is related to anergy, poor skin healing, and lymphoid atrophy. Although appropriate nutritional replacement is of central importance in severely ill patients, the use of immunostimulant diets is more debatable. There is some evidence that dietary supplementation with glutamine is of benefit, but the role of additional specific fatty acids is yet to be defined.

Many drugs, including steroids, non-steroidal anti-inflammatory drugs, H_2 blockers, and even some antibiotics (tetracyclines), have been implicated as potential immunodepressants. It has been suggested that colloids impair the phagocytic capacity of reticuloendothelial cells, but this result has not been reproduced in at least five studies where starches have been shown not to affect phagocytosis, chemotaxis, or cytokine secretion by macrophages. Steroids, routinely used to treat sepsis in the 1970s, are now known to be associated with a worsening of morbidity and mortality. Cellular studies demonstrate that glucocorticoids promote neutrophil survival but induce eosinophil and thymocyte apoptosis, while pretreatment of macrophages with glucocorticoids blocks their release of proinflammatory cytokines in response to activating agents and similarly glucocorticoids block IL-2 release by T cells.

Anesthetic agents, particularly the inhalational anesthetics, can suppress the immune response. Specifically, depression of reticuloendothelial phagocytic function, neutrophil chemotaxis, and both T- and B-cell mitogenic capacity have been reported, although these effects are relatively short lived.

Infections are frequent and severe in patients with both acute and chronic renal failure. Immune alterations in acute renal failure include reduced neutrophil bactericidal activity, impaired macrophage antigen presentation, inappropriate macrophage activation, and defective T-cell function. The mechanisms underlying these changes are multifactorial, with many relating to the critical illness leading to the renal insult. Uremia also exerts a direct immunosuppressive effect, as do many of the filtration membranes used in renal support. Furthermore, inhibitory proteins that can block neutrophil oxygen metabolism and bacterial killing have been described in renal failure.

Critical illness is associated with hypoxia, which itself can depress cellular function and has been shown to stimulate macrophage prostaglandin E_2 and tumor necrosis factor-α production. Relative hypoxia promotes free-radical formation but also prolongs neutrophil survival. These effects may combine to promote local accumulation of neutrophils and their histotoxic products at sites of tissue injury. All these proinflammatory influences may contribute locally to impaired immune cell function.

Patients with a deficient glucocorticoid response to stress have increased morbidity and mortality; however, the immunosuppressive effects of glucocorticoids given early in sepsis are well known, so that a balance in the appropriate circulating levels of steroid hormones must exist. There is normally a circadian rhythm to the body's release of corticotropin-releasing hormone and glucocorticoids, disruption of which can impair cellular immunity. Thyroid hormones, ephedrine, and glucagon also influence immune function (Mannick 1993). Prolactin has a number of positive effects, promoting antibody formation and T-cell-induced macrophage activation. Critical illness may disrupt normal prolactin circadian rhythms, while dopamine, commonly used to maintain urine output, acts on dopamine-2 receptors to inhibit prolactin release, contributing to anergy.

Apart from the rare primary immunodeficiency syndromes, there is now evidence that a patient's genetic make-up can influence the response to critical illness. Genetic influences have been shown to alter the tumor necrosis factor-α response, with patients homozygous for TNFB2 producing greater amounts of tumor necrosis factor-α and having a worse prognosis than heterozygotes for this gene locus (Stüber *et al.* 1996). Within cells the expression of certain genes can influence survival; expression of the proto-oncogene *bcl-2* protects cells from apoptotic death induced by free radicals, drugs, or ionizing radiation. A number of other proto-oncogenes affect cell survival; the *bcl-2* family members *Bax* and *Bcl-xS* both promote cell death, while expression of *bcr-abl* can protect against apoptosis.

Translocation of bacteria and endotoxin from the gut into the splanchnic circulation is well described in experimental models, and there is a sound basis for believing that the gut mucosal barrier is defective in many critically ill patients, allowing a portal for entry of pathogens or simply endotoxin alone (Mannick 1993). Endotoxin can prime and activate cells, causing cytokine and eicosanoid production, including prostaglandin E_2 release, and can depress lymphocyte proliferation, suppressing IL-2 secretion and contributing to the general immune dysfunction associated with critical illness. However, this cannot be the sole explanation, as a similar degree of immunodepression is seen in LPS-responsive and LPS-resistant mice and in germ-free mice subject to similar degrees of insult.

Inflammatory mediators

Tumor necrosis factor-α, IL-1β, and IL-6 are the archetypal proinflammatory cytokines, and they can all influence immune cell function, inhibit lymphocytes, and alter normal phagocyte function (Mannick 1993). IL-6 has a role in the immune response to infection and is synergistic with platelet activating factor in neutrophil priming, possibly leading to an inappropriate inflammatory response. IL-10, the classic anti-inflammatory cytokine, suppresses T cells. IL-4 promotes T- and B-cell proliferation, macrophage expression of class II molecules, and antigen presentation, but inhibits macrophage expression of the LPS receptor CD14 and the release of nitric oxide and proinflammatory cytokines. IL-2, secreted by helper T cells, promotes lymphocyte survival and activation and is central to the appropriate function of the adaptive immune response. Critical illness is associated with impaired IL-2 production, and serum from critically ill patients inhibits IL-2 production by normal T lymphocytes. Not only is the serum immunosuppressive, but wound fluid similarly decreases lymphocyte proliferative responses *in vitro*.

The nature of the serum suppressive factor(s) remains to be elucidated, but IL-10, transforming growth factor-β, prostaglandin E_2, and nitric oxide all have some of these characteristics. Macrophages release hydrogen peroxide and prostaglandin E_2, both of which reduce B-cell responses. Prostaglandin E_2 is a powerful inhibitor of lymphocyte activation and, together with prostacyclin (prostaglandin I_2), can inhibit phagocyte activation and cytokine release (Mannick 1993). Nitric oxide, a constitutive product of the endothelium, is released in large amounts by activated macrophages and can promote or suppress immune function. Lymphocyte proliferation is inhibited by elevated levels of nitric oxide. However, this is complicated as low levels of nitric oxide production are probably necessary for optimal peripheral blood lymphocyte function and for microbial killing, with augmentation above a threshold progressively impairing immune responses by inhibiting the antigen-specific secondary immune response of primed lymph node cells. Likewise, although nitric oxide is important for normal macrophage antimicrobial action, high levels suppress phagocytosis, free-radical production, and protein synthesis and lead to macrophage apoptosis (Albina and Reichner 1995). Modifications of cytokines, colony-stimulating factors, and growth factors by acute phase reactants, heat shock proteins, and changes in cytokine receptor expression lead to a complex acute phase response (Pannen and Robotham 1995).

Reactive oxygen intermediates (including superoxide, hydrogen peroxide, hydroxyl radical, and peroxynitrite) are essential for appropriate bactericidal activity. They are produced in abundance, and can lead to widespread damage of DNA, proteins, and membranes and cell death if not balanced by antioxidants including albumin, vitamin E, and the enzymes superoxide dismutase, catalase, glutathione synthase, and glutathione reductase. Heat shock proteins and acute phase response proteins also protect cells against endotoxin and reactive oxygen intermediate cell death. For example, serum amyloid-A protein binds to neutrophils, reducing their oxidative burst. These acute phase response products may also mediate immunosuppression; a B-cell-suppressive effect of serum amyloid-A protein has been described. Furthermore, receptor shedding and soluble receptors and adhesion molecules further modulate the balance between pro- and anti-inflammatory responses.

Many inflammatory mediators can directly influence the function of cells of the immune system. However, it is difficult to dissect the exact role of each mediator, as the response of cells cultured *in vitro* in the presence of a single inflammatory mediator will be very different from that of cells *in vivo* in the presence of the whole milieu of pro- and anti-inflammatory agents.

In general it is fair to surmise that the proinflammatory mediators are important natural components of an effective inflammatory response which includes appropriate immune function. Only when activated to excess do they exert detrimental effects, which can include effects on immune function. Likewise, the body naturally produces anti-inflammatory mediators that exert a balancing influence on the inflammatory response; however, overexpression of these factors may also impair immune responses. Both excessive immune cell activation from proinflammatory responses and, at different times, excessive immunosuppression from anti-inflammatory mediators probably contribute to the immune dysfunction seen in critical illness (Bone 1996).

Conclusion

Critical illness can depress normal immune function in many ways, and this is influenced by the premorbid state of the patient, his or her genetic make-up, and drugs used in therapy. Although many endogenous inflammatory mediators are implicated, these agents are important in the normal functioning of the inflammatory cascade and it is only when the balance of pro- and anti-inflammatory responses is tipped too far in one direction that they become detrimental. As our understanding of inflammation advances, useful manipulation of soluble mediators can be attempted. Until then therapy must concentrate on adequate and early resuscitation and appropriate treatment of comorbidities, including renal failure and the nutritional state.

References

Albina, J.E. and Reichner, J.S. (1995). Nitric oxide in inflammation and immunity. *New Horizons,* **3**, 46–64.

Bone, R.C. (1996). Sir Isaac Newton, sepsis, SIRS, and CARS. *Critical Care Medicine,* **24**, 1125–8.

Mannick, J.A. (1993). Trauma, sepsis, and immune defects. In *Host defense dysfunction in trauma, shock and sepsis. Mechanisms and therapeutic approaches* (ed. E. Faist, J.L. Meakins, and F.W. Schildberg), pp. 15–21. Springer-Verlag, Berlin.

Pannen, B.H.J. and Robotham, J.L. (1995). The acute-phase response. *New Horizons,* **3**, 183–97.

Stüber, F., Petersen, M., Bokelmann, F., and Schade, U. (1996). A genomic polymorphism within the tumor necrosis factor locus influences plasma tumor necrosis factor-α concentrations and outcome of patients with severe sepsis. *Critical Care Medicine,* **24**, 381–4.

13.5 HIV-related disease

13.5.1 Infections associated with HIV

M. G. Brook and R. F. Miller

Key messages

- The onset and types of infection associated with HIV are dependent on CD4 lymphocyte counts.
- Respiratory infections at high CD4 counts are usually bacterial infections or tuberculosis; at lower counts pneumocystis and fungal infections become more common.
- Septicemia follows similar principles as for non-HIV patients.
- Neurological involvement includes acute myelopathy, neuropathy, encephalopathy, encephalitis, mass lesions, and meningitis
- Cardiac and renal disease, although later findings, may occur and carry significant morbidity and mortality.

Introduction

The complications of HIV are related to the patient's immune status as indicated by their blood level of CD4-positive T-helper lymphocytes (Table 1). There are also social, demographic, and geographical influences on this disease, such that marked differences in the prevalence of HIV-related complications may be seen between patients attending separate treatment centers even within the same city. Additionally, there are differences in disease patterns related to the route of acquisition of HIV (Table 2) and the local use of antiretroviral drugs and prophylactic agents administered to prevent the more common infections such as *Pneumocystis carinii* pneumonitis, tuberculosis, *Toxoplasma gondii* and *Mycobacterium avium intracellulare* complex. HIV-related diseases that may require critical care are more restricted in range, and it is likely that most intensive care units in developed countries will have a similar case mix (Wachter *et al*. 1992; De Palo *et al*. 1995).

HIV-related diseases that may require critical care are presented in a system-orientated way.

Respiratory disease

At higher CD4 counts (over 300 × 10^6/l) respiratory disease is usually due to acute bacterial infection or, less commonly, tuberculosis. At lower CD4 counts other diseases become more prevalent,

Table 1 Relationship between CD4 count and HIV-related complications

CD4 count (× 10^6/l)	Complication
250–500	Tuberculosis Bacterial pneumonia Oral candidiasis
150–200	Kaposi's sarcoma Lymphoma
75–125	*P. carinii* pneumonitis Cerebral toxoplasmosis Cryptococcal meningitis *M. avium intracellulare*
<50	Cytomegalovirus retinitis

Adapted with permission from McCarthy and Mercey (1994).

Table 2 Differences between homosexual men and intravenous drug users (IVDU) in the prevalence of first AIDS-defining events in Edinburgh, UK, 1993

Disease	Homosexual men* (number (%))	IVDU* (number (%))
P. carinii pneumonitis	32 (57)	36 (55)
Esophageal candidiasis	5 (9)†	16 (25)†
Kaposi's sarcoma	11 (20)†	1 (2)†
Wasting syndrome	3 (5)	4 (6)
Atypical *Mycobacterium*	1 (2)	5 (8)
Toxoplasmosis	3 (5)	1 (2)
Disseminated tuberculosis	1 (2)	3 (5)
Other	3 (7)	9 (13)

*Total is greater than 100% as some patients had two or more diagnoses.
†Difference is significant ($p < 0.05$).
Adapted with permission from Ross *et al.* (1994).

including *P. carinii* pneumonitis, pulmonary Kaposi's sarcoma, pulmonary lymphoma, and fungal infection (*Cryptococcus neoformans*). Multiple concurrent pathologies are common.

Pneumocystis carinii pneumonitis

Etiology and epidemiology

P. carinii pneumonitis is the initial AIDS-defining disease in approximately 50 per cent of patients in developed countries (Table 2), although this proportion is falling with the wider use of co-trimoxazole as primary prophylaxis. As a result *P. carinii* pneumonitis often presents for the first time at lower CD4 counts and as a second or subsequent complication of AIDS.

Clinical features

The most common symptoms are dyspnea, fatigue, fever, and cough. The patient frequently appears to be clinically less ill than the chest radiograph and arterial blood gases suggest, and the only physical signs may be fine inspiratory crackles on auscultation of the chest. Deterioration on treatment may be due to super-added bacterial infection, fluid overload, or pneumothorax.

Investigations

The chest radiograph may be normal in mild or early cases but more commonly shows bilateral interstitial shadowing, most prominently at the midzones. Atypical radiographic changes occasionally seen include apical and unilateral shadowing. The majority (over 90 per cent) of patients will have hypoxemia at rest or after exercise. The diagnosis is confirmed in most cases (over 90 per cent) by Grocott's methenamine silver staining of alveolar lavage fluid obtained at bronchoscopy. Microscopy of sputum induced by the inhalation of nebulized saline has a sensitivity of approximately 40 per cent compared with bronchoscopy, but the diagnostic yield is variable as it is operator dependent.

Treatment is outlined in Table 3. The majority of patients (over 85 per cent) improve on co-trimoxazole therapy. Those who require mechanical ventilation have a poor prognosis, with 25 per cent or less surviving their hospital stay.

Table 3 Management of *P. carinii pneumonitis*

Co-trimoxazole 120 mg/kg/day for 2 days, then 90 mg/kg/day intravenously/orally for 19 days

Alternatives: pentamidine 4 mg/kg/day intravenously or clindamycin 450 mg intravenously/orally four times daily plus primaquine 30 mg/day orally or trimetrexate 1.2 mg/kg/day plus folinic acid 0.5 mg/kg orally four times daily or dapsone 100 mg/day orally plus trimethoprim 20 mg/kg/day intravenously/orally (mild disease only)

If $Pa\text{O}_2 < 9.3$ kPa (70 mmHg) or $D(\text{A}-\text{a})\text{O}_2 > 4.3$ kPa (33.2 mmHg) add methylprednisolone 1 g intravenously on days 1–3 (the United States National Institutes of Health recommend methylprednisolone 80 mg three times daily initially followed by reducing doses)

If response to initial steroids is poor, give methylprednisolone 500 mg intravenously on days 4 and 5, and then reduce the prednisolone on days 6–16 starting at 60 mg orally. Give with an H_2 antagonist

Respiratory support

Initial supplemental oxygen by Venturi mask to maximum $F\text{IO}_2$ of 0.6 to maintain $Pa\text{O}_2 > 8$ kPa (60 mmHg) or $Sa\text{O}_2 > 90\%$

If $Sa\text{O}_2 < 90\%$ and $Pa\text{O}_2 < 8$ kPa (60 mmHg) on maximum oxygen by mask, initiate nasal or full-face CPAP with PEEP at 5 cmH_2O and $F\text{IO}_2 = 0.6$

Increase PEEP to 7.5 and then 10 cmH_2O over the next 2–4 h until $Sa\text{O}_2 > 90\%$

If $Sa\text{O}_2 < 90\%$ or $Pa\text{CO}_2$ rises on CPAP, intubate and initiate IPPV with intermittent mandatory ventilation at high respiratory rates and pressure-limited ventilation

CPAP, continuous positive airway pressure; PEEP, positive end-expiratory pressure.
Adapted with permission from Jeffrey *et al.* (1993).

Bacterial pneumonia

Etiology and epidemiology

At CD4 counts over 200 × 10^6/l HIV-infected patients are susceptible to common pathogens such as *Streptococcus pneumoniae*, *Staphylococcus aureus*, and *Hemophilus influenzae*. At lower CD4 counts a wider range of pathogens will be found including *Pseudomonas* species and organisms that are uncommon in the immunocompetent (e.g. *Serratia marcescens*).

Clinical features

At higher CD4 counts the illness is similar to community-acquired bacterial pneumonia seen in the immunocompetent, usually presenting as a lobar pneumonia. At lower CD4 counts the clinical features are similar to *P. carinii* penumonitis, although the cough is more likely to be productive. Approximately 20 per cent of *P. carinii* pneumonitis will be complicated by bacterial infection.

Investigations

The chest radiograph will normally show lobar or segmental consolidation at higher CD4 counts, but at lower levels (below 300 × 10^6/l) patchy bilateral consolidation and interstitial shadowing indistinguishable from changes due to *P. carinii* pneumonitis or other infections may be seen. Blood and sputum cultures should be performed and brochoscopic alveolar lavage performed if the etiology is in doubt.

Treatment

A second- or third-generation cephalosporin such as cefuroxime 750 mg intravenously four times daily, with additional cover for atypical organisms such as *Mycoplasma pneumoniae* (e.g. erythromycin 500 mg four times daily) is appropriate in patients with higher CD4 counts. In the more immunocompromised (CD4 < 200 × 10^6/l) a broad-spectrum antibiotic with anti-pseudomonal activity is required (e.g. ceftazidime 1–2 g intravenously twice daily). Supplemental oxygen by mask may be required to maintain $Pa\text{O}_2$ above 8 kPa (60 mmHg).

Critical care

Management is the same as for *P. carinii* pneumonitis.

Tuberculosis

Etiology and epidemiology

Tuberculosis is particularly common (up to 50 per cent) in HIV-infected patients from countries with a high prevalence of *Mycobacterium tuberculosis*, in intravenous drug users, and in localized areas of countries with an otherwise low prevalence (e.g. New York City). Even in low-prevalence countries such as the United Kingdom 5 per cent or more of HIV-infected patients will develop tuberculosis.

Clinical features

At CD4 counts over 300×10^6/l the presentation is similar to that in the immunocompetent, with lung and lymph node disease predominating and the pulmonary disease largely focal. Typical symptoms include fever, weight loss, and cough. As the CD4 count falls the clinical disease becomes less typical; extrapulmonary disease is present in two-thirds of patients, and pulmonary involvement is more widespread and rapidly progressive, often simulating acute bacterial infection and *P. carinii* pneumonitis.

Investigations

At high CD4 counts the chest radiograph will show focal, often apical, consolidation with cavitation or pleural effusion. Hilar and mediastinal lymphadenopathy are seen in 30 per cent of patients. At lower CD4 counts the chest radiograph is often indistinguishable from that with other respiratory diseases such as acute bacterial infection and *P. carinii* pneumonitis. The chest radiograph is normal in up to 20 per cent of patients with pulmonary involvement. Only 60 per cent of those with pulmonary disease will be sputum smear positive for acid-fast bacilli, although microscopy of bronchoscopic alveolar lavage fluid will increase the sensitivity to 80 to 90 per cent, with the remainder being identified on culture.

Treatment

All patients with confirmed or suspected tuberculosis should be isolated. The normal initial therapy is rifampin (rifampicin) 450 to 600 mg/day, isoniazid 300 mg/day, and pyrazinamide 1.5 to 2 g/day, with the addition of ethambutol 15 mg/kg/day if there is a significant (over 2 per cent) risk of resistance to isoniazid. After 2 months of therapy ethambutol and pyrazinamide are stopped, but the other drugs are continued for 6 months. Outbreaks of infection due to multidrug-resistant organisms have been reported. In these cases treatment should be guided by local susceptibility patterns and the results of antibiotic sensitivity testing of the organism.

Critical care

Management is the same as for *P. carinii* pneumonitis. It is important to isolate such patients.

Other respiratory infections

Infection with fungi such as *Cryptococcus* species, *Aspergillus* species, and *Histoplasma capsulatum*, atypical mycobacteria such as the *M. avium* complex and *Mycobacterium kansasii*, and other organisms such as *Nocardia asteroides* and cytomegalovirus are uncommon causes of acute respiratory failure and are normally encountered in patients with low CD4 counts (below 150×10^6/l).

Pulmonary malignancy

The two most common malignancies associated with HIV are Kaposi's sarcoma and non-Hodgkin's lymphoma. Hodgkin's lymphoma is also seen with increased frequency. Pulmonary Kaposi's sarcoma is normally only seen in patients with low CD4 counts (below 200×10^6/l) who have cutaneous and palatal lesions, and it can cause extensive lung disease. Non-Hodgkin's lymphoma and Hodgkin's lymphoma may be seen at any stage of HIV and behave more aggressively at lower CD4 counts; they only rarely involve the lung.

Septicemia

Epidemiology

Septicemia may arise spontaneously, as a secondary event after acute focal infection such as pneumonia, as a complication of infection of a central venous catheter, or as a result of neutropenia complicating cytotoxic therapy or marrow disease. Most cases arise in patients who are severely immunocompromised. A wide range of organisms may be involved.

Clinical features

The features are similar to those seen in HIV-negative patients.

Treatment

The choice of empirical antibiotic therapy is dictated by the situation, but normally would provide cover for all common Gram-positive and Gram-negative bacteria, including *Pseudomonas* species, with an agent such as ceftazidime 2 g intravenously twice daily. Specific agents should be added in the following circumstances: with associated diarrhea give ciprofloxacin 500 mg intravenously twice daily (for *Salmonellae*); with a pre-existing central catheter in place give vancomycin 500 mg intravenously four times daily (for resistant Gram-positive cocci); if there is profound neutropenia ($< 0.1 \times 10^9$/l), give gentamicin 5 mg/kg/day divided into three doses; if abdominal sepsis is suspected, give metronidazole 500 mg intravenously three times daily.

Critical care

Management is the same as for any hypotensive septic patient. Forty per cent of such patients treated in the intensive care unit (ICU) will survive to leave hospital.

Neurological disease

Acute myelopathy and neuropathy

Peripheral neuropathy, radiculopathy, mononeuritis multiplex, Guillain–Barré syndrome, and acute myelopathy may occur at any time in the course of HIV infection. At low CD4 counts (below 100×10^6/l), cytomegalovirus, herpes simplex, and varicella zoster virus may cause myelopathy and radiculopathy. Myelopathy may also be due to infiltration of the spinal cord or meninges by lymphoma or mycobacterial infection. Neuropathy may be caused by drugs such as zalcitabine (DDC) and isoniazid.

Fortunately, respiratory failure due to myelopathy, radiculopathy, or neuropathy is rare, as the outcome is uncertain and it is likely that 50 per cent or more of such patients will not recover neurological function.

Encephalopathy and encephalitis

The three main causes of encephalopathy and encephalitis in patients with advanced HIV disease and CD4 counts below $100 \times 10^6/l$ are HIV encephalopathy, progressive multifocal leukoencephalopathy, and cytomegalovirus-related encephalitis. Less commonly, herpes simplex virus and varicella zoster virus infection may be implicated. Encephalopathy may also be metabolically induced by drugs or by liver or renal failure.

Magnetic resonance imaging (MRI) of the brain will show cerebral atrophy and variable degrees of white matter change in HIV encephalopathy. Similar changes may be seen with cytomegalovirus. Progressive multifocal leukoencephalopathy produces asymmetrical white matter changes without a mass effect on MRI . Herpes simplex virus and varicella zoster virus lesions are usually more localized and may produce a mass effect in patients with encephalitis. A lumbar puncture should be performed to identify cytomegalovirus, herpes simplex virus (I and II), varicella zoster virus, and polyoma JC virus DNA by the polymerase chain reaction.

Cerebral mass lesions

Cerebral toxoplasmosis and primary central nervous system lymphoma are the most common causes of cerebral mass lesions, but other possible etiologies include *Cryptococcus neoformans*, tuberculosis, and pyogenic abscess. Tuberculosis and pyogenic abscess can occur at any CD4 count, whereas the other conditions normally present in patients with advanced HIV disease.

All these conditions are amenable to therapy in the short term, and therefore it is appropriate to admit patients to the ICU at initial presentation for management of respiratory failure due to rapidly progressive weakness or diminished consciousness due to raised intracranial pressure.

Meningitis

The major causes of meningitis are *M. tuberculosis*, which can occur at any stage of HIV disease, and *C. neoformans* at CD4 counts below $125 \times 10^6/l$. The frequency of pyogenic meningitis is not increased in HIV-infected patients, although cases due to *Listeria monocytogenes* may be more common.

Tuberculous and cryptococcal meningitis both have a subacute onset. Initial features include mild to moderate fever and a headache that is not severe. At this stage, neck stiffness and photophobia are absent in 50 per cent or more of the patients. As the meningitis progresses over days or weeks, the classic signs are more likely to be found as are diminished levels of consciousness and focal neurological signs.

Intensive monitoring and management of respiratory failure due to diminished consciousness is appropriate at initial presentation. The expected survival rate is approximately 50 per cent.

Cardiac disease

Symptomatic cardiac disease related to HIV infection is almost invariably confined to late-stage HIV disease ($CD4 < 100 \times 10^6/l$). It is caused directly by HIV or opportunistic infections including cytomegalovirus, toxoplasmosis, and cryptococcus. Heart failure may also complicate septicemia and severe respiratory disease or may be drug induced. HIV-related cardiomyopathy or myocarditis predominantly involves the left ventricle. Mean survival of patients with symptomatic cardiac failure is under 2 months.

Although the long-term prognosis is poor, a 50 per cent survival can be expected for patients admitted to the ICU presenting with acute cardiac failure.

Renal disease

Renal failure may complicate any severe systemic illness in the course of HIV, with therapeutic drugs and dehydration being additional contributing factors. HIV-associated nephropathy presents with both acute (50 per cent) and chronic renal failure. It is much more common in black patients and usually presents at CD4 counts below $200 \times 10^6/l$.

Renal biopsy typically shows focal and segmental glomerulosclerosis, although other histological changes have been described. Renal failure from any cause requiring hemofiltration or hemodialysis in the ICU is associated with a short-term mortality of at least 50 per cent.

References

De Palo, V.A., Millstein, B.H., Mayo, P.H., Salzman, S.H., and Rosen, M.J. (1995). Outcome of intensive care in patients with HIV infection. *Chest*, **107**, 506–10.

Jeffrey, A.A., Bullen, C., and Miller, R.F. (1993). Intensive care management of *Pneumocystis carinii* pneumonia. *Care of the Critically Ill*, **9**, 258–60.

McCarthy, G.A. and Mercey, D. (1994). The changing clinical features of HIV-1 infection in the United Kingdom. *Communicable Disease Report Review*, **4**, R53–8.

Ross, J.D., Brettle, R., Zhu, C., Haydon, G., and Elton, R.A. (1994). A comparison of AIDS-defining events and subsequent CDC Stage IV events in IDUs and gay men. *International Journal of STD and AIDS*, **5**, 419–23.

Wachter, R.M., Luce, J.M., and Hopewell, P.C. (1992). Critical care of patients with AIDS. *Journal of the American Medical Association*, **267**, 541–7.

13.5.2 Role of intensive care in HIV disease

R. F. Miller and M. G. Brook

Key messages

- Respiratory failure is the most common etiology for intensive care unit admission for HIV-infected patients.
- An understanding of HIV infectivity, testing, universal precautions, and the natural history of the disease is crucial for the health care provider.
- Similar principles of ethics apply whether or not a patient is HIV infected.

Introduction

Despite major advances in our understanding of the etiology, pathogenesis, and natural history of the acquired immunodeficiency syndrome (AIDS), it remains a uniformly fatal disease. Significant improvements in survival of HIV-infected patients have occurred consequent upon early intervention with antiretroviral therapy and use of prophylaxis against *Pneumocystis carinii* pneumonia. In addition, better outcomes have been reported in HIV-infected individuals with *P. carinii* pneumonia requiring intensive care and mechanical ventilation. It is against this background that HIV-infected individuals continue to utilize intensive care resources.

The most common reason for HIV-infected patients to be admitted to the intensive care unit (ICU) is respiratory failure secondary to *P. carinii* or bacterial pneumonia; this accounts for two-thirds of all admissions with additional causes listed in Table 1.

Table 1 Reasons for admission of HIV-infected patients to the ICU

	Etiology
HIV-related processes	
Respiratory failure	*P. carinii* pneumonia Bacterial pneumonia Fungal pneumonia (e.g. *C. neoformans*)
Septicemia/hypotension	Enteric organisms (e.g. *Salmonella*) Central-catheter infections Neutropenia/steroid therapy
(Meningo)encephalitis	Herpes simplex, varicella zoster virus, *C. neoformans*
Cerebral abscess	*Toxoplasma gondii* *M. tuberculosis*
Cardiac failure/dysrhythmias	HIV cardiomyopathy
Acute renal failure	Focal segmental glomerulonephritis Drug toxicity (e.g. pentamidine or phosfonoformate)
General processes	
Trauma	Motor vehicle accident Burns Mugging/assault
Medical	Drug overdose Asthma Diabetic ketoacidosis Pancreatitis
Elective surgery	Open-lung biopsy Craniotomy/brain biopsy Coronary artery surgery

HIV testing

Although the status of some HIV-infected patients admitted to the ICU will be known and declared, in the majority it will not (Table 2).

In some patients the possibility of underlying HIV infection will emerge with the acquisition of the knowledge that the patient belongs to an at-risk group (e.g. intravenous drug user) or from information conveyed to the ICU team by the patient's friends, relatives, or partner(s); in others the discovery of physical signs (e.g. oral candidiasis or cytomegalovirus retinitis, or pathology such as *Cryptococcus neoformans* meningitis and fungemia, disseminated *Mycobacterium tuberculosis*, or Kaposi's sarcoma) will prompt the diagnosis.

A conscious patient should not be tested for HIV unless patient counseling occurs and there are facilities available for post-test counseling in the event of a positive result. In the United States consent is required. Unconscious patients or those sedated and mechanically ventilated are clearly unable to give consent for HIV testing, and in the United States consent comes from the next of kin. The need for testing in this situation is rare, for example if organ donation is being considered (Miller *et al.* 1993).

Before performing an HIV test it is important to consider the natural history of HIV infection in order to interpret the result, particularly as an 'HIV test' is often mistakenly called an 'AIDS test' and it is frequently incorrectly stated that a negative 'HIV test' result excludes HIV infection. It is also important that the physician requesting the test poses the question: 'Will a negative or positive HIV test result influence management of the patient?' Following exposure to HIV, through sex, shared needles, or receipt of contaminated blood, antibodies to HIV develop in most patients within 6 months. In the majority antibodies develop within 6 weeks; however, in some cases there is a delay of up to 12 months, and a few patients never develop antibodies. During this window time between exposure and development of antibodies (so-called seroconversion), patients are infected with HIV, they are infectious, and HIV antigen and DNA are detectable in peripheral blood, yet they are HIV negative on testing as the HIV test detects only antibodies to HIV and not the virus itself.

Other patients, in whom HIV infection has been documented by a previous positive HIV test result, lose their HIV antibodies as their disease progresses and the CD4 lymphocyte count falls. They are also infected with HIV and are infectious, yet they are negative on routine HIV testing. Another small group of patients, despite being

Table 2 Status of HIV-infected patients admitted to the ICU

- Declared by patient
- Known by referring team/ICU team from medical records, previous admissions, etc.
- Known by patient but not declared/actively denied
- Suspected by patient (e.g. because of lifestyle) but never tested
- Suspected by ICU team because of presence of coexisting disease (e.g. Kaposi's sarcoma)
- Unsuspected by patient/ICU team because patient has no risk factors for HIV infection and no overt signs of immunodeficiency

Table 3 Universal precautions in the care of HIV-infected and at-risk patients in the ICU

Education of all staff about HIV infection
Barrier precautions: cover all cuts/abrasions; latex gloves and plastic apron for handling body fluids; wash hands after procedure
Additional precautions: visor or goggles and mask in addition to barrier precautions if performing chest physiotherapy/suctioning
Invasive procedures: barrier and additional precautions plus non-absorbing gown

infected with HIV and having detectable HIV antigen and DNA in peripheral blood, never develop antibodies and yet may have AIDS-defining illnesses such as *P. carinii* pneumonia and low CD4 lymphocyte counts with negative HIV test results.

The HIV test is based on detection of antibodies, most often by an antibody capture–enzyme-linked immunosorbent assay (ELISA). A positive result is always confirmed by other laboratory tests on the same sample using different methods to detect HIV antibodies (e.g. particle agglutination assays and Western blotting) and by repeating the HIV antibody detection tests on a second sample of blood.

In the clinical situation where an HIV test is negative but HIV infection is suspected on clinical grounds (e.g. cerebral toxoplasma abscess) or where it is imperative that HIV infection is excluded (e.g. in a patient being worked up for organ transplantation), more refined tests should be carried out for HIV genome detection by polymerase chain reaction or HIV p24 or RT antigen detection.

Control of infection issues

Many physicians, nurses, and other health care workers in the ICU are anxious about caring for HIV-infected patients, given the irreversible nature of AIDS and the fact that most patients with HIV infection, whether or not they know their status, are asymptomatic.

The risk to ICU staff of nosocomial transmission of HIV from a patient is very low but must not be ignored. HIV has been found in saliva, sputum, bronchial aspirate, and urine as well as in blood. Given the difficulties in accurately and rapidly identifying HIV-infected and at-risk patients, it is appropriate to have universal precautions for all staff in the ICU (Table 3) (Layon *et al.* 1991; Miller *et al.* 1993).

All physicians, nurses, and other health care workers must be educated about the modes of transmission, prevention, natural history, and epidemiology of HIV infection.

There is a need for routine use of barrier precautions when interacting with any patient in the ICU. Any health care worker, including doctors, nurses, physiotherapists, and health care assistants, with exudative skin lesions or 'weeping' dermatitis should not perform care duties in the ICU. Simple cuts and abrasions should be covered with a waterproof plaster or dressing. Latex gloves and a single-use disposable plastic apron should be worn to protect against accidental spillage when handling body fluids such as blood, feces, urine, or drain fluid. Clearly, wearing gloves does not reduce the risk of a needlestick or sharps injury, but it does dramatically reduce the risk of a splash exposure.

Additional precautions are necessary if suctioning is performed via an endobacterial tube or tracheostomy, or if chest physiotherapy is carried out. Even with rigorous attention to technique and the use of 'soft' suction catheters at low pressure, suctioning may cause bronchial secretions to become tinged with blood. Both suctioning and physiotherapy cause bronchial secretions to be aerosolized, and so health care workers performing these activities should wear a disposable face mask of the type used by surgeons and eye protection such as goggles or a visor, in addition to basic barrier precautions, in order to prevent the aerosol droplets from coming into contact with mucosal surfaces.

Health care workers performing invasive procedures, such as cardiac catheterization, transfemoral intra-aortic balloon insertion, vaginal or Cesarean section delivery, or endoscopy (including bronchoscopy), should wear a non-absorbing gown in addition to taking the precautions discussed above (Table 3).

If the skin or mucosal surfaces are accidentally contaminated by splash, the area should be washed with water immediately. If a needlestick or sharps injury is sustained, the area should first be washed with water and then rigorously manipulated in order to provoke local bleeding. After the accident the infection status (HIV and hepatitis B/C) of the patient can be explored. The event should also be reported in confidence to the hospital risk management team and/or occupational health department. The United States Centers for Disease Control and Prevention (CDC) has recently demonstrated that there is benefit from zidovudine if it is given to health care workers immediately following needlestick or splash exposure; many centers continue to offer immediate postexposure 'prophylaxis' with zidovudine (with or without didioxyinosine).

Infection control policies will vary from hospital to hospital, although certain basic principles apply. Patients with suspected or proven *M. tuberculosis* infection should be managed in cubicles, preferably with negative-pressure facilities, in respiratory isolation. As airborne pathogens such as *M. tuberculosis* may be exhaled from the ventilator in mechanically ventilated patients and may disseminate within the ICU environment, it is recommended that a filter be placed in the exhalation limb of the ventilator tubing. To reduce the risks of aerosolization of infectious airborne droplets further, closed-circuit suction systems should be used for performing endotracheal physiotherapy. Non-disposable equipment should be cleaned after use using local infection control policies.

An indwelling arterial cannula, while providing on-line information such as pressure and oxygenation, avoids the need for repeated arterial punctures (and the attendant risk of needlestick injury). After use, all needles and syringes should be disposed of in a rigid-walled puncture-resistant container and not discarded by the bedside or placed in clinical waste bags. Needles should never be bent, broken, or resheathed after used.

Although the HIV status of a patient may be known, admission to the ICU may be the first occasion that this diagnosis is made known to his or her partner and family. Identifying samples for laboratory analysis or requests for imaging investigations as 'high risk' or 'HIV/AIDS', although necessary for safe practice, provides an opportunity for a potential breach of confidentiality. Patients infected with HIV have the same rights to confidentiality as any human being, regardless of their ethnicity, color, religion, sexuality, or financial status (Layon *et al.* 1991).

Ethical considerations

The appropriateness of allocating finite resources to the terminally ill is frequently challenged, and as AIDS is a progressive disease and has no cure many health care workers have questioned the validity of decisions to allow HIV-infected patients to receive intensive care. Patients with AIDS have the same right of access to intensive care as non-HIV-infected patients with conditions such as cirrhosis with varices, severe chronic heart failure with angina, or unresectable non-small-cell lung carcinoma, all of which have similar short- and long-term prognoses to AIDS (Wachter *et al.* 1988). Patients with these conditions are often admitted to the ICU even though they may have no guarantee of recovery or return to normal life; in contrast, ICU admission is frequently refused to AIDS patients.

When HIV-infected patients are questioned about their desire for life-sustaining therapy at a time when they are physically well, the majority say that they would opt for active intervention (Steinbrook *et al.* 1986). When asked what treatment they would want if they developed *P. carinii* pneumonia, 95 per cent of HIV-infected men stated that they would want hospital admission and antibiotics, 55 per cent would want intensive care (and mechanical ventilation if respiratory failure supervened), and 48 per cent would want cardiopulmonary resuscitation. If the scenario was one of *P. carinii* pneumonia on the background of HIV-associated dementia, only 19 per cent and 17 per cent respectively wanted transfer to intensive care with mechanical ventilation and cardiopulmonary resuscitation. Seventy-three per cent of HIV-infected men wanted to discuss these issues with their regular out-patient physician; however, only 33 per cent had been able to do this. HIV-infected patients and their relatives deserve the same considerations as any other critically ill patient; they should be informed and consulted about their medical condition and the need for intensive care, mechanical ventilation/intubation, and cardiopulmonary resuscitation should any of these situations arise.

There appear to be few reasons for not offering full intensive care to HIV-infected patients if it is clinically indicated and the prognosis is clearly explained and understood by the patient and his or her physician. If, during the course of treatment, deterioration occurs and continuing or commencing new therapy is not indicated, and this scenario has been discussed with the patient by his or her doctor at a time when he or she was well, for example when seen in the out-patient clinic or first admitted to the hospital, then active support can be discontinued.

Many patients, whether HIV infected or not, now have 'living wills' or 'advance directives' which lay out their wishes for treatment in the event of physical and/or mental incapacity. Others give durable power of attorney to their partner or another family member, so that treatment decisions can be made even if the patient is too ill to partake in discussions.

References

Layon, A.J., Harman, E.M., and Kilroy, R.A. (1991). Acquired immunodeficiency syndrome in the intensive care unit. In *Critical care practice* (ed. J.W. Hoyt, A.S. Tonnesen, and S.J. Allen), pp. 466–76. W.B. Saunders, Philadelphia, PA.

Miller, R.F., Jeffrey, A.A., and Bullen, C. (1993). HIV positive patients in the intensive care unit. *Care of the Critically Ill*, **9**, 113–15.

Steinbrook, R., Lo, B., Moulton, J., Saika, G., Hollander, H., and Volberding, P.A. (1986). Preferences of homosexual men with AIDS for life-sustaining treatment. *New England Journal of Medicine*, **314**, 457–60.

Wachter, R.M., Cooke, M., Hopewell, P.C., and Luce, J.M. (1988). Attitudes of medical residents regarding intensive care for patients with the acquired immunodeficiency syndrome. *Archives of Internal Medicine*, **148**, 149–52.

13.6 Tropical diseases

13.6.1 Malaria

C. J. M. Whitty and P. L. Chiodini

Key messages

- Falciparum malaria is a common cause of severe illness in patients who have returned from the tropics.
- Thick and thin blood films remain the diagnostic method of choice.
- Severe malaria can occur with low parasite counts.
- Quinine or quinidine are currently the drugs of choice in severe falciparum malaria, except when patients come from very restricted geographical areas of Asia where quinine resistance is a significant problem.
- The most urgent priority is giving adequate antimalarial drugs. Debates about route of administration or exchange transfusion should not be allowed to delay this.
- Renal failure, respiratory problems, and multiple organ failure can occur after all the parasites have been cleared from the blood.
- Pregnant women are at particular risk, particularly of hypoglycemia.

Introduction

Malaria is the most common life-threatening infection acquired by travelers to tropical countries. Chemoprophylaxis reduces the risk of infection, but does not remove it. All patients who become critically ill within 12 months of returning from a malarious area must have malaria excluded, even if they took chemoprophylaxis. Falciparum malaria has been reported up to a year following return from an endemic area. Malaria should also be considered in fever not due to other causes months or years after traveling to an area where *Plasmodium vivax* or *Plasmodium ovale* infections are common, as relapses due to liver hypnozoites can occur long after initial exposure.

Diagnosis

Several methods are available for malaria diagnosis. Direct microscopy remains the gold standard for malaria diagnosis. Giemsa- or Field-stained thin and thick blood films enable the *Plasmodium* species and parasitemia to be determined with a high degree of accuracy by skilled microscopists. The sensitivity of detection by the thick blood film is approximately 0.0001 parasites per 100 red cells. The newer methods discussed below are useful as an adjunct to direct microscopy, particularly for those who are not regularly examining malaria films, but are not a substitute.

Fluorescence microscopy uses acridine orange dye which selectively targets parasite DNA and has been incorporated into a commercial test (QBC II). This is as sensitive as thick films, although parasitemia estimation is unreliable. It has the disadvantage of needing a trained microscopist to perform the test and requires an ultraviolet microscope.

HRP2 antigen capture assays provide a qualitative method which uses a monoclonal antibody to a water-soluble protein of *Plasmodium falciparum*. It may show a positive result before and after that seen by microscopy. It is an easy test to perform and does not need trained staff. It is useful in detecting *P. falciparum* in mixed infections. It is commercially available as ParaSight F and as the ICT malaria *P. falciparum* test.

New methodology

Each *Plasmodium* species produces lactate dehydrogenase specific to the species. Monoclonal antibodies to these are undergoing field trials as a diagnostic tool.

Ribosomal DNA polymerase chain reaction has now progressed to sensitive detection and speciation of *Plasmodium* species. It is not currently suitable for the clinical diagnosis of malaria.

Initial assessment of the patient with falciparum malaria

Critically ill patients with malaria will almost invariably have falciparum malaria. If other forms of malaria are reported, the clinician should consider mixed infections, laboratory error in speciation, or other diagnoses with non-falciparum malaria as an incidental finding.

It is easy to underestimate severity in falciparum malaria. The parasite goes through a repeating cycle of development and reproduction, and it is only at certain points of the cycle that significant pathological damage occurs. This makes assessment much more difficult than for other infections. A patient can appear clinically well and then deteriorate rapidly, despite adequate treatment, as parasites mature. A low parasite count does not mean that the infection is trivial, and many of the complications of malaria can occur after all parasites have been cleared from the peripheral blood.

Any or all of the following signs or test results indicate a complicated case of malaria which will need intensive monitoring and treatment, and in which rapid deterioration is likely (World Health Organization 1990):

(1) reduction in conscious level;

(2) neurological signs or fits;

(3) renal impairment (creatinine above 250 µmol/l) or oliguria;

(4) shock or hypotension;

(5) parasite count above 5 per cent on blood film;

(6) anemia (hematocrit below 15 per cent);

(7) signs of bleeding or disseminated intravascular coagulation (DIC);

(8) jaundice;

(9) pulmonary edema, hypoxia, or acidosis;

(10) hypoglycemia (glucose below 2.2 mmol/l).

Where neurological signs or coma are present, hypoglycemia should be immediately excluded since it is caused by both severe malaria and antimalarial drugs. Pregnant women are at particular risk. In the absence of these, the following factors indicate a potentially complicated case:

(1) pregnancy;

(2) hyperpyrexia (above 39 °C);

(3) parasite count above 1 per cent;

(4) mature parasites (schizonts or late trophozoites) on the blood film.

Treatment

Antimalarial chemotherapy

The first priority in treating all cases of malaria is starting an appropriate antimalarial drug. In cases of severe falciparum malaria the choices are limited (White 1996). Parenteral quinine is the current gold standard and should be used in all cases except where there is a high chance of quinine resistance. Significant quinine resistance is currently limited to a very restricted geographical area (see below). Parenteral quinine is not available in the United States, and quinidine should be substituted. Quinidine is a more active antimalarial drug but has a four times greater effect on the heart. Both quinine and quinidine are proarrhythmogenic drugs and should never be given as bolus injections. Quinidine should be administered with ECG monitoring; if the QT interval is prolonged by more than 25 per cent the infusion must be stopped. Both quinine and quinidine can cause hypoglycemia.

Quinine doses are started with a loading dose in severe cases of malaria. The two most commonly used regimens are as follows:

(1) 20 mg/kg quinine dihydrochloride salt over 4 h in 5 per cent glucose or 0.9 per cent sodium chloride;

(2) 7 mg/kg quinine dihydrochloride salt infused over 30 min, followed by 10 mg/kg over 4 h.

A maintenance dose of 10 mg/kg every 8 to 12 h should follow. This should be given by infusion over 4 h. Cinchonism (ringing in the ears, deafness, and vertigo) is not an indication to stop therapy. The therapeutic range for quinine has not been defined, but plasma concentrations between 5 and 10 mg/l (15 to 30 µmol/l) are safe and effective. Trough quinine levels should be checked after the third parenteral dose where available.

The quinidine loading dose is 10 mg/kg infused over 1 h, followed by 1.2 mg/kg/h (0.02 mg/kg/min). The therapeutic range has not been determined, but plasma concentrations of 5 to 8 mg/l are considered effective and are usually safe. If plasma concentration monitoring is not available, reduction of the quinidine dose by a third should be considered on day 3 of treatment.

Once the patient has begun to respond to treatment he or she can be converted to oral therapy if tolerated. Quinine or quinidine should be continued until asexual parasites are eliminated from the peripheral blood. A second-line drug such as sulfadoxine–pyrimethamine (Fansidar) once, or doxycycline for 7 days, should then be administered to eliminate any residual parasites.

Chemotherapy in falciparum malaria resistant to quinine

Severe quinine-resistant malaria is exceptionally rare at present. It is currently (1997) a clinical problem in a very restricted area of Southeast Asia. Patients who come from the Thai–Burmese or Thai–Cambodian borders should be considered potentially to have quinine resistance, and alternative drugs should be considered. Resistance patterns change rapidly, and up-to-date information about them should be sought in these patients. Only artemisinine derivatives can reliably take the place of quinine for parenteral treatment, and currently these are available only on a named-patient basis in Europe and America. The alternatives are as follows.

1. Artesunate (as sodium artesunate) diluted in 5 to 10 ml of 5 per cent glucose by intravenous or intramuscular injection. The initial dose is 2.4 mg/kg, followed by 1.2 mg/kg at 12 and 24 h, and daily thereafter (Hien and White 1993).

2. Artemether by intramuscular injection. The initial dose is 3.2 mg/kg, followed by 1.6 mg/kg daily.

Both should be followed by a second-line drug. Reports of increased incidence of neurotoxicity with mefloquine make other choices of second-line drugs, such as doxycycline, more attractive.

Where artemesinine derivatives are not available, combined therapy using quinine or quinidine as above with doxycycline up to 3 mg/kg/day will reduce failure rates in quinine resistance.

General management

Patients should be monitored for clinical and parasitological response to treatment. Rises of parasite count up to 12 h after instituting treatment may be due to the natural parasite cycle and should not be taken as evidence of treatment failure. There should be a fall of more than 75 per cent by 48 h; if there is not, resistant malaria is likely and treatment should be altered. Conventional clinical measures of response are time to recovery of consciousness, time to fever falling below 37.5 °C, and time to parasite count falling by 50, 90, and 100 per cent.

Fluid balance, renal function, and respiratory function should be monitored closely. Both renal and pulmonary function can deteriorate despite a parasitological and clinical response to treatment.

Optimizing fluid balance is the most difficult part of supportive care. Patients with severe malaria are often dehydrated and require initial rehydration. They may also require blood transfusion. Overhydration is easy, with potentially fatal results. The pulmonary complications of malaria have a mortality of around 50 per cent in intensive care units, and these may be compounded by overvigorous hydration. Restoration of plasma volume with careful monitoring should avoid the problems of both hypovolemia and fluid excess. The renal failure seen in malaria is most often due to sequestration of parasites in the kidney, and will not respond to repeated fluid challenges.

Apart from antimalarial drugs and caution with fluids, identifying and treating the complications of severe malaria early is the key to good management. No trials of potentially disease-modifying drugs (such as anti-tumor necrosis factor) have demonstrated any reduction in mortality to date.

Complications of falciparum malaria

Cerebral malaria

Technically, cerebral malaria is unrousable coma without fits in the preceding 30 min with no cause except for malaria. Hypoglycemia should always be excluded. Mortality from cerebral malaria can be up to 30 per cent even with ideal treatment. Despite this, patients who survive seldom have significant neurological sequelae.

Trials have shown that prophylactic intramuscular phenobarbital (phenobarbitone) improves outcome in cerebral malaria; where this is not available it is likely that other anticonvulsants given prophylactically help. Fits occur commonly in cerebral malaria and should be treated conventionally. There is no contraindication to using benzodiazepines. The role of cerebral edema in the pathology of cerebral malaria is unclear, but it has been clearly demonstrated that steroids and mannitol have no place in the management of adult cases of cerebral malaria.

In practice, any reduction in conscious level or neurological signs in the context of malaria are significant, and patients should ideally be nursed in a high-dependency unit. There is no indication for routine lumbar puncture or CT scanning in cerebral malaria, except where coexisting meningitis is suspected.

Renal problems

Renal failure is one of the common causes of death in severe falciparum malaria. This can occur acutely or after the patient seems to have begun to recover. Hemofiltration and dialysis should be started early in renal impairment due to malaria as this probably improves outcome, although there are currently no good trials to prove this. Extreme care should be taken to avoid overhydration.

Quinine and quinidine are metabolized in the liver, and so dosage seldom has to be adjusted for at least the first 3 days of treatment when patients are in renal failure. Where facilities to monitor drug levels are not available, it is conventional to reduce dosage by a third on day 3. Dialysis marginally reduces quinine levels. Certain antimalarial second-line drugs, particularly Fansidar, are best avoided in renal failure. Doxycycline is probably the second-line drug of choice, at a reduced dose of 100 mg/day.

'Blackwater fever', an old term mainly used for patients with black urine secondary to intravascular hemolysis, should be treated like any other malaria. It is sometimes triggered by antimalarial drugs, particularly in cases of glucose-6-phosphate dehydrogenase deficiency, but the risks of stopping treatment are almost always greater than those of the hemolysis itself.

Pulmonary complications

Pulmonary involvement in malaria is very serious, with a high mortality even with intensive care. Fluid excess should be excluded with the help of a pulmonary artery catheter, and if present corrected with diuretics. In severe cases, however, it may be caused by a form of acute respiratory distress syndrome. Apart from optimizing the fluid balance and standard intensive care support for acute respiratory distress syndrome, no specific treatment has been shown to help.

Hyperparasitemia

The exact role of exchange blood transfusion where high parasite loads are present is controversial (Wilkinson *et al.* 1994). Exchange transfusion involves substituting unparasitized donated blood for the heavily parasitized blood of the patient. It is not indicated for parasitemias of less than 5 per cent. At parasitemias of greater than 10 per cent it has a place in certain cases where complications are present. At confirmed parasitemias of over 20 per cent it is very likely to be indicated. It should never be performed without close nursing and medical supervision to balance blood coming out with blood going in. Laboratory errors in estimation of parasitemias (by up to a factor of 10) are not uncommon, with overestimation more likely than underestimation. Therefore it is advisable to have high parasitemias checked by a reference laboratory.

Anemia

Profound anemia is common in severe malaria, particularly where there has been a prolonged history. Parasitized cells are broken down and the parasite suppresses hemopoiesis, so that there is seldom a reticulocytosis. Transfusion is usually indicated at a hemoglobin concentration below 5 g/dl. Low platelets are almost invariable in malaria and require no specific treatment unless bleeding occurs. There is no association between platelet count and severity.

Disseminated intravascular coagulation

DIC is a rare but serious complication of severe malaria. It can present with profuse bleeding into the gut. Since almost all patients with DIC and malaria will already be anemic, blood transfusion is usually indicated. The role of heparin, fresh frozen plasma, and other treatments is no more clearly worked out than in other causes of DIC.

Shock (algid malaria)

Shock in malaria is often associated with coexisting sepsis, and all patients with this complication should receive broad-spectrum antibiotics with good Gram-negative cover. Otherwise treatment is supportive. Malaria has surprisingly little effect on cardiac function in most cases.

Pregnancy

Pregnant women who contract malaria are at increased risk. Quinine remains the antimalarial drug of choice. Doxycycline is contraindicated. Fansidar should be used with caution, and always given with a folate supplement. Pregnant women should have their quinine treatment extended for at least 48 h beyond the point where they become parasite free on blood film, and treated for a minimum of a week if no second-line agent is used. Recrudescence can occur more than a month after adequate treatment of pregnant women.

Anemia and hypoglycemia are common in pregnancy. Risk of abortion is increased, and congenital malaria is well recognized when malaria occurs near term, although it is more common in *P. vivax* cases. Low-birth-weight babies are common after malaria in pregnancy.

Malaria in childhood

The general principles of managing children with severe malaria are similar to those for adults. Anemia is very common and renal failure very rare in children. Hypoglycemia is common. It is conventional practice to consider lumbar puncture on all children from malaria endemic areas with cerebral signs to exclude meningitis, although there is a risk of raised intracranial pressure. Doxycycline is contraindicated in children.

References

Hien, T.T. and White, N.J. (1993). Quinghaosu. *Lancet*, **341**, 603–8.

White, N.J. (1996). The treatment of malaria. *New England Journal of Medicine*, **335**, 800–6.

Wilkinson, R.J., Brown, J.L., Pasvol, G., Chiodini, P.L., and Davidson, R.N. (1994). Severe falciparum malaria: predicting the effect of exchange transfusion. *Quarterly Journal of Medicine*, **87**, 553–7.

World Health Organization (1990). Severe and complicated malaria. *Transactions of the Royal Society of Tropical Medicine and Hygiene*, **84** (Supplement 2), 1–65.

13.6.2 Other tropical diseases

C. J. M. Whitty and P. L Chiodini

Key messages

- Typhoid should be suspected in all febrile patients returning from an endemic area; if diagnosed and treated early it has a very low mortality. The general principles of management of enteric fevers are similar to those for other Gram-negative infections, with rehydration and supportive care.
- Diphtheria is common in most of the developing world and the Russian republics. It is sporadic in developed countries.
- There are tens of thousands of cases of rabies every year worldwide, and only four claimed recoveries. Correct post-bite prophylaxis is highly effective in preventing rabies cases. Once symptomatic rabies encephalitis occurs death is virtually inevitable, usually within a week without intensive care.
- It is essential that the possibility of viral hemorrhagic fever is considered in patients who have been in a relevant geographical area. It should be noted that taking and handling laboratory specimens is the most common cause of nosocomial viral hemorrhagic fever.

Typhoid and other enteric fevers

Diagnosis

Typhoid or paratyphoid should be suspected in all febrile and seriously ill patients returning from endemic areas where malaria has been excluded. Typhoid is common in almost all of Asia, Africa, and Southern America. Clinical history and examination are non-specific. A history of fever with headache is typical, and cough, meningism, constipation, and diarrhea are all common. Splenic and hepatic enlargement are common, and a tender distended abdomen occurs in most severe cases. Nuchal rigidity and localized chest signs may be misleading. A relative bradycardia is classical but seldom helpful. Crops of pink blanching papules (rose spots) on the trunk occur in up to 50 per cent of cases, but are difficult to detect in dark skin.

Confirmation of typhoid requires isolation of the organism. Blood cultures are usually positive in the early stages of the disease, but fall to less than 50 per cent in cases presenting late in the course of infection. The classical progression of typhoid is of mild to moderate disease in the first week, progressing to severe disease in the second to third weeks. Therefore confirmation of typhoid in critically ill patients may not be possible with blood cultures. Bone marrow culture has a yield of up to 90 per cent, even where patients have had antibiotics, and should be used where the diagnosis is considered in seriously ill patients. Fecal cultures become positive in many cases as the disease progresses. Urine cultures may yield the organism.

The Widal test is seldom helpful in the acute setting. A rise in titers over time is too late to be of practical help in management. New methods of serodiagnosis are under development. The full blood count can show a normal, raised, or low white cell count. Leukopenia is common as the illness progresses.

Management

Typhoid diagnosed and treated early has a very low mortality. It is a safe policy to treat a critically ill patient who might have typhoid as if they do. The general principles of management for enteric fevers are similar to those for other Gram-negative infections, with rehydration and supportive care. However, there are some specific features.

Ciprofloxacin and other quinolone drugs have been demonstrated to be effective in large clinical trials. Resistance to the traditional choices of chloramphenicol or trimethoprim–sulfamethoxazole (co-trimoxazole) is common in Asia, and increasing elsewhere. The third-generation cephalosporins, including cefotaxime and ceftriaxone, are also effective. Therefore, although chloramphenicol is still a first-line agent for the treatment of sensitive strains, quinolones or third-generation cephalosporins are now often required.

Steroid therapy has been shown to improve outcome in some severely ill patients with typhoid. High-dose dexamethasone (3 mg/kg followed by 1 mg/kg every 6 h) reduces mortality where there is reduced consciousness or shock (Hoffman *et al.* 1984).

Typhoid leads to inflammation in Peyer's patches. This leads to sloughing, particularly in the third week of the clinical disease. The result can be either severe gastrointestinal hemorrhage or perforation, with high mortality. The onset of hemorrhage is usually obvious, with a rapid rise in pulse and a drop in blood pressure. Bowel perforation may be difficult to diagnose early, as patients with typhoid already have a tender distended abdomen with scanty bowel sounds. The use of steroids can further complicate diagnosis. Where the patient deteriorates rapidly, abdominal free fluid and gas under the diaphragm should be actively sought.

The bowel in severe typhoid may be very friable. In the absence of perforation the management of gastrointestinal hemorrhage should be conservative. Early simple surgical intervention has replaced conservative management for bowel perforation.

Infection control

Enteric fever is contracted via the fecal–oral route. *Salmonella typhi* or *Salmonella paratyphi* can be cultured from the patient's blood, stool, or urine. Patients with enteric fever should be isolated and strict enteric precautions applied.

Diphtheria

Diphtheria is common in most of the developing world and the Russian republics. It is sporadic in developed countries. It is primarily a disease of children, with a mortality of 5 to 10 per cent, but is seldom fatal in immunized patients. The disease is due to the toxin produced by the bacteria.

Diagnosis

Diphtheria can infect many sites, but almost all severe cases will involve the nose, throat, or larynx. The early presentation is unremarkable, with sore throat and mild fever, or a bloody nasal discharge. Over 24 h there is a worsening of symptoms, and the diagnostic gray-white adherent membrane becomes apparent and spreads. All cases where the diagnosis is likely should be treated as diphtheria. Occasionally the onset may be very rapid.

The diagnosis is confirmed by isolating *Corynebacterium diphtheriae* from a throat or nasal swab.

Treatment

The complications of severe diphtheria will be local (due to respiratory obstruction) and systemic, mainly myocarditis and neuritis (due to toxin). All patients with a strong clinical suspicion of diphtheria should be treated immediately with antibiotics and antitoxin. Erythromycin 500 mg intravenously every 6 h or penicillin G (benzylpenicillin) 600 to 1200 mg intravenously every 6 h are the recommended regimens to prevent further toxin production; very few strains are currently resistant to erythromycin.

Recommendations on dose of antitoxin are reached on the basis of clinical experience and the assumption that the duration of disease and extent of membrane formation are a guide to the toxic burden. The dose range is from 10 000 IU for tonsillar diphtheria of short duration, through 40 000 to 60 000 IU for pharyngeal disease, to 100 000 to 150 000 IU for extensive disease of duration 3 days or more. The antitoxin can be given intramuscularly or intravenously. In severe disease at least some of the dose should be given intravenously. Antitoxin should be preceded by an intracutaneous test dose. Occasional anaphylactic reactions occur, and the drugs necessary to treat anaphylaxis should be drawn up. Conventional wisdom (but no trials) favors bed rest. Steroids do not help to reduce the incidence of myocarditis or neuritis.

Where there is any evidence of respiratory obstruction in laryngeal diphtheria, tracheostomy should be performed as an emergency procedure without delay. The main role of critical care is in managing the cardiac complications which occur in diphtheria myocarditis, which has a 50 per cent mortality. The early indications are ECG abnormalities with ST changes, ectopics, arrhythmias, and heart block. Mortality in heart block is very high, even with cardiac pacing. Congestive cardiac failure and valvular incompetence may occur as the ventricles dilate. Neurological complications, including respiratory and laryngeal paralysis, may occur many weeks after the infection has been eliminated and the patient seems to be recovering.

Once they have recovered, patients should receive active immunization against diphtheria.

Infection control

Diphtheria is spread by droplet infection from the respiratory tract, by close contact with respiratory tract secretions, or with exudate from cutaneous lesions. Patients with this condition should be isolated. At least 5 days after antibiotic therapy has been completed, throat and nose cultures should be taken 24 to 48 h apart. The patient should remain in isolation until three consecutive cultures have been reported negative. Close contacts who are incompletely immunized, or in whom the immunization status is unclear, should be vaccinated against diphtheria. They should also receive prophylactic antibiotic therapy (usually erythromycin) pending the results of their throat swab cultures.

Rabies

Diagnosis

The diagnosis of rabies depends upon a high level of clinical suspicion and may not be considered in the prodrome unless the patient gives a history of being bitten. There is usually itching around the site of the bite. Vague prodromal symptoms such as headache, myalgia, and fatigue precede the onset of rabies encephalitis. The classical furious rabies, with characteristic hydrophobic spasms, can be mistaken for generalized tetanus. Drug intoxication can also be confused with rabies. In the absence of spasms or hydrophobia, rabies can be misdiagnosed as viral encephalitis, postinfectious encephalitis, tetanus, poliomyelitis, or vaccine reactions. Paralytic (dumb) rabies, which generally follows a slower course, consists of ascending flaccid paralysis, usually from the bite site, with loss of autonomic function but not sensation. It may be misdiagnosed as Guillain–Barré syndrome.

In unvaccinated individuals rabies antibody occurs from late in the first week, but its absence does not exclude rabies. In vaccinated individuals very high levels in the cerebrospinal fluid or blood may be helpful. A fluorescent antibody test on punch biopsies from the nape of the neck is rapid, sensitive, and specific in expert hands (Blenden *et al*. 1986). Rabies virus tissue culture takes several days; saliva, cerebrospinal fluid, tracheal aspirates, or throat swabs may be cultured. Definitive diagnosis is usually made at postmortem.

Management

There are tens of thousands of cases of rabies every year worldwide, and only four claimed recoveries. Correct postbite prophylaxis is highly effective in preventing rabies cases. Once symptomatic rabies encephalitis occurs death is virtually inevitable, usually within a week without intensive care. Death from rabies is agonizing and terrifying. Overenthusiastic efforts to prolong life are inappropriate. The priorities must be to relieve suffering and distress in the patient and to protect carers. All patients should be heavily sedated and nursed in dim quiet surroundings. Hydrophobic spasms can be provoked by sudden movement, draughts of air, bright lights, and loud noises, as well as by water (or even the mention of water).

Where clinicians decide to undertake intensive care to prolong life, many cardiac and respiratory complications will occur. Furious rabies is associated with cardiac arrhythmias (usually requiring pacing), myocarditis, blood pressure swings, and eventually flaccid paralysis. Cardiac and respiratory arrest may occur during spasms. The main complication of paralytic rabies is respiratory failure, with inability to swallow, urinary retention, and constipation. Spasms may occur in the terminal phase.

Infection control

Rabies virus may be found in cerbrospinal fluid, saliva, tears, urine, and other body fluids or tissues. Patients should be barrier nursed, including respiratory precautions for tracheal suction, ideally by staff who have had pre-exposure rabies vaccine. Those who have not been previously immunized should begin a course of postexposure vaccination. However, there are no reports of human-to-human transmission of rabies via infected secretions.

The *Strongyloides* hyperinfection syndrome

Strongyloides stercoralis is a nematode parasite widespread in the tropics. Humans are infected via skin penetration by filariform larvae from moist soil. *Strongyloides* larvae have an ability, almost unique among nematodes, to reinfect the same host. This autoinfection cycle enables the parasite to remain in the same host for many years, sometimes for as long as three to four decades. Long-established strongyloidiasis in a normal host is often asymptomatic. The danger occurs when a patient with non-apparent infection becomes severely immunocompromised by way of underlying disease or iatrogenic immunosuppression. This permits massive autoinfection with the development of an enormous worm burden, often complicated by secondary Gram-negative sepsis in the blood and occasionally the cerebrospinal fluid.

The diagnosis of *Strongyloides* hyperinfection should be suspected in any patient who has, at one time or another, visited or lived in an endemic area, has depressed immunity, and develops any of the following complications.

1. Gastrointestinal: steatorrhea, protein-losing enteropathy, ileus, necrotizing jejunitis, small bowel infarction, gastrointestinal bleeding.
2. Respiratory: shortness of breath and wheeze, respiratory failure.
3. Central nervous system: meningitis, cerebral abscess.
4. Gram-negative shock.

Risk factors for disseminated strongyloidiasis are systemic corticosteroid therapy, human T-cell lymphotropic virus I infection, HIV infection, anticancer chemotherapy, leukemia, lymphoma, lepromatous leprosy, visceral leishmaniasis, systemic lupus erythematosus, and malnutrition.

The disease is often missed and at least 50 per cent of cases are diagnosed at postmortem. The intensive care unit (ICU) staff should maintain a high index of suspicion, as it is likely that this disorder will increase in frequency over the years as more people visit tropical areas.

The diagnosis can be confirmed by finding *Strongyloides* larvae in stool, duodenal fluid, sputum, tracheal aspirate, ascitic fluid, or cerebrospinal fluid, depending on the sites invaded. Widespread dissemination of larvae is common. The laboratory must be asked to look specifically for parasites. Peripheral blood eosinophilia is usually absent. Serology may be helpful.

Treatment

Strongyloides responds to ivermectin, thiabendazole, or albendazole. However, these are oral treatments and none are available as a parenteral preparation licensed for human use, which can give rise to therapeutic difficulty in the presence of paralytic ileus. Successful use of rectal thiabendazole has been reported (Boken *et al.* 1993).

Infection control

Filariform larvae of *Strongyloides* are infective to humans via contact with skin and mucous membranes. Feces and all body fluids should be considered as potentially infectious, and skin and mucous membrane contact with them must be avoided. Strict enteric precautions are required.

Viral hemorrhagic fevers

Most of the information in this section is abstracted from the document *Management and control of viral hemorrhagic fevers* (Advisory Committee on Dangerous Pathogens 1997) which gives detailed practical guidance and should be consulted if a suspected case of viral hemorrhagic fever is encountered. (Crown Copyright is reproduced with the permission of Her Majesty's Stationery Office.) The United States guidelines (Centers for Disease Control 1995) are also useful. Those practitioners working in geographical areas other than the United Kingdom or the United States will need to pay close attention to the requirements of their own health departments regarding viral hemorrhagic fevers which are in force at the time.

There is a risk of secondary nosocomial infection with the viral agents responsible, among both clinical and laboratory staff, by way of accidental inoculation or contamination of broken skin or mucous membranes by infected blood or body fluids.

Occasionally, a severely ill patient requires ICU care while the precise diagnosis is unclear, and it is this scenario which presents a danger to the admitting unit. Therefore it is essential that the possibility of viral hemorrhagic fevers is considered in patients who have been in a relevant geographical area so that a risk assessment can be made and proper specialist care provided.

The viral hemorrhagic fevers which cause most concern are Lassa fever, Ebola virus, Marburg virus, and Crimean/Congo hemorrhagic fever.

Lassa fever is endemic in rural West Africa. The incubation period for control purposes is set at 3 to 21 days. There is insidious onset of fever, shivering, malaise, headache, and generalized aching. Sore throat occurs early. The illness may progress to produce edema of the face and neck, pleural effusion, and ascites. Diarrhea and vomiting may occur. Severe cases exhibit bleeding into the skin, mucosae, and deeper tissues.

Ebola fever has been reported from Zaire, Sudan, Côte d'Ivoire, and Gabon, and the virus has been isolated from cynomolgus monkeys originating in the Philippines. The incubation period is 4 to 16 days. There is abrupt onset of shivering, fever, headache, backache, and muscle and joint pains. Anorexia, nausea, vomiting, and watery diarrhea begin around the third day. A morbilliform rash is common, appearing after 3 to 8 days, lasting 4 to 14 days, and followed by desquamation. There may be inflammation of the throat and conjunctivas. Many patients develop spontaneous bleeding. A few days after the start of symptoms, patients exhibit an altered mental state and extreme lethargy. Renal failure commonly occurs in fatal cases.

Marburg virus has been reported from Uganda, Zimbabwe, and Kenya. The reported incubation period is 3 to 9 days and the clinical course is similar to that of Ebola virus infection.

The virus causing Crimean/Congo hemorrhagic fever is transmitted to humans by tick bite. There is evidence of its presence in the former USSR, Bulgaria, former Yugoslavia, East and West Africa, Dubai, Iraq, South Africa, Pakistan, Greece, Turkey, Albania, Afghanistan, and India. The incubation period is 7 to 12 days followed by abrupt onset of fever, chills, malaise, irritability, headaches, and severe pains in the limbs and loins, and then anorexia, nausea, vomiting, and abdominal pain. Fever is continuous or remittent and resolves by crisis after 8 days. There is flushing and edema of the face and neck, injection of the conjunctivas and pharynx, and edema of the soft palate. Most patients develop a petechial rash. A hemorrhagic rash appears on the soft palate and uvula early; other signs of bleeding, including hematemesis and melena, appear on the fourth or fifth day. Gastrointestinal or nasal hemorrhage can be fatal. Central nervous system involvement is associated with a poor prognosis.

Transmission in a hospital setting

Lassa virus is transmitted by accidental inoculation of blood or body fluids, intimate personal contact, and close exposure to pharyngeal secretions. It can be excreted in body fluids up to a few months after recovery.

Ebola virus is transmitted by contact with infected blood or body fluids. Aerosol transmission has been queried, but has not yet been recorded in a clinical setting.

Marburg virus is transmitted by contact with infected blood or body fluids. Although aerosol transmission is not recorded from a clinical setting, the possibility should not be disregarded in a seriously ill patient with lung involvement.

Crimean/Congo hemorrhagic fever is transmitted by contact with infected blood or body fluids. Secondary causes have followed resuscitation.

It should be noted that taking and handling laboratory specimens is the most common cause of nosocomial viral hemorrhagic fevers.

United Kingdom guidelines

Management of viral hemorrhagic fevers should be undertaken in a designated high-security infectious diseases unit (HSIDU) where there is a concentration of appropriate expertise and designated laboratory facilities. Patients known or suspected to be suffering from a viral hemorrhagic fever should not be admitted to a general ICU, but managed in an HSIDU.

Risk assessment

Where a patient has been admitted to an ICU and viral hemorrhagic fever is subsequently thought to be a possibility, the head of the ICU in conjunction with the infection control doctor and a specialist in tropical or infectious diseases should assess the following:

(1) which viral hemorrhagic fever may be responsible;

(2) how likely the patient is to have a viral hemorrhagic fever;

(3) if infection takes place, the severity of the viral hemorrhagic fever;

(4) the likely location of the virus (e.g. on equipment, in waste, in body fluids from the patient, etc.);

(5) the routes by which staff members may become exposed;

(6) an estimate of likely exposure.

Patient categorization

Epidemiological evidence is very important in this context, particularly as a firm diagnosis can seldom be made on clinical grounds alone. The United Kingdom guidelines assess risk as follows.

Minimal risk

This category includes febrile patients who have not been in known endemic areas before the onset of illness or who have been in endemic areas (or in contact with a known or suspected source of a viral hemorrhagic fever), but in whom the onset of illness was definitely more than 21 days after their last contact with any potential source of infection. These patients can be reassured and managed without extra precautions. Most will have malaria.

Moderate risk

This category includes febrile patients who have been in an endemic area during the 21 days before the onset of illness, but who have none of the additional risk factors which would place them in the high-risk category, or who have not been in a known endemic area, but may have been in adjacent areas or countries during the 21 days before the onset of illness, and who have evidence of severe illness with organ failure and/or hemorrhage which could be due to a viral hemorrhagic fever and for which no alternative diagnosis is currently evident. It should be noted that malaria is the most likely alternative diagnosis in patients suspected to have a viral hemorrhagic fever but who prove to have an alternative condition.

High risk

This category includes the following types of febrile patients.

1. They have been in an endemic area during the 3 weeks before illness and satisfy at least one of the following conditions:

    (a) they have lived or stayed for more than 4 h in a house where there were ill feverish persons known or strongly suspected to have a viral hemorrhagic fever;

    (b) they have taken part in nursing or caring for ill feverish patients known or strongly suspected to have a viral hemorrhagic fever, or have had contact with the body fluids, tissue, or dead body of such a patient;

    (c) they are a laboratory, health, or other worker who has, or has been likely to have, come into contact with the body fluids, tissues, or body of a human or animal known or strongly suspected to have a viral hemorrhagic fever;

    (d) they were previously categorized as moderate risk, but have developed organ failure and/or hemorrhage.

2. They have not been in an endemic area but during the 3 weeks before illness they have done at least one of the following:

   (a) they have cared for a patient or animal known or strongly suspected to have a viral hemorrhagic fever, or have come into contact with the body fluids, tissues, or dead body of such a patient or animal;

   (b) they have handled clinical specimens, tissues, or laboratory cultures known or strongly suspected to contain the agent of a viral hemorrhagic fever.

Such patients should be admitted to an HSIDU.

United States guidelines

The United States recommendations on management of patients with suspected viral hemorrhagic fever (Centers for Disease Control 1995) apply to patients who, within 3 weeks before onset of fever, have traveled in the specific local area of a country where viral hemorrhagic fever has recently occurred, had direct contact with blood, other body fluids, secretions, or excretions of a person or animal with viral hemorrhagic fever, or worked in a laboratory or animal facility that handles hemorrhagic fever viruses. The likelihood of acquiring viral hemorrhagic fever is considered extremely low in persons who do not meet any of these criteria.

Expert advice is available from:

Special Pathogens Branch
Division of Viral Diseases
National Center for Infectious Diseases
Centers for Disease Control and Prevention
1600 Clifton Road NE
Atlanta
GA 30333
USA
Telephone: +1 404 639 1511 (from 4:30 p.m to 8 a.m. +1 404 639 2688)

Principles of management

Lassa fever

Intensive supportive therapy is applied plus ribavirin 30 mg/kg intravenously as a loading dose, followed by 16 mg/kg intravenously every 6 h for 4 days, and then 8 mg/kg intravenously every 8 h for 6 days. Prophylactic use (ribavirin 500 mg orally every 6 h for 7 days) may prevent or delay disease in exposed contacts.

Ebola virus and Marburg virus

No specific therapy is currently regarded as effective for either of these infections and management depends upon intensive supportive measures.

Crimean/Congo hemorrhagic fever

Therapy depends upon intensive supportive treatment. However, early ribavirin therapy can reduce the duration of feverish illness and the severity of hemorrhagic manifestations.

References

Advisory Committee on Dangerous Pathogens (1997). *Management and control of viral hemorrhagic fevers*. HMSO, London.

Blenden, D.C., Creech, W., and Torres-Anjel, M.J. (1996). Use of immunofluorescence examination to detect rabies virus antigen in the skin of humans with clinical encephalitis. *Journal of Infectious Diseases*, **154**, 698–701.

Boken, D.J., Leoni, P.A., and Preheim, L.C. (1993). Treatment of *Strongyloides stercoralis* hyperinfection syndrome with thiabendazole administered per rectum. *Clinical Infectious Diseases*, **16**, 123–6.

Centers for Disease Control (1995). Update: management of patients with suspected viral hemorrhagic fever—United States. *Morbidity and Mortality Weekly*, **44**, 475–9.

Hoffman, S.L., *et al.* (1984). Reduction of mortality in chloramphenicol treated severe typhoid fever by high dose dexamethasone. *New England Journal of Medicine*, **310**, 82–7.

13.7 Systemic inflammatory response

13.7.1 Assessment

O. Boyd and P. Newman

Key messages

- An initial insult results in a primary disease process which can lead to a multisystem disturbance (the systemic inflammatory response syndrome) secondary to inflammatory mediators released during shock.
- A history and examination should initiate the search for the cause of the systemic inflammatory response syndrome.
- Regional assessment includes investigation of the cardiovascular, respiratory, renal, gastrointestinal, hematological, and metabolic systems. Suspected infection should be investigated.

Introduction

An uncontrolled primary disease process can result in secondary multisystem involvement (Fig. 1), now known as the systemic inflammatory response syndrome (SIRS) (Bone *et al.* 1992). This name replaces many older terms such as sepsis and sepsis syndrome (Table 1).

SIRS occurs in many patients in intensive care units (ICUs). The underlying pathophysiology is decreased organ perfusion, which can involve whole organ systems or occur regionally within an organ system. Assessment of multisystem involvement can be confusing, showing a mixture of effects of increased and decreased regional blood flow. Furthermore, there is a temporal separation between the primary disease and involvement of other systems, and a range of severity of organ dysfunction.

Although the criteria for SIRS form a useful starting point, assessment of the severity of secondary multisystem involvement is undertaken by history, examination, and investigation of the patient, looking for evidence of individual organ dysfunction and determining the degree of dysfunction.

History

Early in the progression of multisystem involvement there will be few indicators in the history, although it may be noted that the patient is not attaining the expected recovery pathway following the primary illness. Specific complaints are dominated by the primary illness,

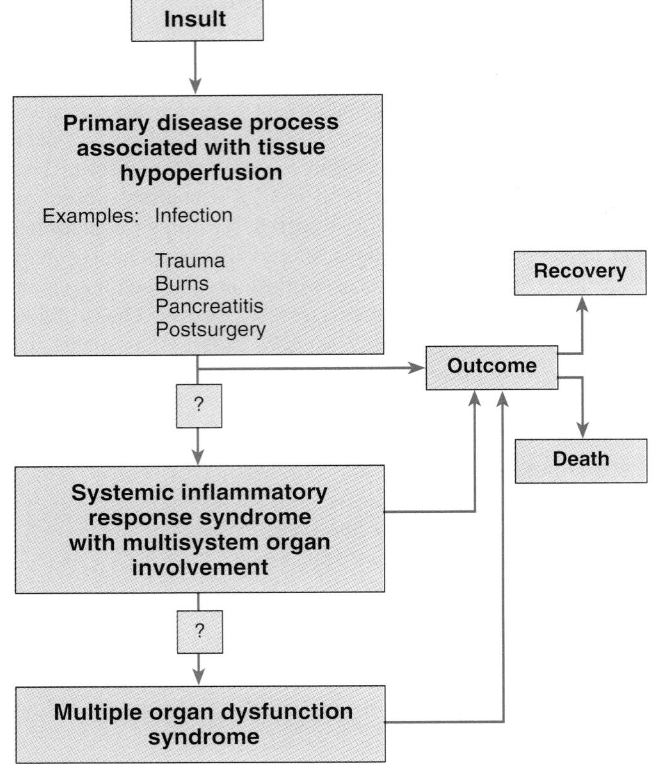

Fig. 1 Schematic representation of the development of multisystem involvement following a primary insult.

Table 1 Criteria for the diagnosis of SIRS in adults
A severe clinical illness and the presence of two or more of the following:
Temperature greater than 38 °C or less than 36 °C
Heart rate greater than 90 beats/min
Tachypnea (>20 breaths/min) or $P\text{CO}_2$ less than 32 mmHg (4.25 kPa)
White blood cell count greater than 12×10^9/l or less than 4×10^9/l, or more than 10% band forms on the peripheral blood film

Data from Bone *et al.* (1992).

with the only clues to more widespread involvement being complaints of general lethargy and tiredness which do not form an easily recognizable pattern.

Enquiry must be made into specific indicators of organ dysfunction. The cardiovascular system may yield a history of palpitations or dizziness. Regarding the respiratory system, the patient may be short of breath, have mild orthopnea, or complain of more florid symptoms of chest infection. Gastrointestinal tract symptoms include nausea or vomiting, abdominal pain, a history of gastrointestinal bleeding, failure to pass flatus, or diarrhea. Symptoms of the central nervous system include confusion and psychosis, particularly in the elderly.

Examination

The hard signs indicating SIRS should be sought (Table 1); however, most patients in ICUs satisfy these criteria and their non-specific nature makes them unhelpful for clinical management. There may be evidence of depleted intravascular volume with dry mucous membranes and low jugular venous pressure. There may be mild icterus, bruising, prolonged bleeding from venous puncture sites, and slow wound healing. Patients may look surprisingly well, with warm extremities, venous distension, and bounding peripheral pulses, or they may look quite ill, with cool and cyanosed peripheries. These two presentations are the extremes of a spectrum and represent the descriptions of 'cold' and 'warm' shock found in older textbooks. Occasionally the patient may be virtually moribund at first presentation, when findings will be of cold poorly perfused peripheries, thready peripheral pulses, and cyanosis; blood pressure will be difficult to measure, respiratory rate is usually raised, and the patient will be comatose.

Examination of the cardiovascular system usually reveals a mild tachycardia; blood pressure, particularly diastolic pressure, is slightly lower than that previously recorded. If there is excessive arterial vasoconstriction, the systolic pressure can occasionally be raised in an attempt to compensate for intravascular volume depletion. The jugular venous pressure is usually low, but if the heart is failing or there is compensatory vasoconstriction it may be raised. There may be added heart sounds, such as a gallop rhythm, and occasionally a flow murmur if the patient has a hyperdynamic circulation. Assessment of the respiratory system demonstrates a raised respiratory rate, and as lung injury and fluid extravasation become more extensive there may be reduced respiratory excursion, basal crackles, and evidence of pleural effusions. It is quite common for the cardiorespiratory findings to be diagnosed erroneously as left ventricular failure, and for the patient to be treated with diuretics; however, if the underlying condition involves volume depletion, this will worsen the perfusion deficit. Further investigations to identify and correct the physiological abnormalities are always necessary.

The abdominal examination may show evidence of reduced gastrointestinal function with distension, absent bowel sounds, and large nasogastric tube output. Signs of central nervous system abnormalities include reduced level of consiousness and confusion or psychosis.

Regional assessment

The urgency and complexity of investigations vary depending on the state of the patient and the facilities available. Although evaluation begins with history and examination, it continues with investigation of the degree and severity of multisystem involvement. Although each region is discussed individually here, they are usually evaluated simultaneously.

Cardiovascular system

The cardiac rhythm is monitored continuously, as dysrhythmia can be unexpected. An ECG should be obtained to look for pre-existing abnormalities and for ST- and T-wave changes indicative of ischemia, which may affect treatment choices. Invasive blood pressure monitoring will continously record extremes of pressure and also allow access for serum chemistries and blood gases. Monitoring central venous pressure and pulmonary artery occlusion pressure provides hemodynamic and oxygen transport data and is useful in precarious situations.

Cardiac output is difficult to assess clinically and can be vital for planning subsequent treatment. Cardiac output is usually measured by thermodilution using a pulmonary artery catheter; suprasternal or esophageal Doppler may also be utilized. Left and right ventricular stroke work and systemic and pulmonary arterial resistance provide information on the status of cardiac and circulatory function. Echocardiography, either surface or transesophageal, is also useful for assessing regional cardiac function, particularly if a new murmur is detected which may be indicative of cardiac dilation, papillary muscle dysfunction, or endocarditis.

Respiratory system

The respiratory system is frequently the first involved in multisystem disease. The chest radiograph may be entirely normal or may exhibit patchy peripheral infiltrates. Blood gases and acid–base abnormalities may be subtle. In early stages of acute lung injury $P\text{O}_2$ may fall slightly. There may be a compensated metabolic acidosis only as the arterial pH may be normal owing to compensatory hyperventilation. As lung injury and oxygenation become worse despite increasing the inspired oxygen concentration, CO_2 elimination may be hindered, leading to a rise in $P\text{CO}_2$ with a consequent mixed metabolic and respiratory acidosis. Blood lactate concentration may rise secondary to tissue hypoperfusion. Owing to the shift of the oxyhemoglobin dissociation curve, oxygen availability is better represented by direct measurement of arterial oxygen saturation rather than partial pressure. A more complete picture of oxygen transport is achieved by measuring mixed venous oxygen saturation and calculating the total body oxygen availability and oxygen consumption.

Renal system

Renal insufficiency may result from an absolute decrease in renal blood flow and/or a change in distribution from cortex to medulla. There is a fall in glomerular filtration rate with an increase in plasma urea and creatinine, and urine output falls below 0.5 ml/kg/h. Repeated biochemical analysis and accurate measurement of urine flow by catheterization are required. Plasma electrolytes, particularly potassium, should be followed closely. Plasma-to-urine ratios of urea, creatinine, sodium, and osmolality can be helpful in distinguishing primarily prerenal failure from renal failure. Because of constantly changing conditions, frequent investigations of creatinine clearance or glomerular filtration rate may not be very helpful. Ultrasound visualization of the kidneys provides information on size, echogenicity, and ureteric dilation; this can suggest previous renal disease, parenchymal disease, or obstructed urine flow.

Gastrointestinal system

The response of the gastrointestinal tract to shock includes decreased peristalsis with gastric dilatation, stress ulceration, and hypoperfusion of the mucosa leading to altered barrier function. A plain abdominal radiograph may indicate ileus or perforation. A left lateral decubitus radiograph may reveal free air more readily. An ultrasound examination or CT scan of the abdomen may show evidence of pancreatitis, ascites, and fluid collections. Intra-abdominal free fluid or collections and fluid from intra-abdominal drains should be sent for culture and biochemistry. If there is suspicion of gastrointestinal bleeding or infection, an esophagogastroduodenoscopy should be performed.

There is increasing interest in direct and indirect assessment of splanchnic perfusion. The mucosa of the gastrointestinal tract may be particularly sensitive to hypoperfusion, and is accessible to monitoring via oral or rectal routes. Gastric tonometry has recently been proposed as a simple non-invasive method of estimating mucosal acidosis ($pH_{intramucosal}$ or pH_i), giving an estimate of the relative state of perfusion of the mucosa. Assessment of pH_i may be of prognostic significance and may be useful in monitoring the gastrointestinal tract in the multisystem involvement following shock. However, its relevance to continued clinical management is as yet unclear.

Hematological system

Assessment of the blood count and clotting status is essential as anemia, alterations in white cell count, and changes in clotting are common. A complete blood count, with indices of red blood cell morphology and white blood cell differential, and film are mandatory. Full investigation of clotting function, including fibrinogen and fibrin degradation product levels, is helpful. Blood should be sent for type and cross-match if anemia or bleeding are suspected or proven.

Metabolic systems (hepatic, pancreas, endocrine)

Minor degrees of hepatic dysfunction appear to occur early in the multisystem involvment of shock. Biochemical assessment of liver function may show elevation of both conjugated and unconjugated bilirubin and transaminases. In more severely ill patients acalculous cholecystitis can occur with marked hyperbilirubinemia. Further investigation of the liver using abdominal ultrasound or CT scanning is often warranted to exclude specific hepatic disease, such as hepatic abscess or biliary obstruction.

Elevation of blood glucose is common in critically ill patients and should be monitored. Hepatic failure may cause a fall in blood glucose. Pancreatitis, which may develop in the shocked patient presumably due to poor pancreatic perfusion, usually demonstrates elevated amylase. Abdominal ultrasound or CT scan may show edema or necrosis of the gland.

Experimental work has suggested that many patients with SIRS may have relative deficiency of cortisol. This is often only demonstrated by a direct challenge with a synacthen (ACTH stimulation) test. Resistant hypotension, elevation of blood potassium, and a fall of blood sodium may also indicate hypoadrenalism.

Infection

Many patients with SIRS have concurrent infection, which may be causing SIRS or may merely be a coexisting finding. All patients with multisystem involvement following a primary disease must be carefully screened for infection, with attempts made to arrive at a microbiological diagnosis. Appropriate skin and wound swabs, as well as specimens of sputum, urine, blood, and drainage fluid for culture and sensitivity, should be taken in the suspected septic patient. Repeat cultures should be taken during pyrexial episodes.

Reference

Bone, R.C., *et al.* (1992). Definitions for sepsis and organ failure and guidelines for the use of innovative therapies in sepsis. *Critical Care Medicine*, **20**, 864–74.

13.7.2 Therapeutic strategy

P. Newman and R. M. Grounds

Key messages

- The fundamental principle of management is the early restoration of organ perfusion and oxygenation with concurrent diagnosis and treatment of the initial insult.
- Restoring perfusion and oxygenation requires resuscitation of the cardiovascular and respiratory systems and transfusion of red cells to restore adequate oxygen-carrying capacity.
- At present the use of supranormal oxygen delivery goals in established shock has not been proved to improve outcome.
- Specific therapies directed at preventing further organ damage should be used to halt the progression to multiple organ failure.

Introduction

The underlying process in the development of multisystem dysfunction in shock is decreased organ perfusion and oxygenation. The initial insult may be multisystem or may be organ specific, but both have the potential to lead to the multiple organ dysfunction syndrome if not diagnosed and treated adequately. The fundamental principle of management is the early restoration of organ perfusion and oxygenation with concurrent treatment of the initial insult. Without diagnosing and then treating the cause of the patient's illness, the outcome of supportive therapies is likely to be poor as the underlying disease process will not be halted and may progress, thus increasing the likelihood of multiple organ failure. For example, 72 h after bowel surgery a patient may show obvious signs of shock, with hypotension and tachycardia and the signs of the systemic inflammatory response syndrome (SIRS), the causes of which may be a chest infection with sepsis, anastomotic breakdown with peritonitis, or a myocardial infarction. Hence the correct diagnosis of the underlying disease process and treatment to halt this is as crucial as the restoration of global perfusion and oxygenation.

Classically, shock has been divided into the four distinct categories: hypovolemic, cardiogenic, septic, and neurogenic. In each of these categories there is a decrease or maldistribution of oxygenated blood supplying the tissues. Tissue oxygen delivery (Do_2) is made up of many components and is calculated using the equation

$$Do_2 = \text{CI} \times \text{Hb} \times Sao_2 \times 0.134 + (0.0031 \times Pao_2)$$

where CI is the cardiac index, Hb is the hemoglobin concentration, Sao_2 is arterial saturation, and Pao_2 is arterial oxygen partial pressure.

The cardiac index is dependent on myocardial contractility and the balance between pulse rate and cardiac filling time. These factors depend on ventricular filling pressure and therefore on optimal intravascular volume status. Restoration of a normal hemoglobin concentration will increase circulating blood volume and restore tissue perfusion. Arterial saturation depends on matching ventilation and perfusion in the lung parenchyma, with sufficient oxygen supply.

Depending on the cause of shock, other specific treatments, such as antibiotics or surgery, may be needed to restore adequate perfusion pressure and correct maldistribution of flow. Organ-specific therapies may also protect and restore function to targeted organs.

Treatment

In order to prevent shock from progressing to multiple organ failure, rapid identification and treatment to minimize the duration and extent of the hemodynamic and metabolic disturbance are essential. Response to therapy involves both non-invasive and invasive monitoring.

Correction of hypovolemia

The aim of fluid resuscitation is to replete the circulation. The primary objective is to identify the source of fluid loss and prevent further loss, whether overt bleeding or fluid sequestration to the extravascular space. An adequate intravascular volume is only part of fluid resuscitation and there are many other considerations.

Measurement of central venous pressure and pulmonary artery occlusion pressure are the most widely used guides to volume status. A pulmonary artery occlusion pressure of 12 to 15 mmHg should be the initial target. Further small fluid boluses of 100 to 200 ml should be given with regular reassessment of pulmonary artery occlusion pressure and cardiac output. Values of stroke volume and left heart work can be derived from hemodynamic information obtained from the pulmonary artery catheter, and fluid therapy may target an optimal stroke volume.

There is an unresolved controversy regarding the use of crystalloid or colloid in the initial phase of shock. If crystalloid is used, more must be given than colloid.

Hemoglobin is the only available oxygen-carrying molecule. There must be a compromise between oxygen-carrying capacity and blood viscosity to allow good capillary blood flow. A hemoglobin concentration of 10 to 12 g/dl and a hematocrit in the range of 30 to 40 per cent are probably optimal, but controversy exists as to whether these goals should pertain in all conditions

Correction of coagulopathy aids volume replacement and reduces further volume loss. Loss, consumption, and dilution of clotting factors during hemorrhage and disseminated intravascular coagulation (DIC) in septic shock can lead to intractable bleeding. Adequate correction with fresh frozen plasma, cryoprecipitate, and platelets may be required to correct this. Hematinics and vitamin K may also be required after massive blood transfusion.

Thus hypovolemia should be corrected by early aggressive fluid resuscitation with adequate crystalloid or colloid, blood, and clotting factors (if required), guided by hemodynamic monitoring to assess whether treatment is effective.

Correction of decreased myocardial contractility (pump failure)

Cardiogenic shock may denote a primary cardiac cause such as myocardial infarction or valvular disease. However, pump failure can occur in a patient with a previously normally functioning heart. The treatment of primary cardiac disease is dealt with elsewhere. The aim of therapy is to optimize cardiac output with a decrease or minimal increase in myocardial work and oxygen consumption.

Tachyarrhythmias are common in shock and need to be treated in order to maximize cardiac function. Fluid resuscitation should be titrated with pulmonary artery occlusion pressure and cardiac output. Nitroglycerin, a venodilator, may be a useful preload reducer.

If there is an adequate preload, a reduction in systemic vascular resistance may improve cardiac function. Nitroprusside and nitroglycerin can be helpful, but the blood pressure may fall sharply. In septic shock there is a loss of peripheral arterial vascular tone, and vasoconstrictors (norepinephrine (noradrenaline), methoxamine) are often required to increase afterload in the presence of an adequate cardiac output. Steal between and within organs may occur, and specific organ vasodilators will be considered later.

Inotropes increase contractility but may also have a deleterious effect on myocardial work, increasing myocardial oxygen consumption. Dopamine, dobutamine, and epinephrine (adrenaline) all have beta effects (chronotropic and inotropic) at lower doses and alpha effects (vasoconstriction) at higher doses. Dopexamine is an inodilator (beta and dopaminergic effects) with no alpha effects and perhaps less effect on myocardial oxygen consumption. The phosphodiesterase inhibitors enoximone and milrinone are inodilators and have been used in cardiogenic shock. Because all the commonly used drugs have a number of potential contradictory effects, their cardiac and circulatory effects must be continuously reassessed.

Assist devices can be used to augment cardiac output and reduce afterload. The intra-aortic balloon counterpulsator has been used in cardiogenic shock and has also been shown to be useful in septic and anaphylactic shock if myocardial depression is a prominent problem.

Correction of low Sao_2

Patients in shock often have a decreased level of consciousness and a patent airway should be a priority. This may involve intubation to secure and protect the airway, particularly in trauma and head injury. Once a patent airway is ensured, oxygen should be administered.

Both intermittent positive-pressure ventilation and positive end-expiratory pressure have adverse effects on hemodynamics by decreasing venous return, necessitating continous monitoring of the vascular status.

Specific therapies for septic shock

The initial aim of management is the identification and aggressive treatment of the source of sepsis, and this is combined with supportive measures.

Septic shock differs from other forms of shock, with a high cardiac output and vasodilated peripheries due to maldistribution of the increased blood flow. There is shunting of blood, a lactic acidosis, and capillary leak. If fluid alone does not restore an adequate blood pressure, vasoconstrictors should be used to provide a perfusion pressure to vital organs. As the disease process worsens, the myocardium fails and inotropic support may be required.

Septic shock may be associated with DIC and consumption of platelets. Transfusion of clotting factors and platelets may be required. Newer agents for treating sepsis, including antibacterial endotoxin antibodies, anti-tumor necrosis factor and IL-1 receptor antagonists, and L-NMMA and aminoguanidine which block nitric oxide production, have not proved to be the cure.

Corticosteroids have been tried as a treatment for shock. There is no evidence that they are of value, and in fact they may worsen the condition by contributing to overwhelming infection. The only use for steroids in shock is immediately after spinal cord injury.

Therapies aimed at preserving organ function

Critically ill patients have an increased metabolic demand requiring nutritional support. Early enteral feeding following a major insult has been advantageous. The gut seems to be involved in the evolution of multiple organ failure, and early feeding may help to prevent this by preventing gut mucosal atrophy, increasing hepatosplanchnic blood flow, maintaining immunocompetence, reducing bacterial translocation, preserving gut flora, and preventing stress ulceration. Special formulas may enhance immunocompetence and protect the mucosa if given early.

If enteral feeding is not tolerated, other measures to protect the gastric mucosa will be required. H_2 antagonists will reduce acid secretion but may allow overgrowth of bacteria in the stomach, increasing the risk of nosocomial pneumonia. Sucralfate, which is a mucosal barrier agent, will not cause bacterial overgrowth. The role of gastric tonometry in assessing gastric and splanchnic perfusion is still being evaluated.

Nosocomial respiratory infection is common in ventilated patients. Therapeutic strategies which reduce the incidence of infection include selective decontamination of the gut (Selective Decontamination of the Digestive Tract Trialists' Collaborative Group 1993). Allowing rest periods during enteral feeding may allow acidity to return to the stomach and kill bacteria, thereby reducing the likelihood of nosocomial infection (Lee *et al.* 1990).

The prevention of acute renal failure when shock occurs revolves around the restoration of adequate oxygen delivery to the kidneys and avoidance of further nephrotoxic insults. Fluid resuscitation alone may not be sufficient to maximize renal blood flow. Historically, dopamine has been used in low-dose infusion (2–4 μg/kg/min) to increase urine output, possibly by increasing cardiac output rather than by a specific dopaminergic vasodilator effect. Maximum fluid loading with concomitant diuretic infusion to prevent the onset of renal failure is also used. This is achieved with colloids, nitrates, norepinephrine and furosemide (frusemide). The diuretics mannitol and furosemide have both been used to increase urine output; however, adequate intravascular volume should be assured.

There is no evidence at present that any specific regimen will prevent acute renal failure. However, avoidance of nephrotoxic insults, such as unmonitored use of aminoglycosides, non-steroidal anti-inflammatory drugs, angiotensin-converting enzyme inhibitors, and contrast agents, is paramount in the patient with perilous renal function.

Supranormal goal-directed therapy in shock

In order to guide therapy in shock, targets must be set. Classical goals have included adequate blood pressure (above 120/80 mmHg), a heart rate within certain limits (60–120 beats/min), a urine output above 0.5 ml/kg/h, a central venous pressure of 4 to 12 cmH_2O, and a cardiac index (cardiac output indexed to body surface area) of 2.8 to 3.5 ml/min/m^2 (Shoemaker *et al.* 1988). Shoemaker and colleagues reviewed a cohort of high-risk surgical patients and found that survival is improved if the following supranormal values can be obtained: cardiac index above 4.5 ml/min/m^2, oxygen delivery above 600 ml/min/m^2, and oxygen consumption above 170 ml/min/m^2. (NB Both oxygen delivery and consumption are indexed to body surface area (Tuchschmidt *et al.* 1992; Boyd *et al.* 1993).) Fluids, blood, and inotropes were used to achieve these goals, and if a patient was unable to achieve them with therapy the prognosis was poor.

Other research has not supported these results. Gattinoni *et al.* (1995) compared three groups: supranormal cardiac index (above 4.5 ml/min/m^2), normal cardiac index (2.5–3.5 ml/min/m^2), and normal mixed venous saturation ($Svo_2 > 70$ per cent). Fluids, blood, and inotropes were used to attain these goals in a heterogeneous group of critically ill shocked patients. It was found that supranormal cardiac index or $Svo_2 > 70$ per cent conferred no benefit to morbidity or mortality in the shocked patient.

Thus, although higher than normal oxygen delivery provides the opportunity for improved oxygen consumption, it cannot ensure that tissue hypoxia will be reversed.

References

Boyd, O., Grounds, R.M., and Bennett, E.D. (1993). A randomized clinical trial of the effect of deliberate perioperative increase of oxygen delivery on mortality in high-risk surgical patients. *Journal of the American Medical Association*, **270**, 2699–707.

Gattinoni, L., *et al.* (1995). A trial of goal-oriented hemodynamic therapy in critically ill patients. *New England Journal of Medicine*, **333**, 1025–32.

Lee, S., Chang, R.W.S., and Jacobs, S. (1990). Intermittent nasogastric feeding: a simple and effective method to reduce pneumonia among ventilated ICU patients. *Clinical Intensive Care*, **1**, 100–2.

Selective Decontamination of the Digestive Tract Trialists' Collaborative Group (1993). Meta-analysis of randomised controlled trials of selective decontamination of the digestive tract. *British Medical Journal*, **307**, 525–32.

Shoemaker, W.C., Appel, P.L., Kram, H.B., Waxman, K., and Lee T.-S. (1988). Prospective trial of supranormal values of survivors as therapeutic goals in high-risk surgical patients. *Chest*, **94**, 1176–86.

Tuchschmidt, J., Fried, J., Astiz, M., and Rackow, E. (1992). Elevation of cardiac output and oxygen delivery improves outcome in septic shock. *Chest*, **102**, 216–20.

13.8 Multiple organ failure

13.8.1 Pathophysiology

J. F. Bion

Key messages

- Multiple organ failure is often preceded by a systemic oxygen debt.
- The splanchnic region is susceptible to ischemia–reperfusion injury.
- Neutrophil activation in the ischemic gut mucosa results in remote organ injury.
- The systemic inflammatory response is both a cause and a consequence of cell injury.
- Systemic oxygen flux differs in the septic and non-septic state
- Multiple physiological insults potentiate each other
- Constitutional factors may determine individual susceptibility to multiple organ failure

Introduction

The syndrome of multiple organ failure was first described in the mid-1970s (Baue 1975). It is the most important reason for admission to intensive care, but it is also a result of sophisticated organ system support for patients who would previously have died rapidly from their acute disease. The mortality from multiple organ failure increases with the number and duration of failing organs (Knaus *et al.* 1985), and the costs of death are high. The challenge for intensive care is to determine the pathogenesis of this diverse syndrome in order to prevent its development.

Epidemiology

The epidemiology of critical illness is not well defined. There are several reasons for this. First, at present there are no universally agreed definitions of organ system failure which adequately describe severity independent of therapy. Second, the risk factors which determine susceptibility to critical illness are poorly understood. Third, multiple organ failure is a syndrome, not a disease, and there are likely to be many mechanisms involved. Fourth, in any individual there is a complex interaction between the degree of homeostatic reserve, the severity, site, and multiplicity of the precipitating insult(s), and the specificity and timing of treatment, all of which will influence the appearance of organ system dysfunction. However, certain characteristic patterns may be observed, and it is the development of secondary, or remote, organ failures following initial resuscitation from an acute physiological disturbance, such as trauma or surgery, which has excited the greatest interest.

A characteristic pattern is the multiple trauma patient who is fluid resuscitated, taken to the operating room for fracture stabilization, and then develops progressive organ system dysfunction during the ensuing days in the intensive care unit (ICU), with a systemic inflammatory response which resembles sepsis and is often associated with microbial colonization or infection at different sites, particularly the lungs. In one study (Sauaia *et al.* 1995), over 60 per cent of late trauma deaths were a consequence of multiple organ failure. This pattern of an initial insult, usually involving hypovolemic or hypodynamic shock ('ebb phase'), followed later by multiple organ failure and a state that looks like the response to infection ('flow phase') encouraged earlier investigators to assume that organ failures were caused in some way by micro-organisms or their products. While this may be so, there is growing evidence that microbial colonization is a consequence, and not necessarily the cause, of organ failures. Two mechanisms at least are considered to be fundamental to this process: tissue hypoxia and activation of the immunoinflammatory cascade, producing a systemic inflammatory response and cellular injury.

Mechanisms of cell injury

Ischemia and oxidants

Multiple organ failure is always preceded by varying degrees of physiological disturbance, the most frequent of which are hypotension, low-cardiac-output states, hypoxemia, sepsis, and direct tissue injury. Common to these conditions are immunoinflammatory cascade activation and an overt or occult reduction in systemic oxygen supply causing cell injury. Ischemia or severe hypoxia results in intracellular ATP depletion, failure of membrane ion pumps, calcium influx activating phospholipases and other enzymes, sodium influx causing cell swelling, and finally cell death. During ischemia, the cellular enzyme xanthine dehydrogenase is converted to xanthine oxidase, while at the same time one of its substrates, hypoxanthine, accumulates. Reperfusion with oxygenated blood results in the generation of highly reactive oxygen species (free radicals) from the action of xanthine oxidase on hypoxanthine. Free radicals cause cell injury by damaging membranes and nuclear DNA by direct oxidation.

The immunoinflammatory cascade

Immunoinflammatory cascade activation may be either a cellular response to injury (the acute phase response) or the cause of cellular injury if the degree of activation is severe and prolonged, as in the systemic inflammatory response syndrome. Leukocyte and endothelial activation and neutrophil migration are central to this process. Cytokines and other cellular chemoattractants activate leukocytes and promote their adhesion to vascular (particularly venous) endothelium, with subsequent migration into the tissues. The specific site of attachment may be determined by the volume of endothelium available, local cytokine signaling from tissue macrophages, and local tissue injury. Neutrophils cannot re-enter the circulation, and remain trapped within the tissues discharging their contents until they are phagocytosed. Not only are neutrophils activated by the cellular response to ischemia–reperfusion and membrane oxidation, but they themselves also cause oxidative damage by generating reactive oxygen species, including nitric oxide. Thus there are considerable similarities between the precipitants of, and the host response to, cell injury. The cytokine network is a complex and very sensitive system for controlling host defenses, but once widely activated it has limited powers of expression and may be poorly targeted.

Cellular defenses

Cells have evolved a number of protective mechanisms against injury and 'premature ageing' which may be impaired by disease. These defense mechanisms include induction of heat shock proteins which preserve cellular protein synthesis and are produced in response to a variety of cellular insults, including thermal injury. Antioxidants such as superoxide dismutase, glutathione, vitamins C and E, and sulfhydryl-containing molecules provide protection against reactive oxygen species, and their depletion potentiates oxidative damage. Several cytokines (e.g. interleukins 10 and 13) have anti-inflammatory properties, and it is possible that constitutional differences in the expression of these mediators might in part determine susceptibility to critical illness.

Cell death and tissue repair

In the normal process of ageing, cell death occurs by an 'intrinsic' programmed gene-mediated sequence of cell shrinkage and nuclear condensation called apoptosis. However, apoptosis may be accelerated by the extrinsic pathological processes described above which result in cell injury and swelling. Apoptosis and extrinsic cell injury have several mechanisms in common, including changes in gene expression and calcium influx, again suggesting the possibility of a link between constitutional susceptibility and acquired disease. Whatever the process, removal of necrotic cell debris by phagocytosis will be impaired if the volume of dead tissue is large or if phagocytic capacity is impaired by immunoincompetence induced by drugs or nutritional depletion. Under these circumstances, tissue remodeling will be delayed with prolongation of organ failures.

The evidence for cellular dysoxia in multiple organ failure: cause or effect?

Systemic oxygen flux

Evidence that cellular dysoxia is an important initiating mechanism for multiple organ failure comes from studies of patients at risk or in the early stages of critical illness, which demonstrate that interventions which increase systemic oxygen supply (mainly fluid resuscitation and inotropic agents) also increase subsequent survival rates (Boyd *et al.* 1993). However, the same techniques do not seem to exert a beneficial effect later in the development of the syndrome, particularly in association with sepsis, even though measures such as base deficit, hyperlactatemia, and tonometry suggest the presence of a tissue oxygen debt. Part of the reason for this is that there are substantial differences in flow distribution and intermediary metabolism between patients presenting in low flow states (e.g. hypovolemia, trauma, cardiac failure) and those with the higher flow states associated with sepsis.

In non-septic patients (the 'ebb phase' of incipient multiple organ failure) oxygen consumption remains stable, despite marked reductions in oxygen delivery, until a critical point is reached below which oxygen consumption becomes dependent on supply, and a tissue oxygen debt accumulates. However, septic patients appear to demonstrate supply dependency over a wide range of values of oxygen delivery, with consumption continuing to increase despite large increases in supply. While this may be explained in part by mathematical coupling and by the thermogenic effect of inotropic agents, it is also likely to be a real effect caused by an increase in metabolic rate, disordered intermediary metabolism, flow maldistribution, increased nitric oxide production with impaired vasoregulation and increased vascular permeability, and a prolonged diffusion distance from red cell to tissues in the septic state.

These two patient populations are evidently very different. However, at some point in the development of multiple organ failure the 'ebb phase' gives way to the later 'flow phase' of the systemic inflammatory response syndrome. How this progression occurs has been the subject of detailed research, much of which has focused on the gut and splanchnic region.

The splanchnic circulation

Tissues differ in their susceptibility to hypoxia, and a minor oxygen debt for some organs may cause considerable dysfunction in others. The splanchnic organs may be at risk of hypoxia-induced cellular dysfunction for several reasons. The gut does not autoregulate flow for pressure, and gut blood flow falls markedly in response to reductions in cardiac output. Splanchnic vasoconstriction persists despite correction of systemic hemodynamics. The gut mucosa and the renal medulla demonstrate approximately the same vascular architecture of arterioles and venules, allowing oxygen shunting and consequent distal hypoxia under normal conditions. The oxygen content in these vessels is also reduced by plasma skimming, giving a hematocrit of around 10 per cent. This potential for dysoxic damage will be aggravated by alterations in systemic oxygen supply, sepsis-induced changes in tissue perfusion, and certain drugs such as non-steroidal anti-inflammatory analgesics. Because of its main function of salt and water retention, the kidney is the only organ required to do more work in the presence of a reduced oxygen supply, and its ability to autoregulate flow for pressure is additionally compromised by prior renal impairment and hypertension. Hepatic blood flow is determined by portal venous flow (affected by changes in cardiac output) and by hepatic arterial flow, which has limited capacity for autoregulation. Given these features, the hypothesis that the splanchnic organs are at particular risk from an acute reduction in systemic oxygen supply seems reasonable. In the case of the gut, this may be due to loss of its barrier function.

Gut barrier function

Determinants of gut permeability

The concept that the gut may cause systemic illness and remote organ injury dates from the end of the nineteenth century, and more recently phrases such as the gut being 'the motor of multiple organ failure' and 'the largest undrained abscess in the body' have established the theory in the medical imagination. Permeability of the gut mucosa has been studied using various molecular probes such as sugars or ^{51}Cr-EDTA. Increased permeability has been demonstrated following many different physiological insults including ischemia–reperfusion injury, endotoxemia, systemic acidoses, glutamine deficiency, and certain cytokines such as γ-interferon. One of several possible mechanisms for hyperpermeability is increased synthesis of nitric oxide by gut epithelium, neutrophils, or vascular endothelium. Nitric oxide may exert different effects on permeability depending on the nature of the stimulus, its timing, and the experimental model being used; these factors may explain why inhibition of nitric oxide synthase has been shown both to reduce and to increase gut permeability. However, the relevance of increased gut leakiness to translocation is uncertain, and it is possible that the phenomenon is merely a local expression of a generalized process.

Translocation and multiple organ failure

Translocation of bacteria and endotoxin to mesenteric lymph nodes or portal and venous blood has been demonstrated in animal models of gut ischemia and hemorrhagic shock, but the situation is more complex in humans. Translocation certainly occurs, but it has been associated with systemic symptoms in only one case report involving deliberate ingestion of *Candida*. In no other study has it been possible to establish a causal link between an acute precipitating illness, translocation, and remote organ injury. Indeed, there is a growing body of evidence which tends to disprove the theory. Studies of patients undergoing laparotomy for blunt trauma have failed to identify portal or systemic endotoxemia or elevation of tumor necrosis factor. Selective antibiotic decontamination of the gut may reduce systemic endotoxemia but has no significant effect on survival. Patients undergoing cardiopulmonary bypass demonstrate gut mucosal ischemia, hyperpermeability, and systemic endotoxemia, but these appear to be independent phenomena. Therefore it is possible that translocation could be a normal process in health or an epiphenomenon in critical illness.

Linking gut mucosal ischemia to multiple organ failure

Interpreting measures of splanchnic blood flow

Several techniques have been developed to measure the adequacy of splanchnic and gut mucosal blood flow. Gastrointestinal tonometry measures intralumenal $P\text{CO}_2$ from which intracellular pH (pH_i) can be derived. Laser Doppler probes and Clark electrodes have been used to measure flow and $P\text{O}_2$ directly. In human and animal studies, gut intramucosal acidosis occurs both in low-flow non-septic states and sepsis, and this is associated with increased gut permeability and a poor clinical outcome. However, in animal models of sepsis a low pH_i occurs despite normal or increased mucosal blood flow and oxygenation. Therefore the low pH_i of sepsis may represent failure of cellular oxygen utilization rather than inadequate oxygen supply, again explaining why maximizing systemic oxygen delivery has not improved outcome from septic shock. The fact that a low pH_i is associated with an increase in mortality in both septic and non-septic patients does not imply a common pathophysiology. In the early phase of critical illness a low pH_i is caused by an inadequate splanchnic oxygen supply, but the associated increase in mortality cannot be attributed to translocation. An alternative explanation is that the ischemic gut releases or activates some other mediator of remote organ injury. One possible mediator is the activated neutrophil.

Leukocyte activation in the splanchnic bed

In laboratory models, ischemia–reperfusion injury results in activation of membrane-bound phospholipase A_2. This enzyme catalyzes the production of the proinflammatory eicosanoids and platelet activating factor from membrane phospholipid. Both these substances can activate neutrophils and enhance their adhesion to endothelium. Platelet activating factor is also a potent mediator of vascular permeability. The gut mucosa contains substantial numbers of neutrophils, has a large surface area for the supply of membrane phospholipids, and has a particularly high concentration of xanthine oxidase which catalyzes the production of reactive oxygen species. This makes the gut a potent source of activated neutrophils. Gut ischemia followed by reperfusion increases vascular (and lung) permeability and causes pulmonary sequestration of neutrophils, and these effects can be abolished by inhibitors of phospholipase A_2 and neutrophil adhesion, and by leukocyte depletion (Biffl and Moore 1996). This process can occur in reverse; acid instillation into the lung produces gut mucosal leak, which can also be minimized by inhibitors of neutrophil adhesion to endothelium.

Multiple physiological insults

Another important finding is that combined or sequential physiological insults produce more injury than a single insult. In animal models, gut ischemia–reperfusion causes pulmonary leak and neutrophil sequestration but little morbidity or mortality; independently, the injection of a low dose of endotoxin causes neutrophil sequestration but no leak. However, the same dose of endotoxin given after reperfusion results in acute lung injury and a high mortality. This type of pattern is much more typical of the clinical situation, where relatively minor disturbances of physiology, many of them iatrogenic, may occur repeatedly following a major precipitant such as trauma, surgery, or cardiopulmonary arrest.

Mechanisms of remote organ injury

Leukocyte-mediated injury

Primary injury of either the gut or the lungs results in a leukocyte-mediated increase in permeability in the other organ, as described above. Multiple or sequential insults translate this into clinically evident organ injury and increased mortality. Other organs will be affected by the same process, although evidently there are many factors which determine the precise pattern of organ involvement and the reasons for certain organs being spared or involved. Once activated neutrophils migrate from the vascular space, they are able to release their intracellular contents, which include hypochlorous acid and other reactive oxygen species, causing cellular injury. Lymphocytes may also be involved either in their role as memory cells, documenting and responding to foreign antigen, or in direct response to intracellular infection. Unlike neutrophils, lymphocytes recirculate via the lymphatics; they also secrete many cytokines which may enhance or inhibit the immune response.

Mechanisms for specific organ involvement

The concept of activated leukocytes adhering to endothelium and then migrating to release their contents in the tissues is evidently a useful model, but it is not the sole explanation of multiple organ failure. Platelets can contribute to cellular or organ injury by the formation of platelet plugs and distal ischemia, and also by promoting neutrophil activation and adhesion. Activated neutrophils become more rigid, and may also cause microvascular occlusion, shunting, and tissue ischemia independent of direct chemical injury. Activated neutrophils will adhere to endothelium, but may not migrate into the tissues unless the endothelium has been stimulated by the additional presence of cytokines such as interleukin 8 (IL-8). Many cytokines promote leukocyte adhesion, and the variability of clinical presentation of multiple organ failure may be a consequence of variable expression by different organs of these chemotactic cytokines, some of which may be produced by injured tissue which then becomes the target of secondary attack. For example, a patient with a prior respiratory infection may be more likely to develop acute respiratory failure following an emergency surgical procedure simply because of pulmonary cytokine signaling.

The flow characteristics in certain organs will also influence the presentation of multiple organ failure, with renovascular disease being an obvious example. Accumulation of toxins, drugs, or active drug metabolites may explain the predominance of certain organ failures such as encephalopathies or renal dysfunction.

Immunoincompetence

Multiple organ failure presents the paradox of widespread activation of the immune system combined with an increased susceptibility to infection which may itself contribute to continuing organ failures. The challenge is to develop therapies which enhance normal host defenses while limiting the self-destructive component. Glutamine is a semi-essential amino acid which is required for lymphocyte proliferation and cytokine expression, macrophage activity, gut mucosal function, muscle homeostasis, and renal function. Deficiency in body stores occurs rapidly in critical illness, and may be associated with increased infection rates; the same phenomenon has been reported following marathon running. Supplementation with intravenous glutamine may reduce mortality from critical illness. Arginine also has an important role in host defenses, and has been shown to reduce infection-related mortality in animal models. Impaired production of anti-inflammatory cytokines such as IL-10 may be as important as overproduction of proinflammatory cytokines such as IL-6.

Conclusions

Multiple organ failure is the consequence of many pathological processes, of which an acute reduction in oxygen supply is one of the most important. The splanchnic region is particularly susceptible to hypoxic injury, and the ischemic gut mucosa may act as a neutrophil activating site. This is followed by leukocyte adhesion to endothelium and migration into the tissues where cytotoxic chemicals are released, producing tissue injury. This process is aggravated by multiple and sequential physiological insults, some of which may not be identified using basic clinical monitoring. The factors which precipitate cell injury and organ failures may be amenable to relatively simple preventative interventions, but once damage has occurred the process becomes increasingly complex and self-sustaining. Natural defense systems may become pathologically hyperactive, or suppressed, at different stages in critical illness, making it difficult to target therapies directed at the immunoinflammatory casade with any degree of accuracy or safety. This may explain in part the failure of immunotherapy for sepsis.

References

Baue, A.E. (1975). Multiple, progressive, or sequential systems failure: a syndrome of the 1970s. *Archives of Surgery*, **110**, 779–81.

Biffl, W.L. and Moore, E.E. (1996). Splanchnic ischaemia/reperfusion and multiple organ failure. *British Journal of Anaesthesia*, **77**, 59–70.

Boyd, O., Grounds, R.M., and Bennett, E.D. (1993). A randomized clinical trial of the effect of deliberate perioperative increase of oxygen delivery on mortality in high-risk surgical patients. *Journal of the American Medical Association*, **270**, 2699–707.

Knaus, W.A., Draper, E.A., Wagner, D.P., and Zimmerman, J.E. (1985). Prognosis in acute organ-system failure. *Annals of Surgery*, **202**, 685–93.

Sauaia, A., *et al.* (1995). Epidemiology of trauma deaths: a reassessment. *Journal of Trauma*, **38**, 185–93.

13.8.2 Assessment

J.-L. Vincent

Key messages

- Organ failure is an ongoing process and assessment requires repeated evaluation over time.
- Many systems available to assess multiple organ failure aim at prediction of mortality. The description of organ failure and the ability to separate out individual organ failures from an overall organ failure score may be more relevant for patient management.
- Effective organ failure assessment systems need to be simple to apply and based on reliable, repeatable, and readily available parameters.

Introduction

The assessment of the patient with multiple organ failure requires the use of a system to score or stage the disease process. There is still no universally accepted method for judging the severity of organ dysfunction, and new systems are continually being developed and adapted for this purpose. Several elements must be considered in the development of a scoring system:

(1) the ability to assess the function of individual organs as well as making a more global patient assessment;

(2) the accessibility of the variables used in the score;

(3) the concept that organ dysfunction is a continuum rather than a single event;

(4) the time factor, since patients dying early may not have had time to develop multiple organ failure, and in those who do, the course over time is often variable, so that regular (daily) assessment is necessary.

The decision as to which organs should be included in any scale of organ dysfunction is largely dependent on the ability to measure organ function objectively. Organ failure can be assessed by the degree of abnormality of a single measure of function, a combination of variables defining a syndrome of abnormality, or the clinical intervention necessary to support the failing organ. The use of a single index offers simplicity but may lack specificity. A combination of variables increases specificity but may lose simplicity. Therapy-based variables are subject to different treatment protocols depending on the center.

Assessment of individual organ failure

The degree of failure of six main organs can be assessed as follows, although the gastrointestinal tract is an elusive organ to evaluate.

1. Respiratory function is best assessed by the evaluation of gas exchange, most commonly using the $Pa\text{O}_2/\text{Fi}\text{O}_2$ ratio. The need for mechanically assisted ventilation may be taken into account.

2. Renal function: urea (or blood urea nitrogen), and creatinine levels can adequately assess the degree of renal dysfunction. The creatinine clearance may be a more precise measure but is not determined daily in many intensive care units (ICUs). Oliguria is another variable that can be employed.

3. Hematological function: coagulation abnormalities are probably the most important indicator of hematological malfunction and are best measured by the platelet count, the prothrombin time, and the activated partial thromboplastin time. A low hematocrit or an abnormal white cell count may be included.

4. Liver function: the degree of hyperbilirubinemia is the most useful index. Levels of the liver enzymes serum glutamic oxaloacetic transaminase and serum glutamic pyruvic transaminase may give useful additional information. Some groups have suggested the use of prothrombin time, but this may reflect abnormalities of the coagulation system and not just liver dysfunction.

5. Cardiovascular function: no specific indicator is a reliable measure of cardiac function, but the presence of significant hypotension is useful. The degree of adrenergic support may be valuable, even though it is clearly a therapeutic measure.

6. Neurological function: the Glasgow Coma Scale is commonly used to assess neurological dysfunction. A major problem with the assessment of neurological function in the intensive care patient is the frequent use of sedatives and analgesics.

7. Gastrointestinal function: it is widely accepted that it would be useful to assess the degree of gastrointestinal dysfunction, particularly with the current interest in the role of the gut in the pathogenesis of multiple organ failure. However, accurate assessment of gastrointestinal function is still virtually impossible. One of the best indicators that could be used is tolerance to enteral feeding, but there is no single objective method of measuring this, and individual interpretation makes it impractical for use in a scoring system.

General severity scoring systems

The Acute Physiological, Age, and Chronic Health Evaluation (APACHE II) score was developed to predict the mortality of severely ill patients. The variables chosen for this assessment are the patient's age and the worst reading of 12 physiological variables in the first 24 h of stay in the ICU. The sum of the 12 individual variables is combined with points for age and chronic health to reach a total APACHE II score. This is then used in combination with a relative weight assigned to the major disease categories to obtain a final risk estimate. Although this score allows some prediction of the risk of mortality, it gives no information on the degree of individual organ failure. It also involves fairly complex calculations and is an initial measure, not generally repeated after the first 24 h of admission. The Simplified Acute Physiology Score (SAPS) similarly uses complex calculations and is again aimed at predicting mortality.

Organ dysfunction scoring systems

In an attempt to improve on currently available systems, a new scale has recently been developed by the European Society of Intensive Care Medicine. The Sepsis-related Organ Failure Assessment (SOFA) score (Table 1) was created to allow easy repeated assessment of the degree of individual organ dysfunction in all critically ill patients. It aims to assess morbidity, and therefore complements other scales which focus on mortality. The score is limited to the assessment of six organs, each being awarded a score between zero (normal) and 4, based on the worst values on each day. The variables necessary for each organ were chosen for their simplicity and are routinely measured in most ICUs. Therefore indices are available on a daily basis, allowing regular reassessment of patient condition. As the mortality is directly related to the degree of organ failure, it is also related to the SOFA score. In retrospective studies, the mortality has been shown to increase with increasing SOFA scores for each organ.

Several other scoring systems have been proposed recently. A major difference between them is in their definition of cardiovascular dysfunction. The Multiple Organ Dysfunction Score (MODS) uses a complex calculation of the pressure-adjusted heart rate averaged over a 24-h period, thus removing the simplicity of the system. The Brussels score uses hypotension and acidemia, but acidemia may be due to factors other than circulatory failure, including renal failure. The SOFA score uses the requirement for adrenergic support, which varies from one hospital to the next, but broad categories are used to avoid major conflicts.

The ability to quantify the severity of dysfunction of each individual organ enables therapy to be appropriately directed, and repeated scores allow the effects of treatment to be assessed

Table 1 The SOFA score

SOFA score	1	2	3	4
Respiration				
$Pa{O_2}/Fi{O_2}$ (mmHg) (kPa)	<400 (<53.3)	<300 (<40)	<200 (<26.7)	<100 (<13.3)
			← with respiratory support	→
Coagulation				
Platelets ($\times 10^9$/l)	<150	<100	<50	<20
Liver				
Bilirubin (mg/dl (μmol/l))	1.2–1.9 (20–32)	2.0–5.9 (33–101)	6.0–11.9 (102–204)	>12.0 (>204)
Cardiovascular				
Hypotension*	MAP <70 mmHg	Dopamine ⩽5 or dobutamine (any dose)	Dopamine >5 or epinephrine ⩽0.1 or norepinephrine ⩽0.1	Dopamine >15 or epinephrine >0.1 or norepinephrine >0.1
Central nervous system				
Glasgow Coma Scale	13–14	10–12	6–9	<6
Renal				
Creatinine (mg/dl (μmol/l)) or	1.2–1.9 (110–170)	2.0–3.4 (171–299)	3.5–4.9 (300–440)	>5.0 (>440)
Urine output (ml/day)			<500	<200

MAP, mean arterial pressure.
*Adrenergic agents administered for at least 1 h (doses are in μg/kg/min).

Conclusions

Any method of assessment must be able to adapt to improved understanding of the pathophysiology of multiple organ failure and progress in developing therapies. Assessment systems must be regularly re-evaluated and adjusted to include new information. Regular patient assessment is essential for monitoring both the effects of therapy and the morbidity and mortality. An effective method of assessment needs to be easy to use, reliable, and based on objective data.

Bibliography

Bernard, G.R., *et al.* (1995). Quantification of organ failure for clinical trials and clinical practice. *American Journal of Respiratory and Critical Care Medicine*, **151**, A323 (abstract).

Knaus, W.A., Draper, E.A., Wagner, D.P., and Zimmerman, J.E. (1985). APACHE II: a severity of disease classification system. *Critical Care Medicine*, **13**, 818–29.

Le Gall, J., Lemeshow, S., and Saulnier, F. (1993). A new simplified acute physiology score (SAPS II) based on a European/North American multicenter study. *Journal of the American Medical Association*, **270**, 2957–63.

Marshall, J.C., Cook, D.J., Christou, N.V., Bernard, G.R., Sprung, C.L., and Sibbald, W.J. (1995). Multiple organ dysfunction score: a reliable descriptor of a complex clinical outcome. *Critical Care Medicine*, **23**, 1638–52.

Vincent, J.-L. (1994). Sepsis and septic shock: update on definitions. In *Update in intensive care and emergency medicine* (ed. K. Eyrich, C. Sprung, and K. Reinhart), pp. 3–15. Springer-Verlag, Berlin.

Vincent, J.-L., *et al.* (1997). The SOFA (Sepsis-Related Organ Failure Assessment) score to describe organ dysfunction/failure. *Intensive Care Medicine*, **22**, 707–10.

13.8.3 Therapeutic strategy

J.-L. Vincent

Key messages

- Multiple organ failure remains a leading cause of mortality in intensive care units.
- Treatment of multiple organ failure is dependent on the removal of the cause and avoidance of tissue hypoxia.
- General therapies include the removal of the septic focus with antibiotics and surgery, and adequate resuscitation with fluids, blood transfusion, and vasoactive agents as necessary.
- Effective resuscitation is vital and can then be followed by organ-specific therapies. There is often no specific treatment for individual organ failure, and support of the organ is all that can be offered.

Introduction

Multiple organ failure remains a leading cause of mortality in intensive care units, with rates as high as 90 per cent, and therefore methods for preventing its development are crucial in the care of the critically ill patient. Major advances have been made in our understanding of the pathophysiology of multiple organ failure, leading to great hopes for new therapies. However, at present the best way of limiting the release of mediators is to identify and remove the cause (e.g. septic focus or devitalized tissue after trauma) and to avoid tissue hypoxia.

The importance of organ, particularly gut, hypoperfusion in the subsequent development of multiple organ failure makes optimal general resuscitation essential. Cornerstones of care include sufficient fluid replacement, adequate oxygenation with respiratory support where indicated, correction of anemia, use of vasopressors where necessary to restore perfusion pressure, and the maintenance of an adequate cardiac output. In addition to these general factors, support for individual failing organs may be required.

Multiple organ failure often starts with an episode of acute circulatory failure, not necessarily frank shock. In these conditions, the function of all organs is altered. Multiple organ failure sometimes appears to affect organs sequentially, starting with the lungs and followed by dysfunction of other organs. However, such a pattern may be related to our ability to diagnose organ dysfunction. For example, lung failure is identifiable early in its course (infiltrates on the chest radiograph and abnormal gas exchange) whereas liver dysfunction is usually recognized by an increase in bilirubin levels, which is a relatively late event. Gastrointestinal dysfunction is even more difficult to define. The likelihood of mortality is related to the number of affected organs and the degree of dysfunction, and can be assessed using various systems such as the Sepsis-related Organ Failure Assessment (SOFA) score.

General therapies

Removal of a septic focus

The persistence of a septic focus is a direct pathway to multiple organ failure, since the continuous release of proinflammatory mediators will alter organ function. Hence the source of infection must be sought actively, identified early, and treated promptly and aggressively. Broad-spectrum antibiotics, based on a knowledge of likely pathogens and local resistance patterns, should be administered until a specific organism is identified. Whenever possible, the antibiotic spectrum should then be narrowed. Any septic focus must be drained and possible sources (e.g. urinary or intravenous catheters) removed or changed.

Resuscitation

Adequate fluid administration is essential to ensure adequate perfusion and to prevent worsening of tissue ischemia. The maintenance of hypovolemia (particularly when associated with vasopressor therapy to maintain arterial pressure) may increase the likelihood of multiple organ failure.

Adequate oxygen delivery to the tissues by the use of oxygen, transfusion of red blood cells to maintain a hemoglobin greater than 10 g/dl, and administration of pharmacological agents to maintain adequate cardiac output is essential in the prevention and treatment of multiple organ failure. A low mixed venous oxygen saturation ($S\text{v}O_2$) should be avoided. Blood lactate levels may be a more valuable guide to the adequacy of oxygen delivery (DO_2). Gastric intramucosal pH values (pH_i) may also give useful additional information.

Nutritional support should be commenced as soon as the cardiovascular status has been stabilized. Whenever possible, the enteral route is preferred to maintain gut integrity and function. If full enteral nutrition is not tolerated, parenteral nutrition or a combination of both should be administered. Nutrients may not be well utilized in the presence of acute circulatory failure (shock), and enteral nutrition may even be harmful if the gut oxygen requirements exceed the oxygen availability.

Organ-directed therapy

After adequate resuscitation further treatment of established multiple organ failure aims at supporting individual organs. Sometimes supportive therapy of one organ may adversely affect the function of another. For example, the use of positive end-expiratory pressure (PEEP) to support lung function may alter hepatosplanchnic blood flow. Regular clinical evaluation, monitoring of hemodynamic and blood gas indices, and the use of blood lactate levels is essential to guide interventions in these patients.

Lung

Respiratory failure occurs when gas exchange is sufficiently impaired to cause hypoxemia. It can vary in severity from acute lung injury to the severe state of the acute respiratory distress syndrome and occurs following a direct or indirect lung insult. Acute lung injury can be defined as the presence of bilateral pulmonary infiltrates on chest radiographs, a pulmonary artery wedge pressure below 18 mmHg (where measured), and a PaO_2/FiO_2 ratio below 40 kPa (300 mmHg). Acute respiratory distress syndrome has similar criteria but with a PaO_2/FiO_2 ratio below 27 kPa (200 mmHg).

There is no established therapy for lung failure; one must aim at optimal support and prevention of complications. The main objectives are to maintain oxygen delivery to the tissues, to ensure carbon dioxide elimination, and to maintain gas exchange with minimal barotrauma. In practice, PaO_2 should be maintained above 8 kPa (60 mmHg) at all times. The early use of respiratory support may improve oxygen delivery, but neither mechanical ventilation nor PEEP prevents the development of respiratory failure. Acute respiratory distress syndrome (and its milder form acute lung injury) is characterized by non-hemodynamic pulmonary edema. In these conditions, the capillary permeability in the lungs is altered so that any increase in capillary pressures can worsen lung edema. Hence it is very important to avoid fluid overload. Since it is equally important to avoid long periods of hypovolemia, invasive hemodynamic monitoring is generally required to monitor the pulmonary artery pressures closely.

Oxygen treatment

The use of oxygen during resuscitation is essential to optimize oxygenation. A face mask or nasal cannulas may be sufficient initially. Although nasal cannulas interfere less with patient feeding and communication, face-mask oxygen is preferred to guarantee a defined FiO_2. Problems of oxygen toxicity can be avoided by using inspired oxygen concentrations of less than 50 per cent for long-term therapy, although higher oxygen concentrations are probably safe when used for less than 24 h.

If the patient's condition fails to improve or deteriorates despite oxygen administration, respiratory support with endotracheal intubation is needed. If this is required for longer than 2 or 3 weeks, tracheostomy may be necessary. Mechanical ventilation not only improves gas exchange, but also removes the work of breathing and thus the total body oxygen requirements. The increase in intrathoracic pressure can also improve left ventricular function by reducing left ventricular afterload. The increased intrathoracic pressure can impede venous return and cause a fall in cardiac output, particularly in hypovolemic patients.

PEEP should be used routinely. A PEEP level of 5 cmH_2O has virtually no risk of hemodynamic imbalance or barotrauma. A higher PEEP level should be used if adequate oxygenation is not achieved with conventional mechanical ventilation at safe inspired oxygen concentrations. PEEP levels in excess of 15 cmH_2O should be used only in selected cases, as they increase the risk of barotrauma. High PEEP levels may also result in a fall in cardiac output so that, although arterial oxygenation may improve, oxygen delivery may actually decrease. In patients with adequate breathing capabilities, the use of continuous positive airway pressure (CPAP) may be sufficient to guarantee adequate blood oxygenation. Placing the patient in the prone position can improve gas exchange in 50 per cent of patients with acute respiratory distress syndrome but poses logistic problems.

Aspiration of tracheal secretions and physiotherapy is essential to control mucus secretions and limit the risk of pulmonary infection. Early recognition and appropriate treatment of infection of pulmonary edema prevents worsening of the failure.

In acute respiratory distress syndrome, administration of nitric oxide (NO) by inhalation produces local vasodilation of well-ventilated units. This intervention diverts blood flow from unventilated units, thus reducing shunting of blood and improving arterial oxygenation. However, NO is a potentially toxic substance and must be administered with caution at the lowest possible effective dose.

Heart

Heart failure can be defined as the inability of the heart to supply sufficient oxygen to satisfy tissue oxygen demand. The primary aim of treatment is to improve cardiac output. This can be achieved by several methods and must be tailored to the individual patient.

In acute heart failure with pulmonary edema, a patient may have a decreased plasma volume and diuretics could worsen the failure and compromise coronary blood flow. A fluid challenge under guidance of cardiac filling pressures should be considered as a first-line intervention.

Improved contractility can be achieved by various agents, each of which have positive and negative aspects and must be selected according to the specific needs of the patient. Dobutamine is our inotropic agent of choice. If there is a low flow state but no cardiovascular collapse, the addition of phosphodiesterase inhibitors can be helpful to enhance the β-adrenergic effects. Digoxin has mild inotropic effects, and so its use should be reserved for rapid atrial fibrillation.

Vasodilators can effectively increase cardiac output by reducing left ventricular afterload. Intravenous agents with a short half-life, such as nitroglycerin or sodium nitroprusside, are preferred. Vasodilating therapy is prohibited in the presence of arterial hypotension. Diuretics are prescribed to reduce the sodium and water retention which aggravates the development of edema. However, diuretics should be used prudently to avoid an excessive reduction in ventricular preload.

Ventilatory support may be required in the management of severe cardiogenic pulmonary edema where hypoxemia is very severe. The increase in intrathoracic pressure can also improve ventricular function by reducing venous return. The use of CPAP by mask may be sufficient to achieve beneficial cardiorespiratory effects in patients with cardiogenic pulmonary edema.

Kidneys

Maintenance of adequate renal blood flow and avoidance of nephrotoxic agents will help to avoid renal failure. Accurate fluid balance charts, serum biochemistry, and urinalysis are important in the assessment of renal function. Low-dose dopamine is still widely used to enhance renal blood flow, but there is no evidence that dopamine has any specific effect on the renal circulation. Low doses of dobutamine may achieve the same effects by increasing global blood flow. Once renal failure is established, hemodialysis or hemofiltration techniques may be required to correct electrolyte disturbances, acidosis, and uremia.

A daily weight chart is valuable for assessing fluid balance in patients with acute renal failure although it can be imprecise. Hence invasive monitoring may be required to guide fluid administration by repeated measurements of pulmonary artery wedge pressures.

Serum potassium can rise rapidly and is cardiotoxic. Mild degrees of hyperkalemia can be treated with cation exchange resins which increase fecal potassium excretion; however, levels greater than 6.0 mmol/l are an indication for hemodialysis. The use of insulin/glucose or sodium bicarbonate is recommended before dialysis can be commenced. Calcium gluconate should be administered in the presence of life-threatening arrhythmias.

In hemodialysis and hemofiltration blood from the patient is pumped through a number of semipermeable membranes allowing contact with a dialysate solution. Diffusion allows removal of certain molecules from the blood and alteration of the biochemistry towards that of the dialysate. The frequency of dialysis is variable, but current thinking encourages the use of more shorter sessions at more frequent intervals. Since dialysis can cause hypotension in patients with already compromised tissue perfusion, continuous treatments, allowing slower less acute removal of uremic metabolites, are often preferred.

In peritoneal dialysis the peritoneal membrane is used as the semipermeable membrane, thus avoiding the need for extracorporeal circulation of blood. A major problem with this technique is the risk of infection and peritonitis, and so it is not generally used following intra-abdominal surgery.

Hemofiltration removes plasma water and substances dissolved in it by convective flow across a high-flux semipermeable membrane and replaces it with a solution of the desired biochemical composition. The potential advantage is that it is a more continuous technique, avoiding the acute hemodynamic alterations frequently associated with hemodialysis. However, it requires continuous anticoagulation and close monitoring of the fluid and electrolyte balance of the patient. Hemofiltration techniques may also enable removal of cytokines which are associated with the inflammatory response, but the amount of material removed is relatively small since many cytokines do not circulate. Removal may be more effective if high flows are used, but this technique removes many elements indiscriminately including those which may be beneficial to the patient such as hormones, drugs, etc. Continuous arteriovenous hemofiltration has the advantage of being simple, but has limited efficacy. Continuous venovenous hemofiltration is more efficacious, but requires the use of a pump. Both techniques can be combined with dialysis.

Coagulation

Maintenance of adequate perfusion will help to prevent coagulation abnormalities. Platelet transfusions may be required for thrombocytopenia with platelet counts below 20×10^9/l, or below 50×10^9/l in the presence of acute bleeding. Once disseminated intravascular coagulation has become established, with a prolonged prothrombin time or activated partial thromboplastin time, and if there is an imminent risk of bleeding, fresh frozen plasma should be administered to replace coagulation factors. Regularly repeated hemoglobin, platelet count, and clotting tests must be carried out to guide further treatment.

Brain

Many patients with multiple organ failure develop altered mental status ranging from disorientation and drowsiness to coma. Midazolam and propofol are useful agents if sedation is required, but they may contribute to the alterations in consciousness. Central nervous system investigations are generally negative in patients with unexplained alteration of mental status: the cerebrospinal fluid is unremarkable and the CT scan is normal. The EEG may show nonspecific diffuse slowing of electrical activity. There is no specific treatment for brain failure.

Gut

A nasogastric tube should be inserted to prevent gastric distension and decrease the risk of aspiration of gastric contents. Gastric residues should also be monitored regularly.

Stress ulceration occurs in critically ill patients but the incidence of associated gastrointestinal bleeding has decreased markedly, probably as a result of generally improved care. Maintaining adequate gastric blood flow can help to limit stress ulcer development. Gastric protective agents are routinely used to prevent stress ulceration, although there is no consensus on the best agent to use. By inhibiting gastric acidity, H_2 blockers may allow overgrowth of pathogenic bacteria from the gastrointestinal tract and thereby increase the risk of nosocomial pulmonary infection. In this respect, the use of barrier agents such as sucralfate may provide a better alternative, although these differences have been de-emphasized recently. The use of early enteral nutrition is important in maintaining gut function and limiting stress ulceration.

Liver

Acute liver failure can be assessed by the degree of alteration of liver biochemistry, particularly the bilirubin level. Liver dysfunction itself influences many other systems including the brain, with encephalopathy and cerebral edema, and the blood with coagulopathies. No specific treatment is available and therapy must be supportive.

Conclusion

For the purposes of treatment multiple organ failure can be viewed as the dysfunction of individual organs, but in reality it is a combination of the dysfunction of many organs with each failing organ having effects on the others. The most important factor in the treatment of multiple organ failure is to prevent its development.

Bibliography

Friedman, G., Berlot, G., Kahn, R.J., and Vincent, J.L. (1995). Combined measurements of blood lactate concentrations and gastric intramucosal pH in patients with severe sepsis. *Critical Care Medicine*, **23**(7), 1184–93.

Kirkpatrick, A.W., Meade, M.O., and Stewart, T.E. (1996). Lung protective ventilatory strategies in ARDS. In *Yearbook of intensive care and emergency medicine* (ed. J.L. Vincent), pp. 398–410. Springer-Verlag, Berlin.

Vincent, J. (1997). Prevention and/or therapy of MOF. *World Journal of Surgery*, **20**, 465–704.

Vincent, J.L. and Tielemans, C. (1995). Continuous hemofiltration in severe sepsis. Is it beneficial? *Journal of Critical Care*, **10**, 27–32.

Zapol, W.M. and Hurford, W.E. (1995). Inhaled nitric oxide: state-of-the art. In *Yearbook of intensive care and emergency medicine* (ed. J.L. Vincent), pp. 323–30. Springer-Verlag, Berlin.

13.9 Anaphylaxis

13.9.1 Pathophysiology

J. H. Levy and M. Fisher

Key messages

- Anaphylaxis is a complex clinical syndrome characterized by acute cardiopulmonary dysfunction.
- Multiple biological pathways can produce a similar clinical syndrome by mediator release, although these reactions are not immmunological in origin.
- There are few human data regarding the acute changes that occur during anaphylactic shock.
- The effects of mediators on the heart, the vascular endothelium and smooth muscle, and the lung produce a complex, enigmatic, and unpredictable adverse response.

Introduction

The word 'anaphylaxis' is derived from Greek and means 'against protection'. The term was first used by Portier and Richet to describe a phenomenon observed when Portuguese man of war venom induced an acute allergic state rather than immunization or prophylaxis in dogs. The first recorded case of anaphylaxis, death due to the sting of a wasp, is described in the tomb of King Menes in Egypt.

Anaphylaxis is a normal inflammatory response designed to render foreign substances inactive and to amplify the response to recruit other immunosurveillance cells. It becomes a life-threatening reaction when patients recognize an antigen as foreign and the host response is exaggerated to produce acute injury. The reason why patients become sensitized and why they respond in such a pathophysiological manner is still unknown.

The study of human anaphylaxis is exceedingly complex. The major reasons for this difficulty are that it occurs infrequently, prior sensitization to a specific antigen is not often recognized, the clinical manifestations may be produced by a number of both immunological and non-immunological mechanisms, and severe anaphylaxis has an explosive, unpredictable, and catastrophic nature with an associated rapid resolution that usually accompanies therapy.

In immunological terms 'anaphylaxis' signifies a Gell and Coombs type I hypersensitivity reaction that is mediated by IgE or IgG antibodies. In a clinical sense 'anaphylaxis' is used to describe reactions which are severe and resemble the classical features of severe type I hypersensitivity irrespective of mechanism. It is generally accepted that 'anaphylaxis' can be due to physical factors such as physical activity (i.e. exercise) that can cause mobilization of inflammatory cells (i.e. mast cells and basophils).

Synonyms for 'anaphylactic' in its immunological sense include 'reaginic', 'allergic', and 'hypersensitive', and non-immunological mechanisms producing the clinical features characteristically associated with anaphylaxis are called 'allergoid', 'anaphylactoid', 'direct histamine release', or 'pseudoallergic'. Here, the expression clinical anaphylaxis will be used to describe the clinical syndrome irrespective of mechanism. Anaphylaxis will be used to refer to type I hypersensitivity, and anaphylactoid will be used to refer to reactions in which there is no possibility of an immune basis.

The mechanism of anaphylaxis: antibodies

The agent triggering an anaphylactic reaction is known as the antigen. A molecule's ability to act as an antigen is called its immunogenicity. Antigens are classically large molecules which stimulate an immune response on their own, or smaller compounds (haptens) that bind to proteins prior to producing a reaction. Most polypeptides can act as complete antigens; however, compounds such as suxamethonium, which have two antigenic groups an appropriate distance apart for bridging cell-bound IgE molecules, may be able to act as antigens without first binding to protein. In some circumstances a metabolic breakdown product of the parent molecule may be the antigenic component.

Antibodies are formed when antigens stimulate B lymphocytes to form plasma cells which secrete protein macromolecules that have an ability to bind to the antigen that stimulated their production, usually rendering the antigen harmless. Antigen–antibody complexes form rapidly, for example after food ingestion, and are rapidly cleared from plasma. If they persist and bridge cell-bound antibodies, or the complex binds to cell walls, allergic symptoms occur, leading to a series of syndromes ranging from minor to life threatening. The antibodies formed by plasma cells are measured as immunoglobulins. The immunoglobulins may be bound to cells, exist free in blood, or be released into secretory systems. They are Y-shaped molecules with Fab (antibody binding) and Fc (cell binding) segments. Five classes are recognized: IgA, IgD, IgE, IgG, and IgM.

IgG has a molecular weight of 150 kDa. When the Fab site combines with the antigen, the Fc region becomes complement activating. IgG4 does not bind complement but binds to mast cells, while IgG and IgG3 readily activate complement. IgG2 is a less active activator of complement. IgG diffuses more readily than other antibodies; it binds to bacteria and adheres to phagocytic cells. IgG4 probably also acts as a reaginic antibody. IgG complement-dependent antibodies

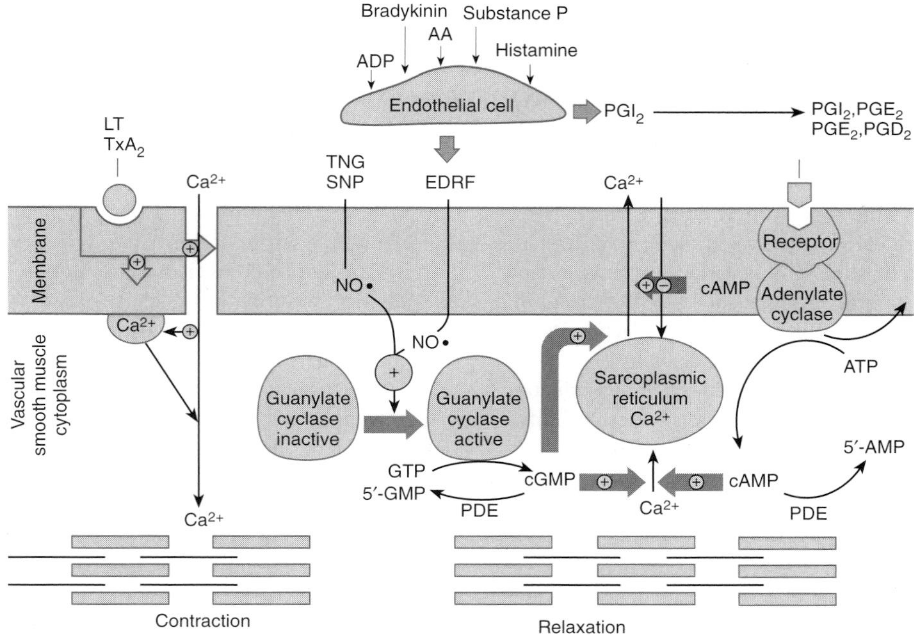

Fig. 1 The action of any mediator on vascular responses is the result of effects on vascular endothelial cells and direct responses on vascular smooth muscle. Leukotrienes (LT), thromboxane A_2 (TxA_2), and histamine (H_1 receptor effects) produce vascular contraction by increasing calcium release or entry in vascular smooth muscle. Bradykinin, substance P, histamine (H_1 receptor effects), and other mediators stimulate endothelial cells to release both endothelium-derived relaxing factor (EDRF) and prostacyclin (PGI_2), which produce vasodilation by vascular relaxation. EDRF, like nitroglycerin (TNG) or sodium nitroprusside (SNP), generates nitric oxide (NO), which activates guanylate cyclase. Guanylate cyclase generates cGMP, which causes calcium uptake into the sarcoplasmic reticulum and extrusion from the cell to produce vascular relaxation and vasodilation. PGI_2, PGE_1, and other prostaglandins directly stimulate receptors on vascular smooth muscle that are coupled to adenylate cyclase. Stimulation of these prostaglandin receptors increases cAMP, which also increases calcium uptake into the sarcoplasmic reticulum and extrusion from the cell to produce vascular relaxation and vasodilation. The cyclic nucleotides in vascular smooth muscle cytoplasm are broken by various phosphodiesterase enzymes (PDE) as shown.

against protamine and cremaphor have been found in the serum of reactors. It is possible that clinical anaphylaxis to cremaphor-based drugs, which usually shows evidence of classical pathway activation, is mediated by IgG antibodies. IgG4 antibodies can be detected by skin tests.

IgE is responsible for the majority of true anaphylaxis. Very small concentrations exist in serum, and the majority of the immunoglobulin is bound to mast cells. The physiological role of IgE is not known. Bridging of IgE molecules by antigen leads to anaphylactic mediator release. Elevated levels of IgE are usually found in patients with a history of allergy, atopy, or asthma.

The mechanism of anaphylaxis: mediator release

A detailed discussion of the mechanism of mediator release is given by Levy (1992).

The mast cells and circulating basophils are the key cells in the allergic response. Mast cells are tissue fixed, residing in the perivascular spaces of the lung, skin, and intestine, i.e. organ systems that function as part of the interaction with our internal and external environments. Basophils comprise approximately 1 per cent of the circulating leukocyte pool. Bridging of cell-bound IgE molecules, anaphylatoxins, and non-specific stimuli all precipitate the release of mediators from some populations of these cells. In the case of IgE bridging, phospholipase C and protein kinase mechanisms are activated in the cell membrane. Membrane phospholipids are further converted to arachidonic acid which is converted to prostaglandins by cyclo-oxygenases and leukotrienes. Mast cell granules which contain preformed histamine, eosinophilic chemotactic factors, neutrophil chemotactic factor, bradykinin, and other mediators move to the periphery of the cell, and the granule and cell membranes fuse causing a pore to form through which the granule contents are extruded. Because the mast cells reside in the perivascular spaces, the mediators rapidly gain access to the central circulation.

Mediators of anaphylaxis

Many other inflammatory mediators of anaphylaxis are also involved, including cytokines, basophil kallikrein, prekallikrein activator, and Hageman factor activator. The exact role of the kinin system is unclear; however, bradykinin is a potent activator of vascular endothelium, releasing both endothelium-derived relaxing factor (nitric oxide) and prostacyclin (Fig. 1).

Multiple mediators produce shock during anaphylaxis via the direct and indirect effects that they have on vascular endothelium, smooth muscle, and the cardiopulmonary system (Fig. 1). However, most of the symptoms and signs of anaphylaxis can be produced by histamine alone. Histamine may be released by a spectrum of non-specific stimuli including drugs, neuropeptides, anaphylatoxins, and antibody-induced release. Allergic and non-allergic histamine release are initiated by different mechanisms.

Histamine exhibits diverse endorgan effects including increased capillary permeability, increased exocrine gland secretion, smooth muscle contraction, increased prostaglandin synthesis, modulation of eosinophil migration, bradycardia, the release of endogenous catecholamines, and the suppression of T-cell function. Capillary permeability is a complex pathophysiological event produced in part by contracting cells on postcapillary venules and the stimulation of terminal arteriolar dilatation which opens intercellular gaps.

The precise role of individual mediators is difficult to assess, as they are usually only able to be studied in isolation. Infusion of any mediator into animal or man will produce some of the effects of clinical anaphylaxis. Possible mediators in anaphylactic shock include platelet activating factor, bradykinin, and cytokines.

The endpoints of mediator release in anaphylaxis are vasodilatation, smooth muscle contraction, increased capillary permeability, and increased exocrine secretion.

The complement system

The complement system is a series of distinct proteins that can be activated through a classic or alternative pathway. The common final pathway leads to generation of the anaphylatoxins C3a and C5a which have direct effects on target organs and cause mediator release. IgG, IgM, and probably IgA antibodies activate complement, producing clinical anaphylaxis. The alternative pathway can be activated by contrast media, drugs, and endotoxin.

Mechanism of clinical anaphylaxis

A detailed discussion of the mechanism of clinical anaphylaxis is given by Fisher (1987). The classic description of types of immune response was given by Coombs and Gell.

Type I sensitivity is anaphylactic hypersensitivity. It is a result of mediator release from mast cells and basophils. The reactions are usually immediate, do not lead to cell lysis, and do not require complement activation. IgE and probably IgG antibodies mediate the reaction.

Type II sensitivity reactions are mediated by IgG or IgM antibodies directed against antigens on the surface of 'foreign cells'. The antigens may be part of the cell wall or haptens absorbed on the cell surface. The reactions are due to complement activation or phagocytosis by killer cells. The most common form are transfusion reactions, hemolytic anemias, idiopathic thrombocytopenic purpura, and Goodpasture's syndrome.

Type III reactions result from soluble antigen–antibody complexes or immune complexes. Complement is activated, and IgG and IgM antibodies participate. Examples include poststreptococcal glomerulonephritis and systemic lupus erythematosus.

Type IV reactions or delayed hypersensitivity are predominantly mononuclear and occur without the involvement of complement or antibodies. They result from the interaction between T lymphocytes and specific antigens. They are characteristically delayed by 24 to 48 h.

The Coombs and Gell classification is an oversimplification. For example, both type I and type IV reactions to the same drug have occurred in the same patient, IgE appears to be involved in some type III reactions, and complement activation has frequently been found in IgE-mediated reactions.

Late-phase reactions involving IgE can occur. The symptoms are not apparent until 2 to 4 h after exposure to antigen. Some reactions have a biphasic pattern characterized by initial resolution of manifestations with a subsequent reappearance. Aggregate anaphylaxis has been primarily described with polypeptides such as gammaglobulin; it has no latent period of sensitization and can be produced experimentally by injection of gammaglobulin from one species into another. Aggregated human gammaglobulin also produces clinical anaphylaxis probably involving complement.

Aspirin and other non-steroidal anti-inflammatory drugs produce clinical anaphylaxis, particularly in patients with respiratory allergy (asthma). The probable mechanism is inhibition of the cyclo-oxygenase pathway shunting arachadonic acid through lipo-oxgenase pathways, thus generating leukotrienes and platelet activating factor.

Kallikrein activation, producing kinins, has been implicated as a cause of reaction to plasma products. The role of the bradykinin system is important; patients using angiotensin-converting enzyme inhibitors for the first time can develop angioedema. Angiotensin-converting enzyme inhibitors block kininase, which promotes the conversion of angiotensin I to angiotensin II but is also responsible for the breakdown of bradykinin. Interactions between angiotensin-converting enzyme inhibitors and plasma products have been described.

Physical factors, exercise, osmolality, and temperature change may all produce clinical anaphylaxis. The mechanisms of exercise-induced anaphylaxis are complex and may represent prior antigen activation followed by mechanical injury to mast cells. Radiocontrast media, glucose solutions, and mannitol are hyperosmotic and can have direct effects on mast cells, producing clinical anaphylaxis.

Pathophysiology of anaphylactic shock

The effects of histamine are shown in Table 1. The presence of bronchospasm does not appear to be related to blood levels.

Equivalent histamine plasma levels have been noted without cardiovascular effects. The state of and quantity of receptors may be important in these responses, and recent work shows that subpopulations of mast cells react differently to different stimuli in terms of histamine release, suggesting that significant reactions may occur without elevating plasma histamine.

Table 1 Plasma histamine levels and symptoms in humans

Situation	Level
Normal	0.8
Tachycardia	<3
Increased gastric secretion	<3
Flush	<3
Headache	<3
Hypotension	5
Shock	>10
Profound shock	>30

Data from Vigorito *et al.* (1983) and Levy *et al.* (1989).

There appears to be no correlation between a drug's ability to release histamine and its ability to cause severe reactions. For example, morphine is a potent histamine releaser but a rare cause of anaphylaxis, whereas penicillin does not release histamine yet causes anaphylaxis.

The role of histamine in the cardiovascular system is complex. Histamine receptors include H_1, H_2, and H_3 receptors. Owing to the large mast cell population in the perivascular spaces and around the heart, histamine may have an unknown physiological role as part of host defense mechanisms. Histamine can also produce either vasoconstriction or vasodilation depending on the state of the endothelium. IgE-mediated hypersensitivity in human cardiac muscle *in vitro* produces increases in inotropic and chronotropic effects. The difficulty with the extrapolation of *in vitro* data to the human heart is the complex interaction of the response to anaphylactic mediators, the sympathetic adrenal response, and the secondary effects of hypoxia and acidosis.

The majority of observations in humans with normal hearts undergoing anaphylactic reactions show that hypotension and reduced cardiac output are primarily related to plasma volume loss and vasodilation, although persistent ventricular dysfunction can occur particularly with pre-existing coronary artery disease.

Observations in non-β-blocked humans with anaphylaxis show reduced left venticular filling pressure and decreased preload and afterload, with no evidence of cardiac dysfunction.

Levy *et al.* (1989) reported anaphylaxis to protamine as part of two studies evaluating the incidence of reactions during cardiac surgery. In an IgE-mediated reaction, profound vasodilation and decreased systemic vascular resistance were found without increases in pulmonary artery pressure. Another reaction was characterized by acute pulmonary vasoconstriction, right ventricular dysfunction, and systemic hypotension.

Etiology

Anaphylaxis in hospital is most commonly caused by drugs. Outside hospital, environmental antigens such as food and venoms are the most common causes.

Anaphylaxis is most commonly related to drug therapy. The Boston Collaborative Drug Surveillance Program (1973) reported a hospital incidence of anaphylaxis of three in 10 000 patients. Incidences of anaphylaxis mediated by various substances are as follows: penicillin, 15 to 40 in 100 000; contrast media, one in 1000; blood volume replacement solutions, one in 400; anaesthetic drugs, one in 20 000. Up to 1 per cent of the population have antibodies to hymenoptera stings, with 60 to 80 deaths occurring annually in the USA.

Protein-based substances, including blood products, vaccines, antivenoms, gammaglobulins, protamine, and desensitizing solutions, cause the majority of reactions. These substances may produce reactions due to specific antibodies or form aggregates which activate complement. Multiple drugs are implicated in anaphylaxis, with antibiotics representing some of the most common parenterally administered agents. Foods (particularly peanuts) and environmental allergens, such as insect stings, are also common causes.

References

Boston Collaborative Drug Surveillance Program (1973). Drug induced anaphylaxis. *Journal of the American Medical Association*, **224**, 613–15.

Fisher, M. (1987). Anaphylaxis. *Disease-a-Month*, **33**.

Levy, J.H. (1992). *Anaphylactic reactions in anaesthesia and intensive care* (2nd edn). Butterworth, Stoneham, MA.

Levy, J.H., Schwieger, I.M., Zaidan, J.R., Faraj, B.A., and Weintraub, W.S. (1989). Evaluation of patients at risk for protamine reactions. *Journal of Thoracic and Cardiovascular Surgery*, **98**, 200–4.

Vigorito, C., Russo, P., Picotti, G.B., Chiariello, M., Poto, S., and Marone, G. (1983). Cardiovascular effects of histamine infusion in man. *Journal of Cardiovascular Pharmacology*, **5**, 531–7.

13.9.2 Assessment and therapeutic strategy

M. Fisher

Key messages

- Factors influencing treatment include the location of the reaction, the rate of progression, the underlying state of the patient, the clinical manifestations, and the mechanism of the reaction.
- Multisystem anaphylaxis with urticaria, asthma, edema, and cardiovascular collapse leads to an obvious diagnosis. However, if symptoms are restricted to a single system such as angio-edema or cardiovascular collapse, the diagnosis may be difficult or not considered.
- The cardinal signs that anaphylaxis is likely to be severe are rapid onset, dyspnea, stridor, facial swelling, and/or hypotension.
- Sympathomimetic drugs are the primary pharmacological agents used in anaphylactic shock. All vasoconstrictor sympathomimetics produce some improvement. The agent of choice is epinephrine (adrenaline).
- In severe anaphylaxis with cardiovascular collapse or pulmonary edema there is a significant volume deficit. Colloid is more efficient in severe cases, since four to eight times the volume of crystalloid is necessary.

Introduction

There have been no randomized controlled trials of the treatment of human anaphylaxis. The unpredictable onset, variable severity, and priority and effectiveness of treatment make such studies impossible. Therefore the treatment of anaphylaxis is based on cellular mechanisms, animal models, and clinical observation. The understanding of the use of epinephrine (adrenaline), the role of steroids, and the appreciation of the need for volume replacement were poor in a review of 105 anaphylactic reactions to bee stings.

Terminology

'Anaphylaxis' is an immunological description of a type I hypersensitivity reaction which is mediated by IgE or IgG. The term is also used clinically to describe a variable group of symptoms produced by a number of mechanisms. Anaphylaxis is an example of a host defense mechanism becoming hostile, similar to septic shock. Many of the same mediators are involved, but their release is more rapid and less well sustained. Histamine is the early major mediator, but plasma levels may normalize within a short period of time, with other mediators being responsible for the maintenance of the clinical state. The term 'clinical anaphylaxis' is used to describe the clinical state irrespective of the mechanism, and in this context the term implies severity.

'Anaphylactoid' is used to describe anaphylaxis-like reactions not involving IgE antibodies. Frequently, the mechanism of reaction eludes detection with current techniques. It is best to assume that all such reactions are immune mediated, and subsequent exposure to the precipitating agent should be avoided.

Factors influencing treatment

The appropriate treatment of anaphylaxis depends on a number of important factors.

The location of the reaction

Although the majority of reactions occur outside the hospital, many progress rapidly. The optimum immediate treatment may vary, particularly if oxygen and intravenous fluid are not available. Epinephrine by intramuscular injection rather than intravenously is effective if given early and is safer than intravenous injection in a patient whose electrocardiograph and blood pressure are not monitored (Fisher 1995).

The rate of progression

Anaphylaxis may progress rapidly or slowly, and may recur after apparently effective treatment. When progress is slow and the early symptoms are minor, antihistamines may be all that is necessary to stop the attack. Rapid progression dictates early and aggressive therapy.

The underlying state of the patient

The patient's health and physical characteristics may both influence the severity and treatment. In particular, patients with asthma are likely to have severe bronchospasm, patients with cardiac disease are more likely to have arrhythmias and heart failure, and patients on β-blockers or who have epidural anesthetics in place are more likely to have severe reactions which are refractory to treatment.

The obese patient may have limited venous access, and intramuscular epinephrine may be preferable to a delay while venous (or intraosseus) access is achieved.

The clinical manifestations

The clinical manifestations of anaphylaxis are variable. Table 1 shows the incidence of clinical features in a series of reactions to intravenous drugs. The nature of the symptoms will have a major impact on the necessary treatment.

Table 1 Clinical features of anaphylaxis in 525 patients who reacted to intravenous drugs perioperatively

	No. of cases	Sole feature	Worst feature
Cardiovascular collapse	463	56	410
Bronchospasm	195	20	94
Cutaneous rash	66		
Erythema	236		
Urticaria	43		
Angio-edema	127	6	18
Generalized edema	37		
Pulmonary edema	13	2	3
Gastrointestinal	36		

The mechanism of the reaction

In its clinical sense 'anaphylaxis' describes a series of symptom complexes which have different etiologies and pathophysiology. Reactions to protamine, for example, may present as classical anaphylactic shock or as pulmonary hypertension and right ventricular failure.

Assessment

Differential diagnosis

Multisystem anaphylaxis with urticaria, asthma, edema, and cardiovascular collapse leads to an obvious diagnosis. However, if symptoms are restricted to a single system such as angio-edema or cardiovascular collapse, the diagnosis may be difficult or not considered. Cases have been described where anaphylaxis was suspected only after the fifth episode of cardiovascular collapse. Anaphylaxis may only be considered as a cause of cardiovascular collapse when removal of drapes or clothing has revealed erythema or urticaria.

Many diseases simulate an allergic response and may mimic anaphylaxis. Distinction from 'cardiac' causes of cardiovascular collapse may be assisted by a lowered venous pressure, vasodilatation, and supraventricular tachycardia as the predominant rhythm in anaphylactic shock. While a volume deficit may occur in hypotension due to cardiac causes, the volume deficit in anaphylaxis is usually greater. Vasovagal reactions after parenteral drug administration, particularly dental block with local anesthesia, are commonly misdiagnosed as anaphylaxis. Initial bradycardia, the rapid return of a strong pulse without treatment, and hyperventilation are important in the diagnosis.

The distinction between anaphylactic bronchospasm and allergic asthma is semantic only. The presence of airway obstruction and tension pneumothorax must be rapidly excluded in the patient who is difficult to inflate after respiratory arrest. 'Factitious' allergic emergencies have been described. Angioneurotic edema may be mimicked by venous or lymphatic obstruction, particularly in head and neck surgery.

The cardinal signs that anaphylaxis is likely to be severe are rapid onset, dyspnea, stridor, facial swelling, and/or hypotension.

Table 2 Clinical features of anaphylaxis

Minor features
Aura
Rhinitis
Cough
Pruritus
Urticaria
Rash
Conjunctivitis
Lacrimation
Flush
Major features
Common
Prolonged unconsciousness
Angioneurotic edema
Laryngeal edema
Generalized edema
Cardiovascular collapse
Bronchospasm
Gastrointestinal cramps
Diarrhea
Vomiting
Uncommon
Hemoptysis
Melena
Pulmonary edema
Cardiac failure
Disseminated intravascular coagulation
Late features
Headaches
Vaginal discharge
Edema

Clinical features

The clinical effects of mediator release have variable clinical expression relating to the mediator released, the mechanisms, the quantity, the timing, and the reactivity, number, and state of the receptors. In general, smooth muscle cells are affected, producing bronchoconstriction in the airway, vasodilatation in peripheral blood vessels, increased capillary permeability, and increased secretion of exocrine glands. This leads to the clinical manifestations shown in Table 2.

Minor features

Aura

A recognizable aura is experienced by some patients who have previously had anaphylactic reactions. This is important in enabling self-treatment or early treatment with epinephrine or antihistamines depending on the usual rate of progression of the disorder.

Skin changes

Three characteristic changes appear, with generalized erythema being most common. Diffuse rashes, often on the legs and lower trunk, and urticarial wheals appear less frequently. In heavily tanned Caucasians the erythema may only be visible on non-tanned areas, and it is extremely difficult to detect in Polynesian, Melanesian, or black patients.

Other features

Rhinorrhoea, conjunctivitis, and swelling of the eyelids may occur.

Major features

Angio-edema

Angio-edema involves the head and upper airway and occurs in 22 per cent of patients. It may progress very slowly; indeed, it may be noticed by the recovering patient or relatives rather than by medical staff. Its slow progression predicates observation of the anaphylactic patient for at least for 12 h.

Bronchospasm

Two patterns of bronchospasm are observed in patients who have anaphylaxis in the operating room. The first is a transient bronchospasm which causes difficulty in ventilating the patient for a short period. The second type is severe protracted bronchospasm which is very difficult to reverse.

Hypotension and shock

Cardiovascular collapse is the most common life-threatening manifestation of anaphylaxis. It is associated with vasodilation and peripheral pooling of blood, reducing venous return and cardiac output. In addition, plasma loss occurs in many patients.

The precise role of cardiac depression in anaphylaxis is controversial, but there is no doubt that myocardial depression is an unusual primary feature.

The situation with respect to 'cardiac anaphylaxis' can be summarized as follows. The human heart possesses histamine receptors and is responsive to histamine, interleukins, and other mediators. Histamine infused into the heart produces an increase in rate and a reduction in filling pressures, systemic blood pressure, left ventricular end-diastolic pressure, pulmonary artery pressure, and systemic and pulmonary vascular resistance (Levi 1988) The human heart *in vitro* has reduced contractility when exposed to histamine. However, depression of cardiac function is rarely described, possibly because the myocardial function is supported by the endogenous catecholamine response. Drugs modifying cardiac function or the sympathoadrenal response make the reaction more severe. The integrity of the myocardium and the myocardial blood supply is a major factor in the response of the heart to both anaphylactic mediators and their effects; patients with cardiac disease are more likely to develop arrhythmias and cardiac failure than those with normal hearts. The cardiac filling pressures in humans rise after type I hypersensitivity, with severe pulmonary vasoconstriction and right ventricular failure occurring in aggregate anaphylaxis and some protamine reactions. The volume deficit in anaphylaxis may be severe; up to half the plasma volume may leak into the periphery within minutes (Fisher 1986).

Pulmonary edema

Pulmonary edema is rarely described in multisystem anaphylaxis, but is a common postmortem finding in death from anaphylactic shock. As a sole clinical feature it has been attributed to delayed reactions to protamine after cardiac bypass and reactions to plasma solutions, particularly in IgA-deficient patients. Acute pulmonary edema during anaphylaxis is usually non-cardiogenic, but heart failure may occur. It is important to remember that a volume deficit occurs in association with severe non-cardiogenic pulmonary edema.

Generalized edema

This is well documented but may be related to crystalloid therapy rather than the disease *per se*.

Gastrointestinal symptoms

The gastrointestinal symptoms of anaphylaxis are a cramping severe abdominal pain which may be associated with vomiting, hematemesis, diarrhea, and melena. They are self-limiting, lasting up to 6 h, and very resistant to treatment.

Other features

Prolonged coagulation indices and disseminated intravascular coagulation have been described. Prolonged unconsciousness has been described as a feature of anaphylaxis, but it is not certain whether this is related to cerebral hypoxemia or anaphylaxis *per se*. Women who have anaphylaxis may develop a profuse watery vaginal discharge, which may be bloodstained, about 3 days after the reaction. It is self-limiting and lasts about 48 h.

Management

The spectrum of severity of anaphylactoid and anaphylactic reactions is broad. Some patients respond to crystalloid infusion and corticosteroids, both of which are regarded as ineffective in severe cases. At the other end of the spectrum are patients who die despite early aggressive management.

Non-specific treatment

When anaphylaxis is suspected a high concentration of oxygen should be administered and intravenous access secured as rapidly as possible. Continuing exposure to the antigen should be stopped. Electrocardiography is the first monitoring priority as it increases the safety of administration of epinephrine. If the patient is unconscious the airway should be secured prior to the development of edema.

If the patient is pulseless external cardiac compression is instituted. Military antishock trousers are helpful in supporting blood pressure in some patients and have been helpful in patients not responding to epinephrine; however, they are not a substitute for fluid and vasopressors.

Pharmacological management

Sympathomimetic drugs, which oppose the effects of mediators on the end-organs and inhibit mediator release, are the primary pharmacological agents used in anaphylactic shock. All vasoconstrictor sympathomimetics produce some improvement (Fisher 1986).

The agent of choice is epinephrine, which has the advantage of being effective in bronchospasm, angio-edema, and hypotension (Fisher 1986; Soreide *et al.* 1988). In adults, the dosage is 0.5 ml of 1:1000 intramuscularly or 3 to 5 ml of 1:10 000 intramuscularly or slowly intravenously. In children, the dosage is 0.01 ml/kg of 1:1000 intramuscularly or 0.1 ml/kg of 1:10 000 intravenously.

There are a few important provisos to the use of epinephrine. When full doses ($>$ 0.3 mg) are given intravenously to patients with cardiac disease and normal blood pressures, arrhythmias and myocardial ischemia may occur. When a patient is monitored, small boluses (2–4 μg) or an infusion should be given. Subcutaneous or intramuscular epinephrine should be reserved for the non-monitored patient. The poorly perfused patient should always receive the drug intravenously.

In severe anaphylaxis with cardiovascular collapse or pulmonary edema there is a significant volume deficit. Although crystalloid solutions are effective in less severe cases, in severe cases colloid is more efficient since four to eight times the volume of crystalloid is necessary, leading to prolonged resuscitation.

In over 90 per cent of patients this treatment will produce a rapid response. Patients on β-blockers or who have epidurals in place may be refractory. Massive doses of epinephrine may be necessary in the β-blocked patient, or consideration should be given to drugs with α-adrenergic effects such as metaraminol or norepinephrine. Glucagon has also been used successfully. If there is no response after 2 liters of colloid in a patient on an epinephrine infusion, a norepinephrine infusion should be tried (Fisher 1986). Angiotensinamide has also been successful in a patient refractory to norepinephrine. If there is still failure to respond to the above therapy, the heart should be imaged. Rarely, anaphylaxis may occur in which the heart is the primary target organ and balloon counterpulsation is required.

Bronchospasm

Severe bronchospasm is the most difficult manifestation to treat. Epinephrine is the drug of first choice. In addition, continuous nebulization of a β_2 agonist such as salbutamol should be commenced and intravenous steroids (1 g methylprednisolone) given. Anesthetic vapors may be effective; isoflurane is the agent of choice when epinephrine is being administered. Ketamine may produce a dramatic response, particularly in children. The major hazard of bronchospasm is barotrauma. Initial ventilation should be at a slow rate (less than 6 breaths/min with the maximum possible expiratory time). Hand ventilation may be necessary if desaturation occurs with mechanical ventilation. There is no need to lower carbon dioxide if the pH is above 7.0 and oxygenation is adequate. Increasing minute volume should be delayed until improvement in airway resistance occurs. An extreme situation occurs when the airway resistance exceeds the elastic recoil of the lungs and chest wall, and expiration does not occur. In these circumstances manually assisting expiration by squeezing the lateral chest wall at the end of expiration may be lifesaving.

Any sudden deterioration should lead to auscultation and insertion of chest drains if pneumothorax is suspected. When all else fails, with unsatisfactory mechanical ventilation leading to barotrauma, and cardiac arrest is imminent, strong consideration should be given to cardiopulmonary bypass.

Angio-edema

Patients with angio-edema should be intubated, and given intramuscular or intravenous epinephrine and an H_1-blocker.

Pulmonary edema

Pulmonary edema is usually controlled by increasing positive end-expiratory pressure (PEEP) until airway 'flooding' is controlled and saturation becomes adequate. The pulmonary edema in anaphylaxis is a membrane edema associated with a volume deficit and thus diuretics are contraindicated. Rarely, at the end of cardiopulmonary bypass there may be severe pulmonary edema, in which losses of high-protein edema fluid (up to 35 liters) may occur over 24 h, and massive replacement and invasive monitoring are required. In these patients, increasing PEEP to stop flooding may lead to the exudation of edema fluid through the lung surface into the thoracic cavity and chest drains or thoracotomy may be necessary to prevent tamponade of the lungs and heart.

Other drugs

Antihistamines

Antihistamines are not drugs of first choice. However, anecdotal cases exist where patients who have been difficult to wean off inotrope infusions or difficult to resuscitate have become 'easy to manage' when treated with large doses of H_1-blockers or H_1- and H_2-blockers.

Steroids

There is no evidence that steroids are of therapeutic benefit in anaphylactic shock. Steroids may be a drug of second choice in anaphylactic bronchospasm as the disease may persist well past the 40 to 50 min required for steroids to be effective. Steroids are theoretically beneficial for stabilizing membranes, potentiating the effects of sympathomimetics, inhibiting calcium uptake, and improving microcirculatory perfusion.

Aminophylline

Aminophylline has been described as dangerous in anaphylaxis, although others have seen no adverse effects (Fisher 1995). In combination with epinephrine, it produces no greater incidence of ventricular fibrillation than either drug alone.

Experimental drugs

Calcium-channel blockers, ketanserin, and naloxone have theoretical possibilities or have been used in animals. Their usefulness remains to be evaluated.

Prostaglandin E has been used in refractory pulmonary hypertension and right ventricular failure after bypass. It has been used in combination with a systemic vasoconstrictor in the treatment of refractory pulmonary hypertension, including that associated with anaphylactic reactions to protamine.

Follow-up

The treatment of acute anaphylaxis does not end with the acute episode. The cause of the reaction should be determined when possible. History is of vital importance. Cutaneous testing and radio-immunoassays for specific IgE should be carried out. The findings should be clearly documented and explained to the patient, who should be instructed to carry a letter and/or a Med-Alert tag at all times. Patients at risk from secondary exposure, particularly when exposure is inadvertent, should be instructed in self-medication or parent medication with injectable or inhaled epinephrine. Patients with recurrent anaphylaxis where a cause cannot be found should be treated with steroids on alternate days. Desensitization should be considered where possible, and pretreatment with antihistamines, sympathomimetics, and steroids may be effective in preventing secondary reactions under some circumstances such as reactions to intravenous contrast media (Lieberman 1990).

References

Fisher, M.M. (1986). Clinical observations on the pathophysiology and treatment of anaphylactic cardiovascular collapse. *Anaesthesia and Intensive Care*, **14**, 17–21.

Fisher, M.M. (1995). Treating anaphylaxis with sympathomimetic drugs. *British Medical Journal*, **305**, 1107–8.

Levi, R. (1988). Cardiac anaphylaxis: models, mediators, mechanisms and clinical considerations. In *Human inflammatory disease* (ed. G. Marone, L.M. Lichtenstein, and M. Conderell), p. 93. Decker, Toronto.

Lieberman, P. (1990). The use of antihistamines in the prevention and treatment of anaphylaxis and anaphylactoid reactions. *Journal of Allergy and Clinical Immunology*, **86**, 684–6.

Soreide, E., Buxrud, T., and Harboe, S. (1988). Severe anaphylactic reactions outside hospital: etiology, symptoms and treatment. *Acta Anaesthesiologica Scandinavica*, **32**, 339–42.

13.10 Connective tissue disorders

13.10.1 Systemic lupus erythematosus

U. B. S. Prakash

- Systemic lupus erythematosus is an autoimmune disease that can affect many organs.
- The thoracic manifestations of systemic lupus erythematosus include pulmonary hemorrhage, pneumonitis, pulmonary thromboembolism, pulmonary hypertension, interstitial pneumonitis, pleurisy, pericarditis, and diaphragmatic dysfunction.

Introduction

Systemic lupus erythematosus (SLE) is a disorder of unknown etiology characterized by immunologically mediated inflammation affecting multiple organs and the presence of autoantibodies to components of the cell nucleus. The clinical criteria necessary to establish the diagnosis of SLE have been published by the American Rheumatologic Association. Almost any organ system can be affected by the disease. Intrathoracic involvement is among the more serious complications. Thoracic complications occur in more than 50 per cent of patients (Table 1). Morphological changes in pleuropulmonary tissues are non-specific. Many of the histopathological lesions are not caused by SLE itself but by secondary factors, such as congestive cardiac failure, infection, aspiration, and immunocompromised status. The thoracic complications with the potential to cause critical illness in patients with SLE include pulmonary alveolar hemorrhage, respiratory infections, pericardial effusion, lupus pneumonitis, pulmonary thromboembolism, pulmonary hypertension, and massive pleural effusion. Other complications, such as interstitial pneumonitis and fibrosis and diaphragmatic weakness, are seldom severe enough to produce respiratory failure. An important clinical 'pearl' is to recognize that almost all the serious thoracic complications occur in the acute (active) phase of SLE; indeed, a thoracic complication may be the presenting manifestation of the disease.

Table 1 Thoracic complications in SLE

Pleurisy with or without pleural effusion
Lupus pneumonitis
Shrinking lung syndrome (atelectasizing pneumonitis)
Interstitial pneumonitis and fibrosis
Lymphocytic interstitial pneumonitis
Pulmonary alveolar hemorrhage
Diaphragmatic dysfunction
Respiratory infections
Pulmonary edema
Obstructive lung disease
Bronchiolitis obliterans with organizing pneumonia (BOOP)
Pulmonary vasculitis including capillaritis
Pulmonary thromboembolism
Pulmonary hypertension
Upper airway disease (epiglottitis, subglottic stenosis, laryngeal edema, vocal cord paralysis, and cricoarytenoid arthritis)
Pericarditis and pericardial effusion
Myocarditis
Endocarditis (Libman–Sacks)

Pulmonary hemorrhage

Pulmonary hemorrhage, characterized by diffuse intra-alveolar bleeding, is an uncommon but life-threatening manifestation of SLE. It is likely that clinically insignificant or undetectable intra-alveolar hemorrhage occurs in many patients. One retrospective study of 76 autopsy cases of SLE collected over a period of 20 years observed that, even though pulmonary hemorrhage was clinically suspected or noted in less than 2 per cent of patients, it was the primary cause of death in 14 per cent. In another series of 57 patients pulmonary bleeding was the cause of death in 10.5 per cent. The mechanism of pulmonary hemorrhage is unclear, although pulmonary vasculitis and capillaritis are probably responsible. Other mechanisms include the activation of the complement system leading to deposition of immune-complex deposits in the capillary basement membrane of the lungs. The end result is a break in the integrity of the alveolar–capillary basement membranes. The presence of uremia, hemorrhagic diathesis, oxygen toxicity, and infection may increase the risk of pulmonary hemorrhage. Viral infection may lead to hemorrhagic pneumonia. Pathological analysis of the lungs in cases of intra-alveolar hemorrhage has shown abnormalities similar to those of lupus pneumonitis (see below) as well as distinctive vasculitis of small vessels.

Pulmonary hemorrhage is almost never encountered in patients whose primary disease is quiescent. As a rule, it is more likely in those with active and progressive disease. The presence of renal failure as a result of SLE increases the risk of pulmonary hemorrhage. A clinical caveat is that pulmonary hemorrhage can be the presenting manifestation of SLE. Pulmonary hemorrhage may vary from subclinical to massive in quantity. Clinically significant hemorrhage will present as cough with hemoptysis and progressive dyspnea. Significant hemoptysis is noted in about 10 per cent of patients with these symptoms. The onset of hemoptysis and progressive respiratory distress can be abrupt, and the clinical presentation frequently resembles that of acute lupus pneumonitis. Chest radiographs usually reveal bibasilar patchy alveolar infiltrates that are predominantly central (perihilar) in location. The degrees of hypoxia, respiratory distress, and respiratory failure are directly proportional to the extent of intra-alveolar bleeding. Bronchoscopy may reveal diffuse oozing of blood from many segmental bronchi, and the effluent from the bronchoalveolar lavage will be bloody throughout the procedure. Chronic subclinical pulmonary alveolar hemorrhage occurs less commonly than the acute variety and may occur intermittently. It may lead to a clinical and pathological picture of pulmonary hemosiderosis. A diagnostic bronchoalveolar lavage is likely to show a large number of hemosiderin-laden macrophages, although this finding alone is not diagnostic.

Pulmonary hemorrhage caused by SLE requires immediate therapy if the bleeding is significant enough to cause respiratory failure. The initial management of the respiratory disease may require supplemental oxygen therapy, endotracheal intubation and ventilation, and diagnostic bronchoscopy. The medical treatment consists of high-dose prednisone therapy (1.5–2.0 mg/kg/day) supplemented by a cytotoxic agent such as cyclophosphamide (2 mg/kg/day). Plasmapheresis has been used successfully to clear the circulating immune complexes. The mortality from SLE-induced pulmonary alveolar hemorrhage is in excess of 50 per cent, with many patients dying within a few days of the onset of hemoptysis.

Lupus pneumonitis

Lupus pneumonitis may present as either an acute or a chronic illness. Lupus pneumonitis is rarer than pulmonary hemorrhage, with an incidence ranging from 1.5 to 9 per cent. In several reported cases, lupus pneumonitis has been the first manifestation of the disease. The pathophysiology of both acute and chronic lupus pneumonitis involves deposition of immune complexes in the blood vessels and alveolar walls with or without associated vasculitis. Acute lupus pneumonitis is reported to occur more frequently during the immediate postpartum period in women who have SLE. It is usually accompanied by other manifestations of SLE. It mimics acute lung infection so that an infectious etiology for respiratory illness should be excluded. Acute lupus pneumonitis is characterized by the acute onset of dyspnea, high fever, and cough, with occasional hemoptysis. Physical findings are minimal unless hypoxia is severe enough to cause cyanosis. A slightly elevated leukocyte count, increased erythrocyte sedimentation rate, and significant hypoxemia may be present. The chest radiograph may exhibit unilateral or bilateral localized, diffuse, or patchy lung infiltrates, predominantly in the lower lung zones, with small pleural effusions. Histologically, acute lupus pneumonitis reveals non-specific changes that may include interstitial pneumonia, edema, and arteriolar thrombi. Vasculitis of major vessels is uncommon. Culture of lung tissue for pathogenic organisms is usually negative. A lung biopsy is rarely indicated and is not usually helpful. Since pulmonary infections and other non-specific pulmonary parenchymal abnormalities are more common in these patients, a diagnostic bronchoalveolar lavage should be performed to exclude an infectious process. The diagnosis of lupus pneumonitis is one of exclusion.

The chronic form of lupus pneumonitis is characterized by diffuse interstitial lung disease. Clinical findings include dyspnea on exertion, non-productive cough, and basal crackles.

Pulmonary thromboembolism

Thrombophlebitis involving the veins of the lower extremities occurs in 5 to 12 per cent of patients with SLE. Predisposing factors include chronic low-grade disseminated intravascular coagulation, angiitis of small vessels, prolonged bed rest, and increased thromboplastin generation. Among the etiologies for thromboembolic phenomenon in patients with SLE is the antiphospholipid antibody syndrome, which refers to a spectrum of autoantibodies, including lupus anticoagulant antibodies and anticardiolipin antibodies, that bind to negatively charged phospholipids. The antiphospholipid antibody syndrome plays a major role in causing the thromboembolic phenomenon in the extremities and the pulmonary vasculature. The syndrome is characterized by a prolonged activated partial thromboplastin time in association with otherwise normal clotting and platelet counts, the presence of anticardiolipin antibody, and a false-positive VDRL test. Recurrent thromboembolism and pulmonary emboli complicated by pulmonary hypertension have been described in patients with the lupus anticoagulant syndrome. Long-term anticoagulant therapy is required in these patients. The adult respiratory distress syndrome, caused by extensive small-vessel thrombosis of multiple organs including the lungs, has been described.

Pulmonary hypertension

Pulmonary hypertension has been described in up to 15 per cent of patients with SLE, although usually it is not clinically significant. The pathology of pulmonary hypertension is not well understood, since vasculitis of the pulmonary vasculature is rarely seen. The role of recurrent pulmonary thromboembolism and antiphospholipid antibody syndrome should be considered among the major etiologies of pulmonary hypertension. Other potential mechanisms responsible include a chronic hypoxic state due to interstitial fibrosis and the chronic alveolar hemorrhage syndrome. Clinically, pulmonary hypertension in SLE is analogous to primary (idiopathic) pulmonary hypertension. Raynaud's phenomenon is common. The secondary form of pulmonary hypertension caused by chronic and recurrent pulmonary emboli can be treated by pulmonary thromboembolectomy. Thus diagnostic tests including pulmonary angiography should be considered. The prognosis in those with persistent pulmonary hypertension is grave. Response to vasodilator therapy is unsatisfactory.

Interstitial pneumonitis

Extensive interstitial fibrosis, as observed in patients with rheumatoid arthritis and scleroderma, is seldom seen in patients with SLE. The prevalence of a diffuse interstitial process in SLE is about 3 per cent. Patchy and irregular areas of non-specific interstitial pneumonitis and fibrosis develop in 15 to 45 per cent. Immunofluorescent staining for IgG may show patchy and 'lumpy-bumpy' staining of the alveolar wall. Histology may include patchy interstitial fibrosis together with hyperplasia of bronchus-associated lymphoid tissue. The most common presentation is the insidious onset of a chronic, non-productive cough, dyspnea on exertion, and a history of recurrent pleuritic pain. Less commonly, a diffuse process may develop following a bout of acute lupus pneumonitis. Clinical manifestations of the diffuse interstitial process are similar to those of rheumatoid arthritis and progressive systemic sclerosis (scleroderma). In a series of 18 patients with SLE-induced diffuse lung disease, the mean age was 45.7 years and the mean duration of the disease was 10.3 years; pulmonary symptoms were present for a mean of 6 years. Pulmonary function tests in patients with diffuse lung disease demonstrate a restrictive pattern and diminished diffusing capacity for carbon monoxide. The clinical course is similar to that of diffuse lung disease associated with rheumatoid arthritis or scleroderma.

Plate-like or discoid atelectatic areas are more common and usually occur in the lower two-thirds of the lung fields. Infectious processes, particularly in patients who are on immunosuppressive therapy for SLE, are the most common cause of pulmonary parenchymal infiltrates. Mortality is more frequently due to sepsis and renal disease than to pulmonary complications.

Pleural involvement

Pleurisy is the most common and often the presenting manifestation of SLE, and is found in up to 83 per cent of patients at necropsy. Pleural effusions are seen more frequently in older patients and in drug-induced SLE. Non-specific pathological changes in the pleura include infiltration by lymphocytes, mononuclear cells, and plasma cells, together with various degrees of fibrosis. Painful pleurisy occurs in 41 to 51 per cent of patients, and pleuritic pain may be the first symptom in many. Young females who present with new-onset pleurisy or pleural effusion should be evaluated for SLE. An important differential diagnosis is pulmonary embolism, since SLE patients are at higher risk of thromboembolic phenomena. Pleuritic pain may be unilateral or bilateral and is usually located at the costophrenic margins. Pleural effusions are usually small to moderate, bilateral in nearly 50 per cent, occasionally associated with a small pericardial effusion, and frequently accompanied by dyspnea, cough, and fever. Hemorrhagic effusions have been infrequently reported. Pleural effusions may also occur as a complication of lupus-induced nephrotic syndrome or an infectious process. Pleural biopsy is rarely required as histological findings are non-specific.

Thoracentesis is usually unnecessary unless the cause of the pleural effusion is uncertain. The pleural fluid may be clear or serosanguinous; grossly bloody effusions are uncommon. The fluid is almost always an exudate with varying numbers of leukocytes, a normal (greater than 60 mg/dl) or high glucose, decreased levels of total hemolytic complement and of C3 and C4 components, and increased immune complexes. Although decreased, and occasionally undetectable, levels of complement components, observed in nearly 80 per cent of lupus effusions, have been used to corroborate the diagnosis of lupus pleuritis, they are not diagnostic because similar findings have been described in rheumatoid pleural effusion and other disorders. The presence of antinuclear antibody in the pleural fluid is non-diagnostic. Likewise, the nature of immune complexes in SLE-induced pleural effusion is unclear. Lupus erythematosus cells have been found in zero to over 85 per cent of lupus effusions. The presence of *in vivo* lupus erythematosus cells in pleural fluid is considered to be characteristic for SLE, and has not been described in other conditions except drug-induced SLE.

Cardiac manifestations

Pericarditis induced by SLE is common and is the most frequent cardiovascular complication. Clinical diagnosis is made in 20 to 30 per cent of patients, with pericardial involvement present in over 60 per cent patients at autopsy. Pericardial involvement may be asymptomatic or manifested by precordial chest pain and a pericardial friction rub. Hemodynamic alterations are uncommon. Pericardiocentesis may be required in the presence of imminent or documented tamponade, and the fluid exhibits characteristics similar to those of pleural fluid.

Other cardiac complications include atypical (non-bacterial) verrucous (Libman–Sacks) endocarditis, myocarditis, and coronary artery disease. The basic histological lesion is the diffuse microvasculitis that leads to the characteristic fibrinoid changes in the connective tissues of the cardiovascular system. Even though the prevalence of cardiac manifestations is reported to be 50 to 60 per cent, clinical manifestations occur only when there is severe involvement.

Diaphragmatic dysfunction

Bilateral elevated hemidiaphragms with diminished diaphragmatic function have been described. The clinical significance of this finding is unclear, although this phenomenon may be responsible for the 'unexplained dyspnea' in the absence of pulmonary parenchymal abnormalities in some patients. The term 'shrinking lung syndrome' has been used to described this complication. The mechanisms responsible for the diaphragmatic weakness are unknown, but may include recurrent episodes of diaphragmatic pleurisy, basal atelectasis, and steroid myopathy. Most patients with lung involvement from SLE will demonstrate a restrictive pattern on pulmonary function testing. Diaphragm dysfunction does not appear to respond to corticosteroid therapy.

Other complications

Obstruction to airflow is rare in SLE, although severe airway obstruction has been reported, possibly from bronchiolitis. Bronchiolitis obliterans with organizing pneumonitis (BOOP), bilateral hilar adenopathy, lymphocytic interstitial pneumonitis, and pulmonary amyloidosis are among the thoracic complications of SLE.

Vasculitis and capillaritis have been demonstrated; however, vasculitis is not characteristic of all SLE-induced pleuropulmonary processes. Nevertheless, vasculitic lesions of large vessels and capillaries of the lung have been observed in nearly 50 per cent of patients.

Upper airway disease in the form of epiglottitis, subglottic stenosis, laryngeal edema or ulceration, inflammatory mass lesions or nodules, vocal cord paralysis, and cricoarytenoid arthritis has been reported.

Acute reversible hypoxemia is a syndrome that occurs in a subpopulation of acutely ill patients with SLE. Aggregation of circulating leukocytes within the pulmonary vasculature as a result of the complement-mediated phenomenon is postulated to cause this syndrome. Corticosteroid therapy results in decreased levels of complement which in turn reverses the hypoxemia induced by leukocyte aggregation.

Pulmonary function testing in patients with SLE usually reveals a restrictive pattern with diminished lung volumes. The diffusing capacity of the lung for carbon monoxide (DLCO) is characteristically reduced and is a reliable method for monitoring patients with pulmonary parenchymal disease. Unless the diaphragmatic weakness is significant, maximal inspiratory and expiratory pressures may not indicate severe abnormalities.

Drug-induced SLE

Drug-induced SLE is a clinically important syndrome, and the number of drugs known to cause it continues to increase. Pleural and pulmonary involvement are common in drug-induced SLE. Pleural disease is more common with certain drugs such as procainamide and hydralazine. The pleural fluid in this group of patients shows biochemical characteristics similar to the classic form of SLE. Withdrawal of the offending chemical usually results in resolution of symptoms and signs within days or weeks.

Treatment and prognosis

Pleural effusion responds favorably to systemic corticosteroid therapy, with an occasional large effusion being refractory. Lupus pneumonitis and alveolar hemorrhage syndrome may respond to high-dose corticosteroid therapy. Cytotoxic agents have been used in patients refractory to corticosteroid therapy, but there is little evidence to suggest improved efficacy of the latter. Sepsis and progressive renal dysfunction are the most frequent causes of mortality in patients with SLE, but acute pulmonary alveolar hemorrhage is also important. Dramatic response can be expected in some patients with lupus pneumonitis following systemic corticosteroid therapy.

Bibliography

Boumpas, D.T., Austin, H.A., Fessler, B.J., Balow, J.E., Klippel, J.H., and Lockshin, M.D. (1995). Systemic lupus erythematosus: emerging concepts. Part 1: Renal, neuropsychiatric, cardiovascular, pulmonary, and hematologic disease. *Annals of Internal Medicine*, **122**, 940–50.

Lynch, J.P. and Hunninghake, G.W. (1992). Pulmonary complications of collagen vascular disease. *Annual Reviews of Medicine*, **43**, 17–35.

Martin, L., Edworthy, S.M., Ryan, J.P., and Fritzler, M.J. (1992). Upper airway disease in systemic lupus erythematosus: a report of 4 cases and a review of the literature. *Journal of Rheumatology*, **19**, 1186–90.

Orens, J.B., Martinez, F.J., and Lynch, J.P. (1994). Pleuropulmonary manifestations of systemic lupus erythematosus. *Rheumatic Disease Clinics of North America*, **20**, 159–93.

Wiedemann, H.P. and Matthay, R.A. (1992). Pulmonary manifestations of systemic lupus erythematosus. *Journal of Thoracic Imaging*, **7**, 1–18.

13.10.2 Vasculitis

U. B. S. Prakash

Key messages

- Wegener's granulomatosis is a necrotizing granulomatous vasculitis which can affect the lungs and kidneys. Treatment includes corticosteroids and cyclophosphamide.
- Takayasu's arteritis is a chronic inflammatory disease of the large arteries and is treated with corticosteroids.
- Behçet's disease is a chronic recurring vasculitis with aphthous stomatitis and life-threatening hemoptysis. Corticosteroid therapy alone is usually inadequate.
- Other vasculitis syndromes include allergic angiitis and granulomatosis, giant cell arteritis, Henoch–Schönlein purpura, hypocomplementemic urticarial vasculitis, polyarteritis nodosa, and microscopic polyangiitis.

Introduction

The vasculitis syndromes are a heterogenous group of disorders characterized by inflammation and necrosis of blood vessels with a wide variety of clinical manifestations. The pathological process may affect the arteries, veins, and capillaries of any organ. The majority of primary vasculitis syndromes have no known etiological factors. Secondary vasculitis can be seen in collagenoses, immunological disorders, and infectious diseases caused by *Aspergillus* and *Mucor* species, hepatitis B and C, and HIV.

There is no satisfactory clinical classification of vasculitis syndromes. However, each of the well-known vasculitis syndromes exhibits certain clinical features that aid in the diagnosis. Even though any vasculitis syndrome can potentially affect any organ system, respiratory involvement is more evident or predominant in certain types. In this chapter we focus on respiratory manifestations in primary vasculitis syndromes.

Wegener's granulomatosis

Wegener's granulomatosis is a systemic vasculitis characterized by necrotizing granulomatous vasculitis of the respiratory tract, kidneys, and other organs. Renal involvement with focal segmental glomerulonephritis is a hallmark of Wegener's granulomatosis. The term 'limited disease' has been used to describe Wegener's granulomatosis involving the respiratory system without glomerulonephritis. Morphological characteristics include necrotizing granulomas, fibrinoid necrosis, microabscesses, focal vasculitis, thrombosis, and fibrous obliteration of the vascular lumen.

Wegener's granulomatosis usually manifests around the age of 45 years; the male-to-female ratio is 2:1. Initially, symptoms are nonspecific and include fever, malaise, weight loss, arthralgias, and myalgias. The disease may manifest in skin (40–50 per cent), eyes (43 per cent), skeletal system (arthralgias in 58 per cent and arthritis in 28 per cent), and nervous system (25 per cent).

Antineutrophil cytoplasmic antibodies, particularly the granular type (c-ANCA), are present in more than 90 per cent of the patients and thus are useful in confirming the diagnosis; the specificity of the test is 98 per cent. Since the titers and sensitivity of the test correlate with disease activity, c-ANCA can be used to monitor this. Other abnormal laboratory findings in patients with Wegener's granulomatosis include mild to moderate normochromic normocytic anemia, leukocytosis, thrombocytosis, elevated erythrocyte sedimentation rate (often above 100 mm/h), positive rheumatoid factor, and elevation of immunoglobulins. When the kidneys are affected, urinalysis shows hematuria, proteinuria, and red cell casts in 80 per cent of patients. Pulmonary function tests, particularly flow–volume curves, are helpful in assessing major airway involvement. Pulmonary parenchymal disease is present in nearly all patients.

Clinical features include dyspnea, cough, and hemoptysis; the latter is present in 98 per cent of patients. Chest pain is observed in 5 to 55 per cent of patients. Clinically, respiratory manifestations can vary from subacute non-specific respiratory illness to rapidly progressive respiratory failure. Pulmonary symptoms are almost always associated with chest radiograph abnormalities such as infiltrates (63 per cent), nodules (31 per cent), infiltrates with cavitation (8 per cent), and nodules with cavitation (10 per cent). The lung nodules range from a few millimeters to several centimeters in size, are commonly bilateral, and are cavitated in one-third of patients. Diffuse alveolar infiltrates may indicate alveolar hemorrhage.

Pulmonary alveolar hemorrhage caused by Wegener's granulomatosis can be life threatening. Diffuse pulmonary hemorrhage from necrotizing capillaritis has been noted in 5 to 45 per cent of biopsy or necropsy cases. Radiograph abnormalities resembling alveolar hemorrhage may be caused by bronchiolitis obliterans with organizing pneumonitis (BOOP). The diagnosis will not be overlooked if additional features, particularly vasculitis, suppurative granulomas, tiny necrotic zones, microabscesses, and multinucleated giant cells, are appreciated.

Upper airway disease, which is present in over 85 per cent of patients, manifests as rhinorrhea with purulent or bloody nasal discharge, nasal mucosal drying and crust formation, and epistaxis. Paranasal sinus involvement leading to deep facial pain, nasal septal perforation, and ulceration of the vomer are important signs. Bacterial infection, particularly from *Staphylococcus aureus*, may complicate the clinical picture. Destruction of the nasal cartilages resulting in 'saddle-nose deformity' is usually a subacute or chronic complication.

Laryngeal and tracheal ulcerated lesions, which are present in 30 per cent of untreated patients, are responsible for hemoptysis. Subglottic stenosis develops in 5 to 8 per cent of treated patients. The symptoms caused by subglottic stenosis are usually insidious in onset. Inspiratory and expiratory flow–volume loops may be useful in assessing the impairment of upper airway function as well as in the follow-up of these patients.

Therapy of Wegener's granulomatosis includes a combination of prednisone (1–1.5 mg/kg/day) and cyclophosphamide (2 mg/kg/day). Complete remission is noted in more than 90 per cent of patients. Corticosteroids alone may be sufficient in the milder form of disease. Trimethoprim–sulfamethoxazole (co-trimoxazole) is effective in maintaining remission. Bronchoscopic balloon dilatation followed by stent insertion has been used to treat tracheobronchial stenosis caused by Wegener's granulomatosis.

Takayasu's arteritis

Takayasu's arteritis is a term synonymous with pulseless disease, aortic arch arteritis or syndrome, or reversed coarctation. It is a chronic inflammatory disease of unknown etiology that affects the large arteries. The aorta and its main branches are primarily involved. Histological features include continuous or patchy panarteritis with granulomatous inflammation and infiltration by lymphocytes, plasma cells, histiocytes, and multinucleated giant cells.

Takayasu's arteritis is more common in Asia, although the disease has been described in all parts of the world. The female-to-male ratio is 8.5:1, and 80 per cent of patients are aged between 11 and 30 years. Initial or acute clinical features include fever, malaise, weight loss, arthralgias, and night sweats lasting 4 to 6 weeks. Chronic disease is the result of chronic ischemia of affected organs.

Pulmonary artery involvement is present in up to 50 per cent of all patients with Takayasu's arteritis. Pathological lesions are generally localized to medium and large pulmonary arteries. The histological abnormalities described above lead to pulmonary arterial occlusion and stenoses. A retrospective study of perfusion lung scans revealed abnormalities in 76 per cent of 120 patients with Takayasu's arteritis; initially, the changes appeared in the upper lobes and then progressed to involve arteries of the middle and lower lobes. In another study of 42 patients, in whom a respiratory problem was not clinically suspected, intravenous digital subtraction angiography showed involvement of the pulmonary artery in 14.3 per cent, even though the chest radiographs were abnormal in only two patients. A comparison study of 59 patients with temporal arteritis showed that, although chest radiographs were abnormal in 68 per cent, pulmonary angiography revealed arterial occlusions in 86 per cent. A poor correlation has been noted between angiographic changes and the results of spirometry and arterial blood gas analysis. Other thoracic complications described include aneurysms of pulmonary arteries. Takayasu arteritis presenting as idiopathic adult respiratory distress syndrome with a pathological diagnosis of acute interstitial pneumonia has been described.

Corticosteroid therapy results in symptomatic remission within days to weeks. Patients with significant pulmonary arterial occlusions may require surgical bypass. Death is usually the result of vascular complications such as rupture of an aneurysm, myocardial infarction, congestive cardiac failure, or cerebrovascular accident. Pulmonary involvement signifies a poor prognosis.

Behcet's disease

Behçet's disease is a chronically recurring vasculitic syndrome characterized by aphthous stomatitis together with two or more of the following: aphthous genital ulcerations, uveitis, cutaneous nodules or pustules, synovitis, or meningoencephalitis. The histological features of mucocutaneous lesions are non-specific and exhibit varying degrees of lymphocytic and plasma cell infiltration and deposition of IgM and C3 in the dermal vessels. Fever, elevated erythrocyte sedimentation rate, and anemia are common. Obstruction of major vessels and aneurysms occurs in 10 to 37 per cent of patients. Thromboses of the superficial and deep veins of both the upper and lower extremities and of the superior and inferior vena cava develop in 7 to 37 per cent of patients.

Massive hemoptysis is the most serious respiratory complication in Behçet's disease. Life-threatening hemoptysis is common and is the cause of death in more than one-third of patients with pulmonary disease. Serious hemoptysis, initially responsive to corticosteroid therapy, demonstrates a propensity to recur. A review of 28 patients with Behçet's disease with respiratory complications noted hemoptysis, fever, chest pain, and dyspnea in all 28 patients, and the lung involvement was associated with extrapulmonary disease. Another report on 49 patients with lung involvement observed that recurrent dyspnea, cough, chest pain, and hemoptysis were the initial features, particularly in young men, and appeared 3.6 years after the onset of Behçet's disease. Chest radiographs may show pulmonary infiltrates, pleural effusions, and prominent pulmonary arteries. The reasons for hemoptysis include aneurysms of the pulmonary artery, pulmonary artery–bronchus fistula, and mucosal ulcerations. Pulmonary angiography may reveal amputation of branches of the pulmonary artery and aneurysmal dilatations. Immunopathological studies have indicated that pulmonary vasculitis is a result of circulating immune complexes.

Pulmonary embolism occurs with greater frequency because of the high incidence of deep vein thrombosis of the extremities and the vena cavas. Ventilation–perfusion scans can be misleading, and anticoagulant therapy for presumed pulmonary embolism can result in catastrophic results. Other complications include formation of fistulas between the coronary and pulmonary arteries and between the bronchial and pulmonary arteries, pulmonary hypertension, pulmonary hemorrhage, thromboembolism of the superior vena cava and/or other mediastinal veins, aneurysms of the aorta, pleural effusion, mediastinal or hilar lymphadenopathy, and interstitial pulmonary fibrosis.

Corticosteroid suppresses but seldom controls all manifestations of the disease. Sudden death in corticosteroid-treated patients has indicated that corticosteroid alone may be inadequate. Other drugs, namely colchicine, chlorambucil, methotrexate, azathioprine, and cyclosporine (cyclosporin), may be needed for chronic therapy. Anticoagulant therapy is contraindicated. Patients with significant hemoptysis or diffuse pulmonary infiltrates carry a poor prognosis; a mortality of 80 per cent within 2 years after the onset of lung disease has been noted.

Allergic angiitis and granulomatosis

Allergic angiitis and granulomatosis, also known as Churg–Strauss syndrome, is characterized by asthma, eosinophilia, non-necrotizing angiitis, and extravascular granulomata, usually with eosinophilic infiltrates. Allergic angiitis and granulomatosis is an uncommon disease. The angiitis is disseminated and involves pulmonary and systemic arteries and veins. Peripheral blood hypereosinophilia and elevation of serum IgE are common.

Pulmonary features in allergic angiitis and granulomatosis are related to underlying asthma which is present in all patients. A phasic pattern of allergic angiitis and granulomatosis syndrome has been described: initial allergic rhinitis, evolving into asthma and followed by peripheral blood eosinophilia, eosinophilic tissue infiltrates, and ultimately vasculitis. A study of 154 patients with allergic angiitis and granulomatosis revealed the following: 84 males, average age of 28 years at onset of allergic rhinitis, average age of 35 years at onset of asthma, and average age of 38 years when vasculitis was diagnosed. Other observations included a mean peak eosinophil count of $12.9 \times 10^9/l$, anemia in 83 per cent, granulomas in 40 per cent, tissue eosinophilia in 50 per cent, and vasculitis in more than 70 per cent. Chest radiograph abnormalities are present in over 60 per cent of patients and include patchy and occasionally diffuse alveolar–interstitial infiltrates in the perihilar area. Massive pulmonary alveolar hemorrhage is rare. Respiratory failure is the cause of death in a small percentage, although death from status asthmaticus has been reported in up to 8 per cent. The main causes of death are heart failure, renal failure, cerebral hemorrhage, and gastrointestinal hemorrhage.

Upper airway disease may manifest as allergic rhinitis, nasal polyps, nasal mucosal crusting, and septal perforation in up to 70 per cent of patients. Chronic paranasal sinus pain or tenderness and seasonal allergy are also present.

Treatment of allergic angiitis and granulomatosis consists of high-dose systemic corticosteroids, which can produce a dramatic response. Therapy should be instituted as soon as the diagnosis is established.

Giant cell arteritis

Giant cell arteritis, also known as temporal arteritis, cranial arteritis, and granulomatous arteritis, is a vasculitis of undetermined origin. Even though the histological findings vary, the characteristic features include lymphocytic infiltration with fragmentation of the internal elastic lamina, granulomatous inflammation, histiocytes, and multinucleated giant cells. These classic histological features are seen in approximately 60 per cent of temporal artery biopsies. Giant cell arteritis is responsible for 26 per cent of all primary vasculitic syndromes in North America. The disease can occur in association with polymyalgia rheumatica, and affects middle-aged or older individuals. There are no specific laboratory tests to diagnose giant cell arteritis, although a majority of patients exhibit an elevated erythrocyte sedimentation rate. Biopsy of the temporal artery is necessary to document the diagnosis.

Pulmonary complications of giant cell arteritis occur in approximately 10 per cent of patients. The respiratory symptoms can be the presenting features, and they include cough, sore throat, and hoarseness. A population-based study of 94 patients with giant cell arteritis found pulmonary symptoms in up to 30 per cent. It should be considered in any older patient with a new cough or throat pain without obvious cause. Disease of the small airways has been observed in 46 per cent of patients with giant cell arteritis; however, the abnormalities have not been significant. Chest radiograph abnormalities consist of pulmonary nodules, interstitial infiltrates, and occlusion and aneurysms of the pulmonary artery.

Virtually all manifestations of the disease respond favorably to systemic corticosteroids at an initial dose of 1 mg/kg/day. Chronic low-dose therapy is required to maintain remission in most patients.

Henoch–Schönlein purpura

Henoch–Schönlein purpura, also known as anaphylactoid purpura or allergic purpura, is manifested by acute purpura, arthritis, colicky abdominal pain, and nephritis. The disease is more common in children than in adults, although an adult form of the disease exists. The mean age is 17 years with equal occurrence in males and females. Morphological characteristics include arteriolitis and venulitis in the superficial dermis and the bowel, and mild proliferative and necrotizing glomerulonephritis. Immunofluorescence histology reveals significant deposits of IgA in the dermis and kidney. Palpable purpura over the buttocks and lower extremities and fever are generally the first signs of the disease. The triad of purpura, arthritis, and abdominal pain is present in approximately 80 per cent of patients.

Joint involvement is typically monoarticular, transient, involves large joints, and causes pain out of proportion to objective evidence of synovitis. Peritonitis and melena are common.

Pulmonary involvement is rare in Henoch–Schönlein purpura. However, several respiratory complications have been observed. Chest radiographs have shown perihilar patchy opacities, reticulonodular changes, and pulmonary alveolar hemorrhage.

Therapy consists of high-dose corticosteroid therapy (0.5–1.0 mg/kg/day). Initial full recovery can be expected in almost all patients. Relapse of purpura, abdominal pain, and arthritis may occur over a 3- to 6-week period before complete resolution is observed.

Hypocomplementemic urticarial vasculitis

Hypocomplementemic urticarial vasculitis is a systemic vasculitic syndrome characterized by hypocomplementemia, cutaneous vasculitis, urticaria, and arthritis. Angio-edema, fever, uveitis, episcleritis, and seizures may also occur. The erythrocyte sedimentation rate is elevated in two-thirds of patients and hypocomplementemia is noted in 38 per cent. Hypocomplementemic urticarial vasculitis has been associated with pulmonary complications.

Pulmonary vasculitis is not characteristic of hypocomplementemic urticarial vasculitis. However, obstructive pulmonary disease occurs in up to 60 per cent of patients with the disease. Obstructive airway disease is thought to result from a combination of smoking and an unknown immunological process, perhaps mediated by the complement cascade. A review of 72 cases of biopsy-proven hypocomplementemic urticarial vasculitis revealed that 32 per cent had hypocomplementemia and 21 per cent had obstructive lung disease. High incidences of angio-edema, ocular inflammation, and glomerulonephritis have been observed. In one report on 18 patients with the disease, pulmonary function was measured in 17 patients, 11 of whom had dyspnea; all dyspneic patients had moderate to severe airflow obstruction, which progressed in all 11 and subsequently improved in only one. Six of these 11 patients died of respiratory failure, one underwent lung transplantation, and three of the remaining four had moderately severe to life-threatening respiratory insufficiency. Treatment did not appear to alter the progression of obstructive lung disease. The pathogenesis of the disease may involve humoral autoimmunity, although it is not clear how autoimmunity would participate in the development of obstructive lung disease. Cigarette smoking appears to be a risk factor for fatal lung disease.

Polyarteritis nodosa

Polyarteritis nodosa is characterized by multiple organ involvement by necrotizing arteritis of small and medium arteries. An important aspect of this syndrome is the rarity with which it affects the respiratory system. Although early postmortem series reported pulmonary involvement as a relatively common manifestation of polyarteritis nodosa, the modern classification of the disease suggests that pulmonary involvement is very uncommon. In a study of 118 patients with polyarteritis nodosa, no pulmonary complications were encountered. However, classic histological findings have been observed in the pulmonary arteries of patients with well-documented classic polyarteritis nodosa. Of the few reported cases with pulmonary manifestations, the majority are autopsy series and the vascular lesion involved the bronchial arteries. Corticosteroids have been used for therapy.

Microscopic polyangiitis

Microscopic polyangiitis or polyarteritis is a separate entity that should be distinguished from polyarteritis nodosa and Wegener's granulomatosis. Microscopic polyangiitis is characterized by pauci-immune necrotizing and crescentic glomerulonephritis and other systemic involvement in the absence of granulomas. Recurrent and diffuse alveolar hemorrhage due to pulmonary capillaritis is the main pulmonary feature of the disease. The presence of perinuclear type ANCA with specificity for myeloperoxidase (MPO-ANCA) is recognized as an indicator of the disease, although it is positive in only 60 per cent of patients. Clinically, microscopic polyangiitis can be confused with Wegener's granulomatosis because of hemoptysis, alveolar hemorrhage, and the presence of c-ANCA in about 40 per cent of patients. Acute respiratory failure caused by pulmonary hemorrhage, acute interstitial processes, and the acute respiratory distress syndrome have been described in this syndrome. Pathologically, interstitial inflammatory cells (alveolitis), hyaline membranes, restructuring, and fibrosis have been observed. The upper airway complications seen in Wegener's granulomatosis are conspicuously absent in microscopic polyangiitis, although progressive airway obstruction has been reported. The treatment of microscopic polyangiitis is similar to that of Wegener's granulomatosis.

Bibliography

Churg, A. (1983). Pulmonary angiitis and granulomatosis revisited. *Human Pathology*, **14**, 868–83.

Doyle, L., McWilliam, L., and Hasleton, P.S. (1988). Giant cell arteritis with pulmonary involvement. *British Journal of Diseases of the Chest*, **82**, 88–92.

Efthimiou, J., Johnston, C., Spiro, S.G., and Turner-Warwick, M. (1986). Pulmonary disease in Behçet's syndrome. *Quarterly Journal of Medicine*, **58**, 259–80.

Schwartz, H.R., McDuffie, F.C., Black, L.F., Schroeter, A.L., and Conn, D.L. (1982). Hypocomplementemic urticarial vasculitis: association with chronic obstructive pulmonary disease. *Mayo Clinic Proceedings*, **57**, 231–8.

Travis, W.D., Hoffman, G.S., Leavitt, R.Y., Pass, H.I., and Fauci, A.S. (1991). Surgical pathology of the lung in Wegener's granulomatosis: review of 87 open lung biopsies from 67 patients. *American Journal of Surgical Pathology*, **15**, 315–33.

13.10.3 Rheumatoid arthritis

U. B. S. Prakash

Key messages

- Rheumatoid arthritis is a chronic inflammatory process involving diarthroidal joints.
- Thoracic complications include pleural effusion, interstitial lung disease, obstructive airway disease, necrotic rheumatoid nodules, and rheumatoid pneumoconiosis.

Table 1 Respiratory complications in rheumatoid arthritis

Pleurisy and pleural effusion
Interstitial pneumonitis and fibrosis
Obliterative bronchiolitis
Necrobiotic nodules
Caplan's syndrome
Cricoarytenoid arthritis
Laryngeal nodules
Pulmonary hypertension
Bronchiectasis
Pulmonary vasculitis

Introduction

Rheumatoid arthritis is a chronic inflammatory process characterized by non-suppurative arthritis involving the diarthroidal joints with frequent involvement of tendons, ligaments, fascia, muscle, and bone. The criteria for the diagnosis of rheumatoid arthritis have been established by the American Rheumatism Association. Extraskeletal organ systems are frequently involved and may be the presenting manifestations. The respiratory system is among the commonly affected organs. Pleuropulmonary manifestations are encountered in 2 to 45 per cent of patients, depending on whether chest radiography or diffusing capacity for carbon monoxide is used to assess the prevalence of lung involvement. The involvement of the respiratory system is peculiar in that while rheumatoid arthritis occurs more frequently in females, the pulmonary complications are more common in males. Furthermore, pleuropulmonary complications may precede the onset of arthritic symptoms. Many of the intrathoracic manifestations have the potential to cause critical illness and respiratory distress (Table 1).

Pleural involvement

Pleural involvement is the most common intrathoracic manifestation of rheumatoid arthritis. Chest radiographic evidence of pleural involvement has been noted in 24 per cent of males and 16 per cent of females. Postmortem analyses have observed pleural involvement in nearly 50 per cent. Clinically, massive pleural effusion with the potential to cause respiratory distress is uncommon. Indeed, one-third of patients with rheumatoid pleural involvement are relatively asymptomatic because of 'painless pleurisy'. The pleural effusion is generally unilateral, small, persistent, or recurrent. Occasionally, effusions become chronic and persist for months to years. Pleural involvement is more common in those with active seropositive rheumatoid arthritis than in those with inactive disease. Other pulmonary complications (Table 1) may occur in association with pleural involvement.

Thoracic effusions may become complicated by empyema. Unusual infections of the pleural space due to *Pseudomonas multocida, Eichinella corrodens*, and anaerobic organisms occur with greater frequency.

Analysis of pleural fluid typically reveals an exudate with a total protein level above 3.0 g/dl and a pH below 7.3. The fluid is usually pale yellow but may occasionally be bloody. It may assume an opalescent green color due to the high cholesterol content ('pseudochylothorax'). Typically, the glucose level is low (below 30 m/dl) because of the selective blockage of glucose transport into the pleural space as well as the increased glucose utilization by the inflamed and metabolically active pleural mesothelial cells. The low pH is the result of localized acidosis due to lactate and CO_2 in the pleural space. Pleural fluid complement (C50, C3, and C4) levels are diminished in over one-third of patients. Chronic rheumatoid pleural effusions usually contain large amounts of cholesterol; some of the highest levels recorded have been in patients with rheumatoid pleural effusions. Pneumothorax secondary to transthoracic needle aspiration may also occur.

Interstitial lung disease

The term 'rheumatoid lung' is used to describe the diffuse interstitial pneumonitis and fibrosis that occurs as a serious complication of rheumatoid arthritis. Although it is detected on 2.0 to 4.5 per cent of chest radiographs, a restrictive pattern of pulmonary dysfunction can be documented by pulmonary function testing in over one-third of patients. The clinical, physiological, and histopathological features of rheumatoid lung disease are analogous to those of idiopathic pulmonary fibrosis. Cough and dyspnea on exertion are common symptoms. Physical examination reveals fine late inspiratory crackles, which are more pronounced in the lung bases. Clubbing occurs in over two-thirds of patients. The chest radiograph usually exhibits a bibasilar interstitial process or micronodules, with progression to other regions as the disease progresses. Late-stage disease may show generalized honeycombing. Predominantly upper-lobe bilateral infiltrates are seen in 6 per cent of patients. High-resolution CT of the chest in the early stages of the disease may show 'ground-glass' infiltrates; as the disease progresses, interstitial patterns and honeycombing will appear.

Pulmonary function testing in patients with rheumatoid lung disease typically demonstrates restrictive lung dysfunction manifested by decreased lung volumes, relatively normal air flow, and diminished diffusing capacity for carbon monoxide; the last mentioned is the earliest physiological abnormality. Exercise-induced hypoxia is common in patients with advanced disease.

Invasive diagnostic procedures are rarely indicated in patients with rheumatoid lung. Diagnostic bronchoalveolar lavage has demonstrated abnormalities identical to those in idiopathic pulmonary fibrosis, namely increased neutrophils as well as lymphocytes. A lung biopsy is rarely required to document rheumatoid lung disease as the morphological features are non-specific without clinical correlation. Histopathological patterns include non-specific interstitial pneumonitis, bronchiolitis obliterans with patchy organizing pneumonia, lymphoid hyperplasia, cellular interstitial infiltrates, and rheumatoid nodules. Diagnostic bronchoalveolar lavage and lung biopsy may be indicated in complicated situations, particularly in patients with immunocompromised status as a result of therapy for rheumatoid arthritis.

The occurrence of rapidly progressive lung disease, respiratory failure, and cor pulmonale as the result of rheumatoid lung is rare. A rapid progression is seen when arthritic features flare, with the respiratory course exhibiting progressive idiopathic pulmonary fibrosis.

Obstructive airway disease

More than 20 per cent of patients with active rheumatoid arthritis develop obstructive airway disease secondary to rheumatoid bronchiolitis, follicular bronchiolitis, and bronchitis. Pulmonary function testing in patients (non-smokers) with rheumatoid arthritis has demonstrated obstructive lung dysfunction in nearly a third. A much higher prevalence of obstructive lung disease has been noted in patients with rheumatoid arthritis who smoke.

Patients with rheumatoid arthritis and respiratory involvement have a 50 per cent incidence of non-PiM (PiMZ and PiMS) phenotypes for α_1-antitrypsin. Deficiency of α_1-antitrypsin is a known cause of panlobular emphysema. Since many patients with rheumatoid arthritis receive penicillamine and gold salts to control joint inflammation, it is essential to recognize that these drugs themselves can cause bronchiolitis obliterans. High-resolution CT of the chest may demonstrate 'ground-glass' infiltrates.

Necrobiotic rheumatoid nodules

Necrobiotic rheumatoid pulmonary nodules represent pulmonary parenchymal rheumatoid nodules that are histologically identical to subcutaneous rheumatoid nodules. These nodules are more common in men and in those with seropositive rheumatoid arthritis. The lung nodules may precede the arthritic symptoms and produce minimal or no pulmonary symptoms; they are usually incidentally identified by chest radiography, when they are usually bilateral and measure from a few millimeters to several centimeters in diameter. The lung nodules tend to occur near pleural surfaces, 65 per cent of them cavitate, and they may rupture into pleural space to produce pneumothorax or empyema. A chronically cavitated nodule may become secondarily infected by *Aspergillus*, resulting in an aspergilloma. Pulmonary rheumatoid necrobiotic nodules appear and disappear with the activity or inactivity of rheumatoid arthritis. Many resolve spontaneously even when rheumatoid arthritis remains active. Respiratory distress and critical illness as the result of rheumatoid lung nodules is very rare.

Rheumatoid pneumoconiosis

The occurrence of rheumatoid nodules in association with pneumoconiosis is known as Caplan's syndrome. The pneumoconiosis may be caused by silicosis, asbestosis, aluminosis, or some other inorganic occupational lung disease. Caplan's syndrome is much more common in Europe than in North America. Pulmonary parenchymal nodules in Caplan's syndrome measure 1 to 5 cm in diameter, evolve rapidly, undergo cavitation, appear in crops, and are often associated with other respiratory lesions. Histopathological examination reveals features similar to those of rheumatoid nodules together with the presence of pneumoconiotic material in the center.

Other pulmonary complications

Cricoarytenoid arthritis is an important complication of rheumatoid arthritis because it may present with inspiratory and expiratory distress due to a fixed obstruction at the level of the larynx. It is recognized clinically in one-third of cases, even though its presence has been noted in most necropsy studies on patients with rheumatoid arthritis. Clinical manifestations include chronic persistent sore throat and hoarseness. Laryngoscopy may show erythema and edema of the vocal apparatus, abnormal motion of the vocal cords, and dysfunction of the arytenoids. Occurrence of rheumatoid nodules in the larynx may present with hoarseness, cough, and dyspnea. It is important for specialists in critical care and anesthesia to recognize that laryngeal nodules may remain asymptomatic and pose problems at the time of tracheal intubation.

Bronchiolitis may predispose to bronchiectasis, particularly in patients with advanced and severe rheumatoid arthritis. Patients with rheumatoid bronchiolitis may have a significantly higher incidence of positive bacterial cultures in their sputum specimens.

Pulmonary hypertension and pulmonary vasculitis, possibly leading to rheumatoid arthritis, have been described in patients with alveolar hemorrhage.

It is important for specialists to recognize that patients with severe rheumatoid arthritis involving the cervical spine may develop instability of the cervical spinal column. This can preclude extension of the neck, necessitating laryngoscopic tracheal intubation. Tracheal intubation in such patients should be accomplished with a flexible bronchoscope.

Treatment and prognosis

Non-steroidal anti-inflammatory therapy in the initial stages of rheumatoid arthritis, if successful in controlling the primary disease, controls most of the pulmonary manifestations. Complicating factors such as large pleural effusions and obstructive airway disease may need to be addressed. Systemic corticosteroid therapy in the earlier stages of rheumatoid interstitial lung disease may reverse the acute inflammatory process and slow or stop the progression to irreversible fibrotic lung disease. As the lung disease becomes chronic, the ability to reverse it with corticosteroids and other pharmacological agents diminishes. Poor prognosis is associated with interstitial disease, with a median survival of 3.5 years and a 5-year survival of 39 per cent. Poor prognosis also follows rheumatoid bronchiolitis. Immunotherapy consisting of corticosteroids and cytotoxic agents has been used, but the response has been unsatisfactory in chronic or advanced cases. It is important to recognize that certain cytotoxic agents used in the treatment of rheumatoid arthritis (e.g. gold, methotrexate, and cyclophosphamide) can also produce pulmonary toxicity. However, methotrexate appears to have no major adverse effect on the pulmonary function of the majority of patients with rheumatoid arthritis, including those with pre-existing pulmonary disease.

Bibliography

Anaya, J.M., *et al.* (1995). Pulmonary involvement in rheumatoid arthritis. *Seminars in Arthritis and Rheumatism*, **24**, 242–54.

Begin, R., Masse, S., Cantin, A., Menard, H.A., and Bureau, M.A. (1982). Airway disease in a subset of nonsmoking rheumatoid patients. *American Journal of Medicine*, **72**, 743–50.

Lawry, G.V., Finerman, M.L., Hanafee, W.N., Mancuso, A.A., Fan, P.T., and Bluestone, R. (1984). Laryngeal involvement in rheumatoid arthritis. *Arthritis and Rheumatism*, **27**, 873–82.

Michalski, J.P., McCombs, C.C., Scopelitis, E., Biundo, J.J., Jr, and Medsger, T.A., Jr (1986). Alpha1-antitrypsin phenotypes, including M subtypes, in pulmonary disease associated with rheumatoid arthritis and systemic sclerosis. *Arthritis and Rheumatism*, **29**, 586–91.

Popp, W., *et al.* (1992). Prediction of interstitial lung involvement in rheumatoid arthritis. The value of clinical data, chest roentgenogram, lung function, and serologic parameters. *Chest*, **102**, 391–4.

14 Perioperative problems

14.1 General postoperative intensive care

14.1.1 Recovery from anesthesia

D. R. Miller and A. D. Baxter

Key messages

- Anesthesia-related problems include postoperative pain, persistent sedation, nausea and vomiting, altered mental status, and hypothermia with shivering.
- The introduction of short-acting anesthetics and pharmacokinetic modeling procedures allow a more predictive pattern of recovery from anesthesia.

Introduction

For most patients, recovery from anesthesia is a smooth and uneventful part of a carefully conducted anesthetic and surgical procedure. However, when patients are at high risk or are unstable, or when complications arise during the conduct of anesthesia or in the early recovery period, prompt intervention by skilled medical and nursing personnel is required. Therefore postanesthesia care units (PACUs) have evolved into highly specialized areas which are able to care simultaneously for patients recovering from routine surgery under general or regional anesthesia, and for critically ill or unstable patients who require intensive monitoring and cardiorespiratory therapy until discharge to a ward or transfer to an intensive care unit (ICU).

Routine recovery

Recovery from anesthesia begins in the operating room following discontinuation of the anesthetic drugs. Times to awakening and the quality of early recovery following general anesthesia may vary greatly, depending upon the properties of selected anesthetic drugs (inhaled anesthetics, sedative–hypnotics, opioid analgesics, and neuromuscular blocking drugs) as well as their additive and/or synergistic interactions (Stanski and Shafer 1995). The speed of recovery from inhaled anesthetics is influenced by four main factors: the solubility coefficient of the drug, the concentration administered and the duration of anesthesia, patient sensitivity, and drug interactions. The recent introduction of two new inhaled anesthetics, desflurane and sevoflurane, may ensure even more rapid and predictable recovery compared with regimens employing halothane, enflurane, or isoflurane (Table 1).

Several recent developments have affected the factors determining speed of recovery from intravenous anesthetics. First, the introduction of new short-acting agents, including propofol, have had an enormous impact on the practice of anesthesia. Propofol is an effective induction agent and is also an effective hypnotic for maintenance of anesthesia when administered by continuous infusion. Propofol is further indicated for both monitored anesthesia care sedation and ICU sedation (titrated to a target level of sedation at rates between 25 and 75 μg/kg/min). The main advantages of this drug include its rapid, predictable, and pleasant recovery, even after prolonged administration (hours to days), and its antiemetic properties. Its disadvantages include insolubility, requiring formulation in intralipid, and a relatively high cost. Midazolam may be less costly when used either by continuous infusion or intermittent boluses for postoperative sedation of mechanically ventilated patients (titrated to a target level of sedation at rates between 0.5 and 1.5 μg/kg/min). However, recovery times are somewhat longer (30 min to several hours) and vary depending upon the duration of drug therapy, the depth of sedation during maintenance, and patient-related factors. Recent trends in cardiovascular anaesthesia have been directed towards early extubation (within 4–6 h of completion of surgery, rather than routinely ventilating patients overnight), a cost-effective practice which tends to be facilitated by the use of the shorter-acting sedatives (Miller 1996).

A second recent advance in intravenous anesthesia has been the development of pharmacokinetic modeling procedures which provide a quantitative description of the relative contributions of distribution and elimination processes during recovery from intravenous anesthetic drugs (Schnider and Shafer 1995). The definitions of these new descriptors of clinical recovery are presented in Table 2.

Recovery from the effects of neuromuscular blocking drugs is influenced by a number of factors, including the duration of action of the drug (Table 1), the dose, and the status of renal and hepatic function. In addition, the dose of anticholinesterase for reversal, patient temperature, acid–base status, and use of drugs that increase the neuromuscular block (e.g. inhaled anesthetics, aminoglycoside antibiotics) are also important determinants of recovery. Careful clinical evaluation of recovery from the effects of neuromuscular blocking drugs is essential, as residual paralysis has been shown to exist in 5 to 40 per cent of patients upon arrival in the PACU. Recovery of neuromuscular function can be confirmed by several simple tests,

Table 1 Duration of the clinical effect of anesthetic drugs

Class/drug	Ultra-short	Short	Intermediate	Long
Inhalational anesthetics	Desflurane	Sevoflurane	Isoflurane Enflurane Halothane	
Sedative–hypnotics	Propofol	—	Midazolam	Thiopental (thiopentone) Diazepam Lorazepam
Opioids	Remifentanil	Alfentanil*	Sufentanil†	Fentanyl Morphine
Muscle relaxants	Succinylcholine (suxamethonium)	Mivacurium	Atracurium Cisatracurium Vecuronium Rocuronium	Pancuronium Curare Doxacurium Pipercuronium Metocurine

This classification scheme must be interpreted in the light of the fact that the duration of action of most drugs increases following repeated boluses or prolonged administration. For most intravenous anesthetic drugs, clinical predictors of recovery are relatively independent of elimination half-lives and are best described by either their context-sensitive half-times or their mean effect times.

*Short duration with intravenous boluses.

†Short duration during infusions lasting for less than 8 h.

Table 2 Clinical predictors of recovery from intravenous anesthetics

Predictor	Relevant effect-site decrement time	Context-sensitive half-time	Mean effect time
Definition	Time for effect-site concentration to decrease from anesthetic to recovery concentration as a function of infusion duration	Time required for a 50% decrease in plasma concentration as a function of infusion duration ('the context')	Area under curve expressing probability of drug effect as a function of time after the agent is discontinued
Sensitive to duration of administration	Yes	Yes	Yes
Modeling assumptions	Yes	Yes	Yes
Clinically validated	No	Yes	No
Advantages concentration	Estimates drug concentration in the brain; most clinically relevant kinetic model	Very useful in comparing kinetics of two drugs	Incorporates both kinetics and dynamics in predicting duration of drug effect
Disadvantages	More dependent on modeling assumptions	50% decrement may be larger than is clinically relevant at the end of surgery; also, IV anesthetics exert effects in brain, not in plasma	Imprecision in quantitating drug effect (binary, response/no-response variables)

IV, intravenous.

Adapted with permission from Miller (1996).

including observation of a sustained muscle contraction (without fade) to tetanic stimulation of the ulnar nerve at the wrist or noting the patient's ability to sustain a head lift for more than 5 s. If signs of persistent residual neuromuscular block persist, an additional dose of anticholinesterase and an antimuscarinic drug should be given immediately (e.g. neostigmine 50 μg/kg plus atropine 20 μg/kg intravenously).

Postoperative anesthesia-related problems

Postoperative pain

Appropriate and aggressive treatment of postoperative pain is one of the primary goals of postanesthesia care, in order to ensure patient comfort and minimize complications related to increased sympathetic nervous system activity (hypertension, tachycardia, myocardial ischemia, agitation). Certain steps are essential in achieving and maintaining adequacy of analgesia in the initial postoperative period. A key principle in the provision of optimal postoperative analgesia is to initiate therapy intraoperatively with appropriate use of one or more of the following modalities:

(1) intermediate and long-acting opioid analgesics titrated towards the end of surgery (e.g. fentanyl 25–50 µg intravenously or morphine 1–2 mg increments intravenously);

(2) non-steroidal anti-inflammatory analgesics (e.g. ketorolac 15–30 mg intravenously);

(3) infiltration of the surgical wound with local anesthetic;

(4) intra-articular infiltration with opioid and/or local anesthetic (joint surgery);

(5) intrathecal or epidural opioid/local anesthetic (thoracic, abdominal, gynecological, and major orthopedic surgery).

Persistent sedation

Delayed recovery from anesthesia is less common, but an organized and thorough approach is required when this problem occurs. First, the times and amounts of all preoperative and intraoperative anesthetic medications should be carefully documented, as recovery profiles based on drug doses, the duration of administration, and clinical recovery times are relatively predictable (Table 1). Second, an appropriate physical examination should be conducted, including a firm tactile stimulus to elicit arousal. Findings should then allow establishment of a differential diagnosis, for which the most common causes of delayed recovery can be established (Table 3). It should be appreciated that, even when using higher concentrations of long-acting anesthetic drugs, all patients should be responsive within a period of 10 to 60 min after surgery. In the rare circumstance where the diagnosis remains elusive, a complete neurological examination should be performed in consultation with a neurologist.

As residual effects of anesthetic drugs represent the single most common cause of delayed recovery, treatment is generally expectant. However, possible contributing factors such as hypothermia, hypercarbia, etc. should be treated aggressively. When delayed recovery from opioids is suspected, naloxone may be administered intravenously in 40-µg increments four times every 2 to 4 min, with due care to avoid reversal of opioid-induced analgesia. In general, unless a patient has received massive doses of opioid analgesic, 0.2 mg of naloxone will induce arousal when unconsciousness is secondary to opioid effects. Nalbuphine, which is an opioid agonist–antagonist, is an effective alternative in doses of 2.5 to 10 mg intravenously. When delayed recovery is thought to be secondary to residual effects of benzodiazepines, flumazenil may be administered in increments of 0.1 mg intravenously to a maximum dose of 0.2 mg. It is important to appreciate that the duration of effect of a single dose of flumazenil is shorter than that of the commonly used benzodiazepine agonists midazolam and diazepam, and that resedation may occur. Furthermore, flumazenil does not specifically reverse the respiratory depressant effects of this class of central nervous system depressants.

Nausea and vomiting

Postoperative nausea and vomiting remain a troublesome and frequent complication of general anesthesia, with an in-hospital incidence varying from 10 to 30 per cent. Predisposing factors include a full stomach (emergency surgery, pregnancy), obesity, gastroschesis, diabetes mellitus, hiatus hernia and/or gastroesophageal reflux, type of surgery, phase of the menstrual cycle in women, and choice of anesthetic (particularly those incorporating higher doses of opioid analgesics and inhalational anesthetics). Vomiting itself may be associated with particular medical risks, as outlined in Table 4. Treatment consists of antiemetic drugs (e.g. droperidol 0.25–0.5 mg intra-

Table 3 Causes of delayed recovery from anesthesia

Cause	Examples
Pre-existing mental dysfunction	Head injury, intoxication
Heavy premedication	Lorazepam, scopolamine
Residual sedation from anesthetics*	Elderly and obese patients, and following prolonged surgery
Persistent paralysis	Neuromuscular blocking drugs not 'reversed' 'Phase II' block following succinylcholine Atypical cholinesterase (prolonged response to mivacurium, succinylcholine) Long-acting drugs (e.g. doxacurium)
Metabolic	Hypoglycemia, hyperglycemia, hyponatremia
Respiratory	CO_2 narcosis, hypoxemia
Primary central nervous system	Delerium tremens, seizure disorder, intraoperative cerebral anoxia or thromboembolism, paradoxical air embolism, intraoperative hemorrhage, postoperative cerebrovascular accident
Hypothermia	Core temperature $<33\,^{\circ}C$

*Most common cause of delayed recovery from anesthesia.

Table 4 Medical risks of postoperative nausea and vomiting

Pulmonary aspiration of gastric contents
Strain on abdominal or inguinal suture lines
Complications from elevated central venous pressure (ocular, tympanic, intracranial)
Sympathetic responsiveness: hypertension, tachycardia, myocardial ischemia in patients at risk
Parasympathetic responsiveness: bradycardia, hypotension
Delayed hospital discharge, overnight admission

venously, dimenhydrinate 12.5–25 mg intravenously, ondansetron 4–8 mg intravenously, or a subhypnotic dose of propofol 0.15 mg/kg intravenously) and/or substitution of alternate opioid or non-opioid analgesics for the treatment of postoperative pain. Undesirable side-effects of many antiemetic drugs include sedation (with the exception of ondansetron) and extrapyramidal side-effects or dysphoria with droperidol.

Altered mental status

Occasionally, patients emerging from general anesthesia exhibit unusual mental reactions, ranging from lethargy to confusion and/or aggressive physical behavior. These phenomena may occur quite suddenly, resulting in the potential for patient injury or disconnection from monitors, airways, and intravenous or arterial cannulas. While treatment is generally supportive, with symptoms usually resolving within 5 to 15 min from the time of onset, due care must be taken to protect and reassure the patient, while at the same time evaluating underlying causes. An adverse psychological response to emergence from general anesthesia is the most common cause of these reactions, but other factors, including premedication with long-acting sedatives, preoperative intoxication, or postoperative withdrawal, can also elicit bizarre emergence behavior. Anesthetic drugs such as ketamine can also cause postoperative dysphoria, although acute reactions are rare. When patient evaluation reveals that altered mental status is secondary to a physiological abnormality such as hypoxemia, acidemia, hypoglycemia, or hypotension, sedative or analgesic medications should not be administered until the underlying problem has been treated and the patient's mental status has been reassessed.

Table 5 Causes of postoperative hypo- and hypertension

Complication	Primary factor	Causes/examples
Hypotension	Spurious	Large blood pressure cuff Improper transducer calibration
	Hypovolemia	Absolute Deficits from operating room Ongoing 'third-space' losses Hemorrhage, coagulopathy Relative Autonomic block (spinal/epidural) Vasodilators (e.g. morphine, droperidol, nitrates) Raised intrathoracic pressure (pneumothorax, inferior vena caval compression)
	Ventricular dysfunction	May occur in patients with pre-existing left ventricular dysfunction
	Myocardial ischemia	Hypotension, tachycardia in patients at risk
	Cardiac dysrhythmias	Sinus or nodal bradycardia, complete heart block Paroxysmal atrial tachycardia, atrial fibrillation/flutter
	Lowered systemic vascular resistance	Regional anesthesia, α-adrenergic blocking drugs Vasoactive blood components, external body warming
Hypertension	Spurious	Small blood pressure cuff Improper transducer calibration
	Preoperative hypertension	Elevated renin–angiotensin activity High preoperative sympathetic nervous system activity
	Enhanced autonomic activity	Noxious stimuli (pain, anxiety, full bladder) Adverse physiological conditions (raised $Pa\text{CO}_2$, lowered $Pa\text{O}_2$, lowered glucose, lowered blood pressure, raised intracranial pressure, myocardial ischemia)
	Medications	Beta-mimetics (ephedrine, epinephrine (adrenaline)) Inotropes (dopamine, dobutamine) Anesthetics (ketamine, desflurane) Bronchodilators (aminophylline, terbutaline)

Table 6 Postoperative respiratory complications

Complication	Primary factor	Causes/examples
Inadequate ventilation	Lowered respiratory drive	Residual effects of inhalational and/or IV anesthetics Postoperative opioids and/or hypnotics CO_2 retainer, morbid obesity
	Increased airways resistance	Obstruction of the upper airway Bronchospasm
	Lowered compliance	Obesity, restrictive lung disease, atelectasis
	Neuromuscular problems	Incomplete reversal of muscle relaxants, thoracic spinal or epidural blockade
	Increased dead-space	Excessive PEEP or CPAP, air embolization, thrombus
	Increased CO_2 production	Shivering, increased work of breathing, infection, sympathetic nervous system hyperactivity
Inadequate oxygenation	Altered distribution of ventilation	Ventilation–perfusion mismatch, lowered FRC, obesity, pulmonary edema, pulmonary aspiration, gastric distension
	Altered distribution of perfusion	Raised pulmonary vascular resistance secondary to pain, hypoxemia, increased airway pressure, effects of anesthetics
	Inadequate $Pa\text{O}_2$	Severe hypoventilation, apnea, airway obstruction, laryngospasm, diffusion hypoxia
	Raised mixed venous $Pa\text{O}_2$	Decreased cardiac output, lowered blood pressure, shivering, hypovolemia

PEEP, positive end-expiratory pressure; CPAP, continuous positive airway pressure; FRC, functional residual capacity; IV, intravenous.

Hypothermia and shivering

Postoperative hypothermia is another relatively frequent occurrence, resulting from the combined effects of radiative and convective heat loss during anesthesia and surgery, in addition to the frequent use of cool irrigating and intravenous solutions. Intraoperative use of dry inspired gases, muscle relaxants which prevent shivering, and altered thermoregulation are additional causal factors. Postoperatively, hypothermia can result in hypoperfusion of peripheral tissues, resulting in local tissue hypoxia and metabolic acidemia which can jeopardize the viability of marginal tissue grafts. Other complications include exacerbation of coagulopathies, hypertension, and tachycardia.

Most patients suffer a loss of 2 to 3 °C in core body temperature but spontaneously rewarm during routine recovery. However, shivering thermogenesis can increase total body oxygen consumption by as much as fivefold. The intensity of postoperative shivering may be increased by inhalational anesthetic-related tremor which has both tonic and clonic components. Myocardial ischemia or ventilatory failure may supervene in patients with coronary artery disease and those with limited ventilatory reserve. Accordingly, patients arriving in the PACU with a core temperature below 35 °C should generally be actively rewarmed using such methods as warmed blankets, radiant heating, and reflective coverings. In addition, all hypothermic patients should receive supplemental oxygen. The resolution of metabolic acidosis usually accompanies rewarming, although bicarbonate may be required following prolonged hypothermia.

Cardiovascular complications

Cardiovascular complications during the postoperative period may be either primary or secondary, and present with different manifestations depending upon the patient's underlying cardiovascular status. While some cardiovascular complications are relatively transient and benign, others, such as ventricular dysrhythmias, are potentially life threatening. Accordingly, the PACU must be equipped with basic cardiovascular monitors for all patients, and with more advanced monitoring facilities (including on-line ST-segment monitoring and invasive pressure monitoring) for those at higher risk. Immediate availability of resuscitation drugs is essential. ST-segment monitoring in selected patients at risk of coronary artery disease or in whom it is documented is well justified, as recent evidence suggests that myocardial ischemia occurring in the first 24 h postoperatively is an independent risk factor for the development of serious cardiovascular complications including fatal and non-fatal myocardial infarction, congestive cardiac failure, and dysrhythmias (Goldman 1995). A description of contributing etiological factors for postoperative hypo- and hypertension is presented in Table 5.

Postoperative respiratory complications

Postoperative respiratory complications may be due to mechanical, hemodynamic, or pharmacological factors related to surgery and anesthesia. The preoperative respiratory status and ventilatory reserves of the patient also play a large role in postoperative outcomes. In order to simplify diagnosis of respiratory problems and direct appropriate therapy, it is useful to categorize respiratory complications as being due to abnormalities of ventilation, arterial oxygenation, or airway patency and protection problems (Mecca 1992). Examples are provided in Table 6.

References

Goldman, L. (1995). Cardiac risk in noncardiac surgery: an update. *Anesthesia and Analgesia*, **80**, 810–20.

Mecca, R.S. (1992). Postoperative recovery. In *Clinical anesthesia* (2nd edn) (ed. P.G Barash, B.F. Cullen, and R.K. Stoelting), pp. 1515–46. J.B. Lippincott, Philadelphia, PA.

Miller, D.R. (1996). Intravenous anaesthesia: new drugs, new concepts and clinical applications. *Canadian Journal of Anaesthesia*, **43**, R142–8.

Schnider, T.W. and Shafer, S.L. (1995). Evolving clinically useful predictors of recovery from intravenous anesthetics. *Anesthesiology*, **83**, 902–5.

Stanski, D.R. and Shafer, S.L. (1995). Quantifying anesthetic drug interaction. *Anesthesiology*, **83**, 1–5.

14.1.2 Postoperative pain relief

A. D. Baxter and D. R. Miller

Key messages

- Use of non-steroidal anti-inflammatory drugs may reduce opioid requirements.
- Opioids are the mainstay for analgesia, but they have side-effects of respiratory depression, urinary retention, and reduced gastrointestinal motility.
- Patient-controlled analgesia and epidural intraspinal therapy have advantages in pain control.

Introduction

The recent recognition of the importance of pain control has led to the development of acute pain services. Inadequate analgesia can lead to increased pulmonary and cardiovascular morbidity. A team approach to analgesia, with preoperative education to reduce anxiety and regular postoperative assessment using various treatment modalities, has ameliorated morbidity.

Analgesic modalities

Non-steroidal anti-inflammatory drugs

Neuronal activation following trauma occurs by direct nociceptor stimulation and the release of chemical mediators from the damaged cells. The latter activate the inflammatory response mediated by local production of prostaglandins, kinins, and other chemicals, resulting in sensitization of nociceptors at the site of injury and primary hyperalgesia. Non-steroidal anti-inflammatory drugs (NSAIDs) are a group of drugs classified as cyclo-oxygenase inhibitors, which inhibit the production of prostaglandins in response to tissue damage. Prostaglandins are involved in the production of secondary hyperalgesia around the injured area and NSAIDs also act here. The peripheral and central effects of NSAIDs result in less activation of nociceptors, less local inflammatory response, perhaps reduced stress response, and less pain.

The clinical properties of NSAIDs include mild to moderate analgesia with a ceiling effect with increasing dose, an anti-inflammatory action, an antipyretic action, reduction in renal blood flow and the glomerular filtration rate, gastric irritation, reduction in platelet adhesiveness, and rare idiosyncratic allergic reactions.

Although less potent than opioids as sole analgesics, many of these drugs reduce opioid requirements in the perioperative period when used as an adjunct to intramuscular, intravenous, or epidural intraspinal opioids, thus yielding improved analgesia, reduced opioid side-effects such as emesis, and faster return of gastrointestinal function (Souter *et al.* 1994). NSAIDs can be given orally, and some can be given rectally (naproxen, indomethacin) or by injection (ketorolac, tenoxicam, diclofenac). Concerns about renal, gastric, and platelet effects limit the use of these drugs for many critical care patients, although the safety of short-term perioperative use in patients having either minor or major surgery make them attractive. The advantage of preoperative loading rather than intra- or postoperative use is not well established and is probably small. Small increases in perioperative blood loss have been noted with preoperative use of NSAIDs (e.g. during total hip or knee arthroplasty), but initiation of therapy immediately postoperatively eliminates this concern. It is prudent to avoid their use in patients with pre-existing risk factors for bleeding and renal dysfunction, and in the elderly who may be more prone to side-effects. NSAIDs continue to be underutilized in the perioperative period, but are becoming more popular as a component of multimodal analgesia and in patients undergoing ambulatory surgery.

Opioid analgesics

Mu opioid receptor agonists continue to be the most effective analgesics in the perioperative period (Cousins 1994). Stimulation of these receptors produces analgesia and euphoria, but also respiratory depression, urinary retention, and reduced gastrointestinal motility; other side-effects include nausea, vomiting, and pruritus.

Morphine is metabolized in the liver to an active potent 6-glucuronide which is excreted by the kidney; this may accumulate in renal failure, causing sedation. Meperidine (pethidine) is shorter acting and is metabolized to normeperidine, which also accumulates in renal failure, causing seizures. Fentanyl in bolus doses is short acting, but with continuous infusion may accumulate with a more prolonged duration of action. The minimum effective blood concentration to produce analgesia may vary up to fivefold between patients.

For rapid and effective pain relief morphine and other opioids are frequently given by intravenous patient-controlled analgesic pump. Patients titrate the medication to their own individual effective analgesic concentrations, varying drug dosing according to analgesic requirements. Morphine in 1- to 2-mg increments or meperidine in 10- to 20-mg increments are adequate for most patients, while basal infusions are best reserved for ventilated or opioid-tolerant patients. In others, analgesia is not improved and the risks of respiratory depression are increased. In general, patient-controlled analgesic pumps provide improved analgesia and patient satisfaction when compared with traditional intermittent intravenous or intramuscular injections. They simplify delivery, record the amount of drug administered, and are less labor intensive for nurses.

Intramuscular injections need to be given regularly rather than 'as required', and close monitoring is necessary with appropriate dose adjustment. This route is also painful and absorption may be erratic (e.g. edema). Continuous intravenous infusions require an appropriate loading dose; otherwise several half-lives are required to reach steady state conditions. Careful monitoring and infusion rate adjustment are also required. Intravenous infusion is effective for the artificially ventilated patient when respiratory depression is not a concern, but during weaning from the ventilator titration of the dose to that required for analgesia is preferable, using either a patient-controlled analgesic pump or a nurse-controlled technique. Transdermal preparations of fentanyl are available; their analgesic and respiratory effects are equivalent to continuous fentanyl infusion. Although useful for chronic cancer pain, they have little or no role in the perioperative period. Iontophoretic modification of these preparations may make them more titratable and suitable for the perioperative patient.

Intraspinal opioid therapy has become increasingly popular in recent years, either using single-dose intrathecal injection during spinal anesthesia, or by repeated bolus or continuous infusion into the epidural space. This route of opioid administration produces selective analgesia with the drugs entering the spinal cord to act at opioid receptors in the substantia gelatinosa in the dorsal horn. A single injection of preservative-free intrathecal morphine may produce analgesia for 24 h or more, and dosage modification according to the site of surgery (e.g. 0.2–0.3 mg after Cesarean section, 0.3–0.4 mg after hip or knee replacement or lower abdominal surgery, 0.4–0.5 mg after upper abdominal surgery, and 0.6–0.7 mg after repair of abdominal aortic aneurysm or thoracotomy) has been found to reduce the frequency of respiratory depression. Fentanyl and meperidine can also be used intrathecally, but the duration of analgesia is only a few hours. Meperidine has mild local anesthetic activity and may cause hypotension and motor system block when injected intrathecally.

Morphine is associated with delayed respiratory depression when used by bolus epidural injection, but continuous infusion of dilute solutions (0.03 mg/ml, 5–10 ml/h) has reduced the frequency of this problem. More lipophilic drugs like fentanyl are readily absorbed from the epidural space, and plasma levels similar to those reported with direct intravenous infusion are found. Although the quality of analgesia may be somewhat better with epidural infusion of fentanyl alone compared with intravenous infusion, the differences are of insufficient magnitude to justify the routine use of the epidural route. Meperidine has been studied less than fentanyl and morphine; it is of intermediate lipophilicity and can provide effective analgesia when infused at 20–30 mg/h.

The addition of a dilute local anesthetic solution (e.g. bupivacaine 1–1.25 mg/ml, 5–10 ml/h) may facilitate reduction of the opioid dose while providing better control of pain with movement and coughing. Occasionally inadequate analgesia may require use of a higher concentration such as 2.5 mg/ml. Tachyphylaxis is rarely a problem with opioid–local anesthetic mixtures, unlike epidural local anesthetics used alone. Hypotension and muscle weakness are problems with the more concentrated solutions, but are uncommon with dilute solutions. Hypotension can be treated with intravenous volume expansion or vasoconstrictors such as phenylephrine or ephedrine.

Segmental analgesia can be obtained with local anesthetics and lipophilic opioids, with the epidural catheter optimally placed in the center of the dermatomal distribution of the incision to minimize dose requirements and side-effects. This is less relevant when using morphine alone, as its greater hydrophilicity results in wider spread within the subarachnoid space. Continuous infusion produces smoother analgesia and less hypotension than repeated bolus injection with local anesthetics and opioids.

Respiratory depression, pruritus, nausea, vomiting, and urinary retention are seen with opioids regardless of the route of administration. Side-effects can be treated with opioid antagonists such as naloxone, while partial agonist–antagonists like nalbuphine may treat the opioid side-effects with less reversal of analgesia. With appropriate patient selection and dosing, clinically significant respiratory depression occurs in less than 1 per cent of patients with all routes of opioid therapy, and its onset is usually preceded by increasing sedation. With the use of small doses of intraspinal opioids, appropriate training of nursing personnel, and frequent monitoring of the level of sedation and respiratory rate, many institutions now manage such patients on routine surgical wards.

Other regional techniques

Peripheral nerve blocks with long-lasting local anesthetics such as bupivacaine can produce postoperative analgesia lasting for several hours in the early postoperative period. Applications include ankle block for foot surgery, ilio-inguinal block for inguinal hernia repair, femoral nerve block for knee surgery, and penile block for circumcision. The patient must be given appropriate instructions to protect the anesthetized body part while it is insensitive and should be warned to inspect the field for bleeding and color change.

After major surgery, catheter techniques may be used to extend the duration of analgesia. Epidural local anesthetic infusions were the earliest example of these, but have problems with hypotension, motor blockade, and tachyphylaxis; these problems are largely avoided by using opioid–local anesthetic mixtures. Following thoracic or upper abdominal surgery, intercostal or paravertebral blocks provide useful analgesia but are limited by the duration of action of the local anesthetic. Intrapleural infusion of bupivacaine has been shown to have an opioid-sparing effect after unilateral flank surgery (e.g. subcostal incision for open cholecystectomy). However, it is less efficacious for analgesia after thoracotomy, where better analgesia may be achieved by infusion of the anesthetic into the chest wall in the extrapleural paravertebral space. Similarly, continuous anesthesia to the brachial or lumbar plexus can provide analgesia after major upper or lower limb surgery or optimize peripheral blood flow by inducing a sympathectomy (e.g. after microvascular digital reimplantation).

Pre-emptive analgesia

Activation of nociceptive pathways leads to peripheral and central sensitization and to hyperalgesia. Peripheral sensitization is mediated by prostaglandins and other inflammatory mediators, while central sensitization involves activation of *N*-methyl-D-aspartate (NMDA) receptors, resulting in 'wind-up' (lowering of thresholds for subsequent noxious stimuli and expansion of neuronal receptive fields). Experimental studies suggest that this sensitization can be prevented or reduced by treatment before the stimulus with antiprostaglandins, intravenous opioids, intraspinal opioids or local anesthetics, or peripheral nerve block. The benefits often last longer than the duration of action of a local anesthetic (Woolf and Chong 1993).

Multimodal analgesia

Advances in our understanding of the nociceptive pathways have led to recent attempts to interrupt these pathways at several sites, rather than relying on a single modality (Fig. 1). Thus peripheral sensitization is reduced by perioperative and/or postoperative use of

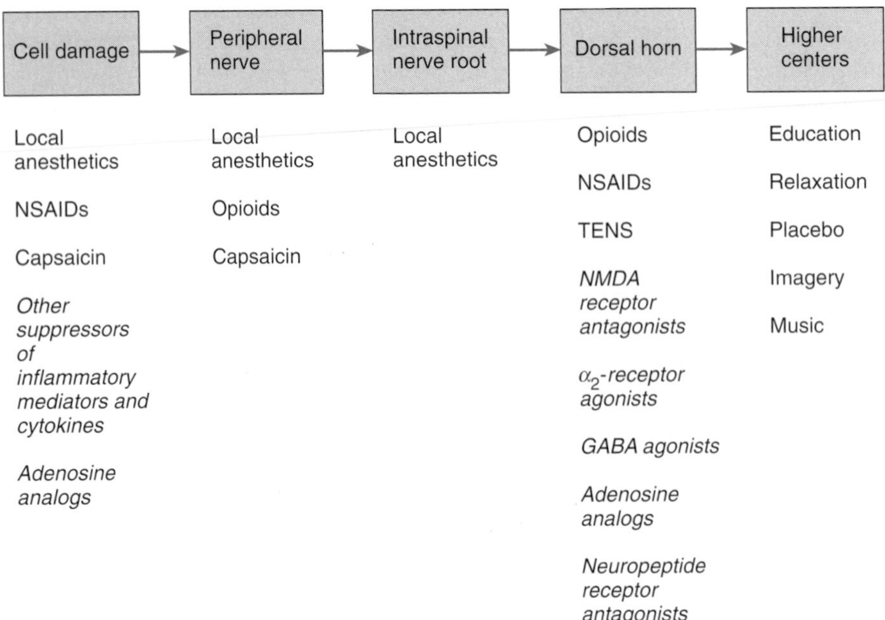

Fig. 1 Present and potential future (italics) modalities for interruption of the nociceptive pathway: TENS, transcutaneous electrical nerve stimulation.

NSAIDs, and central sensitization is reduced by intraoperative infiltration of local anesthetic by the surgeon or regional anesthesia by the anesthetist, resulting in better analgesia and fewer intravenous or epidural opioid requirements and side-effects postoperatively (Kehlet and Dahl 1993). Potential future components of the multimodal approach include the use of drugs for suppression of production of cytokines and other inflammatory mediators at the site of the stimulus, α_2-agonists for intraspinal use, and NMDA antagonists for prevention of 'wind-up'. The use of local anesthetic and opioid mixtures during epidural analgesia is another example of this approach, providing better control of pain with movement than opioids alone, and NSAIDs may also be used to augment intraspinal techniques. Multimodal analgesia may be effective both for patients undergoing major surgery and in the ambulatory care setting, proving improvement in analgesia and reduction of side-effects.

Outcome studies

While early studies claimed a reduction in morbidity with patient-controlled analgesic pumps, more recent studies with better controls show that this advantage is relatively small. Ambulation may be improved, but there is no effect on the stress response. However, epidural analgesia, particularly with opioids and local anesthetics, reduces the hormonal response to stress, oxygen consumption, and cardiac work, especially in high-risk patients (Liu *et al.* 1995). Thoracic epidural analgesia is required to blunt the stress response for abdominal surgery and must be maintained both intra- and postoperatively. This leads to benefits with reduction of cardiac and pulmonary morbidity, a lower incidence of deep vein thromboses, earlier return of gastrointestinal function, and improved graft patency after vascular surgery.

References

Cousins, M.J. (1994). Acute and postoperative pain. In *Textbook of pain* (3rd edn) (ed. P.D. Wall and R. Melzack), pp. 357–85. Churchill Livingstone, Edinburgh.

Kehlet, H. and Dahl, J.B. (1993). The value of 'multimodal' or 'balanced analgesia' in postoperative pain treatment. *Anesthesia and Analgesia*, **77**, 1048–56.

Liu, S., Carpenter, R.L., and Neal, J.M. (1995). Epidural anesthesia and analgesia: their role in postoperative outcome. *Anesthesiology*, **82**, 1474–1506.

Souter, A.J., Fredman, B., and White, P.F. (1994). Controversies in the perioperative use of nonsteroidal antiinflammatory drugs. *Anesthesia and Analgesia*, **79**, 1178–90.

Woolf, C.J. and Chong, M. (1993). Pre-emptive analgesia — treating postoperative pain by preventing the establishment of central sensitization. *Anesthesia and Analgesia*, **77**, 362–79.

14.1.3 Ventilatory function in the postoperative period

J. T. Cooke and P. S. Barie

Key messages

- The vital capacity and functional residual capacity can fall by 50 to 70 per cent during the first 24 h after surgery. These changes occur shortly after induction of anesthesia and can last 7 to 14 days.
- Cough and mucociliary transport are the most important mechanisms for protecting the lung from environmental and infectious agents. Mucociliary clearance is reduced for 2 to 6 days after general anesthesia.
- Management of postoperative pulmonary dysfunction starts before surgery. Risk factors for pulmonary complications include upper abdominal or thoracic surgery, general debility, cigarette smoking, and hypercapnea. Deep breathing exercises and incentive spirometry

significantly reduce the length of stay and pulmonary complications. These exercises should be implemented prior to surgery when possible and continued for at least 3 to 4 days depending on the patient's progress.

Introduction

Anesthesia, the anatomical location of the surgical procedure, and pre-existing conditions impact upon postoperative ventilatory function. These factors alter respiratory muscle function, disrupt lung mechanics, promote atelectasis, and predispose to postoperative pulmonary complications. An understanding of these processes requires a knowledge of lung mechanics and how they are altered by surgical procedures and anesthesia, often resulting in postoperative pulmonary complications.

Lung volumes

Lung volumes are classically described as composed of four components: tidal volume, inspiratory reserve volume, expiratory reserve volume, and residual volume (Fig. 1). Combinations of these volumes form four capacities. All the lung volumes taken together constitute the total lung capacity. The inspiratory reserve volume, tidal volume, and expiratory reserve volume form the vital capacity. The expiratory reserve volume and the residual volume form the functional residual capacity. The functional residual capacity is the volume of gas in the lungs at the end of normal exhalation. It represents a balance between the elastic nature of the lungs to collapse inwards and of the chest wall to spring outwards. The closing capacity approximates the residual volume in normal individuals, representing the volume at which small airways close, inhibiting further flow of gas. The relationship between the closing capacity and the functional residual capacity forms the basis for the pathophysiological ventilatory changes found in the postoperative setting. Normally the functional residual capacity is always greater than the closing capacity. During a tidal breath the small airways remain open throughout inspiration and expiration. If the closing capacity becomes greater than the functional residual capacity, a tidal breath may open small airways only partially or not at all. This results in areas of atelectasis and mismatched ventilation and perfusion (low ventilation-to-perfusion ratio) (Fig. 1).

Thoracic and abdominal surgery produces a restrictive ventilatory defect. All the lung volumes are reduced. The vital capacity and functional residual capacity can be reduced as much as 50 to 70 per cent during the first 24 h after surgery. These changes occur shortly after anesthetic induction and can last as long as 7 to 14 days postoperatively. Pain and splinting can contribute to the reduced lung volumes; although effective pain control with epidural analgesia and nerve blocks may not restore lung volumes to presurgery levels, these can help. General anesthesia induces a number of changes that reduce lung volumes. These include relaxation of the diaphragm, chest wall relaxation, loss of respiratory compliance, and shift of extracellular fluid volume from the chest into the abdomen.

Normal lung volumes and capacities

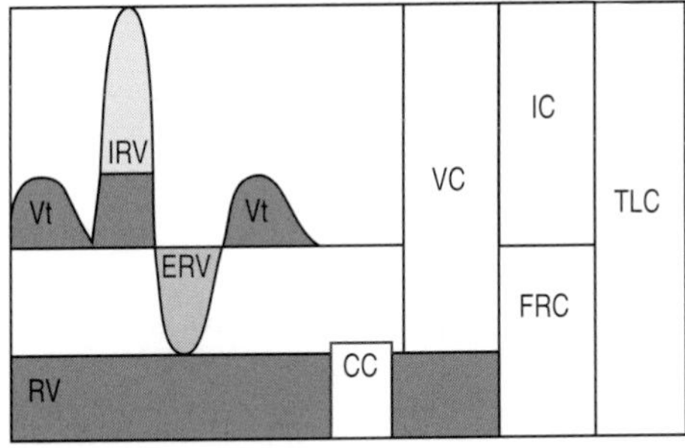

Restrictive lung volumes after surgery

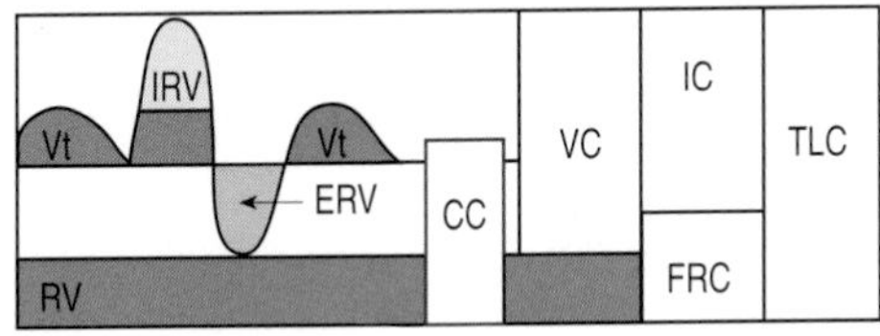

Fig. 1 Normal lung volumes are composed of the tidal volume (Vt), the inspiratory reserve volume (IRV), the expiratory reserve volume (ERV), and the residual volume (RV). The capacities are the vital capacity (VC), the inspiratory capacity (IRC), and the total lung capacity (TLC). The closing capacity (CC) is the volume of gas in the lungs at which the small airways begin to close. Normally it is very close to the residual volume. Surgery produces a restrictive ventilatory defect, reducing all the lung volumes, with the functional residual capacity and the total capacity sometimes becoming less than the closing capacity. As a result, portions of the lung are not ventilated because the airways remain closed throughout inspiration and expiration. The gas in unventilated alveoli is absorbed and atelectasis occurs.

Diaphragm function

Dysfunction of the diaphragm contributes to the postoperative reduction of lung volumes. Studies have shown that, during induction of general anesthesia in supine patients, the diaphragm shifts cephalad, accounting for a loss of 340 to 750 ml of thoracic volume on each side. Intravenous anesthetics depress diaphragmatic tonic activity, and volatile anesthetics depress synaptic transmission affecting the intercostal muscles preferentially (although ketamine preserves respiratory muscle tone). Respiratory muscle tone loss results in perioperative atelectasis. Ninety per cent of anesthesia-induced atelectasis is present 1 h after surgery. Fifty per cent of the anesthesia-induced atelectasis is still present 24 h after surgery. In the aftermath of open cholecystectomy through a subcostal incision, the contribution of the diaphragm to tidal breathing is markedly reduced. Transdiaphragmatic pressure measurements during phrenic nerve stimulation indicate that reduced central nervous system efferent signals to the phrenic nerve are the cause of depressed diaphragm activity. Animal studies suggest that phrenic nerve activity is decreased through the transmission of inhibitory signals from somatic or visceral afferents. Epidural anesthesia ablates the inhibitory signals from the visceral sympathetic receptors; phrenic nerve transmission increases and diaphragm function improves. In contrast, the administration of epidural narcotics ameliorates pain but does not affect phrenic diaphragm function, although intrathecal opioids depress ventilation centrally in a dose-dependent fashion. Pain does not seem to account for the diaphragm dysfunction.

In cardiac and thoracic surgery, late postoperative atelectasis is not explained solely by diaphragmatic dysfunction. Phrenic nerve paralysis is found in fewer than 10 per cent of patients undergoing cardiothoracic surgery. Variables associated with severe atelectasis after cardiac surgery include the number of saphenous vein grafts, the use of internal mammary grafts, the length of cardiac bypass time, and entrance to the pleural space. Mechanisms relate to the low temperature and inadequate blood flow to the alveolar epithelium, resulting in decreased surfactant production or function. Routine thoracic surgery does not involve bypass or cardioplegia, but reduced surfactant activity can occur through the mechanical compression of the lung and the accumulation of lung water.

Gas exchange

Postoperative hypoxemia is common. Initially it is related to the residual effects of general anesthesia. Hypoxemia may last days to weeks after thoracic or upper abdominal surgery. This correlates with a reduced functional residual capacity and an increased closing capacity. Induction of general anesthesia is associated with an increase of dead-space (ventilation of non-perfused lung), shunt (perfusion of non-ventilated lung), and inhibition of hypoxic pulmonary vasoconstriction. There is more ventilation to non-dependent areas of the lung, leading to areas of high ventilation-to-perfusion ratio. In addition to the redistribution of ventilation, the anesthesia circuit can result in a dead-space of up to 50 per cent of the tidal volume. If total ventilation is not increased, alveolar hypoventilation results in hypoxemia as well as hypercapnia. Observation of hypoxemia and hypercapnia after surgery may be due to hypoventilation caused by painful excursion of an incision, narcotic effects upon ventilatory drive, or upper airway obstruction.

Intrapulmonary shunt increases from 5 to about 8 per cent during anesthesia. The increase is greater in patients who are obese or who have pre-existing pulmonary conditions. The mechanism is primarily that of loss of normal respiratory muscle support of the lung and subsequent atelectasis. An additional mechanism may be the administration of 100 per cent oxygen. If the oxygen is entirely absorbed, the air spaces will collapse without the presence of the slowly diffusing nitrogen (absorption atelectasis). Positive end-expiratory airway pressure (PEEP) intraoperatively may not improve hypoxemia. Even though atelectasis is reduced, the shunt fraction does not improve. It is believed that this is the result of reduced cardiac output as PEEP is applied.

Hypoxic vasoconstriction is a point of some controversy. It is the normal constriction of the pulmonary circulation in response to hypoxemia. Pulmonary circulation is shifted from poorly ventilated areas to well-ventilated areas, attempting to preserve normal ventilation–perfusion relationships. Intravenous anesthetics do not affect hypoxic vasoconstriction. Inhalation anesthetics induce hypoxic vasoconstriction inhibition when tested on tissue preparations. However, when they are tested on intact preparations, including the human lung, there is little effect upon hypoxic vasoconstriction.

Postoperatively there is a linear relationship between atelectasis and hypoxemia. After upper abdominal and thoracic surgery the patient takes rapid shallow breaths. The small tidal volumes do not allow the opening of small airways that have closed due to the increased closing capacity and reduced functional residual capacity. Both atelectasis and low ventilation-to-perfusion relationships are promoted, resulting in arterial hypoxemia. Alveolar hypoventilation can occur as the result of increased dead-space ventilation, and decreased mixed venous oxygenation can be found as the result of increased peripheral muscle activity and reduced cardiac output.

Control of ventilation

Inhaled and intravenous anesthetics and narcotic analgesics depress the medullary respiratory center of the brain. With anesthetic induction, minute ventilation is reduced and apnea can occur as the ventilatory response to carbon dioxide is blunted or obliterated entirely. Without ventilatory support the $P\text{aCO}_2$ of the apneic patient will rise by 2 to 3 mmHg/min, causing respiratory acidosis. Anesthetic and narcotic effects on the carotid bodies depress the ventilatory response to hypoxia and the peripheral chemoreceptor response to acidemia. The normal synergistic elevation of minute ventilation in response to hypercapnia and hypoxia is attenuated. Importantly, this phenomenon has been observed even at subanesthetic/analgesic levels of drug administration. Immediately after surgery, a comfortable resting patient can be significantly hypercapnic and hypoxemic. Narcotics, in addition to depressing the normal response to hypercapnia and hypoxia, can change the overall breathing pattern. Sighs are reduced or eliminated, the depth of the breath is shallow, cough is suppressed, and lung volumes may decrease as secretions accumulate.

Defense mechanisms

Cough and mucociliary transport are the most important mechanisms protecting the lung from environmental and infectious agents. Their impairment in the postoperative setting increases the risk of respiratory infection. Coughing is inhibited because of pain and the use of narcotic analgesics. Respiratory muscle dysfunction reduces the expulsive force and the effectiveness of coughing. Mucociliary clearance is reduced for 2 to 6 days after general anesthesia as a result of ciliary damage from dry anesthetic gases, increased mucus viscosity, and reduced clearance from areas of atelectasis.

Upper abdominal surgery

After upper abdominal incisions the patient shifts to a style of breathing with little abdominal volume change and increased rib-cage excursion. The accessory muscles of ventilation assume an increased role in ventilating the postoperative patient. The contraction of the abdominal muscles is also prominent in expiration. The importance of this activity in generating tidal volume is not known. The shift from diaphragm to intercostal muscles is accompanied by redistribution of ventilation and the delivery of less inspiratory gas to the lower lobes. Abdominal muscle activity during expiration decreases the functional residual capacity. Closing volume occurs during a tidal breath, with the net result being atelectasis.

Cardiac surgery

The left hemidiaphragm is often elevated after cardiac surgery. In one study it was found that, if ice slush was used for topical cooling of the heart, the hemidiaphragm was elevated in 26 per cent of patients. If an internal mammary artery graft was used, the incidence increased to 39 per cent. In addition to the risk factors mentioned above, the phrenic nerve may be injured directly by topical cooling. Phrenic nerve injury can also occur during mobilization of the internal mammary artery because of its anatomic proximity. Bilateral phrenic nerve injury can result in the need for prolonged ventilatory assistance and tracheostomy. Another potential mechanism of atelectasis unique to cardiac surgery arises if the cardioplegia solution enters the pulmonary circulation. It is postulated that the high potassium chloride concentration may damage pulmonary endothelial cells.

Recommendations to prevent or minimize atelectasis after cardiac surgery include the use of careful technique in mobilization of the internal mammary artery, use of a pericardial insulating pad, avoidance of entry into the pleural space, and recovery of as much of the cardioplegia solution as possible before it enters the pulmonary circulation.

Thoracotomy and sternotomy

Changes in respiratory function occurring after a thoracotomy may last as long as 3 weeks. In addition to loss of functional lung tissue, the chest wall compliance decreases by as much as 75 per cent. Marked increases in the work of breathing occur. Hypoventilation and respiratory acidosis can occur into the first postoperative day. Hypoxemia can last up to 10 days. The intraoperative collapsing of one lung to facilitate surgery is unique to thoracic surgery. Hypoxic vasoconstriction of the collapsed lung minimizes any increase in the shunt. The collapsed lung should be periodically inflated to prevent prolonged postoperative atelectasis.

Sternotomy results in less postoperative pain and discomfort than a thoracotomy. However, there is no appreciable difference in immediate postoperative pulmonary function. There is no difference in the vital capacity and peak flow of patients undergoing sternotomy compared with those undergoing lateral thoracotomy. After 4 to 7 days the sternotomy patient recovers more rapidly towards preoperative pulmonary function than the lateral thoracotomy patient. Sternotomy for a pulmonary resection may be advantageous to patients with severely impaired preoperative pulmonary function but it is technically more challenging.

During a pneumonectomy, the mediastinum will shift and the remaining lung will distend to fill the available space. Gas exchange efficiency decreases. Early attempts at limiting distension with thoracoplasty resulted in unwanted scoliosis on the operated side and worsened long-term breathing capacity. This problem was solved by using plombage or sponges to fill the space and limiting mediastinal shift, until it was realized that the remaining lung can fill the hemithorax with few complications if the mediastinum is kept in place (neutral position) until fibrosis occurs.

Distension of the remaining lung, movement of the diaphragm, and the production of connective tissue to fill the remaining thoracic space result in reduction of the size of the thoracic cavity. On the operated side, the diaphragm is elevated and the interspaces between the ribs are smaller. As the remaining lung distends to fill the space, only compensatory dilatation occurs.

Interventions

Management of postoperative pulmonary dysfunction starts before surgery. Risk factors for pulmonary complications are identified. These include the anatomical site of surgery, general illness or debility, the presence of chronic obstructive pulmonary disease (COPD), obesity (over 120 kg), cigarette smoking, and hypercapnia. Lung expansion techniques reduce pulmonary complications. Ideally, these are started prior to surgery and then continued through the postoperative period. The goal is to promote alveolar inflation and normalize the functional residual capacity. Expiratory maneuvers reduce lung volumes and promote atelectasis. Deep cough at the end of a maximum inspiration, although an expiratory maneuver, clears inspissated secretions. Inspiratory maneuvers such as deep breathing, intermittent positive-pressure breathing (IPPB), and continuous positive airway pressure (CPAP) promote lung inflation.

Deep breathing to total lung capacity, with emphasis on using the diaphragm, inflates the lung and reverses postoperative hypoxemia. In randomized studies using deep breathing with and without the assistance of an incentive spirometer, there were no significant differences in postoperative complications or length of stay. There was a significant improvement in length of stay and pulmonary complications compared with those patients who did not use either form of respiratory exercise; thus deep breathing exercises or incentive spirometry should be implemented prior to surgery and continued afterwards, although the optimum schedule after surgery has not been developed. Recommendations include performing the exercises as frequently as possible, and maintaining the therapy for at least 3 to 4 days depending on the patient's progress.

IPPB received much early attention with studies showing that it reduced the incidence of complications over controls. However, it was no more effective than deep breathing and incentive spirometry. IPPB requires more equipment and is more expensive than other techniques. Its use can be associated with abdominal distention and discomfort. Substantial insufflation of the stomach, duodenum, and proximal small intestine may be hazardous for hemostasis or anastomotic integrity after splenectomy or upper gastointestinal surgery. At present IPPB is best utilized in those patients where active lung inflation is not possible even with patient co-operation. This would include those patients with kyphoscoliosis, muscular dystrophy, and multiple sclerosis.

Non-invasive ventilatory techniques using mask-delivered CPAP of 7.5 to 15 cmH_2O may promote faster restoration of preoperative pulmonary function compared with incentive spirometry. CPAP requires the use of a tight-fitting nasal or facial mask and the equipment is expensive. Benefit may be achieved in the uncooperative patient. Application of CPAP does not require any effort on the patient's part other than keeping the mask on, which can be a challenge with discomfort due to a poor fit. Aerophagia and gastric distension may occur, leading to aspiration in some circumstances. Where aspiration is a concern, a nasogastric tube should be placed.

The role of bronchoscopy in the management of atelectasis is not fully defined. As a general approach, conservative measures (incentive spirometry, chest physiotherapy) should be tried first. Bronchoscopy should be considered in patients who do not respond or who have inspissated secretions. For complete atelectasis of a lung, therapeutic bronchoscopy can relieve the airway obstruction quickly. In refractory atelectasis, the bronchoscope has been used to insufflate air into the atelectatic segment. The air pressure is maintained for several minutes. Several small studies have demonstrated successful reversal of refractory subsegmental atelectasis using this technique. Bronchoscopy, bronchoalveolar lavage, and protected specimen brushing have been the subjects of multiple studies looking at improving diagnostic accuracy in ventilator-associated pneumonia. At present, the results are inconclusive.

Doxapram is an intravenous respiratory stimulant. In a double-blind randomized control study, Pao_2 was higher and postoperative pulmonary complications were fewer in doxapram-treated patients. However, there was no difference in length of hospital stay. COPD patients with nutritional depletion have improved their maximal inspiratory pressure by 27 per cent when given human growth hormone in addition to nutrition, but other prospective studies have not shown any benefit.

Special situations

Acute upper airway obstruction

Acute airway closure may occur upon extubation as the result of airway edema, laryngospasm, or failure of vocal cord adduction. If no air leak can be detected on deflation of the endotracheal tube cuff or the patient has sustained airway or neck trauma during or before intubation, this may be seen. Intraoperative injury to the recurrent laryngeal nerve or mechanical compression from a cervical hematoma may also be responsible.

Nebulized racemic epinephrine (adrenaline) induces vasoconstriction and decreases mucosal edema, and is useful for non-operative obstruction. Corticosteroids also may reduce laryngeal edema, although there is no role for the routine use of steroids to prevent postextubation laryngeal edema.

A helium–oxygen mixture can be used as a temporizing measure in the management of upper airway obstruction. A mixture of 80 per cent helium and 20 per cent oxygen has a density one-third that of air. The reduced density decreases airway resistance and flow resistive work, providing time for definitive therapy. In cases of rapid respiratory failure, surgical control of the airway is required. Cricothyroidotomy may be performed under emergent conditions; it is quicker than a tracheostomy, can be carried out by non-surgical personnel, and has a lower general complication rate than emergent tracheostomy.

Anaphylaxis

Anaphylactic reactions may occur in the perioperative setting of administering antibiotics, blood products, intravenous contrast media, or almost any pharmaceutical. The reaction is mediated by the release of inflammatory agents from mast cells or basophils. If the reaction is IgE mediated, it is an anaphylactic response with membrane-bound IgE binding the offending agent. Mast cells and basophils release histamine and platelet activating factor, causing bronchospasm, bronchorrhea, laryngeal edema, and increased vascular permeability. Symptoms may occur within seconds to minutes, or they may be delayed for as long as an hour. Anaphylactoid reactions are mediated without IgE. Mechanisms include complement reactions, non-immunological activation of mast cells, and production of arachadonic acid mediators. Antibiotic reactions are generally IgE mediated, whereas radiocontrast reactions cause a non-immunological activation of the mast cells (Table 1).

Avoidance of the offending agents on the basis of a thoughtful history (i.e. after an initial radiocontrast reaction, there is a 35 per cent chance of a reaction upon re-exposure) eliminates this risk. Once a reaction occurs, the first priority is to secure the airway. The approach is similar to the treatment of laryngeal edema. Pharmacological intervention includes epinephrine (both inhaled and parenterally), antihistamines (diphenhydramine), and corticosteroids.

Table 1 Common agents causing anaphylactic and anaphylactoid reactions

Anaphylaxis	Anaphylactoid
Beta-lactams	Blood
Sulfonamides	Serum
Vancomycin	Plasma
Local anesthetics	Plasmanate
Gammaglobulin	Immunoglobulins
Heterologous serum	Narcotics
	Radiocontrast media
	Dextrans
	Non-steroidal anti-inflammatory drugs

COPD and asthma

It is generally believed that COPD increases the risk of postoperative pulmonary complications. There are few or no data estimating the surgical risk in the asthmatic patient. Ideally, medical management is maximized prior to surgery. The patient should be free of wheezes and active infection. In one study, a forced vital capacity in 1 s (FEV_1) of less than 65 per cent of the predicted value and a forced vital capacity of less than 70 per cent of the predicted value resulted in a 100 per cent complication rate. However, there are also studies where patients with severe air-flow obstruction did not suffer any pulmonary complications. Other than in pulmonary resection, there is no level of pulmonary function that is an absolute contraindication to surgery. The preoperative test with the most utility in predicting postoperative complications is a reduced arterial PO_2. In the case of a pulmonary resection, an FEV_1 below 800 ml contraindicates surgery.

COPD and asthma pose special ventilatory problems in the perioperative setting. Air-flow obstruction, inflammation, and increased secretions during acute episodes cause severe hypoxemia and hypercapnia as a result of ventilation–perfusion mismatch, atelectasis, increased work of breathing, and dysfunctional respiratory mechanics. Therapy is directed at correcting hypoxemia, reversing air-flow obstruction, treating infection, reducing inflammation, and in some instances providing ventilatory support. This is accomplished with chest physiotherapy, antibiotics, bronchodilators, corticosteroids, and both invasive and non-invasive mechanical ventilation.

Symptomatic COPD patients should be treated aggressively in the perioperative setting with antibiotics, bronchodilators, and corticosteroids. Bronchodilators and corticosteroids will decrease airway resistance. The reversibility of the air-flow obstruction will be variable, but every small advantage helps. Bronchodilators available include β-agonists, anticholinergics, and theophylline preparations. The β-agonists and the anticholinergics demonstrate greater bronchodilator action than theophylline. Studies of COPD patients suggest that anticholinergics have a slight advantage in both a stable setting and acute exacerbation. Maximal bronchodilation can be achieved with either drug in adequate doses. Adding the other agent does not improve bronchodilation further.

Beta-agonists available include metaproterenol (orciprenaline), albuterol (salbutamol), isoetheraine, pirbuterol, and bitolterol. They can be administered effectively via a nebulizer or metered dose inhaler. In the co-operative patient the metered dose inhaler with a spacer is the most cost-effective method of administering these medications. The spacer traps large particles that would deposit in the mouth and be absorbed, potentially increasing systemic side-effects. These agents can be dosed every 1 to 2 h if needed to achieve a beneficial effect.

Ipratropium is the quarternary salt of atropine. Unlike atropine, it does not readily cross biological membranes. There are few systemic side-effects, and it has a longer duration of action. As with the β-agonists, it can be administered via a nebulizer or metered dose inhaler. During acute exacerbations there can be a benefit to increasing the dosing frequency from every 6 h to every 2 to 4 h. In contrast with COPD, ipratropium is not considered a first line-drug in the therapy of asthma.

The role of theophylline in the management of COPD and asthma has come under scrutiny. In double-blind studies comparing its use with a placebo in patients already receiving β-agonists and corticosteroids, there was no significant improvement in dyspnea, spirometry, or arterial blood gases. More gastrointestinal side-effects occurred in the theophylline groups. Theophylline has a limited role as a bronchodilator in acute exacerbations of asthma and COPD. The beneficial effects seen in theophylline-treated patients are not the result of bronchodilation, but are due to its action as a respiratory stimulant, as a possible inotropic agent for the respiratory muscles, and as a mild diuretic. Dosing of theophylline should be adjusted according to blood levels. Its metabolism is variable, and it interacts with many commonly used drugs.

Corticosteroids are a cornerstone in the treatment of COPD exacerbations. The use of steroids is based on a single randomized study where patients received either methylprednisolone 0.5 mg/kg or placebo every 6 h for 3 days, in addition to standard therapy. The corticosteroid group demonstrated an improvement in FEV_1 after the first 12 h. Of the 40 corticosteroid-treated patients, 22 demonstrated a 40 per cent improvement in FEV_1. There were no differences in the arterial blood gases or the eventual outcome.

Adequate hydration is advocated empirically in both the asthmatic and the COPD patient. Respiratory secretions will be less viscous in the well-hydrated patient. The administration of fluid directly into the airway and the use of mucolytics have not been shown to have consistent benefit. Indeed, the mucolytic agent *N*-acetylcysteine is an airway irritant, and may exacerbate the situation. In the attempt to mobilize secretions and treat atelectasis, chest physiotherapy, deep breathing, and cough offer the greatest benefit.

Mechanical ventilation will be required at times to support the patient with respiratory failure and asthma. The purpose is to support the patient while the disease process runs its course or the pharmacological intervention takes effect. The potential to do harm is great. Intubation predisposes to aspiration and nosocomial pneumonia. Increased work of breathing and respiratory muscle fatigue will occur in the setting of a small endotracheal tube and too little ventilator support. Air-flow obstruction by edema, bronchospasm, and mucus promote air trapping, auto-PEEP, and the risk of barotrauma. Auto-PEEP will also predispose to hypotension and increased work of breathing in triggering a machine breath. Hyperventilation of the patient can result in severe alkalemia. This promotes cardiac irritability and electrolyte disturbances, reduces cerebral blood flow, and suppresses the ventilatory drive. Strategies to minimize these complications include ventilating with tidal volumes of 5 to 10 ml/kg, increasing the peak flow to shorten inspiratory time, and applying extrinsic PEEP of the order of 5 to 10 cmH_2O to offset the effects of auto-PEEP. Additional strategies using non-invasive ventilation (continuous or bi-level positive airways pressure), where support is provided by facial or nasal masks, are being studied.

Bibliography

Forster, R.E., Dubois, A.B., and Fisher, A.B. (1986). *The lung. Physiologic basis of pulmonary function tests* (3rd edn). Year Book, Chicago, IL.

Ingbar, D. (ed.) (1994). Respiratory emergencies II. *Clinics in Chest Medicine*, **15**(3).

Krell, W.S. and Prakash, U.B.S. (1994). Therapeutic bronchoscopy. In *Bronchoscopy* (ed. U.B.S. Prakash), p. 207. Raven Press, New York.

Olsen, G. (ed.) (1993). Perioperative respiratory care. *Clinics in Chest Medicine*, **14**(2).

Owens, G. (1994). Postoperative atelectasis, pneumonia, and aspiration pneumonitis. In *Perioperative medicine* (2nd edn) (ed. D.R. Goldman, F.H. Brown, and D. M. Guarneri), p. 565. McGraw-Hill, New York.

14.1.4 Postoperative fluid and circulatory management

R. M. Grounds and O. Boyd

Key messages

- During the perioperative period there are changes in organ metabolic demand and in nutrient and oxygen availability which can lead to an imbalance between supply and demand, and produce life-threatening organ failure.
- The aim of circulatory management during this perioperative period is to maintain the balance of nutrient and oxygen supply and demand.
- Assessment of each patient is based on history, examination, and investigations.
- Therapy should be directed primarily towards actual or impending cardiovascular and/or circulatory system impairment.
- Management is based on correction of deranged physiology, with particular reference to optimizing blood volume, oxygen-carrying capacity, cardiac output, and organ perfusion.
- Treatment is by continued monitoring of the cardiovascular system and organ function with the judicious use of fluids, blood products, and drugs to improve cardiac function and organ perfusion.

Introduction

During the perioperative period a number of changes occur which alter the normal physiological responses of the cardiorespiratory system. The most obvious of these is the surgery itself, but the physiological changes caused by positive-pressure ventilation, reduction in functional residual capacity, and reduced muscle tone all tend to reduce tissue perfusion. In addition, many drugs given during the perioperative period have the potential to reduce tissue perfusion, redistribute blood flow, and override normal compensatory mechanisms. Conversely, postoperative stress responses, the maintenance of normothermia, the effects of starvation and altered requirement for specific organ function, and the inflammatory response secondary to the trauma of surgery, followed by healing, may all have metabolic consequences requiring an increase in tissue perfusion. By monitoring and, if necessary, manipulating the circulation with intravenous fluid therapy, the goal will be to restore, optimize, and maintain the cardiovascular function in order to prevent tissue hypoperfusion and organ failure.

If the balance of the supply and demand is not maintained, hypoxic tissue damage may result. This is particularly important in those patients with pre-existing organ dysfunction where very mild changes in perfusion cause significant damage to the organ. Clinical trials have shown that if tissue perfusion is maintained during the perioperative period, mortality and morbidity may be reduced (Shoemaker *et al.* 1988; Boyd *et al.* 1993).

Patient assessment

Elucidation of the underlying dysfunction must depend upon the time-honored triad of history, examination, and special investigations.

Emphasis must be placed on determining the limit of physiological reserve, particularly of the cardiorespiratory system. The intention is to identify those patients who will benefit from further investigation and more complex perioperative management in the intensive care unit (ICU) (Shoemaker *et al.* 1988; Boyd *et al.* 1993). The history should also inquire into limitation of physiological reserve in other organ systems.

Examination should specifically attempt to determine the degree of physiological limitation that may have been noted from the history. In particular, physical signs of cardiac failure should be noted, as these have been shown to be good predictors of postoperative morbidity and mortality.

Emphasis must be on defining abnormalities of cardiorespiratory performance and organ perfusion. Selection of the appropriate tests and investigations (Table 1) should be individualized. Preoperative investigations are aimed at defining the extent and nature of any physiological incapacity and identifying potential avenues of treatment. It may not be possible to ameliorate long-standing medical conditions completely, but these should be treated wherever possible. When complete treatment cannot be effected in the time available, investigations should elucidate the degree of outstanding physiological deficit. The actual choice of investigation is difficult, and is a balance between the information accrued by the new investigation and the time taken to perform the test.

Once basic resuscitation has been performed, management should be directed towards optimizing the intravascular volume and cardiac performance to maintain tissue perfusion and tissue oxygen supply.

On the basis of history, examination, and preliminary investigations patients can be classified as those with myocardial impairment, those with intravascular volume depletion, or those with both. Management must be directed toward optimizing the intravascular fluid volume and consequently improving myocardial performance to ensure adequate tissue perfusion. For descriptive purposes

Table 1 Investigative studies which should be considered*

Laboratory tests

Essential
- Complete blood count
- Hematocrit
- Electrolytes
- ECG
- Arterial oxygen saturation
- Blood glucose
- Urea, creatinine

Additional preoperative
- Arterial blood gas determination
- Mixed venous oxygen saturation
- Oxygen delivery and oxygen consumption

Additional postoperative
- Arterial blood gas determination
- Mixed venous oxygen saturation
- Serum lactate
- Oxygen delivery and oxygen consumption

Radiology

Essential
- Chest radiography

Additional preoperative
- Ultrasound of abdomen
- Echocardiogram
- CT scan
- MRI scan
- Thallium scan
- Pulmonary perfusion scan

Additional postoperative
- CT scan
- MRI scan
- Thallium scan
- Echocardiogram
- Pulmonary perfusion scan
- Ultrasound of abdomen

Monitoring

Non-invasive
- Continuous ECG
- Pulse oximetry
- Non-invasive blood pressure
- Esophageal Doppler
- Temperature
- Urine output
- Ventilation indices

Invasive
- Intra-arterial direct pressure monitor
- Central venous pressure monitor
- Pulmonary artery catheter
- Urinary catheter
- Esophageal Doppler
- Core temperature

Derived variables
- Stroke volume
- Systemic vascular resistance
- Pulmonary vascular resistance

MRI, magnetic resonance imaging.
*Not all tests will be required for all patients.

patient management will be discussed under three main headings: optimizing intravascular fluid volume, improving myocardial and circulatory performance, and improving organ perfusion. There is a large overlap between these three areas, and appropriate therapy should be directed to all of them simultaneously whenever possible.

Optimizing intravascular fluid volume

Control of cardiac filling pressure

The principle is to ascend the Frank–Starling curve for each individual by obtaining the optimal filling pressure associated with the best afterload (Fig. 1). It is vital to achieve the optimal filling pressure, thus ensuring an adequate blood volume which will provide complete perfusion for all organs and adequate cardiac filling. The simplest method of assessing the intravascular fluid volume in the perioperative period is by using a central venous pressure line and fluid challenge. With a central venous pressure of 6 to 8 mmHg, set to zero at the mid-axillary line, a fluid challenge (bolus doses of 100 or 200 ml of intravenous fluid) is given (and if necessary repeated) until the central venous pressure maintains a sustained rise in response to the challenge.

In sicker patients the pulmonary artery catheter allows a more sophisticated version of the fluid challenge to be performed. The patient is given intravenous fluid until the pulmonary artery occlusion pressure is adequate (10–14 mmHg). The patient is then given fluid challenges in bolus doses of 100- to 200-ml aliquots of intravenous fluid until the doses result in an increase in pulmonary artery pressure without a corresponding increase in cardiac output.

Intravenous fluid therapy

The aim of intravenous fluid therapy is to increase tissue oxygen delivery by restoring normal intravascular fluid volume with an appropriate hematocrit. This will also increase preload (Fig. 1), and hence cardiac output and oxygen delivery.

In some cases the choice of intravenous fluid is obvious. If the patient is bleeding and losing blood, the fluid replacement will be primarily blood. If the patient has a fistula or large burns, the initial fluid therapy should be matched to replace these losses. However, in many patients the cause of reduced intravascular volume is not immediately obvious. In these cases the intravascular volume should be restored and the hematocrit maintained.

There have been numerous attempts to establish whether intravenous crystalloid or artificial colloidal solutions are most appropriate for resuscitation in these patients and this issue is unresolved.

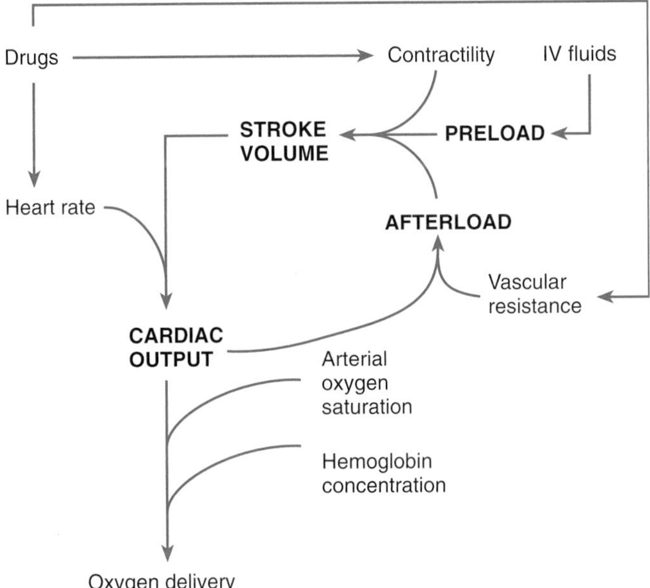

Fig. 1 Summary of the components of management and their effects on cardiac output.

Blood products

When transfusing blood, the improvement in oxygen-carrying capacity must always be balanced against the reduction in flow accompanying the increase in blood viscosity with increasing hemoglobin concentration, the risk of transfusion reaction, and the increasing risk of blood-borne infections.

There is no absolute agreement on the optimal hematocrit; however, the aim should be to maintain the optimum oxygen-carrying capacity while maintaining viscosity as low as possible. Guidelines have been provided, but in the special situation of the perioperative patient maximal oxygen-carrying capacity should be achieved with a hematocrit of between 30 and 40 per cent (Czer and Shoemaker 1978; Malmberg and Woodson 1979; Consensus Conference 1988).

Improving myocardial and circulation performance

It is possible to manipulate cardiovascular performance to enhance tissue oxygen delivery without changing intravascular volume, although this is usually done in tandem with ensuring optimal intravascular volume and oxygen-carrying capacity. When an adequate and appropriate intravascular volume has been ensured, drugs which enhance myocardial performance may be considered if tissue oxygen delivery is not satisfactory (Fig. 1).

Afterload reduction

Afterload reduction may decrease the work of the heart and actually increase cardiac output. However, it will also cause a reduction in blood pressure, and many organs rely on pressure as well as flow for their function and performance. Thus the effects of afterload reduction vary depending on individual circumstances. There is no consensus as to how low blood pressure can be allowed to fall before medical intervention should be initiated, or which drugs should be used to reduce afterload (Table 2). However, as a general guide the patient should not be allowed to have a prolonged episode where the systolic blood pressure is lower than the normal diastolic pressure.

Pump function

The force of contraction of the cardiac muscle depends on a number of factors. Myocardial performance can be affected by intrinsic and extrinsic factors, including catecholamines, inotropes, shortening of refractory period, calcium ions, digitalis, and atrial filling. The principle of therapy should be directed towards treating arrhythmias and

Table 2 Choice of drugs for afterload reduction

Agent	Dose	Arterial dilatation	Venous dilatation
Nitroprusside	0.5–10 μg/kg/min	+++	+++
Nitroglycerin	0.5–5 μg/kg/min	++	+++
Hydralazine	0.5–5 mg every 8 h	++	++
Prazosin	0.5–5 mg every 8 h	++	++
Nifedipine	10 mg orally	++	+
Minoxidil	200 μg–1 mg daily	+++	+
Captopril	6.25–25 mg every 8 h	++	+
Enalapril	2.5–40 mg daily	++	+
Dobutamine	1–10 μg/kg/min	0	++
Dopexamine	0.5–3 μg/kg/min	0	++
Epoprostenol (prostacyclin)	1–5 ng/kg/min	++	?
Phentolamine	1–4 μg/kg/min	+++	0
Trimethepan (trimetaphan)	25–100 μg/kg/min	0	+++
Nitric oxide	1–10 ppm inhaled	+++	++

controlling heart rate. The controlled use of inotropes and vasodilators can then be considered. Vasoactive drugs used in the ICU can have more than one action and should be used selectively (Table 3).

Heart rate

As heart rate increases, the time for ventricular filling in diastole falls and the time for myocardial muscle perfusion is reduced. Thus there is a balance between optimum heart rate and maximum cardiac output. Heart rate is a function of many variables, and cardiac output is the product of stroke volume and heart rate (Fig. 1). In certain circumstances the heart rate may affect the function of the other variables and lead to a reduction in cardiac output.

Arterial oxygen saturation

Supplemental oxygen should be used to maintain arterial oxygen saturation at 95 per cent or above. Oxygen saturation is more important than arterial partial pressure because of variations in the position of the oxygen dissociation curve. In the emergency situation the potential of CO_2 retention in chronic obstructive pulmonary disease has been overstated, since more patients are harmed by lack of sufficient oxygen than by CO_2 retention.

Improving organ perfusion

It is important to consider maintenance or enhancement of specific organ perfusion.

Blood flow and oxygen delivery to the brain are autoregulated. Below a certain systemic blood pressure (usually about 80 mmHg), brain blood flow is directly proportional to systemic blood pressure.

The heart is only perfused during diastole. Thus it is essential not to give agents that increase the heart rate so that it limits myocardial oxygen delivery. There are also specific coronary vasodilators, such as nitrates, which can be used to improve coronary artery blood flow.

If oliguria develops despite adequate filling and perfusion pressure, use of a loop diuretic such as furosemide (frusemide) and/or dopamine should be considered. The combined use of furosemide and dopamine in patients who are normotensive and have adequate intravascular fluid replacement has been shown to reduce acute renal failure and attenuate the severity of established acute renal failure (Lindner 1983).

Splanchnic perfusion may be improved by using specific dopamine receptor agonists such as dopamine or dopexamine. This may reduce

Table 3 Vasoactive drugs used in the ICU

Pressor agent	Dose (μg/kg/min)	Cardiac stimulation (β_1)	Vasodilatation (β_2)	Vasoconstriction	Dopaminergic
Epinephrine (adrenaline)	1–10	++++	++	+++	0
Norepinephrine (noradrenaline)	1–10	+++	0	++++	0
Isoprenaline	0.01–0.1	++++	++++	0	0
Dopamine	1–10	++	+	+	+++
	10–20	+++	+	+++	0
Dobutamine	1–10	+++	+++	+	0
Dopexamine	1–10	+	+++	0	+++
Phenylephrine	20–200	0?	0	++++	0
Metaraminol	1–10	+	0	+++	0
Amrinone	5–15	++	+	0	0
Milrinone	0.375–0.75	++	+	0	0

toxin load and bacterial translocation, which have been implicated in the development of multiple organ failure.

Conclusions

1. Treat the underlying disease process.
2. Maintain and maximize perfusion and oxygenation.
3. Begin basic resuscitation (airway, breathing, circulation) and then proceed to more complex manipulation of the cardiovascular system and intravascular volume:
   (a) maintain adequate oxygenation of the lungs;
   (b) establish venous access;
   (c) rapidily infuse colloid, crystalloid, and/or blood (if hemoglobin is less than 10 g/dl);
   (d) use central venous pressure or pulmonary artery pressure manometry;
   (e) use pulse oximetry;
   (f) once adequate filling has been achieved, consider the use of vasoactive drugs to achieve a specific therapeutic aim.

References

Boyd, O., Grounds, R.M., and Bennett, E.D. (1993). A randomized clinical trial of the effect of deliberate perioperative increase of oxygen delivery on mortality in high-risk surgical patients. *Journal of the American Medical Association*, **270**, 2699–707.

Consensus Conference (1988). Perioperative red blood cell transfusion. *Journal of the American Medical Association*, **260**, 2700–3.

Czer, L.S.C. and Shoemaker, W.C. (1978). Optimal hematocrit value in critically ill postoperative patients. *Surgery, Gynecology and Obstetrics*, **147**, 363–8.

Lindner, A. (1983). Synergism of dopamine and furosemide in diuretic-resistant, oliguric acute renal failure. *Nephron*, **33**, 121–8.

Malmberg, P.O. and Woodson, R D. (1979). Effect of anemia on oxygen transport in hemorrhagic shock. *Journal of Applied Physiology*, **47**, 882–8.

Shoemaker, W.C., Appel, P.L., Kram, H.B., Waxman, K., and Lee, T.-S. (1988). Prospective trial of supranormal values of survivors as therapeutic goals in high-risk surgical patients. *Chest*, **94**, 1176–86.

14.2 Management of specialized surgical patients

14.2.1 Orthopedic surgery

M. Fennelly and G. Edge

Key messages

- Patients with congenital neuromuscular diseases, in addition to the elderly, present many complex conditions perioperatively.
- Scoliosis, rheumatoid arthritis, and ankylosing spondylitis present unusual respiratory and cardiac conditions.
- Malignant hyperthermia and fat embolism syndrome may be seen in patients with bony injury or skeletal disease.

Introduction

Perioperative management of the orthopedic patient depends on the age group involved, the presence of concurrent systemic disease, and the planned surgical procedure. Certain congenital and/or neuromuscular–skeletal problems present in childhood, and these patients may require multiple surgical procedures as they grow older. Patients with major trauma may have multisystem involvement. Elderly patients commonly have intercurrent cardiovascular disease or other conditions, with the main cause of mortality following anesthesia and surgery being due to cardiac disease; outcome is not influenced by choice of either a regional or general anesthetic technique.

In many cases the need for postoperative intensive or high-dependency care can be anticipated and surgery planned accordingly. However, there are a number of patients in whom unexpected complications arise perioperatively, related to the nature of the surgery involved, with often massive blood loss, the use of large quantities of acrylic bone cement, and the occurrence of fat, air, and venous embolism.

Specific orthopedic conditions

Scoliosis

Patients with scoliosis sufficiently severe to warrant surgical intervention usually present in adolescence. These patients often have cardiopulmonary disease. The most common problems are a restrictive ventilatory pattern associated with altered lung volumes, ventilation–perfusion defects, and pulmonary vascular changes. Hypoxemia may occur, and significant ventilation–perfusion mismatch may lead to pulmonary hypertension. The incidence of cardiac anomalies is higher than normal in patients with scoliosis, with mitral valve prolapse occurring in up to 30 per cent. After corrective surgery patients should recover in a high-dependency or intensive care unit as ongoing neurological assessment is mandatory. Patients with known respiratory compromise may require a period of postoperative ventilation.

Allergy

There is a higher incidence of allergy in patients with musculoskeletal disorders, and latex allergy must be considered in patients with spina bifida who present for a wide variety of orthopedic procedures. A careful history is crucial.

Cement

Methyl methacrylate cement is a self-polymerizing acrylic compound used in securing joint prostheses. Insertion of cement into the reamed bone medulla is associated with marked decreases in blood pressure and with decreases in $S\text{ao}_2$ and end-tidal carbon dioxide levels. These have been attributed to three factors:

(1) emboli forced into the circulation from prosthesis insertion;

(2) release of vasoactive substances by the cement;

(3) release of acrylic monomer into the blood stream.

It is essential that patients are not hypovolemic at the time of cement insertion and intraoperative esophageal Doppler monitoring may be useful.

Bone tumors

Most patients with primary bone tumors are managed by tertiary referral hospitals and may present for major resection following chemotherapy and/or radiotherapy. Side-effects from both these treatments vary from minor inconvenience to life threatening. Myelosuppression is common and may result in anemia, leukopenia, and thrombocytopenia. There is an increased association between malignancy and deep vein thrombosis. Cardiomyopathy, pulmonary fibrosis, and renal and liver damage are associated with various chemotherapeutic agents and should be sought.

Malignant hyperthermia

This is a disease of skeletal muscle characterized by massive uncontrolled release of calcium from the sarcoplasmic reticulum. Patients with osteogenesis imperfecta, Duchenne muscular dystrophy, and myotonic disorders have a propensity to develop malignant hyperthermia. Occasionally this may be a problem in patients with scoliosis or talipes where these findings may be an indication of underlying muscle disease.

Spinal cord injury

In the acute phase of high spinal cord injury, airway management, resuscitation, and subsequently anesthesia for decompression or stabilization of the vertebral column may be required. Spinal cord perfusion becomes pressure dependent as autoregulation is lost. Hemodynamic instability is common and response to fluid therapy is unpredictable. Anticholinergic agents, such as atropine or glycopyrrolate, should always be available for patients with high cervical injuries where bradycardia is common. With lesions above C4 voluntary diaphragmatic movement will not occur. Lesions between C5 and C7 are associated with significant respiratory embarrassment due to loss of intercostal and abdominal muscle function and ineffective cough. Temperature should be monitored and carefully managed as thermoregulation is often lost. Life-threatening underlying sepsis may not always be obvious in the patient with spinal cord injury who has a complete lesion. Autonomic hyper-reflexia becomes common with chronicity. Hyperkalemia following the use of succinylcholine (suxamethonium) can become a problem 48 h after spinal cord injury.

Fat embolus

Fat embolus commonly occurs following fractures of the pelvis and long bones, and may occasionally occur during manipulation of fractures or following elective orthopedic surgery. It manifests 2 or 3 days after injury, with a general deterioration in the patient's condition. Pulmonary manifestations are the most common and constitute the major threat to life. Signs and symptoms include dyspnea, cyanosis, and frothy sputum containing fat globules. Chest radiographs show mottling of the lung fields, and blood gas analysis shows low arterial partial pressure of oxygen due to ventilation–perfusion mismatch.

Rheumatoid arthritis

This systemic connective tissue disorder is three times more common in women than in men. It classically involves multiple joints symmetrically, and extremes such as atlantoaxial joint instability with erosion of the odontoid can lead to cord compression. These patients usually present for joint replacement surgery, release of contractures, and repair of ruptured tendons. Extra-articular disease manifestations are common, and include fever, rash, lymphadenopathy, and anemia. Cardiac involvement occurs in a third of patients and presents as pericarditis with pericardial effusion; tamponade may occur. Pulmonary involvement includes fibrosis, which may be diffuse or localized into nodules, and is usually associated with pleural effusions. Renal disease is common and may be due to the disease itself or to drug therapy. Long-standing disease may result in amyloid deposition.

Ankylosing spondylitis

This disease of the sacroiliac joints and spine occurs four times more frequently in men than in women. Vertebral involvement ascends from the lumbar to the cervical regions and results in complete immobility of the spine. Vertebrobasilar insufficiency can occur, and spinal stenosis may produce the cauda equina syndrome.

Complete ankylosis increases the risk of cervical fracture which may be difficult to diagnose or visualize radiographically and can result from a trivial injury. Postural problems are exacerbated by flexion deformity of the hips. The condition is associated with aortic insufficiency and mitral regurgitation, and preoperative echocardiography is useful in assessing the degree of involvement and looking for cardiomegaly. Associated conduction defects and heart block may require use of a pacemaker. Chest wall immobility produces a restrictive respiratory pattern, and occasonally apical pulmonary fibrosis is present. These patients are particularly prone to pulmonary barotrauma during intermittent positive-pressure ventilation. Preoperative lung function and arterial blood gas analysis provide guidance for postoperative management and the need for postoperative ventilatory support.

14.2.2 Vascular surgery

E. P. Wilson and R. A. White

Key messages

- Owing to the physiological and chronological age of most vascular patients, and their associated medical conditions, a thorough preoperative evaluation is required if possible with invasive monitoring as needed perioperatively.
- Prolonged hyper- or hypotension, focal neurological deficits, or wound hematomas after carotid surgery prompt not only re-evaluation but also monitoring in the intensive care unit.
- Abdominal aortic reconstruction can lead to distal embolization, thrombosis, renal failure, and ischemic colitis.
- Prophylaxis against deep venous thrombosis is stratified based on risk and should begin prior to surgery.

Introduction

Patients undergoing vascular procedures represent one of the most challenging populations to be cared for in the intensive care unit (ICU). Since a significant proportion of patients with peripheral vascular disease also have coronary artery disease, hypertension, diabetes mellitus, renal insufficiency, and chronic obstructive pulmonary disease, monitoring and subsequent management of the perioperative course requires an aggressive approach. Coronary disease has long been recognized as the most common cause of postoperative death in patients undergoing major vascular procedures, and the mortality is anticipated to rise as the average age of the population increases. Therefore determination of cardiac risk factors may direct care towards preinterventional invasive treatment of coronary disease, or alternative modalities such as observation with invasive monitoring or proceeding with 'lesser' operations (e.g. extra-anatomic bypasses or endovascular procedures) (Hertzer *et al.* 1984). Although the indication for a given procedure may not allow for extensive preoperative evaluation, as in the case of ruptured aneurysms, crescendo transient cerebrovascular ischemic attacks, or limb-threatening ischemia, stratification based on risk assessment may help guide postprocedural care.

Goldman *et al.* (1977) described a multifactorial index for stratifying patients subjected to non-cardiac surgery based on a variety of easily obtainable data, including history and physical examination. Unfortunately, it appears that, while adequate for general surgery, the predictive value of this classification has not been consistent with vascular patients. Instead, it is recommended that all patients undergoing vascular surgery have a careful cardiac evaluation including an ECG, and if the history, physical examination, or ECG suggest coronary disease, a dipyridamole-thallium scintigram should be performed. If the scan shows either fixed or redistribution defects, or left ventricular dilation with ST segment depression, the postoperative cardiac morbidity is high. Coronary angiography and possibly balloon angioplasty or surgical intervention should be performed on these patients prior to any peripheral vascular reconstruction. If the scan is negative, there is only an associated 2 per cent risk of cardiac death in the case of aortic surgery.

Invasive monitoring is necessary in patients with moderate cardiac risk, or in the case of uncorrectable coronary disease, although monitoring itself is not without risk (Krupski and Bensard 1995). Continuous arterial pressure monitoring is performed, generally using the radial artery. Additionally, pulmonary artery catheterization, particularly oximetry, will provide on-demand cardiac output information together with calculated oxygen delivery and consumption. Optimization of cardiovascular hemodynamics, such as maintaining a cardiac index greater than 2.8 l/min/m^2, a pulmonary artery wedge pressure between 8 and 15 mmHg, and a systemic vascular resistance less than 1100 dyn s cm, or increasing the oxygen delivery index to 600 ml/min can reduce morbidity and mortality, and may reduce early graft failure in peripheral reconstructions. Because these therapeutic manipulations should be instituted preoperatively to obtain their full benefit, admission to the ICU prior to surgery for optimization of cardiovascular indices may be warranted in these high-risk patients. However, there is no specific endpoint for hemodynamic monitoring, and clinical judgment will dictate when to withdraw intensive care in a stable patient.

Most peripheral vascular procedures do not traverse either the thoracic or the abdominal cavity, and therefore excessive fluid shifts, temporary loss of the enteral route, or compromise of pulmonary function are generally only a concern in aortic and visceral reconstructions. In patients with pre-existing pulmonary compromise, continuous oxygen saturation monitoring is necessary, together with aggressive pulmonary physiotherapy, including incentive spirometry, chest percussion, and tracheal suctioning if indicated. The use of a retroperitoneal approach to the aorta and renal vessels may decrease the incidence of postoperative pulmonary complications. In patients with diabetes, frequent blood sugar analysis with appropriate insulin coverage for control will help with fluid balance as well as wound healing. Enteral nutrition should be reinstituted as quickly as possible, although a prolonged ileus (more than 7 days) in mesenteric bypass patients necessitates the use of parenteral formulas. Urine output, serum creatinine, and blood urea nitrogen should be followed for trends in all vascular patients; however, in the case of renal bypass surgery, an improvement over preoperative levels may be anticipated in these variables. An early improvement followed by deterioration or immediate oliguria may signal renal bypass failure, and duplex ultrasonography or limited angiography should be performed urgently.

Considerations in specific vascular procedures

Carotid endarterectomy or bypass

Surgery to treat symptomatic or asymptomatic carotid artery stenosis for the prevention of stroke has become the most commonly performed peripheral vascular operation. This has led to a cost-conscious effort to determine the most effective delivery method for postoperative care. The primary concerns in the period immediately following surgery include neurological status, hypertension, and development of wound hematomas. Morasch *et al.* (1996) designed a protocol to identify patients in the recovery room who will require intensive care monitoring and found that 21 per cent of all endarterectomy patients required further intensive monitoring or intervention. This was identified within a 3-h stay in a postanesthesia recovery unit by the indicators of prolonged hypertension or hypotension following initial pharmacological manipulation, focal neurological deficit or mental status change, or wound hematoma with airway compromise. Systolic blood pressure can be difficult to control; the desired pressure is less than 160 mmHg. Oral nifedipine or intravenous β-blockers may be effective; however, titration with intravenous nitroprusside is indicated in more resistant hypertension. Nitroglycerin may be more beneficial in patients with severe coronary disease.

Although the accepted rate for cerebral events following carotid endarterectomy in an asymptomatic patient is believed to be less than 3 per cent, when a change in mental status not related to anesthesia or a new focal defect is discovered, it can be devastating. If the event occurs in the recovery unit, immediate heparinization (intravenous bolus with 10–15 units/kg) and a transfer to the operating suite is indicated for emergent neck exploration to determine the etiology. Thrombosis is the leading cause in this scenario, although embolization and hypertensive stroke also can occur. If duplex ultrasonography can be obtained expediently, it may rule out vessel thrombosis, and it is particularly useful when the patient has already been transferred to the ICU.

Wound hematomas generally present cosmetic and discomfort problems; however, the primary concern is airway compromise. Evidence of stridor, respiratory compromise, or a new obstructive pattern on arterial blood gas analysis, in the presence of a neck incision hematoma, should prompt opening of the wound and evacuation of the hematoma, with attention directed towards identifying a bleeding source. In the absence of hematoma, vocal cord paresis secondary to recurrent or non-recurrent laryngeal nerve injury should be investigated, particularly in patients with prior neck dissection or contralateral endarterectomy.

Aortic and visceral revascularization

Bleeding is one of the most common complications of vascular surgery. Intraoperatively, venous bleeding is the most likely source because of the proximity and adherence of large veins to the field of dissection (Hunter *et al.* 1993). Postoperatively, however, bleeding from anastamoses or poorly ligated branch vessels, such as lumbar arteries, is more commonly at fault. This is signaled by the continued need for blood replacement postoperatively and possibly by the detection of a retroperitoneal hematoma, although rapid loss will lead to hypovolemic shock. Treatment consists of re-exploration and repair of the bleeding site. Bleeding secondary to coagulopathy should be treated with blood product replacement (e.g. packed red blood cells and fresh frozen plasma) and full reversal of iatrogenically induced anticoagulation. Full screening laboratory tests, including prothrombin time, bleeding time, partial thromboplastin time, and a disseminated intravascular coagulation panel of fibrinogen and fibrin degradation products, should be obtained.

Thrombosis of a graft may be detected by loss of distal pulses, deterioration of renal function, or development of ischemic bowel, depending on the bypass site. Following stabilization of hemodynamics, treatment consists of returning to the operating room for correction of the technical error. If there is no anatomical or hemodynamic cause for the failed graft, or if the thrombosis is recurrent, a search for a hypercoagulable state should be made. Antithrombin III deficiency, protein C or protein S deficiency, and anticardiolipin antibodies are causes of postoperative thrombosis.

Embolization to distal sites, particularly in aortic surgery, can occur at various stages during the procedure, releasing atherosclerotic and thrombotic debris to the distal circulation. This becomes most evident in the foot, with patchy areas of cyanosis or necrosis. Prevention is of paramount importance, and treatment should be expectant, with foot cradles for protection and skin care for prevention of infection. In cases where viability of the foot is a concern, thromboembolectomy with a Fogarty balloon catheter, followed by infusion of urokinase, may be an alternative.

Acute renal failure can be related to various etiologies, and carries a 40 per cent mortality in postoperative elective aortic cases. The most common etiology is decreased renal perfusion secondary to reduced cardiac output. Other causes include suprarenal cross-clamping, ligated left renal vein, and arterial embolization. In each case, prevention is critical. Treament includes maintaining maximum renal perfusion with the use of renal dose dopaminergic agonists (intravenous dopamine 2–4 μg/kg/min) and optimization of cardiac output, together with the judicious use or avoidance of contrast agents and nephrotoxic drugs such as gentamicin.

Bowel and spine ischemia

Intestinal ischemia following aortic surgery involves the colon in almost all cases, and is caused by disruption of collateral blood flow to the bowel with the rectosigmoid region most commonly involved. The incidence in elective cases approaches 10 per cent, while nearly 60 per cent of patients with ruptured aneurysms will experience ischemic colitis. Overall mortality is 50 per cent, with rates as high as 90 per cent when there is transmural involvement. Diagnosis is often difficult because of normal postoperative discomfort; however, progressive distension, worsening peritoneal findings, sepsis, metabolic acidosis, and diarrhea should raise strong suspicion. Limited endoscopy to 40 cm is usually sufficient to establish the diagnosis, and repeat endoscopy can be used to follow mild cases. If there is no evidence of gangrene or full-thickness involvement, treatment consists of bowel rest and antibiotics. Further deterioration should be followed by exploration and resection of non-viable bowel with colostomy.

Spinal cord ischemia following aortic surgery is rare and is primarily caused by interruption of flow to the artery of Adamkiewicz, arising from the posterior division of an intercostal artery between T8 and L2 or from a lumbar branch. The incidence is highest in thoracic aortic reconstructions but can occur in repairing abdominal aortas, particularly when they are ruptured. As in most spinal cord injuries, treatment is supportive.

Infrainguinal bypass

Complications of femoropopliteal and femorotibial arterial reconstructions occurring in the immediate postoperative period are likely to be technical errors. Thrombosis of the graft is signaled by the loss of distal pulses compared with findings at the completion of the procedure. Doppler evaluation at regular intervals may detect impending graft failure. Other signs of a failed graft include recurrent limb pain, parasthesias, paralysis, pallor, and a cool extremity. Operative intervention should be undertaken immediately.

Significant hemorrhage following femoral bypass surgery is uncommon, but may indicate technical errors at the anastamosis, in the graft tunnel, or in the graft itself, such as a non-ligated venous branch. These mechanical failures should be addressed in the operating room once the patient is stabilized, although direct pressure can be applied as needed until that time.

Prophylaxis for deep venous thrombosis

Deep vein thrombosis is an important concern following major gynecological, orthopedic, and general surgery, and, less commonly, vascular surgery. Sequelae include swelling, pain, and limb fatigue, and can lead to pulmonary embolism, postphlebitic syndrome, or phlegmasia cerulea dolens or albicans. Preventive measures can be based on risk stratification.

Low-risk patients, i.e. those less than 40 years of age, without additional risk factors, who are having uncomplicated thoracic or abdominal surgery with general anesthesia of short duration, should wear graduated compression stockings and ambulate early. Moderate-risk patients are aged over 40, undergo general anesthesia for more than 30 min, and have additional risk factors such as obesity or malignancy. Prophylaxis in this population should include either low-dose heparin subcutaneously (5000 units) administered 2 h preoperatively and every 12 h thereafter or intermittent compression devices. A combination of these modalities should be utilized in the high-risk surgical candidate, i.e. patients with a prior history of deep venous thrombosis or pulmonary embolism and patients undergoing orthopedic procedures or surgery for malignant disease.

References

Goldman, L., *et al.* (1977). Multifactorial index of cardiac risk in noncardiac surgical procedures. *New England Journal of Medicine*, **297**, 845–50.

Hertzer, N.R., *et al.* (1984). Coronary artery disease in peripheral vascular patients. A classification of 1000 coronary angiograms and results of surgical management. *Annals of Surgery*, **199**, 223–33.

Hunter, G.C., Bull, D.A., and Bernhard, V.M. (1993). Noninfectious complications in vascular surgery. In *Vascular surgery* (ed. W.S. Moore), pp. 707–35. W.B. Saunders, Philadelphia, PA.

Krupski, W.C. and Bensard, D.D. (1995). Preoperative cardiac risk management. *Surgical Clinics of North America*, **75**, 647–63.

Morasch, M.D., *et al.* (1996): Intensive care after carotid endarterectomy: a prospective evaluation. *Journal of the American College of Surgeons*, **183**, 387–92.

14.2.3 Thoracic surgery

J. A. McGuigan and K. Lowry

Key messages

- Complications following lung resection can be prevented or anticipated by careful preoperative assessment and selection.
- Preoperative use of bronchodilators and energetic treatment of lung infection should be considered.
- Smaller wounds and minimal distraction of ribs lessen postoperative pain.
- A balance needs to be struck between overhydration and dehydration with thickened bronchial secretions.
- An effective method for pain relief after thoracotomy is well-supervised epidural anesthesia.
- Diuretics are often used without good evidence of any benefit.
- The routine use of digoxin to prevent arrhythmias is contoversial.
- While prolonged postoperative positive airways pressure should be avoided following lung resection, a number of patients may benefit from short-term assisted ventilation.

Introduction

Postsurgical general thoracic patients frequently develop complications. The issues in management of thoracic vascular and esophageal surgery are similar to those of general vascular and gastrointestinal tract surgery.

Many patients undergoing thoracic procedures suffer from multiple pathological processes related to age and cigarette smoking. Postoperative problems can be minimized by a thorough assessment of cardiopulmonary status, nutritional state, mental condition, and general fitness preoperatively. It is essential that all patients are encouraged to cease smoking. Smokers with lung cancer are more likely to have cerebrovascular and cardiovascular disease in addition to emphysema and chronic bronchitis in the lung tissue remaining after resection.

Treadmill testing can assess cardiac and respiratory function. Oxygen utilization measurements help to differentiate exercise limitation as a result of pulmonary or cardiological disease. These data may help in patient selection but clinical judgment remains the most important factor.

When time is available, cardiorespiratory function can be 'optimized' with supervised exercise programs, particularly for noncancerous disease. Bronchodilators should be used to improve breathing if reversible obstructive airway disease is present. If steroids are used preoperatively, inhaled drugs are preferable to systemically administered alternatives. Occasionally patients may appear to have relatively poor lung function as a result of pneumonitis, which should be treated.

Subtle changes between pre- and postoperative radiography may help to detect early signs of left ventricular failure or consolidation. Preoperative infections should be energetically treated with antibiotics and surgery should be postponed if possible. When such patients have lung resection there is a risk that their postoperative course may be affected by a recurrence of infection which can have fatal consequences. While efforts must be made to culture organisms, broad-spectrum antibiotics are often used while awaiting bacteriological identification to prevent delay in treatment.

Flexible bronchoscopy not only allows bacterial cultures to be obtained but may also help to clear secretions and recruit unventilated areas of lung.

Major pulmonary surgery shares many of the operative risks of cardiac surgery without the 'safety net' of cardiopulmonary bypass. The heart may be manipulated, the pericardium may be opened, and the left atrium partially clamped or amputated during pulmonary resection. Pulmonary vascular resistance is frequently elevated intraoperatively, and may remain so after resection. Pulmonary resection patients may have similar complications to cardiac patients such as arrhythmias or cardiac failure.

Potential problems following thoracotomy

Potential problems following thoracotomy include the following:

(1) poor respiratory effort and sputum retention (chest wall trauma and pain);

(2) atelectasis, pneumonia, and sepsis;

(3) alveolar and minor bronchial air leaks (pulmonary trauma);

(4) localized or general pulmonary edema;

(5) intercostal, pulmonary, or bronchial vessel hemorrhage;

(6) cardiac arrhythmias, myocardial infarction, and heart failure;

(7) pulmonary and systemic embolism;

(8) chest wall hematoma, wound infection, and dehiscence.

Chest wall trauma is unavoidable during thoracotomies. Smaller wounds with less retraction of ribs produce less pain and wall stiffness. The ultimate reduction in this trend is video-assisted surgery. However, thoracoscopic surgery for cancer does not have universal acceptance. Some degree of atelectasis and sputum retention follows all major thoracotomies. Pneumonia is best prevented by physiotherapy, coughing, and adequate pain relief.

Multiple peripheral air leaks tend to stop after a few days. There is controversy as to whether pleural suction removes air and allows better lung inflation or simply produces unhelpful traction on mediastinal structures with little change in a non-compliant lung. Blood loss into the pleura should decline within 6 to 12 h of surgery. If it increases in volume or persists at more than 200 ml/h, surgery should be considered. Blood loss into the bronchus is usually minimal postoperatively, but if it is persistent or increases in volume, bronchoscopy may be helpful. Heart failure must be excluded in thoracotomy patients developing arrhythmias. Prophylaxis of venous thromboembolism is appropriate for virtually all pulmonary resections. Arterial embolism may be due to dislodgment of lung tumor present in the left atrium.

Potential problems following lung resection

Potential problems following lung resection include the following:

(1) respiratory insufficiency, due to extensive lung resection;

(2) bronchopleural fistula and massive air leak;

(3) early and late mediastinal shift;

(4) torsion of lobe or segment;

(5) residual venous and arterial lung infarction;

(6) blood loss into the pleural space, lung interstitium, or bronchus;

(7) bronchial obstruction from accumulated secretions, blood, and pus;

(8) empyema following air leaks, insufficient lung volume, or overwhelming sepsis;

(9) paradoxical chest wall movement following extensive rib resection;

(10) pulmonary hypertension

Respiratory insufficiency is uncommon when surgeons carry out planned resections but can arise if more lung is resected to deal with intraoperative technical complications. Meticulous attention must be given to prevent dehydration, overtransfusion, and infection in these circumstances. Even those who survive excessive resections will have poor lifestyles.

Bronchopleural fistulas may occur early (usually at 0–3 days) as a result of poor bronchial tissue and poor technique. They may occur later (usually at 8–12 days) as a result of poor vascular supply of the bronchus with or without infection. Following pneumonectomy, fistulas are frequently fatal if pleural fluid is not drained immediately from the side of the resection, since this fluid can enter the remaining lung causing severe respiratory compromise. The patient is often aware of a sudden change in breathing with thin dark brown fluid being coughed up. After complete drainage, surgical repair must be urgently considered.

Abrupt mediastinal shift can cause poor venous return to the heart. This may be seen following pneumonectomy or lobar resection with subsequent collapse of the remaining lung tissue on the same side. Post-pneumonectomy pleural spaces are generally not drained, so that the mediastinum returns to midline. In cases of severe mediastinal shift a drain or needle decompression may be required.

Following right upper or lower lobe resections the middle lobe is prone to torsion. Venous and arterial infarctions are usually identified at postmortem examination. A high index of suspicion has occasionally led to early re-exploration and resection for torsion or infarction. Pulmonary hypertension occurs infrequently and is usually due to inappropriate selection criteria for surgery. When it develops the prognosis is poor.

Failure to clear secretions adequately in the early phase following surgery may lead to bronchial obstruction. Physiotherapy is the first course of action, followed by bronchoscopy if this is unsuccessful. Empyema is infrequent following lung resection and usually requires drainage and antibiotic treatment. Paradoxical chest wall movement follows chest wall resection where repair of the chest wall is inadequate or breaks down afterwards. The diagnosis is clinical and treatment is purely determined by the presence of respiratory insufficiency.

Physiological changes following pneumonectomy

Important physiological changes following pneumonectomy include the following:

(1) reduced compliance;

(2) decreased diffusion capacity;

(3) temporary rise in pulmonary artery pressure;

(4) inconsistent changes in cardiac output;

(5) increase in rate and depth of breathing;

(6) increase in ratio of tidal volume to functional residual volume;

(7) increase of 10 to 30 per cent in residual lung vital capacity and total capacity.

It is important to consider the physiological changes that follow pneumonectomy. These changes are well tolerated by otherwise fit young patients but carry a high morbidity and occasional mortality for older patients. Heart failure and myocardial infarction are common.

Removal of secretions

The most efficient way to remove secretions following thoracic surgery is by coughing. Many patients who have had surgery have impaired coughing due to pain and chest wall and diaphragmatic weakness. Excessive fluid restriction is often used in the management of lung resection cases, and this also leads to difficulty in clearing thickened secretions. Physiotherapy, particularly assisted inspiratory and expiratory therapy, is successful in most patients. A number of patients require nasotracheal suction to remove secretions, and in a small number frequent attempts are necessary. Bronchoscopy may be required. Formal tracheostomy is another major intervention but should be avoided unless other methods have failed. Insertion of an endotracheal tube and ventilation are occasionally successful but are frequently associated with further complications such as nosocomial infection.

Minitracheotomy permits a 4-mm flanged cannula to be inserted through the cricothyroid membrane under local anesthesia, allowing secretions to be adequately removed without delaying recovery (Jackson *et al.* 1991). The patient usually remains conscious and is able to eat and drink. Most importantly, the patient is able to cough and insertion of a catheter through the minitracheotomy elicits a cough reflex. Percutaneous dilation and open tracheostomy are rarely carried out to deal with thicker secretions.

Pain relief

In the past there has been great emphasis in avoiding respiratory depression due to opiate administration. This has led to atelectasis and retained secretions in patients who develop shallow respiration. The best method of pain relief in a conscious breathing patient is well-supervised epidural anesthesia. This system is not always available, and supervision and technical skills are often inadequate in some locations where lung resection is carried out. An alternative system without the risks of epidural anesthesia is the insertion of a catheter between the chest wall and the pleura. In addition, an infusion of bupivacaine can be delivered to the intercostal nerves on the operated side (Carabine *et al.* 1995).

Fluid therapy

Following lung resection pulmonary artery pressure rises, there is an increase in right ventricular stroke work, and an increased risk of developing pulmonary edema. For these reasons post-pneumonectomy patients are often fluid restricted. This policy stems from the mistaken belief that pulmonary edema following lung resection is mainly related to circulating volume. However, such action fails to take into account the principal difference between classical pulmonary edema, which is pressure related (high pulmonary artery wedge pressure), and post-pneumonectomy edema, which is flow related (pulmonary artery wedge pressure normal or low). Post-pneumonectomy patients, whilst not needing to be excessively transfused, do need a sufficiently high right ventricular filling pressure to maintain cardiac output and organ perfusion. Plasma oncotic pressure should be maintained. Attempts to produce hyperosmolar states by protein infusion or the use of diuretics are distinctly unhelpful and are often used without good evidence of any benefit. The former may induce true left ventricular failure, whilst the latter may result in a reduced cardiac output and hypotension.

There is concern regarding the potential combination of relative dehydration and non-steroidal analgesia which is potentially disastrous for renal function, particularly in patients with limited reserve.

Inotropes

The preoperative assessment of patients for lung resection should identify those with impaired cardiac function. Right ventricular failure is an absolute contraindication to elective resection, given the need for increased right ventricular work postresection, while left ventricular failure is a relative contraindication dependent on the response to treatment. Therefore postoperative inotropic support is seldom required, but the agent of choice, in the absence of systemic sepsis, would be dobutamine which increases contractility while reducing systemic and pulmonary vascular resistance.

Tachyarrhythmias are relatively common following thoracotomy, with the most common being atrial fibrillation and atrial flutter. In the past it was routine to digitalize all thoracotomy patients before surgery and to continue digoxin for a period of up to 6 weeks postoperatively. Even in thoracotomies for esophageal resection, pretreatment with digoxin offers no advantage (Ritchie *et al.* 1993); however, there may be a case for pretreatment in those individuals in whom a tachyarrhythmia would be particularly hazardous (e.g. pre-existing severe angina).

The treatment of common atrial arrhythmias following thoracotomy is by conventional therapy with digoxin, verapamil, or amiodarone. Synchronized electrical conversion is used when there is significant hemodynamic instability. Caution should be advised in prescribing amiodarone since it may increase the incidence of acute respiratory distress syndrome following pneumonectomy (Van Mieghem *et al.* 1994).

Ventilation

The accepted rule in some centers is that ventilation is avoided following lung resection since the bronchial suture line may leak. This position is difficult to justify when one considers that lobar bronchi suture lines are routinely tested to be watertight at a pressure of 35–40 cmH_2O intraoperatively. Of more importance is the increased incidence of infection in intubated patients, which may impair stump healing, and overdistention of the contralateral lung with possible mediastinal shift from positive-pressure ventilation. Elective postoperative ventilation should be avoided for pulmonary resections. Short-term planned ventilation may help patients who are frail and likely to become exhausted after complex lung and chest wall resections.

References

Carabine, U.A., Gilliland, H., Johnston, J.R., and McGuigan, J. (1995). Pain relief for thoracotomy. Comparison of morphine requirements using an extrapleural infusion of bupivocaine. *Regional Anesthesia*, **20**, 412–17

Jackson, I.J., Choudhry, A.K., Ryan, D.W., Matthews, H.R., and Corke, C.F. (1991). Minitracheotomy Seldinger—assessment of a new technique. *Anaesthesia*, **46**, 475–7.

Ritchie, A.J., Whiteside, M., Tolan, M., and McGuigan, J.A. (1993). Cardiac dysrhythmia in total thoracic oesophagectomy. A prospective study. *European Journal of Cardio-thoracic Surgery*, **7**, 420–2.

Van Mieghem, W., Coolen, L., Malysse, I., Lacquet, L.M., Deneffe, G.J., and Demedts, M.G. (1994). Amiodarone and the development of ARDS after lung surgery. *Chest*, **105**, 1642–5.

14.3 Postcardiac surgical intensive care

14.3.1 Circulatory management

J. Lloyd and C. Ferguson

Key messages

- Circulatory instability after cardiac surgery is unpredictable in magnitude but is short-lived in the vast majority of patients.
- Early diagnosis of perioperative problems may prevent further complications.
- The preoperative state of the patient and the extent of surgery will influence the postoperative condition.
- Myocardial oxygen supply should be maximized and myocardial oxygen demand minimized

Introduction

Successful management of the circulation after cardiac surgery requires knowlege and understanding of the preoperative condition of the patient and of perioperative events. Preoperative factors include the nature and severity of the cardiac lesion (distribution of coronary artery disease and ventricular function or quantitation of the valve lesion) and associated changes such as pulmonary hypertension. Medication may include β-blockers, calcium-channel antagonists, angiotensin-converting enzyme inhibitors, nitrates, diuretics, and aspirin, and these drugs may exert effects postoperatively by, for example, promoting bleeding (aspirin) or promoting arrhythmias due to electrolyte imbalance (diuretics). Associated conditions include hypertension, peripheral vascular disease, diabetes, and renal dysfunction; patients may have unrelated but important conditions such as hypothyroidism.

The heart suffers inevitable injury during surgery because aortic cross-clamping causes a degree of myocardial ischemia. This may be aggravated by prolonged cross-clamping or suboptimal myocardial protection. Hypothermia, used to reduce oxygen demand on bypass, may have harmful effects, and reperfusion after cross-clamp removal causes a well-defined injury. Revascularization may be incomplete or inadequate, so that surgery may not increase myocardial oxygen delivery sufficiently. Cardiopulmonary bypass also has circulatory effects; the extracorporeal circuit induces a systemic inflammatory response which includes alterations in coagulation and platelet function predisposing to bleeding that may be worsened by inadeqate reversal of systemic heparinization. Rewarming is rarely complete, and the priming volume of the bypass circuit provides an obligatory crystalloid and/or colloid load.

The circulation of the patient after cardiac surgery will be influenced by pre- and postoperative medication, preoperative myocardial and circulatory function, surgical events, and the effects of cardiopulmonary bypass, hypothermia, and the myocardial protection regimen employed.

Assessment of the circulation

Physical examination is valuable in the postoperative period and provides prompt diagnosis of pneumothorax or acute valvular insufficiency, for example, but low cardiac output states are not recognized consistently from clinical signs. Oliguria and metabolic acidosis are also unreliable in this connection. There is a need for direct assessment of myocardial performance by invasive monitoring in some patients.

Routine monitoring includes ECG, direct arterial pressure, right atrial pressure, urine output, and core and peripheral temperatures. More complex monitoring, while providing more information, carries more risk to the patient and should be used only where benefit balances risk. Thus opinions vary widely on the use of the pulmonary artery catheter in these patients. Direct measurement of filling pressures and cardiac output provides information unavailable by any other single intervention and can form the basis for rational use of vasoactive drugs and support of the circulation. However, the complications and costs of these devices are not insignificant, and there is no hard evidence of their use improving outcome.

Cardiac output may be measured less invasively using an esophageal Doppler probe which provides a beat-to-beat display of flow in the descending aorta and calculates analogs for filling pressures and resistances.

Myocardial and valvular performance are increasingly assessed using transesophageal echocardiography which provides accurate estimates of cardiac filling, contraction, and ejection fraction, although it requires considerable expertise and the equipment is costly. Use of this technique to assess regional wall motion abnormalities is more controversial.

Aim of circulatory management

The aim of circulatory management is to allow the heart to recover from the trauma of surgery while providing adequate oxygen delivery to the rest of the body. The circulation is manipulated to minimize myocardial oxygen demand and maximize myocardial oxygen supply while ensuring sufficient cardiac output and systemic blood pressure to provide adequate tissue perfusion.

Although increased heart rate can increase cardiac output up to a point, rapid rates restrict filling and reduce output while reducing diastolic perfusion of the myocardium and thus its oxygen supply. Optimizing preload is the most efficient method of improving cardiac output, at least as far as oxygen demand is concerned. Once this has been achieved, attention can be directed at increasing contractility and reducing afterload.

Oxygen delivery may be increased by blood transfusion, but high blood viscosity reduces tissue oxygen delivery. The optimal hematocrit in humans is unknown and controversial. Regional differences in perfusion caused by vasoconstriction, which may be fixed (atherosclerosis) or functional (vasospasm), may also reduce myocardial and tissue oxygen delivery. Measures to reduce oxygen consumption also reduce the required oxygen delivery. These include analgesia, sedation, and occasionally paralysis to minimize muscular activity, particularly shivering.

Rewarming

Induced hypothermia is employed to varying degrees during cardiopulmonary bypass to reduce oxygen consumption, thus allowing lower hematocrit, lower pump flow, and lower mean arterial pressures to ensure adequate oxygen delivery. Hypothermia also affords a margin of safety should perfusion fail. The extent of hypothermia varys with the technique of myocardial protection employed—cold crystalloid cardioplegia, warm blood cardioplegia, or cross-clamp/fibrillation.

In order to avoid the problems associated with hypothermia, recent interest has focused on 'warm bypass' where the patient is kept normothermic throughout. However, the technique is not widely established and moderate hypothermia of 30 to 32 °C is most commonly used.

Perioperative rewarming, like cooling, is effected by the heat exchanger in the bypass circuit and involves warming the blood above 37 °C, avoiding excessive temperatures that may damage blood components and also have adverse neurological effects. The patient is separated from the bypass at normothermia (measured centrally), but because of uneven rewarming rates different tissues will not be uniformly warm. Further heat is lost during the completion of surgery, so that by the time of arrival in the intensive care unit (ICU) the patient tends to be cold peripherally and centrally, with vasoconstriction. Over the next 2 to 5 h the patient rewarms with progressive vasodilatation, requiring fluid loading to maintain blood pressure and perfusion. Shivering increases oxygen consumption dramatically (by 300–600 per cent) and must be avoided to prevent oxygen demand outstripping oxygen supply with consequent tissue acidosis.

Many techniques are used to aid rewarming, including vasodilators to improve peripheral distribution of heat, prevention of surface loss with blankets, and addition of heat with convective air devices.

Control of hemostasis

After bypass patients have a coagulation defect which has a multifactorial etiology and is understood incompletely. Many of the changes in coagulation cannot be assessed by conventional tests of clotting (e.g. the ability of platelets to stick to von Willebrand factor at the site of injury), and thus specific therapy cannot be ordered.

Heparin and protamine

Anticoagulation during bypass is achieved with heparin which is reversed at the end by protamine. The anticoagulant effect is monitored using the activated clotting time, which is a modification of the whole blood clotting time, and protamine is titrated to return it to the pre-bypass level. The activated clotting time is also affected by low levels of coagulation factors and platelets and by hypothermia, so that protamine reversal of heparin is inexact. Heparin levels can also be used to monitor reversal. A prolonged activated clotting time soon after bypass should be treated with a further dose of protamine.

Uncommonly, heparin rebound can occur following initial correction with protamine; the activated clotting time becomes prolonged some 4 to 6 h later. This is thought to be caused by release of heparin sequestered in the tissues into the circulation and is more likely after large doses of heparin.

Protamine may have an anticoagulant effect in large doses (three to four times those needed to reverse the effects of heparin) and thus may be a rare cause of postoperative bleeding.

Dilutional coagulopathy

The fluids used to prime the bypass circuit together with those given during the procedure dilute the circulating clotting factors. Donor blood may be given and includes few platelets or clotting factors.

Effects of cardiopulmonary bypass

The artificial surface of the bypass circuit is a powerful thrombotic stimulus which is not completely suppressed by heparin anticoagulation. Thrombus is produced and increases progressively during cardiopulmonary bypass, consuming clotting factors, activating platelets, and stimulating fibrinolysis. The activation of coagulation and fibrinolysis is also part of the systemic inflammatory response to cardiopulmonary bypass which includes activation of complement and leukocytes. Attempts to reduce hemostatic activation include improvements in oxygenators (membrane rather than bubble) and the biocompatibility of the materials used in the circuit, the search for drugs that are more effective than heparin in suppressing thrombus formation (e.g. hirudin), and the use of antifibrinolytic agents. Epsilon-aminocaproic acid has been used before and after bypass to reduce bleeding, but has a tendency to produce thrombosis. This is the major concern with all such agents—balancing the risk of thrombosis of grafts and in other vascular beds against that of prolonged bleeding. Tranexamic acid and aprotinin are used before and during bypass to reduce postoperative bleeding. Full risk–benefit analyses are not yet available and thus precise regimens are not defined, with the exception of the use of aprotinin in reoperation.

Hypothermia

The clotting cascade and platelet function are both impaired by hypothermia; thus rewarming is important in restoring normal hemostasis.

Fluid management

The aims of postoperative fluid management are to optimize preload and cardiac output, thus maintaining oxygen delivery to the tissues. Preload may change rapidly because of bleeding, spontaneous diuresis, vasodilatation during rewarming, and the effects of mechanical ventilation (with or without positive end-expiratory pressure). Pressure measurements commonly used as indicators of preload may not reflect end-diastolic volumes after cardiac surgery because ventricular compliance is altered, but are still valuable since a rise in left-sided pressure may precipitate myocardial ischemia or pulmonary edema. Preload is reduced if necessary with nitrates and diuretics.

Capillaries may be 'leaky' post-bypass and the patient receives a crystalloid load perioperatively (the pump prime, with or without cardioplegia), so that a diuresis is promoted, and crystalloid maintenance fluid is restricted to 1 ml/kg/h. Colloids are used to maintain preload. Opinions vary as to the threshold hematocrit for transfusion within the range 0.25 to 0.35. The hazards and expense of hemologous blood transfusion can be avoided to an extent by autotransfusion; the patient's shed blood is collected from the chest drains, filtered, and reinfused at a rate determined by hourly losses. Although this shed blood has a lower hematocrit, increased free hemoglobin, reduced clotting factors and platelets, and increased fibrinolytic breakdown products, the red cells are functional with a normal half-life. This technique is safe in volumes of up to 700 ml and can reduce hemologous blood usage. The risks include increased potential for infection and renal impairment.

Specific conditions

Postoperative hypertension

Postoperative hypertension is common, occurring in 15 to 45 per cent of patients, but is usually short-lived. Its adverse effects include increased myocardial oxygen demand, decreased left ventricular performance, myocardial ischemia or infarction, graft and suture line disruption, and increased bleeding. After adequate analgesia and anxiolysis, optimization of circulating volume, and attention to rewarming, parenteral treatment is with vasodilators (nitrates, sodium nitroprusside, hydralazine, or phentolamine). Calcium-channel blockers and angiotensin-converting enzyme inhibitors may be needed later if hypertension persists beyond the immediate postoperative period.

Bleeding

The most common causes of bleeding are inadequate surgical hemostasis and reduced platelet number or function. Inadequate reversal of heparin and dilutional coagulopathy are less common, and heparin rebound is rare. Definitive diagnosis is difficult because of the limitations of coagulation investigations and because treatment may have to begin before test results are available if bleeding is brisk.

Inadequate hemostasis should be suspected when chest drain losses are excessive (> 300 ml/h) or prolonged and the coagulation profile is normal. Re-exploration should be considered early to prevent the problems of massive transfusion and circulatory instability due to hypovolemia and/or low oxygen-carrying capacity when blood loss is replaced with crystalloid and non-blood colloid.

Initial investigations include complete blood count, activated partial thromboplastin time, and prothrombin time. The activated clotting time may indicate inadequate reversal of heparin. Thromboelastography can also be used to guide therapy. Results of investigations should guide blood component treatment wherever possible and investigations should be repeated at intervals. Intravascular volume must be maintained while investigation and treatment of bleeding is instituted.

Many measures may be taken to conserve blood and reduce the coagulopathy associated with cardiopulmonary bypass. Apart from meticulous surgical hemostasis, hemologous blood use may be reduced by preoperative donation, intraoperative normovolemic hemodilution, and perioperative blood salvage. Coagulopathy is reduced by antifibrinolytics, biocompatible oxygenators and circuits, and accurate heparin reversal. It is possible to identify and, in some cases, treat preoperatively some patients who are at increased risk of bleeding. These include patients with bleeding diatheses, either congenital (hemophilia, von Willebrand's disease) or acquired (uremia, cirrhosis), those on anticoagulants such as warfarin, heparin, or aspirin, those receiving thrombolytics such as streptokinase up to the time of surgery, and those undergoing complex or prolonged surgery (reoperation, combined graft and valve surgery, surgery for aortic dissection). Although no coagulation test can predict which patients will bleed excessively postoperatively, hematological advice is valuable in the management of these complex cases.

Cardiac tamponade

Cardiac tamponade is usually manifest within 24 h of surgery but can develop 10 to 14 days after surgery in association with the postpericardiotomy syndrome or anticoagulant-induced rebleeding. It results in reduced cardiac output with compensatory vasoconstriction and tachycardia, and may lead to myocardial ischemia owing to the combination of reduced coronary perfusion pressure and reduced diastolic time. Classical signs include dyspnea, tachycardia, hypotension, distant heart sounds, pulsus paradoxus, and Kussmaul's sign. The ECG may show electrical alternans.

These signs are not reliable in the postoperative patient, because the pericardium is usually left open and most patients are ventilated, and the diagnosis depends on a high degree of suspicion. It should be considered whenever there is hypotension, tachycardia, and raised filling pressures, particularly if profuse chest drainage markedly reduces or stops suddenly. A chest radiograph typically shows mediastinal widening and, if time permits, echocardiography will confirm the diagnosis.

Treatment is by surgical re-exploration with evacuation of the hematoma. The chest may have to be reopened in the ICU if tamponade is sudden and severe, leading to hemodynamic collapse.

Disorders of rate and rhythm

Although some patients have preoperative arrhythmias, many develop them in the first few hours postoperatively, most commonly ventricular ectopic beats, nodal rhythm, and atrial fibrillation. In those with borderline ventricular function, loss of sinus rhythm may lead to a profound fall in cardiac output and systemic blood pressure. Treatment is required if hemodynamic deterioration occurs or the potential for dangerous rhythms exists (e.g. R on T ventricular ectopy). Attention should be directed to underlying factors such as hypoxia, hypercarbia, acidosis, hypovolemia, pain, anxiety, cardiac chamber distension, and electrolyte disturbances, particularly hypokalemia and hypomagnesemia. Patients on chronic diuretic therapy are at greater risk of the last two. Serum potassium should be maintained above 4.5 mmol/l with frequent measurement and supplements as required. Addressing these predisposing factors will often lead to spontaneous correction of the arrhythmia, but cardioversion, antiarrhythmic drugs, or pacing will be needed if it does not. Arrhythmias, particularly with signs of ischemia, may signal perioperative infarction, inadequate revascularization, or a blocked graft requiring urgent reoperation. Bradycardia predisposes to ventricular overload, ischemia, and heart failure, and is managed with chronotropes (e.g. isoproterenol (isoprenaline)) or pacing.

Left heart dysfunction

After improvement in the first hour after bypass, myocardial function declines to a nadir at 4 to 5 h postoperatively before returning to preoperative levels at 24 h. This depression, which is seen even in those with good preoperative ventricular performance, is more severe in those with poor ventricular function or chronic heart failure, and may result in a low cardiac output syndrome requiring intervention with mechanical or pharmaceutical support.

Risk factors for a low cardiac output syndrome include low ejection fraction, cardiac enlargement and high baseline and postcontrast left ventricular end-diastolic pressure at cardiac catheterization, old age, and female sex. Cardiopulmonary bypass itself impairs ventricular function by a combination of reperfusion injury, the effects of cardioplegia solution, and some degree of ischemic damage which varys with the adequacy of myocardial protection. Myocardial 'stunning' refers to the myocardial dysfunction following ischemia and reperfusion, which is reversible and responsive to inotropic support in contrast with ongoing ischemia where inotropes may cause further injury.

Low cardiac output syndrome can be defined as a cardiac index of 2 l/min/m^2 or less and is associated with increased risk of in-hospital death and multiple organ failure. Poor myocardial function decreases coronary perfusion pressure which reduces coronary blood flow leading to worsening cardiac function. The cycle can only be broken by improving myocardial performance by optimizing myocardial blood flow, increasing contractility, and reducing afterload with inotropic or vasodilator drugs or intra-aortic balloon counterpulsation.

Since inotropes act by raising calcium levels which may be increased already in stunned myocardium, there are good theoretical reasons for using mechanical assist devices rather than excessive doses of inotropes early in the low cardiac output syndrome after cardiac surgery. Inotrope regimens must be tailored individually and guided by invasive monitoring. Myocardial dysfunction is both systolic and diastolic so that the lusitropic effects of inotropic agents are important in this context. Apart from the adrenergic agonists norepinephrine (noradrenaline), epinephrine (adrenaline), dobutamine, and dopamine, the phosphodiesterase II inhibitors milrinone and enoximone may have a place acting as inodilators, increasing contractility, and reducing afterload while avoiding the problems associated with adrenergic receptor downregulation. Dopexamine is also an inodilator, although largely a β_2 agonist.

Myocardial ischemia and infarction

Postoperative ischemia can be detected by ECG, echocardiography, or biochemical tests and the incidence depends on the method used, as there is no agreed gold standard. Echocardiography may be more sensitive than ECG, but it is more expensive and requires greater expertise so that it is not yet considered for routine use. Moreover, detection of ischemia by monitoring regional wall motion abnormalities is controversial since not all such abnormalities are of ischemic origin.

Ischemia is most common in the first 2 days, with a peak incidence 2 h after completion of anastamoses. There is no clear relationship of ischemic episodes to hemodynamic indices. Although tachycardia may be more frequent in patients with postoperative ischemia, approximately half of detected episodes are unrelated to hemodynamic changes.

The etiology of postoperative myocardial ischemia is multifactorial and the relative contributions of global changes in hemodynamics, incomplete revascularization, graft thrombosis or spasm, embolus, kinking, and native coronary artery spasm remain to be defined. The relationship between perioperative ischemia and perioperative infarction is also unclear.

Infarction may be silent or accompanied by hemodynamic changes. More specific markers of myocardial injury, such as the troponin complex proteins troponin I and T, may provide more sensitive diagnosis in the future. Ischemia is treated by reducing myocardial oxygen demand and maximizing supply. The consequences and mangement of infarction will depend on its location and size.

Right heart dysfunction

Right ventricular dysfunction is common after cardiac surgery and may occur in the previously normal ventricle. It may be secondary to left heart dysfunction and pulmonary hypertension or follow right-sided injury. Since the right ventricle has little muscular reserve, small increases in afterload can result in large reductions in stroke volume, leading to dilatation of the ventricle and tricuspid regurgitation.

Right ventricular dysfunction leads to reduced left ventricular preload, and shifts in the intraventricular septum reduce the distensibility of the left ventricle, resulting in reduced cardiac output and hypotension which jeopardize the blood flow to the dilated right ventricle. Ischemia further reduces contractility and right heart failure becomes self-perpetuating, leading to left heart dysfunction.

Depressed postoperative right ventricular contractility may be produced by poor protection of the right heart perioperatively, reperfusion injury, or ischemia. Pulmonary hypertension may result from pulmonary disease, left ventricular failure, microemboli, or acute respiratory distress syndrome, or patients may have pre-existing primary pulmonary hypertension.

Right-sided failure presents as increased right-sided filling pressures, reduced cardiac output, and low systemic blood pressure. The finding of a low cardiac index with a right atrial pressure that increases disproportionately to increases in left atrial or pulmonary artery occlusion pressure is highly suggestive of right ventricular failure. If severe, it will present as cardiogenic shock. Echocardiography or measurement of right ventricular ejection fraction with a fast-response thermistor pulmonary artery catheter can be used to evaluate right heart function. The former technique is more sensitive.

Right ventricular failure is managed by maintaining sinus rhythm (by pacing if necessary), optimizing preload, reducing afterload with pulmonary vasodilators (epoprostenol or inhaled NO), and improving contractility with inotropes or mechanical assist devices. Of course, systemic blood pressure must be maintained.

Bibliography

Gravlee, G.P., Davis, R.F., and Utley, J.R. (ed.) (1993). *Cardiopulmonary bypass: principles and practice*. Williams & Wilkins, Baltimore, MD.

Kapla, J.A. (ed.) (1993). *Cardiac anaesthesia*. W.B. Saunders, Philadelphia, PA.

14.3.2 Ventilatory management

S. J. Brett and B. F. Keogh

Key messages

- The low-risk postoperative cardiac surgical patient has similar ventilatory requirements to any postoperative patient and is generally extubated within hours.
- Coronary revascularization using the internal mammary arteries may lead to violation of the pleural space, pleural collections, and atelectasis, necessitating drainage.
- Valvular surgery and patients with pulmonary hypertension may require prolonged inotropic or vasodilatory support to achieve weaning.
- Cyanotic heart disease correction and assist devices require an understanding of the underlying physiology before weaning is attempted.

Introduction

The range of cardiac surgical interventions has greatly increased in recent years. Postoperative management may be difficult due to either the nature of the surgery or to commonly encountered comorbid disease. Cardiorespiratory stability, adequate sedation and analgesia, and vigilant attention to detail are integral factors in the postoperative management of the high-risk cardiac surgical patient.

A further challenge has recently been posed by the recognition that some cardiac surgical patients can be managed with only the briefest period of postoperative ventilation and intensive care. This controversial 'fast-track' approach, driven substantially by resource considerations, is gaining momentum and is likely to be much more widely applied in the future.

Changes in respiratory function after cardiac surgery

The lungs are exposed to a variety of insults during routine cardiac surgery (Hachenberg 1995). These include loss of functional residual capacity with patchy atelectasis, ischemia–reperfusion injury, and exposure to bypass-activated neutrophils and inflammatory mediators. Increased pulmonary capillary permeability results in increased extravascular lung water, which may be further aggravated by elevated left atrial pressure. Ventilation–perfusion mismatch occurs and pulmonary compliance may markedly decrease. The patient may be hypoxemic, the chest radiograph often demonstrates lobar collapse, and the work of breathing is increased. Pulmonary dysfunction is maximal at around 24 h postoperatively and may persist for several weeks. Despite this, relatively few patients (less than 5 per cent) develop frank postoperative respiratory failure.

Routine ventilatory management of cardiac surgical patients

Initial ventilatory settings are as follows: FiO_2, 0.6 to 0.8; tidal volume, 10 to 12 ml/kg; respiratory rate, 8 to 12 breaths/min; positive end-expiratory pressure (PEEP), 5 cmH_2O; inspiratory-to-expiratory (I:E) ratio, 1:2; peak pressure limit, 30 cmH_2O. These are guidelines only and individual units may vary widely in approach. Flow- or pressure-generated tidal breaths are equally applicable, although mandatory pressure-generated breaths, such as those delivered by the pressure-control mode, are uncomfortable for conscious post-sternotomy patients (because of the high initial inspiratory flow rates and breath characteristics).

Typically, synchronized intermittent mandatory ventilation is established with the addition of pressure support up to 20 cmH_2O. This initiates a smooth weaning process in which respiratory rate is decreased as spontaneous activity increases.

A chest radiograph is obtained to check for pneumo- and hemothoraces, the position of intravascular catheters and the endotracheal tube, and the status of the lung fields. In most patients the aim of the ventilatory strategy is to achieve normal oxygen and CO_2 levels. Initial high inspired oxygen concentration is rapidly reduced once adequate oxygenation is ensured.

In patients with widespread atelectasis, prolongation of the I:E ratio and intermittent gentle hyperinflation may be necessary to recruit alveoli. High levels of PEEP ($> 10\ cmH_2O$) and high mean airway pressures ($> 15\ cmH_2O$) may impair cardiac performance, particularly in the presence of hypovolemia. Transmission of airway pressure to the pleural space is directly related to compliance; hence patients with emphysema are most at risk of ventilator-induced hemodynamic depression. Paradoxically, patients with postoperative pulmonary edema, despite poor cardiac function, may be less adversely affected by lung volume recruitment techniques.

Unstable cardiac patients may tolerate hypercarbia poorly which, in combination with other factors, may contribute to arrhythmia and low output syndromes. Unlike many other ventilated patients, permissive hypercapnia is rarely indicated. Adverse effects on pulmonary hemodynamics may be observed if CO_2 is allowed to rise, and thus the aim is to maintain normal values.

Individual minute volume requirements vary and relate to underlying lung disease, physiological dead-space, the degree of residual hypothermia, the presence of shivering, and the level of consciousness. Violent shivering increases CO_2 production and oxygen requirements. Thermogenic shivering can be treated with active warming blankets, filtered warm air generators, or overhead radiators. Non-thermogenic shivering may respond to small doses of meperidine (25 mg) or doxapram (10–20 mg).

Weaning from ventilation

With the exception of fast-track patients, an expectant approach should be adopted. Guidelines for ventilatory weaning are outlined in Tables 1 and 2. It may be appropriate to start weaning before these criteria are met. For example, some patients may be weaned and extubated prior to removal of an intra-aortic balloon pump.

Adequate analgesia is vital and can be provided by intravenous or extradural opiate infusions, or by small titrated doses of intravenous opiates, commonly administered via patient-controlled analgesia devices. As in all patients, regular clinical assessments are the key to the weaning process. A small group of patients with limited cardiac reserve may decompensate when challenged by the work of spontaneous breathing.

Patient extubation is a clinical decision and guidelines are listed in Table 3. At this stage, an elevated $PaCO_2$ can be accepted in a

Table 1 General criteria for weaning from mechanical ventilation

Hemodynamic stability: low-dose inotropes only
Stable cardiac rhythm
Reasonable hemostasis: chest tube drainage <100 ml/h
Approaching normothermia
No requirement for further interventions (e.g. cardioversion)
Co-operative patient, grossly intact central nervous system
Controllable discomfort

Table 2 Respiratory criteria for weaning from mechanical ventilation

$Pa\text{o}_2 > 10$ kPa with $F\text{io}_2 < 0.5$
$Pa\text{co}_2 < 6.5$ kPa
pH >7.32
PEEP $\leqslant 6$ cmH_2O
No gross abnormality on chest radiograph (e.g. pulmonary edema)
No residual neuromuscular blockade

Table 3 Criteria for extubation

General criteria for weaning still apply (Table 1)
$Pa\text{o}_2 > 10$ kPa with $F\text{io}_2 < 0.5$
$Pa\text{co}_2 < 8$ kPa
Minute volume requirement <150 ml/kg/min
Respiratory rate <20 breaths/min
Tidal volume >8 ml/kg
Comfortable, co-operative, and neurologically intact patient

hemodynamically satisfactory patient, and is indeed very common in a comfortable co-operative postsurgical patient in whom adequate narcotic analgesia has been achieved.

After extubation many patients benefit from additional respiratory support in the form of humidified oxygen by mask or nasal cannula, continuous positive airway pressure by mask, and physiotherapy including the use of incentive spirometry. Nasal positive-pressure ventilation is not usually indicated in the early postoperative phase but may be the logical progression in a patient with limited cardiopulmonary reserve who undergoes a protracted weaning program.

In some patients weaning may be predictably or inadvertently inappropriate in the early postoperative period. Common associated factors include complex and prolonged surgery, re-exploration, perioperative infarction, left ventricular dysfunction, dysrhythmia, pulmonary edema or fluid overload, pulmonary hypertension, obesity, chronic pulmonary disease, neurological injury, and advanced age. Cardiovascular stability should be achieved before weaning is contemplated. In patients with poor left ventricular function, judicious use of diuretics, angiotensin-converting enzyme inhibitors, and phosphodiesterase inhibitors may be necessary during a protracted weaning phase. An early sign of decompensation in such a patient is a sudden rise in right atrial pressure when increased respiratory work demands are imposed. Pleural collections in excess of 250 ml should be drained, sometimes repeatedly and preferably with ultrasound guidance, in patients with limited reserve.

Fast-track approach

In recent years it has become apparent that many 'routine' patients can be extubated early. The use of short-acting anesthetic drugs combined with moderate-dose opiate regimens allows early return of consciousness and spontaneous respiration without obviously compromising patient safety. In addition, the tendency to employ less profound hypothermia for myocardial preservation, and the use of active warming and vasodilators during rewarming, imposes less metabolic stress than previous approaches. Many patients can be extubated within 4 h of surgery and, in some units, in the operating room. Dedicated postcardiac surgery recovery areas, employing the fast-track approach, can provide an efficient service without dependence on the availability of intensive care beds (Chong *et al.* 1992).

The key to a successful fast-track program is proper patient selection, high-level supervision by a disciplined team, and absence of surgical complications.

Specific conditions

Coronary revascularization

Coronary revascularization accounts for approximately 70 per cent of adult cardiac surgery in the Western world, including repeat procedures. Specific ventilatory implications in coronary revascularization relate to the use of arterial conduits. The internal mammary artery is employed in the majority of patients. During harvesting, which may be bilateral, the pleural cavity is often breached. Extrapulmonary collections commonly result and may be associated with lobar collapse. The internal mammary graft usually remains attached proximally at its pedicle and, on some occasions, may be short and under tension with chest closure. Excessive tidal volumes may compromise the graft and even avulse it. Tidal volumes should be limited to 10 to 12 ml/kg and excessive hyperinflation avoided with manual ventilation (e.g. with physiotherapy techniques). Peak pressure should be limited to 30 cmH_2O in patients with chronic obstructive pulmonary disease to avoid hyperinflation; a lower limit may be necessary with severe emphysema.

The gastroepiploic artery and, more recently, the inferior epigastric artery are used for revascularization. Although intra-abdominal manipulation is limited, some patients develop a paralytic ileus which may last for several days and compromise respiratory function. Thus nasogastric drainage should be available in all patients until normal gut function is restored.

Valve and aortic surgery

Valve surgery may be technically difficult and involve several valves or additional coronary grafts. Aortic valve dysfunction may be associated with aortic root abnormalities or aortic dissection. Many patients are now presenting for repeat surgery. All but the most straightforward valve and aortic surgical patients are likely to require significant postoperative ventilatory support.

These patients frequently undergo very long bypass procedures and encounter severe hemostatic derangements and respiratory complications. Phrenic nerve damage may occur during surgery but may only be identified when weaning is attempted. Non-invasive ventilatory support, such as nocturnal nasal positive-pressure ventilation, should be considered.

Pulmonary hypertension

Pulmonary hypertension may be secondary to valve disease, left ventricular failure, congenital heart disease, chronic thromboembolic disease, pulmonary veno-occlusive disease, or a wide range of pulmonary parenchymal pathology. Pulmonary hypertension may be fixed or responsive to some degree. Intravenous pulmonary vasodilators, nitric oxide, nebulized epoprostenol, and phosphodiesterase inhibitors such as milrinone or enoximone may offer benefit. Inotropic agents may decrease pulmonary vascular resistance with net improvement in right ventricular function. Care must be taken not to imbalance the circulation with excessive systemic vasodilation which may cause the right ventricle, with a relatively fixed afterload, to fail. Similarly, in the presence of right-to-left shunts, such as occur in Eisenmenger's syndrome, profound hypoxia may result. Paradoxically, in either case, systemic vasoconstriction by infusion with norepinephrine (noradrenaline) at doses of 0.01 to 0.1 μg/kg/min may be necessary to balance the circulation.

General principles include appropriate hemodynamic support, strict attention to fluid balance, avoidance of acidosis, adequate analgesia, and patience. It is well recognized that patients with mitral valve disease will show improvement in pulmonary hemodynamics a few days after surgery. Although patients with pulmonary hypertension have numerically satisfactory gas exchange in most cases, cardiovascular compromise is the inevitable effect of premature weaning from ventilation.

Cardiomyoplasty and aortomyoplasty

Cardiomyoplasty and aortomyoplasty are increasingly being performed in the management of endstage cardiac failure and may represent an alternative to cardiac transplantation in the absence of available donor organs or if the patients are unsuitable for transplantation.

Both procedures involve the mobilization of the left latissimus dorsi muscle with a pedicle still attached. The muscle is then routed through the left chest wall after one or two ribs are partially resected. In cardiomyoplasty the muscle is wrapped around the ventricle, whereas in aortomyoplasty it is wrapped around the aorta. In the latter case, stimulation eventually results in a counterpulsation phenomenon, analogous to that of an intra-aortic balloon pump, which both decreases left ventricular afterload and improves coronary perfusion. In each case, the muscle is not stimulated for several days and is unlikely to offer any support for several weeks.

Such patients have, by definition, limited cardiac reserve and may have some coexisting respiratory disease. The procedure offers no immediate hemodynamic benefit and, in fact, may be detrimental initially.

Surgery for the correction of congenital heart disease

Patients undergoing correction of simple congenital defects (e.g. atrial and ventricular septal defects), do not normally present particular difficulties in perioperative care. Many different anomalies, with specific requirements related to cardiopulmonary interactions, may be encountered in adolescents and adults with an increase in exposure to corrected cyanotic congential heart disease.

Patients with certain forms of cyanotic congenital heart disease, including tricuspid atresia and univentricular hearts, undergo redirection of systemic venous blood to the pulmonary circulation, bypassing the right ventricle. Surgery may be staged, employing the Glenn shunt (superior vena cava to the right pulmonary artery) initially or a single Fontan-type procedure (right atrium anastomosed to the right pulmonary artery). Such circulations essentially rely on a passive (i.e. venous-pressure-driven) transpulmonary gradient to provide left heart filling. Any increase in pulmonary vascular resistance will reduce flow, and caval flow can reverse during positive-pressure inspiration (Penny *et al.* 1991).

Low cardiac output states are common postoperatively and are usually due to flow reduction related to pulmonary vascular resistance rather than ventricular dysfunction. Ventilation should in principle employ low mean airway pressures, but achieve normocarbia. Low-level PEEP (2–3 cmH_2O) may offer indirect benefit by maintaining functional residual capacity. Early extubation is preferable. Negative-pressure ventilation using external cuirass ventilators increases pulmonary flow by up to 45 per cent and may be a key intervention in postoperative low-output state (Shekerdemian *et al.* 1996). In addition, nitric oxide or nebulized prostacyclin (prostaglandin I_2) may favorably influence pulmonary vascular resistance.

Implantable left ventricular assist devices

Implantable left ventricular assist devices are used as a bridge to transplantation and for long-term support. These devices connect the left ventricle to the ascending aorta and provide no right-sided assistance. Right heart failure due to poor myocardial function and/or elevated pulmonary vascular resistance occurred in 21 per cent of patients in trials of one device, and temporary mechanical support for the right ventricle was required. The ventilation strategy is as for the Fontan circulation; low airway pressures are combined with selective and non-selective pulmonary vasodilators.

Respiratory failure after cardiac surgery

The true incidence of respiratory failure after cardiac surgery is difficult to determine because of the different weaning philosophies adopted and the need for ventilation for non-pulmonary reasons. In a study of 2256 patients in the United Kingdom, 7 per cent remained ventilated at 48 h and a third of these subsequently died (Holmes *et al.* 1994). Many of these patients develop sepsis and subsequent multiple organ failure on a background of the pulmonary insult induced by cardiac surgery. The incidence of acute respiratory distress syndrome after cardiac surgery is 1 to 5 per cent. Particular risk factors include a high level of perioperative intervention, age over 60, and total pumped volume greater than 300 liters (Messent *et al.* 1992).

Conclusions

The majority of cardiac surgical patients do not require prolonged ventilatory support. High-risk patients or those with specific conditions may require prolonged ventilation in concert with hemodynamic and other organ support. Basic principles include lung volume maintenance, avoidance of ventilation-induced damage, and achievement of cardiovascular stability before weaning is undertaken. In individual patients, particularly those with abnormal anatomy, an understanding of cardiopulmonary interactions is vital in constructing respiratory support regimens.

References

Chong, J.L., Pillai, R., Fisher, A., Grebenik, C., Sinclair, M., and Westaby, S. (1992). Cardiac surgery: moving away from intensive care. *British Heart Journal*, **68**, 430–3.

Hachenberg, T. (1995). Pulmonary dysfunction after cardiac surgery. *Current Opinion in Anaesthesiology*, **8**, 565–71

Holmes, L., Loughead, K., Treasure, T., and Gallivan, S. (1994). Which patients will not benefit from further intensive care after cardiac surgery? *Lancet*, **344**, 1200–2.

Messent, M., Sullivan, K., Keogh, B.F., Morgan, C.J., and Evans, T.W. (1992). Adult respiratory distress syndrome following cardiopulmonary bypass: incidence and prediction. *Anaesthesia*, **47**, 267–8.

Penny, D.J., Hayek, Z., and Redington, A.N. (1991). The effects of positive and negative extrathoracic pressure ventilation on pulmonary blood flow after the total cavopulmonary shunt procedure. *International Journal of Cardiology*, **30**, 128–30.

Shekerdemian, L.S., Shore, D.F., Lincoln, C., Bush, A., and Redington, A.N. (1996). The differential effects of negative pressure ventilation on cardiac output in children with congenital heart disease. *Circulation*, **94**, I-651 (abstract).

14.4 Postneurosurgical intensive care

14.4.1 Postoperative protection against cerebral injury

M. Smith

Key messages

- The fundamental aim of postneurosurgical intensive care is to protect the brain from ischemic damage.
- Protection of the brain is achieved by balancing cerebral substrate supply and demand.
- Barbiturates and other anesthetic agents which decrease cerebral metabolic demand have been used for protection of the brain, but there is limited evidence for their efficacy.
- Fluid restriction after brain injury is contraindicated and does not reduce brain swelling.
- Plasma glucose levels should be maintained within normal limits as hyperglycemia is associated with poor outcome after brain injury.
- Moderate hypothermia protects the brain by mechanisms other than a simple reduction in cerebral metabolic rate.
- Specific therapies for protection against mediators of cerebral injury are being investigated.

Introduction

Approaches to the prevention of cerebral ischemia in the postoperative period fall into five broad categories:

(1) balancing cerebral substrate supply to metabolic demand with drugs which reduce cerebral metabolic rate;

(2) maintaining adequate fluid balance while preventing brain swelling;

(3) maintaining normoglycemia because of the adverse effects of hyperglycemia on ischemic brain;

(4) the renewed interest in moderate hypothermia;

(5) the use of specific antagonists against mediators of cerebral injury.

Cerebral ischemia

The brain requires a constant supply of oxygen and glucose, as well as other metabolic substrates, for maintenance of structural integrity and electrophysiological function. This supply must be uninterrupted because the brain has little ability to store nutrients. Therefore an adequate cerebral blood flow must be maintained at all times.

Global cerebral ischemia, such as that occurring after cardiac arrest, results from hypoperfusion of the entire brain. Focal ischemia, resulting in local hypoperfusion, occurs secondary to localized lesions or vessel occlusion distal to the circle of Willis. As collateral flow may develop in focal ischemia, the area of damage is graded with the development of an ischemic core surrounded by a penumbral zone.

As blood flow to cerebral tissue is reduced there is sequential loss of cellular function. First neuronal electrical function ceases, followed by failure of energy-dependent ion pumps within the cell membrane. This is followed by total cessation of neuronal function, which results in uncontrollable movement of ions across the cell membrane and subsequent cell death. Neurons in the ischemic penumbra may continue to receive enough blood to maintain oxygenation above the level required to maintain ion pump function but below that required for electrical, and therefore clinical, function. Metabolic failure is likely to occur in the ischemic penumbra if its precarious oxygen supply is not reversed within 3 to 4 h. However, total cessation of blood flow to the brain leads to an isoelectric electroencephalogram within 30 s and total depletion of high-energy phosphate stores within 2 min. Ischemic brain tissue releases biochemical substances, such as excitatory amino acids and free radicals, which worsen cerebral damage.

Control of metabolic supply and demand

As the supply of metabolic substances to the brain is reduced during ischemia, it should follow that drugs which decrease metabolic demand might favorably influence the supply–demand relationship. Therefore they might be expected to produce prolonged tolerance to, and improved outcome from, a temporary ischemic insult.

Barbiturates have been used in the protection of the brain since 1937, when Horsley first described their effect on cerebrospinal fluid pressure. The exact way in which barbiturates lower intracranial pressure is unknown, but it is almost certainly multifactorial. It is clear that barbiturates lower cerebral metabolic rate by as much as 50 per cent, thereby reducing cerebral blood flow and hence intracranial pressure. This is dose dependent until an isoelectric electroencephalogram is produced, when no further effect can be achieved. The reduction in cerebral metabolic rate is caused by inhibition of energy-dependent synaptic transmission. Drugs which reduce cerebral metabolic rate by an effect on electroencephelographic activity can only be effective in conditions where the ischemic event is mild enough to permit some activity to remain. This situation may occur in the ischemic penumbra. There can be no benefit from drug therapy if the ischemic event is severe enough to cause an isoelectric electroencephelograph. In such circumstances barbiturates can cause no further reduction in electroencephelographic activity and hence no protective effect can be postulated. Clinically therefore barbiturates would be expected to work only in conditions of reversible ischemia such as temporary vascular occlusion.

Although it is logical that barbiturates have been considered as potential candidates for cerebral protection, there is limited evidence for their efficacy. Randomized prospective studies of the use of barbiturates following head injury, cardiac arrest, and stroke have all failed to demonstrate any improvement in outcome. Their use has become limited to the control of intracranial pressure when intracranial hypertension is resistant to other therapeutic modalities, although even in this role they are being superseded by the use of propofol. Barbiturates also cause cerebral vasoconstriction directly, independent of cerebral metabolic rate reduction. They may also block the formation of oxygen free radicals, reduce circulating catecholamines, and decrease central adrenergic neurotransmitter activity.

Barbiturates have many side-effects and infusions are difficult to control because of the marked accumulation of the drug. Hypotension occurs in 50 per cent of patients even when adequate blood volume and cardiac output are maintained. Therefore other drugs which reduce cerebral metabolic rate, including propofol and isoflurane, have been increasingly used in place of barbiturates.

Propofol has effects on the cerebral circulation analogous to those of barbiturates. It causes a dose-dependent fall in cerebral metabolic rate and intracranial pressure, but at modest doses perfusion pressure is maintained if the patient is euvolemic. The reactivity of cerebral vessels to CO_2 and cerebral autoregulation are maintained during propofol infusion. Its favorable pharmacological profile prevents accumulation and sedation levels can easily be controlled.

Isoflurane also causes a dose-dependent fall in cerebral metabolic rate and an isoelectric electroencephelograph at normal clinical doses. Suprisingly, isoflurane has not been shown to offer superior cerebral protection over other inhalational anesthetic agents, even during focal insults. The logistics of administering inhalational anesthetic agents in the intensive care unit (ICU) have prevented their widespread adoption.

It now appears unlikely that the neuroprotective effects of anesthetic agents are related to the decrease in electroencephelograph activity alone, as focal ischemic damage is not reduced in a dose-dependent manner.

Fluid balance

Cerebral edema is the final common pathway of most forms of brain injury and may reduce global or localized blood flow. Brain swelling due to increases in water content may occur because of cellular failure or breakdown of the blood–brain barrier. Unlike in the past, severe fluid restriction does not have a place as it leads to a reduced circulating blood volume, low blood pressure, and a rise in the viscosity of the blood. These may all contribute to inadequate flow to areas of the brain at risk.

Fluid restriction does not decrease the risk of brain swelling because the blood–brain barrier possesses unique permeability properties. It is impermeable to charged particles such as sodium ions, which can therefore form a greater hydrostatic gradient across the blood–brain barrier than oncotically active particles such as albumin. This contrasts with the action of these fluids in other organs. Therefore the influence of intravenous fluids on brain water content is related to how they affect plasma osmolality and not to their oncotic properties. It is reasonable to administer whatever intravenous fluids are necessary to maintain an adequate blood volume provided that plasma osmolality is not reduced in the process. In this way, cerebral blood flow is maintained without adverse effects on brain swelling.

Flow through the cerebral microcirculation and collaterals in ischemic brain is improved if blood viscosity is reduced. Many units aim for a hematocrit of 30 to 35 per cent after intracranial neurosurgery or brain injury and use 0.9 per cent saline or synthetic colloid solutions for volume replacement.

Maintenance of normoglycemia

Hyperglycemia is associated with a worsened outcome after ischemic cerebral injury. In conditions of insufficient oxygen supply, cellular energy requirements may be partially met by anaerobic glycolysis which leads to an accumulation of lactic acid. Since this is virtually all ionized at physiological pH an intracellular acidosis develops, leading to deregulation of ionic membrane pumps and the formation of other mediators of neuronal injury. This intracellular acidosis is the probable mechanism of the adverse effects of hyperglycemia in brain injury, although the evidence in focal ischemia is not as clear. Thus hyperglycemia enhances injury in the ischemic core, whereas damage in the penumbra may be reduced due to collateral flow. Significant hyperglycemia ($>$ 10 mmol/l) should always be corrected with insulin infusion in neurosurgical patients.

Normal brain is almost totally dependent upon a continuous delivery of exogenous glucose for maintenance of cellular energy requirements. Therefore glucose levels should be monitored routinely and maintained between 4 and 8 mmol/l.

The administration of glucose-containing solutions does not produce a significant increase in blood glucose, but may predispose patients to a worsened outcome. Therefore it seems prudent to withhold glucose-containing solutions from neurosurgical patients, except those in whom the onset of hypoglycemia may be rapid and have severe consequences such as diabetics. Hyperglycemia should be corrected prior to a likely ischemic event so as not to exacerbate it.

Hypothermia

There has been renewed interest in the use of hypothermia for cerebral protection since its first use over 30 years ago. Hypothermia protects the brain by mechanisms other than a simple reduction in cerebral metabolic rate.

Hypothermia as a therapy for neuroprotection was abandoned in the neurosurgical population because the adverse effects of cooling to low temperatures, including coagulopathies and hemodynamic disturbances, were found to outweigh the protective benefits. However, there has been a resurgence of interest in its use to attenuate ischemic cerebral damage, and attention has focused on small reductions in temperature. Decreases in brain temperature of only 2 or 3 °C have recently been found to have neuroprotective effects in models of cerebral ischemia. Mild hypothermia has also been found to be advantageous in the treatment of head injury and during certain neurosurgical procedures. It is unlikely that the small reduction in cerebral metabolic rate can explain these findings, and the effect is most likely to be multifactorial. The beneficial effects of hypothermia may in part be mediated by attenuation of the increase in glutamate and aspartate after ischemic injury. Other possible mechanisms include membrane permeability changes, prevention of lipid peroxidation, free-radical reactions, nitric oxide effects, downregulation of protein kinase C activity, and changes in blood–brain barrier permeability.

Active warming should be avoided during neurosurgery and in the management of brain-injured patients in the ICU but therapeutic hypothermia cannot be recommended for routine clinical practice until the results of clinical trials are available. Problems which must be resolved include the degree of hypothermia offering the best risk–benefit ratio, methods of rewarming, and temperature measurement sites.

Specific therapies

Attention has focused on the role of specific antagonists of excitatory neurotransmitters or inhibitors of intracellular events following episodes of loss of ionic homeostasis.

Ischemia results in the loss of normal mechanisms for release and uptake of excitatory neurotransmitters in the central nervous system. Glutamate and aspartate are implicated and mediate neuronal damage and cell death. Energy sources become exhausted during ischemia and the energy-dependent sodium pump fails. This leads to membrane depolarization due a sudden influx of Na^+ and Cl^- and efflux of K^+. Voltage-dependent Ca^{2+} channels are opened, allowing a massive influx of Ca^{2+} which causes a high intracellular Ca^{2+} concentration. This activates enzymes which disrupt receptor function and mitochondrial and cell membrane integrity. Membrane phospholipids can then be broken down to free fatty acids which are metabolized to prostaglandins, leukotrienes, and free radicals. Following this neurotoxic cascade, additional intracellular damage results from second-messenger activation which in turn activates protein kinase C and release of Ca^{2+} from the endoplasmic reticulum.

N-Methyl-D-aspartate (NMDA) antagonists have been observed to reduce neuronal injury under a variety of experimental conditions. One such antagonist, dizocilpine, has been used in therapeutic studies, but initial encouraging results need to be substantiated. Glutamate AMPA-receptor antagonists improve outcome from global ischemia in animals even when given some time after the insult. However, major side-effects have made such compounds slow to enter the clinical field, although human trials are now under way.

Calcium entry into cells can be non-competitively antagonized by dihydropyridines such as nimodipine and nicardipine. These compounds are used for the prevention and treatment of vasospasm after subarachnoid hemorrhage but have little efficacy after global ischemia. Recent studies have failed to show improvement in outcome when nimodipine is given after head injury, except in subgroups of patients with subarachnoid blood. Interestingly, a combination of nimodipine and dizocilpine decreases neuronal damage in animal models of ischemia, presumably because of the overlapping effects of the two drugs on calcium influx.

Oxygen free radicals are generated in excess in ischemic brain and contribute significantly to further neuronal damage and death. Free-radical scavengers, such as superoxide dismutase and dihydrolipoate, may decrease mortality but do not improve outcome. A novel group of compounds have been designed which are potent inhibitors of lipid peroxidation induced by oxygen free radicals. One such drug, tirilazid, has shown some protective benefit in animal models of cerebral ischemia.

Nitric oxide may mediate ischemic cerebral damage. It is produced postsynaptically in response to activation by excitatory amino acids and then diffuses into presynaptic nerve endings, activating guanylate cyclase and thereby increasing levels of cGMP. Nitric oxide is also highly reactive in its own right, reacting with the superoxide anion to produce hydroxyl free radicals in the central nervous system. Nitric oxide synthesis may also play a part in cerebral flow–metabolism coupling and therefore be important in the local regulation of blood flow. Blockade of nitric oxide synthesis with N^G-nitro-L-arginine reduces infarct size in animal models of ischemia.

Conclusion

Although specific antagonists of mediators of cerebral damage are unlikely to be clinically available for some years, there is still much which can be done to minimize ischemic cerebral damage in the neurosurgical patient. Balance of cerebral substrate supply with demand and prevention of hyperglycemia and hyperthermia will provide optimal conditions for cerebral recovery and are measures which can easily be undertaken in the ICU.

Bibliography

Illievich, U.M. and Spiss, C.K. (1994). Hypothermic therapy for the injured brain. *Current Opinion in Anaesthesiology*, 7, 394–400.

Lam, A., Winn, H., Cullen, B., and Sundling, N. (1991). Hyperglycemia and neurological outcome in patients with head injury. *Journal of Neurosurgery*, 75, 545–51.

Siesjo, B.K. (1992). Pathophysiology and treatment of focal cerebral ischemia I. Pathophysiology. *Journal of Neurosurgery*, 77, 169–84.

Verhaegen, M. and Warner, D.S. (1995). Brain protection and brain death. In *Neuroanaesthetic practice* (ed. H. Van Aken), pp. 267–93. BMJ Publishing, London.

Warner, D.S. (1994). Perioperative neuroprotection. *Current Opinion in Anaesthesiology*, 7, 416–20.

14.4.2 Circulatory management

D. K. Menon and A. Swami

Key messages

- Normal cerebrovascular autoregulatory responses may be affected by disease.
- Maintenance of normal plasma osmolality is essential in preventing cerebral edema.
- Cerebral perfusion pressure should be maintained above 60 to 70 mmHg in head-injured patients
- Hypervolemia, hemodilution, and hypertension may be useful in maintaining cerebral perfusion in patients with vasospasm following aneurysmal subarachnoid hemorrhage.

Systemic hemodynamics and cerebral blood flow

A detailed discussion of cerebral circulation is given by Menon (1995). The cranial vault is a closed cavity, and intracranial pressure (ICP) has a major influence on cerebral blood flow (CBF). Consequently, unlike many other organs, the pressure driving cerebral perfusion is not the mean arterial pressure (MAP) but the cerebral perfusion pressure (CPP), which is defined as follows:

$$\text{CPP} = \text{MAP} - \text{ICP}.$$

The cerebrovascular bed is capable of pressure autoregulation: changes in cerebral perfusion pressure are matched by changes in cerebrovascular resistance, so that cerebral blood flow remains constant (Fig. 1(a)). When the capacity for autoregulatory changes in cerebrovascular resistance are exhausted, cerebral blood flow follows cerebral perfusion pressure and is said to be pressure passive. Even within autoregulatory limits, cerebral blood flow is closely coupled to neuronal activity, is exquisitely sensitive to changes in $P\text{aCO}_2$ within the physiological range, and increases with severe hypoxemia (below a $P\text{aO}_2$ of about 6.5 kPa).

These physiological relationships between cerebral perfusion pressure and cerebral blood flow may be seriously disrupted in disease. Experimental data suggest that while cardiac output may only be a minor determinant of cerebral blood flow in health, it may have a marked effect, independent of changes in cerebral perfusion pressure, on cerebral perfusion in disease. The response of the cerebral circulation to systemic hemodynamic changes may also be affected by drugs, including volatile anesthetic agents and systemic vasodilators (nitrates, nitroprusside). Regional flow may be affected by focal ischemia, intra- or extracranial occlusive vascular disease, or large-vessel spasm after subarachnoid hemorrhage. Clearly, many of these effects are focal or regional, and result in heterogeneity in cerebrovascular physiology. Under these circumstances assumptions about uniformity of the cerebrovascular effects of systemic hemodynamic changes may no longer hold true.

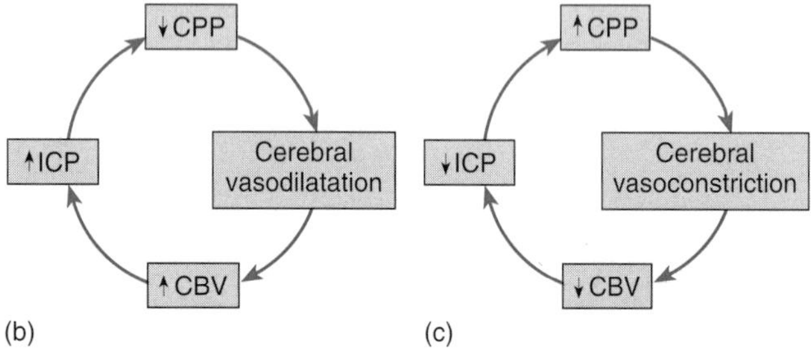

Fig. 1 Schematic representation of the effect of changes in cerebral perfusion pressure (CPP) on cerebral blood flow (CBF), cerebrovascular resistance (CVR), and cerebral blood volume (CBV) within the limits of autoregulation (cerebrovascular resistance and cerebral blood volume curves are non-quantitative). (a) Cerebral blood flow is maintained when cerebral perfusion pressure changes by compensatory changes in cerebrovascular resistance. The changes in cerebrovascular resistance are caused by changes in cerebral vasomotor tone and hence are accompanied by changes in cerebral blood volume, which may have significant effects on intracranial pressure when intracranial compliance is reduced. These dynamic compensatory mechanisms do not operate effectively outside the limits of autoregulation, and reduction of cerebral perfusion pressure below the lower autoregulatory threshold will result in cerebral blood flow reductions and cerebral ischemia. In acute head injury this autoregulatory threshold may be elevated to 70 mmHg. (b) Reductions in cerebral perfusion pressure result in reductions in cerebrovascular resistance. The accompanying increases in cerebral blood volume cause elevation of intracranial pressure and further reduction in cerebral perfusion pressure. (c) Increases in cerebral perfusion pressure initiate vasoconstriction and hence reductions in cerebral blood volume and intracranial pressure.

General issues

The use of systemic dehydration for the management of raised intracranial pressure is inappropriate, since reductions in mean arterial pressure trigger autoregulatory cerebral vasodilatation and increase cerebral blood volume and intracranial pressure (Fig. 1(b)). In general, current practice is to aim for euvolemia and maintain a cerebral perfusion pressure that ensures adequate cerebral perfusion. In many instances mean arterial pressure may need to be increased above normal levels to maintain cerebral perfusion pressure and cerebral blood flow in the face of intracranial hypertension, vasospasm, or regional vascular occlusion.

Fluid therapy

A detailed discussion of fluid therapy is given by Drummond (1995). Fluid therapy is titrated to clinical and laboratory assessment of volume status, but generally involves the administration of about 2.5 to 3 l/day of maintenance in the average adult (30–40 ml/kg/day). Unlike other vascular beds, capillaries in the brain are impermeable to most small molecules including sodium, and fluid flux across an intact blood–brain barrier is primarily governed by osmolality rather than plasma oncotic pressure. Consequently, plasma osmolality is often elevated with mannitol (up to 320 mosmol/l), and it is important to avoid hypotonic fluids, which may reduce plasma osmolality and worsen cerebral edema. Glucose-containing solutions are generally avoided, since the water that accompanies the active transport of dextrose across the blood–brain barrier can worsen cerebral edema, and hyperglycemia can worsen outcome in cerebral ischemia. Disruption of the blood–brain barrier might be expected to reduce the effects of osmolality on fluid flux and increase the relevance of plasma oncotic pressure, but this issue is as yet unresolved. It would seem prudent, however, to avoid marked reductions in colloid oncotic pressure.

Hemodynamic monitoring

Hemodynamic targets are set to optimize cerebral blood flow and oxygen delivery using either data from the literature or bedside multimodality monitoring (cerebral perfusion pressure, jugular oximetry

(Sjo_2), or transcranial Doppler flow measurements) as indices of adequate cerebral perfusion in individual patients.

Thus target levels of mean arterial pressure and cardiac index vary between disease states and individual patients, but their attainment follows generally accepted critical care practice and fluid loading with colloids is used as a first step. In the past this was titrated to central venous pressure. However, myocardial dysfunction may occur even in young neurosurgical patients as a consequence of myocardial contusion in trauma or subendocardial ischemia in subarachnoid hemorrhage, and many of these patients have significant acute lung injury. Therefore right-sided filling pressures may not provide an accurate index of volume status, and nimodipine (which is routinely used in subarachnoid hemorrhage) results in unpredictable hemodynamic consequences. These considerations have prompted the increasing use of pulmonary artery catheters for the measurement of pulmonary capillary wedge pressure, cardiac index, and systemic vascular resistance.

Hemodynamic manipulation

Hemodilution secondary to fluid therapy may reduce blood viscosity and improve microcirculatory characteristics. However, it may also cause a reduction in hemoglobin levels with a net decrement in tissue oxygen delivery. Thus hematocrit levels are generally maintained above 30 per cent.

If volume loading does not achieve the hemodynamic targets, vasoactive agents are administered. Most published work describes the use of dopamine, phenylephrine, or norepinephrine (noradrenaline), with the emphasis being on mean arterial pressure elevation rather than cardiac index enhancement. Cerebral perfusion pressure targets of 60 to 70 mmHg are sought. Guidance regarding cardiac index targets are not available for most conditions, but in our experience isolated cardiac index elevation with dobutamine and dopexamine have not been useful in maintaining cerebral perfusion.

Specific disease states

Acute head injury

Detailed discussions of the management of acute head injury are given by the American Association of Neurological Surgeons (1995) and Drummond (1995).

Direct measurements of cerebral blood flow have demonstrated significant ischemia in the early period after trauma, and postmortem studies in fatal head injury implicate ischemia as an important pathophysiological process. Data from the National Traumatic Coma Data Bank have shown that hypotension at presentation and during emergency surgery (systolic blood pressure below 90 mmHg) is an important correlate of poor outcome following head injury. Several studies have shown that maintenance of a relatively high cerebral perfusion pressure can significantly improve outcome following head injury. The maintenance of cerebral perfusion pressure above 60 to 70 mmHg is associated with an improved outcome. Several methods of assessing the cerebral circulation suggest that, in pooled data in a population of head-injured patients, a cerebral perfusion pressure of 60 to 70 mmHg represents a lower breakpoint for cerebrovascular autoregulation. However, optimal cerebral perfusion pressure levels in some individual patients may be higher, and hyperemia, detected by transcranial Doppler and Sjo_2 monitoring, may suggest the need for hyperventilation or further sedation rather than cerebral perfusion pressure elevation in others. While hemodynamic targets ideally need to be based on specific pathophysiological patterns in individual patients, a cerebral perfusion pressure of 70 mmHg provides a useful starting point for initial management, achieved using adequate hydration (with maintenance of plasma osmolality), colloid loading, and vasoactive agents. Where autoregulation is preserved, further elevation of cerebral perfusion pressure may induce cerebral vasoconstriction and reduce cerebral blood volume and intracranial pressure (Fig. 1(c)).

A complex vasoregulatory protocol involving clonidine, β-blockers, dihydroergotamine, and elevation of plasma oncotic pressure may be helpful, although there are no data directly comparing this regime with the standard cerebral perfusion pressure oriented hemodynamic management protocols used by most neurotrauma centers (Asgiersson *et al.* 1995).

Aneurysmal subarachnoid hemorrhage

Autoregulatory responses are known to be impaired following acute subarachnoid hemorrhage, and hypovolemia and hypotension are poorly tolerated. Disturbances of fluid and electrolyte balance, in the form of cerebral salt wasting, the syndrome of inappropriate ADH secretion, or diabetes insipidus, are also common. Therefore adequate hydration, careful volume assessment including central venous pressure measurement, and maintenance of a normal mean arterial pressure (80–100 mmHg) are mandatory following acute subarachnoid hemorrhage. Specific hemodynamic management following aneurysmal subarachnoid hemorrhage is governed by whether vasospasm is present, whether the aneurysm has been clipped, and the hemodynamic response to nimodipine therapy (Fig. 2).

Many patients with aneurysmal subarachnoid hemorrhage develop delayed neurological deficits, typically between 4 and 10 days following hemorrhage, often associated with angiographic evidence of vasospasm and reductions in cerebral blood flow. These delayed ischemic neurological deficits may be prevented or reduced by Ca^{2+}-channel blockers such as nimodipine. Established delayed ischemic neurological deficits have been successfully treated with a combination of induced hypervolemia, hemodilution, and hypertension (triple-H therapy) (McGrath *et al.* 1995). Hypervolemia is typically achieved by crystalloid and colloid loading to achieve central venous pressure levels of 8 to 12 mmHg, or pulmonary artery wedge pressure levels of 12 to 16 mmHg, and a hematocrit of 30 to 35 per cent. Depending on plasma sodium levels, either desmopressin or fludrocortisone can be used to assist volume loading by preventing diuresis while maintaining normal plasma sodium levels. Phlebotomy may be needed in addition to volume loading to achieve hematocrit targets in some polycythemic patients. Most reports describe systolic blood pressure targets of 160 to 240 mmHg, but hypertension may be titrated to reversal of delayed ischemic neurological deficits (which may not be immediate). Phenylephrine, norepinephrine, and dopamine have commonly been used as adjuncts to volume loading to achieve hypertension (occasionally along with atropine). Dobutamine has also been used to maintain a high cardiac index (approximately 5.0 l/m^2/min).

The presence of an unclipped recently ruptured aneurysm is a self-evident contraindication to aggressive triple-H therapy, and maintenance of a systolic blood pressure of 140 to 150 mmHg is more appropriate in such patients. Indeed, the ability to treat delayed

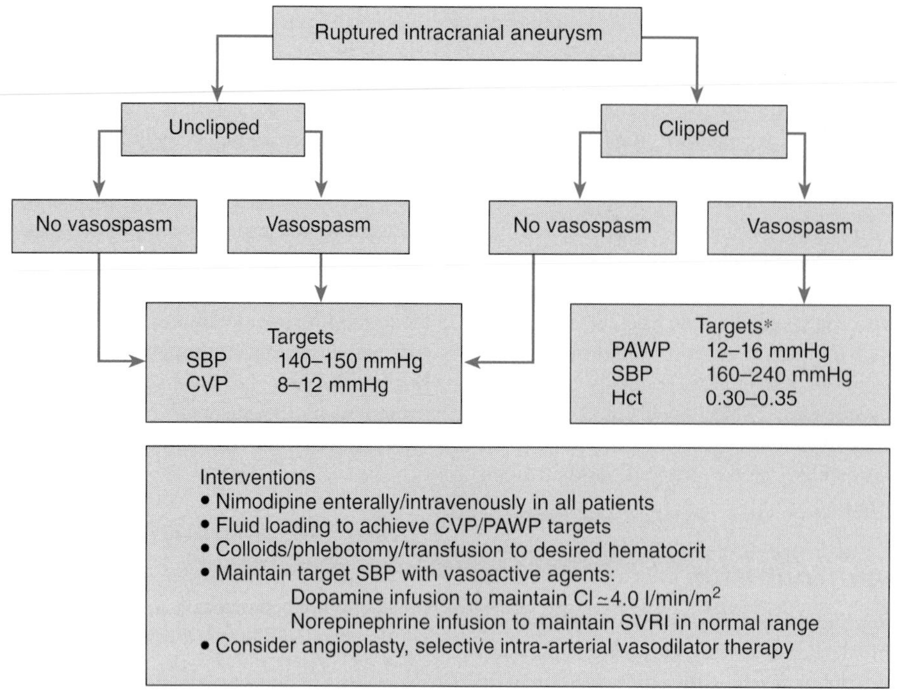

Fig. 2 Flowchart for hemodynamic management in aneurysmal subarachnoid hemorrhage: CVP, central venous pressure; SBP, systolic blood pressure; Hct, hematocrit; PAWP, pulmonary artery wedge pressure; CI, cardiac index.
*Triple-H therapy should be used with caution in the presence of myocardial ischemia or established cerebral infarction, and may need to be withdrawn if fluid overload results in pulmonary edema.

ischemic neurological deficits effectively with triple-H therapy provides one major justification for early clipping in subarachnoid hemorrhage. Complications of aggressive triple-H therapy include pulmonary edema, myocardial ischemia and infarction, and, particularly in patients with an established low-density lesion on CT scanning, intracranial hemorrhage into an established infarction. Therefore special care is required in patients with myocardial ischemia, cardiac decompensation, or established cerebral infarction. Such patients, and those who fail triple-H therapy, are early candidates for balloon angioplasty or selective pharmacological vasodilatation of vasospastic lesions. The use of nimodipine, particularly by the intravenous route, can result in profound vasodilatation, but this response is inconsistent, and comprehensive hemodynamic monitoring is mandatory in these patients before norepinephrine or other vasopressors are employed (Fig. 2).

Non-aneurysmal intracranial hemorrhage

Both rebleeding and vasospasm are thought to be much less common following intracranial hemorrhage due to hypertension or vascular malformations, and therapy is centered on the maintenance of cerebral perfusion and the treatment of mass effect due to the bleed and reactive edema. Typically, this involves maintaining a mean arterial pressure about 10 to 20 per cent above normal values (depending to some extent on the patient's premorbid blood pressure, where known) and avoiding marked hypertension which could precipitate further hemorrhage.

However, these targets may need to be modified in some patients with arteriovenous malformations, in whom the normal brain surrounding the lesion is exposed to lower perfusion pressures due to steal by the malformation. Restoration of higher regional perfusion pressures following excision of the arteriovenous malformations may exceed the upper limits of autoregulation in this previously ischemic tissue, resulting in cerebral edema and hemorrhage which may require sedation, artificial ventilation, and controlled hypotension.

Miscellaneous

Although triple-H therapy has been used in cerebral ischemia other than in the setting of aneurysmal subarachnoid hemorrhage, the available literature does not provide consistent objective evidence for such an approach. However, if intracranial hemorrhage is excluded, maintenance of relatively high mean arterial pressure values may maximize collateral supply and minimize or prevent neuronal loss. This approach must be seen as an adjunct to specific treatment where available, such as thrombolysis for hyperacute ischemic stroke, anticoagulation and/or thrombolysis for vascular occlusion during the course of interventional neuroradiological procedures, and re-exploration of a thrombosed vessel following carotid endarterectomy.

Rarely, patients exhibit persistent cerebral hyperemia (by transcranial Doppler or cerebral blood flow measurement) following carotid endarterectomy, because postoperative increases in perfusion pressure in a previously ischemic cerebrovascular bed exceed the upper limit of autoregulation. Such patients are at high risk of cerebral edema or hemorrhage, and may require a short period of careful blood pressure control at the lower end of their normal range while normal cerebrovascular reactivity is restored.

References

American Association of Neurological Surgeons and the Brain Trauma Foundation—Joint Section on Neurotrauma and Critical Care (1995). *Guidelines for the management of severe head injury*, pp 95–104. Brain Trauma Foundation, New York.

Asgiersson, B., Grande, P.O., and Nordstrom, C.H. (1995). The Lund concept of post-traumatic brain edema therapy. *Acta Anaesthesiologica Scandinavica*, **39** (Supplement 105), 103–6.

Drummond, J.C. (1995). Fluid management of head injured patients. *Acta Anaesthesiologica Scandinavica*, **39** (Supplement 105), 107–11.

McGrath, B.J., Guy, J., Borel, C.O., Friedman, A.H., and Warner, D.S. (1995). Perioperative management of aneurysmal subarachnoid hemorrhage. *Anesthesia and Analgesia*, **81**, 1295–1302.

Menon, D.K. (1995). Cerebral circulation. In *Cardiovascular physiology* (ed. H.J. Priebe and K. Skarvan), pp. 198–223. BMJ Publishing, London.

14.4.3 Ventilatory management

M. Smith

Key messages

- Respiratory complications are an important cause of morbidity and mortality in neurosurgical patients.
- Normal arterial blood gases are a prerequisite for good neurological recovery after brain injury.
- Respiratory failure in neurosurgical patients may occur secondary to intra- and extracranial complications.
- Neurogenic pulmonary edema is caused by an increase in sympathetic activity secondary to a combination of factors.
- Comatose patients and those with bulbar palsies require careful respiratory management and airway protection.
- Therapeutic hyperventilation is an important tool in the acute management of intracranial hypertension but its use over several days has never been shown to improve outcome.
- Overzealous hyperventilation may precipitate cerebral ischemia secondary to cerebral vasoconstriction.
- Cerebral perfusion pressure must be maintained during intensive respiratory support of neurosurgical patients by adjustment of posture and level of positive end-expiratory pressure and by cardiovascular support.

Introduction

Respiratory complications are an important cause of morbidity and mortality in neurosurgical patients and contribute more than any other organ system to postoperative complications. Respiratory failure occurs in 20 per cent of head-injured patients and may be secondary to abnormal respiratory patterns, pulmonary edema, or pulmonary aspiration. Up to 54 per cent of head-injury deaths have been associated with respiratory failure. There is a 17.5 per cent rate of pulmonary complications in the elderly after elective surgery for intracranial malignancy, with 4 per cent of deaths due to respiratory problems.

Normal arterial blood gas tensions are a prerequisite for good neurological recovery after neurosurgery and brain injury. Hypoxemia is associated with poor outcome after brain injury. As the resistance of cerebral vessels is metabolically determined, changes in $Pa\text{CO}_2$, which cause changes in extracellular pH, modulate cerebral blood flow. Rises in $Pa\text{CO}_2$ secondary to hypoventilation lead to rises in cerebral blood flow which may precipitate rises in intracranial pressure. Intracranial hypertension, like hypoxemia, is associated with poor outcome after brain injury. Although moderate hyperventilation is used as a therapeutic tool in the management of acute intracranial hypertension, spontaneous hyperventilation must be avoided as excessive falls in $Pa\text{CO}_2$ may compromise cerebral blood flow and precipitate cerebral ischemia.

Indications for controlled ventilation

It is essential that arterial blood gases are maintained within normal limits. It is accepted that intervention is required to maintain $Pa\text{O}_2$ above 8.5 kPa and $Pa\text{CO}_2$ between 3.8 and 5.5 kPa. Controlled ventilation is instituted in neurosurgical patients for support during respiratory insufficiency or failure, for protection of the airway, for treatment of respiratory depression, or for therapeutic hyperventilation (Table 1).

Respiratory failure

Respiratory insufficiency may occur secondary to aspiration of gastric contents. The incidence of aspiration is highest at the time of the head injury and in the perioperative period, although it can occur at any time. Aspiration may precipitate the well-known sequence of pneumonia, respiratory failure, and acute lung injury.

Basilar atelectasis is prevalent in patients with ineffective cough and gag reflexes, including those with depressed conscious level or bulbar palsies. Pneumonia is a common sequel and a significant cause of mortality in neurosurgical patients. The loss of the normal sighing mechanism secondary to a decreased level of consciousness also predisposes to atelectasis and pneumonia. Early intubation, continuous positive airway pressure, and intermittent mandatory

Table 1 Indications for controlled ventilation in the neurosurgical patient

Respiratory insufficiency/failure
Aspiration
Pneumonia
Neurogenic pulmonary edema
Pulmonary/fat embolus
Acute lung injury
Chest trauma
Head injury
Airway protection/maintenance
Unconscious
Absent gag
Transfer
Respiratory depression
Central
Drug-induced
Therapeutic hyperventilation
Head injury
Postoperative

ventilation, in association with aggressive physiotherapy, may help to prevent pulmonary complications.

Neurogenic pulmonary edema is a rapidly developing protein-rich alveolar edema that develops after any neurological insult, but most commonly following head injury. Neurogenic pulmonary edema is caused by a combination of factors whose common denominator appears to be increased sympathetic activity. This results in rises in left atrial pressure, systemic hypertension, and pulmonary vasoconstriction, which cause increased pressure in the pulmonary capillary bed and direct endothelial damage. Sympathetic nerve stimulation may lead directly to an increase in pulmonary capillary permeability to protein, perhaps via sympathetic innervation of contractile elements in endothelial cells. In this way the tight junctions in the capillary may be opened, allowing protein to leak out. Treatment of neurogenic pulmonary edema involves general cardiovascular support, as indicated by invasive monitoring including pulmonary artery catheterization, and respiratory support which often requires a high inspired oxygen concentration and high levels of positive end-expiratory pressure (PEEP). Adrenergic blockade has also been suggested as a means of attenuating the systemic effects of neurogenic pulmonary edema by blocking hypertensive responses.

Forty per cent of head-injured patients have other injuries, including long-bone and pelvic fractures, which may be the source of fat emboli. These can cause respiratory complications ranging from subclinical effects on gas transfer to acute lung injury.

Neurosurgical patients fall into the high-risk category for thromboembolic disease; they are often immobile for long periods and have undergone prolonged surgical procedures. However, postoperative neurosurgical patients have higher rates of pulmonary embolism than can be explained by these risk factors alone and prophylaxis against deep venous thrombosis and pulmonary emboli should be universal. Out of concern for catastrophic postoperative hemorrhage, many neurosurgeons do not advocate the use of low-dose heparin in the perioperative period. However, physical methods of prophylaxis, such as graduated compression stockings and intermittent calf compression, are equally effective if applied throughout the period of immobility. In patients with prolonged periods of immobility, subcutaneous heparin should be started when the risk of postoperative bleeding has passed.

Neurosurgical patients with chest trauma significant enough to affect gas transfer must be artificially ventilated. Flail chest in the neurosurgical patient should always be managed with controlled ventilation.

Hypoxemia is a common complication of head injury even in patients without clinical evidence of chest wall or pulmonary pathology. The etiology is unclear but may be related to ventilation–perfusion mismatches due to persistent perfusion of underventilated lung. Arteriovenous shunting secondary to autonomic discharge may also be a factor. It has been shown that the functional residual capacity may be reduced by up to a third after severe head injury, and this is directly related to the fall in $P\text{aO}_2$. In these circumstances the application of small amounts of PEEP successfully reverses the hypoxemia.

Airway maintenance and protection

Comatose patients or those with bulbar palsies require careful airway and respiratory management to maintain and protect the airway. The pharynx is the most common site of airway obstruction in comatose patients because of decreased tone in pharyngeal muscles and posterior displacement of soft tissue due to the gravity effects of the supine position. Trauma patients may also have local injury and swelling.

Protective airway reflexes may also be absent in those with a decreased consciousness level or bulbar palsies. The risk of aspiration of gastric contents during passive regurgitation can be minimized by airway protection with an endotracheal tube or tracheostomy. Protective airway reflexes are lost in a predictable manner as the consciousness level decreases; the pharyngeal reflex is lost first, followed by the laryngeal and finally the tracheal reflex.

As a general rule patients with a Glasgow Coma Scale score below 8 require airway control and protection. Postoperative neurosurgical patients may require transfer between departments (e.g. for CT scanning), and it is imperative that the trachea is intubated and ventilation is controlled if the consciousness level is deteriorating or the patient is already comatose.

Respiratory depression

Hypoventilation in the immediate postoperative period may occur because of the residual effects of anesthetic drugs. Central respiratory depression may result from opioids or inhalational anesthetic agents and peripheral effects from muscle relaxants. As hypoxemia and hypercarbia must be avoided, artificial ventilation should be continued into the postoperative period if there is evidence of hypoventilation at the end of surgery.

Respiratory depression and abnormal respiratory patterns may also occur because of central effects related to the intracranial pathology itself. A decreased consciousness level can result in hypoxemia secondary to obstruction of the airway (see above), but rarely causes hypoventilation *per se*. However, specific lesions in the central nervous system are correlated with specific abnormalities of respiratory pattern. Cheyne–Stokes respiration is associated with simultaneous dysfunction of both cerebral hemispheres or the basal ganglia. Lesions in the pons may cause central neurogenic hyperventilation and irregular breathing patterns such as cluster or apneustic breathing. These are often preterminal and may be seen following midbrain compression during tentorial herniation. Ataxic breathing is a sign of medullary failure.

Even in the absence of midbrain compression, spontaneous hyperventilation is common in head-injured patients and can also be seen after intracranial surgery. Ventilation must be controlled to prevent excessive reduction in cerebral blood flow secondary to the fall in $P\text{aCO}_2$.

Therapeutic hyperventilation

The use of moderate hyperventilation is advocated for traumatic coma, but considerable controversy remains as to whether or not it is an effective or safe therapy. The rationale for hyperventilation is based on two premises. It reduces intracranial pressure and reverses cerebral lactic acidosis which, like intracranial hypertension, is correlated with poor outcome after brain injury.

The fall in intracranial pressure induced by hyperventilation occurs secondary to constriction of cerebral vessels and a consequent reduction in cerebral blood flow and volume. Cerebral blood flow decreases by about 15 per cent for each 0.5-kPa reduction in $P\text{aCO}_2$. The reactivity of cerebral vessels to CO_2 appears to be maintained in many patients following head injury, and this argues for the effectiveness of hyperventilation as a therapy for intracranial hypertension. As might be expected, the response to hyperventilation is greatest in those with cerebral hyperemia or 'luxury perfusion'. Since these are the patients who are most likely to develop intracranial hypertension, their relatively greater responsiveness may be of therapeutic significance. However, following closed head injury some patients develop a phase of low cerebral blood flow which may persist for 12 to 24 h. Such patients may be at risk of cerebral ischemia following a further reduction in cerebral blood flow by hyperventilation.

Hyperventilation also reverses the cerebral metabolic acidosis commonly present after head injury. As CO_2 freely passes the blood–brain barrier, the respiratory alkalosis produced by hyperventilation causes cerebral CO_2 to be depleted. The restoration of cerebral autoregulation in patients with global hyperemia is likely to be related to this reversal of cerebral acidosis.

Changes in cerebral blood flow and intracranial pressure may become refractory to hyperventilation because of a normalization of the intracellular and extracellular pH of the central nervous system by the bicarbonate buffering mechanism. In some cases accumulation of lactic acid within the brain may also play a part; this may be due to excessive hyperventilation which can lead to the development of cerebral ischemia and cerebral lactic acidosis. This may explain why patients with cerebral hyperemia are least refractory to prolonged hyperventilation. Maintenance of the reactivity of cerebral vessels to CO_2 does not ensure continued control of intracranial pressure, as progressive brain edema might outweigh the effects of intracranial pressure from therapeutic decreases in cerebral blood volume. It has also been shown that unacceptable rises in intracranial pressure can occur when hyperventilation is withdrawn after many days.

Other beneficial effects of hyperventilation might include a reduction in energy requirement because of the reduced work of breathing and a reduction in cerebrospinal fluid formation, although these are unlikely to be of clinical significance.

An inverse steal (Robin Hood phenomenon) has been reported in severe head injury and brain tumors following hyperventilation. Blood flow to uninjured brain is reduced by hyperventilation, while the cerebral vessels in the injured area do not constrict and blood flow is preferentially diverted to that area.

Hyperventilation can also have deleterious effects. By reducing cerebral blood flow, hyperventilation may change borderline cerebral ischemia into full-blown ischemia. For example, active hyperventilation of awake unsedated subjects to a $Pa\text{CO}_2$ of 2.7 kPa induces electroencephalographic changes compatible with cerebral hypoxia. In some comatose patients treated with prolonged hyperventilation, an improvement in conscious level has been seen when the $Pa\text{CO}_2$ was allowed to return to normal values.

One of the supposed benefits of hyperventilation is the induction of an inverse steal phenomenon as described above. However, recent evidence suggests that, under certain circumstances, preferential blood flow to the injured area may result in progressive edema in this area and a consequent reduction in local oxygen supply.

Hyperventilation causes a leftward shift of the oxygen–hemoglobin dissociation curve and inhibition of oxygen delivery to the tissues. This will reduce oxygen delivery to the brain and may also reduce cardiac output if myocardial oxygen delivery is impaired. Artificial ventilation increases mean intrathoracic pressure, and this may further decrease cardiac output and adversely affect cerebral blood flow. Large increases in intrathoracic pressure can indirectly increase intracranial pressure because of the rise in central venous pressure.

If adaptation to prolonged hyperventilation occurs, rebound intracranial hypertension may result when hyperventilation is discontinued. As the central nervous system pH is decreased via the buffering mechanism during adaptation, a relative central nervous system acidosis and concomitant rise in cerebral blood flow can occur as the $Pa\text{CO}_2$ is allowed to rise. This will persist until the buffering system has had time to recorrect the pH.

No controlled clinical studies have shown an improvement in outcome when prolonged hyperventilation is used as a therapy for intracranial hypertension. There has been a trend away from the use of empirical hyperventilation and towards limiting the drop in $Pa\text{CO}_2$ to moderate levels in those in whom it is indicated. A $Pa\text{CO}_2$ of 4.0 to 4.5 kPa is now accepted as being the endpoint for therapeutic hyperventilation in many units. Monitoring must also be used to titrate hypocapnea, so that reductions in intracranial pressure are achieved while cerebral oligemia is avoided. Monitoring of jugular venous saturation ($Sj\text{O}_2$) has been used to assess the relative adequacy of cerebral oxygen supply.

Although the reduction in cerebral blood volume may be small compared with total brain volume, hypocapnia can be lifesaving in patients with an expanding cerebral lesion.

Problems with controlled ventilation

The problems of controlled ventilation in the neurosurgical patient include those common to other patient groups. They include effects on cardiac output, hormonal changes, risks of barotrauma, and the introduction of infection. Additionally, sedation may be required to facilitate mechanical ventilation, and this may prevent neurological assessment.

PEEP is a necessary adjuvant in the treatment of many patients requiring mechanical ventilation for respiratory failure. However, it can have adverse effects on intracranial pressure which may be disadvantageous in neurosurgical patients. Rises in central venous pressure during PEEP may impair cerebral venous return and increase intracranial pressure. The effect of PEEP on intracranial pressure is variable, but is greatest in those with a low intracranial compliance. The intracranial effects of PEEP can be minimized by positioning the patient in a head-up position (15°–30°), with the neck in a moderate degree of flexion, to encourage maximal cerebral venous drainage. PEEP levels of up to 10 to 12 cmH_2O have little effect on intracranial pressure under such circumstances. PEEP is also less easily transmitted to the intracranial compartment if the pulmonary compliance is low.

Invasive hemodynamic and intracranial pressure monitoring are necessary in all neurosurgical patients requiring more than modest levels of PEEP. In this way the cerebral perfusion pressure (CPP) can be measured and maintained using the following relationship:

$$\text{CPP} = \text{MAP} - \text{ICP}$$

where MAP is the mean arterial pressure and ICP is the intracranial pressure.

The CPP should be maintained at levels between 60 and 70 mmHg by reduction of PEEP, augmentation of blood pressure with inotropes, or reduction in intracranial pressure by changes in the position of the patient or the use of specific therapies.

Conclusion

Respiratory abnormalities and problems of airway management are common in patients with a decreased consciousness level or bulbar palsies. Respiratory insufficiency plays a major role in exacerbating cerebral injury in neurosurgical patients, and careful management of the airway and ventilation is essential.

Bibliography

Baigelman, W. and O'Brien, J.C. (1981). Pulmonary effects of head trauma. *Neurosurgery*, **9**, 729–40.

Demling, R.H. and Riessen, R. (1993). Respiratory failure after cerebral injury. *New Horizons*, **1**, 440–6.

Muizelaar, J.P., *et al.* (1991). Adverse effects of prolonged hyperventilation in patients with severe head injury: a randomized clinical trial. *Journal of Neurosurgery*, **75**, 731–9.

Warters, R.D. and Allen, S.J. (1994). Hyperventilation: new concepts for an old tool. *Current Opinion in Anaesthesiology*, **7**, 391–3.

Wauchob, T.D., Brooks, R.J., and Harrison, K.M. (1984). Neurogenic pulmonary oedema. *Anaesthesia*, **39**, 529–34.

14.5 Post-transplantation surgery

14.5.1 Renal transplantation

I. J. A. Davidson and M. Ulrich

Key messages

- Many measures that can improve the success of renal transplantation are known.
- Albumin administration has a positive effect on post-transplant function.
- Calcium antigonists can compete with cyclosporine A (cyclosporin A) for elimination and increase renal blood flow.
- Although many protocols are available, a drug regimen of steroids, cyclosporine A, verapramil, mycophenolate mofetil, and antithymocyte globulin is useful.

Introduction

The short-term outcome of renal transplantation depends on the status and management of the kidney organ donor, the warm and cold ischemia times, the circumstances surrounding organ recovery, and the recipient's surgery. Up to 10 per cent of transplanted cadaver kidneys have initial non-function, with the recipients requiring temporary post-transplant hemodialysis. Reperfusion injury, precipitated by lack of oxygen, is believed to play a major role, causing microvascular injury in solid organs after transplantation (Ar'Rajab *et al.* 1996)

Long-term outcome is more likely to be determined by immunological factors and the post-transplant immunosuppression protocols employed, as well as being influenced by other perioperative factors, particularly fluid and drug regimens. Although transplant protocols vary widely between centers, the postoperative goals are similar and include early graft function, as evidenced by immediate urine output, avoidance of post-transplant hemodialysis, short hospital stay, prevention of acute and chronic rejection, long-term function and graft survival, and an improved quality of life.

Post-transplant delayed function and acute rejection

The occurrence of the delayed onset of kidney function after transplantation, or delayed function (defined as the need for hemodialysis), is associated with poor outcome including lower graft survival. Therefore it is exceedingly important to identify and institute measures to induce immediate kidney function after surgery. The second major factor leading to graft failure is the occurrence of acute rejection. Designing an immunosuppressive drug protocol that prevents acute rejection but does not lead to increased morbidity in terms of infections and malignancies is a continuing process. Here we describe a perioperative fluid and drug protocol that results in less than 10 per cent incidence of delayed function with excellent renal function, less than 20 per cent biopsy-proven rejection or renal toxicity, and long-term graft survival exceeding 90 per cent.

Perioperative intravascular hemodynamics

The effect of delayed function on outcome is significant. For instance, graft survival at 1 year may be up to 25 per cent worse in patients with delayed function, with a mortality exceeding 10 per cent. Patients with diminished plasma and blood volume are more likely to have delayed onset of urine output. The loss of intravascular albumin during a renal transplant (approximately 1 g/kg body weight) corresponds to about 18 ml/kg body weight, or about 1200 ml of plasma volume in a 70-kg man (Dawidson and Ar'Rajab 1992). Maintaining blood and plasma volume results in earlier and better post-transplant renal function. An important intraoperative factor affecting post-transplant function in cadaver renal transplant recipients is the maintenance of intravascular volume by administration of a colloid such as albumin in a dose of 1.0 to 1.5 g/kg, in addition to about 30 ml/kg of electrolyte solution (Table 1).

Perioperative use of calcium antagonists

Calcium antagonists have a major role in preventing post-transplant delayed function as well as cyclosporine-induced renal toxicity (Loutzenhiser and Epstein 1990). Even though all calcium antagonists exert beneficial hemodynamic renal effects, only verapamil and diltiazam will increase cyclosporine A blood concentration, an effect that can be used to enhance immunosuppression early after transplantation. Verapamil can be given in slow-release formulation (120 mg) preoperatively as part of the premedication, followed by 5 to 10 mg given intraoperatively into the external iliac artery or the aortic patch while the iliac artery distal to the anastomosis is temporarily occluded. Postoperatively, verapamil is started orally (120 mg slow-release formulation) twice daily and continued chronically. Rarely does the verapamil dose have to be decreased in cases of hypotension or peripheral edema. More often, it needs to be increased for treatment of hypertension. Second-line antihypertensive agents include a β-blocker or diuretic. Angiotensin-converting enzyme

Table 1 Beneficial effects of intraoperative administration of albumin in a dose of 1.2–1.5 g/kg in 437 cadaver renal transplants

	Albumin	No albumin	*P*
Urine onset < 30 min (% of cases)	75	39	0.001
Urine volume in first 24 h postoperation (liters)	6.0	3.7	0.0003
Glomerular filtration rate (ml/min)			
Day 1	33	22	0.01
Day 7	47	28	0.003
Delayed function (%)	12	34	0.0001
Graft survival at 1 year (%)	78	49	0.002
Patient survival at 1 year	94	89	0.202

Modified with permission from Dawidson *et al.* (1992).

inhibitors are avoided because of their tendency to induce renal failure in post-transplant patients.

The perioperative use of verapamil or diltiazem significantly decreases the incidence of delayed function. Also, renal blood flow after surgery increases dramatically with verapamil compared with controls. The decrease in renal flow induced by cyclosporine A is prevented by verapamil (Dawidson and Ar'Rajab 1992) and graft survival is significantly improved when compared with controls, with a predicted graft survival at 1 year of 96 per cent. Repeat transplants with verapamil now have a graft survival similar to that of first transplants without verapamil, while repeat transplants without verapamil have a predicted graft survival at 1 year of only 50 per cent. Additionally, the lower incidence of rejection in patients on verapamil may be secondary to the increased cyclosporine A blood levels induced by verapamil through competitive metabolism via the cytochrome P-450 liver enzyme system.

The combined effect of albumin and verapamil

The combined effect of verapamil and albumin has been shown to be additive in decreasing the risk of developing delayed function by a factor of 5.4 (Table 2) (Dawidson *et al.* 1994). Verapamil and albumin

Table 2 Odds ratio estimates for the decreased risk of delayed function after cadaver renal transplant with perioperative verapamil and albumin

	Decreased risk of delayed function	95% confidence limits	*P*
Verapamil	2.4	1.1–5.0	0.028
Albumin	2.3	1.0–5.2	0.018
Verpamil+albumin	5.4	2.9–10.0	0.001

Table 3 Beneficial effects of combined albumin (1.2–1.5 g/kg) and verapamil (10 mg intra-arterially) in cadaver renal transplants (mean±SD)

	Albumin+verapamil	None	*P*
Urine output in first 24 h (liters)	5.7±3.7	2.2±2.1	0.0004
Serum creatinine at day 7 (mg/100 ml)	2.1±1.8	4.3±3.5	0.0004
Delayed function (%)	12	42	0.0001
Graft survival at 1 year (%)	84	62	0.01

Table 4 Summary of a current intraoperative fluid and drug management protocol for renal transplantation

- Intraoperative albumin administration of 1.0–1.5 g/kg. This amounts to approximately 25 g/h intravenous infusion of either 25% or 5% albumin solutions. A central venous pressure of 15–18 mmHg at the time of vascular declamping is sought.
- Electrolyte (i.e. saline) volume infusions are limited to about 30 ml/kg. Usually, this is a slow infusion of 700–1000 ml/h in a 70-kg man based on dialysis history, 'dry weight', and actual body weight.
- Mannitol infusions are avoided since the colloid (albumin) administration provides the short-acting blood volume expansion effect of mannitol (mannitol also induces histological changes in the kidney that may be confused with cyclosporine toxicity).
- Before reperfusion, 125 mg solumedrol is given intravenously.
- At the time of kidney reperfusion 80–100 mg furosemide (frusemide) is given. Intravenously, larger doses have not been proven effective in increasing urine volume. In living related transplants, 20-mg increments may be given.
- After vascular clamps are removed, up to 10 mg (2.5-mg increments) verapamil is injected into the renal artery, observing blood pressure. Blood pressure drop is unusual with albumin blood volume expansion.
- Maintain systolic arterial blood pressure above 100 mmHg.

used together lead to larger urine output, a lower serum creatinine, reduced incidence of delayed function, and improved graft survival (Table 3). One protocol for using albumin and verapamil in renal transplantation is summarized in Table 4. The beneficial effects and possible mechanisms of verapamil in the transplant situation are shown in Table 5.

Table 5 Possible mechanism and beneficial effects of the action of verapamil on cadaver renal transplantation

1. Cellular protection from ischemia
2. Selected vasodilatation of the afferent arteriole preventing decreased blood flow resulting from ischemic injury and cyclosporine-induced vasospasm
3. Induction of elevated cyclosporine A blood levels through competitive metabolism with the liver cytochrome P-450 enzyme system, resulting in better immunosuppression
4. Additive effect with intraoperative volume expansion (albumin and electrolyte administration) reducing the need for hemodialysis
5. Inherent immunosuppressive properties

Early immunosuppression protocols

The immediate post-transplant immunosuppression protocol has a profound impact on the incidence of acute rejection and subsequent graft survival. There is a delicate balance between the desired goal of preventing acute rejection and minimizing side-effects from over-immunosuppression. Complications due to over-immunosuppression are serious and can impact graft as well as patient survival. They include opportunistic infections, malignancies, and organ toxicity, particularly nephrotoxicity from cyclosporine A or tacrolimus (FK 506). Most American transplant immunosuppression protocols are variations of quadruple drug induction therapy. In addition to

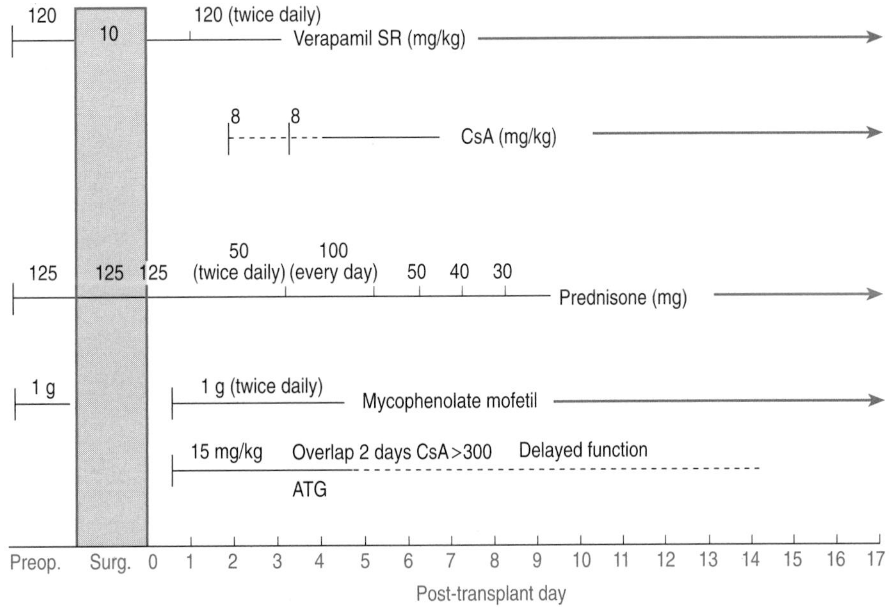

Fig. 1 The early post-transplant immunosuppression protocol currenty used in the Transplant Program at the State University of New York at Buffalo: ATG, antithymocyte globulin; CsA, cyclosporine A; SR, slow release formulation. Perioperatively, methylprednisolone 125 mg is given intravenously, followed by prednisone tapering to 30 mg daily by day 8. Antithymocyte globulin 15 mg/kg, diluted in 500 ml 5 per cent glucose or 0.9 per cent saline, is started within 6 h of surgery and infused over 4 to 6 h daily, overlapping with cyclosporine A (Neoral®) starting at 8 mg/kg orally. When cyclosporine blood levels have exceeded 300 ng/ml for two consecutive days, antithymocyte globulin is stopped. Mycophenolate mofetil (CellCept®) 1 g twice daily is started when the patient can take oral medication, usually within 24 to 48 h. Verapamil is started with 120-mg slow-release formulation as pre-medication; 10 mg verapamil is injected intra-arterially during/after the perfusion and given chronically in a dose of 120 mg twice daily.

corticosteroids in a tapering dose, cyclosporine A is introduced after a few days, overlapping with antithymocyte globulin or the monoclonal antibody OKT-3, until therapeutic cyclosporine A blood concentration is achieved. The introduction of cyclosporine A is delayed to allow the kidney to recover from the ischemic event. The fourth component of the protocol is azathioprine or, more recently, mycophenolate mofetil (RS 61443 CellCept®, Upjohn, Valhalla, NY).

An example of a quadruple induction protocol is given in Fig. 1. The specifics of this protocol include 375 mg of methylprednisolone given perioperatively in three doses, followed by prednisone, tapering the dose to 30 mg daily by day 8. Verapamil is given to protect the kidney from ischemia and cyclosporine-induced vasospasm, and also to increase blood cyclosporine A levels. The old formulation of cyclosporine A (Sandimmune®) has been replaced by a microemulsion formulation (Neoral®, Sandoz Pharmaceuticals, Hanover, NJ) with more consistent uniform absorption characteristics. Cyclosporine A therapy is initiated when serum creatinine is half the preoperative level or below 3.0 mg/dl. Recently, ketoconazole 50 mg daily has also been given with the first dose of cyclosporine A so that therapeutic cyclosporine A blood levels are obtained more rapidly. When cyclosporine A blood levels have reached 300 ng/ml for two consecutive days, antithymocyte globulin and ketoconazole are discontinued and subsequent cyclosporine A is administered at maintenance levels of at least 300 ng/ml for 16 weeks. In cases of delayed function, the introduction of cyclosporine A is delayed and antithymocyte globulin is continued for up to 14 days.

References

Ar'Rajab, A., Dawidson, I.J.A., and Fabia, R. (1996). Reperfusion injury. *New Horizons*, **4**, 224–34.

Dawidson, I. and Ar'Rajab, A. (1992). Perioperative fluid and drug therapy during cadaver kidney transplantation. In *Clinical transplants* (ed. P. Terasaki and M. Cecka), pp. 267–84. UCLA Tissue Typing Laboratory, Los Angeles, CA.

Dawidson, I.J.A., *et al.* (1992). Intraoperative albumin administration affects the outcome of cadaver renal transplantation. *Transplantation*, **53**, 774–82.

Dawidson, I., *et al.* (1994). Perioperative albumin and verapamil improve early outcome after cadaver renal transplantation. *Transplantation Proceedings*, **26**, 3100–1.

Loutzenhiser, R. and Epstein, M. (1990). The renal hemodynamic effects of calcium antagonists. In *Calcium antagonists and the kidney* (ed. M. Epstein and R. Loutzenhiser), pp. 33–73. Hanley & Belfus, Philadelphia, PA.

14.5.2 Cardiac transplantation

G. D. Martich and B. P. Griffith

Key messages

- While postoperative care of the heart transplant recipient remains unique in many ways, it also shares many of the management issues for patients following routine cardiac surgery.
- Hearts used for transplantation have normal ejection fractions and no significant coronary artery disease by donor selection criteria. Therefore an adequate preload should result in adequate cardiac output.
- Because of hypothermic preservation or possible ischemia–reperfusion injury, the rate of the denervated heart may be as slow as 40 to 50 beats/min. In order to ensure an acceptable heart rate, and hence an acceptable cardiac output, patients receive continuous intravenous infusion of β-adrenergic agonists for several days following transplantation.
- The right ventricle may fail as a sequela of acute vascular or cellular rejection. Acute right ventricular failure should be suspected if hypotension, low cardiac output, and elevated central venous pressures exist postoperatively.
- Arrhythmias or new evidence of congestive heart failure should alert the physician to possible acute cellular rejection.

Introduction

Nearly 30 years ago, Christian Barnard performed the first human-to-human heart transplant. With more than 40 000 heart transplants having been performed since 1967, recipients have an almost 90 per cent chance of survival at 1 year post-transplant, and approximately 75 per cent are alive at 5 years. Since the operation has changed very little through the years, the perioperative management of these patients must correlate directly with the observed improvement in survival. Non-operative advances include the advent of cyclosporine (cyclosporin) in the 1980s, more standardized donor and recipient selection criteria, and increased physician experience and expertise in postoperative management of problems peculiar to heart transplant recipients. Table 1 lists general criteria for acceptable donor hearts. These criteria serve as flexible guidelines for the procurement process. The selection criteria for candidates for heart transplantation are listed in Table 2. These criteria serve as general guidelines for the transplant center, but may be altered depending on individual patient presentation.

While postoperative care of the heart transplant recipient remains unique in many ways, it also shares many of the management issues for patients following routine cardiac surgery. The similarities between heart bypass and heart transplant surgery in the immediate postoperative period include optimizing hemodynamics, weaning

Table 1 Heart donor criteria
Age <40 years (variable, oldest donor 65 years)
Negative serologies for HIV and hepatitis B
No active infection or malignancy
Very low probability of coronary artery disease (coronary angiogram may be required to rule out significant coronary disease if risk factors present)
ABO compatibility
Normal ventricular function
Donor weight 80%–120% of recipient's weight

Table 2 Recipient selection criteria for heart transplantation

NYHA class III–IV status
Age <66 years (general rule, oldest recipient 73 years)
No significant pulmonary hypertension defined as:
Transpulmonary gradient (mean pulmonary artery pressure – wedge pressure) >15 mmHg
Pulmonary artery systolic pressure >60 mmHg
Pulmonary vascular resistance >5 Wood units

from mechanical ventilatory support, and achieving adequate hemostasis.

The unique nature of caring for patients following heart transplantation is due, in part, to lack of sympathetic or parasympathetic innervation of the heart. Because of this, special attention needs to be paid to adequacy of heart rate in maintaining cardiac output, need for pacing, and which antiarrhythmic agents are effective. Additionally, issues such as intensity and types of immunosuppression, diagnosis and treatment of rejection, and prevalence of postoperative infectious and general surgical complications make the care of heart transplant recipients potentially much more complicated than that of other cardiac surgical patients. Here we address these subjects as well as some of the long-term sequelae that send heart transplant recipients back to the critical care setting, such as transplant coronary artery disease and post-transplant lymphoproliferative disease.

Hemodynamic support

Blood pressure and cardiac output

Maintaining mean arterial pressure above 65 mmHg and cardiac index above 2.5 l/min/m^2 is one standard for patients following cardiac surgery. This applies equally to patients after coronary artery bypass grafting, valve replacement surgery, and heart transplantation. In order to achieve these hemodynamic goals postoperatively, 250- to 500-ml aliquots of intravenous normal saline are administered up to a total of 3 to 4 liters if the patient has either a low blood pressure or diminished cardiac output. Depressed left ventricular function unresponsive to intravenous fluid loading dictates the need for inotropic and/or vasopressor support.

Hearts used for transplantation have normal ejection fractions and no significant coronary artery disease (or minimal risk factors for coronary disease) by donor selection criteria (Table 1). Therefore an adequate preload (pulmonary artery wedge pressure above 15 mmHg) should result in adequate cardiac output. That all heart transplant recipients have completely denervated hearts potentially complicates the logic that good filling pressure results in good cardiac output. Because of hypothermic preservation or possible ischemia–reperfusion injury, the rate of the denervated heart may be as slow as 40 to 50 beats/min. Despite optimal left ventricular filling, the cardiac output will decrease proportionally to the decrease in heart rate. In order to ensure an acceptable heart rate, and hence an acceptable cardiac output, patients receive continuous intravenous infusion of β-adrenergic agonists for several days following transplantation. Institutional bias varies among dopamine, dobutamine, or isoproterenol (isoprenaline), which are administered to maintain adequate circulating cathecholamine levels. Weaning from the β-agonist takes place slowly over a course of 3 to 7 days. If tachyarrhythmias occur while the patient is still receiving a β-agonist, the first step towards correcting the rhythm problem is stopping the infusion.

Pacing

If the patient develops a low cardiac output as a result of absolute or relative bradycardia, despite pharmocological manipulation, temporary pacing to maintain a heart rate above 90 beats/min is appropriate. Epicardial ventricular pacing wires placed intraoperatively safeguard against bradycardia refractory to medications. The need for temporary pacing following heart transplantation has been reported to be as high 27 per cent. Long-term pacing requirements are much lower (2–11 per cent) and are almost equally divided between sinoatrial and atrioventricular node dysfunction.

Arrhythmias

Postoperative atrial and ventricular arrhythmias are uncommon in our experience. If they occur early in the postoperative period, they can usually be attributed to electrolyte deficiencies and respond to potassium and/or magnesium supplementation. Atrial tachyarrhythmias not secondary to electrolyte imbalance may respond to decreasing the β-agonist infusion. Ventricular ectopy is rare and may respond to a decrease in inotropes, electrolyte supplementation, or removing the pulmonary artery catheter. Ectopy, atrial more often than ventricular, occurring a week or more post-transplant alerts the clinician to the possibility of cardiac rejection.

Tachyarrhythmias and bradyarrhythmias following heart transplantation are treated with drugs that act directly on the heart and not through the sympathetic or parasympathetic systems. Agents from each of the four major antiarrhythmic classifications act, at least in part, directly on the heart. However, both β-blockers and calcium-channel blockers need to be used with caution because of negative inotropic effects. Similarly, adenosine is effective in treating supraventricular arrhythmias, but the starting dose should be a quarter to half normal because of its more pronounced and prolonged action in heart transplant patients. Neither atropine nor digoxin are effective antiarrhythmics in the immediate postoperative period following heart transplantation because their actions are vagally mediated.

Electrocardiogram

Two distinct P waves may complicate interpretation of the immediate postoperative electrocardiogram (Fig. 1). The residual cuffs of native atria contribute one P wave in addition to the transplant-related P wave. A bedside non-invasive test that readily differentiates between the two involves the response elicited by vagal stimulation which affects only the native P wave. A new surgical technique employing bicaval anastomoses eliminates the confusion of two P waves, but is not widely used.

Right ventricular dysfunction

Candidates for heart transplantation undergo thorough evaluation to exclude significant pulmonary hypertension (Table 2). Despite this

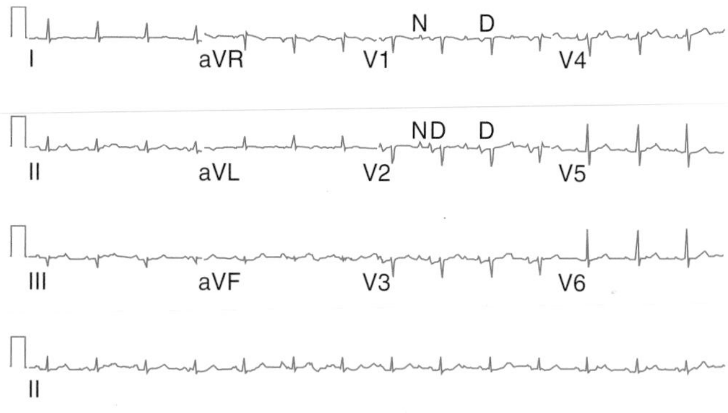

Fig. 1 Twelve-lead electrocardiogram with rhythm strip of a patient after heart transplantation showing two distinct P waves, a native P wave from the right atrial remnant (N) and a donor P wave (D) associated with the QRS complex. (Courtesy of Dr E. Curtiss.)

screening process, even modest pulmonary hypertension may cause the unconditioned right ventricle of the donor to decompensate. The right ventricle may also fail hemodynamically as a sequela of acute vascular or cellular rejection. Acute right ventricular failure should be suspected if hypotension, low cardiac output, and elevated central venous pressures exist postoperatively. Intravenous nitroglycerin or prostaglandin E_1 may be useful for reducing pulmonary vascular resistance while attempting to support the right ventricle with inotropic agents. Possible other treatments include inhaled nitric oxide as well as insertion of a right ventricular assist device if maximal medical therapy fails.

Pericardial tamponade

Pericardial fluid collections are common postoperatively in all cardiac surgical procedures. The pericardium is left open after heart transplantation, but localized collections of fluid can accumulate, causing myocardial compression and subsequently limiting left ventricular filling. The triad of elevation of central venous pressures, low systemic blood pressure, and low cardiac output merits evaluation with transesophageal echocardiography to distinguish cardiac tamponade from right ventricular failure (Table 3). Often, diastolic equalization of pressures is not present.

Table 3 Differential diagnoses of low blood pressure, low cardiac output, and high central venous pressure following heart transplantation

Diagnosis	Diagnostic procedure	Treatment
RV failure secondary to PHTN	TEE	PGE, inotropes, inhaled NO, RVAD
RV failure secondary to VR	TEE	Plasmapheresis, cyclophosphamide, retransplant
Tamponade	TEE	Intravenous fluids until surgical re-exploration

RV, right ventricle; PHTN, pulmonary hypertension; TEE, transesophageal echocardiogram; PGE_1, prostaglandin E_1; NO, nitric oxide; RVAD, right ventricular assist device, VR, vascular rejection.

Immunosuppression and rejection

Immunosuppression

Immunosuppression schemes vary from institution to institution, but most involve a three-drug regimen including cyclosporine, azathioprine, and corticosteroids. Some centers additionally treat heart transplant recipients with lymphocytotoxic agents (i.e. antithymocyte globulin, antilymphocyte globulin, OKT3) to decrease the risk of rejection in the immediate postoperative period.

We use the interleukin 2 (IL-2) inhibitor tacrolimus in place of cyclosporine for primary immunosuppressive therapy in heart transplant recipients. Compared with cyclosporine (also an IL-2 inhibitor), tacrolimus-treated heart transplant recipients have a lower incidence of hypertension and an improved quality of life. However, in a series of liver transplant patients treated with either tacrolimus or cyclosporine, the tacrolimus-treated group displayed a twofold increase in neuro- and nephrotoxicity. Since toxicity and efficacy are dependent on blood levels, and given the wide array of drug interactions seen with both cyclosporine and tacrolimus, adding or withdrawing medications from the transplant patient's regimen requires caution and knowledge of drug interaction. Medications that may alter each other's blood levels are listed in Tables 4 and 5.

Cellular rejection

Acute cellular rejection can be defined as any clinical event requiring treatment by a significant increase in immunosuppression. During the first year following heart transplantation, the average patient will have 1.3 ± 0.7 episodes of acute cellular rejection. Identifiable risk factors for early acute cellular rejection include female sex and young age of the donor. The endomyocardial biopsy has served as the gold standard for diagnosing allograft rejection since its introduction in

Table 4 Drugs known to interact with cyclosporine and their effect on cyclosporine serum levels

Increase cyclosporine levels	Decrease cyclosporine levels
Cimetidine	Nafcillin
Tacrolimus	Octreotide
Diltiazem	Phenobarbital (phenobarbitone)
Verapamil	Phenytoin
Erythromycin	Carbamazepine
Ketoconazole	Rifampin (rifampicin)
Metoclopramide	Sulfamethoxazole
Nicardipine	
Amiodarone	

Reproduced with permission from Martich and Boujoukos (1996).

Table 5 Based on clinical experience these medications may alter tacrolimus levels (no formal drug interaction studies with tacrolimus have been conducted)

Increase tacrolimus levels	Decrease tacrolimus levels
Erythromycin	Phenobarbital
Ciprofloxacin	Phenytoin
Ketoconazole	Carbamazepine
Fluconazole	Rifampin
Diltiazem	
Cyclosporine	
Clotrimazole	

Reproduced with permission from Martich and Boujoukos (1996).

Table 6 Standardized grading system for classifying cellular rejection on endomyocardial biopsy specimens

Grade	New nomenclature	Old nomenclature
0	No rejection	No rejection
IA	Focal (perivascular or interstitial infiltrate) without necrosis	Mild rejection
IB	Diffuse but sparse infiltrate without necrosis	
II	One focus only with aggressive infiltration and/or focal myocyte damage	Focal moderate rejection
IIIA	Multifocal aggressive infiltrates and/or myocyte damage	Low moderate rejection
IIB	Diffuse inflammatory process with necrosis	Borderline/ severe rejection
IV	Diffuse aggressive polymorphous ± infiltrate ± edema ± hemorrhage ± vasculitis, with necrosis	Severe acute rejection

1973. In 1990, a formulation standardizing the grading of endomyocardial biopsies for the degree of cardiac rejection was established (Table 6). Endomyocardial biopsies are generally performed every 1 to 2 months during the first year, when rejection risk is greatest. Arrhythmias or new evidence of congestive heart failure should alert the physician to possible acute cellular rejection. Treatment of significant rejection episodes begins with pulse doses of corticosteroids for 3 days (methylprednisolone 500–1000 mg/day) followed by augmented primary immunosuppression. Steroid refractory acute cellular rejection has been treated with variable success by a variety of methods including lymphocytotoxic therapy and total lymphoid irradiation.

Vascular rejection

A devastating, but fortunately rare, early complication of heart transplantation is vascular rejection, also called antibody-mediated or humoral rejection. Biopsy findings include evidence of endothelial cell proliferation and swelling by light microscopy and vascular deposition of immunoglobulin by immunofluorescence. Vascular rejection usually presents intraoperatively or in the first few days postoperatively. Clinically, vascular rejection manifests first with right ventricular failure followed quickly by left ventricular failure. Therapy for vascular rejection includes aggressive use of inotropes, vasopressors, and in some cases mechanical ventricular support or retransplantation. Medical treatment of vascular rejection may include cyclophosphamide and plasma exchange. Despite aggressive measures, vascular rejection is associated with a much higher mortality than acute cellular rejection.

Complications

Infections

Two peak periods exist for infectious complications following heart transplantation. Nosocomial infections dominate the first postoperative month or early period and include catheter-related infections secondary to staphylococcal species and Gram-negative organisms. Thus all indwelling catheters should be removed early. Despite the need for continued inotropic agents, pulmonary artery and arterial catheters are generally removed on the first postoperative day. Patients receive prophylactic vancomycin for 48 h and third-generation cephalosporin antibiotics in the perioperative period. Infections during the late period (the second to fifth postoperative month) are dominated by viruses, fungi, and bacteria. Treatment is culture and serology based.

Prophylactic treatment of cytomegalovirus-seronegative recipients of cytomegalovirus-seropositive donors remains controversial and institution dependent. Many centers treat cytomegalovirus-negative heart transplant recipients who receive a cytomegalovirus-positive donor heart with ganciclovir for the first several days after transplant; others wait for symptomatic cytomegalovirus disease to manifest before treatment. Cytomegalovirus disease is usually subacute and may cause pneumonia, gastrointestinal symptoms (i.e. anorexia, nausea, epigastric pain, diarrhea), and/or bone marrow suppression (most commonly leukopenia).

Less controversy surrounds treatment of toxoplasma-seronegative recipients who receive a heart from a seropositive donor. Most centers routinely give several weeks of treatment with pyrimethamine and folinic acid. *Toxoplasma gondii* disease can cause a severe myositis with myocardial involvement as well as encephalitis. If disease develops, treatment consists of clindamycin or sulfadiazine plus higher doses of pyrimethamine.

Mediastinitis has traditionally been thought to be a severe and life-threatening infection following heart transplantation. However, two large series totaling over 750 patients reported no increased mortality in heart transplant recipients who developed mediastinitis. The incidence of mediastinitis reported in those reviews was less than 3 per cent.

Other practical infection control issues include the risk of *Pneumocystis carinii* pneumonia and oral herpes, and the need for reverse isolation. *P. carinii* pneumonia has become a much less commonly reported pathogen with the advent of prophylactic trimethoprim–sulfamethoxazole (co-trimoxazole) in all patients at risk. Herpes simplex virus positive recipients all receive a 3-month course of acyclovir (aciclovir) orally to prevent reactivation. Finally, from a practical standpoint, there is no need for routine reverse or protective isolation following heart transplantation.

General surgical complications

The extracardiac complication rate following heart transplantation has been reported to be between 9.5 and 38 per cent. In one series abdominal complications after heart transplantation directly accounted for a 30-day mortality of nearly 10 per cent. Most of the extracardiac complications involve the abdomen and gastrointestinal tract and 4 to 20 per cent may require surgical intervention. The high incidence of surgical abdominal processes may be secondary to the use of immunosuppressive agents masking disease progression until relatively late. Common causes of general surgical intervention following heart transplantation include gastrointestinal bleeding, viscus perforation, pancreatitis, cholecystitis, and *Clostridium difficile* enterocolitis. The prevalence of these complications is far greater than in general cardiac surgical patients in part because of the added side-effects of immunosuppressive agents. Unexplained ileus or abdominal pain/tenderness, particularly with evidence of infection, mandates an aggressive work-up beginning with imaging studies of the abdomen.

Post-transplant lymphoproliferative disease

Abnormal B-cell proliferation following solid organ transplantation results in a peculiar type of lymphoma called post-transplant lymphoproliferative disease. Either primary or secondary Epstein–Barr virus infection plus immunosuppression causes post-transplant lymphoproliferative disease, which has an incidence of approximately 4 per cent following heart transplantation. Higher rates of post-transplant lymphoproliferative disease (up to 11.4 per cent) have been reported following postoperative lymphocytotoxic therapy. The lungs, bone marrow, gastrointestinal tract, central nervous system, and lymph nodes are the most common sites of post-transplant lymphoproliferative disease. Following heart transplantation, post-transplant lymphoproliferative disease may occur early (less than 1 year) or late (more than 1 year) and is divided as such for prognostic reasons. Patients who develop early post-transplant lymphoproliferative disease (peak occurrence at 3 to 4 months post-transplant) have a mortality rate of approximately 25 per cent, but up to 89 per cent respond to the treatment of choice which is a reduction in immunosuppression. Late post-transplant lymphoproliferative disease rarely, if ever, responds to reduction in immunotherapy alone and has an associated mortality of over 70 per cent. If a reduction in the degree of immunosuppression does not induce a remission, post-transplant lymphoproliferative disease may respond to radiation therapy or conventional antilymphoma chemotherapy.

Coronary artery disease

Transplant-associated coronary artery disease replaces infection and rejection as the predominant cause of morbidity and mortality after the first 2 years post-transplant. This accelerated form of coronary disease has an annual incidence of 5 to 10 per cent each year after the first, and a prevalence as high as 50 per cent by 5 years. Transplant coronary disease affects heart transplant recipients regardless of preoperative age or diagnosis. Autopsy examinations have revealed significant transplant-associated coronary artery disease in heart transplant recipients as young as 4 years old.

The cause of transplant-associated coronary artery disease is unknown and probably multifactorial. None of the normal risk factors for developing coronary artery disease, except hyperlipidemia, appear to be independent predictors for developing transplant-associated coronary artery disease. Cytomegalovirus may play an additional role in the development of this disease, but this remains controversial. Recently, much evidence has been presented in support of an immune-mediated etiology for transplant-associated coronary artery disease. The evidence in support of an immune-mediated cause includes a higher incidence of rejection noted in heart transplant recipients with transplant-associated coronary artery disease.

Transplant-associated coronary artery disease may first manifest as sudden death, arrhythmias, or congestive heart failure without endomyocardial biopsy evidence of cardiac rejection. Likewise, transplant-associated coronary artery disease may not be suspected until a routine ECG displays evidence of ischemia or infarction. Transplant-associated coronary artery disease often progresses in the absence of chest pain, although several reports have suggested reinnervation with resulting angina pectoris. Since most heart transplant recipients do not experience chest discomfort, many centers screen for transplant-associated coronary artery disease with yearly coronary angiography. Nonetheless, it is difficult to diagnose. The diffuse concentric nature of transplant-associated coronary artery disease makes coronary angiography of limited benefit and underestimates the degree of vessel narrowing compared with autopsy findings. Several innovative methods for detecting transplant-associated coronary artery disease, such as cine-videodensitometry with nitroglycerin, intracoronary ultrasonography, perfusion scintigraphy, and angioscopy are being investigated.

Minimizing rejection episodes, lowering blood lipids, and controlling hypertension are the aims of medical therapy for transplant-associated coronary artery disease. Diltiazem, a calcium-channel blocker, has been shown to delay the onset of transplant-associated coronary artery disease possibly via enhanced immunosuppressive effect or regulation of lipoprotein-receptor synthesis or a combination of the two. Other more aggressive treatment regimens for transplant-associated coronary artery disease include percutaneous transluminal coronary angioplasty, coronary artery bypass grafting, and retransplantation.

Heterotopic heart transplantation

Severe fixed pulmonary hypertension makes patients ineligible to receive a standard heart transplant because the unconditioned donor right ventricle will quickly fail. Heterotopic heart transplantation offers an alternative to heart–lung transplantation for this group of patients. They obtain their hearts in the same manner (and on the same waiting list) as standard heart transplant candidates, but generally can be given a donor heart from smaller donors. The procedure involves leaving the native heart *in situ* while the donor heart is implanted in the right hemithorax. The donor heart is anastamosed to the native heart in parallel with side-to-side anastomoses between native and donor atria and end-to-side anastomoses between donor and native pulmonary artery and aorta. In this case, the native conditioned right ventricle can help carry the right-side load, preventing acute right ventricular failure of the transplanted heart.

Interpretation of the ECG in heterotopic heart transplant recipients, like standard heart transplant recipients, may be fraught with difficulty. Heterotopic recipients will display two distinct electrocardiograms with different heart rates and frequently different rhythms. The QRS complex associated with the carotid pulse, or that seen better with right chest leads, belongs to the transplanted heart. Major drawbacks of heterotopic heart transplantation are compression of the right lower lobe of the lung and need for chronic anticoagulation because of the risk of thrombus generation in the diseased native cardiac chambers. Hemodynamic monitoring with a pulmonary artery catheter allows measurement of pulmonary artery pressures, pulmonary capillary wedge pressure, and mixed venous oxygen saturation, but placement of the catheter's infusion port within the common right atrium is essential for accurate cardiac output estimations by thermodilution.

Bibliography

Augustine, S.M., Yeo, C.J., Buchman, T.G., Achuff, S.C., and Baumgartner, W.A. (1991). Gastrointestinal complications in heart and in heart–lung transplant patients. *Journal of Heart Lung Transplantation*, **10**, 547–56.

Billingham, M.E., *et al.* (1990). A working formulation for the standardization of nomenclature in the diagnosis of heart and lung rejection: Heart Rejection Study Group. *Journal of Heart Transplantation*, **9**, 587–93.

Martich, G.D. and Boujoukas, A.J. (1996). Adult cardiac transplantation. *Journal of Intensive Care Medicine*, **11**, 23–33.

O'Connell, J.B., *et al.* (1992). Cardiac transplantation: recipient selection, donor procurement, and medical follow-up. *Circulation*, **86**, 1061–79.

Pham, S.M., *et al.* (1994). Cardiac transplantation at the University of Pittsburgh: 1994 update. In *Clinical transplants* (ed. P. Terasaki and M. Cecka). UCLA Tissue Typing Laboratory, Los Angeles, CA.

14.5.3 Hepatic transplantation

M. Manji and J. F. Bion

Key messages

- Elective orthotopic liver transplantation has a 5-year survival of 80 to 90 per cent.
- Multidisciplinary care and teamwork are essential.
- Preoperative severity and acuity are important determinants of outcome.
- Infectious complications significantly increase morbidity and mortality.
- Prothrombin time is the best single measure of graft function.

Introduction

Orthotopic liver transplantation has become a routine treatment for endstage liver disease, and the number of transplants is only limited by the supply of donor organs. Five-year survival rates in excess of 80 per cent are now common for elective procedures (Fig. 1). Factors which have contributed to this success include earlier referral for transplantation, better preoperative preparation, advances in anesthetic and surgical techniques, and skilled perioperative intensive care management, particularly in the case of physiologically unstable patients and those with fulminant hepatic failure.

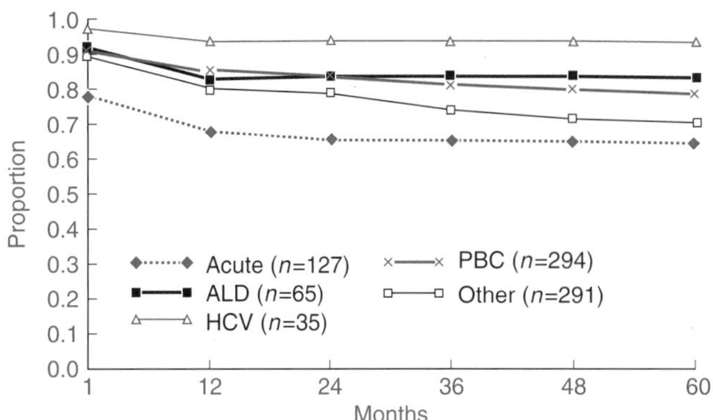

Fig. 1 Five-year survival by diagnosis following orthotopic liver transplantation (Birmingham 1988–1996): ALD, alcoholic liver disease; HCV, hepatitis C virus; PBC, primary biliary cirrhosis. (Data from the Liver Unit, Queen Elizabeth Hospital, Birmingham, UK.)

Preoperative factors

General features

Endstage liver disease affects virtually every other organ system. The pretransplant condition of patients with endstage liver disease makes them more vulnerable to postoperative multiple organ failure. They usually demonstrate a hyperdynamic circulation, with a poor nutritional and immune state. Portal hypertension may contribute to impaired gut barrier function allowing translocation of enteric organisms or their toxic products. Acute complications may occur rapidly, including infections such as spontaneous peritonitis, cerebral depression from the injudicious use of sedative drugs, and gastrointestinal bleeding from varices or ulceration, all of which may result in further hemodynamic and cerebral decompensation.

Short-term outcome following orthotopic liver transplantation depends on the etiology (mortality is higher for hepatitis B and non-A non-B hepatitis) and on preoperative morbidity. In patients with rapidly deteriorating liver failure, adverse prognostic indicators include acidemia, grade III or more encephalopathy, prothrombin time greater than 100 s and serum creatinine greater than 300 μmol/l. Ideally, transplantation should be performed before complications occur, since they reduce the chances of success and may be impossible to reverse even after transplantation.

Cardiovascular dysfunction

The hyperdynamic circulation of endstage liver disease is characterized by increased heart rate, low systemic blood pressure, elevated cardiac output, and depressed systemic vascular resistance combined with varying effects on the pulmonary and renal arterial vasculature. These features may be made worse following graft reperfusion by release of proinflammatory mediators. The pathogenesis is not clearly understood, but may include sympathetic nervous system overactivity, excess of vasoactive mediators such as prostaglandins,

ADH, bile acids, and nitric oxide (NO), and the presence of arteriovenous shunts. Concurrent upregulation of vasoconstrictor endothelins may selectively affect the renal and pulmonary vasculature.

Renal dysfunction

The hepatorenal syndrome is characterized by renal vasoconstriction and a reduction in renal blood flow, glomerular filtration rate, and urine output in the presence of normal renal histology. While it is important to maintain normovolemia and adequate renal perfusion pressure, since this will facilitate renal blood flow and may delay the deterioration in renal function, this may be impossible to achieve in patients with cerebral edema. In this circumstance renal protection takes secondary importance to cerebral protection. The hepatorenal syndrome may be mediated by catecholamines, endothelins, and the renin–angiotensin–aldosterone axis, and may be exacerbated by jaundice, infection, or nephrotoxic agents such as radiocontrast dyes or excessive diuretic therapy (Lerman *et al.* 1991). Spontaneous recovery without hepatic transplantation is rare. Acute tubular necrosis secondary to cardiovascular failure or sepsis may also contribute to, or cause, renal failure during the pretransplant phase.

Pulmonary dysfunction

The hepatopulmonary syndrome is characterized by hypoxemia and dyspnea, which are worse in the erect position in patients with portal hypertension from chronic liver disease. Pulmonary vasodilation results in basal ventilation–perfusion mismatch. Hypoxemia is only partially responsive to supplementary oxygen. This syndrome condition is common (up to 30 per cent), and the prognosis is good. In patients with portocaval shunting, progressive pulmonary vasoconstriction with hypertension, dyspnea, and hemoptysis, and right ventricular failure may supervene. This is a rare (0.5 per cent) but serious complication with a high mortality.

Encephalopathy

In patients with acute fulminant hepatic failure, progression to grade IV encephalopathy is associated with a progressive rise in intracranial pressure, eventually resulting in brainstem herniation. Precipitating causes such as variceal bleeding or occult infection should be sought and promptly corrected. Clinical signs of raised intracranial pressure such as pupillary dilatation or flexor or extensor posturing are unreliable as they may occur too late for treatment, and in any case may be disguised by the use of opioid infusions to obtund autonomic responses and neuromuscular blockade to facilitate mechanical ventilation. Intracranial pressure monitoring is the ideal method for detection and management, but the substantial risk of intracranial hemorrhage associated with these devices means that the coagulopathy must be corrected first, and this can be difficult to achieve. Control of intracranial pressure in the absence of normal renal function is likely to require continuous ultrafiltration.

Cerebral hemorrhage may also occur spontaneously, and will require CT or magnetic resonance imaging for diagnosis. Other causes of impaired cerebral function include hypoglycemia and rapid shifts in sodium balance. Hyponatremia may be induced by excessive use of mannitol, and hypernatremia by sodium-containing colloids or dialysis. Central pontine myelinolysis may be induced by rapid correction of hyponatremia.

Pharmacokinetics

Drug handling varies widely in patients awaiting orthotopic liver transplantation because of their variable hepatic reserve. Sedative drugs to facilitate medical procedures should be given with particular caution, and administered only by those with proper experience in their use.

Postoperative management

The intensity and duration of postoperative intensive care depends on the preoperative state of the patient, the progress of the operative procedure, and the development of postoperative complications. Skilled anesthetic management of cardiovascular variables during graft reperfusion and control of coagulation throughout the operation substantially improve postoperative stability. Prior abdominal operations, adhesions, or technical surgical difficulties are likely to increase blood loss, transfusion requirements, and operative and graft ischemic time. Abdominal bleeding or graft dysfunction will delay progress, as will prior muscle wasting. Graft dysfunction will affect all other organ systems; conversely, impaired function of other organ systems is likely to lead to graft dysfunction.

Monitoring the grafted liver

Survival of the patient is dependent on normal function of the graft. Good clinical prognostic features include evidence of normal cerebral function, the production of bile from the T-tube, normal to high blood sugars, the need for intravenous potassium supplements, and a normalizing prothrombin time. Progressive organ system dysfunction in the absence of another cause indicates graft failure.

Biochemical liver function tests are useful for evaluating graft function. A small early rise in transaminases (serum alanine aminotransferase and aspartate aminotransferase) during the first 24 h post-transplant represents mild injury during organ procurement, storage, and reperfusion. Substantial increases which continue to rise indicate graft ischemic injury, which may be severe and should prompt ultrasound examination of vascular patency and maximization of cardiac output to ensure adequate splanchnic oxygen supply. Alkaline phosphatase and γ-glutamyl transferase are indicators of cholestatic injury. Management is focused on avoiding further injury to the liver graft during the period of recovery by ensuring adequate hemoglobin, circulating volume, and perfusion pressures. A later rise in bilirubin is an indicator of graft rejection.

Duplex ultrasonography is a safe and routine investigation, but must be performed by an experienced radiologist. Absent flow in an otherwise stable patient who is progressing well suggests that there are problems with the investigation, not the patient. Angiography or biopsy may be required for cases that are difficult to interpret. Bile cultures should be performed routinely. Cholangiography should be performed to evaluate the common duct if bile flow is inadequate.

Graft failure

Early graft dysfunction is an important cause of morbidity and mortality. Primary non-function of the graft requires urgent retransplantation. Causes include complete vascular occlusion, ischemia, necrosis or reperfusion injury, or prior disease of the donor liver. It should be distinguished from reversible early graft dysfunction caused by an inadequate vascular anastomosis, acute rejection, and sepsis in the recipient. Clinical features of graft failure include hemodynamic instability similar to sepsis (a low systemic vascular resistance syndrome), reduced bile production, a metabolic (predominantly lactic) acidosis, increasing prothrombin time, persistent hypoglycemia requiring intravenous glucose support, and progressive multiple organ failure.

Graft support

Adequate graft oxygenation is essential. The liver receives its blood supply from both the portal vein and the hepatic artery, and the main determinant of flow is cardiac output. Therefore systemic hemodynamics are important. Prostacylin and *N*-acetylcysteine have been used in fulminant hepatic failure and primary graft non-function to improve the hepatic microcirculation, alter cytokine release, modulate NO production, and enhance resistance to oxidant injury (Harrison *et al.* 1991). Artificial liver support in the form of an extracorporeal liver-assist device containing living hepatocytes is still a research tool. Retransplantation remains the only option for graft failure.

Immunosuppression and prevention of rejection

Cyclosporine A (cyclosporin A) has been fundamental to the success of liver transplantation, as with other solid organs. It inhibits T-cell-mediated production of interleukin 2 (IL-2). T suppressor cells are spared, enhancing natural immunosuppression. Triple therapy (cyclosporine A, azathioprine, prednisolone) is usually started within 12 h postoperatively. Some centers start cyclosporine A preoperatively. Daily cyclosporine A levels are required in the early postoperative period, to avoid both subtherapeutic levels and toxicity such as renal failure, neurological complications, and hepatatoxicity mimicking allograft rejection. However, early rejection remains the most common cause of hepatic dysfunction after orthotopic liver transplantation, with at least one rejection episode occurring in two-thirds of recipients. The dose of cyclosporine A should be reduced or its introduction delayed in patients with evolving renal dysfunction.

Graft rejection

Clinical signs include pyrexia, fluid retention, graft enlargement, and dysfunction of other organ systems. Serum bilirubin is elevated, with subsequent derangement of other liver enzymes. Liver biopsy is conclusive, unless graft failure is multifactorial. Initial therapy is high-dose methylprednisolone. The murine monoclonal antihuman T-cell antibody (OKT3) is used for non-responders or severe cases. It can also be used to allow lower doses of cyclosporine in patients with impaired renal function. The first dose of OKT3 can precipitate life-threatening reaction ranging from serum sickness to acute anaphylaxis. FK506 is a macrolide antibiotic immunosuppressant that may have more immunospecificity and less toxicity than cyclosporine. It suppresses activation and proliferation of T cells as well as inhibiting lymphokines such as IL-2. It may also prove useful as a rescue agent in acute rejection episodes.

Hemodynamics and fluid therapy

A pulmonary artery catheter is required to measure filling pressures, cardiac function, vascular resistance, and oxygen delivery and consumption. Considerable fluid shifts may occur in some patients, and can be difficult to detect; a colloid requirement in excess of 50 ml/kg in the first 24 h is common. Clotting factors must not be given routinely postoperatively as they will interfere with prothrombin times which are an important guide to graft function. Abdominal bleeding, if unchecked, may result in the abdominal compartment syndrome and splanchnic ischemia. Thromboelastography may be helpful in distinguishing between surgical blood loss and coagulopathies. Care must be taken when interpreting filling pressures as a guide to volume status when vasoconstrictor agents such as norepinephrine (noradrenaline) are used. Volume overload may cause hepatic venous congestion, and contribute to pulmonary edema, pleural effusions, and tissue edema.

Most patients will demonstrate a high cardiac output and a systemic vascular resistance sufficiently low to require vasoconstrictors for at least a few hours. This pattern is more common, severe, and prolonged in patients with fulminant hepatic failure. Delayed onset of dependence on vasoconstrictors means that either the graft is failing or microbial sepsis has supervened. It is absolutely essential that vasoconstrictors are not used as a substitute for adequate volume expansion or at a dose which impairs systemic oxygen delivery. It is difficult to give a precise value for adequate oxygen delivery, but an oxygen debt is unlikely in the presence of a decreasing base deficit, normal lactate, and good urine output.

Orthotopic liver transplantation patients rarely need inotropic agents. If they do, either epinephrine (adrenaline) or dopamine may be useful for maintaining a high cardiac output without further vasodilatation; the β_2 agonist effect of adrenaline may increase blood lactate levels or lead to occult ischemic heart disease in older patients. Occasional patients may demonstrate a normal systemic vascular resistance with a high cardiac output and therefore present with hypertension postoperatively, which usually resolves with vasodilators. Hypertension may indicate raised intracranial pressure in encephalopathic patients or those susceptible to intracranial hemorrhage or pain in patients receiving inadequate analgesia, or it may be caused by cyclosporine.

Respiratory management

The duration of mechanical ventilation will be determined by the patient's preoperative condition and perioperative course. Uncomplicated patients should be weaned from mechanical ventilation rapidly, usually within 12 h of surgery. Weaning should start as soon as adequate analgesia has been assured, the graft is producing bile, cerebral function is evidently normal, and the patient is metabolically stable with a resolving base deficit. Caution should be exercised if technical difficulties were experienced during surgery, or there is evidence of continuing blood loss, a metabolic acidosis, a rising prothrombin time, or progressive organ system dysfunction. Patients with fulminant hepatic failure are at risk of cerebral edema for several days, and should be managed appropriately.

Right lower zone opacification on the chest radiograph is a common finding within the first few days after orthotopic liver transplantation. It represents a pleural effusion together with basal collapse, and usually resolves without complications. More substantial fluid collections may require drainage if respiratory function is impaired, particularly in patients with muscle wasting or a poor cough.

Patients with chronic liver disease and those who have undergone orthotopic liver transplantation are susceptible to pulmonary edema. Causes include low intravascular colloid osmotic pressures, endothelial dysfunction leading to acute lung injury, and excessive intravenous fluid loads. Clinical pulmonary deterioration (tachypnea, respiratory distress) often precedes chest radiograph changes, and frequently responds to fluid removal by diuretics or ultrafiltration. Fluid overload is a more common problem in children, and requires meticulous fluid balance charts combined with a high index of suspicion. Positive end-expiratory pressure (PEEP) may be used in intubated patients to reduce shunting, but levels in excess of 10 cmH_2O may reduce cardiac output and graft blood flow.

Late-onset respiratory failure or prolonged ventilator dependence is more common in patients who were severely ill before transplantation and those who have a complicated postoperative course with impaired graft function. The combination of hypoxemia and patchy infiltrates on the radiograph suggests a differential diagnosis of acute lung injury or pulmonary infection. A rapid response to continuous positive airways pressure, PEEP, or inhaled NO in intubated patients suggests fluid overload and pulmonary edema.

Intubated patients are more likely to develop colonization or infection with Gram-negative bacteria or *Candida* during the first 2 weeks after orthotopic liver transplantation. Patients who have been extubated and then relapse are susceptible to a range of organisms, including *Aspergillus*. If fungi are suspected, high-dose liposomal amphotericin should be started immediately. Cytomegalovirus and other opportunistic infections may appear after several weeks. Bronchoscopic lavage and brush biopsy should be performed early and repeated if negative. Selective antibiotic digestive decontamination reduces pulmonary infections in intubated patients after orthotopic liver transplantation but does not reduce endotoxemia or organ system failures (Bion *et al*. 1994). The use of prophylactic antibiotics should be based on written protocols, and therapy should be guided by close consultation with the infectious disease specialist and/or the microbiologist. Failure to find a specific cause for either acute lung injury or pneumonia is not uncommon, even at autopsy.

Hemostasis and rheology

The prothrombin and activated partial thrombin times should normalize by the third postoperative day if the graft is working. Preoperative thrombocytopenia does not always respond to platelet transfusion and usually takes longer (up to a week or more) to recover. Hemoglobin concentration should not exceed 10 g/dl to avoid increasing blood viscosity and the risk of portal or hepatic arterial occlusion. Clotting factors should not be given routinely because the prothrombin time is a valuable guide to graft function. In the presence of continued blood loss, thromboelastography should be used to determine the cause. Aprotinin should be continued if there is evidence of continuing fibrinolysis and cryoprecipitate given if fibrinogen levels are low. Reasonable clotting indices suggest a surgical cause and should prompt re-exploration; intra-abdominal clot may cause the abdominal compartment syndrome, clotting factor consumption, clot expansion, and hyperbilirubinemia.

Renal management

Renal dysfunction following orthotopic liver transplantation usually represents a combination of pre-existing renal injury, acute tubular necrosis, and nephrotoxins. It is associated with increased mortality (Ishitani *et al*. 1993). The most important preventative measures are maintenance of a normal perfusion pressure and an adequate cardiac output, and avoidance of the abdominal compartment syndrome. The role of renal dose dopamine remains controversial. Daily cyclosporine A and aminoglycoside antibiotic monitoring are essential. In the absence of an alternative cause, deteriorating renal function suggests poor graft function, sepsis, or bile leak. Renal replacement therapy should not be delayed, and is best managed using continuous venovenous hemofiltration. Heparin may induce thrombocytopenia; epoprostenol (prostacyclin) may be preferable.

Neurological management

Recovery of consciousness from grade IV encephalopathy after a successful transplant may take more than a week. Daily electroencephalographic recordings are helpful to demonstrate progress and exclude epileptiform activity. New or worsening neurological dysfunction may indicate graft failure, seizures, severe sodium shifts with central pontine myelinolysis, cerebral edema, cerebral hemorrhage, or cyclosporine toxicity, particularly in the presence of severe hypomagnesemia. Late-onset impairment of conscious level or seizures suggests cyclosporine toxicity or cerebral infection with *Aspergillus* or cytomegalovirus.

Infections and their prevention

Early infections are usually caused by Gram-negative bacilli; after a few days these give way to mixed organisms, with *Candida* colonization of the oropharynx and Gram-positive colonization of catheters. Multiple-resistant bacteria may appear, usually in the sicker patients who are still ventilator dependent, with a sensitivity pattern depending on antibiotic utilization. Biliary colonization is common, but should not prompt antibiotic treatment in the absence of symptoms while the T-tube is *in situ*. Late infections include cytomegalovirus, herpesvirus, *Pneumocystis*, and invasive *Aspergillus*.

Viral infection (herpesvirus and cytomegalovirus) of the liver occurs in about 10 per cent of patients between the second and sixth postoperative week; the risk depends on the serological states of the donor and the recipient. Severity ranges from being virtually asymptomatic to a systemic illness with fever, arthralgia, malaise, hepatitis, pneumonitis, and enteritis. Diagnosis is by serology and liver biopsy. Treatment involves a reduction in immunosuppression and the antiviral agent ganciclovir. Ganciclovir produces a leukopenia with suppression of T-cell production.

Broad-spectrum antibiotics are usually prescribed routinely for the first 2 days. Oral nystatin, amphotericin, and low-dose Septrin are used for prophylaxis. Fluconazole is a useful systemic anti-*Candida* prophylactic which mildly inhibits cyclosporine A metabolism. Intravenous liposomal amphotericin should be prescribed (150 mg/day for adults) if *Candida* or *Aspergillus* infection is suspected. There appears to be little role for selective antibiotic digestive decontamination. Antimicrobial agents are only part of a systematic approach to infection prevention, which also inlcudes high-quality nursing care and adherence by medical staff to agreed protocols for prevention of cross-infection. Nutritional factors are also important.

Nutrition

All patients with chronic liver disease are malnourished to varying degrees. Protein catabolism is increased after liver transplantation, and complications such as sepsis and prolonged stay in the intensive care unit increase nitrogen loss and contribute to morbidity (DiCecco *et al.* 1991). Phosphate depletion may cause muscle weakness and delay weaning from ventilation. Enteral nutrition should be started as soon as possible. The role of glutamine and other immunonutrients in this population remains to be determined.

Sedation and analgesia

Adequate analgesia is essential, and is best provided by a continuous infusion of alfentanil. Other opioids may be used, but the kinetics of morphine or fentanyl may be more difficult to predict in this population. It is best to avoid the use of hypnotic agents if the function of the new graft is uncertain; propofol infusions are least likely to cause delayed recovery of consciousness, but in large doses produce cardiovascular depression. Patients with prior fulminant hepatic failure will require neuromuscular blockade for at least 24 h, and atracurium or its isomer cisatracurium are the only relaxants which should be used in this patient population.

References

Bion, J.F., *et al.* (1994). Selective decontamination of the digestive tract reduces Gram-negative pulmonary colonization but not systemic endotoxemia in patients undergoing elective liver transplantation. *Critical Care Medicine*, **22**, 40–9.

DiCecco, S.R., Plevak, D.J., Wiesner, R.H., O'Keefe, S., Hammel, K., and Janzow, D. (1991). Do we accurately estimate caloric and nitrogen requirements in the liver transplant patient? *Critical Care Medicine*, **19**, S55 (abstract).

Harrison, P.M., Wendon, J.A., Gimson, A.E., Alexander, G.J., and Williams, R. (1991). Improvement by acetylcysteine of hemodynamics and oxygen transport in fulminant hepatic failure. *New England Journal of Medicine*, **324**, 1852–7.

Ishitani, M., Wilkowski, M., Stevenson, W., and Pruett, T. (1993). Outcome of patients requiring hemodialysis after liver transplantation. *Transplantation Proceedings*, **25**, 1762–3.

Lerman, A., *et al.* (1991). Elevation of plasma endothelin associated with systemic hypertension in humans following orthotopic liver transplantation. *Transplantation*, **51**, 646–50.

14.5.4 Lung transplantation

R. Ruggiero and C. A. Keller

Key messages

- Lung transplantation requires preparation of the donor and recipient with immunosuppressant agents and antibiotics.
- Early causes of graft failure include reperfusion injury, technical and immunological rejection, and infections. Late causes of graft failure are bronchiolitis obliterans and infections.

Introduction

The first human lung transplant was performed at the University of Mississippi by James Hardy in 1963; the patient survived for 18 days. Twenty years later, the Toronto Lung Transplant Group led by Joel Cooper reported the first long-term survival. Since then, advances in immunosuppression regimens, surgical technique, and postoperative care of lung transplant recipients have allowed remarkable improvement in results and allowed a much larger number of patients with endstage lung disease to benefit from this procedure. Present indications for lung transplantation are outlined in Table 1. The main goals of postoperative care of lung transplant recipients are maintenance of hemodynamic stability, meeting the challenges of early ventilatory management, establishment of immunosuppression, and prevention of infections.

Basics of preoperative and intraoperative care

Once an appropriate donor is allocated and the selection process is concluded, a patient scheduled to receive a lung transplant is rapidly prepared for the operation. This process consists of initiation of immunosuppression and administration of antibiotics. Intravenous infusion of cyclosporine A (cyclosporin A) is started at a dose of 2 to 10 μg/kg/day, based on baseline renal function. Azathioprine 2 mg/kg is given as an intravenous bolus. The first dose of corticosteroids consists of methylprednisolone 500 to 1000 mg intravenous bolus given intraoperatively just before reperfusion of the donor lung graft.

Once taken to the operating room, the patient is intubated with either a double-lumen endotracheal tube or an endotracheal tube

Table 1 Indications for lung transplantation

Candidates for single-lung transplant (age ≤60 years)
Obstructive lung disease
Emphysema
α_1-Antitripsin deficiency
Restrictive lung disease
Idiopathic pulmonary fibrosis
Occupational lung diseases
Sarcoidosis
Pulmonary vascular disease
Primary pulmonary hypertension
Eisenmenger's complex with repairable heart defect
Candidates for double-lung transplant (age ≤ 50 years)
Cystic fibrosis
Bilateral bronchiectasis
Primary pulmonary hypertension
Candidates for heart–lung transplant (age ≤ 50 years)
Cyanotic congenital heart disease
Eisenmenger's complex

with endobronchial blocker; this allows selective single-lung ventilation during the time of implantation of the donor lung. Large-bore intravenous cannulas, a pulmonary artery catheter, a radial arterial cannula, and a Foley catheter are inserted for intraoperative monitoring and management.

The conduct of the operation consists of entrance into the pleural space through a posterolateral fifth intercostal space thoracotomy incision. Once the donor organ has arrived in the operating room, recipient pneumonectomy is carried out, taking care to leave long bronchial, pulmonary artery, and pulmonary vein stumps. Significant bleeding can be encountered at this early stage of the operation, particularly in patients with cystic fibrosis, who often have tenacious inflammatory adhesions, as well as in patients who have undergone previous thoracic surgery (i.e. placement of thoracostomy tube, lung resection, and, more recently, lung reduction surgery). In patients undergoing heart–lung transplantation, injury of pulmonary–systemic collaterals can also be the cause of severe bleeding. The implantation of the donor lung starts with the bronchial anastomosis. In the 1980s the Toronto Lung Transplant Group demonstrated that omental wrapping of the bronchial anastomosis combined with reduction in steroid dosages was beneficial in decreasing the incidence of bronchial dehiscence, which was one of the main problems in the pioneering days of lung transplantation. More recently it has been shown that telescoping the bronchial anastomosis without omental wrapping also allows good bronchial healing. The latter technique is rapidly gaining popularity because it avoids the laparotomy incision necessary for mobilization of the omentum. Reinforcement of the bronchial anastomosis can also be achieved with pericardial, pleural, or intercostal muscle flaps.

The vascular anastomoses are carried out next. Some surgeons prefer to construct the pulmonary artery anastomosis first; others start by reconnecting the donor and recipient left atrial cuffs. Technical guidelines to be followed are avoidance of pulmonary artery kinking or torsion and making the left atrial anastomosis as wide as possible to prevent obstruction of the pulmonary veins. Vascular clamps are partially released before the vascular suture lines are tied to allow air removal. Single-lung transplantation can be carried out almost uniformly using recipient single-lung ventilation. Occasionally cardiopulmonary bypass is required for hemodynamic instability or inadequate gas exchange, usually in patients with primary pulmonary hypertension.

The technique of double-lung transplantation has evolved from *en bloc* simultaneous removal of the recipient lungs during a period of cardiopulmonary bypass through a median sternotomy to the present 'clam-shell' technique, which allows sequential single-lung transplantation through bilateral thoracotomies joined by dividing the sternum transversely; this technique requires no cardiopulmonary bypass. After the donor lung has been reperfused, ventilation is re-established. It is important for the anesthesiologist to carry out the initial period of donor lung ventilation gently to allow complete re-expansion of the transplanted lung. This process can take several minutes, during which the surgeon should avoid manipulation of the atelectatic lung which can cause parenchymal hematomas, hemoptysis, or air leaks.

Early postoperative care

The goals of the early postoperative care of lung transplant recipients are to establish adequate immunosuppression as soon as possible, initiate prophylaxis against infection, and prevent or treat early complications such as bleeding, acute lung graft dysfunction, airway complications, and vascular complications.

Immunosuppression

The standard immunosuppression at present consists of a triple-drug regimen with cyclosporine A, azathioprine, and corticosteroids. Antilymphocytic agents are used to treat refractory rejection and, in some centers, to induce immunosuppression.

Cyclosporine

Cyclosporine A is a polypeptide compound of fungal origin which was identified in 1972. After receiving FDA approval in the early 1980s, it has been part of most immunosuppressive protocols for solid organ transplantation. Its main effect is inhibition of the cellular immune response with overall decreased activation of T lymphocytes by blockage of the release of interleukin 2 (IL-2) from helper T lymphocytes. Nephrotoxicity is the most important adverse side-effect of cyclosporine therapy. Other deleterious side-effects include hypertension, hyperlipidemia, neurotoxicity, gingival hyperplasia, hirsutism, and development of lymphomas. Drugs such as calcium-channel blockers and ketoconazole have been shown to raise cyclosporine levels. Dosages are as follows.

1. Preoperative and intraoperative: continuous intravenous cyclosporine A infusion at 2 to 10 mg/kg/day.
2. Perioperative: continue infusion at 5 to 10 mg/kg/day. When oral intake is well tolerated, the oral dose is calculated by multiplying the total daily intravenous dose by three. This usually amounts to 100 to 300 mg orally every 12 h. The target cyclosporine serum level is 250 to 350 ng/ml.
3. Maintenance: particularly if rejection has been well controlled, cyclosporine A dosage can be reduced to 50 to 200 mg orally every 12 h. The target cyclosporine serum level is 100 to 200 ng/ml.

Over the last 18 months a new cyclosporine preparation, Neoral®, has become available. This new product has the advantage of being absorbed more rapidly, so that more stable serum drug levels are achieved.

Azathioprine

Azathioprine is an antimetabolite purine analog that inhibits cellular DNA and RNA synthesis. Its main effect is depression of T-cell-mediated immunity. Side-effects consist of hepatotoxicity and pancytopenia. Dosages used at our institution are as follows.

1. Preoperative: 2 mg/kg intravenous bolus.
2. Perioperative: 2 mg/kg/day intravenously or orally in two divided doses.
3. Maintenance: 1 mg/kg/day orally in one or two doses.

The white blood cell count should be monitored in patients receiving azathioprine and maintained between 4×10^9/l and 8×10^9/l.

Corticosteroids

Corticosteroids act by reducing production of IL-1 and IL-2 from antigen-presenting cells and activated lymphocytes respectively. Although steroids mainly inhibit T-cell activity, B-cell function is also impaired, although to a lesser degree. Corticosteroids are not only used for induction of immunosuppression, but are also the primary line of treatment in the setting of acute rejection. The adverse side-effects of steroid therapy are numerous and include impaired healing, glucose intolerance, increased susceptibility to infection, hypertension, gastrointestinal bleeding, osteoporosis, avascular bone necrosis, cataracts, and psychological and personality modification. Dosages are as follows.

1. Intraoperative: methylprednisolone 500 to 1000 mg intravenously at the time of graft reperfusion.
2. Perioperative: methylprednisolone 125 mg intravenously every 8 h for three doses, and then methylprednisolone intravenously or prednisone orally 1 mg/kg/day in two divided doses.
3. Maintenance: prednisone 1 mg/kg/day with tapering to 0.3 mg/kg/day.
4. Rescue: methylprednisolone 10 mg/kg/day intravenous bolus for 3 days.

Antilymphocyte and antithymocyte agents

Antilymphocyte and antithymocyte immunoglobulins are produced by injection of human lymphocytes into rabbits or horses. These products have been used in some centers for induction of immunosuppression, but their main role is in rescue therapy after recurrent rejection resistant to corticosteroids. OKT3 is a monoclonal antibody directed against a T-lymphocyte glycoprotein. It is used strictly for rescue therapy of refractory rejection. Methotrexate, cyclophosphamide, FK-506, and mycophenolate mofetil are other immunosuppressive agents available when commonly used drugs are either ineffective or not tolerated.

Bleeding

Bleeding in the immediate postoperative period is detected by excessive drainage from the pleural chest tubes, need for blood transfusion to maintain adequate hemoglobin levels and stable hemodynamics, and chest radiograph changes consistent with hemothorax. Most commonly, bleeding is diffuse in nature and originates from the parietal pleura or the mediastinal soft tissue, and rarely from the vascular anastomoses. Postoperative hemorrhage is more frequent if it is necessary to use cardiopulmonary bypass and in the presence of dense intrapleural inflammatory adhesions. The latter is often the case in cystic fibrosis patients in whom donor pneumonectomy needs to be carried out with the greatest attention to hemostasis. Patients undergoing heart–lung transplantation are also at higher risk of bleeding from pulmonary–systemic collaterals.

Acute graft dysfunction (reperfusion injury)

Acute deterioration of graft function often occurs in lung transplant recipients over the first 24 to 48 h postoperatively. It is attributed to a combination of suboptimal donor lung preservation, ischemic injury, and reperfusion injury. The end result is rapid development of non-cardiogenic pulmonary edema and pulmonary infiltrates. Clinical suspicion should be raised by rapid deterioration in gas exchange with decrease in arterial oxygen saturation, decrease in lung compliance, increase in pulmonary vascular resistance, and unilateral radiological changes resembling the acute respiratory distress syndrome. Fortunately, in the majority of cases, this process responds to aggressive diuretic therapy and adjustment in ventilatory settings by adding or increasing peak end-expiratory pressure. Prevention of acute early graft dysfunction is the main reason for the limit of 4 to 6 h in graft ischemic time recommended in lung transplantation.

Infections

Infections in the first 2 or 3 weeks following lung transplantation are usually bacterial in origin and are due to colonization of the donor lung. The most common pathogens are *Staphylococcus aureus*, *Pseudomonas aeruginosa*, other Gram-negative organisms, and *Candida albicans*. Viral and other fungal infections tend to occur later.

Cytomegalovirus

Cytomegalovirus infections present a unique problem in lung transplantation because of their frequency and severity. They occur more often in the presence of cytomegalovirus mismatches (cytomegalovirus-positive donor–cytomegalovirus-negative recipient, or cytomegalovirus-negative donor–cytomegalovirus-positive recipient). Cytomegalovirus-positive recipients receiving grafts from cytomegalovirus-positive donors are also at a higher risk of developing cytomegalovirus pneumonia or disseminated infection. Cytomegalovirus can be acquired by cytomegalovirus-negative patients receiving a donor lung from a cytomegalovirus-negative donor through transfusion of contaminated blood or through a new infection. When either donor or recipient are cytomegalovirus positive, ganciclovir is started on the seventh postoperative day for cytomegalovirus prophylaxis at a dose of 5 mg/kg intravenously twice a day for 2 weeks, followed by 5 mg/kg/day for as long as 3 months. Acyclovir (aciclovir) 400 mg orally three times daily is given for herpes prophylaxis to patients who do not require ganciclovir for cytomegalovirus prophylaxis.

Antibiotics are given preoperatively and continued postoperatively. We use piperacillin–tazobactam 3.375 mg intravenously every 6 h. Antibiotic coverage is modified as soon as cultures and sensitivities become available from specimens obtained during bronchoscopy performed in the operating room at the completion of the operation. Trimethoprim–sulfamethoxazole (co-trimoxazole), one tablet double-strength three times per week, is started for *Pneumocystis carinii* prophylaxis 1 month post-transplant and continued for as long as a year.

Vascular complications

Vascular complications consist of pulmonary arterial or, more often, pulmonary venous thrombosis. Development of vascular problems is suspected in the presence of deteriorating graft function associated with radiological changes in vascular markings. Diagnosis can be established by transesophageal echocardiography, perfusion scans, or pulmonary angiography. Treatment consists of anticoagulation and/or surgical revision.

Late postoperative care and complications

Bronchiolitis obliterans

Bronchiolitis obliterans is usually a late complication and is the main cause of long-term morbidity and mortality in lung transplant recipients. It presents clinically with progressive dyspnea, coughing, and wheezing. Pulmonary function tests deteriorate rapidly, with a decline in forced expiratory volumes and flow in a pattern similar to that observed with obstructive lung disease. Diagnosis is confirmed histologically with transbronchial biopsy. The etiology of bronchiolitis obliterans is not understood completely. Risk factors for its development seem to be recurrent and severe episodes of acute rejection and cytomegalovirus pneumonia. Once established, bronchiolitis obliterans is difficult to manage, short of retransplantation. The best approach is prevention and aggressive treatment of all episodes of rejection and infection. Surveillance is carried out with spirometric studies. Measurement of lung flows with home-based portable devices has also been found to be reliable.

Conclusions

As of November 1996 the National Transplant Waiting List showed that, in the United States, 2296 patients were listed for lung transplantation and 233 patients for heart–lung transplantation. Over the previous 12 months, 657 patients had undergone either single- or double-lung transplantation, and 65 patients had received heart–lung transplants. Better postoperative care has contributed significantly to the steady improvement in results. One-year survival rates are now approaching 85 per cent for single-lung transplantation, 75 to 80 per cent for double-lung transplantation, and 60 to 65 per cent for heart–lung transplantation. The lack of donor organs remains the main limiting factor preventing wider use of this therapeutic modality in the treatment of patients with endstage lung disease.

Bibliography

Frist, W.H. (1995). Management of complications. In *Thoracic transplantation* (ed. S.J. Shumway and N.E. Shumway), pp. 226–51. Blackwell Science, Cambridge, MA.

Keller, C.A. (1997). Heart–lung, double-lung and single-lung transplantation. In *Critical care* (ed. J.M. Civetta, R.W. Taylor, and R.R. Kirby), pp. 1317–31. Lippincott-Raven, Philadelphia, PA.

McCarthy, P.M. (1995). Immediate postoperative care after thoracic organ transplantation. In *Thoracic transplantation* (ed. S.J. Shumway and N.E. Shumway), pp. 205–18. Blackwell Science, Cambridge, MA.

Maurer, J.R. (1990). Therapeutic challenges following lung transplantation. *Clinics in Chest Medicine*, **11**, 279–90.

Todd, T.R.J. (1990). Early postoperative management following lung transplantation. *Clinics in Chest Medicine*, **11**, 259–67.

15 Management of the critical care department

15.1 The intensive care unit

15.1.1 Design of the department

T. E. Oh

Key messages

- The design and commissioning team should have an appropriate membership.
- The floor plan allocates patient, staff, and storage/support areas.
- A mixture of open wards and single rooms is useful.
- Each bed area requires a floor space of at least 21 m^2 (225 ft^2) and adequate gas and power supplies.
- Windows, hand-washing facilities, and radiograph viewing facilities are essential.
- The space needed for storage and support services is two to three times the total bed area.
- Power, lights, air conditioning, and communications require careful consideration.
- Local safety regulations must be obeyed.

Introduction

The design of an intensive care unit (ICU) must cater for the special needs of managing critically ill patients and comply with the standards of local regulatory bodies. A design and commissioning team should consist of critical care specialists, critical care nurses, administrators, and architects and engineers who are familiar with building hospitals. Input should also be sought from other disciplines (e.g. medicine, surgery, pediatrics, anesthesiology, infectious diseases, biochemistry, radiology, physiotherapy, and pharmacy) and from patient representative groups. The specific function of the ICU must first be determined and established, i.e. what level of care will it provide and whether it is to be a multidisciplinary or single-discipline unit (FICANZCA 1994; Oh 1997). As intensive care medicine continues to evolve, some currently accepted guidelines for the design of ICUs (FICANZCA 1994; ICS 1994; SCCM 1995) may change.

Floor plan

The floor plan of an ICU allocates areas for patients, staff, visitors, storage, and services. In addition to patient bed areas (see below), two to three times more space is required to support the operations of an ICU. The advantages of an open-ward design compared with multiple single rooms are debatable. Single rooms offer isolation facilities and more privacy but demand more nursing staff; a patient will be left unattended if the nurse leaves the room for whatever reason. There is no evidence that either design has a lower nosocomial infection rate than the other (Huebner *et al.* 1989). In practical terms, the debate is largely irrelevant. An open-ward ICU still requires some single rooms (at least 20 per cent of its total beds) for patients who need isolation or extra privacy. An ICU made up entirely of single-room beds will require a high nursing staff level to care for critically ill patients and will be less cost-effective. Thus the ICU design should incorporate open-ward bays (of two to eight beds) as well as single rooms in a ratio appropriate to the defined role and function of the unit. Open-ward bays may be clustered.

Traffic flow patterns should be considered in the floor plan. Traffic patterns should facilitate patient transport and restocking of bedside supplies, alleviate nurse fatigue from moving between the bed and store rooms, and enable rapid staff access in emergencies. Corners and doors must not impede trolley movements. Through traffic to other departments should not be permitted. Access into the ICU should be controlled by siting the reception area at the entry. A separate access for the ICU staff and a service corridor for moving goods and equipment may be desirable if space permits. Of course, the floor plan must conform to local safety regulations such as those pertaining to fire (e.g. fire doors and emergency exits).

The ICU should be located close to the accident and emergency department, operating rooms, acute wards, and radiological and imaging services. If a single-discipline ICU is planned, it is best located adjacent or close to other ICUs to enable sharing of facilities such as a common laboratory.

Patient areas

Spatial dimensions

Each patient bed area in an open ward requires a minimum floor space of 21 m^2 (225 ft^2) (SCCM 1995). There should be a corridor space of at least 2 m (6.6 ft) adjacent to each open-ward bay. Single rooms require a floor space of at least 23 m^2 (250 ft^2), and should be square or rectangular with a minimum length of 4.5 m (15 ft).

Isolation rooms require an additional anteroom of area at least 1.9 m^2 (20 ft^2) for hand-washing and gowning. Doors into single rooms must be large enough to allow easy passage of beds (particularly orthopedic beds) and equipment. Positive/negative-pressure air conditioning for isolation rooms is expensive and of questionable value. This facility should be equipped for only one or two rooms. However, facilities to dispose of body wastes and contaminated items should be available in all isolation rooms.

Bedside services

Bedside service outlets must conform to national standards (including electrical safety and emergency supply). Three oxygen, two air, four suction, and 16 electrical power outlets are appropriate for a high-level ICU. Outlets for portable X-ray machines should be easily accessible from each bed. Provision for gas scavenging should be considered; use of nitric oxide (and possibly other gases) may become more common. Bedside services are usually mounted at the wall, as beds are traditionally placed with the head towards the wall. The design of wall-mounted service outlets must allow for rapid access to the patient's head in emergencies. Other designs offering easier head access provide services through a floor column or overhead pendant; the bed effectively becomes an island. However, space to place extra monitoring equipment can be limited with these alternative designs. Equipment should be kept off the floor as much as possible, but mounted equipment and shelves for portable equipment should not be sited too high for nurses to reach. There must be adequate facilities to hang intravenous and blood containers. Outlets for a telephone, radio, and television at each bed can be planned, even though they may not be commonly used.

Bedside workstation and storage

There should be adequate surface areas at the bedside for frequently used drugs, charts, and computer terminals, and enough storage for other drugs, patient care disposables, intravenous fluids, pillows, linen, and the patient's belongings. A properly designed bed divider for an open-ward bay can offer a good workstation. Lead-lining these dividers will help minimize X-radiation risks to staff and patients.

Hand-wash sinks

An adequate number of hand-wash sinks for every bed area is imperative. Ideally, there should be one sink for every two open-ward beds and one sink in every single room and anteroom. Sinks should be deep and wide so that they are non-splash, and elbow, knee, foot, and infrared-operated taps are recommended to minimize contamination. Proper facilities for hemodialysis, such as filtered water, should be incorporated in selected bed areas.

Radiograph viewers

A distinct area should be designated for storing and viewing radiographic films. At least one multidisplay viewer is needed in each multibed ward. Mechanized carousel viewers will facilitate viewing of serial films. A separate radiology room to site the carousel, to store and develop films, and to park portable machines may be needed in a large ICU. However, location of films away from the bed area is a disadvantage. Consideration could be given to installation of (future) digital technology to retrieve radiographs from the radiology department on bedside computer screens.

Emergency trolleys

Each patient area should have space allocated for parking emergency trolleys (e.g. defibrillators and airway management trolleys). Alcoves furnished with adequate electrical power points for charging battery packs are ideal.

Nursing stations

A central nursing station may be sited in the middle or at one end of an open ward. It is important that the nursing station allows direct visualization of individual patient beds. Hence bed dividers should not be excessively high and room partitions should have windows. Visualization of beds out of the line of direct vision or in single rooms can be achieved by video monitors. Large ICUs will probably be better served by having a nursing station or substation for every cluster of eight to twelve beds. Since critical care nursing is at the bedside, monitoring at a central nursing station is of less relevance except in a coronary care unit. In recent years, the amount of monitoring equipment at an ICU nursing station has been scaled down considerably. Nevertheless, the nursing station should be adequate to house patient and video monitors, telephones, intercoms, computer terminals, a printer, patient records, stationery, and a specimens refrigerator. The nursing station should be designed for necessary staff functions, such as charting, form-filling, tracking personnel, and hand-overs. Comfortable seats, good task lighting, and provision of a clock and calendar are essential. If the ICU secretary or ward clerk is to be sited at the nursing station, space has to be apportioned accordingly.

Support and storage areas

Apart from patient areas, an ICU requires carefully planned areas for supporting services, storage, and staff utilization.

Reception

A reception area at the entrance of the ICU controls access. If the receptionist is also the ICU secretary, files, and indeed the whole reception area, must be securable. The waiting lounge for relatives and visitors should be adjacent to the reception desk. The lounge should be sufficiently large to accommodate the anticipated numbers of visitors. One to two seats for every ICU bed is acceptable. Seats should be comfortable and single-seating arrangements are usually preferable. A television can be placed in one corner. An adjacent interview room is recommended for larger ICUs and for others if the floor plan allows. This will also provide some privacy for grieving relatives. Visitors should have access to a pay-phone, toilets, and refreshments either on-site or nearby. An arrangement whereby a hospital overnight suite can be made available for relatives of dying patients is ideal. Gowning facilities for visitors are not necessary.

Storage

A universal problem with older ICUs is the lack of storage space. Each ICU bed requires an additional 25 per cent floor space for storage of consumable items. Space for equipment is additional, depending on the ICU inventory. An ICU bristling with heavy equipment (e.g. imaging, hemodialysis, and cardiopulmonary bypass machines) may require double the extra space. A crude estimate of the minimum storage space requirement is 30 per cent of total bed areas. Storage rooms should be dedicated (e.g. for linen, intravascular consumables, intravenous fluids, respiratory care, and electronic equipment). Frequently used items should be located closer to patients. Some items, such as linen, can be stored in alcoves with open shelves rather than in rooms or cupboards. The storage area for electronic equipment should have facilities for recharging battery-operated devices. Storage plans should also allow for space to park prepared treatment trolleys (e.g. dressing and monitoring trolleys).

Medication area

A medication area is needed to store and prepare medications. This should consist of clearly labeled drug cabinets, one of which is lockable for controlled drugs, adequate bench space, and a sink with taps. If the medications area is a dedicated room, use of glass windows for visualization of patients and of occupants of the room is recommended.

Meals preparation area

A small room should be made available for the preparation of enteral nutrition solutions and food for those patients able to eat. This room should be furnished with cupboards, benches, a sink with taps, a refrigerator (for food only), and a microwave oven.

Utility rooms

Clean and dirty utility rooms must be separate, each with its own access. A clean utility room is used to prepare and clean items for sterilization and to restock treatment trolleys. An autoclave is usually sited here for in-house sterilization of specific items (sterilization of most items is performed in the central hospital facility). Sinks and taps with bench surfaces that are easy to clean are essential. A dirty utility room is required to clean bedpans and to decontaminate or dispose of used dressings, linen, and monitoring devices. A sluice and sinks with taps are necessary. Systems such as covered containers must be provided to dispose of needles and contaminated substances and fluids.

Procedures room

A room dedicated to special procedures and treatment, such as performing tracheostomies and placing intracranial pressure-monitoring devices, is less important than previously. The undesirability of moving an unstable patient and the introduction of more percutaneous procedural techniques has contributed to a decline in the use of the procedures room. If one is desired, it should be of adequate size and located close to patient beds.

Stat laboratory

ICUs offering high-level care must have access to 24-h laboratory services. However, a stat laboratory is valuable for rapid analysis of arterial blood gases and for simple biochemical and hematological tests, including serum electrolytes, osmolality, lactate, hemoglobin, hematocrit, and oxygen saturation by co-oximeter. The size of the room will depend on the equipment to be housed. Benches, water and gas supplies, and communication links will be required.

Satellite pharmacy

Larger ICUs may require a satellite pharmacy within the unit to store and dispense medications, and to prepare parenteral nutrition solutions and intravenous additives. Drug storage and preparation facilities (e.g. laminar flow cupboards) will be required. The satellite pharmacy must be securable.

Workshop

A room dedicated to uncomplicated repairs, calibrations, assembly, and equipment checks is desirable. Major servicing and repairs are undertaken elsewhere by the hospital engineering department or service contracts. This room requires benches, electrical outlets, piped gases with scavenging, and a sink.

Staff areas

Lounge

A staff rest room with comfortable chairs, a television, and food and drinks facilities is obligatory. Separate lounges for doctors and nurses are a waste of space and money, and erode the team spirit of the unit.

Changing rooms

Every ICU must have staff toilets and showers. Male and female changing rooms are also required, except in small ICUs. These facilities can be incorporated with the staff lounge as a staff amenity area.

Seminar room

A seminar or conference room is extremely useful for educational activities and staff hand-over rounds. If space is lacking, a large staff lounge can double up for these activities.

Offices

The ICU should have a sufficient number of offices for senior ICU staff. Actual numbers depend on the size and function of the ICU.

On-call rooms

An ICU which offers 24-h resident medical services requires on-call rooms inside or adjacent to the ICU. Two or even three rooms may be necessary for a large ICU to cater for the resident medical staff on night and weekend duties. An on-call room should have an *en suite* bathroom and be furnished with a desk, an easy chair, a bed, and full communication links.

Environmental services and control

Time and sensory orientation

It is imperative that all patient and central staff areas should have large clear windows. Lack of natural illumination leads to day–night disorientation, psychosis, and increased stress. Installation of large clocks and calendars visible from patient beds will also help time and sensory orientation. Radios, television sets, and telephones should be available for those less sick patients who are able to use them.

Color and furnishings

Consideration must be given to providing an environment which minimizes stress to patients, visitors, and staff. Choice of warm color schemes and soft furnishings contribute to a calming effect. Blinds, shades, and curtains help to establish patient privacy. Advice from interior designers will be useful. Each bed area will require chairs for the nurse (typist chairs with castors are suitable), visitors (preferably stackable), and the patient to sit out of bed.

Sound

Reducing noise levels will also reduce environmental stress. There may be national standards for minimal sound levels. Noise levels below 45 dB during the day, 40 dB in the evening, and 20 dB at night have been recommended (SCCM 1994). Sound engineers should be consulted in the building and commissioning of the ICU. Sound-absorbing materials and noise filters should be used in walls, ceilings, partitions, and work-tops. Similarly, floors should have coverings that absorb sound, with consideration of maintenance, durability, trolley movement, and infection control factors.

Light and power

General overhead lighting must cater for day and night illumination. Dimmer switches are useful. Task lighting is also necessary at nursing stations and bed workstations for charting and treatment preparations. Bright spotlights are required at the bedside for procedures and emergencies. Each bed area should also be equipped with a night light and a reading light.

Electrical supplies must conform to national safety standards. Outlets close to the head of the bed should be sited well above the floor for easy access, and those at the sides should be sited close to the floor to minimize tripping over cables. A source of emergency power for electricity and light must be available to resupply power immediately if the main source fails.

Air conditioning

The air-conditioning system should be able to perform a minimum of six total air changes per hour, with two changes of outside air; higher rates of changes may be required for single rooms. Control of heating, cooling, and humidification should be possible. Air from dirty utility rooms and isolation rooms should be exhausted outside the ICU system. Air exchange filters should satisfy appropriate standards.

Communications

A good ICU communication network is vital. Within the ICU, it is necessary to have close links between patient bed areas, nursing stations, staff areas, and the reception area. This can be effected by a system of telephones and intercoms. Visual displays of alphanumeric or color codes will not generate noise and can be used for non-emergency communications. An emergency visual and audio call button at the bedside, connected to all areas, will enable immediate help to be summoned. The provision of one or two additional telephones through a different line from the hospital switchboard is recommended. Computer networking between ICU monitors and with other departments (e.g. pathology, radiology, and central admission office) should be implemented. Pagers and mobile voice communicators should be used to track staff outside the ICU. A special code for pagers (e.g. 1111) will enable ICU staff to be summoned immediately in emergencies. Communications between the ICU and other hospital departments may include a pneumatic tube system.

References

FICANZCA (Faculty of Intensive Care, Australian and New Zealand College of Anaesthetists) (1994). *Minimum standards for intensive care units IC-1.* Australian and New Zealand College of Anaesthetists, Melbourne.

Huebner, J., *et al.* (1989). Influence of architectural design on nosocomial infections in intensive care units—a prospective 2-year analysis. *Intensive Care Medicine*, **15**, 179–83.

ICS (Intensive Care Society) (1994). *Standards for intensive care units.* BioMedica, London.

Oh, T.E. (1997). Design and organization of intensive care units. In *Intensive care manual* (4th edn) (ed. T.E. Oh), p. 3. Butterworth-Heinemann, Oxford.

SCCM (Guidelines/Practice Parameters Committee of the American College of Critical Care Medicine, Society of Critical Care Medicine) (1995). Guidelines for intensive care unit design. *Critical Care Medicine*, **23**, 582–8.

15.1.2 Site, size, and interactions with other departments

D. C. Jerwood and E. Major

Key messages

- The direct patient area should occupy no more than about 40 per cent of the total area of an intensive care unit (ICU). Space outside the patient area is no less vital than circulation space around each bed, but storage space and staff space are often reduced in the interests of cost reduction.
- Ideally, the location of an ICU within a particular hospital will be determined by the case mix of patients to be admitted and the situation of other departments with which ICU staff will be in regular contact.
- Sharing engineering services with other departments such as operating theaters might seem advantageous, but this is by no means an overriding factor and may indeed reduce flexibility.
- The normally busy ICU acts as a focal point of the hospital since from time to time every clinician will need to visit a patient, some very frequently.

Introduction

There are a number of published guidelines for the design of intensive care units (ICUs), notably from the European Society of Intensive Care Medicine (1994), the Faculty of Intensive Care of the Australian and New Zealand College of Anaesthetists (1994), and the American College of Critical Care Medicine (1995), but perhaps the most detailed emanate from the Intensive Care Society of the United Kingdom (1996). There is considerable international agreement about design details but little agreement on unit size or the benefits (or otherwise) of regionalization. The following is a distillation of the recommendations.

Site

An ICU should be a geographically distinct area within a hospital. It should function as an autonomous department with controlled access and no through traffic. Ideally, there should be separate access for visitors and staff, and provision should be made for escape from fire.

Ideally, the location of an ICU within a particular hospital will be determined by the case mix of patients to be admitted and the situation of other departments with which ICU staff will be in regular contact. For some hospitals there will be an ideal site. An ICU admitting burn patients from a wide area might best be situated on the ground floor near the ambulance entrance, whereas a cardiothoracic or neurosurgical ICU would be better adjacent to the operating theater suite with easy access to the radiology department. In many cases the location will be a compromise, but most general units will usually be in proximity (either horizontally or vertically) to the operating theater, the accident and emergency department, or the radiology department.

Transport of critically ill patients between units or between departments of the same hospital is known to be associated with risk, but proximity of the radiology department or operating theater to the ICU does not abolish that risk since the time taken to perform the investigation or surgical procedure is usually much longer than the transport time. Obviously time spent in elevators with the attendant risks of breakdown should be minimized, but it seems likely that the additional risks of transporting critically ill patients are more closely related to equipment deficiencies and staff inexperience than to distance between departments (Manji and Bion 1995). Monitoring and supervision of the patient should be no less intense because the patient is being moved; rather, they should be more intense.

There is no specific requirement that an ICU be on a particular level, although evacuation of the unit during a fire hazard would be more readily accomplished in a ground floor unit. Sharing of engineering services with other departments such as operating theaters might seem advantageous, but this is by no means an overriding factor and may indeed reduce flexibility. Patient routes should be separate from transport corridors, and any elevator access required should be under keyed control. Disruption is minimized by separate supply and service access via perimeter corridors, but such corridors will inevitably restrict access to natural light.

Interactions with other departments

Although the ICU is usually self-contained and has its own complement of medical, nursing, and other staff, it certainly is not isolated. The normally busy ICU acts as a focal point of the hospital since from time to time every clinician will need to visit a patient, some very frequently. Doctors in training are often intimidated by equipment that they do not understand and terminology which is new to them, but this should not always be so. Training opportunities abound in intensive care, and all doctors in training would benefit from time spent learning the basics of resuscitation and recognition of serious illness.

During the course of a normal day the ICU has to cope with hundreds of individual visits by doctors, porters, cleaners, maintenance men, professionals allied to medicine, and of course patients' relatives. All too often supplies, beds, staff, visitors, and rubbish share a common entrance corridor cluttered by equipment spilling over from the store room. To cope with all this and retain a relatively calm non-threatening atmosphere for patients is perhaps one of the greatest challenges for designers and ICU staff.

Security is a major problem, and many units have installed entry-phone systems with the consequent need to provide a full-time receptionist to avoid repeated calls into the clinical area. Frequent telephone calls are unavoidable, but personal calls to staff on duty should be discouraged. The person answering the telephone has a responsibility to do so politely and professionally since no-one should be put off calling the ICU for help. Early communication about a patient may be lifesaving, and therefore good telephone etiquette is an important part of the well-run unit.

In the same way that early referral may be lifesaving, postdischarge problems and readmission to ICU may be avoided by provision of detailed information to the medical and nursing teams taking over the care of the patient. This is best done in writing in the case notes, but should be accompanied by a telephone call to inform the doctor or nurse in person. In that way good relationships are constructed. It is also good practice to send a formal discharge summary to the general practitioner who may be unaware that the patient has been seriously ill. This is particularly important because the details of a 2-month stay in the ICU may be glossed over by a busy doctor dictating the hospital discharge summary when the patient leaves the general ward. The pathologist will also appreciate information about the circumstances of a patient's death before performing a postmortem examination.

Good communication improves the image of the ICU and certainly makes it easier for ICU staff to enlist help when needed. The consultant radiologist is much more likely to take an interest in the investigation of an ICU patient if he or she is approached personally by an ICU doctor, and the surgeon performing a laparotomy on an ICU patient will feel more confident of the ICU team's involvement with the patient if they have demonstrated interest by watching the procedure and discussing the findings.

Intensive care medicine may be challenging, stimulating, rewarding, and depressing. Sometimes it may be all these things in one day. None of us can be expert in all aspects of medicine and surgery, but that is precisely the challenge of intensive care. Only regular contact between intensive care doctors and professionals of all specialties can provide the best care for our patients.

Size

Bed numbers

Intensive care bed provision varies widely between countries, but this is partly due to differing interpretations of the term 'intensive care'. Some countries make a distinction between high-dependency care and intensive care, whereas others label all such areas 'intensive care'. In the United States and Scandinavia the provision of intensive care beds may account for up to 10 to 15 per cent of total acute hospital beds, whilst in the United Kingdom the provision has historically been less than 1 per cent. However, this may slowly be changing.

The ideal bed complement of a single ICU is a quite separate issue from the total number of ICU beds required for a population, but there is no international consensus on bed numbers or optimum ICU size. However, there is a clear view within the United Kingdom that 1 per cent of total acute beds represents serious underprovision for intensive care. In some hospitals it has been appropriate to form multiple small units specializing in, for example, neurosurgery, cardiothoracic surgery, burns, or pediatrics, while in other hospitals large multidisciplinary units are favored. To date there is little hard data on which model provides better care, but there appear to be economies of scale and greater nursing flexibility within larger units.

Units containing over ten beds demand full-time specialist medical cover, but this is something which has traditionally not been the case in ICUs in the United Kingdom. In 1994 only a minority of ICUs had a full-time director and the median size was four beds. Perhaps these two findings are linked in that it is clearly not cost effective to provide a full-time director to look after four beds.

Floor space

ICU floor space can be considered under the following headings: patient space, storage space, and staff space.

Patient space

Although separate rooms are ideal for reasons of patient privacy and prevention of cross-infection, staffing costs are high, and a ratio of one single room to every six beds in an open area has been recommended. Not only are single rooms expensive of staff but they also occupy more floor space. United Kingdom recommendations suggest that at least 20 m^2 (215 ft^2) of floor area is required for each open-area bed space and 30 m^2 (323 ft^2) is required for single rooms (Intensive Care Society 1996). This significantly exceeds the recommendation produced by the Society of Critical Care Medicine in the United States which recommends 150 to 200 ft^2 for open bed areas. United Kingdom standards suggest that single rooms also require an anteroom of area 2.5 m^2 with facilities for washing and gowning, whilst beds in an open area should be provided with a 2.5-m wide unobstructed corridor space beyond the working area. However, the provision of services via gantry or stalactite systems may reduce these requirements.

Storage space

Required storage space is directly related to the number of beds. Bulky items such as ventilators, trolleys, dialysis equipment, drip stands, monitoring equipment, infusion pumps, and blood warmers require considerable floor space and 5 m^2 per bed space is recommended. Consumables such as syringes, infusion sets, filters, and items from the central sterile services department occupy another 5 m^2 per bed. Linen must be stored; the space required depends on the frequency of delivery and laundry turn-round, but is estimated to be 2 m^2 per bed space. Thus the total storage per bed space is 12 m^2. Specialized beds, traction frames, portable imaging equipment, and seating for visitors all need storage, and this may add another 20 m^2 to the requirements.

Staff space

Space must be allowed for clean and dirty utility rooms, and a staff base sufficient to accommodate computer terminals, telephones, and stationery. In addition, a blood chemistry laboratory, technical workshop, and receptionist's desk must be accommodated. Medical and nursing staff require office space, and there must be a relatives' waiting room and an additional room for interviewing distressed relatives.

A separate seminar/conference room with projection facilities and a small library space, as well as seating for teaching or multidisciplinary meetings, requires a further 30 m^2. This area may also include a radiology storage/viewing area if not separately accommodated.

A staff rest room of area not less than 21 m^2 (increased by 3 m^2 for every two beds beyond eight beds) should be provided well away from the relatives' room. Changing facilities, lockers, showers, and toilets must be provided, and the specific United Kingdom recommendations are 0.75 m^2 per nurse with a minimum of 15.5 m^2 for female staff and 7.5 m^2 for male staff (assuming fewer male nurses).

Table 1 Space allocation for a 12-bedded ICU

	Area (m^2)
Patient area	
2 single rooms	60
10 beds in open area	200
Circulation space	25
Total bed space	285
Storage	
Clean/dirty utility	30
Pharmacy	15
Waste disposal	2
Equipment	60
Consumables	60
Linen	24
Total storage	191
Staff areas	
Residents' bedroom	15
Offices	45
Kitchen	15
Laboratory	15
Reception area	15
Relatives' waiting room	30
Interview room	15
Seminar room	30
Staff rest room	26
Changing facilities	23
Workshop	15
Cleaners' store	10
Total staff area	254
Total all areas	730

The ICU residents' room should have an area of 15 m^2 and a kitchen facility for use by all staff also requires at least 15 m^2. Cleaning staff require an area for storing equipment and materials, which can be estimated as 6 to 8 m^2 per eight beds.

As with any ward, separate clean and dirty utility areas are needed. The clean utility area for the preparation of sterile equipment and dressing trolleys etc. should be at least 10 m^2, and the dirty utility area should be twice this size. A separate area of 2 m^2 is required for storing bagged clinical waste. Because of the large range and quantity of drugs used, most units require a satellite pharmacy or at least a drug storage area utilizing 15 m^2 of floor space. A satellite pharmacy would, of course, require a bench, sink, refrigerator, and air conditioning.

From the foregoing it can be seen that the direct patient area occupies no more than about 40 per cent of the total area of an ICU (Table 1). Space outside the patient area is no less vital than circulation space around each bed, but storage space and staff space are often reduced in the interests of cost reduction. Few of us are given the opportunity to design a new ICU, but for those fortunate enough to have the funding to do so there is little excuse for repeating the mistakes of past generations since so much has now been written on the subject.

References

European Society of Intensive Care Medicine Task Force (1994). Guidelines for the utilisation of intensive care units. *Intensive Care Medicine*, **20**, 163–4.

Faculty of Intensive Care, Australian and New Zealand College of Anaesthetists (1994). *Minimum standards for intensive care units, IC-1.* Australian and New Zealand College of Anaesthetists, Melbourne.

Guidelines/Practice Parameters Committee of the American College of Critical Care Medicine, Society of Critical Care Medicine (1995). Guidelines for intensive care unit design. *Critical Care Medicine*, **23**, 582–8.

Intensive Care Society (1996). *Standards for intensive care units.* Intensive Care Society, London.

Manji, M. and Bion, J.F. (1995). Transporting critically ill patients. *Intensive Care Medicine*, **21**, 781–3.

15.1.3 Staffing of the department

D. Reis Miranda and L. J. Broerse

Key messages

- The effective and efficient care of patients in the intensive care unit (ICU) is ensured by the complementary knowledge and skills of physicians and nurses.
- The structured integration of the two professions requires that particular attention is paid to those areas where knowledge and skills overlap (organization quality standards such as protocols).
- It is mandatory that all the relevant characteristics of the ICU should be assessed before plans for staffing are made. Fixed staffing rules should not be imposed as they will often lead to severe mismatch between the provision and the demands of resources at ICU level.
- The classification of ICUs into levels of care related to the requirements of nursing staff is the key to appropriate staffing of each unit.
- Attention to non-clinical aspects of organization and management is essential for the effective functioning of the staff in the ICU.

Introduction

Five professional groups should be considered when staffing the intensive care unit (ICU): physicians, nurses, administrative staff, technicians, and ancillary staff.

Physicians include the following:

(1) the medical director who is responsible for the management of the clinical affairs of the ICU;

(2) intensivists who are responsible for the 24-h medical care of the patients in the ICU;

(3) intensive care fellows in training as intensivists;

(4) residents undergoing regular training in intensive care pertaining to the training program of their respective specialties, and who are responsible for the performance of medical care in the ICU under supervision.

Nurses are classified as follows:

(1) the head nurse who is responsible for the management of the nursing affairs of the ICU;

(2) intensive care nurses who are responsible for the 24-h nursing care of the patients in the ICU;

(3) nurses in training;

(4) nursing helpers who are responsible for the performance of nursing tasks under supervision.

With increasing frequency, the tasks of the medical director and the head nurse are integrated in an intensive care directorate which governs both professional groups at ICU level. The functions described above for each professional group aim at covering the tasks generally existing in the ICU: leadership, planning, observation, judgment, decision, action, and support. The titles of such functions, as well their content and diversity, may vary from country to country.

Administrative staff consist of the following:

(1) the ICU secretary who is responsible for the administration of clinical and nursing affairs;

(2) other staff, such as data managers, may be required according to the type of work being done, which usually depends on the size and type of the ICU (e.g. university or non-university).

Technicians are responsible for the maintenance of the equipment in the ICU (see below). Ancillary staff include cleaners and staff to attend to families and visitors to the ICU.

The final composition of the staff is based upon an evaluation which is discussed below.

Organizational aspects

Appropriate staffing of the ICU is based upon a prior assessment of the unit taking the following into account:

(1) number of beds and desired occupancy rate;

(2) the number and case mix of patients to be admitted;

(3) amount and type of equipment and technology;

(4) levels of care in the ICU.

Number and case mix of patients to be admitted

It is generally accepted that the case mix may be characterized by age, severity of illness, diagnosis, type of admission (medical or surgical, scheduled or unscheduled), and origin of the patients (operating theater/recovery room, ward, other ICU, other hospital, etc.). It has been shown that each of these characteristics exerts a significant influence upon the use of resources and the outcome of patients admitted to the ICU.

The type and number of patients admitted, and of the related tasks to be performed, are the major determinants of the type and number of the staff in the ICU. The case mix of the admitted patients (when describing risk of death) should be primarily related to the professional expertise of the staff, whereas the patient turnover (number of admissions) should be related to the number of staff. The two examples below may help to explain these relations.

An ICU with a high patient turnover (more than 80 admissions per bed annually) will use a large proportion of its staff to ensure efficient functioning of the infrastructure associated with frequent admission and discharge of patients. The medical conditions to be dealt with are usually predictable, the overall risk of mortality is low, and the patients can usually be classified into a relatively small number of case-mix groups (usually postoperative diagnoses). The work required is usually the monitoring and treatment of rather short periods of cardiovascular and/or respiratory dysfunction or failure. Therefore the nursing tasks performed in these ICUs usually have a low grade of differentiation, and protocols can easily be designed for many of the tasks performed by physicians. In this ICU leadership, planning, and judgment can be concentrated on training and supervision, so that the variety and number of differentiated professionals required are reduced.

An ICU admitting predominantly patients with a higher risk of mortality (e.g. older, with unscheduled medical and surgical problems, or admitted from the ward, other ICUs, or other hospitals) will necessarily have a lower turnover of patients and therefore require less manpower for administrative tasks. However, this ICU will require more personnel for the clinical care of individual patients. These patients will not only be sicker and need more care, but they will also present a larger variety of often unpredictable critical illnesses (e.g. multiple organ and system dysfunction or failure). Therefore these ICUs will have a high grade of differentiation of tasks performed, so that a higher grade of professional differentiation will be required among the staff. Moreover, centralization will be difficult, except for leadership, and protocols will only be possible for the less differentiated tasks. In other words, the medical and nursing staff providing daily care to the patients will need to have a high degree of professional autonomy.

Patient numbers and case mix can easily be assessed if the ICU is already operative. If the organization of the unit is being planned prior to opening, the survey should be performed in two steps: first, a comparative study of two or three ICUs with characteristics similar to the one being planned which are operating in similar hospitals; second, submission of the original staff planning to revision and approval after the first and second years of operational experience.

Amount and type of equipment and technology

Equipment and technology are usually related to the case-mix characteristics. Depending upon the amount, cost, and sophistication of the equipment, as well as its technical control and maintenance requirements, it may be necessary to establish a technical unit within the ICU. This unit should be staffed by technicians who are present for at least 8 h on 5 working days in the week. Although these technicians are on the pay-roll of the ICU and working there full time, they should be regarded as part of the technical department of the hospital, which will define and control their tasks, be responsible for their periodic professional updating, and ensure their professional career prospects.

Levels of care in the ICU

The case mix of patients admitted to the ICU annually varies very little as it depends on the clinical activities in the hospital. Therefore the resources required to meet the demands of its clinical management are predictable. An important consequence of this is that the use of ICU resources (particularly staff) is almost constant over time, so that it is possible to define staffing levels that will satisfy demand (as measured by severity of illness or risk of death). Because the nursing staff is the most important (about 90 per cent of the total) and the most constantly present in the ICU, it is often used as a measure of level of ICU staffing.

Four levels of care (LOCs) based upon *P*/*N* ratios (the number of patients assisted by one nurse per shift) were defined in 1983 at the Bethesda Consensus Conference. Staffing levels were proposed for only three of these, as the ICUs in the fourth level, where the nursing requirements were difficult to distinguish from those in the general ward, were not considered to be true ICUs. The LOCs were defined on the basis of the average nursing workload in the ICU measured at patient level by means of the Therapeutic Intervention Scoring System (TISS) (Cullen *et al.* 1974). The work of one experienced ICU nurse in each nursing shift is represented by 40 to 45 TISS points daily. Therefore average daily patient requirements are at least 40 TISS points (*P*/*N*=1/1) in the ICU, between 39 and 20 TISS points (*P*/*N* between 1.5 and 2) in high-dependency care, between 19 and 10 TISS points (*P*/*N* between 2 and 4) in the medium- to high-level care, and less than 10 TISS points in medium- to low-level care. The division of ICUs into three levels of care corresponds to a generally experienced need to match the concentration of resources to the intensive care requirements of patients. Therefore medium-care units are generally located next to ICUs in the majority of hospitals. However, the traditional classifications of intensive care, high-dependency care, and medium care should be abandoned, because they are extremely difficult to define, and should be replaced by the calculation of *P*/*N* ratios as the quantitative definition of LOCs.

The relation between workload as measured by TISS and the daily activities of the nursing staff has recently been re-examined (Reis Miranda *et al.* 1995). After simplifying, updating, and validating a new version of TISS, it was determined that one TISS point corresponded to 10.8 min of nursing work and that one nurse could develop the equivalent of 46 TISS points per shift. Therefore it is possible to calculate the *P*/*N* ratio of an ICU by analyzing the nursing workload in a given period of time. Each computed *P*/*N* ratio is itself an LOC. However, previous research has shown that it is possible to aggregate the ICUs into three LOCs, to each of which one-third of the total intensive care population will be admitted when stratifying the patients by severity of illness (Reis Miranda and Gimbrère 1994). ICUs with patients requiring a *P*/*N* ratio below 1.6 are in LOC III, ICUs with patients requiring an average *P*/*N* ratio of 2.5 (range between 1.7 and 2.9) are in LOC II, and ICUs with patients requiring a *P*/*N* ratio above 3.0 are in LOC I. Although the number of patients is similar in each LOC, there are more beds in the higher LOCs because the average stay is longer.

Calculating the staffing of the ICU

Only physicians and nurses are considered in this calculation. As the staffing demands of ICUs are not all the same, the staffing calculations need to be performed according to the LOC at which the ICU is operating or planning to operate. Therefore information on the *P*/*N* ratios (*P*/*P* ratios for physicians) is required. Indications which may be useful for the eventual staffing levels of other professional groups are discussed above.

Every ICU has an appointed medical director, which is a full-time job in LOC II and LOC III ICUs, and a full-time head nurse. The total number of physicians and nurses required to staff the ICU can be calculated using the ratios given in Table 1 and the formula below.

The total number of professionals required in an ICU can be calculated by substituting the appropriate ratios in the following formula:

$$\text{total number of professionals} = \frac{A \times B \times C \times D \times E}{F \times G}$$

where A is the number of shifts per day (usually A=3), B is the total number of beds in the ICU, C is the total number of days that the unit operates per week (usually C=7), D is the desired occupancy rate of the unit (usually D=0.85), E is the staffing allocation to compensate for holidays, sickness, etc. (usually E=1.25), F is the appropriate *P*/*N* or *P*/*P* ratio, and G is the total number of days that each professional works per week (usually G=5). Therefore, in the usual situation, the formula can be simplified to

$$\text{total number of professionals} = \frac{22.3B}{5F}$$

The calculated total number of professionals includes all professional categories indicated above under physicians and nurses. Decisions regarding the grade of professional differentiation among each professional group must be made on the basis of analysis of the tasks required in the ICU.

Education, training, and skills

The knowledge and skills required by intensive care physicians (Guidelines Committee 1992; Thijs *et al.* 1996) and nurses (consultation with various professional organizations) are summarized in Table 2. It is anticipated that the importance of individual requirements may vary with the specific activities of each ICU.

An important feature of Table 2 is that it shows areas where the requirements of knowledge and skills for physicians and nurses overlap. In these overlapping areas the description of duties and responsibilities must be more detailed and supported by the use of protocols whenever possible.

Changing staffing strategies

Staffing the ICU usually follows a traditional approach: (a) defining the professional groups to appoint; (b) calculating the number of professionals in each group; (c) defining the professional qualifications of each (sub)group. In this way appropriate staffing is determined by defining the professional inputs to the department.

However, this staffing strategy is in the process of change as greater importance is attributed to the management of human factors associated with individual performance. These human factors can be classified as follows.

1. Professional career in terms of professional development and future and salary prospects.
2. Organization commitment in terms of identification with the objectives of the ICU and the hospital. A recent European study suggests that organization commitment is a variable with a significant explanatory power *vis-à-vis* mortality in the ICU when controlling for severity of illness.
3. Identification with the culture of the organization which is defined as the collective mental programming of its members. It is important that this culture shifts from the traditional process-oriented culture (bureaucracy based) to a results-oriented culture (effort based). In the European study mentioned above, the results-oriented culture appeared to have a beneficial effect on the performance of the ICU.

Therefore, when planning an ICU, it is important to bear in mind that staffing the unit should be a 'win–win operation', contributing to the achievement of the stated objectives of both the unit and the staff

Table 1 Ratios for calculating the total number of professionals required to staff the ICU

LOC	I	II	III
P/*P* ratio	—*	6.5	5
P/*N* ratio	⩾3	2.9–1.7	⩽1.6
Mean SAPS II score	⩽24	25–37	⩾38
Mean TISS score	⩽15	16–27	⩾28

*A physician (see definition of professional categories in the text) should be present in the ICU or on call in the hospital.

Table 2 Knowledge and skills required by intensive care physicians and nurses

Physicians		Nurses	
Education and training	**Skills**	**Skills**	**Education and training**
Basic and advanced cardiopulmonary resuscitation	Invasive and non-invasive maintenance of open airways including suction, physiotherapy and postural drainage		Basic and advanced cardiopulmonary resuscitation
		Ventilation by bag and mask Cardiopulmonary resuscitation	
Pathophysiology, diagnosis, and therapy of congenital and acquired cardiovascular diseases Cardiovascular monitoring computations and interpretation	Arterial puncture and blood sampling Catheterization of central venous system including pulmonary artery catheterization and transvenous pacemakers Management of cardiovascular emergencies including cardiac tamponade Use of intra-aortic devices	Heart sounds Intravenous access 12-lead ECG Basic interpretation of monitoring Management and interpretation of Swan–Ganz catheter	Heart pump function and monitoring Shock Rhythm disorders Basic indications and pharmacology of cardiovascular drugs Cardiovascular emergencies
Pathophysiology, diagnosis, and therapy of acute and chronic respiratory diseases Mechanical ventilation Respiratory function and monitoring	Implementation and withdrawal of ventilatory modes Management of respiratory emergencies (e.g. pneumothorax) Fiber-optic laryngotracheobronchoscopy BGA: determination and interpretation	Auscultation Oxygenotherapy Pulse oximetry BGA Capnography	Mechanical ventilation ARDS/MOF Acute respiratory failure
	Operation of mechanical ventilators and accessories (e.g. humidifiers)		
		Troubleshooting ventilators	
Pathophysiology, diagnosis, and therapy of acute and chronic renal failure Monitoring and interpretation of renal function		Dialysis techniques	Renal failure Basic indications of dialysis
Pathophysiology, diagnosis, and therapy of life-threatening neurological disorders Psychiatric emergencies Sleep and pain	Lumbar puncture Management of neurological emergencies Monitoring including modified EEG Treatment of pain	Intracranial pressure monitoring Temperature regulation Assessment of level of consciousness	Coma Intracranial pressure Acute intoxications
Pathophysiology, diagnosis, and therapy of metabolic and endocrine disorders Nutrition	Monitoring and assessment of metabolism and nutrition	Tube feeding	Water and electrolyte balance Basic hormonal regulation Nutrition (enteral, parenteral)
Pathophysiology, diagnosis, and therapy of acute GI disorders	Management of GI emergencies such as esophageal bleeding Prophylaxis of upper GI bleeding	Gastric suction Gastric lavage Sengstaken tube management Management of stomas Management of abdominal wounds	Stoma care Abdominal wound care

(*continued*)

Table 2 cont.

Pathophysiology, diagnosis, prevention, and therapy of infections	Antibiotherapy Infection control Sampling, staining, and interpretation of samples for bacteriology studies	Infection control General wound care Bacteriology sampling	Sterility/asepsis Infection control
Pathophysiology, diagnosis, and therapy of hematological disorders	Management of massive transfusion Utilization of blood components	Administration of blood products	Coagulation disorders Blood transfusion
Pathophysiology of trauma Crush injury Burns Drowning	Temporary immobilization of fractures Use of special equipment such as circle beds, G-suit, Roto rest, cervical collar		Cranial trauma Burns Multitrauma
Bio-engineering and biostatistics Principles of monitoring	Utilization, zeroing, and calibration of transducers Use of amplifiers and recorders Basic electrical safety hazards Invasive and non-invasive monitoring		Basic principles of monitoring and electrical safety
Prognostic indices Organization and management Cost effectiveness Budgeting and budget control	Severity scores (e.g. APACHE, SAPS, MPM) Workload scores (e.g. TISS) Record-keeping Reporting		Prognostic indices Cost effectiveness Cost containment
		Ethics regarding: Death and dying Ordinary versus extraordinary life-support mechanisms Withholding life-support measures Organ transplantation Informed consent Patients, families, and next of kin	

BGA, blood gas analysis; EEG, electroencephalography; GI, gastrointestinal; ARDS, acute respiratory distress syndrome; MOF, multiple organ failure.

members. Instead of inputs (e.g. case mix and number of patients to admit), the objectives of the ICU should be the measurable outputs (given the inputs) such as mortality and percentage consumption of manpower (final outputs) and length of stay, occupancy rate, average annual number of admissions per ICU bed, and costs (intermediate outputs). Thus the following factors should be taken into consideration when staffing the ICU.

1. Definitions:
   (a) objectives of the organization;
   (b) staffing strategies enabling these objectives to be achieved;
   (c) analysis of the working process, resulting in the definition of intermediate processes and measurable outputs;
   (d) staffing plan, including detailed descriptions of the jobs and tasks to be performed.
2. Planning the results, i.e. the strategies required to achieve the stated final and intermediate outputs.
3. Staff evaluation: a clear, regular, and effective process of evaluating every staff member must be in place to ensure appropriate monitoring of both individual performance relative to the objectives of the organization and adequate management of the human factors mentioned above.

Therefore changing the strategies for staffing the ICU implies that these strategies should focus on the description and evaluation of the performance and professional output of the staff rather than on the professional input to the department.

References

Cullen, D.J., *et al.* (1974). Therapeutic Intervention Scoring System: a method for quantitative comparison of patient care. *Critical Care Medicine*, **2**, 57–60.

Guidelines Committee (1992). Guidelines for program content for fellowship training in critical care medicine. *Critical Care Medicine*, **20**, 875–82.

Reis Miranda, D. and Gimbrère, J. (1994). The Netherlands. *New Horizons*, **2**, 357–63.

Reis Miranda, D., De Rijk, A., and Schaufeli, W.B. (1995). Simplified therapeutic intervention scoring system. The TISS-28 items: results from a multicenter study. *Critical Care Medicine*, **24**, 64–73.

Thijs, L.G. and the Working Group on Guidelines for Training Programmes in Intensive Care Medicine (1996). Guidelines for a training programme in intensive care medicine. *Intensive Care Medicine*, **22**, 166–72.

15.1.4 Patient safety

S. J. Y. Bard and G. R. Park

Key messages

- Patients are particularly vulnerable in the intensive care unit.
- Patient safety is ultimately dependent on staffing levels and competence.
- Nurses have a crucial role to play, and they must be familiar with infection control procedures and with the instrumentation and drugs used.
- Procedures to monitor staff performance should be in place.

Introduction

Patients in the intensive care unit (ICU) cannot react normally to danger because they are acutely ill. They cannot describe adverse symptoms properly nor indicate pain accurately. In many instances they will be sedated in order to tolerate their treatment. Some may be both sedated and paralysed. Such patients are unable to protect themselves and rely completely on the medical and nursing staff attending them to keep them safe.

Procedures which are designed to benefit intensive care patients also carry risks. Many procedures are invasive, i.e. they break the patient's natural skin barrier, putting him or her at risk of infection. Monitoring and treatment often require the use of numerous pieces of equipment with the consequent danger of malfunction. In addition, drugs have undesirable as well as desirable effects.

Nursing

Staff recruitment and training

Nurses have a crucial role to play in keeping patients safe. Since nurses are the only people who are constantly with critically ill patients, they are in a unique position to observe and protect them. An ICU with insufficient nurses will not be safe.

In order to keep a unit staffed to its funded establishment of nurses, managers must continually take a positive and imaginative approach to nurse recruitment. Before appointing staff there should be careful reference checks and health screening. Suitability to deal with the stresses of intensive care work must be established as far as possible before appointment. Some recommendations as to how this might be done are found in the Allitt Inquiry (Clothier and MacDonald 1994) and are summarized in Table 1.

The competence of established staff at all levels will be improved by further training. A well-planned staff development program will help to achieve and maintain a balanced skill mix, which will contribute significantly to patient safety.

However, it is not sufficient to have a well-balanced establishment. It must also be the aim to achieve a balanced profile of skills throughout each shift. This can be done by good rota planning, careful patient allocation, and planning staff breaks with safety in mind.

Malicious intervention

The possibility of malicious harm to patients was recently demonstrated by the Allitt case. Allitt was a nurse in a pediatric ward who murdered or injured several children in her care over a 3-month period. The principal recommendation of the formal inquiry (Clothier and MacDonald 1994) was that the 'disaster should serve to heighten awareness in all those caring for children of the possibility of malevolent intervention as a cause of unexplained clinical events'. This recommendation could equally apply to those caring for critically ill patients since, like children, they are completely vulnerable.

The inquiry also recommended that 'reports of serious untoward incidents to District and Regional Health Authorities should be in writing and through a single channel which is known to all'. ICUs need a mechanism whereby critical incidents can be analyzed and statistically atypical patterns detected. Good shift-by-shift record keeping is essential. Records should be 'constructed and completed in such a manner as to facilitate monitoring of standards, audit, quality assurance and the investigation of complaints' (UKCC 1993).

Infection control

Infection can kill critically ill patients who might otherwise survive serious medical conditions. The layout of the ICU should incorporate some single rooms which can be used for isolation nursing. There should be sufficient space around the remaining beds to minimize infection spread by droplets and enough well-placed wash basins to facilitate hand-washing by staff. Hand-washing is extremely important in the control of infection, and if the hands are clean the use of alcohol hand rubs before and after procedures is quick and effective.

Close liaison between the ICU and the infection control team is essential if effective infection-control policies are to be established. Regular combined unit rounds by a consultant microbiologist and an intensive care consultant will ensure that antibiotics are used appropriately.

Largely as a result of the indiscriminate use of antibiotics in the last two decades, strains of multiresistant organisms have begun to emerge as a major problem in the 1990s.

Equipment

Highly sophisticated electronic equipment is an integral part of today's ICU, but unless it is competently used it can endanger patients. Staff need training when they encounter new equipment and, if necessary, manufacturers should initially be invited to take on this responsibility. A designated member of staff should take on the continuing training role. As with all nursing/medical practice, staff should decline any duties or responsibilities unless they are able to perform them in a safe and skilled manner.

Alarms, for example on monitoring equipment, ventilators, or dialysis machines, are designed to keep the patient safe. They must be checked and set appropriately by the nurse at the beginning of each span of duty. Dangerous and possibly fatal situations can rapidly occur if it is assumed that the alarms have been set and they have not.

Equipment must be maintained to the best possible standard with agreed maintenance and cleaning routines. However, staff should not

Table 1 Recommendations from the Allitt Inquiry concerning preappointment health screening

- For all those seeking entry to the nursing profession the most recent employer or place of study should be asked to provide a record of time taken off sick
- No candidate for nursing in whom there is evidence of major personality disorder should be employed in the profession
- Nurses should undergo formal health screening when they obtain their first post after qualifying
- The Chairman of the NHS Occupational Physicians suggested that psychiatric profile was more important than psychological testing and recommended that applicants with an adverse psychiatric history should not be accepted for training until they have shown the ability to live an independent life without professional support and have been in stable employment for at least 2 years
- Procedures by which managers can refer nurses to occupational health should make clear the criteria which should trigger such referrals
- Consideration should be given to how general practitioners might, with the candidate's consent, be asked to certify that there is nothing in the medical history of a candidate for employment which would make them unsuitable for the nursing profession

Table 2 Prescription writing in the ICU

- Specific allergies or 'no allergies' should be recorded on the drug chart
- Prescriptions should be written legibly in ink
- There should be a start date and, where appropriate, a review or stop date
- The patient's full name, hospital number, date of birth, and address should be stated on the drug chart
- Dose of drug and frequency of administration should be stated
- If 'as required', a minimum dose interval should be specified
- Approved drug names should be used in full
- Special instructions should be in English without abbreviations
- Micrograms, nanograms, and units should not be abbreviated
- Drugs by infusion should state:
  - the drug
  - the dose
  - the diluent
  - the total volume
  - the rate
  - the route

hesitate to stop using equipment if they suspect that it is not working properly. Where possible, such equipment should be replaced and checked thoroughly before further use. If doubt arises over the correct operation of a ventilator, the patient should be immediately hand-ventilated with 100 per cent oxygen whilst assistance is summoned and the suspected fault investigated.

Drug usage

Drug usage in the ICU is complex. Co-operation between the pharmacy and the ICU increases patient safety. The pharmacy can provide comprehensive information about the correct preparation and administration of intravenous drugs, and about drug compatibility and special dose adjustments (e.g. in renal failure).

Standard solutions for frequently used drug infusions, agreed between ICU and pharmacy staff, can assist less experienced prescribers and reduce risks of error.

A daily review of prescriptions by a visiting pharmacist may eliminate such errors as incorrect frequency of dosing, incorrect dose adjustment during dialysis, synergism of pharmacological action, omission of drugs from a regimen, and transcription errors (Table 2).

Some errors may originate outside the hospital during drug manufacture. Contamination or the mislabeling of a batch of drugs may occur with potentially fatal consequences. Close observation of patients during drug therapy and swift action in the event of unexpected adverse reactions are crucial to patient safety.

Conclusion

A large hospital relies upon efficient and effective intensive care services to support its work. The specialized work of medical and surgical departments, pediatrics, and casualty would be impossible without such a service.

With sufficient competent staff, clear policies and protocols, and close collaboration with other hospital departments, we can be more confident that patients will be safe in the ICU.

References

Clothier, C. and MacDonald, C. (1994) *The Allitt Inquiry: independent inquiry relating to deaths and injuries on the children's ward at Grantham and Kesteven General Hospital during the period February to April 1991.* HMSO, London.

UKCC (1993). *Standards for records and record keeping.* United Kingdom Central Council for Nursing, Midwifery and Health Visiting, London.

15.1.5 Staff safety

T. S. Leary and G. R. Park

Key messages

- Some of the many potential hazards to staff working within an intensive care unit include infection, musculoskeletal injury, exposure to hazardous chemical substances and ionizing radiation, and fatigue.

Introduction

Within the intensive care unit (ICU), there are many areas for concern when attempting to prevent harm to the patients or hospital staff. Although these are often related, it is important that any protective measures for one group do not put another group at increased risk. These measures may rely on the common sense of the individuals concerned, but most are based upon local advice or regulations. The latter may be nationally agreed standards, or may be the result of local risk assessment and management programs. While particular risk areas may be identified, the concept of 'universal precautions' must be outlined. Achieving an entirely risk-free practice is impossible for the foreseeable future, and so staff should approach their work in a way that minimizes exposures without compromising patient care.

The importance of recording accidents cannot be stressed enough since this is often the information source for hazard prediction and accident prevention.

The provision of adequate staff education may prevent many accidents (at least 50 per cent of accidents are avoidable) and so provide the management with a contented workforce that is not adversely affected by sickness and ill-health. This has both ethical and economic implications.

Protecting the staff from infection

Within the ICU, staff are in close proximity to a wide variety of potential pathogens. Some of these and their routes of infection are shown in Table 1. This list is by no means exhaustive, and many organisms can gain entry by a variety of pathways; for example, *Mycobacterium tuberculosis* can be transmitted by inhalation or inoculation.

Needlestick injuries

The risk of transferring infection depends on the virulence of the organisms present, the size of the infecting dose, and the state of the recipient's immune system. Infection with HIV following a 'sharps injury' is less than 0.5 per cent, whereas the risks of acquiring hepatitis B and hepatitis C viruses with a similar injury are 6 to 30 per cent and 2.7 to 10 per cent respectively. These risks are increased (up to five times) in units associated with the treatment of liver disease and transplantation. Techniques to reduce 'sharps exposure' include the following (Berry and Greene 1992).

1. Eliminating non-essential unprotected needle use by the use of needleless or protected-needle devices and by modifying the procedures requiring needles to minimize the risk.
2. Using ports within intravenous cannulas that allow syringe connection without the use of needles.
3. Using plastic ampoules and blunt drawing-up needles to reduce skin perforation.
4. Using self-capping needles and educating staff on the hazards of resheathing needles to reduce the risk. When needles must be resheathed, a one-handed technique should be taught.
5. Continuous intra-arterial blood gas monitoring and closed systems for venous blood sampling should be used to reduce exposure to blood.
6. Potentially hazardous items should be placed, by their user, in separate rigid disposal containers before incineration. Such 'sharps bins' should be readily available.

Table 1 Some of the organisms that may infect staff and their routes of infection

Inoculation	Ingestion	Inhalation
Clostridium tetani	Hepatitis A	*Clostridium diphtheria*
HIV	*Vibrio cholera*	*Mycobacterium tuberculosis*
Cytomegalovirus	*Salmonella typhi*	*Neisseria meningitis*
Staphylococcus aureus	Polio virus	Influenza *Streptococcus pyogenes*

Hand-washing care

Regular hand-washing and the use of disinfectant alcohol solutions, for example before the ingestion of food, reduces not only the interpatient spread of infection but also staff contamination. Gloves provide further protection, since the presence of even small cuts in the natural defense barrier of the skin may allow viruses to enter the bloodstream. Even blood products from a blood bank should be handled with care, because the lifespan of HIV and the hepatitis viruses in stored blood is almost indefinite and screening may fail to detect them. Other viruses survive for differing times: cytomegalovirus survives for only a few days, whereas Epstein–Barr virus may survive for long periods in stored blood.

Protective clothing

The increase in glove usage has been associated with a concomitant increase in glove-related allergic reactions, many of which are potentially debilitating (Hamann and Kick 1994). Communication between users and manufacturers is important in the design of a strong, but flexible and sensitive, hypoallogenic glove.

Special waterproof, but breathable, uniforms are available which provide a further barrier to bacterial infection and limit staff carriage of pathogens outside the ICU.

In contrast with the highly effective barrier of intact skin, mucous membranes are much more permeable to microbial invasion. Unlike the upper respiratory tract, which is protected by mucus, cilia, and macrophages, the open eye presents relatively unprotected mucous membrane surfaces relying on the bactericidal action of lysozyme in tears. While goggles protect the eyes alone, the wearing of a face mask or visors may also protect the upper airway from droplet infection.

The anticipated reduction in the incidence of tuberculosis has not occurred. Indeed, 3 million people die from this infection every year. It may be related to the prevalence of HIV, the development of multidrug resistant strains, and the limited access to health care for those in congregative facilities. When an infectious patient with tuberculosis undergoes a procedure that generates respiratory aerosols, healthcare workers attending the patient may need to wear protective air-purifying respirators with high-efficiency filters.

Immunization

Immunization against common pathogens such as hepatitis B, tuberculosis, tetanus, rubella, and poliomyelitis may be needed and should

Table 2 Potentially hazardous substances found in the ICU

Gases	Vapors	Fluids	Solids
Carbon dioxide	Isoflurane	Glutaraldehyde	Latex
Nitrous oxide	Halothane	Mercury	Methotrexate
Nitric oxide		Formaldehyde	

Table 3 Some of the factors implicated in back injury

Staff	Equipment	Patients	Environment
Lifting alone	Absent lifting aids	Obesity	Rooms too small
Too heavy weights	Broken lifting aids	Unexpected collapse	Area cluttered
Cumulative stress	Poor bed brakes	Uncooperative	
Asymmetrical trunk			

be offered to all staff, at least on a voluntary basis. Comprehensive local policies, for both treatment and counseling, should be available if accidental contamination occurs.

Hazardous chemical substances

Health regulations lay down the essential requirements and a sensible step-by-step approach for the control of hazardous substances and the protection of people exposed to them. Recommendations also exist for the management of accidental exposure. These instructions are based upon information from both the manufacturers and local pharmacy departments, and have been responsible for the prohibition of some agents, such as ethyl chloride, in some ICUs and for the selection of the least toxic alternative. Some of the substances commonly encountered are shown in Table 2.

The evidence that waste anesthetic gases used for sedation significantly affect staff is equivocal. Effects may include an increased risk of spontaneous abortion (Buring *et al.* 1985) and alteration of DNA synthesis by nitrous oxide. Actively replicating cells are most affected, as folic acid metabolism is altered and the synthesis of vitamin B_{12} and DNA is impaired. Long-term exposure may result in several disabilities including megaloblastic anemia, leukopenia, peripheral neuropathy, and possibly fetal damage. ICUs should be properly ventilated and be equipped with scavenging systems. However, given the paucity of evidence linking such agents with staff ill health, these protective measures should not increase patient risk in any way.

Controlled-airflow cupboards may be employed as a means of both protecting staff from harmful fluid splashes and vapors and providing a sterile field.

Musculoskeletal injury

Back injury is a major cause of morbidity among health workers and therefore has implications for financial planning within the establishment of the health provider. Although nurses take fewer days leave through sickness than the general population, they have 30 per cent more absences per year through back pain. The consensus of opinion is that these back injuries result from stresses during patient lifting and transfer (Laflin and Aja 1995). Several factors have been implicated in back injury (Table 3)

The employer has the responsibility for providing in-service education on body mechanics, the means to avoid lifting wherever possible, and the reduction of risks where lifting is absolutely necessary. Uniforms and footwear should allow flexibility while also providing support, as much of staff time is spent standing. Air fluidized beds reduce the need for regular patient 'turning', slides such as the Patslide® simplify patient transfer, and hoists are available. Such back-saving devices are supplied with guidelines for their proper use (often reinforced by local physiotherapy departments). Failure to comply with these instructions, such as their use by appropriate numbers of staff, may put both staff and patients at risk. Beds, chairs, and other furniture should all be adjustable such that work can be completed from a comfortable position.

Protecting the staff from radiation exposure

Ionizing radiation is used in medicine for both imaging and therapeutics. When a radioactive isotope decays, α particles (two protons and two neutrons) or β particles (electrons) may be released. In addition, γ radiation may be emitted. X-rays, like γ-rays, are high-frequency electromagnetic radiation; however, they are produced differently, being emitted when a beam of electrons is accelerated from a cathode to strike an anode, which is usually made of tungsten. The energy from these sources of ionizing radiation is initially dissipated in tissues by the displacement of electrons, thus producing chemically reactive ions. Therefore they may all cause tissue damage and chromosome changes, particularly in dividing cells. Exposure to all forms of radiation should be kept to a minimum.

Distance from the source is the most important factor. Alpha particles can only travel a few centimeters in air, and so the only precautions involve containment and the avoidance of ingestion or inhalation. Beta particles can be shielded by Perspex, which decelerates the particles slowly, thus preventing release of their energy as X-rays. A lead containment of thickness 7 cm will absorb 90 per cent of γ-rays, and much less is needed for protection against X-rays.

Scintillation counters can be used to monitor levels of γ irradiation. In units where high doses of X-ray exposure may be encountered, individuals need to wear photographic film badges to record the total dose absorbed.

Non-ionizing radiation from the lower-frequency range of the electromagnetic spectrum includes ultraviolet (UV) light (wavelength, 10–400 nm). Wavelengths below 200 nm do not pose a hazard as they are absorbed by air, and atmospheric ozone provides protection from wavelengths below 290 nm. Wavelengths of 280 to 320 nm (UV-B) have been associated with skin cancer (these effects are enhanced by UV-A which has wavelengths of 320–400 nm). Although minimal levels are observed from fluorescent lighting or from units where UV is used to limit bacterial contamination, care must be taken in neonatal units where such lights are used in the treatment of jaundice.

There is no evidence that video display unit operators are exposed to electric, magnetic, or radiation levels significantly above ambient levels. Clip-on UV screens are available and may reduce exposure. There is no conclusive evidence of increased eyesight deficits, epilepsy, or facial dermatitis. However, such equipment should be well positioned to avoid muscloskeletal injury.

Protection against fatigue and stress

Common sense suggests that accidents occur when staff are tired. Whereas continuous detailed work in an ICU leads to fatigue, in anesthesia as a whole a clear association has not been shown between the commonly cited human error and levels of tiredness.

The environment in which staff work is also important. Equipment should be chosen, used, and maintained such that noise levels are kept to a minimum. This may not only prevent damage to hearing (damage occurs above 85 dB), but may also avoid the disturbing lower levels which adversely affect health, well being, performance, and safety. Audible alarms should be used where necessary; however, they should be distinguishable, pleasant, and as quiet as safety will allow.

Temperature control within the ICU is of paramount importance. Patient factors often dictate higher than average working temperatures, although the potential deleterious effects on staff, particularly those wearing protective clothing which impedes heat loss by sweat evaporation, should be noted.

The cumulative effects of long hours and irregular food intake have been implicated in causing depression and other stress-related conditions (Firth-Cozens 1989), particularly in an area often associated with periods of acute anxiety among staff about a clinical situation. Much stricter regulations now exist, especially among the medical staff, for controlling hours of work according to intensity. Staff should be provided with a separate rest area where they can relax away from anxious patients and relatives. The use of group meetings during which problems and incidents encountered in the workplace can be freely discussed in an informal atmosphere may help to alleviate stresses.

References

Berry, A.J. and Greene E.S. (1992) The risk of needlestick injuries and needle-stick transmitted diseases in the practice of anesthesiology. *Anesthesiology*, **77**(5), 1007–21.

Buring, J.E., Hennekens, C.H., Magrent, S.L., Rosner, B., Greenberg, E.R., and Cotton, T. (1985). Health experiences of operating room personnel. *Anesthesiology*, **62**, 325–30.

Firth-Cozens, J. (1989) Stress in medical undergraduates and house officers. *British Journal of Hospital Medicine*, **41**, 161–4.

Hamann, C.P. and Kick, S.A. (1994). Allergies associated with medical gloves. Manufacturing issues. *Dermatological Clinics*, **12**(3), 547–59.

Laflin, K. and Aja, D. (1995). Healthcare concerns relating to lifting: an inside look at intervention strategies. *American Journal of Occupational Therapy*, **49**(1), 63–72.

15.1.6 Staff welfare

P. J. Cole and G. R. Park

Key messages

- Staff working with the critically ill are likely to be more stressed than some other health care professionals.
- Stressors include poor communication, dealing with death, excessive workload, bureaucracy, noise and light in the intensive care unit environment, and technology
- The symptoms of stress may be psychological, physiological, or behavioral.
- Unrelieved stress in a high proportion of workers results in a department with low morale and high staff turnover.

Introduction

> Stress is the psychological, physiological and behavioural response by an individual when they perceive a lack of equilibrium between demands placed upon them and their ability to meet those demands. (Palmer 1992, p.39)

The results of stress will depend upon the individual, the type of stress, and its duration. Whilst moderate stress or pressure may improve performance and health, high levels of stress over a period of time may lead to ill health. What constitutes a high level of stress varies from individual to individual.

The medical profession is recognized as being stressed. However, there has been little research on the effects on health of the interaction of physical and psychosocial factors at work, and little hard evidence to show that morbidity or mortality is increased by overwork or stress (Payne and Firth-Cozens 1987). Studies show that doctors have twice as many road accidents, three times the incidence of cirrhosis and suicide, and 30 times the incidence of drug addiction compared with the general population. Nurses have the highest suicide rate among professional groups and top the list of psychiatric out-patient referrals. Some of these statistics are undoubtably caused by occupational stress.

Stress in the intensive care unit

Staff working with the critically ill are likely to be more stressed than some other health care professionals (Table 1). Poor communication between intensive care unit (ICU) staff is a particular stressor. This may be more common in nurses who are qualified but do not have postgraduate critical care training. They may feel that they lack the knowledge or clinical skills to confront doctors and this may lead to feelings of insecurity. However, those with postgraduate qualifications in critical care may be subject to increased stress when faced with conflict in treatment decisions. In addition, frustration is increased when nurses feel that their expertise is not being recognized and that they are undervalued.

Table 1 Potential stressors for those working in the ICU

Interpersonal communication
Between staff
Between staff and relatives
Death and dying
Workload
Staff shortages/staff turnover
Job security
Adequate training and teaching
Shift work (unsocial hours)
Fatigue
Lack of job satisfaction
Management/administration
Financial constraints
Lack of resources and outdated equipment
Increased technology
Lack of recognition

Communication

Inability to communicate with the patient can also cause frustration. Nurses may feel incompetent because their patients cannot communicate seemingly important messages to them (Bergbom-Engberg and Haljamaë 1993).

Talking with relatives is a skill usually learned from experience rather than training. At times this can be stressful since the ICU staff have to remain calm and objective even when relatives or patients are rude, abusive, or aggressive. Death and dying are a common fate of many patients on the ICU. It has been shown to be 'an emotionally charged, highly stressful situation demanding all the skills of the staff'. Difficulties arise with the breaking of bad news to the relatives, in the care of the dying patient, and in supporting the family of the patient and dealing with any reactions that might ensue.

Death

The process of dying and imminent death may affect staff more adversely than the everyday pressures of looking after acutely ill patients. Staff have to come to terms with the realization that no amount of dedication or hard work is going to reverse the situation. Then there is a change in care from an aggressive and invasive approach to one of making the patient comfortable and ignoring adverse physiological changes.

Once the doctors have made the decision for terminal care to take place and appropriate analgesia has been prescribed, the nurses should take control of the patient's care, which becomes palliative rather than curative, thereby allowing nurses the opportunity to deal with the process more effectively.

Often, even before having time to come to terms with the death of one patient, nurses and doctors have to care for another acutely ill patient. This may cause emotional confusion at the sudden turnabout of dealing with death one minute and life the next. There may even be self-criticism at not dealing with the situation in the best possible way.

Workload

One of the most obvious factors of occupational stress is excessive workload. This may be linked to the continual problem of high staff turnover to which critical care units are particularly vulnerable. The high turnover of staff may be associated with the long introductory phase for junior nurses before they can become independent.

The need to spend time orienting new staff may add further stress to the existing staff. Furthermore, once a nurse has achieved independence, delay in promotion to one of the few senior posts may cause frustrated experienced staff to leave.

Bureaucracy

Other stressors include the imposition of unpleasant management decisions, bureaucracy, and the ever increasing amount of administration being done by doctors and clinical nursing staff, together with obvious problems of shift work and fatigue. Legal implications such as accountability or responsibility increase stress.

Environment

The ward environment can also cause stress. Bright lights for 24 h a day are common and are accompanied by the regular noise of equipment. Furthermore, ICU staff are expected to improve their knowledge continuously, and they may not be given time to do this or be paid adequately.

Technology

The influence of technology may be a further cause of stress. Staff may be overwhelmed by the complex monitors that surround the bed space. This may deflect the carers' attention from the patient and dehumanize the person being cared for.

Staff must understand that equipment exists, not to take time away from the patient, but, if used correctly, to increase the time that the nurse spends with the patient. Adequate training allows nurses to

Table 2 Some of the differences in working conditions between doctors and nurses

Doctors	Nurses
Longer hours, more sleep deprivation	Fewer hours per shift with protected breaks, rest period
Responsible for decisions	Required to make fewer decisions
Poor support from colleagues	Good support from colleagues
Distractions (research, examinations, work in other areas)	Continuous exposure to patients and relatives

understand and be competent with the technology. They will then be able to reassure patients and relatives that they, rather than the equipment, are controlling the situation.

Differences between doctors and nurses

Each individual working on the ICU may not be exposed to all the stressors discussed above (Table 1); their job will determine which stressors are more relevant to them. Doctors may find that the major factors include having responsibility for decision-making regarding patient management, longer working hours than their colleagues with subsequent fatigue and sleep deprivation, and other commitments such as research or work in other areas (Table 2).

Nursing staff may work fewer hours with less responsibility for decisions and usually have good support from their colleagues, although they have to deal with problems of continuous exposure to patients and relatives which may be particularly demanding when dealing with death and dying.

The levels of stress to which staff are exposed may cause difficulties with the organization of the ICU, and this will be revealed by organizational symptoms (Table 3). If these are present, then there is the need for action.

Table 3 Organizational symptoms of excessive stress

High staff turnover
High absenteeism
Poor job performance
Increased occurrence of accidents
Low morale
Recruitment/retention failure
Staff burnout

Reproduced with permission from Palmer (1992).

Stress and health

Low or high levels of stress may lead to ill health if the carer is exposed to them for some time. The exact amount of time varies from individual to individual. The symptoms of stress may be psychological, physiological, or behavioral (Table 4).

These symptoms may progress to a syndrome of exhaustion, disillusionment, and withdrawal known as 'burnout'. This is a progressive disorder in staff with high and sustained level of stress. It produces feelings of tension, irritability, and fatigue, ending with a defensive reaction of detachment, apathy, cynicism, or rigidity.

Table 4 Personal symptoms of stress

Psychological	Depression, anxiety, obsessions, indecision, reduced self-esteem, reduced enthusiasm
Physiological	Muscle tension, backache, neckache, eating disturbance, ? hypertension, ? peptic ulceration, ? recurrence of previous diseases
Behavioral	Accident proneness, alcohol and drug abuse, aggressiveness, type A behavior, relationship difficulties, insomnia, withdrawal

Reproduced with permission from Palmer (1992).

Coping with stress

Unrelieved stress in a high proportion of workers results in a department with low morale and high staff turnover. To prevent this, each stressor should be looked at and methods of coping and alleviating the stress should be used. Interpersonal communications can be improved by encouraging staff to attend courses that include interpersonal skills and assertiveness training. Talking about stress-related problems with colleagues, family, partners, or friends may also be an effective coping mechanism. Problems about communicating with awake patients who cannot speak may be reduced by open discussions with other members of the team looking after the patient.

Certain patients are very demanding and different nurses should look after them. This allows nurses to alternate between periods of emotional demand and more relaxed times. It may be useful for staff to meet shortly after having to deal with a difficult patient so that coping mechanisms can be discussed and perhaps the way that staff deal with difficult situations can be improved. Similarly, a meeting after the death of a patient allows the obvious stress to be talked about. Indeed, many American hospitals allow staff to take 'time-out' for informal and confidential discussion to take place. These are usually under the control of a facilitator who was not involved in the death. This results in at least one person being objective.

It is essential that management attempt to minimize staff turnover so that the morale of the remaining staff is maintained. The theory is that if individuals are happy and stimulated within an environment, the decision to leave will not be made. Ways of maintaining job satisfaction have been suggested (Akerman 1993), including the following:

(1) encouraging effective communication between all members of the multidisciplinary team;

(2) providing opportunities for the staff to advance both professionally and academically with sufficient study leave and funding;

(3) arranging research and teaching sessions and encouraging all staff to participate in the running of the ICU.

If workload continues to increase, ways of reducing the immediate problem should be sought by either employing agency nurses or closing beds until it is resolved. If this is not done, the quality of safe consistent care will decrease and nursing staff will attempt to prevent this by working overtime, thus precipitating occupational stress.

If workers are feeling under stress, it is important for them to isolate those stressors that they control themselves from those controlled by senior staff or management. Nurses have used team-building and attempts to control the number of stressors acting at any one time to reduce stress. Self-assessment questionnaires may be useful in the hope that staff can deduce which areas are causing stress and start a self-care strategy (Foxall *et al.* 1990). Methods of preventing stress should be sought before staff start in the ICU. This may include preparing staff for what they may expect and what stressors they may face, establishing peer group support, and teaching techniques to improve interaction with fellow professionals and with patients and their relatives.

If a worker is suffering from symptoms of stress, this should not be ignored but ways of minimizing stress should be sought. These may range from advice about relaxation, to adequate leave for those unable to cope so that they have a period of adjustment, to the availability of professional counselors based either locally or regionally.

References

Akerman, A.D. (1993). Retention of critical care staff. *Critical Care Medicine*, **21**(9), 394–5.

Bergbom-Engberg, I. and Haljamaë, E.H. (1993). The communication process with ventilator patients in the ICU as perceived by nursing staff. *Intensive and Critical Care Nursing*, **9**, 40–7.

Foxall, M.J., Zimmerman, L., Standley, R., and Bené, B. (1990). A comparison of frequency and survey of nursing job stress perceived on intensive care, hospice and medical–surgical nursing. *Journal of Advanced Nursing*, **15**, 577–84.

Palmer, S. (1992). *Stress management: a course reader.* Centre for Stress Management, London.

Payne, R. and Firth-Cozens, J. (1987). *Stress in health professionals.* John Wiley, Chichester.

15.1.7 Fire and electrical safety

J. E. Risdall and G. R. Park

Key messages

- The basic principles of fire safety are fire prevention, early detection, and appropriate responses designed to safeguard life and minimize the damage to property.
- Every intensive care unit should follow the site fire policy and procedure document backed up by its own standing orders covering actions to be taken on discovering a fire, use of fire extinguishers, what assistance to call and from where, the compartmental system of the unit, recognized evacuation routes for personnel and patients, designated holding points and staff rendezvous points, and indications for isolating pipelines.
- Currents as low as 10 A may provoke ventricular fibrillation via intracorporeal electrodes such as saline-filled central venous and pulmonary artery catheters.

Introduction

Although most accidents occur at home, there is no reason to view the intensive care unit (ICU) environment as one of particular safety. The multiplicity of monitors and equipment, all needing electrical power, combined with the dependence, frailty, and relative immobility of the critically ill patient provide the potential for disaster in the event of a fire or electrical emergency. While awareness and good practice will do much to prevent such occurrences, preplanning, training, and rehearsal are essential if the risks posed to patients and staff by such incidents are to be minimized.

Fire

The possibility of fire in an ICU is seldom considered by the staff. Such incidents do occur, although rarely. All ICUs should have a plan for dealing with such an event, and staff should be familiar with their roles. The most serious threat is a fire occurring within the patient area, since, unless it is promptly controlled, the area will rapidly fill with smoke and a co-ordinated evacuation will become impossible. ICUs pose particular fire risks because of stocks of combustible disposable items which, when ignited, are liable to produce smoke and potentially toxic gases. Oxygen and the presence of electrical equipment compound the problem.

Prevention and detection

Design considerations

The design of the ICU can make a major contribution to fire safety. Ideally, the ICU should be divided into compartments so that fire, heat, smoke, and gases are contained and their spread delayed.

Fire and smoke doors are an essential component of fire compartments, but to work they must be closed in the presence of fire. This is best achieved by automatic devices which operate in response to activation of smoke alarms or heat detectors. Permanently closed doors on main thoroughfares are often wedged open. In these circumstances the fire doors will be useless. Smoke detectors should be placed in each compartment of the unit, including staff rooms, offices, and store rooms.

Individual patient rooms should be designed as separate fire compartments. Fire compartments should provide 30 min of containment or protection, increasing to 1 h for high-risk areas such as stores and kitchens. Any opening within the compartment must be protected by fire-resistant materials, and door seals and ventilation grills must be smoke and heat proof. Ideally, two evacuation routes should be included in the design of an ICU. Each should be wide enough to permit easy maneuvering of beds and should not become obstructed by stores or equipment. If the unit is not at ground level, an appropriate escape route to the ground should exist. Lateral movement, at the same level, through fire barriers is acceptable, gaining protection time for each barrier passed.

Air conditioning in the ICU should provide a slight positive pressure, preventing ingress of smoke from fires outside the unit. The exception will be in isolation units which require a negative pressure system. Air conditioning should not entrain or spread smoke around the patient area. The system should be separate from the main hospital system and should not automatically close down in the event of a fire elsewhere in the hospital. However, smoke and heat detectors should be fitted within the ducting and other parts of the system. These should shut the system down if smoke or hot gases enter. This will prevent the spread of smoke and fire within the ICU. The control unit for the air conditioning should be located within the ICU. Service ducts should also be capable of isolation to prevent ingress of smoke.

The generation and spread of fire and smoke within the unit can be limited by ensuring that disposable items are stored within cupboards or bins whose doors and lids are kept closed. Safe storage facilities for oxygen cylinders should also be provided. Furniture and furnishings should be fire resistant and conform to the relevant safety standards.

Oxygen

Oxygen and medical air are usually supplied to the ICU by pipelines. The decision to disconnect the oxygen supply should only be made with the agreement of medical or nursing staff in conjunction with the fire service and hospital engineers. The design of the pipeline supply should be such that it can be isolated from the rest of the hospital. Portable oxygen cylinders should be available as a back-up for each patient.

Damage limitation

Fire extinguishers

An adequate number of fire extinguishers should be provided of appropriate capacity and type to deal with the potential fire hazards, bearing in mind the risk of electrical fires from the equipment in use and potential solvent hazards in laboratory areas. More importantly, staff should be trained in their use without endangering themselves.

Smoke hazard

If 'first-aid' firefighting does not succeed in extinguishing the fire, the area will rapidly fill with smoke. Under such circumstances visibility will be profoundly reduced and staff should evacuate the area. Patients needing mechanical ventilation will be protected from smoke inhalation by their ventilator circuits and the piped gases. Since oxygen and air pipelines are rarely involved in fires, it is usually safer to leave patients connected to mechanical ventilation and the uncontaminated air source. They can be moved subsequently by the fire service with the help of doctors and nurses when the area is safe. Positioning essential equipment at each bed space in the same place will help staff to find emergency equipment more easily in poor visibility.

Evacuation

In an ICU divided into compartments, total evacuation should only be needed in exceptional circumstances. It should be possible to move patients within the unit, away from the seat of the fire, whilst the fire is controlled. Movement of patients on mechanical ventilation and other mechanical support poses problems. A medical officer should decide which infusions and treatments can be stopped. He or she should also triage the patients before evacuation, so that those most able to withstand the stress of evacuation are moved first. A specific area of the hospital, some distance from the ICU but preferably on the same horizontal level, should be designated as the holding area for ventilator-dependent and other critically ill patients.

Joint exercises with the local fire services, using simulated patients, familiarize both the staff and the rescue services with the protocols, technology, and geography of the ICU. Using artificial smoke, from a smoke generator of the type used in theatrical productions, adds realism. A video of the event permits constructive discussion afterwards.

Electricity

The ubiquitous use of electrical equipment in the support, monitoring, and treatment of critically ill patients renders them susceptible to electrical hazards. Some of the risks also affect the staff. These hazards range from power failures to the inadvertent passage of electric current through a patient.

Power failure

Failure of the mains electrical supply is a particular risk to the critically ill patient. All ICUs should have an emergency generator that automatically switches on when the mains supply fails. This system should be tested periodically. This familiarizes staff with any delay while the system switches over and any actions necessary during the changeover. The risks are further minimized by ensuring that as much equipment as possible has an integral battery back-up. These batteries need to be kept fully charged.

Electric shock

Only in medicine is the deliberate connection of current-carrying electrical equipment to living individuals entertained on a regular basis. The situation is worsened in critically ill patients, because the normal resistance of the skin is frequently bypassed either by special treatment of the skin or by the placement of electrodes within the body. In addition, sick patients are likely to be more susceptible to the effects of electrical energy than healthy adults.

Biological effects of electric currents

At fairly low currents, there is a threshold of perception defined as the minimum current that an individual can detect. This is felt as a tingling sensation in an awake patient. As the current is increased, the 'let-go' current is reached. At the 'let-go' current local muscle spasm makes voluntary muscle relaxation at the point of contact impossible. This prevents the victim from releasing a live wire or piece of equipment. Further increases in the current produce respiratory muscle spasm and subsequently ventricular fibrillation. Beyond this level higher currents produce sustained myocardial contraction and tissue burns.

In an unconscious patient receiving mechanical ventilation, the first indication of electrocution may be ventricular fibrillation. This is more likely if the current pathway crosses the heart. The heart is most susceptible during the repolarization phase of the cardiac cycle. The most arrhythmogenic supply frequency is 50 to 60 Hz, ironically the usual frequency of the mains supply.

Microshock

The concept of microshock derives from the fact that, if there is an intracorporeal contact on or close to the heart, currents as low as 10 A may provoke ventricular fibrillation. Intracorporeal electrodes include monitoring devices such as central venous catheters and pulmonary artery catheters, particularly if they are filled with a conductive solution like saline. There is also a risk of microshock in patients with external pacing wires and from temperature monitoring probes placed in the lower third of the esophagus, behind the left atrium.

Protection against electrocution

Any protective cut-out system built into electrical equipment for use in intensive care needs to be very sensitive to prevent electrocution. However, it must not be responsible for unnecessary cut-outs. The use of earth leakage circuit breakers, a common feature in domestic and industrial electrical equipment, is inappropriate in the ICU because low but variable levels of current may be permitted to pass through the patient. Instead, equipment design has concentrated on isolating the patient circuit from the mains supply.

International design standards are published jointly by the International Standards Organization (ISO) and the International Electrotechnical Commission (IEC).

IEC 601 Part 1 classifies equipment according to the means of protection it provides against electric shock. There are three classes for electrical equipment for medical use.

1. Class I equipment has any conducting parts accessible to the user, such as the metal casing, connected to earth. If a fault occurs in the equipment in which an inadvertent connection is made between the live supply and the casing, the circuit is completed to earth. A high current then flows, melting protective fuses and disconnecting the circuit. This system requires that the correct fuses are installed and the earth wire is properly connected.

2. Class II equipment is double insulated. All accessible parts are protected by two layers of insulation, or by reinforced insulation, so that there is no possibility of a person touching any conducting part that may become live through a fault. Therefore an earth wire is not required.

3. Class III equipment has no potentials exceeding 24 V a.c. or 50 V d.c. Normally, it is not possible for these potentials to produce an electric shock in healthy humans. However, there may still be a risk of microshock.

A further classification of equipment is based on the permissible leakage current. This relates to the use for which the equipment is designed. These type classifications include the following.

1. Type B may be Class I, II, or III, or battery powered, but the maximum leakage current to the patient must not exceed 100 A. Type B equipment is not suitable for direct connection to the heart even via a catheter.

2. In type BF equipment, the parts in contact with the patient are electrically isolated.

3. Type CF has a very high degree of isolation with a maximum leakage current of less than 10 A and therefore is suitable for direct connection to the heart or connection via a saline-filled catheter.

All ECG leads, pressure transducers, and cardiac output computers should be rated as type CF.

Mobile telephones

Recent concerns have been expressed about malfunctions occurring in equipment when mobile telephones have been used nearby. Equipment affected includes syringe drivers, infusion pumps, and ventilators containing programmable microchips. At present it is not clear whether this is an effect of all types of mobile telephone or only the later type of satellite telephone. Since many hospital porters and security personnel make use of hand-held radio transmitters, it would seem prudent to prohibit the use of all mobile telephones and other radio-transmitting devices in the ICU.

Bibliography

Bushman, J.A. and Thompson, P.W. (1989). Environmental safety of the patient. In *General anaesthesia* (ed. J.F. Nunn, J.E. Utting, and B.R. Brown), pp. 588–94. Butterworth, London.

Moyle, J.T.B. (1990). Electrical safety. In *Anaesthesia: Review 7* (ed. L. Kaufman), pp. 75–85. Churchill Livingstone, Edinburgh.

Parbrook, G.D., Davis, P.D., and Parbrook, E.O. (1985). *Basic physics and measurement in anaesthesia* (2nd edn). Heinemann Medical, London.

Park, G.R. (1991). *Fire safety in the intensive care unit.* Intensive Care Society, London.

15.1.8 The role of intensive care unit policies

B. Morgan and W. J. Sibbald

Key messages

- Standards, policies, and practice parameters define expectations for clinical practice, create consistency in the method of delivery, and promote continuity of care.
- Policies describe conditions that must exist to facilitate exemplary patient care, and they represent non-negotiable requirements.

- Practice parameters (practice guidelines, clinical protocols, algorithms, practice standards) assist care providers in the definition of an appropriate path for the management of a specific clinical problem.
- Policies and practice parameters, if utilized as intended, are ideal strategies for reducing unnecessary expenditure and may provide a defense in court.

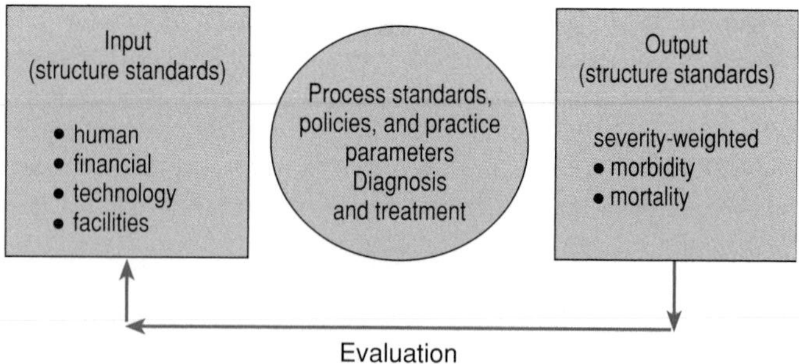

Fig. 1 Definition of inputs and outputs associated with outcomes in process improvement.

Introduction

Health care providers are struggling with the need to manage increases in both utilization and costs. Critical care is disproportionately utilized by our ageing population, new drugs and life-support technologies are expanding available treatment options, and there is an increasing acceptance of organ transplants. Lessons from industry demonstrate that poor quality leads to avoidable increases in costs. An important mechanism for managing the growing imbalance between available resources and patient need is to improve efficiency and therefore deliver better quality services at reduced costs.

Outcomes are clearly altered by the process of health care. Standards, policies, and practice parameters are tools that improve this process. Process improvement involves the definition of activities which transform inputs (patients with health care needs, equipment, etc.) into outputs (including information, diagnostic and treatment decisions) that are associated with outcomes realized by patients and others who are affected by the care (Fig. 1).

Standards, policies, and practice guidelines define expectations for clinical practice, create consistency in the method of delivery, and promote continuity of care. Their use improves care because intensive care units that have well-defined protocols, a medical director to co-ordinate activities, well-educated nurses, and strong collaboration between nurses and physicians demonstrate the best outcomes (Knaus *et al.* 1986).

Standards

Standards describe the desirable and achievable level of performance that can be expected, and are typically presented as structure, process, or outcome standards. Structure standards outline the essential supports and resources necessary to provide safe and adequate care, process standards define the activities consistent with appropriate care, and outcome standards describe a desired patient response to the care provided. While process standards depend in part upon structure standards, outcome standards depend upon both structure and process standards to attain desired responses. For example, examining standards related to infection control in intensive care, a structure standard might state that a sink for hand-washing be available in each patient room. The ability to achieve a process standard such as hand-washing after contact with each patient will depend, in part, upon the structure standard being met. An outcome standard, such as no horizontal transmission of bacteria, requires that both the structure and process standards be achieved.

Standards should be identified prior to developing policies or practice parameters, to establish clearly the goals and purpose for any directives. Most professional bodies have begun to create standards for practice; thus the development of unit-specific standards should be consistent with those set by licensing boards or accrediting bodies.

Policies

Policies describe conditions that must exist to facilitate exemplary patient care, and they represent non-negotiable requirements. Unit policies provide the authority for unit management and decision-making. Policy statements are necessary to ensure consistency related to the utilization of human, environmental, and material resources. Policies should complement the standards of care determined by professional associations.

The development of a policy requires careful balancing of quality and cost, with the objective of promoting optimal quality of care and appropriate allocation of resources. Well-developed policies provide valuable direction for clinical decision-making.

Policy statements are normally limited to issues requiring rigid regulation, and must be developed with input from all appropriate stakeholders. These policy statements must be given authority and support from hospital and unit administration, and physicians and other health care workers must be held accountable for deviations from written policy. Mechanisms for monitoring and revision are essential if policies are to remain useful and current.

Policies that fail to meet the needs of a unit, are not utilized as intended, or bypass intermittent review and revision are wasted efforts. A policy statement that is inconsistent with actual unit practice may increase liability by encouraging practice that is in conflict with expected standards and may promote complacency towards policies in general.

Practice parameters

Practice parameters assist care-providers in the selection of an appropriate path for the management of a specific clinical problem. Also referred to as practice guidelines, clinical protocols, algorithms, or practice standards, practice parameters assist clinical decision-making in improving quality, effectiveness, and appropriateness of care (Cahn 1994). The American College of Critical Care Medicine has adopted the terminology practice parameters and differentiated it from standards and guidelines by the following definitions: parameters represent 'expert consensus on therapeutic issues that identify recommendations upon which appropriate clinical judgments may be based', standards are 'a sociological designation of a level of requirement, excellence, or attainment', and guidelines define 'a statement of policy by an agency having interests in or authority over an activity' (Shapiro 1995).

Practice guidelines can be used to free practitioners from the need to 'weigh the pros and cons of every decision, connect practitioners as a collective consciousness, bring order, direction and consistency and

Table 1 The contrast between standards, policies, and practice parameters

Standards	Policies	Practice parameters
Describe the minimum level of care that should be expected	Define the non-negotiable requirements that must exist within a facility	Evidence-based algorithms used to guide decision-making about patient care
Define structure, process, and outcome expectations	Provide directives for unit operation	Utilize the 'best evidence' to determine the most appropriate intervention

provide an intellectual vehicle for applying research to practice, and pool the expertise of many' (Eddy 1990). This contrast is shown in Table 1.

Why do we need policies and practice parameters?

Health care knowledge and technology have expanded more than health care funding. At the same time, unexplained geographical variations in style, delivery, and costs have been clearly identified. Clearly defined policies and practice parameters, if utilized as intended, are ideal strategies for reducing unnecessary expenditure. Policies and practice parameters provide a defense for limiting diagnostic testing and management strategies to only those which have been proven beneficial.

Outcomes are inversely related to practice variability (Shapiro 1995), i.e. the more variation that exists in practice, the worse the outcomes. Practice policies and parameters are an important part of a total quality management program—by encouraging consistency in practice, reducing unsubstantiated deviations from best clinical evidence, and providing a flow pattern that streamlines the process of monitoring, review, and revision.

Well-developed guidelines do not provide a 'cookbook' approach to health care. While practice parameters promote the use of the current best evidence to guide decision-making, they also require practitioners to assess clinical situations correctly and to use clinical judgment to support actions. When combined with clinical expertise that includes experience, proficiency, and the ability to make clinical judgments, practice parameters promote quality care.

Day-to-day use of practice parameters must be monitored and evaluated, and the process reviewed and revised on a regular basis to ensure consistency with current evidence. The benefit of practice parameters will depend to a large degree upon the quality of the parameter developed. Their quality can be judged according to the Institute of Medicine's eight criteria, which include validity, reliability, applicability, flexibility, clarity, development in a multidisciplinary process, scheduled reviews, and careful documentation (Field and Lohr 1990). Alternatively, parameters can be evaluated according to Eddy's criteria, which state that they should be accurate, accountable, predictable, defensible, and usable (Eddy 1990).

Collaborative practice between health disciplines improves outcomes (Knaus *et al.* 1986). No single health care discipline practices in isolation; therefore careful attention to collaborative development, education, implementation, and evaluation is paramount in the success of any policy or practice parameter. Valid policies and practice parameters should be based on research evidence and expert interpretation. Consequently, policies or guidelines should complement standards for any specific discipline. Inconsistencies in parameters from one discipline to another can render even the most well-designed parameter impotent.

Liability has previously been measured by comparing practice with 'expert opinion'. Unfortunately, expert opinion varies and is predominately subjective; it may also be entrenched in past practices versus best clinical evidence. As a result of increasing support by national organizations to develop practice parameters, complying with published practice parameters may become the standard for measurement of clinical practice in the courts.

As national bodies develop and publish increasing numbers of practice parameters, individual agencies will benefit from already completed projects. The role of the individual units will be to apply practice parameters to their own specific policies and guidelines. Local practice parameters may exceed national standards. For example, if a particular disease were more prevalent in the area serviced by a given unit, it might be appropriate to initiate additional diagnostic screening tests.

As experience and new evidence develop relating to any specific area of practice, one can anticipate that the 'best practice' will be quite different in the future compared with now. Practitioners will need to link closely with organizations developing practice parameters to keep those agencies abreast of new research evidence or practice changes. The need to review practice parameters regularly, to identify barriers to implementation, or to revise outdated practices will be the greatest future challenge. The merits of policies and practice parameters are summarized in Table 2.

Developing policies, protocols, and practice parameters

The quality of any policies or guidelines developed will depend upon the process for policy generation, implementation, monitoring, and revision. The following process can be used in developing any directive intended to improve the quality of care and reduce unnecessary costs.

Assign responsibility

A policy must have authority to be effective. Without authority, a policy cannot be enforced, and health care providers are powerless to demand compliance. Responsibility for unit administration is generally granted to a medical director or administrative leader.

Identify stakeholders

Collaborative practice improves patient outcomes (Knaus *et al.* 1986). Many different medical disciplines and health care professionals participate in the delivery of intensive care and will bring a variety of perspectives to any planning table. These various perspectives must be considered when developing a policy, and their impact on the final decision must be taken into account.

Table 2 The merits and limitations of policies and practice parameters

Benefits	Limitations
Promote evidence-based practice	Potential for misinterpretation of evidence
Guide appropriate allocation of resources	May be insufficient concrete evidence to develop guidelines
Pool expertise to enhance decision-making	Poorly developed parameters may reduce outcomes and increase risk
Control costs	Resources must be available to support practice parameters
Promote consistency in care	Rigid guidelines may restrict judgment
Facilitate collaborative practice	Complex guidelines may be too difficult to apply
Provide education	Value depends upon accurate implementation
Guide orientation of new staff	Compliance must be monitored
Reduce risk	Failure to review and revise may lead to outdated, ineffective, and possibly incorrect guidelines
Promote quality management	

Successful implementation of any policy requires 'buy in' and commitment from each member of the planning team. While unanimous agreement rarely occurs, members need to be able to support the final decision. Failure to include important stakeholders in policy development will adversely impact the degree of clinical support and application of the policy. While inclusion may be more labor intensive than autocratic decision-making, the end result will be a more useful policy with the support of those involved.

Establish the purpose of the policy or protocol

The goals of the policy being developed should be determined. These goals will provide the guiding principles for policy or guideline development.

Review appropriate information to guide decision-making

Any necessary information that will assist with determining policies or guidelines should be collected. This includes any national standards regarding facility requirements or minimum levels of expected practice, relevant research evidence, and benchmark data from similar units. Information regarding standards of care for the population to be cared for is also needed.

Develop policy draft and circulate

Establish a draft policy or guideline, and circulate to any additional stakeholders to provide opportunity for feedback. Establish a date for submission of feedback to ensure timely completion of the protocol.

Determine impact of policy

Any policy or practice change will invariably impact on a number of different practices. While inclusion of all stakeholders from the onset of policy development is important, circulation of a draft to additional individuals may identify issues that the policy development team overlooked. The chain of events that cascades from any policy change must be followed, and the impact of the policy on other patients and care providers considered. For example, is the policy consistent with standards of practice for all disciplines involved, or will a policy change for one population of patients impact on resources for another?

Identify resource implications

Policy changes or practice guidelines have resource implications. These implications might be as simple as establishing an education plan to communicate the policy change. However, expansion of services provided could significantly impact upon personnel, technology, and staff training requirements.

Document and communicate policy or guidelines

Successful policy or guideline implementation will depend upon clear communication of the intent and process of the policies. The key to effective implementation is the need to gain the support of those involved. All policies must be documented to identify expectations and provide a reference for personnel to review.

Identify monitoring indicators/criteria

Even the most appropriate policy is only useful if implemented correctly. Criteria to monitor for variances from the policy, as well as indicators to measure effectiveness, are paramount to quality management.

Review and revise

Monitoring intervals must be determined and the data collected must be reviewed. Once policies become outdated, their value is lost. Outdated policies promote complacency towards policies in general, and failure to revise policies that are no longer appropriate may increase liability.

Conclusion

Critical care is practised by addressing the demands of who should be the recipients of such care and under what circumstances. Even for standard management options, critical care clinicians implicitly identify all available alternatives and possible outcomes, weigh the benefits against the risks and costs, and, in the broadest context, factor in logistic, economic, societal, and legal considerations. Because the critical care unit is a hospital resource with a high level of activity, expenditure, and risk exposure, the benefits of applying evidence-based medicine to the questions 'What should we be doing?' and 'To whom should we be doing it?' are likely to be proportionately great. For the purposes of both maximizing quality improvement and rising to the challenge of most effectively using the available resources for patient care, clinicians and investigators need a common platform upon which they can agree to assemble the facts. Evidence-based clinical guidelines and practice standards are an ideal platform, as they provide the tools required to evaluate and use the medical literature in day-to-day practice.

References

Cahn, M. (1994). Practice guidelines: a piece of the quality puzzle. *Bulletin of Medical Libraries*, **82**, 312–14.

Eddy, D. (1990). Practice policies—what are they? *Journal of the American Medical Association*, **163**, 877–80.

Field, M. and Lohr, K. (1990). *Clinical practice guidelines: directions for a new program*. National Academy Press, Washington, DC.

Knaus, W.A., Draper, E.A., and Wagner, D.P. (1986). An evaluation of outcome from intensive care in major medical centers. *Annals of Internal Medicine*, **104**, 410.

Shapiro, B. (1995). Practice parameters for critical care medicine. *Critical Care Medicine*, **23**, 1458–9.

15.1.9 Resource management in the intensive care unit

M. Jegers and D. Reis Miranda

Key messages

- Effective resource management in the intensive care unit (ICU) is largely dependent on the clear definition of intensive care policies at regional and hospital level.
- Managerial tools such as cost accounting systems and budgets, if correctly implemented, enhance the quality of ICU management, particularly with respect to resource use.
- When comparing ICUs, only the direct costs should be considered as they reflect resource use under the responsibility of the ICU manager.
- When evaluating ICU management, the direct costs should be considered together with ICU output and patient characteristics.
- The costs and the cost structure in the ICU should reflect the number and the type of activities.

Introduction

The intensive care unit (ICU) is a concentration of specific professional expertise and high-technology equipment around a number of beds located in a sophisticated hospital environment. The wealth of the unit, or its richness measured as education and number of personnel, type and quantity of equipment, the use of more expensive medication, etc., may vary widely among ICUs and is largely dependent on the local institutional context of health care organization and financing.

The principal objective of the ICU is to provide monitoring and therapeutic facilities for critically ill patients which cannot be supplied elsewhere in the hospital. Therefore other hospital departments often request the ICU to contribute to the treatment of critically ill patients. Depending upon its professional characteristics and wealth, the ICU may also be requested to contribute to the care of patients in another hospital. In this case, the ICU may assume a kind of central function which was quite probably not considered in its original objectives.

As the activities performed in the ICU can be seen as temporary interventions integrated with the longitudinal process of care in other hospital departments, the ICU is frequently classified in the group of 'cost centers' in the hospital, together with the operating theater, the laboratories, etc. In general, in addition to professional expertise, a cost center is allocated resources for performing activities that are requested 'externally'.

However, because of the large financial impact of ICU activities on the hospital budget, many hospital administrations are beginning to enforce 'resource management' in the ICU. As it becomes clear that effective management of resources requires active contribution from the professionals using the resources, the involvement of ICU physicians and nurses in resource management activities is increasing substantially. This implies that, in addition to their specific professional responsibilities, physicians and nurses in the ICU take responsibility for the whole process of resource consumption in the ICU, from admission to discharge of each patient. Therefore the ICU is becoming a 'responsibility center', negotiating its annual budget for delivery of agreed outputs (number and type of patients, quality of care, etc.) with the hospital administration.

It should be emphasized that the problem does not end here, as prediction of the activities to be developed is the first relevant factor of effective management. The need for prediction for sound resource management requires consideration of health care strategies defined at other levels of decision-making.

Decision-making at the regional level

A health care region is usually defined, from an ICU perspective, as a population served by hospitals and ICUs of different characteristics. Each health care region should make a detailed assessment of the ICU needs in the hospitals of the area: the number and type of patients to be admitted, the general and specific interventions required, the number and type of beds, and the number and differentiation of staff. When this is complete, the general characteristics of each ICU in the area should be defined, with a level of care (LOC) attributed to each unit (Reis Miranda *et al.* 1990). The exchange of expertise and patient care among the various ICUs is then defined and agreed between the hospitals, with the financial implications being taken into consideration.

Decision-making at the hospital level

Every hospital should make a detailed survey of the ICU requirements in the hospital by collecting the data discussed above. After the characteristics of the required ICUs have been determined (one or more ICUs, one or more LOCs), admission and discharge criteria will be established so that the objectives and use of the ICU are understood by the hospital medical and nursing staff. We believe that budgets and management accounting systems are appropriate instruments for establishing and monitoring the resource-related consequences of these decisions.

Management accounting in the ICU

Some basic concepts of management accounting are defined and explained below.

Resources and costs

Resources of various types are used in the ICU: nurses' and physicians' time, equipment time, disposables, medication, etc. In every productive activity some of these inputs or resources are transformed into outputs. An economist defines these inputs as 'costs', i.e. something that has to be foregone in order to perform the productive activity (Sher and Pinola 1986). This definition does not imply that costs are exclusively expressed in monetary terms. However, in a situation where different resources are required, such as in an ICU, they should be quantified using a common expression and 'money' is the most obvious candidate. Therefore the rational resource management of an ICU requires information about costs expressed in monetary terms.

Two monetary cost concepts can be distinguished: financial costs and accounting costs. These two concepts are easily and frequently confused, with deleterious effects on the quality of cost information.

Financial costs

The financial manager of a hospital should, in principle, be concerned with real cash flows: the main expenditures incurred before a new product or process begins to earn a return on its investment. This means that the money effectively paid to obtain equipment or other goods and services is their cost in this frame of reference. To avoid confusion, we will call this cost an 'outlay' or investment.

Accounting costs

Accounting at the hospital level is called 'financial accounting'. Here we shall consider only those concepts that are relevant to our understanding of accounting information at the hospital and the ICU levels.

The cash basis is considered to be unsatisfactory for cost measurement in a large and complex organization such as a hospital. For example, investment in a building would be recorded as a cost in the fiscal year that the payment is effectively made, with no effect whatsoever on the costs in subsequent years. Therefore allocation of the amount invested over the lifetime of the building ('depreciation') is more appropriate. In fact, this method is an application of the more general 'accrual principle'. The accrual-based system records transactions that do not involve cash, such as depreciation of buildings, unpaid bills, unpaid salaries, and uncollected income. Although conceptually clear, accrual accounting has many intricacies and subtleties which go beyond the scope of the present text. The central idea is that costs are considered as a component of the overall costs of the periods in which the resource paid for has an economic effect, which means in an accounting context that it is related to outputs realized in the same periods of time. Therefore this method of allocation does not take the time at which payment is made into consideration. Consequently, outlays and accounting costs for the same period can differ considerably.

ICU costs

Accounting for costs at the unit level is part of 'management accounting' or 'cost accounting'. Accounting costs registered at the hospital level are allocated to the different hospital departments, such as the ICU. Two distinctions are conceptually important.

The first is the distinction between direct and indirect costs. A direct cost of an ICU is a cost that is traceable to ICU activities. Indirect costs are not traceable to ICU activities, although some vague relationship can be assumed. The difference between direct costs and indirect costs is not absolute; the measurement technology used dictates which costs are to be treated as direct and which as indirect. In this sense the activity-based costing technique should be considered as a further refinement of existing techniques, and not as a conceptually new approach to costing.

Examples of indirect costs for an ICU are hospital management, costs of buildings and cleaning, costs of interventions in the operating theater, hospital gardeners, etc. Confusingly, the term 'indirect' cost is used in a different way in some health economic texts; when determining the cost of a specific treatment, costs due to lost earnings, lost production, or lost income are labeled indirect costs (Koopmanschap and Rutten 1994). In our terminology, these are direct costs of the treatment, but not of the ICU; they are not indirect ICU costs, as they are not borne by the hospital. The fact that they cannot be measured easily should not change their definition (Durand-Zaleski 1994).

As other units have similar indirect costs, an important problem is how to allocate them to all the units concerned. Many techniques exist, but none is perfect and all are arbitrary to some extent. This makes cost comparisons between ICUs in different hospitals very difficult, particularly when direct costs and indirect costs are combined. For example, in a given hospital ICU costs themselves might be considered indirect costs for the medical or the surgical departments.

In our view, only direct costs should be taken into consideration when comparing ICUs because they are the only costs that reflect resource use under the responsibility of the ICU manager.

The second distinction is that between fixed costs and variable costs. A fixed cost is a cost that is not influenced, directly or indirectly, by the level of ICU activities in a given period of time, traditionally 1 year. Variable costs are influenced by the level of ICU activities, measured for example by the number of Therapeutic Intervention Scoring System (TISS) points per day, the number of patients, or the number of patient-days.

It must be stressed that direct–indirect and fixed–variable costs are two independent cost characterization scales, resulting in $2 \times 2 = 4$ different possible cost categories.

A final note on cost accounting: revenues

Contrary to what the term 'cost accounting' might suggest, costs and revenues are treated on an equal basis in a cost accounting system. Here also cash inflows can be distinguished from accounting revenues, according to the parallel distinction between cash outflows and accounting costs. Here also a distinction between direct revenues and indirect revenues can be made. As the ICU is a service center for other units in the hospital, it does not generate direct revenues from outside the hospital. Therefore the revenues allocated to the ICU in a traditional cost accounting system are almost exclusively indirect. Awareness of the relation between revenues and resources used by

ICU managers can be increased by implementing a system of transfer pricing or cross-charging: the ICU services are then internally billed to those units who use them. Different transfer-pricing formulas can be devised, for example by taking into consideration the required level of care or the number of TISS points consumed. The advantage of such a system is clear: ICU activity does not have to be capped according to costs, but can be planned with reference to both costs and 'revenues'. A possible disadvantage is the problem of designing a good transfer-pricing system containing the incentives desired by hospital management or health care authorities. Needless to say these incentives should enable both monetary and non-monetary objectives to be achieved.

Cost accounting in European ICUs

In a European study performed between 1994 and 1996 (EURICUS-I), data on cost accounting procedures in 89 ICUs in 12 countries were collected (Reis Miranda *et al.* 1998). Although the participating ICUs were not selected as representative of the respective countries, we assume that the data collected illustrate the present situation in a large number of European ICUs.

A cost accounting system was used in only 14 ICUs in this sample. The costing systems used were diverse, ranging from the very sophisticated to the very simple, such as recording only the costs of disposables and pharmaceuticals or only the costs of materials used and depreciation of equipment, or using a predetermined cost per TISS point.

Direct costs in the ICU

Based on the data collected during the EURICUS-I research, we propose a structure for calculating the direct cost of an ICU over a given period of time (e.g. 1 year) (Table 1). With respect to resource management, the column headed 'units' is the most important. The values it contains must be compared with the output of the ICU in order to assess its overall efficiency.

Comparison of resource use by ICUs should be made by comparing the physical input of resources. When comparing costs, care must be taken to standardize the 'prices' used to transform the number of resource units into cost.

This analysis can be refined by separating fixed costs from variable costs. As fixed costs are not influenced by the ICU's activity level, decision-making about the resources representing fixed costs should be of a different longer-term nature than decision making about variable resource input. Definition of activity level, which can be different for different kinds of resources representing variable costs, is a major problem. For example, TISS points provide a reasonable measure of nursing activity in the ICU, whereas the number of patients or patient-days is probably a better method of quantifying the use of pharmaceuticals.

When the data have been collected as described above, a cost structure for the ICU will emerge. It is also possible to define the cost structure per unit of output, such as patients, nursing-days, etc., or of groups of patients aggregated according to age, diagnosis, whether they are medical or surgical, outcome, etc. The cost of the various activities in the ICU (monitoring, mechanical ventilation, dialysis, etc.) can also be analyzed, so that the costs associated with the care requirements for any new patient or group of patients can readily be ascertained. It should be noted that the cost structure of the ICU should be re-evaluated periodically using a representative sample.

Table 1 Structure for calculating the direct cost of an ICU

	Units	Cost
Personnel		
Physicians		
Nurses		
Others (e.g. secretary, technicians)		
ICU equipment*		
Beds		
Medical equipment (e.g. monitoring, ventilators)		
Other		
Maintenance		
Current expenses		
Pharmaceuticals		
Disposables		
Blood and products		
Laboratory tests		
Radiology		
Particular procedures such as dialysis		
Physiotherapy		
Other		

*Equipment costs can be rents, lease payments, or depreciation.

Budgeting in the ICU

Whereas cost accounting systems record past costs and revenues of the ICU, a budget predicts the financial consequences of future activities. Therefore it is an essential instrument for controlling the use of resources in the unit. A budget is not and should not be a mere forecast of future costs and revenues, such as a weather forecast, but the consequence of strategic decision-making at both the hospital and ICU level. When used properly, it makes a substantial contribution to the rational use of resources, converging to explicitly stated objectives for the hospital and the ICU. In principle, a budget can be produced for any future time period, but in practice budgets are mostly drawn up for 1 year.

An ICU budget should include at least the following components: current wages, other current expenses, and investments. It should be constructed on either a cash-flow or an accrual basis. In practice, these two approaches are frequently mixed in the same budget, leading to confusion and misleading information for ICU managers.

There are three fundamental budgeting procedures: bottom-up, top-down, and just-top. In a bottom-up system the budgeting procedure is initiated by ICU management, and refined and coordinated at the hospital level. In a top-down design, hospital management makes the first budget proposal, which is discussed and possibly changed by unit managers. In a just-top procedure, a given authority (the director, a board, or a council) decides on the ICU budget without any participation by ICU members. This procedure is sometimes disguised as an 'objective' rule, for example last year's budget with some allowance for inflation or a fixed percentage of the hospital budget. Reflection of the hospital and ICU strategy can easily be avoided in such a system. Therefore we believe that it is inadequate for rational resource management.

A budget's usefulness is substantially increased when there is a budget control system. In such a system, previously established budgets are compared with the real situation. When important differences are observed, immediate action may be necessary to preserve the ICU strategy. Without such a system, those differences will go unnoticed for a long time.

Budgeting in European ICUs

EURICUS-I also provided information on budgeting procedures in European ICUs. In a relatively large number of cases, ICU managers stated that there was no specific ICU budget. Only 29 out of a total of 89 ICUs reported the existence of an ICU investment budget. Budget control systems were reported in 37 ICUs. In 14 cases the system was developed by the ICU itself and not at the hospital level. These results suggest that a large number of European ICUs do not have the correct instruments for monitoring, analyzing, and steering the input of resources.

References

Durand-Zaleski, I. (1994). Estimating the cost of intensive care. *Intensive Care Medicine*, **20**, 538–9.

Koopmanschap, M.A. and Rutten, F.F. (1994). The impact of indirect costs on outcomes of health care programs. *Health Economics*, 3, 385–93.

Reis Miranda, D., Williams, A., and Loirat, P. (ed.) (1990). *Management of intensive care. Guidelines for better resource use*. Kluwer Academic, Dordrecht.

Reis Miranda, D., Ryan, D.W., Schaufeli, W.B., and Fidler, V. (ed.) (1998). *Organisation and management of intensive care: a prospective study in 12 European countries*. Springer, Berlin.

Sher, W. and Pinola, R. (1986). *Modern micro-economic theory*. North-Holland, New York.

15.2 Medicolegal and ethical issues

15.2.1 Consent/assent to treatment

S. A. M. McLean

Key messages

- The patient may exercise his or her right to refuse treatment on grounds which can be entirely personal and which, in the case of the competent adult, do not need to be explained to, or understood by, the doctor.
- In the unconscious patient consent to treatment may be impossible. In most countries, treatment in this situation would not be unlawful if there were a need to undertake therapy.
- The older child may well have opinions about treatment which conflict with those of parents and doctors. The capacity to make binding decisions is based on the child's maturity and capacity to understand the information about the treatment and also what would happen if the therapy were refused.
- Therefore discussion protects both patient and doctor and will most likely enhance the relationship of trust between the parties which is so essential to the 'good' practice of medicine.

Introduction

It goes without saying that therapy given without the agreement of a patient is ethically and legally unacceptable. No contemporary practitioner would proceed without any agreement from a competent patient—to do so would be an assault or battery. However, problems may arise when assessing the amount of information which is sufficient to ensure that the patient is adequately informed and able to make a meaningful decision.

Therefore the real problems are not about the need to obtain consent but rather about how real that consent actually is. The doctor who provides insufficient information may be held to have been negligent. In addition, there may be further complications when the patient is not a sane adult. Where the patient is a child, incompetent, or unconscious, different questions are raised which require careful analysis by doctors, based always on the principle that treatment given without the provision of appropriate information, and without an assent from the patient, must be the exception rather than the rule.

For this reason, it is important to remember that obtaining an agreement is much more than a legalistic hurdle to be overcome. It is a reflection of the doctor's respect for the integrity of his or her patient. Even the certainty of a cure is not sufficient to authorize the clinician to intervene without that respect being manifested. The crucial element is the sharing of information and not the outcome. Consent is only one of the possible outcomes of this process of information sharing. The patient may exercise his or her right to refuse treatment on grounds which can be entirely personal and which, in the case of the competent adult, do not need to be explained to, or understood by, the doctor (Re C (adult: refusal of medical treatment) [1994] 1 All ER 819 (England)).

The competent adult

The sane adult patient has a right to accept or refuse treatment without explanation. The doctor's obligation is to provide information of a type which explains the nature, risks, and benefits of the chosen therapy, and also any alteratives. Different legal systems require different standards to be met. In the United Kingdom, the doctor's behavior will be judged by a test which depends effectively on the standard set by his or her professional colleagues, although the court reserves the right (seldom, if ever, exercised) to declare that medical practice is not adequate (*Bolam* v. *Friern Hospital Management Committee* [1957] 2 All ER 118 (England); *Sidaway* v. *Board of Governors, Bethlem Royal Hospital* [1985] 1 All ER 643 (H.L.)(UK)).

In some states in the United States, the test which is applied is that of the 'prudent patient'. In these states the doctor's responsibility is to provide the information which the 'prudent patient' would want to know before reaching a conclusion (*Canterbury* v. *Spence* 464 F 2d 772 (US)).

Most common-law jurisdictions have followed the United Kingdom test, although recently an Australian Court (*Rogers* v. *Whitaker* [1993] 4 Med LR 79 (Australia)) declared this test to be insufficient to meet the requirement that 'Every human being of adult years and sound mind has a right to determine what shall be done with his own body...' (*Schloendorff* v. *Society of New York Hospital* 105 NE 92 (US)). This case may mark the beginning of a move away from the assumption that it is for the doctor to decide what the patient needs to know and start a trend towards recognizing that the decision is one which is personal to the individual patient who may therefore require information which the doctor does not regard as relevant.

The unconscious patient

Although treatment without any consent at all would generally be an assault or a battery, the doctor may confront situations where there is no possibility of agreement being given. In cases of this type, different principles need to be identified in order to decide whether or not treatment can proceed. In most countries, treatment in this situation would not be unlawful if there is a need to undertake therapy. The legal principle of necessity would generally apply, but only as long as the treatment is in fact necessary (*Marshall* v. *Curry* [1933] 3 DLR 260 (Canada)). The doctor would be unwise to undertake any treatment which is optional or which could be postponed. The necessity doctrine authorizes treatment only within a relatively narrow set of constraints—it does not confer *carte blanche* authority on the doctor (*Murray* v. *McMurchy* [1949] 2 DLR 442 (Canada)).

The incompetent patient

This group of patients has recently been the subject of considerable legal attention in the United Kingdom (Law Commission 1993; Scottish Law Commission 1995). Those who are mentally handicapped and those who are mentally ill would fall within this category. Special laws generally exist in respect of the mentally ill, and the requirements of consent are often less problematic, at least with regard to treatment which is specifically related to the mental illness itself. However, the mere status of mental illness will not necessarily mean that the individual should not be given the same opportunity to agree to or refuse therapy for other conditions (Re C (adult: refusal of medical treatment) [1994] 1 All ER 819 (England)). The law, and ethics, would still seek to accord to the mentally ill the same rights as other patients, intervening in a patient's decision only when it is manifestly irrational as a result of the pre-existing condition.

Equally, the mentally handicapped should be given respect. This requires that, whenever possible, the same information should be given to them as would be provided to any other patient. If the patient can understand the information, then his or her choices should be given credence. In this group, however, the problem will sometimes be that the offering of information will serve no purpose since the patient will be unable to understand it. In such cases, where the patient is an adult, the doctor is confronted with real difficulties. If the patient is an adult, there is no one else who can agree to or refuse treatment on his or her behalf. Although it is common to seek the views of relatives, their choices have no legal standing. Therefore in cases of this sort, particularly if the treatment is serious (such as sterilization), the authority of a court will generally be advisable before proceeding.

Courts in the United Kingdom will generally decide on the basis of a 'best interests' test. In other words, they will decide what outcome will provide the most benefit for the patient concerned. This test is often criticized because it seems to follow the clinical recommendation in almost all cases, leading to the suspicion that 'best interests' can be defined as 'best medicine'. However, it is clear that, by analogy with the competent patient, the optimal clinical recommendation is only one part of the equation. Personal and other matters may intrude to lead to a rejection of the recommended treatment. If respect for persons is central to the ethical practice of medicine, then that respect should be extended to all patients and not merely some of them.

In other countries, such as the United States and Canada, courts have sometimes applied an alternative test—that of 'substituted judgment' (*Cruzan* v. *Director, Missouri Department of Health* 110 S Ct 2841 (1990) (United States); Re Eve (1986) 31 DLR (4th) 1 (Canada)). This test requires the court to place itself in the position of the patient and try to work out what the patient would have decided had he or she been able to make a decision. This test is also the subject of concern, since effectively (unless there are individuals who can testify that the patient had previously expressed specific opinions) it involves guesswork. In any event, the outcome of the substituted judgment test may equally be that the clinical recommendation is routinely followed, since most people would imagine that advised or advisable therapy is what the person would have chosen.

Minors

Clearly, when a child is very young, decisions about health care have to be made by others. Parents are given by law the right to care, custody, and control of their young children, and it is their views which will decide whether or not treatment should proceed. However, it should be remembered that children are not possessions and that parents are legally authorized to act only in the best interests of their children. Therefore parental consent to involving children in certain procedures, such as non-therapeutic clinical trials, must be of dubious legal standing.

However, the real dilemmas arise as children begin to mature. The older child may well have opinions about treatment which conflict with those of parents and doctors. The question is whether or not these wishes, rather than those of the parents, should be respected. The age at which a child becomes an adult for legal purposes varies from country to country (in Scotland, for example, the child becomes an adult for these purposes at age 16 (Age of Legal Capacity (Scotland) Act 1991); in England and Wales, partial decision-making powers are given by the Family Law Reform Act 1969. However, even before that age is reached, children begin to be legally entitled to enter into certain transactions without requiring parental authority. These transactions can also include decisions about whether or not to accept health care.

The capacity to make binding decisions is based on the child's maturity and capacity to understand the information about the treatment and also what would happen if the therapy is refused. Lord Scarman, in the Gillick case, put the position in this way: '...I would hold as a matter of law that the parental right to determine whether or not their minor child below the age of 16 will have medical treatment terminates if and when the child achieves a sufficient understanding and intelligence to enable him or her to understand fully what is proposed' (*Gillick* v. *West Norfolk and Wisbech Area Health Authority* [1985] 3 All ER 402 at p. 423 (H.L.)).

Therefore the mature child should be entitled to the same information as an adult and his or her decision should be respected. However, yet again in this situation, there is considerable reluctance in some reported cases to uphold the young person's decision when it does not follow the clinical recommendation (Elliston 1996). Most commonly, this arises when a child refuses treatment. In strict law, and in line with the ethics of medical treatment, this is anomalous and paternalistic. As Hoggett (1986) puts it: '...the capacity to consent must logically include the capacity to dissent: if, then, parental capacity is diminished to the extent that the child herself has acquired capacity, the parents should have no power to insist...a duty to provide adequate medical aid does not necessarily import a power to force it upon a competent child who has rejected it'.

Although societies rightly strive to protect children, protection at the expense of respect is a poor compromise. It must be remembered that the age at which that same individual could make unchallengeable choices is essentially arbitrary and varies from one jurisdiction to another. Therefore there is nothing sacrosanct about it, and the child's decision should be tested against understanding rather than chronological age. Generally speaking, it is the doctor who is charged with deciding on the maturity of the child, and it is essential that this judgment is based on the child's actual capacity rather than on whether or not the doctor approves of the decision which the child makes.

Conclusions

The purpose of discussing risks, benefits, and alternatives with patients is twofold. First, it demonstrates respect for the individual and enhances his or her capacity to act in an autonomous manner in line with his or her own values. This means that the outcome of the process of discussion is for the individual him- or herself. Except in very rare cases, information should not be withheld because it might lead to the patient's rejecting treatment, because if the information resulted in a rejection of treatment then clearly it is relevant to the patient. The therapeutic imperative should not dominate to the detriment of the rights of the patient.

A second purpose of information disclosure is that, if it results in an agreement to proceed with therapy, it renders what would otherwise be an assault lawful. Where adequate information has been given, the doctor may undertake procedures which could otherwise result in criminal charges. Equally, the doctor will not be liable under the civil law of negligence where the required amount of information has been given.

Therefore discussion protects both patient and doctor, and will most likely enhance the relationship of trust between the parties which is so essential to the 'good' practice of medicine. However, few legal systems, if any, are absolutely clear about just how much information needs to be given. In countries where the doctrine of informed consent exists, the amount of information required may be greater than those (such as the United Kingdom) where this doctrine plays no formal role. However, people's moral standing does not change with geography nor with the nature of their illness and, irrespective of the legal tests laid down, each person is entitled to the fullest possible relevant information on which to base his or her decision.

The principles of medical ethics are designed to ensure that patients are not harmed by their involvement with medicine and that they are prioritized in terms of respect. Harm, of course, need not merely be physical. Harm, both ethically and legally, may be no more than disrespect resulting from inadequate discussion. Legally, compensation may be obtained even where the failure to respect the patient's wishes results in a good clinical outcome (*Malette* v. *Shulman* (1990) 67 DLR (4th) 321 (Canada)). This is because the values which underpin medicine are greater than technical capacity. In judging what information a patient should receive, doctors would do well to bear this in mind, since '...it can be concluded that patient autonomy is only protected where there is a meaningful choice made by the patient, on the basis of adequate information, about which of the available therapies is acceptable and as to whether or not to participate in any therapy at all' (McLean 1989).

References

Elliston, S. (1996). If you know what's good for you: refusal of consent to medical treatment by children. In *Contemporary issues in law, medicine and ethics* (ed. S.A.M. McLean), pp. 29–56. Dartmouth, Aldershot.

Hoggett, B. (1986). Parents, children and medical treatment: the legal issues. In *Rights and wrongs in medicine* (ed. P. Byrne), p. 158. Oxford University Press for King's Fund, London.

Law Commission (1993). *Consultation paper no.129. Mentally incapacitated adults and decision-making*. HMSO, London.

McLean, S.A.M. (1989). *A patient's right to know: information disclosure, the doctor and the law*. Dartmouth, Aldershot.

Scottish Law Commission (1995). *Report on incapable adults (Scot Law Com No 151)*. HMSO, Edinburgh.

15.2.2 Medical research

S. L. Cohen

Key messages

- Research in intensive care is difficult owing to complex clinical situations, differing populations, and small numbers.
- Research in emergency life-threatening situations requiring immediate treatment in patients who are unable to communicate and where there is not time to look for relatives or any legal representatives can be conducted without consent under certain conditions.
- The problem of consent remains difficult, but clearly there is a case for carrying out research without consent provided that the proper procedures of a stringent, ethical, and scientific research ethics committee have been followed.

Introduction

Advances in the treatment of patients in intensive care can only occur as a result of research. Care of all patients is guided by the principles of medical ethics:

(1) beneficence, i.e. acting for the patient's benefit;

(2) non-maleficence, i.e. not intentionally doing harm;

(3) autonomy, i.e. respect for a patient's right to self-determination, including to consent or otherwise to any procedure;

(4) justice.

Research in intensive care is difficult for a number of reasons. Clinical situations are complex and it is often hard to compare patients. Each intensive care unit (ICU) attracts a differing population of patients depending on the activities of the hospital and the nature of the area where it is situated (e.g. inner city, suburban, or rural) and varying between countries. The numbers of patients with rare diseases seen in any one unit may be too small for valid generalizations and conclusions to be reached. This has led to the idea of multicenter trials and research which may have deficiencies due to non-homogeneity of patients.

Research workers are often under pressure to produce results, and unfortunately this has led to the occasional occurrence of dishonesty because of the need to be seen to be productive and the need to attract grants to continue research. This underlies the need for good control of research, usually mediated by ethics committees monitoring research projects and, it is hoped, the vigilance of editors of medical journals who ensure that only true, original, and ethical work is published.

Perhaps the greatest problem involved in research in intensive care is the question of consent which by definition must be informed, i.e. the patient must be able to make an informed decision on the risks and benefits of any procedure, research or otherwise. Patients in intensive care are frequently sedated and/or unconscious, either as a result of their disease or because of drug treatment, which makes informed consent impossible. A great deal of thought, time, and energy has gone into the development of means to obviate this problem of informed consent in intensive care, while safeguarding the rights and dignity of the patient.

Principles of research

Research has burgeoned in the last decades. It is recognized that clinical investigation is only acceptable if informed consent to this activity is granted by the patient. Informed consent was first defined in the Nuremberg code as 'the voluntary consent of the human subject is absolutely essential'. This means that the person involved should have the legal capacity to give consent, should be so situated as to exercise free power of choice, without the intervention of an element of force, fraud, deceit, duress, overreaching, or other ulterior form of constraint or coercion, and should have sufficient knowledge and comprehension of the elements of the subject matter involved to enable him or her to make an informed and enlightened decision. Informed consent therefore requires the following:

(1) disclosure of information;

(2) legal competency;

(3) voluntariness and freedom of choice;

(4) understanding by the patient;

(5) decision-making capacity.

These principles grew up as a result of the revulsion felt universally to the practices in so-called 'research' carried out by the Nazi regime on concentration camp victims in the Second World War. This set of principles has been reaffirmed repeatedly in the declaration of the World Medical Association (Tokyo 1964: revised in Helsinki 1975) and in guidelines issued by the World Health Organization. The idea that research could be justified on the grounds of 'good of society' should never prevail over the interests and welfare of any particular human being (Lemaire *et al.* 1997).

These principles have made it essential that all research is scrutinized by a research ethics committee empowered to give or deny consent for research. The need to recognize patient autonomy and to control medical investigation is intended to protect the concerned individuals who, in the case of intensive care, may be in no position to look after their own interests.

The vast majority of patients in intensive care are unable to give consent as defined above. Even when they appear to be conscious, they may be not be competent as a result of either disease causing confusion or the effects of drugs given for sedation. Judgment as to the competency of patients in intensive care to give consent is usually made by the attending intensive care doctors. It is recommended that there be a separation of duties between the responsible ICU physician and the investigator, and that they should not be the same individual. Clearly there are situations where consent cannot be obtained, for example research into treatment of cardiac arrest where consent cannot be asked from the patient. Intensive care specialists strongly believe that there is an ethical obligation to undertake research in these situations to improve patient care and outcome. Various solutions have been proposed to the problem of obtaining consent in the emergency situation and/or intensive care.

Presumed consent or waiver of consent

Research in emergency life-threatening situations requiring immediate treatment in patients who are unable to communicate and where there is not time to look for relatives or any legal representatives can be conducted without consent under certain conditions (Biros *et al.* 1995). The United States Federal Drug Administration allows physicians to carry out research with a waiver of informed consent under certain specified conditions (US Department of Health and Human Services 1995).

1. This situation assumes that the research will benefit the patient him- or herself. In these circumstances it is obviously very important that the research must have been planned in advance and approved by a reputable independent ethics committee who should examine not only the ethics of the research but also critically scrutinize the scientific content.

2. The research should not involve significant risk.

The concept of risk in patients who are critically ill is difficult in that the risk for them is very different to the kind of risk that might be acceptable in a well person.

3. Patients or surrogates will be given full information after recovery and, if needed, proper informed consent will be sought when the patient regains consciousness.

In this kind of critical emergency situation the doctor has to act in the best interest of his or her patient according to a concept of presumed consent. There is a danger with current legislation that it may prove more difficult to carry out research on unconscious patients in intensive care if the general rules for research are overlegislated and thus preclude patients from the benefits that research can give.

Assent from a surrogate

In these situations of emergency research, although according to British law relatives cannot give consent, it is always wise to obtain their assent to the research process if at all possible. The Declaration of Helsinki admitted the possibility of assent being given by a surrogate in a case of physical or mental incapacity. The ethical basis for surrogate consent is a concept of substituted judgment where the surrogate is supposed to act as the patient would have done him- or herself. It is suggested that surrogate consent is better than no consent at all. Perhaps the most important advantage of surrogate consent is that it allows the possibility that the surrogate may refuse consent to the research on the basis of their knowledge of the patient's attitudes. There are obviously problems regarding who can act as a surrogate for patients. In some countries there is a hierarchy of relatives, but in modern life many people may regard their next of kin as somebody who is not a relative at all. Another problem about surrogates is that it may prove very difficult to find the surrogate to allow research under such emergency situations.

Another idea that has been proposed as a solution to this problem of consent of the incapacitated patient in intensive care is the idea of deferred consent (Abramson *et al.* 1986), i.e. obtaining consent to research after the procedure has already started. My own view is that this is not a very satisfactory solution because there is no element of informed consent at the start to either the patient or any surrogate or relative.

In this time of patient autonomy and increasing information patients, particularly in the United States, often have living wills, and a proviso of the living will may be a consent to research or otherwise in the intensive care situation. Perhaps a more useful concept, as practised in the United States although rarely, as yet, in Europe, is the idea of a durable power of attorney which the patient gives to a nominated representative to consent or otherwise to medical procedures if he or she is incapacitated. This would provide a very useful legal basis for consent in the emergency intensive care situation, although again it would suffer from the defect of time delay in finding the person who holds the power of attorney, as in the case of any other surrogate; in these situations time is often of the essence. The Council of Europe, through its Steering Committee on Bioethics, has stated in Article 6, Section 3: 'where according to law an adult does not have the capacity to consent to an intervention because of a mental disability or disease or for similar reasons, the intervention may only be carried out with the authorisation of his or her representative or an authority or a person or body provided by law (Council of Europe 1996). Article 8 of the same document states that when because of an emergency situation the appropriate consent cannot be obtained, any medically necessary intervention may be carried out immediately for the benefit of the health of the individual concerned. Therefore it seems that the European regulations would permit research in intensive care without consent of the patient provided that all the safeguards were maintained.

In the United States a consensus statement from the Coalition Conference of Acute Resuscitation and Critical Care Researchers noted that when the critical nature of the illness or injury, or the need to apply an investigation therapy rapidly, precludes prospective consent for participation in emergency research, patients are vulnerable not only to research risks but also to being denied potentially beneficial therapy for their life-threatening condition. They set out recommendations to resolve some of the problems surrounding research in the emergency situation (Biros *et al.* 1995).

Conclusions

There is a need for research to continue in intensive care for the benefit of patients. The problem of consent remains difficult, but clearly there is a case for carrying out research without consent provided that the proper procedures of a stringent, ethical, and scientific research ethics committee have been followed. These committees should scrutinize proposals and look critically at the scientific basis of the work in order that researchers may retain public confidence, whilst advancing the standards of treatment available in intensive care.

References

Abramson, N.S., Meisel, A., and Safar, P. (1986). Deferred consent: a new approach for resuscitation research on comatose patients. *Journal of the American Medical Association*, **225**, 2466–71.

Biros, M.H., Lewis, R.J., Olson, C.M., Runge, J.W., Cummins, R.O., and Fost, N. (1995). Informed consent in emergency research. *Journal of the American Medical Association*, **273**, 1283–7.

Council of Europe, Directorate of Legal Affairs (1996). *Draft convention for the protection of human rights and dignity of the human being with regard to the application of biology and medicine. Conventional human rights in biomedicine*. Council of Europe, Strasbourg.

Lemaire, F., Blanch, L., Cohen, S.L., and Sprung, C. (1997). ESICM statement on informed consent for research purposes in intensive care patients in Europe. *Intensive Care Medicine*, **23**, 435–9.

US Department of Health and Human Services (1995). Food and Drug Administration. Protection of human subjects: informed consent. *Federal Regulations*, **60**, 49086–103.

15.2.3 Withdrawing treatment

R. Ritz and G. Stratenwerth

Key messages

- There is no difference ethically between primary withholding and secondary withdrawal of life-supporting treatments.
- Local guidelines on withdrawal of treatment, consistent with general ethical principles, must be in written form and strictly followed.
- Any decision based upon the patient's wishes or written will must be compatible with the conscience of the person who has to implement it.
- Following the decision to withhold or withdraw a life-sustaining treatment, responsibility for continuation of the patient's care must be assumed by the whole medical team.

Introduction

The principal task of intensive care medicine is to save lives. This is frequently constrained in practice by borderline situations which may entail withholding or withdrawing life-sustaining measures. It is generally accepted that there is no difference ethically between primary withholding and secondary withdrawal of life-supporting treatments, including mechanical ventilation, hemodialysis, or vasoactive drug administration (American Thoracic Society Bioethics Task Force 1991; World Federation of Societies of Intensive and Critical Care 1992). However, European physicians sometimes have more difficulties in withdrawing life-sustaining therapy that has already been started than in withholding it initially.

'Futility'

In general, opinion as usually presented by newspapers and television, and also by patient directives, is that the physician should be instructed to 'stop all treatment when prognosis is fatal'. Nobody wishes to undertake futile acts, but how can one decide when an act is futile?

Futile acts are described as acts that are ineffective and incapable of achieving a desired result or goal: continuation of treatment could allow the patient to exist in a vegetable state, but would not have the capability of restoring sentient functions and individual personality. In the medical literature, there has been a change from the concept of futile care to the concept of reasonable care; this is an evolutionary, but dangerous, progression within medical ethics (Civetta 1996). This change may reflect an ethical change within society at large, away from the moral authority of physicians and medical paternalism. However, the question remains as to how far one should be allowed to drift from traditional medical ethics. Admittedly, medical ethics is no longer viewed as an immutable framework of absolute truth or the province of physicians alone.

The current state of generally accepted ethical opinion and the view of the general public as to how to proceed in such situations must be respected and integrated in the decision-making process.

In practice, all the physicians and nursing staff in an intensive care unit (ICU) must be capable of abandoning treatment in certain situations. A consensus shoud be established and local guidelines, consistent with general ethical principles, must be available in written form in the ICU and strictly followed when making major decisions in special cases.

Following total acceptance of the decision to stop treatment, life-support measures such as mechanical ventilation, hemodialysis, administration of vasoactive drugs, etc. should be withheld or withdrawn if their continuation only serves to prolong the process of dying for the patient. This also applies to patients with an 'apallic syndrome'.

The ability to make a rational decision about whether or not to stop treatment depends on the ability to make a correct diagnosis, to understand the implications of the diagnosis for the future course, and to communicate this information to the patient's proxies.

Self-determination

Autonomy is one of the four main principles of medical ethics, namely to respect the autonomy of the patient, to cure, not to harm, and to redistribute resources adequately. The increasing tendency to include the patient's wishes in medical decisions is well recognized. In the ICU, where it is frequently impossible to communicate with the patient, the increasing existence of written instructions is welcomed and they must have a strong influence on medical decisions.

The right of self-determination is valid in all patients (World Federation of Societies of Intensive and Critical Care Medicine 1992). In principle, the critically ill patient in the ICU is accounted competent to judge. However, restrictions due to his or her acute illness with severe pain, difficulty in breathing, or an exceptional mental state must be taken into account. If the patient is not fully competent to judge, his or her presumptive wishes are decisive (Smedira *et al.* 1990); his or her own written instructions and information from proxies or pretreating physicians should be respected. Patients' directives must be correctly formulated with the correct date and a hand-written signature. Their liability depends on the precision of the contents of the directive, how long ago it was written, and the putative possibility of foreseeing the present emergency situation at the time of writing.

In rare cases, the medical decision might deviate from that intended by the patient, particularly in the management of crisis (e.g. attempted suicide). There are no general rules for such delicate and individual situations. Any decision based upon the patient's wishes or written will must be compatible with the conscience of the person who has to perform it in practice (Asch *et al.* 1995); withdrawal of life-supporting measures must be carried out by the responsible physician.

Decision process

In addition to the medical indication and the patient's actual or presumed wishes, factors such as the patient's biological age, probable future quality of life, and general state of health, as well as the current situation and prognosis, should be taken into account when making the decision. The question of adequate redistribution of resources should not be crucial in withholding or withdrawing life-sustaining treatments in an individual case.

It is recommended that all medical personnel caring for the patient be included in the decision process; all their judgments must be respected and consensus should be sought. The opinions of proxies should also be integrated in the decision process without burdening these persons with codecision. The final decision has to be made by a physician; he or she will have to take the responsibility for the consequences.

Following the decision to withhold or withdraw a life-sustaining treatment, responsibility for continuation of patient care must be assumed by the whole medical team. After withholding or withdrawing life-support measures, comfort therapy (analgesia etc.) should be applied as required, even if such procedures shorten the patient's life. The patient and his or her family have a right to continuing and adequate human sympathy and respect.

References

American Thoracic Society Bioethics Task Force (1991). Withholding and withdrawing life-sustaining therapy. *American Review of Respiratory Disease*, **144**, 726–31.

Asch, D.A., Hansen, F.J., and Lanken, P.N. (1995). Decision to limit or continue life-sustaining treatment by critical care physicians in the United States: conflicts between physicians' practices and patients' wishes. *American Journal of Respiratory and Critical Care Medicine*, **151**, 288–92.

Civetta, J.M. (1996). Futile care or caregiver frustration? A practical approach. *Critical Care Medicine*, **24**, 346–51.

Smedira, N.G., *et al.* (1990). Withholding and withdrawal of life support from the critically ill. *New England Journal of Medicine*, **322**, 309–15.

World Federation of Societies of Intensive and Critical Care Medicine (1992). Ethical guidelines. *Intensive and Critical Care Digest*, **11**, 40–1.

15.2.4 Law enforcement issues

C. P. H. Heneghan

Key messages

- A deliberate act to kill, intending to kill, is murder.
- Euthanasia is not a defense.
- Withdrawing treatment because it is inappropriate or because consent has been rescinded, resulting in death, is usually not murder, nor is administering dangerous doses of analgesics if necessary to relieve pain.

Introduction

Law enforcement issues in critical care can be divided into issues of life and death and issues of confidentiality. Other criminal issues seldom arise, as doctors, nurses, and other involved professionals set themselves standards of behavior that are more restrictive than the criminal law. These ethical standards are outside the remit of this topic.

The law discussed is the Law of England (and Wales) unless otherwise stated. The law may differ in other jurisdictions. This chapter should not be seen as legal advice. It is intended as an overview of areas of the law which may be of special interest to the intensivist.

Life and death

Murder

What if a patient is or appears to be suffering unbearably? Can we legally (as distinct from morally, where most religions are clear that any taking of human life is wrong) put them out of their misery, say by administering intravenous potassium chloride?

This has recently been addressed in *R.* v. *Cox* [1993]. Dr Cox is a consultant rheumatologist who had an elderly patient immobilized by rheumatoid arthritis, who was in excruciating pain even if lightly touched. No pain killer had useful effect, even in usually lethal doses. He gave her a large dose of potassium chloride intravenously: shortly after, she died. He was open about his actions, which he recorded in the notes. Murder, loosely, is when you kill someone when intending to kill him or her (or intending to do grievous bodily harm). Cox was charged with murder, which was reduced at trial to attempted murder, as it was impossible to show that his injection had actually caused her death, since she was so ill at the time. Dr Cox's intention, freely admitted, was to kill in order to reduce suffering. He was convicted of attempted murder and received a suspended sentence of a year's imprisonment (he was only admonished by the General Medical Council (GMC)).

Intent is a matter for evidence and inference, as are all facts before a court. Those tempted to kill a suffering patient and state a different intention could find their evidence rejected and be convicted of murder, for which the mandatory sentence in England is life imprisonment.

Euthanasia

What if a patient actually asks us to kill them? Is consent, or euthanasia, a defense?

Euthanasia is murder or assisting suicide. The distinction is whether the actions are solely those of someone other than the deceased. Thus the wish of the deceased to die (consent) is no defense to a charge of murder. Although suicide and attempted suicide are no longer criminal offenses, assisting suicide is. Even to explain to someone how to kill themselves might be assisting suicide.

It may be argued that no-one should be prevented from taking his or her own life; therefore he or she should be provided with the information and materials necessary. Counter-arguments include the following.

1. The desire for suicide may be transient, and if postponed by events might not be carried through.
2. If assisting suicide or euthanasia were a defense to murder, it would often be difficult to disprove, with serious consequences.
3. Euthanasia may be abused. Even if restricted to being carried out apparently in good faith by registered medical practitioners, rules may be bent to allow 'euthanasia' of those with conditions like Alzheimer's disease. This has happened with 'therapeutic' abortion, which is only legal in rigidly defined circumstances but is now effectively performed on demand.

The situation in The Netherlands is also interesting. Euthanasia is allowed, also in strict circumstances: the request for euthanasia must be free and voluntary; it must be well considered, durable, and persistent; the patient must be experiencing intolerable (not necessarily physical) suffering; the doctor must consult with an independent medical colleague who has experience in the field; the death must be reported as from euthanasia. However, in 22 per cent of cases there had been only one request, in 30 per cent the interval between requests was 1 h to 1 week, and in 66 per cent the request was purely oral. Although in 56 per cent of cases intolerable suffering was cited as the reason for euthanasia, only 42 per cent of patients had mentioned it, and only 18 per cent as the most important reason, while 24 per cent quoted fear of mental deterioration. Consultation frequently did not occur, and euthanasia was usually not reported; deaths were certified as natural. Possibly, involuntary 'euthanasia' was occurring (Goff 1995). Thus trends are worrying, even in a most 'humane' society.

These are only arguments; in the United Kingdom (as in almost the whole world) euthanasia remains criminal.

Unassisted suicide

What if a patient wants to die, and asks us to turn off the ventilator? Can we do this legally, even if we believe that it will cause his or her death, and life expectancy was otherwise normal?

There has not, as far as the author knows, been such a case in United Kingdom, although there was an analogous one. An insulin-dependent diabetic paraplegic wished to die, and withdrew consent for continued insulin. He died, and no criminal charges were brought. An individual of sound mind may refuse treatment for any or no reason. This applied to the diabetic, is regularly applied to Jehovah's Witnesses who refuse blood transfusion, and would probably apply to withdrawing ventilation. (This conclusion should not be taken to be applicable to children or those of unsound mind. Powers under the Mental Health Act may require continuation of physical treatment if the request to discontinue is a result of mental illness.)

Manslaughter

Manslaughter is relevant to critical care in relation to gross negligence: if someone is killed as a result of really serious negligence, this may result in a charge of manslaughter. This has yet to happen in critical care, although there have been two recent cases of doctors charged with manslaughter, one an anesthetist's failure and the other a serious intrathecal drug error. It should be noted that a jury will decide whether there was gross negligence, advised by the judge and perhaps by medical experts.

Easing the passing

'Easing the passing' was Dr Bodkin Adams' phrase to explain giving morphine to the dying; he was acquitted of murder. What if we give large doses of analgesics to relieve suffering, knowing that it may hasten death? Our intention is not to kill. Is this criminal?

The law normally includes within the definition of 'intention' consequences which are inevitable. If you get up at night to open a window, knowing that this will inevitably wake the baby, in law you intended to wake the baby. Therefore if you give painkillers in doses which could be expected to shorten life, although your primary intention was to relieve suffering, normally your intention would be deemed to include shortening life. However, in the Bodkin Adams case, a general practitioner who administered morphine to sick elderly patients was charged with their murder, and Devlin J directed the jury that when he cannot give a patient back their health, a doctor 'is still entitled to do all that is proper and necessary to relieve pain and suffering even if the measure he takes will incidentally shorten life' (Williams 1983). Therefore this would seem to be legal.

Discontinuing treatment—victims of violence

Victims of assault often come to the intensive care unit. It may happen that further treatment becomes futile and should be discontinued. Would there be any effect on the charge brought against the assailant? Might it prevent a charge of murder, if the victim appeared not to have died directly as a result of the assault?

In one such case (*R.* v. *Malcherek* [1981]), where a ventilator was switched off, it was argued that the original assault was not murder because the victim would still be alive if the doctors had not switched off the ventilator. The judge rejected this argument and was upheld in the Court of Appeal. A similar outcome resulted when a victim of assault who was a Jehovah's Witness refused transfusion and died.

Brain death

Brain death is not a legal concept. The law has not defined death, and has accepted whatever diagnostic method doctors used. Brain death is a medical invention which has been accepted at least once by the courts (*R.* v. *Malcherek* [1981]). No doctor has yet been charged with murder for removing a beating heart from a brain-dead patient in the United Kingdom, and it is not likely to happen.

Coroners and brain death

Some coroners (in the United Kingdom and Australia the coroner is the investigating officer, not the coroner's pathologist, who performs the autopsy) have claimed jurisdiction over brain-dead patients, and asserted a right to refuse organ donation. This is based on the idea that brain death is death, and their jurisdiction commences at time of death. This is probably incorrect, as their jurisdiction commences when the death is reported to them, which may be some time later. Reporting need only occur in certain cases, principally those of violent or unnatural death, or where cause is unknown.

The coroner's duty is to state a cause of death, not as a medical diagnosis, but for example accident or natural causes. Removal of organs for transplant cannot impede that function. If a patient is properly declared brain dead, the cause of death has to be known. If organs are removed for transplantation, they are, by definition, normal.

Although it may be politically advisable to inform a coroner before removal of organs, this cannot be a legal requirement.

Confidentiality

Confidentiality in critical care is unusual. The basic rule is to respect and protect confidential information unless the patient gives permission to disclose it, and the main exceptions, protecting others and preventing crime, do not generally apply in critical care (but see below for AIDS and notifiable diseases). However, in critical care, it is often impossible to disclose any medical information to the patient because he or she is unconscious or sedated. What is the legal position about disclosure to relatives?

GMC guidelines (GMC 1995) allow, in terminal illness, disclosure without consent to a relative when it is judged that disclosure to the patient would be damaging, but disclosure to someone would be in his or her best interest. This does not apply to most critical care patients, and there are no guidelines or cases. Standard practice is to disclose full information, including diagnosis, treatment, prognosis, etc., to relatives, because someone must know what is going on and the obvious candidates are the relatives. The courts and the GMC would probably be guided by this.

But what if there appears to be a dispute between family members? Can one restrict the information? There is no case law, but it would probably be acceptable to restrict information to next of kin who can decide what to do with it. What of an unmarried partner? They seem to be treated as next of kin, which is not unreasonable since we often do not know the true status of claimed next of kin. What if the diagnosis is AIDS? Here the GMC states that a doctor has a duty to see that a sexual partner of an AIDS patient is informed, even if disclosure is refused; this must cover where consent is unavailable. Finally, what of notifiable diseases? Here there is a legal requirement to inform the relevant authority which overrides the duty to the patient.

References

GMC (1995). *Duties of a doctor: confidentiality*, para. 12. General Medical Council, London.

Goff, R. (1995). A matter of life and death. *Medical Law Review*, 3, 1–21.

*R.*v. *Cox*, Winchester Crown Court, Ognall J, 18th September 1992, discussed in *Medical Law Review*, 1, 232–4 (1993).

R. v. *Malcherek* [1981] 1 WLR 690.

Williams, G. (1983). *Textbook of criminal law* (2nd edn), p. 385. Stevens, London.

15.2.5 The role of the coroner in the United Kingdom

M. A. Branthwaite

Key messages

- The coroner has a duty to enquire into violent or unnatural death, sudden death of unknown cause, and death occurring while the deceased was in prison.
- The coroner usually sits alone but must summon a jury in some circumstances.
- The inquest is an enquiry, not an adversarial contest, and the coroner is prohibited from making findings of criminal or civil liability against any person.

- ‘Unnatural’ deaths include those which occur during operation or before recovery from anesthetic as well as events at variance with the expected.
- A finding of ‘lack of care’ does not signify civil negligence but refers to neglect.

Legal background

Coroners in England and Wales are judicial officers with at least 5 years professional experience as medical practitioners, barristers, or solicitors. They are appointed by local authorities but thereafter can only be removed from office by the Lord Chancellor. Their role and responsibilities are defined by Statute, supplemented by the body of decided cases which constitutes common law. Their primary remit is to enquire how, when, and where a person has died when there is reasonable cause to suspect:

(1) the deceased has died a violent or unnatural death;

(2) has died a sudden death of which the cause is unknown;

(3) has died in prison or in such a place or in such circumstances as to require an inquisition under any other Act.

The Coroners Act 1988 confers statutory power which can be invoked to compel a coroner to hold an inquest or, in some circumstances, to quash the findings of an inquest and order a rehearing. The proceedings of a coroner's court are also open to judicial review, a process by which procedural irregularities can be explored and rectified.

Unlike a criminal prosecution or a civil suit for negligence, the coroner's inquest is an enquiry, not an adversarial contest. Indeed, there is a specific requirement to avoid any finding which might appear to determine any question of civil or criminal liability. No witness can be compelled to answer any question tending to incriminate himself, but any person whose conduct is likely, in the opinion of the coroner, to be called into question is either summoned as a witness or must be notified of the date, time, and place of the hearing. Furthermore, the coroner is required to adjourn an inquest if requested to do so by the Chief Officer of Police or by the Director of Public Prosecutions on the grounds that a person may be charged with murder, manslaughter, or infanticide, of causing death by dangerous driving, or of aiding, abetting, counselling, or procuring the suicide of the deceased. If evidence given during an inquest suggests that such a charge might be made, the coroner is required to adjourn the hearing and refer the papers to the Director of Public Prosecutions.

The coroner usually sits alone, assisted by a coroner's officer who is often a former police officer. The coroner may summon a jury for any inquest, but must do so in the following circumstances:

(1) any death occurring in prison, even if entirely natural;

(2) if the death occurred while the deceased was in police custody or resulted from an injury inflicted by a police officer in the purported execution of his duty;

(3) where an inquest is to be held in any event and the death was caused by an accident, poisoning, or disease which requires notification;

(4) if the death occurred in circumstances which, if continued or recurring, would be prejudicial to the health and safety of the public.

Unlike the 12-person jury of the criminal courts, a coroner's jury consists of not less than seven and not more than 11 jurors. Proceedings are usually slower and more formal in the presence of a jury; it is they who will make findings of fact, but only after the coroner has summarized the evidence and directed them upon the law.

There was a time when the coroner or the jury were empowered to add a rider to the verdict making recommendations designed to prevent recurrence of similar fatalities. That power no longer exists, but the coroner is still entitled to report such matters to those in a position to take appropriate action.

The distinction between certifying death and the cause of death

A medical practitioner attending the deceased ‘during his last illness’ is obliged to complete a certificate which states, to the best of his or her knowledge and belief, the cause of death. This certificate is then presented, usually by the next of kin, to the Registrar of Births, Marriages, and Deaths who, if satisfied that it is proper to do so, will issue a Death Certificate; thereafter funeral arrangements can be made. However, the registrar has a duty to report death to the coroner in a number of circumstances which include death from unknown cause, death which the registrar has reason to believe to have been unnatural (including substance abuse, industrial disease, or exposure to toxic substances) or caused by violence, neglect, or abortion, or death which appears to have occurred during an operation or before recovery from the effect of anesthetic. Thus it is not uncommon for death during intensive care to fall within the jurisdiction of the coroner.

What constitutes unnatural death?

The answer to this vexed question will determine whether or not, once informed of the death, the coroner will deem is necessary to hold an inquest. Often, if the chances of death would have been high in any event, the coroner will conclude that the death was ‘natural’. Alternatively, he may reach such a conclusion after considering the results of a postmortem examination which he can order. However, there may be additional circumstances which have the potential to change a natural death into an ‘unnatural’ death, and then an inquest must be held. ‘Unnatural’ is defined in the *Oxford English Dictionary* as ‘at variance with what is natural, usual, or to be expected’. A significant proportion of patients are admitted to intensive care after some unexpected event. An inquest will be necessary if death ensues, whether the event was iatrogenic or an accident occurring despite exemplary care throughout. For example, the development of septicemia and fatal respiratory insufficiency after aspiration of vomit following uneventful gynecological surgery was ‘unnatural’. It has also been held that death from an extremely rare disease should be considered ‘unnatural’, the somewhat surprising examples given being legionnaires' disease and typhoid fever (*R.*v. *HM Coroner for Inner North London* ex parte *Thomas* [1992]). It is important to recognize the distinction between certainty as to cause of death but uncertainty as to whether or not it was ‘natural’. If in doubt, it is wise to inform the coroner, usually possible to explain this further difficulty to the bereaved family without exacerbating their distress, and better for such a requirement to be clear from the outset rather than be met unexpectedly when the registrar declines to issue a death certificate. Once the coroner has been notified, he or she is likely to open an inquest but then adjourn while evidence is gathered and preparations are made for the hearing. In all but exceptional circumstances, the coroner will authorize burial or cremation before the inquest is reopened.

Nature of the inquest

The enquiry is directed exclusively to establishing who the deceased was, and how, when, and where he or she came by his or her death. 'How' is to be interpreted as 'by what means' not 'in what circumstances'. The coroner has power to disallow questions which, in his or her opinion, are not relevant to the enquiry, but defining the boundaries of relevance when standards of care may be deficient is far from easy. It is usual for the coroner to ask questions of each witness followed by 'interested parties' or their legal representatives, often counsel or solicitor acting for the bereaved family. At the end of the hearing, the coroner makes a formal record, the 'inquisition'. It includes the verdict which, strictly, is the entirety of the written report, although the term is often used to signify only the conclusion of the coroner or jury as to the death. A list of suggested conclusions is available which includes natural causes, accident or misadventure, and 'open verdict'; the last of these indicates that the evidence is still inconclusive. The consensus of opinion today is that accident and misadventure should be regarded as synonymous, and it is common for such a term to be chosen when an unexpected complication (iatrogenic or otherwise) occurs during medical care. It has no significance as to civil liability.

'Lack of care'

This distasteful term has generated considerable dispute as to its meaning and relevance. It has nothing to do with the concept of civil negligence and should only be used in the narrow sense to mean lack of physical attention, in other words neglect (*R.* v. *HM Coroner for North Humberside* ex parte *Jamieson* [1994]). The Court of Appeal defined neglect as follows:

> a gross failure to provide adequate nourishment or liquid, or provide or procure basic medical attention or shelter or warmth for someone in a dependent position... who cannot provide it for himself. Failure to provide medical attention for a dependent person whose physical condition is such as to show that he obviously needs it may amount to neglect. So it may be if it is the dependent person's mental condition which obviously calls for medical attention...
>
> ...neglect can rarely, if ever, be an appropriate verdict on its own.

Despite the apparent relevance of this definition to circumstances where intensive care is withheld or withdrawn, medical practitioners need not fear unduly because the circumstances which prompt 'withhold or withdraw' decisions are those where death from natural causes is considered a virtual certainty in any event. The law seeks to establish not only the cause of death in fact, i.e. which system failed, but also causation in law, i.e. the initiating event in a chain of sequential events which culminates in death. Similarly with the advance directive—this is not to be regarded as a variant of suicide or voluntary assisted euthanasia, but as a natural consequence of a valid autonomous decision.

An overlap with unlawful killing?

A coroner's finding of 'unlawful killing' covers all cases of homicide but requires proof to the criminal standard of 'beyond reasonable doubt'. Concerns that decisions to withhold or withdraw intensive care could lead to such a finding are misplaced in the majority of cases. Whatever the factual cause of death, causation in law is likely to be the initiating illness which created the need for intensive care. Special considerations apply to patients who require support of basic biological functions such as nutrition and hydration but whose condition is otherwise stable, a question usually discussed in the context of persistent vegetative state (*Airedale NHS Trust* v. *Bland* [1993]). Here it is more difficult to distinguish causation in fact and in law, and the current position is that application be made to the courts for a ruling that the withdrawal of treatment in an individual case would not be unlawful (House of Lords 1994). It is important to note that the coroner is not empowered to make such a ruling, nor can criticism or even the potential for prosecution be avoided by informing the coroner in advance of an intention to withdraw supportive management from a patient in this condition. It is to be hoped that the current requirement for judicial approval will soon be replaced by accepted guidelines, analogous to the criteria for brainstem death.

References

Airedale NHS Trust v. *Bland* [1993] Appeal Cases 835–899.

House of Lords (1994). *Report of Select Committee on Medical Ethics*. HMSO, London.

R. v. *HM Coroner for Inner North London* ex parte *Thomas* [1993] 2 WLR 547–55.

R. v. *HM Coroner for North Humberside* ex parte *Jamieson* [1994] 3 WLR 82–102.

15.2.6 Key differences between the role of the coroner in the United States and the United Kingdom

M. Graham

Key messages

- Medicolegal death investigations in the United States are carried out by elected or appointed officials possessing various degrees of expertise.
- The major responsibility of the medicolegal official is to certify the cause and manner of death of those persons falling under his or her jurisdiction.
- Only the medicolegal official can certify a death in which the cause or circumstances of death are not entirely natural.

Although medicolegal death investigation in the United States can trace its origins, for the most part, to the British coroner system, there are currently many significant differences in the structures of the systems, the qualifications of the personnel, and their responsibilities.

There is not a uniform system for death investigation in the United States. Some jurisdictions use the coroner system, whereas others use the medical examiner system. Coroners are elected officials whose requirements for office vary by jurisdiction. Some jurisdictions have only age, residency, and non-felon requirements, whereas others specify that the coroner must be a physician. Medical examiners are appointed officials who are required to be physicians. In some jurisdictions further requirements may specify that the medical examiner be a pathologist and, in some places, that he or she must be a forensic pathologist.

The role of the medicolegal official in the United States, whether the coroner or the medical examiner, is primarily to determine and certify the cause and manner of death of the decedents falling under medicolegal jurisdiction as specified by statute and regulation. In general, these deaths include those which are sudden, unexpected, or unattended, occur while in legal custody, are work related, affect the general welfare of the community, or involve trauma (including chemical trauma such as drug overdoses). The medicolegal verdict may be the outcome of an inquest (most commonly in the coroner system) or may be the opinion of the medicolegal official charged with investigating the death (most commonly in the medical examiner system). For death certification purposes, the cause and manner of death are based on the preponderance of the evidence. If significant doubt exists about their true nature, the cause and/or the manner of death may be designated as 'undetermined'. In the United States, only the mediocolegal official is permitted to certify the cause and manner of a death for a case in which jurisdiction has been accepted. The medicolegal investigation of deaths under the jurisdiction of the medicolegal office is not subject to cessation by the police department or the prosecuting attorney. Similar to the British coroner system, it is not the role of the medicolegal official in the United States to assess criminal culpability or civil liability. The cause of death is defined as the disease or injury that starts the lethal train of events resulting in death. The determination of the cause of death is based on medical (antemortem and postmortem) and non-medical (investigative) information.

The medicolegal official is also responsible for certifying the manner of death, which is a determination of how death came about—natural or otherwise. Non-natural deaths include those occurring by accident, homicide, and suicide. Some jurisdictions have an 'unclassified' category which is often used for deaths related to substance abuse or therapeutic complications. If the manner of death cannot reasonably be determined, it is certified as 'undetermined'. For death certification purposes, homicide is the killing of one person through a volitional act of another person. A manner of death of homicide does not imply intent to kill, infer legal culpability, or require a higher standard of proof than other manners of death. If any aspect of a death (cause or manner) is unnatural, the death must be certified by a medicolegal official. Rarity of a disease process does not equate with it being unnatural. A natural cause of death may arise in an unnatural manner (e.g. stress-induced dysrhythmia precipitated by armed robbery) and thus fall under the jurisdiction of the medicolegal official. In contrast, what may appear to be an unnatural component of a death may by custom not remove the death from the 'natural' category. This applies to certification of deaths where therapy may have caused or contributed to death. Modern medical diagnostic and therapeutic techniques have become progressively more aggressive and invasive. Predictably, even in the best of hands some patients will die as a result of medical intervention. Since some of these iatrogenic complications are 'injuries', particularly those related to mechanical procedures, some medicolegal officials certify deaths due to these types of injuries as accidents, whereas others consider reasonably foreseeable complications as an extension of the disease process and thus certify such deaths as natural. Unforeseeable mechanical complications (e.g. explosion of a respirator) lead to unnatural manners of death. Such determinations do not reflect an opinion regarding liability or deviation/adherence to the standard of care.

The foregoing discussion shows that, although there are similarities between death investigation practices in the United States and the United Kingdom, significant differences exist. The clinical practitioner must be familiar with the medicolegal practices in the location in which he or she practices to fulfill the legal requirements. As a general rule, if there is doubt as to whether or not the medicolegal official has jurisdiction in a particular death, the death should be reported to the medicolegal official for an official ruling.

15.2.7 Clinical liability

C. P. H. Heneghan

Key messages

- Clinical liability is liability for acts and omissions in diagnosis and treating of patients.
- Each adult individual of sound mind is liable for his or her own actions. We may also be liable for the acts or omissions of others.
- A duty of care is a duty not to be negligent, and it arises when a doctor–patient relationship arises.
- It is necessary for a plaintiff to prove that alleged negligence has caused (or materially contributed to) alleged damage.

Introduction

Clinical liability is liability for acts and omissions in diagnosis and treating of patients. Who is liable in law, and for what?

The law discussed is the Law of England and Wales, unless otherwise stated. Negligence law in Scotland and Northern Ireland is identical to that in England. The law may differ in other jurisdictions. This chapter should not be seen as legal advice. It is intended as an overview of areas of the law which may be of special interest to the intensive care specialist.

Person

Each adult individual of sound mind is liable for his or her own actions. We may also be liable for the acts or omissions of others. If we employ another, we may be directly or vicariously liable. We may be directly liable if we employ someone who is unqualified or unsuitable for the post, and we knew or ought to have known this, for example by checking qualifications or references. We may be vicariously liable for an employee's acts or omissions in the course of his or her employment. This includes a surprising range of actions, as well as negligence if it occurs in the course of one's employment. The range may include actions specifically forbidden by the contract of employment and by company regulations. This judge-made law is probably policy based; if employers are liable, they must insure against the risk, and this ensures compensation for victims of such negligence.

What of trainees? Is the senior liable for the actions of the junior? Yes, if the junior is actually employed by the senior, who is vicariously liable. Not necessarily, if both are employed by another, such as a hospital, health authority, or trust. This will turn on the facts of the case. If a junior acts on the direct instructions of the senior, or appropriately follows policy guidelines set up by the senior, the senior may then be liable. Otherwise the junior may be personally liable (while the employer will still be vicariously liable).

Vicarious liability of health employers in the United Kingdom has recently been revived under the National Health Service indemnity scheme. This is merely the result of abandoning the agreement that doctors would insure themselves against clinical liability, and that employers would not have to be liable for them. The principal reason for abandonment of this arrangement was that the cost of such insurance became very high, resultant on increased litigation, and this cost was reimbursed by employers; it became cheaper for the employer to be liable. (At least one United Kingdom health authority has tried to exclude liability for negligence in a contract of employment. For many reasons, this attempt is unlikely to succeed.)

In the private sector in the United Kingdom, the doctor is not an employee but is a subcontractor or is independent. This would also apply in many common law jurisdictions, such as Australia, New Zealand, Canada, and most of the United States. The doctor will be liable for his or her own actions and must insure against negligence accordingly. The hospital is not vicariously liable, although it may have a direct responsibility to insure, through accreditation procedures, that only properly qualified practitioners are allowed to practice there.

Duty of care

To whom, and in what circumstances, does a duty of care exist? A duty of care is a duty not to be negligent, and it arises when a doctor–patient relationship arises. This may be by referral by the patient personally or from another doctor, and acceptance of that referral, or by the doctor's being on duty for all patients in certain categories, into which the patient falls, or by the doctor's voluntarily assuming responsibility, for example at a road traffic accident. There is usually no doubt that there is a duty of care, particularly in a critical care unit.

Standard of care

General Medical Council

Contracts of employment will usually not set standards, but contain only descriptive terms such as being employed to undertake diagnosis, treatment, and prevention of disease. The General Medical Council (GMC) is the regulatory body for doctors in the United Kingdom, but until recently it has only set standards of behavior rather than of care. Recently, under the Medical (Professional Performance) Act 1995, the GMC's responsibilities have been extended, requiring the removal of a doctor's name from the Register for a 'seriously deficient standard of professional performance', not defined further. This power will probably usually be applied to a consistent pattern of deficient performance, although it is not thus limited. It remains to be seen how case law will develop, how it will interact with existing law, and whether other regulatory bodies will be given such powers.

Bolam

In existing law, the standard of care has been set by the courts, advised by experts in the relevant field; if a practitioner fails to achieve the required standard in his or her care of the patient, he or she will be found to be negligent. The law was set out by McNair J in *Bolam* v. *Friern Hospital Management Committee* [1957]:

> ...where you get a situation which involves the use of some special skill or competence, then the test as to whether there has been negligence...is the test of the ordinary skilled man exercising and professing to have that special skill. A man need not possess the highest expert skill: it is well established law that it is sufficient if he exercises the ordinary skill of the ordinary competent man exercising that particular art...he is not guilty of negligence if he has acted in accordance with a practice accepted as proper by a responsible body of medical men skilled in that particular art.

Most common law jurisdictions accept almost identical tests (e.g. Ireland, Australia, New Zealand, and South Africa). The United States standard of care has been stated thus: 'A physician has a legal duty to exercise that degree of knowledge, care, and skill that is expected of a reasonably competent practitioner in the same class in which he belongs, acting in the same or similar circumstances' (ATLS 1993). Whilst this is almost identical to *Bolam*, it may be that the phrase 'in the same class in which he belongs' could allow differentiation between trainees and the trained (see below).

Thus it is not required that the standard of care be at the highest, but merely ordinarily competent, and it is a defense that the doctor's actions would be acceptable to a responsible body of practitioners, even if a minority (referred to as the responsible minority defense). Whether the care was actually negligent may be decided by a court, advised by experts. Many experts may advise, differently, and the lawyers or the judge (or jury in some jurisdictions outside the United Kingdom) may have to decide between the expert evidence. (It would be logical to conclude that if a recognized expert supports a course of action, then so does a responsible body of opinion; this is an oversimplification.)

Trainees

Is the standard of care lower for trainees than for senior doctors? Is it an excuse that you have not seen a condition before, and cannot be expected to get it right? No case in medical law has yet addressed this (though see end of last paragraph), although in one case (failing to detect that an umbilical catheter was in the vein rather than an artery as it did not, on radiography, pass down to the groin before going up to the heart) a pediatric registrar was held at fault when his senior house officer was not, which suggests a different standard for different grades. However, a safer principle to adopt is that there is one standard of care in any circumstance, because any other suggestion will be met with the question: 'If you did not know what you were doing, why did you not call someone who did?' The only proper answer, if true, can be that there was not time.

Responsible body of opinion: definitions

What is a responsible body of practitioners? This question can be broken down into two: What is a body of practitioners professing a particular art? What makes it responsible? The first question could be important to critical care specialists, who may not have their own college distinct from anesthetists, physicians, and surgeons. This point has recently been addressed with reference to spinal surgery (*De Freitas* v. *O'Brien & Connolly* [1995]). In defense to a course of action condemned as negligent by orthopedic surgeons, it was argued that there was a subspeciality of spinal surgeons who supported the criticized action. This subspeciality had 11 members in the United Kingdom, of whom one was the defendant and two others were expert witnesses on his behalf. It was decided that size alone was not a determining factor in recognizing a distinct body of practitioners and that spinal surgeons are such a body, and the responsible minority defense was accepted.

The second question (what makes a minority responsible?) has also been addressed recently (*Hepworth* v. *Kerr* [1995]). An anesthetist was accustomed, over many years and in over 1500 patients, to dropping the systolic blood pressure to 35 to 40 mmHg for up to 2 h during middle-ear surgery. A patient developed anterior spinal artery syndrome, and sued. The practitioner was undoubtedly in a minority, perhaps of one. In deciding whether the body of practitioners was responsible and whether the technique was negligent, the judge remarked that there was no published academic support for dropping the blood pressure so low, with the whole weight of publication being against. Although the practitioner himself had published a series of cases on this technique, this was flawed, and there was no scientific validation of what was admittedly experimental, stepping right outside conventional wisdom. Without scientific validation, the use of this experimental technique was not responsible and therefore was negligent. Thus scientific validation is at least one way that a body of opinion achieves responsibility.

After *Bolam*

Lawyers are unhappy with *Bolam* because it appears to allow doctors, and not the courts, to decide the standard of care. This has been modified with regard to consent. Does *Bolam* still apply to treatment? Yes, but also with modification.

In *Joyce* v. *Merton* [1996], a case involving brachial artery catheterization for coronary angiography, postoperative instructions to the patient were found to be negligent although they were standard practice for the time. The Court of Appeal approved the following reformulation of the law:

> In the field of diagnosis and treatment, a defendant is not guilty of negligence if his acts or [decisions not to act] were in accordance with accepted clinical practice, *provided that clinical practice stood up to analysis and was not unreasonable in the light of the state of medical knowledge at the time*. (emphasis added)

The first part of this is *Bolam*; the italicized part is a departure from it. Under *Bolam*, following (even unreasonable) standard practice was a defense. This is no longer so. Although superficially acceptable, as all would expect their practice to be reasonable, there is cause for concern. A (usually medically unqualified) judge, rather than a doctor, may now decide whether medical practice is unreasonable and therefore negligent. It is difficult to predict what effect this will have, but clearly the fullest explanation of the basis for medical actions may be necessary. It remains to be seen whether the House of Lords (the highest appeal court in the United Kingdom) will support this approach, although similar departures from *Bolam* have been adopted in other jurisdictions.

Protocols and guidelines

Management of common conditions may be improved by the application of written protocols or guidelines. Would breach of a protocol (particularly a national consensus protocol) be viewed by the courts as negligent? It may be strongly evidence based, such as the use of thrombolysis in myocardial infarction, less so, like protocols for management of acute severe asthma, or hardly at all, like the recommendation not to administer halothane anesthetics within 3 months of each other. What is the legal status of such guidelines?

Following such guidelines should normally provide a defense to an allegation of negligence, based on the responsible minority defense, subject to *Joyce* above. Failure to follow protocols cannot *per se* be conclusive of negligence, as there may be differing responsible bodies of opinion. However, departure from guidelines could warrant further investigation, and will probably be the subject of cross-examination; the practitioner should expect to need to justify such a departure, and have his or her explanation prepared.

Damage and causation

The tort (a fault which is actionable at civil law) of negligence does not consist only of the negligent act or omission; there must also be damage resulting from the negligence. Mistakes occur every day, but do not damage patients because they are identified and corrected. Therefore legal liability does not arise. Thus it is necessary for the plaintiff to prove that the alleged negligence has caused (or materially contributed to) the alleged damage. Most cases fail at this fence.

References

ATLS (1993). *ATLS student manual* (5th edn). American College of Surgeons, Chicago, IL.

Bolam v. *Friern Hospital Management Committee* [1957] 2 All ER 118.

De Freitas v. *O'Brien and Connolly* (CA) [1995] Med LR 108.

Hepworth v. *Kerr* (QBD) [1995] Med LR 139.

Joyce v. *Merton, Sutton and Wandsworth Health Authority* (CA) [1996] Med LR 1.

15.3 Scoring systems in intensive care

15.3.1 The role and limitations of scoring systems

K. Rowan

Key messages

- Outcomes can be measured in terms of mortality, morbidity, disability, functional health status, and quality of life.
- The outcome of intensive care depends on the inputs available (e.g. equipment, staff), the processes of care (e.g. type of care, skill and timing with which care is administered), and also on the case mix of the patients treated.
- The proposed role of scoring systems has been based on the quantification of case mix and the development of mathematical equations to estimate probabilities of outcome for intensive care patients.
- The roles of scoring systems in intensive care can be divided into three main areas: comparative audit, evaluative research, and the clinical management of individual patients.
- The limitations of scoring systems can be categorized into four main areas: limitations in their application, limitations of the data, limitations of their accuracy, and finally limitations arising from interpretation of their results.

The context for the development of scoring systems

Scoring systems have been developed in response to an increasing emphasis on the evaluation and monitoring of health services, including intensive care.

The desire for high-quality intensive care

The goal of all those involved in the financing, management, and delivery of intensive care is to provide the highest-quality care with the resources available in order to achieve the best outcomes for patients.

There are three aspects to quality: equity, humanity, and clinical effectiveness. In addition, purchasers and providers of health services seek to achieve these goals in as cost effective a way as possible. Equity can be defined in several ways, but most commonly it addresses whether there is equal use for equal need. Humanity addresses the acceptability of care to patients and relatives, and clinical effectiveness addresses the effect of care on the health of patients. Therefore a quality intensive care service would provide equitable, humane, and effective care as cost effectively as possible.

The need to measure the clinical effectiveness of intensive care

While all three aspects of the quality of care are important, it is clear that issues of equity and humanity are less relevant if the care being provided is either ineffective or harmful. Therefore the priority for any type of care, including intensive care, is to evaluate and audit its clinical effectiveness.

Intensive care has developed over the past 50 years with very little rigorous scientific evidence as to what is, or is not, clinically effective. Despite the enormous expenditure of resources on a relatively small number of intensive care patients, those delivering intensive care often have to decide which patients can benefit most in the absence of knowledge derived from rigorous scientific studies regarding the likelihood of successful treatment.

Whereas evaluation of the clinical effectiveness of intensive care through research requires the measurement of outcomes, audit of intensive care can be achieved by measuring either processes or outcomes. However, process audit should only be employed when the clinical effectiveness of the intervention being audited has already been demonstrated through rigorous evaluative research. Process audit can then be useful in assessing the appropriate use of clinically effective interventions in intensive care. Process audit cannot assess the impact of the overall package (combination of interventions) of intensive care. Outcome audit is essential to do this.

Outcomes in intensive care

Outcomes can be measured in terms of mortality, morbidity, disability, functional health status, and quality of life. In many areas of health care, choice of the outcome measure presents considerable methodological and practical difficulties. Deaths are often too rare and therefore too insensitive a measure, and health status is often too complex for routine measurement.

Given the relatively high mortality experienced by intensive care patients, deaths do offer a sensitive, appropriate, and therefore meaningful measure of outcome. However, other outcomes such as morbidity, disability, functional health status, and quality of life should not be ignored.

Evaluating outcomes in intensive care

There is no dispute that, because of their high internal validity, rigorously conducted experiments or randomized controlled trials of sufficient statistical power are the gold-standard research design for detecting the important beneficial or harmful effects of existing or new interventions. In such experiments, individuals are assigned prospectively to one of two (or more) alternative interventions using a random method of allocation.

With rigorous random allocation, the investigator can assume that the two (or more) groups studied have similar distributions of both the known and unknown confounding factors, i.e. the characteristics that could affect both their selection and risk for a particular outcome. Therefore interpretation of the results of randomized controlled trials can assume that the groups being compared were similar in all respects apart from the intervention(s) under study.

In practice, however, there are a number of situations where random allocation is not possible for either ethical or logistical reasons (Black 1996). For example, it is deemed unethical to randomly allocate severely ill patients to receive intensive care or general ward care. Although it might be possible to obtain ethical approval to randomly allocate less severely ill patients, it is likely that neither clinicians nor patients and their relatives would find this acceptable. Low recruitment to such a trial would jeopardize the ability to generalize from the results.

There are also questions about the way that services are organized that preclude evaluation by an experimental method. For example, it is difficult to conceive how the supposed benefits of a regionalized service could be compared with a non-regionalized service within a randomized study design.

In certain situations, even specific medical interventions may not be amenable to a randomized controlled trial. This is because, even though clinicians may vary in how they treat similar patients, each clinician may be quite certain of the correctness of his or her own practice. In other words, although collective equipoise may exist, individual equipoise may be absent (Lilford and Jackson 1995).

The alternative in such a situation is to use observational methods in which the outcome of care that patients receive as part of their 'natural' treatment is studied. In other words, no attempt is made by the investigators to assign patients on a random basis to particular treatments. Instead, naturally occurring variations in how patients are treated form the basis of any comparisons. Before drawing inferences from the outcomes of treatment for such groups of patients, the characteristics of the patients admitted to intensive care have to be taken into account.

The collective term which has been applied to these patient characteristics is case mix, and the term case-mix adjustment has been applied to the process of attempting to account for the presence of these risk factors when comparing hospital death rates following intensive care.

Case-mix measurement and adjustment

Death can result from many factors other than ineffective care. The outcome of intensive care depends not only on the inputs available (e.g. equipment, staff) and the processes of care (e.g. type of care, skill and timing with which care is administered), but also on the case mix of the patients treated.

Patients are admitted for intensive care for a wide range of clinical indications, and it is necessary to consider both the nature of the current crisis and the underlying disease. Some patients are admitted with substantially greater risk of death than others, so that intensive care units (ICUs) admitting greater proportions of high-risk patients would be expected to have higher mortality. An ICU in a large tertiary care center may have a very different patient population from an ICU based in a district general hospital.

Scoring systems

During the past 25 years, through the development of scoring systems, various patient characteristics have been recognized as important in increasing the risk of hospital death following intensive care. Such characteristics include increasing chronological age, greater severity of the acute illness, a past medical history of particular severe conditions, having undergone emergency surgery, and the reason for admission to intensive care. Accounting for such patient characteristics or case mix is essential before any comparisons of outcome are possible.

The proposed role of scoring systems

The proposed role of scoring systems has been based on the quantification of case mix and the development of mathematical equations to estimate probabilities of outcome for intensive care patients. Outcome has usually been measured as death before discharge from hospital following intensive care.

Each scoring system describes the association between independent variables (case mix) and the dependent variable (death before discharge from hospital following intensive care) in the form of a mathematical equation known as a multiple logistic regression equation. APACHE II, APACHE III, and SAPS II sum the weights for some or all of the independent variables into a score before incorporation into the mathematical equation. The mathematical equation describes the strength of the association of each of the different independent variables with the dependent variable, while allowing for the effect of all the other independent variables in the same equation.

The mathematical equation for each scoring system can be applied to a group of intensive care patients, for whom data are available on the independent variables, to estimate the expected hospital death rate. On applying the equation, the probability of death before discharge from hospital following intensive care can be estimated for each patient and summed for all the patients to yield the expected hospital death rate for the whole group of patients. The expected hospital death rate can then be compared with the actual hospital death rate. This is often displayed in the form of a ratio of actual to expected hospital death rates, referred to as the standardized mortality ratio (SMR). Confidence intervals can be calculated to determine whether the value of the SMR is statistically significantly different from 1.0, the value obtained when the expected and actual hospital death rates are the same (Rapoport *et al.* 1994).

Given the above, the proposed roles of scoring systems in intensive care can be divided into three main areas: comparative audit, evaluative research, and the clinical management of individual patients.

Comparative audit

Comparing the actual outcomes with the expected outcomes for groups of patients, calculated using a scoring system, has been proposed as the basis for initial exploratory control comparisons of different providers. The use of case-mix-adjusted outcomes as a measure of the clinical effectiveness of intensive care assumes that an SMR greater than 1.0 may reflect poor care and, conversely, an SMR less than 1.0 may reflect good care.

Local clinical audit could then be implemented to investigate the reasons for any unexpected results. For example, it may be that a high SMR arose because of poor outcomes in one particular group of patients, such as respiratory cases, or at a particular time, such as at night.

Evaluative research

As described above, it is not always possible to evaluate interventions using randomized study designs. When non-randomized or observational methods are used, a valid means of adjusting for case-mix differences between groups of patients is needed. Accurate objective estimates of the probabilities of hospital death, when translated into expected hospital death rates for groups of patients, have been proposed as the basis for research studies to identify those components of intensive care structure and process that are linked to improved patient outcome.

Scoring systems have also been proposed to aid stratification in randomized controlled trials. All interventions used in intensive care should be subject to a randomized controlled trial to demonstrate clinical effectiveness. Given the considerable heterogeneity of the intensive care patient population, it is proposed that stratification using an accurate objective estimate of the risk of hospital death might create a more homogeneous subset of patients to isolate better the effects of the intervention on the outcome.

Clinical management of individual patients

For their simplest use in the clinical management of individual patients, scores from scoring systems have been proposed as a form of clinical shorthand, i.e. a common standard terminology to convey information rapidly about a patient. They have also been proposed for use in triage to classify patients according to severity of illness.

Although the early scoring systems were proposed only as a means for comparing observed and expected outcomes for groups of intensive care patients, some of the subsequent second- and third-generation methods are promoted as methods to guide the clinical care of individual patients. It is proposed that an accurate objective estimate of the risk of hospital death can provide additional information to help make clinical decisions about treatment aims for individual patients. Such decisions might include when to withdraw treatment or when to discharge a patient.

Some of the recent methods have incorporated trend analysis, i.e. the concept of modeling probabilities of outcome over time, into their development in an attempt to improve the ability to predict outcome for individual intensive care patients.

Limitations of scoring systems

The limitations of scoring systems can be categorized into four main areas: limitations in their application, limitations of the data, limitations of their accuracy, and finally limitations arising from interpretation of their results. Each will be addressed in turn.

Limitations in the application of scoring systems

The way in which a scoring system is developed often differs from the way that it is applied. Such differences may arise in the definitions and rules for data collection, the patient inclusion/exclusion criteria, the time period for data collection, the mode of data collection, the precise outcome variable measured, and the handling of data prior to analysis. The assumption is that the method is not affected by such differences. However, it is not known whether they invalidate the method. The differences often arise because the detailed instructions required to apply the scoring system are not fully described in the original scientific paper, possibly because of space limitations. Some examples are given below.

Different patient inclusion criteria

The relationship between the APACHE II score and hospital mortality can be biased by using it to estimate probabilities of death before discharge from hospital for patients selected by substantially different inclusion criteria from those included in its development. APACHE II may not be properly calibrated for such selected patients and may systematically over- or underestimate the probabilities. Many examples of the application of APACHE II, a generic scoring system for intensive care patients, to specific groups of intensive care patients are found in the literature.

Different time period for data collection

The definition of what constitutes the first 24 h of intensive care can vary (Rowan 1996). In addition, observance of the inclusion criteria of a randomized controlled trial might lead to both the inclusion of different patients from those used to develop the scoring system and its application beyond the time period for data collection used in its development (e.g. use of APACHE II scores as inclusion criteria for patients in a trial based on data collected on their fourth day in intensive care). The scoring system may not be valid except for 'the initial 24 h after ICU admission' as stated in the original publication. Considerable variation of the time period for data collection is found in the literature.

Different outcome variable

TISS was developed solely as a proxy measure for the severity of illness of patients by quantifying the type and amount of treatment provided. However, TISS scores have been employed to determine both nursing workload and nurse dependency. These applications may not be valid for the following reasons. A TISS score represents only direct nursing tasks. Labor-intensive nursing tasks are not included in TISS. Nursing tasks, such as consoling bereaved relatives, are not included in TISS. A sedated patient receiving a greater number of TISS interventions may be less nurse dependent than an alert agitated patient receiving far fewer TISS interventions. Such issues may affect the use of TISS as a nurse workload/dependency measure.

Limitations of the data collected for scoring systems

The data collected in studies applying scoring systems can vary in terms of their validity, reliability, and completeness. However, such information is rarely reported in the literature.

Validity and reliability

Poor validity and reliability of data can arise from the lack of objective rules and definitions for data collection for the scoring systems. In addition, it is easier to define and measure age rather than to define and measure the primary reason for admission to intensive care in a valid and reliable way. A lack of objective rules and definitions may lead to the same patients being categorized in different ways in different ICUs.

Completeness

Given that the level of missing data for scoring systems can vary between different ICUs and based on the assumption that missing data are assigned as 'normal' in most scoring systems, incomplete data may reduce the accuracy of scoring systems. It is rare that the level and variation of missing data across ICUs are reported when results of studies employing scoring systems are published.

Limitations of the accuracy of scoring systems

The accuracy of scoring systems must be investigated. Inaccuracy may arise from insufficient adjustment for case mix.

Insufficient adjustment for case mix

The use of case-mix-adjusted mortality as a measure of effectiveness of intensive care assumes that the scoring system can define a baseline rate and that mortality in excess of this baseline rate reflects poor-quality care. If the baseline is set incorrectly, too many or too few cases will be labeled as receiving poor-quality care. Therefore the accuracy of the scoring system used to adjust mortality must be thoroughly tested to avoid under- or overestimation of effectiveness.

If the proportion of variance in outcome explained by the scoring system is high, such statistical adjustment may be extremely potent and yield meaningful comparisons of treatment. However, it must be remembered that, however exhaustive the method employed, some of the residual outcome differences may be due to unmeasured differences in case mix.

The fact that different scoring systems yield different probabilities for the same patients may indicate that other factors, not currently incorporated into the current scoring systems, may be important for estimating outcome (Lemeshow *et al.* 1995).

Limitations arising from interpretation of results

Employing the best descriptor

The lack of a statistically significant difference between the mean APACHE II scores for two groups of intensive care patients used in a comparison is often used to demonstrate that the two groups were similar. However, owing to the differences in the relationship between the APACHE II score and observed hospital mortality for different reasons for admission for intensive care, patients with the same APACHE II score can have very different estimates for the probability of death before discharge from hospital following intensive care. Therefore the APACHE II score is probably not the best descriptor for comparing two groups of intensive care patients, except when the two groups have identical reasons for admission for intensive care.

In such an example, a better descriptor is the estimate of the probability of death before discharge from hospital following intensive care as this encompasses all the identified aspects of the case mix of patients. However, the probabilities are not normally distributed and are skewed towards zero, with many more patients having lower than higher probabilities. The lack of a statistically significant difference between the median probabilities of death before discharge from hospital for two groups of intensive care patients used in a comparison will better demonstrate that the two groups were similar.

Translation of probabilities to predictions

The fact that a given method provides accurate estimates of the probability of death before discharge from hospital following intensive care for groups of intensive care patients does not mean that it will necessarily provide a good prediction of whether an individual patient will live or die. A probability of 61 per cent estimates that 61 out of 100 patients with the same probability would be expected to die.

An arbitrary cutpoint has to be used to translate the estimated probability of death before discharge from hospital into a prediction. Use of these methods in the clinical management of individual patients requires that a probability value is provided above which the aims of treatment for patients will be altered.

Old methods become out of date

The last update of TISS was published in 1983. Since then many new interventions have been introduced into the intensive care setting. Not surprisingly, individuals have modified TISS locally to incorporate the new interventions. For each intervention added, a new weighting is assigned. As a result, there are many variations of TISS in use. It is assumed that, because of the common label, TISS is a common currency. Anecdotal evidence from the United Kingdom suggests that this is a false assumption.

Cautious interpretation

TISS was developed solely as a proxy measure of the severity of illness for a patient by quantifying the type and amount of treatment provided. However, there are at least three false assumptions upon which TISS is based.

First, it assumes that all ICUs treating critically ill patients will manage them in a similar manner. Second, it assumes that every ICU has a reasonably wide and similar range of therapeutic capabilities. Third, it does not measure appropriateness of treatment, i.e. addressing whether a particular intervention given was necessary or a particular intervention omitted was required.

The use of therapy varies greatly between institutions and therefore a system founded on specific therapeutic interventions may not be universally applicable. TISS may reflect certain routine treatment policies such as frequent ordering of arterial blood gas analyses or may vary with the technological level of the ICU.

When interpreting TISS data, it is important to recognize that two identical TISS scores in different ICUs do not necessarily represent the same clinical challenge.

Conclusion

The original development of APACHE I was very influential in promoting research in the area of case-mix adjustment and evaluation of outcome following intensive care. All the direct methods of assessing patients have been based on the fundamental premise that clinical variables measured at or near the time of admission for intensive care are associated with the risk of subsequent hospital mortality. Results to date suggest that this relationship is plausible. Although case mix measured in this way has only been related to in-hospital survival, length and quality of survival also need to be considered.

There is a need to explore the extent to which the theoretical limitations outlined above occur in practice and to investigate the extent to which they affect interpretation of results from scoring systems. However, it should be noted that scoring systems only address one aspect of the quality of care, are only one of the methods that can be used to assess clinical effectiveness, currently only address one of a number of possible outcome measures, and have a number of theoretical methodological limitations.

Despite this, intensive care is in the forefront with regard to both the available scoring systems for case-mix adjustment and the debate of issues around the appropriate uses of such methods. Finally, it should be recognized that obtaining a complete picture of the quality of intensive care will require a number of interrelated quantitative and qualitative audit and research studies.

References

Black, N. (1996). Why do we need observational studies to evaluate the effectiveness of health care. *British Medical Journal*, **312**, 1215–18.

Lemeshow, S., Klar, J., and Teres, D. (1995). Outcome prediction for individual intensive care patients: useful, misused, or abused? *Intensive Care Medicine*, **21**, 770–6.

Lilford, R. and Jackson, J. (1995). Equipoise and the ethics of randomization. *Journal of the Royal Society of Medicine*, **88**, 552–9.

Rapoport, J., Teres, D., Lemeshow, S., and Gehlbach, S. (1994). A method for assessing the clinical performance and cost-effectiveness of intensive care units: a multicenter inception cohort study. *Critical Care Medicine*, **22**, 1385–91.

Rowan, K. (1996). The reliability of case mix measurement in intensive care. *Current Opinion in Critical Care*, **2**, 209–13.

15.3.2 Principles of severity scoring systems

K. Rowan

Key messages

- Scoring systems are methods designed to quantify and reduce a number of discrete but interrelated patient characteristics to a single value.
- Scoring systems can be either specific, relating to only certain types of intensive care patients, or generic, relating to all types of intensive care patients.
- Scoring systems can be developed employing two different approaches: a subjective theoretical approach where, prior to data collection, experts determine the inclusion and weighting of different factors in the score, or an objective empirical approach where, after data collection, the statistical relationship between each factor and the outcome determine its inclusion and weight in the score.
- Analysis of the relationship between the score and outcome (usually death before discharge from hospital following intensive care) in large databases of intensive care patients has led to the development of models to estimate probabilities of outcome.
- The accuracy of a scoring system should be examined by testing the validity and reliability of the score.
- The accuracy of a probability model should be examined by testing the goodness of fit of the model on an independent set of data.

Introduction

Scoring systems are not unique to intensive care; they have been developed and are used in many other areas of health care (Iezzoni 1994). Before discussing the principles of scoring systems in intensive care, it is important to define what is meant by a scoring system, to describe briefly the history of the development of scoring systems in intensive care, to outline the approaches employed in the development of the scoring systems proposed for use, and to relate scoring systems to models derived to estimate probabilities of a particular outcome.

What is a scoring system?

Scoring systems are methods designed to quantify and reduce a number of discrete but interrelated patient characteristics to a single value, with the final value relating to the severity of the condition. This concept is best illustrated by reference to a simple scoring system in general use in intensive care, for example the Glasgow Coma Scale (GCS). Rather than describing a patient's level of neurological function in words and relying on the assumption that colleagues share both common definitions and therefore meaning for those words, the level of neurological function for a patient is expressed as a score. In the case of the GCS, the total score is the sum of discrete scores for interrelated neurological characteristics, namely eye opening, motor response, and verbal response. The lower the GCS, the greater is the impairment of neurological function.

Specific and generic scoring systems

Scoring systems can be either specific or generic. With reference to intensive care, specific scoring systems are used for assessment of only certain types of patients; the Trauma Injury Severity Score (TRISS) method used for assessing trauma patients falls into this category. Generic scoring systems are used for assessment of all, or nearly all, types of intensive care patients; the Acute Physiology and Chronic Health Evaluation (APACHE) method(s) fall into this category.

Anatomical and physiological scoring systems

Scoring systems can be either anatomical or physiological. Anatomical scoring systems assess the extent of injury, whereas physiological scoring systems assess the impact of injury on function. The Injury Severity Score, an anatomical scoring system, divides the body into six major anatomical regions and sums the scores for the type and extent of injury in each region. The TRISS method, which incorporates the Injury Severity Score, also scores the systolic blood pressure, the respiratory rate, and the level of neurological function using the GCS. Once assessed, the Injury Severity Score is a fixed quantity whereas the TRISS score is variable and changes over time as the physiological response to injury varies.

Brief history of scoring systems in intensive care

Specific scoring systems

The early scoring systems developed relating to intensive care were either specific anatomical methods (Abbreviated Injury Scale 1969;

Burns Score 1971; Injury Severity Score 1974) or specific physiological methods (Trauma Index 1971; GCS 1974; Trauma Score 1981; Sepsis Score 1983; TRISS 1987). Most of the early work focused on trauma patients.

Generic scoring systems

There have been two main approaches to the development of generic scoring systems for intensive care; one was aimed at measuring severity by treatment intensity, and the other at measuring severity by patient characteristics and physiological measurements.

The scoring system developed to quantify severity of illness for intensive care patients based on the type and amount of treatment received was the Therapeutic Intervention Scoring System (TISS) published in 1974. The underlying philosophy of the developers of TISS was that the sicker the intensive care patient, the greater the number and complexity of treatments given. Therefore a proxy measure of the severity of illness for a patient could be obtained by quantifying the type and amount of treatment provided. TISS was last updated in 1983.

The first generic physiological scoring system developed to quantify severity of illness by patient characteristics was the APACHE method published in 1981. The underlying philosophy of the developers of APACHE was that the wide variety of physiological measurements routinely obtained on intensive care patients contained precise information on severity of illness. Therefore the severity of illness of a patient could be quantified by selecting and weighting the important physiological measurements.

In response to criticism that the APACHE method was too complex and time consuming to use routinely, two further methods were derived from it, the Simplified Acute Physiology Score (SAPS) published in 1984 and, an update from the original developers, APACHE II, published in 1985. These were both subsequently updated to SAPS II in 1991 and APACHE III in 1993.

Other methods unrelated to the original APACHE method and using a different approach to development (see below), known as the Mortality Prediction Models (MPM), were published in 1988. These were subsequently updated in 1993 to become the Mortality Probability Models (MPM II).

Approaches to the development of generic scoring systems in intensive care

The developers of generic scoring systems in intensive care have adopted either a subjective theoretical approach or a more objective empirical approach to decide which factors (treatments, patient characteristics, physiological measurements) should be included in the score and how they should be weighted.

The subjective theoretical approach involves the prior selection and weighting of factors based on past studies and expert opinion. The developers of TISS and the original APACHE method used this approach in their scoring systems.

In contrast, the developers of MPM adopted the more objective empirical approach by collecting information on many patient factors during treatment and, at the conclusion of treatment, identified those statistically associated with outcome, survival, or non-survival. A combination of univariate and multivariate statistical techniques was then used to reduce the large number of potential factors to a smaller subset and to weight the final factors.

Apart from TISS, an objective empirical approach was used to develop all the scoring systems currently proposed for use in adult intensive care today—APACHE II, APACHE III, SAPS II, MPM II, and TISS.

Scores to probabilities of outcome

Most of the current scoring systems are developed either to yield a score and then an estimate of the probability of a given outcome (TISS, APACHE II, APACHE III, SAPS II) or to yield directly an estimate of the probability of a given outcome (MPM II). The outcome measure which has been the main focus in adult intensive care is hospital mortality, defined as death before discharge from hospital following intensive care.

The weights or coefficients associated with the factors in scoring systems are derived from analysis of large databases of intensive care patients (usually thousands of patients) containing information on both the patient factors required for the scoring system and the outcome, survival, or death before discharge from hospital following intensive care.

For each scoring system, the association between the independent variables (treatments, patient characteristics, physiological measurements) and the dependent variable (death before discharge from hospital following intensive care) is described in the form of a mathematical equation, known as a multiple logistic regression model. APACHE II, APACHE III, and SAPS II sum weights for some or all of the independent variables into a score before incorporation into the model. The model describes the strength of the association of each of the different independent variables with the dependent variable, while allowing for the effect of all the other independent variables in the same model.

Once estimated, the model can be applied to a group of intensive care patients for whom data are available on the independent variables to estimate the expected hospital death rate. By applying the model, the probability of death before discharge from hospital following intensive care can be estimated for each patient and summed for all patients to yield the expected hospital death rate for the whole group of patients.

The expected hospital death rate can then be compared with the actual hospital death rate. This is often displayed in the form of a ratio of actual to expected hospital death rates, referred to as the standardized mortality ratio (SMR). When the actual hospital death rate is greater than the expected the resultant SMR is greater than 1.0, and when the actual hospital death rate is less than expected the resultant SMR is less than 1.0. Confidence intervals can be calculated to determine whether the difference from 1.0 is statistically significant (Rappoport *et al.* 1994).

Principles of scoring systems

For a scoring system to be regarded as a scientific measure, it must be able to yield valid, consistent, and reproducible results. The validity and reliability of the score must be investigated in addition to testing the goodness of fit of the multiple logistic regression model.

Testing of the scoring system and publication of the results in the peer-reviewed scientific literature is an absolute prerequisite before its general acceptance and use. In addition, where possible, testing should be confirmed and published by an independent groups of researchers.

Validity of the score

The validity of a score is investigated to confirm that the score quantifies what it has been designed to measure (Daley 1994). This can be approached in a number of ways: describing the distribution of the score, relating the score to other factors, relating the score to outcome, and comparing the score with other scores developed to measure the same entity. The following serve as examples.

It would be expected that patients would be admitted for intensive care across a wide range of severity as measured by the score, and that patients admitted for intensive care following elective surgery would show a narrower range of scores than patients admitted for intensive care following emergency surgery or as medical admissions.

A severity score would be expected to demonstrate a direct and consistent relationship with outcome, with increasing scores (representing increasing severity of illness) showing a strong association with increasing observed death before discharge from hospital following intensive care. A significant difference between mean severity scores for survivors and non-survivors would be expected.

Comparison of the score with independent clinical assessment of a patient and with other scores developed to measure the same entity should reveal consistent results. If one particular score is regarded as the gold standard, any new score should be tested against it.

Reliability of the score

The reliability of a score is investigated to determine whether the score is consistently reproducible (Hughes and Ash 1994). This can be approached in two ways: intra-rater reliability and inter-rater reliability. The following serve as examples.

The same individual should be able to measure and quantify the severity of illness for the same patient repeatedly yielding the same score (intra-rater reliability). In addition, two different individuals should be able to measure and quantify the severity of illness for the same patient repeatedly yielding the same score (inter-rater reliability).

Goodness of fit of the model

The first application of a model should be to test its fit on a sample of patients independent of the sample used to develop it. The sample of patients used to test the method should be a completely new cohort.

Often, a large sample of patients is randomly split and one portion is used to develop the method and the second portion is used to test the method. However, given that the random split of the sample of patients ensures that the characteristics of the patients in each subset are similar, such testing tends to overestimate the goodness of fit of the model.

The goodness of fit is tested by examining its calibration and discrimination (Hosmer and Lemeshow 1989).

Calibration

Calibration tests the extent of agreement between the expected and actual numbers of hospital deaths across subgroups of patients. For example, patients are grouped by decile of the probability of hospital death. The expected number of hospital deaths within each subgroup (the sum of the individual patient estimates of the probability of hospital death) is compared with the actual number of deaths within each subgroup, and the expected number of hospital survivors is compared with the actual number of hospital survivors. The agreement across the whole range of probabilities can be tested formally using the Hosmer–Lemeshow goodness-of-fit statistic.

Discrimination

Discrimination tests the ability of a method to determine patients who live from patients who die, based on the estimates of the probability of hospital death. A number of arbitrary cut points from zero to 1.0 (zero to 100 per cent) are applied, and for each cutpoint the expected number of deaths (those patients with a probability of hospital death greater than the cutpoint) is compared with the actual number of deaths. The results from each cutpoint applied can be plotted to give a receiver operating characteristic (ROC) curve. The greater the true-positive rate (the proportion of patients predicted to die who actually die) relative to the false-positive rate (the proportion of patients predicted to die who live), the greater the area under the ROC curve. The area may range from 0.5 (the method is no better than chance at determining patients who live from patients who die) to 1.0 (the method is perfect).

Conclusion

Before deciding to use a scoring system, the principles required to ensure that it is a scientific measure, namely the validity and reliability of the score and the goodness of fit of the probability model should always be considered.

References

Daley, J. (1994). Validity of risk-adjustment methods. In *Risk adjustment for health care outcomes* (ed. L.I. Iezzoni), pp. 239–62. Health Administration Press, Ann Arbor, MI.

Hosmer, D.W. and Lemeshow, S. (1989). Assessing the fit of the model. In *Applied logistic regression* (ed. D.W. Hosmer and S. Lemeshow), pp. 135–75. John Wiley, New York.

Hughes, J.S. and Ash, A.S. (1994). Reliability of risk-adjustment methods. In *Risk adjustment for health care outcomes* (ed. L.I. Iezzoni), pp. 263–86. Health Administration Press, Ann Arbor, MI.

Iezzoni, L.I. (1994). Risk and outcomes. In *Risk adjustment for health care outcomes* (ed. L.I. Iezzoni), pp. 1–28. Health Administration Press, Ann Arbor, MI.

Rappoport, J., Teres, D., Lemeshow, S., and Gehlbach, S. (1994). A method for assessing the clinical performance and cost-effectiveness of intensive care units: a multicenter inception cohort study. *Critical Care Medicine*, **22**, 1385–91.

15.3.3 Acute Physiology and Chronic Health Evaluation (APACHE)

J. E. Zimmerman and W. A. Knaus

Key messages

- APACHE II is a useful measure of severity of disease, but mortality predictions are of limited value because they are based on treatment outcomes between 1979 and 1982.

- APACHE III measures severity of disease, and provides an updated standard or benchmark for assessing the effectiveness and efficiency of the intensive care unit.
- Estimates of patient risk can assist in decision-making for individual patients, but will never predict death or survival with 100 per cent accuracy.

The APACHE system

The Acute Physiology and Chronic Health Evaluation (APACHE) system was developed to measure severity of illness for intensive care unit (ICU) patients. APACHE I was developed between 1979 and 1981 and consisted of two parts: an acute physiology score (weights 1–4), which indicated the extent of 34 potential acute physiological abnormalities, and a chronic health evaluation measured using a questionnaire which classified patients (A–D) according to their prior health status. The variables in APACHE I were selected from a review of the literature and weighted by a panel of expert clinicians. The APACHE I score correlated directly with hospital mortality, but the system was complex, required multihospital validation, and did not include probability calculations.

APACHE II – a severity of disease classification system

The APACHE II system was introduced in 1985 and incorporated several important changes in the original APACHE classification system. The number of physiological variables was reduced from 34 to 12 by eliminating variables that were infrequently measured, redundant, or of lower clinical priority. Among the 12 remaining variables, some weights and thresholds were changed. The points assigned to acute renal failure were doubled and the Glasgow Coma Scale (GCS) was given greater weight to reflect the prognostic impact of coma. In addition to changes in the acute physiology score, a score to reflect the prognostic impact of emergency surgery was added to reflect its independent association with mortality risk. The method for chronic health evaluation was also changed. The prior health status questionnaire was abandoned and risk points were awarded to reflect diminished physiological reserve due to ageing, immune deficiency, and chronic cardiac, pulmonary, renal, or liver disease.

APACHE II scores have a theoretical range of zero to 71 and are based on three components:

(1) points for the extent of abnormality of 12 physiological measures (acute physiology score);

(2) points for increasing age;

(3) points for chronic health abnormalities.

APACHE II scoring is performed using worst values for the first ICU day. An increased APACHE II score reflects an increased severity of illness and an escalating risk of hospital death. The APACHE II system also includes an equation to estimate prognosis based on information collected during the first 24 h in intensive care. The APACHE II mortality prediction equation incorporates weights for 40 disease or eight organ system categories, weights reflecting the APACHE II score on the first ICU day, and an additional weight for emergency surgery.

APACHE II was validated using a database consisting of 5815 intensive care admissions at 13 large tertiary care centers. For each increase of 3 points in first-day APACHE II score, there is an increase of approximately 3 per cent in hospital death rate. However, an APACHE II score cannot be directly equated with a specific risk of death. This is because of the prognostic importance of the disease causing intensive care admission. For example, a first-day APACHE II score of 25 for a patient with asthma is associated with a much lower mortality risk than the same score for a patient with septic shock. APACHE II can be used to predict the probability of hospital death for multidiagnostic patient groups based on aggregate individual estimates which account for the prognostic impact of disease, the risk (if any) imposed by emergency surgery, the first-day APACHE II score, and patient outcomes at 13 United States hospitals between 1979 and 1982. The accuracy of these predictions has subsequently been confirmed by other investigators.

APACHE II has been used in clinical research, to evaluate intensive care resource use, and to measure ICU performance. It has been used most frequently to risk-stratify patients entered into research studies. This capability has been used in performing outcome studies among patients with a broad range of diagnoses including acute respiratory distress syndrome, lung neoplasm, granulocytopenia with hematological malignancy, abdominal sepsis, and upper gastrointestinal bleeding. APACHE II has also been used as a case-mix control in therapeutic trials including studies of stress ulcer prophylaxis, open management of the septic abdomen, peritoneal lavage for acute pancreatitis, and monoclonal antibodies in septic shock. In each instance, APACHE II was used at study entry to ensure that treatment and control groups had an equivalent severity of illness. The success and acceptance of APACHE II as a case-mix control and a means for group risk stratification in clinical investigations is attributable to the system's statistical and clinical validity.

A second major use of APACHE II has been to examine and compare the quality of intensive care as reflected by patient outcome. In a series of studies, risk factors for disease were combined with APACHE II scores to estimate group death rates which could then be compared with the actual number of deaths as a measure of performance among ICUs in the United States. Similar performance comparisons have been reported for more than 20 countries including Great Britain and Ireland (Rowan *et al.* 1994) and 12 European and North American countries (Castella *et al.* 1995). In the United States study, uncovering important variations in ICU performance led to specific examination of technical and organizational associations with superior outcomes.

Finally, APACHE II has also been used in studies aimed at improving intensive care utilization. The principles used to identify patients at low and high risk of death during the first day of intensive care were used to assess risk for receiving life-supporting intensive care therapy. Patients admitted to ICUs for monitoring who have minimal physiological abnormalities were at a low risk (< 10 per cent) for subsequently receiving the life-supporting therapies that are unique to intensive care. These low-risk monitoring admissions seldom require unique intensive care services, and units with a substantial number of such admissions may have the ability to reduce them without jeopardizing quality of care.

One of the major limitations of APACHE II is that risk predictions are based on the outcomes of treatment during the period 1979 to 1982. In addition, the system was neither designed nor intended to predict mortality for individual patients and has an error rate of

Eyes open spontaneously or to painful/verbal stimulation

Verbal / Motor	Oriented conversation	Confused conversation	Inappropriate words and incomprehensible sounds	No response
Obeys verbal command	0	3	10	15
Localizes pain	3	8	13	15
Flexion withdrawal/ decorticate rigidity	3	13	24	24
Decerebrate rigidity/ no response	3	13	29	29

Eyes do not open spontaneously or to painful/verbal stimulation

Verbal / Motor	Oriented conversation	Confused conversation	Inappropriate words and incomprehensible sounds	No response
Obeys verbal command				16
Localizes pain				16
Flexion withdrawal/ decorticate rigidity			24	33
Decerebrate rigidity/ no response			29	48

The shaded areas without scores represent unusual and unlikely clinical combinations. There were few or no cases in these cells. For the shaded areas with scores we had data that permit us to extrapolate values. Placing a patient in any of these cells should be done after careful confirmation of clinical findings.

Fig. 1 APACHE III scoring for neurological abnormalities. (Reproduced with permission from Knaus *et al.* (1991).)

approximately 15 per cent when predicting hospital mortality using a 0.50 decision point. However, some investigators have used APACHE II scoring over time to assess prognosis for individual patients. While these authors were able to place patients into ascending risk strata, their predictions were consistently high, i.e. all patients with a greater than 60 per cent risk of death were predicted to die.

APACHE II is also subject to selection and lead-time bias. Selection bias refers to errors in prediction caused by differences in how patients for whom outcomes are to be predicted are actually selected compared with the criteria that were used in creating the reference database. Differences between referral source and selection criteria have led to inaccurate APACHE II outcome estimates. Lead-time bias refers to errors in prediction due to variations in the timing of intensive care admission. Unless all admissions were entered at approximately the same time in the course of their illness, APACHE II predictions have not calibrated well within new databases.

The APACHE III prognostic system

APACHE III was developed in 1991 to expand and improve the prognostic estimates provided by APACHE II (Knaus *et al.* 1991). The APACHE III database was collected between 1988 and 1990 and included data on patient and institutional characteristics within a nationally representative sample of 17 440 intensive care admissions. The study involved 42 ICUs at 40 United States hospitals, including 26 randomly selected to represent institutions with more than 200 beds according to geographic region, size, and teaching status. This process resulted in about 50 per cent of all United States hospitals and 85 per cent of all intensive care beds being eligible to participate.

APACHE III was developed to achieve the following goals:

(1) to re-evaluate the selection and weighting of physiological variables using objective statistical modeling;

(2) to update and expand the size and representativeness of the reference database;

(3) to examine the relationship between outcome and patient selection for and timing of intensive care admission;

(4) to distinguish clearly the use of predictive estimates for patient groups from mortality estimates for individual patients.

To develop APACHE III, the database was randomly split into two. Using the estimation dataset and multivariable logistic regression analysis, weights were developed to capture the prognostic impact of each physiological variable's deviation from normal. During this process, APACHE II was used to control for disease and chronic health status. Next, the significance of 34 items reflecting chronic health status, comorbidities, and age were evaluated, and weights

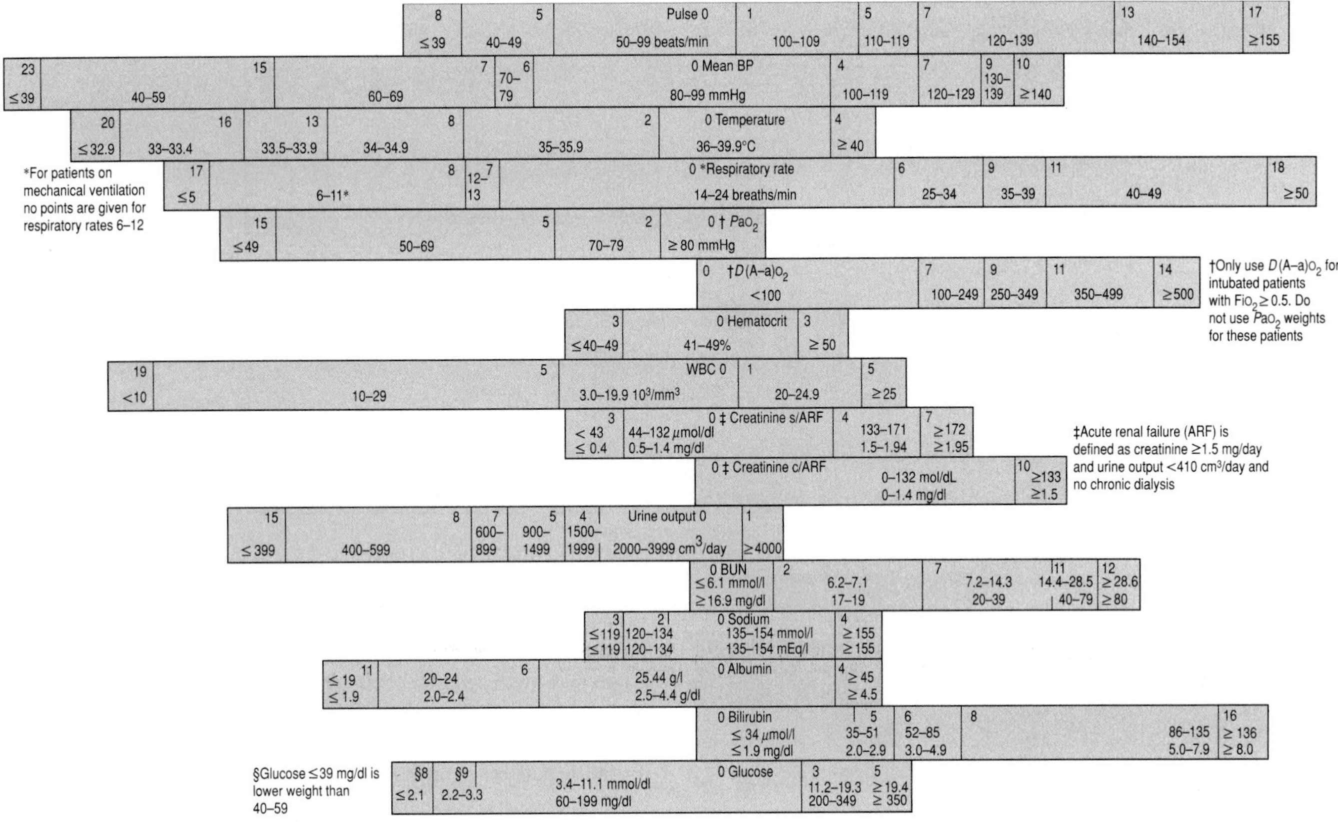

Fig. 2 APACHE III scoring for vital signs and laboratory abnormalities. (Reproduced with permission from Knaus *et al.* (1991).)

were estimated using the entire data file. Validity was then tested by comparing predicted hospital survival with that observed in the validation dataset.

APACHE III is similar to APACHE II in principle but is more complex. The APACHE III score consists of points for 17 physiological abnormalities, points for age, and points for chronic health evaluation. Scoring for the 17 physiological variables (the acute physiology score) reflects the extent of abnormality of five vital signs, 11 laboratory tests, and neurological status. Figure 1 illustrates how GCS variables have been incorporated into APACHE III scoring. In contrast with the GCS, APACHE III scoring for neurological abnormalities is based on the worst rather than the best response during each ICU day. APACHE III scoring for each GCS variable is weighted to reflect the explanatory power of interactions between worst ocular, motor, and verbal responses for hospital survival for the validation half of 17 440 ICU admissions to 42 ICUs at 40 United States hospitals (Knaus *et al.* 1991). Based on reliability testing, the GCS variables were also reformatted by simplifying the evaluation of eye opening and eliminating the distinctions between incomprehensible words and inappropriate sounds, between flexion withdrawal and decorticate rigidity, and between decerebrate rigidity and no response. As shown in Figs 2 and 3, points for the five vital signs and 11 laboratory tests range from zero to 23.

The APACHE III score also includes a score for age and chronic health evaluation (Table 1). Although 34 chronic health items were tested, only seven comorbid conditions had a significant impact on short-term mortality. Each of these seven comorbidities predisposes to infection, the cause of a large number of hospital deaths. We believe that prior cardiac, respiratory, and renal disease were not significant because their prognostic impact is captured by improved physiological measurement. The sum of points for the components of APACHE III yields a numerical score that theoretically ranges from zero to 299; physiological scoring contributes the majority (score 0–252) compared with age (score 0–24) and chronic health evaluation (score 0–23).

The APACHE III system has three major uses. First, the APACHE III score can be used to measure severity of disease and to risk-stratify patients within a single diagnostic category or independently defined patient group. This is because an increased score is associated with an increased risk of hospital death. Second, APACHE III scores can be used to compare patient outcomes, but can only be used for intensive care admissions meeting diagnostic and selection criteria similar to that used in the APACHE III study. Third, APACHE III can be used to predict patient outcomes. However, to do this, the score must be related to diagnosis, selection criteria, and a reference database.

APACHE III predictive equations

An APACHE III equation predicts hospital mortality for groups of intensive care admissions by linking first-day patient characteristics to a 17 440-admission reference database. Although these predictions are based on patient outcomes between 1988 and 1990, they have been updated based on outcomes between 1993 and 1996 for over 37 000 intensive care admissions in the United States. Prognostic estimates use the APACHE III score and additional variables to reflect disease

pH \ P_{CO_2}	<25	25–<30	30–<35	35–<40	40–<45	45–<50	50–<55	55–<60	≥60
<7.15				12				4	
7.15–<7.20									
7.20–<7.25				6		3		2	
7.25–<7.30		9							
7.30–<7.35								1	
7.35–<7.40				0					
7.40–<7.45		5						1	
7.45–<7.50			0	2					
7.50–<7.55									
7.55–<7.60			3				12		
7.60–<7.65	0								
≥7.65									

Fig. 3 APACHE III scoring for acid–base disturbances. (Reproduced with permission from Knaus *et al.* (1991).)

Table 1 APACHE III Scoring for age and chronic health evaluation

Age (years)	
⩽44	0
45–59	5
60–64	11
65–69	13
70–74	16
75–85	17
⩾85	24
Chronic health evaluation (comorbid condition)	
AIDS	23
Hepatic failure	16
Lymphoma	13
Metastatic cancer	11
Leukemia/multiple myeloma	10
Immunosuppression	10
Cirrhosis	4

and selection for intensive care. The database includes 212 potential diagnoses, and there are weights for 66 specific diagnoses and 12 organ-system-related residual categories (Knaus *et al.* 1991). Patient selection is accounted for by weights reflecting whether ICU re-admission or emergency surgery occurred, location prior to ICU admission, and hospital length of stay and location prior to ICU admission. These variables are included to control, at least in part, for differences in patient selection and response (or lack of response) to prior therapy, i.e. for selection and lead-time bias. When individual prognostic estimates are aggregated to the unit level, comparisons of actual and predicted outcomes provide a method for assessing ICU performance (Zimmerman *et al.* 1993). Similar disease-specific predictive models have also been developed to predict hospital mortality and other outcomes for patients with specific conditions such as sepsis and coronary artery bypass surgery.

In addition to hospital mortality, equations incorporating similar but not identical variables are used to predict outcomes which reflect intensive care resource use (Zimmerman *et al.* 1993). These first-day capabilities include predictions of subsequent intensive care and hospital length of stay, therapeutic intervention score, risk for active life-supporting therapy, frequency of laboratory testing, and duration of mechanical ventilation. Each of these outcomes is predicted on the basis of individual patient and institutional characteristics with reference to the APACHE III database. The length of stay predictions have been updated based on data for the period from 1993 to 1996. Comparisons of actual and predicted outcomes aggregated to the unit level provide a means of assessing ICU efficiency.

APACHE III and prognosis for individual patients

Prognostic systems will never predict an individual's death or survival with 100 per cent specificity. Serious diseases and high severity scores do not indicate absolute irreversibility, and a less severe disease and low severity scores do not guarantee against unforeseen complications or chance mortality. Although mortality predictions using the first-day APACHE III equation are as accurate as those based on physician judgment, clinicians recognize that it is rarely possible to estimate prognosis accurately for an individual patient after 1 day of intensive care. The ability to predict an individual patient's probability of survival typically depends on assessment of how he or she responds to therapy over time.

The APACHE III prognostic system uses daily assessment of acute physiology scores as a measure of response to therapy over time (Wagner *et al.* 1994). An individual's probability of hospital mortality is updated daily using a series of equations based on the patient's status on ICU day 1, on physiological status during the current day, and on physiological trends over the most recent 24 h.

These daily prognostic estimates can and are being used to assist in making decisions about individual patients. With the aid of modern clinical information systems, physicians are using these probabilities at the bedside in much the same way they use the results of laboratory tests. However, unlike the results of a blood test which is drawn directly from the patient, a risk probability is derived indirectly from patient characteristics and a reference database. Thus clinicians using this information must consider, as they do in interpreting laboratory tests or other reports, the impact of missing values, data entry errors, and other sources of bias. They must also consider the availability of new therapies and understand that the confidence interval for each estimate widens for each succeeding day, that trends are more important than absolute values, and that some prognostic factors and indicators of response to therapy are not measured by the physiological variables.

References

Castella, X., Artigas, A., Bion, J., Kari, A., and the European/North American Severity Study Group (1995). A comparison of severity of illness scoring systems for ICU patients: results of a multicenter, multinational study. *Critical Care Medicine*, **23**, 1327–35.

Knaus, W.A., *et al*. (1991). The APACHE III prognostic system: risk prediction of hospital mortality for critically ill hospitalized adults. *Chest*, **100**, 1619–36.

Rowan, K.M., Kerr, J.H., Major, E., McPherson, K., Short, A., and Vessey, M.P. (1994). Intensive Care Society's Acute Physiology and Chronic Health Evaluation (APACHE II) study in Britain and Ireland: a prospective, multicenter, cohort study comparing two methods for predicting outcome for adult intensive care patients. *Critical Care Medicine*, **22**, 1392–1401.

Wagner, D.P., Knaus, W.A., Harrell, F.E., Zimmerman, J.E., and Watts, C. (1994). Daily prognostic estimates for critically ill adults in ICUs: results from a prospective multicenter, inception cohort analysis. *Critical Care Medicine*, **22**, 1359–72.

Zimmerman, J.E., *et al*. (1993). Value and cost of teaching hospitals: a prospective, multicenter, inception cohort study. *Critical Care Medicine*, **21**, 1432–42.

15.3.4 Simplified Acute Physiology Score (SAPS)

J.-R. Le Gall

Key messages

- SAPS II was developed by logistic regression analysis of data from a European and North American study.
- All the variables in SAPS II are readily available. No special venous or arterial blood samples are required.
- The probability of mortality is calculated directly from the score using a logistic regression equation, without adding points or any sort of correction for the acute disease.

Introduction

The Second Simplified Acute Physiology Score (SAPS II) was created because published scoring systems were becoming increasingly sophisticated and time and money were needed to collect the data needed to use them. A joint European–North American study was undertaken during 1991 and 1992 to develop, from a large sample of medical and surgical patients, a method for converting the score to a probability of hospital mortality. Logistic regression analysis was used to assist in selecting the variables that would constitute SAPS II, deciding on appropriate groupings and point assignments for each variable, and converting the SAPS II score to a probability of hospital mortality.

Development of SAPS II

Statistical basis and methods

All consecutive admissions (patients aged 18 years or older) to adult ICUs in the participating hospitals were eligible for enrolment, but burn patients, coronary care patients, and cardiac surgery patients were excluded from the statistical analyses. Interoperator quality was controlled by having each site co-ordinator complete a second set of forms for a 5 per cent random sample of the patients studied at that ICU. Kappa statistics (Fleiss 1981) and intraclass correlation coefficients (Fleiss 1986) were used to assess the quality of the data. Data collection included patient demographic information, all variables necessary for computing the original SAPS, a set of new variables that might be included in SAPS II, and vital status at hospital discharge. The physiological variables were recorded by the data collectors as the worst value in the first 24-h period in the ICU. The worst value was defined as the value that would have been assigned the greatest number of SAPS points in the original SAPS score.

To develop SAPS II, 65 per cent of the available patients were randomly selected to form the developmental dataset, while the remaining 35 per cent became the validation dataset. Each of the possible explanatory variables was independently evaluated for its association with hospital mortality. These bivariate analyses were used to screen through the set of independent variables to identify a smaller subset associated with hospital mortality.

Results

Of the 14 745 patients enrolled, 1593 were excluded (burns, coronary, and cardiac surgery patients), leaving 13 152 for model development and validation.

Mortalities varied from 13.8 per cent in Switzerland to 32.4 per cent in the United Kingdom, but it is important to note that these are crude values. The ICUs were not randomly selected and should not be assumed to be representative of all ICUs in any given country. There were more male than female patients in all countries.

Of the 37 variables collected to develop SAPS II, only 17 were included in the final score. Patients with missing information on type of admission (scheduled surgical, unscheduled surgical, medical) could not be included in the final analysis, since they could not be correctly categorized for the assignment of points. Similarly, patients with missing ventilation data for whom the $Pa{O_2}/Fi{O_2}$ ratio could not be calculated were excluded. These exclusions removed 155 cases from the database, so that the final SAPS II was developed and validated on 12 997 patients, of whom 8369 were in the developmental sample and 4628 in the validation sample. Variables

Variable/Points	26	13	12	11	9	7	6	5	4	3	2	0	1	2	3	4	6	7	8	9	10	12	15	16	17	18
Age (years)												<40						40–59				60–69	70–74	75-79		≥80
Heart rate (beats/min)				<40							40–69	70–119				120–159		≥160								
Systolic BP (mmHg)		<70						70–99				100–199		≥200												
Body temperature (°C) (°F)												< 39 < 102.2			≥ 39 ≥102.2											
Only if VENT or CPAP: $P\mathrm{aco}_2$ (mmHg)/ Fio_2 (0.XX) $P\mathrm{aco}_2$ (kPa)/ Fio_2 (0.XX)				<100 <13.3	100–199 13.3–26.5		≥ 200 ≥ 26.6																			
Urinary output (l/day)				<0.500					0.500–0.999			≥ 1.000														
Blood urea (mmol/l) (g/l)												< 10.0 < 0.60					10.0–29.9 0.60–1.79				≥ 30.0 ≥ 1.80					
WBC (10^3/mm^3)			<1.0									1.0–19.9			≥ 20.0											
Serum K (mmol/l)										<3.0		3.0–4.9			≥ 5.0											
Serum Na (mmol/l)								< 125				125–144	≥145													
Serum HCO_3 (mmol/l)							< 15			15–19		≥ 20														
Bilirubin (μmol/l) (mg/l)												<68.4 <4.0				68.4–102.5 40.0–59.9				≥ 102.6 ≥ 60.0						
GCS	<6	6–8				9–10		11–13				14–15														
Chronic diseases																				Met.can.	Hem.mal.				AIDS	
Type of admission												Scheduled surgical					Medical		Unscheduled surgical							
SUM OF POINTS																										

TOTAL SAPS II | pts

RISK OF HOSPITAL DEATH | %

Fig. 1 SAPS II score sheet: BP, blood pressure; VENT, ventilated; CPAP, continuous positive airways pressure; WBC, white blood cells; GCS, Glasgow Coma Scale; Met. can., metastatic cancer; Hem. mal., hematological malignancy.

were excluded because they were unrelated to hospital mortality in the bivariate analyses or because they were not necessary once other variables were included in the multivariate model.

Figure 1 shows the variables, ranges, and points that make up the SAPS II system. The SAPS II score is made up of 17 variables: 12 physiological variables, age, type of admission (scheduled surgical, unscheduled surgical, medical), and three variables related to underlying disease (AIDS, metastatic cancer, and hematological malignancy). Table 1 gives the definitions of the variables constituting the SAPS II system.

The risk of death

From the developmental set of 8369 patients, an equation, based on the multiple logistic regression model, was developed to convert the SAPS II score to a probability of hospital mortality. Since it was observed that the distribution of the SAPS II score was highly skewed, a shrinking power transformation ln(SAPS II score + 1), was incorporated into the model. Thus the model contained two explanatory variables, SAPS II score and ln(SAPS II score + 1).

The first step in calculating the probability of hospital mortality is to compute the logit as follows:

$$\begin{aligned}\text{logit} &= \beta_0 + \beta_1(\text{SAPS II score}) + \beta_2[\ln(\text{SAPS II score} + 1)]\\ &= -7.7631 + 0.0737(\text{SAPS II score})\\ &\quad +0.9972[\ln(\text{SAPS II score} + 1)].\end{aligned}$$

This logit is converted to a probability of hospital mortality with the following equation:

$$\Pr(Y = 1|\text{logit}) = \frac{e^{\text{logit}}}{1 + e^{\text{logit}}}.$$

Evaluation of the SAPS II model

Table 2 presents the goodness-of-fit test performed on the developmental dataset. The p value for this test was 0.883. Since this is large (much greater than 0.05), the model very closely reflects the true mortality experience in the developmental dataset. In the validation dataset, the p value for the goodness-of-fit test was 0.104, suggesting that the model reflected the mortality experience in a group of patients independent of those on whom the model was developed.

The areas under the receiver operating characteristic (ROC) curve for SAPS II were 0.88 (95 per cent confidence interval, 0.87–0.90) in the developmental dataset and 0.86 (95 per cent confidence interval, 0.84–0.88) in the validation dataset. Areas this large are generally acknowledged to constitute excellent discrimination.

The future of the SAPS II system

The selection of variables and the assignment of weights to levels of these variables in SAPS II were accomplished with the assistance of the logistic regression modeling technique. This approach differs from that used in older systems, which were based on clinical judgment alone. Collecting the data necessary to calculate the SAPS II score is very simple and rapid. We estimate that it would take less than 5 min per patient. All the variables in the SAPS II are readily available and do not require special venous or arterial blood samples.

Table 1 Variables and Definitions for SAPS II

Variable	Definition
Age	Use the patient's age at last birthday in years
Heart rate	Use the worst value in 24 h, either low or high heart rate (e.g. if it varies from cardiac arrest (11 points) to extreme tachycardia (3 points), assign 4 points)
Systolic BP	Use the same method as for heart rate (e.g. if it varies from 60 mmHg to 195 mmHg, assign 13 points)
Body temperature	Use the highest temperature in Celsius or Fahrenheit
Pa_{O_2} /Fi_{O_2} ratio	If ventilated or CPAP, use the lowest value of the ratio
Urinary output	If the patient is in the ICU for less than 24 h, make the calculation for 24 h (e.g. 1 liter in 8 h=3 liters in 24 h)
Blood urea	Use the highest value in mmol/l or g/l
WBC	Use the worst (high or low) WBC count according to the score sheet
Serum K	Use the worst (high or low) value in mmol/l according to the score sheet
Serum Na	Use the worst (high or low) value in mmol/l according to the score sheet
Serum HCO_3	Use the lowest value in mmol/l
Bilirubin	Use the highest value in μmol/l or mg/l
GCS	Use the lowest value: if the patient is sedated, record the estimated GCS before sedation
Type of admission	Unscheduled surgical*, scheduled surgical†, or medical‡
AIDS	Yes, if positive HIV with clinical complications such as *Pneumocystis carinii* pneumonia, Kaposi's sarcoma, lymphoma, tuberculosis, toxoplasma infection
Hematological malignancy	Yes, if lymphoma, acute leukemia, multiple myeloma
Metastatic cancer	Yes, if proven metastasis by surgery, CT scan, or any other method

BP, blood pressure; CPAP, continuous positive airways pressure; WBC, white blood cells; GCS, Glasgow Coma Scale.
*Patients added to operating room schedule within 24 h of the operation.
†Patients whose surgery was scheduled at least 24 h in advance.
‡Patients having no surgery within 1 week of ICU admission.

One of the goals of the modeling process was to maintain a pure physiology-based system. However, criteria of calibration and discrimination were improved considerably by including the three underlying chronic clinical conditions. As with any system based on clinical measurements, missing values present a problem which has been identified by others. This problem is generally solved by assuming that values not recorded in the medical record are within normal limits. These rules were followed with SAPS II since, for instance, in some countries serum bilirubin is not systematically measured. Similarly, blood gases are not necessarily measured in all non-ventilated patients.

Some systems require that a single diagnosis be specified for estimating the probability of mortality. In SAPS II the probability of mortality is calculated directly from the score using a logistic regression equation, without adding points or any sort of correction for the acute disease. This decision was made before beginning the study, based on the assumption that selection of a single diagnosis is too difficult for most ICU patients. Although some patients can be categorized according to a specific, simple, and unique diagnosis, such as chronic obstructive pulmonary disease, septic shock, or barbiturate overdose, this is not generally the case. In fact, it has been found possible to categorize only 37 per cent of patients into a single diagnostic category, with the remaining patients having multiple diagnoses (Bahloul *et al.* 1988). While knowledge of diagnosis would certainly have an impact on the estimated probability of mortality, such estimates will be available for only a small percentage of ICU patients. When several diagnoses are possible, it is often difficult to select the most important. For example, if a patient has acute respiratory distress syndrome and associated purulent peritonitis, which is the main diagnosis? Other systems require specification of the principle diagnosis, and risk of death will differ according to the chosen category. Before it can be determined that adjustment for diagnosis has been successful in estimating risk, the calibration of the model must be carefully checked within diagnosis groups.

A more complete discussion can be found in the original article (Le Gall *et al.* 1993).

Customized SAPS II for early severe sepsis

To develop a customized version of SAPS II for ICU patients with early severe sepsis, we selected 1130 patients from the European–North American Study database who had been in intensive care for at least 24 h (Le Gall *et al.* 1996). These patients had severe sepsis as defined by the criteria of the American College of Chest Physicians

Table 2 Goodness-of-fit test for SAPS II in the developmental sample of 8369 patients

	Survived		Died	
Pr (death)	**Observed**	**Expected**	**Observed**	**Expected**
0.00–0.10	4066	4060.4	165	170.7
>0.10–0.20	1164	1174.0	211	201.0
>0.20–0.30	451	442.6	137	145.4
>0.30–0.40	332	338.3	184	177.7
>0.40–0.50	194	187.8	145	151.2
>0.50–0.60	144	146.2	182	179.8
>0.60–0.70	73	71.3	131	132.7
>0.70–0.80	68	67.0	198	199.0
>0.80–0.90	26	31.7	191	185.3
>0.90–1.00	15	13.7	292	293.3
Total	6533		1836	
$\hat{H}=3.70$	df=8		p value=0.883	

and the Society of Critical Care Medicine (systemic inflammatory response syndrome in response to infection plus hypotension, hypoperfusion, or multiple organ dysfunction). In patients with severe sepsis, mortality was higher (48.0 versus 19 per cent among other patients) and 28-day survival was lower. The customized SAPS II was well calibrated ($p = 0.92$ for the goodness-of-fit test) and discriminated well (area under the ROC curve, 0.78). Performance in the validation sample was equally good ($p = 0.85$ for the goodness-of-fit test; area under the ROC curve, 0.79).

The formula for the probability of death for a patient with early severe sepsis is

$$P = \frac{e\beta_0 + \beta_1(\text{SAPS II score})}{1 + e\beta_0 + \beta_1(\text{SAPS II score})}$$

where $\beta_0 = -3.5524$ and $\beta_1 = 0.0694$. For instance with a SAPS II score of 40 the probability of hospital death is 0.315 (the standard formula would have given a probability of 0.247).

Customization provides a simple technique for applying existing models to a subgroup of patients. Accurate assessment of the probability of hospital mortality is a useful adjunct to clinical trials.

References

Bahloul, F., Le Gall, J.-R., Loirat, P., Alperovitch, A., and Patois, E. (1988). Facteurs pronostiques en réanimation. *Presse Médicale*, **17**, 1741–4.

Fleiss, J.L. (1981). *Statistical methods for rates and proportions* (2nd edn). Wiley, New York.

Fleiss, J.L. (1986). *The design and analysis of clinical experiments*. Wiley, New York.

Le Gall, J.-R., Lemeshow, S., and Saulnier, F. (1993). A new simplified acute physiology score (SAPS II) based on a European/North American Multicenter Study. *Journal of the American Medical Association*, **270**, 2957–63.

Le Gall, J.-R., *et al.* for the Intensive Care Unit Scoring Group (1996). Customized probability models for early severe sepsis in adult intensive care patients. *Journal of the American Medical Association*, **276**, 644–50.

15.3.5 Therapeutic Intervention Scoring System (TISS)

D. J. Cullen

Key messages

- TISS is a useful tool for obtaining severity of illness data on intensive care unit (ICU) patients quickly and easily.
- TISS lends itself well to studies of ICU function, personnel and resource utilization, and assessment of the need for ICU beds.
- Small differences in TISS scores are not discriminating and the TISS data should not be overinterpreted.
- The ability to categorize patients makes TISS a very useful tool for ICU administrative and management research.

Introduction

The Therapeutic Intervention Scoring System (TISS) (Cullen and Nemeskal 1989) (Table 1), which is probably the oldest of the widely available methods for classifying critical illness, has been used worldwide for many purposes: determining the severity of illness, establishing nurse-to-patient ratios, assessing utilization of intensive care unit (ICU) facilities, and establishing the number of intensive care beds, particularly in response to a request for a Certificate of Need. By quantitating severity of illness using the TISS system, i.e quantitating therapeutic interventions, the above goals can be achieved rapidly, efficiently, and with reasonable precision.

Table 1 Therapeutic Intervention Scoring System

Four points

(a) Cardiac arrest and/or countershock within past 48 h
Point score for 2 days after most recent cardiac arrest

(b) Controlled ventilation with or without positive end-expiratory pressure
Does not mean intermittent mandatory ventilation which is a 3-point intervention. If the patient's full ventilatory needs are being supplied by the machine, score 4 points

(c) Controlled ventilation with intermittent or continuous muscle relaxants

(d) Balloon tamponade of varices

(e) Continuous arterial infusion
Intra-arterial pitressin infusion does not include standard 3-ml/h heparin flush to maintain catheter patency

(f) Pulmonary artery catheter

(g) Atrial and/or ventricular pacing
Active pacing even if a chronic pacemaker

(h) Hemodialysis in unstable patient
Include first two rounds of an acute dialysis. Include chronic dialysis in patient whose medical situation now renders dialysis unstable

(i) Peritoneal dialysis

(j) Induced hypothermia
Continuous or intermittent cooling to achieve body temperature less than 33°C

(k) Pressure-active blood infusion
Use of a blood pump or manual pumping of blood in the patient who requires rapid blood replacement

(l) G-suit

(m) Intracranial pressure monitoring

(n) Platelet transfusion

(o) Intra-aortic balloon assist

(p) Emergency operative procedures (within past 24 h)
Can include the initial emergency operative procedure, not including diagnostic tests (i.e. angiography, CT scan)

(q) Lavage of acute gastrointestinal bleeding

(r) Emergency endoscopy or bronchoscopy

(s) Vasoactive drug infusion (> one drug)

Three points

(a) Central intravenous hyperalimentation (includes renal, cardiac, hepatic failure fluid)

(b) Pacemaker on standby

(c) Chest tubes

(d) Intermittent mandatory ventilation or assisted ventilation
The patient supplies partial ventilation

(e) Continuous positive airways pressure

(f) Concentrated K^+ infusion via central catheter

(g) Nasotracheal or orotracheal intubation

(h) Blind intratracheal suctioning

(i) Complex metabolic balance (frequent intake and output)
Measurement of intake/output above and beyond the normal 24-h routine. Frequent adjustment of intake according to total output

(j) Multiple blood gas, bleeding, and/or stat studies (>four per shift)

(k) Frequent infusions of blood products (>5 units/24 h)

(l) Bolus intravenous medication (non-scheduled)

(*continued*)

Table 1 *(continued)*

(m) Vasoactive drug infusion (one drug)

(n) Continuous antiarrhythmia infusions

(o) Cardioversion for arrhythmia (not defibrillation)

(p) Hypothermia blanket

(q) Arterial line

(r) Acute digitalization (within 48 h)

(s) Measurement of cardiac output by any method

(t) Active diuresis for fluid overload or cerebral edema

(u) Active therapy for metabolic alkalosis

(v) Active therapy for metabolic acidosis

(w) Emergency thora-, para-, and pericardiocentesis

(x) Acute anticoagulation (initial 48 h)
Includes Rheomacrodex

(y) Phlebotomy for volume overload

(z) Coverage with more than two intravenous antibiotics

(aa) Therapy of seizures or metabolic encephalopathy (within 48 h of onset)

(bb) Complicated orthopedic traction (e.g. Stryker frame, CircOlectric bed)

Two points

(a) Central venous pressure

(b) Two peripheral intravenous catheters

(c) Hemodialysis (stable patient)

(d) Fresh tracheostomy (less than 48 h)

(e) Spontaneous respiration via endotracheal tube or tracheostomy (T-piece or tracheostomy mask)

(f) Gastrointestinal feedings

(g) Replacement of excess fluid loss (over and above the ordered maintenance level)

(h) Parenteral chemotherapy

(i) Hourly neurological vital signs

(j) Multiple dressing changes

(k) Pitressin infusion

One point

(a) ECG monitoring

(b) Hourly vital signs

(c) One peripheral intravenous catheter

(d) Chronic anticoagulation

(e) Standard intake and output (every 24 h)

(f) Stat blood tests

(g) Intermittent scheduled intravenous medications

(h) Routine dressing changes

(i) Standard orthopedic traction

(j) Tracheostomy care

(k) Decubitus ulcer (preventive therapy)

(continued)

Table 1 (*continued*)

(l) Urinary catheter

(m) Supplemental oxygen (nasal or mask)

(n) Intravenous antibiotics (two or less)

(o) Chest physiotherapy

(p) Extensive irrigations, packings, or debridement of wounds, fistulas, or colostomy

(q) Gastrointestinal decompression

(r) Peripheral hyperalimentation/Intralipid therapy

Adapted from Cullen and Nemeskal (1989).

Table 2 Patient classes and TISS points

Class		No. of patients	TISS points (mean ± SE)	Mortality at 1 year
I	Routine wake-up	364	5±0.2	0
II	Close observation	771	11±0.7	15%
III	Intensive nursing care	662	23±1.0	21%
IV	Intensive nursing and physician care	231	43±1.0	73%

Data from Cullen and Nemeskal (1989).

Description of TISS

TISS is based on the premise that, regardless of the diagnosis causing critical illness, the physiological derangements requiring therapeutic support result from the severity of the illness. Further, many critically ill patients lack a diagnosis upon admission and, sometimes, even at the time of their death from multiple organ failure.

Two major assumptions are required when using TISS. First, most or all interventions listed must be available within a particular hospital. Otherwise, the TISS score may be artificially reduced. Second, therapeutic interventions should be initiated for medically justifiable reasons and not because they generate additional income for the physician or hospital.

TISS is a compilation of monitoring and therapeutic modalities, each of which is assigned a weighted score from one to four points depending on their intensity (Table 1). The sum of all the points for all the interventions applied to an individual patient is the TISS score.

Determination of TISS in an individual patient

Data are collected from the previous 24 h, preferably at the same time each day and by the same observer, and the score is tallied. Since some interventions are interrelated, only the highest point intervention is scored. For example, if within a 24-h period a patient received controlled ventilation (4 points) and then was weaned to intermittent mandatory ventilation (3 points), continuous positive airways pressure (3 points), and spontaneous ventilation with supplemental oxygen (1 point), the 4-point intervention (for controlled ventilation) is scored but not the others. As patient status improves, interventions are usually withdrawn and TISS decreases. Conversely, as severity worsens, therapy escalates and TISS points increase.

Although not essential, we recommend that TISS data collectors have a critical care nursing background. Such nurses can easily identify the interventions and evaluate the appropriate associations within and between categories. Once familiar with the tool, an experienced observer can assign interventions and tally points in 3 to 5 min per patient. Table 1 can serve as the data collection instrument.

Stratification of ICU patients

The intensive care population can be stratified into three major groups, with the most critically ill being class IV (Table 2).

Class IV patients are physiologically unstable, requiring intensive nursing and physician care with frequent observations and change of orders by their physicians. These patients usually have multiple organ or multiple system dysfunction. A more objective and quantitative method of measuring the severity of illness in class IV critically ill patients resulted in a further extension of TISS (Cullen *et al.* 1984). We objectively and quantitatively defined severity of illness and demonstrated the extreme nature of class IV critical illness.

Table 3 Comparison of two surgical ICUs (excluding cardiac surgical patients)

	Tertiary hospital	Community hospital
No. of beds	9	4
Average census (%)	7.2 (80%)	3.96 (99%)
Average daily unit TISS (range)	30.3 (26–40)	15 (9.7–25.3)
Average length of stay (range) (days)	4 (1–41)	2.4 (<1–21)
Class I admission	0	6 (16%)
No. of patient days <10 TISS points	0	25/111 (22%)
Patients discharged with TISS <10	1/55 (2%)	26/31 (84%)
Average TISS on discharge	19.3 (9–30)	6.4 (1–11)
Class IV time (no. of patient days/total patient days)	35/201 (17%)	5/111 (4.5%)
Class III time (no. of patient days/total patient days)	162/201 (77%)	18/111 (16%)
Class II time (no. of patient days/total patient days)	14/201 (6%)	63/111 (57%)
Class I time (no. of patient days/total patient days)	0 (0%)	25/111 (22.5%)

Reproduced with permission from Cullen and Nemeskal (1989).

Class III patients are physiologically stable, but require intensive nursing care, invasive monitoring, and predictable medical care. Their condition is expected to remain stable or improve, with early return to general floor care expected. These patients usually have no more than one organ or system requiring major supportive therapy.

Class II patients require intensive monitoring (but no therapy) to detect potential catastrophes.

Class I patients do not require intensive monitoring or therapy and therefore do not need the facilities of an ICU.

The excellent separation of the classes by TISS score suggests that clinical classification of patients into these four groups provides a practical method of describing the ICU population (Table 2). The mortality of class IV patients is very high, with 73 per cent of patients dying within 1 year (Cullen *et al.* 1976).

Intermediate TISS

We developed Intermediate TISS (Cullen *et al.* 1994) to focus on class I patients who, by definition, do not require ICU facilities. Patients believed to be at high risk of developing serious postoperative problems, although not needing intensive care *per se*, are often admitted to the ICU for overnight observation. Similarly, patients are often admitted to medical ICUs or coronary care units to exclude myocardial infarction. Thus we wish to improve identification of medical patients who may not require intensive therapies, but nevertheless need cardiopulmonary monitoring.

Intermediate TISS could also be used to determine the need for additional hospital facilities such as more ICU beds, intermediate care areas, or bedside flexible monitoring. Cost containment may limit or halt the growth of ICU facilities, as less expensive ways of caring for patients are developed. Therefore we modified TISS for intermediate and floor care to emphasize both the medical and surgical aspects of a patient's illness. Intermediate TISS focuses on medical patients with cardiopulmonary problems and diabetes mellitus. Although many of these medical problems could be managed more easily in an ICU, they are increasingly being managed outside the ICU, which is both time-consuming and task intensive for a general care nurse. One criticism of the original TISS was that it was too procedurally oriented and diminished the importance of nursing care. Some of the items added or upweighted for Intermediate TISS recognize the increased nursing effort required for appropriate care of medical patients in the absence of specific procedures. Intermediate TISS can be substituted for the original TISS when assessing the hospital's needs for intermediate care, flexible monitoring, or floor care.

Intermediate TISS adds useful new information for defining the severity of illness in medical patients or surgical patients with medical complications, but does not add significantly to the original TISS for surgical patients (Cullen *et al.* 1994).

Use of TISS to evaluate ICU bed needs

A very important use of TISS is the accurate determination of a hospital's need for ICU beds. The following questions must be answered.

1. How are the ICU facilities currently being utilized?
2. How many additional beds are needed, and in what type of facility should they be located (postanesthetic care unit, intermediate care, or full intensive care)?
3. Were adverse results reported in patients who were denied intensive care because of inadequate bed supply?

We determined the need for intensive care and used the TISS system to follow patients who were triaged from the ICU. This study, which was performed for our own hospital, led us to design methods to evaluate ICU bed needs in other hospitals prior to obtaining a

Certificate of Need for additional ICU beds (Cullen and Nemeskal 1989). These data allow us to suggest a nurse-to-patient ratio based on the severity of illness, to determine if patients are being discharged from the ICU earlier than is acceptable based on their TISS points at the time of discharge and in the follow-up period, to decide if any patients did not need intensive care based on their low admission and in-unit TISS points, and to suggest how many ICU beds are needed to manage each unit's case load.

In Table 3 an extremely busy and overloaded surgical ICU is contrasted with a surgical ICU where there is unnecessary admission and retention of patients is occurring.

References

Cullen, D.J. and Nemeskal, A.R. (1989). Therapeutic Intervention Scoring System (TISS). In *Problems in critical care*, Vol. 3 (ed. J.C. Farmer), pp. 545–62. J.B. Lippincott, Philadelphia, PA.

Cullen, D.J., Ferrara, L.C., Briggs, B.A., Walker P.F., and Gilbert, J. (1976). Survival, hospitalization charges and follow up results in critically ill patients. *New England Journal of Medicine*, **294**, 982–7.

Cullen, D.J., Keene, R., Waternaux, C., Kunsman, J.M., Caldera, D.L., and Peterson, H. (1984). Results, charges, and benefits of intensive care for critically ill patients. Update 1983. *Critical Care Medicine*, **12**, 102–6.

Cullen, D.J., Nemeskal, A.R., and Zaslavsky, A.M. (1994). Intermediate TISS: a new Therapeutic Intervention Scoring System for non-ICU patients. *Critical Care Medicine*, **22**, 1406–11.

15.3.6 Mortality Probability Model (MPM)

A. D. Collins and S. Lemeshow

Key messages

- MPM II is a mortality probability system using the presence or absence of a number of conditions to aid in the process of evaluating patients for actions such as triage, rationing, futility determinations, and quality of care evaluation.
- With the exception of the age variable, all variables have a 'yes' or 'no' response.
- As with any intensive care unit or similar model, the results from an MPM II calculation should be regarded as an estimate and a probability.

Introduction

With the development of the intensive care unit (ICU) for the specialized care of patients who are recovering from complex operations or are otherwise very ill, some clinicians became responsible for large numbers of patients defined not so much by their underlying illness itself as by its severity and complexity. These clinicians noted that patients who died had certain patterns and physiological conditions in common, and that these patterns and conditions appeared to be far more important prognosticators of mortality than the patient's diagnosis. Mortality Probability Model II (MPM II) is a mortality probability system using the presence or absence of a number of these conditions to aid in the process of evaluating patients for actions such as triage, rationing, futility determinations, and quality of care evaluation.

Development of MPM and MPM II

The development of MPM (Lemeshow *et al.* 1985) was distinctly different from that of both APACHE and SAPS. The first difference was that MPM was entirely statistically derived. A large number of variables (137 at admission and 75 at 24 h) were collected on 755 consecutive general medical and surgical ICU patients. Coronary care, cardiac surgery, burn, and pediatric (under 14 years) patients were excluded. Various statistical techniques were used to determine the relative importance and weight of each variable. This process allowed the developers to retain only those variables that were commonly collected, non-ambiguous, and shown statistically to have a strong association with survival status. Finally, a multiple logistic regression model was developed. Using multiple logistic regression, the developers further reduced the number of significant variables and objectively derived weights for each of those remaining. The final multiple logistic regression models directly computed an estimate of the probability of the patient dying during the hospital stay rather than a point score. The results were very compact and simple models with only seven variables at admission and seven variables at 24 h, and very little time was required to collect and record the necessary information.

A second major difference was that, while APACHE II and SAPS were performed 24 h after admission to the ICU and used the 'worst' value during the first day for each of its variables, the MPM system contained models that could be performed both immediately upon admission and at 24 h. Third, the variables were more condition based rather than the physiologically based variables predominating in APACHE II and SAPS. Fourth, the variables collected were generally 'yes' or 'no' answers.

Subsequently, MPM II (Lemeshow *et al.* 1993) was developed with data collected in two separate studies, again using consecutive general medical and surgical ICU admissions excluding cardiac surgery, coronary care, burn, and pediatric (under 18 years) patients. In addition, admissions other than the first were excluded for those with multiple admissions. One of these datasets consisted of approximately 6000 cases collected at six United States hospitals. For the second dataset, the developers of MPM and SAPS (Le Gall *et al.* 1993) joined forces to collect data on over 14 000 patients in 137 hospitals in Europe and North America during 1990 and 1991. Together, the 19 124 cases ultimately included came from a diverse sample of ICUs consisting of 41 per cent university hospitals, 27 per cent university-affiliated hospitals, 16 per cent community teaching hospitals, and 16 per cent community non-teaching hospitals. The data consisted of a large set of variables including all those used in APACHE II, the original MPM, and SAPS, together with others.

For the admission model (MPM_0), the 19 124 patients were randomly separated into two groups: 12 610 cases (65 per cent) were used for model development, and 6514 (35 per cent) were used later for validation purposes. The association of each possible independent variable with hospital mortality was assessed. This list was further analyzed for frequency of inclusion, ease of interpretation, and strength of

Table 1 Variables in MPM_0 (admission model) with definitions and estimated coefficients

Variable	Definition	β
Constant		−5.46836
Physiology		
Coma or deep stupor	At ICU admission, not due to drug overdose. For patients on a paralyzing muscle relaxant, awakening from anesthesia, or heavily sedated, use your best judgment of the level of consciousness prior to sedation Coma: no response to any stimulation, no twitching, no movement in extremities, no response to pain or command; generally corresponds to a GCS of 3 Deep stupor: exhibits decorticate or decerebrate posturing; posturing is spontaneous or in response to stimulation or deep pain, not in response to commands; generally corresponds to GCS of 4 or 5	1.48592
Heart rate	Record whether the heart rate was noted to be greater than or equal to 150 beats/min within 1 h before or after ICU admission	0.45603
Systolic blood pressure	Record whether the systolic blood pressure was noted to be less than or equal to 90 mmHg within 1 h before or after ICU admission	1.06127
Chronic diagnoses		
Chronic renal insufficiency	Should have evidence of an elevation of serum creatinine >176.8 μmol/l (2 mg/dl) and be documented as chronic in the medical history. *Note: If patient has acute and chronic renal failure, only record as for acute*	0.91906
Cirrhosis	'Yes' if history of heavy alcohol use with portal hypertension and varices, other etiologies with evidence of portal hypertension and varices, or biopsy confirmation	1.13681
Metastatic neoplasm	For example, stage IV cancer; excludes regional nodes. 'Yes' if metastases obvious by clinical assessment or confirmed by a pathology report. 'No' if it is not obvious and has to be confirmed by a pathology report that is not available at the time of ICU admission. Acute hematological malignancies are considered in this category. Chronic leukemia is considered only if there are findings attributable to the disease or the patient is under *active* treatment for the leukemia. Findings include sepsis, anemia, stroke caused by clumping of white cells, tumor lysis syndrome (increased uric acid as the result of chemotherapy), pulmonary edema including lymphangeiatic form of ARDS	1.19979
Acute diagnoses		
Acute renal failure	ATN, tubular necrosis, or acute on chronic renal failure. Prerenal failure is not included	1.48210
Cardiac dysrhythmia	Cardiac arrhythmia, paroxysmal tachycardia, fibrillation with rapid ventricular response, second- or third-degree heart block. Do not include chronic and stable arrhythmias	0.28095
Cerebrovascular incident	Cerebral embolism, occlusion, CVA, stroke, brainstem infarction, cerebrovascular arteriovenous malformation (acute stroke or cerebrovascular hemorrhage, not chronic arteriovenous malformation)	0.21338
Gastrointestinal bleeding	Hematemesis, melena. A perforated ulcer is not necessarily a GI bleed; may be identified by obvious 'coffee grounds' in the NG tube. A drop in hemoglobin by itself is not sufficient evidence of acute GI bleeding	0.39653
Intracanial mass effect	Intracranial mass (abscess, tumor, hemorrhage, subdural) as identified by CT or other scan associated with any of the following: (1) midline shift; (2) obliteration or distortion of cerebral ventricles; (3) gross hemorrhage in cerebral ventricles or subarachnoid space; (4) visible mass >4 cm; (5) any mass which enhances with contrast media. If the mass effect is known within 1 h of ICU admission, it can be indicated as 'yes'. CT scanning is not mandated, and is only indicated for patients with major neurological insult	0.86533

(*continued*)

association with mortality. Through multiple logistic regression techniques, variables were further reduced and those remaining were each assigned a weight. Discrimination, or the ability to assign higher probabilities to those who died than those who survived, was assessed using the receiver operating characteristic (ROC) curve (Hanley and McNeil 1982). This compares all possible pairs of a surviving and a dying patient and evaluates the proportion wherein the patient with the highest probability of mortality was the one who actually died. Calibration of the model was assessed using the Hosmer–Lemeshow goodness-of-fit test which tests the degree of correspondence of

Table 1 continued

Variable	Definition	β
Other		
Age	Patient's age at last birthday in years	0.03057
CPR prior to admission	Within 24 h prior to ICU admission CPR includes chest compression, defibrillation, or cardiac massage; record as 'yes' regardless of where the CPR was administered	0.56995
Mechanical ventilation	Patient is on a ventilator at the time of ICU admission or immediately thereafter	0.79105
Non-elective surgery	This question should be answered 'yes' except for (1) elective surgery (i.e. surgery scheduled at least 24 h in advance) patients and (2) preoperative Swan–Ganz insertion if an elective surgery patient	1.19098

GCS, Glasgow Coma Scale; ARDS, acute respiratory distress syndrome; ATN, acute tubular necrosis; CVA, cerebrovascular accident; GI, gastrointestinal; NG, nasogastric; CPR, cardiopulmonary resuscitation.

actual outcome and estimated outcome over groups of patients across the entire range of probabilities. Next, variables were assessed for removal if their elimination improved calibration while not harming discrimination. The resulting model contains 15 variables and directly computes a probability of mortality during the entire hospital stay (Table 1).

In order to test the overall quality of the finished model, discrimination and calibration were tested using the developmental set. The importance of performing and reporting both calibration and discrimination cannot be overstated; it is entirely possible for a model to perform well in one of these calculations and poorly in the other, resulting in a poor model. In the case of MPM_0 the results were an area under the ROC of 0.837 and a Hosmer–Lemeshow goodness-of-fit test showing good calibration with $p = 0.623$. Finally, the model was applied to and computed for all the cases in the validation set. Again, there was excellent discrimination (ROC value, 0.824) and calibration ($p = 0.327$). The result of the MPM II calculation again computes a direct estimate of the probability of mortality for the patient using 15 variables.

A similar process was performed using data available for the 10 357 patients in the development sample still alive at 24 h. The resulting model contains 13 variables: five values already collected in the admission model, two re-evaluated at 24 h, and six new variables (Table 2). This model (MPM_{24}) was tested and demonstrated excellent discrimination (ROC value, 0.844) and good calibration ($p = 0.764$) in the developmental sample, and again for those cases in the validation group who survived at 24 h (ROC value, 0.836; $p = 0.231$ for goodness-of-fit test).

Subsequent to the original publication of the MPM models, models for 48 h (MPM_{48}) and 72 h (MPM_{72}) were developed and published (Lemeshow *et al.* 1994). These models use the same variables, and except for β_0 (or constant term), the same β_k values as MPM_{24}. MPM_{48} and MPM_{72} have their own unique β_0 terms ($\beta_0 = -5.39153$ in MPM_{24}; $\beta_0 = -5.23840$ in MPM_{72}).

Because of the ease of collection, the acceptance and use of MPM_0 and MPM_{24} have been significant. Unlike APACHE III, the MPM II probability of mortality does not depend on a single diagnosis because individual diagnoses, other than those included as part of the model itself, did not seem be significantly related to mortality. This is a significant simplifying factor since the necessity of designating a single overriding diagnosis within the first 24 h of an ICU admission is fraught with obvious difficulties. It is estimated that the collection time for any of the MPM II models is less than 2 min per patient (with some sites reporting less than 1 min) when the data are collected concurrently.

Using MPM II models

The computation of MPM probability is very straightforward. With the exception of the age variable, all variables have a 'yes' or 'no' response. The values for the variables listed for the particular model, the constant term β_0, and the appropriate value(s) of β_k noted for each variable with a 'yes' response, and the age variable obtained by multiplying the age by β_k are entered into the following generic formula:

$$\text{logit} = \beta_0 + \beta_1 x_1 + \beta_2 x_2 + \beta_2 x_2 + \ldots + \beta_k x_k$$

where $\beta_k x_k$ is the value for the kth variable times the value of that variable (note that in MPM II the value of the variable is 1 if 'yes' and 0 if 'no', except for age which uses the actual age). The probability of mortality is obtained from the logit derived above using the following formula:

$$\text{probability of mortality} = \frac{e^{\text{logit}}}{1 + e^{\text{logit}}}.$$

An example of a calculation of the probability of mortality of a patient at admission using MPM_0 follows.

A 60-year-old medical patient is admitted with cirrhosis and gastrointestinal bleeding. There is no history of renal insufficiency, cancer, or cerebral vascular incident. At the time of admission the patient is alert and oriented, and has a good color on room air; the heart rate is 140 beats/min and regular, and the blood pressure is 65/45 mmHg. There is no evidence of intercranial mass effect.

This short history has the following positive findings relevant to MPM_0: age, not a scheduled surgery patient, blood pressure 90 mmHg, history of cirrhosis, and an acute gastrointestinal bleed. Since we are performing an admission calculation, β_0 and the other appropriate values are extracted from Table 1 and inserted into the logit formula as follows:

Table 2 Variables in MPM_{24} with definitions and estimated coefficients

Variable	Definition	β
Constant		−5.64592
Variables at admission		
Age	As in MPM_0	0.32680
Cirrhosis	As in MPM_0	1.08745
Intracranial mass effect	As in MPM_0	0.91314
Metastatic neoplasm	As in MPM_0	1.16109
Non-elective surgery	As in MPM_0	0.83404
Variables reassessed at 24 h		
Coma or deep stupor	At 24 h, whether or not due to drug overdose. Determination is made *at* the 24-h mark. For patients on a paralyzing muscle relaxant, awakening from anesthesia, or heavily sedated, use best judgment of the level of consciousness prior to sedation. *The definition for this variable is the same as for coma/deep stupor in MPM_0, but overdose patients still in coma or deep stupor at 24 h are not an exception to the definition*	1.68790
Mechanical ventilation	Determination applies at any time during the first 24 h	0.80845
New variables		
Creatinine	>176.8 μmol/l (2.0 mg/dl): determination applies at any time during the first 24 h	0.72283
Confirmed infection	*Confirmed* admission infection or *confirmed* new infection. Determination applies at any time during the first 24 h. 'Yes' only if cultures, Gram stains, or radiographs confirm a suspected admission infection or new infection that developed in the first 24 h, or there is evidence of gross purulence. Laboratory confirmation (including verbal or fax-type confirmation) must be obtained by the 24-h mark for 'Confirmed' to be checked. If used to confirm presence of infection, radiological or other imaging evidence must be strongly suggestive	0.49742
Partial pressure of oxygen ($Pa{O_2}$) <7.98 kPa (60 mmHg)	Determination applies at any time during the first 24 h. $P{O_2}$ values in the operating room and recovery room should *not* be considered. Does not count transient drop during tracheal suctioning. A sustained drop in pulse oximetry <90% is acceptable evidence of low $P{O_2}$. Record as 'yes' regardless of $Fi{O_2}$	0.46677
Prothrombin time > 3 s above standard	Determination applies at any time during the first 24 h	0.55352
Urine output <150 ml in 8 h	Determination applies to *any 8-h period* during the first 24 h	0.82286
Vasoactive drug therapy	Continuous IV, defined as *for 1 h or more* Determination applies at any time during the first 24 h Vasoactive drugs include: 1. Any dose of metaraminol (Aramine), epinephrine (adrenaline chloride), norepinephrine (Levophed), dobutamine (Dobutrex), phenylephrine (Neo-Synephrine), amrinone (Inocor) 2. Naloxone (Narcan) if used to elevate blood pressure 3. Dopamine (Intropin) *only if* >5 μg/kg/min	0.71628

$$\text{logit} = -5.46836 + 1.06427 + 1.13681 + 0.39653 + 1.19098 + (0.03057 \times 60) = 0.15143.$$

This result is entered into the probability equation:

$$\text{probability of mortality} = \frac{e^{0.15143}}{1 + e^{0.15143}} = 0.54.$$

As with any ICU or similar model, the results from an MPM II calculation should be regarded as an estimate and a probability. Based upon a similar population of patients, it is expected that 54 per cent of those ICU patients with this constellation of variables will die during their hospital stay. The converse is that 46 per cent of such patients would be expected to survive.

Any multiple logistic regression probability model is developed from a specific population and can only purport to perform in that population. All the ICU models were developed from a diverse population of general medical and surgical patients in ICUs who were receiving intensive care and from data collected at specific times. Therefore their accuracy should only be assumed in similar settings.

In an effort to deal with this, MPM II was designed to allow for customization to account for specific differences. For example, while a model might perform very well across the entire range of hospitals, there might be a benefit in customizing for a certain subgroup such as hospitals in a specific country. In addition, as the MPM II database grows it will be possible and necessary to customize the models to reflect general changes in mortality resulting from changes in treatment and improvements in the quality of care. Similarly, as the database grows, MPM II models for times beyond 72 h will be developed.

References

Hanley, J.A. and McNeil, B.J. (1982) The meaning and use of the area under a receiver operating characteristic (ROC) curve. *Radiology*, **143**, 29–36.

Le Gall, J.-R., Lemeshow, S., and Saulnier, F. (1993) A new Simplified Acute Physiology Score (SAPS II) based on a European/North American multicenter study. *Journal of the American Medical Association*, **270**, 2957–63.

Lemeshow, S., Teres, D., Pastides, H., Avrunin, J.S., and Steingrub, J.S. (1985). A method for predicting survival and mortality of ICU patients using objectively derived weights. *Critical Care Medicine*, **13**, 519–25.

Lemeshow, S., Teres, D., Klar, J., Avrunin, J.S., Gehlbach, S.H., and Rapoport, J. (1993). Mortality Probability Models (MPM II) based on an international cohort of intensive care unit patients. *Journal of the American Medical Association*, **270**, 2478–86.

Lemeshow, S., *et al.* (1994). Mortality probability models for patients in the intensive care unit for 48 or 72 h: a prospective multicenter study. *Critical Care Medicine*, **22**, 1351–8.

15.3.7 Trauma scoring

H. R. Champion and W. J. Sacco

Key messages

- TRISS and ASCOT use anatomical and physiological measures of injury severity, age, and type of injury to relate patient values to survival probability.
- TRISS is now computed using admission values of the Revised Trauma Score, the Injury Severity Score (ISS) based on final diagnoses, patient age, and type of injury (blunt or penetrating).
- ASCOT has been shown to be a better predictor of survival probability than TRISS.

Introduction

A major application of injury severity scales is the control for casemix differences in the asessment of trauma outcomes. TRISS and ASCOT are two scales which are widely used for the study of survival/death outcomes in trauma patients.

TRISS and ASCOT scales

TRISS

The combination of physiological and anatomical scores and other factors has yielded patient characterizations which accurately relate to survival/death outcome. TRISS (Champion *et al.* 1983) combines the Trauma Score, the Injury Severity Score, patient age, and injury type (blunt or penetrating). Logistic function norms relate TRISS values to patient survival probability (P_s). P_s values are used in outcome evaluation methods which statistically compare the number of survivors in an institution with the number expected from norm predictions. In the late 1980s the Trauma Score was replaced in TRISS by the Revised Trauma Score (Champion *et al.* 1989), which demonstrated improved predictive accuracy and reliability.

TRISS is now computed using admission values of the Revised Trauma Score (RTS) (Table 1), the Injury Severity Score (ISS) based on final diagnoses, patient age, and type of injury (blunt or penetrating). TRISS estimates P_s using the logistic function

$$P_s = 1/(1 + e^{-b})$$

where

$$b = b_0 + b_1(\text{RTS}) + b_2(\text{ISS}) + b_3(\text{AGE}).$$

AGE is zero if the patient is less than 55 years old and 1 otherwise, and the b_i are regression coefficients (weights) determined from the data.

Table 2 gives the b_i values for regression coefficients based on the Abbreviated Injury Scale (AIS-85) derived from Major Trauma Outcome Study (MTOS) data for blunt- and penetrating-injured adults. Table 3 gives the Abbreviated Injury Scale (AIS-90) TRISS coefficients based on data from four study sites in the MTOS (Champion *et al.* 1990).

ASCOT

Like TRISS, ASCOT uses anatomical and physiological measures of injury severity, age, and type of injury to relate patient values to survival probability. Patients with extremely poor or good prognoses are divided into four sets and excluded from logistic function modeling. Table 4 defines these 'set-aside' groups and gives their associated ASCOT survival rates. The ASCOT P_s assigned to each patient in a set-aside group equals the survival rate of the group. For the remaining patients (non-set-asides), P_s is estimated with a logistic function whose exponent is

$$k = k_0 + k_1(G) + k_2(S) + k_3(R) + k_4(A) + k_5(B) + k_6(C) + k_7(\text{AGE*})$$

where G (Glasgow Coma Scale), S (systolic blood pressure), and R (respiratory rate) are the coded values of Revised Trauma Score variables on admission to the emergency department (Table 1), and A, B, and C are summary scores of serious injuries (AIS > 2) to various body regions or systems. The value of a component (A, B, or C) is given by the square root of the sum of the squares of the AIS values for a patient's injuries in the component. The body regions, injury severities, and ICD-9-CM codes associated with each component

Table 1 Revised Trauma Score*

Glasgow Coma Scale	Systolic blood pressure (mmHg)	Respiratory rate (breaths/min)	Coded value
13–15	>89	10–29	4
9–12	76–89	>29	3
6–8	50–75	6–9	2
4–5	1–49	1–5	1
3	0	0	0

*RTS=0.9368 GCS_c+0.7326 SBP_c+0.2908 Rr_c where GCS is the Glasgow Coma Scale, SBP is the systolic blood pressure, Rr is the respiratory rate, and *c* is the coded value.

Table 2 TRISS regression weights*

	b_0	b_1(RTS)	b_2(ISS)	b_3(AGE)
Blunt	−1.2470	0.9544	−0.0768	−1.9052
Penetrating	−0.6029	1.1430	−0.1516	−2.6676

*May 1986 MTOS norms.

Table 3 MTOS-controlled site coefficients based on AIS-90

	Constant	RTS	ISS	Age
Blunt	−0.4499	0.8085	−0.0835	−1.7430
Penetrating	−2.5355	0.9934	−0.0851	−1.1383

Table 4 ASCOT set-aside definitions and survival rates

	Blunt		Penetrating	
Set-aside	No.	Survivors (%)	No.	Survivors (%)
(1) AIS=6, RTS=0	48	0	22	0
(2) MAX AIS<6, RTS=0	217	1.4	192	2.6
(3) AIS=6, RTS>0	35	22.9	9	22.2
(4) MAX AIS=1, RTS>0	6262	99.8	1345	99.9

Table 5 ASCOT patient age characterization

AGE*	Age (years)
0	0–54
1	55–64
2	65–74
3	75–84
4	⩾85

AGE* is a variable used to compute patient survival probability.

Table 6 ASCOT model weights (design set)

Variable	Blunt	Penetrating
Constant	−1.1570	−1.1350
G	0.7705	1.0626
S	0.6583	0.3638
R	0.2810	0.3332
A	−0.3002	−0.3702
B	−0.1961	−0.2053
C	−0.2086	−0.3188
AGE*	−0.6355	−0.8365

Table 7 *z* and *W* values for six trauma centers

Center no.	*N*	*z*	*W*
1	530	6.02	5.2
2	42	2.33	8.8
3	234	2.05	2.27
4	158	2.56	3.81
5	291	3.62	3.37
6	2234	10.23	0.58

are defined in Table 4. AGE* is assigned a value between zero and 4 (Table 5). Table 6 gives ASCOT logistic model coefficients based on AIS-85. ASCOT has been shown to be a better predictor of survival probability than TRISS.

Applications

Audit filter

Survivors (non-survivors) whose P_s values are less than (exceed) 0.5 are identified as having statistically 'unexpected outcomes' and as being worthy of peer review.

Survival/death outcome evaluation

z and W statistics

Principally because of the MTOS (1982–1989) endorsed by the American College of Surgeons, z and W statistics have become standards for comparison of survival/death outcomes among patient subsets. z and W compare the actual number (A) of survivors in a patient sample with the expected number (E) based on TRISS or ASCOT norms. W measures the clinical significance and z the statistical significance of the comparison.

W represents the difference between the actual and expected numbers of survivors per 100 patients treated. For example, a W value of +2 signifies that $A > E$ and means that there were two more survivors per 100 patients treated than would have been expected based on national data. Likewise, a W value of −2 signifies that $A < E$ and means that there were two fewer survivors per 100 patients treated than would have been expected.

For W to be statistically different from zero, z must be greater than +1.96 or less than −1.96. If z is between −1.96 and +1.96, the difference between A and E is not statistically different from zero and thus neither is W.

The formulas for W and z are

$$W = 110(A - E)/N$$

where N is the number of patients in the sample and

$$z = (A - E)/S.$$

S is a scale factor that accounts for statistical variation. The following two examples illustrate why both W and z are needed.

Example 1

Suppose that an institution has the same outcome results for five successive years, i.e for each year $N = 500, A = 455, E = 450$, and $S = 1$.

For each year $z = 1.25$ and $W = 1$, and neither is statistically different from zero. However, for all 2500 patients, both z = 2.80 and W = 1 are statistically different from zero. This indicates that, although the yearly outcomes were not statistically better than expected based on MTOS norms, as the sample size increased the results became significant and indicate one more survivor than expected per 100 patients treated.

If the institution had the same results for 20 years, W would still be 1 but z would increase to 10, signifying a stronger statistical certification of the results.

Example 2

The z and W values for six United States trauma centers are given in Table 7. The results for center 5 are better than those for center 3, as N, z, and W are greater for center 5. The results for center 1 are even better than those for center 5. Although center 2 has a striking W value, the small sample size causes one to reserve judgment about its ability to sustain such a large W value as N increases. Center 6 is an example of large z and small W values. The W value of 0.58 indicates a slight improvement in survival rate over the norm; the z value of 10.23 indicates that this slight improvement is strongly statistically significant. However, the large sample size gives the potential for large z values.

The future

Reliable methods of controlling for case-mix differences in injured patients are essential for the scientific study of injury. Without them, epidemiological studies would lack perspective and interventional studies would be severely confounded unless governed by extraordinarily tight case definitions that would seriously restrict patient sample sizes. Severity scales also have potential applications in estimating and prescribing resource needs, and thus are an essential tool in the drive for more cost-effective care. As databases become linked, the combination of automobile crash information, clinical, financial, and rehabilitative databases will allow greater perspective on the impact of injury.

In some respects the widely used scoring systems are primitive, focus on mortality, and are not robust enough for many important applications. Substantial refinement is required to link injury severity with the many outcomes other than death.

References

Abbreviated Injury Scale–1985 Revision. Association for the Advancement of Automotive Medicine, Des Plaines, IL.

Champion, H.R., Sacco, W.J., and Hunt, T.K. (1983). Trauma severity scoring to predict mortality. *World Journal of Surgery*, 7, 4–11.

Champion, H.R., *et al.* (1989). A revision of the Trauma Score. *Journal of Trauma*, **29**, 623.

Champion, H.R., *et al.* (1990). A new characterization of injury severity. *Journal of Trauma*, **30**, 539.

15.3.8 Glasgow Coma Scale

J. E. Zimmerman and W. A. Knaus

Key messages

- The Glasgow Coma Scale (GCS) is the best available measure of the impact of disease or injury on neurological function.
- The GCS score is such an important determinant of patient outcome that it is used in most contemporary severity scoring systems.
- When combined with abnormal brainstem responses, the GCS score identifies individuals at high risk of death or a vegetative state.

The Glasgow Coma Scale (GCS) was described in 1974 as a practical method for assessing coma. Impaired consciousness was classified according to three dimensions of responsiveness: best ocular, best motor, and best verbal responses. Over the past 20 years, the GCS has become a universal language for reliably describing patients with impaired consciousness in a reproducible way. In addition, a numerical score which quantifies the extent of abnormality for the patient's best ocular, motor, and verbal responses permits calculation of a 13-point GCS score ranging from 3 to 15 (Table 1). In providing a summary measure of neurological function, the GCS score assumes that the patient is normotensive, normoxic, and has received no paralytic, narcotic, or other medications that artificially depress neurological status.

Table 1 The GCS score

Best ocular response	Open spontaneously	4
	Open to verbal command	3
	Open to pain	2
	No response	1
Best motor response	Obeys verbal command	6
	Localizes pain	5
	Flexion withdrawal	4
	Abnormal flexion (decorticate posturing)	3
	Abnormal extension (decerebrate posturing)	2
	No response	1
Best verbal response	Oriented and converses	5
	Disoriented and converses	4
	Inappropriate words	3
	Incomprehensible sounds	2
	No response	1
GCS score = ocular + motor + verbal		3–15

Because it can be used to describe impaired consciousness in a wide variety of medical and surgical diseases, the GCS score is probably the most widely used and best known severity scoring system. The ocular, motor, and verbal responses included in the GCS and the associated score have been used independently or in combination with other neurological findings to reflect the severity of neurological damage for patients with head injury, cardiac arrest, intracerebral hemorrhage, cerebral infarction, sepsis, and other causes of non-traumatic coma. The GCS has also been included as part of most contemporary general severity scoring systems including the Mortality Probability Model (MPM II), the Simplified Acute Physiology Score (SAPS II), the Pediatric Risk of Mortality (PRISM), and Acute Physiology and Chronic Health Evaluation II and III (APACHE II and III).

The GCS has been used for group risk stratification in clinical outcome studies, to evaluate intensive care resource use, and to evaluate prognosis for individual patients. In clinical research, the GCS score is the most frequently used measure of severity of head injury. Data from the United States National Traumatic Coma Bank demonstrated a clear association between a decrease in mortality and an increase in the GCS score. For patients in this study with GCS scores below 8, three treatable conditions (hypoxia, shock, and increased intracranial pressure) were found to be significantly correlated with death or severe disability. This information has been used to develop evidence-based guidelines for the management of severe head injury (Chesnut 1995).

The GCS score has also been used to generate computer-based predictions of outcome for individuals with severe head injury and to

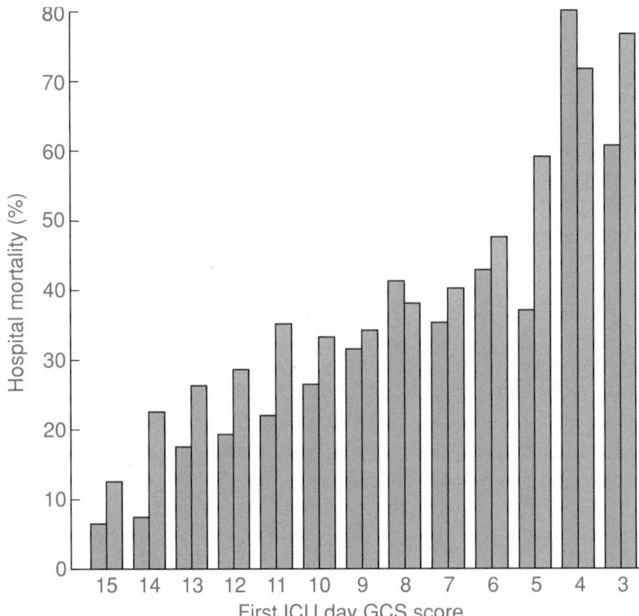

Fig. 1 Relationship between first ICU day GCS score and hospital mortality rate for postoperative patients ($n = 6786$) (gray bar) and non-operative patients ($n = 9187$) (green bar). (Reproduced with permission from Bastos *et al.* (1993).)

measure the impact of these estimates on patient management (Murray *et al.* 1993). There was increased use of intensive monitoring and treatment for patients predicted to have a less than 0.4 probability of a poor outcome (dead or vegetative), and a 39 per cent reduction in the use of the same interventions for patients predicted to have a greater than 0.8 probability of a poor outcome.

The GCS score has been used to assess prognosis for a variety of medical and surgical diseases treated in intensive care units (ICUs). As shown in Fig. 1, a decrease in GCS score is associated with a consistent but non-linear increase in hospital mortality among both non-operative and postoperative patients. The results from both risk stratification and multivariate analysis clearly showed that the GCS score is able to identify patients likely to survive and those likely to die (Bastos *et al.* 1993). However, discriminating ability within intermediate levels of prognosis is poor, with death rates relatively constant between GCS scores of 7 and 11. It is for patients in this intermediate range where the value of factors such as other acute physiological abnormalities, age, comorbidity, and diagnosis are of great prognostic importance.

For 596 patients with non-traumatic coma, most commonly due to cardiac arrest (36 per cent) and cerebral infarction or intracerebral hemorrhage (36 per cent), components of the GCS combined with abnormal brainstem response identified those at high risk of predicted death or severe disability (Hamel *et al.* 1995). For patients with one or more abnormal brainstem responses (absent pupillary or corneal response, absent or dysconjugate roving eye movements) and no motor response to pain, the rate of death or severe disability at 2 months was 96 per cent. These results are similar to those reported by other investigators among patients with intracerebral hemorrhage and comatose survivors of cardiac arrest. For patients with intracerebral hemorrhage, a predictive model using the initial GCS score, hemorrhage size, and pulse pressure provided a valid and easy-to-use model for predicting probability of survival (Tuhrim *et al.* 1995).

Despite its worldwide acceptance and prognostic usefulness, the GCS score has several important limitations. First, initial scoring may be impossible for patients with severe head injury. This is because highly trained emergency medical personnel are able to intubate, sedate, or paralyze these patients before transport to a trauma center. As a result, it has been impossible to make an accurate determination of a true GCS score for almost 50 per cent of comatose head-injured patients in the emergency department. Second, current management for patients with severe head injury often includes sedation, narcotics, and paralysis for control of elevated intracranial pressure. Thus it may also be impossible to determine an accurate daily GCS score for these patients while in the ICU. Third, periorbital swelling hypotension hypoxia and intubation can interfere with scoring. We have encountered similar problems in measuring the GCS score for medical and surgical ICU patients who were included in the APACHE III data collection. Thus, for many critically ill patients, these limitations can interfere with the basic assumption that the GCS score reflects the severity of neurological injury or disease.

Recommendations for overcoming these problems have included the following.

1. Obtain an initial GCS score within 1 to 2 h after injury.
2. Avoid scoring until hypotension or hypoxia have been stabilized.
3. Use an ocular score of 1 for patients with severe periorbital swelling.
4. Carefully adhere to the definitions provided in the original description of the GCS.
5. Delay scoring until 10 to 20 min after the estimated half-life of medications which cause sedation or paralysis.
6. Record a GCS score of 15 if there is no prior accurate determination and sedation, or paralysis cannot be reduced.

At present there is no better measure that summarizes the impact of disease or injury on neurological function. Thus, whether used alone or as an important component of APACHE III or another prognostic system, the GCS score is an important predictor of patient outcome. This is why everything possible should be done to obtain a GCS score on all intensive care admissions. When direct measurement of the GCS score is impossible, a consistent policy of either recording a normal score or using other patient data to impute or estimate the score are the two alternatives. While this means that risk may be underestimated, treating missing data in a systematic manner within a clinical trial will minimize the impact for subsequent analyses.

References

Bastos, P.G., Sun, X., Wagner, D.P., Wu, A.W., and Knaus, W.A. (1993). Glasgow Coma Scale in the evaluation of outcome in the intensive care unit: findings from the Acute Physiology and Chronic Health Evaluation III study. *Critical Care Medicine*, **21**, 1459–65.

Chesnut, R.M. (1995). Medical management of severe head injury: present and future. *New Horizons*, **3**, 581–93.

Hamel, M.B., *et al.* (1995). Identification of comatose patients at high risk for death or severe disability. *Journal of the American Medical Association*, **273**, 1842–8.

Murray, L.S., *et al.* (1993). Does prediction of outcome alter patient management? *Lancet*, **341**, 1487–91.

Tuhrim, S., Horowitz, D.R., Sacher, M., and Godbold, J.H. (1995). Validation and comparison of models predicting survival following intracerebral hemorrhage. *Critical Care Medicine*, **23**, 950–4.

15.3.9 Sedation scoring

A. Hole and P. Klepstad

Key messages

- Sedation scores are valuable in achieving stable and suitable sedation for each individual patient.
- Excessive administration of sedative drugs induces hypotension and reduces respiratory drive.
- Increased depth of sedation must prompt the physician to make a differential diagnosis between oversedation, decreased conscious level due to neurological disease, or depression associated with stay in the intensive care unit.

Why monitor sedation in intensive care unit patients?

Analgesic and sedative drugs are necessary in the intensive care unit (ICU) to reduce pain and discomfort after surgery, trauma, and therapeutic interventions. Self-reports from physicians treated in the ICU describe endotracheal suction as an extremely painful and frightening intervention. Other common bedside procedures, such as intravascular cannulation, wound dressings, and nursing routines, are also painful.

Sedation is necessary in the majority of patients to enable them to tolerate intermittent positive-pressure ventilation. Adequate sedation also reduces the risk of self-extubation, which is a potentially fatal event, particularly in patients with a compromised airway. Critically ill patients with marginal tissue oxygen supply benefit from sedation because of a lower sympathetic activation which reduces oxygen consumption and improves the relationship between oxygen supply and consumption. Although there are no randomized studies in adult critical care other than investigations of surrogate endpoints for outcome, sedation for adult critically ill patients may give a better outcome in terms of mortality and morbidity as well as relief from unnecessary suffering.

However, excessive administration of sedative drugs induces hypotension and reduces respiratory drive. Modern respiratory therapy is based on a synergy between the patient and the machine, and maintenance of respiratory drive will reduce ventilator time and the risk of ventilator-associated complications. Prolonged weaning from respiratory support increases length of ICU stay and costs. The hypotension induced by sedative drugs may compromise tissue perfusion (e.g. cerebral perfusion pressure) or, if the hypotension is corrected, vasoactive drugs may increase the risk of arrhythmias and vasospasm. In addition, oversedation interferes with observation of neurological status and serious central nervous system complications, such as intracerebral hemorrhage or cerebral edema, may occur undetected.

Adaptation of sedation agent dosage to the need of the individual patient is important, as both under- and oversedation are harmful to the patient. The dosage of commonly used sedative drugs to achieve a desired level of sedation varies widely from patient to patient because of large variations in metabolism and pharmacodynamics. A valid method for monitoring sedation is needed to tailor sedation to the individual patient. Also, standardized methods for estimating sedation are often missing in research, so that comparisons between different study populations are obscured.

General demands of scoring systems

The level of sedation measured by a scoring system must relate to clearly defined clinical levels of sedation and be useful for different sedative drugs. A monitoring procedure for clinical use must be simple, non-invasive, and require a minimum of time. Since critically ill patients are treated by many health professionals during their stay in the ICU, the monitoring system must possess good interoperator agreement. The monitored sedation level should be easy to record and read from the patient's chart; it should give a real-time estimate of the sedation and not add unnecessary costs.

Different principles of sedation monitoring

Physiological variables

Blood pressure and heart rate will certainly vary with the degree of wakefulness and discomfort in the ICU setting. However, they are difficult to use as a measure of sedation because of many confounding factors.

Serum concentrations of sedative drugs

Monitoring serum drug concentrations is expensive and invasive, and the results are delayed so that the practical information concerning the level of sedation will be virtually worthless. Also, the development of tolerance to sedative drugs and large interpatient variability (low pharmacokinetic–pharmacodynamic correlation) reduce the value of drug serum concentrations as a measure of sedation.

Neurophysiological methods

Traditionally, electroencephalography forms the basis of electrophysiological monitoring and assessment of the central nervous system function. Routine use of electroencephalography in an ICU is impractical and difficult to interpret. Hence the two-electrode cerebral function monitor and real-time spectral analysis, which is even more advanced, have been used. However, both these methods are expensive, require experienced operators with appropriate technical skills, and should be reserved for scientific use (Avramov and White 1995).

Clinical scoring systems

The methods fulfilling most of the demands of ICU sedation monitoring for routine use are objective clinical sedation scoring systems. These scoring systems are based on the principle that easily recognizable elements of sedation are weighted by a number, and the sedation score is given by this number or a sum of numbers related to several of these elements. Several scores have been developed in the last 25 years. Some are intended for subpopulations of critically ill patients, and sedation scores should not be confused with scores designed for other purposes, such as the Glasgow Coma Scale, whose clinical use should be restricted to describing the degree of consciousness.

Two scoring systems, the Ramsay sedation score and the Cambridge sedation score (also known as the Addenbrooke's sedation score), have survived the test of time and are in widespread use. The Ramsay Score, which was developed in 1974, consists of two parts: the awake levels and the asleep levels (Table 1). The arousability from sleep levels is tested by recording the response to verbal command or, if there is none, the response to a firm glabellar tap (Hansen-Flaschen *et al.* 1994). In the Cambridge sedation score (Table 2) the patient's response to the unstandardized but strong stimulus of tracheal suction is recorded if there is no reaction to verbal stimuli. The observers also register whether the patient is paralyzed or in a normal sleeping state (Shelly and Wang 1992). The UCL Hospitals sedation score (Table 3) is a modification of the Cambridge sedation score (Singer and Webb 1997). To avoid oversedation due to increased sedation in an attempt to abort cough produced by suction, this score replaces response to tracheal suction with response to movement or noxious stimuli. The UCL sedation score also denominates the sedation scores with increasing negative or positive values which indicate over- and undersedation respectively. This method may be of didactic value, as the principle of titrating sedation near a score of zero and avoiding extreme values is easy for nursing staff to understand. The sedation elements of these tests should be interpreted equally easily by doctors and nurses, are not time-consuming, are not drug specific, and give valid information about the sedation level by a single number. However, neither test differentiates between different elements of sedation. This is better achieved by the COMFORT scale developed for use in sedated children which combines the scores of eight clinical variables (Ambuel *et al.* 1992). Adaptation of this score to adult patients is simple.

A common limitation of all clinical sedation scores is that they do not differ between sedation induced by drugs and decreased level of consciousness due the patient's illness or decreased patient response due to major depression. The existing sedation scores are, in effect, an integrated score of drug-induced sedation and somnolence (Hansen-Flaschen *et al.* 1994). Even so the scores are adapted to clinical use but more specific methods for scientific studies of sedating drug efficacy are absent. Another limitation of categorized sedation scores is that it is not possible to assume that the intervals between the scores are equal; therefore, for scientific purposes, the scores must be treated as non-parametric data.

Recommendations

Routine use of sedation scoring and charting in ICU patients is strongly recommended. Sedation scores are valuable tools for achieving stable and suitable sedation for the individual patient. In routine monitoring with sedation scores, increased depth of sedation must prompt the physician to make a differential diagnosis between oversedation, decreased conscious level due to neurological disease (e.g. cerebral insult), or ICU-associated depression. Monitoring of all vital signs in ICU patients, including sedation, includes a legal aspect that should not be underemphasized. Use of one of the simple objective clinical scoring systems is recommended, of which the Ramsey sedation score is currently the most popular.

Table 1 Ramsay sedation score

Awake levels

1. Patient anxious and agitated or restless
2. Patient co-operative, oriented, and tranquil
3. Patient responds to command only

Asleep levels (response to glabellar tap or loud auditory stimulus)

4. Brisk response
5. Sluggish response
6. No response

Table 2 Cambridge sedation score

1. Agitated
2. Awake
3. Roused by voice
4. Roused by tracheal suction
5. Unrousable
6. Paralyzed
7. Asleep

Table 3 UCL Hospitals sedation score*

Agitated and restless	+3
Awake and uncomfortable	+2
Aware but calm	+1
Roused by voice, remains calm	0
Roused by movement	−1
Roused by painful or noxious stimuli	−2
Unrousable	−3
Natural sleep	A
Paralyzed	P

*Sedation doses are adjusted to achieve a score as close as possible to zero. Positive scores require increased sedation doses and negative scores require reduced sedation doses.

References

Ambuel, B., Hamlett, K.W., Marx, C.M., and Blumer, J.L. (1992). Assessing distress in pediatric intensive care environments: the COMFORT scale. *Journal of Pediatric Psychology*, **17**, 95–109.

Avramov, M.N. and White, P.F. (1995). Methods for monitoring the level of sedation. *Critical Care Clinics*, **11**, 803–26.

Hansen-Flaschen, J., Cowen, J., and Polomano, R. (1994). Beyond the Ramsay scale: need for a validated measure of sedating drug efficacy in the intensive care unit. *Critical Care Medicine*, **22**, 732–3.

Shelly, M.P. and Wang, D.Y. (1992). The assessment of sedation: a look at current methods and possible techniques for the future. *British Journal of Intensive Care*, **2**, 195–203.

Singer, M. and Webb, A.R. (1997). *The Oxford handbook of critical care*. Oxford University Press.

15.3.10 Organ failure scoring

J. C. Marshall

Key messages

- Mortality is a function of both the number of failing organ systems and the severity of physiological derangement within each
- The Multiple Organ Dysfunction (MOD) score variables reflect physiological perturbation rather than therapeutic intervention.
- The MOD score is calibrated against mortality and may provide a measure of admission severity, intensity of therapeutic intervention, and global intensive care unit morbidity.
- The MOD score may be useful as an outcome measure in clinical trials.

Introduction

The multiple organ dysfunction syndrome (Bone *et al.* 1992) is an outgrowth of the development of the intensive care unit (ICU). It exists because the ICU permits prolonged survival in the face of otherwise lethal physiological derangement; its conceptual evolution reflects the organ system-based supportive approach that the ICU evolved to provide.

The first reports of specific organ dysfunction in the ICU described states that were either present or absent as determined by arbitrary and variable criteria. Two features of this poorly understood process were apparent. Its development followed overwhelming infection, injury, ischemia, or other stimuli associated with activation of a host inflammatory response, and mortality rose sharply as its severity, reflected in the number of failing systems, increased.

Both the systems studied and the criteria used to define failure differed significantly between studies. A systematic review of 30 reports published between 1969 and 1993 showed that the respiratory, renal, hepatic, cardiovascular, gastrointestinal, hematological, and neurological systems were the most commonly evaluated systems in published reports (Marshall 1995).

The recognition that ICU prognosis is a function not only of the number of failing systems but also of the severity of derangement within each system, and that organ dysfunction is a potentially preventable complication of critical illness whose quantification provides a surrogate measure of clinical morbidity, has stimulated intense interest in the development of valid and objective systems for quantifying organ dysfunction as an outcome measure. Several such systems have emerged (Marshall *et al.* 1995; Le Gall *et al.* 1996; Vincent *et al.* 1996). The similarities of these are more striking than their differences, reflecting an implicit and evolving consensus on descriptive principles. The Multiple Organ Dysfunction (MOD) score (Table 1) (Marshall *et al.* 1995) was structured using a methodologically rigorous approach to maximize the ultimate validity of the measure. Because the description of a complex entity such as the multiple organ dysfunction syndrome involves considerably more than the simple prediction of mortality, this process will be described in greater detail.

The Multiple Organ Dysfunction score

Which organ systems?

A valid measure of organ dysfunction in critical illness must reflect the process as the clinician understands it or, in methodological terms, it must have construct validity. Thus the first step in the development of the MOD score was to understand multiple organ dysfunction syndrome as it is perceived by the clinician. A systematic review of the literature describing multiple organ dysfunction syndrome identified 30 papers published between 1969 and 1993, and is described in detail elsewhere (Marshall 1995).

What descriptive variables?

Selection of optimal descriptive variables for each system required consideration not only of construct validity, but also of content validity, i.e. the extent to which a candidate variable encompassed the entire spectrum of abnormality for the organ system of interest. Thus, in addition to ascertaining what variables have been employed by others in the past, a set of criteria were developed that characterized the 'ideal' descriptor of organ dysfunction (Marshall 1995), and candidate variables were evaluated against these.

Organ dysfunction or failure can be described from one of two perspectives: the physiological perturbation present in a given organ system, or the intervention required to support failing organ function. Only variables that reflected physiological abnormalities were included, because the objective of the score was to describe a biological, rather than a therapeutic, state and because therapeutic decisions vary significantly between centers, thus introducing significant potential for bias.

It was possible to identify acceptable variables for six of the seven organ systems cited in published reviews of multiple organ failure. However, no satisfactory measure of physiological dysfunction could be found for the gastrointestinal system; therefore it was eliminated from further consideration. None of the published descriptors for the cardiovascular system performed well when evaluated against the criteria describing an ideal measure of organ dysfunction. Therefore the pressure-adjusted heart rate (PAR), a physiological measure that corrects for the effects of therapy, was developed:

$$\mathrm{PAR} = \frac{\text{heart rate} \times \mathrm{CVP}}{\mathrm{MAP}} \qquad (1)$$

where CVP is the central venous pressure and MAP is the mean arterial pressure. Increasing degrees of intrinsic cardiovascular dysfunction result in increasing values of the PAR. If the central venous pressure is low (reflecting hypotension resulting from an inadequate circulating volume rather than intrinsic cardiovascular derangement), the value will be similarly low; thus the PAR, like the ratio of Po_2 to Fio_2, partially corrects for the effects of therapeutic intervention.

When and how are the variables measured?

Physiological function in an unstable critically ill patient is in a constant state of flux. Moreover, the frequency with which tests are

Table 1 The MOD score

Organ system	0	1	2	3	4
Respiratory*					
Po_2/Fio_2	>300	226–300	151–225	76–150	≤75
Renal†					
Serum creatinine (μmol/l)	≤100	101–200	201–350	351–500	>500
Hepatic					
Serum bilirubin (μmol/l)	≤20	21–60	61–120	121–240	>240
Cardiovascular‡					
PAR	≤10.0	10.1–15.0	15.1–20.0	20.1–30.0	>30.0
Hematological					
platelet count (platelets/ml)	>120	81–120	51–80	21–50	≤20
Neurological§					
GCS	15	13–14	10–12	7–9	≤6

Selected values are those obtained at a standard time of day (preferably morning) rather than the worst values during the day; missing data receive a score of zero.
*The Po_2/Fio_2 ratio is calculated without reference to the use or mode of mechanical ventilation, and without reference to the use or level of positive end-expiratory pressure.
†The serum creatinine level is measured without reference to the use of dialysis.
‡The pressure-adjusted heart rate (PAR) is calculated as described in the text.
§The GCS is preferably calculated by the patient's nurse, and is scored conservatively (for the patient receiving sedation or muscle relaxants, normal function is assumed unless there is evidence of intrinsically altered mentation).
Reproduced with permission from Marshall *et al.* (1995).

performed varies between centers, introducing a significant potential for sampling bias.

The approach selected to minimize artefactual overestimation of the score comprises three elements.

1. Only post-resuscitation values are used in calculation of the MOD score.
2. The value recorded for a given variable on a particular day is a representative value for that variable, rather than the worst value. This is most readily accomplished by recording the value at a constant time (e.g. the first value of the day).
3. Missing or unobtainable values are presumed to be normal and scored zero. Thus the Glasgow Coma Scale (GCS) in an anesthetized patient is considered normal, as is the PAR in a patient who does not have a central venous pressure line. However, when scores are calculated daily, the previous value of a missing variable is brought forward; thus if the creatinine is not measured on a given day, the first previous value is used.

Finally, to maximize the simplicity of the score, variables are measured without consideration of the therapy employed.

How is the score calibrated?

Criterion validity reflects the extent to which a descriptor accurately measures a phenomenon when compared with an existing gold standard. There is no discrete biochemical or pathophysiological marker of the presence of multiple organ dysfunction syndrome, and no independent method of gauging its severity. ICU mortality was chosen as the criterion against which variables were calibrated since it is accepted that multiple organ dysfunction syndrome is associated with a mortality whose magnitude is a function of the severity of the syndrome. However, this does not imply that a measure of organ dysfunction is a good predictor of mortality, as at least 20 per cent of all deaths in the ICU setting result from disorders other than the multiple organ dysfunction syndrome. ICU mortality, rather than hospital mortality, was selected to acknowledge that a patient capable of short-term survival without ICU support does not have life-threatening multiple organ dysfunction syndrome, even though he or she may have a terminal illness.

Once the methodological approach was defined, optimal descriptors were selected for organ dysfunction through a cohort study of 700 critically ill patients admitted to a surgical ICU in a tertiary care center. Dysfunction within each organ system was measured on a scale ranging from zero to 4 to provide maximal separation of increasing degrees of abnormality. Values of each variable were calibrated so that a score of zero represented normal function and predicted a mortality of 5 per cent or less, whereas a score of 4 reflected profoundly deranged function and correlated with a mortality of 50 per cent or greater. Intervening values of the variable were selected to provide sensible intervals of approximately comparable magnitude. A scale with five incremental increases allowed the

Table 2 Calculation of MOD scores*

Admission MOD score
The sum of the values for the six MOD score variables obtained on the first ICU day

Daily MOD score
The sum of the values for the six MOD score variables calculated separately for each ICU day

Cumulative MOD score
The sum of the worst daily values for each variable up to the current day. The baseline score is the admission MOD score. Further deterioration in any given variable is reflected by increasing the score for that system, while improvement is not recorded. In contrast with the daily score, the cumulative score reflects only worsening of function, and provides a measure of the time course and severity of clinical deterioration following ICU admission

Aggregate MOD score
The sum of the worst day's value for each of the six variables over the entire ICU stay. Thus the value for the respiratory component may be obtained on day 2 and that for the renal component on day 11. This has been the standard approach used to quantify organ failure or dysfunction; it can be appreciated that the aggregate MOD score will be identical with the cumulative score on the last ICU day

ΔMOD score
The difference between the aggregate score, or the cumulative score over a defined period, and the admission MOD score. As a measure of the deterioration in organ function occurring over a set time interval, it provides a measure of morbidity attributable to events occurring following admission to the ICU and therefore potentially amenable to therapeutic intervention

Mortality-adjusted score
The aggregate score, adjusted so that a maximal score of 24 is given to any patient who dies during the ICU stay, regardless of the score at the time of death. The mortality-adjusted score is a combined measure of clinically significant morbidity and mortality during the ICU stay

*Raw data for each of the six systems are recorded daily. For each variable, the value selected is representative, usually the first recorded value of the day. Pre-resuscitation values or obvious laboratory errors are omitted; missing data are assigned the value recorded for the previous day or, if none are available, are scored zero.

final score to be sensitive to relatively small changes in function, yet the increments for each variable were sufficiently large for a one-point change in score to represent a significant change in function.

The quantification of organ dysfunction in critical illness

An organ dysfunction score is a tool that provides an objective quantitative measure of a complex phenomenon such as multiple organ dysfunction syndrome. Such a tool has a number of potential uses (Table 2).

Organ dysfunction as a prognostic indicator

It is a basic truism of intensive care that the sickest patients are those who are most likely to die during their ICU course. Application of this principle to the task of risk prediction in the ICU led to the development of dedicated prognostic scores, each of which was developed using statistical approaches that maximized predictive capability.

A scale that grades increasing degrees of organ system dysfunction will of necessity have some predictive capacity. If 80 per cent of all deaths in the ICU result from multiple organ dysfunction syndrome, then 20 per cent of ICU deaths occur in the absence of significant degrees of organ dysfunction, and a scale that has been designed to maximize construct validity should not have the prognostic power of a scale that has been designed to predict mortality.

Nonetheless, the degree of organ dysfunction early during the ICU stay is an important determinant of the ultimate risk of ICU mortality. Organ dysfunction scores calculated within the first 24 h of ICU admission are highly associated with ultimate risk of mortality, ranging from 2 per cent for patients with a score of zero to 100 per cent for patients with a score of 13 or higher (Marshall *et al.* 1995). Quantification of the degree of organ dysfunction at the time of admission can provide an objective measure of the severity of the syndrome in those patients in whom it is present and therefore could be a useful means of stratifying patients at the outset of clinical trials or of quantifying initial severity of illness in disease processes such as intra-abdominal infection where organ dysfunction is an important outcome.

Organ dysfunction as a point measure of illness severity

Daily organ dysfunction scores provide a measure of both illness severity and intensity of therapeutic intervention. Tracking daily MOD scores over time provides a global evaluation of net clinical improvement or deterioration.

Organ dysfunction as an aggregate measure of ICU morbidity

More commonly, organ dysfunction is described as a process occurring over time. This approach has been almost universally employed in studies of multiple organ failure in which the number of failing systems are counted independent of whether these failures occur concurrently or sequentially. Aggregate scores are calculated by taking the worst values for each variable over a defined time, usually the entire ICU stay. Just as ICU prognosis is a function of the number of failing systems, so there is a strong correlation between aggregate organ dysfunction scores and ultimate ICU or hospital mortality (Fig. 1).

Although it has been customary to evaluate organ dysfunction over the entire ICU stay, the cumulative evolution of organ system dysfunction over any particular interval can also be assessed (Table 2). Such a measure provides an estimate of the time course and magnitude of deteriorating organ function.

ΔMOD scores

ICU prognosis is strongly influenced by the severity of illness at the time that the patient enters the unit. However, from a therapeutic

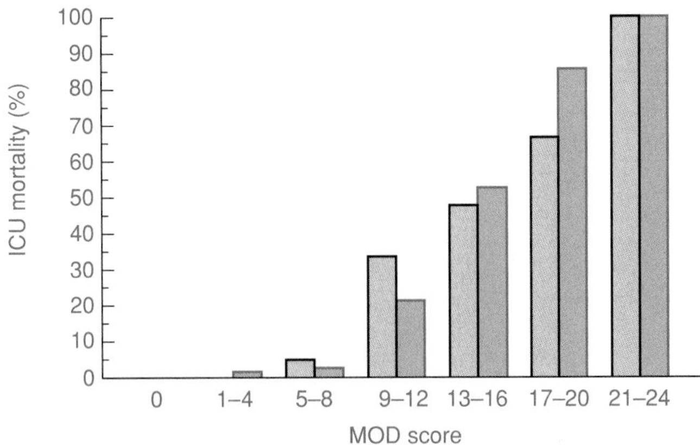

Fig. 1 Incremental increases in aggregate MOD scores are associated with increased ICU mortality: gray bars, patients in the first (developmental) half of the MOD score data set; green bars, patients in the second (validation) half of the MOD score data set. (Reproduced with permission from Marshall *et al.* (1995).)

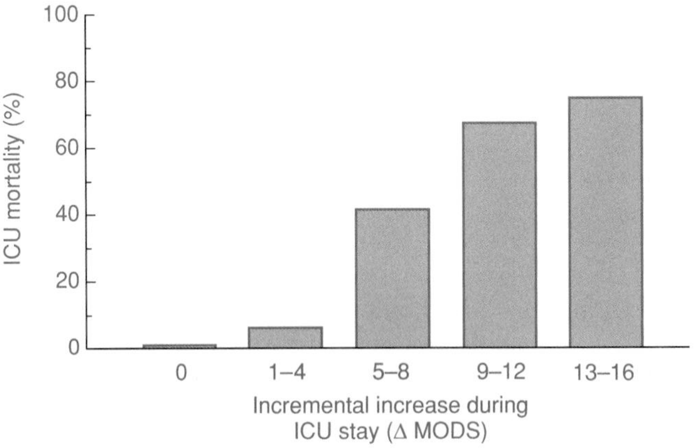

Fig. 2 Increases in MOD scores (the difference between aggregate and admission scores), representing morbidity attributable to events within the ICU, correlated with increasing ICU mortality. (Reproduced with permission from Marshall *et al.* (1995).)

perspective, the clinician is more interested in what happens to the patient after he or she has been admitted. A quantitative measure of the degree of morbidity that develops following ICU admission can be obtained by calculating a ΔMOD score. The ΔMOD score provides an attractive measure of attributable ICU morbidity, and is of potential use in clinical trials or quality assurance activities. Moreover, regardless of the degree of organ dysfunction present when the patient is admitted, mortality increases incrementally with increasing values of the ΔMOD score (Marshall *et al.* 1995) (Fig. 2).

Mortality-adjusted organ dysfunction scores

Finally, organ dysfunction scores can incorporate a measure of both morbidity and mortality by assigning a maximum number of organ dysfunction points to any patient who dies while in the ICU. Since survival despite a high score is clearly a superior outcome to death with a low score, this mortality-adjusted MOD score provides a measure of net clinical benefit that reflects both morbidity and mortality.

Organ dysfunction scoring – the future

A scoring system does not provide new information, but merely combines multiple divergent factors into a single variable.

The greatest potential use of organ dysfunction scores is as an objective measure of clinical outcome that can be used in ICU-based clinical trials. Such scales are more sensitive than mortality to clinically important effects of therapy, and therefore can facilitate the performance of studies by permitting smaller sample sizes. Calculation of severity of illness at the time of admission to the study permits more refined patient stratification. Objective systems for describing organ dysfunction are a prerequisite to studies of the epidemiology and pathophysiology of the syndrome. Moreover, the availability of objective measures of ICU morbidity and mortality has an obvious contribution to make to quality assurance activities.

The clinical utility of such scores has yet to be widely evaluated. Clearly, the science of prognostication is not sufficiently evolved to allow decisions regarding continuation or cessation of therapy on the basis of an organ dysfunction score. However, in patients with a complex disease process, such scales may show evidence of net clinical improvement or deterioration that can provide support for a particular approach to therapy or suggest the need for clinical re-evaluation.

References

Bone, R.C., *et al.* (1992). Definitions for sepsis and organ failure and guidelines for the use of innovative therapies in sepsis. *Chest*, **101**, 1644–55.

Le Gall, J.R., *et al.* (1996). The Logistic Organ Dysfunction System: a new way to assess organ dysfunction in the intensive care unit. *Journal of the American Medical Association*, **276**, 802–10.

Marshall, J.C. (1995). Multiple organ dysfunction syndrome (MODS). In *Clinical trials for the treatment of sepsis* (ed. W.J. Sibbald and J.-L. Vincent), pp. 122–38. Springer-Verlag, Berlin.

Marshall, J.C., Cook, D.J., Christou, N.V., Bernard, G.R., Sprung, C.L., and Sibbald, W.J. (1995). Multiple Organ Dysfunction score: a reliable descriptor of a complex clinical outcome. *Critical Care Medicine*, **23**, 1638–52.

Vincent, J.L., *et al.* (1996). The Sepsis-related Organ Failure Assessment (SOFA) score to describe organ dysfunction/failure. *Intensive Care Medicine*, **22**, 707–10.

16 Monitoring equipment and techniques

16.1 Cardiorespiratory monitoring

16.1.1 ECG monitoring

W. R. Lewis and E. S. Kaufman

Key messages

- Supported cardiac monitoring has been shown to lower mortality in patients with confirmed myocardial infarction.
- Cardiac monitoring equipment is useful in the detection of serious cardiac arrhythmias in other subsets of patients at risk.
- This equipment must be supported by staff able to respond to detected arrhythmias and initiate appropriate treatment.
- The technology has advanced to allow arrhythmia classification, radiotelemetry, and ST-segment monitoring.

Introduction

The routine use of ECG monitoring in the intensive care unit (ICU) setting was initially supported in the management of acute myocardial infarction. Lawrie *et al.* (1967) showed that a 20 per cent reduction in mortality from acute myocardial infarction could be achieved by early defibrillation of ventricular fibrillation detected by an oscilloscope and alarm system. Although monitoring systems have increased in their complexity, the components of successful monitoring remain the same: a reliable ECG display, the ability to detect an arrhythmia (whether human, electronic, or a combination of both), and the personnel and techniques to correct the arrhythmia disturbance.

The routine use of ECG monitoring is most helpful in the management of disease processes in which a cardiac arrhythmia is likely to occur. Economic and space limitations have made it impossible to monitor all individuals at risk for arrhythmias in an intensive care setting. Fortunately, not all such patients require intensive hemodynamic or respiratory monitoring. Coronary care units, high-dependency units, and the advent of radiotelemetry monitors (telemeters) have allowed cardiac monitoring in many patients who do not require intensive nursing care. The cost-efficient nature of telemeters has changed the scope of their use. Initially, they were used solely for protective reasons, i.e. the immediate detection of life-threatening arrhythmias. Subsequently, telemeters have evolved to serve both a protective and a diagnostic purpose; they detect life-threatening arrhythmias and double as in-patient Holter monitors.

Specific indications

Cardiac diagnoses

Monitoring is essential in patients with documented myocardial infarction. The highest risk of ventricular fibrillation after myocardial infarction is in the first 24 h. The typical duration of monitoring is 2 to 3 days following an uncomplicated myocardial infarction. Patients with suspected but not confirmed myocardial infarction can be safely admitted to non-intensive care monitored beds. Patients with unstable angina pectoris should also be admitted to monitored units because they are susceptible to myocardial infarction and malignant cardiac rhythm problems.

ECG monitoring is also beneficial in the setting of latent or suspected cardiac arrhythmias. For example, patients with recent syncope of suspected cardiac origin should be monitored during evaluation. Syncope is more likely to be of cardiac origin if underlying organic heart disease is present. In the case of syncope, monitoring is performed to protect the patient and possibly document the cause.

Non-cardiac diagnoses

Patients with diminished cardiac reserve who have undergone surgery should be monitored in the postoperative period. Myocardial infarction is most likely to occur in the first 6 days after non-cardiac surgery. Patients who have had coronary artery bypass grafting are at risk of ventricular and supraventricular arrhythmias and should have ECG monitoring. If atrial epicardial pacing electrodes are placed in the postbypass patient, another monitor can be used to display the atrial electrogram for early detection and diagnosis of atrial fibrillation and flutter.

Patients with severe neurological problems are susceptible to cardiac rhythm disorders and should have ECG monitoring. ECG monitoring is useful in patients with respiratory distress, including those requiring mechanical ventilation. Patients with acute drug toxicity from agents known to have arrhythmic consequences should be monitored. For example, an overdose of tricyclic antidepressants can result in torsade de pointes. In addition, we admit to telemetry units patients initiating therapy with potentially proarrhythmic drugs. Examples of such drugs are class I antiarrhythmics (such as

Table 1 American College of Cardiology guidelines for in-hospital cardiac monitoring of adults for detection of arrhythmia

Class I: Electrocardiographic monitoring is indicated in most if not all patients (significant risk of life-threatening arrhythmias)

Examples
- Acute myocardial infarction
- Suspected myocardial infarction
- Unstable angina
- Arrhythmogenic drug toxicity
- Severe electrolyte imbalance
- Convalescence from coronary artery bypass surgery
- Initiation of antiarrhythmic drugs
- After coronary angioplasty or catheter ablation
- Severe hemodynamic, respiratory, or neurological illness
- Suspected cardiac syncope
- Documented new arrhythmias (atrioventricular block, ventricular tachycardia)

Class II: Monitoring is indicated in some patients

Examples
- Non-life-threatening arrhythmias undergoing control
- History of life-threatening arrhythmias recently controlled >3 days after uncomplicated myocardial infarction
- Syncope with moderate risk of cardiac origin

Class III: Monitoring is not indicated

Examples
- Postoperative patients at low risk of arrhythmias
- Uncomplicated obstetric patients
- Terminal illness where arrhythmias will not be treated
- Chronic stable atrial fibrillation
- Stable ventricular ectopy
- Uncomplicated cardiac catheterization

Reproduced with permission from Jaffe *et al.* (1991)

quinidine) and class III antiarrhythmics (such as sotalol), which can induce torsade de pointes and other serious arrhythmias. Severe electrolyte abnormalities can cause arrhythmias which have hemodynamic consequences. Hyperkalemia, whether from altered renal function, drug effects, or severe tissue damage (as in burns), can result in severe bradyarrhythmias or ventricular fibrillation. Hypokalemia and hypomagnesemia can result in torsade de pointes. In each of these cases, cardiac monitoring can lead to early diagnosis and effective treatment. A summary of the American College of Cardiology recommendations for ECG monitoring (Jaffe *et al.* 1991) is shown in Table 1.

Safety of telemetry units

When appropriately used, telemetry monitoring units without intensive care nursing are safe. Estrada *et al.* (1994) studied 467 telemetry unit admissions and found that the in-hospital mortality was 1.9 per cent; 8.1 per cent of patients required transfer to an ICU. Their study confirmed that telemetry patients are at lower arrhythmic risk than ICU patients. Only 1 per cent of the total population was transferred to an ICU because of ECG findings; the remainder were transferred because of other symptoms. This study reveals the current trend towards utilizing telemeters for management of lower-risk arrhythmias, diagnostic monitoring, and observation of patients at low risk of life-threatening arrhythmias.

It should be recognized that changes in cardiac rhythm are not sensitive early warning indicators of respiratory or hemodynamic distress. Patients at high risk of hemodynamic or respiratory problems should be admitted to ICUs, where the nurse-to-patient ratio is higher than on cardiac telemetry units.

Monitoring equipment

The technology for cardiac monitoring has improved greatly since 1967. Event detection algorithms are able to alarm for sustained tachycardias and bradycardias. Detection of a cardiac QRS requires not only amplitude recognition but also discrimination between noise and slower upstroke depolarizations. New sophisticated analysis systems are capable of 'learning' the normal QRS morphology and then detecting, displaying, storing, and counting different QRS morphologies, runs, and patterns. Pacing stimulus artifacts can be magnified to allow pacemaker analysis. Computer digitization of the ECG signal has allowed storage of an entire 24-h display in a Holter or full-disclosure format. Reports can be printed which display graphics for heart rate and supraventricular and ventricular ectopy.

Monitoring systems have been developed by several manufacturers. To ensure that different telemeters meet optimum requirements for ECG monitoring, the American Heart Association has developed recommendations for standardized lead placement, instrumentation performance, and personnel training. The American Heart Association recommends that at least two leads be monitored. Multiple-lead display facilitates detection of P waves, QRS-axis changes, and ST-segment changes (ischemia may be manifest in only one plane). Multiple-lead display also minimizes data loss due to noise. Detection algorithms developed by various manufacturers are tested against standardized rhythm databases according to protocols established by the Association for the Advancement of Medical Instrumentation. The American Heart Association Database, The Massachusetts Institute of Technology Database, and the European Society of Cardiology ST-T Database are among the most widely used.

However, we have observed many examples of failure to alarm during life-threatening arrhythmias, no matter how sophisticated the telemeter monitoring system. Automated alarm systems are imperfect. An ECG display is of little value unless the ability to detect and respond to arrhythmias is present. Arrhythmia detection software must be sensitive and specific. If the system is not sufficiently sensitive, arrhythmias will be missed. If the system is very sensitive but not sufficiently specific, frequent alarms for noise may make the staff less apt to respond in a 'boy who cried wolf' syndrome. To overcome some of these problems of instrumentation, we favor using a monitor watcher, i.e. a person trained in arrhythmia recognition who watches the ECG displays in order to alert nursing staff to rhythm abnormalities. In the National Survey of Continuous Electrocardiographic Monitoring in United States Hospitals, 55 per cent of hospitals employed a monitor watcher in their 'step-down' or telemetry unit (Jenkins and George 1995). The monitor watcher need not possess a registered nurse degree. Forty-four per cent of hospitals indicated that they employed monitor technicians. ICU monitoring does not necessarily require a monitor watcher since nursing personnel are usually at the bedside. The National Survey found that only 29.5 per cent of hospitals utilized a monitor watcher in the ICU. Vigilance to the monitors can be physically and mentally challenging. A monitoring station should be quiet and free from frequent interruptions. The space should be comfortable with the monitors at eye level and with desk space for documentation. The number of monitors watched by a single individual should be limited and should be directly proportional to the sophistication of the arrhythmia detection system. Provided that the system has good sensitivity and specificity, 24 tracings can be watched by a single individual.

As previously mentioned, equipment is only one part of an ECG monitoring unit. Appropriate life-support systems including defibrillators, temporary pacing, and medications are required to treat serious arrhythmias. Lifesaving equipment should be accompanied by nurses and physicians fully trained in cardiac resuscitation and advanced cardiac life support. Patients who are placed on cardiac monitors for the detection of life-threatening arrhythmias should remain monitored by escorts when taken to tests at remote sites.

The future

The first three decades of cardiac monitoring have focused on the rhythm of the heart. In the future ECG monitoring will include other aspects of the electrocardiogram. Ischemia and infarction are manifest on the surface electrocardiogram as ST-segment depression or elevation. Current systems are capable of monitoring ST-segment shifts. Such ECG monitors are useful in the setting of acute myocardial infarction and after percutaneous transluminal coronary angioplasty (Gurfinkel *et al.* 1994). Because ischemia can occur in different regions of the heart, the number of leads monitored has increased. The utility of ST-segment monitoring is not limited to cardiac patients. Ischemia is often detected in patients after non-cardiac surgery. The highest incidence of silent ischemia occurs during weaning from the ventilator. Thus monitoring the ST segment may allow for earlier intervention in patients with underlying coronary disease.

Conclusions

ECG monitoring has become the standard of care in patients at risk of cardiac arrhythmias. Many commercial systems are available with a variety of sophistication. Standardized testing of such equipment is performed to assure quality care for the arrhythmia patient. However, equipment is not the only requirement for quality monitoring. Highly trained personnel are required to identify serious arrhythmias and to carry out appropriate corrective interventions. Finally, in the age of optimum utilization, careful patient selection for ICU and telemetry monitoring units is essential.

References

Estrada, C.A., Prasad, J.K., Rosman, H.S., and Young, M.J. (1994). Outcomes of patients hospitalized to a telemetry unit. *American Journal of Cardiology*, **74**, 357–62.

Gurfinkel, E., *et al.* (1994). ST-segment computerized monitoring before and after angioplasty: clinical correlation with recurrent angina during the short-term follow-up. *Clinical Cardiology*, **17**, 433–6.

Jaffe, A.S., *et al.* (1991). Recommended guidelines for in-hospital cardiac monitoring of adults for detection of arrhythmia. *Journal of the American College of Cardiology*, **18**, 1431–3.

Jenkins, L.S. and George, V. (1995). National survey of continuous ECG monitoring in U.S. hospitals. *Journal of Nursing Administration*, **25**, 38–44.

Lawrie, D.M., Greenwood, T.W., Goddard, M., Harvey, A.C., Julian, D.G., and Oliver, M.F. (1967). A coronary-care unit in the routine management of acute myocardial infarction. *Lancet*, **ii**, 109–14.

16.1.2 Pulse oximetry

J. A. Pologe

Key messages

- While the principle upon which pulse oximetry is based is fairly simple, the design and implementation of a working pulse oximeter is difficult.
- Even when completed properly, the finished product falls far short of being able to measure accurately and display arterial oxygen saturation data on all patients under all conditions.
- With proper understanding of its operation and limitations, the pulse oximeter is the single most valuable tool available to the clinician for monitoring arterial oxygen saturation.
- This real-time monitor allows the clinician to detect, with only a few seconds delay, even a small reduction in the arterial oxygen saturation.

Measurement principles

The fundamental physical property that allows the measurement of arterial oxygen saturation by pulse oximetry is that blood changes color with saturation. A pulse oximeter measures the 'color' of the arterial blood and correlates this 'color' with a given oxygen saturation which is then displayed. Hemoglobin, in its reduced or oxygenated state, strongly absorbs light at all wavelengths below approximately 630 nm which includes the entire visible spectrum apart from the red region. This is why the light from a flashlight placed against one side of the hand appears red when viewed from the other side. When blood is well oxygenated it does not absorb much red light, but as it desaturates it absorbs an increasing amount so that the blood takes on a darker appearance. The opposite behavior occurs in the near-infrared region (about 810–1000 nm) where hemoglobin absorbs more light when saturated with oxygen than when it is desaturated. For this reason, current pulse oximeters use two emitters, usually light-emitting diodes; one is designed to generate light in the red region, typically centered around 660 nm, and one which generates light in the near-infrared region, usually centered at 925 or 940 nm.

The saturation of a suspension of pure hemoglobin in a cuvette can be determined by measuring the ratio of light absorbed at a red wavelength (say 660 nm) to that absorbed at an infrared wavelength (say 940 nm); this ratio A_{660}/A_{940} will have a one-to-one correlation with oxygen saturation. The absorption of light, at any given wavelength, is determined by directing a beam of light with a known intensity (the incident intensity) on the hemoglobin solution and measuring its intensity after passage through the solution (the transmitted intensity); absorption is defined as the ratio of the incident intensity to the transmitted intensity. Measurement *in vivo* is not quite so simple. In the body, there is both venous and arterial blood as well as bone and other tissue. Further, the hemoglobin is not homogeneously distributed throughout the blood but confined in red blood cells that are highly effective light scatterers, making it almost impossible to collect all the light that is not absorbed by the sample. These facts make direct measurement of the absorption of the arterial blood quite difficult.

How is it possible to measure the absorption of arterial blood only, so that arterial oxygen saturation can be measured accurately? Instead of measuring the ratio of absorptions, the solution is to measure the ratio of differential absorptions (Fig. 1), or at least an approximation thereof, which is defined as follows:

$$\frac{\Delta A_{660}}{\Delta A_{940}} \approx \frac{dA_{660}}{dA_{940}}$$

where dA is the differential change in absorption and ΔA is the physical measurement made to approximate the differential change in absorption.

Without going through the mathematics involved, suffice it to say that measuring a differential change in absorption rather than absorption itself simplifies the problem significantly. It is no longer necessary to know the incident intensity, so that instrument design is substantially simplified and the effect of scattering is minimized. The measurement now involves only the instantaneous change in absorption which in living tissue is mainly due to the change in path-length of the arterial blood; absorption in venous blood, skin, and bone is almost negligible. The fact that differential rather than direct absorption is measured is what makes pulse oximetry possible.

Limitations

Although the preceding description is only a sketch of how pulse oximetry is actually accomplished, it provides an adequate basis for beginning to understand what pulse oximetry can and cannot measure as well as the limitations inherent in the technique.

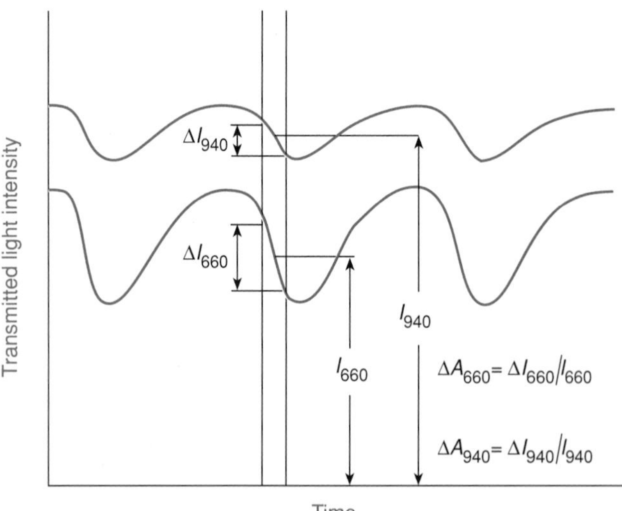

Fig. 1 The light generated by each emitter is modulated by the arterial pulsation as it passes through the tissue. The differential absorption is calculated by dividing the small change in intensity by the total intensity of the output light.

The most obvious limitation of pulse oximetry is based on the fact that it is only a two-channel system (two wavelength bands) and therefore can only resolve two components in the blood. In pulse oximetry it is assumed that only oxyhemoglobin and reduced hemoglobin are present in the arterial blood. Any additional chromophores in the arterial blood which absorb light in the wavelength bands used by the instrument will lead to unpredictable results. These additional chromophores include other hemoglobin species and dyes that may be injected into the blood for diagnostic purposes. The two most common species of hemoglobin, after oxyhemoglobin and reduced hemoglobin, are carboxyhemoglobin and methemoglobin. If either of these dyshemoglobins are present in concentrations above normal, the pulse oximeter will give a falsely elevated reading for the arterial oxygen saturation.

The term 'arterial oxygen saturation' has different meanings in different texts. Here, it is defined as the concentration of oxyhemoglobin divided by the concentration of total hemoglobin, expressed as a percentage. This is also known as fractional oxygen saturation. (Functional saturation is defined as the ratio of the oxyhemoglobin concentration to the sum of the oxyhemoglobin concentration and the reduced hemoglobin concentration, expressed as a percentage.) Thus, if 80 per cent of a patient's arterial hemoglobin is oxygenated and the remaining 20 per cent is deoxygenated (or reduced), the pulse oximeter will read 80 per cent. However, if, instead of deoxygenated hemoglobin, the patient had 20 per cent of his or her hemoglobin bound to carbon monoxide, the pulse oximeter would read almost 97 per cent, thus deceiving the clinician into thinking that the patient's blood was adequately oxygenated. A similar problem would occur in the presence of 20 per cent methemoglobin and 80 per cent oxyhemoglobin; in this case the instrument would read approximately 90 per cent saturation, again indicating a falsely elevated saturation. Note that these pulse oximeter readings in the presence of dyshemoglobins are neither the fractional nor the functional saturation. This is a severe and potentially dangerous limitation of current pulse oximetry.

Some manufacturers calibrate their pulse oximeters to functional saturation and some calibrate to fractional saturation. This is done by slowly desaturating a number of healthy volunteers by having them breathe hypoxic gas mixtures. At various saturation levels, arterial blood samples are drawn for invasive analysis of oxygen saturation. These measurements are paired with the readings made by the pulse oximeter at the same saturation levels, and these pairs of data points are used to generate the pulse oximeter's calibration curves. These are the curves that the instrument will use to relate the measurements made by the pulse oximeter (the ratio of the differential absorbances) to the arterial oxygen saturation to be displayed. It is important to recognize that, while the manufacturer may calibrate to either functional or fractional saturation, the oximeter can read neither. Its measurements are accurate only if the dyshemoglobins present in the blood are at normal levels (1–3 per cent) and all the remaining hemoglobin is either oxygenated or reduced. The difference in the calibration methodologies simply means that an instrument calibrated to functional saturation will always read slightly higher (approximately 1–3 per cent) than the same, or any other, pulse oximeter that is calibrated to fractional saturation.

Another limitation of the pulse oximeter stems from the fact that it needs an arterial pulse in the tissue under test to generate the signal necessary to measure oxygen saturation. When the pulse in the tissue under test, such as a finger in the case of a finger probe, is small or absent, the instrument is apt to give erroneous information. The pulse at the probe site may be small because of poor perfusion of the patient in general or poor perfusion at the probe site (e.g. because of cold hands). When perfusion is low, the pulse oximeter will generally try to increase the amplification of whatever signal it receives in an attempt to continue to provide readings of oxygen saturation. Thus any noise present, say from patient motion or from room lights that may be pulsating, will make a larger contribution to the signal and may cause the oximeter to give readings that have very little to do with the arterial oxygen saturation. Although this is a problem with current oximeters, it need not necessarily mislead the clinician. If the pulse oximeter waveform displays a reasonably clean pressure waveform for approximately 10 s, the oximeter is probably reading correctly. If the instrument does not have a waveform display, the clinician can attempt to correlate the pulse rate with the ECG and observe the stability of the saturation readings as an indication of accuracy. However, these techniques are not as reliable as being able to 'see' the data that the oximeter is actually using to make its calculations.

It would be easy to make a pulse oximeter that rarely reads incorrectly in a low-perfusion situation by not allowing readings at all below some specific trigger level. In fact, oximeters do stop making readings below some preset signal level, but the manufacturers attempt to set this point as low as possible to allow the clinician to obtain saturation readings on patients who are very sick. While this is a good strategy most of the time, it does lead to some additional problems. Sometimes pulse oximeters appear to be reading satisfactorily on a patient whose heart has stopped beating. This phenomenon is due to attempts by the instrument to read the smallest possible signal. Anything that looks like a pulsatile signal will be interpreted by the oximeter as a pulse, and the instrument will make every effort to calculate and display an oxygen saturation reading. This pulsatile signal could be generated by a ventilator inducing a pressure waveform into the venous system, or just by physically moving or slightly jostling the patient and therefore the probe. Any motion of the probe on the patient will introduce a false signal in the pulse oximeter. Again, the pulse oximeter's waveform display allows easy identification of artifactual signals.

A few other potential sources of error should also be mentioned. Excessive patient motion, particularly at the probe site, can generate signals that dwarf those caused by the arterial pulsation. Even when the tissue under test is well perfused, motion can cause false readings which may be either high or low. Electrical noise such as that generated by electrosurgery can also diminish the accuracy of pulse oximeter readings. Because the magnitude of these effects on the pulse oximeter's ability to make accurate readings depends on the perfusion level of the tissue under test, it is always best to try to find a probe site that provides the maximum possible signal level. Perfusion can be enhanced by warming the patient, rubbing the tissue at the probe site, or using vasodilator creams. Finally, high ambient light levels can overwhelm the signal detection circuitry so that the oximeter is unable to make any measurements whatsoever.

Interpretation of readings

It is important to recognize that the pulse oximeter only identifies the percentage of the hemoglobin that is bound to oxygen and not the total amount of oxygen that is being delivered. In other words, if a patient had an oxygen saturation of 100 per cent but was anemic, he or she could still be receiving only half the oxygen that the tissues need. The same situation could occur if the saturation were 100 per cent but the cardiac output was low. Thus with normal levels of dyshemoglobins and adequate signal strength (adequate perfusion at the probe site), we only know for sure that the hemoglobin is being properly loaded with oxygen and therefore that the lungs must also be reasonably well perfused and exchanging gas properly.

On the otherwise healthy patient, arterial oxygen saturation should typically be about 96 per cent and quite stable. If the oximeter is reading less than the normal oxygen saturation levels, the source of the low readings should be investigated (e.g. ventilation–perfusion mismatch or inadequate fraction of inspired oxygen, possibly because of faulty or incorrectly set-up gas mixers; various hemoglobinopathies, such as sickle cell hemoglobin, can reduce the ability of hemoglobin to bind with the available oxygen).

Bibliography

Pologe, J.A. (1987). Pulse oximetry: technical aspects of machine design. *International Anesthesiology Clinics*, **25**, 137–53.

16.1.3 Blood pressure monitoring

R. M. Gardner

Key messages

- Direct vascular pressure measurements can be safe, accurate, and reliable.
- Understanding the interaction of 'plumbing' and monitoring equipment is required if accurate pressure measurements are to be made.
- Zeroing the monitoring system is one of the most important steps taken in pressure monitoring.
- Simple plumbing set-ups, with short lengths of tubing and few stopcocks, provide the most optimal dynamic response and recording mechanism.
- Fast-flush testing to validate adequate dynamic response is essential for deriving accurate systolic and diastolic pressures.

Introduction

Invasive blood pressure monitoring incorporates mechanical and electronic technologies that were formerly only used in the heart catheterization laboratory. With the introduction of disposable devices and modern computerized bedside monitors, clinicians can now conveniently obtain accurate and timely arterial and pulmonary artery pressures. To obtain and maintain the most accurate pressure waveforms and measurements, and to minimize risks to patients, clinicians must understand the principles of operation of these monitoring systems. The two principal purposes for inserting an arterial or pulmonary artery catheter are to measure intravascular pressures and to withdraw blood samples for blood gas or other biochemical analysis.

Continuous and accurate assessment of blood pressures can only be made by inserting vascular catheters. Having pressures displayed continuously permits timely detection of dangerous hemodynamic events and permits display of information to initiate or titrate patient therapy. However, invasive pressure monitoring provides accurate information only when correct equipment and techniques are used.

Equipment

The equipment used to measure blood pressures comprises two components (Fig. 1).

1. The plumbing system consists of a fluid-filled catheter, stopcocks, pressure tubing, a continuous-flush device with its pressurized fluid source, and a transducer.

2. The monitoring system consists of the following electronic and display components: the excitation voltage source for the transducer, a zero control, an oscilloscope display, and a processor to derive measures for the digital display (systolic, diastolic, mean pressures, and heart rate).

Plumbing system

The components comprising the plumbing system must always be kept sterile because the fluid contained within them comes into direct contact with the patient's blood. Most of these components are rather inexpensive and thus are single-use devices.

Catheter

Arterial catheters should be as small in diameter as is practical. The smaller the diameter, the less is the risk of obstruction of the artery. However, the wall must be thick enough to prevent kinking and obstruction of the catheter. The length of the catheter insertion should be sufficient to prevent spontaneous ejection from the artery. The diameter and length of the pulmonary artery catheter are set by the number of lumens required and the length of catheter required to reach the pulmonary artery. The diameter of each lumen is small and, coupled with the long length of the catheter, results in a large resistance which makes it difficult to achieve adequate dynamic response.

Stopcocks

A stopcock is usually located near the catheter connection and is used as a site for blood withdrawal for blood gas or similar analyses. When filling the plumbing system great care should be taken to ensure that all the central cavities of all stopcocks are filled with fluid. Air bubbles can lead to embolism if they are 'flushed' into the patient and can also diminish the fidelity of the pressure recording. In addition, stopcocks are a vulnerable site for patient contamination. Hands should be washed before stopcocks are touched, open stopcock ports should never be touched, and ports not in active use should be covered. Each connection in the plumbing system, particularly those involving stopcocks, can be problematic. The connections present a size discontinuity in the fluid pathway and become sites for entrapment of air bubbles. Unfilled central cavities of stopcocks are a frequent source of air bubbles that distort dynamic response.

Pressure tubing

The catheter and stopcock are normally connected to the pressure transducer with pressure tubing. Since the fluid-filled column between the catheter tip and the transducer must faithfully transmit the patient's pressure waveform, the tubing should be short (typically less than 50 cm) and non-elastic (soft venous tubing should not be used). Long lengths of tubing add resistance to the pressure waveform as it is transmitted from the catheter tip to the pressure transducer. Such resistance distorts the transmission of the pressure waveform. With elastic tubing, the pressure signal is distorted, 'pulsating' the pressure tubing, and therefore the signal detected by the external pressure transducer is distorted.

Continuous-flush device and pressurized fluid source

The continuous-flush device with its attached pressurized fluid source is used to fill the plumbing system. The device contains a tiny controlled 'leak' so that it infuses from 1 to 3 ml/h of fluid through the plumbing system and helps prevent clot formation at the catheter tip. The fluid source is usually an intravenous bag of normal saline pressurized in a cuff to 300 mmHg. The bag of saline should have the air expelled before connecting it to the plumbing system to prevent air embolism as it empties. A microdrip chamber should be attached to the fluid-filled source so that the flow rate can occasionally be monitored to ensure that the bag is pressurized and to detect when it is empty. Since it is impossible to clean and sterilize continuous-flush

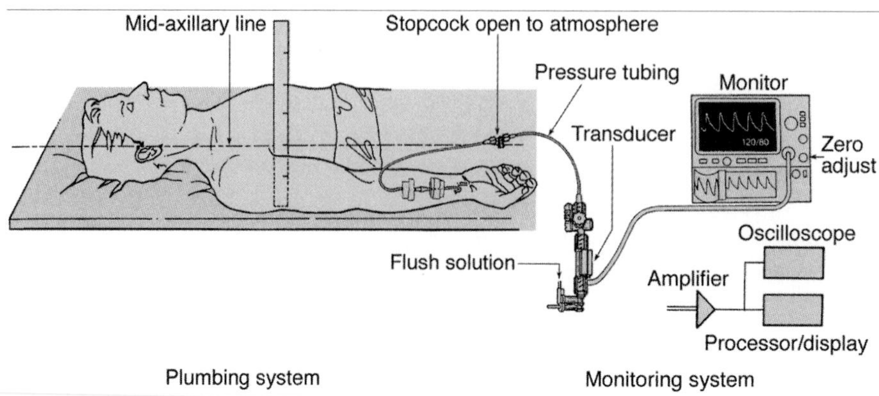

Fig. 1 The plumbing and monitoring systems used for pressure monitoring. The zeroing methodology for a pressure-monitoring system is also shown.

devices thoroughly, and to ensure that the tiny 'leak' functions properly for continuous flushing, they should not be reused.

Pressure transducer

Pressure transducers are used to convert the patient's pressure signal into an electrical signal. Modern blood pressure transducers have resulted from a remarkable set of advances in technology. Semiconductor technology has produced the transistor, processor chips for digital watches, and personal calculators. The same technology is now used to make pressure transducers. These transducers are calibrated to within ± 3 per cent accuracy and maintain their zero set-point. Modern pressure transducers have a sensitivity of 5 μV/mmHg applied per volt of excitation voltage. The transducers are remarkably rugged, accurate over time, temperature stabilized, and relatively inexpensive. Since it is impossible to clean and sterilize disposable transducers thoroughly, they should not be reused.

Monitoring system

Transducer excitation, amplifier, and filter

The sensing elements of most pressure transducers are four resistive elements connected as a Wheatstone bridge. Resistive bridges require the application of an excitation voltage (typically 3–5 V). Modern monitors provide very accurate (± 0.1 per cent) excitation voltages to the transducers. These monitors also contain amplifiers which take the small voltages supplied by the transducer, typically 30 μV/mmHg, and 'magnify' them by about 1000 times. In amplifiers, 'magnification' is called 'gain'. With modern amplifiers, the gain is very stable and is calibrated with an accuracy to about ± 1 per cent. Because of the accuracy, reliability, and stability of both the transducers and amplifiers, it is no longer necessary to calibrate pressure monitors (Cooper and Paulsen 1994; Gardner 1996). Most pressure amplifiers include a low-pass filter to filter out unwanted high-frequency signals.

Zero control

The zero control on a pressure monitor permits the adjustment of the output signal from the monitor to be zero when the fluid–air interface of a designated 'zeroing stopcock' is placed at the mid-axillary position. Once the zeroing process has been completed it is important that the vertical relationship between the patient and the pressure transducer remains fixed. If there is a vertical movement of the transducer or of the patient relative to the transducer, the system must be 're-zeroed.' However, the vertical position of the zeroing stopcock is unimportant once it is closed.

Operational issues

Zeroing

Zeroing the pressure-monitoring system is probably the most important step in establishing the system. Mistakes in zeroing can result in substantial errors in pressure measurement, particularly when measuring pulmonary artery pressures. The zeroing process involves an interaction between the position of the transducer and the stopcock in the plumbing system with the zero control of the monitoring system. For each 10 cm of misalignment of the zeroing point, there is a pressure offset of $10/1.36 = 7.3$ mmHg. Such a misalignment may cause an error of as much as 100 per cent in measurement of central venous pressure. The vertical position of the transducer relative to the patient's mid-axillary line can change, particularly if it is mounted externally. As a result the monitoring system zero should be verified frequently, and always prior to initiation of treatment changes based on pressure data. Proper zeroing is done by opening the appropriate zeroing stopcock to atmosphere and aligning the resulting fluid–air interface point at the patient's mid-axillary line (Fig. 1).

Dynamic response testing

The fluid-filled plumbing system can modify the pressure signal in what is known as a dynamic response distortion (Gardner 1981; Gardner and Hollingsworth 1986). Most pressure-monitoring systems are underdamped (Gardner 1981). Dynamic distortion is caused by modification of the frequency characteristics of the applied pressure waveform as it passes through the plumbing system. The dynamic response of the plumbing system can be tested using a fast-flush test. This is accomplished by first opening the valve of the continuous-flush device and then quickly closing it. For underdamped systems, the rapid closure produces a square wave from which the natural frequency and damping coefficient of the plumbing system can be determined. The natural frequency is determined by measuring the period of a full cycle of oscillation and taking the reciprocal. For example, if the display rate (either oscilloscope or chart

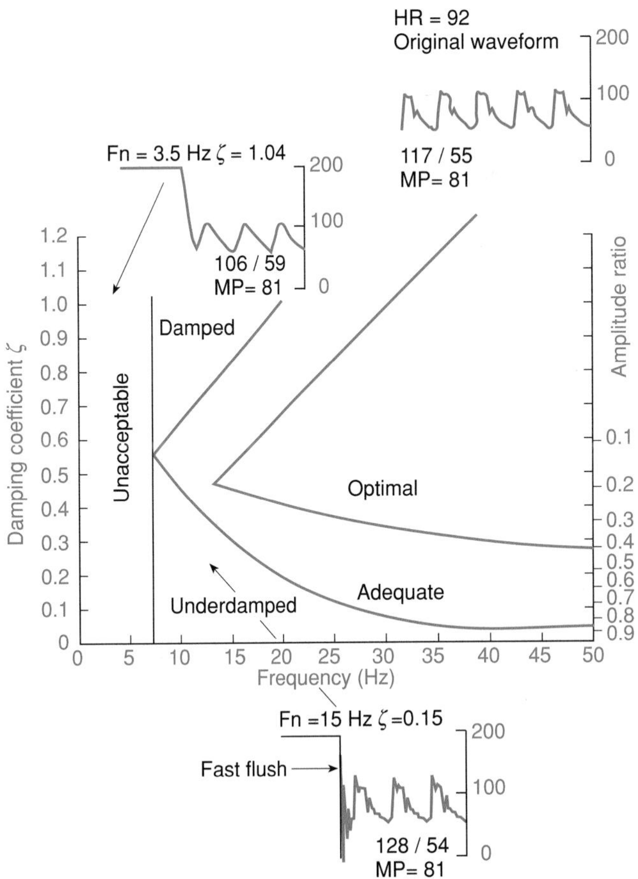

Fig. 2 Dynamic response testing methodology for the pressure-monitoring system. See text for an explanation of how to determine the adequacy of the dynamic response. (Adapted from Gardner (1981).)

Table 1 Relation between arterial pressure waveform and heart condition

Short systolic time	Hypovolemia High peripheral resistance
Marked respiratory swing	Hypovolemia Pericardial effusion Airways obstruction High intrathoracic pressure
Slow systolic upstroke	Poor myocardial contractility High peripheral resistance

recorder) is 25 mm/s and one full cycle of oscillation takes 2 mm, the natural frequency is

$$\frac{25\ \text{mm/s}}{2\ \text{mm/cycle}} = 12.5\ \text{cycles/s} = 12.5\ \text{Hz}.$$

The determination of damping coefficient requires the measurement of the ratio of the amplitude of two succeeding cycles of the resulting oscillating signal following a fast-flush test. The amplitude ratio is then plotted on the non-linear amplitude ratio scale on the far right-hand side of Fig. 2. The damping coefficient ζ is then determined from the linear damping coefficient scale shown in Fig. 2.

Several factors lead to an inadequate dynamic response, including air bubbles in the plumbing system and pressure tubing that is too long or too elastic (Gardner 1981). The best way to enhance the dynamic response of the plumbing system is to maximize the natural frequency. It is clear from Fig. 2 that the higher the natural frequency, the greater the range of damping coefficient can be while still reliably reproducing the pressure waveform. Figure 2 also shows that inadequate dynamic response causes two general types of error. If a system is underdamped, the systolic pressure will tend to be overstated and the diastolic pressure understated. However, if the system is overdamped, the systolic pressure will be understated and diastolic pressure overstated. The pulmonary artery catheter is subjected to 'whip' as it moves in the right ventricle, and this artifact tends to cause major overshoot during the systolic phase of the pressure waveform.

Many clinicians use the shape of the pressure waveform to estimate qualitative information about the heart (Table 1). Since dynamic waveform distortions caused by the plumbing system can mimic many of these conditions, having an adequate dynamic response is essential.

References

Cooper, T. and Paulsen, A.W. (Co-Chairmen) (1994). *American National Standard for blood pressure transducers*. Association for the Advancement of Medical Instrumentation (AAMI), Arlington, VA, and American National Standards Institute (ANSI), New York.

Gardner, R.M. (1981). Direct blood pressure measurement—dynamic response requirements. *Anesthesiology*, **54**, 227–36.

Gardner, R.M. (1996). Accuracy and reliability of disposable pressure transducers coupled with modern pressure monitors. *Critical Care Medicine*, **24**, 879–82.

Gardner, R.M. and Hollingsworth, K.W. (1986). Optimizing ECG and pressure monitoring. *Critical Care Medicine*, **14**, 651–8.

16.1.4 Central venous catheterization

J. L. Peters and R. Moore

Key messages

- A combination of inexperience with technique and an inadequate knowledge of the anatomy of the great veins are important factors in the etiology of procedure-related complications.
- It is imperative that clinical staff receive the appropriate supervision until the various techniques of percutaneous or open surgical procedures have been mastered.
- The need for a rigorous attention to antisepsis and aseptic technique must be appreciated at all times.

Introduction

During the past 40 years there has been an enormous increase in the use of central venous catheterization and cannulation devices for the management of critically ill patients. The introduction of these techniques has allowed the optimal adjustment of circulatory variables by direct hemodynamic measurement and volume replacement, infusion of a wide variety of pharmacologically active agents which cannot be satisfactorily infused via peripheral veins, and the provision of intravenous nutrition for prolonged periods of time. As with all procedures in surgery and medicine, drawbacks and complications have been recorded. Many of these have been iatrogenic in origin.

Indications

A steady expansion in the use of central venous catheters and access procedures has occurred during the past decade and the indications are summarized in Table 1. It is essential to ensure the most appropriate technique, device, and route of insertion is chosen.

Manometry and monitoring

The central venous pressure (CVP) provides an indication of the state of distension of the venous collection system and the relative effectiveness of the pumping action of the heart in emptying the system. Underfilling or overdistension can be recognized by CVP measurements before overt clinical signs have developed. Accordingly, CVP monitoring has been established as an important index to record, for both diagnostic purposes and for assessing the therapeutic effect of volume replacement in critical circulatory problems. The CVP is directly proportional to the venous return and inversely proportional to the myocardial competence or contractility. The venous return is regulated by the blood volume and the capacitance of the venous system and the CVP provides a measure of the effective circulating blood volume relative to the ability of the heart to handle that volume.

It is important to remember, however, that the CVP is not a linear function of blood volume. Isolated CVP readings, unless particularly

Table 1 Indications for central venous catheterization

L	Long-term infusions
	Inotropic drugs
	Hyperosmolar fluids
	Antibiotics
	Cytotoxic drugs
	Prostaglandins
M	Manometry and monitoring
N	Nutrition
O	Oximetry and cardiac output measurement
P	Plasmapheresis and hemodialysis
	Peritoneovenous shunt procedures
	Prevention of pulmonary embolism
	Pacemaker electrode insertion
	Portosystemic shunt insertion and transvenous liver biopsy

high, or low, are of little value. Continuous measurement of the CVP with respect to volume expansion or contraction provides a useful guide in intravenous fluid therapy.

Zero reference level

It is important to establish a common reference level from which comparable values of CVP can be determined. In most clinical units, the suprasternal notch, the manubriosternal junction, or the mid-axillary line are used as convenient zero assessment levels. However, radiographic studies have shown that the zero point which corresponds to the entry of caval veins into the right atrium rests on a plane with the fourth intercostal space, 0.43 times the depth of the chest below the anterior surface of the sternum.

Choice of catheter

It is important that the catheter selected should be applicable to the anatomical route of insertion and due regard should be given to the length of time for which the device will be in place. The selection of the central venous catheter should also take into account the known complications that have been documented.

Central venous catheters introduced using a through-needle technique are obsolete and should not be used since there is a serious risk of catheter embolization occurring during insertion. The safest techniques involve either a Seldinger wire technique or a through-cannula introducer. Over-needle single lumen devices are a satisfactory alternative, particularly for the internal jugular route.

It is wise to avoid catheter types which have 'disconnecting hubs'. Fixed-hub devices provide extra security against the catheter inadvertently snaking into the circulation and disconnection leading to air embolism. The union between catheter and shaft is a potential source of bacterial contamination from the body's surface and can be eliminated by a suitable chemical and mechanical bonding process.

A great deal of research has been performed in respect of the most suitable and least thrombogenic material for use in central venous catheterization. It is likely that there will always be a risk of thrombosis in association with these devices. The advent of silicone rubber and polyurethane polymers has been of great benefit to patients.

In adults, for the approaches using the arm, a 60- to 70-cm catheter is necessary in order to reach the superior vena cava. However, shorter catheters are now being used where the tip rests in the more proximal reaches of the subclavian or innominate veins. For the subclavian route, a 30- to 35-cm catheter is necessary, and for the internal jugular route a 12- to 15-cm length of catheter is required. The correct choice of length is important to avoid malposition. The majority of short Teflon over-needle jugular catheters are inappropriately 'stiff' when introduced into the subclavian vein with the risk of perforation of the superior vena cava or innominate vein. Movement of the patient's shoulder or infusion line will cause a to-and-fro movement of the relatively rigid catheter tip against the endothelium of the vein.

The various tunneling techniques naturally require catheters with different specifications and at the present time these are usually 90 to 110 cm in length. It is vital that the hubs of these devices are more resilient for their intended long-term placement.

Complications

Immediate

The incidence of insertion complications declines with increasing experience of the clinician. It is important to remember that in patients with trauma, a coexistent unilateral hemopneumothorax may already be present. Thus, attempts at cannulation of the contralateral subclavian vein may result in fatal bilateral pneumothoraces. This can also occur following previous open-heart or mediastinal surgery when a communication may have been created between each pleural cavity similar to the anatomy of the 'buffalo' thorax; injury to the pleura in such cases results in bilateral lung collapse. Table 2 lists reported immediate complications of central venous catheterization.

The true incidence of such events is unknown. It would seem sensible and prudent in the future to use ultrasound-guided vascular access techniques (Peters *et al.* 1982).

Malposition

Extravascular

Central venous catheters (or their guidewires) can dissect through the vein wall, pleura, pericardium, heart, or mediastinum, leaving the tip of the catheter resting in a potentially lethal position. This should be suspected if a catheter or guidewire advancement is difficult or if blood cannot be freely aspirated.

Intravascular

Accidental arterial placement of central venous catheters can easily occur in hypotensive patients, but this is usually quickly recognized. Rarely, severe arterial trauma requiring rapid intervention can occur.

It is best to position catheters in the superior vena cava and avoid kinks, loops, or retrograde placement into the internal jugular vein. Should the latter event occur, a conscious patient may complain of the 'ear gurgling' sign as the infusion is commenced. Rarely, the retrograde infusion of lidocaine (lignocaine) has led to coma and

Table 2 Reported immediate complications of central venous catheterization

Site of trauma	Clinical sequelae
Arterial puncture	Hematoma Carotid artery laceration and embolism Subclavian artery perforation Ascending cervical artery laceration Internal mammary artery laceration Pulmonary artery laceration (apical branch) Pulmonary arterial trunk perforation Aortic puncture Aortic dissection Arteriovenous fistula Vertebral artery pseudoaneurysm
Pleura and mediastinum	Pneumothorax Hemothorax Hemomediastinum Hydromediastinum
Venous cannulation	Air embolism Catheter embolism
Heart	Cardiac arrhythmias Cardiac perforation and tamponade
Lymphatic vessels	Thoracic duct laceration
Neurological injury	Phrenic nerve Brachial flexis Recurrent laryngeal nerve Horner's syndrome Hypoglossal nerve Fatal cerebrovascular episodes
Trachea	Deflation of endotracheal tube cuffs
Thyroid	Penetration of thyroid cyst

hyperosmolar solutions have caused cerebral venous thrombosis. Catheters can work themselves into an abnormal aberrant position after a period of time and cause thrombosis. It is wise to avoid over advancement of long-term catheters into the right atrium since endocardial injury and the serious sequelae of arrhythmias, mural thrombosis, perforation, advancement into the coronary sinus, or catheter knotting have been reported.

Delayed events

Air embolism

The need to insert percutaneous central venous catheters with the patient in a 20° to 30° head-down Trendelenburg tilt to prevent this problem is well established, yet episodes continue to be reported because this principal has been ignored. It is vital to remember that accidental disconnection or any defect in the infusion system will create a 'cardio-atmospheric fistula' and air can and will enter rapidly causing cardiovascular collapse, hemiplegia, coma, or death.

Bacterial colonization

Venous access systems can only be kept sterile if a meticulous insertion technique is followed by disciplined maintenance from the nursing staff. Hopefully, in the future, the use of stopcocks or valve side-ports will steadily be eliminated since these increase the risk of bacterial contamination. The role of bacterial adherence to catheter hub plastics and subsequent colonization of the devices is under constant investigation. Recently, the development of antibiotic-impregnated silicone central venous catheters has been achieved and introduced into clinical practice with favorable early results (Raad *et al.* 1997). Care must be taken to exclude patients with known hypersensitivity to the antibiotic. The value of a specialist nurse or team to care for venous access systems has been emphasized and fully reviewed.

If catheter-related sepsis is suspected, early involvement of the microbiology team is mandatory. The appropriate specimens and swabs should be taken; any tunnel track should be examined, and serum expressed and sent for culture. Samples from the infusate, tubing, and bag should be cultured. Recently, the value of using a new endoluminal brush method for *in situ* diagnosis of catheter-related sepsis has been shown to be of great clinical value (Kite *et al.* 1997). Auscultation of the heart for new murmurs should be performed and the fundi examined for evidence of endophthalmitis or emboli. An echocardiogram should also be performed to exclude valvular vegetations. A comprehensive review of complications associated with intravenous access devices has been published (Whitman 1996).

Sites of insertion

Central venous catheters can be inserted via several routes dictated by the presence of local pathology, anatomical abnormality, or venous occlusion due to thrombosis. The principal sites may be subdivided as to whether the procedure involves a percutaneous cannulation technique or direct surgical exposure (cut-down) on a venous tributary. Doubts concerning major central venous occlusion secondary to thrombosis should be resolved with angiography or ultrasound techniques. Occlusive superior central venous thrombosis presents a difficult problem and considerable technical expertise is required to gain venous access in these exacting circumstances. Direct percutaneous catheterization via the inferior vena cava has been performed with ultrasound or CT radiographic guidance. Alternatively, a combined angiographic and operative technique whereby catheters are placed through previously occluded veins using retrograde guidewire and Dormia basket techniques may be used. Fortunately, the need for such techniques is rare.

The preferred sites for central venous access in the absence of complicating factors are listed in Table 3.

The authors prefer to use the infraclavicular subclavian percutaneous route since catheter maintenance is simpler and the device does not traverse the 'beard' area. Another minor but important point to remember is that the cephalic vein is absent in approximately 10 per cent of people and many have been thrombosed following prolonged peripheral venous therapy.

Insertion techniques

The careful assessment and inspection of the selected site in respect of the venous anatomy should precede any procedure. Whenever possible, a chest radiograph should be examined for apical pulmonary bullas and other mediastinal pathology. 'Auscultation' using an ultrasound probe can successfully localize the major veins of the neck and subclavian region.

Table 3 Preferred sites for central venous access

Access procedure	Preferred site
Percutaneous	Internal jugular vein (right) Infraclavicular subclavian vein Antecubital fossa veins of forearm Femoral vein
Surgical exposure	External jugular vein Cephalic vein Internal jugular vein (right) Long saphenous vein and tributaries Antecubital fossa veins

If the technique is carried out with the patient in bed, it is essential that this can be tilted adequately into a head-down situation in order to provide adequate filling of the great veins. Most techniques can be performed using regional local infiltration anesthesia supplemented if necessary by intravenous sedation and anesthesia. Careful antiseptic skin preparation is essential, and the patient should be covered by surgical drapes leaving the operative site adequately exposed. These measures eliminate any accidental contamination of the central venous catheter components during insertion. Clear adhesive incise drapes are also of benefit in the more sophisticated procedures.

Internal jugular vein

This is a popular route for the insertion of single-lumen catheter over-needle devices or multiple-lumen central catheters using a Seldinger technique. A sound knowledge of the anatomy is vital. After emerging from the base of the skull posterior to the internal carotid artery, the vein descends through the neck behind and deep to the sternomastoid muscle, gradually moving to the lateral side of the common carotid artery and ultimately lying anterolateral to this vessel. The relationship of the vein to the deep surface of the sternomastoid muscle is the key to understanding the position of the vein in the neck. It joins the subclavian vein behind the first rib to form the innominate vein. In practice, surface markings are not a reliable means of locating the internal jugular vein as its position, particularly in a lateral plane, tends to vary considerably. Therefore the steady increase in the use of imaging techniques using ultrasound during insertion procedure is a wise development. The main advantage of the internal jugular vein as a cannulation site stems from the fact that the blood flow is relatively rapid. Thrombosis is an unusual complication and usually the same site can be used repeatedly. However, there are disadvantages. Experience and a lot of practice is necessary to obtain the best results for the technique. In the conscious patient, the muscle tone in the sternomastoid can make cannulation more difficult and one must always be aware of the possibility of air embolism during the insertion procedure.

Infraclavicular subclavian vein

Polyurethane elastomer central venous catheters can be easily inserted into the infraclavicular subclavian vein using a Seldinger technique and maintained in place free of bacterial colonization for many weeks. This site is the most comfortable for most patients. This procedure attracted a high incidence of pneumothorax in the early years following its introduction. However, if the correct technique is employed, this complication can be avoided and a major advantage is that there is no risk to the carotid artery and adjacent nerves in the neck. The patient should be relaxed and positioned in a 20° to 30° head-down tilt; it is best to extend the shoulders over a towel or suitable sandbag and turn the head to the contralateral side with the arms lying alongside the thorax. It is wise to have ECG monitoring in place to detect arrhythmias. Right-handed clinicians will find the left subclavian artery the easiest to cannulate. A needle with its attached syringe should be inserted 2 to 3 cm below the clavicle at the junction of its middle and medial thirds with the bevel of the needle facing upwards. The needle should then advance towards the inferior border of the clavicle, aiming for the small space between the clavicle and the first rib. By the time the pectoral muscles are reached, the needle or cannula should be parallel to the coronal plane in order to avoid pleural lacerations. The needle is then advanced slowly along an imaginary line extending from the surgical neck of the humerus, through the junction of the middle and medial thirds of the clavicle to the top of the ipsilateral sternoclavicular joint. The needle should traverse this line in a plane 0.5 to 1 cm posterior to the clavicle and the vein will be entered at the outer border of the first rib. When this technique is used carefully and cautiously, there is no risk of pleural or pulmonary injury since the vein will be entered lateral to the outer edge of the first rib.

Peripheral insertion of central catheters

Veins of forearm

The freedom from the risk of pneumothorax makes this approach of particular value in patients known to have impaired pulmonary function and for those who have had to undergo surgery with a period of intermittent positive-pressure ventilation. It is also of particular value in the very elderly for the same reason.

Care should be taken in selecting the site of insertion because some veins of the antecubital fossa may be obliterated by previous peripheral intravenous cannulations. The best tributary is the median basilic vein on the left, followed by the right median basilic, and finally the median cephalic veins at the elbow. The passage of long central venous catheters through these tributaries leading in to the cephalic vein is not generally preferred because of the difficulty sometimes experienced in negotiating the segment of vein situated in the deltopectoral groove as it penetrates the layers of fascia before entering the subclavian vein. Successful placement is achieved most often by using the simpler and more direct median basilic route.

The venous anatomy should be delineated by a tourniquet. This should be placed high up the arm so that it does not encroach on to the draped area and zone of skin preparation. If the patient is shocked and the veins are collapsed, a cut-down procedure should be considered first rather than destroying a valuable route of venous access by multiple unsuccessful percutaneous puncture attempts. Very recently an ingenious technique has been described (Williams *et al.* 1997). With a tourniquet in place around the upper arm, any small vein present in the arm or hand is identified and the smallest cannula available (e.g. 'blue 22 gauge') inserted. This is then connected to a bag of warm crystalloid solution which is infused under pressure without releasing the upper arm tourniquet. After a few minutes, the venous tree of the arm is filled with crystalloid and it is then easy to identify a suitable large-caliber vein and insert the appropriate central venous catheter or large-bore peripheral catheter under local anesthesia.

The central catheter should be advanced along the vein, slowly and without force. If obstruction of the catheter's progress should be encountered, it may be withdrawn a fraction and a further attempt made. Often, gentle rotation of the catheter around its long access will help to negotiate the valve or tributary. The position of the catheter tip should be checked to ensure that it lies in the superior vena cava or its junction with the innominate vein.

Femoral vein

This route of insertion is only occasionally used because of the obvious risk of the initiation of a significant deep venous thrombosis. However, it is an extremely valuable choice in circumstances due to trauma, burns, or other pathology which may have excluded the neck and upper limb veins. The common femoral vein lies medial to the common femoral artery as it passes deep to the inguinal ligament and the upper aspect of the femoral triangle. The two vessels are closely apposed. The introducing cannula or needle syringe assembly should be inserted through the tissues of the femoral triangle overlying the vein at an angle of approximately 45° and 1 cm medial to the arterial pulsation. There is usually no difficulty in puncturing the vein lumen; the needle should be advanced and rotated as in previous techniques and the guidewire gently passed through into the vena cava.

Catheter removal

When a central venous catheter requires to be removed, either because of suspected bacterial colonization or damage or because it has served its useful purpose, it is important that this should be carried out using an aseptic technique. The catheter tip and a portion of the subcutaneous segment should be sent for culture and the exit wound covered by an occlusive dressing. Unfortunately, air embolism has occurred frequently via the catheter track.

References

Kite, P., *et al.* (1997). Evaluation of a novel endoluminal brush method for *in situ* diagnosis of catheter related sepsis. *Journal of Clinical Pathology*, **50**, 278–82.

Peters, J.L., Belsham, P.A., Garrett, C.P.O., and Kurzer, M. (1982). Doppler ultrasound technique for safer percutaneous catheterization of the infraclavicular subclavian vein. *American Journal of Surgery*, 143, 391.

Raad, I., *et al.* (1997). Central venous catheters coated with minocycline and rifampin for the prevention of catheter-related colonization and bloodstream infections. *Annals of Internal Medicine*, **127**, 267–74.

Whitman, E.D. (1996). Complications associated with the use of central venous access devices. *Current Problems in Surgery*, **33**, 309–88.

Williams, D.J., Baylis, R., and Hinchcliffe, R. (1997). Intravenous access: obtaining large-bore access in the shocked patient. *Annals of the Royal College of Surgeons of England*, **79**, 466.

16.1.5 Pulmonary artery catheterization

A. B. J. Groeneveld

Key messages

- The pulmonary artery wedge pressure (PAWP) only reflects left atrial and thus, in general, left ventricular filling pressure, if there is continuity with the left atrium.
- During spontaneous and artificial ventilation, the appropriate phase in the ventilatory cycle to judge PAWP is at end expiration.
- For clinical purposes, it is recommended that the thermodilution cardiac output in ventilated patients is determined as the mean of at least four thermodilution injections (with valid curves) at random in the ventilatory cycle using a room temperature injectate.
- Complications occur in about 20 per cent of pulmonary artery catheterizations.

Introduction

Hemodynamic changes play a central role in the critically ill patient, so that monitoring of the circulation allows evaluation of the etiology, severity, and course of circulatory instability and its response to treatment, supplementing clinical judgment. The pulmonary artery catheter allows a number of continuous and intermittent measurements to be made from which other variables can be derived (Table 1). However, the relevance of these variables to treatment and prognosis is difficult to prove. In addition, interpretation of the data may be difficult, so that additional bedside imaging techniques may be required. Therefore doubts have been expressed as to whether the benefits of a pulmonary artery catheter outweigh the risks of insertion and whether its use improves outcome. Nevertheless, the catheter may allow evaluation of responses to therapy and assessment of physiological and perhaps prognostically important endpoints for resuscitation, supplementing the clinical indicators of circulatory instability.

Technique

Insertion of a pulmonary artery catheter requires access to a great vein (the internal jugular, subclavian, or femoral). There is no ideal vein for insertion of a pulmonary artery catheter. Puncture is performed with sterile techniques, using a guidewire to introduce a catheter sheath through which the catheter is advanced, with an inflated balloon, into the pulmonary artery and finally into the wedge position. Continuous monitoring of pressures during this procedure allows correct positioning of the catheter without fluoroscopic control.

Table 1 Variables which can be measured or derived using the pulmonary artery catheter

	Formula	Normal range
Measured variables		
Cardiac output (l/min)		4–7
Mean pulmonary artery pressure (mmHg)		11–17
Pulmonary artery wedge pressure (mmHg)		2–12
Central venous pressure (mmHg)		0–6
Mixed venous oxygen saturation (%)		65–80
Derived variables		
Stroke volume (ml)	CO/HR	60–100
SVR (dyn s/cm^5)	[(MAP – CVP) × 80]/CO	900–1400
PVR (dyn s/cm^5)	[(MPAP – PAWP) × 80]/CO	150–250
Left ventricular stroke work (g)	SV × (MAP – PAWP) × 0.0136	77–110
Right ventricular stroke work (g)	SV × (MPAP – CVP) × 0.0136	7–14
Oxygen delivery (ml/min)	$10 \times Ca\text{O}_2 \times \text{CO}$	640–1200
Oxygen uptake (ml/min)	$10 \times (Ca\text{O}_2 - Cv\text{O}_2) \times \text{CO}$	100–180
Oxygen extraction	Oxygen uptake/delivery	0.22–0.30
Indexed variables*		
Cardiac index (l/min/m^2)		2.8–4.2
Stroke volume index (ml/m^2)		30–65
SVR index (dyn s/cm^5/m^2)		1700–2600
PVR index (dyn s/cm^5/m^2)		45–225
Left ventricular stroke work index (g/m^2)		43–61
Right ventricular stroke work index (g/m^2)		7–12
Oxygen delivery index (ml/min/m^2)		500–600
Oxygen uptake index (ml/min/m^2)		120–160

$Ca\text{O}_2$=([Hb] × $Sa\text{O}_2$ × 1.34)+($Pa\text{O}_2$ × 0.003), 18–20 ml/100 ml.
$Cv\text{O}_2$=([Hb] × $Sv\text{O}_2$ × 1.34)+($Pa\text{O}_2$ × 0.003), 13–15 ml/100 ml.
Hb, hemoglobin concentration; $Sa\text{O}_2$, arterial oxygen saturation; $Sv\text{O}_2$, mixed venous oxygen saturation; SVR, systemic vascular resistance; PVR, pulmonary vascular resistance.
*Variables indexed to body surface area.

Measurements and errors

Pulmonary artery pressures

Pressures should be obtained after proper calibration and zeroing to atmospheric pressure, with the mid-chest level as reference. Technical errors are frequent (20 per cent) and may not be detected. The difference between the diastolic pulmonary artery pressure and the pulmonary artery wedge pressure (PAWP) can be regarded as a measure of the pulmonary vascular resistance. The gradient may increase in sepsis and pulmonary embolism.

Pulmonary artery wedge pressure

Proof of a true wedge position of the catheter tip is a high oxygen saturation (higher than in arterial blood) measured in blood aspirated from the distal port of the pulmonary artery catheter with an inflated balloon, but this is not done routinely. Other criteria for the true wedge position are given in Table 2. It is relatively hard to judge the validity of PAWP readings even if the waveform is recorded on paper. Errors in measurements include damping of the curve, 'overwedging', and incomplete wedging.

Table 2 Criteria for the wedge position in West zone III

Normal atrial curve and atrioventricular waves
PAWP <diastolic pulmonary artery pressure
Change in PAWP <half change in PEEP during incremental PEEP
Respiratory variation in PAWP <half airway pressure swings
Catheter tip at or below left atrium
Fully saturated blood after wedging, dead-space removal, and slow aspiration of blood

PAWP only reflects left atrial and thus, in general, left ventricular filling pressure, if there is continuity with the left atrium. Left atrial reflects left ventricular end-diastolic pressure, in the absence of major mitral valve disease (Table 3). PAWP and changes in it may not reflect left ventricular end-diastolic volume, which is a true measure of left ventricular preload, since distensibility of the left ventricle may vary between patients and between clinical situations.

During mechanical ventilation, positive airway pressure is introduced which may be partially transmitted into the alveoli and pulmonary vessels, and may affect PAWP measurements if the catheter tip is in West zones I and II (O'Quinn and Marini 1983). In contrast to zone III, airway pressure exceeds venous (and arterial) pulmonary pressure in zones I and II, so that PAWP may be influenced by intra-alveolar rather than intravascular pressure. Clinically, zone III conditions can be assumed if the tip of the pulmonary artery catheter is below the level of the right atrium (this can be confirmed on a lateral chest radiograph). Sometimes, it is necessary to reposition the catheter tip, although in the majority of cases the catheter tip moves into the dorsal region of the lungs, where blood flow is higher than in the ventral region during insertion in the supine patient (O'Quinn and Marini 1983). Alternatively, changing the patient's position from supine to lateral (catheter tip below the heart) and re-zeroing to atmospheric pressure at the mid-chest level may improve the accuracy of PAWP as a measure of pulmonary venous pressure. Mechanical ventilation with positive end-expiratory pressure (PEEP) may enlarge zones I and II by increasing the intra-alveolar pressure. This effect is greatest during hypovolemia or other conditions associated with pulmonary hypoperfusion. Hence, during incremental PEEP, PAWP may rise more than the pulmonary artery pressure, central venous pressure, and thus left atrial pressure, if the catheter tip is not situated in West zone III.

Table 3 PAWP, left ventricular end-diastolic pressure (LVEDP), and capillary pressure

PAWP >LVEDP	Mitral stenosis, insufficiency and obstruction, pulmonary venous hypertension, false catheter (zone II) position, left–right shunt
PAWP <LVEDP	Reduced left ventricular compliance, aortic insufficiency
PAWP <capillary pressure	Pulmonary veno-occlusive disease

During spontaneous and artificial ventilation, the proper phase in the ventilatory cycle to judge the PAWP is at end-expiration, when air flow is zero and the intrathoracic and atmospheric pressures are equal unless PEEP is applied. This will avoid the need for complicated estimates or measurements of transmural pressures (i.e. intrathoracic intravascular pressure minus intra-alveolar pressure). Conversely, the electronically derived mean pressure (over the respiratory cycle) may not conform to the pressure at end-expiration (O'Quinn and Marini 1983). Hence, manual measurements from graphic recordings at end-expiration are preferable to automated measurements in digital displays of PAWP, which reflect mean values over the respiratory cycle.

With PEEP, measurements of pressures at end-expiration with the atmospheric pressure as reference may not indicate transmural pressures in the thorax (O'Quinn and Marini 1983). This can be corrected by subtracting half the PEEP (in mmHg), reflecting transmitted airway pressure in the presence of normal respiratory compliance, from the measured PAWP at end-expiration (Smiseth *et al.* 1996). Measurement of pleural or esophageal pressure (or pericardial pressure after cardiac surgery) to calculate transmural pressure is not routine. An alternative approach is provided by measuring the intrathoracic intravascular pressure after a sudden discontinuation of PEEP, assuming that the delay between the reduction in PEEP and the increase in venous return prevents changes in cardiac loading. Finally, some authors suggest that central venous pressure (right atrial pressure) may equal pericardial pressure, so that referencing PAWP to central venous pressure may yield a transmural pressure when pericardial pressure is elevated by a high airway pressure (Smiseth *et al.* 1996). However, the validity of this idea has been questioned.

Pulmonary capillary pressure

At the point of inflection between the fast and slow components of the pressure decay curve after balloon inflation and arterial occlusion, the pulmonary capillary pressure downstream from the pulmonary artery pressure but upstream from the PAWP can be derived, provided that superimposed respiratory changes are minimal or absent. This pressure is the microvascular hydrostatic pressure for fluid filtration. The capillary pressure in the lung depends on the (unpredictable) ratio of afferent to efferent pulmonary vascular resistance. The clinical value of this concept has not yet been demonstrated.

Thermodilution cardiac output

The thermodilution method for measuring cardiac output, after injection of the thermal indicator through the central venous port and registration of the temperature change by the thermistor at the distal port of the pulmonary artery catheter, has been extensively validated by comparison with the gold standard, i.e. the cardiac output calculated, using the Fick principle, from measurements of the oxygen uptake (indirect calorimetry, analysis of inspiratory and expiratory breath) and the arterial and mixed venous oxygen contents. In critically ill patients, at low cardiac outputs and with small thermal indicator volumes, the thermodilution method may overestimate the true cardiac output (Renner *et al.* 1993).

There is continuing controversy regarding the optimal phase during which the thermal indicator should be injected for proper measurement of cardiac output during spontaneous and mechanical ventilation. It is generally recommended that the average result for three consecutive thermal indicator injections at either end-expiration or end-inspiration should be taken for a proper cardiac output estimation. However, during mechanical ventilation and relatively large swings in intrathoracic pressure, transmitted to the heart and great vessels, there is a large variation in right-sided thermodilution cardiac output following changes in right ventricular loading during the respiratory cycle. It is difficult to predict the phase and amplitude of these changes in an individual patient because many factors may influence the relation between the cyclic variation in ventilation-induced intrathoracic pressure and thermodilution cardiac output. These factors include the intravascular volume status, the cardiac output itself, and the level of intraalveolar pressure swings which are determined by respiratory rate, pattern, and volume (compliance). Therefore, for clinical purposes, it is recommended that the thermodilution cardiac output in ventilated patients should be determined as the mean of at least four thermodilution injections (with valid curves) made at random in the ventilatory cycle, irrespective of their variation.

By altering the steady-state baseline blood temperature, on which the change in injectate temperature is superimposed and cardiac output is calculated from the area under the curve of the thermal signal, rapid infusion of relatively cold fluids changes the area under the thermodilution curve and thus leads to either under- or over-estimation of cardiac output, depending on the infusion rate. Thermal noise can be reduced by warming infusion fluids to blood temperature.

Another controversial point is the widely recommended use of ice-cold injectates for enhancing signal-to-noise ratio and improving reproducibility (Renner *et al.* 1993). However, many authors have shown that room temperature injectates may be as good as cold injectates, provided that the patient is not severely hypothermic or hyperdynamic. The use of room temperature injectates also avoids the need to warm the injectate before injection and slowing of the heart rate and associated stroke volume changes during cardiac passage of the cold solution. The first error can be avoided by measuring the temperature of the injectate at the injection site. Furthermore, tricuspid regurgitation, which is a relatively frequent phenomenon in critically ill mechanically ventilated patients, may result in systematic and accidental measurement errors of cardiac output compared with measurements using other methods. A systematic underestimation usually results. Finally, the accuracy of thermodilution cardiac output measurements depends on the volume of thermal indicator injectate used. Since reproducibility is improved by using 10 ml rather than 5 ml, the former is widely advocated. The injectate port must be downstream from the introducer sheath of the pulmonary artery catheter, without concomitant infusions via the side-arm, to avoid errors. Otherwise, the side-arm may substitute for the central venous port, if the latter becomes non-functional.

Continuous cardiac output

Thermodilution techniques have been developed for (semi)-continuous monitoring of cardiac output (Haller *et al.* 1995). This technology utilizes a thermal filament in the catheter at the level of the right ventricle. The generated heat (pulse) is detected downstream by a thermistor close to the tip of the catheter. This method yielded results equivalent to those obtained by intermittent thermodilution, even though the respiratory variations in venous return tend to level out with the continuous technique. However, the response time is relatively long. A Doppler pulmonary artery catheter has been developed for intermittent and continuous measurement of blood flow. In some experiments, Doppler blood flow correlated better with electromagnetic blood flow than with thermodilution blood flow.

The pulmonary artery catheter in practice

Hemodynamic monitoring in shock aims at identifying and reversing those hemodynamic and metabolic defects that contribute to the shock syndrome and are associated with a poor outcome. The data obtained with the pulmonary artery catheter may be of considerable help in judging the type, severity, and course of shock, and in choosing and evaluating the effect of therapy, even though the importance of various hemodynamic variables in the reflecting severity of disease has been questioned. Moreover, therapy aimed at increased tissue oxygenation may promote survival, although this has also been questioned (Groeneveld and Thijs 1991).

Insertion of a pulmonary artery catheter is justified in the management of circulatory failure when difficulties in diagnosis arise, i.e. differentiation of obstructive, hypovolemic, cardiogenic, and distributive types of shock (Groeneveld and Thijs 1991). Monitoring of hemodynamic and metabolic variables may also help in therapeutic decisions which aim at adequate tissue oxygenation. The clinical situations that may warrant insertion of a pulmonary artery catheter are listed in Table 4. Pulmonary hypertension is a relatively frequent phenomenon in critically ill patients, and it may have various etiologies. It may be associated with a fatal outcome, but it is unclear whether pulmonary hypertension is central in circulatory compromise or merely reflects severe underlying lung disease carrying an unfavorable prognosis, such as the acute respiratory distress syndrome.

In the normal heart, the left ventricular filling pressure (PAWP) is usually only a few mmHg higher than the right ventricular filling pressure (right atrial or central venous pressure), so that the latter pressure may reflect cardiac preload. Assessment of cardiac filling from the central venous pressure may be difficult in the diseased heart, even after proper measurement. The central venous pressure may not accurately reflect right ventricular end-diastolic volume, and the pressure (changes in pressure) may not predict (changes in) left ventricular filling pressure (PAWP) in the presence of pre-existing or shock-induced left ventricular dysfunction, pulmonary hypertension, and right ventricular dysfunction.

However, variability and changes in left ventricular distensibility may invalidate PAWP as a measure of ventricular preload, i.e. left ventricular end-diastolic volume (O'Quinn and Marini 1983; Groeneveld and Thijs 1991). When volume status is in doubt from PAWP measurements, it may be useful to assess left ventricular ejection fraction and end-diastolic volume at the bedside with the help of two-dimensional echocardiography or radionuclide angiography with a mobile gamma-camera or 'nuclear probe' (O'Quinn and Marini 1983; Groeneveld and Thijs 1991). However, these techniques are not widely available. Irrespective of left ventricular end-diastolic

Table 4 Indications for pulmonary artery catheterization

Shock	High-risk anesthesia and major surgery
	Cardiopulmonary surgery
	Abdominal vascular surgery
	Anesthesia and major surgery in patient with prior cardiac disease
Diagnosis	Unknown etiology when clinical judgment fails
Treatment	Unresponsive to liberal fluid therapy in the absence of a low CVP
	Need for vasopressors
	Therapeutic uncertainties
	Shock and pre-existent cardiac disease
Pulmonary edema	Sepsis, multitrauma
Diagnosis	Differentiation of edema types
Treatment	Shock and pulmonary edema
Miscellaneous	Complicated disease course of myocardial infarction
	Pregnancy-induced hypertension
	Renal insufficiency, to determine prerenal factor (hepatorenal syndrome)
	Pulmonary hypertension (primary and secondary)

CVP, central venous pressure.

volume, PAWP is an important determinant of microvascular filtration in the lung, even in the case of the 'leaky capillaries' associated with the acute respiratory distress syndrome (O'Quinn and Marini 1983).

Pros and cons of pulmonary artery catheterization in the critically ill

It is evident that the pulmonary artery catheter has provided detailed insight into hemodynamic and metabolic disturbances in the critically ill (Groeneveld and Thijs 1991). However, it is less clear whether catheterization should be performed in every patient with cardiopulmonary compromise. The impact of a pulmonary artery catheter and the hemodynamic variables obtained with it on management and outcome are not well defined, and the benefits may not outweigh the risks of catheter insertion. Moreover, understanding of the pulmonary artery catheter varies widely among doctors and nurses, emphasizing the need for continuous education on its role in hemodynamic monitoring.

Investigators favoring liberal use argue that failure of clinical judgment in diagnosing shock or instituting successful treatment is an indication for catheterization, since continuous assessment of essential hemodynamic and metabolic variables reflecting the type, severity, and course of circulatory compromise is needed to ensure immediate adjustment of therapy. In fact, delay in appropriate treatment is one of the most important factors associated with mortality in shock. Proponents further argue that clinical judgment and estimation of circulatory status often fail to predict cardiac filling, output, and function (and derived vascular resistances) in critically ill patients. The introduction of the pulmonary artery catheter by experienced critical care physicians may improve outcome in a variety of specialty units and patient conditions. With careful attention to the technique, insertion of a pulmonary artery catheter is relatively safe and the complication rate decreases with increasing experience.

Investigators who do not favor widespread use of pulmonary artery catheterization claim that clinical judgment is not inferior to invasive assessment of cardiac output. There is controversy concerning the prognostic significance of hemodynamic and metabolic disturbances in shock other than those associated with myocardial infarction. There is little direct (as opposed to circumstantial) evidence that invasive hemodynamic monitoring improves outcome in shock, even when hemodynamic data lead to alterations in therapy.

Complications

Complications occur in about 20 per cent of pulmonary artery catheterizations; they are serious in 4 per cent of patients and fatal in 4 per cent or less. Ventricular arrhythmias may occur in 12 to 70 per cent of patients during pulmonary artery catheterization; they may be more frequent and serious in patients with myocardial infarction than in those with septic shock, but mortality related to this complication is low. Complete heart block, induced by a right bundle branch block during pulmonary artery catheterization in patients with pre-existing left bundle branch block, is extremely rare. Thrombocytopenia may impose an unacceptable bleeding risk. Other complications include pneumothorax, arterial puncture, bleeding at the injection site, ventricular perforation and cardiac tamponade, infection, thrombosis, pulmonary vascular laceration with (fatal) hemorrhage, and pulmonary infarction.

Prevention of infection

Insertion of a pulmonary artery catheter should be performed with meticulous sterile techniques in order to prevent infection. Proper disinfection of pressure transducers is essential in preventing sepsis which may enhance the risk of catheter colonization and infection. As a guideline, pulmonary artery catheters should be removed or, if necessary, replaced after 72 h to prevent catheter sepsis, since prolonged use increases the chance of the development of sepsis, endocardial lesions, and right-sided endocarditis. Replacement over a guidewire increases the risk of bacteremia, but new punctures increase the risk of mechanical complications.

References

Groeneveld, A.B.J. and Thijs, L.G. (1991). Hemodynamic monitoring in septic shock. In *Strategy in hemodynamic monitoring* (ed. J.-F. Dhainaut and D. Payen), pp. 179–96. Springer Verlag, Berlin.

Haller, M., Zöllner, C., Briegel, J., and Forst, H. (1995). Evaluation of a new continuous thermodilution cardiac output monitor in critically ill patients: a prospective criterion standard study. *Critical Care Medicine*, **23**, 860–6.

O'Quinn, R. and Marini, J.J. (1983). Pulmonary artery occlusion pressure: clinical physiology, measurement and interpretation. *American Review of Respiratory Disease*, **128**, 319–26.

Renner, L.A., Morton, M.J., and Sakuma, G.Y. (1993). Indicator amount, temperature, and intrinsic cardiac output affect thermodilution cardiac output accuracy and reproducibility. *Critical Care Medicine*, **21**, 586–97

Smiseth, O.A., Thompson, C.R., Ling, H., Robinson, M., and Miyagishima, R.T. (1996). A potential clinical method for calculating transmural left ventricular filling pressure during positive end-expiratory pressure ventilation: an intraoperative study in humans. *Journal of the American College of Cardiology*, **27**, 155–61.

16.1.6 Mixed venous oxygen saturation monitoring

D. Rosser

Key messages

- Mixed venous oxygen saturation is an accurate direct measurement.
- Mixed venous oxygen saturation provides information about adequacy of supply that is not available from other monitors.
- Mixed venous oxygen saturation requires pulmonary artery catheterization, with attendant risks.
- Mixed venous oxygen saturation obviates the need for inaccurate derived hemodynamic variables.
- There is no clear evidence of outcome benefit to support its use.

Introduction

The fundamental purpose of the cardiovascular and respiratory systems is to supply oxygen to the tissues. In the vast majority of cases of physiological decompensation leading to critical illness it is a reduction of oxygen availability to the tissues, on a regional or global scale, that is the vital defect. Only rarely is a deficit in the supply of other metabolites (e.g. glucose) or a failure of removal of metabolic waste products (e.g. carbon dioxide, lactate) significant at a cellular level in the presence of an adequate oxygen supply. The other situation in which anaerobic metabolism can predominate, in the presence of an adequate oxygen supply, is metabolic poisoning which interferes with the cellular utilization of oxygen (e.g. cyanide, carbon monoxide, or, debatably, sepsis).

The ideal variable to monitor in patients would be oxygen tension at a mitochondrial level; however, this remains impossible in the clinical setting. The closest clinical approach to this ideal is the measurement of interstitial or tissue oxygen tension, which is used clinically in some centers for the early detection of ischemia in free flaps.

It is known that oxygen diffusion occurs in pre- and postcapillary vessels, and that free exchange of oxygen occurs between precapillary arteriole, capillary, postcapillary venule, and tissue. Thus the tissues and the microvasculature have a relatively uniform oxygen tension throughout. Therefore measurement of tissue oxygen tension indicates the oxygen availability at the outer surface of the cells of the tissue studied, which is influenced by the local balance between supply and demand. Thus the technique has the potential to be a very useful indicator of the adequacy of perfusion, but is still limited by the lack of a reliable user-friendly monitoring system. The other major problem with this type of monitoring concept is that it can only look at a small area of tissue which may not be representative of regions that are susceptible to hypoxic injury such as the renal medulla and the liver.

The alternative way of gaining an insight into the balance between oxygen supply and demand is the measurement of venous oxygen content, which is conveniently done by optical measurement of venous hemoglobin oxygen saturation. Unlike arterial saturation, which can be measured transcutaneously by analyzing only that part of the signal which is pulsatile, it is not possible to measure venous saturation non-invasively. In the same way as tissue oxygen tension indicates the balance between supply and demand, a rise in venous saturation indicates a rise in supply relative to demand and a fall in venous saturation indicates the opposite. Venous oxygen saturation can be used to monitor individual organs (e.g. the brain in head injury or during cardiopulmonary bypass), but is most commonly used to monitor mixed venous oxygen saturation ($S\text{vO}_2$) in the pulmonary artery. This global assessment of the adequacy of oxygen delivery to the body is very valuable; however, it does not necessarily reflect the status of individual tissues (Dahn *et al.* 1988).

Clinical techniques

The simplest way to measure $S\text{vO}_2$ is by sampling blood through the distal port of a pulmonary artery catheter. Unfortunately, central venous blood sampling does not correlate well with true mixed venous samples (Martin *et al.* 1992). This is due to the large differences in the supply-to-demand ratios of the organs draining into the major venous system (e.g. resting skin venous saturation exceeds 80 per cent; heart venous saturation is less than 30 per cent); thus passage through the right heart is required to ensure mixing. The measurement of hemoglobin saturation using a modern multi-wavelength co-oximeter is extremely accurate, but the drawing of a mixed venous sample is error prone. Blood needs to be aspirated very slowly as the resistance to flow leads to the generation of high negative pressures which can lower gas tensions (and hence saturation) and predisposes to the introduction of air into the sample. A rapid sampling rate can also lead to drawback of blood which has been exposed to gas-exchanging pulmonary vasculature, giving a falsely high oxygen saturation, as does drawing samples in the wedged position.

There are now a number of commercially available systems which allow continuous monitoring of $S\text{vO}_2$ via an optical fiber built into the pulmonary artery catheter. Two- and three-wavelength systems are available and, although there is some evidence that the three-wavelength system is more accurate and neither are as accurate as a co-oximeter, both are sufficiently accurate for clinical purposes (Scuderi *et al.* 1992). As with all optical methods for measuring hemoglobin saturation, there is a possibility of interference from other optically active compounds (e.g. carboxyhemoglobin, methylene blue, very high bilirubin levels). This is relatively unusual and in practice the major problem with these systems is drift, which may be significant within relatively short periods and is often significant between the preinsertion calibration and the check calibration immediately after insertion. The systems should be recalibrated against a co-oximeter every 4 to 6 h, and should be rechecked before clinical decisions are made primarily on the basis of $S\text{vO}_2$. Migration of the catheter to a permanently wedged position (or failure to deflate the balloon) will lead to a sustained rise in the $S\bar{v}O_2$ reading to arterial levels or above.

Clinical use

The most common indication for measuring SvO_2 is cardiovascular failure with hypotension or otherwise unexplained metabolic acidosis. The current philosophy of hemodynamic management of intensive care patients centers around the maintenance of an adequate cardiac output, although there is still considerable debate about what value should be sought and how it should be achieved. It is too often forgotten that measurement of cardiac output by thermodilution (taken as the clinical gold standard) is far from accurate. Under ideal circumstances thermodilution is accurate to within about 10 per cent, and most centers accept three readings within 10 per cent of each other for averaging. However, there are a number of potential sources of error, such as the presence of even minor tricuspid regurgitation which is common in ventilated intensive care patients (Jullien *et al.* 1995). Esophageal Doppler cardiac output monitoring appears to be as accurate as thermodilution, but it is user dependent to some extent and has its own sources of error.

The simple measurement of cardiac output, even leaving aside the accuracy of the available systems, only tells half the story, i.e. supply, and gives no indication of demand. This, in part, has led to the use of the derived variables oxygen delivery, oxygen consumption, and oxygen extraction ratio. When interpreting these variables it must be remembered that their calculation requires the measurement of a number of primary variables, each with a significant measurement error, creating a large potential error in the derived variable. The practice of indexing the values to body surface area serves to introduce further errors. In clinical practice, rather than in the laboratory, these derived variables add little to the simple measurement of SvO_2 which allows assessment of the adequacy of the cardiac output for the prevailing metabolic conditions.

If the microcirulation is intact and therefore responding appropriately to changes in local demand, a normal or high SvO_2 indicates that the oxygen delivery is adequate, or more than adequate, whereas a low SvO_2 suggests that oxygen delivery is too low. It can be argued that a high SvO_2 may be the result of shunting of blood away from nutrient capillaries and therefore may coexist with areas of hypoperfusion and cellular hypoxia. There is little evidence to support the concept of pathological microvascular shunting of blood in critical illness and no reason to believe that further increasing the cardiac output would improve this situation. Recent clinical studies in multiple organ failure have suggested that raising the cardiac output to supranormal levels increases mortality. The SvO_2 will also be raised if the tissues have stopped using oxygen, either because of metabolic poisoning or significant cell death. In neither of these situations will further increases in oxygen delivery be helpful.

Monitoring SvO_2 in a patient with cardiovascular failure allows a rational approach to be made to managing the circulation without the use of any calculated or derived variables. The patient with a low SvO_2 needs his or her cardiac output to be increased, first by intravascular volume optimization using the response of stroke volume and filling pressures to fluids or vasodilators or both. If SvO_2 remains low once fluid status is optimized, the blood pressure will determine whether cardiac output is best improved by afterload reduction or inotropes. Similarly, the patient with a normal or high SvO_2 who remains hypotensive after intravascular fluid optimization will require an increase in vascular resistance using vasopressors. It is unnecessary to calculate the error-prone systemic vascular resistance in order to make this decision, nor should the systemic vascular resistance itself be treated if the blood pressure and cardiac output are acceptable. It must also be remembered that fluid optimization and the state of the circulation need to be reviewed after each therapeutic maneuver.

Target SvO_2 depends on the clinical situation faced. In high-output hypotensive states requiring vasopressors most would aim to keep SvO_2 at or above normal ($\geqslant$ 70 per cent), but in low-output states a target SvO_2 of 60 per cent is usually accepted. These targets clearly need to be adjusted according to the clinical situation, response to treatment, underlying cardiac function, etc.

The other main indication for the measurement of SvO_2 is severe respiratory failure, as a low SvO_2 will exaggerate the effects of a pulmonary shunt. If some pulmonary blood flow has an oxygen content that is lower than normal, it will mix with the oxygenated blood in the left heart lowering the average arterial saturation proportionately. As the contribution of dissolved oxygen to total oxygen content is insignificant at atmospheric pressure it is not possible to make up for this deficiency by hyperoxygenating the blood flowing through the normal lung. To illustrate simply, a patient with a 25 per cent true shunt and otherwise perfect respiratory function will have an arterial saturation of 91 per cent if the SvO_2 is 75 per cent. If the SvO_2 falls to 40 per cent the arterial saturation will fall to 83 per cent with a consequent further lowering of SvO_2, inducing a self-perpetuating downward spiral. Thus optimizing SvO_2 is a valuable strategy in the support of severe respiratory failure.

Conclusions

SvO_2 monitoring has attained general acceptance and some fervent supporters; however, in common with most monitors there is little hard evidence to support its use. Despite the advantages of the direct and continuous nature of the measurement and the insight given into the adequacy of oxygen delivery, it is clearly a long way from being the perfect monitor. The placement of a pulmonary artery catheter is an invasive procedure with numerous possible complications. Although the systems are intrinsically accurate there remain problems with drift and calibration. It is also possible that regional hypoxia may coexist with a normal or high SvO_2. These problems accepted, SvO_2 provides information not given by any other monitor which is of great value in the logical management of a large number of critically ill patients.

References

Dahn, M.S., Lange, M.P., and Jacobs, L.A. (1988). Central mixed and splanchnic venous oxygen saturation monitoring. *Intensive Care Medicine*, **14**, 373–8.

Jullien, T., Valtier, B., Hongnat, J.M., Dubourg, O., Bourdarias, J.P., and Jardin, F. (1995). Incidence of tricuspid regurgitation and vena caval backward flow in mechanically ventilated patients: a color Doppler and contrast echocardiographic study. *Chest*, **107**, 488–93.

Martin, C., Auffray, J.P., Badetti, C., Perrin, G., Papazian, L., and Gouin, F. (1992). Monitoring of central venous oxygen saturation versus mixed venous oxygen saturation in critically ill patients. *Intensive Care Medicine*, **18**, 101–4.

Scuderi, P.E., Bowton, D.L., Meredith, J.W., Harris, L.C., Evans, J.B., and Anderson, R.L. (1992). A comparison of three pulmonary artery oximetry catheters in intensive care unit patients. *Chest*, **102**, 896–905.

16.1.7 Cardiac output assessment

H. E. Montgomery and M. Singer

Key messages

- Cardiac output can be measured by invasive and non-invasive techniques.
- Non-invasive techniques should be viewed as complementary to, rather than competitive with, invasive methods.
- Additional information is often provided on circulatory status (e.g. filling, contractility, extravascular lung water).
- Reliability and accuracy vary. This may depend as much on the device as on the technique itself. The operator must be aware of the pitfalls and shortcomings of both the technique and the device being utilized, and of specific problems which may arise in individual patients with certain conditions.

Introduction

Measurement of blood flow has lagged behind pressure monitoring because of the inherent risks, complexity, and expense of pulmonary artery catheterization, the benchmark technique for cardiac output measurement in the intensive care unit (ICU). A variety of techniques are now available to assess cardiac output with either complete or relative non-invasiveness. These should be viewed as complementary to, rather than competing against, the invasive methods. While no technique can lay claim to absolute accuracy, technological improvements are continually raising the profile of non-invasive devices because of greater reliability and ease of use. Other than cardiac output, many provide additional information on circulatory status (e.g. preload, contractility, extravascular lung water). All have flaws, and data derived from different methods often correlate poorly. Some techniques may be better applied to monitoring trends rather than to absolute output.

The use of invasive flow monitoring in the ICU tends to be reactive, and it is often turned to when empirical therapy has failed to provide an adequate response to a cardiovascular disturbance. A reliable means of assessing global flow non-invasively would promote early identification of compensated derangements and faster correction.

Prospective users should familiarize themselves with both the theory and practice underlying a particular device, develop sufficient expertise to recognize unreliable or erroneous signals, and be aware of the limitations of both machine and technology. Poor equipment design and inadequate user education have contributed to many techniques being either discredited or used incorrectly.

Thermodilution

Thermodilution cardiac output performed via a pulmonary artery catheter (Royston and Feeley 1993) is the standard technique employed by most ICUs. It is a variant of the dye dilution technique, utilizing cold to create a thermal deficit. The Stewart–Hamilton equation links cardiac output to changes in temperature as follows:

$$\text{cardiac output} = \frac{V(T_\text{B} - T_\text{I})K_1K_2}{T_\text{B}(t)\text{d}t}$$

where V is the volume injected, T_B and T_I are the blood and injectate temperatures respectively, K_1 and K_2 are computational constants, and $T_\text{B}(t)\text{d}t$ is the change in blood temperature as a function of time.

In practice, a 5- or 10-ml bolus of 5 per cent glucose is injected smoothly and rapidly (< 4 s) through the proximal (injectate) lumen of the catheter into the right atrium. A thermistor at the catheter tip detects the temperature change which is computed to produce a curve, the area under which is inversely proportional to the cardiac output. Both technical and physiological errors are commonplace and frequently unrecognized (Gnaegi *et al.* 1997). The thermistor must be freely within the pulmonary arterial lumen rather than abutting the wall. All circuit connections must be leak free to prevent loss of injectate, as volumes injected must be accurate. The injectate may occasionally reflux up the introducer sheath when the proximal port is contained within the sheath; alternatively, rapid infusions through the side port of the sheath may affect the readings. The computer settings must be correct. The injectate may be cooled in ice to increase the temperature difference and the signal-to-noise ratio; however, a 10-ml room temperature injectate is usually adequate. Large variations (up to 50 per cent) in cardiac output can be obtained by performing the measurements at end-inspiration and end-expiration. To improve reproducibility, it is currently recommended that the injections are spaced evenly across the respiratory cycle.

Anatomical or physiological conditions that preclude accurate measurement include intracardiac shunts, tricuspid regurgitation, and cardiac arrhythmias. Notwithstanding errors in the technique itself introduced by inappropriate mixing of the injectate or bidirectional flow of the indicator solution, left heart output may also vary markedly from right heart output in many of these situations.

Modified pulmonary artery catheters can provide continuous monitoring of mixed venous saturation via a fiber-optic probe sited at the catheter tip (e.g. Oximetrix, Abbott Laboratories), right ventricular ejection fraction measurement, or 'continuous' cardiac output monitoring (e.g. Intellicath, Baxter-Edwards). The last technique utilizes pulses of heat emitted every 30 to 60 s from a thermal filament located 14 to 25 cm from the catheter tip. The resulting rise in blood temperature is detected by a thermistor located 4 cm from the tip. A formula cross-correlates the thermistor temperature with the thermal energy input sequence. Cardiac output is computed from the area under the 'washout' curve and a modified Hamilton–Stewart equation. This technique is not strictly continuous, as updated values reflect an average of the previous 3 to 6 min. Correlation against the bolus thermodilution technique has not always been close.

Doppler ultrasound

A more detailed discussion of the Doppler ultrasound technique is given by Royston and Feeley (1993) and Espersen *et al.* (1995).

The Doppler equation enables the velocity of a moving object to be calculated from the shift in reflected frequency of a sound wave of known frequency:

$$V = \frac{\Delta f\, C}{2 f_t \cos\theta}$$

where V is the velocity of moving blood corpuscles, Δf is the Doppler frequency shift, C is the velocity of sound in tissue, f_t is the transmitted frequency, and θ is the angle between the ultrasound beam and the flow direction.

Doppler frequency shift signals can be displayed as velocity–time waveforms (Fig. 1). High-frequency ultrasound waves (usually 2–4 MHz) are used for aortic blood flow measurement. A suprasternal approach, which is easy to perform, non-invasive, and painless, can be used. However, the probe cannot be fixed to enable continuous monitoring and signal acquisition is difficult in 5 per cent of cases (e.g. patients with a short neck, emphysema, or aortic valve disease). With sufficient experience accuracy is good.

Other approaches have been tried, of which descending thoracic aortic blood flow monitoring via the esophagus has proved the most successful. The Esophageal Doppler Monitor (marketed by Abbott Laboratories in Europe and Deltex in the United States) possesses a monitor to verify correct signal measurement and has undergone numerous single- and multicenter validation studies to ensure, with adequate training, reliable and reproducible results within minutes. A probe of diameter 6 mm inserted 35 to 40 cm into the distal esophagus is oriented to obtain a characteristic aortic flow signal. The area (integral) under each velocity–time waveform—the stroke distance—represents stroke volume flowing down the descending thoracic aorta. Applying a correction factor from a nomogram incorporating the patient's age, height, and weight enables an estimate of total left ventricular stroke volume to be determined with 85 to 90 per cent accuracy. Intra- and interobserver variability is low and trend-following is accurate despite wide variations in flow and blood pressure.

Only aortic coarctation and concurrent use of intra-aortic balloon counterpulsation preclude measurement. Aortic regurgitation produces a characteristic reverse flow throughout diastole. Caution should be observed in patients with esophageal varices or other local pathology, and in those with marked coagulopathies; however, no serious adverse event has yet been reported.

The waveform shape provides valuable information on left ventricular preload, afterload, and resistance. There are age-related normal ranges for peak velocity. Values outside this range are indicative of hypo- or hyperdynamic circulations. The flow time can be automatically corrected for heart rate by dividing it by the square root of the cycle time (FTc). This is inversely proportional to systemic vascular resistance. A short FTc is due to hypovolemia, flow obstruction, or excess arterial constriction (e.g. excessive vasopressor dosage). Preload changes predominantly affect the FTc, inotropic changes mainly affect the peak velocity, while afterload changes have an intermediate effect. The effects of therapy can be readily appreciated on a beat-by-beat basis, and Starling-like curves can be constructed to optimize fluid therapy.

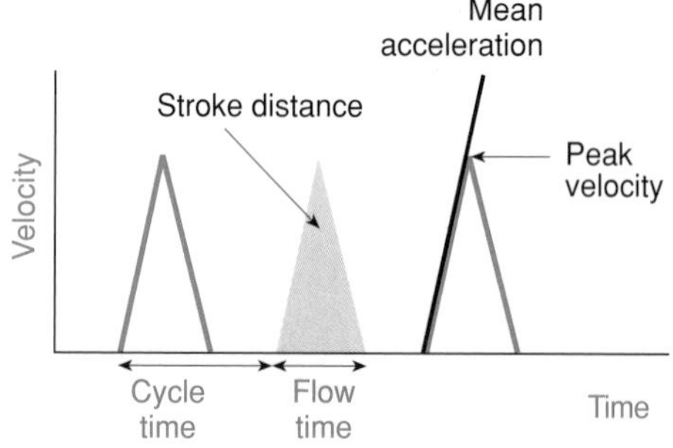

Fig. 1 Doppler flow-velocity waveform.

Echocardiographic evaluation

Echocardiographic techniques (either transthoracic or transesophageal) can estimate cardiac output by either combined use of imaging with Doppler signal derivation or calculation from ventricular volumes.

Combined use of imaging with Doppler signal derivation

With Doppler echocardiography, blood flow velocity is usually measured either within or close to the heart (e.g. aortic root, transmitral). Echocardiographic measurement of cross-sectional area at this point enables calculation of cardiac output as the product of blood velocity averaged over 1 min and cross-sectional area. Non-uniform flow distribution, particularly in the presence of turbulence, may render this measurement unreliable. Further error derives from difficulty in orienting the Doppler signal to the direction of blood flow, and from beat-to-beat variation in stroke volume.

Aortic vessel area at the valve ring (or just distal to the sinuses of Valsalva) is calculated from aortic root diameters determined from two-dimensional images, assuming a circular shape. However, both the aorta and the main pulmonary trunk are more ellipsoidal than circular, and root area varies with timing in the cardiac cycle and with changes in preload, afterload, and cardiac dynamics. Planimetric measurement, enhanced by recently developed edge-detection technology, may improve the accuracy of calculated outputs although it may still prove dificult to obtain accurate short-axis views. Similar methodology has been applied to record flow across the left ventricular outflow tract, the pulmonary artery, and, in particular, the mitral valve. The area of the mitral valve is measured by planimetry or estimated from diameters obtained in two- and four-chamber views assuming an elliptical shape. Transesophageal echocardiographic imaging allows cardiac output estimation in the critically ill more readily than the transthoracic approach.

Calculation from ventricular volumes

Stroke volume can also be calculated as the difference between end-diastolic and end-systolic ventricular volume. Volumes are generally calculated from two-dimensional and M-mode images which rely upon assumptions about left ventricular shape. Recently developed technology allows three-dimensional reconstruction of images of the left ventricle.

1. Cube method: this assumes that wall motion is symmetrical and $V = D^3$, where D is the short-axis diameter of the left ventricle.
2. Teichholz method: this uses the equation $V = D^3[7.0/(2.4 + D)]$ which corrects better for ventricular shape.

3. Bullet method: this assumes that the left ventricular cavity is bullet-shaped and that $V = 5AL/6$ where A is the cross-sectional area of the short axis and L is the cavity length.

4. Area–length calculations: these assume the left ventricle is an elongated ellipse. Images in two planes (best measured by a biplanar transesophageal approach) are derived, and the volume is calculated as $0.524D_1D_2L$ where L is length and D_1 and D_2 are the orthogonal short-axis diameters.

5. Simpson's method: the volume of each 'slice' obtained from multiple short-axis images of known spacing is either calculated from its dimensions or measured by planimetry. The volumes are summed to yield a total ventricular volume. Irregular ventricular shape entails deriving more slices, good endocardial imaging, and accurate timing of images in the cardiac cycle.

6. Automated quantitation of cavity size using sophisticated real-time edge-detection technology has recently been demonstrated. Advantages are continuous measurement with little operator dependence.

Transthoracic impedance

The use of transthoracic impedance is discussed in detail by Preiser *et al.* (1989).

In 1966, Kubicek described the thorax as a cylinder evenly perfused with blood of specific resistivity p, itself related to the hematocrit. Pulsatile thoracic aortic blood flow caused negative impedance changes (measured from a steady state mean base impedance Z, with a maximum rate of change $(dZ/dt)_{max}$ of Z_m during systole between pairs of electrodes placed around the neck and upper abdomen). Thus

$$\text{stroke volume} = p(L^2/Z^2)Z_m$$

where L is the mean separation of the inner pair of electrodes.

The sensing electrodes are usually applied to either side of the neck root and to opposite sides of the body in a midcoronal plain at xiphisternum level. Two pairs of transmitting electrodes are placed 5 cm above the neck sensors and 5 cm below the thoracic sensors respectively. Accurate electrode placement is crucial. A clear ECG trace is required to allow gating of data acquisition; displacement of a single lead will falsely reduce cardiac output.

The original Kubicek equation consistently overestimated stroke volume. As the thorax behaves electrically more like a truncated cone than a cylinder, correction factors have been introduced, notably by Sramek and Bernstein. Computer analysis can also eliminate variations due to gross movement and respiration. Under optimal conditions, bio-impedance cardiac output correlates reasonably with other techniques. However, shivering, valvular regurgitation, pacemaker devices, atrial fibrillation, positive end-expiratory pressure, marked tachycardia, different brands of electrode and, possibly, poor left ventricular function all impede reliability. Cardiac output may be overestimated in low-flow states, and underestimated in hyperdynamic states and soon after cardiac surgery. The frequent presence of one or more of these confounding factors in the ICU patient has precluded its routine adoption.

Pulse contour analysis

Since the rate at which blood flows from arteries to veins is proportional to the rate of fall of arterial pressure, analysis of the contour of the aortic pulse wave enables cardiac output to be determined (Royston and Feeley 1993). The pulse contour cardiac output technique can be performed from an indwelling pressure-transduced radial or femoral cannula. However, it cannot be used for quantitation of cardiac output unless a calibration tool such as thermodilution is used. Various equations have been developed for monitoring stroke volume using the area below the pressure contour, heart rate, and a calibration factor. Sparse data exist regarding validation of the pulse contour cardiac output technique against other techniques. Its major shortcoming is the inconsistent volume–pressure relationship associated with different individuals. This is also influenced by different hemodynamic states, for example changes in blood pressure, posture, body temperature, and vasopressor agents. Further studies are needed before this technique can be recommended for routine usage.

COLD technique

Depending on the desired measurement, the COLD technique (circulation, oxygenation, lung water, and liver function diagnosis) utilizes cannulas placed in either the femoral or radial artery, the pulmonary artery, and/or a central vein. A bolus of cold fluid can be injected into any central vein enabling thermodilution cardiac output to be measured via a 3 French catheter placed in the femoral or radial artery. This catheter has an optional oximetry probe for continuous measurement of arterial oxygen saturation. A similar catheter can be placed in the pulmonary artery for cardiac output measurement and monitoring of mixed venous oxygen saturation. A new development allows continuous measurement of cardiac output by pulse contour analysis with automatic recalibration being performed by intermittent thermodilution injections.

The device is also geared for measurement of indocyanine green in the blood. Following injection of a bolus of indocyanine green into (preferably) a central vein, cardiac output can be computed from the dye dilution curve produced at the arterial catheter site. A fiber-optic reflectance densitometer built into this catheter measures the indocyanine green concentration. An additional advantage is that elimination of this dye is almost entirely via the liver. Thus the plasma disappearance rate can be used to assess hepatic function. With normal liver function only a tiny fraction of the injected indocyanine green is still detectable in the bloodstream after 10 min.

The device also purports to measure (i) global end-diastolic volume, i.e. the sum of the end-diastolic volumes of both left and right atria and ventricles, (ii) intrathoracic blood volume, and (iii) extravascular lung water. It has been suggested that intrathoracic blood volume can be used as an indicator of cardiac preload while cardiac performance can be derived from the ratio of cardiac output to global end-diastolic volume. Extravascular lung water measurement is purported to quantify acute lung injury and is claimed to enable appropriate therapy to shorten ventilation time and reduce complication rates such as pneumonia. Confirmatory studies to support these claims are scanty.

Direct Fick method

The Fick principle states that oxygen consumption ($\dot{V}O_2$) equals the product of cardiac output and the arteriovenous O_2 content difference. Arterial O_2 content is measured from a peripheral arterial blood sample, venous O_2 content is measured from pulmonary arterial blood, and $\dot{V}O_2$ is derived from minute ventilation and inspired and expired gas analysis (Espersen *et al.* 1995). Reliability decreases when the patient is receiving high inspired O_2 concentrations.

CO_2 production ($\dot{V}CO_2$) can be used instead of $\dot{V}O_2$ as cardiac output also equals $\dot{V}CO_2$ divided by ($C\bar{v}CO_2 - CaCO_2$), where $C\bar{v}CO_2$ and $CaCO_2$ are mixed venous and arterial CO_2 contents respectively. The venoarterial difference can be calculated from measured partial presures, and $\dot{V}CO_2$ can be accurately measured using computerized dual-beam infrared absorption analysers.

Lithium dilution

This technique (Linton *et al.* 1997) employs the dye dilution principle using lithium chloride as the indicator. This is injected as a 0.15- to 0.3-ml bolus into a central vein. Arterial plasma [Li^+] levels are measured by a lithium-selective electrode in a flow-through cell connected by a three-way tap to a standard arterial cannula. The blood sampling rate is 4 ml/min. The voltage across the electrode membrane is related to plasma [Li^+]. A correction is applied for plasma sodium concentration as this can cause interference. Cardiac output is computed from the lithium concentration–time curve using the following equation:

$$\text{cardiac output (l/min)} = \frac{\text{LiCl dose} \times 60}{\text{Area} \times (1 - \text{PCV})}$$

where the LiCl dose is in millimoles and Area is the integral of the primary curve (mmol/s). The packed cell volume (PCV) is a necessary correction as lithium is only distributed in plasma and not in erythrocytes. The primary curve is due to the initial circulation of lithium following injection. The integral of the primary curve is obtained from an equation deriving the integral of a lognormal from the first part of the curve. The secondary curve, which is due to recirculation of the indicator, is ignored. Early studies have shown good agreement with cardiac output measured by both continuous and bolus thermodilution techniques.

References

Espersen, K., *et al.* (1995) Comparison of cardiac output measurement techniques: thermodilution, Doppler, CO_2-rebreathing and the direct Fick method. *Acta Anaesthesiologica Scandinavica*, 39, 245–51.

Gnaegi, A., Feihl, F., and Perret, C. (1997). Intensive care physicians' insufficient knowledge of right-heart catheterization at the bedside. Time to act? *Critical Care Medicine*, **25**, 213–20.

Linton, R., Band, D., O'Brien, T., Jonas, M., and Leach, R. (1997). Lithium dilution cardiac output measurement: a comparison with thermodilution. *Critical Care Medicine*, **25**, 1796–1801.

Preiser, J.C., Daper, A., Parquier, J.-N., Contempre, B., and Vincent, J.-L. (1989). Transthoracic electrical bioimpedance versus thermodilution technique for cardiac output measurement during mechanical ventilation *Intensive Care Medicine*, **15**, 221–3.

Royston, D. and Feeley, T.W. (ed.) (1993). *International anesthesiology clinics. Monitoring in anesthesiology: current standards and newer techniques*. Little, Brown, Boston, MA.

16.1.8 Oxygen transport

J. D. Edwards

Key messages

- An absolute reduction in $\dot{D}O_2$ could occur as a result of a reduction in hemoglobin, PaO_2, and hence SaO_2, or from a reduction in cardiac output.
- If $\dot{D}O_2$ falls but oxygen demand continues and $\dot{V}O_2$ is maintained, there will be an increase in the oxygen extraction ratio (OER) and a fall in $S\bar{v}O_2$ if the ability of the tissues to extract oxygen is intact.
- Despite the high values of $\dot{D}O_2$ found in severe sepsis, there is evidence of an inability of peripheral tissues to utilize oxygen, and hence an inappropriately low $\dot{V}O_2$.
- It would seem reasonable to try, wherever possible, to achieve an OER as close to normal as possible. In some situations, such as septic shock, acute respiratory distress syndrome, and trauma, the aim should be a value of about 20 per cent, provided that underlying cardiac disease has been excluded.

Introduction

The relationship between the delivery of oxygen (O_2) to the tissues and its consumption by respiring cells has been the subject of much interest. The various components of this relationship have been termed the oxygen transport system.

Tissue hypoxia can be classified into four common forms: stagnant (low cardiac output), anemic (reduced hemoglobin), hypoxic (reduced arterial oxygen tension (PaO_2)), and cytoxic hypoxia (e.g. due to cyanide intoxication). Oxygen flux is defined as the product of arterial oxygen content and cardiac output.

Current interest in monitoring and manipulating oxygen delivery was stimulated by the work of Shoemaker *et al.* (1988) who demonstrated that certain patterns of oxygen transport variables had predictive prognostic value in high-risk surgical cases and that outcome could be improved by manipulating these variables.

Basic physiology of oxygen transport

As oxygen is carried to the tissues bound to hemoglobin and propelled by the total blood flow (the cardiac output (CO)) oxygen is extracted by respiring cells; at rest this is normally 25 per cent of the available oxygen (Fig. 1).

Definitions, terms, equations, and abbreviations

The total amount of oxygen flowing to the tissues is termed the oxygen delivery ($\dot{D}O_2$) and is calculated (in ml/min/m^2) as follows:

$$\dot{D}O_2 = \text{CI} \times CaO_2 \times 10 \quad (1)$$

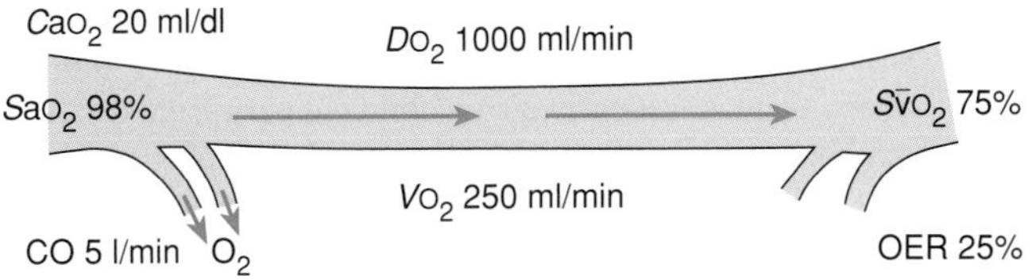

Fig. 1 Normal oxygen transport patterns.

where CI is the cardiac index and Cao_2 is the arterial oxygen content. CI is the ratio of CO to the patient's body surface area which is derived from standard nomograms based on measurements of the patient's height and weight. Cao_2 is given (in ml/dl) by

$$Cao_2 = ([Hb] \times Sao_2 \times K) + \text{dissolved } O_2 \quad (2)$$

where [Hb] is the hemoglobin concentration (in g/dl) and Sao_2 is the percentage arterial oxyhemoglobin saturation. The dissolved oxygen is calculated by multiplying the arterial partial pressure of oxygen (Pao_2) by 0.023 (for kPa) or 0.003 (for mmHg). The mixed venous oxygen content ($C\bar{v}o_2$) is given (in ml/dl) by a similar equation:

$$C\bar{v}o_2 = ([Hb] \times S\bar{v}o_2 \times K) + \text{dissolved } O_2 \quad (3)$$

where $S\bar{v}o_2$ is the percentage mixed venous oxyhemoglobin saturation which is measured in blood taken from the pulmonary artery. Subtracting $C\bar{v}o_2$ from Cao_2 gives the difference $Cao_2 - C\bar{v}o_2$ between the arterial and mixed venous oxygen contents, often inappropriately abbreviated to $D(a-\bar{v})o_2$.

The relationship between supply and consumption of oxygen can be calculated from the oxygen extraction ratio (OER) given (as a percentage) by

$$\text{OER} = \frac{Cao_2 - C\bar{v}o_2}{Cao_2} \times 100. \quad (4)$$

The Fick equation

$$\text{CO} = \frac{Vo_2}{Cao_2 \times Cao_2} \quad (5)$$

was developed in order to calculate CO from measured volumes and concentrations of oxygen in respired gases and from Cao_2 and $C\bar{v}o_2$. Now that CO can be measured routinely at the bedside by the thermodilution technique, an alternative form of the Fick equation (often inaccurately termed the reverse or indirect Fick equation) is used to calculate oxygen consumption (in ml/min/m^2):

$$Vo_2 = \text{CI} \times (Cao_2 \times C\bar{v}o_2) \times 10. \quad (6)$$

Normal values

A hypothetically normal adult patient has a body surface area of 1.84 m^2, CO = 5 l/min, [Hb] = 15 g/dl, Sao_2 approaching but never reaching 100 per cent, and Pao_2 = 13 kPa (depending on age). Thus the following derived variables can be calculated: Cao_2 = 20.73 ml/dl and Do_2 = 563 ml/min/m^2.

The normal $S\bar{v}o_2$ is 75 per cent with a mixed venous oxygen tension ($P\bar{v}o_2$) of 5.3 kPa. Therefore $C\bar{v}o_2$ = 15.76 ml/dl, OER = 24 per cent, and Vo_2 = 132 ml/min/m^2.

Calculating OER from the arterial and mixed oxygen contents obviates any systematic errors in CO determination that may lead to erroneous conclusions about the relationship between Do_2 and Vo_2 which has been thought to be a problem by some authorities.

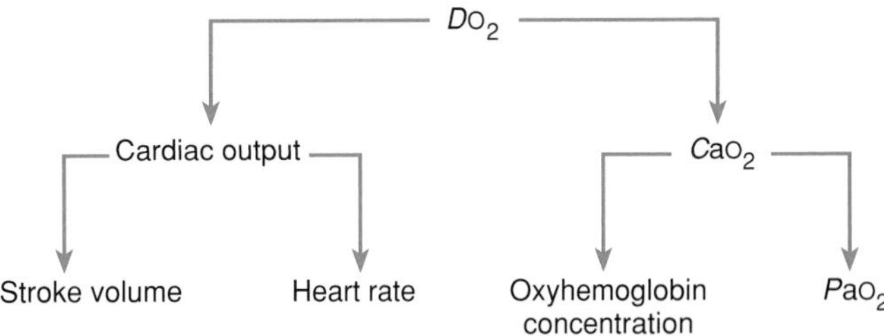

Fig. 2 Determinants of oxygen delivery (Do_2).

Abnormalities of oxygen transport

An absolute reduction in Do_2 could occur as a result of a reduction in hemoglobin, Pao_2, and hence Sao_2, or from a reduction in CO which could be the result of low stroke volume or heart rate. The stroke volume is in turn dependent on cardiac preload, contractility, and afterload. These variables are summarized in Fig. 2.

Hypoxemia should generally be corrected immediately, particularly where Sao_2 is reduced. This is followed by correction of anemia and hypovolemia. There is no universal agreement on the optimal hematocrit. The level of hemoglobin which a patient will tolerate will depend on the underlying cardiac reserve. A patient with a healthy myocardium will tolerate a reduction to 8.0 g/dl for instance, and could maintain Do_2 by a compensatory increase in CI to 5.5 l/min/m^2. If there is pre-existing or acquired myocardial dysfunction this compensation will not occur. If Do_2 falls but oxygen demand continues and Vo_2 is maintained, there will be an increase in OER and a fall in $S\bar{v}o_2$ if the ability of the tissues to extract oxygen is intact. The normal values of OER (25 per cent) and $S\bar{v}o_2$ (75 per cent) are global figures for the whole body. Individual organs may have differing extraction capabilities at rest and under abnormal conditions. For instance, the kidney has an excess of blood flow over oxygen demand with an OER of 15 per cent, whereas the coronary circulation has a resting OER of 75 per cent and this cannot increase with demand. In shock states associated with increased OER survival is unlikely if $S\bar{v}o_2$ falls to less than 40 per cent.

The oxyhemoglobin dissociation curve

The oxyhemoglobin dissociation curve has received scant attention in clinical practice. The circumstances associated with a shift of the dissociation curve (abnormal temperature, $Paco_2$, pH, and 2,3-diphosphoglycerate levels) are almost always present in critically ill patients to varying degrees and in varying combinations. In addition, there may be other hemoglobin ligands such as carboxyhemoglobin or methemoglobin which, if present in significant concentrations, could effectively lower Cao_2 and also lead to errors in oxygen transport calculations.

Shifts of the oxyhemoglobin dissociation curve mainly affect the middle portion, leaving the upper and lower portions relatively unaffected. The concept of consumable or available oxygen has been proposed based on clinical work, notably that of Bryan Brown *et al.* (1973) which strongly suggests that oxygen bound to hemoglobin below a tension of 2.7 kPa (20 mmHg) is unavailable for tissue oxygen consumption because of its inability to diffuse into mitochondria at this partial pressure. It has generally been believed that a shift of the curve to the right is beneficial in that if $P\bar{v}o_2$ remains constant there will be a reduction in $S\bar{v}O_2$ and therefore an automatic increase in $\dot{V}O_2$ due to reduced hemoglobin affinity for oxygen at the mixed venous saturation. This has been referred to as improved 'off-loading' of oxygen. This is a purely teleological point of view and ignores the multiplicity of factors which affect tissue oxygen availability in critically ill patients.

If DO_2 is reduced but the tissues maintain the ability to extract oxygen, $\dot{V}O_2$ can be maintained at a normal or even high value by an increase in OER which produces a reduction in $S\bar{v}O_2$. This is an unsatisfactory mechanism as it may limit oxygen availability towards the venous side of the capillary bed in some tissues, notably the liver. This type of oxygen transport pattern is seen most commonly in poorly resuscitated trauma victims or in cardiogenic shock following acute myocardial infarction.

Oxygen transport patterns in clinical situations

The use of vasoactive drugs (vasopressors, inotropes, or vasodilators) will rapidly change oxygen transport abnormalities in critically ill patients. The following discussion applies to the oxygen transport patterns found in patients before administration of such drugs.

Trauma

There has been much confusion concerning $\dot{V}O_2$ following major trauma in man. Early work suggested that $\dot{V}O_2$ was reduced in the first 24 to 48 h and later increased, the so-called 'ebb phase' followed by the 'flow phase'. More recently, the situation has been reappraised in more severely injured patients and it is now clear that, in the absence of hypothermia, $\dot{V}O_2$ is maintained or even increased within the first 2 to 24 h. Subsequent oxygen transport patterns will depend on the rate of healing of injuries, the use of intubation and anesthesia, surgical procedures, and development of complications such as sepsis and acute respiratory distress syndrome. Even modest increases in OER and reductions in $S\bar{v}O_2$ can be associated with poor outcome in trauma victims.

Cardiogenic shock

Cardiogenic shock continues to affect 5 per cent of all cases of acute myocardial infarction despite use of thrombolysis. The oxygen transport patterns have now been well defined. There is a reduction in CI to 2.2 l/min/m^2 or below, and $\dot{V}O_2$ is maintained at a normal levels by increases in OER and reductions in $S\bar{v}O_2$. If an increase in CI can be achieved using inotropic agents, there are no increases in $\dot{V}O_2$ and significant reductions in OER with increases in $S\bar{v}O_2$. Also, a small but significant number of patients present with the clinical picture of cardiogenic shock but with normal oxygen transport variables. This is an important subgroup as they make a full recovery without any further treatment except basic life support.

Similar oxygen transport patterns have been demonstrated following tension pneumothorax and massive pulmonary embolism.

Sepsis and acute respiratory distress syndrome

Sepsis and acute respiratory distress syndrome are considered together as the oxygen transport abnormalities in these two conditions are very similar.

The cardiac output and DO_2 are usually above the normal range in fully resuscitated patients. Despite the high values of DO_2 found in severe sepsis, there is evidence of an inability of peripheral tissues to utilize oxygen with an inappropriately low $\dot{V}O_2$, frequently in the presence of an increased arterial blood lactate representing anaerobic glycolysis. This oxygen transport pattern has been demonstrated in the earliest and the most recent clinical studies. The causes of this inability to consume oxygen are purely speculative.

Alternative methods of assessing the adequacy of tissue oxygen delivery

Tonometry: monitoring of gastric intramucosal pH

Measurement of gastric PCO_2 by tonometry has been used to calculate the pH of the gastric mucosa (pHi), a reduction of which may predict mortality in critically ill patients who were apparently otherwise well resuscitated. Resuscitating patients to a normal pHi as well as satisfactory global oxygen transport patterns is associated with an improved outcome.

The difference between mixed venous and arterial CO_2 tension

The difference between the CO_2 tension in arterial blood ($PaCO_2$) and mixed venous blood ($P\bar{v}O_2$), termed $P(\bar{v}-a)CO_2$ or more recently PCO_2, is normally less than 1 kPa (7.5 mmHg). PCO_2 may increase to up to 3.0 kPa (22.5 mmHg) in severe shock, and may increase in the presence of apparently adequate oxygen transport patterns and normal lactate levels in septic shock. This variable has an advantage over tonometry in that samples are available from systemic and pulmonary artery catheters which are already *in situ*. However, doubt has been cast on the validity of this variable as the sole indicator of inadequate DO_2.

Jugular bulb saturation

Changes in oxyhemoglobin saturation of jugular venous blood (SjO_2) can provide important information about the relationship between oxygen supply and demand in the brain. Levels below 55 to 50 per cent indicate a critical imbalance between cerebral oxygen consumption and cerebral blood flow. It is generally accepted that the dominant jugular vein, or the ipsilateral vein in the case of a unilateral cerebral lesion, should be cannulated.

Blood lactate levels

Lactic acid is produced as a result of anaerobic metabolism when DO_2 to the tissues is low or inadequate. Currently, blood lactate is measured by bedside analyzers using an enzymatic technique, and a level of 1.5 mmol/l is generally accepted as the upper limit of normal. Lactate levels may be elevated when vital signs and oxygen transport variables appear satisfactory, indicating DO_2 values that are insufficient for the clinical circumstances involved or inability to consume oxygen. It has been suggested that lactate levels may be invalidated by the presence of liver disease. However, even with advanced cirrhosis, patients in shock with a high lactate have a poorer prognosis, and in acute hepatic encephalopathy a high blood lactate is associated with mortality and inadequate oxygen transport patterns.

The 'oxygen flux' test

It has been suggested that if an increase in Do_2 results in an increase in $\dot{V}o_2$, this is evidence of covert tissue hypoxia. This is highly controversial. Concerns have been expressed about complications of increasing cardiac output above normal and the production of toxic O_2 free radicals at the peripheries. Also, some authors have attributed the increase in $\dot{V}o_2$ to a statistical artifact due to mathematical coupling of errors in cardiac output determination by thermodilution and have claimed that this phenomenon is not seen when $\dot{V}o_2$ is calculated independently of cardiac output by indirect calorimetry.

An alternative approach of plotting the OER against CI and ignoring the $\dot{V}o_2$ as a therapeutic goal has been suggested by Silance *et al.* (1994). Indeed, $\dot{V}o_2$ may change independently of tissue hypoxia owing to pain, agitation, or use of sedation and analgesia. It would seem reasonable to try, wherever possible, to achieve an OER as close to normal as possible, and in some situations such as septic shock, acute respiratory distress syndrome, and trauma to aim at a value of 20 per cent provided that underlying cardiac disease has been excluded.

Estimation of $\dot{V}o_2$ from respiratory gases

In theory, the volume of oxygen consumed can be calculated by subtracting the expired volume of oxygen ($\dot{V}eo_2$) from the inspired volume ($\dot{V}io_2$). Fio_2 and Feo_2 must be measured, and expired volume is routinely and easily measured. Because the respiratory ratio is rarely 1.0 there is a difference between inspired and expired gas volumes. Therefore inspired volume is commonly calculated from measurements of N_2 concentrations. As FiO_2 increases above 0.6, errors in measuring N_2 concentrations increase expotentially.

Continuous measurement of $S\bar{v}o_2$

Fiber-optic catheters have been developed which continuously measure and display the value of $S\bar{v}o_2$ using an appropriate monitor. If the ability of the tissues to extract oxygen is preserved, a decrease in $S\bar{v}o_2$ usually indicates a mismatch between supply and demand. $S\bar{v}o_2$ can change because of an increase in oxygen consumption or a decrease in oxygen delivery. Also, in patients with sepsis and/or acute respiratory distress syndrome $S\bar{v}o_2$ could change as a result of variation in the ability of tissues to consume oxygen and could represent either an improvement or a deterioration.

References

Bryan Brown, C., Baek, S., Makbali, G., and Shoemaker, W. (1973). Consumable oxygen: availability of oxygen in relation to oxyhemoglobin dissociation curve. *Critical Care Medicine*, **1**, 17–21.

Shoemaker, W.C., Appel, P.L., Kram, H.B., Waxman, K., and Lee, T.S. (1988). Prospective trial of supranormal values of survivors as therapeutic goals in high risk surgical patients. *Chest*, **94**, 1176–86.

Silance, P.G., Simon, C., and Vincent, J.-L. (1994). The relationship between cardiac index and oxygen extraction in acutely ill patients. *Chest*, **105**, 1190–7.

16.1.9 Right ventricular function

H. E. Montgomery

Key messages

- The right ventricle possesses significant independent systolic function.
- Right ventricular dysfunction is common in the intensive care unit and adversely affects prognosis, cannot be inferred from left ventricular function, and requires specific management.
- Right ventricular assessment should include baseline assessment (history, examination, ECG), assessment of performance (right ventricular volumes, stroke volume, ejection fraction), and assessment of function (load-independent measures).
- Appropriate use of echocardiography and a pulmonary artery flotation thermodilution catheter allows rapid repeatable measurement of cardiac volumes, ejection fraction, pressures, loading conditions, and output, as well as construction of a right ventricular end-systolic pressure–volume graph.

Introduction

The right ventricle exhibits significant systolic function independent of the left ventricle (Oldershaw 1992). Right ventricular dysfunction is common in the intensive care unit (ICU); it complicates cases as diverse as myocarditis, multiple injury, burns, severe mitral regurgitation, septic shock, and ischemic cardiac injury (ischemic cardiac failure, postinfarction ventricular septal defect, and left ventricular infarction) where it also worsens prognosis. Indeed, after inferior myocardial infarction, right ventricular infarction increases mortality and morbidity by a factor of 5 to 7.

Management of right and left ventricular dysfunction differs. For instance, in post-thoracotomy biventricular failure with cardiogenic shock, left ventricular support without right ventricular support worsens prognosis.

Finally, right ventricular function cannot be inferred from that of the left ventricle. Even after myocardial infarction each ventricle should be considered independently. The left anterior descending artery supplies much of the left ventricle, but also parts of the right ventricle (the right ventricular infundibulum, the upper septum, and the anterior right ventricle bordering the septum). The right coronary artery supplies much of the right ventricle (the more proximal and complete a right coronary artery occlusion, the more likely is a right ventricular infarction) and also parts of the left ventricle (e.g. posteromedial papillary muscles, and generally its posterior wall). Thus a variable balance of biventricular injury follows any pattern of coronary occlusion.

Therefore the right ventricle should be considered separately in all ICU patients to provide prognostic information and to guide management strategies. Knowledge of both performance (which is load dependent) and function (which is load independent) is useful. Both should be considered in the light of the probable chronic loading conditions.

Assessment in the ICU demands simple, portable, repeatable, cheap, and non-toxic methodology. Rapid broad classification of function may have advantages over lengthy precise assessment.

Baseline assessment of right ventricular function

Clinical assessment

History

Limitation in exercise capacity correlates with the reduction in right ventricular ejection fraction in those with coronary artery or chronic obstructive airways disease. In the presence of ischemic left ventricular dysfunction, right ventricular dysfunction is more likely amongst those with previous inferior myocardial infarction or chronic lung disease.

Examination

Right ventricular dysfunction elevates right ventricular end-diastolic pressure and hence jugular venous pressure. Sustained elevation leads to dependent edema, ascites, and hepatomegaly. Palpable pulmonary arterial pulsation and a sustained right ventricular heave may be found in patients with chronic right ventricular pressure loading.

Right ventricular compliance is normal with volume loading, and the jugular venous pressure waveform is preserved.

In the hypertrophied pressure-loaded right ventricle, reduced right ventricular compliance causes a more dominant a wave (more than twice the v-wave amplitude measured from the nadir of the x depression), prominent x descent, and shallow y descent.

On auscultation high at the left sternal edge, the sound of aortic valve closure should precede and be louder than that of the pulmonary valve (P2). P2 is delayed if right ventricular ejection is prolonged (as in acute pulmonary arterial hypertension complicating pulmonary embolus), and is louder in cases of pulmonary hypertension. Right ventricular third heart sounds (which increase in intensity during inspiration) may be heard.

Electrocardiography

Right ventricular myocardial infarction

The conventional ECG demonstrates no features specific to right ventricular myocardial infarction, which is best diagnosed using leads V_1R–V_6R (in the mirror image positions of leads V_1–V_6). ST elevation in lead V_4R alone is approximately 90 per cent sensitive and 80 per cent specific for right ventricular myocardial infarction.

Chronic right atrial overload

There is a tall peaked p wave in lead II (P pulmonale), often with a p-wave vector of 60° to 90°.

Right ventricular hypertrophy

This is suggested by a clockwise axis rotation, right axis deviation, or vertical heart position, or by increased anterior lead R-wave amplitude, often with a strain pattern (convex ST depression and T-wave inversion). A large R wave in lead aVR, a deep S wave in the inferior leads, and a small notch, slur, or q wave in lead V_1 are also observed.

Acute right ventricular overload

This may yield the classic S1 Q3 T3 pattern (an S wave in lead I, and a q wave and T-wave inversion in lead III), right bundle branch block, right axis deviation, right ventricular hypertrophy with strain, P pulmonale, atrial fibrillation, or sinus tachycardia.

ST-segment elevation in leads V_3R–V_6R under conditions of cardiovascular stress may suggest reduced right ventricular contractile reserve.

Comparison of end-diastolic pressures

When ventricular end-diastolic pressure (on the x axis) is plotted against stroke volume (on the y axis), the line representing right ventricular function lies somewhat above and to the left of that for the left ventricle. Therefore, at any given cardiac output, right ventricular end-diastolic pressure will be lower than left ventricular end-diastolic pressure, and this discrepancy should increase as the former rises. The ratio of right to left ventricular end-diastolic pressure may be preserved in the event of equivalent biventricular impairment. A rise in this ratio (or the crude ratio of central venous pressure to pulmonary artery wedge pressure) suggests an approximation of the right ventricular and left ventricular function graphs, due to either a specific decline in right ventricular function or to increased left ventricular ejection. The ratio may rise above unity if the right ventricle is selectively or severely damaged, and is often greater than 1.3 in cases of right ventricular infarction. Where peripheral vasodilatation occurs (e.g. sepsis), the curves approximate as the left ventricular curve moves to the left. In chronic airways disease, this approximation is compounded by an additional movement of the right ventricular curve to the right.

Right ventricular volumes as a guide to right ventricular performance

Right ventricular end-diastolic volume is a poor index of right ventricular function. Right ventricular end-systolic volume is better, depending less on preload than on afterload and contractility.

Imaging of right ventricular volumes

Volume assessment by imaging is hampered by the complex crescent-shaped cross-section of the right ventricle, dense free-wall trabeculation disrupting edge detection, overlap of other cardiac chambers, and effects of imaging angle (which can introduce errors of up to 30 per cent). Contrast ventriculography, magnetic resonance imaging, and CT scanning are standard techniques, but are not readily applicable in the ICU (Oldershaw 1992).

Echocardiography

Multiple slices can be used to compute right ventricular volume. Transthoracic imaging is impeded by mechanical ventilation and its position beneath the sternum, and volume assessment is hampered by wall trabeculation, changes in wall reflection in disease, and the complex right ventricular shape. Transesophageal echocardiography and automated border detection may help. Even so, defining the right ventricular boundary by the innermost trabeculae often leads to underestimation. Technological advances may extend the value of echocardiography to the ICU assessment of the right ventricle.

Radionuclide studies

First-pass studies using red corpuscles labeled with technetium-99m and a right anterior oblique projection of 20° to 30° allow temporal separation of the right from the left ventricle, and can assess right ventricular end-systolic and end-diastolic volumes and ejection rates. After tracer equilibration with the intravascular pool (equilibration studies), images gated to different parts of the cardiac cycle can be acquired. Radioactivity in overlapping structures (e.g. the left ventricle, the right ventricular outflow tract, and underlying right atrium) diminishes accuracy. Even with a left anterior oblique projection of 20° and a caudal angulation of 20°, the right atrium may contribute 30 per cent of the signal. Ventricular separation by a 40° left anterior oblique projection is incomplete if the right ventricle is dilated.

Conductance catheter assessment of right ventricular volumes

The conductance catheter, which is placed in the right ventricle, establishes an electric field between an apical and basal electrode by the passage of a high-frequency current. Current is held constant while adjacent pairs of intervening electrodes measure the conductivity of the surrounding blood (which greatly exceeds that of the surrounding myocardium) along this right ventricular long axis (Dickstein *et al.* 1995). However, the technique has limitations.

1. The electrode pairs measure in cross-sectional dimensions. Thus shortening of the longitudinal axis is poorly assessed.
2. Blood resistivity varies with time, and increasing flow velocity may reduce blood conductivity and lead to underestimated values.
3. Structures surrounding the ventricle contribute to the conductance signal.

The last problem might be partly overcome by the proximal injection of 10 ml of cooled saline (2–4 °C). Since

$$G(t) = \frac{\alpha}{L^2 \sigma_b V(t)} + G_p$$

where $G(t)$ is the variation of conductance with time, σ_b is the specific conductivity of blood, α is a dimensionless constant, L is the electrode distance, $V(t)$ is the variation of volume with time, and G_p is the conductance of surrounding structures, such an injection would cause a transient decrease in left ventricular blood conductivity and allow G_p to be calculated. This technique has as yet only been applied to the left ventricle. There are also specific difficulties in applying this technique to the right ventricle. Because of the complex right ventricular geometry, the current density varies between electrode pairs. Right ventricular trabeculation distorts uniform distribution of the excitation current in the right ventricle and leaves some regions unscanned. Finally, the right ventricular wall is thinner than that of the left ventricle, and current leakage is greater. Impedance methods may significantly underestimate the stroke volume, and therefore correction factors are required. Nonetheless, stroke volumes measured by conductance and thermodilution correlate closely.

Ejection fraction as a guide to right ventricular function

The right ventricular ejection fraction, defined as the ratio of stroke volume to right ventricular end-diastolic volume (normal value, 0.52 ± 0.07), is load dependent and thus is a crude measure of right ventricular function, particularly in cases of acutely altered afterload. Nonetheless, at lower values, it may correlate well with load-independent measures although an exponential relationship seems to exist at higher ejection fractions. The use of conductance catheters has distinct advantages over imaging techniques, as outlined above. Two other methods are appropriate to the ICU.

Pulmonary artery thermodilution catheter

A pulmonary artery thermodilution catheter with a rapid-response thermistor is used to plot a thermodilution temperature curve simultaneously with an ECG trace and measured stroke volume. The plateau method originally used to calculate right ventricular ejection fraction from these traces was cumbersome and overestimated the value if it was less than 30 per cent. This method has been modified (Dhainaut *et al.* 1987). The exponential portion of the downslope of the temperature curve begins when temperature has risen by 20 per cent from its nadir, and a computer algorithm yields a single first-order exponential curve to fit this washout curve. Three temperatures (T_1, T_2, and T_3) are recorded at the first three R waves after this 20 per cent rise, and the mean of two residual fractions (T_2/T_3 and T_2/T_1) calculated. The right ventricular ejection fraction is then determined as 1 minus the mean residual fraction, and is little affected by mild tricuspid regurgitation. Commercial automated systems have been developed for this task. The right ventricular end-diastolic volume (RVEDV) and end-systolic volume (RVESV) can now be calculated since

$$\frac{\text{SV}}{\text{EF}} = \frac{\text{SV}}{\text{EDV}} = \frac{\text{SV} \times \text{EDV}}{\text{SV}} = \text{EDV}$$

$$\text{RVESV} = \text{RVEDV} - \text{SV} = \text{SV} \times \left(\frac{1}{\text{EF}} - 1\right)$$

where SV is stroke volume, EF is ejection fraction, EDV is end-diastolic volume, and EDD is end-diastolic diameter.

Echocardiography

On an end-diastolic four-chamber view, an axis from the center of the echocardiographic fan through the right ventricular apex to the junction of the tricuspid valve annulus and right ventricular free wall is defined. The systolic descent of the tricuspid valve ring along this axis (measured where the tricuspid valve plane intersects it at end-systole) is recorded as the tricuspid annular plane systolic excursion. Since both right ventricular walls (septum and free wall) contract longitudinally during right ventricular systole causing pronounced base-to-apex right ventricular shortening, the tricuspid annular plane systolic excursion correlates well with the right ventricular ejection fraction ($r > 0.9$). The systolic tricuspid descent is normally 2 ± 0.2 cm.

Changes in right ventricular longitudinal area are less accurate measures of right ventricular ejection fraction, and, since most right ventricular contraction is longitudinal, changes in cross-sectional dimension and area are poorer indices of right ventricular systolic volume change (Kaul *et al.* 1984).

End-systolic pressure–volume loops

The end-systolic pressure–volume relationship can be derived from right ventricular pressure–volume loops. This is a load-independent measure of function, particularly at higher stroke volumes. The relationship is described by the equation

$$P_{ES} = m(V_{ES} - V_0)$$

where P_{ES} is the end-systolic pressure, V_{ES} is the end-systolic volume, V_0 is the intercept of the x axis (i.e. the end-systolic volume at which the ventricular end-systolic pressure is zero), and m is the gradient of the line. The origin of the graph should not be used as a substitute for V_0. Simple positional changes (semi-upright to flat, or leg-raising), volume loading with colloid or crystalloid, abdominal caval compression, or challenge with pulmonary vasodilators such as inhaled nitric oxide alter loading conditions and allow two graph points to be plotted. Thus m is quantified (normally 0.46 ± 0.07). Estimates of end-systolic pressure–volume relationship might broadly categorize right ventricular function, although large standard errors may limit accurate assessment and monitoring of change. Determination of the right ventricular end-systolic pressure–volume relationship requires two measures: end-systolic volume and right ventricular end-systolic pressure.

End-systolic volume can be measured using any of the techniques described above.

Right ventricular end-systolic pressure is defined as follows:

(1) peak right ventricular pressure;

(2) the pressure in the right ventricle at minimum right ventricular volume;

(3) Pressure at maximal right ventricular elastance (i.e. where the pressure–volume relationship is steepest).

In the ICU setting, mean pulmonary artery dichrotic notch pressure is routinely available from a pulmonary artery catheter, and reflects end-ejection as a measure of end-systole. However, right ventricular ejection may outlast right ventricular mechanical systole. Altered pulmonary vascular resistance influences ejection duration, and hence valve closure and the dichrotic notch timing. Delay causes an underestimate of the right ventricular end-systolic pressure–volume relationship, and early closure causes an overestimate.

Therefore the confidence intervals in describing right ventricular end-systolic pressure–volume relationship curves are greater if defined by right ventricular peak pressure or dichrotic notch, and lower when maximal elastance defines end-systole. Dichrotic notch pressure is lower and peak right ventricular pressure higher than right ventricular end-systolic pressure defined by maximal elastance (Brown and Ditchey 1988).

Combined techniques

In theory, echocardiographic measures of right ventricular volume may be combined with measurement of right ventricular pressure (either direct or from Doppler-derived continuous right ventricular pressure recordings using tricuspid regurgitation jets and the Bernoulli equation) to construct right ventricular pressure–volume relationships.

Combined use of a right ventricular pressure transducer and conductance catheter to measure right ventricular end-systolic volume would allow continuous right ventricular end-systolic pressure–volume relationship monitoring.

Use of the pulmonary artery flotation catheter to calculate right ventricular end-systolic volume and measure dichrotic notch pressure offers simplicity, low cost, easily repeatable measurements, and the use of monitoring equipment already in use.

References

Brown, K.A. and Ditchey, R.V. (1988). Human right-ventricular end-systolic pressure–volume relation defined by maximal elastance. *Circulation*, **78**, 81–91.

Dickstein, M.L., Yano, O., Spotnitz, H.M., and Burkhoff, D. (1995). Assessment of right ventricular contractile state with the conductance catheter technique in the pig. *Cardiovascular Research*, **29**, 820–6.

Dhainaut, J.-F., *et al.* (1987). Bedside evaluation of right ventricular performance using a rapid computerized thermodilution method. *Critical Care Medicine*, **15**, 148–52.

Kaul, S., Tei, C., Hopkins, J.M., and Shah, P.M. (1984). Assessment of right ventricular function using two-dimensional echocardiography. *American Heart Journal*, **107**, 526–31.

Oldershaw, P. (1992). Assessment of right ventricular function and its role in clinical practice. *British Heart Journal*, **68**, 12–15.

16.1.10 Measurement of extravascular lung water

P. Frost and D. Bihari

Key messages

- The gravimetric measurement of extravascular lung water is considered to be the gold standard against which other methods are compared.
- Although chest radiography is the most common technique used in the further assessment of pulmonary edema, there are a variety of factors which reduce its accuracy.
- The indicator dilution technique relies upon the central injection of two indicators and their detection in the femoral artery. One (the non-diffusible indicator) remains confined to the intravascular space whilst the other (the diffusible indicator) distributes between the intravascular and extravascular spaces.
- Fluid management guided by measurement of extravascular lung water may improve outcome.

Introduction

Pulmonary edema is frequently encountered in critically ill patients. Whatever the cause, its presence is an ominous finding and is associated with an increased mortality. Furthermore, in such patients therapeutic intervention is strongly influenced by the desire not to exacerbate any further accumulation of fluid in the lung with subsequent deteriorating gas exchange. In this context it is hardly surprising that the measurement of extravascular lung water is of such great clinical and experimental interest. The ideal method for measuring lung water should be accurate, sensitive, reproducible, non-invasive, convenient, and inexpensive (Staub 1986). However, at present, despite considerable effort, no single method satisfies all these criteria.

Gravimetric measurement of lung water

Gravimetric measurement of extravascular lung water is considered to be the gold standard against which other methods are compared. However, since it is a postmortem technique, it has only experimental rather than clinical value. The technique requires the determination of the wet and dry weights of the lung sample. The normal range of lung water measured in this way is between 4 and 7 ml/kg.

Clinical examination and chest radiography

Bedside assessment of pulmonary edema may be difficult in the intensive care unit (ICU). Clinical examination can be misleading. Furthermore, although lung function tests such as pulmonary compliance and measures of oxygenation may be abnormal, these abnormalities are not specific for pulmonary edema. Although chest radiography is the most common technique used in the further assessment of pulmonary edema, there are a variety of factors which reduce its accuracy. Radiographic diagnostic accuracy depends upon technical consistency. This is often difficult to achieve in the ICU. Furthermore, it is important to appreciate that the characteristic radiographic appearance of pulmonary edema is dependent on the structural integrity of the lungs. In the intact lung, excess water will accumulate in a predictable fashion. If lung architecture is severely distorted by the presence of various pulmonary diseases, the recognition of excess water becomes more difficult. Also, gravity may contribute to an atypical distribution of edema in the supine patient. It has also been shown that, paradoxically, the chest radiograph may show no evidence of edema despite hemodynamic evidence of congestive heart failure. It appears that the radiographic appearance may occasionally lag behind the clinical state (Herman *et al.* 1990).

Despite these limitations chest radiography has several advantages. It is portable, inexpensive, and non-invasive. In most circumstances it is relatively sensitive if somewhat non-specific. This is particularly true of the detection of early pulmonary edema when other techniques such as dye dilution give values for lung water that are still within the normal range. In view of these advantages it is not surprising that considerable effort has been made to standardize this investigation. Numerous scoring systems have been devised to quantify pulmonary edema, and many studies have been carried out to assess their reliability. However, the correlation between radiographic quantification and other techniques such as CT and double-indicator dye dilution have been variable ($r = 0.62$–0.89). Chest radiography is still the most popular technique for the evaluation of extravascular lung water and this is likely to remain the case for the near future.

Indicator dilution measurement of lung water

This technique relies upon the central injection of two indicators and their detection in the femoral artery. One (the non-diffusible indicator) remains confined to the intravascular space, while the other (the diffusible indicator) distributes between the intravascular and extravascular space. The only significant extravascular space between the point of injection and the point of detection is within the lungs. Approximately 20 per cent of this space consists of non-aqueous material and the rest is water. Using appropriate detection equipment, dilution curves can be generated for each indicator within the femoral artery (Fig. 1(a)). The volume of distribution of each indicator can be calculated from these curves, and the extravascular lung volume and hence the lung water can be determined by simple subtraction.

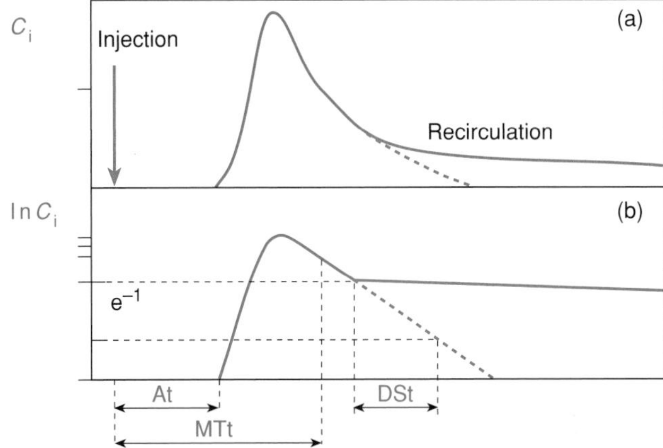

Fig. 1 Schematic representation of the dilution curves and their analysis for transit times; At, appearance time; MTt, mean transit time; DSt, exponential downslope time.

Lewis *et al.* (1979) used negative heat (a cold crystalloid bolus (2–10 °C)) as the diffusible indicator and indocyanine green dye as the non-diffusible indicator. Indocyanine green is suitable as the non-diffusible indicator since it is bound to plasma proteins and therefore remains within the intravascular space. Although indocyanine green is widely accepted as a suitable intravascular marker, the choice of the optimal diffusible indicator remains controversial.

The volume of distribution of an indicator is calculated from the product of the flow and the mean circulation or transit time of that indicator (MTt). Hence the volume of distribution of the thermal indicator (V_{DT}) is given by

$$V_{DT} = CO \times MTt_T$$

where CO is the cardiac output. Similarly, the volume of distribution of the dye indicator (V_{DD}) is given by

$$V_{DD} = CO \times MTt_D.$$

CO or flow can be derived from the familiar Stewart–Hamilton equation utilizing the area under the thermal dilution curve:

$$\mathrm{CO} = \frac{V(T_\mathrm{p} - T_\mathrm{i})k}{\int_{t=0}^{\infty} (T_\mathrm{p} - T_\mathrm{b})\mathrm{d}t}$$

where T_p is the patient's temperature, V is the volume of injectate, T_i is the temperature of the injectate, t is time, and k is a constant.

The mean transit time is actually the sum of the appearance time (At) and the mean time (t) of the indicator dilution curve (Fig. 1(b)). The mean time is defined as follows:

$$t = \int_0^{\infty} C_\mathrm{i} t \mathrm{d}t \Big/ \int_0^{\infty} C_\mathrm{i} \mathrm{d}t$$

where C_i is the concentration of indicator. Hence the extravascular thermal volume (which is virtually equivalent to the extravascular lung water volume) is given by

$$\mathrm{EVTV} = \mathrm{CO}(\mathrm{MTt_T} - \mathrm{MTt_D}).$$

Extravascular lung water can also be calculated in a different way from the exponential decay times, or downslope times (DSt), of semi-logarithmic plots of the two dilution curves using the theory of Newman *et al.* (1951) (Fig. 1(b)). Newman and his colleagues modeled the central circulation of right heart, lungs, and left heart by linking three chambers in series through which fluid was pumped at a constant rate. They noted that when an indicator flowed through these chambers the downslope of the resulting dilution curve was determined by the flow and the volume of the largest chamber in the series only. Hence the volume of the largest mixing chamber between the point of injection and the point of detection is the product of the cardiac output and the downslope time of the indicator.

When this is applied to the central circulation, the largest mixing chamber for the diffusible indicator is the intra- and extravascular space in the lung (the pulmonary thermal volume (PTV)) and that for the non-diffusible indicator is the pulmonary intravascular space (pulmonary blood volume (PBV)):

$$\mathrm{PTV} = \mathrm{CO} \times \mathrm{DSt_T}$$

$$\mathrm{PBV} = \mathrm{CO} \times \mathrm{DSt_D}$$

where subscripts T and D denote thermal and dye respectively. Therefore the extravascular thermal volume will also be given by

$$\mathrm{EVTV} = \mathrm{CO} \times (\mathrm{DSt_T} - \mathrm{DSt_D}).$$

The determination of extravascular thermal volume from the mean circulation time and downslope time is illustrated in Fig. 2.

The Edwards lung water computer was the first system that allowed the bedside computation of extravascular lung water. Unfortunately, the measurements were not reliable and showed reciprocal dependence on cardiac output. More recently, the COLD computer (circulation, oxygenation, lung water, and liver function diagnosis) (Pulsion Medical Systems, Munich) has been developed. This system utilizes a thermistor-tipped fiber-optic catheter to detect the thermal and optical signals in the femoral artery. In addition to the calculation of extravascular lung water, this system can be used for the accurate determination of a number of blood volumes (total blood volume, intrathoracic blood volume, and global heart volume).

EVLW=CO ($\mathrm{MTt_T}$–$\mathrm{MTt_D}$)

Intrathoracic thermal volume

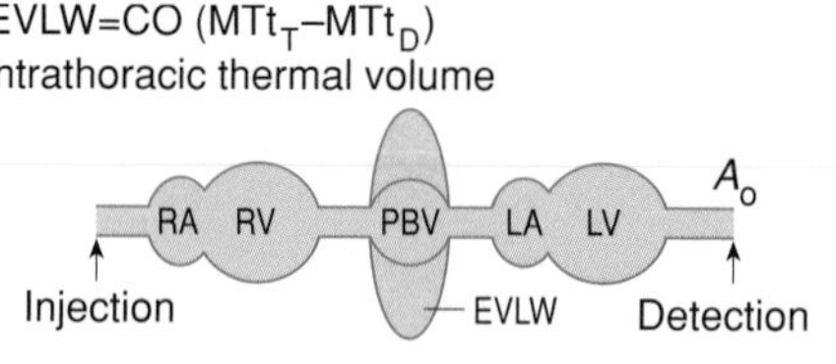

V_DT=CO×$\mathrm{MTt_T}$
The volume of distribution of the thermal indicator between the points of injection and detection

Intrathoracic blood volume

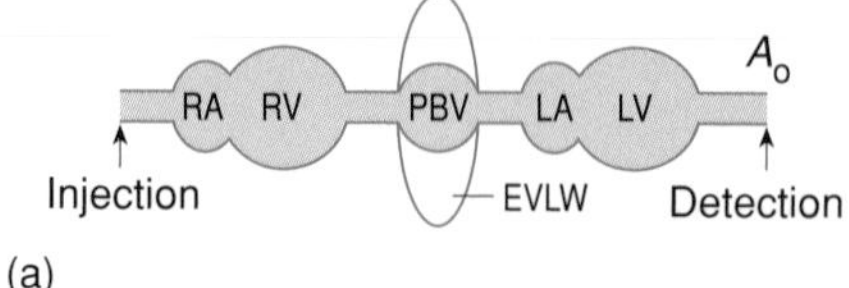

V_DD=CO×$\mathrm{MTt_D}$
The volume of distribution of the dye between the points of injection and detection

(a)

EVLW=CO ($\mathrm{DSt_T}$–$\mathrm{DSt_D}$)

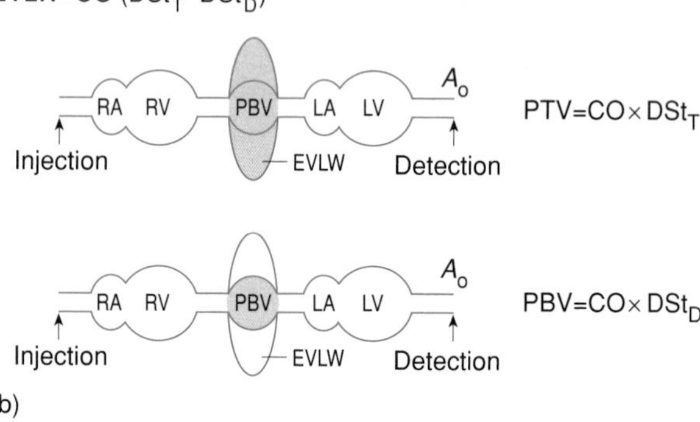

(b)

Fig. 2 (a) Calculation of extravascular lung water (extravascular thermal volume) from mean transit times; (b) calculation of extravascular lung water from downslope times. (Reproduced with permission from McLuckie (1996).)

A number of potential flaws in the use of a thermal indicator have been identified. Thermal volumes will overestimate lung water volume owing to the loss of the heat indicator into non-aqueous structures such as the heart and blood vessel walls as well as the dry tissue mass of the lung. Lewis *et al.* (1982) estimate that this loss is about 9 per cent of the injected indicator. However, given that the dry tissue mass usually remains unchanged, thermal indicator loss will remain constant and can be allowed for in the calculation of lung water.

The establishment of thermal equilibrium between blood and thermistor may artificially prolong the thermal mean transit time. The response time of the fiber-optic system to changing dye concentration is much faster. This inequality in response time may lead to an overestimation of the extravascular thermal volume. This potential source of error could be avoided by using deuterium oxide rather than heat as the diffusible indicator; the advantage is that deuterium oxide can also be detected by a fiber-optic system.

Whatever the diffusible indicator used, accurate lung water measurement must depend logically on diffusion of the indicator throughout the whole extravascular lung space, which in turn depends upon pulmonary perfusion. Interestingly, a study that investigated the effects of pulmonary emboli on the double-indicator dilution measurement of extravascular lung water showed that both over- and underestimation of lung water could occur, depending on the size of the emboli. Despite these theoretical limitations there is excellent correlation between the thermal dye dilution measurement of lung water and gravimetric determination.

Other methods of measurement

Lung water can be measured by a variety of densitometric techniques including X-ray and gamma-ray densitometry, CT, and positron tomography. Nuclear magnetic resonance has also been utilized. In general, the application of these techniques in intensive care has been restricted by a lack of mobility or the need for exposure to ionizing radiation.

Clinical studies of lung water in the critically ill

Although there are a variety of techniques by which lung water can be measured in the critically ill, the important question remains as to whether such measurements lead to improvements in treatment and survival of critically ill patients. Several investigators have tried to establish the role of extravascular lung water measurements in clinical practice. Clinical management based on protocols incorporating extravascular lung water measurement has been associated with reduced mortality in acute respiratory distress syndrome and sepsis patients. Similarly, if patients were managed according to a protocol that emphasized fluid restriction when extravascular lung water was greater than 7 ml/kg, time in the ICU and ventilator days were significantly shorter than for a comparable group of patients whose fluid management was guided by wedge pressure measurements.

More clinical trials are needed if the value of extravascular lung water measurement is to be fully exploited.

References

Herman, P.G., *et al.* (1990). Limited correlation of left ventricular end diastolic pressure with radiographic assessment of pulmonary hemodynamics. *Radiology*, **174**, 721–4.

Lewis, F.R., Elings, V.B., and Storm, J.A. (1979). Bedside measurement of lung water. *Journal of Surgical Research*, **27**, 250–61.

Lewis, F.R., Elings, V.B., Hill, S.L., and Christensen, J.M. (1982). The measurement of extravascular lung water by thermal-green dye indicator dilution. *Annals of the New York Academy of Sciences*, **384**, 394–410.

McLuckie, A. (1996). The COLD system of haemodynamic monitoring. *Intensive Care World*, **13**, 24–48.

Newman, E.V., Merrell, M., Genecin, A., Monge, C., Milnor, W.R., and McKeever, W.P. (1951). The dye dilution method for describing the central circulation. *Circulation*, **4**, 735–45.

Staub, N.C. (1986). Clinical use of lung water measurements: report of a workshop. *Chest*, **90**, 588–94.

16.1.11 Gastrointestinal tonometry

G. Gutierrez

Key messages

- Monitoring increases in tissue CO_2 concentration is useful in establishing the adequacy of aerobic metabolism in the clinical setting.
- Gastrointestinal tonometry is a useful indicator of morbidity and mortality after major surgery and during critical illness.
- Resuscitation guided by pHi may improve outcome in patients admitted to the intensive care unit with normal pHi.
- Air tonometry in conjunction with infrared capnometry obviates the use of blood gas analyzers and markedly decreases the opportunities to introduce errors in the measurement.

Introduction

Tonometry is a relatively non-invasive technique that measures gut mucosal $P\text{CO}_2$ by allowing the equilibration of CO_2 partial pressures between a fluid-filled balloon and the gastrointestinal mucosa (Fig. 1). The physiological basis of gastrointestinal tonometry as a monitor of hypoxia is the increase in tissue CO_2 production that accompanies anaerobic metabolism (Gutierrez and Brown 1995).

Another important aspect governing the response of the gastrointestinal tract to hypoxia is the mucosal microvasculature pattern, which consists of arterioles running in parallel to venules. This microvascular arrangement results in a countercurrent circulation of blood, characterized by the diffusion of oxygen from the arterioles to the adjacent venules. An opposite phenomenon occurs with CO_2, as this gas diffuses from the venules to the arterioles. Therefore tissue concentrations of O_2 are lowest, and those of CO_2 are highest, at the tip of the countercurrent system. Countercurrent vascular systems

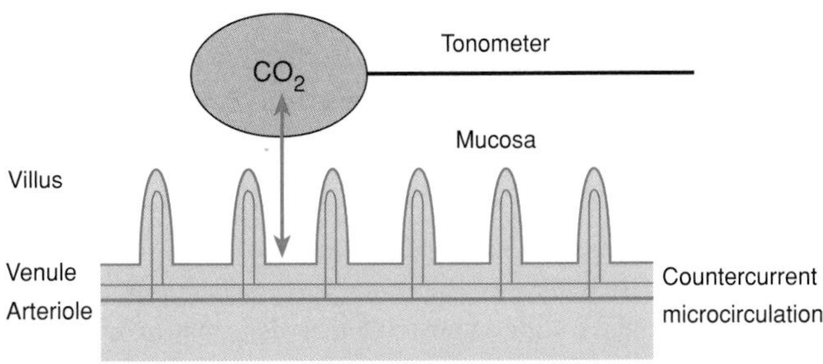

Fig. 1 Schematic representation of a balloon tonometer next to the mucosal surface. After some time, the partial pressures of CO_2 in the mucosa and in the fluid contained in the balloon come into equilibrium. The vulnerability of the mucosa to hypoxic insults is related to the countercurrent microcirculation of the villus.

are characteristic of the renal and splanchnic microcirculation and fare poorly during conditions of hypoperfusion or hypoxia. Thus they manifest the harmful effects of hypoxia or hypoperfusion earlier than other tissues.

The physiological basis of gastrointestinal tonometry

During aerobic metabolism, cellular CO_2 production is proportional to the rate of cellular O_2 consumption. CO_2 production ceases to be proportional to O_2 consumption during hypoxia as the mitochondria fail to dispose of the hydrogen ions (H^+) derived from the hydrolysis of ATP. Excess tissue CO_2 results from the buffering action of bicarbonate on H^+:

$$H^+ + HCO_3^- \rightarrow H_2O + CO_2.$$

Monitoring increases in tissue CO_2 concentration is useful in establishing the adequacy of aerobic metabolism in the clinical setting.

Another possible mechanism to account for increases in tissue CO_2, in addition to cellular hypoxia, is a decrease in local blood flow, or 'flow stagnation'. However, studies in dogs subjected to progressive cardiac tamponade support the notion that CO_2 production in the gut mucosa is the result of anaerobic metabolism (Schlichtig and Bowles 1994).

Technical development of gastrointestinal tonometry

Gastric tonometry was developed in the late 1950s as a method of measure arterial $P\text{CO}_2$ in mechanically ventilated children with poliomyelitis. Clinical studies showed that arterial $P\text{CO}_2$ could be estimated with reasonable accuracy, but that gastric $P\text{CO}_2$ surpassed arterial $P\text{CO}_2$ during severe shock. Although not recognized at the time, it is precisely this property that makes tonometry a useful monitor of regional anaerobic metabolism. For all practical purposes, the concept of tonometry was forgotten until Fiddian-Green *et al.* (1982) advanced the notion that the pH of the gut mucosa could be calculated using this method. This hypothesis was based on the following assumptions: free CO_2 diffusion in tissue; luminal fluid $P\text{CO}_2$ is in equilibrium with mucosal $P\text{CO}_2$; intestinal mucosal bicarbonate concentration $[HCO_3^-]$ is similar to arterial bicarbonate concentration. Making these assumptions, it is possible to estimate the intramucosal pH (pHi) from the Henderson–Hasselbalch equation:

$$\text{pHi} = 6.1 + \log\left(\frac{[\text{HCO}_3^-]}{\alpha \times \text{tonometer } P\text{CO}_2}\right)$$

where α is the solubility of CO_2 in plasma ($\alpha = 0.03$).

The assumptions that underlie the calculation of pHi are applicable to many clinical conditions. Total intestinal ischemia is a notable exception. In this condition mucosal bicarbonate may be lower than arterial bicarbonate. The presence of systemic acidosis also complicates the interpretation of pHi. A low mucosal pH in concert with a low arterial pH does not imply local tissue hypoxia, but most likely reflects generalized tissue acidosis. Given these potential difficulties in the interpretation of pHi, the use of the difference between tonometric and arterial $P\text{CO}_2$ (the $P\text{CO}_2$ gap) is recommended.

Potential sources of error in measuring mucosal $P\text{CO}_2$

Saline tonometry is a labor-intensive technique (Table 1) and its successful use requires the close co-operation of physicians, nurses, and laboratory personnel.

There are ample opportunities to introduce errors at any point during the measurement process, including the following:

(1) inaccurate timing of partially equilibrated samples;

(2) the introduction of air in the saline sample;

(3) erroneous saline $P\text{CO}_2$ measurements by the blood gas analyzer;

(4) high gastric luminal acidity, whose buffering by exogenous or endogenous bicarbonate may elevate tonometer $P\text{CO}_2$.

The last is a common source of error and many investigators recommend the use of H_2-receptor antagonists in patients in whom tonometry is used. The idea is to maintain a relatively alkalotic gastric luminal pH and prevent the spurious generation of CO_2 in the stomach by exogenous or endogenous bicarbonate. The use of sucralfate has been suggested as an alternative to H_2-receptor blockers. Improvements in tonometer technology now permit the measurement of mucosal $P\text{CO}_2$ using air instead of saline.

Applications of gastrointestinal tonometry

Gastrointestinal pHi as a predictor of morbidity

The clinical utility of pHi during the repair of abdominal aneurysm has been established in several studies. There is a significant correlation between decreases in pHi and the development of ischemic colitis in patients undergoing aortic reconstruction, in particular those with pHi values lower than 6.90. Decreases in pHi have also been associated with chronic gastric and mesenteric ischemia. Gastric pHi values less than 7.20 predict major complications after abdominal aortic surgery. Decreases in pHi have also been associated

Table 1 Steps required in the measurement of gastric pHi with the saline balloon system

1. Introduce saline into the tonometer balloon (2.5 ml)
2. Wait for an equilibration period of 30–90 min
3. Draw and discard the saline occupying the tonometer's dead-space (1.0 ml)
4. Sample the balloon's saline using a strict anaerobic technique
5. Obtain a sample of arterial blood and measure blood bicarbonate
6. Measure the $P\text{CO}_2$ of the saline in a blood gas analyzer
7. Compute the steady state $P\text{CO}_2$ using a special-purpose nomogram
8. Compute pHi using the Henderson–Hasselbalch equation

with bleeding from stress ulcers in intensive care unit (ICU) patients and in patients with chronic renal failure undergoing dialysis.

Measures of gastric pHi are predictive of weaning outcome from mechanical ventilation. Patients demonstrating decreases in pHi during the weaning process are likely to require mechanical ventilation. Also, there is an inverse association between changes in gastric pHi and the age of the transfused blood. Patients receiving blood stored for more than 15 days experience decreases in gastric mucosal pHi on a consistent basis, implying that aged red blood cells cause microcirculatory occlusion in some organs. Gastric tonometry also appears to have a useful role in assessing the viability of orthotopic liver transplants, with a greater incidence of major morbidity in patients with pHi < 7.32 at some point during the transplantation procedure.

Mucosal hypoperfusion, as defined by pHi < 7.32, has been associated with longer length of stay in the ICU, greater incidence of major complications, and greater mortality in patients undergoing elective major surgery (Mythen and Webb 1994).

Gastrointestinal pHi as a predictor of mortality

Admission pHi values below 7.35 to 7.32 have been associated with greater ICU mortality. Further, patients with persistently low pHi 12 h after ICU admission appear to have a high mortality compared with those who maintain normal pHi in the ICU.

Gastrointestinal pHi as a therapeutic index in critically ill patients

In a prospective multicenter study Gutierrez *et al.* (1992) tested the efficacy of pHi-guided therapy. A control group of ICU patients were treated according to standard ICU practices, whereas the protocol group received, in addition, treatment to increase systemic O_2 transport whenever pHi fell below 7.35, or by more than 0.10 units from the previous measurements. The results of the study showed that survival for patients admitted with pHi < 7.35 was similar in the protocol and control groups (37 per cent versus 36 per cent), whereas survival for those admitted with normal pHi was significantly greater in the protocol group (58 per cent versus 42 per cent). These findings suggest that pHi-guided resuscitation may improve outcome in patients admitted to the ICU with normal pHi, perhaps by preventing splanchnic ischemia and the development of a systemic oxygen deficit.

Automated gastrointestinal tonometry systems

Gastrointestinal tonometry is perhaps the most important advance in monitoring to appear over the past two decades. Increased gut mucosal $P\text{CO}_2$ is an excellent physiological signal of tissue hypoperfusion or hypoxia. Newer methods are being developed to ease the process of monitoring mucosal $P\text{CO}_2$. One such method uses air in conjunction with an infrared capnometer. Air is pumped in and out of the balloon as frequently as every 10 min. This system obviates the use of blood gas analyzers and markedly decreases the opportunities to introduce errors into the measurement. An example of the use of intermittent gas tonometry is shown in Fig. 2. This anecdotal example serves to illustrate the potential of this technique as a monitor of tissue oxygenation. Prospective trials of air tonometry are required to determine its place among the monitoring techniques in use in the ICU.

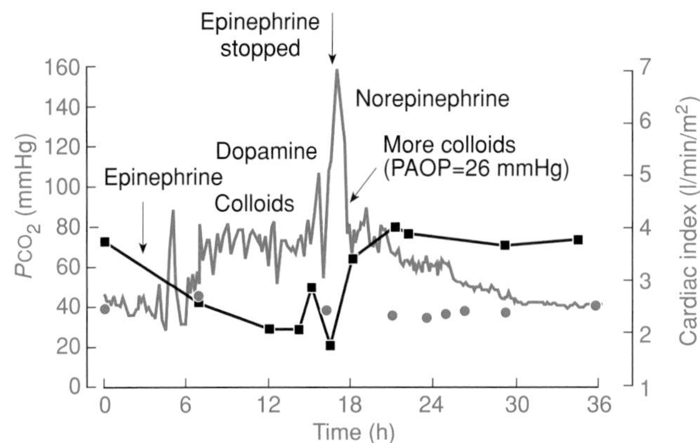

Fig. 2 Changes in mucosal $P\text{CO}_2$ in a patient with sepsis, measured with a closed-loop air tonometer system: ● arterial $P\text{CO}_2$; ■ cardiac index. Gastric mucosal $P\text{CO}_2$ increased following the institution of intravenous epinephrine (adrenaline) to treat hypotension. This was accompanied by a decrease in cardiac index. Mucosal $P\text{CO}_2$ increased to approximately 150 mmHg, a value compatible with the onset of bowel ischemia, when intravenous epinephrine was stopped and a norepinephrine (noradrenaline) infusion started. Tonometer $P\text{CO}_2$ decreased rapidly as the cardiac index increased to baseline values. As regional CO_2 increased, a $P\text{CO}_2$ gap developed that was greater than 20 mmHg and at one point reached 100 mmHg. Over the next hour, regional $P\text{CO}_2$ returned to baseline values and the $P\text{CO}_2$ gap disappeared as the patient improved.

Conclusions

There is sufficient information in the clinical literature to support the prognostic utility of gastric tonometry. Gastric tonometry measures increases in mucosal $P\text{CO}_2$, a very useful physiological signal. This information should help clinicians judge the state of energy balance in a tissue that is exquisitely sensitive to alterations in perfusion and oxygenation.

References

Fiddian-Green, R.G., Pittenger, G., and Whitehouse, W.M. (1982). Back-diffusion of CO_2 and its influence on the intramural pH in gastric mucosa. *Journal of Surgical Research*, **33**, 39–48.

Gutierrez, G. and Brown, S.D. (1995). Gastric tonometry: a new monitoring modality in the intensive care unit. *Journal of Intensive Care Medicine*, **10**, 34–44.

Gutierrez, G., *et al.* (1992). Gastric intramucosal pH as a therapeutic index of tissue oxygenation in critically ill patients. *Lancet*, **339**, 195–9.

Mythen, M.G. and Webb, A.R. (1994). Intra-operative gut mucosal hypoperfusion is associated with increased post-operative complications and cost. *Intensive Care Medicine*, **20**, 99–104.

Schlichtig, R. and Bowles, S.A. (1994). Distinguishing between aerobic and anaerobic appearance of dissolved CO_2 in intestine during low flow. *Journal of Applied Physiology*, **76**, 2443–51.

16.2 Blood gases

16.2.1 Blood gas analyzers

B. A. Shapiro

Key messages

- Blood gas analyzers utilize electrochemical techniques for the measurement of pH, $P\text{CO}_2$, and $P\text{O}_2$.
- Measurements of pH, $P\text{CO}_2$, and $P\text{O}_2$ are traditionally accomplished by blood gas analyzers composed of specially designed electrodes encased in a thermostatically controlled chamber.
- Calibration establishes a relationship between electrode output and the units in which the measured variables are expressed.
- The co-oximeter is the most accurate method available for measuring the four clinically relevant hemoglobin moieties.

Introduction

Techniques for blood analysis of oxygen, carbon dioxide, and hydrogen ion concentration were developed more than a century ago. These techniques were confined to the research laboratory because several hours were required to obtain a single set of measurements (Severinghaus and Astrup 1987). In the 1950s, electrochemical techniques were introduced which took only a few minutes to measure arterial pH, $P\text{CO}_2$, and $P\text{O}_2$. However, the original electrodes required specific laboratory conditions and highly trained technicians. Nearly two decades of technical development were required to evolve blood gas analyzers that could reliably function in hospital clinical laboratories. The last 10 years has provided development of analytical technology that promises to function outside the laboratory environment.

Principles of electrode technology

To evaluate modern blood gas analysis devices, it is essential to have a fundamental understanding of the chemistry and physics associated with electrochemical analysis (Shapiro *et al.* 1994).

Traditional pH electrode

As illustrated in Fig. 1, when two solutions with different pH values are separated by a pH-sensitive glass membrane, a potential difference develops that can be measured by chemical half-cells as a voltage change across the membrane. The reference half-cell, usually composed of mercury/mercurous chloride (calomel) housed in a compartment filled with potassium chloride solution, supplies a constant reference voltage. The measuring half-cell is a Sanz electrode, usually composed of a silver/silver chloride substance embedded within a pH 6.840 buffer chamber. A contact bridge consisting of a potassium chloride solution completes the electronic circuit.

The relationship between the potential difference $E_u - E_k$ and the pH values of the known (pH_k) and unknown (pH_u) solutions is defined by the Nernst equation:

(a)

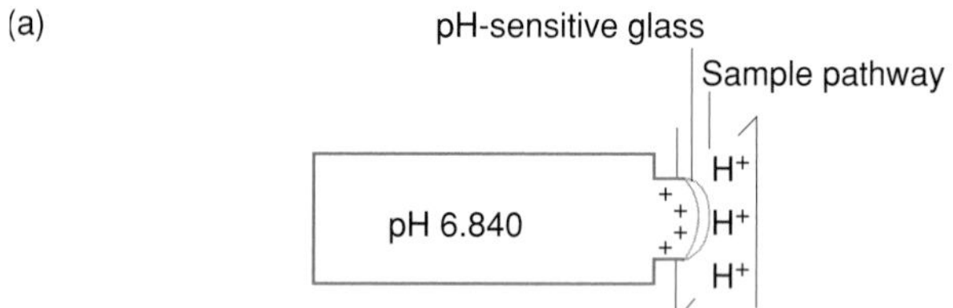

(b)

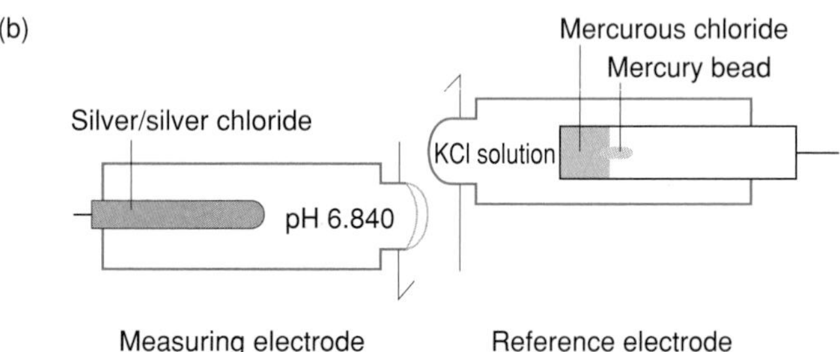

(c)

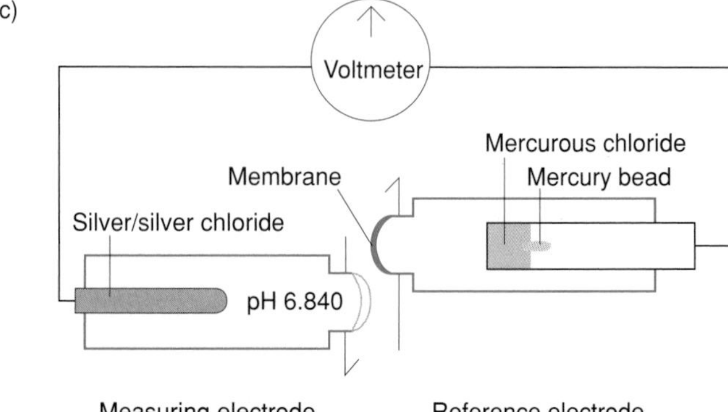

Fig. 1 Basic principles of the pH electrode: (a) a voltage develops across pH-sensitive glass when pH is unequal in the two solutions; (b) chemical half-cells used as measuring and reference electrodes; (c) the pH electrode system completed by a device measuring voltage changes. (Reproduced with permission from Shapiro *et al.* (1994).)

$$pH_u = pH_k + \frac{E_u - E_k}{2.3026T}\frac{F}{R}$$

where R is the molar gas constant, F is the Faraday constant, and T is the absolute temperature. A potential of 61.5 mV will be developed for every pH unit difference between the known solution within the measuring electrode (pH 6.840 buffer) and the sample solution at 37 °C. A potential difference of zero indicates that the pH values of the unknown and known solutions are equal. A potential difference of 33.5 mV will add 0.544 pH units to the known value, resulting in a pH of 7.384 for the sample solution.

Traditional $P\text{CO}_2$ *electrode*

The Severinghaus electrode directly measures CO_2 tensions by allowing the CO_2 gas to undergo a chemical reaction that produces hydrogen ions. Since the amount of CO_2 that diffuses across the permeable membrane is directly proportional to the pressure gradient, the hydrogen ion production will create a proportional difference in potential that is measured by half-cells similar to those of the pH electrode. Both the reference and measuring half-cells of the $P\text{CO}_2$ electrode are silver/silver chloride. The pH-sensitive glass is separated from the blood sample by a silicon elastic membrane and by a nylon spacer saturated with a bicarbonate solution which constantly replenishes the solution at the electrode tip and spacer.

Traditional $P\text{O}_2$ *electrode*

The Clark ($P\text{O}_2$) electrode is a polarographic device that measures oxygen tensions by oxidation–reduction reactions, a chemical process that generates measurable electrical currents. The polarographic electrode consists of a platinum cathode and a silver anode immersed in an electrolyte solution. The Clark electrode is bathed in an electrolyte solution with an oxygen-permeable polypropylene membrane covering the cathode tip. A silver (Ag) anode immersed in an electrolyte solution oxidizes silver producing a constant flow of electrons (current):

$$4Ag \rightarrow 4Ag^+ + 4e^-.$$

An adjacent platinum cathode reacts chemically with oxygen to form hydroxyl ions (OH^-), a reduction reaction that uses electrons:

$$O_2 + 2H_2O + 4e^- \rightarrow 4OH^-.$$

In general, as electrons are consumed at the cathode, the anode reaction is accelerated. The volume of oxygen reduced will be directly proportional to the number of electrons used in the cathode reaction. Thus, by measuring the change in current, the amount of oxygen that diffused across the membrane into the electrolyte solution can be determined.

Modern electrode-based blood gas analyzers

The pH, $P\text{CO}_2$, and $P\text{O}_2$ measurements are traditionally accomplished by blood gas analyzers composed of specially designed electrodes encased in a thermostatically controlled chamber. Although specific features vary among manufacturers, the principles of their respective electrodes and analytical procedures are extremely similar. Modern blood gas analyzers have evolved into automated self-diagnostic instruments, requiring minimal maintenance, that not only calculate bicarbonate, base excess/deficit, and temperature corrections, but also interface with computer systems to allow data storage, trending analysis, blood gas interpretation, and other algorithms. Stability, consistency, and speed have been maximized by the electrode and chamber designs, and by microprocessor control of analysis and calibration processes. Several models have expanded to include ion-selective electrodes for measurements of Na^+, K^+, Cl^-, and Ca^{2+}, as well as electrical conductivity measurements of hematocrit on a single sample whose volume is measured in microliters. Others have combined a blood gas analyzer and co-oximeter into a single unit. Still others have miniaturized and re-formed the traditional electrodes into electrochemical sensors that are imprinted on a computer card and can function reliably outside the traditional laboratory setting, even at the bedside; thus the delay between obtaining the sample and receiving results is significantly reduced. These devices are generally referred to as point-of-care analyzers (Templin-Kozlowski 1995).

Calibration of electrode-based blood gas analyzers

Calibration establishes a relationship between electrode output (in millivolts or nanoamperes) and the units in which the measured variables are expressed (pH units and kilopascals or millimeters of mercury). The calibration process also ensures that the electrode outputs are products of the pH, $P\text{CO}_2$, and $P\text{O}_2$ of the analyzed samples. It is accomplished by introducing calibration standards of known pH, $P\text{CO}_2$, and $P\text{O}_2$ to the respective electrodes and adjusting the analyzer readings to the known values. Calibration standards for most analyzers are buffers for the pH electrode and known gases, corrected for barometric pressure, for the $P\text{CO}_2$ and $P\text{O}_2$ electrodes.

A one-point calibration is an adjustment of the electronic response to a single standard. This should be performed before each sample analysis or every 30 min. A two-point calibration adjusts the electronic responses to two standards. This should be performed every 2 to 8 h and after electrode maintenance and repair measures.

Electrode precision

Modern blood gas analyzers have manufacturer's specifications that state electrode reliability provided the manufacturer's instructions for calibration, maintenance, and sample introduction are followed. Theoretically, proper maintenance and calibration should ensure reliability; however, this is not always the case.

The co-oximeter

The co-oximeter is the most accurate method available for measuring the four clinically relevant hemoglobin moieties and is considered the standard against which other methods must be compared.

Oximetry

Oximetry is a general term pertaining to the various technologies capable of measuring oxyhemoglobin saturation.

All oximetry is based upon spectrophotometric principles that measure the amount of light transmitted and/or absorbed by the hemoglobin moiety. Photoelectric principles allow light intensity to

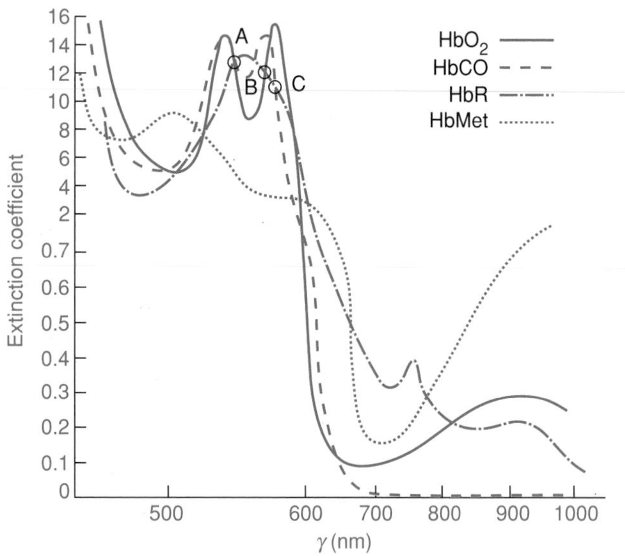

Fig. 2 Spectral analysis of hemoglobin. Points A, B, and C identify three wavelengths where different Hb moieties have the same extinction coefficients (isosbestic points): A, triple isosbestic point at 548 nm (HbO_2, HbCO, and HbR); B, double isosbestic point at 568 nm (HbO_2, HbR); C, double isosbestic point at 578 nm (HBCO, HbR). (Reproduced with permission from Shapiro *et al.* (1994).)

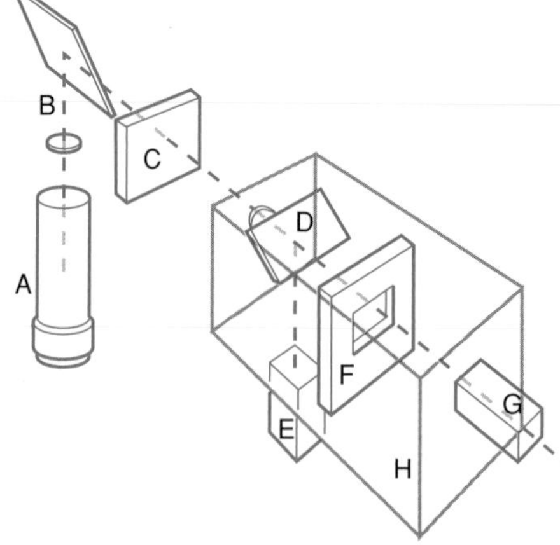

Fig. 3 Basic components of the co-oximeter. A, light source; B, lens and mirror; C, monochromator; D, beam splitter; E, reference detector; F, cuvette; G, sample detector; H, temperature-regulated block. (Reproduced with permission from Shapiro *et al.* (1994).)

be translated to electrical current, which is the basis for modern spectrophotometers. For example, light of a given intensity is passed through a specific substance and some fraction is transmitted to an oxide-coated metallic surface. The resultant current is directly proportional to the transmitted light intensity. These physical principles combined with solid state technology developed in the 1960s made present spectrophotometers possible.

Spectral analysis

The hemoglobin molecule exists in various forms: oxyhemoglobin (HbO_2), reduced hemoglobin (HbR), carboxyhemoglobin (HbCO), sulfhemoglobin (HbS), methemoglobin (HbMet), and fetal hemoglobin (HbF). Each of these forms has its own light spectrum (Fig. 2). The specific wavelengths at which measurements are made are generally chosen so that one wavelength is the maximum difference between the extinction coefficients of the moieties of interest and the other is an isosbestic point, i.e. the extinction coefficients are the same.

Modern co-oximeters are spectrophotometers that simultaneously analyze four hemoglobin moieties: reduced hemoglobin, oxyhemoglobin, carboxyhemoglobin, and methemoglobin. As illustrated in Fig. 3, the basic components of a co-oximeter are a light source, a series of lenses, filters, mirrors that focus the light beam, a cuvette or sample chamber, a monochromator that isolates the wavelengths of interest using filters or gratings, and photodiode detectors that emit electrons in proportion to the amount of light striking their surfaces. The emitted electrons (current) are fed to a circuit whose output is proportional to the relative absorbance of the sample.

Limitations of blood gas analyzers

An analyzer is a device that performs measurements on fluids, excrement, or tissue permanently removed from the body. Blood gas analyzers utilize electrochemical techniques for measurement of pH, $P\text{CO}_2$, and $P\text{O}_2$. Despite impressive technological advances, blood gas analyzers continue to have three major disadvantages.

1. There is a limited frequency of measurements because blood must be permanently removed from the patient.
2. There is delay in receiving the results, which makes the values an isolated reflection of a previous time in a dynamic physiological continuum.
3. The blood sample is subject to preanalytic error.

Despite these impediments associated with blood gas analyzers, arterial pH, $P\text{CO}_2$, and $P\text{O}_2$ measurements are the most frequently ordered laboratory examination in the ICU and the operating room.

References

Severinghaus, J.W. and Astrup, P.B. (1987). History of blood gas analysis. *International Anesthesiology Clinics*, **25**, 1–214.

Shapiro, B.A., Peruzzi, W.T., and Templin-Kozlowski, R. (1994). *Clinical application of blood gases* (5th edn). Mosby–Year Book, Chicago, IL.

Templin-Kozlowski, R. (1995). Blood gas analyzers. *Respiratory Care Clinics of North America*, **1**, 35–46.

16.2.2 Blood gas analysis

B. A. Shapiro

Key messages

- $Pa\mathrm{CO}_2$ falls into one of three categories: respiratory alkalosis (alveolar hyperventilation), normal alveolar ventilation, or respiratory acidosis (ventilatory failure).
- Classification of the arterial pH in relation to $Pa\mathrm{CO}_2$ allows one to determine whether the primary abnormality is ventilatory (respiratory acid–base imbalance) or metabolic acid–base imbalance.
- Complete assessment of the arterial oxygenation status requires measurement of $Pa\mathrm{O}_2$, oxyhemoglobin saturation, and hemoglobin content.
- There is no scientific justification for the routine practice of temperature correcting blood gas measurements when alterations in body temperature are present.

Introduction

Blood gas values require interpretation to be appropriately utilized in clinical care. This makes the clinical application of blood gas data comparable with physiological studies such as electrocardiography and pulmonary function in that they are intended to document a presumptive diagnosis, to specify the insult to allow for prognostic and/or therapeutic assessment, and to quantify the insult so that serial measurements can be used to follow the course of the pathology and/or therapy. With appropriate interpretation, the blood gas measurements permit oxygenation deficiencies to be differentiated from primary ventilatory deficiencies and primary metabolic acid–base abnormalities.

pH and $Pa\mathrm{CO}_2$

Universally accepted normal ranges for pH and $Pa\mathrm{CO}_2$ for adults are listed in Table 1. Since clinical judgments are rarely influenced by minor variations from the normal ranges of arterial pH and $Pa\mathrm{CO}_2$, 'acceptable ranges' are applied to critically ill patients.

$Pa\mathrm{CO}_2$–pH relationship

A predictable and reasonably linear relationship exists between pH and plasma carbonic acid when $Pa\mathrm{CO}_2$ values change acutely (Table 2). These pH changes are due to variations in the concentration of volatile acid (H_2CO_3) and are referred to as respiratory acid–base changes.

$Pa\mathrm{CO}_2$–plasma bicarbonate relationship

The predicted plasma bicarbonate is also calculated from the relationship depicted in Table 2. The difference between the respiratory plasma bicarbonate and the actual plasma bicarbonate provides a rapid and easy assessment of the metabolic component.

Relationship between minute ventilation and $Pa\mathrm{CO}_2$

In normal exercising humans, the minute ventilation increases in proportion to both the metabolic rate and cardiac output. The alveolar ventilation increases because lung perfusion increases in proportion to the ventilation while the anatomical dead-space either decreases (due to larger tidal volumes) or remains unchanged. Thus both CO_2 production and alveolar ventilation increase during exercise, so that $Pa\mathrm{CO}_2$ either remains the same or decreases slightly. In contrast, a normal subject on a positive-pressure ventilator requires an above-normal minute volume to maintain a normal $Pa\mathrm{CO}_2$; this is generally attributed to an increase in dead-space ventilation.

When a normal non-exercising human increases minute ventilation spontaneously, the cardiac output and metabolic rate do not increase in proportion. The predicted relationships between minute volume and $Pa\mathrm{CO}_2$ in non-exercising humans are listed in Table 3. Spontaneously breathing patients with an increased minute ventilation and $Pa\mathrm{CO}_2$ significantly higher than predicted in Table 3 have either significantly increased dead-space ventilation or increased CO_2 production. The existence of a significant disparity between minute volume and $Pa\mathrm{CO}_2$ should alert the clinician to the possibility that an acute dead-space-producing pathology (diminished cardiac output, pulmonary emboli) may be present.

Arterial oxygen tensions

Normal $Pa\mathrm{CO}_2$ is usually stated to be 97 mmHg (12.9 kPa) when breathing room air ($F\mathrm{IO}_2 = 0.21$) at sea level (atmospheric pressure of 760 mmHg or 101 kPa). It is not unusual for $Pa\mathrm{O}_2$ to fall as low as 80 mmHg (10.6 kPa) during normal sleep or rise as high as 105 mmHg (14 kPa) during excitement. Hypoxemia is defined as $Pa\mathrm{O}_2$ below 80 mmHg (10.6 kPa) while breathing room air.

Table 1 Laboratory normal ranges

	Mean	1 SD	2 SD	Acceptable therapeutic range
$Pa\mathrm{CO}_2$ (mmHg)	40	38–42	35–45	30–50
$Pa\mathrm{CO}_2$ (kPa)	5.3	5.0–5.6	4.7–6.0	4.0–6.7
pH	7.40	7.38–7.42	7.35–7.45	7.30–7.50

SD, standard deviation.

Reproduced with permission from Shapiro *et al.* (1994).

Table 2 Approximate $Paco_2$–pH relationship

$Paco_2$ (mmHg)	$Paco_2$ (kPa)	pH	$[HCO_3^-]_p$ (mmol/l)
80	10.7	7.20	28
60	8.0	7.30	26
40	5.3	7.40	24
30	4.0	7.50	22
20	2.7	7.60	20

Assuming an initial $PaCO_2$ of 40 mmHg (5.3 kPa): for every 20 mmHg (2.7 kPa) increase in $PaCO_2$, the pH will decrease by 0.10; for every 10 mmHg (1.3 kPa) decrease in $PaCO_2$, the pH will increase by 0.10.
Assuming an initial pH of 7.40, an arterial $PaCO_2$ of 40 mmHg, and a plasma bicarbonate of 24 mmol/l: an acute $PaCO_2$ increase of 10 mmHg (1.3 kPa) will increase the plasma bicarbonate by 1 mmol/l; an acute $PaCO_2$, decrease of 10 mmHg (1.3 kPa) will decrease the plasma bicarbonate by 2 mmol/L.
$[HCO_3^-]_p$, plasma bicarbonate concentration.
Reproduced with permission from Shapiro *et al.* (1994).

The normal adult range of Pao_2 decreases with age; the general guideline for patients in hospital is to subtract 1 mmHg from the minimal 80-mmHg level for every year over 60 years of age (Cerveri *et al.* 1995). This guideline is not applicable to persons over 90 years of age.

Fio_2–Pao_2 relationship

Oxygen is the drug most commonly administered to patients having blood gas measurements. The degree of hypoxemia can be estimated for the patient receiving oxygen therapy because increasing the inspired oxygen concentration by 10 per cent ($Fio_2 = 0.1$) increases the inspired oxygen tension by approximately 75 mmHg (10 kPa), i.e. 10 per cent of 760 mmHg (101 kPa). In the normal lung, this represents an increase in the ideal alveolar oxygen tension of approximately 50 mmHg (6.7 kPa). Since Pao_2 increases by approximately 50 mmHg for every 10 per cent increment of inspired oxygen concentration, multiplying the percentage inspired oxygen by 5 (or multiplying Fio_2 by 500) will result in the minimal predicted Pao_2 (in mmHg) for that level of oxygen therapy. If the measured Pao_2 value is less than the predicted value, it may be assumed that the patient will be hypoxemic when breathing room air (Shapiro *et al.* 1994).

An interpretive approach

Structured approaches are necessary for the novice and useful for the initiated. Such categorized approaches are accepted in the teaching of radiological, electrocardiographic, and pulmonary function interpretation, and they are no less applicable and necessary in the teaching of arterial blood gas interpretation.

Step 1: assessment of $Paco_2$ and pH

$Paco_2$ must fall into one of three categories:

(1) below 35 mmHg (4.6 kPa) is defined as respiratory alkalosis (alveolar hyperventilation);

(2) between 35 and 45 mmHg (between 4.6 and 6 kPa) is defined as normal alveolar ventilation;

(3) above 45 mmHg (6 kPa) is defined as respiratory acidosis (ventilatory failure).

Classification of the arterial pH relative to the $Paco_2$ allows the clinician to determine whether the primary abnormality is ventilatory (respiratory acid–base imbalance) or metabolic acid–base imbalance.

Acute respiratory alkalosis (alveolar hyperventilation)

In this condition $Paco_2 < 35$ mmHg (4.6 kPa), pH > 7.45, base deficit is less than 3 mmol/l, and bicarbonate is less than 22 mmol/l.

Significant renal compensation has not been elicited. The alveolar hyperventilation is assumed to be recent and probably secondary to either hypoxemia or respiratory center stimulation (pain, fright, anxiety, etc.).

Chronic respiratory alkalosis (alveolar hyperventilation)

In this condition $Paco_2 < 35$ mmHg (4.6 kPa) and pH is between 7.40 and 7.45.

This most probably represents a long-standing hyperventilation (at least 24 h) with renal compensation for the respiratory alkalosis.

Subacute or partly compensated respiratory alkalosis (alveolar hyperventilation)

In this condition $Paco_2 < 35$ mmHg (4.6 kPa), pH is between 7.46 and 7.50, base deficit is greater than 3 mmol/l, and bicarbonate is less than 22 mmol/l.

Table 3 Expected relationships between minute volume and $Paco_2$ in the normal non-exercising human

Minute volume	$Paco_2$ (mmHg)	$Paco_2$ (kPa)	Range (mmHg)	Range (kPa)
Normal	40	5.3	35–45	4.7–6.0
Twice normal	30	4.0	25–35	3.3–4.7
Four times normal	20	2.7	15–25	2.0–3.3

Reproduced with permission from Shapiro *et al.* (1994).

The renal system seldom compensates for an alkalotic pH below 7.45; with long-standing alveolar hyperventilation this is more common.

Completely compensated metabolic acidosis

In this condition $Pa\text{CO}_2 < 35$ mmHg (4.6 kPa), pH is between 7.35 and 7.40, base deficit is greater than 3 mmol/l, and bicarbonate is less than 22 mmol/l.

Alveolar hyperventilation in the presence of a pH between 7.35 and 7.40 reflects a primary metabolic acidosis in which the ventilatory system has normalized the pH. This rarely represents a primary alveolar hyperventilation because it is very unusual for either the renal or the pulmonary system to overcompensate. Further, the renal system seldom compensates for a respiratory alkalosis below pH 7.45.

Partly compensated metabolic acidosis

In this condition $Pa\text{CO}_2 < 35$ mmHg (4.6 kPa), pH < 7.35, base deficit is greater than 5 mmol/l, and bicarbonate is less than 18 mmol/l.

This most probably represents a primary metabolic acidosis to which the ventilatory system has responded with hyperventilation. The ventilatory system is incapable of providing the work of breathing necessary to compensate completely for the metabolic acidosis.

Metabolic alkalosis

In this condition $Pa\text{CO}_2$ is between 35 and 45 mmHg (4.6 and 6 kPa), pH > 7.45, base excess is greater than 3 mmol/l, and bicarbonate is greater than 26 mmol/l.

This represents a primary metabolic alkalosis to which the ventilatory system has not responded. This interpretation may change when accompanied by hypoxemia since that could represent a chronic CO_2 retainer that is relatively hyperventilating in response to an acute arterial oxygenation deficit.

Metabolic acidosis

In this condition $Pa\text{CO}_2$ is between 35 and 45 mmHg (4.6 and 6 kPa), pH < 7.35, base deficit is greater than 3 mmol/l, and bicarbonate is less than 22 mmol/l.

This represents a metabolic acidosis to which the ventilatory system has not responded. It is reasonable to expect some alveolar hyperventilation when pH < 7.30 unless the patient is unable to increase alveolar ventilation.

Partly compensated metabolic alkalosis

In this condition $Pa\text{CO}_2 > 45$ mmHg (6 kPa), pH > 7.45, base excess is greater than 3 mmol/l, and bicarbonate is greater than 26 mmol/l.

This usually represents a primary metabolic alkalosis for which the ventilatory system has partly compensated. It may represent acute alveolar hyperventilation (respiratory alkalosis) superimposed on chronic ventilatory failure. It is rare to see this phenomenon with pH < 7.50 because a conscious person will seldom significantly hypoventilate to compensate for a metabolic alkalosis. It is unusual to see a $Pa\text{CO}_2$ rise higher than 60 mmHg (8 kPa) in a conscious person, but an obtunded or unconscious patient can manifest higher $Pa\text{CO}_2$ levels in response to a severe metabolic alkalosis.

Chronic ventilatory failure (respiratory acidosis)

In this condition $Pa\text{CO}_2 > 45$ mmHg (6 kPa), pH is between 7.35 and 7.45, base excess is greater than 3 mmol/l, and bicarbonate is greater than 26 mmol/l.

This is an exception to the overcompensation rule in that it is common for the chronic hypercapnic patient to have pH > 7.40. This chronic hypercapnia probably reflects an altered intracellular CO_2 concentration because CO_2 excretion has been less than CO_2 production for a considerable period of time. The elevated peripheral (intracellular) CO_2 stores create an increase in intracellular H^+ concentration that is relatively independent of the blood pH. Progressive intracellular adaptation to the elevated $Pa\text{CO}_2$ and H^+ occurs so that essential mitochondrial functions can be maintained.

Acute ventilatory failure (respiratory acidosis)

In this condition $Pa\text{CO}_2 > 45$ mmHg (6 kPa), pH < 7.35, base excess is less than 3 mmol/l, and bicarbonate is greater than 22 mmol/l.

This represents an acute decompensation or failure of the ventilatory mechanism. This is often a life-threatening process and requires appropriate and rapid attention.

Step 2: assessment of arterial oxygenation

Following classification of the $Pa\text{CO}_2$ and pH values (Table 4), complete assessment of the arterial oxygenation status requires measurement of $Pa\text{O}_2$, oxyhemoglobin saturation $Sa\text{O}_2$ (per cent), and hemoglobin content (g/dl).

Hemoglobin content

Although normal hemoglobin content in adults is greater than 12 g/dl, it is reasonable to assume that a content above 8 g/dl provides an adequate oxygen-carrying capacity for patients with adequate myocardial function; patients with heart disease probably require at least 10 g/dl. In the following discussion a hemoglobin content of 10 g/dl is assumed.

Oxyhemoglobin saturation

This measurement provides the best single indication of oxygen content because almost all the oxygen in the blood exists as oxyhemoglobin. As shown in Fig. 1, $Sa\text{O}_2$ has a predictable relationship with $Pa\text{O}_2$ under normal circumstances. Any departure from that relationship implies an alteration in the affinity of hemoglobin for oxygen. Essentially, a shift to the right (decreased affinity) is present when the $Sa\text{O}_2$ is less than predicted for the $Pa\text{O}_2$, and vice versa. When the affinity is decreased, it is prudent to base clinical judgments on $Sa\text{O}_2$ rather than $Pa\text{O}_2$.

Arterial oxygen tension

This is the traditional measurement upon which the clinical assessment of arterial oxygenation is based. Arterial hypoxemia in adults is defined as $Pa\text{O}_2 < 80$ mmHg (10.7 kPa) while breathing room air. Since there is no clinically significant improvement in the oxygen delivery capabilities of blood with 90 per cent $Sa\text{O}_2$ compared with 95 per cent $Sa\text{O}_2$, most practitioners of critical care medicine and respiratory care consider clinically significant hypoxemia to be present at $Pa\text{O}_2 < 60$ mmHg (8 kPa).

Table 4 Seven primary blood gas classifications

Classification	$Paco_2$	pH	$[HCO_3^-]_p$	Base excess
Primary ventilatory				
1. Acute ventilatory failure	↑	↓	N	N
2. Chronic ventilatory failure	↑	N	↑	↑
3. Acute alveolar hyperventilation	↓	↑	N	N
4. Chronic alveolar hyperventilation	↓	N	↓	↓
Primary acid–base				
1. Uncompensated acidosis	N	↓	↓	↓
Uncompensated alkalosis	N	↑	↑	↑
2. Partly compensated acidosis	↓	↓	↓	↓
Partly compensated alkalosis	↑	↑	↑	↑
3. Compensated alkalosis or acidosis	↑ or ↓	N	↑ or ↓	↑ or ↓

Arrows indicate depressed or elevated values; $[HCO_3^-]_p$, plasma bicarbonate; N, normal.
Reproduced with permission from Shapiro *et al.* (1994).

The maintenance of adequate tissue oxygenation with moderate hypoxemia (Pao_2 between 40 and 60 mmHg (5.3 and 8 kPa)) primarily depends upon cardiovascular function and oxygen consumption. However, severe hypoxemia ($Pao_2 < 40$ mmHg (5.3 kPa)) must be considered a direct threat to tissue oxygenation because there is a diminished oxygen 'driving force' in the systemic capillaries and the oxyhemoglobin saturation is less than 75 per cent, which increases the affinity of hemoglobin for oxygen because oxygen does not occupy the fourth heme site. This combination of increased affinity and decreased driving force significantly diminishes the movement of oxygen molecules from the systemic capillaries to the extracellular fluid compartment.

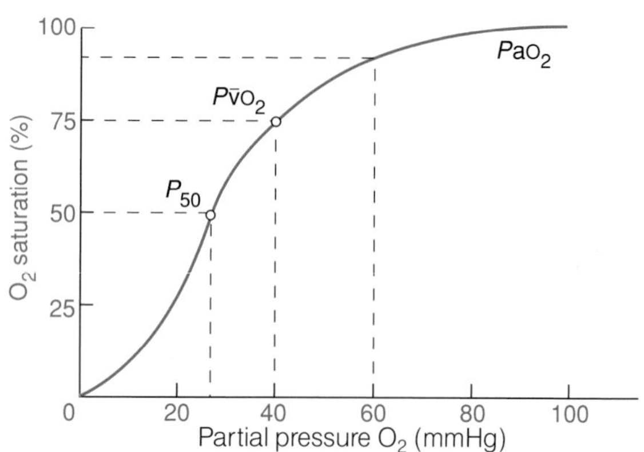

Fig. 1 The hemoglobin dissociation curve for oxygen. The vertical axis represents the portion of the potential oxygen-carrying hemoglobin sites that are chemically combined with oxygen (percentage saturation). The horizontal axis represents the partial pressure of oxygen in the plasma in mmHg (kPa = mmHg × 0.133). The non-linear relationship between percentage saturation and Po_2 accounts for the vast majority of oxygen reserves in the blood. Normally, hemoglobin is 50 per cent saturated with oxygen at a plasma Po_2 of 27 mmHg (3.6 kPa); this is designated P_{50}. Normal mixed venous (pulmonary arterial) blood is 75 per cent saturated with oxygen at a plasma Po_2 of 40 mmHg (5.3 kPa). Arterial blood with a plasma Po_2 of 60 mmHg (8.0 kPa) is normally 90 per cent saturated with oxygen, arterial blood with a plasma Po_2 of 80 mmHg (10.7 kPa) is normally 95 per cent saturated with oxygen, and arterial blood with a plasma Po_2 of 97 mmHg (12.9 kPa) is normally 97 per cent saturated with oxygen. (Reproduced with permission from Shapiro *et al.* (1994).)

Oxygen therapy

Since many patients are already receiving oxygen therapy when blood gas values are obtained, it is essential that the probable hypoxemic state be assessed so that the oxygen therapy is not removed unnecessarily. Apart from special circumstances, oxygen therapy should not be removed to assess hypoxemia.

Uncorrected hypoxemia

In this condition Pao_2 remains below 60 mmHg (8 kPa) despite increased inspired oxygen concentration.

Corrected hypoxemia

On oxygen therapy, Pao_2 is between 60 mmHg (8 kPa) and 100 mmHg (13.3 kPa). Hypoxemia is assumed to exist on room air because Pao_2 is less than the predicted normal level for that oxygen therapy.

Excessively corrected hypoxemia

Pao_2 is less than the minimal predicted normal at room air for that oxygen therapy but greater than 100 mmHg (13.3 kPa). It is assumed that hypoxemia exists on room air so that oxygen therapy is required, but it can be decreased.

When the actual Pao_2 is greater than the theoretically minimal Pao_2 for that Fio_2, the patient may not be hypoxemic on room air. However, it is recommended that the patient is re-evaluated at a significantly lower Fio_2 before discontinuing oxygen therapy completely.

Temperature correction of blood gas values

A blood sample with fixed O_2 and CO_2 contents will manifest different gas tensions when analyzed at various temperatures. Adjustment of the temperature of the measuring electrode to the patient's core temperature would add at least 30 min to the time required for each set of measurements and would severely complicate quality assurance in the laboratory. To avoid these impractical and undesirable factors, the pH, $P\text{CO}_2$, and $P\text{O}_2$ electrodes are encased in a constant 37 °C environment to which the blood sample chamber is also exposed.

The term 'temperature correction' refers to the application of mathematical adjustments to the values measured at 37 °C in order to produce a more accurate reflection of the gas tensions as they existed in the artery from which they were obtained. The availability of microprocessor technology makes the practice of temperature correction convenient and practical. Justification for temperature correction is based on the belief that 'knowing' the precise *in vivo* values is an advantage in patient care. This implies that either the normal ranges for pH, $P\text{CO}_2$, and $P\text{O}_2$ remain constant regardless of body temperature variation, or that the normal ranges are well established for all body temperatures. All available physiological and clinical data suggest that neither of these alternatives is valid.

The following factors (Shapiro 1995) strongly support the contention that there is no scientific basis for the routine practice of applying temperature corrections to blood gas measurements when alterations in body temperature are present.

1. The 37 °C values for pH and $P\text{CO}_2$ reliably reflect the *in vivo* acid–base status at the patient's actual temperature.

2. The pH and $P\text{CO}_2$ changes attributable to temperature variation do not affect the calculated bicarbonate value.

3. Since CO_2 production and $P\text{aCO}_2$ undergo parallel changes with temperature variation, clinical assessment of alveolar ventilation is most reliably reflected by applying the well-established homeostatic reference points at 37 °C.

4. When hypocapnia is desired to minimize the intracranial blood volume, the $P\text{aCO}_2$ measurement at 37 °C is as clinically reliable as the temperature-corrected values.

5. Data reliably quantifying the balance between oxygen delivery and oxygen demand at temperatures other than 37 °C are unavailable; therefore temperature-corrected $P\text{O}_2$ values do not improve our ability to make clinically relevant interpretations.

The popularity of the routine temperature correction of pH, $P\text{CO}_2$, and $P\text{O}_2$ values is based on the observation that large differences in blood gas values are present when the patient's temperature is profoundly hypo- or hyperthermic. This observation leads some clinicians to the unsubstantiated conclusion that the uncorrected 37 °C values are 'wrong'. The danger in this superficial thought process is that one might reach the unfounded conclusion that temperature-corrected values are 'right'.

References

Cerveri, I., *et al.* (1995). Reference values of arterial oxygen tension in the middle-aged and elderly. *American Journal of Respiratory and Critical Care Medicine*, **152**, 934–41.

Shapiro, B.A. (1995). Temperature correction of blood gas values. *Respiratory Care Clinics of North America*, **1**, 69–76.

Shapiro, B.A., Peruzzi, W.T., and Kozelowski-Templin, R (1994). *Clinical application of blood gases* (5th edn). Mosby Year Book, Chicago, IL.

16.3 Ventilatory monitoring

16.3.1 Capnography

L. Blanch and A. Artigas

Key messages

- Currently available systems for breath-by-breath CO_2 analysis include mass spectrometry and infrared light absorption.
- The expiratory capnogram is a technique that provides qualitative information on the waveform patterns associated with mechanical ventilation and quantitative estimation of arterial $P\text{CO}_2$.
- Capnography seems reliable for use as a non-invasive tool for assessing alveolar ventilation during weaning in general intensive care patients.
- Capnography is one of the most reliable methods of detecting esophageal intubation.

Introduction

Capnography permits recognition of CO_2 concentration changes in the patient's airway during the respiratory cycle. The ability to measure a patient's inspired and expired CO_2 has existed for many years. In the past 10 years, this technology has evolved so that it is now available in many intensive care units. However, the understanding and applications of this technique have not been fully tested in clinical practice.

Technical aspects

Currently available systems for breath-by-breath CO_2 analysis include mass spectrometry and infrared light absorption. Mass spectrometry has a faster response time (about 100 ms) and the ability to analyze several gas concentrations simultaneously. However, its main drawback is its high price and costly maintenance. Infrared CO_2 analyzers are cheaper and easier to use; each one is dedicated to a single patient, and the range of response times oscillates between 100 and 200 ms which is sufficient for breath-by-breath monitoring purposes. Two types of CO_2 analyzer based on the infrared radiation principle are available. Mainstream analyzers have the advantage of being part of the ventilatory circuit and exhibiting a very rapid response time. Sidestream analyzers pump the gas sample from the airway at a flow rate of 150 to 500 ml/min through a small tube to a remote analyzer. Their major disadvantages are obstruction of the tubing by pulmonary secretions or condensation and a slower response time than the mainstream analyzers (Hess 1990).

Physiology

Correct analysis of a capnogram requires systematic evaluation of the following steps. First, it is important to determine whether or not there is exhaled CO_2. This is very important for early recognition of esophageal intubation. Second, the four phases of the waveform

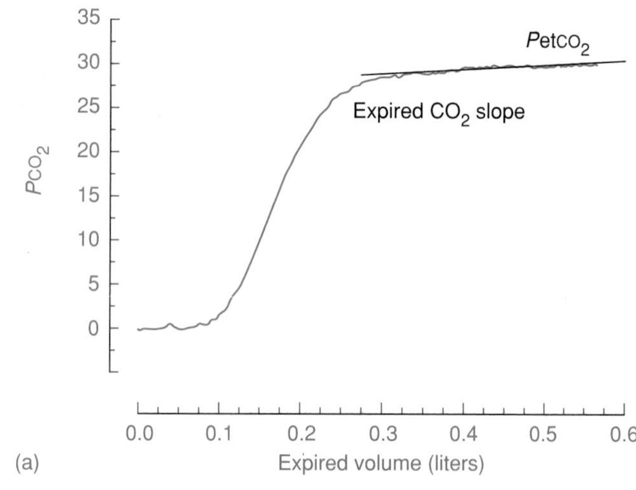

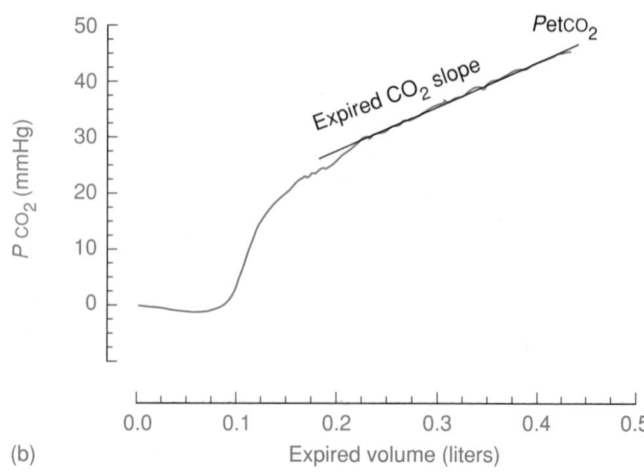

Fig. 1 Expired capnograms from patients receiving mechanical ventilation: (a) a normal capnogram; (b) a capnogram produced during an episode of airway obstruction in a patient with chronic obstructive pulmonary disease.

should be analyzed. The capnogram represents the total CO_2 eliminated by the lungs given that no gas exchange occurs in the airways. Expired gas contains CO_2 from three sequential compartments: phase I contains gas from the apparatus and anatomical dead-space, phase II represents increasing CO_2 concentration resulting from progressive emptying of alveoli, and phase III represents essentially alveolar gas. Phase III is often referred to as the plateau, as it is flat or has a small positive slope (Fig. 1). The highest point is the end-tidal $P\text{CO}_2$ ($P\text{etCO}_2$). Therefore the expiratory capnogram is a technique that provides qualitative information on the waveform patterns associated with mechanical ventilation and a quantitative estimation of arterial $P\text{CO}_2$ ($P\text{aCO}_2$) from $P\text{etCO}_2$. Additionally, the minimum $P\text{CO}_2$ should be determined and the $P\text{aCO}_2 - P\text{etCO}_2$ gradient should be calculated. When all this information is analyzed, we can search for the causes of hypo- or hypercapnia, if present, such as inadequate alveolar ventilation, CO_2 rebreathing, or excessive CO_2 output.

The shape of the capnogram can be characteristic in many clinical situations. The absence of an alveolar plateau is the result of the presence of alveoli with long time constants that continuously contribute to increasing the concentration of CO_2 in the airways (Fig. 1). This usually occurs in patients with chronic obstructive pulmonary disease or asthma. A cleft just before end-expiration may be seen in patients fighting the ventilator or recovering from neuromuscular blockade. Oscillations during phase III are caused by beating of the heart against the lungs and are generally observed in patients ventilated at lower respiratory rates. Moreover, continuous capnographic monitoring at low-speed recording provides information about disturbances in the pulmonary circulation, and exhaled CO_2 tracings are also useful for detecting episodes of hypoventilation during mechanical ventilation (Hess 1990).

Clinical aspects

Relationship between $P\text{aCO}_2$ and $P\text{etCO}_2$ during total ventilatory support

The relationship between $P\text{aCO}_2$ and $P\text{etCO}_2$ has been the subject of a large number of investigations and controversy exists regarding its utility as a non-invasive tool for monitoring alveolar ventilation at the bedside. Several studies performed on patients under total ventilatory support have reported that the values of $P\text{etCO}_2$ at a single point in time usually correlate well with the corresponding value of $P\text{aCO}_2$. However, when the ventilatory pattern was changed, there was a weaker correlation between changes in $P\text{etCO}_2$ and changes in $P\text{aCO}_2$ measured simultaneously for a given patient, indicating that the $P\text{aCO}_2 - P\text{etCO}_2$ gradient varied within individual patients during changes in ventilation. These contradictory data merely reflect the strong influence of pathophysiological states and circulatory and respiratory variables on $P\text{aCO}_2$ and $P\text{etCO}_2$.

Relationship between $P\text{aCO}_2$ and $P\text{etCO}_2$ during weaning from mechanical ventilation

The decision to start a weaning trial in a critically ill patient often implies that patients have correct arterial blood gases. The fact that capnography could provide a non-invasive tool to assess alveolar ventilation, and thereby $P\text{aCO}_2$, induced some investigators to evaluate the utility of capnography during weaning periods. However, these studies have yielded controversial results. Although some authors reported that patients monitored with pulse oximetry and capnography required less blood gas sampling, others found that variations in $P\text{etCO}_2$ did not correctly indicate changes in $P\text{aCO}_2$ in patients with parenchymal lung disease, particularly emphysema. To analyze this clinical problem, we evaluated the relationship between $P\text{aCO}_2$ and $P\text{etCO}_2$ before weaning and during a weaning trial to determine the ability of $P\text{etCO}_2$ to identify clinically relevant episodes of hypercapnia. Interestingly, we found that monitoring $P\text{etCO}_2$ and pulse oximetry provided good assessment of hypercapnic episodes and oxygenation impairment during weaning from mechanical ventilation. Although capnography does not detect minor changes in $P\text{aCO}_2$ with 100 per cent accuracy, continuous $P\text{etCO}_2$ monitoring reduces the need for arterial blood sampling. Accordingly, although not replacing invasive arterial determinations entirely, capnography seems reliable for use as a non-invasive tool for assessing alveolar ventilation during weaning in general intensive care patients (Saura *et al.* 1996).

Assessment of optimal positive end-expiratory pressure using capnography

The difference between $P\text{aCO}_2$ and $P\text{etCO}_2$ can be used to determine the appropiate level of positive end-expiratory pressure (PEEP) in acute respiratory failure. In both experimental animals and patients, the smallest $P\text{aCO}_2 - P\text{etCO}_2$ gradient coincided with the level of PEEP providing the best arterial oxygenation and the least intrapulmonary shunt. When lung overdistension was caused by higher levels of PEEP, the $P\text{aCO}_2 - P\text{etCO}_2$ gradient increased due to an increase in alveolar dead-space with no further improvement in arterial oxygenation or intrapulmonary shunt. Nevertheless, the utility of the $P\text{aCO}_2 - P\text{etCO}_2$ gradient for setting optimal PEEP levels seems to be limited to those patients who are able to recruit collapsed lung units with PEEP (Blanch *et al.* 1987).

Detection of esophageal intubation using capnography

Undetected esophageal intubation is a serious problem in intensive care patients. In patients in whom the vocal cords cannot be visualized, an esophageal intubation may be difficult to recognize. Capnography is one of the most reliable of the methods used to detect esophageal intubation, and it has been suggested that $P\text{etCO}_2$ should be used routinely to verify correct endotracheal tube placement. The limitations on the use of $P\text{etCO}_2$ to detect esophageal intubation are the presence of CO_2 or a carbonated beverage in the stomach and the presence of low pulmonary perfusion or a cardiac arrest (Hess 1990).

Capnography during cardiopulmonary resuscitation

Several investigators have shown CO_2 monitoring to be useful during cardiopulmonary resuscitation. Sudden increases in $P\text{etCO}_2$ during cardiopulmonary resuscitation indicate a return of spontaneous circulation, and it has been found that $P\text{etCO}_2$ values from patients who were likely to be resuscitated were twice as high as those from patients who could not be resuscitated. Moreover, a sudden interruption of pulmonary perfusion due to acute embolism causes an increase in physiological dead-space which reduces $P\text{ETCO}_2$ without markedly affecting $P\text{aCO}_2$ (Falk *et al.* 1987).

Capnography in airway obstruction

The expiratory capnogram represents the elimination of gas from the alveolar units and the resulting waveform may be influenced by lung abnormalities. In a homogeneous normal lung, the $\dot{V}/Q$ ratio, the time constant, and the CO_2 partial pressure are the same for all respiratory units, and there is a well-defined transition point between phase II (steep concentration change) and phase III (alveolar plateau). However, when the mixing of inspired gas with lung resident gas is incomplete, mainly because of regional inequalities of ventilation and perfusion, CO_2 concentration increases in the airways from proximal to distal. Therefore the increasing expiratory CO_2 slope may be due to serial mixing of gas flowing from units with different time constants. As a result, faster units empty more rapidly at first while gas from slower units, presumably with very low $\dot{V}/Q$ ratios and high $P\text{CO}_2$, forms a proportionately greater part of the end-expiratory gas. Consequently, the CO_2 concentration in the exhaled gas progressively increases and contributes to an increase in the slope during phase III of the capnogram. Thus, given a relatively uniform perfusion distribution, slow hypercapnic units, and sequential emptying, the CO_2 elimination waveform does reflect the $\dot{V}/Q$ relationship in many patients.

Bronchial obstruction is associated with regional decreases in airflow with consequent reductions in alveolar ventilation responsible for the heterogeneity of the ventilation–perfusion relationship. Therefore the increasing expired CO_2 slope observed in some patients may be due to the serial mixing of gas flowing from units with different time constants. Several studies conducted in asthmatic patients showed the existence of a significant correlation between indices describing the shape of the capnogram and the usual spirometric parameters.

Critically ill patients on mechanical ventilation may often exhibit high respiratory system resistance (R_{rs}) for a variety of reasons. The expiratory capnogram is correlated with spirometric indices. In ventilated critically ill patients regression analysis revealed a close correlation between R_{rs} and expired CO_2 slope ($r = 0.86$; $p < 0.001$) with 95 per cent confidence intervals for R_{rs} of $\pm 7.39\ \text{cmH}_2\text{O/l/s}$ from the predicted value obtained by the regression equation, where

$$R_{rs} = 11.42 + (2.28 \times \text{expired CO}_2 \text{ slope}).$$

These observations suggest that CO_2 elimination is impaired by a flow-resistive mechanism and that the degree of airway obstruction modulates the rate at which $P\text{CO}_2$ rises during expiration. However, despite the strong correlation between R_{rs} and the expired CO_2 slope, the confidence intervals were wide. Therefore, the expired CO_2 slope has limited clinical applicability in accurately predicting R_{rs} at the bedside (Blanch *et al.* 1994).

In the same study, a weak correlation was found between $P\text{aCO}_2 - P\text{ETCO}_2$ gradients and R_{rs}. Because the $P\text{aCO}_2 - P\text{ETCO}_2$ gradient is influenced by structural airway abnormalities as well as by high and low $\dot{V}/Q$ regions, this may explain why some patients exhibit a large $P\text{aCO}_2 - P\text{ETCO}_2$ gradient with normal R_{rs} and expired CO_2 slope. Finally, low cardiac output states may increase the $P\text{aCO}_2 - P\text{ETCO}_2$ gradient as pulmonary perfusion decreases while lung mechanical properties remain unchanged (Blanch *et al.* 1987; Hess 1990). These observations suggest that CO_2 elimination in critically ill patients is strongly modulated by lung, airway, endotracheal tube, and ventilator equipment resistances.

References

Blanch, L., Fernandez, R., Benito, S., Mancebo, J., and Net, A. (1987). Effect of PEEP on the arterial minus end-tidal carbon dioxide gradient. *Chest*, **92**, 451–4.

Blanch, L., Fernandez, R., Saura, P., Baigorri, F., and Artigas, A. (1994). Relationship between expired capnogram and respiratory system resistance in critically ill patients during total ventilatory support. *Chest*, **105**, 219–23.

Falk, J.L., Rackow, E.C., and Weil, M.H. (1988). End-tidal carbon dioxide concentration during cardiopulmonary resuscitation. *New England Journal of Medicine*, **318**, 607–11.

Hess, D. (1990). Capnometry and capnography: technical aspects, physiologic aspects, and clinical applications. *Respiratory Care*, **35**, 557–76.

Saura, P., Blanch, L., Lucangelo, U., Fernndez, R., Mestre, J., and Artigas, A. (1996). Utility of capnography to detect hypercapnic episodes during weaning from mechanical ventilation. *Intensive Care Medicine*, **22**, 374–81.

16.3.2 Flow–volume and pressure–volume loops

J. C. Goldstone

Key messages

- Pressure–volume curves describe the stiffness of the respiratory system.
- It is important to titrate external positive end-expiratory pressure to prevent airway closure and to avoid overdistension in patients with stiff lungs.
- Flow–volume curves describe the flow limitation of the respiratory system.
- Bronchodilator therapy can be titrated to increase airflow for any given driving pressure across the lungs.
- Approximations to pressure–volume curves can be made during tidal ventilation without disconnecting the patient from the ventilator.

Introduction

Volume, pressure, and flow are measured by many intensive care ventilators and continuous-waveform displays are commonplace. The pressure–volume relationship of the lung and chest wall demonstrates the stiffness of the lung, whereas flow–volume curves demonstrate changes in airways resistance. Many ventilators are able to store waveforms, enabling trends to be established. Great care must be taken to ensure that each waveform is recorded under exactly the same ventilator settings if a comparison is to be made (Tobin and Van de Graaff 1994).

Pressure–volume (*PV*) curves

The stiffness of the lungs and chest wall can be assessed during mechanical ventilation or from small stepped inflations over a larger volume range. Pressure is measured when flow has ceased and the system is at equilibrium. Stepped inflation demands specialist, albeit simple, equipment and demands that the patient can be disconnected from respiratory support for a period while the measurement is made. This may not be possible in unstable patients.

Measurements during mechanical ventilation

Constant-flow ventilation is provided by most intensive care unit ventilators when the machine breath is volume controlled. A constant volume is delivered for every time increment, and so the horizontal time axis becomes a linear analog for volume. It is important that there is no inspiratory pause and that the inspiratory waveform is a square wave. The shape of the inflation pressure–volume curve can be determined from a machine breath when the patient is relaxed. With care it is possible to make this measurement when the patient is not completely relaxed (Fig. 1).

As inspiration begins, pressure is required to overcome the flow resistance of the branching airways and the ventilatory apparatus. This can be seen as a sudden upstroke of pressure (Fig. 2(a)). At this point little gas has entered the chest. If the inspiratory flow is known or set by the ventilator, resistance at this point of the respiratory cycle can be estimated. As flow occurs, the volume of the chest increases (Figs 2(b) and 2(c)). Further pressure is required to overcome the expansion of the elastic components of the lung and chest wall. This portion of the waveform is the pressure–volume curve of the lung and chest wall. The gradient of the line is equal to the effective dynamic compliance, which is calculated as

$$\frac{\text{tidal volume}}{\text{peak airway pressure} - \text{PEEP}}.$$

This method of assessing the pressure–volume relationship has its limitations. Firstly, although pressure may be linearly related to volume over the volume range, this may not be the case for larger volumes where pressure may be markedly non-linear. In addition, the opening pressure of the system may change with disease, and it may not be accurately assessed as the initial upstroke of this waveform is related to the resistance of the system rather than the tissue elasticity which needs to be overcome to prevent airway closure.

Static measurements

The term 'static' refers to the conditions of no flow where the respiratory system is not moving. Increments of lung volume are delivered from a large (1.5 liters) syringe which inflates the lungs at the end of expiration. During deflation, the lung is emptied in similar steps and the pressure is recorded.

The pressure–volume curve (Fig. 3) obtained using the super-syringe technique displays two general characteristics which are amplified when the lungs are stiff. Firstly, pressure must be applied to the lungs to open them and this opening pressure can be clearly seen as the first inflexion point of the graph. Inflation from this point is generally linearly related to the pressure difference. During inflation there is a point where lung volume reaches its maximum. When this happens, high pressures are needed to change volume further. This is the second (upper) inflexion point. Patterns of ventilation which allow the patient to fall below the lower inflexion point would entail opening and closing of alveoli. If ventilation is set such that the upper inflexion point is exceeded, high-pressure damage would be expected. Ideal ventilation would entail the application of external positive end-expiratory pressure (PEEP) to prevent alveolar collapse together with ventilation over the linear pressure–volume range, avoiding high inflation pressures and alveolar distension.

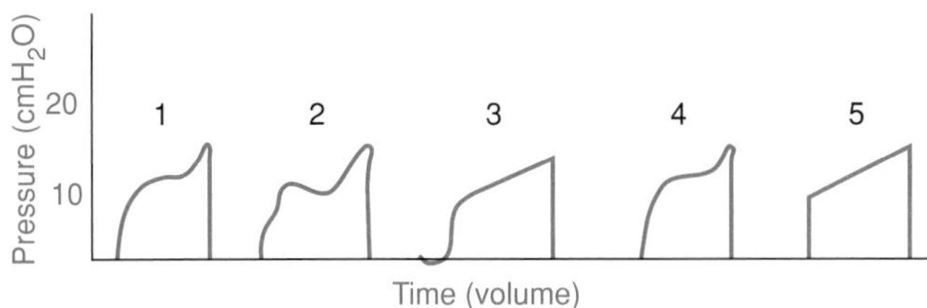

Fig. 1 An example of a sequence of pressure–time curves. Many are irregular, particularly when the patient is interacting with the ventilated breath. Only when the curve is close to the idealized shape (example 5) is it possible to measure the pulmonary mechanics.

Pressure–volume curves of the lung are very sensitive to changes in lung stiffness. This is seen frequently when the lung tissue is damaged, particularly when the response is fibrosis. The changes seen in the acute respiratory distress syndrome provide a good example of the use of pressure–volume curves. In health, the lungs are held open at the relaxed end-expiratory point (functional residual capacity) and inflation occurs at transpleural pressure changes of 2 to 5 cmH_2O. The first change related to increasing lung stiffness in the acute respiratory distress syndrome is the appearance of a lower inflexion point, suggesting that alveolar collapse has occurred at the relaxed end-expiratory point. As the syndrome progresses the slope of the pressure–volume curve changes, suggesting that compliance has decreased. Furthermore, the gap between the inflation and deflation curves (hysteresis) increases (Matamis *et al.* 1984; Beydon *et al.* 1991).

Several therapeutic strategies are available, and the conceptual type of ventilation has been described as ventilation for a 'baby lung'. As

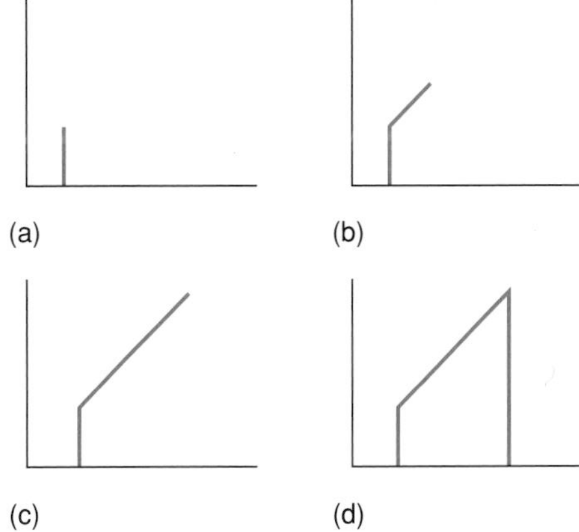

Fig. 2 Constituent parts of an individual pressure–time diagram: (a) initially little gas has entered the chest, no volume expansion of the lung has occurred, and the rapid pressure upstroke overcomes airway resistance; (b) as time progresses expansion of the lungs and chest wall occurs and the pressure curve increases; (c) at the end of the breath the pressure curve is linear, and the pressure–volume curve can be assessed from this line; (d) a completed breath.

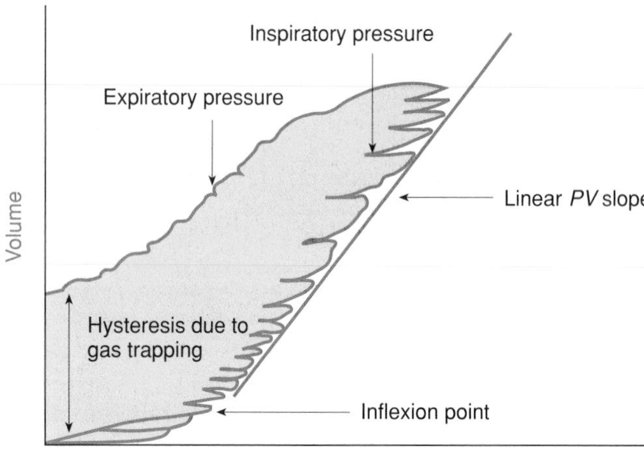

Fig. 3 A pressure–volume tracing obtained from a patient with stiff lungs by means of the supersyringe technique. The inflation curve has two clearly separate components and the inflexion point is indicated. The pressure at the inflexion point is used to assess the external PEEP required to prevent airway closure. The inspiratory and expiratory curves are separated due to hysteresis. This difference in volume is related to gas trapping.

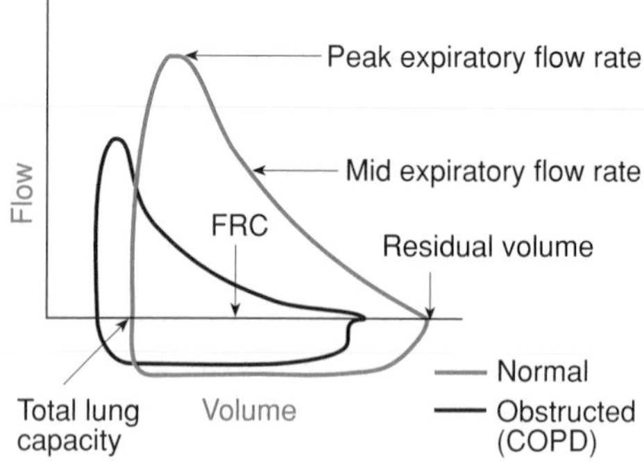

Fig. 4 Flow–volume loops from residual volume to total lung capacity for a normal subject and a patient with chronic obstructive pulmonary disease (COPD). In the normal subject, airflow into the chest is fixed and the subject exhales forcibly, quickly achieving peak flow. This contrasts with the COPD patient's tracing which starts at an inflated lung volume (offset to the left), exceeds the total lung capacity of the normal subject, and cannot reach the same expiratory flow. Note the shallower gradient of the mid-expiratory section of the curve.

with neonates and infants, alveolar collapse occurs readily. Recruitment of partially ventilated or collapsed lungs is difficult, but this problem can be minimized by titrating external PEEP appropriately. Overexpansion of the shrinking 'normal lung' in adults can be avoided if the compliance is measured frequently. During recovery, pressure and volume limitations imposed by lung mechanics can be normalized in a timely manner.

A quasistatic pressure–volume curve can be obtained without ventilator disconnection. If the ventilator can be adjusted so that ventilation pauses and the pressure in the airway can be measured, the compliance of the lung and chest wall can be plotted. The volume of successive breaths can be adjusted so that higher volumes are achieved. The agreement between this approach and the supersyringe technique is good, although care must be taken with all ventilator measurements that the calibration of pressure and volume sensors is accurate (Sydow *et al.* 1991).

Flow–volume curves

In a respiratory function laboratory the effect of changes in resistance is commonly assessed by measuring flow–volume loops (Fig. 4). Both inspiratory and expiratory flow are recorded and the subject is asked to breathe in to total lung capacity from a mouthpiece. Expiration is normally performed in a forced manner back to residual volume and a loop is created. Several loops can be superimposed to obtain a representative example. It is important to note that volume recordings start from residual volume and increase as the trace is read from right to left. A variety of measurements can be read from the tracing, although it is as easy to assess the shape of the graph. For comparison a tracing of a patient with mild obstructive lung disease is shown. In absolute terms, residual volume (and functional residual capacity) are greater, indicating that residual gas is trapped in the lungs. The characteristic shape of the curve is offset to the left with scalloping of the expiratory curve. Peak and all other expiratory flows are reduced from normal, and the ratio of flow to volume indicates an obstructed picture.

In intubated patients similar tracings are obtained but with important differences. Patients rarely, if ever, perform flow–volume loops over the total lung volume range. In addition, they do not perform forced maneuvers and tend to exhale passively. Nevertheless it is possible to measure flow–volume loops in patients over the tidal volume range.

A flow–volume loop for a spontaneously breathing patient is shown in Fig. 5. The loop is smaller, and expiratory flow is limited and equal to that of inspiration. In addition, expiratory flow comes to a sudden abrupt end typical of obstructive lung disease.

Expiratory flow is dependent on effort, the elastic recoil of the lung and chest wall, and the resistance of the airways. If the patient is relaxed the expiratory flow is dependent on airways resistance, although it is important to scrutinize the flow–volume loop to ensure consistent results.

Simple visual inspection of flow–volume loops can be an effective method of demonstrating bronchodilation. Further evidence that treatment has been successful include a reduction in auto-PEEP and an increase in flow for any given pressure difference.

The application of PEEP to patients who exhibit flow limitation has previously been avoided. Expiratory flow will only be decreased and hyperinflation worsened if the external PEEP exceeds the level

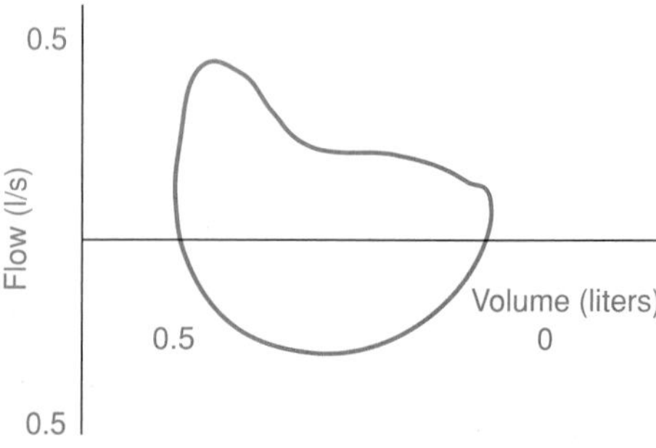

Fig. 5 A flow–volume loop during tidal ventilation. Although the loop is smaller, the same basic components as in Fig. 4 are present. However, in this example expiratory flow is passive and related to the recoil pressure of the lung and chest wall.

of auto-PEEP. As auto-PEEP is difficult to measure in the spontaneously breathing patient, the effect of changes in external PEEP are unknown. By monitoring flow–volume loops, external PEEP can be applied until the flow–volume curves move along the volume axis (Tobin and Lodato 1989).

References

Beydon, L., Lemaire, F., and Jonson, B. (1991). Lung mechanics in ARDS: compliance and pressure–volume curves. In *Adult respiratory distress syndrome* (ed. W.M. Zapol and F. Lemaire), pp. 139–61. Marcel Dekker, New York.

Matamis, D., Lemaire, F., Harf, A., Brun-Buisson, C., Ansquer, J.C., and Atlan, G. (1984). Total respiratory pressure–volume curves in the adult respiratory distress syndrome. *Chest*, **86**, 58–66.

Sydow, M., *et al.* (1991). Improved determination of static compliance by automated single volume steps in ventilated patients. *Intensive Care Medicine*, **17**, 108–14.

Tobin, M.J. and Lodato, R.F. (1989). PEEP, auto-PEEP, and waterfalls. *Chest*, **96**, 449–51.

Tobin, M.J. and Van de Graaff, W.B. (1994). Monitoring of lung mechanics and work of breathing. In *Principles and practice of mechanical ventilation* (ed. M.J. Tobin), pp. 967–1003. McGraw-Hill, New York.

16.3.3 Monitoring gas exchange and ventilation–perfusion mismatching

R. Rodriguez-Roisin and J. Roca

Key messages

- The components that distinctively determine $P\text{O}_2$ and $P\text{CO}_2$ in any single gas exchange unit of the lung are the $\dot{V}_A/\dot{Q}$ ratio, the composition of the inspired gas, and the composition of the mixed venous blood.
- The multiple inert gas elimination technique gives both quantitative and qualitative estimates of the distributions of $\dot{V}_A/\dot{Q}$ ratios.
- If the measured $P\text{aO}_2$ is significantly different from the values estimated using the multiple inert gas elimination technique, other potential mechanisms of hypoxemia, such as diffusion limitation to O_2, increased intrapulmonary parenchymal O_2 consumption, or increased postpulmonary shunt are implicated.

Introduction

The major function of the lung is to exchange respiratory gases (O_2 and CO_2). Arterial $P\text{O}_2$ and $P\text{CO}_2$ are the directly measurable endpoint variables used in practice to interpret respiratory failure properly. The calculated variables $P(\text{A–a})\text{O}_2$ (difference between alveolar and arterial $P\text{O}_2$), venous admixture ratio, and physiological dead-space, which are viewed as simple functional markers of the overall status of pulmonary gas exchange, are also used. However, these indices consider not only the state of the lung, i.e. ventilation–perfusion ($\dot{V}_A/\dot{Q}$) mismatching, intrapulmonary shunt, and limitation on O_2 diffusion, but also the external conditions under which the lung is operating. The components that distinctively determine $P\text{O}_2$ and $P\text{CO}_2$ in any single gas exchange unit of the lung are the $\dot{V}_A/\dot{Q}$ ratio, the composition of the inspired gas, and the composition of the mixed venous blood.

The multiple inert gas elimination technique gives both quantitative and qualitative estimates of the distributions of $\dot{V}_A/\dot{Q}$ ratios, without producing changes in either the pulmonary vascular caliber or the airway tone. In addition, it aids in the interpretation of the complex interplay between the intrapulmonary components ($\dot{V}_A/\dot{Q}$ imbalance, intrapulmonary shunt, and O_2 diffusion limitation) and extrapulmonary components (inspired O_2 concentration, overall ventilation, cardiac output, and O_2 consumption) of gas exchange.

Multiple inert gas elimination technique

The arterial, mixed venous, and mixed expired concentrations of six infused inert gases, measured by gas chromatography, are used to calculate the ratio of arterial pressure (*P*a) to mixed venous pressure (*P*v) (retention) and the ratio of alveolar pressure (*P*A) or mixed expired pressure (*P*E) to mixed venous pressure (excretion). Retention and excretion are then used to compute multicompartment $\dot{V}_A/\dot{Q}$ distributions. Six inert gases (sulfur hexafluoride (partition coefficient, 0.005), ethane, cyclopropane, enfluorane or halothane, ether, and acetone (partition coefficient, 300), listed in ascending order of solubility) are used to characterize the distribution of the $\dot{V}_A/\dot{Q}$ ratios in the whole lung (Wagner *et al.* 1974).

The principle modulating inert gas elimination within the lung is based on the simple concept that the uptake (retention or *P*a/(*P*v) and the elimination (excretion or *P*E/(*P*v) of an inert gas in any ideal homogeneous area of the lung under steady state conditions obey the following equation:

$$\frac{Pa}{Pv}=\frac{PE}{Pv}=\frac{\lambda}{\lambda+\dot{V}_A/\dot{Q}} \tag{1}$$

where λ defines the solubility (partition coefficient). According to this equation, both the retention and the excretion of an inert gas depend only on the solubility of the six inert gases and the $\dot{V}_A/\dot{Q}$ ratio. For each inert gas, the retentions are calculated as the ratio of arterial partial pressure to mixed venous partial pressure and the excretions as the ratio of mixed expired partial pressure to mixed venous partial pressure. By using a multicompartimental approach with enforced smoothing, the retentions of the six inert gases allow the estimation of a continuous distribution of the pulmonary blood flow against $\dot{V}_A/\dot{Q}$ ratio on a logarithmic scale (Evans and Wagner 1977). Similarly, the excretions of the six inert gases provide an estimation of the distribution of the alveolar ventilation against $\dot{V}_A/\dot{Q}$ ratio. Because the term $\lambda/(\lambda+\dot{V}_A/\dot{Q})$ is common to both retention and excretion, the blood flow and the ventilation distributions are mathematically interdependent, reflecting the same distribution of the $\dot{V}_A/\dot{Q}$ ratios seen from opposite sides of the blood–gas barrier. The use of enforced smoothing requires compartments with similar $\dot{V}_A/\dot{Q}$ ratio to have similar perfusion. In summary, the key variables required for determining $\dot{V}_A/\dot{Q}$ distributions for real lungs, in either normal subjects or patients with lung disease, are *P*v, *P*a, and *P*E for the six inert gases, their partition coefficients, and minute ventilation. If mixed venous and arterial blood samples

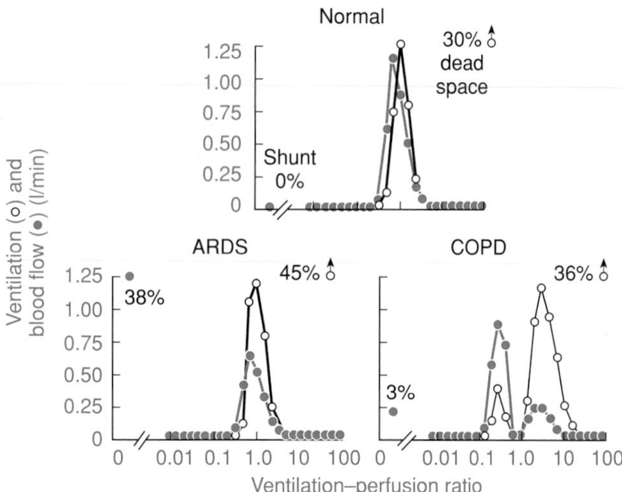

Fig. 1 Representative distributions of $\dot{V}_A/Q$ ratios in a healthy individual breathing spontaneously and in patients needing mechanical ventilation, one with acute respiratory distress syndrome (ARDS) and one with chronic obstructive pulmonary disease (COPD). Note the different amounts of intrapulmonary shunt and $\dot{V}_A/Q$ inequalities in the latter two conditions. Although the main body of each distribution is only slightly broadened in ARDS, shunt is conspicuously increased. In COPD, both distributions are predominantly bimodal but shunt is small. See text for further details.

are available, cardiac output is computed using steady state mass balance equations. The lung is assumed to be composed of a number of homogeneous compartments arranged in parallel, each with constant and continuous ventilation and perfusion. The mathematical model assumes that the ventilation and perfusion distributions have regular contours and do not present sudden irregularities.

Figure 1 shows a typical distribution of $\dot{V}_A/\dot{Q}$ ratio in a young healthy individual at rest in a semirecumbent position and breathing room air. The amounts (distributions) of alveolar ventilation and pulmonary perfusion (*y* axis) are plotted against 50 $\dot{V}_A/\dot{Q}$ ratios, ranging from zero to infinity, on a logarithmic scale (*x* axis). Each data point represents a specific amount of alveolar ventilation or pulmonary blood flow; the lines are drawn to facilitate visual comprehension. Overall blood flow or total alveolar ventilation corresponds to the sum of all data points of their respective distributions. The use of a logarithmic rather than a linear axis for $\dot{V}_A/\dot{Q}$ ratios is based on established practice in the field of pulmonary gas exchange. The two distributions are characteristically symmetric, positioned around a mean $\dot{V}_A/\dot{Q}$ ratio of 1.0, and narrow. In healthy individuals, no blood flow is diverted to the left to the zone of low $\dot{V}_A/\dot{Q}$ ratios (poorly ventilated lung units) nor is ventilation distributed to the right to the zone of high $\dot{V}_A/\dot{Q}$ ratios (incompletely perfused, but still finite, lung units). Intrapulmonary shunt is defined as areas with zero $\dot{V}_A/\dot{Q}$ ratio (less than 0.005), as postpulmonary shunt (bronchial and Thebesian circulations) is insensitive to inert gases. Consequently, intrapulmonary shunt measured by the multiple inert gas elimination technique is lower than the conventional venous admixture ratio (1–2 per cent of cardiac output) when breathing room air as well as the postpulmonary shunt. When breathing 100 per cent O_2 the difference measured by multiple inert gas elimination technique and 100 per cent O_2 is greatly reduced. The normal value of inert physiological dead-space (infinite $\dot{V}_A/\dot{Q}$ ratio (> 100); approximately 30 per cent of overall alveolar ventilation) is also slightly less than that computed using the traditional Bohr formula. While the latter variable includes the dead-space-like effects of all lung units whose alveolar $P\text{CO}_2$ values are less than the arterial $P\text{CO}_2$, the inert gas approach represents only the dead-space-like effects of those alveoli whose $\dot{V}_A/\dot{Q}$ ratios are greater than 100.

The first moment of each distribution, namely the mean $\dot{V}_A/\dot{Q}$ ratio of each distribution, and the second moment (or dispersion) log SD where SD is the standard deviation, are commonly used to calculate the degree of $\dot{V}_A/\dot{Q}$ mismatch. The second moments (square roots) of the pulmonary blood flow and the alveolar ventilation distributions (log SD$\dot{Q}$ and log SD$\dot{V}$ respectively) reflect the variance (standard deviation) of $\dot{V}_A/\dot{Q}$ ratios about the mean. In an ideal homogenous lung, both dispersions should be zero; in practice, in a normal healthy individual the upper 95 per cent confidence limits are 0.60 and 0.65 (Cardús *et al.* 1997). The degree of $\dot{V}_A/\dot{Q}$ inequality can also be expressed as the total percentage of ventilation and perfusion in defined regions of the $\dot{V}_A/\dot{Q}$ spectrum. Other approaches, which are equally valid, have used related parameters such as the root mean square of the difference between retention and excretion over the six inert gases, thus giving an additional quantitative assessment of $\dot{V}_A/\dot{Q}$ mismatch (Gale *et al.* 1985). Finally, the $\dot{V}_A/\dot{Q}$ mismatch can be assessed qualitatively by describing the morphological pattern of each distribution, which can be narrowly or broadly unimodal, or clearly bimodal.

The multiple inert gas elimination technique is such a robust tool that it can also be used to assess the potential presence of O_2 diffusion limitation because, in practice, equilibration of inert gases is not diffusion limited. Accordingly, the technique can be used to compute the $P\text{aO}_2$ predicted from the degree of both $\dot{V}_A/\dot{Q}$ mismatch and shunting and compare it with the measured $P\text{aO}_2$. If the measured $P\text{aO}_2$ is not significantly different from the estimated value, this indicates that other potential mechanisms of hypoxemia, such as O_2 diffusion limitation, increased intrapulmonary parenchymal O_2 consumption, or increased postpulmonary shunt, are absent.

Practical features

The original technique was described using inert gas measurements determined from mixed venous blood (*P*v), arterial blood (*P*a), and mixed expired gas (*P*E). An alternative approach without mixed venous sampling can be used with similar accuracy. In this instance, cardiac output needs to be obtained directly. However, the mixed venous variables can be computed from the mass balance equation:

$$\dot{Q}_T \times \lambda \times P\bar{v} = (\dot{Q}_T \times \lambda \times Pa) + (\dot{V}_E \times PE) \quad (2)$$

$$P\bar{v} = Pa + \frac{\dot{V}_E \times PE}{\lambda \times \dot{Q}_T} \quad (3)$$

where $\dot{V}_E$ is minute ventilation and $\dot{Q}_T$ is cardiac output.

Finally, a third modality of the multiple inert gas elimination technique requiring only mixed expiratory and peripheral venous sampling is available, based on the assumption that inert gases are not metabolized in the peripheral tissues. Virtual equilibration between blood and tissues is achieved and the peripheral venous blood reflects the inert gas concentrations of the inflowing arterial blood. Using this approach, it has been shown that partial pressures of the inert gases in a peripheral vein are 95 per cent of *P*a and sufficiently reproducible to permit substitution of peripheral venous for arterial sampling. However, this approach cannot be used to estimate the first moments of the distributions, the amount of intrapulmonary shunt and dead-space, or the percentage perfusion or ventilation for a given range of $\dot{V}_A/\dot{Q}$ ratios (Wagner *et al.* 1985).

Conclusions

Over the last 20 years the multiple inert gas elimination technique has been widely used in clinical research in pulmonary and critical care medicine. It has been shown to be useful for improving understanding of the mechanisms of abnormal arterial blood gases in disease states and the effects of different interventions, and also in identifying the various intra- and extrapulmonary determinants influencing $P\text{ao}_2$. In patients with acute respiratory distress syndrome and life-threatening bacterial pneumonia the predominant mechanism of low $P\text{ao}_2$ is increased intrapulmonary shunt together with mild to moderate $\dot{V}_A/\dot{Q}$ abnormalities, which are more conspicuous in the latter condition (Fig. 1). In contrast, in patients with respiratory failure secondary to chronic obstructive pulmonary disease or acute severe asthma the major component of abnormal blood gases is $\dot{V}_A/\dot{Q}$ mismatching and intrapulmonary shunt is marginal; in all these conditions inert dead-space is mildly to moderately increased. The role of O_2 diffusion limitation is negligible in all these clinical conditions.

References

Cardús, J., *et al.* (1997). Increase in pulmonary ventilation–perfusion ratio with age in healthy individuals. *American Journal of Respiratory and Critical Care Medicine*, **156**, 648–53.

Evans, J.W. and Wagner, P.D. (1977). Limits on VA/Q distributions from analysis of experimental inert gas elimination. *Journal of Applied Physiology*, **42**, 889–98.

Gale, G.E., Torre-Bueno, J., Moon, R.E., Saltzman, H.A., and Wagner, P.D. (1985). Ventilation–perfusion inequality in normal humans during exercise at sea level and simulated altitude. *Journal of Applied Physiology*, **58**, 978–88.

Wagner, P.D., Saltzman, H.A., and West, J.B. (1974). Measurements of continuous distributions of ventilation–perfusion ratios: theory. *Journal of Applied Physiology*, **36**, 588–99.

Wagner, P.D., Smith, C.M., Davies, N.J.H., McEvoy, R.D., and Gale, G.E. (1985). Estimation of ventilation–perfusion inequality by inert gas elimination without arterial sampling. *Journal of Applied Physiology*, **59**, 376–83.

16.3.4 Respiratory muscle function

M. Aubier

Key messages

- The diaphragm is a unique skeletal muscle that resembles cardiac muscle.
- Respiratory muscle function, and particularly diaphragmatic function, are linked tightly at the level of muscle blood flow.
- Impairment in diaphragmatic contractility reduces peak transdiaphragmatic pressure during maximum static inspiratory effort.
- When the time constant of the exponential phase of relaxation exceeds 75 ms, diaphragmatic dysfunction may be suspected.
- Diaphragmatic muscle fatigue is associated with electromyographic redistribution from high to low frequencies.

Introduction

When an attempt is made to sustain a strong contraction in a human muscle for a period of time, the force inevitably falls. The failure to sustain the force indicates that the muscle has become fatigued. In this mechanistic approach, fatigue can be defined as the inability to maintain or continue to develop the force required to perform a particular task. However, the metabolic changes leading to loss of force presumably started at the beginning of a fatiguing contraction before the development of fatigue as we have defined it. A subject performing a fatiguing task will become progressively more aware of the effort he or she is making. Eventually, the muscle becomes painful (particularly in continuous contractions) and finally the subject is unable to maintain the force. Relaxation rapidly reverses the symptoms.

Fatigue can be local, affecting a single muscle, or generalized, after exhausting the whole body as in exercise. Presumably, the factors limiting performance are different with the two types of fatigue. Almost all studies of skeletal muscle fatigue are acute, with fatigue having developed over the period of an hour or less in a previously normal (fresh) muscle. Virtually nothing is known about chronic fatigue, i.e. a state in which the muscle has not yet recovered from previous effort. Despite our incomplete understanding, chronic fatigue of the inspiratory muscles may be common. As with skeletal muscle fatigue, respiratory muscle fatigue can be defined as an inability of the respiratory muscles to continue to develop sufficient respiratory pressure swings to maintain normal alveolar ventilation. In contrast with skeletal muscle fatigue, respiratory muscle fatigue is characterized by two unique features.

1. Because the respiratory muscles are essential for life, fatigue is life-threatening.
2. The respiratory muscles, particularly the diaphragm, are different from other skeletal muscles. Indeed, the diaphragm must contract phasically during life and thus functionally appears to have more similarities with the myocardium than with limb muscles.

Elements of respiratory muscle contraction

Excitation–contraction coupling

In view of the functional similarity between the heart and the diaphragm, there should be some similarities in their cellular contraction mechanisms, particularly their excitation–contraction coupling processes. This hypothesis has been supported by recent experimental evidence. Although the central role of calcium in the excitation–contraction coupling of skeletal muscle is well established, the cellular mechanisms by which calcium is made available to the contractile proteins during excitation to elicit contraction of the muscle remain unclear. It is generally believed that the activation of the contractile elements is essentially independent of extracellular calcium; the cyclic nature of Ca^{2+} fluxes in skeletal muscle during excitation–contraction coupling are regarded as strictly intracellular. In cardiac muscle, however, the passage of calcium across the cell membrane is of major importance; without it, there would be insufficient calcium from the sarcoplasmic reticulum to initiate contraction.

These classical notions have been challenged by the observation that isolated diaphragmatic fibers obtained from rats behave similarly to single papillary muscle fibers from the same animal. Indeed, a striking similarity has been observed between the contractile responses of diaphragmatic and papillary muscle fibers to extracellular calcium deprivation; in both cases there was complete abolition of twitch. However, no twitch abolition was observed with single soleus or extensor digitorum longus fibers (slow and fast peripheral skeletal muscles) when they were exposed to a calcium-free medium for 3 h. These data have been confirmed *in vivo* in hypocalcemic dogs (Aubier *et al.* 1985). In these animals, a marked depression in diaphragmatic contractility was noted with hypocalcemia, whereas the sartorius, a peripheral skeletal muscle with a similar histochemical profile, was not affected. Therefore it appears that the diaphragm is a unique skeletal muscle that resembles cardiac muscle.

These results provide new insight into the mechanisms of diaphragmatic contraction, emphasizing the peculiarities of the diaphragm with respect to the other striated skeletal muscles. They also have important pharmacological implications. Owing to the similarities between the reliance of diaphragmatic and cardiac muscle fibers on extracellular calcium, it is conceivable that drugs or agents that exert a direct inotropic positive or negative effect on the myocardium may have a similar action on the diaphragm.

Energy supply

In addition to the fact that respiratory muscle dysfunction may result from impairment of calcium movement at the respiratory muscle cell level, respiratory muscle dysfunction also occurs when the supply of nutrients to the respiratory muscles is insufficient to meet metabolic demand. During breathing, the oxygen requirements of the respiratory muscles, particularly the diaphragm, are met primarily by increased diaphragmatic blood flow, which is related to cardiac output during unobstructed hyperventilation, hypoxia, and exercise. It is possible that, when blood flow to the muscle is insufficient, lack of oxygen leads to a shift in the anaerobic glycolytic pathways of high-energy phosphate generation. This shift would increase the acid accumulation associated with a given level of ATP turnover, impairing sarcoplasmic reticulum function and reducing the force-generating capability of the diaphragm.

Blood flow may play a role in 'washing out' the by-products of cellular metabolism. Thus it is possible that increased diaphragmatic blood flow may cause an increase in the washout of toxic by-products of cellular metabolism concomitant with an improvement in diaphragmatic energetics. Whatever the mechanism, because respiratory muscle function, and particularly diaphragmatic function, are tightly linked to the level of muscle blood flow, it may be possible to improve diaphragmatic function by administering pharmacological agents that augment diaphragmatic blood flow in patients in whom such flow is diminished or not high enough to meet the demands of breathing.

Detection of respiratory muscle fatigue

Diaphragmatic fatigue can be detected in various ways (Dureuil and Aubier 1988). Assessment of fatigue for the diaphragm is more difficult than for other skeletal muscles because the tension generated cannot be measured directly; only the mechanical transformation of that tension into pressure can be measured. The transdiaphragmatic pressure (P_{di}), defined as the difference between the gastric pressure and the esophageal pressure, closely reflects the tension produced by the contracting muscle. The clinical manifestations of inspiratory muscle fatigue, which are a paradoxical inward abdominal motion during inspiration (abdominal paradox) and an alteration between abdominal and rib cage breathing (respiratory alternans), have been described by Cohen *et al.* (1982).

These investigators studied 12 patients intubated and ventilated for acute respiratory failure who exhibited difficulties during discontinuation of artificial ventilation. Ventilatory support was discontinued and the investigators searched for clinical signs of respiratory muscle fatigue that could be correlated with electromyographic signs of fatigue. Electromyographic evidence of inspiratory muscle fatigue was described in six patients with the following sequence of events leading to respiratory acidemia. Electromyographic fatigue followed or was accompanied by an increased respiratory rate, which was, in turn, followed by abnormalities of thoracoabdominal mechanics (respiratory alternans, abdominal paradox). These clinical manifestations were followed by an increase in arterial carbon dioxide levels ($Pa\text{CO}_2$). It was concluded that these abnormal respiratory movements may be a reliable index of inspiratory muscle fatigue. Apart from diaphragmatic fatigue, abdominal paradox has been described only in diaphragmatic paralysis. Tobin *et al.* (1987) recently studied normal subjects breathing against severe resistive loads and concluded that rib cage-abdominal paradox is due predominantly to an increase in respiratory load rather than to respiratory muscle fatigue, because the paradox was observed in this study for a level of exercise that can be sustained indefinitely.

Diaphragmatic function can he assessed by recording P_{di} and analyzing two indices: peak P_{di} and the rate of relaxation of P_{di}. Impairment in diaphragmatic contractility reduces peak P_{di} during maximum static inspiratory effort (P_{di}(max)) or during maximum sniff (P_{di}(sniff)). However, the values of peak P_{di} depend not only on diaphragmatic contractility, but also on the subject's ability and motivation to co-operate. Furthermore, comparison of peak P_{di} obtained at two different times is possible only if the abdominal compliance and muscle length and geometry (lung volumes) remain unchanged.

Changes in diaphragmatic contractility are also accompanied by changes in muscle relaxation time (fatigued muscles relax more slowly than fresh muscles). This change in diaphragmatic relaxation time, which appears before the decrease in muscular tension, can be quantified by measuring the maximum P_{di} relaxation rate or time constant (T) of the monoexponential phase of relaxation. The latter index is interesting because it is not influenced by the amplitude of the peak P_{di}, but only by lung volume (T decreases when lung volume increases). When T exceeds 75 ms, diaphragmatic dysfunction may be suspected.

Changes in the frequency content of the diaphragmatic electromyogram during spontaneous breathing also provide information concerning diaphragmatic fatigue. Diaphragmatic muscle fatigue is associated with electromyographic redistribution from high to low frequencies. The technique is to ascertain the ratio of power in a high-frequency band to that in a low-frequency band (high-to-low ratio) of the diaphragmatic electromyogram. The high-to-low ratio of the diaphragm decreases with fatigue with a time course and extent that closely matches the prolongation of relaxation. As T becomes longer, modification of the high-to-low ratio precedes fatigue. It is then possible to detect mechanical impairment of the diaphragm before the muscle fails as a pressure generator. The other ways of assessing diaphragmatic function, i.e. the force–frequency curve and bilateral phrenic stimulation with single pulses, are difficult to use routinely in the intensive care unit (Aubier *et al.* 1981).

References

Aubier, M., *et al.* (1981). Detection of diaphragmatic fatigue in man by phrenic stimulation. *Journal of Applied Physiology*, **50**, 538–44.

Aubier, M., *et al.* (1985). Effects of hypocalcemia on diaphragmatic strength generation. *Journal of Applied Physiology*, **58**, 2053–61.

Cohen, C.A., *et al.* (1982). Clinical manifestations of inspiratory muscle fatigue. *American Journal of Medicine*, **73**, 308–16.

Dureuil, B. and Aubier, M. (1988). Assessment of diaphragmatic function in the intensive care unit. *Intensive Care Medicine*, **14**, 83–5.

Tobin, M.J., *et al.* (1987). Does rib cage abdominal paradox signify respiratory muscle fatigue. *Journal of Applied Physiology*, **63**, 851–60.

16.3.5 Respiratory system compliance and resistance

J. C. Goldstone

Key messages

- The term compliance can be quantified as the slope of the volume–pressure curve (volume change per unit change in pressure).
- If volume is delivered during inspiration by a constant (square-wave) flow, a tracing of pressure against time can be used to measure compliance.
- Resistance to gas flow occurs as a result of frictional forces within the airway and is measured in terms of a pressure drop per unit flow rate.
- Airways resistance can be estimated simply from the pressure–time curves obtained from the ventilated patient. The pressure difference related to resistance is often calculated at the end of inspiration during an end-inspiratory pause.

Introduction

When positive pressure is used to inflate the lungs, the forces that must be overcome are primarily the elasticity of the lungs and chest wall and airflow resistance. From the resting expiratory point (functional residual capacity (FRC)), the lungs (whose normal tendency is to collapse) are held open by the chest wall (whose tendency is to expand); at this point there is no gas flow and the system is in balance. As the respiratory system expands and gas enters the chest, the movement of the system can be described mathematically by the equation of motion

$$P_{ao} = \frac{1}{C_{rs}}V + R_{rs}\frac{dV}{dt} + I_{rs}\frac{d^2V}{dt^2}$$

where the pressure at the airway opening (P_{ao}) is related firstly to the compliance (C_{rs}) of the system, secondly to the resistance to airflow (R_{rs}), and thirdly to the inertia of the gas (I_{rs}). Inertia is a small contributor to the force required to generate lung inflation and is often omitted.

Compliance

In a ventilated patient, the relationship between the volume of the respiratory system and the distending pressure can be described easily. A large calibrated syringe can be used to change volume in small steps and the pressure in the airway is measured when the system is in equilibrium at the end of each inflation (Beydon *et al.* 1991). Several pressure measurements are made as lung volume gradually changes. With stiff lungs, little volume change will occur compared with normal lungs at the same distending pressure. This is obvious when pressure and volume are displayed graphically, as the slope of the volume–pressure curve is steeper in the stiffer system.

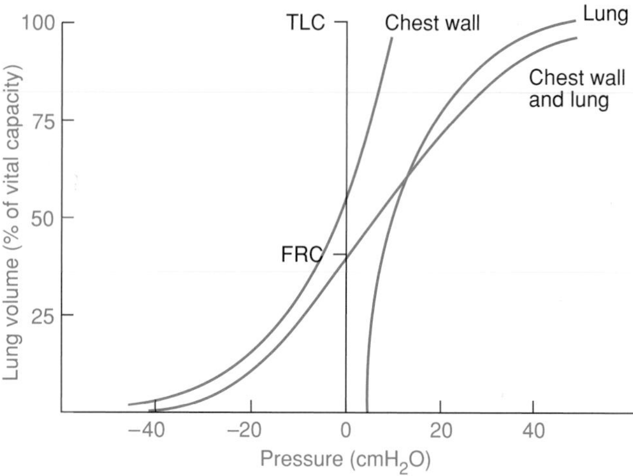

Fig. 1 The pressure–volume curve of the lung, the chest wall, and the lung and chest wall combined. As the lung volume increases, more pressure is required because the lung reaches the limit of distension. When the recoil of the lung equals the outward force of the chest wall, the system is in balance and this is the resting position of the respiratory system (FRC).

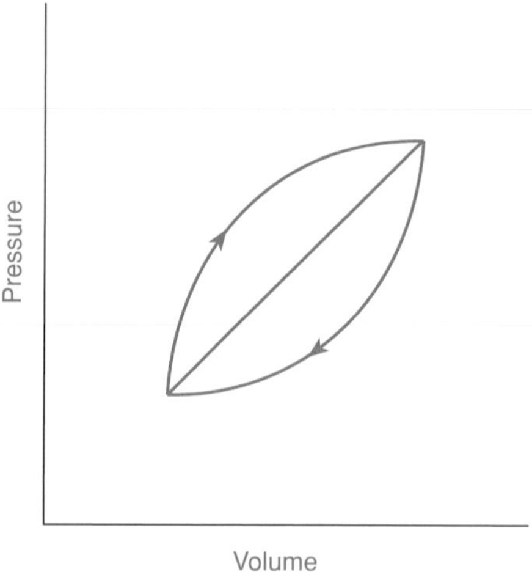

Fig. 2 If the intrapleural pressure is plotted against volume simultaneously, a hysteresis loop is constructed. The degree of hysteresis is related to airways resistance. The gradient of the line drawn between the points of zero flow is the dynamic compliance of the respiratory system.

The combined compliance curve of the respiratory system is comprised of a chest wall component and a lung component (Fig. 1). In a healthy subject, the least pressure required to inflate the lung is found when the volume is in the mid-range. At total lung capacity (TLC), little change in volume occurs for unit change in pressure.

Compliance can be quantified as the slope of the volume–pressure curve (volume change per unit change in pressure). When measured from the FRC, the slope of both the lung and chest wall volume–pressure curves are equal and normal values are around 100 ml/cmH_2O. If the whole system is inflated, the pressure required would be doubled and the total compliance is 50 ml/cmH_2O.

The measurement of compliance is dependent on the lung volume from which the measurement is made. The term specific compliance allows for this by referencing it to lung volume. When lung volume is at or around FRC, the volume–pressure curve is linear and therefore compliance is constant. Changes in compliance can occur when the lung parenchyma is damaged; for example, destruction of the lung parenchyma (emphysema) reduces the 'springiness' of the lung, thus increasing compliance, whereas fibrosis has the opposite effect and reduces compliance.

'Static' compliance (C_{st}) is that measured when no flow occurs and the respiratory system is stationary. In the laboratory total static compliance is measured with the aid of a shutter and a mouthpiece. Pressure is measured at the mouth and the subject relaxes at lung volumes above and below FRC. If the subject is not experienced, the measured pressure can be affected by persistent ventilation or inadvertent contraction of the respiratory muscles.

'Dynamic' compliance (C_{dyn}) is that measured while the system is in motion, i.e. during respiration. In this circumstance the pressure between the airway and the pleural space is measured (transpulmonary pressure). Pleural pressure is transmitted to other mediastinal structures, such as the esophagus, and can be measured with a thin-walled air-filled balloon catheter passed transnasally. The pressure detected by the esophageal balloon is referenced to the pressure at the mouth using a differential pressure transducer and plotted simultaneously with lung volume (Fig. 2).

With a constant distending pressure, the time taken for pressure equilibration will be related to the time constant of the system and the lungs will tend to fill exponentially. Dynamic compliance is measured between points of zero flow from the slope of the transpleural pressure–volume curve (Fig. 2). It should be noted that the hysteresis does not itself contribute to the slope of the curve. Rather, as the time taken for the lung to inflate increases, some parts of the lung may not fill adequately. Under these circumstances the dynamic compliance will be reduced.

Compliance is most easily measured when the patient is fully ventilated, as the effect of continuing respiratory muscle activity is absent. The observation that muscle activity can either increase or decrease intrathoracic pressure underlines the artifacts that can occur during any spontaneous breathing. This is particularly relevant when measurements are automated (Sydow *et al.* 1991).

Resistance

Resistance to gas flow (R) occurs as a result of frictional forces within the airway and is measured as a pressure drop per unit flow rate (cmH_2O/l/s). Resistors in series are additive and a drop in pressure occurs between mouth and the branching airways, across the lung tissue itself, and across the chest wall when gas is moving at a particular flow rate. If it is possible to measure the pressure drop between all these elements during gas flow, then each component of resistance can be isolated. In practice, however, airways resistance (R_{aw}) is commonly measured. R_{aw} is the resistance of the branching airways, and the pressure drop is measured between the airway (P_{ao}) and the alveolus (P_A):

$$R_{aw} = \frac{P_{ao} - P_A}{dV/dt}.$$

Although the peripheral airways are narrow, their combined cross-sectional area results in a small contribution to airways resistance. Most of the pressure drop across the airways occurs at the level of the small bronchi.

More pressure is required for a given flow rate if the flow is turbulent. Although it is possible for airflow to be laminar within the large airways, the internal surface must be smooth and the length of the airway must be more than 30 times the radius. As many of the upper airways are irregular, short, or branched, a mixture of turbulent and laminar airflow results. For the same reason airflow within endotracheal tubes is rarely laminar, particularly when the tube is kinked or coated with secretions.

Resistance changes with lung volume. As the lung deflates, so the caliber of the branching airways decreases. If the measurement of resistance is volume dependent, standardization is required before interpretation of resistance data. This problem can be solved if the hyperbolic volume:resistance relationship is converted to a straight line by using the reciprocal of resistance, i.e. conductance (Sykes 1995).

Measurements in intubated patients

If the patient is fully relaxed measurements of pressure at the airway can be made easily. Flow is commonly measured using a pneumotachograph which consists of a resistor and equipment to measure the small pressure drop across it. The resistor can have a number of different designs and is placed in the gas flow close to the subject's or patient's airway. By introducing a linear resistance, the pressure drop across the resistance is proportional to the flow rate. If the flow signal derived from the pneumotachograph is integrated, volume can be obtained simultaneously. Measurement errors are easily introduced; for example, if the flow past the pneumotachograph is not linear, at a given flow rate the pressure drop will be higher and a greater flow rate will be recorded. This circumstance can occur if dust or secretions occlude the resistor. Therefore it is important to calibrate the flow-measuring equipment accurately before and after use.

Compliance measurements

If volume is delivered during inspiration by a constant (square-wave) flow, a tracing of pressure against time can be used to measure compliance. As no account is taken of the effect of airways resistance on peak airway pressure, or of the effect of intrinsic positive end-expiratory pressure (PEEP), this measurement can be thought of as 'effective' compliance. As flow is constant, the same volume is added for all time periods and the time access can be regarded as a linear analog of volume. The tracing can be regarded as a pressure–volume tracing (Fig. 3).

At the onset of inspiration no flow occurs, and this is the starting point for pressure measurement. The pressure is measured again at the end of the square-wave flow. The slope of the line is the 'effective dynamic compliance' (as the system is in motion). Care should be taken to ensure that the upstroke is linear when resistance or intrinsic PEEP is high.

The 'effective dynamic compliance' can be verified from the airway pressure–time tracings. If a pause is inserted at the end of inspiration, airway pressure declines and, if sufficient time is allowed, the pressure in the airway will reflect the compliance of the lung. As the system is now stationary, this value is related to the total static compliance of the respiratory system. The slopes of alveolar pressure and 'effective dynamic compliance' often match (Fig. 3). Errors can occur if the

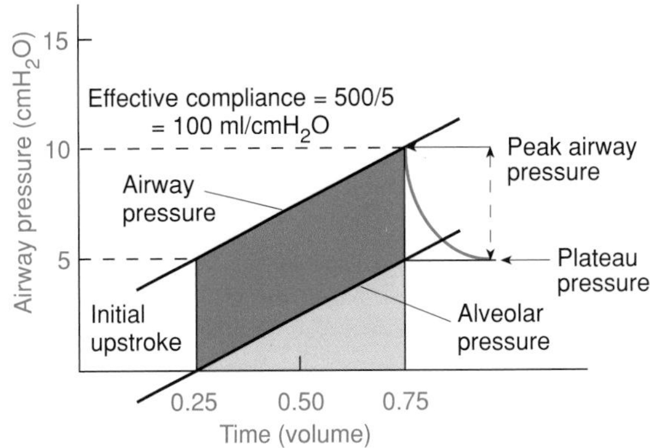

Fig. 3 Airway pressure of a mechanical breath under conditions of constant flow. With an inspiratory pause and occlusion, airway pressure declines to a plateau. If no respiratory effort occurs, this is the static recoil pressure of the lung and chest wall. The gradient of the airway pressure–time curve is termed the effective dynamic compliance. It is offset from alveolar pressure by the effect of airflow resistance. Ideally, the gradients of airway and alveolar pressure are parallel. This can be checked when the peak plateau pressure equals the initial upstroke of the curve.

inspiratory pause is brief relative to the time constants of the lung. When expiratory time is delayed, airway pressure will slowly equilibrate with alveolar pressure and care must be taken to establish a plateau.

Many ventilators compute compliance automatically. Although there is a danger that such systems can be misinterpreted, good agreement has been found when compliance is measured automatically (Sydow *et al.* 1991).

The pressure–volume curve of the lung and chest wall can be constructed formally by measuring the inflation pressure when the lung is inflated with small aliquots (volume 200 ml) from a large calibrated syringe (1–2000 ml) (Matamis *et al.* 1984). Pressure is measured during inflation and deflation and several slopes can be

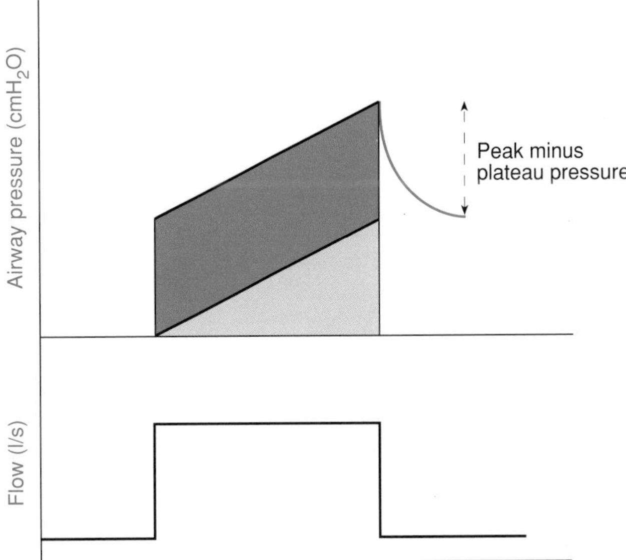

Fig. 4 Airway resistance can be measured from airway pressure curves; resistance is equal to pressure difference divided by flow. In this diagram, pressure difference is estimated from the peak plateau pressure, and flow can be measured directly from the ventilator.

determined. This technique has been used to demonstrate the changes in compliance in acute respiratory distress syndrome, and has been applied to determine the inspiratory inflection point indicative of the point where the airways start to open.

Although dynamic compliance is best measured as the transpulmonary pressure ($P_{aw} - P_{es}$), the technique requires an esophageal balloon catheter. When the patient is relaxed and supine, the weight of the mediastinal contents creates an offset increase in pressure. The absolute measurement of C_{dyn} would also be affected by the presence of esophageal contractions which can be detected as an intermittent positive-pressure deflection.

Resistance measurement

It is important that the site of pressure measurement and the technique used to measure resistance are stated, as differences can occur solely on the basis of the technique. The basis of measurement is that resistance is proportional to the pressure difference and inversely related to the flow rate. If the pressure drop between the airway and alveolus is known, airway resistance (R_{aw}) can be calculated.

Airway resistance can be estimated simply from the pressure–time curves obtained from the ventilated patient. As previously described, the airway pressure tracing during constant-flow ventilation has an initial rapid upstroke. At the start of inspiration little gas has entered the chest and this pressure rise is due to airway resistance. If the inspiratory flow rate does not change (the measurements should be made during a square-wave constant-flow breath) and airway resistance is constant, there will be a constant pressure difference attributable to resistance during inspiration (Fig. 4). The pressure difference related to resistance is often calculated at the end of inspiration during an end-inspiratory pause (Suter *et al.* 1975).

A graph of volume, transthoracic pressure, and flow demonstrates the problems of obtaining a measurement of resistance in a subject. As resistance is related to lung volume, it is clearly important to ensure that this variable, which is changing, is standardized. Furthermore, flow changes during the respiratory cycle, and if we merely divided pressure by flow a wide range of values would be obtained. Which would represent the true measurement of resistance?

Several methods have been used to calculate resistance. To avoid the problem of standardization of lung volume, this can be fixed during inspiration and expiration and the pressure and flow differences between these two states used to measure resistance. An alternative approach is to use a sinusoidal oscillator which delivers a rapidly changing small-volume waveform to the respiratory system. Yet more techniques use other ways of averaging the lung compliance and measure resistance at a particular instantaneous flow rate. When these techniques are compared, little difference is found in absolute terms. Most are also able to detect stepwise changes in resistance.

References

Beydon, L., Lemaire, F., and Jonson, B. (1991). Lung mechanics in ARDS: compliance and pressure volume curves. In *Adult respiratory distress syndrome* (ed. W.M. Zapol and F. Lemaire), pp. 139–61. Marcel Dekker, New York.

Matamis, D., Lemaire, F., Harf, A., Brun-Buisson, C., Ansquer, J.C., and Atlan, G. (1984). Total respiratory pressure–volume curves in the adult respiratory distress syndrome. *Chest*, **86**, 58–66.

Suter, P.M., Fairley, H.B., and Isenberg, M.D. (1975). Optimum end-expiratory pressure in patients with acute pulmonary failure. *New England Journal of Medicine*, **292**, 284–9.

Sydow, M., *et al.* (1991). Improved determination of static compliance by automated single volume steps in ventilated patients. *Intensive Care Medicine*, **17**, 108–14.

Sykes, K. (1995). *Respiratory support*. BMJ, London.

16.4 Imaging

16.4.1 Imaging the respiratory system

J. A. Worrell

Key messages

- All the lung fields should be included. The depth of inspiration should be adequate. The patient should be centered and not significantly rotated.
- Exposure should allow visualization of the peripheral vasculature yet be adequate to visualize the intervertebral disk spaces through the heart.
- All life-support equipment should be accounted for and in proper position, particularly endotracheal tubes, central catheters, and nasogastric tubes.
- The mediastinum should be assessed for heart size, evidence of hematoma, vascular volume status, barotrauma, and patent central airways.
- The lung fields should be examined for evidence of pneumothorax (including tension pneumothorax), aspiration, contusion, atelectasis and pneumonia, effusion, edema, and pulmonary vascular volume status.
- The vertebral column, paraspinal lines, ribs, and shoulder girdle should be searched for evidence of fracture or hematoma. The soft tissues give evidence of local injury as well as systemic interstitial volume. Evidence of free intraperitoneal or retroperitoneal air and hepatosplenomegaly should be sought.

Introduction

Patients in the intensive care unit (ICU) require complex life-support devices and monitors which add to the difficulty of radiographic imaging. At the same time, such imaging is often required to verify proper positioning of these devices as well as to evaluate the results of therapy. Diagnostic imaging in these patients poses challenges that are difficult but not uncommon, with bedside radiography accounting for nearly half the chest radiographs in many acute care hospitals.

The technology of bedside radiographic imaging

The supine anteroposterior bedside chest radiograph, performed using conventional film-screen systems, has been the standard examination of the thorax in the critical care patient. The details of these examinations are beyond the scope of this chapter, but the importance of the delivery of consistently high quality examinations to the clinical management of these patients cannot be overstated.

Film digitization and storage phosphor computed radiography are now challenging conventional film as the gold standard in bedside radiography. The advantages of transmitting images directly to the ICU as well as the increased reliability of image library management are enabling these systems, despite their cost, to be strong challengers to conventional film.

However, digital systems also bring practical challenges. The video monitors commonly available in the ICU are adequate for clinical use, such as monitoring life-support apparatus, but the ICU team should be aware of the limits of these systems for detecting pneumothorax, interstitial lung disease, etc. Clinicians should also realize that video monitors are not as bright as conventional light boxes and that excess ambient light can seriously degrade the images.

The redundancy, reliability, and capacity of any digital system must be assured before full implementation is accomplished. The system should be capable of handling very large amounts of data, remembering that a single chest radiograph may require up to 10 megabytes of data, and that up to a week's worth of images should be immediately available for review (MacMahon and Giger 1996).

Accuracy and efficacy of the chest radiograph

The accuracy and efficacy of the bedside radiograph in assessing cardiopulmonary abnormalities and monitoring life-support apparatus continue to be debated. Routine radiographs in patients with pulmonary disease or hemodynamically unstable cardiac disease have revealed that 57 per cent of these patients had unsuspected abnormalities in contrast with only 3 per cent of hemodynamically stable cardiac patients or those with miscellaneous conditions in the medical ICU (Strain *et al.* 1985). The efficacy of routine radiography has also been studied in the surgical ICU. One study found that only 1.3 per cent of life-support devices were significantly malpositioned, and the rate of cardiopulmonary findings requiring immediate action was less than 1 per cent (Silverstein *et al.* 1993).

Atelectasis and pneumonia

Atelectasis is a very common condition seen in the ICU, with an appearance varying from normal to patchy areas of linear opacity to segmental and lobar collapse. When air bronchograms are absent, mucous plugging or other endobronchial lesions should be considered (Fig. 1). Otherwise, other etiologies, including pneumonia, aspiration, and contusion, should be considered. The distinction between atelectasis and pneumonia can be difficult, but an associated pneumonia should be considered after 3 to 4 days in the postoperative patient, and even sooner in the immune-compromised patient.

The diagnosis of infectious pneumonia is difficult and requires the correlation of clinical and laboratory findings with the chest radiograph appearance, often of a segmental or larger pattern of parenchymal opacification with air bronchograms present. The diagnosis of aspiration pneumonia, unless observed, is often radiographic based on the appearance of a new localized parenchymal opacification that is not due to atelectasis alone (Silverstein *et al.* 1993) (Fig. 2).

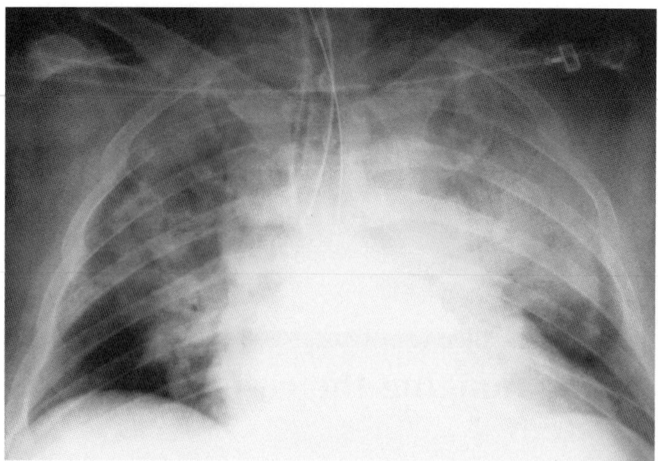

Fig. 2 Aspiration pneumonia. The patient is an alcoholic with witnessed aspiration and bilateral parenchymal opacification in the middle and upper lung fields. Aspiration into the superior segments of the lower lobes and the apical and posterior segments of the upper lobes is common in the supine patient. Notice also the endotracheal tube near the right mainstem bronchial orifice.

Edema

Pulmonary edema can be broadly categorized into hydrostatic (cardiogenic) edema and injury (non-cardiogenic) edema. While there is controversy over how accurately the plain chest radiograph can distinguish types of edema, the correct type of underlying pattern is often recognizable on the bedside radiograph, and valuable contributions to the patient's fluid and respiratory management can be made. The plain film also helps to differentiate pulmonary interstitial hydrostatic edema, systemic interstitial edema, central circulating vascular volume, and pulmonary vascular volume (Milne and Pistolesi 1993).

Hydrostatic edema

Hydrostatic interstitial pulmonary edema is manifest by perihilar or lower lobe blurring of vascular detail and thickening of the bronchial walls, and eventually septal lines, pleural effusions, and air bronchograms as alveolar flooding with edema fluid occurs. Often hydrostatic edema is accompanied by an increase in the pulmonary vascular volume, as manifest by increased size and peripheral appearance of pulmonary arteries and veins. The central circulating vascular volume may be estimated by the vascular pedicle width and the prominence of the right heart, azygos vein, and superior vena caval shadows, and is often increased in hydrostatic edema (Fig. 3). The appearance of the chest radiograph characteristically precedes and recovery afterwards postdates the clinical manifestations of hydrostatic edema. Changes in systemic interstitial fluid may be monitored by comparing the width of the soft tissues of the chest wall on serial radiographs.

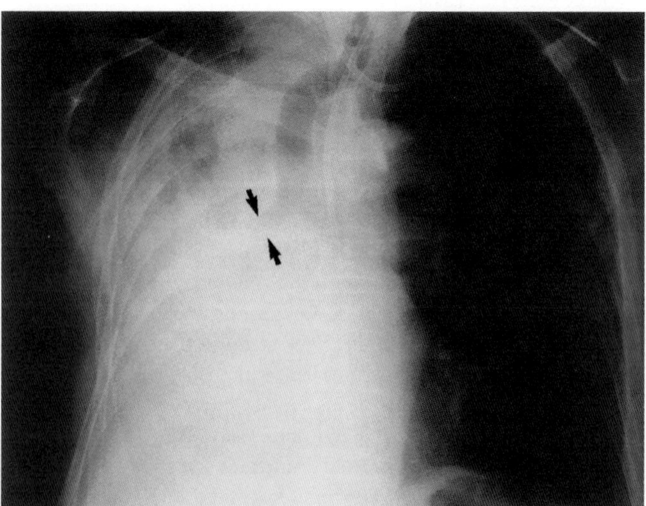

Fig. 1 Mucous plug. The right mainstem bronchus is occluded (arrows), the trachea is shifted to the right, and there is almost complete atelectasis of the right lung. (Reproduced with permission from Worrell (1994).)

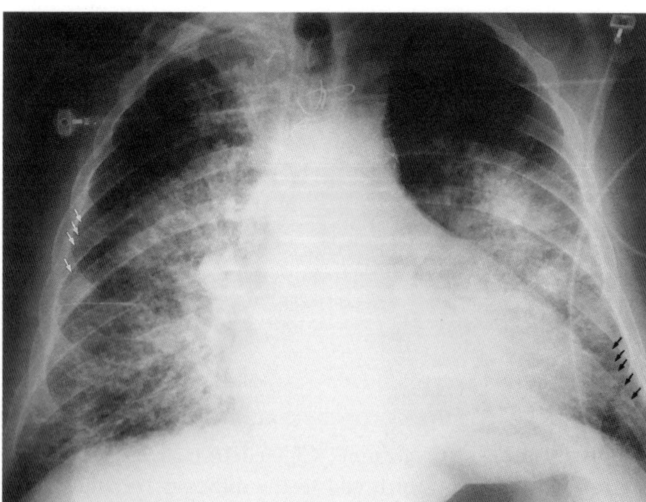

Fig. 3 Hydrostatic pulmonary edema. This patient had renal failure and central vascular volume overload. The heart and vascular pedicle are increased, with a predominantly perihilar pattern of parenchymal opacification, bronchial cuffing, and vascular indistinctness. Notice the prominent septal (Kerley) lines (white arrowheads on the right and black arrowheads on the left).

Injury edema

The radiographic findings of injury edema are a patchy, often peripheral, pattern of pulmonary opacification, with air bronchograms present and a normal central vascular volume. The radiographic appearance of injury edema often lags the clinical manifestations, and diagnosis of the acute respiratory distress syndrome requires correlation of the clinical findings of severe unrelenting hypoxemia which is refractory to oxygen therapy, with the radiographic appearance of injury edema in an appropriate clinical setting (Fig. 4). Accurate interpretation of serial radiographs in the patient with acute respiratory distress syndrome requires a knowledge of the clinical and particularly the ventilatory status in order to differentiate changes in pulmonary aeration due to altered ventilatory support from changes in the underlying state of the injured lung (Milne and Pistolesi 1993).

Pulmonary embolism and infarction

Pulmonary embolic phenomena are frequent complications of prolonged bed rest and are often a diagnostic conundrum. The chest radiograph is neither sensitive nor specific in the diagnosis of pulmonary embolism, but Doppler ultrasound may be used in evaluation of thrombosis of the veins of the lower extremities. Ventilation–perfusion nuclear medicine lung scans are useful in the evaluation of possible pulmonary embolism, and, where available, perfusion scans can be obtained at the bedside. CT also can detect proximal pulmonary emboli and may prove useful in some high-risk patients. However, conventional selective pulmonary angiography remains the gold standard in the diagnosis of pulmonary embolism.

Pleural effusions

Pleural effusions are another common condition in ICU patients, particularly after upper abdominal surgery. There is some disagreement on the effectiveness of the supine chest radiograph in the diagnosis of pleural effusions, but those larger than about 250 ml are reliably diagnosed. Bedside sonography or lateral decubitus views may be performed to confirm suspected effusions. Ultrasound may also be used to guide thoracentesis of small or loculated effusions.

Abnormal intrathoracic air collections

Many patients in the surgical ICU receive ventilatory support that includes positive end-expiratory pressure and are at increased risk of developing barotrauma. Abnormal intrathoracic air may also be the result of penetrating or non-penetrating injury, asthma, or illicit drug use (particularly crack cocaine and marijuana), as well as iatrogenic etiologies such as esophageal dilatation or neck surgery. The basic radiographic sign of a pneumothorax is the thin white line of visceral pleura as it is separated from the parietal pleura which is often seen in the lung apices. In the supine patient, the abnormal air may collect in the lateral costophrenic sulci (giving a deep sulcus sign) or anteromedially, causing abnormal lucencies near the cardiophrenic angle or superior mediastinum. The size of the pneumothorax is only poorly estimated by the supine chest radiograph, and its significance should be interpreted in the light of the clinical circumstances, particularly when a tension pneumothorax is considered.

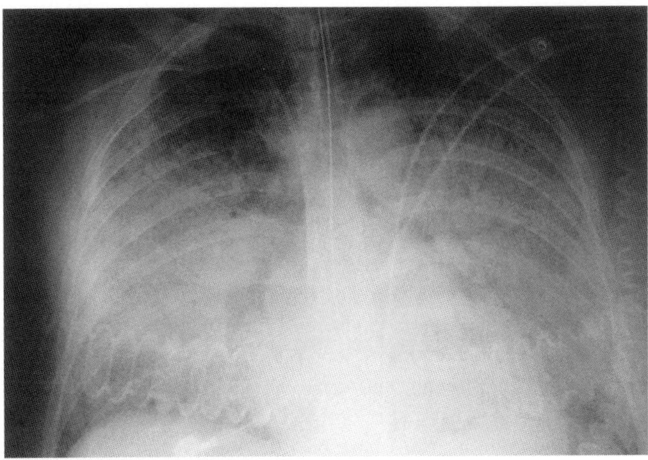

Fig. 4 Acute respiratory distress syndrome. This patient experienced acute respiratory failure following a normal vaginal delivery, complicated by a subsequent ruptured appendix. Notice the normal heart and vascular pedicle size, prominent air bronchograms, and extensive bilateral parenchymal opacification with relative sparing of the apices and costophrenic angles.

The diagnosis of pneumomediastinum depends on the observation of abnormal air collections in the mediastinum, often outlining the inferior mediastinal border (the continuous diaphragm sign) or appearing as a thin white line of mediastinal pleura outlining the lateral borders of the mediastinum, and frequently associated with subcutaneous emphysema in the neck. Pneumopericardium may be diagnosed when there is definite visualization of intrapericardiac structures such as the roots of the great vessels. Although rarely seen, intracardiac air may be identified when intracardiac structures such as the trabeculae of the right ventricle are recognized.

Endotracheal and tracheostomy tubes

A search for proper positioning of all life-support apparatus should be a requisite part of the assessment of all chest radiographs. Proper positioning of the endotracheal tube within the trachea and approximately 5 cm from the carina should be confirmed with routine postintubation radiographs (Henschke *et al.* 1996) (Fig. 5). The cuff of the endotracheal tube should also be observed for overdistention, as evidenced by bulging instead of parallel walls. The tip of the tracheostomy tube should lie between half and two-thirds of the distance between the stoma and the carina. Chest radiographs should be obtained immediately after tracheostomy tube placement to search for complications such as malpositioning, pneumothorax, pneumomediastinum, or atelectasis from aspirated mucus or blood clot (Henschke *et al.* 1996).

Central venous and pulmonary artery catheters

Central venous catheters are often inserted to monitor ICU patients. Ideal positioning should place the tip between the right atrium and the last venous valve, approximately 2.5 cm from the confluence of the internal jugular and subclavian veins, between the medial portion of the first rib and the level of the carina (Henschke *et al.* 1996). Peripherally inserted central catheters are becoming more popular for long-term therapy but are thin and often difficult to visualize. These catheters can almost always be visualized with properly exposed and positioned bedside radiographs and only very rarely require oblique positioning or contrast injection to confirm their placement. Pulmonary artery (Swan–Ganz) catheters are also commonly placed in ICU patients. Typical positioning is in the distal main pulmonary arteries on either side, more commonly on the right (Henschke *et al.* 1996). While ideal radiographic positioning is near the bifurcation of the right pulmonary artery, adequacy of placement should also be correlated with the waveforms obtained. Complications of pulmonary artery catheter placement include arrhythmias, pulmonary infarction, pulmonary artery perforation, knotting of the catheter, and sepsis. Radiographs should be obtained after pulmonary artery catheter placement, but once positioning and complications have been excluded, routine follow-up films need not be obtained (Henschke *et al.* 1996).

Nasogastric tubes

Nasogastric tubes are frequently placed in ICU patients for gastric decompression. Some nasogastric tubes have side-ports to facilitate suction, and these should be located below the gastroesophageal junction. Small-bore feeding tubes help to meet the patient's nutritional requirements, with the tip ideally placed beyond the pylorus in the duodenum. However, the weighted tip tends to remain in the fundus in the supine patient. Positioning the patient in the right lateral decubitus position often helps advance the tube tip into the duodenum, if his or her condition will allow. Radiographs should be obtained before use, particularly in the unresponsive patient, to confirm proper positioning.

Sengstaken–Blakemore or Minnesota tubes are placed in patients with bleeding esophageal varices. Radiographs to confirm the proper placement of these tubes are particularly important prior to inflation of the esophageal and gastric balloons because of the potential for injury when these large-capacity balloons are inflated. Complications include pneumomediastinum, subcutaneous emphysema, pleural effusion, pneumothorax, pneumonia, and aspiration (Henschke *et al.* 1996).

Thoracostomy tubes

Thoracostomy tubes are used to drain fluid, empyemas, blood, and pneumothoraces. Tubes placed for pneumothorax drainage in supine patients work best when placed anterosuperiorly, whereas fluid collections drain best when placed posteroinferiorly. Small-bore chest tubes may be placed with radiographic guidance and are increasingly used to drain small or loculated pneumothoraces. Complications of thoracostomy tubes include injury to the mediastinal vessels or nerves, subcutaneous emphysema (particularly if the side-port is not within the rib cage), contusion of the underlying lung, and malfunction of the tube due to malposition either in a fissure or with the tube in an extrapleural location. After tube removal, a tube track is commonly seen at the site of the placement and should not be confused with a residual pneumothorax. A radiograph should be obtained immediately after thoracostomy tube placement, both to verify the location of the tube and to assess its therapeutic efficacy. Routine follow-up radiographs are not warranted (Henschke *et al.* 1996).

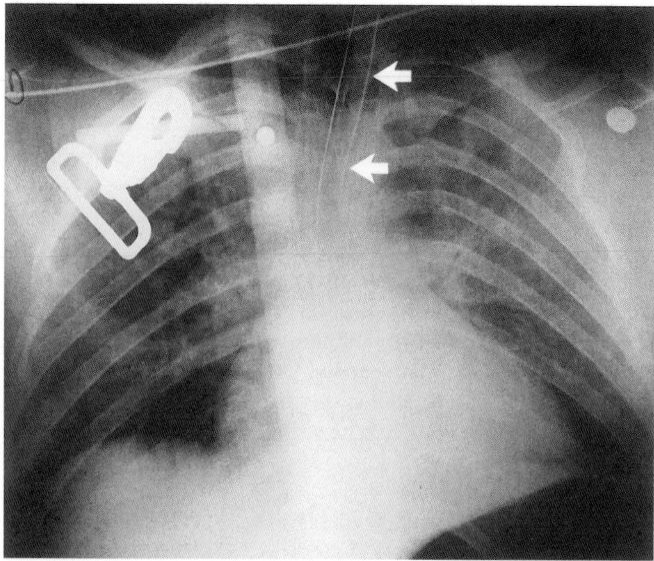

Fig. 5 Inadvertent esophageal intubation. This patient had been unsuccessfully resuscitated elsewhere for over 2 h. Notice that the stomach is distended with gas and that the endotracheal tube (arrows) is located to the left of the tracheal air column in the esophagus. (Reproduced with permission from Worrell (1994).)

Other devices

An increasing number of monitoring and therapeutic devices are visualized on the chest radiograph of the ICU patient. A partial list would include left ventricular bypass devices, intra-aortic balloon pumps, thermistors, intraesophageal pressure monitors, dual-lumen endotracheal tubes, cardiac and diaphragmatic pacing devices, neurostimulators, ventriculoperitoneal shunts, intrathecal catheters, automatic defibrillators, and artificial cardiac valves. It is a challenge for both the radiologist and the clinician to know and understand the radiographic appearance and proper positioning of each of these many devices.

The clinical and imaging team in the ICU

Patient care in the ICU is optimized when the clinical and imaging departments function as a team. Co-operation should begin at the design stage for the purchase and deployment of new imaging equipment and carried through to where and when the clinical and imaging teams meet in consultation. The simple task of properly noting pertinent clinical information with the radiological request can enormously benefit the patient's care as well as reduce the cost of these services. Co-ordination of the routine morning portable examination times so that the images and interpretation are available for morning clinical rounds should be standard. The radiological technologist should understand enough about patient care to co-operate with the nursing staff in consistently obtaining high-quality radiographs in these often difficult patients, and the nursing staff should understand enough of the technologist's duties to be able to help, particularly in stabilizing multiple catheters and monitors, so that the vital clinical function of these devices is not interrupted as the patient is positioned for radiography. The technologists should be trained in the basics of film and processor quality assurance. Each film should be reviewed as it is developed, not only for quality but also for any medical problems requiring immediate attention. Not only should the radiologist be prepared to contact the clinical team immediately with verbal reports of such findings, but the clinical team should be readily available to receive such reports.

One of the great advantages and, paradoxically, one of the great dangers inherent in digital systems is the ability to obtain and simultaneously to deliver diagnostic images to the ICU as well as other diagnostic nodes. This dispersion of the medical image may inadvertently lead to the loss of vital consultation between the radiologist and the clinician, with the result that optimal patient care is jeopardized.

Conclusions

The modern ICU has proved to be a life-saving innovation for the critically ill patient. An understanding of the imaging diagnosis of common clinical problems, as well as how to apply the basics of radiological physics, physiology, clinical medicine, and pathology to diagnose and treat the more uncommon problems, is essential for optimal patient care. The complexities of monitoring and treatment demand that the clinical and imaging teams co-operate so that the extensive expertise of each team member can be brought to bear in optimizing patient care in these, the sickest patients in the hospital.

References

Henschke, C.I., Yankelevitz, D.F., Wand, A., Davis, S.D., and Shiau, M. (1996). Accuracy and efficacy of chest radiography in the intensive care unit. *Radiologic Clinics of North America*, **34**, 21–31.

MacMahon, H. and Giger, M. (1996). Portable chest radiography techniques and teleradiology. *Radiologic Clinics of North America*, **34**, 1–20.

Milne, E.N.C. and Pistolesi, M. (1993). *Reading the chest radiograph. A physiologic approach*. C.V. Mosby, St Louis, MO.

Silverstein, D.S., Livingston, D.H., Elcavage, J., Kovar, L., and Kelly, K.M. (1993). The utility of routine daily chest radiography in the surgical intensive care unit. *Journal of Trauma*, **35**, 643–6.

Strain, D.S., Kinasewitz, G.T., Vereen, L.E., and George, R.B. (1985). Value of routine daily chest X-rays in the medical intensive care unit. *Critical Care Medicine*, **13**, 534–6.

Worrell, J.A. (1994). The plain chest radiograph in the intensive care unit. *Contemporary Diagnostic Radiology*, **17**, 1–5, 16.

16.4.2 Imaging the cardiovascular system

G. Porenta and G. Maurer

Key messages

- Echocardiography can be performed in real time at the bedside, permits qualitative and quantitative image analysis, and can easily be repeated for monitoring purposes.
- Transthoracic and transesophageal echocardiography frequently provide essential or even unique information for the clinical management of critical care patients.
- Quantitative coronary angiography uses digital image processing to measure vessel diameter at stenotic and unobstructed coronary segments and to quantify the severity of coronary stenoses.
- Cardiovascular nuclear imaging includes the assessment of regional myocardial perfusion by perfusion scintigraphy, measurement of ventricular function by multigated radionuclide ventriculography, detection of acute myocardial necrosis by infarct avid imaging, and delineation of myocardial viability and cardiac metabolism by positron emission tomography.
- Magnetic resonance imaging permits exquisite delineation of cardiac structures and the great vessels with excellent resolution.

Introduction

Different imaging modalities permit a qualitative and quantitative assessment of the anatomy, morphology, and function of the cardiovascular system (Marcus *et al.* 1991). Non-invasive imaging methods, which can be performed quickly at the bedside with little discomfort to the patient and with a low risk of morbidity and mortality, are particularly relevant in the critical care setting.

It is very important, yet sometimes difficult, for physicians in critical care medicine to select from many available imaging techniques a single modality that maximizes the expected gain in diagnostic accuracy and minimizes the time required to establish a diagnosis, discomfort to the patient, and costs.

Imaging modalities

Echocardiography

Echocardiography can be performed in real time at the bedside, permits qualitative and quantitative image analysis, can easily be repeated for monitoring purposes, carries no risk to the patient, and is available at a low cost. It is an operator-dependent technique requiring well-trained personnel. The major prerequisite for an echocardiographic examination is an acoustic window that offers adequate access to the region of interest.

Echocardiography has a prominent role as a diagnostic procedure for cardiac and critical care patients. Transthoracic echocardiography with its different imaging modalities, including two-dimensional (B-mode) imaging, M-mode, and Doppler and color Doppler flow mapping, provides unique information for many clinical problems (Weyman 1994).

Two-dimensional echocardiography generates real-time tomographic images of cardiac and extracardiac structures by rapidly sweeping the ultrasound beam through the acoustic window and recording the intensity of the backscattered signal for every spatial location within the sector sweep. Images can be frozen at selected time points, such as end-systole and end-diastole, to permit measurements of distances and areas. Transthoracic imaging probes use frequencies between 2.5 and 3.5 MHz with better penetration but less resolution. In contrast, transesophageal probes are located very near the target structures and can employ higher frequencies (5 or 7 MHz) to generate images with a higher resolution.

M-mode images show the ultrasound signal along a particular scan line versus time and are used to obtain geometric measurements at selected time points during the cardiac cycle (e.g. wall thickness and ventricular dimensions at end-systolic and end-diastolic time periods). Estimates of myocardial function such as the shortening fraction or wall thickening can then be calculated from these M-mode measurements.

Doppler imaging provides information about the velocity of blood flow within cardiac chambers, across cardiac valves, and within vessels. Continous Doppler imaging records all velocities that are detected along an ultrasound beam irrespective of their location. Continuous Doppler imaging can measure high-velocity jets but is unable to localize the source of the velocity within the Doppler beam. In contrast, pulsed Doppler imaging constrains the signal acquisition to a small sample volume within the ultrasound beam and thus is able to determine precisely the location of the Doppler signal. However, pulsed Doppler imaging is unable to quantify high-velocity jets (> 2 m/s) which typically occur across stenotic cardiac valves.

Color Doppler flow mapping extends conventional pulsed Doppler imaging from one sampling point to an entire sector image and provides a two-dimensional sector map of Doppler velocities. For each spatial location in the two-dimensional sector, the Doppler signal is recorded and displayed as a colored pixel value which encodes the velocity and the direction of flow. With this color coding, the observer can assess the direction of flow (e.g. red towards the transducer, blue away from the transducer) and the characteristics of blood flow (yellow for turbulent flow). Color Doppler imaging can be used to assess the severity of valvular regurgitant flow, the location and direction of pathological flow channels, such as septal defects, aneurysms, or shunts, and the hemodynamic impact of intraluminal obstructions.

Transesophageal echocardiography

The transthoracic acoustic window is limited in patients on mechanical ventilation and also in patients with lung disease, skeletal deformities, or extreme obesity, and diagnostic information may be inadequate. Moreover, anatomical structures that are located deeper within the thoracic cavity can often be difficult to visualize with a transthoracic approach. In these cases, transesophageal echocardiography can be employed to image the heart through a posterior acoustic window and provides a more detailed view of many anatomical areas, including the left and right atria, the intra-atrial septum, and native or artificial valves. Transesophageal echocardiographic imaging is particularly helpful in hemodynamically unstable patients when a prompt diagnosis needs to be established for therapeutic decision-making.

Radiographic techniques

Chest radiography

In critical care, the chest radiograph is extremely helpful in the assessment of many cardiac disorders including heart failure with pulmonary congestion, pleural effusions, and pulmonary embolism. It is also used to assess the position and integrity of catheters, pacing electrodes, or artifical heart valves. Fluoroscopy, i.e. the real-time display of X-ray images on image intensifiers, may be used to confirm floating electrodes as a cause of pacemaker dysfunction and to assess the dynamic motion of prosthetic valves when prosthetic dysfunction is suspected.

X-ray computed tomography

X-ray CT generates transaxial images of X-ray absorption and provides better spatial resolution than standard X-ray imaging. In cardiovascular imaging, CT imaging permits assessment of large vascular structures and has the capability of measuring vessel size and detecting dissecting membranes, calcific plaques, or large thrombi.

Angiography

Angiography involves the selective or subselective intravascular injection of a contrast medium to generate single images or image sequences of intraluminal dimensions. With the use of thin catheters and non-ionic contrast media, arterial angiography, including coronary angiography, is a safe procedure with very few severe complications. Quantitative coronary angiography uses digital image processing to measure vessel diameter at stenotic and unobstructed coronary segments and to quantify the severity of coronary stenoses.

Cardiovascular nuclear imaging

Scintigraphic imaging procedures record the spatial or temporal distribution of radioactive isotopes within the body using a gamma camera for the detection of radiation.

Image acquisition can be performed in the planar or tomographic mode. In planar imaging, projection images of the tracer distribution are recorded from selected angles to image the heart from different views. Planar imaging is a simple technique that can be performed with a mobile gamma camera and thus is available at the bedside. In tomographic or single-photon-emission CT (SPECT) imaging, the gamma camera rotates around the patient to record a sequence of 30 to 60 planar projection images. CT is then used to reconstruct and display tomographic transaxial images of the heart. ECG-gated image acquisition can be used with both planar and SPECT imaging to show the distribution of radioactivity at selected time periods within the cardiac cycle (e.g. at end-systole or end-diastole).

The main applications of cardiovascular nuclear imaging include the assessment of regional myocardial perfusion by perfusion scintigraphy, the measurement of left and right ventricular function by multigated radionuclide ventriculography, the detection of acute myocardial necrosis by infarct avid imaging, and the delineation of myocardial viability and cardiac metabolism by positron emission tomography (Zaret and Wackers 1993).

Myocardial perfusion scintigraphy

Myocardial perfusion scintigraphy provides images of the relative distribution of regional myocardial perfusion. After perfusion tracers have been injected intravenously, the tracer is extracted by normal myocardial tissue in relation to myocardial perfusion. Thus the myocardial activity of the tracer substance reflects relative myocardial perfusion. However, myocardial tracer concentration depends not only on perfusion but also on the presence of viable tissue that is able to extract and retain the tracer. Tracer will not be retained in necrotic or fibrotic areas, and perfusion scintigraphy will be abnormal.

At rest, coronary blood flow in a stenotic artery remains unaltered until more than 80 to 90 per cent of the luminal area is obstructed. Thus, in the absence of a prior myocardial infarction, resting perfusion scintigraphy may be normal even though severe coronary disease is present. This situation changes when myocardial perfusion is increased by either stress techniques (e.g. treadmill testing or dobutamine infusion) or pharmacological vasodilatation (e.g. dipyridamole or adenosine). Under hyperemic conditions, perfusion in normal myocardium will increase two- to fivefold while little or no change of flow will occur distal to a stenosis of 50 per cent or more of diameter. The flow inhomogeneity between unobstructed and stenotic vessels will be reflected in a difference of relative tracer uptake on myocardial scintigrams. Thus, in myocardial scintigraphy, stress imaging is employed to assess the functional significance of coronary stenoses, while rest scintigrams are used to depict the extent of myocardial damage and delineate myocardial viability.

Radionuclide ventriculography

In radionuclide ventriculography, red blood cells are labeled and blood pool imaging is performed using a gamma camera. Right and left ventricular contractile function can be quantified by first-pass or equilibrium imaging. Both methods use ECG-gated image acquisition to measure the amount of radioactivity within each cardiac chamber at different time periods during the cardiac cycle. The amount of radioactivity is proportional to the ventricular volume, so that ventricular ejection fractions can be calculated from end-systolic and end-diastolic count rates independent of geometric assumptions. Gated images can be displayed on the display screen as a continuous loop in video mode, giving the visual impression of a beating heart. Regional cardiac wall motion can be assessed visually. Quantitative image analysis can be used to calculate global right and left ventricular ejection fractions, regional wall motion and synchronicity of contraction, and indices of diastolic function such as peak filling rate or time to peak filling rate.

Infarct avid imaging

Localization and extent of myocardial necrosis can be characterized by nuclear medicine imaging using tracers with an affinity for acutely necrotic myocardial tissue which accumulate in such areas over a period of hours following tracer injection. Scintigraphic images display an increased count activity in locations of recent myocardial damage.

Positron emission tomography

Positron emission tomography (PET) is a nuclear imaging technique that records the distribution of positron-emitting isotopes in the body. PET tracer substances can be synthesized that are chemically identical with naturally occurring compounds such as water, acetate, or ammonia. Therefore PET imaging can be used to study metabolic pathways directly rather than by tracing the fate of foreign substances. PET imaging permits quantitation of myocardial perfusion (in ml/min/g) and metabolic pathways (in μmol/g/min). Most PET isotopes have short half-lives in the range of minutes so that an on-site cyclotron is necessary for isotope production. As PET imaging puts considerable demands on financial and technical resources, it is currently available at specialized centers only.

Cardiac applications of PET imaging include the quantitative measurement of myocardial perfusion and assessment of myocardial glucose utilization. In particular, the presence of maintained glucose metabolism in dysfunctional myocardial tissue is considered to be the most established diagnostic method to identify residual myocardial viability.

Magnetic resonance imaging

Magnetic resonance imaging (MRI) permits exquisite delineation of cardiac structures and the great vessels with excellent resolution. Also, ECG-gated acquisition can be performed to monitor cardiac wall motion, wall thickness, and flow phenomena such as turbulent flow. In recent years, magnetic resonance angiography which is capable of visualizing the luminal characteristics of larger vessels has become available.

Clinical applications

Coronary artery disease

The definitive diagnosis of coronary artery disease is established by coronary angiography which is considered the reference method for the diagnosis of coronary artery disease and is required before decisions regarding coronary revascularization can be made. Myocardial perfusion scintigraphy or stress echocardiography are non-invasive methods that can be used to screen for the presence and extent of coronary disease. Both methods have approximately 80 to 90 per cent sensitivity and specificity to detect significant coronary artery disease. Their diagnostic contribution should be sought when there is an intermediate (20–70 per cent) pretest likelihood of coronary disease and coronary angiography is not absolutely indicated for other reasons. These functional tests offer information about the physiological significance of coronary stenoses and thus can complement angiography, particularly when the significance of a given stenosis or the effectiveness of collaterals is uncertain.

Complications of acute and chronic coronary artery disease can be diagnosed by echocardiography. The extent and severity of wall motion abnormalities, the echogenicity and thickness of myocardial segments, and the presence of abnormal flow patterns can serve as valuable indicators characterizing the severity of potentially life-threatening sequelae of coronary disease. In acute myocardial infarction, echocardiography can be used to assess global and regional left ventricular function, to detect the presence of ventricular thrombi, to assess the presence, severity, and etiology of mitral regurgitation, such as 'papillary muscle dysfunction' or ventricular dilatation, to confirm the presence of a ventricular septal defect or rupture of the myocardial wall, and to monitor the development of left ventricular remodeling and cardiac aneurysms. Resting perfusion scintigraphy can be used to assess the extent of myocardial damage and identify the culprit vascular territory in patients with chronic or acute infarction. Infarct avid imaging permits differentiation between chronic and acute myocardial damage.

Demonstration of myocardial viability is frequently required before coronary revascularization by transluminal angioplasty or coronary bypass surgery can be recommended. While a normal wall motion is indicative of maintained myocardial viability, myocardial dysfunction does not necessarily exclude it. Asynergic, yet viable (hibernating), segments will improve wall motion subsequent to adequate revascularization. The most established method to predict residual myocardial viability is PET metabolic imaging. However, this technique is not widely available. Alternatively, demonstration of residual tracer uptake on myocardial resting perfusion scintigraphy and improvement in segmental wall motion during low-dose dobutamine echocardiography are valuable techniques for identifying myocardial viability in asynergic segments.

Valvular and congenital heart disease

Although for many years cardiac catheterization was the method of choice to establish a diagnosis of valvular or congenital heart disease and assess its severity, echocardiography is probably the most important imaging modality in this setting at present. Two-dimensional echocardiography can characterize the morphology of cardiac valves and the etiology of valvular disease. Doppler imaging can measure flow velocity across stenotic or regurgitant valves and estimate corresponding pressure gradients using the modified Bernoulli equation. Doppler studies have also been shown to quantify stenotic valve areas reliably using the continuity equation and pressure half-time methods. Color Doppler imaging permits identification and characterization of regurgitant flow and thus helps to assess the severity of regurgitant lesions or shunt flow.

Cardiomyopathies

Cardiomyopathies are cardiac disorders that affect the vigor or pattern of the contractile function of the heart. Regional myocardial wall motion and wall thickness can be studied by echocardiography which helps to differentiate dilated, hypertrophic, and ischemic cardiomyopathy. Quantitative methods can be employed to assess ventricular function and to document the progression or regression of disease under different therapeutic strategies. Doppler echocardiography can offer important information about diastolic function using the mitral inflow signal and pulmonary venous flow information. Radionuclide ventriculography should be initiated when measurements of ejection fraction independent of geometric assumptions are needed to quantitate or monitor ventricular function. Coronary angiography may need to be employed to document or exclude an ischemic etiology of cardiomyopathy.

Endocarditis

Echocardiography frequently provides unique and essential information which, in addition to blood culture, constitutes a major criterion for establishing a diagnosis of endocarditis. Valvular vegetations are pathognomonic signs of endocarditis and can be identified with either transthoracic or, more reliably, transesophageal echocardiography. Vegetation size, as determined by echocardiography, has been shown to be a predictor of embolic events and death. Complications of endocarditis such as abscess formation, valvular perforations, or the rupture of a sinus Valsalvas are best diagnosed by echocardiography. Moreover, echocardiography can be repeated to monitor the response to antibiotic therapy or determine the need for surgical intervention. A normal echocardiogram does not entirely exclude the presence of endocarditis.

Pericardial disease, cardiac masses, and tumors

Pericardial diseases, including effusions, calcifications, or cysts, can be diagnosed and assessed by echocardiography or MRI. Echocardiography is particularly valuable for defining the location, extent, and hemodynamic significance of pericardial effusions and guiding pericardiocentesis. Echocardiography and MRI are also useful imaging tools for defining the structure and dimensions of masses or tumors and their relation to cardiac anatomy. Ventricular or atrial thrombi, myxomas, cysts, or myocardial tumors can be identified and their composition characterized.

Aortic dissection

The diagnosis of aortic dissection can be established by transesophageal echocardiography, CT imaging, and MRI. While these diagnostic modalities are comparable in their accuracy of detecting aortic dissection, transesophageal echocardiography appears to be developing into the method of choice as it is available at the bedside and offers the capability for an immediate diagnosis in this life-threatening disorder with a very high early mortality.

Intensive care patients

Echocardiography can quickly determine whether a cardiac disorder is the underlying cause of hemodynamic instability and thus guide further patient management. Transesophageal echocardiography is frequently used in patients on mechanical ventilation to monitor left ventricular function and its response to inotropic therapy, to assess the valvular function and characterize its cause and severity, and to identify pericardial effusions. Also, echocardiography provides an immediate assessment of the filling status of the left ventricle and permits diagnosis of a dynamic outflow tract obstruction that may be aggravated by volume depletion. Thus echocardiography offers important decision support regarding the requirement for and the appropriateness of inotropic therapy and volume substitution.

References

Marcus, M., Schelbert, H., Skorton, D.J., and Wolf, G. (ed.) (1991). *Cardiac imaging*. W.B. Saunders, Philadelphia, PA.

Weyman, A.E. (ed.) (1994). *Principles and practice of echocardiography.* Lea & Febiger, Philadelphia, PA.

Zaret, B.L. and Wackers, F.J. (1993). Nuclear cardiology, Parts 1 and 2. *New England Journal of Medicine*, **329**, 775–83, 855–63.

16.4.3 Doppler echocardiography in intensive care

J. L. Fellahi and F. Jardin

Key messages

- Echocardiography allows easy, prompt, and accurate identification of most of the critical cardiac injuries that occur in the intensive care unit.
- Transesophageal echocardiography is technically easy and generally safe to perform in intubated critically ill patients.
- Doppler echocardiography allows assessment of the complications of myocardial infarction, acute valve disease, endocarditis, massive pulmonary embolism, and aortic diseases.
- Examples of useful applications of Doppler echocardiography include diagnosis of right-to-left shunting in unexplained hypoxemia, assessment of chest trauma, and assessment of brain-dead patients as possible organ donors.

Introduction

Doppler echocardiography is routinely used in some intensive care units (ICUs) (Jardin and Dubourg 1986). Moreover, the introduction of transesophageal echocardiography has enabled reliable high-resolution acoustic images of cardiac morphological features, function, and hemodynamics to be obtained for mechanically ventilated patients in whom transthoracic images are frequently suboptimal (Jardin 1995).

Echocardiographic investigation is of great interest for the critical care specialist because it allows easy, prompt, and accurate identification of most of the critical cardiac injuries that occur in ICUs and it is a non-invasive hemodynamic technique which can be repeated frequently, introducing the concept of non-invasive hemodynamic monitoring.

Performance of echocardiography

Echocardiographic examination in a sedated and mechanically ventilated critically ill patient is decribed. It includes transesophageal echocardiography which, in our experience, is superior to transthoracic echocardiography in this setting.

Performance of transesophageal echocardiography

Transesophageal echocardiography is technically easy and generally safe to perform in intubated critically ill patients. Nevertheless, it should be done only by specially trained physicians. Guidelines for appropriate training have been recommended by the American Society of Echocardiography (Pearlman *et al.* 1992). In the ICU, the examination is usually undertaken with the patient in the supine position. Before examination, the history is taken with emphasis on gastrointestinal-tract-related symptoms, such as dysphagia, hematemesis, and esophageal varices, and upper gastrointestinal surgical procedures. The mean duration of a transesophageal echocardiography examination is about 20 min. One transesophageal echocardiography daily is generally sufficient, but the procedure can be repeated if necessary. Echocardiographic images are recorded on videotape, together with an ECG lead and respiratory flow obtained using a pneumotachograph, the signal of which is amplified and displayed on the video screen. Thus precise timing of cardiac beats within the respiratory cycle can be obtained. Because non-invasive hemodynamic monitoring is often the main objective of transesophageal echocardiography in the ICU and the study may have to be interrupted early, we focus on the transgastric short-axis view first. The probe is raised to the mid-esophageal position to obtain a four-chamber view and to the upper esophageal position which allows imaging of basal short-axis views of the heart. Finally, the ascending arch and descending thoracic aorta are systematically explored.

The transgastric short-axis view

This view, which is the cornerstone of non-invasive hemodynamic monitoring, allows an accurate two-dimensional and M-mode echographic assessment of the left ventricle giving areas, diameters, and thicknesses (Fig. 1). It also allows qualitative evaluation of the regional wall motion of the left ventricle. Transesophageal echocardiography is of greater sensitivity in this setting, permitting a more complete examination of the left ventricular free wall.

The four-chamber view

The four-chamber view, which is readily obtained 30 to 35 cm from the dental arches, is ideal for estimating both right and left atrial and ventricular dimensions and kinetics (Fig. 2). Unfortunately, left ventricular dimensions are consistently underestimated and accurate measurements on this view are not recommended. The morphological features of native or prosthetic mitral valves are best visualized using transesophageal echocardiography. In the absence of tachycardia, color and pulse-wave Doppler are extremely useful in searching for valvular heart diseases, abnormalities of left ventricular diastolic function, and hypovolemia. Tricuspid regurgitation, which is frequent in mechanically ventilated patients, is also revealed on the four-chamber view and can be quantified by color Doppler.

The basal short-axis views

Several anatomical structures (ascending aorta, main pulmonary artery, right and left atria, atrial appendages, pulmonary veins, and aortic and pulmonary valves) are readily accessible on these views (Fig. 3). Pathological phenomena, such as small atrial thrombi (particularly in the left atrial appendage) and left atrium spontaneous echo contrast, both associated with an increased risk of thromboemboli, patent foramen ovale during mechanical ventilation with or without positive end-expiratory pressure, and central pulmonary artery thrombi in massive pulmonary embolism, are identified more clearly and more frequently by transesophageal echocardiography than by transthoracic echocardiography.

The thoracic aorta

Transesophageal echocardiography has become an established technique (Fig. 4) for the rapid and reliable detection of traumatic and medical abnormalities associated with the ascending and descending aorta, particularly aortic dissection. This technique compares favorably with other imaging methods such as aortography, CT scanning,

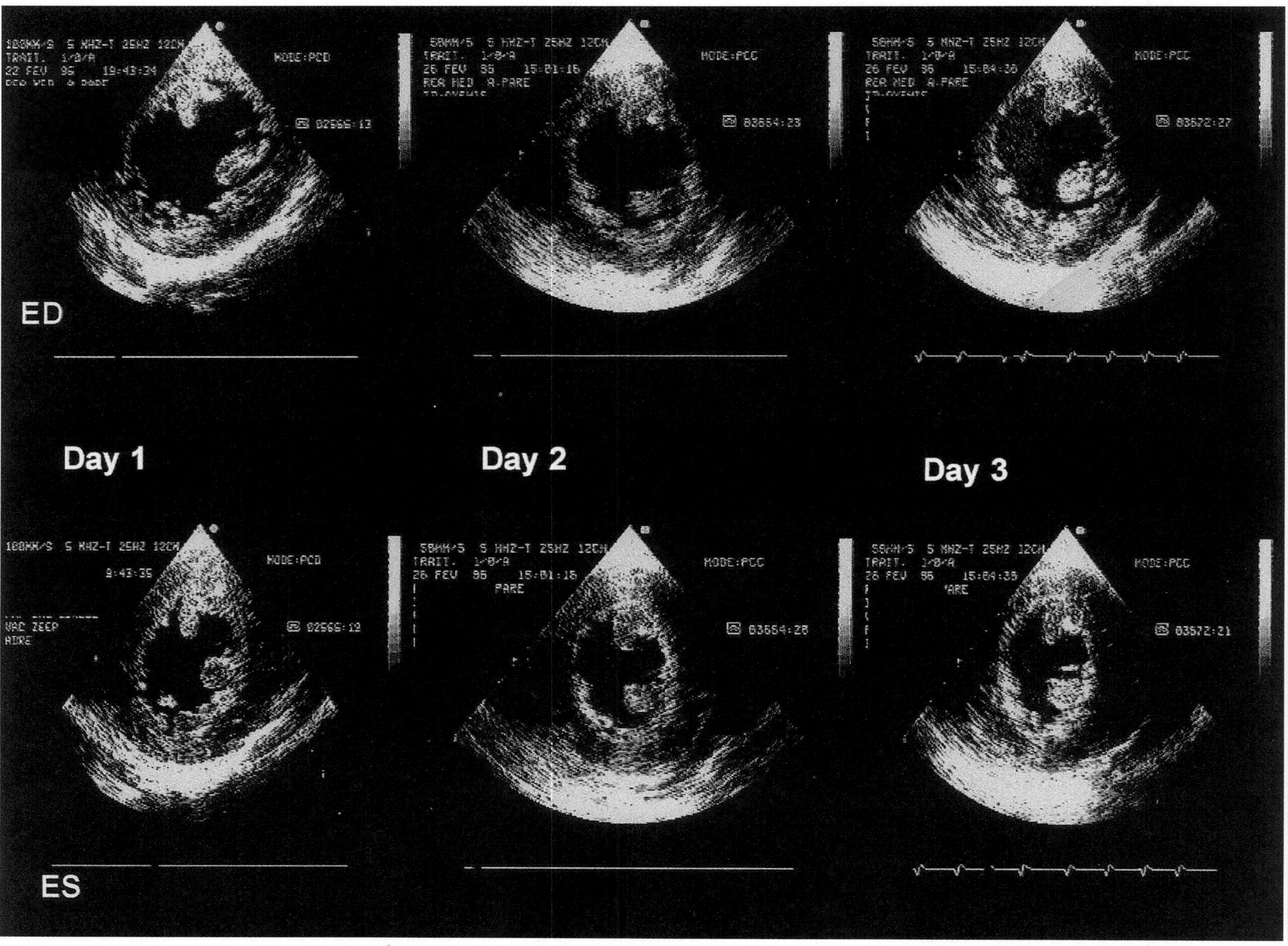

Fig. 1 Short-axis view of the left ventricle obtained using a transgastric approach. In this patient, who was being managed for septic shock, repeated measurement of the area of the left ventricle at end-diastole (ED) and end-systole (ES) allowed the adequacy of left ventricular filling to be checked and assessment of the beneficial effect of hemodynamic support (epinephrine (adrenaline) infusion on day 1, replaced by dobutamine infusion on days 2 and 3, after some improvement).

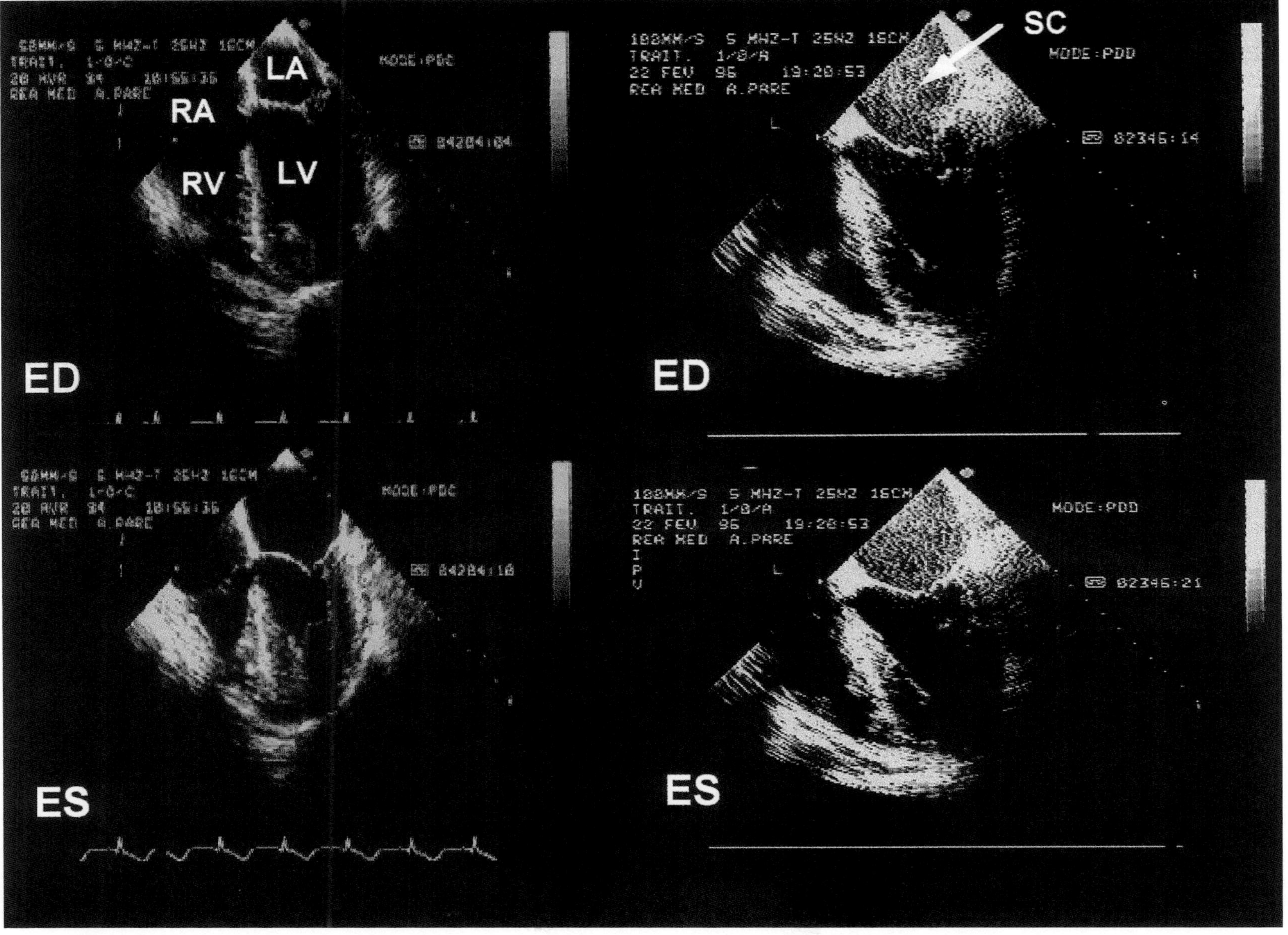

Fig. 2 Long-axis four-chamber views (RA, right atrium; LA, left atrium; RV, right ventricle; LV, left ventricle) at end-diastole (ED) and end-systole (ES) obtained by transesophageal echocardiography in two different patients. Left panel, a patient with some degree of hypertrophic cardiomyopathy and normal LV systolic function who was managed by mechanical ventilation for an acute pulmonary edema. Right panel, a patient, with previously normal cardiac status, who was managed for a septic shock with severe left ventricular systolic dysfunction and extremely low cardiac output resulting in spontaneous contrast (SC) in the left atrium.

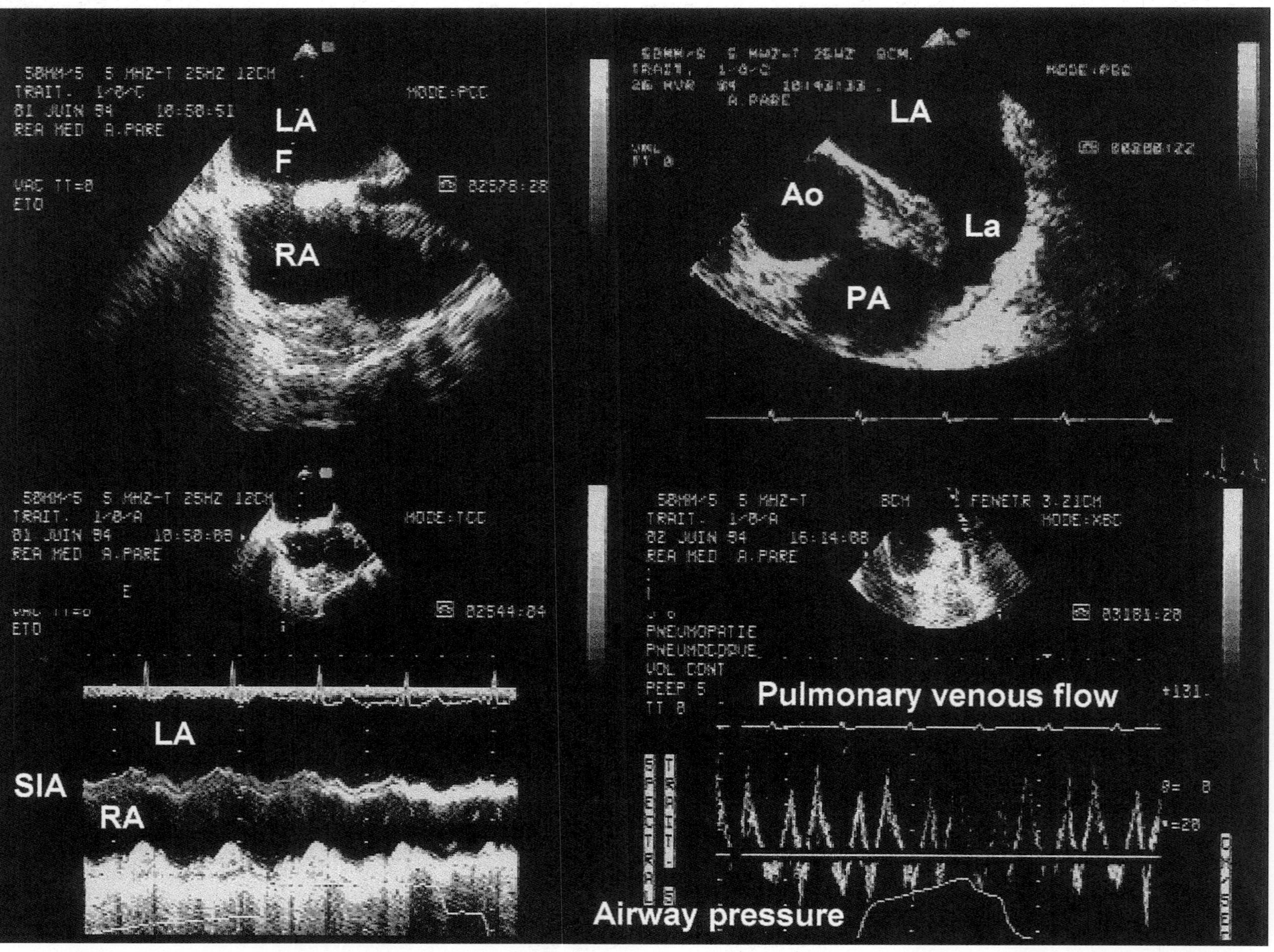

Fig. 3 Basal short-axis view, showing the interatrial septum (SIA) (left upper panel), with the foramen ovale (F). The left lower panel shows the kinetics of the SIA recorded in M-mode; the profile is the same as that of a left atrial pressure recording. The left atrial appendage (La) is shown in the right upper panel, and the pulmonary venous flow velocity from the left superior pulmonary vein is recorded, together with airway pressure, in the right lower panel. RA, right atrium; LA, left atrium; Ao, aorta.

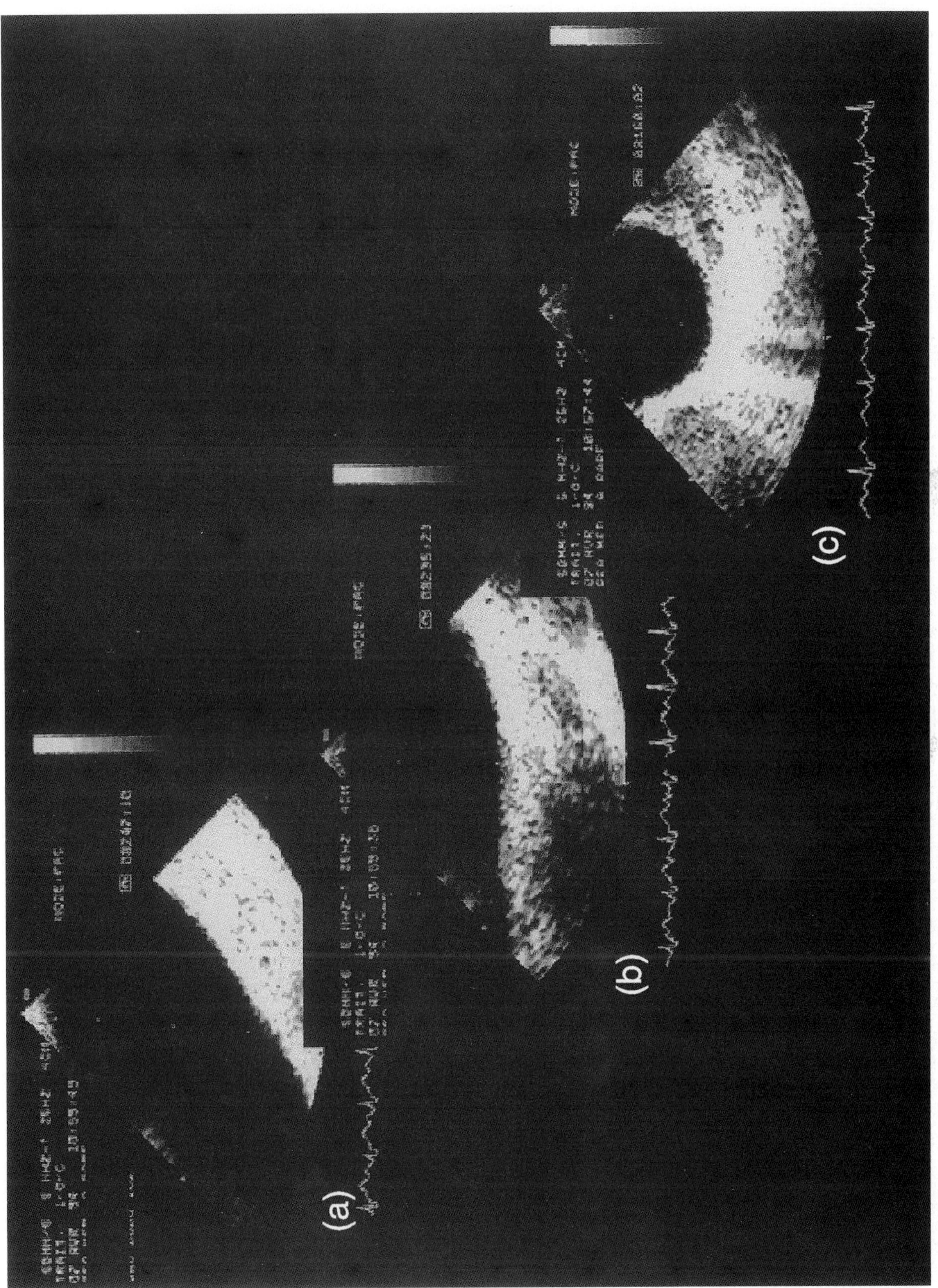

Fig. 4 The thoracic aorta: (a) ascending part; (b) arch; (c) descending part.

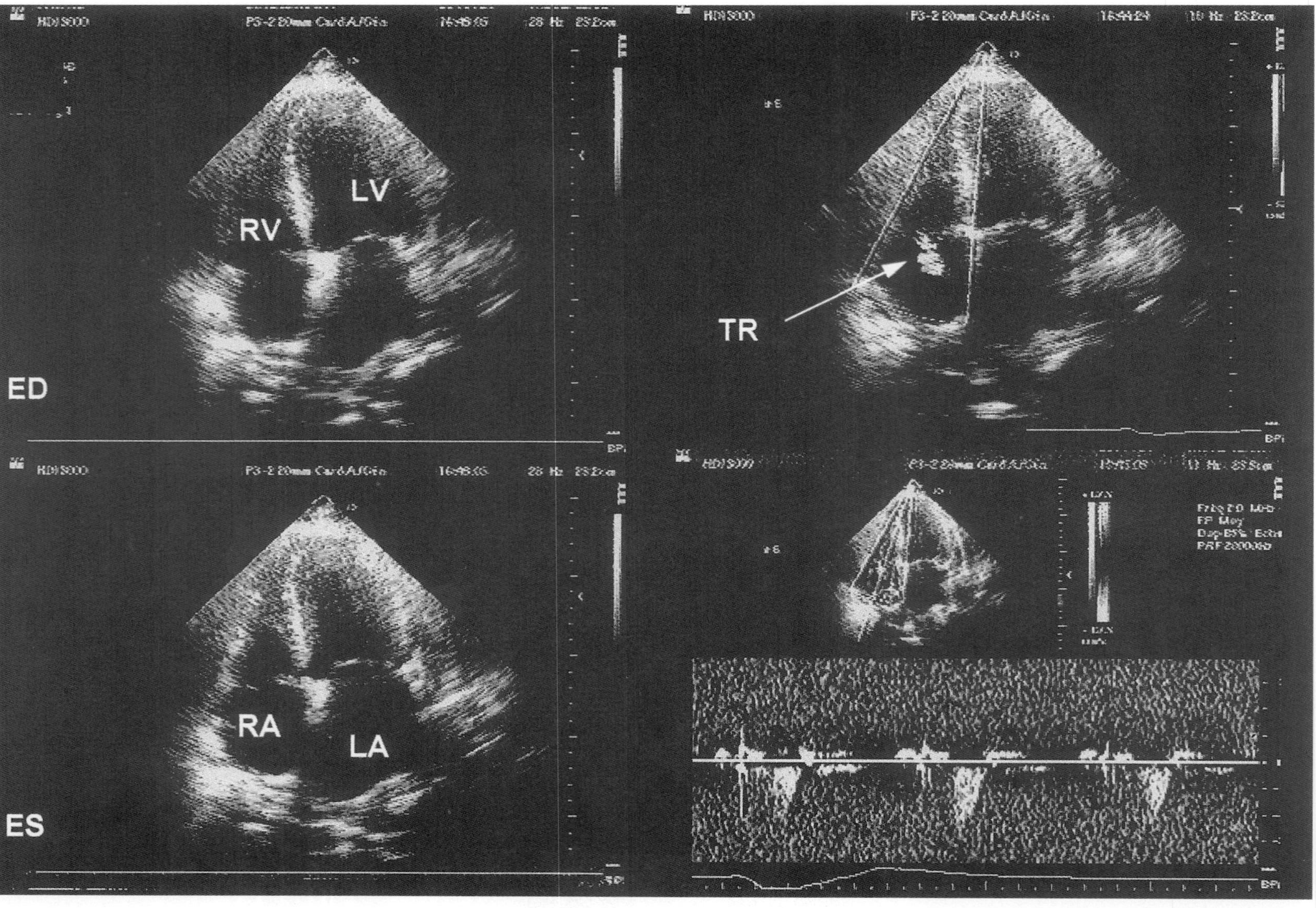

Fig. 5 Accurate measurements of right ventricular (RV) dimensions at end-diastole (ED) and end-systole (ES) and of right atrial (RA) dimensions at ES are best obtained from an apical transthoracic echocardiography four-chamber view (left). This approach can also be used to detect tricuspid regurgitation (TR) by color Doppler (dense cloud of echoes) (right upper panel) and to record backward flow velocity by continuous Doppler (right lower panel), permitting calculation of pulmonary artery systolic pressure.

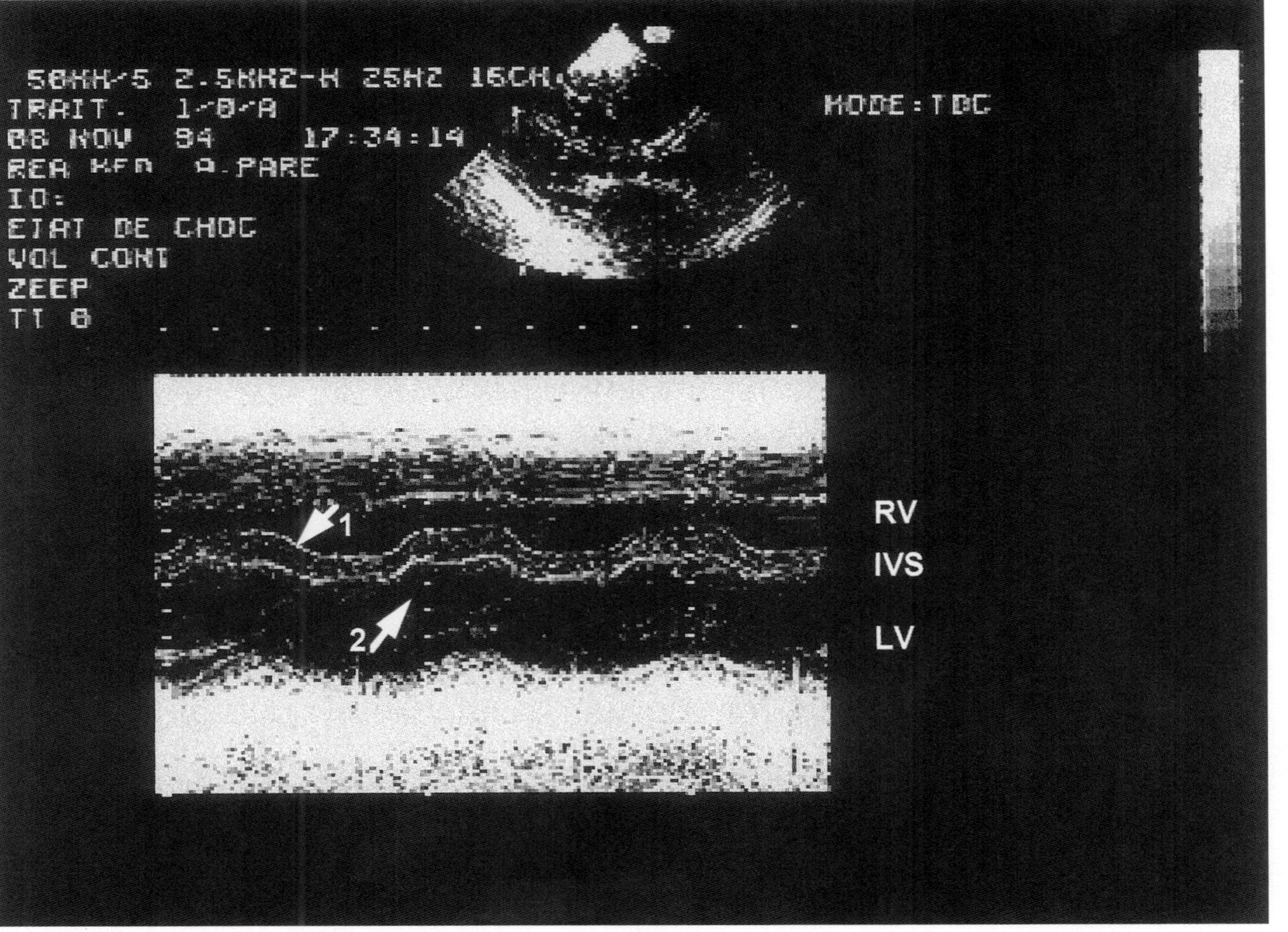

Fig. 6 Parasternal long-axis view of the interventricular septum (IVS). In this patient presenting with massive pulmonary embolism, M-mode recording visualizes paradoxical IVS motion, with its displacement towards the left ventricular (LV) cavity at the onset of diastole (first arrow), and then towards the right ventricular (RV) cavity at the onset of systole (second arrow). The result is a parallel motion of the IVS and the LV posterior wall instead of the normal pattern consisting of an alternating convergent (systole) and divergent (diastole) motion.

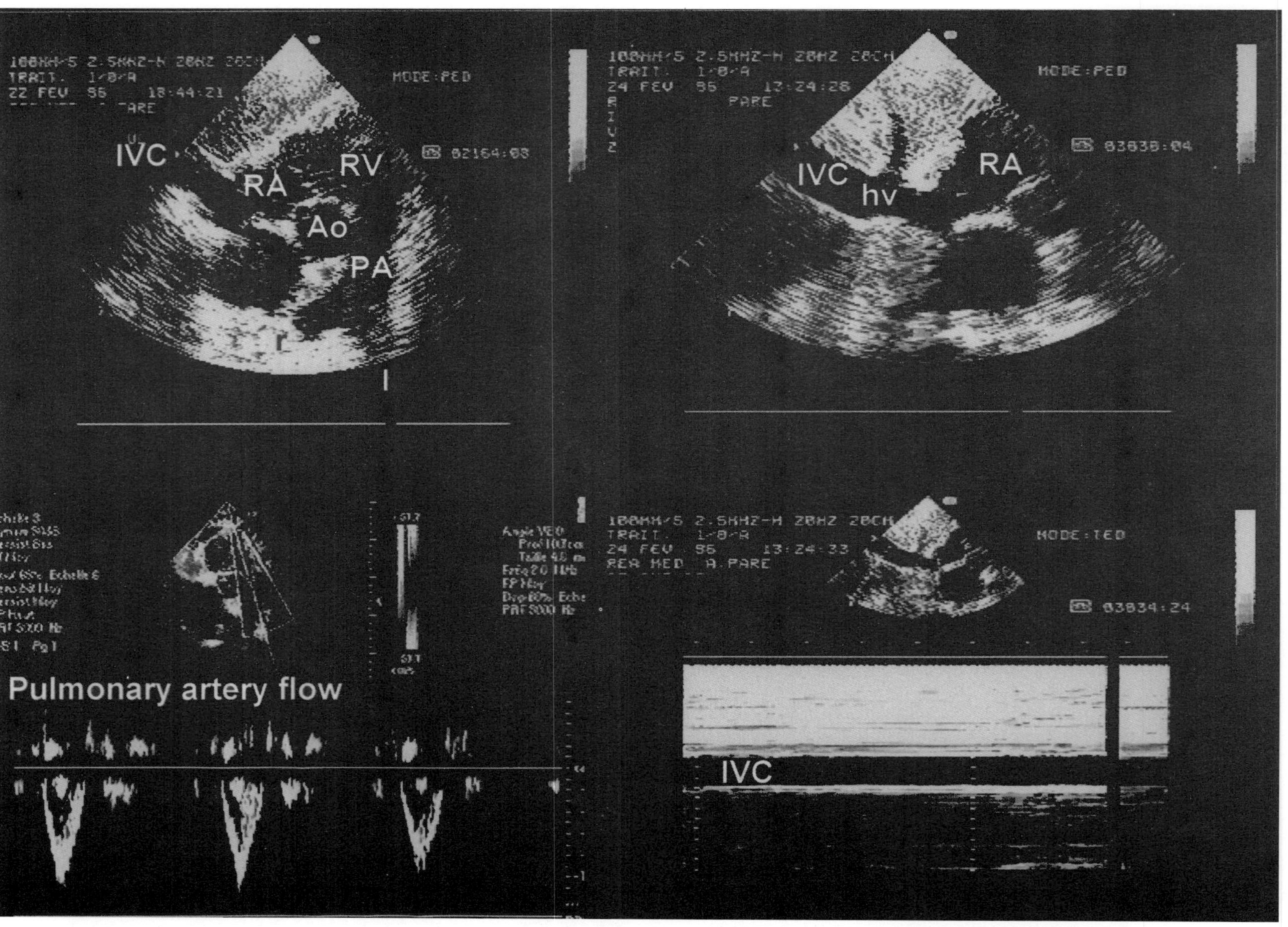

Fig. 7 Transesophageal echocardiography in subcostal view, showing the main pulmonary artery (PA) dividing into right (r) and left (l) branches (left upper panel). This approach allows the pulmonary artery flow velocity to be recorded by pulsed Doppler (left-hand bottom corner) and pulmonary hypertension to be detected. The right upper panel shows the inferior vena cava (IVC) behind the liver, which receives the hepatic veins (hv) and anastomoses to the right atrium (RA). Use of M-mode recording allows accurate measurement of the IVC diameter above the level of hepatic veins (right lower panel). RV, right ventricle; Ao, aorta.

and magnetic resonance imaging. When single-plane transesophageal echocardiography is used, the upper portion of the ascending aorta is masked by the trachea and the proximal bronchus. Dual-plane and multiplane transesophageal echocardiography, combined with color Doppler, partly overcome this limitation.

Performance of transthoracic echocardiography

Transthoracic echocardiography supplements transesophageal echocardiography and provides useful additional information. Unfortunately, it is technically difficult to perform in critically ill patients who are frequently intubated, are in the supine position, and have injuries or surgical incisions, which limit the transducer access, or underlying pathological conditions, such as chronic obstructive pulmonary disease, that result in suboptimal images. Nevertheless, the apical four-chamber view can often be used (Fig. 5), at least during the exhalation part of the respiratory cycle. This approach is useful for detecting right ventricular enlargement, and to estimate it by measuring the ratio of right to left ventricular area at end-diastole. This ratio is normally below 0.6. A ratio ranging from 0.6 to 1 indicates mild right ventricular dilatation, whereas a ratio greater than 1 denotes severe right ventricular dilatation. This view also provides a suitable angle for Doppler analysis of tricuspid backward flow. A small motion of the transducer also permits Doppler analysis of the left ventricular outflow tract. The parasternal short- or long-axis views enable visualization of a paradoxical motion of the interventricular septum, reflecting right ventricular systolic overload (Fig. 6). The subcostal long- and short-axis views are useful for examining patients with chronic obstructive pulmonary disease and emphysema. The subcostal long- and short-axis views are also the best way of accurately determining the diameter of the inferior vena cava (Fig. 7) which closely correlates with central venous pressure in mechanically ventilated patients. Color Doppler and contrast echocardiography in the plane of the inferior vena cava and the hepatic vein can also be informative. The subcostal basal short-axis view allows Doppler analysis of the pulmonary arterial flow (Fig. 7), which can also be obtained by transesophageal echocardiography.

Clinical applications

Doppler echocardiography has many useful applications to diagnosis procedures and hemodynamic monitoring in the ICU.

Examples of diagnostic applications include assessment of the complications of myocardial infarction, cardiac sources of embolism, cardiac tamponade, acute disease of native or prosthetic valves, suspicion of endocarditis, massive pulmonary embolism, aortic diseases, right-to-left shunting in unexplained hypoxemia, and chest trauma. Doppler echocardiography can also be used to assess brain-dead patients as possible organ donors. This technique has been particularly useful for elucidating some of the hemodynamic consequences of mechanical ventilation, and it is an excellent method of detecting detrimental effects of excessive respiratory support.

Hemodynamic monitoring includes unexplained hypotension, hemodynamic instability, patients in shock, acute right or left ventricular dysfunctions (systolic and diastolic), and systemic and pulmonary venous return assessment in the ICU or the operating room.

References

Jardin, F. (1995). *L'évaluation hémodynamique non invasive a u lit du patient par l'écho-cardiographie-Doppler*. Masson, Paris.

Jardin, F. and Dubourg, O. (1986). *L'exploration échocardiographique en médecine d'urgence*. Masson, Paris.

Pearlman, A.S., *et al.* (1992). Guidelines for physician training in transesophageal echocardiography: recommendations of the American Society of Echocardiography Comittee for Physician Training in Echocardiography. *Journal of the American Society of Echocardiography*, 5, 187–94.

16.4.4 Imaging the upper urinary tract

U. Patel and M. J. Kellett

Key messages

- Ultrasound and, less frequently, CT are the preferred investigations in the critical care setting.
- The main questions being asked of diagnostic imaging in the critically ill patient with impaired renal function are as follows. What size are the kidneys? Can obstruction be detected or excluded with confidence?
- Renal failure due to bilateral renal obstruction requires an urgent CT scan because of its superior resolution of the retroperitoneum.
- In renal trauma ultrasound will demonstrate perirenal hematoma, urinoma, and the integrity of the renal artery and vein. However, spiral CT is more informative and allows staging of injury.

Renal imaging

The kidneys can be imaged by plain radiography, ultrasound (gray scale and spectral or color Doppler), intravenous urography, nuclear medicine scans (using a variety of radionuclides tailored to assessing overall renal function or excretion and to searching for a source of sepsis), computed tomography (spiral CT is an informative and rapid scanning method), and magnetic resonance imaging (MRI). Ultrasound and, less frequently, CT are preferred in the critical care setting (Levine 1994). The rest are inconvenient (intravenous urography, nuclear medicine scans, and particularly MRI), imprecise (the plain radiograph), or both (nuclear medicine scans and intravenous urography). The portability and ease of ultrasound makes it the method of choice in most situations.

Table 1 lists the most common indications for imaging the upper urinary tract and the relative value of each modality. It is impossible to quote meaningful figures for the reliability of any given imaging technique in the critically ill as patient status (the presence of multiple catheters or orthopedic metalwork and the often poor renal reserve precluding liberal, or even any, use of intravenous contrast agents) compromises image quality. For this reason, coupled with the dynamic nature of critical illness, repeated and frequent studies may be necessary to achieve a confident diagnosis.

Table 1 Common indications for renal imaging in the ICU

	AXR	IVU	Ultrasound	CT	Angiography	MRI	NM scan
Renal failure	++	+	+++	++	–	+	+
Renal obstruction	++	++	+++	++	–	+	+
Renal artery stenosis	–	+	++	++	+++	+	–
Renal artery thrombosis	–	+	++	++	+++	+	–
Renal vein thrombosis	–	+	++	++	+++	+	–
Acute pyelonephritis	+	–	++	+++	–	–	–
Renal abscess	–	+	+++	+++	–	+	–
Pyonephrosis	–	+	+++	++	–	+	–
Renal trauma	+	++	++	+++	+++	+	+

The relative diagnostic value of currently available imaging modalities, with – being of no value and +++ being the investigation of first choice. In compiling this table both the intrinsic diagnostic ability of the test and its convenience in the ICU were taken into account. Thus MRI, although a very informative modality, is too inconvenient in the critically ill.

AXR, plain abdominal radiograph; IVU, intravenous urography; NM, nuclear medicine.

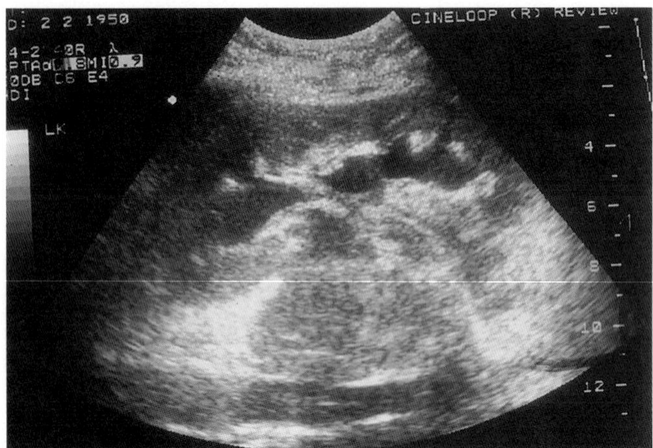

Fig. 1 Moderately severe pelvicalyceal dilatation on sonography. The dilated renal pelvis and calyces are represented by the central echo-free (fluid-containing) zone with separation of the high echoes from the echogenic renal sinus (fat-containing). The renal parenchyma has intermediate echoes.

Renal failure

The main questions being asked of diagnostic imaging in the critically ill patient with impaired renal function are as follows. What size are the kidneys? Can obstruction be detected or excluded with confidence? The size of the kidneys and the depth of the parenchyma will give an indication of either acute, or acute on chronic, renal failure. Ultrasound will provide this information with simple accuracy. Since the apparatus is mobile, the examination can be performed in the intensive care unit (ICU).

The exclusion of obstruction is usually, but not always, straightforward and again ultrasound is the first imaging tool to use. If the collecting system is dilated because of obstruction, it will show up clearly as an echo-free area within the parenchyma (Fig. 1). However, an acute obstruction may present with only mild dilatation, particularly when there is a small intrarenal pelvis. Whenever there is the slightest dilatation of the pelvicalyceal system in a patient with impaired renal function, obstruction must be suspected until proved otherwise. Conversely, renal pelvic cysts can easily mimic dilated pelvicalyceal systems and obstruction can be suggested by the ultrasound image (Fig. 2). The limitation of ultrasound is its operator dependence. The correct settings for any given patient (obese or thin) are vital, and interpretation of the real-time images can be quite taxing.

Spectral Doppler ultrasound studies of obstructed kidneys were originally pursued enthusiastically. The resistive index

$$\mathrm{RI} = \frac{\text{peak systolic} - \text{minimal diastolic}}{\text{peak systolic}}$$

calculated from the velocity waveform of an intrarenal artery, was reported to exceed 0.7 if obstruction was present; however, a more recent study achieved a sensitivity of only 44 per cent in patients with colic (Tubin *et al.* 1994). Doppler study is not always practical in the critically ill, and these more sophisticated machines are not really mobile.

Patients with solitary kidneys (congenital or previous nephrectomy) can have many causes of obstruction which will produce renal failure. The most common cause is calculus disease, and ultrasound may detect renal pelvic stones or stones in the proximal few centimeters of the ureter. Distal ureteric stones lodged at the ureteric orifice can be seen by scanning through the bladder. In any case a plain abdominal radiograph (portable in the ICU if necessary) will be required to exclude mid-ureteric stones, which are easily missed by ultrasound because of overlying bowel.

Patients presenting with renal failure due to bilateral renal obstruction will probably have pelvic or retroperitoneal pathology. It may be important to discover whether pelvic malignancy is the culprit and an urgent CT scan may be indicated because of its superior resolution of the retroperitoneum.

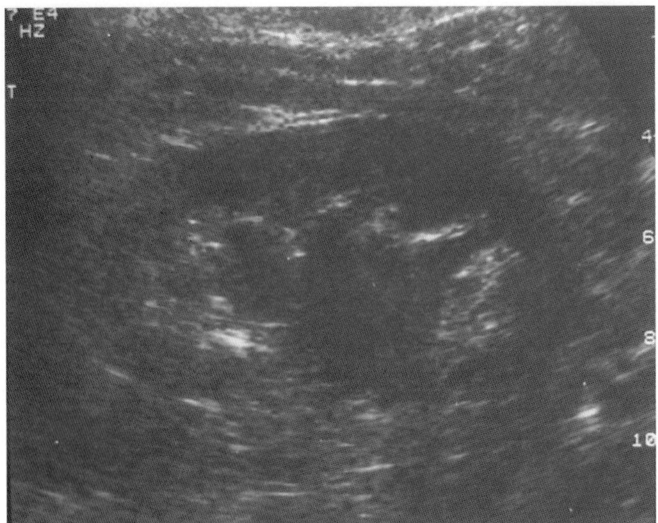

Fig. 2 The central echo-free zone in this case is caused by pelvic cysts and not a dilated renal pelvis or calyces. Renal cysts are an infrequent cause of a false-positive diagnosis of renal obstruction on ultrasound.

Currently, biopsy is almost always necessary in renal parenchymal disease to reach a specific tissue diagnosis, and it can be carried out in the ICU using ultrasound guidance. The possibility of renal artery stenosis is seldom of immediate concern in critical care; its recognition is still challenging and ultrasound has a limited reliability. Exclusion of renal artery or vein thrombosis, or injury (see below), may be more urgent, particularly in the transplant patient. Color Doppler ultrasound is the preferred modality. Nuclear medicine scans are cumbersome, and CT, MRI, and angiography, which is the most specific, are inconvenient. Nuclear medicine scans are uniquely informative in the assessment of the contribution of each kidney to overall renal reserve. This information can be essential in guiding the necessity and timing of intervention.

Antegrade pyelography and percutanous nephrostomy

When there is any suspicion of obstruction, an antegrade pyelogram, with or without percutaneous nephrostomy, is the next step. If the clinical and ultrasound evidence of obstruction is strong with a high serum potassium, an urgent nephrostomy should be performed. If possible, it is preferable to move the patient to the imaging department. An antegrade pyelogram can then be performed first, if merely wanting to exclude obstruction, with fluoroscopy and spot films showing free flow of contrast down the ureter to the bladder. Fluoroscopy is also helpful when placing the nephrostomy to ensure that the catheter is well sited in the collecting system (Kellett and Günther 1992), but if necessary nephrostomy can be undertaken in the ICU with ultrasound to guide the puncture.

Renal sepsis

Acute pyelonephritis is a clinical diagnosis and, if uncomplicated, imaging is not required except to disclose an underlying cause such as calculus disease, for which an ultrasound scan or plain radiography will suffice. Acute focal pyelonephritis may progress to abscess formation or, particularly with diabetes, to emphysematous pyelonephritis, which can be detected by ultrasound (Levine 1994).

Pyonephrosis

Kidneys obstructed by calculi are at a high risk because of infection in the calculi. Urine is an excellent culture medium for Gram-negative bacilli and an obstructed kidney can rapidly be damaged by a pyonephrosis. The danger of a Gram-negative septicemia should prompt urgent imaging and, if necessary, a percutaneous nephrostomy (Fig. 3).

As with obstructive renal failure, ultrasound is the first-line imaging modality (Levine 1994). It may detect a perirenal abscess as a complication of the pyonephrosis, which can be treated with a percutaneous drain inserted at the time of the nephrostomy. In very obese patients, retroperitoneal collection may be difficult to exclude with certainty on ultrasound and a CT scan may be indicated. If access to an abscess is then thought to be awkward, a catheter drain can be inserted under CT guidance.

Renal trauma

The kidneys are vulnerable to blunt, penetrating, and deceleration injury (Lang *et al.* 1985; Federle 1990). Microscopic hematuria alone is a poor indicator of renal trauma, but heavier hematuria and the presence of hemodynamic disturbance warrants renal imaging. Ultrasound will demonstrate perirenal hematoma, urinoma, and the integrity of the renal artery and vein. However, dynamic CT, ideally in the spiral mode, is more informative and allows staging of injury (Fig. 4). Staging is of importance as it can influence management. Renal arteriography may also be necessary and can be followed with embolization of any bleeding points.

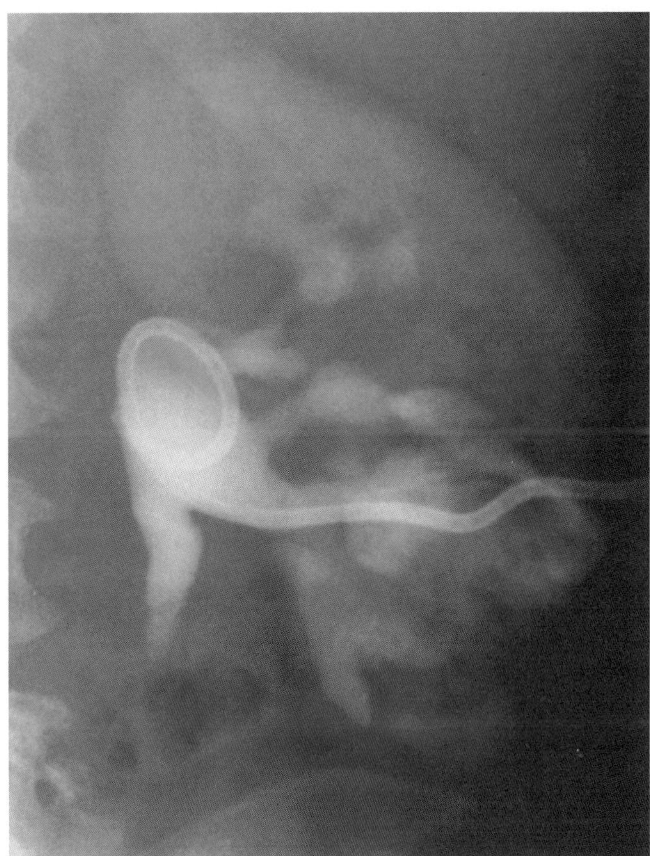

Fig. 3 Nephrostomy inserted to drain an infected and obstructed kidney. The cause of obstruction was a ureteric calculus.

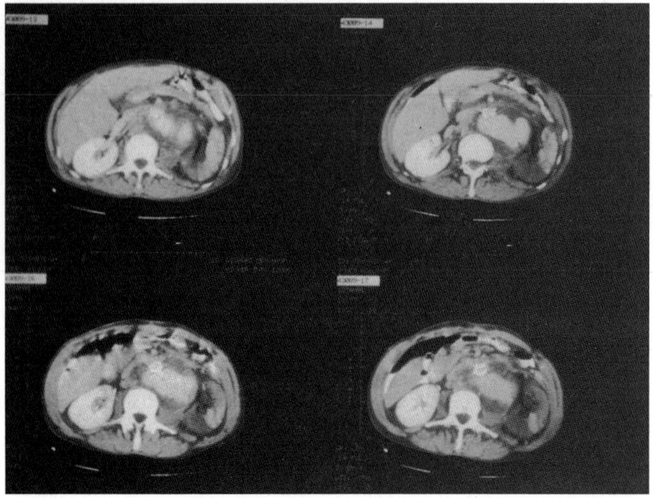

Fig. 4 Photomontage of sequential CT cuts showing a bleeding left renal artery with contrast leakage. The associated renal infarction is seen as an area of non-enhancement in the posteromedial portion of the left kidney. This scan follows the normal layout of cross-sectional images and the left kidney is on the right side of each scan.

Fistulas

Bilateral needle nephrostomies may need to be inserted to divert urine away from a lower-tract fistula. It is imperative that these catheters are made of a material that will not kink, and nephrostomies should be well sited in the collecting system. If necessary, a parallel balloon catheter may be inserted in the ureter to block flow to the fistula completely.

References

Federle, M.P. (1990). Evaluation of renal trauma. In *Clinical urography: an atlas and textbook of urological imaging* (ed. H.M. Pollock), pp. 1472–94. W.B. Saunders, Philadelphia, PA.

Kellett, M.J. and Günther, R.W. (1992). Interventional uroradiology. In *Diagnostic radiology* (2nd edn) (ed. R.G. Grainger and D.J. Allison), pp. 1417–39. Churchill Livingstone, London.

Lang, E.K., Sullivan, J., and Frentz, G. (1985). Renal trauma: radiological studies. Comparison of urography, CT, angiography and radionuclide studies. *Radiology*, **154**, 1–6

Levine, E. (1994). Acute renal and urinary tract disease. *Radiological Clinics of North America*, **32**, 989–1004.

Tubin, M.E., Dodd, G.D., and Verdile, V.P. (1994). Acute renal colic: diagnosis with duplex Doppler ultrasound. *Radiology*, **193**, 697–701.

16.4.5 Imaging the central nervous system

J. M. Stevens and C. Mandel

Key messages

- The most significant recent development in CT scanning has been volumetric data acquisition in times compatible with a single breath hold.
- Although logistically more difficult than CT in the critically ill patient, magnetic resonance imaging is generally preferred because it is more accurate and avoids the high X-ray dose delivered by CT.
- Cranial CT is preferred as the first and only examination in head injury.
- Plain radiography of the cervical spine is indicated only in the unconscious or non-assessable patient, or when there is clinical evidence of neck injury.
- Magnetic resonance imaging may demonstrate potentially treatable causes of traumatic cord compression and is of value in predicting recovery.

Introduction

The widespread availability of CT and magnetic resonance imaging (MRI) has greatly expanded the role of imaging in all areas, but had the earliest and greatest impact in disorders of the central nervous system. Plain radiographs, once the basis of medical imaging, are almost obsolete in the context of central nervous system disease, but still have a limited role in spinal trauma.

Imaging methods

Computed tomography

The most significant recent development has been volumetric data acquisition in times compatible with a single breath hold. This is achieved by a slip ring design allowing the patient to move through the gantry during scanning, often referred to as helical or spiral CT. The data can be presented as two-dimensional images (slices) in any desired plane and at any chosen slice interval, or in less familiar ways such as maximum intensity pixel projections or shaded surface renderings. The quality of the images depends upon, in particular, the tightness of the spiral and the number of data points collected. There is a time penalty for increasing image quality and also a reduction in the area covered; a major constraint is overheating of the X-ray tube. Whole regions such as the head or thoracic spine can now be imaged in seconds, and virtually the whole body can be imaged in a few minutes.

CT still has advantages over MRI. There are no contraindications, as there are to MRI, and the configuration of the gantry is generally far less claustrophobic. Life-support systems are not incompatible, provided that they are portable. CT is better than conventional tomography at delineating most fractures and bone destruction, and may be superior to MRI in correctly diagnosing recently extravasated clotted blood and subarachnoid hemorrhage. The disadvantage of CT is relatively low soft tissue contrast, particularly in areas enclosed by dense bone, such as the vertebral canal.

Special techniques using CT

CT angiography is capable of demonstrating the cervical and intracranial arteries and veins, is widely available, and has been shown reliably to demonstrate carotid stenoses and intracranial aneurysms larger than 3 to 4 mm. Data acquisition takes about 30 s (i.e. faster than magnetic resonance angiography), but the technique requires the rapid injection of about 90 ml of intravenous contrast medium.

CT myelography requires the introduction of contrast medium into the subarachnoid space via a spinal needle. In supine immobile patients this can be achieved via a fluoroscopy-guided lateral cervical puncture at C1–2, rather than a lumbar puncture. CT myelography may require little or no patient manipulation and imaging is much faster and of more reliable quality than conventional myelography.

Magnetic resonance imaging

Although MRI-compatible life-support equipment is available, it is generally more expensive and may be underused. This, combined with poor patient access, makes MRI logistically more difficult than CT in the critically ill patient. MRI is generally preferred to CT not only because it is more accurate, but also to avoid the high X-ray dose delivered by CT (Royal College of Radiologists Working Party 1995); however, considerations of radiation dose are usually far outweighed by other factors in critically ill patients.

The main contraindications to MRI are cardiac pacemakers and other implanted metallic devices such as cochlear implants, heart valves, and intracranial aneurysm clips. Not every device in each of these classes is an absolute contraindication. In practice, however, the necessary MRI compatibility details are often not available in time and, because the potential consequences can themselves be fatal, care is required. A common misconception is that orthopedic metal implants and hemostatic vascular clips in the soft tissues are a contraindication. Most implants do not move in the magnetic fields and, if they do, they are sufficiently embedded in fibrous or osseous tissue not to loosen, although generally a period of 6 weeks after insertion is preferred before exposure to MRI. Heating effects induced by electric currents are negligible.

The advantage of MRI over CT is its far greater sensitivity to soft tissue damage (Soulen *et al.* 1994). It has replaced myelography as the test of choice to demonstrate or exclude spinal cord compression; however, it may not demonstrate fractures. Hematomas that are less than 1 or 2 days old, although usually shown, may present a non-specific appearance.

Special techniques using MRI

Many novel magnetic resonance contrast mechanisms such as diffusion, perfusion, and magnetization transfer are available, as are many methods of image analysis. At present most are too constrained by limited computing power and other design features to be applied in routine clinical practice. Two which have been used are magnetic resonance angiography and, thus far to a more limited extent, diffusion-weighted imaging.

Magnetic resonance angiography can be relied upon to demonstrate carotid stenoses, most carotid and major intracranial branch occlusions, virtually all aneurysms greater than 3 or 4 mm in size, and the patency of dural sinuses and cerebral veins. It is usually not sufficient to display vascular malformations adequately. Magnetic resonance angiography is prone to artefacts, and is sensitive even to slight patient movement and magnetic field distortion.

Diffusion-weighted imaging requires the application of a strong magnetic field gradient, usually in one of three possible directions. Diffusion causes loss of signal, so that areas of restricted diffusion appear brighter. Diffusion-weighted imaging is used to identify ischemic brain damage at an earlier stage than is possible with routine MRI, but early signal changes detected by this technique can be reversible.

Other imaging modalities

Ultrasound

This technique, which is useful in other areas, has found only limited application in the central nervous system. A major advantage is that it can be brought to the bedside. It can be used to detect intracranial hemorrhage and hydrocephalus in the postoperative patient, but in practice the views are usually too restricted to be sufficiently reliable and CT is preferred. Doppler ultrasound reliably detects carotid stenosis and occlusion. Transcranial Doppler can provide information about flow in major intracranial arteries and has been used to cue use of spasmolytic drugs in subarachnoid hemorrhage and exchange transfusion in sickle cell disease, but its reliability cannot be accepted without question.

Nuclear medicine

Outside medical research the roles of nuclear medicine in patient care are continuing to contract rather than expand. Some specialized units with the appropriate facilities may use techniques such as single-photon-emission CT (SPECT) to make clinical observations on cerebral blood flow, which may cue various interventions, or to establish brain death, but in practice little information is provided that cannot be obtained or deduced, often more reliably, in other ways.

Cerebral angiography

Conventional or digital subtraction angiography is virtually never indicated in the critically ill patient. It is still needed to diagnose some aneurysms and arteriovenous malformations, and the site of torrential bleeding, particularly when endovascular treatments are contemplated.

Head injury

Indications for imaging

It is possible to predict reliably, on clinical criteria alone, patients at moderate or high risk of having intracranial hemorrhage. Cranial CT is preferred as the first and only examination for this purpose, and skull radiography should be dispensed with (Moseley 1995). CT is likely to show depressed or compound fractures and the presence of foreign bodies better than skull radiographs, and may adequately define facial and other injuries as well. CT should be available within 4 h for the following indications (Royal College of Radiologists Working Party 1995): impaired or fluctuating consciousness, the unassessable (intoxication, young children), patients at special risk (e.g. those with previous craniotomy, clinical deterioration, or failure to improve), neurological signs, or seizures. Most orbital and facial injuries do not require urgent investigation.

Plain radiography of the cervical spine is indicated only in the unconscious or unassessable patient, or when there is clinical evidence of neck injury; however, up to 30 per cent of injuries occur at C7–T1 level, which may not be well visualized. In patients with severe head injuries, fractures of C1, C2, or the occipital condyles may be shown by cranial CT; only about a third of these are recognizable on plain radiographs, suggesting that the initial cranial CT should include the spine down to the body of C3.

The digitally acquired scout scan (or scanogram) used to plan CT can be very useful in demonstrating regions such as the cervicothoracic junction and will exclude most major injuries detectable by plain radiography.

Serial CT in the absence of clinical deterioration is not required; however, an examination at discharge or transfer is desirable. The main abnormality detected on planned delayed CT is hydrocephalus, usually from 2 to 12 weeks after injury.

MRI is not generally recommended in the early management of patients; however, it may be helpful as a planned delayed investigation to assess prognosis (Gentry 1994).

Some form of angiography may be needed in rare situations, such as diagnosing and guiding treatment of complications such as severe bleeding, false aneurysms, or arteriovenous fistulas.

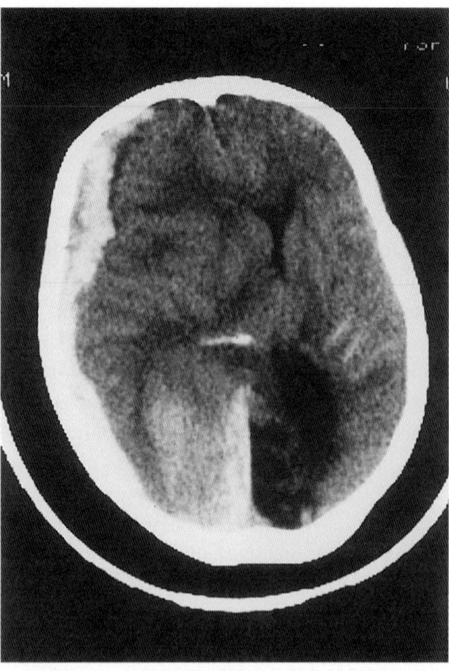

Fig. 2 CT of a patient's head 6 h after a road traffic accident; the patient is drowsy with a right hemiparesis and hemianopia. A large right subdural hematoma is shown with considerable associated mass effect, resulting in transtentorial coning and a left posterior cerebral artery territory infarct.

Intracranial hemorrhage

Extradural hematomas usually present a convex cerebral surface on axial and coronal images. They nearly always contain clotted blood which appears white on CT; a central darker area suggests active bleeding (Fig. 1). Subdural hematomas are generally more extensive and of more variable density (Fig. 2). Cerebral contusions usually are, or soon become, hemorrhagic and local mass effect increases over the first 2 or 3 days before diminishing (Fig. 3). Isolated intracerebral hematomas can first occur or appear after a few days. Massive intraventricular hemorrhage is rare and is generally seen only in patients with severe head injury, a group with a high mortality. Small amounts of intraventricular blood are common and have no prognostic significance.

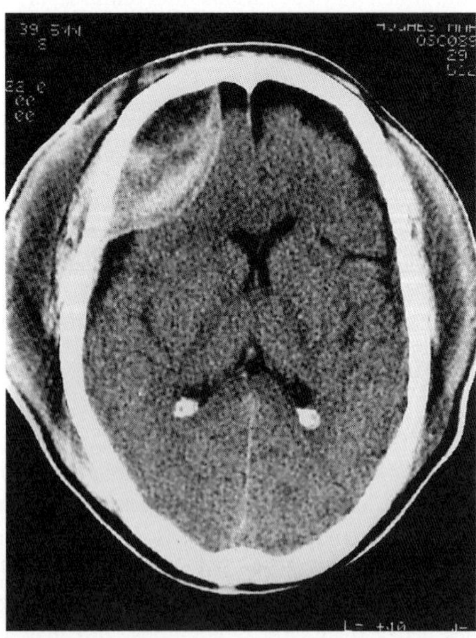

Fig. 1 CT of a patient's head after a fall, showing an acute extradural hematoma under the right side of the frontal bone. Areas of low density within it are suggestive of active bleeding.

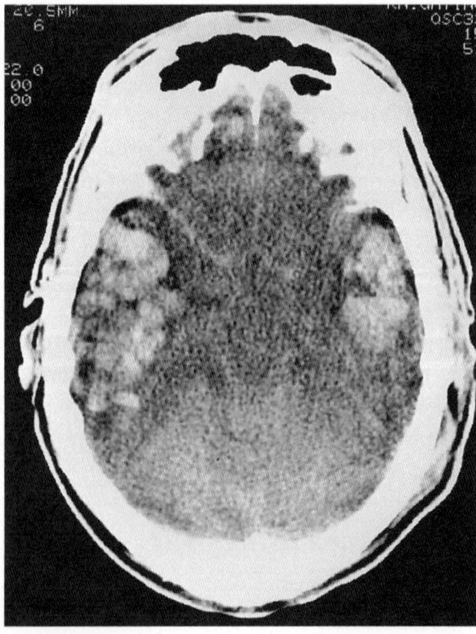

Fig. 3 CT of a patient's head after a severe road traffic accident; the Glasgow Coma Score on admission was 9. Bilateral hemorrhagic contusions are present in the temporal and frontal lobes, with obliteration of the basal cisterns. Low density in the brainstem relative to the cerebellum is a poor prognostic sign. The patient's condition deteriorated and he died 3 days later.

Diffuse axonal injury

CT and MRI show only the so-called marker lesions. They are typically small and occur in characteristic sites in the cerebral white matter including the corpus callosum and upper brainstem; only about 30 per cent appear hemorrhagic. Diffuse vascular injury may manifest as bilateral hemorrhages in the basal ganglia. A hematoma at the pontomedullary junction may indicate a tear in the lower brainstem. MRI is more sensitive than CT in detecting marker lesions.

Post-traumatic cerebral infarction

About one-third of post-traumatic cerebral infarctions are visible within 24 h, and the rest in up to 14 days.

Outcomes

The size of intracranial hematomas on CT is of predictive value, but not the location or degree of midline shift. A normal CT is a very good prognostic sign. The predictive value of delayed MRI over CT is still being evaluated (Gentry 1994).

Suspected raised intracranial pressure

A normal CT or MRI means that intracranial pressure will be normal in the great majority of cases, although not all. The interpretation of 'generalized brain edema' is usually incorrect when the brain substance appears normal, and diffuse hyperemic brain swelling is more common. A mass inducing a midline shift, particularly when associated with dilatation of the contralateral ventricle, is an indication of raised intracranial pressure, as is progressive obliteration of the basal cisterns and compression of the brainstem (Fig. 3). Imaging has shown that pressure cones (herniations) at either the tentorial hiatus or the foramen magnum usually are not lethal in themselves, and may even be asymptomatic (Fisher 1995).

Lumbar puncture and prior brain imaging

CT or MRI generally are not recommended prior to lumbar puncture unless consciousness is depressed or there are neurological signs. In most other cases the main reason for considering cranial imaging is that it may demonstrate lumbar puncture to be unnecessary or inappropriate.

Coma

In the two main causes, head injury and disturbances of cerebral circulation, both cranial CT and MRI may contribute significantly to diagnosis. In other conditions, such as liver disease, infections, and diabetes, they do not usually have a major role.

Diffuse hypoxic brain damage is usually recognizable on initial CT or MRI, but the signs may be subtle. The most common appearance on CT after prolonged cardiac arrest is diffuse bilateral loss of superficial gray–white matter differentiation (Fig. 4). Sulci, ventricles, and basal cisterns often, but not always, appear small or effaced. The basal ganglia may be spared or, less commonly in hypoxic states induced, for example, by some poisonings, they may be selectively involved.

Similar changes, although usually much less symmetrical, may be seen in some forms of encephalitis and can be reversible, sometimes rapidly so. MRI may show abnormalities not revealed by CT, particularly brainstem involvement, but appearances are generally nonspecific and do not make the diagnosis in themselves. In herpes simplex encephalitis both CT and MRI can be normal in the first few days, and the abnormalities shown are less specific than is sometimes implied.

(a)
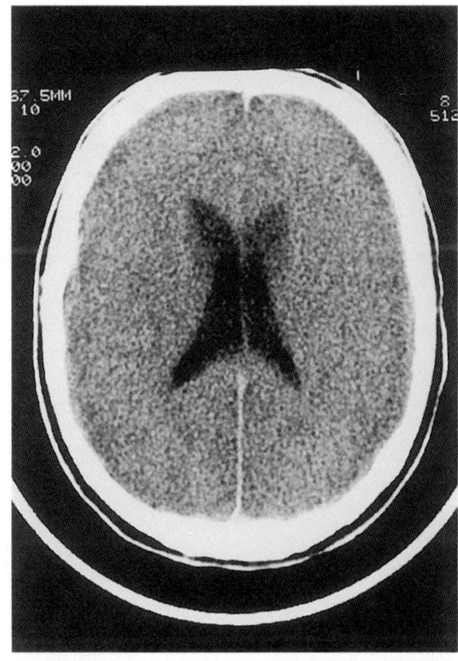

(b)
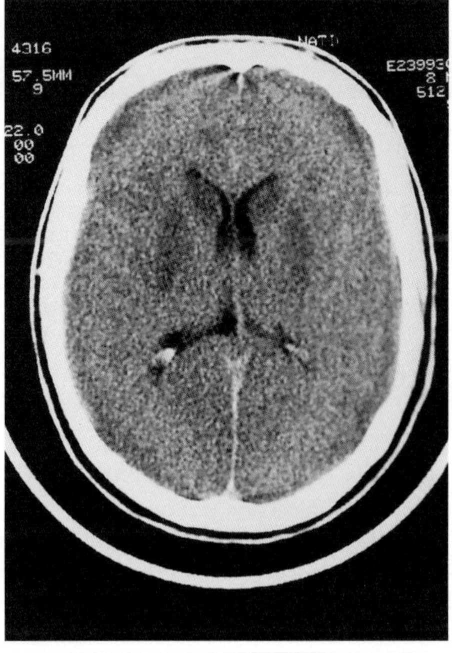

Fig. 4 CT of a patient's head 8 h after a cardiac arrest. Severe diffuse hypoxic brain damage is present with loss of differentiation of gray and white matter because of low density in the damaged cortical gray matter. This case also shows the so called 'CT reversal sign', namely relatively low density of the damaged basal ganglia when compared with the more normal white matter.

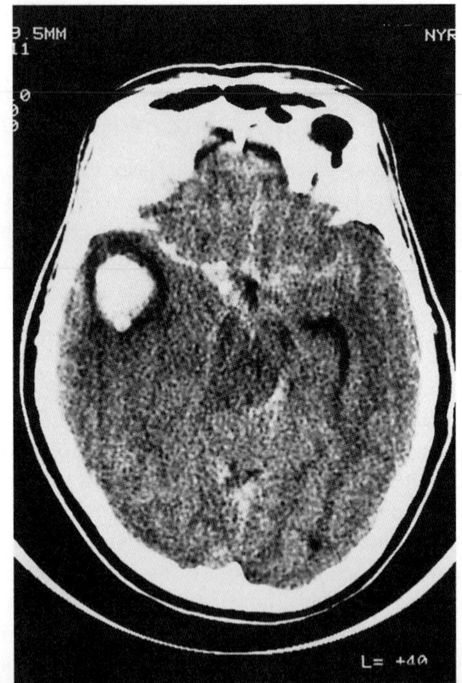

(a)

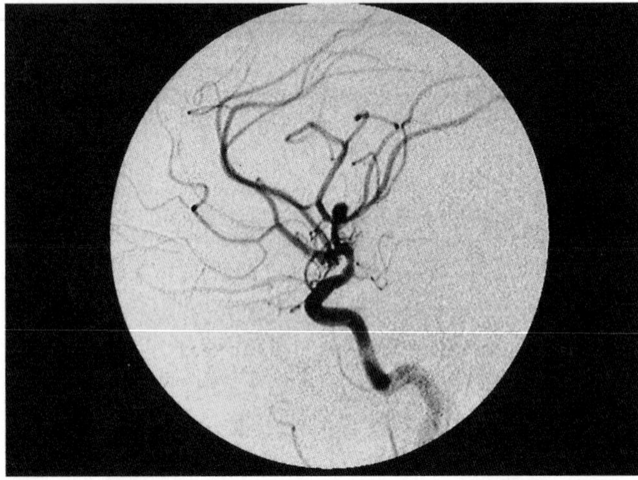

(b)

Fig. 5 (a) Plain CT of a patient's head 12 h after the onset of severe headache showing extensive subarachnoid blood. A localized partly intracerebral hematoma in the right temporal lobe suggests that the cause of hemorrhage is a middle cerebral artery aneurysm. (b) Cerebral digital subtraction angiography: selective injection into the right internal carotid artery demonstrates a right middle cerebral artery aneurysm.

Suspected subarachnoid hemorrhage

Subarachnoid blood will be seen on cranial CT (Fig. 5) in 90 per cent of cases if performed within 3 days of aneurysmal hemorrhage, but in less than 40 per cent after 5 days. In contrast, MRI may detect aneurysmal hemorrhage in about 50 per cent of cases within 3 days, but in over 80 per cent after 3 days and perhaps more when non-routine protocols are used. Nevertheless the specificity of MRI remains questionable; CT is the preferred initial investigation and should precede lumbar puncture. MR or CT angiography have not yet replaced invasive angiography for the detection of small aneurysms that have bled. Subarachnoid clot confined to the midline basal cisterns is very unlikely to be due to an angiographically definable cause and is benign. The development of delayed cerebral infarction after subarachnoid hemhorrage is associated with the overall extent of the blood shown by CT, not its location.

Acute dissection of the wall of the internal carotid artery, which can simulate subarachnoid hemhorrage clinically, is best diagnosed by axial MRI which shows changes in the vessel wall where the hemorrhage is located.

Acute stroke

Both CT and MRI will, with careful interpretation, demonstrate almost all detectable cerebral infarcts within the first 12 h. They will show the early appearance of hemorrhage, although in some cases CT may be easier to interpret. Primary hemorrhages are often distinguishable from hemorrhagic stroke; the former usually show surrounding brain damage confined to the periphery of the hematoma and not extending in the territory of a major artery.

Current trends in the diagnosis of arteriosclerotic carotid stenosis or occlusion, or intracranial branch occlusions, are towards non-invasive vascular imaging (CT angiography, magnetic resonance angiography, Doppler ultrasound).

Seizures

Seizures should be regarded as a significant sign of neurological dysfunction, and early brain imaging by CT or MRI certainly is warranted. MRI will detect more structural lesions than CT, but the majority of additional lesions shown will not have an impact on initial management.

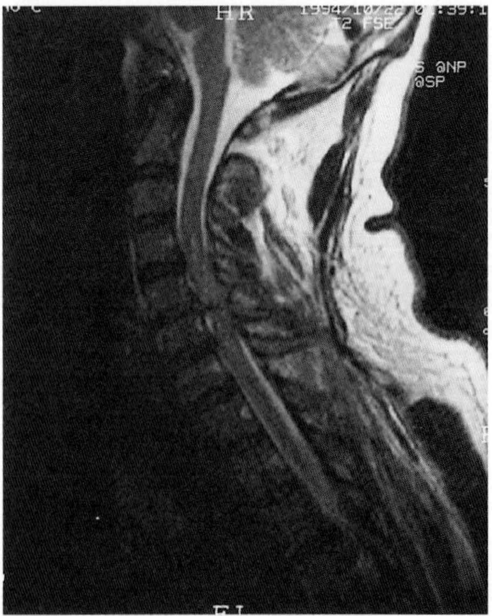

Fig. 6 MRI of the cervical spine in a pedestrian hit by a motor car, resulting in acute paraplegia. An acute disk protrusion at C5–6 results in a spinal cord compression. There is also diffuse increased signal in the spinal cord for one and a half segments above and below, indicating acute spinal cord damage and perhaps edema.

Spinal conditions

Trauma

Spinal imaging is indicated in the alert traumatized patient only when there is pain or signs of neural injury. With spinal cord injury, imaging the entire spine should be considered because a significant number of such patients have fractures at more than one level. High-quality plain radiographs are the single most useful test (Royal College of Radiologists Working Party 1995), although, once detected, fractures are more completely evaluated by CT. Flexion and extension views may be helpful in revealing instability after pain and muscle spasm have subsided.

Imaging the spinal cord is often indicated. MRI may demonstrate potentially treatable causes of traumatic cord compression (Fig. 6), and is of value in predicting recovery, showing areas of cord damage not contiguous with spinal fracture, and diagnosing the cause of delayed deterioration.

Acute non-traumatic myelopathy

The main causes of acute non-traumatic myelopathy are spinal cord compression and transverse myelitis. Although it remains accepted that urgent imaging is required to exclude compression, in practice most patients referred are already paraplegic and have little prospect of useful recovery.

MRI is by far the preferred imaging modality because it is more sensitive and much more specific than myelography for the range of pathologies, both inside and outside the spinal cord, that may be shown.

References

Fisher, C.M. (1995). Brain herniation: a revision of concepts. *Canadian Journal of Neurological Science*, **22**, 83–91.

Gentry, L.R. (1994). Imaging of closed head injury. *Radiology*, **191**, 1–17.

Moseley, I. (1995). Imaging the adult brain. *Journal of Neurology, Neurosurgery and Psychiatry*, **58**, 7–21.

Royal College of Radiologists Working Party (1995). *Making the best use of a department of clinical radiology: guidelines for doctors* (3rd edn), pp. 26–33, 66–80. Royal College of Radiologists, London.

Soulen, R.L., Duman, R.J., and Hoeffner, E. (1994). Magnetic resonance imaging in the critical care setting. *Critical Care Clinics*, **10**, 401–16.

16.5 Laboratory monitoring

16.5.1 Biochemical monitoring

D. C. Borgstrom and J. A. Weigelt

Key messages

- Current technology has allowed the development of various biochemical monitors which work through either electrochemical potential gradient differences or optical changes related to solute concentrations.
- Unfortunately, continuing problems with biocompatibility have limited the clinical application of these monitors.
- Further investigation will undoubtedly define improved sensing techniques which maintain reliability *in vivo* through improved efforts at biocompatibility while minimizing risk to the biological system.
- Clinical situations which warrant instantaneous interpretation of electrolyte, metabolite, and solute concentrations are few. Bedside analyzers or intensive care unit stat laboratories have also improved greatly and often give results for clinical use very rapidly.

Introduction

Biochemical analysis is becoming increasingly sophisticated. The number of metabolites that can be rapidly quantified in blood has grown tremendously in the last 20 years. Many of these tests can now be performed at the bedside, on line with indwelling catheters, or ultimately *in vivo*.

The clinical need for these analyses is clear. Many clinical situations arise in the critical care setting that warrant immediate blood gas analysis, blood glucose levels, or serum potassium levels. Even with an expeditious stat laboratory in the intensive care unit, repeat tests are frequently required before the first result is returned because the clinical situation has changed. Ever-increasing concern for blood-borne infectious diseases has led to the identification of methods of evaluating electrolyte and metabolite levels which minimize the risk of contamination of the health care provider. A variety of chemical sensing techniques have been developed ranging from transcutaneous analyses to on-line *ex vivo* techniques, and ultimately on-line *in vivo* biochemical analysis technology. However, it must be remembered that, although these technological advances are astounding and have many theoretical benefits, the practical clinical benefits may not be as great. The technological advances necessary to make *in vivo* sensors biocompatible will probably be extremely expensive; consequently a cost–benefit analysis in the increasingly cost-conscious medical environment may ultimately determine the success or failure of *in vivo* biochemical analysis.

In vitro biochemical analysis

Even the most expeditious stat laboratory cannot provide the immediate feedback of on-line biochemical sensing technology, although significant improvements have been made in *in vitro* biochemical analysis. The greatest improvements have been in reducing the turn-around time for results to become available and the amount of blood used for testing. Frankel *et al.* (1994) have recently evaluated the use of a bedside microanalyzer in the acute trauma setting. They found excellent correlation between test results from the microanalyzer and from a large batch analyzer used in the central laboratory. More importantly, the mean turn-around time was reduced tenfold by using the microanalyzer which also required less than 1 ml of blood. Phlebotomy for arterial blood gas or electrolyte analysis is quite common in the critical care setting. Consequently there is significant blood loss in this critically ill population as a result of attempts to understand their physiology better. Microanalyzers may play a significant role in improving our ability to care for the critically ill. They improve the turn-around time for test results and, more importantly, they minimize blood loss. In addition, since they are not as comprehensive as central laboratory tests, we may learn to be more judicious in ordering provocative tests which frequently have little impact on patient outcome.

Ex vivo optical sensing techniques

Oximetry, or more specifically pulse oximetry, has become a standard for critical care monitoring. On-line analyzers are capable of determining blood gases, electrolytes, and hematocrit on undiluted whole blood on line. This allows for rapid interpretation of these data; however, it still represents only a snapshot of the clinical situation. Current systems are much less bulky than previous generations; however, their clinical applicability is limited for the vast majority of patients.

In vivo biochemical analysis

Current investigation has focused on *in vivo* analysis, i.e. sensors or probes which can be placed into the arterial blood stream and continuously monitor various metabolite or electrolyte concentrations. Continuous measurement and interpretation of changes can

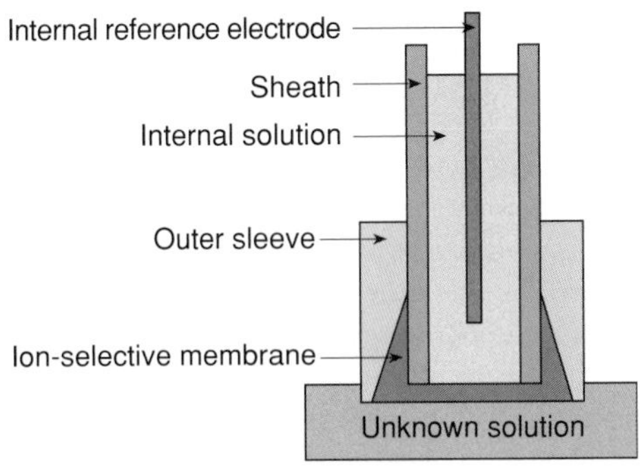

Fig. 1 Ion-selective electrode. The ion-selective membrane allows only the chemical of interest to contact the electrode. (Reproduced with permission from Koryta (1980).)

be made similarly to the current use of pulse oximetry. Electrochemical sensors and optical sensors have been developed in order to achieve this on-line continuous-analysis function. However, biocompatibility has limited their clinical efficacy.

Electrochemical sensors

Electrochemical sensors are of two types: ion-selective electrodes and ion-sensitive field-effect transistors. Both are designed to measure the electrochemical potential of a specific solute. They differ only in their method of analysis.

Ion-selective electrodes

Ion-selective electrodes (Fig. 1) are found most commonly in large batch analyzers in the clinical chemistry laboratory. The sample is diluted to minimize interference from other solutes. The ion-selective electrode allows a particular solute or metabolite through a semipermeable membrane, while excluding others, so that a potential gradient is generated at the electrode surface. This potential difference is then used to calculate the specific ion or solute concentration (Eberhart and Weigelt 1989).

Ion-sensitive field-effect transistor

The ion-sensitive field-effect transistor (Fig. 2) differs from the ion-selective electrode in the method by which it measures the electrochemical potential produced by the ionic species in question. An electrochemical potential is generated at the substrate–gate insulator interface. This potential acts as an effective gate voltage which modulates electrocurrent drain. The circuit is completed by a reference electrode in contact with the solution. Thus a major difference between the ion-selective electrode and the ion-sensitive field-effect transistor is that the latter requires a reference electrode to be in contact with the solution of interest to complete the circuit (Eberhart and Weigelt 1989). Both types of sensor have shown reliability *in vitro*; however, *in vivo* clinical application is still minimal.

Optical sensors

Optical sensors are the latest generation of biochemical monitors. Their benefit over electrochemical sensors is a simple design which improves reliability. Optical sensors generally use extrinsic indicators, typically either absorbance or luminescence, for the purposes of

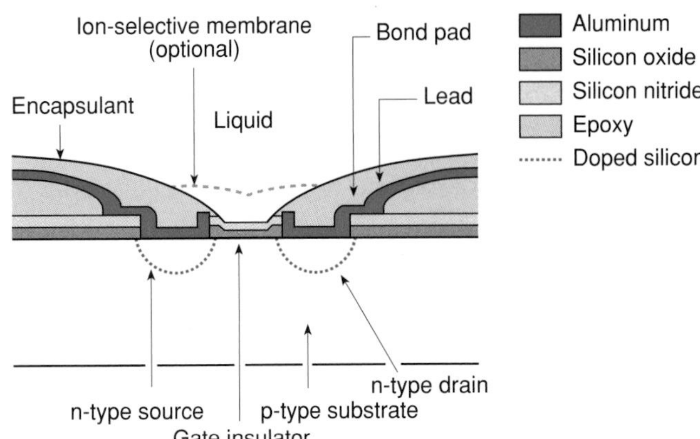

Fig. 2 Diagram of the pH-sensing ion-sensitive field-effect transistor (p-type is the solution or plasma; n-type is the electrochemical current source). Hydrogen ions accumulate at the liquid–gate insulator interface, or are sequestered in the ion-selective membrane, if used. An electric field is produced across the silicon oxide–nitride gate insulator, modulating the current flow between source and drain. (Reproduced with permission from Eberhart *et al.* (1982).)

biochemical monitoring. Absorbent or luminescent materials are fixed to the optical fiber. Light of a given wavelength is transmitted through the optical fiber, and the change in wavelength returned for analysis corresponds to the magnitude of interaction between the solute of interested and the working matrix affixed to the tip of the optical sensor (Lubbers 1995).

Biocompatibility

All the biochemical monitoring techniques described above have shown reliability and efficacy *in vitro*. Unfortunately, the complexities of human plasma, with platelets, plasma proteins, and the clotting cascade, and problems with biocompatibility have made their application *in vivo* unreliable. This has limited application to research designs or short (hours) clinical utility. Plasma protein adherence, platelet adherence, and coagulation protein adherence have all led to ineffective sensor function. Extensive investigation is now focusing on ways of improving the hemocompatibility of invasive sensors.

Exposure of an irregular polymer or fiber-optic surface leads to rapid absorption of various proteins and platelet adherence. Placing a negative charge on the surface will tend to diminish adherence of the large negatively charged plasma proteins. Covalent bonding of heparin limits platelet adhesion and deactivates coagulation enzymes such as thrombin. Albumin coating will also reduce platelet adhesion (Benmakroha *et al.* 1995).

All these approaches have been attempted with mixed success. None provides the protection needed to maintain proper sensor function and minimize risk to the biological system over a suitable length of time. The latest attempts at improving biocompatibility have been in cell membrane mimicry. Synthesized biomembrane copolymers have been developed which mimic the thrombosis-resisting properties of the relatively inert red blood cell membrane (Rolfe 1995). However, this technology is still in its infancy, and will require much further improvement before it is clinically applicable.

References

Benmakroha, Y., Zhang, S., and Rolfe, P. (1995). Haemocompatibility of invasive sensors. *Medical and Biological Engineering and Computing*, 33, 811–21.

Eberhart, R.C. and Weigelt, J.A. (1989). New chemical sensing techniques for critical care medicine. In *Textbook of critical care* (2nd edn) (ed. W.C. Shoemaker), pp. 223–30. W.B. Saunders, Philadelphia, PA.

Eberhart, R.C., *et al.* (1982). Development of indwelling sensors for continuous electrolyte monitoring. In *Applied physiology in clinical respiratory care* (ed. O. Prakash), pp. 269–78. Martinus Nijhoff, Hingham, MA.

Frankel, H.L., *et al.* (1994). Minimizing admission laboratory testing in trauma patients: use of a microanalyzer. *Journal of Trauma*, **37**, 728.

Koryta, J. (1980). *Medical and biological applications of electrochemical devices.* John Wiley, New York.

Lubbers, D.W. (1995). Optical sensors for clinical monitoring. *Acta Anaesthesiologica Scandinavica*, **39** (Supplement 104), 37–54.

Rolfe, P. (1995). Experience with invasive and non-invasive sensors for anaesthesia. *Acta Anaesthesiologica Scandinavica*, **39** (Supplement 104), 61–8.

16.5.2 Lactate

J.-L. Vincent

Key messages

- Lactate production is increased when oxygen availability in the tissues is reduced, but increased cellular metabolism or altered pyruvate metabolism can also be implicated.
- The measurement of blood lactate levels can serve as an indicator of the presence of impaired tissue oxygenation, once other possible causes have been ruled out.
- A normal lactate level can be considered to be about 1 mmol/l, and levels above 1.5 to 2 mmol/l should alert the clinician to the possibility of tissue hypoxia.
- Lactate elimination occurs primarily in the liver.
- Blood lactate levels may reflect generalized poor tissue oxygenation but give no information on regional hypoxia.

Lactate metabolism

Lactate is produced exclusively as the result of glucose metabolism as the normal endpoint of tissue glycolysis. It is produced from pyruvate, mainly by skeletal muscle, erythrocytes, the brain, the gut, and skin, in a reaction catalyzed by lactate dehydrogenase (LDH):

$$\text{pyruvate} + \text{NADH} + \text{H}^{+} \overset{\text{LDH}}{\longleftrightarrow} \text{lactate} + \text{NAD}^{+}.$$

The lactate produced is a dead-end product and is simply converted back to pyruvate, which is then utilized primarily by the liver (about 50 per cent) and the kidneys (about 25 per cent). Lactate is used for gluconeogenesis, or is oxidized to carbon dioxide and water via the tricarboxylic acid cycle in the mitochondria, producing adenosine triphosphate (ATP). In the presence of cellular hypoxia, the ratio of NADH to NAD increases so that the ratio of lactate to pyruvate increases (type A lactic acidosis). When glycolysis is stimulated or pyruvate metabolism is impaired by depressed activity of pyruvate dehydrogenase, the formation of both pyruvate and lactate can be increased, but the ratio of lactate to pyruvate remains normal (type B lactic acidosis).

Differential diagnoses of hyperlactatemia

Increased lactate production occurs when cellular metabolism is abnormal owing to either a lack of oxygen in the tissues (acute circulatory failure) or, more rarely, a failure of cellular oxygen utilization despite the presence of adequate oxygen delivery (Table 1). Clinically, the most important causes are those due to hypoxia, but let us briefly consider the other conditions.

Biguanides accelerate the conversion of glucose into lactate; at high levels they inhibit gluconeogenesis and pyruvate oxidation, thus raising lactate levels. This complication has been described primarily with phenformin, which is no longer available, so that this problem has become extremely rare.

In alcohol intoxication ethanol is oxidized to acetoacetic acid, producing large amounts of NADH and an increase in the ratio of NADH to NAD.

Thiamine is involved in the pyruvate dehydrogenase complex and biotin is involved in pyruvate carboxylase. Therefore deficiency of either substance can increase pyruvate and lactate levels.

Tumors are known to have high glycolytic and lactate production rates, probably due to increased lactate production by the tumor. Liver metastases leading to diminished hepatic lactate metabolism may represent a contributory factor.

Fructose infusion causes a rise in phosphofructokinase levels which reduces intracellular phosphate levels and hence ATP. Glycolysis is thus stimulated and gluconeogenesis inhibited, raising lactate levels. Today glucose has become preferable to fructose and sorbitol as an infused nutrient, and so this problem has virtually disappeared.

Diabetic ketoacidosis impairs lactate transport, and diabetic patients also have reduced muscle pyruvate dehydrogenase activity. However, the degree of lactic acidosis is usually fairly mild in such cases, unless associated with severe hypovolemia when some degree of tissue hypoxia may develop.

Table 1 Causes of lactic acidosis

Type A Imbalance between oxygen supply and demand
Circulatory shock
Cardiac arrest
Severe anemia
Severe hypoxemia
Seizures, intense shivering, exercise
Type B Decreased cellular oxygen utilization
Drugs and toxins, e.g. phenformin, CO, alcohol
Thiamine or biotin deficiency
Neoplastic diseases
Fructose or sorbitol infusion
Uncompensated diabetes
Inborn metabolic errors

Inborn errors of metabolism, including glucose-6-phosphate deficiency and pyruvate dehydrogenase deficiency, can also cause lactate levels that are higher than normal owing to interference with the normal metabolic pathways.

All the above are fairly easily recognized and can be distinguished from hyperlactatemia due to cellular hypoxia.

Clinical applications

The measurement of blood lactate levels can serve as an indicator of the presence of impaired tissue oxygenation, once other possible causes have been ruled out. Many experimental and clinical studies have found raised lactate levels in circulatory failure which correlate well with the severity of the shock state. Raised levels have also been associated with the oxygen uptake/supply dependency phenomenon. Hyperlactatemia has been linked with increased mortality, and falling levels have been shown to indicate a better prognosis.

The measurement of blood lactate levels from a simple blood test using automated analyzers is straightforward and readily available, and repeated sequential levels are easily obtained. The enzymatic conversion of lactate to pyruvate is an oxygen-utilizing reaction. The extraction of oxygen from the sample can be detected by a sensitive oxygen fuel cell sensor and is directly proportional to the sample lactate concentration. A whole-blood sample can be collected in a heparin fluoride tube to prevent coagulation and glycolysis. Nitrite can be used in the sample tube to convert hemoglobin to the met-form, thus avoiding uptake of oxygen during the enzyme reaction. The enzymatic method is specific for the L-isomer and therefore will not detect D-lactate (e.g. in short-bowel syndrome). Levels are identical in the central venous blood, the pulmonary artery, and the arterial blood.

Limitations

Non-hypoxemic causes

As mentioned above, blood lactate levels are raised not only in tissue hypoxia associated with acute circulatory failure but also in certain other conditions (Table 1). Fortunately, these are relatively rare, except for seizures, and all are fairly easily recognized.

Overproduction/underelimination

The lactate level at any time is a balance between its production and its elimination. Overproduction of pyruvate, and hence of lactate, occurs primarily as the result of inadequate oxygenation as outlined above, although some can be produced from protein catabolism. Therefore excessive muscle breakdown, as seen in critical illness, could contribute to raised lactate levels. Since elimination occurs primarily in the liver, liver failure might be expected to cause abnormally high lactate levels. Patients with stable cirrhosis maintain relatively normal levels, but the presence of liver disease can certainly magnify levels already raised due to circulatory failure and slow down their return to normal once resuscitation is complete.

Washout phenomenon

Low blood flow during early hypovolemic shock may cause sequestration of lactate in tissues, so that increased flow following fluid replacement could then cause an apparent sudden rise in blood lactate by a 'washout' phenomenon. Such a phenomenon is of relatively limited amplitude and should not be overestimated.

Endotoxin

Some animal studies have shown increased lactate production in the presence of endotoxin, even in the absence of tissue hypoxia. This may be due to inactivation of pyruvate dehydrogenase, the enzyme mediating pyruvate transformation to acetyl-coA for entry into the tricarboxylic acid cycle. The clinical application of these observations is limited, as patients with severe sepsis but hemodynamic stability typically have normal lactate levels.

Global not regional

Blood lactate levels may reflect generalized poor tissue oxygenation but give no information on regional hypoxia. Lactate is produced by a wide variety of tissues and overproduction by just one organ is uncommon.

Conclusions

Despite the drawbacks, blood lactate levels remain a useful means of assessing tissue oxygenation. The importance of tissue hypoxia in the development of multiple organ failure makes any such measure of tissue oxygenation valuable. A normal value can be considered to be about 1 mmol/l, and levels above 1.5 to 2 mmol/l should alert the clinician to the possibility of reduced tissue oxygenation. Raised levels must be interpreted cautiously in the light of clinical patient assessment and other available indices. Changes in blood lactate levels occur too slowly to guide therapy directly and therefore can never replace established hemodynamic variables. Evaluation of blood lactate levels over time are of most use in following the course of hypoxia and enabling the effects of treatment to be monitored. A combination of blood lactate with a more regional measure of tissue perfusion, such as gastric pHi, may give a more valuable assessment of the overall situation.

Bibliography

Aduen, J., *et al.* (1994). The use and clinical importance of a substrate-specific electrode for rapid determination of blood lactate concentrations. *Journal of the American Medical Association*, **272**, 1678–85.

Bakker, J., Gris, P., Coffernils, M., Kahn, R.J., and Vincent, J.L. (1996). Serial blood lactate levels can predict the development of multiple organ failure following septic shock. *American Journal of Surgery*, **171**, 221–6.

Friedman, G., Berlot, G., Kahn, R.J., and Vincent, J.L. (1995). Combined measurements of blood lactate concentrations and gastric intramucosal pH in patients with severe sepsis. *Critical Care Medicine*, **23**, 1184–93.

Gutierrez, G. and Wulf, M.E. (1996). Lactic acidosis in sepsis: a commentary. *Intensive Care Medicine*, **22**, 6–16.

Vincent, J.L., Dufaye, P., Berre, J., Leeman, M., Degaute, J.P., and Kahn, R.J. (1983). Serial lactate determinations during circulatory shock. *Critical Care Medicine*, **11**, 449–51.

16.5.3 Colloid osmotic pressure

A. R. Webb

Key messages

- The colloid osmotic pressure is that pressure which must be exerted on a colloid solution to prevent any net movement of solvent across a semipermeable membrane.
- Colloid osmotic pressure is dependent on the concentration of solution, the effects of electrically charged colloid molecules on the small ions in the solvent, the effective increase in concentration of solution by the space taken up by large molecules, and other intermolecular interactions.
- In practice, any patient with edema and intravascular hypovolemia needs colloid; raising plasma colloid osmotic pressure is desirable but not always attainable.
- A high plasma colloid osmotic pressure may defer recovery from acute renal failure.

Introduction

The need for urgent and rapid restoration of circulating volume in states of hypovolemia is well recognized but the choice of solution is less clear. To restore and maintain circulating volume a colloid solution must improve and maintain intravascular colloid osmotic pressure (COP) without excessive contribution to interstitial COP.

What is COP?

Osmosis is the process by which a solvent moves from a solution with a low solute concentration to a solution with a higher solute concentration. The osmotic pressure can be defined as that pressure which must be exerted on the solution to prevent any net movement of solvent across a semipermeable membrane. When the semipermeable membrane allows free transfer of salts and small ions, whilst restraining the passage of colloid molecules, the osmotic pressure of this system is termed the colloid osmotic pressure.

The capillary wall impedes transfer of protein (colloid) molecules which are dissolved in a solvent of water and electrolytes such that they are in much greater concentration on the intravascular side than on the interstitial side; therefore COP is greater in the intravascular compartment than in the interstitial compartment. This imbalance of COP across the capillary membrane counteracts the imbalance in hydrostatic pressures and, together, these pressures determine the bulk flow of water and electrolytes across the capillary wall. These Starling forces are related by

$$J_v \propto [(P_c - P_i) - \sigma(\pi_c - \pi_i)]$$

where J_v is the net fluid transfer across the capillary membrane, P_c is the capillary hydrostatic pressure, P_i is the interstitial hydrostatic pressure, π_c is the intravascular COP, π_i is the interstitial COP, and σ is the reflection coefficient which has a value between 0 and 1. If the transport passages across the capillary totally impede transfer of colloid molecules, $\sigma = 1$. A low reflection coefficient means that a high COP difference will be less effective at counteracting the hydrostatic pressure difference.

How do we measure COP?

COP is most commonly measured using a membrane transducer system (Fig. 1). The method requires detection of flow across a semipermeable membrane from a closed reference chamber to the sample chamber. As fluid moves from the closed reference chamber into the sample the hydrostatic pressure in the reference chamber becomes progressively more negative, tending to counterbalance the COP of the sample. When the hydrostatic pressure in the reference chamber equals the osmotic pressure in the sample chamber, an equilibrium is reached. Using such devices, it typically takes less than 3 min to measure the COP of a sample of heparinized blood, plasma, or serum which may be as small as 200 µl. However, when samples are small, it is often necessary to inject several aliquots into the oncometer. This is because the initial high pressure gradient between the COP of the sample and the zero-pressure conditions of the reference chamber causes significant sample dilution. With several repeated injections of sample and as the hydrostatic pressure in the reference chamber approaches the COP of the sample, dilution becomes insignificant. Although COP measurement is easy, there are several potential artifacts which may lead to erroneous results: free hemoglobin in the sample due to hemolysis has a significant COP, liquid heparin reduces COP by dilution, excess dry heparin molecules contribute to COP, a reduction in pH reduces the negative charge on protein, and fibrinogen molecules contribute to COP.

Calculation of COP

By analogy with the ideal gas laws, van't Hoff stated that the COP of an ideal solution (infinitely dilute) is proportional to the concentration of the colloid (Webb 1992).

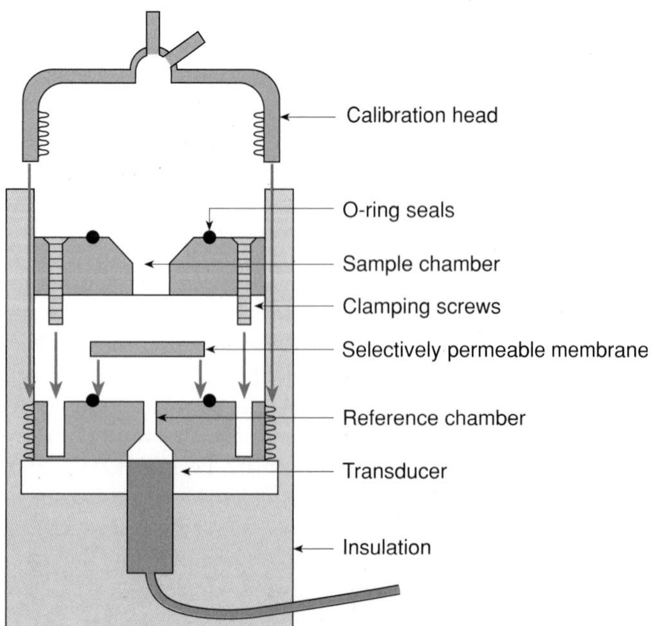

Fig. 1 A membrane transducer oncometer. The COP is the negative pressure in the reference chamber which prevents net movement of salt and water across the semipermeable membrane.

Deviation from van't Hoff behavior

The COP of normal plasma is between 25 and 30 mmHg. According to the van't Hoff analogy and assuming a protein concentration of 70 g/l and an albumin concentration of 40 g/l, COP for normal plasma becomes 15.1 mmHg at 37 °C (Webb 1992). This is a 40 to 50 per cent underestimate of the measured colloid osmotic pressure, representing considerable deviation from ideal behavior. If van't Hoff's law applied to the behavior of plasma solutions, and assuming an albumin-to-globulin ratio of 1.33 as above, the concentrations of albumin and globulin required to give a colloid osmotic pressure of 25 mmHg would be 67 g/l and 50 g/l respectively. Thus the total protein concentration would be 117 g/l, leading to a considerably higher plasma viscosity (Webb 1992).

Gibbs–Donnan equilibrium

The effect of electrical charge on protein molecules separated from a protein-free ultrafiltrate by a semipermeable membrane is to cause an unequal distribution of electrolytes across the membrane. Human albumin is known to carry a net negative charge of 17 at the physiological pH of 7.4 when titrated against a 0.15-M sodium chloride solution.

The Gibbs–Donnan equilibrium requires that the products of diffusible ions on each side of the membrane must be equal. To achieve this equilibrium while maintaining electrical neutrality, an increase in albumin concentration attracts cations and reduces the concentrations of ions in the ultrafiltrate. An osmotic gradient is generated across the membrane which is due to the imbalance in concentration of diffusible ions in addition to the effect of the albumin. This imbalance contributes significantly to the measured osmotic pressure across the membrane.

The excluded-volume effect

Deviation from van't Hoff behavior increases with increasing colloid concentration for all the commonly available plasma substitutes, whether they are electrically charged or not, with the exception of polygeline (Webb 1992). In a polydisperse solution of macromolecules, the shape and size of the larger molecules may be such that they occupy a significant portion of the total volume of the solution. Thus the effective concentration of the smaller molecules is greater, since the volume available for molecular motion is reduced by the space taken up by the larger molecules.

This so-called excluded-volume effect is clearly not apparent for solutions of infinite dilution, since the solute will take up no space under these theoretical conditions.

Human albumin and modified fluid gelatin are the only two plasma substitutes containing charged colloid molecules; the remainder are neutral polymer mixtures. The non-linear relationship between COP and concentration is best explained by the excluded-volume effect for the neutral colloids, and this effect also makes an appreciable contribution for charged colloids. The molecular weight range of polygeline is well below that of the other neutral colloids; hence less space is taken up by these small molecules.

Other solute and solvent interactions

Some colloidal molecules may interact with the salts in the solvent. In particular, albumin is known to bind chloride ions and the fraction of salt available to take part in the Gibbs–Donnan equilibrium is reduced. This leads to a further increase in COP at physiological pH.

Other forces altering the behavior of solutions of polymers include intermolecular attractive and repulsive forces between molecules. These effects are likely to be small, as far as COP is concerned, for solutions used at clinical concentrations.

The physiological significance of the non-linear relationship between COP and concentration.

Deviation from van't Hoff behavior as colloid concentration increases is physiologically advantageous, reducing the viscosity of solution required to generate an adequate COP. If COP were dependent only on molecular numbers (linearly related to concentration), the provision of an adequate COP would require molecules too small to be retained within the circulation or molecules so large that the viscosity of the solution would be far in excess of that of plasma.

COP measurement in the critically ill

The ability of the capillary to impede transfer of plasma proteins is impaired in the critically ill (Fleck *et al.* 1985). The reflection coefficient of the capillary wall is reduced and the COP difference becomes less effective. The COP is commonly less than 20 mmHg in the critically ill as albumin leaks out of the circulation.

Formulas to calculate COP are reasonably accurate in health. However, formulas based upon albumin concentration severely underestimate measured COP in the critically ill (Barclay and Bennett 1987). There are two predominant reasons for this inaccuracy.

1. Globulin molecules are much larger than albumin molecules and the ratio of albumin to globulin is usually reduced in the critically ill; this reduces the number of molecules contributing to COP at a given total protein concentration.
2. Artificial plasma substitutes, which are not accounted for in the formulas, contribute to COP.

Although the clinical measurement of COP provides us with the only accurate measure of this Starling force, it is not as useful as it might appear at first sight. The importance of maintaining the difference between COP and pulmonary artery wedge pressure above 3 mmHg has been stated as a method of avoiding pulmonary edema and allowing reabsorption of established pulmonary edema (Weil *et al.* 1979). While such an achievement undoubtedly helps, it does not take into account the ability of the pulmonary lymphatics to clear interstitial fluid, the ability of the pulmonary capillary endothelium to retain colloid molecules on the intravascular side, or the ability of the pulmonary lymphatics to clear leaked colloid molecules (Civetta 1979). We must be clear in our fluid management of critically ill patients that edema formation is not simply an imbalance of Starling forces allowing excessive transfer of capillary fluid to the interstitium, but is also a failure of the lymphatic drainage system to cope with the additional fluid transfer. Furthermore, the fluid-balance relationships found in one part of the circulation do not necessarily apply to others.

With the currently available plasma substitutes, it remains very difficult to normalize plasma COP in the critically ill, particularly if they have capillary leak. We cannot routinely measure the reflection coefficient, which varies amongst different capillary beds. Although we may be able to increase circulating colloid concentration, if the reflection coefficient is low any difference between plasma and interstitial COP will have less effect against the hydrostatic pressures. In practice, any patient with edema and intravascular hypovolemia needs colloid.

A high plasma COP is a risk if high-molecular-weight plasma substitutes are used in renal failure. Even if dialysis or hemofiltration are used, the pore sizes of typical dialyzers and filters do not allow adequate clearance of larger colloid molecules. The high plasma COP prevents glomerular filtration which is governed by the Starling equation. Measurement of plasma COP and appropriate use of crystalloid or colloids of lower molecular weight may prevent deferred recovery of acute renal failure.

References

Barclay, S.A. and Bennett, D. (1987). The direct measurement of plasma colloid osmotic pressure is superior to colloid osmotic pressure derived from albumin or total protein. *Intensive Care Medicine*, 13, 114–18.

Civetta, J.M. (1979). A new look at the Starling equation. *Critical Care Medicine*, 7, 84–91.

Fleck, A., *et al.* (1985). Increased vascular permeability: a major cause of hypoalbuminaemia in disease and injury. *Lancet*, **i**, 781–3.

Webb, A.R. (1992). Colloid osmotic pressure measurement in clinical practice. In *Plasma volume expansion* (ed. J.F. Baron), pp. 33–52. Arnette Blackwell, Paris.

Weil, M.H., Henning, R.J., and Puri, V.K. (1979). Colloid osmotic pressure: clinical significance. *Critical Care Medicine*, 7, 113–16.

16.5.4 Full (complete) blood count

I. Cavill and C. A. J. Wardrop

Key messages

- The request for a full blood count implies, in addition to routine screening, a request for evaluation of the blood for any unexpected diagnosis.
- The reliable production of red cell absolute values and white cell and platelet indices depends on expert hematological quality control and standard laboratory operating procedures.
- The main indices of diagnostic usefulness in the full blood count are hemoglobin concentration (or hematocrit), mean red cell hemoglobin content and volume, total and differential white blood cell counts, and platelet count.
- Hemoglobin concentration (and hematocrit) do not predict total red cell mass or blood volume in intensive care patients. The development of widely applicable methods for these estimates is urgently needed in the assessment of such patients.

Hematological screening

When a physician requests hematological screening of the critically ill, the implicit requirement is evidence of any unexpected hematological diagnosis or complication, including possible drug toxicity; 'full blood count and film (smear)' allows this evaluation.

There are two reasons for measuring a full blood count on a venous (or arterial) sample which are particularly relevant to the critically ill. The first is simply to measure the hemoglobin concentration (or hematocrit) as an indicator, albeit imperfect (see below), of the total red cell mass in the body. The second is to monitor changes in the red cell, white cell, and platelet populations. Because marrow cell production is a highly dynamic process (new red cells are formed at a rate of 2×10^6/s), it is a sensitive biological indicator of adverse changes at the cellular level.

There is only one way to obtain the correct hemoglobin concentration, which is to take a venous (or arterial) blood sample and measure the hemoglobin concentration with a standardized and properly controlled instrument (Thom 1990). Hemoglobin concentration in a capillary sample is entirely dependent upon the uncontrollable degree of hemoconcentration involved in the process. Most blood gas analyzers do not measure hemoglobin; rather, they infer it. They will be fooled by denatured or aberrant forms of hemoglobin and thus usually use diluted samples.

How the laboratory makes its measurements is important and relevant to the way in which results are viewed. The basic analysis is the full blood count (hemoglobin, red blood cells, packed cell volume, mean red cell volume and hemoglobin content, neutrophils, lymphocytes, monocytes, eosinophils, basophils, and platelets) which is produced as a package by automated cell counters in about 50 s. Reticulocyte counts may become increasingly useful. These results must be available on a time scale which meets intensive care needs. Near-patient testing is only one option, and the benefits and penalties of this must be fully considered. When it is necessary, it should be delivered on site by the laboratory.

Counting blood cells appears to be very easy, but it is not. The technologies that are used to overcome the problems involved can differ substantially between different counters, and it is important to appreciate the strengths and limitations of each type. No result is better than the technique and quality control that has gone into producing it. It is even more important for the clinician to appreciate that the major cause of wrong results is misidentification of the sample. A total quality strategy begins with, and is entirely dependent on, the correct sample from the correct patient going into a bottle that is correctly identified.

It is no longer appropriate to refer to the laboratory and its 'Coulter counter'. Cell counters are not all the same and they have differing solutions to the problem of counting cells. These flexible entities come in a wide range of shapes and sizes. The first way in which technologies differ relates to the way in which they stabilize the cells before they can be counted. The second difference is in the way in which the separate blood cell populations are distinguished from one another.

Some technologies rely heavily on an assessment of size to identify blood cells. This is rarely clear cut, and these systems must employ sophisticated software to try and set cut-off points between the different populations. The problem is particularly difficult when the red cells are microcytic. Although the mean volume of the red cell population is clearly distinct from that of the platelets, the wide distribution of cell sizes contained with the red cell population means that a significant number may be of very similar size to the larger platelets. At the other end of the platelet distribution, i.e. for the smallest fragments in the population, the problem is to distinguish these from other small particles (dust) and the electronic noise in the sensing systems.

The problems involved in counting and classifying white cells (Bentley 1990) are even more complex and revolve around how to define the white cell populations in terms which can be appreciated by a machine rather than the human eye. Traditionally, division of the total white cell population into its constituent elements has been based on visual assessment of a blood smear treated with appropriate dyes. This was never a reliable quantitative measure. Older machines attempt to distinguish white cell populations on the basis of an analysis of the size distribution alone. More recent versions may use other characteristics related to cell conductivity as well as size. The best of the modern machines use size and cytochemical characteristics to define the white cells. While each of the technologies will give similar results in normal samples, they can differ considerably when pathological samples are analysed. This means that analyses carried out on one machine cannot always be translated to another clinical laboratory setting. Reference ranges for any given laboratory must reflect the technology that is in use.

It has been historical practice to express the white cell differential in terms of the percentage of the total white cell count. This has probably contributed substantially to misunderstandings of the significance of the differential count data. It should have been clear that when the total white cell count either increases or decreases, the absolute numbers of each of the white cell populations will change significantly even with a normal percentage differential. All white cell differential count data must be expressed and understood as an absolute count (in units of 10^9/l) for each of the populations. If an automated differential count is not available, a total white blood cell count of less than 1×10^9/l represents severe neutropenia.

In the measurement of red cells there are significant differences in the analyses carried out by different automated cell counters. While the red cell count will often be similar, the red cell indices, particularly the mean cell volume, may vary substantially. As a measure of cell size, the mean cell volume has considerable drawbacks. These arise from the essentially elastic nature of the red cell. *In vivo*, red cells do not have a fixed volume. *Ex vivo*, their volume will vary in relation to storage conditions and oxygenation. A highly agitated and oxygenated sample will have a lower mean cell volume than an unoxygenated sample, but as time passes the cells in an EDTA anticoagulated sample will swell as they take up water. In addition, the red cell can function as an osmometer and will reflect, for example, changes in D-glucose plasma levels. In contrast, the mean cell hemoglobin is unaffected by different technologies and totally unaffected by the storage changes which afflict the mean cell volume.

Changes in erythropoietic activity can be monitored by a quantitative reticulocyte count. Because reticulocytes only exist in the circulation for about 12 to 60 h, changes in the reticulocyte count will reflect current erythropoietic output. Inflammatory cytokines will suppress erythropoiesis, and the earliest indication of this may be obtained by quantitative reticulocyte counting. This should not be confused with the semiquantitative reticulocyte estimation produced as a result of a visual microscope assessment of the reticulocytes on a spread blood film. The latter has a coefficient of variation of 30 to 50 per cent in normal subjects, and may be even higher in cases where the reticulocyte count is diminished. The newest counters now have automated reticulocyte counting built in as part of the regular full blood count.

Iron supply to the erythropoietic tissue has an immediate effect on the adequacy of hemoglobin synthesis. When the rate of iron supply is insufficient to meet the demands of the erythroid tissue, hypochromic red cells will be produced. This is functional, as opposed to storage, iron deficiency. The measurement of the percentage of cells which are hypochromic, i.e. that have an individual red cell hemoglobin concentration below 28 g/dl, is the single most effective variable for monitoring this condition. This measurement should not be confused with the mean corpuscular hemoglobin concentration which reflects the mean hemoglobin concentration in all the cells and is relatively stable and unchanging. Functional iron deficiency is a major limiting factor in the response to erythropoietin therapy. It is most common in patients with chronic renal failure treated with recombinant erythropoietin, but may be equally significant in patients with decreased oxygen supply which stimulates erythropoietin production.

At present only one family of instruments (Technicon H®, Bayer Diagnostics) is capable of making all these analyses. These counters are widely available in hematology laboratories in the United Kingdom, but it is important to ascertain from the laboratory which technologies are being used. When quantitative reticulocyte counting and quantitative percentage hypochromic estimates are required, these needs should be made known to the laboratory. However, interpretation and applications of these indices have yet to be established in intensive care patients.

Near-patient testing

The incentive for near-patient testing should be the clinical need to return the results of investigations to the clinician as speedily as possible. This is a laudable, acceptable, and achievable aim.

There are many ways of addressing this problem (England *et al.* 1995; Machin *et al.* 1996). Increasingly, laboratories are working extended-day or shift systems, as well as providing the traditional on-call services, to meet a 24-h demand. Similarly, rapid sample transport systems can eliminate the delays associated with portering. Any laboratory supporting an intensive care unit should be able to display or print the results of an investigation anywhere in the hospital within microseconds of completion of the analysis. These avenues should be explored and exploited to the full before alternatives are proposed.

Where these systems do not provide a sufficiently rapid return of results, it is appropriate to bring the equipment nearer to the patient. This equipment must give results of equal quality to those provided by the main laboratory. This excludes the superficially attractive, quick, and simple systems that may be appropriate to field medicine. The basic principles that should be followed have been outlined by England *et al.* (1995). They include the strong recommendation that only instrumentation which employs primary sampling of whole blood should be employed for near-patient blood counting. Anything which involves dilution of whole blood in a preanalytical phase should be avoided. The reasons for this have nothing to do with the wish to maintain a professional closed shop and everything to do with the delivery of a service which will stand up to scrutiny.

The philosophy behind establishing near-patient testing must be based on a partnership between the clinical service and the laboratory. A service-level agreement which defines the range of services, operational details, and responsibilities of staff from both sides must be the basis on which this is established. The management of this service will need to involve staff from the laboratory and the clinicians concerned, but the professional head of the laboratory service must take the final responsibility for the quality of the service that is delivered. Training, equipment, safety, handling and logging of results, and quality control (both internal and external) must all be considered when such a service is established.

The point about defining the technological basis for something as simple as blood cell counting should be clear. It should not be undertaken lightly or ill-advisedly. It is not just a question of picking the right machine and plugging it in. After choosing the right machine to meet the specified clinical demand, the next step is to operate it properly. In general, the simpler or more basic (and cheaper) the counter, the more difficult this is. Conversely, the more sophisticated (and expensive) the machine, the more its operation will have been deskilled. All counters will require active quality control and must be operated in an environment in which detailed standard operating procedures have been laid down and where there is active quality control, interlaboratory quality assessment, and accreditation of the laboratory or operator.

Anyone can present a sample to an automated counter and obtain some results. Any competent lawyer can ask who is responsible for the procedure, where are the standard operating procedures, how are they maintained, how is their distribution controlled, who maintains the instrument, where are the instrument maintenance logs, where are the quality control results, what was the 'discriminant interval' for the last quality assessment, where are the quality assessment reports, where is the record of the laboratory procedure, and where is the report filed. A really good lawyer might start to ask who changed the sheath fluid container 6 months ago and what was the stated expiry date on the container. If you are not willing to bear the responsibility for being able to answer these questions and to spend the same time on this that the laboratory does, then do-it-yourself near-patient testing may not be for you.

Blood volume and red cell mass (volume) in intensive care patients

The hemoglobin concentration (and hematocrit) have a very variable relationship with blood volume parameters in critically ill patients, whose plasma volume tends to be low and fluctuating. Ideally, when widely applicable methods are available, blood volume measurement will become almost as frequent a request as 'full blood count' for monitoring intensive care patients and determining their blood transfusion requirements (Wardrop *et al.* 1992).

References

Bentley, S.A. (1990). Automated differential white cell counts: a critical appraisal. *Baillière's Clinical Haematology*, 3, 851–69.

England, J.M., Hyde, K., Lewis, S.M., Mackie, I.J., and Rowen, R.M. (1995). Guide-lines for near patient testing: haematology. *Clinical and Laboratory Haematology*, 17, 301–10.

Machin, S.J., Mackie, I.J., Chitolie, A., and Lawrie, A.S. (1996). Near patient testing (NPT) in haemostasis—a synoptic review. *Clinical and Laboratory Haematology*, 18, 69–74.

Thom, R. (1990). Automated red cell analysis. *Baillière's Clinical Haematology*, **3**, 837–50.

Wardrop, C.A.J., Holland, B.M., Jacobs, S., and Jones, J.G. (1992). Optimization of the blood for oxygen transport and tissue perfusion in critical care. *Postgraduate Medical Journal*, **68** (Supplement 2), S2–6.

16.5.5 Coagulation monitoring

E. C. M. van Gorp and H. ten Cate

Key messages

- Tests assessing platelet function determine either overall platelet reactivity (bleeding time) or specific functions such as adhesion, aggregation, or release.
- The activated partial thromboplastin time determines the classical contact activation pathway.
- The prothrombin time determines the classical extrinsic pathway.
- The thrombin (clotting) time specifically measures the final conversion of fibrinogen to fibrin.

Hemostatic mechanism and its monitoring

Primary hemostasis

Primary hemostasis is maintained by the action of platelets, which interact with the damaged vessel wall, and with each other, to form a platelet plug, in collaboration with coagulation factors. Tests assessing platelet function determine either overall platelet reactivity (bleeding time) or specific functions such as adhesion, aggregation, or release. Most of these latter tests are used to screen patients suspected of having a specific hemostatic defect.

Simple devices for measuring bleeding time are commercially available. However, there are only a few indications for and considerable limitations on their use.

Plasma coagulation

Plasma coagulation, the second part of the hemostatic defense mechanism, involves a series of linked coagulation protease–zymogen reactions, ultimately resulting in the formation of fibrin molecules. In the current hypothetical model of coagulation, thrombin generation is induced by the assembly of the tissue factor–factor VIIa complex, previously identified as the starting point of the extrinsic route of coagulation (Davie *et al.* 1991). Tissue factor is a glycoprotein which is not normally present in the circulation in detectable quantities, but can be induced in monocytes or endothelial cells by a variety of stimuli such as cytokines. Factor VII is bound upon exposure of tissue factor at the cell surface, and after its activation, the complex catalyzes the activation of factors IX and X, resulting in a rapid onset of factor X activation and subsequent conversion of prothrombin to thrombin. Thrombin is a key enzyme in the clotting cascade, enhancing its formation rate by several feedback loops. Thrombin eventually cleaves fibrinogen, yielding fibrin molecules which may polymerize via the action of activated factor XIII to form insoluble fibrin, and vascular occlusion may occur.

Assessment of the functional integrity of the coagulation cascade is classically performed with clotting assays. The activated partial thromboplastin time determines the classical contact activation pathway by activating factor XII, followed by conversion of factors XI, IX, VIII, X, and V, and prothrombin. Clotting is started by adding an activator of the contact route and calcium to citrated plasma, and the clotting time is recorded. The assay is suitable for detecting either functional (inhibitors) or absolute deficiencies in any of the intrinsic route factors, or for measuring the effects of anticoagulant treatment, particularly heparin.

The classical extrinsic route, i.e. the tissue factor–factor VII pathway activates factor X, prothrombin, and fibrinogen, and is monitored by the prothrombin time. In this assay clotting is started by adding thromboplastin (a tissue-factor-containing preparation) and calcium to citrated plasma, and recording the clotting time. This assay is sensitive to functional abnormalities of primarily vitamin-K-dependent clotting factors (factors VII and X, prothrombin), and measures functional or absolute deficiencies of any of these clotting proteins. Furthermore, the test is particularly sensitive to the effects of oral anticoagulant agents (coumarins).

The thrombin (clotting) time specifically measures the final conversion of fibrinogen to fibrin, induced by adding thrombin and calcium to citrated plasma, and recording the clotting time. It is sensitive to reduced levels of fibrinogen or to abnormal clotting activity of this molecule.

By employing a combination of one or more of these clotting assays, a general indication can be obtained of the presence of any abnormality in the clotting mechanism or any effect induced by anticoagulant treatment.

In addition to these clotting assays, more sensitive and specific assays have been developed during the past decade. These assays are aimed at detecting products of the actual clotting process, i.e. activation peptide fragments, which are generated upon conversion of some of the clotting zymogens, free enzymes, or enzyme–inhibitor complexes (Bauer and Weitz 1994). By developing specific antibodies against these reaction products of coagulation (and fibrinolysis), immunoassays were developed for quantitating the plasma levels of the activation markers. This methodology provided a markedly increased level of detection of coagulation activation compared with the less sensitive clotting assays, and showed that even normal individuals had low basal levels of coagulation activity. This enabled coagulation derangements in disease processes to be studied at a very early stage.

Assays for measuring activation fragments, released upon activation of coagulation zymogens such as prothrombin fragment F1+2, fibrin monomers, and fibrinopeptide A, or enzyme inhibitor complexes (thrombin–antithrombin) are commercially available, and others are being developed for clinical use. Most are enzyme-linked immunosorbent assays (ELISA) and radio-immunoassays, and are laborious and expensive to perform. Therefore most of these assays are still unsuitable for routine clinical application.

Fibrinolysis

The process of fibrinolysis involves the enzymatic cascade which helps to break down cross-linked fibrin molecules. The most relevant factor is tissue type plasminogen activator, which converts plasminogen to plasmin which cleaves fibrin into various cleavage products. Tests for measuring overall fibrinolytic activity, such as the previously utilized fibrin clot lysis methods, are of no value in critical care management. Assays exist which quantitate breakdown products of fibrin (fibrin degradation products) and fibrinogen, according to the principle described for clotting cleavage fragments. More specific and very sensitive assays are now being developed, mostly aimed at the D-dimer fragment of fibrin, and these promise to be a valuable addition to the laboratory armory. Very simple bedside applications of the D-dimer test are currently being evaluated in thromboembolic disorders. Again, the main drawback is the high sensitivity and limited specificity; it appears that in many disorders D-dimer plasma levels are elevated, such that these assays are of no value in establishing the diagnosis of thromboembolism or disseminated intravascular coagulation (DIC). However, D-dimer assays may be useful for excluding the presence of thromboembolism, provided that comorbidity that may cause coagulation and fibrinolysis derangements is virtually absent. Unfortunately, this will not be the case in critical care patients.

Assessment of disorders of coagulation in critical care medicine

Bleeding, thrombosis, or a combination of bleeding and thrombosis, as seen in DIC, are the three major presentations of coagulation disorders in critical care medicine.

Clinical history is the mainstay for recognition and diagnosis of a clinically significant bleeding state or predisposition for developing thrombosis. History should be verified for bleeding complications after previous surgical procedures, trauma, or tooth extraction. Family history may raise suspicion of a hereditary coagulation disorder (hemophilia, von Willebrand's disease, rare thrombopathies) or thrombophilia. The patient may be known to have an underlying disease, such as chronic parenchymal liver disease, renal failure (uremia, proteinuria), a myeloproliferative syndrome, or a malignancy, which may lead to coagulation disorders.

Ingestion of drugs affecting hemostasis (e.g. aspirin, non-steroidal anti-inflammatory drugs, heparin, coumarin, chemotherapy) should be investigated thoroughly.

If, based on the above assessment, there is suspicion of a specific and classified diagnosis, appropriate laboratory tests should be performed. If no diagnosis has been made after a specific medical history and corresponding laboratory investigation, a general coagulation screening should be instituted.

General screening should include at least a platelet count, prothrombin time, and activated partial thromboplastin time. The outcome of these tests may help to establish the cause of bleeding (Table 1). It should be stressed that severe profuse bleeding may be the first sign of overt DIC (Table 2).

Bleeding

Bleeding tendency may be caused by a defect in primary hemostasis, caused by thrombocytopenia (Table 1) or thrombocytopathia (e.g. von Willebrand's disease, non-steroidal anti-inflammatory drugs, uremia). Defects in secondary hemostasis result from deficiency of coagulation factors or from the development of inhibitors against coagulation proteins (Table 1). There is no evidence in the literature

Table 1 General coagulation screening

Primary hemostasis	Thrombocytopenia	Idiopathic thrombocytopenic purpura Thrombotic thrombocytopenic purpura Hemolytic uremic syndrome Microangiopathic hemolytic anemia Drug-induced thrombocytopenia
Secondary hemostasis	PTT prolonged with normal PT	Factor XI deficiency Factor IX deficiency Factor VIII deficiency von Willebrand's disease Influenced by heparin
	PT prolonged with normal PTT	Factor VII deficiency Influenced by inhibitors of fibrinogen/fibrin, heparin/coumarin
	PTT and PT prolonged	Vitamin K deficiency (warfarin, coumarin) Factor I (fibrinogen) deficiency Factor II (prothrombin) deficiency Factor V deficiency Factor X deficiency DIC (usually with low platelets) Fibrinolysis (fibrinolytic agents, plasminogen activator) Liver disease

PTT, activated partial thromboplastin time (cephalin time); PT, prothrombin time.

Table 2 Coagulation tests in thrombohemorrhagic disorders

Suspicion of DIC	Suspicion of fibrinolysis
Conventional tests	
Low platelets	Normal or low platelets
Decreased fibrinogen	Abnormal clot lysis time
Thrombin time prolonged	FDP and D-dimer elevated
PTT and PT prolonged	
FDP elevated	
Blood smear: fragmented cells, helmet cells	
Confirmatory tests	
AT lowered	
TAT elevated	
D-dimer elevated	
Clot lysis time normal	
Protein C lowered	

PTT, activated partial thromboplastin time; PT, prothrombin time; FDP, fibrin degradation products; AT, antithrombin; TAT, thrombin–antithrombin complex.

for a predictive value of the bleeding time for the occurrence of bleeding. For this reason the test should be used with great caution. However, in a case of a documented defect in primary hemostasis, the bleeding time may be useful for monitoring the effect of a therapeutic intervention, e.g. administration of deamino-D-arginine vasopressin or von Willebrand factor. If the general coagulation screening tests remain normal in a bleeding patient, a factor XIII deficiency should be considered.

Thromboembolic disease

Patients with serious underlying condition such as malignancy, recent surgical procedures, or major trauma are predisposed to developing thromboembolic disorders. Immobilization, previous deep venous thrombosis, a primary defect in fibrinolysis (plasminogen deficiency), a hypercoagulable state caused by protein deficiencies and abnormalities (protein C or S deficiency, antithrombin deficiency, factor V mutant), and hyperhomocysteinemia are also associated with an increased risk of developing venous thrombosis.

At this time there is no coagulation test available for confirming the presence of active thrombotic disease. Some tests (e.g. D-dimer) are useful for excluding a thrombotic disorder outside the critical care setting (van Beek *et al.* 1996).

Thrombohemorrhagic disorders

Severe illness such as Gram-negative sepsis, obstetric pathology, trauma, (intra)vascular disorders, or malignancy may lead to DIC (ten Cate *et al.* 1993). DIC is a pathological condition in which large amounts of thrombin are generated due to activation of the coagulation system. This may lead to generalized thrombosis and multiple organ failure. In addition life-threatening hemorrhage may occur due to consumption of coagulation factors and activation of the fibrinolytic system (Table 2).

Given the current spectrum of very sensitive assays for hemostatic activation (e.g. thrombin–antithrombin complexes, D-dimer, prothrombin fragment 1+2), it is likely that these assays will give abnormal results when used in a critical care patient.

For this reason we recommend the initial use of several conventional tests to diagnose DIC (in the absence of a gold-standard test) (Table 2). Specific additional tests may be useful to support the suspicion and confirm the diagnosis, but should not be used as a single test until major clinical trials have established the relationship between a given level of an activation marker and clinical outcome. In the future, measurement of activation markers may provide a tool for determining the efficacy of anticoagulant therapy.

Specific entities

Heparin-induced thrombocytopenia is a specific thrombotic disorder associated with arterial and venous thrombosis. A specific laboratory test to confirm the diagnosis of heparin-induced thrombocytopenia is the serotonin release test, which is performed on a plasma or serum sample from the patient. However, this test can only be performed in specialized laboratories.

References

Bauer, K.A. and Weitz, J.I. (1994). Laboratory markers of coagulation and fibrinolysis. In *Hemostasis and thrombosis: basic principles and clinical practice* (3rd edn) (ed. R.W. Colman, J. Hirsh, V.J. Marder, and E.W. Salzman), pp. 1197–1210. J.B. Lippincott, Philadelphia, PA.

Davie, E.W., Fujikawa, K., and Kisiel, W. (1991). The coagulation cascade: initiation, maintenance and regulation. *Biochemistry*, **30**, 10 363–70.

ten Cate, H., Brandjes, D.P.M., Wolters, H.J., and van Deventer, S.J.H. (1993). Disseminated intravascular coagulation: pathophysiology, diagnosis and treatment. *New Horizons*, **1**, 312–23.

van Beek, E.J.R., *et al.* (1996). The role of plasma D-dimer concentration in the exclusion of pulmonary embolism. *British Journal of Haematology*, **92**, 725–32.

16.5.6 Microbiological monitoring

M. R. Jacobs

Key messages

- In critically ill patients with suspected infections, broad-spectrum therapy can be empirically instituted, guided by Gram-stain results if applicable, until culture and susceptibility results are known.
- Adequate collection of appropriate specimens is crucial for the generation of meaningful laboratory results and should not be neglected in the rush to initiate antimicrobial therapy in a critically ill patient.
- Microbiology laboratories can provide valuable information about the presence and antimicrobial susceptibility of pathogens.
- Improvements in rapid diagnostic procedures will be a major advance in the diagnosis and management of infections in critically ill patients.

Introduction

Microbiological monitoring in critical care medicine includes collection and direct examination of specimens by microscopic, antigenic, and molecular methods, culture for bacterial, viral, and fungal pathogens, antimicrobial susceptibility testing, and serological diagnosis.

Specimen collection

The results of investigations performed on a specimen are valid only if the specimen is correctly collected, transported, and processed. Whenever possible, specimens for bacteriological culture should be obtained before institution of antimicrobial therapy. Use of topical anesthetic or preservative-containing solutions should be avoided. Procedure guides for collecting and transporting specimens and information about facilities available, routine laboratory practices, and interpretation of results should be obtained from laboratory staff (Murray 1995).

Respiratory specimens

Respiratory tract infections, including acute primary pneumonias as well as pneumonias secondary to endotracheal intubation, impaired host defenses (resulting from malignancies, immunosuppressive drugs, or chemotherapy), diabetes, alcoholism, and congestive heart failure, are common in critically ill patients. Expectorated sputum is often difficult to obtain, and material for examination should be obtained by bronchoscopy, transtracheal aspiration, or open-lung biopsy. Bronchoscopy with double-lumen catheters minimizes oropharyngeal contamination of specimens; however, the local anesthetic agents possess antibacterial properties.

Open-lung biopsy is an alternative to bronchoscopy but is usually necessary only in immunocompromised hosts. The procedure has a low mortality and morbidity but is contraindicated in patients with hypoxia and thrombocytopenia (Donowitz and Mandell 1985).

Blood specimens

Prompt, adequate, and aseptic collection of blood for culture before starting antimicrobial therapy is essential to identify the causative pathogen in septic shock and to guide antimicrobial therapy. Blood is best collected by direct venipuncture with a needle and syringe or a sterile transfer set. Specimens collected from arterial or venous access catheters may only demonstrate colonization of the catheter and not septicemia. Vascular catheter tips should be submitted for culture when removed.

Bacteremia is best demonstrated by detecting the same organism in multiple blood cultures collected over several hours from different venipuncture sites (Murray 1995). If it is not possible to delay antimicrobial therapy beyond the initial set of cultures, two culture sets from different venipuncture sites can be collected. Collection times should be clearly noted to assess the relationship between clinical condition, antimicrobial therapy, procedures performed, and culture results.

Urine specimens

Collection of urine as clean-catch midstream specimens is usually not possible in critically ill patients, and suprapubic aspiration may be considered. Suprapubic aspirates should be clearly identified because any growth from these specimens is significant.

In catheterized patients, urine should be collected with a needle and syringe directly from the clamped catheter. Such specimens should be clearly identified as conventional quantitation is not used. Urinary catheter tips should not be submitted for culture as they reflect only what is colonizing the catheter and not necessarily the patient.

Other specimens

Body fluids and exudates, pus, and swabs of mucous membranes should be submitted to the laboratory as rapidly as possible. If anaerobic culture is required, an anaerobic transport medium should be used.

Direct examination of specimens

Direct microscopic examination of specimens, either as wet mounts or stained smears, is important in the diagnosis of many infections and remains the major rapid method currently available (Murray 1995). Specimens can be examined for various microbial antigens, molecular components, and other products, as well as for inflammatory cells and chemical composition.

Wet mounts

Direct examination of a fluid specimen on a slide and under a coverslip can be extremely useful for demonstrating protozoa and inflammatory cells. Dark-ground or phase-contrast illumination reveals bacteria, including spirochetes, and other organisms. *Cryptococcus neoformans* can be demonstrated by mixing a drop of India ink with a drop of cerebrospinal or other body fluid; the India ink is a negative stain that outlines the capsule of the organism.

Stained smears

Gram stains

Gram-stained smears of body fluids and exudates can show inflammatory cells as well as bacteria. Clumps of Gram-positive cocci suggest staphylococci or peptococci, chains of Gram-positive cocci suggest streptococci or peptostreptococci, large Gram-positive bacilli suggest *Clostridia*, large Gram-negative bacilli suggest *Enterobacteriaceae*, and small Gram-negative bacilli suggest *Pseudomonas*, *Hemophilus*, or *Bacteroides* species. However, bacterial morphology can be highly variable, particularly in material from abscesses or from patients on antimicrobial therapy.

Gram stains of sputum are helpful when inflammatory cells suggest pneumonia; however, if there are oral squamous epithelial cells and a wide variety of organisms, the specimen includes saliva and should not be cultured. Pneumonia caused by *Streptococcus pneumoniae*, *Hemophilus influenzae*, *Staphylococcus aureus*, and *Enterobacteriaceae* can often be diagnosed with a high degree of accuracy from a sputum Gram stain. Although a Gram stain is easily and rapidly performed, considerable experience and expertise is required as the decolorization time depends on the nature of the material being stained. Adequate quality control of Gram-stained smears must be performed, including staining control smears containing a mixture of Gram-positive cocci and Gram-negative bacilli.

Acid-fast and fluorescent stains

Mycobacteria do not stain with the Gram method, but they resist acid decolorization and therefore can be identified by various acid-fast stains of sputum and other specimens. Both the Ziehl–Neelsen stain using hot carbolfuchsin and the Kinyoun stain using cold concentrated carbolfuchsin will stain mycobacteria red. Auramine will stain mycobacteria a bright yellow under fluorescent illumination.

Fluorescein-tagged fluorescent antibody stains are useful in demonstrating various pathogens, such as *Legionella pneumophila* and common respiratory viruses. Other staining methods include silver stains of biopsies for fungi and *Pneumocystis carinii* and Giemsa stains of blood smears for malarial parasites.

All staining procedures, particularly fluorescent stains, require adequate controls.

Detection of microbial products

The only commonly available techniques for detecting microbial products directly in specimens are various immunological methods used to identify microbial antigens and direct gene probes to detect various viruses, *Neisseria gonorrhoeae*, and *Chlamydia trachomatis*. Enzyme-linked immunosorbent assays (ELISAs), using enzyme-labeled antibodies, have been developed for the detection of bacterial antigens.

Considerable advances have recently been made in the development of amplified gene probes, for example by the polymerase chain reaction (Murray 1995). However, ready availability of commercial products has not yet been attained, and these tests are currently very labor intensive, have a considerable turn-round time, and are more suitable for batch testing than for testing individual specimens.

In patients with meningitis, capsular polysaccharide antigens of *H. influenzae*, *Strep. pneumoniae*, *Neisseria meningitidis*, and *C. neoformans* can be detected in cerebrospinal fluid and sometimes in serum, sputum, urine, and effusion fluids. Antigens can be demonstrated by counterimmunoelectrophoresis, latex agglutination, or staphylococcal coagglutination, but false-positive and false-negative results limit the application of these procedures, which are generally only as sensitive as direct microscopy.

Culture methods

The wide variety of pathogenic bacteria, fungi, viruses, and protozoa require different isolation techniques. Physicians should be aware of which procedures are routine and which are available on request, either on site or at a referral laboratory. Laboratory tests should be specifically requested, particularly when unusual infections are suspected.

Respiratory specimens

Bacterial pneumonia can be diagnosed by culturing sputum, transtracheal aspirates, or biopsy material. Mycobacteria and fungi require prolonged aerobic incubation. *L. pneumophila* and other *Legionella* species can be cultured from respiratory specimens on charcoal–yeast extract agar under CO_2; sputum is the least suitable specimen for isolating these organisms. Viral isolation can be attempted from all respiratory specimens.

Blood specimens

The diagnosis of septicemia is best made by demonstrating sustained bacteremia. After direct inoculation of blood into liquid culture media, culture broths are incubated at 35 °C and examined several times daily. Growth is detected by macroscopic turbidity, blind subculture, or measurement of CO_2 released from ^{14}C-labeled substrates; other techniques are also available.

Most bacteria and yeasts will grow in common commercial blood culture media, but prolonged incubation is required to isolate some yeasts and *Brucella* species. Biphasic broth–agar media can also be used for prolonged incubation. Antimicrobial-removing resin media or lysis-centrifugation systems may be used in addition to these media when patients are receiving antimicrobial therapy or are suspected of having fungal infections.

Urine specimens

Quantitation of organisms in voided urine is used to differentiate infection (more than 100 000 organisms per milliliter) from urethral and perineal contamination (less than 10 000 mixed organisms per milliliter). Urine is cultured quantitatively on blood and MacConkey agar using a calibrated 1-μl loop. Any growth may be significant in catheter and suprapubic specimens, and culture results in patients on antimicrobial therapy need to be interpreted with caution.

Other specimens

Specimens from normally sterile sites are usually cultured for bacteria aerobically and anaërobically. Specimens from mucous membranes or the gastrointestinal tract often require selective and enrichment media to allow isolation of specific pathogens (e.g. *Salmonella* from feces and *N. gonorrhoeae* from cervix, urethra, throat, or rectum).

Identification of isolates

Most common bacteria, fungi, and mycobacteria are identified by their ability to grow on cell-free media, colonial morphology on solid media, rate of growth, temperature requirements, stained microscopic morphology, and a variety of tests to differentiate biochemical or antigenic properties. Atypical isolates, unusual organisms, and differentiation of phage types or serotypes of an organism are generally referred to a reference laboratory for investigation. Identification of epidemic strains can be achieved by many molecular methods such as digested DNA fragment fingerprinting, plasmid profiles, and multilocus enzyme electrophoresis.

Growth of viruses and *Chlamydia* can be detected in cell cultures by cytopathic effects such as cell rounding, cell clustering, syncytial formation, and intranuclear or intracytoplasmic inclusions. Hemadsorption, direct immunofluorescence, and interference can also indicate viral growth. Final identification of a virus requires specific neutralization of viral activity by homologous antiserum.

Antimicrobial susceptibility testing

Determining the susceptibility of isolated pathogens to appropriate antimicrobial agents is an important function of clinical microbiology laboratories. The development and spread of resistant bacteria, including nosocomial pathogens, as well as the increasing numbers of immunocompromised patients and new antimicrobial agents, emphasize the importance of rapid and accurate susceptibility testing of isolates from critically ill patients.

In vivo susceptibility is influenced by several factors: host defense mechanisms, the concentrations of antimicrobial agents at the site of infection, the natural course, nature, and severity of the infection, any delay in starting therapy, and the effects of other therapeutic measures such as surgery. *In vitro* susceptibility depends on the organism, growth medium, atmosphere, inoculum size, and length and temperature of incubation; because these factors have been standardized for common rapidly growing bacteria, the outcome of infection is mainly affected by *in vivo* factors.

Organisms are susceptible to an antimicrobial agent if they are inhibited *in vitro* by a concentration of the agent that is lower than serum concentrations achievable with the usual drug dosage (National Committee for Clinical Laboratory Standards 1993). Resistant organisms are not inhibited or are inhibited only at concentrations above attainable levels. Susceptibility is expressed as the minimum inhibitory concentration (MIC) of an antimicrobial agent required to inhibit growth of a defined population of organisms or as categories based on antimicrobial levels in various body sites. All susceptibility tests are based directly or indirectly on MIC determination and require strict standardization and quality control.

Determination of minimum inhibitory concentration

Serial twofold dilutions of antimicrobial agents in a suitable growth medium, such as Mueller–Hinton broth, are prepared in test tubes (macrodilution method) or microdilution wells (microdilution method), with an inoculum of 10^5 to 10^6 organisms per milliliter. MICs can also be performed by the agar dilution method; multiple inocula of 10^4 organisms per milliliter are placed on antimicrobial-containing plates with a replicating device. After overnight incubation at 35 °C, the lowest antimicrobial concentration completely inhibiting growth is the MIC. Minimum bactericidal concentrations (MBCs) can be determined by quantitative subculture; the MBC is the lowest concentration of antimicrobial agent producing at least a 99.9 per cent reduction in the original inoculum. However, MBCs are rarely determined because of a lack of reproducibility of results and absence of quality control strains.

As the macrodilution method is time consuming and expensive, the microdilution method is preferred and is commercially available in the form of frozen or lyophilized trays. A new method for MIC determination is the E-test, which is performed with a plastic carrier strip that releases an antimicrobial gradient when placed on an agar plate and combines the simplicity of disk diffusion with the accuracy of MIC determination. The E-test is particularly useful for susceptibility testing of fastidious organisms such as *Strep. pneumoniae*, *H. influenzae,* and anaerobes.

MICs are expressed in micrograms per milliliter and should be at least two- to fourfold lower than the mean achievable drug levels to produce an adequate therapeutic response. MICs can also be interpreted in categories based on achievable serum drug concentrations (National Committee for Clinical Laboratory Standards 1995). An organism is susceptible if it is inhibited by antimicrobial levels readily attained in the blood on usual dosage, including oral when applicable, moderately susceptible or intermediate if inhibited only by blood levels achieved with high dosage or when the drug is concentrated (e.g. in urine), and resistant if inhibitory levels are above clinically achievable levels. These general categories are influenced by the patient's condition, drug pharmacokinetics, renal and hepatic disease, the patient's age, presence of shock, and effects of other drugs. They may need to be modified for infections in sites of poor drug penetration such as cerebrospinal fluid.

Disk-diffusion testing

This method is based on the principle that the zone of growth inhibition around an antimicrobial-containing disk is inversely proportional to the MIC of the antimicrobial agent. Up to 12 disks can be placed on a 15-cm diameter Petri dish containing Mueller–Hinton agar seeded with a test organism. Zones of inhibition are recorded after overnight incubation at 35 °C. The degree of microbial growth inhibition is categorized as susceptible, intermediate (moderately susceptible), or resistant on the basis of the diameter of the zone of inhibition around each disk. The disk-diffusion method usually used is the Kirby–Bauer procedure, which is limited by the range of drugs and organisms that can be tested, but is well standardized (National Committee for Clinical Laboratory Standards 1995). Anaerobes should not be tested by disk diffusion.

Automated tests

The MIC and disk-diffusion methods require overnight incubation in addition to the 24 h usually required to isolate the organism. The obvious need for faster turn-round times in the care of critically ill patients led to commercial systems that produce susceptibility results in 3 to 7 h. Staphylococci, enterococci, and enteric Gram-negative bacilli can be tested by several commercially available systems that read and interpret results photometrically. These automated systems perform adequately in comparison with MIC results and can test between nine and 20 drugs at one time. However, they should be used with caution, and results for important isolates should be confirmed by conventional methods.

Beta-lactamase production

Penicillin- and ampicillin-resistant strains of *H. influenzae* and *Staph. aureus* produce β-lactamases that can be demonstrated by various techniques, such as the acidimetric and chromogenic cephalosporin methods.

Choice of susceptibility test method and agents

Choice of method depends on the organism to be tested, the techniques available, and the need for speed and accuracy. The choice of antimicrobial agents to be tested is based on the organism to be tested, its site of isolation, and local patterns of susceptibility and antimicrobial usage. Some antimicrobials, such as amikacin and the newer β-lactams, are tested only if resistance to older aminoglycosides or β-lactams is detected.

Because critically ill patients are often treated with combinations of antimicrobial agents, the interaction of these agents is important. Effective antimicrobial combinations are either synergistic or additive. These effects can be assessed *in vitro* by testing the rate of killing of an organism with drugs alone and in combination. Many drug combinations are known to be synergistic *in vivo*, particularly β-lactams and vancomycin with aminoglycosides. Combinations known to be antagonistic *in vivo* include penicillin G (benzylpenicillin) with tetracycline in pneumococcal meningitis and chloramphenicol with a β-lactam agent in meningitis caused by enteric Gram-negative bacilli.

Susceptibility of yeasts and molds to amphotericin B and itraconazole, and of yeasts to ketoconazole and 5-fluorocytosine, can be determined using techniques similar to MIC determinations for bacteria. However, these techniques are difficult to perform and are best done by specialist reference laboratories.

Monitoring antimicrobial levels

Concentrations of antimicrobial agents in serum and other body fluids can confirm that a therapeutic drug level has been attained, and that levels are not approaching toxicity. Although most drugs do not usually require monitoring, aminoglycosides, chloramphenicol, and vancomycin produce dose-related toxicity. Gentamicin and tobramycin trough values greater than 2 μg/ml, and amikacin values greater than 10 μg/ml, are associated with nephrotoxicity. Reversible hematopoietic toxicity of chloramphenicol occurs with serum concentrations greater than 25 μg/ml, particularly in patients with liver and renal disease. Renal disease, obesity, dehydration, edema, major burns, and fever can all alter aminoglycoside distribution, resulting in potentially toxic drug levels.

Serological diagnosis of infections

Although not often useful for the management of critically ill patients, serology can be important in providing a diagnosis during convalescence. However, the diagnosis of HIV infections is usually made serologically. Acute-phase sera should be collected whenever possible in patients with febrile illnesses, and convalescent samples should be drawn 1 to 6 weeks later. There are many techniques for detecting antibodies, and commercial reagents are increasingly available. Examples of infections causing severe illness that can be diagnosed serologically include typhoid fever, brucellosis, legionnaires' disease, many acute viral diseases, histoplasmosis, rickettsial typhus and spotted fevers, toxoplasmosis, and amoebiasis.

References

Donowitz, G.R. and Mandell, G.L. (1985). Acute pneumonia. In *Principles and practices of infectious diseases* (2nd edn) (ed. G.L. Mandell, R.G. Douglas, and J.E. Bennett). John Wiley, New York.

Murray, P.R. (ed.) (1995). *Manual of clinical microbiology* (6th edn). American Society for Microbiology Press, Washington, DC.

National Committee for Clinical Laboratory Standards (1993). *Methods for dilution antimicrobial susceptibility tests for bacteria which grow aerobically. Approved Standard M7-A3* (3rd edn). National Committee for Clinical Laboratory Standards, Villanova, PA.

National Committee for Clinical Laboratory Standards (1995). *Performance standard for antimicrobic susceptibility testing: sixth informational supplement M100-S6*. National Committee for Clinical Laboratory Standards, Villanova, PA.

16.5.7 Sputum sampling and bronchoalveolar lavage

G. J. Bellingan

Key messages

- Microbiological cultures of sputum and tracheal aspirates are commonly contaminated with colonizing bacteria and thus are an insensitive method of diagnosing ventilator-associated pneumonia.
- The absence of Gram-negative organisms in sputum and tracheal aspirates excludes them as a cause of ventilator-associated pneumonia.
- The false-negative and false-positive rates from both bronchial lavage and brushings are sufficiently variable, even in experienced hands, for them still to be regarded as investigational tools.
- In institutions routinely handling them, Gram staining of bronchoalveolar lavage and quantitative cultures of protected catheter brushings aid in the diagnosis of ventilator-associated pneumonia, but only in patients who are not on antibiotics.

Clinical diagnosis of pneumonia

The clinical diagnosis of pneumonia in ventilated patients is difficult. The usual criteria include the following:

(1) new or progressive chest radiograph infiltrates;

(2) new fever;

(3) leukocytosis or leukopenia;

(4) new or increasing purulent tracheobronchial secretions.

Using six variables may increase the prognostic power (Niederman *et al.* 1994). Mechanically ventilated patients are rapidly colonized with Gram-negative organisms, making interpretation of tracheobronchial cultures difficult. Non-infectious causes of chest radiograph infiltrates are common and there are many causes of fever other than pneumonia (Meduri 1990). Thus clinical criteria alone are insufficient to diagnose ventilator-associated pneumonia, with as few as one-third of clinical diagnoses being confirmed.

Methods of diagnosis

Diagnostic gold standards

Confirmation of recently suspected pneumonia can be provided at postmortem when inflammation involves several airspaces and cultures confirm a pathogen. Cultures from pleural fluid or blood, or appropriate serology in the correct clinical setting without evidence of another focus, can be taken as diagnostic, as can new pulmonary cavitation, although these can be misleading. A positive response to antibiotics may be misleading; recovery in the absence of antibiotics is a clearer finding (Meduri 1990).

Sputum and tracheal aspirates

Sputum and tracheal aspirates are of only limited diagnostic value because of the frequency of bacterial colonization. Patients with suspected ventilator-associated pneumonia commonly have Gram-negative organisms in their sputum, but not in samples from which pneumonia can be reliably diagnosed (postmortem or pleural aspirates). However, the isolation of *Staphylococcus aureus* or anaerobic Gram-negative organisms in sputum is mirrored in 80 per cent of cases by similar isolation in diagnostic cultures. The absence of Gram-negative organisms from purulent sputum excludes them as a cause of pneumonia. Similarly, the culture of tracheal aspirates has an almost 100 per cent sensitivity but only a 29 per cent specificity for the diagnosis of pneumonia. When pneumonia is confirmed, the same organism has usually been cultured in tracheal aspirates but, despite the polymicrobial nature of nosocomial pneumonia, there is a higher incidence of polymicrobial cultures from tracheal aspirates (Chastre and Fagon 1994).

Quantitative cultures of sputum and tracheal aspirates

Quantitative sputum cultures provide few advantages, with one study demonstrating no benefit and a second showed poor specificity (86 per cent of patients with pneumonia had more that 10^5 colony forming units (CFU) per milliliter but so did 40 per cent of those without pneumonia). There is a significant correlation between patients with pneumonia and aspirates with more than 10 bacteria per oil immersion field.

Elastin fibers

Examining for the presence of elastin fibers improves the specificity of tracheal aspirates in patients who have pneumonia. This is a simple process in which a drop of 40 per cent potassium hydroxide (KOH) is mixed with purulent tracheal aspirate on a microscope slide and examined for the KOH-resistant elastin fibers using standard light microscopy. Although not very sensitive (52 per cent), it is highly specific for patients with necrotizing pneumonia; false-positive results occur in patients with lung damage from non-infectious causes such as acute respiratory distress syndrome (Shepherd *et al.* 1995).

Fiber-optic bronchoscopy

Fiber-optic bronchoscopy in the intensive care unit is usually performed down an endotracheal tube of diameter 7 mm or more in ventilated patients. Patients should be preoxygenated, sedated, and paralysed. Non-intubated patients should have topical anesthetic applied to the upper respiratory tract; nebulized lidocaine (lignocaine) is particularly effective. They may be sedated if appropriate. The complication rate in non-intubated patients is less than 0.05 per cent, but this will be higher in ventilated patients where there is a risk of hemorrhage and deterioration in oxygenation. Bronchoscopic washings are often contaminated with multiple upper respiratory tract pathogens, and using a suction channel to introduce topical anesthetic increases this. Two main approaches have been adopted to reduce contamination of lower respiratory tract secretions obtained via the fiber-optic bronchoscope yet still retain the advantage of directed sampling.

Plugged telescoping catheter brush sampling

In 1979 Wimberley described a plugged telescoping catheter for use in sampling respiratory secretions (Fagon *et al.* 1988). This system was the most effective at resisting contamination *in vitro* when passed through a bronchoscope heavily contaminated with saliva. The bronchoscope is positioned at the appropriate bronchial orifice, the plugged telescoping catheter (e.g. Medi-Tech, Key-Med, United Kingdom) is advanced 2 cm into the segmental bronchus, and then the inner catheter is advanced a further 3 cm beyond the protecting plug. The catheter is then wedged and rotated to obtain the maximum innoculum of secretions. When sampling is completed, the brush and then the inner catheter are retracted and finally the whole catheter is removed. The outer and inner portions are cleaned, the brush is clipped off into 1 ml of sterile saline and transported rapidly to the bacteriology laboratory. The brush samples 0.001 ml; thus precise laboratory work is needed to obtain accurate quantitative results. In the absence of antibiotics, pathogens and contaminants can be distinguished by quantitative cultures containing more than 10^3 CFU/ml, although very recent infections may not yet have reached this pathogen load. Gram staining is not sufficiently accurate (Niederman *et al.* 1994).

Animal models of pneumonia using quantitative cultures show a sensitivity of 70 to 90 per cent and a specificity of up to 100 per cent. In non-intubated patients protected brushing demonstrated organisms at a count above 10^3 CFU/ml in 95 per cent of those with definite pneumonia and in only 5 per cent of controls. False-positive results occur in patients with obstructing endobronchial lesions where distal growth of mixed flora occurs commonly without pneumonia.

Fagon *et al.* (1988) obtained protected brushings and immediate postmortem lung biopsies from 26 patients with suspected nosocomial pneumonia. Of these, six had pneumonia histologically from which a total of 19 organisms were cultured, and brushing demonstrated 15 of these. In a second study by the same group only 45 of 147 patients with suspected pneumonia had more than 10^3 CFU/ml on quantitative culture of brush samples. Pneumonia was definitely excluded in only 9 per cent of these 45 patients. Brush samples from the remaining 102 patients had less than 10^3 CFU/ml and none of the patients were shown to have pneumonia. Pooled results of 18 studies of protected brushing show a sensitivity of 89 per cent (95 per cent confidence interval, 87–93 per cent) and a specificity of 94 per cent (95 per cent confidence interval, 92–97 per cent). Drawbacks include the persistent but low false-positive rate and the time between sampling and culture results. Potential false-negative results can occur if samples are taken from patients currently or recently on antibiotics, processed incorrectly, or taken very early in the course of the pneumonia when less than 10^3 but more than 10^2 CFU/ml may be cultured (Niederman *et al.* 1994). Unlike lavage, brushing is not useful for diagnosing pulmonary pathogens such as *Pneumocystis carinii* or mycobacteria.

Bronchoalveolar lavage

The fiber-optic bronchoscope is wedged in the segment implicated on the chest radiograph and 150 ml of sterile saline (less in those requiring a high F_{IO_2}) is instilled. The return from the first 25 ml is discarded. The diagnostic yield can be increased by lavaging bilaterally. Complications include transient worsening of radiograph infiltrates and deterioration in oxygenation, with P_{aO_2} falling by a mean of 1.07 kPa.

Quantitative cultures are required to discriminate pathogens from contaminants; a cut-off of 10^5 CFU/ml lavage fluid is commonly used but this is disputed. Varying volumes of instilled saline and dilution of return may account for some discrepancies. A comparison of protected brushing and bronchoalveolar lavage in eight normal volunteers showed that brushing was sterile in seven but lavage was sterile in only one, although on quantitative culture none had more than 10^4 CFU/ml. A method for 'protected' bronchoalveolar lavage has recently been reported with excellent sensitivity and specificity.

Bronchoalveolar lavage is the standard method for diagnosing pneumonia in the immunocompromised; here the sensitivity is up to 88 per cent and the specificity is nearly 100 per cent. If lavage fails, the gold standard remains open-lung biopsy.

The precision of bronchoalveolar lavage in the diagnosis of pneumonia has been compared with that of tracheal aspirates and protected brushing in mechanically ventilated animals. Of the bacterial species yielded by lung homogenates at postmortem, bronchoalveolar lavage recovered 74 per cent, brushing recovered 41 per cent, and needle aspirates 56 per cent. False-positive results were obtained in 8 per cent of bronchoalveolar lavage cultures but in 40 per cent of tracheal aspirates.

Bronchoalveolar lavage has a sensitivity of 53 to 93 per cent and a specificity of 60 to 100 per cent in the diagnosis of ventilator-associated pneumonia. The lack of consensus on a cut-off for a positive lavage and the variability in sensitivity and specificity limit the usefulness of quantitative cultures obtained using this technique. However, immediate Gram staining of lavage fluid has shown intracellular organisms in more than 7 per cent of leukocytes in 86 per cent of patients with pneumonia and only in 4 per cent of those without, with Gram staining and later cultures agreeing closely. Of those without pneumonia, 91 per cent had less than 2 per cent leukocytes with intracellular organisms (Chastre and Fagon 1994). This sensitivity is lost in those recently or currently on antibiotics.

Blind sampling techniques

Most ventilator-associated pneumonias are multifocal; hence blind sampling, which is cheaper and quicker, may be appropriate. Blind protected brushing can be performed in ventilated patients using a Metras catheter normally used for bronchography. In one study brushing directed by a fiber-optic bronchoscope had a sensitivity of 71 per cent whilst that from blind brushing was 64 per cent, with close agreement in the organisms cultured. Likewise, non-bronchoscopic bronchoalveolar lavage has a reported sensitivity of 75 per cent, even though 86 per cent of samples are right-sided. Patients are preoxygenated and briefly disconnected from the ventilator, and a 50-cm sterile plugged telescoping catheter (Combicath) is blindly inserted until wedged. The plug is expelled by injecting air and the inner catheter advanced at least 3 cm distally until wedged. Lavage is performed with 20 ml of saline and at least 1 ml is recovered for examination and culture.

Transtracheal and transthoracic needle aspirations

Transtracheal aspiration through the cricothyroid membrane, like tracheal aspiration, is sensitive but not specific. Transthoracic needle aspirations performed under radiological guidance require near-normal clotting. A fine-needle aspirate is suitable for microbiology and cytology and has a lower complication rate, but even in non-ventilated patients this is still high; 5 to 15 per cent hemorrhage and 17 to 25 per cent have a pneumothorax. Neither procedure is appropriate in ventilated patients.

Antibody-coated bacteria

Antibody coating of bacteria is used to help diagnose lower respiratory tract infections; it is based on the postulation that invasive bacteria stimulate an immune response whereas colonizing bacteria do not. An aspirate sample is washed, pelleted, plated into shallow wells on microscope slides, and heat fixed, after which fluorescein-conjugated anti-human immunoglobulin is added with appropriate controls. Of patients with pneumonia, 54 to 73 per cent have antibody-coated bacteria compared with none to 2 per cent of the controls. There is no advantage in using specific immunoglobulin classes (Meduri 1990).

Conclusions

Results from both bronchoalveolar lavage and brushings are sufficiently variable, even in experienced hands, for them to be regarded as investigational tools only. Where there is close liaison between the microbiology department and the intensive care unit, these techniques will be useful. They require further evaluation with regard to both their applicability in different subgroups and their effect on survival.

Many physicians will continue to manage patients on the basis of clinical findings and microbiological cultures obtained non-invasively. This remains appropriate until invasive diagnostic sampling with quantitative cultures has been shown to produce a definite improvement in outcome.

References

Chastre, J. and Fagon, J.Y. (1994). Invasive diagnostic testing should be routinely used to manage ventilated patients with suspected pneumonia. *American Journal of Respiratory and Critical Care Medicine*, **150**, 570–4.

Fagon, J.-Y., *et al.* (1988). Detection of nosocomial lung infection in ventilated patients. Use of a protected specimen brush and quantitative culture techniques in 147 patients. *American Review of Respiratory Diseases*, **138**, 110–16.

Meduri, G.U. (1990). Ventilator-associated pneumonia in patients with respiratory failure. A diagnostic approach. *Chest*, **97**, 1209–19.

Niederman, M.S., Torres, A., and Summer, W. (1994). Invasive diagnostic testing is not needed routinely to manage suspected ventilator-associated pneumonia. *American Journal of Respiratory and Critical Care Medicine*, **150**, 565–9.

Shepherd, K.E., Lynch, K.E., Wain, J.C., Brown, E.N., and Wilson, R.S. (1995). Elastin fibers and the diagnosis of bacterial pneumonia in the adult respiratory distress syndrome. *Critical Care Medicine*, **23**, 1829–34.

16.6 Neurological monitoring

16.6.1 Electroencephalogram monitoring

N. J. Smith

Key messages

- Changes in neuronal function produce recognizable electroencephalogram (EEG) patterns, which have been analyzed and classified.
- Progressive reduction of cerebral blood flow is reflected in the EEG and evoked potentials by a sequence of changes which appear before irreversible damage has occurred.
- Fits are clinically unrecognizable in the presence of neuromuscular blockade; therefore if paralysis is used in status epilepticus, EEG monitoring is essential to prevent ischemic neuronal damage due to persistent seizure activity.
- EEG and evoked potential monitoring can affect clinical decision-making in a high proportion of comatose patients.
- In countries where the legal definition of brain death includes an isoelectric EEG, a multichannel EEG should be recorded, conforming to stringent technical standards.

Cortical function and the EEG

The electroencephalogram (EEG) is the continuously varying difference in electrical potential between two points on the scalp. This potential difference is produced by postsynaptic potentials (not action potentials) in the underlying cortical neurons, principally large pyramidal cells in cortical layers 3, 4, and 5. Changes in neuronal function produce recognizable EEG patterns which have been analyzed and classified (Prior and Maynard 1986; Grundy 1995). An evoked potential is the reproducible change in electrical potential recorded from the appropriate sensory pathway, or sensory cortex, following a stimulus to sensory receptors (e.g. the eye for the visual evoked potential) or part of the pathway (e.g. a peripheral nerve for the somatosensory evoked potential). Therefore cortical evoked potentials are affected by lesions anywhere between the site of stimulation and the sensory cortex, and lesions can sometimes be localized by recording subcortical responses to the same stimulus.

EEG and evoked potential effects of cerebral ischemia

In seriously ill patients a number of variables (blood pressure, arterial oxygen and carbon dioxide levels, intracranial pressure, etc.) are routinely monitored with the underlying aim of preventing brain damage. This objective might be achieved more effectively by monitoring the function of the cerebral cortex itself (Prior 1985; Grundy 1995). Both EEG and evoked potentials are sensitive to reduction in cerebral blood flow, and two important thresholds have been identified (Prior 1985).

The first threshold is the blood flow level at which EEG alteration first appears during gradual flow reduction. This occurs when regional cortical blood flow falls to 20 ml/min/100 g, and is manifest in the EEG by an increase in the proportion of low frequencies (below 8 Hz) and the disappearance of higher frequencies. Alternatively, if the reduction in cerebral blood flow is rapid, a burst-suppression pattern dominates the EEG at flow levels below 20 ml/min/100 g. At about the same flow level the waveform of the cortical somatosensory evoked potential changes, with an increase in latency.

The second threshold is the disappearance of spontaneous EEG activity, which occurs when regional cerebral blood flow falls to approximately 15 ml/min/100 g. If flow remains below this level for more than 45 min, neuronal activity does not return with adequate reperfusion and histology demonstrates selective neuronal loss. Evoked potentials disappear when flow falls to about 12 ml/min/100 g, and infarction results if flow is maintained at this level for 2 to 3 h.

Therefore progressive reduction of cerebral blood flow is reflected in the EEG and evoked potentials by a sequence of changes which appear before irreversible damage has occurred. EEG recognition of ischemia is particularly significant in patients in whom the lower limit of cerebral autoregulation is elevated above its usual level of 60 to 70 mmHg, so that arterial blood pressure is no longer a reliable guide to adequate cerebral perfusion.

EEG recording, data reduction, and display

The EEG is conventionally recorded from a number of electrodes applied to the scalp in positions determined by measurement according to the International 10–20 System. In an intensive care unit there is inevitably considerable electrical interference from other equipment, and so EEG signals of a few microvolts can only be recorded satisfactorily if the correct electrodes are selected and applied to appropriate areas of the scalp with low contact impedance (less than 5 k at 10 Hz), and all leads between patient and preamplifier are kept as short as possible (Prior 1985; Prior and Maynard 1986).

Modern EEG machines record 16 or more channels of data at a paper speed of 3 cm/s. If recording is continued for more than a few hours, an excessive quantity of data is generated, much of which is redundant, and so some form of data reduction is essential for long-term monitoring. Spatial information is usually sacrificed or reduced to a simple comparison between corresponding areas of the right and left hemispheres.

Many methods of data reduction have been developed, each having its own advantages (Prior and Maynard 1986). Several popular methods depend upon frequency analysis by the fast Fourier transform, with the results being displayed as a graph of amplitude (or power) against frequency, often as a compressed spectral array in which successive spectra are plotted one above the other, with suppression of those parts of later spectra which would overwrite earlier ones. This unpredictable loss of data means that the sudden disappearance or abrupt attenuation of a frequency component is not detected until several more spectra have been plotted, resulting in potentially dangerous delay. To avoid this disadvantage of the compressed spectral array, the same data can be displayed as a density-modulated spectral array, in which the darkness (density) of each point along the frequency axis is proportional to the amplitude at that frequency. Such a display is difficult to calibrate and has limited dynamic range, but abrupt attenuation is visible immediately in the density-modulated spectral array, unlike the compressed spectral array. The ultimate data reduction is achieved by extracting from the Fourier transform a single number, such as mean frequency or spectral edge frequency. Excessive data reduction of this sort is potentially dangerous and is not recommended for clinical purposes.

Methods based upon frequency analysis cannot reliably detect the onset of a burst-suppression pattern in the EEG or the occurrence of brief high-amplitude events associated with epileptic seizures. These important EEG patterns are clearly demonstrated by data reduction in the time domain to give an indication of EEG amplitude. This has the additional advantage of retaining the variability of the EEG. The simplest display of this type is a graph of amplitude integrated over a short epoch. One device of this type, the cerebral function monitor, plots the amplitude on a logarithmic scale, thereby ensuring a large dynamic range. A further development, the cerebral function analyzing monitor, also displays the percentage of activity in the conventional EEG frequency bands (Prior and Maynard 1986).

Clinical applications of EEG monitoring in critical care

Since the EEG is intimately related to the function of the cerebral cortex, EEG monitoring would appear to be desirable in all critically ill patients with impaired consciousness (Prior 1985; Grundy 1995). However, EEG monitoring is usually reserved for certain clearly defined indications.

Refractory status epilepticus

Fits are unrecognizable clinically in the presence of neuromuscular blockade, and persistent seizure activity leads to ischemic neuronal damage. Therefore EEG monitoring is mandatory to identify seizure activity if paralysis is used in status epilepticus. Many different protocols have been devised for the management of refractory status epilepticus; all require elimination of paroxysmal activity from the EEG, and some specify that sedation should be deep enough to produce a burst-suppression pattern in the EEG. Conventional EEG recording may be used, but a device such as the cerebral function monitor or cerebral function analyzing monitor is ideal for EEG monitoring in status epilepticus, since the marked increase in amplitude associated with seizure activity is clearly displayed and the burst-suppression pattern produced by deep sedation is readily identified (Fig. 1). Nursing staff must mark the EEG trace whenever the patient is turned, endotracheal suction applied, or any other nursing procedure is undertaken, so that the associated artifacts are not mistaken for fits. Seizure activity may occasionally appear without a marked increase in the EEG amplitude; therefore if monitoring is by cerebral function monitor or cerebral function analyzing monitor, the raw EEG should be inspected occasionally and full EEGs recorded at intervals.

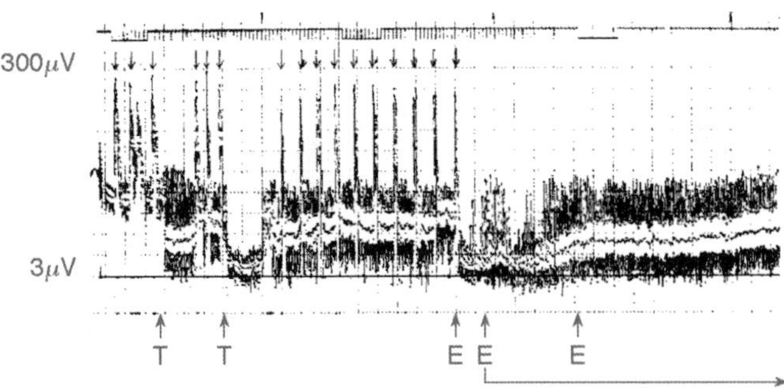

Fig. 1 Refractory status epilepticus monitored by the cerebral function monitor. Repeated fits are associated with marked increase in amplitude. Bolus doses of thiopentone (T) suppressed the fits briefly, but could not be repeated because of hypotension. The fits were finally brought under control permanently by etomidate (E). Bolus doses were administered at the vertical arrows, and a continuous infusion is shown by the horizontal line. Calibration: vertical time markers every minute, and time trace reversed for 10 min every hour. Vertical lines on the graph paper are repeated every 5 min. The amplitude is logarithmic from 3 to 300 μV.

EEG monitoring sometimes fails to show any evidence of paroxysmal activity in a patient undergoing treatment for presumed convulsive status epilepticus. This indicates that the diagnosis is not status epilepticus, and some other condition should be considered. Important possibilities are decerebrate spasms, particularly after head injury, myoclonus status after cardiac arrest, or psychogenic ‘pseudostatus’, a potentially life-threatening condition which is being increasingly recognized.

Coma

Routine EEG monitoring in all comatose patients is not yet generally accepted practice. However, three significant findings have emerged from EEG studies of coma: firstly, the presence of cyclical changes in EEG amplitude and frequency, reminiscent of sleep–wake cycles, and the better prognosis in patients showing these EEG features compared with an unvarying pattern; secondly, the prevalence of unsuspected epileptiform activity, particularly in post-traumatic and post-hypoxic coma; thirdly, the association between inadequate sedation and poor control of intracranial pressure, with EEG arousal often coinciding with lability of blood pressure and intracranial pressure. In addition, patients with brainstem lesions show abnormal EEG slow-wave responses to stimulation, while extensive brainstem hemorrhage or infarction is associated with an unresponsive pattern of widespread rhythmical activity at a frequency of about 8 Hz (‘alpha coma’).

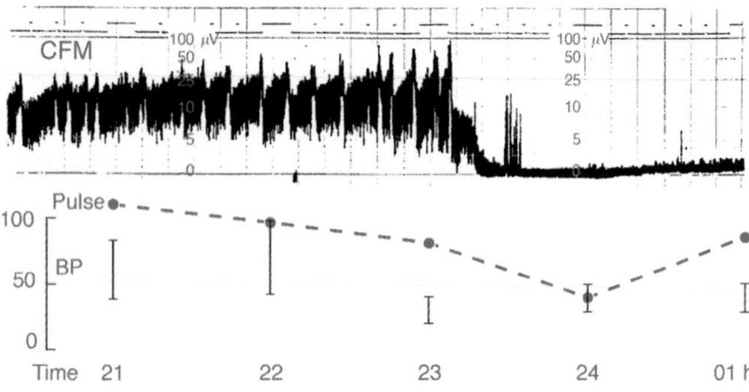

Fig. 2 Cerebral function monitor (CFM) trace in a patient who had been resuscitated after cardiac arrest and was having repeated fits, shown by the marked increase in amplitude. During a period of hypotension, beginning at about 23.00, the EEG amplitude fell below 5 μV, indicating severe cortical ischemia, which persisted until a further cardiac arrest 12 h later. Calibration for cerebral function monitor: time, vertical lines every 10 min. Amplitude as marked (linear from 0 to 10 μV, logarithmic from 10 to 100 μV). BP, blood pressure.

In addition to the general applications of EEG monitoring mentioned above, specific indications include the detection of cerebral hypoxia and ischemia in the first few days after head injury, particularly in patients with multiple injuries, and monitoring progress in patients remaining in coma after resuscitation (Fig. 2) and during recovery from sedative drug overdose or profound hypothermia. In these situations EEG activity often reappears before there is any clinical evidence of improvement.

EEG and evoked potential monitoring can affect clinical decision-making in a high proportion of comatose patients (Jordan 1993), but whether routine neurophysiological monitoring in coma (regardless of the cause) improves the outcome has yet to be determined.

Brain death

Disappearance of EEG activity indicates cessation of cortical function, but this does not necessarily occur at the same time as failure of brainstem function. Following sedative drug overdose or profound hypothermia, the EEG may remain isoelectric for more than 24 h in patients who subsequently recover. Conversely, low-amplitude EEG activity may persist for several hours after brain death has been confidently diagnosed by other methods. In countries where the legal definition of brain death includes an isoelectric EEG, reliance cannot be placed upon a monitoring device; a multichannel EEG should be recorded, conforming to stringent technical standards such as those recommended by the American EEG Society (1994).

References

American EEG Society (1994). Guideline three: minimum technical standards for EEG recording in suspected cerebral death. *Journal of Clinical Neurophysiology*, **11**, 10–13.

Grundy, B.L. (1995). The electroencephalogram and evoked potential monitoring. In *Monitoring in anesthesia and critical care medicine* (3rd edn) (ed. C.D. Blitt and R.L. Hines), pp. 423–89. Churchill Livingstone, New York.

Jordan, K.G. (1993). Continuous EEG and evoked potential monitoring in the neuroscience intensive care unit. *Journal of Clinical Neurophysiology*, **10**, 445–75.

Prior, P.F. (1985). EEG monitoring and evoked potentials in brain ischemia. *British Journal of Anaesthesia*, **57**, 63–81.

Prior, P.F. and Maynard, D.E. (1986). *Monitoring cerebral function: long-term monitoring of EEG and evoked potentials*. Elsevier, Amsterdam.

16.6.2 Intracranial pressure monitoring

R. S. Graham and J. D. Ward

Key messages

- The normal range for intracranial pressure is zero to 10 mmHg. The threshold for treatment of increased intracranial pressure is generally accepted to be 20 mmHg or greater.
- Indications for intracranial pressure monitoring include severe head injury, subarachnoid hemorrhage, comatose meningoencephalitis, hepatic encephalopathy, Reye's syndrome, and following craniotomy.
- Intracranial pressure can be monitored from the intraventricular, intraparenchymal, subarachnoid, subdural, and epidural compartments.
- Fiber-optic catheter devices have been shown to drift during prolonged monitoring. Miniature strain gauge devices have less tendency to drift. Subarachnoid or subdural fluid-coupled devices and epidural devices are less accurate.
- Complications associated with intracranial pressure monitoring include infection, hemorrhage, malfunction, and malposition.

Introduction

Intracranial pressure monitoring has progressed from an investigational measurement to an essential tool in the care of patients with a variety of neurological diseases.

Intracranial physiology

The Monroe–Kelley doctrine is the basis for understanding intracranial pressure–volume interactions. It states that an increase in the volume of one intracranial compartment must be compensated by a decrease in the volume of one or more of the other compartments, so that the total volume remains fixed. The intracranial compartments include intracellular space, extracellular space, cerebral blood volume, and the cerebrospinal fluid volume. The normal range for intracranial pressure is zero to 10 mmHg. The threshold for treatment of increased intracranial pressure is generally accepted to be 20 mmHg or greater.

An intracranial mass lesion or brain swelling forces cerebrospinal fluid out of the intracranial compartment, resulting in obliteration of the ventricular system and basal cisterns with no initial rise in intracranial pressure. A non-contrast head CT scan will readily demonstrate these changes (Fig. 1). If there are further increases in the volume of the lesion or swelling, intracranial pressure begins to rise in an exponential fashion (Fig. 2). Cerebral perfusion pressure is defined as the difference between the mean arterial pressure and the intracranial pressure. It represents the pressure gradient that drives cerebral blood flow. Cerebral ischemia is the most important

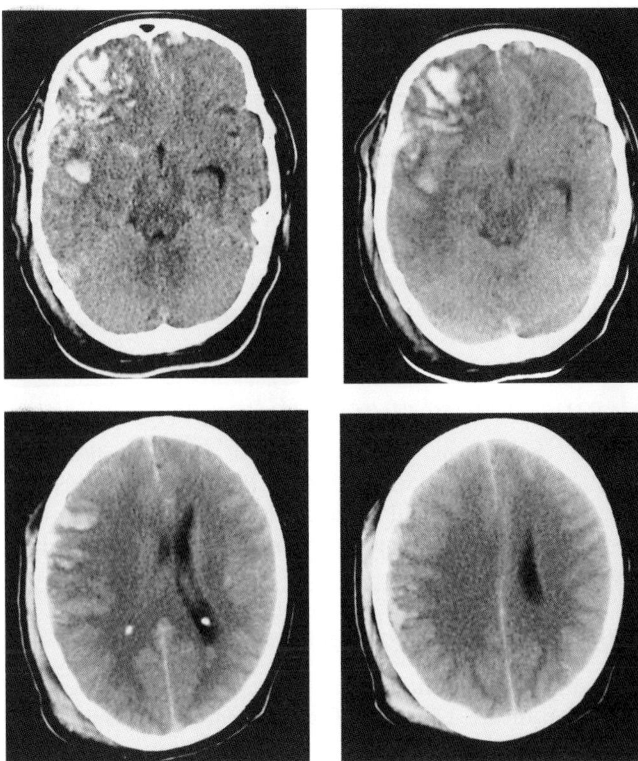

Fig. 1 Non-contrast head CT scan of a severely head-injured patient. A large right frontal hemorrhagic contusion is present with shifting of the midline structures. Additional hallmarks of elevated intracranial pressure are present, including effacement of the basal cisterns, marked right frontal lobe cerebral edema, effacement of the right lateral ventricle, and obliteration of the right temporal horn of the lateral ventricle.

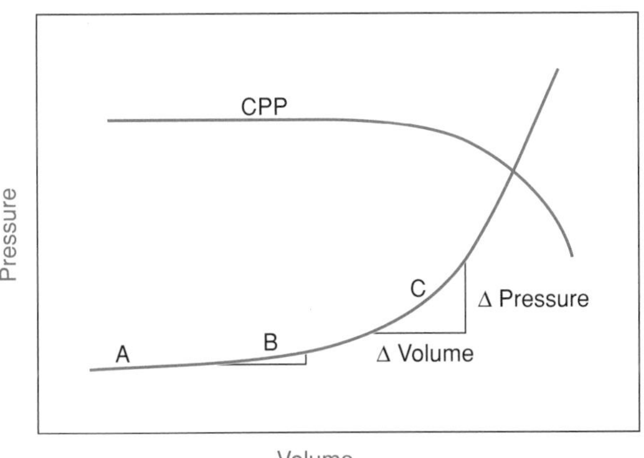

Fig. 2 Idealized intracranial pressure–volume curve. Point A represents the position of a normal brain on the compliance curve. Point B represents an increase in the intracranial volume which has been partially compensated for by the decrease in cerebrospinal fluid and contraction of the cerebral blood volume. Point C represents a further increase in the intracranial volume. In this region of the curve no further compensation is possible and pressure begins to rise exponentially. Cerebral perfusion pressure may drop below the critical levels necessary to maintain neuronal metabolism and result in 'secondary ischemic injury'.

secondary injury occurring in patients with elevated intracranial pressure. Therefore maintenance of a suitable cerebral perfusion pressure is considered by many experts to be of equal or greater importance than strict control of intracranial pressure. The current

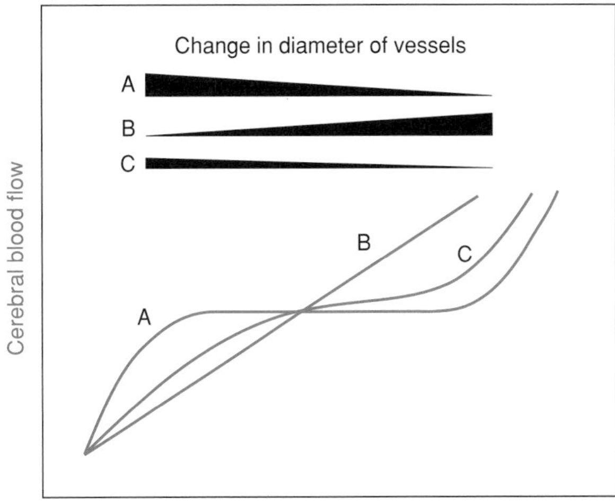

Fig. 3 Idealized autoregulation curve. Curve A represents intact autoregulation. Constant cerebral blood flow is maintained over a wide range of perfusion pressures. Curve B represents complete loss of autoregulation. In this case the cerebral vessels are pressure passive, and cerebral perfusion pressure is directly proportional to cerebral blood flow and cerebral blood volume. Consequently, patients with disturbed autoregulation are caught in a delicate balance between adequate cerebral perfusion pressure and avoidance of increased intracranial pressure from vasodilatation. Curve C represents partial loss of autoregulation. In this case the brain is vulnerable to hypotension, but maintains autoregulation at higher pressures. Increases in cerebral perfusion pressure may result in decreased intracranial pressure through vasoconstriction and contraction of the cerebral blood volume.

consensus for cerebral perfusion pressure management is that a minimum pressure of 70 mmHg should be maintained (Bullock *et al.* 1996).

Autoregulation is the mechanism by which cerebral blood flow is maintained constant despite variations in cerebral perfusion pressure. Autoregulation of flow occurs by dilatation and constriction of the cerebral vessels and subsequent changes in cerebrovascular resistance. Stimuli for autoregulation include mechanical changes in pressure or chemical changes in the local concentration of metabolic products, for example oxygen tension, partial pressure of carbon dioxide, increased potassium, and fluctuations in local pH. Functionally, autoregulation serves to maintain homeostasis. It buffers changes in pressure and maintains precise control of the delivery of oxygen and nutrients to match the metabolic needs of local tissues (Fig. 3). Autoregulation is frequently perturbed to varying degrees in cases of severe head injury and other processes with marked elevations in intracranial pressure. In these situations, the cerebral vessels become pressure passive. The cerebral vascular bed loses its ability to compensate for decreases in cerebral perfusion pressure with arterial vasodilatation; thus hypoperfusion and ischemia result. Conversely, when there is loss of autoregulation in a hypertensive patient, cerebral blood flow exceeds the metabolic requirements, resulting in hyperemia. Additionally, in a pressure-passive vascular bed, cerebral blood volume increases in direct proportion to the cerebral perfusion pressure because of vasodilatation (Fig. 3). Hypertensive therapy in patients with disrupted autoregulation can result in further elevations of intracranial pressure.

Indications for intracranial pressure monitoring

Mild and moderate head injuries

Mild and moderate head-injured patients with Glasgow Coma Scale (GCS) scores of 9 to 15 are usually capable of following commands and may be followed with serial neurological examinations. Patients with GCS declining to 8 or less are reclassified as severe head injuries, and intracranial pressure monitoring is instituted.

Severe head injuries

Severe head injury patients presenting with GCS scores of 8 or less (comatose) are at risk of development of intracranial hypertension. In those patients with abnormal CT scans, the incidence of intracranial hypertension is 53 to 63 per cent. Additionally, 10 to 15 per cent of patients with normal CT scans experience elevated intracranial pressure. The magnitude and duration of intracranial hypertension correlate with poor outcome in severely head-injured patients. Intracranial pressure monitoring allows early detection of developing intracranial mass lesions, rational treatment of intracranial hypertension, and drainage of cerebrospinal fluid to reduce intracranial pressure, and may improve outcome. All severely head-injured patients admitted with abnormalities on CT scan should have intracranial pressure monitoring. In comatose patients with normal admission CT scans, intracranial pressure monitoring is suggested if two or more of the following risk factors are present: age greater than 40, presence of motor posturing, or systolic blood pressure less than 90 mmHg (Bullock *et al.* 1996).

Subarachnoid hemorrhage

Subarachnoid hemorrhage commonly results from rupture of a cerebral aneurysm. Patients presenting with large subarachnoid hemorrhages or poor clinical grades are at high risk of development of intracranial hypertension. The benefits of ventricular catheter placement in these patients is twofold. It allows accurate medical treatment of increased intracranial pressure and drainage of cerebrospinal fluid. The latter is important because of the high incidence of acute hydrocephalus associated with large blood loads and intraventricular blood clots. It is important not to overdrain the cerebrospinal fluid and acutely lower intracranial pressure since, rarely, these actions may precipitate rebleeding from unclipped cerebral aneurysms. Catheters are often left in place for several days postoperatively until normal cerebrospinal fluid circulation and absorption has been re-established.

Spontaneous intracerebral hematoma

Hypertension, arteriovenous malformations, aneurysms, cavernous malformations, amyloid angiopathy, and anticoagulation are common conditions associated with the formation of spontaneous intracerebral hematomas. Insertion of an intracranial pressure monitoring device may be undertaken in certain cases, particularly comatose patients or patients with hematomas which have ruptured into the ventricular system. In the latter case, a ventriculostomy often relieves pressure from life-threatening acute hydrocephalus. However, routine use of intracranial pressure monitoring in patients with spontaneous intracerebral hematomas has not shown proven benefit.

Large ischemic infarcts

Occlusions of the internal carotid artery or middle cerebral artery are responsible for the majority of large ischemic infarctions. These infarctions are associated with a predictable pattern of brain swelling, which may be overwhelming. Brain edema and increased intracranial pressure are the most common causes of death in the first week following an ischemic stroke. Insertion of an intracranial pressure monitor may be useful in guiding therapy through the period of peak brain edema. However, proven benefits of intracranial pressure monitoring and control in this group of patients has not yet been established.

Meningitis and encephalitis

In both children and adults, meningitis and encephalitis are commonly associated with marked intracranial hypertension. Monitoring may be useful for the identification and treatment of life-threatening elevations in intracranial pressure, particularly in comatose patients.

Hepatic encephalopathy

Patients with fulminant hepatic failure progressing to grade 4 encephalopathy are at high risk of development of intracranial hypertension. Intracranial pressure monitoring in this patient population has been shown to increase the duration of survival from the onset of grade 4 encephalopathy and results in altered treatment compared with patients who are not monitored (Keays *et al.* 1993). Additionally, intracranial pressure monitoring provides important prognostic information in this group. Patients with prolonged uncontrollable elevations in intracranial pressure and reduced cerebral perfusion pressure have an increased chance of permanent neurological deficit and may not be appropriate candidates for transplantation. Special consideration should be given to the profound coagulopathy that is present in all patients with fulminant hepatic failure. Epidural devices are recommended in these cases because of the high rate of hemorrhagic complications associated with the use of intraventricular, intraparenchymal, and subdural monitors.

Reye's syndrome

Reye's syndrome is a rare disorder which results in non-icteric hepatic encephalopathy. It occurs primarily in children and adolescents; predisposing factors include a variety of viral infections and the use of aspirin. Pathological changes include acute brain swelling and fatty infiltration of the liver and other visceral organs. In cases where acute cerebral edema is adequately controlled, complete recovery is usually attained. Continuous intracranial pressure monitoring is indicated in cases with neurological deterioration or high serum ammonia levels (above 300 μg per cent (175 μmol/l)) (Chi *et al.* 1990).

Postoperative intracranial pressure monitoring

Insertion of an intracranial pressure monitoring device following intracranial surgery is easily accomplished in sterile fashion. Definite indications for intracranial pressure monitoring in postoperative patients include all patients status postevacuation of acute traumatic intradural hematomas or contusions. Constantini *et al.* (1988)

determined that resection of glioblastoma, repeat surgery, and protracted surgery (greater than 6 h) were risk factors for intracranial hypertension following elective surgery. Intracranial pressure monitoring should be employed following elective craniotomies, particularly when risk factors can be identified prior to the end of surgery. Additionally, intracranial pressure monitoring may provide an early indicator of postoperative complications.

Brain tumors

Brain compression which develops in association with intracranial tumors has a slowly progressive course. Consequently, accommodations are made and intracranial pressure monitoring is rarely necessary. Corticosteriods and resection of tumor are the mainstays of treatment in this situation.

Methods of intracranial pressure monitoring

Intracranial pressure can be monitored from the intraventricular, intraparenchymal, subarachnoid, subdural, and epidural compartments. Intraventricular monitors include ventriculostomy catheters and ventriculostomy catheters equipped with catheter-tip transducers. Ventriculostomy catheters are fluid coupled to external strain gauge pressure transducers. They are maintained at the level of the external auditory meatus for accurate pressure readings. They offer the ability for recalibration after insertion and replacement of pressure transducers if a malfunction occurs. Importantly, therapeutic cerebrospinal fluid drainage is possible with these devices. Ventriculostomy with catheter-tip pressure transducers employs advanced fiber-optics which is significantly more expensive. Ventricular canalization may be difficult in the case of 'slit ventricles', and catheters are occasionally obstructed with blood clot or tissue.

Intraparenchymal catheter-tip pressure transducers include fiber-optic devices such as the Camino device and recently introduced miniature strain gauge transducers (Gopinath *et al.* 1995). These devices are zeroed to atmospheric pressure prior to insertion and recalibration is not possible. Fiber-optic devices have been shown to drift significantly during prolonged monitoring. Miniature strain gauge devices have less tendency to drift (Gopinath *et al.* 1995). Additionally, the fiber-optic catheters are brittle and have a tendency for decreased longevity in the intensive care setting. Subarachnoid or subdural fluid-coupled devices and epidural devices are less accurate (Bullock *et al.* 1996). The primary indication for their use is in patients with a prohibitively high risk of hemorrhage complications such as liver transplant candidates (Keays *et al.* 1993). In summary, the ventriculostomy catheter connected to an external strain gauge transducer is relatively inexpensive and provides the most accurate and reliable method for intracranial pressure monitoring. Additionally, it allows therapeutic drainage of cerebrospinal fluid. Parenchymal pressure tip transducers provide similar intracranial pressure data but are not as reliable owing to drift and the inability to recalibrate.

Complications of intracranial pressure monitoring

Complications associated with intracranial pressure monitoring include infection, hemorrhage, malfunction, and malposition. The vast majority of these complications are minor and do not produce long-term morbidity. Infection is the most common complication. It usually takes the form of bacterial colonization of the catheter or positive surveillance cultures of cerebrospinal fluid. Grossly purulent ventriculitis is rare. Risk factors for infection include duration of monitoring, associated intracranial injuries, concurrent infection, external irrigation, and serial replacement of monitors. Several groups have reported correlation between the duration of monitoring and the frequency of infection. Others have reported that the chance of infection decreases after the monitor has been in place for 6 days or more, suggesting that the primary source of infection was contamination at the time of insertion. The overall infection rate for intracranial pressure monitors is approximately 5 to 10 per cent. At our institution, ventriculostomy catheters are left in place until neurological improvement has occurred, intracranial pressure is stable and has not required treatment for 24 h, or daily surveillance cerebrospinal fluid studies become suspicious for infection (catheters are reinserted through a new site).

Hemorrhage is associated with insertion of intracranial pressure devices in 1 to 2 per cent of cases. The majority of hemorrhages are small and require no treatment. Significant hematomas which require surgical drainage occur in approximately 0.5 per cent of cases. Disseminated intravascular coagulation, fibrinolysis, anticoagulation, and impaired liver synthetic function increase the risk of hemorrhagic complications. At our institution, all patients are screened with routine laboratory tests to detect coagulation abnormalities prior to insertion of intracranial pressure devices. Extraparenchymal devices have lower risk of hemorrhage.

References

Bullock, R., *et al.* (1996). *Guidelines for the management of severe head injury.* Brain Trauma Foundation.

Chi, C.S., Law, K.L., Wong, T.T., Su, G.Y., and Lin, N. (1990). Continuous monitoring of intracranial pressure in Reye's syndrome—5 years experience. *Acta Paediatrica Japonica*, **32**, 426–34.

Constantini, S., Cotev, S., Rappaport, Z.H., Pomeranz, S., and Shalit, M.N. (1988). Intracranial pressure monitoring after elective intracranial surgery: a retrospective study of 514 consecutive patients. *Journal of Neurosurgery*, **69**, 540–4.

Gopinath, S.P., Robertson, C.S., Contant, C.F., Narayan, R.K., and Grossman, R.G. (1995). Clinical evaluation of a miniature strain-gauge transducer for monitoring intracranial pressure. *Neurosurgery*, **36**, 1137–40.

Keays, R.T., Alexander, G.J., and Williams, R. (1993). The safety and value of extradural intracranial pressure monitors in fulminant hepatic failure. *Journal of Hepatology*, **18**, 205–9.

16.6.3 Jugular venous bulb monitoring

K. Ahmed and N. M. Dearden

Key messages

- Development of fiber-optic technology has allowed continuous monitoring of jugular venous oxygen saturation during intensive care in patients at significant risk from cerebral ischemia.
- Previously undetected episodes of cerebral hypoperfusion that affect outcome have been identified following severe brain trauma.
- The technique allows early detection and correction of cerebral hypoperfusion, which is usually due to low blood pressure and the use of excessive hyperventilation.
- During intensive care periods of erroneous recording occur in as many as 60 per cent of patients.
- At the present time jugular venous oxygen saturation monitoring remains the most practical method of continuous monitoring for cerebral hypoperfusion in critical neurosurgical patients.

Introduction

Arteriojugular venous oxygen content difference ($AJDo_2$) and arteriojugular venous lactate content difference are used to monitor neurosurgical patients at risk from cerebral ischemia. According to the Fick principle, the relationship between the global cerebral metabolic rate for oxygen ($CMRo_2$) and the cerebral blood flow (CBF) can be described by measurements of $AJDo_2$ (normal range 4–8 ml oxygen per 100 ml blood):

$$AJDo_2 = CMRo_2/CBF.$$

$AJDo_2$ can be calculated from the formula

$$AJDo_2 = \frac{[Hb] \times 1.39 \times AJSo_2}{100} + \frac{(Pao_2 - Pjo_2) \times 0.3}{100}$$

where [Hb] is the hemoglobin concentration, $AJSo_2$ is the arteriojugular venous oxygen saturation difference, Pao_2 is the arterial O_2 partial pressure (in mmHg), and Pjo_2 is the jugular venous O_2 partial pressure (in mmHg). If the arterial saturation, [Hb], and the position of the Hb dissociation curve remain constant, the jugular bulb saturation (Sjo_2) is proportional to the ratio of global cerebral blood flow to cerebral metabolism.

Normal values for cerebral blood flow, $CMRo_2$, and Sjo_2 are 40 to 50 ml /100 g/min, 3 to 3.5 ml /100 g/min, and 54 to 75 per cent respectively.

If arterial oxygen saturation (Sao_2) is estimated using pulse oximetry, the global cerebral oxygen extraction ratio (OER) can be derived:

$$OER = \frac{Sao_2 - Sjo_2}{Sao_2}.$$

Continuous monitoring of arterial and jugular venous oxygen saturation enables the adequacy of global cerebral perfusion to be assessed. With a normal Sao_2 of 97 per cent and a hemoglobin concentration greater than 12 g/dl, an Sjo_2 of less than 50 per cent indicates global relative hypoperfusion. Once Sjo_2 falls below 40 per cent ($AJDo_2 > 9.0$ ml per cent, OER > 57 per cent), global cerebral ischemia is likely and will be associated with the production of lactic acid.

The lactate oxygen index (LOI) can also be used to identify cerebral ischemia:

$$LOI = \frac{\text{jugular venous [lactate]} - \text{arterial [lactate]}}{AJDo_2}.$$

LOI is normally less than 0.03. In the event of significant regional or global cerebral ischemia the lactate oxygen index will become elevated above 0.08. However, regional cerebral ischemia may be associated with increased lactate production in the presence of a normal or even elevated Sjo_2. Reduced Sjo_2 indicates global hypoperfusion and requires urgent correction, but normal or increased Sjo_2 should not be interpreted as indicating adequacy of cerebral perfusion to all areas of the brain. Proper management of patients at risk of cerebral ischemia requires monitoring of both arteriojugular lactate levels and oxygen delivery.

Anatomical considerations for jugular venous oxygen saturation monitoring

In health the majority of venous drainage from both cerebral hemispheres is via venous sinuses through the sigmoid sinus to the right internal jugular vein, although significant drainage may occur through the left. In 75 per cent of subjects there is free venous crossflow between the transverse sinuses, and therefore there is little gradient in Sjo_2 between right and left jugular bulbs. When cerebral blood flow is normal, about 3 to 5 per cent of jugular bulb blood arises from extracranial venous sources.

If focal intracranial pathology is present, the patterns of venous drainage may change and differences in Sjo_2 between right and left become evident. Some investigators choose to monitor Sjo_2 from the side of the focal pathology, while others suggest using the effects of sequential unilateral compression of the internal jugular veins on intracranial pressure to guide the side for cannulation. In those patients with cerebral aneurysms undergoing angiography, the side of predominant venous drainage can be determined angiographically. If no difference is detected and in patients with diffuse brain injury, the right side is used for ease of insertion.

The jugular venous bulb is found immediately inferior to the jugular foramen and is the dilated initial part of the internal jugular vein. The internal jugular vein descends adjacent to the lateral aspect of the internal and common carotid arteries together with the vagus nerve, and all these structures are enclosed within the carotid sheath.

Blood from extracranial sources may lead to contamination and erroneously high Sjo_2 readings; thus the correct position of the catheter tip, i.e. above the top of the C2 vertebra, should be confirmed on a lateral cervical radiograph.

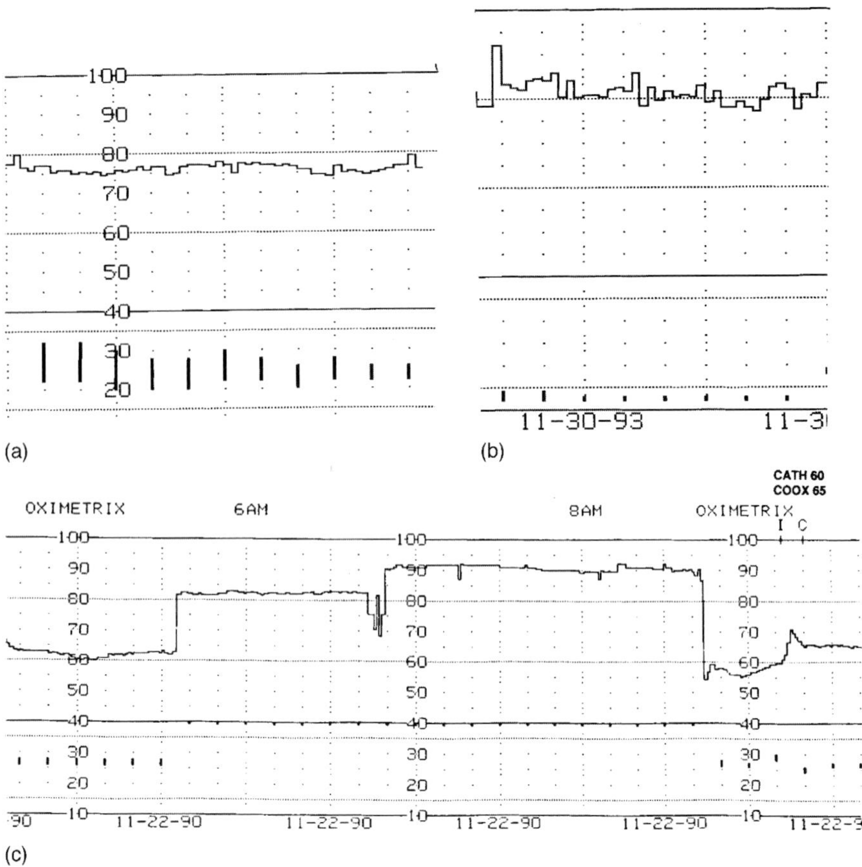

Fig. 1 Displays of Sjo_2 using the Oximetrix system. (a) Acceptable Sjo_2 recording: upper trace, Sjo_2 between 75 and 80 per cent showing characteristic fluctuations; lower trace, vertical bars between two dotted lines indicate adequate light intensity. (b) Unacceptable Sjo_2 recording: the upper trace is reasonable but the lower trace shows the light intensity below the lower dotted lines, indicating catheter obstruction which requires either flushing or catheter replacement. (c) A period of acceptable recordings is followed by a sudden increase in the Sjo_2 trace accompanied by a rise in the light intensity display to above the upper dotted lines. This was due to the patient being turned, resulting in a vessel wall artifact. Subsequent repositioning of the patient restored accuracy as confirmed by an *in vivo* calibration (I, C).

Table 1 Technique for inserting the Oximetrix Sjo_2 catheter

1.	Prepare, sterilize, and drape the neck. Gown and gloves are mandatory.
2.	Perform venipuncture at or above the level of the thyroid cartilage just lateral to the carotid pulsation with the head in the neutral position and, if the intracranial pressure allows, with the patient flat. Aim towards the external auditory meatus. Advance the needle slowly through the carotid sheath into the vein which may be deeper than expected.
3.	Once blood can be aspirated freely use the Seldinger technique for gentle insertion of a 5 or 6 French percutaneous sheath introducer into the jugular vein. If the introducer catheter abuts the skull base (approximately 10–15 cm), withdraw about 1 cm.
4.	Perform the *in vitro* calibration.
5.	Attach a hemostatic valve and contamination sleeve to the introducer.
6.	Prime and then continuously flush the catheter system with heparinized saline.
7.	Advance the fiber-optic catheter, via a protective sleeve, hemostatic valve, and introducer, until the tip abuts the skull base. Withdraw the fiber-optic catheter about 0.5 cm, leaving the introducer *in situ* and therefore maintaining 0.5–1 cm of optical catheter protruding from the introducer tip.
8.	Confirm free aspiration of blood from the catheter (if necessary retract the catheter slightly) and verify satisfactory light intensity display.
9.	Secure with a sterile transparent dressing, noting the length of catheter inserted to allow detection of displacement.
10.	Attach a pressure transducer to the catheter system to monitor for catheter obstruction.
11.	Obtain a lateral cervical radiograph to confirm correct position (catheter tip above C2).
12.	Perform *in vivo* calibration. Aspirate blood slowly maintaining constant Sjo_2 and light intensity readings to minimize sample contamination from extracranial sources. Adjust the machine reading to match the laboratory co-oximeter value.

Contraindications to this technique include local infection or trauma, a bleeding diathesis, or pre-existing impairment to cerebral venous drainage. The effect of insertion of an Sjo_2 catheter on intracranial pressure is minimal if monitoring is confined to 7 days or less, and the risk of cerebral venous thrombosis is below 5 per cent when the catheter is continuously flushed with neutral saline (3 ml/h).

Equipment

Continuous *in vivo* venous oxygen saturation is achieved by reflectance oximetry from a fiber-optic catheter in the jugular venous bulb, and improves detection of transient global cerebral ischemic insults which may be missed by periodic *in vitro* analysis (Andrews *et al.* 1991).

Red and near-infrared light is transmitted at 1-ms intervals down a transmitting fiber and reflected along a receiving fiber to a photoelectric sensor. Reflected signals are averaged over 5 s and updated every second. The Oximetrix 3 system displays trended and current oxygen saturation.

Reflected light intensity is also displayed and when normal appears as a bar between two dotted lines. The reflected light intensity should always be examined before interpreting data. High reflected light intensity indicates that the catheter tip is abutting too closely against the vessel wall, while low intensity suggests catheter-tip obstruction which should be rectified by aspiration and flushing (Fig. 1). Inability to display a light intensity signal larger than a dot between the upper and lower dotted lines may signify damage to the fiber-optics.

A technique for inserting the Oximetrix 3 jugular venous catheter is described in Table 1. The system requires calibration before insertion to confirm its integrity and compares the reference signal with a standard optical reference. Andrews *et al.* (1991) found that use of the preinsertion calibration for jugular oximetry led to over-reading by 7 per cent. Therefore it was necessary to perfom an *in vivo* calibration against a laboratory calibrated oximeter and to repeat this every 8 to 12 h to ensure accurate measurements.

Clinical uses of Sjo_2 monitoring

Continuous monitoring of jugular venous oximetry has been performed in intensive care, the operating room, and the neuroradiology suite to monitor patients with brain trauma, intracranial aneurysms, and arteriovenous malformations.

Episodes of jugular venous desaturation to below 50 per cent have been reported in neurosurgical patients in association with hyperventilation, vasospasm, and compromised cerebral perfusion pressure during intensive care (Chan *et al.* 1992). In a study of 102 patients with severe brain injury, 76 episodes of desaturation below 50 per cent for more than 10 min were recorded and verified in 41 patients, mostly during the first 24 h of intensive care. When outcome was evaluated in 95 patients at 3 months using the Glasgow outcome scale against the number of desaturation episodes, mortalities for none, one, and more than one episode of desaturation were 18 per cent, 46 per cent, and 71 per cent respectively (Robertson 1992). Recent work by Deyne (personal communication) has shown that of 150 patients with severe brain trauma and extremely early Sjo_2 monitoring (mean 4.8 h post-trauma), 85 had episodes of jugular desaturation to below 55 per cent within the first 6 to 12 h of management. In 42 of these patients the cause was a low cerebral perfusion pressure secondary to low blood pressure (< 70 mmHg), while in 29 it was associated with excessive hyperventilation ($Paco_2$ < 30 mmHg).

The potential value of continuously monitoring Sjo_2 is early detection of cerebral hypoperfusion (Sjo_2 < 50 per cent), both intraoperatively and in intensive care, so that appropriate action can be taken more readily. Clinically relevant perturbations of Sjo_2 occur in about 60 per cent of neurological operations. A protocol for use when Sjo_2 falls below 50 per cent in patients with severe brain trauma is shown in Table 2.

Table 2 Protocol for use when Sjo_2 is less than 50 per cent

1.	Check light intensity display
	If high, adjust catheter position
	If low, establish catheter patency
2.	Verify Sao_2
	If <95%, correct hypoxemia
3.	Confirm Sjo_2 with laboratory co-oximeter
	If >5% discrepancy, perform *in vivo* calibration
4.	Measure arterial and jugular lactate concentrations
	Calculate LOI to identify presence of ischemia
5.	Measure blood hemoglobin concentration
	If <10 g/dl, consider transfusion
6.	Measure $Paco_2$
	If <4 kPa (30 mmHg), correct hypocapnia
7.	Check CPP
	If CPP <70 mmHg, correct low BP or high ICP
8.	Correct any remaining cerebral hypoperfusion
	Increase $Paco_2$ further and/or raise CPP (reduce ICP/raise BP)

CPP, cerebral perfusion pressure; BP, blood pressure; ICP, intracranial pressure. Modified from Sheinberg *et al.* (1992).

References

Andrews, P.J.D., Dearden, N.M., and Miller, J.D. (1991). Jugular bulb cannulation: description of a technique and validation of a new continuous monitor. *British Journal of Anaesthesia*, 67, 553–8.

Chan, K.H., Miller, J.D., Dearden, N.M., Andrews, P.J.D., and Midgley, S. (1992). The effect of changes in cerebral perfusion pressure upon middle cerebral artery blood flow velocity and jugular bulb venous saturation after severe brain injury. *Journal of Neurosurgery*, 77, 55–61.

Robertson, C.S. (1992). Treatment of cerebral ischemia in severely head injured patients. *British Journal of Intensive Care*, 1, 12–15.

Sheinberg, M., Kante, M.J., Robertson, C.S., Contant, C.F., Narayan, K., and Grossman, R.G. (1992). Continuous monitoring of jugular venous oxygen saturation in head-injured patients. *Journal of Neurosurgery*, 76, 212–17.

16.6.4 Cerebral blood flow monitoring

P. J. D. Andrews and W. Fitch

Key messages

- The brain is an obligate aerobe.
- The brain depends fundamentally on an adequate supply of substrate (principally oxygen) and the ability to utilize it effectively.
- The monitoring of blood flow to the brain can provide an estimate of the delivery of substrate under physiological or pathophysiological situations.
- Estimates of cerebral blood flow can be made with sufficient precision and reproducibility to aid clinical decision-making.

Introduction

The consumption of energy by the brain is substantial (25 per cent of total glucose and 20 per cent of total oxygen consumed by the resting body). In contrast, the brain has negligible stores of oxygen and only a small content of energy-generating substrates. Therefore it is not surprising that the brain is uniquely susceptible to critical decreases in the delivery of substrate. As a result, monitoring the balance between the supply (cerebral blood flow, hemoglobin concentration, oxygen saturation, glucose concentration) and demand (cerebral metabolic rate for oxygen, cerebral metabolic rate for glucose) is of interest. Many therapeutic measures, such as control of intracranial pressure, modulation of systemic arterial pressure, and control of $Pa\text{CO}_2$, are undertaken in the belief that they may modify this balance beneficially. However, there is increasing evidence that so-called 'ischemic' brain injury may be due more to failure of the utilization of oxygen in a manner similar to that seen in association with the systemic inflammatory response syndrome rather than to failure of the delivery of oxygen *per se*.

Measurement of cerebral blood flow

The theoretical bases of all methods which utilize diffusible tracers are similar, and are based on the Fick principle which can be expressed mathematically as follows:

$$\text{CBF} = \frac{Q_t}{\int_0^{\infty}(C_\text{a} - C_\text{v})\,\text{d}t} \tag{1}$$

where CBF is the cerebral blood flow and has the dimensions of mass or volume per unit time, Q_t is the quantity of the tracer which is taken up in the tissue in unit time t, and C_a and C_v are the arterial and venous concentrations respectively over a short period of time dt.

Unlike the kidney, the brain does not specifically and selectively remove foreign substances from the blood and excrete them—a process which would permit accurate measurement. What the brain can do, however, is absorb (in physical solution) an inert gas when it is presented to the brain via the lungs and/or circulation. Thus the amount of the tracer absorbed by the brain will be independent of the state of mental activity and, since the tracer is inert, will not itself alter the metabolic rate (and hence the blood flow) of the brain. The numerator of the Fick equation (Q_t) is determined by the concentration of the tracer in the brain, the weight of the brain, and the blood–brain partition coefficient (λ) for the tracer:

$$\lambda = \frac{Q_t/W}{C_\text{v}(t)}. \tag{2}$$

In practice, the concentration of tracer can be measured directly or calculated from arteriovenous differences over a timed period. The mass of tissue (brain) can rarely be defined; thus perfusion is calculated as a flow per unit mass (W) of tissue (traditionally 100 g). As a result, it is customary to express flow (F) as milliliters of blood per 100 g brain tissue per minute, such that

$$\text{perfusion} = \frac{100C_\text{v}(t)\,\lambda}{\int_0^t (C_\text{a} - C_\text{v})\,\text{d}t}. \tag{3}$$

This equation is valid only if the partition coefficient is the same throughout the tissue to be measured, the blood flow is uniform in the tissue, and diffusion equilibrium is achieved between the arterial supply and the tissue. Thus the ideal freely diffusible tracer (Table 1) would equilibrate with the entire volume of tissue under investigation and have a 100 per cent extraction efficiency, such that by the end of the first circulation the arterial input would be in diffusion equilibrium with the tissue. Subsequently, there should be a linear relationship between the washout of the tracer and perfusion throughout the range of physiological measurement. Unfortunately, no such ideal tracer exists. Moreover, as far as the brain is concerned, gray matter, white matter, skull, and scalp have different partition coefficients and blood flows, as do normal brain, edematous brain, ischemic brain, and compressed brain. Furthermore, estimations of cerebral perfusion are complicated further by the recirculation of tracer and scatter due to extra- and intracranial activity. However,

Table 1 Characteristics of an ideal tracer

Inert metabolically, and biologically stable over the period of measurement
Not converted into any other substance with different properties during the period of measurement
No effect on metabolism, blood flow, or any other relevant index over the period of study
Suitable for use with available detecting/measuring devices
Measurable to an adequate level of precision and accuracy
Readily available, affordable, and non-toxic

despite these valid considerations, estimates of cerebral blood flow can be made in clinical practice with sufficient reproducibility to permit decisions in relation to treatment and, as alluded to earlier, can be of value in the management of specific patients.

Kety–Schmidt technique

The first quantitative method for the estimation of cerebral blood flow (and cerebral metabolism) in humans was the Kety–Schmidt technique (Kety and Schmidt 1948), in which an inert gas (originally 15 per cent nitrous oxide) was inhaled and global cerebral blood flow was calculated from the difference between the saturation curves obtained from arterial and cerebral venous (jugular bulb) blood.

Technique

Nitrous oxide (3–10 per cent) is administered through the breathing circuit such that the patient's lungs are ventilated with a mixture of oxygen, nitrogen, and nitrous oxide throughout the period of measurement (usually 10–15 min, although a longer period may be required if cerebral blood flow is low). Diffusion equilibrium must be achieved by the end of the period of measurement such that the final arterial and venous concentrations are within 10 per cent of each other. Timed paired samples of arterial and jugular venous blood are withdrawn during the period of nitrous oxide administration. These can be aspirated intermittently (usually six pairs in 10 min) or continuously to give 'integrated' arterial and cerebral venous concentrations. The nitrous oxide concentrations in the blood are determined either spectrophotometrically or chemically.

The basic Kety–Schmidt technique has a number of limitations. First, it is invasive; it is necessary to cannulate an artery (any artery) and the vein which drains the organ being studied. In the case of the brain the appropriate sampling site is the jugular bulb. This introduces the further problem of what is, and what is not, the true venous drainage from the brain itself. The third limitation is that the measurement of flow obtained is the mean of the flows through all the tissues draining into the vein from which the blood samples have been withdrawn; regional variations in flow cannot be detected. Finally, nitrous oxide is not totally inert; in higher concentrations it can alter cerebral metabolism and hence influence cerebral blood flow.

Radioactive inert gases

The most widely used tracer at the present time is the noble gas xenon which has a suitable isotope (xenon-133). Although xenon is not totally inert (at high concentrations it has an anesthetic effect), this is not a significant problem since the isotope has a high specific activity and is administered in low physical concentrations. Importantly, the rate at which the isotope leaves the brain can be determined by monitoring the decay in radioactivity using appropriate externally placed detectors. This technique has two distinct advantages over the Kety–Schmidt technique described earlier. First, no blood samples are required; the flow can be determined from graphical analyses of the clearance curves (ideally by computer) and the result obtained rapidly in real time. Second, it is possible to obtain estimates of regional perfusion by increasing the number of suitably collimated detectors. However, as the number of detectors increases, their size and sensitivity decreases such that distinguishing the true washout curve from statistical noise becomes more difficult. A compromise between sensitivity and resolution must be made, and each clinician must decide the most appropriate approach to the situation under investigation.

Techniques

Intra-arterial injection

When the tracer is given as a single rapid bolus injection (via the carotid artery), which is effectively instantaneous, there is a rapid maximum uptake on the first pass through the brain because of the large concentration gradient between the blood and the brain. Recirculation is minimal since a substantial proportion (95 per cent) of the radioactivity administered is excreted into the alveoli on its first passage through the lungs. The requirement for intracarotid puncture has severely limited the applicability of this technique and has led to the development of other means by which the isotope can be presented to the brain.

Intravenous administration

Xenon-133 (in physiological saline) is administered as a rapid intravenous injection. However, since 80 to 90 per cent of the administered activity is lost on its first passage through the lungs, a significantly greater initial dose is required. Exhaled air must be collected, either into activated charcoal or vented to the atmosphere, and the activity in the exhaled gas (as well as the clearance from the brain) must be measured (the end-expired activity is assumed to be in equilibrium with the arterial input to the cerebral circulation). This estimate of recirculating activity is important since, with this technique (as with the inhalation technique described below), the isotope is distributed not only to the brain but throughout the body. The cranial clearance is subjected to a computerized correction, which takes account of the recirculation of the xenon, and a single washout curve is constructed. As well as the potential for statistical noise in the convolution process, activity in the airways during the early part of the measurement can be greater than that in the brain and results in substantial artifact.

Inhalation technique

The subject breathes, or is ventilated, via a circle breathing system (into which xenon-133 (20 mBq/l) has been introduced) for 2 min. An adequate supply of oxygen is provided, and carbon dioxide is eliminated via soda lime. The circuit is then converted into an open breathing system and the exhaled gas is treated as described above. To minimize error due to the initially high activity in the airways, the first minute of the washout curve is rejected, or a decrease of 20 per cent from the initial maximum count is accepted, before the actual flow is estimated.

Summary

One of the fundamental tenets of any technique of measuring cerebral blood flow is that the flow itself should not be influenced by changes in other variables during the acquisition of the data. In other words, stable physiological conditions (arterial pressure, carbon dioxide tension, mental activity, sympathomimetic activity) should be maintained. Each of the above techniques can be carried out at the bedside. All can be used in the conscious, anesthetized, or ventilated patient, and measurements have been obtained in the operating room (Cook *et al.* 1995) or intensive care unit (Bowton *et al.* 1989). However, many believe that the required stability is difficult to achieve, particularly in the intensive care environment, and/or if the flow is low and the time of the clearance of isotope is prolonged.

Tomographic methods

Although the rates of uptake and clearance of xenon-133 have been recorded satisfactorily for many years using stationary detectors, the advent of the inhalation and intravenous techniques prompted a reappraisal because of the unsatisfactory degree of resolution obtained. As a result, equipment capable of producing tomographic images of the brain has been harnessed to provide tomographic representations of regional cerebral blood flow.

Xenon-enhanced computed tomography

This technique employs relatively high concentrations (15–35 per cent) of stable xenon in association with computed X-ray transmission tomography. The resolution is higher than that obtained with isotope techniques, and if the dose is increased further and the period of monitoring prolonged it is possible to calculate the partition coefficients of various tissues on a point-to-point basis. However, xenon is not totally inert at higher concentrations. Its anesthetic properties will decrease cerebral metabolism and hence modify cerebral blood flow.

Positron emission tomography

Currently, positron emission tomography is the most accurate method for the quantitative measurement of regional cerebral oxygen extraction, glucose uptake, and blood flow. A map giving spatial and depth resolution is produced which has many advantages over xenon-133 planar estimations. However, the technique is very expensive and is available in only a few centers. The interested reader is referred to the review by Myers *et al.* (1992).

Near-infrared spectroscopy

This method is based on the direct Fick principle and utilizes a sudden increase in the arterial saturation of oxygen as the tracer. The amount which accumulates in cerebral tissue is measured over time (t). Cerebral blood flow is calculated from the ratio of the amount of tracer accumulated to the amount introduced during time t:

$$\mathrm{CBF} = \frac{K\Delta[\mathrm{Hbo_2}]}{[\mathrm{tHb}]\int_0^t(\Delta S\mathrm{ao_2})\mathrm{d}t} \tag{4}$$

where $K = 0.89$ is a constant which reflects the ratio of cerebral to large-vessel hematocrit (0.69), the molecular weight of hemoglobin (64 500), and cerebral tissue density.

When there is a small sudden increase in arterial saturation, the initial increase in cerebral $\mathrm{Hbo_2}$ concentration $[\mathrm{Hbo_2}]$ represents the accumulation of tracer. The quantity introduced is given by the product of the integral of $\Delta S\mathrm{ao_2}$ with respect to time and the arterial total hemoglobin concentration [tHb]. It is clear that a number of assumptions must be made during measurement of cerebral blood flow using this method, including constant cerebral blood flow, cerebral blood volume, and oxygen extraction. Early work in adults suggests that cerebral blood flow changes minimally for $P\mathrm{ao_2}$ between 6 and 13 kPa. Similarly, oxygen extraction and blood volume have been observed to vary little (the latter is also measurable by near-infrared spectroscopy).

The equipment required to make this measurement includes a near-infrared spectroscopy monitor and a pulse oximeter capable of beat-by-beat measurement of $S\mathrm{ao_2}$. Even if an ear probe is used, the lung–brain circulation time is 1 to 2 s faster than the lung–ear circulation time and an adjustment must be made. Early work in pediatric practice (transmission near-infrared spectroscopy) (Edwards *et al.* 1988) has been extended to adults (reflectance near-infrared spectroscopy), but further development of these emerging technologies is required before the widespread use of this method can be recommended.

Conclusions

Clearly, monitoring cannot independently improve outcome. However, it can contribute to decreases in morbidity and mortality when integrated into an effective treatment protocol since, in addition to providing warnings, monitoring may guide the clinician in making the correct decision. The monitoring of cerebral blood flow probably falls into the latter category and, although not widely practised, has a defined role in the intensive care environment.

References

Bowton, D.L., Bertels, N.H., Prough, D.S., and Stump, D.A. (1989). Cerebral blood flow is reduced in patients with sepsis syndrome. *Critical Care Medicine*, **17**, 399–403.

Cook, D.J., *et al.* (1995). Cerebral blood flow during cardiac operations: comparison of Kety–Schmidt and xenon 133 clearance methods. *Annals of Thoracic Surgery*, **59**, 614–20.

Edwards, A.D., Richardson, C., Cope, M., Wyatt, J.S., Delpy, D.T., and Reynolds, E.O.R. (1988). Cotside measurement of cerebral blood flow in ill newborn infants by near infrared spectroscopy. *Lancet*, 1 October, 770–1.

Kety, S.S. and Schmidt, C.F. (1948). The nitrous oxide method for the quantitative determination of cerebral blood flow in man: theory, procedure, and normal values. *Journal of Clinical Investigation*, **27**, 476–83.

Myers, R., Spinks, T.J., Luthra, S.K., and Brooks, D.J. (1992). Positron-emission tomography. In *Quantitative methods in neuroanatomy* (ed. M.G. Stewart), pp. 117–61. John Wiley, Chichester.

16.7 Nutritional monitoring

16.7.1 Indirect calorimetry

P. G. Gauger and R. H. Bartlett

Key messages

- Energy expenditure can be measured by direct or indirect calorimetry. Indirect calorimetry, which is more practical and common in critical care, estimates energy expenditure by measuring the amount of oxygen consumed ($\dot{V}O_2$).
- $\dot{V}O_2$ may be measured by open- or closed-circuit variants of indirect calorimetry, depending on the specific clinical situation.
- $\dot{V}O_2$ can also be estimated using the Fick relationship if arterial and mixed venous oxygen contents and cardiac output can be measured.
- $\dot{V}O_2$ is a factor in an equation that converts the amount of substrate consumed (oxygen) to the amount of energy expended under 'resting conditions.'
- In the critically ill patient, resting energy expenditure should be measured frequently under standardized conditions to guide appropriate and sufficient nutritional provision.

Introduction

Ever since Lavoisier discovered *oxygène* and constructed the first calorimeter in the 1780s, scientists have striven to understand better the 'fire of life' that is metabolism. It is clear that critical illness is associated with profound and dynamic abnormalities of metabolic rate and substrate utilization. It is also clear that meeting nutritional demands positively affects the outcome of many serious illnesses. In an effort to optimize nutritional supply and demand, frequent and accurate assessment of patient resting energy expenditure has become common in modern intensive care.

Methods of measuring energy expenditure

Energy expended, in the form of heat produced, could be measured by placing the entire patient in an insulated chamber and measuring temperature change. This process (direct calorimetry) is obviously impractical for critical care. Indirect calorimetry, developed and validated using direct calorimetry, is the calculation of energy utilization based on measurement of O_2 consumption ($\dot{V}O_2$). $\dot{V}O_2$ can be measured directly or calculated using the Fick equation. Once $\dot{V}O_2$ has been measured, the metabolic rate can be described as either milliliters of O_2 consumed per minute (normal value, 100–120 ml O_2/min/m^2) or kilocalories of foodstuffs metabolized per day (normal value, 25–30 kcal/kg/day). Accurate measurement of $\dot{V}O_2$ is the basis for indirect calorimetry. Minor corrections and further calculations can be carried out if $\dot{V}CO_2$ and urinary nitrogen are also measured.

Open-circuit measurement of $\dot{V}O_2$

Measurements of $\dot{V}O_2$, $\dot{V}CO_2$, and gas flow and volume are known collectively as spirometry. Open-circuit variants, such as breath-by-breath, mixing chamber, and dilution systems, all rely on the measurement of O_2 and CO_2 concentrations in both the inspired gas (FiO_2, $FiCO_2$) and the expired gas (FeO_2, $FeCO_2$) together with the exhaled minute ventilation ($\dot{V}_e$). The inhaled minute ventilation ($\dot{V}_i$) can be assumed to be the same as $\dot{V}_e$, directly measured or indirectly calculated. During open-circuit measurements in an ideal situation, $\dot{V}O_2$ can be calculated using the equation

$$\dot{V}O_2 = \dot{V}_i(FiO_2) - \dot{V}_e(FeO_2)$$

and $\dot{V}CO_2$ can be derived similarly from

$$\dot{V}CO_2 = \dot{V}_e(FeCO_2) - \dot{V}_i(FiCO_2).$$

$\dot{V}_e$ is measured either by collection of exhaled gas into a device such as a Douglas bag or by an in-line flow detector. $\dot{V}_i$ is usually assumed to be equivalent to $\dot{V}_e$, although there is a slight difference between the two because of the addition of water vapor from the lungs and the difference between O_2 consumed and CO_2 produced. As it is technically difficult to measure differences between $\dot{V}_i$ and $\dot{V}_e$, $\dot{V}_i$ is often calculated using the Haldane equation in which it is assumed that insoluble nitrogen is not consumed:

$$\dot{V}_i = \frac{FeN_2}{FiN_2}\dot{V}_e.$$

If there are no other gases present (i.e. anesthetic vapors), FiN_2 and FeN_2 can be calculated from the O_2 and CO_2 concentrations alone:

$$FiN_2 = 1 - FiO_2 - FiCO_2$$

and

$$FeN_2 = 1 - FeO_2 - FeCO_2.$$

Substituting, the equation for $\dot{V}O_2$ becomes

$$\dot{V}o_2 = \frac{1 - Feo_2 - FeCo_2}{1 - Fio_2}.$$

Since $Fico_2$ is zero during mechanical ventilation and usually no greater than 0.03 per cent when breathing room air, the equation for $\dot{V}co_2$ can be simplified to

$$\dot{V}co_2 = \dot{V}_e(Feco_2).$$

The key components of all open-circuit systems are accurate volume-measurement systems and the individual O_2 and CO_2 analyzers. These systems must be capable of measuring small changes in gas concentrations in normoxic and hyperoxic environments. The simplest method is to analyze aliquots of mixed inhaled and mixed exhaled gases in a conventional blood gas machine. Many other factors must be considered to ensure accuracy of measurements in the clinical setting. Elevated levels of Fio_2 have an undesirable effect on open-circuit measurements. Errors occur when Fio_2 approaches 1.0 and the denominator of the Haldane equation approaches zero. Also, any error in the measurement of gas concentrations is amplified as Fio_2 increases because $\dot{V}o_2$ becomes a small difference between two large numbers, and because of variations in Fio_2. In most ventilators, Fio_2 can vary slightly during a single breath. This is often the case with ventilators that utilize on-line mixing of air and O_2 with proportional valves instead of an external blender and/or reservoir to dampen fluctuations in Fio_2. Therefore Fio_2 must be measured in mixed inspired gas at the same time as Feo_2 is measured in exhaled gas.

The presence of anesthetic gases or other vapors can complicate open-circuit techniques, specifically by negating the assumptions made in the Haldane equation. Also, water vapor that is not removed by desiccants, temperature manipulations, or special tubing can adversely affect the function of O_2 and CO_2 sensors. During measurements, the ventilator or breathing system circuit must be free from leaks. Loss of gas to the environment or entry of room air into the system will obviously invalidate measurements of $\dot{V}_e$. This is also the case in patients with leaking endotracheal tube cuffs, leaking chest tubes, and bronchopleural fistulas, in which $\dot{V}_e$ and $Feco_2$ are decreased and subsequently $\dot{V}o_2$, $\dot{V}co_2$, and resting energy expenditure are underestimated. The $\dot{V}o_2$ measured by open-circuit technique is the volume at ambient temperature and pressure saturated with water vapor (ATPS), which must be converted to standard temperature and pressure under dry conditions (STPD) for metabolic calculations.

Closed-circuit measurement of $\dot{V}o_2$

Closed-circuit spirometry excludes the patient from the outside air. The volume-loss type uses a volumetric reservoir of O_2. The exhaled CO_2 is absorbed from the system with soda lime so that the loss of volume over time is a measure of oxygen consumption. The volume-replenishment variant uses a similar system which maintains a constant volume in an oxygen reservoir by adding measured increments of oxygen to the system. As the system is similarly scrubbed of CO_2 and water vapor, the volume of oxygen needed to replenish the reservoir correlates with the patient's oxygen consumption. The spirometer is usually filled with 100 per cent O_2 saturated with water vapor (ATPS). Of course, there is no effect of Fio_2 on metabolic rate.

Because analysis of the inspired or expired O_2 concentration is not necessary, closed-circuit calorimetry is not plagued with difficulties of breath-to-breath variation in Fio_2. A potential difficulty with the volume-loss system is that the volume in the spirometer decreases during the measurement, which renders it impractical for prolonged measurement periods. This is not an issue with the volume-replenishment system. System leaks can greatly alter $\dot{V}o_2$ values, as loss of gas to the environment will overestimate $\dot{V}o_2$ while entrainment of room air into the system will cause underestimation. Just as with open-circuit spirometry, air leaks through chest tubes or around endotracheal tube cuffs prevent accurate quantification of exhaled gas and invalidate the measurement of $\dot{V}o_2$ or $\dot{V}co_2$. Similarly, during hemodialysis, CO_2 is exchanged through the dialysis coil, thus decreasing $\dot{V}co_2$ values. Conversely, colonic fermentation of non-digested carbohydrates can lead to the production of short-chain fatty acids, which are then metabolized to CO_2. When this is absorbed and excreted by the patient, the $\dot{V}co_2$ value is falsely elevated (Weissman and Kemper 1995). Interposition of the closed-circuit spirometer between the ventilator and the patient can confound that relationship. The internal volume of the closed-circuit device greatly increases the compressible volume of the system. The ventilator-driven tidal volume must be increased to account for this inefficiency. Patients with poor pulmonary compliance and high inflation pressure requirements are most affected by this problem.

Estimation of $\dot{V}o_2$ using the Fick equation

The Fick equation is frequently used to calculate $\dot{V}o_2$ in the intensive care unit (ICU), but is rarely used to derive $\dot{V}co_2$. The method requires the use of a pulmonary artery catheter to obtain mixed venous blood and to measure cardiac output. Ideally, the oxygen saturations should be measured using a co-oximeter, rather than calculated from the Pao_2 (Weissman and Kemper 1995). $\dot{V}o_2$ is then calculated using the familiar formula

$$\dot{V}o_2 = (Cao_2 - C\bar{v}o_2) \times CO$$

where Cao_2 is the arterial oxygen content, $C\bar{v}o_2$ is the mixed venous oxygen content, and CO is the cardiac output. Because the oxygen content is calculated at STPD, $\dot{V}o_2$ is expressed as the STPD volume.

The Fick method does not allow for continuous measurement, and as the different components are measured at different times, it is actually a composite rather than a real-time measurement (Weissman and Kemper 1995). Also, the determination depends on the accuracy of thermodilution cardiac output which is usually around ± 15 per cent. Under ideal conditions, the values for $\dot{V}o_2$ obtained by the Fick method and the gas exchange methods discussed below are similar, but $\dot{V}co_2$ and the respiratory quotient (RQ) cannot be accurately determined by the Fick method. When $\dot{V}o_2$ measured by gas exchange methods exceeds $\dot{V}o_2$ obtained using the Fick method, this may be due to the contribution of the metabolic activity of the lung.

The specific advantages and difficulties of the different types of $\dot{V}o_2$ measurement should be used to decide which method is appropriate for particular applications. A closed-circuit system is best suited for a patient on control or assist-control modes of mechanical ventilation and requiring elevated Fio_2 (above 0.6) (Branson 1990). Open-circuit methods are suitable for any mechanically ventilated patient with

$F_{IO_2} < 0.6$ and spontaneously breathing patients since the system can be adapted to interface with the patient through a canopy hood instead of via a mask or direct airway access. The Fick method is reserved for patients with major air leaks.

Conversion of $\dot{V}O_2$ to energy expenditure

Measured values of $\dot{V}O_2$ and $\dot{V}CO_2$ are converted to energy expenditure (kcal/day) by using the Weir equation

$$\text{energy expenditure} = (3.941\dot{V}O_2 + 1.11\dot{V}CO_2) \times 1440$$

where $\dot{V}O_2$ and $\dot{V}CO_2$ are expressed in liters per minute and 1440 is the number of minutes per day. This equation assumes that RQ = 0.8. A version of the Weir equation exists that uses the value of excreted 24-h urinary nitrogen to determine the exact RQ, but this is not often used because the measurement is not usually available at the same time as $\dot{V}O_2$ and $\dot{V}CO_2$. Another value derived from spirometry is RQ = $\dot{V}CO_2/\dot{V}O_2$, which can guide interpretation of substrate utilization and also serve as an indicator of quality control for the measurements since it extends over a fairly narrow range (0.67–1.3) (Branson 1990). An even simpler conversion is provided by the equation

$$\text{energy expenditure (kcal/day)} = \dot{V}O_2\ (\text{l/min}) \times 60\ (\text{min/h}) \times 24\ (\text{h/day}) \times 5\ (\text{kcal/l}\ O_2)$$

in which it is assumed that the RQ = 1.0 and the caloric value of all substrates consumed is 5 kcal/l O_2 consumed. In practice, the errors associated with ignoring corrections for RQ, V_i versus V_e, and urinary nitrogen are minor, and the simplified conversion factor given above can be used. It is important to use STPD values for $\dot{V}O_2$ and $\dot{V}CO_2$ in all calorimetry calculations.

Resting energy expenditure is defined as the energy expended when the patient is lying in bed, awake, and aware of his or her surroundings. This is not often the case with critically ill patients, and therefore the goal is to complete measurement at a time when the patient has been resting undisturbed for at least 30 min in a thermoneutral environment. In the strictest sense, there should have been no food intake for the previous 5 h, but it is perhaps more realistic that there should have been no changes in the rate of nutrition. These measurements of resting energy expenditure are actually about 10 per cent higher than the basal metabolic rate as they include the specific dynamic action of foods (diet-induced thermogenesis) (Weissman and Kemper 1995). Measurements should be obtained when this requisite resting state is maintained for at least 10–15 min. Subsequent measurements should be performed under conditions that are as similar as possible to aid in meaningful interpretation of the trends. The difference between actual or total energy expenditure and resting energy expenditure is accounted for by activity. In mechanically ventilated patients, the total energy expenditure is commonly about 10 per cent greater than the resting energy expenditure (Swinamer *et al.* 1987). Similarly, in hospital patients with more activity, such as ambulation and physical therapy, total energy expenditure may exceed resting energy expenditure by 20 to 30 per cent. These factors should be considered when designing individualized nutritional regimens.

Interpretation of metabolic measurements

Measurement of energy expenditure is clearly beneficial in striving for optimal management of critically ill patients. Inadequate nutrition contributes to complications in patients with renal failure, sepsis, or burns, and in mechanically ventilated patients. Conversely, overaggressive nutrition in the metabolically altered patient may result in detrimental metabolic changes. Many predictive equations, such as the Harris–Benedict formula, have been used to estimate energy requirements. Although adequate in normal and minimally stressed patients, formula predictions are ± 40 per cent of the measured value in critically ill patients. Similarly, metabolic rate changes from day to day (but rarely from hour to hour) in ICU patients. Those patients with sepsis, severe respiratory failure, or hemodynamic instability should be assessed daily. Two studies have shown that ICU survival in patients with multiple organ failure was improved when nutritional management was based on indirect calorimetry data (Bartlett *et al.* 1982; Kresowik *et al.* 1984).

Optimal patient care clearly involves provision of appropriate and adequate levels of nutrition. In the critically stressed patient, this cannot be consistently estimated and should be frequently measured. It is important to understand the different methods of $\dot{V}O_2$ measurement, and the conversion of $\dot{V}O_2$ to energy expenditure. Understanding the applications and limitations of the different methods will aid in making accurate and appropriate measurements for different patients.

References

Bartlett, R.H., Dechert, R.E., Mault, J.R., Ferguson, S.K., Kaiser, A.N., and Erlandson, E.E. (1982). Measurement of metabolism in multiple organ failure. *Surgery*, **92**, 771–9.

Branson, R.D. (1990). The measurement of energy expenditure: instrumentation, practical considerations, and clinical application. *Respiratory Care*, **35**, 640–59.

Kresowik, T.F., Dechert, R.E., Mault, J.R., Arnoldi, D.K., Whitehouse, W.M., and Bartlett, R.H. (1984). Does nutritional support affect survival in critically ill patients? *Surgery Forum*, **35**, 108–10.

Swinamer, D.L., Phang, P.T., Jones, R.L., Grace, M., and King, E.G. (1987). Twenty-four hour energy expenditure in critically ill patients. *Critical Care Medicine*, **15**, 637–43.

Weissman, C. and Kemper, M. (1995). Metabolic measurements in the critically ill. *Critical Care Clinics*, **11**, 169–97.

16.8 Specific organ function

16.8.1 Hepatic function

A. Rhodes and J. Wendon

Key messages

- Traditionally, biochemical tests of liver function give relatively little information about how well the liver is performing, as they are essentially indicators of hepatic dysfunction.
- Unconjugated hyperbilirubinemia occurs when the production of bilirubin exceeds the ability of the liver to remove and conjugate it.
- Conjugated hyperbilirubinemia occurs when there is leakage of bilirubin into the blood system, at or distal to the hepatocytes, when its normal route of excretion is blocked.
- The main value of the aminotransferases is in detecting hepatocellular damage and in monitoring the patient's progress. Alanine aminotransferase has greater hepatic specificity than aspartate aminotransferase.
- Serum levels of alkaline phosphatase enzymes rise in many types of liver disease, particularly where there is intra- or extrahepatic obstruction to the flow of bile and in the presence of intrahepatic space-occupying lesions such as tumors.

Introduction

The liver is a complex organ with a diverse array of functions. Thus the biochemical assessment of hepatic function is difficult, as demonstrated by the numerous tests of function which exist. In order to solve this problem, most clinical laboratories provide a battery of liver function tests in an attempt to evaluate some of its function. Traditionally these have included bilirubin, alkaline phosphatase, and the transaminases, which give relatively little information about how well the liver is performing as they are essentially indicators of hepatic dysfunction. What these tests offer is a cheap and simple screening tool. Recently additional investigations have been described which provide a better description of hepatic function in the diseased state.

Bilirubin

Bilirubin is the by-product of heme metabolism when red cells are degraded by the reticuloendothelial system. The iron in heme is reutilized but the tetrapyrrole ring is degraded to bilirubin. Unconjugated bilirubin is insoluble in water and is transferred in blood bound to albumin. It is taken up by the hepatocytes in the liver, transported to the smooth endoplasmic reticulum, and undergoes conjugation to form a diglucuronide. Conjugated bilirubin is water soluble and is excreted via biliary canaliculi to the gut. Little conjugated bilirubin normally appears in the blood unless there is impaired hepatocyte excretion or biliary obstruction.

The normal range for total bilirubin is 3 to 15 mmol/l (0.16–0.88 mg/100 ml), with jaundice becoming clinically apparent above 35 mmol/l (2.0 mg/100 ml). Men have a slightly greater mean total bilirubin than women, with levels 3 to 4 mmol/l (0.16–0.23 mg/100 ml) higher.

Unconjugated hyperbilirubinemia occurs when the production of bilirubin exceeds the ability of the liver to remove and conjugate it (Table 1). The concentration of unconjugated hyperbilirubinemia in adults rarely exceeds 100 mmol/l (5.8 mg/100 ml).

Conjugated hyperbilirubinemia occurs when there is leakage of bilirubin into the blood system, at or distal to the hepatocytes, when its normal route of excretion is blocked (Table 2). The water-soluble bilirubin then enters the blood system and is excreted in the urine. However, some of the conjugated bilirubin becomes covalently bound to serum albumin and other proteins; this portion is not filtered by the kidney and thus bilirubinuria is not always present with conjugated hyperbilirubinemia. With total biliary obstruction, bilirubin levels can reach up to 500 mmol/l (29.2 mg/100 ml) with the levels stabilizing due to renal excretion. The presence of renal failure or hemolysis with obstruction can lead to levels greater than 1000 mmol/l (58.5 mg/100ml).

Table 1 Causes of unconjugated hyperbilirubinemia

Excessive bilirubin production
Ineffective hematopoiesis (e.g. pernicious anemia)
Hemolytic disorders (e.g. transfusion disorders, G6PD deficiency, drugs, etc.)
Abnormal bilirubin metabolism
Congenital inherited defects (e.g. Gilbert's syndrome, Crigler–Najjar syndrome)
Drug effects

G6PD, glucose-6-phosphate dehydrogenase.

Table 2 Causes of conjugated hyperbilirubinemia

Hepatocellular abnormality
Primary hepatocyte disease (e.g. hepatitis, cirrhosis, drugs, neoplasm)
Intrahepatic cholestasis (e.g. pregnancy, drugs)
Congenital conjugated hyperbilirubinemia (e.g. Dubin–Johnson syndrome, Rotor syndrome)
Benign postoperative jaundice
Biliary obstruction
Intrahepatic (e.g. primary biliary cirrhosis, sclerosing cholangitis, infantile obstructive cholangiopathy)
Extrahepatic obstruction (e.g. atresia, calculus, neoplasm, stricture)

Urinary bilirubin

When bilirubinuria is present the urine is usually a dark brown color. The presence of bilirubin in the urine is easily tested for by a test strip impregnated with a diazo reagent. This test is easy to perform at the bedside and is sensitive to bilirubin levels of 1 to 2 mmol/l (0.06–0.12 mg/100 ml). It can establish the presence of liver damage before frank jaundice is clinically apparent. In jaundiced patients, the absence of bilirubinuria usually implies that the hyperbilirubinemia is unconjugated and thus suggests a hemolytic cause (Table 1).

Aspartate and alanine aminotransferases

These enzymes are released when there is cell damage causing an increase in permeability of the cell membrane or cell necrosis. The main value of the aminotransferases is in detecting hepatocellular damage and monitoring the patient's progress. The levels rapidly return to normal following resolution of the factors causing hepatocellular damage. Because they are so non-specific, raised levels of these enzymes are of little value alone in generating a differential diagnosis; however, a very high level (greater than 1000 IU/L) is strongly suggestive of acute hepatitis of viral, drug, or ischemic origin.

Aspartate aminotransferase (AST) is released into blood when there is damage to any one of a number of organs, including the liver, heart muscle, skeletal muscle, kidney, pancreas, and red blood cells. There is no method available for differentiating the organ source of a raised AST and therefore it is a relatively non-specific test.

Alanine aminotransferase (ALT) is far more specific to the liver than AST, but is still present in low concentrations in other tissues and thus can still be raised in conditions such as myositis. ALT tends to rise and fall in tandem with AST; however, a small increase in ALT can often be found prior to, or without, a rise in AST.

Glutathione-*S*-transferase

The glutathione-*S*-transferase enzymes have attracted much attention recently as a more sensitive and specific marker of hepatocellular damage and necrosis than the aminotransferases. The cationic B forms are found only in the liver as bilirubin-binding proteins. They have a shorter half-life than the aminotransferases and thus are better able to follow acute cellular damage and recovery.

Alkaline phosphatases

Alkaline phosphatases are enzymes which release inorganic phosphate from a variety of orthophosphates. They are present in nearly all tissues, particularly the liver, biliary system, kidney, bone, intestine, placenta, and tumors. In the liver, alkaline phosphatase is located in the microvilli of the bile canaliculus and on the sinusoidal surface of the hepatocytes. Serum alkaline phosphatase levels vary depending on age. Levels are raised until about 20 years of age and then again in old age, with increased contributions of liver alkaline phosphatase in males and bone alkaline phosphatase in postmenopausal females. Serum levels of alkaline phosphatase enzymes rise in many types of liver disease, particularly where there is either intra- or extrahepatic obstruction to the flow of bile and in the presence of intrahepatic space-occupying lesions such as tumors. The raised levels of the enzyme seem to be due to increased synthesis in the canalicular membrane rather than failure of excretion. Specific assays are available to differentiate between the different isoenzymes of alkaline phosphatase, which makes specific diagnosis simpler and more accurate. When these specific isoenzyme assays are not available, other markers of biliary obstruction such as γ-glutamyl transferase should be monitored.

Gamma-glutamyl transferase

Gamma-glutamyl transferase is a membrane-bound glycoprotein which is found in cells with a high secretory or absorptive activity, such as the liver, kidney, pancreas, intestine, and prostate. Serum levels of γ-glutamyl transferase are high in almost all forms of liver disease, with particularly high levels in intrahepatic biliary obstruction. Ninety per cent of patients with hepatobiliary disease have raised levels of γ-glutamyl transferase, but the test has a poor specificity because of the many other conditions that also give raised levels (Table 3).

Albumin

Albumin is the most abundant plasma protein and is primarily synthesized in the liver, which produces approximately 15 g/day in an average 70-kg adult. Therefore measurement of albumin could provide an estimate of hepatic synthetic function. Unfortunately,

Table 3 Causes of increased γ-glutamyl transferase

Hepatobiliary disease
Pancreatic disease
Alcohol
Drugs (particularly enzyme inducers)
Miscellaneous (e.g. anorexia nervosa, dystrophia myotonica, Guillain–Barré syndrome, hyperthyroidism, porphyria cutanea tarda, malignant disease)

alterations in serum albumin are determined by a variety of factors other than hepatic synthesis. As well as decreased synthesis in liver disease, malnutrition, malabsorption, and malignancy, there are increased losses in protein-losing enteropathies, the nephrotic syndrome, and increased vascular permeability. Sepsis and hypercatabolic states lead to increased breakdown, and changes in plasma volume due to disease or overhydration will cause a fall in serum level.

Prothrombin time

Coagulation factors II, VII, IX, and X are synthesized by the liver in a vitamin-K-dependent process. The prothrombin time assesses the function of factors II, VII, and X, and thus provides a marker of the synthetic function of the liver. Owing to the short half-life of these factors (5–72 h) this test can provide a sensitive indicator of hepatic function. Interpretation must be related to vitamin K levels as deficiency can occur in cholestasis where vitamin K absorption is poor.

Galactose elimination capacity

Galactose is eliminated from the liver via a phosphorylation pathway involving galactokinase. When this enzyme is saturated, the hepatic galactose clearance follows zero-order kinetics and can be used as a determinant of hepatic function. It has been shown in animals to be proportional to the functioning mass of hepatocytes. In this test 0.5 g/kg of galactose is injected intravenously (Tygstrup 1966) and galactose levels are then measured in the serum to determine clearance between 20 and 50 min following injection. Values below 2 mg/kg/min are incompatible with survival and probably relate to extrahapatic galactose clearance. This test has been shown to be able to predict survival in patients with primary biliary cirrhosis and to correlate with outcome following interferon treatment for hepatitis B.

Indocyanine green clearance

The clearance of indocyanine green can provide an estimate of hepatic perfusion as uptake is almost exclusively from hepatocytes and the extraction ratio is high. As the extraction ratio of this compound across the liver approximates to almost unity, its clearance can also be used as a marker of function (Caesar *et al.* 1961). Therefore interpretation of this test must take into account the varying effects of both liver blood flow and hepatic extraction. In order to quantify the extraction ratio in patients with liver dysfunction, it is necessary to sample indocyanine green concentrations across the liver from the hepatic vein and a peripheral artery. When high levels of indocyanine green are given, the uptake process becomes saturated and the maximal clearance, which equates to the functioning mass of hepatocytes, can be calculated. Reports have suggested that this test can be used as an independent variable in the assessment of prognosis for patients with cirrhosis, in patients awaiting liver transplantation, and in the assessment of potential donor grafts.

Formation of monoethylglycine xylidide from lidocaine

The formation of monoethylglycine xylidide from lidocaine (lignocaine) (1 mg/kg infused slowly over 3 min) is able to quantify the activity of the cytochrome P-450 enzyme system (Oellerich *et al.* 1987). A single sample taken 15 to 30 min after the lidocaine injection is analyzed using a fluorescence polarization immunoassay. Monoethylglycine xylidide is metabolized from lidocaine by multiple enzymes which are primarily found in the liver, and results have proved to be a reliable predictor of 1 year survival in both adult and pediatric patients with chronic liver disease. Caution must be advised with this test in severe liver dysfunction, as repeated doses can lead to lidocaine toxicity.

Arterial ketone body ratio

The ratio of acetoacetate to 3-hydroxybutyrate in arterial blood has been proposed as an indicator of hepatic mitochondrial function and hepatic redox state (Ozawa *et al.* 1983). This test is being increasingly used as it is relatively simple and a dedicated spectrophotometer and diagnostic kit are available. Decreases in the ratio have been associated with increases in morbidity and mortality following hepatic surgery, and it is of prognostic value in patients with fulminant hepatic failure.

Miscellaneous other tests

A variety of other compounds which are utilized by the liver can be used as a test of hepatic function. Glucose homeostasis is closely controlled by the liver and is an ominous and early sign of fulminant hepatic failure. Although non-specific this can be used as a test of function, particularly in the acute phase following liver transplantation. The liver is also the major organ involved in lactate metabolism. A rise in serum lactate from any cause will be compounded by an abnormally performing liver, as metabolism will decrease and levels will rise. There are several other tests that can quantify hepatic function including the aminopyrine breath test and clearance of caffeine, antipyrine, sulfobromophthalein, or sorbitol. Most of these tests are rarely used outside the research laboratory and have little or no place in regular clinical practice.

References

Caesar, J., Shaldon, S., Chiandussi, L., Guevara, L., and Sherlock, S. (1961). The use of indocyanine green in measurement of hepatic blood flow and as a test of hepatic function. *Clinical Science*, **21**, 43–57.

Oellerich, M., *et al.* (1987). Monoethylglycinexylidide formation kinetics: a novel approach to the assessment of liver function. *Journal of Clinical Chemistry and Clinical Biochemistry*, **25**, 845–53.

Ozawa, K., *et al.* (1983). Metabolic abnormalities associated with postoperative organ failure: a redox theory. *Archives of Surgery*, **118**, 1245–51.

Tygstrup, N. (1966). Determination of the hepatic elimination capacity (Lm) of galactose by single injection. *Scandinavian Journal of Clinical Investigation*, **18**, 118–25.

16.8.2 Renal function

P. M. J. M. de Vries

Key messages

- Glomerular filtration rate is the best test of renal function.
- Urinalysis is often the most useful test available for detecting early renal abnormalities.
- Serum creatinine level is probably the most widely used indirect measure of glomerular filtration rate. Unfortunately, correct interpretation in the critical care setting is a problem.
- A normal urinalysis in the setting of acute renal failure suggests the presence of either prerenal or obstructive renal failure.
- Red blood cells and red blood cell casts suggest the presence of glomerular or vascular inflammatory disease and are rarely seen in acute interstitial nephritis.

Introduction

Tests that best detect abnormalities in renal function are those that measure glomerular filtration rate. However, these tests many not be useful for screening purposes in the critical care setting. Apart from practical disadvantages, patients with early renal disease may have a normal or even increased glomerular filtration rate. Urinalysis is often the most useful test available for detecting early renal abnormalities. Measuring urinary protein level or examining the urinary sediment can also help in establishing a specific diagnosis.

Serum creatinine

Serum creatinine level is probably the most widely used indirect measure of glomerular filtration rate. Unfortunately, correct interpretation in the critical care setting is a problem. It is important to consider variation in creatinine production from differences in muscle mass. Muscle mass may decline rapidly in a catabolic state, resulting in an underestimation of the amount of deteroration of renal function. Another confusing factor is the use of standard normal ranges for creatinine levels. In an elderly individual a normal serum creatinine concentration can indicate a twofold reduction in glomerular filtration rate as a result of a comparable reduction in muscle mass (Rowe *et al.* 1976). Variations in creatinine production due to age- and sex-related differences in muscle mass have been measured and incorporated in formulas to improve the ability of serum creatinine to estimate glomerular filtration rate. The most widely used equation is

$$\text{creatinine clearance} = \frac{1.22 \times (140 - \text{age}) \times \text{weight}}{S_{cr}}$$

where age is in years, weight is in kilograms, and S_{cr} is the serum creatinine level in micromoles per liter.

Furthermore, failure to remember the potential effects of tubule secretion on serum creatinine, particularly in patients with reduced renal function, may lead to overestimation of the glomerular filtration rate.

Creatinine clearance

In principle, the renal clearance of creatinine is the urinary creatinine excretion divided by the area under the serum creatinine concentration–time curve over the period in which the urine was sampled. In practice, it is usually calculated as the ratio of the urinary creatinine excretion to a single serum creatinine value:

$$\text{creatinine clearance} = UV/S$$

where U (μmol/ml) is the urine creatinine concentration, V (ml/min) is the urinary volume, and S (μmol/ml) is the serum creatinine level.

Measuring creatinine clearance obviates some of the problems of using serum creatinine as a marker of glomerular filtration rate, but creates others. The above equation assumes that serum creatinine is constant over 24 h. This may not be valid under clinical conditions. Differences in muscle mass do not affect clearance, but variability in tubule secretion of creatinine does. Tubule secretion causes creatinine clearance to overestimate the true glomerular filtration rate.

It can be questioned whether measurement of clearance gives more reliable information than serum creatinine level alone.

Urinalysis

Urinary volume is of very little diagnostic value. The sudden onset of anuria or the fluctuation in urine volume are typical of obstructive uropathy. Oliguria (urinary volume less than 400 ml/day) is seen as pathognomonic of renal failure. However, non-oliguric renal failure is seen in roughly half of patients with renal failure (Anderson and Schrier 1980) and with increasing frequency over the years. This is probably because the non-oliguric state is normally less severe. Early diagnosis and improved supportive management of seriously ill and traumatized patients in the intensive care unit ensure that most patients never reach the oliguric state.

Examination of urine by microscopy remains a useful procedure. A number of conditions affect formed elements in the urine and, when possible, should be optimized. Contamination with bacteria should be avoided by sampling a midstream specimen. A high urine concentration and a low urine pH help to preserve formed

Table 1 Urine indices used in the differential diagnosis of prerenal and intrinsic renal azotemia

Diagnostic index	Prerenal azotemia	Renal azotemia
Urine specific gravity	>1.018	<1.012
Urine osmolality (mosmol/kg)	>500	<350
Urine sodium (mmol/l)	<10	>20
Urine-to-plasma creatinine ratio	>40	<20
FE_{Na}	<1	>1

elements. Thus a first-void morning urine specimen, which normally is most likely to be acid and concentrated, should be used. Urine should be examined as soon as possible after collection and should not be refrigerated.

A normal urinalysis in the setting of acute renal failure suggests the presence of either prerenal or obstructive renal failure. Red blood cells and red blood cell casts suggest the presence of glomerular or vascular inflammatory disease and are rarely seen in acute interstitial nephritis. Hematuria accompanied by red blood cell casts or marked proteinuria is most likely to be glomerular in origin. It has been reported that red blood cells originating in glomeruli have a distinctive dysmorphic appearance that is most readily appreciated by use of phase-contrast microscopy.

Large numbers of polymorphonuclear leukocytes are typical of acute diffuse pyelonephritis or papillary necrosis, while eosinophils are commonly found in acute allergic interstitial nephritis.

Casts are cylindrical bodies several times larger than leukocytes and red blood cells. They form in distal tubules and collecting ducts. Hyaline or fine granular casts can be seen in normal individuals and provide little useful diagnostic information. Red blood cell casts usually indicate glomerular disease, whereas white blood cell casts are most commonly associated with interstitial nephritis but can also be seen in glomerulonephritis. Casts made up of renal tubular cells are always indicative of tubule damage. Coarse granular casts often result from degeneration of a variety of cellular casts. Thus their presence is usually pathological but non-specific.

Urinary protein

Total protein concentration in urine can be estimated at the bedside using chemically impregnated plastic strips. However, these reagent strips are unable to react to many of the heterogeneous proteins found in human urine. All these semiquantitative screening tests measure only total protein or albumin concentration. Their sensitivity and specificity can be markedly influenced by fluid intake, the state of diuresis, and the resulting urinary concentration. Laboratory quantitative protein tests are best performed in 24-h urine. There is a high degree of correlation between 24-h urinary protein excretion and protein-to-creatinine ratios in random single voided urine samples in patients with a variety of renal diseases (Ginsberg *et al.* 1983), suggesting that this ratio is useful as a screening test for renal disease. It has been suggested that protein-to-creatinine ratios above 3.0 mg/mg or below 0.2 mg/mg indicate protein excretion rates of more than 3.0 g/24 h or less than 0.2 g/24 h respectively.

Only a small amount of protein appears in the urine as a result of normal tubule secretion. Two renal mechanisms can give rise to a higher urine protein level. First, a disruption in the capillary wall barrier can lead to a glomerular origin of proteinuria. Second, tubule damage or dysfunction can inhibit the normal absorptive capacity of the proximal tubule and lead to tubule proteinuria. Therefore proteinuria may suggest the presence of glomerulonephritis, vasculitis, or tubule necrosis. Quantifying urine protein excretion may help to distinguish glomerular from tubular proteinuria. Since protein excretion is in the nephrotic range (> 3 g/24 h), a glomerular source is almost certain. Quantification of urine protein excretion may also provide useful prognostic information and assist in monitoring the response to therapy.

Urine chemistry

The kidney responds to hypoperfusion by avidly retaining salt and water with production of urine that may reach a specific gravity of 1.030. Measurement of the specific gravity may be included as part of the standard urinalysis. Specific gravity is a convenient and rapidly obtained indicator of urine osmolality. It can be measured accurately with a refractometer or a hygrometer. The normal range is 1.003 to 1.030, but values fall with age as the kidney's ability to concentrate urine decreases. Specific gravity can be used to obtain a crude estimate of how the concentration of other urine constituents may reflect their total excretion because it correlates inversely with 24-h urine volume (McCormack *et al.* 1991).

In acute intrinsic renal failure, when the kidney is unable to concentrate or dilute the urine, specific gravity is 1.010 (like that of serum). A prerenal cause of failure will give rise to high specific gravity values. There are a few other tests that help to distinguish prerenal from renal conditions (Table 1). Urine osmolality is a reliable index of concentrating ability; values greater than 500 mosmol/l suggest prerenal causes, while values less than 350 mosmol/l point to a parenchymal origin of renal failure. Urinary sodium concentrations below 10 mmol/l are typical of prerenal azotemia. In order to enhance the sensitivity of diagnostic tests, urine-to-plasma ratios have been developed. For example, the urine-to-plasma ratio for creatinine concentrations is a useful diagnostic index. In practice, however, some patients may have intermediate values and the diagnosis cannot be made on the basis of this index alone.

The fractional excretion of sodium (FE_{Na}) is the most sensitive index for distinguishing between prerenal and parenchymal causes of renal failure (Espinel 1976). This index relates sodium clearance to creatinine clearance:

$$FE_{Na} = \frac{U_{Na}P_{Cr}}{P_{Na}U_{Cr}} \times 100 \text{ per cent.}$$

Sodium is reabsorbed avidly from glomerular filtrate in patients with prerenal azotemia as a consequence of suppression of atrial natriuretic peptide secretion, activation of renal nerves and the renin–angiotensin–aldosterone axis, and local changes in peritubular hemodynamics. In contrast, sodium reabsorption is inhibited in acute tubular necrosis as a result of tubule cell injury. Creatinine is reabsorbed to a much smaller extent than sodium in both conditions. Consequently, FE_{Na} is typically less than 1 per cent in patients with prerenal azotemia, whereas it is usually greater than 1 per cent in those with ischemic or nephrotoxic acute renal failure.

However, this diagnostic index is also of limited discriminatory value. It is frequently greater than 1 per cent in patients with prerenal azotemia who are receiving diuretics or have bicarbonaturia or adrenal insufficiency. In contrast, FE_{Na} below 1 per cent is found in 15 per cent of patients with non-oliguric ischemic or nephrotoxic acute renal failure. This has been described in patients with renal failure of a variety of causes, including ischemia, aminoglycosides, radiocontrast agents, rhabdomyolysis, hemolysis, burns, sepsis, and hepotorenal syndrome.

References

Anderson, R.J. and Schrier, R.W. (1980). Clinical spectrum of oliguric and non-oliguric acute renal failure. In *Acute renal failure* (ed. B.M. Brenner and J.H. Stein). Churchill Livingstone, New York.

Espinel, C.H. (1976). The FE_{Na} test: use in the differential diagnosis of acute renal failure. *Journal of the American Medical Association*, **236**, 579–81.

Ginsberg, J.M., Chang, B.S., Matarese, R.A., and Garella, S. (1983). Use of single voided urine samples to estimate quantitative proteinuria. *New England Journal of Medicine*, **309**, 1543–6.

McCormack, M., Dessureault, J., and Guitard, M. (1991). The urine specific gravity dipstick: a useful tool to increase fluid intake in stone forming patients. *Journal of Urology*, **146**, 1475–7.

Rowe, J.W., Andres, R., Tobin, J.D., Norris, A.H., and Shock, N.W. (1976). The effect of age on creatinine clearance in men: a cross-sectional and longitudinal study. *Journal of Gerontology*, **31**, 155–63.

17 Pharmacotherapeutics

17.1 Drugs in critical illness

17.1.1 Analgesics

M. Tryba

Key messages

- Analgesia and sedation are primarily independent goals of intensive care therapy.
- Inadequate analgesia may result in increased morbidity.
- Parenteral opioids are the mainstay of effective pain therapy.
- The different pharmacological properties of the opioids have to be considered during long-term administration.
- Ketamine offers advantages in hemodynamically unstable patients and in those with opioid-induced constipation.
- The synergistic action of clonidine with most central analgesics and sedatives is increasingly used to facilitate analgesia and sedation and to prevent the development of tolerance.
- Although central neuraxial blocks provide powerful analgesia, the possible benefits in ventilated patients have to be carefully weighed against the possible risks.

Table 1 Adverse effects of under- and overtreatment with analgesics

Undertreatment	Overtreatment
Elevated catecholamines	Hypotension
Hypertension	Bradycardia
Tachycardia	PVR ↓ (morphine)
Oxygen consumption ↑	Respiratory depression
Myocardial ischemia	Constipation, ileus
Pulmonary complications (FRC ↓, atelectasis, shunt volume ↑)	Nausea, vomiting
	Immunosuppression
Tracheal tube intolerance	Renal failure (morphine, NSAIDs)
Gastric stasis	Tolerance
ADH ↑	
Sodium and water retention	
Hypergylcemia	
Infection	

FRC, functional residual capacity; ADH, antidiuretic hormone; PVR, peripheral vascular resistance; NSAIDs, non-steroidal anti-inflammatory drugs.

General aims

Anxiolysis, analgesia, sedation, and amnesia are accepted goals of intensive care. Inadequate pain therapy and sedation not only restrict the comfort of critically ill patients but may also cause specific complications. Elevated catecholamine plasma levels may lead to disturbances of the microcirculation, myocardial ischemia and infarction, cerebral disturbances, and renal failure. Patients with cardiovascular diseases are at particular risk. Thus adequate analgesia and sedation should be integral elements in the management of critically ill patients. Sympathetic hyperactivity, due to severe pain as well as analgesic overtreatment, may result in increased morbidity (Table 1). Therefore complete stress reduction should be restricted to the most critical phases of intensive care therapy.

The large variety of recommended drugs and drug combinations demonstrates that the optimal analgesic and sedation regimen does not exist. Instead, the various substances and measures should be adjusted individually considering the underlying diseases and the specific situation, the accompanying therapeutic measures, the expected duration of treatment, and the treatment phase. Specific problems arise in patients during withdrawal from alcohol, opioids, or benzodiazepines, in patients with systemic inflammatory response syndrome, and during weaning from long-term mechanical ventilation.

Consideration of the continual changes in the analgesic and sedative demands of critically ill patients is of particular importance for optimal therapy. Thus, during the early phase after surgery or trauma, pain therapy is of utmost importance. Later in intensive

Table 2 Physiological effects of analgesic drugs

	Morphine	Fentanyl	Sufentanil	Alfentanil	Meperidine
Sedation	+	+	++	0	+
Cough suppression	+	+	?	?	0
Thoracic wall rigidity	0	+	+	++	+
Increase in splanchnic perfusion	+	?	?	?	?
Constipation	++	+/++	++	+/++	+
Biliary duct pressure	++	+++	?	++	+
Arterial hypotension	+	0	0	0	+
Histamine release	+	0	0	0	+
Cholinergic action	+	+	+	+	0
Anticholinergic action	0	0	0	0	+

care, the importance of pain therapy decreases while sedation becomes increasingly difficult. Tolerance to opioids and benzodiazepines may occur after only a few days of treatment and may be facilitated by continuous administration.

Analgesia and sedation are primarily independent goals of intensive care. Thus, at least in long-term ventilated patients, fixed drug combinations of analgesics and hypnotics are not useful. Even experienced intensivists may frequently underestimate analgesic requirements and treat agitation and/or sympathetic hyperactivity syndromes with excessive doses of sedatives, when administration of effective analgesic measures, such as epidural opioids (3–5 mg morphine), would eliminate the need for sedatives and systemic opioids almost entirely. If analgesia is insufficient, the analgesic dosage, the analgesic itself, or the analgesic method should be adjusted to the patient's demands. Simultaneous administration of hypnotics and analgesics requires a knowledge of possible synergistic and antagonistic interactions between the various substances.

Analgesia

After cardiovascular stabilization and during an uncomplicated course minor pain may be well tolerated. In these patients effective pain treatment can easily be achieved during the later course.

There are no validated objective indices for evaluating the quality of analgesia in mechanically ventilated patients. In non-ventilated patients visual or verbal analog scales are useful tools for guiding pain therapy. Clinically, a five-point verbal scale gives sufficient information. Such a scale may also be used in intubated but co-operative patients.

Problems

Prolonged administration of opioids and high dosages are frequently associated with tolerance, which often becomes noticeable after about a week. Drug-specific differences appear to exist. Typical predisposing factors for the development of tolerance are systemic inflammatory response syndrome and tetanus. The underlying pathophysiological and pharmacological mechanisms have not been clarified in detail. During weaning from long-term mechanical ventilation, withdrawal symptoms due to dosage reduction of analgesics and sedatives become dominant, complicating and extending the weaning phase.

Analgesic drugs

Parenteral opioids are the mainstay of effective pain therapy during the early phase of intensive care. There is only a very limited place for non-opioid-analgesics in the intensive care unit (ICU), perhaps with the exception of ketamine and clonidine. Central neuraxial blocks with local anesthetics and/or opioids are highly effective measures for the control of severe pain. However, their use in the ICU is limited because of possible complications (hemorrhage, neurological damage, spinal infection) and contraindications (sepsis, anticoagulants). Specific therapeutic measures such as transcutaneous electric nerve stimulation have shown proven efficacy after cardiothoracic surgery, but have not become routine in most hospitals.

Opioids

In general, all opioids are suitable as analgesics in the ICU. This is particularly true in short-term ICU patients with an uncomplicated course. However, the longer the treatment lasts and the more complicated the postoperative course, the more important are the pharmacological differences between the various opioids (Table 2).

Of particular relevance is the significant cumulation after prolonged administration of opioids. For example, while fentanyl presents as a short-acting opioid during anesthesia, the duration of action approaches that of morphine during prolonged continuous infusion (Table 3). The loss of action of fentanyl initially depends on the distribution of the drug in the body and not on elimination or metabolism. In contrast, the duration of action of sufentanil increases only slightly even after prolonged administration. Furthermore, development of tolerance seems to be less pronounced with sufentanil than with fentanyl. Whether this phenomenon is caused by the high receptor affinity of sufentanil has not been clearly demonstrated. However, the high receptor affinity and and lipophilicity are probably

Table 3 Pharmacological summary of analgesic drugs

	Morphine	Fentanyl	Sufentanil	Alfentanil	Meperidine
Clearance (ml/min/kg)	12–18	10–22	10–15	2–6	10–22
Elimination half-life (h)	2–4	1.5–6	2–3	1–2	3.5–6
Steady state context-sensitive half-time (min)		306	100	60	
Protein binding (%)	30–60	80–85	92	90	64
Elimination/metabolism	Liver≫kidney	Liver	Liver	Liver	Liver⋙kidney
Elimination prolonged if :	Liver failure, septic shock	Prolonged administration	Liver failure	Liver failure	Liver failure, septic shock
Active metabolites	M6G				Normeperidine
Duration of action after a single bolus dose (h)	2–3	0.5–1	1.5–2	0.25–1	1–2
Bolus dose (mg)	2–7	0.05–0.2	0.01–0.07	0.25–1	25–50
Infusion rate (mg/70 kg/h)	2–7	0.1–0.2	0.01–0.07	1–7	10–50

M6G, morphine-6-glucuronide.

the reasons for the superior analgesic efficacy of sufentanil after prolonged administration of opioids. In critically ill patients the analgesic efficacy of alfentanil is often insufficient after only a few days of administration, probably because of the low affinity to the opioid receptor. Furthermore, the duration of action after prolonged administration of alfentanil (context-sensitive half-time) increases similarly to fentanyl (Table 3). Meperidine (pethidine) may release histamine, may result in hypotension, and has the major disadvantage that the active neurotoxic metabolite normeperidine (norpethidine) cumulates during prolonged administration.

Piritramide has about the same duration of action as morphine, is almost free from cardiovascular side-effects, and can be used as an alternative to morphine, although it is significantly more expensive. The mixed opioid agonist–antagonist buprenorphine acts as an agonist in low dosages and as an antagonist in higher dosages. Dose increases are frequently necessary in ICU patients, and therefore mixed agonist–antagonists are only rarely useful in this setting. However, they may offer advantages in individual cases, such as patients with severe constipation. Owing to their limited analgesic power there is no role in the ICU for mixed agonists–antagonists like pentacozine, nalbuphine, or tramadol.

In general, the spectrum of opioids in the ICU should be limited to only a few drugs. Opioids with high analgesic power should be preferred. In patients with an expected short duration of treatment fentanyl can be used without problems during the immediate post-operative period. Sufentanil may have advantages in patients receiving long-term treatment because of the improved quality of analgesia, improved sedation, less tolerance, and earlier recovery. The great advantage of morphine is its unrivalled low price.

Ketamine

In subanesthetic concentrations ketamine has analgesic and sedative properties. It is not suitable for monotherapy because of its excitatory and hallucinogenic side-effects. However, these can be avoided if ketamine is administered in combination with benzodiazepines or propofol. Ketamine acts as an antagonist at the *N*-methyl-D-aspartate (NMDA) receptor, thus preventing and counteracting neurophysiological mechanisms after prolonged nociceptive stimulation (e.g. wind-up) resulting in hyperexcitatory nociceptive spinal neurons. This beneficial effect of ketamine is of clinical relevance if tolerance has developed after prolonged opioid administration. Ketamine doses of 1 to 2 mg/kg/h restore analgesia and simultaneously allow significant reduction of the opioid dosages. Ketamine offers advantages in hemodynamically unstable patients (e.g. septic shock). Because of its catecholamine-releasing properties, exogenous catecholamaines may be significantly reduced in these patients. However, these actions, which may also result in tachycardia and hypertension, should be considered carefully in patients with coronary heart disease. In patients with significant constipation due to high opioid dosage, ketamine should be seriously considered as an alternative analgesic.

Other non-opioid analgesics

This group of drugs includes non-steroidal anti-inflammatory drugs, aspirin, acetaminophen (paracetamol), and metamizol. Their analgesic potency is limited. There is no proof that a combination of opioids and non-opioid analgesics offers any advantages in ICU patients. Furthermore, even short-term inhibition of prostaglandin synthesis may result in significant side-effects in critically ill patients, particularly renal insufficiency or gastrointestinal bleeding. However, acetaminophen or metamizol may have a limited role as an anti-pyretic, particularly after cardiac surgery.

Regional anesthesia

Regional anesthetic techniques provide powerful analgesia during the postoperative period. Evidence is increasing that epidural local anesthetics and/or opioids may even improve cardiovascular and pulmonary morbidity in some patient subgroups (e.g. major abdominal surgery, vascular surgery, morbidly obese). The most promising results have been achieved in patients with thoracic epidural anesthesia and combined local anesthetics and opioids (bupivacaine plus morphine). The greatest concerns about epidural anesthesia in the ICU are associated with an increased risk of hemorrhage in anticoagulated patients or those with coagulation disorders and an increased risk of spinal infections. Experience with prolonged epidural analgesia in long-term ventilated patients is limited. Therefore the possible beneficial effects of central neuraxial blocks in ventilated patients should be carefully weighed against the possible risks on an individual basis.

Adjuvants

Alpha-2-adrenoceptor agonists (clonidine)

Alpha-2-adrenoceptor agonists display numerous different actions in the body. Their inhibition of catecholamine release, in particular norepinephrine (noradrenaline), and synergistic interactions with opioids (and sedatives) are of clinical relevance to analgesia (and sedation) in the ICU. The locus ceruleus has been identified as an essential site of action in the central nervous system. Furthermore, α_2-adrenoceptor agonists interact with the spinal noradrenergic system, resulting in inhibition of nociceptive inputs. Administration of α_2-adrenoceptor agonists functionally results in central sympathicolysis. Short-lasting α-mimetic effects can only be observed after bolus administration of α_2-adrenoceptor agonists. The only α_2-adrenoceptor agonist in clinical use at present is clonidine.

Clonidine was initially introduced into the management of critically ill patients because of its apparent sympathicolytic activity, for example in patients with withdrawal syndrome after chronic alcohol abuse or in patients with tetanus. It has also been shown to be effective in the prophylaxis of delirium tremens and during weaning from long-term mechanical ventilation, which in principle can be considered as iatrogenic withdrawal from opioids and benzodiazepines. Prophylactic dosages range from 0.9 to 1.5 mg/day, but therapeutic dosages must be titrated and guided by cardiovascular variables (heart rate, blood pressure) and may sometimes exceed 15 mg/day. Significant hypotension is a clear sign of (masked) hypovolemia and requires aggressive fluid replacement.

Numerous studies have shown that clonidine acts synergistically with almost all sedatives and analgesics. There is increasing evidence that adjuvant administration of clonidine during long-term mechanical ventilation may facilitate sedation and analgesia, prevent development of tolerance, and simultaneously reduce the dose requirements of sedatives and analgesics. Clonidine dosages usually range from 0.9 to 1.5 mg/day.

Contraindications for clonidine are hypovolemia and severe arrhythmias. Although high clonidine doses may lead to constipation, this effect may be counteracted in most cases because of the reduced dosage requirements of opioids.

Recommended procedures

Non-ventilated patients

Pain should be individually titrated with opioids to the desired level. Co-operative patients may benefit from patient-controlled analgesia in the ICU. Because of its sedative power, sufentanil can often be administered without additional sedatives. Regional anesthetic techniques, in particular central neuraxial blocks, may provide effective analgesia in selected patient groups (e.g. major upper abdominal surgery, major vascular surgery, amputation, major soft tissue injury, multiple rib fractures) with possible outcome benefits.

Mechanically ventilated patients

Postoperative ventilation

In general, postoperative ventilation for up to 24 h does not cause major problems. Any opioid can be used with no significant advantages or disadvantages.

Short-term mechanical ventilation

Short-acting opioids and benzodiazepones are the drugs of choice in patients with an expected duration of mechanical ventilation of up to 5 days. In these patients sufentanil may have pharmacological advantages over other opioids because of its lower cumulation tendency. Tolerance development and withdrawal symptoms may be less pronounced than with fentanyl. The weaning phase may be facilitated by administering clonidine. Regional analgesia is rarely used except in selected patients (e.g. severe pancreatitis).

Long-term mechanical ventilation

The main problems during analgesia and sedation occur in patients who are ventilated for more than 5 days. Sepsis, acute respiratory distress syndrome, renal and liver failure, specific procedures such as the prone position, weaning problems, and sympathetic hyperactivity syndromes (e.g. withdrawal from chronic alcohol abuse) may significantly complicate the control of adequate analgesia and sedation. In general, there seems little sense in using ultra-short-acting substances in patients with an expected duration of treatment of several days or even weeks. Sufentanil may be an exception due to its higher affinity with the opioid receptor, lower tolerance development, and less pronounced withdrawal symptoms. Clonidine may be added if initial signs of tolerance development occur or even earlier to facilitate the control of analgesia and sedation. It should be also considered if sympathetic hyperactivity syndromes (agitation, tachycardia, hypertension) develop during the weaning phase. Administration of catecholamines is not a contraindication. If analgesia and sedation are insufficient despite high opioid and benzodiazepine doses, ketamine may be considered as an alternative in hemodynamically instable patients. However, despite neurophysiological advantages (NMDA antagonist) the high costs of ketamine may restrict its use in many ICUs.

During the weaning phase the analgesics and sedatives should be reduced slowly and continued into the postextubation phase. Hyperventilation may be titrated with low doses of opioids to the desired respiratory frequency.

Conclusions

Sufficient analgesia and sedation can be achieved in most critically ill patients if the benefits and risks of the various drugs and drug combinations are carefully and individually considered. Systemic opioids are the mainstay of analgesia in the critically ill. However, excessive escalations of dosage should be avoided. Instead, alternative drugs (e.g. ketamine, clonidine) or measures (e.g. regional analgesia) may be indicated.

Bibliography

Hughes, M.A., Glass, P.S.A., and Jacobs, J.R. (1992). Context-sensitive half-time in multicompartment pharmacokinetic models for intravenous anesthetic drugs. *Anesthesiology*, **76**, 334–41.

Mirski, M.A., Muffelman, B., Ulatowski, J.A., and Hanley, D.F. (1995). Sedation for the critically ill neurologic patient. *Critical Care Medicine*, **23**, 2038–52.

Murray, M.J. and Plevak, D.J. (1994). Analgesia in the critically ill patient. *New Horizons*, **2**, 56–63.

Park, G.R. and Gempeler, F. (1993). *Sedation and analgesia*. W.B. Saunders, London.

Tryba, M., Kulka, P., and Zenz, M. (1993). Sympathetic hyperactivity syndromes in ventilated patients. In *Yearbook of intensive care and emergency medicine* (ed. J.L. Vincent), pp. 529–39. Springer Verlag, Berlin.

17.1.2 Antiarrhythmics

A. K. B. Slade and A. J. Camm

Key messages

- Antiarrhythmic drugs are still the first therapeutic modality in the treatment of cardiac arrhythmias despite impressive advances in non-pharmacological approaches such as radiofrequency catheter ablation and the implantable cardioverter defibrillator for long-term arrhythmia management.
- Cardiac arrhythmias in critically ill patients are often the consequence of non-cardiac states such as sepsis, electrolyte imbalance, and hypoxia, and any therapy designed to restore sinus rhythm must address such factors.
- Vaughan Williams class I drugs are not commonly used because of adverse effects on myocardial contractility and significant risk of proarrhythmia.
- Beta-blockers are useful in certain situations, and the action of ultra-short-acting agents such as esmolol has enabled β-blocker therapy to be attempted in a wider range of clinical situations.
- Amiodarone is perhaps the most useful acute agent for both ventricular and supraventricular arrhythmias, but there is considerable concern about long-term usage when side-effects are common and potentially serious.
- Magnesium in pharmacological doses is useful in the treatment of torsade de pointes (a variant of ventricular tachycardia).

Introduction

Antiarrhythmic drugs, when used chronically for the prophylaxis of cardiac arrhythmia, have received increasing criticism. Although often efficacious, they may also exacerbate arrhythmia in certain subgroups of patients, resulting in excess mortality. Alternative therapies such as catheter ablation, sophisticated pacing techniques, the implantable cardioverter defibrillator, and arrhythmia surgery have an increasing role in the long-term management of cardiac arrhythmia.

Drugs are still most often used as the first therapeutic modality in the acute management of cardiac arrhythmia. It should be remembered that the ideal antiarrhythmic agent does not exist. Antiarrhythmics are seldom reliably efficacious, have a narrow therapeutic range, often have toxic side-effects, and usually have adverse effects on myocardial contractility and normal impulse formation and conduction.

Cardiac arrhythmias occurring in critically ill patients are often the consequence of non-cardiac states such as sepsis, hypoxia, acid–base imbalance, and electrolyte imbalance, particularly hypokalemia and hypomagnesemia, rather than primary cardiac conditions. Initial management of cardiac arrhythmias should be guided toward correction of underlying abnormalities predisposing to arrhythmia. Furthermore, failure to correct such abnormalities may render attempts to restore sinus rhythm futile. The role of direct current cardioversion should not be forgotten in a patient population likely to be intolerant to antiarrhythmic drugs.

Antiarrhythmic drugs should always be used with caution, particularly in patients with impaired hemodynamics and organ dysfunction. Nonetheless, pharmacological control of pathological arrhythmia may be necessary if the rapid ventricular rate itself is causing hemodymanic compromise. In selecting appropriate antiarrhythmic drugs a number of classifications are available, of which the most widely known is the Vaughan Williams classification based on cellular electrophysiological effects (Table 1). Such classifications are seldom useful to the clinician, particularly when faced with a patient with unstable hemodynamics and impairment of drug excretion as a consequence of organ dysfunction.

Class Ia drugs: quinidine, procainamide, disopyramide

Quinidine used to be popular as an effective antiarrhythmic agent against both supraventricular and ventricular arrhythmias, but has fallen into disuse after a large meta-analysis suggested excess mortality due to drug-induced torsade de pointes. Quinidine directly suppresses sinus node and atrioventricular node conduction and increases the refractory period of atrial, ventricular, and His–Purkinje tissue, although an anticholinergic action may increase atrioventricular nodal conduction. The most important side-effect is torsade de pointes associated with QT prolongation. Other side-effects are seen in the gastrointestinal system, together with tinnitus, thrombocytopenia, and hypotension. The drug may precipitate digoxin toxicity. The drug is metabolized hepatically and dose reduction is necessary in patients with hepatic disease.

Procainamide has a similar mode of action to quinidine but prolongs the QT interval less. Long-term administration may result in a lupus syndrome. It is generally safe and well tolerated. Torsade de pointes is less common. It is useful in the treatment of sustained monomorphic ventricular tachycardia where it has been shown to be superior to lidocaine (lignocaine) in a small randomized cross-over trial (Gorgels *et al.* 1996). The usual dose is 10 mg/kg body weight at a rate of 100 mg/min. A maintenance infusion of 2 to 6 mg/min may be required. Oral use is limited by the short plasma half-life, such that even slow-release preparations have to be administered several times daily to maintain antiarrhythmic efficacy. It is useful as a predictor of

Table 1 The Vaughan Williams classification: the major electrophysiological actions of commonly used antiarrhythmic drugs

Class	Mode of action	Site of action	Examples
I	Membrane-stabilizing agents (fast sodium-channel inhibition)		
	(a) Delayed repolarization Broadened action potential	Atrium Ventricle Accessory pathway	Quinidine Disopyramide Procainamide
	(b) Decreased action potential duration Accelerated repolarization	Ventricle	Lidocaine Mexiletine
	(c) QRS prolongation No change in action potential	Atrium Ventricle His–Purkinje tissue Accessory pathway	Flecainide Propafenone Encainide
II	Beta-blocking agents	Sinus node AV node	Propranolol Atenolol
III	Prolongation of repolarization	Atrium AV node Accessory pathway Ventricle His–Purkinje tissue	Amiodarone Bretylium Sotalol
IV	Calcium-channel antagonists	AV node	Diltiazem Verapamil

AV, atrioventricular.

antiarrhythmic efficacy. Failure of procainamide to suppress inducibilty of ventricular tachycardia by electrophysiological testing indicates that no other single agent is likely to be efficacious.

Disopyramide differs markedly from quinidine in a number of respects. It is profoundly negatively inotropic which makes it particularly unsuitable for use in critically ill patients with impaired myocardial function. Gastrointestinal side-effects are generally less, but anticholinergic side-effects are much more prominent. Such effects may cause acceleration of atrioventricular nodal conduction and may require coadministration of atrioventricular nodal blocking drugs when used for the treatment of supraventricular arrhythmias.

Class Ib drugs: lidocaine, mexiletine

Lidocaine remains the first-line treatment for ventricular arrhythmias despite lack of clinical evidence for its efficacy. It is most appropriately used to treat ventricular arrhythmias in the post-myocardial infarction setting where acute ischemia is thought to be important in the pathogenesis of ventricular arrhythmias. It can no longer be advocated for primary prophylaxis of ventricular fibrillation. The definitive study of paramedic-administered lidocaine showed a statistically significant reduction in ventricular fibrillation at the expense of an increase in asystole (Koster and Dunning 1985). It is usually administered as a bolus injection (100 mg over 1–2 min) followed by a maintenance infusion of 1 to 4 mg/min if required, as plasma levels fall rapidly after a single bolus. It should be remembered that 10 ml of 1 per cent lidocaine contains 100 mg of the drug. If a bolus is ineffective, it is unlikely that an infusion will succeed. Lidocaine cannot be given orally as it undergoes extensive first-pass hepatic metabolism and thus is unsuitable for long-term prophylaxis. Drug metabolism occurs in the liver, and dosage should be reduced in primary hepatic impairment or where hepatic blood flow is reduced. Toxicity manifests with confusion, twitching, and epileptiform fits.

Mexiletine is similar to lidocaine but is available as an oral agent for long-term prophylaxis. It depresses myocardial contractility less than lidocaine. It is usually administered acutely as an intravenous bolus of 100 to 250 mg over 5 to 10 min with subsequent infusion at 0.5 to 1 mg/min. The drug is principally metabolized by the liver, although approximately 15 per cent is excreted unchanged by the kidney.

Class Ic drugs: flecainide

This class of antiarrhythmic drug has recently emerged from the cloud caused by a trial which demonstrated excess mortality in postmyocardial infarction patients with ventricular premature contractions. Intense analysis of this trial has allowed a reappraisal of the use of such agents. Although these drugs were initially licensed for the treatment of ventricular arrhythmias, it is now probably inappropriate to use them in patients with ischemic heart disease or myocardial fibrosis from other causes.

Class Ic drugs are undoubtedly useful in patients with supraventricular arrhythmias, particularly in the cardioversion of atrial fibrillation and subsequent maintenance of sinus rhythm. This is particularly true with acute-onset atrial fibrillation, where return to sinus rhythm may be seen in 70 per cent of patients. It is less effective in well-established atrial fibrillation. Bolus administration of 1 to 2 mg/kg (maximum 150 mg) over 30 min is usual, although care must be taken to watch for hypotension as a consequence of myocardial depression. Side-effects occur as a consequence of its effects on myocardial contractility and conduction, with precipitation of overt heart failure in previously well patients being recognized. Monitoring of the QRS duration offers a means of following therapy. This class also increases pacing thresholds by up to 200 per cent, necessitating reprogramming of permanent units and high outputs in temporary pacing systems.

Beta-blocking agents

Beta-blocking agents act by antagonizing the effects of catecholamines on the heart. Such levels of catecholamines may be appropriate in critically ill patients either endogenously or as a consequence of inotropic support for the failing myocardium. Therefore such agents must be used cautiously in critically ill patients. Conventional agents such as propranolol, atenolol, or metoprolol have comparatively long half-lives, and adverse hemodynamic effects from administration may persist. Beta-blocking agents are particularly useful in arrhythmias with a significant sympathetic component such as is seen in some supraventricular arrhythmias and ventricular tachycardia arising from the right ventricular outflow tract. Beta-blocking agents may also be useful in the acute control of ventricular rate in atrial fibrillation.

The recent introduction of esmolol, an ultra-short-acting β-blocker with complete elimination within 30 min by blood esterases, enables transient use of β-blockade in situations such as postoperative hypertension and where β-blockade may be beneficial but hemodynamic effect is uncertain. The usual dosage range is 50 to 400 g/min.

Xamoterol is a β-blocker with a high degree of intrinsic sympathomimetic activity. It may have a role as an adjuvant agent in patients with refractory ventricular arrhythmias who do not settle on amiodarone alone.

Amiodarone

Amiodarone is an extremely attractive antiarrhythmic drug in the setting of the critically ill patient. It is effective against a wide spectrum of both supraventricular and ventricular arrhythmias. Its effects on the inotropic state of the myocardium are significantly less than those of the vast majority of antiarrhythmic drugs. There is less concern in the acutely ill patient about the extensive long-term toxicity that is seen with chronic oral usage.

Amiodarone has a complex mode of action with activity across the Vaughan Williams classification. It has a delayed onset of action which necessitates aggressive intravenous loading (usually 300 mg over 30 min followed by a maintenance infusion of 900 to 1200 mg/24 h) preferably via central venous access to avoid phlebitis. Acute infusion at high rates may lead to significant hypotension due to an α-receptor-blocking effect. Even with such loading, effective antiarrhythmic action may not be seen for 1 to 24 h. Metabolism is principally hepatic to desethylamiodarone, which has antiarrhythmic action in its own right. Amiodarone potentiates oral anticoagulation and increases blood levels of digoxin, quinidine, and flecainide. Although the drug prolongs the QT interval, drug-induced torsade de pointes is less common than with other class III antiarrhythmic drugs. Despite its delayed onset of action, it has recently proven efficacious and better tolerated in trials comparing it with bretylium in patients with malignant ventricular arrhythmias (Kowey *et al.* 1995). It is also useful in pediatric patients, particularly in the context of postsurgical junctional ectopic tachycardia, an arrhythmia with a very high mortality. It is currently being assessed in the setting of the post-myocardial infarction patient perceived to be at high risk of death in large placebo-controlled randomized trials as small pilot studies have shown potential benefit.

Bretylium

Bretylium tosylate has a unique mode of action. It concentrates in the terminal sympathetic neurons, causing transient release of stored norepinephrine (noradrenaline) before blocking further release. The effect is of a chemical sympathectomy which also gives rise to its major side-effect of hypotension. It is used in patients with ventricular tachycardia refractory to other antiarrhythmic drugs, although it is being superseded by amiodarone which causes significantly less hypotension. Bretylium also has a unique action in that it may cause chemical cardioversion of ventricular fibrillation or facilitate cardioversion of ventricular fibrillation refractory to d.c. cardioversion. It is administered as an intravenous bolus of 5 to 10 mg/kg in emergency situations but is preferably given as an infusion over 30 min which can be repeated at intervals of 1 to 2 h or followed by an infusion of 1 to 2 mg/min. The drug is excreted renally with a half-life of 8 h.

Other class III agents

Because of the increased use of amiodarone as a broad-spectrum antiarrhythmic agent there has been much interest in developing new class III agents without the side-effect profile of amiodarone in chronic use. Sotalol, a racemic mixture of D- and L-sotalol, is perhaps the most widely used but is limited by significant β-blocking activity. D-Sotalol has minimal β-blocking activity but has proved unsuccessful in the setting of high-risk postmyocardial infarction patients, with excess mortality in those taking active drug. Other agents such as dofetilide and ibutilide remain in development, but their ultimate clinical role remains to be determined.

Calcium-channel antagonists

Calcium-channel antagonists, principally verapamil and diltiazem, have important antiarrhythmic action by virtue of their effects on the atrioventricular node. By slowing conduction these agents may terminate junctional re-entrant arrhythmias and slow the ventricular rate in atrial fibrillation or atrial arrhythmias. The drugs can be administered by intravenous bolus: verapamil 5–10 mg over 30–60 s, diltiazem 10–20 mg over a similar time period for acute rate control but care should be taken in patients with impaired myocardial function as precipitate drops in blood pressure can occur due to a significant vasodilating effect. Such hypotension may be prevented by immediate prior administration of a bolus of calcium chloride. Caution should also be used in patients with known sinus or atrioventricular nodal disease and especially if the patient is already taking β-blocking agents when administration of a calcium-channel antagonist may produce profound bradycardia and hypotension.

Digoxin

Digoxin and associated preparations are frequently used to control ventricular rate in patients with rapidly conducted atrial arrhythmias, particularly atrial fibrillation. The drug may be ineffective alone and

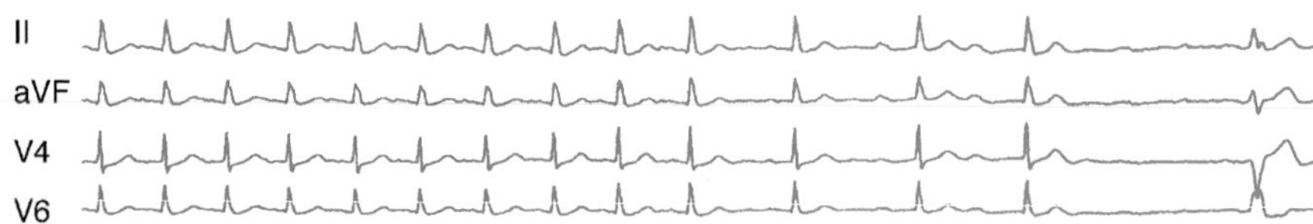

Fig. 1 Termination of atrial tachycardia with 12 mg of intravenous adenosine. Note impaired atrioventricular conduction in beats after successful termination.

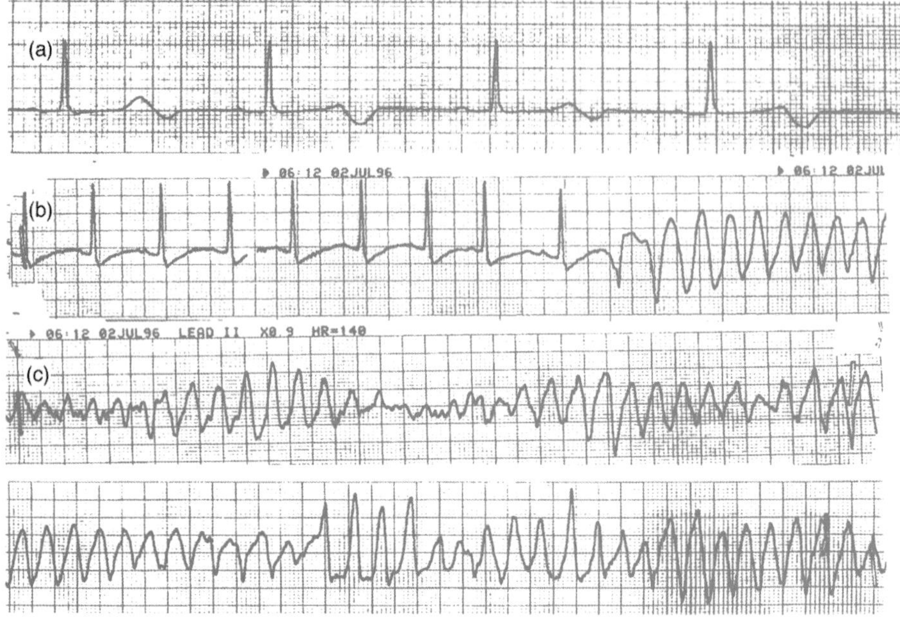

Fig. 2 Ventricular proarrhythmia from class III antiarrhythmic drugs. (a) Sinus bradycardia with a QT interval of 650 ms in an elderly patient taking amiodarone 400 mg daily and atenolol 50 mg daily for rate control of atrial fibrillation. (b) Initiation of torsade de pointes with a single ventricular premature beat. (c) Rhythm strip from the same episode showing characteristic twisting of the electrical axis about the baseline. This episode stopped spontaneously and the proarrhythmia was treated by drug withdrawal and temporary pacing.

may require addition of other agents, such as β-blocking agents or calcium-channel antagonists, to gain adequate rate control.

In this setting intravenous loading with 0.5 to 1 mg over 24 h followed by maintenance doses of 250 to 375 µg/day is usual. The major problem with digoxin is the narrow therapeutic range with considerable potential for exacerbation of arrhythmia when toxic levels are present. Arrhythmias seen with toxicity include atrial tachycardia with block, junctional tachycardia, ventricular premature beats, ventricular tachycardia, sinus bradycardia, and varying degrees of atrioventricular block. Toxicity may be exacerbated by renal impairment, dehydration, hypokalemia, and old age. Concomitant administration of quinidine, amiodarone, and verapamil all increase digoxin levels. In cases of severe overdose, ventricular arrhythmias may require lidocaine or phenytoin, bradycardia may require temporary pacing, and in extreme cases digoxin-specific antibodies can be administered.

Although digoxin is by far the most widely used cardiac glycoside, its relatively prolonged onset of action (1–5 h) may make it unsuitable in the setting of the critically ill patient. Ouabain and deslanoside are only available as intravenous preparations, but both have a rapid onset of action within 10 to 30 min.

Magnesium

There has been increasing interest in the use of intravenous magnesium in a number of clinical settings.

Torsade de pointes is a rare form of polymorphic ventricular tachycardia with a characteristic changing morphology of the QRS complex with the axis twisting around an imaginary baseline. It can occur in the setting of the critically ill patient for a number of reasons: concomitant administration of a number of drugs including antiarrhythmic drugs, metabolic abnormalities such as hypokalemia, hypocalcemia, and hypomagnesemia, and bradycardia. Withdrawal of offending drugs, correction of metabolic abnormalities, and pacing, together with intravenous administration of magnesium sulfate (2–3 g), form the cornerstone of therapy. Magnesium may work in the presence of normal plasma magnesium levels. It may also be of use following cardiac surgery where significant magnesium depletion can occur in the immediate postoperative period. Its role in other settings such as acute myocardial infarction remains uncertain.

Adenosine

Adenosine is an endogenous nucleoside that is capable of causing atrioventricular block when administered in pharmacological concentrations. It has the considerable advantage of having an ultra-short plasma half-life of only a few seconds. Side-effects are usually transient, and include flushing, chest tightness, and bronchospasm, although sustained acceleration of atrial flutter from 2:1 to 1:1 atrioventricular conduction has been reported. Atrioventricular block may persist for several seconds. It has superseded verapamil as the

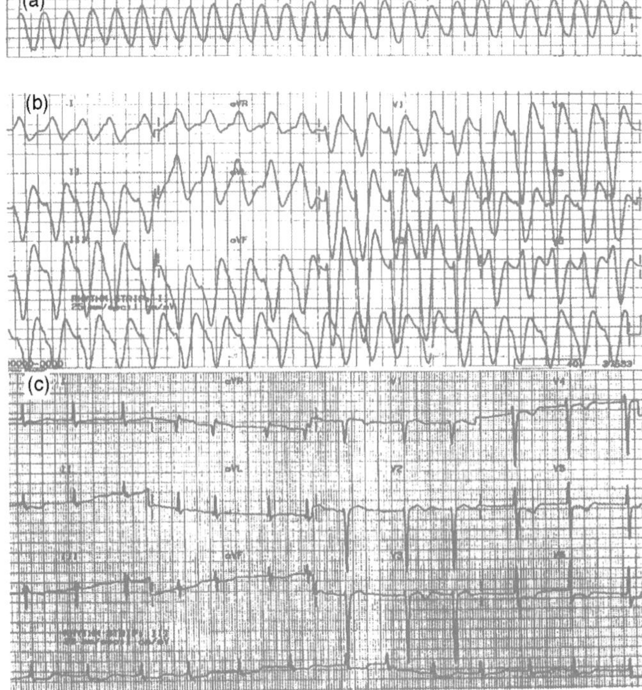

Fig. 3 Proarrhythmia secondary to flecainide. (a) Sinusoidal ventricular tachycardia in a patient taking oral flecainide 100 mg twice daily for suppression of atrial fibrillation. (b) Twelve-lead ECG of a patient on oral flecainide showing atrial fibrillation with very broad (> 200 ms) QRS complexes due to the effect of flecainide on intraventricular conduction. (c) Twelve-lead ECG of the same patient taken 4 days later after withdrawal of flecainide. Note the normal QRS duration.

agent of choice for the acute termination of junctional re-entrant arrhythmias. It is administered as an extremely rapid intravenous bolus of between 3 and 20 mg in an escalating regime until therapeutic or diagnostic effect has been noted (Camm and Garratt 1991) (Fig. 1). Adenosine also has a role in the diagnosis of arrhythmias, particularly broad complex arrhythmias where the non-cardiologist may not easily distinguish ventricular tachycardia from supraventricular arrhythmias conducted with aberration.

Proarrhythmia

There is increasing awareness that the arrhythmia may be worsened in 5 to 10 per cent of patients taking antiarrhythmic drugs, sometimes with lethal consequences (Roden 1994). Such proarrhythmic events are more likely with simultaneous or near-simultaneous administration of different antiarrhythmic drugs and in the presence of metabolic abnormalities, particularly hypokalemia. The capacity for digoxin to precipitate serious arrhythmias when present in excess is widely known, but it took many years for the potential for quinidine to provoke torsade de pointes ('quinidine syncope') to be established.

Two major patterns of ventricular proarrhythmia are seen with antiarrhythmic drugs. Class Ia and class III agents cause QT prolongation and may precipitate torsade de pointes (Fig. 2), whereas class Ic agents may give rise to a sinusoidal ventricular tachycardia refractory to both pharmacological and electrical therapy (Fig. 3).

Conclusion

The wide range of antiarrhythmic drugs confirms that no ideal agent exists. Considering the diverse mechanisms underlying cardiac arrhythmias, this is perhaps not surprising. Such agents almost always have adverse effects on hemodynamics and their use in patients who are critically ill must be carefully considered, with correction of underlying disease states likely to be just as important.

References

Camm, A.J. and Garratt, C.J. (1991). Adenosine and supraventricular tachycardia. *New England Journal of Medicine*, **325**, 1621–9.

Gorgels, A.P., van den Dool, A., Hofs, A., Mulleneers, R., Smeets, J.L., and Vos, M.A. (1996). Comparison of procainamide and lidocaine in terminating sustained monomorphic ventricular tachycardia. *American Journal of Cardiology*, **78**, 43–6.

Koster, R.W. and Dunning, A.J. (1985). Intramuscular lidocaine for prevention of lethal arrhythmias in the prehospitalization phase of acute myocardial infarction. *New England Journal of Medicine*, **313**, 1105–10.

Kowey, P.R., *et al.* (1995). Randomized, double-blind comparison of intravenous amiodarone and bretylium in the treatment of patients with recurrent, hemodynamically destabilizing ventricular tachycardia or fibrillation. *Circulation*, **92**, 3255–63.

Roden, D.M. (1994). Risks and benefits of antiarrhythmic therapy. *New England Journal of Medicine*, **331**, 785–91.

17.1.3 Anticoagulants

A. R. Webb

Key messages

- Heparin is known to increase clotting time by affecting several coagulation factors simultaneously via potentiation of antithrombin III.
- Low-molecular-weight heparin appears to influence factor Xa activity specifically via potentiation of antithrombin III.
- The effects of the prostanoids depend on the balance between thromboxane and prostacyclin.
- Sodium citrate chelates ionized calcium necessary for the coagulation cascade.
- Warfarin produces a controlled deficiency of vitamin-K-dependent coagulation factors.

Introduction

Anticoagulants used in the critically ill include heparin, low-molecular-weight heparin, anticoagulant prostanoids (e.g. epoprostenol, alprostadil), sodium citrate, and warfarin.

Pharmacology

Heparins

Heparin, which was discovered by McLean in 1916, has been used widely for anticoagulation. The predominant mode of action is potentiation of naturally occurring antithrombin III, although platelets and proteins in the cell wall are also affected. Unfractionated heparin reduces the adhesion of platelets to injured arterial walls, probably by maintaining vessel-wall electronegativity. Heparin also binds to platelets and promotes *in vitro* aggregation. In contrast, heparin has been shown to reduce platelet aggregation to collagen.

Heparin is known to increase clotting time by affecting several coagulation factors simultaneously, leading to occult blood loss and hemorrhagic complications. Inhibition of β-factor XIIa by C1 inhibitor is reduced by heparin and varies depending on heparin type. Therefore heparins may potentiate contact activation.

The half-life of heparin is known to be dose dependent and increases with prolonged administration.

Low-molecular-weight heparin

Low-molecular-weight heparin (molecular weight, 4000–6000 Da) appears to influence factor Xa activity specifically via potentiation of antithrombin III; its simpler pharmacokinetics allow for a smaller (around two-thirds) dose to be administered to the same effect (Schrader *et al.* 1990), although it cannot be adequately neutralized by protamine. The half-life is typically 2 to 4 h.

Anticoagulant prostanoids

The effects of the prostanoids depend on the balance between thromboxane and epoprostenol (prostacyclin). At a wounded endothelial surface the effects of thromboxane predominate, allowing platelet activation, aggregation, and plugging of the vessel wall. There is also local vasoconstriction to reduce blood flow into the wounded area. Epoprostenol is the most potent inhibitor of platelet aggregation known and is a potent vasodilator. Anticoagulation may be achieved with epoprostenol by ablating local thromboxane effects or if infused into an extracorporeal circulation where the thromboxane levels are extremely low. Effects are dose dependent and last for up to 30 min after discontinuing an infusion. At high doses epoprostenol may disperse circulating platelet aggregates.

Alprostadil (prostaglandin E_1) has similar effects to epoprostenol, but is up to five times less potent. It is also metabolized in the lungs so that systemic vasodilatation effects should be minimal. This may be an important advantage in the shocked patient.

Sodium citrate

Sodium citrate chelates ionized calcium necessary for the coagulation cascade. Citrate has advantages over heparin for extracorporeal use in that it has no known antiplatelet activity, is readily filtered by a hemofilter (reducing systemic anticoagulation), and is overwhelmed and neutralized when returned to central venous blood.

Warfarin

Warfarin produces a controlled deficiency of vitamin-K-dependent coagulation factors (II, VII, IX, and X).

Uses

Anticoagulants are used in critically ill patients for maintenance of an extracorporeal circulation (Sinclair 1988), treatment of thromboembolism, and prevention of thromboembolism.

Heparin

Heparin forms the mainstay of anticoagulation therapy for the critically ill as a continuous infusion for prevention of coagulation in the extracorporeal circulation or treatment of thromboembolism. Intermittent subcutaneous heparin is more commonly used for prevention of thromboembolism in the immobile patient or following myocardial infarction.

Often, when used for the extracorporeal circulation, systemic anticoagulation is not required. Alternative methods of giving heparin include the following.

1. Regional anticoagulation: heparin infused into the blood leaving the patient may be neutralized by protamine sulfate on return. It is assumed that 1 mg protamine neutralizes about 100 IU of unfractionated heparin. However, protamine sulfate has a number of side-effects which would be particularly deleterious in the critically ill. These include reduced cardiac output, decreased systemic vascular resistance, increased pulmonary vascular resistance, decreased platelet function, and bronchoconstriction.
2. Covalent bonding of heparin to surfaces containing amine groups: the technique is used to produce a biologically active surface on circuits, catheters, and membranes (Carmeda, Stockholm, Sweden). The surface heparin coating acts as a catalyst, accelerating the active inhibition of thrombin binding to the surface by circulating antithrombin III. The limitations of surface heparin coating are similar to those of the vascular endothelium: blood flow must be maintained since blood will clot if stagnant and a high ratio of surface area to volume is required to clear thrombin adequately.

Low-molecular-weight heparin

The indications for low-molecular-weight heparin are the same as those for heparin, including patients with heparin-induced thrombocytopenia. Low-molecular-weight heparin has a less adverse effect on platelet function than unfractionated heparin.

Anticoagulant prostanoids

Epoprostenol is generally used with heparin for anticoagulation of extracorporeal circulations, although heparin interferes with the inhibitory effect of epoprostenol on platelet aggregation *in vitro*. Heparin dosages can be reduced to a minimum or in some cases prostanoids can completely replace heparin. A reduced risk of bleeding with epoprostenol therapy is a particular advantage in the critically ill (Weston 1990). The major disadvantage of epoprostenol is the potential for hypotension as a consequence of peripheral vasodilatation. This effect can be minimized by careful attention to volume status. The less potent prostanoid alprostadil has similar effects on anticoagulation, although there is less systemic vasodilatation at equivalent doses.

Sodium citrate

Trisodium citrate may be used as a regional anticoagulant in at-risk patients undergoing extracorporeal renal replacement therapy. It is a safe and effective alternative to conventional heparinization in the critically ill (Mehta *et al.* 1990). Citrate has advantages over heparin in that it has no known antiplatelet activity, is readily filtered by a hemofilter (reducing systemic anticoagulation), and is overwhelmed and neutralized when returned to central venous blood.

Warfarin

Warfarin is given orally and requires 48 to 72 h to develop its effect. It is used in the treatment and prevention of thromboembolism and is intended for long-term maintenance of anticoagulation. Effects may continue for several days after cessation of treatment; it is usually more appropriate to use short-acting intravenous agents in the critically ill.

Dosages

Heparin

Dose requirement is variable to produce an activated partial thromboplastin time of 1.5 to 3 times control. This usually requires 500 to 2000 IU/h intravenously with an initial loading dose of 3000 to 5000 IU. In treatment of deep vein thrombosis, intermittent subcutaneous injections may be used (250 IU/kg every 12 h).

Heparin may be monitored by global tests or heparin assays, depending on the dosage and type given. Global tests for heparin include the whole-blood clotting time, activated coagulation time, and activated partial thromboplastin time. In patients not receiving heparin, a typical activated coagulation time might be 100 to 140 s; during cardiopulmonary bypass the safe range of heparin anticoagulation, as monitored by the activated coagulation time, is 400 to 500 s. In hemodialysis or hemofiltration an activated coagulation time of 180 to 200 s is usually acceptable.

However, other drugs or coexisting coagulation defects may interfere with coagulation monitoring. For instance, aprotinin inhibits contact activation and therefore can prolong the whole-blood clotting time, activated coagulation time, and activated partial thromboplastin time, thus giving an overestimation of the heparin level. In these cases heparin assays are required, the most specific of which are the anti-Xa assays.

The thrombin clotting time may be used for heparin control, and heparin levels may be titrated using various doses of protamine sulfate to find the dose that neutralizes heparin and normalizes the thrombin time. Typically, a thrombin clotting time of 1.5 times to twice control is acceptable.

For prevention of thromboembolism heparin is usually used subcutaneously in a dose of 5000 IU every 12 h. Control by monitoring coagulation tests is not necessary in these cases.

Low-molecular-weight heparin

Low-molecular-weight heparins cannot be monitored adequately by activated partial thromboplastin time, activated coagulation time, whole-blood clotting time, or thrombin time because of their high ratio of anti-Xa to anti-IIa activities. A prolonged activated partial thromboplastin time is indication of overdose; therefore an anti-Xa assay should be employed.

For deep vein thrombosis prophylaxis 2500 IU is given subcutaneously every 12 h. For anticoagulation of an extracorporeal circuit a bolus of 35 IU/kg is given intravenously followed by an infusion of 13 IU/kg. The dose is adjusted to maintain anti-Xa activity at 0.5 to 1 IU/ml (or 0.2 to 0.4 IU/ml if there is a high risk of hemorrhage).

Anticoagulant prostanoids

It is usual to infuse these drugs at 2.5 to 10 ng/kg/min. If used for an extracorporeal circulation, the infusion should be started 30 min prior to commencement. Although alprostadil is less potent than epoprostenol, similar doses are used for anticoagulation.

Sodium citrate

Sodium citrate is infused at 5 mmol per liter of extracorporeal blood flow.

Warfarin

Warfarin is started at 10 mg/day orally for 2 days, and then at 1 to 10 mg/day according to the international normalized ratio (INR). For deep vein thrombosis prophylaxis, pulmonary embolus, mitral stenosis, atrial fibrillation, and tissue valve replacements, INR should be maintained between 2 and 3. For recurrent deep vein thrombosis or pulmonary embolus and mechanical valve replacements, INR should be kept between 3 and 4.5.

Side-effects

Bleeding is the major side-effect of all anticoagulants and may be particularly problematic in the critically ill because of coexistent abnormalities of the coagulation mechanisms or effects of other drug treatment. Particular caution is required where antiplatelet drugs (e.g. aspirin or dextran) are used. Bleeding can be minimized by careful attention to coagulation monitoring and maintenance of platelet numbers.

Heparin-induced bleeding can be treated by reversal of heparin with protamine sulfate. It is assumed that 1 mg protamine neutralizes about 100 IU of unfractionated heparin. Single doses of protamine should not exceed 50 mg and should be given by slow intravenous injection.

Heparin-induced thrombocytopenia (Bell 1988) assumes two forms. In the non-immune variant heparin binds to the surface of platelets producing a mild transient thrombocytopenia. In the immune variant a severe progressive thrombocytopenia occurs after approximately 7 days of heparin therapy. All heparin must be stopped, including heparin in flush sets, since the syndrome may be precipitated by very small amounts. Immune complex microvascular thrombosis is a risk, and platelet transfusions should be avoided while heparin is present.

Low-molecular-weight heparins may cause moderate elevation of liver enzymes and the dose should be reduced in patients with liver dysfunction.

Epoprostenol infusion may be associated with spontaneous bleeding in the critically ill if there are coexisting abnormalities of platelet function or antiplatelet drugs. The main side-effect of epoprostenol is hypotension due to peripheral vasodilatation. The high pH of the solution may cause necrosis if there is extravasation. Alprostadil has similar anticoagulant side-effects to epoprostenol but hypotension is less severe because of pulmonary metabolism.

Sodium citrate is associated with metabolic problems. The large sodium load requires a reduction of the sodium content of hemofiltration replacement fluids. In addition, citrate is metabolized to bicarbonate and ensuing metabolic alkalosis has been described.

Careful assessment of serum ionized calcium levels is vital to avoid myocardial depression. Additional infusion of calcium may be necessary.

The effects of warfarin may be potentiated by several mechanisms which increase the risk of bleeding:

(1) displacement from plasma protein binding sites (e.g. diuretics, amiodarone);

(2) increased affinity for hepatic receptor sites (e.g. thyroxine);

(3) inhibition of liver enzymes (e.g. alcohol);

(4) other drug interactions (e.g. aminoglycosides, cimetidine, aspirin, acetaminophen (paracetamol));

(5) vitamin K deficiency (e.g. critical illness, liver dysfunction).

Treatment of warfarin overdose requires fresh frozen plasma infusions and vitamin K replacement controlled by INR measurements.

Warfarin may cause necrosis due to intravascular thrombosis in patients with protein C deficiency. This uncommon side-effect occurs when warfarin is used in large initial doses in the absence of heparin.

References

Bell, W.R. (1988). Heparin-associated thrombocytopenia and thrombosis. *Journal of Laboratory and Clinical Medicine*, **111**, 600–5.

Mehta, R.L., McDonald, B.R., Aguilar, M.M., and Ward, D.M. (1990). Regional citrate anticoagulation for continuous arteriovenous hemodialysis in critically ill patients. *Kidney International*, **38**, 976–81.

Schrader, J., *et al.* (1990). Low molecular weight heparin versus standard heparin. A long-term study in hemodialysis and hemofiltration patients. *ASAIO Transactions*, **36**, 28–32.

Sinclair, M.E. (1988). Anticoagulation for extracorporeal systems. In *Update in intensive care and emergency medicine*, No. 5 (ed. J.-L. Vincent), pp. 693–5. Springer-Verlag, Berlin.

Weston, M.J. (1990). Anticoagulation and extracorporeal circuits: the role of prostacyclin. In *Acute renal failure in the intensive care unit* (ed. D. Bihari and G. Nield), pp. 337–47. Springer-Verlag, Berlin.

17.1.4 Antidepressants

S. Heckers and T. Stern

Key messages

- Lack of therapeutic response or development of side-effects with one selective serotonin reuptake inhibitor does not necessarily predict the same outcome with another.
- The selection of an antidepressant should be based on the side-effect profile of the drug and the type of depressive disorder.
- When the patient recovers from a depressive episode, maintenance treatment should be considered to reduce the significant risk of relapse associated with major depression.

Introduction

Pharmacological treatment of depression began in the late 1950s with the introduction of imipramine and iproniazid, the prototypical tricyclic compound and monoamine oxidase inhibitor respectively. With the introduction of fluoxetine (in the United States in 1988), the selective serotonin reuptake inhibitors became the third class of drugs with a proven effect in the treatment of depression. In addition, there are several atypical compounds that share some but not all of the pharmacological features with cyclic antidepressant or selective serotonin reuptake inhibitors.

Tricyclic and tetracyclic antidepressants

Maprotiline, a tetracylic compound, and the large group of tricylic antidepressants (Table 1) effectively treat the neurovegetative signs of depression. Potential sites of action are at the level of receptors (desensitization of presynaptic α_2-receptors, subsensitivity of β-adrenergic receptors), reuptake transporters (block of monoamine reuptake), and intracellular second-messenger systems.

The cyclic compounds do not have significantly different efficacy in treating depression but do show a variety of side-effect profiles. Adverse effects are closely related to their chemical structure and their affinity for postsynaptic receptors. Amitriptyline, clomipramine, and doxepin have a high affinity for muscarinic cholinergic receptors and cause more anticholinergic side-effects (dryness of mouth, blurred vision, constipation, urinary retention) than the other cyclic compounds. Sedation and orthostatic hypotension are two other side-effects that occur frequently with some, but not all, cyclic antidepressants. The ability of cyclic antidepressants to interfere with normal cardiac conduction and potentially lead to lethal cardiac arrhythmias makes them dangerous in the hands of suicidal patients.

Use of clomipramine improves both depression and obsessive–compulsive disorder. Amoxapine, which blocks dopamine D_2 receptors and has moderate neuroleptic activity, can be used to treat patients with psychotic depression, but its use carries the risk of extrapyramidal side-effects.

Monoamine oxidase inhibitors

Phenelzine and tranylcypromine are the two most frequently used monoamine oxidase inhibitors (Table 1). These drugs inhibit both monoamine oxidase isozymes (monoamine oxidase A and monoamine oxidase B), which raises the level of norepinephrine (noradrenaline), serotonin, and dopamine in the synaptic cleft as well as the level of exogenous monoamines taken up by the gastrointestinal tract. The therapeutic effect of monoamine oxidase inhibitors is believed to be due to the inhibition of the monoamine oxidase isozymes, leading to increased aminergic activity. Potential effects on intracellular second-messenger systems are also considered. The sympathomimetic pressor amine tyramine, delivered by food to the gastrointestinal tract, can rise to high levels during monoamine oxidase inhibitor therapy and lead to potentially lethal hypertensive crises. A tyramine-low diet is recommended in patients taking monoamine oxidase inhibitors and a washout period is necessary before starting other antidepressants, since

Table 1 Treatment indices and side-effect profiles of common antidepressants

Drug	Rank order	Daily dose range (mg/day)	Sedative potency	Anticholinergic effects	Blood pressure effects	Other side-effects
Selective serotonin reuptake inhibitors						
Fluoxetine	1	5–80	Low	Absent	Very low	Insomnia, gastrointestinal discomfort, sexual dysfunction, nervousness/akathisia, headache
Fluvoxamine	1	50–300	Low	Absent	Very low	
Paroxetine	1	10–50	Low	Very low	Very low	
Sertraline	1	50–200	Low	Absent	Very low	
Venlafaxine	1	75–375	Low	Absent	Low	
Atypical antidepressants						
Bupropion	1	100–450	Low	Absent	Very low	Agitation, headache, nausea, tremor, seizure
Nefazodone	1	100–600	Moderate	Absent	Low	Headache, nausea
Trazodone	2	100–600	High	Absent	Moderate	Dizziness, headache, nausea, priapism
Cyclic antidepressants						
Amitriptyline	3	50–300	High	High	High	Weight gain, tachycardia
Amoxapine	3	50–300	Low	Low	Moderate	Stimulation, extrapyramidal side-effects
Clomipramine	2	50–250	High	High	High	Tremor, stomach upset, sexual dysfunction
Desipramine	1–2	50–300	Low	Low	High	Jitteriness, palpitations, tachycardia
Doxepin	2	25–300	High	High	Moderate	Tachycardia
Imipramine	2	50–300	Moderate	Moderate	High	Tachycardia
Maprotiline	2	50–200	Moderate	Low	Moderate	Tremor, grand mal seizure
Nortriptyline	1–2	25–150	Low	Low	Low	Sweating, tremor, headache
Monoamine oxidase inhibitors						
Phenelzine	3*	15–90	Low	Very low	High	Insomnia, hypertension, sexual dysfunction
Tranylcypromine	3*	10–90	Low	Very low	High	Insomnia, hypertension, tremor
Stimulants						
Dextroamphetamine	3*	2.5–30	Absent	Very low	Low	Insomnia, tremor, appetite changes
Methylphenidate	3*	5–40	Absent	Very low	Low	Insomnia, tremor, appetite changes

* First choice for subgroup of patients.

monoamine oxidase levels return to baseline only after approximately 10 days.

Monoamine oxidase inhibitors have a very low affinity for postsynaptic receptors and therefore many of the side-effects seen with cyclic antidepressants are absent. However, insomnia is seen frequently and often requires treatment (e.g. with trazodone).

The dietary restrictions and the risk of potentially lethal hypertensive crises preclude the use of monoamine oxidase inhibitors as a first-line drug. However, depressed patients with atypical features (hypersomnia, hyperphagia, and severe psychomotor retardation to the degree of 'leaden paralysis'), anxiety, and hostility may respond better to monoamine oxidase inhibitors than to cyclic antidepressants.

Selective serotonin reuptake inhibitors

All five selective serotonin reuptake inhibitors currently available in the United States (Table 1) inhibit the presynaptic reuptake of serotonin. Fluvoxamine and sertraline demonstrate a relative specificity for the serotonin uptake transporter, whereas venlafaxine, fluoxetine, and paroxetine also inhibit norepinephrine uptake. The proposed mechanism of antidepressant action is the enhancement of serotonergic neurotransmission, but other mechanisms have been proposed.

All selective serotonin reuptake inhibitors effectively treat depression and some lead to improvement of obsessive–compulsive disorder and bulimia nervosa. Selective serotonin reuptake inhibitors do not cause anticholinergic side-effects, cardiac conduction abnormalities, weight gain, or orthostatic hypotension. They are far safer than cyclic antidepressants when taken in overdose. The major side-effects are insomnia, gastrointestinal discomfort, sexual dysfunction, and headache. Lack of therapeutic response or development of side-effects with one selective serotonin reuptake inhibitor does not necessarily predict the same outcome with another.

Atypical antidepressants

Bupropion is a relatively selective dopamine-uptake blocker that is effective in treating depression (Table 1). Its efficacy and side-effect profile are similar to those of the selective serotonin reuptake inhibitors, but bupropion does not cause sexual dysfunction. High doses are associated with an increased risk of seizures.

Trazodone and nefazodone, which are chemically related compounds with an affinity for several receptors and uptake transporters, also treat depression (Table 1). Although trazodone lacks anticholinergic side-effects, it is highly sedative.

Stimulants

Dextroamphetamine (dexamphetamine) and methylphenidate have not shown a significant effect in the treatment of major depression. However, stimulants can be used safely and effectively in medically ill patients who present with apathy and loss of appetite. Their onset of effect is much faster than that of selective serotonin reuptake inhibitors and tricyclic antidepressants.

Treatment strategies

The selection of an antidepressant should be based on the side-effect profile of the drug and the type of depressive disorder. Table 1 gives a rank order for the various antidepressants that takes into account the likelihood of adverse effects as well as potential differences in efficacy. The selective serotonin reuptake inhibitors, bupropion, and nefazodone are recommended as first-line drugs. Nortriptyline and desipramine are preferred over the other cyclic antidepressants because of their favorable side-effect profile. Amitriptyline and amoxapine are considered third-line drugs because of their propensity to induce significant side-effects. The monoamine oxidase inhibitors are also third-line drugs in the treatment of depression, but are often the first choice in depressed patients with atypical features, anxiety, or hostility. Stimulants have a fast onset of effect and can be used effectively in medically ill patients with prominent apathy and lack of appetite.

Monotherapy is typically the initial treatment strategy. Table 1 lists the daily dose ranges of antidepressants. Initiating treatment at low doses and increasing the dose to the usual therapeutic drug levels reduces the risk of adverse effects. It is helpful to check the level of cyclic antidepressants in the blood, but this is not necessary for the monoamine oxidase inhibitors, selective serotonin reuptake inhibitors, and atypical antidepressants.

If the first antidepressant has not shown a therapeutic effect within 4 weeks, alternative strategies should be considered. Switching to another antidepressant (even of the same class) is reasonable. Augmentation of an insufficient treatment response is often effective. This can be achieved by adding a second antidepressant, lithium, a stimulant, or thyroid hormone to the antidepressant. Combining pharmacological treatment with psychotherapy is also helpful.

Patients who do not respond to adequate trials of antidepressants should be considered for electroconvulsive therapy. If the patient recovers from a depressive episode, maintenance treatment should be considered to reduce the significant risk of relapse associated with major depression.

Bibliography

Chiarello, R.L. and Cole, J.O. (1987) The use of psychostimulants in general psychiatry: a reconsideration. *Archives of General Psychiatry*, **44**, 286–95.

Hyman, S.E., Arana, G.W., and Rosenbaum, J.F. (1995). *Handbook of psychiatric drug therapy*. Little, Brown, Boston, MA.

Lipson, R.E. and Stern, T.A. (1991). Management of monoamine oxidase inhibitor-treated patients in the emergency and critical care setting. *Journal of Intensive Care Medicine*, **6**, 117–25.

Preskorn, S.H. (1994). Antidepressant drug selection: criteria and options. *Journal of Clinical Psychiatry*, **55** (Supplement A), 6–22.

Richelson, E. (1994). Pharmacology of antidepressants—characteristics of the ideal drug. *Mayo Clinic Proceedings*, **69**, 1069–81.

17.1.5 Antidiarrheals

G. J. Dobb

Key messages

- Correction of dehydration, electrolyte imbalance, and the underlying cause are more important than antidiarrheals.
- Antidiarrheals can relieve symptoms.
- Contraindications must be excluded before antidiarrheals are given.
- Loperamide is the preferred symptomatic treatment for diarrhea in intensive care patients.

Introduction

Most episodes of diarrhea are self-limiting or will stop when the underlying cause (e.g. medications, lactose in patients with lactase deficiency) is removed. Dehydration and electrolyte abnormalities should be corrected and malnutrition minimized by continuing enteral nutrition or using parenteral nutrition when this proves impossible. Antidiarrheals, other than oral rehydration solutions, should not be given alone to patients with pseudomembranous colitis or acute diarrhea associated with enterotoxigenic *Escherichia coli*, *Salmonella*, or *Shigella* infections. When given together with appropriate antibiotics they usually shorten the period of symptomatic diarrhea, but antidiarrheals should still not be given to patients with high fever or bloody diarrhea. Toxic megacolon can be induced in patients with ulcerative colitis.

Oral rehydration solutions

Oral rehydration solutions are the principle treatment of severe diarrhea in developing countries. The standard World Health Organization (WHO) solution contains 90 mmol/l sodium, 20 mmol/l potassium, 80 mmol/l chloride, 30 mmol/l citrate, and 111 mmol/l (20 g/l) glucose. In controlled trials oral rehydration solutions with a sodium content of 50 mmol/l are as effective as those with a greater sodium content but cause less edema and hypernatremia. Commercially available solutions contain varying amounts of sodium when made up according to manufacturers' directions, for example Pedialyte (Ross, Canada and United States; Abbott, Italy) 45 mmol/l, Gastrolyte (Rhone-Poulenc Rorer, Australia) 60 mmol/l, or Rehydralyte (Ross, United States) 75 mmol/l.

The WHO recommends a carbohydrate content of approximately 2 per cent for oral rehydration solutions. The American Academy of Pediatrics recommends a maximum of 2.5 per cent. Contraindications to the use of oral rehydration solution are shock, persistent vomiting, and impaired consciousness.

The glucose and sodium contents of oral rehydration solutions enhance water absorption at the brush border of the ileal mucosa, thus reducing the volume of diarrhea. When glucose and sodium are present in appropriate ratios, they enhance each other's transport across the brush border, with the movement of solute being associated with the iso-osmotic movement of water.

In children, the use of oral rehydration solution is associated with up to 97 per cent successful rehydration. There is only limited experience of its use in adults developing diarrhea during intensive care (Gallagher and Phelan 1992). Use of an oral rehydration solution maintained hydration and was associated with hemodynamic and metabolic stability and improvement in renal function, but the effect on diarrhea volume was unclear.

Loperamide

Loperamide (Imodium, Janssen-Cilag; Gastro-Stop, Rhone-Poulenc Rorer) is the author's preferred drug for the symptomatic treatment of diarrhea. It binds to peripheral opioid receptors in the gastrointestinal tract and inhibits release of secretory neurotransmitters such as acetylcholine. It also has calmodulin-binding and calcium-channel-blocking effects which may have additional antisecretory and motility-altering actions. In therapeutic doses loperamide has no central effects and approximately 80 per cent remains in the gut wall where it is taken up by the myenteric plexus and longitudinal muscle. Some is also found in the mucosa, probably bound to μ- and δ-opioid receptors (Awouters *et al.* 1993).

The effects of loperamide on the gastrointestinal tract are summarized in Table 1. Its effect on the mucosal transport of water and electrolytes in the non-stimulated state is minimal, but when secretion is stimulated its action on the μ- and δ-opioid receptors decreases secretion of sodium, potassium, and chloride, with this effect being maximal in the jejunum and less marked in the colon. Slowing of intestinal transit increases mucosal contact time and promotes reabsorption of water.

Table 1 Effects of loperamide on the gastrointestinal tract

Minimal effect on gastric motor function
Proabsorption effect on secreting intestine
Slowing of intestinal transit
Decreased propulsive activity
Increased non-propulsive segmenting contractions
Increased rectal volume for threshold sensation

Loperamide is given orally or through a feeding tube with an initial dose of 4 mg followed by 2 mg with every diarrheal bowel evacuation to a maximum of 16 mg/day.

Lomotil

Lomotil (Searle Pharmaceuticals), which is a compound drug containing 2.5 mg diphenoxylate and 25 μg atropine, is widely used. The diphenoxylate has similar effects to loperamide on myenteric plexus opioid receptors. It has more systemic side-effects and no advantage over loperamide. Use of Lomotil is contraindicated in patients with jaundice or impaired liver function, and it can interact with monoamine oxidase inhibitor antidepressants.

Lomotil is given orally as two tablets (5 mg) two to four times daily.

Codeine phosphate

Codeine phosphate was widely used in the past for diarrhea in a dose of 30 mg two to four times daily. It also acts on opioid receptors. Systemic absorption and side-effects limit its usefulness in critically ill patients as an antidiarrheal unless the analgesic properties are needed. Systemic side-effects include drowsiness, respiratory depression, and cough suppression. Larger doses can induce nausea or vomiting.

Clonidine

Clonidine is an α_2-adrenergic receptor agonist which acts both centrally and on α_2-adrenergic receptors on enterocytes. Stimulation of the enterocyte receptors decreases sodium secretion and increases sodium and water absorption; α_2-adrenergic stimulation slows gastrointestinal transit.

Clonidine has been used in the treatment of diarrhea associated with diabetic neuropathy in a dose of up to 0.3 mg/day. It has also been used in cryptosporidosis and appears to be useful in diarrhea associated with drug withdrawal syndromes. Side-effects include hypotension, sedation, and depression. The tendency to induce hypotension limits its usefulness in critically ill patients.

Octreotide

Octreotide is a long-acting somatostatin analog. It is useful in the treatment of diarrhea associated with intestinal endocrine tumors (e.g. carcinoid, vasoactive intestinal peptide) (Kvols *et al.* 1986) and in patients with diarrhea associated with AIDS. A starting dose of 50 μg subcutaneously daily may be increased to 100 μg three to four times daily to control symptoms. Side-effects include transient nausea, bloating, and hyperglycemia.

There is no information on its effect on diarrhea associated with critical illness.

Bismuth subsalicylate

Bismuth subsalicylate is available as Pepto-Bismol (Procter and Gamble, United States, Canada, United Kingdom). The salicylate dissociates in acid pH in the stomach and is absorbed, leaving 99 per cent of the bismuth to pass through the bowel unabsorbed. Bismuth has some antimicrobial activity, although the significance in diarrhea is uncertain, and it may have antisecretory effects.

Non-absorbed bismuth turns the stools black and absorbed bismuth can cause encephalopathy. Bismuth compounds should not be used in patients with impaired renal function.

Other treatments

Other symptomatic treatments for diarrhea include kaolin, which may act as a fluid adsorbent, and fiber, although there is no evidence that added fiber affects the frequency of diarrhea in patients needing intensive care (Hart and Dobb 1988).

Indomethacin can be useful in diarrhea associated with radiation enteritis, cholestyramine can be used when diarrhea is caused by bile acids, and pancreatic enzymes have a specific effect on diarrhea caused by pancreatic insufficiency and malabsorption.

The biotherapeutic agents *Enterococcus fecium* SF68, *Lactobacillus casei* GG, and *Saccharomyces boulardii* have reduced diarrhea in infectious or antibiotic-associated diarrhea. *S. bouldardii* reduces the frequency of diarrhea in critically ill tube-fed patients (Bleichner *et al.* 1997) and appears to be useful in recurrent *Clostridium difficile* infection.

References

Awouters, F., Megens, A., Verlinden, M., Schuurkes, J., Niemegeers, C., and Janssen, P.A.J. (1993). Loperamide: survey of studies on mechanism of its antidiarrheal activity. *Digestive Diseases and Sciences*, 38, 977–95.

Bleichner, G., Bléhaut, H., Mentec, H., and Moyse, D. (1997). *Saccharomyces boulardii* prevents diarrhea in critically ill tube-fed patients. *Intensive Care Medicine*, **23**, 517–23.

Gallagher, H.G. and Phelan, D.M. (1992). Oral rehydration therapy: a Third World solution applied to intensive care. *Intensive Care Medicine*, **18**, 53–5.

Hart, G.K. and Dobb, G.J. (1988). Effect of a fecal bulking agent on diarrhea during enteral feeding in the critically ill. *Journal of Parenteral and Enteral Nutrition*, **12**, 465–8.

Kvols, L.K., Moertel, C.G., O'Connell, M.J., Schutt, A.J., Rubin, J., and Hahn, R.G. (1986). Treatment of the malignant carcinoid syndrome. *New England Journal of Medicine*, **315**, 663–6.

17.1.6 Antiseizure agents

T. P. Bleck

Key messages

- The agents currently available for the prevention of seizures do not interfere with the development of epilepsy.
- The major categories of antiseizure agents include hydantoins, benzodiazepines, barbiturates, and propofol.
- Benzodiazepines are frequently preferred for active seizure termination, while hydantoins are most useful for prophylaxis against subsequent seizures.
- Propofol, high-dose benzodiazepines, and high-dose barbiturates are most useful for the management of refractory status epilepticus.

Introduction

The drugs employed for the prevention and management of seizures lack an appropriate name. Most of these agents only suppress seizures and do not treat epilepsy itself, and for that reason ought not be called antiepileptics. The term anticonvulsant ignores their effects on seizure types other than convulsions. For lack of a better choice, they are described here as antiseizure agents. The important pharmacological properties of these drugs are summarized in Table 1.

In general, patients receiving antiseizure agents prior to developing a critical illness unrelated to their epilepsy should continue with the same drug(s). Patients controlled on an agent only available for oral administration (e.g. carbamazepine) whose illness prevents enteral medication should either be switched to phenytoin or covered with a benzodiazepine (e.g. lorazepam or clonazepam) until enteral medication is possible. Abrupt termination of $GABA_A$ agonists (e.g. benzodiazepines and barbiturates) is associated with withdrawal convulsions even in patients without a history of seizures, and so these drugs should either be tapered slowly or, in the rare case of allergy or a serious adverse reaction, substituted with another $GABA_A$ agonist. It is less certain whether phenytoin, valproate, and other drugs acting by different mechanisms also produce withdrawal convulsions, but it is prudent to assume that patients receiving these drugs need them and therefore to continue treatment.

Only drugs available for parenteral administration are discussed here. The reader is referred to standard texts on antiseizure agents and epilepsy for more detail or for information about enterally administered drugs.

One major source of confusion in the use of antiseizure agents is the rational use of plasma concentrations ('levels'). In out-patient practice, the appropriate concentration of an agent is one in which seizures are controlled without adverse effects. Measurement of serum concentrations is undertaken to ensure compliance and absorption, and to guide changes in the regimen dictated by breakthrough seizures; they are also useful in determining the cause of a patient's apparent intoxication when more than one antiseizure agent is being used. Since seizures are often infrequent events, it is difficult to know what the appropriate concentration of an antiseizure

Table 1 Pharmacological properties of selected antiseizure agents in critical care practice

Agent	Loading dose	Usual maintenance dose	Elimination half-time*	Targeted plasma concentration†	Common adverse effects
Phenytoin	20 mg/kg	5–8 mg/kg/day	20+ h	10–20 μg/ml	Fever, hepatic enzyme elevation
Fosphenytoin	20 mg/kg‡	5–8 mg/kg/day‡	20+ h	10–20 μg/ml	Pruritis during IV loading
Lorazepam	0.1 mg/kg	0.1–0.3 mg/kg/day	4–14 h	Not useful	Sedation, respiratory depression, paradoxical agitation
Clonazepam	1–8 mg	10 mg/24 h	20–80 h	Not useful	Sedation, respiratory depression, paradoxical agitation
Midazolam	0.2 mg/kg	0.1–4.0 mg/kg/h	30 min	Not useful	Sedation, respiratory depression, paradoxical agitation
Phenobarbital	20 mg/kg	1.0–3.0 mg/kg/day	96+ h	15–40 μg/ml	Sedation, respiratory depression, paradoxical agitation
Pentobarbital	5–12 mg/kg	1.0–8.0 mg/kg/h	48 h	Not useful	Sedation, respiratory depression, paradoxical agitation
Propofol	3 mg/kg	1–15 mg/kg/h	15 min	Not useful	Sedation, respiratory depression, paradoxical agitation
Valproate	20 mg/kg	10–20 mg/kg/day	12 h	50–100 μg/ml§	Hepatic dysfunction, bone marrow suppression
Chlormethiazole	320–800 mg	4–160 mg/min	3–12 h	Not useful	Sedation, respiratory depression, paradoxical agitation

IV, intravenous.
* Under normal circumstances.
† For drugs where this is useful.
§ Phenytoin equivalent.
‡ For maintenance therapy; effective level in humans for control of status epilepticus is unknown.

agent will be in a critically ill patient. The laboratory's 'therapeutic range' is generally used in such circumstances, despite the lack of evidence that this is useful in the critically ill patient.

Standard anticonvulsants

Hydantoins

Hydantoins exert their antiseizure action primarily by impairing the ability of neuronal sodium channels to repolarize during abnormally rapid firing. The two agents in this class that are of value in the critically ill patient are phenytoin and fosphenytoin. The important differences between them are confined to the pharmacokinetics of their administration. Because of the long excretion half-time of phenytoin, these drugs require loading doses at the beginning of treatment and whenever an increase in serum concentration is desired.

One advantage of hydantoins over most of the other antiseizure agents is their lack of sedation at the relevant plasma concentrations.

Phenytoin

Phenytoin can be administered enterally or parenterally. The enteral formulations should be given twice daily (except for the extended-release form, which can be given once daily). If the patient is receiving tube feedings, these should be discontinued for 1 h before and after the phenytoin dose to allow absorption of the drug. The parenteral formulation of phenytoin consists of sodium hydroxide (at pH 12) and propylene glycol to keep the otherwise insoluble phenytoin in solution. Extravasation of this material can severely damage the skin and surrounding tissue. The maximum rate of administration is 50 mg/min, but many patients experience hypotension at this rate, mandating slower delivery. Phenytoin can increase atrioventricular block; if it is necessary in a patient with conduction abnormalities, it should be given with an external pacemaker available.

Under normal circumstances, about 90 per cent of phenytoin is bound to plasma proteins, yielding a 'free' phenytoin concentration of 1 to 2 µg/ml as the 'therapeutic range'. In uremic patients, phenytoin is displaced from proteins and the free fraction rises. If necessary, the free phenytoin concentration can be measured directly. Since only the free fraction of phenytoin is metabolized, the maintenance dose is unchanged in renal failure. Patients with liver disease or other causes of hypoproteinemia should receive a smaller loading dose. Those with altered synthetic function will usually need a smaller maintenance dose as well.

Near the upper end of the 'therapeutic range', the clearance mechanisms for phenytoin become saturated and drug metabolism shifts from first-order to zero-order kinetics. As a consequence, small increases in the maintenance dose may translate over time into large increases in the plasma phenytoin concentration.

Fosphenytoin

The major advantages of fosphenytoin, which is a phosphate prodrug of phenytoin, stem from its solubility in water. It can be administered at up to 150 mg/min. However, because of its conversion half-time, the free phenytoin levels produced rise only marginally faster than those produced by phenytoin at 50 mg/min. Extravasation does not damage tissue, but would result in a slower rise in plasma concentrations.

Benzodiazepines

The anticonvulsant effects of benzodiazepines are mediated by $GABA_A$ receptors; these drugs increase the affinity of the receptor for its natural ligand. The differences among the benzodiazepines appear to be pharmacokinetic. The major adverse effect of this class of drugs is respiratory depression with intravenous administration.

Lorazepam

Lorazepam is the most useful parenteral benzodiazepine for acute seizure and status epilepticus termination in North America, where parenteral clonazepam is not available. In this setting, it is typically given at a dose of 0.1 mg/kg. Larger doses are unlikely to be more effective in terminating status epilepticus.

The United States Department of Veterans Affairs randomized trial of therapy for status epilepticus compared lorazepam, diazepam followed by phenytoin, phenytoin alone, and phenobarbital (phenobarbitone) as first-line agents. Lorazepam was the most effective treatment for generalized convulsive status epilepticus.

Clonazepam

Intravenous clonazepam is often used outside North America in a similar manner to lorazepam, except that the initial dose may need to be followed by an infusion.

Midazolam

Because of its short duration of action, midazolam is of limited utility as an antiseizure agent. Its two major roles are as a diagnostic challenge for patients whose electroencephalograms suggest the presence of non-convulsive status epilepticus, and as an infusion in the management of refractory status epilepticus. Although its half-time is very short after a single dose, the drug will accumulate with prolonged administration. Midazolam is effective in the majority of cases of refractory status epilepticus, but the development of tachyphylaxis limits its utility to a few days.

Barbiturates

Barbiturates prolong the open time of chloride channels controlled by $GABA_A$ receptors. Thus their actions are similar to and additive with those of benzodiazepines. High doses of barbiturates decrease venous return as well as impairing myocardial contractility.

Phenobarbital

Phenobarbital remains a useful anticonvulsant because of its broad range of activity, ease of administration, and low incidence of allergic effects. Its major drawback is sedation. In the critical care setting, it is commonly used as a third-line agent in the control of status epilepticus. It will probably be supplanted in this role by propofol. It remains an important drug for intensive care patients with seizures not controlled by a single agent.

Pentobarbital (pentobarbitone) and thiopental (thiopentone)

These barbiturates have a more rapid onset of action than phenobarbital. At hypnotic doses pentobarbital does not exhibit anticonvulsant properties, but at anesthetic doses both pentobarbital and thiopental are useful in the management of status epilepticus. Although thiopental has an onset of action within minutes, approximately 40 per cent of the drug is converted to pentobarbital which has a much longer half-time.

At high doses, these barbiturates are profound myocardial depressants and also decrease venous return. They are also potent immunosuppressants.

Other agents

Propofol

Propofol is a novel anesthetic agent. It acts primarily as a $GABA_A$ agonist, but also has effects at several other receptors. This drug is rapidly becoming the agent of choice for the management of refractory status epilepticus. At the doses required to control refractory status, it appears to produce less myocardial depression than equivalent doses of pentobarbital. The major problem with long infusions at high doses stems from the lipid vehicle in which the drug is dissolved. The vehicle provides 1.1 kcal/ml as fat and may produce elevated plasma triglyceride concentrations, which on rare occasions lead to pancreatitis.

Chlormethiazole

Chlormethiazole is unusual among anesthetic agents in that the antiseizure effect of the drug occurs at concentrations which are frequently below those needed to abolish consciousness. As a consequence, it is frequently used as an initial treatment for status epilepticus in those countries where it is available. As the dose is increased, coma and respiratory depression may occur unpredictably. These effects occur synergistically with other hypnosedative agents.

This drug binds to plastic, decreasing its bioavailability, and may damage some multilumen catheter systems.

Valproate

An intravenous formulation of sodium valproate has recently been approved in the United States as a substitute for oral valproate or divalproex in situations where the patient is unable to take the oral form of the drug. Although effective in some animal models, its role in human status epilepticus remains to be defined.

Isoflurane

Isoflurane is a hydrocarbon anesthetic gas which has been used successfully in the past for the management of refractory status epilepticus, but it has no apparent advantage over other therapies. The requirement for an anesthesia machine to administer this agent, and a gas-scavenging system to clear it from the environment, makes isoflurane difficult to use in the intensive care setting. Even in the operating theater, more convenient agents are usually at hand.

Lidocaine (lignocaine)

Lidocaine (1–2 mg/kg) has been described as an effective antistatus agent in a few studies. It should be remembered that this drug also has the potential to produce seizures.

References

Bebin, M. and Bleck, T.P. (1994). New anticonvulsant drugs. *Drugs*, **48**, 153–71.

Levy, R.H., Mattson, R.H., and Meldrum, B.S. (ed.) (1995). *Antiepileptic drugs* (4th edn). Raven Press, New York.

Parent, J.M. and Lowenstein, D.H. (1994). Treatment of refractory generalized status epilepticus with continuous infusion of midazolam. *Neurology*, **44**, 1837–40.

Payne, T. and Bleck, T.P. (1996). Status epilepticus. *Critical Care Clinics*, **13**, 17–38.

Shorvon, S. (1994). *Status epilepticus: its clinical features in children and adults*. Cambridge University Press.

17.1.7 Antifibrinolytics

S. J. Machin

Key messages

- Antifibrinolytic therapy is useful to control bleeding associated with excess systemic fibrinolysis, localized fibrinolysis, or defective fibrin formation.
- Tranexamic acid (oral or intravenous) is the most useful inhibitor of fibrinolysis.
- Additional therapeutic measures are usually required to control excessive clinical bleeding.

Introduction

The fibrinolytic system is biologically designed to limit the extent of fibrin formation, to repair the injury, and to re-establish blood flow. Excessive local or systemic fibrinolysis may result in uncontrollable clinical bleeding. Antifibrinolytic therapy should rationally be considered in three groups of clinical situations.

1. When systemic fibrinolysis exists (or potentially may develop) due to primary hyperfibrinolysis such as that occurring in acute promyelocytic leukemia following initiation of chemotherapy or after specific thrombolytic therapy for an acute thrombotic event when bleeding complications develop.
2. Patients who have a decreased potential for hemostatic fibrin plug formation (such as hemophiliacs), so that any fibrin formed is not continually being lysed.
3. Certain circumstances when localized bleeding may be influenced by increased amounts of local fibrinolysis at potentially dangerous anatomical sites (i.e. pathological uterine bleeding, subarachnoid hemorrhage).

Antifibrinolytic agents overall limit fibrin degradation and can be categorized into the following groups of compounds.

1. The natural plasmin inhibitors, such as α_2-antiplasmin, which to date have not been developed for human use.
2. Aprotinin and related inhibitory compounds for plasmin and other serine proteases.
3. The synthetic lysine analogs ε-aminocaproic acid and trans-*p*-amino-methyl-cyclohexanecarboxylic acid (tranexamic acid). Both these compounds paradoxically accelerate plasminogen activation by inducing a conformational change in the molecule. However, they block the lysine binding site that is necessary for the binding of plasmin to fibrin (and fibrinogen), thus inhibiting overall fibrinolysis.

Pharmacology of ε-aminocaproic acid and tranexamic acid

Tranexamic acid has a six- to tenfold higher molar potency compared with ε-aminocaproic acid and is generally preferred for clinical usage, although routes of administration, effectiveness, limitations, and side-effects of the two compounds are similar. Tranexamic acid is available as 500-mg tablets (Cyklokapron, Pharmacia, United Kingdom), a syrup, or a colorless solution (5 ml contains 500 mg tranexamic acid) for intravenous infusion. Tranexamic acid is rapidly absorbed from the gastrointestinal tract with a serum half-life of 2 h, and is rapidly excreted unchanged in the urine with greater than 90 per cent elimination within 24 h. The inhibitory effect on fibrinolysis lasts for 7 to 8 h compared with less than 3 h for ε-aminocaproic acid. As urine levels are considerably higher than plasma levels, reduced doses provide effective control of urinary tract bleeding. Recommended adult dosages are 500 to 1000 mg three times daily orally or a similar dose by slow intravenous injection (1 ml/min) three times daily. General side-effects are relatively few, even with prolonged administration, but include abdominal pain, nausea, vomiting, skin rash, and conjunctival suffusion. Importantly, serious thrombotic complications have been reported and care must be taken in certain clinical conditions. These agents do not promote clot formation but only delay or prevent clot dissolution.

Problems have arisen when these agents are given in the presence of an underlying prothrombotic state or ongoing disseminated intravascular coagulation (DIC) resulting in generalized microvascular thrombosis, particularly in the renal glomeruli. The second type of thrombotic complication may develop in patients with upper genitourinary tract bleeding. Naturally occurring urokinase is neutralized by the inhibitor excreted in the urine, and so lysis-resistant clots may form in the renal pelvis causing pain, oliguria, and even renal failure.

Table 1 Clinical indications for antifibrinolytic therapy

	Clinical state	Comment
Systemic fibrinolysis	Carcinoma of prostate with secondaries	May control generalized bleeding
	Acute promyelocytic leukemia	Particularly during initial induction chemotherapy
	Congenital α_2-antiplasmin deficiency	Rare; autosomal recessive condition
	Bleeding during or after thrombolytic therapy (streptokinase, tissue plasminogen activator)	Benefit not proven, may be helpful
	Liver disease with portal hypertension	Exclude DIC first
	Liver transplant surgery	Useful in immediate postoperative period
Defective clot formation	Hemophilia A and B	After extensive surgery, in addition to specific factor concentrates
Localized fibrinolysis	Uterine menorrhagia	Variable benefit
	Subarachnoid hemorrhage	Variable benefit
	Gastrointestinal bleeding	May help in unresponsive cases
	Following prostatectomy	Controls excessive bleeding

Clinical indications for intensive care medicine

The pathogenic situations in which antifibrinolytic drug therapy should be considered are listed in Table 1. When systemic fibrinolysis is suspected, it is important to obtain confirmatory laboratory evidence of hyperfibrinolysis by a screening test before starting therapy. In these situations the euglobin lysis time will be appreciably shortened (the normal range is usually approximately 70 to 300 min) or excessive lysis can be demonstrated by a fibrin plate assay. Acute DIC must also be excluded, as antifibrinolytic therapy is usually contraindicated in this condition. It is also useful to monitor the fibrinogen level, as a reduced level (below 1.0 g/l) due to fibrinogenolysis may require additional specific replacement therapy. In most of the clinical situations listed, antifibrinolytic therapy alone is unlikely to control bleeding completely and additional therapeutic hemostatic measures will be required.

17.1.8 Antimicrobials

A. P. R. Wilson

Key messages

- Meningitis should be treated with cefotaxime or ceftriaxone.
- Ceftazidime or piptazobactam are appropriate for treatment of ventilator-associated pneumonia more than 2 days after intubation.
- Teicoplanin is less likely to cause nephrotoxicity than vancomycin when used in combination with aminoglycosides.
- Imipenem and meropenem should not be used in combination with other β-lactams.
- Clindamycin is the drug of choice in aspiration pneumonia.

Introduction

In critically ill patients, the pharmacokinetic behavior of many antibiotics is unpredictable and clinical trials are difficult to perform and interpret. Antibiotic dose has to be modified repeatedly according to the observed renal or hepatic function. Pharmacodynamic factors are important, i.e. the rapidity of bactericidal effect, the postantibiotic effect (the length of suppression of bacterial growth after antibiotic concentrations have fallen below inhibitory levels), and the serum half-life and tissue penetration of the antibiotic. The intravenous route is used because intramuscular administration is affected by changes in tissue perfusion.

The penicillins, glycopeptides, and cephalosporins act against the cell wall of the organism and bacterial killing depends more on the time during which the minimum inhibitory concentration (MIC) is exceeded than the absolute tissue concentration, provided that the lowest concentration is at least four times the MIC. The aminoglycosides and quinolones require high peaks because bacterial killing depends on concentration rather than duration of exposure. The area under the curve of serum concentration (AUC) versus time and over the MIC (area under the inhibitory curve AUIC) can be used to predict efficacy:

$$\mathrm{AUIC} = (\mathrm{AUC}_{0\text{–}24\,\mathrm{h}})/\mathrm{MIC}.$$

The AUIC is predictive of bacterial eradication *in vivo* and has resulted in the recommendation that pseudomonal pneumonia be treated with a dose of 400 mg of ciprofloxacin every 8 h and not 200 mg every 12 h (Forrest *et al.* 1993).

Vancomycin, teicoplanin, the aminoglycosides, the cephalosporins, and the penicillins are all excreted by the kidneys and will accumulate when creatinine clearance falls below 30 ml/min. Erythromycin, clindamycin, chloramphenicol, fusidic acid, and rifampin (rifampicin) are metabolized in the liver and, although safe in renal failure, should be avoided in liver failure.

Clearance by hemodialysis or hemofiltration depends on the type of machine, the flow rates, and the duration (Table 1) (Cotterill 1995). Aminoglycoside dosage must be determined by serum assay. The cephalosporins are eliminated slowly but dosage is usually reduced by at least half. The glycopeptides need only be given every 3 to 7 days after loading.

Penicillins

The penicillins interrupt bacterial cell wall synthesis and are bactericidal. They penetrate well and reach high concentrations in the urine and the lung. Penicillin G (benzylpenicillin) is highly active against most *Streptococcus pneumoniae*, *Streptococcus* group A, and α-hemolytic streptococci. *Neisseria meningitidis*, *Clostridium perfringens*, and *Treponema pallidum* are susceptible. It is quickly excreted and another dose is needed after 4 h. Allergic reactions develop in up to 10 per cent of patients and can include drug-induced fever. Interstitial nephritis is suggested by a raised eosinophil count. High doses given to patients with meningitis and severe renal failure risk the development of seizures.

Isoxazolyl penicillins (oxacillin, nafcillin, flucloxacillin, cloxacillin) are resistant to the penicillinase of *Staphylococcus aureus* and should be used empirically when infection with this organism is suspected. They have a synergistic interaction with aminoglycosides, but penicillin remains the drug of choice for the few penicillin-susceptible strains of *Staph. aureus*. Methicillin-resistant *Staph. aureus* (MRSA) infections must be treated with glycopeptides. Doses do not need to be reduced in renal failure because there is partial hepatic excretion. Hepatitis or cholestatic jaundice can occur during prolonged use.

Ampicillin, amoxicillin (amoxycyllin), azlocillin, piperacillin, amoxicillin–clavulanate (co-amoxiclav) and piperacillin–tazobactam are active against Gram-negative bacteria. Ampicillin is more active against enterococci than penicillin and is effective against *Listeria* species. Most *Enterobacteriaceae* acquired in the intensive care unit (ICU) will be resistant to ampicillin. Supplemental doses are required after hemodialysis. Amoxicillin differs in its longer half-life. Allergic reactions, particularly rash, are common. Amoxicillin–clavulanate is a combination of amoxicillin and clavulanic acid which prevents the action of β-lactamase. It is active against *Escherichia coli*, *Klebsiella* species, *Staph. aureus*, *Hemophilus influenzae*, *Neisseria gonorrheae*, and *Moraxella catarrhalis*. Other *Enterobacteriaceae* and *Pseudomonas* species are resistant. It can be used as an alternative to cefuroxime in most infections and produces high urinary concentrations (Neu *et al.* 1993). Ticarcillin–clavulanic acid is a similar combination agent with activity against some strains of *Pseudomonas* species.

Piperacillin and azlocillin are active against many Gram-negative species, including some *Pseudomonas* and *Klebsiella* species, but they are susceptible to β-lactamase. They are usually given in combination with an aminoglycoside. Piperacillin–tazobactam contains a β-lactamase inhibitor which allows it to be used as monotherapy. These agents are active against enterococci. They are used in intra-abdominal (e.g. biliary) infections, respiratory or urinary infections, and septicemia.

Cephalosporins

The cephalosporins have relatively low toxicity and a broad spectrum. They inhibit bacterial cell wall synthesis and are active against *Staph. aureus* and many Gram-negative bacteria. They are not effective against enterococci or methicillin-resistant staphylococci. Cephalosporins are frequently used in surgical prophylaxis and may be given in combination with gentamicin in bacteremic shock. Pseudomonal bacteremia is treated with ceftazidime. Cefuroxime can be used in abdominal wound infections (with metronidazole) or orbital cellulitis. In community-acquired respiratory infections in which the pathogen is unknown, cefuroxime is often used with erythromycin to ensure activity against *Strep. pneumoniae*, *H. influenzae*, and atypical causes. Pyelonephritis can be effectively treated by cefuroxime, cefotaxime, or ceftriaxone without an aminoglycoside. Vomiting and allergic reactions are the most common adverse effects.

Cephalexin, cefazolin (cephazolin), cephradine, cefadroxil, and cefaclor are active against the staphylococci, but activity against Gram-negative bacteria is poor except in the urine where high concentrations are achieved. They are not oral equivalents of cefuroxime and should only be used for infections that do not respond to cheaper agents.

Cefuroxime is active against *E. coli*, *Proteus* species, *Klebsiella* species, streptococci, and most strains of *Staph. aureus* (not MRSA). *Pseudomonas* species are resistant. It is used in a wide range of community-acquired or postsurgical infections. It is not recommended in meningitis. Thrombophlebitis and hypersensitivity are the only common adverse effects. Cefotaxime and ceftriaxone are less active against Gram-positive bacteria, but streptococci (not enterococci) are highly susceptible. They are active against pneumococci with a modestly reduced susceptibility to penicillin. Most *Enterobacteriaceae* and *H. influenzae* are susceptible, including nosocomial strains. *Enterobacter cloacae* and *Citrobacter freundii* are often resistant. Polymicrobial infections, bacteremia, and pneumonia are suitable target infections unless *Pseudomonas* is a likely pathogen. They are given as empirical treatment of meningitis in children or for Gram-negative meningitis in adults. Ceftriaxone is reduced in dosage only in severe renal failure or if there is liver and renal impairment. It is highly concentrated in bile and can produce biliary sludge causing abdominal pain and vomiting.

Ceftazidime is the most active agent against *Pseudomonas aeruginosa* but resistance can develop during treatment. It is active against the *Enterobacteriacae* but has poor activity against staphylococci and streptococci. It penetrates the lung well and is commonly used in the treatment of ventilator-associated pneumonia, provided that staphylococcal pneumonia is excluded. Some advise combination with an aminoglycoside in severe cases. It is used to treat febrile episodes in neutropenic patients and pseudomonal urinary infections.

Glycopeptides

Vancomycin and teicoplanin are active only against Gram-positive bacteria but methicillin-resistant staphylococci are susceptible. The range of susceptibility of coagulase-negative staphylococci to teicoplanin is wider than to vancomycin. Teicoplanin is more active than

Table 1 Dosage of parenteral antibiotics in critically ill patients and those with renal or hepatic failure

Antibiotic	Dosage	Renal failure	Hepatic failure
Penicillin G	Meningitis and fasciitis: 2.4 g, 4–6 h, IV Endocarditis: 1.2 g, 4 h, IV Pneumonia: 0.6–1.2 g, 6 h, IV	Reduce (e.g. 0.6 g, 6 h) if anuric Max. 6 g/day if severe (CH reduce 25% 4–6 h)	
Cloxacillin, flucloxacillin	1–2 g, 4–6 h, IV or up to 12 g/day	Reduce if severe	Causes cholestatic jaundice
Amoxicillin	1 g, 6–8 h, IV	Reduce if severe	
Ampicillin	1 g, 4–6 h, IV Meningitis: 2 g, 6 h	Reduce if severe: 1 g, 12 h, IV (CH 1–2 g, 6–12 h)	
Amoxicillin–clavulanate	1.2 g, 6 h, IV	0.6 g, 24 h if severe	Monitor liver function
Azlocillin	5 g, 8 h, IV	5 g, 12 h, IV	Reduce if renal and liver failure
Piperacillin	4 g, 6–8 h, IV	4 g, 8–12 h (CH 4 g, 8 h)	
Piperacillin–tazobactam	4.5 g, 8 h, IV	4.5 g, 8–12 h	Reduce if renal and liver failure
Cefuroxime	0.75–1.5 g, 8 h, IV	0.75–1.5 g, 12–24 h if severe (CH 0.75 g,12 h)	
Cefotaxime	1–2 g, 8 h	Halve dose if severe	
Ceftriaxone	2–4 g, 24 h, IV	Reduce	Reduce
Ceftazidime	2 g, 8–12 h, IV	1–2 g, 12–24 h (CH 1–2 g, 24–48 h)	
Vancomycin	1 g, 12 h, maintain trough 5–10 mg/l Colitis 125 mg, 6 h, oral	Assay: 1 g, 36–96 h Nephrotoxic (CH 1 g, 48 h)	
Teicoplanin	400 mg, 12 h × 3, then daily; can double Keep trough >20 mg/l in staphylococcal endocarditis	400 mg, 48–72 h after loading (CH same)	
Gentamicin	1–1.5 mg/kg, 8 h, but maintain trough <2 mg/l, peak 5–10 mg/l or use nomogram for once daily dosing (7 mg/kg)	Increase interval using assays: 80 mg, 12–48 h (CH 30%–70% of normal dose, 12 h)	
Imipenem	0.5–1 g, 6 h, IV	0.5 g 12 h (CH 500 mg, 6–78 h)	
Meropenem	0.5–1 g, 8 h, IV	Reduce	Monitor liver function
Ciprofloxacin	20 mg, 12 h, IV except pseudomonal pneumonia 400 mg, 8 h, IV/oral	Caution (CH 200 mg, 12 h)	Caution hepatitis
Ofloxacin	200–400 mg, 12 h, IV/oral	100 mg, 24 h	Reduce
Erythromycin	1–4 g/day, IV	Max. 1.5 g/day if severe	Avoid
Clarithromycin	0.5 g, 12 h, IV/oral	Reduce	Avoid
Clindamycin	600 mg, 8 h, IV/oral		Reduce
Trimethoprim–sulfamethoxazole	*Pneumocystis* 120 mg/kg, 14 days	Dose interval to 24 h	
Chloramphenicol	50 mg/kg/day, IV/oral	Avoid, no change	Avoid
Fusidic acid	0.5 h, 8 h, IV/oral		Avoid
Tetracycline	0.5–1 g, 12 h, IV		
Metronidazole	0.5 h, 8 h, IV; 1 g, 8 h, as required	0.5 g, 12 h	
Rifampin	450–600 mg, 24 h, 6 months oral/IV		Avoid
Isoniazid	300 mg, 24 h, 6 months oral/IV	Max. 200 mg/day if severe	Avoid

(continued)

Table 1 continued

Antibiotic	Dosage	Renal failure	Hepatic failure
Pyrazinamide	1.5–2 g, 24 h, 2 months oral		
Ethambutol	25 mg/kg/day oral	Reduce	
Streptomycin	0.75–1 g, 24 h, IM	Assay levels	
Amphotericin B	1 mg/kg/day over 6 h, Abelcet 5 mg/kg/day, Ambisome 3 mg/kg/day, Amphocil 3 mg/kg/day, IV	Use liposomal or colloidal forms, no change	
Fluconazole	400 mg × 1 then 200 mg, 24 h, IV	Reduce (CH 400 mg, 48 h)	
Itraconazole	200 mg, 24 h, oral	(CH 100 mg, 12–24 h)	Avoid
Flucytosine	200 mg/kg/day	50 mg/kg, 24 h (CH 38 mg/kg, 16 h)	Avoid
Acyclovir	5–10 mg/kg, 8 h (>10 days in encephalitis)	2.5–5 mg/kg, 12–24 h (CH 3.5 mg/kg, 24 h)	
Ganciclovir	5 mg/kg, 12 h, 14 days	Reduce	

CH, continuous hemofiltration; IV, intravenous; IM, intramuscular.
Adapted from Cotterill (1995).

vancomycin against enterococci and some vancomycin-resistant enterococci (vanB type). Streptococci, including those resistant to penicillin, are susceptible. Both agents are excreted by the kidneys and dosage reduction is necessary. Oral vancomycin (125 mg every 6 h for 10 days) is used to treat antibiotic-associated colitis. Teicoplanin can be given intravenously as a bolus dose, but vancomycin requires a slow infusion over 1 h to avoid flushing and hypotension. Teicoplanin has a longer half-life, allowing once-daily doses. Neither reliably penetrate the cerebrospinal fluid.

The glycopeptides are used to treat infections caused by MRSA (and other staphylococcal infections in penicillin-allergic patients) and catheter- or prosthesis-related infections caused by coagulase-negative staphylococci. Fever in neutropenic patients is treated with glycopeptides when there is evidence of catheter infection or failure of empirical treatment. Vancomycin concentrations in serum can be monitored to maintain troughs of 5 to 10 mg/l and peaks of 20 to 50 mg/l in order to avoid nephrotoxicity, particularly when used in combination with aminoglycosides. Teicoplanin does not require monitoring for toxicity but to ensure therapeutic levels, for example in staphylococcal endocarditis, when the trough should be over 20 mg/l, in renal failure, and in drug abusers. Ototoxicity and nephrotoxicity are rare using vancomycin alone, but there is a synergistic effect when it is combined with an aminoglycoside. Both glycopeptides can cause rash.

Aminoglycosides

Aminoglycosides are bactericidal agents which bind to bacterial ribosomes. They are active against the *Enterobacteriaceae* and staphylococci. They are synergistic in combination with penicillins except when there is high-level resistance. Penetration into the lung is poor but they are well concentrated in the urine. Their use is limited by nephrotoxicity and ototoxicity, and other agents should be used where possible. Serum concentrations have to be assayed daily in the ICU to avoid potential toxicity. Individual dosing based on daily assays is best practice. A course should be up to 3 to 5 days, and longer courses should only be used exceptionally (e.g. endocarditis). In patients with normal renal function, administration of a large single daily dose (7 mg/kg) has been found to reduce toxicity without affecting efficacy (Nicolau *et al.* 1995). A single assay is taken 6 to 14 h after the dose to determine the next dosage interval. Aminoglycosides are used in severe infections, particularly those acquired in hospital, when resistance to penicillins or cephalosporins is likely. Amikacin can be used in gentamicin-resistant infections.

Carbapenems

Imipenem is a carbapenem antibiotic with an extremely broad spectrum of activity. It is combined with cilastatin, which is an inhibitor of dehydropeptidase, an inactivating enzyme of the renal brush border which ensures adequate urinary concentrations. Meropenem has similar activity but does not require an enzyme inhibitor. Both antibiotics are highly active against streptococci, but enterococci are less susceptible and methicillin-resistant staphylococci are resistant. *Neisseria* and *Hemophilus* species are susceptible, as are *E. coli*, *Klebsiella* species, and *Salmonella* species. *Enterobacter* species, *Serratia* species, *Proteus* species, and *C. freundi* are less susceptible. They have moderate activity against *Pseudomonas* species but *Stenotrophomonas maltophilia* is resistant. *Bacteroides* species, *Fusobacterium* species, and to a lesser extent *Clostridium* species are also susceptible. The carbapenems are inducers of β-lactamase and should not be combined with other β-lactams. Penetration is good into most tissues except the meninges. Excretion is predominantly renal.

The carbapenems are used when several antibiotics would otherwise have to be given. Bacteremia, severe respiratory or urinary infections, febrile neutropenic episodes, and abdominal sepsis are common indications. Pseudomonal infections are better treated with other agents. Nausea and vomiting are common and fits develop in 1 per cent of patients given imipenem. Patients allergic to penicillins may be allergic to carbapenems. Superinfection with resistant pseudomonads or fungi can develop.

Fluoroquinolones

Ciprofloxacin and ofloxacin are available for intravenous use. Their broad spectrum of activity, oral or intravenous administration, and safety have encouraged inappropriate use, and resistance is increasing, particularly in *Ps. aeruginosa* and MRSA. They inhibit bacterial DNA gyrase and are bactericidal. They are active against staphylococci and streptococci, but should not be used in preference to penicillins. *Neisseria* species, *Hemophilus* species, and a high proportion of the *Enterobacteriaceae* are susceptible. *Pseudomonas* species are moderately susceptible, but high doses are needed to inhibit pseudomonads in the lungs. Excretion is by the kidney. The quinolones are used to treat complicated urinary infections and prostatitis. Other agents should be used for community-acquired pneumonia. They are first-line agents in severe *Salmonella* and *Shigella* infections. Osteomyelitis and soft tissue infections caused by Gram-negative bacteria and gonorrhea are other indications. Gastrointestinal symptoms, rash, and rarely convulsions occur. Quinolones are not recommended in children, unless there is no alternative, because of their potential effect on growing cartilage.

Macrolides

Erythromycin acts on protein synthesis at the ribosome and is active against streptococci, *Corynebacterium diphtheriae*, and most *Staph. aureus*. Most Gram-negative species are resistant. *Mycoplasma* species, *Chlamydia* species, and *Legionella* species are susceptible. Erythromycin is excreted by the liver and should be avoided in severe liver disease. It is used in the treatment of mycoplasmal and chlamydial pneumonia, legionnaires' disease, whooping cough, diphtheria, and *Campylobacter* infections. Nausea and vomiting are common adverse effects. Cholestatic jaudice (with estolate) and reversible ototoxicity are reported. Clarithromycin is more active against *Hemophilus* species. It is given twice daily and has good tissue penetration. It is used in respiratory tract infections and gastrointestinal symptoms are less common than with erythromycin.

Clindamycin

Clindamycin inhibits protein synthesis by the ribosome. It has important antianaerobic activity, including 90 per cent of strains of *Bacteroides* species and *Clostridium* species, and it is active against streptococci and staphylococci susceptible to erythromycin and methicillin. *Enterobacteriacae* are not susceptible. Clindamycin penetrates tissues well, particularly bone and pleural fluid (but not the meninges) and is excreted by the liver. The dose should be reduced in liver failure. It is an alternative to metronidazole for abdominal sepsis and is used to treat osteomyelitis caused by susceptible organisms. Its most common use in the ICU is for aspiration pneumonia or lung abscess. Necrotizing fasciitis and other severe infections caused by *Streptococcus pyogenes* can be treated with clindamycin. Clindamycin plus primaquine is an alternative treatment for *Pneumocystis* pneumonia. Diarrhea is a well-recognized adverse effect.

Chloramphenicol

Chloramphenicol has a very wide spectrum of activity, including most Gram-positive and Gram-negative species (not *Pseudomonas* species), anaerobes, and *Rickettsia*, but because of increasing resistance among *H. influenzae* and *Strep. pneumoniae*, it is no longer the drug of choice for empirical treatment of meningitis. It penetrates well into most tissues and is metabolized in the liver. Reversible bone marrow suppression can develop at dose of 4 g/day or on long courses. Serum concentrations should be measured in the very young and should not rise above 20 mg/l to avoid gray baby syndrome (vomiting, cyanosis, collapse). In occasional cases (1 in 30 000) a fatal aplastic anemia develops. The principle uses of chloramphenicol are treatment of meningitis in patients allergic to cephalosporins, brain abscess, and rickettsial infection.

Trimethoprim and trimethoprim–sulfamethoxazole (co-trimoxazole)

Trimethoprim and sulfamethoxazole act at different parts of the same pathway of synthesis of bacterial folate. Their long half-lives allow both to be given at 12-h intervals. Trimethoprim is active against most staphylococci, streptococci (not enterococci), *Enterobacteriaceae*, and *Hemophilus* species, but not *Ps. aeruginosa* or anaerobes. Trimethoprim can be used for urinary tract infection but in hospital-acquired infections susceptibility needs first to be demonstrated. Sulfamethoxazole adds little to trimethoprim for most infections, but trimethoprim–sulfamethoxazole is commonly used to treat *Pneumocystis* pneumonia in AIDS. Adverse effects are common at the high dose used (120 mg/kg/day intravenously in four divided doses) and are due mostly to the sulfonamide. Rash, nausea, vomiting, and diarrhea are common, and occasionally Stevens–Johnson syndrome occurs. Administration in pregnancy is contraindicated.

Tetracyclines

Tetracyclines are bacteriostatic antibiotics with a wide bacterial spectrum but resistance is common. They are the agents of choice in Lyme disease and rickettsial infections, and they are an alternative treatment for atypical pneumonia. Doxycycline has the longest half-life and is the only one that can be used in renal failure. Tetracyclines interfere with growth and development of bone, and must not be given in pregnancy or to children under 8 years old.

Metronidazole

Metronidazole is active against anaerobes, including *Bacteroides* species and *Clostridia*, and against parasites. It is as well absorbed by the oral or rectal route as intravenously and penetrates all tissues. Metronidazole should not be used in pregnancy. It is used to treat anaerobic bacteremia, intra-abdominal sepsis, and brain abscess. Aspiration pneumonia can be treated by metronidazole with a cephalosporin, but clindamycin might be more effective. Oral metronidazole is used to treat mild to moderate antibiotic-associated colitis. Giardiasis and amebiasis are other indications. The dose should be reduced in severe renal or hepatic failure. Peripheral neuropathy may develop during prolonged courses.

Rifampin

Rifampin is a bactericidal antibiotic which inhibits protein synthesis by binding RNA polymerase. It is highly active against staphylococci, *Corynebacteria*, *Moraxella* species, *Legionella* species, and most strains of *Mycobacterium tuberculosis*. Resistance will emerge rapidly if it is used in isolation. Rifampin is distributed throughout the body and is excreted through the liver. It is a first-choice agent in treating tuberculosis and *Mycobacterium avium intracellulare* complex disease. In staphylococcal endocarditis, rifampin may be used in combination with a glycopeptide with or without an aminoglycoside. Fever, rash, and hepatotoxicity are the most important adverse effects. The antibiotic antagonizes the effects of warfarin, cyclosporin, verapamil, and oral hypoglycemics.

Other antituberculosis agents

Isoniazid is given orally at a dose of 5 mg/kg, which is doubled in tuberculous meningitis. It is metabolized in the liver and dosage is not reduced in renal failure. Hepatitis occurs in 1 per cent of patients, but peripheral neuropathy can be prevented by giving pyridoxine. Pyrazinamide is often combined with rifampin and isoniazid. It is given orally and can be used to treat meningitis. Hepatotoxicity can occur. Ethambutol is bacteriostatic and is widely distributed after oral administration. The dose is reduced in renal failure. Optic neuritis (loss of color vision) is an important adverse effect which prevents its use in the unconscious. Streptomycin, like other aminoglycosides, is ototoxic and nephrotoxic, and serum levels must be assayed. However, it can be given parenterally with rifampin and isoniazid.

Antifungals

Amphotericin B

Amphotericin B remains the most common choice for the treatment of systemic fungal infection. It is a polyene which acts by increasing the permeability of the fungal cell membrane. It is highly tissue bound, particularly in the liver, spleen, kidney, and lung. No dose modification is necessary in renal or hepatic failure. Amphotericin B is not removed by hemodialysis. The first 1 mg of a dose may be given over 1 h to assess adverse effects, followed by 49 mg over 6 h. The dose should then be 1 mg/kg/day from the following day.

Treatment of aspergillosis in the profoundly neutropenic patient is often unsuccessful. Amphotericin is often used to treat acute cryptococcal meningitis. Adverse effects are renal failure, fever, chills, headache, hypotension, phlebitis, hypokalemia, anemia, and hepatitis. Liposomal, lipid complex, and colloidal dispersion amphotericin B have been produced to reduce toxicity. High doses can be given safely, but they may be less potent and are only indicated in a patient with previous adverse reactions or renal impairment. *Candida tropicalis* may be resistant.

Imidazoles and triazoles

Miconazole, ketoconazole, fluconazole, and itraconazole are active against most fungi except *Aspergillus* species. They act by inhibiting the production of the fungal cell membrane. Miconazole is not reliable in systemic infections, and ketoconazole is not effective against deep candidiasis. Fluconazole is well absorbed orally and can be given intravenously. Its long half-life allows once-daily dosing. It penetrates most tissues and is excreted renally, largely unchanged. Dosage must be reduced in renal failure. Adverse effects are rare. It is highly effective in mucosal candidiasis, but efficacy in other systemic infections is less well established.

Itraconazole is given orally and is distributed throughout the body except the cerebrospinal fluid. It is used to treat histoplasmosis, paracoccidiodomycosis, and blastomycosis, and can be given to patients with invasive aspergillosis in the presence of immune suppression.

Antivirals

Acyclovir (aciclovir)

Acyclovir is an analog of guanosine. It is active against herpes simplex viruses 1 and 2 and varicella-zoster virus, but poorly active against cytomegalovirus. The intravenous route is most reliable and it is distributed to all tissues including the meninges. Acyclovir is excreted by the kidneys and dose reduction is necessary in renal failure, although it is removed by hemodialysis. For the immune-suppressed patient with herpes zoster or varicella, acyclovir prevents dissemination and reduces shedding. It is effective in the treatment of herpes simplex encephalitis. Chickenpox in adults should be treated with acyclovir to reduce the morbidity associated with pneumonia. Phlebitis, reversible renal impairment, and elevation of transaminases are minor side-effects.

Ganciclovir

Ganciclovir is more active than acyclovir against cytomegalovirus. It is usually administered parenterally and penetrates the lung and liver. Dose should be reduced in renal failure, but it is removed by hemodialysis. Ganciclovir is used to treat cytomegalovirus retinitis in AIDS patients and is also effective in cytomegalovirus gastrointestinal disease. The presence of cytomegalovirus in the lungs of AIDS patients may not be the cause of clinical disease and is usually not treated. Ganciclovir does not affect outcome in cytomegalovirus pneumonitis in transplant recipients unless given with immunotherapy. Myelosuppression is common, and mucositis, fits, and rash can occur.

References

Cotterill, S. (1995). Antimicrobial prescribing in patients on haemofiltration. *Journal of Antimicrobial Chemotherapy*, **36**, 773–80.

Forrest, A., Nix, D.E., Ballow, C.H., Goss, T.F., Birmingham, M.C., and Schentag, J.J. (1993). Pharmacodynamics of intravenous ciprofloxacin in seriously ill patients. *Antimicrobial Agents and Chemotherapy*, **37**, 1073–81.

Neu, H., Wilson, A.P.R., and Grüneberg, R.N. (1993). Amoxicillin/clavulanic acid—a review of its efficacy in over 38 500 patients from 1979 to 1992. *Journal of Chemotherapy*, **5**, 67–93.

Nicolau, D.P., Freeman, C.D., Belliveau, P.P., Nightingale, C.H., Ross, J.W., and Quintiliani, R. (1995). Experience with a once daily aminoglycoside program administered to 2184 adult patients. *Antimicrobial Agents and Chemotherapy*, **39**, 650–5.

17.1.9 Antispasmodics, antiemetics, and drugs enhancing motility

J. J. Cordingley and M. G. A. Palazzo

Key messages

- Antispasmodics are mainly used during gastrointestinal endoscopy and to treat biliary and renal colic.
- Phenothiazines are the most commonly used antiemetics, but their hypotensive effects may be exaggerated in the critically ill.
- Prokinetic agents are frequently used but, although some have been shown to hasten gastric emptying, there are no studies reporting their effect on the establishment of full enteral feeding.

Antispasmodics

The majority of gastrointestinal antispasmodic drugs are synthetic antimuscarinics or atropine-related alkaloids. The most widely used antispasmodic is hyoscine butylbromide because of its efficacy and excellent side-effect profile.

Hyoscine-N-butylbromide (Buscopan)

Buscopan acts as a smooth muscle antispasmodic by inhibiting the action of acetylcholine at postganglionic parasympathetic neuromuscular junctions. It does not cross the blood–brain barrier and therefore is devoid of central nervous system effects.

Dosage and pharmacokinetics

Buscopan is poorly absorbed from the gastrointestinal tract ($<$ 10 per cent) and in the acute situation is normally given parenterally (intravenous, intramuscular, or subcutaneous) at a dose of 20 mg (for an adult), repeated as required. It has low plasma protein binding and is metabolized by the liver.

Adverse effects and drug interactions

Side-effects are mainly those of peripheral parasympathetic blockade and, although these are less marked than with atropine, Buscopan should not be administered to patients with paralytic ileus, glaucoma, or prostatic hypertrophy.

Use in critical care

Buscopan is commonly used to reduce smooth muscle spasm particularly during upper or lower gastrointestinal tract endoscopy. It is also used to treat the acute smooth muscle spasm of biliary and renal colic.

Antiemetics

Physiology

Two areas in the medulla are primarily responsible for the central control of vomiting. The vomiting center is responsible for the muscular control of vomiting and is closely related to the nearby chemoreceptor trigger zone which responds to blood-borne emetic stimuli from outside the blood–brain barrier. Stimuli and the neurotransmitters involved in these pathways are illustrated in Fig. 1. Common etiological factors for vomiting in the critically ill are listed in Table 1.

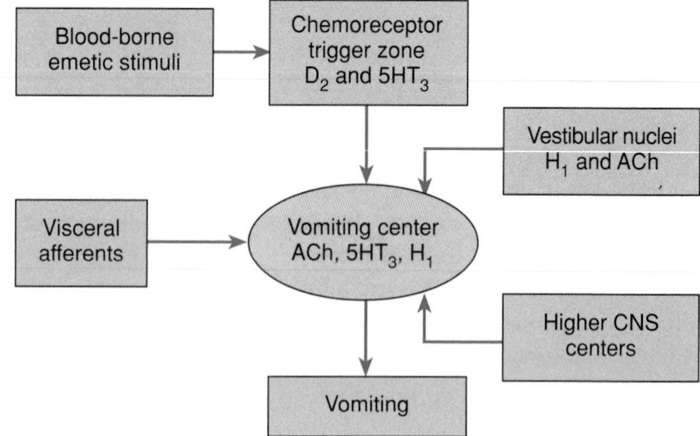

Fig. 1 Pathways involved in the control of vomiting. Receptor types are indicated as follows: Ach, acetylcholine; D_2, dopamine type 2; H_1, histamine type 1; $5HT_3$, 5-hydroxytryptamine type 3.

Antiemetic drugs

The most commonly used antiemetics are phenothiazines, metoclopramide, domperidone, and 5-hydroxytryptamine ($5HT_3$) antagonists. Antimuscarinic agents and H_1-receptor antagonists are mainly used in the management of motion sickness. Details of the most frequently used antiemetics are listed in Table 2.

Phenothiazines

Phenothiazines are antagonists at D_2 receptors in the chemoreceptor trigger zone and at muscarinic, α_1-adrenergic, and H_1 receptors. The most useful agents in this class are prochlorperazine, perphenazine, and trifluoperazine.

Prochlorperazine

After oral administration prochlorperazine undergoes high first-pass metabolism and bioavailability is low. Prochlorperazine has high plasma protein binding and a plasma elimination half-life of approximately 7 h.

Adverse effects and drug interactions

The most common adverse effects are extrapyramidal which reverse after drug withdrawal. Other adverse actions include anticholinergic effects, hypotension, and more rarely cholestatic jaundice, cardiac arrhythmias, and blood dyscrasias. Temperature regulation can be affected, leading to hypothermia.

Use in critical care

In the critical care setting antiemetics are most likely to be given to postoperative patients who may be sedated from the effects of anesthesia and opioid analgesia or be hypovolemic. Adverse effects such as hypotension and sedation are more common in these patients, particularly the elderly. Prochlorperazine is not recommended for use in children less than 10 years of age because of an increased likelihood of extrapyramidal effects.

Table 1 Etiological factors related to vomiting in critical care patients

Patient factors
Hypotension
Pain
Gastric or abdominal distension (mechanical obstruction/ileus)
Gastrointestinal hemorrhage
Raised intracranial pressure
Uremia
Drugs
Opioids
Anesthetic agents
Dopamine agonists
Antibiotics (e.g. erythromycin)
Anticancer chemotherapy
Other interventions
Pharyngeal stimulation
Abdominal or pelvic surgery
Radiotherapy

Perphenazine

Perphenazine has a similar antiemetic action and adverse effects profile to prochlorperazine.

Trifluoperazine

Trifluoperazine administered orally is used as an antiemetic but is more sedative and more likely to produce extrapyramidal reactions than prochlorperazine.

Butyrophenones

Drugs in this class are potent dopamine antagonists and are used primarily as major tranquilizers. The most commonly used butyrophenone for control of emesis is droperidol.

Droperidol

The antiemetic action of droperidol is the result of D_2 receptor antagonism in the chemoreceptor trigger zone.

Dosage and pharmacokinetics

Droperidol may be given orally. It is well absorbed but undergoes significant first-pass metabolism. The adult parenteral dose is 1.25 to 2.5 mg, but this should be reduced in the elderly or debilitated. It is highly plasma protein bound and metabolized in the liver with little excreted unchanged. The elimination half-life is 2 h.

Adverse effects and drug interactions

Droperidol can cause severe hypotension through α_1-adrenoreceptor antagonism and may prolong the QT interval. Other adverse effects include sedation, particularly when combined with other central nervous system depressants or narcotic analgesics. As with all central antidopaminergic agents, movement disorders can occur. Droperidol may cause unpleasant feelings, including agitation and excitement, even at low doses.

Use in critical care

Droperidol is uncommonly used in critical care patients because of its neuroleptic side-effects.

Metoclopramide

Metoclopramide has a dual mechanism of action; an antidopaminergic action at central nervous and peripheral dopamine receptors enhances acetylcholine activity at muscarinic receptors. Metoclopramide causes a rise in esophageal sphincter pressure and faster gastric emptying.

Pharmacokinetics

Metoclopramide is available as both enteral and parenteral preparations. Following intravenous administration, it is rapidly redistributed. It is metabolized in the liver and excreted mainly by the kidneys with an elimination half-life of 3 to 4 h. Dosage should be reduced in patients with significant renal or hepatic impairment.

Adverse effects and drug interactions

As with other antidopaminergic agents, extrapyramidal reactions can occur. Because metoclopramide causes enhanced gastrointestinal contractility it is not recommended for use in the first few days following bowel surgery. Anticholinergic drugs antagonize the peripheral actions of metoclopramide.

Table 2 Routes of administration and adult dosages of antiemetic drugs

	Metoclopramide	Domperidone	Prochlorperazine	Ondansetron
Oral/NG	10 mg, 8 h	10–20 mg, 4–8 h	10 mg at once	16 mg
Rectal	—	30–60 mg, 4–8 h	25 mg at once	—
Intramuscular	10 mg, 8 h	—	12.5 mg, 8 h	4 mg
Intravenous	10 mg, 8 h	—	—	4 mg
IV infusion	Up to 0.5 mg/kg/day	—	—	—

NG, nasogastric; IV, intravenous.

Table 3 Routes of administration and adult dosages of prokinetic drugs

Route	Metoclopramide	Domperidone	Cisapride	Erythromycin
Oral/NG	10 mg, 8 h	10–20 mg, 4–8 h	10 mg, 6–8 h	500 mg, 6 h
Rectal	—	30–60 mg, 4–8 h	30 mg daily	—
Intramuscular	10 mg, 8 h	—	—	—
Intravenous	10 mg, 8 h	—	—	250–500 mg, 6 h
IV infusion	1 mg/h	—	—	—

NG, nasogastric; IV, intravenous.

Use in critical care

Metoclopramide is used as both a first-choice antiemetic and a prokinetic agent (see below). There has been some concern about antagonism of administered dopamine.

Domperidone

Domperidone antagonizes the action of dopamine at D_2 receptors. It has poor central nervous system penetration and its main actions are at receptors in the stomach and chemoreceptor trigger zone. In the bowel this leads to faster gastric emptying and a reduced small bowel transit time.

Dosage and pharmacokinetics

Domperidone is no longer available for intravenous use because of reports of cardiac arrhythmias. Given enterally it has low bioavailability due to high first-pass metabolism. It is highly protein bound and is excreted mainly by the liver. Dosage reduction is not required in renal failure.

Adverse effects

Domperidone is much less likely than other dopamine antagonists to cause central extrapyramidal effects. Minor adverse effects are dry mouth, diarrhea, headache, and thirst.

Use in critical care

The use of domperidone is limited by the non-availability of a parenteral preparation.

$5HT_3$ antagonists

These drugs act by antagonizing $5HT_3$ receptors both centrally in the chemoreceptor trigger zone and peripherally in the bowel. As well as having an antiemetic effect they cause faster gastric emptying, but overall there is an increase in total gut transit time.

Ondansetron

Pharmacokinetics

Ondansetron is available as both oral and parenteral preparations. The usual dose in adults is either 16 mg orally or 4 mg intravenously prior to surgery or 4 mg intravenously or intramuscularly for postoperative control. The oral route has good bioavailability. Ondansetron is approximately 75 per cent bound to plasma proteins and is nearly all metabolized in the liver with a plasma half-life of approximately 3 h. Total daily dosage should not be more than 8 mg in patients with hepatic impairment.

Adverse effects and drug interactions

The adverse effects of ondansetron include headaches and constipation. No significant drug interactions are known.

Use in critical care

The primary indication for $5HT_3$ antagonists is control of nausea and vomiting associated with anticancer chemotherapy and radiotherapy. They are also used for postoperative nausea and vomiting when other agents have not been effective.

Other $5HT_3$ antagonists

Granisetron and tropisetron are alternative $5HT_3$ antagonists with similar actions and adverse effects.

Drugs enhancing motility (prokinesis)

Over recent years there has been an increasing awareness of the importance of enteral feeding as the preferred route of nutrition in critically ill patients. In this group gastrointestinal motility is often impaired due to the effects of surgery, drugs, metabolic derangement, and sepsis (Chang *et al.* 1987). Therefore it is common practice to attempt to hasten the establishment of enteral feeding with prokinetic agents. Although prokinetic agents have been shown to improve gastrointestinal motility, there is no evidence reporting their effects on the establishment of full enteral feeding.

Metoclopramide, domperidone, cisapride, and erythromycin are the prokinetic agents in general use. The routes of administration and dosage of these drugs are listed in Table 3.

Metoclopramide

Metoclopramide is commonly used as a first-line agent to enhance gastric emptying and to assist the establishment of enteral feeding, but there have been no randomized studies in the critically ill.

Domperidone

There are few data on the efficacy of domperidone as a prokinetic agent in critical care patients.

Cisapride

Cisapride stimulates gastrointestinal motility, from the lower esophagus to the colon, by increasing acetylcholine release in the myenteric plexus. This leads to faster gastric emptying and reduced gut transit time.

Pharmacokinetics

There is no parenteral preparation of cisapride, but it can be administered rectally or as a nasogastric suspension. It is absorbed quickly and is highly plasma protein bound. There is high first-pass hepatic metabolism with about half the metabolites excreted by the kidney. The elimination half-life is about 10 h and may be prolonged in renal or hepatic impairment.

Adverse effects and drug interactions

The most likely adverse effects are those related to increased gastrointestinal motility such as abdominal cramps and diarrhea; consequently cisapride is contraindicated in patients with mechanical obstruction or bowel perforation. Enhancement of motility is impaired by the concurrent use of antimuscarinic agents.

Because of its effects on gastric emptying, uptake of drugs which are primarily absorbed in the stomach may be reduced, but those absorbed mainly in the small bowel can be increased.

There have been rare reports of a prolonged QT interval and ventricular arrhythmias in patients taking other medications or with risk factors for arrhythmias such as hypokalemia or hypomagnesemia.

Some drugs inhibit cisapride metabolism and may increase the incidence of QT-interval prolongation. Therefore it is recommended that cisapride should not be used concurrently with erythromycin, clarithromycin, fluconazole, ketoconazole, itraconazole, and miconazole. Cimetidine also inhibits cisapride metabolism.

Use in critical care

Use of cisapride for treatment of postoperative ileus may improve intestinal motility in some patients, but its effects are variable.

Randomized controlled studies of the effect of cisapride in critical care patients have confirmed a significant improvement in gastric emptying compared with controls (Spapen *et al.* 1995; Heyland *et al.* 1996). However, further studies are required to investigate the efficacy of cisapride in promoting the establishment of full enteral feeding.

Erythromycin

Erythromycin is a macrolide antibiotic which has marked stimulatory effects on gastrointestinal motility, believed to be in part due to agonism of motilin receptors.

Erythromycin causes induction of migratory motor complexes with faster gastric emptying and enhanced duodenal contractility. Although colonic motor activity is also increased, there is no stimulation of colonic peristalsis in normal subjects and so the prokinetic effects of erythromycin are in the upper gastrointestinal tract. In an awake patient these are unpleasant and cause nausea, vomiting, abdominal cramps, and diarrhea. However, following encouraging results in patients with diabetic gastroparesis, there has been interest in the use of erythromycin to stimulate motility in sedated intubated intensive care patients.

Dosage and pharmacokinetics

For its prokinetic effect erythromycin is less effective orally than intravenously. It is approximately 80 per cent bound to plasma proteins, and is mainly metabolized and excreted by the liver with only about 10 per cent appearing unchanged in the urine. Plasma half-life is approximately 75 min but is prolonged in hepatic or renal impairment.

Adverse effects and drug interactions

Peripheral intravenous administration frequently causes thrombophlebitis. Reversible cholestatic jaundice is seen rarely, but erythromycin should not be given to patients who already have hepatic dysfunction.

Erythromycin may prolong the QT interval and so should not be used with other drugs which can cause the same effect, particularly astemizole and terfenadine. Theophylline concentrations may rise in patients given erythromycin, and concentrations should be monitored carefully. Because of faster gastric emptying, erythromycin can increase the bioavailability of orally administered drugs (e.g. digoxin) which may result in toxicity. Use of erythromycin will alter gastrointestinal tract flora.

Use in critical care

Erythromycin had no significant effect on clinical markers of gut motility in patients with postoperative ileus following abdominal operations (Bonacini *et al.* 1993). In ventilated patients, erythromycin has been shown to increase acetaminophen (paracetamol) absorption compared with controls (Dive *et al.* 1995).

The prokinetic effects of erythromycin are mainly confined to the upper gastrointestinal tract, and further studies are required to establish whether it is of benefit in establishing enteral feeding in critical care patients.

References

Bonacini, M., Quiason, S., Reynolds, M., Gaddis, M., Pemberton, B., and Smith, O. (1993). Effect of intravenous erythromycin on postoperative ileus. *American Journal of Gastroenterology*, **88**, 208–11.

Chang, R.W.S., Jacobs, S., and Lee, B. (1987). Gastrointestinal dysfunction among intensive care patients. *Critical Care Medicine*, **15**, 909–14.

Dive, A., *et al.* (1995). Effect of erythromycin on gastric motility in mechanically ventilated critically ill patients. A double-blind randomized, placebo-controlled study. *Critical Care Medicine*, **23**, 1356–62.

Heyland, D.K., Tougas, G., Cook, D.J., and Guyatt, G.H. (1996). Cisapride improves gastric emptying in mechanically ventilated, critically ill patients. A randomized double-blind trial. *American Journal of Respiratory and Critical Care Medicine*, **154**, 1678–83.

Spapen, H.D., Duinslaeger, L., Diltoer, M., Gillet, R., Bossuyt, A., and Huyghens, L.P. (1995). Gastric emptying in critically ill patients is accelerated by adding cisapride to a standard enteral feeding protocol: results of a prospective, randomized, controlled trial. *Critical Care Medicine*, **23**, 481–5.

17.1.10 Antithrombotic agents

J. P. Isbister

Key messages

- Urokinase has a higher affinity for fibrin than streptokinase and causes direct activation of plasminogen to form plasmin and less systemic activation of plasminogen.
- Pulse fibrinolytic therapy is generally used in the short term, but longer infusion with lower doses may be used in venous thromboembolic disease and atypical venous thrombotic disorders.
- Intracerebral hemorrhage is the most feared complication of fibrinolytic therapy; it occurs in 0.5 to 1 per cent of cases and is likely to be fatal.

Introduction

Therapeutic thrombolysis (fibrinolysis), although introduced into clinical medicine over 25 years ago, has only become standard care treatment for the majority of patients with an acute transmural myocardial infarction since 1986. Fibrinolytic agents also have a role in a range of other thrombotic states including venous thrombosis, peripheral vascular occlusion, thrombotic strokes, and thrombosis related to indwelling devices (catheters, prosthetic valves, stents, and shunts).

The fibrinolytic system

The plasminogen system plays a central role in the maintenance of vascular patency and cell migration. The binding of plasminogen to surfaces (fibrin or cells) is of importance in regulating the system. Plasmin(ogen) binds to cells with low affinity and high capacity via its lysine binding sites associated with kringle domains and also recognizes carboxy-terminal lysines of cell surface proteins. Plasminogen circulates in inactive forms as glu-plasminogen. This is converted by proteolytic activation to lys-plasminogen which binds with fibrin via lysine binding sites (antagonized by ε-aminocaproic acid). The final conversion to plasmin, which has fibrinolytic and other activities, is mediated by tissue plasminogen activator and/or urokinase.

Tissue plasminogen activator, which is produced by endothelial cells and other tissues, contains similar kringle domains to plasmin which slowly converts plasminogen to plasmin; this action is dramatically potentiated with binding of tissue plasminogen activator to fibrin. This fibrin specificity for plasmin and tissue plasminogen activator results in an ideal complex of fibrin, tissue plasminogen activator, and plasminogen, in which fibrin-bound tissue plasminogen activator converts fibrin-associated plasminogen to plasmin 'on site'. The fibrin-associated plasmin generated is protected from the inhibitory action of α_2-antiplasmin. In the urokinase system, pro-urokinase can exist as either the single-chain urinary plasminogen activator, which is proteolysed to urokinase which has specificity for fibrin, or the two-chain urinary plasminogen activator, which lacks fibrin specificity in converting plasminogen to plasmin.

There are natural inhibitors of fibrinolysis which block the plasminogen activators or plasmin. Free plasmin is inhibited by interaction of its lysine site α_2-antiplasmin, but while bound to fibrin its lysine site is blocked, allowing partial protection against the action of α_2-antiplasmin. Plasminogen activator inhibitors 1, 2, and 3 inhibit tissue plasminogen activator and urokinase. Plasminogen activator inhibitor is the most important of these agents.

Fibrinolytic agents

The action of therapeutric fibrinolytic agents is summarized in Fig. 1.

Streptokinase

Streptokinase, a single-chain polypeptide enzyme derived from β-hemolytic streptococcal cultures, binds to circulating plasminogen in a ratio of 1:1, causing molecular-conformational changes that lead to exposure of plasminogen's active site resulting in partial plasminogen conversion to plasmin. The streptokinase–plasminogen complex, which is not blocked by α_2-antiplasmin, is subsequently converted to streptokinase–plasmin complex which rapidly activates plasminogen to plasmin. Streptokinase–plasmin activity is not fibrin specific, but can also attack fibrinogen and other procoagulants. The half-life of the streptokinase–plasminogen complex is about 23 min; it is inactivated, in part, by antistreptococcal antibodies. Streptokinase decreases plasma fibrinogen levels for 24 to 36 h and the fibrinolytic effect resolves within a few hours; however, a prolonged thrombin time may persist for up to 24 h because of the decrease in fibrinogen levels and the presence of circulating fibrinogen degradation products. Circulating antistreptokinase antibody may be stimulated by recent streptococcal infection and a higher dose may be needed to ensure antibody neutralization. Treatment with strepokinase would be unlikely to induce an effective thrombolytic state when antibody titers are high, as may be seen within 2 years of therapeutic use.

Alteplase

Alteplase (recombinant tissue plasminogen activator) was developed with the aim of providing a fibrin-clot-specific thrombolytic agent that would be less likely to induce systemic fibrinolysis or fibrinogenolysis. It is a purified fibrinolytic glycoprotein of 527 amino acids, synthesized using the complementary DNA for natural human tissue-type plasminogen activator. Alteplase is cleared rapidly from circulating plasma, mainly by the liver, with approximately 80 per cent cleared within 10 min (slightly slower in people with thrombosis). It binds to fibrin in a thrombus and converts the bound plasminogen to plasmin, initiating local fibrinolysis with minimal systemic effects. There is a decrease of 20 to 30 per cent in circulating fibrinogen and associated decreases in plasminogen and α_2-antiplasmin. Alteplase has proved slightly more effective at reopening thrombosed vessels, particularly if administration has been delayed for more than 3 h after the onset of occlusion. This benefit appears to be offset by a higher reocclusion rate if heparin does not produce adequate prolongation of activated partial thromboplastin time. The incidence of cerebral hemorrhage is also greater with tissue plasminogen activator than with streptokinase.

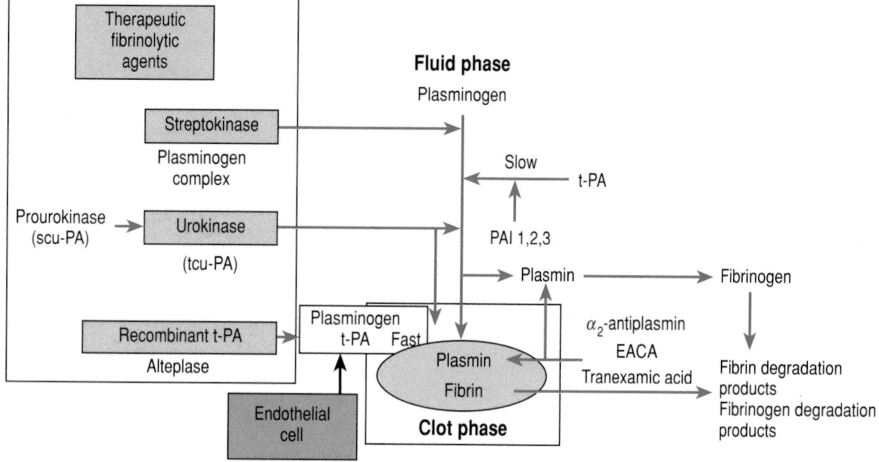

Fig. 1 Action of therapeutic fibrinolytic agents: t-PA, tissue plasminogen activator; PAI, plasminogen activator inhibitor; scu-PA, single-chain urinary plasminogen inhibitor; tcu-PA, two-chain urinary plasminogen inhibitor; EACA, ε-aminocaproic acid.

Urokinase

Urokinase is an enzyme extracted from human urine or produced by culture of fetal kidney cells. In contrast with streptokinase, it is non-antigenic and its use is not associated with allergic reactions. Owing to its greater affinity for fibrin, it should lead to less systemic activation of plasminogen. Urokinase causes direct activation of plasminogen to form plasmin and has a half-life of 14 ± 6 min with 2 to 5 per cent of the dose recovered in the urine and 5 to 8 per cent in the bile. Urokinase forms a urokinase–serum protein complex which retains the biological activity of urokinase. Cost is the main reason limiting its use for clearing thrombosed vascular lines.

Antistreplase

Antistreplase (anisoylated plasminogen streptokinase activator complex (APSAC)) has greater fibrin specificity than streptokinase and relies on the fibrin-binding action of the lys-plasminogen site to localize the complex specifically to the clot. Local tissue enzymes split off the acyl group of APSAC which shields the plasminogen catalytic site, allowing the streptokinase–plasminogen complex to convert plasminogen to plasmin in a more fibrin-specific manner. Antistreplase therapy has the advantage of ease of administration, as it is given as a single intravenous injection. Antistreplase has not been directly compared with alteplase or streptokinase with regard to mortality in large clinical trials, but it is probably equivalent to alteplase in most respects.

Adjunctive therapy

As thrombus formation and its dissolution is a complex process, fibrinolytic treatment should not be seen as ‘isolated’ therapy. As fibrinolytic therapy is increasingly important in the acute management of thrombotic disorders, when a fresh clot is present it is usually combined with, or quickly followed up by, adjunctive or conjunctive therapy. In particular, anticoagulants and antiplatelet agents may be used.

Administration of fibrinolytic therapy

The following general statements can be made about the practical aspects of thrombolytic therapy.

1. The ‘younger’ the thrombus, the greater is the likelihood of success.
2. Arterial clots are generally easier to lyse than venous clots.
3. Local infusion onto or into the clot is more effective.
4. Continuous rather than bolus therapy is more effective, but has a higher risk of systemic fibrinogenolysis and bleeding.
5. Adjunctive therapy is usually needed after successful thrombolysis.
6. Baseline prothrombin time, activated partial thromboplastin time, and thrombin time are indicated. Thrombin time should be checked 4 to 6 h after infusion, and the time should be prolonged by 1.5 to 5 times the baseline.

Treatment regimens

A variety of regimens are used for the administration of fibrinolytic therapy depending on the nature of the clinical problem and whether local or systemic therapy is being administered. In general, short-term pulse therapy is used for most therapy, but longer infusion with lower doses may be used in venous thromboembolic disease and atypical venous thrombotic disorders.

Myocardial infarctions

Dosage regimens for myocardial infarctions are summarized in Table 1.

Table 1 Dosage regimens for fibrinolytic therapy for myocardial infarction

	Streptokinase	Alteplase	Urokinase
Dose	1.5×10^6 IU, infused over 1 h Intracoronary infusion, 20 000 IU bolus followed by 2000 IU/min for 1 h (total dose 140 000 IU)	100 mg for patients ⩾70 kg: 15 mg bolus, 50 mg over 30 min, 35 mg over 60 min 1.5 mg/kg for patients <70 kg: 15 mg bolus, 0.75 mg/kg over 30 min, 0.50 mg/kg over 60 min	Priming dose of 4000 IU/kg over 10 min; subsequently infused at 4000 IU/kg/h for 12 h
Half-life (min)	c.23	c.8	c.14
Hypotensive effect (0–4)	2	1	1
Allergic reactions	Common	Rare	Rare
Cost	Low	High	High

Peripheral arterial occlusion

Systemic administration of thrombolytic agents for peripheral arterial occlusion has been abandoned for local catheter-directed and intra-operative intra-arterial repeated bolus or short-term infusions. Urokinase 250 000 IU diluted in 250 ml of normal saline solution is instilled at the end of mechanical thromboembolectomy over a 30-min period with the arterial inflow occluded.

Pulmonary embolism and venous thrombosis

Regimens similar to those for myocardial infarction can be used. However, it is generally recommended that longer infusion therapy is more appropriate. Streptokinase 250 000 IU is usually administered to neutralize resistant antibodies followed by 100 000 IU/h for 24 to 72 h. Local infusion into the pulmonary artery may also be considered. Alteplase can be used as a single bolus of 1 mg/kg over 10 min (maximum dose, 100 mg).

Occluded arteriovenous shunts

Urokinase 5000 to 25 000 IU in 2 to 3 ml of normal saline is instilled into the affected limb of the shunt which is clamped off for 2 to 4 h. On the venous side, an infusion of 5000 IU in 200 ml of normal saline is run in over 30 min.

Occluded central venous catheters

A solution of urokinase 5000 IU in 1 ml of saline is used for clearance of central intravenous cannulas.

Monitoring fibrinolytic therapy

Thrombolytic agents produce their effects by inducing and amplifying activation of the plasminogen–plasmin proteolytic enzyme system and thus are a potentially dangerous group of pharmaceutical agents. As it is difficult to assay plasmin directly, the detection of circulating fibrin degradation products confirms activation of the fibrinolytic system. Baseline levels are necessary for reference as natural activation occurs in association with arterial and venous thrombosis. A degree of cleavage of fibrinogen also may occur with fibrinolytic therapy, resulting in the generation of fibrinogen degradation products. These degradation products inhibit the thrombin conversion of fibrinogen to fibrin, as can be measured using the thrombin clotting time. Fibrinogen levels may also fall during therapy. Collection of blood samples on aprotinin (150 to 200 IU/ml) minimizes continued *in vitro* lysis which can result in spurious results.

When using fibrinolytic therapy the following questions usually arise.

1. Are there clinical or other objective measures of thrombus dissolution?
2. Which laboratory indices are useful in confirming activation of the fibrinolytic system?
   (a) thrombin time
   (b) fibrin degradation products
   (c) fibrinogen level
3. Which laboratory measurements may be predictive of hemorrhagic complications? In general, tests are not of value in this respect and have a greater role after the event in determining the cause and appropriate therapy.

Contraindications

There are several conditions in which fibrinolysis presents problems and careful analysis of the risk–benefit equation is essential. It is difficult to be dogmatic as to which contraindications are absolute and which are relative.

Contraindications are as follows:

(1) known bleeding disorders and anticoagulant therapy;

(2) when diagnosis is uncertain, particularly if there is suspicion of aortic dissecting aneurysm, esophagitis, or pericarditis;

(3) manifest or recent severe bleeding;

(4) recent history of stroke (within the previous 6 months);

(5) hypertension (diastolic blood pressure above 110 mmHg and/or systolic blood pressure above 180 mmHg despite antihypertensive therapy);

(6) within 10 days of major surgery;

(7) esophageal varicosis;

(8) intracranial neoplasm, arteriovenous malformation, aneurysm;

(9) acute pericarditis;

(10) acute pancreatitis;

(11) diabetic hemorrhagic retinopathy or other hemorrhagic ophthalmic conditions.

Complications

Bleeding

Thrombolytic therapy is associated with a major risk of bleeding complications. Bleeding may be related to known or occult pathology (e.g. peptic ulceration, ulcerative colitis). Physical handling of the patient should be restricted and intramuscular injections avoided. Essential arterial punctures should be confined to the upper extremity vessels, with prolonged pressure and regular checking of the site. Arterial punctures prior to fibrinolytic therapy must be closely monitored, particularly sites through which arteriograms have been performed. Compression devices can be used before thrombolytic therapy is given. Spontaneous soft tissue hemorrhage, particularly retroperitoneal, may occur and patients should avoid strain on their iliopsoas muscles (i.e. sitting up and down in bed). The most feared complication is intracerebral hemorrhage, which may occur in 0.5 to 1 per cent of cases and is likely to be fatal. Risk factors for intracerebral hemorrhage include advanced age, female sex, low body weight, hypertension, and history of cerebrovascular disease. Patients with diabetes and concurrent oral anticoagulant therapy may also have a higher risk of intracerebral hemorrhage.

During strepokinase therapy, circulating fibrinogen levels are reduced to 10 to 20 per cent of normal for up to 24 h. Although alteplase lowers fibrinogen levels less than streptokinase, the incidence of hemorrhagic complications with alteplase is not lower. Intracerebral hemorrhage is more common after tissue plasminogen activator than after streptokinase. Hemorrhagic complications of thrombolytic therapy do not correlate well with the degree of reduction in circulating fibrinogen levels, but correlate with the presence of localized vascular injury. Management of established bleeding complications of thrombolytic therapy includes cessation of therapy; including heparin (protamine sulfate may be indicated) and aspirin. If coagulation tests indicate factor deficiencies, cryoprecipitate and fresh frozen plasma may be necessary. If bleeding persists, administration of antifibrinolytic agents should be considered.

Allergic reactions

Anaphylactic and anaphylactoid reactions, ranging in severity from mild dyspnea to bronchospasm, from periorbital swelling to angioedema, and rarely shock, may sometimes be seen with streptokinase. Urticaria, itching, flushing, nausea, headache, and musculoskeletal pain have also been observed, as have delayed reactions such as vasculitis and interstitial nephritis. Most reactions can be managed with antihistamines and/or corticosteroids, but severe reactions mandate immediate cessation of therapy and treatment with adrenergic agents, antihistamine, and/or corticosteroid. Minor allergic reactions may occur to a lesser extent with the other fibrinolytic agents.

Fever

Fever can occur with streptokinase, but is rarely a major clinical problem and can be treated symptomatically. Occasionally, corticosteroids may be indicated.

Hypotension

Hyotension may occur with any of the fibrinolytic agents and can be of concern in view of the adverse effects of hypoperfusion in the context in which the agent is usually being administered. Hypotension is most common with streptokinase, and slowing the infusion is usually all that is necessary. Transient hypotension is the most common adverse event and is managed by increasing venous return, volume infusion, and a brief reduction of streptokinase infusion.

Bibliography

Goldhaber, S.Z. (1995). Contemporary pulmonary embolism thrombolysis. *Chest*, **107**, 45S–51S.

Gonzalez Fajardo, J.A., Perez-Burkhardt, J.L., and Mateo, A.M. (1995). Intraoperative fibrinolytic therapy for salvage of limbs with acute arterial ischemia: an adjunct to thromboembolectomy. *Annals of Vascular Surgery*, **9**, 179–86.

Habib, G.B. (1995). Current status of thrombolysis in acute myocardial infarction. I. Optimal selection and delivery of a thrombolytic drug. *Chest*, **107**, 225–32.

Lauer, J.E., Heger, J.J., and Mirro, M.J. (1995). Hemorrhagic complications of thrombolytic therapy. *Chest*, **108**, 1520–3.

Verstraete, M. (1995). Thrombolytic therapy of non-cardiac disorders. *Baillière's Clinical Haematology*, **8**, 413–24.

17.1.11 Aprotinin

A. R. Webb

Key messages

- Aprotinin is a naturally occurring non-specific serine protease inhibitor.
- The major role of aprotinin in critical care practice is the prevention of bleeding associated with extracorporeal circulations.
- Life-threatening bleeding due to thrombolytic therapy may be arrested by the plasmin inactivation effect of low-dose aprotinin.
- In septic shock and hepatic cirrhosis, high-dose aprotinin may improve hypotension and organ perfusion.

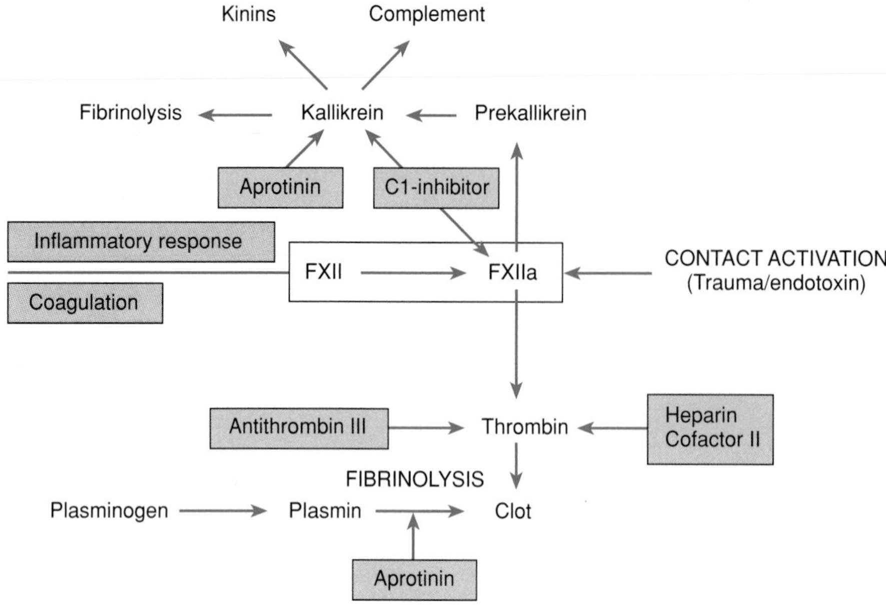

Fig. 1 Schematic representation of the contact pathway. Inhibitors, including aprotinin, are shown in the shaded boxes.

Introduction

Serine protease inhibitors, particularly antithrombin III, are commonly depleted in critically ill patients. The role of serine protease inhibitors in coagulation and anticoagulation is complicated because of their effects at various points in the coagulation pathway (Fig. 1).

Pharmacology

Aprotinin is a naturally occurring non-specific serine protease inhibitor with an elimination half-life of about 2 h (Royston 1990). It is a basic polypeptide of 58 amino acid residues and has a molecular weight of 6512 Da.

The major role of aprotinin in critical care practice is the prevention of bleeding associated with extracorporeal circulations. Prevention of systemic bleeding with aprotinin does not promote coagulation within the extracorporeal circulation and may even contribute to the maintenance of extracorporeal anticoagulation by prevention of contact-associated coagulation. In addition, the non-specific serine protease inhibition may inhibit components of the inflammatory response; this effect has been utilized in shock states.

Modes of action

The effects of aprotinin on the contact pathway are dependent on the circulating plasma concentrations (expressed as kallikrein inactivation units (kIU) per milliliter) since the affinity of aprotinin for plasmin is significantly greater than that for plasma kallikrein (Fritz and Wunderer 1983). At a plasma level of 125 kIU/ml aprotinin inhibits fibrinolysis and complement activation. Inhibition of plasma kallikrein requires higher doses to provide plasma levels of 250 to 500 kIU/ml.

1. Plasma kallikrein inhibition reduces blood coagulation mediated via contact with anionic surfaces and, in the critically ill patient, improves circulatory stability via reduced kinin activation.
2. Prevention of inappropriate platelet activation: neutrophil activation (mediated by complement or kallikrein) causes a secondary activation of platelets. The release of cathepsin G from neutrophil granules is important in this platelet–neutrophil interaction. It has recently been demonstrated that aprotinin can significantly inhibit the platelet activation due to purified cathepsin G; this mechanism forms a direct inhibition of inappropriate neutrophil-mediated platelet activation.

Uses

1. The main role of aprotinin has been to prevent perioperative bleeding. High-dose aprotinin given during cardiopulmonary bypass procedures has been shown to reduce postoperative blood loss dramatically (Royston 1990). In addition, aprotinin has been used to reduce blood loss in neurosurgery and in vascular, orthopedic, and urological surgery.
2. More recently, the same dose regimen of aprotinin has been used to arrest bleeding associated with prolonged extracorporeal CO_2 removal (Brunet *et al.* 1992). This effect may be utilized in patients with bleeding associated with any extracorporeal circulation, including hemofiltration and hemodialysis.
3. Life-threatening bleeding due to thrombolytic therapy may be arrested by the plasmin inactivation effect of aprotinin. Lower doses are required for this effect.
4. Blockade of the kallikrein–kinin system has been utilized in the management of shock states associated with pathological peripheral vasodilatation. High-dose aprotinin may improve hypotension and organ perfusion in septic shock and hepatic cirrhosis.

Dosage

High-dose aprotinin to prevent postoperative bleeding in cardiopulmonary bypass or to arrest bleeding associated with prolonged extracorporeal circulation requires a loading dose of 2×10^6 kIU (over 20 min) followed by 500 000 kIU/h intravenously. As a polypeptide, aprotinin is not active if given orally.

Low-dose aprotinin to reverse the action of thrombolytics requires a loading dose of 500 000 kIU (over 10 min) followed by 200 000 kIU/h intravenously until bleeding stops.

Side-effects

Aprotinin has few side-effects, but allergic and pseudoallergic reactions have been reported. It should be noted that aprotinin prolongs the activated clotting time which cannot, therefore, be used to assess heparin therapy.

References

Brunet, F., *et al.* (1992). Effects of aprotinin on hemorrhagic complications in ARDS patients during prolonged extracorporeal CO_2 removal. *Intensive Care Medicine*, **18**, 364–7.

Fritz, H. and Wunderer, G. (1983). Biochemistry and applications of aprotinin, the kallikrein inhibitor from bovine organs. *Arzneimittelforschung*, **33**, 479–94.

Royston, D. (1990). The serine antiprotease aprotinin (Trasylol): a novel approach to reducing postoperative bleeding. *Blood Coagulation and Fibrinolysis*, **1**, 55–69.

17.1.12 Blood product therapy

R. Liesner

Key messages

- Blood products are expensive to produce and all carry a risk of viral transmission. Therefore they should only be used when there is a valid clinical indication.
- Platelets, fresh frozen plasma, cryoprecipitate, and granulocytes do not undergo any viral inactivation steps, but adult recipients are exposed to only four to twenty donors during each transfusion.
- Albumin, factor concentrates, and immunoglobulin are all made from pooled donor plasma (pools are usually made from more than 1000 donor units) and are then virally inactivated, which substantially reduces but does not abolish infection risk.
- All blood products can cause allergic reactions.

Introduction

Over the last 30 years major advances have been made in transfusion medicine and nowadays virtually all collected blood is separated into its components. There is now the potential for prolonged periods of storage of relatively pure components without functional detriment. Techniques for fractionating plasma have been refined and increasingly sophisticated methods of viral inactivation have been developed in attempts to render all plasma products safe. Advances in molecular technology have resulted in the production of a factor VIII concentrate made by recombinant technology, and there are other recombinant products to follow.

In most hospitals all products described below will be issued to the user by the hospital blood bank, and nowadays it is obligatory that there is adequate documentation detailing what product is given to which patient.

Platelet transfusion

Preparation and administration

Platelets for transfusion can be harvested from whole blood as platelet-rich plasma or buffy coat, or by platelet-pheresis techniques (Fig. 1).

The decision on when to transfuse platelets will depend on the clinical condition of the patient, the platelet count, and the procedure that the patient is to undergo. The usual threshold for transfusion in bone marrow failure syndromes is a platelet count of 5×10^9/l to 15×10^9/l (higher if septic or bleeding), but the threshold level may be higher in other conditions.

An adult dose of platelets is 2×10^{11} to 4×10^{11} which is contained in either a single apheresis pack or a 'pool' of four to six random donor units. Recovery of platelets 1 h after transfusion is 50 to 80 per cent, and the effective half-life in the circulation is about 4 days. Ideally, platelets should be of the same ABO and Rh group as the patient, but in an emergency incompatible groups can be given although this may result in mild immune hemolysis or the development of anti-D.

The normal shelf-life of platelet concentrates is 5 days and they should be kept at room temperature ($22 \pm 2\,^\circ$C), ideally on an agitator. They should be infused through a blood- or platelet-giving set and, if indicated, they can be filtered at the transfusion center or at the bedside and/or gamma-irradiated. Indications for filtration include the prevention of HLA alloimmunization or non-hemolytic transfusion reactions in multiply transfused patients and newly diagnosed patients with aplastic anemia, and also to reduce the chance of cytomegalovirus transmission when cytomegalovirus-negative blood products are not available.

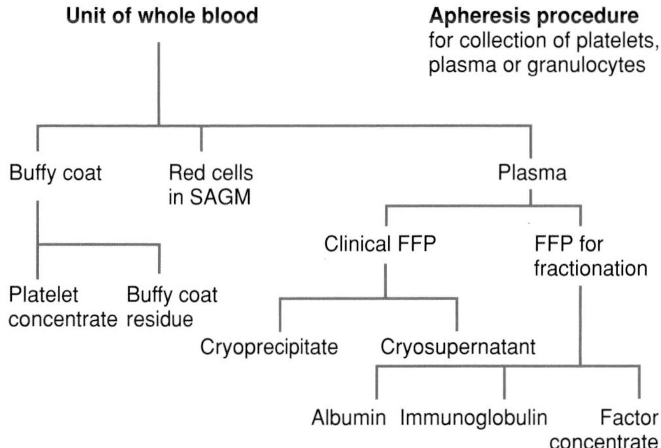

Fig. 1 Donor blood collection protocols indicating processing pathways for the manufacture of blood products: SAGM, saline adenine glucose mannitol; FFP, fresh frozen plasma.

Indications for platelet transfusion

1. Thrombocytopenia:
    (a) bone marrow failure (hematological malignancy, post-chemotherapy, aplastic anemia);
    (b) disseminated intravascular coagulation (DIC);
    (c) massive transfusion;
    (d) immune thrombocytopenia (only in severe bleeding).
2. Platelet dysfunction:
    (a) congenital (rare);
    (b) acquired (cardiopulmonary bypass surgery, drug induced (aspirin), uremia).

Platelet transfusions are contraindicated in thrombotic thrombocytopenic purpura, hemolytic uremic syndrome, and heparin-induced thrombocytopenia because the underlying tendency is for thrombosis rather than bleeding.

Adverse effects

1. Allergic reaction: this may cause fever, urticaria rigor, or rarely anaphylaxis. It is usually due to contaminating white cells or plasma proteins and can be prevented by premedication with intravenous hydrocortisone and chlorpheniramine and filtration.
2. Platelet refractoriness due to HLA alloimmunization, or non-HLA antibodies. This can also be prevented by filtration. If it occurs the patient may require HLA-matched platelets, but this should be discussed with the hematologist.
3. Transmission of infection.
4. After transfusion graft versus host disease following engraftment of donor lymphocytes in immunocompromised recipients. This can only be prevented by gamma irradiation (2500 cGy) of units prior to transfusion.

Fresh frozen plasma

Preparation and administration

Fresh frozen plasma contains all the non-cellular elements of whole blood. It is either separated from anticoagulated whole blood or obtained from plasmapheresis, and it is frozen to a core temperature of –30 °C or below within 6 h of collection. It remains stable if stored for up to 12 months. To transfuse, the plasma is thawed at 37 °C and should be infused within 2 h to avoid loss of potency of clotting factors. Many transfusion centers now supply fresh frozen plasma in several different pack volumes, but the dose given is 10 to 15 ml/kg initially and its effect should always be monitored with coagulation screens pre- and post-transfusion.

ABO-compatible fresh frozen plasma should be used for group A, B, and AB patients, but compatibility testing is not required. Group O patients may be given A or B plasma. Group O plasma, if available, may only be given to group O patients. It is advisable to give females of child-bearing age RhD-compatible fresh frozen plasma, but if RhD-positive fresh frozen plasma has to be given to an RhD-negative female, anti-D immunoglobulin should be given as cover.

A solvent-detergent-treated product manufactured from pooled plasma is widely available in many European countries, and 'standard fresh frozen plasma' is no longer licensed in some (Germany, France, Austria, Belgium). The manufacturing process is highly effective at viral inactivation and also appears to remove the higher-molecular-weight von Willebrand multimers, which seems to be of therapeutic benefit when used for plasma exchange in thrombotic thrombocytopenic purpura.

Indications for administration of fresh frozen plasma

These have been summarized as follows by the British Committee for Standards in Hematology.

(1) Definite indications:
    (a) replacement of factor deficiency where specific factor concentrate is not available (e.g. factors V, X);
    (b) immediate reversal of warfarin effect;
    (c) vitamin K deficiency;
    (d) acute DIC;
    (e) thrombotic thrombocytopenic purpura.

2. Conditional uses—fresh frozen plasma indicated if there is bleeding and disturbed coagulation (prothrombin time or activated partial thromboplastin time more than 1.5 times control value):
    (a) massive transfusion;
    (b) liver disease;
    (c) cardiopulmonary bypass surgery;
    (d) special pediatric indications.

There is no justification for the use of fresh frozen plasma in hypovolemia, plasma exchange procedures (except in thrombotic thrombocytopenic purpura), formula replacement, nutritional support, or treatment of immunodeficiency states.

Adverse effects

1. Allergic reactions: urticaria occurs in 1 to 3 per cent; anaphylaxis is very rare, particularly if flow rates are kept below 30 ml/min.
2. Transmission of infection.
3. Hemolysis if ABO-incompatible fresh frozen plasma is transfused and potent anti-A or anti-B causes lysis of red cells.
4. Fluid overload.
5. Transfusion-related lung injury (very rare). This results from potent antigranulocyte antibodies in donor plasma which cause leukocyte aggregation in pulmonary vessels and acute lung injury.

Cryoprecipitate

Preparation and administration

Cryoprecipitate is prepared from fresh frozen plasma by slow controlled thawing at 4 to 6 °C. The resulting precipitate is separated from the supernatant and refrozen for storage. Cryoprecipitate contains factor VIII, von Willebrand factor, fibrinogen, factor XIII, and fibronectin in higher concentrations than is found in fresh frozen plasma. For these reasons it used to be the mainstay of treatment for hemophilia A and von Willebrand's disease, but it has been superseded by virally inactivated pooled factor concentrates. The same guidelines as for administration of fresh frozen plasma apply for cryoprecipitate and it carries a similar risk of transmission of infection. The dose for one transfusion is 10 to 30 units depending on the circumstances.

Indication for administration of cryoprecipitate

1. Congenital hypofibrinogenemia.
2. DIC with hypofibrinogenemia.

Adverse effects

These are similar to those with fresh frozen plasma.

Granulocytes

Preparation and administration

Granulocytes survive for a very short time *in vitro* and therefore have to be prepared and transfused within 24 h of collection. An effective adult dose is 2×10^{10} to 3×10^{10} cells, and these are either obtained by leukopheresis or 10 to 15 buffy coats. This dose has to be transfused daily. They also contain some platelets and red cells and they should be ABO compatible and cross-matched with the recipient. Hydrocortisone 100 mg and chlorpheniramine 10 mg intravenously must be given before infusion to prevent an immediate febrile reaction.

Indications for the use of granulocyte transfusions

Since the introduction of specific growth factors, the indications are very few because, with the exception of specific rare indications, the hazards associated with granulocyte transfusions outweigh the possible benefits and their use has never been definitely proven in randomized clinical trials. When they are used, it is mainly in immunosuppressed neutropenic patients with uncontrolled bacterial or fungal sepsis.

Adverse effects

1. Alloimmunization: granulocytes provide a powerful antigenic stimulus resulting in the formation of HLA- and granulocyte-specific antibodies which cause platelet refractoriness and febrile transfusion reactions.
2. Transmission of infection, particularly cytomegalovirus and toxoplasmosis.
3. Severe pulmonary infiltration with pulmonary edema, white cell sequestration, and exacerbation of pre-existing pulmonary infection. This is particularly severe if the patient is being treated concurrently with amphotericin B.
4. Graft versus host disease is a rare possibility.

Albumin

Preparation and administration

Albumin accounts for 60 to 80 per cent of normal plasma colloid osmotic pressure and therefore human albumin solutions were originally manufactured from pooled plasma to be used in the resuscitation of patients with acute blood loss. They are prepared by fractionation, filtration, and heat treatment to inactivate any contaminating viruses. However, compared with the artificial colloids that are now available, they are expensive and therefore there are few indications for the use of albumin nowadays. It is supplied in strengths between 4.5 per cent and 20 per cent; the latter is used for edema in hypoproteinemic states.

Indications for albumin

1. Replacement of plasma proteins during therapeutic plasma exchange.
2. Replacement of plasma proteins in burn injuries covering more than 15 per cent surface area (practice varies).
3. Treatment of diuretic-resistant edema in hypoproteinemia.

Adverse effects

1. Pyrogenic or generalized reactions with fever, gastrointestinal disturbance, urticaria, and collapse.
2. Circulatory overload with 20 per cent preparations.
3. Isolated hypotension.
4. Transmission of infection (extremely rare compared with other products prepared from pooled plasma).

Factor concentrates

Preparation and administration

Factor concentrates are the treatment of choice in many inherited coagulation factor deficiencies. They are prepared from pooled donor plasma by fractionation ('intermediate-purity' products) or by chromatographic methods ('high-purity' products). All then undergo viricidal procedures such as pasteurization or solvent-detergent sterilization to reduce the risk of viral transmission to an absolute minimum. They are manufactured in lyophilized form and are reconstituted with small volumes of water for administration. Factor concentrates are currently available for most of the inherited factor deficiencies (fibrinogen, II, VII, VIII, IX, X, XI, XIII) and also for deficiencies of natural anticoagulants (protein C and antithrombin III concentrates), but treatment options should always be discussed with a hematologist as some concentrates may precipitate thrombosis. A recombinant form of factor VIII which avoids the need to give products from donor plasma is now available for hemophilia A but is very expensive.

Indications for use of factor concentrates

1. Inherited or acquired factor deficiencies or inhibitors or von Willebrand's disease.

2. Other congenital deficiencies for which a concentrate is available (antithrombin III, protein C, α_1-antitrypsin, CI esterase inhibitor).

3. Reversal of warfarin (a combination of prothrombin complex concentrate and factor VII concentrate can be used).

Adverse effects

1. Allergic reactions are rare with human-derived products, but are more common with porcine factor VIII which may be given for factor VIII inhibitors.

2. Transmission of infection.

3. Thrombosis can occur, particularly following the use of intermediate-purity factor IX and factor XI concentrates.

Immunoglobulin

Preparation and administration

Intravenous immunoglobulin

Immunoglobulin preparations have been available for therapeutic parenteral use for over four decades, although those suitable for intravenous use (IVIg) have only been manufactured for the last 10 years. These are therapeutic preparations of normal polyspecific immunoglobulin G (IgG) extracted from pooled donor plasma by cold ethanol fractionation. Pepsin treatment at low pH, polyethyleneglycol precipitation, or chromatographic methods can then be used to purify the IgG which is further stabilized to prevent the formation of IgG aggregates. In order to minimize the risk of viral transmission, IgG preparations also undergo a viral inactivation step. The antibody profile reflects the serological responses to endemic pathogens and immunization status of the donor population. They are expensive to produce and this is reflected in the cost. A 2-g/kg course can cost at least \$3000 (£2000) depending on the product.

The mechanism of immunomodulatory action is poorly understood in most conditions for which IVIg has been used, but it alters the immune system in a variety of ways, any or all which may be responsible for its efficacy. Its main effects are reticuloendothelial blockade, suppression of antibody synthesis, and via anti-idiotype antibodies.

Recommended doses for antibody replacements vary from low doses of 0.4 g/kg every 2 to 3 weeks to the high dose of 1 g/kg for 2 days for the treatment of cytomegalovirus infection following bone marrow transplantation. The high dose is used for immunomodulation, although historically in immune thrombocytic purpura the total of 2 g/kg is usually given as 0.4 g/kg for 5 days.

Intramuscular immunoglobulin

Intramuscular immunoglobulin (IMIg) is mainly used when a specific antibody is required to provide short-term protection against infection or as immunotherapy (e.g. in the prevention of Rh isoimmunization). Therefore it is produced as hyperimmune preparations fractionated from donors with high titers of the specific antibody.

Indications for use of intravenous immunoglobulin

Antibody replacement

1. Primary immunodeficiency: X-linked gammaglobulinemia, common variable immunodeficiency, Wiskott–Aldrich syndrome, etc.

2. Secondary antibody deficiency:

    (a) low-grade B-cell malignancies with hypogammaglobulinemia (chronic lymphocytic leukemia, myeloma (selected patients only));
    (b) bone marrow transplantation (prophylaxis and treatment of cytomegalovirus infection, prophylaxis of early and late bacterial infections, prophylaxis or modification of acute graft versus host disease);
    (c) possibly following heart and lung transplantation and for surgical patients at high risk of infection;
    (d) HIV infection in children.

Immune modulation

There are now many proven and possibly immunologically mediated diseases in which there is clinical evidence for a beneficial role of IVIg, but its use in randomized controlled trials has been proved in only a few (conditions marked with an asterisk in the following list).

1. Immune cytopenias: immune thrombocytopenic purpura (including HIV infection and pregnancy and post-transfusion purpura)*, autoimmune neutropenia, hemolytic anemia, parvovirus-associated aplasia, and other red cell aplasias.

2. Acquired factor VIII inhibitors*.

3. Neurological and muscular diseases: Guillain–Barré syndrome*, chronic inflammatory demyelinating polyneuropathy, myasthenia gravis, polymyositis, multiple sclerosis, and intractable childhood epilepsy.

4. Connective tissue diseases: dermatomyositis*, systemic lupus erythematosus, and juvenile rheumatoid arthritis.

5. Vasculitides: Kawasaki disease*, ANCA-positive systemic vasculitides, and primary antiphospholipid syndrome in pregnancy.

6. Chronic fatigue syndrome

Adverse effects

Adverse effects are few and usually mild enough to allow home treatment for patients on regular IVIg.

1. Immune-mediated reactions with vasomotor disturbance and muscle pains. Anaphylaxis is very rare, but can occur in IgA-deficient patients who have IgA antibodies.

2. Hemolysis occurs rarely owing to the presence of blood group antibodies in IVIg.

3. Virus transmission: this is a small but real risk, particularly in IVIg preparations that have not undergone an inactivation step,

and there have been outbreaks of hepatitis viruses in the recent past.

4. Renal function can be further compromised in patients with pre-existing renal impairment.

Indications for the use of intramuscular immunoglobulin

1. Normal IMIg is effective against hepatitis A and measles.
2. Hepatitis B IMIg.
3. Rabies IMIg.
4. Tetanus IMIg.
5. Varicella zoster IMIg.
6. Anti-RhD IMIg.

Bibliography

British Committee for Standards in Haematology, Working Party of the Blood Transfusion Task Force (1992). Guidelines for the use of fresh frozen plasma. *Transfusion Medicine*, 2, 57–63.

Machin, S.J., Kelsey, H., Seghatchian, M.J., Warwick, R., and Mackie, I.J. (1995). Platelet transfusion. *Thrombosis and Haemostasis*, 74, 246–52.

Pamphilion, D. (ed.) (1995). *Modern transfusion medicine*. CRC Press, Boca Raton, FL.

17.1.13 Blood substitutes

O. P. Habler and K. F. Messmer

Key messages

- The increasing number of transfusion recipients and the shortfall of blood donors calls for the development of artificial oxygen carriers as part of perioperative transfusion-sparing strategies.
- Hemoglobin-based oxygen carriers are suitable for 1:1 replacement of lost blood, but have shortcomings due to pharmacokinetic (short half-life) and pharmacodynamic (vasoactivity) properties as well as logistical (hemoglobin supply) problems.
- Perfluorocarbon emulsions transport oxygen in physically dissolved form. They have to be administered in low doses (risk of reticuloendothelial system overload), and intubation and hyperoxic ventilation of the recipient is required; therefore they are suitable mainly for elective surgical procedures allowing for preoperative acute normovolemic hemodilution.

Introduction

Artificial blood substitutes are designed to replace the oxygen-carrying capacity of lost red blood cells in the case of major bleeding (on the battlefield, major intraoperative blood loss). The declared aim is reduction of homologous blood transfusion with its associated costs and risks (virus transmission, immunosuppression, and hemolytic reaction).

To date, approximately 12 million units of red blood cells are transfused each year in the United States, half of them to patients aged 65 years or more. As the number in this age group increases and blood donation decreases, a shortfall of about 4 million units of red blood cells is projected for the year 2030 and an explosion in costs has to be taken into account (actual cost $52 to $64 per unit, expected cost above $200 per unit) (Dietz *et al.* 1996).

The risks of transmission of HIV and non-A non-B hepatitis are currently estimated as 1 in 225 000 transfusions and 1 in 3000 transfusions respectively. Although rare, once virus transmission and outbreak of the disease have occurred, the costs of treatment are enormous (AuBuchon and Birkmeyer 1994).

Major homologous blood transfusion acts as an immunodepressant and increases the incidence of infections, even sepsis syndrome, in the postoperative period. Prolonged stay in hospital and, in the worst case, the necessity for treatment in an intensive care unit represent an important cost factor for the health care system.

Therefore a safe and effective blood substitute could help to prevent a shortfall in the red blood cell supply as well as the increment in costs due to homologous blood transfusion. Several types of products are under preclinical and clinical investigation: solutions of free hemoglobin, liposome-encapsulated hemoglobin, and perfluorocarbons.

Hemoglobin solutions

If hemoglobin solutions were used to replace the oxygen-carrying capacity of 10 per cent of the red blood cell units transfused annually in the United States, 60 000 to 70 000 kg of hemoglobin would be required each year. Possible sources able to provide this amount of hemoglobin are human hemoglobin derived from outdated bank blood, bovine hemoglobin, and genetically engineered recombinant hemoglobin produced by micro-organisms (bacteria, yeasts) (Dietz *et al.* 1996).

The following problems must be overcome to create a safe and effective hemoglobin solution.

1. Intravascular half-life: when hemoglobin is freed from its intraerythrocytic environment, its tetrameric structure of two α chains and two β chains dissociates into α and β dimers or single α and β monomers. Both dimers and monomers are rapidly filtered by the kidneys, reducing their intravascular retention time.
2. Oxygen affinity: removal of hemoglobin from red blood cells is associated with loss of the regulatory function of 2,3-diphosphoglycerate. At low oxygen partial pressures ($P\text{O}_2$), oxygen liberation from hemoglobin is hindered. The P_{50} value ($P\text{O}_2$ at which 50 per cent of hemoglobin is saturated with oxygen) is lowered and the oxygen–hemoglobin dissociation curve is shifted to the left. Because of the higher oxygen affinity of free hemoglobin in the absence of 2,3-diphosphoglycerate, there is less diffusion of oxygen to the tissues, despite an adequate $P\text{O}_2$

gradient (arterial Po_2, 90–100 mmHg; tissue Po_2, 30–40 mmHg). This results in reduced availability of hemoglobin-bound oxygen at the tissue level.

3. Renal toxicity: renal damage occurs when filtered hemoglobin dimers or monomers precipitate in the ascending limb of the loop of Henle. Insufficient purification of the hemoglobin solution from red cell stromal debris may contribute to the renal damage.

4. Vasoconstriction: this phenomenon was observed in the first animal experiments dating from the early 1930s, when a solution of bovine hemoglobin in saline was used for exchange transfusion. Vasoconstriction and the consecutive rise in systemic vascular resistance, mean aortic pressure, and left ventricular oxygen requirements remain uncontrolled side-effects of hemoglobin solution. The most probable explanation of the immediate onset of vasoconstriction upon infusion of hemoglobin solutions is scavenging of nitric oxide (a potent endogenous vasodilator) by the hemoglobin molecule. Upregulation of endothelin and sympathoadrenergic receptors has also been suggested. Additionally, the hemoglobin molecule itself may possess vasoconstricting properties.

Great efforts have been made by pharmaceutical manufacturers to develop new hemoglobin preparations which do not have these problems. The intravascular persistence of hemoglobin has been successfully prolonged by chemical approaches. It is possible to cross-link, polymerize, or conjugate hemoglobin dimers or monomers, thus creating larger molecules with an intravascular half-life prolonged to 6 to 8 h. At the same time renal damage due to filtered monomers and dimers is avoided. Production of hemoglobin solutions based on bovine hemoglobin takes adventage of the fact that the P_{50} value of bovine hemoglobin is similar to that of human hemoglobin and is not controlled by 2,3-diphosphoglycerate but by ionic chloride, which is present in large concentrations in plasma. It is expected that genetic engineering will allow structural modifications of the hemoglobin molecule. The tertiary structure of the binding sites for the nitric oxide molecule might be changed and the vasoconstrictive effect of hemoglobin molecules modified in this way might be reduced. Liposome encapsulation of modified hemoglobin molecules imitates the membrane structure of the red blood cells, allowing incorporation of 2,3-diphosphoglycerate (physiological P_{50}) and prolongation of intravascular half-life.

Despite definite progress in producing effective and safe oxygen-carrying blood substitutes based on hemoglobin, serious limitations on the general use of such compounds still remain. The use of bovine hemoglobin may be seriously endangered by possible interspecies transmission of infectious diseases such as bovine spongiform encephalitis. With better organization of blood banks and a shrinking donor pool, fewer red blood cell units become outdated; hence this source of hemoglobin may become unrealistic. It is unclear at present whether recombinant DNA technology will allow cost-effective provision of the required amounts of hemoglobin and whether complete elimination of bacterial components from the hemoglobin will be achieved. Finally, the use of hemoglobin solutions for the treatment of a broad range of patients requires control of the vasoconstrictor effect of these compounds to satisfy concerns about safety.

Perfluorocarbons

Perfluorocarbons (Riess 1991) are simply constructed molecules derived from hydrocarbons by replacing all the hydrogen atoms with fluorine. Some of these compounds also contain a bromide atom to provide radio-opacity.

Perfluorocarbons are insoluble in water and therefore have to be emulsified. They are not metabolized in the organism but are rapidly (within a few hours) phagocytized by cells of the reticuloendothelial system. In a second slow phase (within a few weeks), they are redistributed to the blood, transported to the lungs, and finally exhaled.

The most important physicochemical feature of perfluorocarbons is that oxygen is highly soluble in them. The relationship between Po_2 and oxygen content in perfluorocarbon solutions is linear, rather than sigmoidal as is the case in blood. At high oxygen partial pressures, achieved by 100 per cent oxygen ventilation, the oxygen content is higher in whole blood than in perfluorocarbons; in the presence of a high Po_2 gradient between arterial blood and tissues, oxygen release to tissues is comparable for blood and perfluorocarbons. For example, 100 g of perfluoro-octyl bromide (perflubron), a second-generation perfluorocarbon, can deliver 15 ml of oxygen in the presence of an arterial Po_2 of 600 mmHg and an estimated tissue Po_2 of 38 mmHg (Po_2 gradient $\Delta Po_2 = 562$ mmHg). This amount of oxygen is approximately equivalent to the amount of oxygen delivered by one unit (450 ml) of blood with a hemoglobin concentration of 14 g/dl.

In theory, intravenously administered perfluorocarbon emulsion should effectively replace the oxygen transport function of red blood cells lost during anemia, thereby preserving tissue oxygenation and organ function. However, the first clinical trials using a 20 per cent weight per volume (w/v) emulsion (Fluosol-DA) have failed. Despite a high intravenous dose, Fluosol-DA 20 per cent was incapable of preserving adequate tissue oxygenation in severly anemic surgical patients who had refused homologous blood transfusion. Six of eight patients died (Gould *et al.* 1986).

One reason for the lack of efficacy of Fluosol-DA 20 per cent is shown schematically in Fig. 1. The oxygen transport capacity of a perfluorocarbon emulsion depends on the concentration of the perfluorocarbon compound in the emulsion and is reflected by the slope of the corresponding straight line. The higher the slope, the more oxygen is dissolved for the same increase in Po_2 and the higher is the oxygen transport capacity of the emulsion. As Fig. 1 indicates, the oxygen transport capacity of Fluosol-DA 20 per cent is low because of its low perfluorocarbon concentration. This is the main reason for the failure of the emulsion in clinical studies.

Production of perfluorocarbon emulsions at higher concentrations was not feasible for many years because of technical difficulties. However, new preparations containing up to 90 per cent w/v perfluorocarbon (perflubron) are being tested experimentally and clinically, and the initial results appear promising (Wahr *et al.* 1996). Low dosages of perflubron do not seem to cause serious side-effects. Animal experiments have shown that the oxygen transport capacity is high enough to preserve tissue oxygenation and organ function even at very low hematocrits (8 per cent).

Conclusions

Artificial blood substitutes which can replace the oxygen transport capacity of lost red blood cells in anemic patients are being investi-

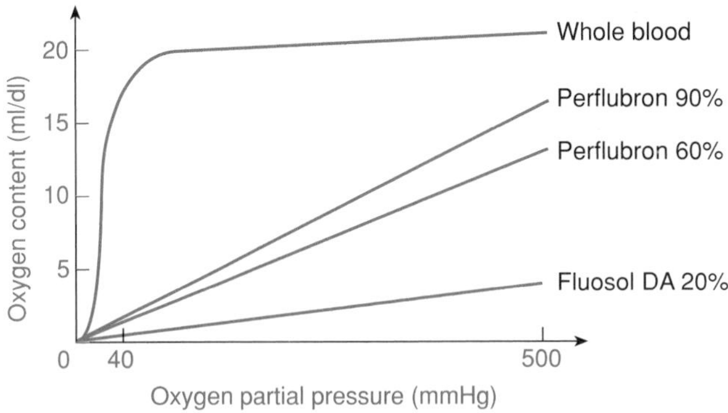

Fig. 1 Relationship between Po_2 and oxygen content in whole blood (sigmoidal relationship) and perfluorocarbon emulsions (linear relationship). The oxygen transport capacity of the perfluorocarbon emulsions (slopes of the straight lines) increases with increasing perfluorocarbon concentration.

gated with the aim of reducing the costs and risks of homologous transfusion. Free hemoglobin solutions increase the hemoglobin-bound fraction of the arterial oxygen content, and perfluorocarbons increase the physically dissolved fraction. Because of the special properties of these two types of oxygen-carrying blood substitute (short intravascular half-life, requirement of 100 per cent oxygen ventilation in the case of perfluorocarbons), they are presently being evaluated in controlled clinical studies of intentional intraoperative dilutional anemia.

References

AuBuchon, J.P. and Birkmeyer, J.D. (1994). Controversies in transfusion medicine. *Transfusion*, **34**, 79–83.

Dietz, N.M., Joyner, M.J., and Warner, M.A. (1996). Blood substitutes: fluids, drugs, or miracle solutions? *Anesthesia and Analgesia*, **82**, 390–405.

Gould, S.A., *et al.* (1986). Fluosol-DA as a red-cell substitute in acute anemia. *New England Journal of Medicine*, **314**, 1653–6.

Riess, J.G. (1991). Fluorocarbon-based *in vivo* oxygen transport and delivery systems. *Vox Sanguinis*, **61**, 225–39.

Wahr, J.A., *et al.* (1996). A pilot study of the effects of a perflubron emulsion, AF 0104, on mixed venous oxygen tension in anesthetized surgical patients. *Anesthesia and Analgesia*, **82**, 103–7.

17.1.14 Bronchodilators

S. W. Benham and C. S. Garrard

Key messages

- There is a selection of β-adrenergic agonists commonly used as bronchodilators that differ in their length of action and β-receptor selectivity.
- The preferred method of administering β-adrenergic agonists is the inhaled route. The intravenous route is reserved for those with a suboptimal response to the inhaled route. The most common cause of suboptimal response to therapy is inadequate dosing.
- Anticholinergic drugs are best employed with β-adrenergic agonists for acute bronchospasm, and are particularly effective in patients with chronic obstructive airways disease.
- Theophylline has a synergistic action with β-adrenergic agonists, and can be delivered intravenously as aminophylline.
- Inhalational anesthetic agents (halothane, enflurane, isoflurane, sevoflurane) have demonstrable bronchodilating activity. Halothane is the most potent, but has more side-effects in the presence of high levels of background sympathomimetic activity.
- Ketamine has been used as an intravenous infusion with good effect in a small number of case reports.

Introduction

Bronchodilators find applications in the treatment of a wide range of disorders characterized by airways obstruction. These include bronchial asthma, chronic obstructive pulmonary disease, and cystic fibrosis. Bronchospasm associated with anaphylactic reactions and cardiogenic pulmonary edema may also benefit from bronchodilators. Drugs with bronchodilator activity are listed in Table 1. Alternative therapies for patients unresponsive to standard therapy are also considered, although they may not be of proven efficacy.

Beta-adrenergic agonists

Several β-adrenergic agonists, which can be administered as inhaled, oral, subcutaneous, and intravenous preparations, are available for use as bronchodilators. These agents act via membrane receptors to stimulate adenyl cyclase, which converts ATP to cyclic AMP. This intracellular messenger is required to mediate relaxation of bronchial smooth muscle. Albuterol (salbutamol) and terbutaline are the most common selective β_2-adrenergic agonists in use; rimiterol has a shorter duration of action and salmeterol a longer duration of action than albuterol. Fenoterol may have less β_2 selectivity than the other drugs. Two related issues should be addressed in the context of bronchodilators in critically ill patients—the methods of administration and the dosage.

There is uncertainty over the amount of drug that reaches the distal airways of patients who are mechanically ventilated when delivery is by metered-dose inhaler. One study demonstrated no physiological response from as many as 100 puffs from a metered-dose inhaler (Manthous *et al.* 1993). More drug reaches the lung when a spacer device is used with the inhaler, or if the drug is delivered by a nebulizer. The inhaled route of administration has the best benefit versus side-effect profile. Intravenous administration should be reserved for the young asthmatic when tidal volume is critically small, when airway resistance is very high, and where the conventional inhaled route has proved unsuccessful (Corbridge and Hall 1995).

Drug dosage should be titrated against response, with due vigilance to detect side-effects. Peak-to-plateau pressure gradient is a useful indicator of bronchodilating effect in patients who are paralyzed and ventilated and delivered a constant tidal volume (Tobin 1988). Mechanically ventilated patients may require much higher doses of

Table 1 Bronchodilators

Drug	Dose	Preferred route	Special points
Beta-adrenergic agonists Albuterol	0.5 ml of 0.5% solution (2.5 mg) diluted with 2.5 ml normal saline via nebulizer initially given every hour, then every 2–4 h or Four puffs (90–100 μg per puff) via spacer device given every 20 min, then every 2–4 h or 3 μg/kg as slow IV injection, followed by an infusion 3–20 μg/min	Nebulized or MDI (with spacer device)	Initial dosing is well above the normal range. Maximum of 25 mg over 2–3 h. Dose should be titrated against response. Presence of side-effects without satisfactory clinical response should prompt the introduction of other drugs
Parasympatholytics Ipratropium bromide	2 ml of 0.025% (500 μg) nebulized solution, every hour for first 6 h, then every 4–6 h or Four to ten puffs of MDI with spacer device every 20 min over first hour, then given every 4–6 h	Nebulized or MDI with spacer device. Not licensed for IV administration	
Methylated xanthine derivatives Theophylline Aminophylline	5 mg/kg as a loading dose in patients not receiving oral theophylline preparations. IV maintenance dose of 0.4 mg/kg. If patient on theophylline, serum level should be obtained, and 1 mg/kg will produce approximately 2 μg/ml rise in serum concentration, if serum level subtherapeutic	IV route most reliable in the acute phase	Target serum concentration 10–20 μg/ml. Side-effects can occur within this range
Corticosteroids Hydrocortisone Prednisolone Beclomethasone	Acute asthma attack: hydrocortisone 200 mg every 6 h IV for 48 h or until oral therapy established; prednisolone 40–60 mg every 6–8 h, reducing to every 24 h after 48 h. Chronic asthma: beclomethasone 200–400 mg twice daily via MDI	IV, oral, inhaled	Steroid dose can be reduced once clinically improved but should be continued for a minimum of 7 days
Inhalational agents Halothane Isoflurane Enflurane Sevoflurane	Halothane 0.25%–1.25% delivered by vaporizer if patient awake. If patient already sedated isoflurane 0.25%–1.25% may be an alternative. Ventilator must be equipped to be able to deliver accurate concentrations of anesthetic vapors	Inhaled	Appropriate monitoring (ECG, blood pressure, inspired vapor concentration) required. With prolonged use, measure halothane metabolites
Ketamine	Initial dose 1–2 mg/kg IV, followed by infusion 10–40 μg/k/min. IM initial dose 5–10 mg/kg	IV, IM	This should not be given without facilities for assisted ventilation being available. Monitoring appropriate for any anesthetic should be provided
Magnesium sulfate	1 g (4 mmol) IV over 15–20 min, followed by 1–2 g (4–8 mmol) infusion every 4 h	IV	Drowsiness and loss of patellar reflex may indicate hypermagnesemia

IV, intravenous; MDI, metered-dose inhaler; IM, intramuscular.

inhaled or nebulized bronchodilators than spontaneously ventilating patients. With careful cardiovascular monitoring, nebulized albuterol can be administered continuously to total doses approaching 25 mg over 2 to 3 h. The most common cause of failure to respond to medical therapy in acute asthma is inadequate dosing. Systemic side-effects should be minimized using selective β_2 agonists given, whenever possible, via the inhaled route. Dose regimens for β-adrenergic agonists are shown in Table 1.

Table 2 Equivalent anti-inflammatory doses of glucocorticoids

Drug	Equivalent anti-inflammatory dose (mg)
Cortisone acetate	25
Hydrocortisone	20
Prednisolone	5
Prednisone	5
Methylprednisolone	4
Triamcinolone	4
Dexamethasone	0.75
Betamethasone	0.75

This table takes no account of mineralocorticoid effects or of variations in duration of action.

The onset of action of albuterol occurs 5 to 15 min after administration, with the peak effect between 30 min and 2 h. Action persists for 3 to 6 h. Side-effects include tremor, insomnia, palpitations, flushing, headaches, and tachyarrhythmias. Relative contraindications are ischemic heart disease, untreated thyrotoxicosis, hypertension, and pregnancy. In such circumstances the relative benefit versus risk has to be evaluated.

Anticholinergics

Anticholinergics produce bronchodilatation by their vagolytic effects. They work by competitively inhibiting acetylcholine at the muscarinic cholinergic receptor. Acetylcholine stimulates the enzyme guanyl cyclase, which is responsible for converting guanosine triphosphate to cyclic GMP which mediates bronchoconstriction. In asthma and chronic obstructive pulmonary disease, improvements in forced expiratory volume in 1 s and hospital admission rates have been demonstrated with a combination of an anticholinergic and a β-agonist compared with using β-agonists alone. Certain patient subgroups seem to benefit, including those with chronic obstructive airways disease, those with severe airflow obstruction, or those receiving β-blockers. Inhaled ipratropium bromide is the agent of choice as its quaternary ammonium structure minimizes its systemic absorption and thus its side-effects.

The onset of action is 15 to 30 min after inhalation, with the peak effect at 30 to 60 min. Action persists for 4 to 6 h. There are almost no side-effects when the inhaled route is used, although higher doses may produce dry mouth, urinary retention, and constipation. Cautions include glaucoma, prostatic hypertrophy, and pregnancy.

Methylated xanthines (theophylline and aminophylline)

Theophylline is inferior to β-agonists as first-line treatment of acute asthmatic attacks, although it may have a synergistic action with β-agonists. It can be combined with ethylenediamine (aminophylline) to become 20 times more soluble than theophylline alone.

The mode of action of theophylline is unclear, but it is known to inhibit the enzyme phosphodiesterase to decrease the metabolism of cyclic AMP. The dose necessary for this action is much higher than that used clinically. Theophylline has a direct relaxant effect on bronchial smooth muscle, a mild inotropic effect, and a diuretic effect, and it delays the onset of diaphragmatic muscle fatigue. It is available intravenously (aminophylline) and orally (theophylline) in preparations with different lengths of action.

The onset of action by the intravenous route is slower than that of the β-agonists because of the need to deliver it over 20 to 30 min. Side-effects include nausea and vomiting, central nervous system irritability and seizures, flushing, hypotension, and a variety of tachyarrhythmias including cardiac arrest. Seizures due to theophylline toxicity are often difficult to treat. The ethlyenediamine component of aminophylline can cause urticaria, erythema, and exfoliative dermatitis. Cautions include liver disease, epilepsy, cardiac disease, and pregnancy.

Therapeutic levels of theophylline are quoted as 10 to 20 μg/ml, and these should be monitored in the critically ill. Rarely, serious side-effects do still occur in this range, and levels should be monitored in those who have previously received theophylline preparations or who are given a maintenance infusion.

Corticosteroids

Corticosteroids are not primarily bronchodilators but reduce airway wall inflammation, decrease mucus production, and have a synergistic effect with β-adrenergic agonists on bronchial smooth muscle relaxation. They are usually prescribed in severe asthma (British Thoracic Society 1990). In some patients with chronic obstructive airways disease, corticosteroids given by either the oral or the inhaled route can hasten an improvement in pulmonary function tests (Hudson and Monti 1990). A reluctance to use steroids or failure to prescribe adequate amounts in acute asthma results in increased hospital admissions, prolonged hospital stay, and increased mortality. The route of administration seems less important than ensuring adequate dosing, and due attention must be paid when changing from intravenous to oral preparations so that equivalent doses are prescribed (Table 2). The advantage of intravenous administration in the acute stage is that the problems of vomiting and delayed absorption are bypassed and the amount of active drug administered is known to have reached the systemic circulation. Inhaled corticosteroids are now accepted as long-term treatment for asthmatics; the inhaled route is inadequate for an acute attack of bronchospasm.

The onset of action depends on the route of administration and the type of steroid, but the peak effect in the asthmatic patient is likely to be 3 to 6 h after administration. Treatment should be continued for at least 7 days to suppress the inflammatory response. Side-effects include hyperglycemia, water retention, euphoria or depression, peptic ulceration, myopathy, increased susceptibility to infection, and adrenal suppression.

Alternative therapeutic agents

Before introducing the following agents the clinician must be convinced that initial therapy has been optimal, with particular emphasis on ensuring that adequate β-adrenergic agonist and corticosteroid has been given.

Anesthetic inhalation agents

Halothane, enflurane, sevoflurane, and isoflurane are anesthetic agents that have bronchodilating properties. Sevoflurane is the least irritating to the airways in awake patients, but its bronchodilating properties are weaker than those of the other agents. Halothane is the next least irritating to inhale and has the most potent bronchodilating properties of all the agents. It has been used from subanesthetic concentrations (0.25 per cent) to anesthetic concentrations (1.25 per cent), and has been beneficial in this group of patients who are unresponsive to conventional therapy.

Halothane can cause serious ventricular arrhythmias, particularly in the presence of high levels of circulating catecholamines. Side-effects include hypotension, bradycardia, and rarely hepatitis. If halothane is used for prolonged periods, accumulation of its metabolites can occur and this should be monitored.

It is imperative that inhalational anesthetics are administered by a clinician who has experience of using these agents and that appropriate cardiorespiratory support and resuscitation facilities are available.

Ketamine

This intravenous or intramuscular anesthetic agent has bronchodilating properties that do not appear to act via the β-adrenergic receptor. There is evidence from animal studies and from case reports in humans that it is a useful bronchodilator. The adult dose ranges from 1 to 2 mg/kg initially, followed by an infusion of 10 to 40 μg/kg/min.

Side-effects include tachycardias, hypertension, hallucinations, and raised intracranial pressure. As with the inhalational agents, this is not a drug for the inexperienced clinician but should be administered by those experienced in its use.

Magnesium sulfate

Both the mechanism of action of magnesium sulfate and conclusive evidence of its efficacy in the severe asthmatic remain unproven in randomized studies. It may benefit subgroups of patients such as women or those being mechanically ventilated. The initial dose is 1 g intravenously over 15 to 20 min, followed by an infusion of 1 to 2 g over 4 h. Side-effects include flushing and hypotension.

Heliox

Although not a bronchodilator, this mixture of helium and oxygen (60 per cent helium and 40 per cent oxygen, or 80 per cent helium and 20 per cent oxygen) has been given to patients with severe bronchospasm who have been unresponsive to conventional therapy. The low density of helium improves the gas flow characteristics in narrowed airways. A reduction in airways pressure in mechanically ventilated patients and an improvement in arterial pH and $Pa\text{CO}_2$ have been demonstrated, but there is not a consistent or predictable response in all patients. Further evaluation is required to identify the patients who would benefit from heliox, and the dose and timing of its use in patients with bronchospasm.

References

British Thoracic Society (1990). Guidelines for management of asthma in adults: II—acute severe asthma. *British Medical Journal*, **301**, 797–800.

Corbridge, T.C. and Hall, J.B. (1995). The assessment and management of adults with status asthmaticus. *American Journal of Respiratory and Critical Care Medicine*, **151**, 1296–1316.

Hudson, L.D. and Monti, C.M. (1990). Rationale and use of corticosteroids in chronic obstructive pulmonary disease. *Medical Clinics of North America*, **74**, 661–90.

Manthous, C.A., Hall, J.B., Schmidt, G.A., and Wood, L.D.H. (1993). Metered-dose inhaler versus nebulised albuterol in mechanically ventilated patients. *American Review of Respiratory Diseases*, **148**, 1567–70.

Tobin, M.J. (1988). Respiratory monitoring in the intensive care unit. *American Review of Respiratory Diseases*, **138**, 1625–42.

17.1.15 Diuretics

J.-A. Romand

Key messages

- Loop diuretics (furosemide (frusemide) and bumetanide) are the most powerful natriuretic drugs. They increase sodium excretion and urine output even in patients with greatly reduced glomerular filtration rate.
- Loop diuretics are prescribed to prevent oliguric acute renal failure or to render it non-oliguric.
- Osmotic diuretics (mannitol) are pharmacologically inert drugs freely filtered in the glomerulus but poorly reabsorbed by the tubules. The mechanism of action resides in the osmotic activity that mannitol produces along the tubular segments, counteracting the normal reabsorption of sodium and water.
- Osmotic diuretics are used as prophylactic therapy when nephron toxicity is anticipated, for example when diagnostic studies or therapeutic intervention require intravenous contrast agents, and during vascular surgery or when increased intracranial pressure is diagnosed.
- Because of its ability to increase renal blood flow selectively through the activation of dopaminergic renal receptors, dopamine has gained in popularity for the prevention and treatment of acute renal failure. Dogma, rather than scientific proof, motivates the use of renal-dose dopamine.

Loop diuretics

Pharmacology

Loop diuretics (furosemide (frusemide) and bumetanide) are the most powerful natriuretic drugs. They increase sodium excretion and urine output even in patients with greatly reduced glomerular filtration rate. The principal site of action of this class of diuretics is the thick ascending limb of the loop of Henle after secretion in the proximal tubules, hence the term 'loop diuretics'. A reversible inhibition of the salt reabsorption transporter is induced when loop diuretics are present in the tubular lumen of the nephrons. The inhibition of Na^+ and Cl^- reabsorption increases the load of the distal tubules, eventually inducing natriuresis and secondary diuresis.

Thus delivery of the diuretic to the tubular lumen is essential for these pharmacological effects (Wittner *et al.* 1991). Following active secretion in the proximal tubule, loop diuretics gain access to this site along the non-specific transport route for organic acid. Calciuria is also induced.

Therapeutic use

The rationale for the use of loop diuretics in acute renal failure is as follows. The inhibition of tubular reabsorption in the thick ascending limb of the loop of Henle produces a cellular resting state and a decrease of oxygen consumption, thereby improving cell tolerance to hypoxia. A protective effect also occurs at the mitochondrial level and prevents the influx of calcium. Furthermore, following inhibition of salt reabsorption at the macula densa level, glomerular vasodilatation occurs secondary to a fall in afferent arteriolar resistance caused by inhibition of vasoconstrictive feedback of the tubular glomeruli. Another possibility is that the hydraulic pressure generated by the induced diuresis leads to dislodgement of obstructive tubular casts in the damaged nephrons. Finally, loop diuretics also increase the secretion of renal vasodilatory prostaglandins, thus counteracting the renal vascular constrictive feedback mechanism. Although loop diuretics are potent drugs, resistance may develop. The etiological factors responsible for the decreased efficacy of natriuretic medication are numerous and represent a mechanism of adaptation to chronic overload of Na^+ at the distal part of the nephron. When this situation occurs, an association with other diuretics acting principally on another segment of the nephron is recommended; for example, metolazone is often used to circumvent this resistance (Channer *et al.* 1994).

Loop diuretics are prescribed to prevent oliguric acute renal failure or to render it non-oliguric. The rationale is that non-oliguric acute renal failure is easier to manage and may have a better prognosis than oliguric acute renal failure. Even if easier management is widely accepted, alteration of the course of acute renal failure with diuretics still needs to be proved. Indeed, spontaneously non-oliguric patients and those converting from oliguric to non-oliguric acute renal failure probably have less severe damage than oliguric patients, and thus a different outcome is expected. Other loop diuretics (e.g. ethacrynic acid, piretanide) have similar pharmacological effects, but are used less or not at all in the critically ill.

Dosage and route of use

Loop diuretics are administered parenterally in critically ill adult patients, and an initial intravenous dosage of 20 mg of furosemide is recommended. If no response is seen after 1 h, an incremental regimen is prescribed which consists of doubling the previous dose every 2 h until satisfactory response is obtained. Alternatively, a daily maximum dose of 2 g of furosemide (10 mg of bumetamide) is given.

If diuresis does not occur after an intravenous bolus of loop diuretics, a continuous infusion can be used. The rationale, based on a number of studies, is that a controlled infusion of furosemide or bumetamide results in greater natriuresis and diuresis than an equivalent dosage administered intermittently. A better efficacy at the diuretic receptor level may explain this improvement. The starting dose is around 2 mg/h of furosemide, titrated every hour until the response is satisfactory, with a maximum of 40 mg/h (Martin and Danziger 1994). The advantages of initiating therapy with a loading dose still need to be determined. A continuous-infusion strategy may also be used when a normal response to bolus injection is seen, when less hemodynamic variation is important, or when a lower total dose of diuretic is essential.

Side-effects

Loop diuretics can induce azotemia when blood pressure decreases following a vigorous diuresis. Furthermore, electrolyte disturbances (hyponatremia, hypokalemia, hypomagnesemia, and metabolic alkalosis) may be seen. Ototoxicity is known to occur when high doses of loop diuretics are given. Finally, gastrointestinal and allergic reactions have been described. Special care must be taken in situations where renal elimination of furosemide is not possible. Indeed, hepatotoxicity related to hepatotoxic metabolism has been reported in experimental studies. However, such adverse effects have not been diagnosed in humans.

Pharmacodynamics in the critically ill

Furosemide is a weak water-soluble acid. Fifty per cent is eliminated unchanged by the kidneys and the remainder is conjugated and excreted as glucoronide; 91 to 99 per cent of furosemide is bound to albumin for which it competes with other acidic drugs. The half-lives of both furosemide and bumetamide are short (1–2 h) and the duration of action is between 3 and 6 h.

Osmotic diuretics

Pharmacology

Osmotic diuretics (mannitol) are pharmacologically inert drugs freely filtered in the glomerulus but poorly reabsorbed by the tubules. The mechanism of action resides in the osmotic activity that mannitol produces along the tubular segments, counteracting the normal reabsorption of sodium and water. Three mechanisms are described. First, the effective intraluminal Na^+ is lower secondary to the increased water in the tubular lumen producing a decrease in the reabsorption force (gradient) for the Na^+ ion. Second, there is an increased flux of Na^+ from the peritubular (extraluminal) fluid back to the tubular lumen; in summary, extraluminal Na^+ is greater than luminal Na^+. Finally, mannitol increases the intravascular volume, decreases the viscosity of the blood, thus increasing the medullary blood flow, and decreases the medullary solute gradient (washout of the normal medullary hypertonicity), preventing concentration of the urine. Another effect lies in the scavenging property of mannitol for oxygen radicals, which may be a beneficial therapeutic tool in the case of reperfusion injuries.

Therapeutic use

Osmotic diuretics are used as prophylactic therapy when nephron toxicity is anticipated, for example when diagnostic studies or therapeutic intervention require intravenous contrast agents, and during vascular surgery or when increased intracranial pressure is diagnosed. The rationale for administering mannitol in the first clinical situation is to decrease the concentration capacity of the urine which inhibits the concentration of the contrast agent within the tubular lumen, thus preventing its toxicity. However, when the nephron becomes impermeable to mannitol, it becomes ineffective.

In vascular (aortic) surgery, the superior efficacy of mannitol over methods which maintain a hyperhydration state is controversial

(Puschett 1994). Furthermore, when aortic cross-clamping takes place above the renal arteries, no preventive effects have been documented.

In cases of increased intracranial pressure, migration of the fluid from the intracellular to the extracellular space allows shrinkage of the brain, which ultimately decreases the swelling of the brain and relieves intracranial hypertension. Intraocular pressure is reduced in a similar way.

Dosage and route of use

Mannitol is available for intravenous use in concentrations of 5 to 25 per cent in volumes ranging from 50 to 1000 ml of water. The adult dose for inducing diuresis ranges from 50 to 200 g adjusted to a predefined urine output level. When euvolemia is necessary, compensatory intravenous fluid must be given. When rapid response is desired, as in intracranial hypertension, the medication is infused over 15 to 30 min at a dose of 1 to 2 g/kg.

Side-effects

As mannitol distributes relatively rapidly into the extracellular space, the administration of hypertonic solution will inevitably be accompanied by a mobilization of fluid from the intracellular space with resulting hyponatremia, hyperkalemia, and metabolic acidosis. These adverse events mainly occur in patients who are unable to increase their urine output. Pulmonary edema may follow because the heart is unable to cope with the increased intravascular volume. Finally, toxic vacuolization of epithelial cells of the late tubule and collecting system has been described following mannitol.

Pharmacodynamics in the critically ill

Diuresis usually begins 30 min after infusion of osmotic diuretics and urine output remains increased for 2 to 3 h. Plasma clearance is normally 30 min to 1 h depending on the glomerular filtration rate.

Renal-dose dopamine (1–3 μg/kg/min)

Because of its ability to increase renal blood flow selectively through the activation of dopaminergic renal receptors, dopamine has gained in popularity for the prevention and treatment of acute renal failure although a definite demonstration of its efficacy has never been obtained. Dopamine has a range of dose responses which have been attributed to activation of different types of endothelial receptor (dopaminergic receptors, β- and α-adrenergic receptors). Dogma, rather than scientific proof, motivates the use of renal-dose dopamine. Indeed, it is known that, even at low doses, patients may have β-adrenergic receptor activation with diuresis secondary to the increase in cardiac output. Many questions concerning the use of renal doses of dopamine remain unanswered. Indeed, what is the clinical relevance of increasing cardiac output? The usefulness of this therapy in acute renal failure still needs to be demonstrated. For all these reasons, in recent years routine renal-dose dopamine administration has been challenged and further carefully designed studies have been called for. In the interval, the possible benefits or advantages of administration of dopamine for renal protection or improvement of diuresis must be weighed against potential adverse effects (Segal *et al.* 1992).

References

Channer, K.S., McLean, K.A., Lawson-Matthew, P., and Richardson, M. (1994). Combination diuretic treatment in severe heart failure: a randomised controlled trial. *British Heart Journal*, **71**, 146–50.

Martin, S.J. and Danziger, L.H. (1994). Continuous infusion of loop diuretics in the critically ill: a review of the literature. *Critical Care Medicine*, **22**, 1323–9.

Puschett, J.B. (1994). Pharmacological classification and renal actions of diuretics. *Cardiology*, **84** (Supplement), 4–13.

Segal, J.M., Phang, P.T., and Walley, K.R. (1992). Low-dose dopamine hastens onset of gut ischemia in a porcine model of hemorrhagic shock. *Journal of Applied Physiology*, **73**, 1159–64.

Wittner, M., Di Stefano, A., Wangemann, P., and Greger, R. (1991). How do loop diuretics act? *Drugs*, **41** (Supplement 3), 1–13.

17.1.16 Fluids

A. R. Webb

Key messages

- The pharmacology of crystalloid solutions depends primarily on their salt content and osmolality.
- Approximately 20 to 25 per cent of infused 0.9 per cent saline will remain intravascular, whereas approximately 10 per cent of infused 5 per cent glucose will remain intravascular.
- Crystalloid fluids are used to provide the daily requirements of water and electrolytes.
- Smaller volumes of colloid are required for resuscitation with less contribution to edema.
- The volume required for correction of hypovolemia is determined by the hemodynamic effect of the fluid given.

Introduction

Intravenous fluids used for volume replenishment in critical care practice encompass the range of crystalloids and colloids.

Crystalloids include:

(1) saline (e.g. 0.9 per cent saline, Ringer's lactate (Hartmann's solution), and 0.18 per cent saline in 4 per cent glucose);

(2) glucose (e.g. 5, 10, and 20 per cent glucose).

Colloids include:

(1) albumin (e.g. 4.5 to 5 per cent and 20 to 25 per cent human albumin solution);

(2) dextran (e.g. 6 per cent dextran 70);

(3) gelatin (e.g. 3.5 per cent polygeline (Hemaccel), 4 per cent succinylated gelatin (Gelofusin));

(4) hydroxyethyl starch (e.g. 6 per cent hetastarch (eloHAES, Hespan), 6 and 10 per cent pentastarch (Pentaspan, HAES-steril)).

Table 1 Ion content, pH, and osmolality of crystalloids

	Na^+ (mmol/l)	K^+ (mmol/l)	HCO_3^- (mmol/l)	Cl^- (mmol/l)	Ca^{2+} (mmol/l)	pH	Osmolality (mosmol/kg)
Saline 0.9%	150			150		5.0	300
Ringer's lactate	131	5	29*	111	2	6.5	280
Saline 0.18% in glucose 4%	31			31		4.0	300

* HCO_3^- provided as lactate.

Pharmacology

Crystalloids

The pharmacology of crystalloid solutions depends primarily on their salt content and osmolality (Table 1). The sodium content of 0.9 per cent saline is equivalent to that of extracellular fluid with which it is isotonic. It follows that infusion of isotonic saline solutions (0.9 per cent saline, Hartmann's solution) will distribute within both the intravascular and interstitial spaces in proportion to their respective volumes. Approximately 20 to 25 per cent of infused 0.9 per cent saline will remain intravascular.

Glucose is provided in non-salt-containing solutions to maintain isotonicity only. Thus 5 per cent glucose is a method of providing intravenous water and will distribute throughout the fluid-containing spaces in proportion to their respective volumes. Only approximately 10 per cent of infused 5 per cent glucose will remain intravascular.

Colloids

A detailed discussion of colloids is given by Mishler (1984) and Baron (1992).

Albumin

Albumin is the fraction of plasma which provides the major part of the circulation's colloid osmotic pressure, and therefore it has been used as a plasma substitute. Although it is naturally occurring, there is a certain amount of heterogeneity in circulating albumin. Albumin for infusion is virtually free from disease transmission since it is heated to 60 °C for approximately 10 h following fractionation. This fractionation process contributes further heterogeneity since the final product contains some dimers and longer-chain polymers. Storage of albumin solutions may produce further heterogeneity through the formation of unstable polymers which may affect *in vivo* characteristics.

Albumin is the main provider of colloid osmotic pressure in the plasma and has a number of other functions:

(1) transport of various molecules;

(2) free-radical scavenging;

(3) binding of toxins;

(4) inhibition of platelet aggregation.

Human albumin solutions have been used successfully as plasma substitutes, and in view of their natural occurrence are considered by many as the gold standard with which synthetic plasma substitutes are compared. The major limitations to the use of human albumin solutions are their high production costs and limited supplies.

Gelatin solutions

Gelatin is a degradation product of animal collagen and therefore is inexpensive and readily available. Gelatin polypeptides are chemically modified to reduce the gel melting point while retaining sufficient molecular size for intravascular retention. In the manufacture of urea-bridged gelatin (polygeline) polypeptides of molecular weight 12 000 to 15 000 Da are formed by thermal degradation of cattle bone gelatin and subsequently cross-linked by hexamethyl di-isocyanate. In the manufacture of succinylated gelatin polypeptides of molecular weight approximately 23 000 Da are produced by thermal degradation of calf skin collagen. These polypeptides are reacted with succinic acid anhydride to replace amino groups with acid carboxyl groups. No cross-links are formed, but the increased net negative charge on the molecule produces a conformational change to open coils. Although little increase in molecular weight is produced by this reaction, molecular size is increased allowing better intravascular retention.

Dextran solutions

Dextrans are high-molecular-weight polysaccharides. They are natural substances produced by the action of the enzyme dextran sucrase during the growth of various strains of the bacteria *Leuconostoc* in media containing sucrose. After partial hydrolysis of raw dextran (molecular weight between 10^7 and 10^8 Da), the resulting hydrolysate is fractionated to produce dextran molecules of average molecular weight 75 000 Da. Since 1953, *Leuconostoc mesenteroides B512* has been the strain used for the manufacture of clinical dextrans. Other strains had been shown to produce dextran molecules with greater degrees of branching, and these molecules were associated with more immunological reactions. Over 90 per cent of the branches in dextran molecules produced by *L. mesenteroides B512* are α1–6 glucosidic bonds, giving relatively few side-chains. The original clinical dextran has undergone changes in the fractionation procedure such that the average molecular weight is now 70 000 Da (dextran 70).

Hydroxyethyl starch solutions

Unmodified starch is unsuitable as a plasma substitute since it is broken down rapidly by amylase. The hydroxyethylation of starch protects the polymer against breakdown by amylase. Waxy starches consisting of 98 per cent amylopectin are used in the manufacture of hydroxyethyl starch. Amylopectin, which is a highly branched polysaccharide resembling natural glycogen, is suspended in water and dilute hydrochloric acid and hydrolysis is allowed to continue until the viscosity of the solution approximates that of dextran 75. The solution is neutralized and treated with ethylene oxide in the presence of sodium hydroxide as a catalyst. Thus glucose units are substituted with hydroxyethyl groups at positions C2, C3, and C6 (Mishler 1984). The characteristics of hydroxyethyl starch solutions are dependent on the range of molecular weights, the degree of substitution of glucose units by hydroxyethyl groups, and the ratio of C2 to C6 substitution. In general, higher molecular weights, higher degrees of substitution, and a high ratio of C2 to C6 substitution are associated with more prolonged effects.

Table 2 Ion content of gastrointestinal fluids

	H^+ (mmol/l)	Na^+ (mmol/l)	K^+ (mmol/l)	HCO_3^- (mmol/l)	Cl^- (mmol/l)
Gastric	40–60	20–80	5–20	150	100–150
Biliary		120–140	5–15	30–50	80–120
Pancreatic		120–140	5–15	70–110	40–80
Small bowel		120–140	5–15	20–40	90–130
Large bowel		100–120	5–15	20–40	90–130

Uses

Provision of daily fluid requirement

Crystalloid fluids are used to provide the daily requirements of water and electrolytes. They should be administered to critically ill patients as a continuous background infusion to supplement fluids given during feeding or to carry drugs. For those who are not taking any nutritional fluid, the total body fluid requirement is usually provided as crystalloid. Higher-concentration glucose infusions may be used to prevent hypoglycemia.

Ringer's lactate and Hartmann's solution have no practical advantage over 0.9 per cent saline for fluid maintenance. Generally local preference dictates the choice of saline solution. The volume required for maintenance fluids should provide the daily requirement of 70 to 80 mmol sodium, although there may be excess loss in sweat and from the gastrointestinal tract. These solutions are not generally required for maintenance if nutritional intake is adequate.

Five per cent glucose is used to supply intravenous water requirements, with 50 g/l glucose being present to ensure an isotonic solution. The normal requirement is 1.5 to 2.0 l/day. Water loss in excess of electrolytes is uncommon, but occurs in excess sweating, fever, hyperthyroidism, diabetes insipidus, and hypercalcemia.

Replacement of excess fluid losses

Fluids lost are invariably salt-containing body fluids, including the plasma component of blood. It is usual to replace excess losses of blood with blood and losses of plasma with plasma or plasma substitutes. However, saline solutions are often used as a temporizing measure. If there are excess losses from the gastrointestinal tract (Table 2) or via sweating, saline solutions are again appropriate. Ringer's lactate is often used for these losses but only offers advantages over 0.9 per cent saline (buffering capacity) in cases of massive replacement.

Ringer's lactate may be useful to maintain acid–base balance if large volumes of crystalloid are exchanged (e.g. during continuous hemofiltration).

Correction of hypovolemia

Colloid solutions are generally used for maintenance of plasma volume and acute replacement of plasma volume deficit. There are two schools of thought regarding the appropriate solutions to correct hypovolemia. Some believe that plasma volume should be maintained or replaced with colloid solutions since crystalloids are rapidly lost from the plasma. Others believe that larger volumes of crystalloid may be used to maintain plasma volume at a lower cost.

Smaller volumes of colloid are required for resuscitation with less contribution to edema. Resuscitation by smaller volumes is quicker and easier. Plasma substitutes maintain plasma colloid osmotic pressure, a useful effect not seen with crystalloids, but they contain no clotting factors or other plasma enzyme systems. Crystalloids are short acting compared with plasma substitutes so that treatment must be repeated more frequently.

It should be noted that plasma substitutes are carried in 0.9 per cent saline so that the majority of critically ill patients will not require 0.9 per cent saline infusions if colloid solutions are used.

The rational choice of colloid solution depends on the length of effect required and knowledge of the persistence of effect of available solutions (Webb 1991) (Table 3):

(1) short-term volume expansion (gelatin, dextran);

(2) medium-term volume expansion (albumin, pentastarch);

(3) long-term volume expansion (hetastarch).

There is no evidence that maintenance of plasma albumin levels, as opposed to maintenance of plasma colloid osmotic pressure with artificial plasma substitutes, is advantageous.

Albumin 20 to 25 per cent and hydroxyethyl starch 10 per cent are hyperoncotic and used to provide colloid where salt restriction is necessary. This use is rarely necessary in intensive care where it has been shown that plasma volume expansion with a particular colloid is related to the weight of colloid infused rather than the concentration (Lamke and Liljedahl 1976). Artificial colloids used with ultrafiltration or diuresis are just as effective in edema states.

Urea-bridged gelatin (polygeline) is a 3.5 per cent solution and contains calcium (6.25 mmol/l). The calcium content prevents the use of the same administration set for blood transfusions. Succinylated gelatin is a 4.0 per cent solution with a larger molecular weight range than polygeline, giving a slightly longer effect. This and the lack of calcium in solution make this a more useful solution than polygeline for short-term plasma volume expansion.

In patients with capillary leak there is considerable leak of albumin and lower-molecular-weight colloids to the interstitium. In these cases it is probably better to use larger-molecular-weight colloids such as hydroxyethyl starch.

Table 3 Relative persistence of colloid effect

Albumin		
Albumin	6–12 h	MW=69 000 Da and negative charge repelled by endothelium
Dextrans		
Dextran 70	4–6 h	Low MW (MW_n=39 000 Da, MW_w=70 000 Da)
Gelatins		
Gelofusin	4–6 h	Low MW (MW_n=22 600 Da, MW_w=35 000 Da) but negative charge repelled by endothelium and increases molecular size
Hemaccel	2–4 h	Low MW (MW_n=24 500 Da, MW_w=35 000 Da)
Hydroxyethyl starches		
Hespan	12–24 h	High MW (MW_n=70 000 Da, MW_w=450 000 Da) and high degree of substitution (0.7)
Pentaspan	4–6 h	Low MW (MW_n=63 000 Da, MW_w=264 000 Da) and low degree of substitution (0.45)
EloHAES	12–24 h	Low MW (MW_n=64 000 Da, MW_w=220 000 Da), medium degree of substitution (0.64), and high C2-to-C6 ratio
HAES-steril	4–6 h	Low MW (MW_n=60 000 Da, MW_w=200 000 Da), low degree of substitution (0.5)

MW, molecular weight; MW_n, number average molecular weight; MW_w, weight average molecular weight.
Persistence is dependent on molecular size and protection from metabolism.
All artificial colloids are polydisperse (i.e. there is a range of molecular sizes).

Hetastarch is usually a 6 per cent solution with a high degree of protection from metabolism. The molecular weight ranges vary, but molecular sizes are large enough to ensure a prolonged effect. These are the most useful colloids in capillary leak. Pentastarch has a lower degree of protection from metabolism and therefore a shorter effect.

The volume required for correction of hypovolemia is determined by the hemodynamic effect of the fluid given. Most colloid solutions are subject to maximum limits suggested by the manufacturers on the basis of available research data. These limits have been exceeded when necessary although side-effects are more likely.

Side-effects

Adverse reactions include the infusion of excess volume which may be associated with edema formation (saline solutions, colloid solutions) or water intoxication (5 per cent glucose). Edema formation may occur despite infusion of apparently appropriate volumes. Conversely, the consequences of inadequate treatment include organ dysfunction and death. Thus the acceptance of the minor side-effect of edema may be necessary to ensure adequate resuscitation.

Allergic phenomena are associated with all colloid fluids (Watkins 1991). These may range from urticaria (the most common effect seen with gelatins) to severe anaphylaxis with cardiovascular collapse (more commonly seen with dextrans). Reactions may be classical IgE mediated or complement mediated and delayed in onset. The incidence of allergic reactions is low with any of the colloid solutions (0.01–0.15 per cent). With longer-acting colloid solutions (e.g. high-molecular-weight hydroxyethyl starch), anaphylactoid reactions may require prolonged support.

Hemostatic defects include dilution coagulopathy associated with any plasma substitute. In addition, dextrans reduce the activities of factors V and VIII, fibrinogen, and prothrombin, decrease platelet adhesion, and prolong bleeding time. These effects are more severe with higher-molecular-weight fractions. Hydroxyethyl starch solutions also reduce factor VIII activity and platelet aggregation (rather than adhesion), again particularly with higher-molecular-weight fractions. These effects are rarely associated with clinical bleeding.

References

Baron, J.-F. (1992). *Plasma volume expansion.* Arnette Blackwell, Paris.

Lamke, L.O. and Liljedahl, S.O. (1976). Plasma volume expansion after infusion of 5 per cent, 20 per cent and 25 per cent albumin solutions. *Resuscitation*, **5**, 85–92.

Mishler, J.M. (1984). Synthetic plasma volume expanders—their pharmacology, safety and efficacy. *Clinics in Haematology*, **13**, 75–92.

Watkins, J. (1991). Allergic and pseudoallergic reactions to colloid plasma substitutes: which colloid? *Care of the Critically Ill*, **7**, 213–17.

Webb, A.R. (1991). The physical properties of plasma substitutes. *Clinical Intensive Care*, **1**, 58–61.

17.1.17 Immunotherapy

S. L. Peake and M. Fisher

Key messages

- The therapeutic use of intravenous immunoglobulin is limited to certain imunological disorders such as immune thrombocytopenic purpura and Guillain–Barré syndrome. No conclusive benefit has been demonstrated in patients with sepsis and septic shock.
- The side-effects of intravenous immunoglobulin are immune complex formation, increased immunoglobulin levels, and infection transmission.
- Monoclonal antibodies against lipid A and tumor necrosis factor-α, interleukin 1 receptor antagonist, and platelet activating factor antagonist have not demonstrated a survival benefit in phase III clinical trials of patients with septic shock.
- The administration of murine monoclonal antibodies leads to the production of human anti-mouse antibodies.

- Granulocyte colony-stimulating factor augments neutrophil number and function and reduces infectious complications during myeloablative chemotherapy.
- There are no convincing data for the use of colony-stimulating factors in critically ill patients with pneumonia and septic shock.

Introduction

In the past decade an exponential growth in therapies which manipulate the immune system and host defense mechanisms has produced immunotherapeutic agents such as immune globulin and monoclonal antibodies with efficacy in fields such as oncology, transplantation, and autoimmune diseases. In the critical care setting, potential immunotherapies include recombinant human granulocyte colony-stimulating factor and interleukin 1 (IL-1) receptor antagonist. Although more traditional pharmacological agents are also able to modulate the immune system, the following discussion will be confined to these novel biological immunotherapies.

Intravenous immunoglobulin

Pharmacology

The pharmacology of immunoglobulins is discussed in detail by Dwyer (1992).

Structure and function of immunoglobulins

The five immunoglobulins classes (IgA, IgD, IgE, IgG, IgM) produced during the humoral immune response have a similar Y-shaped structure with two heavy and two light chains. Each chain consists of a variable and a constant region. Within the variable region is an idiotypic hypervariable or complementarity-determining region (CDR) that binds a unique antigenic determinant. The constant region binds to complement components and Fc receptors on the surface of macrophages, lymphocytes, neutrophils, and eosinophils.

Immunoglobulin preparations

Standard non-specific intravenous immunoglobulin preparations prepared from pooled normal human plasma have been available since the early 1980s. Each manufacturer uses its own donor pool and every batch consists of at least 2000 donors. IgG is the principal class, although IgM and IgA may be present. Various preparations differ in the method of processing, formulation, pH, osmolarity, IgG subclass composition, and antibody titers. All contain antibodies against a variety of micro-organisms. Hyperimmune IgG preparations from donors with high specific neutralizing antibody titers (naturally acquired or following immunization) and immunoglobulin specifically enriched with IgM are also available.

Mechanism of action

In immunosuppressed critically ill patients at high risk of infection, human intravenous immunoglobulin containing a diversity of antibodies that neutralize specific bacterial organisms or products such as lipopolysaccharide has been postulated to enhance the immune response to infection. Antimicrobial effects include enhanced opsonization, complement activation, and antibody-dependent cell cytotoxicity.

The acute inflammatory response may also be attenuated by the selective modulation of cytokine synthesis. An increased ratio of anti-inflammatory mediators (soluble tumor necrosis factor-α receptor and IL-1 receptor antagonist) to proinflammatory mediators (IL-1β, tumor necrosis factor-α, IL-6, and IL 8) has been demonstrated.

In primary humoral immunodeficiencies, intravenous immunoglobulin replaces deficient IgG. In other immunohematological disorders suggested mechanisms of action include the following:

(1) competitive blocking of macrophage IgG Fc receptors by the immunoglobulin Fc portion (prevents autoantibody binding and ingestion of IgG-coated cells);

(2) inhibition of active C3 fragment binding to target cells and complement-dependent clearance;

(3) binding of infused anti-idiotypic antibodies to the autoantibody idiotypic region (increases complex clearance by the reticuloendothelial system) and to the autoantibody-producing B cell (downregulates autoantibody production).

Therapeutic use

Intravenous immunoglobulin therapy is well established in primary humoral immunodeficiencies, immune thrombocytopenic purpura, and prevention of infection in chronic lymphocytic leukemia. Benefits are reported in systemic lupus erythematosus, rheumatoid arthritis, myasthenia gravis, chronic inflammatory demyelinating polyneuropathy, and acute graft versus host disease. A combination of ganciclovir and cytomegalovirus hyperimmune immunoglobulin has also been shown to improve survival in bone marrow transplant patients with cytomegalovirus pneumonitis.

In Guillain–Barré syndrome, high-dose intravenous immunoglobulin produces a functional improvement, a decreased requirement for mechanical ventilation, and fewer complications than plasma exchange. A randomized study of 150 patients has demonstrated it to be more effective. A synergistic effect with high-dose methylprednisolone has been suggested.

No conclusive benefit with either prophylactic or therapeutic intravenous immunoglobulin has been demonstrated in sepsis and septic shock. Both standard non-specific immunoglobulin and hyperimmune preparations enriched with antibodies against specific bacteria or bacterial products have been used. Prophylactic standard and hyperimmune anti-core lipopolysaccharide immunoglobulins have variably shown reduced mortality and infection in neonates, patients with Gram-negative bacteremia, and surgical patients at high risk of Gram-negative infection.

Therapeutic hyperimmune immunoglobulin with high IgG or IgM anti-lipopolysaccharide antibody titers has also shown variable efficacy in patients with suspected Gram-negative bacteremia and septic shock. Observed disparities may relate to differences in protective antibody titers and immunoglobulin class in the various preparations. The lack of standardization, donor availability, unproven efficacy, and risk of disease transmission limit the role of immunoglobulin in the management of septic shock.

Dosage

The optimum dosage and scheduling of intravenous immunoglobulin depends on the clinical indication and specific preparation.

Table 1 Side-effects of intravenous immunoglobulin

Immune complex formation*	Increased immunoglobulins†	Infection transmission
Headache	Hemolysis	HIV
Fever and chills	Neutropenia	Hepatitis B virus
Nausea	Increased plasma viscosity	Hepatitis C virus
Bronchospasm	Thrombosis	Hepatitis δ virus
Hypotension	Aseptic meningitis	Human parvovirus
Anaphylaxis	Renal dysfunction	

*Generally with the first or second dose. The incidence of mild rate-related reactions is 2% to 6%. Severe reactions are more common in patients with IgA deficiency or anti-IgA antibodies.

†Particularly with high doses.

The recommended replacement dose for primary immunodeficiency is 0.1 to 0.4 g/kg/month depending on the degree of antibody production failure and the risk of infection. Larger doses are required for autoimmune diseases such as acute immune thrombocytopenic purpura, and these vary from 0.4 to 1 g/kg/day for 2 to 5 days depending on the clinical response. In demyelinating neuropathies, 0.4 to 1 g/kg/day is administered for 5 days.

Side-effects

Adverse reactions are related to immune complex formation and complement activation, increased immunoglobulin levels and passive transmission of antibodies, and infection transmission (Table 1). Although donor screening and viral inactivation during the manufacturing process minimize the risk of transmission, it is not completely eliminated.

Monoclonal antibodies

Pharmacology

The pharmacology of monoclonal antibodies is discussed by Peake (1993).

Murine antibodies

Kohler and Millstein first described the production of monoclonal-antibody-secreting cell lines in 1975. Immortal non-antibody-secreting murine myeloma cells were fused with antibody-secreting splenocytes from mice immunized with a particular antigen. Following the isolation of hybrid cells, the preferred hybridoma producing the desired antibody was identified and cloned. The technique yielded a perpetually reproducing cell line secreting unlimited quantities of murine monoclonal antibody of a single immunoglobulin class and subclass with identical structure, affinity, and specificity for the antigenic determinant.

Human antibodies

In 1977, the potential immunogenicity of murine antibodies stimulated the production of human antibodies from human myeloma cells and B lymphocytes. However, human antibody production has numerous technical limitations. To overcome these problems, heteromyeloma monoclonal antibodies such as HA-1A (Centocor) have been produced by fusing murine myeloma and human B cells to form non-secreting human–mouse fusion partners which are then fused with human B lymphocytes.

Humanized antibodies

Molecular biology techniques have permitted the development of humanized murine monoclonal antibodies which reduce the immunogenicity of murine antibodies and overcome the limitations of human hybridoma technology. Chimeric human–mouse antibodies are produced by inserting the genetic sequence encoding the mouse variable region of the immunoglobulin next to the human constant region (the most immunogenic portion). The recombinant DNA is then transfected into the murine myeloma cell line and the humanized antibody (approximately 66 per cent human DNA) is secreted. CDR-grafted antibodies (approximately 90 per cent human DNA) can also be produced by inserting only the mouse CDR DNA sequence into the human variable and constant regions.

Therapeutic use

Monoclonal antibody therapy has been successful in acute transplant rejection (murine anti-CD3 antibody), cancer chemotherapy (tumor-specific antibodies coupled to cytotoxic compounds), and rheumatoid arthritis (antitumor necrosis factor-α antibodies).

A role for monoclonal antibody immunotherapy in critical care is not proven, although experimental and limited clinical studies suggest potential benefits (Eidelman *et al.* 1995).

Antibacterial antibodies

Clinical applicability of monoclonal antibodies directed against specific bacterial serotypes is limited since therapy is restricted to documented infections caused by the homologous bacteria. In contrast, antibodies reacting with a diversity of bacteria or their common components, such as the inner-core polysaccharide and lipid A region of lipopolysaccharide, have a wider therapeutic potential.

Two IgM monoclonal antibodies raised against the lipid A region of the J5 mutant, HA-1A (Centocor) and E5 (XOMA), have been clinically evaluated in four large multicenter randomized double-blind placebo-controlled studies. Neither antibody has demonstrated reduced mortality in patients with suspected Gram-negative sepsis or septic shock, and in the second HA-1A study (CHESS trial) a suggestion of increased mortality in non-Gram-negative bacteremic patients resulted in premature study termination.

Anti-cytokine antibodies

Clinically, two large randomized double-blind placebo-controlled phase III trials of the murine anti-tumor necrosis factor-α monoclonal antibody Bay X 1351 (Bayer/Miles) in patients with sepsis, with or without shock, have not demonstrated an overall 28-day survival benefit, although a 17 per cent non-significant reduction in mortality was noted in the subgroup of patients with shock on enrolment in the NORASEPT study. An increased mortality in non-shocked patients was also suggested in the INTERSEPT study. Although studies are ongoing, there is currently no evidence to support the use of anti-cytokine antibodies in patients with sepsis, particularly without shock.

To date, there are no clinical studies involving other anti-cytokine monoclonal antibodies. Experimentally, antibodies to the proinflammatory cytokines IL-1, IL-6, and interferon-γ have been shown to reduce mortality in various models of infection and septic shock.

Antineutrophil adhesion molecule antibodies

Neutrophil adhesion depends on the lipopolysaccharide and cytokine-induced upregulation and binding of adhesion molecules on the neutrophil (CD11/18, L-selectin) and endothelial cell surfaces (intercellular adhesion molecules, E- and P-selectin). Experimentally, monoclonal antibodies against various adhesion molecules have been efficacious in models of ischemia–reperfusion injury (hemorrhagic shock, myocardial ischemia), sepsis, meningitis, allograft rejection, and asthma. However, many studies involve small animals and, to date, there have been no clinical studies.

Dosage

Clinical trials of antibodies against endotoxin and tumor necrosis factor-α have generally employed one or two intravenous doses only: HA-1A, single 100-mg infusion; E5, two 2-mg/kg doses 24 h apart; anti-tumor necrosis factor antibodies, single 3- to 15-mg/kg infusion. However, human pharmacokinetic data are limited, and numerous antibody- and host-related factors must be considered when determining the optimum dosing regimen, including preclinical dose–response studies, phase I and II clinical studies, monoclonal antibody class (IgG versus IgM), circulating half-life, immunogenicity, and the effects of critical illness.

The production of neutralizing human anti-mouse antibodies is reduced with more human antibodies and repeated administration is more likely to be associated with immune complex clearance and decreased clinical efficacy. Murine antibodies are eliminated at the greatest rate. For example, the CDR-grafted anti-tumor necrosis factor-α antibody CDP571(Celltech Inc.) has an elimination half-life of 40 to 90 h, which is two to three times longer than its murine counterpart CB0010 (Celltech Inc.).

Critical illness may also alter the serum half-life. Initial phase I pharmacokinetic studies of HA-1A in stable cancer patients demonstrated a half-life of 29.1 ± 11.3 h. In patients with septic shock, multiple factors such as increased catabolism and reticuloendothelial function decrease the half-life to 15.9 ± 8.8 h.

Side-effects

Adverse effects are uncommon and include headache, fever, chills, gastrointestinal symptoms, and occasionally hypersensitivity reactions and anaphylaxis. Anti-T-lymphocyte antibodies (e.g. anti-CD3) may be associated with T-cell activation and cytokine release with symptoms ranging from an influenza-like illness to shock and cardiac arrest. Murine, chimeric, and heteromyeloma antibodies can also stimulate the development of anti-mouse antibodies (10 to 30 days after exposure) which, although not generally a life-threatening problem, reduce clinical efficacy.

Recombinant cytokines

IL-1 receptor antagonist

Pharmacology

The pharmacology of the IL-1 receptor antagonist is discussed by Dinarello and Wolff (1993). IL-1 consists of two structurally related polypeptides, IL-1α and IL-1β, which bind to the type I and II IL-1 cell surface receptors and have similar actions. The third structurally related interleukin family member is the IL-1 receptor antagonist, which binds to both type I and type II receptors but exerts no agonistic activity. Following endotoxin challenge, plasma levels of IL-1 receptor antagonist are 100 to 1000 times greater than those of IL-1. Such high concentrations are needed to block the IL-1 receptor, since less than 5 per cent occupancy by IL-1 is sufficient to produce an inflammatory response.

Human IL-1 receptor antagonist has been produced by recombinant DNA technology in bacteria. The recombinant form is identical with the naturally occurring form, except for the addition of an N-terminal methionine.

Therapeutic use

IL-1 blockade with recombinant IL-1 receptor antagonist is potentially beneficial in several conditions including inflammatory bowel disease, atherosclerosis, rheumatoid arthritis, graft versus host disease, and myelogenous leukemia. Two large randomized double-blind placebo-controlled phase III studies have been performed in patients with sepsis. No overall reduction in 28-day hospital mortality was demonstrated in the first study, although retrospective analysis suggested decreased mortality in patients with a risk of death greater than 24 per cent (by APACHE score) or organ dysfunction. The second study, designed to evaluate this premise, failed to demonstrate any benefit.

Dosage

In the phase III trials, 1 or 2 mg/kg/h was administered intravenously for 72 h. Similar doses in preliminary pharmacokinetic studies are associated with plasma concentrations of IL-1 receptor antagonist which are 10 000 times greater than IL-1 levels in patients with sepsis.

Side-effects

Since IL-1 receptor antagonist has no agonistic activity, no change in clinical or laboratory parameters has been associated with infusion in volunteers. However, as IL-1 is considered to be a pivotal mediator in the host response to infection and injury, total blockade with IL-1 receptor antagonist would not be desirable.

Colony-stimulating factors

Pharmacology

The pharmacology of colony-stimulating factors is discussed by Lieschke and Burgess (1992). Four colony-stimulating factors promote growth and differentiation of myeloid hemopoietic cells: macrophage colony-stimulating factor, granulocyte colony-stimulating factor, granulocyte–macrophage colony-stimulating factor, and IL-3. Recombinant human forms of granulocyte colony-stimulating factor and granulocyte–macrophage colony-stimulating factor are available for clinical use.

The effects of intravenous recombinant human granulocyte colony-stimulating factor administration include an immediate transient leukopenia (nadir 5 to 15 min), a sustained dose-dependent increase in circulating neutrophils (plateau at 1 week), a leftward shift towards more immature forms, and enhanced neutrophil function (chemotaxis, superoxide generation, phagocytic activity). Neutrophil counts return to baseline 4 to 7 days after therapy ceases. Administration of granulocyte–macrophage colony-stimulating factor is associated with a lesser neutrophil response, but the circulating neutrophil half-life is increased from 8 to 48 h. Prolonged high-dose courses of both factors (2 weeks) is associated with a transient platelet reduction.

Therapeutic use

To date, the principal clinical application has been the treatment of acute neutropenia after myelotoxic chemotherapy. Prophylactic granulocyte colony-stimulating factor reduces the incidence of febrile neutropenic events, infections, hospital admission, and antibiotic use. Neutropenia is not prevented during myeloablative therapy for bone marrow transplantation, but the time to recovery is reduced.

Early studies suggest that augmentation of neutrophil number and function with granulocyte colony-stimulating factor may also be beneficial in non-neutropenic patients with severe pneumonia and sepsis. Clinically, neutrophil counts and free-radical production are increased and L-selectin is downregulated in surgical and trauma patients at risk of sepsis. Increased levels of IL-1 receptor antagonist and soluble tumor necrosis factor-α are also observed. However, preliminary trials in patients with pneumonia do not show a convincing improvement in outcome, and a role for granulocyte colony-stimulating factor in critically ill patients with infection has yet to be established.

Dosage

The recommended dose of granulocyte colony-stimulating factor in hematological patients is 5 μg/kg/day subcutaneously, but higher doses (30–60 μg/kg) have been used. To prevent a rebound leukocytosis, therapy is ceased when the neutrophil count rises to 5000 to 7000/mm^3. The dose of granulocyte–macrophage colony-stimulating factor generally administered is 0.3 to 10 μg/kg/day subcutaneously. Both colony-stimulating factors can also be administered intravenously.

Side-effects

Granulocyte colony-stimulating factor is generally well tolerated. Dose-dependent side-effects include bone pain, rash, increased lactate dehydrogenase and alkaline phosphatase, and, infrequently, vasculitis and anaphylaxis. A large number of adverse effects have limited the clinical use of granulocyte–macrophage colony-stimulating factor. The first dose may be associated with flushing, nausea, vomiting, hypotension, dyspnea, and hypoxia, particularly with intravenous doses of 1 μg/kg or more. With daily doses of 0.3 to 10 μg/kg it causes bone pain, lethargy, anorexia, weight loss, rash, fever and chills (> 3 μg/kg), increased liver enzymes, and decreased albumin. These effects are probably mediated by enhanced cytokine production.

Conclusions

Beneficial effects of immunotherapy in human sepsis and septic shock have not been demonstrated despite animal evidence and immediate hemodyamic improvements suggesting that benefits would occur. This may be related to the complex interaction of cytokines with each other and with other drugs such as sympathomimetics, or real benefits may be concealed by the limitations of clinical trials in human multisystem disease of variable severity. The high cost of such drugs means that approval for use in multisystem disease is unlikely in the absence of demonstration of improved survival. Although these novel biological immunotherapies do have a role to play in conditions other than infection and sepsis, the approved indications are currently limited.

References

Dinarello, C.A. and Wolff, S.M. (1993). The role of IL-1 in disease. *NewEngland Journal of Medicine*, **328**, 106–13.

Dwyer, J.M. (1992). Manipulating the immune system with immune globulin. *New England Journal of Medicine*, **327**, 107–16.

Eidelman, L.A., Pizov, R., and Sprung, C.L. (1995). New therapeutic approaches in sepsis: a critical review. *Intensive Care Medicine*, **21**, S269–72.

Lieschke, G.J. and Burgess, A.W. (1992). Granulocyte colony stimulating-factor and granulocyte–macrophage colony-stimulating factor. *New England Journal of Medicine*, **327**, 28–35, 99–106.

Peake, S.L. (1993). Monoclonal antibodies — immunotherapy for the critically ill. *Anaesthesia and Intensive Care*, **21**, 739–51.

17.1.18 Inhalational anesthetic agents

E. M. Spencer and S. M. Willatts

Key messages

- The main advantage of inhalational sedation is pulmonary excretion and rapid recovery.
- The effective sedative dose of isoflurane is 0.1 to 0.4 per cent.
- Isoflurane is only 0.2 per cent metabolized and can be used safely in patients with renal or hepatic failure.
- Isoflurane causes bronchial relaxation in patients with bronchospasm and is an ideal sedative for ventilated patients with severe asthma and chronic obstructive pulmonary disease.

Introduction

Inhalational gases have been used in the intensive care unit (ICU) to provide sedation intermittently over the last 40 years. Their main advantage over intravenous agents is that they are excreted via the lungs, allowing a rapid and predictable recovery. As the agents are minimally metabolized and quickly eliminated, the depth of sedation can be precisely controlled.

Isoflurane is the only agent that can be recommended for use in critically ill patients in the ICU (Willatts and Spencer 1994). Ether and cyclopropane are too explosive for routine use, although ether has been used to reduce bronchospasm in life-threatening acute asthma. Trichloroethylene and methoxyflurane have now been withdrawn. None of the other agents can be advocated for prolonged use, as nitrous oxide causes bone marrow depression, halothane is hepatotoxic, and enflurane is potentially nephrotoxic. Newer agents such as desflurane and sevoflurane are currently being used during anesthesia, but it is doubtful that they will be of value in the critically ill because of the potential fluoride nephrotoxicity of sevoflurane and the cost of desflurane.

Nitrous oxide

Nitrous oxide is an anesthetic gas with very low blood–gas and oil–gas solubility coefficients, and therefore rapid onset and recovery. It has potent analgesic properties and is used extensively for analgesia in the emergency room and in obstetrics. It was used in the early 1950s to sedate patients with severe tetanus and has since been used for sedation in the critically ill. Its usefulness is limited by myocardial depression, the risk of pulmonary hypertension, and inhibition of methionine synthase activity, leading to bone marrow suppression and peripheral neuropathy with prolonged use.

Nitrous oxide should be avoided in patients with cerebral swelling as it increases cerebral blood flow and intracranial pressure. In patients with pre-existing pulmonary hypertension it increases pulmonary vascular resistance. As nitrous oxide is 34 times more soluble than nitrogen, it will diffuse into air-containing spaces more rapidly than nitrogen can diffuse out causing an increase in size of any air-containing space. Hence it should be administered with great care to patients with pneumothoraces, gut ileus because of abdominal distension, and atrial or ventricular septal defects because of the risk of air embolism from intravenous lines.

Entonox®, a 50 per cent mixture of nitrous oxide in oxygen, is sometimes used in the ICU in situations where pain is predictable and of short duration (e.g. physiotherapy, removal of drains, or changes of dressings).

Halothane

Halothane (CF_3–CHClBr) is a potent inhalational anesthetic which has been used to sedate patients with tetanus. It has a direct action on bronchial smooth muscle, causing bronchodilatation, and it depresses the airway reflexes and so has been useful for the treatment of severe asthma. However, halothane is more than 20 per cent metabolized and some of its metabolites are toxic to the liver; therefore it can no longer be used as a long-term sedative agent.

Enflurane

Enflurane (CHF_2–O–CF_2–CHFCl) is a halogenated ether which has never been used for sedation in the ICU because of the risk of nephrotoxicity. Its metabolism (2 per cent) releases fluoride ions. Prolonged enflurane anesthesia has led to elevated fluoride concentrations and evidence of impaired urine-concentrating ability, and so it cannot be recommended for use in the ICU.

Isoflurane

Isoflurane (CF_3–CHCl–O–CHF_2) is a methyl ethyl ether and an isomer of enflurane. It has a lower blood–gas partition coefficient than halothane or enflurane which allows the alveolar concentration to rise rapidly towards the inspired concentration, leading to a rapid induction of sedation. The level of consciousness can be controlled with precision. The effective dose for sedating ventilated patients is confined to a narrow range (0.1–0.4 per cent). Recovery after prolonged isoflurane sedation is rapid, allowing early extubation and mobilization (Spencer and Willatts 1992).

Respiratory system

Isoflurane depresses ventilation in a dose-related fashion in spontaneously breathing volunteers. Subanesthetic concentrations of 0.1 times the minimal alveolar concentration (MAC) cause a 50 per cent depression of CO_2 response, and concentrations well below this impair responsiveness to hypoxia. These effects are less important in the ICU because isoflurane is only used in patients supported by mechanical ventilation. As isoflurane is eliminated quickly, this ventilatory depression should not be a problem in the weaning period.

In normal human airways there is no effect on bronchomotor tone, but isoflurane may prevent the development of bronchospasm and has been used to treat patients with severe asthma or chronic obstructive pulmonary disease.

Cardiovascular system

All inhalational anesthetic agents depress the myocardium, but cardiac output is maintained with isoflurane because of the substantial reduction in systemic vascular resistance and slight increase in heart rate. Isoflurane decreases vascular tone in all vascular beds, and the decrease in arterial pressure may reduce perfusion to vital organs that already have a compromised circulation. No decrease in arterial pressure or increase in heart rate has been reported at the low concentrations required for sedation. Isoflurane, like all other sedative agents, must be administered carefully in patients who are hypovolemic or have a very unstable cardiovascular system.

Isofurane is a coronary vasodilator, and redistribution of coronary blood flow will occur in the presence of fixed stenotic lesions of the coronary arteries. It has been suggested that this may result in poor endocardial perfusion, the 'coronary steal effect'. However the incidence of myocardial ischemia is not more frequent after isoflurane anesthesia and it is now accepted that inspired concentrations of less than 0.5 per cent isoflurane are safe even in patients with a history of coronary artery disease (Slogoff *et al.* 1991).

Isoflurane does not predispose the heart to arrhythmias and unlike halothane does not increase the probability of ventricular arrhythmias following administration of either epinephrine (adrenaline) or aminophylline. However, it is possible that interactions may occur with other drugs affecting the heart.

Central nervous system effects

All inhalational anesthetics are cerebral vasodilators tending to increase cerebral blood flow and thus intracranial pressure. However, there is no change in cerebral blood flow in normocapnic normotensive volunteers given less than 1 MAC of isoflurane, and isoflurane does not increase the production or prevent the absorption of cerebrospinal fluid. Autoregulation of cerebral blood flow is retained with low doses of isoflurane, and a decrease in $P\text{aCO}_2$ produces a nearly proportional decrease in cerebral blood flow. Isoflurane causes a dose-related decrease in cerebral oxygen consumption up to 1 MAC while maintaining a normal cerebral energy state. This may account for the reports that isoflurane provides more protection from brain ischemia than other agents and has become the agent of choice for neuroanesthesia.

Although isoflurane has been shown to lower intracranial pressure in patients with cerebral disease, some animal studies have shown that it rises in the presence of acute head injury when isoflurane is administered. Under conditions of controlled ventilation and low sedative doses isoflurane is unlikely to increase intracranial pressure, but the effect of isoflurane sedation in patients with head injury is still uncertain.

Renal effects

The proposed metabolic pathway yields two fluoride ions for each isoflurane molecule undergoing metabolism. Plasma inorganic fluoride concentrations above 50 μmol/l have been associated with a reduction in the kidney's ability to concentrate urine.

Isoflurane does not cause any fluoride-induced nephrotoxicity after normal anesthesia, but these findings cannot be extrapolated to more prolonged exposure.

There have been several studies and reports of elevated fluoride concentrations after sedation with isoflurane in the ICU. Despite the high plasma fluoride concentrations seen in some of these ICU patients, there was no effect on renal function as determined by creatinine clearance and urine osmolality. Although there is a relationship between MAC hours and the maximum fluoride concentration seen in each patient, it is not possible to predict how many MAC hours of isoflurane could be administered before a potentially nephrotoxic level of fluoride is reached (Spencer *et al.* 1991).

There appears to be a greater mean plasma concentration after isoflurane sedation than would be expected after the equivalent MAC hours of anesthesia. The metabolism of isoflurane in severely ill patients may vary considerably, like that of other sedatives, owing to variations in hepatic blood flow, cellular oxygenation, and concurrent medications which may interact with liver enzyme systems and interfere with isoflurane metabolism. There may be a more complex biotransformation of isoflurane following chronic exposure as prolonged exposure leads to its accumulation in adipose tissue and continuing metabolism.

Isoflurane should be used with caution for sedation lasting for more than 3 to 4 days and renal function should be monitored closely.

Hepatic effects

Because less than 1 per cent of isoflurane is metabolized and there is no reductive metabolism or free-radical production, it is difficult to implicate isoflurane as a cause of hepatitis. Numerous animal and clinical studies have been performed, but none has produced evidence of hepatic dysfunction following isoflurane anesthesia. This has been ascribed to the chemical stability of isoflurane and the rapidity with which it is eliminated.

Studies of the effect of anesthetic agents on the liver are difficult to interpret as liver function tests are affected by many factors, such as surgical stress or hepatic arterial blood flow, as well as by the agent itself. Glutathione-*S*-transferase has been shown to be a more sensitive indicator of hepatocellular damage than the aminotransferases, and increases in this enzyme have been seen after halothane anesthesia but not following isoflurane anesthesia or sedation (Howie *et al.* 1992).

Clinical use of isoflurane in the ICU

Isoflurane can be used to achieve a desired sedation level which can be easily and rapidly controlled by changing the inspired isoflurane concentration. Age has very little effect on wash-in of isoflurane in healthy patients, although significant pulmonary pathology might alter this finding. Critically ill patients rarely have normal lungs, and an increased dead-space or pulmonary shunt may cause the alveolar concentration to be slightly less than the inspired concentration. However, a stable easily controllable level of sedation can be achieved in these patients and if necessary reversed for assessment. Cardiac output and regional blood flow are important in determining how quickly the effects of the agent are seen. The effective dose range for sedation is very narrow and increased doses should not be attempted to obviate the need for perfect analgesia.

Recovery from isoflurane sedation is very rapid, and procedures requiring disconnection from the ventilator, such as physiotherapy, may require administration of a supplementary agent such as morphine or alfentanil. There is no significant evidence for tolerance to the agent, although this has been reported in one child. There is no evidence that use of isoflurane impairs response to infection. Isoflurane reduces plasma catecholamine concentrations, but prolonged low-dose isoflurane does not inhibit cortisol production.

As isoflurane causes bronchial relaxation in patients with bronchospasm, it is an ideal sedative agent for patients with severe asthma and chronic obstructive pulmonary disease. Isoflurane has been successfully used to control status epilepticus refractory to conventional drugs, but the doses required (0.8–2.0 per cent) were such that the patients needed inotropic support. As only 0.2 per cent of isoflurane is metabolized, it can be used safely in patients with renal or hepatic failure.

Isoflurane has been used successfully for sedation after heart surgery where the desired level of sedation was rapidly attained and the time to endotracheal extubation was shorter than after propofol. Although most of the studies have been performed in adults, isoflurane has been used for sedation of pediatric patients. However, there have been reports of reversible ataxia, agitation, hallucinations, and neuropathies after isoflurane in children. Many other factors could have been associated with these complications, such as concurrent administration of other drugs. Both fentanyl and midazolam have been reported to produce reversible neurological abnormalities.

Sedation in children, particularly those with upper airway infections, can often be a problem and any agent should be used with caution. Deaths due to myocardial dysfunction after propofol sedation have been reported. Isoflurane can effectively provide sedation to pediatric patients but, at present, experience of its use in children is limited and so it should be administered with care.

Equipment

Isoflurane is a good sedative and avoids the problems associated with multiple intravenous infusions such as venous access, incompatibility of solutions, risk of infection, and the dangers associated with bolus adminstration. It requires a vaporizer which is accurate at low inspired concentrations, obviating the need for an expensive monitor to measure the inspired and expired concentrations. For safety the maximum inspired concentration should be 1 per cent with increments of 0.1 per cent. For ease of use and convenience the vaporizer should be portable and compatible with any ventilator and breathing circuit. A Penlon OMV drawover vaporizer with all these features has now been developed specifically for ICU use and is accurate down to an inspired concentration of 0.1 per cent.

Scavenging

Scavenging will be perceived as important as continuing anxiety remains that trace concentrations of inhalational agents may be carcinogenic or fetotoxic. In a review of many large studies in several countries, no clear evidence was obtained for an adverse effect on health produced by the working environment. However, there are now strict regulations about the amount of vapor or gas allowed in the air breathed by hospital staff. Most ICUs do not have piped scavenging, and the most convenient means of scavenging the isoflurane is with an Aldasorber. This contains activated charcoal which adsorbs the anesthetic agent. At the low concentrations of isoflurane used, this does not require changing more than once in 24 h. Used canisters should be sealed in polythene bags until incineration to avoid re-release of the vapor.

Exposure of ICU personnel to isoflurane has been significantly reduced by the development of closed suction systems which do not require disconnection of the patient from the ventilator for tracheal suction.

The cost of isoflurane is about $120 (£80) per day which is comparable with alternative short-acting agents such as propofol.

References

Howie, A.F., Spencer, E.M., and Beckett, G.J. (1992). Aspartate aminotransferase, alanine aminotransferase and glutathione transferase in plasma during and after sedation by low-dose isoflurane and midazolam. *Clinical Chemistry*, **38**, 476–9.

Slogoff, S., *et al.* (1991). Steal-prone coronary anatomy and myocardial ischemia associated with four primary anesthetic agents in humans. *Anesthesia and Analgesia*, **72**, 22–7.

Spencer, E.M. and Willatts, S.M. (1992). Isoflurane for prolonged sedation in the intensive care unit: efficacy and safety. *Intensive Care Medicine*, **18**, 415–21.

Spencer, E.M., Willatts, S.M., and Prys-Roberts, C. (1991). Plasma inorganic fluoride concentrations during and after prolonged isoflurane sedation: effect on renal function. *Anesthesia and Analgesia*, **73**, 731–7.

Willatts, S.M. and Spencer, E.M. (1994). Sedation for ventilation in the critically ill. A role for isoflurane? *Anaesthesia*, **49**, 422–8.

17.1.19 Inotropic agents

R. N. Sladen

Key messages

- The primary indication for an inoconstrictor is a situation where inotropic support is required in the face of low systemic vascular resistance. Inodilators are suited to situations where low cardiac output is associated with high systemic vascular resistance.
- Catecholamines have an agonist effect at the β_1-adrenergic receptor in cardiac muscle, increasing cytoplasmic cAMP which is the fulcrum of all inotropic activity.
- Phosphodiesterase inhibitors exert their inotropic action by inhibiting the breakdown of cAMP by its specific enzyme phosphodiesterase III.

Classification of inotropic agents

The parenteral inotropic agents currently in use include two pharmacological classes, catecholamines and phosphodiesterase inhibitors. However, a more pragmatic classification subdivides inotropic agents on the basis of their actions on the peripheral circulation. Thus those agents which have a predominantly vasoconstrictor effect may be termed inoconstrictors. Endogenous catecholamines, for example norepinephrine (noradrenaline), epinephrine (adrenaline), and dopamine, all have peripheral vasoconstrictor effects mediated by the α_1-receptor; the effect is greatest in the highest dose ranges. Agents whose predominant effect on the peripheral circulation is vasodilation may be termed inodilators. The synthetic derivatives of dopamine, for example dobutamine, dopexamine, and isoproterenol (isoprenaline), have predominant β_2-agonist activity and uniformly induce peripheral vasodilation. Phosphodiesterase inhibitors (amrinone, milrinone, enoximone) are actually more potent vasodilators than inotropic agents.

The primary indication for an inoconstrictor is a situation where inotropic support is required in the face of low systemic vascular resistance, for example septic shock. Inodilators are suited to situations where low cardiac output is associated with high systemic vascular resistance, for example congestive heart failure or severe cardiomyopathy. The vasodilator effect decreases systemic vascular resistance and hence afterload, which further enhances the inotropic action in improving cardiac function.

It may be advantageous to combine an inotropic agent with a vasodilator or vasoconstrictor drug to achieve the optimal balance between inotropic effect, afterload reduction, and perfusion pressure.

In short, the rational choice of an inotropic agent or combination of agents depends on careful evaluation of global hemodynamic indices including cardiac output, mean arterial pressure, and central venous pressure, as well as the effect of the agent on heart rate.

Mechanisms of inotropic action

Inotropic agents exert their action through one of two major mechanisms. Catecholamines have an agonist effect at the β_1-adrenergic

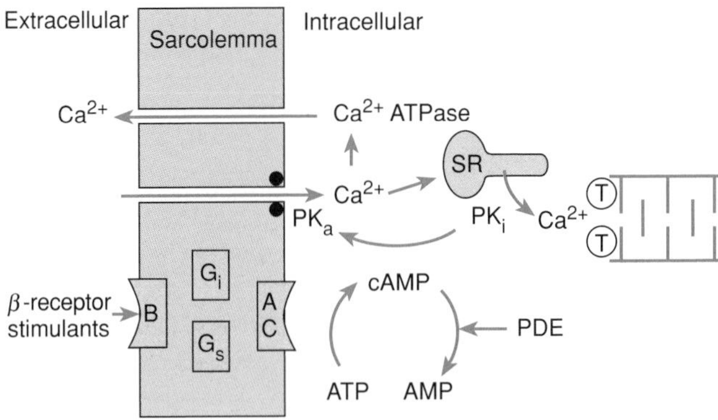

Fig. 1 The β-adrenergic receptor. The β-receptor (B) is coupled to adenylyl cyclase (AC) by G proteins (G_s, stimulatory; G_i, inhibitory). Activation of adenylyl cyclase catalyzes the transformation of ATP to cAMP, which in turn converts protein kinase from its inactive form (PK_i) to its active form (PK_a). Protein in the calcium channel is phosphorylated by PK_a, which opens the 'gate' so that calcium (Ca^{2+}) moves into the cell. Calcium enters the sarcoplasmic reticulum (SR), inducing it to release calcium which binds to troponin C (T) and allows activation of the actin–myosin complex. cAMP is inactivated by conversion to AMP by phosphodiesterase (PDE). Calcium is cycled out of the cell by calcium adenosine triphosphatase (Ca^{2+}ATPase). In contrast, the α-adrenergic receptor is coupled to phospholipase C, generates cGMP, and forms diacyl glycerol and inositol triphosphate. The net effect is similar—calcium entry via the sarcolemma and calcium release from the sarcoplasmic reticulum. However, the α-adrenergic receptor is independent of cAMP. (Reproduced with permission from Royster (1990).)

receptor in cardiac muscle (Fig. 1). The β_1-receptor is located on the outer surface of the sarcolemma, where it is coupled via a G protein to adenylyl cyclase, an enzyme on the inner surface. Within the cytoplasm, ATP is catalyzed by adenylate cyclase to cAMP. The latter is the well-known second messenger, which is the fulcrum of all inotropic activity. The phosphodiesterase inhibitors exert their inotropic action by inhibiting the breakdown of cAMP by its specific enzyme phosphodiesterase III. Thus, whether the mechanism of inotropic effect is β-adrenergic stimulation or phosphodiesterase inhibition, the end result is elevation of cAMP levels within the cardiac muscle cell.

The activation of cAMP induces a protein kinase which opens a calcium gate in the sarcolemma and allows the ingress of calcium into the cardiac muscle cytoplasm. This in turn induces the release of large amounts of calcium from the sarcoplasmic reticulum. Released calcium binds to troponin and forms tropomyosin C which releases tropomyosin from its binding to actin and myosin. Thus the lateral movement of actin and myosin is facilitated, culminating in cardiac muscle contraction.

In vascular smooth muscle, cAMP activation has exactly the opposite effect; it induces an uptake of calcium into the sarcoplasmic reticulum. This results in smooth muscle relaxation. The contrasting effects of cAMP activation in cardiac muscle and smooth muscle explain the inodilator action of phosphodiesterase inhibitors, i.e. a cardiac inotropic effect combined with systemic and pulmonary vasodilation.

Adrenergic receptor agonism

Inotropic activity is mediated through cardiac β_1-adrenergic receptors. In addition to an enhanced force of contraction (inotropy), stimulation of the β_1-receptor also results in a variable degree of increase in sinus node firing rate (chronotropy), an increase in atrioventricular conduction velocity (dromotropy), and a decrease in the excitation threshold (bathmotropy).

Beta-2-receptors and α_1-adrenergic receptors also exist in the myocardium and may mediate an important component of inotropic effect in the presence of chronic congestive heart failure and cardiomyopathy. Normally, about 80 per cent of inotropic action is mediated through the β_1-receptor. However, in chronic congestive heart failure there is a persistent elevation of circulating catecholamines, culminating in downregulation of the cardiac β_1-receptor. The relative proportion of β_2-receptor-induced inotropic effect doubles from 20 to 40 per cent of total inotropic activity.

A small component of inotropic effect is mediated via the α_1-adrenergic receptor, which also becomes more important in congestive heart failure and endstage cardiomyopathy. The α-receptor is coupled to phospholipase C on the inner layer of the sarcolemma by G_q protein. Stimulation induces guanidosine triphosphate to cleave phosphatidyl inositol into inositol triphosphate, which releases calcium from the sarcoplasmic reticulum, and diacylglycerol, which activates protein kinase C and opens the calcium channel in the sarcolemma. The subsequent chain of events is similar to that of β-receptor activation. Importantly, however, the inotropic effect is independent of cAMP.

Phosphodiesterase inhibition

The action of cAMP is terminated when it is cleaved to AMP by phosphodiesterase. Thus phosphodiesterase inhibition increases the concentration of cAMP and enhances β-receptor-mediated inotropic activity. Although phosphodiesterase inhibition depends on cAMP, it is independent of the β-receptor and is effective in the presence of β-receptor downregulation. In addition to mediating positive inotropic action in the heart and vasodilation in vascular smooth muscle, phosphodiesterase inhibition increases cAMP in bronchial smooth muscle (bronchodilation), platelets (decreased platelet aggregation), and liver (gluconeogenesis).

There are three major subgroups of phosphodiesterase. Type I cleaves cGMP, type II cleaves both cGMP and cAMP, and type III is specific for cAMP. Non-selective phosphodiesterase inhibitors have a wide spectrum of clinical effects: aminophylline promotes bronchodilation, cardiac inotropy, and vasodilation; dipyridamole enhances platelet aggregation and coronary vasodilation; papaverine induces profound vasodilation. Thus only those phosphodiesterase inhibitors which are specific for phosphodiesterase III are used as cardiac inotropic agents: the bipyridine derivatives amrinone and milrinone, and the imidazole derivatives enoximone and piroximone. The latter two remain investigational agents in the United States.

Pharmacology of inotropic agents

Inoconstrictors

Norepinephrine

Norepinephrine is an extremely potent β_1- and α_1-adrenergic agonist. Its peripheral action is to induce intense vasoconstriction, particularly in the splanchnic and pulmonary vascular beds. The resultant increase in systemic vascular resistance opposes its inotropic effect. The clinical manifestation is an increase in blood pressure, with a mild increase in heart rate, but little or no change in cardiac output.

To some degree, these effects are dose related, so that in the low-dose range norepinephrine has predominantly β-adrenergic action, whereas in the high-dose range its α-adrenergic vasoconstrictor effects are most prominent.

The dose for norepinephrine is usually in the range from 0.02 to 0.2 g/kg/min, with the very clear endpoint of achieving a mean arterial pressure of about 65 mmHg. Excessive doses of norepinephrine induce splanchnic and renal vasoconstriction, and tissue ischemia that may culminate in progressive metabolic acidosis and disseminated intravascular coagulation (DIC). Norepinephrine should be dosed to a specific clinical endpoint and weaned as soon as the clinical situation permits.

Renal autoregulation is lost in sepsis. Use of norepinephrine to restore normal perfusion pressure in septic patients with profound hypotension and oliguria has been shown to reverse acute renal dysfunction.

Alpha-receptor downregulation

As a potent direct-acting α_1-adrenergic agonist, norepinephrine is indicated to maintain blood pressure in the presence of α_1-receptor downregulation. This may occur after resection of a norepinephrine-secreting pheochromocytoma, long-term therapy (more than 3 weeks) with tricyclic antidepressants and monoamine oxidase inhibitors, and acutely during extracorporeal circulation which may in part explain low systemic vascular resistance states following cardiopulmonary bypass. In acute renal failure the renal arteries are much less sensitive to the vasoconstrictor effects of norepinephrine. This may be used to advantage in maintaining renal perfusion pressure during intermittent hemodialysis to avoid repeated hypotensive insults to the injured kidney.

Epinephrine

Epinephrine is an endogenous catecholamine which is formed by the methylation of norepinephrine ('nor' is from the German *N ohne Radikal*, i.e. the nitrogen atom is without a methyl radical). Like norepinephrine, it acts on the cardiac β_1-receptor and peripheral α_1-receptor to induce a combination of inotropy and vasoconstriction. Its ability to constrict venous beds augments venous return and cardiac preload. However, epinephrine also has substantial peripheral β_2-receptor-mediated vasodilator activity. Thus, although the cutaneous, splanchnic, and pulmonary beds are constricted, skeletal muscle beds (which may receive up to 40 per cent of the cardiac output) are dilated. Therefore the effect of epinephrine on the systemic vascular resistance is more unpredictable than that of norepinephrine. When epinephrine is given in a low cardiac output state with high systemic vascular resistance, cardiac output increases and calculated systemic vascular resistance decreases. In contrast, when epinephrine is administered during anaphylaxis (high cardiac output, low systemic vascular resistance), it increases the systemic vascular resistance.

Like norepinephrine, the effect of epinephrine is predominantly β-adrenergic at lower doses but with increasing dosage vasoconstriction becomes more prominent. Epinephrine has potent chronotropic, dromotropic, and bathmotropic actions, increases myocardial work and oxygen consumption, and may exacerbate myocardial ischemia and induce dangerous tachyarrhythmias. Epinephrine is primarily indicated for low cardiac output states in the dosage range of 0.02 to 0.2 g/kg/min (20–200 ng/kg/min).

Epinephrine stimulates the Emden–Meyerhof pathway, and pyruvate production may actually increase lactic acid levels if the citrate acid cycle is impaired.

Dopamine

Dopamine is the naturally occurring precursor of norepinephrine. It is unique amongst endogenous catecholamines in that it has an agonist action on the dopaminergic (DA_1 and DA_2) receptors. DA_1 receptors are situated in the blood vessels of the kidney and splanchnic bed, where they mediate vasodilation. There are also DA_1 receptors in the lumen of the proximal renal tubule, where they promote sodium and water excretion. DA_2 receptors are located in the presynaptic membrane. When stimulated by free dopamine in the synaptic cleft, they suppress norepinephrine release from presynaptic storage vesicles. This promotes peripheral vasodilation, an action analogous to that of the presynaptic α_2-adrenergic receptors. The dopaminergic effect of dopamine appears to be predominate in the low-dose range (0.5–3 g/kg/min).

At higher doses (3–10 g/kg/min), dopamine has modest β_1-agonist activity, with minimal β_2 peripheral vasodilator action, and maintains venous capacitance, preload, and blood pressure. Dopamine exerts about half its inotropic action indirectly through its conversion to or release of norepinephrine. At doses above 10 g/kg/min, dopamine induces progressive vasoconstriction via α_1-adrenergic mechanisms and norepinephrine release. In fact, there is a direct relationship between the dopamine dose and plasma norepinephrine levels. Excessive doses result in tachyarrhythmias, oliguria, and cutaneous vasoconstriction (skin necrosis and sloughing may occur when dopamine extravasates from a peripheral intravenous catheter).

Inodilators

Catecholamines

Dobutamine

Dobutamine is a synthetic derivative of dopamine and has been widely used for the past decade. It is a very potent β_1- and β_2-agonist which provides direct inotropic action coupled with systemic vasodilation and afterload reduction. However, it is devoid of dopaminergic activity. In hypovolemic patients dobutamine may induce brisk hypotension and tachycardia, and tachyarrhythmias occur in the high-dose range. Dobutamine is generally used in the dose range from 5 to 20 g/kg/min.

Dobutamine is very effective in patients with cardiomyopathy and congestive heart failure, both because it is a direct-acting β_1-adrenergic agent and because of its action on cardiac β_2-adrenergic receptors. In a comparison between dobutamine and epinephrine it was demonstrated that cardiac output was increased more with dobutamine, but mostly because of a greater increase in heart rate; the effects on stroke volume were similar.

Dopexamine

Dopexamine is a relatively new synthetic derivative of dopamine with a unique spectrum of activity. It is a very potent β_2-agonist, a rather weak β_1-agonist, and a dopaminergic agonist at about one-third the potency of dopamine. Its predominant effect appears to be vasodilation and afterload reduction. Increases in contractility are due to the reflex sympathetic response to vasodilation, as well as inhibition of norepinephrine uptake.

Given its vasodilator and dopaminergic activity, it is ideally suited to the patient with congestive heart failure or low cardiac output associated with high systemic vascular resistance, moderately elevated blood pressure, and impaired renal function. Although widely available in Europe, dopexamine remains an investigational agent in the United States.

Studies thus far suggest that dopexamine provides effective afterload reduction and increases in cardiac output in the dose range from 1 to 5 g/kg/min, although with a significant increase in heart rate.

Isoproterenol

Isoproterenol was the first synthetic derivative of dopamine. Like dobutamine, it is a very potent β_1- and β_2-adrenergic agonist. However, its chronotropic effect is excessive and has curtailed its usefulness as an inotropic agent because of its propensity to cause tachyarrhythmias. Today isoproterenol is restricted to occasional use to provide pharmacological pacing in patients with sinus bradycardia with hypotension and/or escape rhythms, or as a chronotropic agent for the denervated heart after cardiac transplantation.

Phosphodiesterase inhibitors

Amrinone

Amrinone is a selective phosphodiesterase III inhibitor, and was the first to be used extensively. It is a yellow bipyridine compound which should be diluted in saline (not glucose) and protected from light to prevent gradual dissociation. Amrinone causes vasodilation and decreases both systemic and pulmonary vascular resistance, providing substantial left and right ventricular afterload reduction. Its modest inotropic action may in part be due to reflex responses to vasodilation. It also has a positive lusitropic action, i.e. it enhances diastolic relaxation and thereby decreases wall stress. Increases in heart rate are relatively benign and overt tachyarrhythmias are unusual. The net effect is improved myocardial efficiency and oxygen balance.

Like all phosphodiesterase III inhibitors, amrinone is effective in the presence of β-receptor downregulation and enhances stroke volume in patients in chronic congestive heart failure or who have been on β-adrenergic agonists for several days. Its adverse effects include excessive hypotension, prolonged duration of action, and thrombocytopenia (if used for long periods or administered orally).

Intracardiac cAMP levels are elevated by increased production (adrenergic stimulation) and decreased breakdown (phosphodiesterase inhibition). The increase in stroke volume is considerably greater when amrinone is used in combination with an inoconstrictor than when either agent is used alone.

The catecholamines are rapidly inactivated by the liver with an elimination half-life of 1 to 2 min. In contrast, amrinone undergoes slow hepatic biotransformation and about 25 per cent is excreted unchanged by the kidney. Its pharmacokinetics have been studied using a two-compartment model, and indicate a short redistribution phase ($(t_{1/2})_\alpha = 2$–4 min) and a long elimination phase ($(t_{1/2})_\beta = 2.5$–4 h). To achieve therapeutic plasma levels (2–9 g/ml) rapidly, amrinone must be loaded, followed by a maintenance infusion. It is recommended that a loading dose of between 1.5 and 3.0 mg/kg be administered over 10 to 15 min. The duration of action of amrinone depends on the size of the loading dose and the duration of the infusion. For example, after a 24-h infusion it may take 2 to 3 h for its effects to dissipate. In contrast with the catecholamines, amrinone does not need to be weaned slowly prior to removal. In most cases it is sufficient to halve the dose or cut it off abruptly and allow it to 'auto-wean'.

Milrinone

Milrinone has largely replaced amrinone in clinical practice, probably because of its greater convenience of use and lesser expense than any great pharmacological superiority. It is about 15 times as potent as amrinone and does not require protection from light. Milrinone is even more dependent on renal clearance, with about 40 per cent being excreted unchanged by the kidney. Therefore it is recommended that the maintenance infusion be progressively decreased with impaired renal function, to about 50 per cent of the standard infusion rate when the creatinine clearance is less than 20 ml/min.

The recommended loading dose on cardiopulmonary bypass is 50 g/kg, followed by a maintenance infusion of between 0.375 and 0.75 g/kg/min.

Chronic oral administration of milrinone in patients with congestive heart failure has actually been associated with increased overall mortality.

Enoximone

Enoximone is an imidazole derivative phosphodiesterase inhibitor with a similar pharmacological profile to amrinone and milrinone. However, it has a much longer duration of action, with an elimination half-life of 6 to 7 h. In addition, it undergoes biotransformation in the liver to an active sulfoxide metabolite which itself has an elimination half-life of 6 to 11 h.

The usual loading dose is 0.5 mg/kg, followed by a maintenance infusion of 2.5 to 10 g/kg/min. It has been suggested that enoximone may be more effective than dobutamine in enhancing inotropic force and heart rate in the β-blocked heart. As with amrinone and milrinone, the combination of enoximone with epinephrine increases stroke volume more than either agent used alone.

Piroximone

Piroximone is an imidazole phosphodiesterase inhibitor with similar actions to enoximone, although without its sustained duration of action. It remains an investigational agent.

Bibliography

Bristow, M.R., Ginsburg, R., and Minobe, W. (1982). Decreased catecholamine sensitivity and β-adrenergic-receptor density in failing human hearts. *New England Journal of Medicine*, **307**, 205–11.

Feneck, R.O. and the European Milrinone Multicentre Trial Group (1992). Intravenous milrinone following cardiac surgery. I. Effects of bolus infusion followed by variable dose maintenance infusion. *Journal of Cardiothoracic and Vascular Anesthesia*, **6**, 554–62.

Ghosh, S., Gray, B., Oduro, A., and Latimer, R. (1991). Dopexamine hydrochloride: pharmacology and use in low cardiac output states. *Journal of Cardiothoracic and Vascular Anesthesia*, **5**, 382–9.

Meier-Hellman, A. and Reinhart, K. (1994). Influence of catecholamines on regional perfusion and tissue oxygenation in septic patients. In *Sepsis: current perspectives in pathophysiology and therapy* (ed. K. Reinhart, K. Eyrich, and C. Sprung), pp. 274–91. Springer-Verlag, Berlin.

Royster, R.L. (1990). Intraoperative administration of inotropes in cardiac surgical patients. *Journal of Cardiothoracic Anesthesia*, **4** (Supplement 5), 17–28.

17.1.20 Muscle relaxants

B. J. Pollard

Key messages

- Relaxants antagonize the action of acetylcholine.
- The principal differences are in onset, duration, elimination, and side-effects. Important factors which modify the action are acid–base and electrolyte changes, muscular diseases, and drug interactions.
- Succinylcholine (suxamethonium) is used for rapid tracheal intubation. Pancuronium, atracurium, vecuronium, or cisatracurium are used to maintain a block.
- Train of four is the most convenient method for monitoring the block following stimulation of the ulnar nerve using a handheld stimulator.
- Indications for the use of a relaxant are tracheal intubation, facilitation of procedures, to assist ventilation, critical gas exchange, tetanus, and transfer of patients.
- Disadvantages of using a relaxant are neurological assessment, disconnection, awareness, disuse atrophy, increased risk of thromboembolism, and increased risk of infection

Physiology

The neuromuscular junction lies between motor nerves and striated muscle. The transmitter acetylcholine is synthesized in the cytoplasm of the prejunctional nerve ending, stored in vesicles, and released in response to a nerve impulse. The receptors for acetylcholine, which are integral with their ion channels, lie on the postjunctional membrane. The simultaneous binding of two acetylcholine molecules is required for a channel to open, whereas occupation of either or both sites by an antagonist molecule will inactivate the channel. The ion channels are selective for small cations, and it is the sudden transmembrane flux of sodium ions through a large number of simultaneously opened channels which is responsible for the generation of an endplate potential and consequently the onward propagation of the impulse. Receptors also exist on prejunctional nerve endings, and these are involved in modulation of impulse transmission. Stimulation of the prejunctional nicotinic receptors by acetylcholine increases transmitter mobilization at higher frequencies.

Pharmacology

Since the introduction of the muscle relaxants in 1942 many potential relaxants have been synthesized, although only a few have become established in clinical practice. Although every agent has been administered to intensive care unit (ICU) patients at some time, we only need to consider five in any detail: succinylcholine (suxamethonium), pancuronium, atracurium, vecuronium, and cisatracurium. The other agents currently available possess a number of undesirable properties which severely limit their use.

The principal differences between the available drugs are speed of onset, duration of action, route of elimination, and side-effects. A rapid onset is of importance when tracheal intubation must be secured without delay, a situation which does not often pertain in the ICU. An intermediate duration of action is convenient because it offers better control than a long-acting drug. The rate of consumption of a short-acting drug would be excessive. Impairment of renal and/or hepatic function are common in the critically ill patient, and ideally any drug should rely as little as possible on either of these systems for elimination. Finally, many critically ill patients have significant cardiovascular instability and drugs should possess minimal side-effects.

A number of pathophysiological changes which may modify the action of a relaxant commonly occur in the critically ill patient. Acidosis (respiratory or metabolic), hypothermia, hypokalemia, hypocalcemia, hypernatremia, and hypermagnesemia may all potentiate a block. Pre-existing muscular or neuromuscular disorders (e.g. myasthenia gravis, myotonias, and muscular dystrophies) prolong a block. Many other drugs may modify the action of muscle relaxants. Important interactions include potentiation of succinylcholine by neostigmine, ecothiopate, alkylating agents, and trimetaphan, and potentiation of the non-depolarizing agents by calcium-channel blockers, aminoglycoside antibiotics, immunosuppressants, furosemide (frusemide), local analgesics, and lithium.

Choice of muscle relaxant

Summary information concerning the four non-depolarizing agents most commonly used in the ICU is given in Tables 1 and 2.

Succinylcholine

Succinylcholine, which is the only depolarizing agent in current clinical use, has the fastest onset and shortest action duration. Its principal use is to secure the airway rapidly, and therefore it is often used in new admissions to the ICU when the time of last food intake is unknown or gastric stasis and a full stomach are suspected. Once the trachea is intubated, it is unusual to continue using succinylcholine.

A number of unwanted side-effects exist including myalgia, masseter spasm, precipitation of malignant hyperthermia, and fasciculations. The side-effect of most potential hazard to the ICU patient is an increase in the serum potassium concentration. Although this increase is only 0.5 to 1.0 mmol/l in normal patients, it is accentuated in patients with burns, muscle trauma, prolonged immobility, and spinal cord injuries. The use of succinylcholine in these patients, or in the presence of an already raised serum potassium, may precipitate a cardiac arrest.

Pancuronium

This synthetic steroid-based non-depolarizing agent has negligible ganglion-blocking activity and is devoid of histamine-releasing potential. It is potent and has a relatively long duration of action. There is a small rise in heart rate (by a combination of vagal inhibition and sympathetic stimulation) and the blood pressure is maintained. It has been a popular mainstay in the ICU for many years. Since it is metabolized in the liver and eliminated via the kidneys, its action is prolonged in patients with liver and/or renal impairment. It is safer to use pancuronium by intermittent bolus doses than by infusion.

Table 1 Basic pharmacodynamic data for four non-depolarizing relaxants

	ED95* (mg/kg)	Approximate infusion rate to maintain steady state (mg/kg/h)	Approximate duration of one ED95 dose (min)
Atracurium	0.23	0.25	20–30
Vecuronium	0.04	0.08	20–30
Pancuronium	0.06	0.03	40–60
Cisatracurium	0.05	0.1	25–35

Data are averages or ranges taken from multiple sources.
*Effective dose to produce 95% response.

Table 2 Elimination (β) half-lives of four non-depolarizing relaxants

	Elimination (β) half-life (min)		
	Normal patients	Patients with hepatic cirrhosis	Patients with renal disease
Atracurium	20	25	22
Vecuronium	54	68	76
Pancuronium	123	208	257
Cisatracurium	29	24	32

Data are averages or ranges taken from multiple sources.

The level of block should be measured regularly in order to guard against cumulation.

Vecuronium

This steroid-based molecule has a similar potency to pancuronium. It possesses good cardiovascular stability and does not release histamine. As a drug with an intermediate duration of action, it is suited to administration by infusion and is popular in many ICUs. Vecuronium is metabolized in the liver and, together with its metabolites (one of which, 3-hydroxy vecuronium, is also a neuromuscular blocker with a potency about 70 per cent of that of vecuronium), is eliminated in the urine. Therefore its action is extended in patients with renal insufficiency and there are reports of cases where recovery from a block has taken many hours (or even days) following an infusion of vecuronium in the presence of renal failure.

Evidence has been accumulating with respect to a possible deleterious effect of vecuronium when given by prolonged infusion in the ICU. Critically ill patients often develop a myopathy as a result of their disease process. This seems to be made worse by the use of a steroid relaxant. However, this so-called 'steroid myopathy' is not clearly understood, and some doubt its existence or disagree over its significance.

Atracurium

Atracurium is an intermediate acting relaxant, is virtually devoid of side-effects, and is suitable for administration by continuous infusion. It releases histamine only at higher doses and this does not appear to be a problem in ICU patients. Its novel mode of metabolism makes it popular for use in the ICU. In the normal healthy individual, about 60 per cent is metabolized by esterases in the liver, while the remainder undergoes spontaneous degradation in the plasma by the Hofmann elimination reaction. Because the Hofmann reaction is completely independent of all organ systems, minimal prolongation of action is seen in the presence of liver and/or renal impairment. Recovery is predictable and it can be reliably expected that the effect will have worn off about 1 h after an infusion is terminated.

The metabolic products of atracurium include laudanosine and acrylates. Although acrylates are potentially hepatotoxic, they are produced in insignificant amounts. More interest has centered around laudanosine, a drug which produces cerebral excitation in certain laboratory animals. Laudanosine accumulates in patients who have received a prolonged infusion of atracurium, particularly in the presence of impaired renal function. It is not known whether laudanosine is toxic in humans and, if so, the plasma concentration at which this might occur. It appears unlikely, on present evidence, that laudanosine is of clinical relevance.

An interesting phenomenon has been observed when atracurium is administered by continuous infusion for several days. After a delay of about 2 to 4 days, resistance to its action begins to develop and an increased infusion rate is required in order to maintain a constant degree of paralysis. This is not observed in every patient and the mechanism is unclear.

Cisatracurium

This substance is one of the component isomers of atracurium, which is itself a racemate. The cardiovascular stability of cisatracurium is equal to that of atracurium and its propensity to release histamine seems to be less. It is approximately four times more potent than atracurium; therefore less is required to secure the same degree of block. This will result in a lower concentration of all metabolites. Clinical experience with cisatracurium in the ICU is limited, but early studies appear promising.

Monitoring neuromuscular blockade in the ICU

Whenever a muscle relaxant is used, its effect should be monitored. The infusion rate (Table 1) can then be adjusted in order to optimize the level of block, thereby avoiding inadequate or excessive block. The optimum technique is yet to be determined. A train of four supramaximal stimuli, delivered from a handheld nerve stimulator attached over the ulnar nerve at the wrist with self-adhesive surface electrodes, is used in the author's ICU. The movements of the thumb are observed by either visual or tactile means. A useful general rule is to maintain the level of block where the first twitch of the train of four is present. If there is no response, the infusion rate of relaxant is decreased, and if there are two or more twitches, the infusion rate is increased. The use of complex equipment designed either to quantify the movements of the thumb using a force or acceleration transducer or to measure the electromyogram of the adductor pollucis muscle are unnecessary for routine clinical purposes.

The use of surface (ECG type) electrodes is convenient, but the response may vary from day to day. It is important to change electrodes at least daily, clean the skin carefully, and maintain a constant position if possible. Accumulation of tissue edema is common in ICU patients, and this may reduce the intensity of what was a supramaximal stimulus. It is usually possible to reduce this problem by exerting firm pressure over the electrodes for 1 to 2 min before stimulating.

Indications for the use of a muscle relaxant

The frequency of use of muscle relaxants fell dramatically from over 90 per cent of ICUs in 1981 to about 16 per cent in 1986. The original perception was that muscle relaxants were an essential part of the sedation regimen of most patients, but this has subsequently been shown not to be the case. At present, muscle relaxants are only used where there is a specific indication.

Tracheal intubation

Tracheal intubation is possible without the use of a relaxant when the patient is weak, has depressed airway protective reflexes, or has received a generous dose of a sedative or anesthetic agent. Intubation may also be performed following local analgesia to the pharynx and larynx. However, the use of a muscle relaxant facilitates intubation and reduces the potential for trauma. It is also particularly useful when the airway has to be secured with the minimum of delay, and in these cases succinylcholine is the drug of choice unless otherwise contraindicated. It must always be remembered that a muscle relaxant must never be administered to a patient unless the operator is certain that the airway can be maintained by some other means (e.g. face mask and oral or nasopharyngeal airway) if tracheal intubation should prove to be difficult.

Facilitation of procedures

The use of a muscle relaxant is highly desirable when certain invasive practical procedures (e.g. rigid bronchoscopy) are to be undertaken.

Assisting ventilation

It is occasionally impossible for a patient to tolerate intermittent positive-pressure ventilation with sedatives alone. The result is marked swings in intrathoracic, intracranial, and intra-abdominal pressures as the patient attempts to breathe out of phase with the ventilator. Many consider the use of a muscle relaxant in patients with a raised intracranial pressure to be mandatory.

Critical gas exchange

In patients with a poor lung compliance, there may be difficulty in achieving satisfactory gas exchange with sedative agents alone. The addition of a muscle relaxant often improves oxygenation.

Tetanus

A muscle relaxant will reduce muscle spasms and facilitate intermittent positive-pressure ventilation.

Transfer of patients

Many clinicians feel that it is appropriate to paralyze patients during transfer between departments and between hospitals.

Disadvantages of using a muscle relaxant

Neurological assessment

When a patient has received a muscle relaxant, it is difficult or even impossible to perform a meaningful neurological assessment. Focal or localizing neurological signs may be missed, leading to a delay in treatment with potentially serious consequences. Epileptiform seizures may also be masked. If it is decided to use a relaxant, it must be discontinued at regular intervals to allow assessments to be made.

Disconnection

In the presence of a relaxant, spontaneous ventilation is not possible in the event of accidental disconnection. However, the high nurse-to-patient ratio and comprehensive ventilator alarms make this a very unlikely situation.

Awareness

Inadequate sedation may be masked by the presence of a relaxant. It is essential to be certain that the patient is adequately sedated before administering a relaxant.

Disuse atrophy

Muscle weakness is present in many critically ill patients as a consequence of their disease process. It is possible that long-term use of a muscle relaxant (particularly a steroid-based drug) may exacerbate this phenomenon.

Increased risk of thromboembolism

The incidence of deep venous thrombosis and pulmonary embolism may be more common in the paralyzed patient. However, the evidence is conflicting.

Increased incidence of infection

Although immobility may predispose to infection, regular physiotherapy and postural care reduces this risk to a minimum.

Bibliography

Hunter, J.M. (1994). Neuromuscular blocking agents in intensive therapy. In *Applied neuromuscular pharmacology* (ed. B.J. Pollard). Oxford University Press.

Pollard, B.J. (1995). Muscle relaxants in the intensive care unit. In *Muscle relaxants in anesthesia* (ed. N.J.N. Harper and B.J. Pollard). Edward Arnold, London.

Sunder, N. and Martyn, J.A.J. (1994). Muscle relaxants. In *Essentials of critical care pharmacology* (2nd edn) (ed. B.Chernow), pp. 212–23. Williams & Wilkins, Baltimore, MD.

17.1.21 Nitric oxide

A. Mebazaa and D. M. Payen

Key messages

- In adults, NO is indicated for patients with acute respiratory distress syndrome with severe hypoxia, patients with severe pulmonary hypertension, or patients undergoing heart transplantation as a test of the reactivity of the pulmonary vascular bed.
- NO is generally administered to mechanically ventilated patients on the inspiratory limb of the respirator either continuously or during the inspiratory phase.
- Inhaled NO is used at 5 to 15 ppm which can be accurately monitored by bedside chemiluminescence apparatus. At these concentrations of NO, nitrogen dioxide (NO_2) and methemoglobin are usually well below toxic levels.

Introduction

Nitric oxide (NO), which is normally released by endothelial cells including those of pulmonary bed, serves to modulate vascular tone by relaxing the subjacent smooth muscle cells. Recently, administration of NO by inhalation was found to produce selective vasodilatation of the pulmonary circulation (Higgenbottam *et al.* 1988). Once in the alveoli, inhaled NO diffuses to the smooth muscle cells of the pulmonary vessels and to the blood perfusing the lung where it is inactivated. Several mechanisms of inactivation have been suggested, including binding of hemoglobin in red cells or other blood proteins by NO.

Low fractions of inhaled NO (18–36 ppm) have been reported to lower pulmonary arterial pressure and improve arterial oxygenation in adult respiratory distress syndrome patients in the intensive care unit (ICU) (Rossaint *et al.* 1993). Several authors have suggested that inhaled NO is distributed predominantly to well-ventilated alveoli and is not delivered to collapsed or fluid-filled areas of the lung. Consequently, local vasodilation of well-ventilated lung regions induces a gradient for regional conductance, leading to diversion of pulmonary arterial blood flow towards well-ventilated alveoli and improving ventilation–perfusion ($\dot{V}/\dot{Q}$) mismatch and arterial oxygenation. We have recently shown that such a gradient can be enhanced by using almitrine bismesylate, a vasoconstrictive agent that acts preferentially on the vessels of poorly ventilated lung regions, reinforcing hypoxic vasoconstriction (Payen *et al.* 1993, 1996).

Administration of inhaled NO

Inhaled NO is generally administered to mechanically ventilated patients and can be added before the ventilator at a fresh gas inlet. The source is a tank containing a known fraction of NO in nitrogen which ranges from 25 to 2000 ppm (1 ppm = 40 nM). A gas-mixing system dilutes NO in additional pure nitrogen from another tank. The NO–N_2 mixture is connected to the gas inlet of the ventilator. A home-made version is provided on the Siemens Servo 900C ventilator. Another prototype NO mixing method was recently described with the new Siemens Servo 300 ventilator.

When NO is delivered at the fresh gas inlet, a soda lime canister may be added to the inspiratory limb downstream of the ventilator to minimize the fraction of nitrogen dioxide (NO_2). NO_2 is produced by oxidation of NO in the presence of a high oxygen fraction, particularly if there is prolonged contact between these two gases. The final acceptable NO_2 fraction delivered to the ventilator must be less than 1 ppm.

Another delivery system has been described where NO in N_2 is directly administered to the inspiratory limb either continuously or during the inspiratory phase. NO is delivered via a nitrogen flowmeter which delivers flows between 10 and 1500 ml/min.

Because NO reacts rapidly with oxygen, we have chosen a delivery system that limits the time of contact between the two gases (Payen *et al.* 1993). At the majority of centers in France, NO is administered continuously via a fenestrated silicone catheter of diameter 1.3 mm (Vygon, Paris, France) positioned within the tracheal tube so that the distal tip of the catheter is 2 cm above the distal end of the tube. Thus the tracheal tube protects the tracheal wall from direct contact with NO which could induce local inflammation and damage. The tube and catheter positions are easily checked by chest radiography. NO is driven into the tracheal tube by continuous flow from a cylinder containing 225 ppm NO in nitrogen (CFPO, Meudon La Fort, France). The flow rate is adapted to provide the therapeutic fraction (5–15 ppm) according to the following formula:

$$\text{NO flow rate (l/min)} = \frac{15\ \text{ppm}}{225\ \text{ppm}} \times \text{minute ventilation (l/min)}.$$

For example, if the minute ventilation is 9 l/min, the NO flow rate must be 0.6 l/min to ensure delivery of 15 ppm NO. Tidal volume is reduced to compensate for the added volume of NO and nitrogen, and to keep the minute ventilation constant. An additional advantage of this delivery system is that, if the ventilator fails, hand ventilation does not interrupt NO administration. However, ventilator breakdown will result in the provision of the tank fraction of NO (225 ppm), which might be toxic.

NO can be administered to non-intubated patients through a facemask connected to a non-rebreathing circuit. Such a system consists of a 5-liter reservoir bag and a one-way valve to separate inspired and expired gas. The fraction of NO–N_2 mixture delivered to the mixing reservoir ranges from 5 to 40 ppm.

Dosage

An NO fraction as low as 60 to 250 ppb may be adequate for some adult patients with acute respiratory distress syndrome. However, this patient population has a wide variation in sensitivity to NO. We use 5 to 15 ppm NO to improve hypoxia in acute respiratory distress syndrome and to reduce pulmonary hypertension.

The absence of tachyphylaxis with inhaled NO, even after administration for 2 months, is a pharmacological advantage. Nevertheless, sudden discontinuation of NO delivery may be followed by severe pulmonary hypertension rebound which appears to be prevented if a slow weaning process (12–48 h) is used.

Clinical uses of inhaled NO

Since 1996, European regulations have allowed administration of inhaled NO to adults for treatment of the following diseases only.

NO is indicated for patients having acute respiratory distress syndrome with severe hypoxia, patients with severe pulmonary hypertension, or patients undergoing heart transplantation as a test of the reactivity of pulmonary vascular bed. In patients with severe acute respiratory distress syndrome, inhaled NO decreases pulmonary hypertension and vascular resistance but has no effect on systemic arterial pressure and vascular resistance. Arterial oxygenation often appears to be improved. We have recently shown that a combination of inhaled NO and intravenous administration of almitrine improves $\dot{V}/\dot{Q}$ matching and has a much greater effect on arterial oxygenation than either agent alone (Payen *et al.* 1993). This beneficial effect of almitrine is probably related to predominant vasoconstriction of pulmonary vessels in non-ventilated lung areas.

In addition to the effects on gas exchange in acute lung injury, inhaled NO may have beneficial effects on lung inflammation in acute respiratory distress syndrome. After 4 days of NO inhalation (18 ppm), polymorphonuclear neutrophils extracted from the bronchoalveolar lavage fluid of patients with acute respiratory distress syndrome were less markedly activated than those from similar patients treated without NO. In addition, the high levels of interleukins 6 and 8 in the bronchoalveolar lavage fluid supernatants fell only in the group of patients receiving NO. Moreover, NO inhalation seemed to reduce the oxidative burst of bronchoalveolar lavage polymorphonuclear neutrophils.

Inhaled NO is also indicated in patients with severe pulmonary hypertension and right ventricular failure. It reduces the pulmonary vascular resistance and pulmonary hypertension without altering the systemic vascular resistance and mean arterial pressure. Several authors have reported that, in patients with either chronic pulmonary hypertension or cardiac disease, NO inhalation at 10 to 40 ppm decreased pulmonary vascular resistance by 5 to 70 per cent without affecting systemic vascular resistance. Inhaled NO may also improve right ventricular function by reducing its afterload with an improvement in right coronary perfusion pressure. This may be observed in patients whose pulmonary hypertension is aggravated following mitral valve replacement for mitral stenosis. Administration of inhaled NO for 12 to 24 h reduces pulmonary hypertension and improves right ventricular function.

NO measurements

The most widely employed methods for monitoring NO and NO_2 in respiratory gas are chemiluminescence and electrochemical detection. Clinicians usually attempt to measure a very small fraction of NO_2 in an oxygen-rich gas mixture with a substantial level of NO. In addition, this mixture undergoes cyclic pressure changes and may have a high humidity.

Chemiluminescence monitors in use at present measure NO and NOx, and the NO_2 fraction is given by the difference between these two values. Fractions of NO and NOx are measured continuously using a fast-response-time chemiluminescence apparatus (Eco Physic, Massy, France). Intratracheal gas is continuously aspirated through the endotracheal tube. At an aspiration flow rate of 150 ml/min, the response time is approximately 30 s and only mean fractions of NO and NOx can be measured accurately. When an aspiration flow rate of 1000 ml/min is selected, the response time is approximately 200 ms and inspiratory and expiratory NO fractions can be measured.

Electrochemical techniques for the measurement of NO and NO_2 levels are rapidly gaining popularity as they are not as costly as the chemiluminescence technique. Although measurements of NO levels have an acceptable accuracy, there have been difficulties with measurement of NO_2 levels.

The frequency of measurement is a subject of much debate. Some authors measure NO and NO_2 levels continuously for days and weeks, so that each patient requires an NO monitor. In our institution, the NO fraction is regularly checked (two to six times daily) on the ventilator expiratory limb and in the tracheal tree by a chemiluminsecence method. This allows us to monitor the NO and NOx fractions of several patients.

Potential toxic effects of inhaled NO

Inhaled NO may react in the bronchial tree with either O_2 or superoxide (O_2^-) to produce agents that are potentially toxic for the lung (Gaston *et al.* 1994). The reaction of NO with O_2 (at $FiO_2 = 0.21$) to produce NO_2 is relatively slow. However, in therapeutic settings and in ventilated patients requiring FiO_2 close to 1, NO conversion to toxic NOx occurs very rapidly (1 s) and it is even quicker in the presence of humidified gas. NO also reacts rapidly with O_2^-, which is present at high levels in acute lung injury, to form the peroxynitrite anion ($OONO^-$). At high fractions, NOx and the peroxynitrite anion are toxic to the lung. At low fractions, recent evidence shows that peroxynitrite anion may play a role in defending the lung against microbial invasion.

During inhalation, the NO that reaches blood binds hemoglobin and produces methemoglobin. The balance between oxidation and reduction of methemoglobin results in a blood level of less than 3 per cent in normal subjects. It should be remembered that, in patients with congenital methemoglobinemia, 10 to 20 per cent of methemoglobin is tolerated without ill effects.

Potential toxic effects in patients treated with inhaled NO were recently investigated by increasing fractions of inhaled NO (100–5000 ppm) in patients with acute respiratory distress syndrome. It was shown that, for the highest fraction of NO, although FiO_2 was 1.0, the intratracheal fraction of inhaled NO_2 was 100 ppb and methemoglobin concentrations always remained below 1.4 per cent. In our experience, in 250 patients treated with inhaled NO (5–15 ppm) the measured NO_2 fraction was always 1 to 2 ppm and the methemoglobin concentration was below 2.5 per cent. Thus it can be concluded that, in patients treated with inhaled NO at doses below 15 ppm, NO_2 and methemoglobin are usually much below toxic levels.

Conclusions

Although early investigations clearly demonstrate the beneficial effect of inhaled NO on pulmonary hypertension and arterial oxygenation, the long-term impact on morbidity and mortality remains unknown. Prospective and randomized trials are examining the clinical effectiveness and safety of this novel therapy.

References

Gaston, B., Drazen, J.M., Loscalzo, J., and Stamler, J.S. (1994). The biology of nitrogen oxides in the airways. *American Journal of Respiratory Critical Care Medicine*, **149**, 538–51.

Higgenbottam, T., Pepke-Zaba, J., Scott, J., Woolman, P., Coutts, C., and Wallwork, J. (1988). Inhaled endothelium derived-relaxing factor (EDRF) in primary hypertension (PPH). *American Review of Respiratory Disease*, **137** (Supplement), 107 (abstract).

Payen, D.M., Gatecel, C., and Plaisance, P. (1993). Almitrine effect on nitric oxide inhalation in adult respiratory distress syndrome. *Lancet*, **341**, 1664–5.

Payen, D.M., Gatecel, C., Kermarrec, N., and Beloucif, S. (1996). Almitrine and inhaled nitric oxide in acute respiratory failure. In *Nitric oxide and the lung* (ed. W.M. Zapol and K.D. Bloch), pp. 313–32. Marcel Dekker, New York.

Rossaint, R., Falke, K., Lopez, F., Slama, K., Pison, U., and Zapol, W.M. (1993). Inhaled nitric oxide for the adult respiratory distress syndrome. *New England Journal of Medicine*, **328**, 399–405.

17.1.22 Oxygen

R. M. Jackson

Key messages

- Oxygen toxicity is mediated by reactive oxygen species, including superoxide, hydrogen peroxide, and hydroxyl radicals.
- Enzymatic and low-molecular-weight antioxidants protect cells from reactive oxygen species, but can be overwhelmed in hyperoxia.
- Arterial hypoxemia due to extreme $\dot{V}/Q$ mismatching (intrapulmonary shunt) requires high levels of inspired oxygen.
- Pulmonary oxygen toxicity resembles acute respiratory distress syndrome in its pathophysiology and morphology.
- Many drugs used clinically, including chemotherapeutic agents and corticosteroids, may exacerbate pulmonary oxygen toxicity.
- No clinically available agents (with the possible unproved exception of surfactants) mitigate human manifestations of oxygen toxicity.

Pharmacology

Mechanisms of oxygen toxicity

Normal cellular respiration produces partially reduced reactive oxygen metabolites. Mitochondria reduce most molecular oxygen completely to water, with the addition of four electrons by the mitochondrial respiratory chain. A fraction of oxygen used metabolically (usually less than 5 per cent) undergoes the addition of a single electron to form the superoxide anion (O_2^-); this process occurs mainly in the ubiquinone–cytochrome B and NADH dehydrogenase regions of the electron transport chain. Production of incompletely reduced oxygen increases proportionately with the increase in intracellular PO_2. At physiological pH, superoxide is protonated to form the perhydroxyl radical ($O_2H^·$), a stronger oxidant that diffuses through cell membranes. Superoxide formed within mitochondria may interact with cellular constituents directly, but most reacts further to form hydrogen peroxide and reactive hydroxyl radicals ($OH^·$) by redox cycling in the presence of catalytic iron or copper. In hyperoxia, the elevated intracellular PO_2 and the law of mass action dictate that mitochondria and other intracellular organelles will form excess superoxide. Iron is normally present in low-molecular-weight complexes, and the production of excess superoxide results in the iron-catalyzed formation of hydroxyl radicals. These highly reactive radicals engage in destructive oxidation and reduction reactions with cell components.

Superoxide can also react rapidly (rate constant greater than 10^9 l/mol/s) with ·NO formed by inflammatory cells (e.g. alveolar macrophages) or lung parenchymal cells. This mechanism may be particularly important during sepsis and the reoxygenation phase of ischemia–reperfusion injury. Excess ·NO reacts with superoxide to form peroxynitrite ($ONOO^-$), which mediates several cytotoxic effects of ·NO such as oxidation of Fe–S centers in respiratory enzymes. Peroxynitrite is a strong long-lived oxidant that decomposes at physiological pH to form a species with the reactivity of hydroxyl radicals ('$OH^· + NO_2$'). Important proteins that contain tyrosine residues, including surfactant protein A and manganese superoxide dismutase, may be nitrated and functionally inhibited by peroxynitrite.

Antioxidant systems

Production of free radicals and hydrogen peroxide is a normal aspect of cellular metabolism, and intracellular enzymes usually eliminate the most toxic products. The prototypes of these antioxidant enzymes are metalloproteins termed superoxide dismutases that catabolize superoxide by extremely rapid (rate constant of 10^9 l/mol/s) conversion to hydrogen peroxide. Catalase (found in peroxisomes and mitochondria) and glutathione peroxidase (found in cytosol and mitochondria) sequentially degrade hydrogen peroxide to water. Glutathione peroxidase removes the products of lipid peroxidation and hydrogen peroxide. Two related cytoplasmic enzymes, glutathione reductase (NADPH dependent) and glucose-6-phosphate dehydrogenase (NADPH producing) participate in antioxidant defense by generating intracellular reduced glutathione. Glutathione is the primary non-protein sulfhydryl intracellular antioxidant. Other non-enzymatic antioxidants, including vitamin A, α-tocopherol, ascorbate, cysteine, ceruloplasmin, urate, and hemoglobin, are present within cells and extracellular fluid. ·NO also functions as an antioxidant by quenching lipid peroxidation chain reactions.

Therapeutic use

Effects of oxygen in disordered gas exchange

Oxygen delivery to the tissues depends on the oxygen content of the blood, the oxygen–hemoglobin binding characteristics, the cardiac output, and undefined factors controlling distribution and diffusion of oxygen from capillary blood. Until it becomes possible to monitor intracellular Po_2 in cells perfused by the vascular bed of clinical interest, dose (Fio_2)–response (intracellular Po_2) relationships will remain assumptions. General relationships exist between inspired Po_2 and changes in arterial Po_2 that are predictable based on knowledge of the intrapulmonary mechanisms of hypoxemia.

If V/Q relationships remain normal during alveolar hypoventilation due to sedation or neuromuscular disease, alveolar Po_2 will increase directly with Pio_2. The alveolar gas equation, which is given in simplified form by

$$PAo_2 = [Fio_2 \times (BP - 47\,\text{mmHg})] - Paco_2/0.8$$

where BP is barometric pressure, then predicts the arterial Po_2 (in mmHg). Patients with decompensated chronic bronchitis and emphysema are often admitted to the intensive care unit (ICU) with respiratory failure characterized by both hypoxemia and hypercapnia. Relatively small increases in Fio_2, often to less than 0.4, substantially correct alveolar Po_2 and permit maintenance of 90 per cent oxygen saturation of hemoglobin. The arterial Po_2 response varies greatly from patient to patient depending on the pattern of breathing and the underlying disorder causing $\dot{V}/Q$ mismatching. Arterial Pco_2 may increase after starting oxygen therapy because of increased physiological dead-space (improved ventilation of poorly perfused regions), although neither minute ventilation nor respiratory drive decreases significantly. Increased Fio_2 overcomes diffusion impairment with improvement in Pao_2, because the rate of oxygen transfer from the lung to the blood varies with alveolar Po_2.

Arterial Po_2 is less responsive to increased Fio_2 when anatomical dead-space is present or when portions of the lung have extreme $\dot{V}/Q$ mismatching, which is the functional equivalent of shunt in acute respiratory distress syndrome. Application of supplemental oxygen therapy may also improve systemic oxygen delivery by increasing the oxygen content of arterial blood. Small increases in Pao_2 may be amplified functionally as increases in oxygen content, which favorably affect mixed venous Po_2 and tissue oxygenation. Oxygen therapy may also be of benefit in disease states characterized by systemic hypoperfusion (e.g. hemorrhagic or cardiogenic shock) even without hypoxemia. However, pharmacologically increased oxygen delivery does not improve survival in septic shock.

Table 1 Oxygen delivery equipment for spontaneous breathing

Fio_2*	Device	O_2 flow (l/min)
0.21–0.40	Nasal cannula	1–6
0.35–0.50	Face mask	5–10
0.50–0.90	Reservoir mask (rebreathing)	5–15
0.50–0.90	Reservoir mask (non-rebreathing)	5–15
0.24–0.60	Venturi	4–15
0.50–1.00	CPAP†	10–15

CPAP, continuous positive airways pressure.
*Fio_2 is approximate and depends on mask fit and ventilatory pattern.
†Tight mask seal, high gas flow rate, and expiratory resistance are required to maintain positive airways pressure.

Oxygen delivery equipment

Equipment currently in use in the ICU allows delivery of increased levels of inspired oxygen ranging from a few per cent above that in room air to pure oxygen at atmospheric pressure. Table 1 lists oxygen therapy equipment available for non-intubated patients and the indications for each type. Although less commonly used for hypoxemia without ventilatory failure, mechanical ventilators with which the Fio_2 can be varied from 0.21 to 1.00 may be required to treat hypoxemic states refractory to oxygen therapy by face mask.

Physiological responses to oxygen

Physiological responses to oxygen are completely reversible on resumption of air breathing. They differ in this functional character from those resulting in the irreversible pathological changes of oxygen toxicity. The most evident physiological responses to oxygen occur in the cardiovascular system.

Oxygen breathing results in decreased pulse rate and pressure, while diastolic blood pressure increases significantly. Significant decreases in cardiac index accompany increases in peripheral vascular resistance and systolic and diastolic blood pressures. Atropine abolishes the heart rate response to oxygen, suggesting vagal control of the bradycardia. Regional circulations appear to respond variably to hyperoxia, but in general blood flow decreases with the rise in venous oxygen saturation. Patients with acute myocardial infarction breathing 40 per cent oxygen had increases in systemic blood pressure and detectable decreases in cardiac output. In acute myocardial infarction beneficial effects of oxygen appear due to increasing coronary arterial oxygen tension and decreasing myocardial oxygen consumption.

Side-effects

Symptoms and physiological manifestations

Four normal men remained in 98 per cent oxygen for periods ranging from 30 to 74 h. During this time oxygen toxicity became clinically evident as symptoms affecting both the respiratory and nervous systems. Cough and chest pain occurred in all four, and paresthesias and anorexia occurred in three. There was a progressive statistically significant fall in vital capacity in all subjects. In all but one, vital capacity returned to normal quickly; however, the remaining subject had a delayed recovery lasting for several weeks. Diffusing capacity decreased during the oxygen exposure, but returned to normal in the follow-up period.

The volume–pressure curves of oxygen-breathing subjects were altered in the direction of decreased volume for a given pressure (i.e. less compliant) compared with control subjects breathing air. Some subjects could not inhale to total lung capacity and seemed unable to generate maximal transpulmonary pressure. Non-uniform behavior of the lung after atelectasis led to disproportionate receptor firing and an early inhibition of inflation.

Clinical manifestations

Five hospitalized patients with irreversible brain damage who were ventilated with pure oxygen developed significantly worse lung function. Gas exchange worsened and a decrease in arterial oxygen tension occurred. Both intrapulmonary shunt and the ratio of dead-space to tidal volume (reflecting abnormalities in $\dot{V}/\dot{Q}$ ratio) increased significantly in the group exposed to pure oxygen. Their lungs showed radiographic evidence of edema and were heavier at autopsy.

Patients who had undergone cardiac surgery were randomized to receive either pure oxygen or a gas mixture sufficient to maintain an arterial Po_2 between 80 and 120 mmHg (10.7 and 16 kPa) for 24 to 48 h. During this time, no difference in intrapulmonary shunt, lung or respiratory system compliance, or the ratio of dead-space to tidal volume could be detected.

Detection of pulmonary oxygen toxicity by traditional physiological testing is insensitive. Decreased tracheal mucous velocity, which can be assessed by direct observation through a fiber-optic bronchoscope, is a very early manifestation of oxygen toxicity. Investigators have detected changes in bronchoalveolar lavage constituents in normal subjects who were exposed to oxygen for approximately 17 h. Albumin and transferrin concentrations increased in lavage fluid from these subjects. Cultured alveolar macrophages from subjects exposed to oxygen released increased amounts of fibronectin and alveolar macrophage-derived growth factor for fibroblasts. The changes disappeared when the subjects were studied 2 weeks later, but they suggested that functional abnormalities may occur very early in oxygen exposure that later initiate fibrosis of the alveolar wall. Table 2 summarizes the approximate time course of oxygen toxicity in humans.

Human studies

The pulmonary function of many patients supported by early mechanical ventilators underwent gradual deterioration. Most patients were ventilated with pressure-limited oxygen-driven ventilators, which resulted in a mean oxygen concentration of over 90 per cent. Patients exposed to the highest concentrations of oxygen for long periods had heavier and more consolidated lungs. Microscopic studies showed an early exudative phase characterized by congestion, alveolar edema, intra-alveolar hemorrhage, and a fibrin exudate together with prominent alveolar membranes. Later, a proliferative phase characterized by alveolar and septal edema and fibroblast proliferation occurred, together with fibrosis and hyperplasia of alveolar lining cells.

Table 2 Time course of pulmonary oxygen toxicity in humans

Manifestation	Time (h)
Decreased tracheal mucous velocity	6
Tracheobronchitis	12
Physiological changes (FVC, C_{lung})	24–48
Gas exchange abnormalities	30
Edema	72–96
Fibrosis	>96

FVC, forced vital capacity; C_{lung}, lung compliance.

Patients exposed to 40 to 100 per cent oxygen for periods of up to 30 days have undergone autopsy. The earliest pathological changes involved alveolar type I cells and endothelial cells. Septal edema was noted, and endothelial sloughing followed by fibrin thrombus formation occurred. Hyaline membranes were found after exposure for about 7 days, and re-epithelialization of the alveoli occurred by proliferation of granular pneumocytes. Septal proliferation and deposition of interstitial collagen, elastin, and fibrosis occurred after 10 days.

More recently, detailed studies of the morphology and time course of oxygen toxicity in baboons have clarified the time course of injury in primates (Fig. 1). The earliest detectable injury includes endothelial injury followed by neutrophil aggregation and thickening of the alveolar interstitium. Later phases of the injury are characterized by loss of alveolar type I cells, denudation of basement membranes, and proliferation of alveolar type II cells.

Similar acute changes that progress to fibrosis occur after cutaneous burns and following smoke inhalation, influenza pneumonia, thoracic irradiation, and ingestion of toxins such as paraquat. The histological resemblance of pulmonary oxygen toxicity to diffuse alveolar damage from other causes (like sepsis and trauma) has led some investigators to propose that oxygen toxicity is a major cause of the acute respiratory distress syndrome. However, near uniformity of oxygen administration to patients with respiratory failure of diverse etiologies makes isolation of oxygen as the primary cause of acute respiratory distress syndrome problematic.

Altered pharmacodynamics

Drug interactions

Many drugs used clinically affect oxygen tolerance, frequently enhancing oxygen toxicity. These fall into several broad categories: drugs that increase tissue oxygen consumption, drugs that undergo redox cycling or produce free radicals during metabolism, and drugs that affect endogenous antioxidant systems.

Epinephrine (adrenaline), norepinephrine (noradrenaline), thyroid hormone, and hyperthermia all increase sensitivity to pulmonary oxygen poisoning. Increased cellular metabolic rate is a common mechanism. There have been no studies investigating the effects of more commonly used adrenergic agents such as dopamine or dobutamine in adult patients receiving hyperoxic mechanical ventilation. As

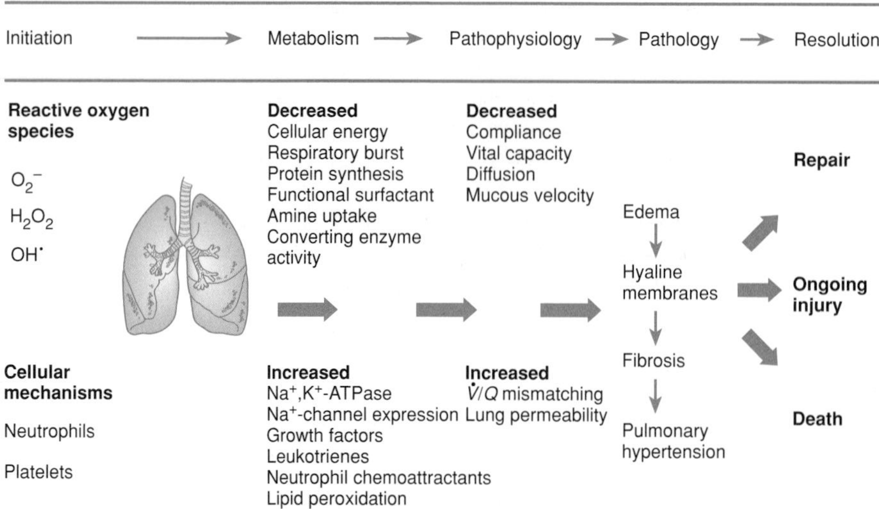

Fig. 1 Pathogenesis of pulmonary oxygen toxicity. Oxygen toxicity is initiated by reactive oxygen metabolites and inflammatory mediators derived from granulocytes and platelets. Initially detectable as impaired cellular metabolism, oxygen toxicity evolves through stages in which physiological or pathological manifestations predominate. The lung injury may resolve or be perpetuated as fibrosis.

well as increasing cellular oxygen utilization, thyroxine lowers lung glutathione content. In contrast, dimethylthiourea, a sulfhydryl compound, protects rats from oxygen toxicity.

Corticosteroids are frequently administered to patients with acute respiratory failure. Mice pretreated with methylprednisolone (10 mg/kg/day for 7 days) suffer significantly greater mortality in hyperoxia. Steroid pretreatment inhibited phospholipid synthesis in lung slices, suggesting that methylprednisolone has an unfavorable effect on the surfactant system of mature lung. The timing of corticosteroid administration in acute lung injury appears to be critical. Methylprednisolone decreases the survival time of adult rats in hyperoxia and minimizes adaptive increases in lung superoxide dismutase. In contrast, when dexamethasone is administered to rats near the end of a 72-h exposure to hyperoxia (when lung inflammation is most pronounced) survival may be improved and lung damage minimized.

Bleomycin, a chemotherapeutic agent, causes acute and chronic lung damage attributed in part to the formation of hydroxyl radicals in the vicinity of cell nuclei. Lung damage due to bleomycin is synergistic with many other chemotherapeutic agents. Simultaneous administration of bleomycin and high Fio_2 produces accelerated lung injury and progressive pulmonary fibrosis. Despite isolated clinical reports to the contrary, in animal models administration of bleomycin well before oxygen exposure results in less rather than more lung injury. This discrepancy is due partially to differences in the acute and chronic effects of bleomycin on lung antioxidant defenses. Lung antioxidant enzymes initially decrease after intratracheal bleomycin administration but increase later.

Cyclophosphamide and 1,3-bis-(2-chloroethyl)-1-nitrosourea (BCNU) similarly increase hyperoxic lung injury while decreasing antioxidant defenses. Nitrofurantoin increases hyperoxic lung injury because reduction of the nitro group of the antibiotic in the presence of oxygen leads to intracellular formation of both superoxide and hydrogen peroxide.

Many other agents used clinically have been investigated in experimental animal models of oxygen toxicity, which may not apply directly to human patients. It is unlikely that similar investigations will ever be performed in humans, and so the animal studies form the only available database.

Pharmacological modulation of oxygen toxicity

To date no effective therapeutic modality has been introduced to reduce oxygen toxicity in adult humans ventilated with high Fio_2. However, recent laboratory studies suggest several potentially useful approaches. Pretreatment of rats with low-dose endotoxin improved survival in above 95 per cent oxygen, while reducing both acute and chronic lung injury. Interstitial edema and neutrophil infiltration occurred in rats pretreated with endotoxin in 100 per cent oxygen, but was much less than in non-pretreated controls. Progressive stepwise increases in Fio_2, which are common clinically, result in accelerated mortality of rats in 95 per cent oxygen, an effect that can also be ameliorated by endotoxin.

After endotoxin pretreatment, lung superoxide dismutase, glutathione peroxidase, and catalase activities increase rapidly during hyperoxic exposure. It also stimulates monocyte production of interleukin 1 (IL-1) and tumor necrosis factor-α, which both individually protect rats from oxygen toxicity. Increases in lung antioxidant enzymes, analogous to those observed in endotoxin-pretreated rats, occur after IL-1 and tumor necrosis factor-α pretreatment in hyperoxia.

Liposome-encapsulated antioxidant enzymes increase intracellular superoxide dismutase and/or catalase activities, and they protect endothelial cells from oxygen toxicity. Intravenous injection of liposome-encapsulated superoxide dismutase and catalase improves survival of adult rats maintained in 100 per cent oxygen, but antioxidant enzyme treatment does not eliminate lung injury. A protective effect is also apparent when liposome-encapsulated superoxide dismutase and catalase are administered intratracheally.

Conjugation of antioxidant enzyme proteins to polyethylene glycol, a high-molecular-weight alcohol, increases circulating lifetime and cell permeability. Exposure of cultured endothelial cells to polyethylene glycol–superoxide dismutase and polyethylene glycol–catalase increases intracellular superoxide dismutase and catalase activities, and makes cells more resistant to damage from reactive oxygen species. Rats pretreated with polyethylene glycol–superoxide dismutase and polyethylene glycol–catalase are protected from pulmonary oxygen toxicity. These rats maintained more normal lung glutathione levels and had less pulmonary edema in 100 per cent oxygen than did rats pretreated with inactivated antioxidant enzymes conjugated with polyethylene glycol. Decreased immunogenicity, increased circulating half-lives, and cell permeability could make such enzymes useful clinical anti-inflammatory agents.

Because oxygen toxicity impairs the surfactant system, attempts have been made to reduce oxygen toxicity by administration of exogenous surfactant. Intratracheal administration of exogenous surfactant significantly protects rabbits from oxygen toxicity. This therapy may prove very useful in neonatal respiratory distress syndrome characterized by endogenous surfactant deficiency and requires prolonged high Fio_2. Trials of its efficacy are also under way in acute respiratory distress syndrome patients who commonly require hyperoxic therapy.

Bibliography

Crapo, J.D. (1986). Morphologic changes in pulmonary oxygen toxicity. *Annual Review of Physiology*, **48**, 721–31.

Fridovich, I. (1978). The biology of oxygen radicals. *Science*, **201**, 875–80.

Jackson, R.M. (1990). Oxygen-induced lung injury: molecular, pharmacologic, and therapeutic aspects. *Clinics in Chest Medicine*, **11**, 73–86.

17.1.23 Prostanoids

A. Brinkmann and P. L. Radermacher

Key messages

- Prostanoids are potent regulators of platelet aggregation and smooth muscle contraction with vasoconstrictive or vasodilatory effects.
- The mechanism of action is via cell surface receptors to increase intracellular cAMP concentrations.
- Therapeutic use in critical care includes pulmonary vasodilatation in acute respiratory distress syndrome and sepsis and as an anticoagulant for the extracorporeal circulation.

Introduction

Prostanoids are among the most prevailing of the naturally occurring physiologically active endogenous substances and have been detected in almost every tissue and body fluid. Prostanoids play an important role as local mediators in the regulation of regional perfusion and metabolism.

Pharmacology

Biochemistry and synthesis

The nomenclature 'prostanoids' encompasses both prostaglandins and thromboxane. These substances are produced by partial oxygenation of arachidonic acid, a natural ubiquitous constituent of the phospholipid in cell membranes. The liberation of arachidonate from the cell membranes is catalyzed by phospholipase A_2. This is the rate-limiting step of prostanoid synthesis.

Prostanoids can be formed by virtually every cell in the cyclo-oxygenase pathway, with minute amounts producing a vast array of potent effects. Thromboxane A_2 (TXA_2) is the main cyclo-oxygenase product of the platelets. Epoprostenol (prostacyclin, PGI_2) is primarily synthesized by endothelial cells. The main biological effects of the most common prostanoids are summarized in Table 1. The letters PG denote the word prostaglandin, and the third letter signifies the structure of the cyclopentane ring. The subscript that follows the third letter indicates the number of double bounds in the structure as well as the fatty acid precursor (Vane and O'Grady 1993).

Mechanism of actions

Prostanoids act on specific cell-surface receptors to stimulate the adenylate cyclase or the phosphate–diacylglycerol pathway resulting in elevated intracellular cAMP concentrations (Vane and O'Grady, 1993).

Metabolism

In general, metabolism leads to a reduction or inactivation of the biological effects of the prostanoid, whilst at the same time facilitating renal clearance.

Metabolism can occur in a number of tissues, particularly in the lung, the liver, and the kidney. The inactivation of prostanoids is rapid and mostly limited to the local site of their release. The lungs occupy a strategic position between the venous and arterial circulation. About 90 per cent of prostanoids escaping the local circulation are removed by a single passage through the lung via enzymatic degradation. Epoprostenol represents an exception of this rule because its pulmonary metabolism is negligible. Epoprostenol and TXA_2 are chemically unstable and hydrolyze spontaneously to the stable metabolites 6-keto-$PGF_{1\alpha}$ and thromboxane B_2. The elimination half-life of prostanoids ranges from 30 s to about 5 min (Vane and O'Grady 1993).

Physiological effects

Cardiovascular system

A local generation of prostanoids is involved in the control of vascular tone in humans.

PGE_2, PGE_1, and epoprostenol represent potent vasodilators in almost every vascular bed studied because of a relaxation of vascular smooth muscles in large and small arterial vessels. Both vasodilation and enhancement of thromboresistance by inhibiting the clumping and sticking of platelets and leukocytes to the vascular endothelium result in improved microcirculatory function.

TXA_2, which is a strong vasoconstrictor, causes biological effects that are generally the opposite of those produced by vasodilatory prostaglandins (Vane and O'Grady 1993)

Table 1 The main biological effects of prostanoids

	Target cells	Biological effect
Thromboxane A_2 (TXA_2)	Blood vessels	Vasoconstriction
	Platelets	Aggregation
Epoprostenol	Blood vessels	Vasodilation
	Endothelium	Inhibition of leukocyte adhesion
	Platelets	Inhibition of activation and aggregation
Prostaglandin E (PGE_1, PGE_2)	Arteries	Vasodilation
	Veins	Vasoconstriction
	Bronchus	Constriction
	Sympathetic nerves	Inhibition of norepinephrine (noradrenaline) release
	Leukolymphocytes	Inhibition of activation
	Uterus	Contraction
Prostaglandin F (PGF_2)	Blood vessels	Vasoconstriction
	Bronchus	Constriction
	Uterus	Contraction

Hematological system

A normal ratio of TXA_2 to epoprostenol is important in maintaining the physiological role of platelets and coagulation. TXA_2 most powerfully activates platelet aggregation, whereas epoprostenol represents its natural opponent. In this respect epoprostenol is much more potent than either PGE_1 or PGE_2, and its effects last longer.

Immune system

Epoprostenol and the E series prostaglandins manifest both pro- and anti-inflammatory properties.

When released at the local side of inflammation the vasodilating prostaglandins tend to act as proinflammatory 'modulators' rather than as true inflammatory mediators, since these substances alone are not capable of causing extravasation of plasma. However, prostanoids contribute to the signs and symptoms of inflammation by accentuating the pain and edema initially produced by histamine, serotonin, and bradykinin.

Epoprostenol and the PGEs inhibit macrophage as well as polymorphonuclear leukocyte activation and adhesion to damaged vascular endothelium. In particular, PGE_2 decreases T-lymphocyte activation by inhibiting the production of the regulatory cytokine interleukin 2 and also interferes with B-cell functions. Moreover, both epoprostenol and the PGEs inhibit the release of the proinflammatory cytokine, i.e. tumor necrosis factor-α and interleukin 1, from activated mononuclear leukocytes.

Gastrointestinal tract

Vasodilating prostaglandins, in particular the E series, exert cytoprotective properties within the gastrointestinal tract. A number of mechanisms are potentially responsible for this protection, including augmentation of mucosal blood flow, inhibition of acid secretion, stimulation of mucus and alkaline secretion, stabilization of tissue lysosomes, and inhibition of proinflammatory cytokine release (Vane and O'Grady 1993).

Endogenous epoprostenol release reduces ischemia–reperfusion injury in experimental animals and limits perioperative endotoxemia and bacterial translocation in patients undergoing major surgery (Brinkmann *et al.* 1996).

Renal system

Integrity of renal prostaglandin synthesis is necessary to preserve renal function. In the kidney epoprostenol and PGE_2 exert five major actions: increased renal blood flow; increased glomerular filtration rate; natriuresis, chloruresis, and water diuresis; stimulation of renin secretion; stimulation of erythropoietin secretion. The local synthesis of prostaglandins is augmented in a compensatory effort to maintain renal function, in particular under vasoconstrictive states, i.e. after volume depletion or ineffective circulating volume (Vane and O'Grady 1993).

Therapeutic use

Usually, epoprostenol and PGE_1 are used as therapeutic agents in critical care medicine. Occasionally, $PGF_{2\alpha}$ and PGE_2 are employed following obstetric complications because of their uterus-contracting properties. The principal side-effects of these prostaglandins are shown in Table 2.

Sepsis and acute respiratory distress syndrome

Prostaglandins and TXA_2 have been demonstrated to be part of the complex network of mediators determining the pathophysiology of sepsis and acute respiratory distress syndrome.

TXA_2 is a major mediator of ischemia and shock, whereas epoprostenol and the PGEs exert potentially beneficial effects. In general, infusing epoprostenol and PGE_1 into patients with sepsis and/or acute respiratory distress syndrome improves cardiac output, right ventricular function, systemic oxygen delivery, and/or consumption, and decreases both mean arterial and pulmonary

Table 2 Principal side-effects of therapeutic prostaglandins

Prostaglandin	Adverse effects
Epoprostenol	Diaphoresis, restlessness Hypotension, tachycardia Increase in pulmonary shunt Facial flushing Nausea, vomiting
Iloprost (stable analog)	Usually mild side-effects Transient flush and headache
PGE_1, PGE_2 (dinoprostone)	Hypotension, tachycardia Increase in pulmonary shunt Facial flushing Diarrhea Uterine contractions
Misoprostol (stable analog)	Headache, altered mental status Abdominal pain, diarrhea
$PGF_{2\alpha}$ (dinoprost)	Hypertension, hypotension Bronchospasm (contraindication asthma) Diarrhea, vomiting Fever

artery pressure, without affecting overall survival in multicenter trials. Moreover, recent studies show improved glucose metabolism and splanchnic oxygenation with epoprostenol infusion in patients with sepsis and septic shock (Scheeren and Radermacher 1997).

The intravenous application of epoprostenol and PGE_1 is limited by potential adverse effects such as hypotension and hypoxemia caused by increased pulmonary shunt (Scheeren and Radermacher 1997).

Inhaled epoprostenol

Aerosolized epoprostenol may provide an alternative to inhaled nitric oxide as a selective pulmonary vasodilator.

Despite its potential toxicity the use of inhaled nitric oxide has been advocated because it only exerts vasodilatory properties in ventilated lung regions and allows selective reduction of pulmonary artery pressure concomitant with improved arterial oxygenation. Owing to its rapid inactivation, it does not have any deleterious effects on systemic hemodynamics. Given the nearly identical pharmacodynamic profile of nitric oxide and epoprostenol, the latter has been administered as an aerosol. Indeed, inhaled epoprostenol also allows reduction of pulmonary artery pressure and an increase in arterial Po_2 with almost no effect on systemic hemodynamics. The efficacy is virtually identical with that of inhaled nitric oxide (Scheeren and Radermacher 1997). The doses of nebulized substance required to obtain this effect range from 1 to 100 ng/kg/min; the individual dosage largely depends on the patient, the underlying pathology, and the type of nebulizer used. Dosages only represent the amount of drug nebulized; because of variable droplet size and unknown drug distribution in the airways, it is not known how much of the drug reaches the alveoli.

Aerosolization of epoprostenol may result in a dose-dependent spillover of the drug into the systemic circulation. This spillover mimics the systemic effects of the intravenous application, including improved splanchnic perfusion and oxygenation (Scheeren and Radermacher 1997).

Aerosolized epoprostenol will probably remain an experimental approach unless nebulizers for continuous long-term use become available.

Pulmonary hypertension unrelated to sepsis and acute respiratory distress syndrome

Epoprostenol and PGE_1 allow treatment of pulmonary hypertension of primary or secondary etiology (mitral stenosis, congestive heart failure, and chronic lung disease).

Epoprostenol is more powerful and has a shorter half-life. Within 30 min of cessation of intravenous infusion, most if not all the clinical effects will dissipate (Scheeren and Radermacher 1997). In patients with primary pulmonary hypertension, epoprostenol is a fundamental element of therapeutic strategy which improves quality of life and survival time, particularly in severely ill patients awaiting transplantation (Barst *et al.* 1996). In addition to long-term intravenous administration of epoprostenol (Barst *et al.* 1996), aerosolization of epoprostenol or its stable analog iloprost offers a new strategy for treatment of patients with primary pulmonary hypertension which appears to be more effective than nitric oxide and oxygen therapy (Olschewski *et al.* 1996).

Extracorporeal circulation

Epoprostenol alone or combined with heparin therapy has been employed to improve platelet count as well as to diminish bleeding complication, during cardiopulmonary bypass, charcoal hemoperfusion, hemodialysis, and hemofiltration (Vane and O'Grady 1993).

During hemodialysis epoprostenol enhances the biological activity of heparin and prevents consumption and activation of platelets without clinically relevant side-effects. During spontaneous and pump-driven hemofiltration, epoprostenol prolongs hemofilter life particularly when combined with low-dose heparin (Scheeren and Radermacher 1997). In patients with combined acute renal and hepatic failure epoprostenol therapy reduces the episodes of hemorrhage compared with conventional heparin anticoagulation for hemofiltration. Potential hemodynamic side-effects and the dosage of epoprostenol can be reduced by directly infusing into the extracorporeal circuit. A combination of low-dose heparin (5000 IU) and epoprostenol (3–5 ng/kg/min) can be recommended as a safe anticoagulation regimen during hemofiltration in critically ill patients (Scheeren and Radermacher 1997).

Acute hemorrhage following uterine atony

$PGF_{2\alpha}$ can be administered intravenously as a pharmacological last resort in patients with severe postpartum hemorrhage due to uterine atony.

Dosage and route of use

Epoprostenol

Epoprostenol is available as a stable freeze-dried sodium salt for intravenous administration in humans. Because of its short elimination half-life of 2 to 3 min, it must be infused continuously, thus permitting precise titration and controllability. Epoprostenol is diluted in a glycine buffer of pH 10 to 11. Therefore it must be infused through a separate line to avoid inactivation by acid drugs (e.g. catecholamines) (Scheeren and Radermacher 1997). The stable analog iloprost may also be orally administered and has minor side-effects (Vane and O'Grady 1993).

PGE series

The naturally occurring prostanoids PGE_1 and PGE_2 (dinoprostone) are also available for intravenous application (Vane and O'Grady 1993). PGE_1 and PGE_2 are rapidly metabolized (half-life < 1 min). However, the active metabolite accounts for part of the vasodilatory activity. The recommended dose for the longer-acting stable analog misoprostol is 100 to 200 g four times daily (Vane and O'Grady 1993).

$PGF_{2\alpha}$

$PGF_{2\alpha}$ (dinoprost) has a very short biological half-life due to rapid metabolism in the circulation (Vane and O'Grady 1993). It may be administered intravenously and intramuscularly. The recommended dose for uterine atony is 250 g up to a total dose of 12 mg.

References

Barst, R.J., *et al.* (1996). A comparison of continuous intravenous epoprostenol (prostacyclin) with conventional therapy for primary pulmonary hypertension. *New England Journal of Medicine*, **334**, 296–302.

Brinkmann, A., *et al.* (1996). Perioperative endotoxemia and bacterial translocation during major abdominal surgery. Evidence for the protective effect of endogenous prostacyclin? *Critical Care Medicine*, **24**, 1293–1301.

Olschewski, H., Walmrath, D., Schermuly, R., Ghofrani, A., Grimminger, F., and Seeger, W. (1996). Aerosolized prostacyclin and iloprost in severe pulmonary hypertension. *Annals of Internal Medicine*, **124**, 820–4.

Scheeren, T. and Radermacher, P. (1997). Prostacyclin (PGI_2): new aspects of an old substance in the treatment of critically ill patients. *Intensive Care Medicine*, **23**, 146–58.

Vane, J.R. and O'Grady, J. (1993). *Therapeutic applications of prostaglandins*. Edward Arnold, London.

17.1.24 Respiratory stimulants

A. A. Jeffrey

Key messages

- Respiratory stimulant effects are exhibited by a number of different classes of drugs, the most significant of which are the analeptics (doxapram and nikethamide), theophylline, and the triazine derivatives (almitrine bismesylate).
- The place of doxapram in respiratory failure remains contentious, but it may have a role where non-invasive mechanical support is impractical.
- Theophyllines have a theoretical role in the treatment of respiratory failure but await large-scale clinical trials in this context.
- Almitrine bismesylate is an orally active respiratory stimulant with potential in the treatment of both acute and chronic respiratory failure.

Introduction

Respiratory stimulant drugs fall into six categories as listed in Table 1, each of which is chemically distinct with different modes of action and clinical application. Only those of significant use within the practice of adult intensive care medicine are discussed in detail here.

Clinical use

Doxapram as a continuous infusion is used to support patients in acute on chronic respiratory failure with hypercapnia and to reduce postoperative complications of anesthesia. These uses, particularly the former, will be discussed in greater detail below.

Nikethamide is a short-acting analeptic agent for reversal of respiratory depression. It is a less specific respiratory stimulant than doxapram and has to be given as repeated doses. The ratio of convulsant to respiratory stimulant doses is approximately 15:1 compared with approximately 40:1 for doxapram. As a result of these disadvantages it is now little used.

Methylxanthines, amongst their other effects, act directly on brainstem respiratory centers, increasing resting ventilation in a dose-dependent manner; respiratory response to hypoxia and hypercapnia are increased. These drugs are most often used for their respiratory stimulant activity in the management of idiopathic apnea of prematurity, although this effect may also be of relevance for their use in acute exacerbations of chronic obstructive pulmonary disease.

Hormones, tricyclic antidepressants, and acetazolamide have respiratory stimulant activity, but are rarely used for their respiratory stimulant effects in the context of adult critical care and will not be discussed further.

Almitrine bismesylate has been shown to be effective in chronic respiratory failure and in supporting patients in acute respiratory failure, but it is still undergoing clinical trials. It acts on peripheral chemoreceptors in the carotid body and can be given orally or intravenously. This drug will be discussed in greater detail below.

Thus only three of these drugs, namely doxapram, almitrene bismesylate, and theophylline, are used routinely or have the prospect of significant clinical application within the context of adult intensive care.

Doxapram

Structure and mode of action

Doxapram is a monohydrated pyrolidinone derivative that acts through both stimulation of peripheral chemoreceptors and direct action on the respiratory center. Its pharmacological effects are summarized in Table 2.

Table 1 Respiratory stimulant drugs

Analeptics
Doxapram
Nikethamide
Methylxanthines
Theophylline
Hormones
Medroxyprogesterone
Tricyclic antidepressants
Protriptyline
Carbonic anhydrase inhibitors
Acetazolamide
Triazine derivative
Almitrine bismesylate

Table 2 Pharmacological effects of doxapram

Respiratory system
Increases tidal volume and hence minute volume at low dose and respiratory rate at higher doses
Cardiovascular system
Increases cardiac output by increasing stroke volume
Central nervous system
Increases cerebral blood flow
Not much convulsant activity; approximately 40-fold difference between the dose that stimulates ventilation and that which induces convulsions
Genitourinary system
Increases urine output and ureteric motility

Clinical use

Although doxapram was originally introduced in the mid-1960s, its therapeutic role is still a matter of debate. Its earliest application was in the reversal of drug-induced respiratory depression, and subsequently in reducing the pulmonary complications following anesthesia. Doxapram has been shown to reduce the effects of opiates on respiratory drive without altering their analgesic effect. Early studies showed that, when doxapram is administered postoperatively as a single dose combined with morphine, the incidence of postoperative cough and sputum expectoration is reduced compared with administration of morphine alone. These findings have not been repeated in more recent studies and doxapram is now used infrequently for this purpose.

Doxapram is now used more commonly to support patients with acute on chronic hypercapnic respiratory failure. Doxapram increases minute volume, at low doses through increased tidal volume and in higher doses by increased respiratory rate, and may support such patients by reducing the need for mechanical support of ventilation while the underlying cause of the exacerbation is treated. Theoretical objections have been raised to the use of doxapram in this circumstance. Experimental work has shown that, in normal subjects, doxapram increases oxygen consumption concomitant with the increase in minute volume. In ventilated patients in whom doxapram has been used during weaning, diaphragmatic dysfunction has been shown to increase as a result of an increase in end-expiratory volume. Some authors cite this as evidence contraindicating the use of doxapram in patients with chronic lung disease. The recent evidence of the role of respiratory muscle fatigue in the context of acute on chronic respiratory failure again argues against the use of a drug that may further increase muscle dysfunction. However, at lower blood levels (1.6–3.0 μg/ml) than those used in these studies, ventilation has been found to increase without changes in O_2 consumption or CO_2 production.

In clinical studies of patients in acute respiratory failure who have not received mechanical ventilation (Moser *et al.* 1973; Riordan *et al.* 1975; Jeffrey *et al.* 1993), doxapram has been shown to be of benefit in improving arterial pH and reducing $P\text{aCO}_2$. Little or no effect on $P\text{aO}_2$ has been found. Although not a controlled trial of doxapram therapy, in the study by Jeffrey *et al.* (1993) doxapram was used as a continuous infusion when the pH could not be kept above 7.25 while on controlled oxygen therapy in 37 of 139 patients all presenting with $P\text{aO}_2 \leqslant 6.5$ kPa and $P\text{aCO}_2 \geqslant 6.5$ kPa. Doxapram was introduced into the regimen at a time determined by a protocol based on these blood gas measurements. Seventy per cent (26/37) survived, with only three receiving mechanical ventilation. It is of note that those patients in whom pH continued to fall while on doxapram had a poor prognosis. This was most apparent 12 to 18 h after starting treatment and may be an indication to consider mechanical support. The overall episode survival of patients in this study was good at 88 per cent. In the last few years, a similar effect has been shown with non-invasive intermitent positive-pressure ventilation, but this facility is not available in many centers and may not be tolerated by patients in severe respiratory distress. Therefore doxapram retains a useful role in the support of such patients.

Pharmacokinetics

The pharmacokinetics of doxapram are such that steady state plasma concentrations are not achieved for many hours with constant-rate infusions. The mean half-life is 2 to 4 h with an apparent volume of distribution (V_D) of 1.5 l/kg. The whole-body clearance is 370 ml/min with less than 5 per cent of an intravenous dose excreted unchanged in the urine in 24 h. As a result, there is a rapid fall in plasma concentration on cessation of an infusion only if this infusion is of short duration. Regimens in which infusion rates start high and then decrease have been recommended. A starting dose of 4 mg/min, gradually reducing over 1 h to 1.5 mg/min, has been shown to produce steady state plasma drug concentrations.

The most commonly noted side-effects of doxapram are cerebral. Marked agitation is common and requires close nursing supervision, particularly if oxygen delivery devices are to be kept in place. Further toxic effects are restlessness, dizziness, hallucinations, excessive sweating, and a sensation of perineal warmth. The metabolic consequences of doxapram include increased catecholamine and steroid secretion with a rise in metabolic rate of up to 30 per cent.

Almitrine bismesylate

Structure and mode of action

Almitrine bismesylate is a triazine derivative which stimulates the carotid body, thus increasing the response to hypoxia with little effect on the response to hypercapnia. Improvement in $\dot{V}/\dot{Q}$ relationships have also been recorded in patients with hypoxemic chronic obstructive airways disease (Powles *et al.* 1983). The mechanism of the latter effect is unclear, but may be due to increased pulmonary vascular resistance and pulmonary artery pressure associated with a small decrease in cardiac output.

Clinical use

Almitrine is active as both an intravenous and an oral preparation. In oral form it has potential in the longer-term management of chronic respiratory failure in patients with chronic obstructive airways disease, either alone or with long-term domiciliary oxygen therapy. Several studies have reported benefit when almitrine is given in daily doses of 100 to 200 mg. There have been reports of the occurrence of peripheral sensory neuropathy between the third and seventh month of treatment in such patients, although whether this is a direct result of the drug or represents the unmasking of the neuropathy reported in association with chronic obstructive airways disease itself is unclear.

In the setting of adult intensive care medicine, almitrine has potential in the management of acute respiratory failure secondary to chronic obstructive airways disease, and there has been interest in a possible role in the management of acute respiratory distress syndrome. The effect of almitrine depends on the degree of hypoxemia present. Almitrine treatment results in a significant improvement in $Pa\text{O}_2$ over that achieved by oxygen alone until normoxia is attained. There have been small-scale trials of almitrine in acute respiratory failure and, as with doxapram, it appears to have been successful in preventing the need for mechanical respiratory support in a number of cases. However, the results of clinical trials have been inconsistent, perhaps because of variable drug absorption from the gut in such acutely ill patients resulting in suboptimal blood levels. The lack of large-scale trials and the limited availability of the drug itself mean that almitrine cannot be recommended for routine use in acute respiratory failure at this stage.

The observed effect of almitrine on ventilation–perfusion ratios in hypoxic patients has stimulated research into its potential use in acute respiratory distress syndrome. In one small-scale study, almitrine was shown to be as effective as 10 cmH_2O of positive end-expiratory pressure in increasing $Pa\text{O}_2$ and reducing venous admixture in 10 patients with acute respiratory distress syndrome who required continuous mechanical ventilation. Once again, larger studies are required before recommending its use in this circumstance.

Almitrene bismesylate is currently licensed for clinical use in a number of countries worldwide, but not in the United Kingdom or the United States.

Pharmacokinetics

Almitrine bismesylate is highly lipid soluble and is absorbed from the gut in 2 to 3 h and highly bound to albumin in the circulation. The plasma half-life is about 2 days. Almitrine is mostly lost through bile and is not excreted in the urine. The physiological effect of a single dose lasts for approximately 6 h. Few adverse reactions have been reported in short-term use; they include headache, urticaria, diarrhea, chest pain, nausea, and vomiting. The most significant side-effect is possible sensory neuropathy in long-term use, already noted above.

Methylxanthines

Methylxanthines, particularly theophylline, have been exciting renewed interest recently in the context of acute respiratory failure secondary to chronic obstructive airways disease. Theophylline is traditionally used as an intravenous infusion because of its (minor) bronchodilator effect in obstructive airways disease, but it is its reported ability to increase diaphragmatic contractility and its potential respiratory stimulant effect that are arousing present interest. The potential to reduce airways obstruction, improve diaphragmatic contractility, and stimulate respiration with a single agent has led some to advocate the routine use of intravenous theophyllines in patients developing acute on chronic respiratory failure as a result of exacerbations of chronic obstructive airways disease to reduce the requirement for mechanical ventilation. The effectiveness of this use has not yet been confirmed in large-scale trials and, given the difficulty of maintaining therapeutic blood levels and the increase in the side-effects of theophylline in acutely ill patients, it should be used with caution in such circumstances.

Conclusions

The use of respiratory stimulants in the context of adult intensive care medicine remains contentious but, given the undoubted problems of the mechanical ventilation of patients with chronic airways obstruction, they retain a place in attempting to avoid the necessity of mechanical intervention in such patients. Future research may clarify their place in the treatment of other conditions associated with hypoxemia.

References

Jeffrey, A.A., Warren, P.M., and Flenley, D.C. (1993). Acute hypercapnic respiratory failure in patients with chronic obstructive lung disease: risk factors and guidelines for management. *Thorax*, **47**, 34–40.

Moser, K.M., *et al.* (1973). Respiratory stimulation with intravenous doxapram in respiratory failure. A double blind co-operative study. *New England Journal of Medicine*, **288**, 427–31.

Powles, A.C., Toxem, D.V., Mahood, C.B., Pugsley, S.O., and Campbell, E.J. (1983). The effect of intravenously administered almitrine, a peripheral chemoreceptor agonist, on patients with chronic airflow obstruction. *American Review of Respiratory Disease*, **127**, 284–9.

Riordan, J.F., Sillett, R.W., and McNicol, M.W. (1975). A controlled trial of doxapram in acute respiratory failure. *British Journal of Diseases of the Chest*, **69**, 57–62.

17.1.25 Sedatives

W. Strobel and R. Ritz

Key messages

- The accurate assessment of the depth of sedation is crucial in any intensive care patient.
- Benzodiazepines are generally among the safest central nervous system active medications.
- Normally, because of its rapid elimination, full recovery to awareness is reached within minutes to a few hours after discontinuation of propofol.

- Indications for haloperidol and other butyrophenone neuroleptics are delirium and withdrawal syndromes.
- Clonidine is a centrally acting imidazoline and α_2-agonist. It can be used as an alternative to benzodiazepines in the treatment of severe alcohol withdrawal symptoms or delirium, which are accompanied by profound activation of the sympathetic nervous system.

Introduction

All patients in the intensive care unit (ICU) are subjected to a variety of unaccustomed stimuli as a result of both their underlying diseases and their therapies. In addition to adequate analgesic treatment, they receive sedative drugs to alleviate anxiety, to ensure co-operation, and to promote sedation and sleep. Drugs used for this purpose are called sedatives or hypnotics; they include the benzodiazepine receptor agonists, propofol, barbiturates, butyrophenones, central-acting α_2-agonists, and the volatile anesthetic isoflurane. Patients receiving artificial ventilation almost invariably require sedatives to enable them to tolerate the endotracheal tube and positive-pressure ventilation. However, before sedatives are given to an anxious or agitated patient, every attempt should be made to exclude other treatable causes such as pain, hypoxemia, metabolic abnormalities, and the toxic effects of drugs.

Assessment of sedation

The desired level of sedation has changed in the last two decades. Deep sedation, with the patient virtually anesthetized, is reserved for a few indications (e.g. severe head injuries and tetanus). In the majority of patients the aim is for them to be sleepy, although easily rousable and hence co-operative. However, endpoints in ICU sedation vary considerably between different institutions, which makes dose recommendations difficult. Generally, there is a tendency towards oversedation, justified by concern for the patient's comfort.

The accurate assessment of the depth of sedation is crucial in any ICU patient. Clinical examinations should be performed at least twice daily using a standardized procedure and the results should be recorded. A score system, for example the 'Ramsay score' (Ramsay *et al.* 1974), can be used for better comparability and reliability of the results. The Glasgow Coma Scale, although frequently used, is not validated for the assessment of drug-induced sedation and is not feasible in intubated patients.

Objective investigator-independent methods to assess the depth of sedation are not established in the clinical routine. Electrophysiological approaches, such as electroencephalograms and brainstem-evoked response audiometry, allow rough assessment of sedation, but these methods are limited by their cost and moderate correlation with clinical presentation. Plasma concentration monitoring of sedatives and their active metabolites is rarely available and, since the intra- and interindividual variability in drug efficacy is very large, 'therapeutic' ranges are poorly defined, particularly for benzodiazepine metabolites.

Benzodiazepines

Benzodiazepines are by far the most frequently used sedatives. All benzodiazepines share powerful anticonvulsant, muscle-relaxant, hypnotic, sedative, and tranquilizing properties. Their mechanism of action is identical; they interact with a specific binding site, termed the benzodiazepine receptor, which is located on the receptor complex of the important endogenous inhibitory neurotransmitter GABA. The benzodiazepine binding augments or facilitates GABA binding, and vice versa, opening chloride channels and thereby producing sedation and an anticonvulsant activity. Some other nervous system effects, such as muscle relaxation and anxiolysis, are probably due to augmentation of glycinergic neurotransmission in the brainstem and the spinal cord.

Parenterally available benzodiazepines commonly used in the ICU are diazepam, lorazepam, and midazolam. Diazepam and lorazepam are insoluble in aqueous fluids and therefore the parenteral preparation contains propylene glycol and benzyl alcohol. These solvents may cause local irritation at the injection site and sometimes subsequent phlebitis. The commercial formulation of midazolam is soluble in water and causes less local discomfort.

All benzodiazepines are extensively metabolized in the liver. The metabolites of diazepam (desmethyldiazepam and oxazepam) are pharmacologically active and accumulate when repeated doses of diazepam are administered. The active metabolite of midazolam (α-hydroxy-midazolam) makes an important contribution to the effects of midazolam in long-term sedation. It has recently been shown that even glucuronidated α-hydroxy-midazolam has benzodiazepine-like properties; hence renal failure may lead to accumulation of this active metabolite (Bauer *et al.* 1995). Lorazepam is directly glucuronidated, and this metabolite has not yet been shown to be pharmacologically active.

Pharmacokinetic interactions between benzodiazepines and other drugs play no role in clinical practice. Except for additive effects with other central nervous system depressants such as neuroleptics, antidepressants, opioid analgesics, and alcohol, relevant pharmacodynamic interactions do not occur.

Owing to the wide interindividual variability in the potency, efficacy, and pharmacokinetics of benzodiazepines (Table 1), standard dose schemes are not established and the dose should be determined by the sedative effect. Patients with hepatic or renal insufficiency require lower doses, whereas previous benzodiazepine abuse makes the necessity for higher doses likely. The time to the maximum sedative effect of an intravenous benzodiazepine bolus varies considerably betwen different benodiazepines (Table 1). A prolonged recovery time, sometimes lasting for several days, may occasionally occur after long-term sedation, even in the absence of liver or kidney dysfunction. Although the elimination half-life of lorazepam is much longer than that of midazolam, lorazepam does not require more recovery time after discontinuation (Pohlmann *et al.* 1994). Owing to its slow elimination and the accumulation of even longer-lasting metabolites, diazepam should not be used for long-term sedation.

Benzodiazepines are generally among the safest central nervous system active medications. They cause severe central nervous system depression and marked respiratory and cardiovascular dysfunction only when they are heavily overdosed. All these symptoms are reversible. Partly due to reduced metabolism, benzodiazepines can induce or augment symptoms of hepatic encephalopathy in patients with

Table 1 Pharmacokinetics, dosage, and costs of commonly used sedatives

	Diazepam	Midazolam	Lorazepam	Propofol	Haloperidol
Volume of distribution (l/kg)*	1.1	1.6	1.6	1.3	18
Albumin binding (%)†	99	95	90	99	92
Clearance (ml/min/kg)	0.4	6.6	1.1	1700	12
Urinary excretion of unchanged drug (%)	<1	<5	<1	<1	1
Plasma half-life (h)	30–60‡	1.5–4‡	10–20‡	<1§	10–24‡
Active metabolites	Desmethyl-diazepam, oxazepam	α-hydroxy-midazolam and its glucuronate	?	None	Reduced haloperidol (weak effects)
Recovery time after long-term sedation	Several days¶	Several hours to a few days¶	Several hours to a few days¶	Minutes to a few hours	Several hours to a few days
Time to maximum sedative effect after intravenous bolus (min)	1–2	5–10	15–30	<1	5–30
Usual doses required for sedation** (mg/kg/h)	0.007–0.15	0.03–0.4	0.002–0.05	0.5–8	0.02–0.25
Mode of administration	Bolus	Bolus/continuous infusion	Bolus	Continuous infusion	Bolus/continuous infusion
Costs	Low	High	High	High††	Medium

*May be markedly reduced in the elderly.
†Reduced in hypalbuminemia and kidney failure.
‡Increased in liver failure.
§Terminal plasma half-life ($t_{1/2\gamma}$) up to 30 h.
¶Depends on previous depth of sedation, markedly increased in liver or kidney failure.
**Dosage depends on desired depth of sedation and impairment of elimination organs.
††Possible reduction of length of stay in the ICU (due to rapid recovery) not taken into account.

liver cirrhosis and therefore should be avoided in such cases. Anterograde amnesia is a commonly seen and sometimes even opportune side-effect, particularly with midazolam. Withdrawal symptoms including insomnia, anxiety, dysphoria, unpleasant dreams, and sweating may occur after long-term adminstration. Hence it is prudent not to discontinue benzodiazepine medication abruptly but to reduce the dosage gradually over some days if possible.

Since a specific benzodiazepine antagonist (flumazenil) is available, all effects of benzodiazepines can readily be reversed. Because the effect of a single dose of flumazenil lasts for only about 1 h, repeated injections or a continuous infusion may be required in benzodiazepine overdose. Because flumazenil may precipitate the onset of seizures, repeated small injections are preferred to a single large bolus injection.

Propofol

Propofol is an alkylphenol and is chemically unrelated to any other sedative drug. It was originally developed as a surgical anesthetic drug. Its exact mechanism of action is still unknown. Propofol is not soluble in aqueous fluids, and therefore the commercial formulation consists of 1 per cent propofol in an oil-in-water emulsion containing 10 per cent soybean oil, 1.2 per cent egg phosphatide, and 2.25 per cent glycerol. As this formulation may cause local irritation and phlebitis, propofol should be delivered via a central catheter separated from other drugs.

After intra- and extrahepatic glucuronidation and sulfation the inactive metabolites are excreted in the urine. The pharmacokinetics of propofol (Table 1) are barely changed by hepatic and renal dysfunction. Concomitant use of fentanyl may reduce the volume of distribution and clearance of propofol and therefore reduce the required dose.

Normally, because of its rapid elimination, full recovery to awareness is reached within minutes to a few hours after discontinuation of propofol, even after long-term administration (Beller *et al.* 1988). Propofol plasma concentrations required for adequate sedation are substantially lower than those needed for surgical anesthesia (about 1.0 μg/ml). These concentrations are achieved with dose rates between 0.3 and 8 mg/kg/h.

The majority of data concerning the side-effects of propofol are derived from its use as an intravenous anesthetic after bolus administration in the operating room and are not necessarily transferable to patients in the ICU. Propofol often causes mild to moderate hypotension, probably due to a decrease in systemic vascular resistance. Severe bradycardia and conduction disturbances are seen sporadically. Although propofol causes marked respiratory depression only in anesthetic doses, it should be used only in patients with mechanical ventilatory support. Data on the effects of propofol on intracranial hemodynamics, particularly in patients with severe head trauma, are limited, but since it might occasionally reduce cerebral perfusion pressure, it should be used with caution in these patients.

Other sedatives

Barbiturates are not often administered for sedation in critically ill patients because they depress the cardiovascular and respiratory systems. In some special indications, for example treatment of elevated intracranial pressure, barbiturates are used with varying degrees of success.

Indications for haloperidol and other butyrophenone neuroleptics are delirium, psychoses, and withdrawal syndromes. In addition to the well-known neuroleptic side-effects, haloperidol may cause QT prolongation on the ECG and therefore should be used with caution, particularly when other drugs known to prolong the QT interval are used. Haloperidol dose rates may reach up to 25 mg/h.

Clonidine is a centrally acting imidazoline and α_2-agonist. It can be used as an alternative or addition to benzodiazepines in the treatment of severe alcohol withdrawal symptoms or delirium, which are accompanied by profound activation of the sympathetic nervous system. In these patients, even very high doses of clonidine (up to 5000 μg/day) do not cause relevant hypotension.

Preliminary reports suggest that isoflurane, a volatile anesthetic agent, can be used for short- and long-term sedation in patients requiring mechanical ventilation. The main advantages of this approach are rapid onset of action, easy titration, and fast dissipation of the effect. Major disadvantages are the need for expensive additional equipment and unavoidable temporary interruptions of the ventilator circuit, for example when the patient is transferred.

Selection of sedatives and modes of administration

The selection of an appropriate sedative drug in ICU patients depends on the comorbidity of the patient, on wanted and unwanted effects of the drug, and on the expected duration of therapy. Benzodiazepines should be chosen for short- and intermediate-term sedation, particularly if anxiolytic and anticonvulsive effects are desired. Haloperidol or other butyrophenones are preferable in psychotic patients and in delirium. Owing to its pharmacokinetic properties, propofol may be favored in liver failure and in situations when rapid recovery from sedation is desired. Finally, the remarkable costs of sedative drugs should be taken into account.

Continuous administration, preceded as required by a bolus, is superior to bolus administration for achieving a constant depth of sedation. However, continuous administration always requires higher cumulative doses than bolus adminstration, which may result in a longer period of sedation after discontinuation of the sedative drug. Because of its short duration of action propofol should be administered continuously, whereas lorazepam, as an intermediate-lasting benzodiazepine, is well suited for bolus administration. Although often used in a continuous infusion, midazolam can also be delivered in repeated bolus injections.

Problems: persistent sedation after drug discontinuation and brain death

Approximately 20 per cent of ICU patients fail to awake in the expected time after discontinuation of sedatives. Different causes have to be excluded: prior overdosage of sedatives, accumulation of the sedative or its active metabolites due to failure of the elimination organ, additive effects of other central nervous system depressants, coma unrelated to administration of sedatives (e.g. hypoxic brain damage, intracerebral hemorrhage, and metabolic disorders). Focal neurological deficits and lack of brainstem reflexes cannot be explained by the previous administration of sedatives and make other causes of coma more likely. In case of doubt, when benzodiazepines have been used for sedation, flumazenil may be administered cautiously in incremental doses of 0.2 to 0.5 mg. The lack of any response after a cumulative dose of 5 mg administered over 10 to 15 min largely excludes benzodiazepine overdose as a cause of persistent coma.

In most countries, the diagnosis of brain death has been prohibited in the presence of any amount of sedative drugs. With the increasing demand for donor organs, this inflexible attitude has been changed and the diagnosis of brain death has become possible if the effect of sedatives is clinically irrelevant. As this term is poorly defined, it is not clear which additional tests are required to meet the definition.

References

Bauer, T., *et al.* (1995). Prolonged sedation due to accumulation of conjugated metabolites of midazolam. *Lancet*, **346**, 145–7.

Beller, J.P., Pottecher, T., Lugnier, A., Mangin, P., and Otteni, J.C. (1988). Prolonged sedation with propofol in ICU patients: recovery and blood concentration changes during periodic interruptions in infusion. *British Journal of Anaesthesia*, **61**, 583–8.

Pohlmann, A.S., Simpson, K.P., and Hall, J.B. (1994). Continuous intravenous infusion of lorazepam versus midazolam for sedation during mechanical ventilatory support: a prospective, randomized study. *Critical Care Medicine*, **22**, 1241–7.

Ramsay, M.A.E., Savege, T.M., Simpson, B.R.J., and Goodwin, R. (1974). Controlled sedation with alphaxalone–alphadolone. *British Medical Journal*, **2**, 656–9.

17.1.26 Stress ulcer prophylaxis and antiulcer drugs

D. J. Cook

Key messages

- The incidence of gastrointestinal bleeding from stress ulceration in critically ill patients has decreased over the last decade.
- Patients at highest risk of bleeding are those who have been mechanically ventilated for 48 h or more and those with a coagulopathy (risk approximately 4 per cent).
- Other critically ill patients have a very low risk of bleeding, and so it may be reasonable to withhold stress ulcer prophylaxis (risk less than 0.01 per cent).
- Antacids, H_2-receptor antagonists, and sucralfate all appear to decrease the rate of bleeding.

- Sucralfate is associated with a lower rate of nosocomial pneumonia than drugs that increase gastric pH.
- No stress ulcer prophylactic agent has been shown definitively to reduce mortality in critically ill patients.

The clinical problem

Upper gastrointestinal stress ulceration was once viewed as a potentially life-threatening complication of critical illness, for which effective prophylactic measures were endorsed. However, the incidence of gastrointestinal bleeding in critically ill patients varies depending on the population studied, the definitions employed, and the methods used to diagnose the bleeding.

Microscopic gastrointestinal bleeding is common in critically ill patients and may be due to nasogastric trauma more often than stress ulceration. This type of bleeding rarely progresses to bleeding that is clinically relevant. Overt bleeding is defined as hematemesis, bloody gastric aspirate, melena, or hematochezia; this is less common, occurring in approximately 5 per cent of a heterogeneous group of intensive care unit (ICU) patients. Clinically important bleeding can be defined as overt bleeding and one of the following: drop in blood pressure of 20 mmHg within 24 h of bleeding, blood pressure drop of 10 mmHg and heart rate increase of 20 beats/min on orthostatic change, or hemoglobin decrease of 20 g/l and transfusion of 2 units of blood within 24 h (Cook *et al.* 1994). Recent reports have suggested that the incidence of clinically important gastrointestinal bleeding is extremely low, and question the widespread use of stress ulcer prophylaxis for all critically ill patients (Cook *et al.* 1994; Zandstra and Stoutenbeek 1994). This decline in clinically important bleeding rates may be due to improvements in critical care over the last decade (e.g. more aggressive hemodynamic support, earlier attention to enteral nutrition, avoidance of corticosteroids in sepsis, etc).

One of these studies, demonstrating a decreased incidence, is a prospective multicenter natural history study evaluating the incidence of, and risk factors for, clinically important bleeding (Cook *et al.* 1994). The study population consisted of consecutive patients more than 16 years old who were admitted to four mixed medical–surgical ICUs. Patients were followed throughout their stay in the ICU for the development of clinically important gastrointestinal bleeding. Of the 2252 patients followed in this study, 33 developed clinically important gastrointestinal bleeding (1.5 per cent; 95 per cent confidence interval, 1.0–2.1 per cent). Gastrointestinal bleeding occurred a mean of 14 $\pm$ 12 days after admission to the ICU.

Who should receive stress ulcer prophylaxis?

Patients with clinically important gastrointestinal bleeding are more likely to be those requiring mechanical ventilation (odds ratio, 16) or those who have a coagulopathy (odds ratio, 4). In patients with one or both of these risk factors, the incidence of clinically important bleeding in one natural history study was 3.7 per cent (95 per cent confidence interval, 2.5–5.2 per cent) (Cook *et al.* 1994). In patients without these risk factors, less than 0.1 per cent (95 per cent confidence interval, 0.02–0.5 per cent) had clinically important bleeding. Also, the mortality was 50 per cent; most patients died with rather than from gastrointestinal bleeding.

Therefore stress ulcer prophylaxis may be warranted most in critically ill patients who are mechanically ventilated or have a coagulopathy.

Costs of clinically important gastrointestinal bleeding

To determine the prolongation of ICU stay and excess mortality attributable to clinically important bleeding, patients who bled were matched with control patients. Cost estimates for resources consumed were derived from hospital costing data. On average, each event resulted in an excess of 6.6 hematology tests, 10.8 units of transfusion products, and 23.6 days of anti-ulcer medication. Our matching procedure estimated the length of stay attributable to clinically important bleeding to be 11.4 days (95 per cent confidence interval, 6.0–28.8 days) in those patients who survived and 0.9 days (95 per cent confidence interval, 5.62–7.48 days) in patients who died, with an average of 6.5 days in all patients (95 per cent confidence interval, 12.3–25.27 days). We determined the mortality attributable to clinically important bleeding to be 12.5 per cent (95 per cent confidence interval, 38.0–63.0 per cent). The total cost of a clinically important bleeding episode was $12 215.60 (1994 Canadian dollars) (Heyland *et al.* 1996). Therefore clinically important bleeding results in significant morbidity and increased health care costs. To the extent that the costing system and management of critically ill patients are similar in other hospitals, these results should be generalizable to other settings.

Options for stress ulcer prophylaxis

Three common classes of drugs are used to prevent stress ulcer bleeding: antacids, H_2-receptor antagonists, and sucralfate. Antacids and H_2-receptor antagonists mediate their beneficial effect through raising gastric pH. Gastric alkalinization has been shown to promote colonization of the stomach with potentially pathogenic bacteria, and this provides a biologically plausible mechanism for a higher incidence of pneumonia. Pneumonia is the most common infection complicating the course of patients in the ICU and leads to a prolonged stay, higher mortality, and increased cost to the health care system. Thus, while preventing stress ulcer bleeding, pH-altering drugs may affect the incidence of pneumonia. The cytoprotective agent sucralfate is a weak base, but has no effect on gastric acidity and may be bacteriostatic (Tryba and Mantey-Stiers 1987). These differences have promoted a lively debate about which is the most effective and least potentially harmful agent for the prevention of stress ulcer bleeding.

Systematic review of prophylaxis

The positive results of randomized trials in stress ulcer prophylaxis have previously led to recommendations that prophylaxis be administered to a large proportion of critically ill patients. However, individually, these studies have not established whether these agents decrease clinically important bleeding, nor have they consistently elucidated the relative merits of different prophylactic regimens.

Several systematic reviews have evaluated drugs for stress ulcer prophylaxis drugs. By gaining adequate power through combining results of different studies, one type of systematic review involving a statistical synthesis (meta-analysis) can resolve issues of therapeutic effectiveness. A recent meta-analysis has resolved discordances between these reviews, providing the current best estimates of the effect of various prophylactic regimens on bleeding, pneumonia, and mortality (Cook *et al.* 1996). The 61 articles identified reported studies of over 7000 critically ill patients. The following inclusion criteria were used to select studies:

(1) randomized trials comparing one or more prophylactic drugs with each other or with an untreated control group;

(2) the target population was critically ill patients;

(3) the outcome measure was gastrointestinal bleeding, pneumonia, or death.

The common odds ratio, which is the weighted average of the individual study odds ratios calculated using the Mantel-Haenszel χ^2 test, was used to interpret the results of this meta-analysis.

Overt bleeding

The common odds ratios demonstrated clear superiority of one regimen over another in two comparisons (Table 1). H_2-receptor antagonists (common odds ratio, 0.42; 95 per cent confidence interval, 0.27–0.60) reduce overt bleeding compared with placebo or no therapy. Three comparisons showed substantial trends which, nevertheless, may still be attributable to chance. Antacids appeared to reduce overt bleeding compared with placebo or no therapy (common odds ratio, 0.63; 95 per cent confidence interval, 0.35–1.11). H_2-receptor antagonists tended to be more effective in reduction of overt bleeding than did antacids (common odds ratio, 0.69; 95 per cent confidence interval, 0.44–1.06). The pooled analysis showed a trend of similar magnitude in favor of sucralfate over H_2-receptor antagonists (common odds ratio, 0.81; 95 per cent confidence interval, 0.44–1.47). A final analysis suggested the equivalence of sucralfate and antacids in the prevention of overt bleeding (common odds ratio, 0.96; 95 per cent confidence interval, 0.54–1.70).

Clinically important bleeding

While the direction of the effect remains the same, there are some differences in results between overt and clinically important bleeding (Table 1). H_2-receptor antagonists remain clearly superior to placebo or no therapy (common odds ratio 0.42, 95 per cent confidence interval 0.21–0.79). Antacids appear to be more efficacious than placebo or no therapy (common odds ratio 0.35, 95 per cent confidence interval 0.08–1.33), although chance must be considered as a possible explanation of the difference.

The trend in favor of H_2-receptor antagonists over antacids in overt bleeding also appeared in the analysis of clinically important bleeding (common odds ratio, 0.91; 95 per cent confidence interval, 0.52–1.61). However, the trend in favor of sucralfate over H_2-receptor antagonists in overt bleeding, disappeared for clinically important bleeding (common odds ratio, 1.35; 95 per cent confidence interval, 0.21–9.35). Finally, while antacids and sucralfate appeared equivalent with respect to overt bleeding, a trend in favor of antacids appeared in the pooled analysis of clinically important bleeding (common odds ratio, 1.54; 95 per cent confidence interval, 0.45–5.25). However, the confidence intervals for these comparisons are very wide.

Pneumonia

H_2-receptor antagonists, when compared with either placebo or no therapy or with antacids, had no impact on the incidence of pneumonia (Table 1). In contrast, sucralfate was associated with a lower incidence of pneumonia compared with both H_2-receptor antagonists (common odds ratio, 0.51; 95 per cent confidence interval, 0.33–0.77) and antacids (common odds ratio, 0.52; 95 per cent confidence interval, 0.34–0.79).

Table 1 Randomized trials of stress ulcer prophylaxis

Comparison	*N*	COR (95% CI)
Antacids versus placebo/control		
Overt bleeding	8	0.63 (0.35–1.11)
Clin. Imp.	3	0.35 (0.08–1.33)
Mortality	5	1.36 (0.83–2.24)
H_2-receptor antagonist versus placebo/control		
Overt bleeding	14	0.42 (0.27–0.60)
Clin. Imp.	9	0.42 (0.21–0.79)
Pneumonia	5	0.92 (0.45–1.89)
Mortality	11	1.03 (0.73–1.48)
H_2-receptor antagonists versus antacids		
Overt bleeding	17	0.69 (0.44–1.06)
Clin. Imp.	10	0.91 (0.52–1.61)
Pneumonia	2	0.91 (0.40–2.05)
Mortality	15	1.06 (0.78–1.46)
Sucralfate versus antacids		
Overt bleeding	10	0.96 (0.54–1.70)
Clin. Imp.	5	1.54 (0.45–5.25)
Pneumonia	6	0.52 (0.34–0.79)
Mortality	11	0.70 (0.52–0.94)
Sucralfate versus H_2-receptor antagonists		
Overt bleeding	7	0.81 (0.44–1.47)
Clin. Imp.	3	1.35 (0.21–9.35)
Pneumonia	7	0.51 (0.33–0.77)
Mortality	7	0.71 (0.49–1.04)

N, number of trials included in the analysis; COR, common odds ratio; 95% CI, 95% confidence interval; Clin. Imp., clinically important gastrointestinal bleeding.

Mortality

The mortality analysis showed a trend in favor of sucralfate over H_2-receptor antagonists (common odds ratio, 0.71; 95 per cent confidence interval, 0.49–1.04), although this fails to reach conventional levels of statistical significance (Table 1). The comparison of sucralfate with antacids yields a common odds ratio of 0.70 (95 per cent confidence interval, 0.52–0.94), indicating that sucralfate is associated with a lower mortality than antacid therapy.

Conclusions

According to current data, clinicians should not be concerned about increasing the risk of nosocomial pneumonia above baseline with the use of any prophylactic agent. However, current evidence supports the use of sucralfate over other agents, both because of a reduction in pneumonia and a possible reduction in mortality. In most settings, this agent is cheaper than alternative prophylactic drugs. Given the inaccuracy of the usual clinical criteria used to diagnose lung infection in previous trials, further trials, in which sucralfate is compared with gastric-pH-altering prophylaxis using sensitive and specific diagnostic approaches such as bronchoalveolar lavage and protected brush catheters, are needed to confirm the apparent reduction in pneumonia associated with sucralfate administration.

Enrolment of patients at risk of both stress ulcer bleeding and nosocomial pneumonia (those ventilated for at least 48 h and those likely to be exposed to study medication for at least 7 days) is also necessary. Whether the effect of sucralfate on ICU mortality is due to its impact on ventilator-associated lung infection or different rates of bacterial translocation remains to be determined. However, examination of both costs and consequences is necessary, and a full-scale economic evaluation is needed.

References

Cook, D.J., Fuller, H., and Guyatt, G.H. for the Canadian Critical Care Trials Group (1994). Risk factors for gastrointestinal bleeding in the critically ill. *New England Journal of Medicine*, **330**, 377–81.

Cook, D.J., *et al.* (1996). Stress ulcer prophylaxis in critically ill patients: resolving discordant meta-analyses. *Journal of the American Medical Association*, **275**, 308–14.

Heyland, D., *et al.* (1996). The clinical and economic consequences of clinically important gastrointestinal bleeding in critically ill patients. *Clinical Intensive Care*, **7**, 121–5.

Tryba, M. and Mantey-Stiers, F. (1987). Antibacterial activity of sucralfate in human gastric juice. *American Journal of Medicine*, **83**, 125–7.

Zandstra, D.F. and Stoutenbeek, C.P. The virtual absence of stress-ulceration related bleeding in ICU patients receiving prolonged mechanical ventilation without any prophylaxis. *Intensive Care Medicine*, **20**, 335–40.

17.1.27 Pulmonary surfactant

W. Seeger and A. Günther

Key messages

- Pulmonary surfactant lowers the alveolar surface tension and makes breathing possible at normal transpulmonary pressures.
- In acute respiratory distress syndrome (ARDS) both composition and surface activity of pulmonary surfactant are markedly altered, yielding elevated alveolar surface tension values.
- In pilot studies, application of 300 to 800 mg/kg of natural surfactant resulted in an acute improvement of gas exchange and slightly reduced 28-day mortality in ARDS.

Pulmonary surfactant is a lipoprotein complex covering the alveolar surface. By profoundly reducing the surface tension at the air–liquid interface, it prevents alveoli from collapse, particularly during expiration. It consists of approximately 90 per cent lipids (mostly phospholipids and a small amount of neutral lipids) and 10 per cent proteins. The phospholipid fraction comprises approximately 80 per cent phosphatidylcholine (mostly dipalmitoylated) and approximately 10 per cent phosphatidylglycerol. Four surfactant-specific apoproteins (surfactant proteins A, B, C, and D) have been identified. Triggered by inspiratory stretching of the alveolar cell layer, alveolar type II cells secrete surfactant-containing lamellar bodies which are extracellularly reorganized into tubular myelin and large multilamellar vesicles, termed large surfactant aggregates. Adsorption of the phospholipids to the air–liquid interface results in the formation of a stable phospholipid film, which, upon compression, can reduce the surface tension to almost zero. In addition to dipalmitoylated phosphatidylcholine and phosphatidylglycerol, surfactant proteins B and C appear to play an essential role in this process. The periodic compression and re-expansion of the interfacial phospholipid film provokes permanent refinement, with the large surfactant aggregates being converted into unilamellar vesicles (small surfactant aggregates). Additional functions of the alveolar surfactant system include prevention of the formation of alveolar edema fluid and, although currently at best fragmentary, a considerable impact on the alveolar host-defense mechanisms.

A lack of surface-active material is the primary cause of the infant respiratory distress syndrome (IRDS), and surfactant therapy (50–100 mg/kg) is now the gold standard. In contrast, surfactant deficiency does not seem to be of major importance in the acute respiratory distress syndrome (ARDS), which is characterized by overhelming inflammatory processes at the gas exchange unit. Rather, a broad pattern of biochemical and biophysical abnormalities of the pulmonary surfactant system is observed, which favors alveolar collapse with ventilation–perfusion mismatch and, in particular, shunt flow. Analysis of bronchoalveolar lavage fluids from ARDS patients consistently demonstrated a decrease in the surface-tension-reducing properties, with minimum surface tension values being increased to above 15 to 20 mN/m compared with near zero in healthy controls (Günther *et al.* 1996). Several biochemical changes have been observed in these patients, including the following.

1. The phospholipid profile is altered, with a reduction in the relative percentages of phosphatidylcholine and phosphatidylglycerol and an increase in the percentages of phosphatidylinositol, phosphatidylethanolamine, and sphingomyelin.
2. The fatty acid composition is altered, with a marked reduction in the relative content of saturated fatty acids, particularly palmitic acid species, in the phospholipid fraction.
3. The levels of surfactant apoproteins are decreased, as shown for the hydrophilic surfactant protein A in native bronchoalveolar lavage fluid and surfactant protein B in the large surfactant aggregate fraction.

Table 1 Gas exchange and surfactant properties following surfactant administration in ARDS

		Gas exchange		Surfactant analysis	
Study	**Dosage/regimen**	**Before treatment**	**After treatment**	**Before treatment**	**After treatment**
Spragg *et al.* (1994)	50 mg/kg Curosurf, via bronchoscope		Increase in PaO_2: 49±16%*	γ_{min}=31.8±5.2 mN/m	γ_{min}=25.7±8.2 mN/m
Gregory *et al.* (1997)	No treatment (control) 8×50 mg/kg 4×100 mg/kg 8×100 mg/kg Survanta via intratracheal instillation	PaO_2/FiO_2 (mmHg) Control: 128 8×50: 98 4×100: 124 8×100: 133	PaO_2/FiO_2 (mmHg) Control: 133 8×50: 161 4×100: 182 8×100: 162	γ_{min} Control: 22.7 8×50: 23.0 4×100: 22.1 8×100: 22.8	γ_{min} Control: 22.9 8×50: 21.8 4×100: 22.4 8×100: 21.6
Anzueto *et al.* (1996)	Exosurf aerosol for 5 days; estimated dose about 5 mg/day	PaO_2/FiO_2 (mmHg) Control: 140±64 Exosurf: 145±82	PaO_2/FiO_2 increase Control: about 35% Exosurf: about 45%	NA	NA
Walmrath *et al.* (1996)	300–500 mg/kg Alveofact via bronchoscope	PaO_2/FiO_2 (mmHg): 85±7 Shunt flow (%): 41.7±3.7	PaO_2/FiO_2 (mmHg): 200±20*** Shunt flow (%): 19.8±2.1*	PPR=0.02±0.01 LSA=27.7±5.3 PC=72.6±2.1 γ_{min}=21.6±9.2 $P\gamma_{min}$=34.4±2.2	PPR=0.23±0.1*** LSA=68.6±8.8* PC=85.3±1.0*** γ_{min}=9.2±2.8** $P\gamma_{min}$=24.2±3.6

PaO_2/FiO_2 is given in mmHg (Gregory *et al.*, Anzueto *et al.*, Walmrath *et al.*) or in percentage change compared with placebo (Spragg *et al.*, only PaO_2 available). Shunt flow was measured using the multiple inert gas elimination technique and is given as the percentage of blood perfusing non-ventilated lung regions.

PPR, ratio of phospholipid to protein in bronchoalveolar lavage fluid; LSA, relative content of large surfactant aggregates as a percentage of the total phospholipids in bronchoalveolar lavage fluid; PC, relative amount of phosphatidylcholine as a percentage of the total phospholipids; γ_{min} ($P\gamma_{min}$), minimum surface tension (mN/m) of lyophilized bronchoalveolar lavage fluid (Spragg *et al.*) or large surfactant aggregates (Gregory *et al.*, Walmrath *et al.*) at 2 mg/ml phospholipid (measured using a pulsating bubble surfactometer) in the absence (γ_{min}: Gregory *et al.*, Walmrath *et al.*) or the presence ($P\gamma_{min}$: Walmrath *et al.*) of proteins derived from bronchoalveolar lavage fluid. Bronchoalveolar lavage was performed 120 h (Gregory *et al.*) or 72 h (Walmrath *et al.*) after treatment.

Data are given as mean±SEM (Spragg *et al.*, Anzueto *et al.*, Walmrath *et al.*) or as median (Gregory *et al.*), and the level of significance is indicated by *($p<0.05$), **($p<0.01$), and ***($p<0.001$).

NA, not available.

4. The content of large surfactant aggregates is reduced. Several experimental studies have suggested that induction of an acute lung injury will result in a higher abundance of the small surfactant aggregates at the expense of the large aggregates. Accordingly, a reduction in the relative content of large surfactant aggregates was demonstrated in ARDS.

5. Surfactant function is inhibited by leaked plasma proteins. Leakage of plasma proteins into the alveolar space is a common finding in ARDS. The ratio of phospholipid to protein in bronchoalveolar lavage fluid is normally about 0.5 (by weight), but reaches mean values as low as 0.05 in ARDS. Strong surfactant-inhibitory capacity has been demonstrated for albumin, hemoglobin, fibrinogen, and particularly fibrin which is known to be easily generated in the alveolar space of ARDS patients because of increased procoagulant and antifibrinolytic activities.

6. Surfactant function is inhibited by inflammatory mediators, particularly proteases and oxygen radicals released by inflammatory cells, which may primarily attack the functionally important hydrophobic surfactant proteins B and C. A corresponding scenario of complex surfactant disturbances, as described for ARDS primarily triggered by diseases remote from the lung (e.g. sepsis, polytrauma, pancreatitis), has also been observed under conditions of severe pneumonia necessitating mechanical ventilation.

Against this background, restoration of alveolar surfactant function appears to be a reasonable approach to improving gas exchange in ARDS patients. Such attempts may include pharmacological approaches to stimulate the secretion of intact surfactant material from type II pneumocytes, but there is no clear evidence that this approach can be effectively used under conditions of acute respiratory failure. In addition, transbronchial administration of exogenous (natural) surfactant preparations may also be employed in ARDS, but large quantities of material will be required to overcome the surfactant-inhibitory capacities in the alveolar space under these conditions. Gregory *et al.* (1997) noted some improvement in gas exchange upon intratracheal instillation of 400 to 800 mg/kg of Survanta (Table 1), and increased survival was suggested in the group receiving 400 mg/kg. Walmrath *et al.* (1996) investigated the safety and efficacy of a bronchoscopic application of another natural surfactant extract (Alveofact) in patients with severe ARDS mostly due to sepsis and severe pneumonia. All patients (currently 26) fulfilled extracorporeal membrane oxygenation criteria (mean Murray lung injury score approximately 3.3) and were treated within the first 5 days of disease, i.e. before the onset of major fibrotic processes. As is evident from Table 1, administration of surfactant resulted in a rapid increase in the mean $Pa{O_2}/Fi{O_2}$ ratio, with a concomitant reduction in the intrapulmonary shunt flow. More than two-thirds of the patients 'responded' with a $Pa{O_2}/Fi{O_2}$ increase of at least 25 per cent. The effect was partially lost within the following hours in some of the responders, but was restored with prolonged improvement of arterial oxygenation by a second administration (data not given in detail). Analysis of the surfactant properties in bronchoalveolar lavage fluid obtained before and after surfactant administration suggested impressive, but still incomplete, restoration of surfactant properties. In contrast, Anzueto *et al.* (1996) found no clinical effect following aerosol application of Exosurf for 5 days in ARDS patients compared with a group receiving saline aerosol. However, there were major drawbacks to this study. Firstly, the authors estimated a pulmonary deposition of 5 mg DPPC/day in their patients. Even if a higher relative efficacy of nebulized versus instilled surfactant material is assumed, this amount of surfactant is clearly far too low to exert a significant clinical response. Secondly, Exosurf, a surfactant-apoprotein-free synthetic surfactant preparation, has repeatedly been shown to possess high sensitivity towards inhibition by plasma proteins and thus might have a lower efficacy than natural surfactant preparations in ARDS.

In conclusion, profound alterations of the alveolar surfactant system are encountered in ARDS, which contribute to the severe impairment in gas exchange. Transbronchial administration of surfactant by bronchoscopy may offer a feasible and safe approach to improving the biochemical and biophysical properties of the endogenous surfactant pool and hence the gas exchange conditions in most severe early-stage ARDS. However, a high and/or repetitive dosage regimen appears to be necessary to overcome inhibitory capacities in the alveolar space of these patients and to achieve sustained alveolar recruitment. Forthcoming studies will have to identify the optimum timing and dosage regimen of this intervention and will have to address the question of whether this therapy is capable of reducing the high mortality of patients with severe ARDS.

References

Anzueto, A., *et al.* (1996). Aerosolized surfactant in adults with sepsis-induced acute respiratory distress syndrome. *New England Journal of Medicine*, **334**, 1417–21.

Gregory, T., *et al.* (1997). Bovine surfactant therapy for patients with acute respiratory distress syndrome. *American Journal of Respiratory and Critical Care Medicine*, **155**, 1309–15.

Günther, A., *et al.* (1996). Surfactant alterations in severe pneumonia, acute respiratory distress syndrome, and cardiogenic lung edema. *American Journal of Respiratory and Critical Care Medicine*, **153**, 176–84.

Spragg, R.G., *et al.* (1994). Acute effects of a single dose of porcine surfactant on patients with the adult respiratory distress syndrome. *Chest*, **105**, 195–202.

Walmrath, D., *et al.* (1996). Bronchoscopic surfactant administration in patients with severe adult respiratory distress syndrome (ARDS) and sepsis. *American Journal of Respiratory and Critical Care Medicine*, **154**, 57–62.

17.1.28 Vasoconstrictors

J.-L. Vincent

Key messages

- In circulatory shock, fluid administration must be optimized before the introduction of any pharmaceutical agent.
- Excessive vasoconstriction can further compromise tissue perfusion and patients must be carefully monitored.
- If hypotension persists despite adequate fluid resuscitation, dopamine administration should be commenced, followed by norepinephrine (noradrenaline) or epinephrine (adrenaline) if required.
- Vasopressors have little positive effect on regional blood flow and the concurrent adminsitration of an inotrope may be considered.

Introduction

Arterial hypotension associated with circulatory shock, and hence altered tissue perfusion, threatens organ function. In all critically ill patients, a priority must be to restore and maintain an adequate tissue perfusion pressure. Fluids must be administered first, but if hypotension persists despite sufficient fluid loading, vasopressor support is indeed required with the aim of restoring a minimal perfusion pressure. Vasoconstrictors can also help to redistribute blood from non-vital organs such as the skin and muscle, where vasoconstriction effects are maximal, to vital organs such as the heart and brain. However, one must be aware of the risks associated with excessive vasoconstriction.

All vasopressors increase cardiac filling pressures as well as systemic pressure, thus favoring the development of pulmonary edema. Therefore it may be misleading to assume that administration of vasopressor agents instead of further fluids will avoid the risk of pulmonary edema. Before any pharmacological intervention, fluid administration must be optimized to ensure that hypovolemia is corrected.

Particular care must be taken about possible impairments in blood flow associated with excessive vasoconstriction. By increasing peripheral resistance, vasoconstriction increases ventricular afterload and thus can limit cardiac output. Its peripheral effects can also limit capillary blood flow and may increase tissue hypoxia, often already compromised. Thus such vasoconstriction may increase the risk of development of multiple organ failure. Hence the additional use of inotropic agents such as dobutamine may be required to maintain oxygen delivery to the tissues.

The effects of agents in common use today, as well as those of some more experimental vasoconstrictors, on common hemodynamic indices are summarized in Table 1.

Adrenergic agents

Adrenergic vasoactive agents exert their actions by activation of α-adrenergic, β-adrenergic, and dopamine receptors (Rudis *et al.*

Table 1 Relative effects of current and experimental vasoconstrictors on common hemodynamic indices

	Blood pressure	Cardiac output	Systemic vascular resistance
Epinephrine	↑↑	↑	↑
Norepinephrine	↑↑	↔	↑↑
Dopamine			
Low dose	↑	↑	↔
High dose	↑↑	↑	↑↑
Ephedrine/phenylephrine	↑↑	↔	↑↑
Angiotensin	↑↑	↓	↑↑
L-NMMA	↑↑	↓	↑↑
Methylene blue	↑↑	↔	↑↑

↑ Increase; ↓ decrease; ↔ no change.

1996). They all have a half-life of 2 to 3 min, enabling a steady state to be reached within 10 to 15 min, and are rapidly metabolized to inactive metabolites. The benefits and side-effects of any particular drug will depend on which receptors they act on (Table 2). The degree of response to adrenergic agents varies widely as it depends on many factors, and so the patient's response should be carefully monitored. Chronic sympathetic stimulation can lead to a decrease in the number and function of adrenergic receptors, which may limit a patient's response to catecholamines. Inflammatory mediators can also result in a decreased response of adrenergic receptors to stimulation. The dose ranges of some vasoconstrictor agents are given in Table 3.

Epinephrine (adrenaline)

Epinephrine is an endogenous catecholamine secreted by the adrenal medulla. It is a potent α- and β_1-agonist and a moderate β_2-agonist. Its effects are dose dependent, with β effects predominating at lower doses (0.04–0.1 μg/kg/min), resulting in increased heart rate, cardiac output, and contractility. The increased contractility may result in myocardial ischemia and arrhythmias, and this limits its use. At higher doses α effects become more prominent, raising peripheral resistance and blood pressure. Recent studies in critically ill patients suggest that epinephrine may reduce splanchnic blood flow more than other adrenergic agents (Meier-Hellmann and Reinhart 1994).

Indications

Cardiac arrest

During cardiopulmonary resuscitation, epinephrine can be administered as a repeated intravenous bolus of 1 mg every 3 to 5 min. As it is absorbed well through the airway mucosa, it can be administered through the endotracheal tube (diluted in 10 ml of sterile water or saline) if venous access is difficult.

Acute anaphylaxis

Epinephrine is the agent of choice as it antagonizes the bronchoconstriction and vasodilation associated with anaphylaxis. It can be administered as a subcutaneous bolus (0.1–0.5 mg) or as a continuous infusion (0.01–0.3 mg/min).

Cardiogenic shock

Epinephrine can be added as a continuous infusion when dopamine or dobutamine have failed to restore minimal perfusion pressure. This occurs in the most severe situations close to cardiopulmonary resuscitation. In some intensive care units (ICUs), epinephrine is administered in other forms of acute circulatory failure as a substitute for dopamine, primarily because of its lower cost, but dopamine is probably preferable.

Table 2 Effects of stimulation of adrenergic or dopamine receptors on various organ systems

	Alpha-receptors	Beta-receptors	Dopamine receptors
Heart	Increase in contractility	Increase in heart rate, contractility, conduction, and irritability	
Blood vessels	Arterial and venous constriction	Dilatation of coronary arteries (β_1) Dilation of arteries (β_2)	Vasodilation, particularly of renal and mesenteric beds
Lungs	Bronchoconstriction	Bronchodilation	
Muscles		Intestinal muscle relaxation (β_1) Uterine muscle relaxation (β_2)	
Eyes	Mydriasis	Increased intraocular tension	
Metabolism	Increased glyconeogenesis and glycogenolysis Decreased insulin secretion	Increased glycogenolysis and lipolysis Increased secretion of insulin, renin, glucagon, parathyroid hormone	
Hematological cells	Stimulation of mastocytes Enhanced platelet aggregation	Inhibition of mastocyte stimulation Enhanced platelet aggregation	

Table 3 Dose ranges of vasoconstrictor agents

Drug	Normal dose range (μg/kg/min)
Epinephrine	0.3–1.5
Norepinephrine	0.1–3
Dopamine	<2 (predominant dopaminergic and β effects)
	5–10 (predominant β effects)
	>10 (predominant α effects)

Norepinephrine (noradrenaline)

Norepinephrine is also an endogenous catecholamine secreted by the adrenal medulla and the terminal endings of postganglionic nerve fibers. It is a potent α-agonist with moderate β_1- and minimal β_2-activity. Therefore it induces marked peripheral vasoconstriction with a resultant rise in blood pressure and often a fall in cardiac output. Importantly, the fall in cardiac output does not always occur early, but slowly over time. Hence norepinephrine can limit tissue oxygen availability. Concurrent use of dobutamine may prevent this. Norepinephrine may also be combined with phosphodiesterase inhibitors (e.g. milrinone) to achieve an additive inotropic effect of the adrenergic and non-adrenergic agents, while the vasodilator and vasoconstrictor effects counterbalance each other. Norepinephrine can also be combined with epoprostenol (prostacyclin) or prostaglandin E_1 in the treatment of severe right heart failure.

An attractive effect of norepinephrine is that the predominant α-adrenergic effects result in limited tachycardia. Hence, the addition of norepinephrine can be considered in the presence of hyperkinetic septic shock associated with severe tachycardia unresponsive to fluids. Nevertheless, the increased afterload related to the vasopressor effects leads to increased myocardial workload and can precipitate acute cardiac failure, myocardial ischemia, and pulmonary edema.

Indications

Severe circulatory shock

Norepinephrine is indicated when arterial hypotension persists despite the administration of dopamine at doses of 20 to 25 μg/kg/min.

Right ventricular failure

Pulmonary hypertension with arterial hypotension can compromise right ventricular function by decreasing right ventricular coronary perfusion. Norepinephrine can be beneficial in restoring coronary perfusion pressure in such circumstances, but this pattern is rare and, when it occurs, is manifest by profound cardiovascular collapse. In right ventricular failure it may be combined with a pulmonary vasodilator such as prostaglandin E_1.

Septic shock

Norepinephrine may be useful for maintaining tissue perfusion pressure in severe circulatory collapse associated with sepsis (Marik and Mohedin 1994).

Dopamine

Dopamine has become a very popular drug as it induces fewer arrhythmias and increases myocardial oxygen consumption less than epinephrine and norepinephrine. It is a naturally occurring amine and is the natural precursor of norepinephrine. Its effects are very dose dependent.

At doses below 2 μg/kg/min dopamine predominantly activates dopamine receptors, increasing renal blood flow. At 5 to 10 μg/kg/min β-adrenergic effects on the myocardium lead to increased contractility and cardiac output, and at doses above 10 μg/kg/min α effects come into play with vasoconstriction causing increased blood pressure and negating the dopaminergic inotropic effects.

Indications

Circulatory shock

Dopamine is the drug of choice for shock when hypotension persists despite fluid therapy. In low cardiac output states, although dobutamine is the agent of choice, the administration of dopamine is usually safer as an initial measure. Dopamine increases blood pressure while maintaining myocardial inotropy, so that cardiac output is better preserved than with norepinephrine.

Augmentation of renal or hepatosplanchnic blood flow

At low doses (above 2 μg/kg/min), the dopaminergic effects of dopamine have been thought to increase renal and hepatosplanchnic flow more than blood flow to other regions. However, there is little clinical evidence to support this function (Vincent and Preiser 1993). The routine use of low-dose dopamine does not reduce the incidence of renal failure and its administration does not consistently improve gastric intramucosal pH.

Phenylephrine and ephedrine

Phenylephrine is a strong vasopressor, primarily affecting α receptors, with some β activity at very high doses. Thus phenylephrine may increase the heart rate even less than norepinephrine.

Ephedrine is a potent non-selective vasopressor with clinical effects similar to those of epinephrine. It has direct adrenoceptor action and also enhances the release of endogenous norepinephrine from sympathetic nerve endings. It has a longer duration of action than the catecholamines and has central nervous system stimulant effects which can be disturbing.

In small studies on septic patients these agents have been shown to increase arterial pressure, but they have little to offer over other vasopressors. These very strong vasopressors have little place in ICU therapy except perhaps transiently in very severe refractory hypotension.

Non-adrenergic agents

Angiotensin

Angiotensin is a naturally occurring octapeptide with a half-life of less than 2 min. It is a potent vasoconstrictor, acting on specific receptors in arteriolar smooth muscle, at presynaptic receptors in peripheral sympathetic nerves and the adrenal medulla to increase catecholamine synthesis and release, and by centrally mediated stimulation of the sympathetic nervous system. It has no myocardial effects, and so it may cause fewer arrhythmias than the adrenergic agents, but it causes a fall in cardiac output with a resultant fall in tissue perfusion despite the rise in blood pressure. Therefore its use is not recommended, even in extreme forms of hyperkinetic shock.

Experimental agents

Nitric oxide (NO) is released in large amounts in response to a septic challenge. It is a potent vasodilator and therefore has been implicated in the hypotension associated with septic shock. NO also plays a role in the maintenance of peripheral tissue perfusion and in intercellular adhesion, and has microbicidal effects. Therefore blocking NO production may restore arterial blood pressure by limiting vasodilation, but could have severe detrimental effects. Two experimental NO blockers which have undergone clinical trials are N^G-monomethyl-L-arginine (L-NMMA) and methylene blue.

L-NMMA is an inhibitor of NO synthase. In septic patients, administration of L-NMMA may increase arterial blood pressure, although with an associated fall in cardiac output. The correct dosage is important, as the use of high doses may increase mortality, and timing of administration also seems to be critical with late intervention being more effective. The degree of inhibition may also be important, with specific inhibition of inducible NOS being most beneficial.

Methylene blue is an inhibitor of guanylate cyclase and therefore may specifically attenuate the hemodynamic effects of NO, which are associated with guanylate cyclase activation. Administration of methylene blue to patients with septic shock increases arterial blood pressure and left ventricular work, with no significant effect on cardiac output, and in animal models methylene blue administration has been shown to improve mesenteric blood flow selectively (Preiser *et al.* 1994). Further work needs to be done to to establish whether there is a place for methylene blue or L-NMMA in the treatment of septic patients.

Conclusions

Adrenergic vasoconstrictor therapy can be used to restore tissue perfusion pressure, but has little capacity to improve the peripheral distribution of blood flow or cellular oxygen availability. Clinically, after adequate fluid resuscitation, it is generally advisable to commence with dopamine and add norepinephrine or epinephrine if doses of 20 to 25 μg/kg/min fail to raise mean arterial pressure sufficiently. The concurrent administration of dobutamine when using vasopressor agents should be considered to maintain an adequate cardiac output.

References

Marik, P.E. and Mohedin, J. (1994). The contrasting effects of dopamine and norepinephrine on systemic and splanchnic oxygen utilization in hyperdynamic sepsis. *Journal of the American Medical Association*, **272**, 1354–7.

Meier-Hellmann, A. and Reinhart, K. (1994). Influence of catecholamines on regional perfusion and tissue oxygenation in septic shock patients. In *Sepsis: current perspectives in pathophysiology and therapy* (ed. K. Reinhart, K. Eyrich, and C. Sprung), pp. 274–91. Springer-Verlag, Berlin.

Preiser, J.C., Zhang, H., and Vincent, J.L. (1994). Administration of methylene blue in septic shock. Is it beneficial? In *Yearbook of intensive care and emergency medicine — 1994* (ed. J.L. Vincent), pp. 90–100. Springer-Verlag, Berlin.

Rudis, M.I., Basha, M.A., and Zarowitz, B.J. (1996). Is it time to reposition vasopressors and inotropes in sepsis? *Critical Care Medicine*, **24**, 525–37.

Vincent, J.L. and Preiser, J.C. (1993). Inotropic agents. *New Horizons*, **1**, 137–44.

17.1.29 Vasodilators

A. B. J. Groeneveld

Key messages

- Vasodilators are the treatment of choice for hypertensive emergencies and cardiogenic pulmonary edema.
- Vasodilating therapy may alleviate pulmonary hypertension, but often at the cost of an increased venous admixture of oxygen in the lungs.
- Vasodilators usually do not benefit the patient in non-cardiogenic shock.
- Invasive monitoring via a pulmonary artery catheter may be indicated for proper institution and guidance of vasodilator therapy.

Introduction

In some conditions requiring intensive care, vasodilator drugs may play a major therapeutic role. These conditions include hypertensive emergencies and primary and secondary pulmonary hypertension. Apart from their blood-pressure-lowering effects, vasodilator drugs are also used to unload the heart, for instance in the case of acute left heart failure or the low cardiac output syndrome associated with various types of circulatory shock. The intravenous route for administration of vasodilators is preferred for treatment in the intensive care unit (ICU), as it allows small starting doses which can be increased over relatively short periods of time guided by the hemodynamic response and which can be followed by continuous maintenance dosing provided that the drugs used have a relatively short half-life. In contrast with the generally prolonged action of orally administered drugs, the intravenous route allows blood pressure to rise rapidly once the infusion has been stopped. Nevertheless, oral therapy is indicated when the patient is prepared for transfer to the general ward. There are large differences between individuals and drugs in dose–response relationships. As a rule, the vasodilating effect is greatest in patients with severe vasoconstriction.

Vasodilator therapy to control arterial blood pressure and to unload the left heart

Hypertensive emergencies

Hypertensive emergencies can be defined as a progression of arterial hypertension associated with signs of acute organ dysfunction, primarily of heart, eye, and brain. Obviously, these conditions should be carefully differentiated from intracranial abnormalities leading to reflex arterial hypertension (Tietjen *et al.* 1996) since a (sudden) reduction in arterial blood pressure in the latter situation may have a detrimental effect on cerebral perfusion. For instance, calcium antagonists may effectively lower arterial blood pressure but may increase cerebral blood volume and decrease cerebral perfusion, whereas adrenergic receptor blockers may lack these adverse effects (Tietjen *et al.* 1996).

Severe arterial hypertension can be effectively treated with any type of antihypertensive drug, and the choice of drug should be guided by the relative advantages and side-effects in the individual patient. However, antihypertensive responses may be unpredictable after both intravenous and non-parenteral administration. Intravenous injections are preferred, with close monitoring of arterial blood pressure, and the infusion pump rate for drugs with a continous maintenance dose should be adjusted according to the blood pressure response. Transient hypotension usually responds to fluid administration. Closed-loop automatic systems for adjusting continuous infusion rates of antihypertensive drugs have been designed.

Drugs used for the treatment of hypertensive emergencies in the ICU include (continuously) intravenously administered sodium nitroprusside, phentolamine, diazoxide, hydralazine, and angiotensin-converting enzyme inhibitors. These drugs effectively lower arterial blood pressure in a relatively short period of time. Intravenously administered calcium antagonists, such as the first-generation dihydropyridine analog nifedipine or the second-generation analogs isradipine and nicardipine, are also used. However, oral administration of calcium blockers, including nifedipine, isradipine, nicardipine, and others, is used more often as a primary or adjunctive therapy for severe hypertension, particularly in milder cases. The oral route is almost as effective as the sublingual route, and administration of both nifedipine and captopril via this route has been shown to be effective in hypertensive emergencies.

Dissecting aneurysm

Acute dissection of the aorta is often associated with severe arterial hypertension. The increased wall stress may invoke (further) intimal tearing. The aim of medical therapy is to diminish wall stress, so that not only arterial blood pressure but also pulse pressure should be reduced. In fact, conservative medical therapy may be as successful as surgical therapy, at least for distal aortic dissection. This is why vasodilating therapy, such as administration of a calcium antagonist which is often accompanied by reflex tachycardia and a rise in cardiac output and pulse pressure, should be combined with β-blocking agents to prevent a rise in wall stress. Alternatively, the combined α- and β-receptor blocker labetalol can be used effectively. Although trimetaphan, a sympatholytic agent, has been the drug of choice in the past for the acute management of aortic dissection because of a lack of reflex tachycardia and rise in pulse pressure, intravenous nitroprusside is now commonly used, together with β-blockers, in the treatment of dissecting aneurysms.

Accelerated hypertension in pregnancy

Some patients with pregnancy-induced hypertensive disease require ICU admission to control blood pressure. Angiotensin-converting enzyme inhibitors are relatively contraindicated because of the risk of fetal death and nitroprusside because of fear of toxicity. Preferred drugs are labetalol, hydralazine, ketanserin, and calcium blockers. The acute effects of orally administered nifedipine (10 mg) is equivalent to that obtained during continuous infusion of hydralazine (1–3 mg), but the former may have fetal side-effects.

Postoperative hypertension

Particularly after cardiopulmonary surgery, arterial hypertension complicating the postoperative course may be dangerous for the newly revascularized heart and therefore should be treated. The drugs used for postoperative hypertension in the recovery room or the ICU include nitroglycerin (glyceryl trinitrate), nitroprusside, ketanserin, and calcium antagonists.

Acute congestive heart failure

This may occur in the course of acute myocardial infarction, valvular disease, cardiomyopathy, and other conditions. It may consist of pulmonary edema, pump failure (cardiogenic shock), or both. In this situation, rapidly active vasodilating drugs have been used to lower preload (pulmonary edema) and afterload (pump failure) of the heart to varying degrees. Vasodilating therapy in acute mitral or aortic regurgitation and septal rupture in the course of myocardial infarction may also unload the heart. Since the treatment of acute congestive heart failure by manipulating pre- and afterload of the heart is often complex, invasive hemodynamic monitoring via a pulmonary artery catheter is usually indicated.

Vasodilating therapy in acute congestive heart failure usually consists of continuously intravenously administered nitroglycerin in normotensive patients and nitroprusside in hypertensive patients. In fact, acute pulmonary edema of cardiac origin is often accompanied by (stress-induced) arterial hypertension, so that the control of arterial blood pressure may help to unload the heart and ameliorate pulmonary congestion. It is claimed that, for a given reduction in arterial blood pressure, nitroglycerin lowers the left heart filling pressure more than nitroprusside because of the greater venodilating effect of the former. Unloading can also be performed by rapidly acting intravenous bolus doses of α-receptor blocking agents (such as phentolamine and urapidil) or enalaprilat (1.25 mg), and by sublingual or intravenous doses of nifedipine and other first-generation calcium antagonists, even though the latter have negative inotropic properties. Adjunctive therapy may consist of a diuretic, morphine, oxygen, and in some cases digitalis and other inotropic drugs such as dobutamine or dopamine.

The use of vasodilators during congestive heart failure may have a varying effect on arterial oxygenation, although, because of the increase in cardiac output achieved with most vasodilators, tissue oxygen delivery is usually enhanced. A potential lowering of $Pa\text{O}_2$ due to the increased pulmonary venous admixture of oxygen caused by the vasodilators may be offset by decreased pulmonary edema.

Vasodilator therapy to control pulmonary artery pressure and to unload the right heart

Pulmonary hypertension may be primary or secondary to lung or heart disease (Prewitt 1987). Also, various modalities of ventilatory support, including positive end-expiratory pressure, may increase transmural pulmonary artery pressure and right ventricular afterload in some patients by 'squeezing' small intrapulmonary blood vessels via the associated rise in intrathoracic pressure and lung volume. During acute pulmonary hypertension, right ventricular dysfunction (cor pulmonale) may limit cardiac output or its response to preload augmentation (fluid loading), and may contribute to arterial hypotension and insufficient peripheral oxygen delivery in the critically ill. Hypotension, in turn, may aggravate right coronary hypoperfusion, thereby further compromising, in light of the elevated oxygen needs associated with increased pulmonary blood pressure and right ventricular afterload, the oxygen supply–demand balance for the right ventricle, resulting in dilatation and dysfunction (Prewitt 1987).

Attempts have been made to decrease pulmonary vascular resistance selectively, although no systemically administered vasodilator has a completely selective vasodilator effect on the pulmonary circulation. The aim is to unload the right heart and thereby increase tissue oxygen delivery and uptake (Prewitt 1987). In cases of shock complicated by pulmonary hypertension, systemic arterial hypotension aggravated by vasodilators may lead to (right) coronary hypoperfusion which further diminishes the right ventricular oxygen supply–demand balance and prevents a rise in cardiac output (Prewitt 1987). Hence it is better to treat pulmonary hypertension complicated by shock with vasoconstrictors rather than with vasodilators (Prewitt 1987). Pulmonary vasodilator therapy, for instance with intravenously administered prostaglandin analogs that are taken up by the lungs with relatively little spill-over into the systemic circulation and subsequent arterial hypotension, may improve right ventricular function.

Primary pulmonary hypertension

The treatment of primary pulmonary hypertension is often a matter of trial and error, since reactivity to one vasodilator does not imply reactivity to another. Most clinicians will admit patients with a decompensated cor pulmonale during disease to the ICU for hemodynamic monitoring. Vasoreactivity can be tested with the help of oxygen, epoprostenol (prostacyclin), prostaglandin E_1, nifedipine, angiotensin-converting enzyme inhibitors, hydralazine, ketanserin, phentolamine, isoproterenol, nitrates, and other vasodilating drugs, and this may help in the subsequent choice of vasodilators that can be administered orally on a long-term basis (Raffy *et al.* 1996). Testing should be done carefully, on the basis of repeated invasive measurements, because of the variability of acute responses and their poor predictive value for chronic responses (Raffy *et al.* 1996). Acute responsiveness of the pulmonary circulation and long-term lowering of pulmonary artery pressure may improve right ventricular function and thereby ultimate survival (Raffy *et al.* 1996).

Secondary pulmonary arterial hypertension

Secondary pulmonary arterial hypertension may occur after major surgery and during chronic obstructive lung disease, pulmonary thromboembolism, acute respiratory distress syndrome, and septic shock. Occasionally, endstage cystic fibrosis or interstitial lung disease is responsible for decompensation of a cor pulmonale. In selected cases, pulmonary hypertension during decompensated chronic obstructive lung disease and acute respiratory failure may limit the circulatory changes adaptive for hypoxemia. In fact, hypoxic vasoconstriction in the lung is a mechanism for maintaining arterial oxygenation in poorly ventilated areas of diseased lungs. Nevertheless, this may lead to right ventricular afterload mismatch and a limitation of cardiac output reserve, whereas, in contrast, a rise in cardiac output may be necessary to maintain tissue O_2 delivery in face of hypoxemia. Hence vasodilators, including nitrates, phosphodiesterase inhibitors, prostaglandins, hydralazine, ketanserin, calcium antagonists, and α-blockers, have been used in attempts to attenuate pulmonary hypertension selectively and unload the right heart. Epoprostenol appears to be a non-selective vasodilator that may lower pulmonary artery pressure and arterial oxygenation and may not increase tissue oxygen delivery. In contrast, prostaglandin E_1 may have greater pulmonary vascular selectivity and less inhibitory effect on hypoxic vasoconstriction. Its administration in decompensated chronic obstructive lung disease may result in amelioration of pulmonary hypertension and increased cardiac output and tissue oxygen delivery.

Pulmonary embolism

Vasodilator therapy has been studied in animals with acute pulmonary embolism, but is not routine in patients (Prewitt 1987). In contrast, vasodilators have been tried , with varying responses, in patients with chronic pulmonary hypertension on the basis of recurrent emboli.

Septic shock and acute respiratory distress syndrome

During sepsis and acute respiratory distress syndrome, pulmonary vascular changes are thought to contribute to the diminished ability of the right heart, in the face of increased afterload, to generate a sufficiently high cardiac output, particularly during fluid loading, to meet increased tissue requirements in some patients (Prewitt 1987). In animal studies, vasodilators have been used in attempts to lower the elevated pulmonary artery pressure during sepsis or endotoxemia (Prewitt 1987). In humans, pulmonary hypertension can be ameliorated to some extent with the serotonin antagonist ketanserin, even without a rise in the venous admixture of oxygen, suggesting a role for serotonin in the pulmonary vascular changes during human acute respiratory distress syndrome. The use of (non-selective) nitrovasodilators may only partly benefit the patient, even though they may effectively ameliorate pulmonary hypertension and unload the right heart, since tissue oxygen delivery may not increase by a limited rise in cardiac output and impaired arterial oxygenation.

Vasodilating prostaglandins have been used in attempts at selective dilatation of the pulmonary vessels (Prewitt 1987). Prostaglandins may increase cardiac output, dilate pulmonary blood vessels, and maintain or increase $P\text{ao}_2$. Side-effects of this type of drug include arterial hypotension, possibly associated with relatively poor pulmonary extraction of these drugs in critically ill patients with a damaged lung vasculature during sepsis and acute respiratory distress syndrome. Although prostaglandin E_1 may have greater selectivity than epoprostenol in lowering pulmonary vascular pressures in chronic obstructive lung disease, this selectivity may be lost when the pulmonary (micro)vasculature has been damaged during sepsis and the acute respiratory distress syndrome. Other major side-effects of vasodilating prostaglandins are their immunomodulating properties, which may limit the inflammatory response in the lungs and thereby increase pulmonary gas exchange and patient survival. These immunomodulating properties mainly consist of inhibition of platelet and neutrophil activation and aggregation. Moreover, lowering pulmonary vascular pressure may help to limit microvascular fluid filtration and edema formation. For these reasons, clinical trials have been performed in patients with acute respiratory distress syndrome to evaluate the effect of prostaglandin E_1 on pulmonary variables and outcome. The effect of prostaglandin E_1 on survival in acute respiratory distress syndrome remains controversial, however, since the lifesaving effects reported in early studies could not be reproduced. Later studies showed no beneficial effect of the drug on hemodynamics, oxygen balance, organ function, and survival in acute respiratory distress syndrome.

Postoperative problems

Major cardiopulmonary and vascular surgery may result in pulmonary vascular damage and hypertension. Various drugs to ameliorate this response have been evaluated, including nitroprusside, ketanserin, and prostaglandin E_1, but none of them has reached routine status (Prewitt 1987). The pulmonary venous admixture of oxygen may increase with nitroprusside and decrease with ketanserin during similar vasodilation after major valvular surgery.

Vasodilator therapy to improve the blood pressure–flow relationship

In the past investigators have tried to increase forward blood flow in animal models of shock by using α-receptor blockers such as phentolamine, hydralazine, calcium antagonists, etc., even though amelioration of sympathetic vasoconstriction during hypoperfusion may be associated with maldistribution of blood flow and decreased oxygen extraction. Sometimes vasodilator therapy has been used as an adjunct to the treatment of low-flow shock of cardiogenic, obstructive, or septic origin in humans, in attempts to improve blood flow and tissue oxygenation. The main disadvantage of this treatment is that arterial blood pressure often falls further, even if cardiac output increases, and arterial oxygenation is reduced so that oxygen delivery may not increase. Hence, close monitoring of the hemodynamic and clinical response of an individual patient to the infusion of these drugs is needed to decide on their potential benefit.

Nitroprusside, nitroglycerin, phosphodiesterase inhibitors, and others, together with inotropic drugs, have been used to increase cardiac output during cardiogenic shock or after extensive cardiopulmonary surgery. Vasodilators are more often used to supplement standard therapy than as primary agents. Moreover, nitroprusside, rather than hydralazine, has been propagated as an adjunct for the emergency treatment of pericardial tamponade (before pericardiocentesis can be performed) after completion of fluid loading. Intravenous infusion of nitroglycerin has been studied as therapy for septic shock complicated by a relatively low cardiac output. However, the results of these studies suggest that the increase in cardiac output obtained with these drugs is modest at best.

Vasodilating prostaglandins such as epoprostenol and prostaglandin E_1 have been used in attempts to increase cardiac output, oxygen delivery, and oxygen uptake by the tissues during septic or traumatic shock, although the effects proved variable. A supply-dependent rise in oxygen uptake with a trial of these drugs may be associated with survival in some studies, but with mortality in others. This may primarily relate to differences in patient populations. A rise in oxygen delivery and uptake may sometimes be accompanied by a rise in oxygen extraction, suggesting improved microcirculatory blood flow with this type of drug (Silverman *et al.* 1990). Although some studies report favorable oxygenation and survival results with these drugs, others have found unchanged tissue oxygenation and outcome (Silverman *et al.* 1990). This has limited the widespread use of these drugs.

During reperfusion following ischemia, cells are overloaded with calcium and this is believed to contribute to reperfusion injury. Hence calcium antagonists have been studied in various models of shock and resuscitation for potentially protective effects on organ function. In clinical practice, there is no place for the routine use of calcium antagonists in situations associated with ischemia and reperfusion, even though the (cerebral) outcome of cardiac arrest, an extreme form of ischemia and reperfusion, was shown to be improved by the administration of the calcium antagonist nimodipine during prolonged cardiopulmonary resuscitation. Calcium antagonists are also used in solutions for preservation of organ transplants and in cardioplegic solutions for cardiac surgery. It has been suggested that 'controlled hypotension' decreases blood losses during anesthesia and surgery.

Finally, vasodilators are used in the management of subarachnoid hemorrhage. The aims is to reduce cerebral vasospasms and subsequent ischemic necrosis rather than lower arterial blood pressure, since the latter may adversely affect cerebral perfusion pressure. The drugs most commonly used for this purpose are the calcium antagonists such as nimodipine that can selectively dilate cerebral vessels. Adjunctive treatment with such drugs has been shown to improve neurological status and survival following subarachnoid hemorrhage.

Specific drugs

Vasodilators used in the ICU and their common side-effects are shown in Tables 1 and 2.

Nitrovasodilators

These drugs are the most frequently used vasodilators in the ICU (Anderson *et al.* 1994). Although sodium nitroprusside has mainly arterial vasodilating properties, nitroglycerin and analogous compounds predominantly dilate veins (Anderson *et al.* 1994). At higher doses, they also have an arterial dilating effect and lower arterial blood pressure. The compounds release nitric oxide (NO) which stimulates production of cyclic guanosine monophosphate in the vessel wall, leading to relaxation (Anderson *et al.* 1994). When administered intravenously, they have a rapid onset of action which also rapidly disappears after discontinuation of the drug. The major limitation is the development of nitrate tolerance, but this can be circumvented by nitrate-free intervals (asymmetric dosing). Indications for administration include severe forms of hypertension (nitroprusside), acute left heart failure with pulmonary edema (nitroglycerin), and unstable angina pectoris (nitroglycerin) (Anderson *et al.* 1994).

The drugs are occasionally used as adjunctive therapy in the treatment of shock, particularly if caused by a low cardiac output (e.g. in cardiogenic shock, after cardiopulmonary surgery, etc.). Similarly, intravenous nitroglycerin has been used to protect patients from the excessive vasoconstriction used in the treatment of esophageal variceal bleeding. The major side-effect of nitroprusside is thiocyanate and cyanide toxicity, a complication that is usually avoided if treatment is limited to 3 days at doses below 3 μg/kg/min with normal renal function. Signs of toxicity include central nervous system symptoms. Thus monitoring of thiocyanate levels in blood is mandatory during prolonged treatment with nitroprusside. The treatment of choice for toxicity is intravenous hydroxycobalamine (5 mg).

Table 1 Vasodilators in use in the ICU

	Route	Dose	Onset	Duration	Side-effects
Nitrates					
Nitroglycerin	IV Sublingual	0.07–5 μg/kg/min 0.4–1 mg	Minutes 20–30 s	5–10 min 1 h	Tachyphylaxis, flush, headache, collapse, confusion
Isosorbide dinitrate	IV Oral	0.5–3.5 μg/kg/min 5–40 mg four times daily	Minutes 15–40 min	Minutes 4–6 h	Tachyphylaxis, flush, headache, collapse, confusion
Sodium nitroprusside	IV	0.5–50 μg/kg/min	30 s	2–3 min	Must be shielded from light, thiocyanate and cyanide toxicity, methemoglobinemia, lactic acidosis, vitamin B_{12} deficiency
Alpha-blockers					
Phentolamine	IV IV	0.5–15 μg/kg/min 1–10 mg bolus	1–2 min	1 h	Flush, vertigo, angina
Labetalol	IV Oral	3–35 μg/kg/min 100–1000 mg twice daily	5–10 min 1 h	3–6 h 6–12 h	Bronchoconstriction, heart block
Calcium antagonists					
Nifedipine	IV IV Oral/sublingual	0.1–0.3 μg/kg/min 5–10 mg bolus 10–20 mg at 1 h or 10–40 mg three times daily	Minutes Minutes 2–5 min 2–5 min	Minutes Minutes 3–6 h 3–6 h	Headache, ankle edema, hyperglycemia
Isradipine	IV IV Oral	0.15–0.6 μg/kg/min 0.5 mg bolus 5 mg twice daily	Minutes 8 min 2 h	Minutes 1–2 h 5–7 h	Headache, ankle edema, hyperglycemia
Nicardipine	IV Oral	1–5 μg/kg/min 20–40 mg three times daily	5–15 min 1–20 min	30 min 3 h	Headache, ankle edema, hyperglycemia
Nimodipine	IV Oral	0.2–0.5 μg/kg/min 30–90 mg four times daily	Minutes	Hours	Headache, ankle edema, hyperglycemia
Angiotensin-converting enzyme inhibitors					
Captopril	Oral/sublingual	12.5–50 mg four times daily	10–15 min	6–12 h	Renal failure, cough
Enalapril(at)	IV Oral	0.2–0.4 μg/kg/min 5–40 mg/day	15 min 1 h	6 h 4–6 h	Headache, ankle edema, hyperglycemia
Prostaglandins					
Epoprostenol	IV	5–35 ng/kg/min	Minutes	Minutes	Flush, hypoxemia
Prostaglandin E_1	IV	20–100 ng/kg/min	Minutes	Minutes	Hypotension, diarrhea, flush, headache

(continued)

Table 1 continued

	Route	Dose	Onset	Duration	Side-effects
Phosphodiesterase inhibitors					
Enoximone	IV	Loading 0.1–2 mg/kg Maintenance 5–20 μg/kg/min	10–30 min	3–6 h	Arrhythmias, headache, gastrointestinal
Amrinone	IV	Loading 0.75 mg/kg Maintenance 5–10 μg/kg/min	10 min	3–6 h	Arrhythmias, headache, gastrointestinal, thrombocytopenia
Milrinone	IV	Loading 50 μg/kg Maintenance 0.5–0.75 μg/kg/min	5 min	6–8 h	Arrhythmias, headache, gastrointestinal
Miscellaneous					
Hydralazine	IV IV Oral	0.2–5 μg/kg/min 5–10 mg bolus 25–100 mg twice daily	15 min 45 min	3 h 4–24 h	
Ketanserin	IV Oral	0.7–1.5 μg/kg/min 20–40 mg twice daily	Minutes	1 h	Arrhythmias, headache
Trimetaphan	IV	5–50 μg/kg/min	1–5 min	10–15 min	Blurred vision, dry mouth, bladder and gut paresis
Diazoxide	IV IV	0.7 μg/kg/min 1–3 mg/kg bolus	5 min	6 h	Hyperglycemia

Table 2 Common side-effects of vasodilators

Hypotension
Hypoxemia
Vertigo
Flush
Nausea and vomiting
Rebound hypertension
Headache
Myocardial/cerebral ischemia at low blood pressure
Sympathetic stimulation during hypotension
Tachycardia
Increased cardiac contractility

Another potential specific side-effect of nitrovasodilators is methemoglobinemia. Continuous intravenous infusion is preferred.

Alpha-blockers

Some α-receptor blockers are non-selective, such as phentolamine (α_1- and α_2-blockade), and some are selective, such as labetalol (α_1- and β-blocker). Stimulation of the α_1 receptor causes arteriolar and venous constriction, and stimulation of the α_2 receptor causes decreased central sympathetic outflow. This type of drug is particularly useful in the treatment of pheochromocytoma crises, hypertension following intracranial hypertension, and pregnancy-induced hypertension (Tietjen *et al.* 1996).

Calcium antagonists

The first-generation calcium antagonists nifedipine, verapamil, and diltiazem have negative inotropic properties, in decreasing order of severity, rendering these drugs less suitable for patients with (borderline) heart failure or for combination therapy with β-blockers. Verapamil and, to a lesser extent, diltiazem are also used to control the ventricular rate during atrial fibrillation or other supraventricular arrhythmias. Preferably, they should not be combined with digitalis. The second-generation drugs, which include isradipine, nicardipine, and nimodipine, have fewer side-effects. These drugs have powerful vasodilating properties and lack significant effects on cardiac conduction and contractility.

These compounds are commonly used in the ICU for control of arterial blood pressure and vasospasm. Intravenous forms can be used to control arterial blood pressure and to unload the left heart during accelerated arterial hypertension. Oral and intravenous nimodipine can be used to prevent cerebral vasospasm during subarachnoid hemorrhage.

Angiotensin-converting enzyme inhibitors

Angiotensin-converting enzyme inhibitors can be used both intravenously and orally. They unload the heart by lowering systemic vascular resistance, and can be useful in the acute (and chronic) treatment of arterial hypertension and pump failure. They can be administered intravenously or sublingually/orally for the rapid lowering of arterial blood pressure. Contraindications for their use include serious disturbances of renal function and the presence of a renal artery stenosis, since administration of these drugs in these situations can lead to severe renal dysfunction.

Phosphodiesterase inhibitors

The mechanism of action of these drugs is inhibition of phosphodiesterase in the heart and vascular wall with accumulation of cAMP. These 'inodilators' increase myocardial contractility without a major rise in heart rate or myocardial oxygen demands, and produce a fall in peripheral vascular resistance without a major blood pressure lowering effect. They increase cardiac compliance and have a prolonged duration of action. They can be used judiciously, mainly as an adjunct to standard therapy in patients in whom cardiac output is (relatively) low, for instance during postmyocardial infarction shock and after cardiopulmonary bypass surgery. Severe hypotension does not usually occur but can be a major limitation in some patients.

Miscellaneous drugs

Hydralazine, diazoxide, and phentolamine can be administered continuously and also as a bolus, since the blood-pressure-lowering effect may be relatively prolonged. The ganglion blocker trimetaphan is used for the treatment of aortic dissection and for controlled hypotension during surgery. This drug, which is continuously intravenously infused, acts rapidly and has a short duration of action. It competitively inhibits the action of acetylcholine at autonomic ganglia, thereby blocking sympathetic and parasympathetic pathways. Thus the hypotensive effect is not accompanied by a baroreceptor-mediated reflex tachycardia.

Conclusions

Vasodilators are commonly used in the ICU. Monitoring responses and guiding treatment towards preset hemodynamic goals is important for documenting the benefit of a vasodilating drug in an individual patient.

References

Anderson, T.J., Meredith, I.T., Canz, P., Selwyn, A.F., and Yeung, A.C. (1994). Nitric oxide and nitrovasodilators: similarities, differences and potential interactions. *Journal of the American College of Cardiology*, **24**, 555–66.

Prewitt, R.M. (1987). Pathophysiology and treatment of pulmonary hypertension in acute respiratory failure. *Journal of Critical Care*, **2**, 206–18.

Raffy, O., *et al.* (1996). Clinical significance of the pulmonary vasodilator response during short-term infusion of epoprostenol in primary pulmonary hypertension. *Circulation*, **93**, 484–8.

Silverman, H.J., *et al.* (1990). Effects of prostaglandin E_1 on oxygen delivery and consumption in patients with the adult respiratory distress syndrome. Results from the prostaglandin E_1 multicenter study. *Chest*, **98**, 405–10.

Tietjen, G.S., Hurn, P.D., Ulatowski, J.A., and Kirsch, J.R. (1996). Treatment modalities for hypertensive patients with intracranial pathology: options and risks. *Critical Care Medicine*, **24**, 311–22.

18 Therapeutic equipment and techniques

18.1 Airway access

18.1.1 Physiology of the intubated airway

G. Hedenstierna

Key messages

- Intubation of the airway increases resistance to gas flow and may increase functional residual capacity, thus increasing respiratory work.
- The anatomical dead-space is reduced by the intubation, but the connecting tubing and any humidifier applied to the respiratory circuit result in an overall increase in respiratory dead-space. This increases the ventilatory demand.
- Bypassing of the nose and the upper airway by the artificial airway interferes with the conditioning of the inspired gas (humidification, heating, and removal of larger particles). The distal end of the tube and the sealing cuff may damage the airway wall and cause infection and stenosis.
- Intubation of the airway will prevent autoinhalation of endogenously produced nasal nitric oxide with possible negative effects on ventilation–perfusion matching.

Introduction

Although the artificial airway establishes an essentially airtight connection between the ventilatory device (or atmosphere) and the lung, it also has a number of side-effects including the following:

(1) increasing resistance to gas flow (with some exceptions);

(2) increasing respiratory dead-space (again, with some exceptions);

(3) interference with conditioning of the inspired gas;

(4) possible effects on functional residual capacity (FRC);

(5) prevention of autoinhalation of endogenously produced nasal nitric oxide (NO) with possible negative effects on ventilation–perfusion matching;

(6) damage to and infection of the airway wall with subsequent stenosis.

Resistance to gas flow

Intubation of the airway, whether via a tracheostomy or by oral or nasal intubation, causes a narrowing of the upper airway. This increases the resistance to gas flow (Habib 1989). Figure 1 demonstrates how the increase in resistance is critically dependent on the size of the airway tube. The use of an airway tube with an internal diameter of 9 mm (French 39) raises the resistance only modestly at flow rates that are commonly used during artificial ventilation. Thus the resistance does not exceed 2 cmH_2O/l/s at flow rates up to 1 l/s. However, there is a rapid increase in resistance with increasing flow rate. If the inside diameter of the airway tube is reduced to 6 mm (French 24), the resistance rapidly increases to typically 6 cmH_2O/l/s at a flow rate of 1 l/s. This compares with the airway resistance seen in moderate asthma. Moreover, the addition of a humidifier adds further to the resistance, so that very high values can be reached (Fig. 1). While this is of little importance during mechanical ventilation, it may make weaning off the ventilator more difficult. Moreover, expiratory resistance will also be increased with a similar rise in expiratory work of breathing. Another aspect is the possible effect of increased expiratory resistance on end-expiratory lung volume.

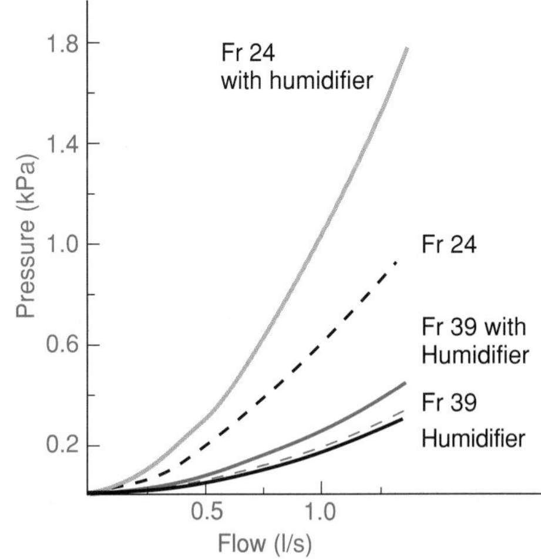

Fig. 1 Pressure–flow curves for French 24 and 39 tracheostomy tubes (internal diameters, 6 mm and 9 mm) with and without a humidifier. Note the considerable increase in pressure required for a given flow when the tube size is decreased or the humidifier is added. Recordings were made in model experiments with flow in the 'inspiratory' direction. (Reproduced with permission from Holst *et al.* (1985).)

Although the tracheal tube may exert an increased resistance to gas flow compared with a normal airway, it is important to point out that inspiratory resistance in the spontaneously breathing non-intubated, but sedated or anesthetized, subject may be increased because of intraction of the upper airways during inspiration (Wheatley *et al.* 1991). During this condition the muscle tone in the pharynx and upper airway seems to be reduced; an upper airway obstruction may cause a reduction of the anteroposterior diameter of the upper airways to 20 per cent of baseline at an intraluminal pressure of -15 cmH_2O. This imposes a dramatic increase in resistance to inspiratory gas flow. Treatment directed towards increasing the tone of the tongue and upper airways, including the pharynx, may be the proper approach, but in the absence of such selective treatment intubation of the airway appears to be the only alternative.

The recording of flow resistance of the tracheal tube may appear to be a simple task, but it is more complicated than generally considered. This is due to difficulty in measuring the correct pressure at the distal end of the airway tube. The introduction of a catheter into the airway lumen will itself exert a certain resistance, both by reducing the airway lumen and by promoting turbulent flow. More important is the fact that a catheter positioned so that its tip is below the distal end of the tube may measure a pressure that is affected not only by the pressure drop in the tube but also by dilatation of the airway distal to the tube. This dilatation reduces the pressure further and causes an overestimation of the calculated tube resistance. The ideal solution is to have a separate channel for recording pressure in the tube wall, with the opening inside the tracheal tube close to the distal end. Three to four openings around the distal end of the tube will compensate for any inhomogeneity in gas flow propagation.

Apparatus dead-space

Intubation of the airway is commonly said to reduce anatomical dead-space by half, i.e. from 100–150 ml to 50–75 ml, or approximately 125 ml to 60 ml. However, the tube needs to be connected to the ventilatory circuit, and the volume of additional tubing and Y-piece, as well as a possible heat–moisture exchanger, will result in an apparatus dead-space that is greater than the bypassed anatomical dead-space. Even the use of a small exchanger will result in a total apparatus dead-space of 120 to 140 ml. Although the increased dead-space volume may seem modest, it easily increases the ventilatory demand by another liter per minute, depending on the ventilatory pattern of the patient. This has implications for both mechanical ventilation and spontaneous breathing. During mechanical ventilation the increased ventilatory demand raises mean airway pressure and may promote volutrauma/barotrauma. During spontaneous breathing the increased ventilatory demand may cause or contribute to respiratory fatigue. The additional dead-space volume has often been neglected but, more recently, there has been increased interest in reducing its effect on gas exchange and ventilation. The technique of tracheal gas insufflation has been developed, based on introducing a catheter through the tracheal tube to its distal end and using it to flush end-expiratory gas out of the tube towards the expiratory port of the ventilatory circuit. This decreases rebreathing, so that the minute ventilation needed to maintain $Pa\text{CO}_2$ is reduced.

Conditioning of the inspired gas

The upper airways, including the nose, condition the inspired gas so that it is heated to body temperature, humidified so that the water saturation is 100 per cent, and filtered to remove larger particles. This will not occur in the intubated patient unless a heat–moisture exchanger or humidifier is connected to the breathing circuit. If the patient is receiving an unconditioned gas, the airways will dry and be more susceptible to infection. A heat–moisture exchanger will reduce fluid loss by condensation of water vapor on the surfaces in the exchanger and evaporation of fluid with the next inspiration. The loss can be completely compensated by more efficient humidifiers based on ultrasonic or heat evaporation principles, and a net inflow of water or water vapor can occur to the extent that interstitial edema may develop.

Functional residual capacity

The decrease in FRC that is regularly seen in the anesthetized intubated patient has been ascribed to the intubation process, with the argument being that the vocal cords and pharyngeal muscle can no longer modulate expiratory flow. However, many observations indicate that FRC is reduced early in the induction of anesthesia, during spontaneous breathing before intubation of the airway, as evidenced by continuous recording of the end-expiratory level (Nunn 1993). The decrease seems more likely to be due to loss of respiratory muscle tone, allowing the elastic recoil of the lung to move the FRC towards lower lung volume.

Rather than lower FRC, the tracheal tube may increase it. This is because the flow resistance of the tube slows down expiratory flow and prolongs expiration. If the expiration is interrupted by the succeeding inspiration, FRC will be elevated as a consequence of the production of intrinsic positive end-expiratory pressure (PEEP), or auto-PEEP. It should be noted that intrinsic PEEP levels of up to 10 cmH_2O have been seen in patients with acute respiratory failure without any primary involvement of the airways. Furthermore, the tube resistance may raise the FRC during spontaneous breathing. This is the normal response to increased expiratory resistance. An example is the air trapping that occurs during hyperventilation in chronic obstructive lung disease. Under certain circumstances the expiratory resistance of the tube may be helpful rather than harmful. The increase in lung volume will also dilate airways and the slower expiratory flow will reduce the pressure drop in the airway tree and move the point where pressure inside and outside the airway is equal (equal pressure point) towards the airway opening. This prevents the airway from undergoing dynamic compression during expiration and is a helpful tool in obstructive lung disease. It should be compared with the artificial devices that such patients breathe through in order to facilitate ventilation.

Bypassing nasal NO

NO has been shown to be a potent vasodilator. When given by inhalation it dilates pulmonary vessels selectively; the systemic circulation is unaffected because of the binding of NO to hemoglobin. NO can be measured in expired gas as an indication of endogenous production in the lung. Only small amounts of NO are expired from the lower airways (2–4 ppb) and most seems to come from the nose. Very high concentrations (around 9000 ppb) have been measured in the paranasal sinuses, which also seem to be the source of NO that is measured in the nose. Whether nasal NO is part of the immunodefense system or whether it modulates ventilation–perfusion matching via autoinhalation is a subject of discussion. Recently, arterial oxygenation has been shown to improve in patients mechanically ventilated for neurological disorders when NO is added to the inspired gas (Lundberg *et al.* 1995). These patients were considered to have essentially healthy lungs; however, an effect of 'autoinhaled' NO was still seen. This supports the hypothesis that nasal NO has a role in the regulation of ventilation–perfusion matching. If this is true, intubation of the trachea eliminates the potentially beneficial effect of nasal NO and may require replacement therapy. However, NO concentrations are extremely small, well below those used for inhalational NO treatment in acute respiratory failure (Lundberg *et al.* 1995).

Tracheal infection and stenosis

The tracheal tube may cause infection and stenosis of the airway. These effects may not be considered to be physiological and therefore will only be dealt with briefly.

Secretions can not only occlude the airway lumen but can also be found along the outside of the tube and may pool above the cuff of the tube. They are frequently contaminated by various bacteria and may cause inflammation. Recently, attempts have been made to eliminate these secretions by introducing a catheter outside the tube and down to the cuff for intermittent aspiration. Tubes are also available with a channel in the wall of the tube opening on the outside just above the cuff.

Tracheal stenosis is a frequent complication of intubation with detectable reductions in tracheal area in more than 50 per cent of patients investigated in a number of studies (Holst *et al.* 1985). However, only a few develop stenosis that is so severe that it will interfere with physical activities. Still, the occurrence of infections and stenosis point to the need for further refinement of the tracheal tube.

References

Habib, M.P. (1989). Physiologic implications of artificial airways. *Chest*, **96**, 180–4.

Holst, M., Hedenstierna, G., Kumlien, J.A., and Shiratzki, H. (1985). Five years experience of coniotomy. *Intensive Care Medicine*, **11**, 202–6.

Lundberg, J.O.N., Lundberg, J.M., Settergren, G., Alving, K., and Weitzberg, E. (1995). Nitric oxide, produced in the upper airways, may act in an 'aerocrine' fashion to enhance pulmonary oxygen uptake in humans. *Acta Physiologica Scandinavica*, **155**, 467–8.

Nunn, J.E. (1993). Respiratory aspects of anesthesia: change in functional residual capacity. In *Nunns applied respiratory physiology* (4th edn), pp. 393–4. Butterworth Heinemann, Oxford.

Wheatley, J.R., Kelly, W.T., Tully, A., and Engel, L.A. (1991). Pressure–diameter relationships of the upper airway in awake supine subjects. *Journal of Applied Physiology*, **70**, 2242–51.

18.1.2 Preparing the patient for endotracheal intubation

R. Whitta

Key messages

- Begin resuscitation of patient before induction of anesthesia.
- Protect the cervical spine.
- Summon experienced help if difficulty is anticipated.
- Life-threatening hyperkalemia is probably the only situation in which the combination of etomidate and succinylcholine (suxamethonium) for anesthesia and muscle relaxation is not adequate.

Introduction

The indications and contraindications for endotracheal intubation are given in Table 1. The contraindications are relative and must be weighed against the result of doing nothing at all. However, basic life support or hand ventilation with oral airway, face mask and rebreathing bag, and oxygen may be more beneficial than an inexperienced person attempting intubation or dealing with the consequences of a failed intubation.

Preparation of the patient is summarized in Table 2.

The patient should be examined as carefully as possible, and resuscitated with volume and inotropes as necessary to improve cardiovascular status. This is important as induction of anesthesia often leads to myocardial depression and hypotension, which are made worse if there is pre-existing volume depletion. While preparations are being made for intubation, the airway should be cleared, using artificial airways if necessary, and oxygenation optimized.

A brief assessment of the likelihood of a difficult intubation should be made so that either more experienced help can be summoned or

Table 1 Indications and contraindications for endotracheal intubation

Indications
Airway protection
Requiring artificial ventilation
Tracheal toilet
Contraindications
Operator inexperience
Inadequate facilities and personnel

Table 2 Preparation of the patient

Optimize cardiovascular status
Make airway and oxygenation as good as possible
Assess likelihood of difficult intubation
Take precautions to prevent pulmonary aspiration of gastric contents
Protect cervical spine
Select position most appropriate for situation, preferably supine with neck flexed and head extended

appropriate equipment can be made available. Precautions against pulmonary aspiration of gastric contents should be taken by turning the unconscious patient into the lateral position, initially at least, and having an assistant apply cricoid pressure when anesthesia is induced. Insertion of a gastric sump tube prior to anesthesia is inappropriate.

The cervical spine should be protected by a hard collar, sandbags, or an assistant's hands if there is suspicion of injury or pre-existing disease such as rheumatoid arthritis.

The patient should be placed in the most appropriate position for intubation, preferably supine, with the head slightly elevated, flexing the cervical spine, and extended on the neck (Fig. 1). The sitting position may be preferable in those who are too dyspneic to lie flat, or who are at major risk of gastric regurgitation (e.g. achalasia of the esophagus), or to prevent contamination of normal lung by infected secretions (e.g. bronchopleural fistula). The lateral position should be used when there is bleeding in the upper airway or to prevent spread of infected pulmonary secretions, with the infected side downwards.

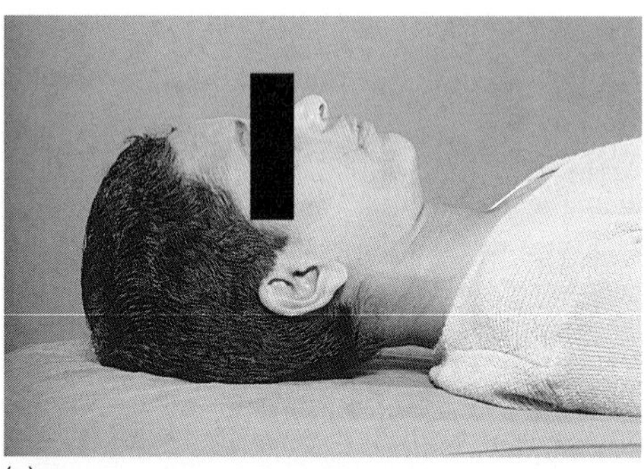

(a)

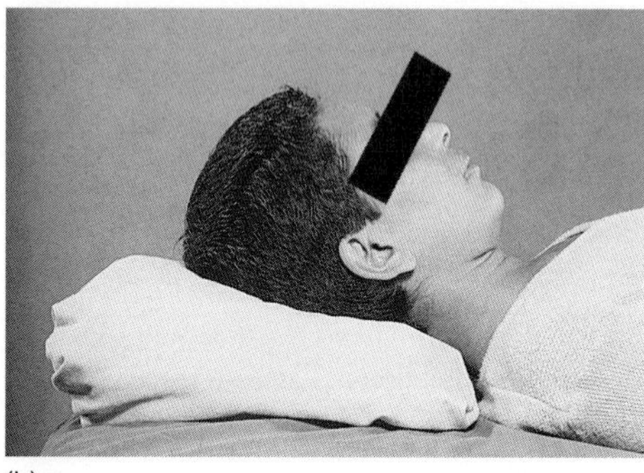

(b)

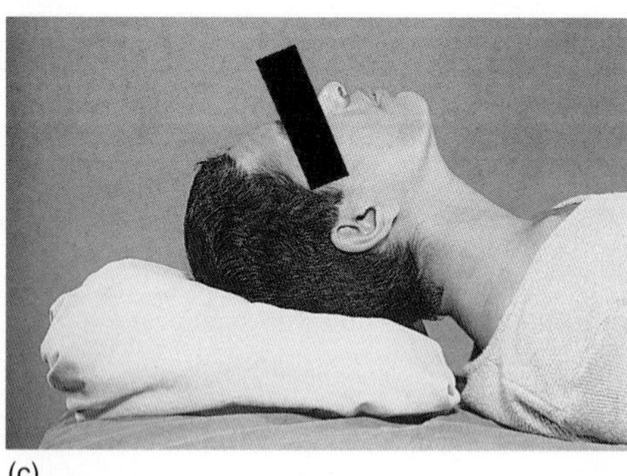

(c)

Fig. 1 (a) When the patient is lying supine with the neck and cervical spine in the neutral position, the angle between the line of the oral cavity and the laryngeal inlet is almost 90°. (b) When the cervical spine is flexed slightly, by providing a pillow, the angle is more obtuse. (c) With head extension they are almost in the same plane.

Intubation under local anesthesia

If the patient is moribund and has no gag reflex, no further drug administration may be required to facilitate intubation. On occasion, intubation under local anesthesia is indicated. This should only be performed by an experienced operator.

Indications

1. Anticipated difficult intubation.
2. Anticipated cardiovascular collapse on induction of sedation or general anesthesia.
3. To isolate a lung leak or prevent contamination of lung segments by infected material (e.g. bronchopleural fistula or tracheo-esophageal fistula).

Local anesthesia for orotracheal intubation

1. Local anesthetic spray (e.g. lidocaine (lignocaine) 4 per cent) to oropharynx and posterior tongue, in combination with advancing laryngoscope and 'spray as you go'. (Alternatively, a 100-mg benzocaine lozenge can be sucked 30 min before intubation, if this is feasible.)
2. Block the superior laryngeal nerve, which provides afferent innervation to the base of tongue, epiglottis, and laryngeal mucosa, as it pierces the thyrohyoid membrane. Displace the hyoid bone towards the side to be injected; a loss of resistance is felt once the thyrohyoid membrane is pierced. Inject 2 ml of local anesthetic (e.g. lidocaine 1 per cent) bilaterally just inferior to the greater cornu of the hyoid bone.
3. Puncture the cricothyroid membrane and inject 2 to 3 ml of 4 per cent lidocaine intratracheally. Correct placement is confirmed by aspiration of air before injection. Remove the needle as soon as injection is made, as the patient will cough and spread local anesthetic up into larynx.

Note that the maximum dose of lidocaine without epinephrine is 3 mg/kg.

Local anesthesia for nasotracheal intubation

1. Apply cocaine paste (10 per cent) to the nostril after it has been checked for patency. This acts as a local anesthetic as well as a vasoconstrictor, and will decrease bleeding.
2. Proceed as above with preferred local anesthetic technique.

Note that the maximum dose of cocaine is 3 mg/kg.

Intubation under general anesthesia

Requirements

1. The operator must have experience of the drugs to be used.

2. Full resuscitation facilities, including medical suction, must be available.
3. An experienced assistant must be present, particularly for the application of cricoid pressure.
4. Ancillary equipment must be available.

Drugs for general anesthesia

A wide range of drugs are available for inducing anesthesia and providing neuromuscular blockade. All intravenous anesthetics and sedatives (except etomidate and ketamine) have the potential for causing cardiovascular collapse by either myocardial depression or vasodilatation (direct or by histamine release). This likelihood is reduced by ensuring adequate cardiovascular resuscitation, using as small a dose as is feasible, and being aware of the possibility of a slow circulation time in the critically ill and therefore a slow onset of anesthesia.

Sedatives

Sedatives can be used as an adjunct to intubation under local anesthesia. However, even small doses in the very ill can lead to unconsciousness, a compromised airway, or hypotension. If used as the sole agent for inducing anesthesia, the onset of action is slower than the general anesthetic agents. Midazolam (dose range 0.1–0.35 mg/kg) acts appreciably faster than diazepam (dose range 0.1–0.6 mg/kg).

Analgesics

Any of the natural or synthetic opioids can be used, but the newer synthetic derivatives such as fentanyl and alfentanil have a more rapid onset and shorter duration of action, together with negligible cardiovascular effects. Therefore, if used in conjunction with an intravenous anesthetic agent, they allow a reduction in dose of that agent and hence greater cardiovascular stability. Morphine can cause hypotension due to histamine release and meperidine in high doses causes myocardial depression, but fentanyl and alfentanil in high doses have minimal effects on the heart and vasculature.

Opiates in high doses can be used as sole agents for anesthetic induction, but the onset is slower than when they are combined with an intravenous anesthetic agent. A good combination is to use a moderate dose of opioid prior to giving an anesthetic agent. This will achieve both a rapid onset of anesthesia and obtund the hypertensive response to laryngoscopy and intubation, if this is undesirable in a particular patient.

Opiate dose ranges are shown in Table 3.

Table 3 Opiate doses

	Moderate	High
Fentanyl	1–2 μg/kg	2–20 μg/kg
Alfentanil	5–15 μg/kg	20–50 μg/kg
Morphine	0.1–0.2 mg/kg	0.5–3 mg/kg

Anesthetic induction agents

All anesthetic induction agents have the potential to produce cardiovascular collapse in the critically ill, particularly if the dose and speed of injection are not modified to take account of the severity of illness. A slow circulation time means that further incremental doses may be given before observing the effect of a dose that has already been administered. Thus there may be an unnecessary accumulation of both dose and adverse effects.

Ketamine (by sympathetic nervous system stimulation) and etomidate (by lack of depressant effect) are least likely to cause hypotension, but the onset of effect of intravenous ketamine is slightly slower than that of the other agents. The neuropsychiatric side-effects may be minimized by the concurrent administration of a benzodiazepine such as diazepam 0.1 to 0.2 mg/kg intravenously. Ketamine may also be given by the intramuscular route, although this takes several minutes to be effective. However, it has the additional advantages of providing analgesia and retaining some degree of muscle tone, thereby aiding maintenance of a patent airway in the unconscious patient. Ketamine and etomidate may also be preferred in the patient with bronchospasm as neither cause histamine release. Etomidate causes pain on injection and myoclonia, while ketamine should not be used where there is pulmonary hypertension or when any increase in systemic vascular resistance is contraindicated.

Thiopental (thiopentone) and propofol can both lead to potentially severe hypotension in the critically ill patient by direct myocardial depression or vasodilatation. Again, this risk is minimized by slow injection, the use of the smallest dose feasible, and the use of adjuncts such as opiates and sedatives. Propofol causes pain on injection and should not be used if there is a history of convulsions. Thiopental is an effective anticonvulsant but is contraindicated in porphyria.

Dose ranges of anesthetic inuction agents are shown in Table 4.

Table 4 Anesthetic induction agent doses

Thiopental 2.5%	1–5 mg/kg
Propofol 1%	1–2 mg/kg
Ketamine	2 mg/kg IV or up to 10 mg/kg IM
Etomidate 0.2%	0.1–0.3 mg/kg

IV, intravenously; IM, intramuscularly.

Neuromuscular blocking agents

Succinylcholine (suxamethonium) is the only agent currently available which provides good intubating conditions in less than a minute. At a dose of 1 mg/kg it causes initial neuromuscular excitation exhibited as muscle fasciculation, followed by flaccid neuromuscular blockade within one arm–brain circulation time. Structurally, it has two acetylcholine molecules linked through the acetate and methyl groups. Succinylcholine has important cholinergic side-effects such as bradycardia. There is the possibility of prolonged paralysis in those with a deficiency in plasma cholinesterase which normally rapidly metabolizes succinylcholine; variants of this deficiency occur in up to 4 per cent of the population. Patients with liver disease may also be deficient in this enzyme. Malignant hyperpyrexia is another important adverse effect. Its effect may last for 3 to 4 min and repeat doses may be used if necessary, although cholinergic effects are more likely and atropine 0.3 to 0.6 mg should be given concurrently.

Once intubation is performed, longer-duration neuromuscular blockade can be achieved with a variety of non-depolarizing agents (e.g. atracurium, vecuronium, mivacurium, pancuronium). The choice depends on concurrent disease (e.g. delayed metabolism and excretion of vecuronium and pancuronium in hepatic and renal disease) and histamine release leading to bronchospasm and hypotension (atracurium and, less so, mivacurium). Important drug interactions, such as prolonged blockade with the aminoglycoside antibiotics, some antiarrhythmic drugs, and local anesthetics, may occur in the intensive care setting.

18.1.3 Selection of tube and route

R. Whitta

Key messages

- Oral intubation is the easiest to perform in the emergency situation.
- Nasal or other routes, although better tolerated, can be planned later.
- Low-pressure high-volume cuffs on polyvinylchloride endotracheal tubes should be used in those in whom intubation is likely to be prolonged (> 24 h).

Types of endotracheal tube

Most endotracheal tubes currently in use are made of polyvinylchloride (PVC) and are for single use only. They are provided in presterilized packs of unmodified length with a 15-mm male connector at the proximal or ventilator end. The tubes can be cut to the desired length when needed; length in centimeters is indicated along the tube, and internal and external diameters are indicated in millimeters. A guide to diameters and lengths is given in Table 1. A radio-opaque strip runs along the length of the tube. The distal end is bevel shaped to minimize the likelihood of complete occlusion if the end abuts against the tracheal wall. Other markings are the inscriptions IT and Z79 which indicate that the material has been tested and approved as non-irritant and non-toxic.

Table 1 Endotracheal tube sizes and lengths

Adult females	Normally 8 mm internal diameter and 21 cm long
Adult males	Normally 9 mm internal diameter and 23 cm long
Add 3 cm to the length if considering nasotracheal intubation, and decrease the internal diameter by 1.5–2 mm	

The distal end also has an inflatable balloon to seal the trachea and prevent the escape of ventilating gases and inhalation of gastric and upper airway secretions. This is connected by a pilot tube which runs within the substance of the tube towards the proximal end where it emerges into an inflatable balloon. The balloon has a one-way valve to allow inflation of the cuff by an air-filled syringe; the inflation of the balloon indicates the state of inflation of the cuff. When cuffs are inflated, this should be to the minimum volume that prevents escape of ventilating gases.

Endotracheal tubes have either high-pressure low-volume cuffs, which are used in anesthesia for short-term intubations, or low-pressure high-volume cuffs, which should be used when intubation is likely to be prolonged. This decreases the likelihood of pressure necrosis of the tracheal mucosa and subsequent tracheal stenosis. Intracuff pressures of 17 to 23 mmHg should prevent gas leak and cause little disturbance to tracheal mucosal blood flow. The pressure can be checked and modified if necessary throughout the patient's period of intubation.

Several other types of tracheal tubes are available and have specific indications in anesthesia (Fig. 1).

1. The original Ring–Adair–Elwin (RAE) tube is a C-shaped preformed PVC oral tube which is used when it is desirable to have ventilating circuit attachments away from the face, as in otolaryngological and maxillofacial procedures.
2. The nasal RAE or 'north-facing' tube is used for nasotracheal intubation; the proximal end is preformed to curve up towards the forehead to allow unimpeded surgical access to the mouth and lower face.
3. Latex rubber tubes which are reinforced with a wire spiral and are longer than normal are used when the patient is operated on in positions other than supine to allow access to the connection and to prevent kinking of the tube.
4. Double-lumen tubes are used when surgery is contemplated in one hemithorax. They allow the deflation of one lung to improve surgical access or to facilitate differential lung ventilation. The main lumen still remains tracheal with a cuff proximal to it, but an extension intubates either the left or right main bronchus, again with a proximal cuff of its own which allows isolated ventilation of that lung. If possible, the position should be checked with a fiber-optic laryngoscope. All plastic tubes soften when left *in situ* and thus are prone to movement; therefore their location should be checked after each change of patient position or at regular intervals. These tubes are still commonly of the red rubber variety and there are several different types.
   (a) The Carlens tube is a left-sided endobronchial lumen tube with a hook at the bifurcation to catch on to the carina to aid placement. However, this may make it more difficult to pass through the larynx.
   (b) The White tube, which is used for right main bronchus intubation, has an aperture in the bronchial cuff to allow ventilation of the right upper lobe and a carinal hook.
   (c) The Robertshaw tube is available in both right- and left-sided versions with bronchial extensions at different angles allowing for anatomical differences. As there is no carinal hook, it is easier to pass through the larynx, when there is intubating difficulty, than the other types. Newer

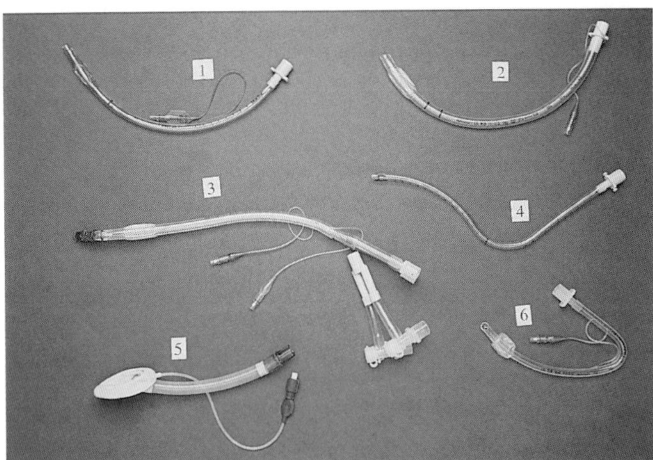

Fig. 1 (1) Standard PVC endotracheal tube (unmodified length) with a 15-mm male connector. (2) Armored tube. Note the wire spiral within the substance of the tube. The length should not be modified. (3) Double-lumen bronchocath. The double catheter mount is provided. (4) Nasal or 'north facing' RAE. (5) Laryngeal mask. (6) Standard RAE tube.

PVC versions called bronchocaths, which are based on Robertshaw tubes, are also available.

Route of intubation

Oral intubation

The oral route is used most commonly when intubating a patient because access to the larynx is easier and the skill is more easily learned. This should be the route of first choice to establish and secure the airway; alternative routes can be planned later.

Advantages

1. Anatomically, oral intubation is the easiest route to visualize the larynx and pass an endotracheal tube.
2. There is minimal tissue trauma on insertion.

Disadvantages

1. Oral intubation can be poorly tolerated, particularly in confused patients. Therefore it may require more sedation and lead to slower weaning.
2. There is less stable fixation where uncuffed tubes are used.
3. Oral and labial sores may develop after prolonged intubation.

Nasotracheal intubation

Advantages

1. Fixation is more stable.
2. Tolerance is better in long-term intubation.
3. There is less likelihood of tube kinking with patient movement.
4. Blind nasal intubation can be performed if laryngoscopy is difficult or impossible.

Disadvantages

1. A more skilled operator is required.
2. Nasal hemorrhage may occur.
3. There is increased incidence of maxillary sinus infection.
4. Nasal intubation is contraindicated in basal skull fracture.

18.1.4 Techniques of tracheal intubation

R. Whitta

Key messages

- Ensure that the appropriate equipment is available and checked.
- Ensure that head positioning is correct.
- Rapid-sequence induction is almost always indicated in the critically ill patient.

Equipment

1. Endotracheal tube: choose the appropriate size, but have one tube each of a size larger and smaller than calculated. Cut to the calculated length and insert a 15-mm male connector.
2. A 10-ml syringe is required to inflate the tracheal cuff. Check the function of cuffs before use.
3. A catheter mount is required to connect the endotracheal tube to the ventilating circuit.
4. A heat and moisture exchanger/bacterial filter is interposed between the catheter mount and the ventilating circuit, preventing contamination of the ventilator by any pathogens in the expired gases or respiratory secretions.
5. Magill's forceps allow manipulation of the distal endotracheal tube into the larynx, if necessary, and removal of foreign bodies.
6. Gum elastic bougie: if placement of the endotracheal tube proves difficult, this flexible introducer can be passed through the cords and a tube 'railroaded' over it.
7. Oral airways (Guedel airways) in a range of sizes.
8. Laryngoscope: in adults a Macintosh blade is appropriate (both right- and left-handed instruments are available).
9. Face mask: select a moulded face mask of appropriate size for preintubation oxygenation and manual ventilation of the lungs if intubation proves prolonged and difficult, or impossible. These should be of the contoured type with a cushioned edge to allow a leak-free seal around the mouth and nose.
10. Ventilating bag and oxygen connection: a high-flow oxygen circuit with a non-rebreathing ventilating bag (2-liter capacity) and an adjustable pressure-limiting valve.

11. Nasal airways are required if there is intra-oral injury or a good airway is not obtained with the use of an oral airway.
12. Medical suction for aspiration of vomitus, blood, and airway secretions.
13. Resuscitation drugs and equipment.
14. A laryngeal mask can be used as a temporary aid to ventilating a patient if intubation with an endotracheal tube proves impossible.

Head position

In order to minimize the angle difference between the oral cavity and the larynx, the ideal position is for the cervical spine to be slightly flexed by placing the head on a low pillow or in a head ring, and for the head to be extended on the neck by tilting it back towards the intubator. Care must be taken for those with limited cervical mobility or cervical spine injury. While the laryngoscope is held in one hand, the free hand can maintain the head position and can also be used to keep the lips clear of the laryngoscope blade and so prevent damage to them.

Laryngoscopy

If the laryngoscope is held in the left hand, it is then passed into the right side of the mouth. The right-angled blade of the Macintosh laryngoscope then aids in laryngoscopy by keeping the tongue out the way to the left-hand side of the mouth. The tip of the blade is passed back in the mouth over the posterior tongue, and directed toward the midline posteriorly. To improve the view the laryngoscope is lifted vertically, not pivoted. If the laryngoscope is pivoted, the view will not be as good and there is more likely to be damage to the patient's upper teeth.

Once the laryngoscope is fully advanced into the vallecula, anterior to the epiglottis, the epiglottis will be seen clearly. As the laryngoscope is elevated, directly behind (below) the epiglottis is the glottic opening surrounded by the aryepiglottic folds laterally and the arytenoid cartilages, covered by the fold, posteriorly. Deep to this are the vocal cords, which are whitish in appearance and are abducted and immobile if the patient has been given neuromuscular blockade.

The endotracheal tube is passed between the cords until the cuff has disappeared below them. The cuff should then be inflated, and the tube secured and connected to the ventilating circuit. The laryngoscope is removed, while the endotracheal tube is held to prevent its accidental displacement.

If laryngoscopy is adequate but there is difficulty in introducing the endotracheal tube into the glottis, the gum elastic bougie may be used. This is advanced into the trachea, held firmly while the endotracheal tube is threaded over it into the trachea, and then removed carefully while the tube is held in place.

Rapid-sequence induction

Otherwise known as 'crash induction', rapid-sequence induction is the method employed to intubate a patient rapidly while protecting against the possibility of aspiration of gastric contents.

The patient is prepared as described above, adopting the most favorable position and preparing drugs and equipment. An experienced assistant is required for this maneuver, to help make equipment available when required and to apply cricoid pressure. This consists of application of moderate pressure to the cricoid cartilage with thumb and index finger. It is important that the correct pressure is applied as the aim is to prevent only passive reflux of gastric fluid. If the patient actively vomits on induction, the cricoid pressure should be released and the patient turned on his or her side, as the continuing application of excessive cricoid pressure could result in esophageal rupture. Overenthusiastic cricoid pressure can also distort the anatomy and make an intubation more difficult than it need be.

The patient is preoxygenated by firmly applying the oxygen mask and administering high-flow oxygen. Oxygen is administered over 2 to 3 min, with the aim of replacing the air occupying the 2 liters of functional residual capacity in the lungs with approaching 100 per cent oxygen. This provides almost 2 liters of oxygen as a metabolic reserve, so that when the patient is rendered apneic for intubation, hypoxemia does not occur for several minutes, allowing more time for intubation, particularly if difficulty is encountered.

When the patient has been preoxygenated, the intravenous induction agent is administered. As the patient becomes unconscious, cricoid pressure is applied by the assistant and succinylcholine (suxamethonium) is administered. Its effect is usually observed within 1 min by peripheral muscular fasciculation. The operator should then proceed to laryngoscopy and intubation. As soon as the endotracheal tube is in place, the tracheal cuff is inflated, the ventilating circuit is attached, and the patient's lungs are inflated. The intubator should not ask for the cricoid pressure to be released until lung inflation is confirmed and there is no audible leak of gas around the cuff.

Correct placement of the endotracheal tube should be confirmed by observing bilateral chest wall movement, auscultating both axillas, negative auscultation of the epigastrium, and capnography of the expired gases if available.

Other positions

If it is necessary to intubate the patient in a position other than supine, i.e. lateral or sitting, the procedure is the same although more difficulty may be encountered with the laryngoscopy. However, as soon as the endotracheal tube is in place and secured, the patient should be placed in the supine position. This will make any further maneuvers, such as resuscitation, easier.

Nasotracheal intubation

Preparation, patient position, and equipment are the same as for endotracheal intubation. Cocaine paste may be applied to the nasal mucosa beforehand if desired. After induction and neuromuscular blockade, the tube is passed carefully through the nostril. There may be some obstruction to the passage in the nasopharynx. This can usually be overcome by rotating the tube as it is gently advanced. Laryngoscopy is then performed and the tube advanced under direct vision. It may pass into the trachea unaided. If not, Magill's forceps can be used to grasp the distal end of the tube and direct it into the trachea. The cuff is then inflated and the procedure continued as before.

For blind nasal intubation the tube is passed though the nares as before and, while it is advanced, the intubator should observe the anterior neck. If the tube impinges in either pyriform fossa, the indentation can be seen. The tube is withdrawn, rotated to the opposite side, and advanced again. If the tube passes easily it will be in either the trachea or the esophagus; this can be differentiated by auscultation and/or capnography. If the tube appears to catch anteriorly in the midline, it is probably against the anterior part of the cricoid or in the vallecula; the patient's head should then be extended further if possible, and the tube withdrawn and advanced again.

Failed intubation

There must be a drill if intubation proves impossible. This can be predicted by examining the patient. The principles are as follows:

(1) recognize early that there is difficulty;

(2) maintain oxygenation;

(3) prevent pulmonary aspiration of gastric contents;

(4) call for more experienced help;

(5) proceed to alternative methods or routes of oxygenation and ventilation.

It is important to recognize when there is likely to be difficulty as prolonged attempts at unsuccessful intubation may cause more trauma to the airway and the patient may become hypoxemic. Inserting an oral airway and gentle hand ventilation with 100 per cent oxygen with cricoid pressure still applied may keep the patient safe while alternative strategies are explored.

The laryngeal mask is available to assist in ventilating patients in these circumstances. If employed, it should not be relied upon to protect the patient against pulmonary aspiration, but should be used as a temporary measure to allow ventilation.

If intubation fails, there are three possible procedures.

1. Call for more experienced help.
2. If help is not available, abandon further attempts at intubation, turn the patient on to his or her side, and maintain ventilation, oxygenation, and airway until the patient recovers from neuromuscular blockade and anesthesia.
3. If the patient's clinical state before the attempted intubation was so poor that this last alternative is impracticable, tracheostomy, minitracheostomy, or cricothyroid puncture with an intravenous cannula, at least to maintain oxygenation, could be lifesaving.

Complications of endotracheal intubation

1. Failed intubation and its sequelae, i.e. hypoxemia.
2. Trauma to eyes, neck, jaw, teeth, lips, nasal mucosa, mouth, tongue, pharynx, larynx, laryngeal nerves, and trachea.
3. Infections of mouth, lips, and maxillary sinuses.
4. Surgical emphysema due to inadvertent puncture of the trachea with the bougie.
5. Myocardial ischemia or cerebral hemorrhage as a result of the hypertensive response to laryngoscopy and intubation.
6. Adverse reactions to drugs.

18.1.5 The difficult intubation

M. E. Faymonville and M. Lamy

Key messages

- A difficult intubation should always be anticipated.
- Fiber-optic techniques represent a dramatic advance in the management of such patients, and are particularly suited to the awake patient in the elective setting.
- If mask ventilation and/or laryngeal mask ventilation prove impossible and the patient still cannot be intubated, transtracheal jet ventilation can be initiated. The patient can then be either awakened, and a semi-elective tracheostomy or cricothyroidotomy performed, or intubated with a special endotracheal intubation technique.

For clinicians involved in airway management, anticipating a difficult intubation may be the first step in preventing major problems. Difficult intubation is defined as inadequate visualization of the glottis, and failed orotracheal intubation as an inability to insert a tracheal tube from the oropharynx into the trachea. Physical limitations to mask ventilation and endotracheal intubation may be accurately identified by thorough observation. All patients should be examined for orofacial anatomical features and for oral structure visibility. A relatively simple grading system which involves preoperative ability to visualize the faucial pillars, soft palate, and the base of the uvula has been designed as a means of predicting the degree of difficulty in laryngeal exposure (Mallampati classes I, II, III). The size of the mandibular space, retrognathia, micrognathia, head–neck configuration, obesity, and the degree of neck mobilization (the ability to assume a 'sniffing' position), as well as prominent central maxillary dentition and limitation of mouth opening, are important factors determining the degree of difficulty of direct laryngoscopy.

Several procedures are available for overcoming the problems of difficult intubation. Fiber-optic techniques represent a dramatic advance in the management of these patients, and are particularly suited to the awake patient in the elective setting. They can also be useful in selected emergency situations. When a difficult intubation is suspected, the patient should be informed about the procedures to be used and brief and reassuring explanations should be given. If the patient agrees to co-operate, premedication is adapted to produce a conscious but relaxed patient. Atropine sulfate is given intramuscularly to decrease nasopharyngeal secretions. Prerequisites for performing a difficult intubation are shown in Table 1.

Anesthetic protocols appropriate for sedation for fiber-optic tracheal intubation must always avoid respiratory depression. We use very light conscious sedation by titration of midazolam and alfentanil upon request to maintain the patient both conscious and co-operative. This has provided good results. In awake patients, fiber-optic intubation maintains a wide margin of safety while producing minimal patient discomfort. However, adequate local anesthesia of the nasopharyngeal and laryngeal mucosa is required. This is initiated with a spray of 10 per cent lidocaine (lignocaine), followed

Table 1 Prerequisites for performing a difficult intubation

Proper selection of the size of the fiberscope in relation to the size of the endotracheal tube
An endotracheal tube size appropriate for passing through the nostrils
Heating the tube in warmed water to avoid trauma to the tissues of the nasopharynx and larynx
Adequate lubrication of the fiberscope and adequate suction
Careful passage of the fiberscope through the distal opening of the tracheal tube
Monitoring of oxygen saturation, ECG, arterial pressure, and end-tidal CO_2
Laryngeal mask
Oxygen supply

by a 10 per cent viscous solution of cocaine for the nasal mucosa by topical application to the anterior ethmoidal nerve and the sensory nerves of the posterior nasal cavity as they emerge from the spheno-palatine ganglion. The cocaine solution is soaked into cotton-tipped applicators that are held in position for 5 to 10 min. The internal branch of the superior laryngeal nerve is blocked by the external approach or by direct application of 1 per cent lidocaine sprayed through the vocal cords by the fiber-optic device. Intimate familiarity with the fiber-optic bronchoscope and the anatomy of the upper airway is essential. The main cause of failure of intubation is lack of expertise in maneuvering the fiberscope.

A more challenging situation occurs when the difficult airway is confronted unexpectedly. When a difficult airway is not recognized and anesthesia is induced, the airway will usually first be controlled by mask ventilation prior to conventional laryngoscopy. If conventional laryngoscopy fails, help should be summoned and the airway controlled by mask ventilation. If endotracheal intubation by conventional laryngoscopy is still unsuccessful after further attempts (perhaps using a different blade or head position) and special alternative techniques fail, the patient should either be awakened, and a laryngeal mask inserted (if an emergency), or a semielective surgical airway (cricothyroidotomy) should be performed. If, at any point, mask ventilation and/or laryngeal mask ventilation prove impossible and the patient still cannot be intubated, transtracheal jet ventilation (Manujet) through a percutaneous intravenous catheter or a transtracheal catheter (acc Ravussin, VBM Medizintechnik GmbH, Germany) should be initiated. Once life-sustaining gas exchange is achieved by transtracheal jet ventilation, the patient should either be awakened, and a semielective tracheostomy or cricothyroidotomy performed, or intubated with a special endotracheal tube intubation technique. Retrograde intubation, a technique consisting of intubation of the trachea over a wire or cannula that has been placed percutaneously through the cricothyroid membrane and threaded up through the vocal cords, remains a viable alternative in this situation.

Complications of intubation include trauma to the tissues of the oronasopharynx and larynx; this can be avoided by proper care and expertise. Prolonged contact of the tube or its cuff with the airway mucosa can produce, sequentially, edema, ulceration, fibrous scarring, and stenotic lesions of the larynx, or trachea.

Endotracheal intubation is frequently used as the initial method of securing a patient's airway in anesthesia and critical care. Respiratory catastrophes are the most common cause of anesthesia-related brain injury and death. Therefore all staff undertaking patient intubations must be adequately trained in both the recognition of difficult intubation and the application of accepted algorithms for management of the compromised airway.

Bibliography

Block, C. and Brechner, V.L. (1971). Unusual problems in airway management. The influence of the temporomandibular joint, the mandible, and associated structures on endotracheal intubation. *Anesthesia and Analgesia*, **50**, 114–23.

Caplan, R.A., *et al.* (1993). Practice guidelines for management of the difficult airway. Report by the American Society of Anesthesiologists' Task Force on management of the difficult intubation. *Anesthesiology*, **78**, 597–602.

Mallampati, S.R., *et al.* (1985). A clinical sign to predict difficult tracheal intubation: a prospective study. *Canadian Anaesthetic Society Journal*, **32**, 429–34.

Mills, P. (1992). Local anaesthesia of the airways. *Today's European Anaesthetist*, **1**, 36–9.

Morris, I.R. (1994). Continuing medical education: fiberoptic intubation. *Canadian Journal of Anaesthesia*, **41**, 996–1008.

18.1.6 Minitracheostomy

D. W. Ryan

Key messages

- Minitracheostomy is a simple method of stimulating coughing and avoiding sputum removal.
- Complications are relatively few and are related to insertion. Bleeding (3 per cent) and surgical emphysema (2 per cent) are the two main problems.

Introduction

Minitracheostomy is used primarily to treat sputum retention but has also been described as a method to overcome acute airway obstruction in laryngeal cancer or trauma using high-frequency ventilation. Less frequent indications include provision of additional oxygen therapy, delivering continuous positive airway pressure in immunocompromised patients, treating sleep apnea, and as a step down in removal of a tracheostomy cannula which is difficult to judge. The technique permits the insertion of a tube of internal diameter 4 mm and external diameter 5.4 mm via the cricothyroid membrane. The patient retains his or her voice and can eat and drink, but the airway is vulnerable due to the lack of a cuff on the tube (Ryan 1990).

Sputum retention

Chest complications are common after major surgery and medical illness, and the confused or uncooperative patient may be unable to assist in expectoration. Additional options to stimulate coughing vary from nasopharyngeal suction, which is often poorly tolerated, direct sucker insertion by direct laryngoscopy, bronchoscopy, and finally tracheostomy. The unacceptable alternative is to allow secretions to build up which will lead to atelectasis with subsequent hypoxia, loss of consciousness, and inevitable death.

Minitracheostomy can prevent this downhill spiral if the clinical situation is appreciated and an early intervention is achieved before airway reflexes are lost, after which intubation is the only solution. Therefore this procedure should be considered after major surgery at the time of extubation or in the early stages of deterioration when the situation is reversible. There have been three small prospective randomized studies of minitracheostomy versus conventional care, all in postoperative thoracotomy cases, with favorable results ($p < 0.05$) (Randell *et al.* 1991).

Insertion techniques

There are three main methods of insertion. The standard technique has been superseded by a Seldinger insertion which reduces the problems of misplacement and repeated insertion attempts which can result in bleeding and surgical emphysema. In certain situations, when location is difficult, a direct insertion with blunt dissection of the neck should be considered; this is easiest when the patient is intubated. The key points to be remembered in an insertion are given in Table 1.

The patient is placed in the tracheostomy position with a bag of saline behind the neck. The skin is cleaned with antiseptic and toweled up. The neck is palpated and infiltrated with local anesthetic (currently bupivacaine 0.5 per cent with 1:2000 ephedrine) along the line of insertion. The operator should stand at the head of the patient, not the side. A mounted needle is then stabbed into the midline of the cricothyroid membrane and air is aspirated into the local anesthetic solution, after which 2 ml are injected into the trachea. This prevents a dangerous coughing attack which encourages bleeding and emphysema. A wire is then threaded down the needle, which is angled to direct the wire down rather than across the trachea. The needle is removed. A dilator is inserted and removed. A moist swab placed over the operation site will be required to stop the air leak and spraying of blood which can be alarming. The mounted minitracheostomy tube is now threaded over the wire and should pass into the trachea (Fig. 1). The step in size between the tube and the dilator sometimes causes a failure to thread easily. A slight screwing action will usually overcome this. The inner dilator is now removed, the tube is in position, air is aspirated again into a fluid-filled syringe, and a suction catheter is passed into the trachea to demonstrate that there is no obstruction and secretions can be removed. If this cannot be done, the tube is incorrectly placed. If the patient has a bulky neck, it is possible to miss the cricothyroid membrane altogether which can result in a high tracheal insertion, but this does not cause any additional difficulties. A failure rate of 1 per cent can be expected. Calcification of the membrane or a very short fat neck are likely to make insertion impossible. Studies of the anatomy of the adult larynx show that a small cricothyroid membrane, normally between 5 and 9 mm, may prevent tube insertion. Insertion using undue force can result in late subglottic stenosis.

Table 1 Insertion key points

Tracheostomy position
Local anesthetic with ephedrine
Stand at head end to insert
Seldinger technique

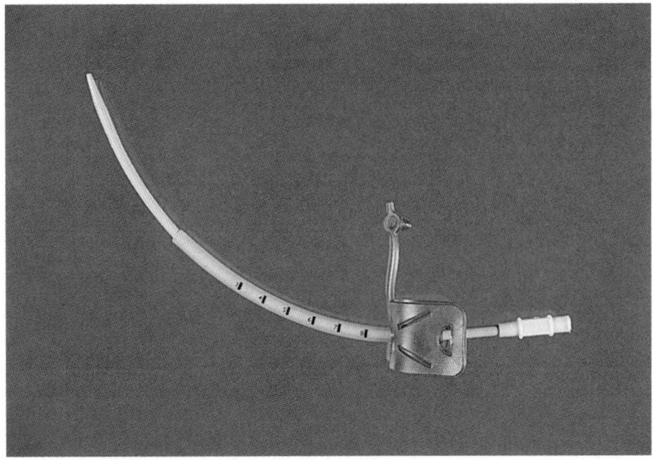

Fig. 1 Minitracheostomy tube, with neck flange folded, mounted on an internal dilator.

General management

The minitracheostomy tube is secured by a neck tie or a Velcro neck strap to avoid accidental removal. A foam dressing between skin and tube prevents abrasion. The plug at the front of the tube can be closed between suction periods or used to entrain oxygen by attaching the appropriate connection. When there is a copious sputum, the plug is left open and the patient is encouraged to cough continuously onto a laryngeal bib. The tube may sometimes become blocked, in which case a replacement is required. If no longer needed, remove the tube by simply pulling it out and cover the site with a non-porous adherent dressing. Healing, except after steroid usage, is quick and the scar is barely visible. Scarring is unusual. Minitracheostomy tubes are rarely required for more than a week.

Practical considerations

Minitracheostomy can be a very quick procedure, but the confused or hypoxic patient will still find it unacceptable, and it is potentially difficult and dangerous if the patient is restless. It is better to intubate such borderline cases and consider minitracheostomy at the time of extubation when the airway is safe. A bleeding diathesis or concomitant dialysis is a contraindication to this method without correction of coagulation; one life-threatening hemorrhage of a patient on anticoagulants has been described. Patients with chronic obstructive pulmonary disease can often only manage sitting up and are prone to coughing bouts; the introduction of blood in the airway can rapidly induce bronchospasm. Each case needs careful appraisal to balance this risk; basically, if the patient can cough, minitracheostomy should be avoided.

What can go wrong?

The principal complications of minitracheostomy are summarized in Table 2.

Misplacement is probably due to inexperience. There are many structures in the neck, and failure to place the patient in the correct position or rushing the insertion can contribute to mistakes; the trick of insertion from above helps to orientate the clinician into the midline. Complications that are described include the loss of the introducer into the pleura, insertion into the mediastinum or the esophagus, and even placement upside down. In one unreported case a pneumothorax was produced. The tube can sometimes kink if the plastic is deformed by too much force on insertion. The use of fiber-optic bronchoscopy ensures that placement is correct, but in practice the tube can easily be seen at direct laryngoscopy if there are any doubts. A lateral neck radiograph may also be helpful.

Surgical emphysema (2 per cent) can be the result of misplacement, an insertion hole which is too large, or repeated attempts at insertion. This was more likely when a scalpel stab was used to insert the tube rather than the Seldinger dilator method. Displacement of the tube during high-frequency ventilation has resulted in emphysema and respiratory distress, leading to recommended safeguards in such usage. Inadvertent penetration of the mucosa by the wire used in the Seldinger technique, producing surgical emphysema, has also been described. The authors concluded that a J-wire would be safer.

Bleeding necessitating surgical intervention (3 per cent) is the other major problem. The thinking behind the midline approach through the cricothyroid membrane is to enter a bloodless field, but this is rarely the case, and it can be frightening to the inexperienced operator when the insertion site is masked by venous bleeding coupled with air bubbling. If there is delay, coughing bouts will ensue as blood is inhaled. Most bleeding stops with compression, but sometimes undersewing or simply large stitches adjacent to the insertion site are required. The bleeding here is from the skin edge. Dramatic life-threatening bleeding necessitating intubation has been described, and complete airway obstruction has been documented in five cases (one fatal).

Table 2 Principal complications
Misplacement
Surgical emphysema
Bleeding

Discussion

Major indications for minitracheostomy are listed in Table 3.

The role of minitracheostomy is principally to provide cough stimulation, enhance secretion removal, improve oxygenation, and prevent atelectasis. In this respect there is good evidence of efficacy (70–85 per cent) in prospective randomized trials. The technique is not without risk. In one prospective study of 152 cases, serious complications were reported in 5.4 per cent. Au *et al.* (1989) studied 144 minitracheostomies using a 20 French silver tube and recorded complications in 5.5 per cent. One study used surgical cricothyroidotomy and a larger tube than currently available as an alternative to tracheostomy and had a complication rate of 6.1 per cent. Pederson *et al.* (1991) found one serious bleed, five less serious bleeds and two cases of surgical emphysema in 73 patients, giving a complication rate of 11 per cent. Tracheostomy, by comparison, is a more major undertaking and is used for a much more varied range of serious conditions. Immediate problems are similar, with severe bleeding at around 5 per cent and surgical emphysema at 1 per cent, and an overall complication rate varying from 6 to 51 per cent.

Table 3 Major indications for insertion
Sputum retention
Airway obstruction
Oxygen entrainment
Weaning from tracheostomy

Minitracheostomy has been used as a means of overcoming significant airway obstruction and to help treatment of acute respiratory failure. It is clearly not a technique for the inexperienced in these circumstances, but it can provide an alternative short-term solution to upper airway obstruction. The use of positive-pressure ventilation is also not without potential danger, but the narrowness of the tube means that either high-frequency ventilation or continuous positive airway pressure will be necessary to ensure that an adequate airflow is achieved in the obstructed upper airway. A recent case describes the management of weaning a flail chest injury after 8 days of ventilation using positive controlled ventilation with a minitracheostomy tube over the next 12 days. Because of the dangers of misplacement of the tube, resulting in serious surgical emphysema and respiratory obstruction, careful supervision of the method is required, notably to ensure placement and patency of the tube.

The administration of longer-term oxygen via a percutaneous tracheal catheter has been described in patients with hypoxemia and continuous positive airway pressure. In appropriate cases, notably intensive care patients where combined sputum removal and oxygenation are necessary, minitracheostomy can provide a solution.

Arguably, the development of percutaneous tracheostomy to enable the intensive care clinician to provide an earlier and safer insertion technique could make minitracheostomy redundant. This would be the case if the complication rate of a tracheostomy was much reduced and the indications for insertion overlapped. Minitracheostomy has very specific indications and limitations which can only benefit the patient if the insertion occurs at an early phase of deterioration.

References

Au, J., Walker, W.S., Inglis, D., and Cameron, E.W.J. (1989). Percutaneous cricothyroidectomy (minitracheostomy) for bronchial toilet: results of therapeutic and prophylactic use. *Annals of Thoracic Surgery*, **48**, 850–2.

Pederson, J., Schuriezek, B.A., Melson, N.C., and Juhl, B. (1991). Is minitracheostomy a simple safe procedure? A prospective investigation in the intensive care unit. *Intensive Care Medicine*, **17**, 333–5.

Randell, T., Tierala, E.K., Lepantalo, M.J., and Lindgren, L. (1991). Prophylactic minitracheostomy after thoracotomy: a prospective, random control, clinical trial. *European Journal of Surgery*, **157**, 501–4.

Ryan, D.W. (1990). Minitracheotomy. *British Medical Journal*, **300**, 958–9.

18.1.7 Tracheostomy

P. B. Hazard

Key messages

- In critical care, the most common role for tracheostomy is to provide a stable interface between mechanically ventilated patients and their respirators.
- The major contraindication to tracheostomy in critically ill patients is local inflammation at the proposed tracheostomy site.
- Conscious patients who are obviously experiencing severe and constant discomfort from translaryngeal intubation should be considered for tracheostomy relatively early in their courses.
- A number of techniques for tracheostomy may be employed. These include the conventional open surgical procedure, cricothyroidotomy, and percutaneous dilatation tracheostomy.
- Percutaneous tracheostomy is more economical and is associated with significantly less morbidity than the operative approach. It is increasingly considered the technique of choice in critically ill patients.

Indications

In the critical care setting, tracheostomy may be required to relieve upper airway obstruction, to expedite bronchopulmonary toilet, and to facilitate mechanical ventilation, particularly when prolonged (Heffner 1988). A fourth indication often cited is to establish airway access in emergencies but, in the absence of upper airway obstruction, emergency airway access can usually be gained more efficiently by endotracheal intubation. Upper airway obstruction in critically ill patients may arise from malignancies, trauma, vocal cord paralysis, or laryngeal edema or spasm. Patients with excessive bronchopulmonary secretions or impaired clearance mechanisms, such as those with neurological conditions, may require tracheostomy in order to provide access for suctioning and therapeutic aerosol delivery. Probably the most common role for tracheostomy is to provide a stable interface between mechanically ventilated patients and their respirators.

Generally, acute airway management in intensive care patients initially involves the use of a translaryngeal endotracheal tube. The endotracheal tube is easily inserted, is less invasive for most patients, and provides an acceptable short-term airway. However, conversion to tracheostomy results in several benefits (Table 1) (Heffner 1988; Hazard *et al.* 1991). For the vast majority of conscious patients, tracheostomy is much more comfortable than translaryngeal intubation. Indeed, it can be argued that intubation is the greatest misery that is routinely visited on such patients, and few would contest the fact that tracheostomy is well tolerated. Few endotracheally intubated patients, alert and unrestrained, will fail to extubate themselves at the first opportunity; however, patients with tracheostomies rarely attempt to remove their airways. The ability to communicate is improved, oral feeding is often possible, and positioning out of bed into a chair, even ambulation in some patients, is facilitated. Apart from the elimination of the continuous discomfort and practical limitations of the translaryngeal tube, patients with tracheostomy are more easily managed from a nursing perspective. Clearance of bronchopulmonary secretions is facilitated. Patient cough efforts and suctioning by nursing staff are likely to be more effective with the shorter more direct airway access. Airway resistance and work of breathing are diminished, and this may facilitate weaning from the ventilator. Earlier tracheostomy may reduce the potential for injury to the larynx that endotracheal intubation provokes.

Table 1 Benefits of tracheostomy

Reduced upper airway injury
Oropharyngeal injury
Laryngeal injury
Improved patient comfort
Elimination of endotracheal tube
Improvement in communication
Potential for oral nutrition
Facilitation of chair positioning and ambulation
Improved pulmonary toilet
Better access to trachea for suctioning
Facilitation of effective cough
Reduced airway resistance and work of breathing
May facilitate ventilator weaning in difficult patients
Secure long-term airway

Contraindications and complications

The major contraindication to tracheostomy in critically ill patients is local infection or inflammation at the proposed tracheostomy site. Patients with extensive burns or soft tissue infection involving the anterior neck should not undergo the procedure. Tracheostomy is relatively contraindicated in patients with cardiovascular or respiratory instability, such as those requiring high-dose vasopressor support or positive end-expiratory pressure at levels greater than 15 to 20 cmH_2O. Severe coagulopathy, uncorrectable by platelet or component transfusion, represents another relative contraindication.

Even in properly selected patients, a large number of complications have been reported with tracheostomy (Stauffer *et al.* 1981; Heffner 1988; Hazard *et al.* 1991). The incidence of these adverse events varies widely in different reports. Table 2 lists the most widely reported complications. Tracheostomy has been associated with a small but definite perioperative mortality, sometimes as high as 5 per cent, although usually much lower. Potentially lethal intraoperative crises include a variety of cardiac arrhythmias and sometimes striking falls in Pao_2 or blood pressure. Perioperative pneumothorax or pneumomediastinum may occur and may be fatal. Hemorrhage may occur during the procedure, or may be delayed for hours or days,

Table 2 Principal complications of tracheostomy

Perioperative death	0%–5%
Pneumothorax	1%–5%
Subcutaneous emphysema	5%–9%
Hemorrhage	3%–40%
Tube displacement	4%–10%
Stomal infection	9%–36%
Pneumonia	11%–66%
Tracheal stenosis	2%–65%
Tracheo-innominate fistula	0%–2%
Tracheo-esophageal fistula	0%–1%
Cosmetic deformity	9%–25%

owing to venous or arterial disruption during the procedure or to delayed erosion into the innominate vein, a condition termed tracheo-innominate fistula which is rare but nearly always fatal. Misplacement of the tracheostomy tube into the anterior pretracheal space may lead to mediastinal emphysema, although mild subcutaneous emphysema may occur more frequently and be of much less importance. Tracheostomy tube misplacement may precipitate acute hypoxemia, cardiac arrhythmias, or shock, and may be a rare cause of mortality. Even in patients who endure the tracheostomy procedure without intraoperative misadventure, problems may develop later. Local infection at the site may be very troublesome to manage, and may lead to extensive necrosis of the anterior neck, sepsis, and, rarely, septic shock. After removal of the tracheostomy tube, healing of the stoma may be delayed, and considerable cosmetic deformity is a common result even when healing progresses well. Perhaps of greater concern, tracheostomy may result in tracheal narrowing with resulting respiratory impairment. Tracheal narrowing may occur at the stomal site, at the cuff site within the trachea, or in the subglottic region, although it is possible that this is partly the effect of antecedent endotracheal intubation.

Timing of tracheostomy

The optimal timing for tracheostomy, a subject surrounded by much controversy, depends on a number of considerations. These include consideration of the relative risk–benefit ratio of prolonged translaryngeal intubation versus tracheostomy, the presence of absolute or relative contraindications to tracheostomy, the degree of discomfort experienced by the endotracheally intubated patient, the extent of problems relating to pulmonary toilet, and the anticipated duration of endotracheal intubation. On the whole, conscious patients who are obviously experiencing severe and constant discomfort from translaryngeal intubation should be considered for tracheostomy relatively early in their courses, whereas those in deep coma might be expected to endure the endotracheal tube longer. Between these extremes, much clinical judgment is called for. Patients with thick secretions, poor airway guarding, or weak cough might be considered candidates for early tracheostomy, while those in whom secretions are minimal or strong cough reflexes are evident might be assigned to later tracheostomy if extubation cannot be accomplished expeditiously. In many patients, obvious oral or labial breakdown arising from the endotracheal tube might dispose toward tracheostomy. Generally speaking, patients for whom prolonged dependence upon an artificial airway can be anticipated, such as those with traumatic cervical spine injuries, should be considered as candidates for early tracheostomy.

Such concerns have led to widely varying recommendations regarding timing of tracheostomy, with some authorities recommending tracheostomy within the first few days of intubation, and others advising delays of 3 to 4 weeks in anticipation of extubation. The emergence of percutaneous tracheostomy, with its simplicity and markedly reduced morbidity, may affect these recommendations. Our own approach is to perform tracheostomy, usually by the percutaneous dilatation method, within 7 days in patients who have not been and show no signs of being able to be weaned from the ventilator. In patients in whom conditions exist that would obviously preclude extubation within this time frame, such as those with high spinal cord injury or many with acute respiratory distress syndrome, we elect to perform tracheostomy within 3 to 5 days of intubation.

Techniques

A number of methods for tracheostomy may be employed. These include the conventional open surgical procedure, cricothyroidotomy, and percutaneous dilatation tracheostomy.

Conventional operative tracheostomy

Most surgeons prefer to perform standard elective tracheostomy in an operating room where sterility is more easily maintained and specialized lighting is readily available. Most also prefer to use general anesthesia, although local anesthesia is sometimes used. The patient is positioned supine, usually with a roll beneath the shoulders to extend the neck. After the anterior neck has been cleaned and painted with antiseptic, a 4-cm transverse incision is made over the second, third, or fourth tracheal ring. The subcutaneous tissues are divided, and the anterior neck muscles and vessels are retracted. If the thyroid isthmus overlies the region selected for tracheostomy, it is ligated and transected. After the trachea has been exposed, one or more cartilages are incised to create an opening into which to place the tube. In some cases, a plug of cartilage is excised from one or more rings. The tracheostomy tube is inserted into the opening, and the incision is closed with sutures or staples. Some surgeons elect to suture the flanges of the tube to the skin, whereas others are content to secure the tube with cloth ties.

Cricothyroidotomy

Cricothyroidotomy is a technique usually reserved for emergency airway access. The cricothyroid space, easily palpated in most patients, marks a level where the trachea is quite close to the skin surface. Thus it is quickly accessible in the occasional acute airway crisis, for example acute upper airway obstruction in a patient who cannot be intubated successfully. The skin overlying the cricothyroid membrane is anesthetized (if time permits) with lidocaine (lignocaine), and a 2- to 3-cm vertical incision is made through the skin over the membrane. The membrane is palpated and a direct stab incision is made though it. The opening is spread with a hemostat, and a small tracheostomy tube (internal diameter of 4–6 mm) is inserted. The hemostat is removed, the cuff inflated, and the tube secured.

The simplicity and speed of the procedure are attractive, but the technique is not widely used in non-emergency situations because of the high morbidity attributed to it. In particular, a high incidence of subglottic stenosis has been ascribed to it by some authorities. Furthermore, because the space is small, the tube used must be of relatively small diameter, and this may pose problems in patients with difficult pulmonary toilet. However, the cricothyroid space has no adjacent important blood vessels and serious bleeding is rarely encountered. Moreover, owing to its high position in the neck, cricothyroidotomy is seldom complicated by pneumothorax. It seems that the potential for injury to the subglottic region relates in large part to previous translaryngeal intubation, and that subglottic stenosis is uncommon when cricothyroidotomy is undertaken in patients who have been intubated for less than 72 h. Electively, the technique may be advantageous in post-sternotomy patients, in whom standard low tracheostomy may predispose to sternotomy wound infection.

Percutaneous dilatation tracheostomy

Percutaneous tracheostomy, a derivation of the Seldinger wire-guided technique for vascular cannulation, was introduced in 1969 as a single-dilatation technique (Toye and Weinstein 1986). Since then it has been modified and refined, and is now used widely throughout the world. The most commonly used method involves the insertion of a needle through the neck into the trachea, followed by passage of a guide-wire through the needle. The needle is removed and its tract gradually enlarged by inserting a series of progressively larger dilators over the wire. After a sufficiently large opening has been created, a tracheostomy tube is inserted. The technique is illustrated in Fig. 1. Bronchoscopic guidance is often used to ensure proper tube placement and promote safety (Barba *et al.* 1995).

The percutaneous technique is attractive because of its simplicity and speed. It can be completed in less than 5 min, whereas conventional operative tracheostomy usually requires more than 30 min (Hazard *et al.* 1991). Furthermore, percutaneous tracheostomy can be safely performed under local anesthesia at the bedside in the intensive care unit, obviating the sometimes precarious exercise of transporting the critically ill patient to and from the operating room. In part because of this fact, the percutaneous approach is far more economical, with an average cost less than one-third that of the conventional procedure (Barba *et al.* 1995). Most importantly, percutaneous tracheostomy is associated with significantly less morbidity than the operative approach, particularly in terms of infection and tracheal stenosis (Hazard *et al.* 1991). For these reasons, it is increasingly considered the technique of choice in critically ill patients.

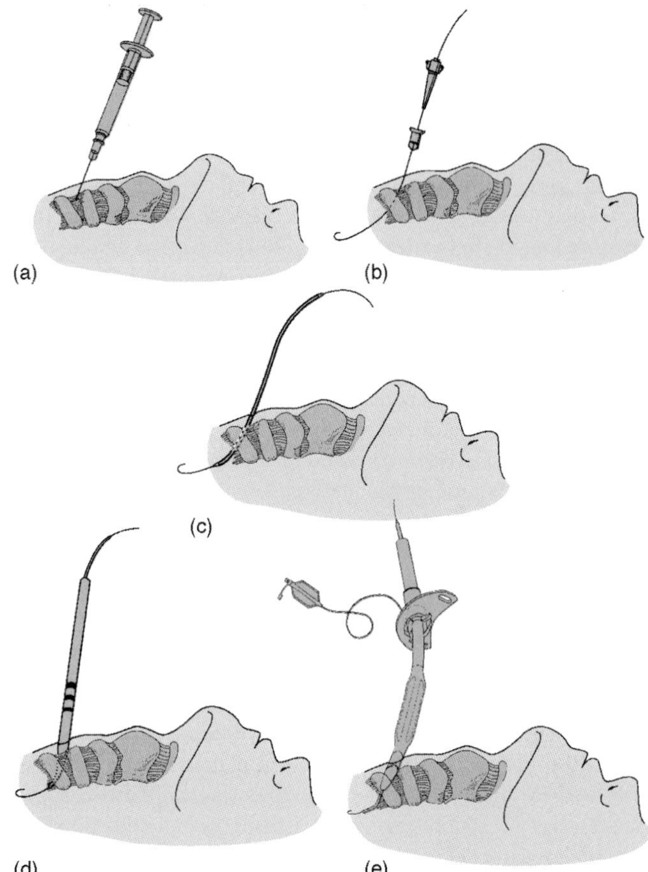

Fig. 1 Technique of percutaneous dilatation tracheostomy. The patient is positioned with a roll beneath the shoulders to extend the neck, and the anterior neck is cleansed and painted with antiseptic. The skin overlying the space between the first and second, or second and third, tracheal cartilages is infiltrated with lidocaine. (a) A 16-gauge needle is inserted through the selected space and into the tracheal lumen. (b) A J-tipped guidewire is passed through the 16-gauge needle, and the needle is removed. A 2-cm vertical superficial incision is made through the skin in such a way as to include the tract of the guide-wire. (c) Dilatation begins by insertion of an 8 French tapered plastic dilator, which is then removed and replaced by a plastic guiding catheter of similar size. (d) The opening is gradually enlarged by passing progressively larger dilators over the guiding catheter. Two to six dilatations may be required to create an opening large enough to accommodate the selected tracheostomy tube. (e) A dilator or dilating obturator is inserted into the tracheostomy tube, and this is passed over the guiding catheter and into the stoma. The guidewire, guiding catheter, and dilator are removed, leaving the tracheostomy tube in place. The cuff is inflated and the tube is secured with cloth ties.

Management of the patient with tracheostomy

Postoperative management

Immediately post-tracheostomy, bilateral breath sounds should be identified and stability of cardiac rhythm, blood pressure, and oxygen saturation should be assured. A chest radiograph is performed to exclude the presence of pneumothorax.

Meticulous care of the tracheostomy tube is necessary, particularly during the first few hours and days. Ideally, the tracheostomy tube should not be changed during the first 5 or 6 days to permit maturation of the tract. If accidental decannulation occurs before such a tract has developed, it may be extremely difficult to reinsert the tube through the opening. In such cases, it is usually best to intubate the patient orally to secure the airway and then address the problem of replacing the tracheostomy tube in a less harried fashion. The tracheostomy site should be covered with a simple split-gauze dressing, and it should be inspected regularly for bleeding from the stoma, which may develop even many hours after the procedure.

Over the ensuing days, regular inspection should be undertaken to detect any signs of wound breakdown, cellulitis, or purulent drainage. If signs of infection are noted, systemic antibiotics after appropriate cultures are indicated. Local cleansing with antiseptic solution may also be of benefit in such cases.

General care

High cuff pressures may cause injury to the tracheal wall, later promoting the development of tracheal stenosis. The tracheostomy tube cuff should be inflated only to the extent necessary to seal the airway, and high inflation pressures should be avoided.

Many patients will be able to take oral feedings soon after tracheostomy. Our practice is to permit sips of water the day after the procedure in alert patients with no gastrointestinal dysfunction. If they tolerate these well, their intakes are advanced to full liquid and then solid food diets.

The inability of intubated patients to communicate with caregivers and family members is a frequent source of anxiety and frustration. Although critically ill patients with cuffed tracheostomy tubes are unable to speak aloud, conversion from endotracheal tube to tracheostomy often improves communication nonetheless. Alert patients can usually mouth the words to describe their simpler wants sufficiently well that bystanders can read their lips. In some cases, the cuff can be partially deflated to allow a small air leak which the patient can use to speak aloud. Spontaneously breathing patients can usually speak when the cuff is deflated and the tube occluded. Fenestrated tracheostomy tubes may further facilitate this, but may promote the growth of excessive granulation tissue. Various specialized devices exist that may improve communication in individual cases of spontaneously breathing patients who continue to require tracheostomy.

Once the tracheostomy has matured, the patient may be able to sit at the bedside or in a chair if he or she is stable otherwise. Some patients may be able to take short assisted walks, utilizing small portable ventilators.

Removal of the tube

A variety of approaches to decannulation can be used, depending on individual situations. In some patients, recovery from the problem that precipitated the need for tracheostomy immediately eliminates the need for the tube and it can simply be removed. However, in most critically ill patients it is difficult to be certain when the tube will no longer be needed, even after apparent recovery, and several approaches to weaning from tracheostomy have been used. An uncuffed fenestrated tube can be inserted and occluded for progressively longer intervals until the patient's ability to breathe spontaneously and maintain adequate pulmonary toilet without an open tracheostomy tube is established. Alternatively, a tracheostomy 'button' can be inserted into the stoma. The button is actually a tube that is just long enough to reach the inner opening of the stoma in the trachea without extending into the lumen. This plugs the tracheostomy, and requires the patient to breathe entirely through the native airway. However, if the patient is unable to maintain secretion clearance or ventilation without the tracheostomy tube, it can quickly be removed and replaced by a standard tracheostomy tube.

After the tracheostomy tube has been removed the stoma should be covered with a simple gauze dressing. Percutaneous tracheostomies usually close completely within 24 h, and conventional tracheostomies within a few days, although on occasion closure may be delayed for several weeks. In rare cases, surgical closure is necessary.

References

Barba, C.A., *et al.* (1995). Bronchoscopic guidance makes percutaneous tracheostomy a safe, cost-effective, and easy-to-teach procedure. *Surgery*, **118**, 879–83.

Hazard, P., Jones, C., and Benitone, J. (1991). Comparative clinical trial of standard operative tracheostomy with percutaneous tracheostomy. *Critical Care Medicine*, **19**, 1018–24.

Heffner, J.E. (1988). Tracheal intubation in mechanically ventilated patients. *Clinics in Chest Medicine*, **9**, 23–35.

Stauffer, J. L., Olson, D.E., and Petty, T.L. (1981). Complications and consequences of endotracheal intubation and tracheostomy. A prospective study of 150 critically ill adult patients. *American Journal of Medicine*, **70**, 65–76.

Toye, F.J. and Weinstein, J.D. (1986). Clinical experience with percutaneous tracheostomy and cricothyroidotomy in 100 patients. *Journal of Trauma*, **26**, 1034–40.

18.2 Ventilatory support

18.2.1 Physiology of positive-pressure ventilation

J. C. Goldstone

Key messages

- Respiratory mechanics can easily be deduced from pressure–time curves measured at the mouth.
- Oxygenation is often dependent on ventilation–perfusion ratios which can be manipulated therapeutically.
- Major hemodynamic interactions occur during mechanical ventilation, offsetting gains in oxygenation.

Introduction

There are many physiological differences between mechanical and spontaneous breaths. During a mechanical breath the ventilator controls many of the primary variables responsible for breathing, for example inspiratory flow rate, inspiratory time, and the length of the respiratory cycle. Inspiratory gas is pushed into the chest and the ventilator performs all the work required to inflate lungs and chest wall. The tidal volume and pattern of gas flow will then be related to the ventilator and the mechanical characteristics of the lungs and chest wall. Gas exchange occurs during each breath and is principally related to ventilation–perfusion abnormalities and to true shunt. As intrathoracic pressure will be greater than atmospheric pressure during a mechanical breath, extrathoracic effects of this positive pressure occur, affecting blood flow to other organs.

The physiology of mechanical ventilation differs from spontaneous breathing in terms of lung mechanics, control of breathing, and gas exchange. The physiological effects will be more obviously altered by disease. Few deleterious effects occur in healthy subjects, but adverse effects may be pronounced in the critically ill.

Lung mechanics during mechanical ventilation

Muscle contraction uses energy. If the muscle moves, external work is performed. This can be measured as the force multiplied by the distance that the muscle has shortened. In many biological situations the length of the muscle is fixed; when this occurs the internal work (within the muscle cell) is difficult to measure. The power that the muscle produces is the product of work and its duration.

In the ventilated patient, work is performed during each breath by the ventilator and energy is dissipated. The external work causes gas to move and this can be measured. Internal work may be performed if the respiratory muscles contract, but no gas flows and this is often not measured. Clearly, considerable amounts of internal work can occur in an occult manner.

The respiratory muscles form a complex geometrical arrangement around the chest wall and the force exerted during contractions cannot be measured directly. However, as the transpleural pressure difference causes gas to move, pressure is used in place of force to calculate work. For similar reasons, the distance that the respiratory muscles move is substituted by the volume of gas which moves when calculating the external work performed during breathing. The external work of breathing is measured by integrating simultaneous measurements of transpleural pressure and the volume change at the mouth.

If the patient is fully ventilated, the machine is performing all the work of breathing. Work is performed principally in expanding the lung against the elastic recoil of the lung and chest wall and, when gas is flowing, in overcoming resistance. Further work is dissipated due to gas compression and rarefaction within the breathing circuit. Under these circumstances, ventilator work can easily be calculated. If pressure is measured at the mouth and inspiratory flow is constant, the work is represented by the area under the pressure–time graph. If this technique is to be used to assess inspiratory work sequentially, the measurement must be made under standardized flow and volume conditions. Additionally, the method assumes that the ventilator is performing all inspiratory work and that the patient is making no effort. Inspiratory effort will reduce the pressure measured at the mouth and the airway pressure–time diagrams will be irregular and convex inwards (Figs 1 and 2).

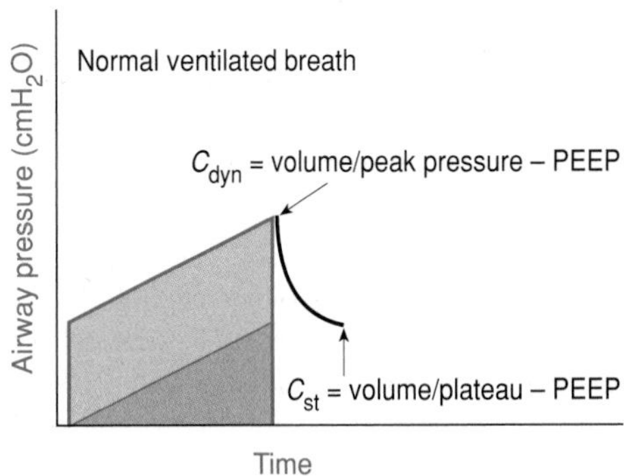

Fig. 1 Airway pressure is measured during constant flow lung inflation. Note the straight upstroke suggestive of a linear relationship between volume and pressure. Compliance can be measured during gas flow (dynamic) or after an inspiratory pause (static).

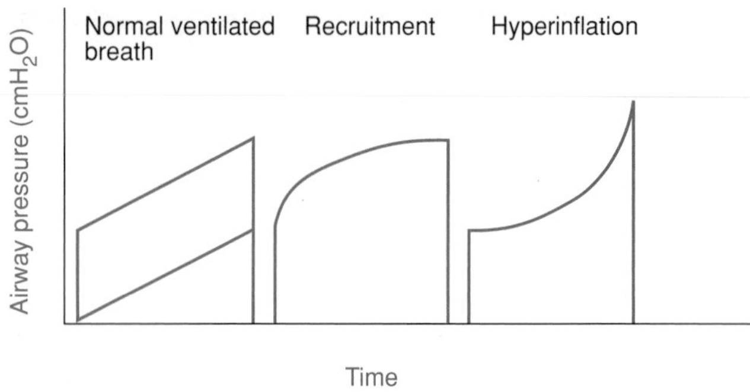

Fig. 2 The upstroke of the pressure–time curves should be linear. If the patient is hyperinflated, pressure rise per unit volume change will accelerate. If pressure changes decelerate, recruitment must be occurring.

The shape of the pressure curve depends on the mechanics of the lung and chest wall. From the relaxed end-expiratory point, the volume–pressure relationship is linear and the slope of the line is related to lung compliance. Dynamic compliance can be calculated by dividing delivered volume by peak end-expiratory pressure.

Hyperinflation of the lung and recruitment of alveoli alter the shape of the pressure–time curves. As lung volume increases, more pressure is required to inflate the chest and the pressure–time curve is non-linear. If lung volume is reduced, recruitment of alveoli may occur during inflation and less pressure is required at the end of inflation.

Gas exchange during mechanical ventilation

While many factors affect gas exchange, oxygenation is primarily dependent upon the composition of the inspired gas, the ventilation–perfusion ratio ($\dot{V}/Q$), and the saturation of mixed venous blood returning to the lungs.

For any given alveolus, the $\dot{V}/Q$ varies between 0.01 and infinity. Low levels of $\dot{V}/Q$ imply shunt (where perfused alveoli are not ventilated), whereas high ratios imply a large dead-space (where ventilated alveoli are not perfused). In the ideal state, $\dot{V}/Q$ tends to 1.0 (equal amounts of ventilation to perfusion) and the scatter of $\dot{V}/Q$ ratios around the mean is low. When $\dot{V}/Q$ is low (0.1) hypoxemia occurs. As $\dot{V}/Q$ rises, the initial response is a rise in Pa_{O_2}. In contrast, Pa_{CO_2} begins to fall when $\dot{V}/Q$ exceeds 1.0.

The effect of mechanical ventilation on this important physiological concept is seen when the effect of posture and loss of muscle tone are considered. When supine, dependent areas of the lung show a decrease in compliance; therefore these lung units are further down the pressure–volume relationship. For a given change in pressure, lung expansion is reduced and ventilation declines. Therefore such dependent areas of the mechanically ventilated lung have a negative effect on the overall $\dot{V}/Q$ state of the lung, and both a higher inspired oxygen tension and increased ventilation are required.

The effect of common disease states can be seen clearly in terms of $\dot{V}/Q$. Intrapulmonary shunt predominates in acute respiratory distress syndrome, although two types of gas exchange have been demonstrated. There are areas of low $\dot{V}/Q$, suggesting very poorly ventilated areas of lung, and also other lung units with essentially normal $\dot{V}/Q$ levels. In contrast, in airflow limitation shunt is typically low and dead-space increased. When $\dot{V}/Q$ relationships are measured, four different patterns of distribution occur and the net effect is one of bimodal changes in $\dot{V}/Q$. Clearly, ventilatory strategies to optimize mechanical ventilation will be fundamentally different in patients with small-volume lungs compared with patients whose primary pathology is hyperinflation.

Control of breathing during mechanical ventilation

Ventilation is often completely controlled. In this circumstance the duration of the inspiratory cycle and the manipulation of pressure and flow rates are preset by the user. Sedation and the use of neuromuscular blocking drugs are often required.

Spontaneous breathing occurs in many ventilated patients and may form part of the ventilatory strategy. The use of inspiratory muscles has been highlighted during conventional ventilation, demonstrating that, far from being silent, respiratory muscle activity can continue throughout inspiration. When the respiratory muscles contract asynchronously, spontaneous work may be comparable with that measured in the non-ventilated state. A study of the timing and duration of inspiratory muscle activity during mechanical ventilation demonstrated the importance of inspiratory flow rate in turning respiratory muscles off. When inspiratory flow is low, respiratory muscle activity persists throughout a mechanical breath.

If patients can perform significant levels of work despite apparently adequate mechanical ventilation, how is rest achieved and which parameters should be adjusted to achieve rest? It may be possible to elicit a history of discomfort and discoordination. Clinical examination may yield several signs of inadequate ventilation. Inspiratory activity may be present (diaphragmatic, sternocleidomastoid, and alae nasa activity); this is particularly significant when it occurs without mechanical assistance. During constant-flow ventilation, inspiratory muscle activity can be detected from irregular asymmetric pressure tracings when these are displayed continuously. If intrathoracic pressure is not measured, an estimate can be made by observing the changes in central venous pressure. During a positive pressure breath, intrathoracic pressure should rise. Conversely, if the central venous pressure falls markedly during a mechanical breath, respiratory muscle activity is likely.

Although hypocapnia has traditionally been considered to be a mechanism underlying suppression of respiratory muscle activity in the mechanically ventilated, cessation of triggering can also be achieved by adjusting tidal ventilation in eucapnic patients. This is particularly evident in patients with airflow limitation ventilated with large-volume breaths.

When the interface between ventilator and patient is inefficient, asynchrony occurs. Triggering or detection of the onset of a spontaneous breath is generally achieved by detecting pressure or flow changes. This is generally measured within the ventilator, although sensitivity increases when the measurement is nearer to the patient. If the lung offers a high resistance to airflow, both pressure and flow signaling is delayed. If the trigger is moved even closer to the respiratory muscles, time delays and inspiratory work can be reduced considerably.

Cardiac effects of mechanical ventilation

Cardiovascular fluctuations during spontaneous breathing are well recognized; it is not surprising that mechanical ventilation can have profound effects on the circulation. Mechanical ventilation, through its effects on both lung volume and intrathoracic pressure, has an influence on the determinants of stroke volume for both right and left ventricles. It is now appreciated that intermittent positive-pressure ventilation has a more complex effect on cardiac output rather than simply decreasing venous blood flow as suggested originally.

When lung volume increases, pulmonary vascular resistance changes. Pulmonary blood vessels are of two major anatomical types. Alveolar blood vessels are closely related to the alveoli and are affected by changes in alveolar pressure. During inflation, the resistance of alveolar blood vessels increases and their capacity decreases. In contrast, extra-alveolar vessels are exposed to intrathoracic pressure changes and volumes. During inflation, the caliber of these vessels increases, their resistance falls, and their capacitance rises. Clearly, the net effect on pulmonary vascular resistance during changes in lung volume is a balance; in health pulmonary vascular resistance increases with lung inflation. In lung diseases characterized by hyperinflation, pulmonary vascular resistance is often elevated.

With lung deflation, there is little change in pulmonary vascular resistance in the alveolar vessels, whereas the reduced lung volume compresses and reduces the caliber of the extrathoracic vessels. The net effect is to increase pulmonary vascular resistance at low lung volumes. In the critically ill, diseases which lead to a reduced lung volume (e.g. acute respiratory distress syndrome) will tend to increase pulmonary vascular resistance. This effect will be exacerbated with other changes in the pulmonary vasculature. One goal of ventilatory therapy is to restore lung volume to functional residual capacity to normalize pulmonary vascular resistance in hyper- and hypoinflated lungs.

Lung volume will also impact on the cardiac system through direct mechanical compression of the heart. This may cause a restrictive effect similar to tamponade when pre-existing lung expansion occurs. As inflation pressures are transmitted to the cardiac chambers, the measurement of cardiac filling pressures becomes inaccurate. Additionally, the cardiac septum may become deviated. This shift in the position of the septum may occur towards or away from the left ventricle, impairing the function of either the right or the left ventricle. As the volume of the lungs increases, so intrathoracic pressure changes. During mechanical ventilation, intrathoracic pressure impedes blood flow to the right atrium and decreases right ventricular diastolic filling.

Bibliography

Marini, J.J. (1989). Should PEEP be used in airflow limitation? *American Review of Respiratory Diseases*, **140**, 1–3.

Marini, J.J., Rodriguez, R.M., and Lamb, V. (1986). Bedside estimation of the inspiratory work of breathing during mechanical ventilation. *Chest*, **89**, 56–63.

Pinsky, M.R. (1990). *The effects of mechanical ventilation on the cardiovascular system*. W.B. Saunders, Philadelphia, PA.

Sykes, K. (1995). *Physiological background to mechanical ventilation*. BMJ Medical, London.

Tobin, M.J. and Lodato, R.F. (1989). PEEP, auto-PEEP, and waterfalls. *Chest*, **96**, 449–51.

18.2.2 Continuous positive airways pressure

J.-A. Romand and P. M. Suter

Key messages

- Continuous positive airways pressure (CPAP) prevents or helps to reopen collapsed alveoli and atelectatic lung zones.
- CPAP is an efficient chest therapy after abdominal or thoracic surgery.
- Non-invasive application can be achieved using a face mask (or nasal mask).
- CPAP improves respiratory and cardiac function in acute cardiogenic pulmonary edema.
- CPAP can be used to facilitate weaning from mechanical ventilation.

Introduction

In the healthy lung, peripheral airways and alveoli are prevented from collapsing by physical means such as alveolar interdependence, the static recoil force of the thoracic wall, and surfactant in the alveoli. In addition, physiological mechanisms, such as periodic sighs and cough, allow reopening of unstable units. Both medical interventions (e.g. anesthesia) and disease states (e.g. abdominal distension, respiratory muscle dysfunction) disturb these physiological forces. Furthermore, different types of acute respiratory failure are also associated with a loss of gas exchange surface area owing to collapsing peripheral airways and alveoli. Typical examples include acute pulmonary edema, atelectasis, pneumonia, and postoperative respiratory dysfunction after abdominal or thoracic surgery. For instance, acute congestive left heart failure results in an increase in extravascular lung water, reductions in both lung volume and respiratory system compliance, and an increase in airway resistance. The work of breathing and the oxygen cost of breathing are both increased, with the potential for an imbalance between oxygen consumption and delivery. In severe cases, the combination of respiratory insufficiency and left ventricular failure constitutes a life-threatening vicious cycle unless appropriate measures are taken promptly to treat both respiratory dysfunction and heart failure. Recruitment of collapsed but potentially functional lung regions is possible in this situation by positive end-expiratory pressure when the patient requires intubation and mechanical ventilation. However, if sufficient alveolar ventilation can be achieved by spontaneous respiration, continuous positive airways pressure (CPAP) is applied with a similar goal, i.e. improvement of gas exchange by recruitment of relatively healthy lung regions for alveolar entry of fresh gas. CPAP is frequently used in different situations although for similar goals.

Respiratory dysfunction in cardiac failure

In acute cardiogenic pulmonary edema secondary to acute myocardial infarction, CPAP can improve left heart function and arterial blood gases, and can also decrease respiratory work (Katz and Marks 1985; Bersten *et al.* 1991; Lenique *et al.* 1997). The relative importance of improved oxygenation, decreased respiratory work, and enhanced circulatory function in the clinical improvement of patients with cardiogenic pulmonary edema has yet to be determined.

Postoperative CPAP administration

Upper abdominal surgery or median sternotomy often results in restrictive pulmonary dysfunction accompanied by atelectasis. This can be observed on more than 60 per cent of postoperative chest radiographs. This regional hypoventilation is complicated by pneumonia in 20 per cent of all patients. The postoperative acute restrictive syndrome is characterized by a 50 per cent fall in vital capacity and a 30 per cent decrease in functional residual capacity. Both recover spontaneously to 40 to 50 per cent of preoperative values by day 5 after abdominal or cardiac surgery. These changes in lung volume increase the alveolo-arterial oxygen gradient $P(A-a)O_2$ postoperatively, producing a decrease in PaO_2.

CPAP delivered during the postoperative period increases lung volumes and systemic oxygenation. Even intermittent CPAP allows as good a recovery of vital capacity and inspiratory reserve volume as other types of chest physiotherapy. Postoperative atelectasis formation and pneumonia can be decreased.

Non-invasive face-mask CPAP to prevent intubation

In some patients with acute respiratory insufficiency (Meduri *et al.* 1989), mask CPAP therapy has been used as the sole mode of ventilatory support (Fig. 1). This type of therapy is also effective in some trauma patients presenting with hypoxemia, thereby avoiding endotracheal intubation. Furthermore, spontaneous breathing through a CPAP face mask results in a better PaO_2 after coronary artery bypass surgery. CPAP applied to patients presenting with postextubation hypoxemia persisting after chest physiotherapy can also be treated satisfactorily with CPAP alone at a level of 5 to 10 cmH_2O. Carbon dioxide retention producing hypercarbia can be seen during mask CPAP ventilation and has been regarded as a relative contraindication to its use. This should not prevent a therapeutic trial. However, this is acceptable only in situations where adequate staffing and capability to measure arterial blood gases are available so that these patients can be monitored very closely. Indeed, non-invasive ventilatory support requires very close supervision by intensive care unit staff; perhaps more than in tracheally intubated patients.

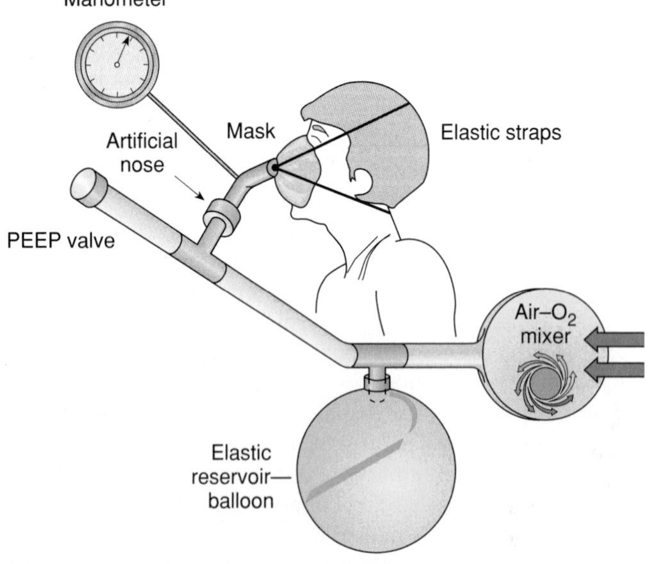

Fig. 1 Face-mask CPAP circuit.

CPAP and intracranial pressure regulation

Some concern may arise when CPAP is delivered to patients with increased intracranial pressure. Positive intrathoracic pressure may indeed decrease venous return and act as a resistive component to venous return from the head, thus increasing intracranial venous pressure and eventually parenchymal intracranial pressure. Decreased systemic venous return may also decrease cardiac output and blood pressure which will decrease oxygen delivery and perfusion to the already jeopardized cerebral tissues. A CPAP level of 10 to 15 cmH_2O increases cerebrospinal fluid pressure by 3 to 5 cmH_2O. The use of CPAP greater than 5 cmH_2O in this type of patient must

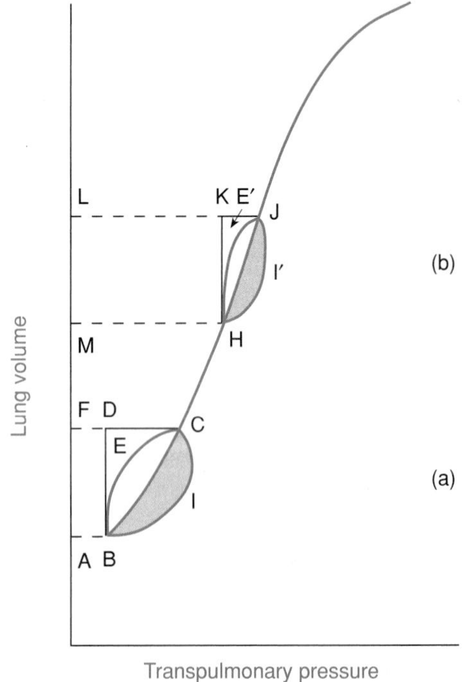

Fig. 2 Pressure–volume diagram of elastic work (curves BCD and HJK) and resistive (non-elastic) work (curves BIC and HI′J) done by the lung during breathing at (a) ambient airway pressure and (b) with CPAP. CPAP decreased both elastic and non-elastic work loads as expressed by the area under the curves HJK and HI′J compared with BCD and BIC respectively. (Reproduced with permission from Katz and Marks (1985).)

be accompanied by frequent monitoring of the patient's neurological status. McGuire *et al.* (1997) demonstrated that, compared with T-piece breathing, CPAP reduces the total inspiratory work without negative effects on cerebral perfusion pressure. The reduction in the work of breathing appears to be related to the minimization of the variation in airway pressure between inspiration and expiration. Thus the equipment used to deliver CPAP should be able to generate an almost constant airway pressure independently of the inspiratory flow of the patient. Demand-valve CPAP systems require more external inspiratory work than high-flow valve-free systems. However, the success or failure of weaning from mechanical ventilation with CPAP depends not only on the characteristics of CPAP systems, but also on patient characteristics such as muscle strength and compliance of the respiratory system (Fig. 2).

Conclusions

CPAP is a mode of ventilation which is used in many different situations to restore pulmonary function. A wide range of physiological effects, implicating almost all systems, are seen when it is administered. When prescribed in non-intubated patients, great care must be taken to prevent complications.

References

Bersten, A.D., Holt, A.W., Vedig, A.E., Skowronski, G.A., and Baggoley, C.J. (1991). Treatment of severe cardiogenic pulmonary edema with continuous positive airway pressure delivered by facemask. *New England Journal of Medicine*, **325**, 1825–30.

Katz, J.A. and Marks, J.D. (1985). Inspiratory work with and without continuous positive airway pressure in patients with acute respiratory failure. *Anesthesiology*, **63**, 598–607.

Lenique, F., Habis, M., Lofaso, F., Dubois-Rand, J.L., Harf, A., and Brochard, L. (1997). Ventilatory and hemodynamic effects of continuous positive airway pressure in left heart failure. *American Journal of Respiratory and Critical Care Medicine*, **155**, 500–5.

McGuire, G., Crossley, D., Richards, J., and Wong, D. (1997). Effects of varying levels of positive end expiratory pressure on intracranial pressure and cerebral perfusion pressure. *Critical Care Medicine*, **25**,1059–62.

Meduri, G.U., Conoscenti, C.C., Menashe, P., and Nair, S. (1989). Noninvasive face-mask ventilation in patients with acute respiratory failure. *Chest*, **95**, 865–70.

18.2.3 Non-invasive positive-pressure ventilation

G. U. Meduri

Key messages

- Non-invasive positive-pressure ventilation (NIPPV) via face mask or nasal mask can be a safe and effective means of improving the pathophysiology of respiratory failure and correcting gas exchange abnormalities.
- Avoiding complications of endotracheal intubation and decreasing the incidence of nosocomial infections are factors involved in improving outcome of patients supported with NIPPV.
- Patient selection, correct implementation, and monitoring are critical to its success.
- Pressure support ventilation is preferred because it minimizes peak inspiratory mask pressure and air leakage and is better tolerated.
- With pressure support ventilation, tidal volume, gas exchange, respiratory rate, and diaphragmatic activity are improved in proportion to the amount of pressure supplied.

In many patients with acute respiratory failure, non-invasive positive-pressure ventilation (NIPPV) via a mask can be as effective in improving the pathophysiology of respiratory failure and correcting gas exchange abnormalities as conventional ventilation via an endotracheal tube (Meduri 1996). When effective, NIPPV avoids the complications associated with endotracheal intubation, improves patient comfort, and preserves airway defense mechanisms, speech, and swallowing. Furthermore, NIPPV provides greater flexibility in instituting and removing mechanical ventilation.

NIPPV via a face mask or nasal mask (Fig. 1) can be a safe and effective means of recruiting alveoli and augmenting ventilation in patients with acute respiratory failure resulting from chronic obstructive pulmonary disease, cardiogenic pulmonary edema, post-extubation, cystic fibrosis, endstage lung disease, postoperative complications, post-traumatic injury, community-acquired pneumonia, or AIDS. Randomized studies have shown it to be superior to conventional management in patients with chronic obstructive pulmonary disease; the procedure significantly decreased the need for intubation, rate of complications, incidence of nosocomial infections, length of stay, mortality, and cost.

The duration of mechanical ventilation with continuous NIPPV is shorter than with conventional mechanical ventilation. In our group experience of 210 patients, the mean duration of mechanical ventilation was 25 h, similar to findings from other studies. The short duration of continuous successful NIPPV is consistently seen across all types of acute respiratory failure, indicating an intrinsic advantage of this procedure (Meduri *et al.* 1996). Factors that may be involved in shortening the duration of mechanical ventilation include earlier intervention, avoiding sedation and paralysis, reduced incidence of

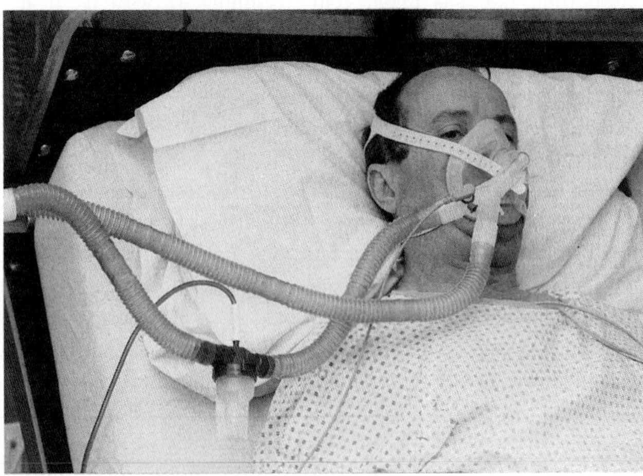

Fig. 1 Patient receiving non-invasive positive pressure ventilation.

Table 1 Criteria for selecting patients for NIPPV

Alert and co-operative patient*
Hemodynamic stability
No need for endotracheal intubation to: Protect the airways[†] Remove excessive secretions
No acute facial trauma

*Patients with chronic obstructive pulmonary disease and CO_2 narcosis are an exception (see text).
[†]Mental obtundation, impaired swallowing, or active upper gastrointestinal bleeding.

mechanical-ventilation-induced respiratory muscle atrophy, eliminating the imposed work by the endotracheal tube, a lower rate of complications (particularly infections), and earlier removal.

Avoiding complications of endotracheal intubation and decreasing the incidence of nosocomial infections are factors involved in improving outcome of patients supported with NIPPV. With NIPPV, the trachea is not invaded by the endotracheal tube or by suction catheters (patients can remove the face mask and expectorate) and the vocal cords are not kept open, thereby lowering the risk of aspiration, which is the leading mechanism of nosocomial pneumonia. Furthermore, patients rarely require invasion by nasogastric, urinary, or central intravenous catheters, which are significant contributors to nosocomial sepsis. Monitoring for clinical and laboratory signs of infection, our group has identified only two cases of pneumonia (and no other infection) among 210 NIPPV-treated patients. A randomized study of chronic obstructive pulmonary disease patients showed a significantly lower rate of intubation (26 per cent) among 43 NIPPV-treated patients and a lower incidence of nosocomial infections (Brochard *et al.* 1995). The standard treatment group had seven pneumonia patients and three episodes of sepsis among intubated patients, while the NIPPV-treated group had two pneumonia patients and two episodes of sepsis (all but one occurring after intubation in patients failing NIPPV). Death related to a nosocomial infection occurred in 10 per cent of the standard treatment group and in only 2 per cent of NIPPV-treated patients. Among 16 studies involving 462 patients treated with NIPPV, the overall complication rate was 15 per cent, including 42 with facial skin necrosis, six with gastric distention, four with pneumonia, nine with conjunctivitis, and five with other complications.

Because correct implementation and monitoring are critical to the success of NIPPV, physicians, hospital respiratory technicians, and nurses should develop familiarity with this methodology. Patient selection is also essential in avoiding complications (Table 1). Patients with morbid obesity (over 200 per cent of ideal body weight) or with unstable angina or acute myocardial infarction should be closely managed, but only by experienced personnel. The response to treatment and duration of mechanical ventilation cannot be predicted by the severity of the underlying lung disease or by the arterial blood gas values obtained before initiating mechanical ventilation.

The methodology for NIPPV in patients with acute respiratory failure used at the University of Tennessee is shown in Table 2. Although no study has directly compared efficacy, one group has reported a higher success rate with face-mask NIPPV versus their institutional historical controls using a nasal mask. Improvement in arterial blood gases appears to be slower in some studies using a nasal mask. Because patient tolerance is essential to the success of NIPPV, a tight uncomfortable fit should be avoided when possible. Even in patients with hypoxemic respiratory failure who are receiving continuous positive airway pressure (CPAP) by mask alone, a small leak will not cause the airway pressure to drop. Placing a nasogastric tube is indicated only for patients developing gastric distension or to provide access for enteral feeding. Gastric distention rarely occurs (less than 2 per cent) in patients treated with NIPPV or mask CPAP. Furthermore, in our experience, none of the patients developing pneumonia on NIPPV had prior gastric distention. In normal patients, the resting upper esophageal sphincter pressure is 33 ± 12 mmHg. When air enters the stomach during NIPPV, a gurgling noise can be heard over the epigastrium with the stethoscope, and a vibration can be felt by palpation. Nurses and therapists should be instructed to observe patients for signs of abdominal distension.

In either hypercapnic or hypoxemic acute respiratory failure, we utilize both CPAP and intermittent positive-pressure ventilation. Pressure support ventilation is the preferred form of intermittent positive-pressure ventilation because it minimizes mask peak inspiratory pressure and air leakage and is better tolerated. The

Table 2 Methodology for NIPPV in patients with acute respiratory failure (University of Tennessee, Memphis)

Position the head of the bed at an angle of 45°
Choose the correct size mask and connect it to the ventilator
Turn the ventilator on and silence the alarms. The initial ventilatory settings are CPAP 0 cmH_2O with pressure support 10 cmH_2O. FiO_2 is titrated to achieve an oxygen saturation of over 90%
Explain the modality to the patient and provide reassurance
Hold the mask gently on the patient's face until the patient is comfortable and in full synchrony with the ventilator
Apply wound care dressing on the nasal bridge and other pressure points
Secure the mask with the headgear, avoiding a tight fit. Allow enough space to pass two fingers beneath the head straps
Slowly increase CPAP to $\geqslant 5$ cmH_2O
Increase pressure support to obtain an exhaled tidal volume of 7 ml/kg, a respiratory rate $\leqslant 25$ breaths/min, and patient comfort
In hypoxic patients, increase CPAP in increments of 2–3 cmH_2O until FiO_2 is 0.6
Avoid peak mask pressure above 30 cmH_2O. Allow minimal air leaks if exhaled tidal volume is adequate
Set the ventilator alarms and apnea backup parameters
Ask the patient to call for needs (repositioning the mask, pain or discomfort, expectoration) or if complications occur (respiratory difficulties, abdominal distention, nausea, vomiting)
Monitor with oximetry and adjust ventilator settings following arterial blood gas results

Table 3 Effects of CPAP and pressure support ventilation delivered by a mask, alone and in combination, on gas exchange and transdiaphragmatic pressure in patients with obstructive and restrictive lung disease

	Obstructive lung disease		Restrictive lung disease	
	Gas exchange	Transdiaphragmatic pressure	Gas exchange	Transdiaphragmatic pressure
CPAP	—↑	↓	↑	↓
PSV	↑	↓	↑	↓
CPAP+PSV	↑↑	↓↓	↑↑	↓↓

PSV, pressure support ventilation.
— no effect; ↑ improved or increased; ↓ worse or decreased.

effects of CPAP and pressure support ventilation delivered by a mask, either alone or in combination, on gas exchange and transdiaphragmatic pressure in patients with obstructive and restrictive lung disease are shown in Table 3. With pressure support ventilation, tidal volume, gas exchange, respiratory rate and diaphragmatic activity are improved in proportion to the amount of pressure applied (Carrey *et al.* 1990; Appendini *et al.* 1994). CO_2 rebreathing can occur during bi-level positive airway pressure (BiPAP) ventilatory assistance using the standard exhalation device (Whisper Swivel) and can be eliminated with a new plateau exhalation device or a non-rebreather valve. Applying expiratory positive airway pressure ($\geqslant$ 4 cmH_2O) decreases inhaled CO_2 which is eliminated at a level of 8 cmH_2O. If air leakage is not improved, despite manipulations of the mask and application of a skin patch, first CPAP and then applied pressure (or tidal volume) are decreased to reduce the peak mask pressure. Portable units, such as the Respironics Bi-PAP device and the Puritan Bennett PB-335, automatically compensate for mask leaks and mouth opening.

In our experience, after the first hour of uncomplicated NIPPV, most patients do not require bedside observation; ventilator and oximetry alarms provide warnings for early intervention if necessary. We have found that time involvement with NIPPV is proportional to the level of experience. In two randomized studies, bedside time commitment by nurses and therapists was similar for patients receiving NIPPV or conventional treatment. After the initial stabilization period on NIPPV (4–6 h), patients with hypercapnic respiratory failure or with hypoxemia on low-level CPAP ($\leqslant$ 5 cmH_2O) can safely remove the mask for 5 to 15 min, during which time they can talk, drink small amounts of liquid, expectorate, or receive nebulized bronchodilator therapy. Because mask ventilation provides a large degree of flexibility, it can be adjusted to meet a patient's individual needs. Continuous oximetry with alarms should be provided, and in patients with severe hypercapnia it is advisable to have an arterial catheter in place. Ventilator settings should be adjusted based on results of arterial blood gases obtained within 1 h and, as necessary, at intervals of 2 to 6 h. Providing reassurance and adequate explanation to the patient about what to expect is of the utmost importance. Patients are instructed to call the nurse if they have needs or develop complications.

Criteria to discontinue NIPPV are shown in Table 4. Following improvement in acute respiratory failure, patients are weaned from mechanical ventilation either by lowering the amount of delivered pressure or by titrating periods off mechanical ventilation to patient tolerance and objective findings, similar to a T-piece weaning trial. At a low level of pressure support (5–8 cmH_2O), the patient is disconnected from the ventilator while receiving supplemental oxygen by nasal cannula or face mask. During weaning, NIPPV eliminates the reintubation factor associated with prematurely removing conventional ventilation, and this may contribute to the shorter duration of ventilation.

Table 4 Criteria for discontinuing NIPPV

- Inability to tolerate the mask due to discomfort or pain
- Inability to improve gas exchange or dyspnea
- Need for endotracheal intubation to manage secretions or protect the airways
- Hemodynamic instability
- ECG instability with evidence of ischemia or significant ventricular arrhythmias
- Failure to improve mental status, within 30 min of initiating NIPPV, of patients who are lethargic from CO_2 retention or agitated from hypoxemia

In patients with hypercapnic acute respiratory failure receiving NIPPV, most studies have found that the arterial blood gas response and need for intubation cannot be predicted by the severity of the underlying lung disease (forced expiratory volume in 1 s and arterial blood gases) or by the arterial blood gas values ($Pa\text{CO}_2$, pH) obtained before implementing NIPPV. Two recent studies involving 125 patients with chronic obstructive pulmonary disease and acute respiratory failure found that non-responders had a higher $Pa\text{CO}_2$ at initiation of NIPPV (Meduri *et al.* 1996). In other studies, the underlying cause of acute respiratory failure did not predict the outcome of NIPPV. Patients with chronic obstructive pulmonary disease and pneumonia or congestive heart failure as the cause of acute respiratory failure have a higher intubation rate. We have consistently found that a reduction in $Pa\text{CO}_2$ or an increase in pH within 1 to 2 h of NIPPV predicted a sustained improvement in gas exchange and requirement for shorter duration of ventilatory support (mean 26 h of NIPPV versus 323 h of mechanical ventilation with endotracheal intubation). This observation has been confirmed by others.

An understanding of this method of mechanical ventilation by physicians and hospital respiratory therapy teams is an extremely important adjunct to ensuring the safety of this technique and its proper use on patients with acute respiratory failure. A dedicated team (therapist and intensivist) should provide in-service training to respiratory therapists, critical care nurses, and house staff prior to implementing this method of ventilation. Continuous supervision is required until providers are fully comfortable in all aspects of non-invasive ventilation. With proper training, NIPPV can be transferred to clinical practice with results similar to those of small studies conducted by a research team. We have produced an instructional videotape , which is available upon request, and developed a comprehensive course to instruct therapists and physicians on correct NIPPV methodology.

References

Appendini, L., *et al.* (1994). Physiologic effects of positive end-expiratory pressure and mask pressure support during exacerbations of chronic obstructive pulmonary disease. *American Journal of Respiratory and Critical Care Medicine*, **149**, 1069–76.

Brochard, L., *et al.* (1995). Noninvasive ventilation for acute exacerbations of chronic obstructive pulmonary disease. *New England Journal of Medicine*, **333**, 817–22.

Carrey, Z., Gottfried, S.B., and Levy, R.D. (1990). Ventilatory muscle support in respiratory failure with nasal positive pressure ventilation. *Chest*, **97**, 150–8.

Meduri, G.U. (1996). Noninvasive positive-pressure ventilation in patients with acute respiratory failure. *Clinics in Chest Medicine*, **17**, 513–53.

Meduri, G.U., Turner, R.E., Abou-Shala, N., Tolley, E., and Wunderink, R.G. (1996). Noninvasive positive pressure ventilation via face mask: first-line intervention in patients with acute hypercapnic and hypoxemic respiratory failure. *Chest*, **109**, 179–93.

18.2.4 Indications for mechanical ventilation

R. F. Armstrong

Key messages

- Indications for mechanical ventilation are as follows:
  - respiratory rate above 30 breaths/min
  - Pa_{O_2} below 11 kPa (83 mmHg) on Fi_{O_2} 0.4
  - Pa_{CO_2} high enough to cause respiratory acidosis (pH < 7.2)
  - shunt fraction above 15 per cent
  - ratio of dead-space to tidal volume above 0.6
  - maximum inspiratory pressure ($P_i$max) less than −20 cmH_2O.

Introduction

The need for mechanical ventilation arises in the patient with respiratory failure. Causes of respiratory failure include inadequate alveolar ventilation (e.g. coma, neuromuscular disorders, cord injury), inadequate oxygenation (e.g. pneumonia, pulmonary edema, acute respiratory distress syndrome), disorders of the thoracic wall (e.g. multiple rib fractures, scoliosis, obesity), airway obstruction (e.g. edema, foreign body, trauma, infection), and cardiac failure (e.g. left ventricular failure, cardiopulmonary resuscitation).

Decisions to introduce respiratory support are based on evidence obtained from clinical history and examination, the Pa_{O_2} measurement in relation to the inspired oxygen fraction (Fi_{O_2}), the Pa_{CO_2} in relation to the arterial pH, the vital capacity, the shunt fraction, the ratio of dead-space to tidal volume, and inspiratory pressure measurement.

Clinical history and examination

The clinical signs of respiratory failure have been described in a classic paper by Gilston (1976) and are summarized in Table 1.

Table 1 Clinical signs of respiratory failure
Respiratory signs
Sweating
Eyebrow lifting
Nasal flaring
Use of accessory muscles of respiration
Cyanosis, difficulty in talking
Expiratory grunting
Tracheal tug
Ineffective cough
Noisy wheezing respiration
Tachypnea
Loss of expiratory pause
Dyscoordinate respiration
Abnormal inspiratory-to-expiratory ratio
Central nervous system signs
Nasal lines
Anxiety and apathy
Drowsiness
Disorientation
Weakness
Picking at bedclothes
Cardiovascular system signs
Cool extremities
Falling blood pressure
Rising central venous pressure
Oliguria

Arterial blood gases

Arterial oxygen tension

The arterial oxygen tension ($P\text{ao}_2$) in a normal subject breathing air at atmospheric pressure ranges from 12.5 kPa (94 mmHg) at age 20 years to 10.8 kPa (81 mmHg) at age 60 to 69 years. The alveolar partial pressure of oxygen $P\text{A}\text{o}_2$ varies with the inspired oxygen concentration as shown by the simplified alveolar air equation

$$P\text{A}\text{o}_2 = \text{Fio}_2\,(P\text{B} - P\text{H}_2\text{O}) - P\text{aco}_2/\text{RQ}$$

where Fio_2 is the inspired oxygen fraction, P_B is the atmospheric pressure (100 kPa), $P\text{H}_2\text{O}$ is the water vapor pressure, $P\text{aco}_2$ is the alveolar partial pressure of carbon dioxide, and RQ is the respiratory quotient (RQ = 0.8).

Thus a healthy patient with a normal $P\text{aco}_2$ breathing 60 per cent oxygen should have a calculated $P\text{ao}_2$ of about 50 kPa (375 mmHg). A $P\text{ao}_2$ of less than a third of the predicted value represents a very significant impairment of respiratory function, and mechanical ventilation should be considered. Use of the $P\text{ao}_2/\text{Fio}_2$ ratio relieves the clinician from calculations of this type. At the American–European Consensus Conference (Bernard *et al.* 1992), acute lung injury was defined as a $P\text{ao}_2/\text{Fio}_2 < 300$ mmHg (40 kPa) and acute respiratory distress syndrome as $P\text{ao}_2/\text{Fio}_2 < 200$ mmHg (26.7 kPa).

Falling oxygen saturation can also be a useful indicator of respiratory failure. However, this can be a pitfall for the inexperienced. A saturation of 100 per cent corresponds to any $P\text{ao}_2$ over 12 kPa (90 mmHg). Thus a patient with severe acute respiratory distress syndrome on 60 per cent oxygen may have a $P\text{ao}_2$ of only 12 kPa (90 mmHg) and yet a saturation of 100 per cent. On this inspired mixture the patient should have a $P\text{ao}_2$ of 50 kPa (375 mmHg) and is clearly in respiratory failure. Misinterpretation of measurements in this way may slow the referral of critically ill patients to the intensive care unit.

Arterial CO_2 tension

Both chronic and acute respiratory failure are accompanied by elevation of $P\text{aco}_2$ above its normal level of 5.3 kPa (40 mmHg). A feature of patients with deteriorating chronic obstructive airways disease is the slow rise in $P\text{aco}_2$ compensated by renal retention of bicarbonate. Usually, there is a rise in bicarbonate (normal range 23–28 mmol/l) of about 4 mmol/l for every 1.3-kPa (10-mmHg) rise in $P\text{aco}_2$. With acute exacerbations secondary to infection and increased sputum production, sharp increases in $P\text{aco}_2$ occur which exceed the ability of the kidney to compensate and precipitate severe respiratory acidosis.

Where expertise allows, nasal ventilation may avert the need for intubation (Udwadia *et al.* 1992), but mechanical ventilation should be introduced in appropriately chosen patients with arterial pH levels below 7.2 and evidence of exhaustion. In acute respiratory failure, elevation of $P\text{aco}_2$ is also a useful marker for the introduction of respiratory support; an irreversible respiratory acidosis with an arterial pH below 7.2 is usually considered intolerable. However, severely asthmatic patients provide an exception to this rule, with ventilation being introduced at lower or even normal levels of $P\text{aco}_2$ if other indications are present (e.g. obtunded, fatigued, falling $P\text{ao}_2$).

Vital capacity measurement

Normal vital capacities range from 65 to 75 ml/kg. Reduction occurs in a variety of conditions associated with weakness, for example cervical cord injury, myasthenia gravis, Guillain–Barré syndrome, persistence of muscle relaxants, poliomyelitis, botulism, and organophosphate poisoning. Rising $P\text{aco}_2$ and falling $P\text{ao}_2$ are also features of these conditions. However, normal blood gases may be maintained in the early stages of these disorders. Vital capacity recordings provide useful insights into the rate of development of weakness. Measurements of less than 10 to 15 ml/kg are generally considered incompatible with spontaneous ventilation and are an indication for mechanical ventilation.

Shunt fraction and dead-space measurements

Intrapulmonary shunt fraction as a percentage of cardiac output can be calculated from the shunt equation:

$$Q_\text{s}/Q_\text{t} = C\text{co}_2 - C\text{ao}_2/C\text{co}_2 - C\text{vo}_2$$

where Q_s is the flow-through shunt, Q_t is the total cardiac output, $C\text{co}_2$ is the oxygen content of pulmonary capillary blood, $C\text{vo}_2$ is the oxygen content of mixed venous blood, and $C\text{ao}_2$ is the oxygen content of arterial blood. The normal range is 1 to 4 per cent. Mechanical ventilation is usually deemed necessary at shunt levels of 15 to 20 per cent. However, the complexity of this calculation, the need for mixed venous blood samples, and the difficulty in accurately measuring saturation in a given patient have made this approach less popular.

The same can be said for the calculation of dead-space as a percentage of the tidal volume (V_d/V_t ratio):

$$V_\text{d} = V_\text{t}\ (P\text{aco}_2 - P\text{eco}_2/P\text{aco}_2)$$

where V_d is the dead-space, V_t is the tidal volume, $P\text{aco}_2$ is the arterial CO_2 tension, and $P\text{eco}_2$ is the mixed expired CO_2 tension. This is normally 30 per cent, and ventilation is recommended at ratios above 60 per cent. Although the calculation is straightforward, collecting mixed expired air renders the process laborious for everyday use.

Inspiratory muscle strength

Inspiratory muscle strength is measured by performing a maximal inspiratory maneuver against a closed airway. The normal range is between −40 and −100 cmH_2O. Weak patients can be identified by failure to achieve these levels. This measurement is difficult to make in the distressed failing patient and is more useful in assessment of weaning potential.

Medical conditions associated with the need for intermittent positive-pressure ventilation

Early ventilation should be considered in severe heart failure and the sepsis syndrome. Possible mechanisms of benefit are a reduction in the work of breathing, a reduction in metabolic rate, relief of diaphragmatic failure associated with an inadequate cardiac output, reduction in cardiac distension by reducing venous return, enabling the use of sedatives and anxiolytics, and reduction of left ventricular afterload.

References

Bernard, G., Artigas, A., and Brigham K. (1994). The American–European Consensus Conference on ARDS. *American Journal of Respiratory and Critical Care Medicine*, **149**, 818–24.

Gilston, A. (1976). Facial signs of respiratory distress after cardiac surgery. *Anaesthesia*, **31**, 385–97.

Schuster, D.P. (1990). A physiologic approach to initiating, maintaining and withdrawing mechanical ventilatory support during acute respiratory failure. *American Journal of Medicine*, **88**, 268–78.

Udwadia, Z., Santis, G.K., Steven, M.H., and Simonds, A. (1992). Nasal ventilation to facilitate weaning in patients with chronic respiratory insufficiency. *Thorax*, **47**, 715–18.

18.2.5 Design and function of mechanical ventilators

R. A. Smith

Key messages

- A microprocessor facilitates high-speed and accurate control of inspiratory and expiratory valve events.
- Program control of tidal volume is designed to regulate gas flow from the inspiratory valve independent of the impedance of the patient's respiratory system.
- Factors contributing to the delay in achieving 'target pressure' in the breathing circuit include inspiratory valve response time, design and function of inspiratory and expiratory valves, inspiratory valve triggering mechanism, and the electronic logic (algorithm design, sensor signal transmission and processing, and speed of computation).
- Appreciation of the design and function of ventilators will enable clinicians to balance the potentially detrimental effects of ventilatory support with the desired physiological goals.

Introduction

Modern mechanical ventilators provide many modes and control options for producing full or partial ventilatory support. An understanding of the major design features and the functional operation of mechanical ventilators enables clinicians to tailor ventilatory support to the desired physiological goals.

Basic design and function

Microprocessor control

Features of mechanical ventilation are selected by input devices such as touch pads and potentiometers on the control panel (Fig. 1). Analog signals from the control panel and from gas pressure and

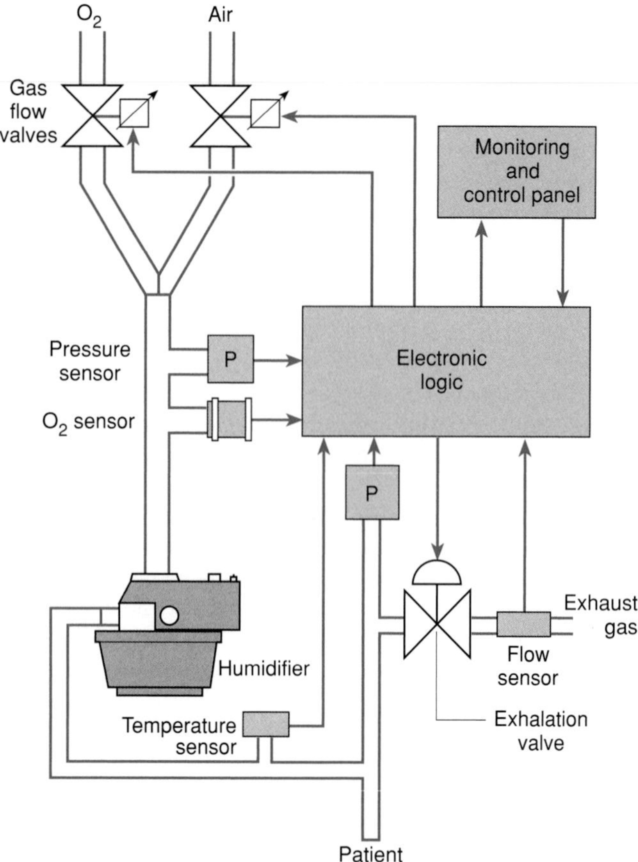

Fig. 1 Schematic illustration of the basic elements of a mechanical ventilator: P, pressure.

flow sensors are 'digitized' by means of an analog-to-digital converter. Digitized signals are transmitted to the microprocessor, which interprets and executes program instructions and performs arithmetical and relational operations. Digital output from the microprocessor is converted to analog signals by means of a digital-to-analog converter in order to control various operations of the mechanical ventilator.

A microprocessor offers the ability to provide rapid and precise control of inspiratory and expiratory valve events while running the user interface (control panel), acquiring data, and monitoring ventilator performance. Microprocessor control is designed as either an open- or a closed-circuit system. In an open-circuit system, input is not influenced by output. In a closed-circuit system, commonly called servocontrol, input is affected by output. Servosystems are designed to prevent any difference (error) between the reference value, represented by control panel settings, and the quantified output, represented by sensor measurements. Strategies in mechanical and software design contribute greatly to the optimal performance of a mechanical ventilator.

Inspiratory and expiratory valves

Most modern mechanical ventilators have program-controlled gas delivery systems with either one inspiratory valve or a pair of inspiratory valves. These valves are almost always operated by an electromagnetic motor (Fig. 2). Ventilators with one electromagnetic valve are supplied with gas from a reservoir which contains

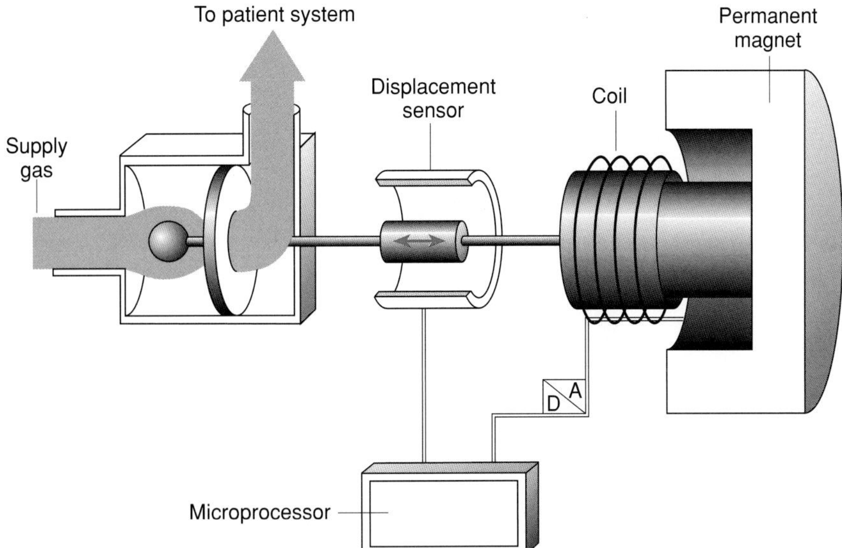

Fig. 2 Schematic illustration of an inspiratory valve of a mechanical ventilator. A microprocessor directs electric current into the coil, via a digital-to-analog converter (D/A), to produce an electromagnetic field with the same polarity as the permanent magnet. Variation in the magnitude of the field causes the armature and ball to be positioned so as to produce the desired rate of gas flow. (Modified with permission from product literature for the Evita, Dragerwerk AG, Lubeck, Germany.)

compressed gas at a specified oxygen concentration. Ventilators with a pair of inspiratory valves proportion compressed air and compressed oxygen independently to produce a selected concentration of oxygen. The function of the inspiratory valve(s) relative to the type and degree of ventilatory support selected is discussed later.

The function of the expiratory valve is to seal the breathing circuit during the mechanical inspiratory phase and to vent gas to ambient during the mechanical exhalation phase (Fig. 3). In most ventilators, the expiratory valves are program controlled. The functional operation of the expiratory valve relative to the type of ventilatory support selected is described later.

Design and function during mechanical inspiratory phase

Volume-controlled mechanical insufflation

Program control of volume is designed to modulate the flow of gas from the inspiratory valve(s) into the patient circuit independent of the impedance of the patient's respiratory system. Variables which can be selected for volume-controlled ventilation include the initiation and termination mechanism and the amount of insufflation pressure.

Initiation of mechanical insufflation

A mechanical breath is initiated (triggered) either automatically by the machine or in synchrony with spontaneous inspiration by the patient. In the first option, a timing mechanism automatically initiates a mandatory breath at intervals independent of spontaneous breathing. A control device regulates the number of mechanical breaths occurring per minute, and this rate establishes the duration of the ventilatory cycle. Alternatively, a mechanical breath is triggered when a gas flow or pressure sensor signals the electronic logic indicating that spontaneous inspiration has occurred.

Gas flow triggering system

The inspiratory valve(s) generate a selected baseline gas flow (usually 5–20 l/min) through the patient circuit. The baseline gas flow vents to ambient via the expiratory valve and is quantified by a sensor. The microprocessor calculates the disparity between baseline gas flow and gas flow exhausted from the patient circuit. The disparity in gas flow necessary to initiate a mechanical breath may be set from 1 l/min to 50 per cent of the baseline rate.

Pressure triggering system

The sensing site for the pressure signal feedback to the microprocessor is near the gas outflow port within the ventilator chassis, or the interface between patient circuit and artificial airway, or the exhalation valve. Regardless of sensing site, the gas pressure transducer is inside the ventilator chassis and the 'pneumatic' signal travels through the sampling tubing at the speed of sound (0.33 m/s at sea level). During pressure-triggered events, the inspiratory and expiratory valves are closed; thus a nearly equivalent pneumatic signal is transmitted regardless of sensor location. Trigger delay includes the time between the start of inspiratory effort and detection of effort by electronic logic and the time from detection of inspiratory effort to initiation of gas flow by the inspiratory valve(s).

Volume-cycled versus time-cycled mechanical insufflation

Mechanical insufflation is terminated either after a selected gas volume has been generated (volume-cycled) or a selected time has elapsed (time-cycled). Activating an inspiratory-hold mechanism extends the inspiratory phase by a selected time after a volume- or time-cycled mechanical insufflation.

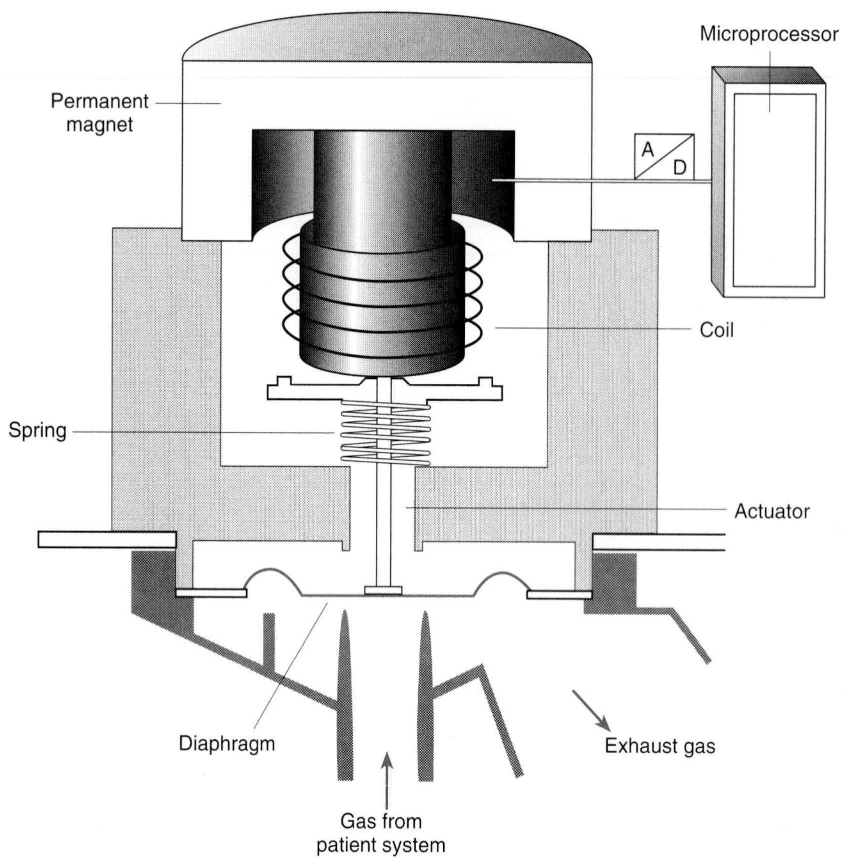

Fig. 3 Schematic illustration of an expiratory valve of a mechanical ventilator. During mechanical insufflation the microprocessor directs electric current into the coil, creating an electromagnetic field with the same polarity as the permanent magnet. Thus the membrane is 'seated' by the downward motion of the actuating rod, which also compresses a metal spring. When the mechanical inspiratory phase is terminated, the electric current is stopped, eliminating the electromagnetic field. The weight of the actuator rod is matched by the tension of the metal spring, thus causing the membrane to unseat and allowing gas to be vented to ambient. Clinician-defined PEEP is produced when the electronic logic directs an electric current through the coil and the membrane is seated with the appropriate pressure. (Modified with permission from product literature for the Veolar, Hamilton Medical Corporation, Reno, NV, USA.)

Gas flow pattern during mechanical insufflation

Gas flow during volume-controlled insufflation can be set as either constant or non-constant (e.g. accelerating, decelerating, or positive-phase sinusoidal). Depending on the design of the ventilator, the peak gas flow or the inspiratory time will change when the gas flow pattern is changed for a given tidal volume. During time-cycled ventilation, in which both tidal volume and inspiratory time are selected, the mean gas flow required to produce a constant pattern is computed by the microprocessor as

$$\dot{V}_{\text{mean}} = V_T/T_I$$

where $\dot{V}_{\text{mean}}$ is the mean gas flow in liters per second, V_T is tidal volume in liters, and T_I is the inspiratory time in seconds. Therefore, to preserve inspiratory time, peak gas flow will be higher when non-constant flow patterns are used. During volume-cycled ventilation, the peak gas flow is selected. Thus the inspiratory time required for insufflation varies with the mean gas rate and tidal volume:

$$T_I = V_T/\dot{V}_{\text{mean}}.$$

Mean gas flow is less during non-constant flow than during constant flow. Therefore inspiratory time is longer when tidal volume is delivered by means of a non-constant gas flow.

Some ventilators are designed to permit gradual increases in inspiratory gas flow to a selected or computed final value during volume-controlled ventilation. The time during which the gas flow increases from the onset of mechanical insufflation to the final gas flow can be set from zero to 10 per cent of the total respiratory cycle. At zero per cent the mean gas flow computed by the electronic logic (V_T/T_I) is immediately generated and sustained throughout mechanical insufflation, thus producing a constant gas flow. As the rise time is increased to 10 per cent, the gas flow is accelerated to its final value.

Pressure-cycled mechanical insufflation

As a safety precaution, an upper limit of acceptable pressure associated with mechanical insufflation is selected. If this is reached, gas flow from the inspiratory valve(s) automatically terminates, the expiratory valve opens, and an alarm activates. The expiratory valve remains open until gas pressure in the breathing circuit decreases to the selected baseline pressure (i.e. atmospheric pressure or a selected level of positive end-expiratory pressure (PEEP)) or until a subsequent machine breath is initiated.

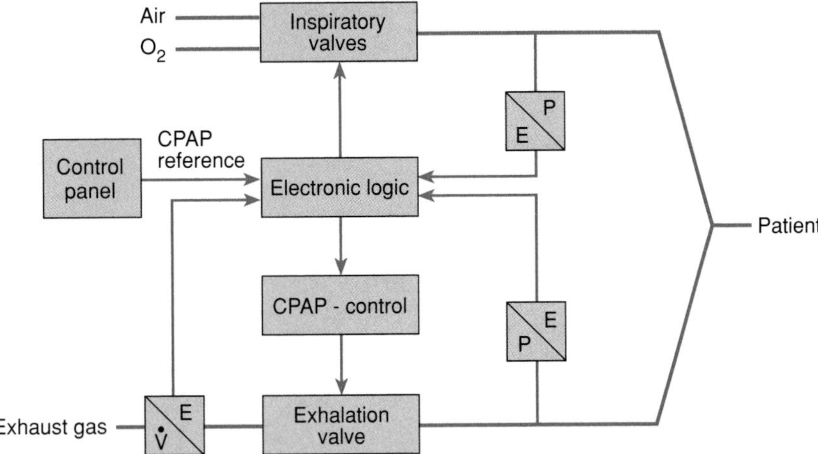

Fig. 4 Block diagram of the operation of a mechanical ventilator during CPAP. To maintain CPAP, the electronic logic first acquires input from the control panel (i.e. the reference value for CPAP) and directs the CPAP. The microprocessor acquires the level of CPAP from the control panel and directs the CPAP control valve to generate pressure on the expiratory valve that is approximately equal to the reference pressure. A transducer (P) in the patient circuit senses the pressure and the signal is transmitted to the electronic logic via an amplifier (E). If the pressure in the patient circuit is lower than the reference pressure, the electronic logic closes the exhalation valve and opens the inspiratory valves, which provides gas flow in order to bring the differential between the reference and patient circuit pressures to zero. If the pressure in the breathing circuit is higher than the reference pressure, the electronic logic closes the inspiratory valves and opens the exhalation valve in order to bring the differential between the reference and patient circuit to zero. (Modified with permission from product literature for the Evita, Dragerwerk AG, Lubeck, Germany.)

Pressure-controlled mechanical insufflation

Program control of pressure is designed to modulate gas flow from the inspiratory valve(s) in order to control pressure at the site where the pressure sensor is located. Some of the variables that can be selected for pressure-controlled ventilation include the initiation and termination mechanisms, and the amount of insufflation pressure.

Initiation of mechanical insufflation

As with volume-controlled ventilation, pressure-controlled ventilation may be initiated automatically or synchronized with spontaneous inspiration. However, after the triggering event, the microprocessor is programmed to sustain the selected pressure at the site of measurement. In contrast with the trigger event, the site of pressure measurement during pressure-controlled insufflation may affect ventilator performance.

Manufacturers typically cite the 'fast' response of their inspiratory valve(s) as performance evidence to promote their mechanical ventilator. However, the response of an isolated valve is of the order of several milliseconds, whereas the response of the ventilator is in the order of several hundred milliseconds. Therefore factors other than the response time of an isolated inspiratory valve determine the general performance of the mechanical ventilator. Other factors contributing to delay in achieving desired pressurization of the patient circuit and inability to sustain a selected pressure include design and function of the inspiratory valve(s) and expiratory valve, the triggering mechanism, and the electronic logic (algorithm design, sensor signal transmission and processing, and speed of computation).

Time-cycled or flow-cycled mechanical insufflation

Pressure-controlled ventilation may be terminated in a time- or flow-cycled manner, i.e. after a selected interval or when the gas flow decreases to a factory-selected level. When time cycled, termination of insufflation occurs as described earlier for volume-controlled ventilation. In contrast with time-cycled volume-controlled ventilation, some ventilators are designed to permit spontaneous breathing during pressure-controlled ventilation.

Flow-cycled pressure-controlled ventilation is terminated when the gas flow decreases to a fixed programmed value, for example 25 per cent of peak gas flow or 5 l/min. As a safety measure, mechanical insufflation ends when a factory-set or clinician-defined upper limit of circuit pressure is reached, or when a factory-programmed inspiratory time elapses.

Design and function during the mechanical expiratory phase

When mechanical insufflation is terminated, expired gas is vented to ambient through the expiration valve. These valves are designed to permit rapid egress of gas with minimal resistance to facilitate reduction of pressure in the patient circuit to atmospheric or a selected level of PEEP. Selected PEEP is generated when the electronic logic closes the expiration valve and interrupts relief of pressure in the patient circuit according to data acquired from a pressure sensor and the control panel.

Some ventilators permit selection of an expiratory hold to determine end-expiratory lung pressure. End-expiratory lung pressure should equal baseline pressure, i.e. either atmospheric pressure or selected PEEP to avoid or increase air trapping respectively.

Design and function during spontaneous breathing

During normal spontaneous breathing, airway opening pressure is constant (atmospheric pressure) and availability of gas is virtually infinite (atmospheric air). When a patient is attached to a mechanical ventilator, gas availability is not infinite. However, ventilators should be designed to deliver as much gas as required to keep the pressure in the patient circuit constant during spontaneous breathing.

Inspiratory valves

Most mechanical ventilators do not incorporate a mechanism dedicated to providing gas for spontaneous breathing, i.e. a 'demand valve'. Typically, a single gas delivery system is used for all modes of ventilatory support. The difference between the various modes is the algorithmic control of the inspiratory valve(s). Therefore performance of mechanical ventilators during spontaneous breathing is determined by factors similar to those discussed for patient-triggered pressure-controlled ventilation, except that the pressure is program controlled during both inspiration and expiration.

During spontaneous breathing, the opening pressure of the inspiratory valve is less than the reference pressure (e.g. continuous positive airway pressure (CPAP)) and the opening pressure of the exhalation valve is greater than the reference pressure (Fig. 4). End-expiratory pressure determines the starting point for the subsequent inspiration. The reduction in pressure necessary to initiate gas delivery is equal to the gradient between the closing pressure of the exhalation valve and the opening pressure of the inspiratory valve. Therefore the design and function of the exhalation valve are also important determinants of ventilator performance during spontaneous breathing modes.

Expiratory valves

Expiratory valve events during spontaneous breathing are controlled by two basic design strategies: the valve is seated by a pressure equal to the reference pressure throughout inspiration (open design), or it is closed coincidently with initiation of inspiration and opened when the end of inspiration is detected by the electronic logic (closed design). Since the patient's demand for gas is unpredictable, a 'trial-and-error' algorithmic control strategy must be employed. If the inspiratory valve opens too slowly, the duration of disparity between actual and reference pressure may be unacceptable. If it opens too rapidly, gas flow delivery may exceed the patient's requirements, producing a circuit pressure greater than reference pressure in closed design systems, i.e. a pressure overshoot or inadvertent pressure support. The presence of inadvertent pressure support may be undesirable, and should be considered when evaluating the patient's breathing efficiency and interpreting the results of bench and clinical studies comparing the work of breathing among various mechanical ventilators. In contrast, in an open-design system, gas delivered in excess of patient demand is vented to ambient, thus preventing inadvertent pressure support. Thus, from a design standpoint, the strategy for control of inspiratory and expiratory valve events involves an engineering compromise.

Conclusions

Mechanical ventilation has not been observed to blunt significantly or reverse the pathophysiology of acute respiratory insufficiency in humans and may even damage healthy or already injured lung. Appreciation of the design and functional characteristics of mechanical ventilators will enable clinicians to balance the potentially detrimental effects of ventilatory support with the desired physiological goals.

Bibliography

Giuliani, R., Mascia, L., Recchia, F., Caracciolo, A., Fiore, T., and Ranieri, V.M. (1995). Patient–ventilator interaction during synchronized intermittent mandatory ventilation. Effects of flow triggering. *American Journal of Respiratory and Critical Care Medicine*, **151**, 1–9

MacIntyre, N., Nishimura, M., Usada, Y., Tokioka, H., Takezawa, J., and Shimada, Y. (1990). The Nagoya Conference on system design and patient–ventilator interactions during pressure support ventilation. *Chest*, **97**, 1463–6.

Mancebo, J., Amaro, P., Mollo, J.L., Lorino, H., Lamaire, F., and Brochard, L. (1995). Comparison of the effects of pressure support ventilation delivered by three different ventilators during weaning from mechanical ventilation. *Intensive Care Medicine*, **21**, 913–19.

Ranieri, V.M., Mascia, L., Petruzzelli, V., Bruno, F., Brienza, A., and Giuliani, R. (1995). Inspiratory effort and measurement of dynamic intrinsic PEEP in COPD patients: effects of ventilator triggering systems. *Intensive Care Medicine*, **21**, 896–903.

Sassoon, C.S., Del Rosario, N., Fei, R., Rheeman, C.H., Gruer, S.E., and Mahutte, C.K. (1994). Influence of pressure- and flow-triggered synchronized intermittent mandatory ventilation on inspiratory muscle work. *Critical Care Medicine*, **12**, 1933–41.

18.2.6 Respiratory rate

G. J. Wrathall

Key messages

- Many patients can be safely mechanically ventilated at respiratory rates similar to their spontaneous rate, but alteration may be required depending upon the following.
- Acceptable $Pa\text{CO}_2$: eucapnia may require high respiratory rates and so care must be taken to avoid ventilator-induced lung injury.
- Metabolic rate: an increase will require a higher alveolar minute ventilation, possibly by respiratory rate adjustment.
- Physiological dead-space: this governs the amount of the tidal breath that results in alveolar ventilation.
- Tidal volume: together with physiological dead-space and respiratory rate, this determines alveolar minute ventilation.
- Mode of ventilation and level of spontaneous respiration: increasing spontaneous ventilation will often allow a reduction in the respiratory rate set.
- Respiratory time constant: this will indicate the risk of the occurrence of auto-PEEP which can be particularly troublesome at higher respiratory rates with shortened expiratory times.

Spontaneous ventilation

The resting spontaneous respiratory rate depends on age, height, weight, airway resistance, and total thoracic compliance. Respiratory frequency falls as puberty is approached, and then rises again in old age (Kacmarek and Venegas 1987).

In healthy adults the respiratory rate is normally 12 to 16 breaths/min as this results in the least work of breathing.

Mechanical ventilation

When a patient is mechanically ventilated, the process of adjusting the respiratory rate (and other variables) is both continuous and repetitive. Alterations in arterial blood gas analysis indicating a respiratory acidosis or alkalosis will often require a change in ventilator settings (Slutsky 1994). Spontaneous tachypnea, agitation, panic, excessive perspiration, tachycardia, and hypertension are all signs that the ventilation of the patient requires reassessing, although there are other causes that may require attention.

A large proportion of adults can be effectively ventilated with a respiratory rate of between 8 and 12 breaths/min, assuming that an appropriate tidal volume is being delivered (Kacmarek and Venegas 1987). However, this simplistic approach will require alteration in many critically ill patients, and the factors that guide the clinician in these changes are discussed below.

Factors affecting respiratory rate

Acceptable $P\text{aCO}_2$

Many clinicians aim for a normal CO_2 tension (4.3–6.5 kPa (32–49 mmHg)) in the arterial blood. A reduction in respiratory rate at constant tidal volume, physiological dead-space, and CO_2 production leads to a rise in $P\text{aCO}_2$. If the CO_2 tension is too low, this helps to restore it to normal. A further reduction in respiratory rate at this point leads to hypercapnia, which may be acceptable if maintaining eucapnia would result in a high possibility of ventilator-induced lung injury. This approach ('permissive hypercapnia') is sometimes used in patients with acute lung injury (Slutsky 1994). Care is required to avoid a significant respiratory acidosis.

In patients with acute on chronic respiratory failure who have a metabolic compensation for hypercapnia, mechanical ventilation should aim to maintain $P\text{aCO}_2$ at the normal level (prior to acute deterioration) for the patient (Kacmarek and Venegas 1987). This often means adjusting the respiratory rate until the pH of the arterial blood is within normal limits.

Metabolic rate

An increased metabolic rate (e.g. pyrexia or thyrotoxicosis) leads to increased oxygen consumption and CO_2 production. Maintaining eucapnia requires an increase in alveolar ventilation. If tidal volume and physiological dead-space remain the same, an increase in respiratory rate will be required.

Physiological dead-space

If physiological dead-space increases (e.g. chronic pulmonary disease), the tidal volume available for gas exchange will be reduced. An increase in respiratory rate, by increasing alveolar ventilation, may restore eucapnia.

Tidal volume

Alterations in tidal volume will require a change in respiratory rate to maintain constant arterial CO_2 tension. A reduction in tidal volume will necessitate an increase in respiratory rate, assuming that the physiological dead-space is not reduced by a similar amount.

Relation between factors

These factors are interrelated, as can be seen by the explanatory text. Their precise relationship can be seen from the following equations (Kacmarek and Venegas 1987):

$$V\text{CO}_2 = (\dot{V}_A \times P\text{aCO}_2)/P_{atm}$$

where $V\text{CO}_2$ is CO_2 production, $\dot{V}_A$ is alveolar ventilation, $P\text{aCO}_2$ is arterial CO_2 tension, and P_{atm} is atmospheric pressure, and

$$\dot{V}_A = f(V_T - V_D)$$

where f is respiratory rate, V_T is tidal volume, and V_D is physiological dead-space.

Mode of ventilation and level of spontaneous respiration

Control mode

The respiratory rate set on the ventilator is that which the patient receives. The rate should not be reduced to 'encourage' the patient to start breathing, as spontaneous respiration is not possible. A change of mode is required.

Assist–control mode

The patient receives a tidal volume when he or she attempts to take ('trigger') a spontaneous breath. As the patient is triggering every breath, theoretically there is no requirement to set a respiratory rate. However, many ventilators provide back-up ventilation if the patient becomes apneic for a set time period. This back-up rate should be set at two to five breaths less than the spontaneous rate to allow adequate ventilation if the patient remains apneic, and also to allow the opportunity for spontaneous respiration to recommence.

Intermittent mandatory ventilation

A set number of mandatory breaths are delivered each minute. In addition, the possibility of spontaneous breaths is ensured. Synchronized intermittent mandatory ventilation prevents 'stacking' of patient and ventilator breaths on top of each other. With both these modes, the respiratory rate is normally set at a level that would provide full ventilatory support if the patient were making no respiratory effort, and then reduced to a level at which the patient is able to respire spontaneously, appears comfortable, and does not develop a respiratory acidosis. The rate is then gradually reduced as the weaning process is continued, so that the patient is required to take an increasing proportion of spontaneous to mandatory breaths. At less than 6 breaths/min, the work of breathing may increase because the majority of ventilation is spontaneous through the resistance of a demand valve and circuit (Brown 1994). Therefore the mandatory breath rate is not usually reduced beyond this point.

Pressure support ventilation

Normally, expiration is cycled when the inspiratory flow rate falls below a set level (typically 25 per cent of peak inspiratory flow) and therefore a respiratory rate is not set on the ventilator. However, some ventilators will not allow a lengthening of inspiratory time beyond an inspiration-to-expiration (I:E) ratio of 4:1 (which theoretically could occur if the patient were to inspire at a slow and steady flow rate). Under these circumstances, inspiration is terminated. Therefore setting a high respiratory rate can lead to cycling to expiration before the patient has inspired fully.

Respiratory time constant

The lung and thorax should return to their resting position at the end of expiration or alveolar pressure will be higher than end-expiratory pressure; this difference is termed 'auto-PEEP'. This increases the risk of ventilator-induced lung injury (Marcy and Marini 1991) and may make triggering of the ventilator more difficult (Slutsky 1994). The time taken for an alveolus to empty is reflected in the 'respiratory time constant', which is equal to resistance multiplied by compliance and is measured in seconds. An increase in resistance (e.g. chronic obstructive pulmonary disease or acute asthma) or compliance (e.g. emphysema) makes the development of auto-PEEP more likely. A reduction in expiratory time, which would occur if the respiratory rate were increased without altering the I:E ratio, also increases the possibility of the development of auto-PEEP.

In patients with conditions that may prolong the time constant, a reduction in respiratory rate to below 8 breaths/min may be required to minimize auto-PEEP (Kacmarek and Venegas 1987). Restrictive lung disorders, with low compliance and normal resistance, may be ventilated at rates higher than this (sometimes more than 20 breaths/min) with a low risk of development of auto-PEEP (Kacmarek and Venegas 1987).

References

Brown, B.R. (1994). Understanding mechanical ventilation: indications for and initiation of therapy. *Journal of the Oklahoma State Medical Association*, **87**, 353–7.

Kacmarek, R.M. and Venegas, J. (1987). Mechanical ventilatory rates and tidal volumes. *Respiratory Care*, **32**, 466–75.

Marcy, T.W. and Marini, J.J. (1991). Inverse ratio ventilation in ARDS. Rationale and implementation. *Chest*, **100**, 494–504.

Slutsky, A.S. (Chairman) (1994). Consensus conference on mechanical ventilation—January 28–30, 1993 at Northbrook, Illinois, USA. Parts 1 and 2. *Intensive Care Medicine*, **20**, 64–79, 150–62.

18.2.7 Tidal and minute volume

G. J. Wrathall

Key messages

- Most critically ill patients should be ventilated with a tidal volume of 5 to 15 ml/kg.
- Acceptable $P\text{aCO}_2$: the risk of ventilator-induced lung injury due to a large tidal volume may alter the acceptable $P\text{aCO}_2$ range.
- Metabolic rate: alterations in alveolar minute ventilation to adjust for metabolic rate changes can often be accomplished by tidal volume changes.
- Physiological dead-space: this changes on commencement of positive-pressure ventilation, and its relationship with tidal volume is not constant.
- Respiratory rate: this also governs alveolar minute ventilation.
- Risk of ventilator-induced lung injury: both plateau pressure and auto-PEEP should be minimized.
- Oxygenation: atelectasis due to inadequate tidal volume should be avoided.
- Circuit compressible volume: this will reduce the delivered tidal volume.
- Cardiovascular stability: increases in intrathoracic pressure with increasing tidal volume may lead to hemodynamic instability.
- Mode of ventilation: tidal volume cannot be set directly in pressure-targeted or spontaneous ventilation.
- Patient comfort: some patients feel more comfortable with a tidal volume at the upper end of the aforementioned range.

Spontaneous ventilation

Normal resting tidal volume is dependent on age, height (to which it is directly related), weight, total thoracic compliance, and airways resistance. Tidal volume increases as puberty approaches, and then falls again in old age (Kacmarek and Venegas 1987).

Healthy adults have a resting tidal volume of approximately 5 ml/kg of lean body weight.

Mechanical ventilation

Constant reassessment of a mechanically ventilated patient is required to select appropriate ventilator settings. Minute volume is the product of respiratory rate and tidal volume. Evidence of inadequate ventilation includes arterial blood gas derangement and clinical signs such as agitation, tachypnea, tachycardia, hypertension, or a poor ventilator–patient interplay ('fighting the ventilator').

A tidal volume of 10 to 15 ml/kg is commonly used, and this is adequate for many patients with normal lungs. However, this tidal volume is not always appropriate, and volumes of 5 ml/kg or even less are sometimes employed. The reasons for this are discussed below.

Factors affecting tidal volume

Acceptable $P\text{aCO}_2$

Maintenance of eucapnia is sometimes only possible with strategies that increase the risk of ventilator-induced lung injury, including the use of a tidal volume that causes alveolar overdistension. In these situations (e.g. acute lung injury), the risk can be reduced by decreasing tidal volume and therefore peak alveolar pressure. If the respiratory rate, physiological dead-space, and CO_2 production remain constant, $P\text{aCO}_2$ will rise. This may be appropriate if there is no significant respiratory acidosis (Slutsky 1994).

Metabolic rate

If the metabolic rate decreases, $Paco_2$ will decrease if alveolar ventilation remains constant. A reduction in tidal volume (if respiratory rate and physiological dead-space remain the same) may return CO_2 tension to normal.

Physiological dead-space

In normal lungs respiring spontaneously, the physiological dead-space (V_D) is approximately 2.2 ml/kg. The tidal volume (V_T) is significantly larger such that the ratio V_D/V_T is between 0.2 and 0.4. Once the lungs become ventilated under positive pressure, deterioration of ventilation–perfusion matching results in an increase in V_D/V_T to approximately 0.5. In pulmonary disease this ratio can be as high as 0.8 (Slutsky 1994). As V_D/V_T increases, the effectiveness of ventilation reduces. An increase in tidal volume may restore alveolar ventilation. However, the relationship between V_D/V_T and tidal volume is not constant. At small tidal volumes, V_D/V_T is high and it falls as tidal volume is increased. At the point where increases in tidal volume cause airway distension, V_D/V_T remains constant until the size of the tidal volume causes a rapid increase in alveolar pressure (when the flat segment of the compliance curve is reached). This causes a reduction in venous return, deterioration in ventilation–perfusion matching, and an increase in V_D/V_T (Kacmarek and Venegas 1987). In normal ventilated lungs, V_D/V_T is lowest when the tidal volume is 10 to 15 ml/kg.

Respiratory rate

If metabolic rate and physiological dead-space remain constant, a reduction in respiratory rate requires an increase in tidal volume to maintain eucapnia.

Relation between factors

The former points are interrelated, as can be seen by the explanatory text. Their precise relationship is shown by the following equations (Kacmarek and Venegas 1987):

$$Vco_2 = (\dot{V}_A \times Paco_2)/P_{atm}$$

where Vco_2 is the CO_2 production, $\dot{V}_A$ is alveolar ventilation, $Paco_2$ is arterial CO_2 tension, and P_{atm} is atmospheric pressure, and

$$\dot{V}_A = f(V_T - V_D)$$

where f is the respiratory rate, V_T is the tidal volume, and V_D is the physiological dead-space.

Ventilator-induced lung injury

Alveolar overdistension increases the likelihood of ventilator-induced lung injury and should be avoided. Limiting the tidal volume to within the steep segment of the compliance curve reduces this possibility (Slutsky 1994), as alveolar volume is difficult to measure. The plateau pressure reflects alveolar pressure, and many clinicians attempt to limit it to below approximately 35 cmH_2O (Slutsky 1994). However, if the compliance of the chest wall is reduced (e.g. severe kyphoscoliosis), a plateau pressure higher than this value is less likely to cause ventilator-induced lung injury.

The lung should be allowed to return to its resting position between each breath. Failure for this to occur leads to higher alveolar end-expiratory pressure than that selected. This auto-PEEP increases the risk of ventilator-induced lung injury (Marcy and Marini 1991). The gas volume V_t above functional residual capacity left in the lung at time t after the start of expiration is calculated by the formula

$$V_t = V_0 \exp(-t/RC)$$

where V_0 is the volume at time zero and RC is the respiratory time constant (the product of resistance and compliance). Therefore a rise in tidal volume increases the chance of development of auto-PEEP if expiratory time is unaltered.

Oxygenation

Selecting a tidal volume of 12 to 15 ml/kg may reduce atelectasis (Shapiro 1994), improving ventilation–perfusion matching and enhancing oxygenation. However, an alternative method may be to use positive end-expiratory pressure (PEEP) to prevent alveolar collapse.

Circuit compressible volume

The volume of gas within the ventilator circuit, which may include a humidifier, will be compressed by the positive pressure during inspiration. Therefore part of the inspiratory tidal volume set on the ventilator is 'lost', and this loss may be increased by compliant circuit tubing. The amount of tidal volume that is lost to this compression is proportional to the change in pressure during the respiratory cycle (peak inspiratory pressure minus positive end-expiratory pressure), and is approximately 3 to 5 ml for a pressure change of 1 cmH_2O (Kacmarek and Venegas 1987). Increases in circuit volume (e.g. magnetic resonance imaging suites) can increase the volume loss. Apart from the reduction in ventilation, this volume loss makes auto-PEEP measurement using an expiratory hold inaccurate.

In circumstances where a large compressible volume loss is possible, exhaled tidal volume should be measured as close to the patient's airway as possible to obtain accurate tidal and minute volumes.

Cardiovascular stability

Intrathoracic pressure rises as tidal volume increases. As the mean airway pressure rises, venous return is impaired and cardiac output falls (Slutsky 1994). Mean airway pressure also depends on respiratory rate, inspiratory-to-expiratory (I:E) ratio, and positive end-expiratory pressure. Therefore strategies to reduce mean airway pressure (including tidal volume reduction) may be required in hemodynamically unstable patients.

Mode of ventilation

With volume-targeted ventilation, tidal volume is reliably controlled. However, with pressure-targeted ventilation tidal volume depends on respiratory impedance and the patient's spontaneous respiratory pattern as well as ventilatory pressures (Slutsky 1994). Therefore it cannot be adjusted directly.

In ventilatory modes which allow spontaneous breaths, the size of these breaths is also not directly controllable, except in the assist–control mode where each spontaneous breath is of a set tidal volume (if the ventilator is volume targeted).

Patient comfort

Mechanically ventilated patients may experience dyspnea despite apparently adequate ventilation. This is not uncommon in patients with neuromuscular disorders (e.g. myasthenia gravis) and can sometimes be overcome by the use of tidal volumes at the upper end of the 10- to 15-ml/kg range (Slutsky 1994).

References

Kacmarek, R.M. and Venegas, J. (1987). Mechanical ventilatory rates and tidal volumes. *Respiratory Care*, **32**, 466–75.

Marcy, T.W. and Marini, J.J. (1991). Inverse ratio ventilation in ARDS. Rationale and implementation. *Chest*, **100**, 494–504.

Shapiro, B.A. (1994). A historical perspective on ventilator management. *New Horizons*, **2**, 8–18

Slutsky, A.S. (Chairman) (1994). Consensus conference on mechanical ventilation—January 28–30, 1993 at Northbrook, Illinois, USA. Parts 1 and 2. *Intensive Care Medicine*, **20**, 64–79, 150–62.

18.2.8 Inspiratory time

G. J. Wrathall

Key messages

- Inspiratory time governs the inspiratory-to-expiratory (I:E) ratio and inspiratory flow rate. In mechanically ventilated patients, a 'standard' inspiratory time of 0.8 to 1.5 s should be adjusted according to the following.

  Oxygenation: improvement in oxygenation may follow lengthening of the inspiratory time to increase mean airway pressure.

  Risk of ventilator-induced lung injury: shortening the expiratory time increases the possibility that auto-PEEP will develop.

  Cardiovascular stability: prolonged inspiration can cause hemodynamic instability by impeding venous return.

  Level of spontaneous respiration and mode of ventilation: lengthened inspiratory time is usually poorly tolerated by patients who are breathing spontaneously or lightly sedated.

- Inspiratory waveform can only be adjusted in volume-targeted ventilation. The choice depends upon the following.

  Patient comfort: square or sinusoidal waveforms are often well tolerated.

  Oxygenation: a descending waveform may improve $Pa\text{O}_2$.

Spontaneous ventilation

The normal resting inspiratory time in adults is 1 to 1.4 s. Thus, with a respiratory rate of 12 to 16 breaths/min, the inspiratory-to-expiratory (I:E) ratio is normally between 1:1.5 and 1:3. Longer expiration is required to avoid using the accessory muscles of respiration for expiration. The possibility of 'gas trapping' in alveoli with long respiratory time constants is also reduced if these alveoli have longer to empty.

Mechanical ventilation

At a constant respiratory rate and tidal volume, the adjustment of inspiratory time governs the I:E ratio and the inspiratory flow rate (the latter also depends on the inspiratory waveform). Inspiratory time may be controlled directly or indirectly depending on the type of ventilator. In indirect methods the inspiratory flow rate or I:E ratio is adjusted. Some ventilators allow direct control of one of these variables, showing the corresponding change in the value of the variable that is not adjusted. If an inspiratory hold, pause, or plateau is required, the time that this takes is part of the inspiratory time.

Many patients undergoing mechanical ventilation can be adequately managed using an inspiratory time similar to their resting value, i.e. 0.8 to 1.5 s. However, there are factors that may result in changes to this inspiratory time being necessary for optimal ventilation.

Oxygenation

Improvement in oxygenation can sometimes be achieved by increasing the mean airway pressure. This is often the case where collapsed alveoli may be recruited by this increased pressure (e.g. acute lung injury) (Marcy and Marini 1991). Prolonging the inspiratory time elevates the mean airway pressure and is often used for this effect. If the inspiratory time is lengthened sufficiently, the I:E ratio is reversed. This is termed 'inverse ratio ventilation'. Where the ventilation is pressure targeted, the square pressure waveform maintains mean airway pressure but limits peak alveolar pressure. With volume-targeted ventilation the mean airway pressure will be lower as the pressure waveform is not square, but it can be elevated by increasing the inspiratory plateau although care must be taken as the peak inspiratory pressure will rise (Marcy and Marini 1991). Although inverse ratio ventilation often increases $Pa\text{O}_2$, there is no conclusive evidence that it improves outcome (Marcy and Marini 1991; Slutsky 1994). In addition, it has significant risks (see below).

Ventilator-induced lung injury

If the lung is unable to return to its resting state during expiration, the alveolar pressure remains higher than the airway pressure measured by the ventilator. This auto-PEEP due to gas trapping increases the risk of alveolar overdistension and ventilator-induced lung injury (Marcy and Marini 1991). The gas volume (V_t) above functional residual capacity left in the lung at time t after the start of expiration is calculated using the formula

$$V_t = V_0 \exp(-t/RC)$$

where V_0 is the volume at time zero and RC is the respiratory time constant (the product of resistance and compliance). Therefore as t is reduced, i.e. inspiratory time is lengthened, the possibility of gas trapping increases. Thus inverse ratio ventilation is associated with a high risk of auto-PEEP development. The above formula also shows that as the respiratory time constant increases, the risk of gas trapping also increases. Conditions with increased compliance (e.g. emphysema) and/or increased resistance (e.g. chronic obstructive pulmonary disease or acute asthma) lead to a prolonged respiratory time constant. A reduction in inspiratory time to allow adequate time for alveolar emptying during expiration may reduce auto-PEEP (Slutsky 1994). However, this reduction in inspiratory time may lead to increased peak inspiratory pressure in volume-targeted ventilation or reduced tidal volume in pressure-targeted ventilation.

Cardiovascular stability

Lengthening inspiratory time to increase mean airway pressure can lead to a fall in cardiac output due to a reduction in venous return. Although $Pa\text{O}_2$ may increase, the fall in cardiac output results in an unpredictable change in oxygen delivery to the tissues (Slutsky 1994). The inspiratory time may need to be reduced again to give an acceptable balance between oxygenation and hemodynamic stability.

Level of spontaneous respiration and mode of ventilation

Prolonging the inspiratory time makes ventilated patients increasingly uncomfortable. To tolerate this, they commonly require sedation and sometimes paralysis if spontaneous respiratory efforts are interfering with adequate ventilation (Marcy and Marini 1991). Therefore prolonged inspiratory times are not usually employed when significant spontaneous ventilation is occurring (e.g. assist–control mode or synchronized intermittent mandatory ventilation).

Inspiratory waveform

Pressure-targeted ventilation

The waveform of the inspiratory flow in this mode of ventilation follows an exponential decelerating pattern. This may help to recruit alveoli with long time constants, particularly if the inspiratory time is lengthened. However, the onset of inspiration may be associated with large shear forces acting on the alveoli owing to the rapid increase in pressure (Marcy and Marini 1991).

Volume-targeted ventilation

On many ventilators in current use, the waveform of the inspiratory flow can be altered. There are four alternatives.

1. Square waveform: the flow rises rapidly to a peak that is maintained during the length of inspiration, and the volume–time waveform is an ascending ramp. Patients who are breathing spontaneously (e.g. assist–control mode or synchronized intermittent mandatory ventilation) find this waveform comfortable.
2. Sinusoidal waveform: the flow–time and volume–time waveforms are both sinusoidal but are out of phase with each other. In clinical use, this waveform is very similar to the square waveform.
3. Ascending waveform: the flow rises constantly through inspiration and the volume–time waveform ascends exponentially. There is no evidence to support the use of this waveform (Slutsky 1994).
4. Descending waveform: the flow rises rapidly to a peak and then falls constantly or exponentially through inspiration. The mean airway pressure is higher and the peak airway pressure lower than a corresponding cycle using the ascending waveform. Therefore it can be used (sometimes together with inverse ratio ventilation) to improve oxygenation by alveolar recruitment.

References

Marcy, T.W. and Marini, J.J. (1991). Inverse ratio ventilation in ARDS. Rationale and implementation. *Chest*, **100**, 494–504.

Slutsky, A.S. (Chairman) (1994). Consensus conference on mechanical ventilation—January 28–30, 1993 at Northbrook, Illinois, USA. Parts 1 and 2. *Intensive Care Medicine*, **20**, 64–79, 150–62.

18.2.9 Positive end-expiratory pressure

P. M. Suter

Key messages

- A certain amount of PEEP (or CPAP in the spontaneously breathing patient) can be considered 'physiological' up to the time of extubation in all forms of acute lung failure.
- In severe lung edema most alveolar spaces are collapsed in paravertebral areas and there is a biphasic *P–V* relationship, corresponding to an important recruitment occurring at a certain level of PEEP in these regions. This 'critical' level of PEEP corresponds to the lower inflection point of the *P–V* curve (P_{flex}).
- PEEP is not solely responsible for the recruitment of alveolar space; tidal volume and end-inspiratory airway pressure also contribute to this effect.
- At zero or low levels of PEEP, mechanical tidal ventilation is preferentially distributed to non-dependent lung zones. With increasing PEEP, more tidal volume is shifted to dependent regions in the supine patient.
- When alveolar collapse occurs during expiration, the next positive pressure breath must 'snap open' this alveolar unit before gas can enter. Repetitive collapsing and snapping open appears to potentiate the shear stress and parenchymal injury from mechanical ventilation. PEEP seems to decrease the severity of ventilator-induced lung damage.

Introduction

Positive end-expiratory pressure (PEEP) is frequently used during mechanical ventilation to improve pulmonary gas exchange. The mechanism of its effects are similar to those obtained by using continuous positive airway pressure (CPAP) during spontaneous breathing, i.e. providing an intrapulmonary pressure above atmospheric to allow opening and/or prevention of closure of terminal bronchioli and alveoli. Thus the functional gas exchange area in the lung is increased, systemic arterial oxygenation is improved, and alveolar dead-space fraction is decreased. Many different types of acute pulmonary failure, resulting in impaired pulmonary gas exchange to such an extent that mechanical ventilatory support is required, are associated with a decrease in functional residual capacity (FRC). The most frequent causes of loss of FRC are the formation of micro- or macroatelectasis due to absence of periodic deep breathing and coughing, and interstitial lung edema compromising small airway and alveolar stability. These phenomena are not distributed evenly throughout the lung tissue, but are clearly gravity dependent.

How does PEEP improve systemic oxygenation?

Prevention and treatment of small airway and alveolar instability leading to atelectasis formation

Anesthesia, intubation, and muscle paralysis cause atelectasis in dependent lung zones. Intermittent tidal volumes that are larger than normal (sigh), and/or the use of PEEP, and/or postural changes can prevent or reverse closure of peripheral airways and collapse of alveoli. This results in an improvement of regional ventilation–perfusion ratios ($\dot{V}/\dot{Q}$) in dependent lung zones and hence improvement of systemic arterial oxygenation. However, when too much PEEP is used, overdistension of non-dependent lung zones and areas of high $\dot{V}/\dot{Q}$ can be created, thereby increasing alveolar dead-space. The most appropriate level of PEEP results in an acceptable balance between sufficient recruitment and tolerable overdistension in other lung areas. The dead-space effect of overdistended areas can be enhanced by PEEP-induced decreases in cardiac output, creating very high $\dot{V}/\dot{Q}$ ratios by increasing $\dot{V}$ and lowering $\dot{Q}$ by lung inflation.

In acute lung injury such as acute respiratory distress syndrome, small airway and alveolar instability are magnified by the important additional effects of interstitial edema, gravitational characteristics of ventilation distribution, and global or regional inflammatory changes.

For all the above reasons, a certain amount of PEEP (or CPAP in the spontaneously breathing patient) can be considered 'physiological' up to the time of extubation in all forms of acute lung failure. Breathing through a T-piece during weaning is associated with a lower FRC than unassisted breathing after extubation; this is because of the impossibility of closing the vocal cords, owing to the presence of the endotracheal tube, and producing an efficient cough.

Massive edema creating areas of low $\dot{V}/\dot{Q}$

Pulmonary edema, acute respiratory distress syndrome, severe pneumonia, and lung contusion are characterized by gravity-dependent massive parenchymal edema with or without additional tissue consolidation. FRC is severely impaired and, for many years, its increase by PEEP has been considered essential for improved lung function (Falke *et al.* 1972). PEEP increases lung volume and the gas exchange surface in two different ways: firstly, by recruitment of collapsed areas and, secondly, by distension of already open regions. The relative importance of both these effects can be assessed by the pressure–volume (*P–V*) curve. Recruitment of collapsed regions results in a curvilinear *P–V* relationship (concave towards the volume axis) with a lower inflection point and an increasing compliance, i.e. a biphasic curve. When open units are distended by PEEP (or tidal volume), the *P–V* curve is linear, at least up to the point where overdistension begins. This latter area is characterized by the upper inflection point (convex towards the volume axis).

Typical *P–V* curves for a normal subject, an emphysematous patient, and a patient with acute respiratory distress syndrome are presented in Fig. 1 (Suter *et al.* 1975). In non-dependent lung areas, i.e. in the parasternal regions in the supine position, PEEP produces essentially a linear increase in gas volume and $\dot{V}/\dot{Q}$; most units are open at zero PEEP in these regions. In severe lung edema or acute respiratory distress syndrome, most alveolar spaces are collapsed in paravertebral areas and there is a biphasic *P–V* relationship, corresponding to an important recruitment occurring at a certain level of PEEP in these regions. This 'critical' level of PEEP corresponds to the lower inflection point of the *P–V* curve (P_{flex}). This recruitment phenomenon increases progressively from the parasternal to the paravertebral areas, and the 'opening' pressure P_{flex} increases linearly along this axis, being highest in dependent lung zones. These effects of PEEP are most striking in the early phases of acute respiratory distress syndrome when lung edema is predominant. In pneumonia or later phases of acute respiratory distress syndrome, when fibrotic changes and remodeling of lung structures are more important, other mechanisms play essential roles in determining the distribution of ventilation with and without PEEP. In addition, PEEP is not solely responsible for the recruitment of alveolar space; tidal volume and end-inspiratory airway pressure also contribute to this effect. Furthermore, postural changes, which are frequently used in the current management of such patients, also influence gravity-dependent mechanisms and their localization, and thus the effects of PEEP and the level of P_{flex} in different lung regions. Lung size, i.e. the distance between the sternum and the vertebrae, is another important variable influencing regional lung weight and the pressure required to counteract compression airway closure and atelectasis.

PEEP redistributes tidal volume

At zero or low levels of PEEP, mechanical tidal ventilation is preferentially distributed to non-dependent lung zones (Gattinoni *et al.* 1995). With increasing PEEP, more tidal volume is shifted to dependent (i.e. paravertebral) regions in the supine patient. In acute lung injury due to edema, pneumonia, or acute respiratory distress syndrome, low levels of PEEP (0–10 cmH_2O) produces less recruitment of collapsed alveolar space than tidal volume. However, at higher PEEP (12–20 cmH_2O) recruitment becomes important.

PEEP and lung tissue injury by mechanical ventilation

When alveolar collapse occurs during expiration, the next positive pressure breath must 'snap open' this alveolar unit before gas can enter. Repetitive collapsing and snapping open appears to potentiate the shear stress and parenchymal injury from mechanical ventilation. PEEP seems to decrease the severity of ventilator-induced lung damage (Dreyfuss and Saumon 1993) at both low and high tidal volumes. However, PEEP can increase end-inspiratory lung distension and thus the potential danger of alveolar rupture and barotrauma. Outcome in patients with acute respiratory failure is

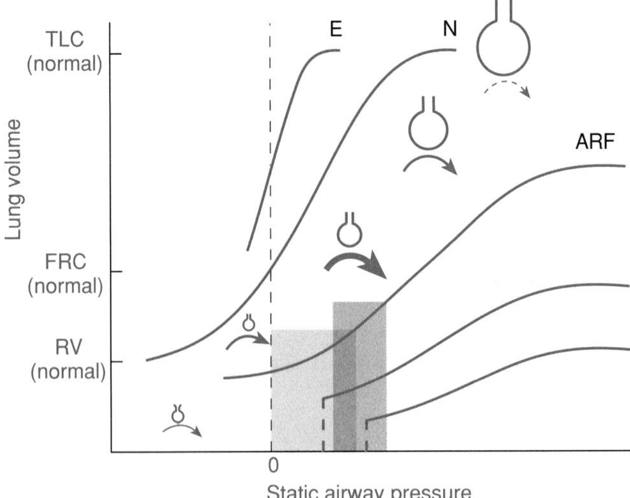

Fig. 1 *P–V* curves for the respiratory system in normal lungs (N) and in those of patients with emphysema (E) and acute respiratory failure (ARF) of the early acute respiratory distress syndrome type: FRC, functional residual capacity; RV, residual volume; TLC, total lung capacity.The shaded areas represent the *P–V* relation during an identical tidal volume at two different PEEP levels. Lower and upper infection zones can be recognized on the *P–V* curve corresponding to acute pulmonary failure.

potentially improved when low tidal volumes and PEEP at a sufficient level to keep the lung open are used (Slutsky 1994).

Guidelines for clinical practice

Choosing the level of PEEP and monitoring

The level of PEEP should be chosen according to the clinical situation and its effects on gas exchange (Table 1). Blood gas measurements show a significant improvement in arterial oxygenation when functional gas exchange units are opened and participate in gas exchange. However, alveolar overdistension of functional alveolar units is more difficult to recognize. Indeed, overdistension typically produces an increase in dead-space ventilation by an excessively high $\dot{V}/\dot{Q}$ ratio but rarely causes any fall in $P\text{a}O_2$. The dead-space fraction of tidal ventilation (V_D/V_T) also depends on tidal volume and cardiac output.

Alveolar recruitment and overdistension can be assessed more precisely by recording a static *P–V* curve of the respiratory system (Suter *et al.* 1975). In general this allows identification of the lower and upper inflection points (P_{flex}). The lower inflection point corresponding to alveolar recruitment can easily be identified in early phases of acute lung injury and in lung edema. This inflection point is generally not found in later phases of acute respiratory distress syndrome, in pneumonia, or during acute exacerbations of chronic pulmonary dysfunction. PEEP should be set above the lower inflection point to improve pulmonary gas exchange and mechanics, and also to avoid potential lung damage by stretch injury exacerbated by opening and closing of lung units during tidal ventilation. The upper inflection area of the *P–V* curve corresponds to the limit above which additional volume or pressure application results in overdistension and possible (high-pressure) lung damage. Recording a static *P–V* curve requires a heavily sedated or paralyzed patient, but it also allows determination of the compliance of the total respiratory system.

Table 1 Clinical guidelines for setting PEEP

1. For optimizing pulmonary gas exchange and systemic oxygenation, set PEEP above the lower inflection point of the pressure curve of the respiratory system
   - To recruit a majority of collapsed gas exchange units
   - But avoid significant falls in cardiac output
   - And adapt tidal volume (lower at high PEEP, higher at low PEEP)
2. For minimizing ventilator-induced lung injury, set PEEP above lower inflection point
   - To decrease stretch injury due to tidal opening and collapsing
   - But avoid reaching upper inflection point
   - To prevent overdistension by lowering PEEP or tidal volume (Fig. 1)
3. Frequently used levels
   - Severe lung edema, early ARDS, 7–10 cmH_2O
   - Lung contusion, severe early ARDS, 10–20 cmH_2O
   - Pneumonia, acute exacerbation of chronic lung failure, late ARDS, 0–10 cmH_2O
4. Frequent reassessment is mandatory, i.e. once or twice daily
5. Monitoring
   - $P\text{a}O_2$, $S\text{a}O_2$
   - Plateau airway pressure
   - $P\text{a}CO_2$, minute ventilation (as an estimate of changes in V_D/V_T)
6. Alternative to increasing (or high) PEEP: prolonged inspiratory time (high I : E ratio)

ARDS, acute respiratory distress syndrome; I : E, inspiratory-to-expiratory.

The concept of 'optimal compliance' (Suter *et al.* 1975) is similar to choosing PEEP above the lower inflection point; tidal ventilation in this range ensures recruitment of collapsed areas and avoids overdistension. Assessments of respiratory mechanics must be repeated regularly as they change due to differences in tidal volume, the amount and location of interstitial edema, changes in tissue elasticity and consolidation, and parenchymal remodeling.

Trade-offs between PEEP, tidal volume, and FiO_2

In a given clinical situation, the potentially harmful effects of tidal volume, FiO_2, and PEEP at their respective values must be considered (Slutsky 1994). For instance, when a tidal volume of 900 ml and a

PEEP of 15 cmH_2O at a fractional inspired oxygen (FiO_2) of 1.0 result in arterial blood gas values of pH 7.43, $Paco_2$ = 5.7 kPa (43 mmHg), and Pao_2 = 7.5 kPa (56 mmHg), and an airway plateau pressure of 53 cmH_2O, the tidal volume should first be lowered to decrease plateau pressure at the (acceptable) risk of an increasing $Paco_2$ and a decrease in arterial pH.

The trade-offs between PEEP and FiO_2 may be more difficult to justify and to decide upon. In a situation such as that described above, a test at an FiO_2 of 0.9 could be performed, with careful continuous monitoring of pulse oximetry and a further blood gas analysis after 15 min. It must be stressed that a less than normal Pao_2, pH, and $Paco_2$ are usually well tolerated by patients with normal reserves in other vital organ functions, and that potential damage to lung tissue by mechanical ventilation and a high FiO_2 may be as important as discrete tissue hypoxia for ultimate survival.

References

Dreyfuss, D. and Saumon, G. (1993). Role of tidal volume, FRC, and end-inspiratory volume in the development of pulmonary edema following mechanical ventilation. *American Review of Respiratory Diseases*, **148**, 1194–1203.

Falke, K., Pontoppidan, H., Kumar, A., Leith, D.E., Geffin, B., and Laver, M.B. (1972). Ventilation with end-expiratory pressure in acute lung disease. *Journal of Clinical Investigation*, **51**, 2315–23.

Gattinoni, L., Pelosi, P., Crotti, S., and Valenza, F. (1995). Effects of positive end-expiratory pressure on regional distribution of tidal volume and recruitment in adult respiratory distress syndrome. *American Journal of Respiratory and Critical Care Medicine*, **151**, 1807–14.

Slutsky, A.S. (Chairman) (1994). Consensus conference on mechanical ventilation—January 28–30, 1993 at Northbrook, Illinois, USA. Parts 1 and 2. *Intensive Care Medicine*, **20**, 64–79, 150–62.

Suter, P.M., Fairley, H.B., and Isenberg, M.D. (1975). Optimum end expiratory airway pressure in patients with acute pulmonary failure. *New England Journal of Medicine*, **292**, 284–9.

18.2.10 Volume-controlled mechanical ventilation

J. J. Marini

Key messages

- Volume-targeted modes guarantee flow, and consequently the volume provided to the circuit in the allowed inspiratory time (tidal volume), at the expense of variable airway pressures.
- If volume-cycled ventilation is used, the physician may select either tidal volume and flow delivery pattern or flow delivery pattern and minimum minute ventilation.
- As a general rule, the ventilator's average inspiratory flow should be approximately four times the minute ventilation. Peak flow should be set 20 to 30 per cent above this average value when a decelerating waveform is used.

Introduction

Positive-pressure inflation can be achieved with machines which control either of the two determinants of ventilating power, i.e. pressure or flow, and terminate inspiration according to pressure, flow, volume, or time limits. However, the flow and pressure waveforms cannot be controlled simultaneously because pressure is developed as a function of flow and the impedance to breathing. The latter is determined by the uncontrolled resistance and compliance. Traditionally, the machine-aided cycles of assist–control and synchronized intermittent mandatory ventilation have been flow controlled and volume targeted.

Volume-targeted versus pressure-targeted ventilation

On initiating mechanical ventilation, the physician must choose between pressure-targeted ventilation ('pressure control') or flow-controlled volume-cycled ventilation ('volume control') (Fig. 1). The fundamental difference is implicit in their names; pressure-targeted modes guarantee pressure at the expense of variable tidal volumes; volume-targeted modes guarantee flow, and consequently the volume provided to the circuit in the allowed inspiratory time (tidal volume), at the expense of variable airway pressures (Table 1). In a well-monitored apneic patient, pressure- and volume-targeted modes can be used with virtually identical effect. If pressure control is used, the targeted inspiratory pressure (above positive end-expiratory pressure (PEEP)) and inspiratory time must be selected, usually with consideration of the desired tidal volume. If volume-cycled ventilation is used, the physician may select (depending on the ventilator) either tidal volume and flow delivery pattern (waveform and peak flow) or flow delivery pattern and minimum minute ventilation (with tidal volume the resulting ratio of minute ventilation to back-up frequency). Once chosen, flow is inflexible to changing inspiratory flow demands.

Flow and tidal volumes are important variables to monitor in pressure targeting, while pressure is the monitored variable of parallel importance in volume targeting. Volume-targeted modes provide a preset volume unless a specified pressure limit is exceeded. The major advantages of volume targeting are the capacity to deliver non-varying tidal volumes (except in the presence of a gas leak or when the pressure limit is exceeded), flexibility of flow and volume adjustments, and ability to ventilate difficult patients. Unlike pressure-targeted ventilation, volume-targeted modes do not ventilate consistently unless the airway is well sealed. Furthermore, once the flow rate is set, the inflation time of the machine remains unresponsive to the patient's native cycling rhythm.

Controlled mechanical ventilation

During controlled mechanical ventilation, the machine provides a fixed number of breaths per minute and remains totally unresponsive to the patient's efforts to alter frequency. This 'lock-out' mode demands constant vigilance to make appropriate adjustments for changes in ventilatory requirements and is used only in situations where pH and/or $Paco_2$ must be tightly controlled. Most patients require sedation to ensure comfort and ablate breathing efforts. Under these conditions, assist–control has similar capability and offers additional advantages.

Table 1 Pressure-controlled versus volume-controlled ventilation

	Pressure control	Volume control
Settings	Pressure target	Tidal volume target
	Inspiratory duration	Flow rate
	Inspiratory rise rate	Flow waveform
Outcome variables	1. Tidal volume	1. Airway pressure
	2. Auto-PEEP	2. Auto-PEEP
Common variables	FiO_2	FiO_2
	PEEP	PEEP
	Mode	Mode
	Frequency	Frequency

When the respiratory system is inflated passively, the mean airway pressure (an easily measured analog of mean alveolar pressure) becomes a valuable indicator of average lung volume. Increments in mean airway pressure tend to raise pleural pressure and impede venous return. Moreover, in patients with acute lung injury, mean airway pressure tends to reflect lung size and the efficiency of arterial oxygenation.

Assist–control ventilation

During assist–control ventilation (or assisted mechanical ventilation), each inspiration triggered by the patient is powered by the ventilator. As a safety mechanism, a back-up rate is set so that a machine cycle is provided automatically if the patient does not initiate a breath within the number of seconds dictated by that frequency. Sensitivity to inspiratory effort can be adjusted to require a small or large negative-pressure deflection below the set level of end-expiratory pressure to initiate the machine's inspiratory phase. Alternatively, many of the latest generation machines can be flow triggered, initiating a cycle when a flow deficit of sufficient magnitude is sensed in the expiratory limb of the circuit relative to the inspiratory limb during the exhalation period. Unlike controlled mechanical ventilation or synchronized intermittent mandatory ventilation, changes in set machine frequency during assisted mechanical ventilation have no effect on V_E unless this back-up frequency is set sufficiently high to supersede or terminate the patient's own respiratory efforts.

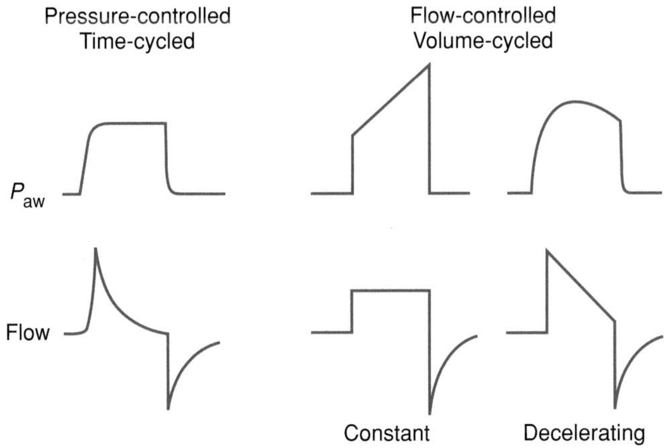

Fig. 1 Airway pressure and flow waveforms during pressure-controlled time-cycled ventilation and during flow-controlled volume-cycled ventilation delivered with constant and decelerating flow profiles. (Reproduced with permission from Marini and Wheeler (1997).)

Ventilator settings for volume-controlled ventilation

Tidal volume

In otherwise healthy individuals, large tidal volumes can be given without generating high pressures. Therefore delivered tidal volumes of 10 to 12 ml/kg of lean body weight are usually appropriate. Obese patients do not have larger lungs, and the tidal volume delivered to a patient with a reduced number of available units must be reduced accordingly (e.g. acute repiratory distress syndrome, pneumonectomy, interstitial fibrosis). Even higher tidal volumes may be needed to satisfy the demands of a hyperpneic subject with normal ventilatory mechanics. Monotonous shallow breaths (< 6 ml/kg) encourage microatelectasis unless interrupted periodically by larger inflations (sighs) or offset by PEEP. As pressure builds during inspiration, a fraction of the inspired gas is stored in tubing and other compressible elements of the ventilator circuit (internal reservoirs, filters, humidifiers, etc.). Although a typical value for such compression losses is 3 to 4 ml/cmH_2O of peak system pressure, compression volume varies with ventilator type and the length, diameter, and composition of the tubing. Under conditions of controlled ventilation, the discrepancy between set or measured inspiratory tidal volume and exhaled tidal volume can quantitate the severity of a bronchopleural fistula.

Frequency

The back-up frequency should be chosen in conjunction with the tidal volume to provide a minimum minute ventilation (V_E) adequate to maintain pH and patient comfort. In the assist mode, the back-up rate should be adjusted to a frequency sufficient to provide 70 to 80 per cent of the usual V_E in case of complete failure of the patient to trigger. In the assist–control mode, any adjustments in the set frequency (up or down) have no effect on V_E or the level of machine support, provided that the patient triggers each breath.

Flow and flow waveform

Controlled cycles

Volume-cycled ventilators allow selection of rate and contour of inspiratory flow. Inappropriately rapid inspiratory flow rates may worsen the distribution of ventilation; the longer exhalation time may reduce air trapping in patients with airflow obstruction. Although peak airway pressure rises as flow rate increases, the mean airway pressure averaged over the entire ventilatory cycle may remain unchanged or even fall as flow rate increases.

On most volume-cycled ventilators, the inspiratory time percentage is usually set indirectly by specifying tidal volume, frequency, and inspiratory flow rate. In general, shorter inspiratory-to-expiratory (I:E) ratios reduce mean intrathoracic pressure. To avoid gas trapping, many ventilators provide a visual warning or auditory alarm when the I:E ratio exceeds 1:1 (duty cycle > 0.5). This threshold defines inverse ratio ventilation.

Triggered cycles

The extent to which the ventilator takes up the inspiratory work of breathing is a function of the triggering sensitivity and the margin by which flow delivery exceeds flow demand (Fig. 2). Triggering sensitivity is generally set to the lowest value that does not result in inadvertent premature 'auto-cycling'. The flow metered by the ventilator must always satisfy the patient's inspiratory drive to breathe. Otherwise, the ventilator not only fails to reduce the work of breathing, but may also force the patient to overcome the resistance of the ventilator circuitry as well as his or her own internal impedance to airflow and chest expansion.

Comfortably rapid inspiratory flow rates are also desirable to ensure that the machine completes inflation before the patient's own ventilatory rhythm cycles into its exhalation phase. Delayed opening of the exhalation valve causes the patient to 'fight the ventilator'. As a general rule, the ventilator's average inspiratory flow should be approximately four times the minute ventilation. Peak flow should be set 20 to 30 per cent above this average value when a decelerating waveform is used. Because the patient's greatest demand for flow occurs in the very first part of the machine cycle, it makes sense to employ a decelerating-flow rather than a constant-flow delivery profile for the same tidal volume and inspiratory time settings. Decelerating-flow profiles also tend to improve the distribution of ventilation in a lung with heterogeneous mechanical properties (widely varying time constants).

To minimize the risk of barotrauma, maximum pressure (alarm and 'pop-off' pressure) should be set no more than 15 to 20 cmH_2O above the peak dynamic cycling pressure observed in a typical breath during constant flow. Unless there is severe airflow obstruction, the pop-off alarm should be set more closely than this (5–10 cmH_2O) if a decelerating flow waveform or pressure control is used because end-inspiratory dynamic and static (plateau) pressures are not as widely separated.

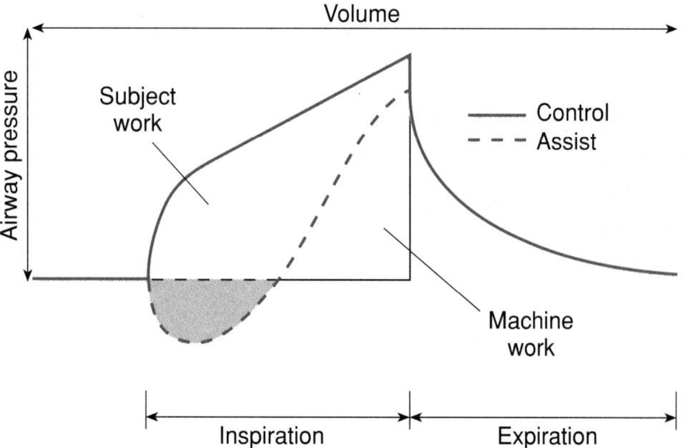

Fig. 2 Airway pressure tracing during volume-cycled ventilation with constant inspiratory flow. With inspiratory flow profile and tidal volume unchanged, the difference in machine work during assisted and controlled cycles quantifies the subject's work of breathing. Vigorous efforts may cause the subject to work against the external circuitry of the ventilator (green shaded region).

Bibliography

McKibben, A. and Ravenscraft, S.A. (1996). Pressure-controlled and volume-cycled mechanical ventilation. *Clinics in Chest Medicine*, **17**, 395–410.

Marini, J.J. and Ravenscraft, S.A. (1992). Mean airway pressure: physiologic determinants and clinical importance. Parts 1 and 2. *Critical Care Medicine*, **20**, 1461–72, 1604–16.

Marini, J.J. and Wheeler, A.P. (1997). *Critical care medicine—the essentials* (2nd edn). Williams & Wilkins, Baltimore, MD.

Ravenscraft, S.A., Burke, W.C., and Marini, J.J. (1992). Volume-cycled decelerating flow. An alternative form of mechanical ventilation. *Chest*, **101**, 1342–51.

Slutsky, A.S. (1993). Mechanical ventilation. American College of Chest Physicians' consensus conference. *Chest*, **104**, 1833–59.

Tobin, M. (1994). Mechanical ventilation. *New England Journal of Medicine*, **330**, 1056–61.

18.2.11 Pressure-controlled mechanical ventilation

S. E. Lapinsky and A. S. Slutsky

Key messages

- Pressure-controlled ventilation delivers a preset pressure with the tidal volume being dependent on lung compliance and airways resistance.
- This mode is well suited to a pressure-targeted lung protective strategy.
- Improved alveolar recruitment and oxygenation may occur in some patients when used with a prolonged inspiratory time (inverse ratio ventilation).
- No outcome benefit of this mode of ventilation has been demonstrated.

Introduction

Pressure-controlled ventilation is a pressure-preset time-cycled ventilatory mode in which a set pressure is delivered by the ventilator. This contrasts with the preset tidal volume of volume-controlled ventilation. Pressure-controlled ventilation is one of a group of pressure-preset modes which includes pressure support ventilation and airway pressure release ventilation. The pressure responsible for delivery of the tidal volume is the gradient between peak alveolar pressure (determined by the preset ventilator pressure) and end-expiratory alveolar pressure (total positive end-expiratory pressure (PEEP)). The tidal volume delivered is influenced by the length of time that this pressure is applied which, in turn, is determined by ventilatory rate and the inspiratory-to-expiratory (I:E) ratio. The tidal volume also varies according to a number of patient factors, including total respiratory system compliance, airways resistance, and intrinsic PEEP (Blanch *et al.* 1993*a*), which interact in a complex way.

The dynamics of pressure-controlled ventilation have been determined by mathematical modeling and validated experimentally (Burke *et al.* 1993). For a given preset pressure and I:E ratio, the tidal volume falls as the ventilatory rate increases (Fig. 1). In the normal lung, increasing ventilatory rate results in a decreased tidal volume. Changes in compliance affect tidal volume, with a decrease in thoracic compliance resulting in a proportional decrease in tidal volume. The tidal volume achieved with non-compliant lungs (e.g. in acute respiratory distress syndrome) is small, with little change as ventilatory rate increases. Tidal volume decreases markedly at higher ventilatory rates in patients with increased airways resistance (e.g. chronic obstructive lung disease). Increased expiratory resistance may produce intrinsic PEEP, reducing the tidal ventilatory pressure and therefore the tidal volume. The decrease in the delivered tidal volume may be overcome by increasing the inspiratory time fraction to allow for pressure equilibration. As a result of the limitations in tidal volume delivery with increasing respiratory rates, minute ventilation increases to an upper limit determined by lung mechanics (Fig. 2). An optimum value for alveolar ventilation occurs at some level of ventilatory rate owing to the relative increase in dead-space fraction as tidal volume decreases.

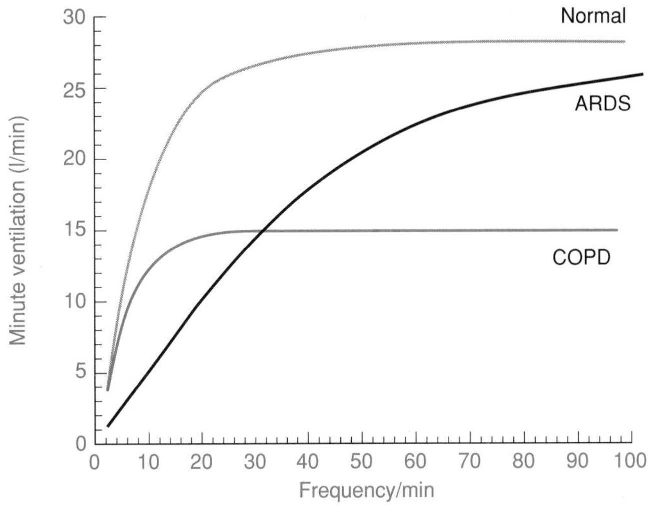

Fig. 2 The relationship of minute ventilation to breathing frequency at a fixed preset pressure and I:E ratio. Minute ventilation increases to a plateau determined by airways resistance. Lung compliance determines the rate of rise of the curve. ARDS = reduced lung compliance; COPD = increased airways resistance. (Adapted with permission from Marini *et al.* (1989).)

Why use pressure-controlled ventilation?

Volume-controlled modes of ventilation would appear to be ideal since they guarantee the total volume delivered to the patient, ensuring the level of alveolar ventilation (Table 1). However, in recent years, there has been increasing focus on the potential detrimental consequences of mechanical ventilation and specifically on ventilation-induced lung injury. A key factor augmenting lung injury is likely to be the degree of end-inspiratory lung stretch. In patients with acute respiratory distress syndrome there is marked regional heterogeneity in compliance, with preferential ventilation to the more compliant regions. In volume-controlled ventilation there is no specific control of regional inflation and the more compliant regions may become overdistended, causing lung injury. This may be avoided by limiting airway pressures or decreasing tidal volumes. The use of pressure-controlled ventilation provides an upper limit to alveolar pressures even if the respiratory system mechanics change (Slutsky 1993). Pressure-controlled ventilation appears to be well suited to this pressure-targeted lung-protective strategy.

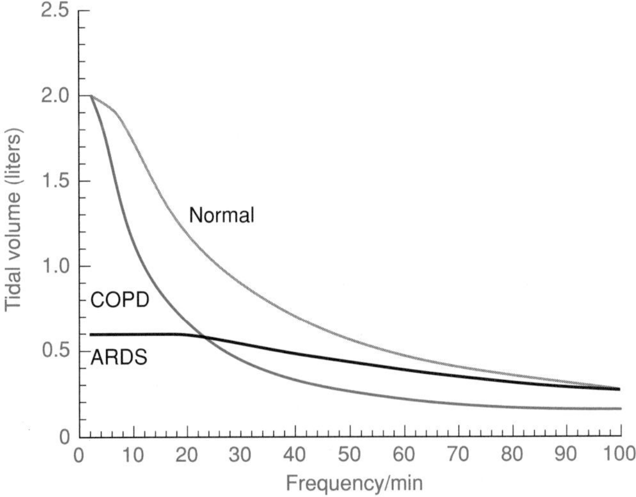

Fig. 1 The relationship of tidal volume to breathing frequency at a fixed preset pressure and I:E ratio. Tidal volume decreases with reduced compliance or with increased airways resistance at higher respiratory rates. ARDS, reduced lung compliance; COPD, increased airways resistance. (Adapted with permission from Marini *et al.* (1989).)

Pressure-controlled ventilation has also been used with a prolonged inspiratory time (see discussion of inverse ratio ventilation below) in an attempt to improve oxygenation. The rationale underlying this approach is that with inverse ratio ventilation it is possible to increase mean airway pressure, while minimizing end-inspiratory pressures. However, not all patients benefit and identifying those who do is difficult. No controlled study has demonstrated an advantage, and therefore no clear indications exist for the use of this mode of ventilation. It is recommended in situations of inadequate oxygenation prior to considering extracorporeal support. Inverse ratio ventilation is most likely to be beneficial early in the disease process when there are still recruitable alveoli. Common criteria include the inability to provide acceptable oxygenation despite a PEEP greater than 10 cmH_2O and Fio_2 greater than 0.80. Some patients do not respond to pressure-controlled inverse ratio ventilation; it has been suggested that this may occur when static thoracic compliance is less than 25 ml/cmH_2O. It is not clear that inverse ratio ventilation offers any distinct advantages over other approaches that increase mean airway pressure. Pressure-preset ventilation is used almost uniformly in neonatology, but is employed far less commonly in adult patients.

Methods

Initiation of respiration may be ventilator triggered ('control breaths') or patient triggered ('assisted breaths'). The inspiratory time and I:E ratio remain constant during ventilator-triggered ventilation but in an assist mode the inspiratory time usually remains fixed, resulting

Table 1 Comparison of volume-controlled and pressure-controlled ventilation

	Volume-controlled	Pressure-controlled
Preset functions	Tidal volume	Pressure
	Flow rate/waveform	I : E ratio
	Rate	Rate
	PEEP	PEEP
	Pressure limit	
Cycling	Volume	Time
Advantages	Guaranteed volume	Control of peak pressures
	Available on all ventilators	Potentially improved distribution of ventilation
	Familiar	
Disadvantages	Pressure varies	Tidal volume varies
Inverse ratio	Often need sedation	May tolerate without sedation

in a variable I:E ratio. Some ventilators allow only ventilator-initiated pressure-controlled breaths, while others provide an assist–control mode and even pressure-controlled synchronized intermittent mandatory ventilation. Owing to the dependence of tidal volume on the I:E ratio, controlled ventilation is usually performed. While sedation and even paralysis are often required, particularly with a prolonged inspiratory time, some patients find this ventilatory mode comfortable. The appropriate inspiratory time may be determined by slowly increasing it while monitoring the increasing tidal volume until no further tidal volume increase occurs (Blanch *et al.* 1993*b*). Further prolongation of the inspiratory time may be necessary to improve oxygenation (see discussion of inverse ratio ventilation below).

Each breath is delivered as a constant pressure maintained for a defined inspiratory time, with flow decelerating exponentially at a rate determined by the patient's thoracic compliance and airways resistance (Fig. 3). This square-wave pressure delivery with decelerating flow appears to have certain physiological advantages, allowing maximal mean airway pressure with minimal peak airway pressure. This may improve oxygenation and dead-space while minimizing lung injury. Unlike volume-preset ventilation, the maximal flow rate is not preset but is a function of the preset pressure and airways resistance. However, the actual maximal flow rate delivered in pressure-control ventilation is limited by the flow capabilities and internal resistance of the ventilator. Pressure and flow patterns similar to those characterizing pressure-controlled ventilation may be emulated to some degree with volume-controlled ventilation when a decelerating flow rather than the usual constant or sinusoidal flow pattern is used. Pressure-controlled ventilation is time cycled, differing from volume-controlled ventilation which is volume cycled and pressure support ventilation which is flow cycled. Exhalation occurs passively with opening of the exhalation valve at the end of the inspiratory time. PEEP may be used to prevent derecruitment of alveoli or, alternatively, prolongation of the inspiratory time (at the expense of expiratory time) may produce intrinsic PEEP with similar physiological effects.

Inverse ratio ventilation

Each respiratory cycle in conventional ventilation is characterized by a short inspiratory phase and longer expiratory phase (e.g. an I:E ratio of 1:2), usually allowing for complete exhalation. Extending the inspiratory time to a ratio greater than unity improves oxygenation, lung compliance, and dead-space ratio in some patients with acute lung injury. Histopathological evidence of reduced lung injury may also be demonstrated. While this benefit of inverse ratio ventilation has been shown in animal studies and small clinical series, no randomized prospective trial has been performed; whether there is benefit remains unclear. Ventilation at I:E ratios ranging from 1:1 to 4:1 is used, but adverse hemodynamic effects and the potential for barotrauma are more common at very high I:E ratios.

A prolonged inspiratory time has the beneficial effect of increasing mean airway pressure and allowing adequate time for ventilation of alveoli with long inspiratory time constants. However, from a theoretical point of view, auto-PEEP generated during inverse ratio ventilation will produce a relatively smaller increase in the regional functional residual capacity of lung units with short expiratory time constants, and hence is potentially not as effective as applied PEEP. The short expiratory time can produce dynamic hyperinflation or intrinsic PEEP. This may be responsible in part for improved oxygenation. Pressure-controlled inverse ratio ventilation may have advantages over similar effects produced by volume-controlled ventilation with an inspiratory pause (Blanch *et al.* 1993*b*). The maintenance of pressure during the prolonged inspiratory phase

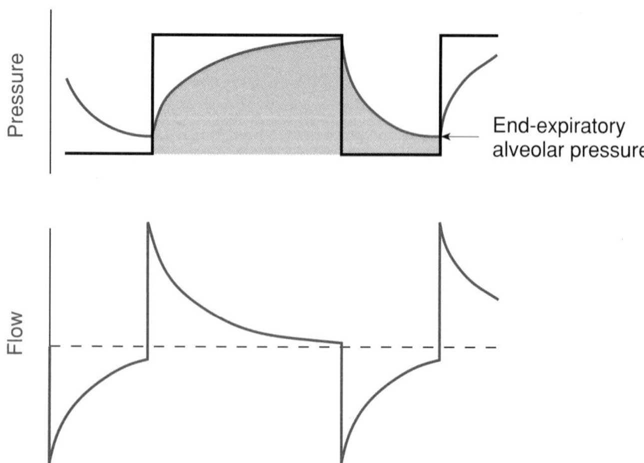

Fig. 3 Schematic diagram of airway and alveolar pressure (upper trace) and flow (lower trace) during pressure-controlled inverse ratio ventilation. Note the square-wave pressure curve (black line) with shaded alveolar pressure. Flow decelerates exponentially during inspiration. Note that at end-expiration there is still expiratory flow and positive alveolar pressure (auto-PEEP). (Reproduced with permission from Marcy and Marini (1991).)

overcomes loss of airway pressure which may occur in the volume-controlled situation owing to leaks and alveolar recruitment.

Close monitoring is necessary after initiation of pressure-controlled inverse ratio ventilation to ensure adequate oxygenation, minute ventilation, and hemodynamic stability. The development of excessive intrinsic PEEP may cause hemodynamic compromise and reduce tidal pressure and, therefore, minute ventilation. Deep sedation or paralysis may be required to prevent dyssynchronous breathing. Inverse ratio ventilation should not be used in the presence of airflow obstruction where exhalation time needs to be maximized.

Limitations

A number of limitations make pressure-controlled ventilation less commonly used than volume control. This mode is not available on all ventilators, although most newer models offer pressure-preset ventilation. Because of the complexities of the interactions between ventilator settings (preset pressure, I:E ratio, rate) and patient factors (thoracic compliance, airways resistance, intrinsic PEEP) with this mode of ventilatory support, significant experience on the part of the clinician is required.

The physician should be aware of specific potential complications which require monitoring. Tidal volume, and therefore minute volume, may vary with changes in ventilator settings, clinical condition, or patient–ventilator dyssynchrony. Thus monitoring of these variables is essential. Pressure-regulated volume-controlled ventilation is a mode which attempts to overcome this concern by adjusting pressure to maintain adequate tidal volume. The development of intrinsic PEEP, particularly in inverse ratio ventilation, may have significant adverse hemodynamic effects, reducing filling pressures and cardiac output. In addition, high intrinsic PEEP may reduce tidal pressure and thus minute ventilation, as well as predisposing to barotrauma. Hypercapnia may be an unavoidable consequence of pressure-controlled ventilation, but this is usually well tolerated. As some patients do not tolerate inverse ratio ventilation, deep sedation or even paralysis may be required with their own inherent complications.

References

Blanch, P.B., Jones, M., Layon, A.J., and Camner, N. (1993*a*). Pressure-preset ventilation. Part 1: Physiologic and mechanical considerations. *Chest*, **104**, 590–9.

Blanch, P.B., Jones, M., Layon, A.J., and Camner, N. (1993*b*). Pressure-preset ventilation. Part 2: Mechanics and safety. *Chest*, **104**, 904–12.

Burke, W.C., Crooke, P.S., Marcy, T.W., Adams, A.B., and Marini, J.J. (1993). Comparison of mathematical and mechanical models of pressure-controlled ventilation. *Journal of Applied Physiology*, **74**, 922–33.

Marini, J.J., Crooke, P.S., and Truwit, J.D. (1989). Determinants and limits of pressure-preset ventilation: a mathematical model of pressure control. *Journal of Applied Physiology*, **67**, 1081–92.

Slutsky, A.S. (1993). Mechanical ventilation. American College of Chest Physicians' consensus conference. *Chest*, **104**, 1833–59.

18.2.12 Pressure support ventilation

J. Mancebo and L. Brochard

Key messages

- Pressure support ventilation is a patient-triggered, pressure-limited, and flow-cycled ventilatory mode. Its most important physiological effects are reduction in the work of breathing of the respiratory muscles, reduction in respiratory rate, and increase in tidal volume.
- The most important clinical aspect of pressure support ventilation is its use in weaning patients from mechanical ventilation and as a non-invasive ventilatory support method in selected patients with an acute decompensation of chronic obstructive pulmonary disease.

Definition and general characteristics

Pressure support ventilation is an assisted ventilatory mode, which is patient triggered (by pressure, airflow, or both), pressure limited, and flow cycled. Thus inspiratory assistance is only provided when the patient makes a breathing effort. The machine then provides a certain amount of inspiratory flow (a decelerated and servoregulated airflow pattern) so as to maintain the preset level of airway pressure (the pressure support level) almost constant throughout the inspiration. The velocity of pressurization, which depends on the shape of the inspiratory flow waveform, can be adjusted in some ventilators. Inspiration ends when the inspiratory airflow reaches a certain threshold value, which theoretically coincides with the end of the inspiratory muscle effort. This flow value is variable and depends on the manufacturer; while some ventilators use a percentage of peak flow (i.e. 25 per cent of peak flow), others use a fixed level (i.e. 5 l/min). Some modern ventilators even allow modification of the flow threshold value to cycle from inspiration to expiration. Additionally, cycling from inspiration to expiration occurs when an increment in airway pressure is detected (i.e. there is an active expiration to stop inspiratory flow) or when a certain time has elapsed. This is important, for example, in case of leaks and non-invasive ventilation. Tracings obtained during pressure support ventilation are shown in Fig. 1.

This ventilatory mode allows the patient to breathe at his or her own respiratory rate. It also provides considerable flexibility regarding the duration of inspiratory and expiratory time, inspiratory flow, and tidal volume. Although these features may render this ventilatory mode more comfortable and provide better synchronization between ventilator and patient compared with volume-controlled modes, these theoretical advantages have not been demonstrated.

Tidal volume depends on the preset level of pressure support, the magnitude of inspiratory effort, and the mechanical characteristics (resistance and compliance) of the respiratory system. Although patients have more flexibility with regard to their own breathing pattern than during assist–control ventilation, they certainly do not have full control of volumes and flows. This

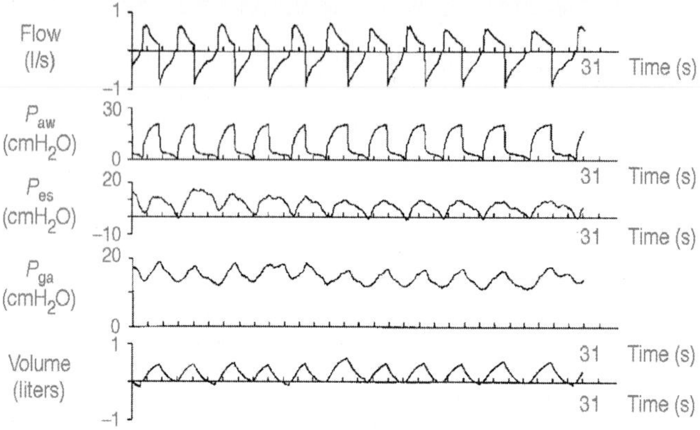

Fig. 1 From top to bottom, tracings of air flow, airway pressure (P_{aw}), esophageal pressure (P_{eso}), gastric pressure (P_{ga}), and tidal volume obtained in a patient ventilated with pressure support mode. The breath-by-breath variability in breathing pattern can be clearly seen.

emphasizes the importance of how the pressure support is actually set by the clinician.

Physiological effects

Work of breathing and breathing pattern

The most important effects are those related to the reduction in the work of breathing, and the changes induced in the characteristics of this work (Brochard *et al.* 1989). Pressure support tends to change the high-pressure and low-tidal-volume inspiratory efforts (which are usually observed in situations of acute respiratory failure) to inspiratory efforts producing lower pressures and higher tidal volumes (as usually occurs during normal breathing). Indeed, pressure support is not primarily directed to improving gas exchange. However, on increasing the levels of support, a decrease in $Pa\text{CO}_2$ is usually observed.

Typical modifications in the breathing pattern induced by pressure support ventilation include an increase in tidal volume, a decrease in respiratory rate, and no major change in minute ventilation. Pressure support ventilation also induces a diminution in transdiaphragmatic pressure swings and in the pressure–time index of the diaphragm when compared with spontaneous breathing. Moreover, in patients who are difficult to wean, progressive increments of pressure support lead to parallel decreases in the work of breathing and oxygen consumption of the respiratory muscles, with disappearance of the electromyographic signs of impending diaphragmatic fatigue. These levels of support also allow accessory respiratory muscles (e.g. the sternocleidomastoid) to rest (Brochard *et al.* 1989).

Another important feature related to pressure support ventilation is its ability to overcome the extra work of breathing induced by the presence of the endotracheal tube and ventilator circuit and their related flow-resistive properties. The levels of pressure necessary to compensate for this extra work are variable and depend on several factors: the diameter of the endotracheal tube, the mean inspiratory drive, and the presence or absence of an intrinsic pulmonary disease. From a clinical point of view, levels of pressure support of about 8 cmH_2O appear useful in overcoming this extra work of breathing.

Cardiovascular effects

Contrary to the hemodynamic consequences induced by conventional positive pressure ventilation, no untoward cardiovascular effects have been attributed to pressure support ventilation. This is probably related to the fact that, during pressure support ventilation, spontaneous inspiratory efforts help maintain a lower intrathoracic pressure; thus venous return and right ventricular afterload are enhanced.

Importance of different working principles

The influence of the different working principles of pressure support ventilation on respiratory muscle unloading should be considered. Different forms of inspiratory flow ramp, cycling criteria, and the flow-impeding characteristics of expiratory circuits induce important clinical changes in the external work of breathing. A ventilator algorithm providing a rapidly ascending inspiratory flow, thus generating a rapid velocity of pressurization, with a cycling criteria at 25 per cent of peak inspiratory flow has been proved more useful in terms of respiratory muscle unloading than those ventilator algorithms which provide a slowly increasing inspiratory flow ramp with cycling criteria near zero flow (Mancebo *et al.* 1995). To provide an optimal reduction in inspiratory muscle effort, high inspiratory airflow rates should probably be mandatory to match adequately the patient's demands for flow.

Many factors related to pressure support ventilation tend to improve the synchrony between the patient's effort and ventilator support. However, this aspect is far from optimal; episodes of dyssynchrony have also been described. These include ineffective efforts at high levels of support, dyssynchrony at the end of inspiration, or insufficient flow during inspiration.

Clinical usefulness

Although pressure support may be used as a total ventilatory assistance method, it should be remembered that neither tidal volume nor respiratory rate are preset in the ventilator. Consequently, pressure support ventilation requires an intact spontaneous inspiratory activity. To the extent that the volume delivered to the patient depends not only on the preset level of pressure but also on the mechanical characteristics of the respiratory system, the efficacy of this method may not be optimal in certain unstable patients. A level of pressure support which may be clinically appropriate for a period of time may be inefficient later if changes in inspiratory demands or respiratory system mechanics occur. As a consequence, close monitoring of respiratory rate and minute ventilation is advised when this method is used in critically ill patients.

Weaning from mechanical ventilation

The most important clinical application of pressure support ventilation is probably the weaning of patients from mechanical ventilation. In one study, the efficacy of pressure support was greater than with other techniques such as T-piece trials or synchronized intermittent mandatory ventilation. This multicenter prospective randomized trial showed that the probability of complete withdrawal of mechanical ventilation 21 days after commencement of weaning was twice as high with pressure support ventilation than with synchronized intermittent mandatory ventilation. The period of weaning was also significantly decreased from a mean duration of 9.3 days for the other two techniques to 5.7 days using pressure support ventilation (Brochard *et al.* 1994).

The above results were not reproduced in another prospective multicenter randomized trial published some months later (Esteban *et al.* 1995). This second study showed that a single daily trial of spontaneous breathing with a T-piece was able to shorten the period of weaning significantly (median of 3 days compared with a median of 4 days with pressure support and 5 days with synchronized intermittent mandatory ventilation).

The contradictory results found in these two studies could be accounted for by the different methodologies used. In particular, modification of the pressure support level and criteria to extubate the patient under pressure support ventilation differed between the two studies. These criteria were more conservative in the second study. However, these investigations do indicate that the way a technique is used is as important as the technique itself; too rigid criteria for extubation probably lengthen the period of weaning.

Adjusting pressure support levels

Another important aspect related to the clinical application of pressure support is the adjustment of pressure levels. Modifications in the breathing pattern (increase in tidal volume and decrease in respiratory rate) induced by pressure support are very useful in adjusting ventilatory settings. Recent data suggest that, at least during the weaning period, levels of pressure support allowing a respiratory rate between 25 and 30 breaths/min would be adequate (Jubran *et al.* 1995). This goal is frequently achieved with pressure support of 15 to 20 cmH_2O, but large interindividual variations in muscle unloading exist.

No additional beneficial physiological effects are obtained by increasing pressure levels more than required to unload the respiratory muscles (Brochard *et al.* 1989). Higher pressure support levels will probably lead to an abnormal breathing pattern with excessively high tidal volumes, appearance of wasted inspiratory efforts, and even some periods of apnea. In the particular case of patients with chronic obstructive pulmonary disease who are difficult to wean, it should be remembered that addition of low levels of external positive end-expiratory pressure (PEEP) (about 5 cmH_2O) may be beneficial because this decreases the extra work of breathing induced by intrinsic PEEP which is due to dynamic hyperinflation with airflow limitation.

Once weaning has begun, reduction of support should be progressive, depending on the clinical tolerance of the patient (cardiovascular and neuropsychological status, gas exchange, and breathing pattern are the most important parameters to be assessed). The reduction is usually accomplished in one or two steps daily with a decrease in pressure support level of 2 to 4 cmH_2O in each step. When the patient tolerates levels of pressure support of approximately 10 cmH_2O (or lower in cases of tracheostomy), extubation is recommended.

Non-invasive pressure support

Finally, another important clinical application of pressure support is non-invasive ventilation. This technique has been used for the supportive treatment of selected patients with acute decompensations of chronic obstructive pulmonary disease and without comorbidities. The use of face-mask pressure support has proved extremely useful in these patients; it improves gas exchange, avoids intubation and mechanical ventilation in approximately 75 per cent of these patients, and also decreases mortality rate and hospital stay when compared with conventional treatment.

References

Brochard, L., Harf, A., Lorino, H., and Lemaire, F. (1989). Inspiratory pressure support prevents diaphragmatic fatigue during weaning from mechanical ventilation. *American Review of Respiratory Disease*, **139**, 513–21.

Brochard, L., *et al.* (1994). Comparison of three methods of gradual withdrawal from ventilatory support during weaning from mechanical ventilation. *American Journal of Respiratory and Critical Care Medicine*, **150**, 896–903.

Esteban, A., *et al.* (1995). A comparison of four methods of weaning patients from mechanical ventilation. *New England Journal of Medicine*, **332**, 345–50.

Jubran, A., Van de Graaff, W.B., and Tobin, M.J. (1995). Variability of patient-ventilator interaction with pressure support ventilation in patients with chronic obstructive pulmonary disease. *American Journal of Respiratory and Critical Care Medicine*, **152**, 129–36.

Mancebo, J., Amaro, P., Mollo, J.L., Lorino, H., Lemaire, F., and Brochard, L. (1995). Comparison of the effects of pressure support ventilation delivered by three different ventilators during weaning from mechanical ventilation. *Intensive Care Medicine*, **21**, 913–19.

18.2.13 Effects of mechanical ventilation on the circulation

M. R. Pinsky

Key messages

- Increasing lung volume and not increasing airway pressure accounts for the hemodynamic effects of mechanical ventilation.
- Hyperinflation increases pulmonary vascular resistance by compressing the alveolar vessels.
- Lung collapse, as occurs during acute lung injury, increases pulmonary vascular resistance by inducing hypoxic pulmonary vasoconstriction.
- Ventilatory strategies that restore end-expiratory lung volumes to basal levels will be associated with the lowest levels of pulmonary vascular resistance.
- Increasing intrathoracic pressure decreases venous return by increasing right atrial pressure.
- Increases in intrathoracic pressure decrease left ventricular afterload, whereas decreases in intrathoracic pressure increase left ventricular afterload.
- All forms of mechanical ventilation share similar hemodynamic outcomes based on their effects on lung volume, intrathoracic pressure, and work of breathing. For the same mean airway pressure and work of breathing, all forms of mechanical ventilation should have similar hemodynamic effects.
- Weaning from mechanical ventilatory support is exercise.

Introduction

Positive-pressure ventilation (PPV) may impair or improve tissue oxygen delivery depending on its overall effects on cardiac output, ventilation–perfusion matching, and alveolar oxygenation. Clinically relevant heart–lung interactions occur because of associated changes in lung volume, intrathoracic pressure (ITP), or metabolic load. All forms of ventilation increase lung volume: spontaneous ventilation decreases ITP while PPV increases ITP, and spontaneous ventilatory efforts constitute exercise. While lung volume and ITP are rarely measured, airway pressures and tidal volume are routinely defined.

Relation between airway pressure, intrathoracic pressure, and lung volume

The degree to which both ITP and lung volume increase during PPV is a function of airways resistance and lung compliance. With reduced compliance, fixed tidal volumes generate greater mean and plateau airway pressures than when compliance is normal. However, for the same tidal increase in lung volume, ITP and pericardial pressure increase similarly in both normal and acute lung injury states despite varying levels of compliance. If lung inflation is sustained and ventricular function normal, as may occur with positive end-expiratory pressure (PEEP), ITP increases more than pericardial pressure, presumably because cardiac volume decreases. Accordingly, ITP may overestimate pericardial pressure during PEEP when left ventricular function is normal. In congestive heart failure, pericardial pressure often exceeds ITP because the pericardium functions as a limiting membrane. However, with progressive increases in PEEP, ITP rises until it approximates pericardial pressure; both pressures will then increase in parallel as PEEP increases further. In heart failure states, hemodynamic effects of PEEP are only seen once ITP and pericardial pressure equalize.

The proportion of airway pressure transmitted to the pleural surface is highly variable. Even the percentage of the increase in airway pressure transmitted to the pericardial surface as PEEP is increased is not constant between individuals. The amount by which either ITP or pericardial pressure will increase cannot be predicted.

Finally, PPV does not increase ITP by similar amounts in all regions of the thorax. Diaphragmatic ITP increases least, while juxtacardiac ITP increases most. Dependent regional ITP exceeds non-dependent ITP by an amount equal to the vertical hydrostatic pressure gradient. Thus, where ITP is measured, if at all, determines both the mean ITP and its swings during PPV.

When tidal volume is kept constant, changes in peak and mean airway pressure reflect changes in the mechanical properties of the lungs and the degree of patient co-operation, but may not reflect changes in ITP. Accordingly, changes in airway pressure should not be used to assess changes in global cardiovascular dynamics in response to PPV under conditions in which lung mechanics vary rapidly.

Hemodynamic effects of changes in lung volume

The major determinant of the hemodynamic response to increases in lung volume are mechanical in nature. Increasing lung volume can decrease pulmonary vascular resistance if the inflation opens collapsed alveoli by reversing hypoxic pulmonary vasoconstriction. Both recruitment of collapsed alveoli and delivery of enriched inspired oxygen to the alveoli reverse hypoxic pulmonary vasoconstriction. Thus patients with muscle weakness (e.g. myasthenia gravis or quadriplegia), alveolar instability (e.g. acute lung injury), pulmonary edema, or alveolar hypoventilation (e.g. asthma, chronic obstructive lung disease, central hypopnea) may decrease pulmonary arterial pressure by the institution of PPV if alveolar recruitment occurs and hypoventilation is corrected. However, lung overdistension will usually increase pulmonary vascular resistance by compressing alveolar pulmonary vessels. In patients prone to hyperinflation (e.g. with chronic obstructive lung disease) mechanical ventilation can increase pulmonary vascular resistance, precipitating cardiovascular collapse secondary to acute cor pulmonale. Thus PEEP may either increase or decrease pulmonary vascular resistance, depending on the initial and final volume status of the lungs. In patients with otherwise normal lung mechanics, a goal of ventilatory therapy is to prevent lung volumes from either collapsing further or overdistending more.

Clearly, left ventricular preload is directly altered by changes in right ventricular output because the two ventricles pump in series. Left ventricular preload can also be indirectly altered by changes in right ventricular end-diastolic volume via the mechanism of ventricular interdependence. If right ventricular volumes decrease, left ventricular diastolic compliance increases. Lung overdistension will compress the heart, increasing pericardial pressure and decreasing cardiac volumes analogous to cardiac tamponade.

Hemodynamic effects of changes in intrathoracic pressure

The heart in the thorax is a pressure chamber within a pressure chamber. Changes in ITP affect pressure gradients for both venous return to the heart and systemic outflow from the heart, independent of the heart itself. PPV, by increasing ITP, will decrease pressure gradients for both venous return and left ventricular ejection, thereby decreasing intrathoracic blood volume. In normal patients who are preload dependent this combination usually decreases cardiac output. However, in patients with congestive heart failure, cardiac output may actually increase upon institution of PPV. This process is analogous to systemic vasodilator therapy which decreases intrathoracic blood volume in all subjects, decreasing cardiac output in normals but increasing cardiac output in heart failure patients.

The primary determinant of cardiac output is venous return. Variations in right atrial pressure represent the major factor determining fluctuations in systemic venous return during ventilation. PPV increases ITP, increasing right atrial pressure and decreasing venous blood flow.

The detrimental effect of PPV on venous return can be minimized by minimizing both mean ITP and swings in lung volume. Prolonging expiratory time, decreasing tidal volume, and avoiding high levels of PEEP all minimize ventilator-induced decreases in cardiac output. Similarly, intravascular fluid infusion or increasing autonomic tone, either endogenously or by vasopressor infusion, increases venous pressure and maintains cardiac output constant despite PPV. Output usually decreases slightly in patients receiving PPV. This decrease is considered to be a form of 'functional hypovolemia'. However, if PPV induces profound hyperinflation, as may occur with acute severe bronchospasm, chronic obstructive lung disease, or the delivery of large rapid tidal breaths, venous return can rapidly decrease to zero, leading to total circulatory arrest.

The decrease in venous return during PPV may be less than predicted based solely on increases in right atrial pressure. Since increasing lung volume also depresses the diaphragm, intra-abdominal pressure also increases, tending to maintain upstream venous pressure. Thus abdominal pressurization by diaphragmatic descent and abdominal muscle contraction minimizes the decrease in venous return during PPV. Recent interest in inverse ratio ventilation has raised questions about its hemodynamic effect because its application includes a large component of unquantified hyperinflation or intrinsic PEEP. However, for the same total PEEP (intrinsic plus extrinsic PEEP), no hemodynamic difference between conventional and inverse ratio ventilation is seen.

As right ventricular end-diastolic volume increases, right atrial pressure also increases. However, both pericardial and right atrial pressure increase during fluid resuscitation, such that right ventricular distending pressure is unchanged. Thus, under normal conditions, right ventricular diastolic compliance is greater than pericardial compliance, or the right ventricle fills below its unstressed volume, or both. This right ventricular filling profile has important and practical clinical implications. A right ventricle filling below its unstressed volume would accommodate increases in venous blood flow without decreasing the pressure gradient for venous return and not inhibiting venous flow. However, if right ventricular diastolic compliance decreased or right atrial pressure increased independent of changes in right ventricular end-diastolic volume, venous return would not be reduced, decreasing cardiac output. Right ventricular diastolic compliance can decrease with acute right ventricular volume overload, i.e. acute cor pulmonale (pulmonary embolism, hyperinflation, and right ventricular infarction). A dissociation occurs between right atrial pressure and right ventricular volume during either tamponade or PPV. Thus PPV, by dissociating right atrial pressure from right ventricular filling pressure, may be maladaptive. The mitigating effects of abdominal pressure in this process probably minimize this dissociation between right ventricular filling and right atrial pressure by matching changes in abdominal pressure with changes in right atrial pressure. Interestingly, even if 'normal' right ventricular interactions are restored by allowing spontaneous ventilatory efforts to occur during PPV, cardiac output will only increase if the right ventricle can transfer the associated increase in venous return into forward pulmonary blood flow. Failure to improve cardiac output in a hemodynamically unstable patient when transiting from PPV to spontaneous ventilation may potentially reflect isolated right ventricular failure.

By increasing ITP, PPV may reduce right ventricular volume and increase left ventricular diastolic compliance through the process of ventricular interdependence. Unlike spontaneous inspiration, which induces intraventricular septal shift, positive-pressure inspiration compresses both ventricles into each other. Since spontaneous inspiratory efforts can and often do occur during PPV and represent the mechanism of ventilation during continuous positive airways pressure and partial ventilatory-assist modes of ventilation, the hemodynamic consequences of spontaneous ventilation are often seen in mechanically ventilated patients.

Changes in ITP alter left ventricular afterload. Increases in ITP unload the left ventricle, whereas decreases will increase left ventricular afterload. This is because maximal left ventricular afterload is proportional to transmural left ventricular systolic pressure which equals the difference between intraluminal left ventricular pressure (or arterial pressure) and ITP. Thus, if arterial pressure were to remain constant as ITP increased, left ventricular afterload would decrease. Similarly, since wall stress is afterload and both transmural systolic pressure and left ventricular end-diastolic volume determine wall stress, by the Laplace equation, if transmural arterial pressure were to remain constant as ITP increased but left ventricular end-diastolic volume were to decrease, then left ventricular afterload would also decrease. This is the more common clinical occurrence, as PPV usually decreases venous return, decreasing left ventricular end-diastolic volume while decreasing left ventricular afterload.

Decreases in ITP increase left ventricular afterload, impeding left ventricular ejection. Spontaneous inspiratory efforts against a resistance decrease left ventricular stroke volume. Increases in ITP proportionally increase arterial pressure but do not alter transmural arterial pressure or aortic blood flow. If the increased ITP is sustained, as may occur with the application of high levels of PEEP, marked hyperinflation, or large-tidal-volume ventilation, venous return and cardiac output will also decrease, thus decreasing arterial pressure. Reflex vasoconstriction tends to keep systemic pressure and flow constant. Therefore the peripheral vasculature will vasodilate reflexively to maintain a constant arterial pressure–flow relationship. Since the coronary circulation both starts and ends in the thorax, coronary perfusion pressure is not altered by ITP-induced increases in arterial pressure. However, both reflex systemic vasodilation and mechanical compression of the coronary arteries by the expanding lungs may induce coronary hypoperfusion.

Mechanically speaking, left ventricular afterload is reduced not only by increases in ITP but also by abolishing negative swings in ITP. Removing negative swings in ITP is more clinically relevant than increasing ITP. There are several reasons for this. First, since many pulmonary disease states are associated with exaggerated negative swings in ITP during inspiration, left ventricular afterload is increased. Second, these exaggerated inspiratory efforts increase the work of breathing, putting yet another load on the circulation. Finally, decreases in ITP can only increase venous return so far before transthoracic venous collapse limits maximal venous blood flow. Thus markedly negative swings in ITP selectively increase left ventricular afterload. Abolishing these swings in ITP should selectively reduce left ventricular afterload and myocardial oxygen demand without altering venous return, and thus not altering steady state cardiac output.

Endotracheal intubation and ventilation of patients with markedly negative swings in ITP should diminish myocardial workload without reducing cardiac output. This cardiac supportive effect of PPV is most noticeable when baseline cardiac performance is impaired but circulating blood volume is adequate. Left ventricular function often improves after institution of PPV in patients with combined cardiovascular insufficiency and respiratory distress. Similarly, weaning patients from PPV by allowing a return of the spontaneous decreases in ITP, may precipitate acute left ventricular failure and pulmonary edema in patients with borderline left ventricular function. Finally, institution of PEEP in patients with congestive heart failure may further augment left ventricular output by reducing left ventricular afterload, despite an associated decrease in left ventricular preload.

Comparisons of different forms of mechanical ventilation

In patients with a markedly increased work of breathing, volume overload, or impaired left ventricular function, the application of mechanical ventilation can be lifesaving because of its ability to unload a weakened cardiovascular system while reducing both global and myocardial oxygen consumption. These effects will be independent of any beneficial effects that mechanical ventilation may have on gas exchange.

In patients with decreased pulmonary elastic recoil, increased pulmonary vascular resistance, hypovolemia, or airflow obstruction, the institution of mechanical ventilatory support may induce cardiovascular instability. This can lead to total cardiovascular collapse. Similarly, withdrawal of ventilatory support invariably increases intrathoracic blood volume and left ventricular afterload, and can be thought of as a cardiovascular stress test. Patients who pass this test without developing worse cardiovascular function can usually be weaned successfully from mechanical ventilatory support, whereas those who fail often are not ready to be weaned or require supplemental inotropic support to sustain spontaneous ventilation. Initiation and withdrawal of ventilatory support can be seen as a ventilatory probe into the determinants of cardiovascular homeostasis in the ventilator-dependent patient. In fact, the left ventricular ejection fraction decreased in patients on a T-tube trial, although not when partially supported by pressure support ventilation.

Although it is not known what percentage of patients fail weaning trials because of cardiovascular insufficiency rather than gas exchange abnormalities, cardiovascular dysfunction appears to play an important role in defining which patients will fail. Identification of such patients early in the course of care may improve their outcome by directing supportive therapies toward improving cardiovascular as well as ventilatory status.

Different ventilatory support modes may also be associated with differing degrees of patient comfort despite similar levels of external work and minute ventilation. If the ventilatory pattern varies between modes, differences in ventilation efficiency between these modes may reflect differing metabolic demands rather than actual weaning effectiveness. Some cardiovascular differences between different modes of ventilation during both maintenance ventilatory support and weaning may reflect differences in patient effort and matching of effort with the ventilator. To the extent that this matching is poor, the metabolic demand of that ventilatory support will be greater.

PPV decreases intrathoracic blood volume compared with spontaneous ventilation. PEEP decreases intrathoracic blood volume even further without altering left ventricular contractile function. The degree of hyperinflation, and not airway pressure, determines the decrease in cardiac output in hemodynamically stable patients. To the extent that decreases in left ventricular end-diastolic volume determine stroke volume, these modes of ventilation will also decrease cardiac output. However, fluid resuscitation can restore intrathoracic blood volume to basal levels despite the continued application of a positive airway pressure. When left ventricular end-diastolic volume is returned to basal levels, cardiac output also returns to its basal level despite the continued application of PEEP. Furthermore, lower-body compression minimizes the decrease in cardiac output in ventilator-dependent septic patients. To the extent that cardiac output does not return to basal levels, other processes such as cor pulmonale or cardiac compression may also influence output. To the extent that cardiac output exceeds basal levels as PEEP is applied, processes other than venous return, such as pulmonary vascular resistance and left ventricular afterload, determine cardiac output.

Most hemodynamic differences between different modes of mechanical ventilation at a constant level of PEEP can be explained by their differential effects on ITP. When two different ventilation modes produce similar changes in ITP and ventilatory effort, their hemodynamic effects are also similar. Thus, in patients with acute lung injury, no significant hemodynamic differences were seen between volume-controlled conventional ventilation and both pressure-controlled conventional and inverse ratio ventilation. Similarly, if pressure-controlled ventilation is delivered with a lower tidal volume and total PEEP, cardiac output will be higher than during conventional ventilation. Furthermore, partial ventilatory support with either intermittent mandatory ventilation or pressure support ventilation gave similar hemodynamic responses and tissue oxygenation when matched for similar tidal volumes.

Increasing levels of PEEP improve cardiac output in patients with impaired left ventricular function and/or an expanded intravascular blood volume. Conversely, in heart failure states, withdrawal of ventilatory support can induce severe cardiac decompensation in otherwise stable patients with acute respiratory failure secondary to chronic airways obstruction. Similarly, in patients with myocardial ischemia and acute left ventricular failure, decreasing levels of ventilatory support worsen ischemia. This detrimental effect can be minimized by preventing negative swings in ITP induced by spontaneous inspiratory effort or by adding exogenous vasopressor therapy.

The beneficial cardiovascular effects of PPV are also seen in patients receiving non-invasive ventilation. As little as 6 h of continuous positive airway pressure daily induces a significant reduction in shunt fraction and the need for subsequent intubation, as well as increasing cardiac output and mixed venous oxygen saturation.

The above concepts can also be used to explain the hemodynamic effects of PPV in patients with chronic obstructive lung disease. Such patients may be even more compromised by dynamic hyperinflation (intrinsic PEEP). Matching extrinsic with intrinsic PEEP induces no measurable hemodynamic effects. There is little hemodynamic difference between increasing airway pressure to generate a breath and decreasing extrathoracic pressure with an iron lung–negative pressure ventilator.

Bibliography

Baigorri, F., *et al.* (1994). Hemodynamic response to external counterbalancing of auto-positive end-expiratory pressure in mechanically ventilated patients with chronic obstructive lung disease. *Critical Care Medicine*, **22**, 1782–91.

Huemer, G., Kolev, N., Kurz, A., and Zimpfer, M. (1994). Influence of positive end-expiratory pressure on right and left ventricular performance assessed by Doppler two-dimensional echocardiography. *Chest*, **106**, 67–73.

Lessard, M.R., Guerot, E., Lorini, H., Lemaire, F., and Brochard, L. (1994). Effects of pressure-controlled with different I:E ratios versus volume-controlled ventilation on respiratory mechanics, gas exchange and hemodynamics in patients with adult respiratory distress syndrome. *Anesthesiology*, **80**, 983–91.

Mang, H., Kacmarek, R.M., Ritz, R., Wilson, R.S., and Kimball, W.P. (1995). Cardiorespiratory effects of volume- and pressure-controlled ventilation at various i/e ratios in an acute lung injury model. *American Journal of Respiratory and Critical Care Medicine*, **151**, 731–6.

Romand, J.A., Shi, W., and Pinsky, M.R. (1995). Cardiopulmonary effects of positive pressure ventilation during acute lung injury. *Chest*, **108**, 1041–8.

18.2.14 Effects of mechanical ventilation on other organ function

M. Hemmer

Key messages

- Positive-pressure ventilation may alter regional hemodynamics and function of liver, gut, kidney, and brain, particularly in hypovolemic and hypotensive patients.
- Preventive measures include careful evaluation of volemic status and maintenance of regional perfuson pressure.

Introduction

Mechanical respiratory support with positive-pressure ventilation is commonly used in critically ill patients with associated acute lung injury. Elevated intrathoracic pressure affects regional hemodynamics and may lead to dysfunction of other organs. The interactions of lung, gut, liver, kidney, and brain are of particular importance in critical illness.

The mechanisms by which elevated intrathoracic pressure may affect local circulations are as follows:

(1) a global reduction in cardiac output and systemic arterial pressure;

(2) impairment of venous return and elevation of venous pressure;

(3) modification of local vasoactive status;

(4) alterations in the distribution of microvascular flow.

The overall effect of elevated intrathoracic pressure on other organs depends on the magnitude of pressure transmission, the patient's volemic status, and the integrity of autoregulatory mechanisms in the regional circulation. Improvements in pulmonary function and arterial oxygenation due to positive-pressure ventilation should counteract the deleterious effects on local hemodynamics. However, the lowest possible peak inspiratory and end-expiratory pressures should be employed in patients with severely impaired autoregulatory mechanisms.

The effect on hepatosplanchnic circulation and function

This effect is difficult to evaluate because of the complex interactions between the forces generated by elevated intrathoracic pressure and the autoregulatory responses of the hepatic vasculature. Although understanding of hepatic circulatory physiology has considerably increased, mechanisms involved in the regulation of liver blood flow have yet to be fully elucidated.

Mechanical ventilation with positive end-expiratory pressure (PEEP) decreases cardiac output by reducing venous return. The liver, which is an important regulator of venous return, plays an essential role in this reponse. Early studies reported that continuous positive-pressure ventilation decreased total hepatic blood flow mainly by reductions in portal venous flow. Recent experimental studies in different animal models (Matuschak and Pinsky 1987, 1989; Brienza *et al.* 1995) have contributed to the understanding of the mechanisms by which positive-pressure ventilation and PEEP alter the characteristics of hepatic pressures and flows (Fig. 1).

PEEP decreases both hepatic arterial and portal flow in parallel with the reduction in cardiac output (Brienza *et al.* 1995). Increased intra-abdominal pressure will decrease transmural pressure in the portal vein and the upstream driving pressure for portal inflow. An increase in right atrial pressure, which represents the effective back-pressure to portal vein flow when the portosinusoidal waterfall is exceeded, decreases the downstream pressure gradient for hepatic venous outflow. Moreover, diaphragmatic descent, which occurs during mechanical ventilation with PEEP, produces mechanical compression of the liver surface, reduction in the diameter and number of perfused sinusoids, and alterations in filtration rate. This explains the significant increase in liver venous resistance observed during ventilation with PEEP, which results in a further decrease in portal flow and total venous return.

The fall in hepatic artery flow is also associated with an increase in back-pressure, consistent with an increase in surrounding tissue pressure. Arterial resistance is not significantly increased, probably because of an intact compensatory hepatic arterial buffer response (Brienza *et al.* 1995).

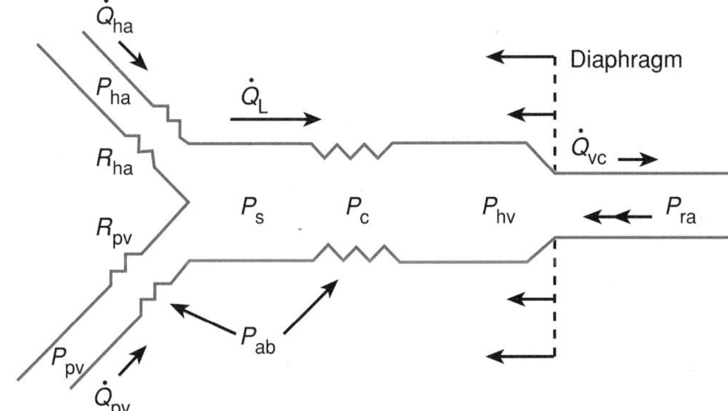

Fig. 1 Schematic representation of the forces interacting to determine the characteristics of hepatic blood flow ($\dot{Q}_L$) during phasic increases in intrathoracic pressure during positive-pressure ventilation. Positive-pressure ventilation can alter $\dot{Q}_L$ and its distribution by (a) decreasing cardiac output and thus hepatic arterial flow ($\dot{Q}_{ha}$) and portal venous flow ($\dot{Q}_{pv}$), (b) diaphragmatic compression, which increases intrahepatic closing pressure (P_c) and creates a postsinusoidal flow-limiting segment, (c) transmission of a peak-inspiratory right atrial pressure wave (P_{ra}) towards the liver, decreasing the pressure gradient for $\dot{Q}_L$, and (d) increasing abdominal pressure (P_{ab}), which influences transmural portal venous pressure (P_{pv}) and P_c. P_{ha}, hepatic arterial pressure; R_{ha}, hepatic arterial resistance; *Rpv*, portal venous resistance; P_s, sinusoidal pressure; P_{hv}, hepatic venous pressure; $\dot{Q}_{vc}$, inferior vena caval flow. (Reproduced with permission from Matuschak and Pinsky (1989).)

Volume expansion, which is commonly used to attenuate the hemodynamic effects of PEEP, restores total cardiac output and liver blood flow to pre-PEEP levels (Matuschak and Pinsky 1987, 1989). Brienza *et al.* (1995) reported that volume expansion during ventilation with PEEP significantly decreased portal venous resistance (owing to the distension of the portal vein by transmural pressure) and restored portal vein flow. However, hepatic artery flow was incompletely restored.

Altered hepatic function, producing a decrease in hepatocyte extraction and intrinsic clearance abilities, impairment of hepatic storage, and reduced biliary excretion, has been observed during mechanical ventilation with and without PEEP. This was not confirmed by Matuschak *et al.* (1987) who showed that addition of PEEP did not influence indocyanide green extraction and clearance. They postulated that despite a decrease in hepatic blood flow, the increase in hepatic back-pressure may increase trans-sinusoidal passage of diffusable substances into the space of Disse and enhance their uptake. However, it is conceivable that the hemodynamic consequences of PEEP could induce liver dysfunction in the presence of hypoxemia, hypotension, or any other condition in which the hepatic arterial buffer response is abolished and oxygen supply to the liver is compromised.

Significant falls in mesenteric blood flow may occur during use of PEEP; this is independent of raised intra-abdominal pressure and cannot be reversed by volume expansion, although it may be reversible by early gut feeding. The role of drugs with β-adrenergic properties, particularly dopexamine, which may compensate for the decrease in splanchnic and portal inflow induced by PEEP should be further investigated. The effect of PEEP-induced hemodynamic changes on the intrinsic metabolic regulation of hepatic circulation by endothelial and Kuppfer cells in steady state, shock, sepsis, and multiple organ dysfunction also remains to be evaluated.

The effect on renal function

Decreased urinary output with lowered renal sodium excretion and a positive water and sodium balance were first described in patients treated with prolonged mechanical ventilation in 1968. The addition of PEEP was shown to enhance this effect. Increases in intrathoracic pressure and decreases in venous return, transmural right atrial pressure, and mean arterial pressure initiate a cascade of interconnected regulatory mechanisms which ultimately reduce both diuresis and natriuresis. Mediation of the renal actions of PEEP is postulated to be by the following mechanisms:

(1) a direct effect of decreased cardiac output and blood pressure on renal perfusion pressure, glomerular filtration rate, and tubular sodium reabsorption;

(2) stimulation of neural mechanisms (stimulation of central sympathetic activity in response to decreased circulatory pressure; unloading of atrial and arterial baroreceptors with attenuation of tonic vagal activity as a consequence of elevated intrathoracic pressure);

(3) hormonal changes induced by the hemodynamic and neural alterations mentioned above.

Direct effects of hemodynamic changes on renal function

Although frequently mentioned, the role of PEEP-induced decreases in renal blood flow seems to be less important than the decrease in renal perfusion pressure. Restoration of renal blood flow by volume expansion does not abolish the decrease in diuresis and natriuresis (Farge *et al.* 1995).

A recent study of human renal allograft recipients showed that renal handling of water and sodium during ventilation with PEEP was dependent on perfusion pressure, despite denervation and renal blood flow stability. Animal work demonstrated that ventilation-independent rises in inferior vena caval and renal vein pressure increased water and sodium retention, and the addition of mechanical ventilation with PEEP enhanced this effect. This phenomenon, which may depend on increases in renal tissue pressure and local neurogenic-reflex-mediated changes, needs further elucidation.

Hormonal changes

Vasopressin secretion

An important increase in plasma arginine vasopressin concentrations during ventilation with PEEP has been described in both animal and human studies. Vasopressin release probably occurs as a consequence of deactivation of atrial and aortic baroreceptors by the increase in intrathoracic pressure and decrease in aortic pressure. However, the quantitative role of vasopressin in PEEP-induced antidiuresis has yet to be clearly evaluated. Extremely high levels of vasopressin were observed when no volume restoration was undertaken during ventilation with PEEP, while normal levels were found with concurrent volume expansion. This could be due to the role of vasopressin as a systemic vasoconstrictor involved in circulatory homeostasis, rather than to its direct effects on renal proximal tubules.

Renin–angiotensin–aldosterone system

Plasma renin activity is elevated during ventilation with PEEP. Decreases in renal perfusion pressure and stimulation of sympathetic activity stimulate the release of renin from the kidney. Renin stimulates the production of angiotensin I from its substrate angiotensinogen by the action of angiotensin-converting enzyme. Angiotensin II increases the mean arterial blood pressure and systemic vascular resistance and decreases the glomerular filtration rate by constricting afferent glomerular vessels and mesangial cells. In addition, angiotensin II is a direct stimulator of tubular renal sodium reabsorption. It also stimulates the adrenal cortex to release aldosterone, which causes water retention and increased sodium absorbtion in the distal tubule of the nephrons.

Atrial natriuretic factor

Atrial natriuretic factor is a peptide hormone that is synthethized, stored, and secreted by the cardiac atria. It acts on the kidney to increase urine flow and sodium excretion, and may also enhance the glomerular filtration rate. Atrial natriuretic factor also antagonizes both the release and endorgan effects of vasopressin, renin, and aldosterone. Atrial distension is a major mechanism of atrial natriuretic factor secretion and, conversely, atrial compression can diminish the production of this hormone. Decreased atrial natriuretic factor secretion during ventilation with PEEP has been described in several human and animal studies. However, the absence of any modification in atrial natriuretic factor secretion during ventilation with 15 cmH_2O PEEP has recently been reported. These conflicting results may be due to differences in atrial transmural pressures and sampling sites (arterial versus atrial).

Neural changes

Stimulation of sympathetic nerves present in renal glomeruli and tubules not only increases renal renin release but also directly decreases the glomerular filtration rate and the reabsorption of water and sodium (Farge *et al.* 1995) The effect of the deactivation of low- and high-pressure intrathoracic baroreceptors has also been previously described.

All these neural and hormonal responses, which could be regarded as potential homeostatic mechanisms protecting circulatory stability during mechanical ventilation with PEEP, may lead to excessive fluid retention in severely ill intensive care patients.

Correction of systemic circulatory depression by volume expansion and increasing water and sodium excretion by low-dose dopamine administration may attenuate the effects of elevated intrathoracic pressure on renal function. The potential therapeutic benefit of angiotensin-converting enzyme inhibitors in this setting is under investigation.

Effect on cerebral hemodynamics and brain function

Elevated intrathoracic pressure caused by mechanical ventilation with PEEP may influence cerebral hemodynamics by several mechanisms, particularly if cerebral blood flow regulation is impaired and pressure dependent. PEEP-related changes in the systemic circulation (decreased venous return, decreased mean arterial pressure (MAP)) will decrease cerebral arterial inflow. Moreover, application of PEEP may increase cerebral venous pressure and intracranial pressure (ICP), and thus further decrease cerebral perfusion pressure (CPP = MAP – ICP) and cerebral venous outflow.

PEEP increases intracranial pressure by direct transmission of elevated pleural pressure through the intervertebral foramina to the spinal and cerebral subarachnoid spaces, and, in some cases, by $P\text{a}CO_2$ elevation as the ratio of physiological dead-space to tidal volume may be increased with the application of PEEP. However, the main transmission of elevated intrathoracic pressure occurs via the systemic and cerebral venous system.

PEEP increases right atrial, internal jugular, and vertebral venous plexus pressures. This is transmitted to the dural sinuses and, at least partially, to cerebral veins with a subsequent rise in intracranial pressure. The magnitude of these changes depends on the amount of PEEP applied and on pulmonary compliance (the stiffer the lung, the less pressure is transmitted to the brain). The transmission of raised atrial pressure to the intracranial contents could be particularly dangerous in patients with pre-existing intracranial hypertension. Abrupt increases of intracranial pressure and decreases in cerebral compliance were observed during application of PEEP in dogs and rabbits with mass lesions artificially produced by inflation of an intracranial balloon.

Conversely, while studying the effect of PEEP in a similar dog model, Huseby *et al.* (1981) found that intracranial hypertension diminished the increase in intracranial pressure for a given level of PEEP. They postulated that there is a compensatory waterfall mechanism with a significant pressure gradient between the cortical veins, which are compressible by the intracranial pressure, and the sagittal venous sinus which is not influenced by intracranial pressure changes. When the intracranial pressure is low, the cortical veins are largely open so that increases in the sagittal sinus pressure are easily transmitted to the intracranial contents, limiting cerebral venous outflow and increasing cerebral blood volume. In contrast, when intracranial pressure is elevated, cortical veins become compressed and the increased pressure gradient between the cortical veins and the sagittal sinus prevents the pressure rise in the sagittal sinus from affecting intracranial pressure (Fig. 2). Another possible compensatory mechanism for elevated cerebral venous pressure is the opening of additional anastomotic channels with a redistribution of cerebral venous outflow.

The waterfall concept was recently contested by Asgeirsson and Grande (1994) who postulated that cerebral venous outflow is regulated by a venous resistance situated at the outflow orifice and directly related to the variations of interstitial tissue pressure; furthermore, they claimed that cortical veins were neither compressed nor collapsed, but remained dilated during increases in intracranial pressure.

The evidence regarding the potential effects of PEEP in patients with brain lesions is conflicting. Some authors have found no clinically significant rise in intracranial pressure on use of PEEP, although others have reported increases in intracranial pressure and neurological deterioration following application of PEEP in patients with decreased cerebral compliance.

The overall effects of elevated intrathoracic pressure on cerebral hemodynamics depend on the interactions of pulmonary and cerebral compliance, the status of cerebral and systemic circulation, and any improvement in respiratory function obtained by ventilation with PEEP. In patients with impaired autoregulation, high intracranial pressure, and reduced cerebral perfusion pressure, the lowest possible intrathoracic pressure should be employed. Moderate head elevation (15–30 per cent), which may decrease the jugular venous

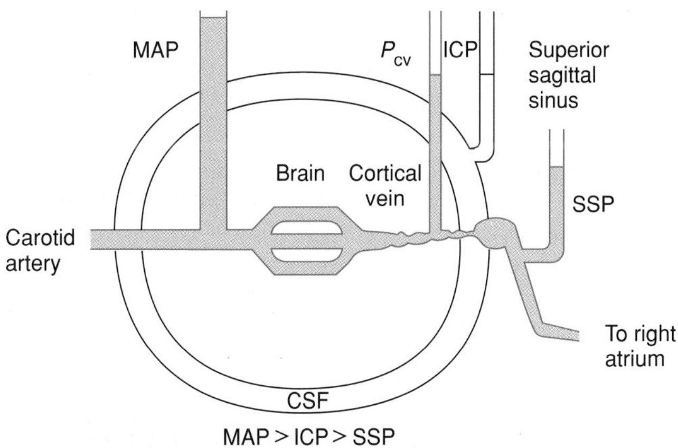

Fig. 2 Schematic illustration of the intracranial space during raised intracranial pressure (ICP). The arrows indicate the position of the hypothesized Starling resistor. Here, the mean arterial pressure (MAP) is greater than intracranial pressure, which is greater than the sagittal sinus pressure (SSP). The cortical vein pressure (P_{cv}) cannot fall below the intracranial pressure, and thus flow is dependent on MAP – ICP and independent of small changes in the sagittal sinus pressure. (Reproduced with permission from Huseby *et al.* (1981).)

pressure, was proposed for the management of these patients. However, conflicting results regarding the beneficial effect of this maneuver and its influence on cerebral perfusion pressure have recently been presented (Asgeirsson and Grande 1994).

References

Asgeirsson, B. and Grande, P.O. (1994). Effects of arterial and venous pressure alterations on transcapillary fluid exchange during raised tissue pressure. *Intensive Care Medicine*, **20**, 567–72.

Brienza, N., Revelly, J.P., Ayuse, T., and Robotham, J.L. (1995). Effects of PEEP on liver arterial and venous blood flows. *American Journal of Respiratory and Critical Care Medicine*, **152**, 504–10.

Farge, D., de la Coussaye, J.E., Beloucif, S., Fratacci, M.D., and Payen, D.M. (1995). Interactions between hemodynamic and hormonal modifications during PEEP induced antidiuresis and antinatriuresis. *Chest*, **107**, 1095–1100.

Huseby, J.S., Luce, J.M., Cary, J.M., Pavlin, E.G., and Butler, M.P. (1981). Effects of positive end expiratory pressure on intracranial pressure in dogs with intracranial hypertension. *Journal of Neurosurgery*, **55**, 704–7.

Matuschak, G.M. and Pinsky, M.R. (1987). Effect of positive end expiratory pressure on hepatic blood flow and performance. *Journal of Applied Physiology*, **62**, 1377–83.

Matuschak, G.M. and Pinsky, M.R. (1989). Effect of positive pressure ventilatory frequency on hepatic blood flow and performance. *Journal of Critical Care*, **4**, 153–65.

18.2.15 Ventilator trauma

J. J. Marini

Key messages

- The rapid development of cystic barotrauma usually presages the occurrence of tension pneumothorax. Effective secretion clearance, treatment of infection, and, most importantly, reduction of airway pressure are fundamental to management.
- In patients with acute respiratory distress syndrome ventilated with high tidal pressures and maintained with low left ventricular filling pressures, alveolar pressures may exceed those in the pulmonary veins. If alveolar rupture opens a communication pathway to the vascular system, this pressure gradient may drive air into the systemic circulation.
- Despite the importance of peak tidal alveolar pressure, there are many contributing predisposing variables: necrotizing and heterogeneous lung pathology, youth, copious retained airway secretions, and extended duration of positive-pressure ventilation.
- As a general rule, high peak pressures applied to a stiff lung cause less alveolar stretch than the same pressures applied to a compliant lung. Pneumothorax becomes much more likely at peak ventilator cycling pressures above 40 cmH_2O.
- Failure to maintain a certain minimum end-expiratory transalveolar pressure (i.e. total PEEP) in the early phase of acute respiratory distress syndrome may intensify pre-existing alveolar damage, particularly when high tidal volumes and inflation pressures are used.

Pneumothorax and other forms of lung rupture

Pathogenesis of alveolar rupture

The various forms of pulmonary barotrauma (interstitial emphysema, pneumomediastinum, pneumoperitoneum, subcutaneous emphysema, lung cyst formation, and pneumothorax) are prominent among the iatrogenic causes of critical illness. Patients with acute respiratory distress syndrome appear to be at highest risk. Although direct rupture of the visceral pleura occasionally occurs, the barotrauma that complicates mechanical ventilation most frequently develops more circuitously. Alveolar rupture is most likely to occur in 'non-partitional' or 'marginal' alveoli, which have bases contiguous with relatively immobile structures (vessels, bronchioles, or fibrous septae). Once excessive pressure gradients between marginal alveoli and contiguous perivascular connective tissues have caused alveolar rupture, extra-alveolar gas follows a pressure gradient along the perivascular sheaths toward the hilum.

In the absence of mediastinal pathology, gas usually dissects along fascial planes into the mediastinum and soft tissues of the neck (subcutaneous emphysema) or retroperitoneum (pneumoperitoneum). Pneumothorax develops in a minority of such cases (perhaps 20–30 per cent) when soft tissue gas ruptures into the pleural space via an interrupted or weakened mediastinal pleural membrane. Unlike pneumothorax, interstitial emphysema, pneumomediastinum, and subcutaneous emphysema have little hemodynamic significance and seldom influence gas exchange in adult patients.

Bronchopulmonary injury

Until recently, ventilator-induced bronchial damage was thought to occur only rarely in adults. However, in the heterogeneously injured lung, high airway pressures can inflict considerable damage in small airways unsupported by cartilage. Airway distortion predisposes to cystic parenchymal damage, disordered gas exchange, and impaired secretion clearance.

Cystic barotrauma

When normal bronchovascular channels are blocked, gas accumulates locally or migrates distally to produce subpleural air cysts. It is not uncommon for a cyst created by positive airway pressure to enlarge quickly, even to a diameter exceeding 10 cm. These cysts compress normal lung tissue, increasing the airway pressure needed for effective ventilation and creating dead-space that raises the ventilatory requirement and mean alveolar pressure. Once under way, cystic barotrauma tends to be self-reinforcing. The rapid development of cystic barotrauma usually presages the occurrence of tension pneumothorax. Effective secretion clearance, treatment of infection, and, most importantly, reduction of airway pressure are fundamental to management.

Systemic gas embolism

In patients with acute respiratory distress syndrome ventilated with high tidal pressures and maintained with low left ventricular filling pressures, alveolar pressures may exceed those in the pulmonary veins. If alveolar rupture opens a communication pathway to the vascular system, this pressure gradient may drive air into the systemic circulation. Irritating microbubbles can then cause vasospasm, with resulting neurological damage or myocardial infarction.

Tension pneumothorax

A sustained increase of pleural pressure ('tension') often complicates a pneumothorax arising during mechanical ventilation. A tension component occasionally develops when a ball-valve mechanism pumps air into the pleural cavity during spontaneous breathing, but is much more common when positive pressure provides the ventilatory power. Positive intrapleural pressure expands the ipsilateral chest cage, impairing inspiratory muscular efficiency. Contralateral pleural pressure tends to be maintained at near-normal levels until late in the process. However, a shifting mediastinum may encroach upon the contralateral hemithorax, compromising lung expansion. Eventually, rising pleural and central venous pressures impede venous return sufficiently to cause hemodynamic deterioration. Vigorous inspiratory efforts tend to maintain intrapleural pressure (averaged for both lungs over the entire respiratory cycle) at near-normal levels until the patient fatigues, is sedated, or receives ventilator assistance. Abrupt hemodynamic deterioration may then occur as the mean pleural pressure rises. Ventilated patients with pneumothorax frequently experience abrupt hemodynamic deterioration immediately after receiving sedating or paralytic drugs. Tension can develop without lung collapse or even major volume loss when the lung is heavily infiltrated, or when air is trapped or bound locally by pleural adhesions.

Risk factors for barotrauma

Alveolar rupture appears to require sustained hyperexpansion of fragile alveoli. Despite the presumed importance of peak tidal alveolar pressure, there are many contributing predisposing variables: necrotizing and heterogeneous lung pathology, youth, copious retained airway secretions, and extended duration of positive-pressure ventilation (Table 1). As major determinants of peak and mean transalveolar pressures, minute ventilation requirements and high levels of PEEP (positive end-expiratory pressure) contribute to the hazard. Peak dynamic and static airway pressures (P_D and P_S) appear to contribute most to the multivariate risk equation; when considered separately from its influence on inspiratory pressure, PEEP contributes little to the risk of barotrauma.

Table 1 Predispositions to barotrauma

Necrotizing lung pathology	High peak cycling pressure
Secretion retention	High mean alveolar pressure
Young age	Minute ventilation requirement
Duration of ventilation	Non-homogeneous parenchymal disease

As a general rule, high peak pressures applied to a stiff lung cause less alveolar stretch than the same pressures applied to a compliant lung. Pneumothorax becomes much more likely at peak ventilator cycling pressures above 40 cmH_2O. However, the inherent susceptibility of lung tissue to rupture, chest wall compliance, and the degree of regional inhomogeneity also play crucial roles. Largely for these reasons, there is no sharp threshold value of peak ventilator cycling pressure below which lung rupture does not occur. A peak static (plateau) pressure above 35 cmH_2O usually achieves or exceeds the alveolar volume corresponding to total lung capacity in a patient with a normal chest wall. However, when the chest wall stiffens, higher plateau pressures may be well tolerated. Secretions, blood clots, or foreign objects in the airway can give rise to heterogeneity or ball-valve mechanisms that encourage barotrauma. The crucial roles of parenchymal inhomogeneity and inflammation in producing barotrauma may explain why pneumothorax tends to develop 1 to 3 weeks after the onset of diffuse lung injury, when some regions are healing while others remain actively inflamed.

Diagnosis of barotrauma

Clinical features

Pneumothorax frequently presents with tachypnea, respiratory distress, tachycardia, diaphoresis, cyanosis, or agitation. In patients receiving volume-cycled ventilation, peak inspiratory (and peak static) airway pressures typically increase, and compliance falls as pneumothorax develops, particularly if tension is present. Volume-cycled ventilators may 'pressure limit' or 'pop off', resulting in ineffective ventilation. During pressure-controlled ventilation, tidal volume and/or minute ventilation decrease, while airway pressures remain unaffected.

Massive gas trapping and auto-PEEP effectively mimic tension pneumothorax, particularly if hyperinflation or infiltration is asymmetrically distributed. Tension is reflected in elevations of central venous, right atrial, and pulmonary arterial pressures.

Radiographic signs of barotrauma

Two useful markers of occult pneumothorax visible on supine films are the deep sulcus sign and hyperlucency centered over the ipsilateral abdominal upper quadrant. In bedridden patients, a lateral decubitus view facilitates visualization by allowing air to collect along the upper margin of the hemithorax. An expiratory chest film thickens the stripe of pleural gas. Pneumothorax under tension can be strongly suspected on a single film when diaphragmatic inversion or extreme mediastinal shift occurs. Obtaining a sequence of films to demonstrate progressive migration of mediastinal contents into the contralateral hemithorax delays diagnosis but confirms validity. Life-threatening tension can exist without complete lung collapse or mediastinal displacement if the lung adheres to the pleura or is densely infiltrated. Tension without mediastinal shift may also develop if the airway becomes obstructed, or the mediastinum is immobilized by infection, fibrosis, neoplasm, or previous surgery.

Value of CT imaging

Thoracic CT scanning is an invaluable aid in determining whether a lucency represents parenchymal or pleural air. Indeed, accurate chest

Table 2 Preventing ventilator-related lung rupture

Minimize minute ventilation	Normalize lung compliance
Minimize PEEP	Improve chest wall compliance
Use lower tidal volume	Encourage spontaneous breathing
Decrease inspiratory-to-expiratory ratio	
Decrease bronchial obstruction	
Use newer modes of ventilation (?)	

tube placement into a loculated pocket of gas or fluid may require CT-guided insertion.

Management of pneumothorax

General principles

The key interventions aimed at avoiding barotrauma (Table 2) are as follows.

1. Treat the underlying disease, particularly suppurative processes.
2. Relieve bronchospasm and maintain effective bronchial hygiene, but minimize unnecessary coughing.
3. Reduce minute ventilation requirement by treating agitation, fever, metabolic acidosis, and bronchospasm. Consider sedation (or paralysis) to enforce hypoventilation and permissive hypercapnia.
4. Reduce peak and mean alveolar pressures by limiting PEEP and auto-PEEP, using small tidal volumes, and increasing the percentage of spontaneous versus machine-aided breaths.
5. Drain the pleural space of air using the least effective suction.

Ventilator-induced pulmonary edema, lung injury, and volutrauma

Pathogenesis

Effect of excessive peak pressures

In excess of 20 000 tidal ventilatory cycles are undertaken daily. Ventilatory patterns that apply high transalveolar stretching forces cause or extend tissue edema and alveolar damage in experimental animals, even when alveolar rupture does not occur. Maximum alveolar pressures above 30 cmH_2O commonly produce regional overdistension in patients with acute respiratory distress syndrome. Peak tidal pressures of this magnitude cause tissue damage in experimental animals when ventilation is sustained for more than 12 to 24 h. The lung may be susceptible to edematous injury produced by high inflating pressures in both the early and late stages of acute respiratory distress syndrome. As the strong collagen infrastructure of the lung degrades, such pressures become increasingly likely to result in overt alveolar disruption (pneumothorax, pneumomediastinum, gas cyst formation) rather than edema.

Importance of end-expiratory lung volume and PEEP

Failure to maintain a certain minimum end-expiratory transalveolar pressure (i.e. total PEEP) in the early phase of acute respiratory distress syndrome may intensify pre-existing alveolar damage, particularly when high tidal volumes and inflation pressures are used. Indeed, shear forces associated with tidal collapse and reinflation of injured alveolar tissues may be responsible for an important component of ventilator-induced lung damage. The end-expiratory pressure required to avert widespread alveolar collapse varies with the regional forces applied to the lung; consequently, a higher end-expiratory pressure is required to prevent atelectasis in dependent compared with more superior regions. Similarly, a higher total PEEP is required when the chest wall is poorly compliant. Gravitational factors help to explain the strikingly dependent distribution of radiographic infiltrates at the onset of lung injury, as well as the reversal of these infiltrates and improvement of arterial oxygenation in the prone position. Total PEEP sufficient to place the tidal volume above the initial low-compliance region (P_{flex}) of the static pressure–volume relationship of the respiratory system appears to attenuate the severe hemorrhagic edema otherwise induced by high ventilating pressure.

Experimentally, damaging the lung requires both high tidal pressures and failure to maintain patency of lung units tending to collapse. What recruits one area of the lung (e.g. dependent regions) is likely to overdistend others (e.g. non-dependent regions) (Fig. 1). Lung recruitment is likely to occur throughout the tidal cycle whenever alveolar pressure reaches a sufficient opening pressure during inspiration and insufficient PEEP is used to keep the same units from closing.

Stress failure of the pulmonary capillaries with resulting extravasation of formed blood elements into the lung tissue may occur at transvascular pressures that exceed 40 to 90 mmHg, depending on animal species. Theoretically, transcapillary mechanical forces of comparable magnitude could be generated when high tidal volumes and peak static tidal pressures are used without sufficient PEEP in the setting of heterogeneous lung disease. High vascular pressure and blood flows may also be important determinants of lung injury.

Ventilatory management

No detailed clinical information is available for guidance regarding the maximum peak and mean alveolar pressures that can be applied safely for extended periods without inducing alveolar damage or retarding lung healing. Clearly, the answer differs between disease stages and among individual patients. The stiffness of the chest wall influences the airway pressure that can be tolerated without overdistension. Moreover, alveolar volumes and stresses undoubtedly vary from site to site within the damaged lung. If avoiding ventilator-induced lung damage is the goal, the pressure applied to the endotracheal tube should allow for the distensibility and vulnerability of all lung units. Although failure to preserve a certain minimum end-expiratory transalveolar pressure intensifies pre-existing alveolar damage in the laboratory setting, this phenomenon has not yet been

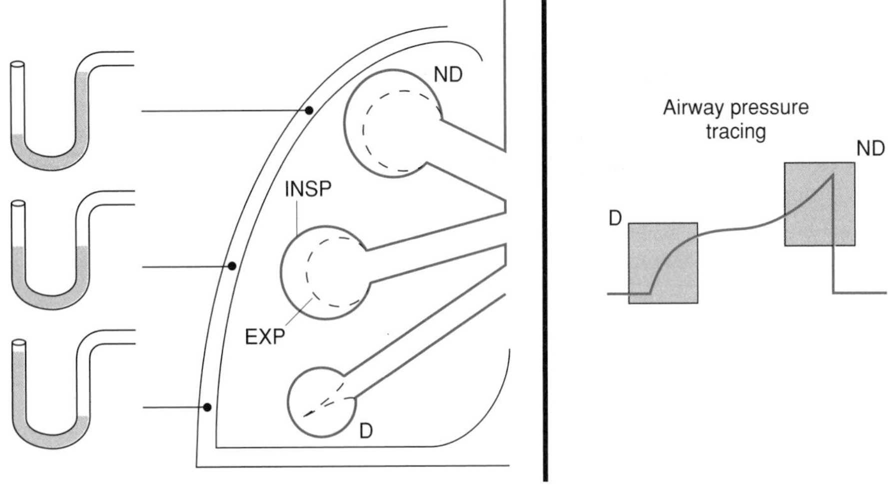

Fig. 1 Regional alveolar mechanics during the tidal breath in acute respiratory distress syndrome. Dependent lung units (D) may be surrounded by sufficiently high pressure to collapse at end-expiration and re-expand at some pressure achieved during the tidal cycle. Non-dependent units (ND) that are exposed to high extra-alveolar pressures may overdistend at end-inspiration, risking 'over-stretch' injury. During controlled ventilation with constant inspiratory flow, the airway pressure tracing may show indirect evidence of these phenomena if it displays segments of rapidly improving and rapidly deteriorating respiratory system compliance. (Reproduced with permission from Marini and Wheeler (1997).)

clearly demonstrated in humans. Moreover, once recruitment has been completed, additional PEEP is probably ineffectual or damaging. Consequently, expert opinion differs on whether applying the least PEEP that accomplishes adequate gas exchange or, alternatively, guaranteeing some minimum value of end-expiratory alveolar pressure is the best course to follow within the first few days of onset of illness. Periodic application of sustained high inflating pressures to recruit unstable lung units (recruiting maneuvers) continues to be advocated by some knowledgeable investigators, particularly when small tidal volumes (< 4–5 ml/kg) are used or high-frequency ventilation is employed. PEEP should be withdrawn later in the disease process, particularly if no inflection region can be identified on the static pressure–volume curve of the respiratory system. Because tidal compliance depends on tidal volume, the appropriate tidal volume to select undoubtedly varies with the level of PEEP and vice versa.

There is no firm consensus regarding the contribution of vascular pressures, position changes, infections, inspired oxygen concentration, and other clinical variables on the incidence or intensity of ventilator-induced lung edema. Furthermore, detailed information is lacking regarding ventilation pressures and patterns of inflation which are safe to apply for extended periods. At fractional inspired oxygen (FiO_2) levels below 0.7, limiting airway pressure to 'safe' levels generally takes precedence over limiting FiO_2. In the absence of definitive data obtained in a clinical context, some practitioners increase end-expiratory lung volume in an attempt to minimize FiO_2, whereas others prefer to use higher FiO_2 rather than increase peak, mean, and end-expiratory airway pressures. Whether methods for achieving a similar mean airway pressure (such as PEEP and inverse ratio ventilation) differ with respect to risks and benefits has not been adequately clarified.

Bibliography

Dreyfuss, D. and Saumon, G. (1993). Role of tidal volume, FRC, and end-inspiratory volume in the development of pulmonary edema following mechanical ventilation. *American Review of Respiratory Disease*, **148**, 1194–1203.

Gammon, B.R., Shin, M.S., and Buchalter, S.E. (1992). Pulmonary barotrauma in mechanical ventilation: patterns and risk factors. *Chest*, **102**, 568–72.

Marini, J.J. (1996). Evolving concepts in the ventilatory management of acute respiratory distress syndrome. *Clinics in Chest Medicine*, **17**, 555–75.

Marini, J.J., and Wheeler, A.P. (1997). *Critical care medicine — the essentials* (2nd edn). Williams & Wilkins, Baltimore, MD.

Pierson, D.J. (1994). Barotrauma and bronchopleural fistula. In *Principles and practice of mechanical ventilation* (ed. M.J. Tobin), pp. 813–36. McGraw-Hill, New York.

Slutsky, A.S. (1993). Barotrauma and alveolar recruitment (editorial). *Intensive Care Medicine*, **19**, 369–71.

18.2.16 High-frequency ventilation

B. F. Keogh and E. H. Gluck

Key messages

- High-frequency ventilation has been use in adult critical care for over 20 years.
- Early devices provided limited ventilatory capacity, particularly in patients with poor pulmonary compliance, and produced indifferent results.
- Ventilatory performance, monitoring and associated safety are greatly enhanced in the new generation of high-frequency ventilation devices.
- Specific indications include airway trauma and persistent pulmonary air leaks (where a low pressure philosophy is applied) and acute respiratory distress syndrome (where a higher pressure approach with attention to lung volume maintenance is employed).

Introduction

High-frequency ventilation (HFV) is a broad term describing ventilation at frequencies greater than four times the natural respiratory frequency of the ventilated subject. First described in 1969, it has attracted considerable scientific interest but has failed to establish a definitive or widely accepted niche in clinical practice. HFV continues to evolve; the availability of more effective technology, allied to enhanced understanding of its principles, has rendered HFV a realistic option in the critical care management of certain forms of respiratory failure.

Gas exchange in high-frequency ventilation

How HFV achieves effective gas exchange despite generating tidal breaths less than the anatomical dead-space remains a mystery. Many theories exist and involve extremely complex physiology and fluid mechanics. Involved processes include augmented gas diffusion in the distal airways, Taylor (lateral) dispersion of gas molecules in more proximal small airways, and high-flow CO_2 washout of central compartments. In addition, *Pendelluft* (cross-ventilation between lung units of different time constants) undoubtedly contributes to effective gas exchange.

The distribution of inspired gas in HFV varies significantly from conventional ventilation. In contrast with low-frequency techniques, where the lower lobes may be preferentially ventilated due to regional lung capacitance, HFV results in more uniform distribution of gas throughout all lobes. Gas distribution in HFV is influenced more by airway resistance and results in more homogeneous gas distribution in both healthy and diseased lungs.

Principles of oxygenation with HFV do not vary significantly from conventional techniques. Effective oxygenation in normal lungs is easily achieved with HFV at low airway pressures but is unpredictable in the presence of pneumonic consolidation or chronic obstructive airways disease. In contrast, in pulmonary edema from any cause, intrapulmonary shunt can be manipulated, as in conventional ventilation, by titration of mean airway pressure, the clinical measure which most usefully reflects lung volume.

Design features of high-frequency jet ventilation devices

Although several HFV devices have been described, the only form currently relevant to adult practice is high-frequency jet ventilation (HFJV) in which short pulses of gas are accelerated through a jet nozzle into the airway (Fig. 1). A negative lateral pressure, which enhances the gas pulse volume, is generated around the jet. Such ventilators operate at frequencies of 60 to 600 breaths/min and employ a passive expiratory phase. High-frequency oscillation, which employs an active expiratory phase and is widely applied in the infant respiratory distress syndrome, has not been effectively translated from pediatric to adult practice.

Different HFJV devices vary markedly in their performance; small design modifications may result in major changes in ventilatory capacity. This variation, coupled with varying, and sometimes inappropriate, HFJV ventilatory philosophies, have led to inconsistent reports of efficacy. In addition, while less powerful devices may prove effective in mildly diseased lungs, they cannot be expected to support patients with severe disease and markedly reduced pulmonary compliance.

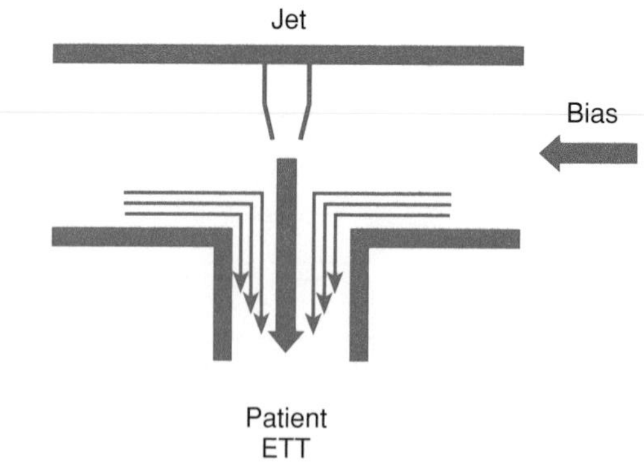

Fig. 1 The jet nozzle sits within the entrainment chamber at the proximal end of the endotracheal tube (ETT), providing an entrainment ratio of up to 200 per cent (in contrast, the entrainment ratio with distal positioning is as low as 25 per cent). The bias gas provides both humidification and variation in fractional inspired oxygen (FiO_2). (Reproduced with permission from Evans and Haslett (1986).)

New HFJV devices employ very rapid response solenoid valves so that the gas pulse is delivered with a square pressure wave, allowing optimal entrainment and maintenance of mean airway pressure if required. Earlier models, which generate pseudosinusoidal pressure waves, are unable to provide the ventilatory capacity now available. New machines provide continuous analysis of ventilator function and safety as well as airway pressure profiles which provide clinically relevant assessments of patient–ventilator interaction.

Controls on HFJV devices are frequency, inspiratory time, and driving pressure. Initial settings depend on the device applied, although general principles exist. Airway pressure and arterial oxygenation are influenced by inspiratory time and driving pressure. Inspiratory time has little effect on CO_2 clearance which is most influenced by driving pressure and frequency. Algorithms for intervention when gas exchange is unacceptable are easily constructed, based on an understanding of the influence of these control parameters. Typical settings when ventilating normal lungs are frequency 60 to 120 breaths/min, driving pressure 1 to 1.5 atm (15–22 psi), and inspiratory time 15 to 20 per cent. In contrast, in acute respiratory distress syndrome (ARDS) frequency is set from 250 to 350 breaths/min, driving pressure 2 to 3 atm (30–40 psi), and inspiratory time 36 to 40 per cent. The considerable difference in these initial settings reflects the need for volume maintenance in the poorly compliant lung. In hypoxic patients, external positive end-expiratory pressure (PEEP) valves are necessary with older devices, but newer more powerful technology relies on the effective generation of intrinsic PEEP. Weaning from HFJV can be smoothly achieved by a gradual reduction in all three controls. The availability of high fresh gas flow in an open system, combined with inspiratory support, should favorably influence the work of breathing.

High-frequency jet ventilation in adult critical care

Some intensive care units have reported their experience in using HFJV as the standard ventilatory mode. This approach is unusual; in most units HFJV is selectively applied in patients considered most likely to benefit. Currently, these patients fall into two major categories in which the philosophy behind application of HFJV is somewhat different.

The management of air leaks—a low-pressure approach

HFJV has been used in the management of bronchopleural fistula for many years. At low pressure, HFJV is thought to decrease gas flow through the fistula, enhancing the possibility of closure. Inspired gas distribution with HFJV is determined by airways resistance (rather than compliance); thus effective ventilation of lung units juxtaposed to an area of infinite compliance (i.e. a bronchopleural fistula), is possible. In this approach, the volume of gas pulses delivered is small. Peak and mean airway pressures are reduced to those which allow effective ventilation, while airflow through the bronchopleural fistula is minimized. This low-pressure philosophy does not invoke aggressive lung volume maintenance techniques or any attempt to generate intrinsic PEEP.

A similar low-pressure philosophy can be employed following tracheal trauma or reconstructive surgery. Ventilation without inflation of the endotracheal tube cuff is desirable following some tracheal surgical procedures and is most efficiently administered with the jet nozzle placed at the proximal end of the endotracheal tube. The technology can be modified to jet distal to surgical anastomoses via a catheter or a stomal airway such as a T-piece. HFJV has been described as the management of choice in traumatic tracheal rupture or following its surgical repair (Brimioulle *et al.* 1990).

HFJV in ARDS—maintaining the mean airway pressure

Ventilatory strategies in ARDS have evolved considerably over the last decade. Lung-protective ventilatory strategies, in which high peak inspiratory pressures and alveolar overdistension are avoided, are now universally recommended. In ARDS specifically, lung volume must be maintained and end-expiratory collapse avoided, not least because low levels of end-expiratory pressure may be associated with ventilator-induced lung damage. Because of its small pulse volumes and short cycle times, HFV theoretically provides an ideal method of achieving effective ventilation while avoiding overdistension and maintaining alveolar volume.

HFJV was first recommended in 1982 as the ideal available ventilatory mode for patients with ARDS. In 1983 a prospective randomized cross-over comparison of HFJV versus conventional ventilation in 309 patients with acute lung injury reported that, while safe, predictable, and effective, HFJV appeared to offer no particular advantage (Carlon *et al.* 1983). This frequently quoted study considerably dampened enthusiasm for HFJV. However, interpretation at the time did not take into account the lack of lung volume recruitment techniques or attempts to maintain lung volume in either treatment arm.

In 1986 a new HFJV device was developed specifically for ARDS with much greater ventilatory capacity to overcome the inability of older technology to achieve adequate gas exchange in poorly compliant lungs. Features included a rapid-response low-inertia solenoid switching valve which generated an effective square wave as described above. The T-piece entrainment chamber was placed at the proximal end of the endotracheal tube, providing an entrainment ratio of up to 200 per cent, thus greatly enhancing ventilatory capacity as well as allowing effective humidification (Fig. 1). Animal experiments indicated that the amplitude of airway pressure changes was rapidly attenuated along the divisions of the respiratory tree, and that, at the 2-mm airway, airway pressure fluctuations were minimal around a value equivalent to the centrally measured mean airway pressure.

The response to this device in 90 patients with severe ARDS was reported by Gluck *et al.* (1993). Lung volume, and hence oxygenation, was maintained by application of the appropriate mean airway pressure in each patient. Significant improvements in arterial oxygenation and reductions in peak and mean airway pressures were observed in the context of excellent CO_2 clearance. The study was a non-randomized prospective assessment in which each patient served as his or her control prior to switching to HFJV. Survival data in such a study are difficult to interpret, but in this group, considered as failing conventional support, survival was 58 per cent. With the addition of further patients, it was observed that 19 of 25 patients (76 per cent) who were switched to HFJV within 48 h of ventilation survived. This suggests that early intervention with HFJV may favorably influence outcome. A recent report in 29 pediatric patients with ARDS also noted that survivors (69 per cent) spent significantly less time on conventional ventilation, again suggesting that early introduction of HFJV may limit ventilator-induced lung damage (Smith *et al.* 1993).

This potent HFJV ventilator now has market clearance from the United States Food and Drugs Administration regulatory body for use in ARDS and is commercially available (Adult Star 1010, Nellcor Puritan Bennett, California, USA). It is currently used enthusiastically in some centers; with wider availability, its use in severe ARDS is likely to increase.

The use of an effective HFJV device whereby mean airway pressure is maintained is based on similar principles to that of pressure-controlled inverse ratio ventilation. Both modes harness the lung volume recruitment capabilities of intrinsic PEEP while operating at different ends of the frequency spectrum (Fig. 2). Peak pressure limitation is inherent in pressure-controlled inverse ratio ventilation and results indirectly from equilibration of directional gas flows with HFJV. Theoretical advantages which can be proposed for HFJV over pressure-controlled inverse ratio ventilation include the following.

1. The minimum alveolar pressure for a given central airway pressure is higher, hence enhancing arterial oxygenation.

2. The peak alveolar pressure for a given airway pressure is lower, hence limiting alveolar volutrauma (in pressure-controlled inverse ratio ventilation (PC-IRV) the alveolus is exposed to the central limiting peak pressure, whereas in HFJV the peak inspiratory pressure is considerably attenuated at alveolar level).

3. With HFJV, physiotherapy and airway toilet procedures can be performed without disconnection from ventilation and subsequent lung volume de-recruitment (relevant in many patients in whom optimal lung volume recruitment may take many hours, or in those returned to supine following prone repositioning as an alveolar recruitment maneuver).

In addition, aggressive lung volume recruitment with PC-IRV commonly mandates the practice of permissive hypercapnia. While hypercapnia is now widely applied, concerns remain about its

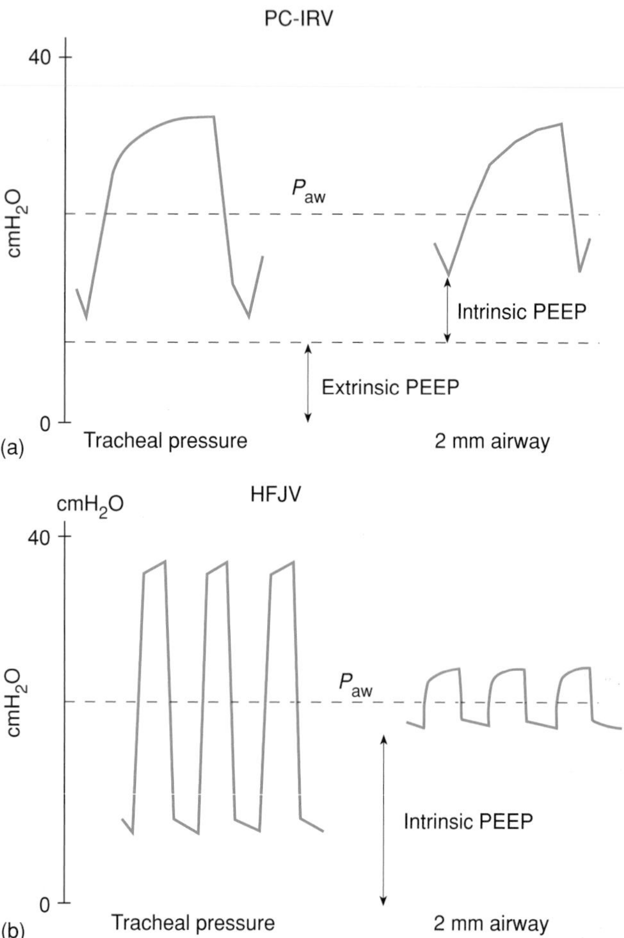

Fig. 2 Schematic representation of airway pressure (P_{aw}) changes with (a) PC-IRV and (b) HFJV in central and peripheral airways. In pressure-controlled inverse ratio ventilation, intrinsic PEEP can be harnessed with or without extrinsic PEEP. The increase in minimium alveolar pressure (by extrapolation) for a given P_{aw} will depend on the level of intrinsic PEEP. Of note is the difference in maximum alveolar pressures; essentially the alveolus is exposed to the central peak inspiratory pressure in pressure-controlled inverse ratio ventilation, whereas peak inspiratory pressure is markedly attenuated in HFJV. (Reproduced with permission from Keogh (1996).)

adverse effects, particularly in certain patient groups. The application of effective HFJV, in which hypercapnia is not necessary and $P\text{aCO}_2$ levels can be effectively controlled, may be more appropriate than pressure-controlled inverse ratio ventilation in patients with renal failure, metabolic acidosis, cardiac insufficiency, or neurological pathology.

Combined high-frequency ventilation

Combined high-frequency ventilation (CHFV), applying features of both conventional and high-frequency ventilation, was first reported in the early 1980s. Favorable gas exchange improvements were reported in refractory ARDS patients, although mortality figures were disappointing. Its future is uncertain and, as better HFJV devices are developed, it is likely that hybrid ventilatory modes will no longer be used.

High-frequency ventilation and inhaled medications

The effective delivery of inhaled medications poses technical difficulties in both conventional ventilators and HFV. Nitric oxide (NO) delivery is particularly controversial and questions of safety and ideal delivery systems persist. NO may be delivered with HFJV devices although, owing to high total gas flows, delivery is cost inefficient. In addition, efficient humidification may sequester active NO and theoretically predispose to the formation of acidic products in ventilator circuits. The higher total NO volume required compared with conventional systems mandates effective expiratory scavenging. In practice, lung volume recruitment with effective HFJV renders the use of NO unnecessary in many patients. If uncertainty exists, HFJV, with its beneficial effects on pulmonary structure, represents a more favorable option than NO application which has cosmetic effects only and does not influence the integrity of lung architecture. Nebulized epoprostenol is an alternative to NO; although costly, it can be effectively delivered via HFJV circuits.

Such technical issues can be overcome by design modifications. The incorporation of atomizers into HFV technology may also offer much more effective delivery of other inhaled medications such as surfactant, α_1-antitrypsin, and antibiotics.

Conclusion

HFV is a realistic ventilatory option in adult intensive care. Delivered at low pressure, it is an established management for airway trauma and persistent bronchopulmonary leaks. Considerable advances in technology now render it an effective option in ARDS, particularly in refractory patients, and in other conditions characterized by poor compliance. Enthusiasm remains sporadic but, with greater availability of the new technology, better understanding of underlying principles, and the further technical improvements envisaged, its application in adult practice is likely to increase considerably in the next few years.

References

Brimioulle, S., Rocmans, P., de Rood, M., Berre, J., and Kahn, R.J. (1990). High-frequency jet ventilation in the management of tracheal laceration. *Critical Care Medicine*, **18**, 338–9.

Carlon, G.C., Howland, W.S., Ray, C., Miodownik, S., Griffin, J.P., and Groeger, J.S. (1983). High-frequency jet ventilation. A prospective randomized evaluation. *Chest*, **84**, 551–59.

Evans, T.W. and Haslett, C. (1986). *Acute respiratory distress in adults*, p. 413. Chapman & Hall, London.

Gluck, E., Heard, S., Patel, C., Mohr, J., and Calkins, J. (1993). Use of ultrahigh frequency ventilation in patients with ARDS. A preliminary report. *Chest*, **103**, 1413–20.

Keogh, B.F. (1996). *Current Anaesthesia and Critical Care*, **7**, 230.

Smith, D.W., *et al.* (1993). High-frequency jet ventilation in children with the adult respiratory distress syndrome complicated by pulmonary barotrauma. *Pediatric Pulmonology*, **15**, 279–86.

18.2.17 External high-frequency oscillation

M. Nevin

Key messages

- External high-frequency oscillation (EHFO) differs from negative-pressure ventilation in that ventilation occurs around a variable baseline and both inspiration and expiration are active.
- EHFO can be used for successful ventilation of normal lungs as well as those demonstrating abnormal compliance.
- EHFO improves sputum clearance and may reduce weaning times from ventilatory support.
- Combination ventilation (EHFO plus pressure-controlled ventilation) may improve outcome in acute respiratory distress syndrome (ARDS) secondary to acute lung injury, provided that treatment is instituted before structural lung changes occur.
- EHFO appears to offer no benefit in patients with infective causes of ARDS.

Introduction

External negative-pressure ventilation was introduced by Dalziel in 1843. Alfred Jones manufactured the first 'body-enclosing lung', but almost 100 years elapsed before the first clinically practical motorized device was developed. Drinker in 1929 and Emerson 3 years later introduced versions of the famous tank respirator known as the 'iron lung'; this was used successfully between 1930 and 1960, and greatly reduced mortality during the poliomyelitis epidemics in the 1950s and 1960s.

Despite this success the iron lung had many limitations, in particular size and limited access to the immobilized patient. A further disadvantage was that only one phase of the respiratory cycle, inspiration, was 'active'. Expiration remained passive, being achieved by elastic recoil of the lungs and chest wall. This limited the frequency of breathing to less than 30 cycles/min. These limitations, as well as the increasing demand for more sophisticated methods of artificial ventilatory support, led to the introduction of positive-pressure ventilation in the 1950s. Over the next 40 years developments in ventilator technology were focused on positive-pressure ventilation, almost to the total exclusion of negative-pressure ventilation.

External high-frequency oscillation

Positive-pressure ventilation is not without its own major problems, in particular the increased risk of pulmonary barotrauma. In an attempt to overcome many of the non-physiological limitations of positive-pressure ventilation, Hayek and Schonfeld (1990) developed a lightweight flexible chest enclosure for human use (Fig. 1). It is connected to a high-frequency oscillator which oscillates the chest around a variable subatmospheric (negative) pressure. Both inspiratory and expiratory phases are active and therefore fully controllable (via chamber pressure and variable inspiratory-to expiratory (I:E) ratio). As a result, respiratory rate is not limited and external high-frequency oscillation (EHFO) can be achieved. This concept of negative-pressure ventilatory support equates closely to the natural physiology of spontaneous respiration.

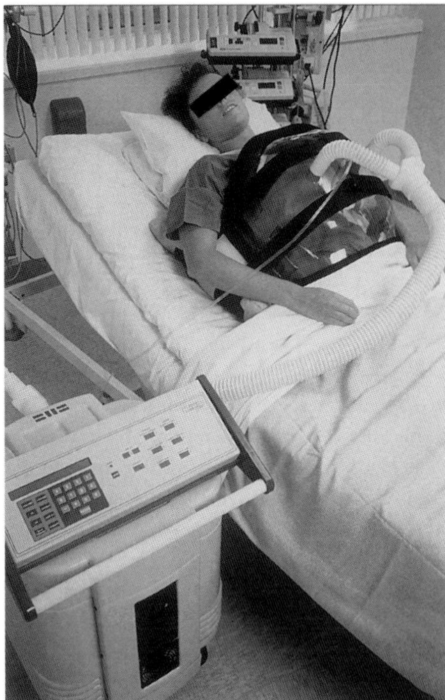

Fig. 1 EHFO: chest cuirass, wide-bore hose, and power unit.

In EHFO both peak inspiratory chamber pressure and end-expiratory chamber pressure (EECP) are controllable. While inspiratory pressures are almost always negative, EECP can be positive, atmospheric, or negative. This allows control over end-expiratory lung volume so that ventilation can proceed below, at, or above functional residual capacity in conditions where lungs are hyperinflated, normal, or underinflated respectively.

As the airway is open to atmospheric pressure, the airway pressure will fluctuate around zero provided that the I:E ratio is 1:1. Altering the I:E ratio will produce predictable variations in both peak inspiratory pressure and EECP. Hayek suggested that a positive EECP is conceptually the same as auto-PEEP; increasing frequency and/or I:E ratio will increase the degree of auto-PEEP present.

Positive end-expiratory chamber pressure (PEECP) will reduce lung volume by compression; the effect of this on oxygenation is balanced by the beneficial effect of the auto-PEEP generated. Another potential advantage of maintaining PEECP is the limiting effect on chronic hyperinflation, further reducing the risk of pulmonary barotrauma. In studies on humans with normal lungs, adequate ventilation was achieved with chamber pressures of +5 to –15 cmH_2O or +10 to –22 cmH_2O. At these pressures, mean chamber pressure is always negative (–5 to –10 cmH_2O).

Tidal volume is determined by 'span', i.e. the pressure difference between end-expiratory and peak inspiratory pressures, and by frequency. Increasing span increases tidal volume, while increasing frequency decreases it. Since minute ventilation is the product of frequency and tidal volume, minute volume will increase provided that the incremental rise in frequency is greater than any associated decrease in tidal volume. CO_2 elimination increases with increasing minute ventilation up to a frequency of 180 cycles/min. Any further increase in frequency will not increase CO_2 clearance. By increasing peak and mean chamber pressures, good CO_2 elimination can be achieved at lower frequencies. Experiments on animals with 'ARDS-like' lungs showed that increasing mean chamber pressure increased lung volume and improved oxygenation.

How does EHFO work?

The auto-PEEP that EHFO generates may well improve resting lung volume and hence improve oxygenation for any given fractional inspired oxygen (Fio_2). It is unlikely that this is the complete explanation of why improvements in oxygenation in many patients are both very rapid and well maintained. A possible solution may be in the positioning of the cuirass, lying as it does over the lower half of the chest and the upper half of the abdomen. The inspiratory phase of EHFO may be capable of initiating the opening of collapsed lung units in dependent areas of high ventilation–perfusion mismatch. This could then allow conventional ventilation to expand the lung units further, but from a non-collapsed (i.e. more compliant) starting point. In this way, EHFO may be looked upon as a 'director' of ventilation, allowing conventional ventilation to recruit lung units which would otherwise be impossible to expand within safe inflation pressure limits.

Clinical applications

In trials on volunteers with normal lungs, EHFO allowed excellent ventilation in all subjects. At constant chamber pressures, the optimal frequency was found to be 60 to 90 cycles/min. EHFO has also been shown to provide effective ventilation in healthy adults in the presence or absence of spontaneous breathing.

Chronic airflow limitation and acute respiratory failure

At frequencies of 60 to 140 cycles/min, EHFO was able to decrease CO_2 and improve oxygenation in all of a group of 10 patients with chronic airflow limitation in incipient respiratory failure (Al-Saady *et al.* 1992). Thus EHFO can provide ventilatory support in patients with severe chronic airflow limitation and respiratory failure, avoiding the need for intubation while eliminating CO_2 and improving sputum clearance. Another study showed that EHFO produced greater improvement in gas exchange and hemodynamic measurements when compared with positive-pressure ventilation in patients with acute respiratory failure.

Organophosphate poisoning

Organophosphate poisoning causes severe respiratory disturbances, including muscle paralysis and bronchial obstruction by copious secretions. If these are left untreated, rapid respiratory distress and demise follows. In cats given organophosphates, EHFO maintained good oxygenation and CO_2 clearance, and improved secretion clearance. Thankfully, this particular application remains untested in the human setting.

Weaning from respiratory support

Patients requiring extended periods of respiratory support may often present major problems as their requirement for positive-pressure ventilation reduces. The patient often needs to be sedated to tolerate an endotracheal tube. The transition from intermittent positive-pressure ventilation to extubation is often a calculated risk, and reintubation is frequently associated with cardiovascular instability. Extubation is often deferred because of this doubt. Modern modes of ventilatory support have reduced both doubt and risk, but have not removed them altogether.

EHFO does not require intubation and is often well tolerated by the awake patient, provided that the frequency is maintained around the patient's own respiratory rate and the span is no greater than 40 cmH_2O. Therefore the decision to stop ventilation is easier as it can be restarted with minimal delay if respiratory embarrassment becomes evident. A further benefit is improvement in sputum clearance. A study of patients in acute respiratory failure showed reductions in both duration of ventilation and overall hospital stay (Gaitini *et al.* 1991).

In our own practice, weaning problems were the original stimulus for assessing the Hayek oscillator. A familiar scenario is illustrated by a patient who, after presenting with staphylococcal pneumonia associated with renal failure, required a protracted period of intermittent positive-pressure ventilation from which he proved difficult to wean. Over a period of weeks ventilatory assistance was reduced to moderate levels of pressure support ventilation, but could not progress further without significant CO_2 retention and associated hypoxemia. The introduction of EHFO was associated with rapid reductions in pressure support requirement and rapid extubation. After a further period of intermittent EHFO lasting for 6 to 18 h/day, the patient made a complete recovery. Subsequent to this case, we have documented five further cases of successful weaning from positive-pressure ventilation after failure of conventional techniques. Provided that the individual is able to tolerate EHFO, sedation can be stopped immediately. Again, this is likely to be associated with improved respiratory effort.

Acute lung injury

Over a 2-year period we identified 40 patients with Pao_2/Fio_2 ratios below 150 mmHg (20 kPa) owing to either lung trauma or acute pneumonic lung disorders. Three-quarters showed maintained improvements in oxygenation with 'combination ventilation' (i.e. EHFO plus positive-pressure ventilation). Overall mortality was 46 per cent, a figure only marginally better than the quoted mortality of around 60 per cent for similar case-mix groups receiving optimal conventional ventilation alone. In most of the patients studied, optimal settings were a frequency of 90 cycles/min, an I:E ratio of 1:1, an inspiratory pressure of −35 to −45 cmH_2O, and an expiratory pressure of +5 cmH_2O (Campbell and Nevin 1993).

A similar study by a group at St George's Hospital, London, examined 'combination ventilation' in ARDS patients (Taylor *et al.* 1994). They showed an initial improvement in 60 per cent of patients, but this was not sustained at 5 h. Overall mortality was 75 per cent.

The differences in outcome from these two initial assessments of 'combination ventilation' warrant an attempt at explanation. Although all patients in both studies met the entry criteria for extracorporeal support, there were major differences which may provide valuable information regarding future applications of the

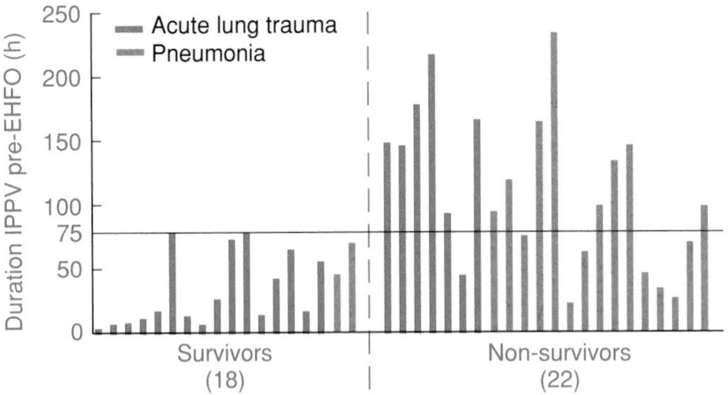

Fig. 2 Pre-EHFO period and outcome.

technique. In the St George's study, many of the patients had undergone extensive periods of conventional ventilation prior to transfer, whereas our study, albeit containing some patients (invariably pneumonias) who had received extensive conventional ventilatory support before entering the trial, consisted mainly of patients who commenced combination ventilation within 72 h of their initial insult. This 'rapid entry' was achieved by setting up a rapid response team which took the equipment to the primary hospital site. A total of 16 hospitals provided patients for the study over the 2-year period. Many patients showed a fluctuating $Pa{O_2}/Fi{O_2}$ ratio during treatment with initial improvement followed by subsequent deterioration and a later recovery. These patients, many of whom were eventual survivors, would have been withdrawn from the St George's study after the measured deterioration.

Analysis of outcome related to the duration of the pre-EHFO period (Fig. 2) strongly suggests that early treatment with combination ventilation is associated with a greater chance of survival than in patients who have received extensive periods of conventional ventilation before proceeding to combination therapy. There appears to be little advantage in using the technique in patients with ARDS secondary to a pneumonic process.

The future

As with all new developments in intensive care practice, there is a need for a prospective randomized assessment to compare outcomes against conventional treatment. While this may be theoretically ideal, it presents many practical problems, not the least of which is the paucity of appropriate cases and the difficulty in standardizing case mix for associated multiple organ failure, as well as the duration and type of pre-entry conventional ventilation. Despite this, an assessment is under way, although several years may be required to produce any clinically significant results.

The Hayek oscillator is not without its problems. Despite considerable efforts by the manufacturers, it remains noisier than conventional ventilators. In addition, the continuous movement of the cuirass can easily cause skin burns. This complication can be avoided if adequate protection is taken by covering the skin contact points with a shock-absorbing protective material such as Granuflex. Although the majority of patients rapidly become accustomed to the oscillator, some spontaneously breathing patients may not tolerate the ventilatory mode. This problem can often be circumvented by starting oscillation at both low frequency and short span. There may be a requirement for sedation in some circumstances. Nevertheless, all the above problems should be overcome with sound sense and a sensitive approach to the feelings of both the patient and attendant nursing staff.

Conclusions

External negative pressure ventilation, as supplied by the Hayek oscillator, appears to be a useful addition to conventional ventilatory techniques in the management of both adult and pediatric intensive care patients. This may be particularly true in patients with a compromised cardiac output or in protracted weaning cases.

Early evidence regarding the use of combination ventilation in ARDS appears to be extremely promising, although further investigation is required before judgment can be passed on its true place in the acute management of such cases. In this context, it would seem sensible to limit the technique to patients with acute lung trauma and to attempt to institute a rapid response environment, allowing the ventilator to be transported to the patient as soon as the appropriate entry criteria have been satisfied.

References

Al-Saady, N.M., Fernando, S.S.D., Singer, M., and Bennett, E.D. (1992). External high frequency oscillation can replace intermittent pressure ventilation in patients with acute respiratory failure. *Intensive Care Medicine*, **18** (Supplement 2), S54.

Campbell, J.M. and Nevin, M. (1993). Hayek oscillator: experience in intensive care. *Clinical Intensive Care*, **4** (Supplement 1), 92.

Gaitini, L., Krimerman, S., Smorgik, Y., Gruber, A., and Werzberger, A. (1991). External high frequency ventilation for weaning from mechanical ventilation. *APICE-Congress, Trieste, 1991*, p. 137.

Hayek, Z. and Schonfeld, T. (1990). External high frequency oscillation around a negative baseline (EHFO-NB). Preliminary trials in humans. In *Intensive and critical care medicine* (ed. O. Aochi, K. Amana, and H. Takashita), pp. 761–2. Excerpta Medica, Amsterdam.

Taylor, N., *et al.* (1994). External high frequency oscillation in patients with acute respiratory distress syndrome and pneumonia. *Proceedings of the Intensive Care Society*, October.

18.2.18 Other novel modes of ventilation

J. Mancebo and V. M. Ranieri

Key messages

- Inverse ratio ventilation has not shown any benefit when compared with conventional volume-controlled ventilation, at the same PEEP levels and minute ventilation, in patients with acute lung injury.
- Airway pressure release ventilation is a mode which allows spontaneous breathing at any time, during either the inspiratory or expiratory phase. One study performed in acute lung injury patients showed an improvement in arterial oxygenation when airway pressure release ventilation was compared with volume-controlled inverse ratio ventilation.

- Proportional assist ventilation is a mode which provides airway pressure in proportion to patient effort. It seems to be very useful for patients with stable respiratory mechanics and changing ventilatory demands.

Inverse ratio ventilation

In inverse ratio ventilation the inspiratory-to-expiratory (I:E) ratio set on the ventilator is greater than 1:1, i.e. inspiratory time is longer than expiratory time. The short expiratory time used in this mode, despite the absence of external positive end-expiratory pressure (PEEP), prevents the collapse of alveoli during expiration due to gas trapping and generation of intrinsic PEEP (auto-PEEP).

In single-compartment systems, the time needed to exhale 96 per cent of the inspired tidal volume is equal to three time constants (the product of resistance and compliance). Consequently, if expiratory time is too short, dynamic hyperinflation will be generated and some level of intrinsic PEEP will be present. Inverse ratio ventilation can be used in volume- or pressure-controlled modes. Typical tracings obtained during inverse ratio ventilation are shown in Figs 1 and 2.

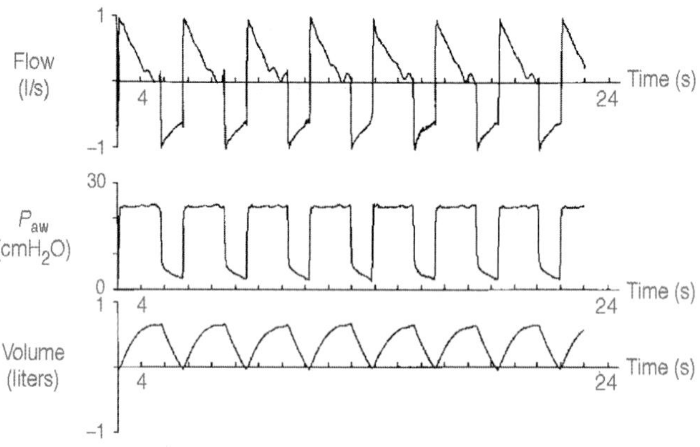

Fig. 1 Tracings of airflow (flow), airway pressure (P_{aw}), and tidal volume obtained in a patient ventilated with pressure-controlled inverse ratio ventilation (I:E ratio, 2:1).

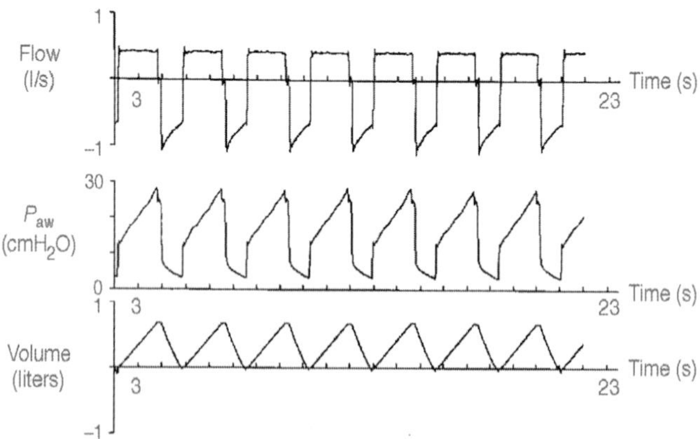

Fig. 2 Tracings of airflow (flow), airway pressure (P_{aw}), and tidal volume obtained in a patient ventilated with volume-controlled inverse ratio ventilation (I:E ratio, 2:1).

Rationale of inverse ratio ventilation

A low inspiratory flow, particularly with a decelerating flow pattern, will theoretically improve intrapulmonary distribution of gas and reduce dead-space, improve the ventilation–perfusion relationship, and enhance alveolar recruitment. Thus intrapulmonary shunting of blood is reduced and gas exchange improved while maintaining the same tidal volume as during conventional ventilation with an I:E ratio of 1:2 without increasing the excursion of volume and pressure (Shantholz and Brower 1994).

Intrapulmonary gas distribution depends not only on inspiratory flow rate but also on the heterogeneity of the time constants of different alveolar units. The distribution of ventilation during inverse ratio ventilation may preferentially reach those alveolar units with slow time constants (slow lung units), where the product of resistance and compliance is high (Shantholz and Brower 1994). An experimental lung model has shown that, when applying the same level of total PEEP by using external PEEP or by inducing intrinsic PEEP (inverting the I:E ratio), the local lung unit end-expiratory pressure and volume differed. Slow lung units inflated more with intrinsic PEEP, while fast lung units inflated more with extrinsic PEEP.

Apart from the theoretical benefits to intrapulmonary gas distribution, the reduction in inspiratory flow rate during inverse ratio ventilation may reduce the stress on lung tissue when there are heterogeneous time constants. Although it has been claimed that pressure-controlled inverse ratio ventilation may reduce the risk of barotrauma compared with volume-controlled ventilation by diminishing peak airway pressure, the alveolar pressure is very similar in both modes when the total PEEP and minute ventilation are kept constant. Moreover, peak airway pressure is clearly irrelevant in this context because it mainly reflects the pressure dissipated to overcome the resistance of the endotracheal tube and central airways, and is unlikely to reflect alveolar distending pressure. To evaluate alveolar distending pressures, one should measure static end-inspiratory airway pressure during a zero flow period.

Mean airway pressure during inverse ratio ventilation

Mean alveolar pressure influences gas exchange and cardiovascular function. It is usually assumed that mean airway pressure is equivalent to, or a good approximation of, mean alveolar pressure. One of the mechanisms advocated to explain the improvement in oxygenation during inverse ratio ventilation is an increase in mean airway pressure; a major physiological determinant seems to be the presence of intrinsic PEEP.

Clinical usefulness of inverse ratio ventilation

Patients are not comfortable during inverse ratio ventilation as it is not a physiological pattern. In consequence, patients ventilated in this particular mode usually require heavy sedation and sometimes muscle paralysis.

Many recent studies have compared the short-term effects of inverse ratio ventilation (commonly used in pressure-controlled modes) on gas exchange, lung mechanics, and hemodynamics in patients with acute lung injury with those obtained when conventional volume-controlled ventilation is used. When maintaining the same total PEEP and minute ventilation for both modes, these studies did not find major differences in either arterial oxygenation or hemodynamics (Lessard *et al.* 1994). Additionally, the static lung mechanics remained unchanged.

Airway pressure release ventilation

First described by Stock and coworkers in 1987, airway pressure release ventilation allows spontaneous breathing at two different levels of continuous positive airway pressure (CPAP), with airway pressure periodically decreasing from the upper level to the lower level (pressure release). It can be patient or time triggered (in the latter it works exactly as pressure-controlled ventilation with PEEP), and is pressure limited and time cycled.

Airway pressure release ventilation allows spontaneous breathing at any time, either during the inspiratory or the expiratory phase. The intermittent liberation of positive pressure produces a decrease in functional residual capacity, thus assisting alveolar ventilation. The usual settings which should be adjusted in this mode are trigger sensitivity, upper and lower levels of CPAP, and the frequency and duration of pressure liberation. Typical tracings are shown in Fig. 3.

Another very similar method, intermittent mandatory pressure release ventilation, was developed to provide better synchronization between the patient's expiration and pressure release, and also to provide a certain degree of assistance during each inspiration by means of pressure support (Rouby 1994). Settings to be adjusted are trigger sensitivity, upper and lower levels of CPAP, the frequency of pressure liberation (adjusted according to the spontaneous respiratory rate, i.e. pressure will be released every two, three, four, or more spontaneous inspiratory cycles), and the level of pressure support. In this mode, ventilatory assistance depends on the patient's spontaneous respiratory rate; the higher the respiratory rate, the higher will be the ventilatory assistance.

Determinants of ventilatory assistance

During airway pressure release ventilation there is a breath-by-breath variation in tidal volume. This depends on the magnitude of the inspiratory effort, the mechanical characteristics of the respiratory system, and the gradient of pressure liberation. Ventilatory assistance will also depend on the duration and frequency of pressure liberation, and the upper and lower levels of CPAP.

According to its working principles, the ventilatory assistance generated by a certain pressure liberation is reduced when respiratory system compliance is decreased as there are small variations in functional residual capacity. This is one explanation of the poor performance of this method in such patients.

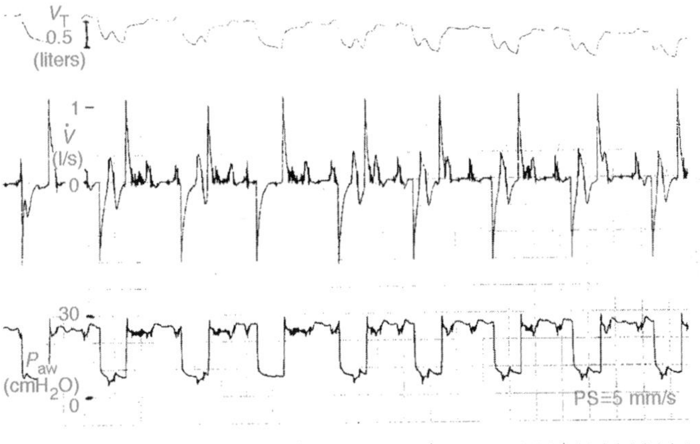

Fig. 3 Tracings of tidal volume (V_T), airflow ($\dot{V}$), and airway pressure (P_{aw}) obtained in a patient ventilated with airway pressure release ventilation. The high-pressure level was set at 25 cmH$_2$O and a 4-s duration, and the low-pressure level was set at 7 cmH$_2$O and a 2-s duration. These settings indicate a back-up rate of 10 cycles/min with an I:E ratio of 2:1.

Clinical usefulness of airway pressure release ventilation

Patients with acute respiratory failure in whom spontaneous ventilation is insufficient to maintain an adequate level of alveolar ventilation because of central or peripheral respiratory depression (i.e. neuromuscular abnormalities, flail chest, etc.) are good candidates for this technique (Rouby 1994). In such situations airway pressure release ventilation can normalize alveolar ventilation without increasing peak airway pressure.

A cross-over design study in patients with acute lung injury (Sydow *et al.* 1994) compared volume-controlled inverse ratio ventilation and airway pressure release ventilation on a long-term basis. After 24 h, an improvement in the ratio of $P(A-a)O_2$ to FiO_2 was observed during airway pressure release ventilation but not during volume-controlled inverse ratio ventilation. These results were attributed to progressive alveolar recruitment and the persistence of spontaneous inspiratory activity.

Proportional assist ventilation

Considerable interest has developed in pressure-assisted methods of ventilatory support. Two of these have been described: pressure support ventilation and the more recent proportional assist ventilation described by Magdy Younes in 1992. These methods assist each spontaneous breath by providing positive pressure at the airway during the period of spontaneous inspiration. During these ventilatory modes, the ventilator-controlled variable is airway pressure, while inspiratory flow and tidal volume are determined by the combined action of the pressure generated by the respiratory muscles (muscular pressure) and the pressure generated by the ventilator (airway pressure). Because the ventilatory consequences of these methods are essentially the result of patient–ventilator interaction, the patient retains considerable control over flow and breathing pattern.

Working principles

Proportional assist ventilation is provided through an experimental prototype—the Winnipeg ventilator (University of Manitoba, Winnipeg, Canada). The gas delivery system consists of a freely moving piston coupled to a motor. As the patient pulls, the piston moves freely into the cylinder and generates force proportional to the current applied to it. This current is proportional to the signals of volume or flow. These signals are amplified by gain controls and are then summed to give a certain amount of current to the motor.

The flow gain signal determines how much pressure (resistive pressure or resistance units) will be given per unit flow. The volume gain signal determines how much pressure (elastic pressure or elastance units) will be given per unit volume. When the piston is activated it

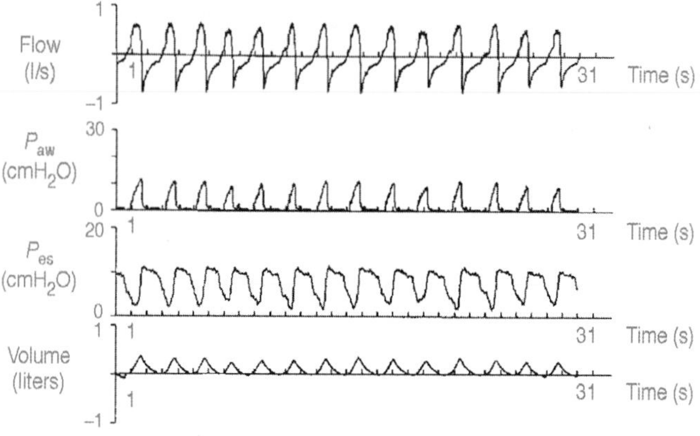

Fig. 4 Tracings of airflow (flow), airway pressure (P_{aw}), esophageal pressure (P_{es}), and tidal volume obtained in a patient ventilated with proportional assist ventilation. There is a typical breath-by-breath variability in P_{aw}, and a proportionality between patient effort (esophageal pressure swings) and P_{aw}. The stronger the patient effort, the higher is P_{aw}.

creates a pressure in the piston chamber and the forward movement of the piston produces airflow. According to the equation of motion of the respiratory system, the total pressure generated by the ventilator during proportional assist ventilation (the airway pressure (P_{aw})) will be a function of the instantaneous airflow and volume, and the respective gains (resistive units and elastic units):

$$P_{aw} = (\text{airflow} \times \text{KR}) + (\text{volume} \times \text{KE})$$

where KR and KE are respectively the gain values for resistance (in $cmH_2O/l/s$) and elastance (in cmH_2O/l) set on the ventilator.

The difference between chamber pressure and proximal airway pressure creates a trigger or assist signal. Gas transfer is terminated when the patient makes an expiratory effort or there is no longer muscular pressure and flow stops. This creates a positive pressure relative to the chamber pressure. When patient pressure exceeds chamber pressure the trigger signal causes the exhalation valve to open, allowing passive deflation. The piston then returns to the starting position. Typical tracings obtained during proportional assist ventilation are shown in Fig. 4.

Potential advantages and disadvantages

Proportional assist ventilation allows complete control of ventilation when demands change and respiratory system mechanics are stable (i.e. sepsis, acidosis, etc.). Additionally, there are no preset limits to the degree of assist (the harder the patient pulls, the harder the machine pushes). These characteristics may render this ventilatory mode very comfortable for the patient.

Disadvantages are related to the lack of assistance when intrinsic PEEP is present. This has to be overcome by the respiratory muscles before initiating inspiratory flow, and thus is not 'seen' by the ventilator as a muscular pressure. For a correct set-up of proportional assist ventilation, the elastance and resistance of the respiratory system must be measured to adjust the ventilator gain settings adequately. Furthermore, during proportional assist ventilation it is mandatory that patients maintain a normal respiratory output. Changes in breathing pattern may be difficult to evaluate. Finally, if inspiratory leaks exist, airway pressure will increase because there is an overestimation of the volume and flow received by the patient.

Differences between pressure support ventilation and proportional assist ventilation

The main theoretical difference is the function that airway pressure is designed to follow during inspiration. With pressure support ventilation, pressure rises to a preset level that remains constant until the cycle-off criterion is reached. Therefore the magnitude of ventilator-applied pressure should be unaffected by patient effort. With proportional assist ventilation, the time course of airway pressure is linked to the time course of patient effort. Airway pressure rises as long as inspiratory muscle effort is produced by the patient. What is preset is not a target pressure, but the proportionality between airway pressure and inspiratory muscle effort, i.e. how much airway pressure will rise for a given increase in muscular pressure during inspiration.

Clinical studies

Substantial differences are expected between patient–ventilator interactions in pressure support ventilation and proportional assist ventilation as a consequence of variations in ventilatory requirements. In mechanically ventilated patients during weaning, the strategy used to compensate for increased ventilatory requirements due to CO_2 stimulation substantially differed during pressure support ventilation and proportional assist ventilation (Ranieri *et al.* 1996). An increase in minute ventilation is obtained by increasing respiratory rate in the former and by increasing tidal volume in the latter. When the energy cost, evaluated as the product of esophageal pressure and time, of such different strategies is compared, a larger amount of inspiratory muscle effort is required to increase minute ventilation in response to hypercapnic stimulation during pressure support ventilation than during proportional assist ventilation. Patient discomfort also appears lower during proportional assist ventilation. Finally, non-invasive proportional assist ventilation has been used to alleviate distress and avoid intubation in patients with acute respiratory failure; early use of proportional assist ventilation resulted in the avoidance of intubation in eight of eleven patients with acute respiratory failure of different etiologies.

References

Lessard, M., Guerot, E., Lorino, H., Lemaire, F., and Brochard, L. (1994). Effects of pressure-controlled with different I:E ratios versus volume-controlled ventilation on respiratory mechanics, gas exchange, and hemodynamics in patients with adult respiratory distress syndrome. *Anesthesiology*, **80**, 983–91.

Ranieri V.M., *et al.* (1996). Patient–ventilator interaction during acute hypercapnia: pressure support vs proportional assist ventilation. *Journal of Applied Physiology*, **81**, 426–36.

Rouby, J.J. (1994). La ventilation en pression positive variable. In *Ventilation artificielle. Principes et applications* (ed. L. Brochard and J. Mancebo), pp. 215–26. Arnette Blackwell, Paris.

Shanholtz, C. and Brower, R. (1994). Should inverse ratio ventilation be used in adult respiratory distress syndrome? *American Journal of Respiratory and Critical Care Medicine*, **149**, 1354–8.

Sydow, M., Burchardi, H., Ephraim, E., Zielmann, S., and Crozier, T. (1994). Long-term effects of two different ventilatory modes on oxygenation in acute lung injury. *American Journal of Respiratory and Critical Care Medicine*, **149**, 1550–6.

18.3 Extracorporeal and non-ventilatory support

18.3.1 Physiology of failure of mechanical ventilation to support gas exchange

K. P. Kelly and R. Rossaint

Key messages

- The iatrogenic factors contributing to failure of mechanical ventilation are as follows:
  - high peak airway pressure and tidal volumes;
  - inappropriate level of positive end-expiratory pressure;
  - shear stress;
  - toxicity of high inspired oxygen concentrations;
  - nosocomial infection.
- Disease factors contributing to worsening of gas exchange are as follows:
  - insufficient viable lung tissue for the demands of the patient;
  - an initiating cause of the acute respiratory distress syndrome remains inadequately treated.

Introduction

The reasons why mechanical ventilation fails to improve gas exchange can be divided into two broad categories: iatrogenic factors that in themselves damage the lung, and the nature and severity of the underlying disease itself.

Iatrogenic factors

Iatrogenic factors contributing to the worsening of gas exchange include high tidal volumes and therefore high airway pressures, positive end-expiratory pressure (PEEP), shear stress, high inspired oxygen concentrations, and nosocomial infection (Table 1).

Previous ventilator strategies used in treating acute respiratory failure have included the following:

(1) striving to achieve arterial blood gas partial pressures in the normal range by utilizing intermittent positive-pressure ventilation settings that are commonly used in the operating room (i.e. relatively large tidal volumes of 10 to 15 ml/kg, and respiratory rates of 10 to 12 breaths/min), thus generating high airway pressures as well as high shear stresses (see below);

(2) PEEP;

(3) increasing the fractional inspired oxygen (FIO_2);

(4) breaching the integrity of the trachea by tracheal intubation, which is an infection risk.

High peak airway pressure and high tidal volumes

If the intermittent positive pressure ventilator is set to deliver 12 breaths of 700 ml on a continuous mandatory ventilation mode, this means that the patient will receive 700-ml breaths, 12 times per minute, regardless of the pressure generated in the airways. This pressure will depend upon the degree of distensibility of the lung (compliance) and the resistance to flow in the airways. Where compliance is reduced, the pressure in the airway may become dangerously high, subjecting the patient not only to the risks of pneumothorax, pneumomediastinum, pneumopericardium, and bronchial rupture, but also to more subtle iatrogenic damage such as pulmonary interstitial emphysema, perivascular hemorrhage, and perialveolar hemorrhage (Swami and Keogh 1992).

In a non-homogeneous disease such as acute respiratory distress syndrome, there are areas of badly affected lung as well as areas of essentially normal tissue which have a normal compliance. These areas of normal lung will receive most of the set tidal volume intended for the whole lung and become overdistended. This volutrauma is probably more damaging than barotrauma (Dreyfuss and Saumon 1992).

In an attempt to limit the amount of volutrauma/barotrauma caused by positive-pressure ventilation, clinicians have instituted pressure-controlled ventilation where the upper pressure limit is restricted (e.g. 30–35 cmH_2O).

PEEP

PEEP is used to recruit collapsed airways or to help keep open airways that would otherwise collapse at the end of expiration. It will increase the functional residual capacity, thereby increasing the number of lung units available for gas exchange. Hence the fraction of the

Table 1 Effects of iatrogenic factors worsening gas exchange during mechanical ventilation

High peak airway pressures	Side-effects of inappropriately high PEEP	Shear stress	Oxygen toxicity	Nosocomial infection
Barotrauma, i.e. pressure-induced damage	Elevation of mean intrathoracic pressure	Shear-stress-induced release of inflammatory mediators	Atelectasis	Pneumonia
Volutrauma, i.e. overdistension	Decreased cardiac output	Loss of surfactant due to failure to keep airway open	Inflammation	
Pulmonary interstitial emphysema	Overdistension of 'good' lung, resulting in diversion of blood from healthy alveoli		Edema	
Perialveolar hemorrhage			Fibrin deposition and hyaline membrane formation	
Pneumothorax			Alveolar hemorrhage	
Pneumopericardium				
Pneumomediastinum				
Bronchial rupture				

cardiac output passing through non-ventilated lung units (the shunt fraction) is reduced.

The price paid for this benefit is an elevation in mean intrathoracic pressure which not only increases the risk of barotrauma/volutrauma, but can also result in a diminution of cardiac output. As the amount of oxygen delivered to the tissues depends upon arterial oxygen content and cardiac output, a decreased cardiac output could result in decreased oxygen delivery even if $Pa\text{O}_2$ is improved.

PEEP may decrease the cardiac output not only by increasing the mean intrathoracic pressure (which may consequently decrease venous return) but also by pressure transmitted directly to the heart from the adjacent lung tissue. In addition, PEEP is transmitted preferentially to normal lung regions. This means that PEEP applied at an inappropriately high level will inhibit blood flow through well-ventilated lung units and will cause a redistribution of blood flow to less well-ventilated areas.

Criteria used to determine 'best PEEP' have included the following:

(1) lung mechanics;

(2) the PEEP setting that reduces shunt fraction to less than 15 per cent of the cardiac output;

(3) the setting giving the greatest oxygen delivery for the lowest FiO_2;

(4) the level showing the greatest recruitment of lung units on CT imaging.

However, the optimal method for determining 'best PEEP' is still controversial, and is usually a compromise of balancing the advantages with the disadvantages of PEEP. In most patients there is arguably little to be gained in choosing a PEEP level higher than 15 cmH_2O.

If we use a combination of limitation of airway pressure and a limited amount of PEEP, this will result in a low tidal volume in patients who have a reduced lung compliance. This accepts that the CO_2 levels will be allowed to rise (permissive hypercapnia). Pressure limitation and accepting smaller tidal volumes may, in addition to the advantages mentioned above, be instrumental in reducing shear stress.

Shear stress

Expanding an atelectatic area from a state where it is completely collapsed to the fully open state requires a much greater change in pressure (pressure amplitude) and higher peak inspiratory pressure than is necessary for inflating a lung area that is already partially held open by an appropriate amount of PEEP. This is similar to inflating a balloon, and follows logically from the the LaPlace law $P = 2\gamma/r$, where P is the pressure to stabilize the alveolus, γ is the surface tension at the air–liquid interface, and r is the radius of the alveolus. The shearing forces generated by repeatedly opening up an atelectatic area and then allowing it to collapse again may lead to structural damage (in particular to the bronchiolar and alveolar epithelia and the capillary endothelium). This disrupted parenchyma may be a source of mediators triggering the continuation of acute respiratory distress syndrome (Lachmann 1992).

Further, if the lung is not held open, it may have a deleterious effect on pulmonary surfactant. In health, surfactant molecules are compressed at end-expiration. If the volume of the alveolus is smaller than the surface area of the surfactant, the excess molecules are squeezed out towards the airways. At the next inspiration, the surfactant in this hypophase rejoins the alveolar surface. However, if the lung is forced open with large tidal volumes or high respiratory rates, rather than judicious use of PEEP, the surfactant is lost into the airways and lost completely from the alveolus. To minimize this phenomenon, as much of the lung as possible should be ventilated without either overdistention of the lung or allowing lung collapse.

Elevated $Fi\text{O}_2$

In addition to mechanical factors which damage the lung, high inspired concentrations of oxygen are believed to be toxic. Most of the evidence for this comes from animal studies, as pathological samples from patients may be difficult to differentiate from the underlying disease. However, it is possible that animal models of oxygen toxicity do not reflect pathology in humans.

A prospective study of 10 patients with irreversible brain damage (five ventilated with an $Fi\text{O}_2$ of 0.21, and the rest at an $Fi\text{O}_2$ of 1.0) found that after 40 h the group ventilated with pure oxygen had significantly poorer oxygenation compared with the 'air' group (temporarily measured at $Fi\text{O}_2$ 1.0 for comparison). The intrapulmonary shunt and ratio of dead-space to tidal volume were also significantly increased in the 'oxygen' group.

Acute tracheobronchitis with substernal distress may begin within 4 to 22 h of breathing 100 per cent oxygen in normal subjects; in contrast, others have found that short exposure (24–48 h) to 100 per cent oxygen appeared to do little harm, with no change in intrapulmonary shunt, dead-space, or clinical outcome following cardiac surgery. However, normal subjects may not reflect the situation in patients with diseased lungs. The individual patient's liability to oxygen toxicity may depend on his or her age, nutritional status, and endocrine status, as well as previous exposure to oxygen or other oxidants (Deneke and Fanburg 1980).

The changes of oxygen toxicity tend to be rather non-specific, i.e. atelectasis, inflammation, edema, alveolar hemorrhage, fibrin deposition, and thickening and hyalinization of alveolar membranes.

High $Fi\text{O}_2$ may cause atelectasis principally by two mechanisms. Firstly, in partially obstructed airways, oxygen is absorbed from the alveoli faster than it is replenished from the upper airways. This tendency will be exaggerated if $Fi\text{O}_2$ is 1.0 and no nitrogen is present to splint the alveoli. Secondly, in the presence of high concentrations of oxygen, surfactant may be deficient in amount and function.

Further toxic effects of oxygen are believed to be mediated through what are loosely termed oxygen radicals, the most important of which are superoxide (O_2^-), hydrogen peroxide (H_2O_2), singlet oxygen ($^1O_2$), and hydroxyl radical (OH^-). These may cause lipid peroxidation, protein sulfhydryl oxidation, and DNA damage, leading to cell death. The protective mechanisms against these radicals under normal circumstances (e.g. superoxide dismutase, catalase, and reduced glutathione) may be overwhelmed in the diseased state. It is not yet clear whether giving reducing agents such as *N*-acetylcysteine or superoxide dismutase has any therapeutic benefit. These toxic effects may lead to an inflammatory response, possibly resulting in edema, fibrin deposition with hyalinization of alveolar membranes, and alveolar hemorrhage.

The 'safe $Fi\text{O}_2$' in humans is unclear. Empirical animal evidence suggests that the dose of oxygen may be a function of partial pressure and length of exposure. In humans, $Fi\text{O}_2$ values of 50 to 60 per cent are generally regarded to be the upper limit of 'safety'. However, in view of the lack of unequivocal evidence about the significance of oxygen toxicity (possibly rarely of practical importance), it is perhaps best to adopt a common-sense approach, i.e. fear of causing oxygen toxicity should not take precedence over the relief of arterial hypoxemia in mechanically ventilated patients, while using the minimum $Fi\text{O}_2$ to provide an acceptable $Pa\text{O}_2$ of 8 kPa.

Nosocomial infection

Mechanical ventilation usually implies intubating the trachea. This instrumentation breaches the defenses of the airways, allowing infection, and thereby increases the risk of a superadded pneumonia in addition to the underlying disease which caused the initial respiratory failure. In order to avoid these drawbacks of mechanical ventilation, non-invasive techniques such as nasal continuous positive airway pressure are under evaluation.

Disease factors

In addition to the iatrogenic factors mentioned above, the disease process itself may be a major reason why mechanical ventilation fails to maintain gas exchange.

Insufficient viable lung tissue

The average basal oxygen requirement in the resting healthy state is 250 ml/min. At a respiratory quotient of 0.8, which is typical for a Western diet, the average basal production of CO_2 in the resting state is 200 ml. Basal oxygen demands may more than double in the presence of fever or infection, and CO_2 production may also increase under these circumstances.

In simple terms, if the lung is acutely unable to cope with these demands, the patient becomes hypoxemic and/or hypercapnic. As the treatment of acute respiratory distress syndrome is basically supportive, in some cases standard mechanical ventilation may not be enough for these increased demands.

Gattinoni *et al.* (1987) postulated that the injured lung could be thought of as comprised of three zones:

(1) an area of normal lung (zone H, healthy);

(2) an area where the lung function is impaired, but may be improved by therapeutic maneuvers (zone R, recruitable);

(3) an area which at present contributes no useful gas exchange, despite therapy (zone D, diseased).

Extrapolating this theory, it may be that the amount of viable lung tissue in zone H and that recruited in zone R, in combination with the extra oxygen demand and CO_2 production due to disease, are insufficient to compensate for zone D. In this case, in order to provide the patient with sufficient $Pa\text{O}_2$ and to remove an adequate amount of CO_2, a non-ventilatory support (extracorporeal respiratory support) must be used.

These techniques do not replace mechanical ventilation but are complementary to it. The technique of extracorporeal CO_2 removal (ECCO$_2$R) is a half-way house, allowing the amount of respiratory support given by a mechanical ventilator to be drastically reduced. This in turn allows airway (distending) pressures applied to the lung to be reduced, as well as reducing PEEP and inspired oxygen concentration.

In comparison, extracorporeal membrane oxygenation (ECMO) transfers a larger amount of oxygen but requires a greater proportion of the cardiac output. Thus the lung is 'rested', and the O_2 demands and CO_2 removal requirements are met during the bridging period while the lung heals or, in a limited number of cases, until lung transplantation.

Inadequately treated underlying disease

It must be emphasized that any initiating cause of acute respiratory distress syndrome should be treated, otherwise this acts as a constant source of 'fuel' stoking the boiler of inflammation (e.g. inadequately treated pneumonia or continuing intra-abdominal sepsis). In the presence of these stimuli, mechanical ventilation is unlikely to be successful.

Conclusion

Mechanical ventilation can be lifesaving. However, there are circumstances where mechanical ventilation fails due to progression of the lung disease in its own right and/or therapy-induced lung damage. It may then be necessary for the patient to be supported by some other means, such as extracorporeal respiratory support, in order to survive.

References

Deneke, S.M. and Fanburg, B.L. (1980). Normobaric oxygen toxicity of the lung. *New England Journal of Medicine*, **303**, 76–86.

Dreyfuss, D. and Saumon, G. (1992). Barotrauma is volutrauma, but which volume is the one responsible? *Intensive Care Medicine*, **18**, 139–41.

Gattinoni, L., Pesenti, A., Avalli, L., Rossi, F., and Bombino, M. (1987). Pressure–volume curve of total respiratory system in acute respiratory failure. *American Review of Respiratory Disease*, **136**, 730–6.

Lachmann, B. (1992). Open up the lung and keep the lung open. *Intensive Care Medicine*, **18**, 319–21.

Swami, A. and Keogh, B.F. (1992). The injured lung: conventional and novel respiratory therapy. *Thorax*, **47**, 555–62.

18.3.2 Extracorporeal respiratory support

L. Gattinoni and L. Brazzi

Key messages

- All conventional methods of maintaining adequate gas exchange in acute respiratory distress syndrome are potentially harmful to the residual healthy lung.
- The rationale for extracorporeal respiratory support is based on the assumption that lung rest provides a better environment for healing.
- Extracorporeal respiratory support should still be considered as an experimental technique which should be concentrated in a few referral centers with sufficient motivation, technical skills, and staff to test new technical solutions safely.

Introduction

Extracorporeal support comprises a variety of techniques. Various acronyms are adopted in the literature to emphasize the main aim of the use of the artificial lung.

1. Extracorporeal membrane oxygenation (ECMO) focuses on the oxygenation aspects of the bypass: it consists of a high-flow venoarterial bypass performed to provide an immediate improvement in the oxygenation of the arterial blood.
2. Extracorporeal CO_2 removal (ECCO_2R) refers mainly to venovenous bypass at relatively low flow (20–30 per cent of the cardiac output). ECCO_2R is intended to provide lung rest by reducing the ventilation of the diseased lung.
3. Partial extracorporeal CO_2 removal (PECO_2R) indicates a very low venovenous bypass with partial removal (30–60 per cent) of the CO_2 produced.
4. Extracorporeal lung assist (ECLA) has been proposed in Japan to emphasize both the oxygenation and ventilatory functions of the extracorporeal support.

Whatever terminology is used, a correct description of the system should include the type of bypass (venoarterial or venovenous), the ratio of extracorporeal blood flow to cardiac output, and the ventilatory management of the natural lungs (low-frequency high-frequency jet ventilation, continuous positive-pressure ventilation, continuous positive airway pressure, etc.).

Here, we focus on the technique of low-frequency positive-pressure ventilation with extracorporeal CO_2 removal (LFPPV–ECCO_2R) , covering the rationale, method, and clinical experience gained by our group over the last 20 years.

Rationale

Since its original description, the acute respiratory distress syndrome (ARDS) has been considered to be a syndrome diffusely affecting the lung parenchyma, as usually seen on anteroposterior chest radiographs. However, CT studies have shown that lesions (seen as densities) are primarily located in the dependent lung, suggesting a non-homogeneous distribution of lung changes. Quantitative analysis of the CT scan images led us to model the ARDS lung into three compartments: the first shows normal inflation, the second is poorly inflated, and the third is non-inflated. The dimensions of the normally inflated lung, which are directly estimated from measurement of lung compliance, may be as small as 20 to 30 per cent of the dimension of a normal adult lung, and may assume the dimensions of a 'baby lung'. Ventilation–perfusion studies in ARDS are consistent with the model described; hence gas exchange is characterized by two physiological compartments, with true shunt in the non-inflated compartment and normal ventilation–perfusion in the baby lung.

Conventionally, two variables are subjected to manipulation by the clinician in order to maintain oxygenation in severe ARDS, namely fractional inspired oxygen (F_{IO_2}) and airway pressure. Oxygen toxicity is an old concept, first described almost a century ago. It is still not known how much oxygen is safe for the diseased lung and for how long, but it is common clinical practice to keep F_{IO_2} as low as possible. The introduction of positive end-expiratory pressure (PEEP) allowed improvement in oxygenation in some patients, without any increase in F_{IO_2}. In 1975 the indiscriminate use of super-PEEP revealed the ominous evidence of barotrauma as a side-effect of positive-pressure ventilation on either specific organ function or on global hemodynamics. The term barotrauma now refers not only to the occurrence of major extrapulmonary air collection, but also to all the structural changes induced by high-peak pressure and high-volume ventilation.

The $Pa\text{CO}_2$ values are typically normal in the early stages of ARDS, but the cost of breathing to maintain these values is greatly increased. Attempts to ventilate the stiff ARDS lung with normal tidal volumes will result in high inflation pressures and specific hyperventilation of the residual healthy zones, factors that have been demonstrated to be deleterious to lung structures.

In summary, all the conventional means of maintaining adequate gas exchange are potentially harmful to the residual normal baby lung. With this in mind, the rationale for extracorporeal support is to avoid further damage to the relatively small part of the lung still available for ventilation, avoiding the need for high-pressure and high-volume ventilation by extracorporeal removal of metabolically produced CO_2.

The LFPPV–ECCO$_2$R technique

LFPPV–ECCO$_2$R is performed using a venovenous bypass. After a bolus of intravenous heparin (100 IU/kg), catheters are usually inserted in femorofemoral sites, even if a jugulofemoral approach is possible. These catheters, which are reinforced with spring-wire and up to 34 French in size, are positioned percutaneously using a modified Seldinger technique.

The extracorporeal apparatus consists of a blood section (artificial lung, blood pump, blood circuit), a gas section (flowmeter, gas lines, humidifiers), and a monitoring apparatus (blood flowmeter, differential pressures across the membrane lung, temperature). A water heater is used to warm the blood flow.

The extracorporeal circuit is assembled and primed with Ringer's lactate after a CO_2 flush to prevent bubble formation. Immediately before connection, the priming solution is substituted with heparinized whole blood at 37 °C. When the patient has been connected to the extracorporeal circuit, the blood flow is started slowly, paying close attention to body temperature and hemodynamic indices. After 20 to 30 min it is usually possible to set the extracorporeal blood flow at the maintenance rate. Clearance of the total CO_2 minute production (200–400 ml/min) occurs mainly through the artificial lung and requires a blood flow of 1.5 to 2.5 l/min with a ventilation of 10–20 l/min in the artificial lung. While oxygen transfer by the artificial lung is strictly dependent on extracorporeal blood flow, the CO_2 clearance is less dependent (ECCO$_2$R increases linearly with the logarithm of extracorporeal blood flow). Consequently, the amount of ECCO$_2$R depends on the surface area of the artificial lung and its level of ventilation. Initially, the ventilator $F\text{IO}_2$ is kept unchanged from the pre-bypass value, while the artificial lungs are ventilated with the same $F\text{IO}_2$. The ventilation of the natural lungs is then decreased to 2 to 4 breaths/min, while PEEP is progressively increased to maintain mean airway pressure at the same level as during the previous period of mechanical ventilation. Oxygen (100 per cent) is then delivered into the carina through a small catheter to provide for the oxygen consumed during the long expiratory pause (apneic oxygenation).

Bleeding is the main complication during long-term bypass, even though heparin-coated circuits with heparinized artificial lungs, introduced in 1992, have reduced the need for systemic anticoagulation. Heparinized circuits can be used without systemic heparinization; however, we routinely maintain an intravenous infusion of heparin (15 000–30 000 IU/day). The activated clotting time (Hemocron, International Technidine, New Jersey, USA) is monitored hourly and the heparin infusion is modified throughout bypass time to maintain activated clotting times between 180 and 210 s. The range of heparin infusion is usually 10 to 30 IU/kg/h. Systemic heparin is discontinued only if major bleeding occurs or when surgical maneuvers are required.

The weaning process, initially from $F\text{IO}_2$ and then from pressures, starts as soon as the patient's gas exchange improves. When oxygenation increases, the $F\text{IO}_2$ of the ventilator is decreased first using a target $Pa\text{O}_2$ of 80 to 100 mmHg (10.7–13.3 kPa). When the target $Pa\text{O}_2$ is maintained with a ventilatory $F\text{IO}_2$ setting of 40 per cent, the $F\text{IO}_2$ of the gas mixture ventilating the artificial lung is decreased stepwise to 21 per cent. If oxygenation is still maintained at the target value, the PEEP level is decreased. When $Pa\text{O}_2$ remains stable at 80 to 100 mmHg (10.7–13.3 kPa), with the PEEP level set at 5 to 10 cmH$_2$O and the $F\text{IO}_2$ set at 40 per cent on the ventilator and 21 per cent on the membrane lung, the patient's ventilatory mode is shifted to spontaneous–assisted ventilation.

Disconnection from bypass is considered when the patient is able to tolerate spontaneous–assisted ventilation for 6 to 12 h with the same level of $F\text{IO}_2$ in the ventilator and the membrane lung. Firstly, gas flow through the artificial lung is stopped and then, if the patient is able to maintain a viable gas exchange without any extracorporeal support, the catheters are removed. Percutaneous cannulation does not require any surgical repair.

The artificial lung routinely used is the Maxima-Carmeda microporous heparinized lung (Medtronic, Anaheim, CA, USA) which allows a decrease of systemic anticoagulation but tends to leak plasma after a few days' use. This leads to frequent changes of the artificial lung. However, the problem now appears close to being solved as a new model, the Maxima Plus PRF, has not shown any plasma leakage (up to 3 weeks) in the most recent patients treated.

A Biomedicus centrifugal pump (Medtronic, Eden Prairie, MN, USA) with a heparinized pump head is used for pumping blood. The main advantage of this device is that it can be used without a reservoir. The reservoir, which is mandatory with a roller pump, may be considered a weak point of the circuit, particularly when heparinized circuits are used, owing to the low blood flow rate and consequent formation of clots, thrombi, and, possibly, emboli. In contrast, the major disadvantage of the heparinized centrifugal pump is the tendency for deposition of small clots, which may cause hemolysis, near the rotor shaft. To prevent this problem the pump head is changed every 5 days.

Clinical experience

We initially selected patients according to the ECMO criteria which are based on oxygenation values in standardized ventilatory conditions. We later realized that a total static lung compliance above 30 ml/cmH$_2$O was usually compatible with spontaneous breathing through a continuous positive airway pressure system. Patients with a total static lung compliance below 30 ml/cmH$_2$O require a more invasive mode such as pressure-controlled inverse ratio ventilation or, if this technique is failing, LFPPV–ECCO$_2$R. Therefore we added the further entry criterion of a total static lung compliance below 30 ml/cmH_2O measured at a tidal volume of 10 ml/kg under both anesthesia and paralysis. We have also recently added the criterion of a negative PEEP response; no patient undergoes extracorporeal support with a clinically important increase in oxygenation and a consistent clearing of CT scan densities when PEEP is increased from 5 to 15 cmH_2O.

Since 1979 we have treated 103 patients with LFPPV–$ECCO_2R$ with an overall survival rate of 42 per cent. The most common indication for $ECCO_2R$ has been pneumonia, either viral or bacterial. This was the underlying etiology in 65 of our 98 ARDS patients. Nineteen had post-traumatic ARDS, and the reason for $ECCO_2R$ in the other 14 patients was thromboembolism, fat embolism, amniotic fluid pulmonary embolism, or septic shock. Three patients were treated with pulmonary lavage during extracorporeal support for alveolar proteinosis. Two patients were treated with partial extracorporeal support; one had decompensated chronic obstructive pulmonary disease, and the other was a patient with recurrent pneumothoraces secondary to congenital bullous emphysema.

Role of extracorporeal support

After 20 years of experience with extracorporeal support in adults, we have little doubt that $ECCO_2R$ is a logical and rational form of therapy for severe ARDS. However, $ECCO_2R$ is not an ideal form of treatment since the technique carries a significant incidence of bleeding, which is a potentially lethal complication. However, the damage induced by mechanical ventilation in severe ARDS is well documented and justifies the search for alternative therapeutic strategies. Permissive hypercapnia is a treatment along these lines and may decrease the need for extracorporeal support. Currently, we still consider extracorporeal support to be an experimental technique for use in referral centers with sufficient motivation, technical skills, and personnel, where new technical solutions may be tested safely. This technique should only be proposed for widespread utilization when technical improvement and better biocompatibility of the components of the artificial lung assure the safety of long-term bypass.

Bibliography

Gattinoni, L., *et al.* (1980). Treatment of acute respiratory failure with low frequency positive pressure ventilation and extracorporeal removal of CO_2. *Lancet*, **ii**, 292–5.

Gattinoni, L., *et al.* (1986). Low frequency positive pressure ventilation with extracorporeal CO_2 removal in severe acute respiratory failure. *Journal of the American Medical Association*, **256**, 881–6.

Kolobow, T., Gattinoni, L., Tomlison, T., and Pierce, J. (1977). Control of breathing using an extracorporeal membrane lung. *Anesthesiology*, **46**, 138–41.

Pesenti, A., Rossi, G.P., Pelosi, P., Brazzi, L., and Gattinoni, L. (1990). Percutaneous extracorporeal CO_2 removal in a patient with bullous emphysema with recurrent bilateral pneumothoraces and respiratory failure. *Anesthesiology*, **72**, 571–3.

Zapol, W.M., *et al.* (1979). Extracorporeal membrane oxygenation in severe acute respiratory failure. A randomized prospective study. *Journal of the American Medical Association*, **242**, 2193–6.

18.4 Cardiovascular system

18.4.1 Defibrillation

R. F. Armstrong and G. Maughan

Key messages

- Early defibrillation is essential in ventricular fibrillation.
- Transthoracic impedance affects outcome of defibrillation.
- Familiarity with equipment improves performance.

Introduction

Many deaths are caused by ventricular fibrillation occurring as a primary electrical event or as a sequel to myocardial infarction.

Early defibrillation has been shown to be the only effective means of correcting ventricular fibrillation. It is now considered the therapy of highest priority in cardiopulmonary resuscitation (Bossaert and Koster 1992).

Apparatus

Modern defibrillators incorporate an a.c.–d.c. converter and a capacitor to store energy provided by the mains a.c. electricity or by battery (Kerber 1993). A transformer allows selection of voltage potentials. The energy is discharged as a high voltage d.c. shock via electrodes placed on the patient's chest. The electrodes may be in the form of hand-held paddles or large adhesive pre-gelled pads. ECG monitoring via the paddles or electrode leads is used with a manual pulse check to verify the diagnosis and result of defibrillation attempts.

Semi-automatic defibrillators interpret the ECG and advise the operator who activates the discharge button. Automatic external defibrillators interpret the cardiac rhythm before delivering a warning and shock without operator action. The automatic external defibrillator is effective, and their use by lay personnel is being encouraged to overcome the delays inherent in prehospital cardiac arrests. Surgically implanted cardioverter–defibrillators are used for high-risk patients as treatment for recurrent ventricular dysrhythmias. They incorporate a pulse generator inserted under the skin connected to a transvenous lead positioned endocardially. The cardioverter–defibrillator senses, paces, and defibrillates if necessary. Energies delivered range from 5 to 34 J.

Energy delivery

The energy delivered by a defibrillator is measured in joules (the product of power (in watts) and duration (in seconds)). Power is the product of circuit voltage and current. Delivered energy may be up to 40 per cent less than actual stored energy because of losses between apparatus and electrodes. Defibrillation results from the passage of current through the myocardium, causing depolarization of a critical mass of myocardial tissue. The size of the current flowing depends on the transthoracic impedance which is measured in ohms:

$$\text{current (amperes)} = \text{potential (volts)}/\text{impedance (ohms)}.$$

Typical impedance values for the human thorax range from 15 to 150 Ω (average 70–80 Ω). Impedance increases if the quality of contact between paddle and skin is poor (low pressure and small contact area), if conductive gels are not used, if the distance between the electrodes increases, and during inspiration (Sirna *et al.* 1988). It doubles when circulation stops. Impedance decreases with successive shocks if contact is maintained between paddle and chest. The optimal defibrillation current is believed to be in the region of 30 to 40 A over 4 to 12 ms.

Because individual impedance levels are unknown, the energy first chosen (200 J) may deliver inadequate current to defibrillate (high impedance) or too much current so that the myocardium is damaged (low impedance). Failure to defibrillate also occurs when transthoracic current follows alternative pathways in the chest, thus reducing current flow through the myocardium. Current-based defibrillators are being developed which measure individual impedance levels and then deliver the appropriate energy level to produce a chosen current. There is some evidence that a biphasic waveform may reduce defibrillation thresholds.

Practical external defibrillation

The electrodes are usually applied below the right clavicle to the right of the sternum (bone is a poor conductor) and apical (fifth left intercostal space in the mid-clavicular to mid-axillary line). Electrodes or pads of diameter 13 cm are suitable for adults; electrodes of diameters 8 cm and 4.5 cm are suitable for children and infants respectively. Energy requirements depend on patient size. An initial shock of 200 J is recommended for a 70-kg adult; 2 J/kg is a useful estimate for smaller patients. Therefore a 10-kg child (about 1 year old) would need 20 J. Pediatric paddles are probably too small for a patient of this size, producing high impedances and inadequate defibrillation current. An electrode of larger diameter should be selected (Atkins *et al.* 1988).

The success of the cardiopulmonary resuscitation sequence depends on basic life support and then early defibrillation at 200,

200, and 360 J. If defibrillation is unsuccessful, intubation, ventilation, intravenous access, adrenaline (epinephrine), and external chest compressions are resumed. Initial cardiac output may be low after successful defibrillation and may only improve slowly.

Cardioversion

Initial energy levels recommended are 50 J for atrial flutter and 100 J for atrial fibrillation. These shocks are synchronized to deliver the shock on the R wave of the ECG. Synchronization in ventricular tachycardia may be difficult or impossible. Recommended energy levels are 100 J in monomorphic and 200 J in polymorphic ventricular tachycardia (Kerber 1993). Anesthesia is necessary as the process is painful.

Dangers of defibrillation

Onlookers and assistants should be asked to stand back during defibrillation to avoid electrocution. Smearing electrode jelly on the chest can result in surface current arcing between electrodes, burning the skin as well as failing to defibrillate. Burns may also occur if there is inadequate conductive material between paddle and skin. Nitrate patches under electrodes and oxygen administration have been reported as causes of fire.

Failure to defibrillate may be due to inadequate current flow and mandates changing to an anteroposterior position of the electrodes (left precordium and right infrascapular region), change of defibrillator, or techniques such as intracardiac or transesophageal defibrillation. Excessive current flow may cause myocardial necrosis.

In the presence of indwelling pacemakers, defibrillation may cause damage to the myocardium as current traverses the pacing electrode causing burns at the tip of the pacing wire. This may cause a rise in pacing threshold. The generator itself may need reprogramming after defibrillation. To minimize damage, the defibrillator paddles should be placed at least 12.5 cm (one paddle width) from the generator.

Finally, defibrillators may not work owing to poor maintenance, inadequate checks, or unfamiliarity with the instrument (Cummins *et al.* 1990). The solutions are obvious but can be difficult to implement.

References

Atkins, D.L., Sirna, S., and Kieso, R. (1988). Pediatric defibrillation: importance of paddle size in determining transthoracic impedance. *Paediatrics*, **82**, 914–18.

Bossaert, L. and Koster, R. (1992). Defibrillation: methods and strategies. *Resuscitation*, **24**, 211–25.

Cummins, R.O., Chesemore, K., and White, R.D. (1990). Defibrillator failures: causes of problems and recommendations for improvement. *Journal of the American Medical Association*, **264**, 1019–25.

Kerber, R.E. (1993). Electrical treatment of cardiac arrythmias: defibrillation and cardioversion. *Annals of Emergency Medicine*, **22**, 296–301.

Sirna, S.J., Ferguson, D.W., Charbonnier, F., and Kerber, R.E. (1988). Factors affecting transthoracic impedance during electrical cardioversion. *American Journal of Cardiology*, **62**, 1048–52.

18.4.2 Pacing and the intensive care unit

H. Montgomery

Key messages

- Cardiac pacing equipment may be used in the diagnosis and treatment of arrhythmias in the intensive care patient.
- Both tachyarrhythmias and bradycardias can be treated by pacing.
- Newer technologies allow safer and more rapid pacing of the right and left atria without the need for fluoroscopy.
- Pacing (dual chamber pacing in particular) is probably underused in the intensive care unit (ICU).

Pacing modalities

A right atrial or ventricular electrode may sense intrinsic electrical activity or provide electrical stimulation (pacing). Pacing in one chamber may occur at a fixed rate, i.e. independent of sensing in either chamber (and hence of intrinsic activity). Alternatively, demand pacing allows sensing to regulate the pacing function. An intrinsic rate (e.g. 70 beats/min) is set, with an associated R–R interval of 857 ms. If a gap of more than 857 ms is detected after the R-wave, i.e. the rate is below 70 beats/min, a pacing spike will be delivered.

The combinations of different chamber sensing and pacing interactions have been coded using five symbols. The first three describe the chamber(s) paced and those sensed (O, none; A, atrium; V, ventricle; D, atrium and ventricle), and the response to sensing (O, none; T, triggered; I, inhibited; D, triggered and inhibited). The fourth and fifth symbols describe programmability and antitachyarrhythmia functions respectively. Common modes encountered are shown in Table 1.

Indications for pacing

The following questions should be addressed.

1. Are correctable factors initiating/sustaining the dysrhythmia (e.g. cardiac ischemia, electrolyte imbalance, thyroid function disturbance, hypoxemia, drug therapy, cardiac irritation by central venous/pulmonary arterial catheters)?
2. Is the patient sufficiently compromised to warrant intervention?
3. Might this dysrhythmia cause cardiovascular compromise either *per se* or through degeneration into a more serious rhythm abnormality?
4. Would pharmacotherapy be more appropriate? Pacing may spare potential proarrhythmic or negatively inotropic effects of antidysrhythmic therapy, particularly when combination therapy is given.
5. Which pacing mode is most appropriate?

Table 1 Common pacing modes

AOO	Fixed-rate atrial pacing (useful in atrial standstill)
VOO	Fixed-rate ventricular pacing (useful in emergencies if ventricular sensing proves unreliable)
AAI	Senses low atrial rates and triggers atrial pacing (useful in sick sinus syndrome)
VVI	Ventricular sensing over a set rate inhibits ventricular pacing
VDD	'Atrial tracking' where atrial sensing triggers a ventricular pacing spike (used in complete heart block with normal sinus node activity)
DVI	Physical atrioventricular sequential pacing if ventricular rate slows below a threshold. Lack of atrial sensing makes it useful only if the pacing rate is greater than the atrial rate; therefore DDD or AAI modes may be more appropriate
DDD	Atrioventricular sequential pacing if intrinsic rate slows below a set threshold

Bradyarrhythmias

Prophylaxis

Pacing may be required in the absence (primary prophylaxis) or presence (secondary prophylaxis) of bradycardia.

Primary prophylaxis is warranted when the occurrence of a symptomatic bradycardia is likely, i.e. under the following circumstances.

1. Sick sinus syndrome, particularly if antitachycardic medications are to be used.
2. Conduction is only occurring through one diseased fascicle. ECG shows first-degree heart block with left bundle branch block or first-degree heart block with right bundle branch block and left axis deviation (signifying left anterior hemiblock). Progression to complete heart block may be precipitated by other factors in the sick ICU patient as described above.
3. First-degree heart block complicating endocarditis (suggesting aortic root abscess) demands immediate pacing. Progression to complete heart block is often associated with sudden asystolic cardiac arrest.

Secondary prophylactic pacing is employed under the following circumstances.

1. Further slowing may prove detrimental.
2. There is a risk of escape tachyarrhythmias such as escape ventricular tachycardia.

Treatment

Any slow rhythm causing cardiovascular compromise should be treated. A poorly functioning ventricle may have a relatively fixed stroke volume; only an increase in heart rate might raise cardiac output without reducing blood pressure.

Tachyarrhythmias

Prophylaxis

Long R–R intervals (prevented by pacing) may be associated with escape ventricular tachycardia (often heralded by frequent ventricular ectopic beats).

Treatment

Pacing may treat re-entrant tachyarrhythmias, involving adjacent pathways. Insertion of an electrical stimulus into the circuit ('capture') may extinguish the arrhythmia. Capture requires the cardiac cycle to be scanned with electrical impulses in one of three ways.

1. Underdrive pacing, which is best employed when cycle frequency is below 150 beats/min. With pacing slightly slower than the native tachycardia, a single extra stimulus may 'break into' the tachycardia in 50 per cent of cases.
2. Single stimulus: the pacing box inserts (on demand) a single stimulus 10 per cent earlier than the anticipated next beat (i.e. 90 per cent of the R–R interval or 'cycle length'), and at progressively earlier stages if unsuccessful.
3. Overdrive pacing: this is generally better with tachycardias over 150 beats/min. Pacing slightly faster than the native tachycardia is more successful than underdriving by reducing the refractory period of the tissue between the pacing wire and the re-entry circuit, thus helping the impulse to 'break into' the circuit. However, it is more likely to accelerate the tachycardia. If specialist equipment is used, each burst may terminate with an extra beat of slightly shorter cycle length.

Antitachycardia pacing, for which an external defibrillator must be readily available, may be used to treat ventricular tachycardia, atrioventricular nodal re-entrant tachycardia, atrioventricular re-entrant tachycardia, and atrial flutter. Pacing is applied to the affected chamber, although overdrive atrial pacing can be used to treat 'slow' ventricular tachycardia, with burst pacing delivered until a ventricular capture beat is seen. Incremental rates or repetition after atropine (to improve atrioventricular conduction if high pacing rates are needed) can be used. In terminating supraventricular tachyarrhythmias, rapid pacing (400–800 beats/min) can be used to cause atrial fibrillation which usually reverts rapidly to sinus rhythm.

Antitachycardia pacing is indicated under the following circumstances.

1. The arrhythmia is due to a drug effect (e.g. antidepressants/antidysrhythmics), and adding further pharmacological agents may have unpredictable effects.
2. The arrhythmia is recurrent, necessitating repeated external cardioversion.
3. The arrhythmia is torsade de pointes.
4. The arrhythmia is still present despite pharmcotherapy.

Improved ventricular dynamics

Pacing may improve cardiac output independently of effects on heart rate. Lone ventricular pacing dissociates the atrioventricular contraction sequence, in which atria contract against closed atrioventricular valves (pacemaker syndrome), reducing output by more than 20 per cent. Atrioventricular sequential pacing maintains the late diastolic atrial 'kick' and improves early diastolic inflow (the atrioventricular valves are open when the atria are full). Thus ventricular preload and cardiac output increase without a rise in venous pressures, particularly where ventricular filling is obstructed (e.g. mitral stenosis) or compliance is reduced. Such ventricular 'stiffness' may complicate myocardial hypertrophy, ischemia, fibrosis, or infiltration. Maintained atrioventricular synchrony may have advantages even in sinus bradycardia, as well as in dysrhythmias complicated by a lack of atrial contraction or a lack of co-ordination between atrial and ventricular contractions (e.g. complete heart block). Atrial activity can be sensed and the ventricle paced in response ('physiological pacing' or 'tracking'), or the chambers can be sequentially paced. The interval between atrial and ventricular activation is usually set at 150 to 200 ms.

Atrioventricular sequential pacing may improve ventricular performance in right ventricular infarction. A short atrioventricular delay may benefit the patient with poor left ventricular compliance, hypertrophic cardiomyopathy, severe dilated cardiomyopathy, or if an interatrial conduction time delay is causing late left atrial activation and contraction against a closed mitral valve.

Diagnosis

Atrial electrodes (*in situ* after cardiac surgery) may aid diagnosis in broad complex tachycardia. Each of the two upper limb leads of a four-lead ECG monitor are connected to the atrial wires, the two lower limb leads are left *in situ*, and recordings are made at 25 mm/s. Lead I records a bipolar electrogram (across the two atrial leads), while leads II and III record unipolar (atrial to lower limb) electrograms. An independent P-wave activity rate lower than the ventricular rate confirms ventricular tachycardia. A higher atrial dissociated rate suggests supraventricular tachyarrhythmia with aberrant conduction; this is confirmed by comparison of the timing of the atrial and ventricular recordings.

Temporary cardiac pacing techniques

Endocardial ventricular pacing leads

Pacing electrodes (5–7 French) are passed through a central venous valved sheath inserted through the subclavian, the femoral, or preferably the right internal jugular vein. The left subclavian system is best avoided if permanent pacemaker insertion is likely. The wire tip is positioned in the right ventricular apex (pointing anteriorly and inferiorly, left and lateral to the spine, and near the cardiac border) in a stable position and with a gentle curve. Advancing the wire by approximately 3 cm once the wire tip is in position improves stability.

Pacing threshold (the minimum output current producing consistent ventricular electrical capture) and sensitivity (the minumum intrinsic R-wave voltage which can be sensed by the pacing box) are assessed sequentially. Sensitivity is set at 3 mV, the rate is set at 20 beats/min above the intrinsic rate, the output is set at 5 mA, and the pacing box is placed in demand mode. Consistent pacing of the heart should be seen on the monitor. Output is steadily reduced until the pacing threshold, i.e. failure to capture, is reached. Ideally, this should be 1 mA. The pacing current is then set at two to three times the threshold. A rapidly rising threshold or a threshold above 6 mA are indications for wire repositioning (using a new wire if the patient is pacemaker dependent). The sensing threshold is now tested, with the box in demand mode, and the rate is set at 20 beats/min below the intrinsic rate. The indicator system of the box should detect 'sensing' with each native beat. The sensing level is steadily reduced from 3 mV until sensing failure occurs. Sensitivity should be as close to 1 mV as possible. Setting the threshold at very low levels may cause failure to sense, while too high a threshold permits detection of artifactual electrical activity such as voluntary muscle movement.

Pacing electrodes may be positioned using fluoroscopic guidance, electrocardiographic guidance, or blind insertion.

Fluoroscopic guidance

Stiff bipolar wires, which can be 'shaped' by hand into a gentle curve before introduction, are generally used. The use of balloon flotation wires is faster, more reliable, and associated with lower morbidity. If placement is difficult, the wire is advanced along the lateral wall of the atrium until the tip catches on the tricuspid valve annulus, advanced to produce a loop of wire in the right ventricle, and then rotated anticlockwise causing the tip to flip inferiorly and leftwards into the right ventricular apex.

Electrocardiographic guidance

ECG limb leads are connected, and the negative-pole electrode is attached to a chest (V) lead. The P-wave of the internal ECG complex will be inverted in the high right atrium, biphasic in the mid-atrium, and upright lower down. The P-wave amplitude normally exceeds that of the QRS complex, with a reduction compared with that of the QRS complex as the right ventricle is entered. All voltages decrease if the inferior vena cava is entered, and all increase upon entry into the coronary sinus.

Blind insertion

This technique has a low risk of perforation and a high success rate. A 1- to 2-ml inflatable balloon is located proximal to the bipolar tip electrodes. With surface ECG monitoring, the pacing wire is connected to the pacing box, the rate is set at 20 beats/min above the intrinsic rate, and a high output is chosen (below 10 mV when immediate capture is essential for lifesaving). The wire is quickly measured against the patient in order to determine the approximate distance from the insertion point to the tip of the right ventricle. Once through the sheath, the balloon is inflated and the wire is advanced smoothly. The lead tip should sweep naturally into and lodge in the right ventricular apex. When ventricular (or, more rarely, accidental atrial) capture is seen on the monitor, the balloon is deflated and the catheter is advanced a further 2 cm. Prior wire 'shaping' (as described above), cooling (to stiffen), or altered patient position (left side up, head-down tilt) may aid placement. In significant tricuspid regurgitation, balloon deflation with blind passage may be needed.

Atrial pacing leads

These are passed through a valved central venous sheath. In emergencies, a soft wire (e.g. a balloon flotation wire) may be used before substitution with a more stable atrial J-wire. The J-wire contains a central stylet which stiffens and straightens the lead to allow passage through the sheath. Further positioning may be guided by fluoroscopy (or internal ECG recording in emergencies). The stylet is slowly withdrawn while the lead is rotated to position the tip inferior to the right atrial appendage. Pulling back on the lead causes it to lodge in the appendage (pointing anteriorly and towards the left shoulder tip). The position is confirmed by lateral radiography. The J should be allowed to flex incompletely on itself, but not open up to an angle of more than 50° to 60°. In difficult cases, the atrium may be paced briefly via the coronary sinus. The optimal atrial pacing threshold is less than 2 mA.

Complications associated with the use of temporary endocardial wires

Wires should be fixed to the skin using sutures and dressings to prevent dislodging during patient movement. Thresholds should be checked with each nursing shift change and entry sites inspected for evidence of infection. After prolonged use, consideration should be given to the insertion of permanent pacing systems. Complications include those of central venous cannulation, sepsis, loss of sensing (leading to inappropriate pacing, and risk of R-on-T arrhythmia), or primary failure to capture (particularly after inferior infarction, where finding an area to pace can be difficult). If pacing is lost, the VOO mode should be selected, the pacemaker output steadily increased to maximum, all connections and power sources checked (or changed), or the pacing wire repositioned. Myocardial perforation may be detected as a stethoscopically audible pericardial rub or complaint of pericardial and/or shoulder pain, and may cause tamponade. Septal perforation leads to a right bundle branch block pattern instead of the usual left bundle branch block in right ventricular pacing. Diaphragmatic pacing may occur, remedied by a reduction in output current or wire repositioning.

Table 2 Points of caution regarding temporary pacemakers

1.	Take care to avoid snaring or dislodging leads when inserting and withdrawing pulmonary artery flotation catheters
2.	Magnetic resonance imaging is contraindicated. Permanent pacemaker damage may result
3.	Pacemakers may have programmed 'hysteresis', i.e. the pacemaker will sense a low rate (e.g. <50 beats/min) before triggering faster pacing (e.g. 70 beats/min). Thus the presence of a bradycardia does not necessarily mean that the pacemaker has malfunctioned

Transesophageal pacing

Transesophageal atrial pacing electrodes are quick, simple, and safe to use. A standard bipolar ventricular wire with an esophageal stethoscope, or a specially developed 'pill' or balloon electrode can be used. If advanced distally, these can also pace the ventricle.

The electrode is inserted to a depth of more than 40 cm and then withdrawn until a maximum atrial electrogram is recorded. Precise positioning is unnecessary, with pacing possible over a 10-cm range. Thresholds may be up to 10 mA.

Transesophageal atrial pacing can be used, like any other atrial electrode, to pace or sense. It can be used to treat atrial arrhythmias by burst pacing. In atrial flutter, pacing at rates 40 per cent higher than the flutter rate may restore sinus rhythm in over 50 per cent of patients. Success is increased by prior digitalization.

Table 3 Common problems encountered with temporary pacing

1.	Failure to sense or pace: temporary pacing wires can be inserted in an emergency, with a magnet applied over the original pacing box. The cardiology/pacing team should then be asked to review the pacemaker function
2.	Failure to deliver pacing spikes: this usually suggests loss of power
3.	Oversensing: the box may detect musculoskeletal activity (e.g. in the shivering patient) and interpret this as intrinsic activity. Magnet application will remove sensing capability in the short term
4.	Tachycardia in the presence of a DDD pacemaker: magnet application over the pacing box causes DOO or VOO pacing, and may reveal P waves or atrial flutter or fibrillation which is being tracked. Pacemaker-mediated tachycardia, in the presence of a DDD pacemaker, may also be the case. A ventricular ectopic beat is retrogradely conducted to the atria, where it is sensed by the atrial wire and a ventricular pacing spike is triggered, leading to another retrogradely conducted beat. This causes a tachycardia in which the atria are contracting against closed atrioventricular valves. Magnet application (conversion to DOO or VOO mode) will terminate the tachycardia. Normal pacing function is resumed upon removal of the magnet
5.	Non-re-entrant ventriculoatrial synchrony: a DDD pacemaker is *in situ*, and a ventricular ectopic is retrogradely conducted and goes unsensed in the atria. The atrial pacing then occurs during the atrial refractory period, followed by a ventricular pacing spike which is again retrogradely conducted. This again causes atrial contraction against closed atrioventricular valves
6.	Bipolar electrocautery is generally safe. Unipolar (cutting) cautery should be avoided. If used, a magnet should be applied (to prevent sensing misinterpretation). Pacemaker damage may still occur
7.	Cardioversion may lead to: program or component malfunction increased pacing threshold undersensing These problems are minimized by using anteroposterior defibrillation, placing paddles at least 10 cm from the box or leads if possible, and using low energies. Back-up temporary pacing facilities should be available

Transesophageal atrial pacing is also used to diagnose tachyarrhythmias or to treat ventricular tachycardia, with success rates of up to 70 per cent at lower ventricular rates.

Epicardial wires

Temporary Teflon-coated epicardial wires, commonly inserted at the time of cardiac surgery, can be utilized for diagnosis and treatment. In the absence of a 'neutral' electrode, a venepuncture needle can be passed through a fold of skin and the connecting cable attached directly to its tip.

External cardiac pacing

Large (> 8 cm) self-adhesive plate electrodes are placed on the anterior and posterior left chest. The pacing unit can usually sense ventricular complexes and pace with outputs below 150 mA. Output is steadily increased to the minimum associated with a detectable output/radial pulse. This may be uncomfortable, and adequate sedation is important. Substitution with endocardial pacing should usually be undertaken as soon as possible.

Problems with pre-existing permanent systems

Medical notes, a patient record card, or contact with the appropriate pacing clinic will usually determine the variety and programming of a permanent pacemaker. Plain chest radiography shows which chambers have leads inserted, while an ECG taken during pacing activity will confirm which chambers are being paced. The pacing service can use transcutaneous devices to interrogate and reprogram the pacemaker. This may give information about how the pacemaker is programmed, how often it has been used, and even what sort of dysrhythmias have recently occurred. Automatic implantable cardiac defibrillator boxes with pacing potential are becoming more common.

Table 2 lists points of caution regarding temporary pacemakers, while Table 3 shows pacemaker-related problems encountered in the ICU.

Specific clinical situations

Cardiac surgery

One to three per cent of cardiac surgical patients operated upon for acquired heart disease will require permanent cardiac pacemaker insertion postoperatively. Risk factors include aortic valve surgery, absence of preoperative sinus rhythm, postoperative myocardial infarction, and female sex (odds ratios 8, 6, 3, and 2 respectively). Repeat surgery, tricuspid valve surgery, preoperative use of digitalis, and advancing age may also increase risk. Approximately 10 per cent of patients undergoing heart transplantation will require temporary pacing, rising to nearly 50 per cent of pediatric heart or heart–lung transplant patients.

Poisoning

Cardiac pacing for management of digitalis-induced cardiac dysrrhythmias may have been superseded by Fab fragment therapy. Its use in cases of overdose of calcium-channel blockers or β-blockers may be advantageous.

Bibliography

Donovan, K.D., Dobb, G.J., and Lee, K.Y. (1991). Hemodynamic benefit in maintaining atrioventricular synchrony during cardiac pacing in critically ill patients. *Critical Care Medicine*, **19**, 320–6.

Jafri, S.M. and Kruse, J.A. (1992). Temporary transvenous cardiac pacing. *Critical Care Clinics*, **8**, 713–25.

Singer, I. and Kupersmith, J. (ed.) (1993). *Clinical manual of electrophysiology*. Williams & Wilkins, Baltimore, MD.

18.4.3 Intra-aortic balloon counterpulsation

P. Kallis

Key messages

- The use of the intra-aortic balloon counterpulsation pump (IABP) can improve the function of the failing heart and may be lifesaving.
- IABP carries a significant mortality and morbidity.
- IABP is precisely timed with the onset of diastole and therefore augments both the diastolic pressure and the coronary perfusion pressure.
- The deflation of the balloon is precisely timed just prior to the onset of systole, thus creating a relative 'vacuum' in the aorta and reducing the afterload.
- Hemodynamically significant aortic regurgitation is a contraindication.

Introduction

Intra-aortic balloon counterpulsation (IABP) utilizes the principle of diastolic counterpulsation, increasing diastolic coronary perfusion pressure (and oxygen supply), reducing systolic afterload (and oxygen demand), and thus improving the cardiac index. Experimental evidence in the 1950s demonstrated that there was an increase in coronary blood flow if the arrival of the arterial pressure wave to the coronary circulation was delayed to coincide with diastole (Kantrowitz and Kantrowitz 1953). The principles of IABP were elucidated in the 1960s, and the first clinical procedure was performed in 1968 (Kantrowitz *et al.* 1968). The safety of IABP has continued to improve, but, although its efficacy has been clearly demonstrated, it should only be used when clearly indicated as it carries a definite morbidity and mortality.

Circulatory effects

Approximately three-quarters of the left ventricular coronary flow occurs during diastole. The IABP is precisely timed with the onset of diastole and therefore augments both the diastolic pressure and the coronary perfusion pressure.

The deflation of the balloon is precisely timed just prior to the onset of systole, thus creating a relative 'vacuum' in the aorta and reducing the afterload (Weber and Janicki 1974). The left ventricular wall tension developed is decreased and the myocardial oxygen demand is reduced. Therefore IABP improves the function of the failing heart by improving the myocardial oxygen supply-to-demand ratio.

The synergistic effects of diastolic augmentation and afterload reduction exert beneficial effects on the failing circulation (Bardet *et al.* 1977). IABP improves the cardiac output and as a result reduces the left atrial filling pressure. The stroke volume increases significantly but the heart rate is not usually altered significantly. The systemic blood pressure drops and the diastolic pressure is increased, but the effect on the mean systemic arterial pressure is variable. The systemic vascular resistance is usually reduced, but this is also manipulated by pharmacological intervention.

Indications

Inability to wean from cardiopulmonary bypass following cardiac surgery despite adequate inotropic support and left ventricular filling, particularly after myocardial revascularization, is one of the most common indications for IABP. Other indications include unstable angina that is not responsive to medical treatment, acute myocardial infarction, particularly when associated with post-infarction complications such as left ventricular failure, papillary muscle rupture, and ventricular septal rupture. IABP is also indicated in the failing heart following transplantation and in patients with viral myocarditis. Occasionally IABP is used prophylactically in cardiac surgery for patients with very poor left ventricular ejection fraction prior to myocardial revascularization. It is sometimes used prophylactically in patients about to undergo high-risk angioplasty.

Contraindications

Untreated aortic dissection and transection are contraindications to IABP, although the more recent method of insertion of the IABP catheter through a Dacron graft on the ascending aorta may overcome this problem. Hemodynamically significant aortic regurgitation is also a contraindication.

IABP insertion

The common femoral artery is the most frequent and easiest route of insertion. A meticulously sterile technique is used and the percutaneous method is attempted first. The artery is cannulated with a needle and a Seldinger technique is used. The IABP is passed along a guidewire either through a sheath or, more recently, without a sheath, particularly in small-caliber femoral arteries. If the percutaneous approach is unsuccessful, the common femoral artery is exposed and the IABP catheter is inserted either using a modified Seldinger technique or by direct cannulation through a purse string. The length of the IABP catheter to be inserted is measured from the entry point of the femoral artery to the angle between the sternum and the manubrium. Fluoroscopy makes the insertion safer and can also check the final position of the catheter, which should be just distal to the left subclavian artery. If fluoroscopy is not available, the position can be verified with a chest radiograph.

If the femoral approach is unsuccessful, the catheter can be inserted through the ascending aorta secured by either a purse string or a long Dacron graft. This approach is possible only following cardiac surgery, and in other situations the axillary or subclavian arteries can be used. These approaches should not be used without careful assessment of benefit versus risk, because the greatest source of morbidity and mortality is related to the arterial access for the insertion of the IABP catheter.

IABP management

In order to minimize the chance of development of thrombus on the balloon, the patients should be anticoagulated with heparin to a partial thromboplastin ratio of 1.5 times normal and the balloon should be kept in motion constantly. The platelet count should be carefully monitored as there is a tendency for the development of thrombocytopenia.

The timing of IABP inflation and deflation is critical in order to achieve optimal diastolic augmentation and reduction of aortic end-diastolic pressure. Inappropriately early inflation of the balloon leads to premature closure of the aortic valve, aortic regurgitation, increased afterload, and myocardial oxygen demand. Premature deflation of the IABP balloon during diastole leads to suboptimal diastolic augmentation (and coronary perfusion) and suboptimal afterload reduction (and increased myocardial oxygen demand).

If the balloon is deflated too late in diastole, the left ventricle will attempt to eject against the inflated balloon. Both the afterload and the myocardial oxygen demand are increased. Late inflation of the balloon after closure of the aortic valve leads to suboptimal diastolic augmentation and coronary perfusion.

Complications

Major complications occur in 8.5 per cent of cases and IABP-related mortality is about 0.8 per cent (McEnany *et al.* 1978). The majority of complications are related to vascular problems and thrombosis, in relation to the site of insertion, is the most common complication. Less common, but very serious, complications include embolism, arterial dissection, and laceration. Pseudoaneurysm formation and femoral artery stenosis can also occur after balloon removal.

Limb ischemia is a serious complication. The majority require embolectomy, but a minority may require a vascular repair or a bypass procedure or even amputation. If the relevant limb becomes ischemic, a very careful assessment of the overall state of the patient must be made. If the patient is judged not to require IABP support any longer, the catheter should be removed. If the relevant limb is subsequently adequately perfused, vascular assessment can be delayed until the patient's condition improves. However, if limb perfusion is inadequate, embolectomy should be attempted with a Fogarty catheter. If this is also inadequate, formal vascular assessment is required with a view to a vascular bypass procedure to re-establish flow. If there has been significant limb ischemia, a prophylactic fasciotomy may also be required in order to prevent compartment syndrome. If the patient is still dependent on IABP, the catheter should be inserted on the opposite side or in one of the other sites as mentioned previously.

Infection of the IABP catheter requires removal as well as antibiotic therapy. Balloon rupture is rare but leads to massive and fatal gas embolism. This is probably related to difficulty in insertion or damage of the balloon by atheroma at the site of insertion. Renal failure may result from renal artery embolization or because of renal hypoperfusion secondary to the poor circulation of the patient. It is controversial whether a properly functioning balloon placed close to the renal arteries can lead to renal failure.

Initiation of and weaning from IABP

IABP therapy must not be unnecessarily delayed, particularly following myocardial revascularization of the already impaired left ventricle. Inotropic support consisting of more than 10 g/kg/min of dopamine or equivalent, despite adequate left atrial pressure, will produce a significant increase in myocardial oxygen demand without an increase in myocardial oxygen supply. Initiation of IABP therapy before any significant myocardial necrosis occurs is more likely to result in a favorable outcome than further escalation of the inotropic support.

The exact timing of weaning from IABP therapy depends on many factors. However, as a rough guide, a patient with a mean arterial pressure greater than 70 mmHg, a cardiac index greater than 2.2 l/m^2/min, and a pulmonary artery pressure below 18 mm Hg without any significant inotropic support should be considered for weaning. Weaning can be achieved by reducing either the inflation ratio or the balloon volume.

The method of removal depends on which method was used to insert the catheter. Heparin is stopped 1 h prior to removal. With the percutaneous method the IABP is turned off and a 50-ml syringe is used to evacuate the balloon gas. An assistant presses on the artery distal to the insertion site to minimize the chance of distal embolization. The catheter is pulled back until the balloon meets the sheath and the two are then pulled together. The artery is allowed to bleed for a few seconds to expel any thrombus, and then firm pressure is applied for at least 30 min or until the bleeding stops completely. A pressure dressing is applied and the patient is kept supine for 24 h.

Open removal is used if the open technique was used for the insertion or when problems are anticipated. The main advantage of this technique is that it allows embolectomy at the same time and a definitive repair of the artery.

References

Bardet, J., *et al.* (1977). Clinical and hemodynamic results of intraaortic balloon counterpulsation and surgery for cardiogenic shock. *American Heart Journal*, **93**, 280–8.

Kantrowitz, A. and Kantrowitz, A. (1953). Experimental augmentation of coronary flow by retardation of the arterial pressure pulse. *Surgery*, **34**, 678–87.

Kantrowitz, A., Tjønneland, S., Freed, P.S., Phillips, S.J., Butner, A.N., and Sherman, J.L., Jr (1968). Initial clinical experience with intraaortic balloon pumping in cardiogenic shock. *Journal of the American Medical Association*, **203**, 113.

McEnany, M.T., *et al.* (1978). Clinical experience with intraaortic balloon pump support in 728 patients. *Circulation*, **58** (Supplement 1), 124–32.

Weber, K.T. and Janicki, J.S. (1974). Intraaortic balloon counterpulsation. A review of physiologic principles, clinical results, and device safety. *Annals of Thoracic Surgery*, **17**, 602–36.

18.4.4 Sengstaken tube

J. Wendon

Key messages

- Balloon tamponade is a lifesaving procedure in patients with torrential variceal hemorrhage.
- It is a procedure which should be undertaken by skilled personnel, and the patient should be nursed in an intensive care or high-dependency environment by experienced staff.
- Tamponade, although highly efficacious in obtaining hemostasis, has a complication rate of up to 20 per cent and can carry a mortality if complicated by aspiration pneumonia, sepsis, and multiple organ failure or by esophageal rupture.

Variceal hemorrhage is a frequent and potentially life-threatening complication in patients with portal hypertension. It has a mortality of approximately 30 per cent, related to the complications of hypovolemia, multiple organ failure, and decompensation of liver disease. Prognosis of the bleeding episode is determined by the magnitude of blood loss and degree of liver failure, with a mortality of approximately 70 per cent for those in Childs–Pugh stage C and 15 per cent in those in Childs–Pugh stage A.

Acute variceal hemorrhage is a medical emergency that should be managed in an intensive care or high-dependency environment by an experienced multidisciplinary team working to well-understood protocols. The aim of therapy should be to correct hypovolemia and achieve hemostasis at the bleeding site. All patients require large-bore venous access, central venous catheter insertion, and appropriate colloid infusion. Coagulation factors, fresh frozen plasma, and platelet transfusions are also likely to be required. Altered conscious level should be acted upon to prevent aspiration, and such patients will frequently benefit from early elective intubation and ventilation. Therapies should be directed to controlling hemorrhage; pharmacological therapy to decrease portal inflow with somatostatin analogs is frequently advocated, and the patient should undergo endoscopy by an experienced practioner with skills in the management of acute variceal hemorrhage. If such an individual is not available and/or if there is massive exsanguinating hemorrhage, balloon tamponade should be considered. This controls hemorrhage by compression of the varices within the stomach and limiting inflow of blood into the esophageal varices by compression just below the gastroesophageal junction.

Tamponade is a highly effective form of therapy, controlling hemorrhage in 85 to 95 per cent of cases in experienced centers, but carries several potential drawbacks. These include the risk of aspiration and pulmonary sepsis, mucosal damage due to direct compression, and incorrect placement with the risk of esophageal rupture. In view of the latter, the tube should ideally only be inflated for 24 h prior to removal and endoscopic therapy or further definitive therapy in the form of surgical shunt or a transjugular intrahepatic portosystemic stent shunt (TIPSS) procedure. Very few patients will safely tolerate a Sengstaken tube; the majority, and certainly all those who are encephalopathic, require intubation to protect their airway and allow safe management of the tube.

Prior to inserting the tube it is essential that the operator is aware of the type of balloon tamponade system being inserted (e.g. a Sengstaken–Blackmore three- or four-lumen tube, a Minnesota tube, or a Linton tube). The practioner should be aware of which lumens are for aspiration and which are for inflation, and furthermore the volumes required to inflate the gastric balloon. Guidelines are listed in Table 1, but these are for guidance only. It should always be ensured that the volumes are appropriate for the system used.

Table 1 Guidelines for volume and pressure requirements in balloon tamponade

Type	Size	Age	Esophageal pressure (mmHg)	Sterile water in gastric balloon (ml)	Omnipaque in gastric balloon (ml)	Total (ml)
Minnesota (4 lumen)	One size	Adult	40	>200	40	>240
Sengstaken (3 lumen)	22	Adult	40	>100	40	140
SB (4 lumen)	20	Adolescent	40	60	20	80
SB (4 lumen)	18	8–12 years	40	40	20	60
SB (4 lumen)		<8 years	40	40	20	60
Linton (single intragastric balloon)	One size	Adolescent and adult		500–700	40	500–700

SB, Sengstaken–Blakemore.

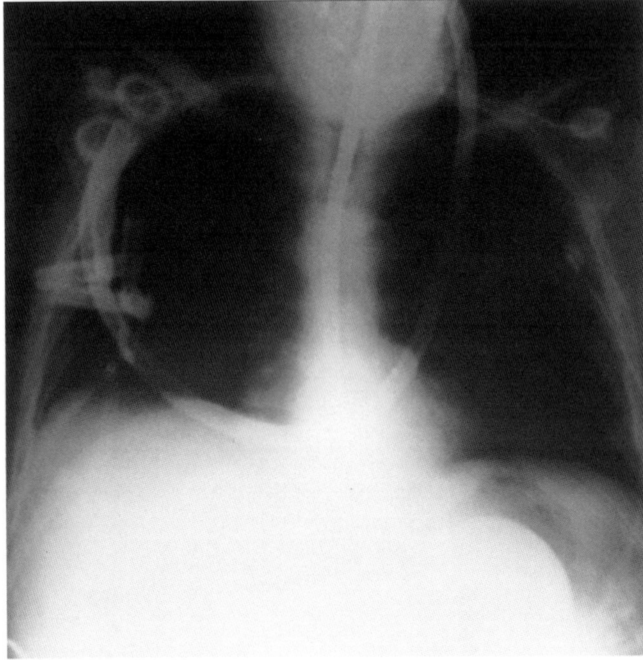

Fig. 1 Appropriately placed Sengstaken–Blackmore tube.

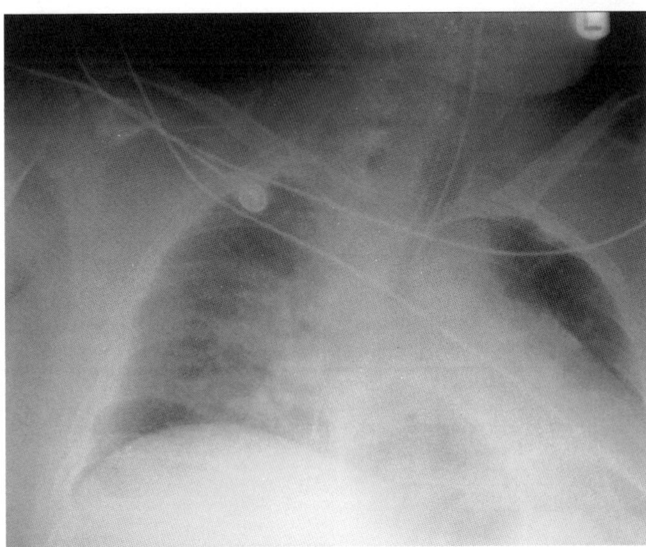

Fig. 2 Incorrectly placed Sengstaken–Blackmore tube.

The balloons should be test inflated prior to insertion; they should then be evacuated and plugs should be inserted into the pressure-monitoring balloon. In normal circumstances it is almost never necessary to inflate the esophageal balloon to obtain hemostasis provided that the gastric balloon has been appropriately inserted and distended. The end of the tube should be lubricated and passed either nasally or orally to the posterior pharyngeal wall and then down the esophagus and into the stomach. The gastric balloon is then inflated with an appropriate volume of sterile water and Omnipaque (~~in~~ order to facilitate visualization on the chest radiograph). It is essential that the operator is certain that the tube has been passed into the stomach prior to inflating the gastric balloon. Firm traction is then applied to the tube until resistance is felt and the balloon is firm against the diaphragm. Aspiration is commenced on both the gastric aspiration lumen, in order to keep the stomach clear of blood, and on the esophageal aspiration port to decrease the risk of aspiration. The tube is then fixed in position either with a tube fixation system or by attaching it to the patient's skin using a dressing such as Elastoplast or Sleek.

The traction should be checked every 2 to 4 h and the skin fixation system moved from right to left on a regular basis to prevent pressure damage to the skin. It is not advisable to apply traction to the tube by the method of attaching a 1-kg weight, such as a bag of intravenous fluid, to the tube. There are two reasons for this: the amount of traction applied may be excessive, and it is also highly variable, changing whenever the patient moves and thus potentially resulting in excessive or subtherapeutic amounts of traction on the gastroesophageal junction.

If the tube is correctly placed, hemostasis is almost always achieved. If bleeding continues consideration should be given to misplacement of the tube, inadequate insufflation of the gastric

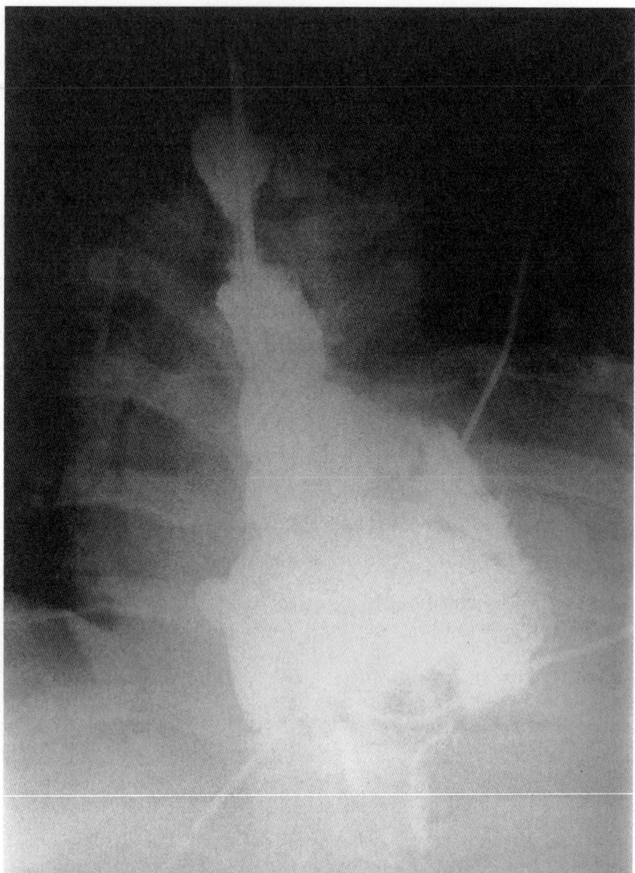

Fig. 3 Gastrografin study showing a lower esophageal tear caused by incorrect tube placement.

balloon, and incorrect application of traction on the gastroesophageal junction. In addition, the possibility that another lesion is accounting for the bleeding or that there are gastric varices in a position that does not allow for compression with the gastric balloon should be considered. As mentioned above, it is rarely necessary to inflate the esophageal balloon but, if this is required, it should be inflated with air, the pressure should be maintained at an appropriate level (25–40 mmHg), and it should be deflated intermittently. It should probably never be inflated except after review by experienced staff. After 24 h or earlier, the gastric balloon should be deflated (the amount in the balloon should be documented clearly in the patient's notes) and the tube gently removed immediately prior to endoscopic review.

In summary, balloon tamponade is highly efficacious in controlling life-threatening bleeding from gastric and esophageal varices, but it should be recognized that it is a specialist technique that carries a complication rate of up to 20 per cent. Although this can be limited by appropriate medical and nursing care of patients requiring balloon tamponade, complications can still occur. If there is any concern that the gastric balloon has been inflated in the esophagus or the gastroesophageal junction has been damaged, the balloons should be deflated immediately. It is then essential not to undertake endoscopy, but to perform appropriate imaging to define the problem and transfer the patient to a unit allowing both cardiothoracic and hepatobiliary support to facilitate repair of the esophagus and surgical shunt to control the portal hypertension. Figure 1 shows an appropriately placed Sengstaken–Blackmore tube. Figure 2 shows an incorrectly placed tube with the gastric balloon filled with air, and Fig. 3 shows the subsequent Gastrografin study confirming a significant lower esophageal tear.

18.5 Renal replacement therapy

18.5.1 Continuous hemofiltration techniques

M. Schetz

Key messages

- The major advantage of continuous renal replacement therapy (CRRT) is the excellent hemodynamic tolerance.
- Continuous pump-driven (venovenous) renal replacement therapy has largely replaced arteriovenous techniques because of the access-related complications and limited efficacy of the latter.
- The need for continuous anticoagulation is the weak point of CRRT.
- Indications for CRRT are acute renal failure with severe fluid overload, hemodynamic instability, cerebral edema, or the need for adequate nutritional input.

Introduction

Continuous renal replacement therapy (CRRT) is an established treatment for acute renal failure in critically ill patients. It was first described by Kramer in 1977 as arteriovenous hemofiltration. The need to increase the efficiency and safety of the procedure led to subsequent modifications of this originally simple technique with remarkable hemodynamic tolerance. A recent survey revealed that CRRT is the preferred method of renal replacement therapy in intensive care units (ICUs) in Europe and Australia.

Methods

Solute transport

Hemofiltration uses convective solute transport; a pressure gradient drives an ultrafiltrate, of composition similar to plasma water, through a high-flux membrane. Blood purification is achieved by partial or complete substitution of the ultrafiltrate with a replacement solution that is administered into the venous (postdilution) or arterial (predilution) limb of the circuit. The clearance of all solutes with molecular weight below the cut-off of the membrane is identical and only limited by the filtration rate.

During hemodialysis the dialysate is directed in countercurrent through the outer side of a hemodialyzer, while solute transport is based on diffusion driven by the concentration difference between blood and dialysate. In contrast with convection, diffusion depends on molecular weight and provides superior elimination of low-molecular-weight solutes but limited elimination of medium-molecular-weight compounds.

Diffusion and convection can be combined in hemodiafiltration.

Vascular access

Arteriovenous methods, using the patient's blood pressure to drive the blood through the extracorporeal circuit, require access to a large artery and vein. Percutaneous cannulation of the femoral artery and vein is the usual method (Fig. 1). Catheters must be short (8–11 cm) and large bore (> 2 mm) in order to reduce resistance. Blood flows of 50 to 120 ml/min are generally observed.

Most complications of arteriovenous hemofiltration are associated with the arterial access. Vessel perforation may lead to retroperitoneal hemorrhage. Other access-related complications include local bleeding, thromboembolic phenomena, aneurysm formation, and infection. The need for arterial access precludes application of arteriovenous techniques in patients with severe arteriosclerosis or a vascular prosthesis.

In venovenous techniques the driving pressure is provided by a blood pump (rate 100–200 ml/min) and a single venous access with a double-lumen catheter suffices (Fig. 2).

Membrane

In order to achieve adequate solute transport, hemofiltration requires membranes with high hydraulic permeability. Synthetic non-cellulose membranes are most suitable and have the additional advantage of improved biocompatibility. The most commonly used membranes

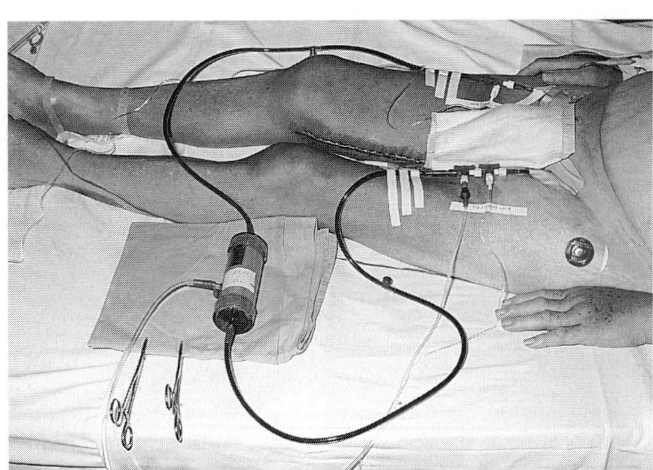

Fig. 1 Technique utilizing arteriovenous vascular access.

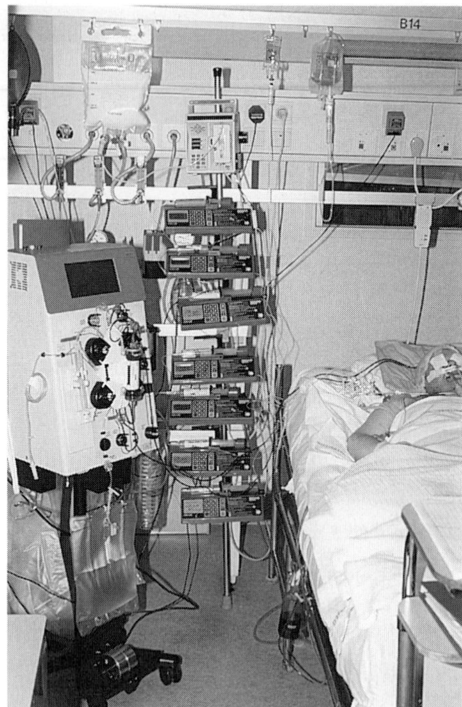

Fig. 2 Technique utilizing a double-lumen venous catheter (venovenous) and blood pump.

in CRRT are the Hospal polyacrylonitrile (AN69), Amicon polysulfone, Gambro polyamide, and Fresenius polysulfone membranes. The cut-off point of these membranes is in the range of 30 000 Da.

Anticoagulation

The need for continuous anticoagulation is the weak point of CRRT. The low-flow low-pressure conditions of the arteriovenous techniques increase the risk of filter clotting. Procedures to reduce the anticoagulant requirements are thorough rinsing of the filter before use in order to avoid all blood–air contact, ensuring maximal blood flow by reducing all resistances in the extracorporeal circuit, use of predilution, and frequent filter flushing with saline.

Extracorporeal anticoagulation can be achieved by prefilter administration of standard heparin (5–10 IU/kg/h), low-molecular-weight heparin, epoprostenol (prostacyclin) or its analogs (5–10 ng/kg/min), citrate, or nafamostate mesylate. Each of these methods has advantages and drawbacks.

When titrating the anticoagulant the primary goal should be avoidance of hemorrhagic complications and not prolongation of filter life. If bleeding risk is high, anticoagulant administration should be minimized or even omitted and frequent filter clotting must be accepted. In patients with severe coagulopathy, a satisfactory filter lifespan can be achieved without anticoagulation.

Replacement solution and dialysate

Volume

The amount of replacement solution is dictated by the filtration rate, the fluid input for nutrition and drug administration, and the fluid balance required. Because of the danger of balancing errors, the manual regulation of the replacement rate proposed originally is now increasingly being replaced by automated balancing devices.

Table 1 Operational characteristics of CRRT

	Q_b (ml/min)	Q_f (ml/h)	Q_d (ml/h)	Cl (ml/min)
CAVH	50–80	±600	—	±10
CAVHD(F)	50–80	(±200)	1000	17 (20)
CVVH	100–200	±1200	—	±20
CVVHD(F)	100–200	(±500)	1000–2000	16–32 (24–40)

Q_b, blood flow; Q_f, filtration rate; Q_c, dialysate flow; Cl, small-solute clearance.

Composition

Commercially available replacement solutions are modified Ringer's solution with an electrolyte composition similar to plasma water: Na^+, 135 to 145 mmol/l; K^+, 0 to 2 mmol/l; Cl^-,105 to 120 mmol/l; Ca^{2+}, 1.75 to 1.88 mmol/l; Mg^{2+}, 0 to 0.75 mmol/l. Dialysis may utilize the same solutions or a commercial dialysis solution. Solutions for peritoneal dialysis have also been recommended, but their high glucose content represents a problem in critically ill patients with limited glucose tolerance.

Lactate and bicarbonate (30–40 mmol/l) are the most commonly used buffers in CRRT. Lactate solutions are stable during manufacture and storage, and are usually well tolerated. In some patients the capacity to metabolize lactate is insufficient because of liver dysfunction or increased endogenous load (lactic acidosis), with continuing bicarbonate loss in the filtrate leading to an increase of acidosis.

Bicarbonate is the most physiological buffer, but bicarbonate solutions are difficult to prepare because of CO_2 loss and precipitation of $CaCO_3$ and $MgCO_3$. Possible solutions are the addition of bicarbonate to the replacement fluid directly before use or the separate administration of calcium and magnesium. Although the use of bicarbonate is controversial in lactic acidosis, many of the problems (volume overload, hypertonicity, etc.) can be prevented during CRRT.

Operational characteristics

The operational characteristics are summarized in Table 1.

Continuous arteriovenous hemofiltration (CAVH)

Convective clearance of small unbound solutes is equal to the filtration rate (Q_f). Determinants of Q_f are the membrane surface area (A), the membrane hydraulic permeability (K_f), and the transmembrane pressure (TMP):

$$Q_f = A \times K_f \times \text{TMP}.$$

TMP is given by

$$\text{TMP} = P_b + P_{uf} - \pi$$

where P_b is the hydrostatic pressure on the blood side, P_{uf} is the negative pressure in the filtrate compartment, and π is the oncotic pressure of the blood.

Arteriovenous techniques operate under conditions of low flow and low pressure. Within the filter there is a progressive decline in hydrostatic pressure accompanied by an increase in osmotic pressure, leading to a point of filtration pressure equilibrium where ultrafiltration ceases and the risk of filter clotting increases. This phenomenon is attenuated by predilution, which decreases oncotic pressure in the filter. Resistance in the filter also depends on blood viscosity which is inversely proportional to the hematocrit.

Satisfactory filtration rates can only be achieved with arteriovenous techniques if the systolic blood pressure is over 80 mmHg and resistance in the extracorporeal circuit (catheters, lines, connections) is kept to a minimum. Attempts to increase filtration rate by lowering the filtrate collector or applying suction to the filtrate line (suction-assisted hemofiltration) augment the risk of reaching filtration pressure equilibrium. Although predilution increases filtration rate, it does not lead to a proportional increase of clearance because of lower solute concentrations in blood and filtrate.

Continuous venovenous hemofiltration (CVVH)

The limited efficiency of CAVH resulted in the development of venovenous pump-driven techniques where transmembrane pressure is higher and primarily determined by the pump performance. Some hemofiltration machines also use a pump on the filtrate line. The use of a pump substantially enhances the complexity of the procedure and requires a variety of control and safety devices (air traps, automatic clamps, pressure monitors, and alarms).

Continuous hemodialysis (CAVHD, CVVHD)

The use of diffusive solute transport is another method of compensating for the limited azotemia control achieved with CAVH. With continuous dialysis, small-solute clearance equals dialysate flow (Q_d) provided that it remains considerably smaller than blood flow, enabling complete equilibration inside the dialyzer.

Combined diffusion and convection

Critically ill patients probably benefit from a combination of diffusion and convection that provides sufficient elimination of small and larger toxins. This combination may be achieved with the following.

1. Hemodiafiltration or continuous hemodialysis with high flux membranes, producing a considerable amount of ultrafiltrate that requires replacement. Diffusive and convective clearance of small solutes are additive.

2. High-flux dialysis or continuous hemodialysis with high-flux membranes and ultrafiltration control, leading to ultrafiltration in the first half of the filter and back-filtration of dialysate in the second part. This backfiltration can be seen as a substitute for the replacement solution.

Efficiency of metabolic control

Electrolytes

Since the electrolyte composition of the filtrate is similar to plasma water, adaptation of the composition of the replacement solution can correct almost every electrolyte disturbance. The sole restriction is the correction of severe hyperkalemia with arteriovenous hemofiltration; in a patient with a plasma potassium above 6 mmol/l, a daily filtration rate of 10 l eliminates only 60 mmol which represents a small fraction of the total body content. Commercially available replacement solutions do not contain phosphate. Large sodium losses can occur if filtrate replacement is partial.

Acid–base equilibrium

Correction of metabolic acidosis may present a problem when lactate is used in patients whose capacity to metabolize lactate is not high enough to compensate for bicarbonate losses in the ultrafiltrate.

Azotemia control

A filtration rate of 10 ml/min is considered necessary to compensate uremia. In hypercatabolic patients these relatively low filtration rates may be inadequate. Efficiency can be increased by addition of dialysis or by the use of a blood pump.

Advantages of CRRT

The most striking advantage of CRRT (Table 2) is its excellent hemodynamic tolerance. Continuous and gradual fluid removal leading to adequate fluid recruitment from the interstitium is the major explanation for this hemodynamic stability. Convective (iso-osmotic) solute transport also contributes to efficient plasma refilling by limiting intracellular fluid shifts. The use of synthetic membranes avoids hypotension induced by bio-incompatibility. Absence of abrupt fluctuations in fluids and solutes prevents the increase of intracranial pressure that is frequently associated with intermittent techniques. Intermittent hemodialysis has been suspected of contributing to the prolongation of acute renal failure through either repetitive periods of hypotension or the use of low-flux bio-incompatible membranes with insufficient removal or generation of the mediators promoting renal failure. Therefore CRRT with biocompatible membranes could lead to a better preservation of residual renal function. Arrhythmias due to rapid changes in electrolyte concentrations are prevented. The tolerance to fluid removal allows unlimited nutritional support; this should take into account the eventual glucose content of the dialysate and a 10 to 15 per cent loss of infused amino acids in the ultrafiltrate.

The continuous nature of CRRT permits equilibration between body compartments and therefore access to the total distribution volume. This improves solute removal, particularly if equilibration is slow. CRRT is flexible, allowing adaptation to the patient's situation at any time. It is managed by the intensive care nurse and thus obviates the need for a dialysis nurse.

Table 2 Advantages of CRRT

Hemodynamic tolerance
Stable intracranial pressure
Renal recovery
Unlimited nutrition
Flexibility
Improved solute removal
No dialysis nurse
Removal of inflammatory mediators?

At the time of its introduction the simplicity of CRRT was considered a significant advantage. The addition of blood pumps, combined diffusion and convection, automated balancing devices, ultrafiltration control, etc. has increased the complexity of these procedures but has greatly contributed to their efficiency and safety.

The potential continuous elimination of mediators of sepsis and multiple organ failure may represent a means of controlling not only the consequences of renal failure but also the underlying disease.

Limitations and disadvantages of CRRT

Most complications of arteriovenous hemofiltration (Table 3) are attributable to the arterial access. Likewise, inadequate azotemia control only applies to the spontaneous arteriovenous techniques. Azotemia control with pump-driven continuous hemofiltration compares favorably with intermittent hemodialysis and, for the same amount of therapy (product of clearance and time), continuous techniques remove more urea because of their access to the total distribution volume. Balancing errors are reduced by the use of automated balancing devices. Extracorporeal losses have to be taken into account when planning nutritional support and drug dosing.

One of the major disadvantages of CRRT is the need for continuous anticoagulation. CRRT leads to patient immobilization and hypothermia. Prevention of disconnection with disastrous bleeding requires the use of Luer lock connections and sensible pressure monitoring. CRRT requires the presence of medical and nursing expertise within the ICU and increases the workload for intensive care nurses.

An improved outcome compared with conventional dialysis has yet to be conclusively demonstrated.

Table 3 Disadvantages and limitations of CRRT

Complications related to arterial access (only if arteriovenous)
Limited efficiency (only if CAVH)
Balancing errors
Loss of nutrients and drugs
Continuous anticoagulation
Immobilization
Hypothermia
Disconnection
Requirement for medical and nursing expertise within the ICU

Table 4 Indications for CRRT

Acute renal failure with
Severe fluid overload
Hemodynamic instability
Cerebral edema
Limited nutritional input
Severe electrolyte and acid–base disturbances
Inflammatory syndromes ?

Indications

Uncomplicated acute renal failure can be treated with intermittent dialysis, but almost 90 per cent of intensive care patients with acute renal failure have associated failure of other organs (Table 4). In these patients continuous therapy should be used, particularly when fluid removal is the primary indication or when there is cerebral edema, hemodynamic instability, or the need for aggressive nutritional support.

Most of these patients present with a systemic inflammatory response syndrome where circulating mediators are involved in the pathogenesis. Although many of these mediators have been shown to be present in the ultrafiltrate, a reduction of plasma level has only been demonstrated for some secondary mediators (complement activation products, arachidonic acid metabolites, interleukin 6, interleukin 8, elastase, myocardial depressant factor) and not for the proinflammatory cytokines tumor necrosis factor and interleukin 1. However, experimental and (to a lesser extent) clinical studies have shown a beneficial effect of hemofiltration on hemodynamic and respiratory status and even suggest an improved outcome in septic subjects.

Finally, other possible indications of CRRT are severe electrolyte and acid–base disturbances.

Bibliography

Bellomo, R. (1993). Recent advances in acute renal replacement therapy. In *Yearbook of intensive care and emergency medicine* (ed. J.-L. Vincent), pp. 311–19. Springer-Verlag, Berlin.

Kierdorf, H. and Sieberth, H.G. (1995). Continuous hemofiltration in acute renal failure. In *Acute renal failure in the critically ill* (ed. R. Bellomo and C. Ronco), pp. 298–310. Springer-Verlag, Berlin.

Mehta, R.L., Dobos, G.J., and Ward, D.M. (1992). Anticoagulation in continuous renal replacement therapy. *Seminars in Dialysis*, **5**, 61–8.

Ronco, C. (1993). Continuous renal replacement therapies for the treatment of acute renal failure in intensive care medicine. *Clinical Nephrology*, **40**, 187–98.

Schetz, M., Ferdinande, P., Van den Berghe, G., Verwaest, C., and Lauwers, P. (1995). Removal of pro-inflammatory cytokines with renal replacement therapy: sense or nonsense? *Intensive Care Medicine*, **21**, 169–76.

18.5.2 Hemodialysis

M. Schetz

Key messages

- Treatment-induced hypotension is a frequent complication of intermittent hemodialysis in critically ill patients.
- Bicarbonate is the buffer of choice in hemodialysis for critically ill patients.
- The use of bio-incompatible membranes may compromise vital organ function in critically ill patients.

- Uncomplicated acute renal failure, life-threatening hyperkalemia, and drug poisoning are indications for conventional hemodialysis.

Introduction

Intermittent hemodialysis was introduced in clinical practice by Kolff in 1944 and for several decades remained the mainstay for the treatment of endstage renal disease and acute renal failure. The goals of dialytic therapy are to replace the homeostatic renal functions (volume control, electrolyte homeostasis, acid–base balance, and clearance of uremic toxins) without additional damage to the kidney and other vital organs.

Timing the initiation of dialysis

Absolute indications for renal replacement therapy include intravascular volume overload unresponsive to diuretics, hyperkalemia, acidemia, and uremic symptoms (encephalopathy, pericarditis, bleeding diathesis). Although there is no consensus on the ideal timing to initiate dialysis in acute renal failure, it is generally accepted that treatment is initiated when a reversible component to the renal insufficiency has been excluded and well before the occurrence of complications or uremic symptoms. Failure of other organs, rate of increase of urea and creatinine, anticipated catabolism, and urine output should be taken into account. Benefits of early dialysis have to be weighed against the effect of dialysis on residual renal function.

Methods

Solute transport

In conventional dialysis blood is circulated through a dialyzer (flow rate, 250–350 ml/min) in the opposite direction to the dialysate (flow rate, 500–800 ml/min). The two flows are separated by a semi-permeable membrane and most solute transport is achieved by diffusion based on a concentration gradient. Fluid removal is achieved by ultrafiltration driven by a hydrostatic pressure gradient (Fig. 1).

Diffusive solute clearance is determined by several variables: time on dialysis (duration and frequency), membrane surface area, membrane diffusive permeability (thickness and porosity), solute molecular weight, blood flow, and dialysate flow. Diffusion is very efficient for the elimination of small solutes, but the removal of larger solutes depends largely on the convective component (ultrafiltration).

Fluid removal is controlled by calculating the transmembrane pressure required to achieve the desired fluid balance, given the ultrafiltration coefficient of the membrane and the time on dialysis. Inaccuracies can be overcome by ultrafiltration control systems which automatically adjust transmembrane pressure and allow more uniform weight reduction over time.

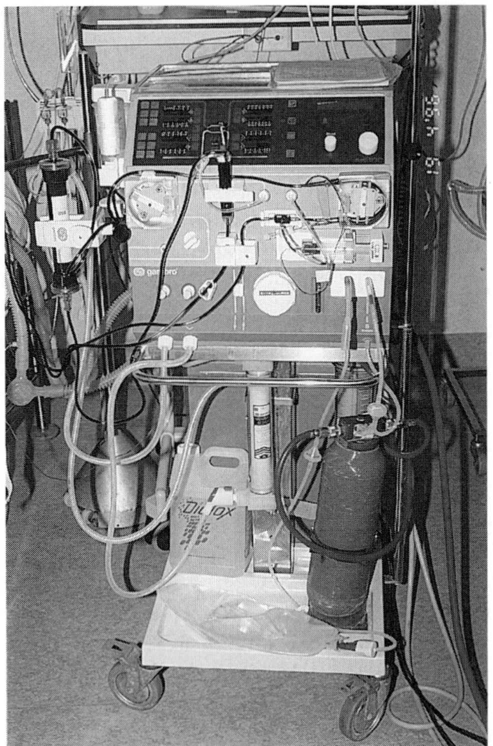

Fig. 1 Dialysis machine.

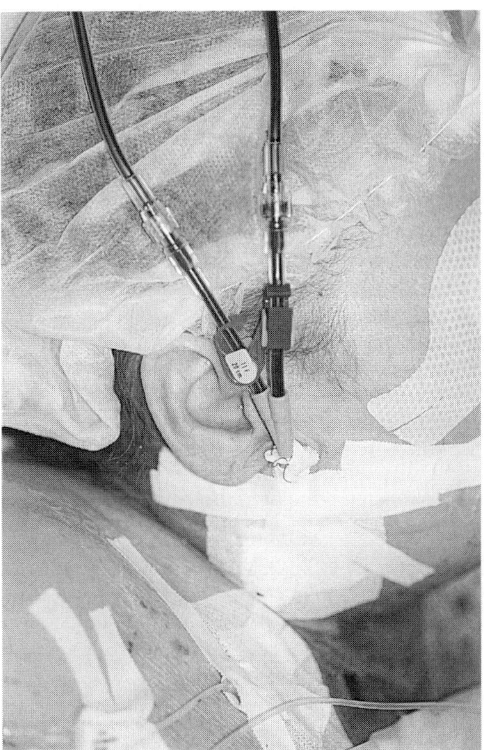

Fig. 2 Double-lumen catheter in internal jugular vein.

Vascular access

Vascular access for acute dialysis is accomplished by percutaneous insertion of a double-lumen catheter in the femoral, subclavian, or jugular vein (Fig. 2). The choice of site is often dictated by the underlying disease (risk of pneumothorax in respiratory compromised patients, 'traffic jam' in the superior vena cava because of the need for invasive monitoring with a pulmonary artery catheter and administration of drugs and nutrition via a triple-lumen catheter, contaminated abdominal wounds, etc.).

Compared with the internal jugular vein, cannulation of the subclavian vein is associated with more acute complications (pneumothorax, hemothorax) as well as an increased incidence of late stenosis. Femoral catheters carry an increased infection risk and limit the patient's mobility. Scribner shunts provide a lower blood flow and have an increased risk of clotting.

Membranes

Most contemporary dialysis devices contain membranes in hollow-fiber format. These membranes differ with regard to composition (cellulose-based versus synthetic polymers), porosity (high-flux versus low-flux), thickness, biocompatibility, and cost.

High-flux membranes have a higher hydraulic permeability and a greater ability to transport larger toxins. There is no consensus on the importance of larger uremic toxins (e.g. β_2-microglobulin) in acute renal failure. Medium-molecular-weight inflammatory mediators such as anaphylatoxins, arachidonic acid metabolites, and cytokines may have deleterious effects in critically ill patients.

Bio-incompatibility is defined as the consequences of blood–membrane contact and includes activation of blood proteins (complement system, coagulation cascade, contact system) and cells (white blood cells and thrombocytes). It may induce several clinical problems (Table 1). Hypoxemia and pulmonary hypertension result from complement activation with pulmonary leukosequestration and release of neutrophil activation products. Generation of anaphylatoxins (complement activation), bradykinin (contact activation), and inflammatory cytokines (monocyte activation) may lead to hypotension. Other sequelae of bio-incompatibility are delayed recovery of renal function due to hypotension and/or generation or decreased removal of nephrotoxic mediators, increased susceptibility to infection resulting from decreased neutrophil function and altered T-lymphocyte function, increased catabolism, activation of coagulation, and thrombocytopenia.

Cellulosic membranes (cuprophane, hemophan, cellulose acetate) are mostly low-flux, thin, bio-incompatible, and cheap, whereas synthetic membranes (polysulfone, polyacrylonitrile, polyamide, polymethylmethacrylate) are mostly high-flux, thicker, biocompatible, and expensive, although this distinction cannot be generalized. Two recent studies reported a decreased morbidity and mortality when biocompatible membranes were used during intermittent dialysis in acute renal failure.

Anticoagulation

Because of the thrombogenicity of the extracorporeal circuit, some form of anticoagulation is required. In acute renal failure it is virtually impossible to devise generally valid schemes of anticoagulation because of the large variability of patient-specific variables. The bleeding risk is estimated on the presence of active bleeding, the time elapsed since surgery, trauma, or invasive procedures, and the location of a potential bleeding source (intracranial, intraspinal). The presence of a coagulopathy reduces anticoagulant requirements, whereas an activated coagulation, such as in severe sepsis, may require more intensive anticoagulation.

Heparin is still the most widely used anticoagulant. Full-dose systemic unfractionated heparin is infrequently used in acute renal failure. The intensity of anticoagulation can be diminished by using a reduced dose of heparin to maintain an activated clotting time 15 to 25 per cent above baseline. Minimal systemic anticoagulation occurs with regional heparinization, where heparin is infused into the arterial catheter of the dialysis circuit and neutralized by infusing protamine into the venous catheter, with maintenance of the patient's activated clotting time at baseline and prolongation of the circuit activated clotting time by 10 s. An additional bolus of protamine at the end of the procedure may be necessary to prevent heparin rebound. Low-molecular-weight heparins provide a superior antithrombotic-to-anticoagulant ratio. Dosage regimens should take into account their predominant renal elimination. Monitoring with routine coagulation tests is inadequate and neutralization by protamine is incomplete. Epoprostenol (prostacyclin) or its analogs prevent platelet activation. The absence of an effect on the clotting cascade usually requires the addition of a reduced dose of heparin. Epoprostenol is very expensive and may induce hypotension. Strict regional anticoagulation can be achieved with prefilter infusion of sodium citrate resulting in calcium depletion. The anticoagulant effect is antagonized by administering calcium to the patient. The citrate infusion is adapted to maintain the circuit-activated clotting time 25 per cent above baseline. Most of the infused citrate is lost in the dialysate, whereas the citrate that returns to the circulation is metabolized to bicarbonate. Therefore the use of citrate may alleviate the need for a buffer in the dialysate. Possible side-effects are hypernatremia, hypocalcemia, and metabolic alkalosis. Citrate anticoagulation is labor intensive and can only be used with diffusive techniques because of insufficient removal of citrate with convective transport. A very high bleeding risk may necessitate omission of anticoagulants with high blood flows and frequent saline flushing to reduce the risk of filter clotting. A future development in this field is the use of heparin-bound surfaces within the circuit.

Table 1 Possible consequences of bio-incompatibility

Hypoxia
Pulmonary hypertension
Hypotension
Delayed recovery of renal function
Increased susceptibility to infection
Catabolism
Thrombin generation
Thrombocytopenia

Dialysate composition

Electrolytes

The sodium content of the dialysate is an important determinant of intradialytic symptoms. Diffusive removal of small solutes entails a reduction of extracellular osmolality, leading to the loss of extracellular fluid into the intracellular compartment. These fluid shifts are reduced by maintaining dialysate sodium at or above the serum concentration. Dialysate sodium concentrations below 140 mmol/l should be avoided in hemodynamically unstable patients.

Rapid and efficient potassium removal represents an advantage of intermittent dialysis; however, it may lead to life-threatening arrhythmias. Progressive reduction of dialysate potassium from 2 to 3 mmol/l to zero to 1 mmol/l allows smoother correction of severe hyperkalemia. A potassium level of 2 to 3 mmol/l is optimal for most intensive care patients, although higher concentrations may be required in patients on digitalis.

Calcium and magnesium concentrations should be ± 3.5 mmol/l and ± 1 mmol/l respectively.

Glucose

In order to prevent major glucose losses, a glucose concentration of 100 to 250 mg/dl is advisable.

Buffer

Bicarbonate was originally used as the buffer but was later replaced by acetate because of difficulties in preparing a stable solution. These practical problems result from the instability of bicarbonate solutions at neutral pH, particularly in the presence of calcium and magnesium. Precipitation of $CaCO_3$ and $MgCO_3$ is prevented by using a two-step proportioning system where a solution containing electrolytes and an acid is mixed with a bicarbonate solution immediately before use. Reaction of bicarbonate with the acid produces CO_2, thereby preventing precipitation of calcium and magnesium salts. Promotion of bacterial growth in bicarbonate solutions is prevented by using powdered concentrates. Bicarbonate concentration should be no greater than 35 mmol/l in order to limit hypoventilation due to rapid correction of acidosis.

Acetate solutions are stable during manufacture and storage. Acetate metabolism takes place in liver and muscles and yields 1 mole of bicarbonate for each mole of acetate. The maintenance of the bicarbonate level depends on the rate of acetate metabolism. Adverse effects may be related to insufficient metabolism (hepatic insufficiency and reduced muscle mass or perfusion), leading to increased acetate levels and a decrease in arterial pH secondary to bicarbonate loss in the dialysate. Acetate metabolism itself may induce symptoms related to increased O_2 consumption or accumulation of intermediates and breakdown products. Acetate-induced symptoms include hypoxemia secondary to hypoventilation (CO_2 loss in the dialysate and CO_2 consumption in acetate metabolism) and hypotension resulting from acetate-induced vasodilation and impaired plasma refilling. The existence of a direct cardiodepressive effect remains controversial. In particular, critically ill patients and patients with impaired left ventricular function are prone to cardiovascular instability during acetate dialysis.

In summary, although bicarbonate dialysis is associated with more work, worry, and cost, it is the buffer of choice in most centers treating patients with acute renal failure.

Efficiency of metabolic control

Dialysis dose is expressed as Kt/V where K is the urea clearance during treatment, t is the time on dialysis, and V is the urea distribution volume. In endstage renal disease Kt/V has been shown to correlate with morbidity and mortality. Dialysis is considered adequate with a Kt/V of 3.6/week when protein intake and catabolic rate is 1 g/kg/day.

The optimal level of azotemia control in acute renal failure is unknown. Most patients with acute renal failure in the ICU are hypercatabolic and require a higher Kt/V to achieve the same level of control. Since critically ill patients often do not tolerate an increase in K (high-efficiency dialysis), the safest way to increase adequacy is to increase t (daily dialysis). Quantitation of dialysis should also take into account a gap between the prescribed and the delivered dose resulting from premature treatment interruptions because of hemodynamic instability, reduced blood flow because of suboptimal vascular access, recirculation, and urea rebound.

Advantages of conventional dialysis

Dialysis treatment is widely available. Greater effectiveness compared with continuous techniques has recently been challenged and only applies to continuous arteriovenous hemofiltration. Because of the rapid and efficient removal of small solutes, conventional dialysis remains the therapy of choice for life-threatening hyperkalemia and drug poisoning. Other major advantages include anticoagulation restricted to the time of treatment and the possibility of mobilizing the patient.

Limitations of conventional dialysis

The major disadvantage of intermittent dialysis is the frequent occurrence of hypotension, sometimes requiring premature discontinuation of the treatment. Dialysis hypotension can usually be attributed to preload reduction because rapid ultrafiltration exceeds fluid recruitment from the interstitium. In turn, the plasma refilling rate is adversely affected by rapid decreases of serum osmolality with intracellular fluid shifts. Acetate and bio-incompatible membranes also contribute to hypotension by induction of inappropriate vasodilatation. Reduction of myocardial contractility is mostly attributable to hypoxemia or acidosis. Convective solute removal (regardless of the ultrafiltration rate, small solute clearance, buffer, and membrane) appears to be hemodynamically better tolerated than diffusion. The mechanism underlying diffusion-related hypotension has not been elucidated. Possible explanations are increased diffusive removal of a small-molecular-weight substance (e.g. an inhibitor of NO synthase leading to increased NO synthesis and vasodilatation) or decreased diffusive removal of a vasodilating middle molecule (e.g. calcitonin gene-related peptide, atrial natriuretic peptide, parathyroid hormone).

In hemodynamically unstable patients, dialysis-induced hypotension can be attenuated by slow ultrafiltration (ultrafiltration controllers), promotion of plasma refilling by infusion of human albumin or the use of high-sodium dialysate, sequential ultrafiltration and dialysis (temporal segregation of ultrafiltration and solute removal), and using biocompatible membranes and bicarbonate (Table 2). Intermittent hemofiltration uses convective solute

Table 2 Prevention of dialysis-related hypotension

Ultrafiltration control
High-sodium dialysate
Human albumin
Sequential ultrafiltration and dialysis
Biocompatible membrane
Bicarbonate
Convective transport

transport resulting in improved hemodynamic tolerance (no diffusion-related hypotension). However, despite these modifications, intermittent techniques will probably never reach the excellent hemodynamic tolerance of the continuous renal replacement techniques.

Rapid removal of osmotically active solutes from the extracellular compartment leads to an increase in cerebral water content (dialysis disequilibrium with headache, nausea, vomiting, and hypertension, eventually progressing to arrhythmias, confusion, tremor, seizures, coma, and death). The disequilibrium syndrome can be attenuated by adding solute (sodium, glucose, mannitol) to the dialysate and by initiation of dialysis before the urea has reached too high a level. In patients with pre-existing cerebral edema, intermittent dialysis can lead to life-threatening increases in intracranial pressure.

Hypoxemia during dialysis results from acetate-induced hypoventilation (only in spontaneously breathing patients) or from the use of bio-incompatible membranes. Intermittent hemodialysis may delay the recovery of renal function because of recurrent periods of hypotension (loss of autoregulation in the injured kidney) or because of the insufficient removal or generation of nephrotoxic mediators by low-flux bio-incompatible membranes. The hemodynamic intolerance to fluid removal often limits nutritional support. The requirement for specialized equipment and trained nursing staff is another limitation of intermittent dialysis.

Indications

Intermittent dialysis is the treatment of choice for uncomplicated acute renal failure since continuous treatment would hamper the mobilization of these otherwise stable patients and expose them to an unnecessary risk of continuous anticoagulation. Life-threatening hyperkalemia and drug poisoning require rapid and efficient solute removal that can only be provided by intermittent hemodialysis.

References

Chertow, G.M., Owen, W.F., and Lazarus, J.M. (1995). Conventional hemodialysis for acute renal failure. In *Acute renal failure in the critically ill* (ed. R. Bellomo and C. Ronco), pp. 265–97. Springer-Verlag, Berlin.

Daugirdas, J.T. (1991). Dialysis hypotension: a hemodynamic analysis. *Kidney International*, 39, 233–46.

Lang, S., Küchle, C., Fricke, H., and Schiffl, H. (1995). Biocompatible intermittent hemodialysis. *New Horizons*, 3, 680–7.

Ledebo, I. (1993). Bicarbonate in high-efficiency hemodialysis. In *Hemodialysis: high-efficiency treatments* (ed. J.B. Bosch and J.H. Stein), pp. 9–25. Churchill Livingstone, New York.

Mehta, R.L. (1994). Therapeutic alternatives to renal replacement for critically ill patients in acute renal failure. *Seminars in Nephrology*, 14, 64–82.

18.5.3 Peritoneal dialysis

F. D. Thompson

Key messages

- Although superseded by hemodialysis and hemofiltration techniques, peritoneal dialysis still has a place in the management of acute renal failure.
- This technique is particularly valuable in children in whom the peritoneal surface area is relatively large.
- Peritoneal dialysis is useful in the removal of potassium and water which may buy valuable time, allowing the patient to be referred to a specialist center for conventional dialysis or filtration.

Introduction

As hemodialysis and filtration have become more generally available, the use of peritoneal dialysis in the management of acute renal failure has diminished. However, there is still a place for this technique, particularly in those units not familiar with vascular dialysis techniques. This technique can still be used in children where the ratio of the peritoneal surface area to the body mass is advantageous, and adequate clearances can often be obtained. Water and potassium can readily be removed from patients, and peritoneal dialysis can still be regarded as an effective means of treatment in those whom circulatory overload and hyperkalemia are a problem.

Procedure

The peritoneal catheter is inserted via a trocar into the abdominal cavity using an entry site in the midline just below the umbilicus. The catheter is directed down towards the pelvis. A purse-string suture secures the catheter and the procedure is carried out using local anesthetic. About 1 to 2 liters of dialysis solution is run in quickly, left to dwell in the abdomen for 30 to 40 min, and then drained. Once dialysis is established, the dwell time and volume used can be increased; 2 liters are often used with the whole cycle of running in, dwell time, and draining taking 2 to 3 h. The dialysis solutions are warmed to body temperature and, in the initial stages, 2000 IU of heparin are added to each liter of fluid to prevent blockage by fibrin deposition within the catheter. The volume of fluid infused and drained should be accurately measured and recorded, allowing the overall balance of dialysate to be calculated. If the patient is overloaded with fluid, a hypertonic solution containing 6.36 per cent dextrose can be used and a negative balance achieved. The isotonic solution containing 1.36 per cent glucose is used when no fluid removal is required. Care has to be taken, as a large negative balance can be achieved even with isotonic solutions. When using a hypertonic solution, the blood glucose should be measured to ensure that significant hyperglycemia does not develop. In the early stages no additional potassium is added to the dialysate but, after several cycles, excess potassium is often removed and so potassium chloride is added to give a potassium concentration of 2 to 3 mmol/l in the dialysate.

Complications

Dialysate leakage

The dialysate can leak from the catheter entry site. Replacing the purse-string suture securing the catheter in place is often all that is required. Dialysate can also leak into the pleural cavity, either as a result of surgery where the diaphragm is breached or via naturally occurring foramina. There is little that can be done to correct this, and alternative forms of dialysis must be used.

Hemorrhage

Occasionally, a blood vessel in the abdominal wall is damaged, causing blood loss, but this blood staining of the dialysate is often short lived and the technique rarely has to be discontinued.

Sepsis

Sepsis is commonly of the peritoneum itself and is often caused by staphylococci or gut intestinal flora. Intraperitoneal vancomycin 50 mg and gentamicin 8 mg can be added empirically to the peritoneal fluid until the results of fluid culture are obtained. There is no need to discontinue the dialysis.

Hypotension

A fall in blood pressure immediately dialysis is commenced is due to stretching of the peritoneum; elevation of the foot of the bed is all that is required. More profound hypotension occurring after several days of dialysis is usually associated with hypovolemia resulting from achievement of too large a negative balance with hypertonic solutions. Protein can also be lost via the dialysate, compounding the picture. Fluid and plasma replacement are then required.

Respiratory difficulty

Splinting of the diaphragm by intraperitoneal fluid can lead to respiratory embarrassment. This is only a problem when respiratory function is already compromised before dialysis commences.

Perforation of a hollow viscus during catheter insertion

Occasionally, the intestine or bladder are damaged by catheter insertion. If there is any suspicion that insertion may be difficult, the abdominal cavity can be filled with dialysate using a fine-bore needle before the large trocar and cannula are inserted.

The major disadvantage of this technique is that patients with severe trauma and sepsis are very catabolic, and the efficacy of peritoneal dialysis is unlikely to achieve adequate and safe biochemical homeostasis. The presence of previous abdominal surgery often prevents its use. On the positive side, peritoneal dialysis can be used effectively in children where the relatively large surface area of the peritoneum often achieves adequate clearances. This technique is very efficient in removing water and potassium which are often life-threatening indications for dialysis. When peritoneal dialysis is established and works well, it is a gentle way of correcting the biochemical changes associated with acute renal failure without causing major changes in the circulation. When first introduced into clinical practice, peritoneal dialysis was occasionally used to remove drugs and toxins in cases of overdose and poisoning. The development of hemodialysis, hemofiltration, and hemoperfusion through adsorbent columns means that this technique is no longer used for this purpose. Similarly, the use of cold or warm dialysate to correct hyper- or hypothermia has been superseded by the use of extracorporeal blood circuits with heat exchangers. Therefore peritoneal dialysis, although overtaken by the modern technology of hemodialysis and filtration, still has a role, albeit small, in the management of acute renal failure. In the presence of intra-abdominal sepsis, peritoneal lavage may also be beneficial.

Bibliography

Daugirdas, J.T. and Ing, T.S. (ed.) (1988). *Handbook of dialysis*, pp. 167–219. Little, Brown, Boston, MA.

18.5.4 Plasmapheresis

L. Y. Vitou and M. Sternbach

Key messages

- Plasmapheresis is usually adjunctive rather than definitive therapy for intensive care conditions requiring immune modulation.
- Several different technological approaches are available.
- Adverse effects depend on underlying disease severity and the frequency, method, and rapidity of volume exchange.
- The strongest evidence for the use of plasmapheresis is in the Guillain–Barré syndrome and thrombotic thrombocytopenic purpura.

Principles

Plasmapheresis is an effective form of definitive or adjunctive therapy for a number of conditions which the intensive care specialist may encounter. Most conditions for which the effect has been shown are associated with circulating antibodies or immune complexes. However, there is evidence to support the concept of immunomodulation through other processes, for example, an increase in splenic clearance of heat-inactivated red blood cells, suggesting improved endogenous clearance of antibodies. Other substances are removed, including intact complement, activated complement products, fibrinogen, and various inflammatory mediators.

Substances removed are distributed primarily in the extravascular space. Therefore plasmapheresis kinetics involve those forces which govern intravascular depletion, catabolism, and synthesis as well as shifting substances from the extravascular to the intravascular compartment. Substance removal from the intravascular space follows the exponential relationship

$$X_t = X_0 \exp(-rt)$$

where X_0 is the initial concentration of substance X, X_t is its concentration at time t, and r is the fractional rate of exchange defined by

$$r = \frac{\text{volume exchanged per hour}}{\text{plasma volume}}.$$

If there is no new synthesis or redistribution of substance X during plasmapheresis, removal of the first plasma volume will produce an initial 63 per cent reduction in plasma concentration of substance X. Additional removal of a second plasma volume produces a further 25 per cent reduction, while removal of a third plasma volume results in only a further 9 per cent reduction.

Endogenous synthesis and shifting of substance X from extravascular to intravascular space are the predominant modes of reaccumulation of the macromolecule between treatments. The magnitude of this reaccumulation is both variable and unpredictable, depending on the substance being considered and individual patient characteristics. Nevertheless, most authors consider four or five single plasma volume exchanges over a period of 7 to 10 days to be adequate short-term therapy, as this would result in a plasma concentration of a given substance of less than 10 per cent of its prepheresis concentration. Serial clinical assessment and measurement of the offending substance (when known) allow decisions to be made regarding the frequency and volume of exchanges, so that treatment protocols can be tailored to specific conditions and patients.

Replacement solutions

The nature and amount of replacement fluid is important. Although specific guidelines have not been developed, except for thrombotic thrombocytopenic purpura–hemolytic uremic syndrome in which fresh frozen plasma is the replacement fluid of choice, general guidelines are useful and can be modified to suit specific clinical situations. Colloid such as albumin must often be administered to maintain hemodynamic stability and to assure a steady colloid osmotic pressure.

Drawbacks of fresh frozen plasma include a small but significant risk of transmission of hepatitis B, hepatitis C, and HIV. Furthermore, administration of fresh frozen plasma may be replenishing the very factor(s) one is endeavoring to remove. A more economical approach is to replace the first third of the removed volume with crystalloid and the last two-thirds with albumin since most of the albumin initially infused will be exchanged. This replacement technique results in a postpheresis serum albumin concentration in the normal range, but can lead to a coagulopathy because albumin solution lacks coagulation factors. Pre-exchange monitoring of coagulation (international normalization ratio and partial thromboplastin time) is recommended, and infusion of fresh frozen plasma as part of the replacement solution may be necessary.

Technical considerations

Circulatory access

This is predicated on the separation technique employed. A blood flow of 40 to 50 ml/min is required for a centrifugal device and can often be achieved via a large peripheral vein. Conversely, when a hollow-fiber technique is used, a blood flow of 100 ml/min is necessary; this can only be sustained by a suitable large-caliber double-lumen central venous catheter. The femoral position may reduce patient mobility, but has the theoretical advantage of allowing relatively hypocalcemic blood to mix with normocalcemic blood before reaching the arteriovenous node, thus reducing the risk of arrhythmias.

Separation technology

Centrifugal devices separate blood into its components according to size, density, and function either continuously or discontinuously.

The intermittent-flow instrument intermittently removes small volumes of blood (200–300 ml) which are then anticoagulated and centrifuged, separating the cellular elements from plasma. The plasma fraction is eliminated, and the cellular elements are returned to the patient together with replacement fluid. This is repeated until the desired volume has been exchanged. The continuous-flow apparatus performs the same task while simultaneously withdrawing blood from the patient, centrifuging it, and returning the cellular fraction to the blood stream. This is accomplished using a loop-shaped ring with sampling ports for plasma and individual cellular elements. Hollow-fiber devices similar to hemodialyzers but with larger pore sizes cannot perform white cell, red cell, or platelet pheresis, and thus are limited to plasmapheresis. However, they offer several advantages. The plasma fraction can be passed through successive devices with decreasing molecular weight cut-offs, allowing different plasma fractions to be separated out (cascade plasmapheresis). The albumin fraction can be returned to the patient, and other fractions (immunoglobulins) selected out. A relatively high blood flow is required with hollow-fiber devices.

Anticoagulation

Traditionally, heparin is used with hollow-fiber hemodialysis and citrate with the centrifugation technique. With heparin, individual dosing is based on monitoring the activated clotting time every 30 min. However, an initial dose of 50 IU/kg and infusions of 25 IU/kg as necessary to maintain the activated clotting time at 1.5 times to twice normal is a conservative approach. In some patients, substantial amounts of heparin can be extracted by the procedure, obviating frequent reinfusion of heparin.

Calculation of plasma volume

This can be performed by nomograms based on height, weight, and hematocrit. However, plasma volume can be approximated to 35 ml/kg for patients with normal hematocrit and to 40 ml/kg in those with a subnormal hematocrit.

Side-effects

Adverse effects are related to the extracorporeal nature of the procedure, necessitating vascular access and anticoagulation, and to the replacement fluid. In a review of 5235 procedures undertaken in 627 patients, side-effects were encountered in 12 per cent of treatment sessions while 40 per cent of patients experienced adverse reactions at least once. Most reactions were minor and transient, and included fever, chills, urticaria, nausea, hyper- and hypotension, paresthesias, and hives. These reactions were more common when fresh frozen plasma was used as replacement fluid. Serious adverse effects, including cardiovascular distress, anaphylactic shock, and disease-specific exacerbation of the underlying condition, occur in seven in 1000 procedures. A fatality rate of three in 10 000 procedures has been estimated, with cardiovascular and respiratory causes most often implicated.

Citrate, which depletes calcium (a necessary cofactor in the coagulation cascade) from plasma, can cause hypocalcemia (paresthesias, carpopedal spasm, or arrhythmias). Arrhythmias are more common when plasma electrolytes are concurrently deranged. Because of removal of albumin and other proteins, drugs which are highly protein bound (e.g. aminoglycosides, Coumadin (warfarin)) should be closely monitored and dosages adjusted.

Indications in the intensive care unit

Diseases exist for which plasmapheresis is considered definitive or, more commonly, useful adjunctive therapy (Table 1). Randomized trials have been undertaken in a few disease states and have shown the procedure to be ineffective (Table 2). Those disorders which may be encountered in the intensive care setting and for which plasmapheresis may be beneficial are reviewed below.

Antiglomerular basement membrane antibody disease

Antiglomerular basement membrane antibody disease, which is one cause of pulmonary–renal syndrome, is characterized by rapidly progressive glomerulonephritis and pulmonary hemorrhage. While there have been a few cases of spontaneous remission, the overwhelming majority succumb to pulmonary hemorrhage or are left with dialysis-dependent renal failure. Plasmapheresis is effective at removing antiglomerular basement membrane antibody, the diagnostic and putative pathogenic factor in antiglomerular basement membrane disease. In conjunction with immunosuppressive therapy, it has become the accepted treatment of choice.

Table 1 Diseases for which plasmapheresis is accepted as definitive or adjunctive therapy

Acute Guillain–Barré syndrome*
Myasthenia gravis
Eaton–Lambert syndrome
Antiglomerular basement membrane antibody disease*
Thrombotic thrombocytopenic purpura–hemolytic uremic syndrome*
Plasma hyperviscosity syndrome
Coagulation factor inhibitors
Autoimmune hemolytic anemia (cold antibody type)
Symptomatic cryoglobinemia
Myeloma kidney (light-chain nephropathy)
Paraprotein-associated peripheral neuropathy
Post-transfusion purpura
Refsum's disease
Pemphigus vulgaris

*Supported by at least one randomized trial.

Table 2 Diseases for which plasmapheresis is not accepted as definitive or adjunctive therapy

Rheumatoid arthritis*
Idiopathic rapidly progressive glomerulonephritis*
Multiple sclerosis*
Polymyositis/dermatomyositis*
Amyotrophic lateral sclerosis*
Fulminant hepatic failure*

*Supported by at least one randomized trial.

Thrombotic thrombocytopenic purpura and hemolytic uremic syndrome

Thrombotic thrombocytopenic purpura is an uncommon disease with an untreated mortality approaching 100 per cent. It is characterized clinically by fever, thrombocytopenic purpura, microangiopathic hemolytic anemia, and ischemically mediated neurological and renal impairment. These findings are variably identified at initial presentation, but the presence of all five signs implies more severe disease and generally portends a worse prognosis. A large multicenter Canadian randomized unblinded trial involving 102 patients demonstrated that plasmapheresis was superior to plasma infusion alone. At 6 months, 15 per cent more patients survived in the plasmapheresis group ($p = 0.036$) with a survival rate of 78 per cent. However, patients with severe renal failure were excluded from the study, less than 65 per cent had neurological findings, and less than 25 per cent had fever at the time of presentation.

No trials have been performed for hemolytic uremic syndrome which shares common pathophysiological features with thrombotic thrombocytopenic purpura. Patients presenting with thrombocytopenic purpura, microangiopathic hemolytic anemia, and neurological signs, fever, and oliguria severe enough to preclude randomization in the Canadian study represented 18 per cent of the total number of patients presenting with thrombotic thrombocytopenic purpura–hemolytic uremic syndrome. They had a survival rate similar to those patients treated with plasmapheresis in the randomized study, suggesting that patients with the most severe form of the disease also benefit. An elevated von Willebrand factor antigen level was found in all patients in whom it was measured; some evidence suggests that cryosupernatant plasma (plasma devoid of cryoprecipitate and thus large multimers of von Willebrand factor) may be more beneficial than fresh frozen plasma in the treatment of thrombotic thrombocytopenic purpura.

Acute inflammatory demyelinating polyneuropathy or Guillain–Barré syndrome

Approximately 20 per cent of patients require mechanical ventilation while 15 per cent remain with residual deficits. It appears to be mediated by an antimyelin antibody. The efficacy of plasmapheresis is well supported by two multicenter randomized trials which demonstrated significant clinical improvement in terms of reduced mortality, shorter functional recovery time, and decreased requirement for mechanical ventilation. Patients receiving plasmapheresis early in the course of their disease benefited more, particularly if they required mechanical ventilation at the time of initiation of therapy, but also if they required mechanical ventilation subsequent to the initiation of plasmapheresis. In a randomized trial comparing plasmapheresis and intravenous immunoglobulin, the two modes of treatment were comparable, with a slight advantage to immunoglobulin. However, the plasmapheresis schedule used in the study was less aggressive than that usually recommended for Guillain–Barré syndrome.

Myasthenia gravis

Myasthenia gravis is caused by an antibody to the acetylcholine receptor. Although the therapeutic rationale for using plasmapheresis is clear, no randomized controlled trials have examined its efficacy. Nevertheless, it is believed to have a role during an acute crisis, as adjunctive therapy at the initiation of immunosuppressive treatment, and pre- and post-thymectomy. Currently, although widely used and accepted, it remains experimental.

Hematological malignancies

A double-blind randomized cross-over study noted hematological improvement in patients with IgA and IgG monoclonal gammopathy, although clinically important outcomes were unaffected. Although there are no randomized trials to support its use in cryoglobulinemia and Waldenstrom's macroglobulinemia, plasmapheresis may provide valuable ancillary therapy to immunosuppression by rapidly depleting the offending cryoglobulins or macroglobulin, particularly in the setting of a crisis or hyperviscosity syndrome. Other hematological conditions in which plasmapheresis may be helpful include cold agglutinin autoimmune hemolytic anemia, acquired coagulant factor inhibitors, and coagulant factor deficiencies which require infusion of large volumes of fresh frozen plasma.

Sepsis-induced multiple organ dysfunction syndrome

A retrospective comparative study of continuous arteriovenous hemofiltration (CAVH), CAVH dialysis, and CAVH sequential plasma filter dialysis in patients with multiple organ dysfunction syndrome showed lower mortality in the latter group. Plasmapheresis may remove mediators inducing multiple organ dysfunction syndrome in such patients, and should be studied in randomized clinical trials.

Bibliography

Barzilay, E., Kessler, D., Berlot, G., Gullo, A., Geber, D., and Ben Zeev, I. (1989). Use of extracorporeal supportive techniques as additional treatment for septic-induced multiple organ failure. *Critical Care Medicine*, **17**, 634–9.

Lewis, J.H., Hasiba, U., and Spero, J.A. (1982). Changes in coagulation factors, complement, immunoglobulins, and immune complex concentrations with plasma exchange. *Transfusion*, **22**, 54–8.

Rock, G.A., *et al.* (1991) Comparison of plasma exchange with plasma infusion in the treatment of thrombotic thrombocytopenic purpura. *New England Journal of Medicine*, **325**, 393–7.

Sutton, D.M.C., *et al.* (1989). Complications of plasma exchange. *Transfusion*, **29**, 124–7.

van der Meché, F.G., *et al.* (1992). A randomized trial comparing intravenous immune globulin and plasma exchange in Guillain–Barré syndrome. *New England Journal of Medicine*, **326**, 1123–9.

Index

Note: Page numbers in *italic* refer to figures and/or tables

B

C

T